Handbook of
STRUCTURAL ENGINEERING

Handbook of
STRUCTURAL ENGINEERING

Edited by

WAI-FAH CHEN
ERIC M. LUI

CRC Press
Taylor & Francis Group
Boca Raton London New York

CRC Press is an imprint of the
Taylor & Francis Group, an **informa** business

SECOND INDIAN REPRINT, 2014

Library of Congress Cataloging-in-Publication Data

Handbook of structural engineering/edited by Wai-Fah Chen,
Eric M. Lui. — 2nd ed.
    p. cm.
    Includes bibliographical references and index.
    ISBN 0-8493-1569-7 (alk. paper)
    1. Structural engineering. I. Chen, Wai-Fah, 1936- II. Lui, E. M. III. Title.

TA633.H36 2004
624.1 — dc22   2004054550

Visit the CRC Press Web site at www.crcpress.com

© 2005 by CRC Press

No claim to original U.S. Government works
International Standard Book Number 0-8493-1569-7
Library of Congress Card Number 2004054550

Printed and bound in India by Replika Press Pvt. Ltd.

FOR SALE IN SOUTH ASIA ONLY.

Abstract

This book is an encapsulation of a myriad of topics of interest to engineers working in the structural analysis, design, and rehabilitation fields. It is a comprehensive reference work and resource book written for advanced students and practicing engineers who wish to review standard practices as well as to keep abreast of new techniques and practices in the field of structural engineering. The *Handbook* stresses professional applications and includes materials that are presented in an easy-to-read and ready-to-use form. It contains many formulas, tables, and charts that give immediate answers to questions arising from practical work. The book covers not only traditional but also novel and innovative approaches to analysis, design, and rehabilitation problems.

Preface

The primary objective of this new edition of the CRC *Handbook of Structural Engineering* is to provide advanced students and practicing engineers with a useful reference to gain knowledge from and seek solutions to a broad spectrum of structural engineering problems. The myriad of topics covered in this handbook will serve as a good resource for readers to review standard practice and to keep abreast of new developments in the field.

Since the publication of the first edition, a number of new and exciting developments have emerged in the field of structural engineering. Advanced analysis for structural design, performance-based design of earthquake resistant structures, life cycle evaluation, and condition assessment of existing structures, the use of high-performance materials for construction, and design for fire safety are some examples. Likewise, a number of design specifications and codes have been revised by the respective codification committees to reflect our increased understanding of structural behavior. All these developments and changes have been implemented in this new edition. In addition to updating, expanding, and rearranging some of the existing chapters to make the book more informative and cohesive, the following topics have been added to the new edition: fundamental theories of structural dynamics; advanced analysis; wind and earthquake resistant design; design of prestressed concrete, masonry, timber, and glass structures; properties, behavior, and use of high-performance steel, concrete, and fiber-reinforced polymers; semirigid frame structures; life cycle evaluation and condition assessment of existing structures; structural bracing; and structural design for fire safety. The inclusion of these new chapters should enhance the comprehensiveness of the handbook.

For ease of reading, the chapters are divided into six sections. Section I presents fundamental principles of structural analysis for static and dynamic loads. Section II addresses deterministic and probabilistic design theories and describes their applications for the design of structures using different construction materials. Section III discusses high-performance materials and their applications for structural design and rehabilitation. Section IV introduces the principles and practice of seismic and performance-based design of buildings and bridges. Section V is a collection of chapters that address the behavior, analysis, and design of various special structures such as multistory rigid and semirigid frames, short- and long-span bridges, cooling towers, as well as tunnel and glass structures. Section VI is a miscellany of topics of interest to structural engineers. In this section are included materials related to connections, effective length factors, bracing, floor system, fatigue, fracture, passive and active control, life cycle evaluation, condition assessment, and fire safety.

Like its previous edition, this handbook stresses practical applications and emphasizes easy implementations of the materials presented. To avoid lengthy and tedious derivations, many equations, tables, and charts are given in passing without much substantiation. Nevertheless, a succinct discussion of the essential elements is often given to allow readers to gain a better understanding of the underlying theory, and many chapters have extensive reference and reading lists and websites appended at the end for engineers and designers who seek additional or more in-depth information. While all chapters in this handbook are meant to be sufficiently independent of one another, and can be perused without first having proficiency in the materials presented in other chapters, some prerequisite knowledge of the fundamentals of structures is presupposed.

This handbook is the product of a cumulative effort from an international group of academicians and practitioners, who are authorities in their fields, graciously sharing their extensive knowledge and invaluable expertise with the structural engineering profession. The authors of the various chapters in this handbook hail from North America, Europe, and Asia. Their scientific thinking and engineering practice are reflective of the global nature of engineering in general, and structural engineering in particular. Their participation in this project is greatly appreciated. Thanks are also due to Cindy Carelli (acquisitions editor), Jessica Vakili (project coordinator), and the entire production staff of CRC Press for making the process of producing this handbook more enjoyable.

Wai-Fah Chen
Honolulu, HI

Eric M. Lui
Syracuse, NY

The Editors

Wai-Fah Chen is presently dean of the College of Engineering at University of Hawaii at Manoa. He was a George E. Goodwin Distinguished Professor of Civil Engineering and head of the Department of Structural Engineering at Purdue University from 1976 to 1999.

He received his B.S. in civil engineering from the National Cheng-Kung University, Taiwan, in 1959, M.S. in structural engineering from Lehigh University, Pennsylvania, in 1963, and Ph.D. in solid mechanics from Brown University, Rhode Island, in 1966.

Dr. Chen received the Distinguished Alumnus Award from National Cheng-Kung University in 1988 and the Distinguished Engineering Alumnus Medal from Brown University in 1999.

Dr. Chen is the recipient of numerous national engineering awards. Most notably, he was elected to the U.S. National Academy of Engineering in 1995, was awarded the Honorary Membership in the American Society of Civil Engineers in 1997, and was elected to the Academia Sinica (National Academy of Science) in Taiwan in 1998.

A widely respected author, Dr. Chen has authored and coauthored more than 20 engineering books and 500 technical papers. He currently serves on the editorial boards of more than 10 technical journals. He has been listed in more than 30 *Who's Who* publications.

Dr. Chen is the editor-in-chief for the popular 1995 *Civil Engineering Handbook*, the 1997 *Structural Engineering Handbook*, the 1999 *Bridge Engineering Handbook*, and the 2002 *Earthquake Engineering Handbook*. He currently serves as the consulting editor for the McGraw-Hill's *Encyclopedia of Science and Technology*.

He has worked as a consultant for Exxon Production Research on offshore structures, for Skidmore, Owings and Merrill in Chicago on tall steel buildings, for the World Bank on the Chinese University Development Projects, and for many other groups.

Eric M. Lui is currently chair of the Department of Civil and Environmental Engineering at Syracuse University. He received his B.S. in civil and environmental engineering with high honors from the University of Wisconsin at Madison in 1980 and his M.S. and Ph.D. in civil engineering (majoring in structural engineering) from Purdue University, Indiana, in 1982 and 1985, respectively.

Dr. Lui's research interests are in the areas of structural stability, structural dynamics, structural materials, numerical modeling, engineering computations, and computer-aided analysis and design of building and bridge structures. He has authored and coauthored numerous journal papers, conference proceedings, special publications, and research reports in these areas. He is also a contributing author to a number of engineering monographs and handbooks, and is the coauthor of two books on the subject of structural stability. In addition to conducting research, Dr. Lui teaches a variety of undergraduate and graduate courses at Syracuse University. He was a recipient of the College of Engineering and Computer Science Crouse Hinds Award for Excellence in Teaching in 1997. Furthermore, he has served as the faculty advisor of Syracuse University's chapter of the American Society of Civil Engineers (ASCE) for more than a decade and was recipient of the ASCE Faculty Advisor Reward Program from 2001 to 2003.

Dr. Lui has been a longtime member of the ASCE and has served on a number of ASCE publication, technical, and educational committees. He was the associate editor (from 1994 to 1997) and later the book editor (from 1997 to 2000) for the ASCE *Journal of Structural Engineering*. He is also a member of many other professional organizations such as the American Institute of Steel Construction, American Concrete Institute, American Society of Engineering Education, American Academy of Mechanics, and Sigma Xi.

He has been listed in more than 10 *Who's Who* publications and has served as a consultant for a number of state and local engineering firms.

Contributors

T. Balendra
Department of Civil Engineering
National University of Singapore
Singapore

Lawrence C. Bank
Department of Civil and Environmental
  Engineering
University of Wisconsin
Madison, Wisconsin

Reidar Bjorhovde
The Bjorhovde Group
Tucson, Arizona

Brian Brenner
Department of Civil and Environmental
  Engineering
Tufts University
Medford, Massachusetts

Siu-Lai Chan
Department of Civil and Structural
  Engineering
Hong Kong Polytechnic University
Kowloon, Hong Kong

Brian Chen
Wiss, Janney, Elstner Associates, Inc.
Irving, Texas

Wai-Fah Chen
College of Engineering
University of Hawaii at Manoa
Honolulu, Hawaii

Franklin Y. Cheng
Department of Civil Engineering
University of Missouri
Rolla, Missouri

G. F. Dargush
Department of Civil Engineering
State University of New York
Buffalo, New York

Robert J. Dexter
Department of Civil Engineering
University of Minnesota
Minneapolis, Minnesota

J. Daniel Dolan
Department of Civil and Environmental
  Engineering
Washington State University
Pullman, Washington

Lian Duan
Division of Engineering Services
California Department of Transportation
Sacramento, California

Allen C. Estes
Department of Civil and Mechanical Engineering
United States Military Academy
West Point, New York

Dan M. Frangopol
Department of Civil, Environmental, and
  Architectural Engineering
University of Colorado
Boulder, Colorado

Phillip L. Gould
Department of Civil Engineering
Washington University
St. Louis, Missouri

Achintya Haldar
Department of Civil Engineering and
  Engineering Mechanics
The University of Arizona
Tucson, Arizona

Ronald O. Hamburger
Simpson Gumpertz & Heger, Inc.
San Francisco, California

Christian Ingerslev
Parsons Brinckerhoff, Inc.
New York, New York

Manabu Ito
University of Tokyo
Tokyo, Japan

S. E. Kim
Department of Civil Engineering
Sejong University
Seoul, South Korea

Richard E. Klingner
Department of Civil Engineering
University of Texas
Austin, Texas

Wilfried B. Krätzig
Department of Civil Engineering
Ruhr-University Bochum
Bochum, Germany

Yoshinobu Kubo
Department of Civil Engineering
Kyushu Institute of Technology
Tobata, Kitakyushu, Japan

Sashi K. Kunnath
Department of Civil and Environmental
   Engineering
University of California
Davis, California

Tien T. Lan
Institute of Building Structures
Chinese Academy of Building Research
Beijing, China

Andy Lee
Ove Arrup & Partners
   Hong Kong Ltd.
Kowloon, Hong Kong

Zongjin Li
Department of Civil Engineering
Hong Kong University of Science and
   Technology
Kowloon, Hong Kong

J. Y. Richard Liew
Department of Civil Engineering
National University of Singapore
Singapore

Eric M. Lui
Department of Civil and
   Environmental Engineering
Syracuse University
Syracuse, New York

Peter W. Marshall
MHP Systems Engineering
Houston, Texas

Edward G. Nawy
Department of Civil and
   Environmental Engineering
Rutgers University — The State University
   of New Jersey
Piscataway, New Jersey

Austin Pan
T.Y. Lin International
San Francisco, California

Mark Reno
Quincy Engineering
Sacramento, California

Phil Rice
Parsons Brinckerhoff, Inc.
New York, New York

Charles Scawthorn
Department of Urban Management
Kyoto University
Kyoto, Japan

Birger Schmidt (deceased)
Parsons Brinckerhoff, Inc.
New York, New York

N. E. Shanmugam
Department of Civil Engineering
National University of Singapore
Singapore

Maurice L. Sharp
Consultant — Aluminum Structures
Avonmore, Pennsylvania

A. K. W. So
Research Engineering Development
   Façade and Fire Testing
   Consultants Ltd.
Yuen Long, Hong Kong

T. T. Soong
Department of Civil Engineering
State University of New York
Buffalo, New York

Shouji Toma
Department of Civil Engineering
Hokkai-Gakuen University
Sapporo, Japan

Shigeki Unjoh
Ministry of Construction
Public Works Research Institute
Tsukuba, Ibaraki, Japan

Jaw-Nan Wang
Parsons Brinckerhoff, Inc.
New York, New York

Yong C. Wang
School of Aerospace,
    Mechanical and Civil Engineering
The University of Manchester
Manchester, United Kingdom

Lei Xu
Department of Civil Engineering
University of Waterloo
Waterloo, Ontario, Canada

Mark Yashinsky
Division of Structures Design
California Department of Transportation
Sacramento, California

Wei-Wen Yu
Department of Civil Engineering
University of Missouri
Rolla, Missouri

Joseph Yura
Department of Civil Engineering
University of Texas
Austin, Texas

Yunsheng Zhang
Department of Materials Science and Engineering
Southeast University
Nanjing, China

List of Abbreviations

2D	two-dimensional
AASHTO	American Association of State Highway and Transportation Officials
ACI	American Concrete Institute
ACMA	American Composites Manufacturers Association
ADAS	Added damping and stiffness
ADRS	Acceleration-displacement response spectrum
AISC	American Institute of Steel Construction
AISI	American Iron and Steel Institute
ANSI	American National Standards Institute
APA	American Plywood Association
AREMA	American Railway Engineering and Maintenance-of-way Association
ARS	Acceleration response spectra
AS	Aerial spinning
ASCE	American Society of Civil Engineers
ASD	Allowable stress design
ASME	American Society of Mechanical Engineers
ASTM	American Society of Testing and Materials
ATC	Applied Technology Council
AWS	American Welding Society
BBC	Basic Building Code
BIA	Brick Industry Association
BOCA	Building Officials and Code Administrators
BOEF	Beam on elastic foundation approach
BSI	British Standards Institution
BSO	Basic safety objective
BSSC	Building Seismic Safety Council
CABO	Council of American Building
CAFL	Constant-amplitude fatigue limit
CALREL	CAL-RELiability
CBF	Concentrically braced frames
CDF	Cumulative distribution function
CEB	Comité Eurointernationale du Béton
CFA	Composite Fabricators Association

CFM	Continuous filament materials
CFRP	Carbon fiber-reinforced plastic
CGSB	Canadian General Standards Board
CHS	Circular hollow section
CIB	Conseil International du Batiment
CIDECT	Comité International pour le Developement et l'Etude de la Construction Tubulaire
CIDH	Cast-in-drilled-hole
CLT	Classical lamination theory
COV	Coefficient of variation
CQC	Complete-quadratic-combination
CRC	Column Research Council
CS	Condition state
CSA	Canadian Standards Association
CSM	Capacity spectrum method
CTOD	The crack tip opening displacement test
CUREE	Consortium of Universities for Research in Earthquake Engineering
CVN	Charpy V-Notch
DBE	Design basis earthquake
DE	Design earthquake
DEn	Department of Energy
DMM	Deep Mixing Method
DOF	Degree-of-freedom
DOT	Department of Transportation
DSP	Densified small particle
EBF	Eccentrically braced frame
EC3	Eurocode 3
ECCS	European Coal and Steel Community
ECS	European Committee for Standardization
ECSSI	Expanded Clay, Shale and Slate Institute
EDA	Elastic dynamic analysis
EDCH	Eurocomp Design Code and Handbook
EDP	Engineering demand parameter
EDR	Energy dissipating restraint
EDWG	Energy Dissipation Working Group
EERI	Earthquake Engineering Research Institute

ELF	Equivalent lateral force	IMF	Intermediate moment frame
EMC	Equilibrium moisture content	IMI	International Masonry Institute
EMS	European Macroseismic Scale	IO	Immediate occupancy
EOF	End one-flange	IOF	Interior one-flange
EPA	Effective peak acceleration	IRC	Institute for Research in
EPB	Earth pressure balance		Construction
EPTA	European Pultrusion Technology	ISA	Inelastic static analysis
	Association	ISO	International Standard Organization
EPV	Effective peak velocity	ITF	Interior two-flange
ERS	Earthquake resisting system	JMA	Japan Meteorological Agency
ERSA	Elastic response spectrum analysis	JRA	Japan Road Association
ESA	Equivalent static analysis	JSME	Japan Society of Mechanical
ESDU	Engineering Sciences Data Unit		Engineers
ETF	End two-flange	LA	Linear analysis
FCAW	Flux-cored arc welding	LAST	Lowest anticipated service
FCAW-S	Self-shielded flux-cored		temperature
	arc welding	LCADS	Life-Cycle Analysis of Deteriorating
FEE	Functional evaluation earthquake		Structures
FEM	Finite element model	LCR	Locked-coil rope
FEMA	Federal Emergency Management	LDP	Linear dynamic procedure
	Agency	LFRS	Lateral force resisting system
FHWA	Federal Highway Administration	LRFD	Load and resistance factor design
FIP	Federation Internationale de la	LSD	Limit states design
	précontrainte	LSP	Linear static procedure
FORM	First-order reliability method	LVDT	Linear Variable Differential
FOSM	First-order second-moment		Transformer
FPF	First-ply-failure	LVL	Laminated veneer lumber
FRC	Fiber-reinforced concrete	MAE	Mid-America Earthquake Center
FRP	Fiber-reinforced polymer	MCAA	Mason Contractors' Association of
FVD	Fluid viscous damper		America
GMAW	Gas metal arc welding	MCE	Maximum considered earthquake
HAZ	Heat-affected zone	MDA	Market Development Association
HDPE	High-density polyethylene	MDOF	Multi-degree-of-freedom
HOG	House over garage	ME	Maximum earthquake
HPC	High-performance concrete	MIG	Metal arc inert gas welding
HPS	High-performance steel	MLIT	Ministry of Land, Infrastructure and
HSLA	High-strength low-alloy		Transport
HSS	Hollow structural section	MMI	Modified Mercalli Intensity
HVAC	Heating, ventilating, and air	MR	Magnetorheological
	conditioning	MRF	Moment-resisting frame
IBC	International Building Code	MSE	Mechanically stabilized earth
ICBO	International Conference of Building	MSJC	Masonry Standards Joint Committee
	Officials	MVFOSM	Mean value first-order
ICC	International Code Council		second-moment
IDA	Incremental dynamic analysis	NA	Nonlinear analysis
IDARC	Inelastic damage analysis of	NAMC	North American Masonry
	reinforced concrete structure		Conference
IDR	Interstory drift ratios	NCMA	National Concrete Masonry
IIW	International Institute of Welding		Association
ILSS	Interlamina shear strength	NDA	Nonlinear dynamic analysis

NDE	Nondestructive evaluation	SBCC	Southern Building Code Congress
NDP	Non-linear dynamic procedure		
NDS	National design specification	SBCCI	Southern Building Code Congress International
NEHRP	National Earthquake Hazard Reduction Program	SCBF	Special concentrically braced frames
NESSUS	Numerical Evaluation of Stochastic Structures Under Stress	SCC	Self-consolidation concrete
NFPA	National Fire Prevention Association	SCF	Stress concentration factor
NLA	National Lime Association	SCL	Structural composite lumber
NSM	Near-surface-mounted	SDAP	Seismic design and analysis procedure
NSP	Non-linear static procedure		
OCBF	Ordinary concentrically braced frames	SDC	Seismic design category
		SDOF	Single degree-of-freedom
OMF	Ordinary moment frame	SDR	Seismic design requirement
OSB	Oriental strand board	SE	Serviceable earthquake
PAAP	Practical advanced analysis program	SEAOC	Structural Engineers Association of California
PBD	Performance-based design		
PBSE	Performance-based seismic engineering	SEAONC	Structural Engineers Association of Northern California
PCA	Portland Cement Association	SEE	Safety evaluation earthquake
PCI	Prestressed Concrete Institute	SFOBB	San Francisco-Oakland Bay Bridge
PD	Plastic design		
PDF	Probability density function	SHRP	Strategic Highway Research Program
PE	Probability of exceedance		
PEER	Pacific Earthquake Engineering Research Center	SLS	Serviceable limit state
		SMAW	Shielded metal arc welding
PEM	Pseudo-excitation method	SMF	Special moment frame
PGA	Peak ground acceleration	SOE	Support of excavation
PGD	Peak ground displacement	SORM	Second-order reliability method
PGV	Peak ground velocity	SPDM	Structural Plastics Design Manual
PI	Point of inflection	SPL	Seismic performance level
POF	Probability of failure	SRC	Steel and reinforced concrete
PPWS	Prefabricated parallel-wire strand	SRF	Stiffness reduction factor
PROBAN	PROBability ANalysis	SRSS	Square-root-of-the-sum-of-the-squares
PSV	Pseudospectral velocity		
PTI	Post-Tensioning Institute	SSI	Soil-structure interaction
PVC	Polyvinyl chloride	SSRC	Structured Stability Research Council
PWS	Parallel wire strand		
Q&T	Quenching and tempering	STMF	Special truss moment frame
QST	Quenching and self-tempering process	SUG	Seismic use group
		TBM	Tunnel boring machine
RBS	Reduced beam section	TCCMAR	Technical Coordinating Committee for Masonry Research
RBSO	Reliability Based Structural Optimization		
		TERECO	TEaching REliability COncepts
RC	Reinforced concrete	TIG	Tungsten arc inert gas welding
RHS	Rectangular hollow section	TLD	Tuned liquid damper
RMS	Root-mean-square	TMCP	Thermal-mechanical controlled processing
SAW	Submerged arc welding		
SBC	Slotted bolted connection	TMD	Tuned mass damper
SBC	Standard Building Code	TMS	The Masonry Society

TT	Through the thickness	VF	Viscous fluid
UBC	Uniform Building Code	VRT	Variance reduction technique
UDL	Uniformed distributed load	WF	Wide flange
ULS	Ultimate limit state	WRF	Wave reflection factor
URM	Unreinforced masonry	WSMF	Welded special moment-frame
USDA	US Department of Agriculture	WUF-W	Welded-unreinforced flange, welded web
USGS	US Geological Survey		
VE	Viscoelastic	ZPA	Zero period acceleration

Contents

SECTION IV Earthquake Engineering and Design

SECTION V Special Structures

SECTION VI Special Topics

Structural Analysis

I

I

Structural Analysis

1

Structural Fundamentals

1.1 Stresses 1-1

Eric M. Lui
*Department of Civil and
Environmental Engineering,
Syracuse University,
Syracuse, NY*

1.1   Stresses

1.1.1   Stress Components and Tractions

Consider an infinitesimal parallelepiped element shown in Figure 1.1. The state of stress of this element is defined by nine stress components or tensors ($\sigma_{11}, \sigma_{12}, \sigma_{13}, \sigma_{21}, \sigma_{22}, \sigma_{23}, \sigma_{31}, \sigma_{32}$, and σ_{33}), of which six ($\sigma_{11}, \sigma_{22}, \sigma_{33}, \sigma_{12} = \sigma_{21}, \sigma_{23} = \sigma_{32}$, and $\sigma_{13} = \sigma_{31}$) are independent. The stress components that act normal to the planes of the parallelepiped ($\sigma_{11}, \sigma_{22}, \sigma_{33}$) are called normal stresses, and the stress components that act tangential to the planes of the parallelepiped ($\sigma_{12} = \sigma_{21}, \sigma_{23} = \sigma_{32}, \sigma_{13} = \sigma_{31}$) are called shear stresses. The first subscript of each stress component refers to the face on which the stress acts, and the second subscript refers to the direction in which the stress acts. Thus, σ_{ij} represents a stress acting on the i face in the j direction. A face is considered positive if a unit vector drawn *perpendicular* to the face directing outward from the inside of the element is pointing in the positive direction as defined

0-8493-1569-7/05/$0.00+$1.50
© 2005 by CRC Press

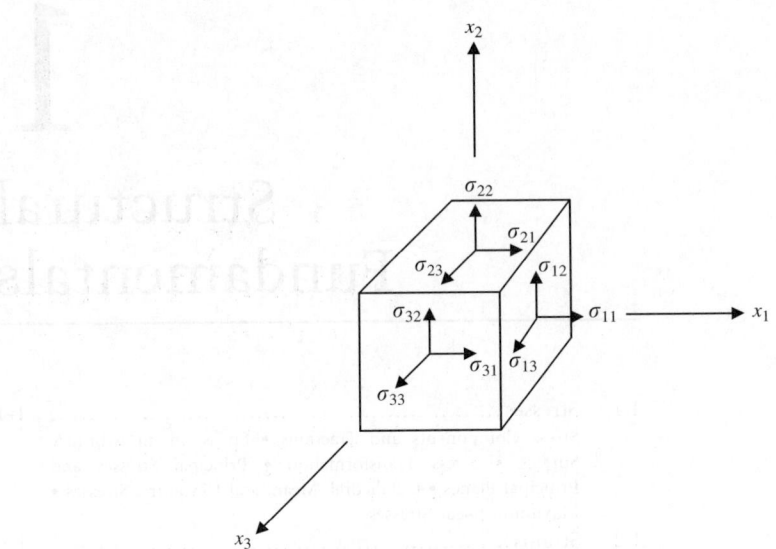

FIGURE 1.1   Stress components acting on the positive faces of a parallelepiped element.

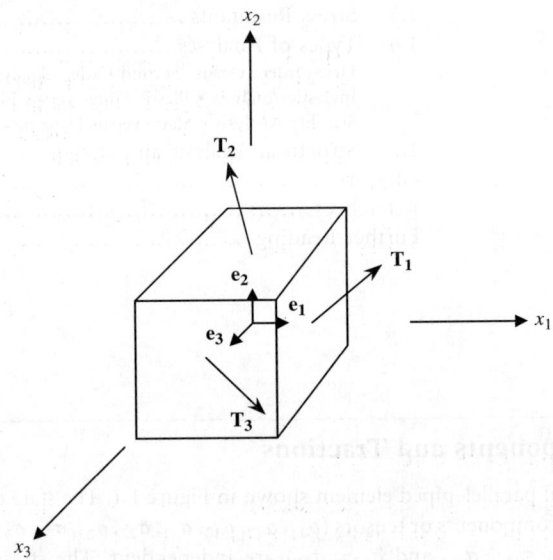

FIGURE 1.2   Tractions acting on the positive faces of a parallelepiped element.

by the Cartesian coordinate system (x_1, x_2, x_3). A stress is considered positive if it acts on a positive face in the positive direction or if it acts on a negative face in the negative direction. It is considered negative if it acts on a positive face in the negative direction or if it acts on a negative face in the positive direction.

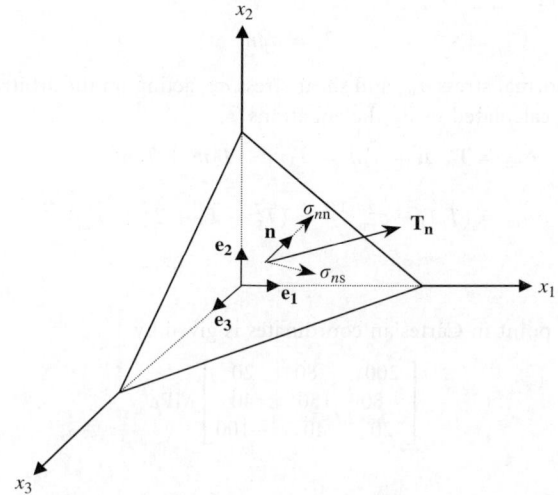

FIGURE 1.3 Traction and stresses acting on an arbitrary plane.

The vectorial sum of the three stress components acting on each face of the parallelepiped produces a traction **T**. Thus, the tractions acting on the three positive faces of the element shown in Figure 1.2 are given by

$$\mathbf{T_1} = \sigma_{11}\mathbf{e_1} + \sigma_{12}\mathbf{e_2} + \sigma_{13}\mathbf{e_3}$$
$$\mathbf{T_2} = \sigma_{21}\mathbf{e_1} + \sigma_{22}\mathbf{e_2} + \sigma_{23}\mathbf{e_3} \tag{1.1}$$
$$\mathbf{T_3} = \sigma_{31}\mathbf{e_1} + \sigma_{32}\mathbf{e_2} + \sigma_{33}\mathbf{e_3}$$

where $\mathbf{e_1}$, $\mathbf{e_2}$, and $\mathbf{e_3}$ are unit vectors corresponding to the x_1, x_2, and x_3 axes, respectively.

Equations 1.1 can be written in tensor or indicial notation as

$$\mathbf{T}_i = \sigma_{ij}\mathbf{e}_j \tag{1.2}$$

Note that both indices (i and j) range from 1 to 3. The dummy index (j in the above equation) denotes summation.

Using Cauchy's definition (Bathe 1982), traction is regarded as the intensity of a force resultant acting on an infinitesimal area. Mathematically, it is expressed as

$$T_i = \frac{dF_i}{dA_i} \tag{1.3}$$

1.1.2 Stress on an Arbitrary Surface

If the tractions acting on three orthogonal faces of a volume element are known, or calculated using Equations 1.1, the traction $\mathbf{T_n}$ acting on any arbitrary surface as defined by a unit normal vector **n** ($= n_1\mathbf{e_1} + n_2\mathbf{e_2} + n_3\mathbf{e_3}$) as shown in Figure 1.3 can be written as

$$\mathbf{T_n} = T_1\mathbf{e_1} + T_2\mathbf{e_2} + T_3\mathbf{e_3} \tag{1.4}$$

where T_1, T_2, and T_3 are the components of $\mathbf{T_n}$ acting in the 1, 2, and 3 directions, respectively, of the Cartesian coordinate system shown. They can be calculated using Cauchy's formulas:

$$T_1 = \sigma_{11}n_1 + \sigma_{21}n_2 + \sigma_{31}n_3$$
$$T_2 = \sigma_{12}n_1 + \sigma_{22}n_2 + \sigma_{32}n_3 \tag{1.5}$$
$$T_3 = \sigma_{13}n_1 + \sigma_{23}n_2 + \sigma_{33}n_3$$

or using indicial notation:

$$T_i = \sigma_{ji} n_j \qquad (1.6)$$

Once $\mathbf{T_n}$ is known, the normal stress σ_{nn} and shear stress σ_{ns} acting on the arbitrary plane as defined by the unit vector $\mathbf{n}$ can be calculated using the equations

$$\sigma_{nn} = \mathbf{T_n} \cdot \mathbf{n} = T_i n_i = T_1 n_1 + T_2 n_2 + T_3 n_3 \qquad (1.7)$$

$$\sigma_{ns} = (T_i T_i - \sigma_{nn}^2)^{1/2} = (T_1^2 + T_2^2 + T_3^2 - \sigma_{nn}^2)^{1/2} \qquad (1.8)$$

EXAMPLE 1.1

If the state of stress at a point in Cartesian coordinates is given by

$$\begin{bmatrix} 200 & -80 & 20 \\ -80 & 150 & 40 \\ 20 & 40 & -100 \end{bmatrix} \text{MPa}$$

Determine:

1. The traction that acts on a plane with unit normal vector $\mathbf{n} = \frac{1}{2}\mathbf{e}_1 + \frac{1}{2}\mathbf{e}_2 + \frac{1}{\sqrt{2}}\mathbf{e}_3$
2. The normal stress and shear stress that act on this plane

   Solution

1. The components of traction that act on the specified plane can be calculated using Equation 1.6:

$$
\begin{aligned}
T_1 &= \sigma_{11} n_1 + \sigma_{21} n_2 + \sigma_{31} n_3 \\
&= (200)\left(\tfrac{1}{2}\right) + (-80)\left(\tfrac{1}{2}\right) + (20)\left(\tfrac{1}{\sqrt{2}}\right) \\
&= 74.1 \text{ MPa} \\
T_2 &= \sigma_{12} n_1 + \sigma_{22} n_2 + \sigma_{32} n_3 \\
&= (-80)\left(\tfrac{1}{2}\right) + (150)\left(\tfrac{1}{2}\right) + (40)\left(\tfrac{1}{\sqrt{2}}\right) \\
&= 63.3 \text{ MPa} \\
T_3 &= \sigma_{13} n_1 + \sigma_{23} n_2 + \sigma_{33} n_3 \\
&= (20)\left(\tfrac{1}{2}\right) + (40)\left(\tfrac{1}{2}\right) + (-100)\left(\tfrac{1}{\sqrt{2}}\right) \\
&= -40.7 \text{ MPa}
\end{aligned}
$$

   From Equation 1.4, the traction acting on the specified plane is

$$\mathbf{T_n} = 74.1\mathbf{e}_1 + 63.3\mathbf{e}_2 - 40.7\mathbf{e}_3$$

2. The normal and shear stresses acting on the plane can be calculated from Equations 1.7 and 1.8, respectively,

$$
\begin{aligned}
\sigma_{nn} &= T_1 n_1 + T_2 n_2 + T_3 n_3 \\
&= (74.1)\left(\tfrac{1}{2}\right) + (63.3)\left(\tfrac{1}{2}\right) + (-40.7)\left(\tfrac{1}{\sqrt{2}}\right) \\
&= 40 \text{ MPa} \\
\sigma_{ns} &= (T_1^2 + T_2^2 + T_3^2 - \sigma_{nn}^2)^{1/2} \\
&= \sqrt{(74.1)^2 + (63.3)^2 + (-40.7)^2 - (40)^2} \\
&= 97.7 \text{ MPa}
\end{aligned}
$$

1.1.3 Stress Transformation

If the state of stress acting on an infinitesimal volume element corresponding to a Cartesian coordinate system $(x_1 - x_2 - x_3)$ as shown in Figure 1.1 is known, the state of stress on the element with respect to another Cartesian coordinate system $(x_1' - x_2' - x_3')$ can be calculated using the tensor equation

$$\sigma_{ij}' = l_{ik} l_{jl} \sigma_{kl} \tag{1.9}$$

where l is the direction cosine of two axes (one corresponding to the new and the other corresponding to the original). For instance,

$$l_{ik} = \cos(i', k), \quad l_{jl} = \cos(j', l) \tag{1.10}$$

represent the cosine of the angle formed by the new $(i'$ or $j')$ and the original $(k$ or $l)$ axes.

1.1.4 Principal Stresses and Principal Planes

Principal stresses are normal stresses that act on planes where the shear stresses are zero. Principal planes are planes on which principal stresses act. Principal stresses are calculated from the equation

$$\det \begin{vmatrix} \sigma_{11} - \sigma & \sigma_{12} & \sigma_{13} \\ \sigma_{12} & \sigma_{22} - \sigma & \sigma_{23} \\ \sigma_{13} & \sigma_{23} & \sigma_{33} - \sigma \end{vmatrix} \tag{1.11}$$

which, upon expansion, gives a cubic equation in σ:

$$\sigma^3 - I_1 \sigma^2 - I_2 \sigma - I_3 = 0 \tag{1.12}$$

where I_1, I_2, and I_3 are the first, second, and third stress invariants (their magnitudes remain unchanged regardless of the choice of the Cartesian coordinate axes) given by

$$\begin{aligned} I_1 &= \sigma_{11} + \sigma_{22} + \sigma_{33} \\ I_2 &= -\det \begin{vmatrix} \sigma_{11} & \sigma_{12} \\ \sigma_{12} & \sigma_{22} \end{vmatrix} - \det \begin{vmatrix} \sigma_{11} & \sigma_{13} \\ \sigma_{13} & \sigma_{33} \end{vmatrix} - \det \begin{vmatrix} \sigma_{22} & \sigma_{23} \\ \sigma_{23} & \sigma_{33} \end{vmatrix} \\ I_3 &= \det \begin{vmatrix} \sigma_{11} & \sigma_{12} & \sigma_{13} \\ \sigma_{12} & \sigma_{22} & \sigma_{23} \\ \sigma_{13} & \sigma_{23} & \sigma_{33} \end{vmatrix} \end{aligned} \tag{1.13}$$

The three roots of Equation 1.12, herein denoted as σ_{P1}, σ_{P2}, and σ_{P3}, are the principal stresses acting on the three orthogonal planes. The components of a unit vector that defines the principal plane (i.e., n_{1Pi}, n_{2Pi}, n_{3Pi}) corresponding to a specific principal stress σ_{Pi} (with $i = 1, 2, 3$) can be evaluated using any two of the following equations:

$$\begin{aligned} n_{1Pi}(\sigma_{11} - \sigma_{Pi}) + n_{2Pi}\sigma_{12} + n_{3Pi}\sigma_{13} &= 0 \\ n_{1Pi}\sigma_{12} + n_{2Pi}(\sigma_{22} - \sigma_{Pi}) + n_{3Pi}\sigma_{23} &= 0 \\ n_{1Pi}\sigma_{13} + n_{2Pi}\sigma_{23} + n_{3Pi}(\sigma_{33} - \sigma_{Pi}) &= 0 \end{aligned} \tag{1.14}$$

and

$$n_{1Pi}^2 + n_{2Pi}^2 + n_{3Pi}^2 = 1 \tag{1.15}$$

The unit vector calculated for each value of σ_{Pi} represents the direction of a *principal axis*. Thus, three principal axes that correspond to the three principal planes can be identified.

Note that the three stress invariants in Equations 1.13 can also be written in terms of the principal stresses:

$$I_1 = \sigma_{P1} + \sigma_{P2} + \sigma_{P3}$$
$$I_2 = -\sigma_{P1}\sigma_{P2} - \sigma_{P2}\sigma_{P3} - \sigma_{P1}\sigma_{P3} \qquad (1.16)$$
$$I_3 = \sigma_{P1}\sigma_{P2}\sigma_{P3}$$

EXAMPLE 1.2

Suppose a plane stress condition exists, derive the equations for (1) stress transformation, (2) principal stresses, and (3) principal planes for this condition.

Solution

1. *Stress transformation.* With reference to Figure 1.4, a direct application of Equation 1.9, with the condition $\sigma_{33} = \sigma_{23} = \sigma_{13} = 0$ applying to a plane stress condition, gives the following stress transformation equations:

$$\sigma'_{11} = \sigma_{11}\cos^2\theta + \sigma_{22}\cos^2(90° - \theta) + \sigma_{12}\cos\theta\cos(90° - \theta) + \sigma_{21}\cos(90° - \theta)\cos\theta$$
$$\sigma'_{22} = \sigma_{11}\cos^2(90° + \theta) + \sigma_{22}\cos^2\theta + \sigma_{12}\cos(90° + \theta)\cos\theta + \sigma_{21}\cos\theta\cos(90° + \theta)$$
$$\sigma'_{12} = \sigma_{11}\cos\theta\cos(90° + \theta) + \sigma_{22}\cos(90° - \theta)\cos\theta + \sigma_{12}\cos^2\theta + \sigma_{21}\cos(90° - \theta)\cos(90° + \theta)$$

Using the trigonometric identities

$$\cos(90° - \theta) = \sin\theta, \quad \cos(90° + \theta) = -\sin\theta,$$
$$\sin^2\theta = \frac{1 - \cos 2\theta}{2}, \quad \cos^2\theta = \frac{1 + \cos 2\theta}{2}, \quad \sin\theta\cos\theta = \frac{\sin 2\theta}{2}$$

the stress transformation equations can be expressed as

$$\sigma'_{11} = \left(\frac{\sigma_{11} + \sigma_{22}}{2}\right) + \left(\frac{\sigma_{11} - \sigma_{22}}{2}\right)\cos 2\theta + \sigma_{12}\sin 2\theta$$
$$\sigma'_{22} = \left(\frac{\sigma_{11} + \sigma_{22}}{2}\right) - \left(\frac{\sigma_{11} - \sigma_{22}}{2}\right)\cos 2\theta - \sigma_{12}\sin 2\theta$$
$$\sigma'_{12} = -\left(\frac{\sigma_{11} - \sigma_{22}}{2}\right)\sin 2\theta + \sigma_{12}\cos 2\theta$$

which are the familiar two-dimensional (2-D) stress transformation equations found in a number of introductory mechanics of materials books (see, e.g., Beer et al. 2001; Gere 2004).

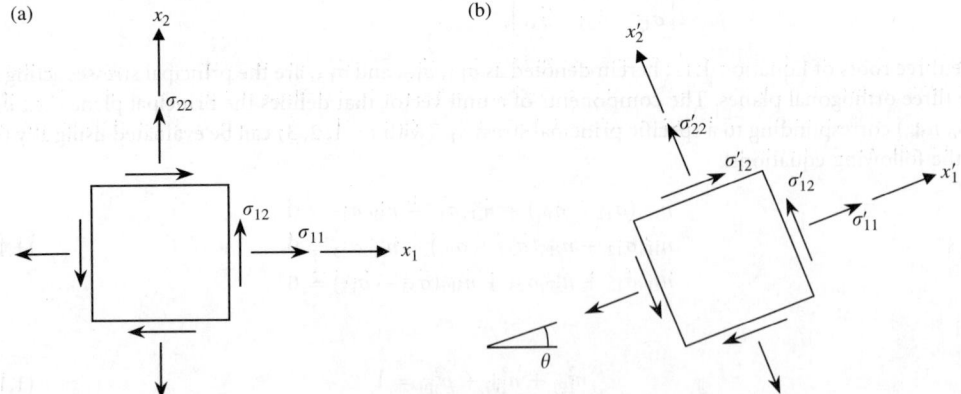

FIGURE 1.4 Two-dimensional (2-D) stress transformation: (a) original state of stress acting on a 2-D infinitesimal element and (b) transformed state of stress acting on a 2-D infinitesimal element.

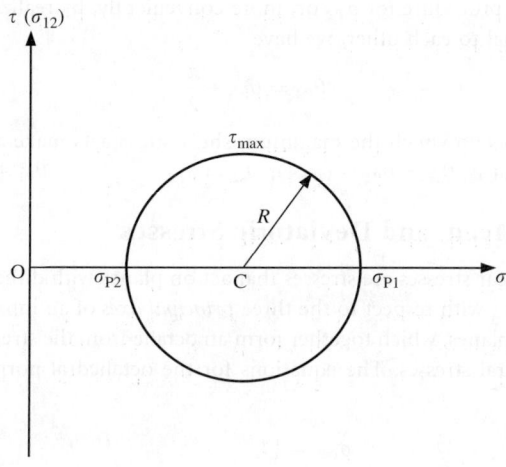

FIGURE 1.5 Mohr's circle.

2. *Principal stresses.* For plane stress condition, Equation 1.11 becomes

$$\det \begin{vmatrix} \sigma_{11} - \sigma & \sigma_{12} \\ \sigma_{12} & \sigma_{22} - \sigma \end{vmatrix}$$

which, upon expansion, gives the quadratic equation

$$\sigma^2 - (\sigma_{11} + \sigma_{22})\sigma + (\sigma_{11}\sigma_{22} - \sigma_{12}^2) = 0$$

The two solutions to the above equation give the two principal stresses as

$$\sigma_{P1} = \frac{\sigma_{11} + \sigma_{22}}{2} + \sqrt{\left(\frac{\sigma_{11} - \sigma_{22}}{2}\right)^2 + \sigma_{12}^2}$$

$$\sigma_{P2} = \frac{\sigma_{11} + \sigma_{22}}{2} - \sqrt{\left(\frac{\sigma_{11} - \sigma_{22}}{2}\right)^2 + \sigma_{12}^2}$$

Note that these stresses represent the rightmost and leftmost points on a *Mohr circle* (Beer et al. 2001), shown in Figure 1.5, with $OC = (\sigma_{11} + \sigma_{22})/2$ and $R = \sqrt{[((\sigma_{11} - \sigma_{22})/2)^2 + \sigma_{12}^2]}$. (Although not asked for in this example, it can readily be seen that the maximum shear stress is the uppermost point on the Mohr circle given by $\tau_{max} = (\sigma_{12})_{max} = R = \sqrt{[((\sigma_{11} - \sigma_{22})/2)^2 + \sigma_{12}^2]}$.)

3. *Principal planes.* Substituting the equation for σ_{P1} into

$$n_{1P1}(\sigma_{11} - \sigma) + n_{2P1}\sigma_{12} = 0$$

and recognizing that

$$n_{1P1}^2 + n_{2P1}^2 = 1$$

it can be shown that the principal plane on which σ_{P1} acts forms an angle $\theta_{P1} = \tan^{-1}(n_{2P1}/n_{1P1})$ with the x_1 (or $x-$) axis and is given by

$$\theta_{P1} = \frac{1}{2}\tan^{-1}\left[\frac{\sigma_{12}}{(\sigma_{11} - \sigma_{22})/2}\right]$$

Following the same procedure for σ_{P2} or, more conveniently, by realizing that the two principal planes are orthogonal to each other, we have

$$\theta_{P2} = \theta_{P1} + \frac{\pi}{2}$$

(Note that the planes on which the maximum shear stress acts make an angle cf $\pm 45°$ with the principal planes, that is, $\theta_{s1} = \theta_{P1} - (\pi/4)$, $\theta_{s2} = \theta_{P2} - (\pi/4) = \theta_{P1} + (\pi/4)$.)

1.1.5 Octahedral, Mean, and Deviatoric Stresses

Octahedral normal and shear stresses are stresses that act on planes with direction indices satisfying the condition $n_1^2 = n_2^2 = n_3^2 = \frac{1}{3}$ with respect to the three *principal axes* of an infinitesimal volume element. Since there are eight such planes, which together form an octahedron, the stresses acting on these planes are referred to as octahedral stresses. The equations for the octahedral normal and shear stresses are given by

$$\sigma_{oct} = \tfrac{1}{3} I_1$$
$$\tau_{oct} = \tfrac{1}{3}\sqrt{2I_1^2 + 6I_2} \qquad (1.17)$$

where I_1 and I_2 are the first and second stress invariants defined in Equations 1.13 or in Equations 1.16. Octahedral stresses are used to define certain failure criteria (e.g., von Mises) for ductile materials.

Mean stress is obtained as the arithmetic average of three normal stresses (or the three principal stresses):

$$\sigma_m = \tfrac{1}{3}(\sigma_{11} + \sigma_{22} + \sigma_{33}) = \tfrac{1}{3}(\sigma_{P1} + \sigma_{P2} + \sigma_{P3}) = \tfrac{1}{3}I_1 \qquad (1.18)$$

Deviatoric stress is defined by the stress tensor

$$\begin{bmatrix} \dfrac{2\sigma_{11} - \sigma_{22} - \sigma_{33}}{3} & \sigma_{12} & \sigma_{13} \\[2mm] \sigma_{12} & \dfrac{2\sigma_{22} - \sigma_{11} - \sigma_{33}}{3} & \sigma_{23} \\[2mm] \sigma_{13} & \sigma_{23} & \dfrac{2\sigma_{33} - \sigma_{11} - \sigma_{22}}{3} \end{bmatrix} \qquad (1.19)$$

The deviatoric stress tensor represents a state of *pure shear*. It is obtained by subtracting the mean stress from the three normal stresses (σ_{11}, σ_{22}, and σ_{33}) in a stress tensor. It is important from the viewpoint of inelastic analysis because experiments have shown that inelastic behavior of most ductile materials is independent of the mean normal stress, but is related primarily to the deviatoric stress.

If the indicial notation s_{ij} is used to represent the nine deviatoric stress components given in Equation 1.19, the maximum deviatoric stress acting on each of the three orthogonal planes (which are the same as the principal planes) can be computed from the cubic equation

$$s^3 - J_1 s^2 - J_2 s - J_3 = 0 \qquad (1.20)$$

where J_1, J_2, and J_3 are the first, second, and third deviatoric stress invariants given by

$$J_1 = s_{ii} = s_{11} + s_{22} + s_{33} = 0$$
$$J_2 = \tfrac{1}{2} s_{ij} s_{ji} = \tfrac{1}{2}(s_{11}^2 + s_{22}^2 + s_{33}^2 + 2s_{12}^2 + 2s_{23}^2 + 2s_{13}^2)$$
$$J_3 = \tfrac{1}{3} s_{ij} s_{jk} s_{ki} = \det \begin{vmatrix} s_{11} & s_{12} & s_{13} \\ s_{12} & s_{22} & s_{23} \\ s_{13} & s_{23} & s_{33} \end{vmatrix} \qquad (1.21)$$

Alternatively, if the principal stresses are known, the three maximum deviatoric stresses can be calculated using the equations

$$s_{P1} = \frac{(\sigma_{P1} - \sigma_{P3}) + (\sigma_{P1} - \sigma_{P2})}{3}$$

$$s_{P2} = \frac{(\sigma_{P2} - \sigma_{P3}) + (\sigma_{P2} - \sigma_{P1})}{3}$$ (1.22)

$$s_{P3} = \frac{(\sigma_{P3} - \sigma_{P1}) + (\sigma_{P3} - \sigma_{P2})}{3}$$

Note that J_1, J_2, and J_3 can also be expressed in terms of I_1, I_2, and I_3, or the three maximum deviatoric stresses, as follows:

$$J_1 = s_{P1} + s_{P2} + s_{P3} = 0$$

$$J_2 = I_2 + \tfrac{1}{3}I_1^2 = \tfrac{1}{2}(s_{P1}^2 + s_{P2}^2 + s_{P3}^2)$$ (1.23)

$$J_3 = I_3 + \tfrac{1}{3}I_1 I_2 + \tfrac{2}{27}I_1^3 = \tfrac{1}{3}(s_{P1}^3 + s_{P2}^3 + s_{P3}^3) = s_{P1}s_{P2}s_{P3}$$

1.1.6 Maximum Shear Stresses

If the principal stresses are known, the maximum shear stresses that act on each of the three orthogonal planes, which bisect the angle between the principal planes with direction indices ($n_1 = \pm 1/\sqrt{2}$, $n_2 = \pm 1/\sqrt{2}$, $n_3 = 0$), ($n_1 = 0$, $n_2 = \pm 1/\sqrt{2}$, $n_3 = \pm 1/\sqrt{2}$), ($n_1 = \pm 1/\sqrt{2}$, $n_2 = 0$, $n_3 = \pm 1/\sqrt{2}$) with respect to the principal axes, are given by

$$\tau_{\max 1} = \tfrac{1}{2}|\sigma_{P1} - \sigma_{P2}|$$

$$\tau_{\max 2} = \tfrac{1}{2}|\sigma_{P2} - \sigma_{P3}|$$ (1.24)

$$\tau_{\max 3} = \tfrac{1}{2}|\sigma_{P1} - \sigma_{P3}|$$

Note that the planes (called principal shear planes) on which these stresses act are *not* pure shear planes. The corresponding normal stresses that act on these principal shear planes are $(\sigma_{P1} + \sigma_{P2})/2$, $(\sigma_{P2} + \sigma_{P3})/2$, and $(\sigma_{P1} + \sigma_{P3})/2$, respectively.

1.2 Strains

1.2.1 Strain Components

Corresponding to the six stress components described in the preceding section are six strain components. With reference to a Cartesian coordinate system with axes labeled 1, 2, and 3 as in Figure 1.1, these strains are denoted as ε_{11}, ε_{22}, ε_{33}, $\gamma_{12} = 2\varepsilon_{12}$, $\gamma_{23} = 2\varepsilon_{23}$, and $\gamma_{31} = 2\varepsilon_{31}$. ε_{11}, ε_{22}, and ε_{33} are called normal strains and γ_{12}, γ_{23}, and γ_{31} are called shear strains. Using the definitions for *engineering* strains (Bathe 1982), normal strain is defined as the ratio of the change in length to the original length of a straight line element, and shear strain is defined as the change in angle (when the element is in a strained state) from an originally right angle (when the element is in an unstrained state).

1.2.2 Strain–Displacement Relationships

If we denote u, v, and w as the translational displacements in the 1, 2, and 3 (or x, y, and z) directions, respectively, then according to the small-displacement theory the six engineering strain components can be written in terms of these displacements as

$$\varepsilon_{11} = \frac{\partial u}{\partial x}, \quad \varepsilon_{22} = \frac{\partial v}{\partial y}, \quad \varepsilon_{33} = \frac{\partial w}{\partial z},$$

$$\gamma_{12} = 2\varepsilon_{12} = \frac{\partial u}{\partial y} + \frac{\partial v}{\partial x}, \quad \gamma_{23} = 2\varepsilon_{23} = \frac{\partial v}{\partial z} + \frac{\partial w}{\partial y}, \quad \gamma_{31} = 2\varepsilon_{31} = \frac{\partial w}{\partial x} + \frac{\partial u}{\partial z}$$ (1.25)

1.2.3 Strain Analysis

Like stresses, strains can be transformed from one Cartesian coordinate system to another. This can be done by replacing σ by ε in Equation 1.9. In addition, one can calculate the three principal strains (ε_{P1}, ε_{P2}, and ε_{P3}) and the maximum shear strains ($\varepsilon_{max\,1}$, $\varepsilon_{max\,2}$, and $\varepsilon_{max\,3}$) acting on the three orthogonal planes by following the same procedure outlined in the preceding section for stresses simply by replacing all occurrences of σ by ε in Equations 1.12 and 1.24. However, it should be noted that except for isotropic elastic materials, principal planes for stresses and principal planes for strains do not necessarily coincide, nor do the planes of maximum shear stresses and maximum shear strains.

1.3 Equilibrium and Compatibility

By using an infinitesimal parallelepiped element subject to a system of positive three-dimensional stresses, equilibrium of the element requires that the following three equations relating the stresses be satisfied (Wang 1953; Timoshenko and Goodier 1970):

$$\frac{\partial \sigma_{11}}{\partial x} + \frac{\partial \sigma_{12}}{\partial y} + \frac{\partial \sigma_{13}}{\partial z} + B_x = 0$$

$$\frac{\partial \sigma_{12}}{\partial x} + \frac{\partial \sigma_{22}}{\partial y} + \frac{\partial \sigma_{23}}{\partial z} + B_y = 0 \qquad (1.26)$$

$$\frac{\partial \sigma_{13}}{\partial x} + \frac{\partial \sigma_{23}}{\partial y} + \frac{\partial \sigma_{33}}{\partial z} + B_z = 0$$

where B_x, B_y, and B_z are the body forces per unit volume acting in the 1, 2, and 3 (or x, y, and z) directions, respectively.

According to Equations 1.25, the six strain components can be expressed in terms of just three displacement variables (u, v, and w). To obtain a unique solution for the displacements for a given loading condition, these strains must be related. By manipulating Equations 1.25, it can be shown (Wang 1953; Timoshenko and Goodier 1970) that the strains are related by the following compatibility equations:

$$\frac{\partial^2 \varepsilon_{11}}{\partial y^2} + \frac{\partial^2 \varepsilon_{22}}{\partial x^2} = 2\frac{\partial^2 \varepsilon_{12}}{\partial x \partial y}$$

$$\frac{\partial^2 \varepsilon_{22}}{\partial z^2} + \frac{\partial^2 \varepsilon_{33}}{\partial y^2} = 2\frac{\partial^2 \varepsilon_{23}}{\partial y \partial z}$$

$$\frac{\partial^2 \varepsilon_{11}}{\partial z^2} + \frac{\partial^2 \varepsilon_{33}}{\partial x^2} = 2\frac{\partial^2 \varepsilon_{13}}{\partial x \partial z}$$

$$\frac{\partial^2 \varepsilon_{11}}{\partial y \partial z} = -\frac{\partial^2 \varepsilon_{23}}{\partial x^2} + \frac{\partial^2 \varepsilon_{13}}{\partial x \partial y} + \frac{\partial^2 \varepsilon_{12}}{\partial x \partial z} \qquad (1.27)$$

$$\frac{\partial^2 \varepsilon_{22}}{\partial x \partial z} = \frac{\partial^2 \varepsilon_{23}}{\partial x \partial y} - \frac{\partial^2 \varepsilon_{13}}{\partial y^2} + \frac{\partial^2 \varepsilon_{12}}{\partial y \partial z}$$

$$\frac{\partial^2 \varepsilon_{33}}{\partial x \partial y} = \frac{\partial^2 \varepsilon_{23}}{\partial x \partial z} + \frac{\partial^2 \varepsilon_{13}}{\partial y \partial z} - \frac{\partial^2 \varepsilon_{12}}{\partial z^2}$$

Since Equations 1.27 were derived from Equations 1.25, they should not be regarded as an independent set of equations. The six stress components (σ_{11}, σ_{22}, σ_{33}, σ_{12}, σ_{23}, and σ_{13}), the six strain components (ε_{11}, ε_{22}, ε_{33}, ε_{12}, ε_{23}, and ε_{13}), and the three displacement components (u, v, and w) constitute a total of 15 unknowns, which cannot be solved using the three equilibrium equations (Equations 1.26) and the six compatibility equations (Equations 1.27). To do so, six additional equations are needed. These equations, which relate stresses with strains, are described in the next section.

1.4 Stress–Strain Relationship

Stress–strain (or constitutive) relationship defines how a material behaves when subjected to applied loads. Depending on the type of material and the magnitude of the applied loads, a material may behave elastically or inelastically. A material is said to behave elastically when loading and unloading follow the same path and no permanent deformation occurs upon full unloading (see Figure 1.6, Paths 1 and 2). A material is said to behave inelastically when loading and unloading do not follow the same path and permanent deformation results upon full unloading (see Figure 1.6, Paths 1 and 3). A material that behaves elastically may be further classified as linear or nonlinear, depending on whether Paths 1 and 2 in Figure 1.6 are linear or nonlinear. If the properties of a material are independent of location in the material, the material is said to be homogenous. Moreover, depending on the directional effect of the mechanical properties exhibited by a material, terms such as isotropic, orthotropic, monoclinic, or anisotropic can also be used to describe a material.

1.4.1 Linear Elastic Behavior

If the material is anisotropic (i.e., no plane of symmetry exists for the material properties), the six stress components are related to the six strain components by 21 independent material constants (D_{ij} in the following matrix equation):

$$
\begin{Bmatrix} \sigma_{11} \\ \sigma_{22} \\ \sigma_{33} \\ \sigma_{12} \\ \sigma_{23} \\ \sigma_{13} \end{Bmatrix} =
\begin{bmatrix}
D_{11} & D_{12} & D_{13} & D_{14} & D_{15} & D_{16} \\
 & D_{22} & D_{23} & D_{24} & D_{25} & D_{26} \\
 & & D_{33} & D_{34} & D_{35} & D_{36} \\
 & & & D_{44} & D_{45} & D_{46} \\
 & \text{sym.} & & & D_{55} & D_{56} \\
 & & & & & D_{66}
\end{bmatrix}
\begin{Bmatrix} \varepsilon_{11} \\ \varepsilon_{22} \\ \varepsilon_{33} \\ \varepsilon_{12} \\ \varepsilon_{23} \\ \varepsilon_{13} \end{Bmatrix}
\tag{1.28}
$$

If the material is monoclinic (i.e., material properties are symmetric about one plane), the number of independent material constants reduces to 13. For instance, if the plane defined by the x_2–x_3 (or y–z)

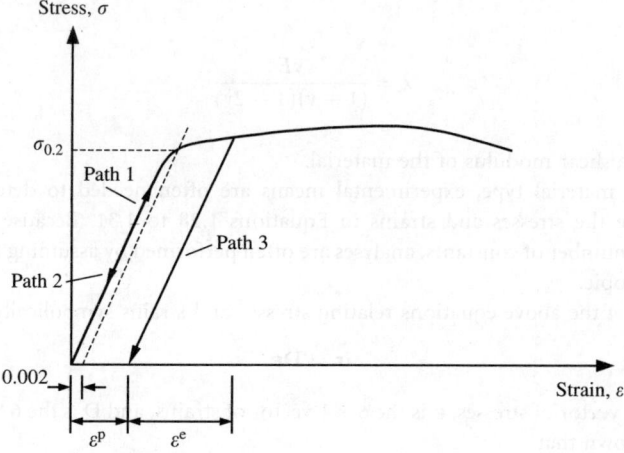

FIGURE 1.6 Uniaxial stress–strain curve.

axes is the plane of symmetry, the stress–strain relationship takes the form

$$
\begin{Bmatrix} \sigma_{11} \\ \sigma_{22} \\ \sigma_{33} \\ \sigma_{12} \\ \sigma_{23} \\ \sigma_{13} \end{Bmatrix} =
\begin{bmatrix}
D_{11} & D_{12} & D_{13} & 0 & D_{15} & 0 \\
 & D_{22} & D_{23} & 0 & D_{25} & 0 \\
 & & D_{33} & 0 & D_{35} & 0 \\
 & & & D_{44} & 0 & D_{46} \\
 & \text{sym.} & & & D_{55} & 0 \\
 & & & & & D_{66}
\end{bmatrix}
\begin{Bmatrix} \varepsilon_{11} \\ \varepsilon_{22} \\ \varepsilon_{33} \\ \varepsilon_{12} \\ \varepsilon_{23} \\ \varepsilon_{13} \end{Bmatrix}
\tag{1.29}
$$

If the material is orthotropic (i.e., material properties are symmetric about two planes), the number of independent material constants further reduces to 9, and the stress–strain relationship takes the form

$$
\begin{Bmatrix} \sigma_{11} \\ \sigma_{22} \\ \sigma_{33} \\ \sigma_{12} \\ \sigma_{23} \\ \sigma_{13} \end{Bmatrix} =
\begin{bmatrix}
D_{11} & D_{12} & D_{13} & 0 & 0 & 0 \\
 & D_{22} & D_{23} & 0 & 0 & 0 \\
 & & D_{33} & 0 & 0 & 0 \\
 & & & D_{44} & 0 & 0 \\
 & \text{sym.} & & & D_{55} & 0 \\
 & & & & & D_{66}
\end{bmatrix}
\begin{Bmatrix} \varepsilon_{11} \\ \varepsilon_{22} \\ \varepsilon_{33} \\ \varepsilon_{12} \\ \varepsilon_{23} \\ \varepsilon_{13} \end{Bmatrix}
\tag{1.30}
$$

If the material is isotropic (i.e., material properties are independent of the direction), the number of independent material constants becomes 2:

$$
\begin{Bmatrix} \sigma_{11} \\ \sigma_{22} \\ \sigma_{33} \\ \sigma_{12} \\ \sigma_{23} \\ \sigma_{13} \end{Bmatrix} =
\begin{bmatrix}
2\mu + \lambda & \lambda & \lambda & 0 & 0 & 0 \\
 & 2\mu + \lambda & \lambda & 0 & 0 & 0 \\
 & & 2\mu + \lambda & 0 & 0 & 0 \\
 & & & 2\mu & 0 & 0 \\
 & \text{sym.} & & & 2\mu & 0 \\
 & & & & & 2\mu
\end{bmatrix}
\begin{Bmatrix} \varepsilon_{11} \\ \varepsilon_{22} \\ \varepsilon_{33} \\ \varepsilon_{12} \\ \varepsilon_{23} \\ \varepsilon_{13} \end{Bmatrix}
\tag{1.31}
$$

where μ and λ are called Lamé constants. They are related to the elastic modulus E and Poisson's ratio v of the material by the following equations:

$$
\mu = \frac{E}{2(1 + v)}
\tag{1.32}
$$

$$
\lambda = \frac{vE}{(1 + v)(1 - 2v)}
\tag{1.33}
$$

Note that $\mu = G$, the shear modulus of the material.

Regardless of the material type, experimental means are often needed to determine the material constants that relate the stresses and strains in Equations 1.28 to 1.31. Because of the difficulty in determining a large number of constants, analyses are often performed by assuming the material is either isotropic or orthotropic.

If we denote any of the above equations relating stresses and strains symbolically as

$$
\boldsymbol{\sigma} = \mathbf{D}\boldsymbol{\varepsilon}
\tag{1.34}
$$

where $\boldsymbol{\sigma}$ is the 6×1 vector of stresses, $\boldsymbol{\varepsilon}$ is the 6×1 vector of strains, and $\mathbf{D}$ is the 6×6 material stiffness matrix, it can be shown that

$$
\boldsymbol{\varepsilon} = \mathbf{D}^{-1}\boldsymbol{\sigma} = \mathbf{C}\boldsymbol{\sigma}
\tag{1.35}
$$

where $\mathbf{C}$ is the material compliance matrix. For an orthotropic material, the expanded form of Equation 1.35 is

$$
\begin{Bmatrix} \varepsilon_{11} \\ \varepsilon_{22} \\ \varepsilon_{33} \\ \varepsilon_{12} \\ \varepsilon_{23} \\ \varepsilon_{13} \end{Bmatrix} =
\begin{bmatrix}
\dfrac{1}{E_{11}} & \dfrac{-v_{21}}{E_{22}} & \dfrac{-v_{31}}{E_{33}} & 0 & 0 & 0 \\
 & \dfrac{1}{E_{22}} & \dfrac{-v_{32}}{E_{33}} & 0 & 0 & 0 \\
 & & \dfrac{1}{E_{33}} & 0 & 0 & 0 \\
 & & & \dfrac{1}{2G_{12}} & 0 & 0 \\
 & & & & \dfrac{1}{2G_{23}} & 0 \\
 & \text{sym.} & & & & \dfrac{1}{2G_{13}}
\end{bmatrix}
\begin{Bmatrix} \sigma_{11} \\ \sigma_{22} \\ \sigma_{33} \\ \sigma_{12} \\ \sigma_{23} \\ \sigma_{13} \end{Bmatrix}
\tag{1.36}
$$

where E_{11}, E_{22}, and E_{33} denote the orthotropic moduli of elasticity measured in three orthogonal directions, G_{12}, G_{23}, and G_{13} denote the orthotropic shear moduli, and v_{ij} denotes the Poisson's ratio obtained by dividing the negative value of the strain induced in the j direction by the strain produced in the i direction by a stress applied in the i direction.

For an isotropic material, the expanded form of Equation 1.35 is

$$
\begin{Bmatrix} \varepsilon_{11} \\ \varepsilon_{22} \\ \varepsilon_{33} \\ \varepsilon_{12} \\ \varepsilon_{23} \\ \varepsilon_{13} \end{Bmatrix} =
\begin{bmatrix}
\dfrac{1}{E} & \dfrac{-v}{E} & \dfrac{-v}{E} & 0 & 0 & 0 \\
 & \dfrac{1}{E} & \dfrac{-v}{E} & 0 & 0 & 0 \\
 & & \dfrac{1}{E} & 0 & 0 & 0 \\
 & & & \dfrac{(1+v)}{E} & 0 & 0 \\
 & & & & \dfrac{(1+v)}{E} & 0 \\
 & \text{sym.} & & & & \dfrac{(1+v)}{E}
\end{bmatrix}
\begin{Bmatrix} \sigma_{11} \\ \sigma_{22} \\ \sigma_{33} \\ \sigma_{12} \\ \sigma_{23} \\ \sigma_{13} \end{Bmatrix}
\tag{1.37}
$$

where the first three of the above matrix equation are often referred to as generalized Hooke's Law for linear elastic, homogeneous, and isotropic materials, respectively.

EXAMPLE 1.3

Determine the stress–strain relationship for a homogeneous isotropic material assuming (1) plane stress condition and (2) plane strain condition.

Solution

1. *Plane stress condition.* If the stresses are acting on the x_1–x_2 (or x–y) plane, plane stress condition implies that $\sigma_{33} = \sigma_{23} = \sigma_{13} = 0$. Substituting this condition into Equation 1.37, we have

$$
\begin{Bmatrix} \varepsilon_{11} \\ \varepsilon_{22} \\ \varepsilon_{12} \end{Bmatrix} =
\begin{bmatrix}
\dfrac{1}{E} & \dfrac{-v}{E} & 0 \\
 & \dfrac{1}{E} & 0 \\
 \text{sym.} & & \dfrac{1+v}{E}
\end{bmatrix}
\begin{Bmatrix} \sigma_{11} \\ \sigma_{22} \\ \sigma_{12} \end{Bmatrix}
\quad \text{and} \quad \varepsilon_{33} = -\frac{v}{E}(\sigma_{11} + \sigma_{22})
$$

or

$$\left\{ \begin{array}{c} \sigma_{11} \\ \sigma_{22} \\ \sigma_{12} \end{array} \right\} = \begin{bmatrix} \dfrac{E}{1-v^2} & \dfrac{vE}{1-v^2} & 0 \\ & \dfrac{E}{1-v^2} & 0 \\ \text{sym.} & & \dfrac{E}{1+v} \end{bmatrix} \left\{ \begin{array}{c} \varepsilon_{11} \\ \varepsilon_{22} \\ \varepsilon_{12} \end{array} \right\} \quad \text{and} \quad \varepsilon_{33} = \dfrac{-v}{1-v}(\varepsilon_{11} + \varepsilon_{22})$$

Note that $\varepsilon_{33} \neq 0$ even though $\sigma_{33} = 0$ (i.e., a biaxial state of stress gives rise to a triaxial state of strain) because of the *Poisson's effect*.

2. *Plane strain condition.* If the strain is negligible in the x_3 (or z) direction, plane strain condition implies that $\varepsilon_{33} = \varepsilon_{23} = \varepsilon_{13} = 0$. Substituting this condition into Equation 1.31, we have

$$\left\{ \begin{array}{c} \sigma_{11} \\ \sigma_{22} \\ \sigma_{12} \end{array} \right\} = \begin{bmatrix} \dfrac{(1-v)E}{1-v-2v^2} & \dfrac{vE}{1-v-2v^2} & 0 \\ & \dfrac{(1-v)E}{1-v-2v^2} & 0 \\ \text{sym.} & & \dfrac{E}{1+v} \end{bmatrix} \left\{ \begin{array}{c} \varepsilon_{11} \\ \varepsilon_{22} \\ \varepsilon_{12} \end{array} \right\} \quad \text{and} \quad \sigma_{33} = \dfrac{vE}{1-v-2v^2}(\varepsilon_{11} + \varepsilon_{22})$$

or

$$\left\{ \begin{array}{c} \varepsilon_{11} \\ \varepsilon_{22} \\ \varepsilon_{12} \end{array} \right\} = \begin{bmatrix} \dfrac{1-v^2}{E} & \dfrac{-v(1+v)}{E} & 0 \\ & \dfrac{1-v^2}{E} & 0 \\ \text{sym.} & & \dfrac{1+v}{E} \end{bmatrix} \left\{ \begin{array}{c} \sigma_{11} \\ \sigma_{22} \\ \sigma_{12} \end{array} \right\} \quad \text{and} \quad \sigma_{33} = v(\sigma_{11} + \sigma_{22})$$

Note that $\sigma_{33} \neq 0$ even though $\varepsilon_{33} = 0$.

1.4.2 Nonlinear Elastic Behavior

If an elastic material exhibits nonlinear behavior, the stress–strain relationship is often cast in *incremental form* relating some increments of strains to stress, or vice versa

$$\mathbf{d\sigma} = \mathbf{D_I}\, \mathbf{d\varepsilon} \tag{1.38}$$

or

$$\mathbf{d\varepsilon} = \mathbf{C_I}\, \mathbf{d\sigma} \tag{1.39}$$

where $\mathbf{d\sigma}$ is the incremental vector of stresses, $\mathbf{d\varepsilon}$ is the incremental vector of strains, $\mathbf{D_I}$ is the incremental material stiffness matrix, and $\mathbf{C_I}$ is the incremental material compliance matrix. If the experimental stress–strain curves of a material are known, the terms in these matrices can be taken as the values of the tangential or secant slopes of these curves. The analysis of structures made of materials that exhibit nonlinear elastic behavior has to be performed numerically in incremental steps as well.

Alternatively, if the nonlinear relationship between any given components of stress (or strain) can be expressed as a mathematical function of strains (or stresses) and material constants k_1, k_2, k_3, etc., as follows:

$$\sigma_{ij} = f_{ij}(\varepsilon_{11}, \varepsilon_{22}, \varepsilon_{33}, \varepsilon_{12}, \varepsilon_{23}, \varepsilon_{13}, k_1, k_2, k_3, \ldots) \tag{1.40}$$

$$\varepsilon_{ij} = g_{ij}(\sigma_{11}, \sigma_{22}, \sigma_{33}, \sigma_{12}, \sigma_{23}, \sigma_{13}, k_1, k_2, k_3, \ldots) \tag{1.41}$$

such relationships can be incorporated directly into the analysis to obtain closed-form solutions. However, this type of analysis can be performed only if both the structure and the loading conditions are very simple.

EXAMPLE 1.4

Derive the load–deflection equation for the axially loaded member shown in Figure 1.7. The member is made from a material with a uniaxial stress–strain relationship described by the equation $\varepsilon = B(\sigma/BnE_0)^n$, where B and n are material constants and E_0 is the initial slope of the stress–strain curve (i.e., the slope at $\sigma = 0$).

The deflection (which for this problem is equal to the elongation) of the axially loaded member can be obtained by integrating the strain over the length of the member; that is,

$$\delta = \int_0^L \varepsilon \, dx = \int_0^L B\left(\frac{\sigma}{BnE_0}\right)^n dx = \int_0^L B\left[\frac{P}{BnE_0 A_0 \left(1 - \dfrac{x}{2L}\right)}\right]^n dx$$

$$= \left(\frac{P}{nE_0 A_0}\right)^n \left(\frac{2^n - 2}{n - 1}\right) B^{1-n} L$$

1.4.3 Inelastic Behavior

For structures subject to uniaxial loading, inelastic behavior occurs once the stress in the structure exceeds the yield stress, σ_y, of the material. The yield stress is defined as the stress beyond which inelastic or permanent strain is induced, as shown in Figure 1.6. While some materials (e.g., structural steel) exhibit a definitive yield point on the uniaxial stress–strain curve, others do not. For such cases, the yield stress is often determined graphically using the 0.2% offset method. In this method, a line parallel to the initial slope of the uniaxial stress–strain curve is drawn from the 0.2% strain point. The 0.2% yield stress is obtained as the stress at which this line intersects the stress–strain curve.

For structures subject to biaxial or triaxial loading, inelastic behavior is assumed to occur when some combined stress state reaches a yield envelope (for a 2-D problem) or a yield surface (for a 3-D problem). Mathematically, the yield condition can be expressed as

$$f(\sigma_{ij}, k_1, k_2, k_3, \ldots) = 0 \tag{1.42}$$

where $k_1, k_2, k_3, \ldots$ are (experimentally determined) material constants.

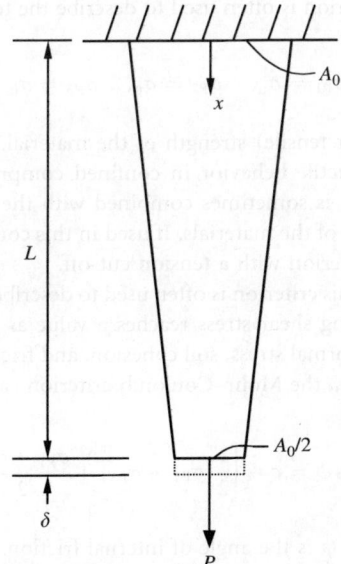

FIGURE 1.7 Tapered axially loaded member.

Over the years, various yield functions f have been proposed to describe the yield condition of a variety of materials (see, e.g., Chen 1982; Chen and Baladi 1985; Chakrabarty 1987; Chen and Han 1988). For ductile materials (e.g., most metals), the Tresca and von Mises yield criteria are often used. A brief discussion of these two criteria is given below:

1. *Tresca criterion.* According to the Tresca yield criterion, yielding occurs when the maximum shear stress at a point calculated using Equations 1.24 reaches a critical value equal to $\sigma_y/2$, where σ_y is the yield stress of the material obtained from a simple tension test. Mathematically, the Tresca yield criterion is expressed as

$$\max \left\{ \begin{array}{c} \frac{1}{2}|\sigma_{P1} - \sigma_{P2}| \\ \frac{1}{2}|\sigma_{P2} - \sigma_{P3}| \\ \frac{1}{2}|\sigma_{P1} - \sigma_{P3}| \end{array} \right\} = \frac{\sigma_y}{2} \tag{1.43}$$

2. *von Mises criterion.* Despite its simplicity, one drawback of the Tresca yield criterion is that it does not take into consideration the effect of the intermediate principal stress. One method to include the effect of this principal stress in the yield function is to use the octahedral shearing stress (or the strain energy of distortion) as the key parameter to describe yielding in the materials. The von Mises yield criterion is one example. The von Mises yield criterion has the form

$$\left[\frac{(\sigma_{P1} - \sigma_{P2})^2 + (\sigma_{P2} - \sigma_{P3})^2 + (\sigma_{P1} - \sigma_{P3})^2}{6} \right]^{1/2} = \frac{\sigma_y}{\sqrt{3}} \tag{1.44}$$

where σ_y is the yield stress obtained from a simple tension test.

It should be noted that both the Tresca and the von Mises yield criteria are independent of hydrostatic pressure effect. As a result, they should be used only for materials that are pressure insensitive. For pressure dependent materials (e.g., soils), other yield (or failure) criteria should be used. A few of these criteria are given below:

1. *Rankine criterion.* This criterion is often used to describe the tensile (fracture) failure of a brittle material. It has the form

$$\sigma_{P1} = \sigma_u, \quad \sigma_{P2} = \sigma_u, \quad \sigma_{P3} = \sigma_u \tag{1.45}$$

where σ_u is the ultimate (or tensile) strength of the material. For materials that exhibit brittle behavior in tension, but ductile behavior in confined compression (e.g., concrete, rocks, and soils), the Rankine criterion is sometimes combined with the Tresca or von Mises criterion to describe the failure behavior of the materials. If used in this context, the criterion is referred to as the Tresca or von Mises criterion with a tension cut-off.

2. *Mohr–Coulomb criterion.* This criterion is often used to describe the shear failure of soil. Failure is said to occur when a limiting shear stress reaches a value as defined by an envelope, which is expressed as a function of normal stress, soil cohesion, and friction angle. If the principal stresses are such that $\sigma_{P1} > \sigma_{P2} > \sigma_{P3}$, the Mohr–Coulomb criterion can be written as

$$\frac{1}{2}(\sigma_{P1} - \sigma_{P3})\cos\phi = c - \left[\frac{1}{2}(\sigma_{P1} + \sigma_{P3}) + \frac{\sigma_{P1} - \sigma_{P3}}{2}\sin\phi \right]\tan\phi \tag{1.46}$$

where c is the cohesion and ϕ is the angle of internal friction.

3. *Drucker–Prager criterion.* This criterion is an extension of the von Mises criterion, where the influence of hydrostatic stress on failure is incorporated by the addition of the term αI_1, where

I_1 is the first stress invariant as defined in Equations 1.13 (note that $\sigma_{11} + \sigma_{22} + \sigma_{33} = \sigma_{P1} + \sigma_{P2} + \sigma_{P3}$)

$$\alpha(\sigma_{P1} + \sigma_{P2} + \sigma_{P3}) + \left[\frac{(\sigma_{P1} - \sigma_{P2})^2 + (\sigma_{P2} - \sigma_{P3})^2 + (\sigma_{P1} - \sigma_{P3})^2}{6}\right]^{1/2} = k \qquad (1.47)$$

where α and k are material constants to be determined by curve-fitting of the above equation to experimental data.

If yielding does not signify failure of a material (which is often the case for ductile materials), the postyield behavior of the material is described by the use of a flow rule. A flow rule establishes the relative magnitudes of the components of plastic strain increment $d\varepsilon_{ij}^P$ and the direction of the plastic strain increment in the strain space. It is written as

$$d\varepsilon_{ij}^P = d\lambda \frac{\partial g}{\partial \sigma_{ij}} \qquad (1.48)$$

where $d\lambda$ is a positive scalar factor of proportionality, g is a plastic potential in stress space, and $\partial g/\partial \sigma_{ij}$ is the gradient, which represents the direction of a normal vector to the surface defined by the plastic potential at point σ_{ij}. Equation 1.48 implies that $d\varepsilon_{ij}^P$ is directed along the normal to the surface of the plastic potential. If the plastic potential g is equal to the yield function f, Equation 1.48 is called the associated flow rule. Otherwise, it is called the nonassociated flow rule.

Using the elastic stress–strain relationship expressed in Equation 1.39, the flow rule expressed in Equation 1.48 with $g = f$ (i.e., associated flow rule), the *consistency condition* for an elastic–perfectly plastic material given by

$$df = \frac{\partial f}{\partial \sigma_{ij}} d\sigma_{ij} = 0 \qquad (1.49)$$

and the following relationship among total, elastic, and inelastic (plastic) strains,

$$d\varepsilon_{ij} = d\varepsilon_{ij}^e + d\varepsilon_{ij}^P \qquad (1.50)$$

it has been shown (Chen and Han 1988) that an incremental stress–strain relationship for an elastic–perfectly plastic material that follows the associated flow rule can be written as

$$d\sigma_{ij} = D_{ijkl}^{ep} d\varepsilon_{kl} \qquad (1.51)$$

where D_{ijkl}^{ep} is the incremental elastic–perfectly plastic material stiffness matrix given by

$$D_{ijkl}^{ep} = D_{ijkl} - \frac{D_{ijmn}(\partial f/\partial \sigma_{mn})(\partial f/\partial \sigma_{pq})D_{pqkl}}{(\partial f/\partial \sigma_{rs})D_{rstu}(\partial f/\partial \sigma_{tu})} \qquad (1.52)$$

where D_{ijkl} (or D_{ijmn}, D_{pqkl} etc.) is the indicial form of $\mathbf{D_I}$ given in Equation 1.38.

1.4.4 Hardening Rules

If a material exhibits *work-hardening* behavior in which a state of stress beyond yield can exist, then in addition to the initial yield surface f a new yield surface, called subsequent yield or loading surface F, needs to be defined. Like the initial yield surface, the loading surface demarcates elastic behavior from inelastic behavior. If the stress point moves on or within the loading surface, no additional plastic strain will be induced. If the stress point is on the loading surface and the loading condition is such that it pushes the stress point out of the loading surface, additional plastic deformations will occur. When this happens, the configuration of the loading surface will change. The condition of loading and unloading for a multiaxial stress state is mathematically defined as follows.

If the stress point is on the loading surface (i.e., if $F = 0$), loading occurs if

$$n_{ij}^F d\sigma_{ij} > 0 \qquad (1.53)$$

and unloading occurs if

$$n_{ij}^F d\sigma_{ij} < 0 \tag{1.54}$$

where n_{ij}^F represents a component of a unit vector that is normal to the loading surface F, that is,

$$n_{ij}^F = \frac{\partial F / \partial \sigma_{ij}}{\sqrt{(\partial F / \partial \sigma_{kl})(\partial F / \partial \sigma_{kl})}} \tag{1.55}$$

For the special case when $n_{ij}^F d\sigma_{ij} = 0$, that is, the loading vector $d\sigma_{ij}$ is perpendicular to the corresponding component of the unit normal vector n_{ij}^F, a state of neutral loading is said to have occurred. Note that additional plastic strain is induced only during loading, but not during neutral loading or unloading.

According to the incremental or flow theory of plasticity, the configuration of the loading surface when loading occurs can be described by the use of a hardening rule. A hardening rule establishes a relationship between the subsequent yield stress of a material and the inelastic deformation accumulated during prior excursion into the inelastic regime. A number of hardening rules have been proposed over the years. They can often be classified into or associated with one of the following:

1. *Isotropic hardening.* This hardening rule assumes that during plastic deformations, the loading surface is merely an expansion, without distortion, of the initial yield surface. Mathematically, this surface is represented by the equation

$$F(\sigma_{ij}) = k^2(\varepsilon_p) \tag{1.56}$$

   where k is a constant, which is a function of the total (i.e., cumulated) plastic strain ε_p. Although this is one of the simplest hardening rules, it has a serious drawback in that it cannot be used to account for the *Bauschinger effect*, which states that the occurrence of an initial plastic deformation in one direction (e.g., in tension) will cause a reduction in material resistance to a subsequent plastic deformation in the opposite direction (e.g., in compression). Since the Bauschinger effect is present in most structural materials, the use of isotropic hardening should be limited to problems that involve only *monotonic loading* in which no stress reversals will occur.

2. *Kinematic hardening.* This hardening rule (Prager 1955, 1956) assumes that during plastic deformation, the loading surface is formed by a simple rigid body translation (with no change in size, shape, and orientation) of the initial yield surface in stress space. Thus, the equation of the loading surface takes the form

$$F(\sigma_{ij} - \eta_{ij}) = k^2 \tag{1.57}$$

   where k is a constant to be determined experimentally and η_{ij} are the coordinates of the centroid of the loading surface, which changes continuously throughout plastic deformation. It should be noted that contrary to isotropic hardening, kinematic hardening takes full account of the Bauschinger effect, so much so that the amount of "loss" of material resistance in one direction during subsequent plastic deformation is exactly equal to the amount of initial plastic deformation the material experiences in the opposite direction, which may or may not be truly reflective of real material behavior.

3. *Mixed hardening.* As the name implies, this hardening rule (Hodge 1957) contains features of both the isotropic and the kinematic hardening rules described above. It has the form

$$F(\sigma_{ij} - \eta_{ij}) = k^2(\varepsilon_p) \tag{1.58}$$

   where η_{ij} and k are as defined in Equations 1.56 and 1.57. In mixed hardening, the loading surface is defined by a translation (as described by the term η_{ij}) and expansion (as measured by the term $k(\varepsilon_p)$), but no change in shape, of the initial yield surface. The advantage of using the mixed hardening rule is that one can conveniently simulate different degrees of the Bauschinger effect by adjusting the two hardening parameters (η_{ij} and k) of the model.

1.4.5 Effective Stress and Effective Plastic Strain

Effective stress and effective plastic strain are variables that allow the hardening parameters contained in the above hardening models to be correlated with an experimentally obtained uniaxial stress–strain curve of the material. The effective stress has unit of stress, and it should reduce to the stress σ_{11} in a uniaxial stress condition. Table 1.1 summarizes the equations for the effective stress and hardening parameter for two materials modeled using the isotropic hardening rule. The equations shown in Table 1.1 can also be used for materials modeled using the kinematic or mixed hardening rule provided that the effective stress σ_e is replaced by a reduced effective stress σ_e^r, computed using a reduced stress tensor given by

$$\sigma_{ij}^r = \sigma_{ij} - \eta_{ij} \tag{1.59}$$

Effective plastic strain increment $d\varepsilon_e^p$ can be defined in the context of plastic work per unit volume in the form

$$dW_p = \sigma_e \, d\varepsilon_e^p \tag{1.60}$$

By using Equation 1.48 in conjunction with a material model, it can be shown (Chen and Han 1988) that for a von Mises material

$$d\varepsilon_e^p = \sqrt{\tfrac{2}{3} d\varepsilon_{ij}^p \, d\varepsilon_{ij}^p} \tag{1.61}$$

and for a Drucker–Prager material

$$d\varepsilon_e^p = \frac{\alpha + \left(1/\sqrt{3}\right)}{\sqrt{3\alpha^2 + (1/2)}} \sqrt{d\varepsilon_{ij}^p \, d\varepsilon_{ij}^p} \tag{1.62}$$

The effective stress and effective plastic strain are related by the incremental stress–strain equation

$$d\sigma_e = H_p \, d\varepsilon_e^p \tag{1.63}$$

where H_p is the plastic modulus, which is obtained as the slope of the uniaxial stress–plastic strain curve at the current value of σ_e.

Using the concept of effective plastic strain, flow rule, consistency condition, relationship between total, elastic, and plastic strains, elastic stress–strain relationship, and a hardening rule, it can be shown (Chen and Han 1988) that an incremental stress–strain relationship for an elastic–work-hardening material can be written in the form of Equation 1.51 with

$$D_{ijkl}^{ep} = D_{ijkl} - \frac{D_{ijmn}(\partial g/\partial \sigma_{mn})(\partial F/\partial \sigma_{pq})D_{pqkl}}{\kappa + (\partial F/\partial \sigma_{rs})D_{rstu}(\partial g/\partial \sigma_{tu})} \tag{1.64}$$

TABLE 1.1 Effective Stress

Material model	Effective stress, σ_e	Hardening parameter, k
von Mises	$\sqrt{3J_2}$	$\sigma_e/\sqrt{3}$
Drucker–Prager	$\left(\sqrt{3}\alpha I_1 + \sqrt{3J_2}\right)/\left(1 + \sqrt{3}\alpha\right)$	$\left(\alpha + \left(1/\sqrt{3}\right)\right)\sigma_e$

Note: J_2 is the second deviatoric stress invariant defined in Equations 1.21, I_1 is the first stress invariant defined in Equations 1.13, and α is a material constant defined in Equation 1.47.

where

$$\kappa = -\frac{\partial F}{\partial \varepsilon_{ij}^p}\frac{\partial g}{\partial \sigma_{ij}} - \frac{\partial F}{\partial k}\frac{dk}{d\varepsilon_e^p}C\sqrt{\frac{\partial g}{\partial \sigma_{ij}}\frac{\partial g}{\partial \sigma_{ij}}} \tag{1.65}$$

where C is a material constant, which is equal to

$$\sqrt{\frac{2}{3}} \tag{1.66a}$$

for a von Mises material and

$$\frac{\alpha + (1/\sqrt{3})}{\sqrt{3\alpha^2 + (1/2)}} \tag{1.66b}$$

for a Drucker–Prager material. From Equation 1.64 it can be seen that D_{ijkl}^{ep} is not necessarily symmetric unless the associated flow rule (i.e., $g = F$) is used in the formulation.

1.5 Stress Resultants

Structural analysis can be performed and results represented in terms of stresses and strains, or forces and displacements. For skeletal structures (i.e., structures that are made up of line elements such as trusses, beams, frames, arches, grillages, etc.), the internal forces and moments, or stress resultants, acting on a given cross-section as shown in Figure 1.8 are related to the stresses acting over the cross-section by the following equations:

$$F_x = \int_A \sigma_{11}\,dA, \quad F_y = \int_A \sigma_{12}\,dA, \quad F_z = \int_A \sigma_{13}\,dA$$

$$M_x = \int_A (-\sigma_{12}z + \sigma_{13}y)\,dA, \quad M_y = \int_A \sigma_{11}z\,dA, \quad M_z = -\int_A \sigma_{11}y\,dA \tag{1.67}$$

where F_x is the axial force, F_y and F_z are the shear forces, M_x is the torque, and M_y and M_z are the bending moments about the y (or x_2) and z (or x_3) axes, respectively. Note that the value of some of these terms

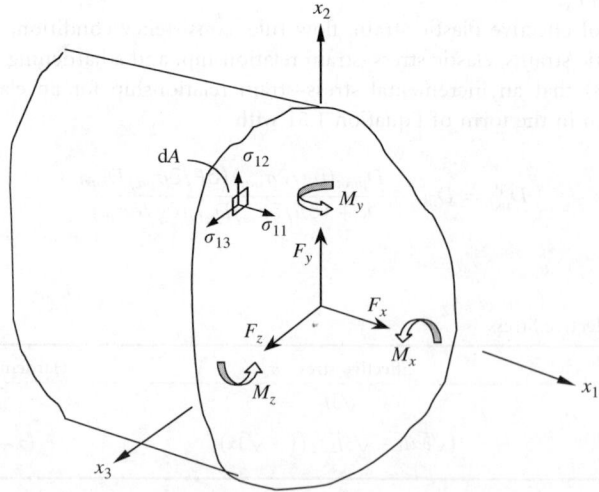

FIGURE 1.8 Stress resultants.

is 0 depending on the structure and the assumptions used in the modeling. For instance, in a truss analysis, it is often assumed that $F_y = F_z = M_x = M_y = M_z = 0$. In a 2-D beam–frame analysis in which the structure is modeled on the x–y (or x_1–x_2) plane, it is often assumed that $F_z = M_x = M_y = 0$. In a 2-D grillage analysis in which the structure is modeled on the x–z (or x_1–x_3) plane, it is often assumed that $F_x = F_z = M_y = 0$.

1.6 Types of Analyses

Depending on the magnitude of the applied loads, the type of structure under consideration, the purpose of performing the analysis, and the degree of accuracy desired, different types of analyses can be performed to determine the force–displacement or stress–strain response of a structural system. Given below is a succinct discussion of some salient features associated with several types of analyses that one can perform depending on the objectives of the analysis and the expectations of the analyst. A more detailed discussion of some of these methods of analysis can be found in later chapters of this handbook.

1.6.1 First-Order versus Second-Order Analysis

A first-order analysis is one in which all equilibrium and kinematic equations are written with respect to the initial or undeformed configuration of the structure. A second-order analysis is one in which equilibrium and kinematic equations are written with respect to the current or deformed geometry of the structure. Because all structures deform under loads, a method of analysis that takes into consideration structural deformation in its formulation will provide a more realistic representation of the structure. However, because of its simplicity, a first-order analysis is often performed in lieu of a second-order analysis. Although the results obtained lack the precision of a second-order analysis, they are sufficiently accurate for design purpose if deflections or deformations of the structure are small.

1.6.2 Elastic versus Inelastic Analysis

An elastic analysis is one in which the effect of yielding is ignored in the analysis. Thus, the stress–strain relationships discussed in Section 1.4.1 (for linear elastic material behavior) or Section 1.4.2 (for nonlinear elastic material behavior) will be used in the analysis. Because all strains (and deformations) are recoverable in an elastic analysis, no consideration is given to the loading history or loading path dependent effect (which is very important in an inelastic analysis) during the analysis. Elastic analysis is therefore much easier to perform than inelastic analysis. However, if yielding does occur, a behavioral model that is capable of capturing the inelastic response of the structure should be used.

1.6.3 Plastic Hinge versus Plastic Zone Analysis

For framed structures, if the applied loads are proportional and monotonic, the loading history effect is inconsequential, and a plastic hinge (also called concentrated plasticity) or plastic zone (also called distributed plasticity) analysis can be performed to capture the inelastic behavior of the system. In the plastic hinge method (ASCE-WRC 1971) of analysis, inelasticity is assumed to concentrate in regions of plastic hinges. A plastic hinge is a zero-length element where the moment is equal to the cross-section plastic moment capacity M_p. If the effects of shear and axial force are ignored, M_p is given by

$$M_p = Z\sigma_y \tag{1.68}$$

where Z is the plastic section modulus (AISC 2001) and σ_y is the material yield stress.

In a simple plastic hinge analysis, once the moment in a cross-section reaches M_p, a hinge is inserted at that location and no additional moment is assumed to be carried by that cross-section. Cross-sections that have moments below M_p are assumed to behave elastically. Because the formation of a plastic hinge is a gradual process in which yielding spreads slowly from the neutral axis toward the extreme fiber of the cross-section (i.e., cross-section plastification effect) as well as along the length of the member (i.e., member plastification effect) as the applied load increases, more realistic models such as the modified plastic hinge approach (White and Chen 1993) and the plastic zone approach (Vogel 1984, 1985; Lui and Zhang 1990; Clarke et al. 1992) have been proposed to capture this spread of plasticity effect. While the modified plastic hinge approach only accounts for cross-section plastification, the plastic zone approach accounts for both cross-section and member plastification as well as for the effect of residual stresses, and is therefore considered the most accurate method of frame analysis. Unfortunately, to achieve this high degree of accuracy, very careful and detailed modeling is required. For practical reasons, plastic zone analysis is rarely performed on a routine basis. It is mostly used as a research tool to calibrate or verify the accuracy of advanced in-house structural analysis programs.

1.6.4 Stability Analysis

Stability analysis is a special type of second-order analysis in which the system under consideration is subjected to compressive force or stress (Allen and Bulson 1980; Chen and Lui 1987, 1991; Bazant and Cedolin 1991). If the force or stress is high enough, a phenomenon known as instability or buckling may occur. At the buckling or critical load, the structural system loses it stiffness, changes its deformation pattern, and loses its ability to carry the applied loads. The mathematics used for the computation of this critical load is called an eigenvalue problem. The system critical load and buckled mode shape are obtained as the lowest eigenvalue and the corresponding eigenvector of the equation

$$\mathbf{KU} = \lambda \mathbf{K_G U} \tag{1.69}$$

where $\mathbf{K}$ is the first-order system stiffness matrix, $\mathbf{K_G}$ is the system geometrical stiffness matrix, λ is the eigenvalue of the system, and $\mathbf{U}$ is the system displacement vector.

Stability analysis can be elastic or inelastic, depending on whether the stiffness matrices in Equation 1.69 are formulated assuming elastic or inelastic material behavior (McGuire et al. 2000). In addition, it should be noted that not all systems experience instability in the form of sudden buckling. Structural systems that are geometrically imperfect (which is often the case for real structures) undergo deformations that may resemble the buckled mode shapes at the outset of loading. The critical load for these geometrically imperfect systems is called the limit load. It is obtained as the peak point of the load–deflection curve generated using a second-order analysis.

1.6.5 Static versus Dynamics Analysis

A static analysis is one in which the effects of damping and inertia are not important and are therefore ignored. It is used when the loads acting on the structure are stationary or applied very slowly over time. A dynamic analysis is performed if the applied loads are time dependent, or if the effects of damping and

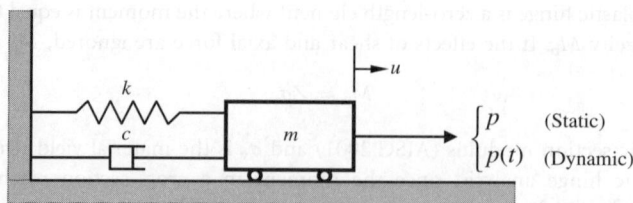

FIGURE 1.9 A simple spring–mass–damper system.

inertia are important. As shown in Figure 1.9, if a static analysis is used, the equilibrium equation has the form

$$ku = p \tag{1.70}$$

where k is the spring stiffness, u is the horizontal displacement of the mass, and p is the applied force. However, if a dynamic analysis is used, the equilibrium equation has the form

$$m\ddot{u} + c\dot{u} + ku = p(t) \tag{1.71}$$

where m is the mass, c is the damping coefficient, k is the spring stiffness, u is the displacement, $\dot{u}$ is the velocity, $\ddot{u}$ is the acceleration of the mass, and $p(t)$ is the time-varying applied load. $m\ddot{u}$ is called the inertia force, $c\dot{u}$ is the damping force from a viscous damper, and ku is the spring force. Note that inertia force and damping force are not present in the static equation, but they are present in the dynamic equation. It is also noteworthy to observe that the static equation is an algebraic equation, but the dynamic equation is a differential equation. A dynamic analysis is therefore more difficult and time consuming to perform than a static analysis, and depending on the form and complexity of the excitation function $p(t)$, recourse to numerical methods is often needed (Cheng 2001; Chopra 2001).

The system shown in Figure 1.9 is referred to as a single degree-of-freedom (dof) system because one displacement variable u is all that is needed to define the displaced configuration of the system. For a multiple dof system, Equations 1.70 and 1.71 need to be written in matrix form as

$$\mathbf{KU} = \mathbf{P} \tag{1.72}$$

$$\mathbf{M\ddot{U}} + \mathbf{C\dot{U}} + \mathbf{KU} = \mathbf{P(t)} \tag{1.73}$$

where $\mathbf{K}$, $\mathbf{M}$, and $\mathbf{C}$ are the system stiffness, system mass, and system damping matrices, respectively, $\mathbf{U}$, $\dot{\mathbf{U}}$, and $\ddot{\mathbf{U}}$ are the system displacement, velocity, and acceleration vectors, respectively, and $\mathbf{P(t)}$ is the time dependent system excitation force vector.

1.7 Structural Analysis and Design

Structural analysis refers to the computation of internal forces, displacements, stresses, and strains of a structure with known geometry, arrangement of components as well as component and material properties under a set of applied loads. Structural design refers to the determination of the proper material, geometry, arrangement of components and component properties to carry a predefined set of applied loads. In general, analysis and design are intertwined, and have to be performed iteratively in sequence. Using a preliminary set of structural and component geometry determined based on experience or the use of simplified behavioral models, an analysis is performed from which internal forces, displacements, stresses, and strains are calculated. These computed quantities are then used (often in conjunction with a design specification) to modify the preliminary design. The basic condition to satisfy in a strength based design is that

$$\text{capacity} \geq \text{demand} \tag{1.74}$$

Another analysis (called reanalysis) is then performed to obtain a more refined set of design quantities. The process is repeated until Equation 1.74 is satisfied in every part of the structure. Very often, different load combinations and different patterns of load applications have to be investigated to identify the worst possible scenario for design. As a result, the use of computers becomes indispensable for the design of complex structures.

Glossary

Mohr's circle — When plotted in a Cartesian coordinate system with the normal stress σ as the abscissa and the shear stress τ as the ordinate, a Mohr circle is a graphical representation of the state of stress at a point. Each pair of coordinates on a Mohr circle represents the magnitude of a pair of normal and shear stresses that exist on a plane with a certain orientation.

Monotonic loading — A loading that does not change direction during the course of the load history.

Poisson's effect — An effect in which an increase (or decrease) of strain in one direction causes a decrease (or increase) of strains in other directions. It is quantified by what is referred to as Poisson's ratio v, which is defined as the ratio of the minus value of the lateral strain to the longitudinal strain. Most materials have Poisson's ratios that fall in the range $0 < v \leq 0.5$.

Principal axes — The three orthogonal axes that are collinear with the unit vectors used to define the three principal planes of a parallelepiped volume element. Principal axes can also be defined as axes about which the product of inertia I_{ij} (when $i \neq j$) vanishes.

References

AISC. (2001). *AISC Manual of Steel Construction — Load and Resistance Factor Design*, 3rd edition, American Institute of Steel Construction, Chicago, IL.

Allen, H.G. and Bulson, P.S. (1980). *Background to Buckling*, McGraw-Hill (U.K.), Maidenhead, Berkshire, England.

ASCE-WRC. (1971). *Plastic Design in Steel — A Guide and Commentary*, 2nd edition, ASCE, New York.

Bathe, K.J. (1982). *Finite Element Procedures in Engineering Analysis*, Prentice Hall, Englewood Cliffs, NJ.

Bazant, Z.P. and Cedolin, L. (1991). *Stability of Structures — Elastic, Inelastic, Fracture, and Damage Theories*, Oxford University Press, New York.

Beer, F.P., Johnston, E.R., Jr., and DeWolf, J.T. (2001). *Mechanics of Materials*, 3rd edition, McGraw-Hill, New York.

Chakrabarty, J. (1987). *Theory of Plasticity*, McGraw-Hill, New York.

Chen, W.F. (1982). *Plasticity in Reinforced Concrete*, McGraw-Hill, New York.

Chen, W.F. and Baladi, G.Y. (1985). *Soil Plasticity: Theory and Implementation*, Elsevier, Amsterdam, The Netherlands.

Chen, W.F. and Han, D.J. (1988). *Plasticity for Structural Engineers*, Springer-Verlag, New York.

Chen, W.F. and Lui, E.M. (1987). *Structural Stability — Theory and Implementation*, Elsevier, New York.

Chen, W.F. and Lui, E.M. (1991). *Stability Design of Steel Frames*, CRC Press, Boca Raton, FL.

Cheng, F.Y. (2001). *Matrix Analysis of Structural Dynamics — Applications and Earthquake Engineering*, Marcel Dekker, New York.

Chopra, A.K. (2001). *Dynamics of Structures — Theory and Applications to Earthquake Engineering*, Prentice Hall, Upper Saddle River, NJ.

Clarke, M.J., Bridge, R.Q., Hancock, G.J., and Trahair, N.S. (1992). Advanced Analysis of Steel Building Frames, *J. Constr. Steel Res.*, 23(1–3), 1–29.

Gere, J.M. (2004). *Mechanics of Materials*, 6th edition, Brooks/Cole, Belmont, CA.

Hodge, P.G., Jr. (1957). Discussion of Prager's (1956) Paper on "A New Method of Analyzing Stress and Strains in Work-Hardening Solids," *J. Appl. Mech.*, 23, 482–484.

Lui, E.M. and Zhang, C.Y. (1990). Nonlinear Frame Analysis by the Pseudo Load Method, *Comput. Struct.*, 37(5), 707–716.

McGuire, W., Gallagher, R.H., and Ziemian, R.D. (2000). *Matrix Structural Analysis*, 2nd edition, John Wiley & Sons, New York.

Prager, W. (1955). The Theory of Plasticity: A Survey of Recent Achievements, *Inst. Mech. Eng.* 169, 41–57.

Prager, W. (1956). A New Method of Analyzing Stress and Strains in Work-Hardening Solids, *J. Appl. Mech., ASME*, 23, 493–496.

Timoshenko, S.P. and Goodier, J.N. (1970). *Theory of Elasticity*, McGraw-Hill, New York.

Vogel, U. (1984). *Ultimate Limit State Calculation of Sway Frames with Rigid Joints*, ECCS Publication No. 33, 1st edition, Rotterdam, The Netherlands.

Vogel, U. (1985). Some Comments on the ECCS Publication No. 33 — Ultimate Limit State Calculation of Sway Frames with Rigid Joints, *Construzioni Metalliche H.I. anno XXXVII*, 35–39.

Wang, C.-T. (1953). *Applied Elasticity*, McGraw-Hill, New York.

White, D.W. and Chen, W.F. (editors). (1993). *Plastic Hinge Based Methods for Advanced Analysis and Design of Steel Frames*, Structural Stability Research Council, Bethlehem, PA.

Further Reading

Bathe, K.J. (1996). *Finite Element Procedures*, Prentice Hall, Upper Saddle River, NJ.

Boresi, A.P. and Schmidt, R.J. (2002). *Advanced Mechanics of Materials*, 6th edition, John Wiley & Sons, New York.

Budynas, R.G. (1999). *Advanced Strength and Applied Stress Analysis*, McGraw-Hill, Boston, MA.

Cook, R.D. and Young, W.C. (1998). *Advanced Mechanics of Materials*, 2nd edition, Prentice Hall, Upper Saddle River, NJ.

Doltsinis, I. (2000). *Elements of Plasticity — Theory and Computation*, WIT Press, Southampton, U.K.

Fung, Y.C. and Tong, P. (2001). *Classical and Computational Solid Mechanics*, World Scientific, Singapore.

Solecki, R. and Conant, R.J. (2003). *Advanced Mechanics of Materials*, Oxford University Press, New York.

Ugural, A.C. and Fenster, S.K. (2003). *Advanced Strength and Applied Elasticity*, 4th edition, Prentice Hall, Upper Saddle River, NJ.

2

Structural Analysis

2.1 Fundamental Principles 2-2
Boundary Conditions • Loads and Reactions • Principle of Superposition

2.2 Beams 2-5
Relation among Load, Shear Force, and Bending Moment • Shear Force and Bending Moment Diagrams • Fixed-Ended Beams • Continuous Beams • Beam Deflection • Curved Beams

2.3 Trusses 2-21
Method of Joints • Method of Sections • Compound Trusses

2.4 Frames 2-24
Slope Deflection Method • Frame Analysis Using Slope Deflection Method • Moment Distribution Method • Method of Consistent Deformations

2.5 Plates 2-38
Bending of Thin Plates • Boundary Conditions • Bending of Rectangular Plates • Bending of Circular Plates • Strain Energy of Simple Plates • Plates of Various Shapes and Boundary Conditions • Orthotropic Plates

2.6 Shells 2-53
Stress Resultants in the Shell Element • Shells of Revolution • Spherical Dome • Conical Shells • Shells of Revolution Subjected to Unsymmetrical Loading • Cylindrical Shells • Symmetrically Loaded Circular Cylindrical Shells

2.7 Influence Lines 2-62
Influence Lines for Shear in Simple Beams • Influence Lines for Bending Moment in Simple Beams • Influence Lines for Trusses • Qualitative Influence Lines • Influence Lines for Continuous Beams

2.8 Energy Methods 2-65
Strain Energy Due to Uniaxial Stress • Strain Energy in Bending • Strain Energy in Shear • Energy Relations in Structural Analysis • Unit Load Method

2.9 Matrix Methods 2-74
Flexibility Method • Stiffness Method • Element Stiffness Matrix • Structure Stiffness Matrix • Loading between Nodes • Semirigid End Connection

2.10 The Finite Element Method 2-87
Basic Principle • Elastic Formulation • Plane Stress • Plane Strain • Choice of Element Shapes and Sizes • Choice of Displacement Function • Nodal Degrees of Freedom • Isoparametric Elements • Isoparametric Families of Elements • Element Shape Functions • Formulation of Stiffness Matrix • Plates Subjected to In-Plane Forces • Beam Element • Plate Element

0-8493-1569-7/05/$0.00+$1.50
© 2005 by CRC Press

2-1

J. Y. Richard Liew
*Department of Civil Engineering,
National University of Singapore,
Singapore*

N. E. Shanmugam
*Department of Civil Engineering,
National University of Singapore,
Singapore*

2.1   Fundamental Principles

The main purpose of *structural analysis* is to determine forces and deformations of the structure due to applied loads. *Structural design* involves form finding, determination of loadings, and proportioning of structural members and components in such a way that the assembled structure is capable of supporting the loads within the design limit states. The analytical model is an idealization of the actual structure. The structural model should relate the actual behavior to material properties, structural details, and loading and boundary conditions as accurately as is practicable.

Structures often appear in three-dimensional form. It is possible to idealize structures that have a regular layout, are rectangular in shape, and are subjected to symmetric loads into two-dimensional frames arranged in orthogonal directions. A structure is said to be two-dimensional or planar if all the members lie in the same plane. *Joints* in a structure are those points where two or more members are connected. Beams are members subjected to loading acting transverse to their longitudinal axis and creating flexural bending only. *Ties* are members that are subjected to axial tension only, while struts (columns or posts) are members subjected to axial compression only. A *truss* is a structural system consisting of members that are designed to resist only axial forces. A structural system in which joints are capable of transferring end moments is called a *frame*. Members in this system are assumed to be capable of resisting bending moments, axial force, and shear force.

2.1.1   Boundary Conditions

A *hinge* or *pinned joint* does not allow translational movements (Figure 2.1a). It is assumed to be frictionless and to allow rotation of a member with respect to the others. A *roller* permits the attached structural part to rotate freely with respect to the rigid surface and to translate freely in the direction parallel to the surface (Figure 2.1b). Translational movement in any other direction is not allowed. A *fixed support* (Figure 2.1c) does not allow rotation or translation in any direction. A *rotational spring* provides some rotational restraint but does not provide any translational restraint (Figure 2.1d). A *translational spring* can provide partial restraints along the direction of deformation (Figure 2.1e).

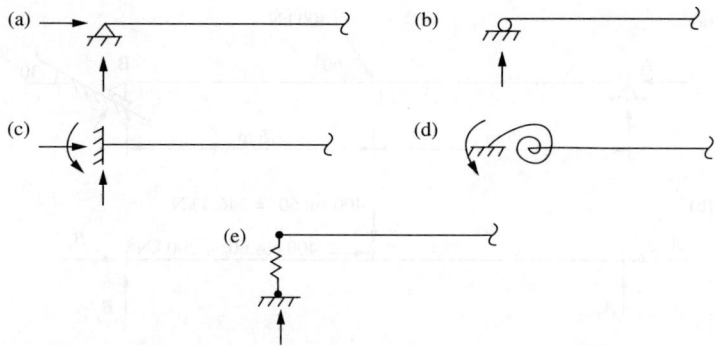

FIGURE 2.1 Various boundary conditions: (a) hinge support, (b) roller support, (c) fixed support, (d) rotational support, and (e) translational spring.

2.1.2 Loads and Reactions

Loads that are of constant magnitude and remain in the original position are called *permanent loads*. They are also referred to as *dead loads*, which may include the self weight of the structure and other loads such as walls, floors, roof, plumbing, and fixtures that are permanently attached to the structure. Loads that may change in position and magnitude are called *variable loads*. They are commonly referred to as live or imposed loads, which may include those caused by construction operations, wind, rain, earthquakes, snow, blasts, and temperature changes in addition to those objects that are movable, such as furniture and warehouse materials.

Ponding loads are due to water or snow on a flat roof that accumulates faster than it runs off. *Wind loads* act as pressures on windward surfaces and pressures or suctions on leeward surfaces. *Impact loads* are caused by suddenly applied loads or by the vibration of moving or movable loads. They are usually taken as a fraction of the live loads. *Earthquake loads* are those forces caused by the acceleration of the ground surface during an earthquake.

A structure that is initially at rest and remains at rest when acted upon by applied loads is said to be in a state of *equilibrium*. The resultant of the external loads on the body and the supporting forces or reactions is zero. If a structure is to be in equilibrium under the action of a system of loads, it must satisfy the six static equilibrium equations:

$$\sum F_x = 0, \quad \sum F_y = 0, \quad \sum F_z = 0$$
$$\sum M_x = 0, \quad \sum M_y = 0, \quad \sum M_z = 0 \tag{2.1}$$

The summation in these equations is for all the components of the forces (F) and of the moments (M) about each of the three axes x, y, and z. If a structure is subjected to forces that lie in one plane, say x–y, the above equations are reduced to

$$\sum F_x = 0, \quad \sum F_y = 0, \quad \sum M_z = 0 \tag{2.2}$$

Consider a beam under the action of the applied loads as shown in Figure 2.2a. The reaction at support B must act perpendicular to the surface on which the rollers are constrained to roll. The support reactions and the applied loads, which are resolved in vertical and horizontal directions, are shown in Figure 2.2b.

With geometry, it can be calculated that $B_y = \sqrt{3}B_x$. Equation 2.2 can be used to determine the magnitude of the support reactions. Taking the moment about B gives

$$10A_y - 346.4 \times 5 = 0$$

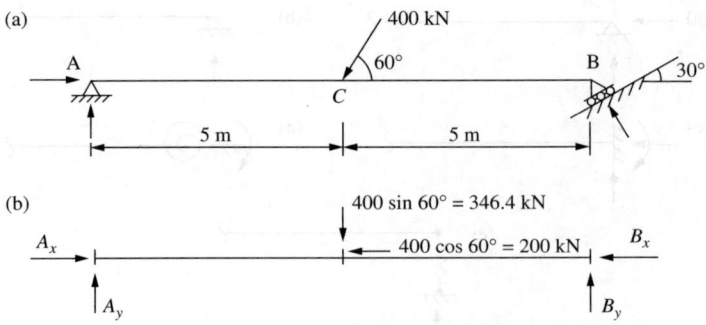

FIGURE 2.2   Beam in equilibrium: (a) applied load and (b) support reactions.

from which we get

$$A_y = 173.2 \text{ kN}$$

Equating the sum of vertical forces, $\sum F_y$, to 0 gives

$$173.2 + B_y - 346.4 = 0$$

and hence we get

$$B_y = 173.2 \text{ kN}$$

Therefore,

$$B_x = B_y/\sqrt{3} = 100 \text{ kN}$$

Equilibrium in the horizontal direction, $\sum F_x = 0$, gives

$$A_x - 200 - 100 = 0$$

and hence,

$$A_x = 300 \text{ kN}$$

There are three unknown reaction components at a fixed end, two at a hinge, and one at a roller. If, for a particular structure, the total number of unknown reaction components equals the number of equations available, the unknowns may be calculated from the equilibrium equations, and the structure is then said to be *statically determinate externally*. Should the number of unknowns be greater than the number of equations available, the structure is *statically indeterminate externally*; if less, it is *unstable externally*. The ability of a structure to support adequately the loads applied to it is dependent not only on the number of reaction components but also on the arrangement of those components. It is possible for a structure to have as many or more reaction components than there are equations available and yet be unstable. This condition is referred to as *geometric instability*.

2.1.3 Principle of Superposition

The principle states that if the structural behavior is linearly elastic, the forces acting on a structure may be separated or divided in any convenient fashion and the structure analyzed for the separate cases. The final results can be obtained by adding up the individual results. This is applicable to the computation of structural responses such as moment, shear, and deflection.

However, there are two situations where the principle of superposition cannot be applied. The first case is associated with instances where the geometry of the structure is appreciably altered under load. The second case is in situations where the structure is composed of a material in which the stress is not linearly related to the strain.

2.2 Beams

One of the most common structural elements is a *beam*; it bends when subjected to loads acting transverse to its centroidal axis or sometimes to loads acting both transverse and parallel to this axis. The discussions in the following subsections are limited to straight beams in which the centroidal axis is a straight line with shear center coinciding with the centroid of the cross-section. It is also assumed that all the loads and reactions lie in a simple plane that also contains the centroidal axis of the flexural member and the principal axis of every cross-section. If these conditions are satisfied, the beam will simply bend in the plane of loading without twisting.

2.2.1 Relation among Load, Shear Force, and Bending Moment

Shear force at any transverse cross-section of a straight beam is the algebraic sum of the components acting transverse to the axis of the beam of all the loads and reactions applied to the portion of the beam on either side of the cross-section. *Bending moment* at any transverse cross-section of a straight beam is the algebraic sum of the moments, taken about an axis passing through the centroid of the cross-section. The axis about which the moments are taken is normal to the plane of loading.

When a beam is subjected to transverse loads, there exist certain relationships among load, shear force, and bending moment. Let us consider the beam shown in Figure 2.3 subjected to some arbitrary loading, p. Let S and M be the shear and bending moment, respectively, for any point m at a distance x, which is measured from A, being positive when measured to the right. Corresponding values of shear and bending moment at point n at a differential distance dx to the right of m are $S + dS$ and $M + dM$, respectively. It can be shown, neglecting the second-order quantities, that

$$p = \frac{dS}{dx} \tag{2.3}$$

and

$$S = \frac{dM}{dx} \tag{2.4}$$

Equation 2.3 shows that the rate of change of shear at any point is equal to the intensity of load applied to the beam at that point. Therefore, the difference in shear at two cross-sections C and D is

$$S_D - S_C = \int_{x_C}^{x_D} p\,dx \tag{2.5}$$

We can write this in the same way for moment as

$$M_D - M_C = \int_{x_C}^{x_D} S\,dx \tag{2.6}$$

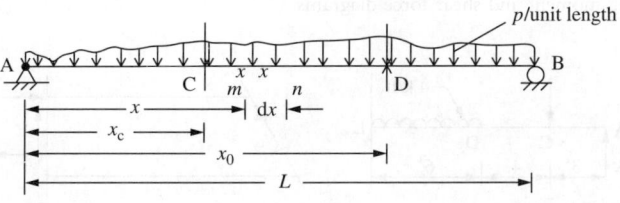

FIGURE 2.3   A beam under arbitrary loading.

2.2.2 Shear Force and Bending Moment Diagrams

To plot the shear force and bending moment diagrams it is necessary to adopt a sign convention for these responses. A shear force is considered to be positive if it produces a clockwise moment about a point in the free body on which it acts. A negative shear force produces a counterclockwise moment about the point. The bending moment is taken as positive if it causes compression in the upper fibers of the beam and tension in the lower fiber. In other words, a sagging moment is positive and a hogging moment negative. The construction of these diagrams is explained with an example given in Figure 2.4.

The section at E of the beam is in equilibrium under the action of applied loads and internal forces acting at E as shown in Figure 2.5. There must be an internal vertical force and internal bending moment to maintain equilibrium at section E. The vertical force or the moment can be obtained as the algebraic sum of all forces or the algebraic sum of the moment of all forces that lie on either side of the section E.

The shear on a cross-section — an infinitesimal distance to the right of point A is +55 and therefore the shear diagram rises abruptly from 0 to +55 at this point. In the portion AC, since there is no additional load, the shear remains at +55 on any cross-section throughout this interval, and the diagram is a horizontal as shown in Figure 2.4. At an infinitesimal distance to the left of C the shear is +55, but at an infinitesimal distance to the right of this point the concentrated load of magnitude 30 has caused the shear to be reduced to +25. Therefore, at point C, there is an abrupt change in the shear force from +55 to +25. In the same manner, the shear force diagram for the portion CD of the beam remains a rectangle. In the portion DE, the shear on any cross-section a distance x from point D is

$$S = 55 - 30 - 4x = 25 - 4x$$

which indicates that the shear diagram in this portion is a straight line decreasing from an ordinate of +25 at D to +1 at E. The remainder of the shear force diagram can easily be verified in the same way. It should be noted that, in effect, a concentrated load is assumed to be applied at a point, and hence at such a point the ordinate to the shear diagram changes abruptly by an amount equal to the load.

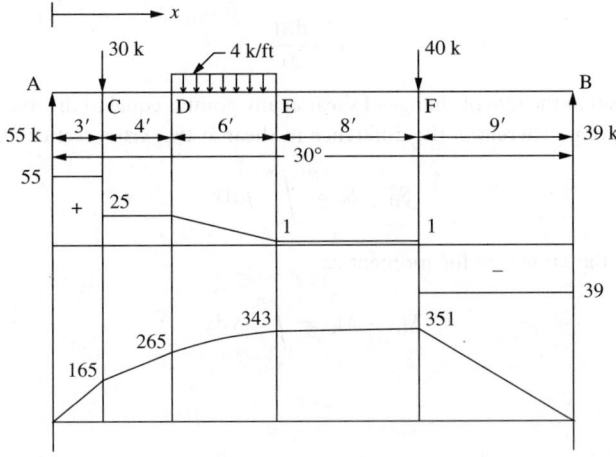

FIGURE 2.4   Bending moment and shear force diagrams.

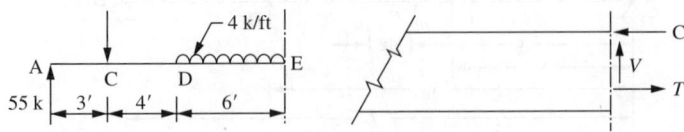

FIGURE 2.5   Internal forces.

In the portion AC, the bending moment at a cross-section a distance x from point A is $M = 55x$. Therefore, the bending moment diagram starts at 0 at A and increases along a straight line to an ordinate of $+165$ at point C. In the portion CD, the bending moment at any point a distance x from C is $M = 55(x+3) - 30x$. Hence, the bending moment diagram in this portion is a straight line increasing from $+165$ at C to $+265$ at D. In the portion DE, the bending moment at any point a distance x from D is $M = 55(x+7) - 30(x+4) - 4x^2/22$. Hence, the bending moment diagram in this portion is a curve with an ordinate of $+265$ at D and $+343$ at E. In an analogous manner, the remainder of the bending moment diagram can easily be constructed.

Bending moment and shear force diagrams for beams with simple boundary conditions and subjected to some selected load cases are given in Figure 2.6.

2.2.3 Fixed-Ended Beams

When the ends of a beam are held so firmly that they are not free to rotate under the action of applied loads, the beam is known as a built-in or fixed-ended beam and it is statically indeterminate. The bending moment diagram for such a beam can be considered to consist of two parts, namely, the free bending moment diagram obtained by treating the beam as if the ends are simply supported and the fixing moment diagram resulting from the restraints imposed at the ends of the beam. The solution of a fixed beam is greatly simplified by considering Mohr's principles, which state that

1. The area of the fixing bending moment diagram is equal to that of the free bending moment diagram.
2. The centers of gravity of the two diagrams lie in the same vertical line; that is, they are equidistant from a given end of the beam.

The construction of the bending moment diagram for a fixed beam is explained with an example shown in Figure 2.7. **PQUT** is the free bending moment diagram, M_s and **PQRS** is the fixing moment diagram, M_i. The net bending moment diagram, M, is shown shaded. If A_s is the area of the free bending moment diagram and A_i the area of the fixing moment diagram then, from the first Mohr principle we have $A_s = A_i$ and

$$\frac{1}{2} \times \frac{Wab}{L} \times L = \frac{1}{2}(M_A + M_B) \times L$$
$$M_A + M_B = \frac{Wab}{L} \tag{2.7}$$

From the second principle, equating the moment about A of A_s and A_i, we have

$$M_A + 2M_B = \frac{Wab}{L^3}(2a^2 + 3ab + b^2) \tag{2.8}$$

Solving Equations 2.7 and 2.8 for M_A and M_B, we get

$$M_A = \frac{Wab^2}{L^2}$$
$$M_B = \frac{Wa^2b}{L^2}$$

Shear force can be determined once the bending moment is known. The shear force at the ends of the beam, that is, at A and B, will be

$$S_A = \frac{M_A - M_B}{L} + \frac{Wb}{L}$$
$$S_B = \frac{M_B - M_A}{L} + \frac{Wa}{L}$$

Bending moment and shear force diagrams for fixed-ended beams subjected to some typical loading cases are shown in Figure 2.8.

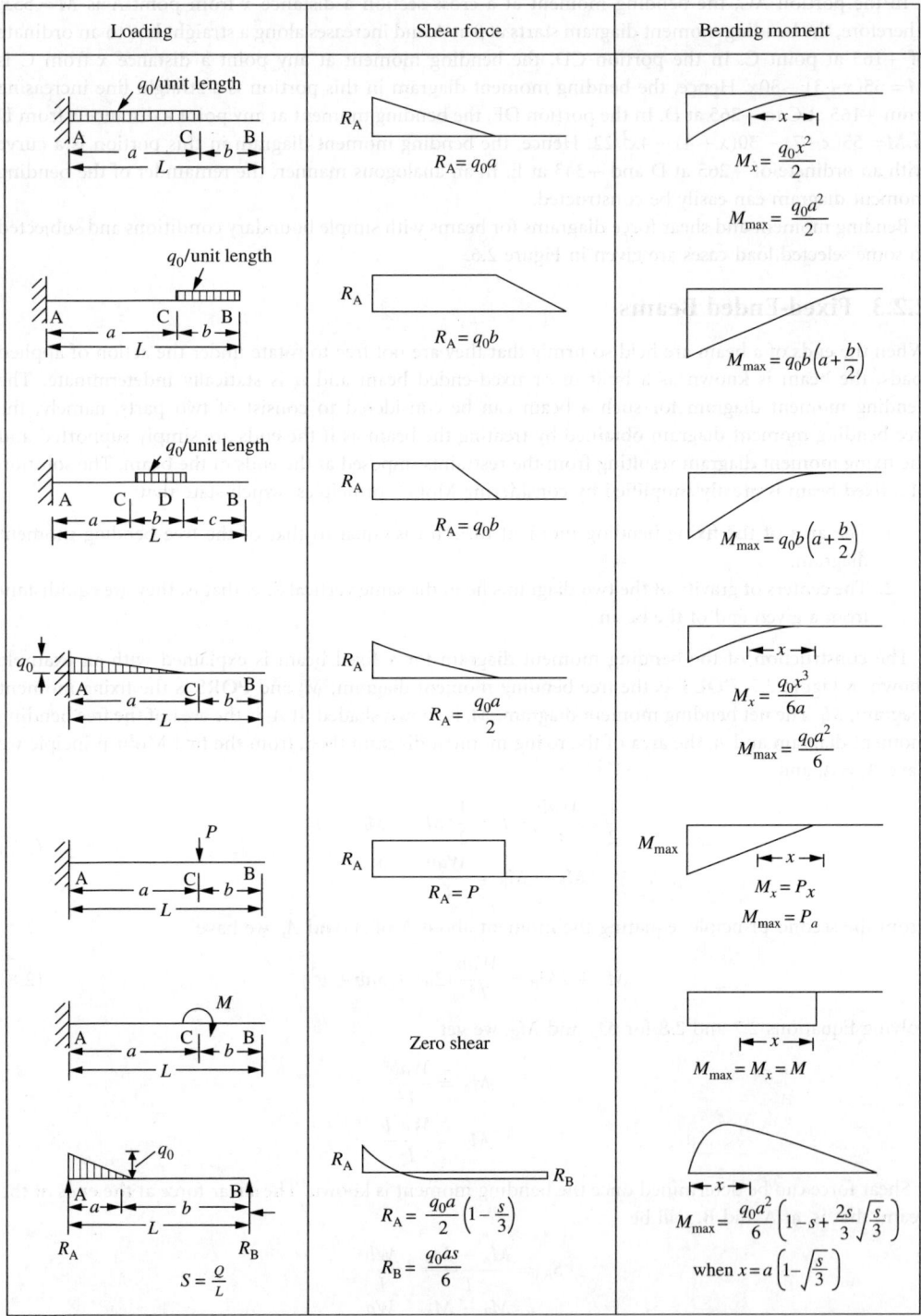

FIGURE 2.6   Shear force and bending moment diagrams for beams with simple boundary conditions subjected to selected loading cases.

Loading	Shear force	Bending moment

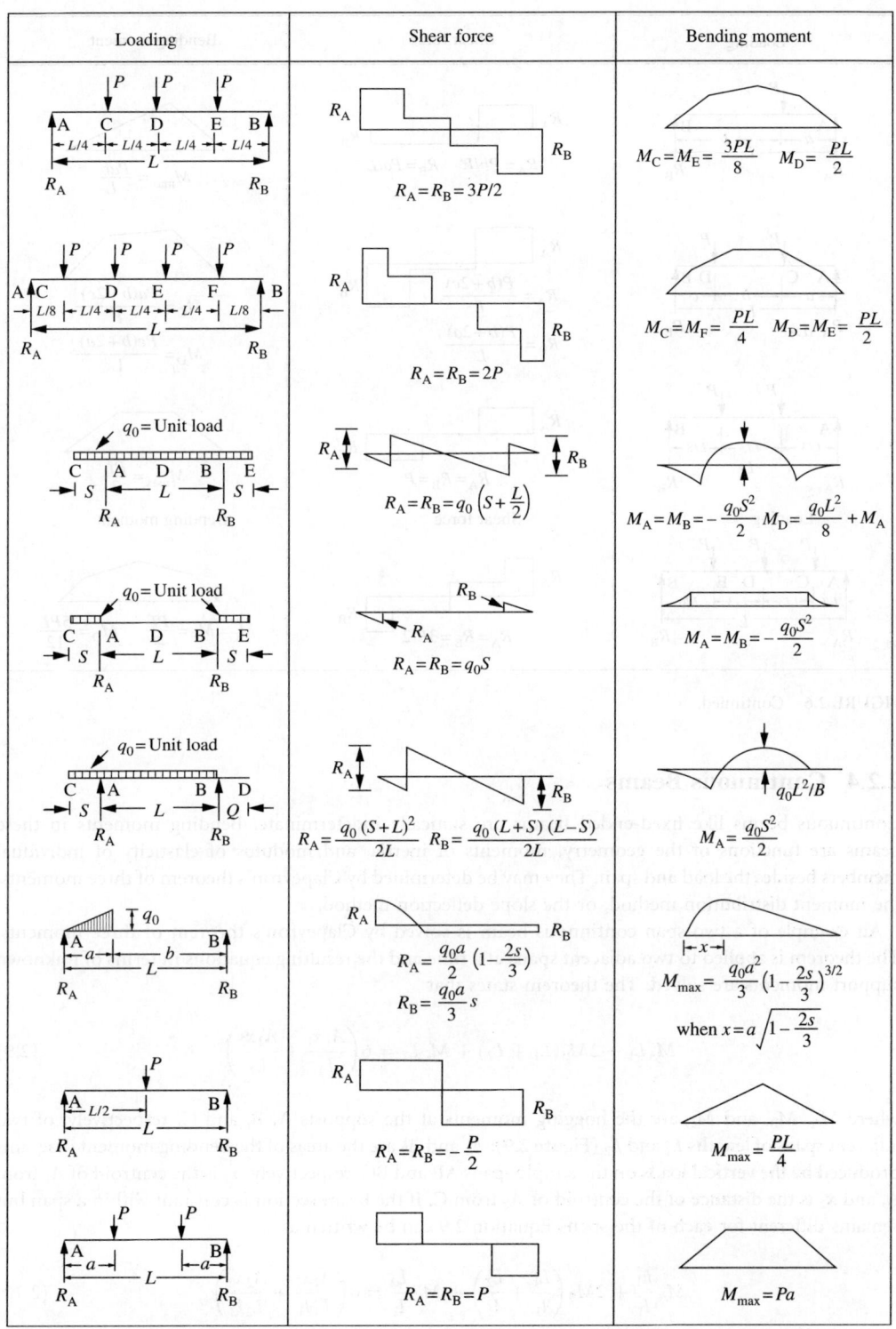

In the table rows:

$$M_C = M_E = \frac{3PL}{8} \quad M_D = \frac{PL}{2}$$

$$R_A = R_B = 3P/2$$

$$R_A = R_B = 2P$$

$$M_C = M_F = \frac{PL}{4} \quad M_D = M_E = \frac{PL}{2}$$

$q_0 = \text{Unit load}$

$$R_A = R_B = q_0\left(S + \frac{L}{2}\right)$$

$$M_A = M_B = -\frac{q_0 S^2}{2} \quad M_D = \frac{q_0 L^2}{8} + M_A$$

$q_0 = \text{Unit load}$

$$R_A = R_B = q_0 S$$

$$M_A = M_B = -\frac{q_0 S^2}{2}$$

$q_0 = \text{Unit load}$

$$R_A = \frac{q_0(S+L)^2}{2L} \quad R_B = \frac{q_0(L+S)(L-S)}{2L}$$

$$M_A = \frac{q_0 S^2}{2}$$

$$q_0 L^2/B$$

$$R_A = \frac{q_0 a}{2}\left(1 - \frac{2s}{3}\right)$$

$$R_B = \frac{q_0 a}{3}s$$

$$M_{max} = \frac{q_0 a^2}{3}\left(1 - \frac{2s}{3}\right)^{3/2}$$

$$\text{when } x = a\sqrt{1 - \frac{2s}{3}}$$

$$R_A = R_B = -\frac{P}{2}$$

$$M_{max} = \frac{PL}{4}$$

$$R_A = R_B = P$$

$$M_{max} = Pa$$

FIGURE 2.6 Continued.

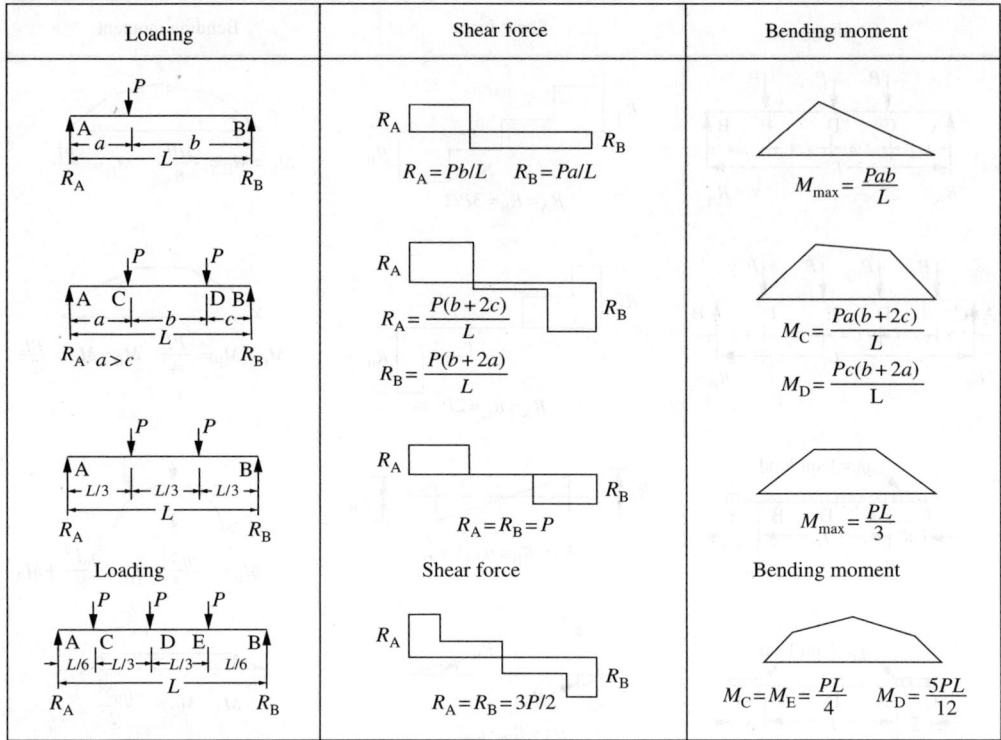

FIGURE 2.6   Continued.

2.2.4 Continuous Beams

Continuous beams like fixed-ended beams are statically indeterminate. Bending moments in these beams are functions of the geometry, moments of inertia, and modulus of elasticity of individual members besides the load and span. They may be determined by Clapeyron's theorem of three moments, the moment distribution method, or the slope deflection method.

An example of a two-span continuous beam is solved by Clapeyron's theorem of three moments. The theorem is applied to two adjacent spans at a time and the resulting equations in terms of unknown support moments are solved. The theorem states that

$$M_A L_1 + 2M_B(L_1 + L_2) + M_C L_2 = 6\left(\frac{A_1 x_1}{L_1} + \frac{A_2 x_2}{L_2}\right) \tag{2.9}$$

where M_A, M_B, and M_C are the hogging moments at the supports A, B, and C, respectively, of two adjacent spans of lengths L_1 and L_2 (Figure 2.9); A_1 and A_2 are the areas of the bending moment diagrams produced by the vertical loads on the simple spans AB and BC, respectively; x_1 is the centroid of A_1 from A, and x_2 is the distance of the centroid of A_2 from C. If the beam section is constant within a span but remains different for each of the spans Equation 2.9 can be written as

$$M_A \frac{L_1}{I_1} + 2M_B\left(\frac{L_1}{I_1} + \frac{L_2}{I_2}\right) + M_C \frac{L_2}{I_2} = 6\left(\frac{A_1 x_1}{L_1 I_1} + \frac{A_2 x_2}{L_2 I_2}\right) \tag{2.10}$$

where I_1 and I_2 are the moments of inertia of the beam section in spans L_1 and L_2, respectively.

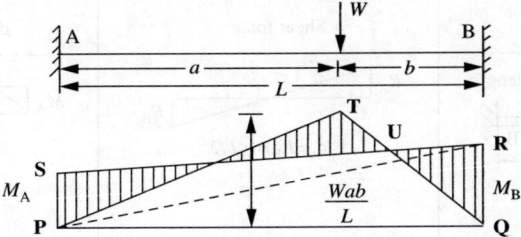

FIGURE 2.7   Fixed-ended beam.

EXAMPLE 2.1

The example in Figure 2.10 shows the application of this theorem.
For spans AC and BC

$$M_A \times 10 + 2M_C(10+10) + M_B \times 10 = 6\left[\frac{(1/2) \times 500 \times 10 \times 5}{10} + \frac{(2/3) \times 250 \times 10 \times 5}{10}\right]$$

Since the support at A is simply supported, $M_A = 0$. Therefore,

$$4M_C + M_B = 1250 \tag{2.11}$$

Considering an imaginary span BD on the right side of B and applying the theorem for spans CB and BD,

$$M_C \times 10 + 2M_B(10) + M_D \times 10 = 6 \times \frac{(2/3) \times 10 \times 5}{10} \times 2 \tag{2.12}$$

$$M_C + 2M_B = 500 \quad (\text{because } M_C = M_D)$$

Solving Equations 2.11 and 2.12, we get

$$M_B = 107.2 \text{ kN m}$$

$$M_C = 285.7 \text{ kN m}$$

Shear force at A is

$$S_A = \frac{M_A - M_C}{L} + 100 = -28.6 + 100 = 71.4 \text{ kN}$$

Shear force at C is

$$S_C = \left(\frac{M_C - M_A}{L} + 100\right) + \left(\frac{M_C - M_B}{L} + 100\right)$$

$$= (28.6 + 100) + (17.9 + 100) = 246.5 \text{ kN}$$

Shear force at B is

$$S_B = \left(\frac{M_B - M_C}{L} + 100\right) = -17.9 + 100 = 82.1 \text{ kN}$$

The bending moment and shear force diagrams are shown in Figure 2.10.

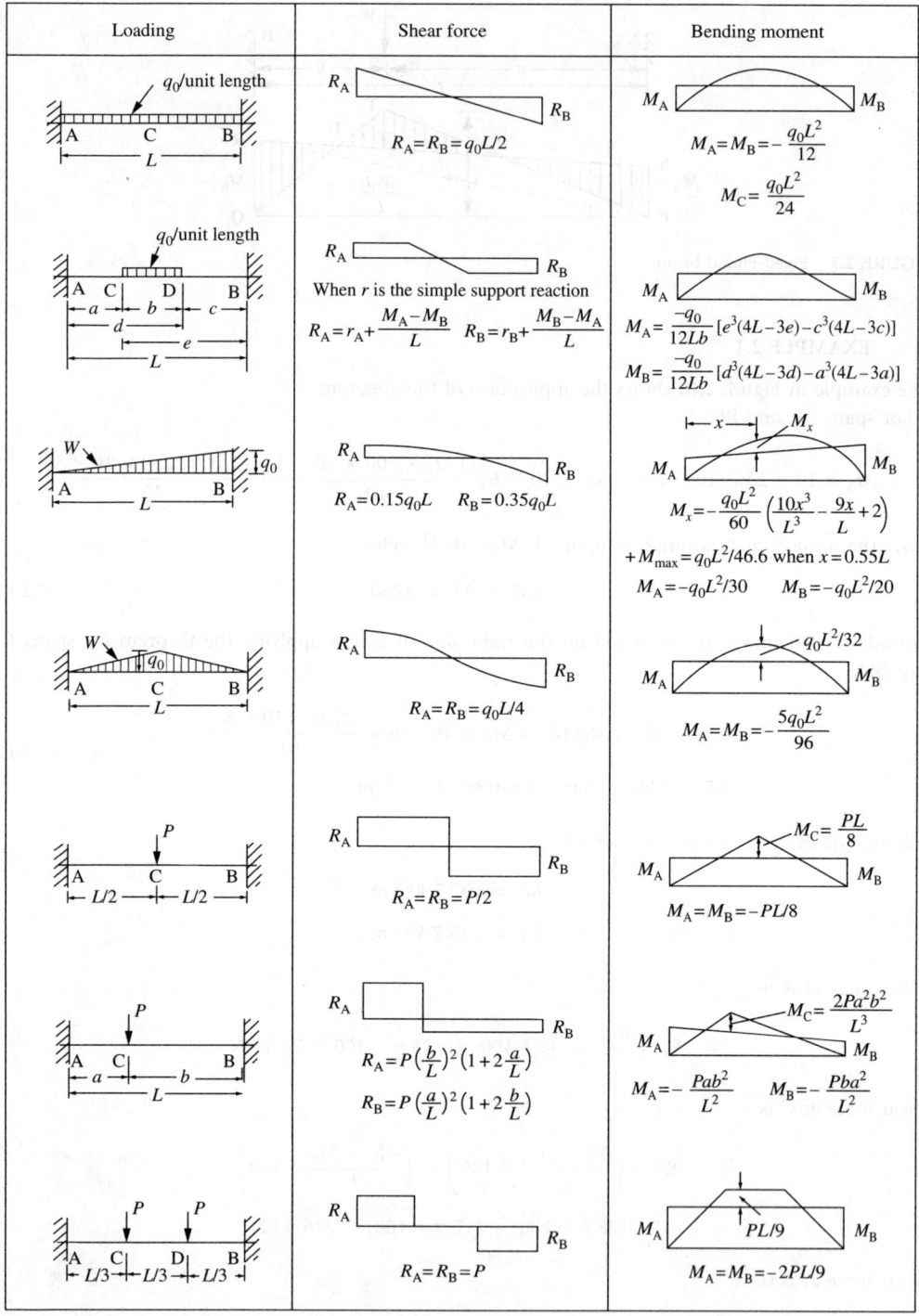

FIGURE 2.8 Shear force and bending moment diagrams for built-up beams subjected to typical loading cases.

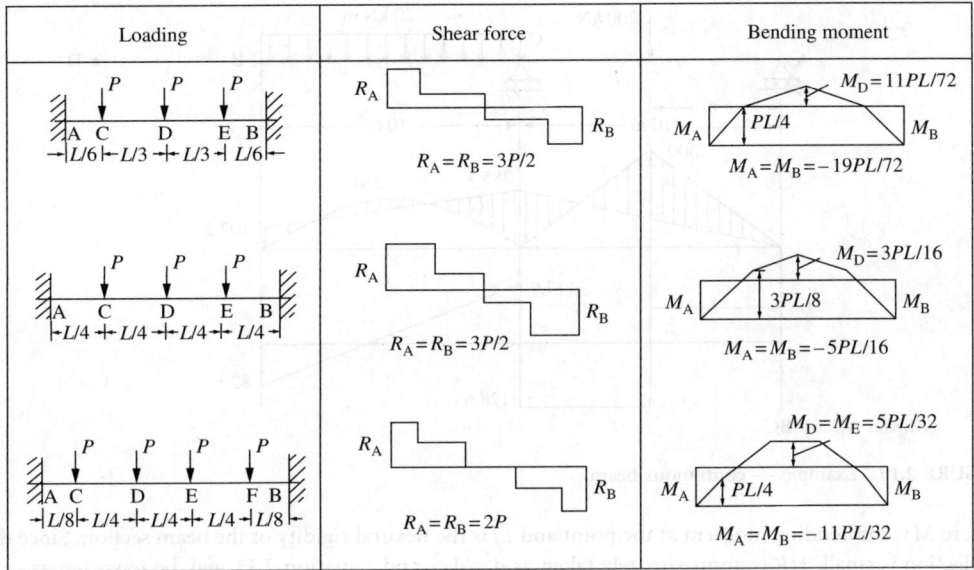

Loading	Shear force	Bending moment

FIGURE 2.8 Continued.

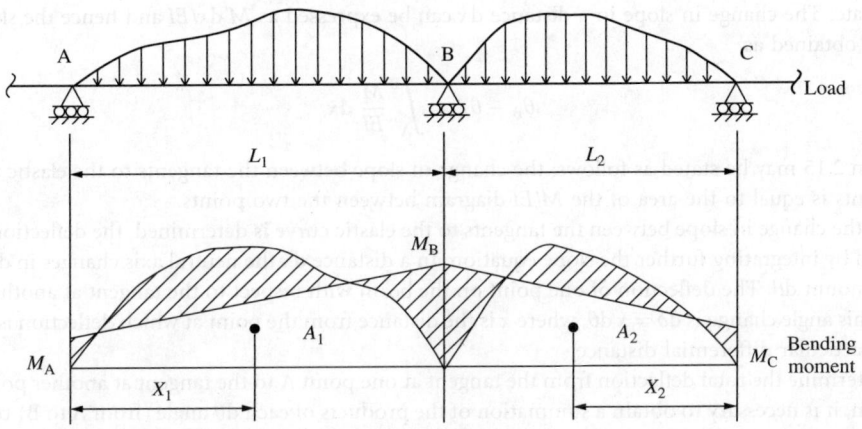

FIGURE 2.9 Continuous beams.

2.2.5 Beam Deflection

There are several methods for determining beam deflections: (i) moment area method, (ii) conjugate-beam method, (iii) virtual work, and (iv) Castigliano's second theorem, among others.

The elastic curve of a member is the shape the neutral axis takes when the member deflects under a load. The inverse of the radius of curvature at any point of this curve is obtained as

$$\frac{1}{R} = \frac{M}{EI} \qquad (2.13)$$

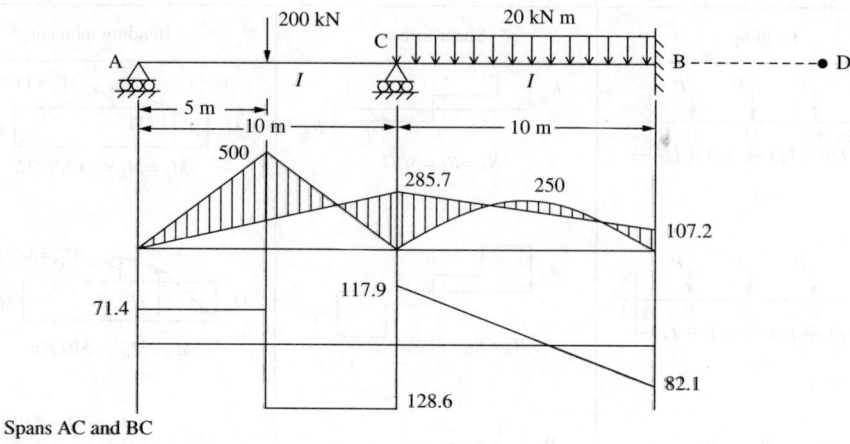

Spans AC and BC

FIGURE 2.10 Example — continuous beam.

where M is the bending moment at the point and EI is the flexural rigidity of the beam section. Since the deflection is small, $1/R$ is approximately taken as d^2y/dx^2, and Equation 2.13 may be rewritten as

$$M = EI \frac{d^2y}{dx^2} \qquad (2.14)$$

In Equation 2.14, y is the deflection of the beam at distance x measured from the origin of the coordinate. The change in slope in a distance dx can be expressed as $M\,dx/EI$ and hence the slope in a beam is obtained as

$$\theta_B - \theta_A = \int_A^B \frac{M}{EI}\,dx \qquad (2.15)$$

Equation 2.15 may be stated as follows: the change in slope between the tangents to the elastic curve at two points is equal to the area of the M/EI diagram between the two points.

Once the change in slope between the tangents to the elastic curve is determined, the deflection can be obtained by integrating further the slope equation. In a distance dx the neutral axis changes in direction by an amount $d\theta$. The deflection of one point on the beam with respect to the tangent at another point due to this angle change is $d\delta = x\,d\theta$, where x is the distance from the point at which deflection is desired to the particular differential distance.

To determine the total deflection from the tangent at one point A to the tangent at another point B on the beam, it is necessary to obtain a summation of the products of each $d\theta$ angle (from A to B) times the distance to the point where deflection is desired or

$$\delta_B - \delta_A = \int_A^B \frac{Mx\,dx}{EI} \qquad (2.16)$$

The deflection of a tangent to the elastic curve of a beam with respect to a tangent at another point is equal to the moment of M/EI diagram between the two points, taken about the point at which deflection is desired.

2.2.5.1 Moment Area Method

The moment area method is most conveniently used for determining slopes and deflections for beams in which the direction of the tangent to the elastic curve at one or more points is known, such as cantilever beams, where the tangent at the fixed end does not change in slope. The method is applied easily to beams loaded with concentrated loads, because the moment diagrams consist of straight lines. These

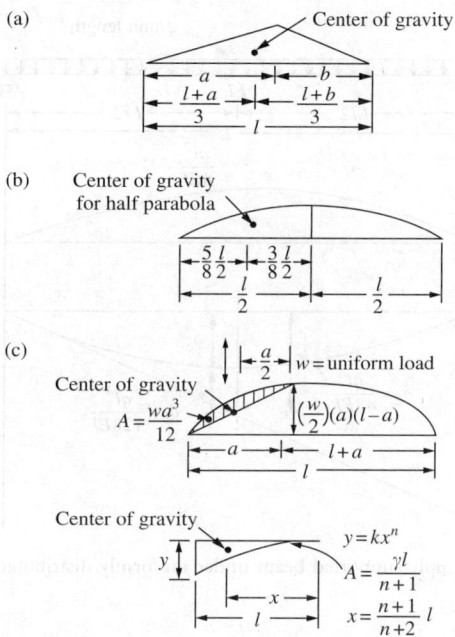

FIGURE 2.11 Typical M/EI diagram.

diagrams can be broken down into single triangles and rectangles. Beams supporting uniform loads or uniformly varying loads may be handled by integration. Properties of some of the shapes of M/EI diagrams that designers usually come across are given in Figure 2.11.

It should be understood that the slopes and deflections that are obtained using the moment area theorems are with respect to tangents to the elastic curve at the points being considered. The theorems do not directly give the slope or deflection at a point in the beam as compared to the horizontal axis (except in one or two special cases); they give the change in slope of the elastic curve from one point to another or the deflection of the tangent at one point with respect to the tangent at another point. There are some special cases in which beams are subjected to several concentrated loads or the combined action of concentrated and uniformly distributed loads. In such cases it is advisable to separate the concentrated loads and uniformly distributed loads and the moment area method can be applied separately to each of these loads. The final responses are obtained by the principle of superposition.

For example, consider a simply supported beam subjected to a uniformly distributed load q as shown in Figure 2.12. The tangents to the elastic curve at each end of the beam are inclined. The deflection δ_1 of the tangent at the left end from the tangent at the right end is found to be $ql^4/24EI$. The distance from the original chord between the supports and the tangent at the right end, δ_2, can be computed as $ql^4/48EI$. The deflection of a tangent at the center from a tangent at the right end, δ_3, is determined in this step as $ql^4/128EI$. The difference between δ_2 and δ_3 gives the centerline deflection as $(5/384)(ql^4/EI)$.

2.2.6 Curved Beams

The beam formulas derived in the previous section are based on the assumption that the member to which a bending moment is applied is initially straight. Many members, however, are curved before a bending moment is applied to them. Such members are called curved beams. In the following discussion all the conditions applicable to the straight-beam formula are assumed valid except that the beam is initially curved.

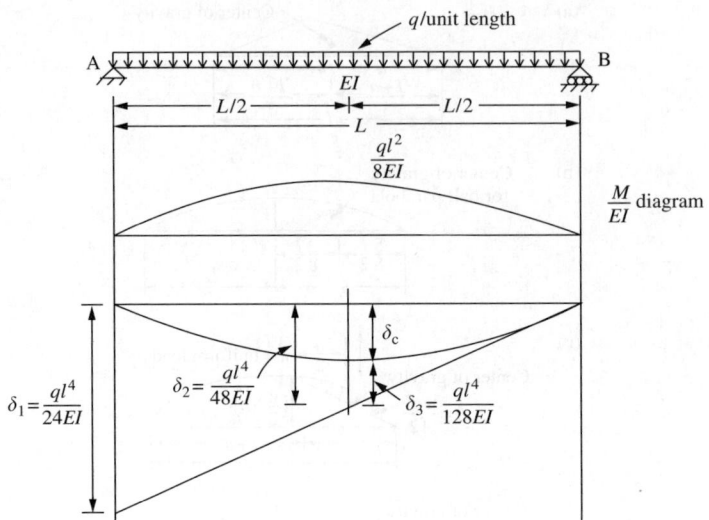

FIGURE 2.12   Deflection — simply supported beam under uniformly distributed load.

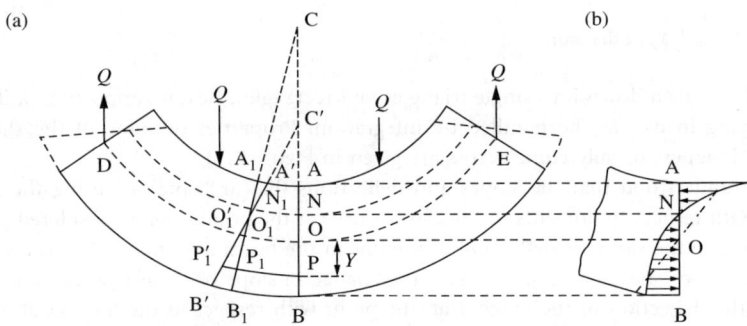

FIGURE 2.13   Bending of curved beams.

Let the curved beam DOE shown in Figure 2.13 be subjected to the loads Q. The surface in which the fibers do not change in length is called the neutral surface. The total deformations of the fibers between two normal sections such as AB and A_1B_1 are assumed to vary proportionally with the distances of the fibers from the neutral surface. The top fibers are compressed while those at the bottom are stretched; that is, the plane section before bending remains a plane after bending.

In Figure 2.13 the two lines AB and A_1B_1 are two normal sections of the beam before the loads are applied. The change in the length of any fiber between these two normal sections after bending is represented by the distance along the fiber between the lines A_1B_1 and $A'B'$; the neutral surface is represented by NN_1, and the stretch of fiber PP_1 is P_1P_1', etc. For convenience, it will be assumed that the line AB is a line of symmetry and does not change direction.

The total deformations of the fibers in the curved beam are proportional to the distances of the fibers from the neutral surface. However, the strains of the fibers are not proportional to these distances because the fibers are not of equal length. Within the elastic limit the stress on any fiber in the beam is proportional to the strain in the fiber, and hence the elastic stresses on the fibers of a curved beam are not proportional to the distances of the fibers from the neutral surface. The resisting moment in a curved beam, therefore, is not given by the expression $\sigma I/c$. Hence, the neutral axis in a curved beam does not pass through the centroid of the section. The distribution of stress over the section and the relative position of the neutral axis are shown in Figure 2.13b; if the beam were straight, the stress would be zero at the centroidal axis and would vary proportionally with the distance from the centroidal axis as indicated by the dot–dash line in the figure. The stress on a normal section such as AB is called the circumferential stress.

2.2.6.1 Sign Conventions

The bending moment M is positive when it decreases the radius of curvature and negative when it increases the radius of curvature; y is positive when measured toward the convex side of the beam and negative when measured toward the concave side, that is, toward the center of curvature. With these sign conventions, σ is positive when it is a tensile stress.

2.2.6.2 Circumferential Stresses

Figure 2.14 shows a free-body diagram of the portion of the body on one side of the section; the equations of equilibrium are applied to the forces acting on this portion. The equations obtained are

$$\sum F_z = 0 \quad \text{or} \quad \int \sigma \, da = 0 \tag{2.17}$$

$$\sum M_z = 0 \quad \text{or} \quad M = \int y\sigma \, da \tag{2.18}$$

Figure 2.15 represents the part ABB_1A_1 of Figure 2.13a enlarged; the angle between the two sections AB and A_1B_1 is $d\theta$. The bending moment causes the plane A_1B_1 to rotate through an angle $\Delta d\theta$, thereby changing the angle this plane makes with the plane BAC from $d\theta$ to $(d\theta + \Delta d\theta)$; the center of curvature is changed from C to C', and the distance of the centroidal axis from the center of curvature is changed from R to ρ. It should be noted that y, R, and ρ at any section are measured from the centroidal axis and not from the neutral axis.

It can be shown that the bending stress σ is given by the relation

$$\sigma = \frac{M}{aR}\left(1 + \frac{1}{Z}\frac{y}{R+y}\right) \tag{2.19}$$

where

$$Z = -\frac{1}{a}\int \frac{y}{R+y} \, da$$

σ is the tensile or compressive (circumferential) stress at a point at a distance y from the centroidal axis of a transverse section at which the bending moment is M; R is the distance from the centroidal axis of the section to the center of curvature of the central axis of the unstressed beam; a is the area of the cross-section; Z is a property of the cross-section, the values of which can be obtained from the expressions for various areas given in Table 2.1. (Detailed information can be obtained from Seely and Smith 1952.)

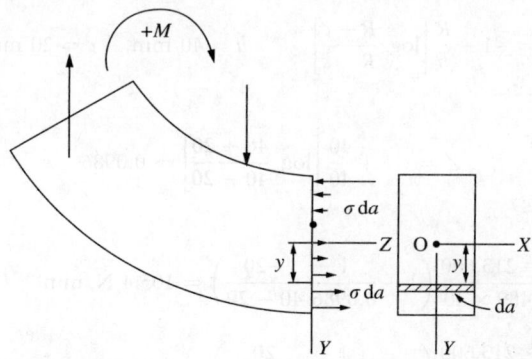

FIGURE 2.14   Free-body diagram of a curved beam segment.

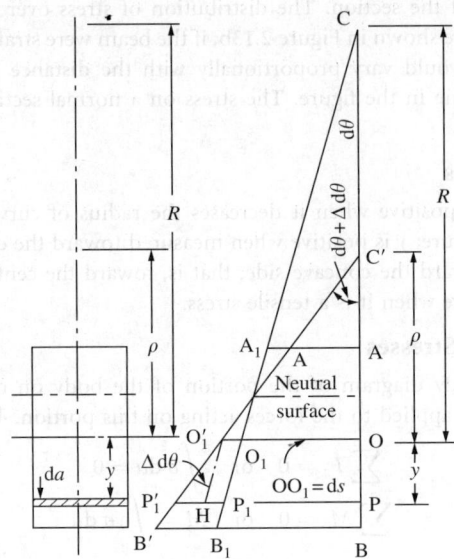

FIGURE 2.15 Curvature in a curved beam.

EXAMPLE 2.2

The bent bar shown in Figure 2.16 is subjected to a load $P = 1780$ N. Calculate the circumferential stress at A and B assuming that the elastic strength of the material is not exceeded.

We know from Equation 2.19,

$$\sigma = \frac{P}{a} + \frac{M}{aR}\left(1 + \frac{1}{Z}\frac{y}{R+y}\right)$$

where

a = area of rectangular section = $40 \times 12 = 480$ mm^2
$R = 40$ mm
$y_A = -20$
$y_B = +20$
$P = 1780$ N
$M = -1780 \times 120 = -213{,}600$ N mm

From Table 2.1, for rectangular section

$$Z = -1 + \frac{R}{h}\left[\log_e\frac{R+c}{R-c}\right], \qquad h = 40 \text{ mm}, \quad c = 20 \text{ mm}$$

Hence,

$$Z = -1 + \frac{40}{40}\left[\log_e\frac{40+20}{40-20}\right] = 0.0986$$

Therefore,

$$\sigma_A = \frac{1780}{480} + \frac{-213{,}600}{480 \times 40}\left(1 + \frac{1}{0.0986}\frac{-20}{40-20}\right) = 105.4 \text{ N/mm}^2 \quad \text{(tensile)}$$

$$\sigma_B = \frac{1780}{480} + \frac{-213{,}600}{480 \times 40}\left(1 + \frac{1}{0.0986}\frac{20}{40+20}\right) = -45 \text{ N/mm}^2 \quad \text{(compressive)}$$

TABLE 2.1 Analytical Expressions for Z

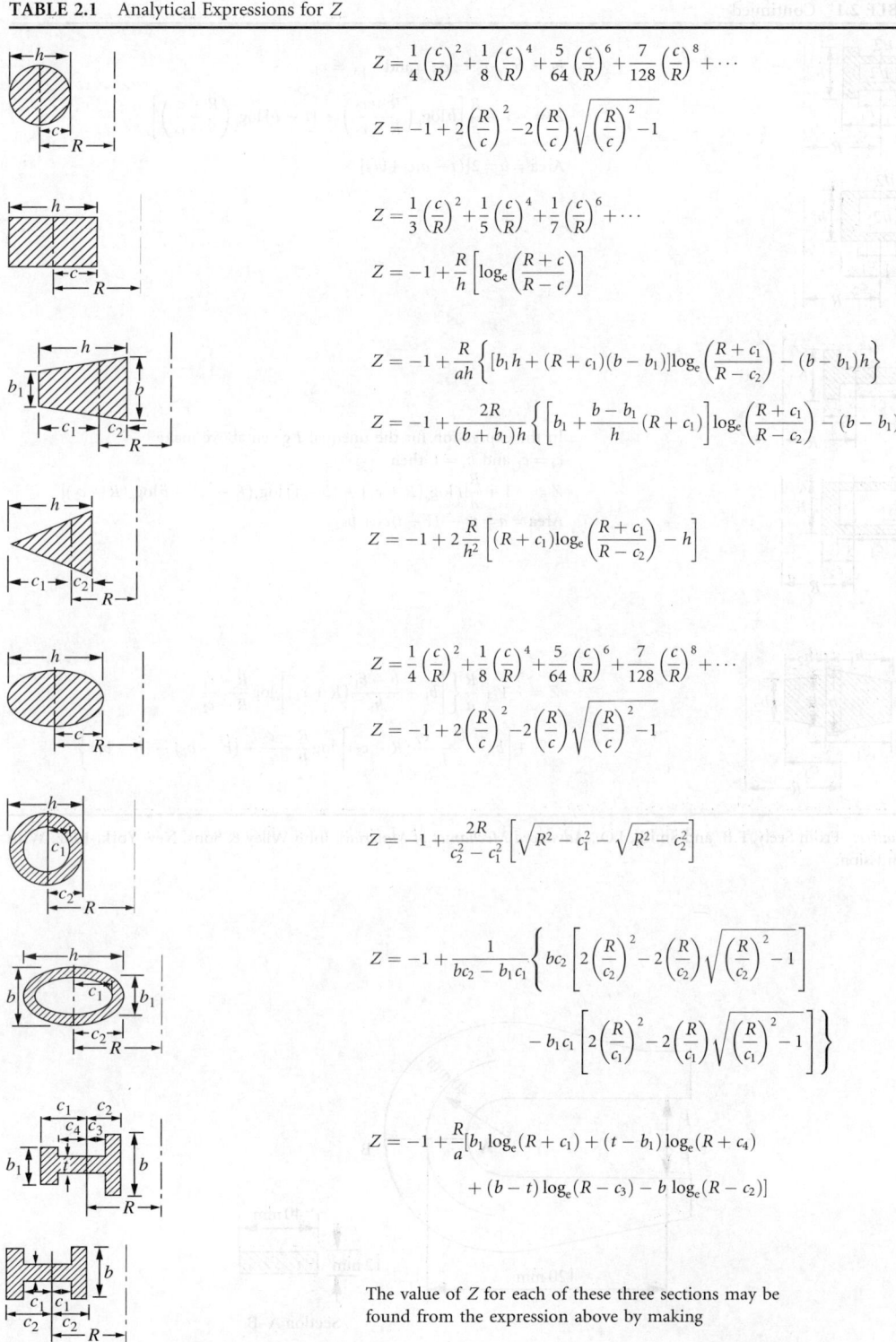

$$Z = \frac{1}{4}\left(\frac{c}{R}\right)^2 + \frac{1}{8}\left(\frac{c}{R}\right)^4 + \frac{5}{64}\left(\frac{c}{R}\right)^6 + \frac{7}{128}\left(\frac{c}{R}\right)^8 + \cdots$$

$$Z = -1 + 2\left(\frac{R}{c}\right)^2 - 2\left(\frac{R}{c}\right)\sqrt{\left(\frac{R}{c}\right)^2 - 1}$$

$$Z = \frac{1}{3}\left(\frac{c}{R}\right)^2 + \frac{1}{5}\left(\frac{c}{R}\right)^4 + \frac{1}{7}\left(\frac{c}{R}\right)^6 + \cdots$$

$$Z = -1 + \frac{R}{h}\left[\log_e\left(\frac{R+c}{R-c}\right)\right]$$

$$Z = -1 + \frac{R}{ah}\left\{[b_1 h + (R + c_1)(b - b_1)]\log_e\left(\frac{R+c_1}{R-c_2}\right) - (b-b_1)h\right\}$$

$$Z = -1 + \frac{2R}{(b+b_1)h}\left\{\left[b_1 + \frac{b-b_1}{h}(R+c_1)\right]\log_e\left(\frac{R+c_1}{R-c_2}\right) - (b-b_1)\right\}$$

$$Z = -1 + 2\frac{R}{h^2}\left[(R+c_1)\log_e\left(\frac{R+c_1}{R-c_2}\right) - h\right]$$

$$Z = \frac{1}{4}\left(\frac{c}{R}\right)^2 + \frac{1}{8}\left(\frac{c}{R}\right)^4 + \frac{5}{64}\left(\frac{c}{R}\right)^6 + \frac{7}{128}\left(\frac{c}{R}\right)^8 + \cdots$$

$$Z = -1 + 2\left(\frac{R}{c}\right)^2 - 2\left(\frac{R}{c}\right)\sqrt{\left(\frac{R}{c}\right)^2 - 1}$$

$$Z = -1 + \frac{2R}{c_2^2 - c_1^2}\left[\sqrt{R^2 - c_1^2} - \sqrt{R^2 - c_2^2}\right]$$

$$Z = -1 + \frac{1}{bc_2 - b_1 c_1}\left\{bc_2\left[2\left(\frac{R}{c_2}\right)^2 - 2\left(\frac{R}{c_2}\right)\sqrt{\left(\frac{R}{c_2}\right)^2 - 1}\right]\right.$$

$$\left. - b_1 c_1\left[2\left(\frac{R}{c_1}\right)^2 - 2\left(\frac{R}{c_1}\right)\sqrt{\left(\frac{R}{c_1}\right)^2 - 1}\right]\right\}$$

$$Z = -1 + \frac{R}{a}[b_1 \log_e(R+c_1) + (t-b_1)\log_e(R+c_4)$$

$$+ (b-t)\log_e(R-c_3) - b\log_e(R-c_2)]$$

The value of Z for each of these three sections may be found from the expression above by making

TABLE 2.1   Continued

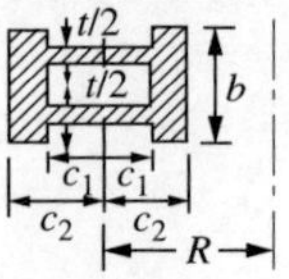

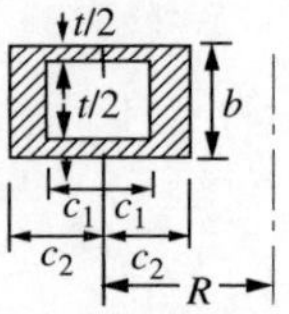

$b_1 = b, \quad c_2 = c_1, \quad \text{and} \quad c_3 = c_4$

$$Z = -1 + \frac{R}{a}\left[b\log_e\left(\frac{R+c_2}{R-c_2}\right) + (t-b)\log_e\left(\frac{R+c_1}{R-c_1}\right)\right]$$

$\text{Area} = a = 2[(t-b)c_1 + bc_2]$

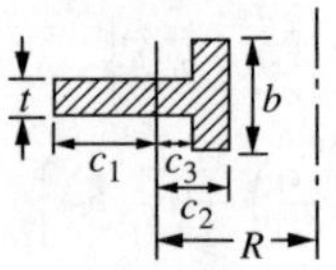

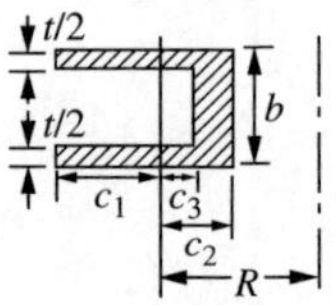

In the expression for the unequal I given above make
$c_4 = c_1$ and $b_1 = t$, then

$$Z = -1 + \frac{R}{a}[t\log_e(R+c_1) + (b-t)\log_e(R-c_3) - b\log_e(R-c_2)]$$

$\text{Area} = a = tc_1 - (b-t)c_3 + bc_2$

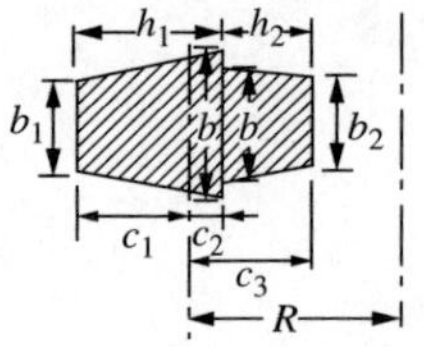

$$Z = -1 + \frac{R}{a}\left\{\left[b_1 + \frac{b-b_1}{h_1}(R+c_1)\right]\log\frac{R+c_1}{R-c_2}\right.$$

$$\left. + \left[b_2 - \frac{b'-b_2}{h_2}(R-c_3)\right]\log\frac{R-c_2}{R-c_3} + (b'-b_2) - (b-b_1)\right\}$$

Source: From Seely, F.B. and Smith, J.O., *Advanced Mechanics of Materials*, John Wiley & Sons, New York, 1952. With permission.

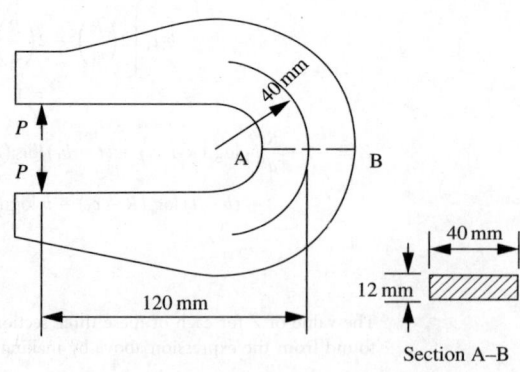

FIGURE 2.16   Bent bar.

2.3 Trusses

A structure that is composed of a number of members pin-connected at their ends to form a stable framework is called a truss. If all the members lie in a plane, it is a planar truss. It is generally assumed that loads and reactions are applied to the truss only at the joints. The centroidal axis of each member is straight, coincides with the line connecting the joint centers at each end of the member, and lies in a plane that also contains the lines of action of all the loads and reactions. Many truss structures are three-dimensional in nature. However, in many cases, such as bridge structures and simple roof systems, the three-dimensional framework can be subdivided into planar components for analysis as planar trusses without seriously compromising the accuracy of the results. Figure 2.17 shows some typical idealized planar truss structures.

There exists a relation among the number of members, m, number of joints, j, and reaction components, r. The expression is

$$m = 2j - r \qquad (2.20)$$

which must be satisfied if it is to be statically determinate internally. r is the least number of reaction components required for external stability. If m exceeds $(2j - r)$, then the excess members are called redundant members and the truss is said to be statically indeterminate.

For a statically determinate truss, member forces can be found by using the method of equilibrium. The process requires repeated use of free-body diagrams from which individual member forces are determined. The method of joints is a technique of truss analysis in which the member forces are determined by the sequential isolation of joints — the unknown member forces at one joint are solved and become known for the subsequent joints. The other method is known as the method of sections in which equilibrium of a part of the truss is considered.

2.3.1 Method of Joints

An imaginary section may be completely passed around a joint in a truss. The joint has become a free body in equilibrium under the forces applied to it. The equations $\sum H = 0$ and $\sum V = 0$ may be applied to the joint to determine the unknown forces in members meeting there. It is evident that no more than two unknowns can be determined at a joint with these two equations.

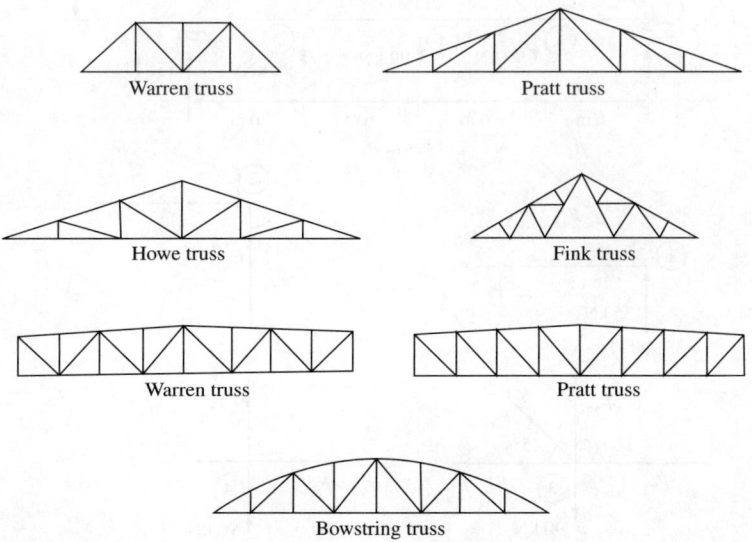

FIGURE 2.17 Typical planar trusses.

EXAMPLE 2.3

A truss shown in Figure 2.18 is symmetrically loaded, and it is sufficient to solve half the truss by considering the joints 1 to 5. At joint 1, there are two unknown forces. Summation of the vertical components of all forces at joint 1 gives

$$135 - F_{12} \sin 45° = 0$$

which in turn gives the force in the member 1–2, $F_{12} = 190$ kN (compressive). Similarly, summation of the horizontal components gives

$$F_{13} - F_{12} \cos 45° = 0$$

Substituting for F_{12} gives the force in the member 1–3 as

$$F_{13} = 135 \text{ kN} \quad \text{(tensile)}$$

Now, joint 2 is cut completely and it is found that there are two unknown forces, F_{25} and F_{23}. Summation of the vertical components gives

$$F_{12} \cos 45° - F_{23} = 0$$

Therefore,

$$F_{23} = 135 \text{ kN} \quad \text{(tensile)}$$

Summation of the horizontal components gives

$$F_{12} \sin 45° - F_{25} = 0$$

and hence,

$$F_{25} = 135 \text{ kN} \quad \text{(compressive)}$$

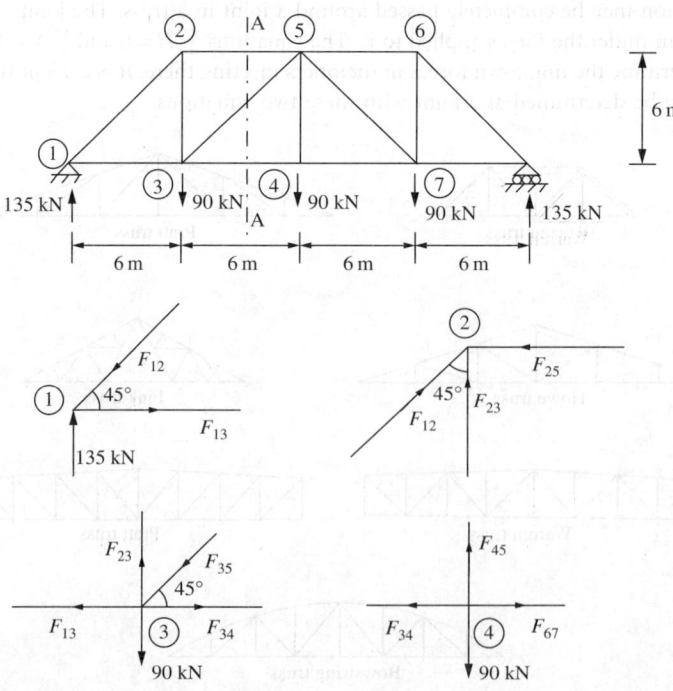

FIGURE 2.18   Example — method of joints, planar truss.

After solving for joints 1 and 2 one proceeds to take a section around joint 3 at which there are now two unknown forces, namely, F_{34} and F_{35}. Summation of the vertical components at joint 3 gives

$$F_{23} - F_{35} \sin 45° - 90 = 0$$

Substituting for F_{23}, one obtains $F_{35} = 63.6$ kN (compressive). Summing the horizontal components and substituting for F_{13} one gets

$$-135 - 45 + F_{34} = 0$$

Therefore,

$$F_{34} = 180 \text{ kN} \quad \text{(tensile)}$$

The next joint involving two unknowns is joint 4. When we consider a section around it, the summation of the vertical components at joint 4 gives

$$F_{45} = 90 \text{ kN} \quad \text{(tensile)}$$

Now, the forces in all the members on the left half of the truss are known and by symmetry the forces in the remaining members can be determined. The forces in all the members of a truss can also be determined by making use of the method of sections.

2.3.2 Method of Sections

In this method, an imaginary cutting line called a section is drawn through a stable and determinate truss. Thus, a section subdivides the truss into two separate parts. Since the entire truss is in equilibrium, any part of it must also be in equilibrium. Either of the two parts of the truss can be considered and the three equations of equilibrium, $\sum F_x = 0$, $\sum F_y = 0$, and $\sum M = 0$, can be applied to solve for member forces.

The example considered in Section 2.3.1 is once again considered (Figure 2.19). To calculate the force in the member 3–5, F_{35}, a section AA should be run to cut the member 3–5 as shown in the figure. It is only required to consider the equilibrium of one of the two parts of the truss. In this case, the portion of the truss on the left of the section is considered. The left portion of the truss as shown in Figure 2.19 is in equilibrium under the action of the forces namely, the external and internal forces. Considering the equilibrium of forces in the vertical direction one can obtain

$$135 - 90 + F_{35} \sin 45° = 0$$

Therefore, F_{35} is obtained as

$$F_{35} = -45\sqrt{2} \text{ kN}$$

The negative sign indicates that the member force is compressive. The other member forces cut by the section can be obtained by considering the other equilibrium equations, namely, $\sum M = 0$. More sections can be taken in the same way to solve for other member forces in the truss. The most important advantage of this method is that one can obtain the required member force without solving for the other member forces.

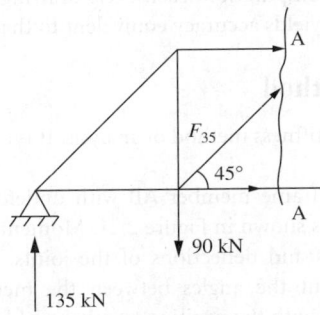

FIGURE 2.19 Example — method of sections, planar truss.

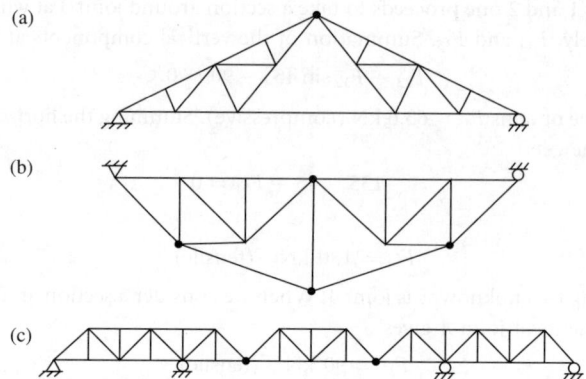

FIGURE 2.20 Compound trusses: (a) compound roof truss, (b) compound bridge truss, and (c) cantilevered construction.

2.3.3 Compound Trusses

A compound truss is formed by interconnecting two or more simple trusses. Examples of compound trusses are shown in Figure 2.20. A typical compound roof truss is shown in Figure 2.20a in which two simple trusses are interconnected by means of a single member and a common joint. The compound truss shown in Figure 2.20b is commonly used in bridge construction and, in this case, three members are used to interconnect two simple trusses at a common joint. There are three simple trusses interconnected at their common joints as shown in Figure 2.20c.

   The method of sections may be used to determine the member forces in the interconnecting members of compound trusses similar to those shown in Figure 2.20a and Figure 2.20b. However, in the case of a cantilevered truss the middle simple truss is isolated as a free-body diagram to find its reactions. These reactions are reversed and applied to the interconnecting joints of the other two simple trusses. After the interconnecting forces between the simple trusses are found, the simple trusses are analyzed by the method of joints or the method of sections.

2.4 Frames

Frames are statically indeterminate in general; special methods are required for their analysis. The slope deflection and moment distribution methods are two such methods commonly employed. Slope deflection is a method that takes into account the flexural displacements such as rotations and deflections and involves solutions of simultaneous equations. Moment distribution on the other hand involves successive cycles of computation, each cycle drawing closer to the "exact" answers. The method is more labor intensive but yields accuracy equivalent to that obtained from the exact methods.

2.4.1 Slope Deflection Method

This method is a special case of the stiffness method of analysis. It is a convenient method for performing hand analysis of small structures.

   Let us consider that a prismatic frame member AB with undeformed position along the x axis is deformed into the configuration P as shown in Figure 2.21. Moments at the ends of frame members are expressed in terms of the rotations and deflections of the joints. It is assumed that the joints in a structure may rotate or deflect, but the angles between the members meeting at a joint remain unchanged. The positive axes, along with the positive member end force components and displacement components, are shown in the figure.

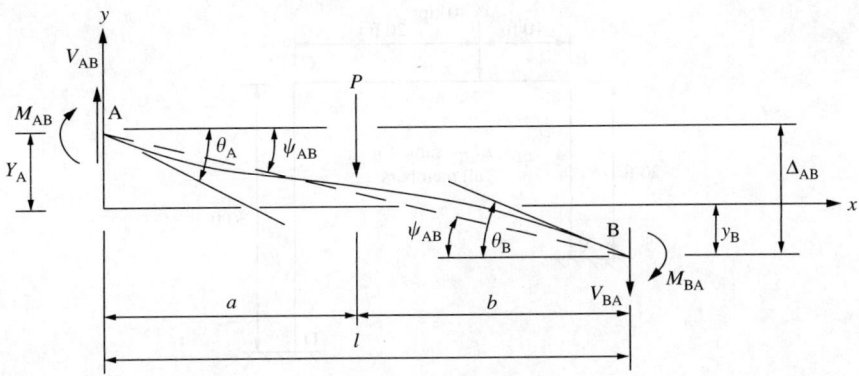

FIGURE 2.21 Deformed configuration of a beam.

The equations for end moments may be written as

$$M_{AB} = \frac{2EI}{l}(2\theta_A + \theta_B - 3\psi_{AB}) + M_{FAB}$$

$$M_{BA} = \frac{2EI}{l}(2\theta_B + \theta_A - 3\psi_{AB}) + M_{FBA}$$

(2.21)

where M_{FAB} and M_{FBA} are fixed-end moments at supports A and B, respectively, due to the applied load. ψ_{AB} is the rotation as a result of the relative displacement between the member ends A and B given as

$$\psi_{AB} = \frac{\Delta_{AB}}{l} = \frac{y_A + y_B}{l}$$

(2.22)

where Δ_{AB} is the relative deflection of the beam ends and y_A and y_B are the vertical displacements at ends A and B. Fixed-end moments for some loading cases may be obtained from Figure 2.8. The slope deflection equations (Equations 2.21) show that the moment at the end of a member is dependent on member properties EI, length l, and displacement quantities. The fixed-end moments reflect the transverse loading on the member.

2.4.2 Frame Analysis Using Slope Deflection Method

The slope deflection equations may be applied to statically indeterminate frames with or without side sway. A frame may be subjected to side sway if the loads, member properties, and dimensions of the frame are not symmetrical about the centerline. Application of the slope deflection method can be illustrated with the following example.

EXAMPLE 2.4

Consider the frame shown in Figure 2.22 subjected to side sway Δ to the right of the frame. Equations 2.21 can be applied to each of the members of the frame as follows:

Member AB:

$$M_{AB} = \frac{2EI}{6}\left(2\theta_A + \theta_B - \frac{3\Delta}{20}\right) + M_{FAB}$$

$$M_{BA} = \frac{2EI}{6}\left(2\theta_B + \theta_A - \frac{3\Delta}{20}\right) + M_{FBA}$$

$$\theta_A = 0, \quad M_{FAB} = M_{FBA} = 0$$

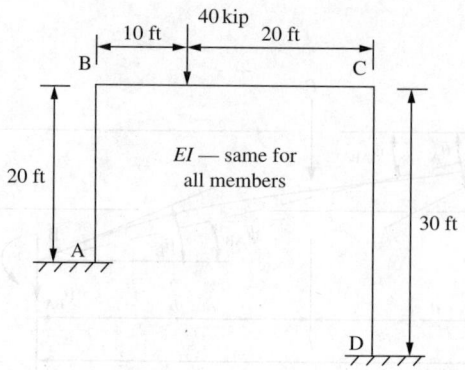

FIGURE 2.22 Example — slope deflection method.

Hence,

$$M_{AB} = \frac{2EI}{6}(\theta_B - 3\psi) \tag{2.23}$$

$$M_{BA} = \frac{2EI}{6}(2\theta_B - 3\psi) \tag{2.24}$$

where $\psi = \Delta/6$.

Member BC:

$$M_{BC} = \frac{2EI}{9}(2\theta_B + \theta_C - 3 \times 0) + M_{FBC}$$

$$M_{CB} = \frac{2EI}{9}(2\theta_C + \theta_B - 3 \times 0) + M_{FCB}$$

$$M_{FBC} = -\frac{180 \times 3 \times 6^2}{9^2} = -240 \text{ kN m}$$

$$M_{FCB} = \frac{180 \times 3^2 \times 6}{9^2} = 120 \text{ kN m}$$

Hence,

$$M_{BC} = \frac{2EI}{9}(2\theta_B + \theta_C) - 240 \tag{2.25}$$

$$M_{CB} = \frac{2EI}{9}(2\theta_C + \theta_B) + 120 \tag{2.26}$$

Member CD:

$$M_{CD} = \frac{2EI}{9}\left(2\theta_C + \theta_D - \frac{3\Delta}{9}\right) + M_{FCD}$$

$$M_{DC} = \frac{2EI}{9}\left(2\theta_D + \theta_C - \frac{3\Delta}{9}\right) + M_{FDC}$$

$$\theta_D = 0, \quad M_{FCD} = M_{FDC} = 0$$

Hence,

$$M_{CD} = \frac{2EI}{9}\left(2\theta_C - \frac{1}{3} \times 6\psi\right) = \frac{2EI}{9}(2\theta_C - 2\psi) \qquad (2.27)$$

$$M_{DC} = \frac{2EI}{9}\left(\theta_C - \frac{1}{3} \times 6\psi\right) = \frac{2EI}{9}(\theta_C - 2\psi) \qquad (2.28)$$

Considering the moment equilibrium at joint B,

$$\sum M_B = M_{BA} + M_{BC} = 0$$

Substituting for M_{BA} and M_{BC} one obtains

$$\frac{EI}{9}(10\theta_B + 2\theta_C - 9\psi) = 240 \quad \text{or} \quad 110\theta_B + 2\theta_C - 9\psi = \frac{2160}{EI} \qquad (2.29)$$

Considering the moment equilibrium at joint C,

$$\sum M_C = M_{CB} + M_{CD} = 0$$

Substituting for M_{CB} and M_{CD} we get

$$\frac{2EI}{9}(4\theta_C + \theta_B - 2\psi) = -120 \quad \text{or} \quad \theta_B + 4\theta_C - 2\psi = -\frac{540}{EI} \qquad (2.30)$$

If summation of the base shear equals zero, we have

$$\sum H = H_A + H_D = 0$$

or

$$\frac{M_{AB} + M_{BA}}{6} + \frac{M_{CD} + M_{DC}}{9} = 0$$

Substituting for M_{AB}, M_{BA}, M_{CD}, and M_{DC} and simplifying

$$27\theta_B + 12\theta_C - 70\psi = 0 \qquad (2.31)$$

solution of Equations 2.29 to 2.31 results in

$$\theta_B = \frac{342.7}{EI}$$

$$\theta_C = \frac{-169.1}{EI} \qquad (2.32)$$

$$\psi = \frac{103.2}{EI}$$

Substituting for θ_B, θ_C, and ψ from Equations 2.32 into Equations 2.23 to 2.28, we get

$M_{AB} = 11.03$ kN m

$M_{BA} = 125.3$ kN m

$M_{BC} = -125.3$ kN m

$M_{CB} = 121$ kN m

$M_{CD} = -121$ kN m

$M_{DC} = -83$ kN m

2.4.3 Moment Distribution Method

The moment distribution method involves successive cycles of computation, each cycle drawing closer to the exact answers. The calculations may be stopped after two or three cycles, giving a very good approximate analysis, or they may be carried on to whatever degree of accuracy is desired. Moment distribution remains the most important hand-calculation method for the analysis of continuous beams and frames and it may be solely used for the analysis of small structures. Unlike the slope deflection method, this method does require the solution to simultaneous equations.

The terms constantly used in moment distribution are fixed-end moment, unbalanced moment, distributed moment, and carry-over moment. When all of the joints of a structure are clamped to prevent any joint rotation, the external loads produce certain moments at the ends of the members to which they are applied. These moments are referred to as *fixed-end moments*. Initially, the joints in a structure are considered to be clamped. When the joint is released, it rotates if the sum of the fixed-end moments at the joint is not equal to zero. The difference between zero and the actual sum of the end moments is the *unbalanced moment*. The unbalanced moment causes the joint to rotate. The rotation twists the ends of the members at the joint and changes their moments. In other words, rotation of the joint is resisted by the members and resisting moments are built up in the members as they are twisted. Rotation continues until equilibrium is reached — when the resisting moments equal the unbalanced moment — at which time the sum of the moments at the joint is equal to zero. The moments developed in the members resisting rotation are the *distributed moments*. The distributed moments in the ends of the member cause moments in the other ends, which are assumed to be fixed, and these are the *carry-over moments*.

2.4.3.1 Sign Convention

The moments at the end of a member are assumed to be positive when they tend to rotate the member clockwise about the joint. This implies that the resisting moment of the joint would be counterclockwise. Accordingly, under gravity loading conditions the fixed-end moment at the left end is assumed to be counterclockwise (negative) and at the right end, clockwise (positive).

2.4.3.2 Fixed-End Moments

Fixed-end moments for several cases of loading may be found in Figure 2.8. Application of moment distribution may be explained with reference to a continuous beam example as shown in Figure 2.23. Fixed-end moments are computed for each of the three spans. At joint B the unbalanced moment is obtained and the clamp is removed. The joint rotates, thus distributing the unbalanced moment to the

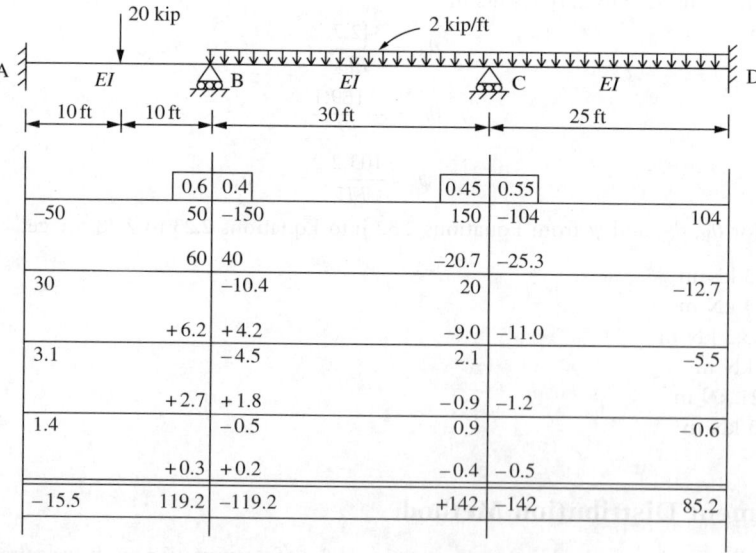

A	0.6	0.4	0.45	0.55	D
−50	50	−150	150	−104	104
	60	40	−20.7	−25.3	
30		−10.4	20		−12.7
	+6.2	+4.2	−9.0	−11.0	
3.1		−4.5	2.1		−5.5
	+2.7	+1.8	−0.9	−1.2	
1.4		−0.5	0.9		−0.6
	+0.3	+0.2	−0.4	−0.5	
−15.5	119.2	−119.2	+142	−142	85.2

FIGURE 2.23 Example — continuous beam by moment distribution.

B-ends of spans BA and BC in proportion to their distribution factors. The values of these distributed moments are carried over at one-half rate to the other ends of the members. When equilibrium is reached, joint B is clamped in its new rotated position and joint C is released afterward. Joint C rotates under its unbalanced moment until it reaches equilibrium, the rotation causing distributed moments in the C-ends of members CB and CD and the resulting carry-over moments. Joint C is now clamped and joint B is released. This procedure is repeated again and again for joints B and C, the amount of unbalanced moments quickly diminishing, until the release of a joint causes negligible rotation. This process is called moment distribution.

The stiffness factors and distribution factors are computed as follows:

$$\text{DF}_{\text{BA}} = \frac{K_{\text{BA}}}{\sum K} = \frac{I/20}{I/20 + I/30} = 0.6$$

$$\text{DF}_{\text{BC}} = \frac{K_{\text{BC}}}{\sum K} = \frac{I/30}{I/20 + I/30} = 0.4$$

$$\text{DF}_{\text{CB}} = \frac{K_{\text{CB}}}{\sum K} = \frac{I/30}{I/30 + I/25} = 0.45$$

$$\text{DF}_{\text{CD}} = \frac{K_{\text{CD}}}{\sum K} = \frac{I/25}{I/30 + I/25} = 0.55$$

The fixed-end moments are as follows:

$$M_{\text{FAB}} = -50, \qquad M_{\text{FBC}} = -150, \qquad M_{\text{FCD}} = -104$$
$$M_{\text{FBA}} = 50, \qquad M_{\text{FCB}} = 150, \qquad M_{\text{FDC}} = 104$$

When a clockwise couple is applied at the near end of a beam, a clockwise couple of half the magnitude is set up at the far end of the beam. The ratio of the moments at the far and near ends is defined as the *carry-over factor* and it is 0.5 in the case of a straight prismatic member. The carry-over factor was developed for carrying over to fixed ends, but it is applicable to simply supported ends, which must have final moments of zero. It can be shown that the beam simply supported at the far end is only three fourths as stiff as the one that is fixed. If the stiffness factors for end spans that are simply supported are modified by three fourths, the simple end is initially balanced to zero, no carry-overs are made to the end afterward. This simplifies the moment distribution process significantly.

2.4.3.3 Moment Distribution for Frames

Moment distribution for frames without side sway is similar to that for continuous beams. The example shown in Figure 2.24 illustrates the applications of moment distribution for a frame without side sway:

$$\text{DF}_{\text{BA}} = \frac{EI/20}{(EI/20) + (EI/20) + (2EI/20)} = 0.25$$

Similarly,

$$\text{DF}_{\text{BE}} = 0.5, \qquad \text{DF}_{\text{BC}} = 0.25$$
$$M_{\text{FBC}} = -100, \qquad M_{\text{FCB}} = 100$$
$$M_{\text{FBE}} = 50, \qquad M_{\text{FEB}} = -50$$

Structural frames are usually subjected to side sway in one direction or the other due to asymmetry of the structure and eccentricity of loading. The sway deflections affect the moments, resulting in an unbalanced moment. These moments could be obtained for the deflections computed and added to the originally distributed fixed-end moments. The sway moments are distributed to columns. Should a frame have columns all of the same length and the same stiffness, the side sway moments will be the same for each column. However, should the columns have differing lengths and stiffness, this will not be the case. The side sway moments should vary from column to column in proportion to their I/l^2 values.

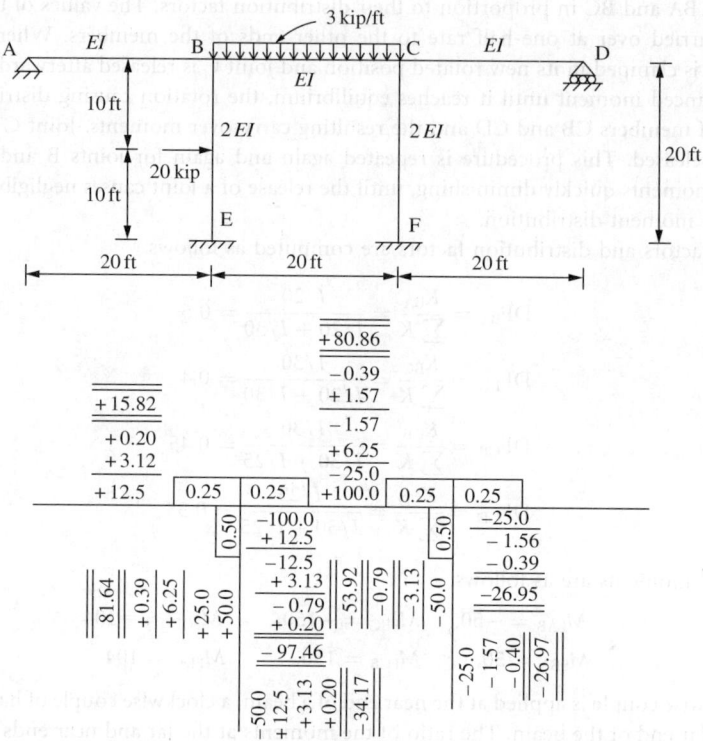

FIGURE 2.24 Example — nonsway frame by moment distribution.

The frame in Figure 2.25 shows a frame subjected to sway. The process of obtaining the final moments is illustrated for this frame.

The frame sways to the right and the side sway moment can be assumed in the ratio

$$\frac{400}{20^2} : \frac{300}{20^2} \quad \text{or} \quad 1:0.75$$

Final moments are obtained by adding distributed fixed-end moments and 13.06/2.99 times the distributed assumed side sway moments.

2.4.4 Method of Consistent Deformations

This method makes use of the principle of deformation compatibility to analyze indeterminate structures. It employs equations that relate the forces acting on the structure to the deformations of the structure. These relations are formed so that the deformations are expressed in terms of the forces and the forces become the unknowns in the analysis.

Let us consider the beam shown in Figure 2.26a. The first step in this method is to determine the degree of indeterminacy or the number of redundants that the structure possesses. As shown in the figure, the beam has three unknown reactions, R_A, R_C, and M_A. Since there are only two equations of equilibrium available for calculating the reactio ns, the beam is said to be indeterminate to the first degree. Restraints that can be removed without impairing the load-supporting capacity of the structure are referred to as *redundants*.

Once the number of redundants is known, the next step is to decide which reaction is to be removed in order to form a determinate structure. Any one of the reactions may be chosen to be the redundant provided that a stable structure remains after the removal of that reaction. For example, let us take the reaction R_C as the redundant. The determinate structure obtained by removing this restraint is the cantilever beam shown in Figure 2.26b. We denote the deflection at end C of this beam, due to P, by Δ_{CP}. The first subscript indicates that the deflection is measured at C and the second subscript that the deflection is due to the applied load P. Using the moment–area method, it can be shown that $\Delta_{CP} = 5PL^3/48EI$. The redundant R_C is then applied to the determinate cantilever beam, as shown in Figure 2.26c. This gives rise to a deflection Δ_{CR} at point C the magnitude of which can be shown to be $R_C L^3/3EI$.

In the actual indeterminate structure, which is subjected to the combined effects of the load P and the redundant R_C the deflection at C is zero. Hence, the algebraic sum of the deflection Δ_{CP} in Figure 2.26b and the deflection Δ_{CR} in Figure 2.26c must vanish. Assuming downward deflections to be positive, we write

$$\Delta_{CP} - \Delta_{CR} = 0 \qquad (2.33)$$

or

$$\frac{5PL^3}{48EI} - \frac{R_C L^3}{3EI} = 0$$

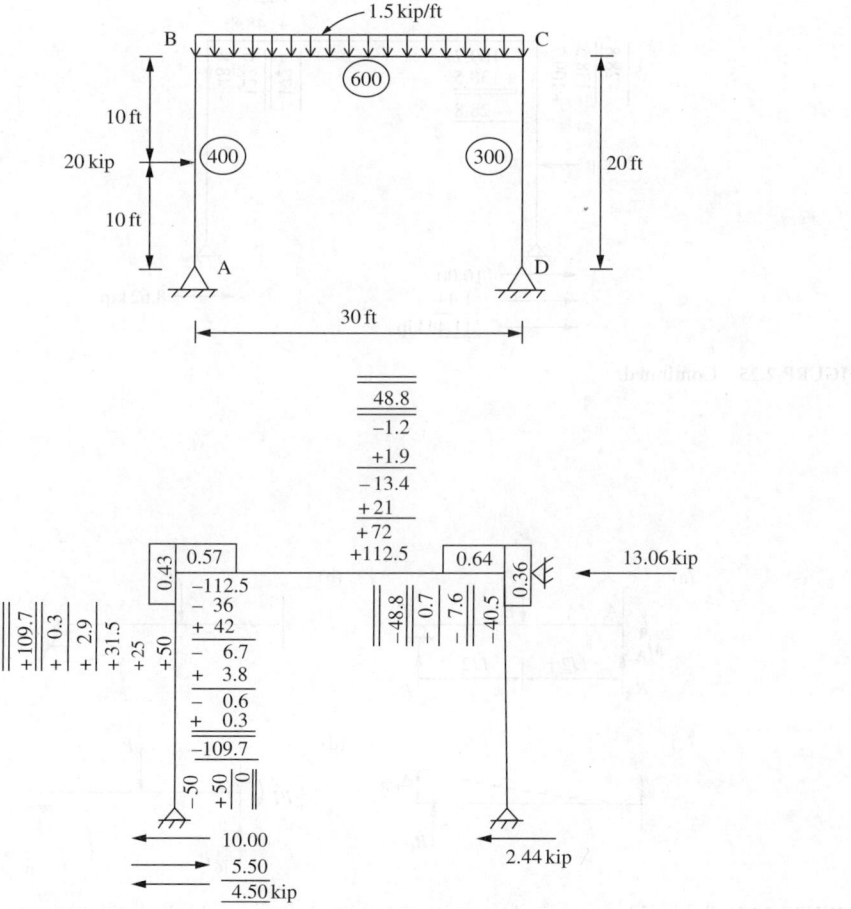

FIGURE 2.25 Example — sway frame by moment distribution.

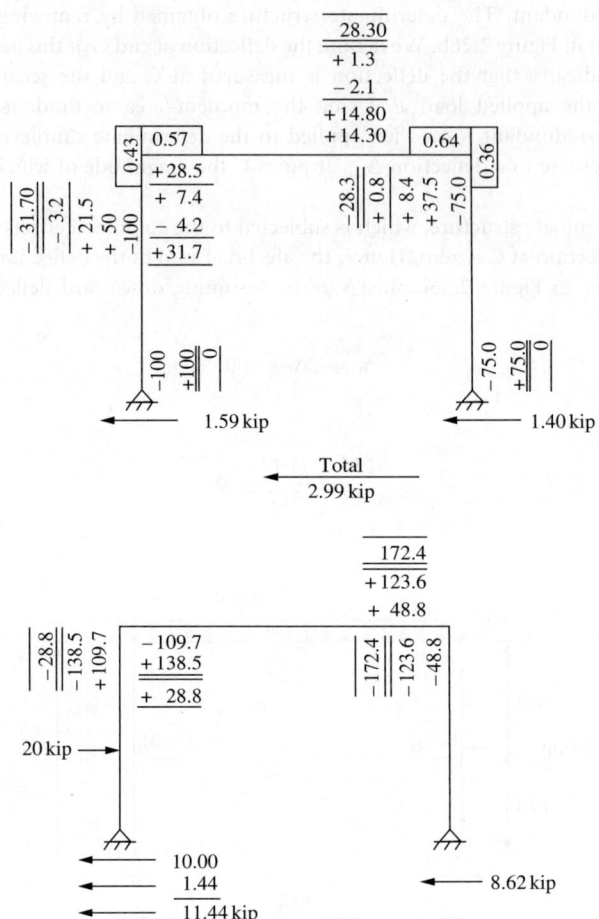

FIGURE 2.25 Continued.

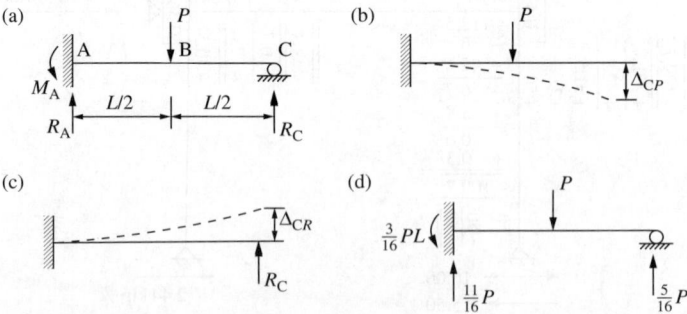

FIGURE 2.26 Beam with one redundant reaction: (a) actual structure; (b) determinate structure subject to actual loads; (c) determinate structure subject to redundant; and (d) beam with all the forces acting on it.

from which

$$R_C = \frac{5}{16}P$$

Equation 2.33, which is used to solve for the redundant, is referred to as an equation of consistent deformations.

Once the redundant R_C has been evaluated, the remaining reactions can be determined by applying the equations of equilibrium to the structure in Figure 2.26a. Thus, $\sum F_y = 0$ leads to

$$R_A = P - \frac{5}{16}P = \frac{11}{16}P$$

and $\sum M_A = 0$ gives

$$M_A = \frac{PL}{2} - \frac{5}{16}PL = \frac{3}{16}PL$$

A free body of the beam, showing all the forces acting on it, is shown in Figure 2.26d.

The steps involved in the method of consistent deformations are

1. The number of redundants in the structure is determined.
2. The redundants required to form a determinate structure are removed.
3. The displacements that the applied loads cause in the determinate structure at the points where the redundants have been removed are then calculated.
4. The displacements at these points in the determinate structure due to the redundants are obtained.
5. At each point where a redundant has been removed, the sum of the displacements calculated in steps (3) and (4) must be equal to the displacement that exists at that point in the actual indeterminate structure. The redundants are evaluated using these relationships.
6. Once the redundants are known the remaining reactions are determined using the equations of equilibrium.

2.4.4.1 Structures with Several Redundants

The method of consistent deformations can be applied to structures with two or more redundants. For example, the beam in Figure 2.27a is indeterminate to the second degree and has two redundant reactions. If the reactions at B and C are selected to be the redundants, then the determinate structure obtained by removing these supports is the cantilever beam shown in Figure 2.27b. To this determinate structure we apply separately the given load (Figure 2.27c) and the redundants R_B and R_C one at a time (Figure 2.27d and e).

Since the deflections at B and C in the original beam are zero, the algebraic sum of the deflections in Figure 2.27c, d, and e at the same points must also vanish. Thus,

$$\Delta_{BP} - \Delta_{BB} - \Delta_{BC} = 0$$
$$\Delta_{CP} - \Delta_{CB} - \Delta_{CC} = 0$$

(2.34)

It is useful in the case of complex structures to write the equations of consistent deformations in the form

$$\Delta_{BP} - \delta_{BB}R_B - \delta_{BC}R_C = 0$$
$$\Delta_{CP} - \delta_{CB}R_B - \delta_{CC}R_C = 0$$

(2.35)

where δ_{BC}, for example, denotes the deflection at B due to a unit load at C in the direction of R_C. Solution of Equations 2.35 gives the redundant reactions R_B and R_C.

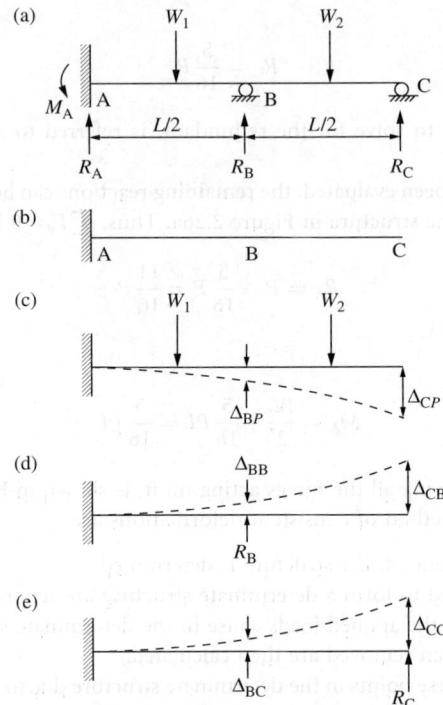

FIGURE 2.27 Beam with two redundant reactions.

EXAMPLE 2.5

Determine the reactions for the beam shown in Figure 2.28 and draw its shear force and bending moment diagrams.

It can be seen from the figure that there are three reactions, namely, M_A, R_A, and R_C, one more than that required for a stable structure. The reaction R_C can be removed to make the structure determinate. We know that the deflection at support C of the beam is zero. One can determine the deflection δ_{CP} at C due to the applied load on the cantilever in Figure 2.28b. In the same way the deflection δ_{CR} at C due to the redundant reaction on the cantilever (Figure 2.28c) can be determined. The compatibility equation gives

$$\delta_{CP} - \delta_{CR} = 0$$

By moment area method,

$$\delta_{CP} = \frac{20}{EI} \times 2 \times 1 + \frac{1}{2} \times \frac{20}{EI} \times 2 \times \frac{2}{3} \times 2 + \frac{40}{EI} \times 2 \times 3 + \frac{1}{2} \times \frac{60}{EI} \times 2 \times \left(\frac{2}{3} \times 2 + 2\right) = \frac{1520}{3EI}$$

$$\delta_{CR} = \frac{1}{2} \times \frac{4R_C}{EI} \times 4 \times \frac{2}{3} \times 4 = \frac{64R_C}{3EI}$$

Substituting for δ_{CP} and δ_{CR} in the compatibility equation, one obtains

$$\frac{1520}{3EI} - \frac{64R_C}{3EI} = 0$$

from which

$$R_C = 23.75 \text{ kN} \uparrow$$

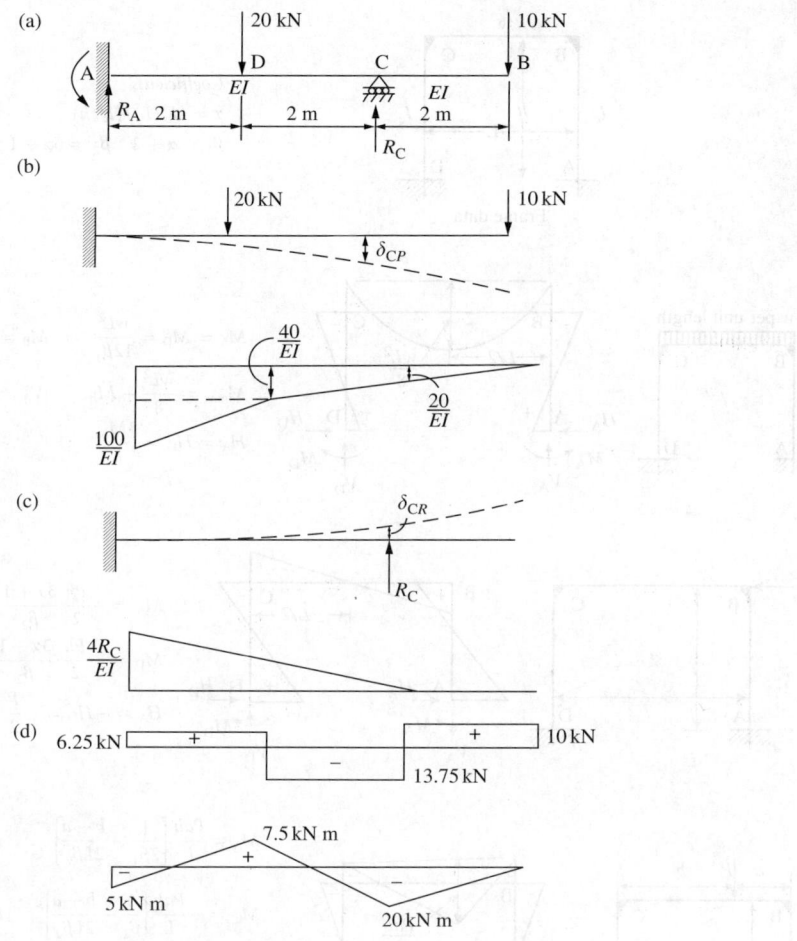

FIGURE 2.28 Example 2.5.

By using statical equilibrium equations we get

$$R_A = 6.25 \text{ kN} \uparrow$$

and

$$M_A = 5 \text{ kN m}$$

The shear force and bending moment diagrams are shown in Figure 2.28d.

1. *Solutions to fix-based portal frames subjected to various types of loading.* Figure 2.29 shows the bending moment diagram and reaction forces of fix-based portal frames subjected to loading typically encountered in practice. Closed form solutions are provided for moments and end forces to facilitate a quick solution to the simple frame problem.
2. *Solutions to pin-based portal frames subjected to various types of loading.* Figure 2.30 shows the bending moment diagram and reaction forces of pin-based portal frames subjected to loading typically encountered in practice. Closed form solutions are provided for moments and end forces to facilitate a quick solution to the simple frame problem.

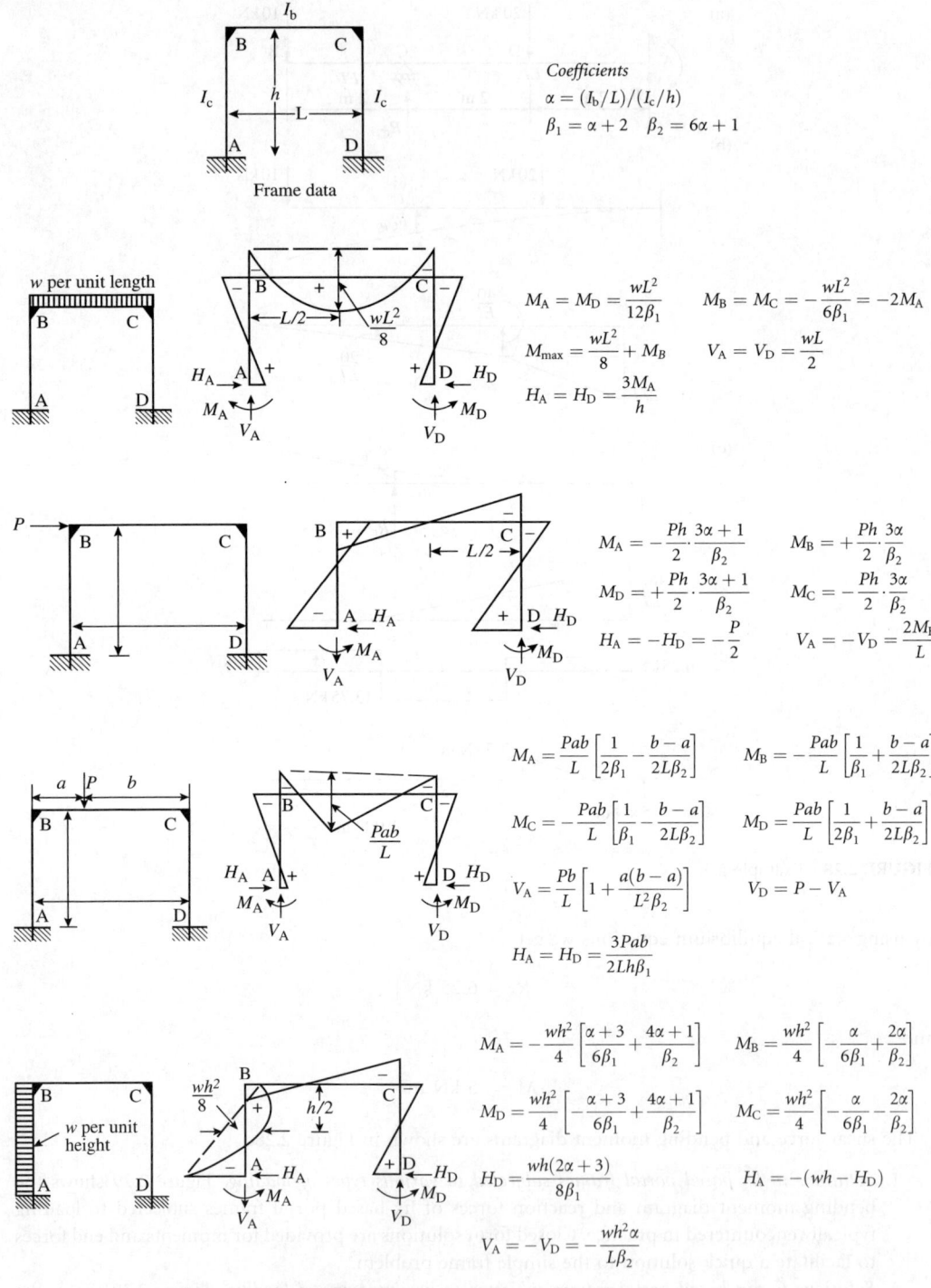

$$\alpha = (I_b/L)/(I_c/h)$$
$$\beta_1 = \alpha + 2 \qquad \beta_2 = 6\alpha + 1$$

$$M_A = M_D = \frac{wL^2}{12\beta_1} \qquad M_B = M_C = -\frac{wL^2}{6\beta_1} = -2M_A$$

$$M_{max} = \frac{wL^2}{8} + M_B \qquad V_A = V_D = \frac{wL}{2}$$

$$H_A = H_D = \frac{3M_A}{h}$$

$$M_A = -\frac{Ph}{2}\cdot\frac{3\alpha+1}{\beta_2} \qquad M_B = +\frac{Ph}{2}\cdot\frac{3\alpha}{\beta_2}$$

$$M_D = +\frac{Ph}{2}\cdot\frac{3\alpha+1}{\beta_2} \qquad M_C = -\frac{Ph}{2}\cdot\frac{3\alpha}{\beta_2}$$

$$H_A = -H_D = -\frac{P}{2} \qquad V_A = -V_D = \frac{2M_B}{L}$$

$$M_A = \frac{Pab}{L}\left[\frac{1}{2\beta_1} - \frac{b-a}{2L\beta_2}\right] \qquad M_B = -\frac{Pab}{L}\left[\frac{1}{\beta_1} + \frac{b-a}{2L\beta_2}\right]$$

$$M_C = -\frac{Pab}{L}\left[\frac{1}{\beta_1} - \frac{b-a}{2L\beta_2}\right] \qquad M_D = \frac{Pab}{L}\left[\frac{1}{2\beta_1} + \frac{b-a}{2L\beta_2}\right]$$

$$V_A = \frac{Pb}{L}\left[1 + \frac{a(b-a)}{L^2\beta_2}\right] \qquad V_D = P - V_A$$

$$H_A = H_D = \frac{3Pab}{2Lh\beta_1}$$

$$M_A = -\frac{wh^2}{4}\left[\frac{\alpha+3}{6\beta_1} + \frac{4\alpha+1}{\beta_2}\right] \qquad M_B = \frac{wh^2}{4}\left[-\frac{\alpha}{6\beta_1} + \frac{2\alpha}{\beta_2}\right]$$

$$M_D = \frac{wh^2}{4}\left[-\frac{\alpha+3}{6\beta_1} + \frac{4\alpha+1}{\beta_2}\right] \qquad M_C = \frac{wh^2}{4}\left[-\frac{\alpha}{6\beta_1} - \frac{2\alpha}{\beta_2}\right]$$

$$H_D = \frac{wh(2\alpha+3)}{8\beta_1} \qquad H_A = -(wh - H_D)$$

$$V_A = -V_D = -\frac{wh^2\alpha}{L\beta_2}$$

FIGURE 2.29 Rigid frames with fixed supports.

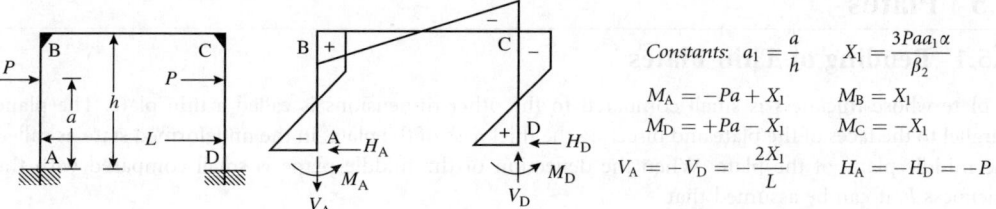

Constants: $a_1 = \dfrac{a}{h}$ $X_1 = \dfrac{3Paa_1\alpha}{\beta_2}$

$M_A = -Pa + X_1$ $M_B = X_1$

$M_D = +Pa - X_1$ $M_C = -X_1$

$V_A = -V_D = -\dfrac{2X_1}{L}$ $H_A = -H_D = -P$

FIGURE 2.29 Continued.

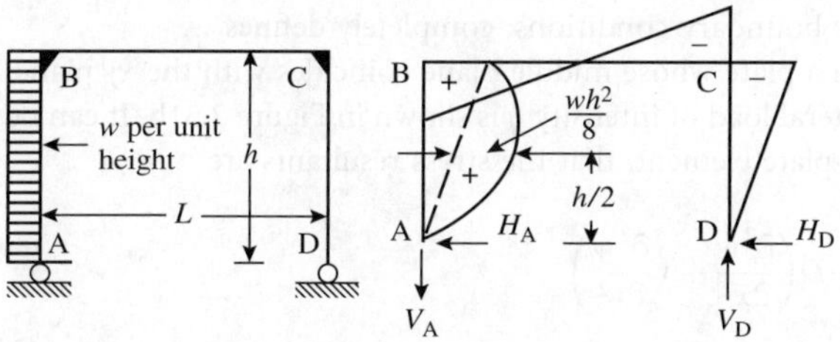

$M_A = \dfrac{wh^2}{4}\left[-\dfrac{\alpha}{2\beta}+1\right]$ $H_D = -\dfrac{M_C}{h}$

$M_C = \dfrac{wh^2}{4}\left[-\dfrac{\alpha}{2\beta}-1\right]$ $H_A = -(wh - H_D)$

$V_A = -V_D = -\dfrac{wh^2}{2L}$

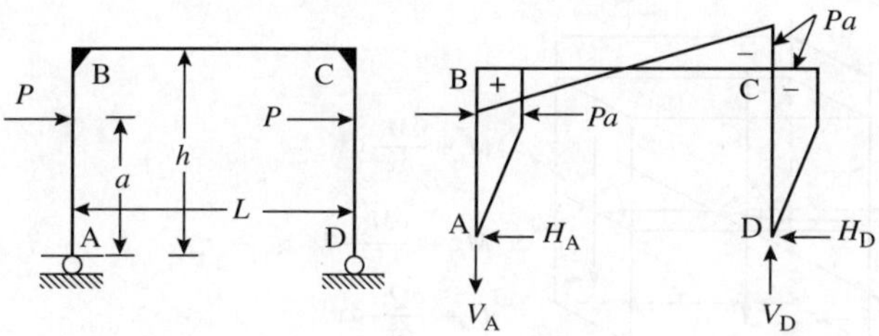

$M_B = -M_C = Pa$ $H_A = H_D = P$

$V_A = -V_D = -\dfrac{2Pa}{L}$

Moment at Loads $= \pm Pa$

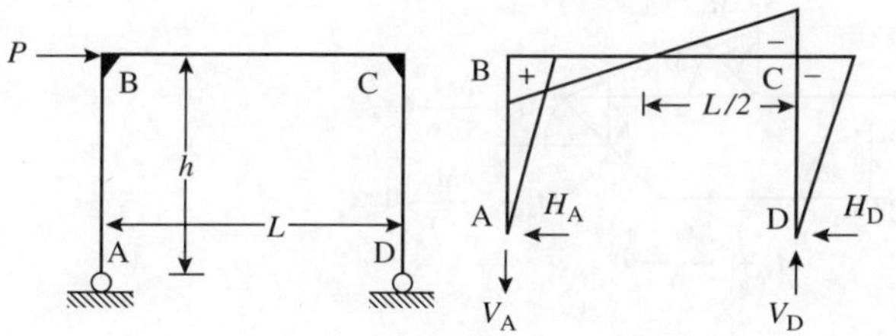

$M_B = -M_C = +\dfrac{Ph}{2}$

$V_A = -V_D = -\dfrac{Ph}{L}$ $H_A = -H_D = -\dfrac{P}{2}$

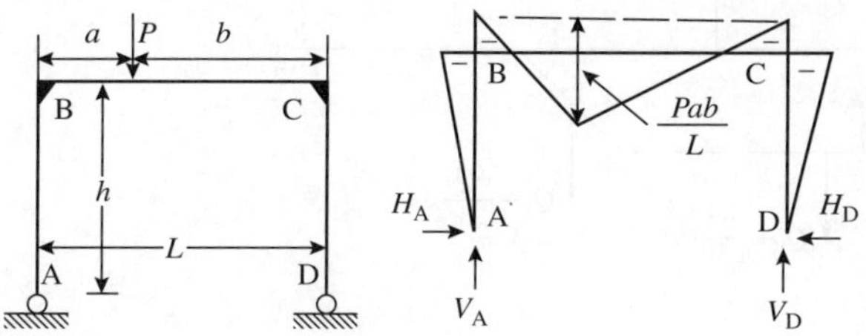

$M_B = M_C = -\dfrac{Pab}{L}\cdot\dfrac{3}{2\beta}$

$V_A = \dfrac{Pb}{L}$ $V_D = \dfrac{Pa}{L}$

$H_A = H_D = -\dfrac{M_B}{h}$

FIGURE 2.30 Rigid frames with pinned supports.

2.5 Plates

2.5.1 Bending of Thin Plates

A plate whose thickness is small compared to the other dimensions is called a thin plate. The plane parallel to the faces of the plate and bisecting the thickness of the plate, in the undeformed state, is called the middle plane of the plate. When the deflection of the middle plane is small compared with the thickness h, it can be assumed that

1. There is no deformation in the middle plane.
2. The normals of the middle plane before bending are deformed into the normals of the middle plane after bending.
3. The normal stresses in the direction transverse to the plate can be neglected.

Based on these assumptions, all stress components can be expressed by deflection w of the plate. w is a function of the two coordinates (x, y) in the plane of the plate. This function has to satisfy a linear partial differential equation, which, together with the boundary conditions, completely defines w.

Figure 2.31a shows a plate element cut from a plate whose middle plane coincides with the xy plane. The middle plane of the plate subjected to a lateral load of intensity q is shown in Figure 2.31b. It can be shown, by considering the equilibrium of the plate element, that the stress resultants are

$$M_x = -D\left(\frac{\partial^2 w}{\partial x^2} + v\frac{\partial^2 w}{\partial y^2}\right)$$

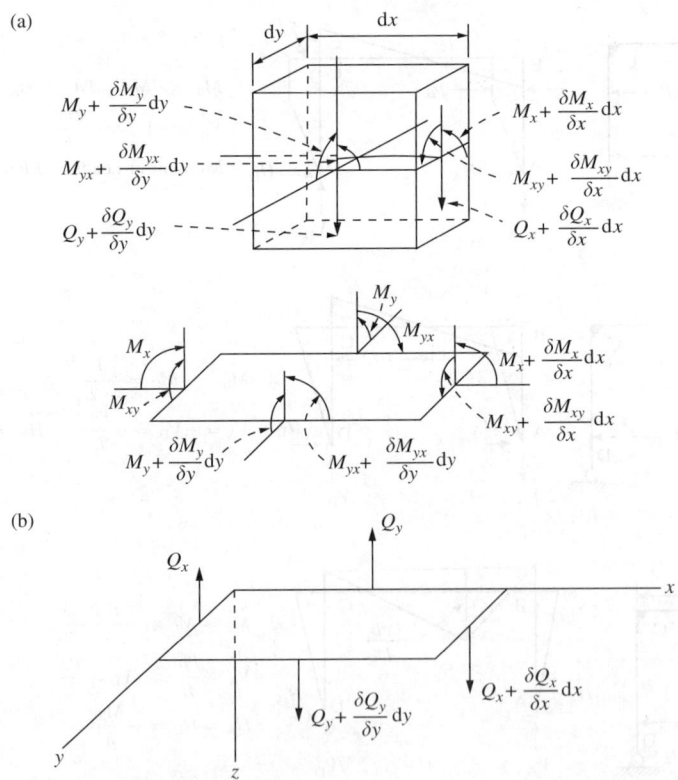

FIGURE 2.31   (a) Plate element and (b) stress resultants.

$$M_y = -D\left(\frac{\partial^2 w}{\partial y^2} + v\frac{\partial^2 w}{\partial x^2}\right) \tag{2.36}$$

$$M_{xy} = -M_{yx} = D(1-v)\frac{\partial^2 w}{\partial x\,\partial y}$$

$$V_y = \frac{\partial^3 w}{\partial y^3} + (2-v)\frac{\partial^3 w}{\partial y\,\partial x^2} \tag{2.37}$$

$$V_x = \frac{\partial^3 w}{\partial x^3} + (2-v)\frac{\partial^3 w}{\partial x\,\partial y^2} \tag{2.38}$$

$$Q_y = -D\frac{\partial}{\partial y}\left(\frac{\partial^2 w}{\partial x^2} + \frac{\partial^2 w}{\partial y^2}\right) \tag{2.39}$$

$$R = 2D(1-v)\frac{\partial^2 w}{\partial x\,\partial y} \tag{2.40}$$

where

M_x, M_y = bending moments per unit length in the x and y directions, respectively
M_{xy}, M_{yx} = twisting moments per unit length
Q_x, Q_y = shearing forces per unit length in the x and y directions, respectively
V_x, V_y = supplementary shear forces in the x and y directions, respectively
R = corner force
D = $Eh^3/(12(1-v^2))$, flexural rigidity of the plate per unit length
v = Poisson's ratio

The governing equation for the plate is obtained as

$$\frac{\partial^4 w}{\partial x^4} + 2\frac{\partial^4 w}{\partial x^2\,\partial y^2} + \frac{\partial^4 w}{\partial y^4} = \frac{q}{D} \tag{2.41}$$

Any plate problem should satisfy the governing Equation 2.41 and boundary conditions of the plate.

2.5.2 Boundary Conditions

There are three basic boundary conditions for plates. These are the clamped edge, simply supported edge, and free edge.

2.5.2.1 Clamped Edge

For this boundary condition, the edge is restrained such that the deflection and slope are zero along the edge. If we consider the edge $x = a$ to be clamped, we have

$$(w)_{x=a} = 0, \quad \left(\frac{\partial w}{\partial x}\right)_{x=a} = 0 \tag{2.42}$$

2.5.2.2 Simply Supported Edge

If the edge $x = a$ of the plate is simply supported, the deflection w along this edge must be zero. At the same time this edge can rotate freely with respect to the edge line. This means that

$$(w)_{x=a} = 0, \quad \left(\frac{\partial^2 w}{\partial x^2}\right)_{x=a} = 0 \tag{2.43}$$

2.5.2.3 Free Edge

If the edge $x = a$ of the plate is entirely free, there are no bending and twisting moments and also vertical shearing forces. This can be written in terms of w, the deflection, as

$$\left(\frac{\partial^2 w}{\partial x^2} + v\frac{\partial^2 w}{\partial y^2}\right)_{x=a} = 0$$

$$\left(\frac{\partial^3 w}{\partial x^3} + (2-v)\frac{\partial^3 w}{\partial x\,\partial y^2}\right)_{x=a} = 0 \tag{2.44}$$

2.5.3 Bending of Rectangular Plates

A plate bending problem may be solved by referring to the differential Equation 2.41. The solution, however, depends on the loading and boundary conditions. Consider a simply supported plate subjected to a sinusoidal loading as shown in Figure 2.32. The differential Equation 2.41 in this case becomes

$$\frac{\partial^4 w}{\partial x^4} + 2\frac{\partial^4 w}{\partial x^2\,\partial y^2} + \frac{\partial^4 w}{\partial y^4} = \frac{q_0}{D}\sin\frac{\pi x}{a}\sin\frac{\pi y}{b} \tag{2.45}$$

The boundary conditions for the simply supported edges are

$$w = 0, \quad \frac{\partial^2 w}{\partial x^2} = 0, \quad \text{for } x = 0 \text{ and } x = a$$

$$w = 0, \quad \frac{\partial^2 w}{\partial y^2} = 0, \quad \text{for } y = 0 \text{ and } y = b \tag{2.46}$$

The deflection function becomes

$$w = w_0 \sin\frac{\pi x}{a}\sin\frac{\pi y}{b} \tag{2.47}$$

which satisfies the boundary conditions in Equations 2.46. w_0 must be chosen to satisfy Equation 2.45. Substitution of Equation 2.47 into Equation 2.45 gives

$$\pi^4\left(\frac{1}{a^2} + \frac{1}{b^2}\right)^2 w_0 = \frac{q_0}{D}$$

The deflection surface for the plate can, therefore, be found to be

$$w = \frac{q_0}{\pi^4 D((1/a^2) + (1/b^2))}\sin\frac{\pi x}{a}\sin\frac{\pi y}{b} \tag{2.48}$$

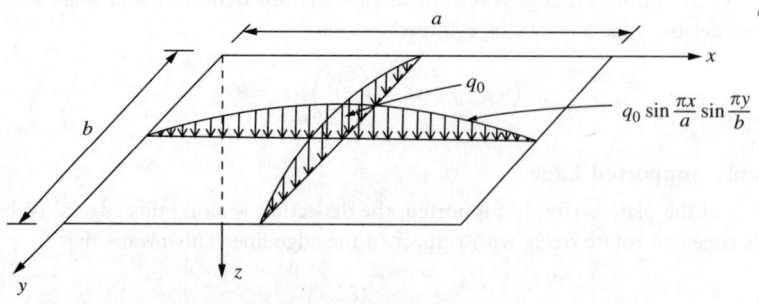

FIGURE 2.32 Rectangular plate under sinusoidal loading.

Using Equations 2.48 and 2.36 we find expression for moments as

$$M_x = \frac{q_0}{\pi^2((1/a^2) + (1/b^2))^2}\left(\frac{1}{a^2} + \frac{v}{b^2}\right)\sin\frac{\pi x}{a}\sin\frac{\pi y}{b}$$

$$M_y = \frac{q_0}{\pi^2((1/a^2) + (1/b^2))^2}\left(\frac{v}{a^2} + \frac{1}{b^2}\right)\sin\frac{\pi x}{a}\sin\frac{\pi y}{b} \qquad (2.49)$$

$$M_{xy} = \frac{q_0(1-v)}{\pi^2((1/a^2) + (1/b^2))^2 ab}\cos\frac{\pi x}{a}\cos\frac{\pi y}{b}$$

Maximum deflection and maximum bending moments that occur at the center of the plate can be written by substituting $x = a/2$ and $y = b/2$ in Equations 2.49 as

$$w_{max} = \frac{q_0}{\pi^4 D((1/a^2) + (1/b^2))^2}$$

$$(M_x)_{max} = \frac{q_0}{\pi^2((1/a^2) + (1/b^2))^2}\left(\frac{1}{a^2} + \frac{v}{b^2}\right) \qquad (2.50)$$

$$(M_y)_{max} = \frac{q_0}{\pi^2((1/a^2) + (1/b^2))^2}\left(\frac{v}{a^2} + \frac{1}{b^2}\right)$$

If the plate is square, then $a = b$ and Equations 2.50 become

$$w_{max} = \frac{q_0 a^4}{4\pi^4 D}$$

$$(M_x)_{max} = (M_y)_{max} = \frac{(1+v)}{4\pi^2}q_0 a^2 \qquad (2.51)$$

If the simply supported rectangular plate is subjected to any kind of loading given by

$$q = q(x, y) \qquad (2.52)$$

the function $q(x, y)$ should be represented in the form of a double trigonometric series as

$$q(x, y) = \sum_{m=1}^{\infty}\sum_{n=1}^{\infty} q_{mn}\sin\frac{m\pi x}{a}\sin\frac{n\pi y}{b} \qquad (2.53)$$

where q_{mn} is given by

$$q_{mn} = \frac{4}{ab}\int_0^a\int_0^b q(x, y)\sin\frac{m\pi x}{a}\sin\frac{n\pi y}{b}\,dx\,dy \qquad (2.54)$$

From Equations 2.45 and 2.52 to 2.54 we can obtain the expression for deflection as

$$w = \frac{1}{\pi^4 D}\sum_{m=1}^{\infty}\sum_{n=1}^{\infty}\frac{q_{mn}}{((m^2/a^2) + (n^2/b^2))^2}\sin\frac{m\pi x}{a}\sin\frac{n\pi y}{b} \qquad (2.55)$$

If the applied load is uniformly distributed with intensity q_0, we have

$$q(x, y) = q_0$$

and from Equation 2.54 we obtain

$$q_{mn} = \frac{4q_0}{ab}\int_0^a\int_0^b \sin\frac{m\pi x}{a}\sin\frac{n\pi y}{b}\,dx\,dy = \frac{16q_0}{\pi^2 mn} \qquad (2.56)$$

where m and n are odd integers. $q_{mn} = 0$ if m or n or both are even numbers. Finally, the deflection of a simply supported plate subjected to uniformly distributed load can be expressed as

$$w = \frac{16q_0}{\pi^6 D} \sum_{m=1}^{\infty} \sum_{n=1}^{\infty} \frac{\sin(m\pi x/a)\sin(n\pi y/b)}{mn((m^2/a^2) + (n^2/b^2))^2} \tag{2.57}$$

where $m = 1, 3, 5, \ldots$ and $n = 1, 3, 5, \ldots$.

The maximum deflection occurs at the center. Its magnitude can be evaluated by substituting $x = a/2$ and $y = b/2$ in Equation 2.57 as

$$w_{\max} = \frac{16q_0}{\pi^6 D} \sum_{m=1}^{\infty} \sum_{n=1}^{\infty} \frac{(-1)^{(m+n)/2-1}}{mn((m^2/a^2) + (n^2/b^2))^2} \tag{2.58}$$

Equation 2.58 is a rapid converging series. A satisfactory approximation can be obtained by taking only the first term of the series; for example, in the case of a square plate,

$$w_{\max} = \frac{4q_0 a^4}{\pi^6 D} = 0.00416\frac{q_0 a^4}{D}$$

Assuming $\nu = 0.3$, the maximum deflection can be calculated as

$$w_{\max} = 0.0454\frac{q_0 a^4}{Eh^3}$$

The expressions for bending and twisting moments can be obtained by substituting Equation 2.57 into Equations 2.36. Figure 2.33 shows some loading cases and the corresponding loading functions.

If the opposite edges at $x = 0$ and $x = a$ of a rectangular plate are simply supported, the solution taking the deflection function as

$$w = \sum_{m=1}^{\infty} Y_m \sin\frac{m\pi x}{a} \tag{2.59}$$

can be adopted. Equation 2.59 satisfies the boundary conditions $w = 0$ and $\partial^2 w/\partial x^2 = 0$ on the two simply supported edges. Y_m should be determined such that it satisfies the boundary conditions along the edges $y = \pm(b/2)$ of the plate shown in Figure 2.34 and also the equation of the deflection surface

$$\frac{\partial^4 w}{\partial x^4} + 2\frac{\partial^4 w}{\partial x^2 \partial y^2} + \frac{\partial^4 w}{\partial y^4} = \frac{q_0}{D} \tag{2.60}$$

q_0 being the intensity of uniformly distributed load.

The solution for Equation 2.60 can be taken in the form

$$w = w_1 + w_2 \tag{2.61}$$

for a uniformly loaded simply supported plate. w_1 can be taken in the form

$$w_1 = \frac{q_0}{24D}(x^4 - 2ax^3 + a^3 x) \tag{2.62}$$

representing the deflection of a uniformly loaded strip parallel to the x axis. It satisfies Equation 2.60 and also the boundary conditions along $x = 0$ and $x = a$.

The expression w_2 has to satisfy the equation

$$\frac{\partial^4 w_2}{\partial x^4} + 2\frac{\partial^4 w_2}{\partial x^2 \partial y^2} + \frac{\partial^4 w_2}{\partial y^4} = 0 \tag{2.63}$$

No.	Load $q(x, y) = \sum_m \sum_n q_{mn} \sin \frac{m\pi x}{a} \sin \frac{n\pi y}{b}$	Expansion coefficients q_{mn}
1	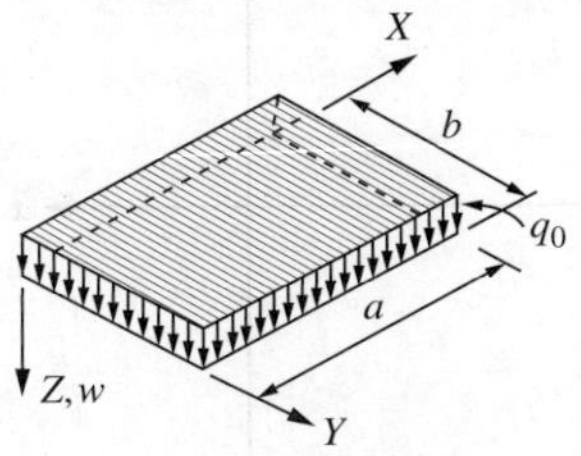	$q_{mn} = \dfrac{16q_0}{\pi^2 mn}$ $(m, n = 1, 3, 5, \ldots)$
2	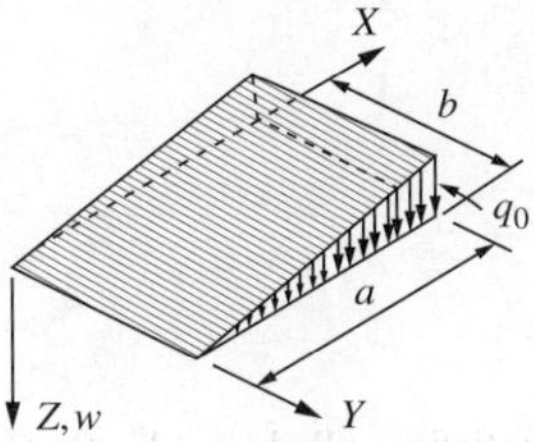	$q_{mn} = \dfrac{-8q_0 \cos m\pi}{\pi^2 mn}$ $(m, n = 1, 3, 5, \ldots)$
3	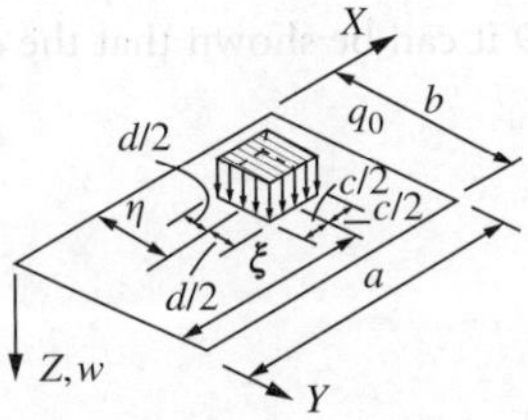	$P_{mn} = \dfrac{16q_0}{\pi^2 mn} \sin \dfrac{m\pi\xi}{a} \sin \dfrac{n\pi\eta}{b}$ $\times \sin \dfrac{m\pi c}{2a} \sin \dfrac{n\pi d}{2b}$ $(m, n = 1, 3, 5, \ldots)$
4	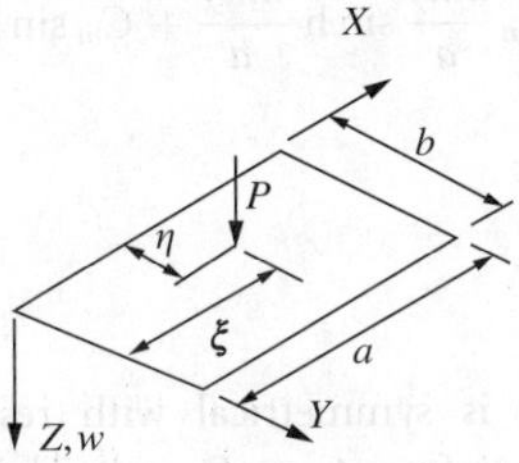	$q_{mn} = \dfrac{4q_0}{ab} \sin \dfrac{m\pi\xi}{a} \sin \dfrac{n\pi\eta}{b}$ $(m, n = 1, 2, 3, \ldots)$
5	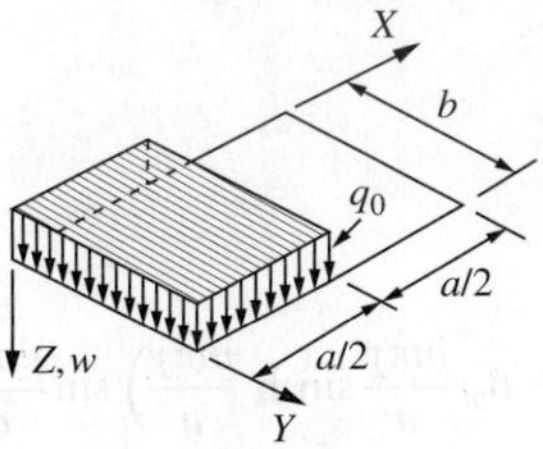	$q_{mn} = \dfrac{8q_0}{\pi^2 mn}$ for $m, n = 1, 3, 5, \ldots$ $q_{mn} = \dfrac{16q_0}{\pi^2 mn}$ for $\begin{cases} m = 2, 6, 10, \ldots \\ n = 1, 3, 5, \ldots \end{cases}$
6	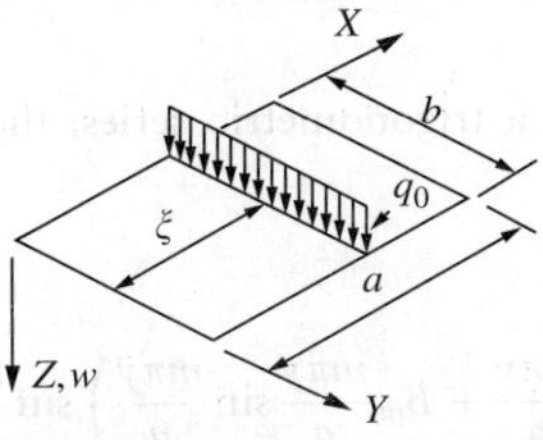	$q_{mn} = \dfrac{4q_0}{\pi an} \sin \dfrac{m\pi\xi}{a}$ $(m, n = 1, 2, 3, \ldots)$

FIGURE 2.33   Typical loading on plates and loading functions.

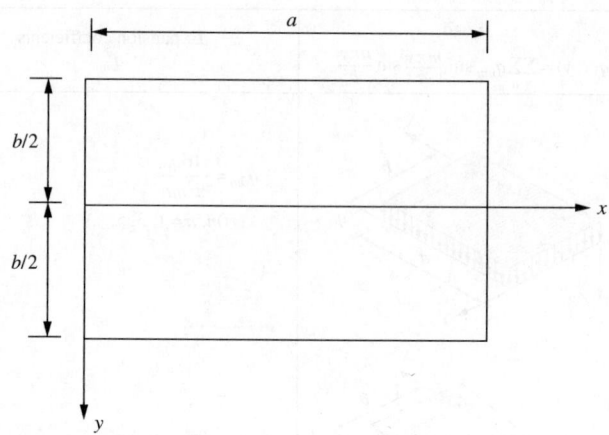

FIGURE 2.34   Rectangular plate.

and must be chosen such that Equation 2.61 satisfies all boundary conditions of the plate. Taking w_2 in the form of series given in Equation 2.59 it can be shown that the deflection surface takes the form

$$w = \frac{q_0}{24D}(x^4 - 2ax^3 + a^3 x)$$

$$+ \frac{q_0 a^4}{24D} \sum_{m=1}^{\infty} \left(A_m \cos h \, \frac{m\pi y}{a} + B_m \frac{m\pi y}{a} \sin h \, \frac{m\pi y}{a} + C_m \sin h \, \frac{m\pi y}{a} \right.$$

$$\left. + D_m \frac{m\pi y}{a} \cos h \, \frac{m\pi y}{a} \right) \sin \frac{m\pi x}{a} \tag{2.64}$$

Observing that the deflection surface of the plate is symmetrical with respect to the x axis we keep in Equation 2.65 only the even function of y; therefore, $C_m = D_m = 0$. The deflection surface takes the form

$$w = \frac{q_0}{24D}(x^4 - 2ax^3 + a^3 x)$$

$$+ \frac{q_0 a^4}{24D} \sum_{m=1}^{\infty} \left(A_m \cos h \, \frac{m\pi y}{a} + B_m \frac{m\pi y}{a} \sin h \, \frac{m\pi y}{a} \right) \sin \frac{m\pi x}{a} \tag{2.65}$$

Developing the expression in Equation 2.62 into a trigonometric series, the deflection surface in Equation 2.65 is written as

$$w = \frac{q_0 a^4}{D} \sum_{m=1}^{\infty} \left(\frac{4}{\pi^5 m^5} + A_m \cos h \, \frac{m\pi y}{a} + B_m \frac{m\pi y}{a} \sin \frac{m\pi y}{a} \right) \sin \frac{m\pi x}{a} \tag{2.66}$$

Substituting Equation 2.66 in the boundary conditions

$$w = 0, \quad \frac{\partial^2 w}{\partial y^2} = 0 \tag{2.67}$$

one obtains the constants of integration A_m and B_m and the expression for deflection may be written as

$$
\begin{aligned}
w = \frac{4q_0 a^4}{\pi^5 D} \sum_{m=1,3,5,\ldots}^{\infty} \frac{1}{m^5} \Bigg(& 1 - \frac{\alpha_m \tan h\, \alpha_m + 2}{2 \cos h\, \alpha_m} \cos h\, \frac{2\alpha_m y}{b} \\
& + \frac{\alpha_m}{2 \cos h\, \alpha_m} \frac{2y}{b} \sin h\, \frac{2\alpha_m y}{b} \Bigg) \sin \frac{m\pi x}{a}
\end{aligned}
\tag{2.68}
$$

where $\alpha_m = (m\pi b)/2a$. Maximum deflection occurs at the middle of the plate, $x = a/2$, $y = 0$, and is given by

$$w_{\max} = \frac{4q_0 a^4}{\pi^5 D} \sum_{m=1,3,5,\ldots}^{\infty} \frac{(-1)^{((m-1)/2)}}{m^5} \left(1 - \frac{\alpha_m \tan h\, \alpha_m + 2}{2 \cos h\, \alpha_m} \right) \tag{2.69}$$

Solution of plates with arbitrary boundary conditions is complicated. It is possible to make some simplifying assumptions for plates with the same boundary conditions along two parallel edges to obtain the desired solution. Alternatively, energy method can be applied more efficiently to solve plates with complex boundary conditions. However, it should be noted that the accuracy of results depends on the deflection function chosen. These functions must be so chosen that they satisfy at least the kinematics boundary conditions.

Figure 2.35 gives the formulas for deflection and bending moments of rectangular plates with typical boundary and loading conditions.

2.5.4 Bending of Circular Plates

In the case of a symmetrically loaded circular plate, the loading is distributed symmetrically about the axis perpendicular to the plate through its center. In such cases, the deflection surface to which the middle plane of the plate is bent will also be symmetrical. The solution of circular plates can be conveniently carried out by using polar coordinates.

Stress resultants in a circular plate element are shown in Figure 2.36. The governing differential equation is expressed in polar coordinates as

$$\frac{1}{r} \frac{d}{dr} \left\{ r \frac{d}{dr} \left[\frac{1}{r} \frac{d}{dr} \left(r \frac{dw}{dr} \right) \right] \right\} = \frac{q}{D} \tag{2.70}$$

where q is the intensity of loading.

In the case of a uniformly loaded circular plate Equation 2.70 can be integrated successively and the deflection at any point at a distance r from the center can be expressed as

$$w = \frac{q_0 r^4}{64D} + \frac{C_1 r^2}{4} + C_2 \log \frac{r}{a} + C_3 \tag{2.71}$$

where q_0 is the intensity of loading and a is the radius of the plate. C_1, C_2, and C_3 are constants of integration to be determined using the boundary conditions.

For a plate with clamped edges under uniformly distributed load q_0 the deflection surface reduces to

$$w = \frac{q_0}{64D} (a^2 - r^2)^2 \tag{2.72}$$

The maximum deflection occurs at the center where $r = 0$ and is given by

$$w = \frac{q_0 a^4}{64D} \tag{2.73}$$

Case no.	Structural system and static loading	Deflection and intensity
1	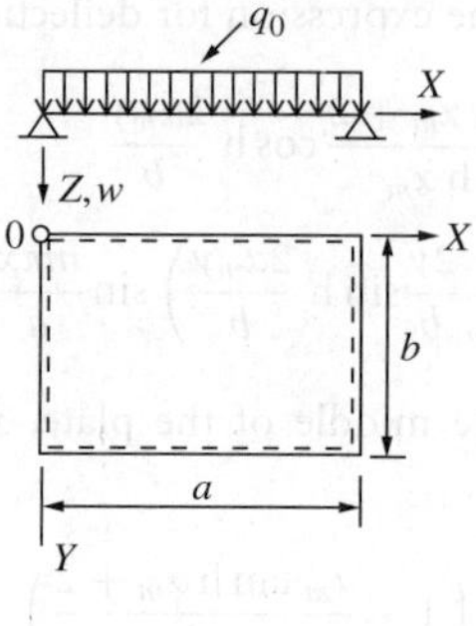 q_0 (uniformly distributed load over simply supported rectangular plate)	$w = \dfrac{16q_0}{\pi^6 D} \sum_m \sum_n \dfrac{\sin\frac{m\pi x}{a} \sin\frac{n\pi y}{b}}{mn\left(\frac{m^2}{a^2}+\frac{n^2}{b^2}\right)}$ $m_x = \dfrac{16q_0 a^2}{\pi^4} \sum_m \sum_n \dfrac{\left(m^2+v\frac{n^2}{\varepsilon^2}\right)\sin\frac{m\pi x}{a}\sin\frac{n\pi y}{b}}{mn\left(m^2+\frac{n^2}{\varepsilon^2}\right)^2}$ $m_y = \dfrac{16q_0 a^2}{\pi^4} \sum_m \sum_n \dfrac{\left(\frac{n^2}{\varepsilon^2}+vm^2\right)\sin\frac{m\pi x}{a}\sin\frac{n\pi y}{b}}{mn\left(m^2+\frac{n^2}{\varepsilon^2}\right)^2}$ $\varepsilon = \dfrac{b}{a}$ $m=1,3,5,\dots,\infty$ $n=1,3,5,\dots,\infty$
2	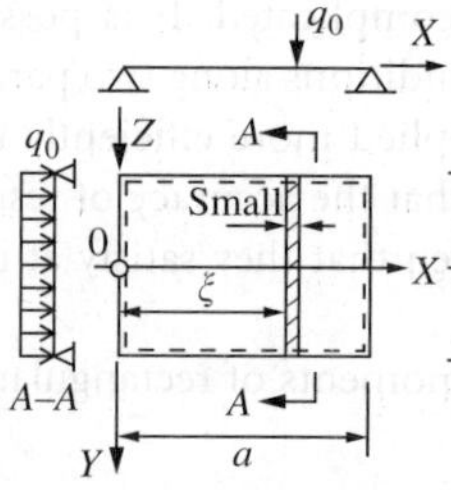 q_0 (line/strip load, small, at distance ξ)	$w = \dfrac{a^4}{D\pi^4}\sum_{m=1}^{\infty}\dfrac{P_m}{m^4}\left(1-\dfrac{2+\alpha_m\tanh\alpha_m}{2\cosh\alpha_m}\cos\lambda_m y\right.$ $\left.+\dfrac{\lambda_m y \sinh\lambda_m y}{2\cosh\alpha_m}\right)\sin\lambda_m x$ where $P_m = \dfrac{2q_0}{a}\sin\dfrac{m\pi\xi}{a}$ $\lambda_m = \dfrac{m\pi}{a}$ $m=1,2,3,\dots$ $\alpha_m = \dfrac{m\pi b}{2a}$
3	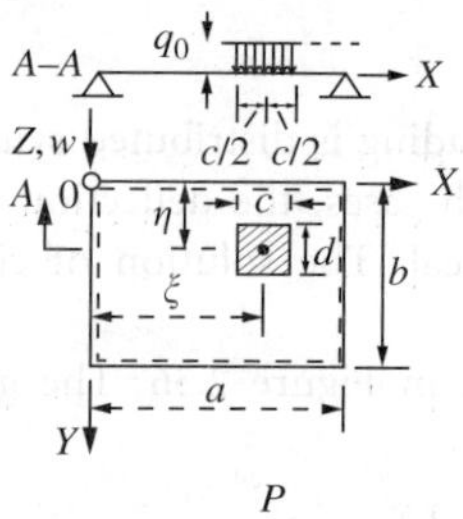 q_0 (patch load $c \times d$ at ξ,η)	$w = \dfrac{16q_0}{D\pi^6}\sum_m\sum_n \dfrac{\sin\frac{m\pi\xi}{a}\sin\frac{n\pi\eta}{b}\sin\frac{m\pi c}{b}\sin\frac{n\pi d}{2b}}{mn\left(\frac{m^2}{a^2}+\frac{n^2}{b^2}\right)}$ $\times \sin\dfrac{m\pi x}{a}\sin\dfrac{n\pi y}{b}$ $m=1,2,3,\dots$ $n=1,2,3,\dots$
4	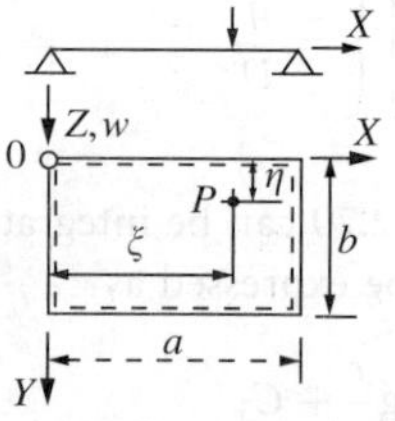 P (concentrated load at ξ,η)	$w = \dfrac{4P}{D\pi^4 ab}\sum_m\sum_n \dfrac{\sin\frac{m\pi\xi}{a}\sin\frac{n\pi\eta}{b}\sin\frac{m\pi x}{a}\sin\frac{n\pi y}{b}}{\left(\frac{m^2}{a^2}+\frac{n^2}{b^2}\right)^2}$ $m=1,2,3,\dots$ $n=1,2,3,\dots$

FIGURE 2.35 Typical loading and boundary conditions for rectangular plates.

Bending moments in the radial and tangential directions are given by

$$M_r = \frac{q_0}{16}\left[a^2(1+v) - r^2(3+v)\right]$$

$$M_t = \frac{q_0}{16}\left[a^2(1+v) - r^2(1+3v)\right]$$

(2.74)

respectively.

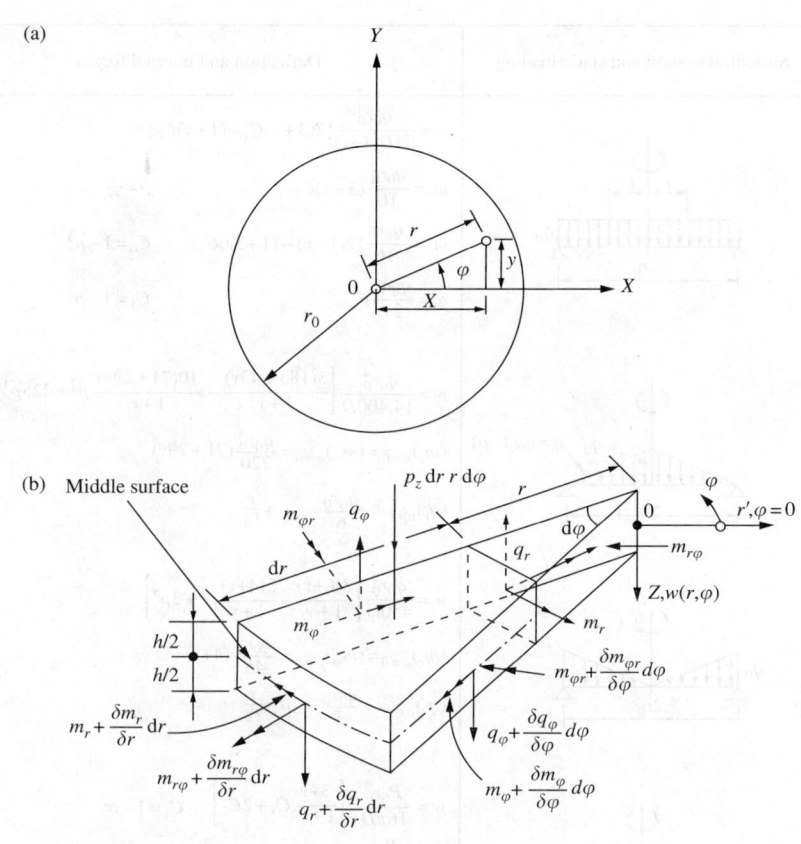

FIGURE 2.36   (a) Circular plate and (b) stress resultants.

The method of superposition can be applied in calculating the deflections for circular plates with simply supported edges. The expressions for deflection and bending moment are given as follows:

$$w = \frac{q_0(a^2 - r^2)}{64D}\left(\frac{5+v}{1+v}a^2 - r^2\right)$$

$$w_{\max} = \frac{5+v}{64(1+v)}\frac{q_0 a^4}{D} \tag{2.75}$$

$$M_r = \frac{q_0}{16}(3+v)(a^2 - r^2)$$

$$M_t = \frac{q_0}{16}[a^2(3+v) - r^2(1+3v)] \tag{2.76}$$

This solution can be used to deal with plates with a circular hole at the center and subjected to concentric moment and shearing forces. Plates subjected to concentric loading and concentrated loading also can be solved by this method. More rigorous solutions are available to deal with irregular loading on circular plates. Once again, the energy method can be employed advantageously to solve circular plate problems. Figure 2.37 gives deflection and bending moment expressions for typical cases of loading and boundary conditions on circular plates.

Case no.	Structural system and static loading	Deflection and internal forces
1	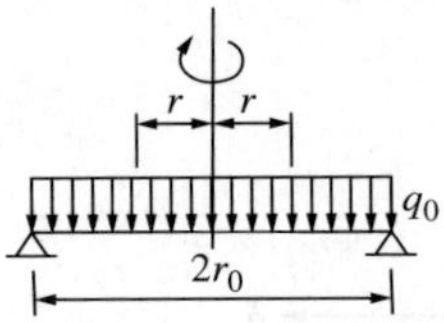	$w = \dfrac{q_0 r_0^4}{64D(1+v)}[2(3+v)C_1-(1+v)C_0]$ $m_r = \dfrac{q_0 r_0^2}{16}(3+v)C_1 \qquad \rho = \dfrac{r}{r_0}$ $m_e = \dfrac{q_0 r_0^2}{16}[2(1-v)-(1+3v)C_1] \qquad C_0 = 1-\rho^4$ $q_r = \dfrac{q_0 r_0}{2}\rho \qquad\qquad\qquad\qquad C_1 = 1-\rho^2$
2	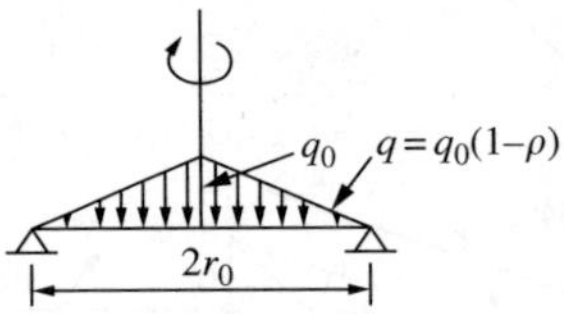	$w = \dfrac{q_0 r_0^4}{14{,}400D}\left[\dfrac{3(183+43v)}{1+v}-\dfrac{10(71+29v)}{1+v}\rho^2+225\rho^4-64\rho^5\right]$ $(m_r)_{\rho=0}=(m_\varphi)_{\rho=0}=\dfrac{q_0 r_0^4}{720}(71+29v)$ $(q_r)_{\rho=1}=-\dfrac{q_0 r_0}{6} \qquad \rho = \dfrac{r}{r_0}$
3	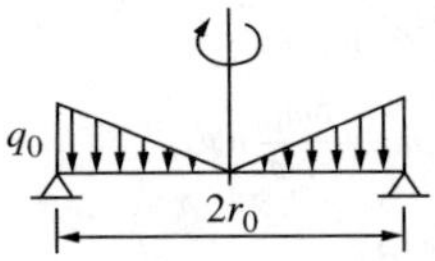	$w = \dfrac{q_0 r_0^4}{450D}\left[\dfrac{3(6+v)}{1+v}-\dfrac{5(4+v)}{1+v}\rho^2+2\rho^5\right]$ $(m_r)_{\rho=0}=(m_\varphi)_{\rho=0}=\dfrac{q_0 r_0^2}{45}(4+v)$ $(q_r)_{\rho=1}=-\dfrac{q_0 r_0}{3} \qquad \rho = \dfrac{r}{r_0}$
4	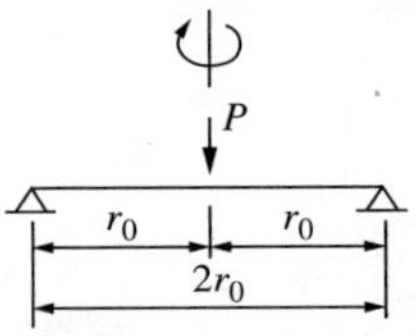	$w = \dfrac{P_0 r_0^2}{16\pi D}\left[\dfrac{3+v}{1+v}C_1+2C_2\right] \qquad C_1 = 1-\rho^2$ $m_r = \dfrac{P}{4\pi}(1+v)C_3 \qquad\qquad C_2 = \rho^2 \ln\rho$ $m_\varphi = \dfrac{P}{4\pi}\left[(1-v)-(1+v)C_3\right] \qquad C_3 = \ln\rho$ $q_r = \dfrac{P}{2\pi r_0 \rho} \qquad\qquad\qquad\qquad \rho = \dfrac{r}{r_0}$
5	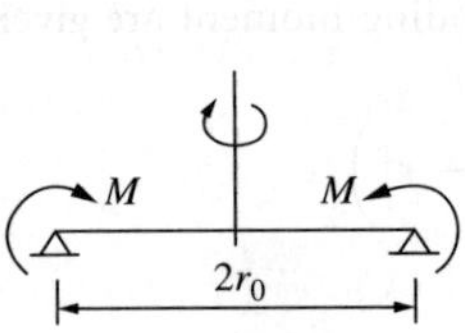	$w = \dfrac{M r_0^2}{2D(1+v)C_1}$ $m_r = m_\varphi = M$ $q_r = 0$ $C_1 = 1-\rho^2, \quad \rho = \dfrac{r}{r_0}$
6	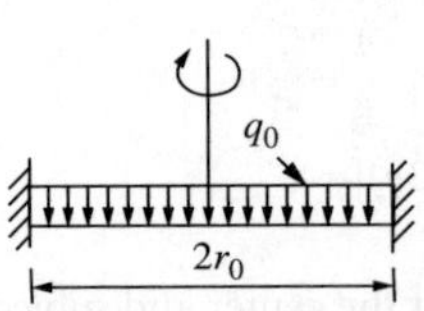	$w = \dfrac{q_0 r_0^4}{64D}(1-\rho^2)^2 \qquad q_r = -\dfrac{q_0 r_0}{2}\rho$ $m_r = \dfrac{q_0 r_0^2}{16}[1+v-(3+v)\rho^2] \qquad \rho = \dfrac{r}{r_0}$ $m_\varphi = \dfrac{q_0 r_0^2}{16}[1+v-(1+3v)\rho^2]$

FIGURE 2.37 Typical loading and boundary conditions for circular plates.

Case no.	Structural system and static loading	Deflection and internal forces
7	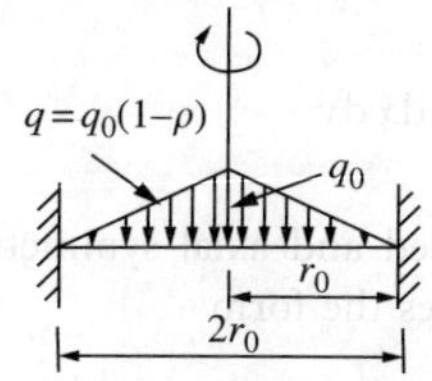 $q = q_0(1-\rho)$ q_0 r_0 $2r_0$	$w = \dfrac{q_0 r_0^4}{14,400D}(129 - 290\rho^2 + 225\rho^4 - 64\rho^5)$ $(m_r)_{\rho=0} = (m_\varphi)_{\rho=0} = \dfrac{29 q_0 r_0^2}{720}(1+\nu)$ $(q_r)_{\rho=1} = -\dfrac{q_0 r_0}{6}$ $(m_r)_{\rho=1} = (m_\varphi)_{\rho=1} = -\dfrac{7 q_0 r_0^2}{120}$ $\rho = \dfrac{r}{r_0}$
8	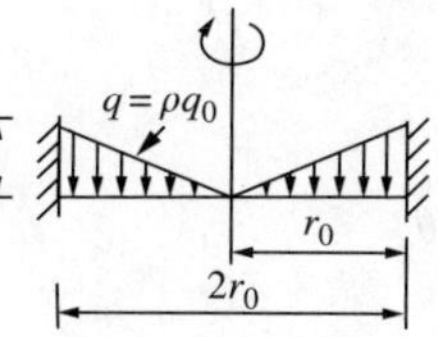 $q = \rho q_0$ q_0 r_0 $2r_0$	$w = \dfrac{q_0 r_0^4}{450D}(3 - 5\rho^2 + 2\rho^5)$ $q_r = -\dfrac{q_0 r_0}{3}\rho^2$ $m_r = \dfrac{q_0 r_0^2}{45}[1 + \nu - (4+\nu)\rho^3]$ $\rho = \dfrac{r}{r_0}$ $m_\varphi = \dfrac{q_0 r_0^2}{45}[1 + \nu - (1+4\nu)\rho^3]$
9	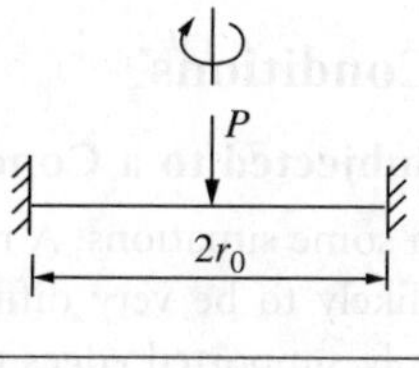 P $2r_0$	$w = \dfrac{P r_0^2}{16\pi D}(1 - \rho^2 + 2\rho^2 \ln\rho)$ $q_r = -\dfrac{P}{2\pi r_0 \rho}$ $m_r = -\dfrac{P}{4\pi}[1 + (1+\nu)\ln\rho]$ $\rho = \dfrac{r}{r_0}$ $m_\varphi = -\dfrac{P}{4\pi}[\nu + (1+\nu)\ln\rho]$

FIGURE 2.37 Continued.

2.5.5 Strain Energy of Simple Plates

The strain energy expression for a simple rectangular plate is given by

$$U = \frac{D}{2}\iint_{area}\left\{\left(\frac{\partial^2 w}{\partial x^2} + \frac{\partial^2 w}{\partial y^2}\right)^2 \right.$$

$$\left. - 2(1-\nu)\left[\frac{\partial^2 w}{\partial x^2}\frac{\partial^2 w}{\partial y^2} - \left(\frac{\partial^2 w}{\partial x \partial y}\right)^2\right]\right\}dx\,dy$$

(2.77)

A suitable deflection function $w(x, y)$ satisfying the boundary conditions of the given plate may be chosen. The strain energy U and the work done by the given load, $q(x, y)$,

$$W = -\iint_{area} q(x,y)w(x,y)\,dx\,dy$$

can be calculated. The total potential energy is, therefore, given as $V = U + W$. Minimizing the total potential energy the plate problem can be solved:

$$\left[\frac{\partial^2 w}{\partial x^2}\frac{\partial^2 w}{\partial y^2} - \left(\frac{\partial^2 w}{\partial x \partial y}\right)^2\right]$$

The term 18 is known as the Gaussian curvature.

If the function $w(x, y) = f(x)f(y)$ (product of a function of x only and a function of y only) and $w = 0$ at the boundary are assumed, then the integral of the Gaussian curvature over the entire plate equals zero. Under these conditions

$$U = \frac{D}{2} \iint_{\text{area}} \left(\frac{\partial^2 w}{\partial x^2} + \frac{\partial^2 w}{\partial y^2} \right)^2 dx\,dy \tag{2.78}$$

If polar coordinates instead of rectangular coordinates are used and axial symmetry of loading and deformation is assumed, the equation for strain energy, U, takes the form

$$U = \frac{D}{2} \iint_{\text{area}} \left\{ \left(\frac{\partial^2 w}{\partial r^2} + \frac{1}{r}\frac{\partial w}{\partial r} \right)^2 - \frac{2(1-v)}{r} \frac{\partial w}{\partial r} \frac{\partial^2 w}{\partial r^2} \right\} r\,dr\,d\theta \tag{2.79}$$

and the work done, W, is written as

$$W = -\iint_{\text{area}} qwr\,dr\,d\theta \tag{2.80}$$

Detailed treatment of Plate Theory can be found in Timoshenko and Woinowsky-Krieger (1959).

2.5.6 Plates of Various Shapes and Boundary Conditions

2.5.6.1 Simply Supported Isosceles Triangular Plate Subjected to a Concentrated Load

Plates of shapes other than circular and rectangular are used in some situations. A rigorous solution of the deflection for a plate with a more complicated shape is likely to be very difficult. Consider, for example, the bending of an isosceles triangular plate with simply supported edges under concentrated load P acting at an arbitrary point (Figure 2.38). A solution can be obtained for this plate by considering a mirror image of the plate as shown in the figure. The deflection of OBC of the square plate is identical with that of a simply supported triangular plate OBC. The deflection owing to the force P can be written as

$$w_1 = \frac{4Pa^2}{\pi^4 D} \sum_{m=1}^{\infty} \sum_{n=1}^{\infty} \frac{\sin(m\pi x_1/a)\sin(n\pi y_1/a)}{(m^2 + n^2)^2} \sin\frac{m\pi x}{a} \sin\frac{n\pi y}{a} \tag{2.81}$$

Upon substitution of $-P$ for P, $(a - y_1)$ for x_1 and $(a - x_1)$ for y_1 in Equation 2.81 we obtain the deflection due to the force $-P$ at A_i:

$$w_2 = -\frac{4Pa^2}{\pi^4 D} \sum_{m=1}^{\infty} \sum_{n=1}^{\infty} (-1)^{m+n} \frac{\sin(m\pi x_1/a)\sin(n\pi y_1/a)}{(m^2 + n^2)^2} \sin\frac{m\pi x}{a} \sin\frac{n\pi y}{a} \tag{2.82}$$

The deflection surface of the triangular plate is then

$$w = w_1 + w_2 \tag{2.83}$$

2.5.6.2 Equilateral Triangular Plates

The deflection surface of a simply supported plate loaded by uniform moment M_o along its boundary and the surface of a uniformly loaded membrane, uniformly stretched over the same triangular boundary, are identical. The deflection surface for such a case can be obtained as

$$w = \frac{M_o}{4aD} \left[x^3 - 3xy^2 - a(x^2 + y^2) + \frac{4}{27}a^3 \right] \tag{2.84}$$

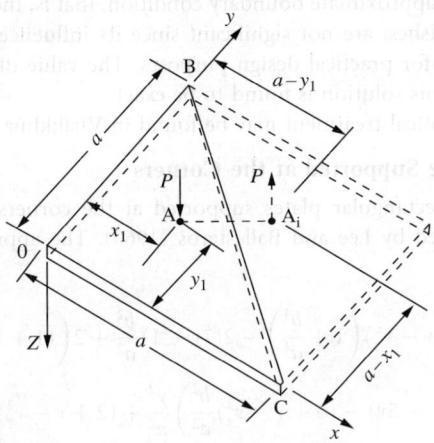

FIGURE 2.38   Isosceles triangular plate.

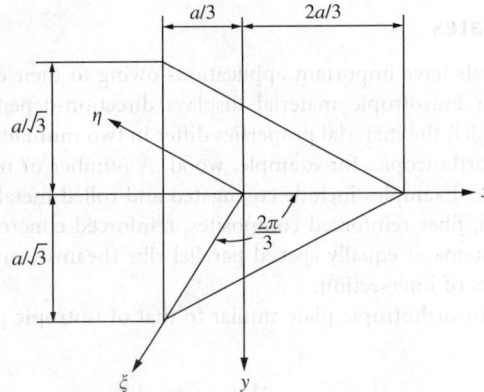

FIGURE 2.39   Equilateral triangular plate with coordinate axes.

If the simply supported plate is subjected to uniform load p_o the deflection surface takes the form

$$w = \frac{p_o}{64aD}\left[x^3 - 3xy^2 - a(x^2 + y^2) + \frac{4}{27}a^3\right]\left(\frac{4}{9}a^2 - x^2 - y^2\right) \tag{2.85}$$

For the equilateral triangular plate (Figure 2.39) subjected to uniform load and supported at the corners, approximate solutions based on the assumption that the total bending moment along each side of the triangle vanishes were obtained by Vijakkhna et al. (1973), who derived the equation for deflection surface as

$$w = \frac{qa^4}{144(1 - v^2)D}\left[\frac{8}{27}(7 + v)(2 - v) - (7 + v)(1 - v)\left(\frac{x^2}{a} + \frac{y^2}{a^2}\right)\right.$$

$$\left. - (5 - v)(1 + v)\left(\frac{x^3}{a^3} - 3\frac{xy^2}{a^3}\right) + \frac{9}{4}(1 - v^2)\left(\frac{x^4}{a^4} + 2\frac{x^2y^2}{a^4} + \frac{y^4}{a^4}\right)\right] \tag{2.86}$$

The errors introduced by the approximate boundary condition, that is, the total bending moment along each side of the triangle vanishes, are not significant since its influence on the maximum deflection and stress resultants is small for practical design purposes. The value of the twisting moment on the edge at the corner given by this solution is found to be exact.

The details of the mathematical treatment may be found in Vijakkhna (1973).

2.5.6.3 Rectangular Plate Supported at the Corners

Approximate solutions for rectangular plates supported at the corners and subjected to uniformly distributed load were obtained by Lee and Ballesteros (1960). The approximate deflection surface is given as

$$
w = \frac{qa^4}{48(1-v^2)D} \left[(10 + v - v^2)\left(1 + \frac{b^4}{a^4}\right) - 2(7v - 1)\frac{b^2}{a^2} + 2\left((1 + 5v)\frac{b^2}{a^2} - (6 + v - v^2)\right)\frac{x}{a} \right.
$$
$$
\left. + 2\left((1 + 5v) - (6 + v - v^2)\frac{b^2}{a^2}\right)\frac{y^2}{a^2} + (2 + v - v^2)\frac{x^4 + y^4}{a^4} - 6(1 + v)\frac{x^2 y^2}{a^4} \right]
$$

$$(2.87)$$

The details of the mathematical treatment may be found in Lee and Ballesteros (1960).

2.5.7 Orthotropic Plates

Plates of anisotropic materials have important applications owing to their exceptionally high bending stiffness. A nonisotropic or anisotropic material displays direction-dependent properties. Simplest among them are those in which the material properties differ in two mutually perpendicular directions. A material so described is orthotropic, for example, wood. A number of manufactured materials are approximated as orthotropic. Examples include corrugated and rolled metal sheets, fillers in sandwich plate construction, plywood, fiber reinforced composites, reinforced concrete, and gridwork. The last example consists of two systems of equally spaced parallel ribs (beams), mutually perpendicular, and attached rigidly at the points of intersection.

The governing equation for orthotropic plate similar to that of isotropic plate (Equation 2.87) takes the form

$$
D_x \frac{\delta^4 w}{\delta x^4} + 2H \frac{\delta^4 w}{\delta x^2 \delta y^2} + D_y \frac{\delta^4 w}{\delta y^4} = q
$$

$$(2.88)$$

where

$$
D_x = \frac{h^3 E_x}{12}, \quad D_y = \frac{h^3 E_y}{12}, \quad H = D_{xy} + 2G_{xy}, \quad D_{xy} = \frac{h^3 E_{xy}}{12}, \quad G_{xy} = \frac{h^3 G}{12}
$$

The expressions for D_x, D_y, D_{xy}, and G_{xy} represent the flexural rigidities and the torsional rigidity of an orthotropic plate, respectively. E_x, E_y, and G are the orthotropic plate moduli. Practical considerations often lead to assumptions, with regard to material properties, resulting in approximate expressions for elastic constants. The accuracy of these approximations is generally the most significant factor in the orthotropic plate problem. Approximate rigidities for some cases that are commonly encountered in practice are given in Figure 2.40.

General solution procedures applicable to the case of isotropic plates are equally applicable to orthotropic plates. Deflections and stress resultants can thus be obtained for orthotropic plates of different shapes with different support and loading conditions. These problems have been researched extensively and solutions concerning plates of various shapes under different boundary and loading conditions may be found in Timoshenko and Woinowsky-Krieger (1959), Tsai and Cheron (1968), Lee et al. (1971), and Shanmugam et al. (1988, 1989).

Geometry	Rigidities
A. Reinforced concrete slab with x and y directed reinforcement steel bars 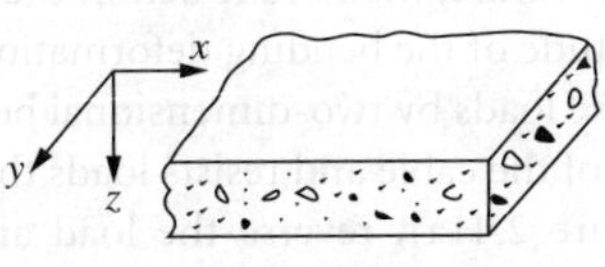 	$D_x = \dfrac{E_c}{1-v_c^2}\left[I_{cx}+\left(\dfrac{E_s}{E_c}-1\right)I_{sx}\right] \qquad D_y = \dfrac{E_c}{1-v_c^2}\left[I_{cy}+\left(\dfrac{E_s}{E_c}-1\right)I_{sy}\right]$ $G_{xy} = \dfrac{1-v_c}{2}\sqrt{D_x D_y} \qquad H = \sqrt{D_x D_y} \qquad D_{xy} = v_c\sqrt{D_x D_y}$ v_c: Poisson's ratio for concrete E_c, E_s: Elastic modulus of concrete and steel, respectively $I_{cx}(I_{sx})$, $I_{cy}(I_{sy})$: Moment of inertia of the slab (steel bars) about neutral axis in the section, $x=$ constant and $y=$ constant, respectively
B. Plate reinforced by equidistant stiffeners 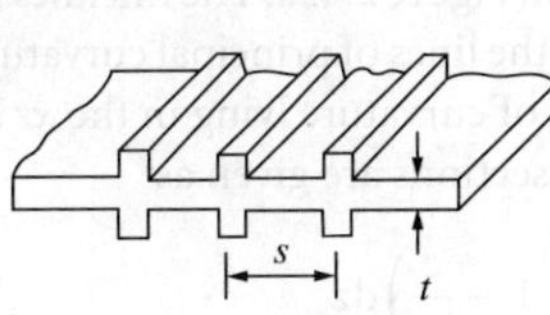 	$D_x = H = \dfrac{Et^3}{12(1-v^2)} \qquad D_y = \dfrac{Et^3}{12(1-v^2)} + \dfrac{E'I}{s}$ E, E': Elastic modulus of plating and stiffeners, respectively v: Poisson's ratio of plating s: Spacing between centerlines of stiffeners I: Moment of inertia of the stiffener cross-section with respect to midplane of plating
C. Plate reinforced by a set of equidistant ribs 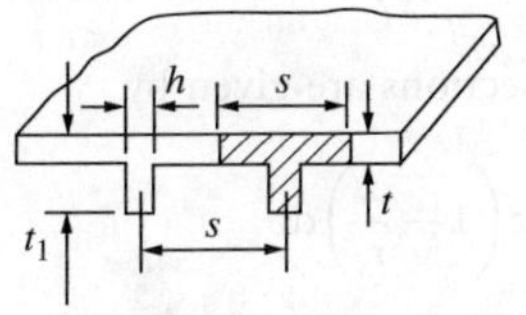 	$D_x = \dfrac{Est^3}{12[s-h+h(t\,t_1)^3]} \qquad D_y = \dfrac{EI}{s}$ $H = 2G'_{xy} + \dfrac{C}{s} \qquad D_{xy} = 0$ C: Torsional rigidity of one rib I: Moment of inertia about neutral axis of a T = section of width s (shown as shaded) G'_{xy}: Torsional rigidity of the plating E: Elastic modulus of the plating
D. Corrugated plate 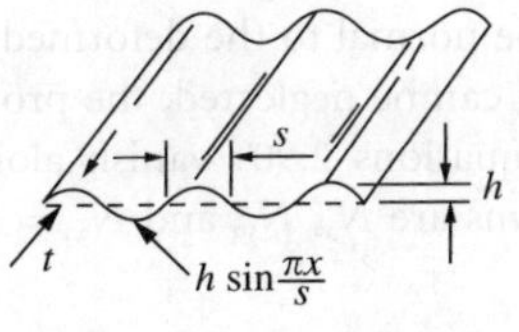 	$D_x = \dfrac{s}{\lambda}\dfrac{Et^3}{12(1-v^2)} \qquad D_y = EI, \ H = \dfrac{\lambda}{a}\dfrac{Et^3}{12(1+v)} \qquad D_{xy} = 0$ where $\lambda = s\left(1+\dfrac{\pi^2 h^2}{4s^2}\right) \qquad I = 0.5h^2 t\left[1-\dfrac{0.81}{1+2.5(h/2s)^2}\right]$

FIGURE 2.40 Various orthotropic plates.

2.6 Shells

2.6.1 Stress Resultants in the Shell Element

A thin shell is defined as a shell with a thickness relatively small compared with its other dimensions. The primary difference between a shell and a plate is that the former has a curvature in the unstressed state, whereas the latter is assumed to be initially flat. The presence of initial curvature is of little consequence as far as flexural behavior is concerned. The membrane behavior, however, is affected significantly by the curvature. Membrane action in a surface is caused by in-plane forces. These forces may be primary forces caused by applied edge loads or edge deformations, or they may be secondary forces resulting from flexural deformations.

In the case of the flat plates, secondary in-plane forces do not give rise to appreciable membrane action unless the bending deformations are large. Membrane action due to secondary forces is, therefore, neglected in small deflection theory. In the case of a shell that has an initial curvature, membrane action caused by secondary in-plane forces will be significant regardless of the magnitude of the bending deformations.

A plate is likened to a two-dimensional beam and resists transverse loads by two-dimensional bending and shear. A membrane is likened to a two-dimensional equivalent of the cable and resists loads through tensile stresses. Imagine a membrane with large deflections (Figure 2.41a), reverse the load and the membrane, and we have the structural shell (Figure 2.41b), provided that the shell is stable for the type of load shown. The membrane resists the load through tensile stresses but the ideal thin shell must be capable of developing both tension and compression.

Consider an infinitely small shell element formed by two pairs of adjacent planes that are normal to the middle surface of the shell and contain its principal curvatures as shown in Figure 2.42a. The thickness of the shell is denoted as h. Coordinate axes x and y are taken as tangent at 0 to the lines of principal curvature and the axis z normal to the middle surface. r_x and r_y are the principal radii of curvature lying in the xz and yz planes, respectively. The resultant forces per unit length of the normal sections are given as

$$N_x = \int_{-h/2}^{h/2} \sigma_x \left(1 - \frac{z}{r_y}\right) dz, \qquad N_y = \int_{-h/2}^{h/2} \sigma_y \left(1 - \frac{z}{r_x}\right) dz$$

$$N_{xy} = \int_{-h/2}^{h/2} \tau_{xy} \left(1 - \frac{z}{r_y}\right) dz, \qquad N_{yx} = \int_{-h/2}^{h/2} \tau_{yx} \left(1 - \frac{z}{r_x}\right) dz \qquad (2.89)$$

$$Q_x = \int_{-h/2}^{h/2} \tau_{xz} \left(1 - \frac{z}{r_y}\right) dz, \qquad Q_y = \int_{-h/2}^{h/2} \tau_{yz} \left(1 - \frac{z}{r_x}\right) dz$$

The bending and twisting moments per unit length of the normal sections are given by

$$M_x = \int_{-h/2}^{h/2} \sigma_x z \left(1 - \frac{z}{r_y}\right) dz, \qquad M_y = \int_{-h/2}^{h/2} \sigma_y z \left(1 - \frac{z}{r_x}\right) dz$$

$$M_{xy} = -\int_{-h/2}^{h/2} \tau_{xy} z \left(1 - \frac{z}{r_y}\right) dz, \qquad M_{yx} = \int_{-h/2}^{h/2} \tau_{yx} z \left(1 - \frac{z}{r_x}\right) dz \qquad (2.90)$$

It is assumed, in bending of the shell, that linear elements as AD and BC (Figure 2.42), which are normal to the middle surface of the shell, remain straight and become normal to the deformed middle surface of the shell. If the conditions of a shell are such that bending can be neglected, the problem of stress analysis is greatly simplified since the resultant moments (Equations 2.90) vanish along with shearing forces Q_x and Q_y in Equations 2.89. Thus, the only unknowns are N_x, N_y, and $N_{xy} = N_{yx}$ and these are called membrane forces.

(a)

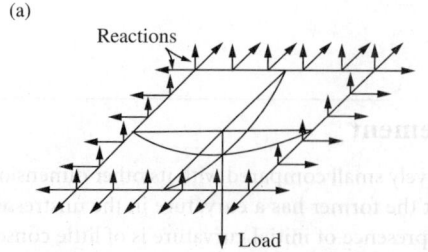

(b)

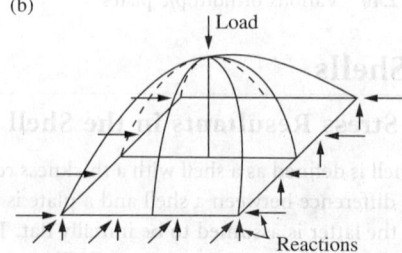

FIGURE 2.41 Membranes with large deflection.

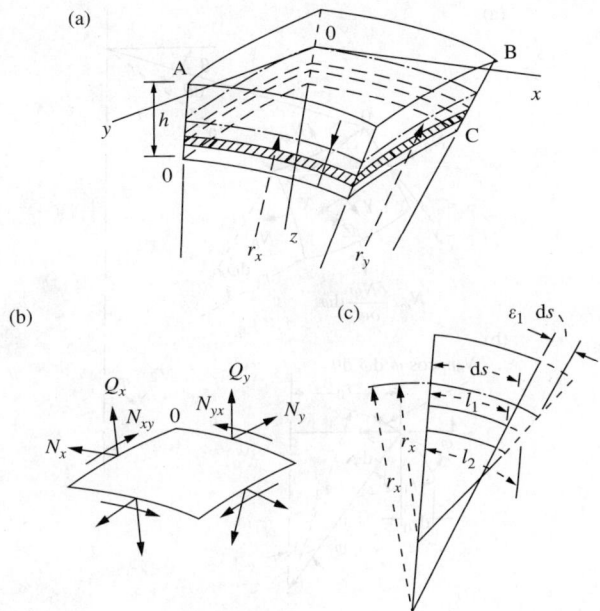

FIGURE 2.42 A shell element.

2.6.2 Shells of Revolution

Shells having the form of surfaces of revolution find extensive application in various kinds of containers, tanks, and domes. Consider an element of a shell cut by two adjacent meridians and two parallel circles as shown in Figure 2.43. There will be no shearing forces on the sides of the element because of the symmetry of loading. By considering the equilibrium in the direction of the tangent to the meridian and z, two equations of equilibrium are written, respectively, as

$$\frac{\mathrm{d}}{\mathrm{d}\varphi}(N_\varphi r_0) - N_\theta r_1 \cos\phi + Y r_1 r_0 = 0$$

$$N_\varphi r_0 + N_\theta r_1 \sin\varphi + Z r_1 r_0 = 0 \tag{2.91}$$

The forces N_θ and N_φ can be calculated from Equations 2.91 if the radii r_0 and r_1 and the components Y and Z of the intensity of the external load are given.

2.6.3 Spherical Dome

The spherical shell shown in Figure 2.44 is assumed to be subjected to its own weight; the intensity of the self weight is assumed as a constant value q_0 per unit area. Considering an element of the shell at an angle φ, the self weight of the portion of the shell above this element is obtained as

$$R = 2\pi \int_0^\phi a^2 q_0 \sin\varphi\, \mathrm{d}\varphi$$

$$= 2\pi a^2 q_0 (1 - \cos\varphi)$$

Considering the equilibrium of the portion of the shell above the parallel circle defined by the angle φ, we can write

$$2\pi r_0 N_\varphi \sin\varphi + R = 0 \tag{2.92}$$

(a)

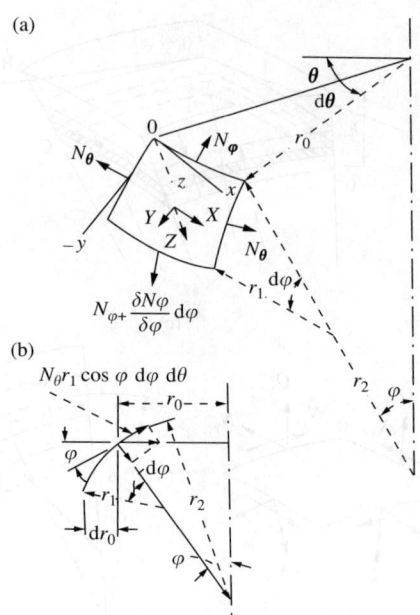

(b)

FIGURE 2.43   An element from shells of revolution — symmetrical loading.

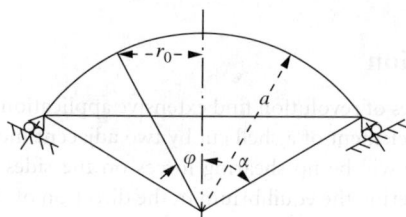

FIGURE 2.44   Spherical dome.

Therefore,

$$N_\varphi = -\frac{aq_0(1 - \cos\varphi)}{\sin^2\varphi} = -\frac{aq_0}{1 + \cos\varphi}$$

We can write from Equations 2.91

$$\frac{N_\varphi}{r_1} + \frac{N_\theta}{r_2} = -Z \tag{2.93}$$

Substituting for N_φ and $z = R$ into Equation 2.93

$$N_\theta = -aq_0\left(\frac{1}{1 + \cos\varphi} - \cos\varphi\right)$$

It is seen that the forces N_φ are always negative. There is thus a compression along the meridians that increases as the angle φ increases. The forces N_θ are also negative for a small angle φ. The stresses as calculated above will represent the actual stresses in the shell with great accuracy if the supports are of such a type that the reactions are tangent to meridians as shown in the figure.

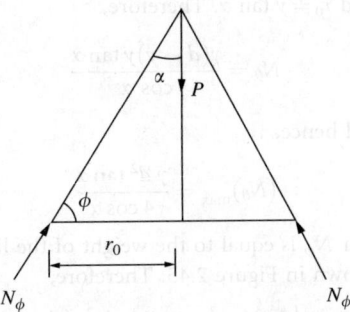

FIGURE 2.45 Conical shell.

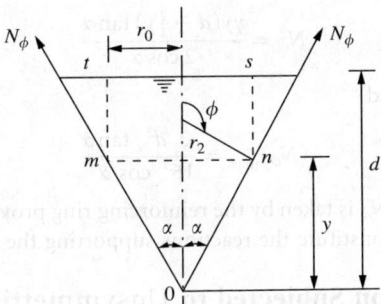

FIGURE 2.46 Inverted conical tank.

2.6.4 Conical Shells

If a force P is applied in the direction of the axis of the cone as shown in Figure 2.45 the stress distribution is symmetrical and we obtain

$$N_\phi = -\frac{P}{2\pi r_0 \cos \alpha}$$

From Equation 2.93, one obtains $N_\theta = 0$.

In the case of a conical surface in which the lateral forces are symmetrically distributed, the membrane stresses can be obtained by using Equations 2.92 and 2.93. The curvature of the meridian in the case of a cone is zero and hence $r_1 = \infty$; Equations 2.92 and 2.93 can therefore be written as

$$N_\phi = -\frac{R}{2\pi r_0 \sin \phi}$$

and

$$N_\theta = -r_2 Z = -\frac{Z r_0}{\sin \phi}$$

If the load distribution is given, N_ϕ and N_θ can be calculated independently.

For example, a conical tank filled with a liquid of specific weight γ is considered as shown in Figure 2.46 The pressure at any parallel circle mn is

$$p = -Z = \gamma(d - y)$$

For the tank, $\phi = \alpha + (\pi/2)$ 19 and $r_0 = y \tan \alpha$. Therefore,

$$N_\theta = \frac{\gamma(d-y)y \tan \alpha}{\cos \alpha}$$

N_θ is maximum when $y = d/2$ and hence,

$$(N_\theta)_{max} = \frac{\gamma\, d^2 \tan \alpha}{4 \cos \alpha}$$

The term R in the expression for N_ϕ is equal to the weight of the liquid in the conical part $mn0$ and the cylindrical part must be as shown in Figure 2.45. Therefore,

$$R = -\left[\tfrac{1}{3}\pi y^3 \tan^2\alpha + \pi y^2 \tan^2\alpha(d-y)\right]\gamma$$

$$= -\pi\gamma y^2 (d - \tfrac{2}{3}y) \tan^2\alpha$$

Hence,

$$N_\phi = \frac{\gamma y(d - \tfrac{2}{3}y) \tan \alpha}{2 \cos \alpha}$$

N_ϕ is maximum when $y = \tfrac{3}{4}d$ and

$$(N_\phi)_{max} = \frac{3}{16}\frac{d^2\gamma \tan \alpha}{\cos \alpha}$$

The horizontal component of N_ϕ is taken by the reinforcing ring provided along the upper edge of the tank. The vertical components constitute the reactions supporting the tank.

2.6.5 Shells of Revolution Subjected to Unsymmetrical Loading

Consider an element cut from a shell by two adjacent meridians and two parallel circles as shown in Figure 2.47. In the general cases shear forces $N_{\varphi\theta} = N_{\theta\varphi}$ and normal forces N_φ and N_θ will act on the sides of the element. Projecting the forces on the element in the y direction, we obtain the governing equation as

$$\frac{\partial}{\partial\varphi}(N_\varphi r_0) + \frac{\partial N_{\theta\varphi}}{\partial\theta}r_1 - N_\theta r_1 \cos\varphi + Y r_1 r_0 = 0 \qquad (2.94)$$

Similarly, the forces in the x direction can be summed up to give

$$\frac{\partial}{\partial\varphi}(r_0 N_{\varphi\theta}) + \frac{\partial N_\theta}{\partial\theta}r_1 + N_{\theta\varphi}r_1 \cos\varphi + X r_0 r_1 = 0 \qquad (2.95)$$

Since the projection of shearing forces on the z axis vanishes, the third equation is the same as Equation 2.93. The problem of determining membrane stresses under unsymmetrical loading reduces to the solution of Equations 2.93 to 2.95 for given values of the components X, Y, and Z of the intensity of the external load.

2.6.6 Cylindrical Shells

It is assumed that the generator of the shell is horizontal and parallel to the x axis. An element is cut from the shell by two adjacent generators and two cross-sections perpendicular to the x axis, and its position is defined by the coordinate x and the angle φ. The forces acting on the sides of the element are shown in Figure 2.48b.

The components of the distributed load over the surface of the element are denoted as X, Y, and Z. Considering the equilibrium of the element and summing up the forces in the x direction, we obtain

$$\frac{\partial N_x}{\partial x}r\, d\varphi\, dx + \frac{\partial N_{\varphi x}}{\partial\varphi}\, d\varphi\, dx + Xr\, d\varphi\, dx = 0$$

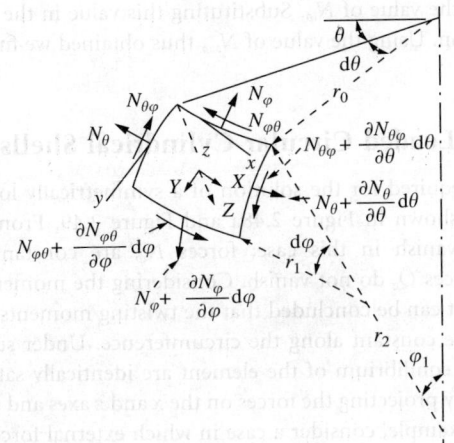

FIGURE 2.47 An element from shells of revolution — unsymmetrical loading.

(a)

(b)

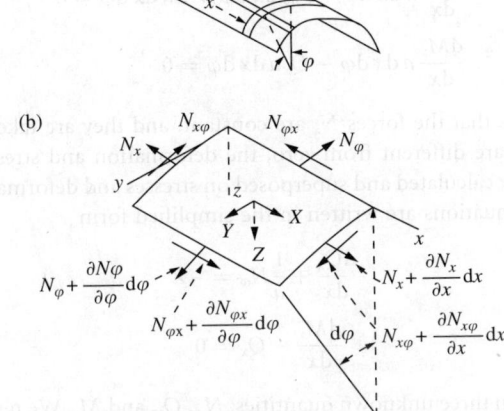

FIGURE 2.48 Membrane forces on a cylindrical shell element.

The corresponding equations of equilibrium in the y and z directions are given, respectively, as

$$\frac{\partial N_{x\varphi}}{\partial x} r\, d\varphi\, dx + \frac{\partial N_{\varphi}}{\partial \varphi}\, d\varphi\, dx + Yr\, d\varphi\, dx = 0$$

$$N_{\varphi}\, d\varphi\, dx + Zr\, d\varphi\, dx = 0$$

The three equations of equilibrium can be simplified and represented in the following form

$$\frac{\partial N_x}{\partial x} + \frac{1}{r}\frac{\partial N_{x\varphi}}{\partial \varphi} = -X$$

$$\frac{\partial N_{x\varphi}}{\partial x} + \frac{1}{r}\frac{\partial N_{\varphi}}{\partial \varphi} = -Y \qquad (2.96)$$

$$N_{\varphi} = -Zr$$

In each case we readily find the value of N_φ. Substituting this value in the second of the equations, we then obtain $N_{x\varphi}$ by integration. Using the value of $N_{x\varphi}$ thus obtained we find N_x by integrating the first equation.

2.6.7 Symmetrically Loaded Circular Cylindrical Shells

To establish the equations required for the solution of a symmetrically loaded circular cylinder shell, we consider an element, as shown in Figure 2.48a and Figure 2.49. From symmetry, the membrane shearing forces $N_{x\varphi} = N_{\varphi x}$ vanish in this case; forces N_φ are constant along the circumference. From symmetry, only the forces Q_z do not vanish. Considering the moments acting on the element in Figure 2.49, from symmetry it can be concluded that the twisting moments $M_{x\varphi} = M_{\varphi x}$ vanish and that the bending moments M_φ are constant along the circumference. Under such conditions of symmetry three of the six equations of equilibrium of the element are identically satisfied. We have to consider only the equations obtained by projecting the forces on the x and z axes and by taking the moment of the forces about the y axis. For example, consider a case in which external forces consist only of a pressure normal to the surface. The three equations of equilibrium are

$$\frac{dN}{dx} a \, dx \, d\varphi = 0$$

$$\frac{dQ_x}{dx} a \, dx \, d\varphi + N_\varphi \, dx \, d\varphi + Za \, dx \, d\varphi = 0 \qquad (2.97)$$

$$\frac{dM_x}{dx} a \, dx \, d\varphi - Q_x a \, dx \, d\varphi = 0$$

The first equation indicates that the forces N_x are constant, and they are taken as equal to zero in the further discussion. If they are different from zero, the deformation and stress corresponding to such constant forces can be easily calculated and superposed on stresses and deformations produced by lateral load. The remaining two equations are written in the simplified form

$$\frac{dQ_x}{dx} + \frac{1}{a} N_\varphi = -Z$$

$$\frac{dM_x}{dx} - Q_x = 0 \qquad (2.98)$$

These two equations contain three unknown quantities: N_φ, Q_x, and M_x. We need, therefore, to consider the displacements of points in the middle surface of the shell.

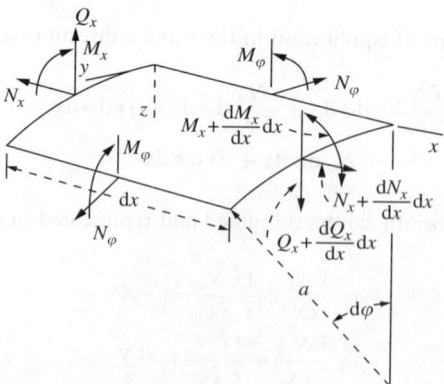

FIGURE 2.49   Stress resultants in a cylindrical shell element.

The component v of the displacement in the circumferential direction vanishes because of symmetry. Only the components u and w in the x and z directions, respectively, are to be considered. The expressions for the strain components then become

$$\varepsilon_x = \frac{du}{dx}, \quad \varepsilon_\varphi = -\frac{w}{a} \tag{2.99}$$

By Hooke's law, we obtain

$$N_x = \frac{Eh}{1-v^2}(\varepsilon_x + v\varepsilon_\varphi) = \frac{Eh}{1-v^2}\left(\frac{du}{dx} - v\frac{w}{a}\right) = 0$$

$$N_\varphi = \frac{Eh}{1-v^2}(\varepsilon_\varphi + v\varepsilon_x) = \frac{Eh}{1-v^2}\left(-\frac{w}{a} + v\frac{du}{dx}\right) = 0 \tag{2.100}$$

From the first of these equation it follows that

$$\frac{du}{dx} = v\frac{w}{a}$$

and the second equation gives

$$N_\varphi = -\frac{Ehw}{a} \tag{2.101}$$

Considering the bending moments, we conclude from symmetry that there is no change in curvature in the circumferential direction. The curvature in the x direction is equal to $-d^2w/dx^2$. Using the same equations as for plates, we obtain

$$M_\varphi = vM_x$$

$$M_x = -D\frac{d^2w}{dx^2} \tag{2.102}$$

where

$$D = \frac{Eh^3}{12(1-v^2)}$$

is the flexural rigidity per unit length of the shell.

Eliminating Q_x from Equations 2.98 we obtain

$$\frac{d^2M_x}{dx^2} + \frac{1}{a}N_\varphi = -Z$$

from which, by using Equations 2.101 and 2.102, we obtain

$$\frac{d^2}{dx^2}\left(D\frac{d^2w}{dx^2}\right) + \frac{Eh}{a^2}w = Z \tag{2.103}$$

All problems of symmetrical deformation of circular cylindrical shells thus reduce to the integration of Equation 2.103.

The simplest application of this equation is obtained when the thickness of the shell is constant. Under such conditions Equation 2.103 becomes

$$D\frac{d^4w}{dx^4} + \frac{Eh}{a^2}w = Z$$

Using the notation

$$\beta^4 = \frac{Eh}{4a^2D} = \frac{3(1-v^2)}{a^2h^2} \qquad (2.104)$$

Equation 2.104 can be represented in the simplified form

$$\frac{d^4w}{dx^4} + 4\beta^4 w = \frac{Z}{D} \qquad (2.105)$$

The general solution of this equation is

$$w = e^{\beta x}(C_1 \cos \beta x + C_2 \sin \beta x) + e^{-\beta x}(C_3 \cos \beta x + C_4 \sin \beta x) + f(x) \qquad (2.106)$$

Detailed treatment of shell theory can be obtained from Timoshenko and Woinowsky-Krieger (1959).

2.7 Influence Lines

Bridges, industrial buildings with traveling cranes, and frames supporting conveyer belts are often subjected to moving loads. Each member of these structures must be designed for the most severe conditions that can possibly be developed in that member. Live loads should be placed at the positions where they will produce these severe conditions. The critical positions for placing live loads will not be the same for every member. On some occasions it is possible by inspection to determine where to place the loads to give the most critical forces, but on many other occasions it is necessary to resort to certain criteria to find the locations. The most useful of these methods is the influence line.

An influence line for a particular response, such as reaction, shear force, bending moment, and axial force, is defined as a diagram, the ordinate to which at any point equals the value of that response attributable to a unit load acting at that point on the structure. Influence lines provide a systematic procedure for determining how the force in a given part of a structure varies as the applied load moves about on the structure. Influence lines of responses of statically determinate structures consist only of straight lines whereas they are curves for statically indeterminate structures. They are primarily used to determine where to place live loads to cause maximum force and to compute the magnitude of those forces. Knowledge of influence lines helps one to study the structural response under different moving load conditions.

2.7.1 Influence Lines for Shear in Simple Beams

Figure 2.50 shows influence lines for shear at two sections of a simply supported beam. It is assumed that positive shear occurs when the sum of the transverse forces to the left of a section is in the upward direction or when the sum of the forces to the right of the section is in the downward direction. A unit force is placed at various locations and the shear force at sections 1–1 and 2–2 are obtained for each position of the unit load. These values give the ordinate of influence line with which the influence line diagrams for shear force at sections 1–1 and 2–2 can be constructed. Note that the slope of the influence line for shear on the left of the section is equal to the slope of the influence line on the right of the section. This information is useful in drawing the shear force influence line in other cases.

2.7.2 Influence Lines for Bending Moment in Simple Beams

Influence lines for bending moment at the same sections 1–1 and 2–2 of the simple beam considered in Figure 2.50 are plotted as shown in Figure 2.51. For a section, when the sum of the moments of all the forces to the left is clockwise or when the sum to the right is counterclockwise, the moment is taken as positive. The values of bending moment at sections 1–1 and 2–2 are obtained for various positions of unit load and plotted as shown in the figure.

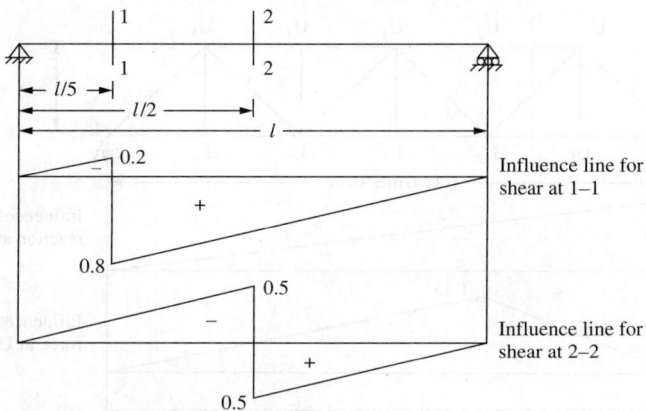

FIGURE 2.50 Influence line for shear force.

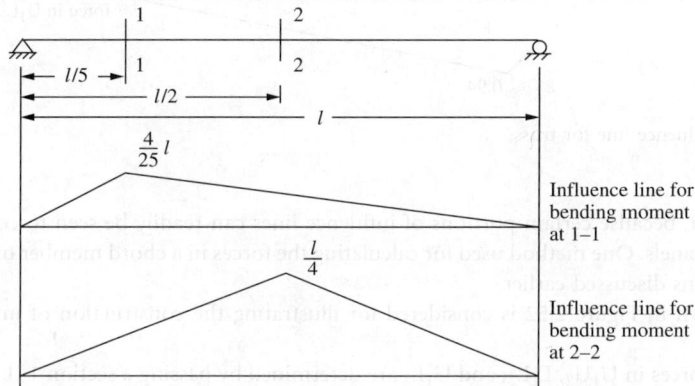

FIGURE 2.51 Influence line for bending moment.

It should be understood that a shear or bending moment diagram shows the variation of shear or moment across an entire structure for loads fixed in one position. On the other hand, an influence line for shear or moment shows the variation of that response at one particular section in the structure caused by the movement of a unit load from one end of the structure to the other.

Influence lines can be used to obtain the value of a particular response for which it is drawn when the beam is subjected to any particular type of loading. If, for example, a uniform load of intensity q_0 per unit length is acting over the entire length of the simple beam shown in Figure 2.50, the shear force at section 1–1 is given by the product of the load intensity, q_0, and the net area under the influence line diagram. The net area is equal to 0.3 and the shear force at section 1–1 is therefore equal to $0.3q_0$. In the same way, the bending moment at the section can be found as the area of the corresponding influence line diagram times the intensity of loading, q_0. The bending moment at the section is equal to $0.08q_0l^2$.

2.7.3 Influence Lines for Trusses

Influence lines for support reactions and member forces may be constructed in the same manner as those for various beam functions. They are useful to determine the maximum load that can be applied to the truss. The unit load moves across the truss, and the ordinates for the responses under consideration may be computed for the load at each panel point. Member force, in most cases, need not be calculated for

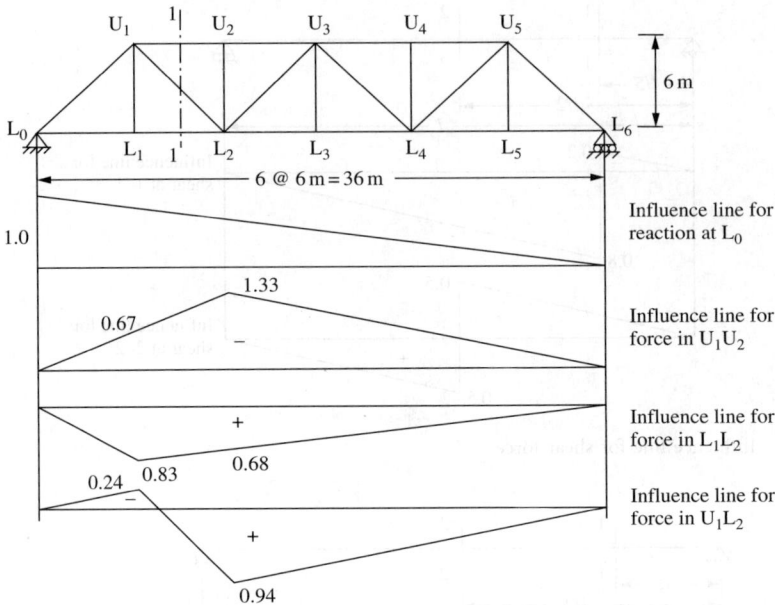

FIGURE 2.52 Influence line for truss.

every panel point, because certain portions of influence lines can readily be seen to consist of straight lines for several panels. One method used for calculating the forces in a chord member of a truss is by the method of sections discussed earlier.

The truss shown in Figure 2.52 is considered for illustrating the construction of influence lines for trusses.

The member forces in U_1U_2, L_1L_2, and U_1L_2 are determined by passing a section 1–1 and considering the equilibrium of the free-body diagram of one of the truss segments. Unit load is placed at L_1 first and the force in U_1U_2 is obtained by taking the moment about L_2 of all the forces acting on the right-hand segment of the truss and dividing the resulting moment by the lever arm (the perpendicular distance of the force in U_1U_2 from L_2). The value thus obtained gives the ordinate of the influence diagram at L_1 in the truss. The ordinate at L_2 obtained similarly represents the force in U_1U_2 for unit load placed at L_2. The influence line can be completed with two other points, one at each of the supports. The force in the member L_1L_2 due to unit load placed at L_1 and L_2 can be obtained in the same manner and the corresponding influence line diagram can be completed. By considering the horizontal component of force in the diagonal of the panel the influence line for force in U_1L_2 can be constructed. Figure 2.52 shows the respective influence diagram for member forces in U_1U_2, L_1L_2, and U_1L_2. Influence line ordinates for the force in a chord member of a "curved-chord" truss may be determined by passing a vertical section through the panel and taking moments at the intersection of the diagonal and the other chord.

2.7.4 Qualitative Influence Lines

One of the most effective methods of obtaining influence lines is the use of Müller-Breslau's principle, which states that the ordinates of the influence line for any response in a structure are equal to those of the deflection curve obtaining by releasing the restraint corresponding to this response and introducing a corresponding unit displacement in the remaining structure. In this way, the shape of the influence lines for both statically determinate and indeterminate structures can be easily obtained, especially for beams.

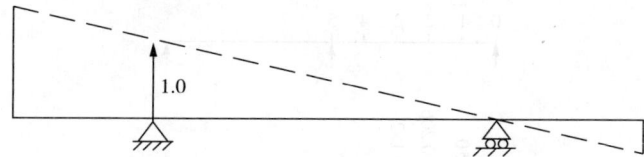

FIGURE 2.53   Influence line for support reaction.

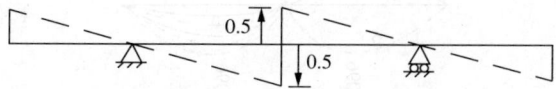

FIGURE 2.54   Influence line for midspan shear force.

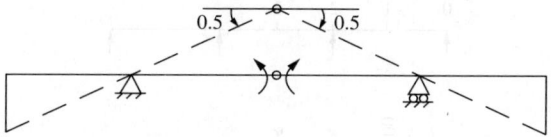

FIGURE 2.55   Influence line for midspan bending moment.

To draw the influence lines of

1. *Support reaction.* Remove the support and introduce a unit displacement in the direction of the corresponding reaction to the remaining structure as shown in Figure 2.53 for a symmetrical overhang beam.
2. *Shear.* Make a cut at the section and introduce a unit relative translation (in the direction of positive shear) without relative rotation of the two ends at the section as shown in Figure 2.54.
3. *Bending moment.* Introduce a hinge at the section (releasing the bending moment) and apply bending (in the direction corresponding to positive moment) to produce a unit relative rotation of the two beam ends at the hinged section as shown in Figure 2.55.

2.7.5 Influence Lines for Continuous Beams

Using Müller-Breslau's principle, the shape of the influence line of any response of a continuous beam can be sketched easily. One of the methods for beam deflection can then be used for determining the ordinates of the influence line at critical points. Figure 2.56 to Figure 2.58 show the influence lines of the bending moment at various points of two-, three-, and four-span continuous beams.

2.8 Energy Methods

Energy methods are a powerful tool in obtaining numerical solutions of statically indeterminate problems. The basic quantity required is the *strain energy*, or work stored due to deformations, of the structure.

2.8.1 Strain Energy Due to Uniaxial Stress

In an axially loaded bar with constant cross-section, the applied load causes normal stress σ_y as shown in Figure 2.59. The tensile stress σ_y increases from 0 to a value σ_y as the load is gradually applied. The original, unstrained position of any section such as C–C will be displaced by an amount dv.

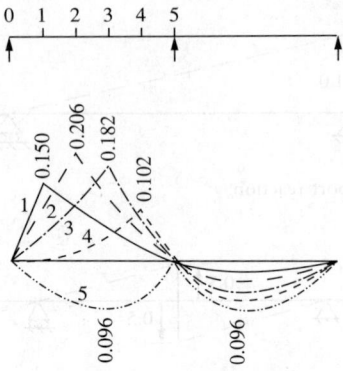

FIGURE 2.56 Influence line for bending moments — two-span beam.

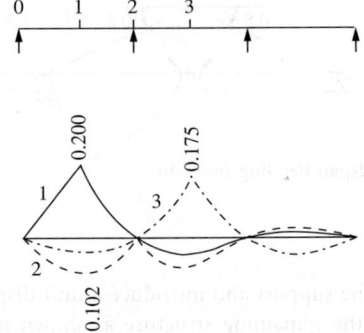

FIGURE 2.57 Influence line for bending moment — three-span beam.

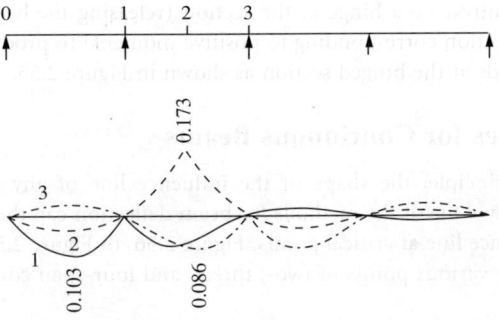

FIGURE 2.58 Influence line for bending moment — four-span beam.

A section D–D located a differential length below C–C will be displaced by an amount $v + (\partial v/\partial y)\,dy$. As σ_y varies with the applied load, from 0 to σ_y, the work done by the forces external to the element can be shown to be

$$dV = \frac{1}{2E}\sigma_y^2 A\,dy = \frac{1}{2}\sigma_y \varepsilon_y A\,dy \qquad (2.107)$$

where A is the area of cross-section of the bar and ε_y is the strain in the direction of σ_y.

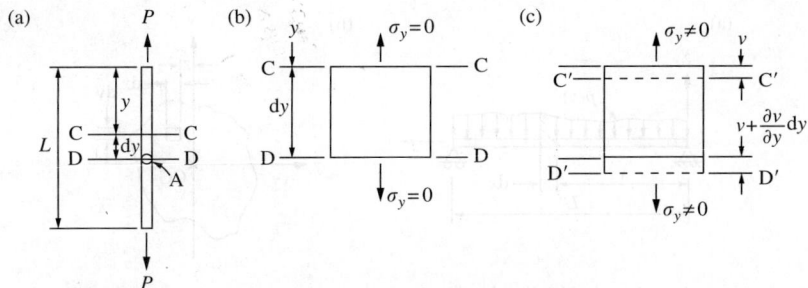

FIGURE 2.59 Axially loaded bar.

2.8.2 Strain Energy in Bending

It can be shown that the strain energy of a differential volume $dx\,dy\,dz$ stressed in tension or compression in the x direction only by a normal stress σ_x will be

$$dV = \frac{1}{2E}\sigma_x^2\,dx\,dy\,dz = \frac{1}{2}\sigma_x\varepsilon_x\,dx\,dy\,dz \tag{2.108}$$

When σ_x is the bending stress given by $\sigma_x = (My)/I$ (see Figure 2.60), then $dV = (1/2E)(M^2y^2)/I^2\,dx\,dy\,dz$, where I is the moment of inertia of the cross-sectional area about the neutral axis.

The total strain energy of bending of a beam is obtained as

$$V = \iiint_{\text{volume}} \frac{1}{2E}\frac{M^2}{I^2}y^2\,dz\,dy\,dx$$

where

$$I = \iint_{\text{area}} y^2\,dz\,dy$$

Therefore,

$$V = \int_{\text{length}} \frac{M^2}{2EI}\,dx \tag{2.109}$$

2.8.3 Strain Energy in Shear

Figure 2.61 shows an element of volume $dx\,dy\,dz$ subjected to shear stress τ_{xy} and τ_{yx}. For static equilibrium, it can readily be shown that

$$\tau_{xy} = \tau_{yx}$$

The shear strain, γ, is defined as AB/AC. For small deformations, it follows that

$$\gamma_{xy} = \frac{\text{AB}}{\text{AC}}$$

Hence, the angle of deformation γ_{xy} is a measure of the shear strain. The strain energy for this differential volume is obtained as

$$dV = \tfrac{1}{2}(\tau_{xy}\,dz\,dx)\gamma_{xy}\,dy = \tfrac{1}{2}\tau_{xy}\gamma_{xy}\,dx\,dy\,dz \tag{2.110}$$

Hooke's Law for shear stress and strain is

$$\gamma_{xy} = \frac{\tau_{xy}}{G} \tag{2.111}$$

(a)   (b)

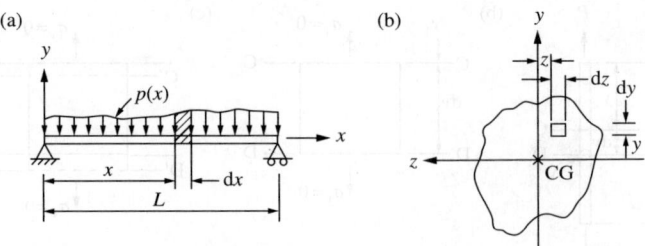

FIGURE 2.60 Beam under arbitrary bending load.

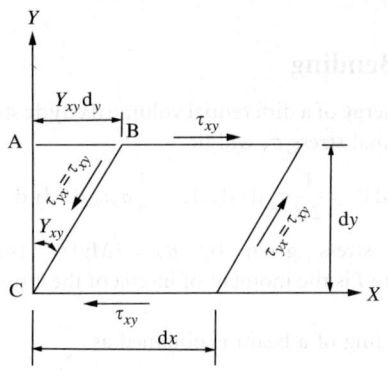

FIGURE 2.61 Shear loading.

where G is the shear modulus of elasticity of the material. The expression for strain energy in shear reduces to

$$dV = \frac{1}{2G}\tau_{xy}^2 \, dx \, dy \, dz \qquad (2.112)$$

2.8.4 Energy Relations in Structural Analysis

The energy relations or laws such as (i) Law of Conservation of Energy, (ii) Theorem of Virtual Work, (iii) Theorem of Minimum Potential Energy, and (iv) Theorem of Complementary Energy are of fundamental importance in structural engineering and are used in various ways in structural analysis.

2.8.4.1 The Law of Conservation of Energy

The Law of Conservation of Energy states that "if a structure and the external loads acting on it are isolated so that these neither receive nor give out energy, then the total energy of this system remains constant."

A typical application of the Law of Conservation of Energy can be made by referring to Figure 2.62, which shows a cantilever beam of constant cross-sections subjected to a concentrated load at its end. If only bending strain energy is considered,

$$\text{external work} = \text{internal work}$$

$$\frac{P\delta}{2} = \int_0^L \frac{M^2}{2EI} \, dx$$

Substituting $M = -Px$ and integrating along the length gives

$$\delta = \frac{PL^3}{3EI} \qquad (2.113)$$

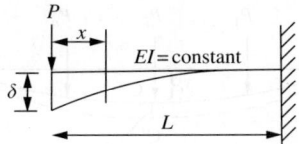

FIGURE 2.62 Cantilever beam.

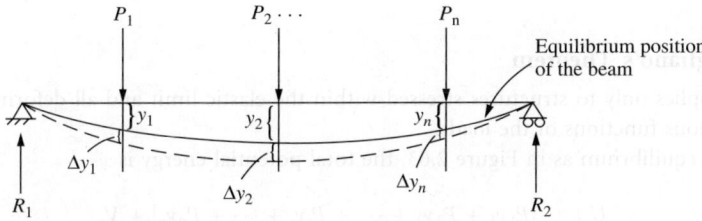

FIGURE 2.63 Equilibrium of a simple supported beam under loading.

2.8.4.2 The Theorem of Virtual Work

The Theorem of Virtual Work can be derived by considering the beam shown in Figure 2.63. The full curved line represents the equilibrium position of the beam under the given loads. Assume the beam to be given an additional small deformation consistent with the boundary conditions. This is called a virtual deformation and corresponds to increments of deflection $\Delta y_1, \Delta y_2, \ldots, \Delta y_n$ at loads $P_1, P_2, \ldots, P_n$ as shown by the dashed line.

The change in potential energy of the loads is given by

$$\Delta(\text{PE}) = \sum_{i=1}^{n} P_i \Delta y_i \tag{2.114}$$

By the Law of Conservation of Energy this must be equal to the internal strain energy stored in the beam. Hence, we may state the Theorem of Virtual Work as: "if a body in equilibrium under the action of a system of external loads is given any small (virtual) deformation, then the work done by the external loads during this deformation is equal to the increase in internal strain energy stored in the body."

2.8.4.3 The Theorem of Minimum Potential Energy

Let us consider the beam shown in Figure 2.64. The beam is in equilibrium under the action of loads, P_1, $P_2, P_3, \ldots, P_i, \ldots, P_n$. The curve ACB defines the equilibrium positions of the loads and reactions. Now apply by some means an additional small displacement to the curve so that it is defined by AC'B. Let y_i be the original equilibrium displacement of the curve beneath a particular load P_i. The additional small displacement is called δy_i. The potential energy of the system while it is in the equilibrium configuration is found by comparing the potential energy of the beam and loads in equilibrium and in the undeflected position. If the change in potential energy of the loads is W and the strain energy of the beam is V, the total energy of the system is

$$U = W + V \tag{2.115}$$

If we neglect the second-order terms, then

$$\delta U = \delta(W + V) = 0 \tag{2.116}$$

The above is expressed as the Principle or Theorem of Minimum Potential Energy, which can be stated as, "if all displacements satisfy the given boundary conditions, those that satisfy the equilibrium conditions make the potential energy a minimum."

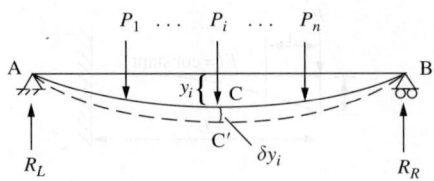

FIGURE 2.64 Simply supported beam under point loading.

2.8.4.4 Castigliano's Theorem

This theorem applies only to structures stressed within the elastic limit and all deformations must be linear homogeneous functions of the loads.

For a beam in equilibrium as in Figure 2.63, the total potential energy is

$$U = -[P_1 y_1 + P_2 y_2 + \cdots + P_j y_j + \cdots + P_n y_n] + V \tag{2.117}$$

For an elastic system, the strain energy, V, turns out to be one half the change in the potential energy of the loads:

$$V = \frac{1}{2} \sum_{i=1}^{i=n} P_i y_i \tag{2.118}$$

Castigliano's Theorem results from studying the variation in the strain energy, V, produced by a differential change in one of the loads, say P_j.

If the load P_j is changed by a differential amount δP_j and if the deflections y are linear functions of the loads, then

$$\frac{\partial V}{\partial P_j} = \frac{1}{2} \sum_{i=1}^{i=n} P_i \frac{\partial y_i}{\partial P_j} + \frac{1}{2} y_j = y_j \tag{2.119}$$

Castigliano's Theorem states that "the partial derivatives of the total strain energy of any structure with respect to any one of the applied forces is equal to the displacement of the point of application of the force in the direction of the force."

To find the deflection of a point in a beam that is not the point of application of a concentrated load, one should apply a load $P = 0$ at that point and carry the term P into the strain energy equation. Finally, introduce the true value of $P = 0$ into the expression for the answer.

EXAMPLE 2.6

It is required to determine the bending deflection at the free end of a cantilever loaded as shown in Figure 2.65.

Solution

$$V = \int_0^L \frac{M^2}{2EI} \, dx$$

$$\Delta = \frac{\partial V}{\partial W_1} = \int_0^L \frac{M}{EI} \frac{\partial M}{\partial W_1} \, dx$$

$$M = W_1 x \qquad\qquad 0 < x < L/2$$
$$= W_1 x + W_2 (x - l/2) \quad L/2 < x < L$$

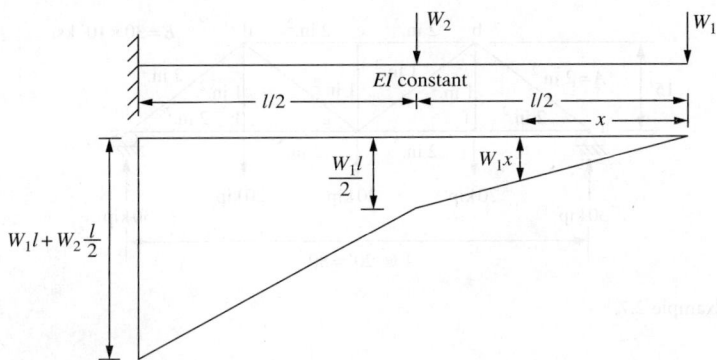

FIGURE 2.65   Example 2.6.

$$\Delta = \frac{1}{EI}\int_0^{l/2} W_1 x \times x\, dx + \frac{1}{EI}\int_{l/2}^{l}\left[W_1 x + W_2\left(x - \frac{P}{2}\right)\right]x\, dx$$

$$= \frac{W_1 l^3}{24EI} + \frac{7W_1 l^3}{24EI} + \frac{5W_2 l^3}{48EI}$$

$$= \frac{W_1 l^3}{3EI} + \frac{5W_2 l^3}{48EI}$$

Castigliano's Theorem can be applied to determine deflection of trusses as follows.

We know that the increment of strain energy for an axially loaded bar is given as

$$dV = \frac{1}{2E}\sigma_y^2 A\, dy$$

Substituting $\sigma_y = S/A$, where S is the axial load in the bar, and integrating over the length of the bar, the total strain energy of the bar is given as

$$V = \frac{S^2 L}{2AE} \tag{2.120}$$

The deflection component Δ_i of the point of application of a load P_i in the direction of P_i is given as

$$\Delta_i = \frac{\partial V}{\partial P_i} = \frac{\partial}{\partial P_i}\sum \frac{S^2 L}{2AE} = \sum \frac{S(\partial S/\partial P_i)L}{AE}$$

EXAMPLE 2.7

Determine the vertical deflection at g of the truss subjected to three-point load as shown in Figure 2.66. Let us first replace a 20-unit load at g by P and carry out the calculations in terms of P. At the end, P will be replaced by the actual load of 20 units.

Member	A	L	S	$\delta S/\delta P$	n	$nS(\delta S/\delta P)(L/A)$
ab	2	25	$-(33.3 + 0.83P)$	-0.83	2	$(691 + 17.2P)$
af	2	20	$(26.7 + 0.67P)$	0.67	2	$(358 + 9P)$
fg	2	20	$(26.7 + 0.67P)$	0.67	2	$(358 + 9P)$
bf	1	15	20	0	2	0
bg	1	25	$0.83P$	0.83	2	$34.4P$
bc	2	20	$-26.7 - 1.33P$	-1.33	2	$(710 + 35.4P)$
cg	1	15	0	0	1	0
"n" indicates the number of similar members				$\sum \frac{S(\delta S/\delta P)L}{A}$		$2117 + 105P$

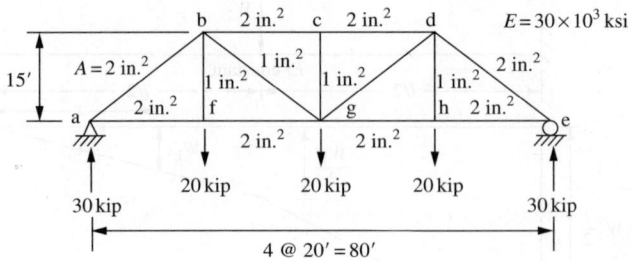

FIGURE 2.66 Example 2.7.

With

$$P = 20$$

$$\Delta_g = \sum \frac{S(\delta S/\delta P)L}{AE} = \frac{(2117 + 105 \times 20) \times 12}{30 \times 10^3} = 1.69$$

2.8.5 Unit Load Method

The unit load method is a versatile tool in the solution of deflections of both trusses and beams. Consider an elastic body in equilibrium under loads P_1, P_2, P_3, P_4, ..., P_n and a load p applied at point O, as shown in Figure 2.67. By Castigliano's Theorem, the component of the deflection of point O in the direction of the applied force p is

$$\delta_{O_p} = \frac{\partial V}{\partial p} \tag{2.121}$$

where V is the strain energy of the body. It has been shown in Equation 2.109, that the strain energy of a beam, neglecting shear effects, is given by

$$V = \int_0^L \frac{M^2}{2EI}\,dx$$

Also, it was shown that if the elastic body is a truss, from Equation 2.120

$$V = \sum \frac{S^2 L}{2AE}$$

For a beam, therefore, from Equation 2.121

$$\delta_{O_p} = \int_L \frac{M(\partial M/\partial p)\,dx}{EI} \tag{2.122}$$

and for a truss

$$\delta_{O_p} = \sum \frac{S(\partial S/\partial p)L}{AE} \tag{2.123}$$

The bending moments M and the axial forces S are functions of the load p as well as of the loads P_1, P_2, ..., P_n. Let a unit load be applied at O on the elastic body and the corresponding moment be m if the body is a beam, and the forces in the members of the body be u if the body is a truss. For the

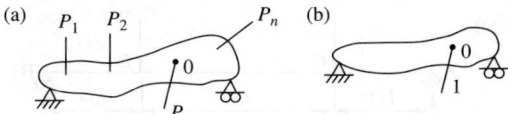

FIGURE 2.67 Elastic body in equilibrium under load.

body in Figure 2.67 the moments M and the forces S due to the system of forces $P_1, P_2, \ldots, P_n$ and p at O applied separately can be obtained by superposition as

$$M = M_P + pm \tag{2.124}$$

$$S = S_P + pu \tag{2.125}$$

where M_P and S_P are, respectively, moments and forces produced by $P_1, P_2, \ldots, P_n$.
   Then

$$\frac{\partial M}{\partial p} = m = \text{moments produced by a unit load at O} \tag{2.126}$$

$$\frac{\partial S}{\partial p} = u = \text{stresses produced by a unit load at O} \tag{2.127}$$

Using Equations 2.126 and 2.127 in Equations 2.122 and 2.123, respectively,

$$\delta_{O_p} = \int_L \frac{Mm\,dx}{EI} \tag{2.128}$$

$$\delta_{O_p} = \sum \frac{SuL}{AE} \tag{2.129}$$

EXAMPLE 2.8

Determine, using the unit load method, the deflection at C of a simple beam of constant cross-section loaded as shown in Figure 2.68a.

Solution

The bending moment diagram for the beam due to the applied loading is shown in Figure 2.68b. A unit load is applied at C where it is required to determine the deflection as shown in Figure 2.68c and the corresponding bending moment diagram is shown in Figure 2.68d. Now, using Equation 2.128, we have

$$
\begin{aligned}
\delta_C &= \int_0^L \frac{Mm\,dx}{EI} \\
&= \frac{1}{EI} \int_0^{L/4} (Wx)\left(\frac{3}{4}x\right) dx + \frac{1}{EI} \int_{L/4}^{3L/4} \left(\frac{WL}{4}\right)\frac{1}{4}(L-x)\,dx \\
&\quad + \frac{1}{EI} \int_{3L/4}^L W(L-x)\frac{1}{4}(L-x)\,dx \\
&= \frac{WL^3}{48EI}
\end{aligned}
$$

Further details on energy methods in structural analysis may be found in Borg and Gennaro (1959).

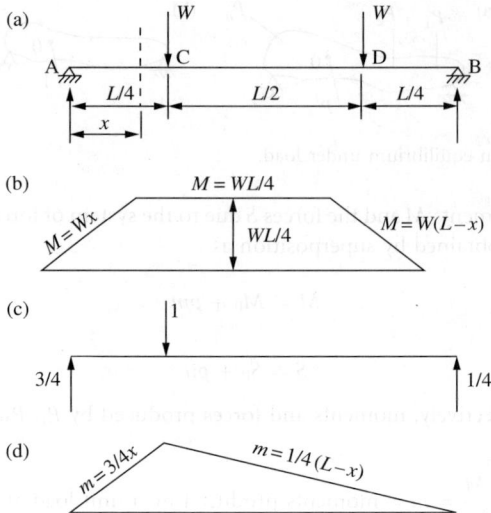

FIGURE 2.68   Example 2.8.

2.9   Matrix Methods

In this method, a set of simultaneous equations that describe the load–deformation characteristics of the structure under consideration are formed. These equations are solved using the matrix algebra to obtain the load–deformation characteristics of discrete or finite elements into which the structure has been subdivided. The matrix method is ideally suited for performing structural analysis using a computer. In general, there are two approaches for structural analysis using the matrix analysis. The first is called the flexibility method in which forces are used as independent variables and the second is called the stiffness method, which employs deformations as the independent variables. The two methods are also called the force method and the displacement method, respectively.

2.9.1   Flexibility Method

In the flexibility method, the forces and displacements are related to one another by using stiffness influence coefficients. Let us consider, for example, a simple beam in which three concentrated loads W_1, W_2, and W_3 are applied at sections 1, 2, and 3, respectively, as shown in Figure 2.69. The deflection at section 1, Δ_1, can be expressed as

$$\Delta_1 = F_{11} W_1 + F_{12} W_2 + F_{13} W_3$$

where F_{11}, F_{12}, and F_{13} are the flexibility coefficients, defined as the deflection at section 1 due to unit loads applied at sections 1, 2, and 3, respectively. Deflections at sections 2 and 3 are similarly given as

$$\Delta_2 = F_{21} W_1 + F_{22} W_2 + F_{23} W_3 \tag{2.130}$$

and

$$\Delta_3 = F_{31} W_1 + F_{32} W_2 + F_{33} W_3$$

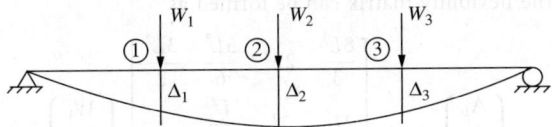

FIGURE 2.69 Simple beam under concentrated loads.

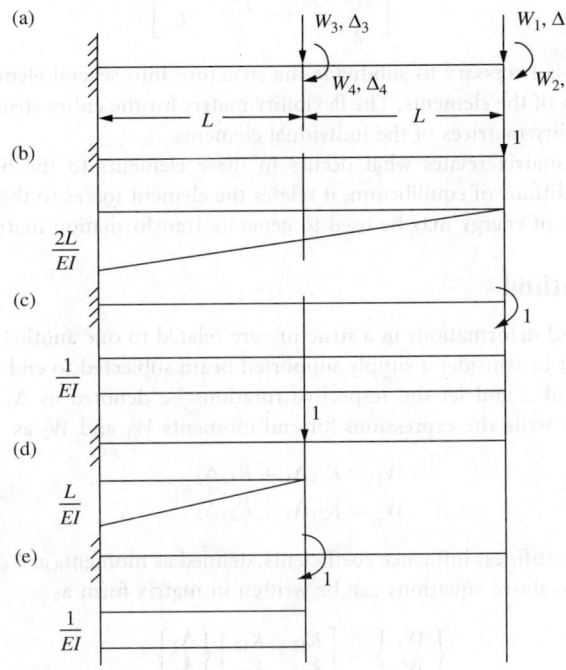

FIGURE 2.70 Cantilever beam.

These expressions are written in the matrix form as

$$\begin{Bmatrix} \Delta_1 \\ \Delta_2 \\ \Delta_3 \end{Bmatrix} = \begin{bmatrix} F_{11} & F_{12} & F_{13} \\ F_{21} & F_{22} & F_{23} \\ F_{31} & F_{32} & F_{33} \end{bmatrix} \begin{Bmatrix} W_1 \\ W_2 \\ W_3 \end{Bmatrix}$$

or

$$\{\Delta\} = [F]\{W\} \tag{2.131}$$

The matrix $[F]$ is called the flexibility matrix. It can be shown, by applying Maxwell's reciprocal theorem (Borg and Gennaro 1959), that the matrix $[F]$ is a symmetric matrix.

Let us consider a cantilever beam loaded as shown in Figure 2.70a. The first column in the flexibility matrix can be generated by applying a unit vertical load at the free end of the cantilever as shown in Figure 2.70b and making use of the moment–area method. We get

$$F_{11} = \frac{8L^3}{3EI}, \quad F_{21} = \frac{2L^2}{EI}, \quad F_{31} = \frac{5L^3}{6EI}, \quad F_{41} = \frac{3L^2}{2EI}$$

Columns 2, 3, and 4 are similarly generated by applying unit moment at the free end and unit force and unit moment at the midspan as shown in Figure 2.70c, Figure 2.70d, and Figure 2.70e, respectively.

Combining the results, the flexibility matrix can be formed as

$$
\begin{Bmatrix} \Delta_1 \\ \Delta_2 \\ \Delta_3 \\ \Delta_4 \end{Bmatrix} = \frac{1}{EI}
\begin{bmatrix}
\dfrac{8L^3}{3} & 2L^2 & \dfrac{5L^3}{6} & \dfrac{3L^2}{2} \\[2mm]
2L^2 & 2L & \dfrac{L^2}{2} & L \\[2mm]
\dfrac{5L^3}{6} & \dfrac{L^2}{2} & \dfrac{L^3}{3} & \dfrac{L^2}{2} \\[2mm]
\dfrac{3L^2}{2} & L & \dfrac{L^2}{2} & L
\end{bmatrix}
\begin{Bmatrix} W_1 \\ W_2 \\ W_3 \\ W_4 \end{Bmatrix}
\tag{2.132}
$$

For a given structure, it is necessary to subdivide the structure into several elements and to form the flexibility matrix for each of the elements. The flexibility matrix for the entire structure is then obtained by combining the flexibility matrices of the individual elements.

Force transformation matrix relates what occurs in these elements to the behavior of the entire structure. Using the conditions of equilibrium, it relates the element forces to the structure forces. The principle of conservation of energy may be used to generate transformation matrices.

2.9.2 Stiffness Method

In this method, forces and deformations in a structure are related to one another by means of stiffness influence coefficients. Let us consider a simply supported beam subjected to end moments W_1 and W_2 applied at supports 1 and 2 and let the respective rotations be denoted as Δ_1 and Δ_2 as shown in Figure 2.71. We can now write the expressions for end moments W_1 and W_2 as

$$
\begin{aligned}
W_1 &= K_{11}\Delta_1 + K_{12}\Delta_2 \\
W_2 &= K_{21}\Delta_1 + K_{22}\Delta_2
\end{aligned}
\tag{2.133}
$$

where K_{11} and K_{12} are the stiffness influence coefficients, defined as moments at 1 due to unit rotation at 1 and 2, respectively. The above equations can be written in matrix form as

$$
\begin{Bmatrix} W_1 \\ W_2 \end{Bmatrix} =
\begin{bmatrix} K_{11} & K_{12} \\ K_{21} & K_{22} \end{bmatrix}
\begin{Bmatrix} \Delta_2 \\ \Delta_2 \end{Bmatrix}
$$

or

$$
\{W\} = [K]\{\Delta\}
\tag{2.134}
$$

The matrix $[K]$ is referred to as stiffness matrix. It can be shown that the flexibility matrix of a structure is the inverse of the stiffness matrix and vice versa. The stiffness matrix of the whole structure is formed by the stiffness matrices of the individual elements that make up the structure.

(a) W_1

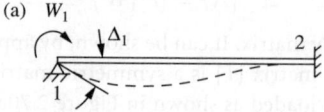

(b) W_1

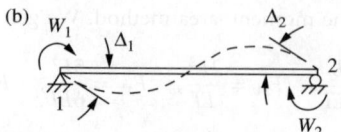

FIGURE 2.71   Simply supported beam.

2.9.3 Element Stiffness Matrix

2.9.3.1 Axially Loaded Member

Figure 2.72 shows an axially loaded member of constant cross-sectional area with element forces q_1 and q_2 and displacements δ_1 and δ_2. They are shown in their respective positive directions. With unit displacement $\delta_1 = 1$ at node 1, as shown in Figure 2.72, axial forces at nodes 1 and 2 are obtained as

$$K_{11} = \frac{EA}{L}, \quad K_{21} = -\frac{EA}{L}$$

In the same way, by setting $\delta_2 = 1$ as shown in Figure 2.72 the corresponding forces are obtained as

$$K_{12} = -\frac{EA}{L}, \quad K_{22} = \frac{EA}{L}$$

The stiffness matrix is written as

$$\begin{Bmatrix} q_1 \\ q_2 \end{Bmatrix} = \begin{bmatrix} K_{11} & K_{12} \\ K_{21} & K_{22} \end{bmatrix} \begin{Bmatrix} \delta_1 \\ \delta_2 \end{Bmatrix}$$

or

$$\begin{Bmatrix} q_1 \\ q_2 \end{Bmatrix} = \frac{EA}{L} \begin{bmatrix} 1 & -1 \\ -1 & 1 \end{bmatrix} \begin{Bmatrix} \delta_1 \\ \delta_2 \end{Bmatrix} \tag{2.135}$$

2.9.3.2 Flexural Member

The stiffness matrix for the flexural element can be constructed by referring to Figure 2.73. The forces and the corresponding displacements, namely, the moments, the shears, and the corresponding rotations and translations at the ends of the member, are defined in the figure. The matrix equation that relates these forces and displacements can be written in the form

$$\begin{bmatrix} q_1 \\ q_2 \\ q_3 \\ q_4 \end{bmatrix} = \begin{bmatrix} K_{11} & K_{12} & K_{13} & K_{14} \\ K_{21} & K_{22} & K_{23} & K_{24} \\ K_{31} & K_{32} & K_{33} & K_{34} \\ K_{41} & K_{42} & K_{43} & K_{44} \end{bmatrix} \begin{bmatrix} \delta_1 \\ \delta_2 \\ \delta_3 \\ \delta_4 \end{bmatrix}$$

The terms in the first column consist of the element forces q_1 through q_4 that result from displacement $\delta_1 = 1$ when $\delta_2 = \delta_3 = \delta_4 = 0$. This means that a unit vertical displacement is imposed at the left end of the member while translation at the right end and rotation at both ends are prevented as shown in Figure 2.73. The four member forces corresponding to this deformation can be obtained using the moment area method.

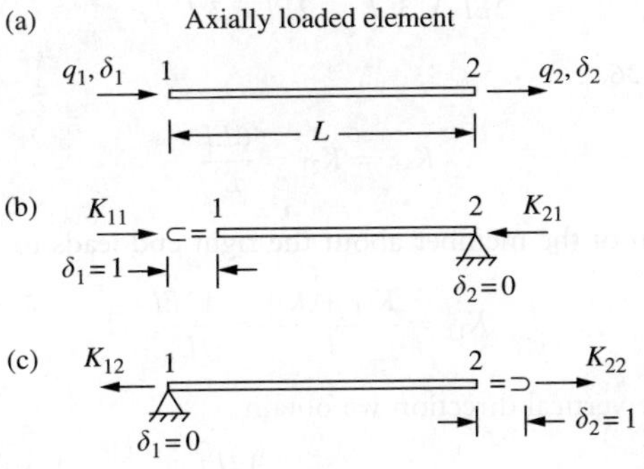

FIGURE 2.72 Axially loaded member.

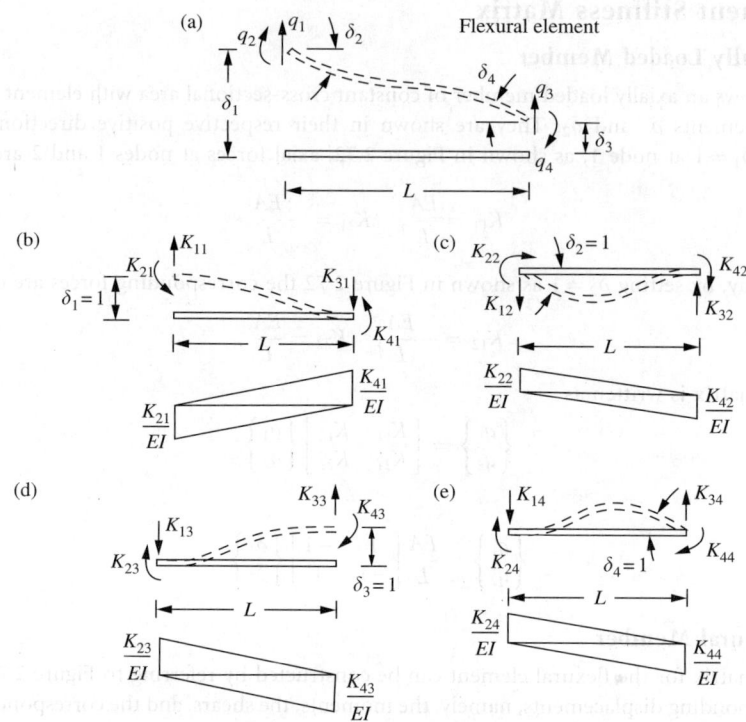

FIGURE 2.73   Beam element — stiffness matrix.

The change in slope between the two ends of the member is zero and the area of the M/EI diagram between these points must therefore vanish. Hence,

$$\frac{K_{41}L}{2EI} - \frac{K_{21}L}{2EI} = 0$$

and

$$K_{21} = K_{41} \tag{2.136}$$

The moment of the M/EI diagram about the left end of the member is equal to unity. Hence,

$$\frac{K_{41}L}{2EI}\left(\frac{2L}{3}\right) - \frac{K_{21}L}{2EI}\left(\frac{L}{3}\right) = 1$$

and in view of Equation 2.136,

$$K_{41} = K_{21} = \frac{6EI}{L^2}$$

Finally, moment equilibrium of the member about the right end leads to

$$K_{11} = \frac{K_{21} + K_{41}}{L} = \frac{12EI}{L^3}$$

and from equilibrium in the vertical direction we obtain

$$K_{31} = K_{11} = \frac{12EI}{L^3}$$

The forces act in the directions indicated in Figure 2.73b. To obtain the correct signs, one must compare the forces with the positive directions defined in Figure 2.73a. Thus,

$$K_{11} = \frac{12EI}{L^3}, \quad K_{21} = -\frac{6EI}{L^2}, \quad K_{31} = -\frac{12EI}{L^3}, \quad K_{41} = \frac{6EI}{L^2}$$

The second column of the stiffness matrix is obtained by letting $\delta_2 = 1$ and setting the remaining three displacements equal to zero as indicated in Figure 2.73c. The area of the M/EI diagram between the ends of the member for this case is equal to unity, and hence,

$$\frac{K_{22}L}{2EI} - \frac{K_{42}L}{2EI} = 1$$

The moment of the M/EI diagram about the left end is zero, so that

$$\frac{K_{22}L}{2EI}\left(\frac{L}{3}\right) - \frac{K_{42}L}{2EI}\left(\frac{2L}{3}\right) = 0$$

Therefore, one obtains

$$K_{22} = \frac{4EI}{L}, \quad K_{42} = \frac{2EI}{L}$$

From vertical equilibrium of the member,

$$K_{12} = K_{32}$$

and moment equilibrium about the right end of the member leads to

$$K_{12} = \frac{K_{22} + K_{42}}{L} = \frac{6EI}{L^2}$$

Comparison of the forces in Figure 2.73c with the positive directions defined in Figure 2.73a indicates that all the influence coefficients except K_{12} are positive. Thus,

$$K_{12} = -\frac{6EI}{L^2}, \quad K_{22} = \frac{4EI}{L}, \quad K_{32} = \frac{6EI}{L^2}, \quad K_{42} = \frac{2EI}{L}$$

Using Figure 2.73d and Figure 2.73e, the influence coefficients for the third and fourth columns can be obtained. The results of these calculations lead to the following element stiffness matrix:

$$\begin{bmatrix} q_1 \\ q_2 \\ q_3 \\ q_4 \end{bmatrix} = \begin{bmatrix} \dfrac{12EI}{L^3} & -\dfrac{6EI}{L^2} & -\dfrac{12EI}{L^3} & -\dfrac{6EI}{L^2} \\[2mm] -\dfrac{6EI}{L^2} & \dfrac{4EI}{L} & \dfrac{6EI}{L^2} & \dfrac{2EI}{L} \\[2mm] -\dfrac{12EI}{L^3} & \dfrac{6EI}{L^2} & \dfrac{12EI}{L^3} & \dfrac{6EI}{L^2} \\[2mm] -\dfrac{6EI}{L^2} & \dfrac{2EI}{L} & \dfrac{6EI}{L^2} & \dfrac{4EI}{L} \end{bmatrix} \begin{bmatrix} \delta_1 \\ \delta_2 \\ \delta_3 \\ \delta_4 \end{bmatrix} \tag{2.137}$$

Note that Equation 2.136 defines the element stiffness matrix for a flexural member with constant flexural rigidity EI.

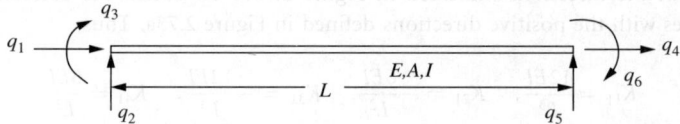

FIGURE 2.74 Beam element with axial force.

If axial load in a frame member is also considered the general form of an element stiffness matrix for an element shown in Figure 2.74 becomes

$$
\begin{bmatrix} q_1 \\ q_2 \\ q_3 \\ q_4 \\ q_5 \\ q_6 \end{bmatrix} =
\begin{bmatrix}
\dfrac{EA}{L} & 0 & 0 & -\dfrac{EA}{L} & 0 & 0 \\[2mm]
0 & \dfrac{12EI}{L^3} & \dfrac{6EI}{L^2} & 0 & -\dfrac{12EI}{L^3} & -\dfrac{6EI}{L^2} \\[2mm]
0 & -\dfrac{6EI}{L^2} & \dfrac{4EI}{L} & 0 & \dfrac{6EI}{L^2} & \dfrac{2EI}{L} \\[2mm]
-\dfrac{EI}{L} & 0 & 0 & \dfrac{EI}{L} & 0 & 0 \\[2mm]
0 & -\dfrac{12EI}{L^3} & \dfrac{6EI}{L^2} & 0 & \dfrac{12EI}{L^3} & \dfrac{6EI}{L^2} \\[2mm]
0 & -\dfrac{6EI}{L^2} & \dfrac{2EI}{L} & 0 & \dfrac{6EI}{L^2} & \dfrac{4EI}{L}
\end{bmatrix}
\begin{bmatrix} \delta_1 \\ \delta_2 \\ \delta_3 \\ \delta_4 \\ \delta_5 \\ \delta_6 \end{bmatrix}
$$

or

$$[q] = [k_c][\delta] \tag{2.138}$$

The member stiffness matrix can be written as

$$
K = 0
\begin{bmatrix}
\dfrac{GJ}{L} & 0 & 0 & -\dfrac{GJ}{L} & 0 & 0 \\[2mm]
0 & \dfrac{12EI_z}{L^3} & \dfrac{6EI_z}{L^2} & 0 & -\dfrac{12EI_z}{L^3} & \dfrac{6EI_z}{L^2} \\[2mm]
0 & \dfrac{6EI_z}{L^2} & \dfrac{4EI_z}{L} & 0 & -\dfrac{6EI_z}{L^2} & \dfrac{2EI_z}{L} \\[2mm]
-\dfrac{GJ}{L} & 0 & 0 & \dfrac{GJ}{L} & 0 & 0 \\[2mm]
0 & -\dfrac{12EI_z}{L^2} & -\dfrac{6EI_z}{L^2} & 0 & \dfrac{12EI_z}{L^3} & -\dfrac{6EI_z}{L^2} \\[2mm]
0 & \dfrac{6EI_z}{L^2} & \dfrac{2EI_z}{L} & 0 & -\dfrac{6EI_z}{L^2} & \dfrac{4EI_z}{L}
\end{bmatrix}
\tag{2.139}
$$

2.9.4 Structure Stiffness Matrix

Equation 2.138 has been expressed in terms of the coordinate system of the individual members. In a structure consisting of many members there would be as many systems of coordinates as the number of members. Before the internal actions in the members of the structure can be related, all forces and deflections must be stated in terms of one single system of axes common to all — the global axes. The transformation from element to global coordinates is carried out separately for each element and the resulting matrices are then combined to form the structure stiffness matrix. A separate transformation matrix $[T]$ is written for each element and a relation of the form

$$[\delta]_n = [T]_n [\Delta]_n \tag{2.140}$$

is written in which $[T]_n$ defines the matrix relating the element deformations of element n to the structure deformations at the ends of that particular element. The element and structure forces are related in the same way as the corresponding deformations as

$$[q]_n = [T]_n [W]_n \tag{2.141}$$

where $[q]_n$ contains the element forces for element n and $[W]_n$ contains the structure forces at the extremities of the element. The transformation matrix $[T]_n$ can be used to transform element n from its local coordinates to structure coordinates. We know, for an element n, the force–deformation relation is given as

$$[q]_n = [k]_n [\delta]_n$$

Substituting for $[q]_n$ and $[\delta]_n$ from Equations 2.140 and 2.141 one obtains

$$[T]_n [W]_n = [k]_n [T]_n [\Delta]_n$$

or

$$[W]_n = [T]_n^{-1} [k]_n [T]_n [\Delta]_n$$
$$= [T]_n^{T} [k]_n [T]_n [\Delta]_n$$
$$= [K]_n [\Delta]_n$$
$$[K]_n = [T]_n^{T} [k]_n [T]_n \tag{2.142}$$

$[K]_n$ is the stiffness matrix that transforms any element n from its local coordinate to structure coordinates. In this way, each element is transformed individually from element coordinate to structure coordinate and the resulting matrices are combined to form the stiffness matrix for the entire structure.

The member stiffness matrix $[K]_n$ in global coordinates for a truss member shown in Figure 2.75, for example, is given as

$$[K]_n = \frac{AE}{L} \begin{bmatrix} \lambda^2 \mu & \lambda\mu & -\lambda^2 & -\lambda\mu \\ \lambda\mu & \mu^2 & -\lambda\mu & -\mu^2 \\ -\lambda^2 & -\lambda\mu & \lambda^2 & \lambda\mu \\ -\lambda\mu & -\mu^2 & \lambda\mu & \mu^2 \end{bmatrix} \begin{matrix} i \\ j \\ k \\ l \end{matrix} \tag{2.143}$$

where $\lambda = \cos\phi$ and $\mu = \sin\phi$.

To construct $[K]_n$ for a given member it is necessary to have the values of λ and μ for the member. In addition, the structure coordinates i, j, k, and l at the extremities of the member must be known.

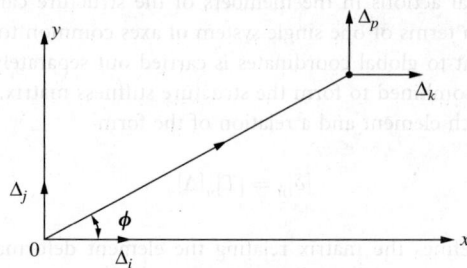

FIGURE 2.75 A grid member.

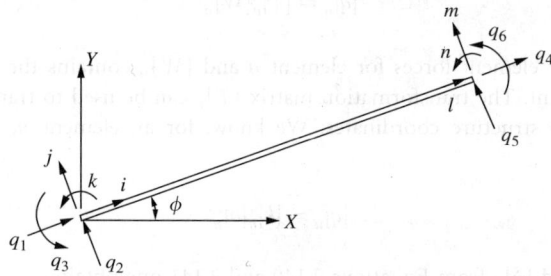

FIGURE 2.76 A flexural member in global coordinate.

The member stiffness matrix $[K]_n$ in structural coordinates for a flexural member shown in Figure 2.76 can be written as

$$
[K]_n = 0 \begin{bmatrix}
\lambda^2 \dfrac{AE}{L} + \mu^2 \dfrac{12EI}{L^3} & & & & & \text{symmetric} \\[2ex]
\mu\lambda\left(\dfrac{AE}{L} - \dfrac{12EI}{L^3}\right) & \mu^2 \dfrac{AE}{L} + \lambda^2 \dfrac{12EI}{L^3} & & & & \\[2ex]
-\mu\left(\dfrac{6EI}{L^2}\right) & \lambda\dfrac{6EI_z}{L^2} & \dfrac{4EI}{L} & 0 & & \\[2ex]
-\lambda^2 \dfrac{AE}{L} - \mu^2 \dfrac{12EI}{L^3} & \mu\lambda\left(\dfrac{AE}{L} - \dfrac{12EI}{L^3}\right) & \mu\left(\dfrac{6EI}{L^2}\right) & \lambda^2 \dfrac{AE}{L} + \mu^2 \dfrac{12EI}{L^3} & 0 & 0 \\[2ex]
-\mu\lambda\left(\dfrac{AE}{L} - \dfrac{12EI}{L^3}\right) & -\left(\mu^2 \dfrac{AE}{L} + \lambda^2 \dfrac{12EI}{L^3}\right) & -\lambda\left(\dfrac{6EI}{L^2}\right) & \mu\lambda\left(\dfrac{AE}{L} - \dfrac{12EI}{L^3}\right) & \left(\mu\dfrac{AE}{L} + \lambda^2 \dfrac{12EI}{L^3}\right) & \\[2ex]
\mu\dfrac{6EI_z}{L^2} & \lambda\dfrac{6EI_z}{L^2} & \dfrac{2EI}{L} & \mu\left(\dfrac{6EI}{L^2}\right) & -\lambda\dfrac{6EI}{L^2} & \dfrac{4EI}{L}
\end{bmatrix}
$$

$$\tag{2.144}$$

where $\lambda = \cos\phi$ and $\mu = \sin\phi$

EXAMPLE 2.9

Determine the displacement at the loaded point of the truss shown in Figure 2.77a. Both members have the same area of cross-section $A = 3$ and $E = 30 \times 10^3$.

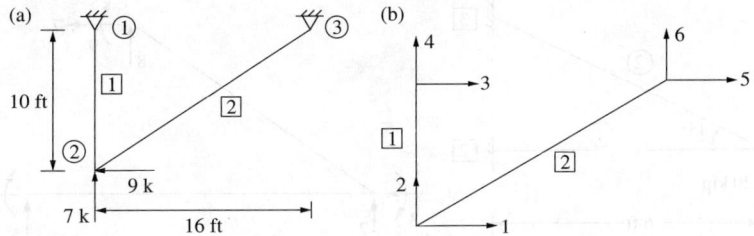

FIGURE 2.77   Example 2.9.

The details required to form the element stiffness matrix with reference to structure coordinate axes are listed below (see Figure 2.77b):

Member	Length	ϕ	λ	μ	i	j	k	l
1	10	90°	0	1	1	2	3	4
2	18.9	32°	0.85	0.53	1	2	5	6

We now use these data in Equation 2.143 to form $[K]_n$ for the two elements.

For member 1

$$\frac{AE}{L} = \frac{3 \times 30 \times 10^3}{120} = 750$$

$$[K]_1 = \begin{array}{c} \\ 1 \\ 2 \\ 3 \\ 4 \end{array} \begin{array}{cccc} 1 & 2 & 3 & 4 \\ 0 & 0 & 0 & 0 \\ 0 & 750 & 0 & -750 \\ 0 & 0 & 0 & 0 \\ 0 & -750 & 0 & 750 \end{array}$$

For member 2

$$\frac{AE}{L} = \frac{3 \times 30 \times 10^3}{18.9 \times 12} = 397$$

$$[K]_2 = \begin{array}{c} \\ 1 \\ 2 \\ 5 \\ 6 \end{array} \begin{array}{cccc} 1 & 2 & 5 & 6 \\ 286 & 179 & -286 & -179 \\ 179 & 111 & -179 & -111 \\ -286 & -179 & 286 & 179 \\ -179 & -111 & 179 & 111 \end{array}$$

Combining the element stiffness matrices, $[K]_1$ and $[K]_2$, one obtains the structure stiffness matrix as follows:

$$\begin{bmatrix} W_1 \\ W_2 \\ W_3 \\ W_4 \\ W_5 \\ W_6 \end{bmatrix} = \begin{bmatrix} 286 & 179 & 0 & 0 & -286 & -179 \\ 179 & 861 & 0 & -750 & -179 & -111 \\ 0 & 0 & 0 & 0 & 0 & 0 \\ 0 & -750 & 0 & 750 & 0 & 0 \\ -286 & -179 & 0 & 0 & 286 & 179 \\ -179 & -111 & 0 & 0 & 179 & 111 \end{bmatrix} \begin{bmatrix} \Delta_1 \\ \Delta_2 \\ \Delta_3 \\ \Delta_4 \\ \Delta_5 \\ \Delta_6 \end{bmatrix}$$

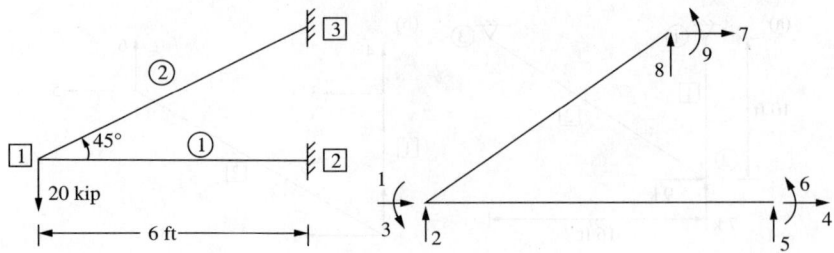

FIGURE 2.78   Example 2.10.

The stiffness matrix can now be subdivided to determine the unknowns. Let us consider Δ_1 and Δ_2, the deflections at joint 2, which can be determined in view of $\Delta_3 = \Delta_4 = \Delta_5 = \Delta_6 = 0$ as follows:

$$\begin{bmatrix} \Delta_1 \\ \Delta_2 \end{bmatrix} = \begin{bmatrix} 286 & 179 \\ 179 & 861 \end{bmatrix}^{-1} \begin{bmatrix} -9 \\ 7 \end{bmatrix}$$

or

$$\Delta_1 = 0.042$$

$$\Delta_2 = 0.0169$$

EXAMPLE 2.10

A simple triangular frame is loaded at the tip by 20 units of force as shown in Figure 2.78. Assemble the structure stiffness matrix and determine the displacements at the loaded node.

Member	Length	A	I	ϕ	λ	μ
1	72	2.4	1037	0	1	0
2	101.8	3.4	2933	45°	0.707	0.707

For members 1 and 2 the stiffness matrices in structure coordinates can be written by making use of Equation 2.144:

$$[K]_1 = 10^3 \times \begin{array}{c c c c c c c} & 1 & 2 & 3 & 4 & 5 & 6 \\ \begin{bmatrix} 1 & 0 & 0 & -1 & 0 & 0 \\ 0 & 1 & 36 & 0 & -1 & 36 \\ 0 & 36 & 1728 & 0 & -36 & 864 \\ -1 & 0 & 0 & 1 & 0 & 0 \\ 0 & -1 & -36 & 0 & 1 & -36 \\ 0 & 36 & 864 & 0 & -36 & 1728 \end{bmatrix} & \begin{matrix} 1 \\ 2 \\ 3 \\ 4 \\ 5 \\ 6 \end{matrix} \end{array}$$

and

$$[K]_2 = 10^3 \times \begin{array}{c c c c c c c} & 1 & 2 & 3 & 7 & 8 & 9 \\ \begin{bmatrix} 1 & 0 & -36 & -1 & 0 & -36 \\ 0 & 1 & 36 & 0 & 1 & 36 \\ -36 & 36 & 3457 & 36 & -36 & 1728 \\ -1 & 0 & 36 & 1 & 0 & 36 \\ 0 & 1 & -36 & 0 & 1 & -36 \\ -36 & 36 & 1728 & 36 & -36 & 3457 \end{bmatrix} & \begin{matrix} 1 \\ 2 \\ 3 \\ 7 \\ 8 \\ 9 \end{matrix} \end{array}$$

Combining the element stiffness matrices $[K]_1$ and $[K]_2$ one obtains the structure stiffness matrix as follows:

$$[K] = 10^3 \times \begin{bmatrix} 2 & 0 & -36 & -1 & 0 & 0 & -1 & 0 & -36 \\ 0 & 2 & 72 & 0 & -1 & 36 & 0 & 1 & 36 \\ -36 & 72 & 5185 & 0 & -36 & 864 & 36 & -36 & 1728 \\ -1 & 0 & 0 & 1 & 0 & 0 & 0 & 0 & 0 \\ 0 & -1 & -36 & 0 & 1 & -36 & 0 & 0 & 0 \\ 0 & 36 & 864 & 0 & -36 & 1728 & 0 & 0 & 0 \\ -1 & 0 & 36 & 0 & 0 & 0 & 1000 & 0 & 36 \\ 0 & 1 & -36 & 0 & 0 & 0 & 0 & 1 & -36 \\ -36 & 36 & 1728 & 0 & 0 & 0 & 36 & 36 & 3457 \end{bmatrix} \begin{matrix} 1 \\ 2 \\ 3 \\ 4 \\ 5 \\ 6 \\ 7 \\ 8 \\ 9 \end{matrix}$$

The deformations at joints 2 and 3 corresponding to Δ_5 to Δ_9 are zero since joints 2 and 4 are restrained in all directions. Cancelling the rows and columns corresponding to zero deformations in the structure stiffness matrix, one obtains the force–deformation relation for the structure:

$$\begin{bmatrix} F_1 \\ F_2 \\ F_3 \end{bmatrix} = \begin{bmatrix} 2 & 0 & -36 \\ 0 & 2 & 72 \\ -36 & 72 & 5185 \end{bmatrix} \times 10^3 \begin{bmatrix} \Delta_1 \\ \Delta_2 \\ \Delta_3 \end{bmatrix}$$

Substituting for the applied load $F_2 = -20$ the deformations are given as

$$\begin{bmatrix} \Delta_1 \\ \Delta_2 \\ \Delta_3 \end{bmatrix} = \begin{bmatrix} 2 & 0 & -36 \\ 0 & 2 & 72 \\ -36 & 72 & 5185 \end{bmatrix}^{-1} \times 10^3 \begin{bmatrix} 0 \\ -20 \\ 0 \end{bmatrix}$$

or

$$\begin{bmatrix} \Delta_1 \\ \Delta_2 \\ \Delta_3 \end{bmatrix} = \begin{bmatrix} 6.66 \\ -23.334 \\ 0.370 \end{bmatrix} \times 10^3$$

2.9.5 Loading between Nodes

The problems discussed so far have involved concentrated forces and moments applied to nodes only. But real structures are subjected to distributed or concentrated loading between nodes, as shown in Figure 2.79. Loading may range from a few concentrated loads to an infinite variety of uniform or nonuniformly distributed loads. The solution method of matrix analysis must be modified to account for such load cases.

One way to treat such loads in the matrix analysis is to insert artificial nodes, such as p and q as shown in Figure 2.79. The degrees of freedom corresponding to the additional nodes are added to the total structure and the necessary additional equations are written by considering the requirements of equilibrium at these nodes. The internal member forces on each side of nodes p and q must equilibrate the

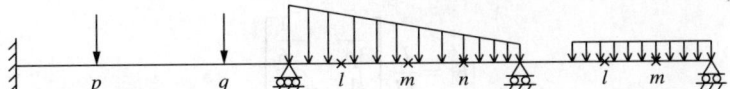

FIGURE 2.79 Loading between nodes.

external loads applied at these points. In the case of distributed loads, suitable nodes, such as *l, m, n* shown in Figure 2.79, are selected arbitrarily and the distributed loads are lumped as concentrated loads at these nodes. The degrees of freedom corresponding to the arbitrary and real nodes are treated as unknowns of the problem. There are different ways of obtaining equivalence between the lumped and the distributed loading. In all cases the lumped loads must be statically equivalent to the distributed loads they replace.

The method of introducing arbitrary nodes is not a very elegant procedure because the number of unknown degrees of freedom makes the solution procedure laborious. The approach that is of most general use with the displacement method is one employing the related concepts of artificial joint restraint, fixed-end forces, and equivalent nodal loads.

2.9.6 Semirigid End Connection

A rigid connection holds unchanged the original angles between interesting members; a simple connection allows the member end to rotate freely; a semirigid connection possesses a moment resistance intermediate between the simple and the rigid. A simplified linear relationship between the moment *M* acting on the connection and the resulting connection rotation ψ in the direction of *M* is assumed giving

$$M = R\frac{EI}{L}\psi \tag{2.145}$$

where *EI* and *L* are the flexural rigidity and length of the member, respectively. The nondimensional quantity *R*, which is a measure of the degree of rigidity of the connection, is called the rigidity index. For a simple connection, *R* is zero and for a rigid connection, *R* is infinity. Considering the semirigidity of joints, the member flexibility matrix for flexure is derived as

$$\begin{bmatrix} \phi_1 \\ \phi_2 \end{bmatrix} = \frac{L}{EI} \begin{bmatrix} \dfrac{1}{3} + \dfrac{1}{R_1} & -\dfrac{1}{6} \\ -\dfrac{1}{6} & \dfrac{1}{3} + \dfrac{1}{R_2} \end{bmatrix} \begin{bmatrix} M_1 \\ M_2 \end{bmatrix} \tag{2.146}$$

or

$$[\phi] = [F][M] \tag{2.147}$$

where ϕ_1 and ϕ_2 are as shown in Figure 2.80.

For convenience, two parameters are introduced as follows:

$$p_1 = \frac{1}{1 + (3/R_1)}$$

and

$$p_2 = \frac{1}{1 + (3/R_2)}$$

where p_1 and p_2 are the fixity factors. For hinged connections, both the fixity factors, *p*, and the rigidity index, *R*, are zero; but for rigid connections, the fixity factor is 1 and the rigidity index is infinity. Since the fixity factor can only vary from 0 to 1.0, it is more convenient to use in the analyses of structures with semirigid connections.

Equation 2.146 can be rewritten to give

$$[F] = \frac{L}{EI} \begin{bmatrix} \dfrac{1}{3p_1} & -\dfrac{1}{6} \\ \dfrac{1}{6} & \dfrac{1}{3p_2} \end{bmatrix} \tag{2.148}$$

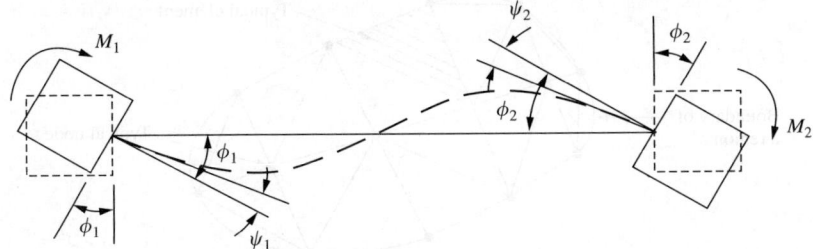

FIGURE 2.80 A flexural member with semirigid end connections.

From Equation 2.148, the modified member stiffness matrix $[K]$ for a member with semirigid end connections expresses the member end moments, M_1 and M_2, in terms of the member end rotations, ϕ_1 and ϕ_2, as

$$[K] = EI \begin{bmatrix} K_{11} & K_{12} \\ K_{21} & K_{22} \end{bmatrix} \qquad (2.149a)$$

Expressions for K_{11} and $K_{12} = K_{21}$ and K_{22} may be obtained by inverting the $[F]$ matrix, thus

$$K_{11} = \frac{12/p_1}{4/(p_1 p_2) - 1} \qquad (2.149b)$$

$$K_{12} = K_{21} \frac{6}{4(p_1 p_2) - 1} \qquad (2.149c)$$

$$K_{22} = \frac{12/p_1}{4/(p_1 p_2) - 1} \qquad (2.149d)$$

The modified member stiffness matrix $[K]$, as expressed by Equations 2.149a to d, will be needed in the stiffness method of analysis of frames in which there are semirigid member end connections.

2.10 The Finite Element Method

For problems involving complex material properties and boundary conditions, numerical methods are employed to provide approximate but acceptable solutions. Of the many numerical methods developed before and after the advent of computers, the finite element method has proven to be a powerful tool. This method can be regarded as a natural extension of the matrix methods of structural analysis. It can accommodate complex and difficult problems such as nonhomogeneity, nonlinear stress–strain behavior, and complicated boundary conditions. The finite element method is applicable to a wide range of boundary value problems in engineering and it dates back to the mid-1950s with the pioneering work of Argyris (1960), Clough (1993), and others. The method was first applied to the solution of plane stress problems and extended subsequently to the solution of plates, shells, and axisymmetric solids.

2.10.1 Basic Principle

The finite element method is based on the representation of a body or a structure by an assemblage of subdivisions called finite elements, as shown in Figure 2.81. These elements are considered to be connected at nodes. Displacement functions are chosen to approximate the variation of displacements over each finite element. Polynomials functions are commonly employed to approximate these displacements. Equilibrium equations for each element are obtained by means of the principle of minimum potential energy. These equations are formulated for the entire body by combining the equations for the individual elements so that the continuity of displacements is preserved at the nodes. The resulting equations are solved by satisfying the boundary conditions to obtain the unknown displacements.

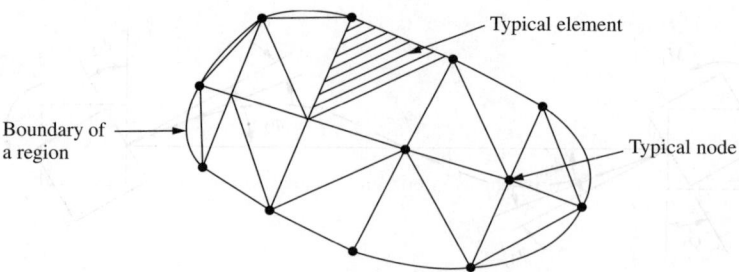

FIGURE 2.81   Assemblage of subdivisions.

The entire procedure of the finite element method involves the following steps:

1. The given body is subdivided into an equivalent system of finite elements.
2. A suitable displacement function is chosen.
3. The element stiffness matrix is derived using a variational principle of mechanics such as the principle of minimum potential energy.
4. The global stiffness matrix for the entire body is formulated.
5. The algebraic equations thus obtained are solved to determine unknown displacements.
6. Element strains and stresses are computed from the nodal displacements.

2.10.2 Elastic Formulation

Figure 2.82 shows the state of stress in an elemental volume of a body under a load. It is defined in terms of three normal stress components σ_x, σ_y, and σ_z and three shear stress components τ_{xy}, τ_{yz}, and τ_{zx}. The corresponding strain components are three normal strains ε_x, ε_y, and ε_z and three shear strains γ_{xy}, γ_{yz}, and γ_{zx}. These strain components are related to the displacement components u, v, and w at a point as follows:

$$\varepsilon_x = \frac{\partial u}{\partial x}, \quad \gamma_{xy} = \frac{\partial v}{\partial x} + \frac{\partial u}{\partial y}$$

$$\varepsilon_y = \frac{\partial v}{\partial y}, \quad \gamma_{yz} = \frac{\partial w}{\partial y} + \frac{\partial v}{\partial z} \tag{2.150}$$

$$\varepsilon_z = \frac{\partial w}{\partial z}, \quad \gamma_{zx} = \frac{\partial u}{\partial z} + \frac{\partial w}{\partial x}$$

The relations given in Equation 2.150 are valid in the case of the body experiencing small deformations. If the body undergoes large or finite deformations, higher-order terms must be retained.

The stress–strain equations for isotropic materials may be written in terms of Young's modulus and Poisson's ratio as

$$\sigma_x = \frac{E}{1 - v^2} [\varepsilon_x + v(\varepsilon_y + \varepsilon_z)]$$

$$\sigma_y = \frac{E}{1 - v^2} [\varepsilon_y + v(\varepsilon_z + \varepsilon_x)]$$

$$\sigma_z = \frac{E}{1 - v^2} [\varepsilon_z + v(\varepsilon_x + \varepsilon_y)] \tag{2.151}$$

$$\tau_{xy} = G\gamma_{xy}, \quad \tau_{yz} = G\gamma_{yz}, \quad \tau_{zx} = G\gamma_{zx}$$

2.10.3 Plane Stress

When the elastic body is very thin and there are no loads applied in the direction parallel to the thickness, the state of stress in the body is said to be plane stress. A thin plate subjected to in-plane loading as

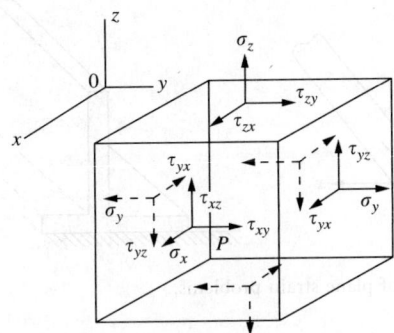

FIGURE 2.82 State of stress in an elemental volume.

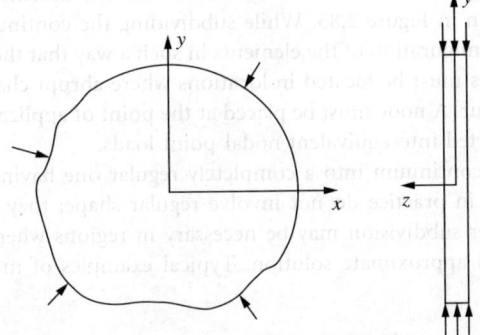

FIGURE 2.83 Plane stress problem.

shown in Figure 2.83 is an example of a plane stress problem. In this case, $\sigma_z = \tau_{yz} = \tau_{zx} = 0$ and the constitutive relation for an isotropic continuum is expressed as

$$
\begin{bmatrix} \sigma_x \\ \sigma_y \\ \sigma_{xy} \end{bmatrix} = \frac{E}{1-v^2} \begin{bmatrix} 1 & v & 0 \\ v & 1 & 0 \\ 0 & 0 & (1-v)/2 \end{bmatrix} \begin{bmatrix} \varepsilon_x \\ \varepsilon_y \\ \gamma_{xy} \end{bmatrix}
\tag{2.152}
$$

2.10.4 Plane Strain

The state of plane strain occurs in members that are not free to expand in the direction perpendicular to the plane of the applied loads. Examples of some plane strain problems are retaining walls, dams, long cylinder, and tunnels, as shown in Figure 2.84. In these problems ε_z, γ_{yz}, and γ_{zx} will vanish and hence,

$$
\sigma_z = v(\sigma_x + \sigma_y)
$$

The constitutive relation for an isotropic material is written as

$$
\begin{bmatrix} \sigma_x \\ \sigma_y \\ \tau_{xy} \end{bmatrix} = \frac{E}{(1+v)(1-2v)} \begin{bmatrix} (1-v) & v & 0 \\ v & (1-v) & 0 \\ 0 & 0 & (1-2v)/2 \end{bmatrix} \begin{bmatrix} \varepsilon_x \\ \varepsilon_y \\ \gamma_{xy} \end{bmatrix}
\tag{2.153}
$$

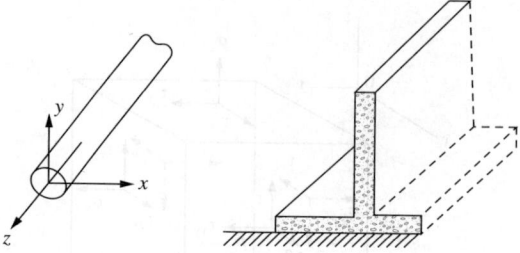

FIGURE 2.84 Practical examples of plane strain problems.

2.10.5 Choice of Element Shapes and Sizes

A finite element generally has a simple one-, two-, or three-dimensional configuration. The boundaries of elements are often straight lines and the elements can be one-dimensional, two-dimensional, or three-dimensional, as shown in Figure 2.85. While subdividing the continuum, one has to decide the number, shape, size, and configuration of the elements in such a way that the original body is simulated as closely as possible. Nodes must be located in locations where abrupt changes in geometry, loading, and material properties occur. A node must be placed at the point of application of a concentrated load because all loads are converted into equivalent nodal-point loads.

It is easy to subdivide a continuum into a completely regular one having the same shape and size. But problems encountered in practice do not involve regular shape; they may have regions of steep gradients of stresses. A finer subdivision may be necessary in regions where stress concentrations are expected to obtain a useful approximate solution. Typical examples of mesh selection are shown in Figure 2.86.

2.10.6 Choice of Displacement Function

Selection of displacement function is an important step in finite element analysis, since it determines the performance of the element in the analysis. Attention must be paid to select a displacement function that

1. Has the number of unknown constants as the total number of degrees of freedom of the element
2. Does not have any preferred directions
3. Allows the element to undergo rigid-body movement without any internal strain
4. Is able to represent states of constant stress or strain
5. Satisfies the compatibility of displacements along the boundaries with adjacent elements

Elements that meet both requirements 3 and 4 are known as *complete elements*

A polynomial is the most common form of displacement function. Mathematics of polynomials are easy to handle in formulating the desired equations for various elements and are convenient in digital computation. The degree of approximation is governed by the stage at which the function is truncated. Solutions closer to exact solutions can be obtained by including a greater number of terms. The polynomials are of the general form

$$w(x) = a_1 + a_2 x + a_3 x^2 + \cdots + a_{n+1} x^n \tag{2.154}$$

The coefficient a is known as a *generalized displacement amplitude*. The general polynomial form for a two-dimensional problem can be given as

$$u(x, y) = a_1 + a_2 x + a_3 y + a_4 x^2 + a_5 xy + a_6 y^2 + \cdots + a_m y^n$$
$$v(x, y) = a_{m+1} + a_{m+2} x + a_{m+3} y + a_{m+4} x^2 + a_{m+5} xy + a_{m+6} y^2 + \cdots + a_{2m} y^n$$

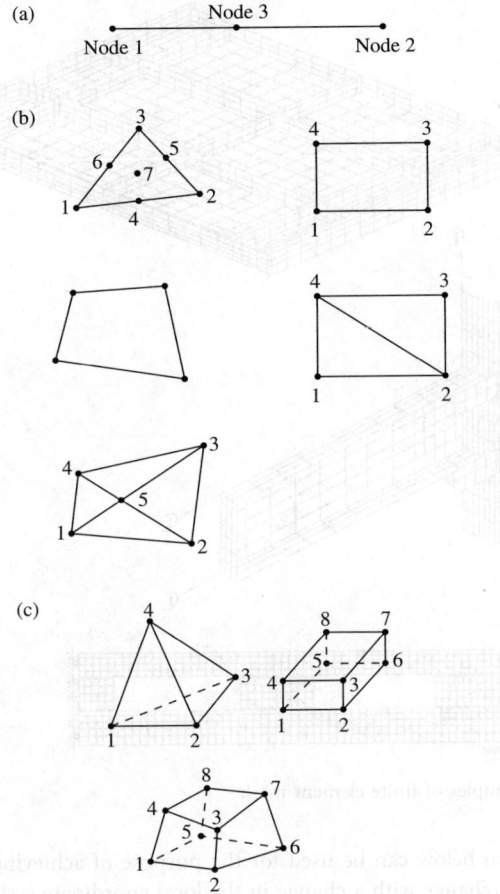

FIGURE 2.85 (a) One-dimensional element, (b) two-dimensional element, and (c) three-dimensional element.

where

$$m = \sum_{i=1}^{n+1} i \tag{2.155}$$

These polynomials can be truncated at any desired degree to give constant, linear, quadratic, or higher-order functions. For example, a linear model in the case of a two-dimensional problem can be given as

$$u = a_1 + a_2 x + a_3 y$$
$$v = a_4 + a_5 x + a_6 y \tag{2.156}$$

A quadratic function is given by

$$u = a_1 + a_2 x + a_3 y + a_4 x^2 + a_5 xy + a_6 y^2$$
$$v = a_7 + a_8 x + a_9 y + a_{10} x^2 + a_{11} xy + a_{12} y^2 \tag{2.157}$$

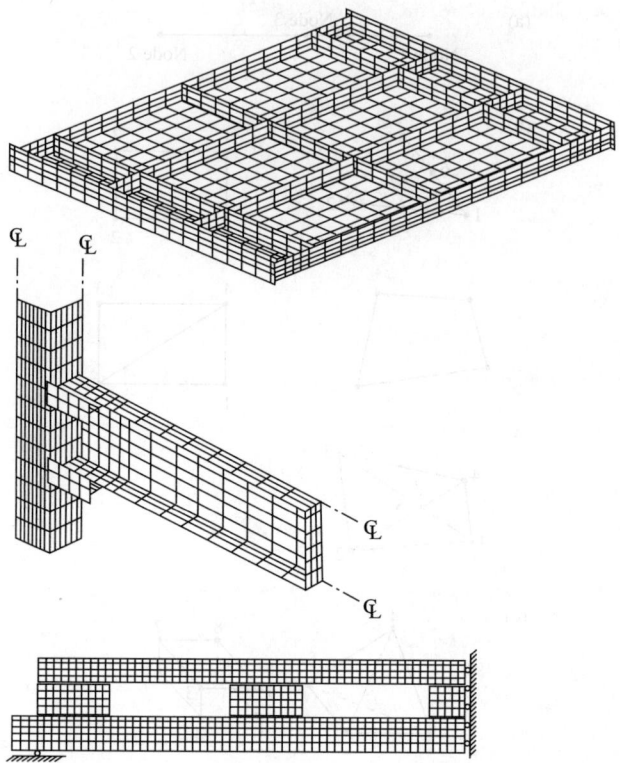

FIGURE 2.86   Typical examples of finite element mesh.

The Pascal triangle shown below can be used for the purpose of achieving isotropy, that is, to avoid displacement shapes that change with a change in the local coordinate system.

											Constant
				1							Constant
			x		y						Linear
		x^2		xy		y^2					Quadratic
	x^3		x^2y		xy^2		y^3				Cubic
x^4		x^3y		x^2y^2		xy^3		y^4			Quantic
x^5	x^4y		x^3y^2		x^2y^3		xy^4		y^5		Quintic

2.10.7 Nodal Degrees of Freedom

The deformation of the finite element is specified completely by the nodal displacement, rotations, or strains, which are referred to as *degrees of freedom*. Convergence, geometric isotropy, and potential energy function are the factors that determine the minimum number of degrees of freedom necessary for a given element. Additional degrees of freedom beyond the minimum number may be included for any element by adding secondary external nodes and such elements with additional degrees of freedom are called higher-order elements. The elements with more additional degrees of freedom become more flexible.

2.10.8 Isoparametric Elements

The scope of finite element analysis is also measured by the variety of element geometries that can be constructed. Formulation of element-stiffness equations requires the selection of displacement expressions with as many parameters as there are node-point displacements. In practice, for planar conditions, only the four-sided (quadrilateral) element finds as wide an application as the triangular element. The simplest form of quadrilateral, the rectangle, has four node points and involves two displacement components at each point for a total of eight degrees of freedom. In this case one would choose four-term expressions for both u and v displacement fields. If the description of the element is expanded to include nodes at the midpoints of the sides, an eight-term expression would be chosen for each displacement component.

The triangle and rectangle can approximate the curved boundaries only as a series of straight line segments. A closer approximation can be achieved by means of *isoparametric* coordinates. These are nondimensionalized curvilinear coordinates whose description is given by the same coefficients as are employed in the displacement expressions. The displacement expressions are chosen to ensure continuity across element interfaces and along supported boundaries, so that geometric continuity is ensured when the same forms of expressions are used as the basis of description of the element boundaries. The elements in which the geometry and displacements are described in terms of the same parameters and are of the same order are called *isoparametric elements*. The isoparametric concept enables one to formulate elements of any order that satisfy the completeness and compatibility requirements and that have isotropic displacement functions.

2.10.9 Isoparametric Families of Elements

2.10.9.1 Definitions and Justifications

For example, let u_i represent nodal displacements and x_i represent nodal x coordinates. The interpolation formulas are

$$u = \sum_{i=1}^{m} N_i u_i, \quad x = \sum_{i=1}^{n} N_i' x_i$$

where N_i and N_i' are shape functions written in terms of the intrinsic coordinates. The value of u and the value of x at a point within the element are obtained in terms of nodal values of u_i and x_i, from the above equations when the (intrinsic) coordinates of the internal point are given. Displacement components v and w in the y and z directions are treated in a similar manner.

The element is *isoparametric* if $m = n, N_i = N_i'$, and the same nodal points are used to define both element geometry and element displacement (Figure 2.87a); the element is *subparametric* if $m > n$, the order of N_i higher than that of N_i' (Figure 2.87b); the element is *superparametric* if $m < n$, the order of N_i lower than that of N_i' (Figure 2.87c). The isoparametric elements can correctly display rigid-body and constant-strain modes.

2.10.10 Element Shape Functions

The finite element method is not restricted to the use of linear elements. Most finite element codes, commercially available, allow the user to select between elements with linear or quadratic interpolation functions. In the case of quadratic elements fewer elements are needed to obtain the same degree of accuracy in the nodal values. Also, the two-dimensional quadratic elements can be shaped to model a curved boundary. Shape functions can be developed based on the following properties: (i) each shape function has a value of 1 at its own node and is 0 at each of the other nodes, (ii) the shape functions for two-dimensional elements are zero along each side that the node does not touch, and (iii) each shape

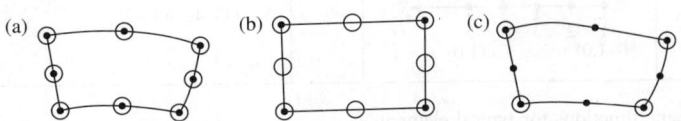

FIGURE 2.87   (a) Isoparametric element, (b) subparametric element, and (c) superparametric element.

function is a polynomial of the same degree as the interpolation equation. Shape functions for typical elements are given in Figure 2.88a and b.

2.10.11 Formulation of Stiffness Matrix

It is possible to obtain all the strains and stresses within the element and to formulate the stiffness matrix and a consistent load matrix once the displacement function has been determined. This consistent load matrix represents the equivalent nodal forces that replace the action of external distributed loads.

As an example, let us consider a linearly elastic element of any of the types shown in Figure 2.89. The displacement function may be written in the form

$$\{f\} = [P]\{A\} \tag{2.158}$$

(a)

Element name	Configuration	DOF	Shape functions
Two-node linear element	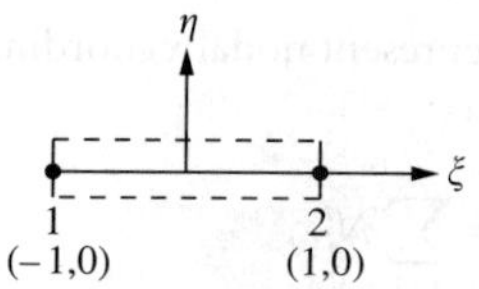	+	$N_i = \frac{1}{2}(1+\xi_0); \quad i=1,2$
Three-node parabolic element	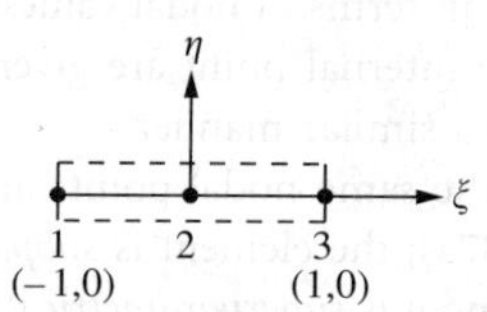	+	$N_i = \frac{1}{2}\xi_0(1+\xi_0); \quad i=1,3$ $N_i = (1-\xi^2); \quad i=2$
Four-node cubic element	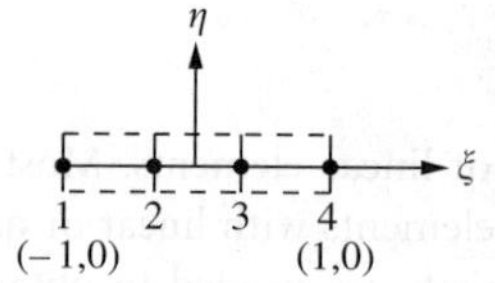	+	$N_i = \frac{1}{16}(1+\xi_0)(9\xi^2-1); \quad i=1,4$ $N_i = \frac{9}{16}(1+9\xi_0)(1-\xi^2); \quad i=2,3$
Five-node quartic element	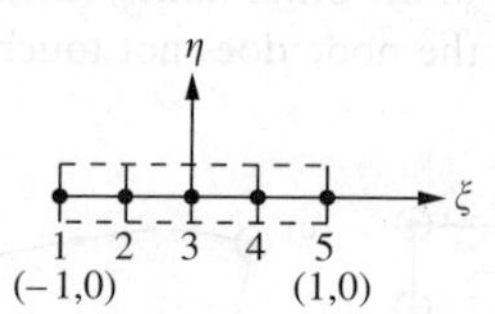	+	$N_i = \frac{1}{16}(1+\xi_0)(4\xi_0(1-\xi^2)+3\xi_0); \quad i=1,5$ $N_i = 4\xi_0(1-\xi^2)(1+4\xi_0); \quad i=2,4$ $N_3 = (1-4\xi^2)(1-\xi^2)$

FIGURE 2.88a Shape functions for typical elements.

(b)

Element name	Configuration	DOF	Shape functions
Four-node plane quadrilateral	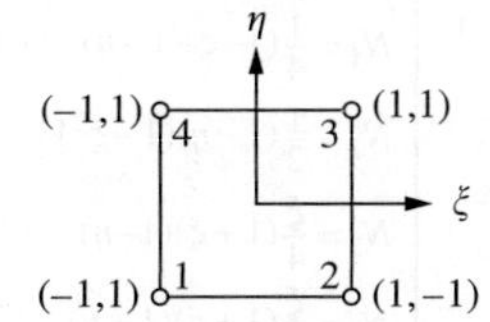	u, v	$N_i = \frac{1}{4}(1+\xi_0)(1+\eta_0); \quad i=1, 2, 3, 4$
Eight-node plane quadrilateral	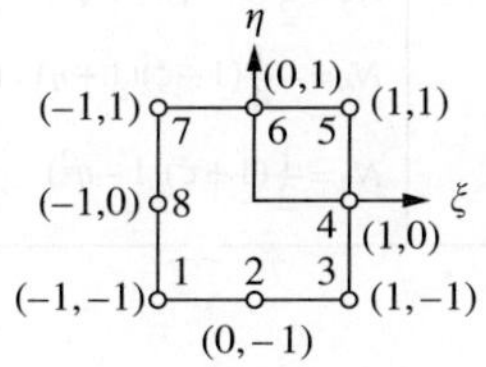	u, v	$N_i = \frac{1}{4}(1+\xi_0)(1+\eta_0)(\xi_0+\eta_0-1);$ $i=1, 3, 5, 7$ $N_i = \frac{1}{2}(1-\xi^2)(1+\eta_0); \quad i=2, 6$ $N_i = \frac{1}{2}(1-\eta^2)(1+\xi_0); \quad i=4, 8$
Twelve-node plane quadrilateral	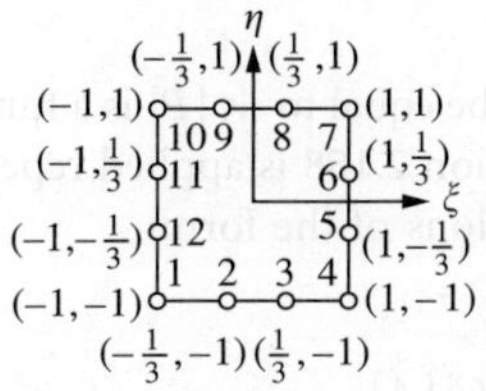	u, v	$N_i = \frac{1}{32}(1+\xi_0)(1+\eta_0)(-10+9(\xi^2+\eta^2))$ $i=1, 4, 7, 10$ $N_i = \frac{9}{32}(1+\xi_0)(1+\eta^2)(1+9\eta_0)$ $i=5, 6, 11, 12$ $N_i = \frac{9}{32}(1+\eta_0)(1-\xi^2)(1+9\xi_0)$ $i=2, 3, 8, 9$
Six-node linear quadrilateral	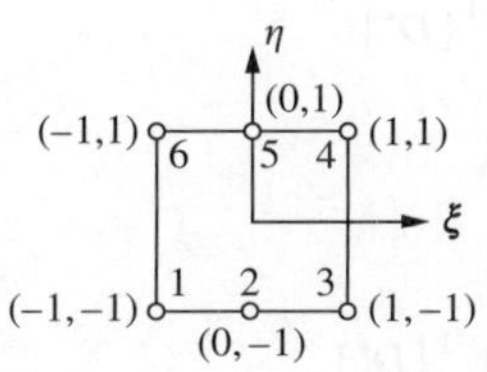	u, v	$N_i = \frac{\xi_0}{4}(1+\xi_0)(1+\eta_0)$ $i=1, 3, 4, 6$ $N_i = \frac{1}{2}(1-\xi^2)(1+\eta_0)$ $i=2, 5$
Eight-node plane quadrilateral	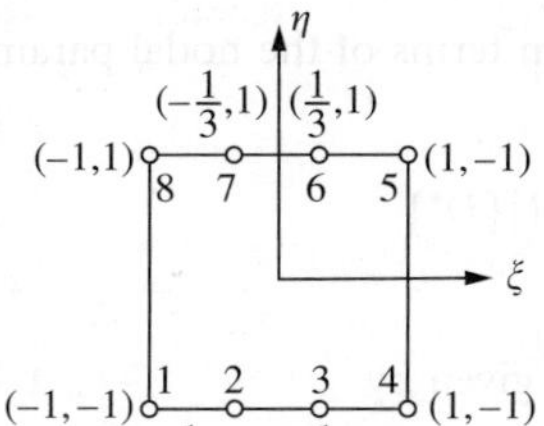	u, v	$N_i = \frac{1}{32}(1+\xi_0)(-1+9\xi^2)(1+\eta_0)$ $i=1, 4, 5, 8$ $N_i = \frac{9}{32}(1-\xi^2)(1+9\xi_0)(1+\eta_0)$ $i=2, 3, 6, 7$

FIGURE 2.88b Continued.

Element name	Configuration	DOF	Shape functions
Seven-node plane quadrilateral	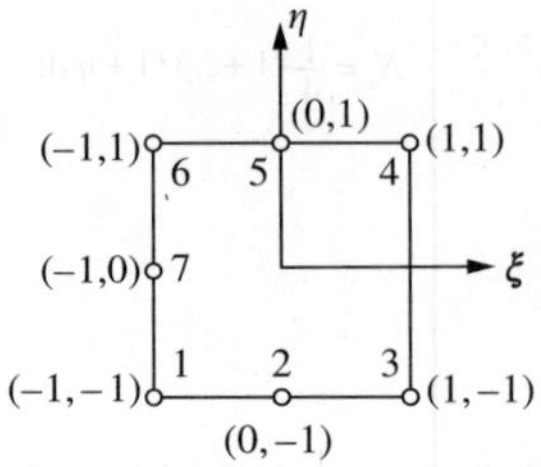	u, v	$N_1 = \frac{1}{4}(1-\xi)(1-\eta)\ (1+\xi+\eta)$ $N_2 = \frac{1}{2}(1-\eta)(1-\xi^2)$ $N_3 = \frac{\xi}{4}(1+\xi)(1-\eta)$ $N_4 = \frac{\xi}{4}(1+\xi)(1+\eta)$ $N_5 = \frac{1}{2}(1+\eta)(1-\xi^2)$ $N_6 = -\frac{1}{4}(1-\xi)(1+\eta)\ (1+\xi-\eta)$ $N_7 = \frac{1}{2}(1+\xi)(1-\eta^2)$

FIGURE 2.88b   Continued.

where $\{f\}$ may have two components $\{u, v\}$ or simply be equal to w, $[P]$ is a function of x and y only, and $\{A\}$ is the vector of undetermined constants. If Equation 2.158 is applied repeatedly to the nodes of the element one after the other, we obtain a set of equations of the form

$$\{D^*\} = [C]\{A\} \tag{2.159}$$

where $\{D^*\}$ refers to the nodal parameters and $[C]$ to the relevant nodal coordinates. The undetermined constants $\{A\}$ can be expressed in terms of the nodal parameters $\{D^*\}$ as

$$\{A\} = [C]^{-1}\{D^*\} \tag{2.160}$$

Substituting Equation 2.160 into Equation 2.158

$$\{f\} = [P][C]^{-1}\{D^*\} \tag{2.161}$$

Constructing the displacement function directly in terms of the nodal parameters, one obtains

$$\{f\} = [L]\{D^*\} \tag{2.162}$$

where $[L]$ is a function of both (x, y) and $(x, y)_{i,j,m}$ given by

$$[L] = [P][C]^{-1} \tag{2.163}$$

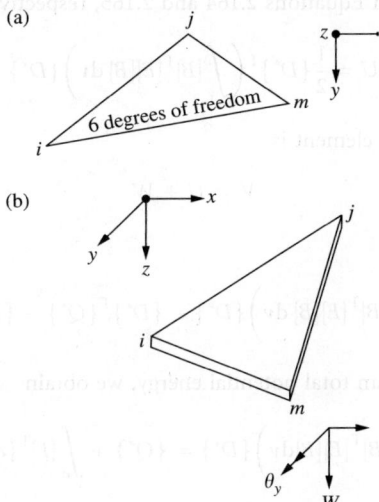

FIGURE 2.89 Degrees of freedom: (a) triangular plane-stress element and (b) triangular bending element.

The various components of strain can be obtained by appropriate differentiation of the displacement function. Thus,

$$\{\varepsilon\} = [B]\{D^*\} \tag{2.164}$$

$[B]$ is derived by differentiating appropriately the elements of $[L]$ with respect to x and y. The stresses $\{\sigma\}$ in a linearly elastic element are given by the product of the strain and a symmetrical elasticity matrix $[E]$. Thus,

$$\{\sigma\} = [E]\{\varepsilon\}$$

or

$$\{\sigma\} = [E][B]\{D^*\} \tag{2.165}$$

The stiffness and the consistent load matrices of an element can be obtained using the principle of minimum total potential energy. The potential energy of the external load in the deformed configuration of the element is written as

$$W = -\{D^*\}^T\{Q^*\} - \int_a \{f\}^T\{q\}\, da \tag{2.166}$$

In Equation 2.166 $\{Q^*\}$ represents concentrated loads at nodes and $\{q\}$ the distributed loads per unit area. Substituting for $\{f\}^T$ from Equation 2.162 one obtains

$$W = -\{D^*\}^T\{Q^*\} - \{D^*\}^T \int_a [L]^T\{q\}\, da \tag{2.167}$$

Note that the integral is taken over the area a of the element. The strain energy of the element integrated over the entire volume v, is given as

$$U = \frac{1}{2} \int_v \{\varepsilon\}^T\{\sigma\}\, dv$$

Substituting for $\{\varepsilon\}$ and $\{\sigma\}$ from Equations 2.164 and 2.165, respectively,

$$U = \frac{1}{2}\{D^*\}^{\mathrm{T}}\left(\int_v [B]^{\mathrm{T}}[E][B]\,\mathrm{d}v\right)\{D^*\} \tag{2.168}$$

The total potential energy of the element is

$$V = U + W$$

or

$$V = \frac{1}{2}\{D^*\}^{\mathrm{T}}\left(\int_v [B]^{\mathrm{T}}[E][B]\,\mathrm{d}v\right)\{D^*\} - \{D^*\}^{\mathrm{T}}\{Q^*\} - \{D^*\}^{\mathrm{T}}\int_a [L]^{\mathrm{T}}\{q\}\mathrm{d}a \tag{2.169}$$

Using the principle of minimum total potential energy, we obtain

$$\left(\int_v [B]^{\mathrm{T}}[E][B]\mathrm{d}v\right)\{D^*\} = \{Q^*\} + \int_a [L]^{\mathrm{T}}\{q\}\,\mathrm{d}a$$

or

$$[K]\{D^*\} = \{F^*\} \tag{2.170}$$

where

$$[K] = \int_v [B]^{\mathrm{T}}[E][B]\,\mathrm{d}v \tag{2.171}$$

and

$$\{F^*\} = \{Q^*\} + \int_a [L]^{\mathrm{T}}\{q\}\,\mathrm{d}a \tag{2.172}$$

2.10.12 Plates Subjected to In-Plane Forces

The simplest element available in two-dimensional stress analysis is the triangular element. The stiffness and consistent load matrices of such an element will now be obtained by applying the equation derived in the previous section.

Consider the triangular element shown in Figure 2.89a. There are two degrees of freedom per node and a total of six degrees of freedom for the entire element. We can write

$$u = A_1 + A_2 x + A_3 y$$

and

$$v = A_4 + A_5 x + A_6 y$$

expressed as

$$\{f\} = \begin{Bmatrix} u \\ v \end{Bmatrix} = \begin{bmatrix} 1 & x & y & 0 & 0 & 0 \\ 0 & 0 & 0 & 1 & x & y \end{bmatrix} \begin{Bmatrix} A_1 \\ A_2 \\ A_3 \\ A_4 \\ A_5 \\ A_6 \end{Bmatrix} \tag{2.173}$$

or

$$\{f\} = [P]\{A\} \tag{2.174}$$

Once the displacement function is available, the strains for a plane problem are obtained from

$$\varepsilon_x = \frac{\partial u}{\partial x}, \quad \varepsilon_y = \frac{\partial v}{\partial y}$$

and

$$\gamma_{xy} = \frac{\partial u}{\partial y}\frac{\partial v}{\partial x}$$

The matrix $[B]$ relating the strains to the nodal displacement $\{D^*\}$ is thus given as

$$[B] = \frac{1}{2\Delta} \begin{bmatrix} b_i & 0 & b_j & 0 & b_m & 0 \\ 0 & c_i & 0 & c_j & 0 & c_m \\ c_i & b_j & c_j & b_j & c_m & b_m \end{bmatrix} \tag{2.175}$$

b_i, c_i, etc., are constants related to the nodal coordinates only. The strains inside the element must all be constant and hence the name of the element.

For derivation of the strain matrix, only isotropic material is considered. The plane stress and plane strain cases can be combined to give the following elasticity matrix, which relates the stresses to the strains:

$$[E] = \begin{bmatrix} C_1 & C_1 C_2 & 0 \\ C_1 C_2 & C_1 & 0 \\ 0 & 0 & C_{12} \end{bmatrix} \tag{2.176}$$

where

$$C_1 = \bar{E}/(1 - v^2) \quad \text{and} \quad C_2 = v$$

for plane stress and

$$C_1 = \frac{\bar{E}(1 - v)}{(1 + v)(1 - 2v)} \quad \text{and} \quad C_2 = \frac{v}{(1 - v)}$$

for plane strain and for both cases

$$C_{12} = C_1(1 - C_2)/2$$

and $\bar{E}$ is the modulus of elasticity.

The stiffness matrix can now be formulated according to Equation 2.171a

$$[E][B] = \frac{1}{2\Delta} \begin{bmatrix} C_1 & C_1 C_2 & 0 \\ C_1 C_2 & C_1 & 0 \\ 0 & 0 & C_{12} \end{bmatrix} \begin{bmatrix} b_i & 0 & b_j & 0 & b_m & 0 \\ 0 & c_i & 0 & c_j & 0 & c_m \\ c_i & b_i & c_j & b_j & c_m & b_m \end{bmatrix}$$

where Δ is the area of the element.

The stiffness matrix is given by Equation 2.177a as

$$[K] = \int_v [B]^{\mathrm{T}}[E][B]\,\mathrm{d}v \tag{2.177a}$$

The stiffness matrix has been worked out algebraically to be

$$
[K] = \frac{h}{4\Delta}
\begin{bmatrix}
C_1 b_i^2 & & & & & \\
+C_{12}c_i^2 & & & & & \\
C_1 C_2 b_i c_i & C_1 c_i^2 & & & & \text{symmetrical} \\
+C_{12}b_i c_i & +C_{12}b_i^2 & & & & \\
C_1 b_i b_j & C_1 C_2 b_j c_i & C_1 b_j^2 & & & \\
+C_{12}c_i c_j & +C_{12}b_i c_j & +C_{12}c_j^2 & & & \\
C_1 C_2 b_i c_j & C_1 c_i c_j & C_1 C_2 b_j c_j & C_1 c_j^2 & & \\
+C_{12}b_j c_i & +C_1 b_i b_j & +C_{12}b_j c_j & +C_{12}b_j^2 & & \\
C_1 b_i b_m & C_1 C_2 b_m c_i & C_1 b_j b_m & C_1 C_2 b_m c_j & C_1 b_m^2 & \\
+C_{12}c_i c_m & +C_{12}b_i c_m & +C_{12}c_j c_m & +C_{12}b_j c_m & +C_{12}c_m^2 & \\
C_1 C_2 b_i c_m & C_1 c_i c_m & C_1 C_2 b_j c_m & C_1 c_j c_m & C_1 C_2 b_m c_m & C_1 c_m^2 \\
+C_{12}b_m c_i & +C_{12}b_i b_m & +C_{12}b_m c_j & +C_{12}b_j b_m & +C_{12}b_m c_m & +C_{12}b_m^2
\end{bmatrix}
\tag{2.177b}
$$

2.10.13 Beam Element

The stiffness matrix for a beam element with two degrees of freedom (one deflection and one rotation) can be derived in the same manner as for other finite elements using Equation 2.171.

The beam element has two nodes, one at each end, and two degrees of freedom at each node, giving it a total of four degrees of freedom. The displacement function can be assumed as

$$
f = w = A_1 + A_2 x + A_3 x^2 + A_4 x^3
$$

that is,

$$
f = [1 \ x \ x^2 \ x^3]
\begin{Bmatrix}
A_1 \\
A_2 \\
A_3 \\
A_4
\end{Bmatrix}
$$

or

$$
f = [P]\{A\}
$$

With the origin of the x and y axis at the left-hand end of the beam, we can express the nodal-displacement parameters as

$$
D_1^* = (w)_{x=0} = A_1 + A_2(0) + A_3(0)^2 + A_4(0)^3
$$

$$
D_2^* = \left(\frac{dw}{dx}\right)_{x=0} = A_2 + 2A_3(0) + 3A_4(0)^2
$$

$$
D_3^* = (w)_{x=1} = A_1 + A_2(l) + A_3(l)^2 + A_4(l)^3
$$

$$
D_4^* = \left(\frac{dw}{dx}\right)_{x=1} = A_2 + 2A_3(l) + 3A_4(l)^2
$$

or

$$
\{D^*\} = [C]\{A\}
$$

where

$$
\{A\} = [C]^{-1}\{D^*\}
$$

and

$$[C]^{-1} = \begin{bmatrix} 1 & 0 & 0 & 0 \\ 0 & 1 & 0 & 0 \\ -\dfrac{3}{l^2} & -\dfrac{2}{l} & \dfrac{3}{l^2} & -\dfrac{1}{l} \\ \dfrac{2}{l^3} & \dfrac{1}{l^2} & -\dfrac{2}{l^3} & \dfrac{1}{l^2} \end{bmatrix}$$

Using Equation 2.163, we obtain

$$[L] = [P][C]^{-1}$$

or

$$[L]^{-1} = \left[\left(1 - \frac{3x^2}{l^2} + \frac{2x^3}{l^3} \right) \left(x - \frac{2x^2}{l} + \frac{x^3}{l^2} \right) \left(\frac{3x^2}{l^2} - \frac{2x^3}{l^3} \right) \left(-\frac{x^2}{l} + \frac{x^3}{l^2} \right) \right] \qquad (2.178)$$

Neglecting shear deformation

$$\{\varepsilon\} = -\frac{d^2 y}{dx^2} \qquad (2.179)$$

Substituting Equation 2.178 into Equation 2.163 and the result into Equation 2.179

$$\{\varepsilon\} = \left[\left| \frac{6}{l^2} - \frac{12x}{l^3} \right| \frac{4}{l} - \frac{6x}{l^2} \left| -\frac{6}{l^2} + \frac{12x}{l^3} \right| \frac{2}{l} - \frac{6x}{l^2} \right] \{D^*\}$$

or

$$\varepsilon = [B]\{D^*\}$$

Moment–curvature relationship is given by

$$M = \bar{E}I \left(-\frac{d^2 y}{dx^2} \right)$$

where $\bar{E}$ is the modulus of elasticity.

We know that $\{\sigma\} = [E]\{\varepsilon\}$, so we have for the beam element

$$[E] = \bar{E}I$$

The stiffness matrix can now be obtained from Equation 2.171a written in the form

$$[K] = \int_0^l [B]^T [d][B] \, dx$$

with the integration over the length of the beam. Substituting for $[B]$ and $[E]$, we obtain

$$[k] = \bar{E}I \int_0^l \begin{bmatrix} \dfrac{36}{l^4} - \dfrac{144x}{l^5} + \dfrac{144x^2}{l^6} & & & \text{symmetrical} \\[2ex] \dfrac{24}{l^3} - \dfrac{84x}{l^4} + \dfrac{72x^2}{l^5} & \dfrac{16}{l^2} - \dfrac{48x}{l^3} + \dfrac{36x^2}{l^4} & & \\[2ex] -\dfrac{36}{l^4} + \dfrac{144x}{l^5} - \dfrac{144x^2}{l^6} & -\dfrac{24}{l^3} + \dfrac{84x}{l^4} - \dfrac{72x^2}{l^5} & \dfrac{36}{l^4} - \dfrac{144x}{l^5} + \dfrac{144x^2}{l^6} & \\[2ex] \dfrac{12}{l^3} - \dfrac{60x}{l^4} + \dfrac{72x^2}{l^5} & \dfrac{8}{l^2} - \dfrac{36x}{l^3} + \dfrac{36x^2}{l^4} & -\dfrac{12}{l^3} + \dfrac{60x}{l^4} - \dfrac{72x^2}{l^5} & \dfrac{4}{l^2} - \dfrac{24x}{l^3} + \dfrac{36x^2}{l^4} \end{bmatrix} dx$$

or

$$[K] = \bar{E}I \begin{bmatrix} \dfrac{12}{l^3} & & \text{symmetrical} & \\ \dfrac{6}{l^2} & \dfrac{4}{l} & & \\ -\dfrac{12}{l^3} & -\dfrac{6}{l^2} & \dfrac{12}{l^3} & \\ \dfrac{6}{l^2} & \dfrac{2}{l} & -\dfrac{6}{l^2} & \dfrac{4}{l} \end{bmatrix} \tag{2.180}$$

2.10.14 Plate Element

For the rectangular bending element shown in Figure 2.90 with three degrees of freedom (one deflection and two rotations) at each node the displacement function can be chosen as a polynomial with 12 undetermined constants as

$$\{f\} = w = A_1 + A_2 x + A_3 y + A_4 x^2 + A_5 xy + A_6 y^2 + A_7 x^3$$
$$+ A_8 x^2 y + A_9 xy^2 + A_{10} y^3 + A_{11} x^3 y + A_{12} xy^3 \tag{2.181}$$

or

$$\{f\} = \{P\}\{A\}$$

The displacement parameter vector is defined as

$$\{D^*\} = \{w_i, \theta_{xi}, \theta_{yi} | w_j, \theta_{xj}, \theta_{yj} | w_k, \theta_{xk}, \theta_{yk} | w_l, \theta_{xl}, \theta_{yl}\}$$

where

$$\theta_x = \frac{\partial w}{\partial y} \quad \text{and} \quad \theta_y = -\frac{\partial w}{\partial x}$$

As in the case of beam it is possible to derive from Equation 2.181 a system of 12 equations relating $\{D^*\}$ to constants $\{A\}$. The last equation is

$$w = \left[[L]_i | [L]_j | [L]_k | [L]_l \right] \{D^*\} \tag{2.182}$$

The curvatures of the plate element at any point (x, y) are given by

$$\{\varepsilon\} = \left\{ \begin{array}{c} \dfrac{-\partial^2 w}{\partial x^2} \\[2mm] \dfrac{-\partial^2 w}{\partial y^2} \\[2mm] \dfrac{2\partial^2 w}{\partial x \partial y} \end{array} \right\}$$

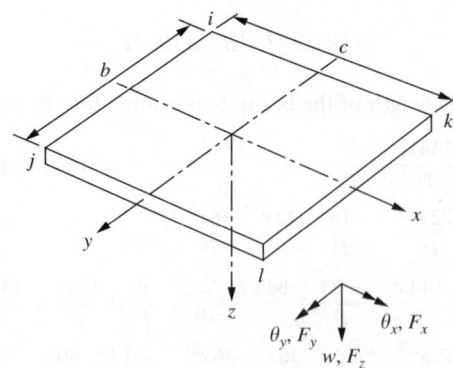

FIGURE 2.90 Rectangular bending element.

By differentiating of Equation 2.182, we obtain

$$\{\varepsilon\} = \left[[B]_i [B]_j [B]_k [B]_l \right] \{D^*\} \tag{2.183}$$

or

$$\{\varepsilon\} = \sum_{r=i,j,k,l} [B]_r \{D^*\}_r \tag{2.184}$$

where

$$[B]_r = \begin{bmatrix} -\dfrac{\partial^2}{\partial x^2} [L]_r \\[1ex] \cdots \\[1ex] -\partial \dfrac{2}{\partial y^2} [L]_r \\[1ex] \cdots \\[1ex] 2\dfrac{\partial^2}{\partial x \, \partial y} [L]_r \end{bmatrix} \tag{2.185}$$

and

$$\{D^*\}_r = \{w_r, \theta_{xr}, \theta_{yr}\} \tag{2.186}$$

For an isotropic slab, the moment–curvature relationship is given by

$$\{\sigma\} = \{M_x \; M_y \; M_{xy}\} \tag{2.187}$$

$$[E] = N \begin{bmatrix} 1 & v & 0 \\ v & 1 & 0 \\ 0 & 0 & \dfrac{1-v}{2} \end{bmatrix} \tag{2.188}$$

and

$$N = \frac{\bar{E}h^3}{12(1-v^2)} \tag{2.189}$$

For orthotropic plates with the principal directions of orthotropy coinciding with the x and y axes, no additional difficulty is experienced. In this case, we have

$$[E] = \begin{bmatrix} D_x & D_1 & 0 \\ D_1 & D_y & 0 \\ 0 & 0 & D_{xy} \end{bmatrix} \tag{2.190}$$

where D_x, D_1, D_y, and D_{xy} are the orthotropic constants used by Timoshenko and Woinowsky-Krieger (1959), and

$$D_x = \frac{E_x h^3}{12(1 - v_x v_y)}$$

$$D_y = \frac{E_y h^3}{12(1 - v_x v_y)}$$

$$D_1 = \frac{v_x E_y h^3}{12(1 - v_x v_y)} = \frac{v_y E_x h^3}{12(1 - v_x v_y)} \tag{2.191}$$

$$D_{xy} = \frac{G h^3}{12}$$

where E_x, E_y, v_x, v_y, and G are the orthotropic material constants and h is the plate thickness.

Unlike the strain matrix for the plane stress triangle (see Equation 2.175), the stress and strain in the present element vary with x and y. In general, we calculate the stresses (moments) at the four corners. These can be expressed in terms of the nodal displacements by Equation 2.165 which, for an isotropic element, takes the form

$$
\begin{Bmatrix} \{\sigma\}_i \\ \{\sigma\}_j \\ \{\sigma\}_k \\ \{\sigma\}_r \end{Bmatrix}
= \frac{N}{cb}
\begin{bmatrix}
6p^{-1}+6vp & 4vc & -4b & -6vp & 2vc & 0 & -6p^{-1} & 0 & -2b & 0 & 0 & 0 \\
6p+6vp^{-1} & 4c & -4vb & -6p & 2c & 0 & -6vp^{-1} & 0 & -2vp & 0 & 0 & 0 \\
-(1-v) & -(1-v)b & (1-v)c & (1-v) & 0 & -(1-v)c & (1-v) & (1-v)b & 0 & -(1-v) & 0 & 0 \\
-6vp & -2vc & 0 & 6p^{-1}+6vp & -4vc & -4b & 0 & 0 & 0 & -6p^{-1} & 0 & -2b \\
-6p & -2c & 0 & 6p+6vp^{-1} & -4c & -4vb & 0 & 0 & 0 & -6vp^{-1} & 0 & -2vb \\
-(1-v) & 0 & (1-v)c & (1-v) & -(1-v)b & (1-v) & 0 & 0 & -(1-v) & (1-v)b & 0 & 0 \\
-6p^{-1} & 0 & 2b & 0 & 0 & 0 & 6p^{-1}+6vp & 4vc & 4b & -6vp & 2vc & 0 \\
-6vp^{-1} & 0 & 2b & 0 & 0 & 0 & 6p+6vp^{-1} & 4c & 4vb & -6p & 2c & 0 \\
-(1-v) & -(1-v)b & 0 & (1-v) & 0 & 0 & (1-v) & (1-v)b & (1-v)c & -(1-v) & 0 & -(1-v)c \\
0 & 0 & 0 & -6p^{-1} & 0 & 2b & -6vp & -2vc & 0 & 6p^{-1}+6vp & -4vc & 4b \\
0 & 0 & 0 & -6vp^{-1} & 0 & 2vb & -6p & -2c & 0 & 6p+6vp^{-1} & -4c & 4vb \\
-(1-v) & 0 & 0 & (1-v) & -(1-v)b & 0 & (1-v) & 0 & (1-v)c & -(1-v) & (1-v)b & -(1-v)c
\end{bmatrix}
$$

$$
\times \begin{Bmatrix} \{D^*\}_i \\ \{D^*\}_j \\ \{D^*\}_k \\ \{D^*\}_r \end{Bmatrix}
\tag{2.192}
$$

The stiffness matrix corresponding to the 12 nodal coordinates can be calculated by

$$
[K] = \int_{-b/2}^{b/2} \int_{-c/2}^{c/2} [B]^T [E][B] \, dx \, dy
\tag{2.193}
$$

For an isotropic element, this gives

$$
[K^*] = \frac{N}{15cb} [T][\bar{k}][T]
\tag{2.194}
$$

where

$$
[T] = \begin{bmatrix} [T_s] & & & \\ & [T_s] & & \\ & & [T_s] & \\ & & & [T_s] \end{bmatrix} \quad \text{(submatrices not shown are zero)}
\tag{2.195}
$$

$$
[T_s] = \begin{bmatrix} 1 & 0 & 0 \\ 0 & b & 0 \\ 0 & 0 & c \end{bmatrix}
\tag{2.196}
$$

and

$$
[\bar{K}] =
\begin{bmatrix}
60p^{-2}+60p^2-12v+42 & & & & & & & & & \\[4pt]
30p^2+12v+3 & 20p^2-4v+4 & & & & & & & & \\[4pt]
-(30p^{-2}-60p^2+3) & 15v & 20p^{-2}-4v+4 & & & & & & & \\[4pt]
30p^2-3v+3 & 5p^2-v+1 & 0 & 20p^2-4v+4 & & & & & & \\[4pt]
-15p^{-2}+12v-3 & 0 & 5p^{-2}-3v+1 & 15v & 20p^{-2}-4v+4 & & & & & \\[4pt]
-(60p^{-2}+30p^2-12v+42) & -15p^{-2}-3v-3 & -30p^2-3v-3 & -15p^{-2}+12v-3 & -30p^2+3v-3 & 60p^{-2}+60p^2-12v+42 & & & & \\[4pt]
15p^{-2}-3v+3 & 10p^2+4v-4 & 0 & 5p^2-v+1 & 0 & -30p^2+12v+3 & 20p^2-4v+4 & & & \\[4pt]
-30p^2+3v-3 & 0 & 10p^{-2}+v-1 & 0 & 10p^2+v-1 & -(30p^{-2}+30p^2+3) & 15v & 20p^{-2}-4v+4 & & \\[4pt]
15p^{-2}-3v-3 & 5p^2-v+1 & 0 & 10p^2+4v-4 & 0 & -15p^{-2}-12v-3 & 5p^2-v+1 & 0 & 20p^2-4v+4 & \\[4pt]
-15p^{-2}-3v-3 & 0 & 10p^{-2}+v-1 & 0 & 10p^{-2}+4v-4 & -(30p^{-2}+12v+3) & 0 & 10p^{-2}+v-1 & -15v & 20p^{-2}-4v+4
\end{bmatrix}
$$

symmetrical

$$\tag{2.197}$$

If the element is subjected to a uniform load in the z direction of intensity q, the consistent load vector becomes

$$\{Q_q^*\} = q \int_{-b/2}^{b/2} \int_{-c/2}^{c/2} [L]^{\mathrm{T}} \, \mathrm{d}x \, \mathrm{d}y \qquad (2.198)$$

where $\{Q_q^*\}$ are 12 forces corresponding to the nodal-displacement parameters.

Evaluating the integrals in this equation gives

$$\{Q_q^*\} = qcb \begin{Bmatrix} 1/4 \\ b/24 \\ -c/24 \\ \cdots \\ 1/4 \\ -b/24 \\ -c/24 \\ \cdots \\ 1/4 \\ b/24 \\ c/24 \\ \cdots \\ 1/4 \\ -b/24 \\ c/24 \end{Bmatrix} \qquad (2.199)$$

More details on the finite element method can be found in Desai and Abel (1972) and Ghali and Neville (1978).

2.11   Inelastic Analysis

2.11.1   An Overall View

Inelastic analyses can be generalized into two main approaches. The first approach is known as *plastic hinge analysis*. The analysis assumes that structural elements remain elastic except at critical regions where plastic hinges are allowed to form. The second approach is known as *spread of plasticity analysis*. This analysis follows explicitly the gradual spread of yielding throughout the structure. Material yielding in the member is modeled by discretization of members into several line elements and subdivision of the cross-sections into many "fibers." Although the spread of plasticity analysis can predict accurately the inelastic response of the structure, the plastic hinge analysis is considered to be computationally more efficient and less expensive to execute.

If geometric nonlinear effect is not considered, the plastic hinge analysis predicts the maximum load of the structure corresponding to the formation of a plastic collapse mechanism (Chen and Sohal 1994). First-order plastic analysis has considerable application in continuous beams and low-rise building frames where members are loaded primarily in flexure. For tall building frames and for frames with slender columns subjected to side sway, the interaction between yielding and instability may lead to collapse prior to the formation of a plastic mechanism (SSRC 1988). If an incremental analysis is carried out based on the updated deformed geometry of the structure, the analysis is termed *second order*. The need for a second-order analysis of steel frame is increasing in view of the modern codes and standards that give explicit permission for the engineer to compute load effects from a direct second-order analysis.

This section presents the virtual work principle to explain the fundamental theorems of plastic hinge analysis. Simple and approximate techniques of practical plastic analysis methods are then introduced.

The concept of hinge-by-hinge analysis is presented. The more advanced topics such as second-order elastic-plastic hinge, refined plastic hinge analysis, and spread of plasticity analysis are covered in Section 2.12.

2.11.2 Ductility

Plastic analysis is strictly applicable for materials that can undergo large deformation without fracture. Steel is one such material with an idealized stress–strain curve as shown in Figure 2.91. When steel is subjected to tensile force, it will elongate elastically until the yield stress is reached. This is followed by an increase in strain without much increase in stress. Fracture will occur at very large deformation. This material idealization is generally known as *elastic–perfectly plastic* behavior. For a compact section, the attainment of initial yielding does not result in failure of the section. The compact section will have reserved plastic strength that depends on the shape of the cross-section. The capability of the material to deform under constant load without decrease in strength is the *ductility* characteristic of the material.

2.11.3 Redistribution of Forces

The benefit of using a ductile material can be demonstrated from an example of a three-bar system shown in Figure 2.92. From the equilibrium condition of the system

$$2T_1 + T_2 = P \tag{2.200}$$

Assuming elastic stress–strain law, the displacement and force relationship of the bars may be written as

$$\delta = \frac{T_1 L_1}{AE} = \frac{T_2 L_2}{AE} \tag{2.201}$$

Since $L_2 = L_1/2 = L/2$, Equation 2.201 can be written as

$$T_1 = \frac{T_2}{2} \tag{2.202}$$

where T_1 and T_2 are the tensile forces in the rods, L_1 and L_2 are lengths of the rods, A is the cross-section area, and E is the elastic modulus. Solving Equations 2.201 and 2.202 for T_2

$$T_2 = \frac{P}{2} \tag{2.203}$$

The load at which the structure reaches the first yield (in Figure 2.92b) is determined by letting $T_2 = \sigma_y A$. From Equations 2.203

$$P_y = 2T_2 = 2\sigma_y A \tag{2.204}$$

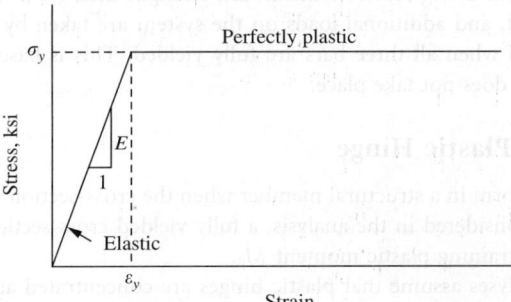

FIGURE 2.91 Idealized stress–strain curve.

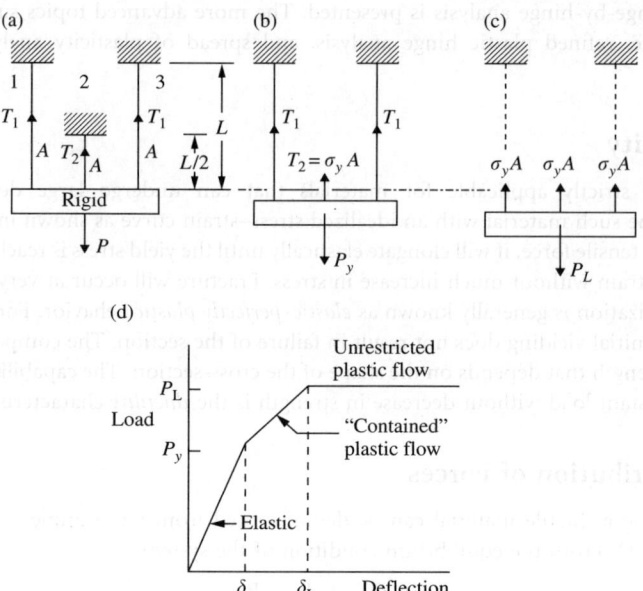

FIGURE 2.92 Force redistribution in a three-bar system: (a) elastic, (b) partially yielded, (c) fully plastic, and (d) load–deflection curve.

The corresponding displacement at first yield is

$$\delta_y = \varepsilon_y L = \frac{\sigma_y L}{2E} \qquad (2.205)$$

After bar 2 is yielded, the system continues to take additional load until all the three bars reach their maximum strength of $\sigma_y A$, as shown in Figure 2.92c. The plastic limit load of the system is thus written as

$$P_L = 3\sigma_y A \qquad (2.206)$$

The process of successive yielding of bars in this system is known as inelastic redistribution of forces. The displacement at the incipient of collapse is

$$\delta_L = \varepsilon_y L = \frac{\sigma_y L}{E} \qquad (2.207)$$

Figure 2.92d shows the load–displacement behavior of the system when subjected to increasing force. As load increases, bar 2 will reach its maximum strength first. As it is yielded, the force in the member remains constant, and additional loads on the system are taken by the less critical bars. The system will eventually fail when all three bars are fully yielded. This is based on an assumption that material strain hardening does not take place.

2.11.4 Concept of Plastic Hinge

A plastic hinge is said to form in a structural member when the cross-section is fully yielded. If material strain hardening is not considered in the analysis, a fully yielded cross-section can undergo indefinite rotation at a constant restraining plastic moment M_p.

Most of the plastic analyses assume that plastic hinges are concentrated at zero-length plasticity. In reality, the yield zone is developed over a certain length, normally called the *plastic hinge length*, depending on the loading, boundary conditions, and geometry of the section. The hinge lengths of beams (ΔL) with

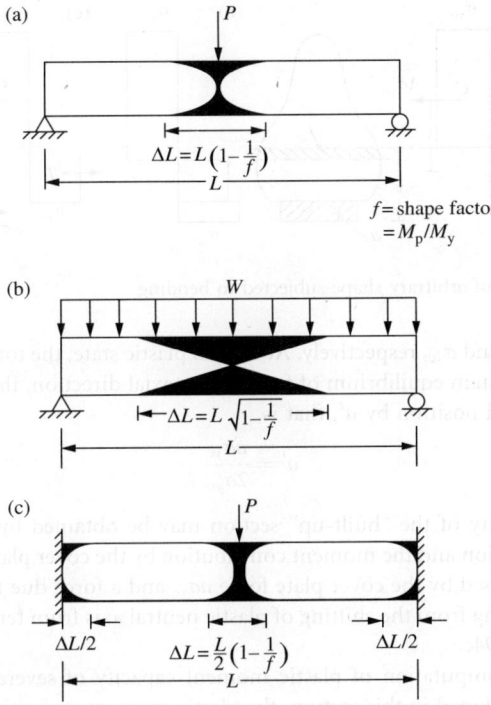

$$\Delta L = L\left(1 - \frac{1}{f}\right)$$

f = shape factor
$= M_p/M_y$

FIGURE 2.93 Hinge lengths of beams with different support and loading conditions.

different support and loading conditions are shown in Figure 2.93. Plastic hinges are developed first at the sections subjected to the greatest moment. The possible locations for plastic hinges to develop are at the points of concentrated loads, at the intersections of members involving a change in geometry, and at the point of zero shear for a member under uniform distributed load.

2.11.5 Plastic Moment Capacity

A knowledge of full plastic moment capacity of a section is important in plastic analysis. It forms the basis for limit load analysis of the system. Plastic moment is the moment resistance of a fully yielded cross-section. The cross-section must be fully compact to develop its plastic strength. The component plates of a section must not buckle prior to the attainment of full moment capacity.

The plastic moment capacity, M_p, of a cross-section depends on the material yield stress and the section geometry. The procedure for the calculation of M_p may be summarized in the following two steps:

1. The plastic neutral axis of a cross-section is located by considering the equilibrium of forces normal to the cross-section. Figure 2.94a shows a cross-section of arbitrary shape subjected to increasing moment. The plastic neutral axis is determined by equating the force in compression (C) to that in tension (T). If the entire cross-section is made of same material, the plastic neutral axis can be determined by dividing the cross-sectional area into two equal parts. If the cross-section is made of more than one type of material, the plastic neutral axis must be determined by summing the normal force and letting the force equal zero.
2. The plastic moment capacity is determined by obtaining the moment generated by the tensile and compressive forces.

Consider an arbitrary section with area $2A$ and one axis of symmetry, which is strengthened by a cover plate of area a as shown in Figure 2.94b. Further, assume that the yield strengths of the original section

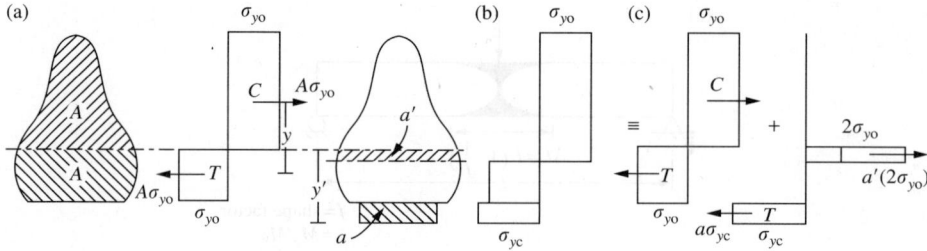

FIGURE 2.94 Cross-section of arbitrary shape subjected to bending.

and the cover plate are σ_{yo} and σ_{yc}, respectively. At the full plastic state, the total axial force acting on the cover plate is $a\sigma_{yc}$. To maintain equilibrium of force in the axial direction, the plastic neutral axis must shift down from its original position by a', that is,

$$a' = \frac{a\sigma_{yc}}{2\sigma_{yo}} \tag{2.208}$$

The resulting plastic capacity of the "built-up" section may be obtained by summing the full plastic moment of the original section and the moment contribution by the cover plate. The additional capacity is equal to the moment caused by the cover plate force $a\sigma_{yc}$ and a force due to the fictitious stress $2\sigma_{yo}$ acting on the area a' resulting from the shifting of plastic neutral axis from tension zone to compression zone as shown in Figure 2.94c.

Figure 2.95 shows the computation of plastic moment capacity of several shapes of cross-section. Based on the principle developed in this section, the plastic moment capacities of typical cross-sections may be generated. Additional information on sections subjected to combined bending, torsion, shear, and axial load can be found in Mrazik et al. (1987).

2.11.6 Theory of Plastic Analysis

There are two main assumptions for first-order plastic analysis:

1. The structure is made of ductile material that can undergo large deformations beyond the elastic limit without fracture or buckling.
2. The deflections of the structure under loading are small so that second-order effects can be ignored.

An "exact" plastic analysis solution must satisfy three basic conditions. They are *equilibrium*, *mechanism*, and *plastic moment* conditions. The plastic analysis disregards the continuity condition as required by the elastic analysis of indeterminate structures. The formation of plastic hinge in members leads to discontinuity of slope. If sufficient plastic hinges are formed to allow the structure to deform into a mechanism, it could constitute a mechanism condition. Since plastic analysis utilizes the limit of resistance of a member's plastic strength, the plastic moment condition is required to ensure that the resistance of the cross-sections is not violated anywhere in the structure. Lastly, the equilibrium condition, which is the same condition to be satisfied in elastic analysis, requires that the sum of all applied forces and reactions be equal to zero and all internal forces be self-balanced.

When all the three conditions are satisfied, the resulting plastic analysis for limiting load is the "correct" limit load. The collapse loads for simple structures such as beams and portal frames can be solved easily using a direct approach or through visualization of the formation of a "correct" collapse mechanism. However, for more complex structures, the exact solution satisfying all the three conditions may be difficult to predict. Thus, simple techniques using approximate methods of analysis are often used to assess these solutions. These techniques, named equilibrium and mechanism methods, will be discussed in the subsequent sections.

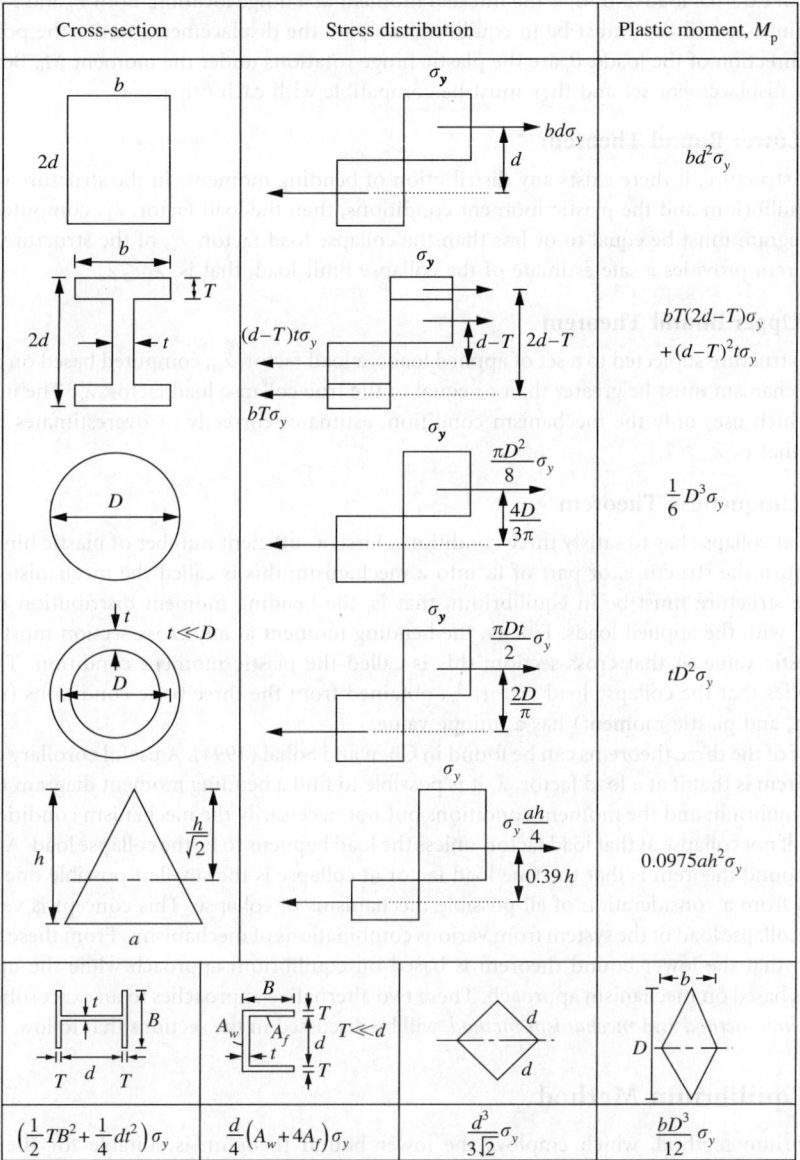

Cross-section	Stress distribution	Plastic moment, M_p
b, $2d$	σ_y, $bd\sigma_y$, d	$bd^2\sigma_y$
b, $2d$, T, t	σ_y, $(d-T)t\sigma_y$, $d-T$, $2d-T$, $bT\sigma_y$	$bT(2d-T)\sigma_y$ $+(d-T)^2t\sigma_y$
D	σ_y, $\dfrac{\pi D^2}{8}\sigma_y$, $\dfrac{4D}{3\pi}$	$\dfrac{1}{6}D^3\sigma_y$
$t \ll D$, D	σ_y, $\dfrac{\pi Dt}{2}\sigma_y$, $\dfrac{2D}{\pi}$	$tD^2\sigma_y$
h, $\dfrac{h}{\sqrt{2}}$, a	σ_y, $\sigma_y\dfrac{ah}{4}$, $0.39h$	$0.0975ah^2\sigma_y$
t, B, T, d, T	B, A_w, A_f, $T \ll d$, d, T, t	d, d	$\vdash b \dashv$, D
$\left(\dfrac{1}{2}TB^2+\dfrac{1}{4}dt^2\right)\sigma_y$	$\dfrac{d}{4}\left(A_w+4A_f\right)\sigma_y$	$\dfrac{d^3}{3\sqrt{2}}\sigma_y$	$\dfrac{bD^3}{12}\sigma_y$

FIGURE 2.95 Plastic moment capacities of sections.

2.11.6.1 Principle of Virtual Work

The virtual work principle may be applied to relate a system of forces in equilibrium to a system of compatible displacements. For example, if a structure in equilibrium is given a set of small compatible displacements, then the work done by the external loads on these external displacements is equal to the work done by the internal forces on the internal deformation. In plastic analysis, internal deformations are assumed to be concentrated at plastic hinges. The virtual work equation for hinged structures can be written in explicit form as

$$\sum P_i \delta_j = \sum M_i \theta_j \qquad (2.209)$$

where P_i is an external load and M_i is the internal moment at a hinge location. Both P_i and M_i constitute an equilibrium set and they must be in equilibrium. δ_j are the displacements under the point loads P_i and in the direction of the loads. θ_j are the plastic hinge rotations under the moment M_i. Both δ_j and θ_j constitute a displacement set and they must be compatible with each other.

2.11.6.2 Lower Bound Theorem

For a given structure, if there exists any distribution of bending moments in the structure that satisfies both the equilibrium and the plastic moment conditions, then the load factor, λ_L, computed from this moment diagram must be equal to or less than the collapse load factor, λ_c, of the structure. The lower bound theorem provides a safe estimate of the collapse limit load, that is, $\lambda_L \leq \lambda_c$.

2.11.6.3 Upper Bound Theorem

For a given structure subjected to a set of applied loads, a load factor, λ_u, computed based on an assumed collapse mechanism must be greater than or equal to the true collapse load factor, λ_c. The upper bound theorem, which uses only the mechanism condition, estimates correctly or overestimates the collapse limit load, that is, $\lambda_u \geq \lambda_c$.

2.11.6.4 Uniqueness Theorem

A structure at collapse has to satisfy three conditions. First, a sufficient number of plastic hinges must be formed to turn the structure, or part of it, into a mechanism; this is called the mechanism condition. Second, the structure must be in equilibrium, that is, the bending moment distribution must satisfy equilibrium with the applied loads. Finally, the bending moment at any cross-section must not exceed the full plastic value of that cross-section; this is called the plastic moment condition. The theorem simply implies that the collapse load factor, λ_c, obtained from the three basic conditions (mechanism, equilibrium, and plastic moment) has a unique value.

The proof of the three theorems can be found in Chen and Sohal (1994). A useful corollary of the lower bound theorem is that if at a load factor, λ, it is possible to find a bending moment diagram that satisfies both the equilibrium and the moment conditions but not necessarily the mechanism condition, then the structure will not collapse at that load factor, unless the load happens to be the collapse load. A corollary of the upper bound theorem is that the true load factor at collapse is the smallest possible one that can be determined from a consideration of all possible mechanisms of collapse. This concept is very useful in finding the collapse load of the system from various combinations of mechanisms. From these theorems, it can be seen that the lower bound theorem is based on equilibrium approach while the upper bound technique is based on mechanism approach. These two alternative approaches to an exact solution, called the *equilibrium method* and *mechanism method*, will be discussed in the sections that follow.

2.11.7 Equilibrium Method

The equilibrium method, which employs the lower bound theorem, is suitable for the analysis of continuous beams and frames in which the structural redundancies do not exceed two. The procedures for obtaining the equilibrium equations of a statically indeterminate structure and evaluating its plastic limit load are as follows:

To obtain the equilibrium equations of a statically indeterminate structure

1. Select the redundant(s).
2. Free the redundants and draw a moment diagram for the determinate structure under the applied loads.
3. Draw a moment diagram for the structure due to the redundant forces.
4. Superimpose the moment diagrams in steps 2 and 3.
5. Obtain maximum moment at critical sections of the structure utilizing the moment diagram in step 4.

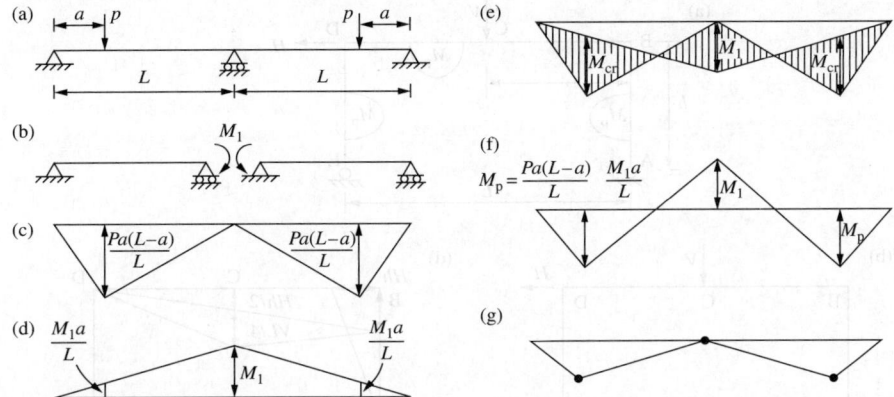

FIGURE 2.96 Analysis of a two-span continuous beam using equilibrium method.

To evaluate the plastic limit load of the structure

6. Select the value(s) of redundant(s) such that the plastic moment condition is not violated at any section in the structure.
7. Determine the load corresponding to the selected redundant(s).
8. Check for the formation of a mechanism. If a collapse mechanism condition is met, then the computed load is the exact plastic limit load. Otherwise, it is a lower bound solution.
9. Adjust the redundant(s) and repeat steps 6 to 9 until the exact plastic limit load is obtained.

EXAMPLE 2.11

Continuous Beam

Figure 2.96a shows a two-span continuous beam analyzed using the equilibrium method. The plastic limit load of the beam is calculated based on the step-by-step procedure described in the previous section as follows:

1. Select the redundant force as M_1 which is the bending moment at the intermediate support, as shown in Figure 2.96b.
2. Free the redundants and draw a moment diagram for the determinate structure under the applied loads, as shown in Figure 2.96c.
3. Draw a moment diagram for the structure due to the redundant moment M_1, as shown in Figure 2.96d.
4. Superimpose the moment diagrams in Figure 2.96c and d and the results are shown in Figure 2.96e.
5. The moment diagram in Figure 2.96e is redrawn on a single straight base line. The critical moment in the beam is

$$M_{cr} = \frac{Pa(L - a)}{L} - \frac{M_1 a}{L} \qquad (2.210)$$

The maximum moment at critical sections of the structure is obtained by using the moment diagram in Figure 2.96e. By letting $M_{cr} = M_p$, the resulting moment distribution is shown in Figure 2.96f.

6. A lower bound solution may be obtained by selecting a value of redundant moment M_1.

For example, if $M_1 = 0$ is selected, the moment diagram is reduced to that shown in Figure 2.96c. By equating the maximum moment in the diagram to the plastic moment, M_p, we have

$$M_{cr} = \frac{Pa(L - a)}{L} = M_p \qquad (2.211)$$

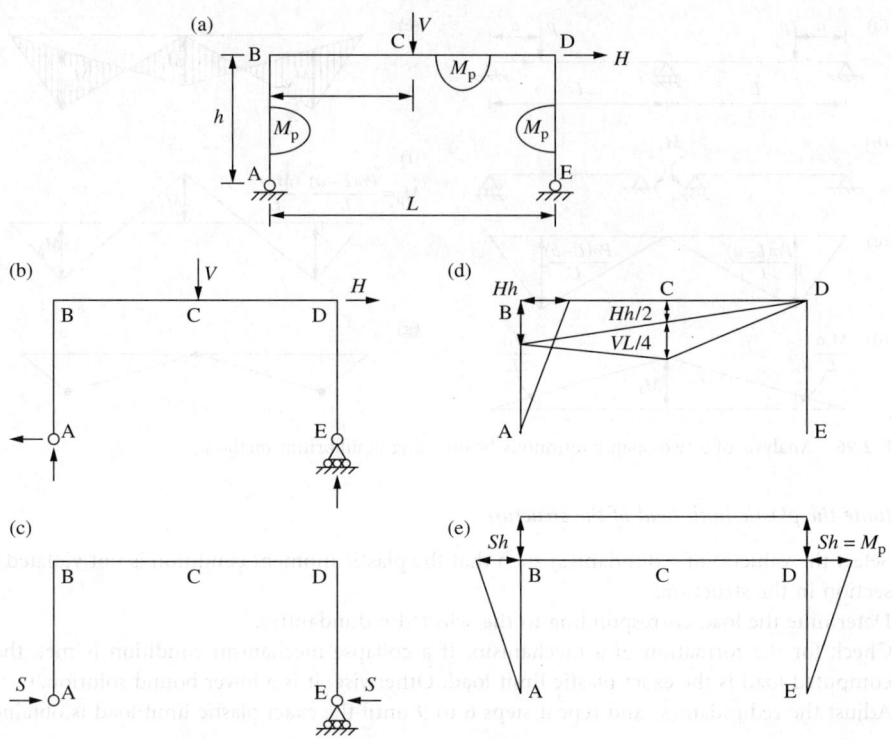

FIGURE 2.97   Analysis of portal frame using equilibrium method.

which gives $P = P_1$ as

$$P_1 = \frac{M_p L}{a(L - a)}$$ (2.212)

The moment diagram in Figure 2.96c shows a plastic hinge formed at each span. Since two plastic hinges in each span are required to form a plastic mechanism, the load P_1 is a lower bound solution. However, if the redundant moment M_1 is set equal to the plastic moment M_p, and by letting the maximum moment in Figure 2.96f equal the plastic moment, we have

$$M_{cr} = \frac{Pa(L - a)}{L} - \frac{M_p a}{L} = M_p$$ (2.213)

which gives $P = P_2$ as

$$P_2 = \frac{M_p(L + a)}{a(L - a)}$$ (2.214)

7. Since a sufficient number of plastic hinges have formed in the beams (Figure 2.96g) to arrive at a collapse mechanism, the computed load, P_2, is the exact plastic limit load.

EXAMPLE 2.12

Portal Frame

A pin-based rectangular frame is subjected to a vertical load V and a horizontal load H as shown in Figure 2.97a. All the members of the frame are made of the same section with moment capacity M_p. The objective is to determine the limit value of H if the frame's width-to-height ratio L/h is 1.0.

Procedure

The frame has one degree of redundancy. The redundancy for this structure can be chosen as the horizontal reaction at E. Figure 2.97b and c show the resulting determinate frame loaded by the applied loads and redundant force. The moment diagrams corresponding to these two loading conditions are shown in Figure 2.97d and e.

The horizontal reaction S should be chosen in such a manner that all three conditions, equilibrium, plastic moment, and mechanism, are satisfied. Formation of two plastic hinges is necessary to form a mechanism. The plastic hinges may be formed at B, C, and D. Assuming that a plastic hinge is formed at D as shown in Figure 2.97e, we have

$$S = \frac{M_p}{h} \tag{2.215}$$

Corresponding to this value of S, the moments at B and C can be expressed as

$$M_B = Hh - M_p \tag{2.216}$$

$$M_C = \frac{Hh}{2} + \frac{VL}{4} - M_p \tag{2.217}$$

The condition for the second plastic hinge to form at B is $|M_B| > |M_C|$. From Equations 2.216 and 2.217 we have

$$Hh - M_p > \frac{Hh}{2} + \frac{VL}{4} - M_p \tag{2.218}$$

and

$$\frac{V}{H} < \frac{h}{L} \tag{2.219}$$

The condition for the second plastic hinge to form at C is $|M_C| > |M_B|$. From Equations 2.216 and 2.217 we have

$$Hh - M_p < \frac{Hh}{2} + \frac{VL}{4} - M_p \tag{2.220}$$

and

$$\frac{V}{H} > \frac{h}{L} \tag{2.221}$$

For a particular combination of V, H, L, and h, the collapse load for H can be calculated.

1. When $L/h = 1$ and $V/H = \frac{1}{3}$, we have

$$M_B = Hh - M_p \tag{2.222}$$

$$M_C = \frac{Hh}{2} + \frac{Hh}{12} - M_p = \frac{7}{12}Hh - M_p \tag{2.223}$$

Since $|M_B| > |M_C|$, the second plastic hinge will form at B and the corresponding value for H is

$$H = \frac{2M_p}{h} \tag{2.224}$$

2. When $l/h = 1$ and $V/H = 3$, we have

$$M_B = Hh - M_p \tag{2.225}$$

$$M_C = \frac{Hh}{2} + \frac{3}{4}Hh - M_p = \frac{5}{4}Hh - M_p \tag{2.226}$$

Since $|M_C| > |M_B|$, the second plastic hinge will form at C and the corresponding value for H is

$$H = \frac{1.6M_p}{h} \tag{2.227}$$

2.11.8 Mechanism Method

This method, which is based on the upper bound theorem, states that the load computed on the basis of an assumed failure mechanism is never less than the exact plastic limit load of a structure. Thus, it always predicts the upper bound solution of the collapse limit load. It can also be shown that the minimum upper bound is the limit load itself. The procedure of using the mechanism method has the following two steps:

1. Assume a failure mechanism and form the corresponding work equation from which an upper bound value of the plastic limit load can be estimated.
2. Write the equilibrium equations for the assumed mechanism and check the moments to see whether the plastic moment condition is met everywhere in the structure.

To obtain the true limit load using the mechanism method, it is necessary to determine every possible collapse mechanism of which some are the combinations of a certain number of independent mechanisms. Once the independent mechanisms have been identified, a work equation may be established for each combination and the corresponding collapse load is determined. The lowest load among those obtained by considering all the possible combinations of independent mechanisms is the correct plastic limit load.

2.11.8.1 Independent Mechanisms

The number of possible independent mechanisms, n, for a structure can be determined from the following equation:

$$n = N - R \tag{2.228}$$

where N is the number of critical sections at which plastic hinges might form and R indicates the degrees of redundancy of the structure.

Critical sections generally occur at the points of concentrated loads, at joints where two or more members meet at different angles, and at sections where there is an abrupt change in section geometries or properties. To determine the number of redundancies (R) of a structure, it is necessary to free sufficient supports or restraining forces in structural members so that the structure becomes an assembly of several determinate substructures.

Figure 2.98 shows two examples. The cuts that are made in each structure reduce the structural members to either cantilevers or simply supported beams. The fixed-end beam requires a shear force and a moment to restore continuity at the cut section, and thus $R = 2$. For the two-store frame, an axial force, shear, and moment are required at each cut section for full continuity and thus $R = 12$.

2.11.8.2 Types of Mechanism

Figure 2.99a shows a frame structure subjected to a set of loading. The frame may fail by different types of collapse mechanisms dependent on the magnitude of loading and the frame's configurations. The collapse mechanisms are

1. *Beam mechanism.* Possible mechanisms of this type are shown in Figure 2.99b.
2. *Panel mechanism.* The collapse mode is associated with side sway, as shown in Figure 2.99c.
3. *Gable mechanism.* The collapse mode is associated with the spreading of column tops with respect to the column bases, as shown in Figure 2.99d.
4. *Joint mechanism.* The collapse mode is associated with the rotation of joints of which the adjoining members developed plastic hinges and deformed under an applied moment, as shown in Figure 2.99e.
5. *Combined mechanism.* It can be a partial collapse mechanism as shown in Figure 2.99f or it may be a complete collapse mechanism as shown in Figure 2.99g.

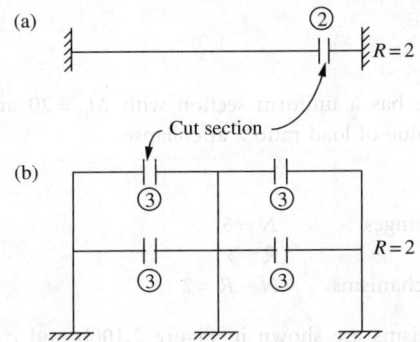

FIGURE 2.98   Number of redundants in a (a) beam and (b) frame.

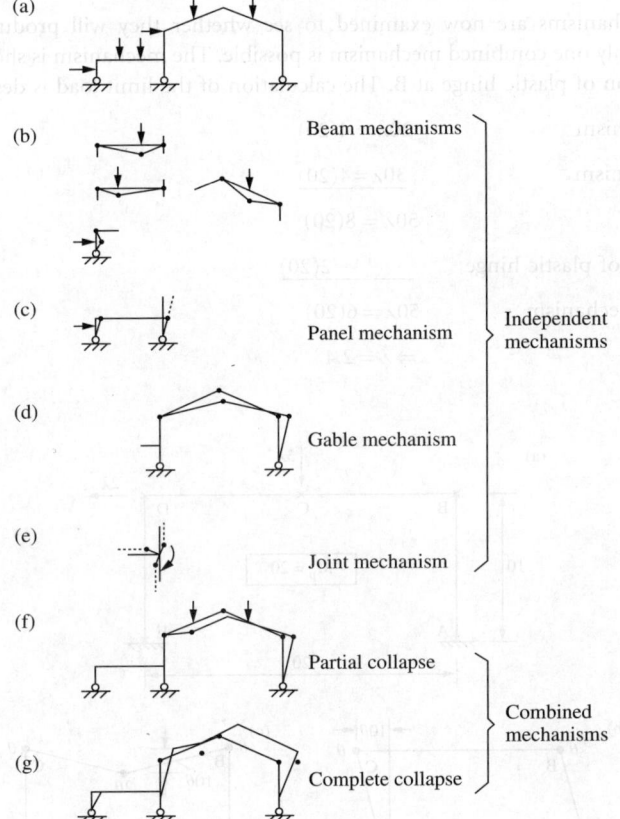

FIGURE 2.99   Typical plastic mechanisms.

The principal rule for combining independent mechanisms is to obtain a lower value of collapse load. The combinations are selected in such a way that the external work becomes a maximum and the internal work becomes a minimum. Thus, the work equation would require that the mechanism involve as many applied loads as possible and at the same time eliminate as many plastic hinges as possible. This procedure will be illustrated in the following example.

EXAMPLE 2.13

Rectangular frame

A fixed-end rectangular frame has a uniform section with $M_p = 20$ and carries the load shown in Figure 2.100. Determine the value of load ratio λ at collapse.

Solution

Number of possible plastic hinges $N = 5$
Number of redundancies $R = 3$
Number of independent mechanisms $N - R = 2$

The two independent mechanisms are shown in Figure 2.100b and c and the corresponding work equations are

Panel mechanism $20\lambda = 4(20) = 80 \;\Rightarrow\; \lambda = 4$
Beam mechanism $30\lambda = 4(20) = 80 \;\Rightarrow\; \lambda = 2.67$

The combined mechanisms are now examined to see whether they will produce a lower λ value. It is observed that only one combined mechanism is possible. The mechanism is shown in Figure 2.100c involving cancellation of plastic hinge at B. The calculation of the limit load is described below:

Panel mechanism $20\lambda = 4(20)$

Beam mechanism $\underline{30\lambda = 4(20)}$

Addition $50\lambda = 8(20)$

Cancellation of plastic hinge $\underline{ -2(20)}$

Combined mechanism $50\lambda = 6(20)$

                                  $\Rightarrow \lambda = 2.4$

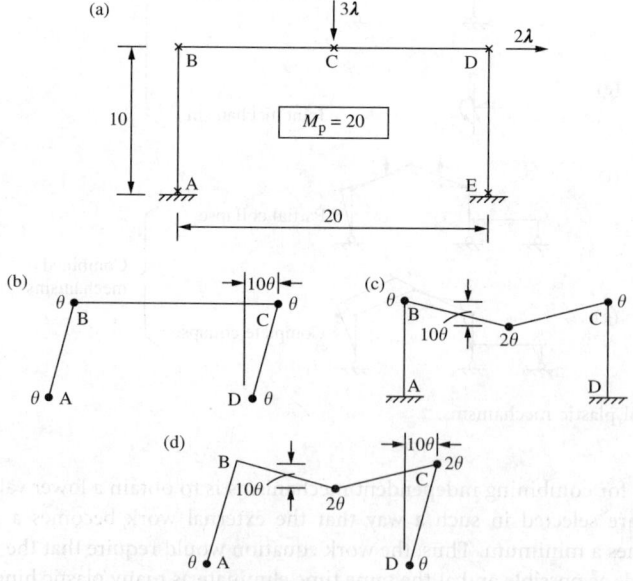

FIGURE 2.100   Collapse mechanisms of a fixed-base portal frame.

The combined mechanism results in a smaller value for λ and no other possible mechanism can produce a lower load. Thus, $\lambda = 2.4$ is the collapse load.

EXAMPLE 2.14

Frame subjected to distributed load

When a frame is subjected to distributed loads, the maximum moment and hence the plastic hinge location are not known in advance. The exact location of the plastic hinge may be determined by writing the work equation in terms of the unknown distance and then maximizing the plastic moment by formal differentiation.

Consider the frame shown in Figure 2.101a. The side sway collapse mode in Figure 2.101b leads to the following work equation:

$$4M_p = 24(10\theta)$$

which gives

$$M_p = 60$$

The beam mechanism of Figure 2.101c gives

$$4M_p\theta = \frac{1}{2}(10\theta)32$$

which gives

$$M_p = 40$$

In fact the correct mechanism is shown in Figure 2.101d, in which the distance Z from the plastic hinge location is unknown. The work equation is

$$24(10\theta) + \frac{1}{2}(1.6)(20)(z\theta) = M_p\left(2 + 2\left(\frac{20}{20-z}\right)\right)\theta$$

which gives

$$M_p = \frac{(240 + 16z)(20 - z)}{80 - 2z}$$

To maximize M_p, the derivative of M_p is set to zero, that is,

$$(80 - 2z)(80 - 32z) + (4800 + 80z - 16z^2)(2) = 0$$

which gives

$$z = 40 - \sqrt{1100} = 6.83$$

and

$$M_p = 69.34$$

In practice, uniform load is often approximated by applying several equivalent point loads to the member under consideration. Plastic hinges thus can be assumed to form only at the concentrated load points, and the calculations become simpler when the structural system becomes more complex.

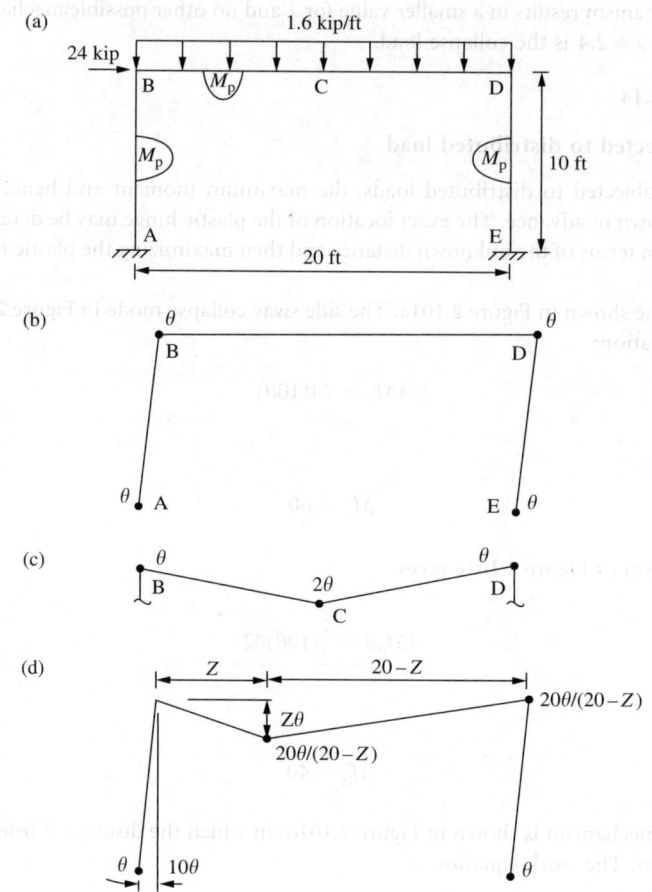

FIGURE 2.101   A portal frame subjected to combined uniform distributed load and horizontal load.

2.11.9 Gable Frames

The mechanism method is used to determine the plastic limit load of the gable frame shown in Figure 2.102. The frame is composed of members with plastic moment capacity of 270 kip ft. The column bases are fixed. The frame is loaded by a horizontal load H and a vertical concentrated load V. A graph from which V and H cause the collapse of the frame is to be produced.

Solution

Consider the three modes of collapse as follows.

2.11.9.1 Plastic Hinges Form at A, C, D, and E

The mechanism is shown in Figure 2.102b. The instantaneous center O for member CD is located at the intersection of AC and ED extended. From similar triangles ACC1 and OCC2, we have

$$\frac{OC_2}{CC_2} = \frac{C_1A}{C_1C}$$

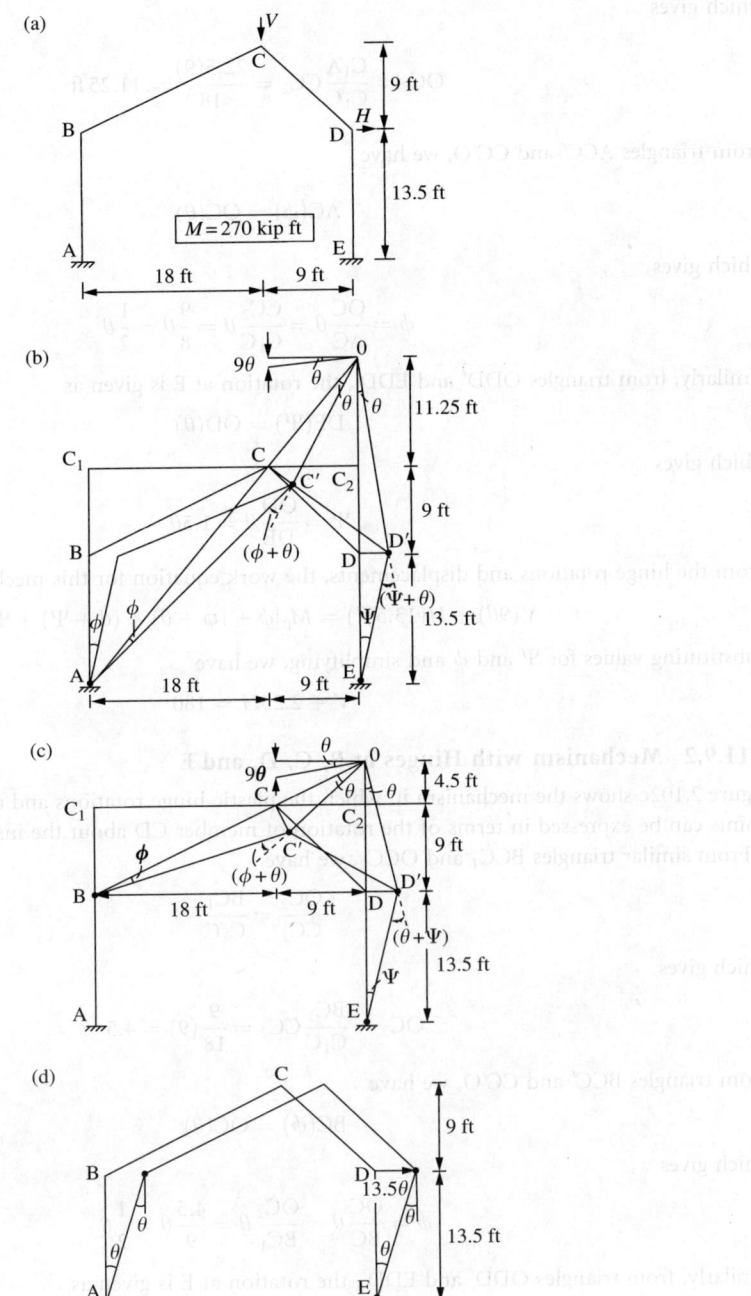

FIGURE 2.102 Collapse mechanisms of a fixed-base gable frame.

which gives

$$OC_2 = \frac{C_1A}{C_1C}CC_2 = \frac{22.5(9)}{18} = 11.25\,\text{ft}$$

From triangles ACC′ and CC′O, we have

$$AC(\phi) = OC(\theta)$$

which gives

$$\phi = \frac{OC}{AC}\theta = \frac{CC_2}{C_1C}\theta = \frac{9}{8}\theta = \frac{1}{2}\theta$$

Similarly, from triangles ODD′ and EDD′, the rotation at E is given as

$$DE(\Psi) = OD(\theta)$$

which gives

$$\Psi = \frac{OD}{DE}\theta = 1.5\theta$$

From the hinge rotations and displacements, the work equation for this mechanism can be written as

$$V(9\theta) + H(13.5\Psi) = M_p[\phi + (\phi + \theta) + (\theta + \Psi) + \Psi]$$

Substituting values for Ψ and ϕ and simplifying, we have

$$V + 2.25H = 180$$

2.11.9.2 Mechanism with Hinges at B, C, D, and E

Figure 2.102c shows the mechanism in which the plastic hinge rotations and displacements at the load points can be expressed in terms of the rotation of member CD about the instantaneous center O.
From similar triangles BCC_1 and OCC_2, we have

$$\frac{OC_2}{CC_2} = \frac{BC_1}{C_1C}$$

which gives

$$OC_2 = \frac{BC_1}{C_1C}CC_2 = \frac{9}{18}(9) = 4.5$$

From triangles BCC′ and CC′O, we have

$$BC(\phi) = OC(\theta)$$

which gives

$$\phi = \frac{OC}{BC}\theta = \frac{OC_2}{BC_1}\theta = \frac{4.5}{9}\theta = \frac{1}{2}\theta$$

Similarly, from triangles ODD′ and EDD′, the rotation at E is given as

$$DE(\Psi) = OD(\theta)$$

which gives

$$\Psi = \frac{OD}{DE}\theta = \theta$$

The work equation for this mechanism can be written as

$$V(9\theta) + H(13.5\Psi) = M_p[\phi + (\phi + \theta) + (\theta + \Psi) + \Psi]$$

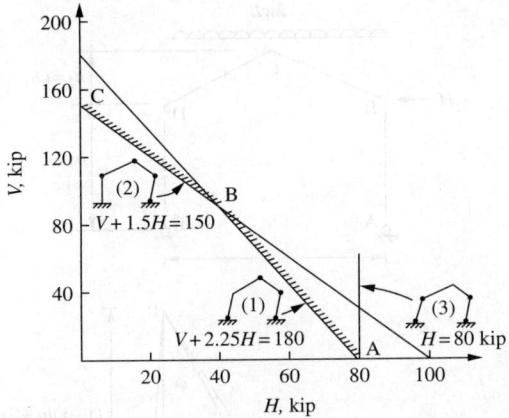

FIGURE 2.103 Vertical load and horizontal force interaction curve for collapse analysis of gable frame.

Substituting values of Ψ and ϕ and simplifying, we have

$$V + 1.5H = 150$$

2.11.9.3 Mechanism with Hinges at A, B, D, and E

The hinge rotations and displacements corresponding to this mechanism are shown in Figure 2.102d. The rotation of all hinges is θ. The horizontal load moves by 13.5θ but the horizontal load has no vertical displacement. The work equation becomes

$$H(13.5\theta) = M_p(\theta + \theta + \theta + \theta)$$

or

$$H = 80$$

The interaction equations corresponding to the three mechanisms are plotted in Figure 2.103. By carrying out moment checks, it can be shown that mechanism 1 is valid for the portion AB of the curve, mechanism 2 for portion BC, and mechanism 3 is valid only when $V=0$.

2.11.10 Analysis Aids for Gable Frames

2.11.10.1 Pin-Based Gable Frames

Figure 2.104a shows a pinned-end gable frame subjected to a uniform gravity load λwL and a horizontal load $\lambda_1 H$ at the column top. The collapse mechanism is shown in Figure 2.104b. The work equation is used to determine the plastic limit load. First, the instantaneous center of rotation, O, is determined by considering similar triangles:

$$\frac{OE}{CF} = \frac{L}{xL} \quad \text{and} \quad \frac{OE}{CF} = \frac{OE}{h_1 + 2xh_2} \tag{2.229}$$

and

$$OD = OE - h_1 = \frac{(1-x)h_1 + 2xh_2}{x} \tag{2.230}$$

From the horizontal displacement of D

$$\theta h_1 = \phi OD \tag{2.231}$$

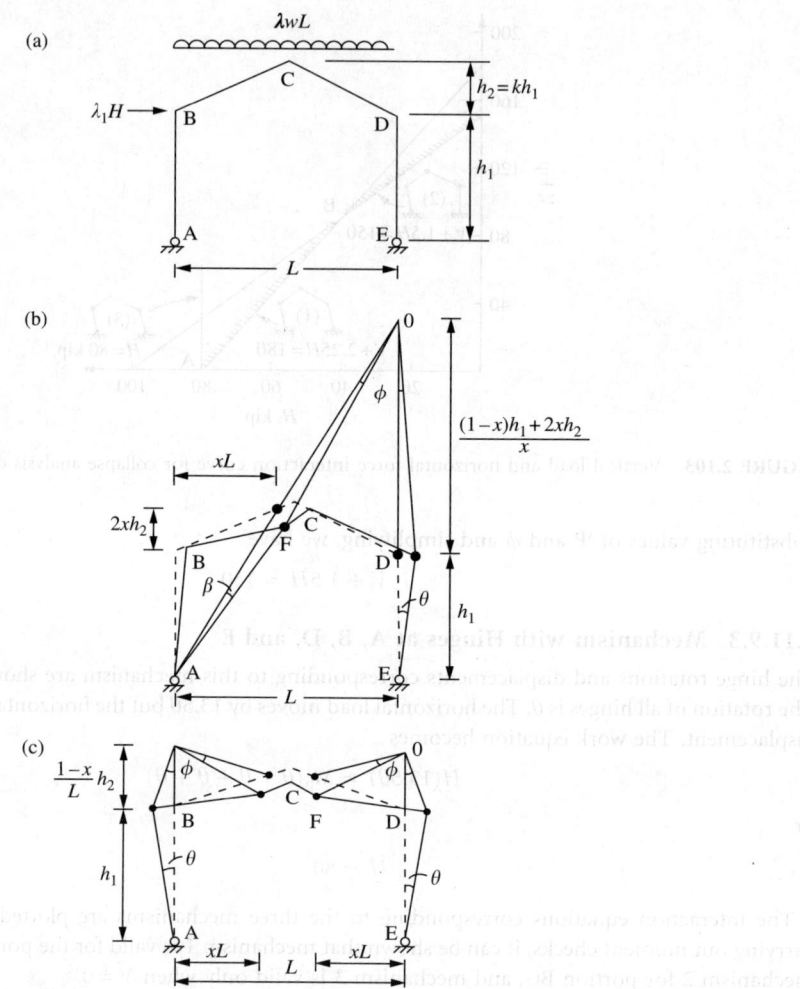

FIGURE 2.104 Pinned-base gable frame subjected to combined uniform distributed load and horizontal load.

where

$$\phi = \frac{x}{(1-x)+2xk}\theta \qquad (2.232)$$

where $k = h_2/h_1$. From the vertical displacement at C,

$$\beta = \frac{1-x}{(1-x)+2xk}\theta \qquad (2.233)$$

The work equation for the assumed mechanism is

$$\lambda_1 H h_1 \beta + \frac{\lambda w L^2}{2}(1-x)\phi = M_p(\beta + 2\phi + \theta) \qquad (2.234)$$

which gives

$$M_p = \frac{(1-x)\lambda_1 H h_1 + (1-x)x\lambda w L^2/2}{2(1+kx)} \qquad (2.235)$$

Differentiating M_p in Equation 2.235 with respect to x and solving for x,

$$x = \frac{A - 1}{k} \tag{2.236}$$

where

$$A = \sqrt{(1+k)(1-Uk)} \quad \text{and} \quad U = \frac{2\lambda_1 H h_1}{\lambda w L^2} \tag{2.237}$$

Substituting for x in the expression for M_p gives

$$M_p = \frac{\lambda w L^2}{8}\left[\frac{U(2+U)}{A^2 + 2A - Uk^2 + 1}\right] \tag{2.238}$$

In the absence of horizontal loading, the gable mechanism, as shown in Figure 2.104c, is the failure mode. In this case, letting $H = 0$ and $U = 0$ gives (Horne 1964)

$$M_p = \frac{\lambda w L^2}{8}\left[\frac{1}{1 + k + \sqrt{1+k}}\right] \tag{2.239}$$

Equation 2.238 can be used to produce a chart as shown in Figure 2.105 by which the value of M_p can be determined quickly by knowing the values of

$$k = \frac{h_2}{h_1} \quad \text{and} \quad U = \frac{2\lambda_1 H h_1}{\lambda w L^2} \tag{2.240}$$

2.11.10.2 Fixed-Base Gable Frames

A similar chart can be generated for a fixed-base gable frame as shown in Figure 2.106. Thus, if the values of loading, λw and $\lambda_1 H$ and frame geometry, h_1, h_2, and L, are known, the parameters k and U can be

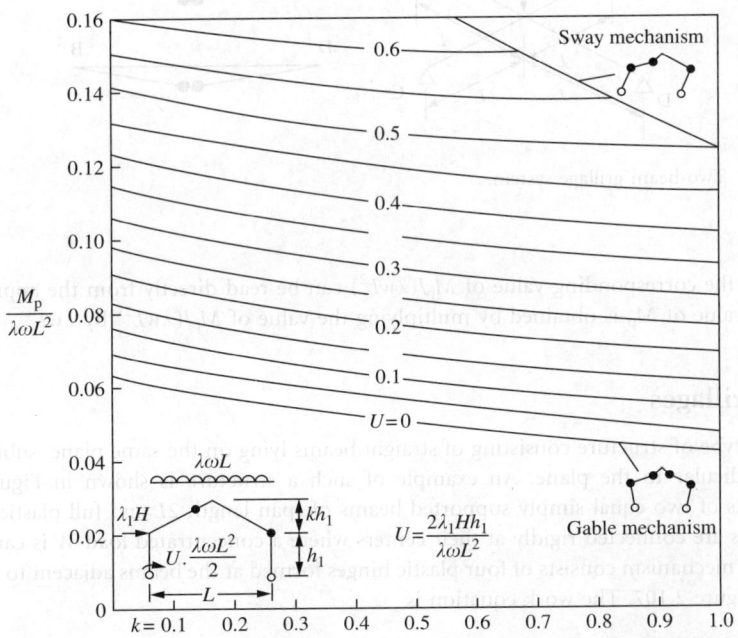

FIGURE 2.105 Analysis chart for pinned-base gable frame.

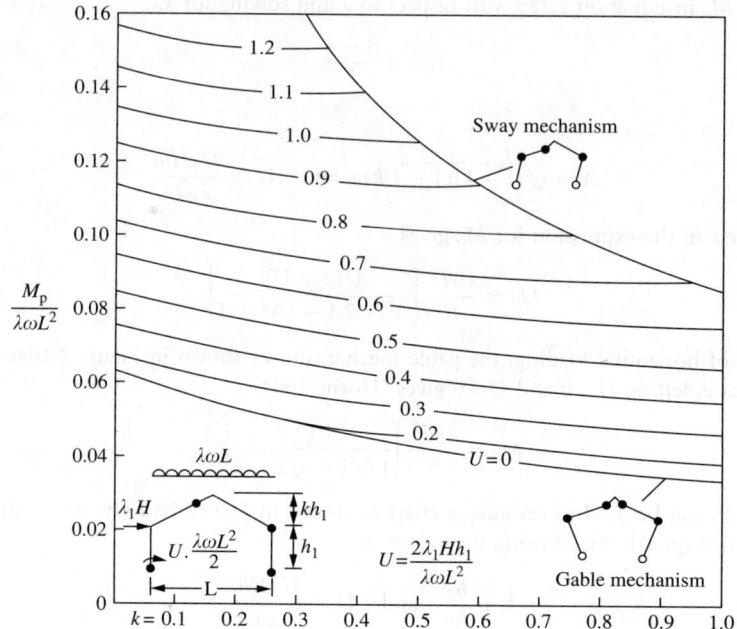

FIGURE 2.106 Analysis chart for fixed-base gable frame.

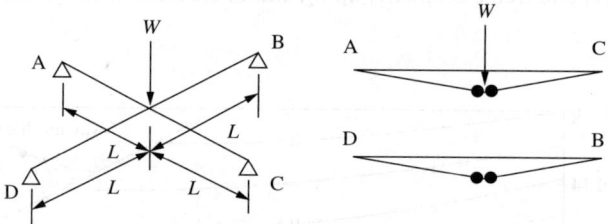

FIGURE 2.107 Two-beam grillage system.

evaluated and the corresponding value of $M_p/(\lambda wL^2)$ can be read directly from the appropriate chart. The required value of M_p is obtained by multiplying the value of $M_p/(\lambda wL^2)$ by λwL^2.

2.11.11 Grillages

A grillage is a type of structure consisting of straight beams lying on the same plane, subjected to loads acting perpendicular to the plane. An example of such a structure is shown in Figure 2.107. The grillage consists of two equal simply supported beams of span length $2L$ and full plastic moment M_p. The two beams are connected rigidly at their centers where a concentrated load W is carried.

The collapse mechanism consists of four plastic hinges formed at the beams adjacent to the point load as shown in Figure 2.107. The work equation is

$$WL\theta = 4M_p\theta$$

where the collapse load is

$$W = \frac{4M_p}{L}$$

2.11.11.1 Six-Beam Grillage

A grillage consisting of six beams of span length $4L$ each and full plastic moment M_p is shown in Figure 2.108. A total load of $9W$ acts on the grillage, splitting into concentrated loads W at the nine nodes. Three collapse mechanisms are possible. Ignoring member twisting due to torsional forces, the work equations associated with the three collapse mechanisms are computed as follows:

Mechanism 1 (Figure 2.109a)

Work equation:

$$9wL\theta = 12M_p\theta$$

where

$$w = \frac{12}{9}\frac{M_p}{L} = \frac{4M_p}{3L}$$

Mechanism 2 (Figure 2.109b)

Work equation:

$$wL\theta = 8M_p\theta$$

where

$$w = \frac{8M_p}{L}$$

Mechanism 3 (Figure 2.109c)

Work equation:

$$w2L2\theta + 4 \times w2L\theta = M_p(4\theta + 8\theta)$$

where

$$w = \frac{M_p}{L}$$

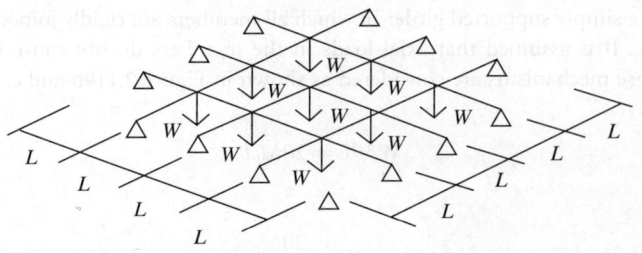

FIGURE 2.108 Six-beam grillage system.

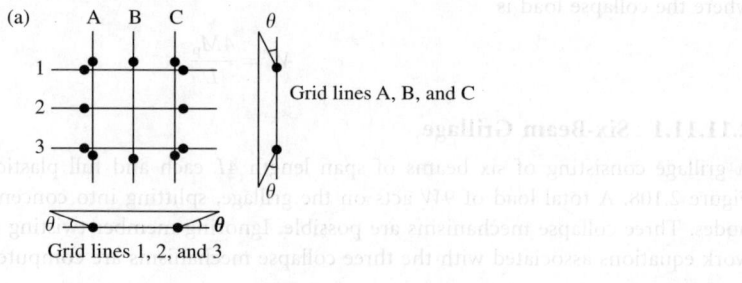

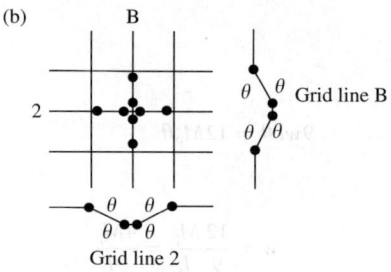

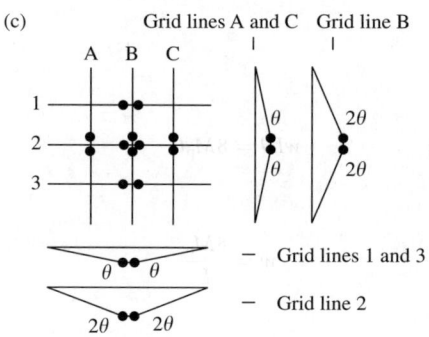

FIGURE 2.109 Six-beam grillage system: (a) mechanism 1, (b) mechanism 2, and (c) mechanism 3.

The lowest upper bound load corresponds to mechanism 3. This can be confirmed by conducting a moment check to ensure that bending moments anywhere are not violating the plastic moment condition. Additional discussion of plastic analysis of grillages can be found in Baker and Heyman (1969) and Heyman (1971).

2.11.12 Vierendeel Girders

Figure 2.110 shows a simply supported girder in which all members are rigidly joined and have the same plastic moment M_p. It is assumed that axial loads in the members do not cause member instability. Two possible collapse mechanisms are considered as shown in Figure 2.110b and c. The work equation for mechanism 1 is

$$W3\theta L = 20M_p\theta$$

so that

$$W = \frac{20M_p}{3L}$$

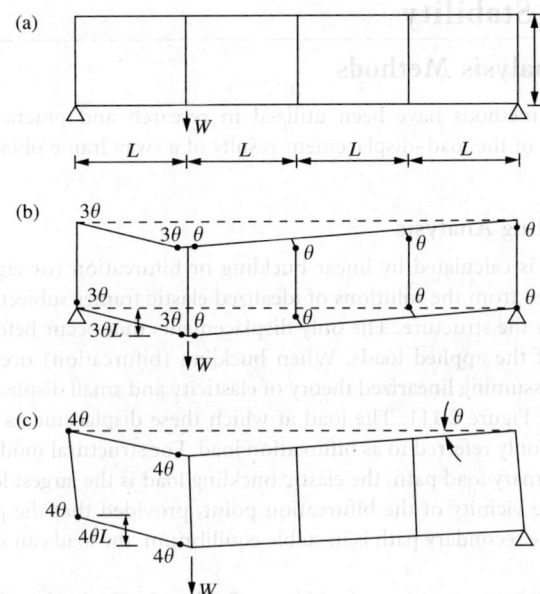

FIGURE 2.110 Collapse mechanism of a Vierendeel girder.

The work equation for mechanism 2 is

$$W3\theta L = 16M_p\theta$$

or

$$W = \frac{16M_p}{3L}$$

It can be easily proved that the collapse load associated with mechanism 2 is the correct limit load. This is done by constructing an equilibrium set of bending moments and checking that they are not violating the plastic moment condition.

2.11.13 Hinge-by-Hinge Analysis

Instead of finding the collapse load of the frame, it may be useful to obtain information about the distribution and redistribution of forces prior to reaching the collapse load. Elastic–plastic hinge analysis (also known as hinge-by-hinge analysis) determines the order of plastic hinge formation, the load factor associated with each plastic hinge formation, and member forces in the frame between each hinge formation. Thus, the state of the frame can be defined at any load factor rather than only at the state of collapse. This allows a more accurate determination of member forces at the design load level.

Educational and commercial software are now available for elastic–plastic hinge analysis (Chen and Sohal 1994). The computations of deflections for simple beams and multistorey frames can be done using the virtual work method (Knudsen et al. 1933; Beedle 1958; ASCE 1971; Chen and Sohal 1994). The basic assumption of first-order elastic–plastic hinge analysis is that the deformations of the structure are insufficient to alter radically the equilibrium equations. This assumption ceases to be true for slender members and structures, and the method gives unsafe predictions of limit loads.

2.12   Structural Stability

2.12.1   Stability Analysis Methods

Several stability analysis methods have been utilized in research and practice. Figure 2.111 shows schematic representations of the load–displacement results of a sway frame obtained from each type of analysis to be considered.

2.12.1.1   Elastic Buckling Analysis

The elastic buckling load is calculated by linear buckling or bifurcation (or eigenvalue) analysis. The buckling loads are obtained from the solutions of idealized elastic frames subjected to loads that do not produce direct bending in the structure. The only displacements that occur before buckling occurs are those in the directions of the applied loads. When buckling (bifurcation) occurs, the displacements increase without bound, assuming linearized theory of elasticity and small displacement as shown by the horizontal straight line in Figure 2.111. The load at which these displacements occur is known as the buckling load, or is commonly referred to as bifurcation load. For structural models that actually exhibit a bifurcation from the primary load path, the elastic buckling load is the largest load that the model can sustain, at least within the vicinity of the bifurcation point, provided that the postbuckling path is in unstable equilibrium. If the secondary path is in stable equilibrium, the load can still increase beyond the critical load value.

Buckling analysis is a common tool for calculations of column effective lengths. The effective length factor of a column member can be calculated using the procedure described in Section 2.12.2.5. The buckling analysis provides useful indices of the stability behavior of structures; however, it does not predict actual behavior of all but idealized structures with gravity loads applied only at the joints.

2.12.1.2   Second-Order Elastic Analysis

The analysis is formulated based on the deformed configuration of the structure. When derived rigorously, a second-order analysis can include both the member curvature (P–δ) and the side sway

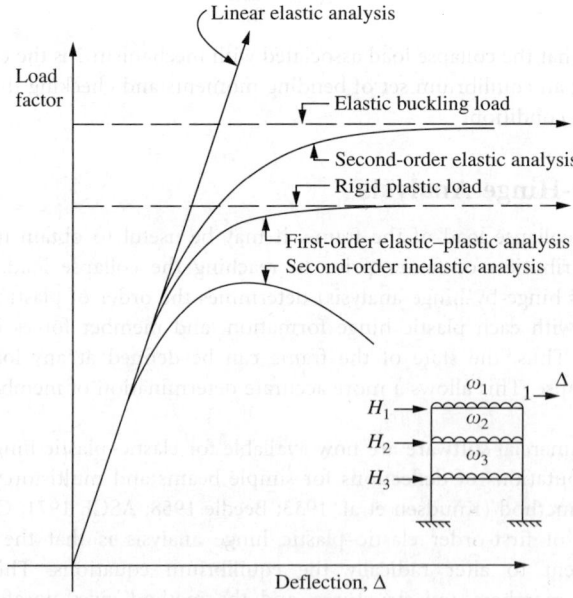

FIGURE 2.111   Categorization of stability analysis methods.

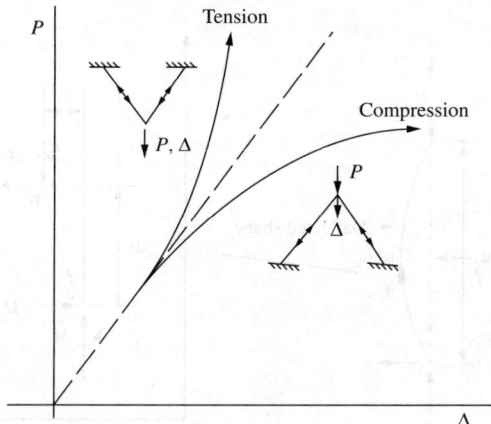

FIGURE 2.112   Behavior of frame in compression and tension.

$(P-\Delta)$ stability effects. The $P-\delta$ effect is associated with the influence of the axial force acting through the member displacement with respect to the rotated chord, whereas the $P-\Delta$ effect is the influence of axial force acting through the relative side sway displacements of the member ends. A structural system will become stiffer when its members are subjected to tension. Conversely, the structure will become softer when its members are in compression. Such a behavior can be illustrated by a simple model shown in Figure 2.112. There is a clear advantage for a designer in making use of the stiffer behavior of tension structures. However, the detrimental effects associated with second-order deformations due to compression forces must be considered in designing structures subjected to predominant gravity loads.

Unlike first-order analysis in which solutions can be obtained in a rather simple and direct manner, a second-order analysis often requires an iterative procedure to obtain solutions. The load–displacement curve generated from a second-order elastic analysis will gradually approach the horizontal straight line that represents the buckling load obtained from the elastic buckling analysis, as shown in Figure 2.111. Differences in the two limit loads may arise from the fact that the elastic stability limit is calculated for equilibrium based on the deformed configuration whereas the elastic critical load is calculated as a bifurcation from equilibrium on the undeformed geometry of the frame.

The load–displacement response of many practical structures usually does not involve any bifurcation of equilibrium path. In some cases, the second-order elastic incremental response may not have yielded any limit. The reader is referred to Chen and Lui (1987) for a basic discussion of these behavioral issues.

Recent work on second-order elastic analysis have been reported by Chen and Lui (1991), Liew et al. (1991), White and Hajjar (1991), and Chen and Toma (1994), among others. Second-order analysis programs that can take into consideration connection flexibility are also available (Chen et al. 1996; Chen and Kim 1997; Faella et al. 2000).

2.12.1.3 Second-Order Inelastic Analysis

Second-order inelastic analysis refers to methods of analysis that can capture geometrical and material nonlinearities of the structures. The most rigorous inelastic analysis method is called spread-of-plasticity analysis. It involves discretization of a member into many line segments and the cross-section of each segment into a number of finite elements. Inelasticity is captured within the cross-sections and along the member length. The calculation of forces and deformations in the structure after yielding requires iterative trial-and-error processes because of the nonlinearity of the load–deformation response, and the change in the effective stiffness of the cross-section at inelastic regions associated with the increase in the

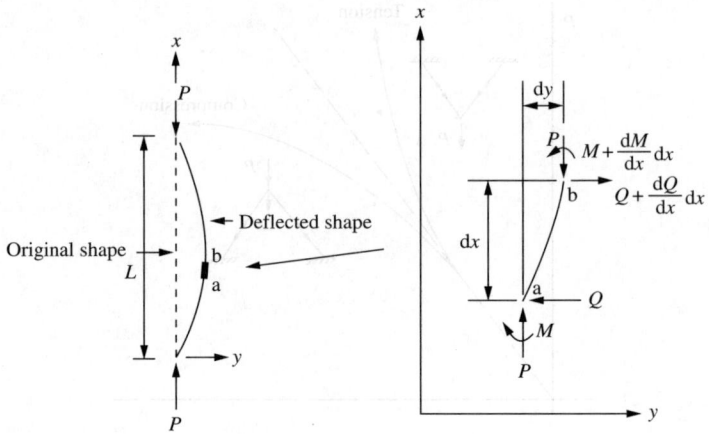

FIGURE 2.113   Stability equations of a column segment.

applied loads and the change in structural geometry. Although most of the spread-of-plasticity analysis methods have been developed for planar analysis (White 1985; Vogel 1985), three-dimensional spread-of-plasticity techniques are also available involving various degrees of refinements (Chen and Atsuta 1977; Clark 1994; Wang 1988; White 1988; Jiang et al. 2002).

The simplest second-order inelastic analysis is the *elastic–plastic hinge* approach. The analysis assumes that the element remains elastic except at its ends where zero-length plastic hinges are allowed to form. Plastic hinge analysis of planar frames can be found in Orbison 1982; Ziemian et al. 1992a,b; Liew et al. 1993; White et al. 1993; Chen and Toma 1994; Chen and Sohal 1994; Chen et al. 1996, among others. Advanced analyses of three-dimensional frames are reported in Chen et al. (2000) and Liew et al. (2000). Second-order plastic hinge analysis allows efficient analysis of large-scale building frames. This is particularly true for structures in which the axial forces in the component members are small and the behavior is predominated by bending actions. Although elastic–plastic hinge approaches can provide essentially the same load–displacement predictions as second-order plastic-zone methods for many frame problems, they cannot be classified as advanced analysis for use in frame design. Some modifications to the elastic–plastic hinge are required to qualify the methods as advanced analysis, and they are discussed in Section 2.12.7.

Figure 2.111 shows the load–displacement curve (a smooth curve with a descending branch) obtained from the second-order inelastic analysis. The computed limit load should be close to that obtained from the plastic-zone analysis.

2.12.2 Column Stability

2.12.2.1 Stability Equations

The stability equation of a column can be obtained by considering an infinitesimal deformed segment of the column as shown in Figure 2.113. Considering the moment equilibrium about point b, we obtain

$$Q\,dx + P\,dy + M - \left(M + \frac{dM}{dx}\,dx\right) = 0$$

or upon simplification

$$Q = \frac{dM}{dx} - P\frac{dy}{dx} \qquad (2.241)$$

Summing force horizontally, we can write

$$-Q + \left(Q + \frac{dQ}{dx}dx\right) = 0$$

or upon simplification,

$$\frac{dQ}{dx} = 0 \tag{2.242}$$

Differentiating Equation 2.241 with respect to x, we obtain

$$\frac{dQ}{dx} = \frac{d^2M}{dx^2} - P\frac{d^2y}{dx^2} \tag{2.243}$$

which, when compared with Equation 2.242, gives

$$\frac{d^2M}{dx^2} - P\frac{d^2y}{dx^2} = 0 \tag{2.244}$$

Since moment $M = -EI(d^2y/dx^2) = 0$, Equation 2.244 can be written as

$$EI\frac{d^4y}{dx^4} + P\frac{d^2y}{dx^2} = 0 \tag{2.245}$$

or

$$y^{IV} + k^2 y'' = 0 \tag{2.246}$$

Equation 2.246 is the general fourth-order differential equation that is valid for all support conditions. The general solution to this equation is

$$y = A\sin kx + B\cos kx + Cx + D \tag{2.247}$$

To determine the critical load, it is necessary to have four boundary conditions, two at each end of the column. In some cases, both geometric and force boundary conditions are required to eliminate the unknown coefficients (A, B, C, D) in Equation 2.247.

2.12.2.2 Column with Pinned Ends

For a column pinned at both ends, as shown in Figure 2.114a, the four boundary conditions are

$$y(x = 0) = 0, \quad M(x = 0) = 0 \tag{2.248}$$

$$y(x = L) = 0, \quad M(x = L) = 0 \tag{2.249}$$

Since $M = -EIy''$, the moment conditions can be written as

$$y''(0) = 0 \quad \text{and} \quad y''(x = L) = 0 \tag{2.250}$$

Using the these conditions, we have

$$B = D = 0 \tag{2.251}$$

The deflection function (Equation 2.247) reduces to

$$y = A\sin Kx + Cx \tag{2.252}$$

Using the conditions $y(L) = y''(L) = 0$, Equation 2.252 gives

$$A\sin KL + CL = 0 \tag{2.253}$$

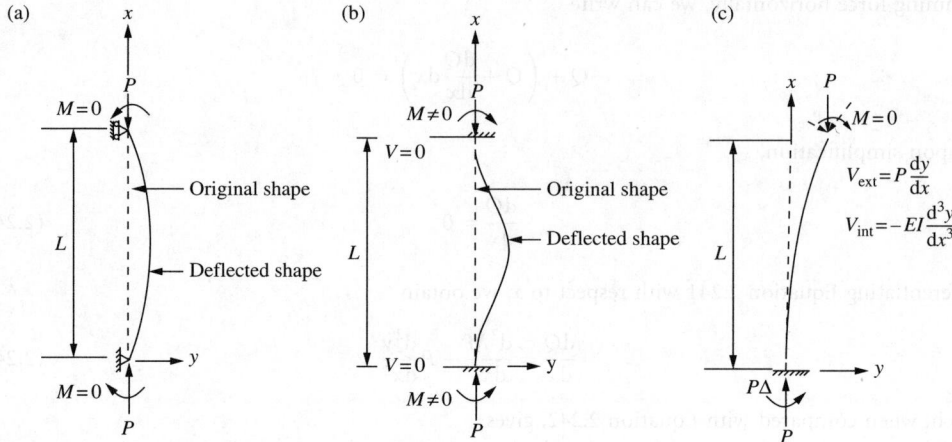

FIGURE 2.114 Column with (a) pinned ends, (b) fixed ends, and (c) fixed–free ends.

and

$$-AK^2 \sin kL = 0 \tag{2.254}$$

$$\begin{bmatrix} \sin kL & L \\ -k^2 \sin kL & 0 \end{bmatrix} \begin{bmatrix} A \\ C \end{bmatrix} = \begin{bmatrix} 0 \\ 0 \end{bmatrix} \tag{2.255}$$

   If $A = C = 0$, the solution is trivial. Therefore, to obtain a nontrivial solution, the determinant of the coefficient matrix of Equation 2.255 must be zero, that is,

$$\det \begin{vmatrix} \sin kL & L \\ -k^2 \sin kL & 0 \end{vmatrix} = 0 \tag{2.256}$$

or

$$k^2 L \sin kL = 0 \tag{2.257}$$

Since $k^2 L$ cannot be zero, we must have

$$\sin kL = 0 \tag{2.258}$$

or

$$kL = n\pi, \quad n = 1, 2, 3, \ldots \tag{2.259}$$

   The lowest buckling load corresponds to the first mode obtained by setting $n = 1$

$$P_{cr} = \frac{\pi^2 EI}{L^2} \tag{2.260}$$

2.12.2.3 Column with Fixed Ends

The four boundary conditions for a fixed end column are (Figure 2.114b)

$$y(x = 0) = y'(x = 0) = 0 \tag{2.261}$$

$$y(x = L) = y'''(x = L) = 0 \tag{2.262}$$

Using the first two boundary conditions, we obtain

$$D = -B, \quad C = -Ak \tag{2.263}$$

The deflection function (Equation 2.247) becomes

$$y = A(\sin kx - kx) + B(\cos kx - 1) \tag{2.264}$$

Using the last two boundary conditions, we have

$$\begin{bmatrix} \sin kL - kL & \cos kL - 1 \\ \cos kL - 1 & -\sin kL \end{bmatrix} \begin{bmatrix} A \\ B \end{bmatrix} = \begin{bmatrix} 0 \\ 0 \end{bmatrix} \tag{2.265}$$

For a nontrivial solution, we must have

$$\det \begin{bmatrix} \sin kL - kL & \cos kL - 1 \\ \cos kL - 1 & -\sin kL \end{bmatrix} = 0 \tag{2.266}$$

or after expanding

$$kL \sin kL + 2\cos kL - 2 = 0 \tag{2.267}$$

Using trigonometrical identities, $\sin kL = 2\sin(kL/2)\cos(kL/2)$ and $\cos kL = 1 - 2\sin^2(kL/2)$, Equation 2.267 can be written as

$$\sin \frac{kL}{2} \left(\frac{kL}{2} \cos \frac{kL}{2} - \sin \frac{kL}{2} \right) = 0 \tag{2.268}$$

The critical load for the symmetric buckling mode is $P_{cr} = 4\pi^2 EI/L^2$ by letting $\sin(kL/2) = 0$. The buckling load for the antisymmetric buckling mode is $P_{cr} = 80.8 EI/L^2$ by letting the bracket term in Equation 2.268 equal zero.

2.12.2.4 Column with One End Fixed and One End Free

The boundary conditions for a fixed–free column are (Figure 2.114c): at the fixed end

$$y(x = 0) = y'(x = 0) = 0 \tag{2.269}$$

and at the free end the moment $M = EIy''$ is equal to zero

$$y''(x = L) = 0 \tag{2.270}$$

and the shear force $V = -dM/dx = -EIy'''$ is equal to Py', which is the transverse component of P acting at the free end of the column.

$$V = -EIy''' = Py' \tag{2.271}$$

It follows that the shear force condition at the free end has the form

$$y''' + k^2 y' = 0 \tag{2.272}$$

Using the boundary conditions at the fixed end, we have

$$B + D = 0 \quad \text{and} \quad Ak + C = 0 \tag{2.273}$$

The boundary conditions at the free end give

$$A \sin kL + B \cos kL = 0 \quad \text{and} \quad C = 0 \tag{2.274}$$

In matrix form, Equations 2.273 and 2.274 can be written as

$$\begin{bmatrix} 0 & 1 & 1 \\ k & 0 & 0 \\ \sin kL & \cos kL & 0 \end{bmatrix} \begin{bmatrix} A \\ B \\ C \end{bmatrix} = \begin{bmatrix} 0 \\ 0 \\ 0 \end{bmatrix} \tag{2.275}$$

TABLE 2.2 Boundary Conditions for Various End Conditions

End conditions	Boundary conditions
Pinned	$y=0$, $y''=0$
Fixed	$y=0$, $y'=0$
Guided	$y'=0$, $y'''=0$
Free	$y''=0$, $y'''+k^2y'=0$

For a nontrivial solution, we must have

$$\det \begin{vmatrix} 0 & 1 & 1 \\ k & 0 & 0 \\ \sin kL & \cos kL & 0 \end{vmatrix} = 0 \qquad (2.276)$$

The characteristic equation becomes

$$k\cos kL = 0 \qquad (2.277)$$

Since k cannot be zero, we must have $\cos kL = 0$ or

$$kL = \frac{n\pi}{2}, \quad n = 1,3,5,\ldots \qquad (2.278)$$

The smallest root ($n=1$) gives the lowest critical load of the column

$$P_{cr} = \frac{\pi^2 EI}{4L^2} \qquad (2.279)$$

The boundary conditions for columns with various end conditions are summarized in Table 2.2.

2.12.2.5 Column Effective Length Factor

The effective length factor, K, of columns with different end boundary conditions can be obtained by equating the P_{cr} load obtained from the buckling analysis with the Euler load of a pinned-end column of effective length KL

$$P_{cr} = \frac{\pi^2 EI}{(KL)^2}$$

The effective length factor can be obtained as

$$K = \sqrt{\frac{\pi^2 EI / L^2}{P_{cr}}} \qquad (2.280)$$

The K factor can be multiplied to the actual length of the end-restrained column to give the length of an equivalent pinned-ended column whose buckling load is the same as that of the end-restrained column. Table 2.3 (AISC 1993) summarizes the theoretical K factors for columns with different boundary conditions. Also shown in the table are the recommended K factors for design applications. The recommended values for design are equal or higher than the theoretical values to account for semirigid effects of the connections used in practice.

2.12.3 Stability of Beam–Columns

Figure 2.115a shows a beam–column subjected to an axial compressive force P at the ends, a lateral load w along the entire length, and end moments M_A and M_B. The stability equation can be derived by considering the equilibrium of an infinitesimal element of length ds as shown in Figure 2.115b. The cross-section forces S and H act in the vertical and horizontal directions.

TABLE 2.3 Comparison of Theoretical and Design *K* Factors

Buckled shape of column is shown by dashed line	(a)	(b)	(c)	(d)	(e)	(f)
Theoretical *k* value	0.5	0.7	1.0	1.0	2.0	2.0
Recommended design value when ideal conditions are approximated	0.65	0.80	1.2	1.0	2.10	2.0

End condition code	
⊥	Rotation fixed and translation fixed
⊽	Rotation free and translation fixed
▨	Rotation fixed and translation free
○	Rotation free and translation free

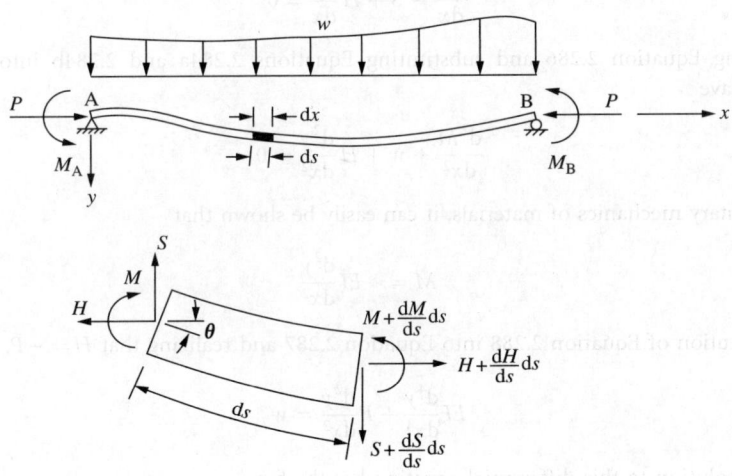

FIGURE 2.115 Basic differential equation of a beam–column.

Considering equilibrium of forces:

1. *Horizontal equilibrium*

$$H + \frac{dH}{ds}\,ds - H = 0 \qquad (2.281)$$

2. *Vertical equilibrium*

$$S + \frac{dS}{ds}ds - S + w\,ds = 0 \tag{2.282}$$

3. *Moment equilibrium*

$$M + \frac{dM}{ds}ds - M - \left(S + \frac{dS}{ds} + S\right)\cos\theta\left(\frac{ds}{2}\right) + \left(H + \frac{dH}{ds}ds + H\right)\sin\theta\left(\frac{ds}{2}\right) = 0 \tag{2.283}$$

Since $(dS/ds)ds$ and $(dH/ds)ds$ are negligibly small compared to S and H, the above equilibrium equations can be reduced to

$$\frac{dH}{ds} = 0 \tag{2.284a}$$

$$\frac{dS}{ds} + w = 0 \tag{2.284b}$$

$$\frac{dM}{ds} - S\cos\theta + H\sin\theta = 0 \tag{2.284c}$$

For small deflections and neglecting shear deformations

$$ds \cong dx, \quad \cos\theta \cong 1, \quad \sin\theta \cong \theta \cong \frac{dy}{dx} \tag{2.285}$$

where y is the lateral displacement of the member. Using the above approximations, Equation 2.280 can be written as

$$\frac{dM}{dx} - S + H\frac{dy}{dx} = 0 \tag{2.286}$$

Differentiating Equation 2.286 and substituting Equations 2.284a and 2.284b into the resulting equation, we have

$$\frac{d^2M}{dx^2} + w + H\frac{d^2y}{dx^2} = 0 \tag{2.287}$$

From elementary mechanics of materials, it can easily be shown that

$$M = -EI\frac{d^2y}{dx^2} \tag{2.288}$$

Upon substitution of Equation 2.288 into Equation 2.287 and realizing that $H = -P$, we obtain

$$EI\frac{d^4y}{dx^4} + P\frac{d^2y}{dx^2} = w \tag{2.289}$$

The general solution to this differential equation has the form

$$y = A\sin kx + B\cos kx + Cx + D + f(x) \tag{2.290}$$

where

$$k = \sqrt{P/EI}$$

and $f(x)$ is a particular solution satisfying the differential equation. The constants A, B, C, and D can be determined from the boundary conditions of the beam–column under investigation.

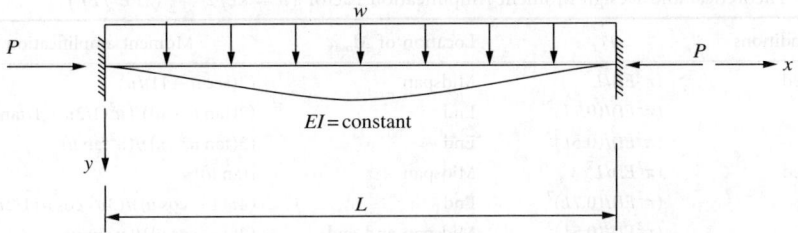

FIGURE 2.116 Beam–column subjected to uniform loading.

2.12.3.1 Beam–Column Subjected to Transverse Loading

Figure 2.116 shows a fixed-ended beam–column with uniformly distributed load w.
The general solution to Equation 2.289 is

$$y = A \sin kx + B \cos kx + Cx + D + \frac{w}{2EIk^2} x^2 \qquad (2.291)$$

Using the boundary conditions

$$y_{x=0} = 0, \quad y'_{x=0} = 0, \quad y_{x=L} = 0, \quad y'_{x=L} = 0 \qquad (2.292)$$

where the prime denotes differentiation with respect to x, it can be shown that

$$A = \frac{wL}{2EIk^3} \qquad (2.293a)$$

$$B = \frac{wL}{2EIk^3 \tan(kL/2)} \qquad (2.293b)$$

$$C = -\frac{wL}{2EIk^2} \qquad (2.293c)$$

$$D = -\frac{wL}{2EIk^3 \tan(kL/2)} \qquad (2.293d)$$

Upon substitution of these constants into Equation 2.291, the deflection function can be written as

$$y = \frac{wL}{2EIk^3} \left[\sin kx + \frac{\cos kx}{\tan(kL/2)} - kx - \frac{1}{\tan(kL/2)} + \frac{kx^2}{L} \right] \qquad (2.294)$$

The maximum moment for this beam–column occurs at the fixed ends and is equal to

$$M_{\max} = -EIy''|_{x=0} = -EIy''|_{x=L} = -\frac{wL^2}{12} \left[\frac{3(\tan u - u)}{u^2 \tan u} \right] \qquad (2.295)$$

where $u = kL/2$.

Since $wL^2/12$ is the maximum first-order moment at the fixed ends, the term in the brackets represents the theoretical moment amplification factor due to the P–δ effect.

For beam–columns with other transverse loading and boundary conditions, a similar approach can be followed to determine the moment amplification factor. Table 2.4 summarizes the expressions for the theoretical and design moment amplification factors for some loading conditions (AISC 1989).

TABLE 2.4   Theoretical and Design Moment Amplification Factor $\left(u = kL/2 = \frac{1}{2}\sqrt{PL^2/EI}\right)$

Boundary conditions	Pc_r	Location of $M_{\max}$	Moment amplification factor
Hinged–hinged	$(\pi^2 EI)/L^2$	Midspan	$(2(\sec u - 1))/u^2$
Hinged–fixed	$(\pi^2 EI)/(0.7L)^2$	End	$(2(\tan u - u))/(u^2(1/2u - 1/\tan 2u))$
Fixed–fixed	$(\pi^2 EI)/(0.5L)^2$	End	$(3(\tan u - u))/(u^2 \tan u)$
Hinged–hinged	$(\pi^2 EI)/L^2$	Midspan	$(\tan u)/u$
Hinged–fixed	$(\pi^2 EI)/(0.7L)^2$	End	$(4u(1 - \cos u))/(3u^2 \cos u (1/2u - 1/\tan 2u))$
Fixed–fixed	$(\pi^2 EI)/(0.5L)^2$	Midspan and end	$(2(1 - \cos u))/(u \sin u)$

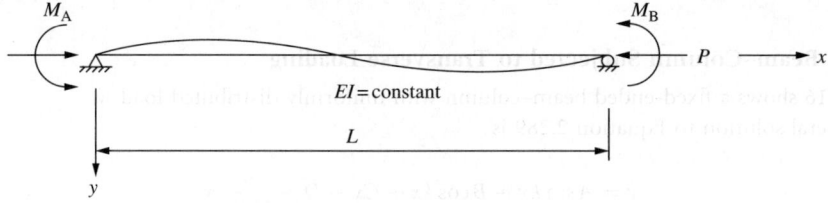

FIGURE 2.117   Beam–column subjected to end moments.

2.12.3.2   Beam–Column Subjected to End Moments

Consider the beam–column shown in Figure 2.117. The member is subjected to an axial force of P and end moments M_A and M_B. The differential equation for this beam–column can be obtained from Equation 2.289 by setting $w = 0$:

$$EI \frac{d^4 y}{dx^4} + P \frac{d^2 y}{dx^2} = 0 \tag{2.296}$$

The general solution is

$$y = A \sin kx + B \cos kx + Cx + D \tag{2.297}$$

The constants A, B, C, and D are determined by enforcing the four boundary conditions

$$y_{x=0} = 0, \quad y''_{x=0} = \frac{M_A}{EI}, \quad y_{x=L} = 0, \quad y''_{x=L} = \frac{-M_B}{EI} \tag{2.298}$$

to give

$$A = \frac{M_A \cos kL + M_B}{EIk^2 \sin kL} \tag{2.299a}$$

$$B = -\frac{M_A}{EIk^2} \tag{2.299b}$$

$$C = -\left(\frac{M_A + M_B}{EIk^2 L}\right) \tag{2.299c}$$

$$D = \frac{M_A}{EIk^2} \tag{2.299d}$$

Substituting Equations 2.299a to d into the deflection function Equation 2.297 and rearranging gives

$$y = \frac{1}{EIk}\left[\frac{\cos kL}{\sin kL}\sin kx - \cos kx - \frac{x}{L} + 1\right]M_A + \frac{1}{EIk^2}\left[\frac{1}{\sin kL}\sin kx - \frac{x}{L}\right]M_B \tag{2.300}$$

The maximum moment can be obtained by first locating its position by setting $dM/dx = 0$ and substituting the result into $M = -EIy''$ to give

$$M_{max} = \frac{\sqrt{(M_A^2 + 2M_A M_B \cos kL + M_B^2)}}{\sin kL} \tag{2.301}$$

Assuming that M_B is the larger of the two end moments, Equation 2.301 can be expressed as

$$M_{max} = M_B \left[\frac{\sqrt{\{(M_A/M_B)^2 + 2(M_A/M_B)\cos kL + 1\}}}{\sin kL} \right] \tag{2.302}$$

Since M_B is the maximum first-order moment, the expression in brackets is therefore the theoretical moment amplification factor. In Equation 2.302, the ratio (M_A/M_B) is positive if the member is bent in double (or reverse) curvature and negative if the member is bent in single curvature. A special case arises when the end moments are equal and opposite (i.e., $M_B = -M_A$). By setting $M_B = -M_A = M_0$ in Equation 2.302, we get

$$M_{max} = M_0 \left[\frac{\sqrt{\{2(1 - \cos kL)\}}}{\sin kL} \right] \tag{2.303}$$

For this special case, the maximum moment always occurs at midspan.

2.12.4 Slope Deflection Equations

The slope deflection equations of a beam–column can be derived by considering the beam–column shown in Figure 2.117. The deflection function for this beam–column can be obtained from Equation 2.300 in terms of M_A and M_B as

$$y = \frac{1}{EIk^2} \left[\frac{\cos kL}{\sin kL} \sin kx - \cos kx - \frac{x}{L} + 1 \right] M_A + \frac{1}{EIk^2} \left[\frac{1}{\sin kL} \sin kx - \frac{x}{L} \right] M_B \tag{2.304}$$

from which

$$y' = \frac{1}{EIk} \left[\frac{\cos kL}{\sin kL} \cos kx + \sin kx - \frac{1}{kL} \right] M_A + \frac{1}{EIk} \left[\frac{\cos kx}{\sin kL} - \frac{1}{kL} \right] M_B \tag{2.305}$$

The end rotations θ_A and θ_B can be obtained from Equation 2.305 as

$$\theta_A = y'(x = 0) = \frac{1}{EIk} \left[\frac{\cos kL}{\sin kL} - \frac{1}{kL} \right] M_A + \frac{1}{EIk} \left[\frac{1}{\sin kL} - \frac{1}{kL} \right] M_B$$

$$= \frac{L}{EI} \left[\frac{kL \cos kL - \sin kL}{(kL)^2 \sin kL} \right] M_A + \frac{L}{EI} \left[\frac{kL - \sin kL}{(kL)^2 \sin kL} \right] M_B \tag{2.306}$$

and

$$\theta_B = y'(x = L) = \frac{1}{EIk} \left[\frac{1}{\sin kL} - \frac{1}{kL} \right] M_A + \frac{1}{EIk} \left[\frac{\cos kL}{\sin kL} - \frac{1}{kL} \right] M_B$$

$$= \frac{L}{EI} \left[\frac{kL - \sin kL}{(kL)^2 \sin kL} \right] M_A + \frac{L}{EI} \left[\frac{KL \cos kL - \sin kL}{(kL)^2 \sin kL} \right] M_B \tag{2.307}$$

The moment–rotation relationship can be obtained from Equations 2.306 and 2.307 by arranging M_A and M_B in terms of θ_A and θ_B as

$$M_A = \frac{EI}{L}\left(s_{ii}\,\theta_A + s_{ij}\,\theta_B\right) \tag{2.308}$$

$$M_B = \frac{EI}{L}\left(s_{ji}\,\theta_A + s_{jj}\,\theta_B\right) \tag{2.309}$$

where

$$s_{ii} = s_{jj} = \frac{kL\sin kL - (kL)^2\cos kL}{2 - 2\cos kL - kL\sin kL} \tag{2.310}$$

and

$$s_{ij} = s_{ji} = \frac{(kL)^2 - kL\sin kL}{2 - 2\cos kL - kL\sin kL} \tag{2.311}$$

are referred to as the *stability functions*.

Equations 2.308 and 2.309 are the slope deflection equations for a beam–column that is not subjected to transverse loading and relative joint translation. When P approaches zero, $kL = (\sqrt{P/EI})L$ approaches zero, and by using L'Hôpital's rule, it can be shown that $s_{ii} = 4$ and $s_{ij} = 2$. Values for s_{ii} and s_{ij} for various values of kL are plotted as shown in Figure 2.118.

Equations 2.309 and 2.310 are valid if the following conditions are satisfied.

1. The beam is prismatic.
2. There is no relative joint displacement between the two ends of the member.
3. The member is continuous, that is, there is no internal hinge or discontinuity in the member.
4. There is no in-span transverse loading on the member.
5. The axial force in the member is compressive.

If these conditions are not satisfied, some modifications to the slope deflection equations are necessary.

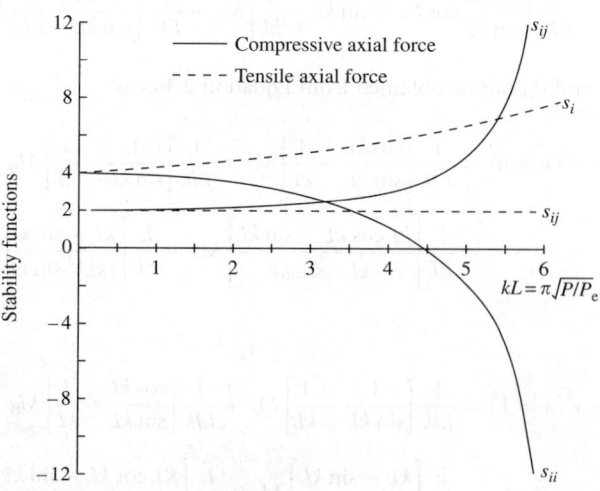

FIGURE 2.118   Plot of stability functions.

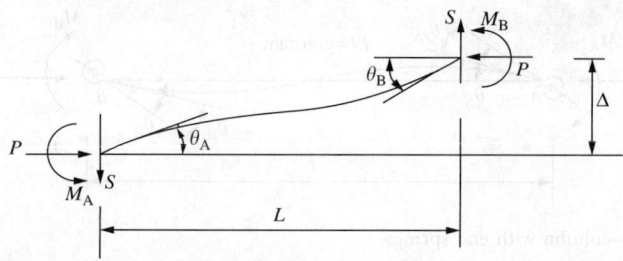

FIGURE 2.119 Beam–column subjected to end moments and side sway.

2.12.4.1 Member Subjected to Side Sway

If there is a relative joint translation, Δ, between the member ends, as shown in Figure 2.119, the slope–deflection equations are modified as

$$M_A = \frac{EI}{L}\left[s_{ii}\left(\theta_A - \frac{\Delta}{L}\right) + s_{ij}\left(\theta_B - \frac{\Delta}{L}\right)\right] = \frac{EI}{L}\left[s_{ii}\theta_A + s_{ij}\theta_B - (s_{ii} + s_{ij})\frac{\Delta}{L}\right] \tag{2.312}$$

$$M_B = \frac{EI}{L}\left[s_{ij}\left(\theta_A - \frac{\Delta}{L}\right) + s_{ii}\left(\theta_B - \frac{\Delta}{L}\right)\right] = \frac{EI}{L}\left[s_{ij}\theta_A + s_{ii}\theta_B - (s_{ii} + s_{ij})\frac{\Delta}{L}\right] \tag{2.313}$$

2.12.4.2 Member with a Hinge at One End

If a hinge is present at the B-end of the member, the end moment there is zero, that is,

$$M_B = \frac{EI}{L}(s_{ij}\theta_A + s_{ii}\theta_B) = 0 \tag{2.314}$$

from which

$$\theta_B = -\frac{s_{ij}}{s_{ii}}\theta_A \tag{2.315}$$

Upon substituting Equation 2.315 into Equation 2.312, we have

$$M_A = \frac{EI}{L}\left(s_{ii} - \frac{s_{ij}^2}{s_{ii}}\right)\theta_A \tag{2.316}$$

If the member is hinged at A rather than at B, Equation 2.316 is still valid provided that the subscript A is changed to B.

2.12.4.3 Member with End Restraints

If the member ends are connected by two linear elastic springs, as in Figure 2.120, with spring constants, R_{kA} and R_{kB} at the A and B ends, respectively, the end rotations of the linear springs are M_A/R_{kA} and M_B/R_{kB}. If we denote the total end rotations at joints A and B by θ_A and θ_B, respectively, then the corresponding member end rotations, with respect to its chord, will be $(\theta_A - M_A/R_{kA})$ and $(\theta_B - M_B/R_{kB})$. As a result, the slope deflection equations are modified to

$$M_A = \frac{EI}{L}\left[s_{ii}\left(\theta_A - \frac{M_A}{R_{kA}}\right) + s_{ij}\left(\theta_B - \frac{M_B}{R_{kB}}\right)\right] \tag{2.317}$$

$$M_B = \frac{EI}{L}\left[s_{ij}\left(\theta_A - \frac{M_A}{R_{kA}}\right) + s_{jj}\left(\theta_B - \frac{M_B}{R_{kB}}\right)\right] \tag{2.318}$$

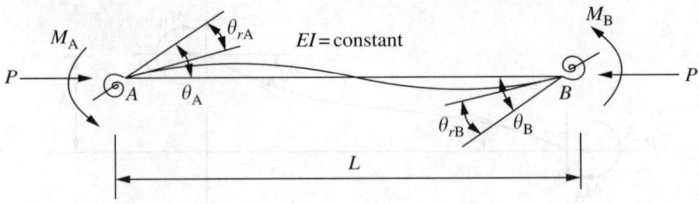

FIGURE 2.120   Beam–column with end springs.

Solving Equations 2.317 and 2.318 simultaneously for M_A and M_B gives

$$M_A = \frac{EI}{LR^*}\left[\left(s_{ii} + \frac{EIs_{ii}^2}{LR_{kB}} - \frac{EIs_{ij}^2}{LR_{kB}}\right)\theta_A + s_{ij}\,\theta_B\right] \qquad (2.319)$$

$$M_B = \frac{EI}{LR^*}\left[s_{ij}\,\theta_A + \left(s_{ii} + \frac{EIs_{ii}^2}{LR_{kA}} - \frac{EIs_{ij}^2}{LR_{kA}}\right)\theta_B\right] \qquad (2.320)$$

where

$$R^* = \left(1 + \frac{EIs_{ii}}{LR_{kA}}\right)\left(1 + \frac{EIs_{ii}}{LR_{kB}}\right) - \left(\frac{EI}{L}\right)^2 \frac{s_{ij}^2}{R_{kA}R_{kB}} \qquad (2.321)$$

In writing Equations 2.319 and 2.320, the equality $s_{jj} = s_{ii}$ has been used. Note that as R_{kA} and R_{kB} approach infinity, Equations 2.319 and 2.320 reduce to Equations 2.308 and 2.309, respectively.

2.12.4.4 Member with Transverse Loading

For members subjected to transverse loading, the slope deflection Equations 2.308 and 2.309 can be modified by adding an extra term for the fixed-ended moment of the member:

$$M_A = \frac{EI}{L}(s_{ii}\theta_A + s_{ij}\theta_B) + M_{FA} \qquad (2.322)$$

$$M_B = \frac{EI}{L}(s_{ij}\theta_A + s_{jj}\theta_B) + M_{FB} \qquad (2.323)$$

Table 2.5 gives the expressions for the fixed-end moments of five commonly encountered cases of transverse loading. Readers are referred to Chen and Lui (1987, 1991) for more details.

2.12.4.5 Member with Tensile Axial Force

For members subjected to tensile force, Equations 2.308 and 2.309 can be used provided that the stability functions are redefined as

$$s_{ii} = s_{jj} = \frac{(kL)^2 \cosh kL - kL \sinh kL}{2 - 2\cosh kL + kL \sinh kL} \qquad (2.324)$$

$$s_{ij} = s_{ji} = \frac{kL \sinh kL - (kL)^2}{2 - 2\cosh kL + kL \sinh kL} \qquad (2.325)$$

TABLE 2.5 Beam–Column Fixed-End Moments (Chen and Lui 1991) $(u = kL/2 = (L/2)\sqrt{P/EI})$

Case	Fixed-end moments

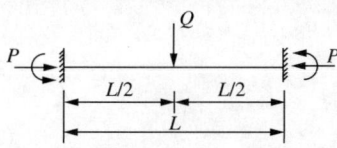

$$M_{FA} = \frac{QL}{8}\left[\frac{2(1 - \cos u)}{u\sin u}\right]$$

$$M_{FB} = -M_{FA}$$

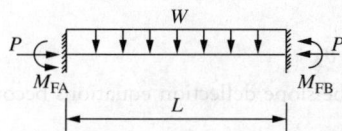

$$M_{FA} = \frac{WL^2}{12}\left[\frac{3(\tan u - u)}{u^2\tan u}\right]$$

$$M_{FB} = -M_{FA}$$

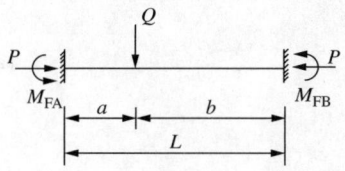

$$M_{FA} = \frac{QL}{d}\left[\frac{2ub}{L}\cos 2u - 2u\cos\frac{2ub}{L} - \sin 2u\right.$$

$$\left. + \sin\frac{2ua}{L} + \sin\frac{2ub}{L} + \frac{2ua}{L}\right]$$

$$M_{FB} = -\frac{QL}{d}\left[\frac{2ua}{L}\cos 2u - 2u\cos\frac{2ua}{L} - \sin 2u\right.$$

$$\left. + \sin\frac{2ub}{L} + \sin\frac{2ua}{L} + \frac{2ub}{L}\right]$$

where

$$d = 2u(2 - 2\cos 2u - 2u\sin 2u)$$

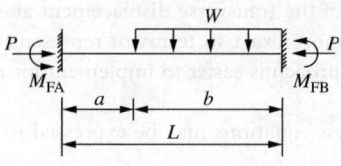

$$M_{FA} = \frac{wL^2}{8u^2 e}\left[(2u\csc 2u - 1)\left(\frac{2ub}{L} - \sin\frac{2ub}{L}\right)\right.$$

$$\left. + (\tan u)\left(1 - \cos\frac{2ub}{L} - \frac{2u^2b^2}{L^2}\right)\right]$$

$$M_{FB} = \frac{-wL^2}{8u^2 e}\left[(2u\cot 2u - 1)\left(\frac{2ub}{L} - \sin\frac{2ub}{L}\right)\right.$$

$$+ (\tan u)\left(1 - \cos\frac{2ub}{L} + \frac{2u^2b^2}{L^2}\right)$$

$$\left. - 2u\left(1 - \cos\frac{2ub}{L}\right)\right]$$

where

$$e = \tan u - u$$

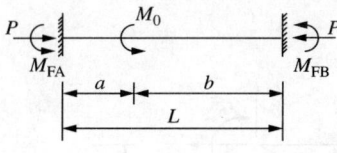

$$M_{FA} = \frac{M_0}{2e}\left[(2u\csc 2u - 1)\sin\frac{2ub}{L}\right.$$

$$\left. - (\tan u)\left(1 - \cos\frac{2ub}{L}\right)\right]$$

$$M_{FB} = \frac{M}{2e}\left[(1 - 2u\cot 2u)\sin\frac{2ub}{L}\right.$$

$$- (\tan u)\left(1 + \cos\frac{2ub}{L}\right)$$

$$\left. + 2u\cos\frac{2ub}{L}\right]$$

where

$$e = \tan u - u$$

2.12.4.6 Member Bent in Single Curvature with $\theta_B = -\theta_A$

For the member bent in single curvature in which $\theta_B = -\theta_A$, the slope deflection equations reduce to

$$M_A = \frac{EI}{L}(s_{ii} - s_{ij})\theta_A \qquad (2.326)$$

$$M_B = -M_A \qquad (2.327)$$

2.12.4.7 Member Bent in Double Curvature with $\theta_B = \theta_A$

For the member bent in double curvature such that $\theta_B = \theta_A$, the slope deflection equations become

$$M_A = \frac{EI}{L}(s_{ii} - s_{ij})\theta_A \qquad (2.328)$$

$$M_B = M_A \qquad (2.329)$$

2.12.5 Second-Order Elastic Analysis

There are two methods for incorporating second-order effects, the stability function approach and the geometric stiffness (or finite element) approach. The stability function approach is based on the governing differential equations of the problem as described in Section 2.12.4, whereas the stiffness approach is based on an assumed cubic polynomial variation of the transverse displacement along the element length. Therefore, the stability function approach is more exact in terms of representing the member stability behavior. However, the geometric stiffness approach is easier to implement for matrix analysis.

For either of these approaches, the linearized element stiffness equations may be expressed in either incremental or total force and displacement forms as

$$[K]\{d\} + \{r_f\} = \{r\} \qquad (2.330)$$

where $[K]$ is the element stiffness matrix, $\{d\} = \{d_1, d_2, \ldots, d_6\}^T$ is the element nodal-displacement vector, $\{r_f\} = \{r_{f1}, r_{f2}, \ldots, r_{f6}\}^T$ is the element fixed-end force vector due to the presence of in-span loading, and $\{r\} = \{r_1, r_2, \ldots, r_6\}^T$ is the nodal force vector as shown in Figure 1.121. If the stability

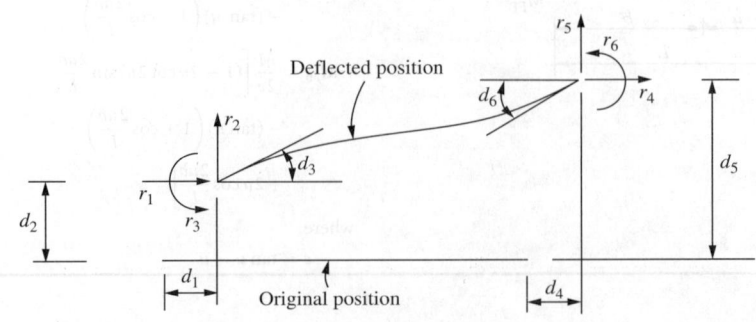

FIGURE 2.121 Nodal displacements and forces of a beam–column element.

function approach is employed, the stiffness matrix of a two-dimensional beam–column element may be written as

$$[K] = \frac{EI}{L} \begin{bmatrix} \dfrac{A}{I} & 0 & 0 & -\dfrac{A}{I} & 0 & 0 \\[2mm] & \dfrac{2(S_{ii}+S_{ij})-(kL)^2}{L^2} & \dfrac{S_{ii}+S_{ij}}{L} & 0 & \dfrac{-2(S_{ii}+S_{ij})+(kL)^2}{L^2} & \dfrac{S_{ii}+S_{ij}}{L} \\[2mm] & & S_{ii} & 0 & \dfrac{-(S_{ii}+S_{ij})}{L} & S_{ij} \\[2mm] & & & \dfrac{A}{I} & 0 & 0 \\[2mm] & & & & \dfrac{2(S_{ii}+S_{ij})-(kL)^2}{L^2} & \dfrac{-(S_{ii}+S_{ij})}{L} \\[2mm] & \text{symmetry} & & & & S_{ii} \end{bmatrix} \tag{2.331}$$

where S_{ii} and S_{ij} are the member stiffness coefficients obtained from the elastic beam–column stability functions (Chen and Lui 1987). These coefficients may be expressed as

$$S_{ii} = \begin{cases} \dfrac{kL\,\sin(kL)-(kL)^2\cos(kL)}{2-2\cos(kL)-kL\,\sin(kL)} & \text{for } P < 0 \\[4mm] \dfrac{(kL)^2\cosh(kL)-kL\,\sinh(kL)}{2-2\cosh(kL)+kL\,\sinh(kL)} & \text{for } P > 0 \end{cases} \tag{2.332}$$

$$S_{ij} = \begin{cases} \dfrac{(kL)^2-kL\,\sin(kL)}{2-2\cos(kL)-\rho\,\sin(kL)} & \text{for } P < 0 \\[4mm] \dfrac{kL\,\sinh(kL)-(kL)^2}{2-2\cosh(kL)+\rho\,\sinh(kL)} & \text{for } P > 0 \end{cases} \tag{2.333}$$

where $kL = L\sqrt{P/EI}$ and P is positive in compression and negative in tension.

The fixed-end force vector r_f is a 6×1 matrix that can be computed from the in-span loading in the beam–column. If curvature shortening is ignored, $r_{f1} = r_{f4} = 0$, $r_{f3} = M_{FA}$, and $r_{f6} = M_{FB}$. M_{FA} and M_{FB} can be obtained from Table 2.5 for different in-span loading conditions. r_{f2} and r_{f5} can be obtained from the equilibrium of forces.

If the axial force in the member is small, Equation 2.331 can be simplified by ignoring the higher-order terms of the power series expansion of the trigonometric functions. The resulting element stiffness matrix becomes

$$[K] = \frac{EI}{L} \begin{bmatrix} \dfrac{A}{I} & 0 & 0 & \dfrac{-A}{I} & 0 & 0 \\[2mm] & \dfrac{12}{L^2} & \dfrac{6}{L} & 0 & \dfrac{-12^2}{L} & \dfrac{6}{L} \\[2mm] & & 4 & 0 & \dfrac{-6}{L} & 2 \\[2mm] & & & \dfrac{A}{I} & 0 & 0 \\[2mm] & & & & \dfrac{12}{L^2} & \dfrac{-6}{L} \\[2mm] & \text{symmetry} & & & & 4 \end{bmatrix} + P \begin{bmatrix} 0 & 0 & 0 & 0 & 0 & 0 \\[2mm] & \dfrac{6}{5L} & \dfrac{1}{10} & 0 & \dfrac{-6}{5L} & \dfrac{1}{10} \\[2mm] & & \dfrac{2L}{15} & 0 & \dfrac{-1}{10} & \dfrac{-L}{30} \\[2mm] & & & 0 & 0 & 0 \\[2mm] & & & & \dfrac{6}{5L} & \dfrac{-1}{10} \\[2mm] & \text{symmetry} & & & & \dfrac{2L}{15} \end{bmatrix}$$

$$\tag{2.334}$$

The first term on the right is the first-order elastic stiffness matrix and the second term is the geometric stiffness matrix, which accounts for the effect of axial force on the bending stiffness of the member.

Detailed discussions on the limitation of the geometric stiffness approach versus the stability function approach are given in Liew et al. (2000).

2.12.6 Modifications to Account for Plastic Hinge Effects

There are two commonly used approaches for representing plastic hinge behavior in a second-order elastic–plastic hinge formulation (Chen et al. 1996). The most basic approach is to model the plastic hinge behavior as "real hinge" for the purpose of calculating the element stiffness. The change in moment capacity due to the change in axial force can be accommodated directly in the numerical formulation. The change in moment is determined in the force recovery at each solution step such that, for continued plastic loading, the new force point is positioned at the strength surface at the current value of axial force. A detailed description of these procedures is given by Lee and Basu (1989), Chen and Lui (1991), and Chen et al. (1996), among others.

Alternatively, the elastic–plastic hinge model may be formulated based on the "extending and contracting" plastic hinge model. The plastic hinge can rotate and extend and contract for plastic loading and axial force. The formulation can follow the force–space plasticity concept using the normality flow rule relative to the cross-section surface strength (Chen and Han 1988). Formal derivations of the beam–column element based on this approach have been presented by Porter and Powell (1971), Orbison (1982), and Liew et al. (2000), among others.

2.12.7 Modification for End Connections

The moment–rotation relationship of the beam–column with end connections at both ends can be expressed as Equations 2.319 and 2.320

$$M_A = \frac{EI}{L}\left[s_{ii}^* \theta_A + s_{ij}^* \theta_B \right] \tag{2.335}$$

$$M_B = \frac{EI}{L}\left[s_{ij}^* \theta_A + s_{jj}^* \theta_B \right] \tag{2.336}$$

where

$$S_{ii}^* = \frac{S_{ii} + \dfrac{EIS_{ii}^2}{LR_{kB}} - \dfrac{EIS_{ij}^2}{LR_{kB}}}{\left[1 + \dfrac{EIS_{ii}}{LR_{kA}}\right]\left[1 + \dfrac{EIS_{jj}}{LR_{kB}}\right] - \left[\dfrac{EI}{L}\right]^2 \dfrac{S_{ij}^2}{R_{kA}R_{kB}}} \tag{2.337}$$

$$S_{jj}^* = \frac{S_{ii} + \dfrac{EIS_{ii}^2}{LR_{kA}} - \dfrac{EIS_{ij}^2}{LR_{kA}}}{\left[1 + \dfrac{EIS_{ii}}{LR_{kA}}\right]\left[1 + \dfrac{EIS_{jj}}{LR_{kB}}\right] - \left[\dfrac{EI}{L}\right]^2 \dfrac{S_{ij}^2}{R_{kA}R_{kB}}} \tag{2.338}$$

and

$$S_{ij}^* = \frac{S_{ij}}{\left[1 + \dfrac{EIS_{ii}}{LR_{kA}}\right]\left[1 + \dfrac{EIS_{jj}}{LR_{kB}}\right] - \left[\dfrac{EI}{L}\right]^2 \dfrac{S_{ij}^2}{R_{kA}R_{kB}}} \tag{2.339}$$

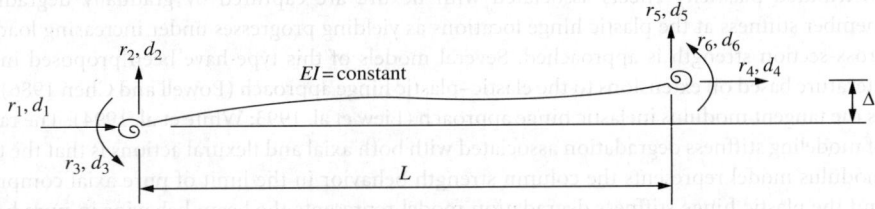

FIGURE 2.122 Nodal displacements and forces of a beam–column with end connections.

The member stiffness relationship can be written in terms of the six degrees of freedom of the beam–column element shown in Figure 2.122 as

$$
\begin{pmatrix} r_1 \\ r_2 \\ r_3 \\ r_4 \\ r_5 \\ r_6 \end{pmatrix} = \frac{EI}{L}
\begin{bmatrix}
\dfrac{A}{I} & 0 & 0 & \dfrac{-A}{I} & 0 & 0 \\[2mm]
 & \dfrac{S_{ii}^{*}+2S_{ij}^{*}+S_{jj}^{*}-(kL)^{2}}{L^{2}} & \dfrac{S_{ii}^{*}+S_{ij}^{*}}{L} & 0 & \dfrac{-\left(S_{ii}^{*}+2S_{ij}^{*}+S_{jj}^{*}\right)+(kL)^{2}}{L^{2}} & \dfrac{S_{ij}^{*}+S_{jj}^{*}}{L} \\[2mm]
 & & S_{ii}^{*} & 0 & \dfrac{-\left(S_{ii}^{*}+S_{ij}^{*}\right)}{L} & S_{ij}^{*} \\[2mm]
 & & & \dfrac{A}{I} & 0 & 0 \\[2mm]
 & & & & \dfrac{S_{ii}^{*}+2S_{ij}^{*}+S_{jj}^{*}-(kL)^{2}}{L^{2}} & \dfrac{-\left(S_{ij}^{*}+S_{jj}^{*}\right)}{L} \\[2mm]
\text{symmetry} & & & & & S_{jj}^{*}
\end{bmatrix}
\begin{pmatrix} d_1 \\ d_2 \\ d_3 \\ d_4 \\ d_5 \\ d_6 \end{pmatrix}
$$

(2.340)

2.12.8 Second-Order Refined Plastic Hinge Analysis

The main limitation of the conventional elastic–plastic hinge approach is that it overpredicts the strength of columns that fail by inelastic flexural buckling. The key reason for this limitation is the modeling of the member by a perfect elastic element between the plastic hinge locations. Furthermore, the elastic–plastic hinge model assumes that material behavior changes abruptly from the elastic state to the fully yielded state. The element under consideration exhibits a sudden stiffness reduction upon formation of a plastic hinge. This approach, therefore, overestimates the stiffness of a member loaded into the inelastic range (White et al. 1991, 1993; Liew et al. 1993). This leads to further research and development of an alternative method called the *refined plastic hinge approach*. This approach is based on the following improvements to the elastic–plastic hinge model:

1. A column tangent modulus model E_t is used in place of the elastic modulus E to represent the distributed plasticity due to axial force effects along the length of a member. The member inelastic stiffness, represented by the member axial and bending rigidities $E_t A$ and $E_t I$, is assumed to be the function of axial load only. In other words, $E_t A$ and $E_t I$ can be thought of as the properties of an effective core of the section, considering column action only. The tangent modulus captures the effect of early yielding in the cross-section due to residual stresses, which was believed to be the cause for the low strength of inelastic column buckling. The tangent modulus approach also has been utilized in previous work by Orbison (1982), Liew (1992), and White et al. (1993) to improve the accuracy of the elastic–plastic hinge approach for structures in which members are subjected to large axial forces.

2. Distributed plasticity effects associated with flexure are captured by gradually degrading the member stiffness at the plastic hinge locations as yielding progresses under increasing load as the cross-section strength is approached. Several models of this type have been proposed in recent literature based on extensions to the elastic–plastic hinge approach (Powell and Chen 1986) as well as the tangent modulus inelastic hinge approach (Liew et al. 1993; White et al. 1994). The rationale of modeling stiffness degradation associated with both axial and flexural actions is that the tangent modulus model represents the column strength behavior in the limit of pure axial compression, and the plastic hinge stiffness degradation model represents the beam behavior in pure bending, thus the combined effects of these two approaches should also satisfy the cases in which the member is subjected to combined axial compression and bending.

It has been shown that with the above two improvements, the refined plastic hinge model can be used with sufficient accuracy to provide a quantitative assessment of a member's performance up to failure. Detailed descriptions of the method and discussion of results generated by the method are given in White et al. (1993) and Chen et al. (1996). Significant works have been done to implement the refined plastic hinge methods for the design of three-dimensional real-size structures (Al-Bermani 1995; Liew et al. 2000).

2.12.9 Second-Order Spread of Plasticity Analysis

Spread of plasticity analyses can be classified into two main types, namely three-dimensional shell element and two-dimensional beam–column approaches. In the three-dimensional spread of plasticity analysis, the structure is modeled using a large number of finite three-dimensional shell elements, and the elastic constitutive matrix, in the usual incremental stress–strain relations, is replaced by an elastic–plastic constitutive matrix once yielding is detected. This analysis approach typically requires numerical integration for the evaluation of the stiffness matrix. Based on a deformation theory of plasticity, the combined effects of normal and shear stresses may be accounted for. The three-dimensional spread-of-plasticity analysis is computational intensive and best suited for analyzing small-scale structures.

The second approach for plastic-zone analysis is based on the use of beam–column theory, in which the member is discretized into many beam–column segments, and the cross-section of each segment is further subdivided into a number of fibers. Inelasticity is typically modeled by the consideration of normal stress only. When the computed stresses at the centroid of any fibers reach the uniaxial normal strength of the material, the fiber is considered as yielded. Compatibility is treated by assuming that full continuity is retained throughout the volume of the structure in the same manner as for elastic range calculations. Most of the plastic-zone analysis methods developed are meant for planar (2-D) analysis (Vogel 1985; White 1985; Chen and Toma 1994) Three-dimensional plastic-zone techniques are also available involving various degrees of refinements (Wang 1988; White 1988).

A plastic-zone analysis, which includes the spread of plasticity, residual stresses, initial geometric imperfections, and any other significant second-order behavioral effects, is often considered to be an exact analysis method. Therefore, when this type of analysis is employed, the checking of member interaction equations is not required. However, in reality, some significant behavioral effects such as the performances of joints and connections tend to defy precise numerical and analytical modeling. In such cases, a simpler method of analysis that adequately captures the inelastic behavior would be sufficient for engineering application. Second-order plastic hinge based analysis is still the preferred method for advanced analysis of large-scale steel frames.

2.12.10 Three-Dimensional Frame Element

The two-dimensional beam–column formulation can be extended to a three-dimensional space frame element by including additional terms due to shear force, bending moment, and torsion. The following

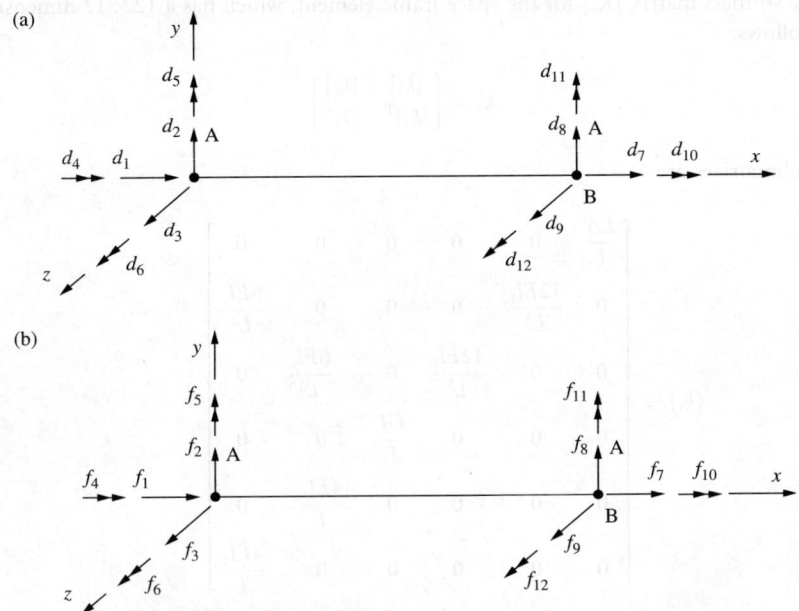

FIGURE 2.123 Three-dimensional frame element: (a) nodal degrees of freedom and (b) nodal forces.

stiffness equation for a space frame element has been derived by Yang and Kuo (1994) by referring to Figure 2.123:

$$[k_e]\{d\} + [k_g]\{d\} = \{{}^2f\} - \{{}^1f\} \tag{2.341}$$

where

$$\{d\}^T = \{d_1, d_2, \ldots, d_{12}\} \tag{2.342}$$

is the displacement vector, which consists of three translations and three rotations at each node and

$$\{{}^i f\}^T = \{{}^i f_1, {}^i f_2, \ldots, {}^i f_{12}\} \quad i = 1, 2 \tag{2.343}$$

are the force vectors, which consist of the corresponding nodal forces at configurations $i = 1$ and $i = 2$, respectively.

The physical interpretation of Equation 2.341 is as follows. By increasing the nodal forces acting on the element from $\{{}^1f\}$ to $\{{}^2f\}$, further deformations $\{d\}$ may occur with the element, resulting in the motion of the element from the configuration associated with the forces $\{{}^1f\}$ to the new configuration associated with $\{{}^2f\}$. During this process of deformation, the increments in the nodal forces, that is, $\{{}^2f\} - \{{}^1f\}$, will be resisted not only by the elastic actions generated by the elastic stiffness matrix $[k_e]$ but also by the forces induced by the change in geometry as represented by the geometric stiffness matrix $[k_g]$.

The only assumption with the incremental stiffness equation is that the strains occurring with each incremental step should be small so that the approximations implied by the incremental constitutive law are not violated.

The elastic stiffness matrix $[K_e]$ for the space frame element, which has a 12×12 dimension, can be derived as follows:

$$[k] = \begin{bmatrix} [k_1] & [k_2] \\ [k_2]^{\mathrm{T}} & [k_3] \end{bmatrix} \quad (2.344)$$

where the submatrices are

$$[k_1] = \begin{bmatrix} \dfrac{EA}{L} & 0 & 0 & 0 & 0 & 0 \\[2mm] 0 & \dfrac{12EI_z}{L^3} & 0 & 0 & 0 & \dfrac{6EI_z}{L^2} \\[2mm] 0 & 0 & \dfrac{12EI_y}{L^3} & 0 & -\dfrac{6EI_y}{L^2} & 0 \\[2mm] 0 & 0 & 0 & \dfrac{GJ}{L} & 0 & 0 \\[2mm] 0 & 0 & 0 & 0 & \dfrac{4EI_y}{L} & 0 \\[2mm] 0 & 0 & 0 & 0 & 0 & \dfrac{4EI_z}{L} \end{bmatrix} \quad (2.345)$$

$$[k_2] = \begin{bmatrix} -\dfrac{EA}{L} & 0 & 0 & 0 & 0 & 0 \\[2mm] 0 & -\dfrac{12EI_z}{L^3} & 0 & 0 & 0 & \dfrac{6EI_z}{L^2} \\[2mm] 0 & 0 & -\dfrac{12EI_y}{L^3} & 0 & -\dfrac{6EI_y}{L^2} & 0 \\[2mm] 0 & 0 & 0 & -\dfrac{GJ}{L} & 0 & 0 \\[2mm] 0 & 0 & \dfrac{6EI_y}{L^2} & 0 & \dfrac{2EI_y}{L} & 0 \\[2mm] 0 & -\dfrac{6EI_z}{L^2} & 0 & 0 & 0 & \dfrac{2EI_z}{L} \end{bmatrix} \quad (2.346)$$

$$[k_3] = \begin{bmatrix} \dfrac{EA}{L} & 0 & 0 & 0 & 0 & 0 \\[2mm] 0 & \dfrac{12EI_z}{L^3} & 0 & 0 & 0 & -\dfrac{6EI_z}{L^2} \\[2mm] 0 & 0 & \dfrac{12EI_y}{L^3} & 0 & \dfrac{6EI_y}{L^2} & 0 \\[2mm] 0 & 0 & 0 & \dfrac{GJ}{L} & 0 & 0 \\[2mm] 0 & 0 & 0 & 0 & \dfrac{4EI_y}{L} & 0 \\[2mm] 0 & 0 & 0 & 0 & 0 & \dfrac{4EI_z}{L} \end{bmatrix} \quad (2.347)$$

where I_x, I_y, and I_z are the moments of inertia about the x, y, and z axes, respectively, L is the member length, E is the modulus of elasticity, A is the cross-sectional area, G is the shear modulus, and J is the torsional stiffness.

The geometric stiffness matrix for a three-dimensional space frame element can be given as follows:

$$[k_g] = \begin{bmatrix} a & 0 & 0 & 0 & -d & -e & -a & 0 & 0 & 0 & -n & -o \\ & b & 0 & d & g & k & 0 & -b & 0 & n & -g & k \\ & & c & e & h & g & 0 & 0 & -c & o & -h & -g \\ & & & f & i & l & 0 & -d & -e & -f & -i & -l \\ & & & & j & 0 & d & -g & h & -i & p & -q \\ & & & & & m & e & -k & -g & -l & q & r \\ & & & & & & a & 0 & 0 & 0 & n & o \\ & & & & & & & b & 0 & -n & g & -k \\ & & & & & & & & c & -o & h & g \\ & \text{sym}^* & & & & & & & & f & i & l \\ & & & & & & & & & & j & o \\ & & & & & & & & & & & m \end{bmatrix} \quad (2.348)$$

where

$$a = -\frac{f_6 + f_{12}}{L^2}, \quad b = \frac{6f_7}{5L}, \quad c = -\frac{f_5 + f_{11}}{L^2}, \quad d = \frac{f_5}{L}, \quad e = \frac{f_6}{L}, \quad f = \frac{f_7 J}{AL}$$

$$g = \frac{f_{10}}{L}, \quad h = -\frac{f_7}{10}, \quad i = \frac{f_6 + f_{12}}{6}, \quad j = \frac{2f_7 L}{15}, \quad k = -\frac{f_5 + f_{11}}{6}, \quad l = \frac{f_{11}}{L}$$

$$m = \frac{f_{12}}{L}, \quad n = -\frac{f_7 L}{30}, \quad o = -\frac{f_{10}}{2}$$

Further details can be obtained from Yang and Kuo (1994).

2.12.11 Buckling of Thin Plates

2.12.11.1 Rectangular Plates

The main difference between columns and plates is that quantities such as deflections and bending moments that are functions of a single independent variable in columns become functions of two independent variables in plates. Consequently, the behavior of plates is described by partial differential equations, whereas ordinary differential equations suffice for describing the behavior of columns. The main difference between column and plate buckling is that column buckling terminates the ability of the member to resist axial load; the same is, however, not true for plates. Upon reaching the critical load, the plate continues to resist increasing axial force, and it does not fail until a load considerably in excess of the elastic buckling load is reached. The critical load of a plate is, therefore, not its failure load. Instead, one must determine the load-carrying capacity of a plate by considering its postbuckling strength.

To determine the critical in-plane loading of a plate, a governing equation in terms of biaxial compressive forces N_x and N_y and constant shear force N_{xy} as shown in Figure 2.124 can be derived as

$$D\left(\frac{\delta^4 w}{\delta x^4} + 2\frac{\delta^4 w}{\delta x^2 \delta y^2} + \frac{\delta^4 w}{\delta y^4}\right) + N_x \frac{\delta^2 w}{\delta x^2} + N_y \frac{\delta^2 w}{\delta y^2} + 2N_{xy}\frac{\delta^2 w}{\delta x \delta y} = 0 \quad (2.349)$$

The critical load for uniaxial compression can be determined from the differential equation

$$D\left(\frac{\delta^4 w}{\delta x^4} + 2\frac{\delta^4 w}{\delta x^2 \delta y^2} + \frac{\delta^4 w}{\delta y^4}\right) + N_x \frac{\delta^2 w}{\delta x^2} = 0 \quad (2.350)$$

---

*The word "sym" in the matrix $[k_g]$ implies symmetrical matrix in future occurrences.

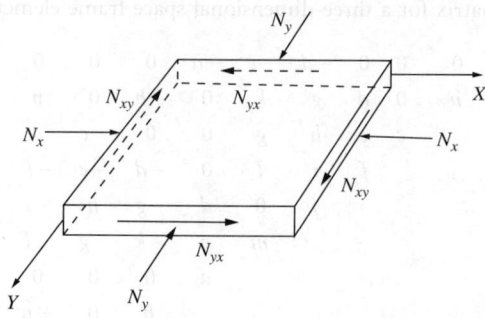

FIGURE 2.124 Plate subjected to in-plane forces.

which is obtained by setting $N_y = N_{xy} = 0$ in Equation 2.349. For example, in the case of a simply supported plate Equation 2.350 can be solved to give

$$N_x = \frac{\pi^2 a^2 D}{m^2} \left(\frac{m^2}{a^2} + \frac{n^2}{b^2} \right)^2 \qquad (2.351)$$

The critical value of N_x (i.e., the smallest value) can be obtained by taking n equal to 1. The physical meaning of this is that a plate buckles in such a way that there can be several half-waves in the direction of compression but only one half-wave in the perpendicular direction. Thus, the expression for the critical value of the compressive force becomes

$$(N_x)_{cr} = \frac{\pi^2 D}{a^2} \left(m + \frac{1}{m} \frac{a^2}{b^2} \right)^2 \qquad (2.352)$$

The first factor in this expression represents the Euler load for a strip of unit width and of length *a*. The second factor indicates in what proportion the stability of the continuous plate is greater than the stability of an isolated strip. The magnitude of this factor depends on the magnitude of the ratio *a/b* and also on *m*, which is the number of half-waves into which the plate buckles. If *a* is smaller than *b*, the second term in the parantheses in Equation 2.347 is always smaller than the first and the minimum value of the expression is obtained by taking $m = 1$, that is, by assuming that the plate buckles in one half-wave. The critical value of N_x can be expressed as

$$N_{cr} = \frac{k\pi^2 D}{b^2} \qquad (2.353)$$

The factor *k* depends on the aspect ratio *a/b* of the plate and *m*. The variation of *k* with *a/b* for different values of *m* can be plotted as shown in Figure 2.125. The critical value of N_x is the smallest value that is obtained for $m = 1$ and the corresponding value of *k* is equal to 4.0. This formula is analogous to Euler's formula for buckling of a column.

In the case where the normal forces N_x and N_y and the shearing forces N_{xy} are acting on the boundary of the plate, the same general method can be used. The critical stress for the case of a uniaxially compressed simply supported plate can be written as

$$\sigma_{cr} = 4 \frac{\pi^2 E}{12(1 - v^2)} \left(\frac{h}{b} \right)^2 \qquad (2.354)$$

The critical stress values for different loading and support conditions can be expressed in the form

$$f_{cr} = k \frac{\pi^2 E}{12(1 - v^2)} \left(\frac{h}{b} \right)^2 \qquad (2.355)$$

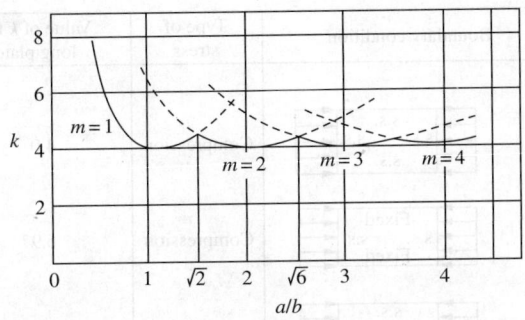

FIGURE 2.125 Buckling stress coefficients for uniaxially compressed plate.

Values of k for plates with several different boundary and loading conditions are given in Figure 2.126.

2.12.11.2 Circular Plates

The critical value of the compressive forces N_r uniformly distributed around the edge of a circular plate of radius r_0 clamped along the edge (Figure 2.127) can be determined by

$$r^2 \frac{d^2\phi}{dr^2} + r \frac{d\phi}{dr} - \phi = -\frac{Qr^2}{D} \tag{2.356}$$

where ϕ is the angle between the axis of revolution of the plate surface and any normal to the plate, r is the distance of any point from the center of the plate, and Q is the shearing force per unit of length. When there are no lateral forces acting on the plate, the solution of Equation 2.356 involves Bessel function of the first order of the first and second kinds and the resulting critical value of N_r is obtained as

$$(N_r)_{cr} = \frac{14.68D}{r_0^2} \tag{2.357}$$

The critical value of N_r for the plate when the edge is simply supported can be obtained in the same way as

$$(N_r)_{cr} = \frac{4.20D}{r_0^2} \tag{2.358}$$

2.12.12 Buckling of Shells

If a circular cylindrical shell is uniformly compressed in the axial direction, buckling symmetrical with respect to the axis of the cylinder (Figure 2.128) may occur at a certain value of the compressive load. The critical value of the compressive force N_{cr} per unit length of the edge of the shell can be obtained by solving the differential equation

$$D \frac{d^4w}{dx^4} + N \frac{d^2w}{dx^2} + Eh \frac{w}{a^2} = 0 \tag{2.359}$$

where a is the radius of the cylinder and h is the wall thickness.

Alternatively, the critical force per unit length may also be obtained by using the energy method. For a cylinder of length L, simply supported at both ends, one obtains

$$N_{cr} = D \left(\frac{m^2 \pi^2}{L^2} + \frac{EhL^2}{Da^2 m^2 \pi^2} \right) \tag{2.360}$$

Case	Boundary condition	Type of stress	Value of k for long plate
(a)	s.s. / s.s. s.s. / s.s.	Compression	4.0
(b)	Fixed / s.s. s.s. / Fixed	Compression	6.97
(c)	s.s. / s.s. s.s. / Free	Compression	0.425
(d)	Fixed / s.s. s.s. / Free	Compression	1.277
(e)	Fixed / s.s. s.s. / s.s.	Compression	5.42
(f)	s.s. / s.s. s.s. / s.s.	Shear	5.34
(g)	Fixed / Fixed Fixed / Fixed	Shear	8.98
(h)	s.s. / s.s. s.s. / s.s.	Bending	23.9
(i)	Fixed / Fixed Fixed / Fixed	Bending	41.8

FIGURE 2.126   Values of K for plate with different boundary and loading conditions.

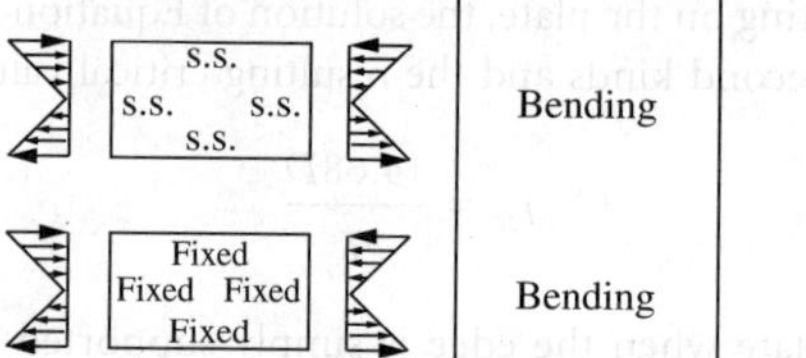

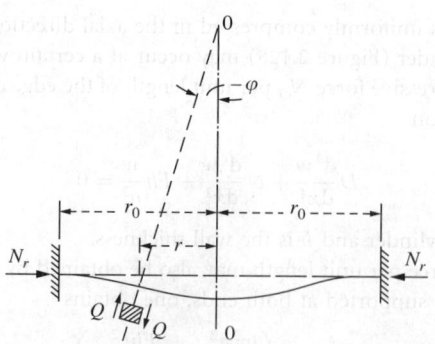

FIGURE 2.127   Circular plate under compressive loading.

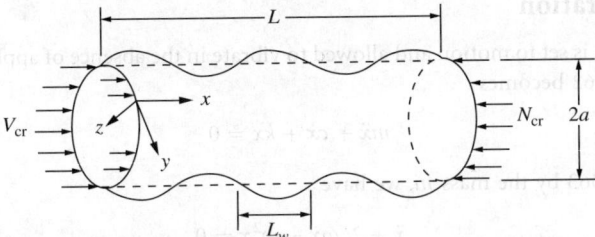

FIGURE 2.128 Buckling of a cylindrical shell.

For each value of m there is a unique buckling mode shape and a unique buckling load. The lowest value is of greatest interest and is thus found by setting the derivative of N_{cr} with respect to L equal to zero for $m = 1$. With Poisson's ratio $= 0.3$, the buckling load is obtained as

$$N_{cr} = 0.605 \frac{Eh^2}{a} \qquad (2.361)$$

It is possible for a cylindrical shell to be subjected to uniform external pressure or to the combined action of axial and uniform lateral pressure. More detailed treatment of such a case may be found in Timoshenko and Gere (1963).

2.13 Structural Dynamic

2.13.1 Equation of Motion

The essential physical properties of a linearly elastic structural system subjected to external dynamic loading are its mass, stiffness properties, and energy absorption capability or damping. The principle of dynamic analysis may be illustrated by considering a simple single-storey structure as shown in Figure 2.129. The structure is subjected to a time-varying force $f(t)$. k is the spring constant that relates the lateral storey deflection x to the storey shear force, and the dash pot relates the damping force to the velocity by a damping coefficient c. If the mass, m, is assumed to concentrate at the beam, the structure becomes a single-degree-of-freedom (SDOF) system. The equation of motion of the system may be written as

$$m\ddot{x} + c\dot{x} + kx = f(t) \qquad (2.362)$$

Various solutions to Equation 2.362 can give an insight into the behavior of the structure under dynamic situation.

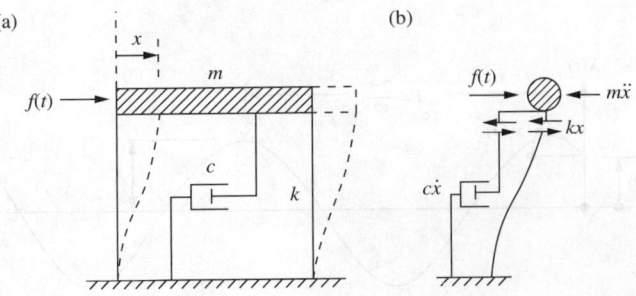

FIGURE 2.129 (a) One DOF structure and (b) forces applied to structures.

2.13.2 Free Vibration

In this case the system is set to motion and allowed to vibrate in the absence of applied force $f(t)$. Letting $f(t) = 0$, Equation 2.362 becomes

$$m\ddot{x} + c\dot{x} + kx = 0 \tag{2.363}$$

Dividing Equation 2.363 by the mass m, we have

$$\ddot{x} + 2\xi\omega\dot{x} + \omega^2 x = 0 \tag{2.364}$$

where

$$2\xi\omega = \frac{c}{m} \quad \text{and} \quad \omega^2 = \frac{k}{m} \tag{2.365}$$

The solution to Equation 2.364 depends on whether the vibration is damped or undamped.

2.13.2.1 Case 1: Undamped Free Vibration

In this case, $c = 0$, and the solution to the equation of motion may be written as

$$x = A\sin\omega t + B\cos\omega t \tag{2.366}$$

where $\omega = \sqrt{k/m}$ is the circular frequency. A and B are constants that can be determined by the initial boundary conditions. In the absence of external forces and damping the system will vibrate indefinitely in a repeated cycle of vibration with an amplitude of

$$X = \sqrt{A^2 + B^2} \tag{2.367}$$

and a natural frequency of

$$f = \frac{\omega}{2\pi} \tag{2.368}$$

The corresponding natural period is

$$T = \frac{2\pi}{\omega} = \frac{1}{f} \tag{2.369}$$

The undamped free vibration motion as described by Equation 2.366 is shown in Figure 2.130.

2.13.2.2 Case 2: Damped Free Vibration

If the system is not subjected to applied force and damping is presented, the corresponding solution becomes

$$x = A\exp(\lambda_1 t) + B\exp(\lambda_2 t) \tag{2.370}$$

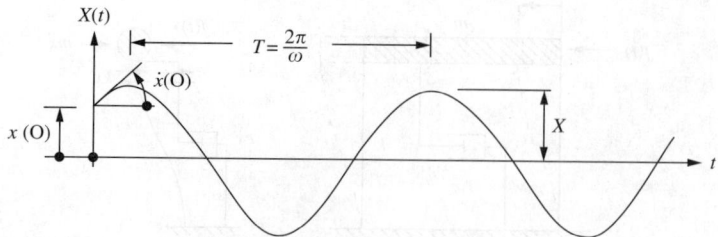

FIGURE 2.130 Response of undamped free vibration.

where

$$\lambda_1 = \omega\left[-\xi + \sqrt{\xi^2 - 1}\right] \qquad (2.371)$$

and

$$\lambda_2 = \omega\left[-\xi - \sqrt{\xi^2 - 1}\right] \qquad (2.372)$$

The solution of Equation 2.370 changes its form with the value of ξ defined as

$$\xi = \frac{c}{2\sqrt{mk}} \qquad (2.373)$$

If $\xi^2 < 1$, the equation of motion becomes

$$x = \exp(-\xi\omega t)(A\cos\omega_d t + B\sin\omega_d t) \qquad (2.374)$$

where ω_d is the damped angular frequency defined as

$$\omega_d = \sqrt{(1 - \xi^2)}\omega \qquad (2.375)$$

For most building structures ξ is very small (about 0.01) and therefore $\omega_d \approx \omega$. The system oscillates about the neutral position at the amplitude decays with time t. Figure 2.131 illustrates an example of such motion. The rate of decay is governed by the amount of damping present.

If the damping is great, then oscillation will be prevented. This happens when $\xi^2 > 1$ and the behavior is referred to as *overdamped*. The motion of such behavior is shown in Figure 2.132.

Damping with $\xi^2 = 1$ is called critical damping. This is the case where minimum damping is required to prevent oscillation and the critical damping coefficient is given as

$$c_{cr} = 2\sqrt{km} \qquad (2.376)$$

where k and m are the stiffness and the mass of the system respectively.

The degree of damping in the structure is often expressed as a proportion of the critical damping value. Referring to Equations 2.373 and 2.376, we have

$$\xi = \frac{c}{c_{cr}} \qquad (2.377)$$

where ξ is called the critical damping ratio.

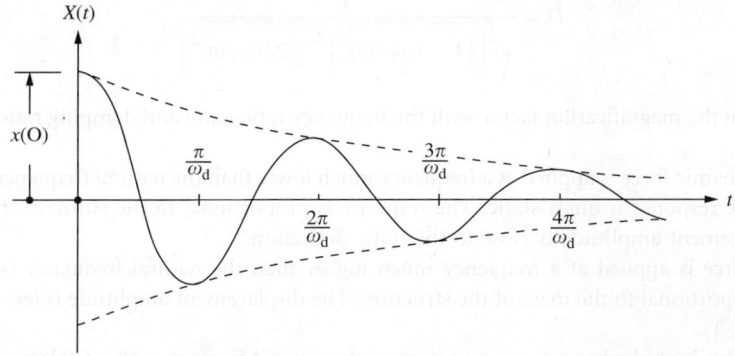

FIGURE 2.131   Response of damped free vibration.

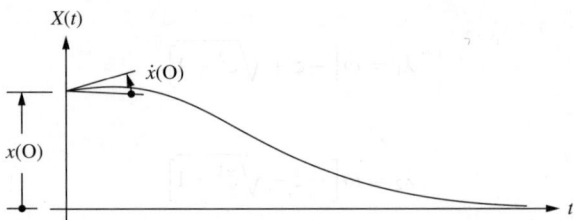

FIGURE 2.132   Response of free vibration with critical damping.

2.13.3 Forced Vibration

If a structure is subjected to a sinusoidal motion such as a ground acceleration of $\ddot{x} = F \sin \omega_f t$, it will oscillate and after some time the motion of the structure will reach a steady state. For example, the equation of motion due to the ground acceleration (from Equation 2.364) is

$$\ddot{x} + 2\xi\omega\dot{x} + \omega^2 x = -F \sin \omega_f t \tag{2.378}$$

The solution to the above equation consists of two parts: the complementary solution given by Equation 2.366 and the particular solution. If the system is damped, oscillation corresponding to the complementary solution will decay with time. After some time the motion will reach a steady state, and the system will vibrate at a constant amplitude and frequency. This motion, which is called force vibration, is described by the particular solution expressed as

$$x = C_1 \sin \omega_f t + C_2 \cos \omega_f t \tag{2.379}$$

It can be observed that the steady force vibration occurs at the frequency of the excited force, ω_f, not the natural frequency of the structure, ω.

Substituting Equation 2.379 into Equation 2.378, the displacement amplitude can be shown to be

$$X = -\frac{F}{\omega^2} \frac{1}{\sqrt{\left[\{1 - (\omega_f/\omega)^2\}^2 + (2\xi\omega_f/\omega)^2\right]}} \tag{2.380}$$

The term $-F/\omega^2$ is the static displacement caused by the force due to the inertia force. The ratio of the response amplitude relative to the static displacement $-F/\omega^2$ is called the dynamic displacement amplification factor, D, given as

$$D = \frac{1}{\sqrt{\left[\{1 - (\omega_f/\omega)^2\}^2 + (2\xi\omega_f/\omega)^2\right]}} \tag{2.381}$$

The variation of the magnification factor with the frequency ratio ω_f/ω and damping ratio ξ is shown in Figure 2.133.

When the dynamic force is applied at a frequency much lower than the natural frequency of the system ($\omega_f/\omega \ll 1$), the response is quasi-static. The response is proportional to the stiffness of the structure, and the displacement amplitude is close to the static deflection.

When the force is applied at a frequency much higher than the natural frequency ($\omega_f/\omega \gg 1$), the response is proportional to the mass of the structure. The displacement amplitude is less than the static deflection ($D < 1$).

When the force is applied at a frequency close to the natural frequency, the displacement amplitude increases significantly. The condition at which $\omega_f/\omega = 1$ is known as resonance.

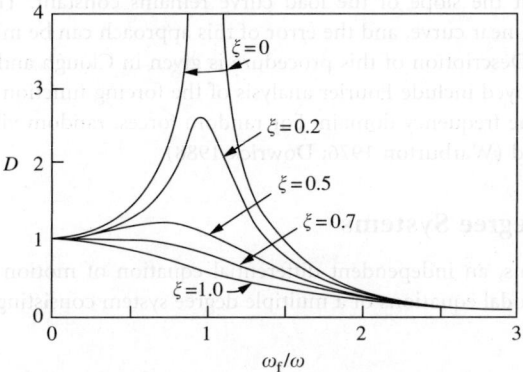

FIGURE 2.133 Vibration of dynamic amplification factor with frequency ratio.

Similarly, the ratio of the acceleration response relative to the ground acceleration may be expressed as

$$D_a = \left| \frac{\ddot{x} + \ddot{x}_g}{\ddot{x}_g} \right| = \sqrt{ \frac{1 + (2\xi \omega_f/\omega)^2}{\left[\{1 - (\omega_f/\omega)^2\}^2 + (2\xi \omega_f/\omega)^2 \right]} } \tag{2.382}$$

D_a is called the dynamic acceleration magnification factor.

2.13.4 Response to Suddenly Applied Load

Consider the spring-mass damper system where a load P_o is applied suddenly. The differential equation is given by

$$M\ddot{x} + c\dot{x} + kx = P_o \tag{2.383}$$

If the system is started at rest, the equation of motion is

$$x = \frac{P_o}{k} \left[1 - \exp(-\xi \omega t) \left\{ \cos \omega_d t + \frac{\xi \omega}{\omega_d} \sin \omega_d t \right\} \right] \tag{2.384}$$

If the system is undamped, then $\xi = 0$ and $\omega_d = \omega$, we have

$$x = \frac{P_o}{k} [1 - \cos \omega_d t] \tag{2.385}$$

The maximum displacement is $2(P_o/k)$ corresponding to $\cos \omega_d t = -1$. Since P_o/k is the maximum static displacement, the dynamic amplification factor is equal to 2. The presence of damping would naturally reduce the dynamic amplification factor and the force in the system.

2.13.5 Response to Time-Varying Loads

Some forces and ground motions that are encountered in practice are rather complex. In general, numerical analysis is required to predict the response of such effects, and the finite element method is one of the most common techniques to be employed in solving such problems.

The evaluation of responses due to time-varying loads can be carried out using the piecewise exact method. In using this method, the loading history is divided into small time intervals. Between these points, it is assumed that the slope of the load curve remains constant. The entire load history is represented by piecewise linear curve, and the error of this approach can be minimized by reducing the length of the time steps. Description of this procedure is given in Clough and Penzien (1993).

Other techniques employed include Fourier analysis of the forcing function followed by solution for Fourier components in the frequency domain. For random forces, random vibration theory and spectrum analysis may be used (Warburton 1976; Dowrick 1988).

2.13.6 Multiple Degree Systems

In multiple degree systems, an independent differential equation of motion can be written for each degree of freedom. The nodal equations of a multiple degree system consisting of n degrees of freedom may be written as

$$[m]\{\ddot{x}\} + [c]\{\dot{x}\} + [k]\{x\} = \{F(t)\} \tag{2.386}$$

where $[m]$ is a symmetrical $n \times n$ matrix of mass, $[c]$ is a symmetrical $n \times n$ matrix of damping coefficient, and $\{F(t)\}$ is the force vector, which is zero in the case of free vibration.

Consider a system under free vibration without damping. The general solution of Equation 2.386 is assumed in the form

$$
\begin{Bmatrix} x_1 \\ x_2 \\ \vdots \\ x_n \end{Bmatrix} =
\begin{bmatrix}
\cos(\omega t - \phi) & 0 & 0 & 0 \\
0 & \cos(\omega t - \phi) & 0 & 0 \\
\vdots & \vdots & \vdots & \vdots \\
0 & 0 & 0 & \cos(\omega t - \phi)
\end{bmatrix}
\begin{Bmatrix} C_1 \\ C_2 \\ \vdots \\ C_n \end{Bmatrix}
\tag{2.387}
$$

where angular frequency ω and phase angle ϕ are common to all values of x. In this assumed solution, ϕ and $C_1, C_2, \ldots, C_n$ are the constants to be determined from the initial boundary conditions of the motion and ω is a characteristic value (eigenvalue) of the system.

Substituting Equation 2.387 into Equation 2.386 yields

$$
\begin{bmatrix}
k_{11} - m_{11}\omega^2 & k_{12} - m_{12}\omega^2 & \cdots & k_{1n} - m_{1n}\omega^2 \\
k_{21} - m_{21}\omega^2 & k_{22} - m_{22}\omega^2 & \cdots & k_{2n} - m_{2n}\omega^2 \\
\vdots & \vdots & \vdots & \vdots \\
k_{n1} - m_{n1}\omega^2 & k_{n2} - m_{n2}\omega^2 & \cdots & k_{nn} - m_{nn}\omega^2
\end{bmatrix}
\begin{Bmatrix} C_1 \\ C_2 \\ \vdots \\ C_n \end{Bmatrix}
\cos(\omega t - \phi) =
\begin{Bmatrix} 0 \\ 0 \\ \vdots \\ 0 \end{Bmatrix}
\tag{2.388}
$$

or

$$[[k] - \omega^2[m]]\{C\} = \{0\} \tag{2.389}$$

where $[k]$ and $[m]$ are the $n \times n$ matrices, ω^2 and $\cos(\omega t - \phi)$ are scalars, and $\{C\}$ is the amplitude vector. For nontrivial solution, $\cos(\omega t - \phi) \neq 0$, thus solution to Equation 2.389 requires the determinant of $[[k] - \omega^2[m]] = 0$. The expansion of the determinant yields a polynomial of nth degree as a function of ω^2, the n roots of which are the *eigenvalues* $\omega_1, \omega_2, \ldots, \omega_n$.

If the eigenvalue ω for a normal mode is substituted in Equation 2.389, the amplitude vector $\{C\}$ for that mode can be obtained. $\{C_1\}, \{C_2\}, \{C_3\}, \ldots, \{C_n\}$ are therefore called the *eigenvectors*, the absolute values of which must be determined through initial boundary conditions. The resulting motion is a sum of n harmonic motions, each governed by the respective natural frequency ω, written as

$$\{x\} = \sum_{i=1}^{n} \{C_i\} \cos(\omega_i t - \phi_i) \tag{2.390}$$

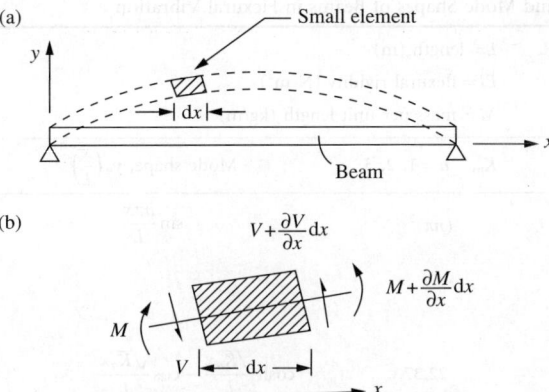

FIGURE 2.134 (a) Beam in flexural vibration and (b) equilibrium of beam segment in vibration.

2.13.7 Distributed Mass Systems

Although many structures may be approximated by lumped mass systems, in practice, all structures are distributed mass systems consisting of an infinite number of particles. Consequently, if the motion is repetitive, the structure has an infinite number of natural frequency and mode shapes. The analysis of a distributed-parameter system is entirely equivalent to that of a discrete system once the mode shapes and frequencies have been determined, because in both cases the amplitudes of the modal response components are used as generalized coordinates in defining the response of the structure.

In principle, an infinite number of these coordinates are available for a distributed-parameter system, but in practice, only a few modes, usually those of lower frequencies, will make a significant contribution to the overall response. Thus, the problem of a distributed-parameter system can be converted to a discrete system form in which only a limited number of modal coordinates are used to describe the response.

2.13.7.1 Flexural Vibration of Beams

The motion of the distributed mass system is best illustrated by a classical example of a uniform beam with of span length L and flexural rigidity EI and a self-weight of m per unit length, as shown in Figure 2.134a. The beam is free to vibrate under its self-weight. From Figure 2.134b, dynamic equilibrium of a small beam segment of length dx requires that

$$\frac{\partial V}{\partial x} dx = m \, dx \frac{\partial^2 y}{\partial t^2} \tag{2.391}$$

where

$$\frac{\partial^2 y}{\partial x^2} = \frac{M}{EI} \tag{2.392}$$

and

$$V = -\frac{\partial M}{\partial x}, \quad \frac{\partial V}{\partial x} = -\frac{\partial^2 M}{\partial x^2} \tag{2.393}$$

Substituting these equations into Equation 2.391 gives the equation of motion of the flexural beam:

$$\frac{\partial^4 y}{\partial x^4} + \frac{m}{EI}\frac{\partial^2 y}{\partial t^2} = 0 \tag{2.394}$$

Equation 2.394 can be solved for beams with given sets of boundary conditions. The solution consists of a family of vibration mode with corresponding natural frequencies. Standard results are available in

TABLE 2.6 Frequencies and Mode Shapes of Beams in Flexural Vibration

$$f_n = \frac{k_n}{2\pi}\sqrt{\frac{EI}{mL^4}}\ \text{Hz}$$

$n = 1, 2, 3, \ldots$

$L = $ length (m)

$EI = $ flexural rigidity (N m^2)

$M = $ mass per unit length (kg/m)

Boundary condition	K_n, $n = 1, 2, 3$	Mode shape, $y_n\left(\frac{x}{L}\right)$	A_n, $n = 1, 2, 3, \ldots$
Pinned–pinned	$(n\pi)^2$	$\sin\dfrac{n\pi x}{L}$	—
Fixed–fixed	22.37	$\cosh\dfrac{\sqrt{K_n}x}{L} - \cos\dfrac{\sqrt{K_n}x}{L}$	0.98250
	61.67		1.00078
	120.90	$-A_n\left(\sinh\dfrac{\sqrt{K_n}x}{L} - \sin\dfrac{\sqrt{K_n}x}{L}\right)$	0.99997
	199.86		1.00000
	298.55		0.99999
	$(2n+1)\dfrac{\pi^2}{4},$ $n > 5$		1.0, $n > 5$
Fixed–pinned	15.42	$\cosh\dfrac{\sqrt{K_n}x}{L} - \cos\dfrac{\sqrt{K_n}x}{L}$	1.00078
	49.96		1.00000
	104.25	$-A_n\left(\sinh\dfrac{\sqrt{K_n}x}{L} - \sin\dfrac{\sqrt{K_n}x}{L}\right)$	1.0, $n > 3$
	178.27		
	272.03		
	$(4n+1)^2\dfrac{\pi^2}{4},$ $n > 5$		
Cantilever	3.52	$\cosh\dfrac{\sqrt{K_n}x}{L} - \cos\dfrac{\sqrt{K_n}x}{L}$	0.73410
	22.03		1.01847
	61.69	$-A_n\left(\sinh\dfrac{\sqrt{K_n}x}{L} - \sin\dfrac{\sqrt{K_n}x}{L}\right)$	0.99922
	120.90		1.00003
	199.86		1.0, $n > 4$
	$(2n-1)^2\dfrac{\pi^2}{4},$ $n > 5$		

Table 2.6 to compute the natural frequencies of uniform flexural beams with different supporting conditions. Methods are also available for dynamic analysis of continuous beams (Clough and Penzien 1993).

2.13.7.2 Shear Vibration of Beams

Beams can deform by flexure or shear. Flexural deformation normally dominates the deformation of slender beams. Shear deformation is important for short beams or in higher modes of slender beams. Table 2.7 gives the natural frequencies of uniform beams in shear, neglecting flexural deformation. The natural frequencies of these beams are inversely proportional to the beam length L rather than L^2, and the frequencies increase linearly with the mode number.

TABLE 2.7 Frequencies and Mode Shapes of Beams in Shear Vibration

$$f_n = \frac{K_n}{2\pi} \sqrt{\frac{KG}{\rho L^2}} \text{ Hz}$$

L = length
K = shear coefficient (Cowper 1966)
G = shear modulus = $E/[2(1+v)]$
ρ = mass density

Boundary condition	K_n, $n = 1, 2, 3, \ldots$	Mode shape, $y_n\left(\frac{x}{L}\right)$

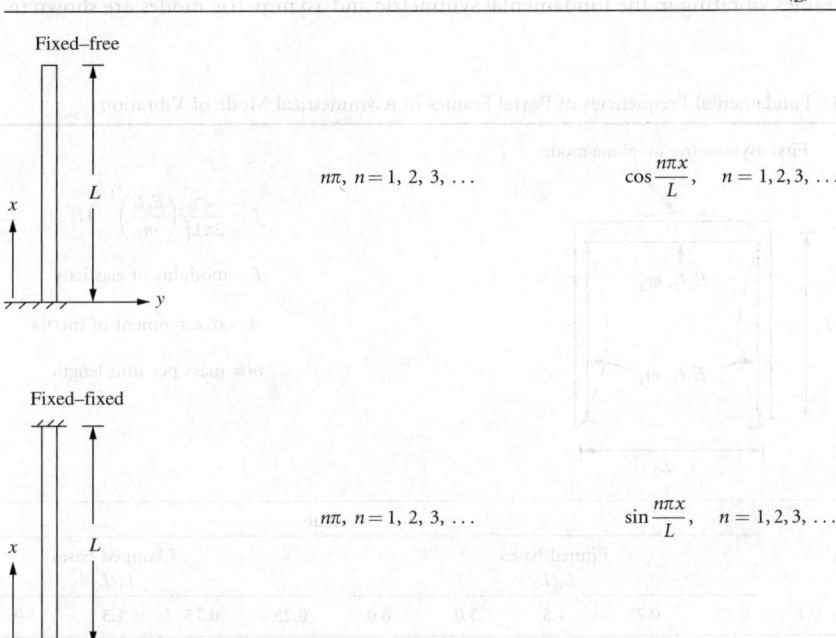

Fixed–free

$n\pi$, $n = 1, 2, 3, \ldots$ $\cos\dfrac{n\pi x}{L}$, $n = 1, 2, 3, \ldots$

Fixed–fixed

$n\pi$, $n = 1, 2, 3, \ldots$ $\sin\dfrac{n\pi x}{L}$, $n = 1, 2, 3, \ldots$

2.13.7.3 Combined Shear and Flexure

The transverse deformation of real beams is the sum of flexure and shear deformations. In general, numerical solutions are required to incorporate both the shear and flexural deformations in the prediction of natural frequency of beams. For beams with comparable shear and flexural deformations, the following simplified formula may be used to estimate the beam's frequency:

$$\frac{1}{f^2} = \frac{1}{f_f^2} + \frac{1}{f_s^2} \tag{2.395}$$

where f is the fundamental frequency of the beam, and f_f and f_s are the fundamental frequencies predicted by the flexure and shear beam theories, respectively (Rutenberg 1975).

2.13.7.4 Natural Frequency of Multistory Building Frames

Tall building frames often deform more in the shear mode than in flexure. The fundamental frequencies of many multistory building frameworks can be approximated by (Rinne 1952; Housner 1963)

$$f = \alpha \frac{\sqrt{B}}{H} \tag{2.396}$$

where α is approximately equal to $11\sqrt{m}/s$, B is the building width in the direction of vibration, and H is the building height. This empirical formula suggests that a shear beam model with f inversely proportional to H is more appropriate than flexural beam for predicting natural frequencies of buildings.

2.13.8 Portal Frames

A portal frame consists of a cap beam rigidly connected to two vertical columns. The natural frequencies of portal frames vibrating in the fundamental symmetric and asymmetric modes are shown in Table 2.8

TABLE 2.8 Fundamental Frequencies of Portal Frames in Asymmetrical Mode of Vibration

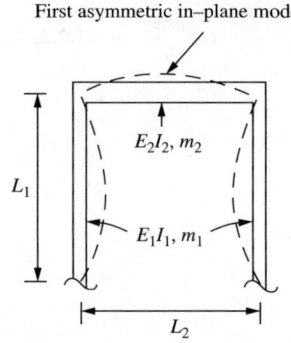

First asymmetric in–plane mode

$$f = \frac{\lambda^2}{2\pi L_1^2}\left(\frac{E_1 I_1}{m_1}\right)^{1/2} \text{ Hz}$$

$E = $ modulus of elasticity

$I = $ area moment of inertia

$m = $ mass per unit length

							λ value				
		Pinned bases						Clamped bases			
		L_1/L_2						L_1/L_2			
$\dfrac{m_1}{m_2}$	$E_1 I_1 / E_2 I_2$	0.25	0.75	1.5	3.0	6.0	0.25	0.75	1.5	3.0	6.0
0.25	0.25	0.6964	0.9520	1.1124	1.2583	1.3759	0.9953	1.3617	1.6003	1.8270	2.0193
	0.75	0.6108	0.8961	1.0764	1.2375	1.3649	0.9030	1.2948	1.5544	1.7999	2.0051
	1.5	0.5414	0.8355	1.0315	1.2093	1.3491	0.8448	1.2323	1.5023	1.7649	1.9853
	3.0	0.4695	0.7562	0.9635	1.1610	1.3201	0.7968	1.1648	1.4329	1.7096	1.9504
	6.0	0.4014	0.6663	0.8737	1.0870	1.2702	0.7547	1.1056	1.3573	1.6350	1.8946
0.75	0.25	0.8947	1.1740	1.3168	1.4210	1.4882	1.2873	1.7014	1.9262	2.0994	2.2156
	0.75	0.7867	1.1088	1.2776	1.3998	1.4773	1.1715	1.6242	1.8779	2.0733	2.2026
	1.5	0.6983	1.0368	1.2281	1.3707	1.4617	1.0979	1.5507	1.8218	2.0390	2.1843
	3.0	0.6061	0.9413	1.1516	1.3203	1.4327	1.0373	1.4698	1.7454	1.9838	2.1516
	6.0	0.5186	0.8314	1.0485	1.2414	1.3822	0.9851	1.3981	1.6601	1.9072	2.0983
1.5	0.25	1.0300	1.2964	1.4103	1.4826	1.5243	1.4941	1.9006	2.0860	2.2090	2.2819
	0.75	0.9085	1.2280	1.3707	1.4616	1.5136	1.3652	1.8214	2.0390	2.1842	2.2695
	1.5	0.8079	1.1514	1.3203	1.4326	1.4982	1.2823	1.7444	1.9837	2.1515	2.2521
	3.0	0.7021	1.0482	1.2414	1.3821	1.4694	1.2141	1.6583	1.9070	2.0983	2.2206
	6.0	0.6011	0.9279	1.1335	1.3024	1.4191	1.1570	1.5808	1.8198	2.0234	2.1693
3.0	0.25	1.1597	1.3898	1.4719	1.5189	1.5442	1.7022	2.0612	2.1963	2.2756	2.3190
	0.75	1.0275	1.3202	1.4326	1.4981	1.5336	1.5649	1.9834	2.1515	2.2520	2.3070
	1.5	0.9161	1.2412	1.3821	1.4694	1.5182	1.4752	1.9063	2.0982	2.2206	2.2899
	3.0	0.7977	1.1333	1.3024	1.4191	1.4896	1.4015	1.8185	2.0233	2.1693	2.2595
	6.0	0.6838	1.0058	1.1921	1.3391	1.4395	1.3425	1.7382	1.9366	2.0964	2.2094
6.0	0.25	1.2691	1.4516	1.5083	1.5388	1.5545	1.8889	2.1727	2.2635	2.3228	2.3385
	0.75	1.1304	1.3821	1.4694	1.5181	1.5440	1.7501	2.0980	2.2206	2.2899	2.3268
	1.5	1.0112	1.3023	1.4191	1.4896	1.5287	1.6576	2.0228	2.1693	2.2595	2.3101
	3.0	0.8827	1.1919	1.3391	1.4395	1.5002	1.5817	1.9358	2.0963	2.2095	2.2802
	6.0	0.7578	1.0601	1.2277	1.3595	1.4502	1.5244	1.8550	2.0110	2.1380	2.2309

and Table 2.9, respectively. The beams in these frames are assumed to be uniform and sufficiently slender so that shear, axial, and torsional deformations can be neglected. The method of analysis of these frames is given in Yang and Sun (1973). The vibration is assumed to be in the plane of the frame, and the results are presented for portal frames with pinned and fixed bases.

If the beam is rigid and the columns are slender and uniform, but not necessarily identical, the natural fundamental frequency of the frame can be approximated using the following formula of Robert (1979):

$$f = \frac{1}{2\pi}\left[\frac{12\sum E_i I_i}{L^3(M + 0.37\sum M_i)}\right]^{1/2} \text{ Hz} \qquad (2.397)$$

where M is the mass of the beam, M_i is the mass of the ith column, and $E_i I_i$ is the flexural rigidity of the ith column. The summation refers to the sum of all columns, and i must be greater than or equal to 2. Additional results for frames with inclined members are discussed in Chang (1978).

TABLE 2.9 Fundamental Frequencies of Portal Frames in Symmetrical Mode of Vibration

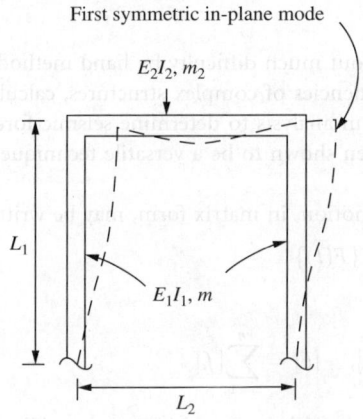

First symmetric in-plane mode

$E_2 I_2, m_2$

$E_1 I_1, m$

$f = \frac{\lambda^2}{2\pi L_1^2}\left(\frac{E_1 I_1}{m_1}\right)^{1/2}$ Hz

E = modulus of elasticity

I = area moment of inertia

m = mass per unit length

λ value

$\left(\dfrac{m_2}{m_1}\right)^{1/4}\left(\dfrac{E_2 I_2}{E_1 I_1}\right)^{3/4}$	$\left(\dfrac{E_1 I_1}{E_2 I_2}\dfrac{m_2}{m_1}\right)^{1/4}\dfrac{L_2}{L_1}$						
	8.0	4.0	2.0	1.0	0.8	0.4	0.2
Pinned bases							
8.0	0.4637	0.8735	1.6676	3.1416	3.5954	3.8355	3.8802
4.0	0.4958	0.9270	1.7394	3.1416	3.4997	3.7637	3.8390
2.0	0.5273	0.9911	1.8411	3.1416	3.4003	3.6578	3.7690
1.0	0.5525	1.0540	1.9633	3.1416	3.3110	3.5275	3.6642
0.8	0.5589	1.0720	2.0037	3.1416	3.2864	3.4845	3.6240
0.4	0.5735	1.1173	2.1214	3.1416	3.2259	3.3622	3.4903
0.2	0.5819	1.1466	2.2150	3.1416	3.1877	3.2706	3.3663
Clamped bases							
8.0	0.4767	0.8941	1.6973	3.2408	3.9269	4.6167	4.6745
4.0	0.5093	0.9532	1.7847	3.3166	3.9268	4.5321	4.6260
2.0	0.5388	1.0185	1.9008	3.4258	3.9268	4.4138	4.5454
1.0	0.5606	1.0773	2.0295	3.5564	3.9267	4.2779	4.4293
0.8	0.5659	1.0932	2.0696	3.5988	3.9267	4.2351	4.3861
0.4	0.5776	1.1316	2.1790	3.7176	3.9267	4.1186	4.2481
0.2	0.5842	1.1551	2.2575	3.8052	3.9266	4.0361	4.1276

2.13.9 Damping

Damping is found to increase with the increasing amplitude of vibration. It arises from the dissipation of energy during vibration. The mechanisms that contribute to energy dissipation are material damping, friction at interfaces between components, and energy dissipation due to the foundation interacting with soil, among others. Material damping arises from the friction at bolted connections and the frictional interaction between structural and nonstructural elements such as partitions and cladding.

The amount of damping in a building can never be predicted precisely, and design values are generally derived based on dynamic measurements of structures of a corresponding type. Damping can be measured based on the rate of decay of free vibration following an impact, by spectral methods based on analysis of response to wind loading, or by force excitation by mechanical vibrator at varying frequency to establish the shape of the steady state resonance curve. However, these methods may not be easily carried out if several modes of vibration close in frequency are presented.

Table 2.10 gives the values of modal damping that are appropriate for use when amplitudes are low. Higher values are appropriate at larger amplitudes where local yielding may develop, for example, in seismic analysis.

2.13.10 Numerical Analysis

Many less complex dynamic problems can be solved without much difficulty by hand methods. More complex problems, such as determination of natural frequencies of complex structures, calculation of response due to time-varying loads, and response spectrum analysis to determine seismic forces, may require numerical analysis. Finite element method has been shown to be a versatile technique for this purpose.

The global equations of an undamped force-vibration motion, in matrix form, may be written as

$$[M]\{\ddot{x}\} + [K]\{x\} = \{F(t)\} \tag{2.398}$$

where

$$[K] = \sum_{i=1}^{n} [k_i], \quad [M] = \sum_{i=1}^{n} [m_i], \quad [F] = \sum_{i=1}^{n} [f_i] \tag{2.399}$$

are the global stiffness, mass, and force matrices, respectively. $[k_i]$, $[m_i]$, and $\{f_i\}$ are the stiffness, mass, and force of the ith element, respectively. The elements are assembled using the direct stiffness method to obtain the global equations such that intermediate continuity of displacements is satisfied at common nodes and, in addition, interelement continuity of acceleration is also satisfied.

Equation 2.398 is the matrix equations discretized in space. To obtain the solution of the equation, discretization in time is also necessary. The general method used is called direct integration. There are two methods for direct integration: *implicit* or *explicit*. The first, and simplest, is an explicit method known as the central difference method (Biggs 1964). The second, more sophisticated but more versatile, is an implicit method known as the Newmark method (Newmark 1959). Other integration methods are also available in Bathe (1982).

TABLE 2.10 Typical Structural Damping Values

Structural type	Damping value, ξ (%)
Unclad welded steel structures	0.3
Unclad bolted steel structures	0.5
Floor, composite and noncomposite	1.5–3.0
Clad buildings subjected to side sway	1

The natural frequencies are determined by solving Equation 2.398 in the absence of force $F(t)$ as

$$[M]\{\ddot{x}\} + [K]\{x\} = 0 \tag{2.400}$$

The standard solution for $x(t)$ is given by the harmonic equation in time

$$\{x(t)\} = \{X\}e^{i\omega t} \tag{2.401}$$

where $\{X\}$ refers to the part of the nodal displacement matrix called natural modes that are assumed to be independent of time, i is the imaginary number, and ω is the natural frequency.

Differentiating Equation 2.401 twice with respect to time, we have

$$\ddot{x}(t) = \{X\}(-\omega^2)e^{i\omega t} \tag{2.402}$$

Substituting Equations 2.401 and 2.402 into Equation 2.400 yields

$$e^{i\omega t}([K] - \omega^2[M])\{X\} = 0 \tag{2.403}$$

Since $e^{i\omega t}$ is not zero, we obtain

$$([K] - \omega^2[M])\{X\} = 0 \tag{2.404}$$

Equation 2.404 is a set of linear homogenous equations in terms of displacement mode $\{X\}$. It has a nontrivial solution if the determinant of the coefficient matrix $\{X\}$ is nonzero, that is,

$$[K] - \omega^2[M] = 0 \tag{2.405}$$

In general, Equation 2.405 is a set of n algebraic equations, where n is the number of degrees of freedom associated with the problem.

References

AISC (1989) *Allowable Stress Design and Plastic Design Specifications for Structural Steel Buildings*, 9th ed., American Institute of Steel Construction, Chicago, IL.

AISC (1993) *Load and Resistance Factor Design Specification for Structural Steel Buildings*, 2nd ed., American Institute of Steel Construction, Chicago, IL.

Al-Bermani, F.G.A., Zhu, K., and Kitipornchai, S. (1995) Bounding-surface plasticity for nonlinear analysis of space structures, *Int. J. Numer. Methods Eng.*, 38(5): 797–808.

Argyris, J.H. (1960) *Energy Theorems and Structural Analysis*, Butterworth, London (Reprinted from *Aircraft Engineering*, Oct. 1954–May 1955).

ASCE (1971) *Plastic Design in Steel — A Guide and Commentary, Manual 41*, American Society of Civil Engineers.

Baker, L. and Heyman, J. (1969) *Plastic Design of Frames: 1. Fundamentals*, Cambridge University Press, Cambridge.

Bathe, K.J. (1982) *Finite Element Procedures in Engineering Analysis*, Prentice Hall, Englewood Cliffs, NJ.

Beedle, L.S. (1958) *Plastic Design of Steel Frames*, John Wiley & Sons, New York.

Biggs, J.M. (1964) *Introduction to Structural Dynamic*, McGraw-Hill, New York.

Borg, S.F. and Gennaro, J.J. (1959) *Advanced Structural Analysis*, D. Van Nostrand Company, Princeton, NJ.

Chang, C.H. (1978) Vibration of frames with inclined members, *J. Sound Vib.*, 56: 201–214.

Chen, H., Liew, J.Y.R., and Shanmugam, N.E. (2000) Nonlinear inelastic analysis of building frames with thin-walled cores, *Thin-Walled Structures*, Vol. 37, Elsevier, U.K., pp. 189–205.

Chen, W.F. and Atsuta, T. (1977) *Theory of Beam-Column, Vol. 2, Space Behavior and Design*, McGraw-Hill, New York.

Chen, W.F., Goto, Y., and Liew, J.Y.R. (1996) *Stability Design of Semi-Rigid Frames*, John Wiley & Sons, New York.

Chen, W.F. and Han, D.J. (1988) *Plasticity for Structural Engineering*, Springer-Verlag, New York.

Chen, W.F. and Kim, S.E. (1997) *LRFD Steel Design Using Advanced Analysis*, CRC Press, Boca Raton, FL.

Chen, W.F. and Lui, E.M. (1987) *Structural Stability — Theory and Implementation*, Prentice Hall, Upper Saddle River, NJ.

Chen, W.F. and Lui, E.M. (1991) *Stability Design of Steel Frames*, CRC Press, Boca Raton, FL.

Chen, W.F. and Sohal, I.S. (1994) *Plastic Design and Advanced Analysis of Steel Frames*, Springer-Verlag, New York.

Chen, W.F. and Toma, S. (1994) *Advanced Analysis in Steel Frames: Theory, Software and Applications*, CRC Press, Boca Raton, FL.

Clark, M.J. (1994) Plastic-zone analysis of frames, in *Advanced Analysis of Steel Frames, Theory, Software, and Applications*, Edited by W.F. Chen and S. Toma, CRC Press, Boca Raton, FL, pp. 195–319.

Clough, R.W. and Penzien, J. (1993) *Dynamics of Structures*, 2nd ed., McGraw-Hill, New York.

Cowper, G.R. (1966) The shear coefficient in Timoshenko's beam theory, *J. Appl. Mech.*, 33: 335–340.

Desai, C.S. and Abel, J.F. (1972) *Introduction to the Finite Element Method*, Van Nostrand Reinhold Company, New York.

Dowrick, D.J. (1988) *Earthquake Resistant Design for Engineers and Architects*, 2nd ed., John Wiley & Sons, New York.

Faella, C., Piluso, V., and Rizzano, G. (2000) *Structural Steel Semi-Rigid Connections — Theory, Design and Software*, CRC Press, Boca Raton, FL.

Ghali, A. and Neville, A.M. (1978) *Structural Analysis*, Chapman and Hall, London.

Heyman, J. (1971) *Plastic Design of Frames: 2. Applications*, Cambridge University Press, Cambridge.

Horne, M.R. (1964) *The Plastic Design of Columns*, BCSA Publication No. 23.

Housner, G.W. and Brody, A.G. (1963) Natural periods of vibration of buildings, *J. Eng. Mech. Div.*, ASCE, 89: 31–65.

Jiang, X.M., Chen, H., and Liew, J.Y.R. (2002) Spread-of-plasticity analysis of 3-D steel frames, *J. Construc. Steel Res.*, Elsevier, U.K., 58(2), 193–212.

Kardestuncer, H. and Norrie, D.H. (1988). *Finite Element Handbook*, McGraw-Hill, New York.

Knudsen, K.E., Yang, C.H., Johnson, B.G., and Beedle, L.S. (1953) Plastic strength and deflections of continuous beams, *Weld. J.*, 32(5), 240–245.

Lee, S.L. and Ballesteros, P. (1960) Uniformly loaded rectangular plate supported at the corners, *Int. J. Mech. Sci.*, 2: 206–211.

Lee, S.L. and Basu, P.K. (1989) Secant method for nonlinear semi-rigid frames, *J. Construct. Steel Res.*, 14(2): 49–67.

Lee, S.L., Karasudhi, P., Zakeria, M., and Chan, K.S. (1971) Uniformly loaded orthotropic rectangula plate supported at the corners, *Civil Eng. Trans.*, 101–106.

Liew, J.Y.R. (1992) *Advanced Analysis for Frame Design*, PhD dissertation, School of Civil Engineering, Purdue University, West Lafayette, IN.

Liew, R.J.Y., Chen, H., Shanmugam, N.E., and Chen, W.F. (2000) Improved nonlinear plastic hinge analysis of space frames, *Eng. Struct.*, 22: 1324–1338.

Liew, J.Y.R., and Chen, W.F. (1994) Trends toward advanced analysis, in *Advanced Analysis in Steel Frames: Theory, Software and Applications*, Edited by W.F. Chen and S. Toma, CRC Press, Boca Raton, FL, pp. 1–45.

Liew, J.Y.R., White, D.W., and Chen, W.F. (1991) Beam-column design in steel frameworks — Insight on current methods and trends, *J. Construct. Steel Res.*, 18: 259–308.

Liew, J.Y.R., White, D.W., and Chen, W.F. (1993) Second-order refined plastic hinge analysis for frame design: parts 1 & 2, *J. Struct. Eng.*, ASCE, 119(11): 3196–3237.

Mrazik, A., Skaloud, M., and Tochacek, M. (1987) *Plastic Design of Steel Structures*, Ellis Horwood, Chichester, England.

Newmark, N.M. (1959) A method of computation for structural dynamic, *J. Eng. Mech.*, ASCE, 85(EM3): 67–94.

Orbison, J.G. (1982) *Nonlinear Static Analysis of Three-Dimensional Steel Frames*, Department of Structural Engineering., Report No. 82-6, Cornell University, Ithaca, NY.

Porter, F.L. and Powell, G.M. (1971) *Static and Dynamic Analysis of Inelastic Frame Structures*, Report No. EERC 71-3, Earthquake Engineering Research Centre, University of California, Berkeley.

Powell, G.H. and Chen, P.F.-S. (1986) 3D beam-column element with generalized plastic hinges, *J. Eng. Mech.*, ASCE, 112(7): 627–641.

Rinne, J.E. (1952) Building code provisions for aseismic design, in *Proceedings of Symposium on Earthquake and Blast Effects on Structures*, Los Angeles, CA, pp. 291–305.

Robert, D.B. (1979) *Formulas for Natural Frequency and Mode Shapes*, Van Nostrand Reinhold Company, New York.

Rutenberg, A. (1975) Approximate natural frequencies for coupled shear walls, *Earthquake Eng. Struct. Dynam.*, 4: 95–100.

Seely, F.B. and Smith, J.O. (1952) *Advanced Mechanics of Materials*, John Wiley & Sons, New York.

Shanmugam, N.E., Huang, R., Yu, C.H., and Lee, S.L. (1988) Uniformly loaded rhombic orthotropic plates supported at corners, *Comput. Struct.*, 30(5): 1037–1045.

Shanmugam, N.E., Huang, R., Yu, C.H., and Lee, S.L. (1989) Corner supported isosceles triangular orthotropic plates, *Comput. Struct.*, 32(5): 963–972.

Structural Stability Research Council (SSRC), (1988) *Guide to Stability Design Criteria for Metal Structures*, 4th ed., Edited by T.V. Galambos, John Wiley & Sons, New York.

Timoshenko, S.P. and Gere, J.M. (1963) *Theory of Elastic Stability*, McGraw-Hill, New York.

Timoshenko, S.P. and Woinowsky-Krieger, S. (1959) *Theory of Plates and Shells*, McGraw-Hill, New York.

Tsai, S.W. and Cheron, T. (1968) *Anisotropic Plates* (translated from the Russian edition by S.G. Lekhnitskii), Gordon and Breach Science Publishers, New York.

Vijakkhna, P., Karasudhi, P., and Lee, S.L. (1997) Corner supported equilateral triangular plates, *Int. J. Mech. Sci.*, 15: 123–128.

Vogel, U. (1985) Calibrating frames, *Stahlbau*, Oct: 295–301.

Wang, Y.C. (1988) *Ultimate Strength Analysis of 3-D Beam Columns and Column Subassemblages with Flexible Connections*, PhD thesis, University of Sheffield, England.

Warburton, G.B. (1976) *The Dynamical Behaviour of Structures*, 2nd ed., Pergamon Press, New York.

White, D.W. (1985) *Material and Geometric Nonlinear Analysis of Local Planar Behavior in Steel Frames Using Iterative Computer Graphics*, M.S. thesis, Cornell University, Ithaca, NY.

White, D.W. (1988) *Analysis of Monotonic and Cyclic Stability of Steel Frame Subassemblages*, Ph.D. dissertation, Cornell University, Ithaca, NY.

White, D.W. and Hajjar, J.F. (1991) Application of second-order elastic analysis in LRFD: Research to practice, *Eng. J.*, AISC, 28(4): 133–148.

White, D.W., Liew, J.Y.R., and Chen, W.F. (1991) *Second-Order Inelastic Analysis for Frame Design: A Report to SSRC Task Group 29 on Recent Research and the Perceived State-of-the-Art*, Structural Engineering Report, CE-STR-91-12, Purdue University, West Lafayette, IN.

White, D.W., Liew, J.Y.R., and Chen, W.F. (1993) Toward advanced analysis in LRFD, in *Plastic Hinge Based Methods for Advanced Analysis and Design of Steel Frames — An Assessment of the State-of-the-Art*, Structural Stability Research Council, Lehigh University, Bethlehem, PA, pp. 95–173.

Yang, Y.B. and Kuo, S.R. (1994) *Theory and Analysis of Nonlinear Framed Structures*, Prentice Hall, Singapore.

Yang, Y.T. and Sun, C.T. (1973) Axial-flexural vibration of frameworks using finite element approach, *J. Acoust. Soc. Am.*, 53: 137–146.

Ziemian, R.D., McGuire, W., and Deierlien, G.G. (1992a) Inelastic limit states design: Part I — planar frame studies, *J. Struct. Eng.*, ASCE, 118(9).
Ziemian, R.D., McGuire, W., and Deierlien, G.G. (1992b) Inelastic limit states design: Part II — three-dimensional frame study, *J. Struct. Eng.*, ASCE, 118(9).

Further Reading

Bath, K.J. (1996) *Finite Element Procedure*, Prentice Hall, Englewood Cliffs, NJ.
Chen, W.F. and Asuta, T. (1977) *Theory of Beam-columns, Vol. 2: Space Behavior and Design*, McGraw-Hill, New York.
Gallagher, R.H. (1975) *Finite Element Analysis Fundamentals*, Prentice Hall, Englewood Cliffs, NJ.
McGuire, W., Gallagher, R.H., and Ziemian, R.D. (2000) *Matrix Structural Analysis*, John Wiley & Sons, New York.
Timishenko, S. and Goodier, J.N. (1970) *Theory of Elasticity*, 3rd ed., McGraw-Hill, New York.
Yang, Y.B. and Kuo, S.-R. (1994) *Theory and Analysis of Nonlinear Framed Structures*, Prentice Hall, Englewood Cliffs, NJ.

3

Structural Dynamics

Franklin Y. Cheng
*Curators' Professor Emeritus of
Civil Engineering,
University of Missouri,
Rolla, MO*

0-8493-1569-7/05/$0.00+$1.50
© 2005 by CRC Press

3.1 Dynamic Forces and Structural Models

3.1.1 Characteristics of Dynamic Forces

The subject presented herein deals with the response of structures subjected to dynamic forces or loads, whose magnitude varies with time. Generally, most of the forces applied to a structure involve, in some manner, time variation; static force may be viewed as a special dynamic case when the force applied is slow enough without causing structural vibration. As structural materials, construction methods, and computer technology rapidly advance, constructed facilities of building structures and nonbuilding structures become taller and slender. Therefore, dynamic behavior of such structures must be included in their design.

The dynamic forces acting on a structure can be categorized in different ways according to (1) the original sources causing vibration, (2) the characteristics of vibration, whether periodic, nonperiodic, or random, or (3) the definite function of time as deterministic or nondeterministic. Rotating machinery, blast, wind, and earthquake are in the first category. Dynamic force due to unbalanced machinery varied repeatedly in magnitude with time is called *periodic force*. Earthquake, wind, and blast, however, do not have any periodicity and are hence called *nonperiodic* or *random forces*. A deterministic force is one where its time function can be specified in regular or irregular variation; for instance, the time variation of rotating machinery can be represented by a mathematical function, blast and impulse may be specified by mathematical curves or lines, and an earthquake may be specified by accelerograms in magnitude with time intervals. These forces may be classified as deterministic. On the contrary, nondeterministic forces cannot be specified as definite functions of time because of the inherent uncertainty in their magnitude and in their variation with time. These types of load should be described through a statistical approach. Wind is in this category and earthquake is also non-deterministic because the magnitude and frequency distribution of any future earthquake cannot be predicted with certainty but can be estimated only in a probabilistic sense. The classification of loading is shown in Figure 3.1. This chapter deals with both deterministic and nondeterministic loadings or forces for which the response analysis methods are presented. Stochastic analysis is illustrated with seismic response.

3.1.2 Mathematical Models of Structural Systems

Analytical accuracy and computational efficiency of dynamics problems depends on several key features: structural modeling, material property idealization, loading assumptions, and numerical techniques. This chapter covers three well-known models:

- Lumped-mass system
- Continuous-mass system
- Finite element system

In fact, the lumped-mass system and the finite element system are similar in modeling and therefore are sometimes classified into one group known as the discrete system.

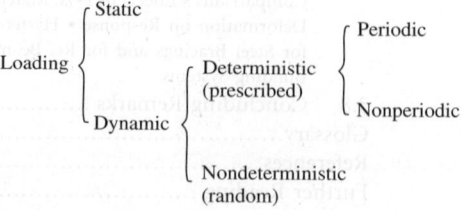

FIGURE 3.1 Loading classification.

3.2 Response Analysis of Single d.o.f. Systems

3.2.1 Definition of d.o.f.

The degree of freedom (d.o.f.) of a structure may be first explained from the nature of loading as statically applied load or dynamic excitation to a structure. In general, d.o.f. represents the independent movement of structural nodes for a static case, but the independent movements of lumped masses for a dynamic case. The number of structural nodes can be more than the number of lumped masses. Furthermore, each lumped mass may have more than one independent motion. For instance, the plane frame shown in Figure 3.2 should have three d.o.f. (x_1, x_2, x_3) for a static case. However, the lumped-mass model may have one (x_3), two (x_3 and x_4), or three (x_3, x_4, and x_5) d.o.f. corresponding to the response analysis of lateral motion (x_3) only, lateral and vertical motions (x_3, x_4) only, or lateral, vertical, and rocking (x_3, x_4, x_5) motions. Therefore, the number of masses and the dynamic d.o.f. are determined by the structural analyst based on the structural configuration and the interest of the analytical results. Note that x_1 and x_2 in Figure 3.2b are not dynamic d.o.f., but they must be given in order to allow the structural joints to rotate during vibration.

3.2.2 Undamped and Damped Free Vibration

3.2.2.1 Undamped with Initial Conditions

Consider the spring-mass model shown in Figure 3.3. This model, which consists of a mass of weight, W, suspended by means of a spring with stiffness, K, is idealized from the accompanying unsymmetrical rigid frame where L and I signify the member's length and its cross-sectional moment of inertia, respectively. The spring stiffness, K, is defined as the force necessary to stretch or compress the spring

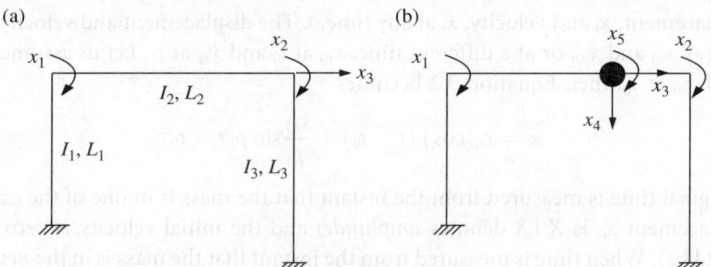

FIGURE 3.2 Plane unsymmetric rigid frame: (a) static d.o.f. and (b) dynamic d.o.f.

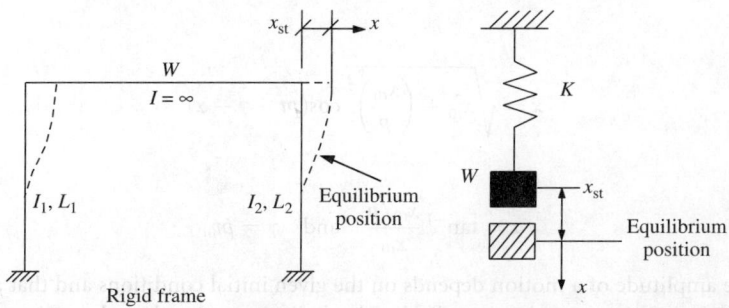

FIGURE 3.3 Structures and spring-mass model.

one unit of length, which, in fact, is the force caused by a unit side sway of the frame. The mass is in equilibrium under the action of two equal and opposite forces: the weight, W, acts downward and the spring force, Kx_{st}, upward. The term x_{st} denotes *static deflection*, which is the amount of movement from the undeformed position to the *equilibrium position* where the displacement of the mass is measured during the vibration.

The motion equation of the structure is

$$M\ddot{x} + Kx = 0 \tag{3.1a}$$

or

$$\ddot{x} + p^2 x = 0 \tag{3.1b}$$

in which $p = \sqrt{K/M}$, where p is the angular frequency with the unit of rad/s; $M = W/g$; and g is the gravity of acceleration. Note that the units of K and M must be consistent. The relationships between frequency and period may be expressed as

$$p = \frac{2\pi}{T} = 2\pi f \quad \text{(angular frequency, rad/s)}$$

$$f = \frac{1}{T} = \frac{p}{2\pi} \quad \text{(natural frequency, cycle/s)}$$

$$T = \frac{1}{f} = \frac{2\pi}{p} \quad \text{(natural period, s)}$$

The solution of Equation 3.1 is

$$x = A \sin pt + B \cos pt \tag{3.2}$$

where the integration constants A and B should be determined by using the information of motion as the known displacement, x, and velocity, $\dot{x}$, at any time, t. The displacement and velocity may be given at the same time, say x_{t0} and $\dot{x}_{t0}$, or at a different time, x_{t0} at t_0 and $\dot{x}_{t1}$ at t_1. Let us assume that x and $\dot{x}$ are given as x_{t0} and $\dot{x}_{t0}$ at t_0, then Equation 3.2 becomes

$$x = x_{t0} \cos p(t - t_0) + \frac{\dot{x}_{t0}}{p} \sin p(t - t_0) \tag{3.3}$$

When the original time is measured from the instant that the mass is in one of the extreme positions, the initial displacement x_0 is X (X denotes *amplitude*) and the initial velocity is zero (as the physical condition should be). When time is measured from the instant that the mass is in the *neutral position*, the initial conditions are $x = 0$ and $\dot{x} = \dot{x}_0$. If the origin is located at t_0 units of time after the mass passes the neutral position with the initial conditions of $x = x_{t0}$. The general expression becomes

$$x = x_{t0} \cos(pt - \gamma) + \frac{\dot{x}_{t0}}{p} \sin(pt - \gamma) \tag{3.4a}$$

or

$$x = \sqrt{x_{t0}^2 + \left(\frac{\dot{x}_{t0}}{p}\right)^2} \cos(pt - \gamma - \alpha) \tag{3.4b}$$

in which

$$\alpha = \tan^{-1} \frac{\dot{x}_{t0}/p}{x_{t0}} \quad \text{and} \quad \gamma = pt_0$$

Note that the amplitude of a motion depends on the given initial conditions and that all the motions are in the same manner except they are displaced relative to each other along time t. The relative magnitude in radians between x, $\dot{x}$, and $\ddot{x}$ is called the *phase angle*.

3.2.2.2 Damped Vibration with Initial Conditions

In the previous discussion, we assumed an ideal vibrating system free from internal and external damping. *Damping* may be defined as a force that resists motion at all times. Therefore, a free undamped vibration continues in motion indefinitely without its amplitude diminishing or its frequency changing. Real systems, however, do not possess perfectly elastic springs nor are they surrounded by a frictionless medium. Various damping agents — such as the frictional forces of structural joints and bearing supports, the resistance of surrounding air, and the internal friction between molecules of the structural materials — always exist.

It is difficult, if not impossible, to derive a mathematical formula for damping resistance that represents the actual behavior of a physical system. A simple yet realistic damping model for mathematical analysis is that the damping force is proportional to velocity. This model can represent structural damping of which the force is produced by the viscous friction and is therefore called *viscous damping*. Figure 3.4 shows a vibration model consisting of an ideal spring and *dashpot* in parallel. The dashpot exerts a damping force, $c\dot{x}$, proportional to the relative velocity, in which c is a proportionality and is called the *coefficient of viscous damping*. The governing differential equation is

$$M\ddot{x} + c\dot{x} + Kx = 0 \tag{3.5}$$

of which the standard solution is

$$x = C_1 e^{\alpha_1 t} + C_2 e^{\alpha_2 t} \tag{3.6}$$

where C_1 and C_2 are integration constants, and α_1 and α_2 may be expressed as

$$\alpha_1 = -\frac{c}{2M} + \sqrt{\frac{c^2}{4M^2} - \frac{K}{M}} \tag{3.7}$$

$$\alpha_2 = -\frac{c}{2M} - \sqrt{\frac{c^2}{4M^2} - \frac{K}{M}} \tag{3.8}$$

After substituting Equations 3.7 and 3.8 for the corresponding terms in Equation 3.6, possible solutions can be obtained for three cases of $c^2/4M^2 = K/M$, $c^2/4M^2 > K/M$, and $c^2/4M^2 < K/M$, corresponding to *critical damping*, *overdamping*, and *underdamping*, respectively.

When $c^2/4M^2 = K/M$, the value of c is called critical damping and takes the form

$$c_{cr} = 2\sqrt{KM} = 2Mp = \frac{2K}{p} \tag{3.9}$$

The ratio of c/c_{cr} is called *viscous damping factor* or simply *damping factor*, ρ, and may be expressed as

$$\rho = \frac{c}{c_{cr}} = \frac{c}{2Mp} = \frac{c}{2\sqrt{(KM)}} = \frac{cp}{2K} \tag{3.10}$$

In most structural and mechanical systems, the assumption of underdamping is justified, that is, $\rho < 1$. For this case, the motion equation is

$$\ddot{x} + 2\rho p\dot{x} + p^2 x = 0 \tag{3.11}$$

The displacement response may be obtained from

$$x = e^{-\rho pt}(A \cos \sqrt{1 - \rho^2}\, pt + B \sin \sqrt{1 - \rho^2}\, pt) \tag{3.12a}$$

or

$$x = C e^{-\rho pt} \cos(\sqrt{1 - \rho^2}\, pt - \alpha) \tag{3.12b}$$

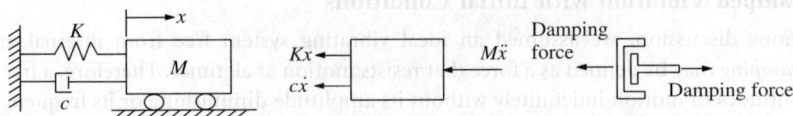

FIGURE 3.4 Spring-mass and viscous damping model.

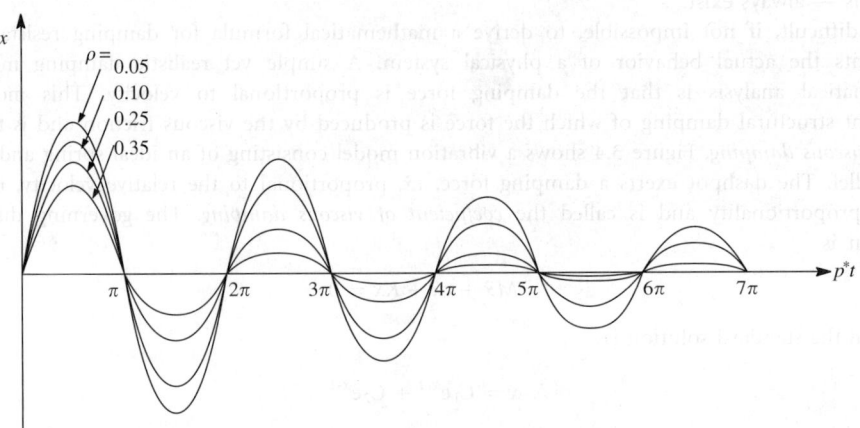

FIGURE 3.5 Motion with underdamping. (Reprinted from Ref. [1, p. 11] by courtesy of Marcel Dekker, Inc.)

Note that the oscillatory motion has the frequency, *damped frequency*, and the associated period, *damped period*, which may be expressed as

$$p^* = p\sqrt{1 - \rho^2} \tag{3.13}$$

$$T^* = \frac{2\pi}{p\sqrt{1 - \rho^2}} = \frac{T}{\sqrt{1 - \rho^2}} \tag{3.14}$$

Actually, the difference between p^*, T^* and p, T is slight. The terms p and T may be used instead of p^* and T^* in damped vibrations without introducing a serious error.

If the given initial conditions of $x_0 = 0$ and $\dot{x} = p$ at $t = 0$ are inserted into Equation 3.12b, we obtain

$$x = \frac{1}{\sqrt{1 - \rho^2}} e^{-\rho p t} \cos\left(\sqrt{1 - \rho^2}\, pt - \frac{\pi}{2}\right) \tag{3.15}$$

which is plotted in Figure 3.5 for various damping factors of 0.05, 0.10, 0.25, and 0.35. It can be seen that the amplitudes of successive cycles are different and the periods of successive cycles are the same; strictly speaking, the motion is not regarded as being periodic but as *time-periodic*. In most engineering structures, ρ may vary from 0.02 to 0.08. Of course, the damping factor for some buildings may be as high as 0.15, depending on the nature of the material used in their construction and the degree of looseness in their connections.

3.2.3 Undamped and Damped Forced Vibration

3.2.3.1 Undamped Forced Vibration with Harmonic Force (Steady-State Response)

Consider the spring-mass model shown in Figure 3.3 where the mass is subjected to a harmonic force $F \sin \omega t$ with *forced frequency* ω. Let $F \sin \omega t$ be considered positive to the right of the equilibrium position from which displacement, x, is measured. The differential equation of motion is

$$M\ddot{x} + Kx = F \sin \omega t \tag{3.16a}$$

or

$$\ddot{x} + p^2 x = p^2 x_{st} \sin \omega t \tag{3.16b}$$

in which $x_{st} = F/K$ as static displacement. The homogeneous solution is $x_h = A \sin pt + B \cos pt$; the term, $p = \sqrt{(K/M)}$, is independent of the forced frequency, ω. The particular solution may be obtained by trying the following:

$$x_p = C_1 \sin \omega t + C_2 \cos \omega t \tag{3.17}$$

Substituting x_p and $\ddot{x}_p$ in Equation 3.17 for C_1 and C_2, we then obtain the following complete solution:

$$x = x_h + x_p = A \sin pt + B \cos pt + \frac{F}{K - M\omega^2} \sin \omega t \tag{3.18a}$$

in which A and B should be determined by using the initial conditions of free vibration. When the force $F \sin \omega t$ is applied from a position of rest, ignore x_h (free vibration due to x_0 and $\dot{x}_0$). The response of forced vibration corresponding to the third term of Equation 3.18a is

$$x = \frac{F/K}{1 - (\omega^2/p^2)} \sin \omega t \tag{3.18b}$$

which indicates that the motion is periodic with the same frequency as that of the force and may endure as long as the force remains on the mass; this is called *steady-state vibration*. Note that F/K is static displacement x_{st}; when $\sin \omega t = 1$, the displacement x is the amplitude X. In general, the application of a disturbing force can produce an additional motion superimposed on steady-state motion. This additional motion is from the homogeneous solution of the free vibration. Consider the initial conditions of $x_0 = 0$ and $\dot{x}_0 = 0$ at $t = 0$ in Equation 3.18a. The result is

$$x = \frac{F/K}{(K - M\omega^2)} \left(\sin \omega t - \frac{\omega}{p} \sin pt \right) \tag{3.19}$$

Comparing Equation 3.18b with Equation 3.19 reveals that there is another term associated with $\sin pt$. This is due to the fact that the application of a disturbing force produces some free vibrations of the system. Thus, the actual motion is a superposition of two harmonic motions with different frequencies, amplitudes, and phase angles. In practical engineering, there is always some damping. So, free vibration is eventually damped out and only forced vibration remains. The early part of a motion consisting of a forced vibration and a few cycles of free vibration is called *transient vibration*, which can be important in aircraft design for landing and for gust loading.

3.2.3.2 Undamped Forced Vibration with Impulses (Shock Spectra)

When the structure is subjected to impulses of duration ζ, the maximum response can be defined in terms of *amplification factor*, A_m, expressed in terms of amplitude (X) and static displacement (x_{st}) as

$$A_m = \frac{X}{x_{st}} = \sqrt{2 \left(1 - \cos \frac{2\pi \zeta}{T} \right)} \tag{3.20}$$

The variation of amplitude in terms of force duration and structural natural period (ζ/T) is expressed in the *shock spectrum* shown in Figure 3.6a. The shock spectra for three other types of impulses are similarly sketched in Figure 3.6b. Note that maximum amplification factor is twice the static displacement and that the amplification factor can be higher after the impulse than that during the impulse (see the case of triangular impulse in sine function shape).

(a)   (b)

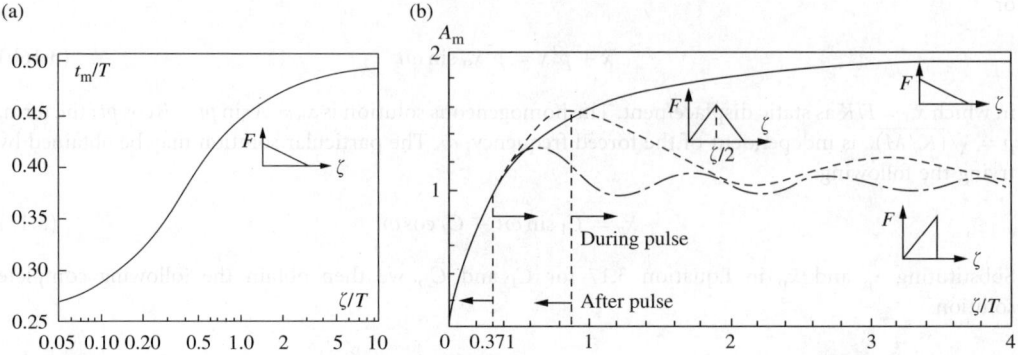

FIGURE 3.6   Shock spectra of impulses. (Reprinted from Ref. [1, pp. 21 and 22] by courtesy of Marcel Dekker, Inc.)

3.2.3.3   Damped Forced Vibration with Harmonic Force (Steady-State Response)

Assuming that the structure is subjected to a harmonic force $F \sin \omega t$, the motion equation is

$$M\ddot{x} + c\dot{x} + Kx = F \sin \omega t \tag{3.21a}$$

Following Equations 3.11 and 3.16b, the motion equation can be rewritten as

$$\ddot{x} + 2\rho p\dot{x} + p^2 x = \frac{F}{M}\sin \omega t = p^2 x_{st} \sin \omega t \tag{3.21b}$$

Similar to Equation 3.18b, the amplification factor for forced damped vibration may be expressed as

$$A_m = \frac{X}{x_{st}} = \frac{1}{\sqrt{[1 - (\omega/p)^2]^2 + (2\rho\omega/p)^2}} \tag{3.22}$$

Equation 3.22 is plotted in Figure 3.7, A_m versus ω/p, for various values of ρ. It is seen that the *peak amplitude*, defined as the amplitude at $d(A_m)/d(\omega/p) = 0$, is greater than the *resonant amplitude*, defined as the amplitude at $\omega/p = 1$. Because they occur practically at the same frequency and it is easy to find the resonant frequency, engineers usually overlook peak amplitude.

3.2.3.4   Damped Forced Vibration with Harmonic Force or Foundation Movement (Transmissibility)

When a harmonic force $F \cos \omega t$ is acting on the mass, the ratio between the amplitude of the force (A_f) transmitted to the foundation and the amplitude of the driving force F is called *transmissibility* and is expressed as follows:

$$T_r = \frac{A_f}{F} = \sqrt{\frac{1 + (2\omega\rho/p)^2}{[1 - (\omega/p)^2]^2 + (2\omega\rho/p)^2}} \tag{3.23}$$

When the foundation is subjected to $X_1 \cos \omega t$, the structural mass is induced to vibrate. The ratio of the amplitude of the mass motion, X_2, to the amplitude of the support motion, X_1, is also the transmissibility shown in Equation 3.23.

The transmissibility versus ω/p for various ρs is shown in Figure 3.8. A few interesting features of *vibration isolation* can be observed from the figure: T_r is always less than one when ω/p is greater than $\sqrt{2}$, regardless of the damping ratio; when ω/p is less than $\sqrt{2}$, T_r, depending on the damping ratio, is always equal to or greater than one; and T_r is equal to one when ω/p equals $\sqrt{2}$ regardless of the amount of damping.

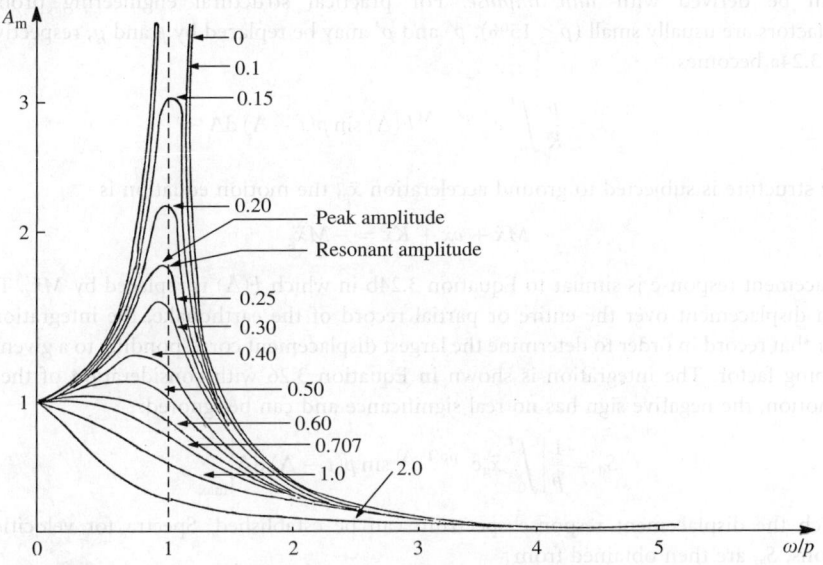

FIGURE 3.7 A_m versus ω/p for various values of ρ. (Reprinted from Ref. [1, p. 31] by courtesy of Marcel Dekker, Inc.)

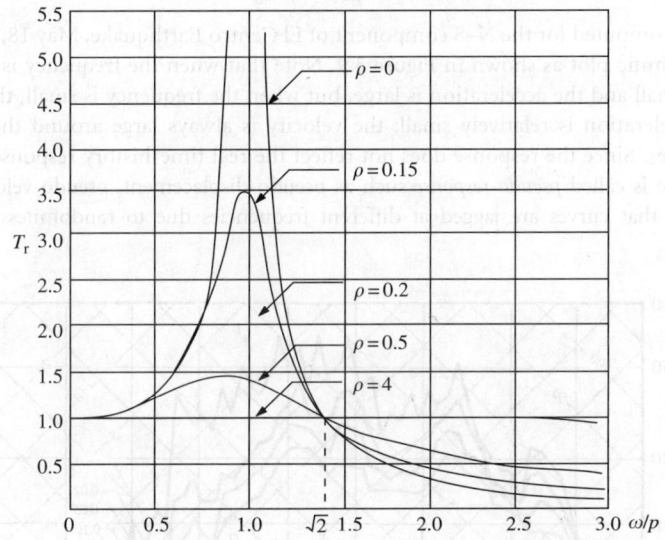

FIGURE 3.8 T_r versus ω/p for various damping factors. (Reprinted from Ref. [1, p. 38] by courtesy of Marcel Dekker, Inc.)

3.2.3.5   Damped Forced Vibration with General Forcing Function or Earthquakes (Duhamel's Integral)

For general forcing function, $F(t)$, the displacement may be obtained from Equation 3.21 by using *Duhamel's integral* (also *convolution integral*) as

$$x = \frac{p}{K\sqrt{(1-\rho^2)}} \int_0^t e^{-\rho^* p^*(t-\Delta)} F(\Delta) \sin p^*(t-\Delta)\, d\Delta \qquad (3.24a)$$

which can be derived with *unit impulse*. For practical structural engineering problems, the damping factors are usually small ($\rho < 15\%$). p^* and ρ^* may be replaced by p and ρ, respectively. Thus, Equation 3.24a becomes

$$x = \frac{p}{K} \int_0^t e^{-\rho p(t-\Delta)} F(\Delta) \sin p(t-\Delta) \, d\Delta \tag{3.24b}$$

When the structure is subjected to ground acceleration $\ddot{x}_g$, the motion equation is

$$M\ddot{x} + c\dot{x} + Kx = -M\ddot{x}_g \tag{3.25}$$

The displacement response is similar to Equation 3.24b in which $F(\Delta)$ is replaced by $M\ddot{x}_g$. To find the maximum displacement over the entire or partial record of the earthquake, the integration must be carried for that record in order to determine the largest displacement corresponding to a given frequency and damping factor. The integration is shown in Equation 3.26 with consideration of the nature of ground motion, the negative sign has no real significance and can be ignored:

$$S_d = \frac{1}{p} \left| \int_0^t \ddot{x}_g e^{-\rho p(t-\Delta)} \sin p(t-\Delta) \, d\Delta \right|_{\max} \tag{3.26}$$

from which the displacement response spectrum can be established. Spectra for velocities, S_v and accelerations, S_a, are then obtained from

$$S_v = pS_d \tag{3.27}$$

$$S_a = pS_v = p^2 S_d \tag{3.28}$$

Response spectra computed for the N–S component of El Centro Earthquake, May 18, 1940, are given in a tripartite logarithmic plot as shown in Figure 3.9. Note that when the frequency is large, the relative displacement is small and the acceleration is large, but when the frequency is small, the displacement is large and the acceleration is relatively small; the velocity is always large around the region of intermediate frequencies. Since the response does not reflect the real time-history response but a maximum value, the response is called *pseudo-response* such as pseudo-displacement, pseudo-velocity, and pseudo-acceleration. Note that curves are jagged at different frequencies due to randomness of seismic input.

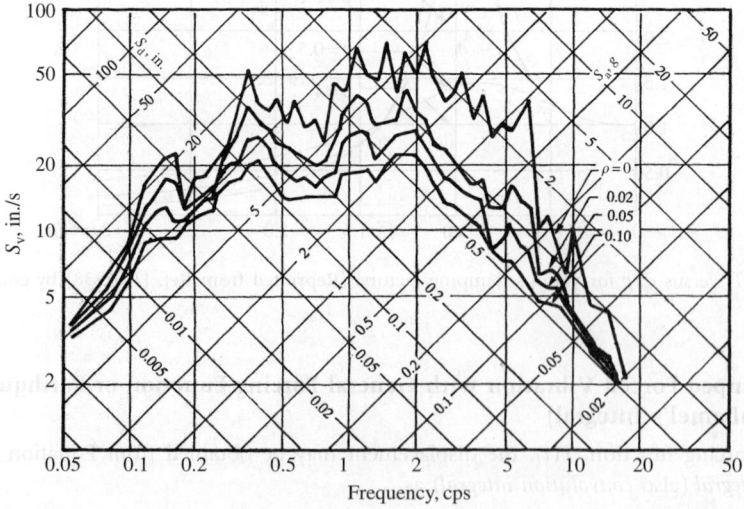

FIGURE 3.9   Response spectra for the N–S component of El Centro Earthquake, May 18, 1940. (Reprinted from Ref. [1, p. 37] by courtesy of Marcel Dekker, Inc.)

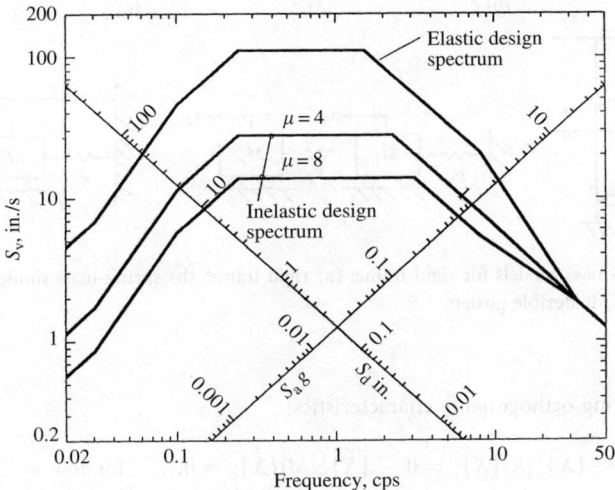

FIGURE 3.10 Elastic and inelastic design spectra. (Reprinted from Ref. [1, p. 376] by courtesy of Marcel Dekker, Inc.)

Smooth *design spectra* for elastic and inelastic responses are available with consideration of various damping ratios, soil profiles, and earthquake records [1, pp. 366–379]. Sample elastic and inelastic design spectra are shown in Figure 3.10 for $\rho = 0.05$. The inelastic spectra given are only for *ductility* $\mu = 4$ and 8. Ductility factor is designed as the ratio of the maximum displacement to yielding displacement.

3.3 Response Analysis of Multiple d.o.f. Systems — Lumped-Mass Formulation

3.3.1 Nature of Spring-Mass Model

A structure that is assumed to be a discrete parameter (lumped mass) system must be conceived of as a model consisting of a finite number of masses connected by massless springs. The *spring-mass model*, depending on the characteristics of the structure, can be established in different ways. An example is shown in Figure 3.11, where M_1 and M_2 are masses lumped from girders and columns, and k_1 and k_2 represent column stiffnesses. When the girder is infinitely rigid, the structure has no joint rotations; this spring-mass model is shown in Figure 3.11b. When the girders are flexible and structural joint rotations exist, the spring-mass model differs as shown in Figure 3.11c. Note the reason for the difference: if x_2 is displaced and the girders are rigid, no force is transmitted to the support. However, with flexible girders, the joints at the first floor rotate, the column below is distorted, and force is transmitted to the support.

3.3.2 Normal Modes, Modal Matrix, and Characteristics of Orthogonality

The motion equations associated with free undamped vibration of a spring-mass system can be expressed in matrix form as

$$[M]\{\ddot{x}\} + [K]\{x\} = 0 \tag{3.29}$$

where $[M]$ and $[K]$ are called the *mass matrix* and *structural stiffness matrix*, respectively. Let the displacement vector be $\{x\} = (\cos pt)\{X\}$; then, Equation 3.29 may be expressed as (since $\cos pt \neq 0$)

$$-p^2[M]\{\ddot{X}\} + [K]\{X\} = 0 \tag{3.30}$$

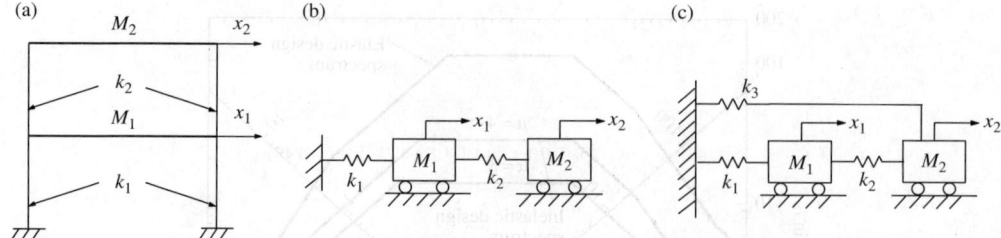

FIGURE 3.11 Spring-mass models for rigid frame: (a) rigid frame; (b) spring-mass model for rigid girders; and (c) spring-mass model for flexible girders.

which has the following orthogonality characteristics:

$$\{X\}_u^T[K]\{X\}_v = 0, \quad \{X\}_u^T[M]\{X\}_v = 0, \qquad \text{for } u \neq v \tag{3.31a}$$

$$\{X\}_u^T[K]\{X\}_v \neq 0, \quad \{X\}_u^T[M]\{X\}_v \neq 0, \qquad \text{for } u = v \tag{3.31b}$$

In Equation 3.30, p is the *eigenvalue* (also angular frequency) and X is the eigenvector (also *normal modes*). u and v in Equation 3.31 represent uth and vth modes associated with frequencies p_u and p_v, respectively. Note that normal modes (or mode shapes) have significant meaning in displacement response, which in fact, results from a combination of the modes of a system.

Let $\{\Phi\}_u$ and $\{\Phi\}_v$ be modal displacements corresponding to uth and vth modes, respectively, such that

$$\{\Phi\}_u^T[M]\{\Phi\}_u = 1 \tag{3.32}$$

When the modal vectors are collected in a single square matrix of order n, corresponding to n modes, the resulting matrix is called *modal matrix*, $[\Phi]$. Using $[\Phi]$ in Equation 3.31 yields Equation 3.33

$$[\Phi]^T[K][\Phi] = [\backslash p_i^2 \backslash], \qquad [\Phi]^T[M][\Phi] = [I] \tag{3.33}$$

Note that Equations 3.31 and 3.33 are derived on the basis that $[M] = [M]^T$, $[K] = [K]^T$, and $p_u \neq p_v$. The orthogonality condition for the unsymmetrical case as well as for zero and repeating eigenvalues is available [1, pp. 98–106].

3.3.3 Response Analysis and Relevant Parameters

3.3.3.1 Response Analysis and Participation Factor

This section covers both undamped and damped vibration analyses. Mathematical formulations are first established in general form with viscous damping and then simplified to the undamped case. Consider the shear building shown in Figure 3.12 subjected to applied force $F(t)$ and viscous damping force expressed as $c\dot{x}_0$. Using the free-body diagrams, the motion equations may be written as

$$M_1\ddot{x}_1 + k_1 x_1 + c_1\dot{x}_1 - k_2 x_2 + k_2 x_1 - c_2\dot{x}_2 + c_2\dot{x}_1 = F_1(t) \tag{3.34a}$$

$$M_2\ddot{x}_2 + k_2 x_2 - k_2 x_1 + c_2\dot{x}_2 - c_2\dot{x}_1 = F_2(t) \tag{3.34b}$$

and in matrix notation as

$$[M]\{\ddot{x}\} + [C]\{\dot{x}\} + [K]\{x\} = \{F(t)\} \tag{3.34c}$$

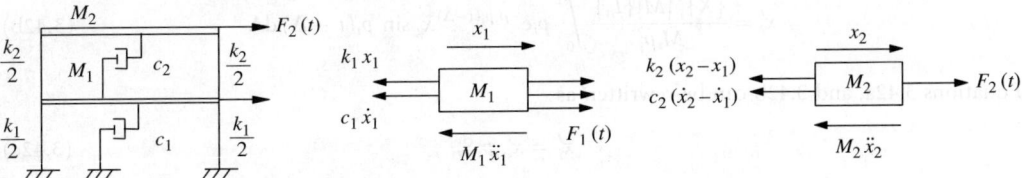

FIGURE 3.12 Structural model with viscous damping.

where the damping matrix is

$$[C] = \begin{bmatrix} c_1 + c_2 & -c_2 \\ -c_2 & c_2 \end{bmatrix} \tag{3.35}$$

The force matrix is $\{F(t)\} = \{F\}f(t)$. For earthquake excitation, $\{F(t)\} = -[M]\{I_n\}\ddot{x}_g$, $\{I_n\} = [\,1 \quad 1\,]^T$ is the *influence factor* to specify the force induced by $\ddot{x}_g$ at each floor. By using the modal matrix $[\Phi]$ or normal mode matrix $[X]$ with the orthogonality conditions, Equation 3.34c may be decoupled as Equation 3.36 or 3.37.

$$\{\ddot{x}'\} + [\Phi]^T[C][\Phi]\{\dot{x}'\} + \lceil p_i^2 \rfloor\{x'\} = [\Phi]^T\{F(t)\} \tag{3.36}$$

$$\lceil \bar{M} \rfloor\{\ddot{x}'\} + [X]^T[C][X]\{\dot{x}'\} + \lceil p_i^2 \rfloor\lceil \bar{M} \rfloor\{x'\} = [X]^T\{F(t)\} \tag{3.37}$$

in which, $\lceil \bar{M} \rfloor = [X]^T[M][X]$, let

$$\{\Phi\}_i^T[C]\{\Phi\}_j = 2\rho_i p_i \delta_{ij} \tag{3.38}$$

$$\{X\}_i^T[C]\{X\}_j = 2\rho_i p_i \bar{M}_i \delta_{ij} \tag{3.39}$$

where δ_{ij} is the *Kronnecker delta*. Note that for given damping coefficients based on physical condition, $[\Phi]^T[C][\Phi]$ need not necessarily be a diagonal matrix. However, one must use the diagonal elements in Equation 3.36 or 3.37 to carry out simple calculations using Duhamel's integral. Consequently, any row of Equation 3.36 or 3.37 is identical to the displacement response equation of the single d.o.f. system as shown in Equation 3.21. For instance, the ith row of Equation 3.36 is

$$\ddot{x}_i' + 2\rho_i p_i \dot{x}_i' + p_i^2 x_i' = \{\Phi\}_i^T\{F(t)\} \tag{3.40}$$

Thus following Equations 3.12a and 3.24b with $p = p^*$, the complete solution of Equation 3.40 is

$$x_i' = e^{-\rho_i p_i t}(A \cos p_i t + B \sin p_i t) + \frac{\{\Phi\}_i^T\{F\}}{p_i} \int_0^t e^{-\rho_i p_i(t-\Delta)}\{f(\Delta)\} \sin p_i(t - \Delta)\, d\Delta \tag{3.41a}$$

Similarly, the complete ith row of Equation 3.37 is

$$x_i' = e^{-\rho_i p_i t}(A \cos p_i t + B \sin p_i t) + \frac{\{X\}_i^T\{F\}}{\bar{M}_i p_i} \int_0^t e^{-\rho_i p_i(t-\Delta)}\{f(\Delta)\} \sin p_i(t - \Delta)\, d\Delta \tag{3.41b}$$

Considering the forced vibration due to an earthquake, we may express Equations 3.41a and 3.41b as Equations 3.42a and 3.42b, respectively

$$x_i' = \frac{\{\Phi\}_i^T[M]\{I_n\}}{p_i^2} \int_0^t p_i e^{-\rho_i p_i(t-\Delta)}\ddot{x}_g \sin p_i(t - \Delta)\, d\Delta \tag{3.42a}$$

$$x_i' = \frac{\{X\}_i^T[M]\{I_n\}}{\bar{M}_i p_i^2} \int_0^t p_i e^{-\rho_i p_i(t-\Delta)} \ddot{x}_g \sin p_i(t-\Delta)\, d\Delta \tag{3.42b}$$

Equations 3.42a, and 3.42b can be rewritten as

$$x_i' = x_{sti}' p_i S_{vi} \tag{3.42c}$$

in which x_{sti}' is equivalent to static displacement and S_{vi} is the velocity response spectrum corresponding to the ith mode. For undamped vibration, we simply let $e^{-\rho_i p_i t}$ and $e^{-\rho_i p_i(t-\Delta)}$ be a unity in Equations 3.40 to 3.42. The actual displacement response is obtained from

$$\{x\} = [\Phi]\{x'\} \tag{3.43a}$$

or

$$\{x\} = [X]\{x'\} \tag{3.43b}$$

As mentioned before, $[\Phi]$ and $[X]$ have significant physical meaning because they show how much contribution is made from each mode to the total displacement.

For example, the structure shown in Figure 3.13 has $p_1 = 2.218$ rad/s, $p_2 = 10.781$ rad/s, and the applied forces $F_1(t) = F_1 f(t)$, $F_2(t) = F_2 f(t)$, where $F_1 = 20$ k, $F_2 = 30$ k, and $f(t) = 1 - (t/\zeta)$ are given in the accompanying figure. Find the displacement contributed from each individual mode and the total response. Assume initial conditions are zero. $[\Phi_{11} \ \Phi_{21}] = [0.340 \ 0.293]$, $[\Phi_{12} \ \Phi_{22}] = [-0.293 \ 0.340]$. Based on Equations 3.41 and 3.43a, the displacement response is

$$\{x\} = [\Phi] \int_0^t [\backslash \sin p(t-\Delta)_\backslash][\backslash p_\backslash]^{-1}[\Phi]^T\{F(\Delta)\}\, d\Delta$$

$$= [\Phi] \sum_{r=1}^n \frac{\Phi_{ri} F_r}{p_i} \int_0^t f(t) \sin p_i(t-\Delta)\, d\Delta$$

$$= \begin{bmatrix} \Phi_{11} x_1' + \Phi_{12} x_2' \\ \Phi_{21} x_1' + \Phi_{22} x_2' \end{bmatrix}$$

$$= \begin{bmatrix} x_1(\text{due to the first mode}) + x_1(\text{due to the second mode}) \\ x_2(\text{due to the first mode}) + x_2(\text{due to the second mode}) \end{bmatrix} \tag{3.44}$$

The results obtained from Equation 3.44 are illustrated in Figure 3.14 and Figure 3.15. This example reveals how the individual modes contribute to the total response. In this case, the first mode is indeed contributing the most. For tall buildings, in general, the first several fundamental modes are essential in affecting response behavior and are practically needed in design. This example also reveals that $\Phi_{ri} F_r$ is measuring how much of the applied force contributing to the rth mode. Thus $\Phi_{ri} F_r$, $X_{ri} F_r$, $\Phi_{ri} M_r$, and $X_{ri} M_r$ in Equations 3.41a, 3.41b and 3.42a, 3.41b are all measuring how much the rth mode participates

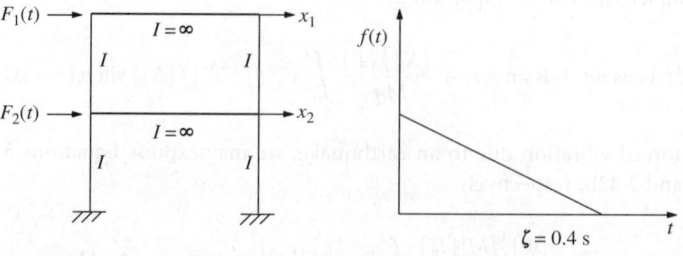

FIGURE 3.13   Undamped forced vibration of a shear building.

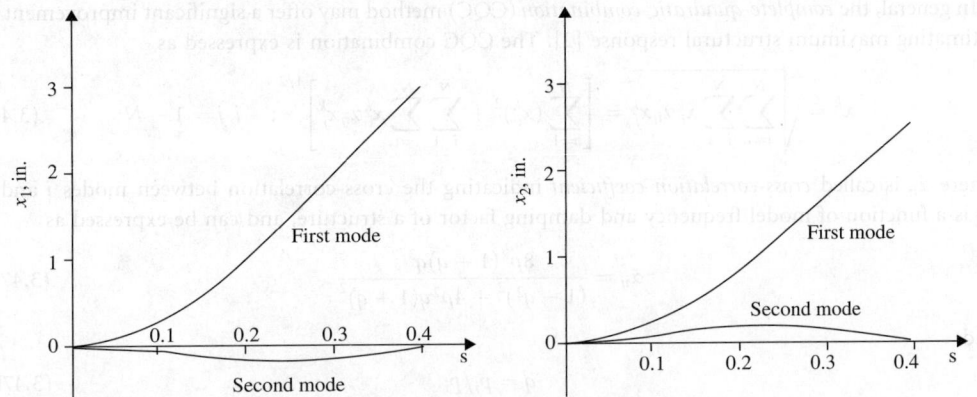

FIGURE 3.14 x_1 and x_2 influenced by individual modes. (Reprinted from Ref. [1, p. 62] by courtesy of Marcel Dekker, Inc.)

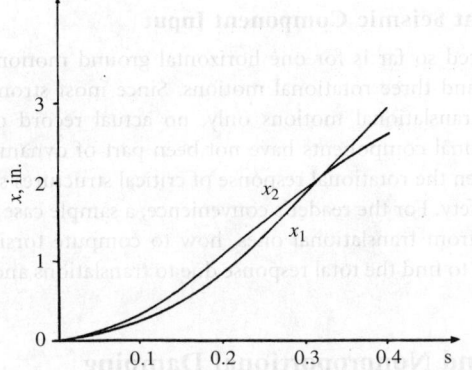

FIGURE 3.15 Displacements x_1 and x_2. (Reprinted from Ref. [1, p. 63] by courtesy of Marcel Dekker, Inc.)

in synthesizing the structure's total load and are called *participation factor(s)*. Note that the time function $f(t)$ in Equations 3.41a, 3.41b and 3.42a, 3.42b is supposed to be the same for all the exciting forces. When $f(t)$ differs among individual forces, the modal matrix equation can be solved by using numerical integration.

3.3.3.2 Modal Combination Methods

For using response spectra, solutions obtained from Equation 3.42c are always the maximum for each mode during the vibration. The displacements are apparently overestimated and conservative. One technique to level off the overestimated response is called the *root-mean-square* method or *square-root-of-the-sum-of-the-squares* (SRSS) and can be expressed as follows:

$$x^k = \sqrt{\sum_{n=1}^{N} (x^k)_n^2}; \quad N = \text{number of modes considered} \tag{3.45}$$

in which superscript k represents the kth d.o.f. of the structural system. This method of combination is known to give a good approximation of the response for frequencies distinctly separated in neighboring modes.

In general, the *complete-quadratic-combination* (CQC) method may offer a significant improvement in estimating maximum structural response [2]. The CQC combination is expressed as

$$x^k = \sqrt{\sum_{i=1}^{N}\sum_{j=1}^{N} x_i^k \alpha_{ij} x_j^k} = \left[\sum_{i=1}^{N} (x_i^k)^2 + \sum_{j=1}^{N}\sum_{i=1}^{N} x_i^k \alpha_{ij} x_j^k \right]^{1/2} ; \quad i,j = 1 - N \tag{3.46}$$

where α_{ij} is called *cross-correlation coefficient* indicating the cross-correlation between modes i and j. α_{ij} is a function of model frequency and damping factor of a structure, and can be expressed as

$$\alpha_{ij} = \frac{8\rho^2(1+q)q^{3/2}}{(1-q^2)^2 + 4\rho^2 q(1+q)^2} \tag{3.47a}$$

and

$$q = p_j/p_i \tag{3.47b}$$

The correlation coefficient diminishes when q is small, that is, p_j and p_i are distinctly separate, particularly when damping is small, such as $\rho = 0.05$ or less. The CQC method is significant only for a narrow range of q. Note that when α_{ij} is small, the second term of Equation 3.46 can be neglected; consequently, CQC is reduced to SRSS given in Equation 3.45. A computer program for CQC is available [1, pp. 871–874].

3.3.3.3 Multicomponent Seismic Component Input

The modal analysis presented so far is for one horizontal ground motion. Actually, earthquakes can induce three translational and three rotational motions. Since most strong motion seismographs are designed to record three translational motions only, no actual record of rotational earthquakes is available. In general, rotational components have not been part of dynamic structural analysis. But if their effect is significant, then the rotational response of critical structures such as nuclear power plants should be considered for safety. For the reader's convenience, a sample case is shown of how to generate seismic rotational records from translational ones, how to compute torsional response spectra from rotational records, and how to find the total response due to translations and rotations [1, pp. 383–414].

3.3.4 Proportional and Nonproportional Damping

There are several kinds of damping, namely, *structural damping*, *coulomb damping*, and *viscous damping*. Structural damping results from internal friction within the material or at connections of a structure. Coulumb damping is due to a body moving on a dry surface. Viscous damping results from a system vibrating in air or liquid, of which examples are shock absorbers, hydraulic dashpots, and a body sliding on a lubricated surface. Viscous damping is commonly used in structural dynamics and is therefore presented in this chapter. The damping matrix can be symmetric or nonsymmetric, proportional or nonproportional. For general structural dynamics, a damping matrix can be treated as symmetric and proportional. To identify whether a damping formulation is proportional or not, one may use Equation 3.10, $\rho = c/(2Mp) = c/[2\sqrt{(KM)}] = cp/(2K)$, which indicates that the damping factor, ρ, can be expressed in terms of mass, stiffness, or a combination of both. For a general expression, let

$$[C] = \alpha[M] + \beta[K] \tag{3.48}$$

which is substituted in Equation 3.38 with consideration of orthogonality condition; then

$$\alpha + \beta p_i^2 = 2\rho_i p_i \tag{3.49}$$

Thus, the damping factor can be determined for a given set of α and β as

$$\rho_i = \frac{\alpha}{2p_i} + \frac{\beta p_i}{2} \tag{3.50a}$$

Examining α and β reveals the physical sense of ρ. If $\alpha = 0$, then

$$\rho_i = \frac{\beta p_i}{2} \tag{3.50b}$$

which means that ρ_i is proportional to p_i. For higher modes of larger p_i, ρ_i will be larger; then, the higher modes of a system will be damped faster than the lower modes. Since $[C] = \beta[K]$, the damping is proportional to stiffness and is called *relative damping* because it is associated with relative velocities of displacement coordinates.

Let $\beta = 0$; then

$$\rho_i = \frac{\alpha}{2p_i} \tag{3.50c}$$

which means that ρ_i is inversely proportional to p_i. Therefore, lower modes will be damped out more quickly than higher modes. Since $[C] = \alpha[M]$, which is associated with absolute velocity of displacement coordinates, the damping is called *absolute damping*. Damping expressed in Equation 3.48 is called *proportional damping*, from which $[\Phi]^T[C][\Phi]$ is always diagonal. Nonproportional damping can be due to different damping factors assigned to different d.o.f. of a system. For instance, engineers may use different construction materials at various floors of a building, employ a concentrated damper (such as a control system) at a certain structural level, or assign larger damping coefficients for the foundation than for the superstructure when foundation–structure interaction is encountered. For response analysis with nonproportional damping, detailed analysis procedures are available including calculation of complex eigenvalues [1, pp. 128–159].

3.3.5 Various Eigensolution Techniques

The methods used in finding natural frequencies and normal modes are referred here as *eigensolution techniques*. Calculation of eigensolution is time consuming and expensive in computation. Most of the computer programs have a few eigensolution subroutines for the user to choose; therefore, a understanding of the numerical procedures of the techniques is essential to use the computer program properly and to interpret the results with confidence. Thus, a few well-known eigensolution techniques along with numerical examples may be found in Ref. [1], including (1) the determinant method, (2) the iteration method, (3) Choleski's decomposition method, (4) the generalized Jacobi method, and (5) the Sturm sequence method. Since formulations and numerical procedures are given in detail in Ref. [1, pp. 72–98], brief comments on the methods' characteristics are presented to facilitate the user's choice:

1. *Determinant method.* The method is convenient for longhand solution of a matrix with small dimension. However, if a singular matrix is modified to a triangular matrix then the method can be used for a larger matrix dimension and can be applied to various eigen formulations [1, pp. 186–187, 225–229].

2. *Iteration method.* The method is convenient for longhand or computer solution. If the stiffness matrix $[K]$ is used in iteration, then the first eigenvalue corresponds to the highest mode, and the subsequent solution is the second highest. However, if the inversion of the stiffness matrix, $[K]^{-1}$, is used in the iteration, then the first solution is the fundamental mode. Since only the first several lower modes are essential in a tall building design, inversion of $[K]$ is necessary. The method is proved to be a converging procedure and can be applied to unsymmetrix matrix as well as complex eigensolution formulation [1, pp. 72–80, 98–105, 137–149].

3. *Choleski's decomposition method.* The method is based on the aforementioned iteration approach. Since the iteration method requires time-consuming stiffness matrix inversion in order to find the fundamental modes, Choleski's decomposition method avoids the matrix inversion and therefore is significant for a structural formulation with large dimension [1, pp. 81–86].

4. *Generalized Jocobi method.* The Jacobi method has several versions. The method presented herein is applicable to a symmetric matrix and is capable of solving negative, zero, or positive eigenvalues.

However, it must solve simultaneously for all the eigenvalues and corresponding eigenvectors. For a large structural system, if only some eigenpairs are needed, this method can be inefficient [1, pp. 87–95].

5. *Sturm sequence method.* The method is suitable for calculating a limited number of eigenpairs. For large structural systems, only several fundamental modes are of practical use; the method is therefore very useful. The method is also useful for determining eigenpairs at a given range of frequencies. Thus, starting from the fundamental mode is not required. The method can be applied to a structure having rigid body motion; that is, [K] can be singular but must be symmetric [1, pp. 95–98].

3.4 Response Analysis of Continuous Systems — Dynamic-Stiffness Formulation

3.4.1 Characteristics and Formulation of the Model

The dynamic-stiffness formulation presented in this section is for the structures composed of prismatic members with and without superimposed masses. It is focused on flexural vibration based on Bernoulli–Euler theory and Timoshenko theory as well as coupling vibration of axial and flexural. The model, however, has been extensively developed for torsional and flexural vibrations with elastic media and P–Δ effect. Coupling vibration was also presented for a structure vibrating in torsional and flexural modes, and longitudinal, torsional, and flexural modes. The method may be considered to be exact because the dynamic stiffness is formulated on the basis of a partial differential equation and consequently yields a lower bound, in comparison with those obtained by using other methods. Note that the model has been used in *transport matrix* formulation for some special structures [3].

3.4.2 Derivation of Dynamic Stiffness Based on Bernoulli–Euler Equation

3.4.2.1 Bernoulli–Euler Equation

Let the typical prismatic beam shown in Figure 3.16 be subjected to a time-dependent load, $w(x, t)$, that is, the magnitude of the load varies continuously from section to section, and the direction varies with time. This load will cause motions of deflection, $y(x, t)$ (assume positive downward), velocity, $\partial y(x, t)/\partial t$, and acceleration, $\partial^2 y(x, t)/\partial t^2$, as well as internal forces of moment, $M(x, t)$ and shear, $V(x, t)$. The equilibrium condition of the free-body diagram yields

$$\frac{\partial^2 M(x, t)}{\partial x^2} = m \frac{\partial^2 y(x, t)}{\partial t^2} - w(x, t) \tag{3.51}$$

where m is the mass per unit length. Considering uniform moment of inertia, I and small deflection, one then obtains

$$EI \frac{\partial^4 y(x, t)}{\partial x^4} + m \frac{\partial^2 y(x, t)}{\partial t^2} = w(x, t) \tag{3.52}$$

where E is the modulus of elasticity. Equation 3.52 is a *partial differential equation* for which the

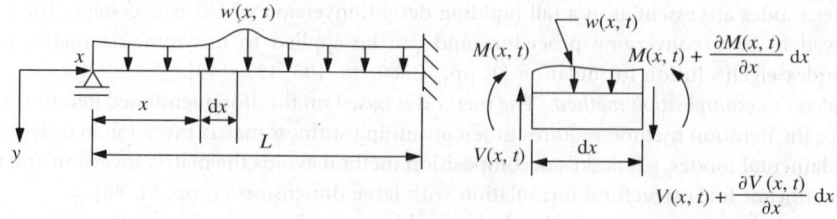

FIGURE 3.16   Prismatic beam.

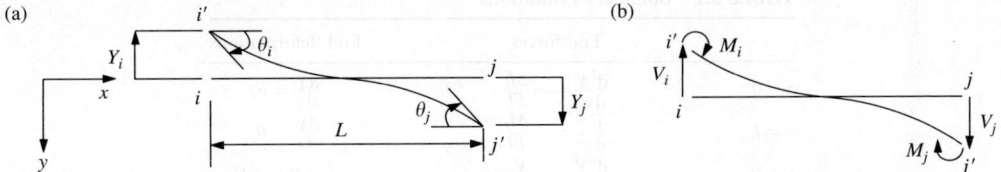

FIGURE 3.17 Typical deformed member: (a) end deformation and (b) end force. (Reprinted from Ref. [1, p. 166] by courtesy of Marcel Dekker, Inc.)

separation of variable technique should be employed for solution. Thus

$$y(x, t) = Y(x) \cdot g(t) \tag{3.53}$$

where $Y(x)$ is called the *shape function*, which is expressed in terms of x along a member and $g(t)$ is the *time function*, which is related to the variable time of motion. The shape function is

$$Y = A \sin \lambda x + B \cos \lambda x + C \sinh \lambda x + D \cosh \lambda x \tag{3.54}$$

where

$$\lambda^4 = \frac{p^2 m}{EI} \tag{3.55}$$

p is a constant independent of time and has unit of rad/s; A, B, C, and D are arbitrary constants and can be determined by using the boundary conditions of a member. The derivation is based on Bernoulli–Euler theory with consideration of bending deformation.

3.4.2.2 Dynamic-Stiffness Coefficients

For the arbitrary member, "ij," of a framework shown in Figure 3.17, let the end moments, M_i, M_j, the end shears, V_i, V_j, and their associated end deflections, Y_i, Y_j, as well as the end slopes, θ_i, θ_j, be considered positive. According to the shape function derived in Table 3.1, the following boundary conditions can be established. Let $\{Q_e\}=[M_i \ M_j \ V_i \ V_j]^T$ and $\{q_e\} = [\theta_i \ \theta_j \ Y_i \ Y_j]^T$, then the flexural dynamic-stiffness coefficients can be expressed as

$$\{Q_e\} = [K_e]\{q_e\} \tag{3.56}$$

$$[K_e] = \begin{bmatrix} \frac{\sinh \phi \cos \phi - \cosh \phi \sin \phi}{G} & \frac{\sin \phi - \sinh \phi}{G} & \frac{\sinh \phi \sin \phi}{G}\left(\frac{\phi}{L}\right) & \frac{\cosh \phi - \cos \phi}{G}\left(\frac{\phi}{L}\right) \\ & \frac{\sinh \phi \cos \phi - \cosh \phi \sin \phi}{G} & \frac{\cosh \phi - \cos \phi}{G}\left(\frac{\phi}{L}\right) & \frac{\sinh \phi \sin \phi}{G}\left(\frac{\phi}{L}\right) \\ & & \frac{(-\cos \phi \sinh \phi - \cosh \phi \sin \phi)}{G}\left(\frac{\phi}{L}\right)^2 & \frac{-\sinh \phi - \sin \phi}{G}\left(\frac{\phi}{L}\right)^2 \\ \text{sym.} & & & \frac{(-\cos \phi \sinh \phi - \cosh \phi \sin \phi)}{G}\left(\frac{\phi}{L}\right)^2 \end{bmatrix}$$

$$\tag{3.57}$$

TABLE 3.1   Boundary Conditions

	End forces	End deformations
$x=0$	$\dfrac{d^2 Y}{dx^2} = -\dfrac{M_i}{EI}$	$\dfrac{dY}{dx} = \theta_i$
$x=L$	$\dfrac{d^2 Y}{dx^2} = -\dfrac{M_j}{EI}$	$\dfrac{dY}{dx} = \theta_j$
$x=0$	$\dfrac{d^3 Y}{dx^3} = -\dfrac{V_i}{EI}$	$Y = -Y_i$
$x=L$	$\dfrac{d^3 Y}{dx^3} = -\dfrac{V_j}{EI}$	$Y = -Y_j$

where $\phi = \lambda L$, $G = [(\cosh \phi \cos \phi - 1)L]/\phi EI$. When ϕ approaches zero, which implies that p must be zero in Equation 3.55, Equation 3.57 becomes

$$[K_e] = \begin{bmatrix} 4 & 2 & \dfrac{-6}{L} & \dfrac{-6}{L} \\ & 4 & \dfrac{-6}{L} & \dfrac{-6}{L} \\ & & \dfrac{12}{L^2} & \dfrac{12}{L^2} \\ \text{sym.} & & & \dfrac{12}{L^2} \end{bmatrix} \dfrac{EI}{L} \tag{3.58}$$

which represents static stiffness coefficients.

3.4.3 Dynamic System Matrix and Eigensolutions

Let a rigid frame have joint rotations $\{X_\theta\}$, sideways $\{X_s\}$, and concentrated masses $[M]$ (associated with sideways such as floor masses), then the dynamic system matrix without externally applied force may be expressed as [1, pp. 175–186]

$$\left(\begin{bmatrix} K_{11} & K_{12} \\ K_{21} & K_{22} \end{bmatrix} - p^2 \begin{bmatrix} 0 & 0 \\ 0 & M \end{bmatrix} \right) \begin{Bmatrix} X_\theta \\ X_s \end{Bmatrix} = 0 \tag{3.59}$$

If the structure has rotations only, the terms corresponding to $\{X_s\}$ should not be included. Eigenvalues can be calculated from the nontrivial solution of Equation 3.59 by making the determinant of the coefficients equal to zero as

$$\left\| \begin{bmatrix} K_{11} & K_{12} \\ K_{21} & K_{22} \end{bmatrix} - p^2 \begin{bmatrix} 0 & 0 \\ 0 & M \end{bmatrix} \right\| = 0 \tag{3.60}$$

The eigen vector of any mode can be determined from the singular matrix in Equation 3.59 after zero determinant [1, pp. 187–194].

3.4.4 Response Analysis

3.4.4.1 Formulation

The dynamic deflection of a member at any point x and time t can be represented as the summation of modal components as

$$y(x, t) = \sum^n Y_i(x) q_i(t) \tag{3.61}$$

where $Y_i(x)$ is the shape function of the member, $q_i(t)$ are generalized coordinates (similar to x_i' in the lumped-mass model), and $i = 1, 2, \ldots, n$, the number of normal modes.

The uncoupled motion equation for the kth mode is

$$\ddot{q}_k + 2\rho p_k \dot{q}_k + p_k^2 q_k = \frac{f(t) \int_0^L F(x) Y_k \, dx}{\int_0^L m Y_k^2 \, dx} \tag{3.62}$$

Let

$$\bar{M}_k = \int_0^L m Y_k^2 \, dx \tag{3.63}$$

$$F_k = \int_0^L F(x) Y_k \, dx \tag{3.64}$$

Then, Equation 3.62 becomes

$$q_k = \frac{F_k}{\bar{M}_k p_k^2} A_k(t) \quad \text{or} \quad q_k = \frac{\sum_{l=1}^{NM} F_{kl}}{p_k^2 \sum_{l=1}^{NM} \bar{M}_{kl}} A_k(t) \tag{3.65}$$

where NM is the total number of members of a structure and

$$A_k(t) = \frac{1}{\sqrt{1-\rho^2}} \int_0^t e^{\rho^* p^*(t-\Delta)} \sin p^*(t-\Delta) F(\Delta) \, d\Delta \tag{3.66}$$

Note that F_k is a general expression, which can represent dynamic force or ground excitation. $A_k(t)$ can be simplified not to include the effect of damping on frequency, then

$$A_k(t) = \int_0^t e^{-\rho p(t-\Delta)} \sin p(t-\Delta) F(\Delta) \, d\Delta \tag{3.67}$$

Let

$$Y_{stk} = \frac{F_k}{\bar{M}_k p_k^2} \quad \text{or} \quad Y_{stk} = \frac{\sum_{l=1}^{NM} F_{kl}}{p_k^2 \sum_{l=1}^{NM} \bar{M}_{kl}} \tag{3.68}$$

Then, the total response is obtained by superimposing the modes shown in Equation 3.68

$$y(x,t) = \sum^n Y_{stk} A_k(t) Y_k(x) \tag{3.69}$$

The moment and shear at time t are

$$M(x,t) = EIy''(x,t) = EI \sum^n Y_{stk} A_k(t) Y_k''(x) \tag{3.70}$$

$$V(x,t) = EIy'''(x,t) = EI \sum^n Y_{stk} A_k(t) Y_k'''(x) \tag{3.71}$$

Note that A_k is the *amplification factor* or *dynamic load factor* for the kth mode, Y_{stk} is the *pseudo-static displacement* at the kth mode, and F_k is the *participation factor*.

3.4.4.2 Numerical Example

A two-span continuous beam shown in Figure 3.18 is subjected to a uniform impulse within the left span as $F(x,t) = 3000[1-(t/0.1)]$ lb/in. $(5.25 \times 10^2[1-(t/0.1)]$ kN/m). Structural properties are $L = 150$ in. (3.81 m), $EI = 6 \times 10^9$ lb/in.^{2} $(4.1 \times 10^{10}$ kN/m^2), and $m = 0.2$ lbs^2/in.^{2} $(1378.9$ Ns^2/m^2). Find the moments at support B and check the equilibrium condition.

(a)

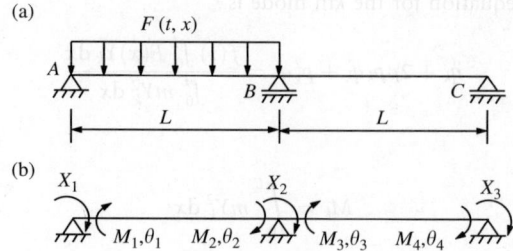

(b)

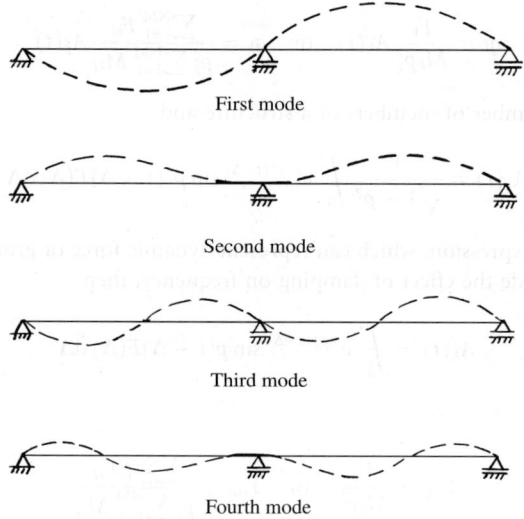

FIGURE 3.18   Two-span continuous beam: (a) given beam and (b) d.o.f., internal moments.

First mode

Second mode

Third mode

Fourth mode

FIGURE 3.19   Normal modes. (Reprinted from Ref. [1, p. 209] by courtesy of Marcel Dekker, Inc.)

Since the beam has rotational d.o.f., Equation 3.60 becomes $\det(K_{11}) = 0$, which yields ϕ_i (also λ_i). The eigen vector $\{X\}_i$ is then calculated. The results are

$$\phi_1 = 3.14160; \qquad \lambda_1 = \frac{\phi_1}{L} = 0.020940; \qquad p_1 = 75.97625 \text{ rad/s}; \qquad \{X\}_1 = [1 \ \ -1 \ \ 1]^{\mathrm{T}}$$

$$\phi_2 = 3.92660; \qquad \lambda_2 = \frac{\phi_2}{L} = 0.026177; \qquad p_2 = 118.68938 \text{ rad/s}; \qquad \{X\}_2 = [-1 \ \ 0 \ \ 1]^{\mathrm{T}}$$

$$\phi_3 = 6.28318; \qquad \lambda_3 = \frac{\phi_3}{L} = 0.0418879; \qquad p_3 = 303.905 \text{ rad/s}; \qquad \{X\}_3 = [1 \ \ 1 \ \ 1]^{\mathrm{T}}$$

$$\phi_4 = 7.06858; \qquad \lambda_4 = \frac{\phi_4}{L} = 0.0471239; \qquad p_4 = 384.6268 \text{ rad/s}; \qquad \{X\}_4 = [-1 \ \ 0 \ \ 1]^{\mathrm{T}}$$

The normal modes are shown in Figure 3.19. Substituting $\{X\}_i$ in Equation 3.54, one can calculate the constants A, B, C, and D for each mode. Consequently, the shape function of these members are obtained as

$Y_{11} = 47.74648 \sin \lambda_1 x$

$Y_{12} = -47.74648 \sin \lambda_1 x$

TABLE 3.2 Calculated Results

	F_{kl}		$\bar{M}_{kl}$			
k	$l=1$	$l=2$	$l=1$	$l=2$	Y_{stk}	$A_k(t)$
1	13,678,350	0	34,195.8	34,195.8	0.034648	1.62095
2	−10,166,310	0	20,702.6	20,702.6	−0.017429	0.75110
3	0	0	8,549.0	8,549.0	0	0.09350
4	558,594	0	6,771.0	7,771.0	0.000279	1.07073

Source: Reprinted from Ref. [1, p. 210], by courtesy of Marcel Dekker, Inc.

$Y_{21} = -37.16499 \sin \lambda_2 x - 1.035983 \sinh \lambda_2 x$
$Y_{22} = -26.28982 \sin \lambda_2 x + 26.26940 \cos \lambda_2 x + 26.28982 \sinh \lambda_2 x - 26.26940 \cosh \lambda_2 x$
$Y_{31} = 23.87324 \sin \lambda_3 x$
$Y_{32} = 23.87324 \sin \lambda_3 x$
$Y_{41} = -21.24624 \sin \lambda_4 x + 0.025582 \sinh \lambda_4 x$
$Y_{42} = 15.02337 \sin \lambda_4 x - 15.02335 \cos \lambda_4 x - 15.02337 \sinh \lambda_4 x + 15.02335 \cosh \lambda_4 x$

Then, from Equation 3.65

$$q_k = \frac{\sum_{l=1}^{2} F_{kl}}{p_k^2 \sum_{l=1}^{2} \bar{M}_{kl}} A_k(t), \quad k = 1, 2, 3, 4 \tag{3.72}$$

in which the amplification factor of impulse load, for $t \leq \zeta$, $\zeta = 0.1$ s, is

$$A_k(t) = 1 - \cos p_k t + \frac{\sin p_k t}{p_k \zeta} - \frac{t}{\zeta} \tag{3.73}$$

The maximum displacement of the first mode occurs during the pulse at 0.0379 s, which is used for the response of other modes. The calculations are summarized in Table 3.2. Substituting the results from Table 3.2 into Equation 3.70 yields the moments at B for $t = 0.0379$ s associated with the first, second, and fourth modes as

$$M_2 = -EIy''(L) = -EI[Y_{st1}A_1 Y_{11}''(L) + Y_{st2}A_2 Y_{21}''(L) + Y_{st4}A_4 Y_{41}''(L)]$$
$$= -EI[0.034648(1.62093)(0) + (-0.017429(0.75101)(-0.036))$$
$$+ 0.000279(1.070783)(0.06672)] = -4.912053(10^{-4})EI \tag{3.74}$$

$$M_3 = EIy''(0) = EI[Y_{st1}A_1 Y_{12}''(0) + Y_{st2}A_2 Y_{22}''(0) + Y_{st4}A_4 Y_{42}''(0)]$$
$$= EI[0.034648(1.62093)(0) + (-0.017429(0.75101)(-0.036))$$
$$+ 0.000279(1.070728)(0.06672)] = 4.912053(10^{-4})EI \tag{3.75}$$

Note that the third mode does not contribute to the structural response because $Y_{st3} = 0$. Note also that the equilibrium check on moments is satisfied at the joint because $\sum M = 0$.

3.4.5 Effects of Rotatory Inertia as well as Bending and Shear Deformation on Frequencies

Bernoulli–Euler theory is derived based on bending deformation only. In fact, a flexural vibration can include bending and *shear deformation* as well as *rotatory inertia* as derived in Timoshenko theory. Consider an element shown in Figure 3.20. Let z be the distance measured at any point from the neutral axis; then the displacement of a fiber located at z is

$$y^a = -z \frac{\partial y}{\partial x} \tag{3.76}$$

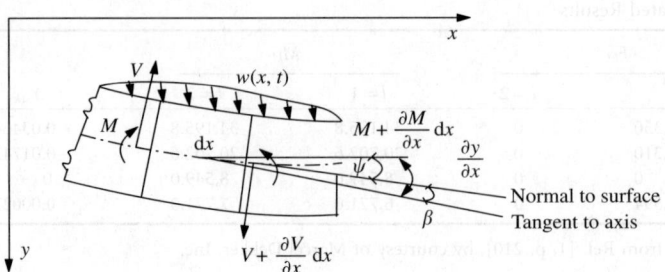

FIGURE 3.20   Element of Timoshenko beam.

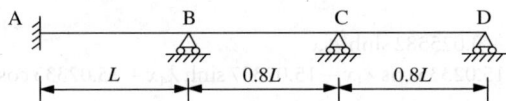

FIGURE 3.21   Three-span beam.

Since y varies with time and distance, y^a-must also vary with time and distance along the axis. Thus, for every dx length of a beam, the cross-section dA has an inertia force

$$\frac{\gamma}{g}(dA)(dx)\ddot{y}^a \tag{3.77}$$

Also shown in the figure, the total slope $\partial y/\partial x$ is a combination of the bending slope, ψ and the shear slope, β

$$\frac{\partial y}{\partial x} = \psi + \beta \tag{3.78}$$

Using the bending and shear slopes, one can write

$$M = -EI\frac{\partial \psi}{\partial x} \quad \text{and} \quad V = \mu AG\beta = \mu AG\left(\frac{\partial y}{\partial x} - \psi\right) \tag{3.79}$$

where G is the *shear modulus*, μ is a constant called the *shear coefficient*, defined as the ratio of average shear stress on a section to the product of shear modulus and shear strain at the neutral axis of the member.

   Using the Timoshenko theory, one can derive dynamic-stiffness coefficients and system matrix for eigensolution and response analysis [1, pp. 240–253]. The presentation focused here is to show the significant effect of Timoshenko theory on natural frequencies. The three-span, continuous, uniform beam shown in Figure 3.21 is analyzed for illustration. The given conditions are $g = 9.8\,\text{m/s}^2$, $E = 2.1 \times 10^7\,\text{kN/m}^2$, γ (beam unit weight) $= 77\,\text{kN/m}^3$, and $G = 8.27 \times 10^8\,\text{kN/m}^2$. Let the slenderness ratio, L/R, of span AB vary from 20, 30, 40, 50, to 60; then find the first five natural frequencies by considering: (A) bending deformation only and (B) bending and shear deformation as well as rotatory inertia. In Case B, the values of μ are assumed to be 0.833 and $\frac{2}{3}$ for showing the effect of the shear factor on natural frequencies.

   Let p be frequencies of Bernoulli–Euler theory and p^* be frequencies of Timoshenko theory; then, the ratio of frequencies, p^*/p, of the first five modes for various slenderness ratios of Cases A and B are shown in Figure 3.22. Observation reveals that (1) Timoshenko theory yields lower frequencies than Bernoulli–Euler theory; (2) reduction is more pronounced for higher modes and lower slenderness ratios; and (3) shear factor has a greater effect on higher modes while smaller μ reduces frequency more than larger μ.

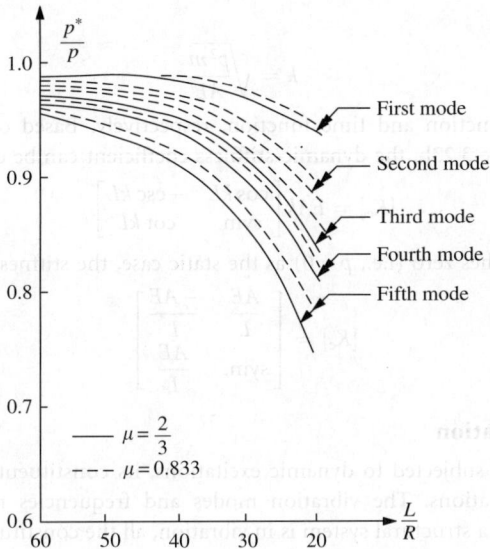

FIGURE 3.22 Comparison of natural frequencies. (Reprinted from Ref. [1, p. 253] by courtesy of Marcel Dekker, Inc.)

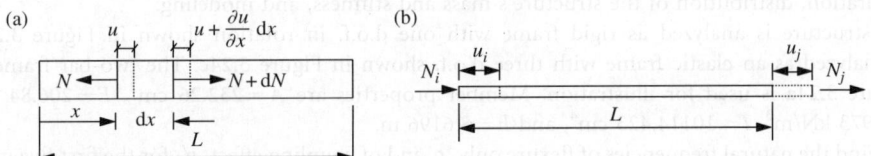

FIGURE 3.23 Longitudinal vibration: (a) element and (b) force–deformation relationship.

3.4.6 Coupling of Longitudinal and Flexural Vibration

3.4.6.1 Longitudinal Vibration and Stiffness Coefficients

Consider element dx of a longitudinal bar shown in Figure 3.23a; the equilibrium equation of the element is

$$(m\,dx)\ddot{u} = N + dN - N \tag{3.80}$$

where u is the longitudinal displacement and N is the axial force in tension. Since the axial force can be expressed in terms of area, as $N = AE\varepsilon = AE\,\partial u/\partial x$, Equation 3.80 becomes

$$\frac{\partial^2 u}{\partial t^2} = a^2 \frac{\partial^2 u}{\partial x^2} \tag{3.81}$$

where $a^2 = AE/m$. Using the separation of variables and substituting $u = X(x)g(t)$ into Equation 3.81

$$\frac{d^2 X}{dx^2} + \frac{p^2}{a^2} X = 0; \qquad \frac{\partial^2 g}{\partial t^2} + p^2 g = 0 \tag{3.82}$$

of which the solutions are

$$X = C_1 \sin kx + C_2 \cos kx; \qquad g = d_1 \sin pt + d_2 \cos pt \tag{3.83}$$

where

$$k = \sqrt{\frac{p^2 m}{AE}} \tag{3.84}$$

X and g are the shape function and time function, respectively. Based on the following boundary conditions shown in Figure 3.23b, the dynamic-stiffness coefficient can be derived as [1, pp. 213–214]

$$[K_e] = EAk \begin{bmatrix} \cos kL & -\csc kL \\ \text{sym.} & \cot kL \end{bmatrix} \tag{3.85a}$$

Note that when k approaches zero (i.e., $p = 0$) as the static case, the stiffness coefficients become

$$[K_e] = \begin{bmatrix} \dfrac{AE}{L} & \dfrac{-AE}{L} \\ \text{sym.} & \dfrac{AE}{L} \end{bmatrix} \tag{3.85b}$$

3.4.6.2 Coupling Vibration

When a plane structure is subjected to dynamic excitations, its constituent members may have longitudinal and flexural vibrations. The vibration modes and frequencies may be coupled. *Coupling vibration* means that when a structural system is in vibration, all the constituent members vibrate in the same frequency for both the longitudinal and flexural motions. On the other hand, *uncoupling vibration* implies that longitudinal and flexural motions are independent of each other so the vibration mode of a system depends on whether the mode is associated with longitudinal or flexural frequency but not affected by both. Whether a structure is in coupling or uncoupling motion depends on the structural configuration, distribution of the structure's mass and stiffness, and modeling.

The structure is analyzed as rigid frame with one d.o.f. in rotation shown in Figure 3.24b and then analyzed as an elastic frame with three d.o.f. shown in Figure 3.24c. The two-bar frame shown in Figure 3.24a is used for illustration. Member properties are $A = 232.26$ cm^2, $E = 206.84$ GN/m^2, $\gamma = 76.973$ kN/m^3, $I = 10114.423$ cm^4, and $h = 8.6196$ m.

(A) Find the natural frequencies of flexure only, p, and of coupling effect, p', for the first five modes by considering a wide range of slenderness ratios for the two identical members: $L/R = 20, 40, 60,$ and 80; where the radius of gyration, R, is constant based on the given cross-section and L is changed. (B) Study the influence of longitudinal frequency parameter on the coupling frequencies by letting the longitudinal dynamic stiffness be replaced by static stiffness as AE/L; find the pseudo-coupling frequencies p'', and compare them with p' obtained in A. p'/p, p''/p', and p''/p versus L/R are plotted in Figure 3.25 to 3.27, respectively. These three figures reveal that (1) p and p' are, respectively, upper and lower bounds of the frequencies; (2) the coupling effect on frequencies becomes more significant for higher modes and smaller slenderness ratios; and (3) the pseudo-coupling approach may be used for lower modes.

3.4.7 Effects of Elastic Media, Torsion, and Axial Force on Vibration

The dynamic-stiffness formulation is also developed for investigating the effect of elastic media on longitudinal and flexural vibrations, axial force on flexural vibration for both Bernoulli–Euler and

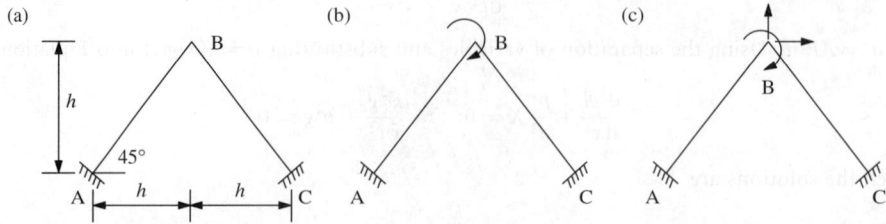

FIGURE 3.24 Coupling vibration: (a) given frame; (b) rigid frame; and (c) elastic frame.

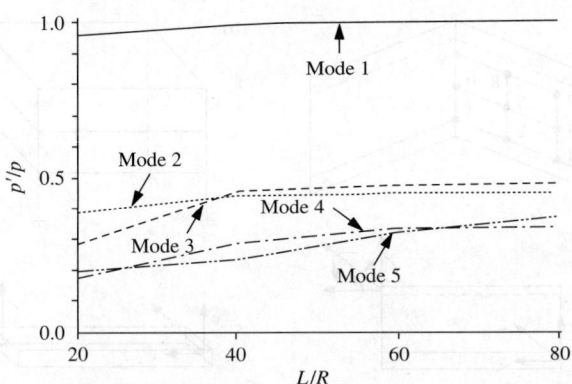

FIGURE 3.25 p'/p versus L/R. (Reprinted from Ref. [1, p. 228] by courtesy of Marcel Dekker, Inc.)

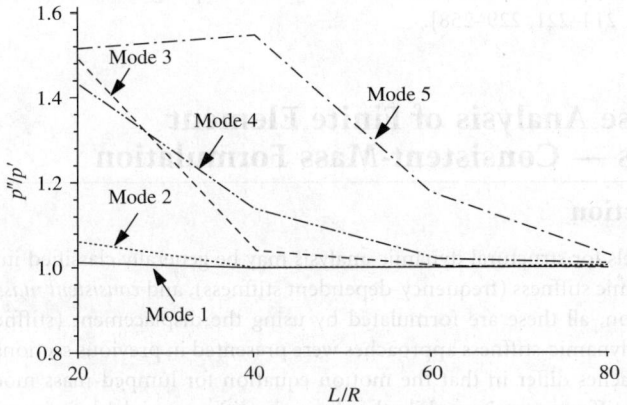

FIGURE 3.26 p''/p' versus L/R. (Reprinted from Ref. [1, p. 228] by courtesy of Marcel Dekker, Inc.)

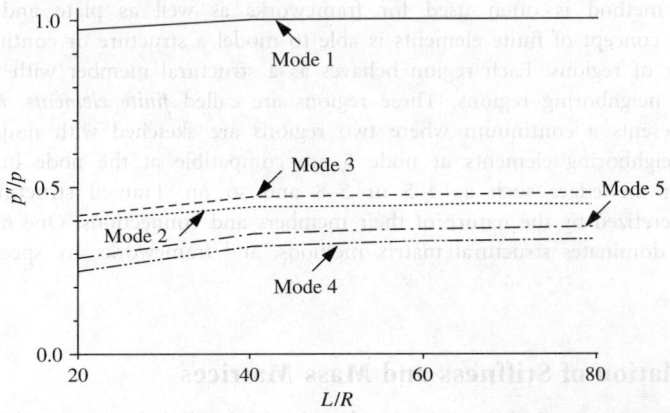

FIGURE 3.27 p''/p versus L/R. (Reprinted from Ref. [1, p. 229] by courtesy of Marcel Dekker, Inc.)

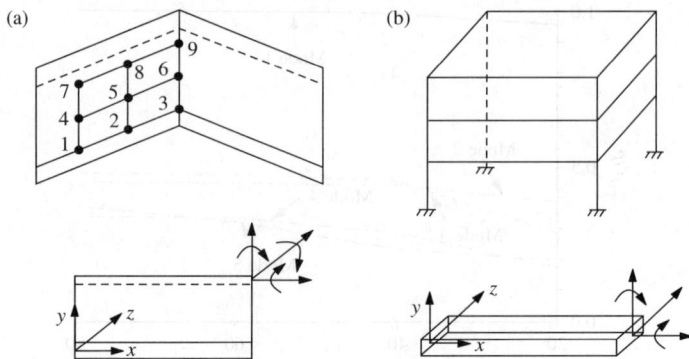

FIGURE 3.28   Discretized elements: (a) flat plate and (b) framework.

Timoshenko theories with and without elastic media, and coupling torsional and flexural vibrations of gird systems [1, pp. 214–221, 229–258].

3.5   Response Analysis of Finite Element Systems — Consistent-Mass Formulation

3.5.1   Introduction

Mathematical models for structural dynamic analysis may be generally classified into three approaches: lumped mass, dynamic stiffness (frequency-dependent stiffness), and *consistent mass (finite element)*. For computer application, all these are formulated by using the displacement (stiffness) matrix method. Lumped-mass and dynamic-stiffness approaches were presented in previous sections. The characteristics of these two approaches differ in that the motion equation for lumped-mass model consists of independent mass and stiffness matrices while the dynamic stiffness model has mass implicitly combined with stiffness. The lumped-mass and consistent-mass approaches are similar in terms of motion equation: both of them have independent mass and stiffness matrices; their mass matrices, however, are not the same.

Consistent mass may be considered as an alliance of finite elements normally used in *continuum mechanics*. This method is often used for frameworks as well as plate and shell structures. The fundamental concept of finite elements is able to model a structure or continuum by dividing it into a number of regions. Each region behaves as a structural member with nodes compatible to the nodes of neighboring regions. These regions are called *finite elements*. A plate shown in Figure 3.28 represents a continuum where two regions are sketched with nodes $1, 2, \ldots, 9$. the boundaries of neighboring elements at node 5 are compatible at the node but not necessarily compatible along the edges such as 4–5 or 5–8 and so on. Framed structures, however, are automatically discretized by the nature of their members and connections. One may say that finite element analysis dominates structural matrix methods, and frameworks are special cases of finite elements.

3.5.2   Formulation of Stiffness and Mass Matrices

Let $y(x, t)$ be the transverse displacement of each point in the direction perpendicular to the axis of a structural element where x denotes the points of the coordinates. If $N_i(x)$ is chosen as *coordinate functions* of the element and $q_e(t)$ represents the element's coordinates, then the dynamic deflection of the element can be expressed as

$$y(x, t) = \sum_{i=1}^{n} N_i(x) q_e(t) \qquad (3.86)$$

where n is the number of generalized coordinates. For a typical element shown in Figure 3.17, Equation 3.86 may be expressed without time variable as

$$Y(x) = [N_1 \quad N_2 \quad N_3 \quad N_4]\{q_e\} \qquad (3.87)$$

in which the shape functions are

$$\left.\begin{array}{ll} N_1 = x - \dfrac{2x^2}{L} + \dfrac{x^3}{L^2}; & N_2 = -\dfrac{x^2}{L} + \dfrac{x^3}{L^2} \\[2mm] N_3 = -1 + \dfrac{3x^2}{L^2} - \dfrac{2x^3}{L^3}; & N_4 = \dfrac{3x^2}{L^2} - \dfrac{2x^3}{L^3} \end{array}\right\} \qquad (3.88)$$

and the stiffness and mass coefficients of the element can be obtained from

$$k_{ij} = \int EI(x) \frac{d^2 N_i}{dx^2} \frac{d^2 N_j}{dx^2} \, dx \qquad (6.89)$$

$$m_{ij} = \int \rho(x) N_i N_j \, dv \qquad (3.90)$$

where moment inertia $I(x)$ and mass density $\rho(x)$, expressed as the function of x, signify that the formulation is applicable to a member with nonuniform cross-section. Equation 3.90 is integration with respect volume dv.

For a prismatic bar, Equations 3.89 and 3.90, respectively, yield Equations 3.91 and 3.92

$$[K_e] = \frac{EI}{L^3} \begin{bmatrix} 4L^2 & 2L^2 & -6L & -6L \\ & 4L^2 & -6L & -6L \\ & \text{sym.} & 12 & 12 \\ & & & 12 \end{bmatrix} \qquad (3.91)$$

$$[M_e] = \rho A L \begin{bmatrix} \dfrac{L^2}{105} & -\dfrac{L^2}{140} & -\dfrac{11L}{210} & \dfrac{13L}{420} \\[2mm] & \dfrac{L^2}{105} & \dfrac{13L}{420} & -\dfrac{11L}{210} \\[2mm] & & \dfrac{13}{35} & -\dfrac{9}{70} \\[2mm] & \text{sym.} & & \dfrac{13}{35} \end{bmatrix} \qquad (3.92)$$

Note that Equation 3.91 is identical to Equation 3.58 when ϕ (or p) approaches zero. The motion equation of a system may be expressed as

$$[M]\{\ddot{x}\} + [C]\{\dot{x}\} + [K]\{x\} = \{F(t)\} \qquad (3.93)$$

where $[M]$ is the mass matrix, $[C]$ is the damping matrix, $[K]$ is the stiffness matrix, and $\{F(t)\}$ represents the matrix involving externally applied forces or ground motion.

Note that Equation 3.93 is identical to Equation 3.34c of lumped-mass formulation. The major difference is that the mass matrix of the lumped-mass model is mostly a diagonal matrix associated with d.o.f. of side sway (the inertia force due to structural joint rotation is relatively small and thus negligible). However, the mass matrix of the consistent-mass model is a full matrix associated with joint rotations and side sways as shown in Equations 3.92 of a typical constituent member of a system. Thus, the computation effort for the lumped-mass model is much less than that for consistent mass because the dimension of the mass and stiffness matrices of a lumped-mass system can be significantly reduced through matrix condensation by eliminating rotational d.o.f.

3.5.3 Frequency Comparison for Lumped-Mass, Dynamic-Stiffness, and Consistent-Mass Models

In general, the computation efforts associated with response analysis for the lumped-mass, consistent-mass, and dynamic-stiffness models are reduced in the order of the individual models for a given problem, and the solution accuracy, however, may also be reduced, respectively. The accuracy sensitivity is illustrated by frequency comparison of a rigid frame shown in Figure 3.29a using these mathematical models: (A) the lumped-mass method, (B) the consistent-mass method, and (C) the dynamic-stiffness method (based on Bernoulli–Euler theory). For "A" and "B" the structural members are divided into three, six, and nine elements, respectively, and the masses of "A" are lumped at the center of the divided segments. Dynamic-stiffness, lumped-mass, and consistent-mass models are shown in Figure 3.29b,c, and d, respectively. Assume that all members are identical with $m = 0.04837$ kgs^2/m^2, $I = 0.00286$ cm^4, $E = 20684.27$ kN/cm^2, and $L = 0.2413$ m.

Eigenvalues of the first three modes are shown in Table 3.3 for comparison with the accurate solution by the dynamic stiffness method. Observation of the solutions reveals that the lumped-mass method needs six elements for the first two modes and nine for the third mode, while the consistent-mass method needs three elements for the first mode and six for the second and third modes. The lumped-mass model can give eigenvalues higher or lower than the dynamic stiffness's solution depending on how much mass is lumped at each node of the structure. The consistent-mass model always gives frequencies higher than the dynamic stiffness.

The consistent-mass model does not yield accurate solutions for higher modes because the shape functions, which are based on four generalized coordinates (see Equation 3.87) in deriving the mass and stiffness coefficients of a typical member, cannot be flexible enough to represent the deformed shape of

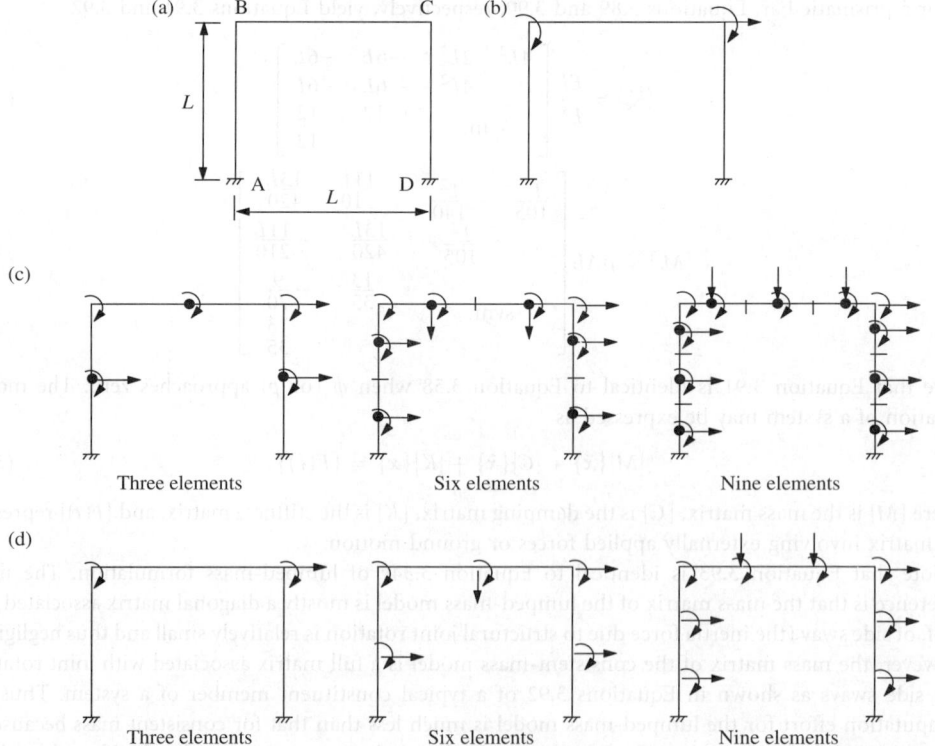

FIGURE 3.29   Modeling for three analysis methods: (a) given structure; (b) dynamic stiffness model; (c) lumped-mass model; and (d) consistent-mass model. (Reprinted from Ref. [1, p. 284] by courtesy of Marcel Dekker, Inc.)

TABLE 3.3 Comparison of Eigenvalues by Lumped-Mass, Consistent-Mass, and Dynamic Stiffness

Methods	Number of elements	First mode (rad/s)	Second mode (rad/s)	Third mode (rad/s)
Lumped-mass	3	210.5	531.2	0840.0
	6	195.8	762.5	1323.6
	9	194.5	765.7	1265.0
Consistent-mass	3	194.5	891.7	1988.6
	6	194.3	771.7	1264.9
	9	194.3	767.8	1254.2
Dynamic stiffness	3	194.3	766.8	1250.7

Source: Reprinted from Ref. [1, p. 284], by courtesy of Marcel Dekker, Inc.

higher modes. For the same reason, the lumped-mass method also yields larger rigidity in structural modeling. It is worthwhile to mention that for large structural systems, such as tall buildings, only the first several fundamental modes are essential for design. Thus, the eigenvalue inaccuracy resulting from the lumped-mass model can be ignored.

3.5.4 Axial, Torsional Vibration, and Flexural Vibration with Timoshenko Theory

Equations 3.86, 3.89, and 3.90 are basics for vibration of finite element systems and frameworks with Timoshenko theory in flexural vibration, axial as well as torsional vibrations. Equation 3.93 can be used for system response analysis. Detailed results are available [1, pp. 265–270, 285–317].

3.5.5 Dynamic Motion Equation with P–Δ Effect

When a member is subjected to a compressive force, P, the force times the member's deflection, Δ, yields an additional moment that is called *second-order moment* due to the P–Δ effect. For the dynamic-stiffness method, the P–Δ is implicitly expressed in the stiffness coefficient [1, pp. 238–240, 253–258]. For consistent-mass and lumped-mass methods, the P–Δ effect is formulated separately from stiffness in a *geometric matrix*. A geometric matrix is important in response analysis for tall buildings because heavy floor load transmitted to supporting columns can affect the vibrating frequencies significantly. The dynamic motion equation is similar to Equations 3.93 with additional term $[K_g]\{q\}$ as follows:

$$[M]\{\ddot{x}\} + [C]\{\dot{x}\} + [K]\{x\} - [K_g]\{x\} = \{F(t)\} \tag{3.94}$$

where $[K_g]$ is called geometric matrix in a consistent-mass model but *string matrix* in a lumped-mass model [1, pp. 303–306, 317–318]. The former involves both rotational and side sway d.o.f. and the latter involves side sway d.o.f. only. The negative sign corresponds to the axial force in compression. If the force in a member is in tension, then the $[k_g^i]$ should have a positive sign. Then, the system matrix is $[K_g] = \sum [k_g^i]$. It worthwhile to point out that compression reduces the stiffness that consequently reduces natural frequencies. It is also worthwhile to note that when the compressive force is a harmonic excitation, such as machinery vibration, then the structural response can be a *dynamic instability* problem. The instability behavior depends on the ratio of the structure's natural frequency to the forcing frequency of the axial force [4].

3.6 Elastic and Inelastic Response Analysis Methods Based on Nature of Exciting Forces

3.6.1 Nature of Exciting Forces

The nature of exciting forces and their categories were discussed in Section 3.1.1. Various response analysis methods are summarized herein for deterministic forces as well as nondeterministic forces. Earthquakes are treated for both deterministic and nondeterministic cases.

3.6.2 Modal Analysis

The method was presented in Sections 3.3.2 and 3.3.3 from which the response analysis can be obtained by using Duhamel's integral or response spectra. Note that this approach is only for an elastic system because it requires natural frequencies and mode shapes.

3.6.3 Direct Integration Methods

Direct integrations are of paramount importance in dynamic response analysis for many reasons including (1) they can be used for structural response at various deformation stages from elastic to inelastic and (2) they can be applied to motion equations for various irregular forcing functions such as earthquake accelerations in digital data. There are a number of numerical integration methods such as Newmark method, Wilson-θ method, Runge–Kutta fourth-order method, constant acceleration method, linear acceleration method, and average acceleration method. These methods have general basic characteristics: (1) determination of response involves computation of displacement, velocity, and acceleration; (2) at the beginning of integration, response parameter values must be given or have been calculated at one or more points proceeding with the specific time intervals of the integration; (3) truncation errors due to finite number of terms in Taylor series expansion for replacing the differential equation by a finite difference equivalent; and (4) propagation error resulting from tendency of error growth from the integration step to the next, wherein the solution can become unbounded and unstable. Two well-known methods are selected to outline integration procedures and their associated truncation as well as propagation errors.

3.6.3.1 Newmark Method (Linear Acceleration Version)

Newmark originally derived a general integration in which two parameters can be modified to change the integration to three cases:

1. *Average acceleration method.* The acceleration of the system remains constant over the time interval Δt and its value is equal to the average values of acceleration at the beginning and end of the interval.
2. *Constant acceleration method.* The acceleration of the system is constant and is equal to its value at the beginning of the time interval.
3. *Linear acceleration method.* The acceleration of the system varies linearly over the time interval.

Without detailed derivation, the end results of the linear acceleration method are expressed as

$$\left(\frac{6}{\Delta t^2}[M] + \frac{3}{\Delta t}[C] + [K]\right)\{x(t+\Delta t)\} = \{F(t+\Delta t)\} - [M]\{A\} - [C]\{B\} \tag{3.95}$$

where

$$\{A\} = -\frac{6}{\Delta t^2}\{x(t)\} - \frac{6}{\Delta t}\{\dot{x}(t)\} - 2\{\ddot{x}(t)\} \tag{3.96a}$$

and

$$\{B\} = -2\{\dot{x}(t)\} - \frac{\Delta t}{2}\{\ddot{x}(t)\} - \frac{3}{\Delta t}\{x(t)\} \tag{3.96b}$$

Note that $\ddot{x}(t)$, $\dot{x}(t)$, and $x(t)$ are supposed to be given or calculated at time t, thus the response parameters are unknown for the next step calculation at $t + \Delta t$. Thus, Equation 3.95 can be simplified to the following expression:

$$[\bar{K}]\{x(t+\Delta t)\} = \{\bar{F}(t+\Delta t)\} \tag{3.97}$$

After $\{x(t+\Delta t)\}$ is calculated, acceleration and velocity are then obtained as

$$\{\ddot{x}(t+\Delta t)\} = \frac{6}{\Delta t^2}\{x(t+\Delta t)\} + \{A\} \tag{3.98}$$

$$\{\dot{x}(t+\Delta t)\} = \frac{3}{\Delta t}\{x(t+\Delta t)\} + \{B\} \tag{3.99}$$

It is important to note that Equation 3.97 has the matrix form for static load; therefore, some computation algorithms such as Gauss elimination for finite element systems of static case can be applied, and the static and dynamic analysis can share the same computer subroutines [1, pp. 329–332, 855–862].

3.6.3.2 Wilson-θ Method

Wilson-θ method is similar to the linear acceleration method with the assumption that acceleration varies linearly over the time interval $\theta\Delta t$ where θ is always greater than 1 and is selected by the analyst to give the desired accuracy and stability. The final form of Wilson's integration has the following form [1, pp. 332–334, 863–870]:

$$[\bar{K}]\{\Delta x_\theta\} = \{\Delta\bar{F}\} \tag{3.100}$$

where $\Delta t_\theta = \theta\Delta t$, Δx_θ is an incremental displacement

$$[\bar{K}] = [K] + \frac{6}{(\Delta t_\theta)^2}[M] + \frac{3}{\Delta t_\theta}[C] \tag{3.101a}$$

$$[\Delta\bar{F}] = \{\Delta F_\theta\} + [M]\{Q\} + [C]\{R\} \tag{3.101b}$$

in which

$$\{\Delta F_\theta\} = \theta[\{F(t+\Delta t)\} - \{F(t)\}] \tag{3.102a}$$

$$\{Q\} = \frac{6}{\Delta t_\theta}\{\dot{x}(t)\} + 3\{\ddot{x}(t)\} \tag{3.102b}$$

$$\{R\} = 3\{\dot{x}(t)\} + \frac{\Delta t_\theta}{2}\{\ddot{x}(t)\} \tag{3.102c}$$

After solving $\{\Delta x_\theta\}$ from Equation 3.100 as

$$\{\Delta x_\theta\} = [\bar{K}]^{-1}\{\Delta F\} \tag{3.103}$$

$\{\Delta\ddot{x}\}$ is determined by the following formula:

$$\{\Delta\ddot{x}\} = \frac{1}{\theta}\{\Delta\ddot{x}_\theta\}$$

The incremental velocity vector, $\{\Delta\dot{x}\}$, and displacement vector, $\{\Delta x\}$, are obtained from

$$\{\Delta\dot{x}\} = \{\ddot{x}(t)\}\Delta t + \frac{\Delta t}{2}\{\Delta\ddot{x}\}$$

$$\{\Delta x\} = \Delta t\{\dot{x}(t)\} + \frac{\Delta t^2}{2}\{\ddot{x}(t)\} + \frac{\Delta t^2}{6}\{\Delta\ddot{x}\}$$

Total displacement, velocity, and acceleration vectors are then determined from

$$\{x(t+\Delta t)\} = \{x(t)\} + \{\Delta x\} \tag{3.104}$$

$$\{\dot{x}(t+\Delta t)\} = \{\dot{x}(t)\} + \{\Delta\dot{x}\} \tag{3.105}$$

$$\{\ddot{x}(t+\Delta t)\} = \{\ddot{x}(t)\} + \{\Delta\ddot{x}\} \tag{3.106}$$

3.6.4 Stability Condition and Selection of Time Interval

3.6.4.1 Stability

Stability of a numerical integration method requires that any error in displacement, velocity, and acceleration at time t does not grow for different incremental time intervals used in the integration. Therefore, the response of an undamped system subjected to an initial condition should be a harmonic motion with constant amplitude that must not be amplified when different Δts are employed in the analysis. Stability of an integration method can thus be determined by examining the behavior of the numerical solution for arbitrary initial conditions based on the following recursive relationship of a single d.o.f. motion:

$$\{X(t + \Delta t)\} = [A]\{X(t)\} \tag{3.107a}$$

or

$$[\ddot{x}(t + \Delta t), \dot{x}(t + \Delta t), x(t + \Delta t)]^{\mathrm{T}} = [A][\ddot{x}(t), \dot{x}(t), x(t)]^{\mathrm{T}} \tag{3.107b}$$

where $[A]$ is an integration approximation matrix. If we start at time t, and take n time steps, Equation 3.107 may be expressed as

$$\{X(t + n\Delta t)\} = [A]^n\{X(t)\} \tag{3.108}$$

To investigate the stability of an integration method, we use the decomposed form of matrix $[A]$ in Equation 3.108 as

$$[A]^n = [\Phi][\lambda^n][\Phi]^{-1} \tag{3.109}$$

where $[\lambda^n]$ is a diagonal matrix with eigenvalues λ_1^n, λ_2^n, λ_3^n in the diagonal position; and is a modal matrix with eigenvectors Φ_1, Φ_2, and Φ_3. Now, define the spectral radius of matrix $[A]$ as

$$r(A) = \max|\lambda_i|; \quad i = 1, 2, 3 \tag{3.110}$$

from which we must have

$$r(A) \leq 1 \tag{3.111}$$

in order to keep $[A]^n$ in Equation 3.108 from growing without bound. The condition of Equation 3.111 is known as the *stability criterion* for a given method.

Numerical results for the Newmark method (linear acceleration and constant acceleration) and the Wilson-θ method from $\Delta t/T = 0.001$ to $\Delta t/T = 100$ are plotted in Figure 3.30. It can be seen that the spectral radius for linear acceleration is stable ($r(A) \leq 1$) at approximately $\Delta t/T < 0.55$ and becomes unstable ($r(A) > 1$) at $\Delta t/T \geq 0.55$. The stability of this method depends on the magnitude of Δt, and is called the *conditional stability* method. However, the spectral radii for the constant acceleration method in the range of $\Delta t/T = 0.001$ to 100 are all less than or equal to 1 ($r(A) \leq 1$); this case is called the *unconditional stability* because it does not depend on the magnitude of Δt. The Wilson-θ method with $\theta = 1.4$ is unconditionally stable and it becomes conditionally stable with $\theta = 1.36$. For unconditional stability, the solution is not divergent even if time increment Δt is large.

3.6.4.2 Selection of Δt

Numerical error, sometimes referred to as *computational error*, is due to the incremental time-step expressed in terms of $\Delta t/T$. Such errors result not from the stability behavior from two other sources: (1) externally applied force or excitation and (2) number of d.o.f. assigned to a vibrating system. A forcing function, particularly an irregular one such as earthquake ground motion, is composed of a number of forcing periods (or frequencies). A larger Δt may exclude a significant part of a forcing function. That part is associated with smaller periods, a forcing function's higher modes. This error may occur for both single- and multiple-d.o.f. vibrating systems. Therefore, Δt must be selected small enough to ensure solution accuracy by including the first several significant vibrating modes in the analysis.

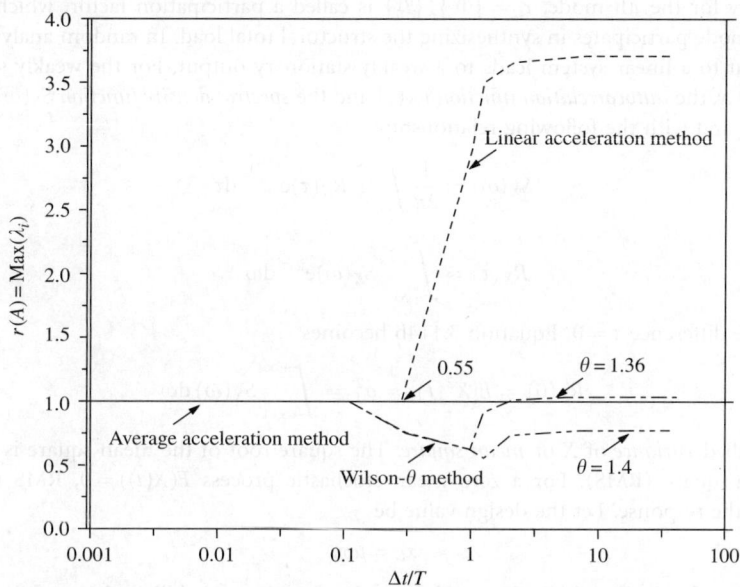

FIGURE 3.30 Spectral radii for Newmark method and Wilson-θ method. (Reprinted from Ref. [1, p. 357] by courtesy of Marcel Dekker, Inc.)

For reasonable accuracy, $\Delta t/T = 0.1$ is recommended, where T is the natural period of highest mode considered in the analysis. Properly selecting the number of modes depends on an individual problem. For the direct integration approach, frequency analysis is not needed. Therefore, the periods of desired methods are not available. A suggested approach is to follow the building code that stipulates the number of modes, and the lower bound of period should be considered. Generally, we can use one Δt and another slightly smaller Δt to find solutions continuously, until two successive solutions are reasonably close. We can also inspect the exciting function to select approximate Δt (for instance, Δt should not be greater than the intervals of the earthquake records).

3.6.5 Nondeterministic Analysis

3.6.5.1 Introduction to Stochastic Seismic Response Analysis

As discussed in Section 3.6.1, when a force cannot be specified as a definite function by time, the response analysis should be determined through a stochastic approach. Earthquake force is a typical example that has inherent uncertainty in magnitude and in time variation. For brief presentation, the seismic response is chosen for discussion in this section. For a structure subjected to an earthquake, the motion equation is similar to Equation 3.94 and is expressed as

$$[M]\{\ddot{x}\} + [C]\{\dot{x}\} + [K]\{x\} = [M]\{I_n\}\ddot{x}_g \qquad (3.112)$$

where $\ddot{x}_g$ is the ground motion acceleration and $\{I_n\}$ is the influence coefficient factor. By the modal analysis method, the motion equation is decomposed to N (d.o.f. of the system) independent second-order differential equations

$$\ddot{x}'_i + 2\rho_i p_i \dot{x}'_i + p_i^2 x'_i = r_i \ddot{x}_g \quad i = 1, 2, \ldots, N \qquad (3.113)$$

where x'_i is the modal displacement for the ith modes and has the relation with the floor displacement as $\{x\} = [\Phi']\{x'\}$ and $[\Phi']$ is the *normalized modal matrix*. ρ_i and p_i are damping coefficient and angle frequency for the *i*th mode. $r_i = \{\Phi'\}_i^{\mathrm{T}}\{I_n\}$ is called a participation factor, which implies how much the *i*th mode participates in synthesizing the structural total load. In random analysis, the *weakly stationary* input to a linear system leads to a weakly stationary output. For the weakly stationary stochastic process X, the *autocorrelation function* $R_X(\tau)$ and the *spectral density function* $S_X(\omega)$ are a Fourier transform pair and with the following relationship:

$$S_X(\omega) = \frac{1}{2\pi} \int_{-\infty}^{+\infty} R_X(\tau) e^{-i\omega\tau}\, d\tau \tag{3.114a}$$

$$R_X(\tau) = \int_{-\infty}^{+\infty} S_X(\omega) e^{i\omega\tau}\, d\omega \tag{3.114b}$$

By setting time difference $\tau = 0$, Equation 3.114b becomes

$$R_X(0) = E[X^2(t)] = \sigma_X^2 = \int_{-\infty}^{+\infty} S_X(\omega)\, d\omega \tag{3.115}$$

where σ_X^2 is called *variance* of X or *mean square*. The square root of the mean square is σ_X, known as the root-mean-square (RMS). For a zero mean stochastic process $E(X(t)) = 0$, RMS represents the magnitude of the response. Let the design value be

$$x_l = \mu \sigma_{x_l} \tag{3.116}$$

where l represents lth d.o.f. and μ is the coefficient related to the probability of the safety. By the sum of the modal contribution, the mean square value of the lth d.o.f. can be obtained as

$$\sigma_{x_l}^2 = \sum_{i=1}^{N} \Phi_{li}'^2 r_i^2 \int_{-\infty}^{+\infty} |H_i(\omega)|^2 S_{\ddot{x}_g}(\omega)\, d\omega \tag{3.117}$$

where Φ_{li}' is the lth element in the ith modal vector, r_i is the participation factor of the ith mode, $S_{\ddot{x}_g}(\omega)$ is the spectral density function for the ground acceleration input, and $H_i(\omega)$ is the frequency response function for the ith mode

$$H_i(\omega) = \frac{1}{p_i^2 - \omega^2 + 2\mathrm{j}\rho_i p_i \omega}; \quad \mathrm{j} = \sqrt{-1} \tag{3.118}$$

in which ρ_i, p_i are structural frequencies and damping coefficient of the ith mode, respectively. The spectral density function for a stationary ground acceleration is

$$S_{\ddot{x}_g}(\omega) = \frac{p_g^4 + 4\rho_g^2 p_g^2 \omega^2}{(p_g^2 - \omega^2)^2 + 4\rho_g^2 p_g^2 \omega^2} S_0 \tag{3.119}$$

which was proposed by Kanai and Tajimi [5,6] based on the study of the frequency content from ground motion records. In Equation 3.119, p_g and ρ_g are prevailing frequency and damping ratio describing soil layer characteristics, S_0 is the intensity of the excitation and can be determined by the strength of the ground motion. By setting the ground acceleration RMS as $\sigma_{\ddot{x}_g}$ on the left-hand side of Equation 3.115 and substituting $S_{\ddot{x}_g}(\omega)$ of Equation 3.119 into the right-hand side of that equation, S_0 can be obtained after integration as

$$S_0 = \frac{2\rho_g \sigma_{\ddot{x}_g}^2}{\pi(4\rho_g^2 + 1)p_g} \tag{3.120}$$

The integral in Equation 3.117 can be found after the substitution of Equation 3.119 by the *theorem of residue* [7], and the results are

$$\int_{-\infty}^{+\infty} |H_i(\omega)|^2 S_{\ddot{x}_g}(\omega)\, d\omega = \frac{\pi S_0}{2} \frac{p_g}{p_i^4} \frac{N_{q_i}}{D_{q_i}} \tag{3.121a}$$

$$N_{q_i} = (1 + 4\rho_g^2)\rho_i + 4\rho_g(\rho_i^2 + \rho_g^2)s + 4\rho_g^2\rho_i s^2 + \rho_g s^3; \quad s = p_g/p_i \quad (3.121b)$$

$$D_{q_i} = \rho_g\rho_i + 4\rho_g^2\rho_i^2 s + 2\rho_g\rho_i(2\rho_g^2 + 2\rho_i^2 - 1)s^2 + 4\rho_g^2\rho_i^2 s^3 + \rho_g\rho_i s^4 \quad (3.121c)$$

Substituting Equation 3.121a into Equation 3.117 yields

$$\sigma_{x_l}^2 = \frac{\pi S_0}{2} \sum_{i=1}^{N} \Phi_{li}'^2 r_i^2 \frac{p_g}{p_i^4} \frac{N_{q_i}}{D_{q_i}} \quad (3.122)$$

Similar to Equation 3.117, the mean square of the velocity $\dot{x}_l$ can be expressed as

$$\sigma_{\dot{x}_l}^2 = \sum_{i=1}^{N} \Phi_{li}'^2 r_i^2 \int_{-\infty}^{+\infty} \omega^2 |H_i(\omega)|^2 S_{\ddot{x}_g}(\omega)\, d\omega \quad (3.123)$$

in which the integral is

$$\int_{-\infty}^{+\infty} \omega^2 |H_i(\omega)|^2 S_{\ddot{x}_g}(\omega)\, d\omega = \frac{\pi S_0}{2} \frac{p_g^3}{p_i^4} \frac{N_{\dot{q}_i}}{D_{\dot{q}_i}} \quad (3.124a)$$

$$N_{\dot{q}_i} = 4\rho_g^3 + \rho_i(1 + 4\rho_g^2)s + \rho_g^2 s^2 \quad (3.124b)$$

$$D_{\dot{q}_i} = s[\rho_g\rho_i + 4\rho_g^2\rho_i^2 s + 2\rho_g\rho_i(2\rho_g^2 + 2\rho_i^2 - 1)s^2 + 4\rho_g^2\rho_i^2 s^3 + \rho_g\rho_i s^4] \quad (3.124c)$$

Thus, Equation 3.123 becomes

$$\sigma_{\dot{x}_l}^2 = \frac{\pi S_0}{2} \sum_{i=1}^{N} \Phi_{li}'^2 r_i^2 \frac{p_g^3}{p_i^4} \frac{N_{\dot{q}_i}}{D_{\dot{q}_i}} \quad (3.125)$$

After the RMS values of the response are obtained, one can calculate the probability of safety in duration τ by

$$P_s(\tau) = P\{X(0) \in D\} \exp(-\nu_D \tau) \quad (3.126)$$

where D is the safety range assigned for the response X and $P\{X(0) \in D\}$ is the probability of the initial X value in the range D. ν_D is called the D-outcrossing rate, expressing the rate of response X going out of range D

$$\nu_D = \nu_a^+ + \nu_b^- \quad (3.127)$$

which is the sum of up-crossing rate ν_a^+ as upper limit, a and down-crossing rate ν_b^- as bottom limit, b, for the safety range $D = [a \ b]$. For the stationary Gaussian process with mean m and variance σ_X^2, the up-crossing and down-crossing rate for $\pm a$ are

$$\nu_a^+ = \nu_{-a}^- = \frac{\sigma_{\dot{X}}}{2\pi\sigma_X} \exp\left[-\frac{(a - m)^2}{2\sigma_X^2}\right] \quad (3.128)$$

If the probability of the initial value X in range D is 1, Equation 3.126 becomes

$$P_s(\tau) = \exp(-\nu_D \tau) \quad (3.129a)$$

The probability of safety for the range $D = [a \ -a]$ can be obtained by substituting Equations 3.127 and 3.128 into Equation 3.129a as

$$P_s(\tau) = \exp\left[-\frac{\sigma_{\dot{X}}\tau}{\pi\sigma_X} \exp\left[-\frac{(a - m)^2}{2\sigma_X^2}\right]\right] \quad (3.129b)$$

In summary, one can determine the RMS value of the response x_l from Equations 3.122 and 3.125 with a given RMS value of the ground acceleration in Equation 3.120, in which the soil layer property is considered with ρ_i and p_i. Finally, the probability of safety can be calculated from Equation 3.129b after the response variance is determined.

3.6.5.2 Numerical Example

Consider the three-story shear building shown in Figure 3.31 with mass $M_1 = M_2 = M_3 = 12{,}304\ \mathrm{N\,s^2/m}$; $l = 4.2672\ \mathrm{m}$; $I_1 = 1.98168 \times 10^{-4}\ \mathrm{m^4}$, $I_2 = 1.4601 \times 10^{-4}\ \mathrm{m^4}$,   and $I_3 = 9.9229 \times 10^{-5}\ \mathrm{m^4}$; $E = 206842.8 \times 10^6\ \mathrm{N/m^2}$. Let the structure be located on the medium firm soil and the ground acceleration be a zero mean stationary process with RMS $\sigma_{\ddot{x}_g} = 0.1g$. Assume the damping factor $\rho_i = 0.02$ for each mode, find the RMS of the floor response and the probability of safety of the third-floor displacement within three times of its RMS.

After performing eigensolution calculation, the structural frequencies are

$$[\,p_1 \quad p_2 \quad p_3\,] = [\,12.82 \quad 32.11 \quad 48.79\,]\ \mathrm{rad/s}$$

and the normalized modal matrix is

$$[\Phi'] = \begin{bmatrix} 0.2545 & -0.5777 & 0.7756 \\ 0.5449 & -0.5769 & -0.6085 \\ 0.7989 & 0.5775 & 0.1680 \end{bmatrix}$$

For horizontal ground motion input, the participation factors are

$$\{r\}^{\mathrm{T}} = [\Phi']^{\mathrm{T}}\{I_n\} = [\,-1.5984 \quad 0.5771 \quad -0.3350\,]$$

where $\{I_n\} = -[1\ 1\ 1]^{\mathrm{T}}$. Since the structure is located on medium firm soil, the prevailing frequency is $p_g = 15.6\ \mathrm{rad/s}$ and the damping factor is $\rho_g = 0.60$. From Equations 3.121b, 3.121c and 3.124b and 3.124c

$$[\,N_{q_1} \quad N_{q_2} \quad N_{q_3}\,] = [\,0.2318 \quad 0.5449 \quad 0.3478\,]$$

$$[\,D_{q_1} \quad D_{q_2} \quad D_{q_3}\,] = [\,0.0303 \quad 0.0114 \quad 0.0116\,]$$

$$[\,N_{\dot{q}_1} \quad N_{\dot{q}_2} \quad N_{\dot{q}_3}\,] = [\,1.8146 \quad 1.0295 \quad 0.9409\,]$$

$$[\,D_{\dot{q}_1} \quad D_{\dot{q}_2} \quad D_{\dot{q}_3}\,] = [\,0.0369 \quad 0.0056 \quad 0.0037\,]$$

For the ground acceleration RMS value assumed as $\sigma_{\ddot{x}_g} = 0.1g$, Equation 3.120 yields $S_0 = 9.6 \times 10^{-3}\ \mathrm{m^2/s^3}$. Then, from Equations 3.122 and 3.125, the RMS values for each floor's displacement and velocity are calculated as

$$[\,\sigma_{x_1} \quad \sigma_{x_2} \quad \sigma_{x_3}\,] = [\,0.0104 \quad 0.0222 \quad 0.0325\,]\ \mathrm{m}$$

$$[\,\sigma_{\dot{x}_1} \quad \sigma_{\dot{x}_2} \quad \sigma_{\dot{x}_3}\,] = [\,0.1366 \quad 0.2842 \quad 0.4150\,]\ \mathrm{m/s}$$

For the third-floor displacement, the range of safety is $D = [3\sigma_{x_3}, -3\sigma_{x_3}] = [0.097 - 0.097]\ \mathrm{m}$, then the probability of safety can be determined from Equation 3.129b with $m = 0$, $\sigma_{x_3} = 0.0325\ \mathrm{m}$, and $\sigma_{\dot{x}_3} = 0.4150\ \mathrm{m/s}$ as

$$P_s(\tau) = \begin{cases} 95.6\% & \text{for duration } \tau = 1\ \mathrm{s} \\ 79.8\% & \text{for duration } \tau = 5\ \mathrm{s} \\ 63.6\% & \text{for duration } \tau = 10\ \mathrm{s} \end{cases}$$

which implies that the probability of safety for the third-floor displacement within $\pm 0.097\ \mathrm{m}$ is 95.6% in 1 s, 79.8% in 5 s, and 63.6% in 10 s.

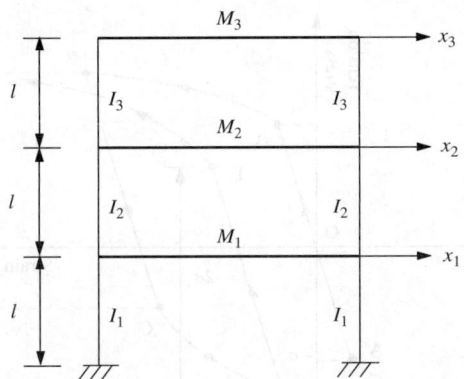

FIGURE 3.31 Three-story shear building.

3.6.5.3 Comments on Nonlinear Systems

The stochastic response of nonlinear structures cannot be obtained by the aforementioned method because of the requirement of superposition. One of the approaches is to generate an ensemble of ground motion accelerograms by the techniques based on *filtered white noise* and then determine deterministically the time-history response of the nonlinear structure to each input accelerogram, and finally examine the output response process using *Monte Carlo methods*. Usually, one is interested primarily in the mean and standard deviation values of the extreme response [8].

3.6.6 Comments on Wave Propagation Analysis

Theoretically, mode superposition is an effective method of obtaining the free or forced vibration response of a continuous system of finite extent. In practice, several difficulties may arise in application of the method. For instance, it may not be possible to determine the mode shapes and frequencies of the continuous system being analyzed. A system of infinite extent has a continuous band of frequencies, and the term "mode shape" loses its meaning. Obviously, modal analysis is not possible in such a case. However, an alternative method known as *wave propagation analysis* may prove to be quite effective in obtaining the response of the system [9].

3.7 Hysteresis Models and Nonlinear Response Analysis

3.7.1 Introduction

The previous sections focused on elastic structures with emphasis on mathematical models, analytical methodologies, and response characteristics. When a structure is subjected to dynamic force or ground motion, its constituent members may deform beyond their elastic limit, such as yielding stress of steel or crack stress of concrete. If we assume that the members continue to behave elastically, then their response behavior is based on *linear* or *elastic analysis* as presented previously. When the stress–strain relationship beyond the elastic stage is considered, the response then results from *nonlinear* or *inelastic analysis*. Naturally, nonlinear analysis always encompasses linear analysis because of elastic material behavior at the early loading stage. When inelastic material behavior is considered in formulating the force–deformation relationship of a structural member, the relationship is called *hysteresis model*. A typical stress–strain relationship of structural steel is shown in Figure 3.32. The linear relationship between O and A is defined as elastic behavior. After initial yielding, σ_y, the slope of the stress–strain curve is not constant and material behavior becomes inelastic. Unloading path B–C and reloading path D–E are elastic and form straight lines parallel to the initial elastic path O–A. Absolute values of the

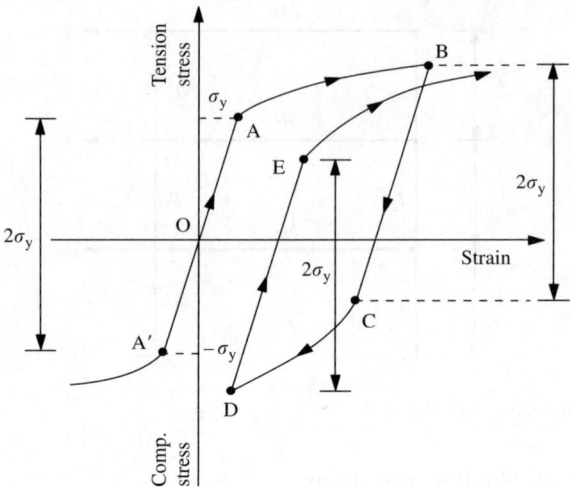

FIGURE 3.32 Material nonlinearity.

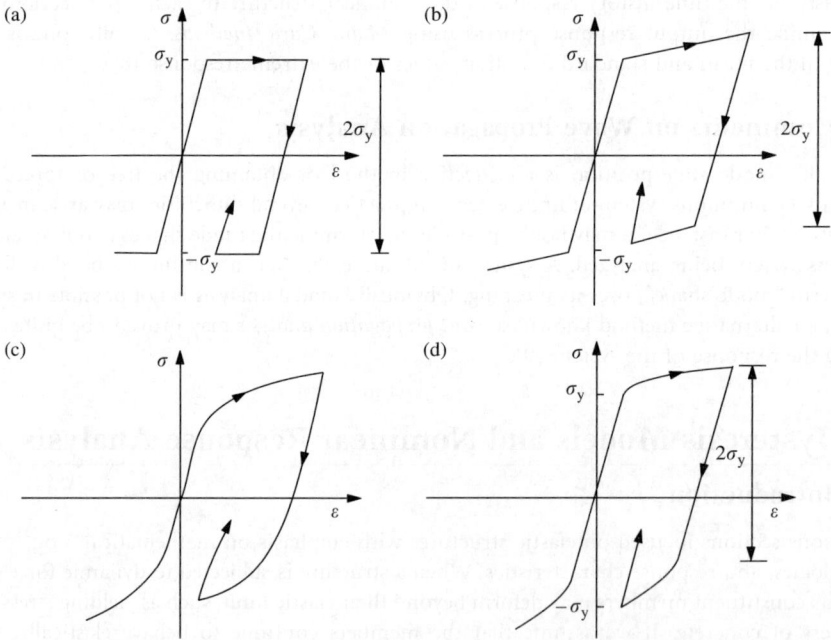

FIGURE 3.33 Simplified stress–strain models: (a) elasto-plastic; (b) bilinear; (c) curvilinear; and (d) Ramberg–Osgood.

initial yield points in tension A and in compression A′ are the same, but the values change for subsequent points B and C or D and E. A stress magnitude of $2\sigma_y$ is observed for $\overline{AA'}$, $\overline{BC}$, and, $\overline{DE}$ which is known as *Bauschinger effect*. For practical purposes, some well-known simplified models are often used. Most typical are these shown in Figure 3.33 as *elasto-plastic, bilinear, curvilinear,* and *Ramberg–Osgood*. For dynamic nonlinear analysis, the main task is the derivation of incremental stiffness coefficients of a typical member. The motion equation is still expressed as Equation 3.34 or 3.93 but in incremental form. The direct integration methods presented in Section 3.6.3 are then employed for response analysis. The incremental stiffness coefficients of elasto-plastic and bilinear models are given in Sections 3.7.2 and 3.7.3, respectively. The derivations of incremental stiffness coefficients of curvilinear and Ramberg–Osgood models are lengthy and are not included here [1, pp. 555–578].

3.7.2 Elasto-Plastic Stiffness Formulation

The elasto-plastic model in Figure 3.34 shows that when the moment reaches the *ultimate moment* capacity of a member, the plastic moment cannot increase but the rotation of the plastic hinge at the cross-section can increase. A plastic hinge develops at the member's end where the magnitude of the moment is greater than at other locations as in the case of dynamic response. The member end behaves like a real center hinge with a constant ultimate moment, M_p. When the member end rotates in reverse, the moment decreases elastically and the plastic hinge disappears. Elastic behavior remains unchanged until the moment reaches ultimate moment capacity. Consequently, a plastic hinge forms again. Plastic hinge formation in a member has three possibilities: a hinge at the *i*-end, the *j*-end, or both ends (see *i* and *j* in Figure 3.35). Force–deformation relationships associated with elastic state and the three states of yield condition are given as follows [1, pp. 529–534].

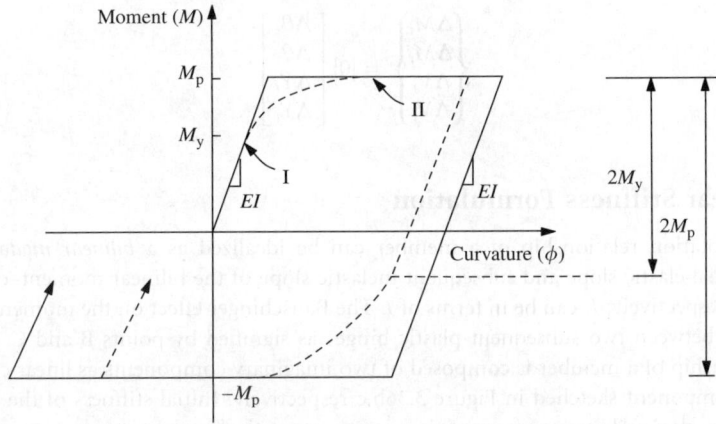

FIGURE 3.34   Elasto-plastic moment–curvature relationship.

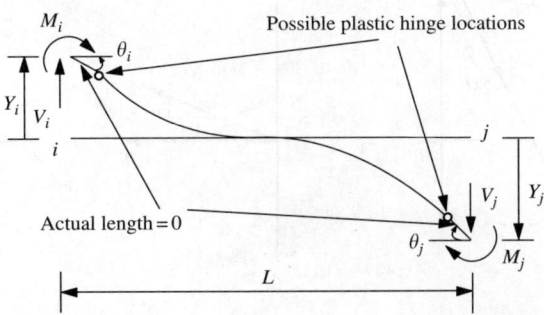

FIGURE 3.35   Elasto-plastic member.

1. Both ends linear

$$
\begin{Bmatrix} \Delta M_i \\ \Delta M_j \\ \Delta V_i \\ \Delta V_j \end{Bmatrix} = 4EI \begin{bmatrix} 1/L & 1/2L & -3/2L^2 & -3/2L^2 \\ & 1/L & -3/2L^2 & -3/2L^2 \\ & & 3/L^3 & 3/L^3 \\ \text{sym.} & & & 3/L^3 \end{bmatrix} \begin{bmatrix} \Delta\theta_i \\ \Delta\theta_j \\ \Delta Y_i \\ \Delta Y_j \end{bmatrix} \tag{3.130}
$$

2. *i*-end nonlinear and *j*-end linear

$$
\begin{Bmatrix} \Delta M_i \\ \Delta M_j \\ \Delta V_i \\ \Delta V_j \end{Bmatrix} = 3EI \begin{bmatrix} 0 & 0 & 0 & 0 \\ & 1/L & -1/L^2 & -1/L^2 \\ & & 1/L^3 & 1/L^3 \\ \text{sym.} & & & 3/L^3 \end{bmatrix} \begin{bmatrix} \Delta\theta_i \\ \Delta\theta_j \\ \Delta Y_i \\ \Delta Y_j \end{bmatrix} \tag{3.131}
$$

3. *i*-end linear and *j*-end nonlinear

$$
\begin{Bmatrix} \Delta M_i \\ \Delta M_j \\ \Delta V_i \\ \Delta V_j \end{Bmatrix} = 3EI \begin{bmatrix} 1/L & 0 & -1/L^2 & -1/L^2 \\ & 0 & 0 & 0 \\ & & 1/L^3 & 1/L^3 \\ \text{sym.} & & & 1/L^3 \end{bmatrix} \begin{bmatrix} \Delta\theta_i \\ \Delta\theta_j \\ \Delta Y_i \\ \Delta Y_j \end{bmatrix} \tag{3.132}
$$

4. Both ends nonlinear

$$
\begin{Bmatrix} \Delta M_i \\ \Delta M_j \\ \Delta V_i \\ \Delta V_j \end{Bmatrix} = [0] \begin{bmatrix} \Delta\theta_i \\ \Delta\theta_j \\ \Delta Y_i \\ \Delta Y_j \end{bmatrix} \tag{3.133}
$$

3.7.3 Bilinear Stiffness Formulation

The moment–rotation relationship of a member can be idealized as a *bilinear model*, as shown in Figure 3.36. Initial elastic slope and subsequent inelastic slope of the bilinear moment–curvature curve are EI and EI_1, respectively; I_1 can be in terms of I. The Bauschinger effect on the moment magnitude of $2M_p$ also exists between two subsequent plastic hinges as signified by points B and C. The moment–rotation relationship of a member is composed of two imaginary components as linear component and elasto-plastic component sketched in Figure 3.36b,c respectively. Initial stiffness of the hysteresis loop and of the elastic, elasto-plastic components is a, a_1, a_2, respectively, where $a = a_1 + a_2$, $a_1 = pa$, $a_2 = qa$,

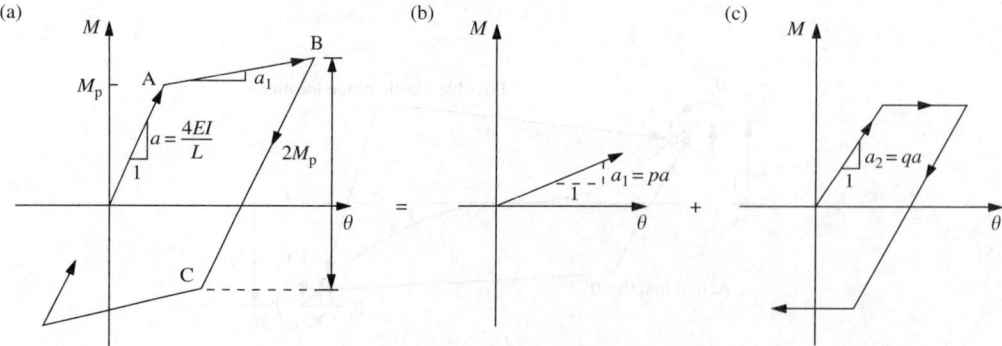

FIGURE 3.36 Bilinear moment–rotation: (a) bilinear hysteresis loop; (b) linear component; and (c) elasto-plastic component.

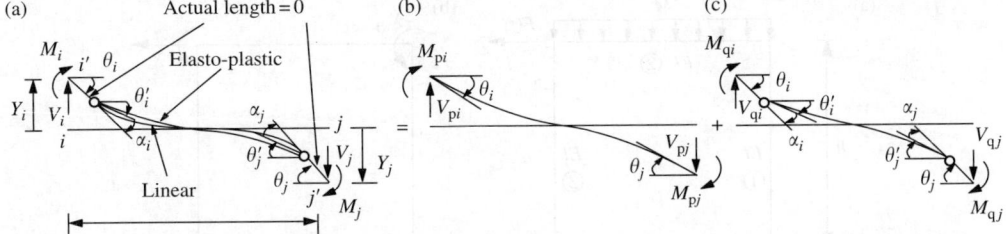

FIGURE 3.37 Bilinear member: (a) nonlinear beam; (b) linear component; and (c) elasto-plastic component.

and $p + q = 1$. p is the fraction of stiffness apportioned to the linear component and q is the fraction of stiffness apportioned to the elasto-plastic component. The second slope a_1 of the hysteresis loop is the same as the initial slope of the linear component. The incremental stiffness coefficients are given as follows [1, pp. 534–538].

1. Both ends linear

$$
\begin{Bmatrix} \Delta M_i \\ \Delta M_j \\ \Delta V_i \\ \Delta V_j \end{Bmatrix} = \begin{bmatrix} a & b & -c & -c \\ & a & -c & -c \\ & & d & d \\ \text{sym.} & & & d \end{bmatrix} \begin{bmatrix} \Delta \theta_i \\ \Delta \theta_j \\ \Delta Y_i \\ \Delta Y_j \end{bmatrix}
\tag{3.134}
$$

in which $a = 4EI/L$, $b = 2EI/L$, $c = 6EI/L^2$, and $d = 12EI/L^3$.

2. i-end nonlinear and j-end linear

$$
\begin{Bmatrix} \Delta M_i \\ \Delta M_j \\ \Delta V_i \\ \Delta V_j \end{Bmatrix} = \begin{bmatrix} pa & pb & -pc & -pc \\ & pa + qe & -pc - qf & -pc - qf \\ & & pd + qg & pd + qg \\ \text{sym.} & & & pd + qg \end{bmatrix} \begin{bmatrix} \Delta \theta_i \\ \Delta \theta_j \\ \Delta Y_i \\ \Delta Y_j \end{bmatrix}
\tag{3.135}
$$

in which $e = 3EI/L$, $f = 3EI/L^2$, and $g = 3EI/L^3$.

3. i-end linear and j-end nonlinear

$$
\begin{Bmatrix} \Delta M_i \\ \Delta M_j \\ \Delta V_i \\ \Delta V_j \end{Bmatrix} = \begin{bmatrix} pa + qe & pb & -pc - qf & -pc - qf \\ & pa & -pc & -pc \\ & & pd + qg & pd + qg \\ \text{sym.} & & & pd + qg \end{bmatrix} \begin{bmatrix} \Delta \theta_i \\ \Delta \theta_j \\ \Delta Y_i \\ \Delta Y_j \end{bmatrix}
\tag{3.136}
$$

4. Both ends nonlinear

$$
\begin{Bmatrix} \Delta M_i \\ \Delta M_j \\ \Delta V_i \\ \Delta V_j \end{Bmatrix} = p \begin{bmatrix} a & b & -c & -c \\ & a & -c & -c \\ & & d & d \\ \text{sym.} & & & d \end{bmatrix} \begin{bmatrix} \Delta \theta_i \\ \Delta \theta_j \\ \Delta Y_i \\ \Delta Y_j \end{bmatrix}
\tag{3.137}
$$

which are actually the incremental forces of the linear component. Note that when $p = 0$, $q = 1$, the bilinear model presented above becomes the elasto-plastic model.

3.7.4 Elasto-Plastic and Bilinear Response Comparisons

The effects of various ps on response behavior are shown for the rigid frame shown in Figure 3.38. Assume $EI = 1000$ kip in.^{2} (2.8697 kN m^2), $M = 2 \times 10^{-4}$ k s^2/in. (35.0236 N s^2/m), $h = 10$ ft (3.048 m), $l = 20$ ft (6.096 m), $F(t) = 0.02 \sin(\pi t)k$ (88.96 $\sin(\pi t)N$), $p = 0.05$, and ultimate moment capacity $M_p = 0.2$ k in. (22.59584 N m). The initial conditions are $x_0 = \dot{x}_0 = \ddot{x}_0 = 0$. The effect of ps on x_s

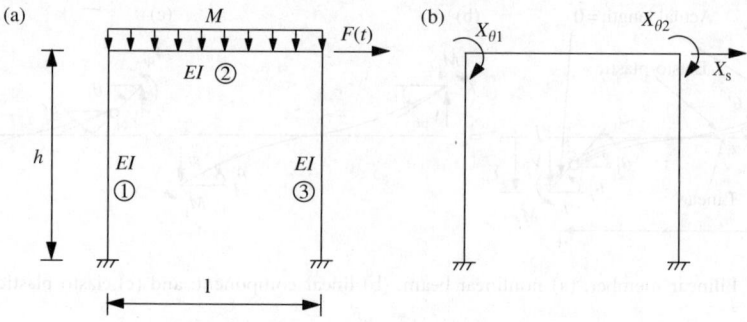

FIGURE 3.38 Response behavior illustration: (a) given structure and (b) rigid frame with joint rotation and side sway.

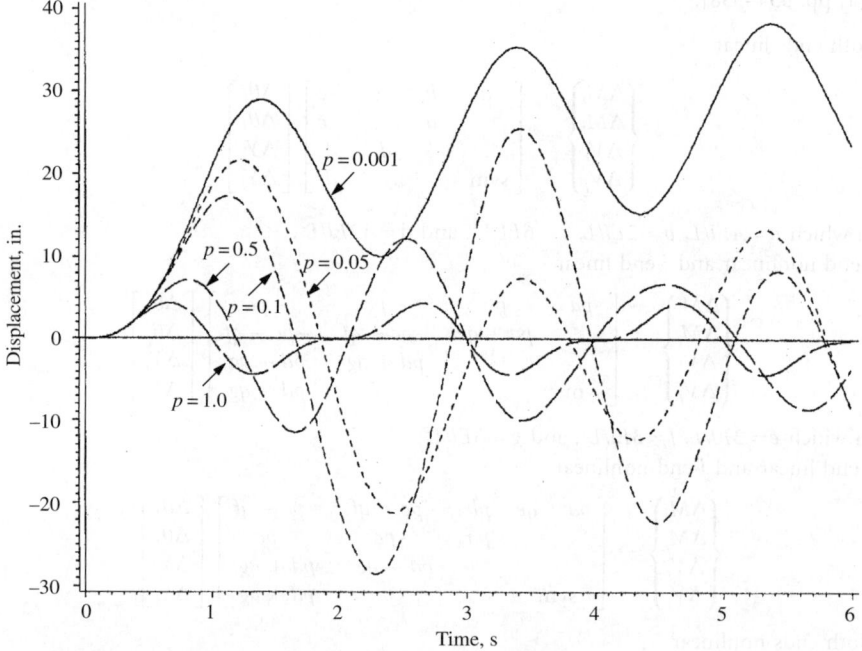

FIGURE 3.39 Effect of ps on x_s versus time. (Reprinted from Ref. [1, p. 547] by courtesy of Marcel Dekker, Inc.)

versus time are shown in Figure 3.39. Note that $p = 1$ is corresponding to elastic response and $p = 0.001$ is close to elasto-plastic case, more decreases in ps and more increases in deflections.

3.7.5 Effects of P–Δ, Material Nonlinearity, and Large Deformation on Response

The material nonlinear behavior was presented in Sections 3.7.2 to 3.7.4, and the P–Δ effect was introduced in Section 3.5.5. Linear and nonlinear analyses can be conducted with consideration of *small* or *large deflection*. Large deflection implies that structural configuration deforms markedly, which results in change of originally assumed directions of forces and displacements. The direction of members' internal forces in a structure after large deformation is not the same as that used in formulation for this structure with small deformation. Consequently, equilibrium equations between internal forces (shears, axial forces, etc.) and external loads (applied force, inertia force, etc.) and the compatibility condition between internal deformations and external displacements at each structural node should be modified at various loading stages. Large deflection formulation is generally considered in inelastic analysis but sometimes used in elastic cases for a material with a small modulus of elasticity. Note that the geometric stiffness matrix or string stiffness matrix due to the P–Δ effect presented previously can be included in either small or large deflection formulation. Naturally, the P–Δ effect becomes more pronounced for large deflections.

The rigid frame given in Figure 3.38 is used to show various P–Δ effects and the influences of material as well as geometric nonlinearities on the displacement response. This study involves the following. First assume each column is subjected to a compression, P, and then formulate $[K_g]$ using the string stiffness model. The buckling load is then determined by finding the singular solution of $([K] - [K_g])\{X\} = 0$. The response analyses are finally conducted for the following cases:

1. Linear analysis (LA) without P–Δ effect or geometric and material nonlinearities.
2. Nonlinear analysis (NA) having material nonlinearity of bilinear model with $p = 0.05$ but not P–Δ effect or large deflection.
3. Nonlinear analysis with $p = 0.05$ combined with 4% of buckling load (NA; $0.04P_{cr}$) but not including large deflection.
4. Same as (3) except 20% of buckling load (NA; $0.2P_{cr}$).
5. Nonlinear analysis with $p = 0.05$, 4% of critical load, and large deflection (NA; $0.04P_{cr}$; LgD).

The displacements versus time are shown in Figure 3.40. Note that material and geometric non-linearity causes large deflection, and the P–Δ effect can induce the structure to be unstable.

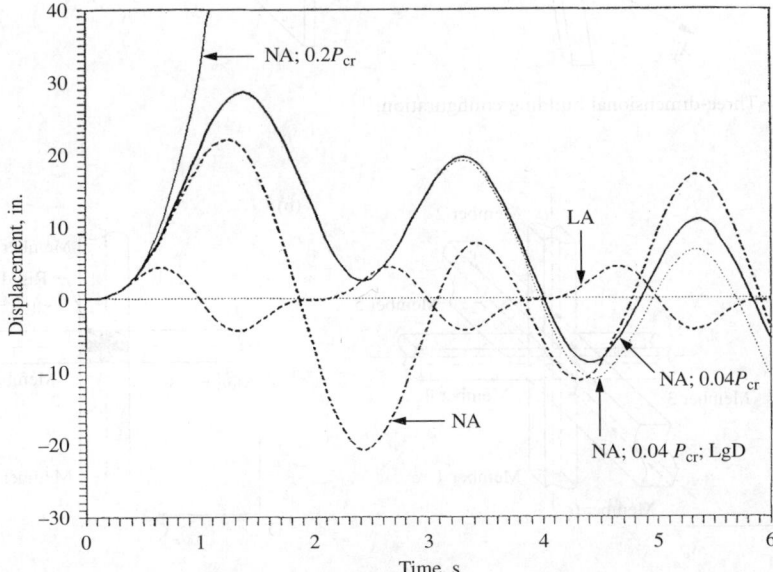

FIGURE 3.40   Lateral displacement versus time. (Reprinted from Ref. [1, p. 588] by courtesy of Marcel Dekker, Inc.)

3.7.6 Hysteresis Models and Stiffnesses for Steel Bracings and for RC Beams, Columns, and Walls of Building Systems

3.7.6.1 Building Systems

Civil engineering building structures usually have beams, beam-columns, bracing elements, shear walls, floor slabs, and rigid zones at structural joints. The three-dimensional building configuration and rigid zone of structural connection are shown in Figure 3.41 and Figure 3.42, respectively.

3.7.6.2 Hysteresis Models

Four hysteresis models are sketched in Figure 3.43 to Figure 3.46. Since the stiffness formulations of these models are too long to list, computer programs are provided from which readers can find detailed calculation procedures [1, pp. 875–977]. Figure 3.43 is for steel bracing (pinned-end truss member) based on the work of Goel and co-workers [10]. Figure 3.44, based on the Takeda model, is for slender RC members in bending such as beams and columns [11]. Figure 3.45 is the Cheng–Mertz model for RC shear walls with consideration of axial deformation, coupling bending, and shear deformation for both

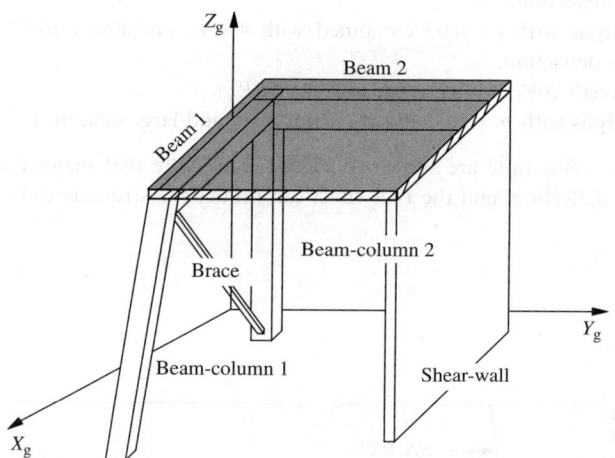

FIGURE 3.41 Three-dimensional building configuration.

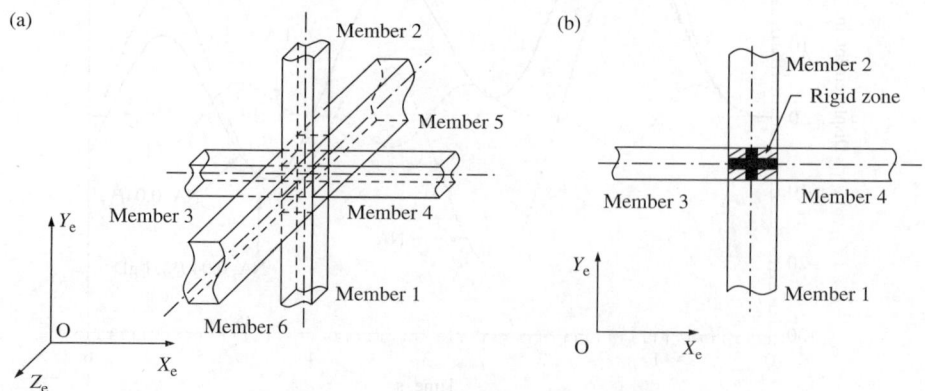

FIGURE 3.42 Rigid zones of structural connection: (a) structural connection and (b) rigid zones in X_e–Y_e plane.

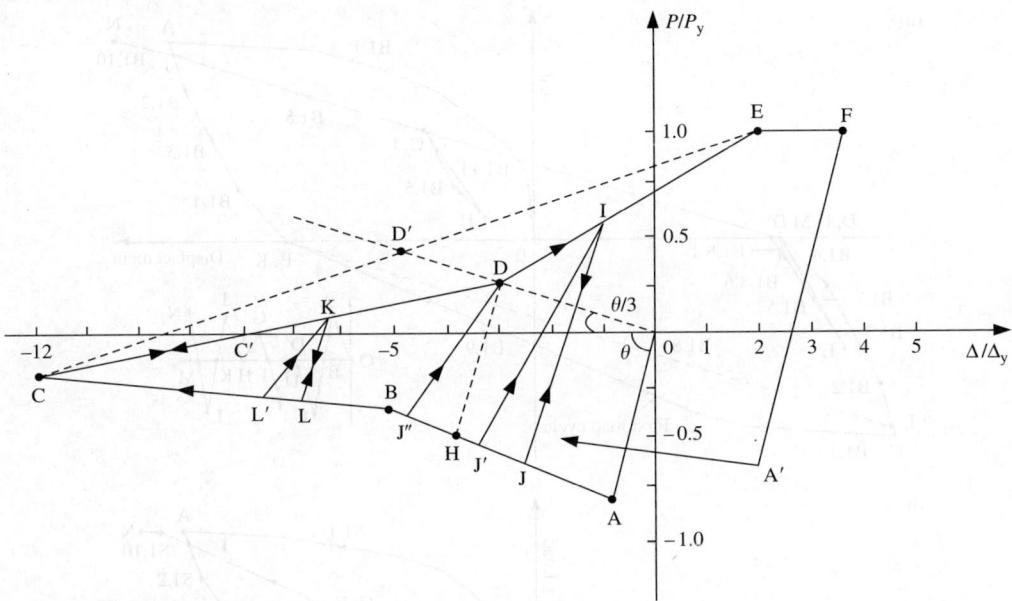

FIGURE 3.43 Jain–Goel–Hanson steel-bracing hysteresis model. (Reprinted from Ref. [1, p. 875] by courtesy of Marcel Dekker, Inc.)

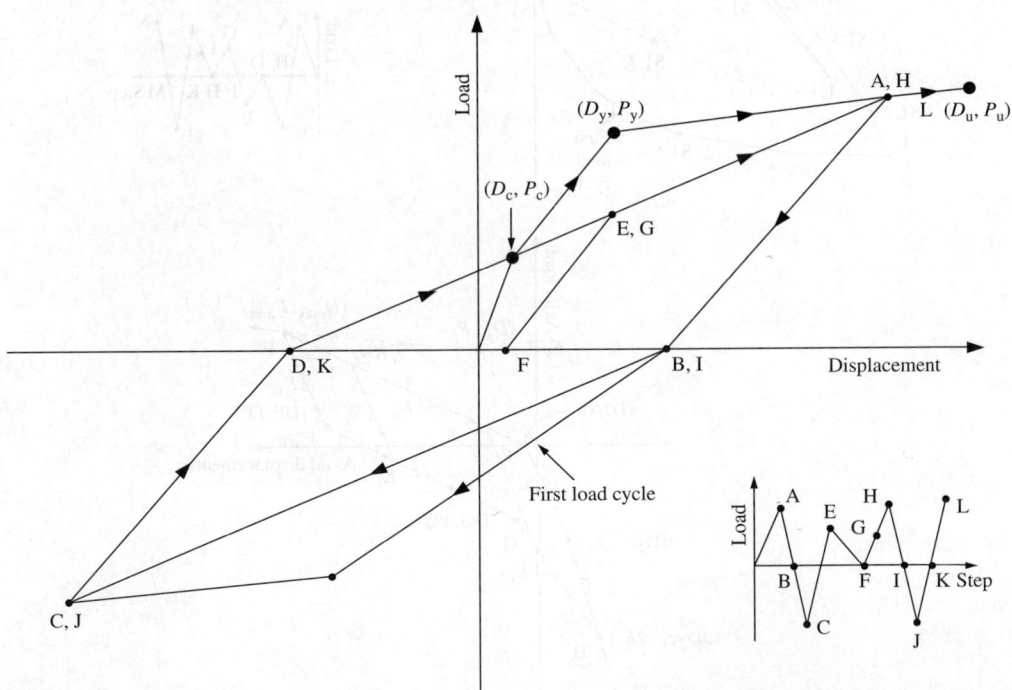

FIGURE 3.44 Takeda model for RC columns and beams. (Reprinted from Ref. [1, p. 895] by courtesy of Marcel Dekker, Inc.)

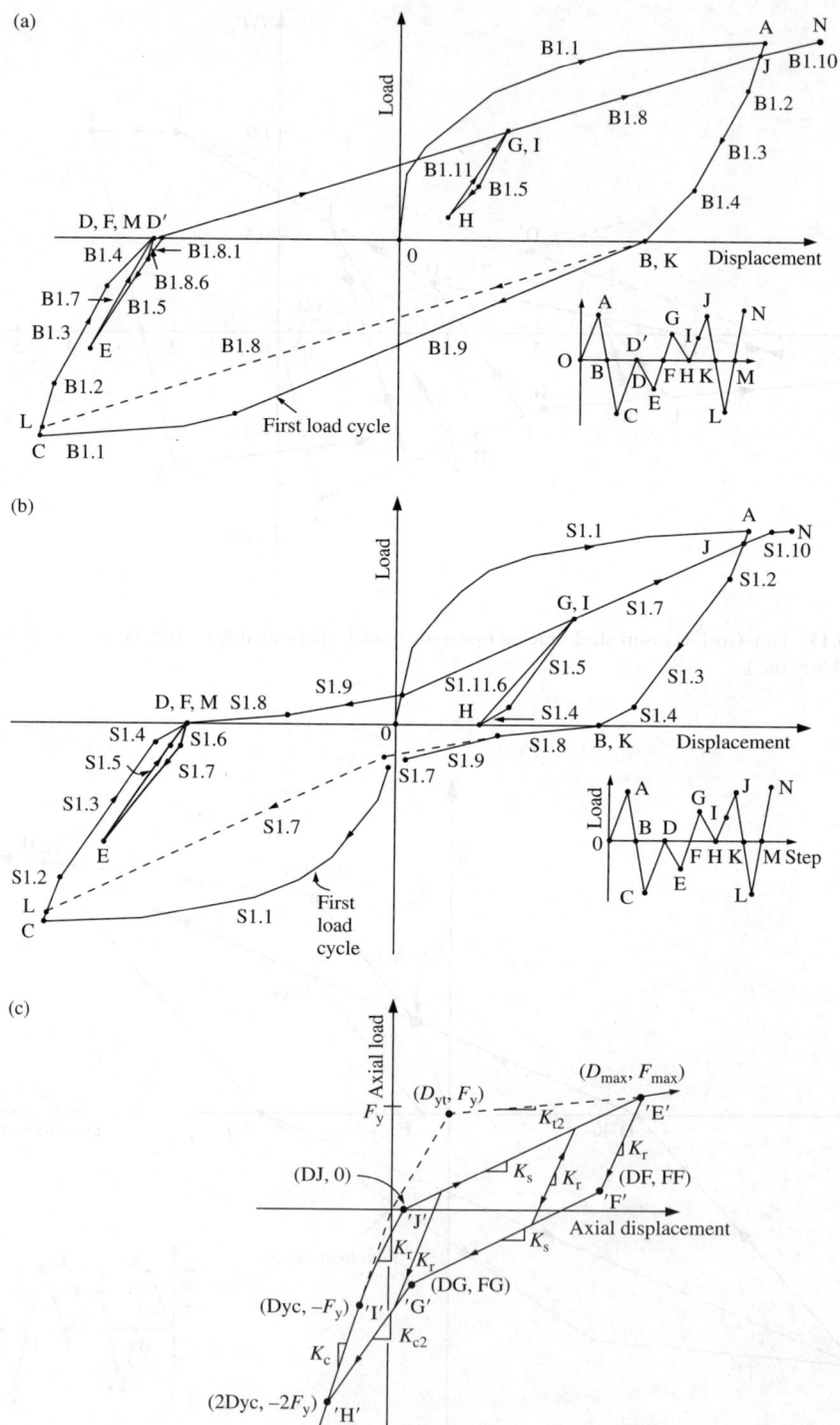

FIGURE 3.45 Cheng–Mertz model for bending coupling with shear of low-rise shear walls: (a) bending; (b) shear; and (c) axial. (Reprinted from Ref. [1, pp. 913, 932, and 952] by courtesy of Marcel Dekker, Inc.)

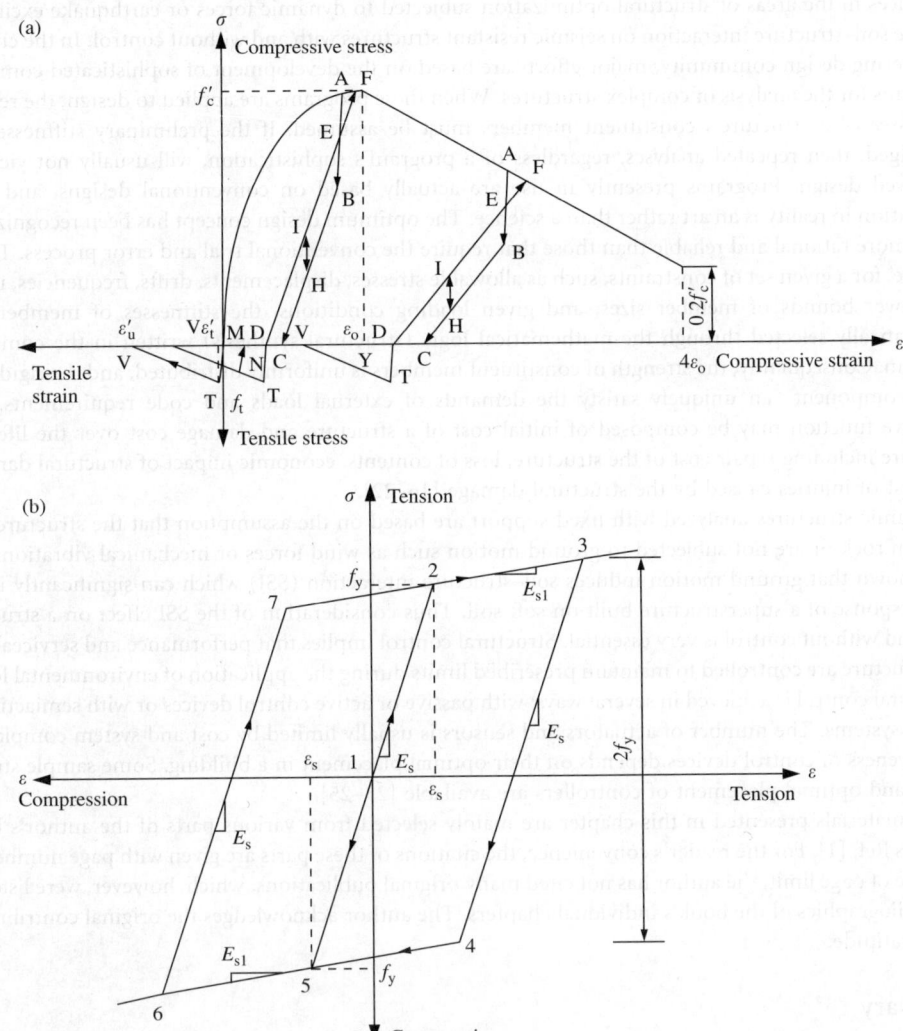

FIGURE 3.46 Cheng–Lou axial hysteresis model for RC members: (a) concrete under cyclic loading and (b) reinforcement under cyclic loading. (Reprinted from Ref. [1, p. 967] by courtesy of Marcel Dekker, Inc.)

low- and high-rise shear walls [12]. Figure 3.46 is the Cheng–Lou model for axial deformation of columns and walls. These models can be used on system analysis. For example, a braced steel frame needs hysteresis models of beams, columns, and bracings. If the steel structure is also composed of trusses, the hysteresis model of trusses must be employed. The latest version of INRESB-3D-II includes the afore-mentioned models for both PC and supercomputer [13,14]. The program has been used for case studies to verify different instances of building damage and collapse induced by strong ground motion [15].

3.8 Concluding Remarks

This chapter has composed most of the essential analytical techniques in structural dynamics. However, some special topics are currently in vogue but not included herein. The reader may find them from the references in the areas of structural optimization subjected to dynamic forces or earthquake excitation and the soil–structure interaction on seismic resistant structures with and without control. In the current engineering design community, major efforts are based on the development of sophisticated computer programs for the analysis of complex structures. When these programs are applied to design, the relative stiffnesses of a structure's constituent members must be assumed. If the preliminary stiffnesses are misjudged, then repeated analyses, regardless of a program's sophistication, will usually not yield an improved design. Programs presently in use are actually based on conventional designs, and their application in reality is an art rather than a science. The optimum design concept has been recognized as being more rational and reliable than those that require the conventional trial and error process. This is because, for a given set of constraints, such as allowable stresses, displacements, drifts, frequencies, upper and lower bounds of member sizes, and given loading conditions, the stiffnesses of members are automatically selected through the mathematical logic (structural synthesis) written in the computer program. Consequently, the strength of constituent members is uniformly distributed, and the rigidity of every component can uniquely satisfy the demands of external loads and code requirements. The objective function may be composed of initial cost of a structure and damage cost over the life of a structure including repair cost of the structure, loss of contents, economic impact of structural damage, and cost of injuries caused by the structural damage [16–22].

Dynamic structures analyzed with fixed support are based on the assumption that the structures are built on rock or are not subjected to ground motion such as wind forces or mechanical vibration. It is well known that ground motion induces soil–structure interaction (SSI) which can significantly influence response of a superstructure built on soft soil. Thus consideration of the SSI effect on a structure with and without control is very essential. Structural control implies that performance and serviceability of a structure are controlled to maintain prescribed limits during the application of environmental loads. Structural control is achieved in several ways: with passive or active control devices or with semiactive or hybrid systems. The number of actuators and sensors is usually limited by cost and system complexity. Effectiveness of control devices depends on their optimal placement in a building. Some sample studies of SSI and optimal placement of controllers are available [23–25].

The materials presented in this chapter are mainly selected from various parts of the author's book listed as Ref. [1]. For the reader's convenience, the citations of these parts are given with page number(s). Because of page limit, the author has not cited many original publications, which, however, were listed in the bibliographies of the book's individual chapters. The author acknowledges the original contribution with gratitude.

Glossary

Continuous system — Structures having constituent members with distributed mass for dynamic stiffness or finite element formulation.

Deterministic — A force's time function can be specified in regular or irregular variation.

Damping factor — Ratio of damping coefficient to the critical damping coefficient.

Ductility factor — Ratio of maximum displacement (or rotation) to the yielding displacement (or rotation).

Eigensolution — Eigenvalues and eigenvectors of a singular matrix that are usually referred to angular frequencies and normal modes in structural dynamics.

Finite element — A continuum is divided into a number of regions; each region behaves as a structural member with nodes compatible to the nodes of neighboring regions.

Hysteresis model — An inelastic force–deformation relationship of a structural member subjected to cyclic loading.

Influence factor — An vector matrix to specify the force induced by ground accelerations at each floor of a building.

Multicomponent seismic components — Earthquake accelerations expressed in more than one direction such as three translations and three rotations.

Participation factor — A factor for measuring how much a given vibrating mode participates in synthesizing the structural total load.

Periodic motion — A motion repeats itself in a certain period. Harmonic motion is periodic but periodic motion is not harmonic. A combination of two harmonic motions is always a periodic motion.

Pseudo-response — The response (displacement, velocity, or acceleration) dose not reflect real time-history response but a maximum value.

Stability criterion — The numerical integration results can be conditionally stable and unconditionally stable. For unconditionally stability, the solution is not divergent even if time increment used in the integration method is large. Conditional stability, however, depends on the time increment.

Transmissibility — A ratio between the amplitude of the force transmitted to the foundation to the amplitude of the driving harmonic force.

References

[1] Cheng, F.Y., *Matrix Analysis of Structural Dynamics — Applications and Earthquake Engineering*, Marcel Dekker, Inc., New York, 2001.

[2] Wilson, E.L., Kiureghian, A.D., and Bayo, E.P., A replacement for the SRSS method in seismic analysis, *J. Earthq. Eng. Struct. Dyn.*, 9, 187–194, 1981.

[3] Tuma, J.J. and Cheng, F.Y., *Dynamics Structural Analysis*, McGraw-Hill, Inc., New York, 1987.

[4] Cheng, F.Y. and Tseng, W.H., Dynamic Instability and Ultimate Capacity of Inelastic Systems Parametrically Excited by Earthquakes, Part I, NSF Report, National Technical Information Service, US Department of Commerce, Virginia, NTIS no. PB261096/AS, 1973 (143 pp.).

[5] Kanai, K., Semi-empirical formula for the seismic characteristics of the ground, *Bull. Earthq. Res. Inst., Univ. Tokyo*, 35, 309–325, 1957.

[6] Tajimi, H., A statistical method of determining the maximum response of a building structure during an earthquake, in *Proceeding of 2nd World Conference on Earthquake Engineering*, Japan, 1960, pp. 781–797.

[7] Mathews, J.H. and Howell, R.W., *Complex Analysis for Mathematics and Engineering*, William C. Brown Publishing, Dubuque, IA, 1996.

[8] Clough, R. and Penzien, J., *Dynamics of Structures*, 2nd ed., McGraw-Hill, Inc., New York, 1993, pp. 713–726.

[9] Humar, J.L., *Dynamics of Structures*, Prentice Hall, Englewood Cliffs, NJ, 1990, p. 728.

[10] Jain, A.K., Goel, S.C., and Hanson, R.D., Hysteretic cycles of axially loaded steel members, *ASCE J. Struct. Div.*, 106, 1775–1795, 1980.

[11] Takeda, T., Sozen, M.A., and Nielsen, N.N., Reinforced concrete response to simulated earthquakes, *ASCE J. Struc. Div.*, 96, 2557–2573, 1970.

[12] Cheng, F.Y. et al., Computed versus observed inelastic seismic response of low-rise RC shear walls, *ASCE J. Struct. Eng.*, 119(11), 3255–3275, 1993.

[13] Cheng, F.Y., Ger, J.F., Li, D., and Yang, J.S., INRESB-3D-SUPII Program Listing for Supercomputer: General Purpose Program for Inelastic Analysis of RC and Steel Building Systems for 3D static and Dynamic Loads and Seismic Excitations, NSF Report, National Technical Information Service, US Department of Commerce, Virginia, NTIS no. PB97-123616, 1996 (114 pp.).

[14] Cheng, F.Y., Ger, J.F., Li, D., and Yang, J.S., INRESB-3D-SUPII Program Listing for PC: Inelastic Analysis of RC and Steel Building Systems for 3D Static and Dynamic Loads and Seismic Excitations, NSF Report, National Technical Information Service, US Department of Commerce, Virginia, NTIS no. PB97-123632, 1996 (109 pp.).

[15] Ger, J.F., Cheng, F.Y., and Lu, L.W., Collapse behavior of Pino-Suarez building during 1985 Mexico Earthquake, *ASCE J. Struct. Eng.*, 119(3), 852–870, 1993.

[16] Cheng, F.Y., *Multiobjective Optimum Design of Seismic-Resistant Structures in Recent Advances in Optimal Structural Design*, S.A. Burns, ed., ASCE, Virginia, 2002, Chap. 9.

[17] Cheng, F.Y., and Li, D., Multiobjective optimization of structures with and without control, *AIAA J. Guid. Control Dyn.*, 19(2), 392–397, 1996.

[18] Cheng, F.Y. and Li, D., Multiobjective optimization design with pareto genetic algorithm, *ASCE J. Struct. Eng.*, 123(9), 1252–1261, 1997.

[19] Cheng, F.Y. and Ang, A.H.-S., Cost-effectiveness optimization for aseismic design criteria of RC buildings, in *Proceedings of Case Studies in Optimal Design and Maintenance Planning of Civil Infrastructure Systems*, D.M. Frangopol, ed., ASCE, 1998, pp. 13–25.

[20] Chang, C.C., Ger, J.R., and Cheng, F.Y., Reliability-based optimum design for UBC and nondeterministic seismic spectra, *ASCE J. Struct. Eng.*, 120(1), 139–160, 1994.

[21] Cheng, F.Y. and Juang, D.S., Assessment of various code provisions based on optimum design of steel structures, *Int. J. Earthquake Eng. Struct. Dyn.*, 16, 45–61, 1988.

[22] Truman, K.Z. and Cheng, F.Y., Optimum assessment of irregular three-dimensional buildings, *ASCE J. Struct. Eng.*, 116(12), 3324–3337, 1989.

[23] Cheng, F.Y. and Suthiwong, S., Active Control for Seismic-Resistant Structures on Embedded Foundation in Layered Half-Space, NSF Report, National Technical Information Service, US Department of Commerce, Virginia, NTIS no. PB97-121345, 1996 (261 pp.).

[24] Cheng, F.Y. and Jiang, H.P., Hybrid control of seismic structures with optimal placement of control devices, *ASCE J. Aerospace Eng.*, 11(2), 52–58, 1998.

[25] Cheng, F.Y. and Zhang, X.Z., Building structures with intelligent hybrid control and soil–structure interaction, in *Proceedings of US–Korea Workshop on Smart Infrastructural Systems*, C.K. Choi, F.Y. Cheng, and C.B. Yun, eds., KAIST, Korea, 2002, pp. 133–142.

Further Reading

Cheng, F.Y. and Ger, J.F., Maximum response of buildings to multi-seismic input, *ASCE Dyn. Struct.* 397–410, 1987.

Cheng, F.Y. and Ger, J.F., Response analysis of 3-D pipeline structures with consideration of six-component seismic input, in *Proceedings of Symposium on Resent Developments in Lifeline Earthquake Engineering, American Society Mechanical Engineers (ASME) and Japan Society of Mechanical Engineers (JSME)*, Vol. I, pp. 257–271, 1989.

Cheng, F.Y. and Ger, J.F., The effect of multicomponent seismic excitation and direction on response behavior of 3-D structures, in *Proceedings of 4th US National Conference on Earthquake Engineering, Earthquake Engineering Research Institute (EERI)*, Vol. 2, pp. 5–14, 1990.

Chopra, A.K., *Dynamics of Structures*, Prentice Hall, Englewood Cliffs, NJ, 1995.

Craig, R.R., *Structural Dynamics*, John Wiley, New York, 1981.

Newmark, N.M., Torsion in symmetrical buildings, in *Proceedings of 4th World Conference on Earthquake Engineering*, Vol. A3, Santiago, pp. 19–32, 1969.

Newmark, N.M. and Hall, W.J., *Earthquake Spectra and Design*, Earthquake Engineering Research Institute (EERI), Oakland, CA, 1982.

Newmark, N.M. and Rosenblueth, R., *Fundamentals of Earthquake Engineering*, Prentice Hall, Englewood Cliffs, NJ, 1971.

Paz, M., *Structural Dynamics*, 2nd ed., Van Nostrand, New York, 1985.

Penzien, J. and Watabe, M., Characteristic of 3-dimensional earthquake ground motion, *J. Earthq. Eng. Struct. Dyn.*, 3, 365–373, 1975.

Soong, T.T. and Grigoriu, M., *Random Vibration of Mechanical and Structural Systems*, Prentice Hall, Englewood Cliffs, NJ, 1993.

Timoshenko, S.P., Young, D.H., and Weaver, W. Jr., *Vibration Problems in Engineering*, 4th ed., John Wiley, New York, 1974.

Structural Design

II

II

Structural Design

4

Steel Structures

Eric M. Lui
*Department of Civil and
Environmental Engineering,
Syracuse University,
Syracuse, NY*

4.1 Materials

4.1.1 Stress–Strain Behavior of Structural Steel

Structural steel is a construction material that possesses attributes such as *strength, stiffness, toughness,* and *ductility* that are desirable in modern constructions. Strength is the ability of a material to resist stress. It is measured in terms of the material's yield strength F_y and ultimate or tensile strength F_u. Steel used in ordinary constructions normally have values of F_y and F_u that range from 36 to 50 ksi (248 to 345 MPa) and from 58 to 70 ksi (400 to 483 MPa), respectively, although higher-strength steels are becoming more common. Stiffness is the ability of a material to resist deformation. It is measured in terms of the modulus of elasticity E and modulus of rigidity G. With reference to Figure 4.1, in which several uniaxial engineering stress–strain curves obtained from coupon tests for various grades of steels are shown, it is seen that the modulus of elasticity E does not vary appreciably for the different steel grades. Therefore, a value of 29,000 ksi (200 GPa) is often used for design. Toughness is the ability of a material to absorb energy before failure. It is measured as the area under the material's stress–strain curve. As shown in Figure 4.1, most (especially the lower grade) steels possess high toughness that made them suitable for both static and seismic applications. Ductility is the ability of a material to undergo large inelastic (or plastic) deformation before failure. It is measured in terms of percent elongation or percent reduction in area of the specimen tested in uniaxial tension. For steel, percent elongation ranges from around 10 to 40 for a 2-in. (5-cm) gage length specimen. Ductility generally decreases with increasing steel strength. Ductility is a very important attribute of steel. The ability of structural steel to deform considerably before failure by fracture allows an indeterminate structure to undergo stress redistribution. Ductility also enhances the energy absorption characteristic of the structure, which is extremely important in seismic design.

4.1.2 Types of Steel

Structural steels used for construction are designated by the American Society of Testing and Materials (ASTM) as follows:

ASTM designation*	Steel type
A36/A36M	Carbon structural steel
A131/A131M	Structural steel for ships
A242/A242M	High-strength low-alloy structural steel
A283/A283M	Low and intermediate tensile strength carbon steel plates
A328/A328M	Steel sheet piling
A514/A514M	High-yield strength, quenched and tempered alloy steel plate suitable for welding
A529/A529M	High-strength carbon–manganese steel of structural quality
A572/A572M	High-strength low-alloy columbium–vanadium steel
A573/A573M	Structural carbon steel plates of improved toughness
A588/A588M	High-strength low-alloy structural steel with 50 ksi (345 MPa) minimum yield point to 4 in. [100 mm] thick
A633/A633M	Normalized high-strength low-alloy structural steel plates
A656/A656M	Hot-rolled structural steel, high-strength low-alloy plate with improved formability
A678/A678M	Quenched and tempered carbon and high-strength low-alloy structural steel plates
A690/A690M	High-strength low-alloy steel H-Piles and sheet piling for use in marine environments
A709/A709M	Carbon and high-strength low-alloy structural steel shapes, plates, and bars and quenched and tempered alloy structural steel plates for bridges

ASTM designation*	Steel type
A710/A710M	Age-hardening low-carbon nickel–copper–chromium–molybdenum–columbium alloy structural steel plates
A769/A769M	Carbon and high-strength electric resistance welded steel structural shapes
A786/A786M	Rolled steel floor plates
A808/A808M	High-strength low-alloy carbon, manganese, columbium, vanadium steel of structural quality with improved notch toughness
A827/A827M	Plates, carbon steel, for forging and similar applications
A829/A829M	Plates, alloy steel, structural quality
A830/A830M	Plates, carbon steel, structural quality, furnished to chemical composition requirements
A852/A852M	Quenched and tempered low-alloy structural steel plate with 70 ksi [485 MPa] minimum yield strength to 4 in. [100 mm] thick
A857/A857M	Steel sheet piling, cold formed, light gage
A871/A871M	High-strength low-alloy structural steel plate with atmospheric corrosion resistance
A913/A913M	High-strength low-alloy steel shapes of structural quality, produced by quenching and self-tempering process (QST)
A945/A945M	High-strength low-alloy structural steel plate with low carbon and restricted sulfur for improved weldability, formability, and toughness
A992/A992M	Steel for structural shapes (W-sections) for use in building framing

* The letter M in the designation stands for Metric.

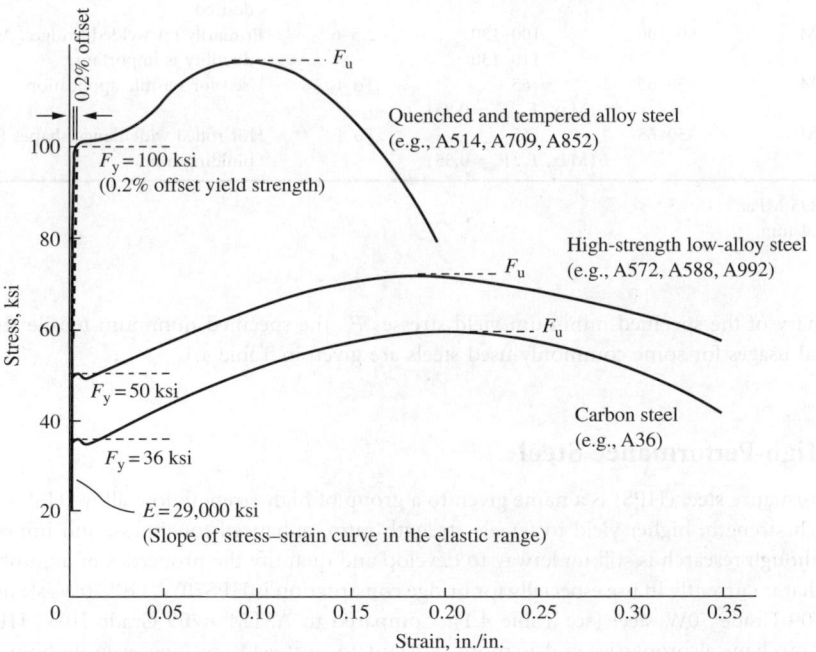

FIGURE 4.1 Uniaxial stress–strain behavior of steel.

TABLE 4.1   Steel Types and General Usages

ASTM designation	F_y (ksi)[a]	F_u (ksi)[a]	Plate thickness (in.)[b]	General usages
A36/A36M	36	58–80	To 8	Riveted, bolted, and welded buildings and bridges
A529/A529M	50	65–100	To 2.5	Similar to A36. The higher yield stress for
	55	70–100	To 1.5	A529 steel allows for savings in weight. A529 supersedes A441
A572/A572M				Grades 60 and 65 not suitable for welded
Grade 42	42	60	To 6	bridges
Grade 50	50	65	To 4	
Grade 55	55	70	To 2	
Grade 60	60	75	To 1.25	
Grade 65	65	80	To 1.25	
A242/A242M	42	63	1.5–5	Riveted, bolted, and welded buildings and
	46	67	0.75–1.5	bridges. Used when weight savings
	50	70	0.5–0.75	and enhanced atmospheric corrosion resistance are desired. Specific instructions must be provided for welding
A588/A588M	42	63	5–8	Similar to A242. Atmospheric corrosion
	46	67	4–5	resistance is about four times that of
	50	70	To 4	A36 steel
A709/A709M				Primarily for use in bridges
Grade 36	36	58–80	To 4	
Grade 50	50	65	To 4	
Grade 50W	50	70	To 4	
Grade 70W	70	90–110	To 4	
Grade 100 and 100W	90	100–130	2.5–4	
Grade 100 and 100W	100	110–130	To 2.5	
A852/A852M	70	90–110	To 4	Plates for welded and bolted construction where atmospheric corrosion resistance is desired
A514/A514M	90–100	100–130	2.5–6	Primarily for welded bridges. Avoid usage if
		110–130		ductility is important
A913/A913M	50–65	65 (Max. $F_y/F_u = 0.85$)	To 4	Used for seismic applications
A992/A992M	50–65	65 (Max. $F_y/F_u = 0.85$)	To 4	Hot-rolled wide flange shapes for use in building frames

[a] 1 ksi = 6.895 MPa.
[b] 1 in. = 25.4 mm.

A summary of the specified minimum yield stresses F_y, the specified minimum tensile strengths F_u, and general usages for some commonly used steels are given in Table 4.1.

4.1.3 High-Performance Steel

High-performance steel (HPS) is a name given to a group of high-strength low-alloy (HSLA) steels that exhibit high strength, higher yield to tensile strength ratio, enhanced toughness, and improved weldability. Although research is still underway to develop and quantify the properties of a number of HPS, one HPS that is currently in use especially for bridge construction is HPS70W. HPS70W is a derivative of ASTM A709 Grade 70W steel (see Table 4.1). Compared to ASTM A709 Grade 70W, HPS70W has improved mechanical properties and is more resistant to postweld cracking even without preheating before welding.

4.1.4 Fireproofing of Steel

Although steel is an incombustible material, its strength (F_y, F_u) and stiffness (E) reduce quite noticeably at temperatures normally reached in fires when other materials in a building burn. Exposed steel members that may be subjected to high temperature in a fire should be fireproofed to conform to the fire ratings set forth in city codes. Fire ratings are expressed in units of time (usually hours) beyond which the structural members under a standard ASTM Specification (E119) fire test will fail under a specific set of criteria. Various approaches are available for fireproofing steel members. Steel members can be fireproofed by encasement in concrete if a minimum cover of 2 in. (5.1 mm) of concrete is provided. If the use of concrete is undesirable (because it adds weight to the structure), a lath and plaster (gypsum) ceiling placed underneath the structural members supporting the floor deck of an upper story can be used. In lieu of such a ceiling, spray-on materials, such as mineral fibers, perlite, vermiculite, gypsum, etc., can also be used for fireproofing. Other means of fireproofing include placing steel members away from the source of heat, circulating liquid coolant inside box or tubular members, and the use of insulative paints. These special paints foam and expand when heated, thus forming a shield for the members (Rains 1976). For a more detailed discussion of structural steel design for fire protection, refer to the latest edition of AISI publication No. FS3, *Fire-Safe Structural Steel — A Design Guide.* Additional information on fire-resistant standards and fire protection can be found in the AISI booklets on *Fire Resistant Steel Frame Construction, Designing Fire Protection for Steel Columns,* and *Designing Fire Protection for Steel Trusses* as well as in the *Uniform Building Code.*

4.1.5 Corrosion Protection of Steel

Atmospheric corrosion occurs when steel is exposed to a continuous supply of water and oxygen. The rate of corrosion can be reduced if a barrier is used to keep water and oxygen from contact with the surface of bare steel. Painting is a practical and cost-effective way to protect steel from corrosion. The Steel Structures Painting Council issues specifications for the surface preparation and the painting of steel structures for corrosion protection of steel. In lieu of painting, the use of other coating materials such as epoxies or other mineral and polymeric compounds can be considered. The use of corrosion resistance steels such as ASTM A242, A588 steel, or galvanized or stainless steel is another alternative. Corrosion resistant steels such as A588 retard corrosion by the formation of a layer of deep reddish-brown to black patina (an oxidized metallic film) on the steel surface after a few wetting–drying cycles, which usually take place within 1 to 3 years. Galvanized steel has a zinc coating. In addition to acting as a protective cover, zinc is anodic to steel. The steel, being cathodic, is therefore protected from corrosion. Stainless steel is more resistant to rusting and staining than ordinary steel primarily because of the presence of chromium as an alloying element.

4.1.6 Structural Steel Shapes

Steel sections used for construction are available in a variety of shapes and sizes. In general, there are three procedures by which steel shapes can be formed: hot rolled, cold formed, and welded. All steel shapes must be manufactured to meet ASTM standards. Commonly used steel shapes include the wide flange (W) sections, the American Standard beam (S) sections, bearing pile (HP) sections, American Standard channel (C) sections, angle (L) sections, tee (WT) sections, as well as bars, plates, pipes, and hollow structural sections (HSS). Sections that, by dimensions, cannot be classified as W or S shapes are designated as miscellaneous (M) sections and C sections that, by dimensions, cannot be classified as American Standard channels are designated as miscellaneous channel (MC) sections.

Hot-rolled shapes are classified in accordance with their tensile property into five size groups by the American Society of Steel Construction (AISC). The groupings are given in the AISC Manuals (1989,

2001). Groups 4 and 5 shapes and group 3 shapes with flange thickness exceeding $1\frac{1}{2}$ in. are generally used for application as *compression members*. When weldings are used, care must be exercised to minimize the possibility of cracking in regions at the vicinity of the welds by carefully reviewing the material specification and fabrication procedures of the pieces to be joined.

4.1.7 Structural Fasteners

Steel sections can be fastened together by rivets, bolts, and welds. While rivets were used quite extensively in the past, their use in modern steel construction has become almost obsolete. Bolts have essentially replaced rivets as the primary means to connect nonwelded structural components.

4.1.7.1 Bolts

Four basic types of bolts are commonly in use. They are designated by ASTM as A307, A325, A490, and A449 (ASTM 2001a–d). A307 bolts are called common, unfinished, machine, or rough. They are made from low-carbon steel. Two grades (A and B) are available. They are available in diameters from $\frac{1}{4}$ to 4 in. (6.4 to 102 mm) in $\frac{1}{8}$ in. (3.2 mm) increments. They are used primarily for low-stress connections and for secondary members. A325 and A490 bolts are called high-strength bolts. A325 bolts are made from a heat-treated medium-carbon steels. They are available in two types: Type 1 — bolts made of medium-carbon steel. Type 3 — bolts having atmospheric corrosion resistance and weathering characteristics comparable to A242 and A588 steels. A490 bolts are made from quenched and tempered alloy steel and thus have higher strength than A325 bolts. Like A325 bolts, two types (Types 1 and 3) are available. Both A325 and A490 bolts are available in diameters from $\frac{1}{2}$ to $1\frac{1}{2}$ in. (13 to 38 mm) in $\frac{1}{8}$ in. (3.2 mm) increments. They are used for general construction purposes. A449 bolts are made from quenched and tempered steels. They are available in diameters from $\frac{1}{4}$ to 3 in. (6.4 to 76 mm). Because A449 bolts are not produced to the same quality requirements nor have the same heavy-hex head and nut dimensions as A325 or A490 bolts, they are not to be used for slip critical connections. A449 bolts are used primarily when diameters over $1\frac{1}{2}$ in. (38 mm) are needed. They are also used for anchor bolts and threaded rod.

High-strength bolts can be tightened to two conditions of tightness: snug tight and fully tight. The snug-tight condition can be attained by a few impacts of an impact wrench or the full effort of a worker using an ordinary spud wrench. The snug-tight condition must be clearly identified in the design drawing and is permitted in bearing-type connections where slip is permitted, or in tension or combined shear and tension applications where loosening or fatigue due to vibration or load fluctuations are not design considerations. Bolts used in slip-critical conditions (i.e., conditions for which the integrity of the connected parts is dependent on the frictional force developed between the interfaces of the joint) and in conditions where the bolts are subjected to direct tension are required to be tightened to develop a pretension force equal to about 70% of the minimum tensile stress F_u of the material from which the bolts are made. This can be accomplished by using the turn-of-the-nut method, the calibrated wrench method, or by the use of alternate design fasteners or direct tension indicator (RCSC 2000).

4.1.7.2 Welds

Welding is a very effective means to connect two or more pieces of materials together. The four most commonly used welding processes are shielded metal arc welding (SMAW), submerged arc welding (SAW), gas metal arc welding (GMAW), and flux core arc welding (FCAW) (AWS 2000). Welding can be done with or without filler materials although most weldings used for construction utilize filler materials. The filler materials used in modern-day welding processes are electrodes. Table 4.2 summarizes the electrode designations used for the aforementioned four most commonly used welding processes. In general, the strength of the electrode used should equal or exceed the strength of the steel being welded (AWS 2000).

TABLE 4.2 Electrode Designations

Welding processes	Electrode designations	Remarks
Shielded metal arc welding (SMAW)	E60XX E70XX E80XX E100XX E110XX	The "E" denotes electrode. The first two digits indicate tensile strength in ksi.[a] The two "X"s represent numbers indicating the electrode usage
Submerged arc welding (SAW)	F6X-EXXX F7X-EXXX F8X-EXXX F10X-EXXX F11X-EXXX	The "F" designates a granular flux material. The digit(s) following the "F" indicate the tensile strength in ksi (6 means 60 ksi, 10 means 100 ksi, etc.). The digit before the hyphen gives the Charpy V-notched impact strength. The "E" and the "X"s that follow represent numbers relating to the electrode usage
Gas metal arc welding (GMAW)	ER70S-X ER80S ER100S ER110S	The digits following the letters "ER" represent the tensile strength of the electrode in ksi
Flux cored arc welding (FCAW)	E6XT-X E7XT-X E8XT E10XT E11XT	The digit(s) following the letter "E" represent the tensile strength of the electrode in ksi (6 means 60 ksi, 10 means 100 ksi, etc.)

[a] 1 ksi = 6.895 MPa.

Finished welds should be inspected to ensure their quality. Inspection should be performed by qualified welding inspectors. A number of inspection methods are available for weld inspections, including visual inspection, the use of liquid penetrants, magnetic particles, ultrasonic equipment, and radiographic methods. Discussion of these and other welding inspection techniques can be found in the *Welding Handbook* (AWS 1987).

4.1.8 Weldability of Steel

Weldability is the capacity of a material to be welded under a specific set of fabrication and design conditions and to perform as expected during its service life. Generally, weldability is considered very good for low-carbon steel (carbon level < 0.15% by weight), good for mild steel (carbon levels 0.15 to 0.30%), fair for medium-carbon steel (carbon levels 0.30 to 0.50%), and questionable for high-carbon steel (carbon levels 0.50 to 1.00%). Because weldability normally decreases with increasing carbon content, special precautions such as preheating, controlling heat input, and post-weld heat treating are normally required for steel with carbon content reaching 0.30%. In addition to carbon content, the presence of other alloying elements will have an effect on weldability. Instead of more accurate data, the table below can be used as a guide to determine the weldability of steel (Blodgett, undated).

Element	Range for satisfactory weldability	Level requiring special care (%)
Carbon	0.06–0.25%	0.35
Manganese	0.35–0.80%	1.40
Silicon	0.10% max.	0.30
Sulfur	0.035% max.	0.050
Phosphorus	0.030% max.	0.040

A quantitative approach for determining weldability of steel is to calculate its *carbon equivalent value.* One definition of the carbon equivalent value C_{eq} is

$$C_{eq} = \text{Carbon} + \frac{(\text{manganese} + \text{silicon})}{6} + \frac{(\text{copper} + \text{nickel})}{15}$$
$$+ \frac{(\text{chromium} + \text{molybdenum} + \text{vanadium} + \text{columbium})}{5} \qquad (4.1)$$

A steel is considered weldable if $C_{eq} \leq 0.50\%$ for steel in which the carbon content does not exceed 0.12% and if $C_{eq} \leq 0.45\%$ for steel in which the carbon content exceeds 0.12%.

Equation 4.1 indicates that the presence of alloying elements decreases the weldability of steel. An example of high-alloy steels is stainless steel. There are three types of stainless steel: austenitic, martensitic, or ferritic. Austenitic stainless steel is the most weldable, but care must be exercised to prevent thermal distortion because heat dissipation is only about one third as fast as in plain carbon steel. Martensitic steel is also weldable but prone to cracking because of its high hardenability. Preheating and maintaining interpass temperature are often needed, especially when the carbon content is above 0.10%. Ferritic steel is weldable but decreased ductility and toughness in the weld area can present a problem. Preheating and postweld annealing may be required to minimize these undesirable effects.

4.2 Design Philosophy and Design Formats

4.2.1 Design Philosophy

Structural design should be performed to satisfy the criteria for strength, serviceability, and economy. *Strength* pertains to the general integrity and safety of the structure under extreme load conditions. The structure is expected to withstand occasional overloads without severe distress and damage during its lifetime. *Serviceability* refers to the proper functioning of the structure as related to its appearance, maintainability, and durability under normal, or service load, conditions. Deflection, vibration, permanent deformation, cracking, and corrosion are some design considerations associated with serviceability. *Economy* concerns with the overall material, construction, and labor costs required for the design, fabrication, erection, and maintenance processes of the structure.

4.2.2 Design Formats

At present, steel design in the United States is being performed in accordance with one of the following three formats.

4.2.2.1 Allowable Stress Design (ASD)

ASD has been in use for decades for steel design of buildings and bridges. It continues to enjoy popularity among structural engineers engaged in steel building design. In allowable stress (or working stress) design, member stresses computed under service (or working) loads are compared to some predesignated stresses called allowable stresses. The allowable stresses are often expressed as a function of the yield stress (F_y) or tensile stress (F_u) of the material divided by a factor of safety. The factor of safety is introduced to account for the effects of overload, understrength, and approximations used in structural analysis. The general format for an allowable stress design has the form

$$\frac{R_n}{\text{FS}} \geq \sum_{i=1}^{m} Q_{ni} \qquad (4.2)$$

where R_n is the nominal resistance of the structural component expressed in unit of stress (i.e., the allowable stress), Q_{ni} is the service or working stresses computed from the applied working load of type i, FS is the factor of safety; i is the load type (dead, live, wind, etc.), and m is the number of load types considered in the design.

4.2.2.2 Plastic Design (PD)

PD makes use of the fact that steel sections have reserved strength beyond the first yield condition. When a section is under flexure, yielding of the cross-section occurs in a progressive manner, commencing with the fibers farthest away from the neutral axis and ending with the fibers nearest the neutral axis. This phenomenon of progressive yielding, referred to as *plastification*, means that the cross-section does not fail at first yield. The additional moment that a cross-section can carry in excess of the moment that corresponds to first yield varies depending on the shape of the cross-section. To quantify such reserved capacity, a quantity called *shape factor*, defined as the ratio of the *plastic moment* (moment that causes the entire cross-section to yield, resulting in the formation of a *plastic hinge*) to the *yield moment* (moment that causes yielding of the extreme fibers only) is used. The shape factor for hot-rolled I-shaped sections bent about the strong axes has a value of about 1.15. The value is about 1.50 when these sections are bent about their weak axes.

For an indeterminate structure, failure of the structure will not occur after the formation of a plastic hinge. After complete yielding of a cross-section, force (or, more precisely, moment) redistribution will occur in which the unyielded portion of the structure continues to carry some additional loadings. Failure will occur only when enough cross-sections have yielded rendering the structure unstable, resulting in the formation of a *plastic collapse mechanism*.

In PD, the factor of safety is applied to the applied loads to obtain *factored loads*. A design is said to have satisfied the strength criterion if the load effects (i.e., forces, shears, and moments) computed using these factored loads do not exceed the nominal plastic strength of the structural component. PD has the form

$$R_n \geq \gamma \sum_{i=1}^{m} Q_{ni} \tag{4.3}$$

where R_n is the nominal plastic strength of the member, Q_{ni} is the nominal load effect from loads of type i, γ is the load factor, i is the load type, and m is the number of load types.

In steel building design, the load factor is given by the AISC Specification as 1.7 if Q_n consists of dead and live gravity loads only, and as 1.3 if Q_n consists of dead and live gravity loads acting in conjunction with wind or earthquake loads.

4.2.2.3 Load and Resistance Factor Design (LRFD)

LRFD is a probability-based limit state design procedure. A *limit state* is defined as a condition in which a structure or structural component becomes unsafe (i.e., a violation of the strength limit state) or unsuitable for its intended function (i.e., a violation of the serviceability limit state). In a limit state design, the structure or structural component is designed in accordance to its limits of usefulness, which may be strength related or serviceability related. In developing the LRFD method, both load effects and resistance are treated as random variables. Their variabilities and uncertainties are represented by frequency distribution curves. A design is considered satisfactory according to the strength criterion if the resistance exceeds the load effects by a comfortable margin. The concept of safety is represented schematically in Figure 4.2. Theoretically, the structure will not fail unless the load effect Q exceeds the resistance R as shown by the shaded portion in the figure. The smaller this shaded area, the less likely that the structure will fail. In actual design, a resistance factor ϕ is applied to the nominal resistance of the structural component to account for any uncertainties associated with the determination of its strength and a load factor γ is applied to each load type to account for the uncertainties and difficulties associated with determining its actual load magnitude. Different load factors are used for different load types to reflect the varying degree of uncertainties associated with the determination of load magnitudes. In general, a lower load factor is used for a load that is more predicable and a higher load factor is used for a load that is less predicable. Mathematically, the LRFD format takes the form

$$\phi R_n \geq \sum_{i=1}^{m} \gamma_i Q_{ni} \tag{4.4}$$

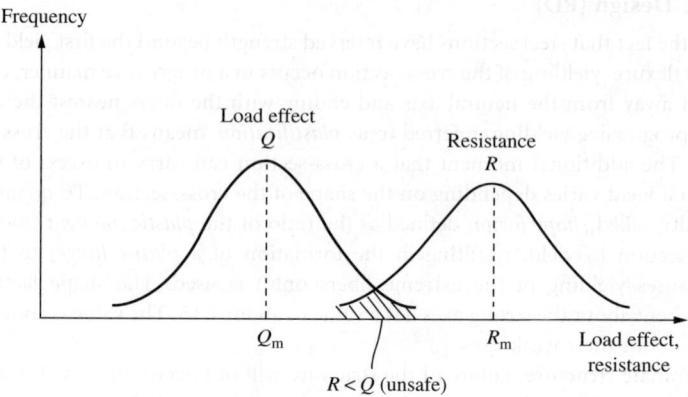

FIGURE 4.2 Frequency distribution of load effect and resistance.

TABLE 4.3 Load Factors and Load Combinations

$1.4(D + F)$

$1.2(D + F + T) + 1.6(L + H) + 0.5(L_r \text{ or } S \text{ or } R)$

$1.2D + 1.6(L_r \text{ or } S \text{ or } R) + (L \text{ or } 0.8W)$

$1.2D + 1.6W + L + 0.5(L_r \text{ or } S \text{ or } R)$

$1.2D + 1.0E + L + 0.2S$

$0.9D + 1.6W + 1.6H$

$0.9D + 1.0E + 1.6H$

Notes: D is the dead load, *E* is the earthquake load, *F* is the load due to fluids with well-defined pressures and maximum heights, *H* is the load due to the weight and lateral pressure of soil and water in soil, *L* is the live load, L_r is the roof live load, *R* is the rain load, *S* is the snow load, *T* is the self-straining force, and *W* is the wind load.

The load factor on *L* in the third, fourth, and fifth load combinations shown above can be set to 0.5 for all occupancies (except for garages or areas occupied as places of public assembly) in which the design live load per square foot of area is less than or equal to 100 psf (4.79 kN/m^2). The load factor on *H* in the sixth and seventh load combinations shall be set to zero if the structural action due to *H* counteracts that due to *W* or *E*.

where ϕR_n represents the design (or usable) strength and $\sum \gamma_i Q_{ni}$ represents the required strength or load effect for a given load combination. Table 4.3 shows examples of load combinations (ASCE 2002) to be used on the right-hand side of Equation 4.4. For a safe design, all load combinations should be investigated and the design is based on the worst-case scenario.

4.3 Tension Members

Tension members are designed to resist tensile forces. Examples of tension members are hangers, truss members, and bracing members that are in tension. Cross-sections that are used most often for tension members are solid and hollow circular rods, bundled bars and cables, rectangular plates, single and double angles, channels, WT- and W-sections, and a variety of built-up shapes.

4.3.1 Tension Member Design

Tension members are to be designed to preclude the following possible failure modes under normal load conditions: yielding in gross section, fracture in effective net section, block shear, shear rupture along plane through the fasteners, bearing on fastener holes, prying (for lap- or hanger-type joints). In addition, the fasteners' strength must be adequate to prevent failure in the fasteners. Also, except for rods in tension, the slenderness of the tension member obtained by dividing the length of the member by its least radius of gyration should preferably not exceed 300.

4.3.1.1 Allowable Stress Design

The computed tensile stress f_t in a tension member shall not exceed the allowable stress for tension, F_t, given by $0.60F_y$ for yielding on the gross area and by $0.50F_u$ for fracture on the effective net area. While the gross area is just the nominal cross-sectional area of the member, the *effective net area* is the smallest cross-sectional area accounting for the presence of fastener holes and the effect of *shear lag*. It is calculated using the equation

$$A_e = UA_n = U\left[A_g - \sum_{i=1}^{m} d_{ni}t_i + \sum_{j=1}^{k}\left(\frac{s^2}{4g}\right)_j t_j\right] \tag{4.5}$$

where U is a reduction coefficient given by (Munse and Chesson 1963)

$$U = 1 - \frac{\bar{x}}{l} \le 0.90 \tag{4.6}$$

in which l is the length of the connection and $\bar{x}$ is the larger of the distance measured from the centroid of the cross-section to the contact plane of the connected pieces or to the fastener lines. In the event that the cross-section has two symmetrically located planes of connection, $\bar{x}$ is measured from the centroid of the nearest one-half the area (Figure 4.3). This reduction coefficient is introduced to account for the shear lag effect that arises when some component elements of the cross-section in a joint are not connected, rendering the connection less effective in transmitting the applied load. The terms in brackets in Equation 4.5 constitute the so-called net section A_n. The various terms are defined as follows: A_g is the gross cross-sectional area, d_n is the nominal diameter of the hole (bolt cutout) taken as the nominal bolt diameter plus $\frac{1}{8}$ in. (3.2 mm), t is the thickness of the component element, s is the longitudinal center-to-center spacing (pitch) of any two consecutive fasteners in a chain of staggered holes, and g is the transverse center-to-center spacing (gage) between two adjacent fasteners gage lines in a chain of staggered holes.

The second term inside the brackets of Equation 4.5 accounts for loss of material due to bolt cutouts; the summation is carried for all bolt cutouts lying on the failure line. The last term inside the brackets of Equation 4.5 indirectly accounts for the effect of the existence of a combined stress state (tensile and shear) along an inclined failure path associated with staggered holes; the summation is carried for all staggered paths along the failure line. This term vanishes if the holes are not staggered. Normally, it is necessary to investigate different failure paths that may occur in a connection; the critical failure path is the one giving the smallest value for A_e.

To prevent block shear failure and shear rupture, the allowable strengths for block shear and shear rupture are specified as follows:

Block shear:

$$R_{BS} = 0.30A_v F_u + 0.50A_t F_u \tag{4.7}$$

Shear rupture:

$$F_v = 0.30F_u \tag{4.8}$$

where A_v is the net area in shear, A_t is the net area in tension, and F_u is the specified minimum tensile strength.

The tension member should also be designed to possess adequate thickness and the fasteners should be placed within a specific range of spacings and edge distances to prevent failure due to bearing and failure by prying action (see Section 4.11).

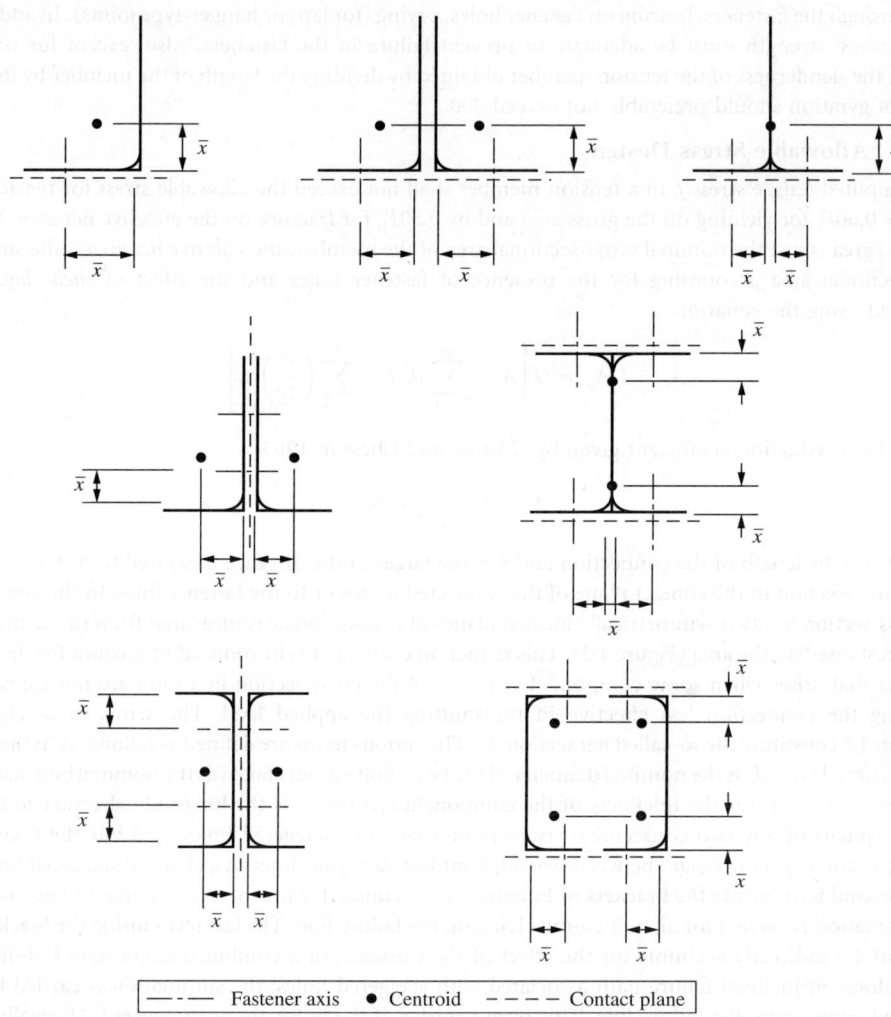

--------- Fastener axis   ● Centroid   - - - - Contact plane

FIGURE 4.3   Definition of $\bar{x}$ for selected cross-sections.

4.3.1.2 Load and Resistance Factor Design

According to the LRFD Specification (AISC 1999), tension members designed to resist a factored axial force of P_u calculated using the load combinations shown in Table 4.3 must satisfy the condition of

$$\phi_t P_n \geq P_u \tag{4.9}$$

The design strength $\phi_t P_n$ is evaluated as follows:

Yielding in gross section:

$$\phi_t P_n = 0.90[F_y A_g] \tag{4.10}$$

where 0.90 is the resistance factor for tension, F_y is the specified minimum yield stress of the material, and A_g is the gross cross-sectional area of the member.

Fracture in effective net section:

$$\phi_t P_n = 0.75[F_u A_e] \tag{4.11}$$

where 0.75 is the resistance factor for fracture in tension, F_u is the specified minimum tensile strength, and A_e is the effective net area given in Equation 4.5.

Block shear: If $F_u A_{nt} \geq 0.6 F_u A_{nv}$ (i.e., shear yield–tension fracture)

$$\phi_t P_n = 0.75[0.60 F_y A_{gv} + F_u A_{nt}] \leq 0.75[0.6 F_u A_{nv} + F_u A_{nt}] \qquad (4.12a)$$

and if $F_u A_{nt} < 0.6 F_u A_{nv}$ (i.e., shear fracture–tension yield)

$$\phi_t P_n = 0.75[0.60 F_u A_{nv} + F_y A_{gt}] \leq 0.75[0.60 F_u A_{nv} + F_u A_{nt}] \qquad (4.12b)$$

where 0.75 is the resistance factor for block shear, F_y, F_u are the specified minimum yield stress and tensile strength, respectively, A_{gv} is the gross shear area, A_{nt} is the net tension area, A_{nv} is the net shear area, and A_{gt} is the gross tension area.

EXAMPLE 4.1

Using LRFD, select a double-channel tension member shown in Figure 4.4a to carry a dead load D of 40 kip and a live load L of 100 kip. The member is 15 ft long. Six 1-in. diameter A325 bolts in standard size holes are used to connect the member to a $\frac{3}{8}$-in. gusset plate. Use A36 steel ($F_y = 36$ ksi, $F_u = 58$ ksi) for all the connected parts.

Load combinations: From Table 4.3, the applicable load combinations are

$$1.4D = 1.4(40) = 56 \text{ kip}$$
$$1.2D + 1.6L = 1.2(40) + 1.6(100) = 208 \text{ kip}$$

The design of the tension member is to be based on the larger of the two, that is, 208 kip and so *each* channel is expected to carry 104 kip.

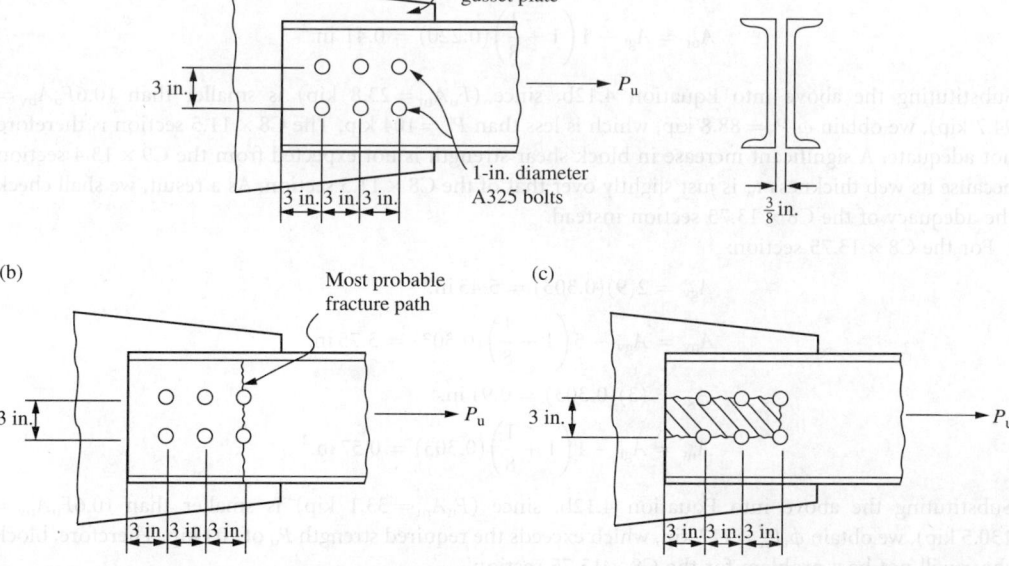

FIGURE 4.4 Design of (a) double-channel tension member (1 in. = 25.4 mm); (b) fracture failure; and (c) block shear failure.

Yielding in gross section: Using Equations 4.9 and 4.10, the gross area required to prevent cross-section yielding is

$$0.90[F_y A_g] \geq P_u$$
$$0.90[(36)(A_g)] \geq 104 \text{ kip}$$
$$(A_g)_{\text{req'd}} \geq 3.21 \text{ in.}^2$$

From the section properties table contained in the AISC-LRFD Manual, one can select the following trial sections: C8 × 11.5 ($A_g = 3.38$ in.²), C9 × 13.4 ($A_g = 3.94$ in.²), and C8 × 13.75 ($A_g = 4.04$ in.²).

Check for the limit state of fracture on effective net area: The above sections are checked for the limiting state of fracture in the following table:

Section	A_g (in.²)	t_w (in.)	$\bar{x}$ (in.)	U^a	A_e^b (in.²)	$\phi_t P_n$ (kip)
C8 × 11.5	3.38	0.220	0.571	0.90	2.6	113.1
C9 × 13.4	3.94	0.233	0.601	0.90	3.07	133.5
C8 × 13.75	4.04	0.303	0.553	0.90	3.02	131.4

[a] Equation 4.6.
[b] Equation 4.5, Figure 4.4b.

From the last column of the above table, it can be seen that fracture is not a problem for any of the trial sections.

Check for the limit state of block shear: Figure 4.4c shows a possible block shear failure mode. To avoid block shear failure the required strength of $P_u = 104$ kip should not exceed the design strength, $\phi_t P_n$, calculated using Equations 4.12a or 4.12b, whichever is applicable.

For the C8 × 11.5 section:

$$A_{gv} = 2(9)(0.220) = 3.96 \text{ in.}^2$$

$$A_{nv} = A_{gv} - 5\left(1 + \frac{1}{8}\right)(0.220) = 2.72 \text{ in.}^2$$

$$A_{gt} = (3)(0.220) = 0.66 \text{ in.}^2$$

$$A_{nt} = A_{gt} - 1\left(1 + \frac{1}{8}\right)(0.220) = 0.41 \text{ in.}^2$$

Substituting the above into Equation 4.12b, since ($F_u A_{nt} = 23.8$ kip) is smaller than ($0.6 F_u A_{nv} = 94.7$ kip), we obtain $\phi_t P_n = 88.8$ kip, which is less than $P_u = 104$ kip. The C8 × 11.5 section is therefore not adequate. A significant increase in block shear strength is not expected from the C9 × 13.4 section because its web thickness t_w is just slightly over that of the C8 × 11.5 section. As a result, we shall check the adequacy of the C8 × 13.75 section instead.

For the C8 × 13.75 section:

$$A_{gv} = 2(9)(0.303) = 5.45 \text{ in.}^2$$

$$A_{nv} = A_{gv} - 5\left(1 + \frac{1}{8}\right)(0.303) = 3.75 \text{ in.}^2$$

$$A_{gt} = (3)(0.303) = 0.91 \text{ in.}^2$$

$$A_{nt} = A_{gt} - 1\left(1 + \frac{1}{8}\right)(0.303) = 0.57 \text{ in.}^2$$

Substituting the above into Equation 4.12b, since ($F_u A_{nt} = 33.1$ kip) is smaller than ($0.6 F_u A_{nv} = 130.5$ kip), we obtain $\phi_t P_n = 122$ kip, which exceeds the required strength P_u of 104 kip. Therefore, block shear will not be a problem for the C8 × 13.75 section.

Check for the limiting slenderness ratio: Using parallel axis theorem, the least radius of gyration of the double-channel cross-section is calculated to be 0.96 in. Therefore, $L/r = (15 \text{ ft})(12 \text{ in./ft})/0.96 \text{ in.} = 187.5$, which is less than the recommended maximum value of 300.

Check for the adequacy of the connection: The calculations are shown in an example in Section 4.11.

Longitudinal spacing of connectors: According to Section J3.5 of the LRFD Specification, the maximum spacing of connectors in built-up tension members shall not exceed:

- Twenty-four times the thickness of the thinner plate or 12 in. (305 mm) for painted members or unpainted members not subject to corrosion.
- Fourteen times the thickness of the thinner plate or 7 in. (180 mm) for unpainted members of weathering steel subject to atmospheric corrosion.

Assuming the first condition applies, a spacing of 6 in. is to be used. Use 2C8 × 13.75 connected intermittently at 6-in. interval.

4.3.2 Pin-Connected Members

Pin-connected members shall be designed to preclude the following failure modes:

- Tension yielding in the gross section
- Tension fracture on the effective net area
- Longitudinal shear on the effective area
- Bearing on the projected pin area (Figure 4.5)

4.3.2.1 Allowable Stress Design

The allowable stresses for tension yield, tension fracture, and shear rupture are $0.60F_y$, $0.45F_y$, and $0.30F_u$, respectively. The allowable stresses for bearing are given in Section 4.11.

4.3.2.2 Load and Resistance Factor Design

The design tensile strength $\phi_t P_n$ for a pin-connected member are given as follows:

Tension on gross area: see Equation 4.10.
Tension on effective net area:

$$\phi_t P_n = 0.75[2tb_{\text{eff}} F_u] \tag{4.13}$$

Shear on effective area:

$$\phi_{sf} P_n = 0.75[0.6A_{sf} F_u] \tag{4.14}$$

Bearing on projected pin area: see Section 4.11.

The terms in Figure 4.5 and the above equations are defined as follows: a is the shortest distance from the edge of the pin hole to the edge of the member measured in the direction of the force, A_{pb} is the projected bearing area $= dt$, $A_{sf} = 2t(a + d/2)$, $b_{\text{eff}} = 2t + 0.63$, in. (or, $2t + 16$, mm) but not more than the actual distance from the edge of the hole to the edge of the part measured in the direction normal to the applied force, d is the pin diameter, and t is the plate thickness.

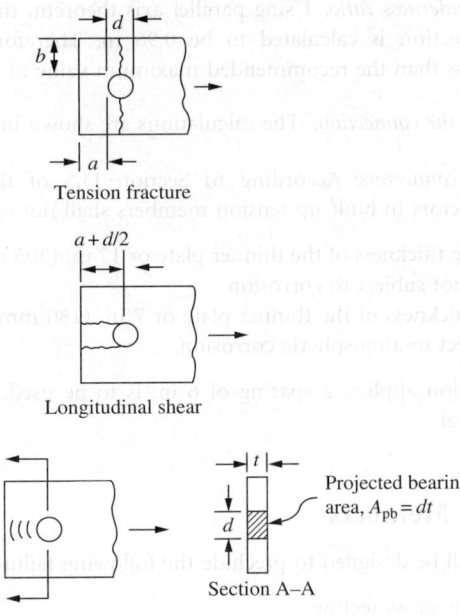

FIGURE 4.5   Failure modes of pin-connected members.

4.3.3 Threaded Rods

4.3.3.1 Allowable Stress Design

Threaded rods under tension are treated as bolts subject to tension in allowable stress design. These allowable stresses are given in Section 4.11.

4.3.3.2 Load and Resistance Factor Design

Threaded rods designed as tension members shall have an gross area A_b given by

$$A_b \geq \frac{P_u}{\phi\, 0.75 F_u} \tag{4.15}$$

where A_b is the gross area of the rod computed using a diameter measured to the outer extremity of the thread, P_u is the factored tensile load, ϕ is the resistance factor given as 0.75, and F_u is the specified minimum tensile strength.

4.4 Compression Members

Members under compression can fail by yielding, inelastic buckling, or elastic buckling depending on the slenderness ratio of the members. Members with low slenderness ratios tend to fail by yielding while members with high slenderness ratio tend to fail by elastic buckling. Most compression members used in construction have intermediate slenderness ratios and so the predominant mode of failure is inelastic buckling. Overall member buckling can occur in one of three different modes: flexural, torsional, and flexural–torsional. Flexural buckling occurs in members with doubly symmetric or doubly antisymmetric cross-sections (e.g., I or Z sections) and in members with singly symmetric sections (e.g., channel, tee, equal-legged angle, double-angle sections) when such sections are buckled about an axis that is *perpendicular* to the axis of symmetry. Torsional buckling occurs in members with doubly symmetric sections such as cruciform or built-up shapes with very thin walls. Flexural–torsional buckling occurs in members with singly symmetric cross-sections (e.g., channel, tee, equal-legged angle, double-angle sections) when such sections are buckled about the axis of symmetry and in members with unsymmetric cross-sections (e.g., unequal-legged L). Normally, torsional buckling of symmetric shapes is not particularly important in the design of hot-rolled compression members. It either does not govern or its buckling strength does not differ significantly from the corresponding weak axis flexural buckling strengths. However, torsional buckling may become important for open sections with relatively thin component plates. It should be noted that for a given cross-sectional area, a closed section is much stiffer torsionally than an open section. Therefore, if torsional deformation is of concern, a closed section should be used. Regardless of the mode of buckling, the governing effective slenderness ratio (Kl/r) of the compression member preferably should not exceed 200.

In addition to the slenderness ratio and cross-sectional shape, the behavior of compression members is affected by the relative thickness of the component elements that constitute the cross-section. The relative thickness of a component element is quantified by the width–thickness ratio (b/t) of the element. The width–thickness ratios of some selected steel shapes are shown in Figure 4.6. If the width–thickness ratio falls within a limiting value [denoted by the LRFD specification (AISC 1999) as λ_r] as shown in Table 4.4, the section will not experience local buckling prior to overall buckling of the member. However, if the width–thickness ratio exceeds this limiting width–thickness value, consideration of local buckling in the design of the compression member is required.

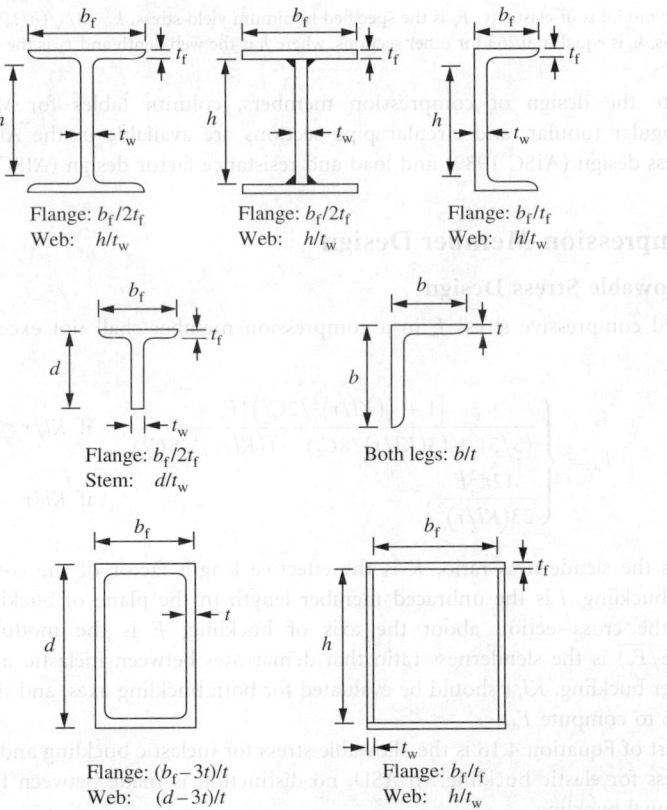

FIGURE 4.6 Definition of width–thickness ratio of selected cross-sections.

TABLE 4.4 Limiting Width–Thickness Ratios for Compression Elements Under Pure Compression

Component element	Width–thickness ratio	Limiting value, λ_r
Flanges of I-shaped sections; plates projecting from compression elements; outstanding legs of pairs of angles in continuous contact; flanges of channels	b/t	$0.56\sqrt{E/F_y}$
Flanges of square and rectangular box and HSS of uniform thickness; flange cover plates and diaphragm plates between lines of fasteners or welds	b/t	$1.40\sqrt{E/F_y}$
Unsupported width of cover plates perforated with a succession of access holes	b/t	$1.86\sqrt{E/F_y}$
Legs of single-angle struts; legs of double-angle struts with separators; unstiffened elements (i.e., elements supported along one edge)	b/t	$0.45\sqrt{E/F_y}$
Flanges projecting from built-up members	b/t	$0.64\sqrt{E/(F_y/k_c)}$
Stems of tees	d/t	$0.75\sqrt{E/F_y}$
All other uniformly compressed stiffened elements (i.e., elements supported along two edges)	b/t h/t_w	$1.49\sqrt{E/F_y}$
Circular hollow sections	D/t	$0.11E/F_y$
	D is the outside diameter and *t* is the wall thickness	

Note: E is the modulus of elasticity, F_y is the specified minimum yield stress, $k_c = 4/\sqrt{(h/t_w)}$, and $0.35 \le k_c \le 0.763$ for I-shaped sections, k_c is equal to 0.763 for other sections, where h is the web depth and t_w is the web thickness.

To facilitate the design of compression members, column tables for W, tee, double angle, square/rectangular tubular, and circular pipe sections are available in the AISC Manuals for both allowable stress design (AISC 1989) and load and resistance factor design (AISC 2001).

4.4.1 Compression Member Design

4.4.1.1 Allowable Stress Design

The computed compressive stress f_a in a compression member shall not exceed its allowable value given by

$$
F_a =
\begin{cases}
\dfrac{\left[1 - \left((Kl/r)^2/2C_c^2\right)\right]F_y}{(5/3) + (3(Kl/r)/8C_c) - ((Kl/r)^3/8C_c^3)}, & \text{if } Kl/r \le C_c \\[2ex]
\dfrac{12\pi^2 E}{23(Kl/r)^2}, & \text{if } Kl/r > C_c
\end{cases}
\tag{4.16}
$$

where Kl/r is the slenderness ratio, K is the effective length factor of the compression member in the plane of buckling, l is the unbraced member length in the plane of buckling, r is the radius of gyration of the cross-section about the axis of buckling, E is the modulus of elasticity, and $C_c = \sqrt{(2\pi^2 E/F_y)}$ is the slenderness ratio that demarcates between inelastic member buckling from elastic member buckling. Kl/r should be evaluated for both buckling axes, and the larger value used in Equation 4.16 to compute F_a.

The first part of Equation 4.16 is the allowable stress for inelastic buckling and the second part is the allowable stress for elastic buckling. In ASD, no distinction is made between flexural, torsional, and flexural–torsional buckling.

4.4.1.2 Load and Resistance Design

Compression members are to be designed so that the design compressive strength $\phi_c P_n$ will exceed the required compressive strength P_u. $\phi_c P_n$ is to be calculated as follows for the different types of overall buckling modes.

Flexural buckling (with width–thickness ratio $\leq \lambda_r$):

$$\phi_c P_n = \begin{cases} 0.85[A_g(0.658^{\lambda_c^2})F_y], & \text{if } \lambda_c \leq 1.5 \\[2mm] 0.85\left[A_g\left(\dfrac{0.877}{\lambda_c^2}\right)F_y\right], & \text{if } \lambda_c > 1.5 \end{cases} \tag{4.17}$$

where $\lambda_c = (KL/r\pi)\sqrt{(F_y/E)}$ is the slenderness parameter, A_g is the gross cross-sectional area, F_y is the specified minimum yield stress, E is the modulus of elasticity, K is the effective length factor, l is the unbraced member length in the plane of buckling, and r is the radius of gyration of the cross-section about the axis of buckling.

The first part of Equation 4.17 is the design strength for inelastic buckling and the second part is the design strength for elastic buckling. The slenderness parameter $\lambda_c = 1.5$ is the slenderness parameter that demarcates between inelastic behavior from elastic behavior.

Torsional buckling (with width–thickness ratio $\leq \lambda_r$): $\phi_c P_n$ is to be calculated from Equation 4.17, but with λ_c replaced by λ_e and given by

$$\lambda_e = \sqrt{\frac{F_y}{F_e}} \tag{4.18}$$

where

$$F_e = \left[\frac{\pi^2 E C_w}{(K_z L)^2} + GJ\right]\frac{1}{I_x + I_y} \tag{4.19}$$

in which C_w is the warping constant, G is the shear modulus $= 11{,}200$ ksi (77,200 MPa), I_x, I_y are the moments of inertia about the major and minor principal axes, respectively, J is the torsional constant, and K_z is the effective length factor for torsional buckling.

The warping constant C_w and the torsional constant J are tabulated for various steel shapes in the AISC-LRFD Manual (AISC 2001). Equations for calculating approximate values for these constants for some commonly used steel shapes are shown in Table 4.5.

Flexural–torsional buckling (with width–thickness ratio $\leq \lambda_r$): Same as for torsional buckling except F_e is now given by

For singly symmetric sections:

$$F_e = \frac{F_{es} + F_{ez}}{2H}\left[1 - \sqrt{1 - \frac{4F_{es}F_{ez}H}{(F_{es} + F_{ez})^2}}\right] \tag{4.20}$$

where

$F_{es} = F_{ex}$ if the x-axis is the axis of symmetry of the cross-section, or
$F_{es} = F_{ey}$ if the y-axis is the axis of symmetry of the cross-section
$F_{ex} = \pi^2 E/(Kl/r)_x^2$
$F_{ey} = \pi^2 E/(Kl/r)_x^2$
$H = 1 - (x_o^2 + y_o^2)/r_o^2$

in which K_x, K_y are the effective length factors for buckling about the x and y axes, respectively, l is the unbraced member length in the plane of buckling, r_x, r_y are the radii of gyration about the x and y axes, respectively, x_o, y_o are shear center coordinates with respect to the centroid (Figure 4.7), and $r_o^2 = x_o^2 + y_o^2 + r_x^2 + r_y^2$.

TABLE 4.5 Approximate Equations for C_w and J

Structural shape	Warping constant, C_w	Torsional constant, J	
I	$h'^2 I_c I_t/(I_c + I_t)$	$\sum C_i(b_i t_i^3/3)$	
C	$(b' - 3E_0)h'^2 b'^2 t_f/6 + E_0^2 I_x$	where	
	where	b_i = width of component element i	
	$E_0 = b'^2 t_f/(2b' t_f + h' t_w/3)$	t_i = thickness of component element i	
		C_i = correction factor for component element i (see values below)	
T	$(b_f^3 t_f^3/4 + h''^3 t_w^3)/36$ (≈ 0 for small t)	b_i/t_i	C_i
L	$(l_1^3 t_1^3 + l_2^3 t_2^3)/36$ (≈ 0 for small t)	1.00	0.423
		1.20	0.500
		1.50	0.588
		1.75	0.642
		2.00	0.687
		2.50	0.747
		3.00	0.789
		4.00	0.843
		5.00	0.873
		6.00	0.894
		8.00	0.921
		10.00	0.936
		∞	1.000

Note: b' is the distance measured from toe of flange to centerline of web, h' is the distance between centerline lines of flanges, h'' is the distance from centerline of flange to tip of stem, l_1, l_2 are the length of the legs of the angle, t_1, t_2 are the thickness of the legs of the angle, b_f is the flange width, t_f is the average thickness of flange, t_w is the thickness of web, I_c is the moment of inertia of compression flange taken about the axis of the web, I_t is the moment of inertia of tension flange taken about the axis of the web, and I_x is the moment of inertia of the cross-section taken about the major principal axis.

Numerical values for r_0 and H are given for hot-rolled W, channel, tee, single-angle, and double-angle sections in the AISC-LRFD Manual (AISC 2001).

For unsymmetric sections: F_e is to be solved from the cubic equation

$$(F_e - F_{ex})(F_e - F_{ey})(F_e - F_{ez}) - F_e^2(F_e - F_{ey})\left(\frac{x_0}{r_0}\right)^2 - F_e^2(F_e - F_{ex})\left(\frac{y_0}{r_0}\right)^2 = 0 \qquad (4.21)$$

The definitions of the terms in the above equation are as in Equation 4.20.

Local Buckling (with width–thickness ratio $\geq \lambda_r$): local buckling in the component element of the cross-section is accounted for in design by introducing a reduction factor Q in Equation 4.17 as follows:

$$\phi_c P_n = \begin{cases} 0.85[A_g Q(0.658^{Q\lambda^2})F_y], & \text{if } \lambda\sqrt{Q} \leq 1.5 \\ 0.85[A_g\left(\dfrac{0.877}{\lambda^2}\right)F_y], & \text{if } \lambda\sqrt{Q} > 1.5 \end{cases} \qquad (4.22)$$

where $\lambda = \lambda_c$ for flexural buckling and $\lambda = \lambda_e$ for flexural–torsional buckling.

The Q factor is given by

$$Q = Q_s Q_a \qquad (4.23)$$

where Q_s is the reduction factor for unstiffened compression elements of the cross-section (see Table 4.6) and Q_a is the reduction factor for stiffened compression elements of the cross-section (see Table 4.7).

4.4.2 Built-Up Compression Members

Built-up members are members made by bolting and/or welding together two or more standard structural shapes. For a built-up member to be fully effective (i.e., if all component structural

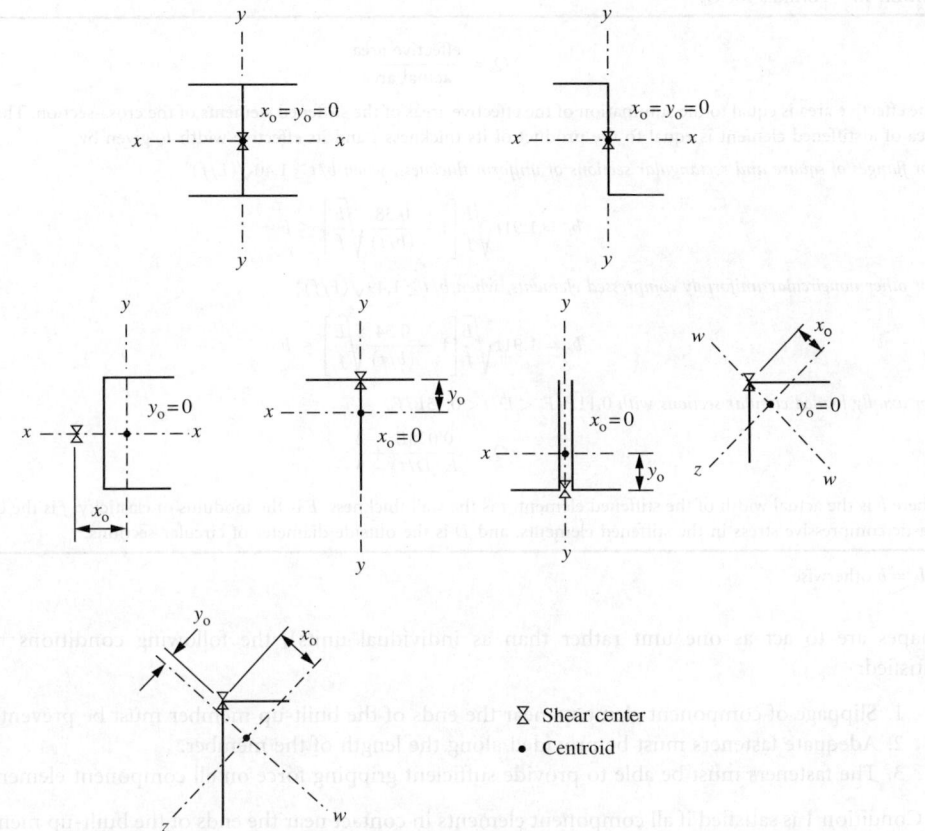

FIGURE 4.7 Location of shear center for selected cross-sections.

TABLE 4.6 Formulas for Q_s

Structural element	Range of b/t or d/t	Q_s
Single-angles	$0.45\sqrt{(E/F_y)} < b/t < 0.91\sqrt{(E/F_y)}$ $b/t \geq 0.91\sqrt{(E/F_y)}$	$1.340 - 0.76(b/t)\sqrt{(F_y/E)}$ $0.53E/[F_y(b/t)^2]$
Flanges, angles, and plates projecting from columns or other compression members	$0.56\sqrt{(E/F_y)} < b/t < 1.03\sqrt{(E/F_y)}$ $b/t \geq 1.03\sqrt{(E/F_y)}$	$1.415 - 0.74(b/t)\sqrt{(F_y/E)}$ $0.69E/[F_y(b/t)^2]$
Flanges, angles, and plates projecting from built-up columns or other compression members	$0.64\sqrt{[E/(F_y/k_c)]} < b/t < 1.17\sqrt{[E/(F_y/k_c)]}$ $b/t \geq 1.17\sqrt{[E/(F_y/k_c)]}$	$1.415 - 0.65(b/t)\sqrt{(F_y/k_c E)}$ $0.90Ek_c/[F_y(b/t)^2]$
Stems of tees	$0.75\sqrt{(E/F_y)} < d/t < 1.03\sqrt{(E/F_y)}$ $d/t \geq 1.03\sqrt{(E/F_y)}$	$1.908 - 1.22(b/t)\sqrt{(F_y/E)}$ $0.69E/[F_y(b/t)^2]$

Notes: k_c is defined in the footnote of Table 4.4, E is the modulus of elasticity, F_y is the specified minimum yield stress, b is the width of the component element, and t is the thickness of the component element.

TABLE 4.7 Formula for Q_a

$$Q_s = \frac{\text{effective area}}{\text{actual area}}$$

The effective area is equal to the summation of the effective areas of the stiffened elements of the cross-section. The effective area of a stiffened element is equal to the product of its thickness t and its effective width b_e given by

For flanges of square and rectangular sections of uniform thickness, when $b/t \geq 1.40\sqrt{(E/f)^a}$

$$b_e = 1.91t\sqrt{\frac{E}{f}}\left[1 - \frac{0.38}{(b/t)}\sqrt{\frac{E}{f}}\right] \leq b$$

For other noncircular uniformly compressed elements, when $b/t \geq 1.49\sqrt{(E/f)^a}$

$$b_e = 1.91t\sqrt{\frac{E}{f}}\left[1 - \frac{0.34}{(b/t)}\sqrt{\frac{E}{f}}\right] \leq b$$

For axially loaded circular sections with $0.11E/F_y < D/t < 0.45E/F_y$

$$Q_a = \frac{0.038E}{F_y(D/t)} + \frac{2}{3}$$

where b is the actual width of the stiffened element, t is the wall thickness, E is the modulus of elasticity, f is the computed elastic compressive stress in the stiffened elements, and D is the outside diameter of circular sections.

[a] $b_e = b$ otherwise.

---

shapes are to act as one unit rather than as individual units), the following conditions must be satisfied:

1. Slippage of component elements near the ends of the built-up member must be prevented.
2. Adequate fasteners must be provided along the length of the member.
3. The fasteners must be able to provide sufficient gripping force on all component elements.

Condition 1 is satisfied if all component elements in contact near the ends of the built-up member are connected by a weld having a length not less than the maximum width of the member or by bolts spaced longitudinally not more than four diameters apart for a distance equal to one and a half times the maximum width of the member. Condition 2 is satisfied if continuous welds are used throughout the length of the built-up compression member. Condition 3 is satisfied if either welds or fully tightened bolts are used as the fasteners. While condition 1 is mandatory, conditions 2 and 3 can be violated in design. If condition 2 or 3 is violated, the built-up member is not fully effective and slight slippage among component elements may occur. To account for the decrease in capacity due to slippage, a modified slenderness ratio is used to compute the design compressive strength when buckling of the built-up member is about an axis *coinciding* or *parallel* to at least one plane of contact for the component shapes. The modified slenderness ratio $(KL/r)_m$ is given as follows:

If condition 2 is violated

$$\left(\frac{KL}{r}\right)_m = \sqrt{\left(\frac{KL}{r}\right)_o^2 + \frac{0.82\,\alpha^2}{(1+\alpha^2)}\left(\frac{a}{r_{ib}}\right)^2} \tag{4.24}$$

If condition 3 is violated

$$\left(\frac{KL}{r}\right)_m = \sqrt{\left(\frac{KL}{r}\right)_o^2 + \left(\frac{a}{r_i}\right)^2} \tag{4.25}$$

In the above equations, $(KL/r)_o = (KL/r)_x$ if the buckling axis is the x-axis and at least one plane of contact between component elements is parallel to that axis; $(KL/r)_o = (KL/r)_y$ if the buckling axis is the

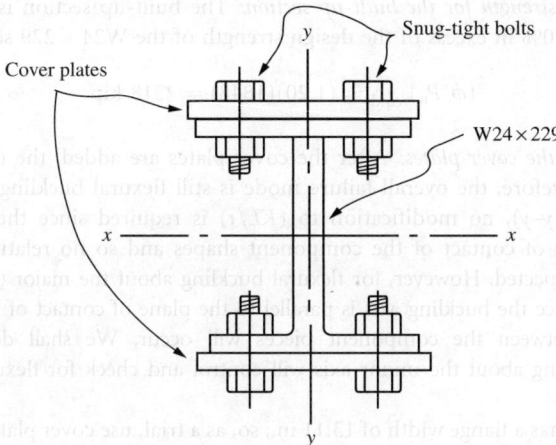

FIGURE 4.8 Design of cover plates for a compression member.

y-axis and at least one plane of contact is parallel to that axis. a is the longitudinal spacing of the fasteners, r_i is the minimum radius of gyration of any component element of the built-up cross-section, r_{ib} is the radius of gyration of individual component relative to its centroidal axis parallel to the axis of buckling of the member, and h is the distance between centroids of components elements measured perpendicularly to the buckling axis of the built-up member.

No modification to (KL/r) is necessary if the buckling axis is perpendicular to the planes of contact of the component shapes. Modifications to both $(KL/r)_x$ and $(KL/r)_y$ are required if the built-up member is so constructed that planes of contact exist in both the x and y directions of the cross-section.

Once the modified slenderness ratio is computed, it is to be used in the appropriate equation to calculate F_a in allowable stress design, or $\phi_c P_n$ in load and resistance factor design.

An additional requirement for the design of built-up members is that the effective slenderness ratio, Ka/r_i, of each component element, where K is the effective length factor of the component element between adjacent fasteners, does not exceed $\frac{3}{4}$ of the governing slenderness ratio of the built-up member. This provision is provided to prevent component element buckling between adjacent fasteners from occurring prior to overall buckling of the built-up member.

EXAMPLE 4.2

Using LRFD, determine the size of a pair of cover plates to be bolted, using fully tightened bolts, to the flanges of a W24 × 229 section as shown in Figure 4.8 so that its design strength, $\phi_c P_n$, will be increased by 20%. Also determine the spacing of the bolts along the longitudinal axis of the built-up column. The effective lengths of the section about the major $(KL)_x$ and minor $(KL)_y$ axes are both equal to 20 ft. A992 steel is to be used.

Determine the design strength for the W24 × 229 section: Since $(KL)_x = (KL)_y$ and $r_x > r_y$, $(KL/r)_y$ will exceed $(KL/r)_x$ and the design strength will be controlled by flexural buckling about the minor axis. Using section properties, $r_y = 3.11$ in. and $A = 67.2$ in.^{2}, obtained from the AISC-LRFD Manual (AISC 2001), the slenderness parameter λ_c about the minor axis can be calculated as follows:

$$(\lambda_c)_y = \frac{1}{\pi}\left(\frac{KL}{r}\right)_y \sqrt{\frac{F_y}{E}} = \frac{1}{3.142}\left(\frac{20 \times 12}{3.11}\right)\sqrt{\frac{50}{29,000}} = 1.02$$

Substituting $\lambda_c = 1.02$ into Equation 4.17, the design strength of the section is

$$\phi_c P_n = 0.85 \lfloor 67.2(0.658^{1.02^2})50 \rfloor = 1848 \text{ kip}$$

Determine the design strength for the built-up section: The built-up section is expected to possess a design strength that is 20% in excess of the design strength of the W24 × 229 section, so

$$(\phi_c P_n)_{\text{req'd}} = (1.20)(1848) = 2218 \text{ kip}$$

Determine the size of the cover plates: After the cover plates are added, the resulting section is still doubly symmetric. Therefore, the overall failure mode is still flexural buckling. For flexural buckling about the minor axis (*y*–*y*), no modification to (*KL*/*r*) is required since the buckling axis is perpendicular to the plane of contact of the component shapes and so no relative movement between the adjoining parts is expected. However, for flexural buckling about the major (*x*–*x*) axis, modification to (*KL*/*r*) is required since the buckling axis is parallel to the plane of contact of the adjoining structural shapes and slippage between the component pieces will occur. We shall design the cover plates assuming flexural buckling about the minor axis will control and check for flexural buckling about the major axis later.

A W24 × 229 section has a flange width of 13.11 in.; so, as a trial, use cover plates with widths of 14 in. as shown in Figure 4.8. Denoting *t* as the thickness of the plates, we have

$$(r_y)_{\text{built-up}} = \sqrt{\frac{(I_y)_{\text{W-shape}} + (I_y)_{\text{plates}}}{A_{\text{W-shape}} + A_{\text{plates}}}} = \sqrt{\frac{651 + 457.3t}{67.2 + 28t}}$$

and

$$(\lambda_c)_{y,\text{built-up}} = \frac{1}{\pi}\left(\frac{KL}{r}\right)_{y,\text{built-up}} \sqrt{\frac{F_y}{E}} = 3.17\sqrt{\frac{67.2 + 28t}{651 + 457.3t}}$$

Assuming $(\lambda)_{y,\text{built-up}}$ is less than 1.5, one can substitute the above expression for λ_c in Equation 4.17. When $\phi_c P_n$ equals 2218, we can solve for *t*. The result is $t \approx \frac{3}{8}$ in. Back-substituting $t = \frac{3}{8}$ into the above expression, we obtain $(\lambda)_{c,\text{built-up}} = 0.975$, which is indeed < 1.5. So, try 14 in. × $\frac{3}{8}$ in. cover plates.

Check for local buckling: For the I-section:

$$\text{Flange:} \quad \left[\frac{b_f}{2t_f} = 3.8\right] < \left[0.56\sqrt{\frac{E}{F_y}} = 0.56\sqrt{\frac{29,000}{50}} = 13.5\right]$$

$$\text{Web:} \quad \left[\frac{h_c}{t_w} = 22.5\right] < \left[1.49\sqrt{\frac{E}{F_y}} = 1.49\sqrt{\frac{29,000}{50}} = 35.9\right]$$

For the cover plates, if $\frac{3}{4}$ in. diameter bolts are used and assuming an edge distance of 2 in., the width of the plate between fasteners will be $13.11 - 4 = 9.11$ in. Therefore, we have

$$\left[\frac{b}{t} = \frac{9.11}{3/8} = 24.3\right] < \left[1.40\sqrt{\frac{E}{F_y}} = 1.40\sqrt{\frac{29,000}{50}} = 33.7\right]$$

Since the width–thickness ratios of all component shapes do not exceed the limiting width–thickness ratio for local buckling, local buckling is not a concern.

*Check for flexural buckling about the major (*x*–*x*) axis:* Since the built-up section is doubly symmetric, the governing buckling mode will be flexural buckling regardless of the axes. Flexural buckling will occur about the major axis if the modified slenderness ratio $(KL/r)_m$ about the major axis exceeds $(KL/r)_y$. Therefore, as long as $(KL/r)_m$ is less than $(KL/r)_y$, buckling will occur about the minor axis and flexural buckling about the major axis will not be controlled. In order to arrive at an optimal design, we shall determine the longitudinal fastener spacing, a, such that the modified slenderness ratio $(KL/r)_m$ about the major axis will be equal to $(KL/r)_y$. That is, we shall solve for a from the equation

$$\left[\left(\frac{KL}{r}\right)_m = \sqrt{\left(\frac{KL}{r}\right)_x^2 + \left(\frac{a}{r_i}\right)^2} = \left(\frac{KL}{r}\right)_y = 73.8\right]$$

In the above equation, $(KL/r)_x$ is the slenderness ratio about the major axis of the built-up section, r_i is the least radius of gyration of the component shapes, which in this case is the cover plate. Substituting $(KL/r)_x = 21.7$, $r_i = r_{\text{cover plate}} = \sqrt{(I/A)_{\text{cover plate}}} = \sqrt{[(\frac{3}{8})^2/12]} = 0.108$ into the above equation, we obtain $a = 7.62$ in. Since $(KL) = 20$ ft, we shall use $a = 6$ in. for the longitudinal spacing of the fasteners.

Check for component element buckling between adjacent fasteners:

$$\left[\frac{Ka}{r_i} = \frac{1 \times 6}{0.108} = 55.6\right] \approx \left[\frac{3}{4}\left(\frac{KL}{r}\right)_y = \frac{3}{4}(73.8) = 55.4\right]$$

so, the component element buckling criterion is not a concern.

Use 14 in. $\times \frac{3}{8}$ in. cover plates bolted to the flanges of the W24 $\times$ 229 section by $\frac{3}{4}$ in. diameter fully tightened bolts spaced 6 in. longitudinally.

4.4.3 Column Bracing

The design strength of a column can be increased if lateral braces are provided at intermediate points along its length in the buckled direction of the column. The AISC-LRFD Specification (1999) identifies two types of bracing systems for columns. A relative bracing system is one in which the movement of a braced point with respect to other adjacent braced points is controlled, for example, the diagonal braces used in buildings. A nodal (or discrete) brace system is one in which the movement of a braced point with respect to some fixed points is controlled, for example, the guy wires of guyed towers. A bracing system is effective only if the braces are designed to satisfy both stiffness and strength requirements. The following equations give the required stiffness and strength for the two bracing systems.

Required braced stiffness:

$$\beta_{cr} = \begin{cases} \dfrac{2.67P_u}{L_{br}} & \text{for relative bracing} \\[2mm] \dfrac{10.7P_u}{L_{br}} & \text{for nodal bracing} \end{cases} \tag{4.26}$$

where P_u is the required compression strength of the column and L_{br} is the distance between braces (for ASD, replace P_u in Equation 4.26 by $1.5P_a$, where P_a is the required compressive strength based on ASD load combinations).

If L_{br} is less than L_q (the maximum unbraced length for P_u), L_{br} can be replaced by L_q in the above equations.

Required braced strength:

$$P_{br} = \begin{cases} 0.004P_u & \text{for relative bracing} \\ 0.01P_u & \text{for nodal bracing} \end{cases} \tag{4.27}$$

where P_u is defined as in Equation 4.26.

4.5 Flexural Members

Depending on the width–thickness ratios of the component elements, steel sections used as *flexural members* are classified as compact, noncompact, and slender element sections. Compact sections are sections that can develop the cross-section plastic moment (M_p) under flexure and sustain that moment through a large hinge rotation without fracture. Noncompact sections are sections that either cannot develop the cross-section full plastic strength or cannot sustain a large hinge rotation at M_p, probably due to local buckling of the flanges or web. Slender elements are sections that fail by local buckling of component elements long before M_p is reached. A section is considered compact if all its component elements have width–thickness ratios less than a limiting value (denoted as λ_p in LRFD). A section is considered noncompact if one or more of its component elements have width–thickness ratios that fall in between λ_p and λ_r. A section is considered a slender element if one or more of its component elements have width–thickness ratios that exceed λ_r. Expressions for λ_p and λ_r are given in the Table 4.8.

In addition to the compactness of the steel section, another important consideration for beam design is the lateral unsupported (unbraced) length of the member. For beams bent about their strong axes, the failure modes, or limit states, vary depending on the number and spacing of lateral supports provided to brace the compression flange of the beam. The compression flange of a beam behaves somewhat like a compression member. It buckles if adequate lateral supports are not provided in a phenomenon called *lateral torsional buckling*. Lateral torsional buckling may or may not be accompanied by yielding, depending on the lateral unsupported length of the beam. Thus, lateral torsional buckling can be inelastic or elastic. If the lateral unsupported length is large, the limit state is elastic lateral torsional buckling. If the lateral unsupported length is smaller, the limit state is inelastic lateral torsional buckling. For compact section beams with adequate lateral supports, the limit state is full yielding of the cross-section (i.e., plastic hinge formation). For noncompact section beams with adequate lateral supports, the limit state is flange or web local buckling. For beams bent about their weak axes, lateral torsional buckling will not occur and so the lateral unsupported length has no bearing on the design. The limit states for such beams will be the formation of the plastic hinge if the section is compact and the limit state will be a flange or web local buckling if the section is noncompact.

Beams subjected to high shear must be checked for possible web shear failure. Depending on the width–thickness ratio of the web, failure by shear yielding or web shear buckling may occur. Short, deep beams with thin webs are particularly susceptible to web shear failure. If web shear is of concern, the use of thicker webs or web reinforcements such as stiffeners is required.

Beams subjected to concentrated loads applied in the plane of the web must be checked for a variety of possible flange and web failures. Failure modes associated with concentrated loads include local flange bending (for tensile concentrated load), local web yielding (for compressive concentrated load), web crippling (for compressive load), side-sway web buckling (for compressive load), and compression buckling of the web (for a compressive load pair). If one or more of these conditions is critical, transverse stiffeners extending at least one-half the beam depth (use full depth for compressive buckling of the web) must be provided adjacent to the concentrated loads.

Long beams can have deflections that may be too excessive, leading to problems in serviceability. If deflection is excessive, the use of intermediate supports or beams with higher flexural rigidity is required.

The design of flexural members should satisfy, at the minimum, the following criteria:

- Flexural strength criterion
- Shear strength criterion
- Criteria for concentrated loads
- Deflection criterion

To facilitate beam design, a number of beam tables and charts are given in the AISC Manuals (1989, 2001) for both allowable stress and load and resistance factor design.

TABLE 4.8 λ_p and λ_r for Members Under Flexural Compression

Component element	Width–thickness ratio[a]	λ_p	λ_r
Flanges of I-shaped rolled beams and channels	b/t	$0.38\sqrt{(E/F_y)}$	$0.83\sqrt{(E/F_L)}$[b] F_L = smaller of $(F_{yf} - F_r)$ or F_{yw} F_{yf} = flange yield strength F_{yw} = web yield strength F_r = flange compressive residual stress (10 ksi for rolled shapes, 16.5 ksi for welded shapes)
Flanges of I-shaped hybrid or welded beams	b/t	$0.38\sqrt{(E/F_{yf})}$ for nonseismic application $0.31\sqrt{(E/F_{yf})}$ for seismic application F_{yf} = flange yield strength	$0.95\sqrt{[E/(F_L/k_c)]}$[c] F_L = as defined above k_c = as defined in the footnote of Table 4.4
Flanges of square and rectangular box and HSS of uniform thickness; flange cover plates and diaphragm plates between lines of fasteners or welds	b/t	$0.939\sqrt{(E/F_y)}$ for plastic analysis	$1.40/\sqrt{(E/F_y)}$
Unsupported width of cover plates perforated with a succession of access holes	b/t	NA	$1.86\sqrt{(E/F_y)}$
Legs of single-angle struts; legs of double-angle struts with separators; unstiffened elements	b/t	NA	$0.45/\sqrt{(E/F_y)}$
Stems of tees	d/t	NA	$0.75\sqrt{(E/F_y)}$
Webs in flexural compression	h_c/t_w	$3.76\sqrt{(E/F_y)}$ for nonseismic application $3.05\sqrt{(E/F_y)}$ for seismic application	$5.70\sqrt{(E/F_y)}$[d]
Webs in combined flexural and axial compression	h_c/t_w	For $P_u/\phi_b P_y \leq 0.125$: $3.76(1 - 2.75P_u/\phi_b P_y)\sqrt{(E/F_y)}$ for nonseismic application $3.05(1 - 1.54P_u/\phi_b P_y)\sqrt{(E/F_y)}$ for seismic application For $P_u/\phi_b P_y > 0.125$: $1.12(2.33 - P_u/\phi_b P_y)\sqrt{(E/F_y)} \geq$ $1.49\sqrt{(E/F_y)}$ $\phi_b = 0.90$ P_u = factored axial force $P_y = A_g F_y$ $0.07E/F_y$	$5.70(1 - 0.74P_u/\phi_b P_y)\sqrt{(E/F_y)}$
Circular hollow sections	D/t D = outside diameter t = wall thickness		$0.31E/F_y$

[a] See Figure 4.5 for definition of b, h_c, and t.
[b] For ASD, this limit is $0.56\sqrt{(E/F_y)}$.
[c] For ASD, this limit is $0.56\sqrt{[E/(F_{yf}/k_c)]}$, where $k_c = 4.05/(h/t)^{0.46}$ if $h/t > 70$, otherwise $k_c = 1.0$.
[d] For ASD, this limit is $4.46\sqrt{(E/F_b)}$, where F_b = allowable bending stress.
Note: In all the equations, E is the modulus of elasticity and is F_y is the minimum specified yield strength.

4.5.1 Flexural Member Design

4.5.1.1 Allowable Stress Design

4.5.1.1.1 Flexural Strength Criterion

The computed flexural stress, f_b, shall not exceed the allowable flexural stress, F_b, as given below (in all equations, the minimum specified yield stress, F_y, can not exceed 65 ksi).

4.5.1.1.1.1 Compact Section Members Bent about Their Major Axes — For $L_b \leq L_c$

$$F_b = 0.66F_y \tag{4.28}$$

where

$$L_c = \text{smaller of } \left\{ 76b_f/\sqrt{F_y}, 20{,}000/(d/A_f)F_y \right\}, \text{ for I and Channel shapes}$$
$$= [1{,}950 + 1{,}200(M_1/M_2)](b/F_y) \geq 1{,}200(b/F_y), \text{ for box sections, rectangular and circular tubes}$$

in which b_f is the flange width (in.), d is the overall depth of the section (ksi), A_f is the area of the compression flange (in.^{2}), b is the width of the cross-section (in.), and M_1/M_2 is the ratio of the smaller to larger moment at the ends of the unbraced length of the beam; M_1/M_2 is positive for reverse curvature bending and negative for single curvature bending.

For the above sections to be considered as compact, in addition to having the width–thickness ratios of their component elements falling within the limiting value of λ_p shown in Table 4.8, the flanges of the sections must be continuously connected to the webs. For box-shaped sections, the following requirements must also be satisfied: the depth-to-width ratio should not exceed six, and the flange-to-web thickness ratio should not exceed two.

For $L_b > L_c$, the allowable flexural stress in tension is given by

$$F_b = 0.60F_y \tag{4.29}$$

and the allowable flexural stress in compression is given by the larger value calculated from Equations 4.30 and 4.31. While Equation 4.30 normally controls for deep, thin-flanged sections where warping restraint torsional resistance dominates, Equation 4.31 normally controls for shallow, thick-flanged sections where St Venant torsional resistance dominates:

$$F_b = \begin{cases} \left[\dfrac{2}{3} - \dfrac{F_y(l/r_T)^2}{1{,}530 \times 10^3 C_b} \right] F_y \leq 0.60F_y, & \text{if } \sqrt{\dfrac{102{,}000 C_b}{F_y}} \leq \dfrac{l}{r_T} < \sqrt{\dfrac{510{,}000 C_b}{F_y}} \\[4ex] \dfrac{170{,}000 C_b}{(l/r_T)^2} \leq 0.60F_y, & \text{if } \dfrac{l}{r_T} \geq \sqrt{\dfrac{510{,}000 C_b}{F_y}} \end{cases} \tag{4.30}$$

$$F_b = \frac{12{,}000 C_b}{ld/A_f} \leq 0.60F_y \tag{4.31}$$

where l is the distance between cross-sections braced against twist or lateral displacement of the compression flange (in.), r_T is the radius of gyration of a section comprising the compression flange plus $\frac{1}{3}$ of the compression web area taken about an axis in the plane of the web (in.), A_f is the compression flange area (in.^{2}), d is the depth of cross-section (in.), $C_b = 12.5M_{max}/(2.5M_{max} + 3M_A + 4M_B + 3M_C)$, where M_{max}, M_A, M_B, M_C are absolute values of the maximum moment, quarter-point moment, midpoint moment, and three-quarter point moment along the unbraced length of the member, respectively. (For simplicity in design, C_b can conservatively be taken as unity.)

It should be cautioned that Equations 4.30 and 4.31 are applicable only to I and Channel shapes with an axis of symmetry in, and loaded in the plane of the web. In addition, Equation 4.31 is applicable only if the compression flange is solid and approximately rectangular in shape and its area is not less than the tension flange.

4.5.1.1.1.2 Compact Section Members Bent about Their Minor Axes — Since lateral torsional buckling will not occur for bending about the minor axes, regardless of the value of L_b, the allowable flexural stress is

$$F_b = 0.75F_y \qquad (4.32)$$

4.5.1.1.1.3 Noncompact Section Members Bent about Their Major Axes — For $L_b \leq L_c$

$$F_b = 0.60F_y \qquad (4.33)$$

where L_c is defined as in Equation 4.28.

For $L_b > L_c$, F_b is given in Equations 4.29 to 4.31.

4.5.1.1.1.4 Noncompact Section Members Bent about Their Minor Axes — Regardless of the value of L_b,

$$F_b = 0.60F_y \qquad (4.34)$$

4.5.1.1.1.5 Slender Element Sections — Refer to Section 4.10.

4.5.1.1.2 Shear Strength Criterion

For practically all structural shapes commonly used in constructions, the shear resistance from the flanges is small compared to the webs. As a result, the shear resistance for flexural members is normally determined on the basis of the webs only. The amount of web shear resistance is dependent on the width–thickness ratio h/t_w of the webs. If h/t_w is small, the failure mode is web yielding. If h/t_w is large, the failure mode is web buckling. To avoid web shear failure, the computed shear stress, f_v, shall not exceed the allowable shear stress, F_v, given by

$$F_v = \begin{cases} 0.40F_y, & \text{if } \dfrac{h}{t_w} \leq \dfrac{380}{\sqrt{F_y}} \\[2mm] \dfrac{C_v}{2.89}F_y \leq 0.40F_y, & \text{if } \dfrac{h}{t_w} > \dfrac{380}{\sqrt{F_y}} \end{cases} \qquad (4.35)$$

where

$C_v = 45{,}000k_v/[F_y(h/t_w)^2]$, if $C_v \leq 0.8$

$\quad = [190/(h/t_w)]\sqrt{(k_v/F_y)}$, if $C_v > 0.8$

$k_v = 4.00 + 5.34/(a/h)^2$, if $a/h \leq 1.0$

$\quad = 5.34 + 4.00/(a/h)^2$, if $a/h > 1.0$

t_w = web thickness (in.)
a = clear distance between transverse stiffeners (in.)
h = clear distance between flanges at section under investigation (in.)

4.5.1.1.3 Criteria for Concentrated Loads

4.5.1.1.3.1 Local Flange Bending — If the concentrated force that acts on the beam flange is tensile, the beam flange may experience excessively bending, leading to failure by fracture. To preclude this type of failure, transverse stiffeners are to be provided opposite the tension flange unless the length of the load when measured across the beam flange is less than 0.15 times the flange width, or if the flange thickness, t_f, exceeds

$$0.4\sqrt{\frac{P_{bf}}{F_y}} \qquad (4.36)$$

where P_{bf} is the computed tensile force multiplied by $\frac{5}{3}$ if the force is due to live and dead loads only or by $\frac{4}{3}$ if the force is due to live and dead loads in conjunction with wind or earthquake loads (kip) and F_y is the specified minimum yield stress (ksi).

4.5.1.1.3.2 Local Web Yielding — To prevent local web yielding, the concentrated compressive force, R, should not exceed $0.66R_n$, where R_n is the web yielding resistance given in Equation 4.54 or 4.55, whichever applies.

4.5.1.1.3.3 Web Crippling — To prevent web crippling, the concentrated compressive force, R, should not exceed $0.50R_n$, where R_n is the web crippling resistance given in Equations 4.56, 4.57, or 4.58, whichever applies.

4.5.1.1.3.4 Sideways Web Buckling — To prevent sideways web buckling, the concentrated compressive force, R, should not exceed R_n, where R_n is the sideways web buckling resistance given in Equation 4.59 or 4.60, whichever applies, except the term $C_r t_w^3 t_f/h^2$ is replaced by $6800 t_w^3/h$.

4.5.1.1.3.5 Compression Buckling of the Web — When the web is subjected to a pair of concentrated force acting on both flanges, buckling of the web may occur if the web depth clear of fillet, d_c, is greater than

$$\frac{4100 t_w^3 \sqrt{F_y}}{P_{bf}} \tag{4.37}$$

where t_w is the web thickness, F_y is the minimum specified yield stress, and P_{bf} is as defined in Equation 4.36.

4.5.1.1.4 Deflection Criterion

Deflection is a serviceability consideration. Since most beams are fabricated with a camber that somewhat offsets the dead load deflection, consideration is often given to deflection due to live load only. For beams supporting plastered ceilings, the service live load deflection preferably should not exceed $L/360$ where L is the beam span. A larger deflection limit can be used if due considerations are given to ensure the proper functioning of the structure.

EXAMPLE 4.3

Using ASD, determine the amount of increase in flexural capacity of a W24 × 55 section bent about its major axis if two 7 in. × $\frac{1}{2}$ in. (178 mm × 13 mm) cover plates are bolted to its flanges as shown in Figure 4.9. The beam is laterally supported at every 5-ft (1.52-m) interval. Use A36 steel. Specify the type, diameter, and longitudinal spacing of the bolts used if the maximum shear to be resisted by the cross-section is 100 kip (445 kN).

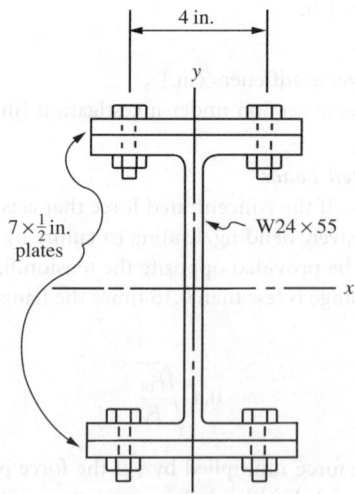

FIGURE 4.9 Beam section with cover plates.

Section properties: A W24 × 55 section has the following section properties:

$$b_f = 7.005 \text{ in.}, \quad t_f = 0.505 \text{ in.}, \quad d = 23.57 \text{ in.}, \quad t_w = 0.395 \text{ in.}, \quad I_x = 1350 \text{ in.}^4, \quad S_x = 114 \text{ in.}^3$$

Check compactness: Refer to Table 4.8, and assuming that the transverse distance between the two bolt lines is 4 in., we have

$$\text{Beam flanges} \quad \left[\frac{b_f}{2t_f} = 6.94\right] < \left[0.38\sqrt{\frac{E}{F_y}} = 10.8\right]$$

$$\text{Beam web} \quad \left[\frac{d}{t_w} = 59.7\right] < \left[3.76\sqrt{\frac{E}{F_y}} = 107\right]$$

$$\text{Cover plates} \quad \left[\frac{4}{1/2} = 8\right] < \left[0.939\sqrt{\frac{E}{F_y}} = 26.7\right]$$

Therefore, the section is compact.

Determine the allowable flexural stress, F_b: Since the section is compact and the lateral unbraced length, $L_b = 60$ in., is less than $L_c = 83.4$ in., the allowable bending stress from Equation 4.28 is $0.66F_y = 24$ ksi.

Determine the section modulus of the beam with cover plates:

$$S_{x,\text{combination section}} = \frac{I_{x,\text{combination section}}}{c}$$

$$= \frac{1350 + 2\left[\left(\frac{1}{12}\right)(7)\left(\frac{1}{2}\right)^3 + (7)\left(\frac{1}{2}\right)(12.035)^2\right]}{\left[\left(\frac{23.57}{2}\right) + \left(\frac{1}{2}\right)\right]}$$

$$= 192 \text{ in.}^3$$

Determine flexural capacity of the beam with cover plates:

$$M_{x,\text{combination section}} = S_{x,\text{combination section}} F_b = (192)(24) = 4608 \text{ k in.}$$

Since the flexural capacity of the beam without cover plates is

$$M_x = S_x F_b = (114)(24) = 2736 \text{ k in.}$$

the increase in flexural capacity is 68.4%.

Determine diameter and longitudinal spacing of bolts: From Mechanics of Materials, the relationship between the shear flow, q, the number of bolts per shear plane, n, the allowable bolt shear stress, F_v, the cross-sectional bolt area, A_b, and the longitudinal bolt spacing, s, at the interface of two component elements of a combination section is given by

$$\frac{nF_v A_b}{s} = q$$

Substituting $n = 2$, $q = VQ/I = (100)[(7)(\frac{1}{2})(12.035)]/2364 = 1.78$ k/in. into the above equation, we have

$$\frac{F_v A_b}{s} = 0.9 \text{ k/in.}$$

If $\frac{1}{2}$ in. diameter A325-N bolts are used, we have $A_b = \pi(\frac{1}{2})^2/4 = 0.196$ in.^{2} and $F_v = 21$ ksi (from Table 4.12), from which s can be solved from the above equation to be 4.57 in. However, for ease of installation, use $s = 4.5$ in.

In calculating the section properties of the combination section, no deduction is made for the bolt holes in the beam flanges nor the cover plates, which is allowed provided that the following condition is satisfied:

$$0.5F_u A_{fn} \geq 0.6F_y A_{fg}$$

where F_y and F_u are the minimum specified yield strength and tensile strength, respectively. A_{fn} is the net flange area and A_{fg} is the gross flange area. For this problem

Beam flanges $\quad [0.5F_u A_{fn} = 0.5(58)(7.005 - 2 \times 1/2)(0.505) = 87.9\,\text{kip}]$
$$>[0.6F_y A_{fg} = 0.6(36)(7.005)(0.505) = 76.4\,\text{kip}]$$

Cover plates $\quad [0.5F_u A_{fn} = 0.5(58)(7 - 2 - 1/2)(1/2) = 87\,\text{kip}]$
$$>[0.6F_y A_{fg} = 0.6(36)(7)(1/2) = 75.6\,\text{kip}]$$

so the use of gross cross-sectional area to compute section properties is justified. In the event that the condition is violated, cross-sectional properties should be evaluated using an effective tension flange area A_{fe} given by

$$A_{fe} = \frac{5}{6}\frac{F_u}{F_y}A_{fn}$$

So, use $\frac{1}{2}$-in. diameter A325-N bolts spaced 4.5 in. apart longitudinally in two lines 4 in. apart to connect the cover plates to the beam flanges.

4.5.1.2 Load and Resistance Factor Design

4.5.1.2.1 Flexural Strength Criterion

Flexural members must be designed to satisfy the flexural strength criterion of

$$\phi_b M_n \geq M_u \tag{4.38}$$

where $\phi_b M_n$ is the design flexural strength and M_u is the required strength. The design flexural strength is determined as given below.

4.5.1.2.1.1 Compact Section Members Bent about Their Major Axes — For $L_b \leq L_p$ (plastic hinge formation)

$$\phi_b M_n = 0.90 M_p \tag{4.39}$$

For $L_p < L_b \leq L_r$ (inelastic lateral torsional buckling)

$$\phi_b M_n = 0.90 C_b \left[M_p - (M_p - M_r)\left(\frac{L_b - L_p}{L_r - L_p}\right) \right] \leq 0.90 M_p \tag{4.40}$$

For $L_b > L_r$ (elastic lateral torsional buckling)

For I-shaped members and channels:

$$\phi_b M_n = 0.90 C_b \left[\frac{\pi}{L_b} \sqrt{EI_y GJ + \left(\frac{\pi E}{L_b}\right)^2 I_y C_w} \right] \leq 0.90 M_p \tag{4.41}$$

For solid rectangular bars and symmetric box sections:

$$\phi_b M_n = 0.90 C_b \frac{57,000\sqrt{JA}}{L_b/r_y} \leq 0.90 M_p \tag{4.42}$$

The variables used in the above equations are defined in the following:

L_b = lateral unsupported length of the member

L_p, L_r = limiting lateral unsupported lengths given in the following table:

Structural shape	L_p	L_r
I-shaped sections, channels	$1.76 r_y / \sqrt{(E/F_{yf})}$ where r_y = radius of gyration about minor axis E = modulus of elasticity F_{yf} = flange yield strength	$[r_y X_1 / F_L]\{\sqrt{[1 + \sqrt{(1 + X_2 F_L^2)}]}\}$ where r_y = radius of gyration about minor axis, in. $X_1 = (\pi/S_x)\sqrt{(EGJA/2)}$ $X_2 = (4C_w/I_y)(S_x/GJ)^2$ F_L = smaller of $(F_{yf} - F_r)$ or F_{yw} F_{yf} = flange yield stress, ksi F_{yw} = web yield stress, ksi F_r = 10 ksi for rolled shapes, 16.5 ksi for welded shapes S_x = elastic section modulus about the major axis, in.^{3} (use S_{xc}, the elastic section modulus about the major axis with respect to the compression flange if the compression flange is larger than the tension flange) I_y = moment of inertia about the minor axis, in.^{4} J = torsional constant, in.^{4} C_w = warping constant, in.^{6} E = modulus of elasticity, ksi G = shear modulus, ksi
Solid rectangular bars, symmetric box sections	$[0.13 r_y E\sqrt{(JA)}]/M_p$ where r_y = radius of gyration about minor axis E = modulus of elasticity J = torsional constant A = cross-sectional area M_p = plastic moment capacity $= F_y Z_x$ F_y = yield stress Z_x = plastic section modulus about the major axis	$[2 r_y E\sqrt{(JA)}]/M_r$ where r_y = radius of gyration about minor axis J = torsional constant A = cross-sectional area $M_r = F_{yf} S_x$ F_y = yield stress F_{yf} = flange yield strength S_x = elastic section modulus about the major axis

Note: L_p given in this table are valid only if the bending coefficient C_b is equal to unity. If $C_b > 1$, the value of L_p can be increased. However, using the L_p expressions given above for $C_b > 1$ will give conservative value for the flexural design strength.

and

$M_p = F_y Z_x$

$M_r = F_L S_x$ for I-shaped sections and channels, $F_{yf} S_x$ for solid rectangular bars and box sections

F_L = smaller of $(F_{yf} - F_r)$ or F_{yw}

F_{yf} = flange yield stress, ksi

F_{yw} = web yield stress, ksi

F_r = 10 ksi for rolled sections, 16.5 ksi for welded sections

F_y = specified minimum yield stress

S_x = elastic section modulus about the major axis
Z_x = plastic section modulus about the major axis
I_y = moment of inertia about the minor axis
J = torsional constant
C_w = warping constant
E = modulus of elasticity
G = shear modulus
C_b = $12.5 M_{max}/(2.5 M_{max} + 3 M_A + 4 M_B + 3 M_C)$
M_{max}, M_A, M_B, M_C = absolute value of maximum moment, quarter-point moment, midpoint moment, and three-quarter point moment along the unbraced length of the member, respectively.

C_b is a factor that accounts for the effect of moment gradient on the lateral torsional buckling strength of the beam. Lateral torsional buckling strength increases for a steep moment gradient. The worst loading case as far as lateral torsional buckling is concerned is when the beam is subjected to a uniform moment resulting in single curvature bending. For this case $C_b = 1$. Therefore, the use of $C_b = 1$ is conservative for the design of beams.

4.5.1.2.1.2 Compact Section Members Bent about Their Minor Axes — Regardless of L_b, the limit state will be plastic hinge formation

$$\phi_b M_n = 0.90 M_{py} = 0.90 F_y Z_y \tag{4.43}$$

4.5.1.2.1.3 Noncompact Section Members Bent about Their Major Axes — For $L_b \leq L'_p$ (flange or web local buckling)

$$\phi_b M_n = \phi_b M'_n = 0.90 \left[M_p - (M_p - M_r)\left(\frac{\lambda - \lambda_p}{\lambda_r - \lambda_p}\right) \right] \tag{4.44}$$

where

$$L'_p = L_p + (L_r - L_p)\left(\frac{M_p - M'_n}{M_p - M_r}\right) \tag{4.45}$$

L_p, L_r, M_p, M_r are defined as before for compact section members, and

For flange local buckling:
 $\lambda = b_f/2t_f$ for I-shaped members, b_f/t_f for channels
 λ_p, λ_r are defined in Table 4.8
For web local buckling:
 $\lambda = h_c/t_w$
 λ_p, λ_r are defined in Table 4.8

in which b_f is the flange width, t_f is the flange thickness, h_c is twice the distance from the neutral axis to the inside face of the compression flange less the fillet or corner radius, and t_w is the web thickness.

For $L'_p < L_b \leq L_r$ (inelastic lateral torsional buckling), $\phi_b M_n$ is given by Equation 4.40 except that the limit $0.90 M_p$ is to be replaced by the limit $0.90 M'_n$.

For $L_b > L_r$ (elastic lateral torsional buckling), $\phi_b M_n$ is the same as for compact section members as given in Equation 4.41 or 4.42.

4.5.1.2.1.4 Noncompact Section Members Bent about Their Minor Axes — Regardless of the value of L_b, the limit state will be either flange or web local buckling, and $\phi_b M_n$ is given by Equation 4.42.

4.5.1.2.1.5 Slender Element Sections — Refer to Section 4.10.

4.5.1.2.1.6 Tees and Double Angle Bent about Their Major Axes — The design flexural strength for tees and double-angle beams with flange and web slenderness ratios less than the corresponding limiting slenderness ratios λ_r shown in Table 4.8 is given by

$$\phi_b M_n = 0.90 \left[\frac{\pi \sqrt{EI_y GJ}}{L_b} \left(B + \sqrt{1 + B^2} \right) \right] \leq 0.90(\beta M_y) \qquad (4.46)$$

where

$$B = \pm 2.3 \left(\frac{d}{L_b} \right) \sqrt{\frac{I_y}{J}} \qquad (4.47)$$

Use the plus sign for B if the *entire* length of the stem along the unbraced length of the member is in tension. Otherwise, use the minus sign. β equals 1.5 for stems in tension and equals 1.0 for stems in compression. The other variables in Equation 4.46 are defined as before in Equation 4.41.

4.5.1.2.2 Shear Strength Criterion
For a satisfactory design, the design shear strength of the webs must exceed the factored shear acting on the cross-section, that is

$$\phi_v V_n \geq V_u \qquad (4.48)$$

Depending on the slenderness ratios of the webs, three limit states can be identified: shear yielding, inelastic shear buckling, and elastic shear buckling. The design shear strength that corresponds to each of these limit states are given as follows:

For $h/t_w \leq 2.45\sqrt{(E/F_{yw})}$ (shear yielding of web)

$$\phi_v V_n = 0.90[0.60 F_{yw} A_w] \qquad (4.49)$$

For $2.45\sqrt{(E/F_{yw})} < h/t_w \leq 3.07\sqrt{(E/F_{yw})}$ (inelastic shear buckling of web)

$$\phi_v V_n = 0.90 \left[0.60 F_{yw} A_w \frac{2.45\sqrt{(E/F_{yw})}}{h/t_w} \right] \qquad (4.50)$$

For $3.07\sqrt{(E/F_{yw})} < h/t_w \leq 260$ (elastic shear buckling of web)

$$\phi_v V_n = 0.90 A_w \left[\frac{4.52 E}{(h/t_w)^2} \right] \qquad (4.51)$$

The variables used in the above equations are defined in the following, where h is the clear distance between flanges less the fillet or corner radius, t_w is the web thickness, F_{yw} is the yield stress of web, $A_w = d t_w$, and d is the overall depth of the section.

4.5.1.2.3 Criteria for Concentrated Loads
When concentrated loads are applied normal to the flanges in planes parallel to the webs of flexural members, the flanges and webs must be checked to ensure that they have sufficient strengths ϕR_n to withstand the concentrated forces R_u, that is

$$\phi R_n \geq R_u \qquad (4.52)$$

The design strengths for a variety of limit states are given below.

4.5.1.2.3.1 Local Flange Bending — The design strength for local flange bending is given by

$$\phi R_n \geq 0.90[6.25 t_f^2 F_{yf}] \tag{4.53}$$

where t_f is the flange thickness of the loaded flange and F_{yf} is the flange yield stress.

The design strength in Equation 4.53 is applicable only if the length of load across the member flange exceeds $0.15b$, where b is the member flange width. If the length of load is less than $0.15b$, the limit state of local flange bending need not be checked. Also, Equation 4.53 shall be reduced by a factor of half if the concentrated force is applied less than $10t_f$ from the beam end.

4.5.1.2.3.2 Local Web Yielding — The design strength for yielding of the beam web at the toe of the fillet under tensile or compressive loads acting on one or both flanges are

If the load acts at a distance from the beam end which exceeds the depth of the member

$$\phi R_n = 1.00[(5k + N)F_{yw} t_w] \tag{4.54}$$

If the load acts at a distance from the beam end which does not exceed the depth of the member

$$\phi R_n = 1.00[(2.5k + N)F_{yw} t_w] \tag{4.55}$$

where k is the distance from the outer face of the flange to the web toe of the fillet, N is the length of bearing on the beam flange, F_{yw} is the web yield stress, and t_w is the web thickness.

4.5.1.2.3.3 Web Crippling — The design strength for crippling of beam web under compressive loads acting on one or both flanges are

If the load acts at a distance from the beam end which exceeds half the depth of the beam

$$\phi R_n = 0.75 \left\{ 0.80 t_w^2 \left[1 + 3 \left(\frac{N}{d} \right) \left(\frac{t_w}{t_f} \right)^{1.5} \right] \sqrt{\frac{EF_{yw} t_f}{t_w}} \right\} \tag{4.56}$$

If the load acts at a distance from the beam end which does not exceed half the depth of the beam and if $N/d \leq 0.2$

$$\phi R_n = 0.75 \left\{ 0.40 t_w^2 \left[1 + 3 \left(\frac{N}{d} \right) \left(\frac{t_w}{t_f} \right)^{1.5} \right] \sqrt{\frac{EF_{yw} t_f}{t_w}} \right\} \tag{4.57}$$

If the load acts at a distance from the beam end which does not exceed half the depth of the beam and if $N/d > 0.2$

$$\phi R_n = 0.75 \left\{ 0.40 t_w^2 \left[1 + \left(\frac{4N}{d} - 0.2 \right) \left(\frac{t_w}{t_f} \right)^{1.5} \right] \sqrt{\frac{EF_{yw} t_f}{t_w}} \right\} \tag{4.58}$$

where d is the overall depth of the section, t_f is the flange thickness, and the other variables are the same as those defined in Equations 4.54 and 4.55.

4.5.1.2.3.4 Sideways Web Buckling — Sideways web buckling may occur in the web of a member if a compressive concentrated load is applied to a flange not restrained against relative movement by stiffeners or lateral bracings. The sideways web buckling design strength for the member is

If the loaded flange is restrained against rotation about the longitudinal member axis and $(h/t_w)(l/b_f)$ is less than 2.3

$$\phi R_n = 0.85 \left\{ \frac{C_r t_w^2 t_f}{h^2} \left[1 + 0.4 \left(\frac{h/t_w}{l/b_f} \right)^3 \right] \right\} \tag{4.59}$$

If the loaded flange is not restrained against rotation about the longitudinal member axis and $(d_c/t_w)(l/b_f)$ is less than 1.7

$$\phi R_n = 0.85 \left\{ \frac{C_r t_w^2 t_f}{h^2} \left[0.4 \left(\frac{h/t_w}{l/b_f} \right)^3 \right] \right\} \qquad (4.60)$$

where t_f is the flange thickness (in.), t_w is the web thickness (in.), h is the clear distance between flanges less the fillet or corner radius for rolled shapes; distance between adjacent lines of fasteners or clear distance between flanges when welds are used for built-up shapes (in.), b_f is the flange width (in.), l is the largest laterally unbraced length along either flange at the point of load (in.), $C_r = 960,000$ ksi if $M_u/M_y < 1$ at the point of load and $C_r = 480,000$ ksi if $M_u/M_y \geq 1$ at the point of load, and M_y is the yield moment.

4.5.1.2.3.5 Compression Buckling of the Web — This limit state may occur in members with unstiffened webs when both flanges are subjected to compressive forces. The design strength for this limit state is

$$\phi R_n = 0.90 \left[\frac{24 t_w^3 \sqrt{EF_{yw}}}{h} \right] \qquad (4.61)$$

This design strength shall be reduced by a factor of half if the concentrated forces are acting at a distance less than half the beam depth from the beam end. The variables in Equation 4.61 are the same as those defined in Equation 4.58 to 4.60.

Stiffeners shall be provided in pairs if any one of the above strength criteria is violated. If the local flange bending or the local web yielding criterion is violated, the stiffener pair to be provided to carry the excess R_u need not extend more than one-half the web depth. The stiffeners shall be welded to the loaded flange if the applied force is tensile. They shall either bear on or be welded to the loaded flange if the applied force is compressive. If the web crippling or the compression web buckling criterion is violated, the stiffener pair to be provided shall extend the full height of the web. They shall be designed as axially loaded compression members (see Section 4.4) with an effective length factor $K = 0.75$, a cross-section A_g composed of the cross-sectional areas of the stiffeners plus $25 t_w^2$ for interior stiffeners and $12 t_w^2$ for stiffeners at member ends.

4.5.1.2.4 Deflection Criterion

The deflection criterion is the same as that for ASD. Since deflection is a serviceability limit state, service (rather than factored) loads are used in deflection computations.

4.5.2 Continuous Beams

Continuous beams shall be designed in accordance with the criteria for flexural members given in the preceding section. However, a 10% reduction in negative moments due to gravity loads is permitted at the supports provided that

- The maximum positive moment between supports is increased by $\frac{1}{10}$ the average of the negative moments at the supports.
- The section is compact.
- The lateral unbraced length does not exceed L_c (for ASD) or L_{pd} (for LRFD) where L_c is as defined in Equation 4.26 and L_{pd} is given by

$$L_{pd} = \begin{cases} \left[0.12 + 0.076 \left(\frac{M_1}{M_2} \right) \right] \left(\frac{E}{F_y} \right) r_y, & \text{for I-shaped members} \\ \left[0.17 + 0.10 \left(\frac{M_1}{M_2} \right) \right] \left(\frac{E}{F_y} \right) r_y \geq 0.10 \left(\frac{E}{F_y} \right) r_y, & \text{for solid rectangular and box sections} \end{cases}$$

$$(4.62)$$

in which, F_y is the specified minimum yield stress of the compression flange, M_1/M_2 is the ratio of smaller to larger moment within the unbraced length, taken as positive if the moments cause reverse curvature and negative if the moments cause single curvature, and r_y is the radius of gyration about the minor axis.

- The beam is not a hybrid member.
- The beam is not made of high-strength steel.
- The beam is continuous over the supports (i.e., not cantilevered).

EXAMPLE 4.4

Using LRFD, select the lightest W-section for the three-span continuous beam shown in Figure 4.10a to support a uniformly distributed dead load of 1.5 k/ft (22 kN/m) and a uniformly distributed live load of 3 k/ft (44 kN/m). The beam is laterally braced at the supports A, B, C, and D. Use A36 steel.

Load combinations: The beam is to be designed based on the worst load combination of Table 4.3. By inspection, the load combination $1.2D + 1.6L$ will control the design. Thus, the beam will be designed to support a factored uniformly distributed dead load of $1.2 \times 1.5 = 1.8$ k/ft and a factored uniformly distributed live load of $1.6 \times 3 = 4.8$ k/ft.

Placement of loads: The uniform dead load is to be applied over the entire length of the beam as shown in Figure 4.10b. The uniform live load is to be applied to spans AB and CD as shown in Figure 4.10c to obtain the maximum positive moment and it is to be applied to spans AB and BC as shown in Figure 4.10d to obtain the maximum negative moment.

Reduction of negative moment at supports: Assuming the beam is compact and $L_b < L_{pd}$ (we shall check these assumptions later), a 10% reduction in support moment due to gravity load is allowed provided that the maximum moment is increased by $\frac{1}{10}$ the average of the negative support moments. This reduction is shown in the moment diagrams as solid lines in Figure 4.10b and d. (The dotted lines in these figures represent the unadjusted moment diagrams.) This provision for support moment reduction takes into consideration the beneficial effect of moment redistribution in continuous beams, and it allows for the selection of a lighter section if the design is governed by negative moments. Note that no reduction in negative moments is made to the case when only spans AB and CD are loaded. This is because for this load case, the negative support moments are less than the positive in-span moments.

Determination of the required flexural strength, M_u: Combining load case 1 and load case 2, the maximum positive moment is found to be 256 kip ft. Combining load case 1 and load case 3, the maximum negative moment is found to be 266 kip ft. Thus, the design will be controlled by the negative moment and so $M_u = 266$ kip ft.

Beam selection: A beam section is to be selected based on Equation 4.38. The critical segment of the beam is span AB. For this span, the lateral unsupported length, $L_b = 20$ ft. For simplicity, the bending coefficient, C_b, is conservatively taken as 1. The selection of a beam section is facilitated by the use of a series of beam charts contained in the AISC-LRFD Manual (AISC 2001). Beam charts are plots of flexural design strength $\phi_b M_n$ of beams as a function of the lateral unsupported length L_b based on Equation 4.39 to 4.41. A beam is considered satisfactory for the limit state of flexure if the beam strength curve envelopes the required flexural strength for a given L_b.

For the present example, $L_b = 20$ ft and $M_u = 266$ kip ft, the lightest section (the first solid curve that envelopes $M_u = 266$ kip ft for $L_b = 20$ ft) obtained from the chart is a W16 × 67 section. Upon adding the factored dead weight of this W16 × 67 section to the specified loads, the required flexural strength increases from 266 to 269 kip ft. Nevertheless, the beam strength curve still envelopes this required strength for $L_b = 20$ ft; therefore, the section is adequate.

Check for compactness: For the W16 × 67 section

$$\text{Flange:} \quad \left[\frac{b_f}{2t_f} = 7.7 \right] < \left[0.38 \sqrt{\frac{E}{F_y}} = 10.8 \right]$$

$$\text{Web:} \quad \left[\frac{h_c}{t_w} = 35.9 \right] < \left[3.76 \sqrt{\frac{E}{F_y}} = 106.7 \right]$$

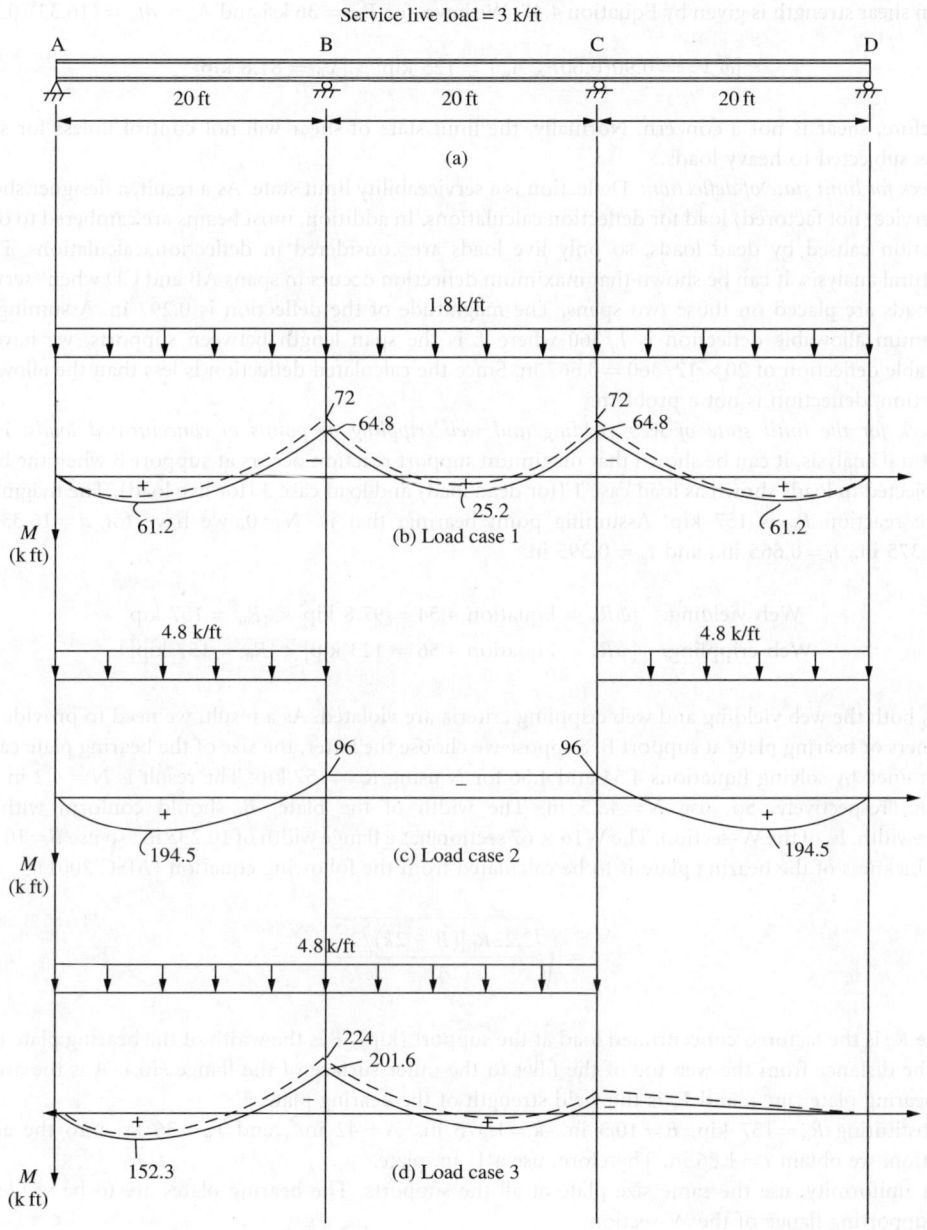

FIGURE 4.10 Design of a three-span continuous beam (1 k = 4.45 kN, 1 ft = 0.305 m).

Therefore, the section is compact.

Check whether $L_b < L_{pd}$: Using Equation 4.62, with $M_1/M_2 = 0$, $r_y = 2.46$ in., and $F_y = 36$ ksi, we have $L_{pd} = 246$ in. (or 20.5 ft). Since $L_b = 20$ ft is less than $L_{pd} = 20.5$ ft, the assumption made earlier is validated.

Check for the limit state of shear: The selected section must satisfy the shear strength criterion of Equation 4.48. From structural analysis, it can be shown that maximum shear occurs just to the left of support B under load case 1 (for dead load) and load case 3 (for live load). It has a magnitude of 81.8 kip. For the W16 × 67 section, $h/t_w = 35.9$, which is less than $2.45\sqrt{(E/F_{yw})} = 69.5$, so the design shear strength is given by Equation 4.47. We have, for $F_{yw} = 36$ ksi and $A_w = dt_w = (16.33)(0.395)$

$$[\phi_v V_n = 0.90(0.60 F_{yw} A_w) = 125 \text{ kip}] > [V_u = 81.8 \text{ kip}]$$

Therefore, shear is not a concern. Normally, the limit state of shear will not control unless for short beams subjected to heavy loads.

Check for limit state of deflection: Deflection is a serviceability limit state. As a result, a designer should use service (not factored) load for deflection calculations. In addition, most beams are cambered to offset deflection caused by dead loads, so only live loads are considered in deflection calculations. From structural analysis, it can be shown that maximum deflection occurs in spans AB and CD when (service) live loads are placed on those two spans. The magnitude of the deflection is 0.297 in. Assuming the maximum allowable deflection is $L/360$ where L is the span length between supports, we have an allowable deflection of $20 \times 12/360 = 0.667$ in. Since the calculated deflection is less than the allowable deflection, deflection is not a problem.

Check for the limit state of web yielding and web crippling at points of concentrated loads: From structural analysis, it can be shown that maximum support reaction occurs at support B when the beam is subjected to loads shown as load case 1 (for dead load) and load case 3 (for live load). The magnitude of the reaction R_u is 157 kip. Assuming point bearing, that is, $N = 0$, we have for $d = 16.33$ in., $k = 1.375$ in., $t_f = 0.665$ in., and $t_w = 0.395$ in.

Web yielding: $[\phi R_n = \text{Equation } 4.54 = 97.8 \text{ kip}] < [R_u = 157 \text{ kip}]$

Web crippling: $[\phi R_n = \text{Equation } 4.56 = 123 \text{ kip}] < [R_u = 157 \text{ kip}]$

Thus, both the web yielding and web crippling criteria are violated. As a result, we need to provide web stiffeners or bearing plate at support B. Suppose we choose the latter, the size of the bearing plate can be determined by solving Equations 4.54 and 4.56 for N using $R_u = 157$ kip. The result is $N = 4.2$ in. and 3.3 in., respectively. So, use $N = 4.25$ in. The width of the plate, B, should conform with the flange width, b_f, of the W-section. The W16 × 67 section has a flange width of 10.235 in., so use $B = 10.5$ in. The thickness of the bearing plate is to be calculated from the following equation (AISC 2001):

$$t = \sqrt{\frac{2.22 R_u [(B - 2k)/2]^2}{A F_y}}$$

where R_u is the factored concentrated load at the support (kip), B is the width of the bearing plate (in.), k is the distance from the web toe of the fillet to the outer surface of the flange (in.), A is the area of the bearing plate (in.²), and F_y is the yield strength of the bearing plate.

Substituting $R_u = 157$ kip, $B = 10.5$ in., $k = 1.375$ in., $A = 42$ in.², and $F_y = 36$ ksi into the above equation, we obtain $t = 1.86$ in. Therefore, use a $1\frac{7}{8}$ in. plate.

For uniformity, use the same size plate at all the supports. The bearing plates are to be welded to the supporting flange of the W-section.

Use a W16 × 67 section. Provide bearing plates of size $1\frac{7}{8}$ in. × 4 in. × $10\frac{1}{2}$ in. at the supports.

4.5.3 Beam Bracing

The design strength of beams that are bent about their major axes depends on their lateral unsupported length L_b. The manner a beam is braced against out-of-plane deformation affects its design. Bracing can be provided by various means such as cross-frames, cross-beams or diaphragms, or encasement of the beam flange in the floor slab (Yura 2001). Two types of bracing systems are identified in the AISC-LRFD Specification — relative and nodal. A relative brace controls the movement of a braced point with respect to adjacent braced points along the span of the beam. A nodal (or discrete) brace controls the movement of a braced point without regard to the movement of adjacent braced points. Regardless of the type of bracing system used, braces must be designed with sufficient strength and stiffness to prevent out-of-plane movement of the beam at the braced points. Out-of-plane movement consists of lateral deformation of the beam and twisting of cross-sections. Lateral stability of beams can be achieved by lateral bracing, torsional bracing, or a combination of the two. For lateral bracing, bracing shall be attached near the compression flange for members bend in single curvature (except cantilevers). For cantilevers, bracing shall be attached to the tension flange at the free end. For members bend in double curvature, bracing shall be attached to both flanges near the inflection point. For torsional bracing, bracing can be attached at any cross-sectional location.

4.5.3.1 Stiffness Requirement for Lateral Bracing

The required brace stiffness of the bracing assembly in a direction perpendicular to the longitudinal axis of the braced member, in the plane of buckling, is given by

$$\beta_{br} = \begin{cases} \dfrac{5.33 M_u C_d}{L_{br} h_o} & \text{for relative bracing} \\[3mm] \dfrac{13.3 M_u C_d}{L_{br} h_o} & \text{for nodal bracing} \end{cases} \tag{4.63}$$

where M_u is the required flexural strength, $C_d = 1.0$ for single curvature bending; 2.0 for double curvature bending near the inflection point, L_{br} is the distance between braces, and h_o is the distance between flange centroids.

L_{br} can be replaced by L_q (the maximum unbraced length for M_u) if $L_{br} < L_q$. (For ASD, replace M_u in Equation 4.63 by $1.5 M_a$, where M_a is the required flexural strength based on ASD load combinations.)

4.5.3.2 Strength Requirement for Lateral Bracing

In addition to the stiffness requirement as stipulated above, braces must be designed for a required brace strength given by

$$P_{br} = \begin{cases} 0.008 \dfrac{M_u C_d}{h_o} & \text{for relative bracing} \\[3mm] 0.02 \dfrac{M_u C_d}{h_o} & \text{for nodal bracing} \end{cases} \tag{4.64}$$

The terms in Equation 4.64 are defined as in Equation 4.63.

4.5.3.3 Stiffness Requirement for Torsional Bracing

The required bracing stiffness is

$$\beta_{Tbr} = \frac{\beta_T}{\left(1 - \dfrac{\beta_T}{\beta_{sec}}\right)} \geq 0 \tag{4.65}$$

where

$$\beta_T = \begin{cases} \dfrac{3.2 L M_u^2}{n E I_y C_b^2} & \text{for nodal bracing} \\[3mm] \dfrac{3.2 M_u^2}{E I_y C_b^2} & \text{for continuous bracing} \end{cases} \tag{4.66}$$

and

$$\beta_{sec} = \begin{cases} \dfrac{3.3 E}{h_o} \left(\dfrac{1.5 h_o t_w^3}{12} + \dfrac{t_s b_s^3}{12} \right) & \text{for nodal bracing} \\[3mm] \dfrac{3.3 E t_w^3}{12 h_o} & \text{for continuous bracing} \end{cases} \tag{4.67}$$

in which L is the span length, M_u is the required moment, n is the number of brace points within the span, E is the modulus of elasticity, I_y is the moment of inertia of the minor axis, C_b is the bending coefficient as defined earlier, h_o is the distance between the flange centroids, t_w is the thickness of the beam web, t_s is the thickness of the web stiffener, and b_s is the width of the stiffener (or, for pairs of stiffeners, b_s = total width of stiffeners). (For ASD, replace M_u in Equation 4.66 by $1.5M_a$, where M_a is the required flexural strength based on ASD load combinations.)

4.5.3.4 Strength Requirement for Torsional Bracing

The connection between a torsional brace and the beam being braced must be able to withstand a moment given by

$$M_{Tbr} = \frac{0.024 M_u L}{n C_b L_{br}} \tag{4.68}$$

where L_{br} is the distance between braces (if $L_{br} < L_q$, where L_q is the maximum unbraced length for M_u, then use L_q). The other terms in Equation 4.68 are defined in Equation 4.66.

EXAMPLE 4.5

Design an I-shaped cross-beam 12 ft (3.7 m) in length to be used as lateral braces to brace a 30-ft (9.1 m) long simply supported W30 × 90 girder at every third point. The girder was designed to carry a moment of 8000 kip in. (904 kN m). A992 steel is used.

Because a brace is provided at every third point, $L_{br} = 10$ ft $= 120$ in., $M_u = 8000$ kip in. as stated. $C_d = 1$ for single curvature bending. $h_o = d - t_f = 29.53 - 0.610 = 28.92$ in. for the W30 × 90 section. Substituting these values into Equations 4.63 and 4.64 for nodal bracing, we obtain $\beta_{br} = 30.7$ kip/in., and $P_{br} = 5.53$ kip.

As the cross-beam will be subject to compression, its slenderness ratio, l/r, should not exceed 200. Let us try the smallest size W-section, a W4 × 13 section, with $A = 3.83$ in.^{2}, $r_y = 1.00$ in., $\phi_c P_n = 25$ kip.

$$\text{Stiffness:} \quad \frac{EA}{l} = \frac{(29,000)(3.83)}{12 \times 12} = 771 \text{ kip/in.} > 30.7 \text{ kip/in.}$$

$$\text{Strength:} \quad \phi_c P_n = 25 \text{ kip} > 5.53 \text{ kip}$$

$$\text{Slenderness:} \quad \frac{l}{r_y} = \frac{12 \times 12}{1.00} = 144 < 200$$

Since all criteria are satisfied, the W4 × 13 section is adequate. Use W4 × 13 as cross-beams to brace the girder.

4.6 Combined Flexure and Axial Force

When a member is subject to the combined action of bending and axial force, it must be designed to resist stresses and forces arising from both bending and axial actions. While a tensile axial force may induce a stiffening effect on the member, a compressive axial force tends to destabilize the member, and the instability effects due to member instability (P–δ effect) and frame instability (P–Δ effect) must be properly accounted for. The P–δ effect arises when the axial force acts through the lateral deflection of the member relative to its chord. The P–Δ effect arises when the axial force acts through the relative displacements of the two ends of the member. Both effects tend to increase member deflection and moment, and so they must be considered in the design. A number of approaches are available in the literature to handle these so-called P–Δ effects (see, e.g., Chen and Lui 1991; Galambos 1998). The design of members subject to combined bending and axial force is facilitated by the use interaction equations. In these equations, the effects of bending and axial actions are combined in a certain manner to reflect the capacity demand on the member.

4.6.1 Design for Combined Flexure and Axial Force

4.6.1.1 Allowable Stress Design

The interaction equations are

If the axial force is tensile:

$$\frac{f_a}{F_t} + \frac{f_{bx}}{F_{bx}} + \frac{f_{by}}{F_{by}} \leq 1.0 \tag{4.69}$$

where f_a is the computed axial tensile stress, f_{bx}, f_{by} are computed bending tensile stresses about the major and minor axes, respectively, F_t is the allowable tensile stress (see Section 4.3), and F_{bx}, F_{by} are allowable bending stresses about the major and minor axes, respectively (see Section 4.5).

If the axial force is compressive:
Stability requirement

$$\frac{f_a}{F_a} + \left[\frac{C_{mx}}{1 - (f_a/F'_{ex})}\right]\frac{f_{bx}}{F_{bx}} + \left[\frac{C_{my}}{1 - (f_a/F'_{ey})}\right]\frac{f_{by}}{F_{by}} \leq 1.0 \tag{4.70}$$

Yield requirement

$$\frac{f_a}{0.66F_y} + \frac{f_{bx}}{F_{bx}} + \frac{f_{by}}{F_{by}} \leq 1.0 \tag{4.71}$$

However, if the axial force is small (when $f_a/F_a \leq 0.15$), the following interaction equation can be used in lieu of the above equations.

$$\frac{f_a}{F_a} + \frac{f_{bx}}{F_{bx}} + \frac{f_{by}}{F_{by}} \leq 1.0 \tag{4.72}$$

The terms in Equations 4.70 to 4.72 are defined as follows:

f_a, f_{bx}, f_{by} = Computed axial compressive stress, computed bending stresses about the major and minor axes, respectively. These stresses are to be computed based on a *first-order analysis*.

F_y = Minimum specified yield stress.

F'_{ex}, F'_{ey} = Euler stresses about the major and minor axes ($\pi^2 E/(Kl/r)_x$, $\pi^2 E/(Kl/r)_y$) divided by a factor of safety of $\frac{23}{12}$.

C_m = a coefficient to account for the effect of moment gradient on member and frame instabilities (C_m is defined in Section 4.6.1.2).

The other terms are defined as in (Equation 4.69).

The terms in brackets in (Equation 4.70) are moment magnification factors. The computed bending stresses f_{bx}, f_{by} are magnified by these magnification factors to account for the P–δ effects in the member.

4.6.1.2 Load and Resistance Factor Design

Doubly or singly symmetric members subject to combined flexure and axial forces shall be designed in accordance with the following interaction equations:

For $P_u/\phi P_n \geq 0.2$

$$\frac{P_u}{\phi P_n} + \frac{8}{9}\left(\frac{M_{ux}}{\phi_b M_{nx}} + \frac{M_{uy}}{\phi_b M_{ny}}\right) \leq 1.0 \tag{4.73}$$

For $P_u/\phi P_n < 0.2$

$$\frac{P_u}{2\phi P_n} + \left(\frac{M_{ux}}{\phi_b M_{nx}} + \frac{M_{uy}}{\phi_b M_{ny}}\right) \leq 1.0 \qquad (4.74)$$

where, if P is tensile,

P_u = factored tensile axial force
P_n = design tensile strength (see Section 4.3)
M_u = factored moment (preferably obtained from a second-order analysis)
M_n = design flexural strength (see Section 4.5)
ϕ = ϕ_t = resistance factor for tension = 0.90
ϕ_b = resistance factor for flexure = 0.90

and, if P is compressive,

P_u = factored compressive axial force
P_n = design compressive strength (see Section 4.4)
M_u = required flexural strength (see discussion below)
M_n = design flexural strength (see Section 4.5)
ϕ = ϕ_c = resistance factor for compression = 0.85
ϕ_b = resistance factor for flexure = 0.90

The required flexural strength M_u shall be determined from a second-order elastic analysis. In lieu of such an analysis, the following equation may be used:

$$M_u = B_1 M_{bt} + B_2 M_{lt} \qquad (4.75)$$

where

M_{nt} = factored moment in member assuming the frame does not undergo lateral translation (see Figure 4.11)
M_{lt} = factored moment in member as a result of lateral translation (see Figure 4.11)
B_1 = $C_m/(1 - P_u/P_e) \geq 1.0$ is the P–δ moment magnification factor
P_e = $\pi^2 EI/(KL)^2$, with $K \leq 1.0$ in the plane of bending
C_m = a coefficient to account for moment gradient (see discussion below)
B_2 = $1/[1 - (\sum P_u \Delta_{oh}/\sum HL)]$ or $B_2 = 1/[1 - (\sum P_u/\sum P_e)]$
$\sum P_u$ = sum of all factored loads acting on and above the story under consideration
Δ_{oh} = first-order interstory translation
$\sum H$ = sum of all lateral loads acting on and above the story under consideration
L = story height
P_e = $\pi^2 EI/(KL)^2$

For end-restrained members that do not undergo relative joint translation and not subject to transverse loading between their supports in the plane of bending, C_m is given by

$$C_m = 0.6 - 0.4\left(\frac{M_1}{M_2}\right)$$

where M_1/M_2 is the ratio of the smaller to larger member end moments. The ratio is positive if the member bends in reverse curvature and negative if the member bends in single curvature.

For members that do not undergo relative joint translation and subject to transverse loading between their supports in the plane of bending, C_m can conservatively be taken as 1.

For purpose of design, Equations 4.73 and 4.74 can be rewritten in the form (Aminmansour 2000):
For $P_u/\phi_c P_n > 0.2$

$$bP_u + mM_{ux} + nM_{uy} \leq 1.0 \qquad (4.76)$$

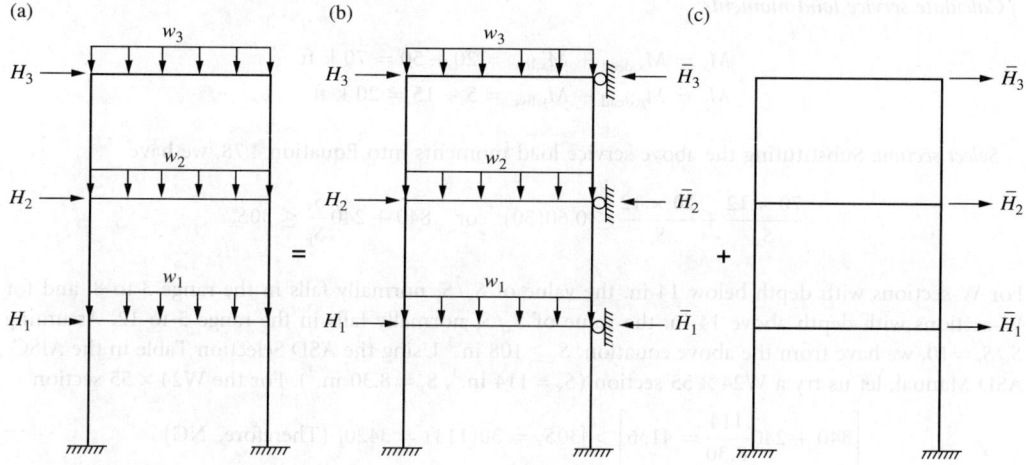

FIGURE 4.11 Calculation of M_{nt} and M_{lt}: (a) original frame; (b) nonsway frame analysis for M_{nt}; and (c) sway frame analysis M_{lt}.

For $P_u/\phi_c P_n \leq 0.2$

$$\frac{b}{2}P_u + \frac{9}{8}mM_{ux} + \frac{9}{8}nM_{uy} \leq 1.0 \qquad (4.77)$$

where

$b = 1/(\phi_c P_n)$
$m = 8/(9\phi_b M_{nx})$
$n = 8/(9\phi_b M_{ny})$

Numerical values for b, m, and n are provided in the AISC Manual (AISC 2001).

4.7 Biaxial Bending

Members subjected to bending about both principal axes (e.g., purlins on an inclined roof) should be designed for *biaxial bending*. Since both moment about the major axis M_{ux} and moment about the minor axis M_{uy} create flexural stresses over the cross-section of the member, the design must take into consideration these stress combinations.

4.7.1 Design for Biaxial Bending
4.7.1.1 Allowable Stress Design

The following interaction equation is often used for the design of beams subject to biaxial bending

$$f_{bx} + f_{by} \leq 0.60F_y \quad \text{or} \quad \frac{M_x}{S_x} + \frac{M_y}{S_y} \leq 0.60F_y \qquad (4.78)$$

where M_x, M_y are service load moments about the major and minor beam axes, respectively, S_x, S_y are elastic section moduli about the major and minor axes, respectively, and F_y is the specified minimum yield stress.

EXAMPLE 4.6

Using ASD, select a W-section to carry dead load moments $M_x = 20$ k ft (27 kN m) and $M_y = 5$ k ft (6.8 kN m) and live load moments $M_x = 50$ k ft (68 kN m) and $M_y = 15$ k ft (20 kN m). Use A992 steel.

Calculate service load moments:

$$M_x = M_{x,\text{dead}} + M_{x,\text{live}} = 20 + 50 = 70 \text{ k ft}$$
$$M_y = M_{y,\text{dead}} + M_{y,\text{live}} = 5 + 15 = 20 \text{ k ft}$$

Select section: Substituting the above service load moments into Equation 4.78, we have

$$\frac{70 \times 12}{S_x} + \frac{20 \times 12}{S_y} \leq 0.60(50) \quad \text{or} \quad 840 + 240\frac{S_x}{S_y} \leq 30S_x$$

For W-sections with depth below 14 in. the value of S_x/S_y normally falls in the range 3 to 8, and for W-sections with depth above 14 in. the value of S_x/S_y normally falls in the range 5 to 12. Assuming $S_x/S_y = 10$, we have from the above equation, $S_x \geq 108$ in.^{3} Using the ASD Selection Table in the AISC-ASD Manual, let us try a W24 × 55 section ($S_x = 114$ in.^{3}, $S_y = 8.30$ in.^{3}). For the W24 × 55 section

$$\left[840 + 240\frac{114}{8.30} = 4136\right] > [30S_x = 30(114) = 3420] \quad \text{(Therefore, NG)}$$

The next lightest section is W21 × 62 ($S_x = 127$ in.^{3}, $S_y = 13.9$ in.^{3}). For this section

$$\left[840 + 240\frac{127}{13.9} = 3033\right] < [30S_x = 30(127) = 3810] \quad \text{(Therefore, Okay)}$$

Therefore, use a W21 × 62 section.

4.7.1.2 Load and Resistance Factor Design

To avoid distress at the most severely stressed point, the following equation for the limit state of yielding must be satisfied:

$$f_{un} \leq \phi_b F_y \tag{4.79}$$

where $f_{un} = (M_{ux}/S_x + M_{uy}/S_y)$ is the flexural stress under factored loads, S_x, S_y are elastic section moduli about the major and minor axes, respectively, $\phi_b = 0.90$, and F_y is the specified minimum yield stress.

In addition, the limit state for lateral torsional buckling about the major axis should also be checked, that is,

$$\phi_b M_{nx} \geq M_{ux} \tag{4.80}$$

$\phi_b M_{nx}$ is the design flexural strength about the major axis (see Section 4.5). To facilitate design for biaxial bending, Equation 4.79 can be rearranged to give

$$S_x \geq \frac{M_{ux}}{\phi_b F_y} + \frac{M_{uy}}{\phi_b F_y}\left(\frac{S_x}{S_y}\right) \approx \frac{M_{ux}}{\phi_b F_y} + \frac{M_{uy}}{\phi_b F_y}\left(3.5\frac{d}{b_f}\right) \tag{4.81}$$

In the above equation, d is the overall depth and b_f the flange width of the section. The approximation $(S_x/S_y) \approx (3.5d/b_f)$ was suggested by Gaylord et al. (1992) for doubly symmetric I-shaped sections.

4.8 Combined Bending, Torsion, and Axial Force

Members subjected to the combined effect of bending, torsion, and axial force should be designed to satisfy the following limit states:

Yielding under normal stress:

$$\phi F_y \geq f_{un} \tag{4.82}$$

where $\phi = 0.90$, F_y is the specified minimum yield stress, and f_{un} is the maximum normal stress determined from an elastic analysis under factored loads.

Yielding under shear stress:

$$\phi(0.6F_y) \geq f_{uv} \tag{4.83}$$

where $\phi = 0.90$, F_y is the specified minimum yield stress, and f_{uv} is the maximum shear stress determined from an elastic analysis under factored loads.

Buckling:

$$\phi_c F_{cr} \geq f_{un} \quad \text{or} \quad \phi_c F_{cr} \geq f_{uv}, \qquad \text{whichever is applicable} \tag{4.84}$$

where $\phi_c F_{cr} = \phi_c P_n/A_g$, in which $\phi_c P_n$ is the design compressive strength of the member (see Section 4.4), A_g is the gross cross-section area, and f_{un}, f_{uv} are normal and shear stresses as defined in Equations 4.82 and 4.83.

4.9 Frames

Frames are designed as a collection of structural components such as beams, beam–columns (columns), and connections. According to the restraint characteristics of the connections used in the construction, frames can be designed as Type I (rigid framing), Type II (simple framing), Type III (semirigid framing) in ASD, or fully restrained (rigid) and partially restrained (semirigid) in LRFD.

- The design of rigid frames necessitates the use of connections capable of transmitting the full or a significant portion of the moment developed between the connecting members. The rigidity of the connections must be such that the angles between intersecting members should remain virtually unchanged under factored loads.
- The design of simple frames is based on the assumption that the connections provide no moment restraint to the beam insofar as gravity loads are concerned, but these connections should have adequate capacity to resist wind moments.
- The design of semirigid frames is permitted upon evidence of the connections to deliver a predicable amount of moment restraint. Over the past two decades, a large body of work has been published in the literature on semirigid connection and frame behavior (see, e.g., Chen 1987, 2000; CTBUH 1993; Chen et al. 1996; Faella et al. 2000). However, because of the vast number of semirigid connections that can exhibit an appreciable difference in joint behavior, no particular approach has been recommended by any building specifications as of this writing. The design of semirigid or partially restrained frames is dependent on the sound engineering judgment by the designer.

Semirigid and simple framings often incur inelastic deformation in the connections. The connections used in these constructions must be proportioned to possess sufficient ductility to avoid overstress of the fasteners or welds.

Regardless of the types of constructions used, due consideration must be given to account for member and frame instability (P–δ and P–Δ) effects either by the use of a second-order analysis or by other means such as moment magnification factors or notional loads (ASCE 1997). The end-restrained effect on the member should also be accounted for by the use of the effective length factor K.

4.9.1 Frame Design

Frames can be designed as *side-sway inhibited* (braced) or *side-sway uninhibited* (unbraced). In side-sway inhibited frames, frame *drift* is controlled by the presence of a bracing system (e.g., shear walls, diagonal, cross, K-braces, etc.). In side-sway uninhibited frames, frame drift is limited by the flexural rigidity of the connected members and the diaphragm action of the floors. Most side-sway uninhibited frames are designed as Type I or Type FR frames using moment connections. Under normal circumstances, the amount of interstory drift under service loads should not exceed $h/500$ to $h/300$ where h is the story height. A higher value of interstory drift is allowed only if it does not create serviceability concerns.

Beams in side-sway inhibited frames are often subject to high axial forces. As a result, they should be designed as beam–columns using beam–column interaction equations. Furthermore, vertical bracing systems should be provided for braced multistory frames to prevent vertical buckling of the frames under gravity loads.

When designing members of a frame, a designer should consider a variety of loading combinations and load patterns, and the members are designed for the most severe load cases. Preliminary sizing of members can be achieved by the use of simple behavioral models such as the simple beam model, cantilever column model, and portal and cantilever method of frame analysis (see, e.g., Rossow 1996).

4.9.2 Frame Bracing

The subject of frame bracing is discussed in a number of references; see, for example, SSRC (1993) and Galambos (1998). According to the LRFD Specification (AISC 1999), the required story or panel bracing shear stiffness in side-sway inhibited frames is

$$\beta_{cr} = \frac{2.67 \sum P_u}{L} \tag{4.85}$$

where $\sum P_u$ is the sum of all factored gravity loads acting on and above the story or panel supported by the bracing and L is the story height or panel spacing (for ASD, replace P_u in Equation 4.85 by $1.5P_a$, where P_a is the required compressive strength based on ASD load combinations).

The required story or panel bracing force is

$$P_{br} = 0.004 \sum P_u \tag{4.86}$$

4.10 Plate Girders

Plate girders are built-up beams. They are used as flexural members to carry extremely large lateral loads. A flexural member is considered a plate girder if the width–thickness ratio of the web, h_c/t_w, exceeds $760/\sqrt{F_b}$ (F_b is the allowable flexural stress) according to ASD, or λ_r (see Table 4.8) according to LRFD. Because of the large web slenderness, plate girders are often designed with transverse stiffeners to reinforce the web and to allow for postbuckling (shear) strength (i.e., *tension field action*) to develop. Table 4.9 summarizes the requirements of transverse stiffeners for plate girders based on the web slenderness ratio h/t_w. Two types of transverse stiffeners are used for plate girders: bearing stiffeners and intermediate stiffeners. Bearing stiffeners are used at unframed girder ends, and at concentrated load points where the web yielding or web crippling criterion is violated. Bearing stiffeners extend the full depth of the web from the bottom of the top flange to the top of the bottom flange. Intermediate stiffeners are used when the width–thickness ratio of the web, h/t_w, exceeds 260, when the shear criterion is violated, or when tension field action is considered in the design. Intermediate stiffeners need not extend to the full depth of the web but must be in contact with the compression flange of the girder.

Normally, the depths of plate girder sections are so large that simple beam theory, which postulates that plane sections before bending remain in plane after bending, does not apply. As a result, a different set of design formulae are required for plate girders.

TABLE 4.9 Web Stiffeners Requirements

Range of web slenderness	Stiffeners requirements
$\dfrac{h}{t_w} \leq 260$	Plate girder can be designed without web stiffeners
$260 < \dfrac{h}{t_w} \leq \dfrac{0.48E}{\sqrt{F_{yf}(F_{yf} + 16.5)}}$	Plate girder must be designed with web stiffeners. The spacing of stiffeners, a, can exceed $1.5h$. The actual spacing is determined by the shear criterion
$\dfrac{0.48E}{\sqrt{F_{yf}(F_{yf} + 16.5)}} < \dfrac{h}{t_w} \leq 11.7\sqrt{\dfrac{E}{F_{yf}}}$	Plate girder must be designed with web stiffeners. The spacing of stiffeners, a, cannot exceed $1.5h$

Note: a is the clear distance between stiffeners, *h* is the clear distance between flanges when welds are used or the distance between adjacent lines of fasteners when bolts are used, t_w is the web thickness, and F_{yf} is the compression flange yield stress (ksi).

4.10.1 Plate Girder Design

4.10.1.1 Allowable Stress Design

4.10.1.1.1 Allowable Bending Stress

The maximum bending stress in the compression flange of the girder computed using the flexure formula shall not exceed the allowable value, F_b', given by

$$F_b' = F_b R_{PG} R_e \tag{4.87}$$

where

F_b = applicable allowable bending stress as discussed in Section 4.5 (ksi)
R_{PG} = plate girder stress reduction factor = $1 - 0.0005(A_w/A_f)(h/t_w - 760/\sqrt{F_b}) \leq 1.0$
R_e = hybrid girder factor = $[12 + (A_w/A_f)(3\alpha - \alpha^3)]/[12 + 2(A_w/A_f)] \leq 1.0$, $R_e = 1$ for a nonhybrid girder
A_w = area of the web
A_f = area of the compression flange
α = $0.60 F_{yw}/F_b \leq 1.0$
F_{yw} = yield stress of the web

4.10.1.1.2 Allowable Shear Stress

Without tension field action: The allowable shear stress is the same as that for beams given in Equation 4.35.

With tension field action: The allowable shear stress is given by

$$F_v = \frac{F_y}{2.89} \left[C_v + \frac{1 - C_v}{1.15\sqrt{1 + (a/h)^2}} \right] \leq 0.40 F_y \tag{4.88}$$

Note that the tension field action can be considered in the design only for nonhybrid girders. If the tension field action is considered, transverse stiffeners must be provided and spaced at a distance so that the computed average web shear stress, f_v, obtained by dividing the total shear by the web area does not exceed the allowable shear stress, F_v, given by Equation 4.88. In addition, the computed bending tensile stress in the panel where tension field action is considered cannot exceed $0.60F_y$, nor $(0.825 - 0.375 f_v/F_v)F_y$ where f_v is the computed average web shear stress and F_v is the allowable web shear stress given in Equation 4.88. The shear transfer criterion given by Equation 4.91 must also be satisfied.

4.10.1.1.3 Transverse Stiffeners

Transverse stiffeners must be designed to satisfy the following criteria.

   Moment of inertia criterion: With reference to an axis in the plane of the web, the moment of inertia of the stiffeners, in in.[4], shall satisfy the condition

$$I_{st} \geq \left(\frac{h}{50} \right)^4 \tag{4.89}$$

where h is the clear distance between flanges, in inches.
   Area criterion: The total area of the stiffeners, in in.[2], shall satisfy the condition

$$A_{st} \geq \frac{1 - C_v}{2} \left[\frac{a}{h} - \frac{(a/h)^2}{\sqrt{1 + (a/h)^2}} \right] Y D h t_w \tag{4.90}$$

where C_v is the shear buckling coefficient as defined in Equation 4.35, a is the stiffeners' spacing, h is the clear distance between flanges, t_w is the web thickness, Y is the ratio of web yield stress to stiffener yield stress, D is equal to 1.0 for stiffeners furnished in pairs, 1.8 for single angle stiffeners, and 2.4 for single plate stiffeners.

Shear transfer criterion: If tension field action is considered, the total shear transfer, in kip/in., of the stiffeners shall not be less than

$$f_{vs} = h\sqrt{\left(\frac{F_{yw}}{340}\right)^3} \tag{4.91}$$

where F_{yw} is the web yield stress (ksi) and h is the clear distance between flanges (in.).

The value of f_{vs} can be reduced proportionally if the computed average web shear stress, f_v, is less than F_v given in Equation 4.88.

4.10.1.2 Load and Resistance Factor Design

4.10.1.2.1 Flexural Strength Criterion

Doubly or singly symmetric, single web plate girders loaded in the plane of the web should satisfy the flexural strength criterion of Equation 4.38. The plate girder design flexural strength is given by

For the limit state of tension flange yielding:

$$\phi_b M_n = 0.90[S_{xt} R_e F_{yt}] \tag{4.92}$$

For the limit state of compression flange buckling:

$$\phi_b M_n = 0.90[S_{xc} R_{PG} R_e F_{cr}] \tag{4.93}$$

where

S_{xt} = section modulus referred to the tension flange = I_x/c_t
S_{xc} = section modulus referred to the compression flange = I_x/c_c
I_x = moment of inertia about the major axis
c_t = distance from neutral axis to extreme fiber of the tension flange
c_c = distance from neutral axis to extreme fiber of the compression flange
R_{PG} = plate girder bending strength reduction factor
   = $1 - a_r[h_c/t_w - 5.70\sqrt{E/F_{cr}})]/[1200 + 300a_r] \leq 1.0$
R_e = hybrid girder factor
   = $[12 + a_r(3m - m^3)]/[12 + 2a_r] \leq 1.0$ ($R_e = 1$ for nonhybrid girder)
a_r = ratio of web area to compression flange area
m = ratio of web yield stress to flange yield stress or ratio of web yield stress to F_{cr}
F_{yt} = tension flange yield stress
F_{cr} = critical compression flange stress calculated as follows:

Limit state	Range of slenderness	F_{cr} (ksi)
Flange local buckling	$\dfrac{b_f}{2t_f} \leq 0.38\sqrt{\dfrac{E}{F_{yf}}}$	F_{yf}
	$0.38\sqrt{\dfrac{E}{F_{yf}}} < \dfrac{b_f}{2t_f} \leq 1.35\sqrt{\dfrac{E}{F_{yf}/k_c}}$	$F_{yf}\left[1 - \dfrac{1}{2}\left(\dfrac{(b_f/2t_f) - 0.38\sqrt{E/F_{yf}}}{1.35\sqrt{E/(F_{yf}/k_c)} - 0.38\sqrt{E/F_{yf}}}\right)\right] \leq F_{yf}$
	$\dfrac{b_f}{2t_f} > 1.35\sqrt{\dfrac{E}{F_{yf}/k_c}}$	$\dfrac{26{,}200k_c}{(b_f/2t_f)^2}$
Lateral torsional buckling	$\dfrac{L_b}{r_T} \leq 1.76\sqrt{\dfrac{E}{F_{yf}}}$	F_{yf}
	$1.76\sqrt{\dfrac{E}{F_{yf}}} < \dfrac{L_b}{r_T} \leq 4.44\sqrt{\dfrac{E}{F_{yf}}}$	$C_b F_{yf}\left[1 - \dfrac{1}{2}\left(\dfrac{(L_b/r_T) - 1.76\sqrt{E/F_{yf}}}{4.44\sqrt{E/F_{yf}} - 1.76\sqrt{E/F_{yf}}}\right)\right] \leq F_{yf}$
	$\dfrac{L_b}{r_T} > 4.44\sqrt{\dfrac{E}{F_{yf}}}$	$\dfrac{286{,}000C_b}{(L_b/r_T)^2}$

$k_c = 4/\sqrt{(h/t_w)}, \ 0.35 \leq k_c \leq 0.763$

b_f = compression flange width

t_f = compression flange thickness

L_b = lateral unbraced length of the girder

$r_T = \sqrt{[(t_f b_f^3/12 + h_c t_w^3/72)/(b_f t_f + h_c t_w/6)]}$

h_c = twice the distance from the neutral axis to the inside face of the compression flange less the fillet

t_w = web thickness

F_{yf} = yield stress of compression flange (ksi)

C_b = Bending coefficient (see Section 4.5)

F_{cr} must be calculated for both flange local buckling and lateral torsional buckling. The smaller value of F_{cr} is used in Equation 4.93.

The plate girder bending strength reduction factor R_{PG} is a factor to account for the nonlinear flexural stress distribution along the depth of the girder. The hybrid girder factor is a reduction factor to account for the lower yield strength of the web when the nominal moment capacity is computed assuming a homogeneous section made entirely of the higher yield stress of the flange.

4.10.1.2.2 Shear Strength Criterion

Plate girders can be designed with or without the consideration of tension field action. If tension field action is considered, intermediate web stiffeners must be provided and spaced at a distance, a, such that a/h is smaller than 3 or $[260/(h/t_w)]^2$, whichever is smaller. Also, one must check the flexure–shear interaction of Equation 4.96, if appropriate. Consideration of tension field action is not allowed if

- The panel is an end panel
- The plate girder is a hybrid girder
- The plate girder is a web tapered girder
- a/h exceeds 3 or $[260/(h/t_w)]^2$, whichever is smaller

The design shear strength, $\phi_v V_n$, of a plate girder is determined as follows:

If tension field action is not considered: $\phi_v V_n$ are the same as those for beams as given in Equations 4.49 to 4.51.

If tension field action is considered and $h/t_w \leq 1.10\sqrt{(k_v E/F_{yw})}$:

$$\phi_v V_n = 0.90[0.60 A_w F_{yw}] \qquad (4.94)$$

and if $h/t_w > 1.10\sqrt{(k_v E/F_{yw})}$:

$$\phi_v V_n = 0.90\left[0.60 A_w F_{yw} \left(C_v + \frac{1 - C_v}{1.15\sqrt{1 + (a/h)^2}} \right) \right] \qquad (4.95)$$

where

$k_v = 5 + 5/(a/h)^2$ (k_v shall be taken as 5.0 if a/h exceeds 3.0 or $[260/(h/t_w)]^2$, whichever is smaller)

$A_w = d t_w$ (where d is the section depth and t_w is the web thickness)

F_{yw} = web yield stress

C_v = shear coefficient, calculated as follows:

Range of h/t_w	C_v
$1.10\sqrt{\dfrac{k_v E}{F_{yw}}} \leq \dfrac{h}{t_w} \leq 1.37\sqrt{\dfrac{k_v E}{F_{yw}}}$	$\dfrac{1.10\sqrt{k_v E/F_{yw}}}{h/t_w}$
$\dfrac{h}{t_w} > 1.37\sqrt{\dfrac{k_v E}{F_{yw}}}$	$\dfrac{1.51 k_v E}{(h/t_w)^2 F_{yw}}$

4.10.1.2.3 Flexure–Shear Interaction

Plate girders designed for tension field action must satisfy the flexure–shear interaction criterion in regions where $0.60\phi V_n \leq V_u \leq \phi V_n$ and $0.75\phi M_n \leq M_u \leq \phi M_n$

$$\frac{M_u}{\phi M_n} + 0.625 \frac{V_u}{\phi V_n} \leq 1.375 \tag{4.96}$$

where $\phi = 0.90$.

4.10.1.2.4 Bearing Stiffeners

Bearing stiffeners must be provided for a plate girder at unframed girder ends and at points of concentrated loads where the web yielding or the web crippling criterion is violated (see Section 4.5.1.1.3). Bearing stiffeners shall be provided in pairs and extend from the upper flange to the lower flange of the girder. Denoting b_{st} as the width of one stiffener and t_{st} as its thickness, bearing stiffeners shall be portioned to satisfy the following limit states:

For the limit state of local buckling:

$$\frac{b_{st}}{t_{st}} \leq 0.56\sqrt{\frac{E}{F_y}} \tag{4.97}$$

For the limit state of compression: The design compressive strength, $\phi_c P_n$, must exceed the required compressive force acting on the stiffeners. $\phi_c P_n$ is to be determined based on an effective length factor K of 0.75 and an effective area, A_{eff}, equal to the area of the bearing stiffeners plus portion of the web. For end bearing, this effective area is equal to $2(b_{st}t_{st}) + 12t_w^2$; and for interior bearing, this effective area is equal to $2(b_{st}t_{st}) + 25t_w^2$, where t_w is the web thickness. The slenderness parameter, λ_c, is to be calculated using a radius of gyration, $r = \sqrt{(I_{st}/A_{eff})}$, where $I_{st} = t_{st}(2b_{st} + t_w)^3/12$.

For the limit state of bearing: The bearing strength, ϕR_n, must exceed the required compression force acting on the stiffeners. ϕR_n is given by

$$\phi R_n \geq 0.75[1.8 F_y A_{pb}] \tag{4.98}$$

where F_y is the yield stress and A_{pb} is the bearing area.

4.10.1.2.5 Intermediate Stiffeners

Intermediate stiffeners shall be provided if

- The shear strength capacity is calculated based on tension field action.
- The shear criterion is violated (i.e., when the V_u exceeds $\phi_v V_n$).
- The web slenderness h/t_w exceeds $2.45(\sqrt{E/F_{yw}})$.

Intermediate stiffeners can be provided in pairs or on one side of the web only in the form of plates or angles. They should be welded to the compression flange and the web but they may be stopped short of the tension flange. The following requirements apply to the design of intermediate stiffeners:

Local buckling: The width–thickness ratio of the stiffener must be proportioned so that Equation 4.97 is satisfied to prevent failure by local buckling.

Stiffener area: The cross-section area of the stiffener must satisfy the following criterion:

$$A_{st} \geq \frac{F_{yw}}{F_y}\left[0.15 Dht_w(1 - C_v)\frac{V_u}{\phi_v V_n} - 18t_w^2\right] \geq 0 \tag{4.99}$$

where F_y is the yield stress of stiffeners, $D = 1.0$ for stiffeners in pairs, $D = 1.8$ for single angle stiffeners, and $D = 2.4$ for single plate stiffeners.

The other terms in Equation 4.99 are defined as before in Equations 4.94 and 4.95.

Stiffener moment of inertia: The moment of inertia for stiffener pairs taken about an axis in the web center or for single stiffeners taken about the face of contact with the web plate must satisfy the following criterion:

$$I_{st} \geq a t_w^3 \left[\frac{2.5}{(a/h)^2} - 2 \right] \geq 0.5 a t_w^3 \tag{4.100}$$

Stiffener length: The length of the stiffeners l_{st}, should fall within the range

$$h - 6 t_w < l_{st} < h - 6 t_w \tag{4.101}$$

where h is the clear distance between the flanges less the widths of the flange-to-web welds and t_w is the web thickness.

If intermittent welds are used to connect the stiffeners to the girder web, the clear distance between welds shall not exceed $16 t_w$ by not more than 10 in. (25.4 cm). If bolts are used, their spacing shall not exceed 12 in. (30.5 cm).

Stiffener spacing: The spacing of the stiffeners, a, shall be determined from the shear criterion $\phi_v V_n \geq V_u$. This spacing shall not exceed the smaller of $3h$ and $[260/(h/t_w)]^2 h$.

EXAMPLE 4.7

Using LRFD, design the cross-section of an I-shaped plate girder shown in Figure 4.12a to support a factored moment M_u of 4600 kip ft (6240 kN m); dead weight of the girder is included. The girder is a 60-ft (18.3-m) long simply supported girder. It is laterally supported at every 20 ft (6.1 m) interval. Use A36 steel.

Proportion of the girder web: Ordinarily, the overall depth to span ratio d/L of a building girder is in the range $\frac{1}{12}$ to $\frac{1}{10}$. So, let us try $h = 70$ in.

Also, because h/t_w of a plate girder is normally in the range $5.70\sqrt{(E/F_{yf})}$ to $11.7\sqrt{(E/F_{yf})}$, using $E = 29,000$ ksi and $F_{yf} = 36$ ksi, let us try $t_w = \frac{5}{16}$ in.

Proportion of the girder flanges: For a preliminary design, the required area of the flange can be determined using the flange area method

$$A_f \approx \frac{M_u}{F_y h} = \frac{4600 \text{ kip ft} \times 12 \text{ in./ft}}{(36 \text{ ksi})(70 \text{ in.})} = 21.7 \text{ in.}^2$$

So, let $b_f = 20$ in. and $t_f = 1\frac{1}{8}$ in. giving $A_f = 22.5$ in.^{2}
Determine the design flexural strength $\phi_b M_n$ of the girder:
Calculate I_x:

$$\begin{aligned} I_x &= \sum [I_i + A_i y_i^2] \\ &= [8,932 + (21.88)(0)^2] + 2[2.37 + (22.5)(35.56)^2] \\ &= 65,840 \text{ in.}^4 \end{aligned}$$

Calculate S_{xt}, S_{xc}:

$$S_{xt} = S_{xc} = \frac{I_x}{c_t} = \frac{I_x}{c_c} = \frac{65,840}{35 + 1.125} = 1,823 \text{ in.}^3$$

Calculate r_T: Refer to Figure 4.12b,

$$r_T = \sqrt{\frac{I_T}{A_f + (1/6)A_w}} = \sqrt{\frac{(1.125)(20)^3/12 + (11.667)(5/16)^3/12}{22.5 + (1/6)(21.88)}} = 5.36 \text{ in.}$$

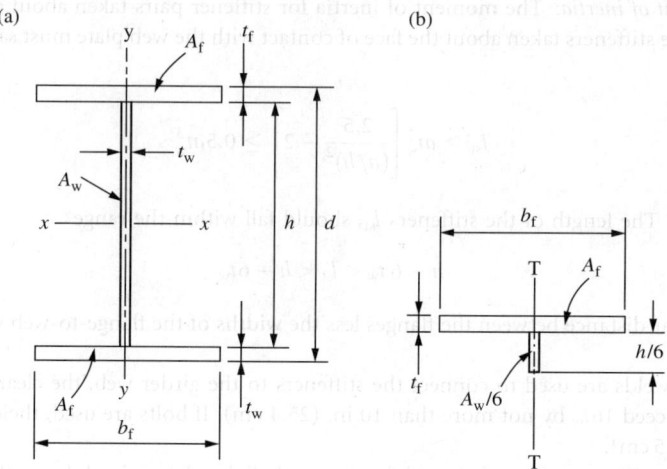

FIGURE 4.12   Design of a plate girder cross-section: (a) plate girder nomenclature and (b) calculation of r_T.

Calculate F_{cr}:

For flange local buckling

$$\left[\frac{b_f}{2t_f} = \frac{20}{2(1.125)} = 8.89\right] < \left[0.38\sqrt{\frac{E}{F_{yf}}} = 10.8\right], \quad \text{so } F_{cr} = F_{yf} = 36 \text{ ksi}$$

For lateral torsional buckling

$$\left[\frac{L_b}{r_T} = \frac{20 \times 12}{5.36} = 44.8\right] < \left[1.76\sqrt{\frac{E}{F_{yf}}} = 50\right], \quad \text{so } F_{cr} = F_{yf} = 36 \text{ ksi}$$

Calculate R_{PG}:

$$R_{PG} = 1 - \frac{a_r(h_c/t_w - 5.70\sqrt{(E/F_{cr})})}{(1{,}200 + 300a_r)} = 1 - \frac{0.972[70/(5/16) - 5.70\sqrt{(29{,}000/36)}]}{[1{,}200 + 300(0.972)]} = 0.96$$

Calculate $\phi_b M_n$:

$$\phi_b M_n = \text{smaller of} \begin{cases} 0.90 S_{xt} R_e F_{yt} & = (0.90)(1{,}823)(1)(36) = 59{,}065 \text{ kip in.} \\ 0.90 S_x R_{PG} R_e F_{cr} & = (0.90)(1{,}823)(0.96)(1)(36) = 56{,}700 \text{ kip in.} \end{cases}$$

$$= 56{,}700 \text{ kip in.}$$
$$= 4{,}725 \text{ kip ft.}$$

Since $[\phi_b M_n = 4{,}725$ kip ft$] > [M_u = 4{,}600$ kip ft$]$, the cross-section is acceptable. Use web plate $\frac{5}{16}$ in. × 70 in. and two flange plates $1\frac{1}{8}$ in. × 20 in. for the girder cross-section.

EXAMPLE 4.8

Design bearing stiffeners for the plate girder of the preceding example for a factored end reaction of 260 kip.

Since the girder end is unframed, bearing stiffeners are required at the supports. The size of the stiffeners must be selected to ensure that the limit states of local buckling, compression, and bearing are not violated.

Limit state of local buckling: Refer to Figure 4.13, try $b_{st} = 8$ in. To avoid problems with local buckling $b_{st}/2t_{st}$ must not exceed $0.56\sqrt{(E/F_y)} = 15.8$. Therefore, try $t_{st} = \frac{1}{2}$ in. So, $b_{st}/2t_{st} = 8$, which is less than 15.8.

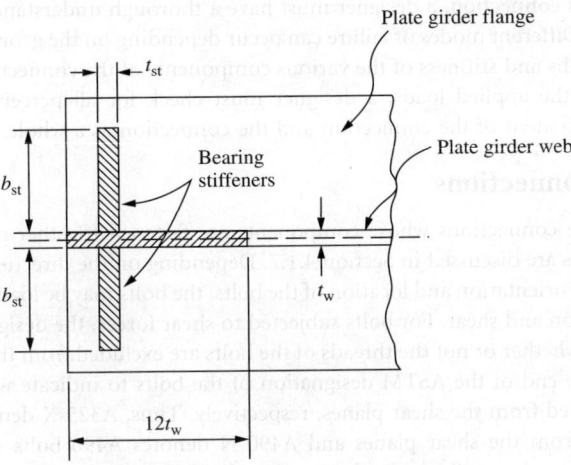

FIGURE 4.13   Design of bearing stiffeners.

Limit state of compression:

$$A_{eff} = 2(b_{st}t_{st}) + 12t_w^2 = 2(8)(0.5) + 12(5/16)^2 = 9.17 \text{ in.}^2$$
$$I_{st} = t_{st}(2b_{st} + t_w)^3/12 = 0.5[2(8) + 5/16]^3/12 = 181 \text{ in.}^4$$
$$r_{st} = \sqrt{(I_{st}/A_{eff})} = \sqrt{(181/9.17)} = 4.44 \text{ in.}$$
$$Kh/r_{st} = 0.75(70)/4.44 = 11.8$$
$$\lambda_c = (Kh/\pi r_{st})\sqrt{(F_y/E)} = (11.8/3.142)\sqrt{(36/29,000)} = 0.132$$

and from Equation 4.17

$$\phi_c P_n = 0.85(0.658^{\lambda^2})F_y A_{st} = 0.85(0.658)^{0.132^2}(36)(9.17) = 279 \text{ kip}$$

Since $\phi_c P_n > 260$ kip, the design is satisfactory for compression.

Limit state of bearing: Assuming there is a $\frac{1}{4}$-in. weld cutout at the corners of the bearing stiffeners at the junction of the stiffeners and the girder flanges, the bearing area for the stiffener pairs is: $A_{pb} = (8 - 0.25)(0.5)(2) = 7.75 \text{ in.}^2$. Substituting this into Equation 4.98, we have $\phi R_n = 0.75(1.8)(36)(7.75) = 377$ kip, which exceeds the factored reaction of 260 kip. So bearing is not a problem.

Use two $\frac{1}{2}$ in. × 8 in. plates for bearing stiffeners.

4.11   Connections

Connections are structural elements used for joining different members of a framework. Connections can be classified according to

- *The type of connecting medium used.* Bolted connections, welded connections, bolted–welded connections, riveted connections.
- *The type of internal forces the connections are expected to transmit.* Shear (semirigid, simple) connections, moment (rigid) connections.
- *The type of structural elements that made up the connections.* Single plate angle connections, double web angle connections, top and seated angle connections, seated beam connections, etc.
- *The type of members the connections are joining.* Beam-to-beam connections (beam splices), column-to-column connections (column splices), beam-to-column connections, hanger connections, etc.

To properly design a connection, a designer must have a thorough understanding of the behavior of the joint under loads. Different modes of failure can occur depending on the geometry of the connection and the relative strengths and stiffness of the various components of the connection. To ensure that the connection can carry the applied loads, a designer must check for all perceivable modes of failure pertinent to each component of the connection and the connection as a whole.

4.11.1 Bolted Connections

Bolted connections are connections whose components are fastened together primarily by bolts. The four basic types of bolts are discussed in Section 4.1.7. Depending on the direction and line of action of the loads relative to the orientation and location of the bolts, the bolts may be loaded in tension, shear, or a combination of tension and shear. For bolts subjected to shear forces, the design shear strength of the bolts also depends on whether or not the threads of the bolts are excluded from the shear planes. A letter X or N is placed at the end of the ASTM designation of the bolts to indicate whether the threads are excluded or not excluded from the shear planes, respectively. Thus, A325-X denotes A325 bolts whose threads are excluded from the shear planes and A490-N denotes A490 bolts whose threads are not excluded from the shear planes. Because of the reduced shear areas for bolts whose threads are not excluded from the shear planes, these bolts have lower design shear strengths than their counterparts whose threads are excluded from the shear planes.

Bolts can be used in both bearing-type connections and slip-critical connections. Bearing-type connections rely on bearing between the bolt shanks and the connecting parts to transmit forces. Some slippage between the connected parts is expected to occur for this type of connections. Slip-critical connections rely on the frictional force that develops between the connecting parts to transmit forces. No slippage between connecting elements is expected for this type of connection. Slip-critical connections are used for structures designed for vibratory or dynamic loads such as bridges, industrial buildings, and buildings in regions of high seismicity. Bolts used in slip-critical connections are denoted by the letter F after their ASTM designation, for example A325-F, A490-F.

Bolt holes. Holes made in the connected parts for bolts may be standard sized, oversized, short slotted, or long slotted. Table 4.10 gives the maximum hole dimension for ordinary construction usage.

Standard holes can be used for both bearing-type and slip-critical connections. Oversized holes shall be used only for slip-critical connections and hardened washers shall be installed over these holes in an outer ply. Short-slotted and long-slotted holes can be used for both bearing-type and slip-critical connections provided that when such holes are used for bearing, the direction of slot is transverse to the direction of loading. While oversized and short-slotted holes are allowed in any or all plies of the connection, long-slotted holes are allowed in only one of the connected parts. In addition, if long-slotted

TABLE 4.10 Nominal Hole Dimensions (in.)

Bolt diameter, d (in.)	Standard (dia.)	Oversize (dia.)	Short slot (width × length)	Long slot (width × length)
$\frac{1}{2}$	$\frac{9}{16}$	$\frac{5}{8}$	$\frac{9}{16} \times \frac{11}{16}$	$\frac{9}{16} \times 1\frac{1}{4}$
$\frac{5}{8}$	$\frac{11}{16}$	$\frac{13}{16}$	$\frac{11}{16} \times \frac{7}{8}$	$\frac{11}{16} \times 1\frac{9}{16}$
$\frac{3}{4}$	$\frac{13}{16}$	$\frac{15}{16}$	$\frac{13}{16} \times 1$	$\frac{13}{16} \times 1\frac{7}{8}$
$\frac{7}{8}$	$\frac{15}{16}$	$1\frac{1}{16}$	$\frac{15}{16} \times 1\frac{1}{8}$	$\frac{15}{16} \times 2\frac{3}{16}$
1	$1\frac{1}{16}$	$1\frac{1}{4}$	$1\frac{1}{16} \times 1\frac{5}{16}$	$1\frac{1}{16} \times 2\frac{1}{2}$
$\geq 1\frac{1}{8}$	$d + \frac{1}{16}$	$d + \frac{5}{16}$	$\left(d + \frac{1}{16}\right) \times \left(d + \frac{3}{8}\right)$	$\left(d + \frac{1}{16}\right) \times (2.5d)$

Note: 1 in. = 25.4 mm.

holes are used in an outer ply, plate washers, or a continuous bar with standard holes having a size sufficient to cover the slot shall be provided.

4.11.1.1 Bolts loaded in tension

If a tensile force is applied to the connection such that the direction of load is parallel to the longitudinal axes of the bolts, the bolts will be subjected to tension. The following conditions must be satisfied for bolts under tensile stresses.

Allowable stress design:

$$f_t \leq F_t \tag{4.102}$$

where f_t is the computed tensile stress in the bolt and F_t is the allowable tensile stress in bolt (see Table 4.11).

Load and resistance factor design:

$$\phi_t F_t \geq f_t \tag{4.103}$$

where $\phi_t = 0.75$, f_t is the tensile stress produced by factored loads (ksi), and F_t is the nominal tensile strength given in Table 4.11.

4.11.1.2 Bolts Loaded in Shear

When the direction of load is perpendicular to the longitudinal axes of the bolts, the bolts will be subjected to shear. The conditions that need to be satisfied for bolts under shear stresses are as follows:

Allowable stress design: For bearing-type and slip-critical connections, the condition is

$$f_v \leq F_v \tag{4.104}$$

where f_v is the computed shear stress in the bolt (ksi) and F_v is the allowable shear stress in bolt (see Table 4.12).

Load and resistance factor design: For bearing-type connections designed at factored loads, and for slip-critical connections designed at service loads, the condition is

$$\phi_v F_v \geq f_v \tag{4.105}$$

where $\phi_v = 0.75$ (for bearing-type connections design for factored loads), $\phi_v = 1.00$ (for slip-critical connections designed at service loads), f_v is the shear stress produced by factored loads (for bearing-type connections) or produced by service loads (for slip-critical connections), (ksi), and F_v is the

TABLE 4.11 F_t of Bolts (ksi)

	ASD		LRFD	
Bolt type	F_t, ksi (static loading)	F_t, ksi (fatigue loading)	F_t, ksi (static loading)	F_t, ksi (fatigue loading)
A307	20	Not allowed	45	Not allowed
A325	44	If $N \leq 20,000$: $F_t =$ same as for	90	F_t is not defined, but the
A490	54	static loading	113	condition $F_t < F_{SR}$ needs
		If $20,000 < N \leq 500,000$:		to be satisfied
		$F_t = 40$ (A325) $= 49$ (A490)		where $f_t =$ tensile stress caused
		If $N > 500,000$:		by service loads calculated
		$F_t = 31$ (A325) $= 38$ (A490)		using a net tensile area
		where $N =$ number of stress		given by
		range fluctuations in design		$A_t = \dfrac{\pi}{4}\left(d_b - \dfrac{0.9743}{n}\right)^2$
		life $F_u =$ minimum specified		F_{SR} is the design stress range
		tensile strength, ksi		given by
				$F_{SR} = \left(\dfrac{3.9 \times 10^8}{N}\right)^{1/3} \geq 7$ ksi
				in which $d_b =$ nominal bolt
				diameter, $n =$ threads per inch,
				$N =$ number of stress range
				fluctuations in the design life

Note: 1 ksi = 6.895 MPa.

TABLE 4.12 F_v or F_n of Bolts (ksi)

Bolt type	F_v (for ASD)	F_v (for LRFD)
A307	10.0[a] (regardless of whether or not threads are excluded from shear planes)	24.0[a] (regardless of whether or not threads are excluded from shear planes)
A325-N	21.0[a]	48.0[a]
A325-X	30.0[a]	60.0[a]
A325-F[b]	17.0 (for standard size holes) 15.0 (for oversized and short-slotted holes) 12.0 (for long-slotted holes when direction of load is transverse to the slots) 10.0 (for long-slotted holes when direction of load is parallel to the slots)	17.0 (for standard size holes) 15.0 (for oversized and short-slotted holes) 12.0 (for long-slotted holes when direction of load is transverse to the slots) 10.0 (for long-slotted holes when direction of load is parallel to the slots)
A490-N	28.0[a]	60.0[a]
A490-X	40.0[a]	75.0[a]
A490-F[b]	21.0 (for standard size holes) 18.0 (for oversized and short-slotted holes) 15.0 (for long-slotted holes when direction of load is transverse to the slots) 13.0 (for long-slotted holes when direction of load is parallel to the slots)	21.0 (for standard size holes) 18.0 (for oversized and short-slotted holes) 15.0 (for long-slotted holes when direction of load is transverse to the slots) 13.0 (for long-slotted holes when direction of load is parallel to the slots)

[a] Tabulated values shall be reduced by 20% if the bolts are used to splice tension members having a fastener pattern whose length, measured parallel to the line of action of the force, exceeds 50 in.
[b] Tabulated values are applicable only to Class A surface, that is, unpainted clean mill surface and blast cleaned surface with Class A coatings (with slip coefficient = 0.33). For design strengths with other coatings, see "Load and Resistance Factor Design Specification to Structural Joints Using ASTM A325 or A490 Bolts" (RCSC 2000).
  Note: 1 ksi = 6.895 MPa.

nominal shear strength given in Table 4.12.

For slip-critical connections designed at factored loads, the condition is

$$\phi r_{str} \geq r_u \tag{4.106}$$

where

ϕ = 1.0 (for standard holes)
  = 0.85 (for oversized and short-slotted holes)
  = 0.70 (for long-slotted holes transverse to the direction of load)
  = 0.60 (for long-slotted holes parallel to the direction of load)
r_{str} = design slip resistance per bolt = $1.13 \mu T_b N_s$
  μ = 0.33 (for Class A surfaces, i.e., unpainted clean mill surfaces or blast-cleaned surfaces with Class A coatings)
  = 0.50 (for Class B surfaces, i.e., unpainted blast-cleaned surfaces or blast-cleaned surfaces with Class B coatings)
  = 0.35 (for Class C surfaces, i.e., hot-dip galvanized and roughened surfaces)
  T_b = minimum fastener tension given in Table 4.13
  N_s = number of slip planes
r_u = required force per bolt due to factored loads

4.11.1.3 Bolts Loaded in Combined Tension and Shear

If a tensile force is applied to a connection such that its line of action is at an angle with the longitudinal axes of the bolts, the bolts will be subjected to combined tension and shear. The conditions that need to be satisfied are given below:

Allowable stress design: The conditions are

$$f_v \leq F_v \quad \text{and} \quad f_t \leq F_t \tag{4.107}$$

TABLE 4.13 Minimum Fastener Tension (kip)

Bolt diameter (in.)	A325 bolts	A490 bolts
$\frac{1}{2}$	012	015
$\frac{5}{8}$	019	024
$\frac{3}{4}$	028	035
$\frac{7}{8}$	039	049
1	051	064
$1\frac{1}{8}$	056	080
$1\frac{1}{4}$	071	102
$1\frac{3}{8}$	085	121
$1\frac{1}{2}$	103	148

Note: 1 kip = 4.45 kN.

TABLE 4.14 F_t for Bolts under Combined Tension and Shear (ksi)

	Bearing-type connections			
	ASD		LRFD	
Bolt type	Threads not excluded from the shear plane	Threads excluded from the shear plane	Threads not excluded from the shear plane	Threads excluded from the shear plane
A307	$26 - 1.8f_v \leq 20$	$26 - 1.8f_v \leq 20$	$59 - 2.5f_v \leq 45$	$59 - 2.5f_v \leq 45$
A325	$\sqrt{(44^2 - 4.39f_v^2)}$	$\sqrt{(44^2 - 2.15f_v^2)}$	$117 - 2.5f_v \leq 90$	$117 - 2.0f_v \leq 90$
A490	$\sqrt{(54^2 - 3.75f_v^2)}$	$\sqrt{(54^2 - 1.82f_v^2)}$	$147 - 2.5f_v \leq 113$	$147 - 2.0f_v \leq 113$

Slip-critical connections

For ASD:
Only $f_v \leq F_v$ needs to be checked
where
f_v = computed shear stress in the bolt, ksi
$F_v = [1 - (f_t A_b / T_b)] \times$ (values of F_v given in Table 4.12)
f_t = computed tensile stress in the bolt, ksi
A_b = nominal cross-sectional area of bolt, in.^{2}
T_b = minimum pretension load given in Table 4.13.

For LRFD:
Only $\phi_v F_v \geq f_v$ needs to be checked
where
$\phi_v = 1.0$
f_v = shear stress produced by service load
$F_v = [1 - (T/0.8T_b N_b)] \times$ (values of F_v given in Table 4.12)
T = service tensile force in the bolt, kip
T_b = minimum pretension load given in Table 4.13
N_b = number of bolts carrying the service-load tension T

Note: 1 ksi = 6.895 MPa.

where f_v, F_v are defined in Equation 4.104, f_t is the computed tensile stress in the bolt (ksi), and F_t is the allowable tensile stress given in Table 4.14.

Load and resistance factor design: For bearing-type connections designed at factored loads and slip-critical connections designed at service loads, the conditions are

$$\phi_v F_v \geq f_v \quad \text{and} \quad \phi_t F_t \geq f_t \qquad (4.108)$$

TABLE 4.15   Bearing Capacity

	ASD	LRFD
Conditions	Allowable bearing stress, F_p, ksi	Design bearing strength, ϕR_n, kip
1. For standard, oversized, or short-slotted holes loaded in any direction	$L_e F_u / 2d \leq 1.2 F_u$	$0.75[1.2 L_c t F_u] \leq 0.75[2.4 dt F_u]^a$
2. For long-slotted holes with direction of slot perpendicular to the direction of bearing	$L_e F_u / 2d \leq 1.0 F_u$	$0.75[1.0 L_c t F_u] \leq 0.75[2.0 dt F_u]$
3. If hole deformation at service load is not a design consideration	$L_e F_u / 2d \leq 1.5 F_u$	$0.75[1.5 L_c t F_u] \leq 0.75[3.0 dt F_u]$

[a] This equation is also applicable to long-slotted holes when the direction of slot is parallel to the direction of bearing force.

  Note: L_e is the distance from free edge to center of the bolt; L_c is the clear distance, in the direction of force, between the edge of the hole and the edge of the adjacent hole or edge of the material; d is the nominal bolt diameter; t is the thickness of the connected part; and F_u is the specified minimum tensile strength of the connected part.

where ϕ_v, F_v, f_v are defined in Equation 4.105, ϕ_t is equal to 0.75, f_t is the tensile stress due to factored loads (for bearing-type connection) or due to service loads (for slip-critical connections) (ksi), and F_t is the nominal tension stress limit for combined tension and shear given in Table 4.14.

For slip-critical connections designed at factored loads, the condition is given in Equation 4.106, except that the design slip resistance per bolt ϕr_{str} shall be multiplied by a reduction factor given by $1 - (T_u/(1.13 T_b N_b))$, where T_u is the factored tensile load on the connection, T_b is given in Table 4.13, and N_b is the number of bolts carrying the factored-load tension T_u.

4.11.1.4 Bearing Strength at Fastener Holes

Connections designed on the basis of bearing rely on the bearing force developed between the fasteners and the holes to transmit forces and moments. The limit state for bearing must therefore be checked to ensure that bearing failure will not occur. Bearing strength is independent of the type of fastener. This is because the bearing stress is more critical on the parts being connected than on the fastener itself. The AISC specification provisions for bearing strength are based on preventing excessive hole deformation. As a result, bearing capacity is expressed as a function of the type of holes (standard, oversized, slotted), bearing area (bolt diameter times the thickness of the connected parts), bolt spacing, edge distance (L_e), strength of the connected parts (F_u), and the number of fasteners in the direction of the bearing force. Table 4.15 summarizes the expressions and conditions used in ASD and LRFD for calculating the bearing strength of both bearing-type and slip-critical connections.

4.11.1.5 Minimum Fastener Spacing

To ensure safety, efficiency, and to maintain clearances between bolt nuts as well as to provide room for wrench sockets, the fastener spacing, s, should not be less than $3d$ where d is the nominal fastener diameter.

4.11.1.6 Minimum Edge Distance

To prevent excessive deformation and shear rupture at the edge of the connected part, a minimum edge distance L_e must be provided in accordance with the values given in Table 4.16 for standard holes. For oversized and slotted hole, the values shown must be incremented by C_2 given in Table 4.17.

4.11.1.7 Maximum Fastener Spacing

A limit is placed on the maximum value for the spacing between adjacent fasteners to prevent the possibility of gaps forming or buckling from occurring in between fasteners when the load to be transmitted by the connection is compressive. The maximum fastener spacing measured in the direction of the force is given as follows:

For painted members or unpainted members not subject to corrosion: smaller of $24t$ where t is the thickness of the thinner plate and 12 in. (305 mm).

TABLE 4.16 Minimum Edge Distance for Standard Holes (in.)

Nominal fastener diameter (in.)	At sheared edges	At rolled edges of plates, shapes, and bars or gas cut edges
$\frac{1}{2}$	$\frac{7}{8}$	$\frac{3}{4}$
$\frac{5}{8}$	$1\frac{1}{8}$	$\frac{7}{8}$
$\frac{3}{4}$	$1\frac{1}{4}$	1
$\frac{7}{8}$	$1\frac{1}{2}$	$1\frac{1}{8}$
1	$1\frac{3}{4}$	$1\frac{1}{4}$
$1\frac{1}{8}$	2	$1\frac{1}{2}$
$1\frac{1}{4}$	$2\frac{1}{4}$	$1\frac{5}{8}$
Over $1\frac{1}{4}$	$1\frac{3}{4}$ × Fastener diameter	$1\frac{1}{4}$ × Fastener diameter

Note: 1 in. = 25.4 mm.

TABLE 4.17 Values of Edge Distance Increment, C_2 (in.)

Nominal diameter of fastener (in.)	Oversized holes	Slotted holes		Slot parallel to edge
		Slot transverse to edge		
		Short slot	Long slot[a]	
$\leq\frac{7}{8}$	$\frac{1}{16}$	$\frac{1}{8}$	$3d/4$	0
1	$\frac{1}{8}$	$\frac{1}{8}$		
$\geq1\frac{1}{8}$	$\frac{1}{8}$	$\frac{3}{16}$		

[a] If the length of the slot is less than the maximum shown in Table 4.10, the value shown may be reduced by one-half the difference between the maximum and the actual slot lengths.

Note: 1 in. = 25.4 mm.

For unpainted members of weathering steel subject to atmospheric corrosion: smaller of 14t where t is the thickness of the thinner plate and 7 in. (178 mm).

4.11.1.8 Maximum Edge Distance

A limit is placed on the maximum value for edge distance to prevent prying action from occurring. The maximum edge distance shall not exceed the smaller of 12t where t is the thickness of the connected part and 6 in. (15 cm).

EXAMPLE 4.9

Check the adequacy of the connection shown in Figure 4.4a. The bolts are 1-in. diameter A325-N bolts in standard holes. The connection is a bearing-type connection.

Check bolt capacity: All bolts are subjected to double shear. Therefore, the design shear strength of the bolts will be twice that shown in Table 4.12. Assuming each bolt carries an equal share of the factored applied load, we have from Equation 4.105

$$[\phi_v F_v = 0.75(2 \times 48) = 72 \text{ ksi}] > \left[f_v = \frac{208}{(6)(\pi d^2/4)} = 44.1 \text{ ksi} \right]$$

The shear capacity of the bolt is therefore adequate.

Check bearing capacity of the connected parts: With reference to Table 4.15, it can be seen that condition 1 applies for the present problem. Therefore, we have

$$[\phi R_n = 0.75(1.2 L_c t F_u) = 0.75(1.2)\left(3 - 1\tfrac{1}{8}\right)\left(\tfrac{3}{8}\right)(58) = 36.7 \text{ kip}$$
$$< 0.75(2.4 d t F_u) = 0.75(2.4)(1)\left(\tfrac{3}{8}\right)(58) = 39.2 \text{ kip}$$
$$> [R_u = \frac{208}{6} = 34.7 \text{ kip}]$$

and so bearing is not a problem. Note that bearing on the gusset plate is more critical than bearing on the webs of the channels because the thickness of the gusset plate is less than the combined thickness of the double channels.

Check bolt spacing: The minimum bolt spacing is $3d = 3(1) = 3$ in. The maximum bolt spacing is the smaller of $14t = 14(0.303) = 4.24$ in. or 7 in. The actual spacing is 3 in., which falls within the range of 3 to 4.24 in., so bolt spacing is adequate.

Check edge distance: From Table 4.16, it can be determined that the minimum edge distance is 1.25 in. The maximum edge distance allowed is the smaller of $12t = 12(0.303) = 3.64$ in. or 6 in. The actual edge distance is 3 in., which falls within the range of 1.25 to 3.64 in., so edge distance is adequate.

The connection is therefore adequate.

4.11.1.9 Bolted Hanger-Type Connections

A typical hanger connection is shown in Figure 4.14. In the design of such connections, the designer must take into account the effect of *prying action*. Prying action results when flexural deformation occurs in the tee flange or angle leg of the connection (Figure 4.15). Prying action tends to increase the tensile force, called prying force, in the bolts. To minimize the effect of prying, the fasteners should be placed as close to the tee stem or outstanding angle leg as the wrench clearance will permit [see Tables on Entering and Tightening Clearances in Volume II — Connections of the AISC-LRFD Manual (AISC 2001)]. In addition, the flange and angle thickness should be proportioned so that the full tensile capacities of the bolts can be developed.

Two failure modes can be identified for hanger-type connections: formation of plastic hinges in the tee flange or angle leg at cross-sections 1 and 2, and tensile failure of the bolts when the tensile force including prying action $B_c\ (= T + Q)$ exceeds the tensile capacity of the bolt B. Since the determination of the actual prying force is rather complex, the design equation for the required thickness for the tee flange or angle leg is semiempirical in nature. It is given by

If ASD is used:

$$t_{\text{req'd}} = \sqrt{\frac{8Tb'}{pF_y(1 + \delta\alpha')}} \qquad (4.109)$$

where T is the tensile force per bolt due to service load exclusive of initial tightening and prying force (kip). The other variables are as defined in Equation 4.110, except that B in the equation for α' is defined as the allowable tensile force per bolt. A design is considered satisfactory if the thickness of the tee flange or angle leg t_f exceeds $t_{\text{req'd}}$ and $B > T$.

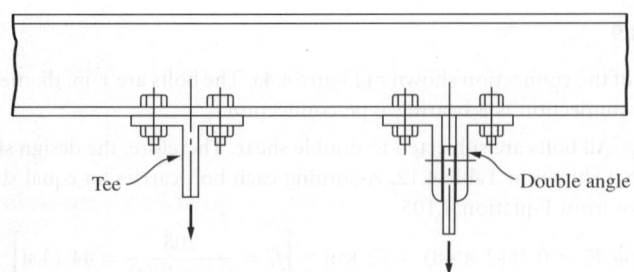

Tee

Double angle

FIGURE 4.14 Hanger connections.

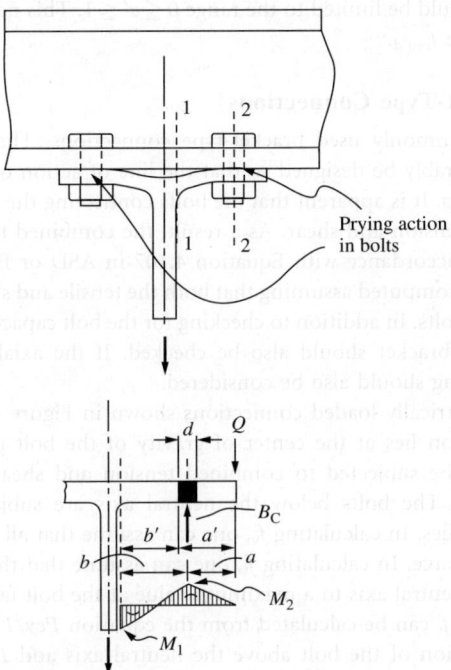

FIGURE 4.15 Prying action in hanger connections.

If LRFD is used:

$$t_{req'd} = \sqrt{\frac{4T_u b'}{\phi_b p F_y (1 + \delta \alpha')}} \tag{4.110}$$

where

$\phi_b = 0.90$

T_u = factored tensile force per bolt exclusive of initial tightening and prying force, kip

p = length of flange tributary to each bolt measured along the longitudinal axis of the tee or double-angle section, in.

δ = ratio of net area at bolt line to gross area at angle leg or stem face = $(p - d')/p$

$\qquad d'$ = diameter of bolt hole = bolt diameter + $\frac{1}{8}$ in.

$\alpha' = \dfrac{(B/T_u - 1)(a'/b')}{\delta[1 - (B/T_u - 1)(a'/b')]} \leq 1$ (if α' is less than zero, use $\alpha' = 1$)

$\qquad B$ = design tensile strength of one bolt = $\phi F_t A_b$ (kip) (ϕF_t is given in Table 4.11 and A_b is the nominal area of the bolt)

$\qquad a' = a + d/2$

$b' = b - d/2$

$\qquad a$ = distance from bolt centerline to edge of tee flange or angle leg but not more than $1.25b$ in.

$\qquad b$ = distance from bolt centerline to face of tee stem or outstanding leg, in.

$\qquad d$ = nominal bolt diameter, in.

A design is considered satisfactory if the thickness of the tee flange or angle leg t_f exceeds $t_{req'd}$ and $B > T_u$.

Note that if t_f is much larger than $t_{req'd}$, the design will be too conservative. In this case α' should be recomputed using the equation

$$\alpha' = \frac{1}{\delta}\left[\frac{4T_u b'}{\phi_b p t_f^2 F_y} - 1\right] \tag{4.111}$$

As before, the value of α' should be limited to the range $0 \le \alpha' \le 1$. This new value of α' is to be used in Equation 4.110 to recalculate $t_{\text{req'd}}$.

4.11.1.10 Bolted Bracket-Type Connections

Figure 4.16 shows three commonly used bracket-type connections. The bracing connection shown in Figure 4.16a should preferably be designed so that the line of action of the force will pass through the centroid of the bolt group. It is apparent that the bolts connecting the bracket to the column flange are subjected to combined tension and shear. As a result, the combined tensile-shear capacities of the bolts should be checked in accordance with Equation 4.107 in ASD or Equation 4.108 in LRFD. For simplicity, f_v and f_t are to be computed assuming that both the tensile and shear components of the force are distributed evenly to all bolts. In addition to checking for the bolt capacities, the bearing capacities of the column flange and the bracket should also be checked. If the axial component of the force is significant, the effect of prying should also be considered.

In the design of the eccentrically loaded connections shown in Figure 4.16b, it is assumed that the neutral axis of the connection lies at the center of gravity of the bolt group. As a result, the bolts above the neutral axis will be subjected to combined tension and shear and so Equation 4.107 or 4.108 needs to be checked. The bolts below the neutral axis are subjected to shear only and so Equation 4.104 or 4.105 applies. In calculating f_v, one can assume that all bolts in the bolt group carry an equal share of the shear force. In calculating f_t, one can assume that the tensile force varies linearly from a value of zero at the neutral axis to a maximum value at the bolt farthest away from the neutral axis. Using this assumption, f_t can be calculated from the equation Pey/I where y is the distance from the neutral axis to the location of the bolt above the neutral axis and $I = \sum A_b y^2$ is the moment of inertia of the bolt areas where A_b is the cross-sectional area of each bolt. The capacity of the connection is determined by the capacities of the bolts and the bearing capacity of the connected parts.

For the eccentrically loaded bracket connection shown in Figure 4.16c, the bolts are subjected to shear. The shear force in each bolt can be obtained by adding vectorally the shear caused by the applied

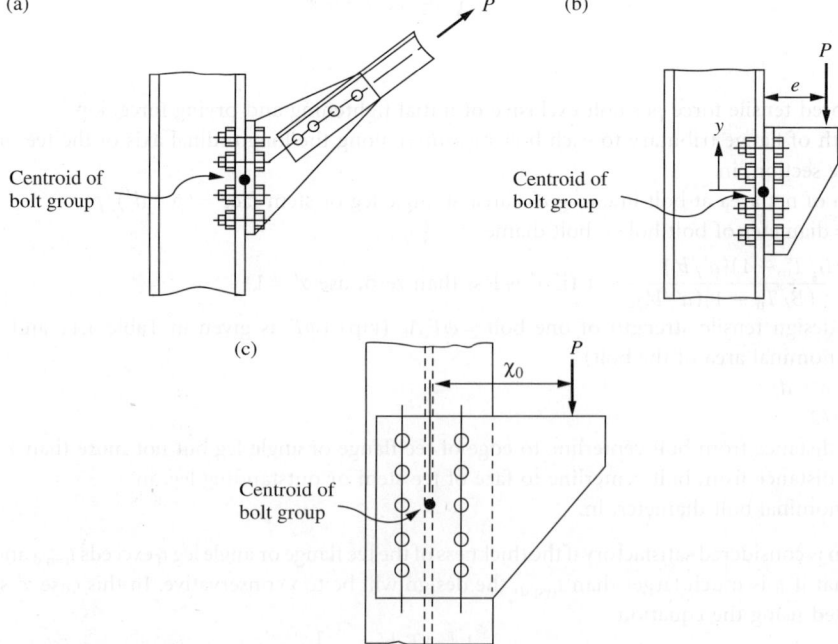

FIGURE 4.16 Bolted bracket-type connections.

load *P* and the moment $P\chi_0$. The design of this type of connections is facilitated by the use of tables contained in the AISC Manuals for Allowable Stress Design and Load and Resistance Factor Design (AISC 1986, 2001).

In addition to checking for bolt shear capacity, one needs to check the bearing and shear rupture capacities of the bracket plate to ensure that failure will not occur in the plate.

4.11.1.11 Bolted Shear Connections

Shear connections are connections designed to resist shear force only. They are used in Type 2 or Type 3 construction in ASD, and Type PR construction in LRFD. These connections are not expected to provide appreciable moment restraint to the connection members. Examples of these connections are shown in Figure 4.17. The framed beam connection shown in Figure 4.17a consists of two web angles that are often

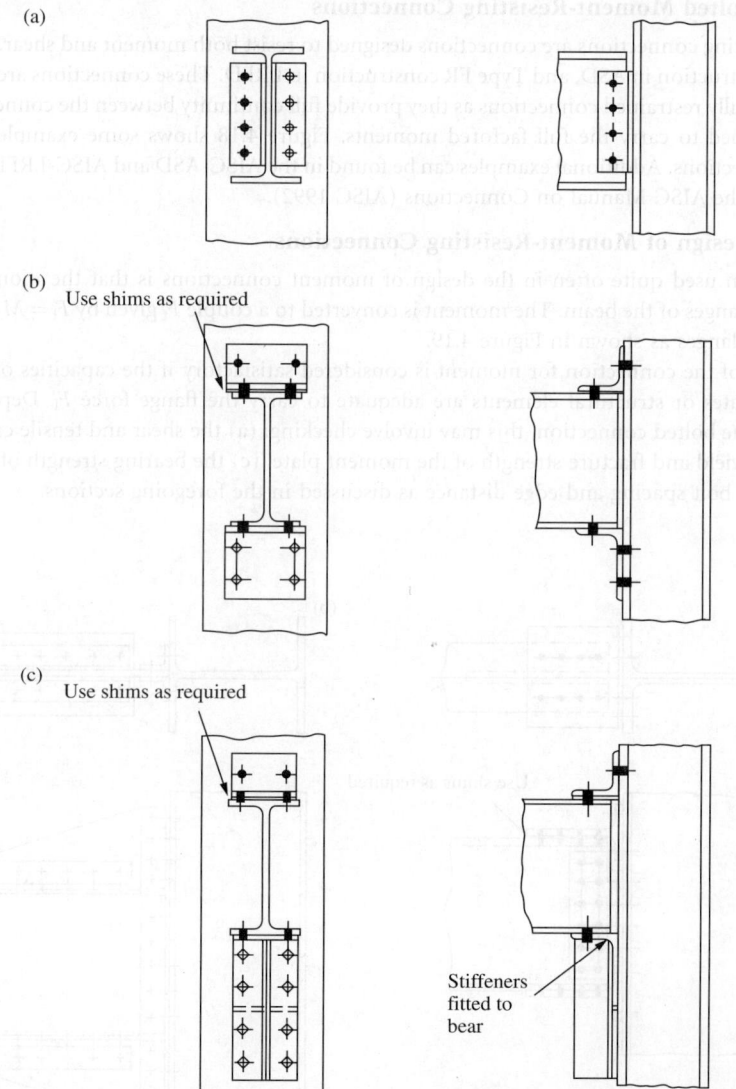

FIGURE 4.17 Bolted shear connections: (a) bolted frame beam connection; (b) bolted seated beam connection; and (c) bolted stiffened seated beam connection.

shop-bolted to the beam web and then field-bolted to the column flange. The seated beam connection shown in Figure 4.17b consists of two flange angles often shop-bolted to the beam flange and field-bolted to the column flange. To enhance the strength and stiffness of the seated beam connection, a stiffened seated beam connection shown in Figure 4.17c is sometimes used to resist large shear force. Shear connections must be designed to sustain appreciable deformation and yielding of the connections is expected. The need for ductility often limits the thickness of the angles that can be used. Most of these connections are designed with angle thickness not exceeding $\frac{5}{8}$ in. (16 mm).

The design of the connections shown in Figure 4.17 is facilitated by the use of design tables contained in the AISC-ASD and AISC-LRFD Manuals. These tables give design loads for the connections with specific dimensions based on the limit states of bolt shear, bearing strength of the connection, bolt bearing with different edge distances, and block shear (for coped beams).

4.11.1.12 Bolted Moment-Resisting Connections

Moment-resisting connections are connections designed to resist both moment and shear. They are used in Type 1 construction in ASD, and Type FR construction in LRFD. These connections are often referred to as rigid or fully restrained connections as they provide full continuity between the connected members and are designed to carry the full factored moments. Figure 4.18 shows some examples of moment-resisting connections. Additional examples can be found in the AISC-ASD and AISC-LRFD Manuals and Chapter 4 of the AISC Manual on Connections (AISC 1992).

4.11.1.13 Design of Moment-Resisting Connections

An assumption used quite often in the design of moment connections is that the moment is carried solely by the flanges of the beam. The moment is converted to a couple F_f given by $F_f = M/(d - t_f)$ acting on the beam flanges as shown in Figure 4.19.

The design of the connection for moment is considered satisfactory if the capacities of the bolts and connecting plates or structural elements are adequate to carry the flange force F_f. Depending on the geometry of the bolted connection, this may involve checking: (a) the shear and tensile capacities of the bolts; (b) the yield and fracture strength of the moment plate; (c) the bearing strength of the connected parts; and (d) bolt spacing and edge distance as discussed in the foregoing sections.

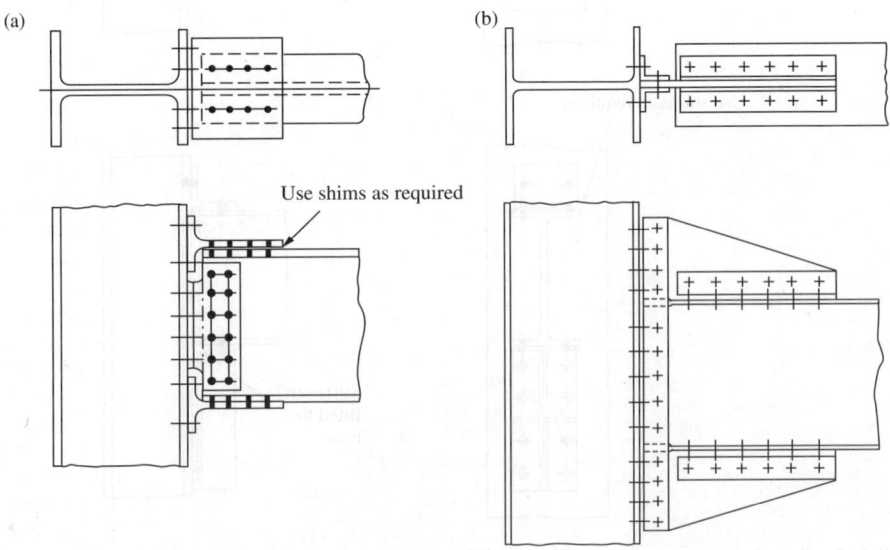

FIGURE 4.18 Bolted moment connections.

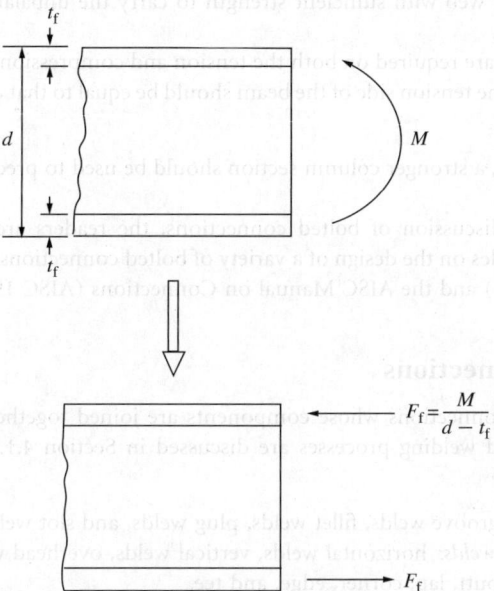

FIGURE 4.19 Flange forces in moment connections.

As for shear, it is common practice to assume that all the shear resistance is provided by the shear plates or angles. The design of the shear plates or angles is governed by the limit states of bolt shear, bearing of the connected parts, and shear rupture.

If the moment to be resisted is large, the flange force may cause bending of the column flange, or local yielding, crippling, or buckling of the column web. To prevent failure due to bending of the column flange or local yielding of the column web (for a tensile F_f) as well as local yielding, crippling, or buckling of the column web (for a compressive F_f), column stiffeners should be provided if any one of the conditions discussed in Section 4.5.1.1.3 is violated.

Following is a set of guidelines for the design of column web stiffeners (AISC 1989, 2001):

1. If local web yielding controls, the area of the stiffeners (provided in pairs) shall be determined based on any excess force beyond that which can be resisted by the web alone. The stiffeners need not extend more than one-half the depth of the column web if the concentrated beam flange force F_f is applied at only one column flange.
2. If web crippling or compression buckling of the web is controlled, the stiffeners shall be designed as axially loaded compression members (see Section 4.4). The stiffeners shall extend the entire depth of the column web.
3. The welds that connect the stiffeners to the column shall be designed to develop the full strength of the stiffeners.

In addition, the following recommendations are given:

1. The width of the stiffener plus one-half of the column web thickness should not be less than one-half the width of the beam flange nor the moment connection plate that applies the force.
2. The stiffener thickness should not be less than one-half the thickness of the beam flange.
3. If only one flange of the column is connected by a moment connection, the length of the stiffener plate does not have to exceed one-half the column depth.
4. If both flanges of the column are connected by moment connections, the stiffener plate should extend through the depth of the column web and welds should be used to connect the stiffener

plate to the column web with sufficient strength to carry the unbalanced moment on opposite sides of the column.

5. If column stiffeners are required on both the tension and compression sides of the beam, the size of the stiffeners on the tension side of the beam should be equal to that on the compression size for ease of construction.

In lieu of stiffener plates, a stronger column section should be used to preclude failure in the column flange and web.

For a more thorough discussion of bolted connections, the readers are referred to the book by Kulak et al. (1987). Examples on the design of a variety of bolted connections can be found in the AISC-LRFD Manual (AISC 2001) and the AISC Manual on Connections (AISC 1992).

4.11.2 Welded Connections

Welded connections are connections whose components are joined together primarily by welds. The four most commonly used welding processes are discussed in Section 4.1.7. Welds can be classified according to

- *The types of welds:* groove welds, fillet welds, plug welds, and slot welds.
- *The positions of the welds:* horizontal welds, vertical welds, overhead welds, and flat welds.
- *The types of joints:* butt, lap, corner, edge, and tee.

Although fillet welds are generally weaker than groove welds, they are used more often because they allow for larger tolerances during erection than groove welds. Plug and slot welds are expensive to make and they do not provide much reliability in transmitting tensile forces perpendicular to the faying surfaces. Furthermore, quality control of such welds is difficult because inspection of the welds is rather arduous. As a result, plug and slot welds are normally used just for stitching different parts of the members together.

4.11.2.1 Welding Symbols

A shorthand notation giving important information on the location, size, length, etc. for the various types of welds was developed by the American Welding Society (AWS 1987) to facilitate the detailing of welds. This system of notation is reproduced in Figure 4.20.

4.11.2.2 Strength of Welds

In ASD, the strength of welds is expressed in terms of allowable stress. In LRFD, the design strength of welds is taken as the smaller of the design strength of the base material ϕF_{BM} (expressed as a function of the yield stress of the material) and the design strength of the weld electrode ϕF_W (expressed as a function of the strength of the electrode F_{EXX}). These allowable stresses and design strengths are summarized in Table 4.18 (ASD 1989; AISC 1999). During design using ASD, the computed stress in the weld shall not exceed its allowable value. During design using LRFD, the design strength of welds should exceed the required strength obtained by dividing the load to be transmitted by the effective area of the welds.

4.11.2.3 Effective Area of Welds

The effective area of groove welds is equal to the product of the width of the part joined and the effective throat thickness. The effective throat thickness of a full-penetration groove weld is taken as the thickness of the thinner part joined. The effective throat thickness of a partial-penetration groove weld is taken as the depth of the chamfer for J, U, bevel, or V (with bevel $\geq 60°$) joints and it is taken as the depth of the chamfer minus $\frac{1}{8}$ in. (3 mm) for bevel or V joints if the bevel is between 45° and 60°. For flare bevel groove welds the effective throat thickness is taken as $5R/16$ and for flare V-groove the effective throat thickness is taken as $R/2$ (or $3R/8$ for GMAW process when $R \geq 1$ in. or 25.4 mm). R is the radius of the bar or bend.

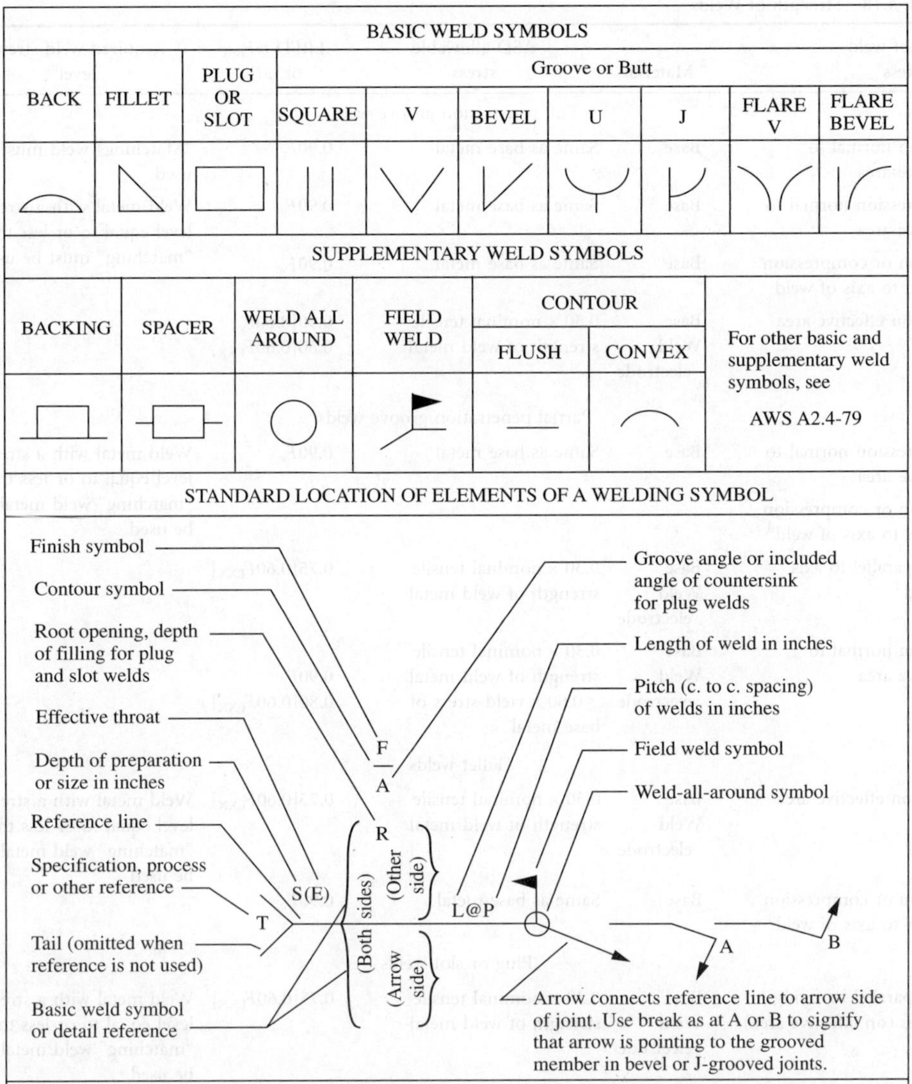

BASIC WELD SYMBOLS									
			Groove or Butt						
BACK	FILLET	PLUG OR SLOT	SQUARE	V	BEVEL	U	J	FLARE V	FLARE BEVEL

SUPPLEMENTARY WELD SYMBOLS						
				CONTOUR		
BACKING	SPACER	WELD ALL AROUND	FIELD WELD	FLUSH	CONVEX	For other basic and supplementary weld symbols, see AWS A2.4-79

STANDARD LOCATION OF ELEMENTS OF A WELDING SYMBOL

Finish symbol

Contour symbol

Root opening, depth of filling for plug and slot welds

Effective throat

Depth of preparation or size in inches

Reference line

Specification, process or other reference

Tail (omitted when reference is not used)

Basic weld symbol or detail reference

Groove angle or included angle of countersink for plug welds

Length of weld in inches

Pitch (c. to c. spacing) of welds in inches

Field weld symbol

Weld-all-around symbol

Arrow connects reference line to arrow side of joint. Use break as at A or B to signify that arrow is pointing to the grooved member in bevel or J-grooved joints.

F, A, R, S(E), T, L@P, (Other side), (Both sides), (Arrow side), A, B

Note:

Size, weld symbol, length of weld and spacing must read in that order from left to right along the reference line. Neither orientation of reference line nor location of the arrow alters this rule.

The perpendicular leg of $\triangle$, $\vee$, $\vdash$, $\vert\hspace{-2pt}\lceil$ weld symbols must be at·left.

Arrow and Other Side welds are of the same size unless otherwise shown. Dimensions of fillet welds must be shown on both the Arrow Side and the Other Side Symbol.

The point of the field weld symbol must point toward the tail.

Symbols apply between abrupt changes in direction of welding unless governed by the "all around" symbol or otherwise dimensioned.

These symbols do not explicitly provide for the case that frequently occurs in structural work, where duplicate material (such as stiffeners) occurs on the far side of a web or gusset plate. The fabricating industry has adopted this convention: that when the billing of the detail material discloses the existence of a member on the far side as well as on the near side, the welding shown for the near side shall be duplicated on the far side.

FIGURE 4.20 Basic weld symbols.

TABLE 4.18 Strength of Welds

Types of weld and stress[a]	Material	ASD allowable stress	LRFD ϕF_{BM} or ϕF_W	Required weld strength level[b,c]
		Full penetration groove weld		
Tension normal to effective area	Base	Same as base metal	$0.90F_y$	"Matching" weld must be used
Compression normal to effective area	Base	Same as base metal	$0.90F_y$	Weld metal with a strength level equal to or less than "matching" must be used
Tension or compression parallel to axis of weld	Base	Same as base metal	$0.90F_y$	
Shear on effective area	Base Weld electrode	$0.30 \times$ nominal tensile strength of weld metal	$0.90[0.60F_y]$ $0.80[0.60F_{EXX}]$	
		Partial penetration groove welds		
Compression normal to effective area	Base	Same as base metal	$0.90F_y$	Weld metal with a strength level equal to or less than "matching" weld metal may be used
Tension or compression parallel to axis of weld[d]				
Shear parallel to axis of weld	Base Weld electrode	$0.30 \times$ nominal tensile strength of weld metal	$0.75[0.60F_{EXX}]$	
Tension normal to effective area	Base Weld electrode	$0.30 \times$ nominal tensile strength of weld metal $\leq 0.60 \times$ yield stress of base metal	$0.90F_y$ $0.80[0.60F_{EXX}]$	
		Fillet welds		
Stress on effective area	Base Weld electrode	$0.30 \times$ nominal tensile strength of weld metal	$0.75[0.60F_{EXX}]$	Weld metal with a strength level equal to or less than "matching" weld metal may be used
Tension or compression parallel to axis of weld[d]	Base	Same as base metal	$0.90F_y$	
		Plug or slot welds		
Shear parallel to faying surfaces (on effective area)	Base Weld electrode	$0.30 \times$ nominal tensile strength of weld metal	$0.75[0.60F_{EXX}]$	Weld metal with a strength level equal to or less than "matching" weld metal may be used

[a] See below for effective area.
[b] See AWS D1.1 for "matching" weld material.
[c] Weld metal one strength level stronger than "matching" weld metal will be permitted.
[d] Fillet welds and partial-penetration groove welds joining component elements of built-up members such as flange-to-web connections may be designed without regard to the tensile or compressive stress in these elements parallel to the axis of the welds.

The effective area of fillet welds is equal to the product of length of the fillets including returns and the effective throat thickness. The effective throat thickness of a fillet weld is the shortest distance from the root of the joint to the face of the diagrammatic weld as shown in Figure 4.21. Thus, for an equal leg fillet weld, the effective throat is given by 0.707 times the leg dimension. For fillet weld made by the SAW process, the effective throat thickness is taken as the leg size (for $\frac{3}{8}$ in. or 9.5 mm and smaller fillet welds) or as the theoretical throat plus 0.11 in. or 3 mm (for fillet weld over $\frac{3}{8}$ in. or 9.5 mm). A larger value for the effective throat thickness is permitted for welds made by the SAW process to account for the inherently superior quality of such welds.

The effective area of plug and slot welds is taken as the nominal cross-sectional area of the hole or slot in the plane of the faying surface.

4.11.2.4 Size and Length Limitations of Welds

To ensure effectiveness, certain size and length limitations are imposed for welds. For partial-penetration groove welds, minimum values for the effective throat thickness are given in the Table 4.19.

For plug welds, the hole diameter shall not be less than the thickness of the part that contains the weld plus $\frac{5}{16}$ in. (8 mm), rounded to the next larger odd $\frac{1}{16}$ in. (or even mm), nor greater than the minimum diameter plus $\frac{1}{8}$ in. (3 mm) or $2\frac{1}{4}$ times the thickness of the weld. The center-to-center spacing of plug welds shall not be less than four times the hole diameter. The thickness of a plug weld in material less than $\frac{5}{8}$ in. (16 mm) thick shall be equal to the thickness of the material. In material over $\frac{5}{8}$ in. (16 mm) thick, the thickness of the weld shall be at least one-half the thickness of the material but not less than $\frac{5}{8}$ in. (16 mm).

For slot welds, the slot length shall not exceed 10 times the thickness of the weld. The slot width shall not be less than the thickness of the part that contains the weld plus $\frac{5}{16}$ in. (8 mm) rounded to the nearest

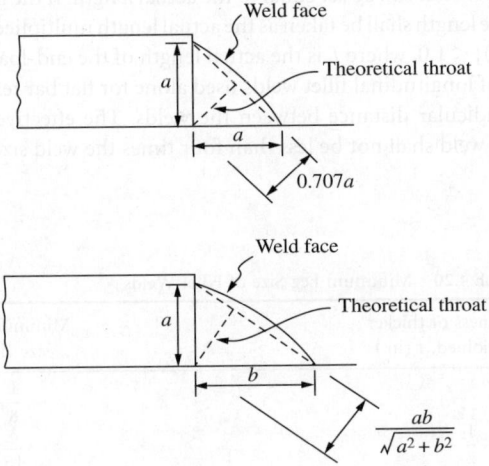

FIGURE 4.21   Effective throat of fillet welds.

TABLE 4.19   Minimum Effective Throat Thickness of Partial-Penetration Groove Welds

Thickness of the thicker part joined, t (in.)	Minimum effective throat thickness (in.)
$t \leq \frac{1}{4}$	$\frac{1}{8}$
$\frac{1}{4} < t \leq \frac{1}{2}$	$\frac{3}{16}$
$\frac{1}{2} < t \leq \frac{3}{4}$	$\frac{1}{4}$
$\frac{3}{4} < t \leq 1\frac{1}{2}$	$\frac{5}{16}$
$1\frac{1}{2} < t \leq 2\frac{1}{4}$	$\frac{3}{8}$
$2\frac{1}{4} < t \leq 6$	$\frac{1}{2}$
>6	$\frac{5}{8}$

Note: 1 in. = 25.4 mm.

larger odd $\frac{1}{16}$ in. (or even mm), nor larger than $2\frac{1}{4}$ times the thickness of the weld. The spacing of lines of slot welds in a direction transverse to their length shall not be less than four times the width of the slot. The center-to-center spacing of two slot welds on any line in the longitudinal direction shall not be less than two times the length of the slot. The thickness of a slot weld in material less than $\frac{5}{8}$ in. (16 mm) thick shall be equal to the thickness of the material. In material over $\frac{5}{8}$ in. (16 mm) thick, the thickness of the weld shall be at least one-half the thickness of the material but not less than $\frac{5}{8}$ in. (16 mm).

For fillet welds, the following size and length limitations apply:

- *Minimum leg size.* The minimum leg size is given in Table 4.20.
- *Maximum leg size.* Along the edge of a connected part less than $\frac{1}{4}$ in. (6 mm) thickness, the maximum leg size is equal to the thickness of the connected part. For thicker parts, the maximum leg size is t minus $\frac{1}{16}$ in. (2 mm), where t is the thickness of the part.
- *Length limitations.* The minimum effective length of a fillet weld is four times its nominal size. If a shorter length is used, the leg size of the weld shall be taken as $\frac{1}{4}$ in. (6 mm) its effective length for the purpose of stress computation. The effective length of end-loaded fillet welds with a length up to 100 times the leg dimension can be set equal to the actual length. If the length exceeds 100 times the weld size, the effective length shall be taken as the actual length multiplied by a reduction factor given by $[1.2 - 0.002(L/w)] \leq 1.0$, where L is the actual length of the end-loaded fillet weld, and w is the leg size. The length of longitudinal fillet welds used alone for flat bar tension members shall not be less than the perpendicular distance between the welds. The effective length of any segment of an intermittent fillet weld shall not be less than four times the weld size or $1\frac{1}{2}$ in. (38 mm).

TABLE 4.20   Minimum Leg Size of Fillet Welds

Thickness of thicker part joined, t (in.)	Minimum leg size (in.)
$t \leq \dfrac{1}{4}$	$\dfrac{1}{8}$
$\dfrac{1}{4} < t \leq \dfrac{1}{2}$	$\dfrac{3}{16}$
$\dfrac{1}{2} < t \leq \dfrac{3}{4}$	$\dfrac{1}{4}$
$t > \dfrac{3}{4}$	$\dfrac{5}{16}$

Note: 1 in. = 25.4 mm.

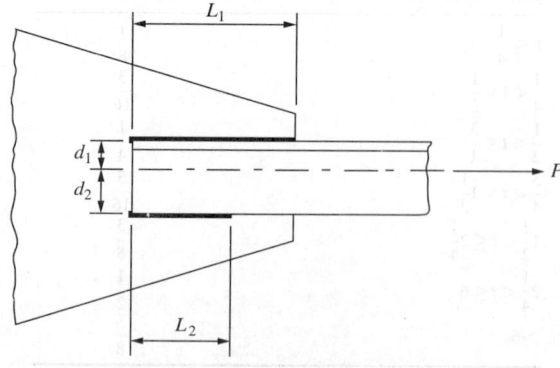

FIGURE 4.22   An eccentrically loaded welded tension connection.

4.11.2.5 Welded Connections for Tension Members

Figure 4.22 shows a tension angle member connected to a gusset plate by fillet welds. The applied tensile force P is assumed to act along the center of gravity of the angle. To avoid eccentricity, the lengths of the two fillet welds must be proportioned so that their resultant will act along the center of gravity of the angle. For example, if LRFD is used, the following equilibrium equations can be written:

Summing force along the axis of the angle:

$$(\phi F_M) t_{eff} L_1 + (\phi F_M) t_{eff} L_2 = P_u \qquad (4.112)$$

Summing moment about the center of gravity of the angle:

$$(\phi F_M) t_{eff} L_1 d_1 = (\phi F_M) t_{eff} L_2 d_2 \qquad (4.113)$$

where P_u is the factored axial force, ϕF_M is the design strength of the welds as given in Table 4.18, t_{eff} is the effective throat thickness, L_1, L_2 are the lengths of the welds, and d_1, d_2 are the transverse distances from the center of gravity of the angle to the welds. The two equations can be used to solve for L_1 and L_2.

4.11.2.6 Welded Bracket-Type Connections

A typical welded bracket connection is shown in Figure 4.23. Because the load is eccentric with respect to the center of gravity of the weld group, the connection is subjected to both moment and shear. The welds must be designed to resist the combined effect of direct shear for the applied load and any additional shear from the induced moment. The design of a welded bracket connection is facilitated by the use of design tables in the AISC-ASD and AISC-LRFD Manuals. In both ASD and LRFD, the load capacity for the connection is given by

$$P = C C_1 D l \qquad (4.114)$$

where P is the allowable load (in ASD), or factored load, P_u (in LRFD), kip; l is the length of the vertical weld, in.; D is the number of sixteenths of an inch in fillet weld size; C are the coefficients tabulated in the AISC-ASD and AISC-LRFD Manuals. In the tables, values of C for a variety of weld geometries and dimensions are given; C_1 are the coefficients for the electrode used (see following table).

Electrode		E60	E70	E80	E90	E100	E110
ASD	F_v (ksi)	18	21	24	27	30	33
	C_1	0.857	1.0	1.14	1.29	1.43	1.57
LRFD	F_{EXX} (ksi)	60	70	80	90	100	110
	C_1	0.857	1.0	1.03	1.16	1.21	1.34

4.11.2.7 Welded Connections With Welds Subjected to Combined Shear and Flexure

Figure 4.24 shows a welded framed connection and a welded seated connection. The welds for these connections are subjected to combined shear and flexure. For purpose of design, it is common practice to assume that the shear force per unit length, R_S, acting on the welds is a constant and is given by

$$R_S = \frac{P}{2l} \qquad (4.115)$$

where P is the allowable load (in ASD), or factored load, P_u (in LRFD), and l is the length of the vertical weld. In addition to shear, the welds are subjected to flexure as a result of load eccentricity. There is no general agreement on how the flexure stress should be distributed on the welds. One approach is to assume that the stress distribution is linear with half the weld subjected to tensile flexure stress and the other half is

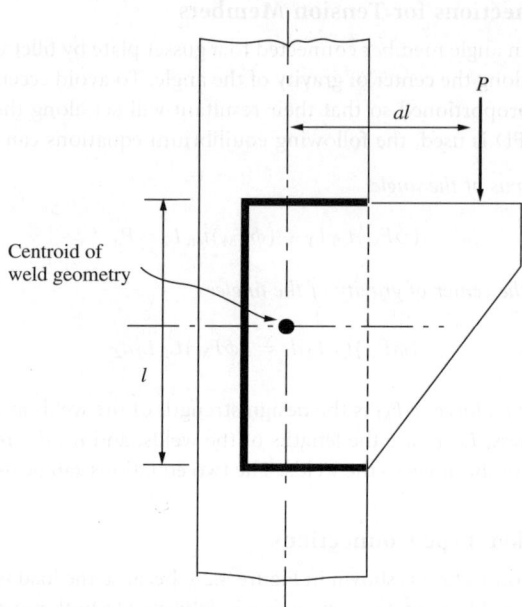

FIGURE 4.23 An eccentrically loaded welded bracket connection.

subjected to compressive flexure stress. Based on this stress distribution and ignoring the returns, the flexure tension force per unit length of weld, R_F, acting at the top of the weld can be written as

$$R_F = \frac{Mc}{I} = \frac{Pe(l/2)}{2l^3/12} = \frac{3Pe}{l^2}$$ (4.116)

where e is the load eccentricity.

The resultant force per unit length acting on the weld, R, is then

$$R = \sqrt{R_S^2 + R_F^2}$$ (4.117)

For a satisfactory design, the value R/t_{eff}, where t_{eff} is the effective throat thickness of the weld should not exceed the allowable values or design strengths given in Table 4.18.

4.11.2.8 Welded Shear Connections

Figure 4.25 shows three commonly used welded shear connections: a framed beam connection, a seated beam connection, and a stiffened seated beam connection. These connections can be designed by using the information presented in the earlier sections on welds subjected to eccentric shear and welds subjected to combined tension and flexure. For example, the welds that connect the angles to the beam web in the framed beam connection can be considered as eccentrically loaded welds and so Equation 4.114 can be used for their design. The welds that connect the angles to the column flange can be considered as welds subjected to combined tension and flexure and so Equation 4.117 can be used for their design. Like bolted shear connections, welded shear connections are expected to exhibit appreciable ductility and so the use of angles with thickness in excess of $\frac{5}{8}$ in. should be avoided. To prevent shear rupture failure, the shear rupture strength of the critically loaded connected parts should be checked.

To facilitate the design of these connections, the AISC-ASD and AISC-LRFD Manuals provide design tables by which the weld capacities and shear rupture strengths for different connection dimensions can be checked readily.

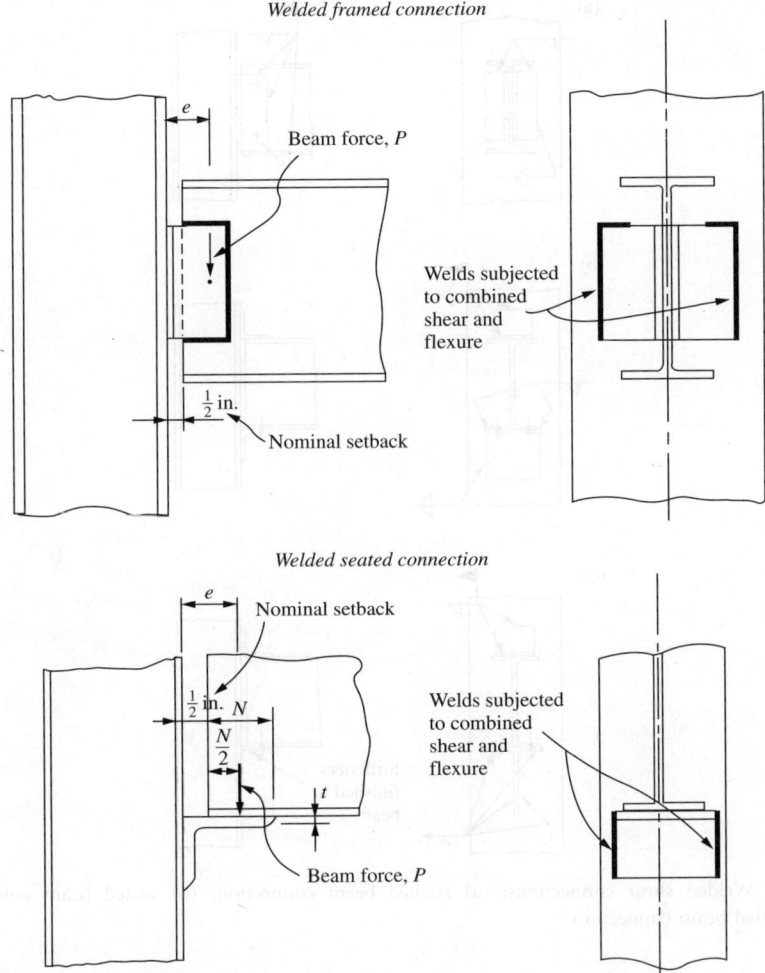

Welded framed connection

e

Beam force, *P*

Welds subjected
to combined
shear and
flexure

$\frac{1}{2}$ in.

Nominal setback

Welded seated connection

e

Nominal setback

$\frac{1}{2}$ in. *N*

$\frac{N}{2}$

t

Welds subjected
to combined
shear and
flexure

Beam force, *P*

FIGURE 4.24 Welds subjected to combined shear and flexure.

4.11.2.9 Welded Moment-Resisting Connections

Welded moment-resisting connections (Figure 4.26), like bolted moment-resisting connections, must be designed to carry both moment and shear. To simplify the design procedure, it is customary to assume that the moment, to be represented by a couple F_f as shown in Figure 4.19, is to be carried by the beam flanges and that the shear is to be carried by the beam web. The connected parts (the moment plates, welds, etc.) are then designed to resist the forces F_f and shear. Depending on the geometry of the welded connection, this may include checking: (a) the yield and fracture strength of the moment plate; (b) the shear and tensile capacity of the welds; and (c) the shear rupture strength of the shear plate.

If the column to which the connection is attached is weak, the designer should consider the use of column stiffeners to prevent failure of the column flange and web due to bending, yielding crippling, or buckling (see section on Design of moment-resisting connections).

Examples on the design of a variety of welded shear and moment-resisting connections can be found in the AISC Manual on Connections (AISC 1992) and the AISC-LRFD Manual (AISC 2001).

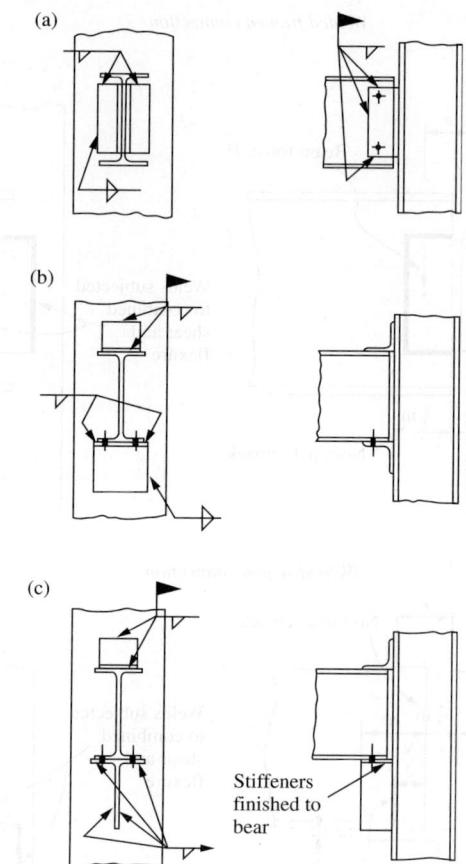

FIGURE 4.25 Welded shear connections: (a) framed beam connection; (b) seated beam connection; and (c) stiffened seated beam connection.

4.11.3 Shop Welded–Field Bolted Connections

A large percentage of connections used for construction are shop welded and field bolted types. These connections are usually more cost effective than fully welded connections and their strength and ductility characteristics often rival those of fully welded connections. Figure 4.27 shows some of these connections. The design of shop welded–field bolted connections is also covered in the AISC Manual on Connections and the AISC-LRFD Manual. In general, the following should be checked: (a) shear/tensile capacities of the bolts and/or welds; (b) bearing strength of the connected parts; (c) yield and/or fracture strength of the moment plate; and (d) shear rupture strength of the shear plate. Also, as for any other type of moment connections, column stiffeners shall be provided if any one of the following criteria — column flange bending, local web yielding, crippling and compression buckling of the column web — is violated.

4.11.4 Beam and Column Splices

Beam and column splices (Figure 4.28) are used to connect beam or column sections of different sizes. They are also used to connect beam or column of the same size if the design calls for extraordinary long span. Splices should be designed for both moment and shear unless it is the intention of the designer to utilize the splices as internal hinges. If splices are used for internal hinges, provisions must be made to ensure that the connections possess adequate ductility to allow for large hinge rotation.

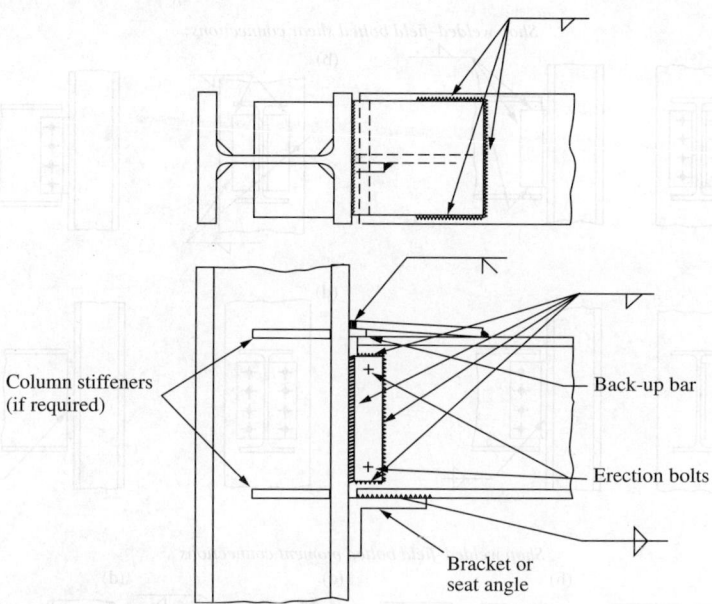

FIGURE 4.26 Welded moment connections.

Splice plates are designed according to their intended functions. Moment splices should be designed to resist the flange force $F_f = M/(d - t_f)$ (Figure 4.19) at the splice location. In particular, the following limit states need to be checked: yielding of gross area of the plate, fracture of net area of the plate (for bolted splices), bearing strengths of connected parts (for bolted splices), shear capacity of bolts (for bolted splices), and weld capacity (for welded splices). Shear splices should be designed to resist the shear forces acting at the locations of the splices. The limit states that are needed to be checked include: shear rupture of the splice plates, shear capacity of bolts under an eccentric load (for bolted splices), bearing capacity of the connected parts (for bolted splices), shear capacity of bolts (for bolted splices), weld capacity under an eccentric load (for welded splices). Design examples of beam and column splices can be found in the AISC Manual of Connections (AISC 1992) and the AISC-LRFD Manuals (AISC 2001).

4.12 Column Base Plates and Beam Bearing Plates (LRFD Approach)

4.12.1 Column Base Plates

Column base plates are steel plates placed at the bottom of columns whose function is to transmit column loads to the concrete pedestal. The design of column base plate involves two major steps:

- Determining the size $N \times B$ of the plate.
- Determining the thickness t_p of the plate.

Generally, the size of the plate is determined based on the limit state of bearing on concrete and the thickness of the plate is determined based on the limit state of plastic bending of critical sections in the plate. Depending on the types of forces (axial force, bending moment, shear force) the plate will be subjected to, the design procedures differ slightly. In all cases, a layer of grout should be placed between the base plate and its support for the purpose of leveling and anchor bolts should be provided to stabilize the column during erection or to prevent uplift for cases involving a large bending moment.

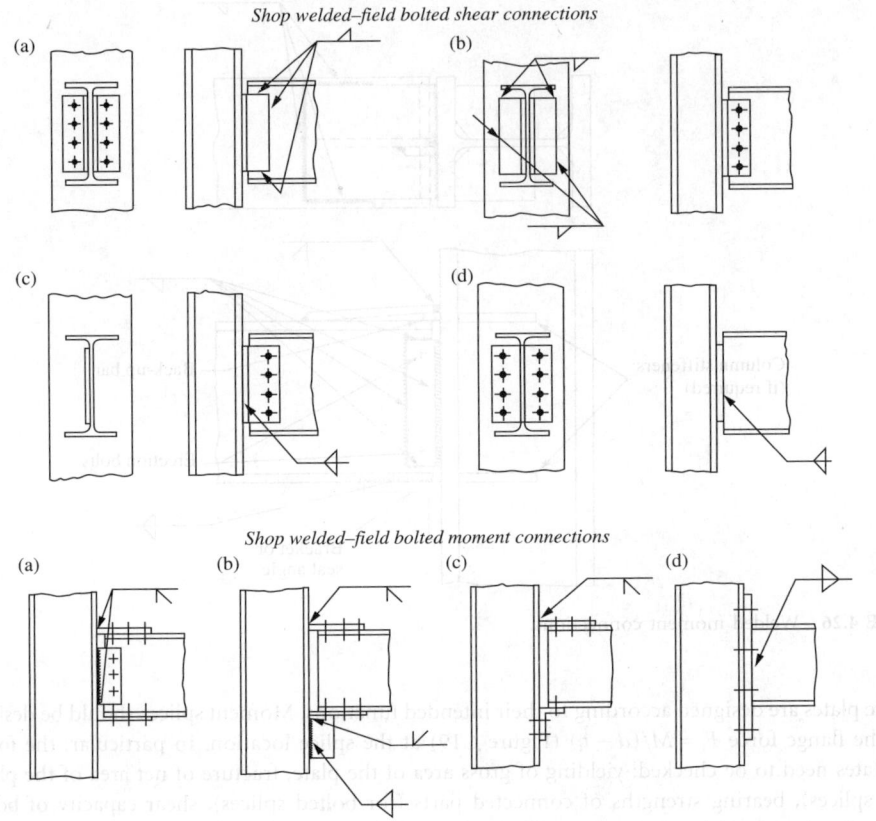

FIGURE 4.27 Shop welded–field bolted connections.

4.12.1.1 Axially Loaded Base Plates

Base plates supporting concentrically loaded columns in frames in which the column bases are assumed pinned are designed with the assumption that the column factored load P_u is distributed uniformly to the area of concrete under the base plate. The size of the base plate is determined from the limit state of bearing on concrete. The design bearing strength of concrete is given by the equation

$$\phi_c \, P_p = 0.60 \left[0.85 f_c' A_1 \sqrt{\frac{A_2}{A_1}} \, \right] \tag{4.118}$$

where f_c' is the compressive strength of concrete, A_1 is the area of the base plate, and A_2 is the area of the concrete pedestal that is geometrically similar to and concentric with the loaded area, $A_1 \leq A_2 \leq 4A_1$.

From Equation 4.118, it can be seen that the bearing capacity increases when the concrete area is greater than the plate area. This accounts for the beneficial effect of confinement. The upper limit of the bearing strength is obtained when $A_2 = 4A_1$. Presumably, the concrete area in excess of $4A_1$ is not effective in resisting the load transferred through the base plate.

Setting the column factored load, P_u, equal to the bearing capacity of the concrete pedestal, $\phi_c P_p$, and solving for A_1 from Equation 4.118, we have

$$A_1 = \frac{1}{A_2} \left[\frac{P_u}{0.6(0.85 f_c')} \right]^2 \tag{4.119}$$

Beam splices

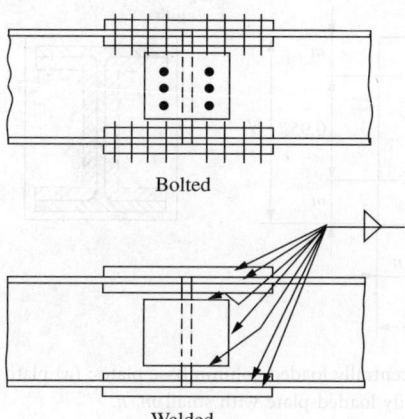

Bolted

Welded

Column splices

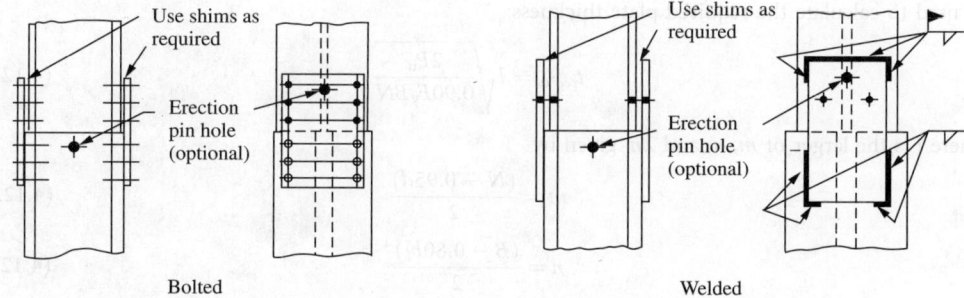

Use shims as required

Erection pin hole (optional)

Use shims as required

Erection pin hole (optional)

Bolted   Welded

FIGURE 4.28 Bolted and welded beam and column splices.

The length, N, and width, B, of the plate should be established so that $N \times B > A_1$. For an efficient design, the length can be determined from the equation

$$N \approx \sqrt{A_1} + 0.50(0.95d - 0.80b_f) \tag{4.120}$$

where $0.95d$ and $0.80b_f$ define the so-called effective load bearing area shown cross-hatched in Figure 4.29a. Once N is obtained, B can be solved from the equation

$$B = \frac{A_1}{N} \tag{4.121}$$

Both N and B should be rounded up to the nearest full inches.

The required plate thickness, $t_{req'd}$, is to be determined from the limit state of yield line formation along the most severely stressed sections. A yield line develops when the cross-section moment capacity is equal to its plastic moment capacity. Depending on the size of the column relative to the plate and the magnitude of the factored axial load, yield lines can form in various patterns on the plate. Figure 4.29 shows three models of plate failure in axially loaded plates. If the plate is large compared to the column, yield lines are assumed to form around the perimeter of the effective load bearing area (the cross-hatched area) as shown in Figure 4.29a. If the plate is small and the column factored load is light, yield lines are assumed to form around the inner perimeter of the I-shaped area as shown in Figure 4.29b. If the plate is small and the column factored load is heavy, yield lines are assumed to form around the inner edge of the

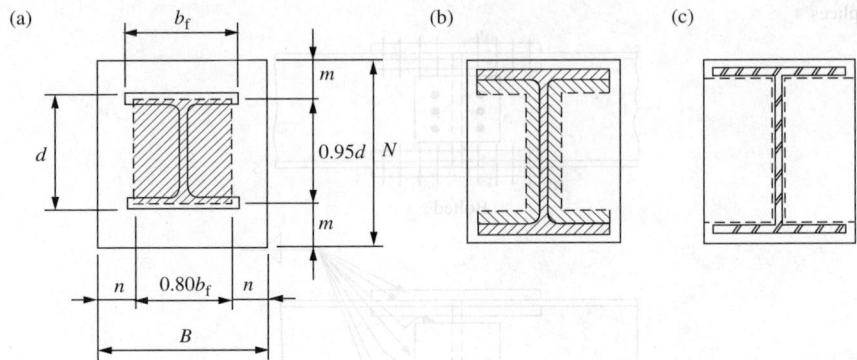

FIGURE 4.29   Failure models for centrally loaded column base plates: (a) plate with large *m*, *n*; (b) lightly loaded plate with small *m*, *n*; and (c) heavily loaded plate with small *m*, *n*.

column flanges and both sides of the column web as shown in Figure 4.29c. The following equation can be used to calculate the required plate thickness:

$$t_{\text{req'd}} = l\sqrt{\frac{2P_{\text{u}}}{0.90F_{\text{y}}BN}} \tag{4.122a}$$

where *l* is the larger of *m*, *n*, and $\lambda n'$ given by

$$m = \frac{(N - 0.95d)}{2} \tag{4.122b}$$

$$n = \frac{(B - 0.80b_{\text{f}})}{2} \tag{4.122c}$$

$$n' = \sqrt{\frac{db_{\text{f}}}{4}} \tag{4.122d}$$

and

$$\lambda = \frac{2\sqrt{X}}{1 + \sqrt{1 - X}} \leq 1 \tag{4.122e}$$

in which

$$X = \left(\frac{4db_{\text{f}}}{(d + b_{\text{f}})^2}\right)\frac{P_{\text{u}}}{\phi_{\text{c}}P_{\text{p}}} \tag{4.122f}$$

4.12.1.2 Base Plates for Tubular and Pipe Columns

The design concept for base plates discussed above for I-shaped sections can be applied to the design of base plates for rectangular tubes and circular pipes. The critical section used to determine the plate thickness should be based on 0.95 times the outside column dimension for rectangular tubes and 0.80 times the outside dimension for circular pipes (Dewolf and Ricker 1990).

4.12.1.3 Base Plates with Moments

For columns in frames designed to carry moments at the base, base plates must be designed to support both axial forces and bending moments. If the moment is small compared to the axial force, the base plate can be designed without consideration of the tensile force that may develop in the anchor bolts. However, if the moment is large, this effect should be considered. To quantify the relative magnitude of this moment, an eccentricity $e = M_u/P_u$ is used. The general procedures for the design of base plates for different values of e will be given in the following (Dewolf and Ricker 1990).

4.12.1.3.1 Small Eccentricity, $e \leq N/6$

If e is small, the bearing stress is assumed to distribute linearly over the entire area of the base plate (Figure 4.30). The maximum bearing stress is given by

$$f_{max} = \frac{P_u}{BN} + \frac{M_u c}{I} \tag{4.123}$$

where c is equal to $N/2$ and I is equal to $BN^3/12$.

The size of the plate is to be determined by a trial and error process. The size of the base plate should be such that the bearing stress calculated using Equation 4.123 does not exceed $\phi_c P_p/A_1$, that is

$$0.60 \left[0.85 f_c' \sqrt{\frac{A_2}{A_1}} \right] \leq 0.60[1.7 f_c'] \tag{4.124}$$

The thickness of the plate is to be determined from

$$t_p = \sqrt{\frac{4 M_{plu}}{0.90 F_y}} \tag{4.125}$$

where M_{plu} is the moment per unit width of critical section in the plate. M_{plu} is to be determined by assuming that the portion of the plate projecting beyond the critical section acts as an inverted cantilever loaded by the bearing pressure. The moment calculated at the critical section divided by the length of the critical section (i.e., B) gives M_{plu}.

4.12.1.3.2 Moderate Eccentricity, $N/6 < e \leq N/2$

For plates subjected to moderate moments, only portion of the plate will be subjected to bearing stress (Figure 4.31). Ignoring the tensile force in the anchor bolt in the region of the plate where no bearing occurs and denoting A as the length of the plate in bearing, the maximum bearing stress can be calculated from force equilibrium consideration as

$$f_{max} = \frac{2 P_u}{AB} \tag{4.126}$$

where A is equal to $3(N/2 - e)$ is determined from moment equilibrium. The plate should be proportioned such that f_{max} does not exceed the value calculated using Equation 4.124. t_p is to be determined from Equation 4.125.

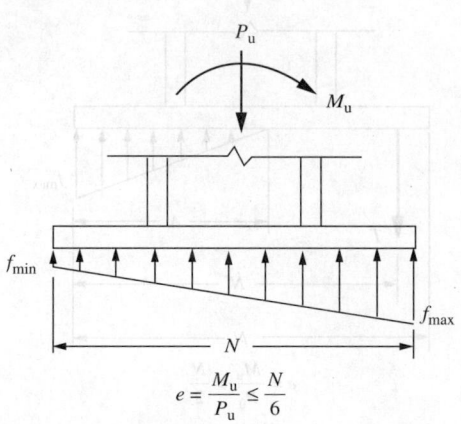

FIGURE 4.30 Eccentrically loaded column base plate (small load eccentricity).

4.12.1.3.3 Large Eccentricity, $e > N/2$

For plates subjected to large bending moments so that $e > N/2$, one needs to take into consideration the tensile force develops in the anchor bolts (Figure 4.32). Denoting T as the resultant force in the anchor bolts, A as the depth of the compressive stress block, N' as the distance from the line of action of the tensile force to the extreme compression edge of the plate, force equilibrium requires that

$$T + P_u = \frac{f_{max}AB}{2} \tag{4.127}$$

and moment equilibrium requires that

$$P_u\left(N' - \frac{N}{2}\right) + M = \frac{f_{max}AB}{2}\left(N' - \frac{A}{3}\right) \tag{4.128}$$

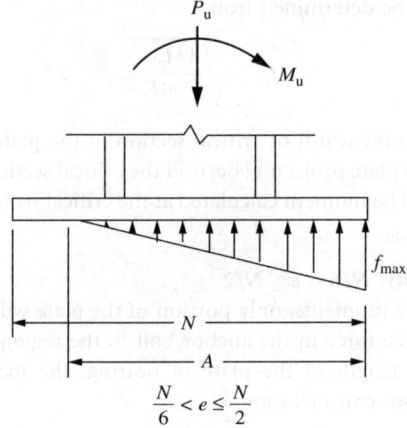

$$\frac{N}{6} < e \leq \frac{N}{2}$$

FIGURE 4.31 Eccentrically loaded column base plate (moderate load eccentricity).

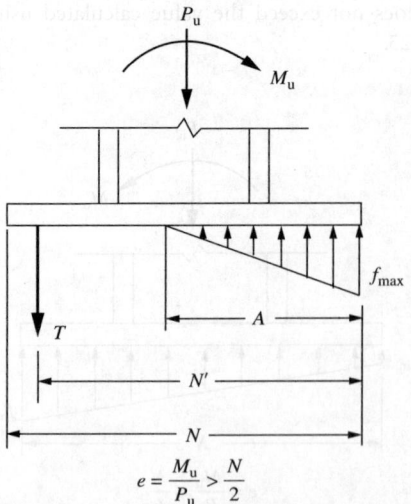

$$e = \frac{M_u}{P_u} > \frac{N}{2}$$

FIGURE 4.32 Eccentrically loaded column base plate (large load eccentricity).

The above equations can be used to solve for A and T. The size of the plate is to be determined using a trial and error process. The size should be chosen such that f_{max} does not exceed the value calculated using Equation 4.124, A should be smaller than N', and T should not exceed the tensile capacity of the bolts.

Once the size of the plate is determined, the plate thickness t_p is to be calculated using Equation 4.125. Note that there are two critical sections on the plate, one on the compression side of the plate and the other on the tension side of the plate. Two values of M_{plu} are to be calculated, and the larger value should be used to calculate t_p.

4.12.1.4 Base Plates with Shear

Under normal circumstances, the factored column base shear is adequately resisted by the frictional force developed between the plate and its support. Additional shear capacity is also provided by the anchor bolts. For cases in which exceptionally high shear force is expected such as in a bracing connection or in which uplift occurs which reduces the frictional resistance, the use of shear lugs may be necessary. Shear lugs can be designed based on the limit states of bearing on concrete and bending of the lugs. The size of the lug should be proportioned such that the bearing stress on concrete does not exceed $0.60(0.85f_c')$. The thickness of the lug can be determined from Equation 4.125. M_{plu} is the moment per unit width at the critical section of the lug. The critical section is taken to be at the junction of the lug and the plate (Figure 4.33).

4.12.2 Anchor Bolts

Anchor bolts are provided to stabilize the column during erection and to prevent uplift for cases involving large moments. Anchor bolts can be cast-in-place bolts or drilled-in bolts. The latter are placed after the concrete is set and are not used often. Their design is governed by the manufacturer's specifications. Cast-in-place bolts are hooked bars, bolts, or threaded rods with nuts (Figure 4.34) placed before the concrete is set. Anchor rods and threaded rods shall conform to one of the following ASTM specifications: A36/A36M, A193/A193M, A354, A572/A572M, A588/A588M, or F1554.

Of the three types of cast-in-place anchors shown in the figure, the hooked bars are recommended for use only in axially loaded base plates. They are not normally relied upon to carry significant tensile force. Bolts and threaded rods with nuts can be used for both axially loaded base plates or base plates with moments. Threaded rods with nuts are used when the length and size required for the specific design exceed those of standard size bolts.

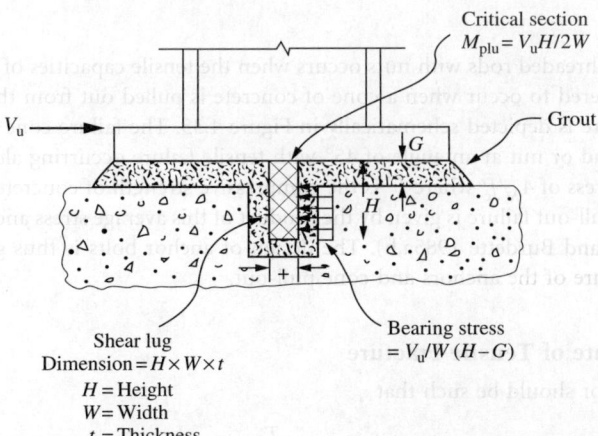

FIGURE 4.33 Column base plate subjected to shear.

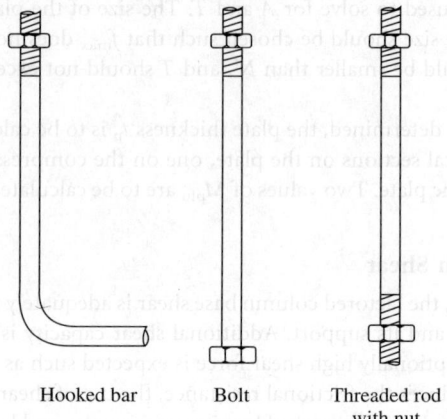

Hooked bar   Bolt   Threaded rod
                          with nut

FIGURE 4.34   Base plate anchors.

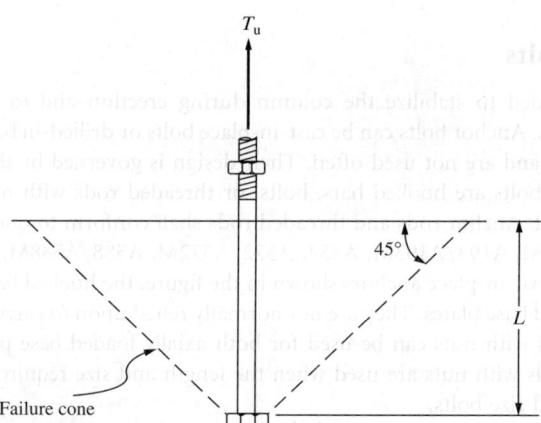

FIGURE 4.35   Cone pull-out failure.

Failure of bolts or threaded rods with nuts occurs when the tensile capacities of the bolts are reached. Failure is also considered to occur when a cone of concrete is pulled out from the pedestal. This cone pull-out type of failure is depicted schematically in Figure 4.35. The failure cone is assumed to radiate out from the bolt head or nut at an angle of 45° with tensile failure occurring along the surface of the cone at an average stress of $4\sqrt{f_c'}$ where f_c' is the compressive strength of concrete in psi. The load that will cause this cone pull-out failure is given by the product of this average stress and the projected area of the cone A_p (Marsh and Burdette 1985a,b). The design of anchor bolts is thus governed by the limit states of tensile fracture of the anchors and cone pull-out.

4.12.2.1 Limit State of Tensile Fracture

The area of the anchor should be such that

$$A_g \geq \frac{T_u}{\phi_t 0.75 F_u} \tag{4.129}$$

where A_g is the required gross area of the anchor, F_u is the minimum specified tensile strength, and $\phi_t = 0.75$ is the resistance factor for tensile fracture.

4.12.2.2 Limit State of Cone Pull-out

From Figure 4.35, it is clear that the size of the cone is a function of the length of the anchor. Provided that there are sufficient edge distance and spacing between adjacent anchors, the amount of tensile force required to cause cone pull-out failure increases with the embedded length of the anchor. This concept can be used to determine the required embedded length of the anchor. Assuming that the failure cone does not intersect with another failure cone or the edge of the pedestal, the required embedded length can be calculated from the equation

$$L \geq \sqrt{\frac{A_p}{\pi}} = \sqrt{\frac{(T_u/\phi_t 4\sqrt{f_c'})}{\pi}} \tag{4.130}$$

where A_p is the projected area of the failure cone, T_u is the required bolt force in pounds, f_c' is the compressive strength of concrete in psi, and ϕ_t is the resistance factor assumed to be equal to 0.75. If failure cones from adjacent anchors overlap one another or intersect with the pedestal edge, the projected area A_p must be adjusted according (see, e.g., Marsh and Burdette 1985a,b).

The length calculated using the above equation should not be less than the recommended values given by Shipp and Haninger (1983). These values are reproduced in the following table. Also shown in the table are the recommended minimum edge distances for the anchors.

Bolt type (material)	Minimum embedded length	Minimum edge distance
A307 (A36)	12d	5d > 4 in.
A325 (A449)	17d	7d > 4 in.

Note: d is the nominal diameter of the anchor.

4.12.3 Beam Bearing Plates

Beam bearing plates are provided between main girders and concrete pedestals to distribute the girder reactions to the concrete supports (Figure 4.36). Beam bearing plates may also be provided between cross-beams and girders if the cross-beams are designed to sit on the girders.

Beam bearing plates are designed based on the limit states of web yielding, web crippling, bearing on concrete, and plastic bending of the plate. The dimension of the plate along the beam axis, that is, N, is determined from the web yielding or web crippling criterion (see Section 4.5.1.1.3), whichever is more critical. The dimension B of the plate is determined from Equation 4.121 with A_1 calculated using Equation 4.119. P_u in Equation 4.119 is to be replaced by R_u, the factored reaction at the girder support. Once the size $B \times N$ is determined, the plate thickness t_p can be calculated using the equation

$$t_p = \sqrt{\frac{2R_u n^2}{0.90 F_y B N}} \tag{4.131}$$

where R_u is the factored girder reaction, F_y is the yield stress of the plate, and $n = (B - 2k)/2$ in which k is the distance from the web toe of the fillet to the outer surface of the flange. The above equation was developed based on the assumption that the critical sections for plastic bending in the plate occur at a distance k from the centerline of the web.

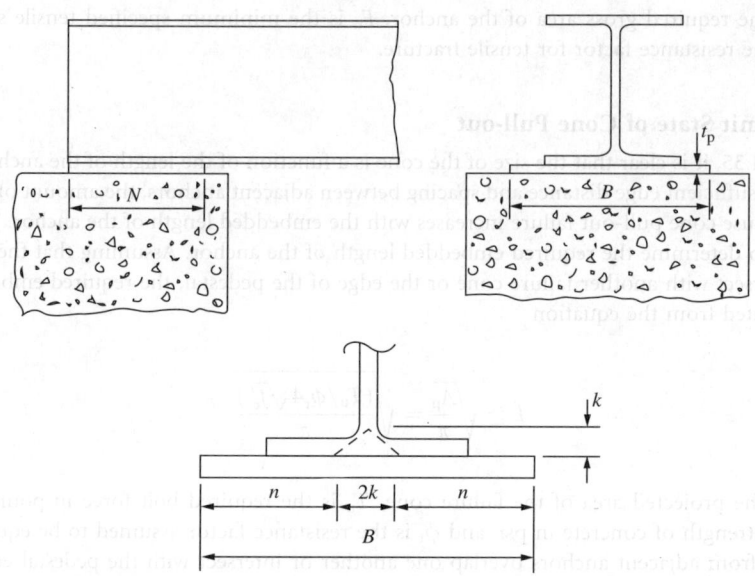

FIGURE 4.36 Beam bearing plate.

4.13 Composite Members (LRFD Approach)

Composite members are structural members made from two or more materials. The majority of composite sections used for building constructions are made from steel and concrete, although in recent years the use of fiber-reinforced polymer has been rising especially in the area of structural rehabilitation. Composite sections made from steel and concrete utilize the strength provided by steel and the rigidity provided by concrete. The combination of the two materials often results in efficient load-carrying members. Composite members may be concrete encased or concrete filled. For concrete encased members (Figure 4.37a), concrete is cast around steel shapes. In addition to enhancing strength and providing rigidity to the steel shapes, the concrete acts as a fire-proofing material to the steel shapes. It also serves as a corrosion barrier shielding the steel from corroding under adverse environmental conditions. For concrete filled members (Figure 4.37b), structural steel tubes are filled with concrete. In both concrete encased and concrete filled sections, the rigidity of the concrete often eliminates the problem of local buckling experienced by some slender elements of the steel sections.

Some disadvantages associated with composite sections are that concrete creeps and shrinks. Furthermore, uncertainties with regard to the mechanical bond developed between the steel shape and the concrete often complicate the design of beam–column joints.

4.13.1 Composite Columns

According to the LRFD Specification (AISC 1999), a compression member is regarded as a composite column if

- The cross-sectional area of the steel section is at least 4% of the total composite area. If this condition is not satisfied, the member should be designed as a reinforced concrete column.
- Longitudinal reinforcements and lateral ties are provided for concrete encased members. The cross-sectional area of the reinforcing bars shall be 0.007 in.^{2}/in. (180 mm^2/m) of bar spacing. To avoid spalling, lateral ties shall be placed at a spacing not greater than $\frac{2}{3}$ the least dimension of

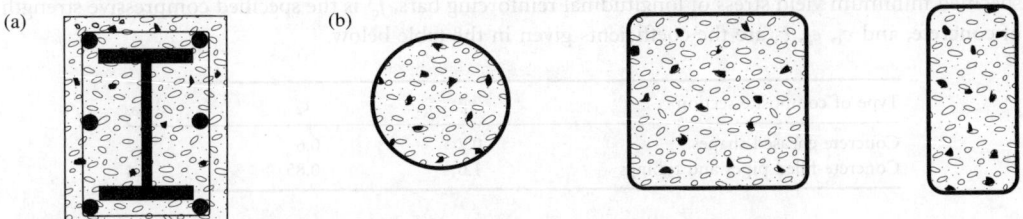

FIGURE 4.37 Composite columns: (a) concrete encased composite section and (b) concrete filled composite sections.

the composite cross-section. For fire and corrosion resistance, a minimum clear cover of 1.5 in. (38 mm) shall be provided.

- The compressive strength of concrete f_c' used for the composite section falls within the range 3 ksi (21 MPa) to 8 ksi (55 MPa) for normal weight concrete and not less than 4 ksi (28 MPa) for light weight concrete. These limits are set because they represent the range of test data available for the development of the design equations.
- The specified minimum yield stress for the steel sections and reinforcing bars used in calculating the strength of the composite columns does not exceed 60 ksi (415 MPa). This limit is set because this stress corresponds to a strain below which the concrete remains unspalled and stable.
- The minimum wall thickness of the steel sections for concrete filled members is equal to $b\sqrt{(F_y/3E)}$ for rectangular sections of width b and $D\sqrt{(F_y/8E)}$ for circular sections of outside diameter D.

14.13.1.1 Design Compressive Strength

The design compressive strength, $\phi_c P_n$, shall exceed the factored compressive force, P_u. The design compressive strength is given as follows:

For $\lambda_c \leq 1.5$:

$$\phi_c P_n = \begin{cases} 0.85\left[(0.658^{\lambda_c^2})A_s F_{my}\right], & \text{if } \lambda_c \leq 1.5 \\ 0.85\left[\left(\dfrac{0.877}{\lambda_c^2}\right)A_s F_{my}\right], & \text{if } \lambda_c > 1.5 \end{cases} \tag{4.132}$$

where

$$\lambda_c = \frac{KL}{r_m \pi}\sqrt{\frac{F_{my}}{E_m}} \tag{4.133}$$

$$F_{my} = F_y + c_1 F_{yr}\left(\frac{A_r}{A_s}\right) + c_2 f_c'\left(\frac{A_c}{A_s}\right) \tag{4.134}$$

$$E_m = E + c_3 E_c\left(\frac{A_c}{A_s}\right) \tag{4.135}$$

where r_m is the radius of gyration of steel section and shall not be less than 0.3 times the overall thickness of the composite cross-section in the plane of buckling, A_c is the area of concrete, A_r is the area of longitudinal reinforcing bars, A_s is the area of steel shape, E is the modulus of elasticity of steel, E_c is the modulus of elasticity of concrete, F_y is the specified minimum yield stress of steel shape, F_{yr} is the specified minimum yield stress of longitudinal reinforcing bars, f_c' is the specified compressive strength of concrete, and c_1, c_2, c_3 are the coefficients given in the table below.

Type of composite section	c_1	c_2	c_3
Concrete encased shapes	0.7	0.6	0.2
Concrete-filled pipes and tubings	1.0	0.85	0.4

In addition to satisfying the condition $\phi_c P_n \geq P_u$, shear connectors spaced no more than 16 in. (405 mm) apart on at least two faces of the steel section in a symmetric pattern about the axes of the steel section shall be provided for concrete encased composite columns to transfer the interface shear force V_u' between steel and concrete. V_u' is given by

$$V_u' = \begin{cases} V_u\left(1 - \dfrac{A_s F_y}{P_n}\right) & \text{when the force is applied to the steel section} \\ \\ V_u\left(\dfrac{A_s F_y}{P_n}\right) & \text{when the force is applied to the concrete encasement} \end{cases} \tag{4.136}$$

where V_u is the axial force in the column, A_s is the area of the steel section, F_y is the yield strength of the steel section, and P_n is the nominal compressive strength of the composite column without consideration of slenderness effect.

If the supporting concrete area in direct bearing is larger than the loaded area, the bearing condition for concrete must also be satisfied. Denoting $\phi_c P_{nc}$ ($= \phi_c P_{n,\text{composite section}} - \phi_c P_{n,\text{steel shape alone}}$) as the portion of compressive strength resisted by the concrete and A_B as the loaded area, the condition that needs to be satisfied is

$$\phi_c P_{nc} \leq 0.65[1.7 f_c' A_B] \tag{4.137}$$

4.13.2 Composite Beams

Composite beams used in construction can often be found in two forms: steel beams connected to a concrete slab by shear connectors and concrete encased steel beams.

4.13.2.1 Steel Beams with Shear Connectors

The design flexure strength for steel beams with shear connectors is $\phi_b M_n$. The resistance factor ϕ_b and nominal moment M_n are determined as follows:

Condition	ϕ_b	M_n
Positive moment region and $h/t_w \leq 3.76\sqrt{(E/F_{yf})}$	0.85	Determined from plastic stress distribution on the composite section
Positive moment region and $h/t_w > 3.76\sqrt{(E/F_{yf})}$	0.90	Determined from elastic stress superposition considering the effects of shoring
Negative moment region	0.90	Determined for the steel section alone using equations presented in Section 4.5

4.13.2.2 Concrete Encased Steel Beams

For steel beams fully encased in concrete, no additional anchorage for shear transfer is required if (a) at least $1\frac{1}{2}$ in. (38 mm) concrete cover is provided on top of the beam and at least 2 in. (51 mm) cover is provided over the sides and at the bottom of the beam and (b) spalling of concrete is prevented by adequate mesh or other reinforcing steel. The design flexural strength $\phi_b M_n$ can be computed using either an elastic analysis or a plastic analysis.

If an elastic analysis is used, ϕ_b shall be taken as 0.90. A linear strain distribution is assumed for the cross-section with zero strain at the neutral axis and maximum strains at the extreme fibers. The stresses are then computed by multiplying the strains by E (for steel) or E_c (for concrete). Maximum stress in steel shall be limited to F_y and maximum stress in concrete shall be limited to $0.85 f_c'$. The tensile strength of concrete shall be neglected. M_n is to be calculated by integrating the resulting stress block about the neutral axis.

If a plastic analysis is used, ϕ_b shall be taken as 0.90 and M_n shall be assumed to be equal to M_p, the plastic moment capacity of the steel section alone.

4.13.3 Composite Beam–Columns

Composite beam–columns shall be designed to satisfy the interaction equation of Equation 4.73 or 4.74 whichever is applicable, with $\phi_c P_n$ calculated based on Equations 4.132 to 4.135, P_e calculated using the equation $P_e = A_s F_{my}/\lambda_c^2$ and $\phi_b M_n$ calculated using the following equation (Galambos and Chapuis 1980):

$$\phi_b M_n = 0.90\left[ZF_y + \frac{1}{3}(h_2 - 2c_r)A_r F_{yr} + \left(\frac{h_2}{2} - \frac{A_w F_y}{1.7 f_c' h_1}\right)A_w F_y\right] \tag{4.138}$$

where Z is the plastic section modulus of the steel section, c_r is the average of the distance measured from the compression face to the longitudinal reinforcement in that face and the distance measured from the tension face to the longitudinal reinforcement in that face, h_1 is the width of the composite section perpendicular to the plane of bending, h_2 is the width of the composite section parallel to the plane of bending, A_r is the cross-sectional area of longitudinal reinforcing bars, A_w is the web area of the encased steel shape ($=0$ for concrete-filled tubes), F_y is the yield stress of the steel section, and F_{yr} is the yield stress of reinforcing bars.

If $0 < (P_u/\phi_c P_n) \le 0.3$, a linear interpolation of $\phi_b M_n$ calculated using the above equation assuming $P_u/\phi_c P_n = 0.3$ and that calculated for beams with $P_u/\phi_c P_n = 0$ (see Section 4.13.2) should be used.

4.13.4 Composite Floor Slabs

Composite floor slabs (Figure 4.38) can be designed as shored or unshored. In shored construction, temporary shores are used during construction to support the dead and accidental live loads until the concrete cures. The supporting beams are designed on the basis of their ability to develop composite action to support all factored loads after the concrete cures. In unshored construction, temporary shores are not used. As a result, the steel beams alone must be designed to support the dead and accidental live loads before the concrete has attained 75% of its specified strength. After the concrete is cured, the composite section should have adequate strength to support all factored loads.

Composite action for the composite floor slabs shown in Figure 4.38 is developed as a result of the presence of shear connectors. If sufficient shear connectors are provided so that the maximum flexural strength of the composite section can be developed, the section is referred to as fully composite. Otherwise, the section is referred to as partially composite. The flexural strength of a partially composite section is governed by the shear strength of the shear connectors. The horizontal shear force V_h that should be designed for at the interface of the steel beam and the concrete slab is given by

In regions of positive moment:

$$V_h = \min\left(0.85 f_c' A_c, A_s F_y, \sum Q_n\right) \tag{4.139}$$

In regions of negative moment:

$$V_h = \min\left(A_r F_{yr}, \sum Q_n\right) \tag{4.140}$$

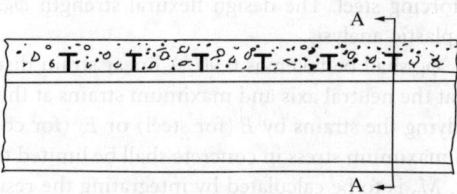

Composite floor slab with stud shear connectors

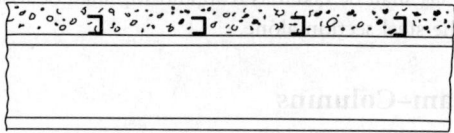

Composite floor slab with channel shear connectors

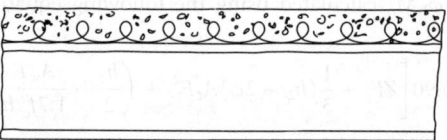

Composite floor slab with spiral shear connectors

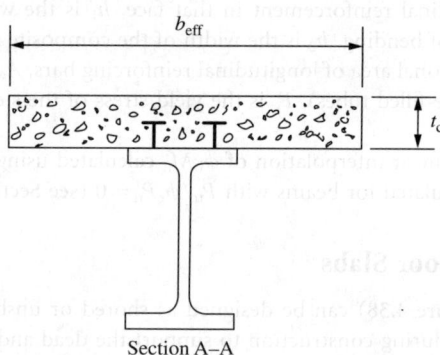

Section A–A

FIGURE 4.38 Composite floor slabs.

where

f_c' = compressive strength of concrete

A_c = effective area of the concrete slab = $t_c b_{eff}$

t_c = thickness of the concrete slab

b_{eff} = effective width of the concrete slab

   = min ($L/4$, s) for an interior beam

   = min ($L/8$ + distance from beam centerline to edge of slab, $s/2$ + distance from beam centerline to edge of slab) for an exterior beam

L = beam span measured from center-to-center of supports

s = spacing between centerline of adjacent beams

A_s = cross-sectional area of the steel beam

F_y = yield stress of the steel beam

A_r = area of reinforcing steel within the effective area of the concrete slab

F_{yr} = yield stress of the reinforcing steel
$\sum Q_n$ = sum of nominal shear strengths of the shear connectors

The nominal shear strength of a shear connector (used without a formed steel deck) is given by

For a stud shear connector:

$$Q_n = 0.5A_{sc}\sqrt{f_c' E_c} \leq A_{sc}F_u \qquad (4.141)$$

For a channel shear connector:

$$Q_n = 0.3(t_f + 0.5t_w)L_c\sqrt{f_c' E_c} \qquad (4.142)$$

where A_{sc} is the cross-sectional area of the shear stud (in.²), f_c' is the compressive strength of concrete (ksi), E_c is the modulus of elasticity of concrete (ksi), F_u is the minimum specified tensile strength of the stud shear connector (ksi), t_f is the flange thickness of the channel shear connector (in.), t_w is the web thickness of the channel shear connector (in.), and L_c is the length of the channel shear connector (in.).

If a formed steel deck is used, Q_n must be reduced by a reduction factor. The reduction factor depends on whether the deck ribs are perpendicular or parallel to the steel beam.

For deck ribs perpendicular to steel beam:

$$\frac{0.85}{\sqrt{N_r}}\left(\frac{w_r}{h_r}\right)\left[\left(\frac{H_s}{h_r}\right) - 1.0\right] \leq 1.0 \qquad (4.143)$$

When only a single stud is present in a rib perpendicular to the steel beam, the reduction factor expressed in Equation 4.143 shall not exceed 0.75.

For deck ribs parallel to steel beam:

$$0.6\left(\frac{w_r}{h_r}\right)\left[\left(\frac{H_s}{h_r}\right) - 1.0\right] \leq 1.0 \qquad (4.144)$$

The reduction factor expressed in Equation 4.144 is applicable only if $(w_r/h_r) < 1.5$. In the above equations, N_r is the number of stud connectors in one rib at a beam intersection, not to exceed three in computations regardless of the actual number of studs installed, w_r is the average width of the concrete rib or haunch, h_r is the nominal rib height, and H_s is the length of stud connector after welding, not to exceed the value $h_r + 3$ in. (75 mm) in computations regardless of the actual length.

For full composite action, the number of connectors required between the *maximum* moment point and the *zero* moment point of the beam is given by

$$N = \frac{V_h}{Q_n} \qquad (4.145)$$

For partial composite action, the number of connectors required is governed by the condition $\phi_b M_n \geq M_u$, where $\phi_b M_n$ is governed by the shear strength of the connectors.

The placement and spacing of the shear connectors should comply with the following guidelines:

- The shear connectors shall be uniformly spaced between the points of maximum moment and zero moment. However, the number of shear connectors placed between a concentrated load point and the nearest zero moment point must be sufficient to resist the factored moment M_u.
- Except for connectors installed in the ribs of formed steel decks, shear connectors shall have at least 1 in. (25.4 mm) of lateral concrete cover. The slab thickness above the formed steel deck shall not be less than 2 in. (50 mm).
- Unless located over the web, the diameter of shear studs must not exceed 2.5 times the thickness of the beam flange. For the formed steel deck, the diameter of stud shear connectors shall not exceed $\frac{3}{4}$ in. (19 mm), and shall extend not less than $1\frac{1}{2}$ in. (38 mm) above the top of the steel deck.

- The longitudinal spacing of the studs should fall in the range six times the stud diameter to eight times the slab thickness if a solid slab is used or four times the stud diameter to eight times the slab thickness or 36 in. (915 mm), whichever is smaller, if a formed steel deck is used. Also, to resist uplift, the steel deck shall be anchored to all supporting members at a spacing not to exceed 18 in. (460 mm).

The design flexural strength $\phi_b M_n$ of the composite beam with shear connectors is determined as follows:

In regions of positive moments: For $h_c/t_w \leq 3.76/\sqrt{(E/F_{yf})}$, $\phi_b = 0.85$, M_n is the moment capacity determined using a plastic stress distribution assuming concrete crushes at a stress of $0.85 f_c'$ and steel yields at a stress of F_y. If a portion of the concrete slab is in tension, the strength contribution of that portion of concrete is ignored. The determination of M_n using this method is very similar to the technique used for computing moment capacity of a reinforced concrete beam according to the ultimate strength method.

For $h_c/t_w > 3.76/\sqrt{(E/F_{yf})}$, $\phi_b = 0.90$, M_n is the moment capacity determined using superposition of elastic stress, considering the effect of shoring. The determination of M_n using this method is quite similar to the technique used for computing the moment capacity of a reinforced concrete beam according to the working stress method.

In regions of negative moment: $\phi_b M_n$ is to be determined for the steel section alone in accordance with the requirements discussed in Section 4.5.

To facilitate design, numerical values of $\phi_b M_n$ for composite beams with shear studs in solid slabs are given in tabulated form in the AISC-LRFD Manual. Values of $\phi_b M_n$ for composite beams with formed steel decks are given in a publication by the Steel Deck Institute (2001).

4.14 Plastic Design

Plastic analysis and design is permitted only for steels with yield stress not exceeding 65 ksi. The reason for this is that steels with high yield stress lack the ductility required for inelastic deformation at hinge locations. Without adequate inelastic deformation, moment redistribution, which is an important characteristic for PD, cannot take place.

In PD, the predominant limit state is the formation of plastic hinges. Failure occurs when sufficient plastic hinges have formed for a collapse mechanism to develop. To ensure that plastic hinges can form and can undergo large inelastic rotation, the following conditions must be satisfied:

- Sections must be compact. That is, the width–thickness ratios of flanges in compression and webs must not exceed λ_p in Table 4.8.
- For columns, the slenderness parameter λ_c (see Section 4.4) shall not exceed $1.5K$ where K is the effective length factor and P_u from gravity and horizontal loads shall not exceed $0.75A_g F_y$.
- For beams, the lateral unbraced length L_b shall not exceed L_{pd} where

For doubly and singly symmetric I-shaped members loaded in the plane of the web

$$L_{pd} = \frac{3600 + 2200(M_1/M_p)}{F_y} r_y \tag{4.146}$$

and *for solid rectangular bars and symmetric box beams*

$$L_{pd} = \frac{5000 + 3000(M_1/M_p)}{F_y} r_y \geq \frac{3000\, r_y}{F_y} \tag{4.147}$$

In the above equations, M_1 is the smaller end moment within the unbraced length of the beam, M_p is the plastic moment ($= Z_x F_y$) of the cross-section, r_y is the radius of gyration about the minor axis, in inches, and F_y is the specified minimum yield stress, in ksi.

L_{pd} is not defined for beams bent about their minor axes, nor for beams with circular and square cross-sections because these beams do not experience lateral torsional bucking when loaded.

4.14.1 Plastic Design of Columns and Beams

Provided that the above limitations are satisfied, the design of columns shall meet the condition $1.7F_aA \geq P_u$ where F_a is the allowable compressive stress given in Equation 4.16, A is the gross cross-sectional area and P_u is the factored axial load.

The design of beams shall satisfy the conditions $M_p \geq M_u$, and $0.55F_yt_wd \geq V_u$ where M_u and V_u are the factored moment and shear, respectively. M_p is the plastic moment capacity, F_y is the minimum specified yield stress, t_w is the beam web thickness, and d is the beam depth. For beams subjected to concentrated loads, all failure modes associated with concentrated loads (see Sections 4.5.1.1.3 and 4.5.1.2.3) should also be prevented.

Except at the location where the last hinge forms, a beam bending about its major axis must be braced to resist lateral and torsional displacements at plastic hinge locations. The distance between adjacent braced points should not exceed l_{cr} given by

$$l_{cr} = \begin{cases} \left(\dfrac{1375}{F_y} + 25 \right) r_y, & \text{if } -0.5 < \dfrac{M}{M_p} < 1.0 \\[4mm] \left(\dfrac{1375}{F_y} \right) r_y, & \text{if } -1.0 < \dfrac{M}{M_p} \leq -0.5 \end{cases} \qquad (4.148)$$

where r_y is the radius of gyration about the weak axis, M is the smaller of the two end moments of the unbraced segment, and M_p is the plastic moment capacity.

M/M_p is taken as positive if the unbraced segment bends in reverse curvature and is taken as negative if the unbraced segment bends in single curvature.

4.14.2 Plastic Design of Beam–Columns

Beam–columns designed on the basis of plastic analysis shall satisfy the following interaction equations for stability (Equation 4.149) and for strength (Equation 4.150):

$$\frac{P_u}{P_{cr}} + \frac{C_m M_u}{\left(1 - \dfrac{P_u}{P_e}\right) M_m} \leq 1.0 \qquad (4.149)$$

$$\frac{P_u}{P_y} + \frac{M_u}{1.18M_p} \leq 1.0 \qquad (4.150)$$

where

P_u = factored axial load
P_{cr} = $1.7F_aA$, F_a is defined in Equation 4.16 and A is the cross-sectional area
P_y = yield load = AF_y
P_e = Euler buckling load = $\pi^2 EI/(Kl)^2$
C_m = coefficient defined in Section 4.4
M_u = factored moment
M_p = plastic moment = ZF_y
M_m = maximum moment that can be resisted by the member in the absence of axial load
      = M_{px} if the member is braced in the weak direction
      = $\{1.07-[(l/r_y)\sqrt{F_y}]/3160\}M_{px} \leq M_{px}$ if the member is unbraced in the weak direction
l = unbraced length of the member
r_y = radius of gyration about the minor axis
M_{px} = plastic moment about the major axis = $Z_x F_y$
F_y = minimum specified yield stress, ksi

4.15 Reduced Beam Section

Reduced beam section (RBS) or dogbone connection is a type of connection in welded steel moment frames in which portions of the bottom beam flange or both top and bottom flanges are cut near the beam-to-column connection thereby reducing the flexural strength of the beam at the RBS region and thus force a plastic hinge to form in a region away from the connection (Engelhardt et al. 1996; Iwankiw and Carter 1996; Plumier 1997). The presence of this reduced section in the beam also tends to decrease the force demand on the beam flange welds and so mitigate the distress that may cause fracture in the connection. RBS can be bottom flange cut only, or both top and bottom flange cuts. Bottom flange RBS is used if it is difficult or impossible to cut the top flange of an existing beam (e.g., if the beam is attached to a concrete floor slab). Figure 4.39 shows some typical cut geometries for RBS. The constant cut offers the advantage of ease of fabrication. The tapered cut has the advantage of matching the beam's flexural strength to the flexural demand on the beam under a gravity load. The radius cut is relatively easy to fabricate and because the change in geometry of the cross-section is rather gradual, it also has the advantage of minimizing stress concentration. Based on experimental investigations (Engelhardt et al. 1998; Moore et al. 1999), the radius cut RBS has been shown to be a reliable connection for welded steel moment frames.

The key dimensions of a radius cut RBS is shown in Figure 4.40. The distance from the face of the column to the start of the cut is designated as a, the length and depth of the cut are denoted as b and c, respectively. Values of a, b, c are given as follows (Engelhardt et al. 1998; Gross et al. 1999):

$$a \approx (0.5 \text{ to } 0.75)b_f \tag{4.151}$$

$$b \approx (0.65 \text{ to } 0.85)d \tag{4.152}$$

$$c \approx 0.25b_f \quad \text{for a bottom flange RBS} \tag{4.153}$$

where b_f is the beam flange width and d is the beam depth. Using geometry, the cut radius R can be calculated as

$$R = \frac{4c^2 + b^2}{8c} \tag{4.154}$$

and the distance from the face of the column to the critical plastic section s_c is given by

$$s_c = a + \frac{b}{2} \tag{4.155}$$

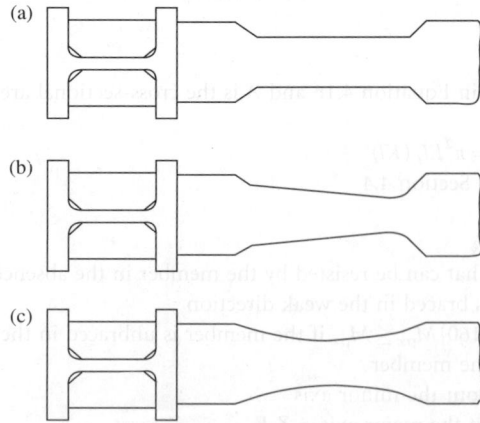

FIGURE 4.39 Reduced beam section cut geometries: (a) constant cut; (b) tapered cut; and (c) radius cut.

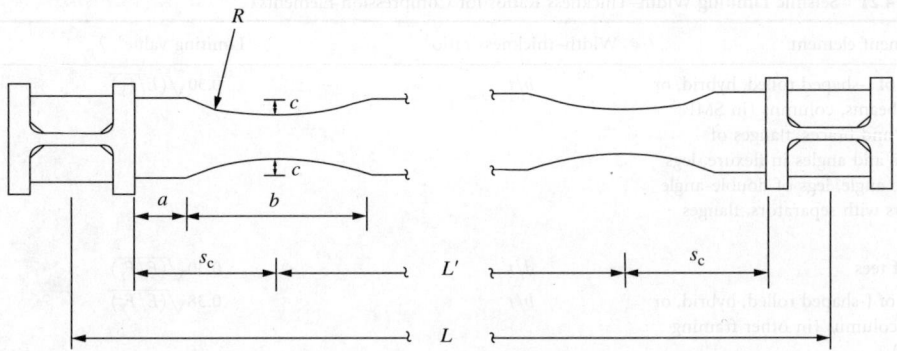

FIGURE 4.40   Key dimensions of a radius cut reduced beam section.

An optimal RBS is one in which the moment at the face of the column will be minimized. To achieve this condition, the following procedure is recommended (Gross et al. 1999):

- Set $c = 0.25b_f$
- Compute the RBS plastic section modulus using the equation

$$Z_{RBS} = Z_b - \frac{(ct_f)^2}{t_w} - ct_f(d - t_f) \tag{4.156}$$

  where Z_b is the plastic section modulus of the full beam cross-section, c is the depth of cut as shown in Figure 4.40, and d, t_f, t_w are the beam depth, beam flange thickness, and web thickness, respectively.

- Compute η, the ratio of moment at the face of the column to plastic moment of the connecting beam, from the equation

$$\eta = 1.1\left(1 + \frac{2s_c}{L'}\right)\frac{Z_{RBS}}{Z_b} + \frac{wL's_c}{2Z_bF_{yf}} \tag{4.157}$$

  where s_c is given in Equation 4.155 and shown in Figure 4.40, L' is the beam span between critical plastic sections (see Figure 4.40), Z_{RBS} and Z_b are defined in Equation 4.156, w is the magnitude of the uniformly distributed load on the beam, and F_{yf} is the beam flange yield strength.

- If $\eta < 1.05$, then the RBS dimensions are satisfactory. Otherwise, use RBS cutouts in both the top and bottom flanges, or consider using other types of moment connections (Gross et al. 1999).

Experimental studies (Uang and Fan 1999; Engelhardt et al. 2000; Gilton et al. 2000; Yu et al. 2000) of a number of radius cut RBS with or without the presence of a concrete slab have shown that the connections perform satisfactorily and exhibit sufficient ductility under cyclic loading. However, the use of RBS beams in a moment resistant frame tends to cause an overall reduction in frame stiffness of around 4 to 7% (Grubbs 1997). If the increase in frame drift due to this reduction in frame stiffness is appreciable, proper allowances must be made in the analysis and design of the frame.

4.16   Seismic Design

Special provisions (AISC 2002) apply for the design of steel structures to withstand earthquake loading. To ensure sufficient ductility, more stringent limiting width–thickness ratios than those shown in Table 4.8 for compression elements are required. These seismic limiting width thickness ratios are shown in Table 4.21.

TABLE 4.21 Seismic Limiting Width–Thickness Ratios for Compression Elements

Component element	Width–thickness ratio[a]	Limiting value[a], λ_p
Flanges of I-shaped rolled, hybrid, or welded beams, columns (in SMF system) and braces, flanges of channels and angles in flexure, legs of single angle, legs of double-angle members with separators, flanges of tees	b/t	$0.30\sqrt{(E/F_y)}$
Webs of tees	d/t	$0.30\sqrt{(E/F_y)}$
Flanges of I-shaped rolled, hybrid, or welded columns (in other framing systems)	b/t	$0.38\sqrt{(E/F_y)}$
Flanges of H-piles	b/t	$0.45\sqrt{(E/F_y)}$
Flat bars	b/t	2.5
Webs in flexural compression for beams in SMF	h_c/t_w	$2.45\sqrt{(E/F_y)}$
Other webs in flexural compression	h_c/t_w	$3.14\sqrt{(E/F_y)}$
Webs in combined flexural and axial compression	h_c/t_w	For $P_u/\phi_b P_y \leq 0.125$: $3.14(1-1.54P_u/\phi_b P_y)\sqrt{(E/F_y)}$ For $P_u/\phi_b P_y > 0.125$: $1.12(2.33-P_u/\phi_b P_y)\sqrt{(E/F_y)}$
Round HSS in axial compression or flexure	D/t	$0.044E/F_y$
Rectangular HSS in axial compression or flexure	b/t or h_c/t	$0.64\sqrt{(E/F_y)}$
Webs of H-pile sections	h/t_w	$0.94\sqrt{(E/F_y)}$

[a] See Table 4.8 for definitions of the terms used, and replace F_y by F_{yf} for hybrid sections.

Moreover, the structure needs to be designed for loads and load combinations that include the effect of earthquakes, and if the applicable building code (e.g., UBC 1997; IBC 2000) requires that amplified seismic loads be used, the horizontal component of the earthquake load shall be multiplied by an overstrength factor Ω_0, as prescribed by the applicable building code. In lieu of a specific definition of Ω_0, the ones shown in Table 4.22 can be used.

The seismic provisions of AISC (2002) identify seven types of frames for seismic resistant design: special moment frames (SMF), intermediate moment frames (IMF), ordinary moment frames (OMF), special truss moment frames (STMF), special concentrically braced frames (SCBF), ordinary concentrically braced frames (OCBF), and eccentrically braced frames (EBF). Examples of these framing systems are shown in Figure 4.41. Specific requirements for the design of these frames are given by AISC (2002). Note that the word "special" is used if the frame is expected to withstand significant inelastic deformation, the word "intermediate" is used if the frame is expected to withstand moderate inelastic deformations, and the word "ordinary" is used if the frame is expected to withstand only limited inelastic deformations. For instance, all connections in an SMF must be designed to sustain an interstory drift angle of at least 0.04 radians at a moment (determined at the column face) equal to at least 80% of the nominal plastic moment of the connecting beam, while in an IMF the ductility value is reduced to 0.02 radians, and in an OMF a value of 0.01 radians is expected at a column face moment no less than 50% of the nominal connecting beam plastic moment. Also, connections used in SMF and IMF must be FR moment connections that are prequalified or tested for conformance with the aforementioned ductility requirement, whereas both FR and PR moment connections are permitted for OMF. To avoid lateral torsional instability, beams in SMF must have both flanges laterally supported at a distance not to exceed $0.086r_y E/F_y$. This limit is more stringent than those (L_p) discussed in Section 4.5. In addition, lateral supports are needed at locations near concentrated forces, changes in cross-section, and regions of plastic hinges.

TABLE 4.22   Value of Ω_0

Seismic load resisting system	Ω_0
Moment frame systems	3
Eccentrically braced frames	2.5
All other systems meeting Part I (AISC 2002) requirements	2

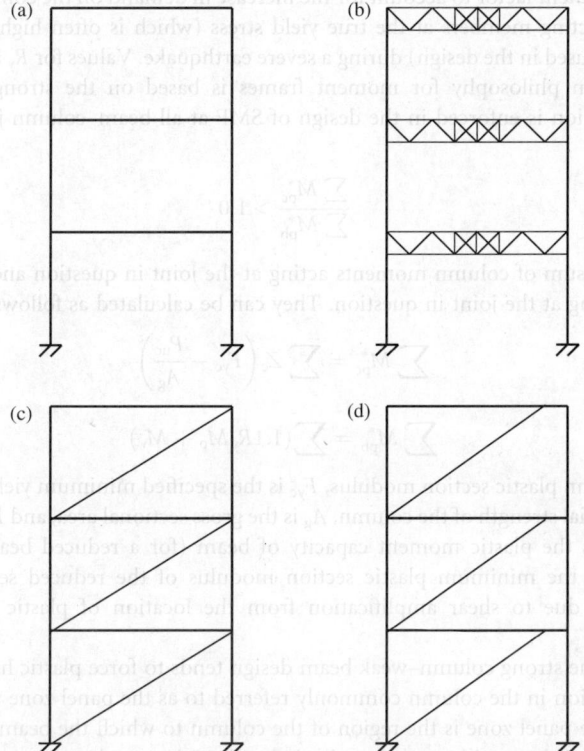

FIGURE 4.41   Types of frames: (a) moment frame; (b) truss moment frame; (c) concentrically braced frame; and (d) eccentrically braced frame.

If bolts are used, they should all be pretensioned high-strength bolts designed for standard or short-slotted (with direction of slot perpendicular to applied force) holes with hole deformation considered in the design (i.e., nominal bearing strength $\leq 2.4dtF_u$). Although design shear strength for the bolts is permitted to be calculated as that of bearing-type joints, all faying surfaces must be prepared as required for Class A or better slip-critical joints. Bolted connections for members that form part of the seismic load resisting system shall be designed to ensure that a ductile limit state (either in the connection or in the member) controls the design. If welds are used, they shall conform to procedures and standards outlined in AWS D1.1 (AWS 2000). If both bolts and welds are used, bolts shall not be designed to share load with welds on the same faying surface.

To avoid brittle fracture of welds, a minimum Charpy V-Notch (CVN) toughness of 20 ft lb (27 J) at $-20°F$ ($-29°C$) is required for all welds in members and connections that are part of the seismic lateral load resisting system. For structures with enclosed SMF or IMF maintained at a temperature of 50°F (10°C) or higher, complete joint penetration welds of beam flanges to columns, column splices, and groove welds of shear tabs and beam webs to columns, shall be made with a filler material with minimum CVN toughness of 20 ft lb (27 J) at $-20°F$ ($-29°C$) and 40 ft lb (54 J) at $70°F$ ($21°C$). If the service temperature is lower than $50°F$ ($10°C$), these minimum CVN toughness values shall be reduced accordingly.

Regardless of the type of framing system used, the required strength of the connections or related components must be designed for an expected yield strength F_{ye} of the connection members given by

$$F_{ye} = R_y F_y \qquad (4.158)$$

where R_y is an adjustment factor to account for the increase in demand on the connections when yielding occurs in the connecting members at the true yield stress (which is often higher than the minimum specified yield stress used in the design) during a severe earthquake. Values for R_y are given in Table 4.23.

The current design philosophy for moment frames is based on the strong column–weak beam concept. This condition is enforced in the design of SMF at all beam–column joints by the following equation

$$\frac{\sum M_{pc}^*}{\sum M_{pb}^*} > 1.0 \qquad (4.159)$$

where $\sum M_{pc}^*$ is the sum of column moments acting at the joint in question and $\sum M_{pb}^*$ is the sum of beam moments acting at the joint in question. They can be calculated as follows:

$$\sum M_{pc}^* = \sum Z_c \left(F_{yc} - \frac{P_{uc}}{A_g} \right) \qquad (4.160)$$

$$\sum M_{pb}^* = \sum (1.1 R_y M_p + M_v) \qquad (4.161)$$

where Z_c is the column plastic section modulus, F_{yc} is the specified minimum yield stress of the column, P_{uc} is the required axial strength of the column, A_g is the gross sectional area, and R_y are the values shown in Table 4.23, M_p is the plastic moment capacity of beam (for a reduced beam section, M_p can be calculated based on the minimum plastic section modulus of the reduced section), and M_v is the additional moment due to shear amplification from the location of plastic hinge to the column centerline.

While the use of the strong column–weak beam design tends to force plastic hinges to develop in the beams, there is a region in the column commonly referred to as the panel zone where high stresses are often developed. The panel zone is the region of the column to which the beams are attached. For the moment frame to be able to withstand the seismic loading, the panel zone must be strong enough to allow the frame to sustain inelastic deformation and dissipate energy before failure. For SMF, the design criterion for a panel zone (when the beam web is parallel to the column web) is

$$\phi_v R_v \geq R_u \qquad (4.162)$$

TABLE 4.23 Value of R_y

Application	R_y
Hot rolled structural steel shapes	
ASTM A36/A36M	1.5
ASTM A572/A572M Grade 42 (290)	1.3
ASTM A992/A992M	1.1
All others	1.1
HSS shapes	
ASTM A500, A501, A618, A847	1.3
Steel pipes	
ASTM A53, A53M	1.4
Plates	1.1
All other products	1.1

where $\phi_v = 0.75$ and R_v is the design shear strength of the panel zone and is given by

$$R_v = \begin{cases} 0.60F_y d_c t_p \left[1 + \dfrac{3b_{cf} t_{cf}^2}{d_b d_c t_p} \right], & \text{if } \dfrac{P_u}{P_y} \le 0.75 \\ 0.60F_y d_c t_p \left[1 + \dfrac{3b_{cf} t_{cf}^2}{d_b d_c t_p} \right] \left[1.9 - \dfrac{1.2P_u}{P_y} \right], & \text{if } \dfrac{P_u}{P_y} > 0.75 \end{cases} \tag{4.163}$$

where b_{cf} is the column flange width, d_b is the overall beam depth, d_c is the overall column depth, t_{cf} is the column flange thickness, t_p is the thickness of the panel zone including the thickness of any doubler plates (if used), F_y is the specified minimum yield strength of the panel zone, P_u is the required strength of the column containing the panel zone, P_y is the yield strength of the column containing the panel zone, and R_u is the required shear strength of the panel zone.

In addition, the individual thickness t of the column web and doubler plates (if used) should satisfy the condition

$$t \ge (d_z + w_z)/90 \tag{4.164}$$

where d_z is the panel zone depth between continuity plates and w_z is the panel zone width between column flanges.

If doubler plates are used, and are placed against the column web, they should be welded across the top and bottom edges to develop the proportion of the total force that is transmitted to the plates. When doubler plates are placed away from the column web, they should be placed symmetrically in pairs and welded to the continuity plates to develop the proportion of total force that is transmitted to the plates. In either case, the doubler plates should be welded to the column flanges using either complete joint penetration groove welds or fillet welds that can develop the design shear strength of the full plate thickness.

Glossary

ASD — Acronym for allowable stress design.

Beam–columns — Structural members whose primary function are to carry loads both along and transverse to their longitudinal axes.

Biaxial bending — Simultaneous bending of a member about two orthogonal axes of the cross-section.

Built-up members — Structural members made of structural elements jointed together by bolts, welds, or rivets.

Composite members — Structural members made of both steel and concrete.

Compression members — Structural members whose primary function is to carry loads along their longitudinal axes.

Design strength — Resistance provided by the structural member obtained by multiplying the nominal strength of the member by a resistance factor.

Drift — Lateral deflection of a building.

Factored load — The product of the nominal load and a load factor.

Flexural members — Structural members whose primary function are to carry loads transverse to their longitudinal axes.

Limit state — A condition in which a structural or structural component becomes unsafe (strength limit state) or unfit for its intended function (serviceability limit state).

Load factor — A factor to account for the unavoidable deviations of the actual load from its nominal value and uncertainties in structural analysis in transforming the applied load into a load effect (axial force, shear, moment, etc.).

LRFD — Acronym for load and resistance factor design.

PD — Acronym for plastic design.

Plastic hinge — A yielded zone of a structural member in which the internal moment is equal to the plastic moment of the cross-section.

Reduced beam section — A beam section with portions of flanges cut out to reduce the section moment capacity.

Resistance factor — A factor to account for the unavoidable deviations of the actual resistance of a member from its nominal value.

Service load — Nominal load expected to be supported by the structure or structural component under normal usage.

Sideways inhibited frames — Frames in which lateral deflections are prevented by a system of bracing.

Sideways uninhibited frames — Frames in which lateral deflections are not prevented by a system of bracing.

Shear lag — The phenomenon in which the stiffer (or more rigid) regions of a structure or structural component attract more stresses than the more flexible regions of the structure or structural component. Shear lag causes stresses to be unevenly distributed over the cross-section of the structure or structural component.

Tension field action — Postbuckling shear strength developed in the web of a plate girder. Tension field action can develop only if sufficient transverse stiffeners are provided to allow the girder to carry the applied load using truss-type action after the web has buckled.

References

AASHTO. 1997. *Standard Specification for Highway Bridges*, 16th ed. American Association of State Highway and Transportation Officials, Washington D.C.

AASHTO. 1998. *LRFD Bridge Design Specification*, 2nd ed. American Association of State Highway and Transportation Officials, Washington D.C.

AISC. 1989. *Manual of Steel Construction — Allowable Stress Design*, 9th ed. American Institute of Steel Construction, Chicago, IL.

AISC. 1992. *Manual of Steel Construction — Volume II Connections*. ASD 1st ed./LRFD 1st ed. American Institute of Steel Construction, Chicago, IL.

AISC. 1999. *Load and Resistance Factor Design Specification for Structural Steel Buildings*. American Institute of Steel Construction, Chicago, IL.

AISC. 2001. *Manual of Steel Construction — Load and Resistance Factor Design*, 3rd ed. American Institute of Steel Construction, Chicago, IL.

AISC. 2002. *Seismic Provisions for Structural Steel Buildings*. American Institute of Steel Construction, Chicago, IL.

Aminmansour, A. 2000. A New Approach for Design of Steel Beam-Columns. *AISC Eng. J.*, **37**(2):41–72.

ASCE. 1997. *Effective Length and Notional Load Approaches for Assessing Frame Stability: Implications for American Steel Design*. American Society of Civil Engineers, New York, NY.

ASCE. 2002. *Minimum Design Loads for Buildings and Other Structures*. SEI/ASCE 7-02. American Society of Civil Engineers, Reston, VA.

ASTM. 2001a. *Standard Specification for Carbon Steel Bolts and Studs, 60000 psi Tensile Strength (A307–00)*. American Society for Testing and Materials, West Conshohocken, PA.

ASTM. 2001b. *Standard Specification for Structural Bolts, Steel, Heat-Treated 120/105 ksi Minimum Tensile Strength (A325-00)*. American Society for Testing and Materials, West Conshohocken, PA.

ASTM. 2001c. *Standard Specification for Heat-Treated Steel Structural Bolts, 150 ksi Minimum Tensile Strength (A490-00)*. American Society for Testing and Materials, West Conshohocken, PA.

ASTM. 2001d. *Standard Specification for Quenched and Tempered Steel Bolts and Studs (A449-00)*. American Society for Testing and Materials, West Conshohocken, PA.

AWS. 1987. *Welding Handbook*, 8th ed., **1**, *Welding Technology*. American Welding Society, Miami, FL.

AWS. 2000. *Structural Welding Code-Steel*. ANSI/AWS D1.1:2002, American Welding Society, Miami, FL.

Blodgett, O.W. Undated. Distortion... How to Minimize It with Sound Design Practices and Controlled Welding Procedures Plus Proven Methods for Straightening Distorted Members. *Bulletin G261*, The Lincoln Electric Company, Cleveland, OH.

Chen, W.F. (editor). 1987. *Joint Flexibility in Steel Frames*. Elsevier, London.

Chen, W.F. (editor). 2000. *Practical Analysis for Semi-Rigid Frame Design*. World Scientific, Singapore.

Chen, W.F. and Lui, E.M. 1991. *Stability Design of Steel Frames*. CRC Press, Boca Raton, FL.

Chen, W.F., Goto, Y., and Liew, J.Y.R. 1996. *Stability Design of Semi-Rigid Frames*. John Wiley & Sons, New York, NY.

CTBUH. 1993. *Semi-Rigid Connections in Steel Frames*. Council on Tall Buildings and Urban Habitat, Committee 43, McGraw-Hill, New York, NY.

Dewolf, J.T. and Ricker, D.T. 1990. *Column Base Plates*. Steel Design Guide Series 1, American Institute of Steel Construction, Chicago, IL.

Disque, R.O. 1973. Inelastic K-Factor in Column Design. *AISC Eng. J.*, **10**(2): 33–35.

Engelhardt, M.D., Winneberger, T., Zekany, A.J., and Potyraj, T.J. 1996. The Dogbone Connection: Part II. *Modern Steel Construction*, **36**(8): 46–55.

Engelhardt, M.D., Winneberger, T., Zekany, A.J., and Potyraj, T.J. 1998. Experimental Investigation of Dogbone Moment Connections. *AISC Eng. J.*, **35**(4): 128–139.

Engelhardt, M.D., Fry, G.T., Jones, S.L., Venti, M., and Holliday, S.D. 2000. Experimental Investigation of Reduced Beam Section Connections with Composite Slabs. Paper presented at the Fourth US–Japan Workshop on Steel Fracture Issues, San Francisco, 11 pp.

Faella, C., Piluso, V., and Rizzano, G. 2000. *Structural Steel Semirigid Connections*. CRC Press, Boca Raton, FL.

Galambos, T.V. (editor). 1998. *Guide to Stability Design Criteria for Metal Structures*, 5th ed., John Wiley & Sons, New York, NY.

Galambos, T.V. and Chapuis, J. 1980. LRFD Criteria for Composite Columns and Beam Columns. Washington University, Department of Civil Engineering, St. Louis, MO.

Gaylord, E.H., Gaylord, C.N., and Stallmeyer, J.E. 1992. *Design of Steel Structures*, 3rd ed. McGraw-Hill, New York, NY.

Gilton, C., Chi, B., and Uang, C.-M. 2000. Cyclic Response of RBS Moment Connections: Weak-Axis Configuration and Deep Column Effects. *Structural Systems Research Project Report No. SSRP-2000/03*, Department of Structural Engineering, University of California, San Diego, La Jolla, CA, 197 pp.

Gross, J.L., Engelhardt, M.D., Uang, C.-M., Kasai, K., and Iwankiw, N.R. 1999. *Modification of Existing Welded Steel Moment Frame Connections for Seismic Resistance*. Steel Design Guide Series 12, American Institute of Steel Construction, Chicago, IL.

Grubbs, K.V. 1997. The Effect of Dogbone Connection on the Elastic Stiffness of Steel Moment Frames. M.S. Thesis, University of Texas at Austin, 54 pp.

International Building Code (IBC). 2000. International Code Council. Falls Church, VA.

Iwankiw, N.R. and Carter, C. 1996. The Dogbone: A New Idea to Chew On. *Modern Steel Construction*, **36**(4): 18–23.

Kulak, G.L., Fisher, J.W., and Struik, J.H.A. 1987. *Guide to Design Criteria for Bolted and Riveted Joints*, 2nd ed. John Wiley & Sons, New York, NY.

Lee, G.C., Morrel, M.L., and Ketter, R.L. 1972. *Design of Tapered Members*. WRC Bulletin No. 173.

Marsh, M.L. and Burdette, E.G. 1985a. Multiple Bolt Anchorages: Method for Determining the Effective Projected Area of Overlapping Stress Cones. *AISC Eng. J.*, **22**(1): 29–32.

Marsh, M.L. and Burdette, E.G. 1985b. Anchorage of Steel Building Components to Concrete. *AISC Eng. J.*, **22**(1): 33–39.

Moore, K.S., Malley, J.O., and Engelhardt, M.D. 1999. *Design of Reduced Beam Section (RBS) Moment Frame Connections*. Steel Tips. Structural Steel Educational Council Technical Information & Product Service, 36 pp.

Munse, W.H. and Chesson, E., Jr. 1963. Riveted and Bolted Joints: Net Section Design. *ASCE J. Struct. Div.*, **89**(1): 107–126.

Plumier, A. 1997. The Dogbone: Back to the Future. *AISC Eng. J.*, **34**(2): 61–67.

Rains, W.A. 1976. A New Era in Fire Protective Coatings for Steel. *Civil Eng.*, ASCE, September: 80–83.

RCSC. 2000. *Load and Resistance Factor Design Specification for Structural Joints Using ASTM A325 or A490 Bolts.* American Institute of Steel Construction, Chicago, IL.

Rossow, E.C. 1996. *Analysis and Behavior of Structures.* Prentice Hall, Upper Saddle River, NJ.

Shipp, J.G. and Haninger, E.R. 1983. Design of Headed Anchor Bolts. *AISC Eng. J.*, **20**(2): 58–69.

SSRC 1993. *Is Your Structure Suitably Braced?* Structural Stability Research Council, Bethlehem, PA.

Steel Deck Institute. 2001. *Design Manual for Composite Decks, Form Decks and Roof Decks, Publication No. 30.* Steel Deck Institute, Fox River Grove, IL.

Uang, C.-M. and Fan, C.-C. 1999. Cyclic Instability of Steel Moment Connections with Reduced Beams Sections. *Structural Systems Research Project Report No. SSRP-99/21*, Department of Structural Engineering, University of California, San Diego, La Jolla, CA, 51 pp.

Uniform Building Code (UBC). 1997. Volume 2 — Structural Engineering Design Provisions, *International Conference of Building Officials*, Whittier, CA.

Yu, Q.S., Gilton, C., and Uang, C.-M. 2000. Cyclic Response of RBS Moment Connections: Loading Sequence and Lateral Bracing Effects. *Structural Systems Research Project Report No. SSRP-99/13*, Department of Structural Engineering, University of California, San Diego, La Jolla, CA, 119 pp.

Yura, J.A. 2001. Fundamentals of Beam Bracing. *AISC Eng. J.*, **38**(1): 11–26.

Further Reading

The following publications provide additional sources of information for the design of steel structures:

General Information

AISC Design Guide Series: Design Guide 1: *Column Base Plates*, Dewolf and Ricker; Design Guide 2: *Design of Steel and Composite Beams with Web Openings*, Darwin; Design Guide 3: *Considerations for Low-Rise Buildings*, Fisher and West; Design Guide 4: *Extended End-Plate Moment Connections*, Murray; Design Guide 5: *Design of Low- and Medium-Rise Steel Buildings*, Allison; Design Guide 6: *Load and Resistance Factor Design of W-Shapes Encased in Concrete*, Griffes; Design Guide 7: *Industrial Buildings — Roofs to Column Anchorage*, Fisher; Design Guide 8: *Partially Restrained Composite Connections*, Leon; Design Guide 9: *Torsional Analysis of Structural Steel Members*, Seaburg and Carter; Design Guide 10: *Erection Bracing of Low-Rise Structural Steel Frames*, Fisher and West; Design Guide 11: *Floor Vibration Due to Human Activity*, Murray, Allen and Ungar; Design Guide 12: *Modification of Existing Steel Welded Moment Frame Connections for Seismic Resistance*, Gross, Engelhardt, Uang, Kasai and Iwankiw (1999); Design Guide 13: *Wide-Flange Column Stiffening at Moment Connections*, Carter. American Institute of Steel Construction, Chicago, IL.

Chen, W.F. and Lui, E.M. 1987. *Structural Stability — Theory and Implementation.* Elsevier, New York, NY.

Chen, W.F. and Kim, S.-E. 1997. *LRFD Steel Design Using Advanced Analysis.* CRC Press, Boca Raton, FL.

Englekirk, R. 1994. *Steel Structures — Controlling Behavior through Design.* John Wiley & Sons, New York, NY.

Fukumoto, Y. and Lee, G. 1992. *Stability and Ductility of Steel Structures under Cyclic Loading.* CRC Press, Boca Raton, FL.

Stability of Metal Structures — A World View. 1991. 2nd ed. Lynn S. Beedle (editor-in-chief), Structural Stability Research Council, Lehigh University, Bethlehem, PA.

Trahair, N.S. 1993. *Flexural–Torsional Buckling of Structures.* CRC Press, Boca Raton, FL.

Allowable Stress Design

Adeli, H. 1988. *Interactive Microcomputer-Aided Structural Steel Design.* Prentice Hall, Englewood Cliffs, NJ.

Cooper, S.E. and Chen, A.C. 1985. *Designing Steel Structures — Methods and Cases.* Prentice Hall, Englewood Cliffs, NJ.

Crawley, S.W. and Dillon, R.M. 1984. *Steel Buildings Analysis and Design*, 3rd ed. John Wiley & Sons, New York, NY.

Fanella, D.A., Amon, R., Knobloch, B., and Mazumder, A. 1992. *Steel Design for Engineers and Architects*, 2nd ed. Van Nostrand Reinhold, New York, NY.

Kuzmanovic, B.O. and Willems, N. 1983. *Steel Design for Structural Engineers*, 2nd ed. Prentice Hall, Englewood Cliffs, NJ.

McCormac, J.C. 1981. *Structural Steel Design*. 3rd ed. Harper & Row, New York, NY.

Segui, W.T. 1989. *Fundamentals of Structural Steel Design*. PWS-KENT, Boston, MA.

Spiegel, L. and Limbrunner, G.F. 2002. *Applied Structural Steel Design*, 4th ed. Prentice Hall, Upper Saddle River, NJ.

Plastic Design

Horne, M.R. and Morris, L.J. 1981. *Plastic Design of Low-Rise Frames*, Constrado Monographs. Collins, London.

Plastic Design in Steel—A Guide and Commentary. 1971. 2nd ed. ASCE Manual No. 41, ASCE-WRC, New York, NY.

Load and Resistance Factor Design

Geschwindner, L.F., Disque, R.O., and Bjorhovde, R. 1994. *Load and Resistance Factor Design of Steel Structures*, Prentice Hall, Englewood Cliffs, NJ.

McCormac, J.C. 1995. *Structural Steel Design — LRFD Method*, 2nd ed. Harper & Row, New York, NY.

Salmon, C.G. and Johnson, J.E. 1996. *Steel Structures — Design and Behavior*, 4th ed. Harper & Row, New York, NY.

Segui, W.T. 1999. *LRFD Steel Design*, 2nd ed. Brooks/Cole, Pacific Grove, CA.

Smith, J.C. 1996. *Structural Steel Design — LRFD Approach*, 2nd ed. John Wiley & Sons, New York, NY.

Tamboli, A.R. 1997. *Steel Design Handbook — LRFD Method*, McGraw-Hill, New York, NY.

Relevant Websites

www.aisc.org
www.aws.org
www.boltcouncil.org
www.icbo.org
www.iccsafe.org
www.sbcci.org
www.steel.org

5

Steel Frame Design Using Advanced Analysis

S. E. Kim
Department of Civil Engineering,
Sejong University,
Seoul, South Korea

Wai-Fah Chen
College of Engineering,
University of Hawaii at Manoa,
Honolulu, HI

5.1   Introduction

The steel design methods used in the United States are allowable stress design (ASD), plastic design (PD), and load and resistance factor design (LRFD). In ASD, the stress computation is based on a first-order elastic analysis, and the geometric nonlinear effects are implicitly accounted for in the member design equations. In PD, a first-order plastic-hinge analysis is used in the structural analysis. PD allows inelastic force redistribution throughout the steel structural system. Since geometric nonlinearity and gradual yielding effects are not accounted for in the analysis of PD, they are approximated in member design equations. In LRFD, a first-order elastic analysis with amplification factors or a direct second-order elastic analysis is used to account for geometric nonlinearity, and the ultimate strength of beam–column members is implicitly reflected in the design interaction equations. All three design

0-8493-1569-7/05/$0.00+$1.50
© 2005 by CRC Press

methods require separate member capacity checks including the calculation of the K-factor. In the following, the characteristics of the LRFD method are briefly described.

The strength and stability of a structural system and its members are related, but the interaction is treated separately in the current American Institute of Steel Construction (AISC)-LRFD Specification [1]. In current practice, the interaction between the structural system and its members is represented by the effective length factor. This aspect is described in the following excerpt from SSRC Technical Memorandum No. 5 [2]:

> Although the maximum strength of frames and the maximum strength of component members are interdependent (but not necessarily coexistent), it is recognized that in many structures it is not practical to take this interdependence into account rigorously. At the same time, it is known that difficulties are encountered in complex frameworks when attempting to compensate automatically in column design for the instability of the entire frame (for example, by adjustment of column effective length). Therefore, SSRC recommends that, in design practice, the two aspects, stability of separate members and elements of the structure and stability of the structure as a whole, be considered separately.

This design approach is marked in Figure 5.1 as the indirect analysis and design method.

In the current AISC-LRFD Specification [1], first- or second-order elastic analysis is used to analyze a structural system. In using first-order elastic analysis, the first-order moment is amplified by B_1 and B_2 factors to account for second-order effects. In the Specification, the members are isolated from a structural system, and they are then designed by the member strength curves and interaction equations as given in the Specifications, which implicitly account for the second-order effects, inelasticity, residual stresses, and geometric imperfections [3]. The column curve and the beam curve were developed by a curve-fit to both theoretical solutions and experimental data, while the beam–column interaction equations were determined by a curve-fit to the so-called "exact" plastic-zone solutions generated by Kanchanalai [4].

In order to account for the influence of a structural system on the strength of individual members, the effective length factor is used as illustrated in Figure 5.2. The effective length method generally provides a good design of framed structures. However, several difficulties are associated with the use of the effective length method, which are as follows:

1. The effective length approach cannot accurately account for the interaction between the structural system and its members. This is because the interaction in a large structural system is too complex to be represented by the simple effective length factor K. As a result, this method cannot accurately predict the actual strengths required of its framed members.

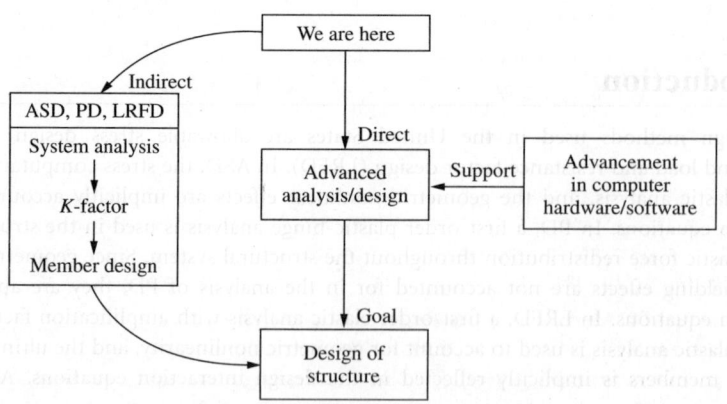

FIGURE 5.1   Analysis and design methods.

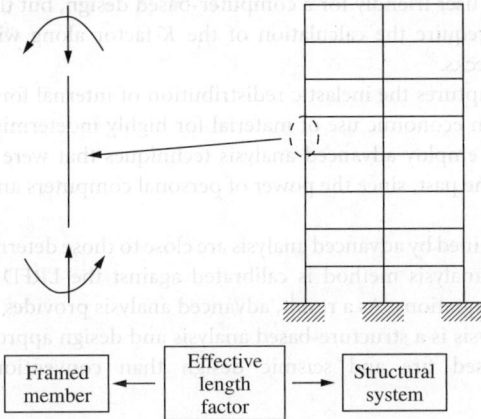

Framed member	←	Effective length factor	→	Structural system

FIGURE 5.2   Interaction between a structural system and its component members.

2. The effective length method cannot capture the inelastic redistributions of internal forces in a structural system, since the first-order elastic analysis with B_1 and B_2 factors accounts only for second-order effects and not the inelastic redistribution of internal forces. The effective length method provides a conservative estimation of the ultimate load-carrying capacity of a large structural system.

3. The effective length method cannot predict the failure modes of a structural system subject to a given load. This is because the LRFD interaction equation does not provide any information about failure modes of a structural system at the factored loads.

4. The effective length method is not user friendly for a computer-based design.

5. The effective length method requires a time-consuming process of separate member capacity checks involving the calculation of K-factors.

With the development of computer technology, two aspects, the stability of separate members and the stability of the structure as a whole, can be treated rigorously for the determination of the maximum strength of the structures. This design approach is shown in Figure 5.1 as the direct analysis and design method. The development of the direct approach to design is called "advanced analysis" or, more specifically, "second-order inelastic analysis for frame design." In this direct approach, there is no need to compute the effective length factor, since separate member capacity checks encompassed by the specification equations are not required. With the current available computing technology, it is feasible to employ advanced analysis techniques for direct frame design. This method was considered impractical for design office use in the past. The purpose of this chapter is to present a practical, direct method of steel frame design, using advanced analysis, that will produce almost identical member sizes as those of the LRFD method.

The advantages of advanced analysis in design use are outlined as follows:

1. Advanced analysis is another tool used by structural engineers in steel design, and its adoption is not mandatory but will provide a flexibility of options to the designer.

2. Advanced analysis captures the limit state strength and stability of a structural system and its individual members directly, so separate member capacity checks encompassed by specification equations are not required.

3. Compared to the LRFD and ASD, advanced analysis provides more information of structural behavior by direct inelastic second-order analysis.

4. Advanced analysis overcomes the difficulties due to incompatibility between the elastic global analysis and the limit state member design in the conventional LRFD method.

5. Advanced analysis is user friendly for a computer-based design, but the LRFD and ASD methods are not, since they require the calculation of the *K*-factor along with the analysis to separate member capacity checks.

6. Advanced analysis captures the inelastic redistribution of internal forces throughout a structural system, and allows an economic use of material for highly indeterminate steel frames.

7. It is now feasible to employ advanced analysis techniques that were considered impractical for design office use in the past, since the power of personal computers and engineering workstations is rapidly increasing.

8. Member sizes determined by advanced analysis are close to those determined by the LRFD method, since the advanced analysis method is calibrated against the LRFD column curve and beam–column interaction equations. As a result, advanced analysis provides an alternative to the LRFD.

9. Since advanced analysis is a structure-based analysis and design approach, it is more appropriate for performance-based fire and seismic design than conventional member-based design approaches [5].

Among various advanced analyses, including plastic zone, quasi-plastic hinge, elastic–plastic hinge, notional-load plastic hinge, and refined plastic-hinge methods, the refined plastic-hinge method is recommended since it retains the efficiency and simplicity of computation and accuracy for practical use. The method is developed by imposing simple modifications on the conventional elastic–plastic hinge method. These include a simple modification to account for the gradual sectional stiffness degradation at the plastic-hinge locations and to include the gradual member stiffness degradation between two plastic hinges.

The key considerations of the conventional LRFD method and the practical advanced analysis method are compared in Table 5.1. While the LRFD method does account for key behavioral effects implicitly in its column strength and beam–column interaction equations, the advanced analysis method accounts for these effects explicitly through stability functions, stiffness degradation functions, and geometric imperfections, which are discussed in detail in Section 5.2.

Advanced analysis holds many answers about the real behavior of steel structures and, as such, the authors recommend the proposed design method to engineers seeking to perform frame design with efficiency and rationality, yet consistent with the present LRFD Specification. In the following sections, we will present a practical advanced analysis method for the design of steel frame structures with LRFD. The validity of the approach will be demonstrated by comparing case studies of actual members and frames with the results of analysis/design based on exact plastic-zone solutions and

TABLE 5.1   Key Considerations of Load and Resistance Factor Design (LRFD) and Proposed Methods

Key considerations	LRFD	Proposed methods
Second-order effects	Column curve B_1, B_2 factors	Stability function
Geometric imperfection	Column curve	Explicit imperfection modeling method
		$\psi = 1/500$ for unbraced frame
		$\delta_c = L_c/1000$ for braced frame
		Equivalent notional load method
		$\alpha = 0.002$ for unbraced frame
		$\alpha = 0.004$ for braced frame
		Further reduced tangent modulus method
		$E'_t = 0.85E_t$
Stiffness degradation associated with residual stresses	Column curve	CRC tangent modulus
Stiffness degradation associated with flexure	Column curve Interaction equations	Parabolic degradation function
Connection nonlinearity	No procedure	Power model/rotational spring

LRFD designs. The wide range of case studies and comparisons should confirm the validity of this advanced method.

5.2 Practical Advanced Analysis

This section presents a practical advanced analysis method for the direct design of steel frames by eliminating separate member capacity checks by the specification. The refined plastic-hinge method was developed and refined by simply modifying the conventional elastic–plastic hinge method to achieve both simplicity and a realistic representation of the actual behavior [6,7]. Verification of the method will be given in the next section to provide the final confirmation of the validity of the method.

Connection flexibility can be accounted for in advanced analysis. Conventional analysis and design of steel structures are usually carried out under the assumption that beam-to-column connections are either fully rigid or ideally pinned. However, most connections in practice are "semirigid" and their behavior lies between these two extreme cases. In the AISC-LRFD Specification [1], two types of construction are designated: Type FR (fully restrained) construction and Type PR (partially restrained) construction. The LRFD Specification permits the evaluation of the flexibility of connections by "rational means."

Connection behavior is represented by its moment–rotation relationship. Extensive experimental work on connections has been performed, and a large body of moment–rotation data collected. With this database, researchers have developed several connection models including: linear, polynomial, *B*-spline, power, and exponential. Herein, the three-parameter power model proposed by Kishi and Chen [8] is adopted.

Geometric imperfections should be modeled in frame members when using advanced analysis. Geometric imperfections result from unavoidable error during fabrication or erection. For structural members in building frames, the types of geometric imperfections are out-of-straightness and out-of-plumbness. Explicit modeling and equivalent notional loads were used in the past to account for geometric imperfections. In this section, a new method based on further reduction of the tangent stiffness of members is developed [6,9]. This method provides a simple means to account for the effect of imperfection without inputting notional loads or explicit geometric imperfections.

The practical advanced analysis method described in this section is limited to two-dimensional braced, unbraced, and semirigid frames subjected to static loads. The spatial behavior of frames is not considered, and lateral torsional buckling is assumed to be prevented by adequate lateral bracing. A compact W-section is assumed so that sections can develop full plastic moment capacity without local buckling. Both strong-axis and weak-axis bending of wide flange sections have been studied using the practical advanced analysis method [6]. In recent developments, several studies have used advanced analysis of structures by including effects such as spatial behavior [10–12], member local buckling [13,14], and lateral torsional buckling [15,16]. The present method may be considered an interim analysis/design procedure between the conventional LRFD method widely used now and a more rigorous advanced analysis/design method such as the plastic-zone method to be developed in the future for practical use.

5.2.1 Second-Order Refined Plastic-Hinge Analysis

In this section, a method called the *refined plastic-hinge* approach is presented. This method is comparable to the elastic–plastic hinge analysis in efficiency and simplicity, but without its limitations. In this analysis, stability functions are used to predict second-order effects. The benefit of stability functions is that they make the analysis method practical by using only one element per beam column. The refined plastic-hinge analysis uses a two-surface yield model and an effective tangent modulus to account for stiffness degradation due to distributed plasticity in framed members. The member stiffness is assumed to degrade gradually as the second-order forces at critical locations approach the cross-section plastic strength. Column tangent modulus is used to represent the effective stiffness of the member when it is loaded with a high axial load. Thus, the refined plastic-hinge model approximates the effect of distributed plasticity along the element length caused by initial imperfections and large bending and axial force actions. In fact, researches by Liew et al. [7,17], Kim and Chen [9], and Kim [6] have shown that refined plastic-hinge analysis captures the interaction of strength and stability of structural systems and that of their component elements. This type of analysis method may, therefore, be classified as an *advanced analysis* and separate specification member capacity checks are not required.

5.2.1.1 Stability Function

To capture second-order effects, stability functions are recommended since they lead to large savings in modeling and solution efforts by using one or two elements per member. The simplified stability functions reported by Chen and Lui [18] or an alternative may be used. Considering the prismatic beam–column element, the incremental force–displacement relationship of this element may be written as

$$\begin{bmatrix} \dot{M}_A \\ \dot{M}_B \\ \dot{P} \end{bmatrix} = \frac{EI}{L} \begin{bmatrix} S_1 & S_2 & 0 \\ S_2 & S_1 & 0 \\ 0 & 0 & A/I \end{bmatrix} \begin{bmatrix} \dot{\theta}_A \\ \dot{\theta}_B \\ \dot{e} \end{bmatrix} \tag{5.1}$$

where S_1, S_2 are stability functions, $\dot{M}_A$, $\dot{M}_B$ are incremental end moments, $\dot{P}$ is the incremental axial force, $\dot{\theta}_A$, $\dot{\theta}_B$ are incremental joint rotations, $\dot{e}$ is the incremental axial displacement, A, I, L are area, moment of inertia, and length of beam–column element, respectively, and E is the modulus of elasticity.

In this formulation, all members are assumed to be adequately braced to prevent out-of-plane buckling and their cross-sections are compact to avoid local buckling.

5.2.1.2 Cross-Section Plastic Strength

Based on the AISC-LRFD bilinear interaction equations [1], the cross-section plastic strength may be expressed as Equation 5.2. These AISC-LRFD cross-section plastic strength curves may be adopted for both strong-axis and weak-axis bendings (Figure 5.3):

$$\frac{P}{P_y} + \frac{8}{9}\frac{M}{M_p} = 1.0 \quad \text{for } \frac{P}{P_y} \geq 0.2 \tag{5.2a}$$

$$\frac{1}{2}\frac{P}{P_y} + \frac{M}{M_p} = 1.0 \quad \text{for } \frac{P}{P_y} \leq 0.2 \tag{5.2b}$$

where P, M are second-order axial force and bending moment, P_y is the squash load, and M_p is the plastic moment capacity.

5.2.1.3 CRC Tangent Modulus

The CRC tangent modulus concept is employed to account for the gradual yielding effect due to residual stresses along the length of members under axial loads between two plastic hinges. In this concept, the elastic modulus E, instead of moment of inertia I, is reduced to account for the reduction of the elastic portion of the cross-section since the reduction of elastic modulus is easier to implement than that of moment of inertia for different sections. The reduction rate in stiffness between weak and strong axes is different, but this is not considered here because rapid degradation in stiffness in the weak-axis strength is compensated well by the stronger weak-axis plastic strength. As a result, this simplicity will make the present methods practical. From Chen and Lui [18], the CRC E_t is written as (Figure 5.4):

$$E_t = 1.0E \quad \text{for } P \leq 0.5P_y \tag{5.3a}$$

$$E_t = 4\frac{P}{P_y}E\left(1 - \frac{P}{P_y}\right) \quad \text{for } P > 0.5P_y \tag{5.3b}$$

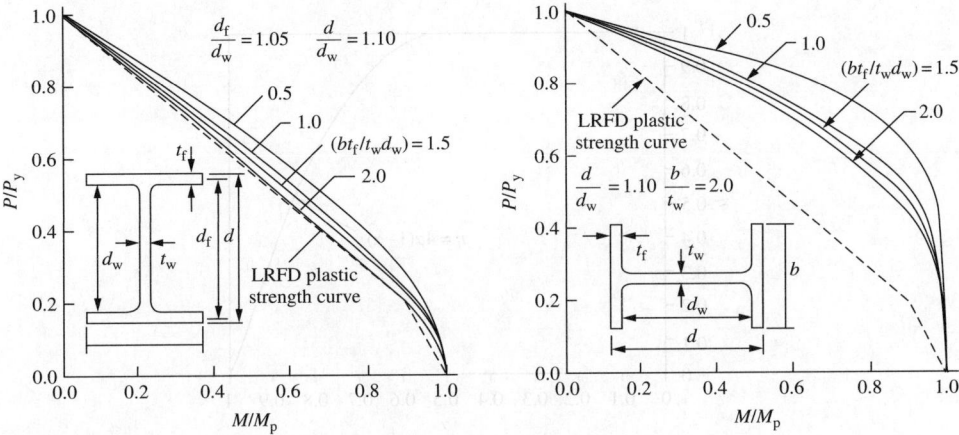

FIGURE 5.3 Strength interaction curves for wide-flange sections.

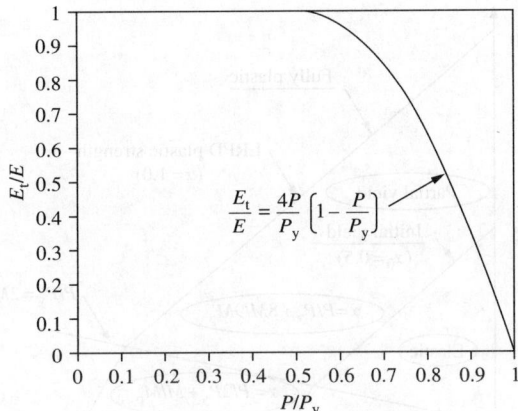

FIGURE 5.4 Member tangent stifness degradation derived from the CRC column curve.

5.2.1.4 Parabolic Function

The tangent modulus model in Equation 5.3 is suitable for $P/P_y > 0.5$, but is not sufficient to represent the stiffness degradation for cases with small axial forces and large bending moments. A gradual stiffness degradation of the plastic hinge is required to represent the distributed plasticity effects associated with bending actions. We shall introduce the hardening plastic-hinge model to represent the gradual transition from elastic stiffness to zero stiffness associated with a fully developed plastic hinge. When the hardening plastic hinges are present at both ends of an element, the incremental force–displacement relationship may be expressed as [19]:

$$
\begin{bmatrix} \dot{M}_A \\ \dot{M}_B \\ \dot{P} \end{bmatrix} = \frac{E_t I}{L} \begin{bmatrix} \eta_A \left[S_1 - \dfrac{S_2^2}{S_1}(1 - \eta_B) \right] & \eta_A \eta_B S_2 & 0 \\ \eta_A \eta_B S_2 & \eta_B \left[S_1 - \dfrac{S_2^2}{S_1}(1 - \eta_A) \right] & 0 \\ 0 & 0 & A/I \end{bmatrix} \begin{bmatrix} \dot{\theta}_A \\ \dot{\theta}_B \\ \dot{e} \end{bmatrix} \tag{5.4}
$$

where $\dot{M}_A$, $\dot{M}_B$, $\dot{P}$ are incremental end moments and axial force, respectively, S_1, S_2 are stability functions, E_t is the tangent modulus, and η_A, η_B are the element stiffness parameters.

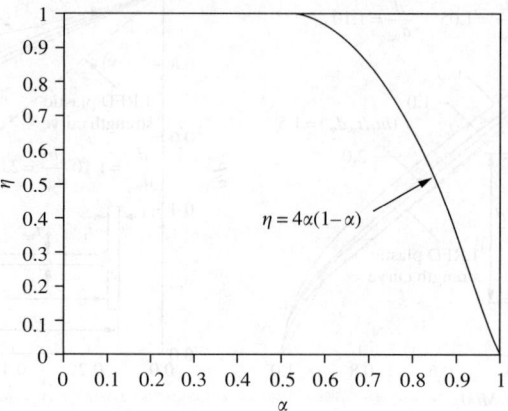

FIGURE 5.5 Parabolic plastic-hinge stiffness degradation function with $\alpha_0 = 0.5$ based on the load and resistance factor design sectional strength equation.

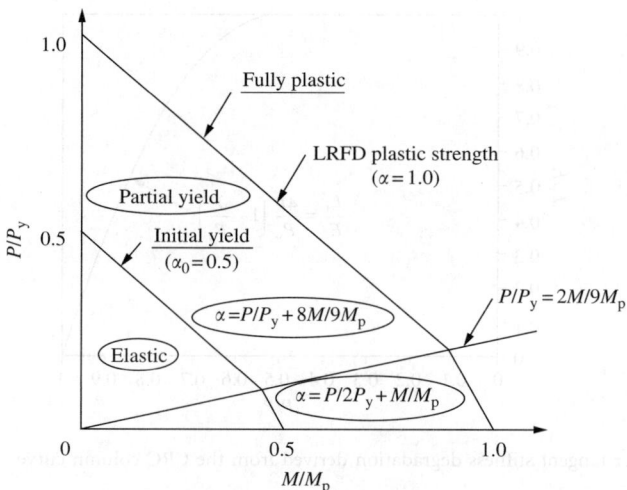

FIGURE 5.6 Smooth stiffness degradation for a work-hardening plastic hinge based on the load and resistance factor design sectional strength curve.

The parameter η represents a gradual stiffness reduction associated with flexure at sections. The partial plastification at cross-sections in the end of elements is denoted by $0 < \eta < 1$. η may be assumed to vary according to the parabolic expression (Figure 5.5):

$$\eta = 4\alpha(1 - \alpha) \quad \text{for } \alpha > 0.5 \tag{5.5}$$

where α is the force-state parameter obtained from the limit state surface corresponding to the element end (Figure 5.6):

$$\alpha = \frac{P}{P_y} + \frac{8}{9}\frac{M}{M_p} \quad \text{for } \frac{P}{P_y} \geq \frac{2}{9}\frac{M}{M_p} \tag{5.6a}$$

$$\alpha = \frac{1}{2}\frac{P}{P_y} + \frac{M}{M_p} \quad \text{for } \frac{P}{P_y} < \frac{2}{9}\frac{M}{M_p} \tag{5.6b}$$

where P, M are second-order axial force and bending moment at the cross-section, respectively and M_p is the plastic moment capacity.

5.2.2 Analysis of Semirigid Frames

5.2.2.1 Practical Connection Modeling

The three-parameter power model contains three parameters: initial connection stiffness R_{ki}, ultimate connection moment capacity M_u, and shape parameter n. The power model may be written as (Figure 5.7):

$$m = \frac{\theta}{(1 + \theta^n)^{1/n}} \quad \text{for } \theta > 0, \; m > 0 \tag{5.7}$$

where $m = M/M_u$, $\theta = \theta_r/\theta_0$, θ_0 is the reference plastic rotation, M_u/R_{ki}, M_u is the ultimate moment capacity of the connection, R_{ki} is the initial connection stiffness, and n is the shape parameter. When the connection is loaded, the connection tangent stiffness, R_{kt}, at an arbitrary rotation, θ_r, can be derived by simply differentiating Equation 5.7 as

$$R_{kt} = \frac{dM}{d|\theta_r|} = \frac{M_u}{\theta_0(1 + \theta^n)^{1+1/n}} \tag{5.8}$$

When the connection is unloaded, the tangent stiffness is equal to the initial stiffness:

$$R_{kt} = \frac{dM}{d|\theta_r|} = \frac{M_u}{\theta_0} = R_{ki} \tag{5.9}$$

It is observed that a small value of the power index, n, makes a smooth transition curve from the initial stiffness, R_{kt}, to the ultimate moment, M_u. On the contrary, a large value of the index, n, makes the transition more abruptly. In the extreme case, when n is infinity, the curve becomes a bilinear line consisting of the initial stiffness, R_{ki} and the ultimate moment capacity, M_u.

5.2.2.2 Practical Estimation of Three Parameters Using a Computer Program

An important task for the practical use of the power model is to determine the three parameters for a given connection configuration. One difficulty in determining the three parameters is the need for numerical iteration, especially to estimate the ultimate moment, M_u. A set of nomographs was

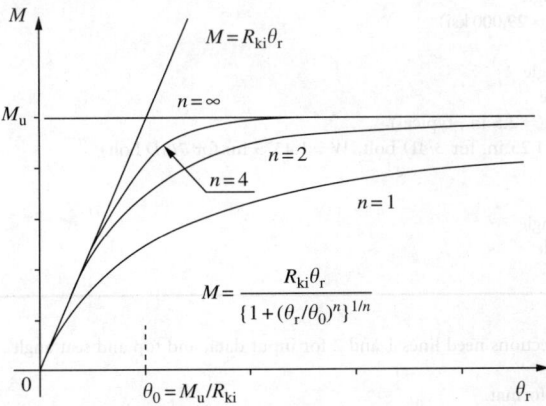

FIGURE 5.7 Moment–rotation behavior of the three-parameter model.

proposed by Kishi et al. [20] to overcome the difficulty. Even though the purpose of these nomographs is to allow the engineer to rapidly determine the three parameters for a given connection configuration, the nomographs require other efforts for engineers to know how to use those, and the values of the nomographs are approximate.

Herein, one simple way to avoid difficulties described above is presented. A direct and easy estimation of the three parameters may be achieved by use of a simple computer program 3PARA.f. The operating procedure of the program is shown in Figure 5.8. The input data CONN.DAT may be easily generated corresponding to the input format listed in Table 5.2.

As for the shape parameter n, the equations developed by Kishi et al. [20] are implemented here. Using a statistical technique for n values, empirical equations of n are determined as a linear function of $\log_{10}\theta_0$ shown in Table 5.3. This n value may be calculated using 3PARA.f.

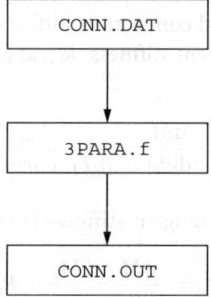

FIGURE 5.8 Operating procedure of computer program estimating the three parameters.

TABLE 5.2 Input Format

Line	Input data	Remark
1	ITYPE F_y E	Connection type and material properties
2	l_t t_t k_t g_t W d	Top/seat-angle data
3	l_a t_a k_a g_a	Web-angle data

ITYPE = Connection type (1 = top- and seat-angle connection, 2 = with web-angle connection).
F_y = yield strength of angle
E = Young's modulus (= 29,000 ksi)
l_t = length of top angle
t_t = thickness of top angle
k_t = k value of top angle
g_t = gauge of top angle (= 2.5 in., typical)
W = width of nut (W = 1.25 in. for 3/4D bolt, W = 1.4375 in. for 7/8D bolt)
d = depth of beam
l_a = length of web angle
t_a = thickness of web angle
k_a = k value of web angle
g_a = gauge of web angle

Notes:
1. Top- and seat-angle connections need lines 1 and 2 for input data, and top and seat angles with web-angle connections need line 1, 2, and 3.
2. All input data are in free format.
3. Top- and seat-angle sizes are assumed to be the same.
4. Bolt sizes of top angle, seat angle, and web angle are assumed to be the same.

5.2.2.3 Load–Displacement Relationship Accounting for Semirigid Connection

The connection may be modeled as a rotational spring in the moment–rotation relationship represented by Equation 5.10. Figure 5.9 shows a beam–column element with semirigid connections at both ends. If the effect of connection flexibility is incorporated into the member stiffness, the incremental element force–displacement relationship of Equation 5.1 is modified as [18,19]

$$
\begin{bmatrix} \dot{M}_A \\ \dot{M}_B \\ \dot{P} \end{bmatrix} = \frac{E_t I}{L} \begin{bmatrix} S_{ii}^* & S_{ij}^* & 0 \\ S_{ij}^* & S_{jj}^* & 0 \\ 0 & 0 & A/I \end{bmatrix} \begin{bmatrix} \dot{\theta}_A \\ \dot{\theta}_B \\ \dot{e} \end{bmatrix}
\tag{5.10}
$$

where

$$
S_{ii}^* = \left(S_{ii} + \frac{E_t I S_{ii} S_{jj}}{L R_{ktB}} - \frac{E_t I S_{ij}^2}{L R_{ktB}} \right) \Big/ R^*
\tag{5.11a}
$$

$$
S_{jj}^* = \left(S_{jj} + \frac{E_t I S_{ii} S_{jj}}{L R_{ktA}} - \frac{E_t I S_{ij}^2}{L R_{ktA}} \right) \Big/ R^*
\tag{5.11b}
$$

$$
S_{ij}^* = S_{ij} / R^*
\tag{5.11c}
$$

$$
R^* = \left(1 + \frac{E_t I S_{ii}}{L R_{ktA}} \right) \left(1 + \frac{E_t I S_{jj}}{L R_{ktB}} \right) - \left(\frac{E_t I}{L} \right)^2 \frac{S_{ij}^2}{R_{ktA} R_{ktB}}
\tag{5.11d}
$$

where R_{ktA}, R_{ktB} are tangent stiffnesses of connections A and B, respectively, S_{ii}, S_{ij} are generalized stability functions, and S_{ii}^*, S_{jj}^* are modified stability functions that account for the presence of end connections. The tangent stiffness (R_{ktA}, R_{ktB}) accounts for the different types of semirigid connections (see Equation 5.8).

TABLE 5.3 Empirical Equations for Shape Parameter n

Connection type	n
Single web-angle connection	$0.520 \log_{10} \theta_0 + 2.291$ for $\log_{10} \theta_0 > -3.073$
	0.695 for $\log_{10} \theta_0 < -3.073$
Double web-angle connection	$1.322 \log_{10} \theta_0 + 3.952$ for $\log_{10} \theta_0 > -2.582$
	0.573 for $\log_{10} \theta_0 < -2.582$
Top- and seat-angle connection	$2.003 \log_{10} \theta_0 + 6.070$ for $\log_{10} \theta_0 > -2.880$
	0.302 for $\log_{10} \theta_0 < -2.880$
Top- and seat-angle connection with double web angle	$1.398 \log_{10} \theta_0 + 4.631$ for $\log_{10} \theta_0 > -2.721$
	0.827 for $\log_{10} \theta_0 < -2.721$

Source: From Kishi, N., Goto, Y., Chen, W.F., and Matsuoka, K.G. 1993. *Eng. J.*, AISC, pp. 90–107. With permission.

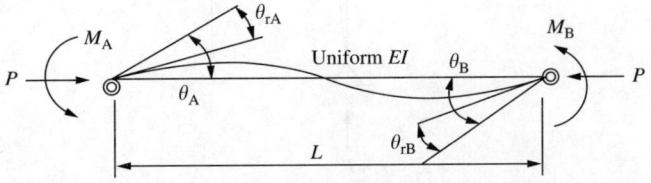

FIGURE 5.9 Beam–column element with semirigid connections.

5.2.3 Geometric Imperfection Methods

Geometric imperfection modeling combined with the Column Research Council (CRC) tangent modulus model is discussed in what follows. There are three methods: the explicit imperfection modeling method, the equivalent notional load method, and the further reduced tangent modulus method.

5.2.3.1 Explicit Imperfection Modeling Method

5.2.3.1.1 Braced Frames

The refined plastic-hinge analysis implicitly accounts for the effects of both residual stresses and spread of yielded zones. To this end, refined plastic-hinge analysis may be regarded as equivalent to the plastic-zone analysis. As a result, geometric imperfections are necessary only to consider fabrication error. For braced frames, member out-of-straightness, rather than frame out-of-plumbness, needs to be used for geometric imperfections. This is because the $P-\Delta$ effect due to the frame out-of-plumbness is diminished by braces. The ECCS [21,22], AS [23], and Canadian Standard Association (CSA) [24,25] specifications recommend an initial crookedness of column equal to 1/1000 times the column length. The AISC Code recommends the same maximum fabrication tolerance of $L_c/1000$ for member out-of-straightness. In this study, a geometric imperfection of $L_c/1000$ is adopted.

The ECCS [21,22], AS [23], and CSA [24,25] specifications recommend the out-of-straightness varying parabolically with a maximum in-plane deflection at the mid-height. They do not, however, describe how the parabolic imperfection should be modeled in analysis. Ideally, many elements are needed to model the parabolic out-of-straightness of a beam–column member, but it is not practical. In this study, two elements with a maximum initial deflection at the mid-height of a member are found adequate for capturing the imperfection. Figure 5.10 shows the out-of-straightness modeling for a braced beam–column member. It may be observed that the out-of-plumbness is equal to 1/500 when the half-segment of the member is considered. This value is identical to that of sway frames as discussed in recent papers by Kim and Chen [9,26,27]. Thus, it may be stated that the imperfection values are essentially identical for both sway and braced frames. It is noted that this explicit modeling method in braced frames requires the inconvenient imperfection modeling at the center of columns although the inconvenience is much lesser than that of the conventional LRFD method for frame design.

5.2.3.1.2 Unbraced Frames

The CSA [23,24] and the AISC Code of Standard Practice [1] set the limit of erection out-of-plumbness at $L_c/500$. The maximum erection tolerances in the AISC are limited to 1 in. toward the exterior

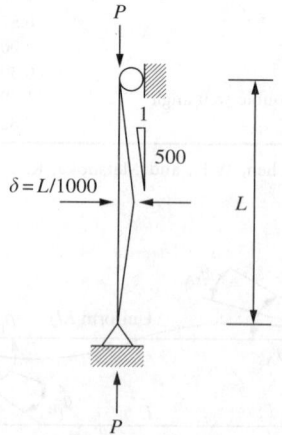

FIGURE 5.10 Explicit imperfection modeling of a braced member.

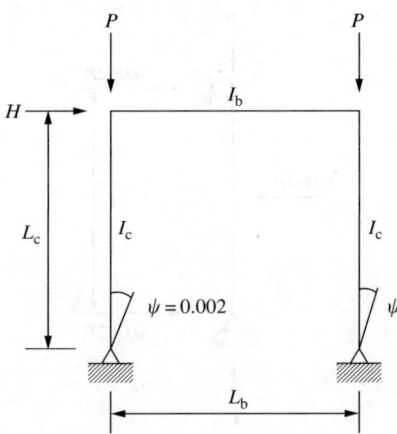

FIGURE 5.11 Explicit imperfection modeling of an unbraced frame.

of buildings and 2 in. toward the interior of buildings less than 20 stories. Considering the maximum permitted average lean of 1.5 in. in the same direction of a story, the geometric imperfection of $L_c/500$ can be used for buildings up to six stories with each story approximately 10 ft high. For taller buildings, this imperfection value of $L_c/500$ is conservative since the accumulated geometric imperfection calculated by 1/500 times building height is greater than the maximum permitted erection tolerance.

In this study, we shall use $L_c/500$ for the out-of-plumbness without any modification because the system strength is often governed by a weak story that has an out-of-plumbness equal to $L_c/500$ [28] and a constant imperfection has the benefit of simplicity in practical design. The explicit geometric imperfection modeling for an unbraced frame is illustrated in Figure 5.11.

5.2.3.2 Equivalent Notional Load Method

5.2.3.2.1 Braced Frames

The ECCS [21,22] and the CSA [23,24] introduced the equivalent load concept, which accounted for the geometric imperfections in an unbraced frame, but not in braced frames. The notional load approach for braced frames is also necessary to use the proposed methods for braced frames.

For braced frames, an equivalent notional load may be applied at mid-height of a column since the ends of the column are braced. An equivalent notional load factor equal to 0.004 is proposed here, and it is equivalent to the out-of-straightness of $L_c/1000$. When the free body of the column shown in Figure 5.12 is considered, the notional load factor, α, results in 0.002 with respect to one-half of the member length. Here, as in explicit imperfection modeling, the equivalent notional load factor is the same in concept for both sway and braced frames.

One drawback of this method for braced frames is that it requires tedious input of notional loads at the center of each column. Another is the axial force in the columns must be known in advance to determine the notional loads before analysis, but these are often difficult to calculate for large structures subject to lateral wind loads. To avoid this difficulty, it is recommended that either the explicit imperfection modeling method or the further reduced tangent modulus method be used.

5.2.3.2.2 Unbraced Frames

The geometric imperfections of a frame may be replaced by the equivalent notional lateral loads expressed as a fraction of the gravity loads acting on the story. Herein, the equivalent notional load factor of 0.002 is used. The notional load should be applied laterally at the top of each story. For sway frames subject to combined gravity and lateral loads, the notional loads should be added to the lateral loads. Figure 5.13 shows an illustration of the equivalent notional load for a portal frame.

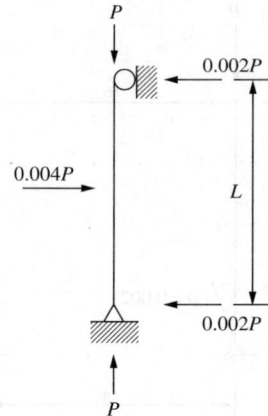

FIGURE 5.12   Equivalent notional load modeling for geometric imperfection of a braced member.

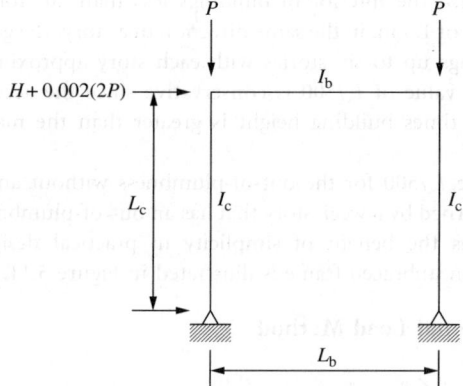

FIGURE 5.13   Equivalent notional load modeling for geometric imperfection of an unbraced frame.

5.2.3.3 Further Reduced Tangent Modulus Method

5.2.3.3.1 Braced Frames

The idea of using the reduced tangent modulus concept is to further reduce the tangent modulus, E_t, to account for further stiffness degradation due to geometrical imperfections. The degradation of member stiffness due to geometric imperfections may be simulated by an equivalent reduction of member stiffness. This may be achieved by a further reduction of tangent modulus [6,9]:

$$E'_t = 4 \frac{P}{P_y} \left(1 - \frac{P}{P_y} \right) E \xi_i \quad \text{for } P > 0.5P_y \tag{5.12a}$$

$$E'_t = E \xi_i \quad \text{for } P \leq 0.5P_y \tag{5.12b}$$

where E'_t is the reduced E_t and ξ_i is the reduction factor for geometric imperfection.

Herein, a reduction factor of 0.85 is used; the further reduced tangent modulus curves for the CRC E_t with geometric imperfections are shown in Figure 5.14. The further reduced tangent modulus concept satisfies one of the requirements for advanced analysis recommended by the SSRC task force report [29], that is, "The geometric imperfections should be accommodated implicitly within the element model.

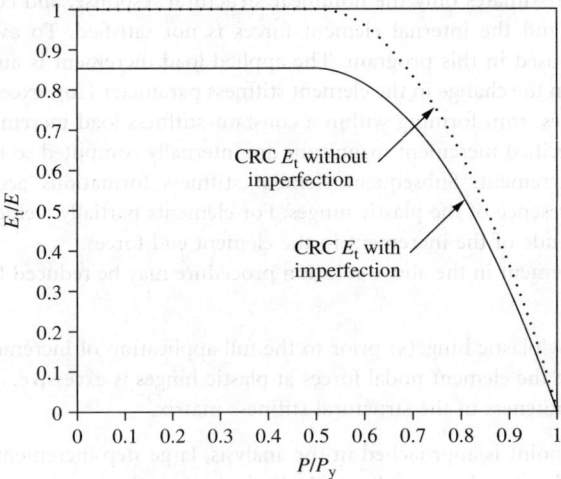

FIGURE 5.14 Further reduced CRC tangent modulus for members with geometric imperfections.

This would parallel the philosophy behind the development of most modern column strength expressions. That is, the column strength expressions in specifications such as the AISC-LRFD implicitly include the effects of residual stresses and out-of-straightness."

The advantage of this method over the other two methods is its convenience for design use, because it eliminates the inconvenience of explicit imperfection modeling or equivalent notional loads. Another benefit of this method is that it does not require the determination of the direction of geometric imperfections, often difficult to determine in a large system. On the other hand, in the other two methods, the direction of geometric imperfections must be taken correctly in coincidence with the deflection direction caused by bending moments, otherwise the wrong direction of geometric imperfection in braced frames may help the bending stiffness of columns rather than reduce it.

5.2.3.3.2 Unbraced Frames
The idea of the further reduced tangent modulus concept may also be used in the analysis of unbraced frames. Herein, as in the braced frame case, an appropriate reduction factor of 0.85 to E_t can be used [27,30,31]. The advantage of this approach over the other two methods is its convenience and simplicity because it completely eliminates the inconvenience of explicit imperfection modeling or the notional load input.

5.2.4 Numerical Implementation

The nonlinear global solution methods may be subdivided into two subgroups: (1) iterative methods and (2) simple incremental method. Iterative methods such as Newton–Raphson method, modified Newton–Raphson method, and quasi-Newton method satisfy equilibrium equations at specific external loads. In these methods, the equilibrium out-of-balance present following linear load step is eliminated (within tolerance) by taking corrective steps. The iterative methods possess the advantage of providing the exact load–displacement frame; however, they are inefficient, especially for practical purposes, in the trace of the hinge-by-hinge formation due to the requirement of the numerical iteration process.

The simple incremental method is a direct nonlinear solution technique. This numerical procedure is straightforward in concept and implementation. The advantage of this method is its computational efficiency. This is especially true when the structure is loaded into the inelastic region since tracing the hinge-by-hinge formation is required in the element stiffness formulation. For a finite increment size, this approach approximates only the nonlinear structural response, and equilibrium between the external applied loads and the internal element forces is not satisfied. To avoid this, an improved incremental method is used in this program. The applied load increment is automatically reduced to minimize the error when the change in the element stiffness parameter ($\Delta\eta$) exceeds a defined tolerance. To prevent plastic hinges from forming within a constant-stiffness load increment, load step sizes less than or equal to the specified increment magnitude are internally computed so that plastic hinges form only after the load increment. Subsequent element stiffness formations account for the stiffness reduction due to the presence of the plastic hinges. For elements partially yielded at their ends, a limit is placed on the magnitude of the increment in the element end forces.

The applied load increment in the above solution procedure may be reduced for any of the following reasons:

1. Formation of new plastic hinge(s) prior to the full application of incremental loads.
2. The increment in the element nodal forces at plastic hinges is excessive.
3. Nonpositive definiteness of the structural stiffness matrix.

As the stability limit point is approached in the analysis, large step increments may overstep a limit point. Therefore, a smaller step size is used near the limit point to obtain accurate collapse displacements and second-order forces.

5.3 Verifications

In the previous section, a practical advanced analysis method was presented for a direct two-dimensional frame design. The practical approach of geometric imperfections and semirigid connections was also discussed together with the advanced analysis method. The practical advanced analysis method was developed using simple modifications to the conventional elastic–plastic hinge analysis.

In this section, the practical advanced analysis method will be verified by the use of several benchmark problems available in the literature. Verification studies are carried out by comparing with the plastic-zone solutions as well as the conventional LRFD solutions. The strength predictions and the load–displacement relationships are checked for a wide range of steel frames including axially loaded columns, portal frame, six-story frame, and semirigid frames [6]. The three imperfection modelings, including explicit imperfection modeling, equivalent notional load modeling, and further reduced tangent modulus modeling, are also verified for a wide range of steel frames [6].

5.3.1 Axially Loaded Columns

The AISC-LRFD column strength curve is used for the calibration since it properly accounts for second-order effects, residual stresses, and geometric imperfections in a practical manner. In this study, the column strength of the proposed methods is evaluated for columns with slenderness parameters, $\left[\lambda_c = (KL/r)\sqrt{F_y/(\pi^2 E)}\right]$, varying from 0 to 2, which is equivalent to slenderness ratios (L/r) from 0 to 180 when the yield stress is equal to 36 ksi.

In explicit imperfection modeling, the two-element column is assumed to have an initial geometric imperfection equal to $L_c/1000$ at column mid-height. The predicted column strengths are compared with the LRFD curve in Figure 5.15. The errors are found to be less than 5% for slenderness ratios up to 140 (or λ_c up to 1.57). This range includes most columns used in engineering practice.

In the equivalent notional load method, notional loads equal to 0.004 times the gravity loads are applied mid-height to the column. The strength predictions are the same as those of the explicit imperfection model (Figure 5.16).

In the further reduced tangent modulus method, the reduced tangent modulus factor equal to 0.85 results in an excellent fit to the LRFD column strengths. The errors are less than 5% for columns of all slenderness ratios. These comparisons are shown in Figure 5.17.

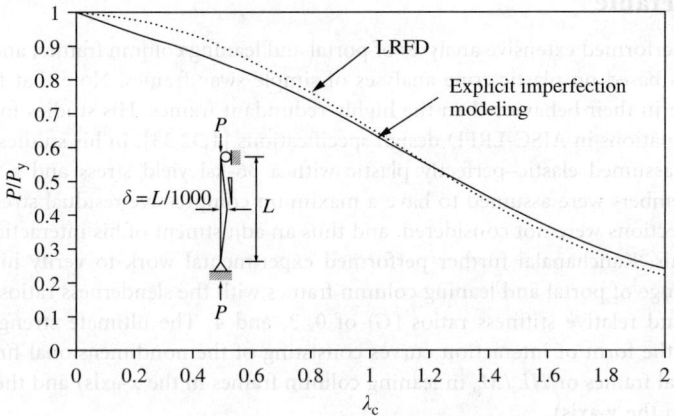

FIGURE 5.15 Comparison of strength curves for an axially loaded pin-ended column (explicit imperfection modeling method).

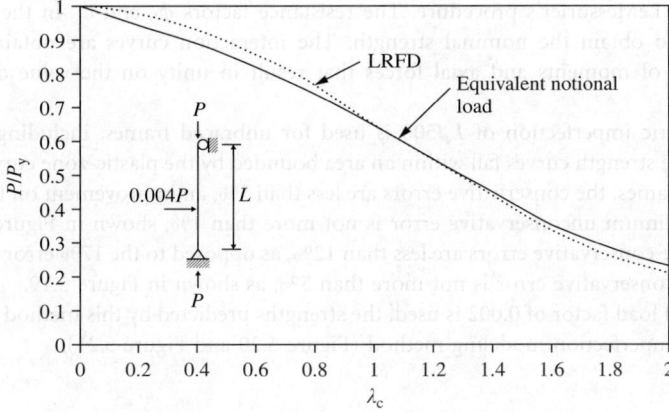

FIGURE 5.16 Comparison of strength curves for an axially loaded pin-ended column (equivalent notional load method).

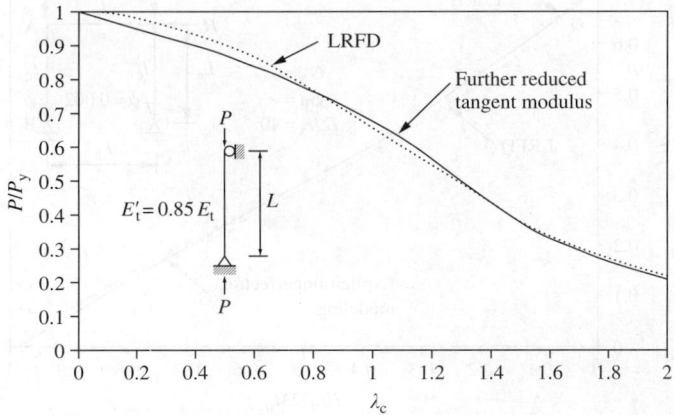

FIGURE 5.17 Comparison of strength curves for an axially loaded pin-ended column (further reduced tangent modulus method).

5.3.2 Portal Frame

Kanchanalai [4] performed extensive analyses of portal and leaning column frames, and developed exact interaction curves based on plastic-zone analyses of simple sway frames. Note that the simple frames are more sensitive in their behavior than the highly redundant frames. His studies formed the basis of the interaction equations in AISC-LRFD design specifications [1,32,33]. In his studies, the stress–strain relationship was assumed elastic–perfectly plastic with a 36-ksi yield stress and a 29,000-ksi elastic modulus. The members were assumed to have a maximum compressive residual stress of $0.3F_y$. Initial geometric imperfections were not considered, and thus an adjustment of his interaction curves is made to account for this. Kanchanalai further performed experimental work to verify his analyses, which covered a wide range of portal and leaning column frames with the slenderness ratios of 20, 30, 40, 50, 60, 70, and 80, and relative stiffness ratios (G) of 0, 3, and 4. The ultimate strength of each frame was presented in the form of interaction curves consisting of the nondimensional first-order moment ($HL_c/2M_p$ in portal frames or HL_c/M_p in leaning column frames in the x-axis) and the nondimensional axial load (P/P_y in the y-axis).

In this study, the AISC-LRFD interaction curves are used for strength comparisons. The strength calculations are based on the LeMessurier K factor method [34] since it accounts for story buckling and results in more accurate predictions. The inelastic stiffness reduction factor, τ [1], is used to calculate K in the LeMessurier's procedure. The resistance factors ϕ_b and ϕ_c in the LRFD equations are taken as 1.0 to obtain the nominal strength. The interaction curves are obtained by the accumulation of a set of moments and axial forces that result in unity on the value of the interaction equation.

When a geometric imperfection of $L_c/500$ is used for unbraced frames, including leaning column frames, most of the strength curves fall within an area bounded by the plastic-zone curves and the LRFD curves. In portal frames, the conservative errors are less than 5%, an improvement on the LRFD error of 11%, and the maximum unconservative error is not more than 1%, shown in Figure 5.18. In leaning column frames, the conservative errors are less than 12%, as opposed to the 17% error of the LRFD, and the maximum unconservative error is not more than 5%, as shown in Figure 5.19.

When a notional load factor of 0.002 is used, the strengths predicted by this method are close to those given by explicit imperfection modeling method (Figure 5.20 and Figure 5.21).

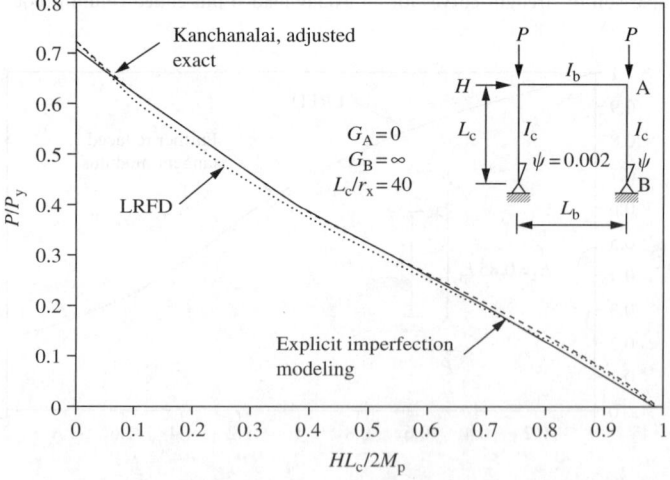

FIGURE 5.18 Comparison of strength curves for a portal frame subject to strong-axis bending with $L_c/r_x = 40$, $G_A = 0$ (explicit imperfection modeling method).

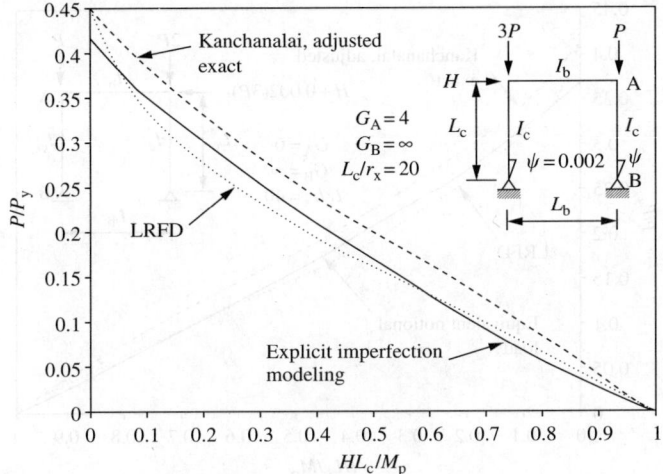

FIGURE 5.19 Comparison of strength curves for a leaning column frame subject to strong-axis bending with $L_c/r_x = 20$, $G_A = 4$ (explicit imperfection modeling method).

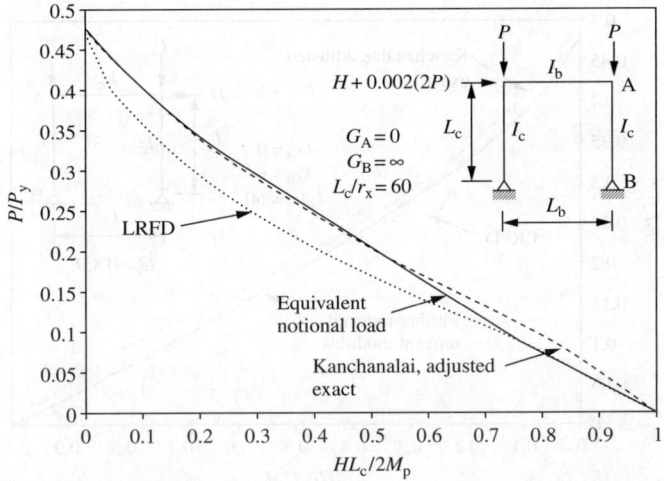

FIGURE 5.20 Comparison of strength curves for a portal frame subject to strong-axis bending with $L_c/r_x = 60$, $G_A = 0$ (equivalent notional load method).

When the reduced tangent modulus factor of 0.85 is used for portal and leaning column frames, the interaction curves generally fall between the plastic zone and LRFD curves. In portal frames, the conservative error is less than 8% (better than 11% error of the LRFD) and the maximum unconservative error is not more than 5% (Figure 5.22). In leaning column frames, the conservative error is less than 7% (better than 17% error of the LRFD), and the maximum unconservative error is not more than 5% (Figure 5.23).

5.3.3 Six-Story Frame

Vogel [35] presented the load–displacement relationships of a six-story frame using plastic-zone analysis. The frame is shown in Figure 5.24. Based on ECCS recommendations, the maximum

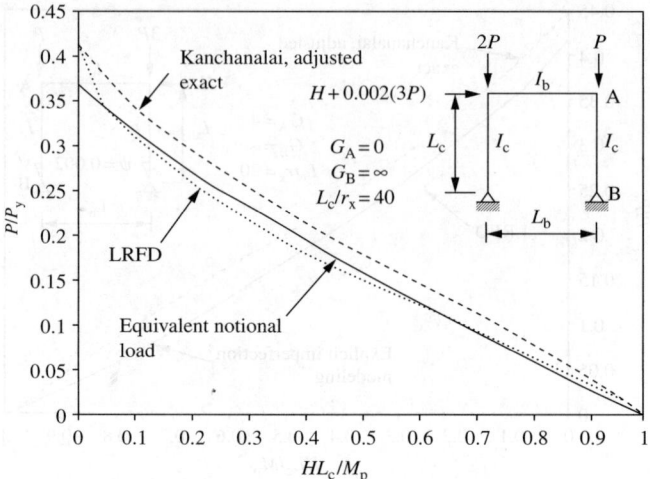

FIGURE 5.21 Comparison of strength curves for a leaning column frame subject to strong-axis bending with $L_c/r_x = 40$, $G_A = 0$ (equivalent notional load method).

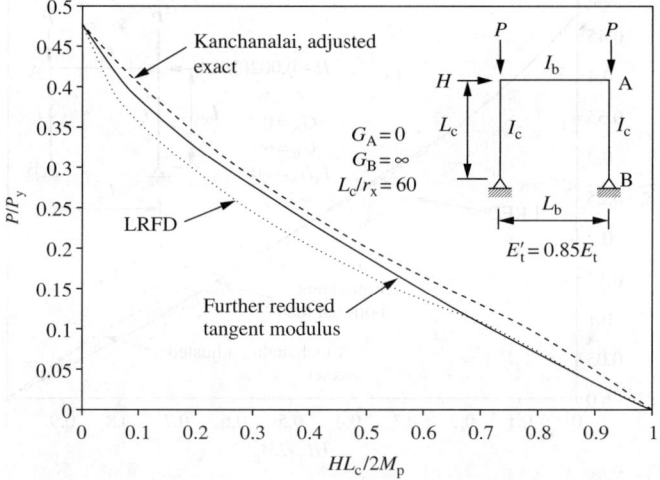

FIGURE 5.22 Comparison of strength curves for a portal frame subject to strong-axis bending with $L_c/r_x = 60$, $G_A = 0$ (further reduced tangent modulus method).

compressive residual stress is $0.3F_y$ when the ratio of depth to width (d/b) is greater than 1.2, and is $0.5F_y$ when the d/b ratio is less than 1.2 (Figure 5.25). The stress–strain relationship is elastic–plastic with strain hardening as shown in Figure 5.26. The geometric imperfections are $L_c/450$.

For comparison, the out-of-plumbness of $L_c/450$ is used in the explicit modeling method. The notional load factor of $1/450$ and the reduced tangent modulus factor of 0.85 are used. The further reduced tangent modulus is equivalent to the geometric imperfection of $L_c/500$. Thus, the geometric imperfection of $L_c/4500$ is additionally modeled in the further reduced tangent modulus method, where $L_c/4500$ is the difference between the Vogel's geometric imperfection of $L_c/450$ and the proposed geometric imperfection of $L_c/500$.

The load–displacement curves in the proposed methods together with Vogel's plastic-zone analysis are compared in Figure 5.27. The errors in strength prediction by the proposed methods

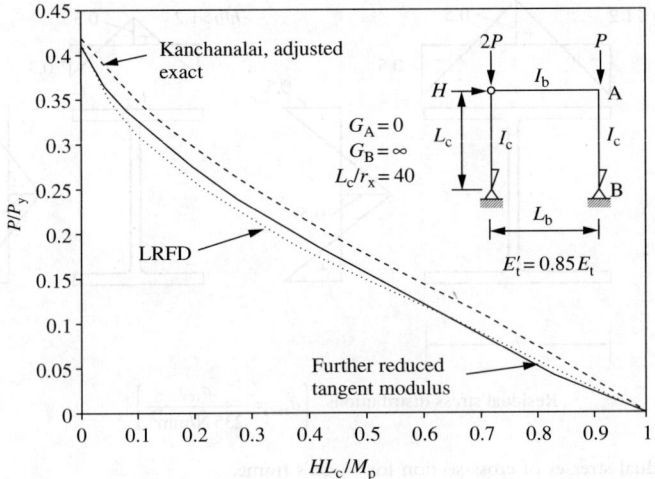

FIGURE 5.23 Comparison of strength curves for a leaning column frame subject to strong-axis bending with $L_c/r_x = 40$, $G_A = 0$ (further reduced tangent modulus method).

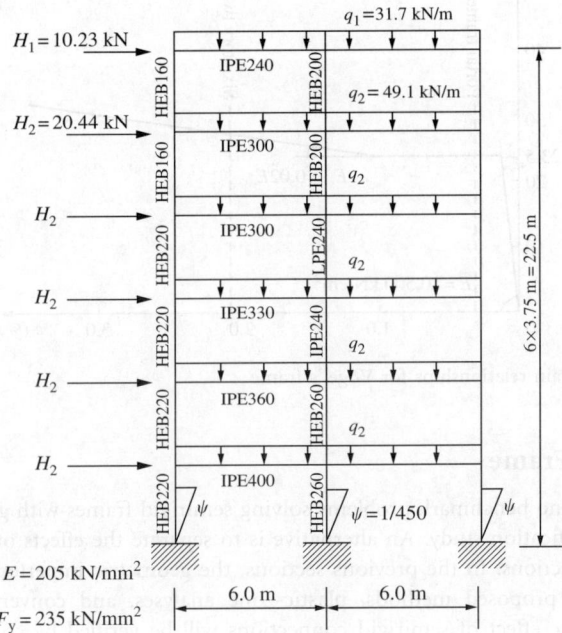

FIGURE 5.24 Configuration and load condition of Vogel's six-story frame for verification study.

are less than 1%. Explicit imperfection modeling and the equivalent notional load method under-predict lateral displacements by 3%, and the further reduced tangent modulus method shows a good agreement in displacement with the Vogel's exact solution. Vogel's frame is a good example of how the reduced tangent modulus method predicts lateral displacement well under reasonable load combinations.

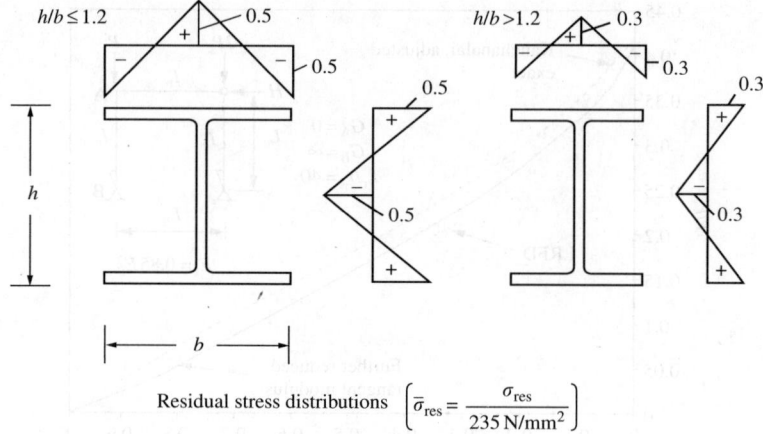

Residual stress distributions $\left[\bar{\sigma}_{\text{res}} = \dfrac{\sigma_{\text{res}}}{235\ \text{N/mm}^2}\right]$

FIGURE 5.25   Residual stresses of cross-section for Vogel's frame.

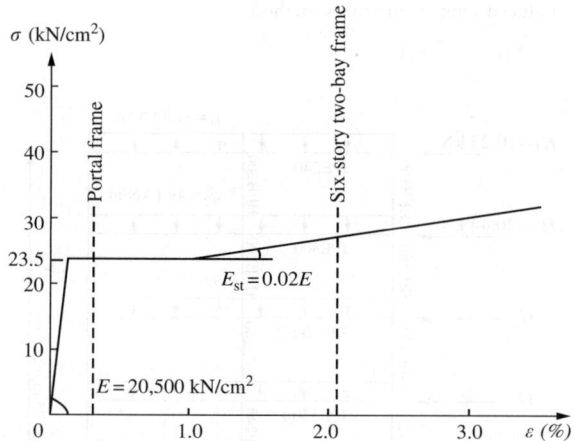

FIGURE 5.26   Stress–strain relationships for Vogel's frame.

5.3.4 Semirigid Frame

In the open literature, no benchmark problems solving semirigid frames with geometric imperfections are available for a verification study. An alternative is to separate the effects of semirigid connections and geometric imperfections. In the previous sections, the geometric imperfections were studied and comparisons between proposed methods, plastic-zone analyses, and conventional LRFD methods were made. Herein, the effect of semirigid connections will be verified by comparing analytical and experimental results.

Stelmack [36] studied the experimental response of two flexibly connected steel frames. A two-story, one-bay frame in his study is selected as a benchmark for the present study. The frame was fabricated from A36 W5 × 16 sections, with pinned base supports (Figure 5.28). The connections were bolted top and seat angles (L4 × 4 × $\frac{1}{2}$) made of A36 steel and A325 $\frac{3}{4}$in.-diameter bolts (Figure 5.29). The experimental moment–rotation relationship is shown in Figure 5.30. A gravity load of 2.4 kip was applied at third points along the beam at the first level, followed by a lateral load application. The lateral load–displacement relationship was provided in Stelmack.

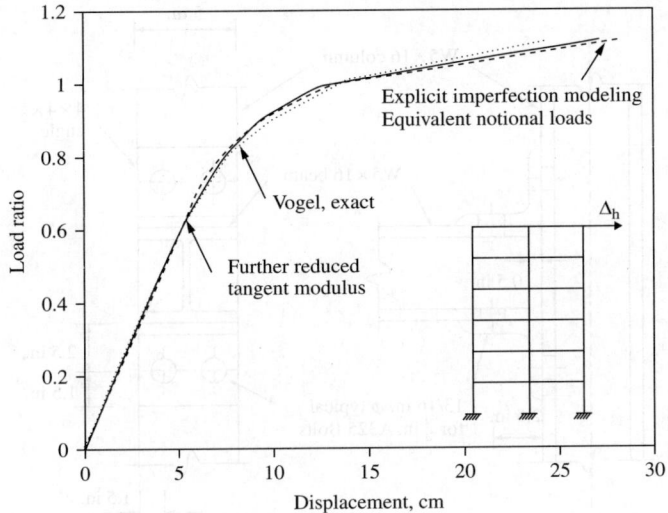

FIGURE 5.27 Comparison of displacements for Vogel's six-story frame.

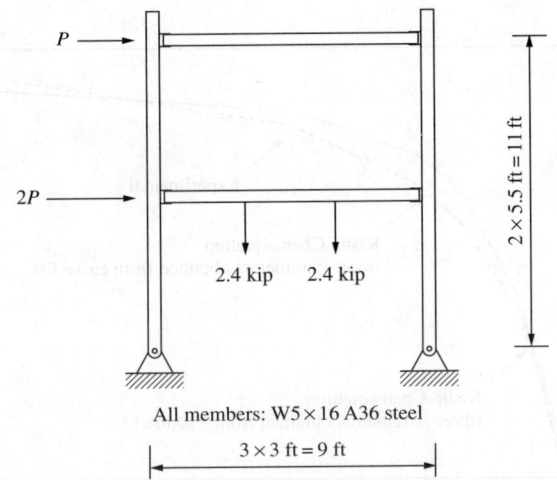

FIGURE 5.28 Configuration and load condition of Stelmack's two-story semirigid frame.

Herein, the three parameters of the power model are determined by curve fitting and the program 3PARA.f presented in Section 2.2. The three parameters obtained by the curve-fit are $R_{ki} = 40,000$ k-in./rad, $M_u = 220$ k-in., and $n = 0.91$. We obtain three parameters of $R_{ki} = 29,855$ kip/rad, $M_u = 185$ k-in, and $n = 1.646$ with 3PARA.f.

The moment–rotation curves given by experiment and curve fitting show good agreement (Figure 5.30). The parameters given by the Kishi–Chen equations and by experiment show some deviation (Figure 5.30). In spite of this difference, the Kishi–Chen equations, using the computer program 3PARA.f, are a more practical alternative in design since experimental moment–rotation curves are not usually available [30]. In the analysis, the gravity load is first applied and then the lateral load. The lateral displacements given by the proposed methods and by the experimental method compare well (Figure 5.31). The proposed method adequately predicts the behavior and strength of semirigid connections.

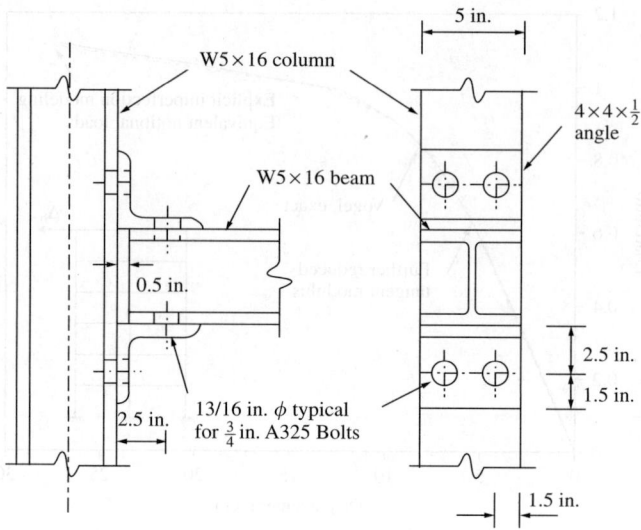

FIGURE 5.29   Top- and seat-angle connection details.

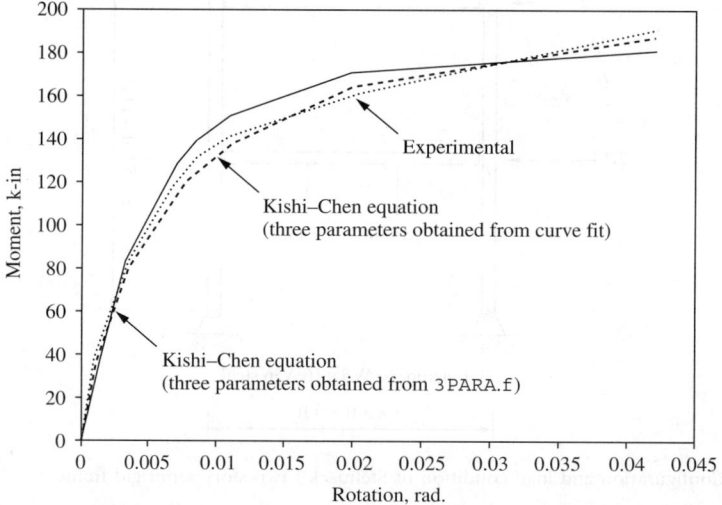

FIGURE 5.30   Comparison of moment–rotation relationships of semirigid connection by experiment and Kishi–Chen equation.

5.4   Analysis and Design Principles

In the preceding section, the proposed advanced analysis method was verified using several benchmark problems available in the literature. Verification studies were carried out by comparing it to the plastic-zone and conventional LRFD solutions. It was shown that practical advanced analysis predicted the behavior and failure mode of a structural system with reliable accuracy.

In this section, analysis and design principles are summarized for the practical application of the advanced analysis method. Step-by-step analysis and design procedures for the method are presented.

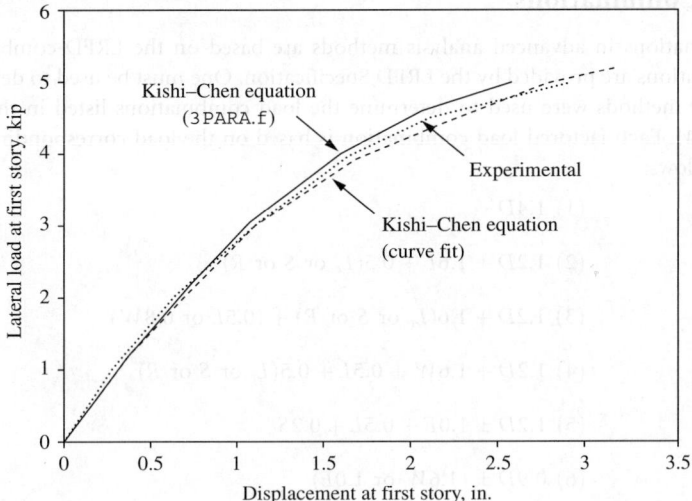

FIGURE 5.31 Comparison of displacements of Stelmack's two-story semirigid frame.

5.4.1 Design Format

Advanced analysis follows the format of LRFD. In LRFD, the factored load effect does not exceed the factored nominal resistance of the structure. Two safety factors are used: one is applied to loads and the other to resistances. This approach is an improvement on other models (e.g., ASD and PD) because both the loads and the resistances have unique factors for unique uncertainties. LRFD has the format

$$\phi R_n \geq \sum_{i=1}^{m} \gamma_i Q_{ni} \qquad (5.13)$$

where R_n is the nominal resistance of the structural member, Q_n is the nominal load effect (e.g., axial force, shear force, bending moment), ϕ is the resistance factor (≤ 1.0) (e.g., 0.9 for beams, 0.85 for columns), γ_i is the load factor (usually > 1.0) corresponding to Q_{ni} (e.g., $1.4D$ and $1.2D + 1.6L + 0.5S$), i is the type of load (e.g., $D =$ dead load, $L =$ live load, $S =$ snow load), and m is the number of load type.

Note that the LRFD [32] uses separate factors for each load and therefore reflects the uncertainty of different loads and combinations of loads. As a result, a relatively uniform reliability is achieved.

The main difference between the conventional LRFD method and advanced analysis methods is that the left-hand side of Equation 5.13 (ϕR_n) in the LRFD method is the resistance or strength of the component of a structural system, but in the advanced analysis method it represents the resistance or the load-carrying capacity of the whole structural system.

5.4.2 Loads

Structures are subjected to various loads including dead, live, impact, snow, rain, wind, and earthquake loads. Structures must be designed to prevent failure and limit excessive deformation; thus, an engineer must anticipate the loads a structure may experience over its service life with reliability.

Loads may be classified as static or dynamic. Dead loads are typical of static loads, and wind or earthquake loads are dynamic. Dynamic loads are usually converted to equivalent static loads in conventional design procedures, and it may be adopted in advanced analysis as well [37].

5.4.3 Load Combinations

The load combinations in advanced analysis methods are based on the LRFD combinations [1]. Six factored combinations are provided by the LRFD Specification. One must be used to determine member sizes. Probability methods were used to determine the load combinations listed in the LRFD Specification (LRFD-A4). Each factored load combination is based on the load corresponding to the 50-year recurrence as follows:

$$(1)\ 1.4D \tag{5.14a}$$

$$(2)\ 1.2D + 1.6L + 0.5(L_r \text{ or } S \text{ or } R) \tag{5.14b}$$

$$(3)\ 1.2D + 1.6(L_r \text{ or } S \text{ or } R) + (0.5L \text{ or } 0.8W) \tag{5.14c}$$

$$(4)\ 1.2D + 1.6W + 0.5L + 0.5(L_r \text{ or } S \text{ or } R) \tag{5.14d}$$

$$(5)\ 1.2D \pm 1.0E + 0.5L + 0.2S \tag{5.14e}$$

$$(6)\ 0.9D \pm (1.6W \text{ or } 1.0E) \tag{5.14f}$$

where D is the dead load (the weight of the structural elements and the permanent features on the structure), L is the live load (occupancy and moveable equipment), L_r is the roof live load, W is the wind load, S is the snow load, E is the earthquake load, and R is the rainwater or ice load.

The LRFD Specification specifies an exception that the load factor on live load, L, in combination (3)–(5) must be 1.0 for garages, areas designated for public assembly, and all areas where the live load is greater than 100 psf.

5.4.4 Resistance Factors

The AISC-LRFD cross-section strength equations may be written as

$$\frac{P}{\phi_c P_y} + \frac{8}{9}\frac{M}{\phi_b M_p} = 1.0 \quad \text{for} \quad \frac{P}{\phi_c P_y} \geq 0.2 \tag{5.15a}$$

$$\frac{1}{2}\frac{P}{\phi_c P_y} + \frac{M}{\phi_b M_p} = 1.0 \quad \text{for} \quad \frac{P}{\phi_c P_y} < 0.2 \tag{5.15b}$$

where P, M are second-order axial force and bending moment, respectively, P_y is the squash load, M_p is the plastic moment capacity, and ϕ_c, ϕ_b are the resistance factors for axial strength and flexural strength, respectively.

Figure 5.32 shows the cross-section strength including the resistance factors ϕ_c and ϕ_b. The reduction factors ϕ_c and ϕ_b are built into the analysis program and are thus automatically included in the calculation of the load-carrying capacity. The reduction factors are 0.85 for axial strength and 0.9 for flexural strength, corresponding to the AISC-LRFD Specification [1]. For connections, the ultimate moment, M_u, is reduced by the reduction factor 0.9.

5.4.5 Section Application

The AISC-LRFD Specification uses only one column curve for rolled and welded sections of W, WT, and HP shapes, pipe, and structural tubing. The specification also uses some interaction equations for doubly and singly symmetric members including W, WT, and HP shapes, pipe, and structural tubing, even though the interaction equations were developed on the basis of W shapes by Kanchanalai [4].

The present advanced analysis method was developed by calibration with the LRFD column curve and interaction equations described in Section 5.3. To this end, it is concluded that the proposed method can be used for various rolled and welded sections, including W, WT, and HP shapes, pipe, and structural tubing without further modifications.

5.4.6 Modeling of Structural Members

Different types of advanced analysis are (1) plastic-zone method, (2) quasi-plastic hinge method, (3) elastic–plastic hinge method, and (4) refined plastic-hinge method. An important consideration in making these advanced analyses practical is the required number of elements for a member in order to predict realistically the behavior of frames.

A sensitivity study of advanced analysis is performed on the required number of elements for a beam member subject to distributed transverse loads. A two-element model adequately predicts the strength of a member. To model a parabolic out-of-straightness in a beam column, a two-element model with a maximum initial deflection at the mid-height of a member adequately captures imperfection effects. The required number of elements in modeling each member to provide accurate predictions of the strengths is summarized in Table 5.4. It is concluded that practical advanced analysis is computationally efficient.

5.4.7 Modeling of Geometric Imperfection

Geometric imperfection modeling is required to account for fabrication and erection tolerances. The imperfection modeling methods used here are the explicit imperfection, the equivalent notional load, and the further reduced tangent modulus models. Users may choose one of these three models in an advanced analysis. The magnitude of geometric imperfections is listed in Table 5.5.

Geometric imperfection modeling is required for a frame but not a truss element, since the program computes the axial strength of a truss member using the LRFD column strength equations, which account for geometric imperfections.

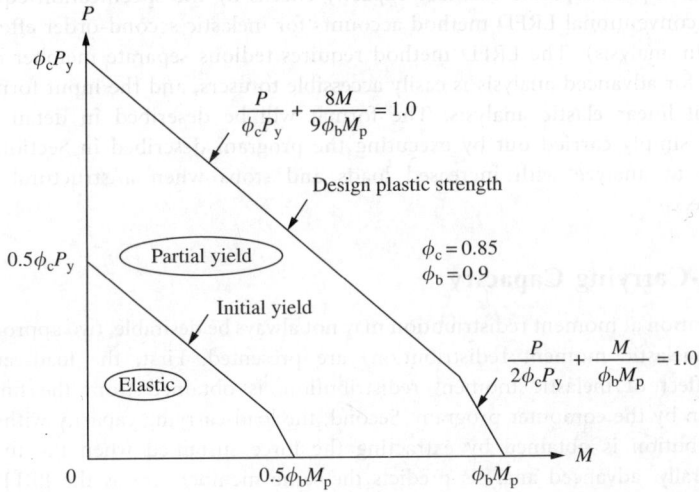

FIGURE 5.32 Stiffness degradation model including reduction factors.

TABLE 5.4 Necessary Number of Elements

Member	Number of elements
Beam member subject to uniform loads	2
Column member of braced frame	2
Column member of unbraced frame	1

TABLE 5.5   Magnitude of Geometric Imperfection

Geometric imperfection method	Magnitude
1. Explicit imperfection modeling method	$\psi = 2/1000$ for unbraced frames
	$\psi = 1/1000$ for braced frames
2. Equivalent notional load method	$\alpha = 2/1000$ for unbraced frames
	$\alpha = 4/1000$ for braced frames
3. Further reduced tangent modulus method	$E_t' = 0.85E_t$

5.4.8 Load Application

It is necessary, in an advanced analysis, to input proportional increment load (not the total loads) to trace nonlinear load–displacement behavior. The incremental loading process can be achieved by scaling down the combined factored loads by a number between 10 and 50. For a highly redundant structure (such as one greater than six stories), dividing by about 10 is recommended, and for a nearly statically determinate structure (such as a portal frame), the incremental load may be factored down by 50. One may choose a number between 10 and 50 to reflect the redundancy of a particular structure. Since a highly redundant structure has the potential to form many plastic hinges and the applied load increment is automatically reduced as new plastic hinges form, the larger incremental load (i.e., the smaller scaling number) may be used.

5.4.9 Analysis

Analysis is important in the proposed design procedures, since the advanced analysis method captures key behaviors including second-order and inelasticity in its analysis program. Advanced analysis does not require separate member capacity checks by the specification equations. On the other hand, the conventional LRFD method accounts for inelastic second-order effects in its design equations (not in analysis). The LRFD method requires tedious separate member capacity checks. Input data used for advanced analysis is easily accessible to users, and the input format is similar to the conventional linear elastic analysis. The format will be described in detail in Section 5.5. Analyses can be simply carried out by executing the program described in Section 5.5. This program continues to analyze with increased loads and stops when a structural system reaches its ultimate state.

5.4.10 Load-Carrying Capacity

Because consideration at moment redistribution may not always be desirable, two approaches (including and excluding inelastic moment redistribution) are presented. First, the load-carrying capacity, including the effect of inelastic moment redistribution, is obtained from the final loading step (limit state) given by the computer program. Second, the load-carrying capacity without the inelastic moment redistribution is obtained by extracting the force sustained when the first plastic hinge is formed. Generally, advanced analysis predicts the same member size as the LRFD method when moment redistribution is not considered. Further illustrations on these two choices will be presented in Section 5.6.

5.4.11 Serviceability Limits

The serviceability conditions specified by the LRFD consist of five limit states: (1) deflection, vibration, and drift; (2) thermal expansion and contraction; (3) connection slip; (4) camber; and (5) corrosion. The most common parameter affecting the design serviceability of steel frames is the deflections.

TABLE 5.6   Deflection Limitations of Frame

Item	Deflection ratio
Floor girder deflection for service live load	$L/360$
Roof girder deflection	$L/240$
Lateral drift for service wind load	$H/400$
Interstory drift for service wind load	$H/300$

Based on the studies by the Ad Hoc Committee [38] and Ellingwood [39], the deflection limits recommended (Table 5.6) were proposed for general use. At service load levels, no plastic hinges are permitted anywhere in the structure to avoid permanent deformation under service loads.

5.4.12 Ductility Requirements

Adequate inelastic rotation capacity is required for members in order to develop their full plastic moment capacity. The required rotation capacity may be achieved when members are adequately braced and their cross-sections are compact. The limitations of compact sections and lateral unbraced length in what follows lead to an inelastic rotation capacity of at least three and seven times the elastic rotation corresponding to the onset of the plastic moment for nonseismic and seismic regions, respectively.

Compact sections are capable of developing the full plastic moment capacity, M_p, and sustaining large hinge rotation before the onset of local buckling. The compact section in the LRFD Specification is defined as:

1. Flange
   - For nonseismic region

$$\frac{b_f}{2t_f} \leq 0.38 \sqrt{\frac{E_s}{F_y}} \tag{5.16}$$

   - For seismic region

$$\frac{b_f}{2t_f} \leq 0.31 \sqrt{\frac{E_s}{F_y}} \tag{5.17}$$

   where E_s is the modulus of elasticity, b_f is the width of flange, t_f is the thickness of flange, and F_y is the yield stress.

2. Web
   - For nonseismic region

$$\frac{h}{t_w} \leq 3.76 \sqrt{\frac{E_s}{F_y}} \left(1 - \frac{2.75 P_u}{\phi_b P_y} \right) \quad \text{for} \quad \frac{P_u}{\phi_b P_y} \leq 0.125 \tag{5.18a}$$

$$\frac{h}{t_w} \leq 1.12 \sqrt{\frac{E_s}{F_y}} \left(2.33 - \frac{P_u}{\phi_b P_y} \right) \geq 1.49 \sqrt{\frac{E_s}{F_y}} \quad \text{for} \quad \frac{P_u}{\phi_b P_y} > 0.125 \tag{5.18b}$$

   - For seismic region

$$\frac{h}{t_w} \leq 3.05 \sqrt{\frac{E_s}{F_y}} \left(1 - \frac{1.54 P_u}{\phi_b P_y} \right) \quad \text{for} \quad \frac{P_u}{\phi_b P_y} \leq 0.125 \tag{5.19a}$$

$$\frac{h}{t_w} \leq 1.12 \sqrt{\frac{E_s}{F_y}} \left(2.33 - \frac{P_u}{\phi_b P_y} \right) \quad \text{for} \quad \frac{P_u}{\phi_b P_y} > 0.125 \tag{5.19b}$$

   where h is the clear distance between flanges, t_w is the thickness of the web, and F_y is the yield strength.

In addition to the compactness of the section, the lateral unbraced length of beam members is also a limiting factor for the development of the full plastic moment capacity of members. The LRFD provisions provide the limit on spacing of braces for beam as:

- For nonseismic region

$$L_{pd} \leq \left[0.12 + 0.076\left(\frac{M_1}{M_2}\right)\right]\left(\frac{E_s}{F_y}\right) r_y \qquad (5.20a)$$

- For seismic region

$$L_{pd} \leq 0.086\left(\frac{E_s}{F_y}\right) r_y \qquad (5.20b)$$

where L_{pd} is the unbraced length, r_y is the radius of gyration about y-axis, F_y is the yield strength, M_1, M_2 are smaller and larger end moments, and M_1/M_2 is the positive in double curvature bending.

The AISC-LRFD Specification explicitly specifies the limitations for beam members as described above, but not for beam–column members. More studies are necessary to determine the reasonable limits leading to adequate rotation capacity of beam–column members. Based on White's study [40], the limitations for beam members seem to be used for beam–column members until the specification provides the specific values for beam–column members.

5.4.13　Adjustment of Member Sizes

If one of the following three conditions — strength, serviceability, or ductility — is not satisfied, appropriate adjustments of the member sizes should be made. This can be done by referring to the sequence of plastic hinge formation shown in the P.OUT. For example, if the load-carrying capacity of a structural system is less than the factored load effect, the member with the first plastic hinge should be replaced with a stronger member. On the other hand, if the load-carrying capacity exceeds the factored load effect significantly, members without plastic hinges may be replaced with lighter members. If lateral drift exceeds drift requirements, columns or beams should be sized up, or a braced structural system should be considered instead to meet this serviceability limit.

In semirigid frames, behavior is influenced by the combined effects of members and connections. As an illustration, if an excessive lateral drift occurs in a structural system, the drift may be reduced by increasing member sizes or using more rigid connections. If the strength of a beam exceeds the required strength, it may be adjusted by reducing the beam size or using more flexible connections. Once the member and connection sizes are adjusted, the iteration leads to an optimum design. Figure 5.33 shows a flow chart of analysis and design procedure in the use of advanced analysis.

5.5　Computer Program

This section describes the practical advanced analysis program (PAAP) for two-dimensional steel frame design [6,19]. The program integrates the methods and techniques developed in Sections 5.2 and 5.3. The names of variables and arrays correspond as closely as possible to those used in theoretical derivations. The main objective of this section is to present an educational version of software to enable engineers and graduate students to perform planar frame analysis for a more realistic prediction of the strength and behavior of structural systems.

The instructions necessary for user input into PAAP are presented in Section 5.5.4. Except for the requirement to input geometric imperfections and incremental loads, the input data format

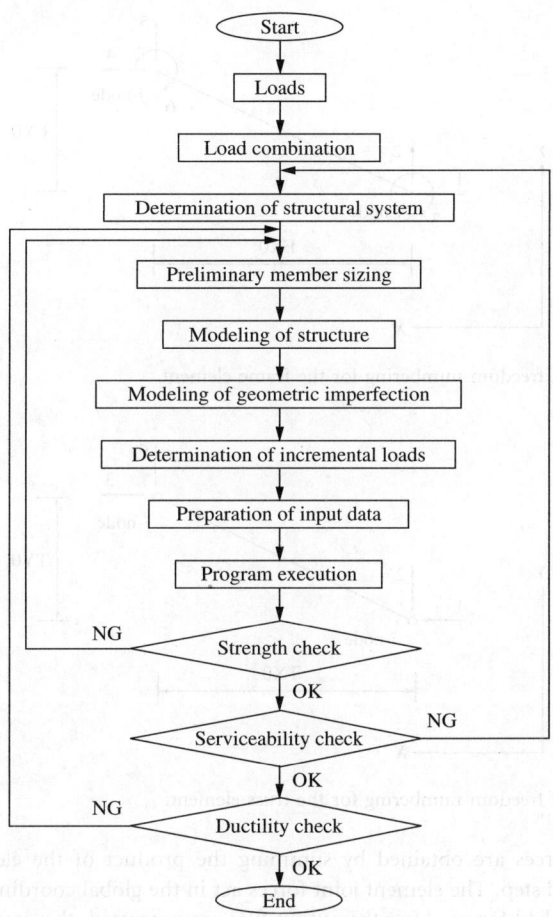

FIGURE 5.33 Analysis and design procedure.

of the program is basically the same as that of the usual linear elastic analysis program. The user is advised to read all the instructions, paying particular attention to the appended notes, to achieve an overall view of the data required for a specific PAAP analysis. The reader should recognize that no system of units is assumed in the program, and take the responsibility to make all units consistent. Mistaken unit conversion and input are a common source of erroneous results.

5.5.1 Program Overview

This FORTRAN program is divided into three parts: DATAGEN, INPUT, and PAAP. The first program, DATAGEN, reads an input data file, P.DAT, and generates a modified data file, INFILE. The second program, INPUT, rearranges INFILE into three working data files, DATA0, DATA1, and DATA2. The third program, PAAP, reads the working data files and provides two output files named P.OUT1 and P.OUT2. P.OUT1 contains an echo of the information from the input data file, P.DAT. This file may be used to check for numerical and incompatibility errors in input data. P.OUT2 contains the load and displacement information for various joints in the structure as well as the element joint forces for all types of elements at every load step. The load–displacement results are presented at the end of every load increment. The sign conventions for loads and displacements should follow the frame degrees of freedom, as shown in Figure 5.34 and Figure 5.35.

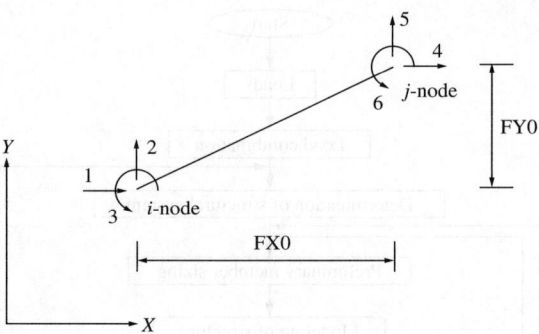

FIGURE 5.34 Degrees of freedom numbering for the frame element.

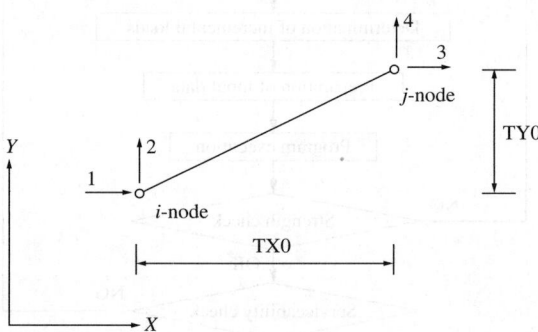

FIGURE 5.35 Degrees of freedom numbering for the truss element.

The element joint forces are obtained by summing the product of the element incremental displacements at every load step. The element joint forces act in the global coordinate system and must be in equilibrium with applied forces. After the output files are generated, the user can view these files on the screen or print them with the MS-DOS PRINT command. The schematic diagram in Figure 5.36 sets out the operation procedure used by PAAP and its supporting programs [6].

5.5.2 Hardware Requirements

This program has been tested in two computer processors. It was first tested on an IBM 486 or equivalent personal computer system using Microsoft's FORTRAN 77 compiler v1.00 and Lahey's FORTRAN 77 compiler v5.01. Then, its performance in the workstation environment was tested on a Sun 5 using a Sun FORTRAN 77 compiler. The program sizes of DATAGEN, INPUT, and PAAP are 8, 9, and 94 kB, respectively. The total size of the three programs is small, 111 kB (= 0.111 MB), and so a 3.5-in. high-density diskette (1.44 MB) can accommodate the three programs and several example problems.

The memory required to run the program depends on the size of the problem. A computer with minimum 640 K of memory and a 30 MB hard disk is generally required. For the PC applications, the array sizes are restricted as follows:

1. Maximum total degrees of freedom, MAXDOF = 300
2. Maximum translational degrees of freedom, MAXTOF = 300
3. Maximum rotational degrees of freedom, MAXROF = 100
4. Maximum number of truss elements, MAXTRS = 150
5. Maximum number of connections, MAXCNT = 150

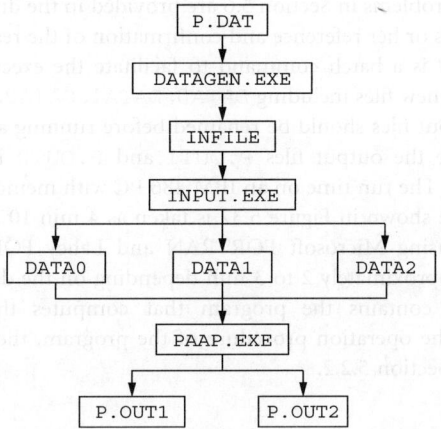

FIGURE 5.36 Operating procedures of the proposed program.

It is possible to run bigger jobs in UNIX workstations by modifying the above values in the PARAMETER and COMMON statements in the source code.

5.5.3 Execution of Program

A computer diskette is provided in *LRFD Steel Design Using Advanced Analysis*, by Chen and Kim [37], containing four directories with following files:

1. Directory PSOURCE
   - DATAGEN.FOR
   - INPUT.FOR
   - PAAP.FOR
2. Directory PTEST
   - DATAGEN.EXE
   - INPUT.EXE
   - PAAP.EXE
   - RUN.BAT (batch file)
   - P.DAT (input data for a test run)
   - P.OUT1 (output for a test run)
   - P.OUT2 (output for a test run)
3. Directory PEXAMPLE
   - All input data for the example problems presented in Section 6
4. Directory CONNECT
   - 3PARA.FOR (program for semirigid connection parameters)
   - 3PARA.EXE
   - CONN.DAT (input data)
   - CONN.OUT (output for three parameters)

To execute the programs, one must first copy them onto the hard disk (i.e., copy DATAGEN.EXE, INPUT.EXE, PAAP.EXE, RUN.BAT, and P.DAT from the directory PTEST on the diskette to the hard disk). Before launching the program, the user should test the system by running the sample example provided in the directory. The programs are executed by issuing the command RUN. The batch file, RUN.BAT executes DATAGEN, INPUT, and PAAP in sequence. The output files produced are P.OUT1 and P.OUT2. When the authors' and the user's compilers are different, the program (PAAP.EXE) may not be executed. This problem may be easily solved by recompiling the source programs in the directory PSOURCE.

The input data for all the problems in Section 5.6 are provided in the directory PEXAMPLE. The user may use the input data for his or her reference and confirmation of the results presented in Section 5.6. It should be noted that RUN is a batch command to facilitate the execution of PAAP. Entering the command RUN will write the new files including DATA0, DATA1, DATA2, P.OUT1, and P.OUT2 over the old ones. Therefore, output files should be renamed before running a new problem.

The program can generate the output files P.OUT1 and P.OUT2 in a reasonable time period, as described in the following. The run time on an IBM 486 PC with memory of 640 K to get the output files for the eight-story frame shown in Figure 5.37 is taken as 4 min 10 s and 2 min 30 s in real time rather than CPU time by using Microsoft FORTRAN and Lahey FORTRAN, respectively. In the Sun 5, the run time varies approximately 2 to 3 min depending on the degree of occupancy by users.

The directory CONNECT contains the program that computes the three parameters needed for semirigid connections. The operation procedure of the program, the input data format, and two examples were presented in Section 5.2.2.

5.5.4 User Manual

5.5.4.1 Analysis Options

PAAP was developed on the basis of the theory presented in Section 5.2. While the purpose of the program is basically for advanced analysis using second-order inelastic concept, the program can also be used for first- and second-order elastic analyses. For a first-order elastic analysis, the total factored load should be applied in one load increment to suppress numerical iteration in the nonlinear analysis algorithm. For a second-order elastic analysis, a yield strength of an arbitrarily large value should be assumed for all members to prevent yielding.

5.5.4.2 Coordinate System

A two-dimensional (x, y) global coordinate system is used for the generation of all the input and output data associated with the joints. The following input and output data are prepared with respect to the global coordinate system:

1. Input data
   • joint coordinates
   • joint restraints
   • joint load

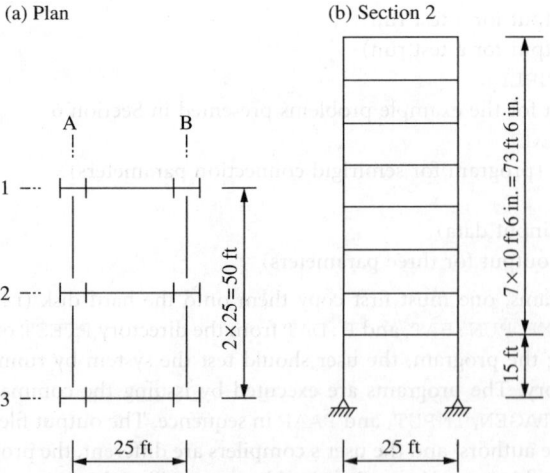

FIGURE 5.37 Configuration of the unbraced eight-story frame.

2. Output data
   - joint displacement
   - member forces

5.5.4.3 Type of Elements

The analysis library consists of three elements: a plane frame, a plane truss, and a connection. The connection is represented by a zero-length rotational spring element with a user-specified nonlinear moment–rotation curve. Loading is allowed only at nodal points. Geometric and material nonlinearities can be accounted for by using an iterative load-increment scheme. Zero-length plastic hinges are lumped at the element ends.

5.5.4.4 Locations of Nodal Points

The geometric dimensions of the structures are established by placing joints (or nodal points) on the structures. Each joint is given an identification number and is located in a plane associated with a global two-dimensional coordinate system. The structural geometry is completed by connecting the predefined joints with structural elements, which may be a frame, a truss, or a connection. Each element also has an identification number.

The following are some of the factors that need to be considered in placing joints in a structure:

1. The number of joints should be sufficient to describe the initial geometry and the response behavior of the structures.
2. Joints need to be located at points and lines of discontinuity (e.g., at changes in material properties or section properties).
3. Joints should be located at points on the structure where forces and displacements need to be evaluated.
4. Joints should be located at points where concentrated loads will be applied. The applied loads should be concentrated and act on the joints.
5. Joints should be located at all support points. Support conditions are represented in the structural model by restricting the movement of the specific joints in specific directions.
6. Second-order inelastic behavior can be captured by the use of one or two elements per member corresponding to the following guidelines:
   - Beam member subjected to uniform loads: two elements
   - Column member of braced frames: two elements
   - Column member of unbraced frames: one element.

5.5.4.5 Degrees of Freedom

A two-joint frame element has six displacement components as shown in Figure 5.34. Each joint can translate in the global *x*- and *y*-directions, and rotate about the global *z*-axis. The directions associated with these displacement components are known as degrees of freedom of the joint. A two-joint truss element has four degrees of freedom as shown in Figure 5.35. Each joint has two translational degrees of freedom and no rotational component.

If the displacement of a joint corresponding to any one of its degrees of freedom is known to be zero (such as at a support), then it is labeled an inactive degree of freedom. Degrees of freedom where the displacements are not known are termed active degree of freedoms. In general, the displacement of an inactive degree of freedom is usually known, and the purpose of the analysis is to find the reaction in that direction. For an active degree of freedom, the applied load is known (it could be zero), and the purpose of the analysis is to find the corresponding displacement.

5.5.4.6 Units

There are no "built-in" units in PAAP. The user must prepare the input in a consistent set of units. The output produced by the program will conform to the same set of units. Therefore, if the user chooses to use kips and inches as the input units, all the dimensions of the structure must be entered in inches and all the loads in kip. The material properties should also conform to these units. The output units will then be in kips and inches, so that the frame member axial force will be in kips, bending moments will be in kip-inches, and displacements will be in inches. Joint rotations, however, are in radians, irrespective of units.

5.5.4.7 Input Instructions

In this section, the input sequence and data structure used to create an input file called `P.DAT` are described. The analysis program, `PAAP`, can analyze any structures with up to 300 degrees of freedom, but it is possible to recompile the source code to accommodate more degrees of freedom by changing the size of the arrays in the `PARAMETER` and `COMMON` statements. The limitation of degree of freedom can be solved by using dynamic storage allocation. This procedure is common in finite element programs [41,42], and will be used in the next release of the program.

The input data file is prepared in a specific format. The input data consists of 13 data sets, including five control data, three section property data, three element data, one boundary condition, and one load data set:

1. Title
2. Analysis and design control
3. Job control
4. Total number of element types
5. Total number of elements
6. Connection properties
7. Frame element properties
8. Truss element properties
9. Connection element data
10. Frame element data
11. Truss element data
12. Boundary conditions
13. Incremental loads

Input of all data sets are mandatory, but some of the data associates with elements (data sets 6–11) may be skipped depending on whether the use of the element. The order of data sets in the input file must be strictly maintained. Instructions for inputting data are summarized in Table 5.7.

5.6 Design Examples

In previous sections, the concept, verifications, and computer program of the practical advanced analysis method for steel frame design have been presented. The present advanced analysis method has been developed and refined to achieve both simplicity in use and, as far as possible, a realistic representation of behavior and strength. The advanced analysis method captures the limit state strength and stability of a structural system and its individual members. As a result, the method can be used for practical frame design without the tedious separate member capacity checks, including the calculation of K factor.

The aim of this section is to provide further confirmation of the validity of the LRFD-based advanced analysis methods for practical frame design. The comparative design examples in this section show the detailed design procedure for advanced and LRFD design procedures [6]. The design procedures conform to those described in Section 5.4 and may be grouped into four basic steps: (1) load condition, (2) structural modeling, (3) analysis, and (4) limit state check. The design examples cover simple structures, truss structures, braced frames, unbraced frames, and semirigid frames. The three practical models — explicit imperfection, equivalent notional load, and further reduced tangent modulus — are used for the design examples. Member sizes determined by advanced procedures are compared with those determined by the LRFD, and good agreement is generally observed.

TABLE 5.7 Input Data Format for the Program PAAP

Data set	Column	Variable	Description
Title	A70	—	Job title and general comments
Analysis and design control	1–5	IGEOIM	Geometric imperfection method
			0: No geometric imperfection (default)
			1: Explicit imperfection modeling
			2: Equivalent notional load
			3: Further reduced tangent modulus
	6–10	ILRFD	Strength reduction factor $\phi_c = 0.85$, $\phi_b = 0.9$
			0: No reduction factors considered (default)
			1: Reduction factors considered
Job control	1–5	NNODE	Total number of nodal points of the structure
	6–10	NBOUND	Total number of supports
	11–15	NINCRE	Allowable number of load increments (default = 100); at least two or three times larger than the scaling number
Total number of element types	1–5	NCTYPE	Number of connection types (1–30)
	6–10	NFTYPE	Number of frame types (1–30)
	11–15	NTTYPE	Number of truss types (1–30)
Total number of elements	1–5	NUMCNT	Number of connection elements (1–150)
	6–10	NUMFRM	Number of frame elements (1–100)
	11–15	NUMTRS	Number of truss elements (1–150)
Connection property	1–5	ICTYPE	Connection type number
	6–15[a]	M_u	Ultimate moment capacity of connection
	16–25[a]	R_{ki}	Initial stiffness of connection
	26–35[a]	N	Shape parameter of connection
Frame element property	1–5	IFTYPE	Frame type number
	6–15[a]	A	Cross-section area
	16–25[a]	I	Moment of inertia
	26–35[a]	Z	Plastic section modulus
	36–45[a]	E	Modulus of elasticity
	46–55[a]	FY	Yield stress
	55–60	IFCOL	Identification of column member, IFCOL = 1 for column (default = 0)
Truss element property	1–5	ITYPE	Truss type number
	6–15[a]	A	Cross-section area
	15–25[a]	I	Moment of inertia
	25–35[a]	E	Modulus of elasticity
	36–45[a]	FY	Yield stress
	46–50	ITCOL	Identification of column member, ITCOL = 1 for column (default = 0)
Connection element data	1–5	LCNT	Connection element number
	6–10	IFMCNT	Frame element number containing the connection
	11–15	IEND	Identification of element ends containing the connection
	16–20	JDCNT	1: Connection attached at element end i
	21–25	NOSMCN	2: Connection attached at element end j
	26–30	NELINC	Connection type number
			Number of same elements for automatic generation (default = 1)
			Element number (IFMCNT) increment of automatically generated elements (default = 1)
Frame element data	1–5	LFRM	Frame element number
	6–15[a]	FXO	Horizontal projected length; positive for i–j direction in global x direction
	16–25[a]	FYO	Vertical projected length; positive for i–j direction in global y direction

TABLE 5.7   Continued

Data set	Column	Variable	Description
	26–30	JDFRM	Frame type number
	31–35	IFNODE	Number of node i
	36–40	JFNODE	Number of node j
	41–45	NOSMFE	Number of same elements for automatic generation (default = 1)
	46–50	NODINC	Node number increment of automatically generated elements (default = 1)
Truss element data	1–5	LTRS	Truss element number
	6–15[a]	TXO	Horizontal projected length; positive for i–j direction in global x direction
	16–25[a]	TYO	Vertical projected length; positive for i–j direction in global y direction
	26–30	JDTRS	Truss type number
	31–35	ITNODE	Number of node i
	36–40	JTNODE	Number of node j
	51–55	NOSMTE	Number of same elements for automatic generation (default = 1)
	56–60	NODINC	Node number increment of automatically generated elements (default = 1)
Boundary condition	1–5	NODE	Node number of support
	6–10	XFIX	XFIX = 1 for restrained in global x-direction (default = 0)
	11–15	YFIX	YFIX = 1 for restrained in global y-direction (default = 0)
	16–20	RFIX	RFIX = 1 for restrained in rotation (default = 0)
	21–25	NOSMBD	Number of same boundary conditions for automatic generation (default = 1)
	26–30	NODINC	Node number increment of automatically generated supports (default = 1)
Incremental loads	1–5	NODE	Node number where a load applied
	6–15[a]	XLOAD	Incremental load in global x direction (default = 0)
	16–25[a]	YLOAD	Incremental load in global y direction (default = 0)
	26–35[a]	RLOAD	Incremental moment in global θ-direction (default = 0)
	36–40	NOSMLD	Number of same loads for automatic generation (default = 1)
	41–45	NODINC	Node number increment of automatically generated loads (default = 1)

[a] Indicates that real value (F or E format) should be entered; otherwise input the integer value (I format).

The design examples are limited to two-dimensional steel frames, so that the spatial behavior is not considered. Lateral torsional buckling is assumed to be prevented by adequate lateral braces. Compact W sections are assumed so that sections can develop their full plastic moment capacity without buckling locally. All loads are statically applied.

5.6.1 Roof Truss

Figure 5.38 shows a hinged-jointed roof truss subject to gravity loads of 201 kip at the joints. A36 steel pipe is used. All member sizes are assumed identical.

5.6.1.1 Design by Advanced Analysis

Step 1: Load condition and preliminary member sizing. The critical factored load condition is shown in Figure 5.38. The member forces of the truss may be obtained (Figure 5.39) using equilibrium conditions. The maximum compressive force is 67.1 kip. The effective length is the same as the actual length (22.4 ft) since K is 1.0. The preliminary member size of steel pipe is 6 in. diameter with 0.28 in. thickness ($\phi P_n = 81$ kip), obtained using the column design table in the LRFD Specification.

Step 2: Structural modeling. Each member is modeled with one truss element without geometric imperfection since the program computes the axial strength of the truss member with the LRFD column strength equations, which indirectly account for geometric imperfections. An incremental load of 0.51 kip is determined by dividing the factored load of 201 kip by a scaling factor of 40 as shown in Figure 5.40.

Step 3: Analysis. Referring to the input instructions described in Section 5.5.4, the input data may be easily generated, as listed in Table 5.8. Note that the total number of supports (NBOUND) in the hinged-jointed truss must be equal to the total number of nodal points, since the nodes of a truss element are restrained against rotation. Programs DATAGEN, INPUT, and PAAP are executed in sequence by entering the batch file command RUN on the screen.

Step 4: Check of load-carrying capacity. Truss elements 10 and 13 fail at load step 48, with loads at nodes 6, 7, and 8 being 241 kip. Since this truss is statically determinant, failure of one member leads to failure of the whole system. Load step 49 shows a sharp increase in displacement and indicates a system failure. The member force of element 10 is 80.41 kip ($F_x = 72.0$ kip, $F_y = 35.7$ kip). Since the load-carrying capacity of 241 kip at nodes 6, 7, and 8 is greater than the applied load of 20 kip, the member size is adequate.

Step 5: Check of serviceability. Referring to P.OUT2, the deflection at node 3 corresponding to load step 1 is equal to 0.02 in. This deflection may be considered elastic since the behavior of the beam

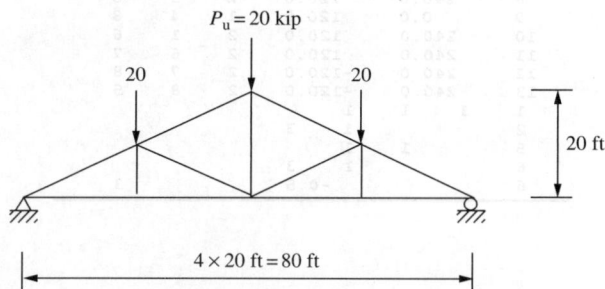

FIGURE 5.38   Configuration and load condition of the hinged-jointed roof truss.

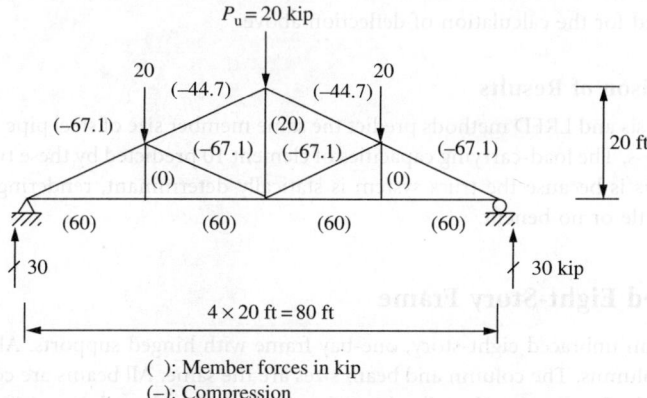

FIGURE 5.39   Member forces of the hinged-jointed roof truss.

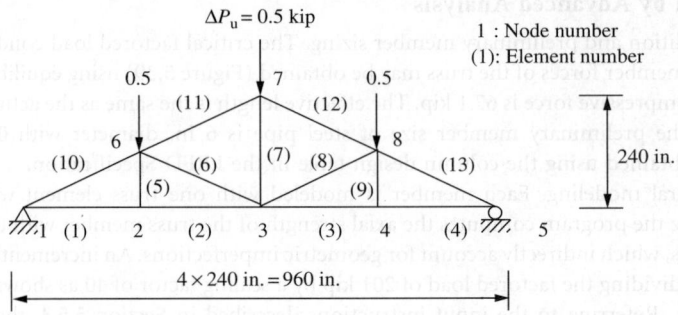

FIGURE 5.40   Modeling of the hinged-jointed roof truss.

TABLE 5.8   Input Data, P.DAT, of the Hinged-Jointed Roof Truss

```
Roof truss
    0    1
    8    8   100
    0    0     1
    0    0    13
    1        5.58        28.1    29000.0      36.0
    1     240.0           0.0       1   1    2    4
    5       0.0         120.0       1   2    6
    6    -240.0         120.0       2   3    6
    7       0.0         240.0       1   3    7
    8     240.0         120.0       2   3    8
    9       0.0         120.0       1   4    8
   10     240.0         120.0       2   1    6
   11     240.0         120.0       2   6    7
   12     240.0        -120.0       2   7    8
   13     240.0        -120.0       2   8    5
    1    1    1          1
    2               1    3
    5         1    1
    6               1    3
    6             -0.5                 3
```

is linear and elastic under small loads. The total deflection of 0.8 in. is obtained by multiplying the deflection of 0.02 in. by the scaling factor of 40. The deflection ratio over the span length is 1/1200, which meets the limitation 1/360. The deflection at the service load will be smaller than that above since the factored load is used for the calculation of deflection above.

5.6.1.2   Comparison of Results

The advanced analysis and LRFD methods predict the same member size of steel pipe with 6 in. diameter and 0.28 in. thickness. The load-carrying capacities of element 10 predicted by these two methods are the same, 80.5 kip. This is because the truss system is statically determinant, rendering inelastic moment redistribution of little or no benefit.

5.6.2   Unbraced Eight-Story Frame

Figure 5.37 shows an unbraced eight-story, one-bay frame with hinged supports. All beams are rigidly connected to the columns. The column and beam sizes are the same. All beams are continuously braced about their weak axis. Bending is primarily about the strong axis at the column. A36 steel is used for all members.

5.6.2.1 Design by Advanced Analysis

Step 1: Load condition and preliminary member sizing. The uniform gravity loads are converted to equivalent concentrated loads, as shown in Figure 5.41. The preliminary column and beam sizes are selected as W33 × 130 and W21 × 50.

Step 2: Structural modeling. Each column is modeled with one element since the frame is unbraced and the maximum moment in the member occurs at the ends. Each beam is modeled with two elements.

The explicit imperfection and the further reduced tangent modulus models are used in this example, since they are easier in preparing the input data compared to equivalent notional load models. Figure 5.42 shows the model for the eight-story frame. The explicit imperfection model uses an out-of-plumbness of 0.2%, and in the further reduced tangent modulus model, $0.85E_t$ is used.

Herein, a scaling factor of 10 is used due to the high indeterminacy. The load increment is automatically reduced if the element stiffness parameter, η, exceeds the predefined value 0.1. The 54 load steps required to converge on the solution are given in P.OUT2.

Step 3: Analysis. The input data may be easily generated, as listed in Table 5.9. Programs are executed in sequence by typing the batch file command RUN.

Step 4: Check of load-carrying capacity. From the output file P.OUT, the ultimate load-carrying capacity of the structure is obtained as 5.24 and 5.18 kip with respect to the lateral load at roof in load combination 2 by the imperfection method and the reduced tangent modulus method, respectively. This load-carrying capacity is 3 and 2% greater, respectively, than the applied factored load of 5.12 kip. As a result, the preliminary member sizes are satisfactory.

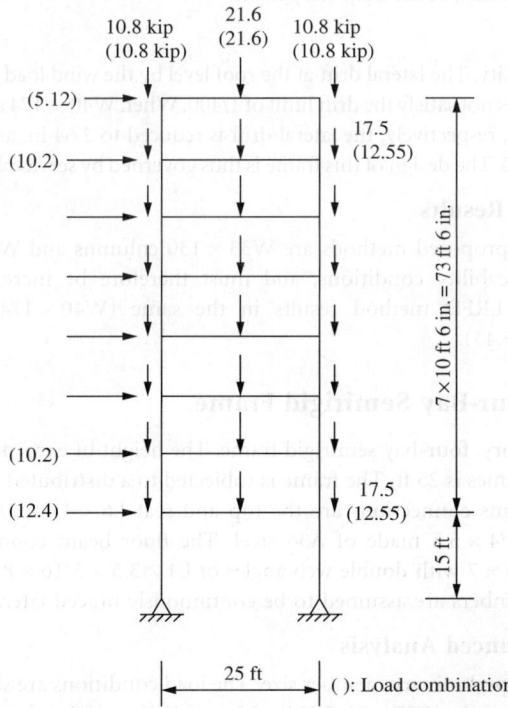

FIGURE 5.41 Concentrated load condition converted from the distributed load for the two-story frame.

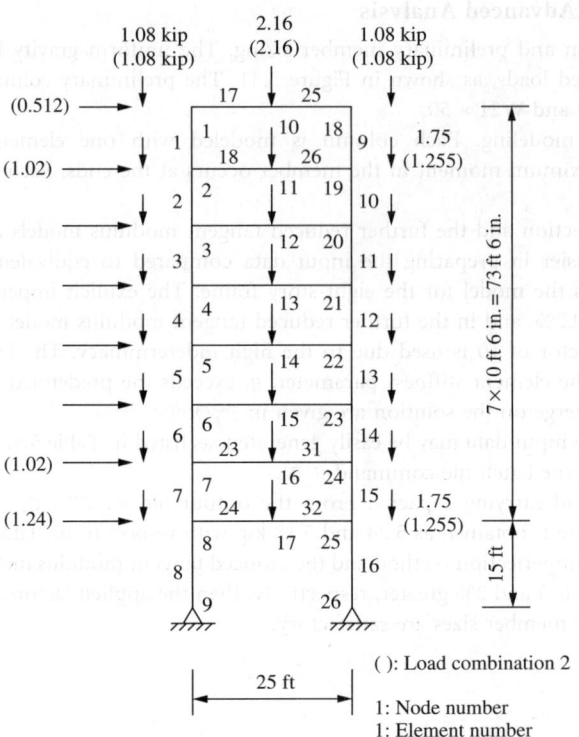

FIGURE 5.42 Structural modeling of the eight-story frame.

Step 5: Check of serviceability. The lateral drift at the roof level by the wind load (1.0W) is 5.37 in. and the drift ratio is 1/198, which does not satisfy the drift limit of 1/400. When W40 × 174 and W24 × 76 are used for column and beam members, respectively, the lateral drift is reduced to 2.64 in. and the drift ratio is 1/402, which satisfies the limit 1/400. The design of this frame is thus governed by serviceability rather than strength.

5.6.2.2 Comparison of Results

The sizes predicted by the proposed methods are W33 × 130 columns and W21 × 50 beams. They do not, however, meet serviceability conditions, and must therefore be increased to W40 × 174 and W24 × 76 members. The LRFD method results in the same (W40 × 174) column but a larger (W27 × 84) beam (Figure 5.43).

5.6.3 Two-Story, Four-Bay Semirigid Frame

Figure 5.44 shows a two-story, four-bay semirigid frame. The height of each story is 12 ft and it is 25 ft wide. The spacing of the frames is 25 ft. The frame is subjected to a distributed gravity and concentrated lateral loads. The roof beams connections are the top and seat L6 × 4.0 × 3/8 × 7 angle with double web angles of L4 × 3.5 × 1/4 × 5.5 made of A36 steel. The floor beam connections are the top and seat angles of L6 × 4 × 9/16 × 7 with double web angles of L4 × 3.5 × 5/16 × 8.5. All fasteners are A325 ¾in.-diameter bolts. All members are assumed to be continuously braced laterally.

5.6.3.1 Design by Advanced Analysis

Step 1: Load condition and preliminary member size. The load conditions are shown in Figure 5.44. The initial member sizes are selected as W8 × 21, W12 × 22, and W16 × 40 for the columns, the roof beams, and the floor beams, respectively.

TABLE 5.9   Input Data, P.DAT, of the Explicit Imperfection Modeling for the Unbraced Eight-Story Frame: (a) Explicit Imperfection Modeling and (b) Further Reduced Tangent Modulus

(a) Unbraced eight-story frame, explicit imperfection modeling

1	1						
26	2						
0	2	0					
0	32	0					
1	38.30	6710.00	467.00	29000.00		36.00	
2	14.70	984.00	110.00	29000.00		36.00	
1	0.252	126.00	1	2	1	7	
8	0.360	180.00	1	9	8		
9	0.252	126.00	1	19	18	7	
16	0.360	180.00	1	26	25		
17	150.00	0.00	2	1	10	8	
25	150.00	0.00	2	10	18	8	
9	1	1					
26	1	1					
1	0.512	-1.080					
2	1.020	-1.255			6		
8	1.240	-1.255					
10		-2.160					
11		-2.510			7		
18		-1.080					
19		-1.255			7		

(b) Unbraced eight-story frame, further reduced tangent modulus

3	1						
26	2						
0	2	0					
0	32	0					
1	38.30	6710.00	467.00	29000.00		36.00	1
2	14.70	984.00	110.00	29000.00		36.00	
1	0.	126.00	1	2	1	7	
8	0.	180.00	1	9	8		
9	0.	126.00	1	19	18	7	
16	0.	180.00	1	26	25		
17	150.00	0.00	2	1	10	8	
25	150.00	0.00	2	10	18	8	
9	1	1					
26	1	1					
1	0.512	-1.080					
2	1.020	-1.255			6		
8	1.240	-1.255					
10		-2.160					
11		-2.510			7		
18		-1.080					
19		-1.255			7		

Step 2: Structural modeling. Each column is modeled with one element and beam with two elements. The distributed gravity loads are converted to equivalent concentrated loads on the beam, as shown in Figure 5.45. In explicit imperfection modeling, the geometric imperfection is obtained by multiplying the column height by 0.002. In the equivalent notional load method, the notional load is 0.002 times the total gravity load plus the lateral load. In the further reduced tangent modulus method, the program automatically accounts for geometric imperfection effects. Although users can choose any of these three models, the further reduced tangent modulus model is the only one presented herein. The incremental loads are computed by dividing the concentrated load by the scaling factor of 20.

Step 3: Analysis. The three parameters of the connections can be computed by the use of the computer program 3PARA. Corresponding to the input format in Table 5.2, the input data, CONN.DAT, may be generated, as shown in Table 5.10.

Referring to the input instructions (Section 5.5.4), the input data are written in the form shown in Table 5.11. Programs DATAGEN, INPUT, and PAAP are executed sequentially by typing "RUN." The program will continue to analyze with increasing load steps up to the ultimate state.

Step 4: Check of load-carrying capacity. As shown in output file, P.OUT2, the ultimate load-carrying capacities of the load combinations 1 and 2 are 46.2 and 42.9 kip, respectively, at nodes 7–13 (Figure 5.45). Compared to the applied loads, 45.5 and 31.75 kip, the initial member sizes are adequate.

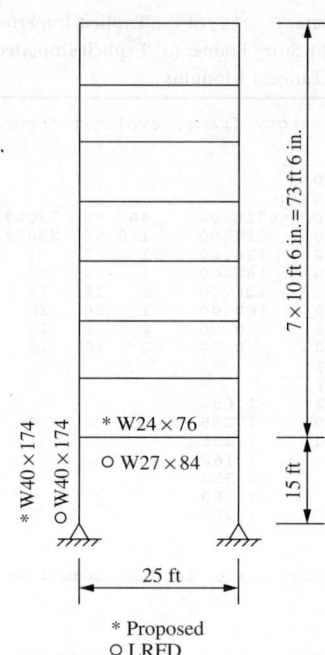

FIGURE 5.43   Comparison of member sizes of the eight-story frame.

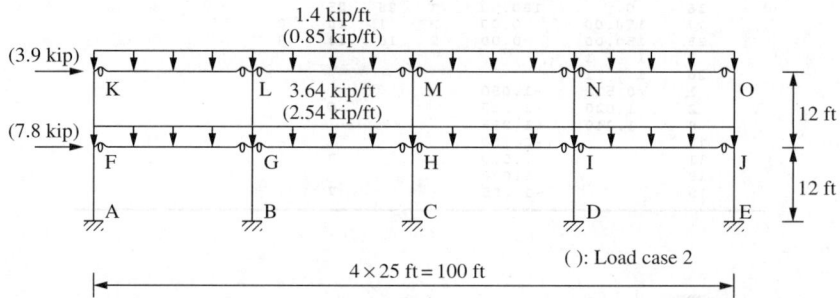

FIGURE 5.44   Configuration and load condition of two-story, four-bay semirigid frame.

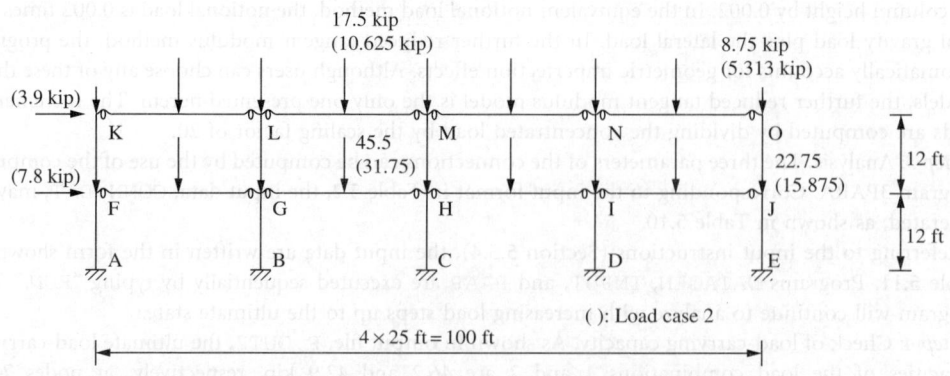

FIGURE 5.45   Concentrated load condition converted form the distributed load for the two-story, one-bay semirigid frame.

Step 5: Check of serviceability. The lateral displacement at roof level corresponding to 1.0W is computed as 0.51 in. from the computer output P.OUT2. The drift ratio is 1/565, which satisfies the limitation 1/400. The preliminary member sizes are satisfactory.

5.6.3.2 Comparison of Results

The member sizes by the advanced analysis and the LRFD method are compared in Figure 5.46. The beam members are one size larger in the advanced analysis method, and the interior columns are one size smaller.

TABLE 5.10 Input Data, CONN.DAT, of Connection for (a) Roof Beam and (b) Floor Beam

		(a) Roof beam			
2	36.0	29,000	2.5	1.25	12
7	0.375	0.875	2.5		
5.5	0.25	0.6875			
		(b) Floor beam			
2	36.0	29.000	2.5	1.25	16
7	0.5625	1.0625	2.5		
8.5	0.3125	0.75			

TABLE 5.11 Input Data, P.DAT, of the Four-Bay, Two-Story Semirigid Frame: (a) Load Case 1 and (b) Load Case 2

```
(a) Four-bay two-story semi-rigid frame (Load case 1)
          3     1
         23     5   100
          2     3     0
         16    26     0
          1  1361.0  607384.0     0.927
          2   446.0   90887.0     1.403
          1     6.16      75.3    20.4    29000.0    36.0    1
          2     6.48     156.0    29.3    29000.0    36.0
          3    11.8      518.0    72.9    29000.0    36.0
          1    11     1     1
          2    12     2     1
          3    13     1     1
          4    14     2     1
          5    15     1     1
          6    16     2     1
          7    17     1     1
          8    18     2     1
          9    19     1     2
         10    20     2     2
         11    21     1     2
         12    22     2     2
         13    23     1     2
         14    24     2     2
         15    25     1     2
         16    26     2     2
          1     0.0    144.0     1     1     6
          2     0.0    144.0     1     2     8
          3     0.0    144.0     1     3    10
          4     0.0    144.0     1     4    12
          5     0.0    144.0     1     5    14
          6     0.0    144.0     1     6    15
          7     0.0    144.0     1     8    17     3     2
         10     0.0    144.0     1    14    23
         13   150.0      0.0     3     6     7     8
         19   150.0      0.0     2    15    16     8
          1     1     1     1     5
          6           -1.1375
          7           -2.2750              7
         14           -1.1375
         15           -0.4375
         16           -0.8750              7
         23           -0.4375
```

TABLE 5.11 Continued

(b) Four-bay two-story semi-rigid frame (Load case 2)

```
  3    1
 23    5  100
  2    3    0
 16   26    0
  1  1361.0  607384.0   0.927
  2   446.0   90887.0   1.403
  1     6.16      75.3     20.4   29000.0   36.0   1
  2     6.48     156.0     29.3   29000.0   36.0
  3    11.8      518.0     72.9   29000.0   36.0
  1   11    1    1
  2   12    2    1
  3   13    1    1
  4   14    2    1
  5   15    1    1
  6   16    2    1
  7   15    1    1
  8   16    2    1
  9   19    1    2
 10   20    2    2
 11   21    1    2
 12   22    2    2
 13   23    1    2
 14   24    2    2
 15   25    1    2
 16   26    2    2
  1    0.0   144.0    1    1    6
  2    0.0   144.0    1    2    8
  3    0.0   144.0    1    3   10
  4    0.0   144.0    1    4   12
  5    0.0   144.0    1    5   14
  6    0.0   144.0    1    6   15
  7    0.0   144.0    1    8   17    3    2
 10    0.0   144.0    1   14   23
 11  150.0     0.0    3    6    7    8
 19  150.0     0.0    2   15   16    8
  1    1    1    5
  6   0.39   -0.7938
  7         -1.5880              7
 14         -0.7938
 15  0.195  -0.2657
 16         -0.5313              7
 23         -0.2657
```

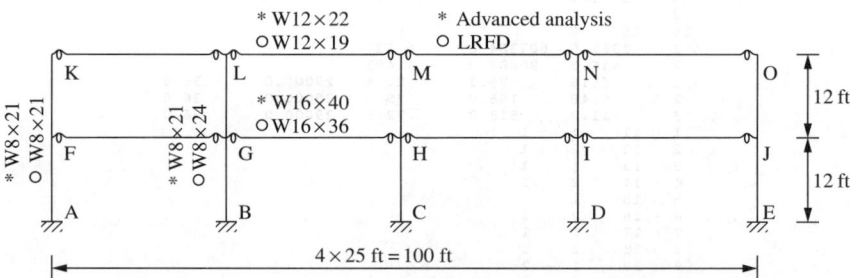

FIGURE 5.46   Comparison of member sizes of the two-story, four-bay semirigid frame.

Glossary

Advanced analysis — Analysis predicting directly the stability of a structural system and its component members and not needing separate member capacity checks.

ASD — Acronym for allowable stress design.

Beam columns — Structural members whose primary function is to carry axial force and bending moment.

Braced frame — Frame in which lateral deflection is prevented by braces or shear walls.

Column — Structural member whose primary function is to carry axial force.

CRC — Acronym for Column Research Council.

Drift — Lateral deflection of a building.

Ductility — Ability of a material to undergo a large deformation without a significant loss in strength.

Factored load — The product of the nominal load and a load factor.

Flexural member — Structural member whose primary function is to carry bending moment.

Geometric imperfection — Unavoidable geometric error during fabrication and erection.

Limit state — A condition in which a structural or structural component becomes unsafe (strength limit state) or unfit for its intended function (serviceability limit state).

Load factor — A factor to account for the unavoidable deviations of the actual load from its nominal value and uncertainties in structural analysis.

LRFD — Acronym for load resistance factor design.

Notional load — Load equivalent to geometric imperfection.

PD — Acronym for plastic design.

Plastic hinge — A yield section of a structural member in which the internal moment is equal to the plastic moment of the cross-section.

Plastic zone — A yield zone of a structural member in which the stress of a fiber is equal to the yield stress.

Refined plastic hinge analysis — Modified plastic hinge analysis accounting for gradual yielding of a structural member.

Resistance factors — A factor to account for the unavoidable deviations of the actual resistance of a member or a structural system from its nominal value.

Second-order analysis — Analysis to use equilibrium equations based on the deformed geometry of a structure under load.

Semirigid connection — Beam-to-column connection whose behavior lies between fully rigid and ideally pinned connection.

Service load — Nominal load under normal usage.

Stability function — Function to account for the bending stiffness reduction due to axial force.

Stiffness — Force required to produce unit displacement.

Unbraced frame — Frame in which lateral deflections are not prevented by braces or shear walls.

References

[1] American Institute of Steel Construction. 2001. *Load and Resistance Factor Design Specification*, 3rd ed., Chicago.

[2] SSRC. 1981. General principles for the stability design of metal structures, Technical Memorandum No. 5, Civil Engineering, ASCE, February, pp. 53–54.

[3] Chen, W.F. and Lui, E.M. 1986. *Structural Stability — Theory and Implementation*, Elsevier, New York.

[4] Kanchanalai, T. 1977. The Design and Behavior of Beam–Columns in Unbraced Steel Frames, AISI Project No. 189, Report No. 2, Civil Engineering/Structures Research Lab., University of Texas at Austin.

[5] Hwa, K. 2003. *Toward Advanced Analysis in Steel Frame Design*, PhD dissertation, Department of Civil and Environmental Engineering, University of Hawaii at Manoa, Honolulu, HI.

[6] Kim, S.E. 1996. *Practical Advanced Analysis for Steel Frame Design*, PhD thesis, School of Civil Engineering, Purdue University, West Lafayette, IN.

[7] Liew, J.Y.R., White, D.W., and Chen, W.F. 1993. Second-order refined plastic hinge analysis of frame design: Part I, *J. Struct. Eng.*, ASCE, 119(11), 3196–3216.

[8] Kishi, N. and Chen, W.F. 1990. Moment–rotation relations of semi-rigid connections with angles, *J. Struct. Eng.*, ASCE, 116(7), 1813–1834.

[9] Kim, S.E. and Chen, W.F. 1996. Practical advanced analysis for steel frame design, *ASCE Structural Congress XIV*, Chicago, Special Proceeding Volume on Analysis and Computation, April, pp. 19–30.

[10] Chen, W.F., Kim, S.E., and Choi, S.H. 2001. Practical second-order inelastic analysis for three-dimensional steel frames, *Steel Struct.*, 1, 213–223.

[11] Kim, S.E. and Lee, D.H. 2002. Second-order distributed plasticity analysis of space steel frames, *Eng. Struct.*, 24, 735–744.

[12] Kim, S.E., Park, M.H., and Choi, S.H. 2001. Direct design of three-dimensional frames using practical advanced analysis, *Eng. Struct.*, 23, 1491–1502.

[13] Avery, P. 1998. *Advanced Analysis of Steel Frames Comprising Non-compact Sections*, PhD thesis, School of Civil Engineering, Queensland University of Technology, Brisbane, Australia.

[14] Kim, S.E. and Lee, J.H. 2001. Improved refined plastic-hinge analysis accounting for local buckling, *Eng. Struct.*, 23, 1031–1042.

[15] Kim, S.E., Lee, J.H., and Park, J.S. 2002. 3D second-order plastic hinge analysis accounting for lateral torsional buckling, *Int. J. Solids Struct.*, 39, 2109–2128.

[16] Wongkeaw, K. and Chen, W.F. 2002. Consideration of out-of-plane buckling in advanced analysis for planar steel frame design, *J. Constr. Steel Res.*, 58, 943–965.

[17] Liew, J.Y.R., White, D.W., and Chen, W.F. 1993. Second-order refined plastic-hinge analysis for frame design: Part 2, *J. Struct. Eng.*, ASCE, 119(11), 3217–3237.

[18] Chen, W.F. and Lui, E.M. 1992. *Stability Design of Steel Frames*, CRC Press, Boca Raton, FL.

[19] Liew, J.Y.R. 1992. *Advanced Analysis for Frame Design*, PhD thesis, School of Civil Engineering, Purdue University, West Lafayette, IN.

[20] Kishi, N., Goto, Y., Chen, W.F., and Matsuoka, K.G. 1993. Design aid of semi-rigid connections for frame analysis, *Eng. J.*, AISC, 4th quarter, 90–107.

[21] ECCS. 1991. *Essentials of Eurocode 3 Design Manual for Steel Structures in Building*, ECCS-Advisory Committee 5, No. 65.

[22] ECCS. 1984. *Ultimate Limit State Calculation of Sway Frames with Rigid Joints*, Technical Committee 8 — Structural stability technical working group 8.2-system, Publication No. 33.

[23] Standards Australia. 1990. *AS4100–1990, Steel Structures*, Sydney, Australia.

[24] Canadian Standard Association. 1994. *Limit States Design of Steel Structures*, CAN/CSA-S16.1-M94.

[25] Canadian Standard Association. 1989. *Limit States Design of Steel Structures*, CAN/CSA-S16.1-M89.

[26] Kim, S.E. and Chen, W.F. 1996. Practical advanced analysis for braced steel frame design, *J. Struct. Eng.*, ASCE, 122(11), 1266–1274.

[27] Kim, S.E. and Chen, W.F. 1996. Practical advanced analysis for unbraced steel frame design, *J. Struct. Eng.*, ASCE, 122(11), 1259–1265.

[28] Maleck, A.E., White, D.W., and Chen, W.F. 1995. Practical application of advanced analysis in steel design, *Proc. 4th Pacific Structural Steel Conf.*, Vol. 1, Steel Structures, pp. 119–126.

[29] White, D.W. and Chen, W.F., Eds. 1993. *Plastic Hinge Based Methods for Advanced Analysis and Design of Steel Frames: An Assessment of the State-of-the-art*, SSRC, Lehigh University, Bethlehem, PA.

[30] Kim, S.E. and Chen, W.F. 1996. Practical advanced analysis for semi-rigid frame design, *Eng. J.*, AISC, 33(4), 129–141.

[31] Kim, S.E. and Chen, W.F. 1996. Practical advanced analysis for frame design — Case study, *SSSS J.*, 6(1), 61–73.

[32] American Institute of Steel Construction. 1994. *Load and Resistance Factor Design Specification*, 2nd ed., Chicago.

[33] American Institute of Steel Construction. 1986. *Load and Resistance Factor Design Specification for Structural Steel Buildings*, Chicago.

[34] LeMessurier, W.J. 1977. A practical method of second order analysis, Part 2 — Rigid Frames. *Eng. J.*, AISC, 2nd quarter, 14(2), 49–67.

[35] Vogel, U. 1985. Calibrating frames, *Stahlbau*, 10, 1–7.

[36] Stelmack, T.W. 1982. *Analytical and Experimental Response of Flexibly-Connected Steel Frames*, MS dissertation, Department of Civil, Environmental, and Architectural Engineering, University of Colorado.

[37] Chen, W.F. and Kim, S.E. 1997. *LRFD Steel Design Using Advanced Analysis*, CRC Press, Boca Raton, FL.

[38] Ad Hoc Committee on Serviceability. 1986. Structural serviceability: a critical appraisal and research needs, *J. Struct. Eng.*, ASCE, 112(12), 2646–2664.

[39] Ellingwood, B.R. 1989. Serviceability Guidelines for Steel Structures, *Eng. J.*, AISC, 26, 1st Quarter, 1–8.

[40] White, D.W. 1993. Plastic hinge methods for advanced analysis of steel frames, *J. Constr. Steel Res.*, 24(2), 121–152.

[41] Cook, R.D., Malkus, D.S., and Plesha, M.E. 1989. *Concepts and Applications of Finite Element Analysis*, 3rd ed., John Wiley & Sons, New York.

[42] Hughes, T.J.R. 1987. *The Finite Element Method: Linear Static and Dynamic Finite Element Analysis*, Prentice Hall, Englewood Cliffs, NJ.

[43] American Institute of Steel Construction. 2002. *Seismic Provisions for Structural Steel Buildings*, Chicago.

6

Cold-Formed Steel Structures

Wei-Wen Yu
Department of Civil Engineering,
University of Missouri,
Rolla, MO

6.1 Introduction

Cold-formed steel members as shown in Figure 6.1 are widely used in building construction, bridge construction, storage racks, highway products, drainage facilities, grain bins, transmission towers, car bodies, railway coaches, and various types of equipment. These sections are cold-formed from carbon,

0-8493-1569-7/05/$0.00+$1.50
© 2005 by CRC Press

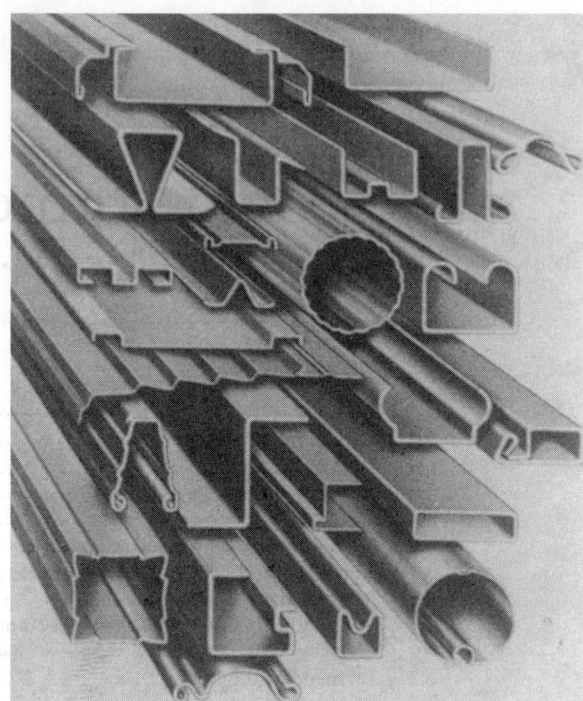

FIGURE 6.1   Various shapes of cold-formed steel sections (courtesy of Yu, W.W. 1991).

low-alloy steel, or stainless steel sheet, strip, plate, or flat bar in cold-rolling machines or by press brake or bending brake operations. The thicknesses of such members usually range from 0.0149 in. (0.378 mm) to about 0.25 in. (6.35 mm) even though steel plates and bars as thick as 1 in. (25.4 mm) can be cold-formed into structural shapes.

The use of cold-formed steel members in building construction began around the 1850s in both the United States and Great Britain. However, such steel members were not widely used in buildings in the United States until the 1940s. At present, cold-formed steel members are widely used as construction materials worldwide.

Compared with other materials such as timber and concrete, cold-formed steel members can offer the following advantages: (1) lightness, (2) high strength and stiffness, (3) ease of prefabrication and mass production, (4) fast and easy erection and installation, and (5) economy in transportation and handling, to name a few.

From the structural design point of view, cold-formed steel members can be classified into two major types: (1) individual structural framing members (Figure 6.2) and (2) panels and decks (Figure 6.3).

In view of the fact that the major function of the individual framing members is to carry load, structural strength and stiffness are the main considerations in design. The sections shown in Figure 6.2 can be used as primary framing members in buildings up to four or five stories in height. In tall multistory buildings, the main framing is typically of heavy hot-rolled shapes and the secondary elements such as wall studs, joists, decks, or panels may be of cold-formed steel members. In this case, the heavy hot-rolled steel shapes and the cold-formed steel sections supplement each other.

The cold-formed steel sections shown in Figure 6.3 are generally used for roof decks, floor decks, wall panels, and siding material in buildings. Steel decks not only provide structural strength to carry loads, but they also provide a surface on which flooring, roofing, or concrete fill can be applied as shown in Figure 6.4. They can also provide space for electrical conduits. The cells of cellular panels can also be used

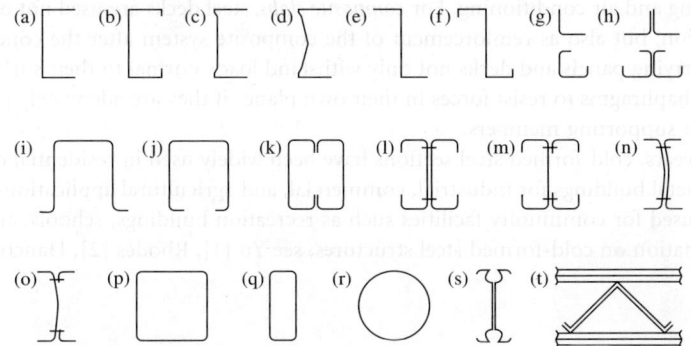

FIGURE 6.2 Cold-formed steel sections used for structural framing (courtesy of Yu, W.W. 1991).

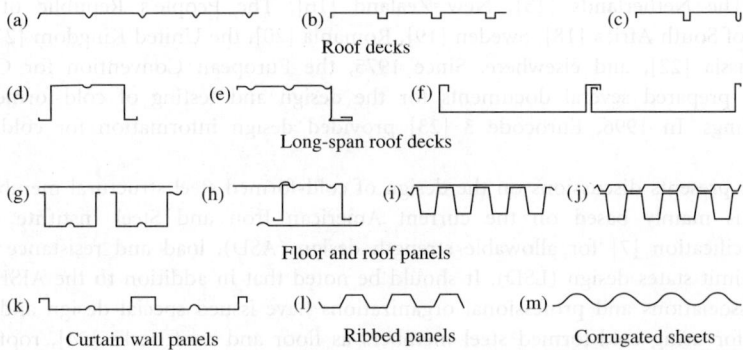

FIGURE 6.3 Decks, panels, and corrugated sheets (courtesy of Yu, W.W. 1991).

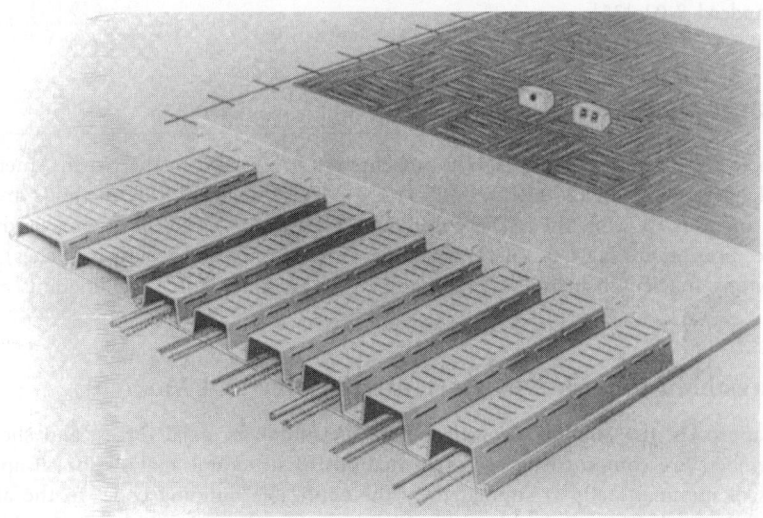

FIGURE 6.4 Cellular floor decks (courtesy of Yu, W.W. 1991).

as ducts for heating and air conditioning. For *composite slabs*, steel decks are used not only as formwork during construction, but also as reinforcement of the composite system after the concrete hardens. In addition, load-carrying panels and decks not only withstand loads normal to their surface, but they can also act as shear diaphragms to resist forces in their own planes if they are adequately interconnected to each other and to supporting members.

During recent years, cold-formed steel sections have been widely used in residential construction and pre-engineered metal buildings for industrial, commercial, and agricultural applications. Metal building systems are also used for community facilities such as recreation buildings, schools, and churches. For additional information on cold-formed steel structures, see Yu [1], Rhodes [2], Hancock et al. [3], and Ghersi et al. [4].

6.2 Design Standards

Design standards and recommendations are now available in Australia [5], Austria [6], Canada [7], Czech Republic [8], Finland [9], France [10], Germany [11], India [12], Italy [13], Japan [14], Mexico [7], The Netherlands [15], New Zealand [16], The People's Republic of China [17], The Republic of South Africa [18], Sweden [19], Romania [20], the United Kingdom [21], the United States [7], Russia [22], and elsewhere. Since 1975, the European Convention for Constructional Steelwork has prepared several documents for the design and testing of cold-formed sheet steel used in buildings. In 1996, Eurocode 3 [23] provided design information for cold-formed steel members.

This chapter presents discussions on the design of cold-formed steel structural members for use in buildings. It is mainly based on the current American Iron and Steel Institute (AISI) North American Specification [7] for allowable strength design (ASD), load and resistance factor design (LRFD), and limit states design (LSD). It should be noted that in addition to the AISI specification, many trade associations and professional organizations have issued special design and construction requirements for using cold-formed steel members as floor and roof decks [24], roof trusses [25], open web steel joists [26], transmission poles and towers [27], storage racks [28], shear diaphragms [7,29], composite slabs [30], metal buildings [31], light framing systems [32–34], guardrails, structural supports for highway signs, luminaries, and traffic signals [36], automotive structural components [37], and others. For the design of cold-formed stainless steel structural members, see SEI/ASCE Standard 8-02 [35].

6.3 Design Bases

For cold-formed steel design, three design approaches are being used in the North American Specification [7]. They are (1) ASD, (2) LRFD, and (3) LSD. The ASD and LRFD methods are used in the United States and Mexico, while the LSD method is used in Canada. The unit systems used in the North American Specification are (1) U.S. customary units (force in kip and length in inches), (2) SI units (force in newtons and length in millimeters), and (3) MKS units (force in kilograms and length in centimeters).

6.3.1 Allowable Strength Design (United States and Mexico)

In the ASD approach, the *required allowable strengths* (moments, axial forces, and shear forces) in structural members are computed by accepted methods of structural analysis for all applicable load combinations of nominal loads as stipulated by the applicable building code. In the absence of an applicable building code, the nominal loads and load combinations should be those stipulated in the American Society of Civil Engineers (ASCE) Standard ASCE 7 [38].

The required allowable strengths should not exceed the allowable *design strengths* permitted by the applicable design standard. The allowable design strength is determined by dividing the *nominal strength* by a factor of safety as follows:

$$R_a = R_n/\Omega \tag{6.1}$$

where R_a is the allowable design strength, R_n is the nominal strength, and Ω is the *factor of safety*. For the design of cold-formed steel structural members using the AISI ASD method, the factors of safety are given in Table 6.1. For details, see the AISI Specification [7].

6.3.2 Load and Resistance Factor Design (United States and Mexico)

Two types of *limit states* are considered in the LRFD method. They are (1) the limit state of strength required to resist the extreme loads during its life and (2) the limit state of serviceability for a structure to perform its intended function.

For the limit state of strength, the general format of the LRFD method is expressed by the following equation:

$$R_u \leq \phi R_n \tag{6.2}$$

where $R_u = \sum \gamma_i Q_i$ is the required strength, ϕR_n is the design strength, γ_i is the load factor, Q_i is the load effect, ϕ is the resistance factor, and R_n is the nominal strength.

The structure and its components should be designed so that the design strengths, ϕR_n, are equal to or greater than the required strengths, $\sum \gamma_i Q_i$, which are computed on the basis of the factored nominal loads and load combinations as stipulated by the applicable building code or, in the absence of an applicable code, as stipulated in ASCE 7 [38].

In addition, the following LRFD criteria can be used for roof and floor composite construction using cold-formed steel:

$$1.2D_s + 1.6C_w + 1.4C \tag{6.3}$$

where D_s is the weight of the steel deck, C_w is the weight of wet concrete during construction, and C is the construction load, including equipment, workmen, and formwork, but excluding the weight of the wet concrete. Table 6.1 lists the ϕ factors, which are used for the LRFD and LSD methods for the design of cold-formed steel members and connections. For details, see the AISI Specification [7].

6.3.3 Limit States Design (Canada)

The methodology for the LSD method is the same as for the LRFD method, except that the load factors, load combinations, target reliability indices, and assumed live-to-dead ratio used in the development of the design criteria are different. In addition, a few different terms are used in the LSD method.

Like the LRFD method, the LSD method specifies that the structural members and connections should be designed to have factored resistance equal to or greater than the effect of factored loads as follows:

$$\phi R_n \geq R_f \tag{6.4}$$

where R_f is the effect of factored loads, R_n is the nominal resistance, ϕ is the resistance factor, and ϕR_n is the factored resistance.

For the LSD method, the load factors and load combinations used for the design of cold-formed steel structures are based on the National Building Code of Canada. For details, see appendix B of the North American Specification [7]. The resistance factors are also listed in Table 6.1.

TABLE 6.1   Factors of Safety, Ω, and Resistance Factors, ϕ, used in the North American Specification [7]

Type of strength	ASD factor of safety, Ω	LRFD resistance factor, ϕ	LSD resistance factor, ϕ
Tension members			
For yielding	1.67	0.90	0.90
For fracture away from the connection	2.00	0.75	0.75
For fracture at the connection (see connections)			
Flexural members			
Bending strength			
For sections with stiffened or partially stiffened compression flanges	1.67	0.95	0.90
For sections with unstiffened compression flanges	1.67	0.90	0.90
Laterally unbraced beams	1.67	0.90	0.90
Beams having one-flange through fastened-to-deck or sheathing (C- or Z-sections)	1.67	0.90	0.90
Beams having one-flange fastened to a standing seam roof system	1.67	0.90	a
Web design			
Shear strength	1.60	0.99	0.80
Web crippling			
Built-up sections	1.65–2.00	0.75–0.90	0.60–0.80
Single-web channel and C-sections	1.65–2.00	0.75–0.90	0.65–0.80
Single-web Z-sections	1.65–2.00	0.75–0.90	0.65–0.80
Single-hat sections	1.70–2.00	0.75–0.90	0.65–0.75
Multiweb deck sections	1.65–2.25	0.65–0.90	0.55–0.80
Stiffeners			
Transverse stiffeners	2.00	0.85	0.80
Concentrically loaded compression members	1.80	0.85	0.80
Combined axial load and bending			
For tension	1.67	0.95	0.90
For compression	1.80	0.85	0.80
For bending	1.67	0.90–0.95	0.90
Closed cylindrical tubular members			
Bending strength	1.67	0.95	0.90
Axial compression	1.80	0.85	0.80
Wall studs and wall assemblies			
Wall studs in compression	1.80	0.85	0.80
Wall studs in bending	1.67	0.90–0.95	0.90
Diaphragm construction	2.00–3.00	0.50–0.65	0.50
Welded connections			
Groove welds			
Tension or compression	1.70	0.90	0.80
Shear (welds)	1.90	0.80	0.70
Shear (base metal)	1.70	0.90	0.80
Arc spot welds			
Welds	2.55	0.60	0.50
Connected part	2.20–3.05	0.50–0.70	0.40–0.60
Minimum edge distance	2.20–2.55	0.60–0.70	0.50–0.60
Tension	2.50–3.00	0.50–0.60	0.40–0.50
Arc seam welds			
Welds	2.55	0.60	0.50
Connected part	2.55	0.60	0.50
Fillet welds			
Longitudinal loading (connected part)	2.55–3.05	0.50–0.60	0.40–0.50
Transverse loading (connected part)	2.35	0.65	0.60
Welds	2.55	0.60	0.50
Flare groove welds			
Transverse loading (connected part)	2.55	0.60	0.50
Longitudinal loading (connected part)	2.80	0.55	0.45
Welds	2.55	0.60	0.50

TABLE 6.1 Continued

Type of strength	ASD factor of safety, Ω	LRFD resistance factor, ϕ	LSD resistance factor, ϕ
Resistance welds	2.35	0.65	0.55
Shear lag effect	2.50	0.60	0.50
Bolted connections			
Shear, spacing, and edge distance	2.00–2.22	0.60–0.70	[a]
Fracture in net section (shear lag)	2.00–2.22	0.55–0.65	0.55
Bearing strength	2.22–2.50	0.60–0.65	0.50–0.55
Shear strength of bolts	2.40	0.65	0.55
Tensile strength of bolts	2.00–2.25	0.75	0.65
Screw connections	3.00	0.50	0.40
Rupture			
Shear rupture	2.00	0.75	[a]
Block shear rupture			
For bolted connection	2.22	0.65	[a]
For welded connections	2.50	0.60	[a]

[a] See appendix B of the North American Specification for the provisions applicable to Canada [7].

6.4 Materials and Mechanical Properties

In the AISI Specification [7], 15 different steels are presently listed for the design of cold-formed steel members. Table 6.2 lists steel designations, American Society for Testing Materials (ASTM) designations, and *yield points*, tensile strengths, and elongations for these steels. From a structural standpoint, the most important properties of steel are as follows:

1. Yield point or yield strength, F_y
2. Tensile strength, F_u
3. Stress–strain relationship
4. Modulus of elasticity, tangent modulus, and shear modulus
5. Ductility
6. Weldability
7. Fatigue strength

In addition, formability, durability, and toughness are also important properties of cold-formed steel.

6.4.1 Yield Point, Tensile Strength, and Stress–Strain Relationship

As listed in Table 6.2, the yield points or yield strengths of all 15 different steels range from 24 to 80 ksi (166 to 552 MPa or 1687 to 5624 kg/cm^2). The tensile strengths of the same steels range from 42 to 100 ksi (290 to 690 MPa or 2953 to 7030 kg/cm^2). The ratios of the tensile strength to yield point vary from 1.08 to 1.88. As far as the stress–strain relationship is concerned, the stress–strain curve can be either the sharp-yielding type (Figure 6.5a) or the gradual-yielding type (Figure 6.5b).

6.4.2 Strength Increase from Cold Work of Forming

The mechanical properties (yield point, tensile strength, and ductility) of cold-formed steel sections, particularly at the corners, are sometimes substantially different from those of the flat steel sheet, strip, plate, or bar before forming. This is because the cold-forming operation increases the yield point and tensile strength and at the same time decreases the ductility. The effects of cold work on the mechanical properties of corners usually depend on several parameters. The ratios of tensile strength to yield point, F_u/F_y, and inside bend radius to thickness, R/t, are considered to be the most important factors to affect the change in mechanical properties of cold-formed steel sections. Design equations are given in the AISI

TABLE 6.2　Mechanical Properties of Steels Referred to in the AISI North American Specification[a,b]

Steel designation	ASTM designation[c]	Yield point, F_y (ksi)	Tensile strength, F_u (ksi)	Elongation (%)	
				In 2-in. gage length	In 8-in. gage length
Structural steel	A36	36	58–80	23	—
High-strength low-alloy structural steel	A242				
$\frac{3}{4}$ in. and below		50	70	—	18
$\frac{3}{4}$ in. to $1\frac{1}{2}$ in.		46	67	21	18
Low and intermediate	A283 Grade A	24	45–60	30	27
tensile strength carbon	B	27	50–65	28	25
plates, shapes, and bars	C	30	55–75	25	22
	D	33	60–80	23	20
Cold-formed welded and	A500				
seamless carbon steel	Round tubing				
structural tubing in	A	33	45	25	—
rounds and shapes	B	42	58	23	—
	C	46	62	21	—
	D	36	58	23	—
	Shaped tubing				—
	A	39	45	25	—
	B	46	58	23	—
	C	50	62	21	—
	D	36	58	23	—
Structural steel with 50-ksi	A529 Grade 50	50	70–100	21	—
minimum yield point	55	55	70–100	20	—
High-strength low-alloy	A572 Grade 42	42	60	24	20
columbium–vanadium	50	50	65	21	18
steels of structural quality	60	60	75	18	16
	65	65	80	17	15
High-strength low-alloy structural steel with 50-ksi minimum yield point	A588	50	70	21	18
Hot-rolled and cold-	A606				
rolled high-strength	Hot-rolled as rolled	45	65	22	—
low-alloy steel sheet and	coils; annealed, or				
strip with improved	normalized; cold				
corrosion resistance	rolled				
	Hot-rolled as rolled cut lengths	50	70	22	—
Zinc-coated steel sheets of	A653				
structural quality	SS Grade 33	33	45	20	—
	37	37	52	18	—
	40	40	55	16	—
	50 Class 1	50	65	12	—
	50 Class 3	50	70	12	—
	HSLAS Grade 40	40	50	22	—
	50	50	60	20	—
	60	60	70	16	—
	70	70	80	12	—
	80	80	90	10	—

TABLE 6.2 Continued

Steel designation	ASTM designation[c]	Yield point, F_y (ksi)	Tensile strength, F_u (ksi)	Elongation (%) In 2-in. gage length	Elongation (%) In 8-in. gage length
Aluminum–zinc alloy coated by the hot-dip process (general requirements)	A792 Grade 33	33	45	20	—
	37	37	52	18	—
	40	40	55	16	—
	50	50	65	12	—
Cold-formed welded and seamless high-strength, low-alloy structural tubing with improved atmospheric corrosion resistance	A847	50	70	19	—
Zinc–5% aluminum alloy-coated steel sheet by the hot-dip process	A875				
	SS Grade 33	33	45	20	—
	37	37	52	18	—
	40	40	55	16	—
	40	40	55	16	—
	50 Class 3	50	70	12	—
	HSLAS Type A Grade 50	50	60	20	—
	60	60	70	16	—
	70	70	80	12	—
	80	80	90	10	—
	HSLAS Type B Grade 50	50	60	22	—
	60	60	70	18	—
	70	70	80	14	—
	80	80	90	12	—
Metal- and nonmetal-coated carbon steel sheet	A1003				
	ST Grade 33H	33	d	10	
	37H	37	d	10	—
	40H	40	d	10	—
	50H	50	d	10	—
Cold-rolled steel sheet, carbon structural, high-strength low-alloy with improved formability	A1008				
	SS Grade 25	25	42	26	—
	30	30	45	24	—
	33 Types 1 and 2	33	48	22	—
	40 Types 1 and 2	40	52	20	—
	HSLAS Grade 45 Class 1	45	60	22	—
	45 Class 2	45	55	22	—
	50 Class 1	50	65	20	—
	50 Class 2	50	60	20	—
	55 Class 1	55	70	18	—
	55 Class 2	55	65	18	—
	60 Class 1	60	75	16	—
	60 Class 2	60	70	16	—
	65 Class 1	65	80	15	—
	65 Class 2	65	75	15	—
	70 Class 1	70	85	14	—
	70 Class 2	70	80	14	—

TABLE 6.2 Continued

Steel designation	ASTM designation[c]	Yield point, F_y (ksi)	Tensile strength, F_u (ksi)	Elongation (%) In 2-in. gage length	Elongation (%) In 8-in. gage length
	HSLAS-F Grade 50	50	60	22	—
	60	60	70	18	—
	70	70	80	16	—
	80	80	90	14	—
Hot-rolled steel sheet and strip, carbon, structural, high-strength low-alloy with improved formability	A1011				
	SS Grade 30	30	49	21–25	19[e]
	33	33	52	18–23	18[e]
	36 Type 1	36	53	17–22	17[e]
	36 Type 2	36	58–80	16–21	16[e]
	40	40	55	15–21	16[e]
	45	45	60	13–19	14[e]
	50	50	65	11–17	12[e]
	55	55	70	9–15	10[e]
	HSLAS Grade 45 Class 1	45	60	23–25	—
	45 Class 2	45	55	23–25	—
	50 Class 1	50	65	20–22	—
	50 Class 2	50	60	20–22	—
	55 Class 1	55	70	18–20	—
	55 Class 2	55	65	18–20	—
	60 Class 1	60	75	16–18	—
	60 Class 2	60	70	16–18	—
	65 Class 1	65	80	14–16	—
	65 Class 2	65	75	14–16	—
	70 Class 1	70	85	12–14	—
	70 Class 2	70	80	12–14	—
	HSLAS-F Grade 50	50	60	22–24	—
	60	60	70	20–22	—
	70	70	80	18–20	—
	80	80	90	16–18	—

[a] The tabulated values are based on ASTM standards.

[b] 1 in. = 25.4 mm; 1 ksi = 6.9 MPa = 70.3 kg/cm^2.

[c] Structural Grade 80 of A653, A875, and A1008 steel and Grade 80 of A792 are allowed in AISI Specification under special conditions. For these grades, $F_y = 80$ ksi, $F_u = 82$ ksi, and elongations are unspecified. See AISI Specification for reduction of yield point and tensile strength.

[d] For type H of A1003 steel, the minimum tensile strength is not specified. The ratio of tensile strength to yield strength should not be less than 1.08.

[e] For A1011 steel, the specified minimum elongation in 2-in. gage length varies with the thickness of the steel sheet and strip. The elongation in 8-in. gage length is for thickness below 0.23 in.

Specification [7] for computing the tensile yield point of corners and the average full-section tensile yield point for design purposes.

6.4.3 Modulus of Elasticity, Tangent Modulus, and Shear Modulus

The strength of cold-formed steel members that are governed by buckling depends not only on the yield point but also on the modulus of elasticity, E, and the tangent modulus, E_t. A value of $E = 29,500$ ksi (203 GPa or 2.07×10^6 kg/cm^2) is used in the AISI Specification for the design of cold-formed steel structural members. This E value is slightly larger than the value of 29,000 ksi (200 GPa or 2.04×10^6 kg/cm^2), which is used in the American Institute of Steel Construction (AISC) Specification for the design of hot-rolled shapes. The tangent modulus is defined by the slope of the stress–strain curve at any given stress level as shown in Figure 6.5b. For sharp-yielding steels, $E_t = E$ up to the yield, but with gradual-yielding steels, $E_t = E$ only up to the proportional limit, f_{pr} (Figure 6.5b). Once the stress exceeds the proportional

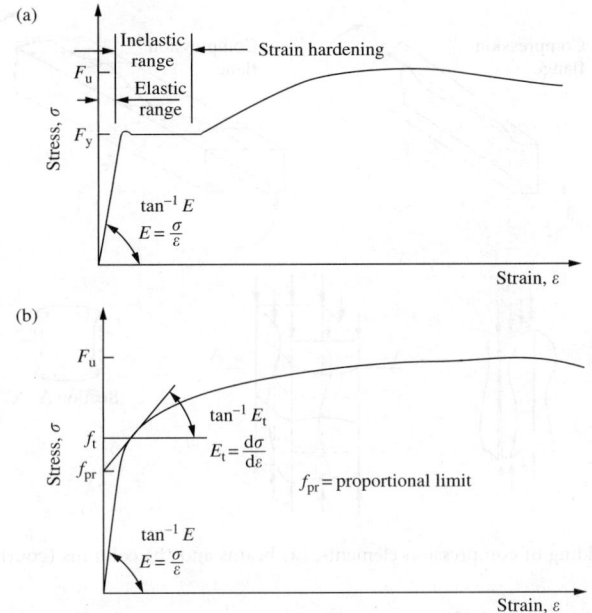

FIGURE 6.5 Stress–strain curves of steel sheet or strip: (a) sharp-yielding and (b) gradual-yielding (courtesy of Yu, W.W. 1991).

limit, the tangent modulus E_t becomes progressively smaller than the initial modulus of elasticity. For cold-formed steel design, the shear modulus is taken as $G = 11,300$ ksi (77.9 GPa or 0.794×10^6 kg/cm^2) according to the AISI Specification.

6.4.4 Ductility

According to the AISI Specification, the ratio F_u/F_y for the steels used for structural framing members should not be less than 1.08, and the total elongation should not be less than 10% for a 2-in. (50.8 mm) gage length. If these requirements cannot be met, the following limitations should be satisfied when such a material is used for purlins and girts: (1) local elongation in a $\frac{1}{2}$-in. (12.7 mm) gage length across the fracture should not be less than 20% and (2) uniform elongation outside the fracture should not be less than 3%. It should be noted that the required ductility for cold-formed steel structural members depends mainly on the type of application and the suitability of the material. The same amount of ductility that is considered necessary for individual framing members may not be needed for roof panels, siding, and similar applications. For this reason, even though Structural Grade 80 of ASTM A653, A875, and 1008 steel, and Grade 80 of A792 steel do not meet the AISI requirements for the F_u/F_y ratio and the elongation, these steels can be used for roofing, siding, and similar applications provided that (1) the yield strength, F_y, used for design is taken as 75% of the specified minimum yield point or 60 ksi (414 MPa or 4218 kg/cm^2), whichever is less, and (2) the tensile strength, F_u, used for design is taken as 75% of the specified minimum tensile stress or 62 ksi (427 MPa or 4359 kg/cm^2). For multiple web configurations a reduced yield point is permitted by the AISI Specification [7] for determining the nominal flexural strength of the section on the basis of the initiation of yielding.

6.5 Element Strength

For cold-formed steel members, the width to thickness ratios of individual elements are usually large. These thin elements may buckle locally at a stress level lower than the yield point of steel when they are

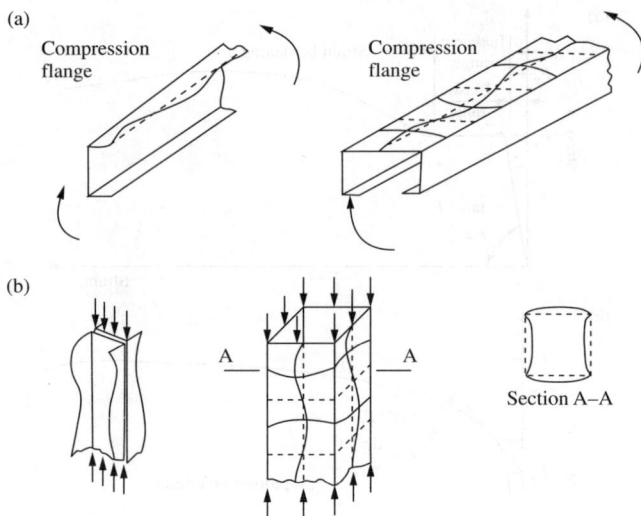

FIGURE 6.6   Local buckling of compression elements: (a) beams and (b) columns (courtesy of Yu, W.W. 1991).

subjected to compression in flexural bending and axial compression as shown in Figure 6.6. Therefore, for the design of such thin-walled sections, *local buckling* and postbuckling strength of thin elements have often been the major design considerations. In addition, shear buckling and web crippling should also be considered in the design of beams.

6.5.1   Maximum Flat Width to Thickness Ratios

In cold-formed steel design, the maximum *flat width to thickness ratio, w/t,* for flanges is limited to the following values in the AISI Specification [7]:

1. *Stiffened compression element* having one longitudinal edge connected to a web or flange element, the other stiffened by
   - Simple lip — 60
   - Any other kind of adequate stiffener — 90
2. *Stiffened compression element* with both longitudinal edges connected to other stiffened element — 500
3. *Unstiffened compression element* — 60

For the design of beams, the maximum depth to thickness ratio, *h/t,* for webs is:

1. Unreinforced webs, $(h/t)_{max} = 200$
2. Webs that are provided with transverse stiffeners:
   - Using bearing stiffeners only, $(h/t)_{max} = 260$
   - Using bearing stiffeners and intermediate stiffeners, $(h/t)_{max} = 300$

6.5.2   Stiffened Elements under Uniform Compression

The strength of a stiffened compression element such as the compression flange of a hat section is governed by yielding if its *w/t* ratio is relatively small. It may be governed by local buckling as shown in Figure 6.7 at a *stress* level less than the yield point if its *w/t* ratio is relatively large.

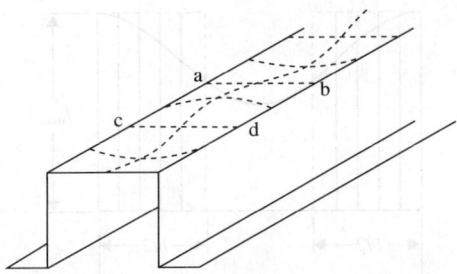

FIGURE 6.7   Local buckling of stiffened compression flange of hat-shaped beam.

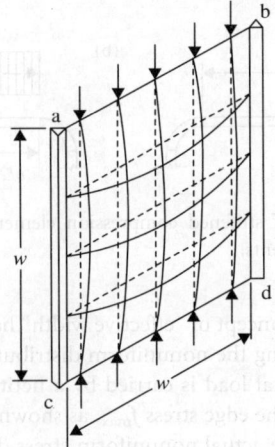

FIGURE 6.8   Postbuckling strength model (courtesy of Yu, W.W. 1991).

The elastic local buckling stress, F_{cr}, of simply supported square plates and long plates can be determined as follows:

$$F_{cr} = \frac{k\pi^2 E}{12(1-\mu^2)(w/t)^2} \tag{6.5}$$

where k is the local buckling coefficient, E is the modulus of elasticity of steel $= 29.5 \times 10^3$ ksi (203 GPa or 2.07×10^6 kg/cm^2), w is the width of the plate, t is the thickness of the plate, and μ is the Poisson's ratio $= 0.3$.

It is well known that stiffened compression elements will not collapse when the local buckling stress is reached. An additional load can be carried by the element after buckling by means of a redistribution of stress. This phenomenon is known as postbuckling strength and is most pronounced for elements with large w/t ratios.

The mechanism of the postbuckling action can easily be visualized from a square plate model as shown in Figure 6.8 [39]. It represents the portion abcd of the compression flange of the hat section illustrated in Figure 6.7. As soon as the plate starts to buckle, the horizontal bars in the grid of the model will act as tie rods to counteract the increasing deflection of the longitudinal struts.

In the plate, the stress distribution is uniform prior to its buckling. After buckling, a portion of the prebuckling load of the center strip transfers to the edge portion of the plate. As a result, a nonuniform stress distribution is developed, as shown in Figure 6.9. The redistribution of stress continues until the stress at the edge reaches the yield point of steel and then the plate begins to fail.

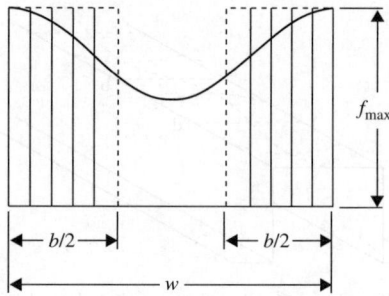

FIGURE 6.9 Stress distribution in stiffened compression elements.

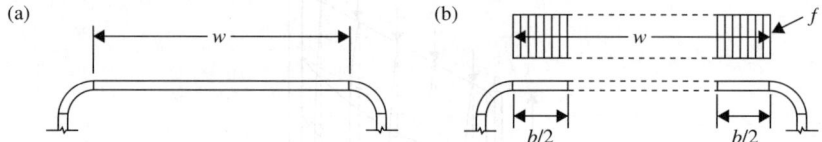

FIGURE 6.10 Effective design width of stiffened compression elements: (a) actual element and (b) effective element, *b* and stress, *f*, on effective elements.

For cold-formed steel members, a concept of "effective width" has long been used for practical design. In this approach, instead of considering the nonuniform distribution of stress over the entire width of the plate *w*, it is assumed that the total load is carried by a fictitious effective width *b*, subjected to a uniformly distributed stress equal to the edge stress f_{max}, as shown in Figure 6.9. The width *b* is selected so that the area under the curve of the actual nonuniform stress distribution is equal to the sum of the two parts of the equivalent rectangular shaded area with a total width *b* and an intensity of stress equal to the edge stress f_{max}. Based on the research findings of von Karman, Sechler, Donnell, and Winter [40], the following equations have been developed in the AISI Specification for computing the *effective design width, b,* for stiffened elements under uniform compression [7].

6.5.2.1 Strength Determination

$$\text{When } \lambda \leq 0.673, \quad b = w \tag{6.6}$$

$$\text{When } \lambda > 0.673, \quad b = \rho w \tag{6.7}$$

where *b* is the effective design width of uniformly compressed element for strength determination (Figure 6.10), *w* is the flat width of compression element, and ρ is the reduction factor determined from Equation 6.8:

$$\rho = (1 - 0.22/\lambda)/\lambda \leq 1 \tag{6.8}$$

where λ is the plate slenderness factor determined from Equation 6.9:

$$\lambda = \sqrt{f/F_{cr}} = (1.052/\sqrt{k})(w/t)(\sqrt{f/E}) \tag{6.9}$$

where *k* is equal to 4.0 is the plate buckling coefficient for stiffened elements supported by a web on each longitudinal edge as shown in Figure 6.10, *t* is the thickness of compression element, *E* is the modulus of elasticity, and *f* is the maximum compressive edge stress in the element without considering the factor of safety.

6.5.2.2 Serviceability Determination

For serviceability determination, Equations 6.6 through 6.9 can also be used for computing the effective design width of compression elements, except that the compressive stress should be computed on the basis of the effective section at the load for which serviceability is determined.

The relationship between ρ and λ according to Equation 6.8 is shown in Figure 6.11.

EXAMPLE 6.1

Calculate the effective width of the compression flange of the box section (Figure 6.12) to be used as a beam bending about the x-axis. Use $F_y = 33$ ksi. Assume that the beam webs are fully effective and that the bending moment is based on initiation of yielding.

Solution

Because the compression flange of the given section is a uniformly compressed stiffened element, which is supported by a web on each longitudinal edge, the effective width of the flange for strength determination can be computed by using Equations 6.6 through 6.9 with $k = 4.0$.

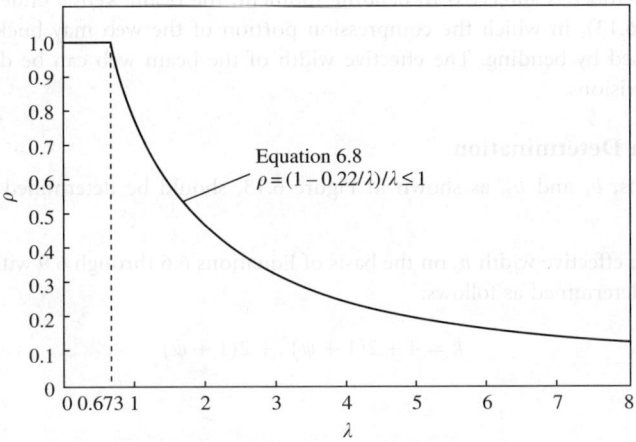

FIGURE 6.11 Reduction factor, ρ, versus slenderness factor, λ (courtesy of Yu, W.W. 1991).

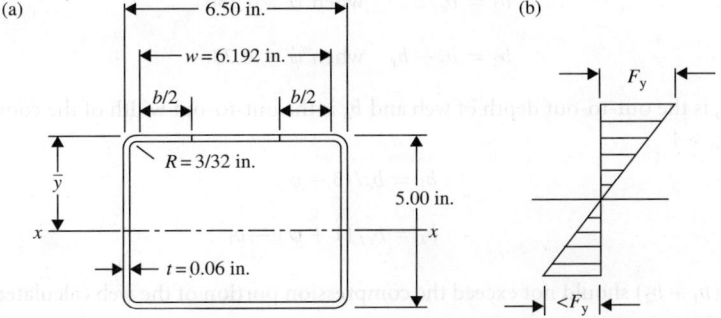

FIGURE 6.12 Example 6.1: (a) tubular section and (b) stress distribution for yield moment (courtesy of Yu, W.W. 1991).

Assume that the bending strength of the section is based on initiation of yielding, $\bar{y} \geq 2.50$ in. Therefore, the slenderness factor λ for $f = F_y$ can be computed from Equation 6.9, that is,

$$
\begin{aligned}
k &= 4.0 \\
w &= 6.50 - 2\,(R + t) = 6.192 \text{ in.} \\
w/t &= 103.2 \\
f &= 33 \text{ ksi} \\
\lambda &= \sqrt{f/F_{\text{cr}}} = (1.052/\sqrt{k})(w/t)\sqrt{f/E} \\
&= (1.052/\sqrt{4.0})(103.2)\sqrt{33/29{,}500} = 1.816
\end{aligned}
$$

Since $\lambda > 0.673$, use Equations 6.7 and 6.8 to compute the effective width, b, as follows:

$$
\begin{aligned}
b = \rho w &= [(1 - 0.22/\lambda)/\lambda]w \\
&= [(1 - 0.22/1.816)/1.816](6.192) = 3.00 \text{ in.}
\end{aligned}
$$

6.5.3 Stiffened Elements with Stress Gradient

When a flexural member is subjected to bending moment, the beam web is under the stress gradient condition (Figure 6.13), in which the compression portion of the web may buckle due to the compressive stress caused by bending. The effective width of the beam web can be determined from the following AISI provisions.

6.5.3.1 Strength Determination

The effective widths, b_1 and b_2, as shown in Figure 6.13, should be determined from the following procedure:

1. Calculate the effective width b_e on the basis of Equations 6.6 through 6.9 with f_1 substituted for f and with k determined as follows:

$$
k = 4 + 2(1 + \psi)^3 + 2(1 + \psi) \tag{6.10}
$$

where

$$
\psi = |f_2/f_1| \quad \text{(absolute value)} \tag{6.11}
$$

2. For $h_o/b_o \leq 4$

$$
b_1 = b_e/(3 + \psi) \tag{6.12}
$$

$$
b_2 = b_e/2 \qquad \text{when } \psi > 0.236 \tag{6.13a}
$$

$$
b_2 = b_e - b_1 \quad \text{when } \psi \leq 0.236 \tag{6.13b}
$$

where h_o is the out-to-out depth of web and b_o is the out-to-out width of the compression flange.

3. For $h_o/b_o > 4$

$$
b_1 = b_e/(3 + \psi) \tag{6.12}
$$

$$
b_2 = b_e/(1 + \psi) - b_1 \tag{6.14}
$$

The value of $(b_1 + b_2)$ should not exceed the compression portion of the web calculated on the basis of effective section.

For the effective width of b_1 and b_2 for C-section webs with holes under stress gradient, see the AISI Specification [7].

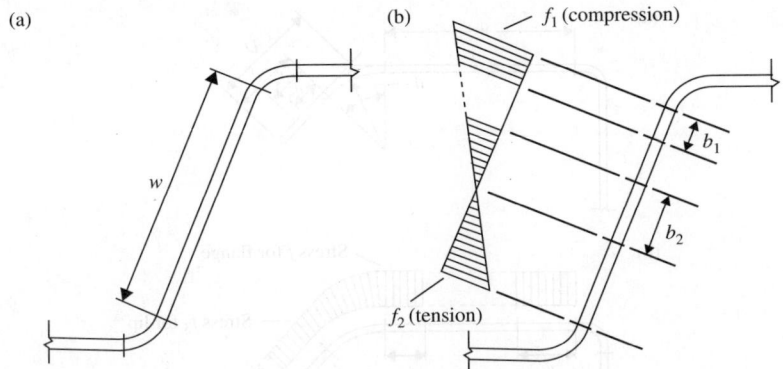

FIGURE 6.13 Stiffened elements with stress gradient: (a) actual element and (b) effective element and stress on effective elements.

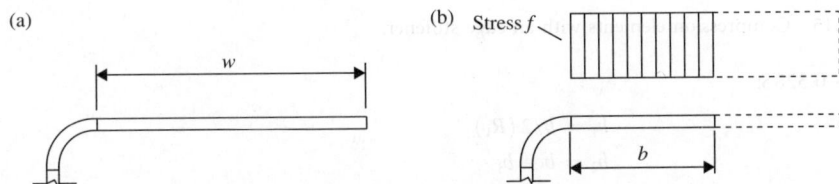

FIGURE 6.14 Effective design width of unstiffened compression elements: (a) actual element and (b) effective element and stress on effective elements.

6.5.3.2 Serviceability Determination

The effective widths used in determining serviceability should be determined as above, except that f_{d1} and f_{d2} are substituted for f_1 and f_2, where f_{d1} and f_{d2} are the computed stresses f_1 and f_2 as shown in Figure 6.13 based on the effective section at the load for which serviceability is determined.

6.5.4 Unstiffened Elements under Uniform Compression

The effective width of unstiffened elements under uniform compression as shown in Figure 6.14 can also be computed by using Equations 6.6 through 6.9, except that the value of k should be taken as 0.43 and the flat width w is measured as shown in Figure 6.14.

6.5.5 Uniformly Compressed Elements with an Edge Stiffener

The following equations can be used to determine the effective width of the uniformly compressed elements with an edge stiffener as shown in Figure 6.15.

6.5.5.1 Strength Determination

For $w/t \leq 0.328S$:

$$I_a = 0 \quad \text{(no edge stiffener needed)} \tag{6.15}$$

$$b = w \tag{6.16}$$

$$b_1 = b_2 = w/2 \tag{6.17}$$

$$d_s = d_s' \quad \text{for simple lip stiffener} \tag{6.18}$$

$$A_s = A_s' \quad \text{for other stiffener shapes} \tag{6.19}$$

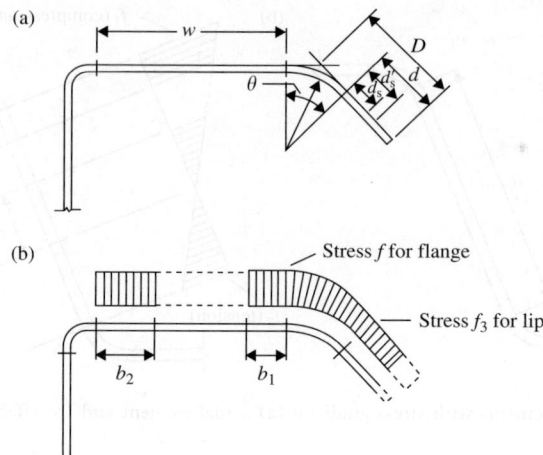

FIGURE 6.15   Compression elements with an edge stiffener.

For $w/t > 0.328S$:

$$b_1 = b/2 \, (R_1) \tag{6.20}$$

$$b_2 = b - b_1 \tag{6.21}$$

$$d_s = d'_s R_I \quad \text{for simple lip stiffeners} \tag{6.22}$$

$$A_s = A'_s R_I \quad \text{for other stiffener shapes} \tag{6.23}$$

where

$$(R_1) = I_s/I_a \le 1 \tag{6.24}$$

$$I_a = 399t^4 \left[\frac{w/t}{S} - 0.328 \right]^3 \le t^4 \left[115 \frac{w/t}{S} + 5 \right] \tag{6.25}$$

$$S = 1.28\sqrt{E/f} \tag{6.26}$$

The effective width, b, is calculated in accordance with Equations 6.6 through 6.9 with k determined as follows.

For simple lip edge stiffener ($140° \ge \theta \ge 40°$)

$D/w \le 0.25$:

$$k = 3.57(R_I)^n + 0.43 \le 4 \tag{6.27}$$

$0.25 < D/w \le 0.8$:

$$k = (4.82 - 5D/w)(R_I)^n + 0.43 \le 4 \tag{6.28}$$

For other edge stiffener shapes

$$k = 3.57(R_I)^n + 0.43 \le 4 \tag{6.29}$$

where

$$n = \left[0.582 - \frac{w/t}{4S} \right] \ge \frac{1}{3} \tag{6.30}$$

In the above equations, k is the plate buckling coefficient, d, w, and D are the dimensions shown in Figure 6.15, d_s is the reduced effective width of the stiffener, d'_s is the effective width of the stiffener calculated as unstiffened element under uniform compression, and b_1 and b_2 are the effective widths shown in Figure 6.15, A_s is the reduced area of the stiffener, I_a is the adequate moment of inertia of the stiffener, so that each component element will behave as a stiffened element, and I_s and A'_s are the moments of inertia of the full-section of the stiffener, about its own centroidal axis parallel to the element to be stiffened, and the effective area of the stiffener, respectively.

For the stiffener shown in Figure 6.15

$$I_s = (d^3 t \sin^2\theta)/12 \tag{6.31}$$

$$A'_s = d'_s t \tag{6.32}$$

6.5.5.2 Serviceability Determination

The effective width, b_d, used in determining serviceability is calculated as in Section 6.5.5.1, except that f_d is substituted for f.

6.5.6 Uniformly Compressed Elements with Intermediate Stiffeners

The effective width of uniformly compressed elements with intermediate stiffeners can also be determined from the AISI Specification, which includes separate design rules for compression elements with only one intermediate stiffener and compression elements with more than one intermediate stiffener.

6.5.6.1 Uniformly Compressed Elements with One Intermediate Stiffener

The following equations can be used to determine the effective width of the uniformly compressed elements with one intermediate stiffener as shown in Figure 6.16.

6.5.6.1.1 Strength Determination

For $b_o/t \le S$

$$I_a = 0 \quad \text{(no intermediate stiffener required)} \tag{6.33}$$

$$b = w \tag{6.34}$$

$$A_s = A'_s \tag{6.35}$$

For $b_o/t > S$

$$A_s = A'_s(R_I) \tag{6.36}$$

where

$$R_I = I_s/I_a \le 1 \tag{6.37}$$

where for $S < b_o/t < 3S$

$$I_a = t^4 \left[50 \frac{b_o/t}{S} - 50 \right] \tag{6.38}$$

and for $b_o/t \ge 3S$

$$I_a = t^4 \left[128 \frac{b_o/t}{S} - 285 \right] \tag{6.39}$$

The effective width, b, is calculated in accordance with Equations 6.6 through 6.9 with k determined as follows:

$$k = 3(R_I)^n + 1 \tag{6.40}$$

where

$$n = \left[0.583 - \frac{b_o/t}{12S} \right] \ge \frac{1}{3} \tag{6.41}$$

All symbols in the above equations have been defined previously and are shown in Figure 6.16.

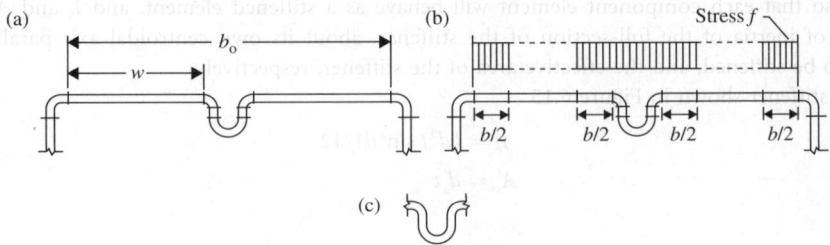

FIGURE 6.16   Compression elements with one intermediate stiffener: (a) actual elements, (b) effective elements and stress on effective elements, and (c) stiffener section.

6.5.6.1.2 *Serviceability Determination*

The effective width, b_d, used in determining serviceability is calculated as in Section 6.5.6.1.1, except that f_d is substituted for f.

6.5.6.2 Uniformly Compressed Elements with More Than One Intermediate Stiffener

In the AISI North American Specification [7], new design equations are provided for determining the effective width of uniformly compressed stiffened elements with multiple intermediate stiffeners.

6.6 Member Design

This chapter deals with the design of the following cold-formed steel structural members: (1) tension members, (2) *flexural members*, (3) concentrically loaded compression members, (4) combined axial load and bending, and (5) closed cylindrical tubular members. The nominal strength equations with factors of safety (Ω) and resistance factors (ϕ) are provided in the Specification [7] for the given limit states.

6.6.1 Sectional Properties

The sectional properties of a member such as area, moment of inertia, section modulus, and radius of gyration are calculated by using the conventional methods of structural design. These properties are based on full cross-section dimensions, effective widths, or net section, as applicable.

   For the design of tension members, the nominal tensile strength is presently based on the gross and net sections. However, for flexural members and axially loaded compression members, the full dimensions are used when calculating the critical moment or load, while the effective dimensions, evaluated at the stress corresponding to the critical moment or load, are used to calculate the nominal strength.

6.6.2 Linear Method for Computing Sectional Properties

Because the thickness of cold-formed steel members is usually uniform, the computation of sectional properties can be simplified by using a "linear" or "midline" method. In this method, the material of each element is considered to be concentrated along the centerline or midline of the steel sheet and the area elements are replaced by straight or curved "line elements." The thickness dimension t is introduced after the linear computations have been completed. Thus, the total area $A = Lt$, and the moment of inertia of the section $I = I't$, where L is the total length of all line elements and I' is the moment of inertia

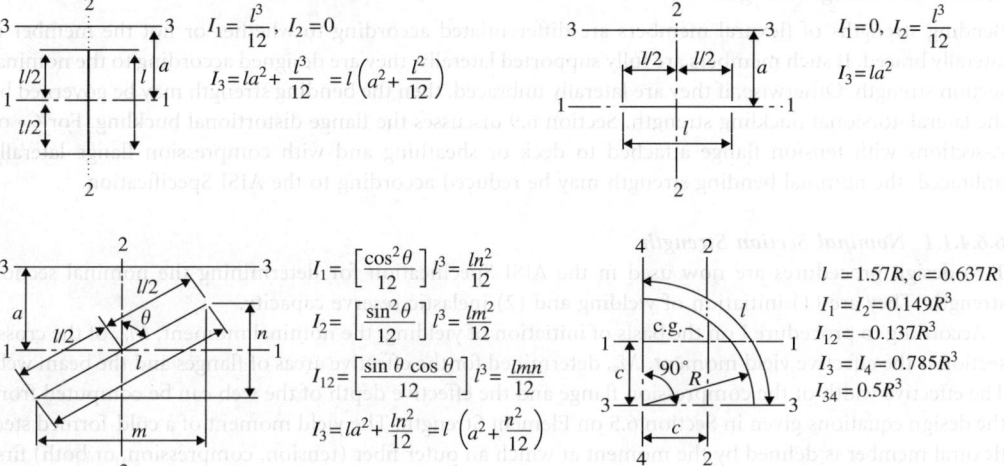

FIGURE 6.17 Properties of line elements.

of the centerline of the steel sheet. The moments of inertia of straight line elements and circular line elements are shown in Figure 6.17.

6.6.3 Tension Members

In the United States and Mexico, the nominal tensile strength of axially loaded cold-formed steel tension members is determined by the following equations:

1. *For yielding:*

$$T_n = A_g F_y \qquad (6.42)$$

2. *For fracture away from connection:*

$$T_n = A_n F_u \qquad (6.43)$$

where T_n is the nominal strength of the member when loaded in tension, A_g is the gross area of cross-section, A_n is the net area of cross-section, F_y is the design yield point of steel, and F_u is the tensile strength of steel.

3. *For fracture at connection:* The nominal tensile strength is also limited by sections E2.7, E3, and E4 of the North American Specification [7] for tension members using welded connections, bolted connections, and screw connections. For details, see chapter E of the Specification under the title "Connections and Joints."

In Canada, the design of tension members is based on appendix B of the AISI North American Specification [7].

6.6.4 Flexural Members

For the design of flexural members, consideration should be given to several design features: (1) bending strength and deflection, (2) shear strength of webs and combined bending and shear, (3) web crippling strength and combined bending and web crippling, and (4) bracing requirements. For some cases, special considerations should also be given to shear lag and flange curling due to the use of thin materials.

6.6.4.1 Bending Strength

Bending strengths of flexural members are differentiated according to whether or not the member is laterally braced. If such members are fully supported laterally, they are designed according to the nominal section strength. Otherwise, if they are laterally unbraced, then the bending strength may be governed by the lateral–torsional buckling strength. Section 6.9 discusses the flange distortional buckling. For C- or Z-sections with tension flange attached to deck or sheathing and with compression flange laterally unbraced, the nominal bending strength may be reduced according to the AISI Specification.

6.6.4.1.1 Nominal Section Strength

Two design procedures are now used in the AISI Specification for determining the nominal section strength. They are (1) initiation of yielding and (2) inelastic reserve capacity.

According to procedure I on the basis of initiation of yielding, the nominal moment, M_n, of the cross-section is the effective yield moment, M_y, determined for the effective areas of flanges and the beam web. The effective width of the compression flange and the effective depth of the web can be computed from the design equations given in Section 6.5 on Element Strength. The yield moment of a cold-formed steel flexural member is defined by the moment at which an outer fiber (tension, compression, or both) first attains the yield point of the steel. Figure 6.18 shows three types of stress distribution for yield moment based on different locations of the neutral axis. Accordingly, the nominal section strength for initiation of yielding can be computed as follows:

$$M_n = M_y = S_e F_y \qquad (6.44)$$

where S_e is the elastic section modulus of the effective section calculated with the extreme compression or tension fiber at F_y and F_y is the design yield stress.

For cold-formed steel design, S_e is usually computed by using one of the following two cases:

1. If the neutral axis is closer to the tension than to the compression flange (case c of Figure 6.18), the maximum stress occurs in the compression flange, and therefore the plate slenderness factor λ (Equation 6.9) and the effective width of the compression flange are determined by the w/t ratio and $f = F_y$. This procedure is also applicable to those beams for which the neutral axis is located at the mid-depth of the section (case a).
2. If the neutral axis is closer to the compression than to the tension flange (case b), the maximum stress of F_y occurs in the tension flange. The stress in the compression flange depends on the location of the neutral axis, which is determined by the effective area of the section. The latter cannot be determined unless the compressive stress is known. The closed-form solution of this type of design is possible but would be a very tedious and complex procedure. It is, therefore, customary to determine the sectional properties of the section by successive approximation.

See Examples 6.2 and 6.3 for the calculation of nominal bending strengths.

EXAMPLE 6.2

Use the ASD and LRFD methods to check the adequacy of the I-section with unstiffened flanges as shown in Figure 6.19. The nominal moment is based on the initiation of yielding using $F_y = 50$ ksi.

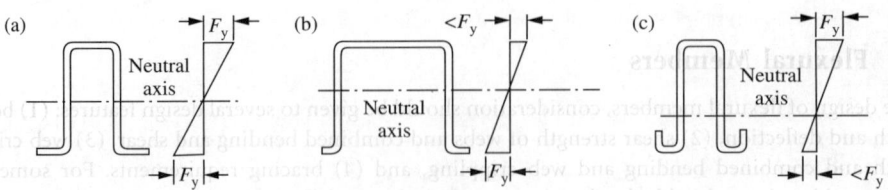

FIGURE 6.18   Stress distribution for yield moment (based on initiation of yielding).

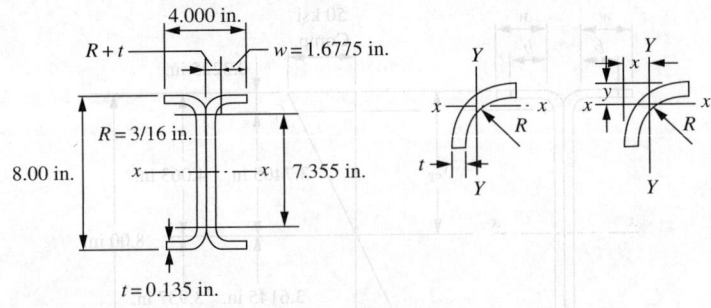

FIGURE 6.19 Example 6.2 (courtesy of Yu, W.W. 1991).

Assume that lateral bracing is adequately provided. The dead load moment $M_D = 30$ in. kip and the live load moment $M_L = 150$ in. kip.

Solution

ASD method

1. Location of neutral axis. For $R = \frac{3}{16}$ in. and $t = 0.135$ in., the sectional properties of the corner element are as follows:

$$I_x = I_y = 0.0003889 \text{ in.}^4$$
$$A = 0.05407 \text{ in.}^2$$
$$x = y = 0.1564 \text{ in.}$$

For the unstiffened compression flange

$$w = 1.6775 \text{ in.}, \quad w/t = 12.426$$

Using $k = 0.43$ and $f = F_y = 50$ ksi,

$$\lambda = \sqrt{f/F_{cr}} = (1.052/\sqrt{k})(w/t)\sqrt{f/E} = 0.821 > 0.673$$
$$b = [(1 - 0.22/0.821)/0.821](1.6775) = 1.496 \text{ in.}$$

Assuming that the web is fully effective, the neutral axis is located at $y_{cg} = 4.063$ in. Since $y_{cg} > d/2$, initial yield occurs in the compression flange. Therefore, $f = F_y$.

2. Check the web for full effectiveness as follows (Figure 6.20):

$$f_1 = 46.03 \text{ ksi (compression)}$$
$$f_2 = 44.48 \text{ ksi (tension)}$$
$$\psi = |f_2/f_1| = 0.966$$

Using Equation 6.10,

$$k = 4 + 2(1 + \psi)^3 + 2(1 + \psi) = 23.13$$
$$h = 7.355 \text{ in.}$$
$$h/t = 54.48$$
$$\lambda = \sqrt{f/F_{cr}} = (1.052/\sqrt{k})\,(54.48)\sqrt{46.03/29,500}$$
$$= 0.471 < 0.673$$
$$b_e = h = 7.355 \text{ in.}$$

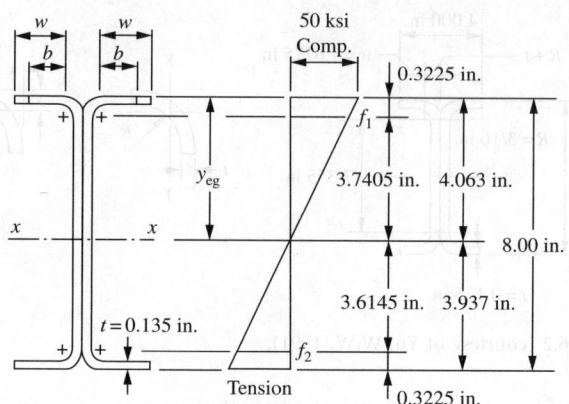

FIGURE 6.20   Stress distribution in webs (courtesy of Yu, W.W. 1991).

Since $h_o = 8.00$ in., $b_o = 2.00$ in., $h_o/b_o = 4$. Using Equations 6.12 and 6.13a to compute b_1 and b_2 as follows:

$$b_1 = b_e/(3 + \psi) = 1.855 \text{ in.}$$
$$b_2 = b_e/2 = 3.6775 \text{ in.}$$

Since $b_1 + b_2 = 5.5325$ in. > 3.7405 in., the web is fully effective.

3. The moment of inertia I_x is

$$I_x = \sum (Ay^2) + 2I_{web} - \left(\sum A \right)(y_{cg})^2$$
$$= 25.382 \text{ in.}^4$$

The section modulus for the top fiber is

$$S_e = I_x/y_{cg} = 6.247 \text{ in.}^3$$

4. Based on initiation of yielding, the nominal moment for section strength is

$$M_n = S_e F_y = 312.35 \text{ in. kip}$$

5. The allowable design moment is

$$M_a = M_n/\Omega = 312.35/1.67 = 187.04 \text{ in. kip}$$

Based on the given data, the required allowable moment is

$$M = M_D + M_L = 30 + 150 = 180 \text{ in. kip}$$

Since $M < M_a$, the I-section is adequate for the ASD method.

LRFD method

1. Based on the nominal moment M_n computed above, the design moment is

$$\phi_b M_n = 0.90(312.35) = 281.12 \text{ in. kip}$$

2. According to the ASCE Standard 7 [38] the required moment for combined dead and live moments is

$$M_u = 1.2M_D + 1.6M_L$$
$$= (1.2 \times 30) + (1.6 \times 150)$$
$$= 276.00 \text{ in. kip}$$

Since $\phi_b M_n > M_u$, the I-section is adequate for bending strength according to the LRFD approach.

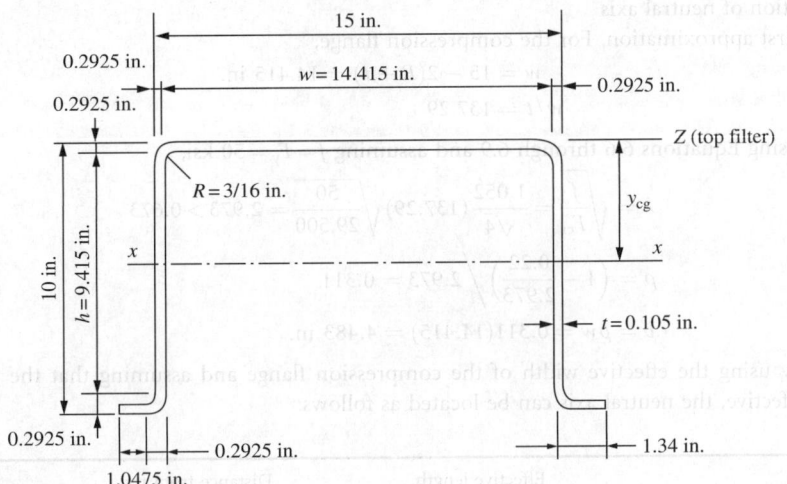

FIGURE 6.21   Example 6.3 (courtesy of Yu, W.W. 1991).

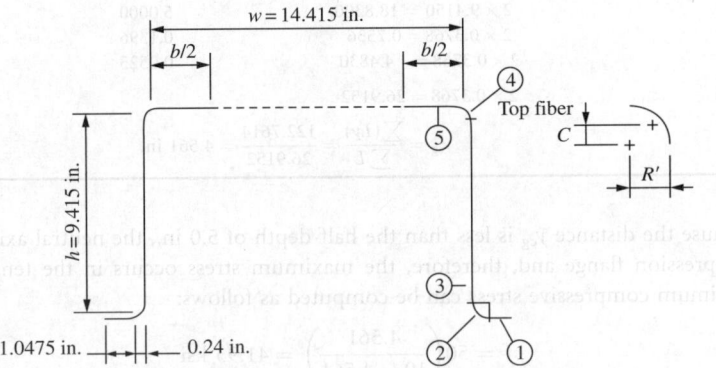

FIGURE 6.22   Line elements (courtesy of Yu, W.W. 1991).

EXAMPLE 6.3

Determine the nominal moment about the *x*-axis for the hat section with stiffened compression flange as shown in Figure 6.21. Assume that the yield point of steel is 50 ksi. Use the *linear method*. The nominal moment is determined by initiation of yielding.

Solution

Calculation of sectional properties. For the linear method, midline dimensions are as shown in Figure 6.22:

  1. Corner element (Figure 6.17 and Figure 6.22)

$$R' = R + t/2 = 0.240 \text{ in.}$$

  Arc length

$$L = 1.57R' = 0.3768 \text{ in.}$$
$$c = 0.637R' = 0.1529 \text{ in.}$$

2. Location of neutral axis
   - First approximation. For the compression flange,
$$w = 15 - 2(R + t) = 14.415 \text{ in.}$$
$$w/t = 137.29$$

Using Equations 6.6 through 6.9 and assuming $f = F_y = 50$ ksi,

$$\lambda = \sqrt{\frac{f}{F_{cr}}} = \frac{1.052}{\sqrt{4}} (137.29) \sqrt{\frac{50}{29,500}} = 2.973 > 0.673$$

$$\rho = \left(1 - \frac{0.22}{2.973}\right) \Big/ 2.973 = 0.311$$

$$b = \rho w = 0.311(14.415) = 4.483 \text{ in.}$$

By using the effective width of the compression flange and assuming that the web is fully effective, the neutral axis can be located as follows:

Element	Effective length, L (in.)	Distance from top fiber, y (in.)	Ly (in.^{2})
1	$2 \times 1.0475 = 2.0950$	9.9475	20.8400
2	$2 \times 0.3768 = 0.7536$	9.8604	7.4308
3	$2 \times 9.4150 = 18.8300$	5.0000	94.1500
4	$2 \times 0.3768 = 0.7536$	0.1396	0.1052
5	$2 \times 0.3768 = 4.4830$	0.0525	0.2354
Total	$2 \times 0.3768 = 26.9152$		122.7614

$$y_{cg} = \frac{\sum(Ly)}{\sum L} = \frac{122.7614}{26.9152} = 4.561 \text{ in.}$$

Because the distance y_{cg} is less than the half-depth of 5.0 in., the neutral axis is closer to the compression flange and, therefore, the maximum stress occurs in the tension flange. The maximum compressive stress can be computed as follows:

$$f = 50 \left(\frac{4.561}{10 - 4.561}\right) = 41.93 \text{ ksi}$$

Since the above computed stress is less than the assumed value, another trial is required.
   - Second approximation. Following several trials, assume that

$$f = 40.70 \text{ ksi}$$
$$\lambda = 2.682 > 0.673$$
$$b = 4.934 \text{ in.}$$

Element	Effective length, L (in.)	Distance from top fiber, y (in.)	Ly (in.^{2})	L_y^2 (in.^{3})
1	22.0950	9.9475	20.8400	207.3059
2	20.7536	9.8604	7.4308	73.2707
3	18.8300	5.0000	94.1500	470.7500
4	20.7536	0.1396	0.1052	0.0147
5	24.9340	0.0525	0.2590	0.0136
Total	27.3662		122.7850	751.3549

$$y_{cg} = \frac{122.7850}{27.3662} = 4.487 \text{ in.}$$

$$f = \left(\frac{4.487}{10 - 4.487} \right) = 40.69 \text{ ksi}$$

Since the above computed stress is close to the assumed value, it is alright.

3. Check the effectiveness of the web. Use the AISI Specification to check the effectiveness of the web element. From Figure 6.23,

$$f_1 = 50(4.1945/5.513) = 38.04 \text{ ksi} \quad \text{(compression)}$$
$$f_2 = 50(5.2205/5.513) = 47.35 \text{ ksi} \quad \text{(tension)}$$
$$\psi = |f_2/f_1| = 1.245.$$

Using Equation 6.10,

$$k = 4 + 2(1 + \psi)^3 + 2(1 + \psi)$$
$$= 4 + 2(2.245)^3 + 2(2.245) = 31.12$$
$$h/t = 9.415/0.105 = 89.67 < 200 \quad \text{OK.}$$
$$\lambda = \sqrt{\frac{f}{F_{cr}}} = \frac{1.052}{\sqrt{31.12}} (89.67) \sqrt{\frac{38.04}{29,500}} = 0.607 < 0.673$$
$$b_e = \quad h = 9.415 \text{ in.}$$

Since $h_o = 10.0$ in., $b_o = 15.0$ in., $h_o/b_o = 0.67 < 4$, use Equation 6.12 to compute b_1:

$$b_1 = b_e/(3 + \psi) = 2.218 \text{ in.}$$

Since $\psi > 0.236$,

$$b_2 = b_e/2 = 4.7075 \text{ in.}$$
$$b_1 + b_2 = 6.9255 \text{ in.}$$

Because the computed value of $(b_1 + b_2)$ is greater than the compression portion of the web (4.1945 in.), the web element is fully effective.

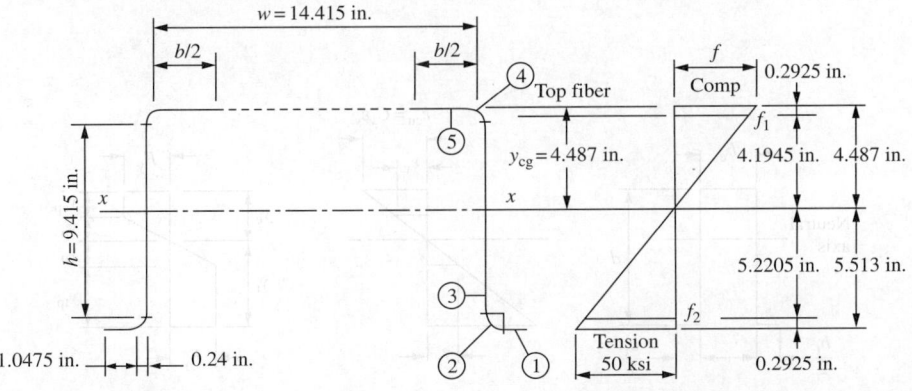

FIGURE 6.23 Effective lengths and stress distribution using fully effective webs (courtesy of Yu, W.W. 1991).

4. Moment of inertia and section modulus. The moment of inertia based on line elements is

$$\sum (Ly^2) = 751.3549$$

$$2I_3' = 2\left(\tfrac{1}{12}\right)(9.415)^3 = 139.0944$$

$$I_z' = 2I_3' + \sum (Ly^2) = 890.4493 \text{ in.}^3$$

$$\left(\sum L\right)(y_{cg})^2 = 27.3662(4.487)^2 = 550.9683 \text{ in.}^3$$

$$I_x' = I_z' - \left(\sum L\right)(y_{cg})^2 = 339.4810 \text{ in.}^3$$

The actual moment of inertia is

$$I_x = I_x't = (339.4810)(0.105) = 35.646 \text{ in.}^4$$

The section modulus relative to the extreme tension fiber is

$$S_x = 35.646/5.513 = 6.466 \text{ in.}^3$$

Nominal moment. The nominal moment for section strength is

$$M_n = S_e F_y = S_x F_y = (6.466)(50) = 323.30 \text{ in.-kip}$$

Once the nominal moment is computed, the design moments for the ASD and LRFD methods can be determined as illustrated in Example 6.2.

According to procedure II of the AISI Specification, the nominal moment, M_n, is the maximum bending capacity of the beam by considering the inelastic reserve strength through partial plastification of the cross-section as shown in Figure 6.24. The inelastic stress distribution in the cross-section depends on the maximum strain in the compression flange, which is limited by the Specification for the given width to thickness ratio of the compression flange. On the basis of the maximum compression strain allowed in the Specification, the neutral axis can be located by Equation 6.45 and the nominal moment, M_n, can be determined by using Equation 6.46:

$$\int \sigma \, dA = 0 \tag{6.45}$$

$$\int \sigma y \, dA = M \tag{6.46}$$

where σ is the stress in the cross-section. For additional information, see Yu [1].

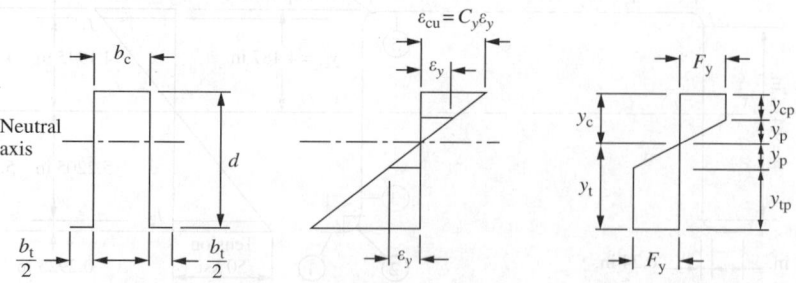

FIGURE 6.24 Stress distribution for maximum moment (inelastic reserve strength) (courtesy of Yu, W.W. 1991).

6.6.4.1.2 Lateral–Torsional Buckling Strength

The nominal lateral–torsional buckling strength of unbraced segments of singly, doubly, and point-symmetric sections subjected to lateral–torsional buckling, M_n, can be determined as follows:

$$M_n = S_c F_c \tag{6.47}$$

where S_c is the elastic section modulus of the effective section calculated at a stress F_c in the extreme compression fiber and F_c is the critical lateral–torsional buckling stress determined as follows:

1. For $F_e \geq 2.78 F_y$

$$F_c = F_y \tag{6.48}$$

2. For $2.78 F_y > F_e > 0.56 F_y$

$$F_c = \frac{10}{9} F_y \left(1 - \frac{10 F_y}{36 F_e} \right) \tag{6.49}$$

3. For $F_e \leq 0.56 F_y$

$$F_c = F_e \tag{6.50}$$

where F_e is the elastic critical lateral–torsional buckling stress.

1. For singly, doubly, and point-symmetric sections

$F_e = C_b r_0 A \sqrt{\sigma_{ey} \sigma_t} / S_f$ for bending about the symmetry axis. For singly symmetric sections, x-axis is the axis of symmetry oriented such that the shear center has a negative x-coordinate. For point-symmetric sections, use $0.5 F_e$. Alternatively, F_e can be calculated using the equation for doubly symmetric I-sections, singly symmetric C-sections, or point-symmetric sections given in (2)

$F_e = C_s A \sigma_{ex} [j + C_s \sqrt{j^2 + r_0^2 (\sigma_t / \sigma_{ex})}] / (C_{TF} S_f)$ for bending about the centroidal axis perpendicular to the symmetry axis for singly symmetric sections

$C_s = +1$ for moment causing compression on shear center side of centroid

$C_s = -1$ for moment causing tension on shear center side of centroid

$\sigma_{ex} = \pi^2 E / (K_x L_x / r_x)^2$

$\sigma_{ey} = \pi^2 E / (K_y L_y / r_y)^2$

$\sigma_t = [GJ + \pi^2 E C_w / (K_t L_t)^2] / (A r_0^2)$

A = full unreduced cross-sectional area

S_f = elastic section modulus of full unreduced section relative to extreme compression fiber

$$C_b = 12.5 M_{max} / (2.5 M_{max} + 3 M_A + 4 M_B + 3 M_C) \tag{6.51}$$

In Equation 6.51

M_{max} = absolute value of maximum moment in unbraced segment

M_A = absolute value of moment at quarter point of unbraced segment

M_B = absolute value of moment at centerline of unbraced segment

M_C = absolute value of moment at three-quarter point at unbraced segment

C_b is permitted to be conservatively taken as unity for all cases. For cantilevers or overhangs where the free end is unbraced, C_b is taken as unity.

E = modulus of elasticity

$C_{TF} = 0.6 - 0.4 \ (M_1 / M_2)$

where M_1 is the smaller and M_2 the larger bending moment at the ends of the unbraced length in the plane of bending, and where M_1/M_2, the ratio of end moments, is positive when M_1 and M_2 have the same sign (reverse curvature bending) and negative when they are of opposite sign (single-curvature bending). When the bending moment at any point within an unbraced length is larger than that at both ends of this length, C_{TF} is taken as unity.

r_0 = Polar radius of gyration of cross-section about the shear center

$$= \sqrt{r_x^2 + r_y^2 + x_0^2} \qquad (6.52)$$

r_x, r_y = radii of gyration of cross-section about centroidal principal axes

G = shear modulus

K_x, K_y, K_t = effective length factors for bending about the x- and y-axes and for twisting

L_x, L_y, L_t = unbraced length of compression member for bending about x- and y-axes, and for twisting

x_0 = distance from shear center to centroid along the principal x-axis, taken as negative

J = St. Venant torsion constant of cross-section

C_w = torsional warping constant of cross-section

$$j = \left[\int_A x^3 \, dA + \int_A xy^2 \, dA \right] \Big/ (2I_y) - x_0 \qquad (6.53)$$

2. For I-sections, singly symmetric C-sections, or Z-sections bent about the centroidal axis perpendicular to the web (x-axis), the following equations are permitted to be used in lieu of (1) to calculate F_e:

$$F_e = C_b \pi^2 E d I_{yc} / [S_f (K_y L_y)^2] \qquad (6.54)$$

for doubly symmetric I-sections and singly symmetric C-sections and

$$F_e = C_b \pi^2 E d I_{yc} / [2S_f (K_y L_y)^2] \qquad (6.55)$$

for point-symmetric Z-sections. In Equations 6.54 and 6.55 d is the depth of the section and I_{yc} is the moment of inertia of compression portion of section about centroidal axis of the entire section parallel to web, using full unreduced section.

EXAMPLE 6.4

Determine the nominal moment for lateral–torsional buckling strength for the I-beam used in Example 6.2. Assume that the beam is braced laterally at both ends and midspan. Use $F_y = 50$ ksi.

Solution

Calculation of sectional properties
Based on the dimensions given in Example 6.2 (Figure 6.19 and Figure 6.20), the moment of inertia, I_x, and the section modulus, S_f, of the full-section can be computed as shown in the following table:

Element	Area, A (in.²)	Distance from mid-depth, y (in.)	Ay^2 (in.⁴)
Flanges	4(1.6775)(0.135) = 0.9059	3.9325	14.0093
Corners	4(0.05407) = 0.2163	3.8436	3.1955
Webs	2(7.355)(0.135) = 1.9859	0	0
Total	3.1081		17.2048

$$2I_{web} = 2(1/12)(0.135)(7.355)^3 = 8.9522$$

$$I_x = 26.1570 \text{ in.}^4$$

$$S_f = I_x / (8/2) = 6.54 \text{ in.}^3$$

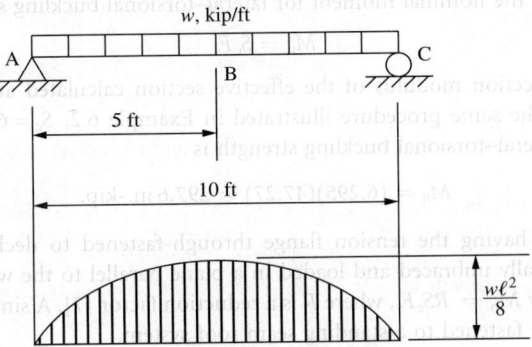

FIGURE 6.25 Example 6.4 (courtesy of Yu, W.W. 1991).

The value of I_{yc} can be computed as shown below:

Element	Area, A (in.^{2})	Distance from y-axis, x-axis (in.)	Ax^2 (in.^{4})
Flanges	$4(1.6775)(0.135) = 0.9059$	1.1613	1.2217
Corners	$4(0.05407) = 0.2163$	0.1564	0.0053
Webs	$2(7.355)(0.135) = 1.9859$	0.0675	0.0090
Total	3.1081	$I_{flanges} = 4(1/12)0.135(1.6775)^3 = 1.2360$	
			0.2124
			$I_y = 1.4484$ in.^{4}

$$I_{yc} = I_y/2 = 0.724 \text{ in.}^4$$

Considering the lateral supports at both ends and midspan, and the moment diagram shown in Figure 6.25, the value of C_b for the segment AB or BC is 1.30 according to Equation 6.51.

Using Equation 6.54,

$$F_e = C_b \pi^2 E \frac{d I_{yc}}{S_f (K_y L_y)^2}$$

$$= (1.30)\pi^2 (29,500) \frac{(8)(0.724)}{(6.54)(1 \times 5 \times 12)^2} = 93.11 \text{ ksi}$$

$$0.56 F_y = 28 \text{ ksi}$$
$$2.78 F_y = 139 \text{ ksi}$$

Since $2.78 F_y > F_e > 0.56 F_y$, from Equation 6.49

$$F_c = \frac{10}{9} F_y \left(1 - \frac{10 F_y}{36 F_e}\right)$$

$$= \frac{10}{9}(50)\left[1 - \frac{10(50)}{36(93.11)}\right]$$

$$= 47.27 \text{ ksi}$$

Based on Equation 6.47, the nominal moment for lateral–torsional buckling strength is

$$M_n = S_c F_c$$

where S_c is the elastic section modulus of the effective section calculated at a compressive stress of $f = 47.27$ ksi. By using the same procedure illustrated in Example 6.2, $S_c = 6.295$ in.^{3}. Therefore, the nominal moment for lateral–torsional buckling strength is

$$M_n = (6.295)(47.27) = 297.6 \text{ in.-kip.}$$

For C- or Z-sections having the tension flange through-fastened to deck or sheathing with the compression flange laterally unbraced and loaded in a plane parallel to the web, the nominal flexural strength is determined by $M_n = RS_e F_y$, where R is a reduction factor [7]. A similar approach is used for beams having one-flange fastened to a standing seam roof system.

For closed box members, the nominal lateral–torsional buckling strength can be determined on the basis of the following elastic critical lateral buckling stress:

$$F_e = \frac{C_b \pi}{K_y L_y S_f} \sqrt{EGJI_y} \tag{6.56}$$

where I_y is the moment of inertia of full unreduced section about centroidal axis parallel to web and J is the torsional constant of box section.

6.6.4.1.3 Unusually Wide Beam Flanges and Short Span Beams

When beam flanges are unusually wide, special consideration should be given to the possible effects of shear lag and flange curling. Shear lag depends on the type of loading and the span to width ratio and is independent of the thickness. Flange curling is independent of span length but depends on the thickness and width of the flange, the depth of the section, and the bending stresses in both tension and compression flanges.

To consider the shear lag effects, the effective widths of both tension and compression flanges should be used according to the AISI Specification.

When a beam with unusually wide and thin flanges is subjected to bending, the portion of the flange most remote from the web tends to deflect toward the neutral axis due to the effect of longitudinal curvature of the beam and the applied bending stresses in both flanges. For the purpose of controlling the excessive flange curling, AISI Specification provides an equation to limit the flange width.

6.6.4.2 Shear Strength

The shear strength of beam webs is governed by either yielding or buckling of the web element, depending on the depth to thickness ratio, h/t, and the mechanical properties of steel. For beam webs having small h/t ratios, the nominal shear strength is governed by shear yielding. When the h/t ratio is large, the nominal shear strength is controlled by elastic shear buckling. For beam webs having moderate h/t ratios, the shear strength is based on inelastic shear buckling.

For the design of beam webs without holes, the AISI Specification provides the following equations for determining the nominal shear strength:

1. For $h/t \le \sqrt{Ek_v/F_y}$

$$V_n = 0.60 F_y ht \tag{6.57}$$

2. For $\sqrt{Ek_v/F_y} < h/t \le 1.51\sqrt{Ek_v/F_y}$

$$V_n = 0.60t^2 \sqrt{k_v F_y E} \tag{6.58}$$

3. For $h/t > 1.51\sqrt{Ek_v/F_y}$

$$V_n = \pi^2 Ek_v t^3 / [12(1 - \mu^2)h] = 0.904 Ek_v t^3 / h \tag{6.59}$$

where V_n is the nominal shear strength of the beam, h is the depth of the flat portion of the web measured along the plane of the web, t is the web thickness, and k_v is the shear buckling coefficient determined as follows:

1. For unreinforced webs, $k_v = 5.34$.
2. For beam webs with transverse stiffeners satisfying the AISI requirements

when $a/h \le 1.0$

$$k_v = 4.00 + \frac{5.34}{(a/h)^2}$$

when $a/h > 1.0$

$$k_v = 5.34 + \frac{4.00}{(a/h)^2}$$

where a is the shear panel length for unreinforced web element, h is the clear distance between transverse stiffeners for reinforced web elements.

For a web consisting of two or more sheets, each sheet should be considered as a separate element carrying its share of the shear force.

For the design of C-section webs with holes, the above nominal shear strength should be multiplied by a factor q_s specified in section C3.2.2 of the AISI Specification [7].

6.6.4.3 Combined Bending and Shear

For continuous beams and cantilever beams, high bending stresses often combine with high shear stresses at the supports. Such beam webs must be safeguarded against buckling due to the combination of bending and shear stresses. Based on the AISI Specification, the moment and shear should satisfy the interaction equations listed in Table 6.3.

6.6.4.4 Web Crippling

For cold-formed steel beams, transverse stiffeners are not frequently used for beam webs. The webs may cripple due to the high local intensity of the load or reaction as shown in Figure 6.26. Because the theoretical analysis of web crippling is rather complex due to the involvement of many factors, the present AISI design equations are based on the extensive experimental investigations conducted at Cornell University, University of Missouri-Rolla, University of Waterloo, and University of Sydney under four loading conditions: (1) end one-flange (EOF) loading, (2) interior one-flange (IOF) loading, (3) end two-flange (ETF) loading, and (4) interior two-flange (ITF) loading [7]. The loading conditions used for the tests are illustrated in Figure 6.27.

TABLE 6.3 Interaction Equations Used for Combined Bending and Shear

	ASD	LRFD and LSD
Beams with unreinforced webs	$M \le M_n/\Omega_b$ and $V \le V_n/\Omega_v$	$\bar{M} \le \phi_b M_n$ and $\bar{V} \le \phi_v V_n$
	$\left(\frac{\Omega_b M}{M_{nxo}}\right)^2 + \left(\frac{\Omega_v V}{V_n}\right)^2 \le 1.0$ (6.60)	$\left(\frac{\bar{M}}{\phi_b M_{nxo}}\right)^2 + \left(\frac{\bar{V}}{\phi_v V_n}\right)^2 \le 1.0$ (6.61)
Beams with transverse web stiffeners	$0.6\left(\frac{\Omega_b M}{M_{nxo}}\right) + \left(\frac{\Omega_v V}{V_n}\right) \le 1.3$ (6.62)	$0.6\left(\frac{\bar{M}}{\phi_b M_{nxo}}\right) + \left(\frac{\bar{V}}{\phi_v V_n}\right) \le 1.3$ (6.63)

Note: M is the bending moment, V is the unfactored shear force, Ω_b and Ω_v are the factors of safety for bending and shearing, respectively, ϕ_b is the resistance factor for bending, ϕ_v is the resistance factor for shear, M_n is the nominal flexural strength when bending alone exists, M_{nxo} is the nominal flexural strength about the centroidal x-axis determined in accordance with the specification excluding the consideration of lateral–torsional buckling, $\bar{M}$ is the required flexural strength (M_u for LRFD and M_f for LSD), V_n is the nominal shear strength when shear alone exists, and $\bar{V}$ is the required shear strength (V_u for LRFD and V_f for LSD).

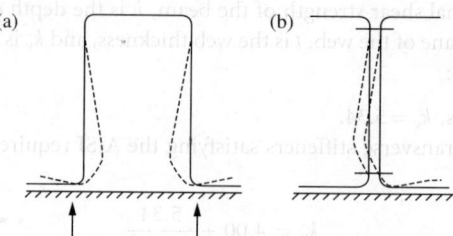

FIGURE 6.26 Web crippling of cold-formed steel beams.

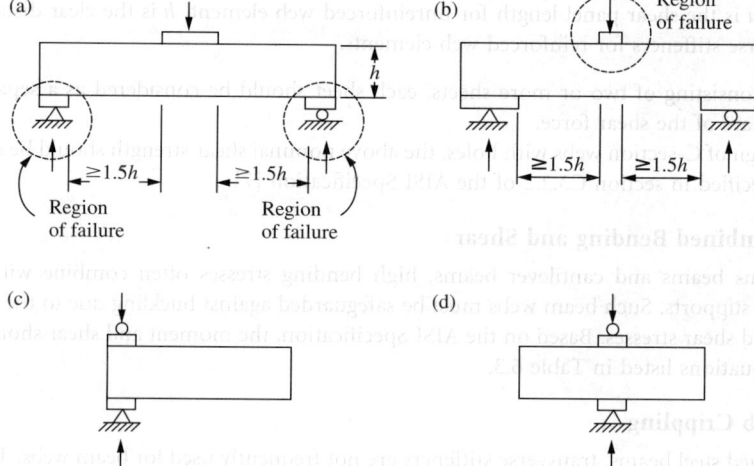

FIGURE 6.27 Loading conditions for web crippling tests: (a) EOF loading, (b) IOF loading, (c) ETF loading, and (d) ITF loading (courtesy of Yu, W.W. 1991).

The nominal web crippling strength of webs without holes for a given loading condition can be determined from the following AISI equation [7] on the basis of the thickness of web element, design yield point, the bend radius to thickness ratio, the depth to thickness ratio, the bearing length to thickness ratio, and the angle between the plane of the web and the plane of the bearing surface:

$$P_n = C t^2 F_y \sin \theta \left[1 - C_R \sqrt{\frac{R}{t}} \right] \left[1 + C_N \sqrt{\frac{N}{t}} \right] \left[1 - C_h \sqrt{\frac{h}{t}} \right] \tag{6.64}$$

where P_n is the nominal web crippling strength, C is the coefficient, C_h is the web slenderness coefficient, C_N is the bearing length coefficient, C_R is the inside bend radius coefficient, F_y is the design yield point, h is the flat portion of web measured in the plane of the web, N is the bearing length ($\frac{3}{4}$ in. [19 mm] minimum), R is the inside bend radius, t is the web thickness, and θ is the angle between the plane of the web and the plane of the bearing surface, $45° \leq \theta \leq 90°$. Values of C, C_R, C_N, C_h, factor of safety, and resistance factor are listed in separate tables of the AISI Specification [7] for built-up sections, single web channel and C-sections, single web Z-sections, single hat sections, and multiweb deck sections.

For C-section webs with holes, the web crippling strength determined in accordance with the above equation should be reduced by using a reduction factor as given in the Specification [7].

6.6.4.5 Combined Bending and Web Crippling

For combined bending and web crippling, the design of beam webs should be based on the interaction equations provided in the AISI Specification [7]. These equations are presented in Table 6.4.

6.6.4.6 Bracing Requirements

In cold-formed steel design, braces should be designed to restrain lateral bending or twisting of a loaded beam and to avoid local crippling at the points of attachment. When C-sections and Z-sections are used as beams and loaded in the plane of the web, the AISI Specification [7] provides design requirements to restrain twisting of the beam under the following two conditions: (1) the top flange is connected to deck or sheathing material in such a manner as to effectively restrain lateral deflection of the connected flange and (2) neither flange is connected to sheathing. In general, braces should be designed to satisfy the strength and stiffness requirements. For beams using symmetrical cross-sections such as I-beams, the AISI Specification does not provide specific requirements for braces. However, the braces may be designed for a capacity of 2% of the force resisted by the compression portion of the beam. This is a frequently used rule of thumb but is a conservative approach, as proven by a rigorous analysis.

6.6.5 Concentrically Loaded Compression Members

Axially loaded cold-formed steel *compression members* should be designed for the following limit states: (1) yielding, (2) overall column buckling (flexural buckling, torsional buckling, or *torsional–flexural buckling*), and (3) local buckling of individual elements. The governing failure mode depends on the configuration of the cross-section, thickness of material, unbraced length, and end restraint. For distorsional buckling of compression members, see Section 6.9 on Direct Strength Method.

TABLE 6.4 Interaction Equations for Combined Bending and Web Crippling

	ASD		LRFD and LSD	
Shapes having single unreinforced webs	$1.2\left(\dfrac{\Omega_w P}{P_n}\right) + \left(\dfrac{\Omega_b M}{M_{nxo}}\right) \leq 1.5$	(6.65)	$1.07\left(\dfrac{\bar{P}}{\phi_w P_n}\right) + \left(\dfrac{\bar{M}}{\phi_b M_{nxo}}\right) \leq 1.42$	(6.66)
Shapes having multiple unreinforced webs such as I-sections	$1.1\left(\dfrac{\Omega_w P}{P_n}\right) + \left(\dfrac{\Omega_b M}{M_{nxo}}\right) \leq 1.5$	(6.67)	$0.82\left(\dfrac{\bar{P}}{\phi_w P_n}\right) + \left(\dfrac{\bar{M}}{\phi_b M_{nxo}}\right) \leq 1.32$	(6.68)
Support point of two nested Z-shapes	$\dfrac{M}{M_{no}} + 0.85\dfrac{P}{P_n} \leq \dfrac{1.65}{\Omega}$	(6.69)	$\dfrac{\bar{M}}{M_{no}} + 0.85\dfrac{\bar{P}}{P_n} \leq 1.65\phi$	(6.70)
	$M \leq M_{no}/\Omega_b$ and $P \leq P_n/\Omega_w$		$\bar{M} \leq \phi_b M_{no}$ and $\bar{P} \leq \phi_w P_n$	

Note: The AISI Specification includes some exception clauses, under which the effect of combined bending and web crippling need not be checked. P is the concentrated load or reaction in presence of bending moment, P_n is the nominal web crippling strength for concentrated load or reaction in the absence of bending moment (for Equations 6.65, 6.66, 6.67, and 6.68), P_n is the nominal web crippling strength assuming single web interior one-flange loading for the nested Z-sections, that is, the sum of the two webs evaluated individually (for Equations 6.69 and 6.70), $\bar{P}$ is the required strength for concentrated load or reaction in the presence of bending moment (P_u for LRFD and P_f for LSD), M is the applied bending moment at, or immediately adjacent to, the point of application of the concentrated load or reaction, M_{no} is the nominal yield moment for nested Z-sections, that is, the sum of two sections evaluated individually, M_{nxo} is the nominal flexural strength about the centroidal x-axis determined in accordance with the specification excluding the consideration of lateral–torsional buckling, $\bar{M}$ is the required flexural strength at, or immediately adjacent to, the point of application of the concentrated load or reaction $\bar{P}$ (M_u for LRFD and M_f for LSD), Ω is the factor of safety = 1.75, Ω_b and Ω_w are the factors of safety for bending and web crippling, respectively, ϕ is the resistance factor (0.9 for LRFD and 0.80 for LSD), ϕ_b is the resistance factor for bending, and ϕ_w is the resistance factor for web crippling.

6.6.5.1 Yielding

A very short, compact column under axial load may fail by yielding. For this case, the nominal axial strength is the yield load, that is,

$$P_n = P_y = A_g F_y \tag{6.71}$$

where A_g is the gross area of the column and F_y is the yield point of steel.

6.6.5.2 Overall Column Buckling

Overall column buckling may be one of the following three types:

1. *Flexural buckling — bending about a principal axis.* The elastic flexural buckling stress is

$$F_e = \frac{\pi^2 E}{(KL/r)^2} \tag{6.72}$$

   where E is the modulus of elasticity, K is the effective length factor for flexural buckling (Figure 6.28), L is the unbraced length of member for flexural buckling, and r is the radius of gyration of the full-section.

2. *Torsional buckling — twisting about shear center.* The elastic torsional buckling stress is

$$F_e = \frac{1}{A r_0^2} \left[GJ + \frac{\pi^2 E C_w}{(K_t L_t)^2} \right] \tag{6.73}$$

   where A is the full cross-sectional area, C_w is the torsional warping constant of cross-section, G is the shear modulus, J is the St. Venant torsion constant of cross-section, K_t is the effective length factor for twisting, L_t is the unbraced length of member for twisting, and r_0 is the polar radius of gyration of cross-section about shear center.

3. *Torsional–flexural buckling — bending and twisting simultaneously.* The elastic torsional–flexural buckling stress is

$$F_e = \left[(\sigma_{ex} + \sigma_t) - \sqrt{(\sigma_{ex} + \sigma_t)^2 - 4\beta \sigma_{ex} \sigma_t} \right] \Big/ (2\beta) \tag{6.74}$$

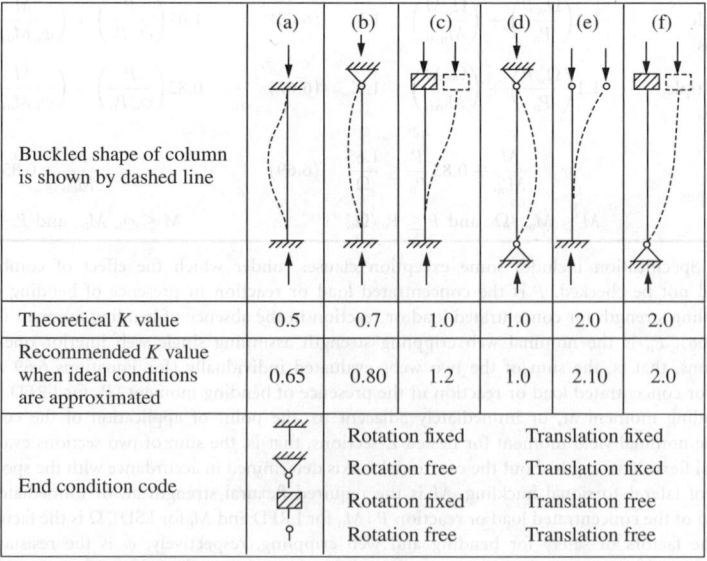

	(a)	(b)	(c)	(d)	(e)	(f)
Buckled shape of column is shown by dashed line						
Theoretical K value	0.5	0.7	1.0	1.0	2.0	2.0
Recommended K value when ideal conditions are approximated	0.65	0.80	1.2	1.0	2.10	2.0
End condition code	Rotation fixed	Translation fixed				
	Rotation free	Translation fixed				
	Rotation fixed	Translation free				
	Rotation free	Translation free				

FIGURE 6.28 Effective length factor K for concentrically loaded compression members.

where $\beta = 1 - (x_0/r_0)^2$, $\sigma_{ex} = \pi^2 E/(K_x L_x/r_x)^2$, σ_t = Equation 6.73, and x_0 is the distance from shear center to centroid along the principal x-axis.

For doubly symmetric and *point-symmetric shapes* (Figure 6.29), the overall column buckling can be either flexural type or torsional type. However, for singly symmetric shapes (Figure 6.30), the overall column buckling can be either flexural buckling or torsional–flexural buckling.

For overall column buckling, the nominal axial strength is determined by Equation 6.75:

$$P_n = A_e F_n \tag{6.75}$$

where A_e is the effective area determined for the stress F_n and F_n is the nominal buckling stress determined as follows:

For $\lambda_c \leq 1.5$

$$F_n = (0.658^{\lambda_c^2})F_y \tag{6.76}$$

For $\lambda_c > 1.5$

$$F_n = \left[\frac{0.877}{\lambda_c^2}\right] F_y \tag{6.77}$$

The use of the effective area A_e in Equation 6.75 is to reflect the effect of local buckling on the reduction of column strength. In Equations 6.76 and 6.77,

$$\lambda_c = \sqrt{F_y/F_e} \tag{6.78}$$

where F_e is the least of elastic flexural buckling stress (Equation 6.72), torsional buckling stress (Equation 6.73), and torsional–flexural buckling stress (Equation 6.74), whichever is applicable.

Concentrically loaded angle sections should be designed for an additional bending moment in accordance with section C5.2 of the Specification [7].

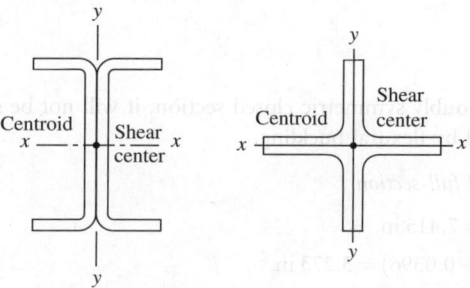

FIGURE 6.29 Doubly symmetric shapes.

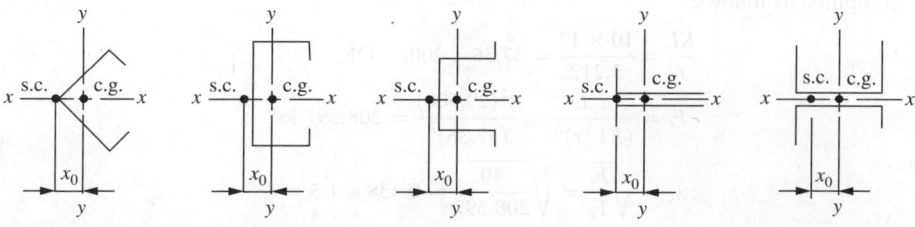

FIGURE 6.30 Singly symmetric shapes (courtesy of Yu, W.W. 1991).

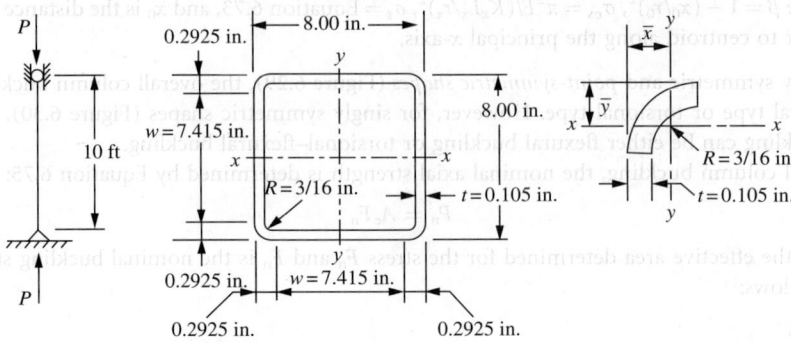

FIGURE 6.31 Example 7.5 (courtesy of Yu, W.W. 1991).

For nonsymmetric shapes whose cross-sections do not have any symmetry, either about an axis or about a point, the elastic torsional–flexural buckling stress should be determined by rational analysis or by tests. See AISI Design Manual [41].

In addition to the above design provisions for the design of axially loaded columns, the AISI Specification also provides design criteria for compression members having one-flange through-fastened to deck or sheathing.

EXAMPLE 6.5

Determine the allowable axial load for the square tubular column shown in Figure 6.31. Assume that $F_y = 40$ ksi, $K_x L_x = K_y L_y = 10$ ft, and the dead to live load ratio is $\frac{1}{5}$. Use the ASD and LRFD methods.

Solution

ASD method

Since the square tube is a doubly symmetric closed section, it will not be subject to torsional–flexural buckling. It can be designed by flexural buckling.

1. *Sectional properties of full-section.*

   $w = 8.00 - 2(R+t) = 7.415$ in.

   $A = 4(7.415 \times 0.105 + 0.0396) = 3.273$ in.^{2}

   $I_x = I_y = 2(0.105)[(1/12)(7.415)^3 + 7.415(4 - 0.105/2)^2] + 4(0.0396)(4.0 - 0.1373)^2 = 33.763$ in.^{4}

   $r_x = r_y = \sqrt{I_x/A} = \sqrt{33.763/3.273} = 3.212$ in.

2. *Nominal buckling stress, F_n.* According to Equation 6.72, the elastic flexural buckling stress, F_e, is computed as follows:

   $$\frac{KL}{r} = \frac{10 \times 12}{3.212} = 37.36 < 200, \quad \text{OK.}$$

   $$F_e = \frac{\pi^2 E}{(KL/r)^2} = \frac{\pi^2 (29,500)}{(37.36)^2} = 208.597 \text{ ksi}$$

   $$\lambda_c = \sqrt{\frac{F_y}{F_e}} = \sqrt{\frac{40}{208.597}} = 0.438 < 1.5$$

   $$F_n = (0.658^{\lambda_c^2})F_y = (0.658^{0.438^2})40 = 36.914 \text{ ksi}$$

3. *Effective area, A_e*. Because the given square tube is composed of four stiffened elements, the effective width of stiffened elements subjected to uniform compression can be computed from Equations 6.6 through 6.9 by using $k = 4.0$:

$$w/t = 7.415/0.105 = 70.619$$

$$\lambda = \sqrt{\frac{f}{F_{cr}}} = \frac{1.052}{\sqrt{k}}\left(\frac{w}{t}\right)\sqrt{\frac{F_n}{E}}$$

$$= 1.052/\sqrt{4}(70.619)\sqrt{36.914/29,500} = 1.314$$

Since $\lambda > 0.673$, from Equation 6.7

$$b = \rho w$$

where

$$\rho = (1 - 0.22/\lambda)/\lambda = (1 - 0.22/1.314)/1.314 = 0.634$$

Therefore,

$$b = (0.634)(7.415) = 4.701 \text{ in.}$$

The effective area is

$$A_e = 3.273 - 4(7.415 - 4.701)(0.105) = 2.133 \text{ in.}^2$$

4. *Nominal and allowable loads.* Using Equation 6.75, the nominal load is

$$P_n = A_e F_n = (2.133)(36.914) = 78.738 \text{ kip}$$

The allowable load is

$$P_a = P_n/\Omega_c = 78.738/1.80 = 43.74 \text{ kip}$$

LRFD method

In the ASD method, the nominal axial load, P_n, was computed to be 78.738 kip. The design axial load for the LRFD method is

$$\phi_c P_n = 0.85(78.738) = 66.93 \text{ kip}$$

Based on the load combination of dead and live loads, the required axial load is

$$P_u = 1.2 P_D + 1.6 P_L = 1.2 P_D + 1.6(5 P_D) = 9.2 P_D$$

where P_D is the axial load due to dead load and P_L is the axial load due to live load.

By using $P_u = \phi_c P_n$, the values of P_D and P_L are computed as follows:

$$P_D = 66.93/9.2 = 7.28 \text{ kip}$$
$$P_L = 5 P_D = 36.40 \text{ kip}$$

Therefore, the allowable axial load is

$$P_a = P_D + P_L = 43.68 \text{ kip}$$

It can be seen that the allowable axial loads determined by the ASD and LRFD methods are practically the same.

6.6.6 Combined Axial Load and Bending

The AISI Specification provides interaction equations for combined axial load and bending.

6.6.6.1 Combined Tensile Axial Load and Bending

For combined tensile axial load and bending, the required strengths should satisfy the interaction equations presented in Table 6.5. These equations are to prevent yielding of the tension flange and to prevent failure of the compression flange of the member.

6.6.6.2 Combined Compressive Axial Load and Bending

Cold-formed steel members under combined compressive axial load and bending are usually referred to as *beam–columns*. Such members are often found in framed structures, trusses, and exterior wall studs. For the design of these members, the required strengths should satisfy the AISI interaction equations presented in Table 6.6.

TABLE 6.5 Interaction Equations for Combined Tensile Axial Load and Bending

	ASD		LRFD and LSD	
Check tension flange	$\dfrac{\Omega_b M_x}{M_{nxt}} + \dfrac{\Omega_b M_y}{M_{nyt}} + \dfrac{\Omega_t T}{T_n} \le 1.0$	(6.79)	$\dfrac{\bar{M}_x}{\phi_b M_{nxt}} + \dfrac{\bar{M}_y}{\phi_b M_{nyt}} + \dfrac{\bar{T}}{\phi_t T_n} \le 1.0$	(6.80)
Check compression flange	$\dfrac{\Omega_b M_x}{M_{nx}} + \dfrac{\Omega_b M_y}{M_{ny}} - \dfrac{\Omega_t T}{T_n} \le 1.0$	(6.81)	$\dfrac{\bar{M}_x}{\phi_b M_{nx}} + \dfrac{\bar{M}_y}{\phi_b M_{ny}} - \dfrac{\bar{T}}{\phi_t T_n} \le 1.0$	(6.82)

Note: M_{nx} and M_{ny} are the nominal flexural strengths about the centroidal x- and y-axes, respectively, M_{nxt}, $M_{nyt} = S_{ft}F_y$, $\bar{M}_x$ and $\bar{M}_y$ are the required flexural strengths with respect to the centroidal axes (M_{ux} and M_{uy} for LRFD, M_{fx} and M_{fy} for LSD), M_x and M_y are the required moments with respect to the centroidal axes of the section, S_{ft} is the section modulus of the full-section for the extreme tension fiber about the appropriate axis, T is the required tensile axial load, T_n is the nominal tensile axial strength, $\bar{T}$ is the required tensile axial strength (T_u for LRFD and T_f for LSD), ϕ_b is the resistance factor for bending, ϕ_t is the resistance factor for tension (0.95 for LRFD and 0.90 for LSD), Ω_b is the safety factor for bending, and Ω_t is the safety factor for tension.

TABLE 6.6 Interaction Equations for Combined Compressive Axial Load and Bending

ASD			LRFD and LSD	
When $\Omega_c P/P_n \le 0.15,$	$\dfrac{\Omega_c P}{P_n} + \dfrac{\Omega_b M_x}{M_{nx}} + \dfrac{\Omega_b M_y}{M_{ny}} \le 1.0$	(6.83)	When $\bar{P}/\phi_c P_n \le 0.15,$ $\quad \dfrac{\bar{P}}{\phi_c P_n} + \dfrac{\bar{M}_x}{\phi_b M_{nx}} + \dfrac{\bar{M}_y}{\phi_b M_{ny}} \le 1.0$	(6.84)
When $\Omega_c P/P_n > 0.15,$	$\dfrac{\Omega_c P}{P_n} + \dfrac{\Omega_b C_{mx} M_x}{M_{nx}\alpha_x} + \dfrac{\Omega_b C_{my} M_y}{M_{ny}\alpha_y} \le 1.0$	(6.85)	When $\bar{P}/\phi_c P_n > 0.15,$ $\quad \dfrac{\bar{P}}{\phi_c P_n} + \dfrac{C_{mx}\bar{M}_x}{\phi_b M_{nx}\alpha_x} + \dfrac{C_{my}\bar{M}_y}{\phi_b M_{ny}\alpha_y} \le 1.0$	(6.86)
	$\dfrac{\Omega_c P}{P_{no}} + \dfrac{\Omega_b M_x}{M_{nx}} + \dfrac{\Omega_b M_y}{M_{ny}} \le 1.0$	(6.87)	$\dfrac{\bar{P}}{\phi_c P_{no}} + \dfrac{\bar{M}_x}{\phi_b M_{nx}} + \dfrac{\bar{M}_y}{\phi_b M_{ny}} \le 1.0$	(6.88)

Note: M_x and M_y are the required moments with respect to the centroidal axes of the effective section determined for the required axial strength alone, M_{nx} and M_{ny} are the nominal flexural strengths about the centroidal axes, $\bar{M}_x$ and $\bar{M}_y$ are the required flexural strengths with respect to the centroidal axes of the effective section determined for the required axial strength alone (M_{ux} and M_{uy} for LRFD, M_{fx} and M_{fy} for LSD), P is the required axial load, P_n is the nominal axial strength determined in accordance with Equation 6.75, P_{no} is the nominal axial strength determined in accordance with Equation 6.75, for $F_n = F_y$, $\bar{P}$ is the required compressive axial strength (P_u for LRFD and P_f for LSD), $\alpha_x = 1 - \Omega_c P/P_{EX}$ (for Equation 6.85), $\alpha_y = 1 - \Omega_c P/P_{EY}$ (for Equation 6.85), $\alpha_x = 1 - \bar{P}/P_{EX}$ (for Equation 6.86), $\alpha_y = 1 - \bar{P}/P_{EY}$ (for Equation 6.86), $P_{EX} = \pi^2 EI_x/(K_x L_x)^2$, $P_{EY} = \pi^2 EI_y/(K_y L_y)^2$, Ω_b is the factor of safety for bending, Ω_c is the factor of safety for concentrically loaded compression, and C_{mx} and C_{my} are the coefficients whose value is taken as follows:

1. For compression members in frames subject to joint translation (side sway), $C_m = 0.85$.
2. For restrained compression members in frames braced against joint translation and not subjected to transverse loading between their supports in the plane of bending, $C_m = 0.6 - 0.4(M_1/M_2)$, where M_1/M_2 is the ratio of the smaller to the larger moment at the ends of that portion of the member under consideration which is unbraced in the plane of bending. M_1/M_2 is positive when the member is bent in reverse curvature and negative when it is bent in single curvature.
3. For compression members in frames braced against joint translation in the plane of loading and subjected to transverse loading between their supports, the value of C_m may be determined by rational analysis. However, in lieu of such analysis, the following values may be used: (1) for members whose ends are restrained, $C_m = 0.85$ and (2) for members whose ends are unrestrained, $C_m = 1.0$.

I_x, I_y, L_x, L_y, K_x, and K_y have been defined previously.

6.6.7 Closed Cylindrical Tubular Members

Thin-walled closed cylindrical tubular members are economical sections for compression and torsional members because of their large ratio of radius of gyration to area, the same radius of gyration in all directions, and the large torsional rigidity. The AISI design provisions are limited to the ratio of outside diameter to wall thickness, D/t, not being greater than $0.441E/F_y$.

6.6.7.1 Bending Strength

For cylindrical tubular members subjected to bending, the nominal flexural strengths are as follows according to the D/t ratio:

$$M_n = F_c S_f$$

1. For $D/t \leq 0.0714\ E/F_y$

$$F_c = 1.25F_y \tag{6.89}$$

2. For $0.0714E/F_y < D/t \leq 0.318E/F_y$

$$F_c = [0.970 + 0.020(E/F_y)/(D/t)]F_y \tag{6.90}$$

3. For $0.318E/F_y < D/t \leq 0.441E/F_y$

$$F_c = 0.328E/(D/t) \tag{6.91}$$

where D is the outside diameter of the cylindrical tube, t is the wall thickness, F_c is the critical buckling stress, and S_f is the elastic section modulus of full, unreduced cross-section. Other symbols have been defined previously.

6.6.7.2 Compressive Strength

When cylindrical tubes are used as concentrically loaded compression members, the nominal axial strength is determined by Equation 6.75, except that (1) the elastic buckling stress, F_e, is determined for flexural buckling by using Equation 6.72 and (2) the effective area, A_e, is calculated by Equation 6.92 given below:

$$A_e = A_0 + R(A - A_0) \tag{6.92}$$

where

$$R = F_y/(2F_e) \leq 1.0$$

$$A_0 = \{0.037/[(DF_y)/(tE)] + 0.667\}A \leq A \quad \text{for } D/t \leq 0.441E/F_y \tag{6.93}$$

$$A = \text{area of unreduced cross-section}$$

In the above equations, the value A_0 is the reduced area due to the effect of local buckling [1,7].

6.7 Connections and Joints

Welds, bolts, screws, rivets, and other special devices such as metal stitching and adhesives are generally used for cold-formed steel connections. The AISI Specification contains only the design provisions for welded connections, bolted connections, and screw connections. These design equations are based primarily on the experimental data obtained from extensive test programs.

6.7.1 Welded Connections

Welds used for cold-formed steel constructions may be classified as arc welds (or fusion welds) and resistance welds. Arc welding is usually used for connecting cold-formed steel members to each other as well as connecting such thin members to heavy, hot-rolled steel framing members. It is used for groove welds, arc spot welds, arc seam welds, fillet welds, and flare groove welds (Figures 6.32 to 6.36). The AISI design provisions for welded connections are applicable only for cold-formed steel structural members, in which the thickness of the thinnest connected part is 0.18 in. (4.57 mm) or less. Otherwise, when the thickness of connected parts is thicker than 0.18 in. (4.57 mm), the welded connection should be designed according to the AISC Specifications [42,43]. Additional design information on structural welding of sheet steels can also be found in the AWS Code [44].

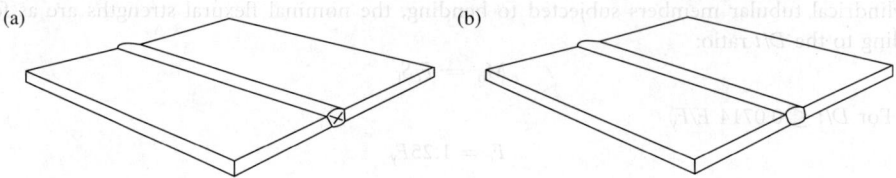

FIGURE 6.32 Groove welds.

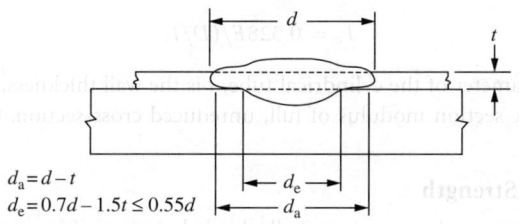

$$d_a = d - t$$
$$d_e = 0.7d - 1.5t \le 0.55d$$

FIGURE 6.33 Arc spot weld — single thickness of sheet.

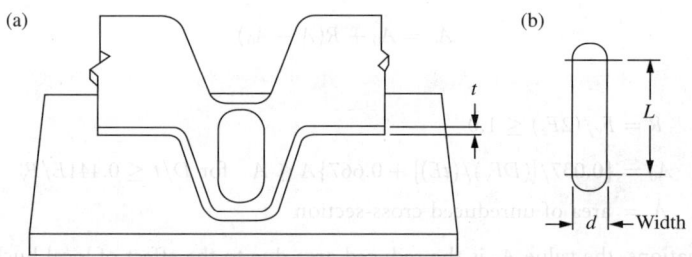

FIGURE 6.34 Arc seam weld.

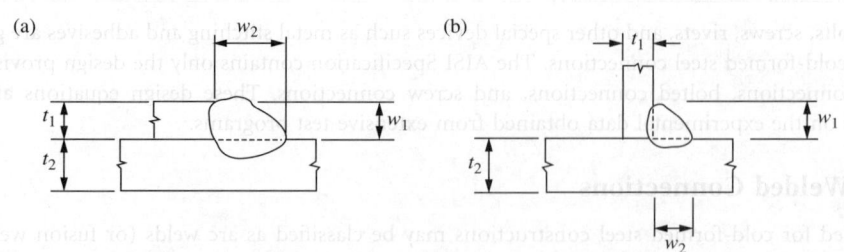

FIGURE 6.35 Fillet welds.

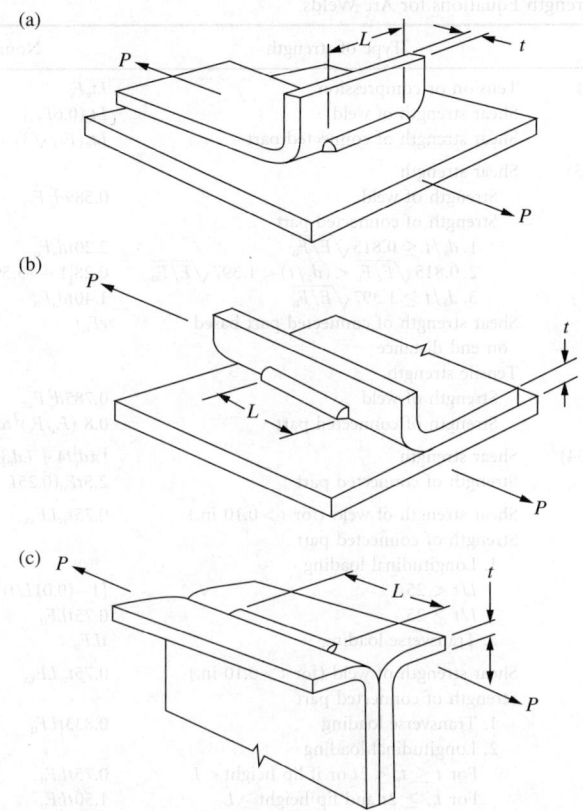

FIGURE 6.36 Flare groove welds: (a, b) flare bevel groove weld and (c) flare V-groove weld.

6.7.1.1 Arc Welds

According to the AISI Specification, the nominal strengths of arc welds can be determined from the equations given in Table 6.7. The design strengths can then be computed by using the factor of safety or resistance factor provided in Table 6.1.

6.7.1.2 Resistance Welds

The nominal shear strengths of resistance spot welds are provided in the AISI Specification [7] according to the thickness of the thinnest outside sheet and the unit used for the nominal shear strength. They are applicable for all structural grades of low-carbon steel, uncoated or galvanized with 0.9 oz/ft² of sheet or less, and medium carbon and low-alloy steels.

6.7.2 Bolted Connections

Due to the thinness of the connected parts, the design of bolted connections in cold-formed steel construction is somewhat different from that in hot-rolled heavy construction. The AISI design provisions are applicable only to cold-formed members or elements less than $\frac{3}{16}$ in. (4.76 mm) in thickness. For materials not less than $\frac{3}{16}$ in. (4.76 mm), the bolted connection should be designed in accordance with the AISC Specifications (42,43).

In the AISI Specification, five types of bolts (A307, A325, A354, A449, and A490) are used for connections in cold-formed steel construction, in which A449 and A354 bolts should be used as

TABLE 6.7 Nominal Strength Equations for Arc Welds

Type of weld	Type of strength	Nominal strength, P_n
Groove welds (Figure 6.32)	Tension or compression	Lt_eF_y
	Shear strength of weld	$Lt_e(0.6F_{xx})$
	Shear strength of connected part	$Lt_e(F_y/\sqrt{3})$
Arc spot welds (Figure 6.33)	Shear strength	
	Strength of weld	$0.589d_e^2F_{xx}$
	Strength of connected part	
	1. $d_a/t \le 0.815\sqrt{E/F_u}$	$2.20td_aF_u$
	2. $0.815\sqrt{E/F_u} < (d_a/t) < 1.397\sqrt{E/F_u}$	$0.28[1 + (5.59\sqrt{E/F_u})/(d_a/t)](td_aF_u)$
	3. $d_a/t \ge 1.397\sqrt{E/F_u}$	$1.40td_aF_u$
	Shear strength of connected part based on end distance	eF_ut
	Tensile strength	
	Strength of weld	$0.785d_e^2F_{xx}$
	Strength of connected part	$0.8\ (F_u/F_y)^2 td_aF_u$
Arc seam welds (Figure 6.34)	Shear strength	$[\pi d_e^2/4 + Ld_e]0.75F_{xx}$
	Strength of connected part	$2.5tF_u(0.25L + 0.96d_a)$
Fillet welds (Figure 6.35)	Shear strength of weld (for $t > 0.10$ in.)	$0.75t_wLF_{xx}$
	Strength of connected part	
	1. Longitudinal loading	
	$\quad L/t < 25$	$[1 - (0.01L/t)]tLF_u$
	$\quad L/t \ge 25$	$0.75tLF_u$
	2. Transverse loading	tLF_u
Flare groove welds (Figure 6.36)	Shear strength of weld (for $t > 0.10$ in.)	$0.75t_wLF_{xx}$
	Strength of connected part	
	1. Transverse loading	$0.833tLF_u$
	2. Longitudinal loading	
	$\quad$ For $t \le t_w < 2t$ or if lip height $< L$	$0.75tLF_u$
	$\quad$ For $t_w \ge 2t$ and lip height $\ge L$	$1.50tLF_u$

Note: d is the visible diameter of outer surface of arc spot weld, d_a is the average diameter of the arc spot weld at midthickness of t, $d_a = (d - t)$ for single sheet or multiple sheets not more than four sheets, d_e is the effective diameter of fused area at the plane of maximum shear transfer, $d_e = 0.7d - 1.5t \le 0.55d$, e is the distance measured in the line of force from the centerline of a weld to the nearest edge of an adjacent weld or to the end of the connected part toward which the force is directed, F_u is the tensile strength of the connected part, F_y is the yield point of steel, F_{xx} is the filler metal strength designation in AWS electrode classification, L is the length of the weld, P_n is the nominal strength of the weld, t is the thickness of the connected sheet, t_e is the effective throat dimension for groove weld, t_w is the effective throat for fillet welds or flare groove weld not filled flush to surface, $t_w = 0.707w_1$ or $0.707w_2$, whichever is smaller, and w_1 and w_2 are the leg of the weld, respectively.

equivalents of A325 and A490 bolts, respectively, whenever a diameter smaller than $\frac{1}{2}$ in. (12.7 mm) is required.

On the basis of the failure modes occurring in the tests of bolted connections, the AISI criteria deal with three major design considerations for the connected parts: (1) longitudinal shear failure, (2) tensile failure, and (3) bearing failure. The nominal strength equations are given in Table 6.8.

In addition, design strength equations are provided for shear and tension in bolts. Accordingly, the AISI nominal strength for shear and tension in bolts can be determined as follows:

$$P_n = A_b F_n \tag{6.94}$$

where A_b is the gross cross-sectional area of bolt and F_n is the nominal shear or tensile stress given in Table 6.9 for ASD and LRFD. For the LSD method, see appendix B of the North American Specification [7].

TABLE 6.8 Nominal Strength Equations for Bolted Connections (ASD and LRFD)[a]

Type of strength	Nominal strength, P_n
Shear strength based on spacing and edge distance	teF_u
Tensile strength in net section (shear lag)	
1. Flat sheet connections not having staggered holes	
With washers under bolt head and nut	
Single bolt in the line of force	$(0.1 + 3d/s)F_uA_n \leq F_uA_n$
Multiple bolts in the line of force	F_uA_n
Without washers under bolt head or nut	
Single bolt in the line of force	$(2.5d/s)F_uA_n \leq F_uA_n$
Multiple bolt in the line of force	F_uA_n
2. Flat sheet connections having staggered hole (see AISI Specification for the modified A_n)	Same as item (1), except for A_n
3. Other than flat sheet (see AISI Specification for the effective net area A_e)	F_uA_e
Bearing strength	
1. Without consideration of bolt hole deformation (see AISI Specification for bearing factor, C, and modification factor, m_f)	m_fCdtF_u
2. With consideration of bolt hole deformation (see AISI Specification for coefficient α)	$(4.64\alpha t + 1.53)dtF_u$

[a] For the LSD method, see appendix B of the AISI North American Specification [7] for shear strength based on spacing and edge distance and for tensile strength in net section.

Note: A_n is the net area of the connected part, d is the nominal diameter of bolt, F_u is the tensile strength of the connected part, s is the sheet width divided by the number of bolt holes in cross-section being analyzed, and t is the thickness of the thinnest connected part.

TABLE 6.9 Nominal Tensile and Shear Stresses for Bolts (ASD and LRFD)[a]

Description of bolts	Nominal tensile stress, F_{nt} (ksi)	Nominal shear stress, F_{nv} (ksi)
A307 bolts, Grade A, $\frac{1}{4}$ in. $\leq d < \frac{1}{2}$ in.	40.5	24.0
A307 bolts, Grade A, $d \geq \frac{1}{2}$ in.	45.0	27.0
A325 bolts, when threads are not excluded from shear planes	90.0	54.0
A325 bolts, when threads are excluded from shear planes	90.0	72.0
A354 Grade BD bolts, $\frac{1}{4}$ in. $\leq d < \frac{1}{2}$ in., when threads are not excluded from shear planes	101.0	59.0
A354 Grade BD bolts, $\frac{1}{4}$ in. $\leq d < \frac{1}{2}$ in., when threads are excluded from shear planes	101.0	90.0
A449 bolts, $\frac{1}{4}$ in. $\leq d < \frac{1}{2}$ in., when threads are not excluded from shear planes	81.0	47.0
A449 bolts, $\frac{1}{4}$ in. $\leq d < \frac{1}{2}$ in., when threads are excluded from shear planes	81.0	72.0
A490 bolts, when threads are not excluded from shear planes	112.5	67.5
A490 bolts, when threads are excluded from shear planes	112.5	90.0

[a] For the LSD method, see appendix B of the AISI North American Specification [7].

Note: 1 in. = 25.4 mm, 1 ksi = 6.9 MPa = 70.3 kg/cm².

For bolts subjected to the combination of shear and tension, the reduced nominal tension stress for the ASD and LRFD is given in Table 6.10. For the LSD method, refer to appendix B of the North American Specification.

6.7.3 Screw Connections

Screws can provide a rapid and effective means to fasten sheet metal siding and roofing to framing members and to connect individual siding and roofing panels. Design equations are presently given in the AISI Specification for determining the nominal shear strength and the nominal tensile strength of connected parts and screws. These design requirements should be used for self-tapping screws with diameters greater than or equal to 0.08 in. (2.03 mm) but not exceeding $\frac{1}{4}$ in. (6.35 mm). The screw can be thread-forming or thread-cutting, with or without drilling point. The spacing between the centers of screws should not be less than $3d$ and the distance from the center of a screw to the edge of any part should not be less than $1.5d$, where d is the diameter of screw. In the direction of applied force, the end distance is also limited by the shear strength of the connected part.

According to the AISI North American Specification, the nominal strength per screw is determined from Table 6.11. See Figure 6.37 and Figure 6.38 for t_1, t_2, F_{u1}, and F_{u2}.

For the convenience of designers, the following table gives the correlation between the common number designation and the nominal diameter for screws.

Number designation	Nominal diameter, d (in.)[a]
0	0.060
1	0.073
2	0.086
3	0.099
4	0.112
5	0.125
6	0.138
7	0.151
8	0.164
10	0.190
12	0.216
1/4	0.250

[a] 1 in. = 25.4 mm.

In addition to the design requirements discussed above, the AISI North American Specification also includes some provisions for built-up compression members composed of two sections in contact or for compression elements joined to other parts of built-up members by intermittent connections.

6.7.4 Rupture

In connection design, due consideration should be given to shear rupture, tension rupture, and block shear rupture. For details, see the AISI Specification [7].

6.8 Structural Systems and Assemblies

In the past, cold-formed steel components have been used in different structural systems and assemblies such as metal buildings, shear diaphragms, shell roof structures, wall stud assemblies, residential construction, and composite construction.

TABLE 6.10 Nominal Tension Stresses, F'_{nt}, for Bolts Subjected to the Combination of Shear and Tension (ASD and LRFD)[a]

Description of bolts	Threads not excluded from shear planes	Threads excluded from shear planes
	(A) ASD method	
A325 bolts	$110 - 3.6f_v \leq 90$	$110 - 2.8f_v \leq 90$
A354 Grade BD bolts	$122 - 3.6f_v \leq 101$	$122 - 2.8f_v \leq 101$
A449 bolts	$100 - 3.6f_v \leq 81$	$100 - 2.8f_v \leq 81$
A490 bolts	$136 - 3.6f_v \leq 112.5$	$136 - 2.8f_v \leq 112.5$
A307 bolts, Grade A		
When $\frac{1}{4}$ in. $\leq d < \frac{1}{2}$ in.	$52 - 4f_v \leq 40.5$	$52 - 4f_v \leq 40.5$
When $d \geq \frac{1}{2}$ in.	$58.5 - 4f_v \leq 45$	$58.5 - 4f_v \leq 45$
	(B) LRFD method	
A325 bolts	$113 - 2.4f_v \leq 90$	$113 - 1.9f_v \leq 90$
A354 Grade BD bolts	$127 - 2.4f_v \leq 101$	$127 - 1.9f_v \leq 101$
A449 bolts	$101 - 2.4f_v \leq 81$	$101 - 1.9f_v \leq 81$
A490 bolts	$141 - 2.4f_v \leq 112.5$	$141 - 1.9f_v \leq 112.5$
A307 bolts, Grade A		
When $\frac{1}{4}$ in. $\leq d < \frac{1}{2}$ in.	$47 - 2.4f_v \leq 40.5$	$47 - 2.4f_v \leq 40.5$
When $d \geq \frac{1}{2}$ in.	$52 - 2.4f_v \leq 45$	$52 - 2.4f_v \leq 45$

[a] For the LSD method, see appendix B of the AISI North American Specification [7].
Note: d is the nominal diameter of bolt and f_v is the shear stress based on gross cross-sectional area of bolt; 1 in. = 25.4 mm, 1 ksi = 6.9 MPa = 70.3 kg/cm^2.

TABLE 6.11 Nominal Strength Equations for Screws

Type of strength	Nominal strength
Shear strength	
1. Connection shear limited by tilting and bearing	
For $t_2/t_1 \leq 1.0$, use smallest of three considerations	1. $P_{ns} = 4.2(t_2^3 d)^{1/2}F_{u2}$
	2. $P_{ns} = 2.7t_1 dF_{u1}$
	3. $P_{ns} = 2.7t_2 dF_{u2}$
For $t_2/t_1 \geq 2.5$, use smallest of two considerations	1. $P_{ns} = 2.7t_1 dF_{u1}$
	2. $P_{ns} = 2.7t_2 dF_{u2}$
For $1.0 < t_2/t_1 < 2.5$, use linear interpolation	
2. Connection shear limited by end distance	$P_{ns} = teF_u^a$
3. Shear in screws	$P_{ns} = 0.8P_{ss}$
Tensile strength	
1. Connection tension	
Pull-out strength	$P_{not} = 0.85t_c dF_{u2}$
Pull-over strength	$P_{nov} = 1.5t_1 d_w F_{u1}$
2. Tension in screws	$P_{nt} = 0.8P_{ts}$

[a] For the LSD method see appendix B of the AISI North American Specification [7].
Note: d is the nominal diameter of screw, d_w is the larger of screw head diameter and washer diameter, F_{u1} is the tensile strength of member in contact with the screw head (Figure 6.38), F_{u2} is the tensile strength of member not in contact with the screw head (Figure 6.37), P_{ns} is the nominal shear strength per screw, P_{ss} is the nominal shear strength per screw as reported by the manufacturer or determined by tests, P_{nt} is the nominal tension strength per screw, P_{not} is the nominal pull-out strength per screw, P_{nov} is the nominal pull-over strength per screw, P_{ts} is the nominal tension strength per screw as reported by the manufacturer or determined by tests, t_1 is the thickness of member in contact with the screw head (Figure 6.38), t_2 is the thickness of member not in contact with the screw head (Figure 6.37), and t_c is the lesser of depth penetration and thickness t_2.

6.8.1 Metal Buildings

Standardized metal buildings have been widely used in industrial, commercial, and agricultural applications. This type of metal building has also been used for community facilities because of

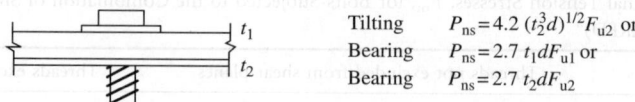

Tilting	$P_{ns} = 4.2\,(t_2^3 d)^{1/2} F_{u2}$ or
Bearing	$P_{ns} = 2.7\,t_1 dF_{u1}$ or
Bearing	$P_{ns} = 2.7\,t_2 dF_{u2}$

FIGURE 6.37   Screw connection for $t_2/t_1 \le 1.0$.

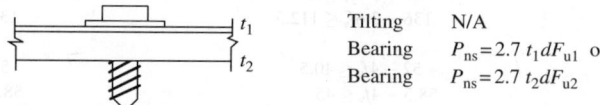

Tilting	N/A
Bearing	$P_{ns} = 2.7\,t_1 dF_{u1}$ or
Bearing	$P_{ns} = 2.7\,t_2 dF_{u2}$

FIGURE 6.38   Screw connection for $t_2/t_1 \ge 2.5$.

its attractive appearance, fast construction, low maintenance, easy extension, and lower long-term cost.

In general, metal buildings are made of welded rigid frames with cold-formed steel sections used for purlins, girts, roofs, and walls. In the United States, the design of standardized metal buildings is often based on the *Metal Building Systems Manual* published by the Metal Building Manufacturers' Association [31]. This manual contains load applications, crane loads, serviceability, common industry practices, guide specifications, AISC-MB certification, wind load commentary, fire protection, wind, snow, and rain data by US county, glossary, appendix and bibliography. In other countries, many design concepts and building systems have also been developed.

6.8.2 Shear Diaphragms

In building construction, it has been a common practice to provide a separate bracing system to resist horizontal loads due to wind load or earthquake. However, steel floor and roof panels, with or without concrete fill, are capable of resisting horizontal in-plane loads in addition to the beam strength for gravity loads if they are adequately interconnected to each other and to the supporting frame. For the same reason, wall panels not only provide enclosure surface and support normal loads, but they can also provide diaphragm action in their own planes.

The structural performance of a diaphragm construction can be evaluated by either calculations or tests. Several analytical procedures exist, and are summarized in the literature [29,45–47]. Tested performance can be measured by the procedures of the *Standard Method for Static Load Testing of Framed Floor, Roof and Wall Diaphragm Construction for Buildings*, ASTM E455 [41,48]. A general discussion of structural diaphragm behavior is given by Yu [1].

Shear diaphragms should be designed for both strength and stiffness. After the nominal shear strength is established by calculations or tests, the design strength can be determined on the basis of the factor of safety or resistance factor given in the Specification [7]. Six cases are currently classified in the AISI Specification for the design of shear diaphragms according to the type of failure mode, type of connections, and type of loading. Because the quality of mechanical connectors is easier to control than that of welded connections, a relatively smaller factor of safety or larger resistance factor is used for mechanical connections. As far as the loading is concerned, the factors of safety for earthquake are slightly larger than those for wind due to the ductility requirements of seismic loading.

6.8.3 Shell Roof Structures

Shell roof structures such as folded-plate and hyperbolic paraboloid roofs have been used in building construction for churches, auditoriums, gymnasiums, schools, restaurants, office buildings, and airplane hangars. This is because the effective use of steel panels in roof construction is not only to provide an economical structure but also to make the building architecturally attractive and flexible for future extension. The design methods used in engineering practice are mainly based on the successful investigation of shear diaphragms and the structural research on shell roof structures.

A folded-plate roof structure consists of three major components. They are (1) steel roof panels, (2) fold line members at ridges and valleys, and (3) end frame or end walls as shown in Figure 6.39. Steel roof panels can be designed as simply supported slabs in the transverse direction between fold lines. The reaction of the panels is then applied to fold lines as a line loading, which can be resolved into two components parallel to the two adjacent plates. These load components are carried by an inclined deep girder spanned between end frames or end walls. These deep girders consist of fold line members as flanges and steel panels as a web element. The longitudinal flange force in fold line members can be obtained by dividing the bending moment of the deep girder by its depth. The shear force is resisted by the diaphragm action of the steel roof panels. In addition to the strength, the deflection characteristics of the folded-plate roof should also be investigated, particularly for long-span structures. In the past, it has been found that a method similar to the Williot diaphragm for determining truss deflections can also be used for the prediction of the deflection of a steel folded-plate roof. The in-plane deflection of each plate should be computed as a sum of the deflections due to flexure, shear, and seam slip, considering the plate temporarily separated from the adjacent plates. The true displacement of the fold line can then be determined analytically or graphically by a Williot diagram. The above discussion deals with a simplified method. The finite-element method can provide a more detailed analysis for various types of loading, support, and material.

The hyperbolic paraboloid roof has also gained popularity due to the economical use of materials and its appearance. This type of roof can be built easily with either single-layer or double-layer standard steel roof deck panels because the hyperbolic paraboloid has straight line generators. Figure 6.40 shows four common types of hyperbolic paraboloid roofs, which may be modified or varied in other ways to achieve a striking appearance. The method of analysis depends on the curvature of the shell used for the roof. If the uniformly loaded shell is deep, the membrane theory may be used. For the case of a shallow shell or a deep shell subjected to unsymmetrical loading, the finite-element method will provide accurate results. Using the membrane theory, the panel shear for a uniformly loaded hyperbolic paraboloid roof can be determined by $wab/2h$, where w is the applied load per unit surface area, a and b are horizontal projections, and h is the amount of corner depression of the surface. This panel shear force should be carried by tension and compression framing members. For additional design information, see Yu [1].

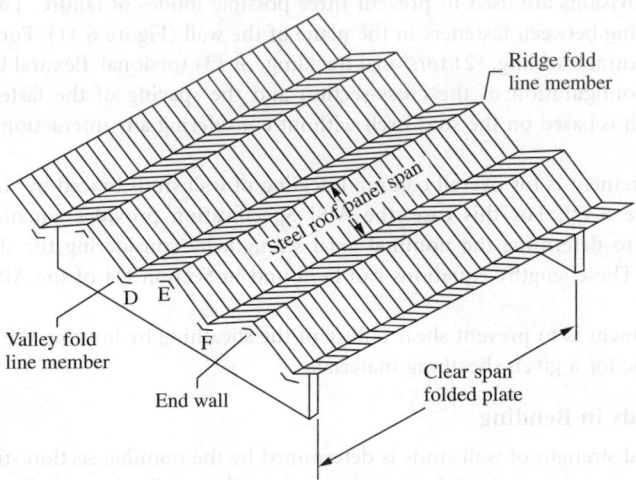

FIGURE 6.39   Folded-plate structure (courtesy of Yu, W.W. 1991).

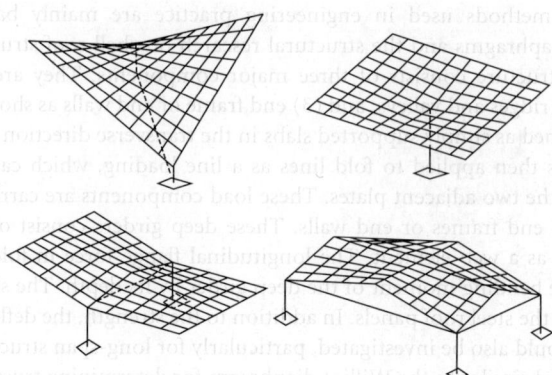

FIGURE 6.40   Types of hyperbolic paraboloid roofs (courtesy of Yu, W.W. 1991).

6.8.4 Wall Stud Assemblies

Cold-formed steel I-, C-, Z-, or box-type studs are widely used in walls with their webs placed perpendicular to the wall surface. The walls may be made of different materials, such as fiber board, lignocellulosic board, plywood, or gypsum board. If the wall material is strong enough and there is adequate attachment provided between wall material and studs for lateral support of the studs, then the wall material can contribute to the structural economy by increasing the usable strength of the studs substantially.

The AISI North American Specification provides the requirements for two types of stud design. The first type is "All Steel Design," in which the wall stud is designed as an individual compression member neglecting the structural contribution of the attached sheathing. The second type is "Sheathing Braced Design," in which consideration is given to the bracing action of the sheathing material due to the shear rigidity and the rotational restraint provided by the sheathing. Both solid and perforated webs are permitted. The subsequent discussion deals with the sheathing braced design of wall studs.

6.8.4.1 Wall Studs in Compression

The AISI design provisions are used to prevent three possible modes of failure. The first requirement is for column buckling between fasteners in the plane of the wall (Figure 6.41). For this case, the limit state may be (1) flexural buckling, (2) torsional buckling, or (3) torsional–flexural buckling depending on the geometric configuration of the cross-section and the spacing of the fasteners. The nominal compressive strength is based on the stud itself without considering any interaction with the sheathing material.

The second requirement is for overall column buckling of wall studs braced by shear diaphragms on both flanges (Figure 6.42). For this case, the AISI Specification provides equations for calculating the critical stresses to determine the nominal axial strength by considering the shear rigidity of the sheathing material. These lengthy equations can be found in Section D4 of the AISI North American Specification [7].

The third requirement is to prevent shear failure of the sheathing by limiting the shear strain within the permissible value for a given sheathing material.

6.8.4.2 Wall Studs in Bending

The nominal flexural strength of wall studs is determined by the nominal section strength by using the "All Steel Design" approach and neglecting the structural contribution of the attached sheathing material.

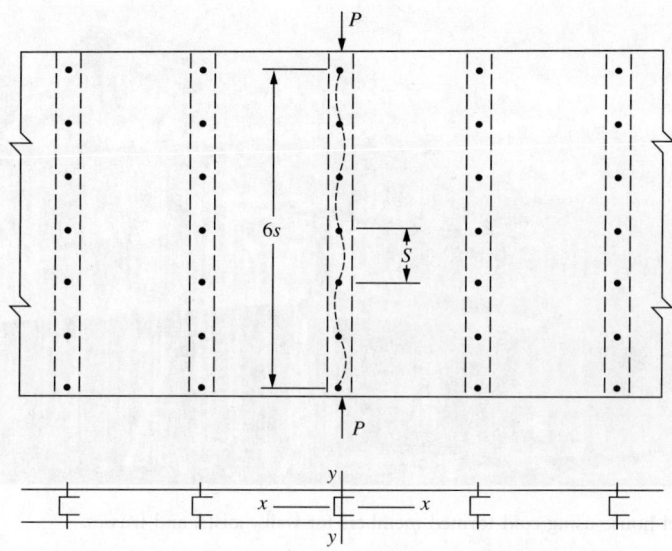

FIGURE 6.41 Buckling of studs between fasteners (courtesy of Yu, W.W. 1991).

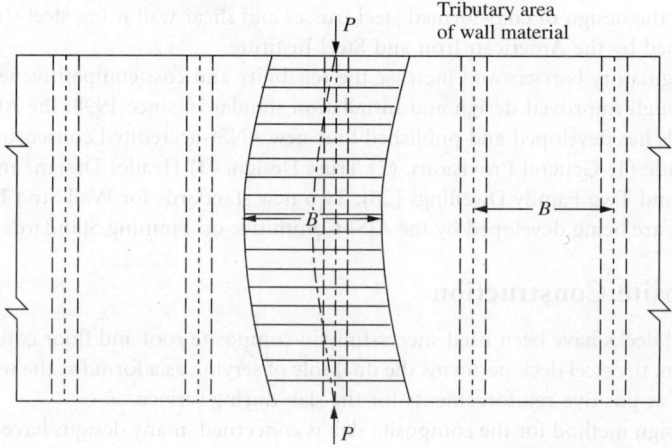

FIGURE 6.42 Overall column buckling of studs (courtesy of Yu, W.W. 1991).

6.8.4.3 Wall Studs with Combined Axial Load and Bending

The AISI interaction equations discussed in Table 6.6 are also applicable to wall studs subjected to combined axial load and bending with the exception that the nominal flexural strength is evaluated by excluding lateral–torsional buckling considerations.

6.8.5 Residential Construction

In recent years, cold-formed steel members have been increasingly used in residential construction as roof trusses, wall framing, and floor systems (Figure 6.43). Because of the lack of standard sections and

FIGURE 6.43   Steel house using cold-formed members for walls, joists, and trusses.

design tables, prescriptive standards have been developed by the National Association of Home Builders Research Center and the Housing and Urban Development. The sectional properties and load-span design tables for a selected group of C-sections have been calculated in accordance with the AISI Specification. For the design of cold-formed steel trusses and shear wall using steel studs, design guides have been published by the American Iron and Steel Institute.

To eliminate regulatory barriers and increase the reliability and cost competitiveness of cold-formed steel framing through improved design and installation standards, since 1998, the AISI Committee on Framing Standards has developed and published four new ANSI-accredited consensus standards. These publications include (1) General Provisions, (2) Truss Design, (3) Header Design, and (4) Prescriptive Method for One and Two Family Dwellings [25]. Two new standards for Wall Stud Design and Lateral Resistance Design are being developed by the AISI Committee on Framing Standards at present (2003).

6.8.6 Composite Construction

Cold-formed steel decks have been used successfully in composite roof and floor construction. For this type of application, the steel deck performs the dual role of serving as a form for the wet concrete during construction and as positive reinforcements for the slab during service.

As far as the design method for the composite slab is concerned, many designs have been based on the SDI Specification for composite steel floor deck [24]. This document contains requirements and recommendations on materials, design, connections, and construction practice. A design handbook is also available from Steel Deck Institute [49]. Since 1984, the American Society of Civil Engineers has published a standard specification for the design and construction of composite slabs [30].

When the composite construction is composed of steel beams or girders with cold-formed steel deck, the design should be based on the AISC Specifications [42,43].

6.9   Computer-Aided Design and Direct Strength Method

The design method discussed in Sections 6.5 and 6.6 deals with the limit states of yielding, local buckling, and overall buckling of structural members. According to the current edition of the AISI North American Specification [7] and previous editions of the AISI Specification, the *effective width design approach* has been and is being used to determine the structural strength of the member. Because hand calculations are excessively lengthy and difficult for solving complicated problems, a large number of computer programs have been prepared by many universities and companies for the analysis and design of cold-formed steel structures. Some of the computer programs can be found on the AISI website at www.steel.org and the website of Wei-Wen Yu Center for Cold-Formed Steel Structures at www.umr.edu/~ccfss.

In recent years, the direct strength method has been developed for determining the member strength [3,50,51]. Instead of using the effective width design approach of individual elements, this method uses the entire cross-section for elastic buckling determination and incorporates local, distortional, and overall buckling in the design process. The cross-section elastic buckling solution insures interelement compatibility and equilibrium at element junctures. Thus, the actual restraining effects of adjoining elements are taken into account. This new method can be used as a rational engineering analysis as permitted by section A1.1(b) of the 2001 edition of the AISI *North American Specification for the Design of Cold-Formed Steel Structural Members*. For details, see the forthcoming new appendix in the North American Specification [7,51]. The computer program for the Direct Strength Method can be obtained from the website at www.ce.jhu.edu/bschafer.

Glossary

ASD (allowable strength design) — A method of proportioning structural components such that the allowable stress, allowable force, or allowable moment is not exceeded by the required allowable strength of the component determined by the load effects of all appropriate combinations of nominal loads. This method is used in the United States and Mexico.

Beam–column — A structural member subjected to combined compressive axial load and bending.

Buckling load — The load at which a compressed element, member, or frame assumes a deflected position.

Cold-formed steel members — Shapes that are manufactured by press-braking blanks sheared from sheets, cut lengths of coils or plates, or by roll forming cold- or hot-rolled coils or sheets.

Composite slab — A slab in which the load-carrying capacity is provided by the composite action of concrete and steel deck (as reinforcement).

Compression members — Structural members whose primary function is to carry concentric loads along their longitudinal axes.

Design strength — R_n/Ω, for ASD or ϕR_n for LRFD and LSD (force, moment, as appropriate), provided by the structural component.

Distorsional buckling — A mode of buckling involving change in cross-sectional shape, excluding local buckling.

Effective design width — Reduced flat width of an element for design purposes. The reduced design width is termed the effective width or effective design width.

Effective length — The equivalent length KL used in design equations.

Factor of safety — A ratio of the stress (or strength) at incipient failure to the computed stress (or strength) at design load (or service load).

Flat width to thickness ratio — The flat width of an element measured along its plane, divided by its thickness.

Flexural members (beams) — Structural members whose primary function is to carry transverse loads or moments.

Limit state — A condition in which a structure or component becomes unsafe (strength limit state) or no longer useful for its intended function (serviceability limit state).

Load factor — A factor that accounts for unavoidable deviations of the actual load from the nominal load.

Local buckling — Buckling of elements only within a section, where the line junctions between elements remain straight and angles between elements do not change.

LRFD (load and resistance factor design) — A method of proportioning structural components such that no applicable limit state is exceeded when the structure is subjected to all appropriate load combinations of factored loads. This method is used in the United States and Mexico.

LSD (limit states design) — A method of proportioning structural components such that no applicable limit state is exceeded when the structure is subjected to all appropriate load combinations of factored loads. This method is used in Canada.

Multiple-stiffened elements — An element that is stiffened between webs, or between a web and a stiffened edge, by means of intermediate stiffeners that are parallel to the direction of stress. A subelement is the portion between adjacent stiffeners or between web and intermediate stiffener or between edge and intermediate stiffener.

Nominal loads — The loads specified by the applicable code not including load factors.

Nominal strength — The capacity of a structure or component to resist the effects of loads, as determined by computations using specified material strengths and dimensions with equations derived from accepted principles of structural mechanics or by tests of scaled models, allowing for modeling effects and differences between laboratory and field conditions.

Point-symmetric section — A point-symmetric section is a section symmetrical about a point (centroid) such as a Z-section having equal flanges.

Required strength — Load effect (force, moment, as appropriate) acting on the structural component determined by structural analysis from the factored loads for LRFD and LSD or nominal loads for ASD (using the most appropriate critical load combinations).

Resistance factor — A factor that accounts for unavoidable deviations of the actual strength from the nominal value.

Stiffened or partially stiffened compression elements — A stiffened or partially stiffened compression element is a flat compression element of which both edges parallel to the direction of stress are stiffened by a web, flange, stiffening lip, intermediate stiffener, or the like.

Stress — Stress as used in this chapter means force per unit area and is expressed in ksi (kips per square inch) for U.S. customary units, MPa for SI units, and kg/cm^2 for the MKS system.

Tensile strength — Maximum stress reached in a tension test.

Thickness — The thickness of any element or section should be the base steel thickness, exclusive of coatings.

Torsional–flexural buckling — A mode of buckling in which compression members can bend and twist simultaneously without change in cross-sectional shape.

Unstiffened compression elements — A flat compression element that is stiffened at only one edge parallel to the direction of stress.

Yield point — Yield point as used in this chapter means either yield point or yield strength of steel.

References

[1] Yu, W.W. 1991. *Cold-Formed Steel Design*, 2nd Edition, John Wiley & Sons, New York, NY (see also 2000, 3rd Edition).

[2] Rhodes, J. 1991. *Design of Cold-Formed Steel Structures*, Elsevier Publishing Co., New York, NY.

[3] Hancock, G.J., Murray, T.M., and Ellifritt, D.S. 2001. *Cold-Formed Steel Structures to the AISI Specification*, Marcel Dekker, Inc., New York, NY.

[4] Ghersi, A., Landolfo, R., and Mazzolani, F.M. 2002. *Design of Metallic Cold-Formed Thin-Walled Members*, Spon, London.

[5] Standards Australia. 1996. *Cold-Formed Steel Structures*, Australia, Suppl 1, 1998.

[6] Lagereinrichtungen. 1974. ONORM B 4901, Dec.

[7] American Iron and Steel Institute. 2001. *North American Specification for the Design of Cold-Formed Steel Structural Members*, Washington, DC.

[8] Czechoslovak State Standard. 1987. *Design of Light Gauge Cold-Formed Profiles in Steel Structures*, CSN 73 1402.

[9] Finnish Ministry of Environment. 1988. *Building Code Series of Finland: Specification for Cold-Formed Steel Structures.*

[10] Centre Technique Industriel de la Construction Metallique. 1978. *Recommendations pourle Calcul des Constructions a Elements Minces en Acier.*

[11] DIN 18807, 1987. *Trapezprofile im Hochbau* (Trapezoidal Profiled Sheeting in Building). Deutsche Norm (German Standard).

[12] Indian Standards Institution. 1975. *Indian Standard Code of Practice for Use of Cold-Formed Light-Gauge Steel Structural Members in General Building Construction,* IS:801-1975.

[13] CNR, 10022. 1984. *Profilati di acciaio formati a freddo. Istruzioni per l' impiego nelle construzioni.*

[14] Architectural Institute of Japan. 1985. *Recommendations for the Design and Fabrication of Light Weight Steel Structures,* Japan.

[15] Groep Stelling Fabrikanten — GSF. 1977. *Richtlijnen Voor de Berekening van Stalen Industriele Magezijnstellingen,* RSM.

[16] Standards New Zealand. 1996. *Cold-Formed Steel Structures,* New Zealand, Suppl 1, 1998.

[17] People's Republic of China National Standard. 2002. *Technical Code for Cold-formed/Thin-Walled Steel Structures,* GB50018-2002, Beijing, China.

[18] South African Institute of Steel Construction. 1995. *Code of Practice for the Design of Structural Steelwork.*

[19] Swedish Institute of Steel Construction. 1982. *Swedish Code for Light Gauge Metal Structures,* Publication 76.

[20] *Romanian Specification for Calculation of Thin-Walled Cold-Formed Steel Members,* STAS 10108/2-83.

[21] British Standards Institution. 1991. *British Standard: Structural Use of Steelwork in Building. Part 5. Code of Practice for Design of Cold-Formed Sections,* BS 5950: Part 5: 1991.

[22] State Building Construction of USSR. 1988. *Building Standards and Rules: Design Standards — Steel Construction,* Part II, Moscow.

[23] Eurocode 3. 1996. *Design of Steel Structures, Part 1–3: General Rules–Supplementary Rules for Cold-Formed Thin Gauge Members and Sheeting.*

[24] Steel Deck Institute. 1995. *Design Manual for Composite Decks, Form Decks, Roof Decks, and Cellular Floor Deck with Electrical Distribution,* Publication No. 29. Fox River Grove, IL.

[25] American Iron and Steel Institute. 2001. *Standards for Cold-Formed Steel Framing: (a) General Provisions, (b) Truss Design, (c) Header Design, and (d) Prescriptive Method for One and Two Family Dwellings,* Washington, DC.

[26] Steel Joist Institute. 1995. *Standard Specification and Load Tables for Open Web Steel Joists,* 40th Edition, Myrtle Beach, SC.

[27] American Society of Civil Engineers. 1988. *Guide for Design of Steel Transmission Towers, Manual 52,* New York, NY.

[28] Rack Manufacturers Institute. 2002. *Specification for the Design, Testing, and Utilization of Industrial Steel Storage Racks,* Charlotte, NC.

[29] Luttrell, L.D. 1987. *Steel Deck Institute Diaphragm Design Manual,* 2nd Edition, Steel Deck Institute, Canton, OH.

[30] American Society of Civil Engineers. 1984. *Specification for the Design and Construction of Composite Steel Deck Slabs,* ASCE Standard, New York, NY.

[31] Metal Building Manufacturers Association. 2002. *Metal Building Systems Manual,* Cleveland, OH.

[32] American Iron and Steel Institute. 1996. *Residential Steel Framing Manual,* including the Cold-Formed Steel Framing Design Guide prepared by T.W.J. Trestain, Washington, DC.

[33] Association of the Wall and Ceiling Industries–International and Metal Lath/Steel Framing Association. 1979. *Steel Framing Systems Manual,* Chicago, IL.

[34] Steel Stud Manufacturers Association. 2001. *Product Technical Information,* Chicago, IL.

[35] American Society of Civil Engineers. 2002. *Specification for the Design of Cold-Formed Stainless Steel Structural Members,* SEI/ASCE-8-02, Reston, VA.

[36] American Iron and Steel Institute. 1983. *Handbook of Steel Drainage and Highway Construction Products*, Washington, DC.

[37] American Iron and Steel Institute. 2000. *Automotive Steel Design Manual*, Washington, DC.

[38] American Society of Civil Engineers. 2002. *Minimum Design Loads for Buildings and Other Structures*, ASCE 7, Washington, DC.

[39] Winter, G. 1970. *Commentary on the Specification for the Design of Cold-Formed Steel Structural Members*, American Iron and Steel Institute, New York, NY.

[40] Winter, G. 1947. Strength of Thin Steel Compression Flanges, *Trans.* ASCE, 112.

[41] American Iron and Steel Institute. 2002. *Cold-Formed Steel Design Manual*, Washington, DC.

[42] American Institute of Steel Construction. 1989. *Specification for Structural Steel Buildings — Allowable Stress Design and Plastic Design*, Chicago, IL.

[43] American Institute of Steel Construction. 2001. *Load and Resistance Factor Design Specification for Structural Steel Buildings*, Chicago, IL.

[44] American Welding Society. 1998. *Structural Welding Code — Sheet Steel*, AWS D1.3-98, Miami, FL.

[45] American Iron and Steel Institute. 1967. *Design of Light-Gage Steel Diaphragms*, New York, NY.

[46] Bryan, E.R. and Davies, J.M. 1981. *Steel Diaphragm Roof Decks — A Design Guide with Tables for Engineers and Architects*, Granada Publishing, New York, NY.

[47] Department of Army. 1992. *Seismic Design for Buildings*, US Army Technical Manual 5-809-10, Washington, DC.

[48] American Society for Testing and Materials. 1993. *Standard Method for Static Load Testing of Framed Floor, Roof and Wall Diaphragm Construction for Buildings*, ASTM E455, Philadelphia, PA.

[49] Heagler, R.B., Luttrell, L.D., and Easterling, W.S. 1997. *Composite Deck Design Handbook*, Steel Deck Institute, Fox River Grove, IL.

[50] Schafer, B.W. and Pekoz, T. 1998. Direct Strength Prediction of Cold-Formed Steel Members using Numerical Elastic Buckling Solutions, *Thin-Walled Structures: Research and Development*, Elsevier, New York, NY.

[51] Schafer, B.W. 2002. Progress on the Direct Strength Method, *Proceedings of the 16th International Specialty Conference on Cold-Formed Steel Structures*, University of Missouri-Rolla, Rolla, MO.

Further Reading

Guide to Stability Design Criteria for Metal Structures, edited by T.V. Galambos, presents general information, interpretation, new ideas, and research results on a full range of structural stability concerns. It was published by John Wiley & Sons in 1998.

Cold-Formed Steel in Tall Buildings, edited by W.W. Yu, R. Baehre, and T. Toma, provides readers with information needed for the design and construction of tall buildings, using cold-formed steel for structural members and architectural components. It was published by McGraw-Hill in 1993.

Thin-Walled Structures, edited by J. Rhodes, J. Loughlan, and K.P. Chong, is an international journal that publishes papers on theory, experiment, design, etc. related to cold-formed steel sections, plate and shell structures, and others. It is published by Elsevier Applied Science. A special issue of the journal on cold-formed steel structures was edited by J. Rhodes and W.W. Yu, guest editors, and published in 1993.

Proceedings of the International Specialty Conference on Cold-Formed Steel Structures, edited by W.W. Yu, J.H. Senne, and R.A. LaBoube, have been published by the University of Missouri–Rolla since 1971. These publications contain technical papers presented at the International Specialty Conferences on Cold-Formed Steel Structures.

"Cold-Formed Steel Structures," by J. Rhodes and N.E. Shanmugan, *The Civil Engineering Handbook* (W.F. Chen and J.Y.R. Liew, Editors-in-Chief), presents discussions on cold-formed steel sections, local buckling of plate elements, and the design of cold-formed steel members and connections. It was published by CRC Press in 2003.

7

Reinforced Concrete Structures

Austin Pan
*T.Y. Lin International,
San Francisco, CA*

7.1 Introduction

Reinforced concrete is a composite material. A lattice or cage of steel bars is embedded in a matrix of Portland cement concrete (see Figure 7.1). The specified compressive strength of the concrete typically ranges from 3,000 to 10,000 psi. The specified yield strength of the reinforcing steel is normally 60,000 psi. Reinforcement bar sizes range from $\frac{3}{8}$ to $2\frac{1}{4}$ in. in diameter (see Table 7.1). The steel reinforcement bars are manufactured with lugs or protrusion to ensure a strong bond between the steel and concrete for composite action. The placement location of the steel reinforcement within the concrete is specified by the concrete *cover*, which is the clear distance between the surface of the concrete and the reinforcement. Steel bars may be bent or hooked.

The construction of a reinforced concrete structural element requires molds or forms usually made of wood or steel supported on temporary shores or falsework (see Photo 7.1). The reinforcement bars are typically cut, bent, and wired together into a mat or cage before they are positioned into the forms. To maintain the specified clear cover, devices such as bar chairs or small blocks are used to support the rebars. Concrete placed into the forms must be vibrated well to remove air pockets. After placement, exposed concrete surfaces are toweled and finished, and sufficient time must be allowed for the concrete to set and cure to reach the desired strength.

The key structural design concept of reinforced concrete is the placement of steel in regions in the concrete where tension is expected. Although concrete is relatively strong in compression, it is weak in tension. Its tensile cracking strength is approximately 10% of its compressive strength. To overcome this weakness, steel reinforcement is used to resist tension; otherwise, the structure will crack excessively and may fail. This strategic combination of steel and concrete results in a composite material that has high strength and retains the versatility and economic advantages of concrete.

To construct concrete structures of even greater structural strength, very high-strength steel, such as Grade 270 strands, may be used instead of Grade 60 reinforcement bars. However, the high strength levels of Grade 270 steel is attained at high strain levels. Therefore, for this type of steel to work effectively with concrete, the high-strength strands must be prestrained or prestressed. This type of structure is

PHOTO 7.1 A 30-story reinforced concrete building under construction. The Pacific Park Plaza is one of the largest reinforced concrete structures in the San Francisco Bay area. It survived the October 17, 1989, Loma Prieta earthquake without damage. Instrumentation in the building recorded peak horizontal accelerations of 0.22g at the base and 0.39g at the top of the building (courtesy of Mr. James Tai, T.Y. International, San Francisco).

behoved ACI 318 for the concepts of reinforced concrete in their design codes. There may be minor changes or additions. The ACI code is incorporated in the International Building Code (IBC), as well as in the bridge design codes of the American Association of State Highway and Transportation Officials (AASHTO). The ACI code is recognized internationally. Design concepts and provisions adopted by other countries are similar to those found in the ACI code.

7.3 Material Properties

With respect to structural design, the most important property of concrete that must be specified by the structural designer is the compressive strength. In the United States, compressive strength, f_c', is expressed in units of psi. For steel reinforcement, the most important property is the yield stress, f_y. Steel reinforcement with specified yield strength of 60,000 psi (i.e., Grade 60) has become the norm (see standard in the United States). Other important material structural properties of concrete relevant for design are given in the following sections.

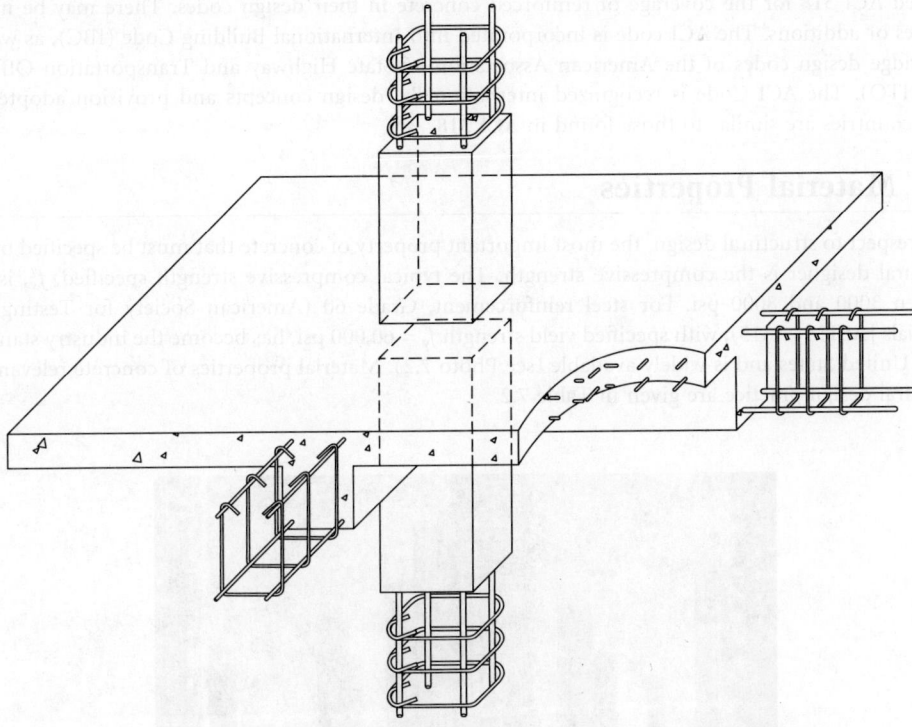

FIGURE 7.1　Reinforced concrete structure.

TABLE 7.1　Reinforcing Bar Properties

| Bar size | Nominal properties | | |
	Diameter (in.)	Area (in.^{2})	Weight (lb/ft)
3	0.375	0.11	0.376
4	0.500	0.20	0.668
5	0.625	0.31	1.043
6	0.750	0.44	1.502
7	0.875	0.60	2.044
8	1.000	0.79	2.670
9	1.128	1.00	3.400
10	1.270	1.27	4.303
11	1.410	1.56	5.313
14	1.693	2.25	7.650
18	2.257	4.00	13.600

Note: Yield stress of ASTM 615 Grade 60 bar = 60,000 psi; modulus of elasticity of reinforcing steel = 29,000,000 psi.

referred to as *prestressed concrete*. Prestressed concrete is considered an extension of reinforced concrete, but it has many distinct features. It is not the subject of this chapter.

7.2 Design Codes

The primary design code for reinforced concrete structures in U.S. design practice is given by the American Concrete Institute (ACI) 318. The latest edition of this code is dated 2002 and is the main reference of this chapter. Most local and state jurisdictions, as well as many national organizations, have adopted ACI 318 for the coverage of reinforced concrete in their design codes. There may be minor changes or additions. The ACI code is incorporated into International Building Code (IBC), as well as the bridge design codes of the American Association of State Highway and Transportation Officials (AASHTO). The ACI Code is recognized internationally; design concepts and provision adopted by other countries are similar to those found in ACI 318.

7.3 Material Properties

With respect to structural design, the most important property of concrete that must be specified by the structural designer is the compressive strength. The typical compressive strength specified, f'_c, is one between 3000 and 8000 psi. For steel reinforcement, Grade 60 (American Society for Testing and Materials [ASTM] A615), with specified yield strength $f_y = 60,000$ psi, has become the industry standard in the United States and is widely available (see Photo 7.2). Material properties of concrete relevant for structural design practice are given in Table 7.2.

PHOTO 7.2 Installation of reinforcing bars in the Pacific Park Plaza building (courtesy of Mr. James Tai, T.Y. International, San Francisco).

TABLE 7.2 Concrete Properties

Concrete strength f'_c (psi)	Modulus of elasticity, $57,000\sqrt{f'_c}$ (psi)	Modulus of rupture, $7.5\sqrt{f'_c}$ (psi)	One-way, shear baseline, $2\sqrt{f'_c}$ (psi)	Two-way, shear baseline, $4\sqrt{f'_c}$ (psi)
3000	3,122,019	411	110	219
4000	3,604,997	474	126	253
5000	4,030,509	530	141	283
6000	4,415,201	581	155	310
7000	4,768,962	627	167	335
8000	5,098,235	671	179	358

Note: Typical range of normal-weight concrete = 145 to 155 pcf; typical range of lightweight concrete = 90 to 120 pcf.

7.4   Design Objectives

For reinforced concrete structures, the design objectives of the structural engineer typically consist of the following:

1. To configure a workable and economical structural system. This involves the selection of the appropriate structural types and laying out the locations and arrangement of structural elements such as columns and beams.
2. To select structural dimensions, depth and width, of individual members, and the concrete cover.
3. To determine the required reinforcement, both longitudinal and transverse.
4. Detailing of reinforcement such as development lengths, hooks, and bends.
5. To satisfy serviceability requirements such as deflections and crack widths.

7.5   Design Criteria

In achieving the design objectives, there are four general design criteria of SAFE that must be satisfied:

1. *Safety, strength, and stability.* Structural systems and member must be designed with sufficient margin of safety against failure.
2. *Aesthetics.* Aesthetics include such considerations as shape, geometrical proportions, symmetry, surface texture, and articulation. These are especially important for structures of high visibility such as signature buildings and bridges. The structural engineer must work in close coordination with planners, architects, other design professionals, and the affected community in guiding them on the structural and construction consequences of decisions derived from aesthetical considerations.
3. *Functional requirements.* A structure must always be designed to serve its intended function as specified by the project requirements. Constructability is a major part of the functional requirement. A structural design must be practical and economical to build.
4. *Economy.* Structures must be designed and built within the target budget of the project. For reinforced concrete structures, economical design is usually not achieved by minimizing the amount of concrete and reinforcement quantities. A large part of the construction cost are the costs of labor, formwork, and falsework. Therefore, designs that replicate member sizes and simplify reinforcement placement to result in easier and faster construction will usually result in being more economical than a design that achieves minimum material quantities.

7.6   Design Process

Reinforced concrete design is often an iterative trial-and-error process and involves the judgment of the designer. Every project is unique. The design process for reinforced concrete structures typically consists of the following steps:

1. Configure the structural system.
2. Determine design data: design loads, design criteria, and specifications. Specify material properties.
3. Make a first estimate of member sizes, for example, based on rule-of-thumb ratios for deflection control in addition to functional or aesthetic requirements.
4. Calculate member cross-sectional properties; perform structural analysis to obtain internal force demands: moment, axial force, shear force, and torsion. Review magnitudes of deflections.
5. Calculate the required longitudinal reinforcement based on moment and axial force demands. Calculate the required transverse reinforcement from the shear and torsional moment demands.

6. If members do not satisfy the SAFE criteria (see previous section), modify the design and make changes to steps 1 and 3.
7. Complete the detailed evaluation of member design to include additional load cases and combinations, and strength and serviceability requirements required by code and specifications.
8. Detail reinforcement. Develop design drawings, notes, and construction specifications.

7.7 Modeling of Reinforced Concrete for Structural Analysis

After a basic structural system is configured, member sizes selected, and loads determined, the structure is analyzed to obtain internal force demands. For simple structures, analysis by hand calculations or approximate methods would suffice (see Section 7.8); otherwise, structural analysis software may be used. For most reinforced concrete structures, a linear elastic analysis, assuming the gross moment of inertia of cross-sections and neglecting the steel reinforcement area, will provide results of sufficient accuracy for design purposes. The final design will generally be conservative even though the analysis does not reflect the actual nonlinear structural behavior because member design is based on ultimate strength design and the ductility of reinforced concrete enables force redistributions (see Sections 7.9 and 7.11). Refined modeling using nonlinear analysis is generally not necessary unless it is a special type of structure under severe loading situations like high seismic forces.

For structural modeling, the concrete modulus E_c given in Table 7.2 can be used for input. When the ends of beam and column members are cast together, the rigid end zone modeling option should be selected since its influence is often significant. Reinforced concrete floor systems should be modeled as rigid diaphragms by master slaving the nodes on a common floor. Tall walls or cores can be modeled as column elements. Squat walls should be modeled as plate or shear wall elements. If foundation conditions and soil conditions are exceptional, then the foundation system will need more refined modeling. Otherwise, the structural model can be assumed to be fixed to the ground. For large reinforced concrete systems or when geometrical control is important, the effects of creep and shrinkage and construction staging should be incorporated in the analysis.

If slender columns are present in the structure, a second-order analysis should be carried out that takes into account cracking by using reduced or effective cross-sectional properties (see Table 7.3 and Section 7.14). If a refined model and nonlinear analysis is called for, then the moment curvature analysis results will be needed for input into the computer analysis (see Section 7.10).

7.8 Approximate Analysis of Continuous Beams and One-Way Slabs

Under typical conditions, for continuous beams and one-way slabs with more than two spans the approximate moment and shear values given in Figure 7.2 may be used in lieu of more accurate analysis methods. These values are from ACI 8.3.3.

TABLE 7.3 Suggested Effective Member Properties for Analysis

Member	Effective moment of inertia for analysis
Beam	$0.35 I_g$
Column	$0.70 I_g$
Wall — uncracked	$0.70 I_g$
Wall — cracked	$0.35 I_g$
Flat plates and flat slabs	$0.25 I_g$

Note: I_g is the gross uncracked moment of inertia. Use gross areas for input of cross-sectional areas.

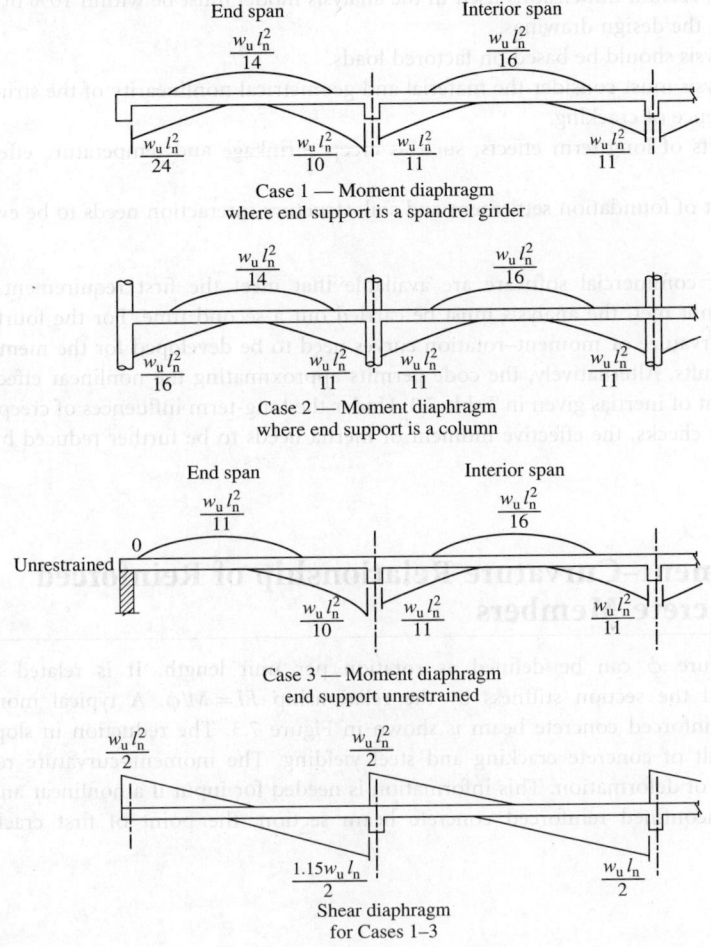

FIGURE 7.2 Approximate moment and shear of continuous beams or one-way slabs (ACI 8.3.3).

7.9 Moment Redistribution

The moment values of a continuous beam obtained from structural analysis may be adjusted or redistributed according to guidelines set by ACI 8.4. Negative moment can be adjusted down or up, but not more than $1000\varepsilon_t$ or 20% (see Notation section for ε_t). After the negative moments are adjusted in a span, the positive moment must also be adjusted to maintain the statical equilibrium of the span (see Section 7.13.12). Redistribution of moment is permitted to account for the ductile behavior of reinforcement concrete members.

7.10 Second-Order Analysis Guidelines

When a refined second-order analysis becomes necessary, as in the case where columns are slender, ACI 10.10.1 places a number of requirements on the analysis.

1. The analysis software should have been validated with test results of indeterminate structures and the predicted ultimate load within 15% of the test results.

2. The cross-section dimensions used in the analysis model must be within 10% of the dimensions shown in the design drawings.
3. The analysis should be based on factored loads.
4. The analysis must consider the material and geometrical nonlinearity of the structure, as well as the influence of cracking.
5. The effects of long-term effects, such as creep shrinkage and temperature effects, need to be assessed.
6. The effect of foundation settlement and soil–structure interaction needs to be evaluated.

A number of commercial software are available that meet the first requirement. If the second requirement is not met, the analysis must be carried out a second time. For the fourth requirement, the moment–curvature or moment–rotation curves need to be developed for the members to provide the accurate results. Alternatively, the code permits approximating the nonlinear effects by using the effective moment of inertias given in Table 7.3. Under the long-term influences of creep and shrinkage, and for stability checks, the effective moment of inertia needs to be further reduced by dividing it by $(1 + \beta_d)$.

7.11 Moment–Curvature Relationship of Reinforced Concrete Members

Member curvature ϕ can be defined as rotation per unit length. It is related to the applied moment M and the section stiffness by the relationship $EI = M/\phi$. A typical moment–curvature diagram of a reinforced concrete beam is shown in Figure 7.3. The reduction in slope of the curve (EI) is the result of concrete cracking and steel yielding. The moment–curvature relationship is a basic parameter of deformation. This information is needed for input if a nonlinear analysis is carried out. For an unconfined reinforced concrete beam section, the point of first cracking is usually

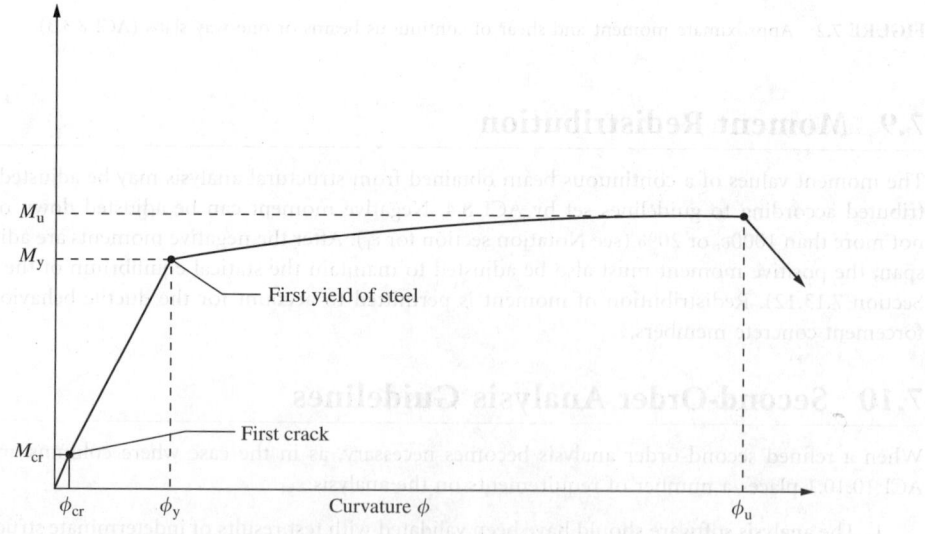

FIGURE 7.3 Typical moment–curvature diagram of a reinforced concrete beam.

neglected for input; the curvature points of first yield ϕ_y and ultimate ϕ_u are calculated from the following formulas:

$$\phi_y = \frac{f_y/E_s}{d(1-k)} \tag{7.1}$$

where

$$k = \left[(\rho + \rho')^2 n^2 + 2 \left(\rho + \frac{\rho'd'}{d} \right) n \right]^{1/2} - (\rho + \rho')n \tag{7.2}$$

At ultimate

$$\phi_u = \frac{0.85\beta_1 E_s f_c'}{f_y^2(\rho - \rho')} \varepsilon_c \left\{ 1 + (\rho + \rho')n - \left[(\rho + \rho')^2 n^2 + 2 \left(\rho + \frac{\rho'd'}{d} \right) n \right]^{1/2} \right\} \tag{7.3}$$

The concrete strain at ultimate ε_c is usually assumed to be a value between 0.003 and 0.004 for unconfined concrete. Software is available to obtain more refined moment–curvature relationships and to include other variables. If the concrete is considered confined, then an enhanced concrete stress–strain relationship may be adopted. For column members, the strain compatibility analysis must consider the axial load.

7.12 Member Design for Strength

7.12.1 Ultimate Strength Design

The main requirement of structural design is for the structural capacity, S_C, to be equal to or greater than the structural demand, S_D:

$$S_C \geq S_D$$

Modification factors are included in each side of the equation. The structural capacity S_C is equal to the nominal strength F_n multiplied by a capacity reduction safety factor ϕ:

$$S_C = \phi F_n$$

The nominal strength F_n is the internal ultimate strength at that section of the member. It is usually calculated by the designer according to formulas derived from the theory of mechanics and strength of materials. These strength formulas have been verified and calibrated with experimental testing. They are generally expressed as a function of the cross-section geometry and specified material strengths. There are four types of internal strengths: nominal moment M_n, shear V_n, axial P_n, and torsional moment T_n.

The capacity reduction safety factor ϕ accounts for uncertainties in the theoretical formulas, empirical data, and construction tolerances. The ϕ factor values specified by ACI are listed in Table 7.4.

The structural demand, S_D, is the internal force (moment, shear, axial, or torsion) at the section of the member resulting from the loads on the structure. The structural demand is usually obtained by carrying out a structural analysis of the structure using hand, approximate methods, or computer software. Loads to be input are specified by the design codes and the project specifications and normally include dead, live, wind, and earthquake loads. Design codes such and ACI, IBC, and AASHTO also specify the values of safety factors that should be multiplied with the specified loads and how different types of loads should be combined (i.e., $S_D = 1.2\text{Dead} + 1.6\text{Live}$). ACI load factors and combinations are listed in Table 7.5.

Combining the two equations above, a direct relationship between the nominal strength F_n and the structural demand S_D can be obtained

$$F_n \geq S_D/\phi \tag{7.4}$$

This relationship is convenient because the main design variables, such as reinforcement area, which are usually expressed in terms F_n, can be related directly to the results of the structural analysis.

TABLE 7.4   ACI Strength Reduction Factors ϕ

Nominal strength condition	Strength reduction factor ϕ
Flexure (tension-controlled)	0.90
Compression-controlled (columns)	
Spiral transverse reinforcement	0.70[a]
Other transverse reinforcement	0.65[a]
Shear and torsion	0.75
Bearing on concrete	0.65
Structural plain concrete	0.55

[a] ϕ is permitted to be linearly increased to 0.90 as the tensile strain in the extreme steel increases from the compression-controlled strain of 0.005.

Note: Under seismic conditions strength reduction factors may require modifications.

TABLE 7.5   ACI Load Factors

Load case	Structurals demand S_D or (required strength U)
1	$1.4(D + F)$
2	$1.2(D + F + T) + 1.6(L + H) + 0.5(L_r \text{ or } S \text{ or } R)$
3	$1.2D + 1.6(L_r \text{ or } S \text{ or } R) + (1.0L \text{ or } 0.8W)$
4	$1.2D + 1.6W + 1.0L + 0.5(L_r \text{ or } S \text{ or } R)$
5	$1.2D + 1.0E + 1.0L + 0.2S$
6	$0.9D + 1.6W + 1.6H$
7	$0.9D + 1.0E + 1.6H$

Note: D is the dead load, or related internal moments and forces, *E* is the seismic load, *F* is the weight and pressure of well-defined fluids, *H* is the weight and pressure of soils, water in soil, or other materials, *L* is the live load, L_r is the roof live load, *R* is the rain load, *S* is the snow load, *T* is the time-dependent load (temperature, creep, shrinkage, differential settlement, etc.), and *W* is the wind load.

7.12.2 Beam Design

The main design steps for beam design and the formulas for determining beam capacity are outlined in the following.

7.12.2.1 Estimate Beam Size and Cover

Table 7.6 may be referenced for selecting a beam thickness. For practical construction, the minimum width of a beam is about 12 in. Economical designs are generally provided when the beam width to thickness ratio falls in the range of $\frac{1}{2}$ to 1. Minimum concrete covers are listed in Table 7.7 and typically should not be less than 1.5 in.

7.12.2.2 Moment Capacity

Taking a beam segment, flexural bending induces a force couple (see Figure 7.4). Internal tension N_T is carried by the reinforcement (the tensile strength of concrete is low and its tension carrying capacity is neglected). Reinforcement at the ultimate state is required to yield, hence

$$N_T = A_s f_y \tag{7.5}$$

At the opposite side of the beam, internal compression force N_C is carried by the concrete. Assuming a simplified rectangular stress block for concrete (uniform stress of $0.85f_c'$),

$$N_C = 0.85f_c' ab \tag{7.6}$$

To satisfy equilibrium, internal tension must be equal to internal compression, $N_C = N_T$. Hence, the depth of the rectangular concrete stress block a can be expressed as

$$a = \frac{A_s f_y}{0.85f_c' b} \tag{7.7}$$

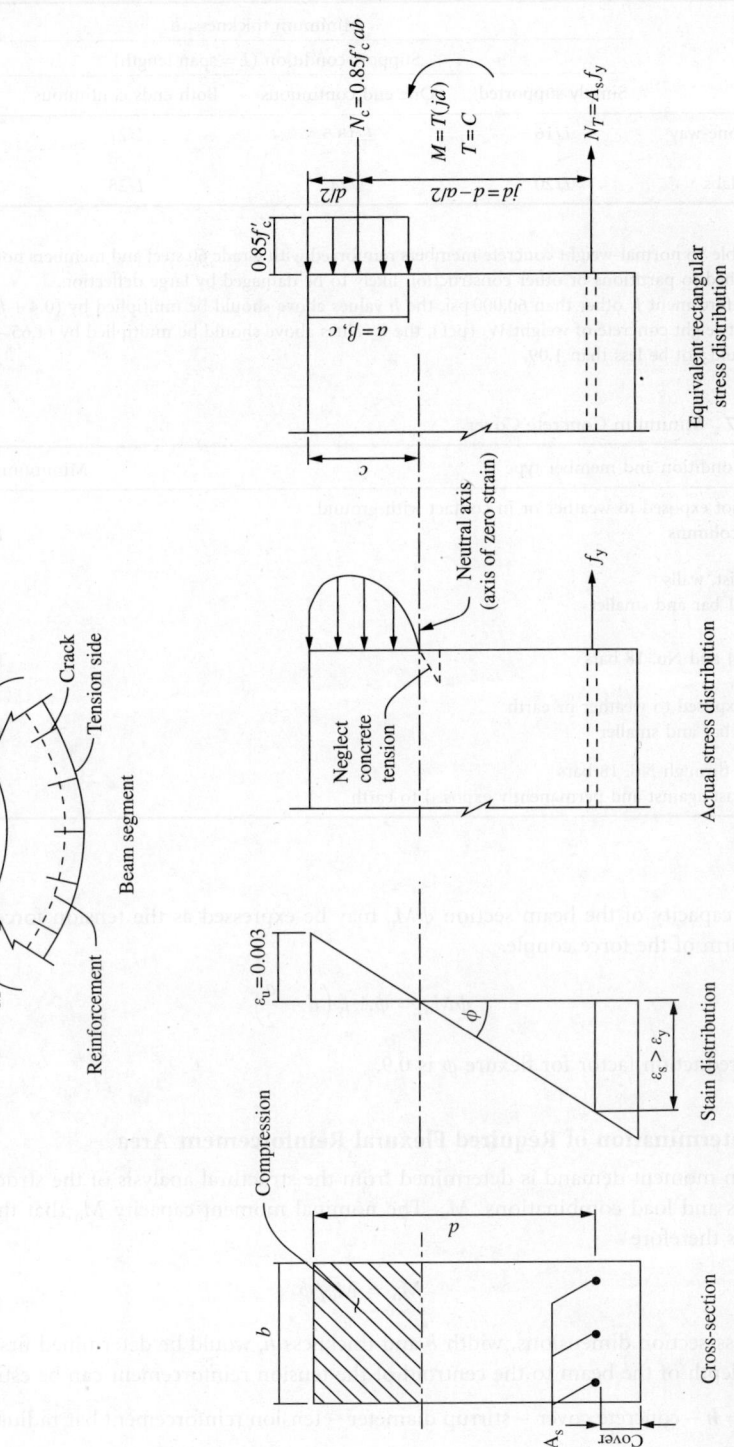

FIGURE 7.4 Mechanics of reinforced concrete beam under flexure.

TABLE 7.6 Minimum Depth of Beams

| | Minimum thickness, h | | | |
| | Support condition (L = span length) | | | |
Member	Simply supported	One end continuous	Both ends continuous	Cantilever
Beams or one-way joists	$L/16$	$L/18.5$	$L/21$	$L/8$
One-way slabs	$L/20$	$L/24$	$L/28$	$L/10$

Notes:
1. Applicable to normal-weight concrete members reinforced with Grade 60 steel and members not supported or attached to partitions or other construction likely to be damaged by large deflection.
2. For reinforcement f_y other than 60,000 psi, the h values above should be multiplied by $(0.4 + f_y/100,000)$.
3. For lightweight concrete of weight W_c (pcf), the h values above should be multiplied by $(1.65 - 0.005 W_c)$, but should not be less than 1.09.

TABLE 7.7 Minimum Concrete Cover

Exposure condition and member type	Minimum cover (in.)
Concrete not exposed to weather or in contact with ground	
Beams, columns	$1\frac{1}{2}$
Slabs, joist, walls	
No. 11 bar and smaller	$\frac{3}{4}$
No. 14 and No. 18 bars	$1\frac{1}{2}$
Concrete exposed to weather or earth	
No. 5 bar and smaller	$1\frac{1}{2}$
No. 6 through No. 18 bars	2
Concrete cast against and permanently exposed to earth	3

The moment capacity of the beam section ϕM_n may be expressed as the tension force multiplied by the moment arm of the force couple.

$$\phi M_n = \phi A_s f_y \left(d - \frac{a}{2} \right) \tag{7.8}$$

The strength reduction factor for flexure ϕ is 0.9.

7.12.2.3 Determination of Required Flexural Reinforcement Area

The maximum moment demand is determined from the structural analysis of the structure under the specified loads and load combinations, M_u. The nominal moment capacity M_n that the cross-section must supply is therefore

$$M_n = M_u/\phi \tag{7.9}$$

The beam cross-section dimensions, width b and thickness h, would be determined first or a first trial selected; the depth of the beam to the centroid of the tension reinforcement can be estimated by

$$d = h - \text{concrete cover} - \text{stirrup diameter} - \text{tension reinforcement bar radius} \tag{7.10}$$

A reasonable size of the stirrup and reinforcement bar can be assumed, if not known (a No. 4 or No. 5 bar size for stirrups is reasonable).

· Rearranging the moment capacity equations presented in the previous section, the required flexural reinforcement is obtained by solving for A_s

$$A_s = \frac{M_n}{f_y \left(d - \frac{1}{2}\left(A_s f_y / 0.85 f_c' b\right)\right)} \qquad (7.11)$$

The required tension reinforcement area A_s is obtained from the quadratic expression

$$A_s = \frac{f_y d \pm \sqrt{\left(f_y d\right)^2 - 4 M_n K_m}}{2 K_m} \qquad (7.12)$$

where K_m is a material constant:

$$K_m = \frac{f_y^2}{1.7 f_c' b} \qquad (7.13)$$

Then, the sizes and quantity of bars are selected. Minimum requirements for reinforcement area and spacing must be satisfied (see the next two sections).

7.12.2.4 Limits on Flexural Reinforcement Area

1. *Minimum reinforcement area for beams:*

$$A_{s,min} = \frac{3\sqrt{f_c'}}{f_y} b_w d \geq 200 b_w d / f_y \qquad (7.14)$$

2. *Maximum reinforcement for beams:* The maximum reinforcement A_s must satisfy the requirement that the net tensile strain ε_t (extreme fiber strain less effects of creep, shrinkage, and temperature) is not less than 0.004. The net tensile strain is solved from the compatibility of strain (see Figure 7.4).

$$\varepsilon_t = 0.003 \frac{d - c}{c} \qquad (7.15)$$

The neutral axis location c is related to the depth of the compression stress block a by the relationship (ACI 10.2.7.3)

$$c = a / \beta_1 \qquad (7.16)$$

The factor β_1 is dependent on the concrete strength as shown in Figure 7.5.

7.12.2.5 Detailing of Longitudinal Reinforcement

Clear spacing between parallel bars should be large enough to permit the coarse aggregate to pass through to avoid honeycombing. The minimum clear spacing should be d_b, but it should not be less than 1 in.
    For crack control, center-to-center spacing of bars should not exceed

$$\frac{540}{f_s} - 2.5 c_c \leq \frac{432}{f_s} \qquad (7.17)$$

where f_s (in ksi) is the stress in the reinforcement at service load, which may be assumed to be 60% of the specified yields strength. Typically, the maximum spacing between bars is about 10 in. The maximum bar spacing rule ensures that crack widths fall below approximately 0.016 in. For very aggressive exposure environments, additional measures should be considered to guard against corrosion, such as reduced concrete permeability, increased cover, or application of sealants.
    If the depth of the beam is large, greater than 36 in., additional reinforcement should be placed at the side faces of the tension zone to control cracking. The amount of skin reinforcement to add need not exceed one half of the flexural tensile reinforcement and it should be spread out for a distance $d/2$. The spacing of the skin reinforcement need not exceed $d/6$, 12 in., and $1000 A_b / (d - 30)$.

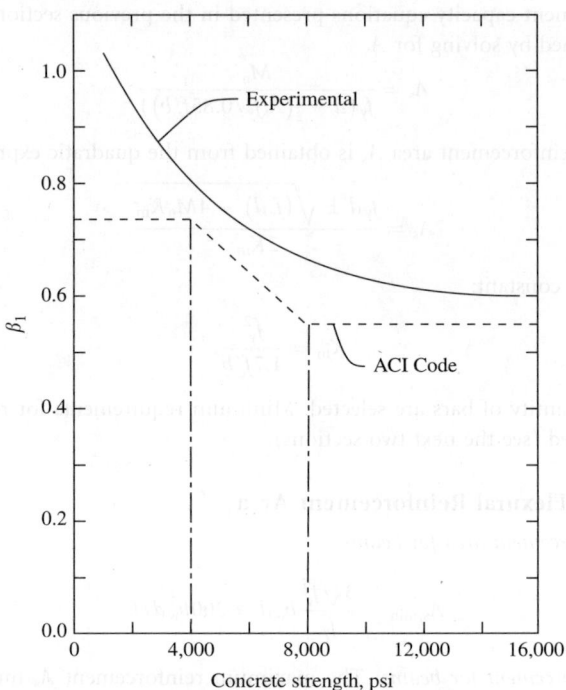

FIGURE 7.5   Relation between β_1 and concrete strength.

To ease reinforcement cage fabrication, a minimum of two top and two bottom bars should run continuously through the span of the beam. These bars hold up the transverse reinforcement (stirrups). At least one fourth of all bottom (positive) reinforcement should run continuously. If moment reversal is expected at the beam–column connection, that is, stress reversal from compression to tension, bottom bars must be adequately anchored into the column support to develop the yield strength.

The remaining top and bottom bars may be cut short. However, it is generally undesirable to cut bars within the tension zone (it causes loss of shear strength and ductility). It is good practice to run bars well into the compression zone, at least a distance d, $12d_b$ or $l_n/16$ beyond the point of inflection (PI) (see Figure 7.6). Cut bars must also be at least one development length l_d in length measured from each side of their critical sections, which are typically the point of peak moment where the yield strength must be developed. See Section 7.17 for development lengths.

To achieve structural integrity of the structural system, beams located at the perimeter of the structure should have minimum continuous reinforcement that ties the structure together to enhance stability, redundancy, and ductile behavior. Around the perimeter at least one sixth of the top (negative) longitudinal reinforcement at the support and one quarter of the bottom (positive) reinforcement should be made continuous and tied with closed stirrups (or open stirrups with minimum 135° hooks). Class A splices may be used to achieve continuity. Top bars should be spliced at the midspan, bottom bars at or near the support.

7.12.2.6 Beams with Compression Reinforcement

Reinforcement on the compression side of the cross-section (see Figure 7.4) usually does not increase in flexural capacity significantly, typically less than 5%, and for most design purposes its contribution to

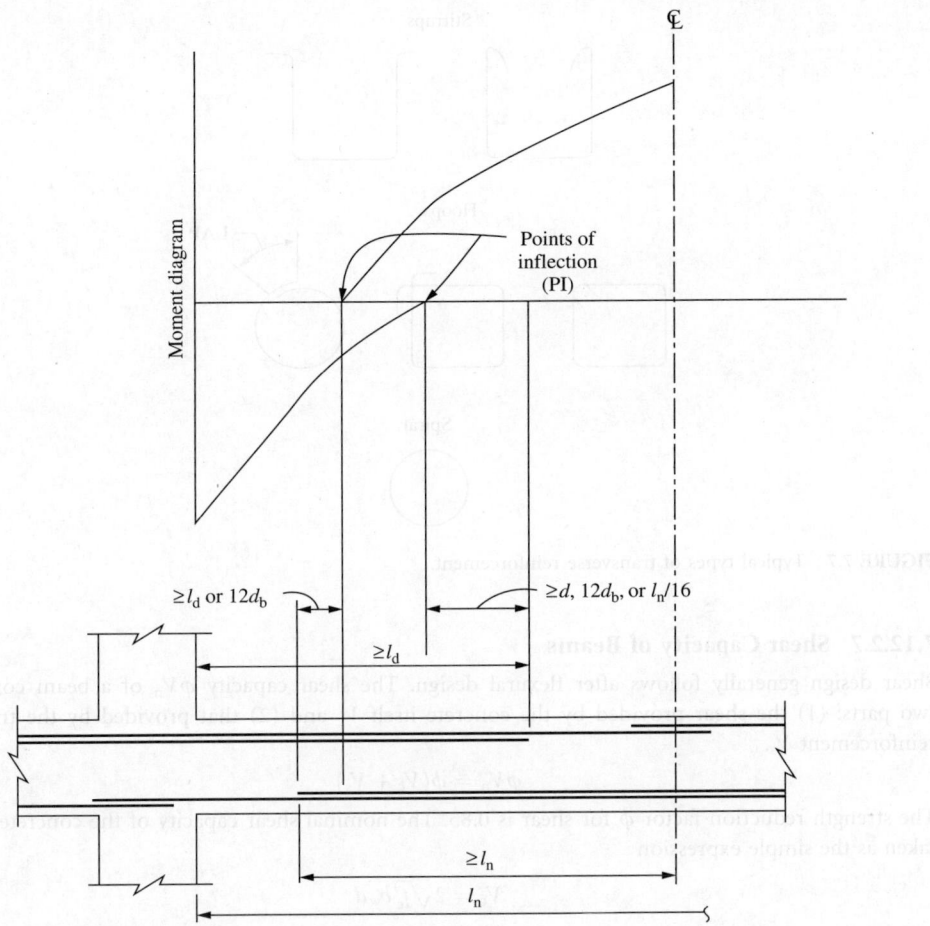

FIGURE 7.6 Typical reinforcement cutoffs for continuous beam.

strength can be neglected. The moment capacity equation considering the compression reinforcement area A_s' located at a distance d' from the compression fiber is

$$\phi M_n = A_s' f_y (d - d') + (A_s - A_s') f_y \left(d - \frac{a}{2} \right) \tag{7.18}$$

where

$$a = \frac{[A_s - A_s'(1 - 0.85 f_c'/f_y)] f_y}{0.85 f_c' b} \tag{7.19}$$

The above expressions assume the compression steel yield, which is typically the case (compression steel quantity is not high). For the nonyielding case, the stress in the steel needs to be determined by a stress–strain compatibility analysis.

Despite its small influence on strength, compression reinforcement serves a number of useful serviceability functions. It is needed for supporting the transverse shear reinforcement in the fabrication of the steel cage. It helps to reduce deflections and long-term creep, and it enhances ductile performance.

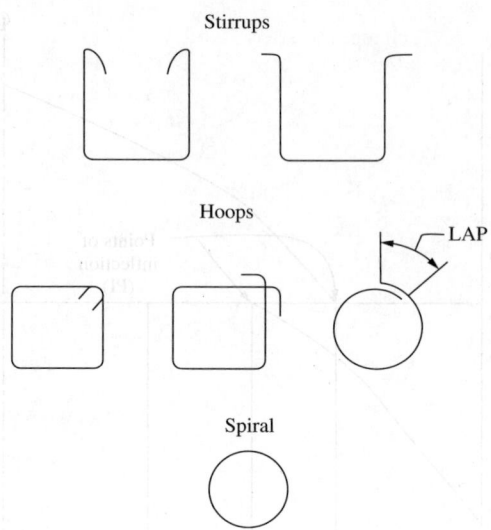

FIGURE 7.7   Typical types of transverse reinforcement.

7.12.2.7 Shear Capacity of Beams

Shear design generally follows after flexural design. The shear capacity ϕV_n of a beam consists of two parts: (1) the shear provided by the concrete itself V_c and (2) that provided by the transverse reinforcement V_s.

$$\phi V_n = \phi(V_c + V_s) \tag{7.20}$$

The strength reduction factor ϕ for shear is 0.85. The nominal shear capacity of the concrete may be taken as the simple expression

$$V_c = 2\sqrt{f_c'}\,b_w d \tag{7.21a}$$

which is in pound and inch units. An alternative empirical formula that allows a higher concrete shear capacity is

$$V_c = \left(1.9\sqrt{f_c'} + 2500\rho_w\frac{V_u d}{M_u}\right)b_w d \le 3.5\sqrt{f_c'}\,b_w d \tag{7.21b}$$

where M_u is the factored moment occurring simultaneously with V_u at the beam section being checked. The quantity $V_u d/M_u$ should not be taken greater than 1.0.

Transverse shear reinforcements are generally of the following types (see Figure 7.7): stirrups, closed hoops, spirals, or circular ties. In addition, welded wire fabric, inclined stirrups, or longitudinal bars bent at an angle may be used. For shear reinforcement aligned perpendicular to the longitudinal reinforcement, the shear capacity provided by transverse reinforcement is

$$V_s = \frac{A_v f_y d}{s} \le 8\sqrt{f_c'}\,b_w d \tag{7.22a}$$

When spirals or circular ties or hoops are used with this formula, d should be taken as 0.8 times the diameter of the concrete cross-section, and A_v should be taken as two times the bar area.

When transverse reinforcement is inclined at an angle α with respect to the longitudinal axis of the beam, the transverse reinforcement shear capacity becomes

$$V_s = \frac{A_v f_y (\sin\alpha_i + \cos\alpha_i) d}{s} \le 8\sqrt{f_c'}\,b_w d \tag{7.22b}$$

The shear formulas presented above were derived empirically, and their validity has also been tested by many years of design practice. A more rational design approach for shear is the strut-and-tie model, which is given as an alternative design method in ACI Appendix A. Shear designs following the strut-and-tie approach, however, often result in designs requiring more transverse reinforcement steel since the shear transfer ability of concrete is neglected.

7.12.2.8 Determination of Required Shear Reinforcement Quantities

The shear capacity must be greater than the shear demand V_u, which is based on the structural analysis results under the specified loads and governing load combination

$$\phi V_n \geq V_u \tag{7.23}$$

Since the beam cross-section dimensions b_w and d would usually have been selected by flexural design beforehand or governed by functional or architectural requirements, the shear capacity provided by the concrete V_c can be calculated by Equations 7.21a or 7.21b. From the above equations, the required shear capacity to be provided by shear reinforcement must satisfy the following:

$$V_s \geq \frac{V_u}{\phi} - V_c \tag{7.24}$$

Inserting V_s from this equation into Equation 7.22a, the required spacing and bar area of the shear reinforcement (aligned perpendicular to the longitudinal reinforcement) must satisfy the following:

$$\frac{s}{A_v} \leq \frac{f_y d}{V_s} \tag{7.25}$$

For ease of fabrication and bending, a bar size in the range of No. 4 to No. 6 is selected, then the required spacing s along the length of the beam is determined, usually rounded down to the nearest $\frac{1}{2}$ in.

In theory, the above shear design procedure can be carried out at every section along the beam. In practice, a conservative approach is taken and shear design is carried out at only one or two locations of maximum shear, typically at the ends of the beam, and the same reinforcement spacing s is adopted for the rest of the beam. Where the beam ends are cast integrally or supported by a column, beam, wall, or support element that introduces a region of concentrated compression, the maximum value of the shear demand need not be taken at the face of the support, but at a distance d away (see Figure 7.8).

Transverse reinforcement in the form of closed stirrups is preferred for better ductile performance and structural integrity. For beams located at the perimeter of the structure, ACI requires closed stirrups (or open stirrups within minimum 135° hooks). In interior beams, if closed stirrups are not provided, at least one quarter of the bottom (positive) longitudinal reinforcement at midspan should be made continuous over the support, or at the end support, detailed with a standard hook.

7.12.2.8.1 Minimum Shear Reinforcement and Spacing Limits

After the shear reinforcement and spacing are selected they should be checked against minimum requirements. The minimum shear reinforcement required is

$$A_{vmin} = 0.75 \sqrt{f_c'} \frac{b_w s}{f_y} \geq \frac{50 b_w s}{f_y} \tag{7.26}$$

This minimum shear area applies in the beam where $V_u \geq \phi V_2/2$. It does not apply to slabs, footings, and concrete joists. The transverse reinforcement spacing s should not exceed $d/2$ nor 24 in. These spacing limits become $d/4$ and 12 in. when V_s exceeds $4\sqrt{f_c'}b_w d$.

When significant torsion exists, additional shear reinforcement may be needed to resist torsion. This is covered in Section 7.16.

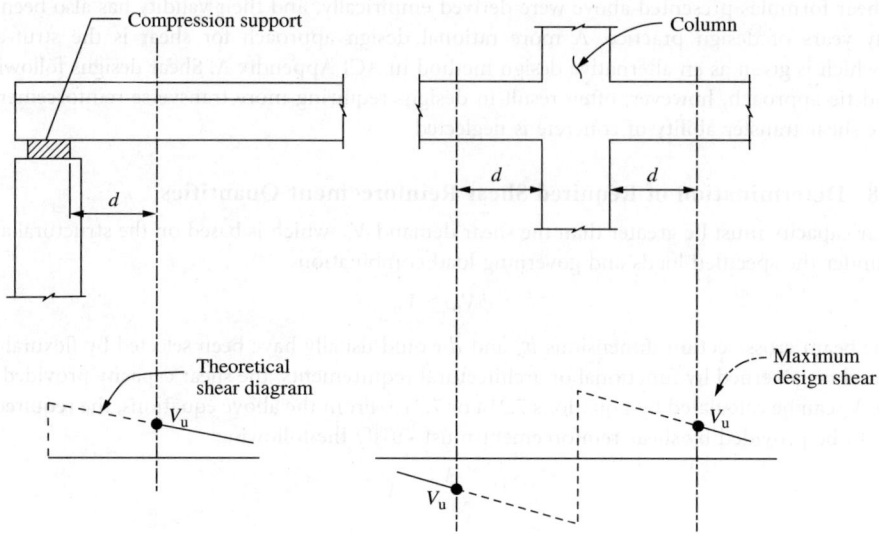

FIGURE 7.8   Typical support conditions for locating factored shear force V_u.

7.12.2.8.2 Modifications for High-Strength and Lightweight Concretes

For concretes with compressive strengths greater than 10,000 psi, the values of $\sqrt{f_c'}$ in all the shear capacity and design equations above should not exceed 100 psi. For lightweight concretes, $\sqrt{f_c'}$ should be multiplied by 0.75 for all-lightweight concrete, or 0.85 for sand-lightweight concrete. If the tensile strength f_{ct} of the concrete is specified, $\sqrt{f_c'}$ may be substituted by $f_{ct}/6.7$, but should not be greater than $\sqrt{f_c'}$.

7.12.2.9   Detailing of Transverse Reinforcement

Transverse reinforcement should extend close to the compression face of a member, as far as cover allows, because at ultimate state deep cracks may cause loss of anchorage. Stirrup should be hooked around a longitudinal bar by a standard stirrup hoop (see Figure 7.9). It is preferable to use transverse reinforcement size No. 5 or smaller. It is more difficult to bend a No. 6 or larger bar tightly around a longitudinal bar. For transverse reinforcement sizes No. 6, No. 7, and No. 8, a standard stirrup hook must be accompanied by a minimum embedment length of $0.014 d_b f_y / \sqrt{f_c'}$ measured between the midheight of the member and the outside end of the hook.

7.12.3   One-Way Slab Design

When the load normal to the surface of a slab is transferred to the supports primarily in one major direction, the slab is referred to as a one-way slab. For a slab panel supported on all four edges, one-way action occurs when the aspect ratio, the ratio of its long-to-short span length, is greater than 2. Under one-way action, the moment diagram remains essentially constant across the width of the slab. Hence, the design procedure of a one-way slab can be approached by visualizing the slab as an assembly of the same beam strip of unit width. This beam strip can be designed using the same design steps and formulas presented in the previous section for regular rectangular beams.

The required cover for one-way slab is less than for beams, typically $\frac{3}{4}$ in. The internal forces in one-way slabs are usually lower, so smaller bar sizes are used. The design may be controlled by the minimum temperature and shrinkage reinforcement. Shear is rarely a controlling factor for one-way slab design. Transverse reinforcement is difficult to install in one-way slabs. It is more economical to thicken or haunch the slab.

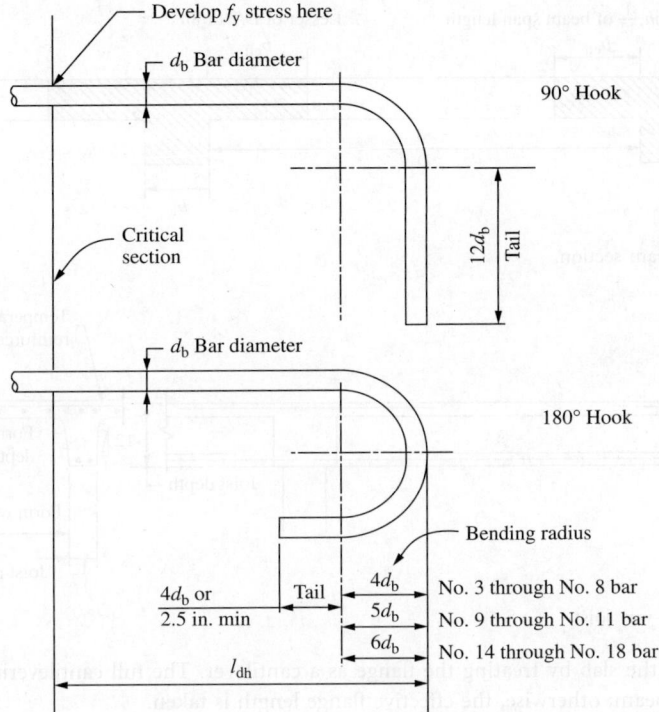

FIGURE 7.9 Standard hooked bar details.

7.12.3.1 Shrinkage and Temperature Reinforcement (ACI 7.12)

For Grade 60 reinforcement, the area of shrinkage and temperature reinforcement should be 0.0018 times the gross concrete area of the slab. Bars should not be spaced farther than five times the slab thickness or 18 in. The shrinkage and temperature requirements apply in both directions of the slab, and the reinforcement must be detailed with adequate development length where yielding is expected.

7.12.4 T-Beam Design

Where a slab is cast integrally with a beam, the combined cross-section acts compositely (see Figure 7.10). The design of T-beam differs from that of a rectangular beam only in the positive moment region, where part of the internal compression force occurs in the slab portion. The design procedures and formulas for T-beam design are the same as for rectangular beams, except for the substitution of b in the equations with an effective width b_{eff} at positive moment sections. The determination of b_{eff} is given in Figure 7.10. The effective width b_{eff} takes into account the participation of the slab in resisting compression. In the rare case where the depth of the compression stress block a exceeds the slab thickness, a general stress–strain compatibility analysis would be required. For shear design the cross-section width should be taken as the width of the web b_w.

7.12.4.1 Requirements for T-Beam Flanges

If the T-beam is an isolated beam and the flanges are used to provide additional compression area, the flange thickness should be not less than one half the width of the web and the effective flange width not more than four times the width of the web. For a slab that forms part of the T-beam flange and if the slab primary flexural reinforcement runs parallel to the T-beam, adequate transverse reinforcement needs

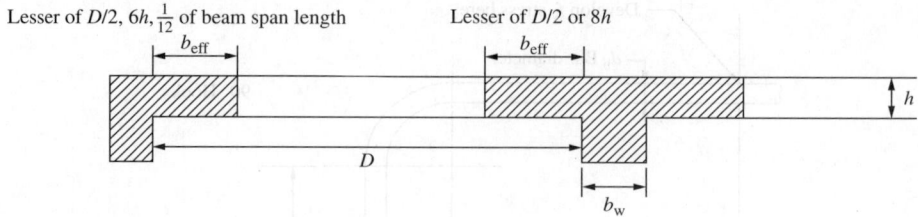

FIGURE 7.10 T-beam section.

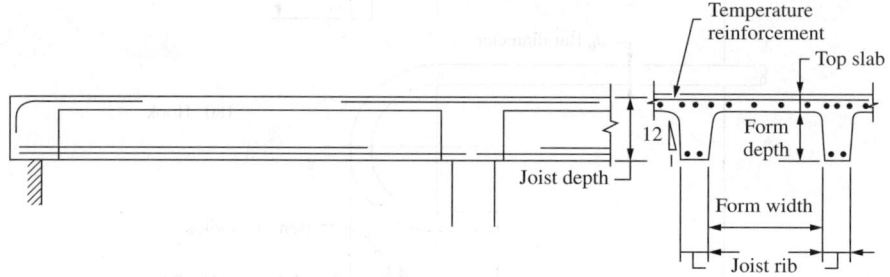

FIGURE 7.11 One-way joist.

to be provided in the slab by treating the flange as a cantilever. The full cantilevering length is taken for an isolated T-beam; otherwise, the effective flange length is taken.

7.12.5 One-Way Joist Design

A one-way joist floor system consists of a series of closely spaced T-beams (see Figure 7.11). The ribs of joists should not be less than 4 in. in width and should have a depth not more than 3.5 times the minimum width of the rib. Flexural reinforcement is determined by T-section design. The concrete ribs normally have sufficient shear capacity so that shear reinforcement is not necessary. A 10% increase is allowed in the concrete shear capacity calculation, V_c, if the clear spacing of the ribs does not exceed 30 in. Alternatively, higher shear capacity can be obtained by thickening the rib at the ends of the joist where the high shear demand occurs. If shear reinforcements are added, they are normally in the form of single-leg stirrups. The concrete forms or fillers that form the joists may be left in place; their vertical stems can be considered part of the permanent joist design if their compressive strength is at least equal to the joist. The slab thickness over the permanent forms should not be less than $\frac{1}{12}$ of the clear distance between ribs or less than 1.5 in. Minimum shrinkage and temperature reinforcement need to be provided in the slab over the joist stems. For structural integrity, at least one bottom bar in the joist should be continuous or spliced with a Class A tension splice (see Section 7.17) over continuous supports. At discontinuous end supports, bars should be terminated with a standard hook.

7.13 Two-Way Floor Systems

Design assuming one-way action is not applicable in many cases, such as when a floor panel is bounded by beams with a long to short aspect ratio of less than 2. Loads on the floor are distributed in both directions, and such a system is referred to as a two-way system (see Figure 7.12). The design approach of two-way floor systems remains in many ways similar to that of the one-way slab, except that the floor slab should now be visualized as being divided into a series of slab strips spanning *both* directions of the floor panel (see Figure 7.12). In the case of one-way slabs, each slab strip carries the same design moment diagram. In two-way systems, the design moment diaphragm varies from one strip to another. Slab strips

(a)

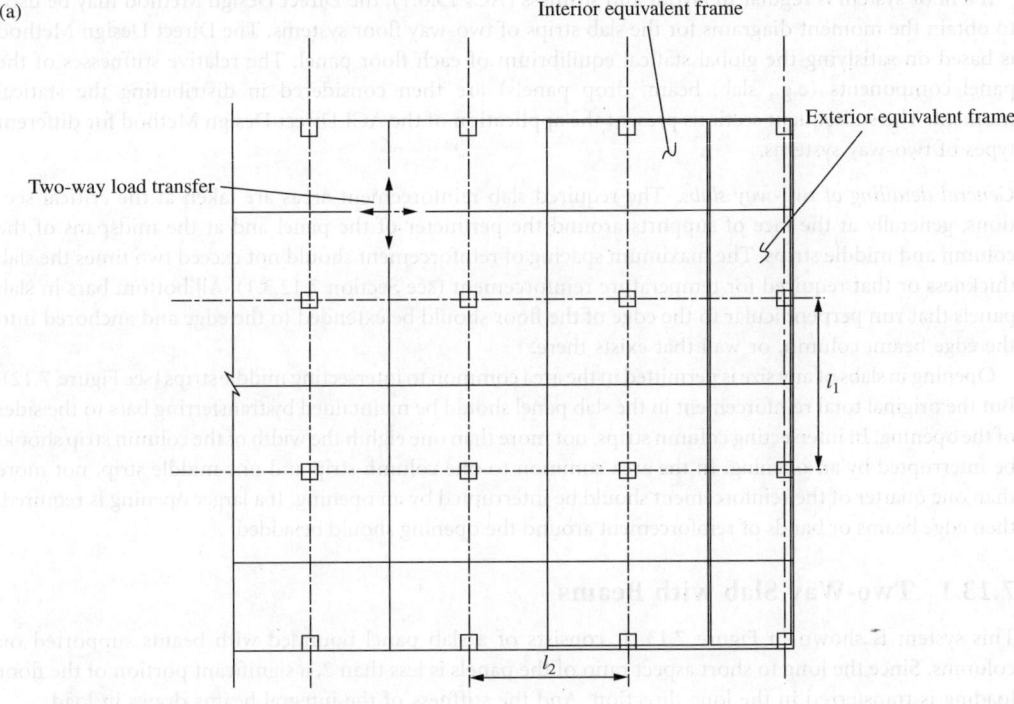

Interior equivalent frame

Exterior equivalent frame

Two-way load transfer

l_1

l_2

(b)

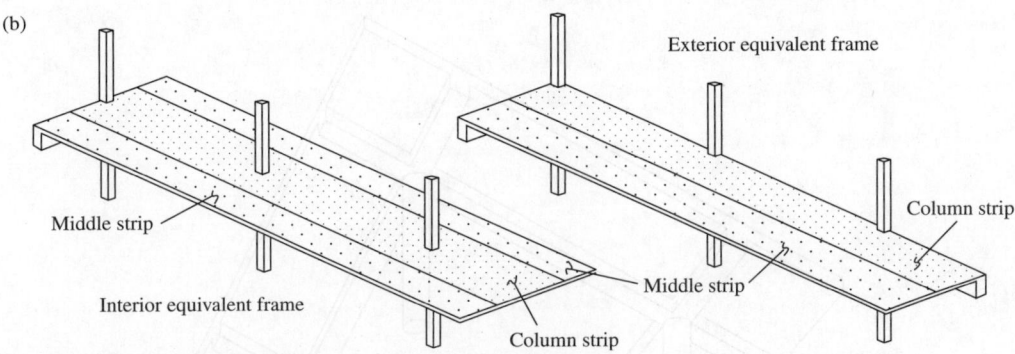

Exterior equivalent frame

Middle strip

Column strip

Interior equivalent frame

Middle strip

Column strip

FIGURE 7.12   (a) Two-way floor system and (b) equivalent frames.

closer to the column support lines would generally carry a higher moment than strips at midspan. Hence, a key design issue for two-way floor design becomes one of analysis, on how to obtain an accurate estimate of internal force distribution among the slab strips. After this issue is resolved, and the moment diagrams of each strip are obtained, the flexural reinforcement design of each slab strip follows the same procedures and formulas as previously presented for one-way slabs and beams. Of course, the analysis of two-way floor systems can also be solved by computer software, using the finite element method, and a number of structural analysis software have customized floor slab analysis modules. The ACI Code contains an approximate manual analysis method, the Direct Design Method, for two-way floors, which is practical for design purposes. A more refined approximate method, the Equivalent Frame Method, is also available in the ACI.

If a floor system is regular in layout and stiffness (ACI 13.6.1), the Direct Design Method may be used to obtain the moment diagrams for the slab strips of two-way floor systems. The Direct Design Method is based on satisfying the global statical equilibrium of each floor panel. The relative stiffnesses of the panel components (e.g., slab, beam, drop panels) are then considered in distributing the statical moment. The subsequent sections present the application of the ACI Direct Design Method for different types of two-way systems.

General detailing of two-way slabs. The required slab reinforcement areas are taken at the critical sections, generally at the face of supports around the perimeter of the panel and at the midspans of the column and middle strips. The maximum spacing of reinforcement should not exceed two times the slab thickness or that required for temperature reinforcement (see Section 7.12.3.1). All bottom bars in slab panels that run perpendicular to the edge of the floor should be extended to the edge and anchored into the edge beam, column, or wall that exists there.

Opening in slabs of any size is permitted in the area common to intersecting middle strips (see Figure 7.12). But the original total reinforcement in the slab panel should be maintained by transferring bars to the sides of the opening. In intersecting column strips, not more than one eighth the width of the column strip should be interrupted by an opening. In the area common to one column strip and one middle strip, not more than one quarter of the reinforcement should be interrupted by an opening. If a larger opening is required, then edge beams or bands of reinforcement around the opening should be added.

7.13.1 Two-Way Slab with Beams

This system is shown in Figure 7.13. It consists of a slab panel bounded with beams supported on columns. Since the long to short aspect ratio of the panels is less than 2, a significant portion of the floor loading is transferred in the long direction. And the stiffness of the integral beams draws in load.

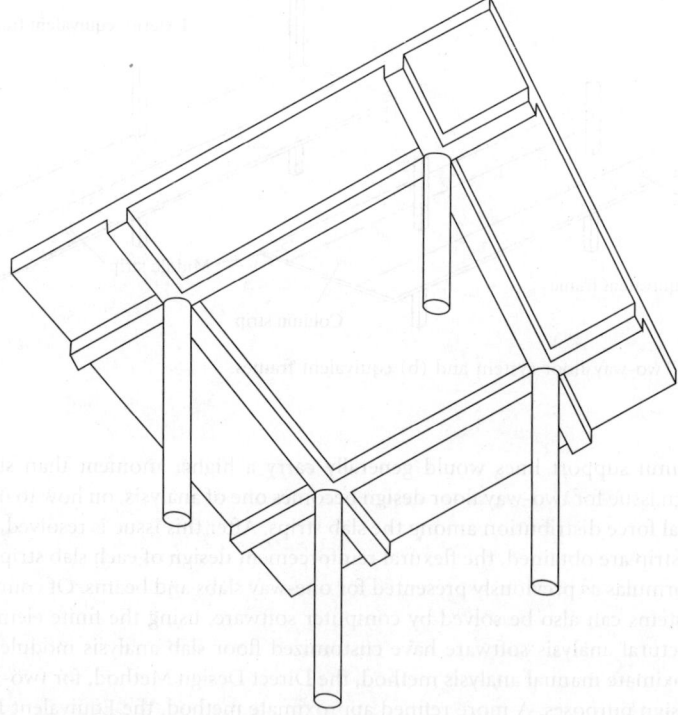

FIGURE 7.13 Two-way slab with beams.

TABLE 7.8 Minimum Thickness of Flat Plates (Two-Way Slabs without Interior Beams)

Yield strength, f_y (psi)	Without drop panels			With drop panels		
	Exterior panels		Interior panels	Exterior panels		Interior panels
	Without edge beams	With edge beams		Without edge beams	With edge beams	
40,000	$l_n/33$	$l_n/36$	$l_n/36$	$l_n/36$	$l_n/40$	$l_n/40$
60,000	$l_n/30$	$l_n/33$	$l_n/33$	$l_n/33$	$l_n/36$	$l_n/36$
75,000	$l_n/28$	$l_n/31$	$l_n/31$	$l_n/31$	$l_n/34$	$l_n/34$

Notes:
1. l_n is length of clear in long direction, face-to-face of support.
2. Minimum thickness if slabs without drop panels should not be less than 5 in.
3. Minimum thickness of slabs with drop panel should not be less than 4 in.

The minimum thickness of two-way slabs is dependent on the relative stiffness of the beams α_m. If $0.2 \leq \alpha_m \leq 2.0$, the slab thickness should not be less than 5 in. or

$$\frac{l_n \left(0.8 + (f_y/200,000)\right)}{35 + 5\beta(\alpha_m - 0.2)} \tag{7.27}$$

If $\alpha_m > 2.0$, the denominator in the above equation should be replaced with $(36 + 9\beta)$, but the thickness should not be less than 3.5 in. When $\alpha_m < 0.2$, the minimum thickness is given by Table 7.8.

7.13.1.1 Column Strips, Middle Strips, and Equivalent Frames

For the Direct Design Method, to take into account the change of the moment across the panel, the floor system is divided into column and middle strips in each direction. The column strip has a width on each side of a column centerline equal to $0.25l_2$ or $0.25l_1$, whichever is less (see Figure 7.12). A middle strip is bounded by two column strips. The moment diagram across each strip is assumed to be constant and the reinforcement is designed for each strip accordingly.

In the next step of the Direct Design Method, equivalent frames are set up. Each equivalent frame consists of the columns and beams that share a common column or grid line. Beams are attached to the slabs that extend to the half-panel division on each side of the grid line, so the width of each equivalent frame consists of one column strip and two half middle strips (see Figure 7.12). Equivalent frames are set up for all the grid lines in both directions of the floor system.

7.13.1.2 Total Factored Static Moment

The first analysis step of the Direct Design Method is determining the total static moment in each span of the equivalent frame

$$M_0 = \frac{w_u l_2 l_n^2}{8} \tag{7.28}$$

Note that w_u is the full, not half, factored floor load per unit area. The clear span l_n is measured from face of column to face of column. The static moment is the absolute sum of the positive midspan moment plus the average negative moment in each span (see Figure 7.14).

The next steps of the Direct Design Method involve procedures for distributing the static moment M_0 into the positive (midspan) and negative moment (end span) regions, and then on to the column and middle strips. The distribution procedures are approximate and reflect the relative stiffnesses of the frame components (Table 7.9).

7.13.1.3 Distribution of Static Moment to Positive and Negative Moment Regions

The assignment of the total factored static moment M_0 to the negative and positive moment regions is given in Figure 7.14. For interior spans, $0.65M_0$ is assigned to each negative moment region and $0.35M_0$

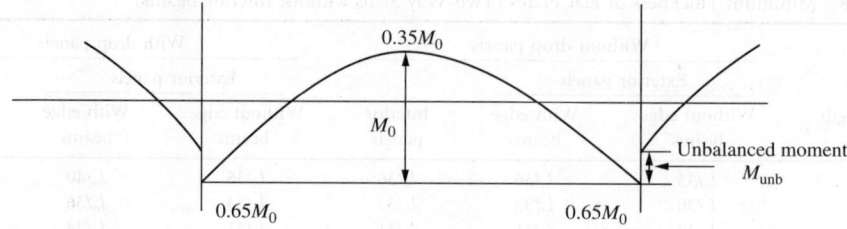

FIGURE 7.14   Static moment in floor panel.

TABLE 7.9   Distribution of Statical Moment for End Span Slab Panels

	Exterior edge unrestrained	Slab with beams between all supports	Slab without beams between interior supports		Exterior edge fully restrained
			Without edge beam	With edge beam	
Interior negative factored moment	0.75	0.70	0.70	0.70	0.65
Positive factored moment	0.63	0.57	0.52	0.50	0.35
Exterior negative factored moment	0	0.16	0.26	0.30	0.65

to the positive moment region. For the exterior span, the percentage of distribution is a function of the degree of restraint, as given in Table 7.9.

After the static moment is proportioned to the negative and positive regions, it is further apportioned on to the column and middle strips. For positive moment regions, the proportion of moment assigned to the column strips is given in Table 7.10. The parameter α_1 is a relative stiffness of the beam to slab, based on the full width of the equivalent frame:

$$\alpha_1 = \frac{E_{cb}I_b}{E_{cs}I_s} \tag{7.29}$$

For interior negative moment regions the proportion of moment assigned to the column strip is given by Table 7.11.

For negative moment regions of an exterior span, the moment assigned to the column follows Table 7.12, which takes into account the torsional stiffness of the edge beam. The parameter β_t is the ratio of torsional stiffness of edge beam section to flexural stiffness of a width of the slab equal to the center-to-center span length of the beam

$$\beta_t = \frac{E_{cb}C}{2E_{cs}I_s} \tag{7.30}$$

The remaining moment, that was not proportioned to the column strips, is assigned to the middle strips.

Column strip moments need to be further divided into their slab and beam. The beam should be proportioned to take 85% of the column strip moment if $\alpha_1 l_1/l_2 \geq 1.0$. Linear interpolation is applied if this parameter is less than 1.0. If the beams are also part of a lateral force resisting system, then moments due to lateral forces should be added to the beams. After the assignment of moments, flexural reinforcement in the beams and slab strips can be determined following the same design procedures presented in Sections 7.12.2 and 7.12.3 for regular beams and one-way slabs.

TABLE 7.10 Distribution of Positive Moment in Column Strip

l_2/l_1	0.5	1.0	2.0
$(\alpha_1 l_2/l_1) = 0$	60	60	60
$(\alpha_1 l_2/l_1) \geq 1.0$	90	75	45

TABLE 7.11 Distribution of Interior Negative Moment in Column Strip

l_2/l_1	0.5	1.0	2.0
$(\alpha_1 l_2/l_1) = 0$	75	75	75
$(\alpha_1 l_2/l_1) \geq 1.0$	90	75	45

TABLE 7.12 Distribution of Negative Moment to Column of an Exterior Span

l_2/l_1		0.5	1.0	2.0
$(\alpha_1 l_2/l_1) = 0$	$\beta_t = 0$	100	100	100
	$\beta_t \geq 2.5$	75	75	75
$(\alpha_1 l_2/l_1) \geq 1.0$	$\beta_t = 0$	100	100	100
	$\beta_t \geq 2.5$	90	75	45

7.13.1.4 Shear Design

The shear in the beam may be obtained by assuming that floor loads act according to the 45° tributary areas of each respective beam. Additional shear from lateral loads and the direct loads on the beam should be added on. The shear design of the beam then follows the procedure presented in Section 7.12. Shear stresses in the floor slab are generally low, but they should be checked. The strip method, which approximates the slab shear by assuming a unit width of slab strip over the panel, may be used to estimate the shear force in the slab.

7.13.2 Flat Plates

Floor systems without beams are commonly referred to as flat plates, (see Figure 7.15). Flat plates are economical and functional because beams are eliminated and floor height clearances are reduced. Minimum thicknesses of flat plates are given in Table 7.8 and should not be less than 5 in. The structural design procedure is the same as for flat slab with beams, presented in the previous sections, except that for flat plates $\alpha_1 = 0$. Refer to Section 7.13.1.2 for the static moment calculation. For the exterior span the distribution of the static moment is given in Figure 7.14. Table 7.10 and Table 7.11 provide the application for moment assignments to column strips.

7.13.2.1 Transfer of Forces in slab–column connections

An important design requirement of the flat plate system is the transfer of forces between the slab and its supporting columns (see Figure 7.14 and Figure 7.16). This transfer mechanism is a complex one. The accepted design approach is to assume that a certain fraction of the unbalanced moment M_{unb} in the slab connection is transferred by direct bending into the column support. This γ_f fraction is estimated to be

$$\gamma_f = \frac{1}{1 + (2/3)\sqrt{b_1/b_2}} \tag{7.31}$$

The moment $\gamma_f M_{unb}$ is transferred over an effective slab width that extends 1.5 times the slab thickness outside each side face of the column or column capital support. The existing reinforcement in the column strip may be concentrated over this effective width or additional bars may be added.

The fraction of unbalanced moment not transferred by flexure γ_v ($\gamma_v = 1 - \gamma_f$) is transferred through eccentricity of shear that acts over an imaginary critical section perimeter located at a

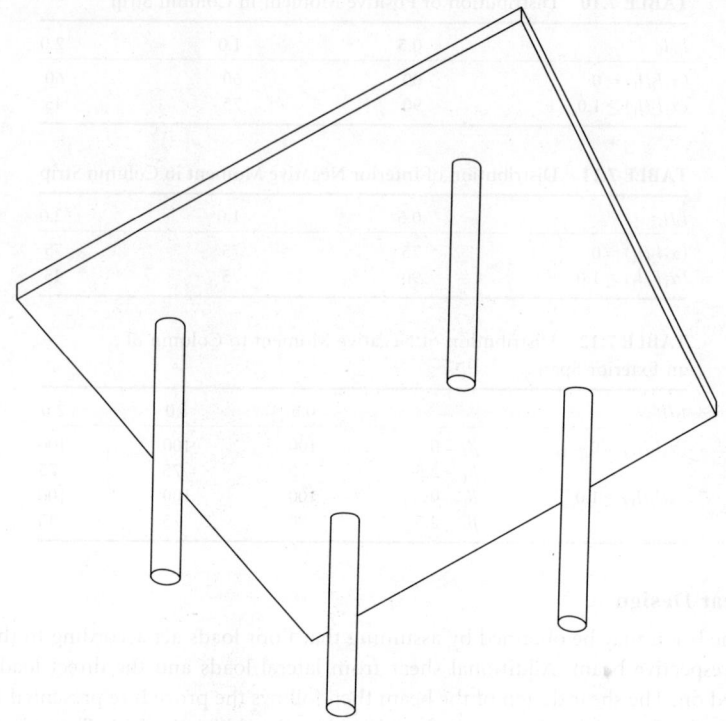

FIGURE 7.15   Flat plate.

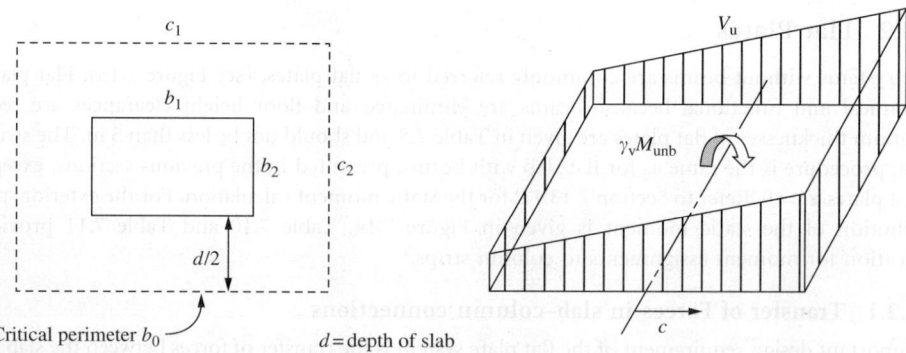

FIGURE 7.16   Transfer of shear in slab–column connections.

distance $d/2$ from the periphery of the column support (see Figure 7.16). Shear stress at the critical section is determined by combining the shear stress due to the direct shear demand V_u (which may be obtained from tributary loading) and that from the eccentricity of shear due to the unbalanced moment:

$$v_u = \frac{V_u}{A_c} \pm \frac{\gamma_v M_{unb} c}{J_c} \qquad (7.32)$$

where the concrete area of the critical section $A_c = b_0 d = 2d(c_1 + c_2 + 2d)$, and J_c is the equivalent polar moment of inertia of the critical section

$$J_c = \frac{d(c_1 + d)^3}{6} + \frac{(c_1 + d)d^3}{6} + \frac{d(c_2 + d)(c_1 + d)^2}{2} \tag{7.33}$$

The maximum shear stress v_u on the critical section must not exceed the shear stress capacity defined by

$$\phi v_n = \phi V_c / b_0 d \tag{7.34}$$

The concrete shear capacity V_c for two-way action is taken to be the lowest of the following three quantities:

$$V_c = 4\sqrt{f_c'} b_0 d \tag{7.35}$$

$$V_c = \left(2 + \frac{4}{\beta_c}\right) 4\sqrt{f_c'} b_0 d \tag{7.36}$$

$$V_c = \left(\frac{\alpha_s d}{b_0} + 2\right) 4\sqrt{f_c'} b_0 d \tag{7.37}$$

where β_c is the ratio of long side to short side of the column. The factor α_s is 40 for interior columns, 30 for edge columns, or 20 for corner columns.

If the maximum shear stress demand exceeds the capacity, the designer should consider using a thicker slab or a larger column, or increasing the column support area with a column capital. Other options include insertion of shear reinforcement or shearhead steel brackets.

7.13.2.2 Detailing of Flat Plates

Refer to Figure 7.17 for minimum extensions for reinforcements. All bottom bars in the column strip should be continuous or spliced with a Class A splice. To prevent progressive collapse, at least two of the column strip bottom bars in each direction should pass within the column core or be anchored at the end supports. This provides catenary action to hold up the slab in the event of punching failure.

7.13.3 Flat Slabs with Drop Panels and/or Column Capitals

The capacity of flat plates may be increased with drop panels. Drop panels increase the slab thickness over the negative moment regions and enhance the force transfer in the slab–column connection. The minimum required configuration of drop panels is given in Figure 7.18. The minimum slab thickness is given in Table 7.8 and should not be less than 4 in.

Alternatively, or in combination with drop panels, column capitals may be provided to increase capacity (see Figure 7.19). The column capital geometry should follow a 45° projection. Column capitals increase the critical section of the slab–column force transfer and reduce the clear span lengths. The design procedure outlined for flat plates in the previous sections are applicable for flat slabs detailed with drop panels or column capitals.

7.13.4 Waffle Slabs

For very heavy floor loads or very long spans, waffle slab floor systems become viable (see Figure 7.20). A waffle slab can be visualized as being a very thick flat plate but with coffers to reduce weight and gain efficiency. The design procedure is therefore the same as for flat plates as presented in Section 7.13.2.

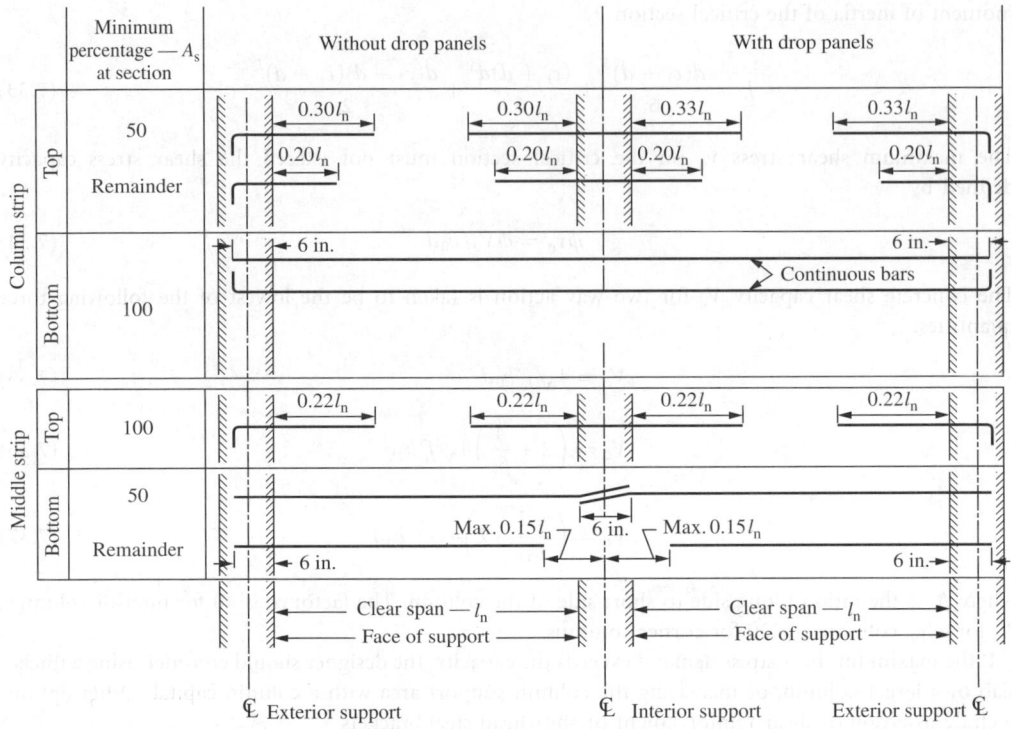

FIGURE 7.17 Detailing of flat plates.

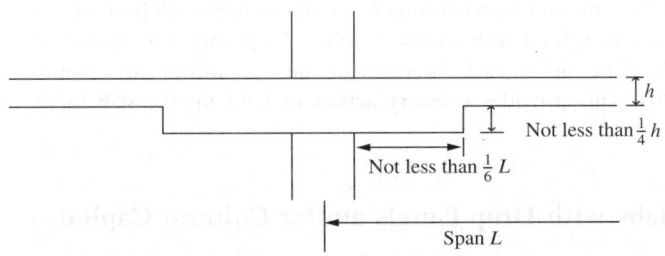

FIGURE 7.18 Drop panel dimensions.

The flexural reinforcement design is based on T-section strips instead of rectangular slab strips. Around column supports, the coffers may be filled in to act as column capitals.

7.14 Columns

Typical reinforcement concrete columns are shown in Figure 7.21. Longitudinal reinforcements in columns are generally distributed uniformly around the perimeter of the column section and run continuously through the height of the column. Transverse reinforcement may be in the form of rectangular hoops, ties, or spirals (Figure 7.21). Tall walls and core elements in buildings (Figure 7.22) are column-like in behavior and the design procedures presented in the following are applicable.

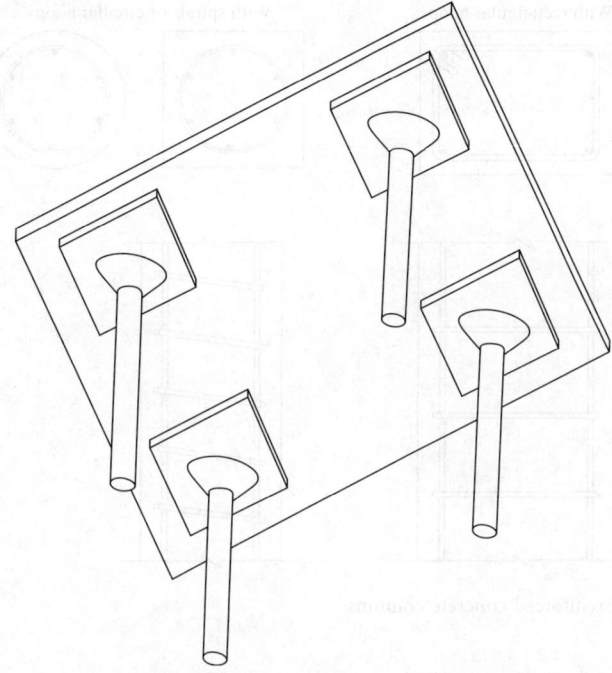

FIGURE 7.19   Flat slab with drop panels and column capital.

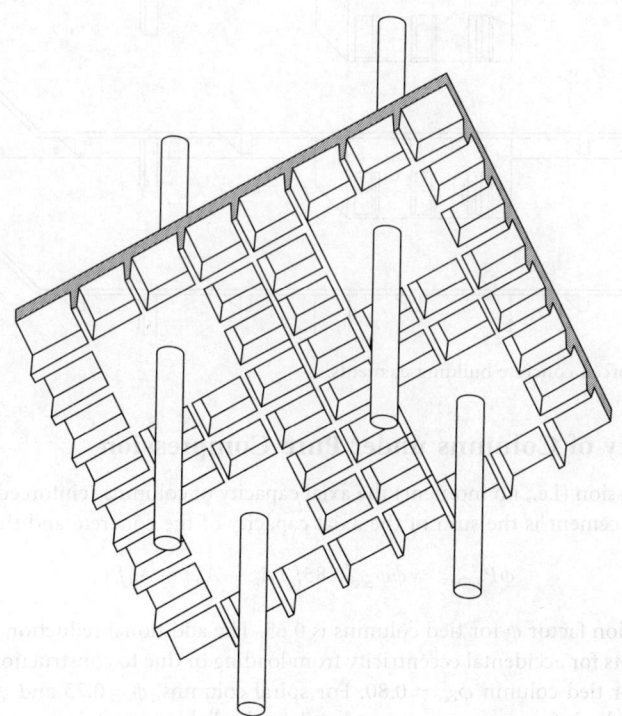

FIGURE 7.20   Waffle slab.

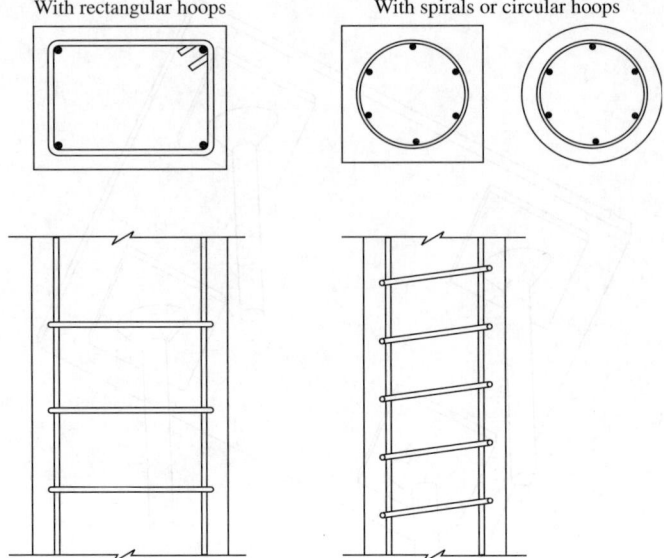

FIGURE 7.21 Typical reinforced concrete columns.

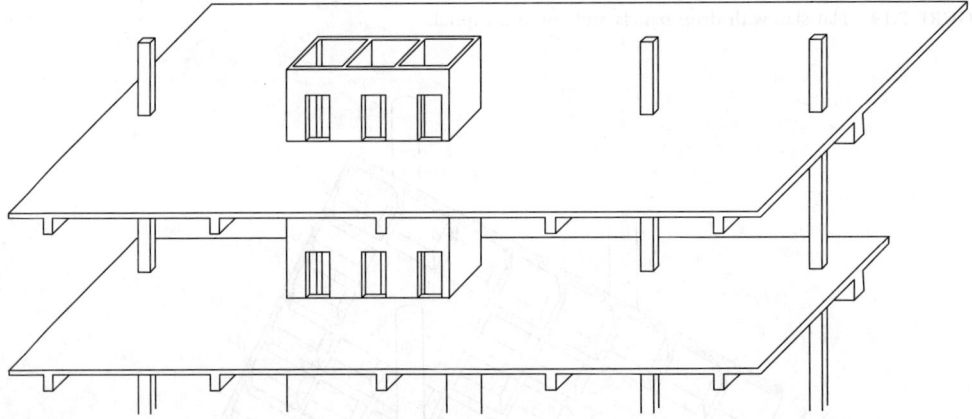

FIGURE 7.22 Reinforced concrete building elements.

7.14.1 Capacity of Columns under Pure Compression

Under pure compression (i.e., no moment) the axial capacity of columns reinforced with hoops and ties as transverse reinforcement is the sum of the axial capacity of the concrete and the steel:

$$\phi P_{n,max} = \phi \phi_{ecc}[0.85 f_c'(A_g - A_{st}) + A_{st} f_y] \tag{7.38}$$

The strength reduction factor ϕ for tied columns is 0.65. The additional reduction factor ϕ_{ecc} shown in the equation accounts for accidental eccentricity from loading or due to construction tolerances that will induce moment. For tied column $\phi_{ecc} = 0.80$. For spiral columns, $\phi = 0.75$ and $\phi_{ecc} = 0.85$. Columns reinforced with spiral reinforcement are more ductile and reliable in sustaining axial load after spalling of concrete cover. Hence, lower reduction factors are assigned by ACI.

7.14.2 Preliminary Sizing of Columns

For columns that are expected to carry no or low moment, the previous equation can be rearranged to estimate the required gross cross-sectional area to resist the axial force demand P_u:

$$A_g > \frac{(P_u/\phi\phi_{ecc}) - A_{st}f_y}{0.85f_c'} \tag{7.39}$$

The ACI Code limits the column reinforcement area A_{st} to 1 to 8% of A_g. Reinforcement percentages less than 4% are usually more practical in terms of avoiding congestion and to ease fabrication. If a column is expected to carry significant moment, the A_g estimated by the above expression would not be adequate. To obtain an initial trial size in that case, the above A_g estimate may be increased by an appropriate factor (e.g., doubling or more).

7.14.3 Capacity of Columns under Combined Axial Force and Moment

Under the combined actions of axial force and moment, the capacity envelope of a column is generally described by an interaction diagram (see Figure 7.23). Load demand points (M_u, P_u) from all load combinations must fall inside the $\phi P_n - \phi M_n$ capacity envelope; otherwise, the column is considered inadequate and should be redesigned. Computer software are typically used in design practice to generate column interaction diagrams.

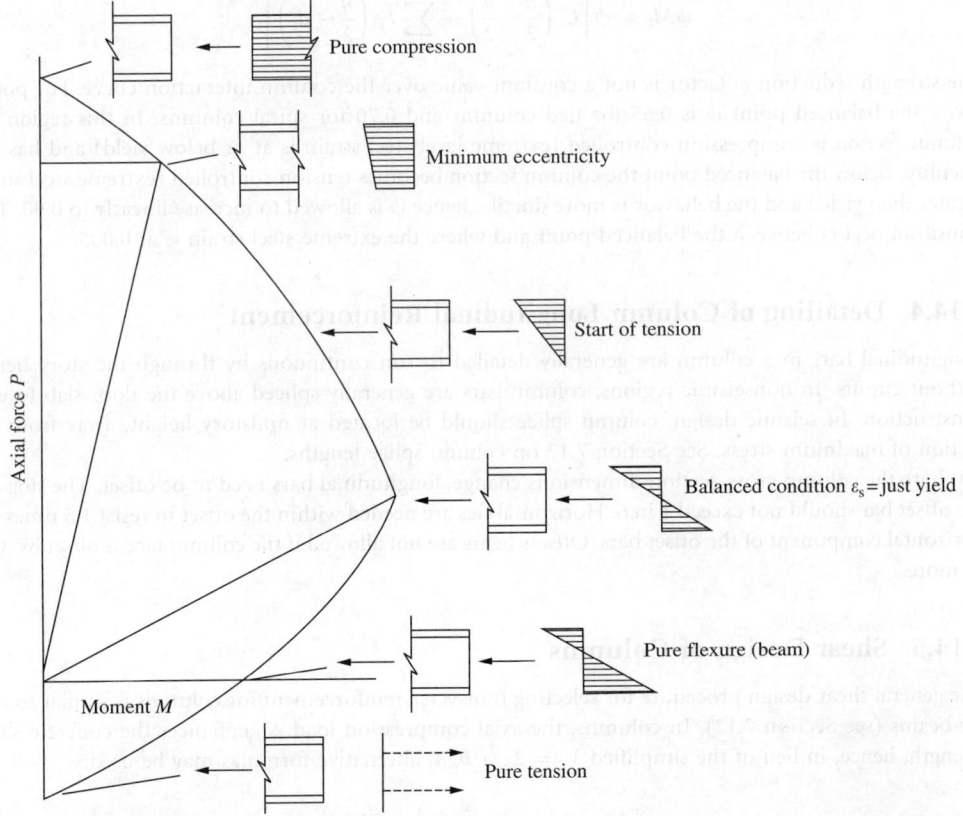

FIGURE 7.23 Column interaction diagram.

The upper point of an interaction curve is the case of pure axial compression. The lowest point is the case of pure axial tension, $\phi P_{n,\text{tension}} = \phi A_{st} f_y$ (it is assumed that the concrete section cracks and supplies no tensile strength). Where the interaction curve intersects with the moment axis, the column is under pure bending, in which case the column behaves like a beam. The point of maximum moment on the interaction diagram coincides with the balanced condition. The extreme concrete fiber strain reaches ultimate strain (0.003) simultaneously with yielding of the extreme layer of steel on the opposite side ($f_y/E_s = 0.002$).

Each point of the column interaction curve represents a unique strain distribution across the column section. The axial force and moment capacity at each point is determined by a strain compatibility analysis, similar to that presented for beams (see Section 7.12) but with an additional axial force component. The strain at each steel level i is obtained from similar triangles $\varepsilon_{si} = 0.003(c - d_i)/c$. Then, the steel stress at each level is $f_{st} = \varepsilon_{si} E_s$, but not greater in magnitude than the yield stress f_y. The steel force at each level is computed by $F_{si} = A_{si} f_{si}$. The depth of the equivalent concrete compressive stress block a is approximated by the relationship $a = \beta_1 c$. β_1 is the concrete stress block factor given in Figure 7.5. Hence, the resultant concrete compression force may be expressed as $C_c = 0.85 f_c' ab$. To satisfy equilibrium, summing forces of the concrete compression and the n levels of the steel, the axial capacity is obtained as

$$\phi P_n = \phi \left(C_c + \sum_{i=1}^{n} F_{si} \right) \tag{7.40}$$

The flexural capacity is obtained from summation of moments about the plastic centroid of the column

$$\phi M_n = \phi \left[C_c \left(\frac{h}{2} - \frac{a}{2} \right) + \sum_{i=1}^{n} F_{si} \left(\frac{h}{2} - d_i \right) \right] \tag{7.41}$$

The strength reduction ϕ factor is not a constant value over the column interaction curve. For points above the balanced point ϕ is 0.65 for tied columns and 0.70 for spiral columns. In this region the column section is compression controlled (extreme level steel strain is at or below yield) and has less ductility. Below the balanced point the column section becomes tension controlled (extreme steel strain greater than yield) and the behavior is more ductile, hence ϕ is allowed to increase linearly to 0.90. This transition occurs between the balanced point and where the extreme steel strain is at 0.005.

7.14.4 Detailing of Column Longitudinal Reinforcement

Longitudinal bars in a column are generally detailed to run continuous by through the story height without cutoffs. In nonseismic regions, column bars are generally spliced above the floor slab to ease construction. In seismic design, column splice should be located at midstory height, away from the section of maximum stress. See Section 7.17 on column splice lengths.

Where the column cross-section dimensions change, longitudinal bars need to be offset. The slope of the offset bar should not exceed 1 in 6. Horizontal ties are needed within the offset to resist 1.5 times the horizontal component of the offset bars. Offsets bents are not allowed if the column face is offset by 3 in. or more.

7.14.5 Shear Design of Columns

The general shear design procedure for selecting transverse reinforcement for columns is similar to that for beams (see Section 7.12). In columns, the axial compression load N_u enhances the concrete shear strength, hence, in lieu of the simplified $V_c = 2\sqrt{f_c'} b_w d$, alternative formulas may be used:

$$V_c = 2 \left(1 + \frac{N_u}{2000 A_g} \right) \sqrt{f_c'} b_w d \tag{7.42a}$$

The quantity N_u/A_g must be in units of pounds per square inch. A second alternative formula for concrete shear strength V_c is

$$V_c = \left(1.9\sqrt{f_c'} + 2500\rho_w \frac{V_u d}{M_m}\right) b_w d \leq 3.5\sqrt{f_c'} b_w d \sqrt{1 + \frac{N_u}{500A_g}} \quad (7.42b)$$

where

$$M_m = M_u - N_u \frac{(4h - d)}{8} \quad (7.43)$$

If M_m is negative, the upper bound expression for V_c is used.

Under seismic conditions, additional transverse reinforcement is required to confine the concrete to enhance ductile behavior. See Section IV of this book on earthquake design.

7.14.6 Detailing of Column Hoops and Ties

The main transverse reinforcement should consist of one or a series of perimeter hoops (see Figure 7.24), which not only serve as shear reinforcement, but also prevent the longitudinal bars from buckling out through the concrete cover. Every corner and alternate longitudinal bar should have a hook support (see Figure 7.24). The angle of the hook must be less than 135°. All bars should be hook supported if the clear spacing between longitudinal bars is more than 6 in. The transverse reinforcement must be at least a No. 3 size if the longitudinal bars are No. 10 or smaller, and at least a No. 4 size if the longitudinal bars are greater than No. 10.

To prevent buckling of longitudinal bars, the vertical spacing of transverse reinforcement in columns should not exceed 16 longitudinal bar diameters, 48 transverse bar diameters, or the least dimension of the column size.

7.14.7 Design of Spiral Columns

Columns reinforced with spirals provide superior confinement for the concrete core. Tests have shown that spiral columns are able to carry their axial load even after spalling of the concrete cover. Adequate confinement is achieved when the center-to-center spacing s of the spiral of diameter d_b and yield strength f_y satisfies the following:

$$s \leq \frac{\pi f_y d_b^2}{0.45 h_c f_c' \left[(A_g/A_c) - 1\right]} \quad (7.44)$$

where h_c is the diameter of the concrete core measured out-to-out of the spiral.

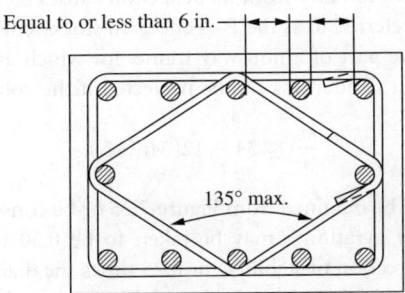

Equal to or less than 6 in.

135° max.

FIGURE 7.24 Column transverse reinforcement detailing.

7.14.8 Detailing of Columns Spirals

Spiral columns require a minimum of six longitudinal bars. Spacers should be used to maintain the design spiral spacing and to prevent distortions. The diameter of the spiral d_b should not be less than $\frac{3}{8}$ in. The clear spacing between spirals should not exceed 3 in. or be less than 1 in. Spirals should be anchored at each column end by providing an extra one and one-half turns of spiral bar. Spirals may be spliced by full mechanical or welded splices or by lap splices with lap lengths not less than 12 in. or $45d_b$ ($72d_b$ if plain bar). While spirals are not required to run through the column-to-floor connection zones, ties should be inserted in those zones to maintain proper confinement, especially if horizontal beams do not frame into these zones.

7.14.9 Detailing of Column to Beam Joints

Joints will perform well if they are well confined. By containing the joint concrete, its structural integrity is ensured under cyclic loading, which allows the internal force capacities, as well as the splices and anchorages detailed within the joint, to develop. Often, confinement around a joint will be provided by the beams or other structural elements that intersect at the joint, if they are of sufficient size. Otherwise, some closed ties, spirals, or stirrups should be provided within the joint to confine the concrete. For nonseismic design, the ACI has no specific requirements on joint confinement.

7.14.10 Columns Subject to Biaxial Bending

If a column is subject to significant moments biaxially, for example, a corner column at the perimeter of a building, the column capacity may be defined by an interaction surface. This surface is essentially an extension of the 2-D interaction diagram described in Figure 7.23 to three coordinate axes $\phi P_n - \phi M_{nx} - \phi M_{ny}$. For rectangular sections under biaxial bending the resultant moment axis may not coincide with the neutral axis. (This is never the case for a circular cross-section because of point symmetry.) An iterative procedure is necessary to determine this angle of deviation. Hence, an accurate generation of the biaxial interaction surface generally requires computer software. Other approximate methods have been proposed. The ACI Code Commentary (R10.3.7) presents the Reciprocal Load Method in which the biaxial capacity of a column ϕP_{ni} is related in a reciprocal manner to its uniaxial capacities, ϕP_{nx} and ϕP_{ny}, and pure axial capacity P_0:

$$\frac{1}{\phi P_{ni}} = \frac{1}{\phi P_{nx}} + \frac{1}{\phi P_{ny}} - \frac{1}{\phi P_0} \tag{7.45}$$

7.14.11 Slender Columns

When columns are slender the internal forces determined by a first-order analysis may not be sufficiently accurate. The change in column geometry from its deflection causes secondary moments to be induced by the column axial force, also referred to as the P–Δ effect. In stocky columns these secondary moments are minor. For columns that are part of a nonsway frame, for which analysis shows limited side-sway deflection, the effects of column slenderness can be neglected if the column slenderness ratio

$$\frac{kl_u}{r} \leq 34 - 12(M_1/M_2) \tag{7.46}$$

The effective length factor k can be obtained from Figure 7.25 or be conservatively assumed to be 1.0 for nonsway frames. The radius of gyration r may be taken to be 0.30 times the overall dimension of a rectangular column (in the direction of stability) or 0.25 times the diameter for circular columns. The ratio of the column end moments (M_1/M_2) is taken as positive if the column is bent in single curvature, and negative in double curvature.

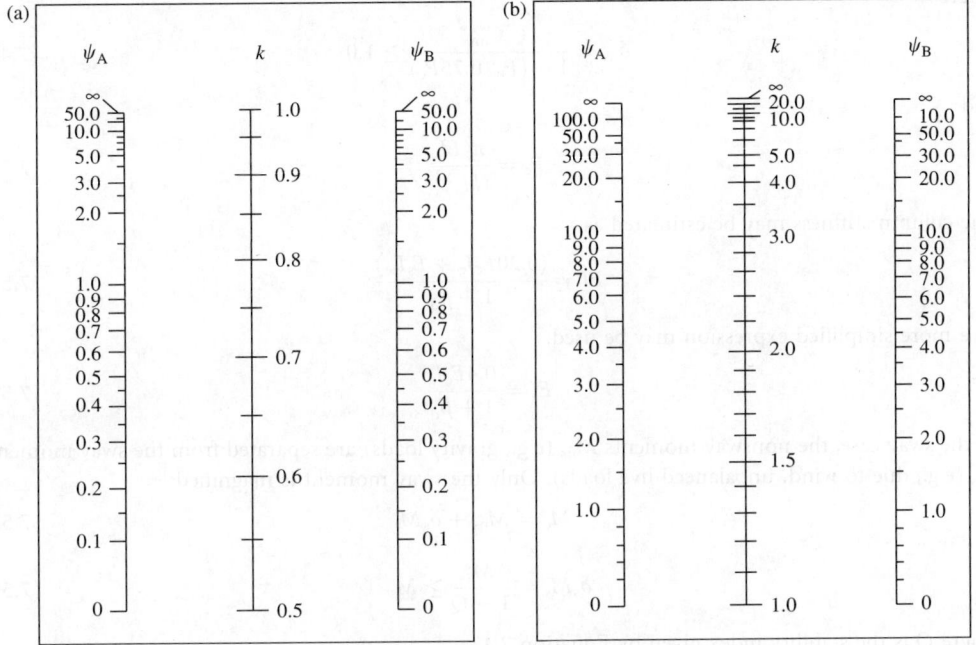

FIGURE 7.25 Effective length factor k: (a) nonsway frames and (b) sway frames.
Note: ψ is the ratio of the summation of column stiffness $[\sum(EI/L)]$ to beam stiffness at the beam–column joint.

For a building story, a frame is considered to be nonsway if its stability index

$$Q = \frac{\sum P_u \Delta_0}{V_u l_c} \leq 0.05 \qquad (7.47)$$

where Δ_0 is the first-order relative deflection between the top and bottom of the story and $\sum P_u$ and V_u are the total vertical load and story shear, respectively.

For sway frames, slenderness may be neglected if the slenderness ratio $kl_u/r \leq 22$. The k factor must be taken as greater than or equal to 1.0 (see Figure 7.25).

For structural design, it is preferable to design reinforced concrete structures as nonsway systems and with stocky columns. Structural systems should be configured with stiff lateral resistant elements such as shear walls to control sway. Column cross-sectional dimensions should be selected with the slenderness criteria in mind.

If slender columns do exist in a design, adopting a computerized second-order analysis should be considered so that the effects of slenderness will be resolved internally by the structural analysis (see Section 7.7). Then, the internal force demands from the computer output can be directly checked against the interaction diagram in like manner as a nonslender column design. Alternatively, the ACI code provides a manual method called the Moment Magnifier Method to adjust the structural analysis results of a first-order analysis.

7.14.12 Moment Magnifier Method

The Moment Magnifier Method estimates the column moment M_c in a slender column by magnifying the moment obtained from a first-order analysis M_2. For the nonsway case, the factor δ_{ns} magnifies the column moment:

$$M_c = \delta_{ns} M_2 \qquad (7.48)$$

where

$$\delta_{ns} = \frac{C_m}{1 - (P_u/0.75P_c)} \geq 1.0 \qquad (7.49)$$

and

$$P_c = \frac{\pi^2 EI}{(kl_u)^2} \qquad (7.50)$$

The column stiffness may be estimated as

$$EI = \frac{(0.20E_c I_g + E_s I_{se})}{1 + \beta_d} \qquad (7.51)$$

or a more simplified expression may be used:

$$EI = \frac{0.4E_c I_g}{1 + \beta_d} \qquad (7.52)$$

In the sway case, the nonsway moments M_{ns} (e.g., gravity loads) are separated from the sway moments M_s (e.g., due to wind, unbalanced live loads). Only the sway moment is magnified:

$$M_c = M_{ns} + \delta_s M_s \qquad (7.53)$$

$$\delta_s M_s = \frac{M_s}{1 - Q} \geq M_s \qquad (7.54)$$

where Q is the stability index given by Equation 7.47.

7.15 Walls

If tall walls (or shear walls) and combined walls (or core walls) subjected to axial load and bending behave like a column, the design procedures and formulas presented in the previous sections are generally applicable. The reinforcement detailing of wall differs from that of columns. Boundary elements, as shown in Figure 7.26, may be attached to the wall ends or corners to enhance moment capacity. The ratio ρ_n of vertical shear reinforcement to gross area of concrete of horizontal section should not be less than

$$\rho_n = 0.0025 + 0.5\left(2.5 - \frac{h_w}{l_w}\right)(\rho_h - 0.0025) \geq 0.0025 \qquad (7.55)$$

The spacing of vertical wall reinforcement should not exceed $l_w/3$, $3h$, or 18 in. To prevent buckling, the vertical bars opposite each other should be tied together with lateral ties if the vertical reinforcement is greater than 0.01 the gross concrete area.

7.15.1 Shear Design of Walls

The general shear design procedure given in Section 7.12 for determining shear reinforcement in columns applies to walls. For walls in compression, the shear strength provided by concrete V_c may be taken as $2\sqrt{f_c'}hd$. Alternatively, V_c may be taken from the lesser of

$$3.3\sqrt{f_c'}hd + \frac{N_u d}{4l_w} \qquad (7.56)$$

and

$$\left[0.6\sqrt{f_c'} + \frac{l_w(1.25\sqrt{f_c'} + 0.2(N_u/l_w h))}{(M_u/V_u) - (l_w/2)}\right]hd \qquad (7.57)$$

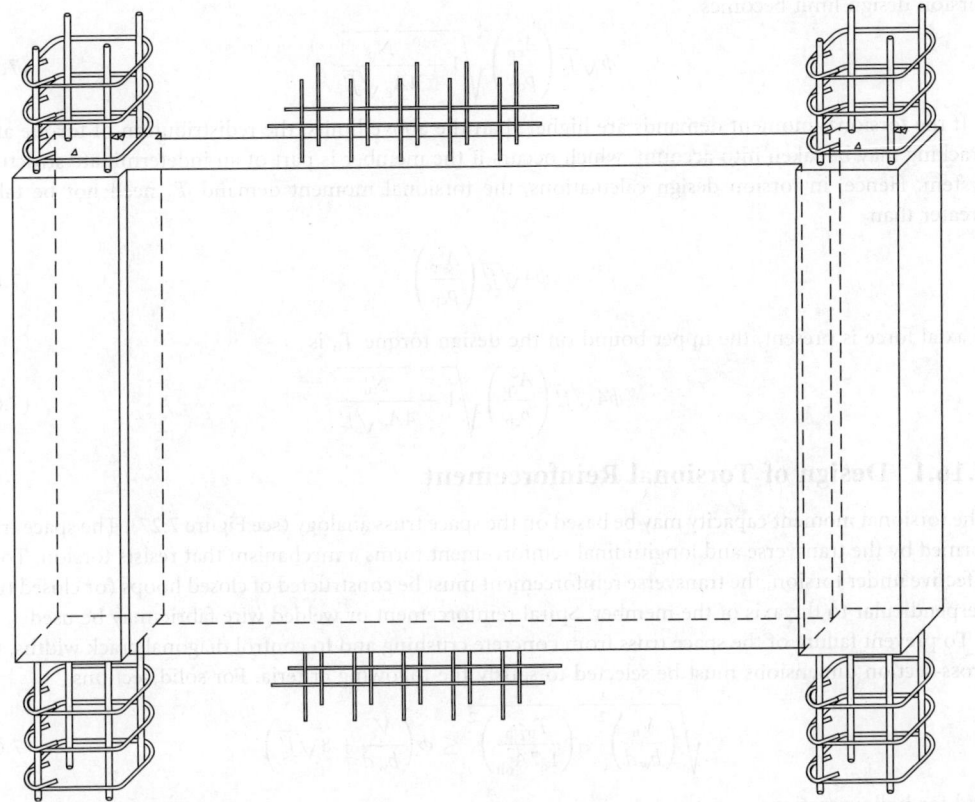

FIGURE 7.26   Reinforced concrete wall with boundary columns.

In lieu of a strain compatibility analysis, the depth of walls d may be assumed to be $0.8l_w$. Shear strength provided by the horizontal reinforcement in walls is also calculated by the equation $V_s = A_v f_y d/s$. The shear capacity of walls $\phi V_n = \phi(V_c + V_s)$ should not be greater than $\phi 10\sqrt{f_c'}hd$.

The spacing of horizontal wall reinforcement should not exceed $l_w/5$, $3h$, or 18 in. The minimum ratio of horizontal wall reinforcement should be more than 0.0025 (or 0.0020 for bars not larger than No. 5). The vertical and horizontal wall bars should be placed as close to the two faces of the wall as cover allows.

7.16   Torsion Design

Torsion will generally not be a serious design issue for reinforced concrete structures if the structural scheme is regular and symmetrical in layout and uses reasonable member sizes. In building floors, torsion may need to be considered for edge beams and members that sustain large unbalanced loading. Concrete members are relatively tolerant of torsion. The ACI permits torsion design to be neglected if the factored torsional moment demand T_u is less than

$$\phi\sqrt{f_c'}\left(\frac{A_{cp}^2}{P_{cp}}\right) \tag{7.58}$$

which corresponds to about one quarter of the torsional cracking capacity. For hollow sections the gross area of section A_g should be used in place of A_{cp}. If an axial compressive or tensile force N_u exists, the torsion design limit becomes

$$\phi\sqrt{f_c'}\left(\frac{A_{cp}^2}{p_{cp}}\right)\sqrt{1+\frac{N_u}{4A_g\sqrt{f_c'}}} \tag{7.59}$$

If the torsional moment demands are higher than the above limits, the redistribution of torque after cracking may be taken into account, which occurs if the member is part of an indeterminate structural system. Hence, in torsion design calculations, the torsional moment demand T_u need not be taken greater than

$$\phi4\sqrt{f_c'}\left(\frac{A_{cp}^2}{p_{cp}}\right) \tag{7.60}$$

If axial force is present, the upper bound on the design torque T_u is

$$\phi4\sqrt{f_c'}\left(\frac{A_{cp}^2}{p_{cp}}\right)\sqrt{1+\frac{N_u}{4A_g\sqrt{f_c'}}} \tag{7.61}$$

7.16.1 Design of Torsional Reinforcement

The torsional moment capacity may be based on the space truss analogy (see Figure 7.27). The space truss formed by the transverse and longitudinal reinforcement forms a mechanism that resists torsion. To be effective under torsion, the transverse reinforcement must be constructed of closed hoops (or closed ties) perpendicular to the axis of the member. Spiral reinforcement or welded wire fabric may be used.

To prevent failure of the space truss from concrete crushing and to control diagonal crack widths, the cross-section dimensions must be selected to satisfy the following criteria. For solid sections

$$\sqrt{\left(\frac{V_u}{b_w d}\right)^2+\left(\frac{T_u p_h}{1.7A_{oh}^2}\right)^2}\le\phi\left(\frac{V_c}{b_w d}+8\sqrt{f_c'}\right) \tag{7.62}$$

and for hollow sections

$$\left(\frac{V_u}{b_w d}\right)+\left(\frac{T_u p_h}{1.7A_{oh}^2}\right)\le\phi\left(\frac{V_c}{b_w d}+8\sqrt{f_c'}\right) \tag{7.63}$$

After satisfying these criteria, the torsional moment capacity is determined by

$$\phi T_n=\phi\frac{2A_o A_t f_{yv}}{s}\cot\theta \tag{7.64}$$

The shear flow area A_o may be taken as $0.85A_{oh}$, where A_{oh} is the area enclosed by the closed hoop (see Figure 7.28). The angle θ may be assumed to be $45°$. More accurate values of A_o and θ may be used from analysis of the space truss analogy.

To determine the additional transverse torsional reinforcement required to satisfy ultimate strength, that is, $\phi T_n\ge T_u$, the transverse reinforcement area A_t and its spacing s must satisfy the following:

$$\frac{A_t}{s}>\frac{T_u}{\phi2A_o f_{yv}\cot\theta} \tag{7.65}$$

The area A_t is for one leg of reinforcement. This torsional reinforcement area should then be combined with the transverse reinforcement required for shear demand A_v (see Section 7.11). The total transverse reinforcement required for the member is thus

$$\frac{A_v}{s}+2\frac{A_t}{s} \tag{7.66}$$

The above expression assumes that the shear reinforcement consists of two legs. If more than two legs are present, only the legs adjacent to the sides of the cross-section are considered effective for torsional resistance.

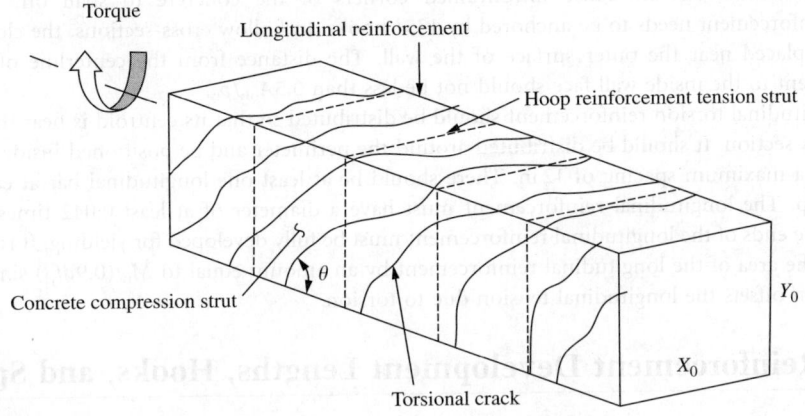

FIGURE 7.27   Truss analogy for torsion.

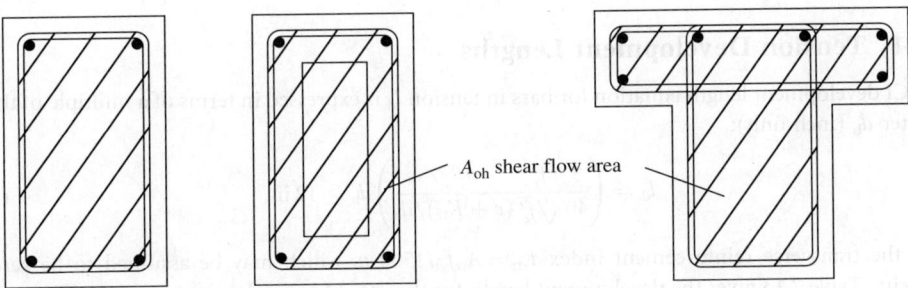

FIGURE 7.28   Torsional reinforcement and shear flow area.

The total transverse reinforcement must exceed the following minimum amounts:

$$0.75\sqrt{f_c'}\frac{b_w}{f_{yv}} \geq \frac{50b_w}{f_{yv}} \tag{7.67}$$

A minimum amount of longitudinal reinforcement is also required:

$$A_l = \frac{A_t}{s}p_h\left(\frac{f_{yv}}{f_{yt}}\right)\cot^2\theta \tag{7.68}$$

The reinforcement area A_l is additional to that required for resisting flexure and axial forces and should not be less than

$$\frac{5\sqrt{f_c'}A_{cp}}{f_{yl}} - \left(\frac{A_t}{s}\right)p_h\frac{f_{yv}}{f_{yl}} \tag{7.69}$$

where A_t/s should not be less than $25b_w/f_{yv}$. The torsional–longitudinal reinforcement should be distributed around the section in a uniform manner.

7.16.2   Detailing of Torsional Reinforcement

The spacing of closed transverse reinforcement under torsion must not exceed $p_h/8$ or 12 in. Torsion reinforcement should be provided for a distance of at least $(b_t + d)$ beyond the point theoretically required. Torsional stresses cause unrestrained corners of the concrete to spall off. Transverse torsion reinforcement needs to be anchored by 135° hooks. In hollow cross-sections, the closed hoops should be placed near the outer surface of the wall. The distance from the centerline of the hoop reinforcement to the inside wall face should not be less than $0.5A_{oh}/p_h$.

The longitudinal torsion reinforcement should be distributed so that its centroid is near the centroid of the cross-section. It should be distributed around the perimeter and be positioned inside the closed hoop with a maximum spacing of 12 in. There should be at least one longitudinal bar at each corner of the hoop. The longitudinal reinforcement must have a diameter of at least 0.042 times the hoop spacing. The ends of the longitudinal reinforcement must be fully developed for yielding. It is permitted to reduce the area of the longitudinal reinforcement by an amount equal to $M_u/(0.9df_{yl})$ since flexural compression offsets the longitudinal tension due to torsion.

7.17 Reinforcement Development Lengths, Hooks, and Splices

The various ultimate capacity formulas presented in the previous sections are premised on the assumption that the reinforcement will reach its yield strength f_y. This is not assured unless the reinforcement has (1) sufficient straight embedment length on each side of the point of yielding, (2) a hook of sufficient anchorage capacity, or (3) a qualified mechanical anchor device.

7.17.1 Tension Development Lengths

The ACI development length equation for bars in tension l_d is expressed in terms of a multiple of the bar diameter d_b (inch unit):

$$l_d = \left(\frac{3}{40} \frac{f_y}{\sqrt{f_c'}} \frac{\alpha\beta\gamma\lambda}{(c + K_{tr})/d_b} \right) d_b \geq 12 \text{ in.} \tag{7.70}$$

where the transverse reinforcement index $K_{tr} = A_{tr}f_{yt}/1500sn$, which may be assumed to be zero for simplicity. Table 7.13 gives the development length for the case of normal weight concrete ($\lambda = 1.0$) and uncoated reinforcement ($\beta = 1.0$). Development lengths need to be increased under these conditions: beam reinforcement positioned near the top surface, epoxy coating, lightweight concrete, and bundling of bars (see ACI Section 12.2.4).

TABLE 7.13 Development Lengths in Tension

| | Tension development length (in.) | |
| | Concrete strength (psi) | |
Bar size	4000	8000
3	12	12
4	12	12
5	15	12
6	21	15
7	36	26
8	47	34
9	60	43
10	77	54
11	94	67
14	136	96
18	242	171

Note: Normal-weight concrete, Grade 60 reinforcement. $\alpha = 1.0$, $\beta = 1.0$, $c = 1.5$ in., and $K_{tr} = 0$.

7.17.2 Compression Development Lengths

For bars under compression, such as in columns, yielding is assured if the development length meets the largest value of $(0.02f_y/\sqrt{f_c'})d_b$, $(0.0003f_y)d_b$, and 8 in. Compression development lengths l_{dc} are given in Table 7.14. Compression development length may be reduced by the factor $(A_s$ required$)/(A_s$ provided$)$ if reinforcement is provided in excess of that required by the load demand. Reinforcement within closely spaced spirals or tie reinforcement may be reduced by the factor 0.75 (spiral not less than $\frac{1}{4}$ in. in diameter and not more than 4 in. in pitch; column ties not less than No. 4 in size and spaced not more than 4 in.).

7.17.3 Standard Hooks

The standard (nonseismic) hook geometry as defined by ACI is shown in Figure 7.9. The required hook length l_{dh} is given in Table 7.15 and is based on the empirical formula $(0.02f_y/\sqrt{f_c'})d_b$. Hook lengths may be reduced by 30% when the side and end covers over the hook exceed 2.5 and 2 in., respectively. A 20% reduction is permitted if the hook is within a confined concrete zone where the transverse

TABLE 7.14 Development Lengths in Compression

	Compression development length (in.)	
	Concrete strength (psi)	
Bar size	4000	8000
3	8	8
4	9	9
5	12	11
6	14	14
7	17	16
8	19	18
9	21	20
10	24	23
11	27	25
14	32	30
18	43	41

Note: Grade 60 reinforcement.

TABLE 7.15 Development Lengths of Hooks in Tension

	Development length of standard hook (in.)	
	Concrete strength (psi)	
Bar size	4000	8000
3	7	6
4	9	7
5	12	8
6	14	10
7	17	12
8	19	13
9	21	15
10	24	17
11	27	19
14	32	23
18	43	30

Note: Grade 60 steel. $\beta = 1.0$, $\lambda = 1.0$, l_{dh} not less than $8d_b$ nor 6 in.

reinforcement spacing is less than three times the diameter of the hooked bar. Note that whether the standard hook is detailed to engage over a longitudinal bar has no influence on the required hook length.

When insufficient hook length is available or in regions of heavy bar congestion, mechanical anchors may be used. There are a number of proprietary devices that have been tested and prequalified. These generally consist of an anchor plate attached to the bar end.

7.17.4 Splices

There are three choices for joining bars together: (1) mechanical device, (2) welding, and (3) lap splices. The mechanical and welded splices must be tested to show the development in tension or compression of at least 125% of the specified yield strength f_y of the bar. Welded splices must conform to ANSI/AWS D1.4, "Structural Welding Code — Reinforcing Steel." Since splices introduce weak leaks into the structure, they should be located as much as possible away from points of maximum force and critical locations.

7.17.4.1 Tension Lap Splices

Generally, bars in tension need to be lapped over a distance of $1.3l_d$ (Class B splice, see Section 7.17.1 for l_d), unless laps are staggered or more than twice the required steel is provided (Class A splice $= 1.0l_d$).

7.17.4.2 Compression Lap Splices and Column Splices

Compression lap splice lengths shall be $0.0005f_y d_b$, but not less than 12 in. If any of the load demand combinations is expected to introduce tension in the column reinforcement, column bars should be lapped as tension splices. Class A splices ($1.0l_d$) are allowed if half or fewer of the bars are spliced at any section and alternate lap splices are staggered by l_d. Column lap lengths may be multiplied by 0.83 if the ties provided through the lap splice length have an effective area not less than $0.0015hs$. Lap lengths within spiral reinforcement may be multiplied by 0.75.

7.18 Deflections

The estimation of deflections for reinforced concrete structures is complicated by the cracking of the concrete and the effects of creep and shrinkage. In lieu of carrying out a refined nonlinear analysis involving the moment curvature analysis of member sections, an elastic analysis may be used to incorporate a reduced or effective moment of inertia for the members. For beam elements an effective moment of inertia may be taken as

$$I_e = \left(\frac{M_{cr}}{M_a}\right)^3 I_g + \left[1 - \left(\frac{M_{cr}}{M_a}\right)^3\right] I_{cr} \le I_g \tag{7.71}$$

where the cracking moment of the section

$$M_{cr} = \frac{f_r I_g}{y_t} \tag{7.72}$$

The cracking stress or modulus of rupture of normal weight concrete is

$$f_r = 7.5\sqrt{f_c'} \tag{7.73}$$

For all-lightweight concrete f_r should be multiplied by 0.75, for sand-lightweight concrete, by 0.85.

For estimating the deflection of prismatic beams, it is generally satisfactory to take I_e at the section at midspan to represent the average stiffness for the whole member. For cantilevers, the I_e at the support should be taken. For nonprismatic beams, an average I_e of the positive and negative moment sections should be used.

Long-term deflections may be estimated by multiplying the immediate deflections of sustained loads (e.g., self-weight, permanent loads) by

$$\lambda = \frac{\xi}{1 + 50\rho'} \tag{7.74}$$

The time-dependent factor ξ is plotted in Figure 7.29. More refined creep and shrinkage deflection models are provided by ACI Committee 209 and the CEP-FIP Model Code (1990).

Deflections of beams and one-way slab systems must not exceed the limits in Table 7.16. Deflection control of two-way floor systems is generally satisfactory by following the minimum slab thickness

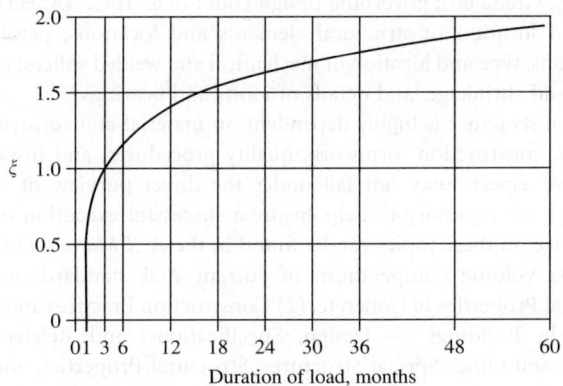

FIGURE 7.29 Time-dependent factor ξ.

TABLE 7.16 Deflection Limits of Beams and One-Way Slab Systems

Type of member	Deflection to be considered	Deflection limitation
Flat roots not supporting or attached to nonstructural elements likely to be damaged by large deflections	Immediate deflection due to live load L	$l/180$[a]
Floors not supporting or attached to nonstructural elements likely to be damaged by large deflections	Immediate deflection due to live load L	$l/360$
Roof or floor construction supporting or attached to nonstructural elements likely to be damaged by large deflections	That part of the total deflection occurring after attachment of nonstructural elements (sum of the long-term deflection due to all sustained loads and the immediate deflection due to any additional live load)[b]	$l/480$[c]
Roof or floor construction supporting or attached to nonstructural elements not likely to be damaged by large deflections		$l/240$[d]

[a] Limit not intended to safeguard against ponding. Ponding should be checked by suitable calculations of deflection, including added deflections due to ponded water, and consideration of long-term effects of all sustained loads, camber, construction tolerances, and reliability of provisions for drainage.

[b] Long-term deflection should be determined in accordance with Equation 7.74, but may be reduced by the amount of deflection calculated to occur before attachment of nonstructural elements. This amount should be determined on the basis of accepted engineering date relating to time deflection characteristics of members similar to those being considered.

[c] Limit may be exceeded if adequate measures are taken to prevent to supported or attached elements.

[d] Limit should be greater than the tolerance provided for nonstructural elements. Limit may be exceeded if camber is provided so that total deflection minus camber does not exceed limit.

requirements (see Table 7.8). Lateral deflections of columns may be a function of occupancy comfort under high wind or seismic drift criteria (e.g., $H/200$).

7.19 Drawings, Specifications, and Construction

Although this chapter has focused mainly on the structural mechanics of design, design procedures and formulas, and rules that apply to reinforced concrete construction, the importance of drawings and specifications as part of the end products for communicating the structural design must not be overlooked. Essential information that should be included in the drawings and specifications are: specified compressive strength of concrete at stated ages (e.g., 28 days) or stage of construction; specified strength or grade of reinforced (e.g., Grade 60); governing design codes (e.g., IBC, AASHTO); live load and other essential loads; size and location of structural elements and locations; development lengths, hook lengths, and their locations; type and location of mechanical and welded splices; provisions for the effects of temperature, creep, and shrinkage; and details of joints and bearings.

The quality of the final structure is highly dependent on material and construction quality measures that improve durability, construction formwork, quality procedures, and inspection of construction. Although many of these aspects may not fall under the direct purview of the structural designer, attention and knowledge are necessary to help ensure a successful execution of the structural design. Information and guidance on these topics can be found in the *ACI Manual of Concrete Practice*, which is a comprehensive five-volume compendium of current ACI standards and committee reports: (1) Materials and General Properties of Concrete, (2) Construction Practices and Inspection, Pavements, (3) Use of Concrete in Buildings — Design, Specifications, and Related Topics, (4) Bridges, Substructures, Sanitary, and Other Special Structures, Structural Properties, and (5) Masonry, Precast Concrete, Special Processes.

Notation

a	= depth of concrete stress block
A_s'	= area of compression reinforcement
A_b	= area of an individual reinforcement
A_c	= area of core of spirally reinforced column measured to outside diameter of spiral
A_c	= area of critical section
A_{cp}	= area enclosed by outside perimeter of concrete cross-section
A_g	= gross area of section
A_l	= area of longitudinal reinforcement to resist torsion
A_o	= gross area enclosed by shear flow path
A_{oh}	= area enclosed by centerline of the outermost closed transverse torsional reinforcement
A_s	= area of tension reinforcement
$A_{s,min}$	= minimum area of tension reinforcement
A_{st}	= total area of longitudinal reinforcement
A_t	= area of one leg of a closed stirrup resisting torsion within a distance s
A_{tr}	= total cross-sectional area of all transverse reinforcement that is within

	the spacing s and that crosses the potential place of splitting through the reinforcement being developed
A_v	= area of shear reinforcement
$A_{v,min}$	= minimum area of shear reinforcement
b	= width of compression face
b_1	= width of critical section in l_1 direction
b_2	= width of critical section in l_2 direction
b_0	= perimeter length of critical section
b_t	= width of that part of the cross-section containing the closed stirrups resisting torsion
b_w	= web width
C	= cross-sectional constant to define torsional properties $= \sum (1 - 0.63(x/y))/(x^3 y/3)$ (total section is divided into separate rectangular parts, where x and y are the shorter and longer dimensions of each part, respectively).
c	= distance from centroid of critical section to its perimeter (Section 7.13.2.1)
c	= spacing or cover dimension
c_1	= dimension of column or capital support in l_1 direction

c_2 = dimension of column or capital support in l_2 direction

c_c = clear cover from the nearest surface in tension to the surface of the flexural reinforcement

C_c = resultant concrete compression force

C_m = factor relating actual moment diagram to an equivalent uniform moment

d = distance from extreme compression fiber to centroid of tension reinforcement

d' = distance from extreme compression fiber to centroid of compression reinforcement

d_b = nominal diameter of bar

d_i = distance from extreme compression fiber to centroid of reinforcement layer i

E_c = modulus of elasticity of concrete

E_{cb} = modulus of elasticity of beam concrete

E_{cs} = modulus of elasticity of slab concrete

EI = flexural stiffness of column

E_s = modulus of elasticity of steel reinforcement

f'_c = specified compressive strength of concrete

F_n = nominal structural strength

f_r = modulus of rupture of concrete

f_s = reinforcement stress

F_{si} = resultant steel force at bar layer i

f_y = specified yield stress of reinforcement

f_{yl} = specified yield strength of longitudinal torsional reinforcement

f_{yt} = specified yield strength of transverse reinforcement

f_{yv} = specified yield strength of closed transverse torsional reinforcement

h = overall thickness of column or wall

h_c = diameter of concrete core measured out-to-out of spiral

h_w = total height of wall

I_b = moment of inertia of gross section of beam

I_{cr} = moment of inertia of cracked section transformed to concrete

I_e = effective moment of inertia

I_s = moment of inertia of gross section of slab

I_{se} = moment of inertia of reinforcement about centroidal axis of cross-section

J_c = equivalent polar moment of inertia of critical section

k = effective length factor for columns

K_m = material constant

K_{tr} = transverse reinforcement index

L = member length

l_1 = center-to-center span length in the direction moments are being determined

l_2 = center-to-center span length transverse to l_1

l_c = center-to-center length of columns

l_d = development length of reinforcement in tension

l_{dc} = development length of reinforcement in compression

l_{dh} = development length of standard hook in tension, measured from critical section to outside end of hook

l_n = clear span length, measured from face-to-face of supports

l_u = unsupported length of columns

l_w = horizontal length of wall

M_1 = smaller factored end moment in a column, negative if bent in double curvature

M_2 = larger factored end moment in a column, negative if bent in double curvature

M_a = maximum moment applied for deflection computation

M_c = factored magnified moment in columns

M_{cr} = cracking moment

M_m = modified moment

M_n = nominal or theoretical moment strength

M_{ns} = factored end moment of column due to loads that do not cause appreciable side sway

M_0 = total factored static moment

M_s = factored end moment of column due to loads that cause appreciable side-ways

M_u = moment demand

M_{unb} = unbalanced moment at slab–column connections

n = modular ratio = E_s/E_c

N_C = resultant compressive force of concrete

N_T = resultant tensile force of reinforcement

N_u = factored axial load occurring simultaneously with V_u or T_u, positive sign for compression

P_c = critical load

p_{cp} = outside perimeter of concrete cross-section

p_h = perimeter of centerline of outermost concrete cross-section

P_n = nominal axial load strength of column

$P_{n,max}$ = maximum nominal axial load strength of column

P_{ni} = nominal biaxial load strength of column

P_{nx} = nominal axial load strength of column about x-axis

P_{ny} = nominal axial load strength of column about y-axis

P_0 = nominal axial load strength of column at zero eccentricity

P_u = axial load demand

Q = stability index

r = radius of gyration of cross-section

s = spacing of shear or torsional reinforcement along longitudinal axis of member

S_C = structural capacity

S_D = structural demand

T_n = nominal torsional moment strength

T_u = torsional moment demand

V_c = nominal shear strength provided by concrete

V_n = nominal shear strength

v_n = nominal shear stress strength of critical section

V_s = nominal shear strength provided by shear reinforcement

V_u = shear demand

v_u = shear stress at critical section

w_u = factored load on slab per unit area

y_t = distance from centroidal axis of gross section to extreme tension fiber

α = ratio of flexural stiffness of beam section to flexural stiffness of width of a slab bounded laterally by centerlines of adjacent panels on each side of beam = $E_{cb}I_b/E_{cs}I_s$

α = reinforcement location factor (Table 7.13)

α_i = angle between inclined shear reinforcement and longitudinal axis of member

α_m = average value of α for all beams on edges of a panel

α_s = shear strength factor

α_1 = α in direction of l_1

β = ratio of clear spans in long to short direction of two-way slabs

β = reinforcement coating factor (Section 7.17.1)

β_c = ratio of long side to short side dimension of column

β_d = ratio of maximum factored sustained axial load to maximum factored axial load

β_t = ratio of torsional stiffness of edge beam section to flexural stiffness of a width of slab equal to span length of beam, center-to-center of supports

β_1 = equivalent concrete stress block factor defined in Figure 7.5

δ_{ns} = nonsway column moment magnification factor

δ_s = sway column moment magnification factor

Δ_0 = first-order relative deflection between the top and bottom of a story

ε_c = concrete strain

ε_t = steel strain

γ = reinforcement size factor = 0.8 for No. 6 and smaller bars; = 1.0 for No. 7 and larger

γ_f = fraction of unbalanced moment transferred by flexure at slab–column connections

γ_v = fraction of unbalanced moment transferred by eccentricity of shear at slab–column connections

λ = lightweight aggregate concrete factor (Section 7.17); = 1.3 for light weight concrete

λ = multiplier for additional long-term deflection

ϕ_{ecc} = strength reduction factor for accidental eccentricity in columns = 1.3 for lightweight concrete

ϕ_u = curvature at ultimate

ϕ_y = curvature at yield

ρ = ratio of tension reinforcement = A_s/bd

ρ' = ratio of compression reinforcement = A_s'/bd

ρ_h = ratio of horizontal wall reinforcement area to gross section area of horizontal section

ρ_n = ratio of vertical wall reinforcement area to gross section area of horizontal section

ρ_w = ratio of reinforcement = $A_s/b_w d$

ξ = time-dependent factor for sustained load

ϕ = strength reduction factor, see Table 7.4

θ = angle of compression diagonals in truss analogy for torsion

Useful Web Sites

American Concrete Institute: www.aci-int.org
Concrete Reinforcing Steel Institute: www.crsi.org
Portland Cement Association: www.portcement.org
International Federation of Concrete Structures: http://fib.epfl.ch
Eurocode 2: www.eurocode2.info
Reinforced Concrete Council: www.rcc-info.org.uk
Japan Concrete Institute: www.jci-net.or.jp
Emerging Construction Technologies: www.new-technologies.org

8

Prestressed Concrete*

Edward G. Nawy
*Department of Civil and Environmental
Engineering
Rutgers University — The State
University of New Jersey,
Piscataway, NJ*

---

*This chapter is a condensation from several chapters of *Prestressed Concrete — A Fundamental Approach*, 4th edition, 2003, 944 pp., by E. G. Nawy, with permission of the publishers, Prentice Hall, Upper Saddle River, NJ.

8.1 Introduction

Concrete is strong in compression, but weak in tension: its tensile strength varies from 8 to 14% of its compressive strength. Due to such a low tensile capacity, flexural cracks develop at early stages of loading. In order to reduce or prevent such cracks from developing, a concentric or eccentric force is imposed in the longitudinal direction of the structural element. This force prevents the cracks from developing by eliminating or considerably reducing the tensile stresses at the critical midspan and support sections at service load, thereby raising the bending, shear, and torsional capacities of the sections. The sections are then able to behave elastically, and almost the full capacity of the concrete in compression can be efficiently utilized across the entire depth of the concrete sections when all loads act on the structure.

Such an imposed longitudinal force is termed a *prestressing force*, that is, a compressive force that prestresses the sections along the span of the structural element prior to the application of the transverse gravity dead and live loads or transient horizontal live loads. The type of prestressing force involved, together with its magnitude, are determined mainly on the basis of the type of system to be constructed and the span length. As a result, permanent stresses in the prestressed structural member are created before the full dead and live loads are applied, in order to eliminate or considerably reduce the net tensile stresses caused by these loads.

With reinforced concrete, it is assumed that the tensile strength of the concrete is negligible and disregarded. This is because the tensile forces resulting from the bending moments are resisted by the bond created in the reinforcement process. Cracking and deflection are therefore essentially irrecoverable in reinforced concrete once the member has reached its limit state at service load. In prestressed concrete elements, cracking can be controlled or totally eliminated at the service load level. The reinforcement required to produce the prestressing force in the prestressed member actively preloads the member, permitting a relatively high controlled recovery of cracking and deflection.

8.2 Concrete for Prestressed Elements

Concrete, particularly high-strength concrete, is a major constituent of all prestressed concrete elements. Hence, its strength and long-term endurance have to be achieved through proper quality control and quality assurance at the production stage. The mechanical properties of hardened concrete can be classified into two categories: short-term or instantaneous properties, and long-term properties. The short-term properties are strength in compression, tension, and shear; and stiffness, as measured by the modulus of elasticity. The long-term properties can be classified in terms of creep and shrinkage. The following subsections present some details on these properties.

8.2.1 Compressive Strength

Depending on the type of mix, the properties of aggregate, and the time and quality of the curing, compressive strengths of concrete can be obtained up to 20,000 psi or more. Commercial production of concrete with ordinary aggregate is usually in the range 4,000 to 12,000 psi, with the most common concrete strengths being in the 6,000 psi level.

The compressive strength f_c' is based on standard 6 in. by 12 in. cylinders cured under standard laboratory conditions and tested at a specified rate of loading at 28 days of age. The standard specifications used in the United States are usually taken from American Society for Testing and Materials (ASTM) C-39. The strength of concrete in the actual structure may not be the same as that of the cylinder because of the difference in compaction and curing conditions.

8.2.2 Tensile Strength

The tensile strength of concrete is relatively low. A good approximation for the tensile strength f_{ct} is $0.10f_c' < f_{ct} < 0.20f_c'$. It is more difficult to measure tensile strength than compressive strength because of the gripping problems with testing machines. A number of methods are available for tension testing, the most commonly used method being the cylinder splitting, or Brazilian, test.

For members subjected to bending, the value of the modulus of rupture f_r rather than the tensile splitting strength f_t' is used in design. The modulus of rupture is measured by testing to failure plain concrete beams 6 in.^{2} in cross-section, having a span of 18 in., and loaded at their third points (ASTM C-78). The modulus of rupture has a higher value than the tensile splitting strength. The American Concrete Institute (ACI) specifies a value of 7.5 for the modulus of rupture of normal-weight concrete.

In most cases, lightweight concrete has a lower tensile strength than does normal-weight concrete. The following are the code stipulations for lightweight concrete:

1. If the splitting tensile strength f_{ct} is specified

$$f_r = 1.09 f_{ct} \leq 7.5\sqrt{f_c'} \tag{8.1}$$

2. If f_{ct} is not specified, use a factor of 0.75 for all-lightweight concrete and 0.85 for sand-lightweight concrete. Linear interpolation may be used for mixtures of natural sand and lightweight fine aggregate. For high-strength concrete, the modulus of rupture can be as high as $11-12\sqrt{f_c'}$.

8.2.3 Shear Strength

Shear strength is more difficult to determine experimentally than the tests discussed previously because of the difficulty in isolating shear from other stresses. This is one of the reasons for the large variation in shear-strength test values reported in the literature, varying from 20% of the compressive strength in normal loading to a considerably higher percentage of up to 85% of the compressive strength in cases where direct shear exists in combination with compression. Control of a structural design by shear strength is significant only in rare cases, since shear stresses must ordinarily be limited to continually lower values in order to protect the concrete from the abrupt and brittle failure in diagonal tension

$$E_c = 57,000\sqrt{f_c'} \text{ psi } (4,700\sqrt{f_c'} \text{ MPa}) \tag{8.2a}$$

or

$$E_c = 0.043 w^{1.5}\sqrt{f_c'} \text{ MPa} \tag{8.2b}$$

8.2.4 High-Strength Concrete

High-strength concrete is termed as such by the ACI 318 Code when the cylinder compressive strength exceeds 6,000 psi (41.4 MPa). For concrete having compressive strengths 6,000 to 12,000 psi (42 to 84 MPa), the expressions for the modulus of concrete are [1–3]

$$E_c \text{ (psi)} = \left[40,000\sqrt{f_c'} + 10^6\right]\left(\frac{w_c}{145}\right)^{1.5} \tag{8.3a}$$

where $f_c' = $ psi and $w_c = $ lb/ft^3

$$E_c \text{ (MPa)} = \left[3.32\sqrt{f_c'} + 6895\right]\left(\frac{w_c}{2320}\right)^{1.5} \tag{8.3b}$$

where $f_c' = $ MPa and $w_c = $ kg/m^3.

Today, concrete strength up to 20,000 psi (138 MPa) is easily achieved using a maximum stone aggregate size of $\frac{3}{8}$ in. (9.5 mm) and pozzolamic cementitious partial replacements for the cement such as silica fume. Such strengths can be obtained in the field under strict quality control and quality assurance conditions. For strengths in the range of 20,000 to 30,000 (138 to 206 MPa), other constituents such as steel or carbon fibers have to be added to the mixture. In all these cases, mixture design has to be made by several field trial batches (five or more), modifying the mixture components for the workability needed in concrete placement. Steel cylinder molds size 4 in. (diameter) × 8 in. length have to be used, applying the appropriate dimensional correction.

PHOTO 8.1 A rendering of the new Maumee River Bridge, Toledo, Ohio. This cable-stayed bridge spans the Maumee River in downtown Toledo as a monument icon for the city. The design includes single pylon, single plane of stays, and a main span with a horizontal clearance of 612 ft in both directions. The main pylon is clad on four of its eight sides with a glass curtain wall system, symbolizing the glass industry and heritage of Toledo. This glass prismatic system and stainless steel clad cables create a sleek and industrial look during the day. At night, the glass becomes very dynamic with the use of LED arrays back-lighting the window wall. Owner: Ohio Department of Transportation (courtesy of the Designer, Figg Engineering Group, Linda Figg, President, Tallahassee, Florida).

8.2.5 Initial Compressive Strength and Modulus

Since prestressing is performed in most cases prior to concrete's achieving its 28-day strength, it is important to determine the concrete compressive strength f'_{ct} at the prestressing stage as well as the concrete modulus E_c at various stages in the loading history of the element. The general expression for the compressive strength as a function of time [4] is

$$f'_{ci} = \frac{t}{\alpha + \beta t} f'_c \qquad (8.4a)$$

where

f_c' = 28-day compressive strength

t = time in days

α = factor depending on type of cement and curing conditions

= 4.00 for moist-cured type-I cement and 2.30 for moist-cured type-III cement

= 1.00 for steam-cured type-I cement and 0.70 for steam-cured type-III cement

β = factor depending on the same parameters for α giving corresponding values of 0.85, 0.92, 0.95, and 0.98, respectively

Hence, for a typical moist-cured type-I cement concrete

$$f_{ci}' = \frac{t}{4.00 + 0.85t} f_c'$$ (8.4b)

8.2.6 Creep

Creep, or lateral material flow, is the increase in strain with time due to a sustained load. The initial deformation due to load is the *elastic strain*, while the additional strain due to the same sustained load is the *creep strain*. This practical assumption is quite acceptable, since the initial recorded deformation includes few time-dependent effects.

The ultimate creep coefficient, C_u, is given by

$$C_u = \rho_u E_c$$ (8.5)

or average $C_u \cong 2.35$.

Branson's model, verified by extensive tests, relates the creep coefficient C_t at any time to the ultimate creep coefficient (for standard conditions) as

$$C_t = \frac{t^{0.6}}{10 + t^{0.6}} C_u$$ (8.6)

or, alternatively,

$$\rho_t = \frac{t^{0.6}}{10 + t^{0.6}}$$ (8.7)

where t is the time in days and ρ_t is the time multiplier. Standard conditions as defined by Branson pertain to concretes of slump 4 in. (10 cm) or less and a relative humidity of 40%.

When conditions are not standard, creep correction factors have to be applied to Equation 8.6 or 8.7 as follows:

1. For moist-cured concrete loaded at an age of 7 days or more

$$k_a = 1.25t^{-0.118}$$ (8.8a)

2. For steam-cured concrete loaded at an age of 1 to 3 days or more

$$k_a = 1.13t^{-0.095}$$ (8.8b)

For greater than 40% relative humidity, a further multiplier correction factor of

$$k_{c_1} = 1.27 - 0.0067H$$ (8.9)

8.2.7 Shrinkage

Basically, there are two types of shrinkage: plastic shrinkage and drying shrinkage. *Plastic shrinkage* occurs during the first few hours after placing fresh concrete in the forms. Exposed surfaces such as floor slabs are more easily affected by exposure to dry air because of their large contact surface. In such cases, moisture evaporates faster from the concrete surface than it is replaced by the bleed water from the lower layers of the concrete elements. *Drying shrinkage*, on the other hand, occurs after the concrete has already attained its final set and a good portion of the chemical hydration process in the cement gel has been accomplished.

Drying shrinkage is the decrease in the volume of a concrete element when it loses moisture by evaporation. The opposite phenomenon, that is, volume increase through water absorption, is termed *swelling*. In other words, shrinkage and swelling represent water movement out of or into the gel structure of a concrete specimen due to the difference in humidity or saturation levels between the specimen and the surroundings irrespective of the external load.

Shrinkage is not a completely reversible process. If a concrete unit is saturated with water after having fully shrunk, it will not expand to its original volume. The rate of the increase in shrinkage strain decreases with time since older concretes are more resistant to stress and consequently undergo less shrinkage, such that the shrinkage strain becomes almost asymptotic with time.

Branson recommends the following relationships for the shrinkage strain as a function of time for standard conditions of humidity ($H \cong 40\%$):

1. For moist-cured concrete any time t after 7 days

$$\varepsilon_{\mathrm{SH},t} = \frac{t}{35 + t}\left(\varepsilon_{\mathrm{SH,u}}\right) \tag{8.10a}$$

where $\varepsilon_{\mathrm{SH,u}} = 800 \times 10^{-6}$ in./in. if local data are not available.

2. For steam-cured concrete after the age of 1 to 3 days

$$\varepsilon_{\mathrm{SH},t} = \frac{t}{55 + t}\left(\varepsilon_{\mathrm{SH,u}}\right) \tag{8.10b}$$

For other than standard humidity, a correction factor has to be applied to Equations 8.10a and 8.10b as follows:

1. For $40 < H \le 80\%$

$$k_{\mathrm{SH}} = 1.40 - 0.010H \tag{8.11a}$$

2. For $80 < H \le 100\%$

$$k_{\mathrm{SH}} = 3.00 - 0.30H \tag{8.11b}$$

8.3 Steel Reinforcement Properties

8.3.1 Non-Prestressing Reinforcement

Steel reinforcement for concrete consists of bars, wires, and welded wire fabric, all of which are manufactured in accordance with ASTM standards (see Table 8.1).

TABLE 8.1   Weight, Area, and Perimeter of Individual Bars

Bar designation number	Weight per foot (lb)	Standard nominal dimensions		
		Diameter, d_b [in. (mm)]	Cross-sectional area, A_b (in.^{2})	Perimeter (in.)
3	0.376	0.375 (10)	0.11	1.178
4	0.668	0.500 (13)	0.2	1.571
5	1.043	0.625 (16)	0.31	1.963
6	1.502	0.750 (19)	0.44	2.356
7	2.044	0.875 (22)	0.6	2.749
8	2.670	1.000 (25)	0.79	3.142
9	3.400	1.128 (29)	1	3.544
10	4.303	1.270 (32)	1.27	3.99
11	5.313	1.410 (36)	1.56	4.43
14	7.65	1.693 (43)	2.25	5.32
18	13.6	2.257 (57)	4	7.09

8.3.2 Prestressing Reinforcement

Because of the high creep and shrinkage losses in concrete, effective prestressing can be achieved by using very high strength steels in the range of 270,000 psi or more (1,862 MPa or higher). Such highly stressed steels are able to counterbalance these losses in the surrounding concrete and have adequate leftover stress levels to sustain in the long term the required prestressing force (see Table 8.2).

A typical reinforcement stress–strain plot is shown in Figure 8.1.

TABLE 8.2 Seven-Wire Standard Strand for Prestressed Concrete

Nominal diameter of strand (in.)	Breaking strength of strand (min. lb)	Nominal steel area of strand (sq in.)	Nominal weight of strands (lb per 1000 ft)[a]	Minimum load at 1% extension (lb)
Grade 250				
0.25	9,000	0.036	122	7,650
0.313	14,500	0.058	197	12,300
0.375	20,000	0.08	272	17,000
0.438	27,000	0.108	367	23,000
0.500	36,000	0.144	490	30,600
0.600	54,000	0.216	737	45,900
Grade 270				
0.375	23,000	0.085	290	19,550
0.438	31,000	0.115	390	26,350
0.500	41,300	0.153	520	35,100
0.600	58,600	0.217	740	49,800

[a] 100,000 psi = 689.5 MPa.
0.1 in. = 2.54 mm; 1 in.^{2} = 645 mm^2.
Weight: multiply by 1.49 to obtain weight in kg per 1,000 m.
1,000 lb = 4,448 Newton.
  Source: Post-tensioning Institute.

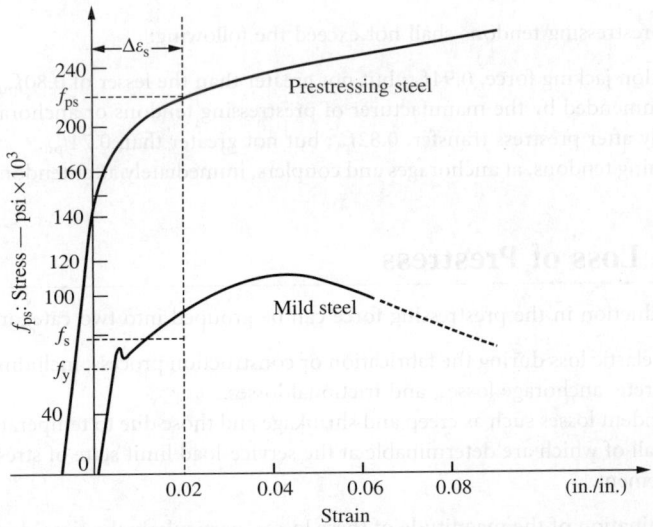

FIGURE 8.1 Stress–strain diagram for steel prestressing strands in comparison with mild steel bar reinforcement [5].

8.4 Maximum Permissible Stresses

The following are definitions of some important mathematical terms used in this section:

f_{py} is the specified yield strength of prestressing tendons, in psi.
f_y is the specified yield strength of nonprestressed reinforcement, in psi.
f_{pu} is the specified tensile strength of prestressing tendons, in psi.
f_c' is the specified compressive strength of concrete, in psi.
f_{ct}' is the compressive strength of concrete at time of initial prestress.

8.4.1 Concrete Stresses in Flexure

Stresses in concrete immediately after prestress transfer (before time-dependent prestress losses) shall not exceed the following:

1. Extreme fiber stress in compression, $0.60 f_{ci}'$.
2. Extreme fiber stress in tension except as permitted in (3), $3\sqrt{f_{ci}'}$.
3. Extreme fiber stress in tension at ends of simply supported members $6\sqrt{f_{ci}'}$.

Where computed tensile stresses exceed these values, bonded auxiliary reinforcement (nonprestressed or prestressed) has to be provided in the tensile zone to resist the total tensile force in concrete computed under the assumption of an uncracked section.

Stresses in concrete at service loads (after allowance for all prestress losses) should not exceed the following:

1. Extreme fiber stress in compression due to prestress plus sustained load, where sustained dead load and live load are a large part of the total service load, $0.45 f_c'$.
2. Extreme fiber stress in compression due to prestress plus total load, if the live load is transient, $0.60 f_c'$.
3. Extreme fiber stress in tension in precompressed tensile zone, $6\sqrt{f_c'}$.
4. Extreme fiber stress in tension in precompressed tensile zone of members (except two-way slab systems), where analysis based on transformed cracked sections and on bilinear moment–deflection relationships shows that immediate and long-time deflections comply with the ACI definition requirements and minimum concrete cover requirements, $12\sqrt{f_c'}$.

8.4.2 Prestressing Steel Stresses

Tensile stress in prestressing tendons shall not exceed the following:

1. Due to tendon jacking force, $0.94 f_{py}$; but not greater than the lesser of $0.80 f_{pu}$ and the maximum value recommended by the manufacturer of prestressing tendons or anchorages.
2. Immediately after prestress transfer, $0.82 f_{py}$; but not greater than $0.74 f_{pu}$.
3. Posttensioning tendons, at anchorages and couplers, immediately after tendon anchorage, $0.70 f_{pu}$.

8.5 Partial Loss of Prestress

Essentially, the reduction in the prestressing force can be grouped into two categories:

- Immediate elastic loss during the fabrication or construction process, including elastic shortening of the concrete, anchorage losses, and frictional losses.
- Time-dependent losses such as creep and shrinkage and those due to temperature effects and steel relaxation, all of which are determinable at the service-load limit state of stress in the prestressed concrete element.

An exact determination of the magnitude of these losses, particularly the time-dependent ones, is not feasible, since they depend on a multiplicity of interrelated factors. Empirical methods of estimating losses differ with the different codes of practice or recommendations, such as those of the Prestressed Concrete Institute, the ACI–ASCE joint committee approach, the AASHTO lump-sum approach, the Comité Eurointernationale du Béton (CEB), and the FIP (Federation Internationale de la Précontrainte). The degree of rigor of these methods depends on the approach chosen and the accepted practice of record (see Table 8.3 to Table 8.5).

In Table 8.5, $\Delta f_{pR} = \Delta f_{pR}(t_0, t_{tr}) + \Delta f_{pR}(t_{tr}, t_s)$, where t_0 is the time at jacking, t_{tr} is the time at transfer, and t_s is the time at stabilized loss. Hence, computations for steel relaxation loss have to be performed for the time interval t_1 through t_2 of the respective loading stages.

As an example, the transfer stage, say, at 18 h, would result in $t_{tr} = t_2 = 18\,h$ and $t_0 = t_1 = 0$. If the next loading stage is between transfer and 5 years (17,520 h), when losses are considered stabilized, then $t_2 = t_s = 17,520\,h$ and $t_1 = 18\,h$. Then, if f_{pi} is the initial prestressing stress that the concrete element is subjected to and f_{pj} is the jacking stress in the tendon.

8.5.1 Steel Stress Relaxation (R)

The magnitude of the decrease in the prestress depends not only on the duration of the sustained prestressing force, but also on the ratio f_{pi}/f_{py} of the initial prestress to the yield strength of the

TABLE 8.3 AASHTO Lump-Sum Losses

	Total loss	
Type of prestressing steel	$f'_c = 4,000$ psi (27.6 N/mm^2)	$f'_c = 5,000$ psi (34.5 N/mm^2)
Pretensioning strand		45,000 psi (310 N/mm^2)
Posttensioning[a] wire or strand	32,000 psi (221 N/mm^2)	33,000 psi (228 N/mm^2)
Bars	22,000 psi (152 N/mm^2)	23,000 psi (159 N/mm^2)

[a] Losses due to friction are excluded. Such losses should be computed according to Section 6.5 of the AASHTO specifications.

TABLE 8.4 Approximate Prestress Loss Values for Posttensioning

	Prestress loss, psi	
Posttensioning tendon material	Slabs	Beams and joists
Stress-relieved 270-K strand and stress-relieved 240-K wire	30,000 (207 N/mm^2)	35,000 (241 N/mm^2)
Bar	20,000 (138 N/mm^2)	25,000 (172 N/mm^2)
Low-relaxation 270-K strand	15,000 (103 N/mm^2)	20,000 (138 N/mm^2)

TABLE 8.5 Types of Prestress Losses

	Stage of occurrence		Tendon stress loss	
Type of prestress loss	Pretensioned members	Posttensioned members	During time interval (t_i, t_j)	Total or during life
Elastic shortening of concrete (ES)	At transfer	At sequential jacking	...	Δf_{pES}
Relaxation of tendons (R)	Before and after transfer	After transfer	$\Delta f_{pR}\,(t_i, t_j)$	Δf_{pR}
Creep of concrete (CR)	After transfer	After transfer	$\Delta f_{pCR}\,(t_i, t_j)$	Δf_{pCR}
Shrinkage of concrete (SH)	After transfer	After transfer	$\Delta f_{pSH}\,(t_i, t_j)$	Δf_{pSH}
Friction (F)	...	At jacking	...	Δf_{pF}
Anchorage seating loss (A)	...	At transfer	...	Δf_{pA}
Total	Life	Life	$\Delta f_{pT}\,(t_i, t_j)$	Δf_{pT}

reinforcement. Such a loss in stress is termed *stress relaxation*. The ACI 318-02 Code limits the tensile stress in the prestressing tendons to the following:

1. For stresses due to the tendon jacking force, $f_{pj} = 0.94 f_{py}$, but not greater than the lesser of $0.80 f_{pu}$ and the maximum value recommended by the manufacturer of the tendons and anchorages.
2. Immediately after prestress transfer, $f_{pi} = 0.82 f_{py}$, but not greater than $0.74 f_{pu}$.
3. In posttensioned tendons, at the anchorages and couplers immediately after force transfer $= 0.70 f_{pu}$.

The range of values of f_{py} is given by the following:

Prestressing bars: $f_{py} = 0.80 f_{pu}$
Stress-relieved tendons: $f_{py} = 0.85 f_{pu}$
Low-relaxation tendons: $f_{py} = 0.90 f_{pu}$

If f_{pR} is the remaining prestressing stress in the steel after relaxation, the following expression defines f_{pR} for stress relieved steel:

$$\frac{f_{pR}}{f_{pi}} = 1 - \left(\frac{\log t_2 - \log t_1}{10}\right)\left(\frac{f_{pi}}{f_{py}} - 0.55\right) \tag{8.12}$$

In this expression, log t in hours is to the base 10, f_{pi}/f_{py} exceeds 0.55, and $t = t_2 - t_1$. Also, for low-relaxation steel, the denominator of the log term in the equation is divided by 45 instead of 10.

An approximation of the term $(\log t_2 - \log t_1)$ can be made in Equation 8.13 so that $\log t = \log(t_2 - t_1)$ without significant loss in accuracy. In that case, the stress-relaxation loss becomes

$$\Delta f_{pR} = f'_{pi} \frac{\log t}{10}\left(\frac{f'_{pi}}{f_{py}} - 0.55\right) \tag{8.13}$$

where f'_{pi} is the initial stress in steel to which the concrete element is subjected.

If a step-by-step loss analysis is necessary, the loss increment at any particular stage can be defined as

$$\Delta f_{pR} = f'_{pi}\left(\frac{\log t_2 - \log t_1}{10}\right)\left(\frac{f_{pi}}{f_{py}} - 0.55\right) \tag{8.14}$$

where t_1 is the time at the beginning of the interval and t_2 is the time at the end of the interval from jacking to the time when the loss is being considered.

For low relaxation steel, change the divider to 45 instead of 10 in Equation 8.14.

8.5.2 Creep Loss (CR)

The creep coefficient at any time t in days can be defined as

$$C_t = \frac{t^{0.60}}{10 + t^{0.60}} C_u \tag{8.15}$$

As discussed earlier, the value of C_u ranges between 2 and 4, with an average of 2.35 for ultimate creep. The loss in prestressed members due to creep can be defined for bonded members as

$$\Delta f_{CR} = C_t \frac{E_{ps}}{E_c} f_{cs} \tag{8.16}$$

where f_{cs} is the stress in the concrete at the level of the centroid of the prestressing tendon. In general, this loss is a function of the stress in the concrete at the section being analyzed. In posttensioned, nonbonded members, the loss can be considered essentially uniform along the whole span. Hence, an average value of the concrete stress f_{cs} between the anchorage points can be used for calculating the creep in posttensioned members.

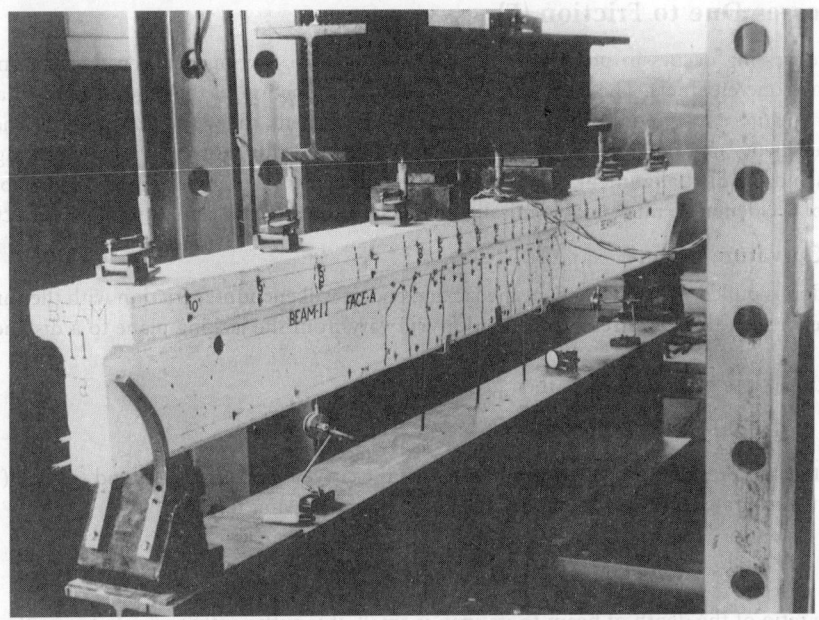

PHOTO 8.2 Prestressed concrete pretensioned T-beam tested to failure at the Rutgers University Concrete Research Laboratory (courtesy E.G. Nawy).

8.5.3 Shrinkage Loss (SH)

As with concrete creep, the magnitude of the shrinkage of concrete is affected by several factors. They include mixture proportions, type of aggregate, type of cement, curing time, time between the end of external curing and the application of prestressing, size of the member, and the environmental conditions. Size and shape of the member also affect shrinkage. Approximately 80% of shrinkage takes place in the first year of life of the structure. The average value of ultimate shrinkage strain in both moist-cured and steam-cured concrete is given as 780×10^{-6} in./in. in the ACI 209 R-92 Report. This average value is affected by the length of initial moist curing, ambient relative humidity, volume–surface ratio, temperature, and concrete composition. To take such effects into account, the average value of shrinkage strain should be multiplied by a correction factor γ_{SH} as follows:

$$\varepsilon_{SH} = 780 \times 10^{-6} \gamma_{SH} \tag{8.17}$$

Components of γ_{SH} are factors for various environmental conditions and tabulated in Ref. [11, Section 2].

The Prestressed Concrete Institute (PCI) stipulates for standard conditions an average value for nominal ultimate shrinkage strain $(\varepsilon_{SH})_u = 820 \times 10^{-6}$ in./in. (mm/mm) [7]. If ε_{SH} is the shrinkage strain after adjusting for relative humidity at volume-to-surface ratio V/S, the loss in prestressing in pretensioned member is

$$\Delta f_{pSH} = \varepsilon_{SH} \times E_{ps} \tag{8.18}$$

For posttensioned members, the loss in prestressing due to shrinkage is somewhat less since some shrinkage has already taken place before posttensioning. If the relative humidity is taken as a percent value and the V/S ratio effect is considered, the PCI general expression for loss in prestressing due to shrinkage becomes

$$\Delta f_{pSH} = 8.2 \times 10^{-6} K_{SH} E_{ps} \left(1 - 0.06 \frac{V}{S}\right)(100 - \text{RH}) \tag{8.19}$$

where RH is the relative humidity (see Table 8.6).

8.5.4 Losses Due to Friction (F)

Loss of prestressing occurs in posttensioning members due to friction between the tendons and the surrounding concrete ducts. The magnitude of this loss is a function of the tendon form or alignment, called the *curvature effect*, and the local deviations in the alignment, called the *wobble effect*. The values of the loss coefficients are often refined while preparations are made for shop drawings by varying the types of tendons and the duct alignment. Whereas the curvature effect is predetermined, the wobble effect is the result of accidental or unavoidable misalignment, since ducts or sheaths cannot be perfectly placed.

8.5.4.1 Curvature Effect

As the tendon is pulled with a force F_1 at the jacking end, it will encounter friction with the surrounding duct or sheath such that the stress in the tendon will vary from the jacking plane to a distance L along the span.

The frictional loss of stress Δf_{pF} is then given by

$$\Delta f_{\mathrm{pF}} = f_1 - f_2 = f_1\left(1 - e^{-\mu\alpha - KL}\right) \tag{8.20}$$

Assuming that the prestress force between the start of the curved portion and its end is small ($\cong 15\%$), it is sufficiently accurate to linearize Equation 8.20 into the following form:

$$\Delta f_{\mathrm{pF}} = -f_1(\mu\alpha + KL) \tag{8.21}$$

where L is in feet.

Since the ratio of the depth of beam to its span is small, it is sufficiently accurate to use the projected length of the tendon for calculating α, giving

$$\alpha = 8y/x \text{ radian} \tag{8.22}$$

Table 8.7 gives the design values of the curvature friction coefficient μ and the wobble or length friction coefficient K adopted from the ACI 318 Code.

TABLE 8.6 Values of K_{SH} for Posttensioned Members

Time from end of moist curing to application of prestress, days	1	3	5	7	10	20	30	60
K_{SH}	0.92	0.85	0.8	0.77	0.73	0.64	0.58	0.45

Source: Prestressed Concrete Institute.

TABLE 8.7 Wobble and Curvature Friction Coefficients

Type of tendon	Wobble coefficient, K per foot	Curvature coefficient, μ
Tendons in flexible metal sheathing		
Wire tendons	0.0010–0.0015	0.15–0.25
Seven-wire strand	0.0005–0.0020	0.15–0.25
High-strength bars	0.0001–0.0006	0.08–0.30
Tendons in rigid metal duct		
Seven-wire strand	0.0002	0.15–0.25
Mastic-coated tendons		
Wire tendons and seven-wire strand	0.0010–0.0020	0.05–0.15
Pregreased tendons		
Wire tendons and seven-wire strand	0.0003–0.0020	0.05–0.15

Source: Prestressed Concrete Institute.

8.5.5 Example 1: Prestress Losses in Beams

A simply supported posttensioned 70-ft-span lightweight steam-cured double T-beam as shown in Figure 8.2 is prestressed by twelve $\frac{1}{2}$-in. diameter (twelve 12.7 mm diameter) 270-K grade stress-relieved strands. The tendons are harped, and the eccentricity at midspan is 18.73 in. (476 mm) and at the end 12.98 in. (330 mm). Compute the prestress loss at the critical section in the beam of 0.40 span due to dead load and superimposed dead load at

1. stage I at transfer
2. stage II after concrete topping is placed
3. two years after concrete topping is placed

Suppose the topping is 2 in. (51 mm) normal-weight concrete cast at 30 days. Suppose also that prestress transfer occurred 18 h after tensioning the strands. Given

$$f_c' = 5000 \text{ psi, lightweight (34.5 MPa)}$$
$$f_{ci}' = 3500 \text{ psi (24.1 MPa)}$$

and the following noncomposite section properties:

$$A_c = 615 \text{ in.}^2 \ (3{,}968 \text{ cm}^2)$$
$$I_c = 59{,}720 \text{ in.}^4 \ (2.49 \times 10^6 \text{ cm}^4)$$
$$c_b = 21.98 \text{ in. (55.8 cm)}$$
$$c^t = 10.02 \text{ in. (25.5 cm)}$$
$$S_b = 2{,}717 \text{ in.}^3 \ (44{,}520 \text{ cm}^3)$$
$$S^t = 5{,}960 \text{ in.}^3 \ (97{,}670 \text{ cm}^3)$$
$$W_D \text{ (no topping)} = 491 \text{ plf (7.2 kN/m)}$$
$$W_{SD} \text{ (2-in. topping)} = 250 \text{ plf (3.65 kN/m)}$$

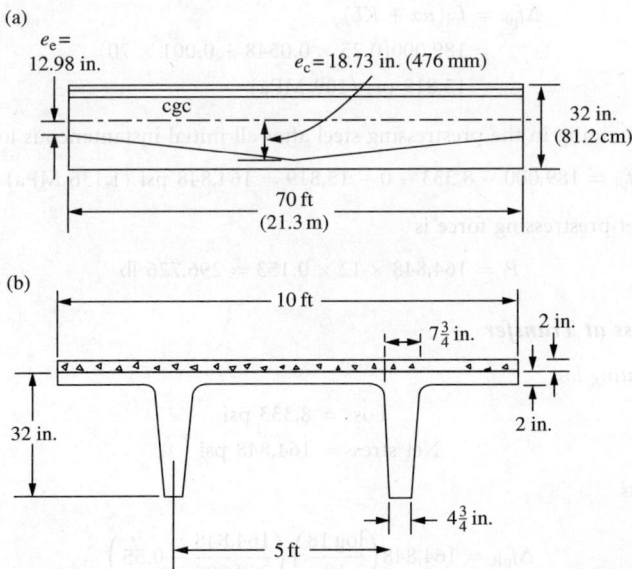

FIGURE 8.2 Double tee pretensioned beam in Example 1: (a) elevation and (b) pretensioned section [5].

$$W_L = 40 \text{ psf } (1{,}915 \text{ Pa}) \text{ --- transient}$$

$$f_{pu} = 270{,}000 \text{ psi } (1{,}862 \text{ MPa})$$

$$f_{py} = 0.85 f_{pu} \approx 230{,}000 \text{ psi } (1{,}589 \text{ MPa})$$

$$f_{pi} = 0.70 f_{pu} = 0.82 f_{py} = 0.82 \times 0.85 f_{pu} \cong 0.70 f_{pu}$$
$$= 189{,}000 \text{ psi } (1{,}303 \text{ MPa})$$

$$E_{ps} = 28 \times 10^6 \text{ psi } (193.1 \times 10^6 \text{ MPa})$$

$$18\text{-day modular ratio} = 9.72$$

Solution

1. *Anchorage seating loss*

$$\Delta_A = \tfrac{1}{4} \text{ in.} = 0.25 \text{ in.,} \quad L = 70 \text{ ft}$$

The anchorage slip stress loss is

$$\Delta f_{pA} = \frac{\Delta_A}{L} E_{ps} = \frac{0.25}{70 \times 12} \times 28 \times 10^6 \cong 8333 \text{ psi } (40.2 \text{ MPa})$$

2. *Elastic shortening.* Since all jacks are simultaneously posttensioned, the elastic shortening will precipitate during jacking. As a result, no elastic shortening stress loss takes place in the tendons. Hence, $\Delta f_{pES} = 0$.
3. *Frictional loss.* Assume that the parabolic tendon approximates the shape of an arc of a circle. Then, from Equation 8.22

$$\alpha = \frac{8y}{x} = \frac{8(18.73 - 12.98)}{70 \times 12} = 0.0548 \text{ radian}$$

From Table 8.7, use $K = 0.001$ and $\mu = 0.25$.
From Equation 8.21, the stress loss in prestress due to friction is

$$\Delta f_{pF} = f_{pi}(\mu\alpha + KL)$$
$$= 189{,}000(0.25 \times 0.0548 + 0.001 \times 70)$$
$$= 15{,}819 \text{ psi } (109 \text{ MPa})$$

The stress remaining in the prestressing steel after all initial instantaneous losses is

$$f_{pi} = 189{,}000 - 8{,}333 - 0 - 15{,}819 = 164{,}848 \text{ psi } (1{,}136 \text{ MPa})$$

Hence, the net prestressing force is

$$P_i = 164{,}848 \times 12 \times 0.153 = 296{,}726 \text{ lb}$$

Stage I: Stress at Transfer

1. *Anchorage seating loss*

$$\text{Loss} = 8{,}333 \text{ psi}$$
$$\text{Net stress} = 164{,}848 \text{ psi}$$

2. *Relaxation loss*

$$\Delta f_{pR} = 164{,}848 \left(\frac{\log 18}{10}\right)\left(\frac{164{,}848}{230{,}000} - 0.55\right)$$
$$\cong 3{,}450 \text{ psi } (23.8 \text{ MPa})$$

3. *Creep loss*

$$\Delta f_{pCR} = 0$$

4. *Shrinkage loss*

$$\Delta f_{pSH} = 0$$

So the tendon stress f_{pi} at the end of stage I is

$$164{,}848 - 3{,}450 = 161{,}398 \text{ psi } (1{,}113 \text{ MPa})$$

Stage II: Transfer to Placement of Topping after 30 Days

1. *Creep loss*

$$P_i = 161{,}398 \times 12 \times 0.153 = 296{,}327 \text{ lb}$$

$$\bar{f}_{cs} = -\frac{P_i}{A_c}\left(1 + \frac{e^2}{r^2}\right) + \frac{M_D e}{I_c}$$

$$= -\frac{296{,}327}{615}\left(1 + \frac{(17.58)^2}{97.11}\right) + \frac{3{,}464{,}496 \times 17.58}{59{,}720}$$

$$= -2016.2 + 1020.0 = 996.2 \text{ psi } (6.94 \text{ MPa})$$

Hence, the creep loss for lightweight concrete, K_{CR} is reduced by 20%, hence $= 1.6 \times 0.80 = 1.28$

$$\Delta f_{pCR} = nK_{CR}\left(\bar{f}_{cs} - \bar{f}_{csd}\right)$$
$$= 9.72 \times 1.28(996.2 - 519.3) \cong 5933 \text{ psi } (41 \text{ MPa})$$

2. *Shrinkage loss.* $K_{SH} = 0.58$ at 30 days, Table 8.6

$$\Delta f_{pSH} = 6190 \times 0.58 = 3590 \text{ psi } (24.8 \text{ MPa})$$

3. *Steel relaxation loss at 30 days*

$$f_{ps} = 161{,}398 \text{ psi}$$

The relaxation loss in stress becomes

$$\Delta f_{pR} = 161{,}398\left(\frac{\log 720 - \log 18}{10}\right)\left(\frac{161{,}398}{230{,}000} - 0.55\right)$$
$$\cong 3{,}923 \text{ psi } (27.0 \text{ MPa})$$

Stage II: Total Losses

$$\Delta f_{pT} = \Delta f_{pCR} + \Delta f_{pSH} + \Delta f_{pR}$$
$$= 5{,}933 + 3{,}590 + 3{,}923 = 13{,}446 \text{ psi } (93 \text{ MPa})$$

The increase f_{SD} in stress in the strands due to the addition of topping is

$$f_{SD} = n\bar{f}_{csd} = n(M_{SD} \times e/I_c) = n \times 519.3 = 9.72 \times 519.3 = 5048 \text{ psi } (34.8 \text{ MPa}) \text{ in tension}$$

$f_{SD} = 5048$ psi (34.8 MPa); hence, the strand stress at the end of stage II is

$$f_{pe} = f_{ps} - \Delta f_{pT} + \Delta f_{SD} = 161{,}398 - 13{,}446 + 5{,}048 = 153{,}000 \text{ psi } (1{,}055 \text{ MPa})$$

Stage III: At the End of 2 Years

$$f_{pe} = 151{,}516 \text{ psi}$$
$$t_1 = 720 \text{ h}$$
$$t_2 = 17{,}520 \text{ h}$$

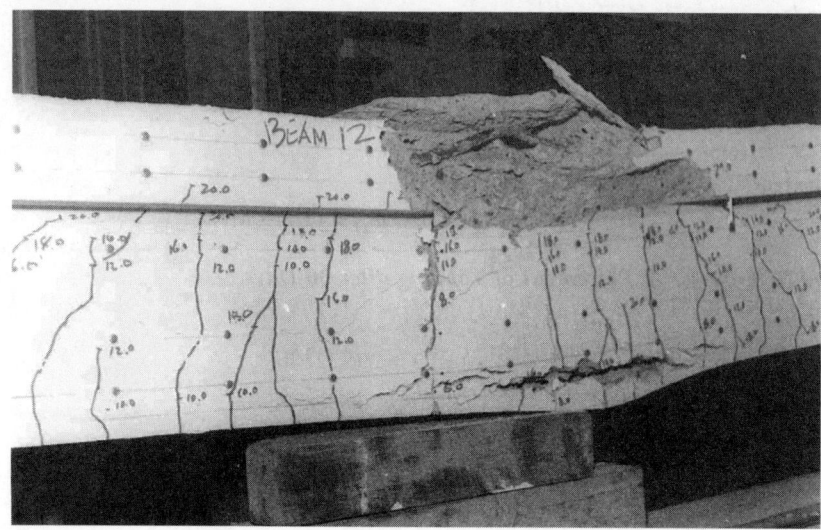

PHOTO 8.3   Crushing of the concrete at the top fibers and yield of reinforcement at the bottom fibers of the prestressed concrete T-beam in Photo 8.2 at Ultimate State of Failure. Tested at the Rutgers University Concrete Research Laboratory (courtesy E.G. Nawy).

The steel relaxation stress loss is

$$\Delta f_{\mathrm{pR}} = 153{,}000 \left(\frac{\log 17{,}520 - \log 720}{10} \right) \left(\frac{153{,}000}{230{,}000} - 0.55 \right)$$

$$\cong 2{,}444 \text{ psi (16.9 MPa)}$$

Hence, the strand stress f_{pe} at the end of stage III is approximately

$$153{,}000 - 2{,}444 = 150{,}536 \text{ psi (1,038 MPa)}$$

8.5.6   Example 2: Prestressing Losses Evaluation Using SI Units

Solve Example 1 using SI units for losses in prestress, considering self-weight and superimposed dead load only.

Data

$$f_c' = 34.5 \text{ MPa}$$
$$f_{ci}' = 24.1 \text{ MPa}$$

$$A_c = 3{,}968 \text{ cm}^2, \qquad S^t = 97{,}670 \text{ cm}^3$$
$$I_c = 2.49 \times 10^6 \text{ cm}^4, \quad S_b = 44{,}520 \text{ cm}^3$$
$$r^2 = I_c / A_c = 626 \text{ cm}^2$$
$$c_b = 55.8 \text{ cm}, \qquad\qquad c^t = 25.5 \text{ cm}$$
$$e_c = 47.6 \text{ cm}, \qquad\qquad e_e = 33.0 \text{ cm}$$
$$f_{\mathrm{pu}} = 1{,}860 \text{ MPa}$$
$$f_{\mathrm{py}} = 0.85 f_{\mathrm{pu}} = 1{,}580 \text{ MPa}$$
$$f_{\mathrm{pi}} = 0.82 f_{\mathrm{py}} = (0.82 \times 0.85) f_{\mathrm{pu}}$$
$$\qquad = 0.7\, f_{\mathrm{pu}} = 1{,}300 \text{ MPa}$$
$$E_{\mathrm{ps}} = 193{,}000 \text{ MPa}$$

Span $l = 21.3$ m

$$A_{ps} = 12 \text{ strands, } 12.7\text{-mm diameter } (99 \text{ mm}^2)$$

$$= 12 \times 99 = 1{,}188 \text{ mm}^2$$

$$M_D = 391 \text{ kN m}, \quad M_{SD} = 199 \text{ kN m}$$

$$\Delta_A = 0.64 \text{ cm}$$

$$V/S = 1.69, \quad RH = 70\%$$

$$E_c = w^{1.5} 0.043 \sqrt{f_c'}, \quad w \text{ (lightweight) } \approx 1830 \text{ kg/m}^3$$

$$E_{ci} = w^{1.5} 0.043 \sqrt{f_{ci}'}$$

$$MPa = 10^6 \text{ N/m}^2 = \text{N/mm}^2$$

$$(\text{psi}) \ 0.006895 = MPa$$

$$(\text{lb/ft}) \ 14.593 = \text{N/m}$$

$$(\text{in. lb}) \ 0.113 = \text{N m}$$

Solution

1. *Anchorage seating loss*

$$\Delta_A = 0.64 \text{ cm}, \quad l = 21.3 \text{ m}$$

$$\Delta f_{PA} = \frac{\Delta_A}{L} E_{ps} = \frac{0.64}{21.3 \times 100} \times 193{,}000 = 58.0 \text{ MPa}$$

2. *Elastic shortening*
Since all jacks are simultaneously tensioned, the elastic shortening will simultaneously precipitate during jacking. As a result, no elastic shortening loss takes place in the tendons.

$$E_c = w^{1.5} 0.043 \sqrt{34.5}$$

$$= 1{,}830^{1.5} \times 0.043 \sqrt{34.5} = 19{,}770 \text{ MPa}$$

$$n = \frac{E_{ps}}{E_c} = \frac{193{,}000}{19{,}770} = 9.76$$

$\bar{f}_{csd} =$ stress in concrete at cgs due to all superimposed dead loads after prestressing is accomplished

$$\bar{f}_{csd} = \frac{M_{SD}e}{I_c} = \frac{1.99 \times 10^7 \text{ N cm} \times 44.7}{2.49 \times 10^6} \times \frac{1}{100} \text{ N/mm}^2$$

$$= 3.57 \text{ MPa}$$

$$K_{CR} = 1.6 \text{ for posttensioned beam}$$

$$\Delta f_{pCR} = n K_{CR}(\bar{f}_{cs} - \bar{f}_{csd})$$

$$= 9.76 \times 1.6(6.90 - 3.57) = 52.0 \text{ MPa}$$

3. *Shrinkage loss at 30 days*
From Equation 8.19

$$\Delta f_{pSH} = 8.2 \times 10^{-6} K_{SH} E_{ps} \left(1 - 0.06 \frac{V}{S}\right)(100 - RH)$$

K_{SH} at 30 days $= 0.58$ (Table 8.6):

$$\Delta f_{pSH} = 8.2 \times 10^{-6} \times 0.58 \times 193,000(1 - 0.06 \times 1.69)(100 - 70)$$
$$= 24.7 \text{ MPa}$$

4. *Relaxation loss at 30 days (720 h)*

$$f_{ps} = 1108 \text{ MPa}$$

$$\Delta f_{pR} = 1108 \left(\frac{\log 720 - \log 18}{10} \right) \left(\frac{1108}{1580} - 0.55 \right)$$
$$= 110.8(2.85 - 1.25)0.151 = 26.8 \text{ MPa}$$

Stage II: Total Losses

$$\Delta f_{pT} = \Delta f_{pCR} + \Delta f_{pSH} + \Delta f_{pR}$$
$$= 52.0 + 24.7 + 26.8 = 104 \text{ MPa}$$

Increase of tensile stress at bottom cgs fibers due to addition of topping is from before

$$\Delta f_{SD} = n f_{CSD} = 9.76 \times 3.57 = 34.8 \text{ MPa}$$
$$f_{pe} = f_{ps} - \Delta f_{pT} + \Delta f_{SD}$$
$$= 1108 - 103.5 + 34.5 = 1039 \text{ MPa}$$

Stage III: At End of 2 Years

$$f_{pe} = 1039 \text{ MPa}$$

$$t_1 = 720 \text{ h}, \qquad t_2 = 17,520 \text{ h}$$

$$\Delta f_{pR} = 1,039 \left(\frac{\log 17,520 - \log 720}{10} \right) \left(\frac{1,039}{1,580} - 0.55 \right)$$
$$= 103.9(4.244 - 2.857)0.108 = 15.6 \text{ MPa}$$

On the assumption that Δf_{pCR} and Δf_{pSH} were stable in this case, the stress in the tendons at end of stage III can approximately be $f_{ps} = 1039 - 15.6 \cong 1020 \text{ MPa}$ (see Figure 8.3).

8.6 Flexural Design of Prestressed Concrete Elements

Unlike the case of reinforced concrete members, the external dead load and partial live load are applied to the prestressed concrete member at varying concrete strengths at various loading stages. These loading stages can be summarized as follows:

- Initial prestress force P_i is applied; then, at transfer, the force is transmitted from the prestressing strands to the concrete.
- The full self-weight W_D acts on the member together with the initial prestressing force, provided that the member is simply supported, that is, there is no intermediate support.
- The full superimposed dead load W_{SD}, including topping for composite action, is applied to the member.
- Most short-term losses in the prestressing force occur, leading to a reduced prestressing force P_{eo}.
- The member is subjected to the full service load, with long-term losses due to creep, shrinkage, and strand relaxation taking place and leading to a net prestressing force P_e.
- Overloading of the member occurs under certain conditions up to the limit state at failure.

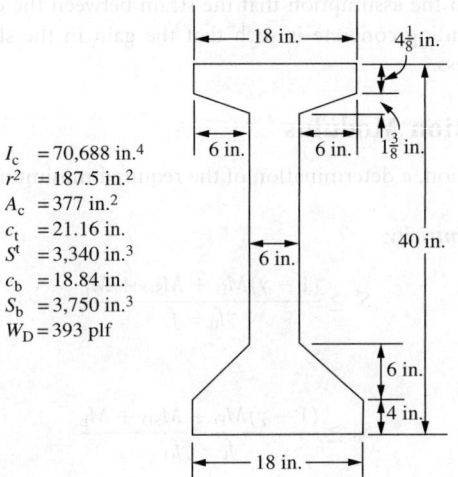

$$I_c = 70,688 \text{ in.}^4$$
$$r^2 = 187.5 \text{ in.}^2$$
$$A_c = 377 \text{ in.}^2$$
$$c_t = 21.16 \text{ in.}$$
$$S^t = 3,340 \text{ in.}^3$$
$$c_b = 18.84 \text{ in.}$$
$$S_b = 3,750 \text{ in.}^3$$
$$W_D = 393 \text{ plf}$$

FIGURE 8.3   I-Beam section in Example 2 [5].

Stress at transfer

$$f^t = -\frac{P_i}{A_c}\left(1 - \frac{ec_t}{r^2}\right) - \frac{M_D}{S^t} \leq f_{ti} \qquad (8.23a)$$

$$f_b = -\frac{P_i}{A_c}\left(1 + \frac{ec_b}{r^2}\right) + \frac{M_D}{S_b} \leq f_{ci} \qquad (8.23b)$$

where P_i is the initial prestressing force. While a more accurate value to use would be the horizontal component of P_i, it is reasonable for all practical purposes to disregard such refinement.

Effective stresses after losses

$$f^t = -\frac{P_e}{A_c}\left(1 - \frac{ec_t}{r^2}\right) - \frac{M_D}{S^t} \leq f_t \qquad (8.24a)$$

$$f_b = -\frac{P_e}{A_c}\left(1 + \frac{ec_b}{r^2}\right) + \frac{M_D}{S_b} \leq f_c \qquad (8.24b)$$

Service-load final stresses

$$f^t = -\frac{P_e}{A_c}\left(1 - \frac{ec_t}{r^2}\right) - \frac{M_T}{S^t} \leq f_c \qquad (8.25a)$$

$$f_b = -\frac{P_e}{A_c}\left(1 + \frac{ec_b}{r^2}\right) + \frac{M_T}{S_b} \leq f_c \qquad (8.25b)$$

where $M_T = M_D + M_{SD} + M_L$; P_i is the initial prestress; P_e is the effective prestress after losses, where t denotes the top and b denotes the bottom fibers; e is the eccentricity of tendons from the concrete section center of gravity, cgc; r^2 is the square of radius of gyration; and S^t/S_b is the top/bottom section modulus value of concrete section.

The *decompression stage* denotes the increase in steel strain due to the increase in load from the stage when the effective prestress P_e acts *alone* to the stage when the additional load causes the compressive stress in the concrete at the cgs level to reduce to zero. At this stage, the *change* in concrete stress due to decompression is

$$f_{decomp} = \frac{P_e}{A_c}\left(1 + \frac{e^2}{r^2}\right) \qquad (8.25c)$$

This relationship is based on the assumption that the strain between the concrete and the prestressing steel bonded to the surrounding concrete is such that the gain in the steel stress is the same as the decrease in the concrete stress.

8.6.1 Minimum Section Modulus

To design or choose the section, a determination of the required minimum section modulus, S_b and S^t, has to be made first.

1. *For variable tendon eccentricity:*

$$S^t \geq \frac{(1 - \gamma)M_D + M_{SD} + M_L}{\gamma f_{ti} - f_c} \qquad (8.26a)$$

and

$$S_b \geq \frac{(1 - \gamma)M_D + M_{SD} + M_L}{f_t - \gamma f_{ci}} \qquad (8.26b)$$

2. *For constant tendon eccentricity:*

$$S^t \geq \frac{M_D + M_{SD} + M_L}{\gamma f_{ti} - f_c} \qquad (8.27a)$$

and

$$S_b \geq \frac{M_D + M_{SD} + M_L}{f_t - \gamma f_{ci}} \qquad (8.27b)$$

The required eccentricity value at the critical section, such as the support for an ideal beam section having properties close to those required by Equations 8.27a and 8.27b, is

$$e_e = \left(f_{ti} - \bar{f}_{ci} \right) \frac{S^t}{P_i} \qquad (8.28)$$

Table 8.8 and Table 8.9 list the properties of standard sections.

TABLE 8.8 Geometrical Outer Dimensions and Section Moduli of Standard AASHTO Bridge Sections

	AASHTO sections					
Designation	Type 1	Type 2	Type 3	Type 4	Type 5	Type 6
Area A_c, in.^{2}	276	369	560	789	1,013	1,085
Moment of inertia						
$I_g(x - x)$, in.^{4}	22,750	50,979	125,390	260,741	521,180	733,320
$I_g(y - y)$, in.^{4}	3,352	5,333	12,217	24,347	61,235	61,619
Top-/bottom-section modulus, in.^{3}	1,476	2,527	5,070	8,908	16,790	20,587
	1,807	3,320	6,186	10,544	16,307	20,157
Top flange width, b_f (in.)	12	12	16	20	42	42
Top flange average thickness, t_f (in.)	6	8	9	11	7	7
Bottom flange width, b_2 (in.)	16	18	22	26	28	28
Bottom flange average thickness, t_2 (in.)	7	9	11	12	13	13
Total depth, h (in.)	28	36	45	54	63	72
Web width, b_w (in.)	6	6	7	8	8	8
c_t/c_b (in.)	15.41	20.17	24.73	29.27	31.04	35.62
	12.59	15.83	20.27	24.73	31.96	36.38
r^2, in.^{2}	82	132	224	330	514	676
Self-weight w_D, lb/ft	287	384	583	822	1,055	1,130

TABLE 8.9 Geometrical Outer Dimensions and Section Moduli of Standard PCI Double T-Sections

Designation	Top-/bottom-section modulus, in.^{3}	Flange width b_f, in.	Flange depth t_f, in.	Total depth h, in.	Web width $2b_w$, in.
8DT12	1,001/315	96	2	12	9.5
8DT14	1,307/429	96	2	14	9.5
8DT16	1,630/556	96	2	16	9.5
8DT20	2,320/860	96	2	20	9.5
8DT24	3,063/1,224	96	2	24	9.5
8DT32	5,140/2,615	96	2	32	9.5
10DT32	5,960/2,717	120	2	32	12.5
12DT34^a	10,458/3,340	144	4	34	12.5
15DT34^a	13,128/4,274	180	4	34	12.5

^{a} Pretopped.

8.6.2 Example 3: Flexural Design of Prestressed Beams at Service Load Level

Design an I-section for a beam having a 65-ft (19.8 m) span to satisfy the following section modulus values:

$$\text{Required } S^t = 3,570 \text{ in.}^3 \ (58,535 \text{ cm}^3)$$

$$\text{Required } S_b = 3,780 \text{ in.}^3 \ (61,940 \text{ cm}^3)$$

Given

$$W_{SD} = 100 \text{ plf}$$

$$W_L = 1,100 \text{ plf}$$

$$f_c' = 5,000 \text{ psi } (34.5 \text{ MPa})$$

$$f_{ci}' \text{ at transfer} = 75\% \text{ of cylinder strength}$$

$$f_{pu} = 270,000 \text{ psi } (1,862 \text{ MPa})$$

$$f_t = 12\sqrt{f_c'}$$

Solution

Since the section moduli at the top and bottom fibers are almost equal, a symmetrical section is adequate. Next, analyze the section in Figure 8.3 chosen by trial and adjustment.

Analysis of Stresses at Transfer

$$\bar{f}_{ci} = f_{ti} - \frac{c_t}{h}(f_{ti} - f_{ci})$$

$$= +184 - \frac{21.16}{40}(+184 + 2,250) \cong -1,104 \text{ psi } (C) \ (7.6 \text{ MPa})$$

$$P_i = A_c \bar{f}_{ci} = 377 \times 1,104 = 416,208 \text{ lb } (1,851 \text{ kN})$$

$$M_D = \frac{393(65)^2}{8} \times 12 = 2,490,638 \text{ in. lb } (281 \text{ kN m})$$

The eccentricity required at the section of maximum moment at midspan is

$$e_c = (f_{ti} - \bar{f}_{ci})\frac{S_t}{P_i} + \frac{M_D}{P_i}$$

$$= (184 + 1,104)\frac{3,572}{416,208} + \frac{2,490,638}{416,208}$$

$$= 11.05 + 5.98 = 17.04 \text{ in. } (433 \text{ mm})$$

Since $c_b = 18.84$ in., and assuming a cover of 3.75 in., try $e_c = 18.84 - 3.75 = 15.0$ in. (381 mm):

$$\text{Required area of strands } A_p = \frac{P_i}{f_{pi}} = \frac{416,208}{189,000} = 2.2 \text{ in.}^2 \ (14.2 \text{ cm}^2)$$

$$\text{Number of strands} = \frac{2.2}{0.153} = 14.38$$

Try thirteen $\frac{1}{2}$-in. strands, $A_p = 1.99$ in.^{2} (12.8 cm^2), and an actual $P_i = 189,000 \times 1.99 = 376,110$ lb (1,673 kN), and check the concrete extreme fiber stresses. From Equation 8.23a

$$f^t = -\frac{P_i}{A_c}\left(1 - \frac{ec_t}{r^2}\right) - \frac{M_D}{S^t}$$

$$= -\frac{376,110}{377}\left(1 - \frac{15.0 \times 21.16}{187.5}\right) - \frac{2,490,638}{3,340}$$

$$= +691.2 - 745.7 = -55 \text{ psi } (C), \text{ no tension at transfer, OK.}$$

From Equation 8.23b

$$f_b = -\frac{P_i}{A_c}\left(1 + \frac{ec_b}{r^2}\right) + \frac{M_D}{S_b}$$

$$= -\frac{376,110}{377}\left(1 + \frac{15.0 \times 18.84}{187.5}\right) + \frac{2,490,638}{3,750}$$

$$= -2501.3 + 664.2 = -1837 \text{ psi } (C) < f_{ci} = 2,250 \text{ psi, OK.}$$

Analysis of stresses at service load. From Equation 8.25a

$$f^t = -\frac{P_e}{A_c}\left(1 - \frac{ec_t}{r^2}\right) - \frac{M_T}{S^t}$$

$$P_e = 13 \times 0.153 \times 154,980 = 308,255 \text{ lb } (1,371 \text{ kN})$$

$$M_{SD} + M_L = (100 + 1,100)(65)^2 \times 12/8 = 7,605,000 \text{ in. lb}$$

$$\text{Total moment } M_T = M_D + M_{SD} + M_L = 2,490,638 + 7,605,000$$

$$= 10,095,638 \text{ in. lb } (1,141 \text{ kN m})$$

$$f^t = -\frac{308,255}{377}\left(1 - \frac{15.0 \times 21.16}{187.5}\right) - \frac{10,095,638}{3,340}$$

$$= +566.5 - 3022.6 = -2,456 \text{ psi } (C) > f_c = -2,250 \text{ psi}$$

Hence, either enlarge the depth of the section or use higher strength concrete. Using $f_c' = 6,000$ psi

$$f_b = -\frac{P_e}{A_c}\left(1 + \frac{ec_b}{r^2}\right) + \frac{M_T}{S_b} = -\frac{308,255}{377}\left(1 + \frac{15.0 \times 18.84}{187.5}\right) + \frac{10,095,638}{3,750}$$

$$= -2,050 + 2692.2 = 642 \text{ psi } (T), \text{ OK.}$$

Check support section stresses. Allowable

$$f_{ci}' = 0.75 \times 6000 = 4500 \text{ psi}$$

$$f_{ci} = 0.60 \times 4500 = 2700 \text{ psi}$$

$$f_{ti} = 3\sqrt{f_{ci}'} = 201 \text{ psi} \quad \text{for midspan}$$

$$f_{ti} = 6\sqrt{f_{ci}'} = 402 \text{ psi} \quad \text{for support}$$

$$f_c = 0.45f_c' = 2700 \text{ psi}$$

$$f_{t1} = 6\sqrt{f'_c} = 465 \text{ psi}$$
$$f_{t2} = 12\sqrt{f'_c} = 930 \text{ psi}$$

1. *At transfer.* Support section compressive fiber stress

$$f_b = -\frac{P_i}{A_c}\left(1 + \frac{ec_b}{r^2}\right) + 0$$

For $P_i = 376{,}110$ lb

$$-2{,}700 = -\frac{376{,}110}{377}\left(1 + \frac{e \times 18.84}{187.5}\right)$$

so that

$$e_e = 16.98 \text{ in.}$$

Accordingly, try $e_e = 12.49$ in.

$$f^t = -\frac{376{,}110}{377}\left(1 - \frac{12.49 \times 21.16}{187.5}\right) - 0$$
$$= 409 \text{ psi } (T) > f_{ti} = 402 \text{ psi}$$
$$f_b = 2{,}250 \text{ psi}$$

Thus, use mild steel at the top fibers at the support section to take all tensile stresses in the concrete, or use a higher strength concrete for the section, or reduce the eccentricity.

2. *At service load*

$$f^t = -\frac{308{,}255}{377}\left(1 - \frac{12.49 \times 21.16}{187.5}\right) - 0 = 335 \text{ psi } (T) < 930 \text{ psi, OK.}$$
$$f_b = -\frac{308{,}255}{377}\left(1 + \frac{12.49 \times 18.84}{187.5}\right) + 0 = -1{,}844 \text{ psi } (C) < -2{,}700 \text{ psi, OK.}$$

Hence, adopt the 40-in. (102-cm)-deep I-section prestressed beam of f'_c equal to 6,000 psi (41.4 MPa) normal-weight concrete with thirteen $\frac{1}{2}$-in. strands tendon having midspan eccentricity $e_c = 15.0$ in. (381 mm) and end section eccentricity $e_e = 12.5$ in. (318 m).

An alternative to this solution is to continue using $f'_c = 5000$ psi, but change the number of strands and eccentricities.

8.6.3 Development and Transfer Length in Pretensioned Members and Design of their Anchorage Reinforcement

As the jacking force is released in pretensioned members, the prestressing force is dynamically transferred through the bond interface to the surrounding concrete. The interlock or adhesion between the prestressing tendon circumference and the concrete over a finite length of the tendon gradually transfers the concentrated prestressing force to the entire concrete section at planes away from the end block and toward the midspan. The length of embedment determines the magnitude of prestress that can be developed along the span: the larger the embedment length, the higher is the prestress developed.

As an example for $\frac{1}{2}$-in. seven-wire strand, an embedment of 40 in. (102 cm) develops a stress of 180,000 psi (1,241 MPa), whereas an embedment of 70 in. (178 cm) develops a stress of 206,000 psi (1,420 MPa). The embedment length l_d that gives the full development of stress is a combination of the transfer length l_t and the flexural bond length l_f. These are given, respectively, by

$$l_t = \frac{1}{1000}\left(\frac{f_{pe}}{3}\right) d_b \tag{8.29a}$$

or

$$l_t = \frac{f_{pe}}{3000} d_b \qquad (8.29b)$$

and

$$l_f = \frac{1}{1000} \left(f_{ps} - f_{pe} \right) d_b \qquad (8.29c)$$

where f_{ps} is the stress in prestressed reinforcement at nominal strength (psi), f_{pe} is the effective prestress after losses (psi), and d_b is the nominal diameter of prestressing tendon (in.).

Combining Equations 8.29b and 8.29c gives

$$\text{Min } l_d = \frac{1}{1000} \left(f_{ps} - \frac{2}{3} f_{pe} \right) d_b \qquad (8.29d)$$

Equation 8.29d gives the minimum required development length for prestressing strands. If part of the tendon is sheathed toward the beam end to reduce the concentration of bond stresses near the end, the stress transfer in that zone is eliminated and an increased adjusted development length l_d is needed.

8.6.3.1 Design of Transfer Zone Reinforcement in Pretensioned Beams

Based on laboratory tests, empirical expressions developed by Mattock et al. give the total stirrup force F as

$$F = 0.0106 \frac{P_i h}{l_t} \qquad (8.30)$$

where h is the pretensioned beam depth and l_t is the transfer length. If the average stress in a stirrup is taken as *half* the maximum permissible steel f_s, then $F = \frac{1}{2} A_t f_s$. Substituting this for F in Equation 8.30 gives

$$A_t = 0.021 \frac{P_i h}{f_s l_t} \qquad (8.31)$$

where A_t is the total area of the stirrups and $f_s \leq 20{,}000$ psi (138 MPa) for crack-control purposes.

8.6.4 Posttensioned Anchorage Zones: Strut-and-Tie Design Method

The anchorage zone can be defined as the volume of concrete through which the concentrated prestressing force at the anchorage device spreads transversely to a linear distribution across the entire cross-section depth along the span. The length of this zone follows St Venant's principle, namely, that the stress becomes uniform at an approximate distance ahead of the anchorage device equal to the depth, h, of the section. The entire prism that would have a transfer length, h, is the total anchorage zone.

This zone is thus composed of two parts:

1. *General zone:* The general extent of the zone is identical to the total anchorage zone. Its length extent along the span is therefore equal to the section depth, h, in standard cases.
2. *Local zone:* This zone is the insert prism of concrete surrounding and immediately ahead of the anchorage device and the confining reinforcement it contains.

After significant cracking is developed, compressive stress trajectories in the concrete tend to congregate into straight lines that can be idealized as straight compressive struts in uniaxial compression. These struts would become part of truss units where the principal tensile stresses are idealized as tension ties in the truss unit with the nodal locations determined by the direction of the idealized compression struts. Figure 8.4 sketches standard strut-and-tie idealized trusses for concentric and eccentric cases both for solid and flanged sections as given in ACI 318-02 Code.

Simplified equations can be used to compute the magnitude of the bursting force, T_{burst}, and its centroid distance, d_{burst}, from the major bearing surface of the anchorage [9]. The member has to have

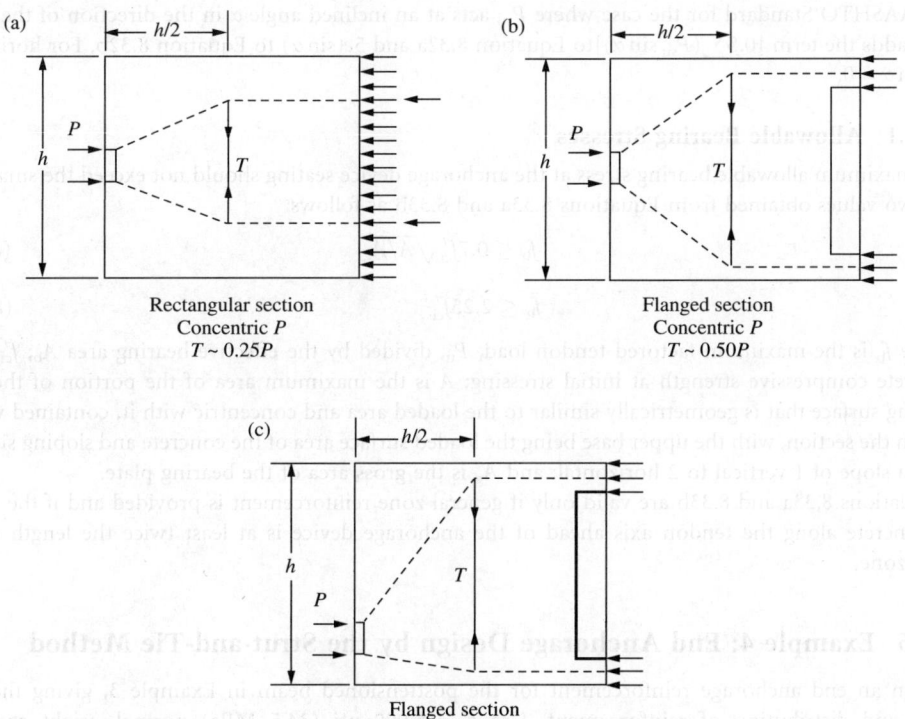

FIGURE 8.4 Strut-and-tie idealized trusses in standard concentric and eccentric cases, ACI 318-02 [8].

a rectangular cross-section with no discontinuities along the span. The bursting force, T_{burst} and its distance, d_{burst}, can be computed from the following expressions:

$$T_{burst} = 0.25 \sum P_{su} \left(1 - \frac{a}{h}\right) \tag{8.32a}$$

$$d_{burst} = 0.5(h - 2e) \tag{8.32b}$$

where $\sum P_{su}$ is the sum of the total factored prestress loads for the stressing arrangement considered (lb), a is the plate width of anchorage device or single group of closely spaced devices in the direction considered (in.), e is the eccentricity (always taken positive) of the anchorage device or group closely spaced devices with respect to the centroid of the cross-section (in.), and h is the depth of the cross-section in the direction considered (in.).

The ACI 318 Code requires that the design of confining reinforcement in the end anchorage block of posttensioned members be based on the *factored* prestressing force P_{su} for both the general and local zones. A load factor of 1.2 is to be applied to an end anchorage stress level $f_{pi} = 0.80 f_{pu}$ for low-relaxation strands at the *short time interval* of jacking, which can reduce to an average value of $0.70 f_{pu}$ for the total group of strands at the completion of the jacking process. For stress-relieved strands, a lower $f_{pi} = 0.70 f_{pu}$ is advised. The maximum force P_{su} stipulated in the ACI 318 Code for designing the confining reinforcement at the end-block zone is as follows for the more widely used low-relaxation strands:

$$P_{su} = 1.2 A_{ps}(0.8 f_{pu}) \tag{8.32c}$$

The AASHTO Standard for the case where P_{su} acts at an inclined angle α in the direction of the beam span adds the term $[0.5\sum(P_{su}\sin\alpha)]$ to Equation 8.32a and $5e(\sin\alpha)$ to Equation 8.32b. For horizontal $P_{su}\sin\alpha = 0$.

8.6.4.1 Allowable Bearing Stresses

The maximum allowable bearing stress at the anchorage device seating should not exceed the smaller of the two values obtained from Equations 8.33a and 8.33b as follows:

$$f_b \leq 0.7 f'_{ci}\sqrt{A/A_g} \qquad (8.33a)$$

$$f_b \leq 2.25 f'_{ci} \qquad (8.33b)$$

where f_b is the maximum factored tendon load, P_u, divided by the effective bearing area A_b; f'_{ci} is the concrete compressive strength at initial stressing; A is the maximum area of the portion of the supporting surface that is geometrically similar to the loaded area and concentric with it, contained wholly within the section, with the upper base being the loaded surface area of the concrete and sloping sideway with a slope of 1 vertical to 2 horizontal; and A_g is the gross area of the bearing plate.

Equations 8.33a and 8.33b are valid only if general zone reinforcement is provided and if the extent of concrete along the tendon axis ahead of the anchorage device is at least twice the length of the local zone.

8.6.5 Example 4: End Anchorage Design by the Strut-and-Tie Method

Design an end anchorage reinforcement for the posttensioned beam in Example 3, giving the size, type, and distribution of reinforcement. Use $f'_{ci} = 5,000$ psi (34.5 MPa) normal-weight concrete. $f_{pu} = 270,000$ psi low-relaxation steel.

Assume that the beam ends are rectangular blocks extending 40 in. (104 cm.) into the span beyond the anchorage devices which then transitionally reduce to the 6-in. thick web.

1. *Establish the configuration of the tendons to give eccentricity $e_e = 12.49$ in. (317 mm).* From Example 3, $c_b = 18.84$ in.; hence, distance from the beam fibers $= c_b - e_e = 6.35$ in. (161 mm). For a centroidal distance of the $13\frac{1}{2}$-in. size strands $= 6.35$ in. from the beam bottom fibers, try the following row arrangement of tendons with the indicated distances from the bottom fibers:
   first row: five tendons at 2.5 in.
   second row: five tendons at 7.0 in.
   third row: three tendons at 11.5 in.

   $$\text{Distance of the centroid of tendons} = \frac{(5 \times 2.5 + 5 \times 7.0 + 3 \times 11.5)}{13} \cong 6.35 \text{ in., OK.}$$

2. *Factored forces in tendon rows and bearing capacity of the concrete.* From Equation 8.32c for low-relaxation strands, $f_{pi} = 0.80 f_{pu}$. The factored jacking force at the short jacking time interval is

   $$P_{su} = 1.2 A_{ps}(0.80 f_{pu}) = 0.96 f_{pu} A_{ps} = 0.96 \times 270,000 = 259,200 \text{ psi.}$$

   If stress-relieved strands are used, it would have been advisable to use $f_{pi} = 0.70 f_{pu}$:
   first row force: $P_u 1 = 5 \times 0.153 \times 259,200 = 198,286$ lb (882 kN)
   second row force: $P_u 2 = 5 \times 0.153 \times 259,200 = 198,286$ lb (882 kN)
   third row force: $P_u 3 = 3 \times 0.153 \times 259,200 = 118,973$ lb (529 kN)
   The total ultimate compressive force $= 198,286 + 198,286 + 118,973 = 515,545$ lb (2,290 kN). The total area of rigid bearing plates supporting the Supreme 13-chucks anchorage devices $= 14 \times 11 + 6 \times 4 = 178$ in.^{2} (113 cm^2). The actual bearing stress

   $$f_b = \frac{515,545}{178} = 2,896 \text{ psi (19.9 MPa)}$$

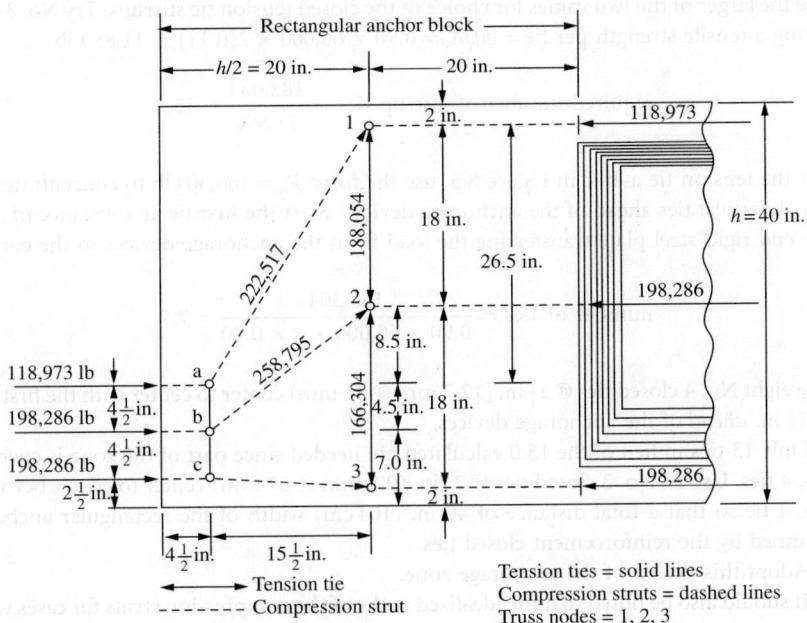

FIGURE 8.5   Struts-and-ties in Example 4 [5].

From Equations 8.33a and 8.33b, the maximum allowable bearing pressure on the concrete is the lesser of

$$f_b \leq 0.7 f'_{ci} \sqrt{A/A_g}$$
$$f_b \leq 2.25 f'_{ci}$$

Assume that the initial concrete strength at stressing is $f'_{ci} = 0.75 f'_c = 0.75 \times 5000 = 3750$ psi. The concentric area, A, of concrete with the bearing plates $\cong (14+4)(11+4) + (6+4)(4+4) = 350$ in.^{2} Allowable bearing stress, $f_b = 0.7 \times 0.90 \times 3750 \sqrt{322/178} = 3178$ psi > 3020 psi, OK. The bearing stress does not control. It should be noted that the area A_g of the rigid steel plate or plates, and the corresponding concrete pyramid base area A within the end block assumed to receive the bearing stress, are purely determined by engineering judgment. The areas are based on the geometry of the web and bottom flange of the section, the rectangular dimension of the beam end, and the arrangement and spacing of the strand anchorages in contact with the supporting steel end bearing plates.

3. *Draw the strut-and-tie model.* Total length of distance a between forces $P_{u1} - P_{u3} = 11.5 - 2.5 = 9.0$ in. Hence depth $a/2$ ahead of the anchorages $= 9.0/2 = 4.5$ in. Construct the strut-and-tie model assuming it to be as shown in Figure 8.5. The geometrical dimensions for finding the horizontal force components from the ties 1–2 and 3–2 have cotangent values of 26.5/15.5 and 13.0/15.5, respectively. From statics, truss analysis in Figure 8.5 gives the member forces as follows:

$$\text{tension tie } 1\text{–}2 = 118{,}973 \times \frac{24.5}{15.5} = 188{,}054 \text{ lb } (836 \text{ kN})$$

$$\text{tension tie } 3\text{–}2 = 198{,}286 \times \frac{13}{15.5} = 166{,}304 \text{ lb } (699 \text{ kN})$$

Use the larger of the two values for choice of the closed tension tie stirrups. Try No. 3 closed ties, giving a tensile strength per tie $= \phi f_y A_v = 0.90 \times 60{,}000 \times 2(0.11) = 11{,}880$ lb

$$\text{required number of stirrup ties} = \frac{188{,}054}{11{,}880} = 15.8$$

For the tension tie a–b–c in Figure 8.5, use the force $P_u = 166{,}304$ lb to concentrate additional No. 4 vertical ties ahead of the anchorage devices. Start the first tie at a distance of $1\frac{1}{2}$ in. from the end rigid steel plate transferring the load from the anchorage devices to the concrete

$$\text{number of ties} = \frac{166{,}304}{0.90 \times 60{,}000 \times 2 \times 0.20} = 7.7$$

Use eight No. 4 closed ties @ $1\frac{1}{4}$ in. (12.7 mm @ 32 mm) center to center with the first tie to start at $1\frac{1}{2}$ in. *ahead* of the anchorage devices.

Only 13 ties in lieu of the 15.0 calculated are needed since part of the zone is covered by the No. 4 ties. Use 13 No. 3 closed ties @ $2\frac{1}{2}$ in. (9.5 mm @ 57 mm) center to center beyond the last No. 4 tie so that a total distance of 40 in. (104 cm) width of the rectangular anchor block is confined by the reinforcement closed ties.

Adopt this design of the anchorage zone.

It should also be noted that the idealized paths of the compression struts for cases where there are several layers of prestressing strands should be such that at each layer level a stress path is assumed in the design.

8.6.6 Ultimate-Strength Flexural Design

8.6.6.1 Cracking-Load Moment

One of the fundamental differences between prestressed and reinforced concrete is the continuous shift in the prestressed beams of the compressive C-line away from the tensile cgs line as the load increases. In other words, the moment arm of the internal couple continues to increase with the load without any appreciable change in the stress f_{pe} in the prestressing steel. As the flexural moment continues to increase when the full superimposed dead load and live load act, a loading stage is reached where the concrete compressive stress at the bottom-fibers reinforcement level of a simply supported beam becomes zero.

This stage of stress is called the limit state of *decompression*. Any additional external load or overload results in cracking at the bottom face, where the modulus of rupture of concrete f_r is reached due to the cracking moment M_{cr} caused by the first cracking load. At this stage, a sudden increase in the steel stress takes place and the tension is dynamically transferred from the concrete to the steel. It is important to evaluate the first cracking load, since the section stiffness is reduced and hence an increase in deflection has to be considered. Also, the crack width has to be controlled in order to prevent reinforcement corrosion or leakage in liquid containers.

The concrete fiber stress at the tension face is

$$f_b = -\frac{P_e}{A_c}\left(1 + \frac{ec_b}{r^2}\right) + \frac{M_{cr}}{S_b} = f_r \tag{8.34}$$

where the modulus of rupture $f_r = 7.5\sqrt{f_c'}$ and the cracking moment M_{cr} is the moment due to all loads at that load level $(M_D + M_{SD} + M_L)$. From Equation 8.34

$$M_{cr} = f_r S_b + P_e\left(e + \frac{r^2}{c_b}\right) \tag{8.35}$$

8.6.6.2 ACI Load Factors Equations

The ACI 318 Building Code for concrete structures is an international code. As such, it has to conform to the International Building Codes, IBC 2000 and IBC 2003 [10] and be consistent with the ASCE-7 Standard on Minimum Design Loads for Buildings and Other Structures. The effect of one or more loads not acting simultaneously has to be investigated. Structures are seldom subjected to dead and live loads alone. The following equations present combinations of loads for situations in which wind, earthquake, or lateral pressures due to earthfill or fluids should be considered:

$$U = 1.4(D + F) \tag{8.36a}$$

$$U = 1.2(D + F + T) + 1.6(L + H) + 0.5(L_r \text{ or } S \text{ or } R) \tag{8.36b}$$

$$U = 1.2D + 1.6(L_r \text{ or } S \text{ or } R) + (1.0L \text{ or } 0.8W) \tag{8.36c}$$

$$U = 1.2D + 1.6W + 0.5L + 1.0(L_r \text{ or } S \text{ or } R) \tag{8.36d}$$

$$U = 1.2D + 1.0E + 1.0L + 0.2S \tag{8.36e}$$

$$U = 0.9D + 1.6W + 1.6H \tag{8.36f}$$

$$U = 0.9D + 1.0E + 1.6H \tag{8.36g}$$

where D is the dead load, E is the earthquake load, F is the lateral fluid pressure load and maximum height; H is the load due to the weight and lateral pressure of soil and water in soil; L is the live load, L_r is the roof load, R is the rain load, S is the snow load; T is the self-straining force such as creep, shrinkage, and temperature effects; and W is the wind load.

It should be noted that the philosophy used for combining the various load components for earthquake loading is essentially similar to that used for wing loading.

8.6.6.2.1 Exceptions to the Values in These Expressions

1. The load factor on L in Equations 8.36c to 8.36e is allowed to be reduced to 0.5 except for garages, areas occupied as places of public assembly, and all areas where the live load L is greater than 100 lb/ft^2.
2. Where wind load W has not been reduced by a directionality factor, the code permits to use $1.3W$ in place of $1.6W$ in Equations 8.36d and 8.36f.
3. Where earthquake load E is based on service-level seismic forces, $1.4E$ shall be used in place of $1.0E$ in Equations 8.36e and 8.36g.
4. The load factor on H is to be set equal to zero in Equations 8.36f and 8.36g if the structural action due to H counteracts that due to W or E. Where lateral earth pressure provides resistance to structural actions from other forces, it should not be included in H but shall be included in the design resistance.

Due regard has to be given to sign in determining U for combinations of loadings, as one type of loading may produce effects of opposite sense to that produced by another type. The load combinations with $0.9D$ are specifically included for the case where a higher dead load reduces the effects of other loads.

8.6.6.3 Design Strength versus Nominal Strength: Strength-Reduction Factor ϕ

The strength of a particular structural unit calculated using the current established procedures is termed *nominal strength*. For example, in the case of a beam, the resisting moment capacity of the section calculated using the equations of equilibrium and the properties of concrete and steel is called the *nominal strength moment* M_n of the section. This nominal strength is reduced using a strength reduction factor ϕ to account for inaccuracies in construction, such as in the dimensions or position of reinforcement or variations in properties. The reduced strength of the member is defined as the design strength of the member.

For a beam, the design moment strength ϕM_n should be at least equal to, or slightly greater than the external factored moment M_u for the worst condition of factored load U. The factor ϕ varies for the

PHOTO 8.4 Paramount Apartments, San Francisco, CA: completed in 2002, it is the first hybrid precast prestressed concrete moment-resistant 39 floor high-rise frame building in a high seismicity zone. Based on tests of large-scale prototype of the moment-resisting connections, it was determined that the performance of the system was superior to cast-in-place concrete, both in cracking behavior and ductility. Hence, as a system, this new development succeeds in enhancing the performance of high-rise buildings and bridges in high seismicity earthquake zones. Since completion of this structure, several other buildings in California have been completed using the same principles (courtesy Charles Pankow Ltd, Design/Build contractors. Structural Engineers: Robert Englekirk Inc.).

different types of behavior and for the different types of structural elements. For beams in flexure, for instance, ϕ is 0.9.

For tied columns that carry dominant compressive loads, the factor ϕ equals 0.65. The smaller strength-reduction factor used for columns is due to the structural importance of the columns in supporting the total structure compared to other members, and to guard against progressive collapse and brittle failure with no advance warning of collapse. Beams, on the other hand, are designed to undergo excessive deflections before failure. Hence, the inherent capability of the beam for advanced warning of failure permits the use of a higher strength reduction factor or resistance factor. Table 8.10 summarizes the resistance factors ϕ for various structural elements as given in the ACI code.

TABLE 8.10 Resistance or Strength Reduction Factor ϕ

Structural element	Factor ϕ
Beam or slab: bending or flexure[a]	0.9
Columns with ties	0.65
Columns with spirals	0.7
Columns carrying very small axial loads (refer to Chapter 5 for more details)	0.65–0.9 or 0.70–0.9
Beam: shear and torsion[b]	0.75

[a] Flexure: for factory-produced precast prestressed concrete members, $\phi = 1.0$. For posttensioned cast-in-place concrete members, $\phi = 0.95$ by AASHTO.

[b] Shear and Torsion: Reduction factor for prestressed members, $\phi = 0.90$ by AASHTO.

PHOTO 8.5 Walt Disney World Monorail, Orlando, Florida: a series of hollow prestressed concrete 100-ft box girders individually posttensioned to provide a six-span continuous structure. Design by ABAM Engineers and owned by Walt Disney World Company (courtesy E.G. Nawy).

8.6.7 Limit States in Bonded Members from Decompression to Ultimate Load

The effective prestress f_{pe} at service load due to all loads results in a strain ε_1 such that

$$\varepsilon_1 = \varepsilon_{pe} = \frac{f_{pe}}{E_{ps}} \tag{8.37a}$$

At decompression, that is, when the compressive stress in the surrounding concrete at the level of the prestressing tendon is neutralized by the tensile stress due to overload, a decompression strain $\varepsilon_{decomp} = \varepsilon_2$ results such that

$$\varepsilon_2 = \varepsilon_{decomp} = \frac{P_e}{A_c E_c}\left(1 + \frac{e^2}{r^2}\right) \tag{8.37b}$$

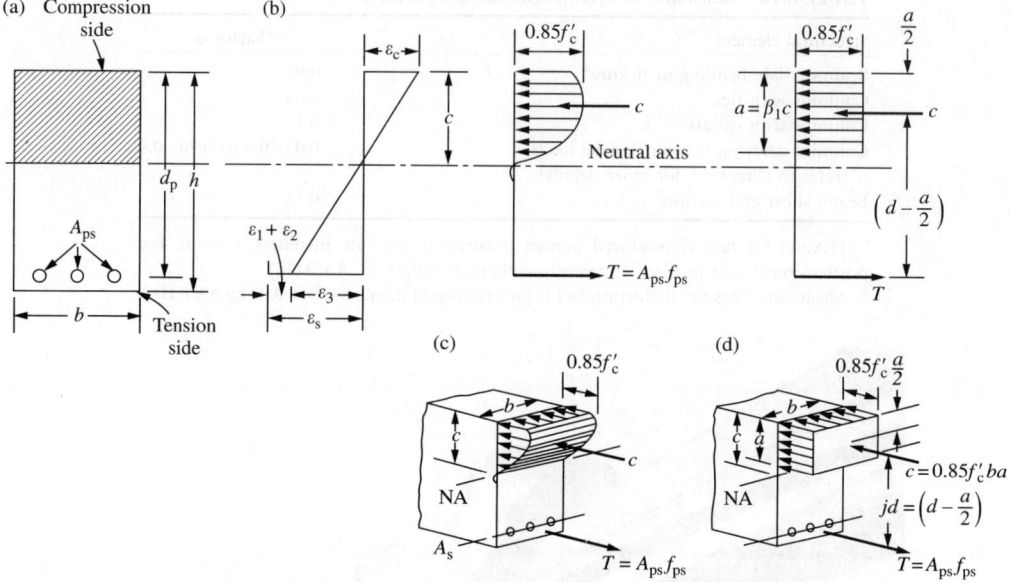

FIGURE 8.6   Stress and strain distribution across beam depth: (a) beam cross-section; (b) strains; (c) actual stress block; and (d) equivalent stress block [11].

Figure 8.6 illustrates the stress distribution in a prestressed concrete member at and after the decompression stage where the behavior of the prestressed beam starts to resemble that of a reinforced concrete beam.

As the load approaches the limit state at ultimate, the additional strain ε_3 in the steel reinforcement follows the linear triangular distribution shown in Figure 8.6b, where the maximum compressive strain at the extreme compression fibers is $\varepsilon_c = 0.003$ in./in. In such a case, the steel strain increment due to overload above the decompression load is

$$\varepsilon_3 = \varepsilon_c \left(\frac{d - c}{c} \right) \tag{8.37c}$$

where c is the depth of the neutral axis. Consequently, the total strain in the prestressing steel at this stage becomes

$$\varepsilon_s = \varepsilon_1 + \varepsilon_2 + \varepsilon_3 \tag{8.37d}$$

The corresponding stress f_{ps} at nominal strength can be easily obtained from the stress–strain diagram of the steel supplied by the producer. If $f_{pe} < 0.50 f_{pu}$, the ACI Code allows computing f_{ps} from the following expression:

$$f_{ps} = f_{pu} \left(1 - \frac{\gamma_p}{\beta_1} \left[\rho_p \frac{f_{pe}}{f_c'} + \frac{d}{d_p} (\omega - \omega') \right] \right) \tag{8.37e}$$

where γ_p can be used as 0.40 for $f_{py}/f_{pu} > 0.85$.

8.6.7.1   The Equivalent Rectangular Block and Nominal Moment Strength

1. The strain distribution is assumed to be linear. This assumption is based on Bernoulli's hypothesis that plane sections remain plane before bending and perpendicular to the neutral axis after bending.

2. The strain in the steel and the surrounding concrete is the same prior to cracking of the concrete or yielding of the steel as after such cracking or yielding.
3. Concrete is weak in tension. It cracks at an early stage of loading at about 10% of its compressive strength limit. Consequently, concrete in the tension zone of the section is neglected in the flexural analysis and design computations and the tension reinforcement is assumed to take the total tensile force.

8.6.7.2 Strain Limits Method for Analysis and Design

8.6.7.2.1 General Principles

In this approach, sometimes referred to as the "unified method," since it is equally applicable to flexural analysis of prestressed concrete elements, the nominal flexural strength of a concrete member is reached when the net compressive strain in the extreme compression fibers reaches the ACI code-assumed limit 0.003 in./in. It also stipulates that when the net tensile strain in the extreme tension steel, ε_t, is sufficiently large, as discussed in the previous section, at a value equal to or greater than 0.005 in./in., the behavior is fully ductile. The concrete beam section is characterized as *tension controlled*, with ample warning of failure as denoted by excessive cracking and deflection.

If the net tensile strain in the extreme tension fibers, ε_t, is small, such as in compression members, being equal to or less than a *compression-controlled* strain limit, a brittle mode of failure is expected, with little warning of such an impending failure. Flexural members are usually tension controlled. Compression members are usually compression controlled. However, some sections, such as those subjected to small axial loads, but large bending moments, the net tensile strain, ε_t, in the extreme tensile fibers, will have an intermediate or transitional value between the two strain limit states, namely, between the compression-controlled strain limit $\varepsilon_t = f_y/E_s = 60{,}000/29 \times 10^6 = 0.002$ in./in., and the tension-controlled strain limit $\varepsilon_t = 0.005$ in./in. Figure 8.7 delineates these three zones as well as the variation in the strength reduction factors applicable to the total range of behavior.

For the tension-controlled state, the strain limit $\varepsilon_t = 0.005$ corresponds to reinforcement ratio $\rho/\rho_b = 0.63$, where ρ_b is the balanced reinforcement ratio for the balanced strain $\varepsilon_t = 0.002$ in the extreme tensile reinforcement. The net tensile strain $\varepsilon_t = 0.005$ for a tension-controlled state is a single value that applies to all types of reinforcement regardless of whether mild steel or prestressing steel. High reinforcement ratios that produce a net tensile strain less than 0.005 result in a ϕ-factor value lower than 0.90, resulting in less economical sections. Therefore, it is more efficient to add compression

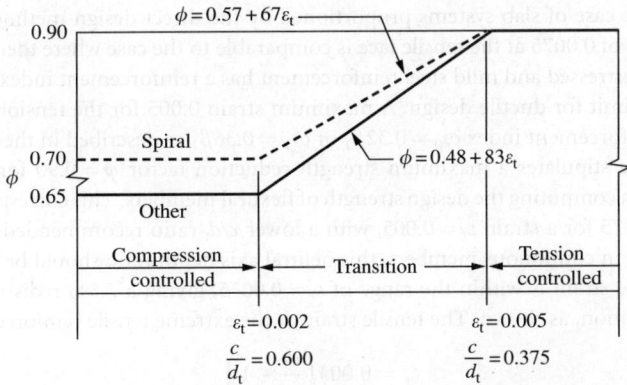

Interpolation on c/d_t: Spiral $\phi = 0.37 + 0.20/(c/d_t)$
Other $\phi = 0.23 + 0.25/(c/d_t)$

FIGURE 8.7 Strain limit zones and variation of strength reduction factor ϕ with the net tensile strain ε_t [1,5,8].

reinforcement if necessary or deepen the section in order to make the strain in the extreme tension reinforcement, $\varepsilon_t \geq 0.005$.

8.6.7.2.1.1 Variation of ϕ as a Function of Strain — Variation of the ϕ value for the range of strain between $\varepsilon_t = 0.002$ and $\varepsilon_t = 0.005$ can be linearly interpolated to give the following expressions:

Tied sections:

$$0.65 \leq [\phi = 0.48 + 83\varepsilon_t] \leq 0.90 \tag{8.38a}$$

Spirally reinforced sections:

$$0.70 \leq [\phi = 0.57 + 67\varepsilon_t] \leq 0.90 \tag{8.38b}$$

8.6.7.2.1.2 Variation of ϕ as a Function of Neutral Axis depth Ratio c/d_t — Equations 8.38a and 8.38b can be expressed in terms of the ratio of the neutral axis depth c to the effective depth d_t of the layer of reinforcement closest to the tensile face of the section as follows:

Tied sections:

$$0.65 \leq \left[\phi = 0.23 + \frac{0.25}{c/d_t}\right] \leq 0.90 \tag{8.39a}$$

Spirally reinforced sections:

$$0.70 \leq \left[\phi = 0.37 + \frac{0.20}{c/d_t}\right] \leq 0.90 \tag{8.39b}$$

For balanced strain, where the reinforcement at the tension side yields at the same time as the concrete crushes at the compression face ($f_s = f_y$), the neutral axis depth ratio for a limit strain $\varepsilon_t = 0.002$ in./in. can be defined as

$$\frac{c_b}{d_t} = \left(\frac{87,000}{87,000 + f_y}\right) \tag{8.40}$$

The Code permits decreasing the negative elastic moment at the supports for continuous members by not more than $[1000\varepsilon_t]\%$, with a maximum of 20%. The reason is that for ductile members plastic hinge regions develop at points of maximum moment and cause a shift in the elastic moment diagram. In many cases, the result is a reduction of the negative moment and a corresponding increase in the positive moment. The redistribution of the negative moment as permitted by the code can only be used when ε_t is equal or greater than about 0.0075 in./in. at the section at which the moment is reduced. Redistribution is inapplicable in the case of slab systems proportioned by the direct design method.

A minimum strain of 0.0075 at the tensile face is comparable to the case where the reinforcement ratio for the combined prestressed and mild steel reinforcement has a reinforcement index of ω not exceeding $0.24\beta_1$ as an upper limit for ductile design. A maximum strain 0.005 for the tension-controlled state is comparable to a reinforcement index $\omega_p = 0.32\beta_1$ or $\omega_T = 0.36\beta_1$ as described in the Code commentary.

The ACI 318 Code stipulates a maximum strength reduction factor $\phi = 0.90$ for tension-controlled bending, to be used in computing the design strength of flexural members. This corresponds to neutral axis depth ratio $c/d_t = 0.375$ for a strain $\varepsilon_t = 0.005$, with a lower c/d_t ratio recommended. For a useful redistribution of moment in continuous members, this neutral axis depth ratio should be considerably lower, so that the net tensile strain is within the range of $\varepsilon_t = 0.0075$, giving a 7.5% redistribution, and 0.020, giving 20% redistribution, as shown. The tensile strain at the extreme tensile reinforcement has the value

$$\varepsilon_t = 0.003\left(\frac{d_t}{c} - 1\right) \tag{8.41}$$

Although the code allows a maximum redistribution of 20% or $1000\varepsilon_t$, it is more reasonable to limit the redistribution percentage to about 10 to 15%. Summarizing, the ACI 318-02 code stipulates that a redistribution (reduction) of the moments at supports of continuous flexural members *not to exceed*

$1000\varepsilon_t\%$, with a maximum of 20% while increasing the positive midspan moment accordingly. But inelastic moment redistribution should only be made when ε_t is equal or greater than 0.0075 at the section for which the moment is reduced. An adjustment in one span should also be applied to all the other spans in flexure, shear, and bar cutoffs.

8.6.7.3 Nominal Moment Strength of Rectangular Sections

The compression force C can be written as $0.85 f_c' ba$ — that is, the *volume* of the compressive block at or near the ultimate limit state when the tension steel has yielded ($\varepsilon_s > \varepsilon_y$). The tensile force T can be written as $A_{ps}f_{ps}$ and equating C to T gives

$$a = \beta_1 c = \frac{A_{ps}f_{ps}}{0.85 f_c' b}$$ (8.42)

where for f_c' range of 4000 to 8000 psi

$$\beta_1 = 0.85 - 0.05\left(\frac{f_c' - 4000}{1000}\right)$$ (8.43)

The maximum value of β_1 is 0.85 and its minimum value is 0.65.

The nominal moment strength is obtained by multiplying C or T by the moment arm $(d_p - a/2)$, yielding

$$M_n = A_{ps}f_{ps}\left(d_p - \frac{a}{2}\right)$$ (8.44)

where d_p is the distance from the compression fibers to the center of the prestressed reinforcement. The steel percentage $\rho_p = A_{ps}/bd_p$ gives the nominal strength of the prestressing steel only as follows if the mild tension and compression steel is accounted for:

$$a = \frac{A_{ps}f_{ps} + A_sf_y - A_s'f_y}{0.85 f_c' b}$$ (8.45)

where b is the section width of the compression face of the beam.

Taking moments about the center of gravity of the compressive block in Figure 8.6, the nominal moment strength becomes

$$M_n = A_{ps}f_{ps}\left(d_p - \frac{a}{2}\right) + A_sf_y\left(d - \frac{a}{2}\right) + A_s'f_y\left(\frac{a}{2} - d'\right)$$ (8.46)

8.6.7.4 Nominal Moment Strength of Flanged Sections

When the compression flange thickness h_f is less than the neutral axis depth c and equivalent rectangular block depth a, the section can be treated as a flanged section

$$T_p + T_s = T_{pw} + T_{pf}$$ (8.47a)

where

T_p = total prestressing force = $A_{ps}f_{ps}$
T_s = ultimate force in the nonprestressed steel = A_sf_y
T_{pw} = part of the total force in the tension reinforcement required to develop the web = $A_{pw}f_{ps}$
A_{pw} = total reinforcement area corresponding to the force T_{pw}
T_{pf} = part of the total force in the tension reinforcement required to develop the flange

On the basis of these definitions

$$A_{pw}f_{ps} = A_{ps}f_{ps} + A_sf_y - 0.85f_c'(b - b_w)h_f$$ (8.47b)

Hence

$$a = \frac{A_{pw}f_{ps}}{0.85f_c' b_w} = \frac{A_{ps}f_{ps} + A_s f_y - 0.85f_c'(b - b_w)h_f}{0.85f_c' b_w} \qquad (8.48)$$

$$M_n = A_{pw}f_{ps}\left(d_p - \frac{a}{2}\right) + A_s f_y(d - d_p) + 0.85f_c'(b - b_w)h_f\left(d_p - \frac{h_f}{2}\right) \qquad (8.49)$$

The design moment in all cases would be

$$M_u = \phi M_n \qquad (8.50)$$

where $\phi = 0.90$ for flexure for the tension zone in Figure 8.7, and reduced in the transition and compression zone.

The value of the stress f_{ps} of the prestressing steel at failure is not readily available. However, it can be determined by *strain compatibility* through the various loading stages up to the limit state at failure. Such a procedure is required if

$$f_{pe} = \frac{P_e}{A_{ps}} < 0.50f_{pu} \qquad (8.51)$$

Approximate determination is allowed by the ACI 318 building code provided that

$$f_{pe} = \frac{P_e}{A_{ps}} \geq 0.50f_{pu} \qquad (8.52)$$

with separate equations for f_{ps} given for bonded and nonbonded members.

In order to ensure ductility of behavior, the percentage of reinforcement such that the reinforcement index, ω_p, does not exceed $0.36\beta_1$, noting that $0.32\beta_1$ is comparable to 0.005 in the strain limits approach. The ACI Code requires instead to compute the zone in Figure 8.7 applicable to the analyzed section so as to determine the appropriate strength reduction factor ϕ to be used for getting the design moment $M_u = \phi M_n$ as an indirect measure for limiting the reinforcement percentage. It also requires for ensuring ductility behavior that the maximum moment M_u should be equal or greater than 1.2 times the cracking moment M_{cr}.

8.6.8 Example 5: Ultimate Limit State Design of Prestressed Concrete Beams

Design the bonded beam in Example 3 by the ultimate-load theory using nonprestressed reinforcement to *partially* carry part of the factored loads. Use strain compatibility to evaluate f_{ps}, given the modified section in Figure 8.8 with a composite 3 in. top slab and

$f_{pu} = 270,000$ psi (1,862 MPa)
$f_{py} = 0.85f_{pu}$ for stress-relieved strands
$f_y = 60,000$ psi (414 MPa)
$f_c' = 5,000$ psi for normal-weight concrete (34.5 MPa)

Use seven-wire $\frac{1}{2}$-in. dia. strands. The nonprestressed partial mild steel is to be placed with a $1\frac{1}{2}$-in. clear cover, and no compression steel is to be accounted for. No wind or earthquake is taken into consideration.

Solution

From Example 3

Service $W_L = 1100$ plf (16.1 kN/m)
Service $W_{SD} = 100$ plf (1.46 kN/m)
Assumed $W_D = 393$ plf (5.74 kN/m)
Beam span $= 65$ ft (19.8 m)

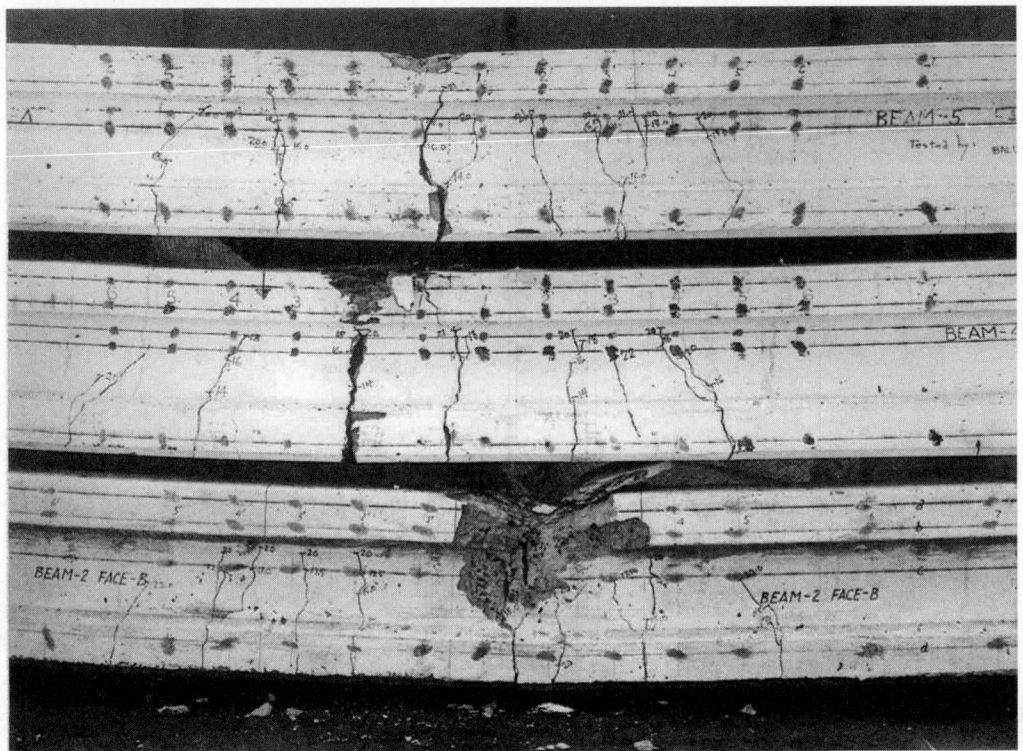

PHOTO 8.6 Different cracking development and patterns of prestressed concrete beams at the limit state at failure as a function of the reinforcement percentage. Tests made at the Rutgers University Concrete Research Laboratory (courtesy E.G. Nawy).

1. *Factored moment*

$$W_u = 1.2(W_D + W_{SD}) + 1.6W_L$$
$$= 1.2(100 + 393) + 1.6(1100) = 2352 \text{ plf } (34.4 \text{ kN/m})$$

The factored moment is given by

$$M_{\hat{u}} = \frac{w_u l^2}{8} = \frac{2,352(65)^2 12}{8} = 14,905,800 \text{ in. lb } (1,684 \text{ kN m})$$

and the required nominal moment strength is

$$M_u = \frac{M_u}{\phi} = \frac{14,905,800}{0.90} = 16,562,000 \text{ in. lb } (1,871 \text{ kN m})$$

2. *Choice of preliminary section.* Assuming a depth of 0.6 in./ft of span as a reasonable practical guideline, we can have a trial section depth $h = 0.6 \times 65 \cong 40$ in. (102 cm). Then assume a mild partial steel $4\#6 = 4 \times 0.44 = 1.76$ in.^{2} (11.4 cm^2). The empirical area of the concrete segment in compression up to the neutral axis can be empirically evaluated from

$$A_c' = \frac{M_n}{0.68 f_c' h} = \frac{16,562,000}{0.68 \times 5,000 \times 40} = 121.8 \text{ in.}^2 \text{ (786 cm}^2)$$

Assume on trial basis a total area of the beam section as $3A_c' = 3 \times 121.8 \approx 362.0$ in.^{2}

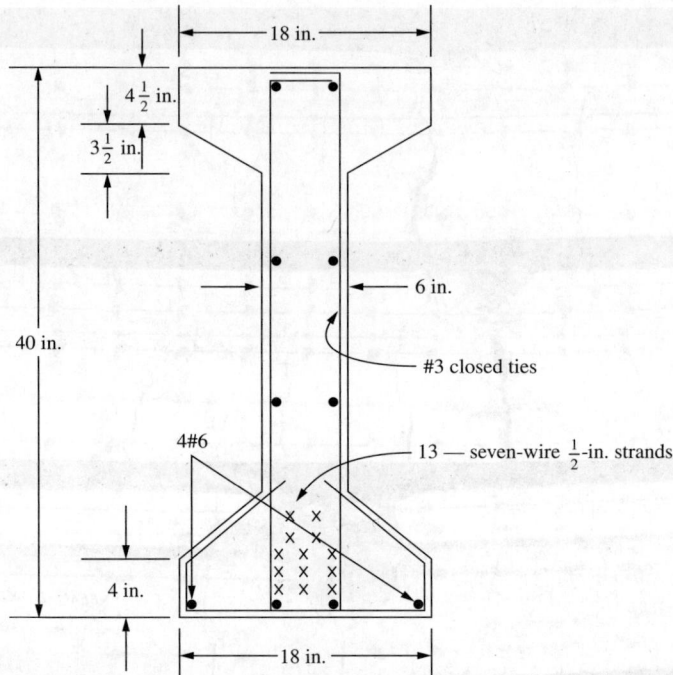

FIGURE 8.8 Midspan section of the beam in Example 5 [5].

Assume a flange width of 18 in. Then the average flange thickness = 121.8/18 ≅ 7.0 in. (178 mm). So suppose the web $b_w = 6$ in. (152 mm), to be subsequently verified for shear requirements. An empirical expression for the area of the prestressing steel can be expressed as

$$A_{ps} = \frac{M_n}{0.72 f_{pu} h} = \frac{16,562,000}{0.72 \times 270,000 \times 40} = 2.13 \text{ in.}^2 \ (13.3 \text{ cm}^2)$$

Number of $\frac{1}{2}$-in. stress-relieved wire strands = 2.13/0.153 = 13.9. So try thirteen $\frac{1}{2}$-in. tendons

$$A_{ps} = 13 \times 0.153 = 1.99 \text{ in.}^2 \ (12.8 \text{ cm}^2)$$

Try the section in Figure 8.8 for analysis.

3. *Calculate the stress f_{ps} in the prestressing tendon at nominal strength using the strain-compatibility approach.* The geometrical properties of the trial section are very close to the assumed dimensions for the depth h and the top flange width b. Hence, use the following data for the purpose of the example:

A_c = 377 in.^{2}

c_t = 21.16 in.

d_p = 15 + c_t = 15 + 21.16 = 36.16 in.

r^2 = 187.5 in.^{2}

e = 15 in. at midspan

e^2 = 225 in.^{2}

e^2/r^2 = 225/187.5 = 1.20

$$E_c = 57,000\sqrt{5,000} = 4.03 \times 10^6 \text{ psi } (27.8 \times 10^3 \text{ MPa})$$
$$E_{ps} = 28 \times 10^6 \text{ psi } (193 \times 10^3 \text{ MPa})$$

The maximum allowable compressive strain ε_c at failure $= 0.003$ in./in. Assume that the effective prestress at service load is $f_{pe} = 155,000$ psi (1,069 MPa).

A.

$$\varepsilon_1 = \varepsilon_{pe} = \frac{f_{pe}}{E_{ps}} = \frac{155,000}{28 \times 10^6} = 0.0055 \text{ in./in.}$$

$$P_e = 13 \times 0.153 \times 155,000 = 308,295 \text{ lb}$$

The increase in prestressing steel strain as the concrete is decompressed by the increased external load (see Figure 8.6 and Equation 8.37b) is given as

$$\varepsilon_2 = \varepsilon_{decomp} = \frac{P_e}{A_c E_c}\left(1 + \frac{e^2}{r^2}\right)$$

$$= \frac{308,295}{377 \times 4.03 \times 10^6}(1 + 1.20) = 0.0004 \text{ in./in.}$$

B. Assume that the stress $f_{ps} = 205,000$ psi as a first trial. Suppose the neutral axis inside the flange is verified on the basis of $h_f = 3 + 4\frac{1}{2} + 3\frac{1}{2}/2 = 9.25$ in. then, from Equation 8.37c

$$a = \frac{A_{ps}f_{ps} + A_s f_y}{0.85 f_c' b} = \frac{1.99 \times 205,000 + 1.76 \times 60,000}{0.85 \times 5,000 \times 18}$$

$$= 6.71 \text{ in. } (17 \text{ cm}) < h_f = 9.25 \text{ in.}$$

Hence, the equivalent compressive block is inside the flange and the section has to be treated as rectangular. Accordingly, for 5000 psi concrete

$$\beta_1 = 0.85 - 0.05 = 0.80$$

$$c = \frac{a}{\beta_1} = \frac{6.71}{0.80} = 8.39 \text{ in. } (22.7 \text{ cm})$$

$$d = 40 - (1.5 + \tfrac{1}{2} \text{ in. for stirrups} + \tfrac{5}{16} \text{ in. for bar}) \cong 37.6 \text{ in.}$$

From Equation 4.37c, the increment of strain due to overload to the ultimate is

$$\varepsilon_3 = \varepsilon_c\left(\frac{d-c}{c}\right) = 0.003\left(\frac{37.6 - 8.39}{8.39}\right) = 0.0104 \text{ in./in.} \gg 0.005 \text{ in./in. OK.}$$

The total strain is

$$\varepsilon_{ps} = \varepsilon_1 + \varepsilon_2 + \varepsilon_3$$
$$= 0.0055 + 0.0004 + 0.0104 = 0.0163 \text{ in./in.}$$

From the stress–strain diagram in Figure 8.1, the f_{ps} corresponding to $\varepsilon_{ps} = 0.0163$ is 230,000 psi.

Second trial for f_{ps} value. Assume

$$f_{ps} = 229,000 \text{ psi}$$

$$a = \frac{1.99 \times 229,000 + 1.76 \times 60,000}{0.85 \times 5,000 \times 18} = 7.34 \text{ in., consider section as a rectangular beam}$$

$$c = \frac{7.34}{0.80} = 9.17 \text{ in.}$$

$$\varepsilon_3 = 0.003\left(\frac{37.6 - 9.17}{9.17}\right) = 0.0093 \text{ in./in.}$$

Then the total strain is $\varepsilon_{ps} = 0.0055 + 0.0004 + 0.0093 = 0.0152$ in./in. From Figure 8.1, $f_{ps} = 229{,}000$ psi (1.579 MPa), OK. Use

$$A_{ps} = 4\#6 = 1.76 \text{ in.}^2$$

4. *Available moment strength.* From Equation 8.46, if the neutral axis were to fall within the flange

$$M_n = 1.99 \times 229{,}000 \left(36.16 - \frac{7.34}{2} \right) + 1.76 \times 60{,}000 \left(37.6 - \frac{7.34}{2} \right)$$

$$= 14{,}806{,}017 + 3{,}583{,}008 = 18{,}389{,}025 \text{ in. lb } (2{,}078 \text{ kN m})$$

$$> \text{required } M_n = 16{,}562{,}000 \text{ in. lb, OK.}$$

5. *Check of the reinforcement allowable limits.*
   A. Min $A_s = 0.004A$, where A is the area of the segment of the concrete section between the tension face and the centroid of the entire section

$$A = 377.0 - 18 \left(4.125 + \frac{1.375}{2} \right) - 6(21.16 - 5.5) = 201.0 \text{ in.}^2$$

   Min $A_s = 0.004 \times 201.0 = 0.80$ in.$^2 < 1.76$ in.^{2} used, hence alright.
   B. Maximum reinforcement index $\omega_T = d/d_p(\omega_p - \omega) \leq 0.36 \, \beta_1 < 0.29$ for $\beta_1 = 0.80$
      Actual

$$\omega_T = \frac{1.99 \times 229{,}000}{18 \times 36.16 \times 5{,}000} + \frac{37.6}{36.16} \left(\frac{1.76 \times 60{,}000}{18 \times 37.6 \times 5{,}000} \right)$$

$$= 0.14 + 0.03 = 0.17 < 0.29, \text{ hence alright.}$$

Alternatively, the ACI Code limit strain provisions as given in Figure 8.7 do not prescribe a maximum percentage of reinforcement. They require that a check be made of the strain ε_t at the level of the extreme tensile reinforcement to determine whether the beam is in the tensile, the transition, or the compression zone for verifying the appropriate ϕ value. In this case, for $c = 9.17$ and $d_t = 37.6$ in., and from similar triangles in the strain distribution across the beam depth

$$\varepsilon_t = 0.003 \times (37.6 - 9.17)/9.17 = 0.0093 > 0.005$$

Hence, the beam is in the tensile zone of Figure 8.7, with $\phi = 0.90$ as used in the solution, and the design is alright in terms of ductility and reinforcement limits.

8.6.9 Example 6: Ultimate Limit State Design of Prestressed Beams in SI Units

Solve Example 5 using SI units. Strands are bonded.

   Data

$$A_c = 5045 \text{ cm}^2, \qquad\qquad b = 45.7 \text{ cm}, \qquad\qquad b_w = 15.2 \text{ cm}$$

$$I_c = 7.04 \times 10^6 \text{ cm}^4$$

$$r^2 = 1394 \text{ cm}^2$$

$$c_b = 89.4 \text{ cm}, \qquad\qquad c^t = 32.5 \text{ cm}$$

$$e_e = 84.2 \text{ cm}, \qquad\qquad e_e = 60.4 \text{ cm}$$

$$S_b = 78{,}707 \text{ cm}^3, \qquad\qquad S^t = 216{,}210 \text{ cm}^3$$

$$w_D = 11.9 \times 10^3 \text{ kN/m}, \qquad w_{SD} = 1{,}459 \text{ N/m}, \qquad w_L = 16.1 \text{ kN/m}$$

$$l = 19.8 \text{ m}$$

$$f_c' = 34.5 \text{ MPa,} \qquad\qquad f_{pi} = 1300 \text{ MPa}$$
$$f_{pu} = 1860 \text{ MPa,} \qquad\qquad f_{py} = 1580 \text{ MPa}$$

Prestress loss $\gamma = 18\%$, $f_y = 414$ MPa

$A_{ps} = 13$ strands, diameter 12.7 mm $(A_{ps} = 99 \text{ mm}^2)$

$\qquad = 13 \times 99 = 1287 \text{ mm}^2$

Required $M_n = 16.5 \times 10^6$ kN m

Solution

Assume $\phi = 0.90$ to be subsequently verified. Hence, $M_u = \phi M_n = 0.9 \times 10^6 = 14.9 \times 10^6$ kN m

1. *Section properties:*
   Flange width $b = 18$ in. $= 45.7$ cm
   Average thickness $h_f = 4.5 + \frac{1}{2}(3.5) \cong 6.25$ in. $= 15.7$ cm

   Try 4 No. 20 M mild steel bars for partial prestressing (diameter = 19.5 mm, $A_s = 300 \text{ mm}^2$)

   $$A_s = 4 \times 300 = 1200 \text{ mm}^2$$

2. *Stress f_{ps} in the prestressing steel at nominal strength and neutral axis position:*

   $$f_{pe} = \gamma f_{pi} = 0.82 \times 1300 \cong 1066 \text{ MPa}$$

   Verify neutral axis position. If outside flange, its depth has to be greater than $a = A_{pw} f_{ps}/ 0.85 f_c' b_w$; $0.5 f_{pu} = 0.50 \times 1860 = 930$ MPa < 1066. Hence, one can use the ACI approximate procedure for determining f_{ps}. From Equation 8.37e.

   $$f_{ps} = f_{pu}\left(1 - \frac{\gamma_p}{\beta_1}\left[\rho_p \frac{f_{pu}}{f_c'} + \frac{d}{d_p}(\omega - \omega')\right]\right)$$

   $d_p = 36.16$ in. $= 91.8$ cm, $d = 37.6$ in. $= 95.5$ cm

   $$\frac{f_{py}}{f_{pu}} = \frac{1580}{1860} = 0.85, \quad \text{use } \gamma_p = 0.40$$

   $$\rho_p = \frac{A_{ps}}{b d_p} = \frac{1287}{457 \times 918} = 0.00306$$

   $$\rho = \frac{A_s}{bd} = \frac{1200}{457 \times 955} = 0.00275$$

   $$\omega_p = \frac{A_{ps}}{b d_p} \times \frac{f_{ps}}{f_c'} = 0.00306 \times \frac{1674}{34.5} = 0.14$$

   $$\omega = \frac{A_s}{bd} \times \frac{f_y}{f_c'} = 0.00275 \times \frac{414}{34.5} = 0.033$$

   $$\omega' = 0$$

   $$f_c' = 34.5 \text{ MPa,} \quad \beta_1 = 0.80$$

   $$f_{ps} = 1860\left(1 - \frac{0.40}{0.80}\left[0.00306 \times \frac{1860}{34.5} + \frac{955}{918} \times 0.033\right]\right)$$

   $$= 1860(1 - 0.1) = 1674 \text{ MPa}$$

From Equation 8.48

$$a = \frac{A_{pw}f_{ps}}{0.85f_c'b_w}$$

where $A_{pw}f_{ps} = A_{ps}f_{ps} + A_sf_y - 0.85f_c'(b-b_w)h_f$

$$A_{pw}f_{ps} = 1287 \times 1674 + 1200 \times 414 - 0.85 \times 34.5 \times (45.7 - 15.2)15.7 \times 10^2$$

$$= 10^6(2.15 + 0.5 - 1.14) \text{ N} = 1240 \text{ kN}$$

$$a = \frac{1240 \times 10}{0.85 \times 34.7 \times 15.2} = 24.7 \text{ cm} > h_f = 15.7 \text{ cm}$$

Hence, the neutral axis is outside the flange and analysis has to be based on a T-section.

3. *Available nominal moment strength.* Checking the limits of reinforcement:

$$\omega_T = \omega_p + \omega = 0.14 + 0.033 = 0.173 < 0.36\beta_1$$

hence, the maximum allowable reinforcement index is not exceeded. Alternatively by the ACI Code, $c/d_t = a/\beta_1 d_t = 15.7/(0.80 \times 91.8) = 0.22 < 0.375$. From Figure 8.7, the beam is in the tensile zone, giving $\phi = 0.90$ for determining the design moment M_u as assumed; hence, the section satisfies the maximum reinforcement limit, as an indirect Code to ensure ductility of the member.

The nominal moment strength is as follows for this flanged section:

$$M_n = A_{pw}f_{ps}\left(d - \frac{a}{2}\right) + A_sf_y(d - d_p) + 0.85f_c'(b - b_w)h_f\left(d_p - \frac{h_f}{2}\right)$$

$$\text{Available } M_n = 1.24 \times 10^6\left(91.8 + \frac{27.7}{2}\right) + 1200 \times 414(95.5 - 91.8)$$

$$+ 0.85 \times 34.5(45.7 - 15.2)15.7\left(91.8 - \frac{15.2}{2}\right) \times 10^2$$

$$= 10^6(96.6 + 1.83 + 118.2)\text{N cm} = 2166 \text{ kN m}$$

$$> \text{Required } M_n = 1871 \text{ kN m}$$

Hence, the section is alright for the design moment M_u on the basis of the $\phi = 0.90$ used for determining the required nominal strength M_n.

8.7 Shear and Torsional Strength Design

Two types of shear control the behavior of prestressed concrete beams: flexure shear (V_{ci}) and web shear (V_{cw}). To design for shear, it is necessary to determine whether flexure shear or web shear controls the choice of concrete shear strength V_c

$$V_{ci} = \frac{M_{cr}}{M/V - d_p/2} + 0.6b_wd_p\sqrt{f_c'} + V_d \tag{8.53}$$

where V_d is the vertical shear due to self-weight. The vertical component V_p of the prestressing force is disregarded in Equation 8.53, since it is small along the span sections where the prestressing tendon is not too steep.

The value of V in Equation 8.53 is the factored shear force V_i at the section under consideration due to externally applied loads occurring simultaneously with the maximum moment M_{max} occurring at that section, that is

$$V_{ci} = 0.6\lambda\sqrt{f_c'}\,b_w d_p + V_d + \frac{V_i}{M_{max}}M_{cr} \geq 1.7\lambda\sqrt{f_c'}\,b_w d_p$$
$$\leq 5.0\lambda\sqrt{f_c'}\,b_w d_p \tag{8.54}$$

where

$\lambda = 1.0$ for normal-weight concrete
$\quad\,\, = 0.85$ for sand-lightweight concrete
$\quad\,\, = 0.75$ for all-lightweight concrete
$V_d =$ shear force at section due to unfactored dead load
$V_{ci} =$ nominal shear strength provided by the concrete when diagonal tension cracking results from combined vertical shear and moment
$V_i =$ factored shear force at section due to externally applied load occurring simultaneously with M_{max}

For lightweight concrete, $\lambda = f_{ct}/6.7\sqrt{f_c'}$ if the value of the tensile splitting strength f_{ct} is known. Note that the value $\sqrt{f_c'}$ should not exceed 100.

The equation for M_{cr}, the moment causing flexural cracking due to external load, is given by

$$M_{cr} = \frac{I_c}{y_t}\left(6\sqrt{f_c'} + f_{ce} - f_d\right) \tag{8.55}$$

where f_{ce} is the concrete compressive stress due to effective prestress after losses at *extreme* fibers of section where tensile stress is caused by external load, psi. At the centroid, $f_{ce} = \bar{f}_c$; f_d is the stress due to unfactored dead load at extreme fiber of section resulting from self-weight only where tensile stress is caused by externally applied load, psi; y_t is the distance from centroidal axis to extreme fibers in tension; and M_{cr} is the portion of the applied *live load* moment that causes cracking. For simplicity the section modulus S_b may be substituted by I_c/y_t.

The web-shear crack in the prestressed beam is caused by an indeterminate stress that can best be evaluated by calculating the principal tensile stress at the critical plane. The shear stress v_c can be defined as the web-shear stress v_{cw} and is maximum near the centroid cgc of the section where the actual diagonal crack develops, as extensive tests to failure have indicated. If v_{cw} is substituted for v_c and $\bar{f}_c$, which denotes the concrete stress f_c due to effective prestress *at the cgc level*, is substituted for f_c in the equation, the expression equating the principal tensile stress in the concrete to the direct tensile strength becomes

$$f_t' = \sqrt{(\bar{f}_c/2) + v_{cw}^2} - \frac{\bar{f}_c}{2} \tag{8.56}$$

where $v_{cw} = V_{cw}/(b_w d_p)$ is the shear stress in the concrete due to all loads causing a nominal strength vertical shear force V_{cw} in the web. Solving for v_{cw} in Equation 8.56 gives

$$v_{cw} = f_t'\sqrt{1 + \bar{f}_c/f_t'} \tag{8.57a}$$

Using $f_t' = 3.5\sqrt{f_c'}$ as a reasonable value of the tensile stress on the basis of extensive tests, Equation 8.57a becomes

$$v_{cw} = 3.5\sqrt{f_c'}\left(\sqrt{1 + \bar{f}_c/3.5\sqrt{f_c'}}\right) \tag{8.57b}$$

which can be further simplified to

$$v_{cw} = 3.5\sqrt{f_c'} + 0.3\bar{f}_c \tag{8.57c}$$

In the ACI code, $\bar{f_c}$ is termed f_{pc}. The notation used herein is intended to emphasize that this is the stress in the concrete, and not the prestressing steel. The nominal shear strength V_{cw} provided by the concrete when diagonal cracking results from *excessive principal tensile stress* in the web becomes

$$V_{cw} = \left(3.5\lambda\sqrt{f_c'} + 0.3\bar{f_c}\right)b_w d_p + V_p \tag{8.58}$$

where V_p is the vertical component of the effective prestress at the particular section contributing to added nominal strength, λ is equal to 1.0 for normal-weight concrete, and less for lightweight concrete, and d_p is the distance from the extreme compression fiber to the centroid of prestressed steel, or $0.8h$, whichever is greater.

The ACI code stipulates the value of $\bar{f_c}$ to be the resultant concrete compressive stress at either the centroid of the section or the junction of the web and the flange when the centroid lies within the flange. In case of composite sections, $\bar{f_c}$ is calculated on the basis of stresses caused by prestress and moments resisted by the precast member acting *alone*.

The spacing of the web shear reinforcement is determined from

$$s = \frac{A_v f_y d}{(V_u/\phi) - V_c} = \frac{A_v \phi f_y d}{V_u - \phi V_c} \tag{8.59}$$

with the following limitations:

1. $s_{max} \le \frac{3}{4}h \le 24$ in., where h is the total depth of the section.
2. If $V_s > 4\lambda\sqrt{f_c'}b_w d_p$, the maximum spacing in (1) shall be reduced by half.
3. If $V_s > 8\lambda\sqrt{f_c'}b_w d_p$, enlarge the section.

8.7.1 Composite-Action Dowel Reinforcement

Ties for horizontal shear may consist of single bars or wires, multiple leg stirrups, or vertical legs of welded wire fabric. The spacing cannot exceed four times the least dimension of the support element or 24 in., whichever is less. If μ is the coefficient of friction, then the nominal horizontal shear force F_h can be defined as

$$F_h = \mu A_{vf} f_y \le V_{nh} \tag{8.60}$$

The ACI values of μ are based on a limit shear-friction strength of 800 psi, a quite conservative value as demonstrated by extensive testing. The Prestressed Concrete Institute recommends, for concrete placed against an intentionally roughened concrete surface, a maximum $\mu_e = 2.9$ instead of $\mu = 1.0\lambda$, and a maximum design shear force

$$V_u \le 0.25\lambda^2 f_c' A_c \le 1000\lambda^2 A_{cc} \tag{8.61a}$$

with a required area of shear-friction steel of

$$A_{vf} = \frac{V_{uh}}{\phi f_y \mu_e} \tag{8.61b}$$

The PCI less conservative approach stipulates

$$\mu_e = \frac{1000\lambda^2 b_v I_{vh}}{F_h} \le 2.9$$

The minimum required reinforcement area is

$$A_{vf} = \frac{50 b_v s}{f_y} = \frac{50 b_v I_{vh}}{f_y} \tag{8.62}$$

where $b_v I_{vh} = A_{cc}$, wherein A_{cc} is the concrete contact surface area.

8.7.2 Example 7: Design of Web Reinforcement for Shear

Design the bonded beam of Example 3 to be safe against shear failure, and proportion the required web reinforcement.

Solution

Data and nominal shear strength determination

$$f_{pu} = 270,000 \text{ psi } (1,862 \text{ MPa})$$
$$f_y = 60,000 \text{ psi } (1,862 \text{ MPa})$$
$$f_{pe} = 155,000 \text{ psi } (1,862 \text{ MPa})$$
$$f_c' = 5,000 \text{ psi normal-weight concrete}$$
$$A_{ps} = 13 \text{ seven-wire } \tfrac{1}{2}\text{-in. strands tendon}$$
$$= 1.99 \text{ in.}^2 (12.8 \text{ cm}^2)$$
$$A_s = 4\#6 \text{ bars} = 1.76 \text{ in.}^2 (11.4 \text{ cm}^2)$$
$$\text{Span} = 65 \text{ ft } (19.8 \text{ m})$$
$$\text{Service } W_L = 1,100 \text{ plf } (16.1 \text{ kN/m})$$
$$\text{Service } W_{SD} = 100 \text{ plf } (1.46 \text{ kN/m})$$
$$\text{Service } W_D = 393 \text{ plf } (5.7 \text{ kN/m})$$

Section properties

$$h = 40 \text{ in. } (101.6 \text{ cm})$$
$$d_p = 36.16 \text{ in. } (91.8 \text{ cm})$$
$$d = 37.6 \text{ in. } (95.5 \text{ cm})$$
$$b_w = 6 \text{ in. } (15 \text{ cm})$$
$$e_c = 15 \text{ in. } (38 \text{ cm})$$
$$e_e = 12.5 \text{ in. } (32 \text{ cm})$$
$$I_c = 70,700 \text{ in.}^4 (18.09 \times 10^6 \text{ cm}^4)$$
$$A_c = 377 \text{ in.}^2 (2,432 \text{ cm}^2)$$
$$r^2 = 187.5 \text{ in.}^2 (1,210 \text{ cm}^2)$$
$$c_b = 18.84 \text{ in. } (48 \text{ cm})$$
$$c_t = 21.16 \text{ in. } (54 \text{ cm})$$
$$p_e = 308,255 \text{ lb } (1,371 \text{ kN})$$
$$\text{Factored load } W_u = 1.2D + 1.6L$$
$$= 1.2(100 + 393) + 1.6 \times 1,100 = 2,352 \text{ plf}$$

Factored shear force at face of support $= V_u = W_u L/2 = (2,352 \times 65)/2 = 76,440 \text{ lb}$

Req. $V_n = V_u/\phi = 76,440/0.75 = 101,920 \text{ lb at support}$

Plane at $\tfrac{1}{2} d_p$ from face of support

Nominal shear strength V_c of web

$$\frac{1}{2} d_p = \frac{36.16}{2 \times 12} \cong 1.5 \text{ ft}$$
$$V_c = 101,920 \times \frac{[(65/2) - 1.5]}{65/2} = 97,216 \text{ lb}$$

Flexure-shear cracking, V_{ci}

From Equation 8.54

$$V_{ci} = 0.6\lambda\sqrt{f_c'}\,b_w d_p + V_d + \frac{V_i}{M_{max}}(M_{cr}) \geq 1.7\lambda\sqrt{f_c'}\,b_w d$$

From Equation 8.55, the cracking moment is

$$M_{cr} = \frac{I_c}{y_t}\left(6\sqrt{f_c'} + f_{ce} - f_d\right)$$

where $I_c/y_t = S_b$, since y_t is the distance from the centroid to the extreme tension fibers. Thus

$$I_c = 70,700 \text{ in.}^4$$
$$c_b = 18.84 \text{ in.}$$
$$P_e = 308,255 \text{ lb}$$
$$S_b = 3,753 \text{ in.}^3$$
$$r^2 = 187.5 \text{ in.}^2$$

The concrete stress at the extreme bottom fibers *due to prestress only* is

$$f_{ce} = -\frac{P_e}{A_c}\left(1 + \frac{ec_b}{r^2}\right)$$

and the tendon eccentricity at $d_p/2 \cong 1.5$ ft from the face of the support is

$$e \cong 12.5 + (15 - 12.5)\frac{1.5}{65/2} = 12.62 \text{ in.}$$

Thus

$$f_{ce} = -\frac{308,255}{377}\left(1 + \frac{12.62 \times 18.84}{187.5}\right) \cong -1,855 \text{ psi (12.8 MPa)}$$

From Example 3, the unfactored dead load due to self-weight $W_D = 393$ plf (5.7 kN/m) is

$$M_{d/2} = \frac{W_D x(l - x)}{2} = \frac{393 \times 1.5(65 - 1.5) \times 12}{2} = 224,600 \text{ in. lb (25.4 kN m)}$$

and the stress due to the unfactored dead load at the extreme concrete fibers where tension is created by the external load is

$$f_d = \frac{M_{d/2}c_b}{I_c} = \frac{224,600 \times 18.84}{70,700} = 60 \text{ psi}$$

Also

$$M_{cr} = 3,753(6 \times 1.0 \times \sqrt{6,000} + 1855 - 60)$$
$$= 8,480,872 \text{ in. lb (958 kN m)}$$

$$V_d = W_D\left(\frac{l}{2} - x\right) = 393\left(\frac{65}{2} - 1.5\right) = 12,183 \text{ lb (54.2 kN)}$$

$$W_{SD} = 100 \text{ plf}$$

$$W_L = 1,100 \text{ plf}$$

$$W_U = 1.2 \times 100 + 1.6 \times 1,100 = 1,880 \text{ plf}$$

The factored shear force at the section due to externally applied loads occurring simultaneously with M_{max} is

$$V_i = W_U\left(\frac{l}{2} - x\right) = 1,880\left(\frac{65}{2} - 1.5\right) = 58,280 \text{ lb (259 kN)}$$

and

$$M_{\max} = \frac{W_U x(l-x)}{2} = \frac{1,880 \times 1.5(65-1.5)}{2} \times 12$$
$$= 1,074,420 \text{ in. lb } (122 \text{ kN m})$$

Hence

$$V_{ci} = 0.6 \times 1.0\sqrt{6,000} \times 6 \times 36.16 + 12,183 + \frac{58,280}{1,074,420}(8,480,872)$$
$$= 482,296 \text{ lb } (54.5 \text{ kN m})$$
$$1.7\lambda\sqrt{f_c'}b_w d_p = 1.7 \times 1.0\sqrt{6,000} \times 6 \times 36.16 = 28,569 \text{ lb } (127 \text{ kN}) < V_{ci} = 482,296 \text{ lb}$$

Hence, $V_{ci} = 482,296$ lb (214.5 kN).

Web-shear cracking, V_{cw}

From Equation 8.58

$$V_{cw} = (3.5\sqrt{f_c'} + 0.3\bar{f_c})b_w d_p + V_p$$
$$\bar{f_c} = \text{compressive stress in concrete at the cgc}$$
$$= \frac{P_e}{A_c} = \frac{308,255}{377} \cong 818 \text{ psi } (5.6 \text{ MPa})$$
$$V_p = \text{vertical component of effective prestress at section}$$
$$= P_e \tan\theta \text{ (more correctly } P_e \sin\theta)$$

where θ is the angle between the inclined tendon and the horizontal. So

$$V_p = 308,255\frac{(15-12.5)}{(65/2) \times 12} = 1,976 \text{ lb } (8.8 \text{ kN})$$

Hence

$$V_{cw} = (3.5\sqrt{6,000} + 0.3 \times 818) \times 6 \times 36.16 + 1,976 = 114,038 \text{ lb } (507 \text{ kN})$$

In this case, web-shear cracking controls [i.e., $V_c = V_{cw} = 114,038$ lb (507 kN)] is used for the design of web reinforcement

$$V_s = \frac{V_u}{\phi = 0.75} - V_c = (97,216 - 114,038) \text{ lb}, \quad \text{namely } V_c > V_n$$

So no web steel is needed unless $V_u/\phi > \frac{1}{2}V_c$. Accordingly, we evaluate the latter:

$$\frac{1}{2}V_c = \frac{114,038}{2} = 57,019 \text{ lb } (254 \text{ kN}) < 97,216 \text{ lb } (432 \text{ kN})$$

Since $V_u/\phi > \frac{1}{2}V_c$ but $< V_c$, use minimum web steel in this case.

Minimum web steel

$$\text{Req. } \frac{A_v}{s} = 0.0077 \text{ in.}^2/\text{in.}$$

So, trying #3 U stirrups, we get $A_v = 2 \times 0.11 = 0.22$ in.^{2}, and it follows that

$$s = \frac{A_v}{\text{Req. } A_v/s} = \frac{0.22}{0.0077} = 28.94 \text{ in. } (73 \text{ cm})$$

We then check for the minimum A_v as the lesser of the two values given by

$$A_v = 0.75\sqrt{f_c'}\left(\frac{b_w s}{f_y}\right), \quad A_v = 50b_w s/f_y, \quad \text{whichever is larger}$$

and

$$A_v = \frac{A_{ps} \, f_{pu} s}{80 \, f_y d_p} \sqrt{\frac{d_p}{b_w}}$$

So the maximum allowable spacing $\leq 0.75\,h \leq 24$ in. Use #3 U stirrups at 22 in. center to center over a stretch length of 84 in. from the face of the support (see Figure 8.8).

8.7.3 SI Expressions for Shear in Prestressed Concrete Beams

$$V_{ci} = \left[\frac{\lambda \sqrt{f_c'}}{20} b_w d + V_d + V_i \left(\frac{M_{cr}}{M_{max}} \right) \right] \geq \left(\frac{\sqrt{f_c'}}{7} \right) b_w d \tag{8.63}$$

$$M_{cr} = S_b \left(0.5 \lambda \sqrt{f_c'} + f_{ce} - f_d \right) \tag{8.64}$$

$$V_{cw} = 0.3 \left(\lambda \sqrt{f_c'} + 0.3 \bar{f_c} \right) b_w d + V_p \tag{8.65}$$

where M_{cr} for shear analysis is equal to the moment causing flexural cracking at section due to *externally* applied load

$$V_c = \left(\frac{\lambda \sqrt{f_c'}}{20} + f \frac{V_u d}{M_u} \right) b_w d; \quad \frac{V_u d}{M_u} \leq 1.0$$

$$\geq \left[\lambda \frac{\sqrt{f_c'}}{5} \right] b_w d$$

$$\leq \left[0.4 \lambda \sqrt{f_c'} b_w d \right] \tag{8.66}$$

$$s = \frac{A_v f_y d}{(V_u/\phi) - V_c} = \frac{A_y f_y d}{V_s} \tag{8.67}$$

Min. A_v: the smaller of

$$A_v \geq \frac{0.35 b_w s}{f_y} \quad \text{or} \quad \frac{A_{ps} \, f_{pu} s}{80 \, f_y d_p} \sqrt{\frac{d}{b_w}}$$

where b_w, s, and d are in millimeters and f_y is in MPa.

8.7.4 Design of Prestressed Concrete Beams Subjected to Combined Torsion, Shear, and Bending in Accordance with the ACI 318-02 Code

8.7.4.1 Compatibility Torsion

In statically indeterminate systems, stiffness assumptions, compatibility of strains at the joints, and redistribution of stresses may affect the stress resultants, leading to a reduction of the resulting torsional shearing stresses. A reduction is permitted in the value of the factored moment used in the design of the member if part of this moment can be redistributed to the intersecting members. The ACI Code permits a maximum factored torsional moment at the critical section $h/2$ from the face of the supports for prestressed concrete members as follows:

A_{cp} = area enclosed by outside perimeter of concrete cross-section = $x_0 y_0$
p_{cp} = outside perimeter of concrete cross-section A_{cp}, in. = $2(x_0 + y_0)$

$$T_u = \phi 4 \sqrt{f_c'} \left(\frac{A_{cp}^2}{p_{pc}} \right) \sqrt{1 + \frac{\bar{f_c}}{4\sqrt{f_c'}}} \tag{8.68}$$

where $\bar{f_c}$ is the average compressive stress in the concrete at the centroidal axis due to effective prestress only after allowing for all losses.

Neglect of the full effect of the total value of external torsional moment in this case does not, in effect, lead to failure of the structure but may result in excessive cracking if $\phi 4\sqrt{f_c'}(A_{cp}^2/p_{pc})$ is considerably smaller in value than the actual factored torque.

If the actual factored torque is less than that given in Equation 8.68, the beam has to be designed for the lesser torsional value. Torsional moments are neglected however if for prestressed concrete

$$T_u < \phi\sqrt{f_c'}\left(\frac{A_{cp}^2}{p_{pc}}\right)\sqrt{1 + \frac{\bar{f_c}}{4\sqrt{f_c'}}} \tag{8.69}$$

8.7.4.2 Torsional Moment Strength

The size of the cross-section is chosen on the basis of reducing unsightly cracking and preventing the crushing of the surface concrete caused by the inclined compressive stresses due to shear and torsion defined by the left-hand side of the expressions in Equation 8.70. The geometrical dimensions for torsional moment strength in both reinforced and prestressed members are limited by the following expressions:

1. *Solid sections*

$$\sqrt{\left(\frac{V_u}{b_w d}\right)^2 + \left(\frac{T_u p_h}{1.7A_{oh}^2}\right)} \leq \phi\left(\frac{V_c}{b_w d} + 8\sqrt{f_c'}\right) \tag{8.70}$$

2. *Hollow sections*

$$\left(\frac{V_u}{b_w d}\right) + \left(\frac{T_u p_h}{1.7A_{oh}^2}\right) \leq \phi\left(\frac{V_c}{b_w d} + 8\sqrt{f_c'}\right) \tag{8.71}$$

where A_{oh} is the area enclosed by the centerline of the outermost closed transverse torsional reinforcement (sq. in.) and p_h is the perimeter of the centerline of the outermost closed transverse torsional reinforcement (in.).

The sum of the stresses at the left-hand side of Equation 8.71 should not exceed the stresses causing shear cracking plus $8\sqrt{f_c'}$. This is similar to the limiting strength for $V_s \leq 8\sqrt{f_c'}$ for shear without torsion. The strength of the plain concrete in the web is taken as

$$V_c = \left(0.6\lambda\sqrt{f_c'} + 700\frac{V_u d_p}{M_u}\right)b_w d_p; \quad \frac{V_u d_p}{M_u} \leq 1.0$$

$$\geq 1.7\lambda\sqrt{f_c'}b_w d_p$$

$$\leq 5.0\lambda\sqrt{f_c'}b_w d_p \tag{8.72}$$

where $f_{pe} > 0.4f_{pu}$.

8.7.4.3 Torsional Web Reinforcement

Meaningful additional torsional strength due to the addition of torsional reinforcement can be achieved only by using both stirrups and longitudinal bars. Ideally, *equal* volumes of steel in both the closed stirrups and the longitudinal bars should be used so that both participate equally in resisting the twisting moments. This principle is the basis of the ACI expressions for proportioning the torsional web steel. If s is the spacing of the stirrups, A_l is the total cross-sectional area of the longitudinal bars, and A_t is the cross-section of one stirrup leg, the transverse reinforcement for torsion has to be based on the full external torsional moment strength value T_n, namely, T_u/ϕ

$$T_n = \frac{2A_0 A_t f_{yv}}{s}\cot\theta \tag{8.73}$$

where A_0 is the gross area enclosed by the shear flow path (sq. in.), A_t is the cross-sectional area of one leg of the transverse closed stirrups (sq. in.), f_{yv} is the yield strength of closed transverse torsional reinforcement not to exceed 60,000 psi, and θ is the angle of the compression diagonals (struts) in the space truss analogy for torsion.

Transposing terms in Equation 8.73, the transverse reinforcement area becomes

$$\frac{A_t}{s} = \frac{T_n}{2A_0 f_{yv} \cot \theta} \tag{8.74}$$

The area A_0 is determined by analysis, except that the ACI 318 Code permits taking $A_0 = 0.85 A_{oh}$ in lieu of the analysis.

The factored torsional resistance ϕT_n must equal or exceed the factored external torsional moment T_u. All the torsional moments are assumed in the ACI 318-02 Code to be resisted by the closed stirrups and the longitudinal steel with the torsional resistance, T_c, of the concrete disregarded, namely $T_c = 0$. The shear V_c resisted by the concrete is assumed to be unchanged by the presence of torsion. The angle θ subtended by the concrete compression diagonals (struts) should not be taken smaller than 30° nor larger than 60°. It can be obtained by analysis as detailed by Hsu [6]. The additional longitudinal reinforcement for torsion should not be less than

$$A_l = \frac{A_t}{s} p_h \left(\frac{f_{yv}}{f_{yl}} \right) \cot^2 \theta \tag{8.75}$$

where f_{yl} is the yield strength of the longitudinal torsional reinforcement, not to exceed 60,000 psi.

The same angle θ should be used in both Equations 8.73 and 8.74. It should be noted that as θ gets smaller, the amount of required stirrups required by Equation 8.74 decreases. At the same time the amount of longitudinal steel required by Equation 8.75 increases.

In lieu of determining the angle θ by analysis, the ACI Code allows a value of θ equal to

1. 45° for nonprestressed members or members with less prestress than in 2,
2. 37.5° for prestressed members with an effective prestressing force larger than 40% of the tensile strength of the longitudinal reinforcement.

The PCI recommends computing the value of θ from the expression:

$$\cot \theta = \frac{T_u/\phi}{1.7 A_{oh} (A_t/S) f_{yv}} \tag{8.76}$$

8.7.4.3.1 Minimum Torsional Reinforcement

It is necessary to provide a minimum area of torsional reinforcement in all regions where the factored torsional moment T_u exceeds the value given by Equation 8.69. In such a case, the minimum area of the required transverse closed stirrups is

$$A_v + 2A_t \geq \frac{50 b_w s}{f_{yv}} \tag{8.77}$$

The maximum spacing should not exceed the smaller of $p_n/8$ or 12 in.

The minimum total area of the additional longitudinal torsional reinforcement should be determined by

$$A_{l,min} = \frac{5\sqrt{f_c'} A_{cp}}{f_{yl}} - \left(\frac{A_t}{s} \right) p_h \frac{f_{yv}}{f_{yl}} \tag{8.78}$$

where A_t/s should not be taken less than $25 b_w/f_{yv}$. The additional longitudinal reinforcement required for torsion should be distributed around the perimeter of the closed stirrups with a maximum spacing of 12 in. The longitudinal bars or tendons should be placed inside the closed stirrups and at least one longitudinal bar or tendon in each corner of the stirrup. The bar diameter should be at least

$\frac{1}{16}$ of the stirrup spacing but not less than a No. 3 bar. Also, the torsional reinforcement should extend for a minimum distance of $(b_t + d)$ beyond the point theoretically required for torsion because torsional diagonal cracks develop in a helical form extending beyond the cracks caused by shear and flexure. b_t is the width of that part of cross-section containing the stirrups resisting torsion. The critical section in beams is at a distance d from the face of the support for reinforced concrete elements and at $h/2$ for prestressed concrete elements, d being the effective depth and h the total depth of the section.

8.7.4.4 SI-Metric Expressions for Torsion Equations

In order to design for combined torsion and shear using the SI (System International) method, the following equations replace the corresponding expressions in the PI (Pound–Inch) method:

$$T_u \leq \frac{\phi \sqrt{f_c'}}{3} \left(\frac{A_{cp}^2}{p_{cp}} \right) \sqrt{1 + \frac{3\bar{f}_c}{\sqrt{f_c'}}} \tag{8.79}$$

$$T_u \leq \frac{\phi \sqrt{f_c'}}{12} \left(\frac{A_{cp}^2}{p_{cp}} \right) \sqrt{1 + \frac{3\bar{f}_c}{\sqrt{f_c'}}} \tag{8.80}$$

$$\sqrt{\left(\frac{V_u}{b_w d} \right)^2 + \left(\frac{T_u p_h}{1.7 A_{oh}^2} \right)^2} \leq \phi \left(\frac{V_c}{b_w d} + \frac{8\sqrt{f_c'}}{12} \right) \tag{8.81}$$

$$\left(\frac{V_u}{b_w d} \right) + \left(\frac{T_u p_h}{1.7 A_{oh}^2} \right) \leq \phi \left(\frac{V_c}{b_w d} + \frac{8\sqrt{f_c'}}{12} \right) \tag{8.82}$$

$$V_c = \left(\frac{\lambda \sqrt{f_c'}}{20} + \frac{5 V_u d_p}{M_u} \right) b_w d_p$$

$$\geq \left(0.17 \lambda \sqrt{f_c'} \right) b_w d_p$$

$$\leq \left(0.4 \lambda \sqrt{f_c'} \right) b_w d_p \tag{8.83}$$

$$\frac{V_u d_p}{M_u} \leq 1.0$$

$$T_n = \frac{2 A_0 A_t f_{yv}}{s} \cot \theta \tag{8.84}$$

where f_{yv} is in MPa, s is in millimeter, A_0, A_t are in mm^2, and T_n is in kN m

$$\frac{A_t}{s} = \frac{T_n}{2 A_0 f_{yv} \cot \theta} \tag{8.85}$$

$$A_l = \frac{A_t}{s} p_h \left(\frac{f_{yv}}{f_{yl}} \right) \cot^2 \theta \tag{8.86}$$

where f_{yv} and f_{yl} are in MPa, p_h and s are in millimeters, and A_l and A_t are in mm^2

$$A_v = \frac{0.35 b_w s}{f_y} \tag{8.87}$$

$$A_{l,min} = \frac{5 \sqrt{f_c'} A_{cp}}{12 f_{yl}} - \left(\frac{A_t}{s} \right) p_h \left(\frac{f_{yv}}{f_{yl}} \right) \tag{8.88}$$

PHOTO 8.7   Sunshine Skyway Bridge, Tampa, Florida. Designed by Figg and Muller Engineers, Inc. The bridge has a 1200 ft cable-stayed main span with a single pylon, 175 ft vertical clearance, a total length of 21,878 ft and twin 40-ft roadways (courtesy Portland Cement Association).

where A_t/s should not be taken less than $0.175b_w/f_{yv}$. Maximum allowable spacing of transverse stirrups is the smaller of $\frac{1}{8}p_h$ or 300 mm, and bars should have a diameter of at least $\frac{1}{16}$ of the stirrups spacing but not less than No. 10 M bar size. Max. f_{yv} or f_{yl} should not exceed 400 MPa. Min. A_{vt} the smaller of

$$\frac{A_{vt}}{s} \geq \frac{0.35b_w}{f_y} \quad \text{or} \quad \frac{A_{vt}}{s} = \frac{1}{16}\sqrt{f_c'}\left(\frac{b_w}{f_y}\right)$$

whichever is larger, where b_w, d_p, and s are in millimeters

$$\geq \frac{A_{ps}f_{pu}}{80f_y d_p}\sqrt{\frac{d_p}{b_w}}$$

Use the lesser of the two sets.

8.8   Camber, Deflection, and Crack Control

8.8.1   Serviceability Considerations

Prestressed concrete members are continuously subjected to sustained eccentric compression due to the prestressing force, which seriously affects their long-term creep deformation performance. Failure to predict and control such deformations can lead to high reverse deflection, that is, camber, which can produce convex surfaces detrimental to proper drainage of roofs of buildings, to uncomfortable ride characteristics in bridges and aqueducts, and to cracking of partitions in apartment buildings, including misalignment of windows and doors.

The difficulty of predicting very accurately the total long-term prestress losses makes it more difficult to give a precise estimate of the magnitude of expected camber. Accuracy is even more difficult in partially prestressed concrete systems, where limited cracking is allowed through the use of additional nonprestressed reinforcement. Creep strain in the concrete increases camber, as it causes a negative increase in curvature that is usually more dominant than the decrease produced by the decrease in prestress losses due to creep, shrinkage, and stress relaxation. A best estimate of camber increase should be based on accumulated experience, span-to-death ratio code limitations, and a correct choice of the modulus E_c of the concrete. Calculation of the moment–curvature relationships at the major incremental stages of loading up to the limit state at failure would also assist in giving a more accurate evaluation of the stress-related load deflection of the structural element.

The cracking aspect of serviceability behavior in prestressed concrete is also critical. Allowance for limited cracking in "partial prestressing" through the additional use of nonprestressed steel is prevalent. Because of the high stress levels in the prestressing steel, corrosion due to cracking can become detrimental to the service life of the structure. Therefore, limitations on the magnitudes of crack widths and their spacing have to be placed, and proper crack width evaluation procedures used. The presented discussion of the state of the art emphasizes the extensive work of the author on cracking in pretensioned and posttensioned prestressed beams.

Prestressed concrete flexural members are classified into three classes in the new ACI 318 Code:

1. *Class U:*

$$f_t \leq 7.5\sqrt{f_c'} \qquad (8.89a)$$

In this class, the gross section is used for section properties when both stress computations at service loads and deflection computations are made. No skin reinforcement needs to be used in the vertical faces.

2. *Class T:*

$$7.5\sqrt{f_c'} \leq f_t \leq 12\sqrt{f_c'} \qquad (8.89b)$$

This class is a transition between uncracked and cracked sections. For stress computations at service T loads, the gross section is used. The cracked bilinear section is used in the deflection computations. No skin reinforcement needs to be used in the vertical faces.

3. *Class C:*

$$f_t > 12\sqrt{f_c'} \qquad (8.89c)$$

This class denotes cracked sections. Hence, a cracked section analysis has to be made for evaluation of the stress level at service and for deflection. Computation of Δf_{ps} or f_s for crack control is necessary, where Δf_{ps} is the stress increase beyond the decompression state and f_s is the stress in the mild reinforcement when mild steel reinforcement is also used. Prestressed two-way slab systems are to be designed as Class U.

Ideally, the load–deflection relationship is trilinear, as shown in Figure 8.9. The three regions prior to rupture are:

Region I — Precracking stage, where a structural member is crack free.

Region II — Postcracking stage, where the structural member develops acceptable controlled cracking in both distribution and width.

Region III — Postserviceability cracking stage, where the stress in the tensile reinforcement reaches the limit state of yielding.

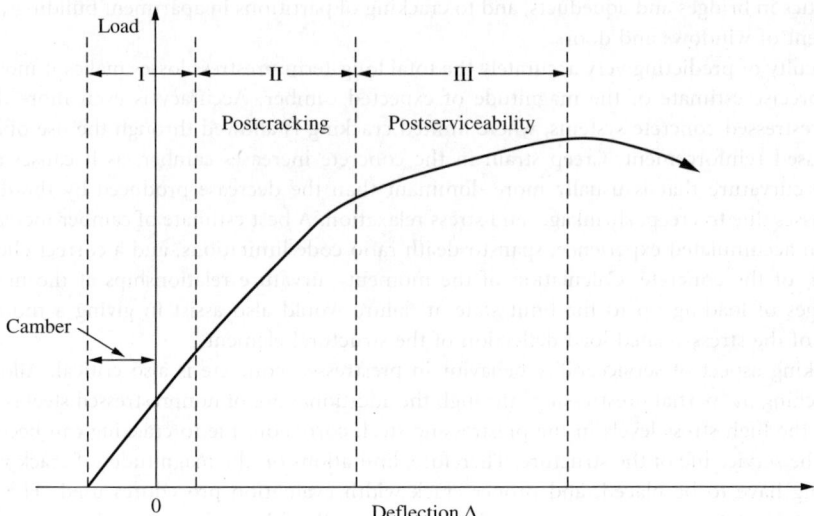

FIGURE 8.9 Beam load–deflection relationship: Region I, precracking stage; Region II, postcracking stage; Region III, postserviceability stage [5,11].

The precracking segment of the load–deflection curve is essentially a straight line defining full elastic behavior, as in Figure 8.9. The maximum tensile stress in the beam in this region is less than its tensile strength in flexure, that is, it is less than the modulus of rupture f_r of concrete. The flexural stiffness EI of the beam can be estimated using Young's modulus E_c of concrete and the moment of inertia of the uncracked concrete cross-section.

The precracking region ends at the initiation of the first crack and moves into region II of the load–deflection diagram in Figure 8.9. Most beams lie in this region at service loads. A beam undergoes varying degrees of cracking along the span corresponding to the stress and deflection levels at each section. Hence, cracks are wider and deeper at midspan, whereas only narrow, minor cracks develop near the supports in a simple beam.

The load–deflection diagram in Figure 8.9 is considerably flatter in region III than in the preceding regions. This is due to substantial loss in stiffness of the section because of extensive cracking and considerable widening of the stabilized cracks throughout the span. As the load continues to increase, the strain ε_s in the steel at the tension side continues to increase beyond the yield strain ε_y with no additional stress. The beam is considered at this stage to have structurally failed by initial yielding of the tension steel. It continues to deflect without additional loading, the cracks continue to open, and the neutral axis continues to rise toward the outer compression fibers. Finally, a secondary compression failure develops, leading to total crushing of the concrete in the maximum moment region followed by rupture. Figure 8.10 gives the deflection expressions for the most common loading cases in terms of both load and curvature.

8.8.1.1 Strain and Curvature Evaluation

The distribution of strain across the depth of the section at the controlling stages of loading is linear, as is shown in Figure 8.11, with the angle of curvature dependent on the top and bottom concrete extreme fiber strains ε_{ct} and ε_{cb}. From the strain distributions, the curvature at the various stages of loading can be expressed as follows:

1. *Initial prestress:*

$$\phi_i = \frac{\varepsilon_{cbi} - \varepsilon_{cti}}{h} \tag{8.90a}$$

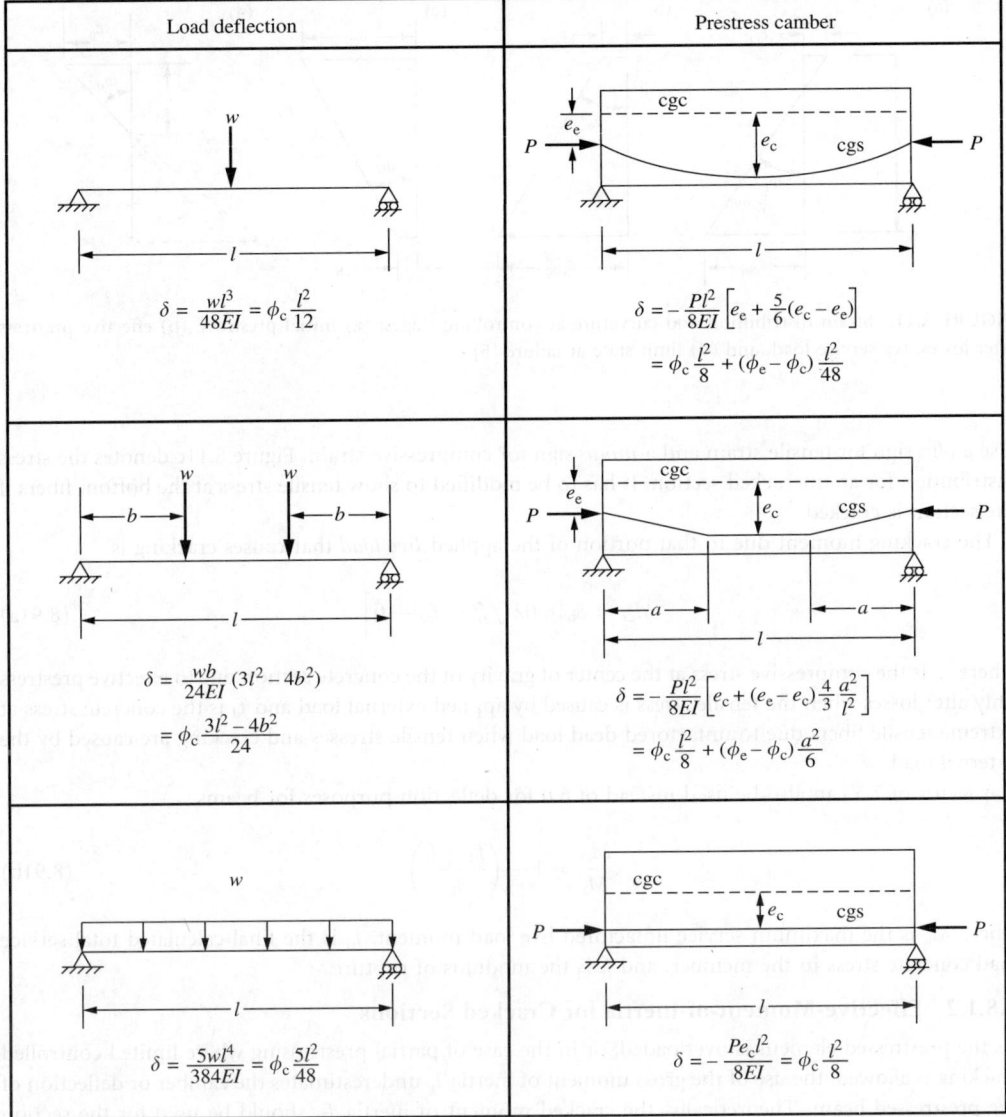

Load deflection	Prestress camber
$\delta = \dfrac{wl^3}{48EI} = \phi_c \dfrac{l^2}{12}$	$\delta = -\dfrac{Pl^2}{8EI}\left[e_e + \dfrac{5}{6}(e_c - e_e)\right]$ $= \phi_c \dfrac{l^2}{8} + (\phi_e - \phi_c)\dfrac{l^2}{48}$
$\delta = \dfrac{wb}{24EI}(3l^2 - 4b^2)$ $= \phi_c \dfrac{3l^2 - 4b^2}{24}$	$\delta = -\dfrac{Pl^2}{8EI}\left[e_c + (e_e - e_c)\dfrac{4}{3}\dfrac{a^2}{l^2}\right]$ $= \phi_c \dfrac{l^2}{8} + (\phi_e - \phi_c)\dfrac{a^2}{6}$
$\delta = \dfrac{5wl^4}{384EI} = \phi_c \dfrac{5l^2}{48}$	$\delta = -\dfrac{Pe_c l^2}{8EI} = \phi_c \dfrac{l^2}{8}$

FIGURE 8.10 Short-term deflection expressions for prestressed concrete beams. Subscript c indicates midspan and subscript e the support [5].

2. *Effective prestress after losses:*

$$\phi_e = \frac{\varepsilon_{cbe} - \varepsilon_{cte}}{h} \qquad (8.90b)$$

3. *Service load:*

$$\phi = \frac{\varepsilon_{cb} - \varepsilon_{ct}}{h} \qquad (8.90c)$$

4. *Failure:*

$$\phi_u = \frac{\varepsilon_u}{c} \qquad (8.90d)$$

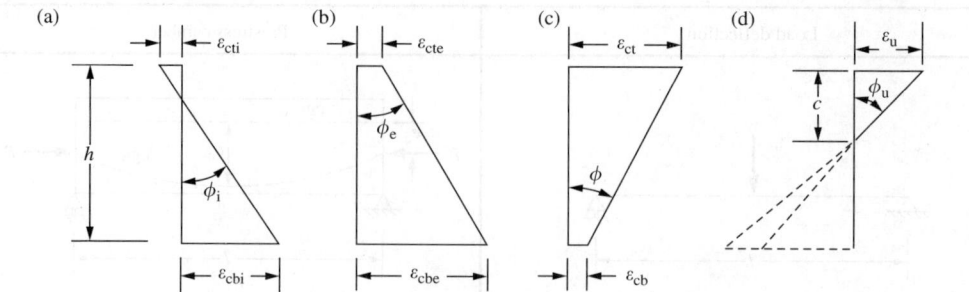

FIGURE 8.11 Strain distribution and curvature at controlling stages: (a) initial prestress, (b) effective prestress after losses, (c) service load, and (d) limit state at failure [5].

Use a *plus* sign for tensile strain and a *minus* sign for compressive strain. Figure 8.11c denotes the stress distribution for an uncracked section. It has to be modified to show tensile stress at the bottom fibers if the section is cracked.

The cracking moment due to that portion of the applied *live load* that causes cracking is

$$M_{cr} = S_b \left[6.0\lambda \sqrt{f_c'} + f_{ce} - f_d \right] \tag{8.91a}$$

where f_{ce} is the compressive stress at the center of gravity of the concrete section due to effective prestress only after losses when the tensile stress is caused by applied external load and f_d is the concrete stress at extreme tensile fibers due to unfactored dead load when tensile stresses and cracking are caused by the external load.

A factor of 7.5 can also be used instead of 6.0 for deflection purposes for beams

$$\frac{M_{cr}}{M_a} = 1 - \left(\frac{f_{tl} - f_r}{f_L} \right) \tag{8.91b}$$

where M_a is the maximum service unfactored live load moment, f_{tl} is the final calculated total service load concrete stress in the member, and f_r is the modulus of rupture.

8.8.1.2 Effective-Moment-of-Inertia for Cracked Sections

As the prestressed element is overloaded, or in the case of partial prestressing where limited controlled cracking is allowed, the use of the gross moment of inertia I_g underestimates the camber or deflection of the prestressed beam. Theoretically, the cracked moment of inertia I_{cr} should be used for the section across which the cracks develop while the gross moment of inertia I_g should be used for the beam sections between the cracks. However, such refinement in the numerical summation of the deflection increases along the beam span is sometimes unwarranted because of the accuracy difficulty of deflection evaluation. Consequently, an effective moment of inertia I_e can be used as an average value along the span of a simply supported bonded tendon beam, a method developed by Branson. According to this method

$$I_e = I_{cr} + \left(\frac{M_{cr}}{M_a} \right)^3 (I_g - I_{cr}) \leq I_g \tag{8.92a}$$

Equation 8.92a can also be written in the form

$$I_e = \left(\frac{M_{cr}}{M_a} \right)^3 I_g + \left[1 - \left(\frac{M_{cr}}{M_a} \right)^3 \right] I_{cr} \leq I_g \tag{8.92b}$$

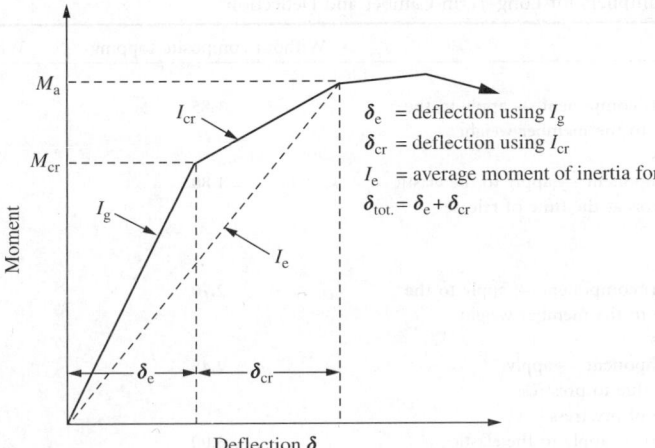

FIGURE 8.12 Moment–deflection relationship [1,5].

The ratio (M_{cr}/M_a) from Equation 8.91b can be substituted into Equations 8.92a and 8.92b to get the effective moment of inertia. Figure 8.12 gives the bilinear moment–deflection relationship for I_g, I_{cr}, and I_e, where

$$I_{cr} = n_p A_{ps} d_p^2 (1 - 1.6\sqrt{n_p \rho_p}) \qquad (8.93a)$$

where $n_p = E_{ps}/E_c$. If nonprestressed reinforcement is used to carry tensile stresses, namely, in "partial prestressing," Equation 8.93a can be modified to give

$$I_{cr} = (n_p A_{ps} d_p^2 + n_s A_s d^2)(1 - 1.6\sqrt{n_p \rho_p + n_s \rho_s}) \qquad (8.93b)$$

8.8.2 Long-Term Effects on Deflection and Camber

8.8.2.1 PCI Multipliers Method

The PCI multipliers method provides a multiplier C_1 that takes account of long-term effects in pre-stressed concrete members as presented in Table 8.11. This table, based on Ref. [7], can provide reasonable multipliers of immediate deflection and camber provided that the upward and downward components of the initial calculated camber are separated in order to take into account the effects of loss of prestress, *which only apply to the upward component*. Substantial reduction can be achieved in long-term camber by the addition of nonprestressed steel. In that case, a reduced multiplier C_2 can be used, given by

$$C_2 = \frac{C_1 + A_s/A_{ps}}{1 + A_s/A_{ps}} \qquad (8.94)$$

where C_1 is the multiplier from Table 8.11, A_s is the area of nonprestressed reinforcement, and A_{ps} is the area of prestressed strands.

8.8.3 Permissible Limits of Calculated Deflection

The ACI Code requires that the calculated deflection has to satisfy the serviceability requirement of maximum permissible deflection for the various structural conditions listed in Table 8.12. Note that long-term effects cause measurable increases in deflection and camber with time and result in excessive overstress in the concrete and the reinforcement, *requiring* computation of deflection and camber.

TABLE 8.11 C_1 Multipliers for Long-Term Camber and Deflection

	Without composite topping	With composite topping
At erection		
Deflection (downward) component — apply to the elastic deflection due to the member weight at release of prestress	1.85	1.85
Camber (upward) component — apply to the elastic camber due to prestress at the time of release of prestress	1.80	1.80
Final		
Deflection (downward) component — apply to the elastic deflection due to the member weight at release of prestress	2.70	2.40
Camber (upward) component — apply to the elastic camber due to prestress at the time of release of prestress	2.45	2.20
Deflection (downward) — apply to the elastic deflection due to the superimposed dead load only	3.00	3.00
Deflection (downward) — apply to the elastic deflection caused by the composite topping	—	2.30

TABLE 8.12   ACI Minimum Permissible Ratios of Span (l) to Deflection (δ) (l = Longer Span)

Type of member	Deflection δ to be considered	$(l/\delta)_{\min}$
Flat roofs not supporting and not attached to nonstructural elements likely to be damaged by large deflections	Immediate deflection due to live load L	180[a]
Floors not supporting and not attached to nonstructural elements likely to be damaged by large deflections	Immediate deflection due to live load L	360
Roof or floor construction supporting or attached to nonstructural elements likely to be damaged by large deflections	That part of total deflection occurring after attachment of nonstructural elements; sum of long-term deflection due to all sustained loads (dead load plus any sustained portion of live load) and immediate deflection due to any additional live load[b]	240[c]
Roof or floor construction supporting or attached to nonstructural elements not likely to be damaged by large deflections		480

[a] Limit not intended to safeguard against ponding. Ponding should be checked by suitably calculating deflection, including added deflections due to ponded water, and considering long-term effects of all sustained loads, camber, construction tolerances, and reliability of provisions for drainage.

[b] Long-term deflection has to be determined, but may be reduced by the amount of deflection calculated to occur before attachment of nonstructural elements. This reduction is made on the basis of accepted engineering data relating to time-deflection characteristics of members similar to those being considered.

[c] Ratio limit may be lower if adequate measures are taken to prevent damage to supported or attached elements, but should not be lower than tolerance of nonstructural elements.

AASHTO permissible deflection requirements, shown in Table 8.13, are more rigorous because of the dynamic impact of moving loads on bridge spans.

8.8.3.1   Approximate Time-Steps Method

The approximate time-steps method is based on a simplified form of summation of constituent deflections due to the various time-dependent factors. If C_u is the long-term creep coefficient, the

TABLE 8.13 AASHTO Maximum Permissible Deflection (l = Longer Span)

Type of member	Deflection considered	Maximum permissible deflection	
		Vehicular traffic only	Vehicular and pedestrian traffic
Simple or continuous spans	Instantaneous due to service live load plus impact	$l/800$	$l/1000$
Cantilever arms		$l/300$	$l/375$

curvature at effective prestress P_e can be defined as

$$\phi_e = \frac{P_i e_x}{E_c I_c} + (P_i - P_e)\frac{e_x}{E_c I_c} - \left(\frac{P_i + P_e}{2}\right)\frac{e_x}{E_c I_c}C_u \tag{8.95}$$

The following expression can predict the time-dependent increase in deflection $\Delta\delta$:

$$\Delta\delta = -\left[\eta + \frac{(1+\eta)}{2}k_r C_t\right]\delta_{i(P_i)} + k_r C_t \delta_{i(D)} + K_a k_r C_t \delta_{i(SD)} \tag{8.96}$$

where
$\eta = P_e/P_i$
C_t = creep coefficient at time t
K_a = factor corresponding to age of concrete at superimposed load application
     = $1.25t^{-0.118}$ for moist-cured concrete
     = $1.13t^{-0.095}$ for steam-cured concrete
t = age, in days, at loading
$k_r = 1/(1 + A_s/A_{ps})$ when $A_s/A_{ps} \ll 1.0$
     $\cong 1$ for all practical purposes

For the final deflection increment, C_u is used in place of C_t in Equation 8.96.
  For noncomposite beams, the total deflection $\delta_{T,t}$ becomes [4]

$$\delta_{T,t} = -\delta_{pi}\left[1 - \frac{\Delta P}{P_0} + \lambda(k_r C_t)\right] + \delta_D[1 + k_r C_t] + \delta_{SD}[1 + K_a k_r C_t] + \delta_L \tag{8.97}$$

where δ_p is the deflection due to prestressing, ΔP is the total loss of prestress excluding initial elastic loss, and $\lambda = 1 - (\Delta P/2P_0)$, in which P_0 is the prestress force at transfer after elastic loss and P_i less than the elastic loss.
  For composite beams, the total deflection is

$$\delta_T = -\delta_{pi}\left[1 - \frac{\Delta P}{P_0} + K_a k_r C_u \lambda\right] + \delta_D[1 + K_a k_r C_u]$$

$$+ \delta_{pi}\frac{I_e}{I_{comp.}}\left[1 - \frac{\Delta P - \Delta P_c}{P_0} + k_r C_u(\lambda - \alpha\lambda')\right]$$

$$+ (1-\alpha)k_r C_u \delta_D \frac{I_c}{I_{comp.}} + \delta_D\left[1 + \alpha k_r C_u \frac{I_c}{I_{comp.}}\right] + \delta_{df} + \delta_L \tag{8.98}$$

where
$\lambda' = 1 - (\Delta P_c/2P_0)$
ΔP_c = loss of prestress at time composite topping slab is cast, excluding initial elastic loss
$I_{comp.}$ = moment of inertia of composite section
δ_{df} = deflection due to differential shrinkage and creep between precast section and composite topping slab
     = $F y_{cs} l^2/8E_{cc}I_{comp.}$ for simply supported beams (for continuous beams, use the appropriate factor in the denominator)

y_{cs} = distance from centroid of composite section to centroid of slab topping
F = force resulting from differential shrinkage and creep
E_{cc} = modulus of composite section
α = creep strain at time t divided by ultimate creep strain
$\quad = t^{0.60}/(10 + t^{0.60})$

8.8.4 Long-Term Deflection of Composite Double-Tee Cracked Beam [2,5]

8.8.4.1 Example 8: Deflection Computation of a Double-Tee Beam

A 72-ft (21.9 m) span simply supported roof normal weight concrete double-T-beam (Figure 8.13) is subjected to a superimposed topping load $W_{SD} = 250$ plf (3.65 kN/m) and a service live load $W_L = 280$ plf (4.08 kN/m). Calculate the short-term (immediate) camber and deflection of this beam by (a) the I_e method and (b) the bilinear method as well as the time-dependent deflections after 2-in. topping is cast (30 days) and the final deflection (5 years), using the PCI multipliers method. Given prestress losses 18%.

	Noncomposite	Composite
A_c, in.^{2}	615 (3,968 cm^2)	855 (5,516 cm^2)
I_c, in.^{4}	59,720 (24.9 × 10^5 cm^4)	77,118 (32.1 × 10^5 cm^4)
r^2, in.^{2}	97 (625 cm^2)	90 (580 cm^2)
c_b, in.	21.98 (558 mm)	24.54 (623 mm)
c_t, in.	10.02 (255 mm)	9.46 (240 mm)
S_b, in.^{3}	2,717 (4.5 × 10^4 cm^3)	3,142 (5.1 × 10^4 cm^3)
S^t, in.^{3}	5,960 (9.8 × 10^4 cm^3)	8,152 (13.4 × 10^4 cm^3)
W_d, plf	641 (9.34 kN/m)	891 (13.0 kN/m)

$$V/S = 615/364 = 1.69 \text{ in. (43 mm)}$$

$$RH = 75\%$$

$$e_c = 18.73 \text{ in. (476 mm)}$$

$$e_{\hat{e}} = 12.81 \text{ in. (325 mm)}$$

$$f_c' = 5000 \text{ psi (34.5 MPa)}$$

$$f_{ci}' = 3750 \text{ psi (25.9 MPa)}$$

Topping $f_c' = 3000$ psi (20.7 MPa)

f_t at bottom fibers $= 12\sqrt{f_c'} = 849$ psi (5.9 MPa)

$$A_{ps} = \text{twelve } \tfrac{1}{2}\text{-in. diameter low-relaxation prestressing steel depressed at midspan only}$$

$$f_{pu} = 270,000 \text{ psi (1,862 MPa), low relaxation}$$

$$f_{pi} = 189,000 \text{ psi (1,303 MPa)}$$

$$f_{pj} = 200,000 \text{ psi (1,380 MPa)}$$

$$f_{py} = 260,000 \text{ psi (1,793 MPa)}$$

$$E_{ps} = 28.5 \times 10^6 \text{ psi (19.65 × 10}^4 \text{ MPa)}$$

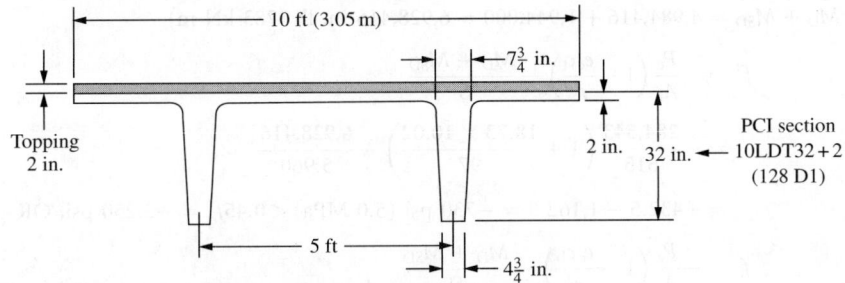

FIGURE 8.13   Double-tee composite beam in Example 8 [5].

Solution by the I_e Method

1. *Midspan section stresses*

$$f_{pj} = 200,000 \text{ psi at jacking}$$

$$f_{pi} \text{ assumed} = 0.945 f_{pj} = 189,000 \text{ psi at transfer}$$

$$e_c = 18.73 \text{ in. (475 mm)}$$

$$P_i = 12 \times 0.153 \times 189,000 = 347,004 \text{ lbs (1,540 kN)}$$

Self-weight moment

$$M_D = \frac{641(72)^2}{8} \times 12 = 4,984,416 \text{ in.-lb}$$

A. *At transfer*

$$
\begin{aligned}
f^t &= -\frac{P_i}{A_c}\left(1 - \frac{e_c c_t}{r^2}\right) - \frac{M_D}{S^t} \\
&= -\frac{347,004}{615}\left(1 - \frac{18.73 \times 10.02}{97}\right) - \frac{4,984,416}{5,960} \\
&= +527.44 - 836.31 \\
&= -308.87 \text{ psi } (C), \text{ say } 310 \text{ psi } (C)(2.1 \text{ MPa}) < 0.60 f'_{ci} = 0.60(3,750) \\
&= 2,250 \text{ psi, OK}
\end{aligned}
$$

$$
\begin{aligned}
f_b &= -\frac{P_i}{A_c}\left(1 - \frac{e_c c_b}{r^2}\right) - \frac{M_D}{S_b} \\
&= -\frac{347,004}{615}\left(1 + \frac{18.73 \times 21.98}{97}\right) + \frac{4,984,416}{2,717} \\
&= -2958.95 + 1834.53 \\
&= -1124.42 \text{ psi } (C), \text{ say } 1125 \text{ psi } (C) < -2,250 \text{ psi, OK}
\end{aligned}
$$

B. *After slab is cast.* At this load level assume 18% prestress loss

$$f_{pe} = 0.82 f_{pi} = 0.82 \times 189,000 = 154,980 \text{ psi}$$
$$P_e = 12 \times 0.153 \times 154,980 = 284,543 \text{ lb}$$

For the 2-in. slab

$$W_{SD} = \frac{2}{12} \times 10 \text{ ft} \times 150 = 250 \text{ plf (3.6 kN/m)}$$

$$M_{SD} = \frac{250(72)^2}{8} \times 12 = 1,944,000 \text{ in.-lb}$$

$$M_D + M_{SD} = 4,984,416 + 1,944,000 = 6,928,416 \text{ in. lb } (783 \text{ kN m})$$

$$f^t = -\frac{P_e}{A_c}\left(1 - \frac{e_c c_t}{r^2}\right) - \frac{M_D + M_{SD}}{S^t}$$

$$= -\frac{284,543}{615}\left(1 + \frac{18.73 \times 10.02}{97}\right) + \frac{6,928,416}{5.960}$$

$$= +432.5 - 1,162.5 = -730 \text{ psi } (5.0 \text{ MPa}) < 0.45 f_c' = -2.250 \text{ psi, OK}$$

$$f_b = -\frac{P_e}{A_c}\left(1 + \frac{e_c c_b}{r^2}\right) - \frac{M_D + M_{SD}}{S^t}$$

$$= -\frac{284,543}{615}\left(1 + \frac{18.73 \times 21.98}{97}\right) + \frac{6,928,416}{2,717}$$

$$= -2426.33 + 2550.02 = +123.7 \ (0.85 \text{ MPa}), \text{ say } 124 \text{ psi } (T), \text{ OK}$$

This is a very low tensile stress when the unshored slab is cast and before the service load is applied, $\ll 12\sqrt{f_c'} = 849$ psi.

C. *At service load for the precast section.* Section modulus for composite section at the top of the precast section is

$$S_c^t = \frac{77,118}{9.46 - 2} = 10,337 \text{ in.}^3$$

$$M_L = \frac{280(72)^2}{8} \times 12 = 2,177,288 \text{ in.-lb } (246 \text{ kN m})$$

$$f^t = -\frac{P_e}{A_c}\left(1 - \frac{e_c c_t}{r^2}\right) - \frac{M_D + M_{SD}}{S^t} - \frac{M_{CSD} + M_L}{S_c^t}$$

$$M_{CSD} = \text{superimposed dead load} = 0 \text{ in this case}$$

$$f^t = -730 - \frac{2,177,288}{10,337}$$

$$= -730 - 210 = -940 \text{ psi } (6.5 \text{ MPa}) \ (C), \text{ OK}$$

$$f_b = +123.7 + \frac{2,177,288}{3,142} = +123.7 + 693.0$$

$$= +816.7, \text{ say } 817 \text{ psi } (T) \ (5.4 \text{ MPa}) < f_t = 849 \text{ psi, OK}$$

D. *Composite slab stresses.* Precast double-T concrete modulus is

$$E_c = 57,000\sqrt{f_c'} = 57,000\sqrt{5,000} = 4.03 \times 10^6 \text{ psi } (3.8 \times 10^4 \text{ MPa})$$

Situ-cast slab concrete modulus is

$$E_c = 57,000\sqrt{3,000} = 3.12 \times 10^6 \text{ psi } (2.2 \times 10^4 \text{ MPa})$$

Modular ratio

$$n_p = \frac{3.12 \times 10^6}{4.03 \times 10^6} = 0.77$$

S_c^t for 2-in. slab top fibers $= 8,152 \text{ in.}^3$ from data.

S_{cb} for 2-in. slab bottom fibers $= 10,337 \text{ in.}^3$ from before for top of precast section.

Stress f_{cs}^t at top slab fibers $= n\dfrac{M_L}{S_c^t}$

$$= -0.77 \times \frac{2{,}177{,}288}{8{,}152} = -207 \text{ psi } (1.4 \text{ MPa}) \ (C)$$

Stress f_{csb} at bottom slab fibers

$$= -0.77 \times \frac{2{,}177{,}288}{10{,}377} = -162 \text{ psi } (1.1 \text{ MPa}) \ (C)$$

2. *Support section stresses*

   Check is made at the support face (a slightly less conservative check can be made at $50d_b$ from end)

   $$e_c = 12.81 \text{ in.}$$

A. *At transfer*

$$f^t = -\frac{347{,}004}{615}\left(1 - \frac{12.81 \times 10.02}{97}\right) - 0$$

$$= +182 \text{ psi } (T) \ (1.26 \text{ MPa}) \ll -2{,}250 \text{ psi, OK}$$

$$f_b = -\frac{347{,}004}{615}\left(1 + \frac{12.81 \times 21.98}{97}\right) + 0$$

$$= -2{,}202 \text{ psi } (C) \ (15.2 \text{ MPa}) < 0.60 f'_{ci} = -2{,}250 \text{ psi, OK}$$

B. After the slab is cast and at service load, the support section stresses both at top and bottom extreme fibers were found to be below the allowable; hence, OK.

Summary of midspan stresses (psi)

	f^t	f^b
Transfer P_e only	$+433$	-2426
W_D at transfer	-1163	$+2550$
Net at transfer	-730	$+124$
External load (W_L)	-210	$+693$
Net total at service	-940	$+817$

3. *Camber and deflection calculation*

   At transfer

   Initial

   $$E_{ci} = 57{,}000\sqrt{3{,}570} = 3.49 \times 10^6 \text{ psi } (2.2 \times 10^4 \text{ MPa})$$

   From before, 28 days

   $$E_c = 4.03 \times 10^6 \text{ psi } (2.8 \times 10^4 \text{ MPa})$$

   Due to initial prestress only, from Figure 8.10

   $$\delta_i = \frac{P_i e_c l^2}{8 E_{ci} I_g} + \frac{P_i (e_e - e_c) l^2}{24 E_{ci} I_g}$$

   $$= \frac{(-347{,}004)(18.73)(72 \times 12)^2}{8(3.49 \times 10^6)59{,}720}$$

   $$+ \frac{(-347{,}004)(12.81 - 18.73)(72 \times 12)^2}{24(3.49 \times 10^6)59{,}720}$$

   $$= -2.90 + 0.30 = -2.6 \text{ in. } (66 \text{ mm}) \uparrow$$

Self-weight intensity $w = 641/12 = 53.42$ lb/in.

$$\text{Self-weight } \delta_D = \frac{5wl^4}{384E_{ci}I_g} \text{ for uncracked section}$$

$$= \frac{5 \times 53.42(72 \times 12)^4}{384(3.49 \times 10^6)59,720} = 1.86 \text{ in. } (47 \text{ mm}) \downarrow$$

Thus, the net camber at transfer is

$$-2.6 + 1.86 = -0.74 \text{ in. } (19 \text{ mm}) \uparrow$$

4. *Immediate service load deflection*
   A. *Effective I_e method*
   Modulus of rupture

$$f_r = 7.5\sqrt{f'_c} = 7.5\sqrt{5000} = 530 \text{ psi}$$

f_b at service load $= 817$ psi (5.4 MPa) in tension (from before). Hence, the section is cracked and the effective I_e from Equation 8.92a or 8.92b should be used

$$d_p = 18.73 + 10.02 + 2(\text{topping}) = 30.75 \text{ in. } (780 \text{ mm})$$

$$\rho_p = \frac{A_{ps}}{bd_p} = \frac{12(0.153)}{120 \times 30.75} = 4.98 \times 10^{-4}$$

From Equation 8.93

$$I_{cr} = n_p A_{ps} d_p^2 (1 - 1.6\sqrt{n_p\rho_p})$$

$$n_p = 28.5 \times 10^6 / 4.03 \times 10^6 = 7 \text{ to be used in Equation 8.93}$$

Equation 8.93a gives $I_{cr} = 11,110$ in.^{4} (4.63×10^5 cm^4), use. From Equation 8.93a and the stress f_{pe} and f_d values already calculated for the bottom fibers at midspan with $f_r = 7.5\sqrt{f'_c} = 530$ psi. Moment M_{cr} due to that portion of live load that causes cracking is

$$M_{cr} = S_b\left(7.5\sqrt{f'_c} + f_{ce} - f_d\right)$$

$$= 3,142(530 + 2,426 - 2,550)$$

$$= 1,275,652 \text{ in.-lb}$$

M_a, unfactored maximum live load moment $= 2,177,288$ in.-lb

$$\frac{M_{cr}}{M_a} = \frac{1,275,652}{2,177,288} = 0.586$$

where M_{cr} is the moment due to that portion of the *live load* that causes cracking and M_a is the maximum service *unfactored live load*.

Using the preferable PCI expression of (M_{cr}/M_a) from Equation 8.91b, and the stress values previously tabulated

$$\frac{M_{cr}}{M_a} = 1 - \frac{f_{tl} - f_r}{f_L} = 1 - \left(\frac{817 - 530}{693}\right) = 0.586$$

$$\left(\frac{M_{cr}}{M_a}\right)^3 = (0.586)^3 = 0.20$$

Hence, from Equation 8.92b

$$I_e = \left(\frac{M_{cr}}{M_a}\right)^3 I_g + \left[1 - \left(\frac{M_{cr}}{M_a}\right)^3\right] I_{cr} \leq I_g$$

$$I_e = 0.2(77,118) + (1 - 0.2)11,110$$

$$= 15,424 + 8,888 = 24,312 \text{ in.}^4$$

$$w_{SD} = \tfrac{1}{12}(891 - 641) = 20.83 \text{ lb/in.}$$

$$w_L = \tfrac{1}{12} \times 280 = 23.33 \text{ lb/in.}$$

$$\delta_L = \frac{5wl^4}{384 E_c I_e} = \frac{5 \times 23.33(72 \times 12)^4}{384(4.03 \times 10^6)24,312}$$

$$= +1.73 \text{ in. (45 mm)} \downarrow \text{ (as an average value)}$$

When the concrete 2-in. topping is placed on the precast section, the resulting topping deflection with $I_g = 59,720 \text{ in.}^4$ becomes

$$\delta_{SD} = \frac{5 \times 20.83(72 \times 12)^4}{384(4.03 \times 10^6)59,720} = +0.63 \text{ in.} \downarrow$$

5. *Long-term deflection (camber) by PCI multipliers*
Using PCI multipliers at slab topping completion stage (30 days) and at the final service load (5 years), the following are the tabulated deflection values:

Load	Transfer δ_p, in. (1)	PCI multipliers	δ_{30}, in. (2)	PCI multiplier (composite)	δ_{Final}, in. (3)
Prestress	−2.60	1.80	−4.68	2.20	−5.71 ↑
w_D	+1.86	1.85	+3.44	2.40	+4.46 ↓
	−0.74 ↑		−1.24 ↑		−1.25 ↑
w_{SD}			+0.63 ↓	2.30	+1.45 ↓
w_L			+1.89 ↓		+1.89 ↓
Final δ	−0.74 ↑		+1.28 ↓		+2.09 ↓

Hence, final deflection ≈ 2.1 in. (53 mm) $\downarrow$

$$\text{Allowable deflection} = \text{span}/180 = \frac{72 \times 12}{180} = 4.8 \text{ in.} > 2.1 \text{ in., OK}$$

8.8.5 Cracking Behavior and Crack Control in Prestressed Beams

If Δf_s is the net stress in the prestressed tendon or the magnitude of the tensile stress in the normal steel at any crack width load level in which the decompression load (decompression here means $f_c = 0$ at the level of the reinforcing steel) is taken as the reference point, then for the prestressed tendon

$$\Delta f_s = f_{nt} - f_d \text{ ksi } (=1000 \text{ psi}) \tag{8.99}$$

where f_{nt} is the stress in the prestressing steel at any load level beyond the decompression load and f_d is the stress in the prestressing steel corresponding to the decompression load.

The unit strain $\varepsilon_s = \Delta f_s / E_s$. Because it is logical to disregard as insignificant the unit strains in the concrete due to the effects of temperature, shrinkage, and elastic shortening, the maximum crack width can be defined as

$$w_{\text{max}} = k a_{\text{cs}} \varepsilon_s^\alpha \tag{8.100}$$

where k and α are constants to be established by tests, a_{cs} is the spacing of cracks, and ε_s is the strain in the reinforcement.

For *pretensioned beams*, the maximum crack width can be evaluated from the maximum crack width at the reinforcing steel level

$$w_{\text{max}} = 5.85 \times 10^{-5} \frac{A_t}{\sum o} (\Delta f_s) \tag{8.101a}$$

and the maximum crack width (in.) at the concrete tension face

$$w'_{\text{max}} = 5.85 \times 10^{-5} R_i \frac{A_t}{\sum o} (\Delta f_s) \tag{8.101b}$$

where R_i is the ratio of distance from neutral axis to tension face to the distance from neutral axis to centroid of reinforcement.

For *posttensioned bonded beams*, the expression for the maximum crack width at the reinforcement level is

$$w_{\text{max}} = 6.51 \times 10^{-5} \frac{A_t}{\sum o} (\Delta f_s) \tag{8.102a}$$

At the tensile face, the crack width is

$$w'_{\text{max}} = 6.51 \times 10^{-5} R_i \frac{A_t}{\sum o} (\Delta f_s) \tag{8.102b}$$

For nonbonded beams, the factor 6.51 in Equations 5.102a and 5.102b becomes 6.83. The crack spacing stabilizes itself beyond an incremental stress Δf_s of 30,000 to 35,000 psi, depending on the *total* reinforcement percent ρ_T of both prestressed and nonprestressed steels.

Recent work by Nawy et al. [13] on the cracking performance of high-strength prestressed concrete beams, both pretensioned and posttensioned, has shown that the factor 5.85 in Equation 8.101a is considerably reduced. For concrete strengths in the range of 9,000 to 14,000 psi (60 to 100 MPa), this factor reduces to 2.75, so that the expression for the maximum crack width at the reinforcement level (inch) becomes

$$w_{\text{max}} = 2.75 \times 10^{-5} \frac{A_t}{\sum o} (\Delta f_s) \tag{8.103a}$$

In SI units, the expression is

$$w_{\text{max}} = 4.0 \times 10^{-5} \frac{A_t}{\sum o} (\Delta f_s) \tag{8.103b}$$

where A_t, cm^2; $\sum o$, cm; Δf_s, MPa.

For more refined values in cases where the concrete cylinder compressive strength ranges between 6,000 and 12,000 psi or higher, a modifying factor for particular f'_c values can be obtained from the following expressions:

$$\lambda_r = \frac{2}{\left(0.75 + 0.06 \sqrt{f'_c}\right) \sqrt{f'_c}} \tag{8.104a}$$

For posttensioned beams, the reduction multiplier λ_0 is

$$\lambda_0 = \frac{1}{0.75 + 0.06\sqrt{f_c'}} \tag{8.104b}$$

where f_c' and the reinforcement stress are in ksi.

8.8.6 ACI Expression for Cracking Mitigation

The ACI expression used for crack control in reinforced concrete structural elements through bar spacing is extended to prestressed concrete bonded beams, on the assumption of the desirability of a "seamless transition" between serviceability requirements for nonprestressed members and fully prestressed members. However, the mechanism of crack generation differs in the prestressed beam from that in reinforced concrete due to initially imposed precompression. Also, effects of environmental conditions are considerably more serious in the case of prestressed concrete elements due to the corrosion risks to the tendons. These provisions stipulate that the spacing of the bonded tendons should not exceed $\frac{2}{3}$ of the maximum spacing permitted for nonprestressed reinforcement. The ACI expression for prestressed members becomes

$$s = \frac{2}{3}\left(\frac{540}{\Delta f_s} - 2.5c_s\right) \tag{8.105}$$

but not to exceed $8(36/\Delta f_{s0})$.

In SI units, the expression becomes

$$s = \frac{2}{3}\left(\frac{95{,}000}{\Delta f_s} - 2.5c_s\right) \tag{8.106}$$

but not to exceed $200(252/\Delta f_s$, where Δf_s is in MPa and c_s is in mm)

Δf_s = difference between the stress computed in the prestressing tendon at service load based on cracked section analysis, and the decompression stress f_{dc} in the prestressing tendon. The code permits using the effective prestress f_{pe} in lieu of f_{dc}, ksi. A limit $\Delta f_s = 36$ ksi, and no check needed if Δf_s is less than 20 ksi.

c_c = clear cover from the nearest surface in tension to the flexural tension reinforcement, in.

While the code follows the author's definition of Δf_s given in Section 8.8.5, Equations 8.105 and 8.106 still lack the practicability of use as a crack control measure and the $\frac{2}{3}$ factor used in the expressions is arbitrary and not substantiated by test results. It should be emphasized that beams have finite web widths. Such spacing provisions as presented in the Code are essentially unworkable, since actual spacing of the tendons in almost all practical cases is *less* than the code equation limits, hence almost all beams satisfy the code, though cracking levels may be detrimental in bridge decks, liquid containment vessels, and other prestressed concrete structures in severe environment or subject to overload. They require additional mild steel reinforcement to control the crack width. Therefore, the expressions presented in Section 8.8.5 in conjunction with Table 8.14 from the ACI 224 Report [14] should be used for safe mitigation of cracking in prestressed concrete members.

TABLE 8.14 Maximum Tolerable Flexural Crack Widths

Exposure condition	Crack width	
	in.	mm
Dry air or protective membrane	0.016	0.41
Humidity, moist air, soil	0.012	0.30
De-icing chemicals	0.007	0.18
Seawater and seawater spray; wetting and drying	0.006	0.15
Water-retaining structures (excluding nonpressure pipes)	0.004	0.10

8.8.7 Long-Term Effects on Crack-Width Development

Limited studies on crack-width development and increase with time show that both sustained and cyclic loadings increase the amount of microcracking in the concrete. Also, microcracks formed at service-load levels in partially prestressed beams do not seem to have a recognizable effect on the strength or serviceability of the concrete element. Macroscopic cracks, however, do have a detrimental effect, particularly in terms of corrosion of the reinforcement and appearance. Hence, an increase of crack width due to sustained loading significantly affects the durability of the prestressed member regardless of whether prestressing is circular, such as in tanks, or linear, such as in beams. Information obtained from sustained load tests of up to 2 years and fatigue tests of up to one million cycles indicates that a *doubling* of crack width with time can be expected. Therefore, engineering judgment has to be exercised as regards the extent of tolerable crack width under long-term loading conditions.

8.8.8 Tolerable Crack Widths

The maximum crack width that a structural element should tolerate depends on the particular function of the element and the environmental conditions to which the structure is liable to be subjected. Table 8.14, from the ACI Committee 224 report on cracking, serves as a reasonable guide on the acceptable crack widths in concrete structures under the various environmental conditions encountered.

8.8.9 Example 9: Crack Control Check

A pretensioned prestressed concrete beam has a T-section as shown in Figure 8.14. It is prestressed with fifteen $\frac{7}{16}$-in. diameter seven-wire strand 270-K grade. The locations of the neutral axis and center of

PHOTO 8.8 West Kowloon Expressway Viaduct, Hong Kong, during construction, comprising 4.2 km dual three-lane causeway connecting Western Harbor Crossing to new airport (courtesy Institution of Civil Engineers, London, and [5]).

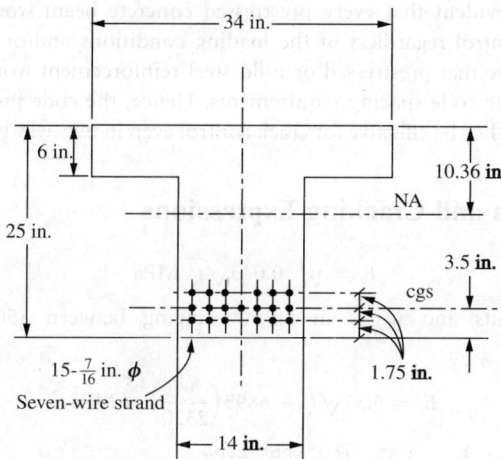

FIGURE 8.14 Beam cross-section in Example 9 [5].

gravity of steel are shown in the figure. $f_c' = 5,000$ psi, $E_c = 57,000\sqrt{f_c'}$, and $E_s = 28 \times 10^6$ psi. Find the mean stabilized crack spacing and the crack widths at the steel level as well as at the tensile face of the beam at $\Delta f_s = 30 \times 10^3$ psi. Assume that no failure in shear or bond takes place.

Solution

$$\Delta f_s = 30,000 \text{ psi} = 30 \text{ ksi}$$

Mean stabilized crack spacing

$$A_t = 7 \times 14 = 98 \text{ sq in.}$$

$$\sum o = 15\pi D = 15\pi\left(\frac{7}{16}\right) = 20.62 \text{ in.}$$

$$a_{cs} = 1.2\left(\frac{A_t}{\sum o}\right) = 1.2\left(\frac{98}{20.62}\right) = 5.7 \text{ in. (145 mm)}$$

Maximum crack width at steel level

$$w_{max} = 5.85 \times 10^{-5}\frac{A_t}{\sum o}(\Delta f_s) = 5.85 \times 10^{-5}\left(\frac{98}{20.62}\right)30$$

$$= 834.1 \times 10^{-5} \text{ in.} \cong 0.0083 \text{ in. (0.21 mm)}$$

Maximum crack width at tensile face of beam

$$R_i = \frac{25 - 10.36}{25 - 10.36 - 3.5} = 1.31$$

$$w'_{max} = w_{max} R_i = 0.0083 \times 1.31 = 0.011 \text{ in. (0.28 mm)}$$

By the ACI method

$$\Delta f_s = 30 \text{ ksi}$$
$$c_c = 1.5 \text{ in.}$$

From Equation 8.105

$$s = \frac{2}{3}\left(\frac{540}{30} - 2 \times 1.5\right) = 10 \text{ in.} < 12 \text{ in., OK}$$

From this solution, it is evident that every prestressed concrete beam would satisfy the ACI Code requirements for crack control regardless of the loading conditions and/or overloading, or environmental conditions. It is rare that prestressed or mild steel reinforcement would ever be spaced within a flange that can violate the code spacing requirements. Hence, the code provisions are not effective, and probably rarely would they be effective for crack control even in two-way prestressed concrete plates.

8.8.10 SI Deflection and Cracking Expressions

$$E_c = w_c^{1.5} 0.043 \sqrt{f'_c} \text{ MPa} \tag{8.107}$$

where f'_c is in MPa units and w_c is in kg/m^3 ranging between 1500 and 2500 kg/m^3. For $f'_c > 35$ MPa, < 80 MPa

$$E_c = 3.32 \sqrt{f'_c} + 6895 \left(\frac{w_c}{2320} \right)^{1.5} \text{ MPa}$$

For normal-weight concrete, $E_c = 3.32 \sqrt{f'_c} + 6895$ MPa

$$f_r = 0.62 \sqrt{f'_c} \tag{8.108}$$

$$I_e = \left(\frac{M_{cr}}{M_a} \right)^3 I_g + \left[1 - \left(\frac{M_{cr}}{M_a} \right)^3 \right] I_{cr} \tag{8.109}$$

$$\left(\frac{M_{cr}}{M_a} \right) = \left[1 + \left(\frac{f_{tl} - f_r}{f_L} \right) \right] \tag{8.110}$$

$$I_{cr} = n_p A_{ps} d_p^2 \left(1 - 1.6 \sqrt{n_p \rho_p} \right) \tag{8.111}$$

$$= \left(n_p A_{ps} d_p^2 + n_s A_s d^2 \right) \left(1 - 1.6 \sqrt{n_p \rho_p + n_s \rho} \right) \tag{8.112}$$

$$w_{max} = \alpha_w \times 10^{-5} \frac{A_t}{\sum o} (\Delta f_s), \text{ mm} \tag{8.113}$$

where A_t, cm^2; $\sum o$, cm; Δf_s, MPa

$$\alpha_w = 8.48 \times 10^{-5} \text{ for pretensioned}$$

$$= 9.44 \times 10^{-5} \text{ for posttensioned}$$

$$= 4.0 \times 10^{-5} \text{ for concretes with } f'_c > 70 \text{ MPa}$$

$$\text{MPa} = \text{N/mm}^2$$

$$\text{(psi) } 0.006895 = \text{MPa}$$

$$\text{(lb/ft) } 14.593 = \text{N/m}$$

$$\text{(in. lb) } 0.113 = \text{N m}$$

Acknowledgments

The author wishes to acknowledge the permission granted by Prentice Hall, Upper Saddle River, New Jersey, and Ms Marcia Horton, its vice president and publications director, to use several chapters of his book, *Prestressed Concrete — A Fundamental Approach*, 4th Edition, 2003, 944 pp. [5] as the basis for extracting the voluminous material that has formed this chapter. Also, grateful thanks are due to

Ms Mayrai Gindy, PhD candidate at Rutgers University, who helped in putting together and reviewing the final manuscript, including processing the large number of equations and computations.

Glossary

ACI Committee 116 Report "Cement and Concrete Terminology" and other definitions pertinent to prestressed concrete are presented:

AASHTO — American Association of State Highway Officials.

ACI — American Concrete Institute.

Allowable stress — Maximum permissible stress used in design of members of a structure and based on a factor of safety against yielding or failure of any type.

Balanced strain — Deformation caused by combination of axial force and bending moment that causes simultaneous crushing of concrete at the compression side and yielding of tension steel at the tension side of concrete members.

Beam — A structural member subjected primarily to flexure due to transverse load.

Beam-column — A structural member that is subjected simultaneously to bending and substantial axial forces.

Bond — Adhesion and grip of concrete or mortar to reinforcement or to other surfaces against which it is placed; to enhance bond strength, ribs or other deformations are added to reinforcing bars.

Camber — Reverse deflection (convex upwards) that is intentionally built into a structural element or form to improve appearance or to offset the deflection of the element under the effects of loads, shrinkage, and creep. It is also reverse deflection (convex upward) in prestressed concrete beams.

Cast-in-place concrete — Concrete placed in its final or permanent location, also called *in situ* concrete, in contrast to precast concrete.

Column — A member that supports primarily axial compressive loads with a height of at least three times its least lateral dimension; the capacity of short columns is controlled by strength; the capacity of long columns is limited by buckling.

Column strip — The portion of a flat slab over a row of columns consisting of a width equal to quarter of the panel dimension on each side of the column centerline.

Composite construction — A type of construction using members made of different materials (e.g., concrete and structural steel), or combining members made of cast-in-place and precast concrete such that the combined components act together as a single member.

Compression member — A member subjected primarily to longitudinal compression; often synonymous with "column."

Compressive strength — Strength typically measured on a standard 6×12 in. cylinder of concrete in an axial compression test, 28 days after casting. For high-strength concrete, 4×8 in. cylinders are used.

Concrete — A composite material that consists essentially of a binding medium within which are embedded coarse and fine aggregates; in portland cement concrete, the binder is a mixture of portland cement and water.

Confined concrete — Concrete enclosed by closely spaced transverse reinforcement to restrain concrete expansion in directions perpendicular to the applied stresses.

Construction joint — The interaction surface between two successive placements of concrete across which it may be desirable to achieve bond, and through which reinforcement may be continuous.

Continuous beam or slab — A beam or slab that extends as a unit over three or more supports in a given direction and is provided with the necessary reinforcement to develop the negative moments over the interior supports; a redundant structure that requires a statically indeterminant analysis (opposite of simple supported beam or slab).

Cover — In reinforced and prestressed concrete, the shortest distance between the surface of the reinforcement and the outer surface of the concrete; minimum values are specified to protect the reinforcement against corrosion and to assure sufficient bond strength with the reinforcement.

Cracks — Fracture in concrete elements when tensile stresses exceed the concrete tensile strength; a design goal is to keep their widths small (hairline cracks well-distributed along a member).

Cracked section — A section designed or analyzed on the assumption that concrete has no resistance to tensile stress.

Cracking load — The load that causes tensile stress in a member equal to or exceeding the modulus of rupture of concrete.

Deformed bar — Reinforcing bar with a manufactured pattern of surface deformations intended to prevent slip when the embedded bar is subjected to tensile stress.

Design strength — Ultimate load and moment capacity of a member multiplied by a strength reduction factor.

Development length — The length of embedded reinforcement to develop the design strength of the reinforcement; a function of bond strength.

Diagonal crack — An inclined crack caused by diagonal tension, usually at about 45° to the neutral axis of a concrete member.

Diagonal tension — The principal tensile stress resulting from the combination of normal and shear stresses acting upon a structural element.

Drop panel — The portion of a flat slab in the area surrounding a column such as column capital that is thicker than the slab in order to reduce the intensity of shear stresses.

Ductility — Capability of a material or structural member to undergo large inelastic deformations without distress; opposite of brittleness; very important material property, especially for earthquake-resistant design; steel is naturally ductile, concrete is brittle but it can be made ductile if well confined.

Durability — The ability of concrete to maintain its design qualities long term while exposed to weather, freeze–thaw cycles, chemical attack, abrasion, and other service load environmental conditions.

Effective depth — Depth of a beam or slab section measured from the compression face to the centroid of the tensile reinforcement.

Effective flange width — Width of slab adjoining a beam stem or web assumed to function as the flange of a T-section or L-section.

Effective prestress — The stress remaining in the prestressing reinforcement after all losses have occurred.

Effective span — The lesser of the distance between centers of supports and the clear distance between supports plus the effective depth of the beam or slab.

End block — End segment of a prestressed concrete beam.

Equivalent lateral force method — Static method for evaluating the horizontal base shear due to seismic forces.

Flat slab — A concrete slab reinforced in two, generally without beams or girders to transfer the loads to supporting members, sometimes with drop panels or column capitals or both.

High-early strength cement — Cement producing strength in mortar or concrete earlier than regular cement.

Hoop — A one-piece closed reinforcing tie or continuously wound tie that encloses the longitudinal reinforcement.

Interaction diagram — Load–moment curve for a member subjected to both axial force and the bending moment, indicating the moment capacity for a given axial load and vice versa; used to develop design charts for reinforced and prestressed concrete compression members.

Lightweight concrete — Concrete of substantially lower unit weight than that made using normal-weight gravel or crushed stone aggregate.

Limit analysis — See **Plastic analysis.**

Limit design — A method of proportioning structural members based on satisfying certain strength and serviceability limit states.

Load and resistance factor design (LRFD) — See **Ultimate strength design.**

Load factor — A factor by which a service load is multiplied to determine the factored load used in ultimate strength design.

LRFD — Load Resistance Factor Design.

Modulus of elasticity — The ratio of normal stress to corresponding strain for tensile or compressive stresses below the proportional limit of the material; for steel typically, $E_s = 29,000$ ksi; for concrete it is a function of the cylinder compressive strength f'_c; for normal-weight concrete, a common approximation is $E_c = 57,000\sqrt{f'_c}$ for concrete having strength not exceeding 6,000 psi.

Modulus of rupture — The tensile strength of concrete at the first fracture load.

Mortar — A mixture of cement paste and fine aggregate; in fresh concrete, the material filling the voids between the coarse aggregate particles.

Nominal strength — The strength of a structural member based on its assumed material properties and sectional dimensions, before application of any strength reduction factor.

Partial loss of prestress — Loss in the reinforcement prestressing due to elastic shortening, creep, shrinkage, relaxation, and friction.

PCI — Precast/Prestressed Concrete Institute.

Plastic analysis — A method of structural analysis to determine the intensity of a specified load distribution at which the structure forms a collapse mechanism.

Plastic hinge — Region in a flexural member where the ultimate moment capacity can be developed and maintained with corresponding significant inelastic rotation, as main tensile steel is stressed beyond the yield point.

Posttensioning — A method of prestressing concrete elements where the tendons are tensioned after the concrete has hardened (opposite of pretensioning).

Precast concrete — Concrete cast at a location different than its final placement, usually in plants or sites close to the final site (contrary to cast-in-place concrete).

Prestressed concrete — Concrete in which longitudinal compressive stresses are induced in the member prior to the placement of external loads through the tensioning of prestressing strands that are placed over the entire length of the member.

Prestressing steel — High-strength steel used to apply the compressive prestressing force in the member, commonly seven-wire strands, single wires, bars, rods, or groups of wires or strands.

Pretensioning — A method of prestressing the concrete element whereby the tendons are tensioned before the concrete has been placed (opposite to posttensioning).

PTI — Posttensioning Institute.

Reinforced concrete — Concrete containing adequate reinforcement and designed on the assumption that the two materials act together in resisting the applied forces.

Reinforcement — Bars, wires, strands, and other slender elements that are embedded in concrete in such a manner that the reinforcement and the concrete act together in resisting forces.

Relaxation — Time loss in the prestressing steel due to creep effect in the reinforcement.

Safety factor — The ratio of a load producing an undesirable state (such as collapse) to an expected service load.

Service loads — Loads on a structure with high probability of occurrence, such as dead weight supported by a member or the live loads specified in building codes and specifications.

Shear span — The distance from a support face of a simply supported beam to the nearest concentrated load for concentrated loads and the clear span for distributed loads.

Shear wall — See **Structural wall.**

Shotcrete — Mortar or concrete pneumatically projected at high velocity onto a surface.

Silica fume — Very fine noncrystalline silica produced in electric arc furnaces as a by-product in the production of metallic silicon and various silicon alloys (also know as condensed silica fume); used as a mineral admixture in concrete.

Slab — A flat, horizontal cast layer of plain or reinforced concrete, usually of uniform thickness, either on the ground or supported by beams, columns, walls, or other frame work. See also **Flat slab**.

Slump — A measure of consistency of freshly mixed concrete equal to the subsidence of the molded specimen immediately after removal of the slump cone, expressed in inches.

Splice — Connection of one reinforcing bar to another by lapping, welding, mechanical couplers, or other means.

Tensile split cylinder test — Test for tensile strength of concrete in which a standard cylinder is loaded to failure in diametral compression applied along the entire length of the cylinder (also called Brazilian test).

Standard cylinder — Cylindrical specimen of 12-in. height and 6-in. diameter, used to determine standard compressive strength and splitting tensile strength of concrete. For high-strength concrete the size is usually 4×8 in.

Stiffness coefficient — The coefficient k_{ij} of stiffness matrix **K** for a multi-degree of freedom structure is the force needed to hold the ith degree of freedom in place, if the jth degree of freedom undergoes a unit of displacement, while all others are locked in place.

Stirrup — A type of reinforcement used to resist shear and diagonal tension stresses in a structural concrete member; typically a steel bar bent into a U or rectangular shape and installed perpendicular to the longitudinal reinforcement and properly anchored; the term "stirrup" is usually applied to lateral reinforcement in flexural members and the term "tie" to lateral reinforcement in compression members. See **Tie**.

Strength design — See **Ultimate strength design**.

Strength reduction factor — Capacity reduction factor (typically designated as ϕ) by which the nominal strength of a member is to be multiplied to obtain the design strength; specified by the ACI Code for different types of members and stresses.

Structural concrete — Concrete used to carry load or to form an integral part of a structure (opposite of, e.g., insulating concrete).

Structural wall — Reinforced or prestressed wall carrying loads and subjected to stress, particularly horizontally due to seismic loading.

Strut-and-tie procedure — Procedure for the design of confining reinforcement in prestressed concrete beam end blocks.

T-beam — A beam composed of a stem and a flange in the form of a "T," with the flange usually provided by the slab part of a floor system.

Tie — Reinforcing bar bent into a loop to enclose the longitudinal steel in columns; tensile bar to hold a form in place while resisting the lateral pressure of unhardened concrete.

Ultimate strength design (USD) — Design principle such that the actual (ultimate) strength of a member or structure, multiplied by a strength factor, is no less than the effects of all service load combinations, multiplied by the respective overload factors.

Unbonded tendon — A tendon that is not bonded to the concrete.

Under-reinforced beam — A beam in which the strain in the extreme tension reinforcement is less than the balanced strain.

Water–cement ratio — Ratio by weight of water to cement in a mixture, inversely proportional to concrete strength.

Water-reducing admixture — An admixture capable of lowering the mix viscosity, thereby allowing a reduction of water (and increase in strength) without lowering the workability (also called superplasticizer).

Whitney stress block — A rectangular area of uniform stress intensity $0.85f_c'$, whose area and centroid are similar to those of the actual stress distribution in a flexural member at failure.

Workability — General property of freshly mixed concrete that defines the ease with which the concrete can be placed into the forms without honeycombing; closely related to slump.

Yield-line theory — Method of structural analysis of concrete plate structures at the collapse load level.

References

[1] Nawy, E. G., *Reinforced Concrete — A Fundamental Approach*, 5th ed. Prentice Hall, Upper Saddle River, NJ, 2003, 864 pp.

[2] ACI Committee 435, *Control of Deflection in Concrete Structures*, ACI Committee Report R435-95, E. G. Nawy, Chairman, American Concrete Institute, Farmington Hills, MI, 1995, 77 pp.

[3] Nawy, E. G., *Fundamentals of High Performance Concrete*, 2nd ed. John Wiley & Sons, New York, 2001, 460 pp.

[4] Branson, D. E., *Deformation of Concrete Structures*, McGraw-Hill, New York, 1977.

[5] Nawy, E. G., *Prestressed Concrete — A Fundamental Approach*, 4th ed. Prentice Hall, Upper Saddle River, NJ, 2003, 944 pp.

[6] Hsu, T. T. C., *Unified Theory of Reinforced Concrete*, CRC Press, Boca Raton, FL, 1993, 313 pp.

[7] Prestressed Concrete Institute. *PCI Design Handbook*, 5th ed. PCI, Chicago, 1999.

[8] ACI Committee 318, *Building Code Requirements for Structural Concrete (ACI 318–02)* and *Commentary (ACI 318 R-02)*. American Concrete Institute, Farmington Hills, MI, 2002, 446 pp.

[9] Breen, J. E., Burdet, O., Roberts, C., Sanders, D., and Wollman, G. "Anchorage Zone Reinforcement for Posttensioned Concrete Girders." NCHRP Report, 356 pp.

[10] International Code Council. *International Building Codes 2000–2003 (IBC)*, Joint UBC, BOCA, SBCCI, Whittier, CA, 2003.

[11] Nawy, E. G., Editor-in-Chief, *Concrete Construction Engineering Handbook*. CRC Press, Boca Raton, FL, 1998, 1250 pp.

[12] Englekirk, R. E., Design-Construction of the Paramount — A 39-Story Precast Prestressed Concrete Apartment Building. *PCI J.*, Precast/Prestressed Concrete Institute, Chicago, IL, July–August 2002, pp. 56–69.

[13] Nawy, E. G., *Design For Crack Control in Reinforced and Prestressed Concrete Beams, Two-Way Slabs and Circular Tanks — A State of the Art*. ACI SP-204, Winner of the 2003 ACI Design Practice Award, American Concrete Institute, Farmington Hills, MI, 2002, pp. 1–42.

[14] ACI Committee 224, *Control of Cracking in Concrete Structures*. ACI Committee Report R224-01, American Concrete Institute, Farmington Hills, MI, 2001, 64 pp.

[15] Posttensioning Institute. *Posttensioning Manual*, 5th ed. PTI, Phoenix, AZ, 2003.

[16] Nawy, E. G. and Chiang, J. Y., "Serviceability Behavior of Posttensioned Beams." *J. Prestressed Concrete Inst.* 25 (1980): 74–85.

[17] Nawy, E. G. and Huang, P. T., "Crack and Deflection Control of Pretensioned Prestressed Beams." *J. Prestressed Concrete Inst.* 22 (1977): 30–47.

[18] Branson, D. E., *The Deformation of Non-Composite and Composite Prestressed Concrete Members*. ACI Special Publication SP-43, *Deflection of Concrete Structures*, American Concrete Institute, Farmington Hills, MI, 1974, pp. 83–127.

[19] PCI, *Prestressed Concrete Bridge Design Handbook*. Precast/Prestressed Concrete Institute, Chicago, 1998.

[20] AASHTO, *Standard Specifications for Highway Bridges*, 17th ed. and 2002 Supplements. American Association of State Highway and Transportation Officials, Washington, DC, 2002.

[21] Nawy, E. G., "Discussion — The Paramount Building." *PCI J.*, Precast/Prestressed Concrete Institute, Chicago, IL, November–December 2002, p. 116.

[22] Freyssinet, E., *The Birth of Prestressing*. Public Translation, Cement and Concrete Association, London, 1954.

[23] Guyon, Y. *Limit State Design of Prestressed Concrete*, vol. 1. Halsted-Wiley, New York, 1972.

[24] Gerwick, B. C., Jr. *Construction of Prestressed Concrete Structures*. Wiley-Interscience, New York, 1997, 591 pp.

[25] Lin, T. Y. and Burns, N. H., *Design of Prestressed Concrete Structures*, 3rd ed. John Wiley & Sons, New York, 1981.

[26] Abeles, P. W. and Bardhan-Roy, B. K., *Prestressed Concrete Designer's Handbook*, 3rd ed. Viewpoint Publications, London, 1981.

[27] American Concrete Institute, *ACI Manual of Concrete Practice*, 2003. *Materials*. American Concrete Institute, Farmington Hills, MI, 2003.

[28] Portland Cement Association. *Design and Control of Concrete Mixtures*, 13th ed. PCA, Skokie, IL, 1994.

[29] Nawy, E. G., Ukadike, M. M., and Sauer, J. A. "High Strength Field Modified concretes." *J. Struct. Div., ASCE* 103 (No. ST12) (1977): 2307–2322.

[30] American Society for Testing and Materials. "Standard Specification for Cold-Drawn Steel Wire for Concrete Reinforcement, A8 2-79." ASTM, Philadelphia, 1980.

[31] Chen, B. and Nawy, E. G., "Structural Behavior Evaluation of High Strength Concrete Reinforced with Prestressed Prisms Using Fiber Optic Sensors." *Proc., ACI Struct. J.*, American Concrete Institute, Farmington Hills, MI, (1994): pp. 708–718.

[32] Posttensioning Institute. *Posttensioning Manual*, 5th ed. PTI, Phoenix, AZ, 1991.

[33] Cohn, M. Z., *Partial Prestressing, From Theory to Practice*. NATO-ASI Applied Science Series, vols. 1 and 2. Martinus Nijhoff, Dordrecht, The Netherlands, 1986.

[34] Yong, Y. K., Gadebeku, C. and Nawy, E. G., "Anchorage Zone Stresses of Posttensioned Prestressed Beams Subjected to Shear Forces." *ASCE Struct. Div. J.* 113 (8) (1987): 1789–1805.

[35] Nawy, E. G., "Flexural Cracking Behavior of Pretensioned and Posttensioned Beams — The State of the Art." *J. Am. Concrete Inst.* December 1985: 890–900.

[36] Federal Highway Administration. "Optimized Sections for High Strength Concrete Bridge Girders." FHWA Publication No. RD-95-180, Washington, DC, August 1997, 156 pp.

[37] Collins, M. P. and Mitchell, D. "Shear and Torsion Design of Prestressed and Non-Prestressed Concrete Beams." *J. Prestressed Concrete Inst.* 25 (1980): 32–100.

[38] Zia, P. and Hsu, T. T. C., *Design for Torsion and Shear in Prestressed Concrete*. ASCE Annual Convention, Reprint No. 3423, 1979.

[39] International Conference of Building Officials, *Uniform Building Code (UBC)*, vol. 2, ICBO, Whittier, CA, 1997.

[40] Naja, W. M., and Barth, F. G., "Seismic Resisting Construction," Chapter 26, in E. G., Nawy, editor-in-chief, *Concrete Construction Engineering Handbook*. CRC Press, Boca Raton, FL, 1998, pp. 26-1–26-69.

9

Masonry Structures

Richard E. Klingner
Department of Civil Engineering,
University of Texas,
Austin, TX

9.1 Introduction

Masonry is traditionally defined as hand-placed units of natural or manufactured material, laid with mortar. In this chapter, the earthquake behavior and design of masonry structures is discussed, extending the traditional definition somewhat to include thin stone cladding.

Masonry makes up approximately 70% of the existing building inventory in the United States (TMS, 1989). U.S. masonry comprises Indian cliff dwellings, constructed of sandstone at Mesa Verde (Colorado); the adobe missions constructed by Spanish settlers in Florida, California, and the southwestern United States; bearing-wall buildings such as the 16-story Monadnock Building, completed in 1891 in Chicago; modern reinforced bearing-wall buildings; and many veneer applications. Clearly, the behavior and design of each type of masonry is distinct. In this chapter, fundamental applications and nomenclature of U.S. masonry are discussed, major construction categories are reviewed, historical seismic performance of masonry is presented, and principal design and retrofitting approaches are noted. Its purpose is to give designers, constructors, and building officials a basic foundation for further study of the behavior and design of masonry.

9.2   Masonry in the United States

9.2.1   Fundamentals of Masonry in the United States

Masonry can be classified according to architectural or structural function. Each is discussed later in this chapter. Regardless of how it is classified, U.S. masonry uses basically the same materials: units, mortar, grout, and accessory materials. In this section, those materials are discussed, with reference to the national consensus specifications of the American Society for Testing and Materials (ASTM). Additional information is available at the web sites of associations such as the National Concrete Masonry Association (NCMA), the Brick Industry Association (BIA), and The Masonry Society (TMS).

9.2.1.1   Masonry Units

Of the more than 20 different classifications of masonry units commercially available in the United States, only the most widely used are discussed here.

9.2.1.1.1   Clay or Shale Masonry Units

The most common structural clay or shale masonry units are Building Brick and Facing Brick. The former are specified using ASTM C62 Building Brick (Solid Masonry Units Made from Clay or Shale). The latter, specifically intended for use when appearance is important, are specified using ASTM C216 Facing Brick (Solid Masonry Units Made from Clay or Shale). Units are usually cored rather than being completely solid. The net cross-sectional area of the unit must be at least 75% of the gross area, that is, the cores occupy less than 25% of the area of the unit.

Many different sizes and shapes of clay or shale masonry units are available, varying widely from region to region of the United States. One common size is probably the "modular" unit, which measures $7\frac{5}{8}$ in. (194 mm) long by $2\frac{1}{4}$ in. (57 mm) high by $3\frac{5}{8}$ in. (92 mm) deep. Using mortar joints $\frac{3}{8}$ in. (9 mm) thick, this unit produces modules 8 in. (203 mm) wide by $2\frac{2}{3}$ in. (68 mm) high. That is, three courses of such units produce modules 8 in. (203 mm) wide by 8 in. high.

Clay or shale masonry units are sampled and tested using ASTM C67 (Methods of Sampling and Testing Brick and Structural Clay Tile). Specified properties include compressive strength and durability. Facing brick can have more restrictive dimensional tolerances and appearance requirements.

9.2.1.1.2   Concrete Masonry Units

The most common concrete masonry units are hollow load-bearing concrete masonry units, specified in ASTM C90 (Loadbearing Concrete Masonry Units). The units are typically made from low- or zero-slump concrete. In the eastern United States, these units are used for unreinforced inner wythes of cavity walls. In the western United States, these units are used for reinforced, fully grouted shear and bearing walls. The net area of the units is usually about 55 to 60% of their gross cross-sectional area. These units are commonly $15\frac{5}{8}$ in. (397 mm) long by $7\frac{5}{8}$ in. (194 mm) high by $7\frac{5}{8}$ in. (194 mm) thick. Using mortar joints $\frac{3}{8}$ in. (9 mm) thick, this unit produces modules 8 in. (203 mm) wide by 8 in. high. These modules are compatible with those of the modular clay brick discussed above. Concrete masonry units are sampled and tested using ASTM C140 (Methods of Sampling and Testing Concrete Masonry Units). Specified properties include shrinkage, compressive strength, and absorption.

9.2.1.2   Mortar

Mortar holds units together, and also compensates for their dimensional tolerances. In the United States, mortar for unit masonry is specified using ASTM C270 (Specification for Mortar for Unit Masonry), which addresses three cementitious systems: portland cement–lime, masonry cement, and mortar cement. These cementitious systems are combined with sand and water to produce mortar.

Portland cement–lime mortar consists of portland cement and other hydraulic cements, hydrated mason's lime, sand, and water. Masonry cement mortar consists of masonry cement, sand, and water. The contents of masonry cement and mortar cement, specified under ASTM C91 and ASTM C1329, respectively, vary from manufacturer to manufacturer, and are not disclosed. They typically include portland cement and other hydraulic cements, finely ground limestone, and air-entraining and water-retention admixtures. Mortar cement has a minimum specified tensile bond strength and a lower maximum air content than masonry cement. Model codes prohibit the use of masonry cement in seismic design categories C and higher. Portland cement–lime mortars and mortar cement mortars are not restricted in this respect.

Within each cementitious system, masonry mortar is also classified according to type. Types are designated as M, S, N, O, and K (derived from every other letter of the phrase "MaSoN wOrK"). These designations refer to the proportion of portland cement in the mixture. Type M has the most, S less, and so on. Higher proportions of portland cement result in faster strength gain, higher compressive strength, and higher tensile bond strength; they also result in lower long-term deformability. Mortar types S and N are typically specified.

Within each cementitious system, mortar can be specified by proportion or by property, with the former being the default. For example, Type S portland cement–lime mortar, specified by proportion, consists of one volume of portland cement, $\frac{1}{2}$ volume of hydrated mason's lime, about $4\frac{1}{2}$ volumes of masons' sand, and sufficient water for good workability. Type S masonry cement mortar or mortar cement mortar is made with one volume of masonry cement or mortar cement, respectively, three volumes of mason's sand, and sufficient water for good workability.

9.2.1.3 Grout

Masonry grout is essentially fluid concrete, used to fill spaces in masonry and to surround reinforcement and connectors. It is specified using ASTM C476 (Grout for Masonry). Grout for masonry is composed of portland cement and other hydraulic cements, sand, and (in the case of coarse grout) pea gravel. It is permitted to contain a small amount of hydrated mason's lime, but usually does not. It is permitted to be specified by proportion or by property, with the former being the default. A coarse grout specified by proportion would typically contain one volume of portland cement or other hydraulic cements, about three volumes of mason's sand, and about two volumes of pea gravel.

Masonry grout is placed with a slump of at least 8 in. (203 mm), so that it will flow freely into the cells of the masonry. Because of its high water–cement (w/c) ratio at the time of grouting, masonry grout undergoes considerable plastic shrinkage as the excess water is absorbed by the surrounding units. To prevent the formation of voids due to this process, the grout is consolidated during placement and reconsolidated after initial plastic shrinkage. Grouting admixtures, which contain plasticizers and water-retention agents, are also useful in the grouting process.

If grout is specified by property (compressive strength), the compressive strength must be verified using permeable molds, duplicating the loss of water and decreased w/c ratio that the grout would experience in actual use.

9.2.1.4 Accessory Materials

Accessory materials for masonry consist of reinforcement, connectors, sealants, flashing, coatings, and vapor barriers. In this section, each is briefly reviewed.

Reinforcement consists of deformed reinforcing bars or joint reinforcement. Deformed reinforcing bars are placed vertically in the cells of hollow units, horizontally in courses of bond-beam units, or vertically and horizontally between wythes of solid units. Model codes require that it be surrounded by grout. Joint reinforcement is placed in the bed (horizontal) joints of masonry and is surrounded by mortar.

Connectors are used to connect the wythes of a masonry wall (ties), to connect a masonry wall to a frame (anchors), or to connect something else to a masonry wall (fasteners).

Sealants are used to prevent the passage of water at places where gaps are intentionally left in masonry walls. Three basic kinds of gaps (joints) are used: expansion joints are used in brick masonry to accommodate expansion, control joints are used in concrete masonry to conceal cracking due to shrinkage, and construction joints are placed between different sections of a structure.

Flashing is a flexible waterproof barrier, intended to permit water that has penetrated the outer wythe to re-exit the wall. It is placed at the bottom of each story level (on shelf angles or foundations), over window and door lintels, and under window and door sills. Flashing should be lapped, and ends of flashing should be defined by end dams (flashing turned up at ends). Directly above the level of the flashing, weepholes should be provided at 24-in. spacing. Flashing is made of metal, polyvinyl chloride (PVC), or rubberized plastic (EPDM). Metallic flashing lasts much longer than plastic flashing. Nonmetallic flashings are subject to tearing. Modern EPDM self-adhering flashing is a good compromise between durability and ease of installation.

9.2.1.5 Masonry Nomenclature by Architectural Function

The architectural functions of masonry include acting as a building envelope to resist liquid water. Masonry walls are classified in terms of this function into barrier walls and drainage walls. Barrier walls act by a combination of thickness, coatings, and integral water-repellent admixtures. Drainage walls act by the above, plus drainage details. Examples of each are shown in Figure 9.1 and Figure 9.2. In drainage walls, an outer wythe (thickness of masonry) is separated from an inner wythe of masonry or from a backup system by a cavity with drainage details.

9.2.1.6 Masonry Nomenclature by Structural Function

From the viewpoint of structural function, U.S. masonry can be broadly classified as nonload-bearing and load-bearing. The former resists gravity loads from self-weight alone, and possibly out-of-plane wind loads or seismic forces from its own mass only. The latter may resist gravity and lateral loads from overlying floors or roof. Both classifications of masonry use the same materials.

Nonload-bearing masonry includes panel walls (an outer wythe of masonry connected to an inner wythe of masonry or a backup system), curtain walls (masonry spanning horizontally between columns), and interior partitions.

Load-bearing masonry walls resist out-of-plane loads by spanning as horizontal or vertical strips, in-plane gravity loads by acting as a shallow beam–column loaded perpendicular to the plane of the wall, and in-plane shear forces by acting as a deep beam–column loaded in the plane of the wall.

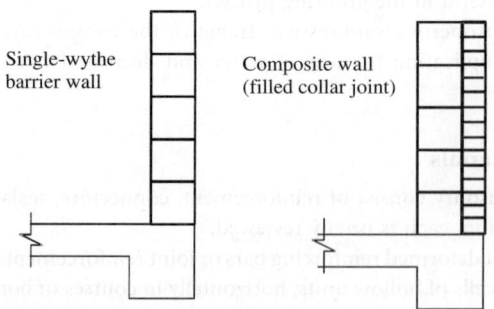

FIGURE 9.1 Examples of barrier walls.

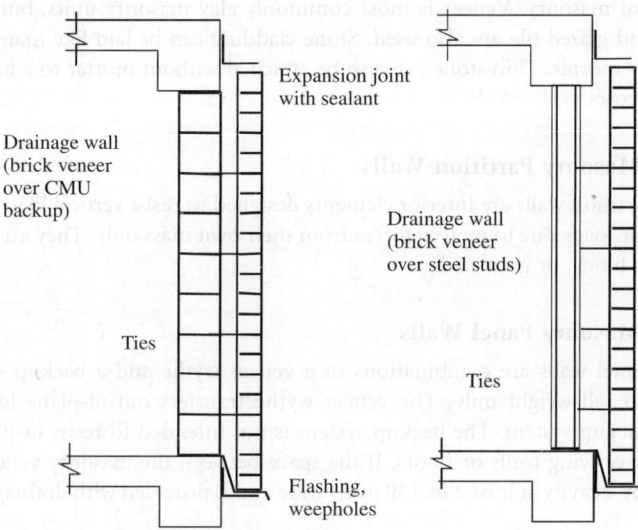

FIGURE 9.2 Examples of drainage walls.

9.2.1.7 Masonry Nomenclature by Design Intent

From the viewpoint of design intent, U.S. masonry can be broadly classified as unreinforced and reinforced. Unreinforced masonry is designed assuming that flexural tension is resisted by masonry alone, and neglecting stresses in reinforcement. Reinforced masonry is designed assuming that flexural tension is resisted by reinforcement alone, and neglecting the flexural tensile resistance of masonry. Both types of masonry are designed assuming that masonry has some diagonal tensile resistance, because both types permit some shear to be resisted without shear reinforcement.

To decipher design intent may be impossible by examination of the masonry alone, with no knowledge of its design process. Masonry elements, no matter how designed, are required to have minimum prescriptive reinforcement whose location and percentage depend on the seismic design category of the structure in which they are located.

Differences in historical tradition have led to potentially confusing differences in nomenclature. For example, in parts of the United States where the *Uniform Building Code* (UBC) has been dominant (roughly speaking, to the west of Denver), "partially reinforced" masonry referred to reinforced masonry whose reinforcement did not comply with UBC requirements for prescriptive reinforcement in zones of highest seismic risk. East of Denver, however, partially reinforced masonry referred to masonry reinforced with wire-type bed-joint reinforcement only, rather than deformed reinforcement placed in grouted cells or bond beams.

9.2.2 Modern Masonry Construction in the United States

A decade ago, it might have been possible to distinguish between modern masonry in the eastern versus the western United States, with the latter being characterized by more emphasis on seismic design. As model codes increasingly adopt the philosophy that almost all regions of the United States have some level of seismic risk, such regional distinctions are disappearing.

9.2.2.1 Modern Masonry Veneer

Modern masonry veneer resists vertical loads due to self-weight only, and transfers out-of-plane loads from wind or earthquake to supporting elements such as wooden stud walls, light-gage steel framing, or a backup wythe of masonry. Veneer is most commonly clay masonry units, but concrete masonry units, glass block, and glazed tile are also used. Stone cladding can be laid like manufactured masonry units, using masonry mortar. Thin stone can also be attached without mortar to a backup frame, using stainless steel connectors.

9.2.2.2 Modern Masonry Partition Walls

Modern masonry partition walls are interior elements designed to resist vertical loads due to self-weight only and out-of-plane loads due to inertial forces from their own mass only. They are of clay or concrete masonry units, glass block, or glazed tile.

9.2.2.3 Modern Masonry Panel Walls

Modern masonry panel walls are combinations of a veneer wythe and a backup system. They resist vertical loads due to self-weight only. The veneer wythe transfers out-of-plane loads from wind or earthquake to the backup system. The backup system is not intended to resist in-plane shear loads or vertical loads from overlying roofs or floors. If the space between the masonry veneer and the backup system is separated by a cavity at least 2 in. (50 mm) wide and is provided with drainage details, the result is a drainage wall.

9.2.2.4 Modern Masonry Curtain Walls

Curtain walls are multistory masonry walls that resist gravity loads from self-weight only and out-of-plane loads from wind or earthquake. Their most common application is for walls of industrial buildings, warehouses, gymnasiums, or theaters. They are most commonly single-wythe walls. Because they occupy multiple stories, curtain walls are generally designed to span horizontally between columns or pilasters. If a single wythe of masonry is used, horizontal reinforcement is often required for resistance to out-of-plane loads. This reinforcement is usually provided in the form of welded wire reinforcement, placed in the horizontal joints of the masonry.

9.2.2.5 Modern Masonry Bearing and Shear Walls

Bearing walls resist gravity loads from self-weight and overlying floor and roof elements, out-of-plane loads from wind or earthquake, and in-plane shears. If bearing walls are composed of hollow units, vertical reinforcement consists of deformed bars placed in vertical cells, and horizontal reinforcement consists of either deformed bars placed in grouted courses (bond beams) or bed-joint reinforcement. Bearing walls, whether designed as unreinforced or reinforced, must have reinforcement satisfying seismic requirements. If bearing walls are composed of solid units, vertical and horizontal reinforcement generally consist of deformed bars placed in a grouted space between two wythes of masonry.

Although model codes sometimes distinguish between bearing walls and shear walls, in practical terms every bearing wall is also a shear wall, because it is impossible in practical terms for a wall to resist gravity loads from overlying floor or roof elements and yet be isolated from in-plane shears transmitted from those same elements.

Reinforced masonry shear walls differ from reinforced concrete shear walls primarily in that their inelastic deformation capacity is lower. They are usually not provided with confined boundary elements because these are difficult or impossible to place. Their vertical reinforcement is generally distributed uniformly over the plane length of the wall. Sections of masonry shear wall that separate window or door openings are commonly referred to as "piers."

Using this type of construction, 30-story masonry bearing-wall buildings have been built in Las Vegas, NV, a region of seismic risk in the United States (Suprenant 1989).

Masonry infills are structural panels placed in a bounding frame of steel or reinforced concrete. This mode of structural action, though common in panel walls, is not addressed directly by design codes. Provisions are under development. Historical masonry infills are addressed later in this chapter.

9.2.2.6 Modern Masonry Beams and Columns

Other modern masonry elements are beams and columns. Masonry beams are most commonly used as lintels over window or door openings, but can also be used as isolated elements. They are reinforced horizontally for flexure. Although shear reinforcement is theoretically possible, it is difficult to install and is rarely used. Instead, masonry beams are designed deep enough so that shear can be resisted by masonry alone.

Isolated masonry columns are rare. The most common form of masonry beam–column is a masonry bearing wall subjected to a combination of axial load from gravity and out-of-plane moment from eccentric axial load or out-of-plane wind or seismic loads.

9.2.2.7 Role of Horizontal Diaphragms in Structural Behavior of Modern Masonry

Horizontal floor and roof diaphragms play a critical role in the structural behavior of modern masonry. In addition to resisting gravity loads, they transfer horizontal forces from wind or earthquake to the lateral force-resisting elements of a masonry building, which are usually shear walls. Modern horizontal diaphragms are usually composed of cast-in-place concrete or of concrete topping overlying hollow-core, prestressed concrete planks or corrugated metal deck supported on open-web joists. Distinctions between rigid and flexible diaphragms, and appropriate analytical approaches for each, are addressed later in this chapter. Performance of horizontal diaphragms in modern masonry is addressed by structural requirements for in-plane flexural and shear resistance and by detailing requirements for continuous chords and other embedded elements.

9.2.3 Historical Structural Masonry in the United States

9.2.3.1 Historical Unreinforced Masonry Bearing Walls in the United States

Unreinforced masonry bearing walls were constructed before 1933 in the western United States and as late as the 1950s elsewhere in the United States. They commonly consisted of two wythes of masonry, bonded by masonry headers, and sometimes also had an interior wythe of rubble masonry (pieces of masonry units surrounded by mortar).

9.2.3.2 Historical Masonry Infills in the United States

Masonry infills are structural panels placed in a bounding frame of steel or reinforced concrete. Before the advent of drywall construction, masonry infills of clay tile were often used to fill interior or exterior bays of steel or reinforced concrete frames. Although sometimes considered nonstructural, they have high elastic stiffness and are usually built tight against the bounding frame. As a result, they can significantly alter the seismic response of the frame in which they are placed.

9.2.3.3 Role of Horizontal Diaphragms in Structural Behavior of Historical Masonry in the United States

Horizontal diaphragms play a crucial role in the seismic resistance of historical as well as modern masonry construction. In contrast to their role in modern construction, however, the behavior of horizontal diaphragms in historical masonry is usually deficient. Historical diaphragms are usually composed of lumber, supported on wooden joists inserted in pockets in the inner wythe of unreinforced masonry walls. Such diaphragms are not strong enough, and not sufficiently well connected, to transfer horizontal seismic forces to the building's shear walls. Out-of-plane deformations of the bearing walls can cause the joists to slip out of their pockets, often resulting in collapse of the entire building. For this reason, horizontal diaphragms are among the elements addressed in the seismic rehabilitation of historical masonry.

9.3 Fundamental Basis for Design of Masonry in the United States

Design of masonry in the United States is based on the premise that reinforced masonry structures can perform well under combinations of gravity and lateral loads, including earthquake loads, provided that they meet the following conditions:

1. They must have engineered lateral-force-resisting systems, generally consisting of reinforced masonry shear walls distributed throughout their plan area and acting in both principal plan directions.

2. Their load–displacement characteristics under cyclic reversed loading must be consistent with the assumptions used to develop their design loadings:
   - If they are intended to respond primarily elastically, they must be provided with sufficient strength to resist elastic forces. Such masonry buildings are typically low-rise, shear-wall structures.
   - If they are intended to respond inelastically, their lateral-force-resisting elements must be proportioned and detailed to be capable of resisting the effects of the reversed cyclic deformations consistent with that inelastic response. They must be proportioned, and must have sufficient shear reinforcement, so that their behavior is dominated by flexure ("capacity design"). The most desirable structural system for such a response is composed of multiple masonry shear walls, designed to act in flexure and loosely coupled by floor slabs.

U.S. masonry has shown good performance under such conditions (TMS Northridge 1994).

Good load–displacement behavior has also been observed under laboratory conditions. This research has been described extensively in U.S. technical literature over the last two decades. A representative sample is given in the proceedings of North American Masonry conferences (NAMC 1985, 1987, 1990, 1993, 1996, 1999).

Of particular relevance is the U.S. Coordinated Program for Masonry Building Research, also known as the TCCMAR Program (Noland 1990). With the support of the National Science Foundation and the masonry industry, the Technical Coordinating Committee for Masonry Research (TCCMAR) was formed in February 1984 for the purpose of defining and performing both analytical and experimental research and development necessary to improve masonry structural technology, and specifically to lay the technical basis for modern, strength-based design provisions for masonry. Under the coordination of TCCMAR, research was carried out in the following areas:

1. Material properties and tests.
2. *Reinforced masonry walls.* In-plane shear and combined in-plane shear and vertical compression.
3. *Reinforced masonry walls.* Out-of-plane forces combined with vertical compression.
4. Floor diaphragms.
5. Bond and splicing of reinforcement in masonry.
6. Limit state design concepts for reinforced masonry.
7. Modeling of masonry components and building systems.
8. Large-scale testing of masonry building systems.
9. Determination of earthquake-induced forces on masonry buildings.

Work began on the initially scheduled research tasks in September 1985, and the program lasted for more than 10 years. Numerous published results include the work of Hamid et al. (1989) and Blondet and Mayes (1991), who studied masonry walls loaded out-of-plane, and He and Priestley (1992), Leiva and Klingner (1994), and Seible et al. (1994a,b), who studied masonry walls loaded in-plane. In all cases, flexural ductility was achieved without the use of confining reinforcement.

Using pseudodynamic testing procedures, Seible et al. (1994a,b) subjected a full-scale, five-story masonry structure to simulated earthquake input. The successful inelastic performance of this structure under global drift ratios exceeding 1% provided additional verification for field observations and previous TCCMAR laboratory testing. With proper proportioning and detailing, reinforced masonry assemblies can exhibit significant ductility.

Limited shaking-table testing has been conducted on reinforced masonry structures built using typical modern U.S. practice:

1. *Gulkan et al. (1990a,b).* A series of single-story, one-third-scale masonry houses were constructed and tested on a shaking table. The principal objective of the testing was to verify prescriptive reinforcing details for masonry in zones of moderate seismic risk.
2. *Abrams and Paulson (1991).* Two 3-story, quarter-scale reinforced masonry buildings were tested to evaluate the validity of small-scale testing.
3. *Cohen (2001).* Two low-rise, half-scale, reinforced masonry buildings with flexible roof diaphragms were subjected to shaking-table testing. Results were compared with the results of static testing and analytical predictions.

Results of these tests have generally supported field observations of satisfactory behavior of modern reinforced masonry structures in earthquakes.

9.3.1 Design Approaches for Modern U.S. Masonry

Three design approaches are used for modern U.S. masonry: allowable-stress design, strength design, and empirical design. In this section, each approach is summarized.

9.3.1.1 Allowable-Stress Design

Allowable-stress design is the traditional approach of building codes for calculated masonry design. Stresses from unfactored loads are compared with allowable stresses, which are failure stresses reduced by a factor of safety that is usually between 2.5 and 4.

9.3.1.2 Strength Design

Within the past decade, strength-design provisions for masonry have been developed within the 1997 UBC, the 1997 and 2000 NEHRP documents, and the 2000 *International Building Code* (IBC). The 2002 edition of the Masonry Standards Joint Committee (MSJC) code includes strength-design provisions, and those provisions will be referenced by the 2003 IBC.

Strength-design provisions for masonry are generally similar to those for concrete. Factored design actions are compared with nominal capacities reduced by capacity reduction factors. Strength-design provisions for masonry differ from those for reinforced concrete, however, in three principal areas, unreinforced masonry, confining reinforcement, and maximum flexural reinforcement:

1. Some masonry can be designed as unreinforced (flexural tension resisted by masonry alone). For this purpose, nominal flexural tensile capacity is computed as the product of the masonry's tensile bond strength (modulus of rupture) and the section modulus of the section under consideration. This nominal strength is then reduced by a capacity reduction factor.
2. Because it is impractical to confine the compressive zones of masonry elements, the inelastic strain capacity of such elements is less than that of confined reinforced concrete elements. The available displacement ductility ratio of masonry shear walls is therefore lower than that of reinforced concrete shear walls with confined boundary elements, and corresponding R factors (response modification factors) are lower.
3. Maximum flexural reinforcement for masonry elements is prescribed in terms of the amount of steel required to equilibrate the compressive stress block of the element under a critical strain

FIGURE 9.3　Monadnock Building, Chicago (1891).

gradient, in which the maximum strain in masonry is the value used in design (0.0025 for concrete masonry and 0.0035 for clay masonry), and the maximum strain in the extreme tensile reinforcement is a multiple of the yield strain. That multiple depends on the ductility expected of the element. under reversed cyclic inelastic deformations. In practical terms, if inelastic response is possible, an element cannot be designed to work above its balanced axial load. The intent of these provisions is to ensure, for inelastic elements, that the flexural reinforcement can yield and begin to strain-harden before the compression toe crushes.

9.3.1.3　Empirical Design

At the end of the 19th century, masonry bearing-wall buildings were designed using empirical rules of thumb, such as using walls 12 in. (305 mm) thick at the top of a building and increasing the wall thickness by 4 in. (102 mm) for every story. The Monadnock Building, built in Chicago in 1891 (Figure 9.3), is 16 stories high, is of unreinforced masonry, and has bearing walls 6 ft (1.83 m) thick at the base. It is still in use today.

Today's empirical design is the descendant of those rules, adapted for the characteristics of modern structures. They involve primarily limitations on the length to thickness ratios of elements, with some rudimentary axial stress checks and limits on the arrangement of lateral-force-resisting elements and the plan aspect ratio of floor diaphragms.

9.4　Masonry Design Codes Used in the United States

9.4.1　Introduction to Masonry Design Codes in the United States

The United States has no national design code, primarily because the U.S. Constitution has been interpreted as delegating building code authority to the states, which in turn delegate it to municipalities

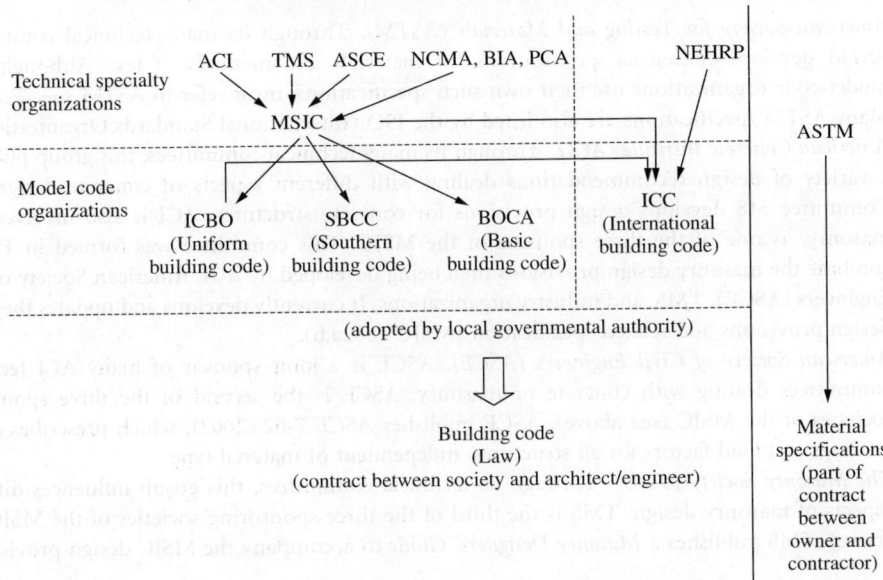

Masonry design provisions in the United States

ANSI rules (balance of interests, letter balloting, resolution of negatives, public comment)

FIGURE 9.4 Schematic of code-development process for masonry in the United States.

and other local governmental agencies. Design codes used in the United States are developed by a complex process involving technical experts, industry representatives, code users, and building officials. As this process applies to the development of design provisions for masonry, it is shown in Figure 9.4 and described herein:

1. *Consensus design provisions* and *specifications for materials or methods of testing* are first drafted in mandatory language by *technical specialty organizations*, operating under consensus rules approved by the American National Standards Institute (ANSI). These consensus rules can vary from organization to organization, but must include the following requirements:
   - Balance of interests (producer, user, and general interest).
   - Written balloting of proposed provisions, with prescribed requirements for a successful ballot.
   - Resolution of negative votes. Negative votes must be discussed and found nonpersuasive before a ballot item can pass. A single negative vote, if found persuasive, can prevent an item from being passed.
   - Public comment. After being approved within the technical specialty organization, the mandatory-language provisions must be published for public comment. If significant public comments are received (usually more than 50 comments on a single item), the organization must respond to the comments.
2. These consensus design provisions and specifications are adopted, sometimes in modified form, by *model-code organizations*, and take the form of *model codes*.
3. These model codes are adopted, sometimes in modified form, by *local governmental agencies* (such as cities or counties). Upon adoption, but not before, they acquire legal standing as *building codes*.

9.4.1.1 Technical Specialty Organizations

Technical specialty organizations are open to designers, contractors, product suppliers, code developers, and end users. Their income (except for Federal Emergency Management Agency [FEMA], a U.S. government agency) is derived from member dues and the sale of publications. Technical specialty organizations active in the general area of masonry include the following:

1. *American Society for Testing and Materials (ASTM)*. Through its many technical committees, ASTM develops consensus specifications for materials and methods of test. Although some model-code organizations use their own such specifications, most refer to ASTM specifications. Many ASTM specifications are also listed by the ISO (International Standards Organization).
2. *American Concrete Institute (ACI)*. Through its many technical committees, this group publishes a variety of design recommendations dealing with different aspects of concrete design. ACI Committee 318 develops design provisions for concrete structures. ACI is also involved with masonry, as one of the three sponsors of the MSJC. This committee was formed in 1982 to combine the masonry design provisions then being developed by ACI, American Society of Civil Engineers (ASCE), TMS, and industry organizations. It currently develops and updates the MSJC design provisions and related specification (MSJC 2002a,b).
3. *American Society of Civil Engineers (ASCE)*. ASCE is a joint sponsor of many ACI technical committees dealing with concrete or masonry. ASCE is the second of the three sponsoring societies of the MSJC (see above). ASCE publishes *ASCE 7-02* (2002), which prescribes design loadings and load factors for all structures, independent of material type.
4. *The Masonry Society (TMS)*. Through its technical committees, this group influences different aspects of masonry design. TMS is the third of the three sponsoring societies of the MSJC (see above). TMS publishes a *Masonry Designers' Guide* to accompany the MSJC design provisions.

9.4.1.2 Industry Organizations

1. *Portland Cement Association (PCA)*. This marketing and technical support organization is composed of cement producers. Its technical staff participates in technical committee work.
2. *National Concrete Masonry Association (NCMA)*. This marketing and technical support organization is composed of producers of concrete masonry units. Its technical staff participates in technical committee work and also produces technical bulletins that can influence consensus design provisions.
3. *Brick Industry Association (BIA)*. This marketing, distributing, and technical support organization consists of clay brick and tile producers. Its technical staff participates in technical committee work and also produces technical bulletins that can influence consensus design provisions.
4. *National Lime Association (NLA)*. This marketing and technical support organization consists of hydrated lime producers. Its technical staff participates in technical committee work.
5. *Expanded Clay, Shale and Slate Institute (ECSSI)*. This marketing and technical support organization consists of producers. Its technical staff participates in technical committee meetings.
6. *International Masonry Institute (IMI)*. This is a labor-management collaborative supported by dues from union masons. Its technical staff participates in technical committee meetings.
7. *Mason Contractors' Association of America (MCAA)*. This organization consists of nonunion mason contractors. Its technical staff participates in technical committee meetings.

9.4.1.3 Governmental Organizations

1. *Federal Emergency Management Agency (FEMA)*. FEMA has jurisdiction over the National Earthquake Hazard Reduction Program (NEHRP) and develops and periodically updates the NEHRP provisions (NEHRP 2000), a set of recommendations for earthquake-resistant design. This document includes provisions for masonry design. The document is published by the Building Seismic Safety Council (BSSC), which operates under a contract with FEMA. BSSC is not an ANSI consensus organization. Its recommended design provisions are intended for consideration and possible adoption by consensus organizations. The 2000 NEHRP *Recommended Provisions* (NEHRP 2000) is the latest of a series of such documents, now issued at 3-year intervals

and pioneered by *ATC 3-06*, which was issued by the ATC in 1978 under contract to the National Bureau of Standards. The 2000 NEHRP *Recommended Provisions* addresses the broad issue of seismic regulations for buildings. It contains chapters dealing with the determination of seismic loadings on structures and with the design of masonry structures for those loadings.

9.4.1.4 Model-Code Organizations

Model-code organizations consist primarily of building officials, although designers, contractors, product suppliers, code developers, and end users can also be members. Their income is derived from dues and the sale of publications. The United States has three model-code organizations:

1. *International Conference of Building Officials (ICBO)*. In the past, this group developed and published the UBC. The latest and final edition is the 1997 UBC.
2. *Southern Building Code Congress (SBCC)*. In the past, this group developed and published the *Standard Building Code* (SBC). The latest and final edition is the 1999 SBC.
3. *Building Officials/Code Administrators (BOCA)*. In the past, this group developed and published the *Basic Building Code* (BBC). The latest and final edition is the 1999 BBC.

In the past, certain model codes were used more in certain areas of the country. The UBC has been used throughout the western United States and in the state of Indiana. The SBC has been used in the southern part of the United States. The BBC has been used in the eastern and northeastern United States.

Since 1996, intensive efforts have been under way in the United States to harmonize the three model building codes. The primary harmonized model building code is called the IBC. It has been developed by the International Code Council (ICC), consisting primarily of building code officials of the three model-code organizations. The first edition of the IBC (2000) was published in May 2000. In most cases, it references consensus design provisions and specifications. It is intended to take effect when adopted by local jurisdictions. It is intended to replace the three current model building codes. Although not all details have been worked out, it is generally understood that the IBC will continue to be administered by the three model-code agencies. Another harmonized model code is being developed by the National Fire Protection Association, consisting primarily of fire-protection officials.

9.4.2 Masonry Design Provisions of Modern Model Codes in the United States

Over the next 3 to 5 years, as the IBC and other harmonized model codes such as that of the National Fire Protection Association are adopted by local jurisdictions, their provisions will become minimum legal design requirements for masonry structures throughout the United States. The 2003 IBC will reference the masonry design provisions of the 2002 MSJC code in essentially direct form.

9.4.2.1 Strength-Design Provisions of the 2002 MSJC Code

Strength-design provisions of the 2002 MSJC code deal with unreinforced as well as reinforced masonry.

Strength-design provisions for unreinforced masonry address failure in flexural tension, in combined flexural and axial compression, and in shear. Nominal capacity in flexural tension is computed as the product of the masonry's flexural tensile strength and its section modulus. Nominal capacity in combined flexural and axial compression is computed using a triangular stress block, with an assumed failure stress of 0.80 f'_m (the specified compressive strength of the masonry). Nominal capacity in shear is computed as the least of several values, corresponding to different possible failure modes (diagonal tension, crushing of compression diagonal, sliding on bed joints). Each nominal capacity is multiplied by a capacity reduction factor.

Strength-design provisions for reinforced masonry address failure under combinations of flexure and axial loads, and in shear. Nominal capacity under combinations of flexure and axial loads is computed using a moment–axial force interaction diagram, determined assuming elasto-plastic behavior of tensile reinforcement and using an equivalent rectangular stress block for the masonry. The diagram also has an upper limit on pure compressive capacity. Shear capacity is computed as the summation of capacity from masonry plus capacity from shear reinforcement, similar to reinforced concrete. Nominal capacity of masonry in shear is computed as the least of several values, corresponding to different possible failure modes (diagonal tension, crushing of compression diagonal, sliding on bed joints). Each nominal capacity is multiplied by a capacity reduction factor.

Strength-design provisions of the 2002 MSJC impose strict upper limits on reinforcement, which are equivalent to requiring that the element remain below the balanced axial load. Nominal flexural and axial strengths are computed neglecting the tensile strength of masonry, using a linear strain variation over the depth of the cross section, a maximum usable strain of 0.0035 for clay masonry and 0.0025 for concrete masonry, and an equivalent rectangular compressive stress block in masonry with a stress of $0.80 f'_m$. These provisions are regarded as too conservative by many elements of the masonry technical community and are currently undergoing extensive review.

9.4.2.2 Allowable-Stress Design Provisions of the 2002 MSJC Code

The allowable-stress provisions of the 2002 MSJC Code are based on linear elastic theory.

Allowable-stress design provisions for unreinforced masonry address failure in flexural tension, in combined flexural and axial compression, and in shear. Flexural tensile stresses are computed elastically, using an uncracked section, and are compared with allowable flexural tensile stresses, which are observed strengths divided by a factor of safety of about 2.5. Allowable flexural tensile stresses for in-plane bending are zero. Flexural and axial compressive stresses are also computed elastically and are compared with allowable values using a so-called "unity equation." Axial stresses divided by allowable axial stresses, plus flexural compressive stresses divided by allowable flexural compressive stresses, must not exceed unity. Allowable axial stresses are one quarter of the specified compressive strength of the masonry, reduced for slenderness effects. Allowable flexural compressive stresses are one third of the specified compressive strength of the masonry. Shear stresses are computed elastically, assuming a parabolic distribution of shear stress based on beam theory. Allowable stresses in shear are computed corresponding to different possible failure modes (diagonal tension, crushing of compression diagonal, sliding on bed joints). The factor of safety for each failure mode is at least 3.

Allowable-stress design provisions for reinforced masonry address failure in combined flexural and axial compression, and in shear. Stresses in masonry and reinforcement are computed using a cracked transformed section. Allowable tensile stresses in deformed reinforcement are the specified yield strength divided by a safety factor of 2.5. Allowable flexural compressive stresses are one third of the specified compressive strength of the masonry. Allowable capacities of sections under combinations of flexure and axial force can be expressed using an allowable-stress moment–axial force interaction diagram, which also has a maximum allowable axial capacity as governed by compressive axial stress. Shear stresses are computed elastically, assuming a uniform distribution of shear stress. Allowable shear stresses in masonry are computed corresponding to different possible failure modes (diagonal tension, crushing of compression diagonal, sliding on bed joints). If those allowable stresses are exceeded, all shear must be resisted by shear reinforcement, and shear stresses in masonry must not exceed a second, higher set of allowable values. The factor of safety for shear is at least 3.

9.4.3 Seismic Design Provisions for Masonry in the 2000 International Building Code

In contrast to wind loads, which are applied forces, earthquake loading derives fundamentally from imposed ground displacements. Although inertial forces are applied to a building as a result, these inertial forces depend on the mass, stiffness, and strength of the building as well as the characteristics of the ground motion itself. This is also true for masonry buildings. In this section, the seismic design provisions of the 2000 IBC are summarized as they apply to masonry elements.

Modern model codes in the United States address the design of masonry for earthquake loads by first prescribing seismic design loads in terms of the building's geographic location, its function, and its underlying soil characteristics. These three characteristics together determine the building's "seismic design category." Seismic Design Category A corresponds to a low level of ground shaking, typical use, and typical underlying soil. Increasing levels of ground shaking, an essential facility, and unknown or undesirable soil types correspond to higher seismic design categories, with Seismic Design Category F being the highest. In addition to being designed for the seismic forces corresponding to their seismic design category, masonry buildings must comply with four types of prescriptive requirements, whose severity increases as the building's seismic design category increases from A to F:

1. Seismic-related restrictions on materials
2. Seismic-related restrictions on design methods
3. Seismic-related requirements for connectors
4. Seismic-related requirements for locations and minimum percentages of reinforcement

The prescriptive requirements are incremental; for example, a building in Seismic Design Category C must also comply with prescriptive requirements for buildings in Seismic Design Categories A and B.

9.4.3.1 Determination of Seismic Design Forces

For structural systems of masonry, as for other materials, seismic design forces are determined based on the structure's location, underlying soil type, and degree of structural redundancy, and the system's expected inelastic deformation capacity. The last characteristic is described indirectly in terms of shear wall types: ordinary plain, ordinary reinforced, detailed plain, intermediate reinforced, and special reinforced. As listed, these types are considered to have increasing inelastic deformation capacity, and a correspondingly increasing response modification coefficient, R, is applied to the structural systems comprising them. Higher values of R correspond, in turn, to lower seismic design forces.

9.4.3.2 Seismic-Related Restrictions on Materials

In Seismic Design Categories A through C, no additional seismic-related restrictions apply beyond those related to design in general. In Seismic Design Categories D and E, Type N mortar and masonry cement are prohibited, due to their relatively low tensile bond strength.

9.4.3.3 Seismic-Related Restrictions on Design Methods

In Seismic Design Category A, masonry structural systems can be designed by strength design, allowable-stress design, or empirical design. They are permitted to be designed including the flexural tensile strength of masonry.

In Seismic Design Category B, elements that are part of the lateral-force-resisting system can be designed by strength design or allowable-stress design only, but not by empirical design. Elements not part of the lateral-force-resisting system, however, can still be designed by empirical design.

No additional seismic-related restrictions on design methods apply in Seismic Design Category C. In Seismic Design Category D, elements that are part of the lateral-force-resisting system must be designed as reinforced, by either strength design or allowable-stress design.

No additional seismic-related restrictions on design methods apply in Seismic Design Categories E and F.

9.4.3.4 Seismic-Related Requirements for Connectors

In Seismic Design Category A, masonry walls are required to be anchored to the roof and floors that support them laterally. This provision is not intended to require a mechanical connection.

No additional seismic-related restrictions on connection forces apply in Seismic Design Category B.

In Seismic Design Category C, connectors for masonry partition walls must be designed to accommodate story drift. Horizontal elements and masonry shear walls must be connected by connectors capable of resisting the forces between those elements, and minimum connector capacity and maximum spacing are also specified.

No additional seismic-related requirements for connectors apply in Seismic Design Categories D and higher.

9.4.3.5 Seismic-Related Requirements for Locations and Minimum Percentages of Reinforcement

In Seismic Design Categories A and B, there are no seismic-related requirements for locations and minimum percentages of reinforcement.

In Seismic Design Category C, masonry partition walls must have reinforcement meeting requirements for minimum percentage and maximum spacing. The reinforcement is not required to be placed parallel to the direction in which the element spans. Masonry walls must have reinforcement with an area of at least 0.2 in.^{2} (129 mm^2) at corners, close to each side of openings, movement joints, and ends of walls, and spaced no farther than 10 ft (3 m) apart.

In Seismic Design Category D, masonry walls that are part of the lateral-force-resisting system must have uniformly distributed reinforcement in the horizontal and vertical directions, with a minimum percentage in each direction of 0.0007 and a minimum summation (in both directions) of 0.002. Maximum spacing in either direction is 48 in. (1.22 m). Closer maximum spacing requirements apply for stack bond masonry. Masonry shear walls have additional requirements for minimum vertical reinforcement and hooking of horizontal shear reinforcement.

In Seismic Design Categories E and F, stack-bonded masonry partition walls have minimum horizontal reinforcement requirements. Stack-bonded masonry walls that are part of the lateral-force-resisting system have additional requirements for spacing and percentage of horizontal reinforcement. Masonry shear walls must be "special reinforced."

9.4.4 Future of Design Codes for Masonry in the United States

The next decade is likely to witness increased harmonization of U.S. model codes, and increasing direct reference to the MSJC Code and Specification. In the MSJC design provisions, the Specification is likely to be augmented by a Code chapter dealing with construction requirements.

9.5 Seismic Retrofitting of Historical Masonry in the United States

The preceding sections of this chapter have dealt primarily with the seismic performance of masonry built under the reinforced masonry provisions that were introduced in the western United States as part of the reaction to the 1933 Long Beach Earthquake. Such masonry generally behaves well. Unreinforced masonry, in contrast, often collapses or experiences heavy damage. This is true whether the masonry was built in the western United States prior to 1933 or in other places either before or after 1933. As a result of this observed poor behavior, recent decades have witnessed significant interest in the seismic retrofitting of historical masonry in the United States. In this section, those retrofitting efforts are briefly reviewed.

9.5.1 Observed Seismic Performance of Historical U.S. Masonry

As noted earlier in this chapter, unreinforced masonry buildings have performed poorly in many U.S. earthquakes. The 1933 Long Beach, California, earthquake severely damaged many unreinforced masonry buildings, particularly schools. One consequence of this damage was the passage of California's Field Act, which prohibited the use of masonry (as it was then used) in all public buildings in the state. When masonry construction was revived in California during the mid-1940s, it was required to comply with newly developed UBC provisions. Those provisions, based on the reinforced concrete design practice of the time, required that minimum seismic lateral forces be considered in the design of masonry buildings, that tensile stresses in masonry be resisted by reinforcement, and that all masonry have at least a minimum percentage of horizontal and vertical reinforcement.

9.5.2 Laboratory Performance of Historical U.S. Masonry

Although laboratory testing of historical U.S. masonry is difficult, some testing has been carried out on masonry specimens cut from existing structures. Much more extensive testing has been carried out on reduced-scale replicas of unreinforced masonry construction. Some of the testing is typical of testing in the United States; some is typical of that found in other countries. That testing includes the following:

1. *Benedetti et al. (1998).* Twenty-four, simple, 2-story, half-scale, unreinforced masonry buildings were tested to varying degrees of damage, repaired and strengthened, and tested again.
2. *Tomazevic and Weiss (1994).* Two, 3-story, reduced-scale, plain and reinforced masonry buildings were tested. The efficacy of reinforcement was confirmed.
3. *Costley and Abrams (1996).* Two reduced-scale brick masonry buildings were tested on a shaking table. Observed and predicted behavior were compared and used to make recommendations for the use of FEMA retrofitting guidelines.

9.5.3 Basic Principles of Masonry Retrofitting

Over the past 15 years, efforts have focused on the seismic response and retrofitting of existing URM buildings. The goals of seismic retrofitting are

1. To correct deficiencies in the overall structural concept.
2. To correct deficiencies in behavior of structural elements.
3. To correct deficiencies in behavior of nonstructural elements.

The most basic elements of seismic retrofitting involve bracing parapets to roofs and connecting floor diaphragms to walls using through anchors (mechanical, grouted, or adhesive).

9.5.4 History of URM Retrofitting in the Los Angeles Area

9.5.4.1 Division 88

In 1949 the city of Los Angeles passed the Parapet Correction Ordinance, which required that unreinforced masonry or concrete parapets above exits, and parapets above public access, be retrofitted to minimize hazards. As a result, such parapets were either laterally braced or removed. Consequently, many unreinforced masonry buildings withstood the 1971 San Fernando earthquake better than previous earthquakes (Lew et al. 1971).

Following the February 1971 San Fernando Earthquake, the city of Los Angeles, the Federal Government, and the Structural Engineers Association of Southern California joined forces in a 10-year investigation. As a result of this investigation, Los Angeles adopted an ordinance known as Division 68 on February 13, 1981. Division 68 required seismic retrofitting of all unreinforced masonry bearing-wall buildings that were built, that were under construction, or for which a permit had been issued prior to October 6, 1933. The ordinance did not include one- or two-family dwellings or detached apartment houses comprising fewer than five dwelling units and used solely for residential purposes.

The 1985 edition of the Los Angeles Building Code revised Division 68 to Division 88 and included provisions for the testing and strengthening of mortar joints to meet minimum values for shear strength.

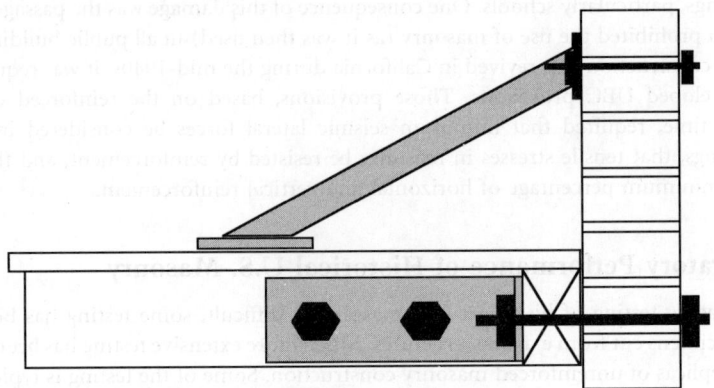

FIGURE 9.5   Schematic of Division 88 URM retrofitting techniques for URM.

Furthermore, Division 88 required that unreinforced masonry be positively anchored to floor and roof diaphragms with anchors spaced not more than 6 ft apart. There were also parapet height limitations, based on wall thickness. Continuous inspection was also required on the retrofitting work. These retrofitting measures are shown schematically in Figure 9.5.

Alternatives to these specific provisions were also possible. Division 88 was renamed Chapter 88 in the 1988 City of Los Angeles Code. In addition to masonry bearing walls, veneer walls constructed before October 6, 1933, were included. This edition also added Section 8811 ("Design Check — Compatibility of Roof Diaphragm Stiffness to Unreinforced Masonry Wall Out-of-Plane Stability").

At the time of the Northridge Earthquake, it is believed that essentially all URM buildings in the city of Los Angeles had their parapets either removed or laterally braced. Unconfirmed reports indicate that in the city of Los Angeles, about 80% of URM buildings had been retrofitted to comply with Division 88; however, the percentage was reported to be considerably lower in other cities in the Los Angeles area.

9.5.4.2 Other Retrofitting Guidelines

The NEHRP, in conjunction with the FEMA, has produced a series of documents dealing with the seismic evaluation and retrofitting of structures, including masonry structures:

1. FEMA 172 (1992), *Handbook of Techniques for the Seismic Rehabilitation of Existing Buildings*, provides a general list of retrofitting techniques.
2. FEMA 178 (1992) presents an overall method for engineers to identify buildings or building components that present unacceptable risks in case of an earthquake.
3. FEMA 273 (1997) and 274 (1997), *NEHRP Guidelines and Commentary* for the seismic rehabilitation of buildings, provide code-type procedures for the assessment, evaluation, analysis, and rehabilitation of existing building structures.
4. FEMA 306 (1998), *Evaluation of Earthquake-Damaged Concrete and Masonry Wall Buildings, Basic Procedures Manual*, provides guidance on evaluation of damage and on performance analysis and includes newly formulated Component Damage Classification Guides and Test and Investigation Guides. The procedures characterize the observed damage caused by an earthquake in terms of the loss in building performance capability.
5. FEMA 307 (1998), *Evaluation of Earthquake-Damaged Concrete and Masonry Wall Buildings, Technical Resources*, contains supplemental information, including results from a theoretical

analysis of the effects of prior damage on single-degree-of-freedom mathematical models, additional background information on the Component Damage Classification Guides, and an example of the application of the basic procedures.

6. FEMA 308 (1998), *The Repair of Earthquake-Damaged Concrete and Masonry Wall Buildings*, discusses the technical and policy issues pertaining to the repair of earthquake-damaged buildings and includes guidance on the specification of individual repair techniques and newly formulated Repair Guides.

7. FEMA 356 (2000), *Prestandard and Commentary for the Seismic Rehabilitation of Buildings*, is an attempt to encourage the use of FEMA 273 and to put the guidelines of that document into mandatory language.

9.6 Future Challenges

This chapter has presented an overview of current issues in the design of masonry for earthquake loads and of the historical process by which those issues have developed. It would not be complete without at least a brief mention of the challenges facing the masonry technical community in this area.

9.6.1 Performance-Based Seismic Design of Masonry Structures

Across the entire spectrum of construction materials, increased attention has been focused on performance-based seismic design, which can be defined as design whose objective is a structure that can satisfy different performance objectives under increasing levels of probable seismic excitation. For example, a structure might be designed to remain operational under a design earthquake with a relatively short recurrence interval, to be capable of immediate occupancy under a design earthquake with a longer recurrence interval, to ensure life safety under a design earthquake with a still longer recurrence interval, and to not collapse under a design earthquake with a long recurrence interval.

This design approach, accepted qualitatively since the 1970s, has been adopted quantitatively in recent documents related to seismic rehabilitation (FEMA 356 2000), and will probably be incorporated into future seismic design provisions for new structures as well. Because masonry structures are inherently composed of walls rather than frames, they tend to be laterally stiff, which is usually a useful characteristic in meeting performance objectives for seismic response.

9.6.2 Increased Consistency of Masonry Design Provisions

The 2003 IBC will reference the 2002 MSJC Code and Commentary essentially in its entirety. Other model codes will probably do the same. As a result, future development of seismic design provisions for masonry structures will take place almost exclusively within the MSJC, rather than in a number of different technical forums. This is expected to lead to increased rationality of design provisions, increased consistency among designs produced by different design methods (e.g., strength versus allowable-stress design), and possibly also simplification of design provisions. As an additional benefit, the emergence of a single set of ANSI-consensus design provisions for masonry is expected to encourage the production of computer-based tools for the analysis and design of masonry structures using those provisions.

References

Abrams, D. and Paulson, T. (1991). "Modeling Earthquake Response of Concrete Masonry Building Structures," *ACI Struct. J.*, 88(4), 475–485.

ACI SP-127 (1991). *Earthquake-Resistant Concrete Structures: Inelastic Response and Design (Special Publication SP-127)*, S.K. Ghosh, editor, 479–503.

ACI 318-02 (2002). ACI Committee 318, *Building Code Requirements for Reinforced Concrete (ACI 318-99)*, American Concrete Institute, Farmington Hills, MI.

ASCE 7-02 (2002). *Minimum Design Loads for Buildings and Other Structures (ASCE 7-02)*, American Society of Civil Engineers.

ATC 3-06 (1978). *Tentative Provisions for the Development of Seismic Regulations for Buildings (ATC 3-06)*, Applied Technology Council, National Bureau of Standards.

BBC (1999). *Basic Building Code*, Building Officials/Code Administrators International.

Benedetti, D., Carydis, P., and Pezzoli, P. (1998). "Shaking-Table Tests on 24 Simple Masonry Buildings," *Earthquake Eng. Struct. Dyn.*, 27(1), 67–90.

Binder, R.W. (1952). "Engineering Aspects of the 1933 Long Beach Earthquake," *Proceedings of the Symposium on Earthquake Blast Effects on Structures*, Berkeley, CA, 186–211.

Blondet, M. and Mayes, R.L. (1991). *The Transverse Response of Clay Masonry Walls Subjected to Strong Motion Earthquakes — Volume 1: General Information*, U.S.–Japan Coordinated Program for Masonry Building Research, TCCMAR Report No. 3.2(b)-2.

Cohen, G.L. (2001). "Seismic Response of Low-Rise Masonry Buildings with Flexible Roof Diaphragms," M.S. Thesis, The University of Texas at Austin, TX.

Costley, A.C. and Abrams, D.P. (1996). "Response of Building Systems with Rocking Piers and Flexible Diaphragms," *Proceedings, Structures Congress*, American Society of Civil Engineers, Chicago, IL, April 15–18, 1996, 135–140.

FEMA 172 (1992). *Handbook of Techniques for the Seismic Rehabilitation of Existing Buildings*, Building Seismic Safety Council.

FEMA 178 (1992). *NEHRP Handbook for the Seismic Evaluation of Existing Buildings*, Building Seismic Safety Council.

FEMA 273 (1997). *NEHRP Guidelines for the Seismic Rehabilitation of Buildings*, Building Seismic Safety Council.

FEMA 274 (1997). *NEHRP Commentary on the Guidelines for the Seismic Rehabilitation of Buildings*, Building Seismic Safety Council.

FEMA 306 (1998). *Evaluation of Earthquake-Damaged Concrete and Masonry Wall Buildings: Basic Procedures Manual*, Federal Emergency Management Agency.

FEMA 307 (1998). *Evaluation of Earthquake-Damaged Concrete and Masonry Wall Buildings: Technical Resources*, Federal Emergency Management Agency.

FEMA 308 (1998). *Repair of Earthquake-Damaged Concrete and Masonry Wall Buildings*, Federal Emergency Management Agency.

FEMA 356 (2000). *Prestandard and Commentary for the Seismic Rehabilitation of Buildings*, Federal Emergency Management Agency.

Gulkan, P., Clough, R.W., Mayes, R.L., and Manos, G. (1990a). "Seismic Testing of Single-Story Masonry Houses, Part I," *J. Struct. Eng.*, American Society of Civil Engineers, 116(1), 235–256.

Gulkan, P., Clough, R.W., Mayes, R.L., and Manos, G. (1990b). "Seismic Testing of Single-Story Masonry Houses, Part II," *J. Struct. Eng.*, American Society of Civil Engineers, 116(1), 257–274.

Hamid, A., Abboud, B., Farah, M., Hatem, K., and Harris, H. (1989). *Response of Reinforced Block Masonry Walls to Out-of-plane State Loads*, U.S.–Japan Coordinated Program for Masonry Building Research, TCCMAR Report No. 3.2(a)-1.

He, L. and Priestley, M.J.N. (1992). *Seismic Behavior of Flanged Masonry Shear Walls* U.S.–Japan Coordinated Program for Masonry Building Research, TCCMAR Report No. 4.1-2.

IBC (2000). *International Building Code*, International Code Council.

Leiva, G. and Klingner, R.E. (1994). "Behavior and Design of Multi-Story Masonry Walls under In-Plane Seismic Loading," *Masonry Soc. J.*, 13(1), 15–24.

Lew, H.S., Leyendecker, E.V., and Dikkers, R.D. (1971). *Engineering Aspects of the 1971 San Fernando Earthquake*, Building Science Series 40, United States Department of Commerce, National Bureau of Standards, 412 pp.

MSJC (2002a). *Building Code Requirements for Masonry Structures (ACI 530-02/ASCE 5-02/TMS402-02)*, American Concrete Institute, American Society of Civil Engineers, and The Masonry Society.

MSJC (2002b). *Specifications for Masonry Structures (ACI 530.1-02/ASCE 6-02/TMS 602-02)*, American Concrete Institute, American Society of Civil Engineers, and The Masonry Society.

NAMC (1985). *Proceedings of the Third North American Masonry Conference*, University of Texas, Arlington, TX, June 3–5, 1985, The Masonry Society, Boulder, CO.

NAMC (1987). *Proceedings of the Fourth North American Masonry Conference*, University of California, Los Angeles, CA, August 16–19, 1987, The Masonry Society, Boulder, CO.

NAMC (1990). *Proceedings of the Fifth North American Masonry Conference*, University of Illinois, Urbana-Champaign, IL, June 3–6, 1990, The Masonry Society, Boulder, CO.

NAMC (1993). *Proceedings of the Sixth North American Masonry Conference*, Drexel University, Philadelphia, PA, June 7–9, 1993, The Masonry Society, Boulder, CO.

NAMC (1996). *Proceedings of the Seventh North American Masonry Conference*, University of Notre Dame, Notre Dame, IN, June 2–5, 1996, The Masonry Society, Boulder, CO.

NAMC (1999). *Proceedings of the Eighth North American Masonry Conference*, University of Texas at Austin, Austin, TX, June 6–9, 1999, The Masonry Society, Boulder, CO.

NEHRP (1997). *NEHRP (National Earthquake Hazards Reduction Program) Recommended Provisions for the Development of Seismic Regulations for New Buildings (FEMA 222)*, Building Seismic Safety Council.

NEHRP (2000). *NEHRP (National Earthquake Hazards Reduction Program) Recommended Provisions for the Development of Seismic Regulations for New Buildings (FEMA 368)*, Building Seismic Safety Council.

Noland, J.L. (1990). "1990 Status Report: US Coordinated Program for Masonry Building Research," *Proceedings, Fifth North American Masonry Conference*, University of Illinois at Urbana-Champaign, IL, June 3–6, 1990.

SBC (1999). *Standard Building Code*, Southern Building Code Congress International.

Seible, F., Hegemier, A., Igarashi, A., and Kingsley, G. (1994a). "Simulated Seismic-Load Tests on Full-Scale Five-Story Masonry Building," *J. Struct. Eng.*, American Society of Civil Engineers, 120(3), 903–924.

Seible, F., Priestley, N., Kingsley, G., and Kurkchubashe, A. (1994b). "Seismic Response of Full-Scale Five-Story Reinforced-Masonry Building," *J. Struct. Eng.*, American Society of Civil Engineers, 120(3), 925–947.

Suprenant, B.A. (1989). "A Floor a Week per Tower," *Masonry Constr.*, 2(11), 478–482.

TMS (1989). "The Masonry Society," *Proceedings of an International Seminar on Evaluating, Strengthening, and Retrofitting Masonry Buildings*, Construction Research Center, The University of Texas at Arlington, TX, October 1989.

TMS Northridge (1994). Klingner, R.E., editor, *Performance of Masonry Structures in the Northridge, California Earthquake of January 17, 1994*, Technical Report 301-94, The Masonry Society, Boulder, CO, 100 pp.

Tomazevic, M. and Weiss, P. (1994). "Seismic Behavior of Plain- and Reinforced-Masonry Buildings," *J. Struct. Eng.*, American Society of Civil Engineers, 120(2), 323–338.

10

Timber
Structures

J. Daniel Dolan
*Department of Civil and
Environmental Engineering,
Washington State University,
Pullman, WA*

10.1 Introduction

Stone and wood were the first materials used by man to build shelter, and in the United States wood continues to be the primary construction material for residential and commercial buildings today. In California, for example, wood accounts for 99% of residential buildings (Schierle 2000). Design and construction methods for wood currently used by the residential construction industry in North America have developed through a process of evolution and tradition. Historically, these construction methods have been sufficient to provide acceptable performance under seismic loading mainly due to the relatively light weight of wood and the historical high redundancy in single-family housing. However, in recent years architectural trends and society's demands for larger rooms, larger windows, and a more open, airy

0-8493-1569-7/05/$0.00+$1.50
© 2005 by CRC Press

feel to the structure have resulted in a reduction in the structural redundancy of the typical house, as well as a reduction in symmetry of stiffness and strength that was inherent in traditional structures.

If one were to review the type of structure that was built in the 1930s, 1940s, and even into the 1950s, one would realize that the average house had a pedestrian door, small double-hung windows, and relatively small rooms with lots of walls. If one compares this typical construction to buildings that are being built in the early 2000s, newer buildings have large windows, if not four walls of primarily glass, large great rooms, and often a single room that pierces the first-story ceiling, becoming two stories high and causing a torsional irregularity in the second-floor structure. If one then includes multifamily construction, which includes apartments, condominiums, and townhouses, the structures become fairly stiff and strong in one direction while in the orthogonal direction (the side with the windows and the doors to the hallways or patio) they become very weak and flexible, due to the lack of structural wall space. In general, modern structures typically have more torsional irregularities, vertical irregularities such as soft stories, and uneven stiffness and strength in orthogonal directions when compared to traditional buildings. As a result, it was observed in the 1994 Northridge earthquake that "most demolished single-family dwellings and multifamily dwellings (as a percentage of existing buildings) were built 1977–1993" (Schierle 2000). The CUREE-Caltech woodframe project also illustrated the inherent soft-story response of light-frame construction.

An additional need for improved understanding of the material and structure used in modern timber buildings is the continued movement toward performance-based design methods and an increased concern over damage. If one considers that the house is the single largest investment that the average person makes in his or her lifetime, it should be no surprise that concern over accumulated damage due to moderate seismic events has begun to be discussed in the context of model building codes. To support the seismic design of timber structures, the wood industry has sponsored the development of the *Standard for Load and Resistance Factor Design (LRFD) for Engineered Wood Construction* (American Forest and Paper Association 1996; American Society of Civil Engineers 1996). However, most designers continue to use the *National Design Specification (NDS$^®$) for Wood Construction* and its supplements (American Forest and Paper Association 2001) for designing wood structures in North America. The NDS is an allowable stress design methodology, while the LRFD is a strength-based design methodology that is intended to provide a better design for seismic concerns. The NDS has been repackaged into the *ASD* (Allowable Stress Design) *Manual for Engineered Wood Construction* (American Forest and Paper Association 2001).

This section reviews the types of wood products available for use in timber construction, the types of structures that are typically designed and built in North America, the design standards that are available for directing the design process, the industry resources that are available, the performance of wood buildings in recent earthquakes, and some of the restrictions that are placed on wood structures due to issues other than seismic concerns. While the rest of the chapter will focus primarily on strength-based design (i.e., LRFD), the NDS will be referred to from time to time where differences are significant. Since the average design firm continues to use ASD for wood structures, it is important that these differences be highlighted.

10.1.1 Types of Wood-Based Products

There are a wide variety and an increasing number of wood-based products available for use in building construction. While the largest volume of wood-based products includes dimensional lumber, plywood, and oriented strand board (OSB), new composites include structural composite lumber (SCL), I-joists, laminated veneer lumber (LVL), and plastic wood (Figure 10.1). In addition, when large sizes are required, glued–laminated lumber (*glulam*) and SCL, such as Paralam, can be used. The current trend is to move toward increased use of wood-based composites in building construction. This is due to the increased difficulty in obtaining large sizes of timber because of restrictions on logging and changes in the economic structure of manufacturing. Therefore, designers should become familiar with SCL, LVL, and glulam when long spans or heavy loads are anticipated. Many of the new composites can be custom manufactured to the size and strength required for a particular application. Designers must obtain the proprietary technical information required to design structures with most of the new composites from the suppliers of the products.

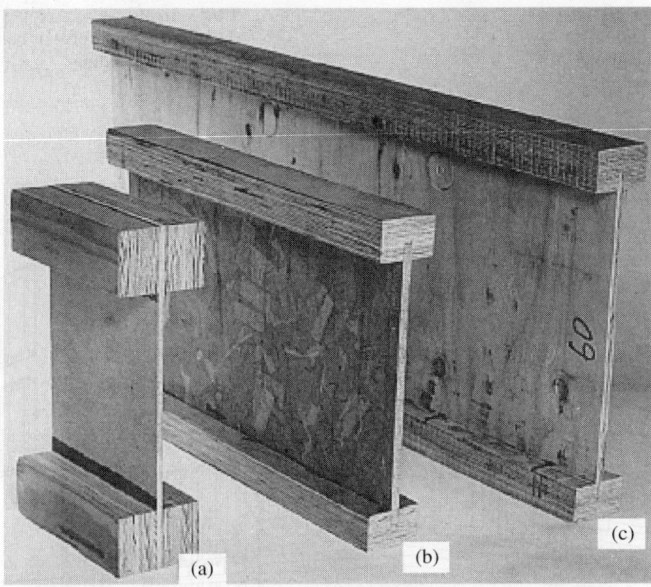

FIGURE 10.1 Prefabricated I-joists with laminated veneer lumber flanges and structural panel webs. One experimental product has (a) a hardboard web; the other two commercial products have (b) oriented strandboard, and (c) plywood webs (courtesy of U.S. Department of Agriculture. 1999. *Wood Handbook: Wood as an Engineering Material*, Agriculture Handbook 72, Forest Products Laboratory, U.S. Department of Agriculture (USDA), Madison, WI).

10.1.2 Types of Structures

Timber structures can be classified into two general categories:

1. *Heavy timber* construction includes buildings such as sports arenas, gymnasiums, concert halls, museums, office buildings, and parking garages. These heavier structures are usually designed to resist higher levels of loading and are therefore designed with the intent of using large section timbers that typically require the use of glulam timber, LVL, structural composite lumber, or similar products. These types of structures require a fairly high level of engineering to ensure the safe performance of the structure due to the lower redundancy of the structure.
2. *Light-frame* construction is by far the largest volume of timber construction in North America. These types of buildings include one- and two-family dwellings (Figure 10.2), apartments, townhouses, hotels, and other light-commercial buildings. These types of structures are highly redundant and indeterminate, and there are currently no computer analysis tools that provide a detailed analysis of these structures.

Light-frame construction consists of 2-in. nominal dimension lumber that ranges in size from 2×4 to 2×12. While the design specifications have geometric parameters that include 2×14, the availability of such large sizes is questionable. The lattice of framing comprising $2\times$ dimensional lumber is then typically sheathed with panel products that include plywood, OSB, fiberboard, gypsum, stucco, or other insulation-type products (Figure 10.2). This system of light-framing sheathed with load-distributing elements then acts to transmit the load horizontally through the roof and floors and vertically through the walls, constituting the lateral-force-resisting system (Figure 10.3). This results in a highly redundant and indeterminate structure that has a good history of performance in seismic loading. With modern architectural trends, the reduction in redundancy results in an increased need to involve structural engineering to ensure good performance.

Roof/floor span systems
1. Wood joist and rafter
2. Diagonal sheathing
3. Straight sheathing

Wall systems
4. Stud wall (platform or balloon frame)
5. Horizontal siding

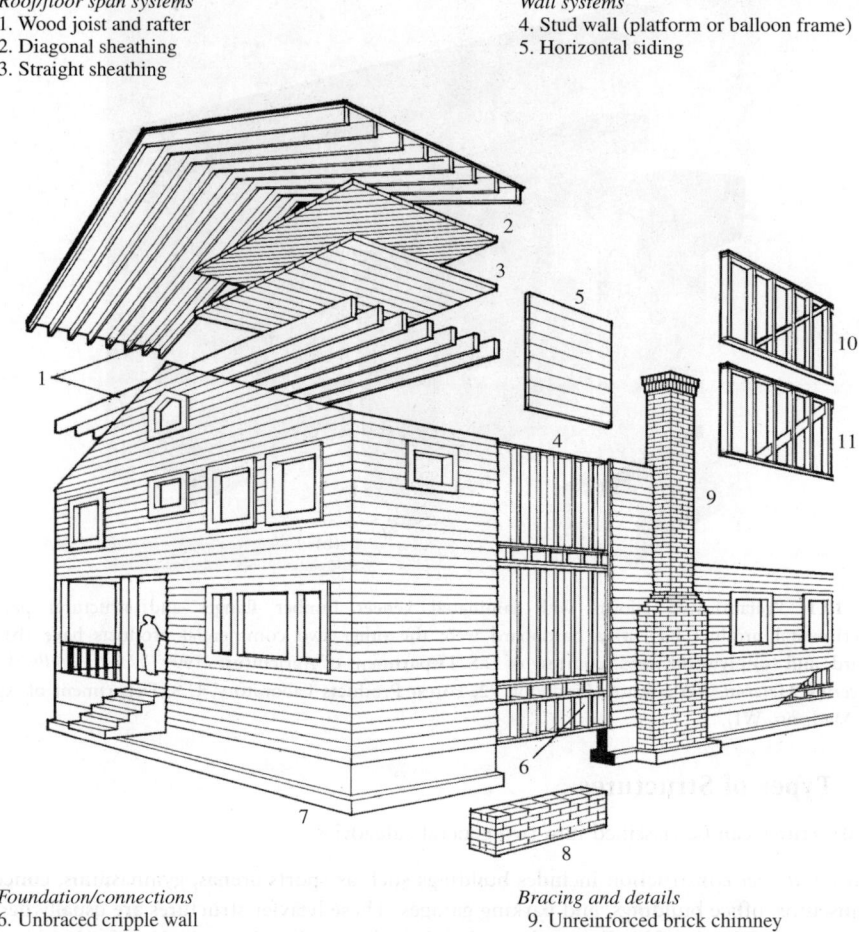

Foundation/connections
6. Unbraced cripple wall
7. Concrete foundation
8. Brick foundation

Bracing and details
9. Unreinforced brick chimney
10. Diagonal blocking
11. Let-in brace (only in vintage)

FIGURE 10.2   Schematic of wood light-frame construction (courtesy of Federal Emergency Management Agency. 1988. *Rapid Visual Screening of Buildings for Potential Seismic Hazards: A Handbook*, FEMA 154 Federal Emergency Management Agency (FEMA), Washington, DC).

Light-frame construction can be broken into two principal categories: fully designed and prescriptive. Currently, the design requirements for using mechanics-based design methods are in the *International Building Code* (IBC) (International Code Council 2003a) or the National Fire Prevention Association *NFPA 5000 Building Code* (National Fire Prevention Association 2002). Requirements for prescriptive construction are contained in the *International Residential Code* (IRC) (International Code Council 2003b), which is a replacement for the Council of American Building Officials' (CABO) *One- and Two-Family Dwelling Code* (Council of American Building Officials 1995). The IRC allows for a mix between a fully designed and rationalized system and the prescriptive systems. In other words, a building that is primarily designed and constructed according to the prescriptive rules of the IRC can have elements that are rationally designed to eliminate such things as irregularities in the structure due to form. The IRC and IBC have consistent seismic provisions as far as the load determination is concerned.

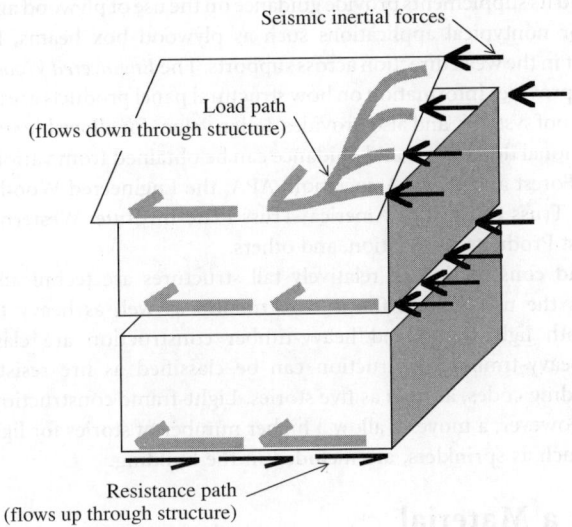

FIGURE 10.3   Lateral-force-resisting system: the load is transmitted horizontally through the roof and floors and vertically through the walls.

10.1.3   Design Standards

Design standards can be divided into two categories: the performance requirements, which are typically covered by the building codes, and the required design methodology to provide that performance, which is included in the design standards. In North America, buildings are typically governed by the IBC or NFPA 5000 for engineered systems. The IRC provides the performance requirements and the methodology to provide the required resistance for one- and two-family residential structures. All of the model building codes available in the United States, plus the *ASCE-7: Minimum Design Load for Buildings and Other Structures* (American Society of Civil Engineers 2002) base their seismic design requirements on the National Earthquake Hazard Reduction Program (*NEHRP*) *Recommended Provisions for Seismic Regulations for New Buildings and Other Structures* (Building Seismic Safety Council 2003a).

All model building codes in effect in the United States recognize either the LRFD standard (American Forest and Paper Association 1996; American Society of Civil Engineers 1996) for strength-based design or the NDS (American Forest and Paper Association 2001) for allowable stress design for engineered wood construction. To parallel the format of the LRFD design manual, the NDS has been incorporated into the *ASD Manual for Wood Construction* (American Forest and Paper Association 2001). In this format, all of the wood-based structural products available for use in the design and construction of timber structures are available to the designer. Most of the provisions of the LRFD standard are similar to those used in the NDS or ASD manual. The difference is that the ASD design methodology bases its requirements in terms of working stresses or allowable stresses, while the LRFD manual bases its design values on the nominal strength values. The two design methodologies also use different load combinations. Since the NEHRP provisions require the use of strength-based design methodology, this chapter will focus on LRFD.

In addition to the ASD and LRFD manuals, the designer may utilize industry documents that are available for most of the trade associations associated with the timber industry. Some of these references include the *Timber Construction Manual* (American Institute of Timber Construction 2004), the *Plywood Design Specification* (APA 1997), the *Engineered Wood Construction Guide* (APA 2003), and the *Wood Frame Construction Manual (WFCM) for One- and Two-Family Dwellings* (American Forest and Paper Association 2001). The *Timber Construction Manual* (American Institute of Timber Construction 2004) provides the designer with guidance on the use of glulam construction and associated issues such as heavy timber connectors, notching and drilling of beams, and the design of arches and other curved members. The Plywood

Design Specification and its supplements provide guidance on the use of plywood and other structural panel products, especially for nontypical applications such as plywood box beams, folded plate roofs, and spanning of the product in the weak direction across supports. The *Engineered Wood Construction Guide* is a general document that provides information on how structural panel products are graded and marked, and the design of floor and roof systems, and also provides design values for allowable stress design of shear walls and diaphragms. Additional documents and guidance can be obtained from various industry associations such as the American Forest and Paper Association, APA, the Engineered Wood Association, Canadian Wood Council, Wood Truss Council of America, Truss Plate Institute, Western Wood Products Association, Southern Forest Products Association, and others.

While the design and construction of relatively tall structures are technically achievable, a typical building code restricts the use of light-frame construction as well as heavy timber construction to low-rise buildings. Both light-frame and heavy-timber construction are classified as combustible materials. However, heavy-timber construction can be classified as fire resistant and can be built, according to most building codes, as high as five stories. Light-frame construction is typically limited to four stories. There is, however, a move to allow a higher number of stories for light-frame, provided fire suppression systems, such as sprinklers, are included in the building.

10.2 Wood as a Material

Wood in general can be considered a relatively brittle material from a structural standpoint. Wood is a natural material that is *viscoelastic* and *anisotropic*, but is generally considered and analyzed as an *orthotropic* material. Wood has a cellular form that enhances the *hygroscopic* or affinity to water response. From a structural engineering point of view, wood can be considered as similar to over-reinforced concrete, and one might consider designing a wood building as building with overreinforced concrete members that are connected with ductile connections.

One needs to differentiate between wood and *timber*. In this chapter, wood is considered as small clear specimens that one might idealize as perfect material. On the other hand, timber comes in the sizes that are typically used in construction and has growth characteristics such as knots, slope of grain, splits and checks, and other characteristics due to the conditions under which the tree grew or the *lumber* was manufactured. These growth characteristics contribute to significant differences in performance between wood and timber. They also provide the inherent weakness in the material that the structural engineer or designer needs to be aware of.

Since wood is a natural material and is hygroscopic and viscoelastic, certain environmental end-use conditions affect the long-term performance of the material. Some of these variables include moisture content, dimensional stability, bending strength, stiffness, and load duration. Each of these variables will be dealt with individually.

Moisture content is one of the most important variables that a designer needs to consider when designing timber structures. Any inspection of wood buildings should include testing of the moisture content with standard moisture meters that are available on the commercial market. Moisture content is determined as the weight of water in a given piece of timber divided by the weight of the woody material within that member in an oven-dried state. Moisture content affects virtually all mechanical properties of timber. One can consider wood as being similar to a sponge; as the water is absorbed from the dry state, it enters the walls of the individual cells of the material. At some point these walls become saturated and any additional water is then stored in the lumen of the cell. The point at which the walls become saturated is known as the fiber saturation point. The fiber saturation point varies for different species, and even within a single species, but ranges somewhere between 23 and 35%. The average fiber saturation point is usually assumed to be 30% for most general applications.

Dimensional stability of timber is affected directly by the moisture content. One can assume a linear variation of shrinkage or swelling with changes of moisture content between the oven-dry state and the fiber saturation point. Once the fiber saturation point is reached, any additional increase in moisture content only adds water to the air space of the member and does not affect the dimensional portions of the member. One can estimate the dimensional stability or dimensional changes of an individual wood member using shrinkage coefficients that are available in resources such as the *Wood Handbook* (U.S. Department of Agriculture 1999). However, one can usually assume that shrinkage in the perpendicular-to-grain directions (either radial or tangential) may be significant, while shrinkage parallel-to-grain can be considered as negligible. This is why connections that are restrained by steel plates through significant depths (even as little as 12-in. members) cause the wood to split, because as the wood dries out it will shrink while the steel and the connectors holding the wood to the steel prevent it from moving.

Moisture content changes have effects in addition to simply changing the perpendicular-to-grain dimension. Restrained wood subjected to the cycling of moisture content, can cause distortion of the members. General issues of warping can include *checking* or *splitting* of the lumber due to restraint of the connections, or wood members with grain oriented perpendicular to each other can cause the member to *bow, warp,* or *cup.* Cup is often seen in lumber deckboards that are laid with the growth rings on the cross-section oriented such that the center of the tree is facing up. All distortional responses can have adverse effects from a structural standpoint, and many cause serviceability concerns. Due to the effects of dimensional change on the straightness and mechanical properties of timber, it is strongly recommended that designers specify, and require without substitution, the use of dry material for timber structures.

Bending strength of timber is affected by moisture content as well. Bending strength may increase as much as 4% for each 1% decrease in moisture content below the fiber saturation point. While the cross-section of the member will also decrease due to shrinkage, the strengthening effect of drying out is significantly larger than the effects of the reduction in cross-section. Stiffness also increases with a decrease in moisture content, and again, the effect of stiffening of the material is greater than the effect of the reduction in cross-section due to shrinkage. This is why it is recommended that timber structures be constructed with dry material that will remain dry during its service life. The less moisture content change within the material at the time it is constructed and used, the less the shrinkage that occurs, and the maximum bending and stiffness values can be used by the designer.

The final two variables to be considered in this section have to do with viscoelastic properties. Because wood, and therefore timber, is viscoelastic, it tends to creep over time and also has a duration of load effect on strength. *Creep* is the continued increase in deflection that occurs in the viscoelastic material that sustains a constant load. In wood that is dry when installed and remains dry during service, the creep effect may cause a 50% increase in elastic deflection over a period of 1 year. However, if the material is unseasoned at the time of installation and allowed to dry out in service, the deflections can increase by 100% over the elastic response. Finally, if the moisture content is cycled between various moisture contents during its lifetime, the deflections can be increased to as much as 200% over the initial elastic response. A clear difference between wood and timber is that wood is affected by moisture content at all times, while timber is only affected by moisture content in the stronger grade levels. This is why Select Structural grade lumber is affected more than Number 3 grade lumber, and the design manuals have a check for minimum strength before the moisture effect or wet service factor is applied.

Duration load is a variable that is directly accounted for in both the LRFD and ASD manuals. Load duration factor, C_D, is based on experimental results from the 1950s at the U.S. Forest Products Laboratory. This curve is often referred to as the Madison curve. It forms the basis of the provisions in the LRFD and ASD manuals. There is one significant difference in how load duration is dealt with in the two design methodologies. In LRFD, the reference time is set at a duration of between 5 and 10 min. This duration is associated with a C_D of 1.0, with longer durations having factors less than 1. The ASD manual uses a reference duration of 10 years, with an associated value of C_D of 1.0. All values of C_D for load durations shorter than 10 years are given values greater than 1.0 for ASD. The designer is responsible for changing the reference time, depending on which design methodology is used.

Finally, ductility and energy dissipation, which are important concepts for seismic design, are beneficial for most wood structures. Provided the structure's connections are detailed properly, timber structures have relatively high values of ductility and energy dissipation. Provided the connections yield, such as would occur for nail and small-diameter bolted connections, ductilities in the range of 4 to 8 are not uncommon. Energy dissipation, on the other hand, from a material standpoint is very low for wood.

Typically, the material damping that occurs in timber structures is less than 1%. However, the hysteretic energy dissipation of structural assemblies such as shear walls is very high. Equivalent viscous damping values for timber assemblies can range from a low of 15% to a high of more than 45% critical damping.

10.3 Seismic Performance of Wood Buildings

10.3.1 General

As noted above, wood frame structures tend to be mostly low rise (one to three stories, occasionally four stories). The following discussion of the seismic performance of wood buildings is drawn from several sources. In the next few paragraphs, we provide an overview of wood construction and performance drawn from FEMA 154 (1988), followed by a limited discussion of performance in specific earthquakes drawn from various sources.

Vertical framing may be of several types: stud wall, braced post and beam, or timber pole. Stud wall structures ("stick-built") are by far the most common type of wood structure in the United States and are typically constructed of 2-in. by 4-in. nominal wood members vertically set about 16 in. apart. These walls are braced by plywood or by diagonals made of wood or steel. Most detached single and low-rise multiple family residences in the United States are of stud wall wood frame construction. Post-and-beam construction is not very common and is found mostly in older buildings. These buildings usually are not residential, but are larger buildings such as warehouses, churches, and theaters. This type of construction consists of larger rectangular (6 in. by 6 in. and larger) or sometimes round wood columns framed together with large wood beams or trusses.

Stud wall buildings have performed well in past earthquakes due to inherent qualities of the structural system and because they are lightweight and low rise. Cracks in the plaster and stucco (if any) may appear, but these seldom degrade the strength of the building and are therefore classified as nonstructural damage. In fact, this type of damage dissipates a lot of the earthquake-induced energy. The most common type of structural damage in older buildings results from a lack of connection between the superstructure and the foundation. Houses can slide off their foundations if they are not properly bolted to the foundation, resulting in major damage to the building as well as to plumbing and electrical connections. Overturning of the entire structure is usually not a problem because of the low-rise geometry. In many municipalities, modern codes require wood structures to be bolted to their foundations. However, the year that this practice was adopted will differ from community to community and should be checked.

Another problem in older buildings is the stability of cripple walls. Cripple walls are short stud walls between the foundation and the first floor level (Figure 10.4). Often these have no bracing and thus may collapse when subjected to lateral earthquake loading (Figure 10.5). If the cripple walls collapse, the house will sustain considerable damage and may also collapse. This type of construction is generally found in older homes. Plywood sheathing nailed to the cripple studs may have been used to strengthen the cripple walls.

Garages often have a very large door opening in one wall with little or no bracing. This wall has almost no resistance to lateral forces, which is a problem if a heavy load such as a second story sits on top of the garage. Homes built over garages have sustained significant amounts of damage in past earthquakes, with many collapses. Therefore the house-over-garage configuration, which is found commonly in low-rise apartment complexes and some newer suburban detached dwellings, should be examined more carefully and perhaps strengthened.

10.3.2 1971 San Fernando Earthquake, California

The San Fernando Earthquake occurred on February 9, 1971 and measured 6.6 on the Richter scale. The following commentary is excerpted from Yancey et al. (1998):

There were approximately 300,000 wood-frame dwellings in the San Fernando Valley of which about 5% were located in the region of heaviest shaking (Steinbrugge et al. 1971). A survey of

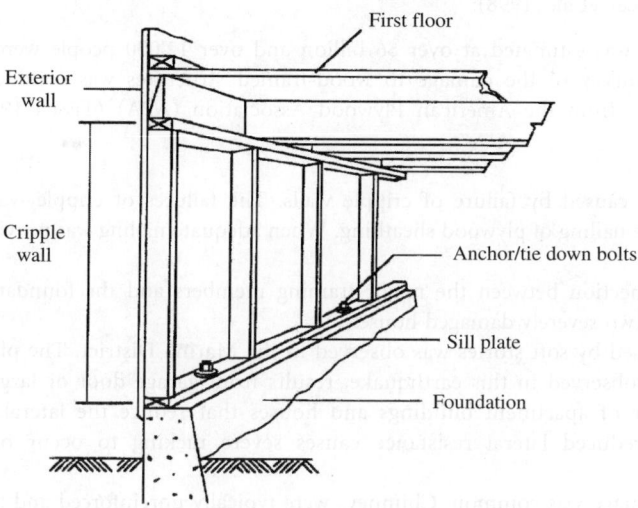

FIGURE 10.4 Cripple wall (courtesy of Benuska, L., Ed. 1990. *Earthquake Spectra*, 6 [Suppl.]).

FIGURE 10.5 Houses damaged due to cripple wall failure, 1983 Coalinga Earthquake (courtesy of EQE International).

12,000 single-family wood-frame houses was conducted by the Pacific Fire Rating Bureau (Steinbrugge et al. 1971). Most of the dwellings were constructed within the two decades prior to the earthquake. Typical types of foundations were either slab on-grade or continuous concrete foundation around the perimeter with concrete piers in the interior, with the former being more common. The majority of the houses were single-story. The survey showed that within the region of most intense shaking, 25% of the wood-frame dwellings sustained losses greater than 5% of the dwelling's value, with the remainder sustaining smaller losses. The number of houses with damage above the 5% threshold is equivalent to 1% of all the wood-frame dwellings in the San Fernando Valley.

10.3.3 1989 Loma Prieta Earthquake, California

The Loma Prieta Earthquake occurred on October 17, 1989, in the San Francisco Bay region and measured 7.1 on the Richter scale (Lew 1990). The following general observation and commentary are excerpted from Yancey et al. (1998):

> Property damage was estimated at over $6 billion and over 12,000 people were displaced from their homes. A survey of the damage to wood-framed structures was conducted by a group of three engineers from the American Plywood Association (APA) (Tissell 1990). Their main findings were

> 1. Damage was caused by failure of cripple walls. The failures of cripple walls were the result of inadequate nailing of plywood sheathing. When adequate nailing was provided, no failure was observed.
> 2. Lack of connection between the major framing members and the foundation was the cause of failure of two severely damaged houses.
> 3. Damage caused by soft stories was observed in the Marina District. The phenomenon of soft stories, first observed in this earthquake, results from garage door or large openings on the ground floor of apartment buildings and houses that reduce the lateral resistance of that story. The reduced lateral resistance causes severe racking to occur or increases lateral instability.
> 4. Chimney damage was common. Chimneys were typically unreinforced and not sufficiently tied to the structure.
> 5. Upward ground movements caused doors to be jammed and damage to basement floors.
> 6. Post-supported buildings were damaged because of inadequate connections of the floor to the post foundation and unequal stiffnesses of the posts due to unequal heights. Houses where the poles were diagonally braced were not damaged.

A particularly noteworthy concentration of damage was in the Marina section of San Francisco, where seven 1920s-era three- to five-story apartment buildings collapsed, and many were severely damaged (Figure 10.6). The excessive damage was due in large part to the man-made fill in the Marina, which liquefied and greatly increased ground motion accelerations and displacements during the shaking. However, the primary cause of the collapse was the soft-story nature of the buildings, due to required off-street parking. The buildings lacked adequate lateral-force-resisting systems and literally were a "house of cards."

10.3.4 1994 Northridge Earthquake, California

An earthquake with a magnitude of 6.8 struck the Northridge community in the San Fernando Valley on January 17, 1994. The effects of this earthquake were felt over the entire Los Angeles region. Approximately 65,000 residential buildings were damaged with 50,000 of those being single-family houses (U.S. Department of Housing and Urban Development 1995). The estimated damage based on insurance payouts was over $10 billion (Holmes and Somers 1996) for single- and multifamily residences.

City and county building inspectors estimated that 82% of all structures rendered uninhabitable by the earthquake were residential. Of these, 77% were apartments and condominiums, and the remaining 23% were single-family dwellings. A week after the earthquake, approximately 14,600 dwelling units were deemed uninhabitable (red or yellow tagged). Severe structural damage to residences was found as far away as the Santa Clarita Valley to the north, south-central Los Angeles to the south, Azusa to the east, and eastern Ventura County to the west.

10.3.4.1 Multifamily Dwellings

Particularly vulnerable were low-rise, multistory, wood-frame apartment structures with a soft (very flexible) first story and an absence of plywood shear walls. The soft first-story condition was most apparent in buildings with parking garages at the first-floor level (Figure 10.7). Such buildings, with large, often continuous, openings for parking, did not have enough wall area and strength to withstand

(a)

(b)

FIGURE 10.6 Collapsed apartment buildings in Marina district of San Francisco, 1989 Loma Prieta Earthquake (courtesy of EQE International).

FIGURE 10.7 Typical soft story "tuck-under" parking, with apartments above (courtesy of EQE International).

the earthquake forces. The lack of first-floor stiffness and strength led to collapse of the first floor of many structures throughout the valley. The main reason for failure was the lack of adequate bracing, such as plywood shear walls. Most older wood-frame structures had poor if any seismic designs and resisted lateral forces with stucco, plaster, and gypsum board wall paneling and diagonal let-in bracing.

10.3.4.2 Single-Family Dwellings

Widespread damage to unbolted houses and to older houses with cripple-stud foundations occurred. Newer houses on slab-on-grade foundations were severely damaged because they were inadequately anchored. Two-story houses without any plywood sheathing typically had extensive cracking of interior sheetrock, particularly on the second floor. Nine hillside houses built on stilts in Sherman Oaks collapsed. All but one of the homes were constructed in the 1960s — predating the major building code revisions made after the 1971 San Fernando Earthquake.

10.4 Design Considerations

Design considerations can be divided into essentially three categories:

1. Material choice
2. Performance requirements
3. Resistance determination

These are governed by the location of the project, the adopted building code regulations, and the design process of choice. Once the choice of designing a wood structure is made, the designer must also consider what types and what grades of wood products are readily available in the region of the project. Both LRFD and ASD include a wide variety of products, product sizes, and grades of products. Local building supply companies rarely stock all of the available products. The economics of building with wood is often the reason for choosing the material for a project, and the choice of products within the broad spectrum of available products that are locally available and stocked greatly improves the economics of a project. If products are specified that must be specially ordered or shipped long distances relative to what is already available in the market, the cost of the project will increase.

Performance requirements for a given building are usually determined by the building code that is enforced for the jurisdiction. However, there are also the local amendments and local conditions that must be considered when determining the performance requirements of a given project. Resistance determination and sizing of lumber for a given project are governed by the choice of design methodology used. If LRFD is used, one set of load combinations from the building code or ASCE 7 is required. If ASD is used, another set of load combinations should be used.

10.4.1 Building Code Loads and Load Combinations

Building codes that are in effect in the United States reference the ASCE load standard (American Society of Civil Engineers 2002) to provide guidance to the engineer on which load combination is to be used. The 2003 IBC and the NFPA 5000 Building Code both reference the ASCE 7–02 load standard. The following load combinations are to be considered when designing wood structures using the LRFD design methodology:

$$
\begin{aligned}
&1.\ 1.4(D+F) \\
&2.\ 1.2(D+F+T) + 1.6(L+H) + 0.5(L_r \text{ or } S \text{ or } R) \\
&3.\ 1.2D + 1.6(L_r \text{ or } S \text{ or } R) + (L \text{ or } 0.8W) \\
&4.\ 1.2D + 1.6W + L + 0.5(L_r \text{ or } S \text{ or } R) \\
&5.\ 1.2D + 1.0E + L + 0.2S \\
&6.\ 0.9D + 1.6W + 1.6H \\
&7.\ 0.9D + 1.0E + 1.6H
\end{aligned}
\qquad (10.1)
$$

where D is the dead load, E is the earthquake load, F is the load due to fluids with well-defined pressures and maximum heights, H is the load due to lateral earth pressure, groundwater pressure, or pressure of bulk materials, L is the live load, L_r is the roof live load, R is the rain load, S is the snow load, T is the self-straining force, and W is the wind load.

If the designer is using ASD, the ASCE 7–02 standard provides different load combinations for use. The following load combinations should be used when designing timber structures following ASD:

$$
\begin{aligned}
&1.\ D + F\\
&2.\ D + L + F + H + T\\
&3.\ D + H + F + (L_r \text{ or } S \text{ or } R)\\
&4.\ D + H + F + 0.75(L + T) + 0.75(L_r \text{ or } S \text{ or } R)\\
&5.\ D + H + F + (W \text{ or } 0.7E)\\
&6.\ D + H + F + 0.75(W \text{ or } 0.7E) + 0.75L + 0.75(L_r \text{ or } S \text{ or } R)\\
&7.\ 0.6D + W + H\\
&8.\ 0.6D + 0.7E + H
\end{aligned}
\tag{10.2}
$$

Since components within a timber structure will be stressed to their capacity during a design seismic event, it is strongly recommended that designers use the LRFD format when considering seismic performance. This is the reason why the NEHRP provisions require that LRFD be used.

10.5 Resistance Determination

The *Load and Resistance Factor Design (LRFD) Manual for Engineered Wood Construction* (American Forest and Paper Association 1996) distinguishes between nominal design values for visually and mechanically graded lumber connections and has supplements to provide guidance for all other wood-based products. This document provides both the reference design mechanical property and the applicable adjustment factors to account for end-use and environmental conditions. The reference properties that are included in the document are F_b, bending strength, F_t, tension parallel-to-grain strength, F_s, shear strength, F_c, the compression parallel-to-grain strength, $F_{c\perp}$, the compression strength perpendicular-to-grain, and E, the modulus of elasticity. Two values for modulus of elasticity are typically provided in the supplements: the mean and fifth percentile values. It should be stressed that the resistance values provided in the LRFD manual should not be mixed or combined with values obtained from the ASD manuals. The two design methodologies use different reference conditions on which to base their design, and mixing the values may result in nonconservative designs. Similar resistance values for the mechanical properties of wood-based materials for ASD are provided in the *ASD Manual for Engineered Wood Construction* (American Forest and Paper Association 2001). The significant difference between ASD and LRFD is that ASD bases the design values for a normal duration of load of 10 years, while LRFD resistance values are based on a 5- to 20-min load duration. Since wood is a viscoelastic material, the duration of load has a significant effect on the strength and deflections of timber structures. Two exceptions for adjustment of mechanical properties for load duration are modulus elasticity and compression perpendicular-to-grain. Both values are based on mean mechanical properties and are not adjusted for load duration since compression perpendicular-to-grain is not associated with fracture-type failure and deflection is considered a serviceability rather than a safety criterion.

Since wood is a natural material that is affected by environmental conditions and has characteristics inherent to the material that cause size or volume effects, several adjustments are made to the design values to account for these variables. Reference strength is adjusted with factors that include the time effect factor (load duration factor), wet service factor, temperature factor, instability factor, size factor, volume factor, flat use factor, incising factor, repetitive member factor, curvature factor, form factor, calm stability factor, shear stress factor, buckling stiffness factor, and bearing area factor. While many of these variables affect virtually all members of design, most factors only affect special situations. Therefore, the more commonly used factors will be covered here, and the reader is directed to either the LRFD or ASD manuals for a full description of the less used factors. In addition to these, the LRFD specification includes adjustment factors for preservative treatments and fire retardant treatments. The values that should be used for these last two variables should be obtained from the supplier of the products, since each product used is proprietary in nature and each treatment affects the performance of the timber differently.

Values of the adjustments are all equal to 1.0 unless the application does not meet the reference conditions. Both LRFD and NDS use the following reference conditions as the basis:

1. Materials are installed having a maximum equilibrium moisture content (EMC) not exceeding 19% for solid wood and 16% for glued products.

2. Materials are new (not recycled or reused).
3. Members are assumed to be single members (not in a structural system, such as a wall or a floor).
4. Materials are untreated (except for poles or piles).
5. The continuous ambient temperature is not higher than 100°F, with occasional temperatures as high as 150°F. (If sustained temperatures are between 100°F and 150°F, adjustments are made. Timber should not be used at sustained temperatures higher than 150°F.)

The effect of load duration will be discussed later in this section.

The reference design values are adjusted for conditions other than the reference conditions using adjustment factors. The adjustment factors are applied in a cumulative fashion by multiplying the published reference design value by the appropriate values of the adjustments. The equation for making the adjustments is

$$R' = R \cdot C_1 \cdot C_2 \cdot C_3 \cdots C_n \tag{10.3}$$

where R' is the adjusted design value for all conditions, R is the tabulated reference design resistance, and $C_1, C_2, C_3, \ldots, C_n$ are the applicable adjustment factors.

Most factors for adjusting for end use are common between the LRFD and ASD methodologies, and many are a function of the species of lumber, strength grade, or width of lumber. Most adjustment factors are provided either by the LRFD or by the ASD manuals. However, some adjustments are associated with proprietary products (chemical treatments) and must be obtained from the product supplier.

Since wood is a viscoelastic material and therefore affected by time, the load duration variable is of particular interest to designers. The LRFD and ASD design methodologies handle the time effects or load duration differently. Regardless of which design methodology is used, the shortest duration load in the load combination will determine the value of the time effect factor. This is because the failure of timber is governed by a creep-rupture mechanism. This implies that timber can sustain higher magnitude loads for short periods of time. In the LRFD design methodology, the time effect factor λ is based on the load combinations considered and the design methodology specifically defines a time effect factor for each combination. These combinations, along with associated time effect factors, are shown in Table 10.1. The reference load duration for LRFD is 5 to 10 min, and this duration is given a value of 1.0. All others follow the Madison curve for load duration, and therefore longer duration loads have time effect factors that are less than 1.0.

The ASD design methodology assumes a load duration of 10 years accumulative, at the design level. Therefore, the load duration factor used in the ASD has a value of 1.0 for normal duration loads, which is 10 years. The shortest-duration load included in a given load combination determines which load duration factor is applicable. All load combinations with individual loads that have a cumulative duration of less than 10 years have load duration factor values greater than 1.0. The loads associated with various load duration factors for ASD are provided in Table 10.2.

Therefore, the adjusted values for either design methodology are determined essentially following the same process, only using the appropriate values. First, the reference design values are obtained from the appropriate table and then adjusted for the end-use conditions, size effects, etc. by multiplying the reference value by a string of adjustment factors. The final adjustment is for the time effect, and

TABLE 10.1 LRFD Load Combinations and Time Effect Factors

LRFD load combination	Time effect factor
$1.4D$	0.6
$1.2D + 1.6L + 0.5(L_r$ or S or $R)$	0.7 for L representing storage
	0.8 for L representing occupancy
	1.25 for L representing impact
$1.2D + 1.6(L_r$ or S or $R) + (0.5L$ or $0.8W)$	0.8
$1.2D + 1.3W + 0.5L + 0.5(L_r$ or S or $R)$	1.0
$1.2D + 1.0E + 0.5L + 0.2S$	1.0
$0.9D - (1.3W$ or $1.0E)$	1.0

TABLE 10.2 ASD Design Loads and Associated Load
Duration Factors

Design load type	Load duration factor value
Dead load	0.9
Occupancy live load	1.0
Snow load	1.15
Construction load	1.25
Wind or seismic load	1.6
Impact load	2.0

the corresponding time effect factor is determined depending on which load combination is being considered.

10.5.1 Bending Members

The most common application for sawn lumber, LVL, glulam timber structural composite lumber, and I-joists is to resist bending forces. In fact, since it is the most common application for dimensional timber members, the visual and machine stress-rated grading rules for lumber are developed around the concept that the members will be placed in bending about their strong axis. In this application, the bending members must account for size effects, duration load, and end-use conditions for the structure. Many times, load distribution elements such as sheathing on floors and walls can provide load sharing between bending members, which can be accounted for in design by use of the repetitive member factor (ASD) or load-sharing factor (LRFD).

Beams must be designed to resist moment, bending shear, bearing, and deflection criteria. The moment, shear, and bearing criteria are all designed using similar formats. The design format for LRFD is

$$L_u \leq \lambda \theta R' \qquad (10.4)$$

where L_u is the resultant of the actions caused by the factored load combination, λ is the time effect factor associated with the load combination being considered, θ is the resistance factor (flexure $= 0.85$, stability $= 0.85$, shear $= 0.75$), and R' is the adjusted resistance.

The moment, shear, or bearing force being considered is determined using engineering mechanics and structural analysis. The assumption of linear elastic behavior is typically made for this analysis and is appropriate considering that timber behaves like a brittle material. Nonlinear analysis is acceptable; however, experimental data will probably be required to support such an analysis.

The resistance value involves consideration of the lateral support conditions and load-sharing conditions, and whether the member is or is not part of a larger assembly. The adjusted resistance is obtained by multiplying the published reference strength values for bending, shear, or compression perpendicular-to-grain by the appropriate adjustment factor. The adjustment factor includes the end use and adjustments based on product. The end-use adjustment factors include the wet service factor (C_M), temperature (C_t), preservative treatment factor (C_{pt}), and fire retardant treatment (C_{rt}). Adjustments for member configuration include composite action (C_E), load-sharing factor (C_r), size factor (C_F), beam stability factor (C_L), bearing area (C_b), and form factor (C_f). Additional adjustments for structural lumber and glulam timber include shear stress (C_H), stress interaction (C_I), buckling stiffness (C_T), volume effect (C_V), curvature (C_c), and flat use (C_{fu}). The form of the equation for determining the adjusted resistance is

$$R' = (G) \times F \times C_1 \cdot C_2 \cdot C_3 \cdots C_n \qquad (10.5)$$

where R' is the adjusted resistance (e.g., adjusted strong-axis moment, adjusted shear), G is a geometry variable consistent with the resistance calculated (e.g., S_X), F is the reference design strength (e.g., F_b for

TABLE 10.3 Load-Sharing Factor, C_r, Values

Bending member product type	C_r
Dimensional lumber	1.15
Structural composite lumber	1.04
Prefabricated I-joists with visually graded lumber	1.07
Prefabricated I-joists with structural composite lumber flanges	1.04

bending strength, F_v for shear strength, F_c for perpendicular-to-grain compression strength), and C_i is the appropriate adjustment factor(s).

The reader is referred to the LRFD or ASD manuals for a full description of the adjustment factors and the values associated with the various conditions. Adjustment factors account for conditions that do not meet the reference conditions. For example, when lumber is being used for bending about its weak axis, the flat-use factor will be used, or if the beam is not fully supported against lateral movement along the compression edge, the beam stability factor C_L will be required.

One of the most commonly used variables for adjusting a bending strength is the load-sharing factor C_r, which is a variable that provides an increase in design bending resistance that accounts for the load sharing that occurs when beams are used in parallel. The condition under which the load-sharing factor is applicable and causes an effective increase in the bending resistance occurs when the spacing between beams is no more than 610 mm (24 in.) at the center and the beams are connected together using a load-distributing element, such as structural wood panel sheathing or lumber decking. This factor essentially accounts for the system effects that an effective connection between parallel beams provides for floor, roof, and wall systems. Because of the strong correlation between the variability of a given product and the magnitude of the system effect, the load-sharing factor has different values depending on the products used for the beams. Table 10.3 provides the load-sharing factors associated with the more common products used as beams in timber construction. The load-sharing factor C_r applies only to the moment resistance and is not applicable to any of the other design resistances.

When shear strength is being checked, the shear stress adjustment factor C_H may be used to increase the shear resistance. However, this is not recommended for general design because at the design stage, the designer is unable to guarantee that any given piece of dimensional lumber used as a beam will not have checks or splits on the wide face of the lumber. In addition, the designer is not able to guarantee that the member will not split during the lifetime of the structure. It is recommended that the shear stress factor be used only when analyzing existing structures, when inspection has been done to determine the length of the splits that are present.

10.5.2 Axial Force Members

Axial force members are considered to be either tension or compression members and are handled with single equations in both cases. Compression members, that is, columns, have been historically classified as short, intermediate, and long columns depending on whether material crushing or Euler elastic buckling was the controlling mechanism of failure. With the widespread use of computers, the ability to program complex equations has eliminated the need for the three classifications of columns.

10.5.2.1 Compression Members

Compression members are usually called columns, although the term may include drag struts and truss members. There are essentially three basic types of columns. The most common one is the solid or traditional column, which consists of a single member, usually dimensional lumber, post and timbers, poles and piles, or glulam timber. The second type of column is the spaced column, which is made of two or more parallel single-member columns that are separated by spacers, located at specific locations along the column and rigidly tied together at the ends of the column. The third type of column is the built-up column, which consists of two or more members that are mechanically fastened together, such as multiple nailed studs within a wall supporting a girder.

The slenderness ratio defines the primary mechanism of failure. Shorter columns obviously will be controlled by the material strength of the wood parallel-to-grain, while longer columns will be controlled by Euler buckling. The slenderness ratio is defined as the ratio of the effective length of the column, l_e, to the radius of gyration, which is

$$r = \frac{I}{A} \tag{10.6}$$

where I is the moment of inertia about the weak axis and A is the cross-sectional area. The effective length, l_e, is determined by multiplying the unbraced length by the buckling length coefficient:

$$l_e = K_e * l \tag{10.7}$$

where K_e is the buckling length coefficient, l is the unbraced length of the column. K_e is dependent on the end support conditions of the column and whether side sway of the top or bottom of the column is restrained or not. Theoretical values for ideal columns and empirical values for the buckling length coefficient, K_e, that are recommended by NDS and LRFD manuals are provided in Table 10.4. The LRFD and NDS specifications require that the maximum permitted slenderness ratio be 175.

The LRFD design equation that must be satisfied for solid columns has a similar form to the other equations:

$$P_u \leq \lambda \phi_c P_c' \tag{10.8}$$

where P_u is the compressive force due to the factored load combination being considered, λ is the applicable time effect factor for the load combination being considered, ϕ_c is the resistance factor for compression parallel-to-grain (0.9), and P_c' is the adjusted compressive resistance parallel-to-grain.

The adjusted compressive resistance P_c' is determined as the gross area times the adjusted compressive strength parallel-to-grain, F_c'. The adjusted compressive strength is determined in the same manner as all mechanical properties of wood, which is to multiply the reference compressive strength parallel-to-grain by all the applicable adjustment factors, such as duration load, temperature, etc. The one additional factor for columns is that the column stability factor C_p must be calculated. C_p accounts for the partial lateral support provided to a column and is determined by

$$C_p = \frac{1 + \alpha_c}{2c} \sqrt{\left(\frac{1 + \alpha_c}{2c}\right)^2 - \frac{\alpha_c}{c}} \tag{10.9}$$

where

$$\alpha_c = \frac{\phi_s P_e}{\lambda \phi_c P_0'} \tag{10.10}$$

$$P_e = \frac{\pi^2 E_{05}' I}{(K_e l)^2} = \frac{\pi^2 E_{05}' A}{[K_e(l/r)]^2} \tag{10.11}$$

and c is a coefficient based on the variability of material being used for the column ($c = 0.8$ for dimensional lumber, 0.85 for round poles and piles, and 9.0 for glulam members and structural

TABLE 10.4 Buckling Length Coefficient for Compression in Wood Column Design

End support conditions	Side-sway restraint	Theoretical coefficients	Empirical recommended coefficients
Fixed–fixed	Restrained	0.5	0.65
Pin–fixed	Restrained	0.7	0.8
Fixed–fixed	Free	1.0	1.2
Pin–pin	Restrained	1.0	1.1
Pinned–fixed	Free	2.0	2.1
Fixed–pinned	Free	2.0	2.4

composite lumber). ϕ_s is the resistance factor for stability, which has a value of 0.85, E'_{05} is the adjusted modulus of elasticity at the fifth percentile level, A is the cross-sectional area, K_e is the effective length factor, and r is the radius of gyration.

If a prismatic column is notched at a critical location, then the factored compressive resistance is determined using the net section rather than the gross cross-section. In addition, C_p should be computed using the properties of the net area if the notches or holes are located in the middle half of the length between inflection points of the column and the net moment inertia is less than 80% of the gross moment of inertia or if the longitudinal dimension of the knot or hole is greater than the larger cross-sectional dimension of the column. If the notch is located in noncritical locations, then the adjusted compressive resistance can be computed using the gross cross-sectional area and C_p, or the net cross-sectional area times the factored compressive strength of the material.

Spaced columns are another common form of compression members that consist of two or more dimensional members connected together at a specific spacing using blocking to form a set of parallel columns that are restrained in a common manner. Figure 10.8 illustrates the concept of a spaced column. The general form of a spaced column has the ability to buckle in more than one manner. First, the overall column could buckle. Second, the individual members making up the column could buckle between the spacing blocks. Therefore, the design specifications restrict the geometry of spaced columns to prevent unexpected element buckling. The following maximum length to width ratios are imposed by the LRFD manual:

1. In a spaced column direction L_1/d_1 should not exceed 80.
2. In a spaced column direction l_3/d_1 should not exceed 40.
3. In solid column direction l_2/d_2 should not exceed 50.

Spaced columns that do not conform to these restrictions must be designed considering that each element within the column is acting as an independent column, unless a rational analysis can be used to account for the restraint conditions used.

Built-up columns are the final type of column. Built-up columns consist of two or more dimensional members that are mechanically fastened together to act as a single unit. Built-up columns can have the form of multiple studs nailed together to form a column to support a girder bearing on top of the wall or

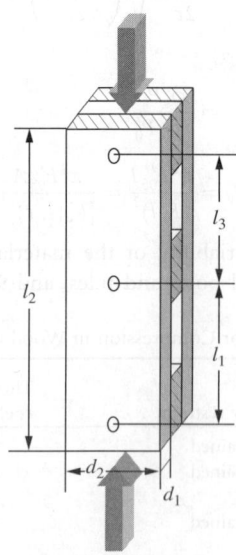

FIGURE 10.8   Illustration of variables associated with the design of spaced columns.

they may have the shape of a hollow column made of members that are nailed together to form the outer circumference of the shape. The capacity of this type of column can conservatively be estimated by considering each of the elements that make up the column as independent columns and adding their collective compressive strengths together. In all cases, the fasteners that connect the members together must be designed and spaced appropriately to transfer the shear and tension forces that occur within and between the members of the built-up column.

10.5.2.2 Tension Members

Compared to compression members, tension members are relatively easy to design and the adjustment factors for the factored resistance are all tabulated values and do not require independent calculations. The basic design equation for designing tension members in the LRFD manual has the form

$$T_u \leq \lambda \phi_t T' \tag{10.12}$$

where T_u is the tension force due to the load combination being considered, λ is the time effect factor corresponding to the load combination being considered, ϕ_t is the resistance factor for tension ($\phi_t = 0.8$), and T' is the adjusted tension resistance parallel-to-grain, which is calculated based on the net section and the adjusted tension strength. The adjusted tension strength parallel-to-grain, F'_t, is determined by multiplying the reference tension strength parallel-to-grain by the appropriate adjustment factors, which include temperature, size effects, and load duration, among others. None of the adjustment factors require independent calculations.

If the tension forces are imparted into the member such that the eccentricity between the connection centroid and the centroid of the member is greater than 5% of the member dimension, then the member must be designed considering combined loading of tension and bending.

10.5.3 Combined Loading

Combined loading for member design in wood structures consists of three different categories:

1. Biaxial bending
2. Bending plus tension
3. Bending plus compression

Biaxial bending is a subset of either bending plus tension or bending plus compression. Essentially, biaxial bending uses the same equations, only setting the axial force being considered in those equations equal to 0. The simplified equation takes the form

$$\frac{M_{ux}}{\lambda \phi_b M'_s} + \frac{M_{uy}}{\lambda \phi_b M'_y} \leq 1.0 \tag{10.13}$$

where M_u is the factored load moment for the load combination being considered, λ is the time effect factor for the load combination being considered, ϕ_b is the resistance factor for bending ($\phi_b = 0.85$), M'_s is the computed bending resistance moment about the x-axis with the beam stability factor $C_L = 1.0$, and any volume factor, C_v (used when designing glulam members) is included. M'_y is the adjusted moment resistance about the weak axis considering the lateral bracing conditions.

Combined bending and axial tension is the second easiest combination to calculate for combined loading. However, there are two conditions that must be considered: first, the condition along the tension face, for which lateral stability is not a concern and second, the condition along the compression face, for which lateral stability is of concern. Therefore, the following two equations must be satisfied for designing members with combined bending and axial tension:

Tension face:

$$\frac{T_u}{\lambda \phi_t T'} + \frac{M_{ux}}{\lambda \phi_b M'_s} + \frac{M_{uy}}{\lambda \phi_b M'_y} \leq 1.0 \tag{10.14}$$

Compression face:

$$\frac{(M_{ux} - (d/6)T_u)}{\lambda\phi_b M'_x} + \frac{M_{uy}}{\lambda\phi_b M'_y(1 - (M_{ux}/\phi_b M_e))^2} \leq 1.0 \tag{10.15}$$

where T_u is the tension force due to the factored load combination being considered, M_e is the elastic lateral buckling moment of the member, and d is the member depth.

If Equation 10.15 is used for nonrectangular cross-sections, the variable $d/6$ should be replaced by the ratio of the strong axis section modulus to the gross cross-sectional area, S_x/A. When the member is being designed for combined bending and compression, the following equation must be satisfied for the design conditions being considered:

$$\left(\frac{P_u}{\lambda\phi_c P'}\right) + \frac{M_{mx}}{\lambda\phi_b M'_x} + \frac{M_{my}}{\lambda\phi_b M'_y} \leq 1.0 \tag{10.16}$$

where P_u is the axial compressive force due to the factored load combination being considered; P' is the adjusted resistance for axial compression parallel-to-grain acting alone for the axis of buckling providing the lower buckling strength; M_{mx} and M_{my} are the factored moment resistances, including any magnification for second-order effects for strong and weak axis bending, respectively; and M'_x and M'_y are the adjusted moment resistances for the strong and weak axes, respectively, from multiplying the section properties by the nominal bending resistance and appropriate adjustment factors.

The moments M_{mx} and M_{my} may be determined using second-order analysis for a simplified magnification method that is outlined in the design manual. The extent of the simplified method compels us to refer the reader to the design specification rather than including it here.

10.6 Diaphragms

When resisting lateral loads such as seismic forces, most light-frame wood buildings can be conceptualized as a box system. Forces are transmitted horizontally through diaphragms (i.e., roofs and floors) to reactions that are provided by the shear walls at the ends of the diaphragms. These forces are in turn transmitted to the lower stories and finally to the foundations. Some designers consider shear walls as vertical diaphragms, since the reaction to loading is similar to half of a diaphragm. However, this chapter will consider shear walls as separate elements.

A *diaphragm* is a structural unit that acts as a deep beam or girder that may or may not be able to transfer torsional loads, depending on the relative stiffness of the diaphragm and supporting shear walls. The analogy to a girder is somewhat more appropriate because girders and diaphragms can be made as assemblies. The sheathing acts as a web that is assumed to resist all of the shear loads applied to the diaphragm, and the framing members at the boundaries of the diaphragm are considered to act as flanges to resist tension and compression forces due to moment. The sheathing is stiffened by intermediate framing members to provide support for gravity loading and to transfer the shear load from one sheathing element to the adjacent sheathing element. Chords act as flanges and often consist of the top plates of the walls, ledgers that attach the diaphragm to concrete or masonry walls, bond beams that are part of the masonry walls, or any other continuous element at the parameter of the diaphragm. The third element of a diaphragm is the struts. Struts provide the load transfer mechanism from the diaphragm to the shear walls at the ends of the diaphragm and act parallel to the loading of the diaphragm. Chords act to resist internal forces acting perpendicular to the general loading in the direction of the diaphragm.

The chord can serve several functions at the same time, providing resistance to loads and forces from different sources and functioning as the tension or compression flange of the diaphragm. It is important that the connection to the sheathing be designed to accomplish the shear transfer since most diaphragm chords consist of many pieces. It is important that splices be designed to transmit the tension or compression occurring at the location of the splice. It is also important to recognize that the direction of application of the seismic forces will reverse. Therefore, it should be recognized that chords need to be designed for equal magnitude torsion and compression forces. When the seismic forces are acting at 90° to the original direction analyzed, the chords act as the struts for the diaphragm to transfer the reaction loads to the shear walls below. If the shear walls are not continuous along the length of the diaphragm, then the strut may act as the drag strut between the segments of the shear wall as well. Diaphragms often have openings to facilitate stairwells, great rooms that are more than one story high, or access to roof systems and ventilation. The transfer of forces around openings can be treated in a manner similar to the transfer around openings in the webs of steel girders. Members at the edges of the openings have forces due to flexure and the high web shear induced in them, and the resulting forces must be transferred into the body of the diaphragm beyond the opening.

In the past, wood sheathed diaphragms have been considered to be flexible by many registered design professionals and most code enforcement agencies. Recent editions of the model building codes recognize that diaphragms have a stiffness relative to the walls that are supporting them, and this relative rigidity determines how the forces will be distributed to the vertical resistance elements.

10.6.1 Stiffness versus Strength

Often in large diaphragms, as an economic measure, the designer will change the nailing or blocking requirements for the diaphragm to remove the high nail schedule or blocking requirements in the central section of the diaphragm where the shear loading is lowest. It is therefore imperative that the designer distinguish between the stiffness of the diaphragm and the strength of the diaphragm. However, at locations where blocking requirements change or the nail schedule is increased, the stiffness of the diaphragm also goes through a significant change. These locations result in potential stress concentrations and when a nail schedule or blocking schedule is changed, the designer should consider ensuring that the change occurs at locations where the loads are not simply equal to the resistance, but sufficiently below the changing resistance at that location. Timber-framed diaphragms and structures with light-frame shear walls are capable of being relatively rigid compared to the vertical resistance system. Added stiffness due to concrete toppings, blocking, and adhesives also makes the relative stiffness greater. Therefore, determination of the stiffness of the diaphragm relative to the vertical resistance system must be considered in the design to determine whether a flexible or rigid analysis is required.

The equation used to estimate the deflection of diaphragms was developed by APA and can be found in several references (Applied Technology Council 1981; APA 1997; National Fire Prevention Association 2002). The midspan deflection of a simple-span, blocked, structural panel sheathed diaphragm, uniformly nailed throughout, can be calculated using

$$\Delta = \frac{5vl^3}{8wEA} + \frac{vl}{4Gt} + 0.188\, le_n + \frac{\sum(\Delta_c X)}{2w} \tag{10.17}$$

where Δ is the calculated deflection, v is the maximum shear due to factored design loads in the direction under consideration, l is the diaphragm length, w is the diaphragm width, E is the elastic modulus of chords, A is the area of chord cross-section, Gt is the panel rigidity through the thickness, e_n is the nail deformation, and $\sum(\Delta_c X)$ is the sum of individual chord splice slip values on both sides of the diaphragm, each multiplied by its distance to the nearest support.

If the diaphragm is not uniformly nailed, the constant 0.188 in the third term must be modified accordingly. Guidance for using this equation can be found in ATC 7 (Applied Technology Council 1981). This formula was developed based on engineering principles and modified by testing. Therefore, it provides an estimate of diaphragm deflection due to loads applied in the factored resistance shear range. The effects of cyclic loading and energy dissipation may alter the values for nail deformation in the third term, as well as the chord splice effects in the fourth term, if mechanically spliced wood chords are used. The formula is not applicable to partially blocked or unblocked diaphragms.

Recent research, part of the wood frame project of the Consortium of Universities for Research in Earthquake Engineering (CUREE), was conducted as part of the shake table tests by Fischer et al. (2001) and Dolan et al. (2002). These new studies provide deflection equations that are broken into two components: one for bending deflection and one for shear deflection. The tests used nailed chord splices. If other types of splices are utilized in a design, additional terms to account for the deformation of the splice effects on the diaphragm need to be added. In addition, the CUREE equations are useful for working stress level loads and have not been validated for deflections approaching those associated with the capacity of the diaphragm. However, the equations are derived based on cyclic tests and the average cyclic stiffness within the reasonable deflection range.

The provisions are based on assemblies having energy dissipation capacities that were recognized in setting the *R*-factors included in the model building codes. For diaphragms utilizing timber framing, the energy dissipation is almost entirely due to nail bending. Fasteners other than nails and staples have not been extensively tested under cyclic load applications. When screws or adhesives have been tested in assemblies subjected to cyclic loading, they have had brittle failures in adhesives and provided minimal energy dissipation. For this reason, adhesives have been prohibited in light-frame shear wall assemblies in high seismic regions. However, the deformation range typically experienced by diaphragms during seismic events has not justified the restriction of using adhesives in the horizontal diaphragms. In fact, the addition of adhesives in most timber diaphragms provides significantly more benefits in the form of higher strength and stiffness to distribute loads to the horizontal members more efficiently. While in the Dolan et al. (2002) diaphragm study, the adhesives by themselves did not have as large an effect on stiffness as blocking, the use of adhesives provided a more uniform deflection pattern and enhanced the behavior of the diaphragm over a nailed-only diaphragm.

10.6.2 Flexible versus Rigid Diaphragms

The purpose of determining whether a diaphragm is flexible or rigid is to determine whether a diaphragm should have the loads proportioned according to the tributary area or the relative stiffness of the supports. For flexible diaphragms, the loads should be distributed according to the tributary area, whereas for rigid diaphragms, the load should be distributed according to the stiffness. The distribution of seismic forces to the vertical elements of the lateral force resistance system is dependent, first, on the relative stiffness of the vertical elements versus the horizontal elements and, second, on the relative stiffness of the vertical elements when they have varying deflection characteristics. The first issue defines when a diaphragm can be considered flexible or rigid. In other words, it sets limits on whether the diaphragms can act to transmit torsional resistance or cantilever. When the relative deflections of the diaphragm and shear walls are determined at the factored load resistance level, and the midspan deflection of the diaphragm is determined to be more than two times the average deflection of the vertical resistant elements, the diaphragms may be considered as being flexible. Conversely, a diaphragm should be considered rigid when the diaphragm deflection is equal to or less than two times the shear wall drift. Obviously, the performance of most diaphragms falls in a broad spectrum between perfectly rigid and flexible. However, at the current time, there are no design tools available to provide for analyzing diaphragms in the intermediate realm. Therefore, model building codes simply differentiate between the two extreme conditions.

The flexible diaphragm seismic forces should be distributed to the vertical resisting elements according to the tributary area and simple beam analysis. Although rotation of the diaphragm may occur because lines of vertical elements have different degrees of stiffness, the diaphragm is not considered sufficiently stiff to redistribute the seismic forces through rotation. The diaphragm may be visualized as a single-span beam supported on rigid supports in this instance.

For diaphragms defined as rigid, rotational or torsional behavior is expected and the action results in a redistribution of shear to the vertical force-resisting elements. Requirements for horizontal shear distribution involve a significantly more detailed analysis of the system than the assumption of flexibility. Torsional response of a structure due to an irregular stiffness at any level within the structure can be the potential cause of failure in the building. As a result, dimensional and diaphragm aspect ratio limitations are imposed for different categories of construction. Also, additional requirements are imposed on the diaphragm when the structure is deemed to have a general torsional irregularity such as when re-entering corners or when diaphragm discontinuities are present.

In an effort to form a frame of reference in which to judge when stiffness of the diaphragm may be critical to the performance of the building, one can consider two different categories of diaphragms. The first category includes rigid diaphragms that must rely on torsional response to distribute the loads to the building. A common example would be an open front structure with shear walls on three sides, such as a strip mall. This structurally critical category has the following limitations:

1. The diaphragm may not be used to resist the forces contributed by masonry or concrete in structures over one story in height.
2. The length of the diaphragm normal to the opening may not exceed 25 ft, and the aspect ratios are limited to being less than 1:1 for one-story structures or 1:1.5 for structures over one story in height. Where calculations show that the diaphragm deflections can be tolerated, the length will be permitted to increase so as to allow aspect ratios not greater than 1.5:1 when the diaphragm is sheathed with structural use panels.

The aspect ratio for diaphragms should not exceed 1:1 when sheathed with diagonal sheathing.

The second category of rigid diaphragms that may be considered is those that are supported by two or more shear walls in each of the two perpendicular directions, but have a center of mass that is not coincident with the center of rigidity, thereby causing a rotation in the diaphragm. This category of diaphragm may be divided into two different categories where category 2a would consist of diaphragms with a minimal eccentricity that may be considered on the order of incidental eccentricities, in which case the following restrictions would apply:

1. The diaphragm may not be used to resist forces contributed by masonry or concrete in structures over one story.
2. The aspect ratio of the diaphragm may not exceed 1:1 for one-story structures or 1:1.5 for structures greater than one story in height.

On the other hand, flexible diaphragms or nonrigid diaphragms have minimal capacity for distributing torsional forces to the shear walls. Therefore, limitations of aspect ratios are used to limit diaphragm deformation such that reasonable behavior will occur. The resulting deformation demand on the structure is also limited such that higher aspect ratios are allowed provided calculations demonstrate that the higher diaphragm deflections can be tolerated by the supporting structure. In this case, it becomes important to determine whether the diaphragm rigidity adversely affects the horizontal distribution and the ability of the other structural elements to withstand the resulting deformations.

Several proposals to prohibit wood diaphragms from acting in rotation have been advanced following the 1994 Northridge earthquake. However, the committees that have reviewed the reports to date have concluded that the collapses that occurred in that event were due in part to a lack of deformation compatibility between the various vertical resisting elements, rather than solely due to the inability of the diaphragm to act in rotation.

Often diaphragms are used to cantilever past the diaphragm's supporting structure. Limitations concerning diaphragms cantilevered horizontally past the outermost shear walls or other vertical support element are related to, but slightly different from, those imposed due to diaphragm rotation. Such diaphragms can be flexible or rigid, and the rigid diaphragms may be categorized in one of the three categories previously discussed. However, both the limitations based on the diaphragm rotation, if they are applicable, and a diaphragm limitation of not exceeding the lesser of 25 ft or two thirds of the diaphragm width must be considered in the design. This is due to the additional demand placed on a structure due to the irregularity resulting from the cantilever configuration of the diaphragm.

Further guidance on the design and detailing of diaphragms can be obtained from Breyer et al. (1998) and Faherty and Williamson (1999), both providing significant details on dealing with torsional irregularities of wood diaphragms.

10.6.3 Connections to Walls

When postevent reports are reviewed for many historical earthquakes, one finds that the principal location of failures in diaphragms occurs at the connection between the diaphragm and the supporting walls. Two of the most prevalent failures are due to cross-grain bending of ledger boards that are used to attach diaphragms to concrete and masonry walls, or to the use of the diaphragm sheathing to make the connection between the wall and the diaphragm. When one uses the ledger board to attach a wood diaphragm to a masonry or concrete wall, the diaphragm will tend to pull away from the wall, which causes cross-grain bending to occur in the ledger board. Cross-grain bending is not allowed in the design specifications for wood construction and is one of the weakest directions in which wood can be loaded. The same mechanism causes splitting of sill plates on shear walls due to the uplift forces of the sheathing.

When diaphragms are attached to walls, regardless of the construction of the wall, the connections must be made directly between the framing or reinforcement elements of the wall and the framing of the diaphragm. Straps should extend into the diaphragm a significant distance to provide adequate length in which to develop the forces experienced. Sheathing rarely provides sufficient capacity to transfer diaphragm forces and the sheathing nailing will inevitably fail. These straps also should extend down along the timber frame wall and be attached to the studs of the wall, or for masonry and concrete walls, be attached to reinforcement in the wall near the location.

10.6.4 Detailing around Openings

Openings in diaphragms may be designed similar to the way openings in steel girders are designed. After the bending and shear forces are determined, connections to transfer those loads can be designed. Construction of such connections often requires the use of blocking between the framing members and straps that extend past the opening, often for multiple joist spacings. It is imperative that connections to transfer such loads be made to the framing members and not simply to the sheathing. This is also an area which a designer should consider as requiring special inspection, to ensure that the nailing used to make the connection is actually located in the framing and not simply through the sheathing into air.

10.6.5 Typical Failure Locations

As stated earlier, the typical locations for failure of diaphragms are the connections between the diaphragm and the vertical force-resisting elements. However, additional areas need to be considered. These include openings in the diaphragm, where shear transfer is being designed, and locations where nail schedules or blocking requirements change. Often openings are placed in diaphragms without the design considering the force transfer required due to the absence of sheathing. This location also results in a difference or an anomaly in the stiffness distribution along the length of the diaphragm. These two actions combine to cause the sheathing nailing, and possibly the framing nailing, to separate, thus causing local failures.

10.7 Shear Walls

Shear walls are typically used as the vertical elements in the lateral-force-resisting system. The forces are distributed to the shear walls of the building by the diaphragms and the shear walls transmit the loads down to the next lower story or foundation. A shear wall can be defined as a vertical structural element that acts as a cantilever beam where the sheathing resists the shear forces and the end studs, posts, or chords act as the flanges of the beam in resisting the induced moment forces associated with the shear applied to the top of the wall. The sheathing is stiffened by the intermediate studs and is therefore braced against buckling. An exploded view of a typical light-frame shear wall is shown in Figure 10.9.

Light-frame shear walls can be divided into two categories: designed and prescriptive. Designed walls are sized and configured using a rational analysis of the loads and design resistance associated with the

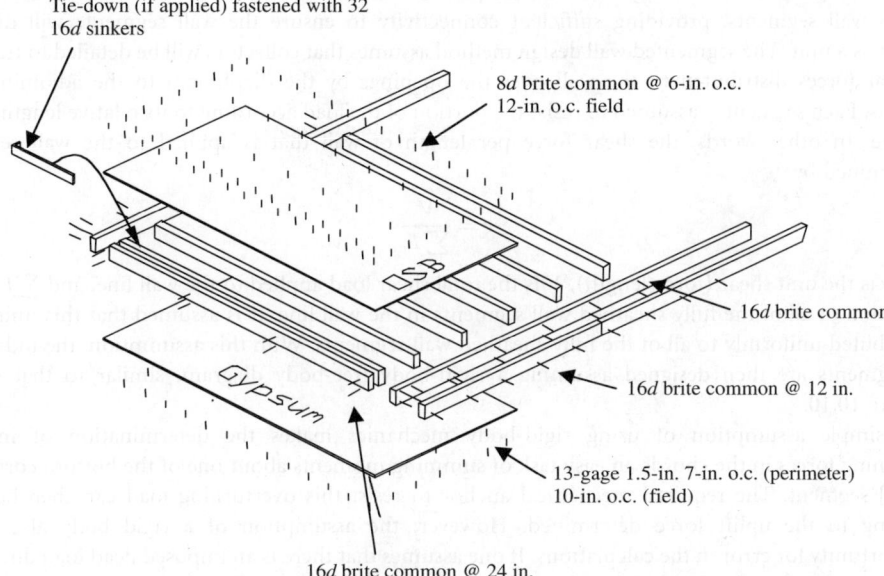

Tie-down (if applied) fastened with 32
16*d* sinkers

8*d* brite common @ 6-in. o.c.
12-in. o.c. field

16*d* brite common

16*d* brite common @ 12 in.

13-gage 1.5-in. 7-in. o.c. (perimeter)
10-in. o.c. (field)

16*d* brite common @ 24 in.

FIGURE 10.9 Typical shear wall assembly (shown lying on its side, rather than vertical) (courtesy of Heine, C.P. 1997. "Effect of Overturning Restraint on the Performance of Fully Sheathed and Perforated Timber Framed Shear Walls," M.S. thesis, Virginia Polytechnic and State University, Blacksburg, VA).

sheathing thickness, and fastener size and schedule. Prescriptive walls are often called shear panels rather than shear walls to differentiate them from the walls designed using rational analysis. Prescriptive walls are constructed according to a set of rules provided in the building code. In all cases, the current building code requires that all shear walls using wood structural panels as the sheathing material be fully blocked.

Adhesives are not allowed for attaching sheathing to the framing in shear walls. Whereas adhesives are advocated for attaching sheathing to roof and floor diaphragms as a method to improve their ability to distribute the loads to the shear walls, the use of adhesives in shear walls changes the ductile system usually associated with light-frame construction to a stiff, brittle system that would have to be designed with a significantly lower *R*-factor. Adhesives are allowed to be used in regions with low seismic hazard since they improve the performance of the building when wind loading is considered. An *R*-factor of 1.5 is recommended for seismic checks even in the low seismic hazard regions (Building Seismic Safety Council 2003a) when adhesives are used.

10.7.1 Rationally Designed Walls

Rationally designed light-frame shear walls can be further divided into two categories. The first is the traditional or segmented shear wall design, which involves the use of tables or mechanics. The second method of design is the perforated shear wall method, which is an empirically based method that accounts for the openings in a wall line. The principal difference between the two methods is the assumptions associated with the free-body diagrams used in each method. Each of these design methods will be discussed independently.

10.7.1.1 Segmented Wall Design

The segmented shear wall design method is the traditional method for designing light-frame shear walls. This method assumes a rigid free-body diagram and a uniform distribution of shear along the top of the wall line. Both this method and the perforated shear wall method assume that the tops of all wall segments in a wall line will displace the same amount. In other words, the assumption is that the tops of the wall segments are tied together with the platform above, the top plate of the wall, or collectors between wall segments, providing sufficient connectivity to ensure the wall segments will displace together as a unit. The segmented wall design method assumes that collectors will be detailed to transmit the shear forces distributed to the wall over the openings by the diaphragm to the adjoining wall segments. Each segment is assumed to resist the portion of the load according to its relative length in the wall line. In other words, the shear force per length of wall that is applied to the wall segment is determined by

$$v = \frac{V}{\sum L_i} \tag{10.18}$$

where v is the unit shear (force/length), V is the total shear load applied to the wall line, and $\sum L_i$ is the summation of all of the fully sheathed wall segments in the wall line. It is assumed that this unit shear is distributed uniformly to all of the fully sheathed wall segments. With this assumption, the individual wall segments are then designed assuming a rigid-body free-body diagram, similar to that shown in Figure 10.10.

The simple assumption of using rigid-body mechanics makes the determination of induced overturning forces in the chords an easy task of summing moments about one of the bottom corners of the wall segment. The required mechanical anchor to resist this overturning load can then be sized according to the uplift force determined. However, the assumption of a rigid body also opens an opportunity for error in the calculations. If one assumes that there is an imposed dead load due to the structure above (say, the floor of the story above), one might assume that the vertical forces due to this dead load may act to resist the imposed overturning action of the lateral load on the individual wall segment. In this case, the size of the mechanical anchor would be determined for the difference in the uplift force due to the overturning moment and the resisting force due to the dead load of the structure above. If the assumption of rigid-body action were valid, this would be an acceptable mechanism of resistance. However, if one investigates the construction of light-frame shear walls, the vertical load is applied to the wall across the top plate as a distributed load. However, the top plate of the wall is usually a double 2 × 4 nominal framing member, which has questionable ability to transmit vertical loads along the length of the wall through bending action. In addition, this top plate is supported by repetitive framing members called studs that would transmit the vertical load to the base of the wall rather than allow the top plate to distribute the load to the end stud or chord for the wall. Currently, there are therefore two schools of thought on how the mechanical anchors to resist overturning forces should be determined. One is to use the full dead load acting on the top of the wall to reduce the uplift forces at the chord. The other is to assume that little or none of the dead load acts to resist the uplift forces. The latter is obviously the more conservative assumption, but it also considers the top plate of the wall as a beam on an elastic foundation, for how vertical loads are transmitted along the length of the wall.

The final step in the design of the segmented shear wall is to determine the thickness of the sheathing and associated nail schedule to be used to attach the sheathing to the framing. This information is usually obtained using design tables available in the building code or design specification. However, it is permitted to use the properties of the individual nail connection and engineering mechanics to determine the resistance of a given sheathing thickness and nailing configuration.

10.7.1.2 Perforated Shear Wall Design

The perforated shear wall method is included in the *NEHRP Provisions* (Building Seismic Safety Council 2003a,b) and the 2003 IBC (International Code Council 2000a) for use in seismic design. It had been adopted earlier for wind design by the Building Officials and Code Administrators (BOCA) and Southern Building Code Congress International (SBCCI) building codes. The method is an empirical design method that accounts for the added resistance provided by the wall segments above and below window and door openings in the wall, if they are sheathed with equivalent sheathing to that used in the fully sheathed segments of the wall. The method was originally developed by Sugiyama and Matsumoto (1994) using reduced-scale light-frame wall specimens, and the method was

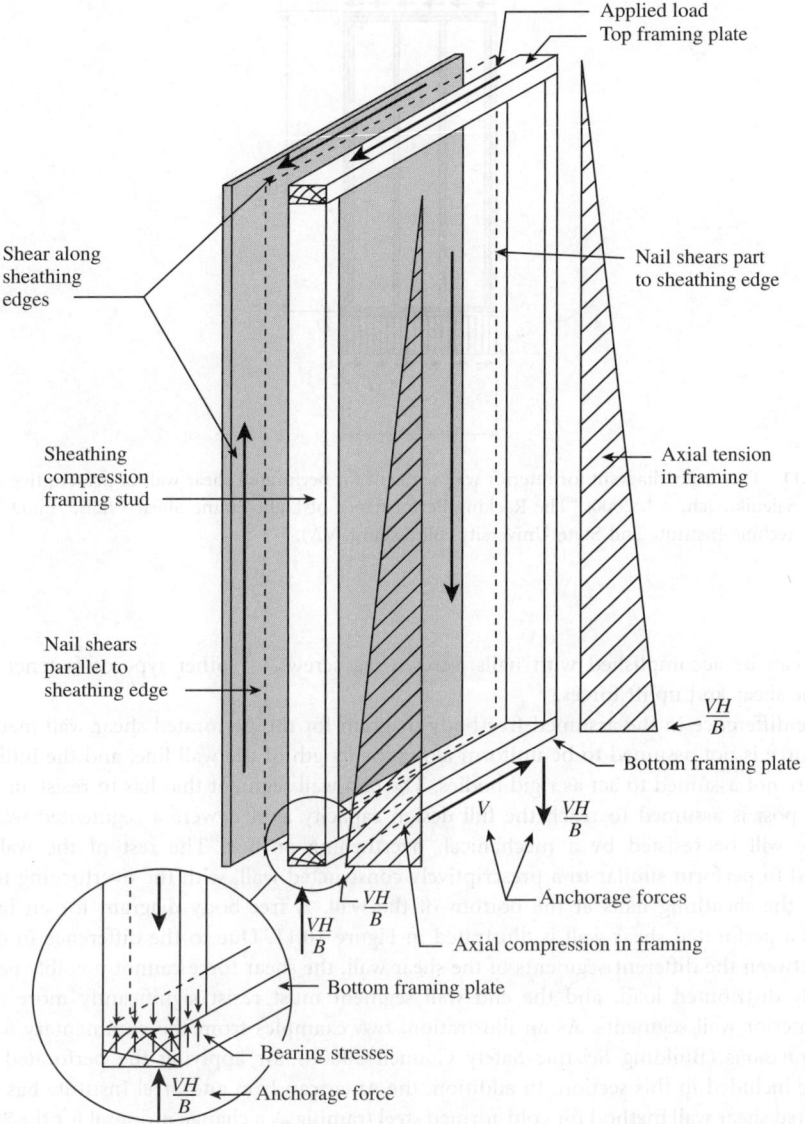

FIGURE 10.10 Rigid free-body diagram assumed for segmented shear wall design (courtesy of Stewart, W.G. 1987. "The Seismic Design of Plywood Sheathed Shearwalls," Ph.D. dissertation, University of Canterbury, New Zealand).

validated for full-scale wall construction under cyclic loads by Dolan and Heine (1997a,b) and Dolan and Johnson (1997a,b).

The perforated shear wall method of design assumes that the segments of wall above and below openings are not specifically designed for force transfer around the opening. Rather, the only assumptions are that the top of the wall line will displace uniformly (i.e., tied together with the top plate of the wall line) and the end full-height sheathed wall segments have mechanical overturning restraint at the extreme ends of the wall line. Other assumptions are that the bottom plate of the wall is attached to the floor platform or foundation sufficiently to resist the distributed shear force applied to the wall and a distributed uplift force equal to the distributed shear force is resisted along the wall length. This

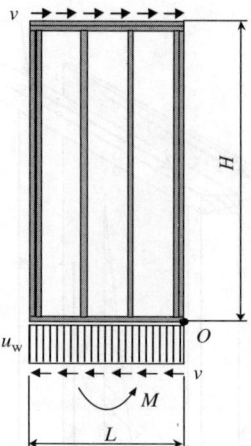

FIGURE 10.11   Free-body diagram for interior wall segment for perforated shear wall or prescriptive wall segment (courtesy of Salenikovich, A.J. 2000. "The Racking Performance of Light-Frame Shear Walls," Ph.D. dissertation, Virginia Polytechnic Institute and State University, Blacksburg, VA).

anchorage can be accomplished with nails, screws, lag screws, or other type of fastener capable of resisting the shear and uplift forces.

The basic difference in the assumed free-body diagram for the perforated shear wall method is that the shear force is not assumed to be uniform along the length of the wall line, and the individual wall segments are not assumed to act as rigid bodies. The end wall segment that has to resist an uplift force in the end post is assumed to reach the full design capacity as if it were a segmented wall since the uplift force will be resisted by a mechanical, overturning anchor. The rest of the wall segments are assumed to perform similar to a prescriptively constructed wall, with the overturning forces being resisted by the sheathing nails at the bottom of the wall. A free-body diagram for an interior wall segment of a perforated shear wall is illustrated in Figure 10.11. Due to the difference in overturning restraint between the different segments of the shear wall, the shear force cannot possibly be resisted as a uniformly distributed load, and the end wall segment must resist significantly more of the load than the interior wall segments. As an illustration, two examples from the commentary for the 2000 *NEHRP Provisions* (Building Seismic Safety Council 2000b) for applying the perforated shear wall method are included in this section. In addition, the American Iron and Steel Institute has introduced the perforated shear wall method for cold-formed steel framing as a change proposal for the 2003 edition of the IBC.

EXAMPLE 10.1: Perforated Shear Wall

Problem description

The perforated shear wall illustrated in Figure 10.12 is sheathed with $\frac{15}{32}$-in. wood structural panel with 10d common nails with 4-in. perimeter spacing. All full-height sheathed sections are 4 ft wide. The window opening is 4 ft high by 8 ft wide. The door opening is 6.67 ft high by 4 ft wide. Sheathing is provided above and below the window and above the door. The wall length and height are 24 and 8 ft, respectively. Hold-downs provide overturning restraint at the ends of the perforated shear wall and anchor bolts are used to restrain the wall against shear and uplift between perforated shear wall ends. Determine the shear resistance adjustment factor for this wall.

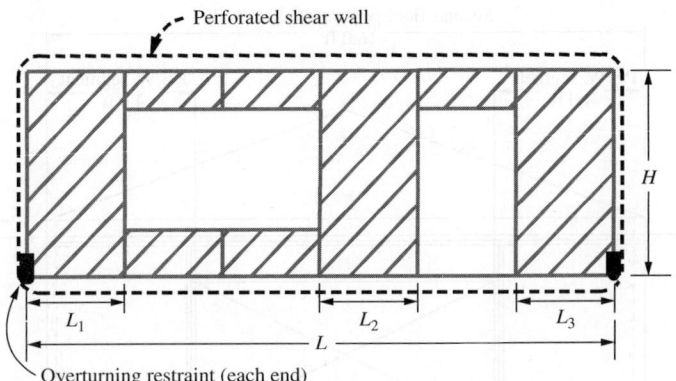

FIGURE 10.12 Perforated shear wall configuration (Example 10.1).

Solution

The wall defined in the problem description meets the application criteria outlined for the perforated shear wall design method. Hold-downs provide overturning restraint at perforated shear wall ends and anchor bolts provide shear and uplift resistance between perforated shear wall ends. Perforated shear wall height, factored shear resistances for the wood structural panel shear wall, and aspect ratio of full-height sheathing at perforated shear wall ends meet requirements of the perforated shear wall method.

The process of determining the shear resistance adjustment factor involves determining percent full-height sheathing and maximum opening height ratio. Once these are known, a shear resistance adjustment factor can be determined from Table 2305.3.7.2 of the 2003 IBC (International Code Council 2003a). From the problem description and Figure 10.12

$$\text{Percent full-height sheathing} = \frac{\textit{sum of perforated shear wall segment widths}, \; \sum L}{\text{Length of perforated shear wall}} = \frac{4 \text{ ft} + 4 \text{ ft} + 4 \text{ ft}}{24 \text{ ft}} \times 100 = 50\%$$

$$\text{Maximum opening height ratio} = \frac{\textit{maximum opening height}}{\text{Wall height}, \; h} = \frac{6.67 \text{ ft}}{8 \text{ ft}} = \frac{5}{6}$$

For a maximum opening height ratio of $\frac{5}{6}$ (or maximum opening height of 6.67 ft when wall height $h = 8$ ft) and percent full-height sheathing equal to 50%, a shear resistance adjustment factor $C_0 = 0.57$ is obtained from Table 2305.3.7.2 of the 2003 IBC (International Code Council 2003a). Note that if wood structural panel sheathing were not provided above and below the window or above the door, the maximum opening height would equal the wall height h.

EXAMPLE 10.2: Perforated Shear Wall

Problem Description

Figure 10.13 illustrates one face of a two-story building with the first- and second-floor walls designed as perforated shear walls. Window heights are 4 ft and door height is 6.67 ft. A trial design is performed in this example based on applied loads V. For simplification, dead load contribution to overturning and uplift restraint is ignored and the effective width for shear in each perforated shear wall segment is assumed to be the sheathed width. Framing is Douglas fir. After basic perforated shear wall resistance and force requirements are calculated, detailing options to provide for adequate shear, v, and uplift, t, transfer between perforated shear wall ends are covered. Method A considers the condition

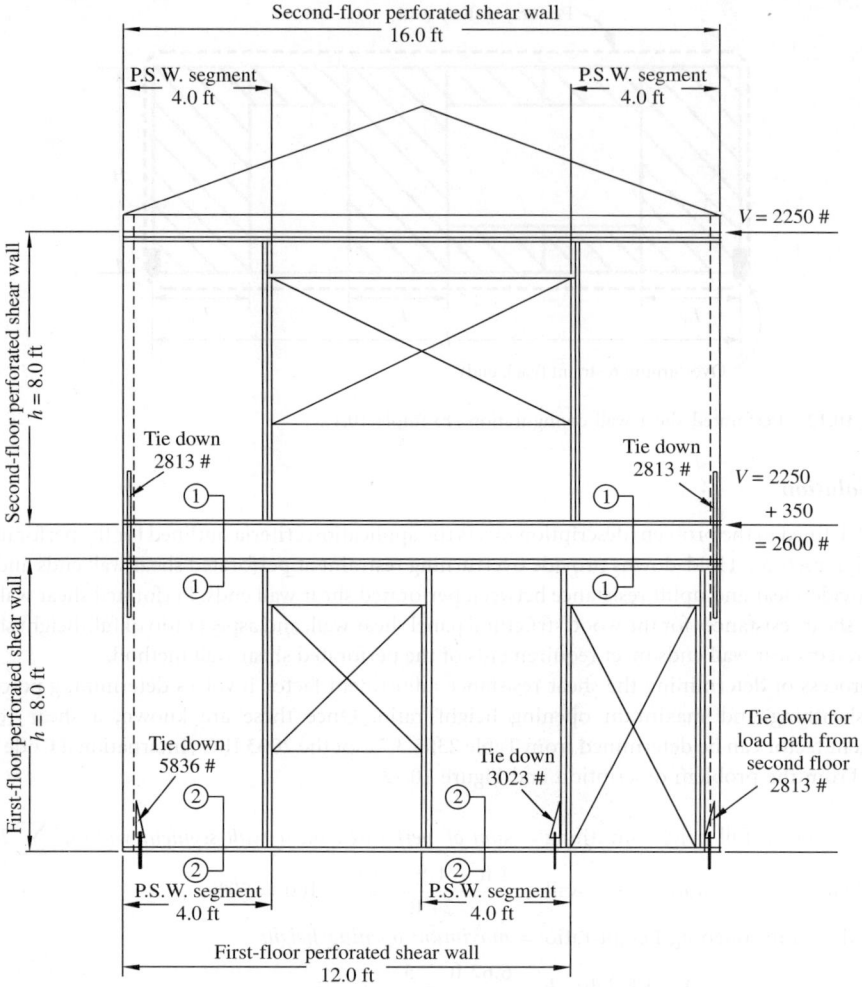

FIGURE 10.13   Perforated shear wall (Example 10.2), two-story building.

where a continuous rim joist is present at the second floor. Method B considers the case where a continuous rim joist is not provided, as when floor framing runs perpendicular to the perforated shear wall with blocking between floor framing joists.

Solution, second-floor wall

Determine the wood structural panel sheathing thickness and fastener schedule needed to resist applied load $V = 2.250$ kip, from the roof diaphragm, such that the shear resistance of the perforated shear wall is greater than the applied force. Also, determine anchorage and load path requirements for uplift force at ends, in plane shear, uplift between wall ends, and compression.

$$\text{Maximum opening height ratio} = \frac{4\,\text{ft}}{8\,\text{ft}} = \frac{1}{2}$$

$$\text{Percent full-height sheathing} = \frac{4\,\text{ft} + 4\,\text{ft}}{16\,\text{ft}} \times 100 = 50\%$$

Shear resistance adjustment factor $C_0 = 0.80$.

Try 15/32 rated sheathing with 8d common nails (0.131 by 2.5 in.) at 6-in. perimeter spacing. Unadjusted shear resistance (Table 5.4 LRFD Structural-Use Panels Supplement) = 0.36 klf (American Forest and Paper Association 1996).

Adjusted shear resistance = (unadjusted shear resistance)(C_0) = (0.36 klf)(0.80) = 0.288 klf.

Perforated shear wall resistance = (adjusted shear resistance)$(\sum L_i)$ = (0.288 klf)(4 ft + 4 ft) = 2.304 kip 2.304 kip > 2.250 kip.

Required resistance due to story shear forces V.

Overturning at shear wall ends

$$T = \frac{Vh}{C_0 \sum L_i} = \frac{2.25 \text{ kip}(8 \text{ ft})}{0.08(4 \text{ ft} + 4 \text{ ft})} = 2.813 \text{ kip}$$

In-plane shear

$$v = \frac{V}{C_0 \sum L_i} = \frac{2.25 \text{ kip}}{0.80(4 \text{ ft} + 4 \text{ ft})} = 0.352 \text{ klf}$$

Uplift t, between wall ends = $v = 0.352$ klf.

Compression chord force C, at each end of each perforated shear wall segment = $T = 2.813$ kip.

Solution, first-floor wall

Determine the wood structural panel sheathing thickness and fastener schedule needed to resist applied load $V = 2.600$ kip, at the second-floor diaphragm, such that the shear resistance of the perforated shear wall is greater than the applied force. Also, determine anchorage and load path requirements for uplift force at ends, in plane shear, uplift between wall ends, and compression.

Percent full-height sheathing = [(4 ft + 4 ft)/12 ft] × 100 = 67%.

Shear resistance adjustment factor, $C_0 = 0.67$.

Unadjusted shear resistance (Table 5.4 LRFD Structural-Use Panels Supplement) = 0.49 klf (American Forest and Paper Association 1996).

Adjusted shear resistance = (unadjusted shear resistance)(C_0) = (0.49 klf)(0.67) = 0.328 klf.

Perforated shear wall resistance = (adjusted shear resistance) $(\sum L_i)$ = (0.328 klf)(4 ft + 4 ft) = 2.626 kip, 2.626 kip > 2.600 kip.

Required resistance due to story shear forces = V.

Overturning at shear wall ends

$$T = \frac{Vh}{C_0 \sum L_i} = \frac{2.600 \text{ kip}(8 \text{ ft})}{0.67(4 \text{ ft} + 4 \text{ ft})} = 3.880 \text{ kip}$$

When maintaining load path from story above, $T = T$ from second floor + T from first floor = 2.813 kip + 3.880 kip = 6.693 kip.

In-plane shear

$$v = \frac{V}{C_0 \sum L_i} = \frac{2.600 \text{ kip}}{0.67(4 \text{ ft} + 4 \text{ ft})} = 0.485 \text{ klf}$$

Uplift t, between wall ends, = $v = 0.485$ klf.

Uplift t, can be cumulative with 0.352 klf from story above to maintain load path. Whether this occurs depends on detailing for transfer of uplift forces between end walls.

Compression chord force C at each end of each perforated shear wall segment $= T = 3.880$ kip.

When maintaining load path from story above, $C = 3.880$ kip $+ 2.813$ kip $= 6.693$ kip.

Hold-downs and posts and the ends of perforated shear wall are sized using calculated force T. The compressive force, C, is used to size compression chords as columns and ensure adequate bearing.

Method A: continuous rim joist
See Figure 10.14.

Second floor: Determine fastener schedule for shear and uplift attachment between perforated shear wall ends. Recall that $v = t = 0.352$ klf.

Wall bottom plate (1.5-in. thickness) to rim joist: Use 20d box nail (0.148 by 4 in.). Lateral resistance $\phi \lambda Z' = 0.254$ kip per nail and withdrawal resistance $\phi \lambda W' = 0.155$ kip per nail.

Nails for shear transfer $=$ (shear force, v)/$\phi \lambda Z' = 0.352$ klf/0.254 kip per nail $= 1.39$ nails per foot.

Nails for uplift transfer $=$ (uplift force, t)/$\phi \lambda W' = 0.352$ klf/0.155 kip per nail $= 2.27$ nails per foot.

Net spacing for shear and uplift $= 3.3$ in. on center.

Rim joist to wall top plate: Use 8d box nails (0.113 by 2.5 in.) toe-nailed to provide shear transfer. Lateral resistance $\phi \lambda Z' = 0.129$ kip per nail.

Nails for shear transfer $=$ (shear force, v)/$\phi \lambda Z' = 0.352$ klf/0.129 kip per nail $= 2.73$ nails per foot.

Net spacing for shear $= 4.4$ in. on center.

See detail in Figure 10.14 for alternate means for shear transfer (e.g., metal angle or plate connector).

Transfer of uplift, t, from the second floor in this example is accomplished through attachment of second-floor wall to the continuous rim joist, which has been designed to provide sufficient strength to resist the induced moments and shears. Continuity of load path is provided by hold-downs at the ends of the perforated shear wall.

First floor: Determine the anchorage for shear and uplift attachment between perforated shear wall ends. Recall that $v = t = 0.485$ klf. Wall bottom plate (1.5-in. thickness) to concrete. Use 0.5-in. anchor bolt with lateral resistance $\phi \lambda Z' = 1.34$ kip.

Bolts for shear transfer $=$ (shear force, v)/$\phi \lambda Z' = 0.485$ klf/1.34 kip per bolt $= 0.36$ bolts per foot.

Net spacing for shear $= 33$ in. on center.

Bolts for uplift transfer: Check axial capacity of bolts for $t = v = 0.485$ klf and size plate washers accordingly. No interaction between axial and lateral load on anchor bolt is assumed (e.g., presence of axial tension does not affect lateral strength).

Method B: blocking between joists
See Figure 10.14.

Second floor: Determine fasteners schedule for shear and uplift attachment between perforated shear wall ends. Recall that $v = t = 0.352$ klf.

Wall bottom plate (1.5-in. thickness) to rim joist: Use 20d box nail (0.148 by 4 in.). Lateral resistance $\phi \lambda Z' = 0.254$ kip per nail.

Nails for shear transfer $=$ (shear force, v)/$\phi \lambda Z' = 0.352$ klf/0.254 kip per nail $= 1.39$ nails per foot.

Net spacing for shear $= 8.63$ in. on center.

Rim joist to wall top plate: Use 8d box nails (0.113 by 2.5 in.) toe-nailed to provide shear transfer. Lateral resistance $\phi \lambda Z' = 0.129$ kip per nail.

Nails for shear transfer $=$ (shear force, v)/$\phi \lambda Z' = 0.352$ klf/0.129 kip per nail $= 2.73$ nails per foot.

Net spacing for shear $= 4.4$ in. on center.

See detail in Figure 10.14 for alternative means for shear transfer (e.g., metal angle or plate connector).

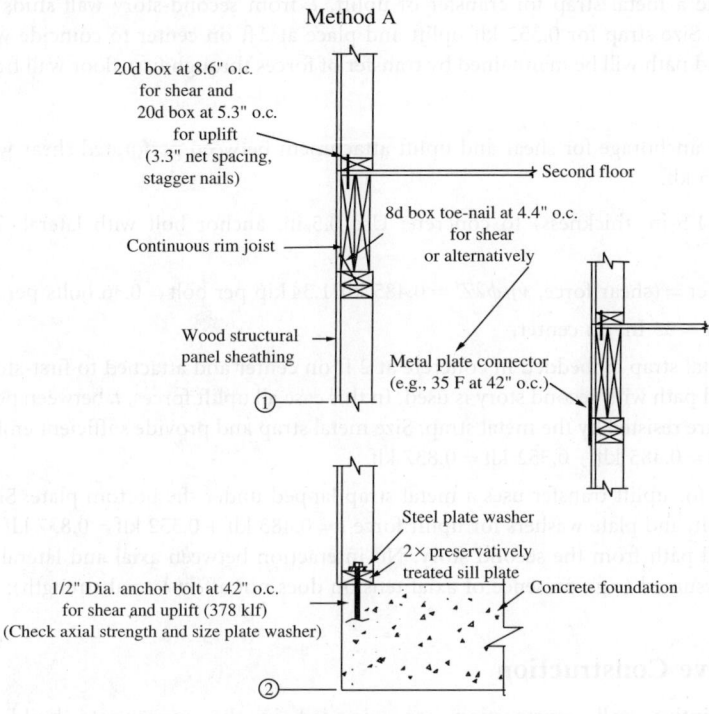

Method A

20d box at 8.6" o.c.
for shear and
20d box at 5.3" o.c.
for uplift
(3.3" net spacing,
stagger nails)

Second floor

Continuous rim joist

8d box toe-nail at 4.4" o.c.
for shear
or alternatively

Wood structural
panel sheathing

Metal plate connector
(e.g., 35 F at 42" o.c.)

①

Steel plate washer

2× preservatively
treated sill plate

1/2" Dia. anchor bolt at 42" o.c.
for shear and uplift (378 klf)
(Check axial strength and size plate washer)

Concrete foundation

②

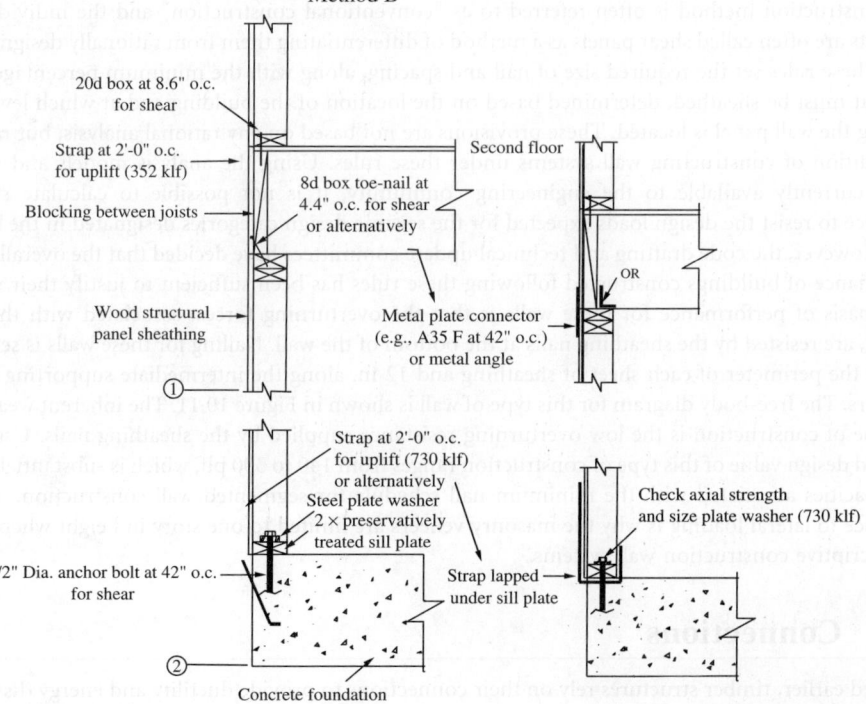

Method B

20d box at 8.6" o.c.
for shear

Strap at 2'-0" o.c.
for uplift (352 klf)

Blocking between joists

Second floor

8d box toe-nail at
4.4" o.c. for shear
or alternatively

Wood structural
panel sheathing

Metal plate connector
(e.g., A35 F at 42" o.c.)
or metal angle

OR

①

Strap at 2'-0" o.c.
for uplift (730 klf)
or alternatively
Steel plate washer
2 × preservatively
treated sill plate

1/2" Dia. anchor bolt at 42" o.c.
for shear

Strap lapped
under sill plate

Check axial strength
and size plate washer (730 klf)

②

Concrete foundation

FIGURE 10.14 Details for perforated shear wall (Example 10.2).

Stud to stud: Provide a metal strap for transfer of uplift, t, from second-story wall studs to first-story wall studs. Size strap for 0.352-klf uplift and place at 2 ft on center to coincide with stud spacing. This load path will be maintained by transfer of forces through first-floor wall framing to the foundation.

***First floor*:** Determine anchorage for shear and uplift attachment between perforated shear wall ends. Recall that $v = t = 0.485$ klf.

Wall bottom plate (1.5-in. thickness) to concrete: Use 0.5-in. anchor bolt with lateral resistance $\phi \lambda Z' = 1.34$ kip.

Bolts for shear transfer = (shear force, v)/$\phi \lambda Z' = 0.485$ klf/1.34 kip per bolt = 0.36 bolts per foot.

Net spacing for shear = 33 in. on center.

Uplift transfer: A metal strap embedded in concrete at 2 ft on center and attached to first-story studs maintaining load path with second story is used. In this case all uplift forces, t, between perforated shear wall ends are resisted by the metal strap. Size metal strap and provide sufficient embedment for uplift force $t = 0.485$ klf + 0.352 klf = 0.837 klf.

An alternative detail for uplift transfer uses a metal strap lapped under the bottom plate. Size metal strap, anchor bolt, and plate washers for uplift force $t = 0.485$ klf + 0.352 klf = 0.837 klf are used to maintain load path from the second story. No interaction between axial and lateral load on anchor bolt is assumed (e.g., presence of axial tension does not affect lateral strength).

10.7.2 Prescriptive Construction

The rules for prescriptive wall construction are provided in the appropriate building code (IRC [International Code Council 2003b] or CABO [Council of American Building Officials 1995]). This construction method is often referred to as "conventional construction" and the individual wall segments are often called shear panels as a method of differentiating them from rationally designed shear walls. These rules set the required size of nail and spacing, along with the minimum percentage of wall area that must be sheathed, determined based on the location of the building and at which level of the building the wall panel is located. These provisions are not based on any rational analysis, but rather on the tradition of constructing wall systems under these rules. Using the analysis models and wall test results currently available to the engineering community, it is not possible to calculate sufficient resistance to resist the design loads expected for the seismic design categories designated in the building code. However, the code drafting and technical update committees have decided that the overall historic performance of buildings constructed following these rules has been sufficient to justify their use.

The basis of performance for these walls is that the overturning forces, associated with the lateral loading, are resisted by the sheathing nails at the bottom of the wall. Nailing for these walls is set at 6 in. around the perimeter of each sheet of sheathing and 12 in. along the intermediate supporting framing members. The free-body diagram for this type of wall is shown in Figure 10.11. The inherent weakness of this type of construction is the low overturning resistance supplied by the sheathing nails. Usually the assumed design value of this type of construction ranges from 140 to 300 plf, which is substantially below the capacities associated with the minimum nail schedule for segmented wall construction. This low resistance to lateral loading is why the masonry veneers are limited to one story in height when applied to prescriptive construction wall systems.

10.8 Connections

As stated earlier, timber structures rely on their connections to provide ductility and energy dissipation. Therefore, it is imperative that connections be given significant consideration when determining how to detail a structure for good seismic performance. In general, the concept for good performance of timber structures is to design connections where the steel yielding in the connection is the governing behavior, rather than the wood crushing. While both steel yielding and wood crushing occur in all connections that are loaded beyond their elastic range, the connection can be designed to favor the steel yielding as the dominant mechanism.

There are many types of connections used in wood. They range from nails, screws, and bolts (referred to as *dowel* connections) to metal plate connectors, shear plates, split rings, and other proprietary connectors. This chapter focuses on nails and bolts, since these are the most widely used connectors in timber structures. Connectors such as metal plate connectors and expanded tubes are proprietary connectors and the designer must contact the suppliers of these products to obtain the necessary information to safely design with them.

Dowel connections can be divided into two principal groups: small and large dowel connections. The designation refers to the relative length of the fastener in the wood member to the diameter, similar to the slenderness ratio for columns. This differentiation of dowel connections can be made because small dowel connections tend to be governed by the yield strength of the dowel and are usually considered to share the load equally. This is because as the individual fasteners yield, the load is redistributed to the other fasteners in the connection and all of the fasteners will yield and bend before the wood fails. On the other hand, large dowel connections tend to be increasingly governed by the crushing strength of the wood. Imperfections in the connection due to construction tolerances and variability of the wood material cause the load to be carried unequally between the individual fasteners in the connection (group action effects). The differentiation diameter for dimensional lumber connections is approximately $\frac{1}{4}$ in.

10.8.1 Design Methodology

In recent years, the design standards for timber construction in the United States changed the lateral design methodology for dowel connections from one with an empirical, restrictive basis to one based on mechanics. The new basis of lateral design is the yield strength of the fastener and the bearing strength of the materials being joined. This results in the ability to configure the connection assembly to provide a connection that is governed by the dowel yielding and bending, the dowel remaining straight and the wood crushing, or a combination of the two. This change also provides flexibility to the designer in choosing connection configurations that are more applicable to special situations, where the previous design methods were restricted to a few typical configurations unless special testing was done.

The new design method is called the yield theory and is best illustrated in Figure 10.15. As can be seen in the figure, there are basically four classes of yielding that are considered by the theory. Mode I is governed by crushing of the wood material by the dowel, and the bolt is held firmly by one member and does not even rotate as the connection displaces. Mode II is also governed by the crushing of the wood, but the dowel rotates as the connection displaces. Mode III is governed by a combination of the dowel yielding and the wood crushing. A single plastic hinge is formed in the dowel as the connection displaces. Finally, mode IV is governed by the dowel yielding, and two plastic hinges are formed in the dowel as the connection displaces.

The designer is referred to either the LRDF or ASD design standards for the complete set of applicable design equations for the particular type of dowel fastener being used. Each type of fastener (i.e., nails, spikes, wood screws, lag screws, bolts, and dowels) has a slightly different set of design equations that predict the yield mode and associated design value, since each type of fastener has different geometries for both the fastener itself and the connection as a whole.

10.8.2 Small-Diameter Dowel Connections

Small dowels, such as nails and screws, are by far the most prevalent fasteners used in wood construction. The framing in light-frame construction is typically nailed together and the sheathing applied to most timber structures is nailed to the framing. The driven fastener has several advantages over other options,

Single shear connections   Double shear connections

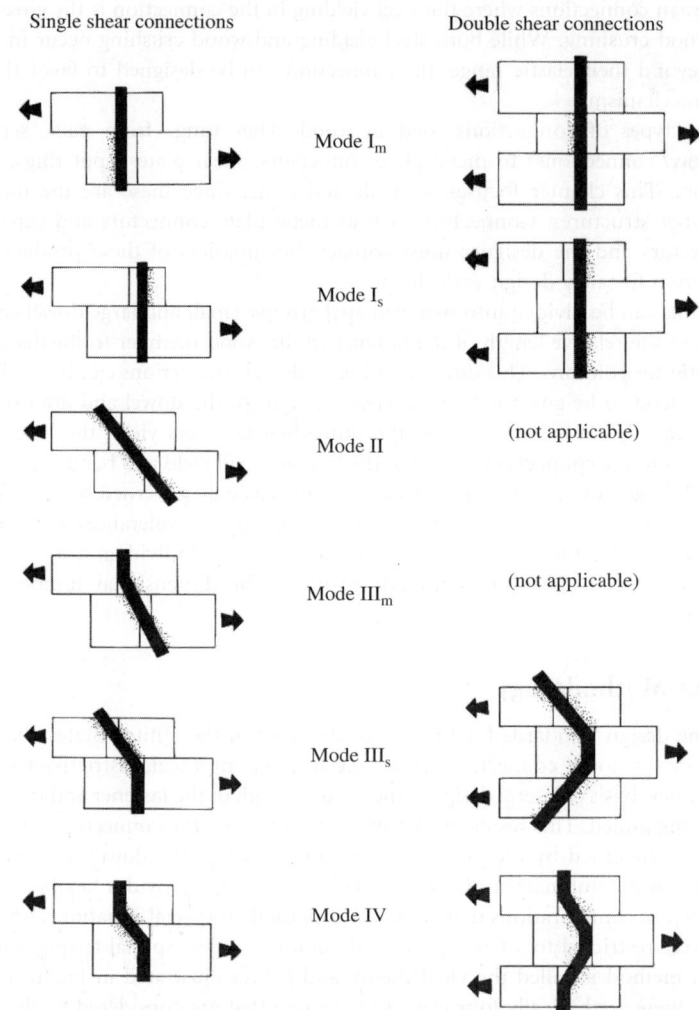

Mode I$_m$

Mode I$_s$

Mode II   (not applicable)

Mode III$_m$ (not applicable)

Mode III$_s$

Mode IV

FIGURE 10.15 Yield modes considered for dowel connections in wood construction (courtesy of American Forest and Paper Association. 2001. *ASD Allowable Stress Design Manual for Engineered Wood Construction and Supplements*, Washington, DC).

in that it is easily installed (typically completed with pneumatic nailing tools and does not require predrilling), provides reasonable resistance to lateral and withdrawal loads, reduces splitting of the timber (unless the spacing is very close), and distributes the resistance over a larger area of the structure. In addition, these types of fasteners easily yield and provide significant levels of damping to the structure.

10.8.2.1 Lateral Resistance of Small-Diameter Fasteners

While these connections can be configured to yield in all of the modes, nails and screws are governed by a reduced set of equations due to the fact that certain subsets of the yield modes are not possible since the fastener does not pass through the entire connection for nails and screws. In dimensional lumber and larger sizes, small-diameter fasteners typically yield in Modes III and IV. These two yield modes provide the highest ductility and energy dissipation since the metal of the dowel yields in plastic yielding and the friction associated with the connection displacement is significant.

Since small dowel connections typically yield in the highly ductile yield modes, the displacement of the connection is such that by the time the assembly load reaches near capacity, all of the fasteners will have yielded and the load is assumed to be shared equally by all of the fasteners. In other words, there is no group action factor and the capacity of a multiple-fastener connection is equal to the sum of all of the connectors. In equation form, the resistance of a multiple nail or screw connection is

$$R' = \lambda \phi_z \sum Z' = \lambda \phi_z n Z' \qquad (10.19)$$

where R' is the total factored resistance, λ is the applicable time effect factor, ϕ_z is the resistance factor for connections (0.65), Z' is the factored resistance for a single fastener, and n is the number of fasteners in the connection.

Small dowels are also not typically affected by grain direction. Nail and screw connections are typically governed almost completely by the yielding of the fastener. This implies that even the perpendicular-to-grain embedment strength of the wood is sufficiently high so that the yielding still is governed by the bending of the dowel.

10.8.2.2 Withdrawal Resistance of Small-Diameter Fasteners

Withdrawal of smooth shank nails can be problematic to the performance of timber structures. In some cases, such as when the nail is driven into the end-grain of a member (i.e., typically how the framing of light-frame timber walls are connected), the design resistance of the fastener is zero. On the other hand, driven into the side-grain of the member, the smooth shank nail may provide sufficient resistance to withdrawal.

Smooth shank fasteners are also significantly affected by changes in the moisture content of the wood during the life of the connection. If the connection is above 19% at any time during its life, and then dries out to below 19%, or vice versa, then 75% of the design withdrawal resistance is lost. If the connection is fabricated at one moisture condition and remains in this condition throughout its life, there is no reduction. Under these same changing moisture content conditions, the lateral design values for the connection are reduced by 30%. This implies that if a designer specifies, or allows, green lumber to be used to construct shear walls and roof systems, the design values used for the shear walls and roof systems should be reduced by a significant amount.

Hardened threaded and ring-shank nails overcome this weakness in performance by the way the deformed nail shank interlocks with the wood fibers. The result is that the design values of the threaded nails are not affected by changes in moisture content. This implies that if a designer allows green lumber to be used in a project, and does not wish to impose the design value reductions associated with moisture content changes, they should specify threaded or ring-shank nails. Helically threaded nails provide the best performance of all of the deformed shank nails available. This is because the deformation pattern allows the nail to withdraw a significant amount without a drop in resistance. This is due to the nail being able to withdraw without tearing the wood fibers around the nail itself. Other types of deformed nails such as ring-shank nails provide higher withdrawal resistance than smooth shank nails, but the resistance drops quickly as the nail withdraws due to the localized damage to the wood fibers.

10.8.3 Large-Diameter Dowel Connections

Large-diameter dowel connections can be designed to yield in any of the four modes shown in Figure 10.15. The diameter relative to the thickness of the timber member will determine the mode of yield that occurs. For a given thickness of timber member, the larger the diameter of the dowel, the lower the yield mode will be, and the less ductile the connection will be. This is the reason why the design standards have been reducing the maximum size of the bolt that is included in the design standard. While there is no restriction against designing with larger-diameter bolts, the LRFD and ASD manuals have 1-in. diameter bolts as the largest diameter included in any of the tables. If one uses the yield equations for bolts, lag screws, or dowels in the design standard, larger sizes can be used.

When determining the size of dowel to use in a given connection, a balancing act of choosing between a few large-diameter fasteners with high capacity and a larger number of fasteners with lower capacity must be performed by the designer. In both cases, the higher number of fasteners used in a given row of fasteners results in a group action effect. This means that the capacity of the connection is less than the sum of capacities of the individual fasteners. Large-diameter bolts do not share the load equally due to placement tolerances and variation in the material properties of the timber. The design specification provides an equation that determines the reduction factor associated with multiple-fastener connections, and should be used for all connections that utilize more than one fastener per row.

If one reviews Figure 10.15 again, it becomes clear that connections that yield in Modes I and II will fail in a brittle manner, since wood as a material fails in a brittle manner. By the same deduction process, connections that yield in Modes III and IV will behave in a more ductile manner and provide higher damping ability. Thus, connections that yield in Modes III and IV should be the preferred connection configurations for seismic design. This is not always possible, due to the fact that this implies that a larger number of smaller-diameter fasteners will be used, and space limitations may not allow the large number of fasteners to be used. If connections yielding in Modes I or II must be used in the design, and these connections are the controlling connections in the structure, then the designer should assume that the structure will remain essentially elastic so that the potential for a brittle failure is minimized.

Recent research by Heine (2001) and Anderson (2002) indicates that the current design specifications provide nonconservative results for multiple-bolt, single-shear connections. There are possibly two problems currently in the design specification. The first is that the minimum spacing requirement of four times the bolt diameter for full design strength may be insufficient to prevent the failure from occurring between the bolts. The second problem is that the current group action factors are derived based on an assumption of elastic response. Together these two issues result in connections that perform as much as 40% below the predicted strength. There is some evidence that the capacity of large connections (100 kip plus design loads) may not be able to achieve the anticipated design values. Figure 10.16 illustrates the experimental and theoretical results for connections made with $\frac{1}{2}$-in. diameter bolts. Note that the 4D spacing has a significant weakening effect on the connection. If one were to consider the LRFD group action factor for this same configuration, the smallest it

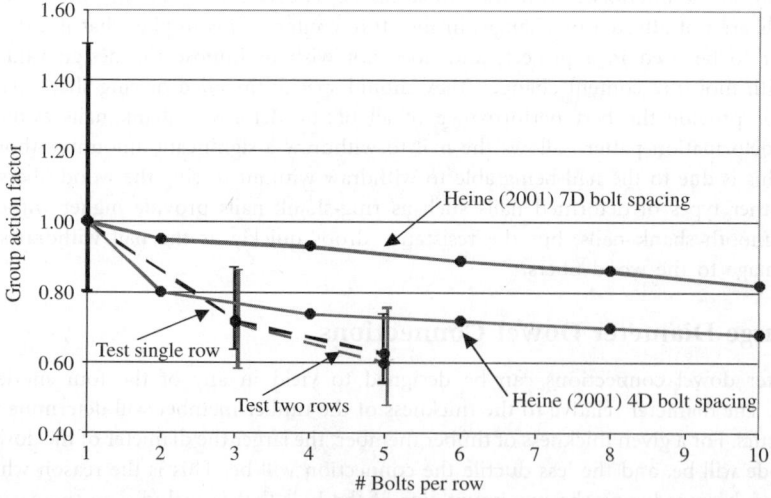

FIGURE 10.16 Experimental and theoretical (Heine) group action factor for $\frac{1}{2}$-in. bolts and variable spacing requirements (courtesy of Anderson, G.T. 2002. "Experimental Investigation of Group Action Factor for Bolted Wood Connections," M.S. thesis, Virginia Polytechnic Institute and State University, Blacksburg, VA).

would be is 0.88. Due to this problem, the 2001 edition of the ASD manual has a new appendix that provides guidance in designing multiple-bolt connections. The new appendix recommends additional checks on the capacity of the wood members to prevent block shear and other potential modes of failure in multiple-bolt connections.

This discussion of the yield modes and group action factor provides some guidance to the designer. First, the designer should try to use, when possible, connections that yield in Modes III and IV to provide the highest possible ductility and energy dissipation for the structure. Second, where possible, the spacing between bolts should be increased to at least seven times the diameter of the bolt (the minimum spacing for full design load that is required by the European design standard). This will minimize the group action effects in multiple-bolt connections. Third, the designer should follow the additional checks outlined in the new appendix for the 2001 ASD manual for preventing unintentional failure modes from occurring in multiple-bolt connections.

10.8.4 Heavy Timber Connectors

Shear plates and split rings are timber connectors that have very high capacities, but also tend to fail in more brittle modes. Historically, these connections have been used in heavy timber and glulam timber structures. Several applications have been made in large timber bow-string truss and glulam timber connections. Design of these connections is covered by the LRDF (American Forest and Paper Association 1996) and ASD (American Forest and Paper Association 2001) manuals, and the reader is referred to either of these documents for a clear description of how to design these connections. Additional guidance on design of these types of connections as well as general heavy timber structure design may be found in the *Timber Construction Manual* (American Institute of Timber Construction 2004). These connections should be considered in a similar class of connection to large-dowel connections. The connections are susceptible to group action and geometry effects in similar ways in which large-dowel connections experience the same phenomena.

While these connectors are good for situations where large numbers of smaller-dowel connections might be required, there are potential problems associated with their use. The first problem is that most contractors are not familiar with the installation of these specialized connectors. There are special tools for drilling and cutting the required surfaces to ensure proper bearing of the connectors in the connection, and the contractor would be advised to manufacture a couple of trial connections before beginning the real connections to learn how to use the cutting tools properly. Second, the connection cannot easily be inspected to ensure proper installation after the members are joined. The main components in these two types of connections need to bear uniformly around their perimeter. If the holes are overdrilled, not drilled circular, or both sides of the connection do not exactly line up for all of the connectors, the connection will not perform properly and will fail at load levels below the intended design level. Also, once the connection is assembled, there is no way to see whether the shear plate or split ring is even present, let alone installed properly. Therefore, these types of connections need to be constructed by a conscientious contractor who can be trusted to perform the installation properly, or continuous inspection of the installation by a responsible party is recommended.

Glossary

Anisotropic — Having different properties in different directions (i.e., not *isotropic*, which is to have the same properties in all directions).

Bow — The distortion of lumber in which there is a deviation, in a direction perpendicular to the flat face, from a straight line from end to end of the piece.

Box system — A type of lateral-force-resisting system (LFRS) in which forces are transmitted via horizontal diaphragms (i.e., roof and floors) to gravity load-bearing walls that act in shear, and thus form the main vertical elements of the LFRS.

Checking — A lengthwise separation of the wood that usually extends across the rings of annual growth and commonly results from stresses set up in wood during seasoning.

Creep — The continued increase in deflection that occurs in the viscoelastic material that sustains a constant load.

Cripple walls — Short stud walls between the foundation and the first-floor level.

Cup — A distortion of a board in which there is a deviation flat-wise from a straight line across the width of the board.

Diaphragm — A nearly horizontal structural unit that acts as a deep beam or girder when flexible relative to supports, and as a plate when its stiffness is higher than the associated stiffness of the walls.

Dowel — A wood connector, such as a nail, screw, or bolt.

Glulam — Glued–laminated lumber is an engineered product made by gluing together 50 mm (2 in.) or thinner pieces of lumber.

Hygroscopic — Readily absorbing moisture, as from the atmosphere.

I-joists — Wood joists that are structural members composed of an oriented strand board (OSB) or plywood web and two laminated vener lumber (LVL), oriented strand lumber (OSB), or solid lumber flanges, the newest and fastest growing engineered wood product, partially due to the declining availability of high-quality, large-dimension lumber for which it substitutes.

Laminated veneer lumber (LVL) — A structural composite lumber product made by adhesively bonding thin sheets of wood veneer oriented with the grain parallel in the long direction. Primary uses include headers, beams, rafters, and flanges for wood I-joists.

Lateral-force-resisting system (LFRS) — Any continuous load path potentially capable of resisting lateral (e.g., seismic) forces.

Lumber — Timber sawed into boards, planks, or other structural members of standard or specified dimensions.

Oriented strand board (OSB) — A structural panel made by adhesively bonding chips of small-diameter softwoods and previously underutilized hardwoods.

Orthotropic — Having different properties at right angles.

Plywood — A structural panel made from wood sheets (typically $\frac{1}{8}$-in. thick) peeled from tree trunks and adhesively laminated so as to orthogonally orient the wood grain in alternate plies.

Spaced column — Compression member made of two or more parallel single-member columns that are separated by spacers, located at specific locations along the column, and are rigidly tied together at the ends of the column.

Structural composite lumber (SCL) — A structural member, made by adhesively bonding thin strips of wood, oriented with the grain parallel to the long axis of the member. Primarily used for headers, heavily loaded beams, girders, and heavy timber construction.

Timber — Trees considered as a source of wood. Also, timbers used in the original round form, such as poles, piling, posts, and mine timbers.

Viscoelastic — A material that exhibits both viscous (time-dependent) and elastic responses to deformation.

Warp — Any variation from a true or plane surface. Warp includes bow, crook, cup, and twist, or any combination thereof.

References

American Forest and Paper Association. 1996. *Load and Resistance Factor Design (LRFD) Manual For Engineered Wood Construction*, American Forest and Paper Association, Washington, DC.

American Forest and Paper Association. 2001. *National Design Specification for Wood Construction*, American Forest and Paper Association, Washington, DC.

American Forest and Paper Association. 2001b. *ASD Allowable Stress Design Manual for Engineered Wood Construction and Supplements*, American Forest and Paper Association, Washington, DC.

American Forest and Paper Association/American Society of Civil Engineers. 2001. *Wood Frame Construction Manual (WFCM) for One- and Two-Family Dwellings*, American Forest and Paper Association/American Society of Civil Engineers, Washington, DC.

American Institute of Timber Construction. 2004. *Timber Construction Manual*, 5th ed., John Wiley & Sons, New York.

American Society of Civil Engineers. 1996. *Load and Resistance Factor Design (LRFD) for Engineered Wood Construction*, AF&PA/ASCE 16–95, American Society of Civil Engineers, New York.

American Society of Civil Engineers. 2002. *Minimum Design Loads for Buildings and Other Structures*, ASCE 7–02, American Society of Civil Engineers, New York.

Anderson, G.T. 2002. "Experimental Investigation of Group Action Factor for Bolted Wood Connections," M.S. thesis, Virginia Polytechnic Institute and State University, Blacksburg, VA.

APA, The Engineered Wood Association. 1997. *The Plywood Design Specification and Supplements*, APA, Tacoma, WA.

APA, The Engineered Wood Association. 2003. *Engineered Wood Construction Guide*, APA, Tacoma, WA.

Applied Technology Council. 1981. *Guidelines for the Design of Horizontal Wood Diaphragms*, ATC-7, Applied Technology Council, Redwood, CA.

Benuska, L., Ed. 1990. "Loma Prieta Earthquake Reconnaissance Report," *Earthquake Spectra*, 6 (Suppl.).

Breyer, D., Fridley, K.J., Pollock, D.G., and Cobeen, K.E. 2003. *Design of Wood Structures*, 5th ed., McGraw-Hill, New York.

Building Seismic Safety Council. 2003a. *NEHRP Recommended Provisions for Seismic Regulations for New Buildings and Other Structures. Part I: Provisions FEMA 368*, Building Seismic Safety Council, Washington, DC.

Building Seismic Safety Council. 2003b. *NEHRP Recommended Provisions for Seismic Regulations for New Buildings and Other Structures. Part 2: Commentary FEMA 369*, Building Seismic Safety Council, Washington, DC.

Council of American Building Officials. 1995. *CABO One- and Two-Family Dwelling Code*, Council of American Building Officials, Falls Church, VA.

Dolan, J.D. and Heine, C.P. 1997a. *Monotonic Tests of Wood-Frame Shear Walls with Various Openings and Base Restraint Configurations*, Timber Engineering Center Report no. TE-1997–001, Virginia Polytechnic Institute and State University, Blacksburg, VA.

Dolan, J.D. and Heine, C.P. 1997b. *Sequential Phased Displacement Cyclic Tests of Wood-Frame Shear Walls with Various Openings and Base Restraint Configurations*, Timber Engineering Center Report no. TE-1997–002, Virginia Polytechnic Institute and State University, Blacksburg, VA.

Dolan, J.D. and Johnson, A.C. 1997a. *Monotonic Performance of Perforated Shear Walls*, Timber Engineering Center Report no. TE-1996–001, Virginia Polytechnic Institute and State University, Blacksburg, VA.

Dolan, J.D. and Johnson, A.C. 1997b. *Sequential Phased Displacement (Cyclic) Performance of Perforated Shear Walls*, Timber Engineering Center Report no. TE-1996–002, Virginia Polytechnic Institute and State University, Blacksburg, VA.

Dolan, J.D., Bott, W., and Easterling, W.S. 2002. *Design Guidelines for Timber Diaphragms*, CUREE Publication W-XX, Consortium of Universities for Research in Earthquake Engineering, Richmond, CA.

Faherty, K. and Williamson, T. 1999. *Wood Engineering and Construction Handbook*, 3rd ed., McGraw-Hill, New York.

Federal Emergency Management Agency (FEMA). 1988. *Rapid Visual Screening of Buildings for Potential Seismic Hazards: A Handbook*, FEMA 154, Federal Emergency Management Agency, Washington, DC.

Fischer, D., Filliatrault, A., Folz, B., Uang, C.-M., and Seible, F. 2001. *Shake Table Tests of a Two-Story Woodframe House*, CUREE Publications W-06, Consortium of Universities for Research in Earthquake Engineering, Richmond, CA.

Heine, C.P. 1997. "Effect of Overturning Restraint on the Performance of Fully Sheathed and Perforated Timber Framed Shear Walls," M.S. thesis, Virginia Polytechnic Institute and State University, Blacksburg, VA.

Heine, C.P. 2001. "Simulated Response of Degrading Hysteretic Joints with Slack Behavior," Ph.D. dissertation, Virginia Polytechnic and State University, Blacksburg, VA.

Holmes, W.T. and Somers, P., Eds. 1996. "Northridge Earthquake Reconnaissance Report, Vol. 2," *Earthquake Spectra*, 11 (Suppl. C), 125–176.

International Code Council (ICC). 2003a. *International Building Code (IBC)*, Falls Church, VA.

International Code Council (ICC). 2003b. *International Residential Code (IRC)*, Falls Church, VA.

Lew, H.S., Ed. 1990. *Performance of Structures during the Loma Prieta Earthquake of October 17, 1989*, NIST Special Publication 778, National Institute of Standards and Technology, Gaithersburg, MD.

National Fire Prevention Association. 2002. *NFPA 5000 Building Code*, National Fire Prevention Association, Boston, MA.

Salenikovich, A.J. 2000. "The Racking Performance of Light-Frame Shear Walls," Ph.D. dissertation, Virginia Polytechnic Institute and State University, Blacksburg, VA.

Schierle, G.G. 2000. *Northridge Earthquake Field Investigations: Statistical Analysis of Woodframe Damage*, CUREE Publication W-02, Consortium of Universities for Research in Earthquake Engineering, Richmond, CA.

Steinbrugge, K.V., Schader, E.E., Bigglestone, H.C., and Weers, C.A. 1971. *San Fernando Earthquake February 9, 1971*, Pacific Fire Rating Bureau, San Francisco, CA.

Stewart, W.G. 1987. "The Seismic Design of Plywood Sheathed Shearwalls," Ph.D. dissertation, University of Canterbury, New Zealand.

Sugiyama, H. and Matsumoto, T. 1994. "Empirical Equations for the Estimation of Racking Strength of a Plywood Sheathed Shear Wall with Openings," *Mokuzai Gakkaishi*, 39, 924–929.

Tissell, J. 1990. "Performance of Wood-Framed Structures in the Loma Prieta Earthquake," in *Wind and Seismic Effects: Proceedings of the Twenty-Second Joint Meeting of the U.S.–Japan Cooperative Program in Natural Resources Panel on Wind and Seismic Effects*, NIST SP 796, National Institute of Standards and Technology, Gaithersburg, MD, September, pp. 324–330.

U.S. Department of Agriculture. 1999. *Wood Handbook: Wood as an Engineering Material*, Agriculture Handbook 72, Forest Products Laboratory, U.S. Department of Agriculture, Madison, WI (available online at http://www.fpl.fs.fed.us/documnts/FPLGTR/fplgtr113/fplgtr113.htm).

U.S. Department of Housing and Urban Development. 1995. *Preparing for the "Big One": Saving Lives through Earthquake Mitigation in Los Angeles, California, HUD-I511-PD&R*, January 17, U.S. Department of Housing and Urban Development, Washington, DC.

Yancey, C.W. et al. 1998. *A Summary of the Structural Performance of Single Family Wood Framed Housing, NISTIR 6224*, Building and Fire Research Laboratory, National Institute of Standards and Technology, Gaithersburg, MD.

Further Reading

The *Wood Handbook* (U.S. Department of Agriculture 1999) is a good introduction to the properties of wood and wood products. The several wood design handbooks (American Institute of Timber Construction 2004; American Forest and Paper Association 1996, 2001; APA 1997, 2003) are all required references for wood structure designers, and contain much useful background information. Breyer et al. (2003) is a good overall text for wood structure design. The Building Seismic Safety Council (2003a,b) provides the current consensus guidelines specific to seismic design of wood buildings. The CUREE project (www.curee.org) is a major research effort to better understand wood building performance and develop improved design data and practices. The CUREE Website provides publications and other information.

11

Aluminum Structures

Maurice L. Sharp
Consultant — Aluminum Structures,
Avonmore, PA

11.1 Introduction

11.1.1 The Material

11.1.1.1 Background

Of the structural materials used in construction, aluminum was the latest to be introduced into the market place even though it is the most abundant of all metals, making up about $\frac{1}{12}$ of the earth's crust. The commercial process was invented simultaneously in the United States and Europe in 1886. Commercial production of the metal started thereafter using an electrolytic process that economically separated aluminum from its oxides. Prior to this time aluminum was a precious metal. The initial uses of aluminum were for cooking utensils and electrical cables. The earliest significant structural use of aluminum was for the skins and members of a dirigible called the *Shenendoah* completed in 1923. The first structural design handbook was developed in 1930 and the first specification was issued by the industry in 1932 (Sharp 1994).

11.1.1.2 Product Forms

Aluminum is available in all the common product forms, flat-rolled, extruded, cast, and forged. Fasteners such as bolts, rivets, screws, and nails are also manufactured. The available thicknesses of flat-rolled products range from 0.006 in. or less for foil to 7.0 in. or more for plate. Widths to 17 ft are possible. Shapes in aluminum are extruded. Some presses can extrude sections up to 31 in. wide. The extrusion process allows the material to be placed in areas that maximize structural properties and joining ease. Because the cost of extrusion dies is relatively low, most extruded shapes are designed for specific applications.

    Castings of various types and forgings are possibilities for three-dimensional shapes and are used in some structural applications. The design of castings is not covered in detail in structural design books and specifications primarily because there can be a wide range of quality depending on the casting process. The quality of the casting affects structural performance.

11.1.1.3 Alloy and Temper Designation

The four-digit number used to designate alloys is based on the main alloying ingredients. For example, magnesium is the principal alloying element in alloys whose designation begins with a 5 (5083, 5456, 5052, etc.).

Cast designations are similar to wrought designations but a decimal is placed between the third and fourth digits (356.0). The second part of the designation is the temper, which defines the fabrication process. If the term starts with T, for example, -T651, the alloy has been subjected to a thermal heat treatment. These alloys are often referred to as heat-treatable alloys. The numbers after the T show the type of treatment and any subsequent mechanical treatment such as a controlled stretch. The temper of alloys that harden with mechanical deformation starts with H, for example, -H116. These alloys are referred to as non-heat-treatable alloys. The type of treatment is defined by the numbers in the temper designation. A 0 temper is the fully annealed temper. The full designation of an alloy has the two parts that define both chemistry and fabrication history, for example, 6061-T651.

11.1.2 Alloy Characteristics

11.1.2.1 Physical Properties

Physical properties usually vary only by a few percent depending on the alloy. Some nominal values are given in Table 11.1. The density of aluminum is low, about one third that of steel, which results in lightweight structures. The modulus of elasticity is also low, about one third of that of steel, which affects design when deflection or buckling controls.

11.1.2.2 Mechanical Properties

Mechanical properties for a few alloys used in general purpose structures are given in Table 11.2. The stress–strain curves for aluminum alloys do not have an abrupt break when yielding but rather have a gradual bend (see Figure 11.1). The yield strength is defined as the stress corresponding to a 0.002 in./in. permanent set. The alloys shown in Table 11.2 have moderate strength, excellent resistance to corrosion in the atmosphere, and are readily joined by mechanical fasteners and welds. These alloys often are employed in outdoor structures without paint or other protection. The higher-strength aerospace alloys are not shown. They usually are not used for general purpose structures because they are not as resistant to corrosion and normally are not welded.

11.1.2.3 Toughness

The accepted measure of toughness of aluminum alloys is fracture toughness. Most high-strength aerospace alloys can be evaluated in this manner; however, the moderate-strength alloys employed for

TABLE 11.1 Some Nominal Properties of Aluminum Alloys

Property	Value
Weight	0.1 lb/in.^{3}
Modulus of elasticity	
Tension and compression	10,000 ksi
Shear	3,750 ksi
Poisson's ratio	1/3
Coefficient of thermal expansion (68 to 212°F)	0.000013 per °F

Source: Gaylord, Gaylord, and Stallmeyer, *Structural Engineering Handbook*, McGraw-Hill, 1997.

TABLE 11.2 Minimum Mechanical Properties

Alloy and temper	Product	Thickness range, in.	Tension TS	Tension YS	Compression YS	Shear US	Shear YS	Bearing US	Bearing YS
3003-H14	Sheet and plate	0.009–1.000	20	17	14	12	10	40	25
5456-H116	Sheet and plate	0.188–1.250	46	33	27	27	19	87	56
6061-T6	Sheet and plate	0.010–4.000	42	35	35	27	20	88	58
6061-T6	Shapes	All	38	35	35	24	20	80	56
6063-T5	Shapes	to 0.500	22	16	16	13	9	46	26
6063-T6	Shapes	All	30	25	25	19	14	63	40

Note: All properties are in ksi. TS is the tensile strength, YS is the yield strength, and US is the ultimate strength.

Source: The Aluminum Association, *Structural Design Manual*, 2000.

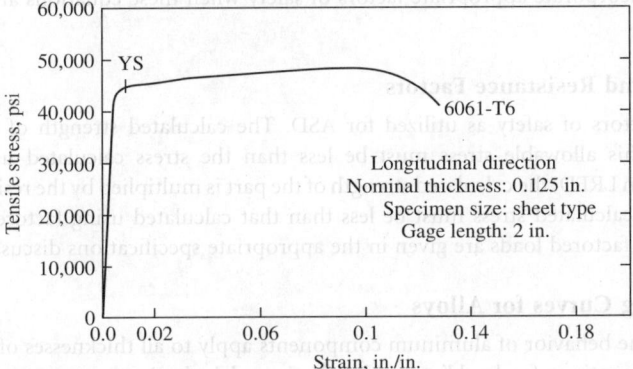

FIGURE 11.1 Stress–strain curve.

general purpose structures cannot be evaluated because they are too tough to get valid results in the test. Aluminum alloys also do not exhibit a transition temperature, their strength and ductility actually increase with decrease in temperature. Some alloys have a high ratio of yield strength to tensile strength (compared to mild steel) and most alloys have a lower elongation than mild steel, perhaps 8–10%, both considered to be negative factors for toughness. However, these alloys do have sufficient ductility to redistribute stresses in joints and in sections in bending to achieve full strength of the components. Their successful use in various types of structures, bridges, bridge decks, tractor trailers, railroad cars, building structures, and automotive frames, has demonstrated that they have adequate toughness. Thus far, there has not been a need to modify design based on toughness of aluminum alloys.

11.1.3 Codes and Specifications

Allowable stress design (ASD) for building, bridge, and other structures, which need the same factor of safety, and load and resistance factor design (LRFD) for building and similar type structures have been published by the Aluminum Association (2000). These specifications are included in a design manual that also has design guidelines, section properties of shapes, design examples, and numerous other aids for the designer.

The American Association of State Highway and Transportation Officials have published LRFD Specifications that cover bridges of aluminum and other materials (AASHTO 2002). The equations for strength and behavior of aluminum components are essentially the same in all of these specifications. The margin of safety for design differs depending on the type of specification and the type of structure.

Codes and standards are available for other types of aluminum structures. Lists and summaries are provided elsewhere (Sharp 1993; Aluminum Association 2000).

11.2 Structural Behavior

11.2.1 General

11.2.1.1 Behavior Compared to Steel

The basic principles of design for aluminum structures are the same as those for other ductile metals such as steel. Equations and analysis techniques for global structural behavior such as load–deflection behavior are the same. Component strength, particularly buckling, postbuckling, and fatigue, are defined specifically for aluminum alloys. The behavior of various types of components are provided in Sections 11.2.1.3–11.2.1.6 and Section 11.2.2. Strength equations are also given. The designer needs to incorporate appropriate factors of safety when these equations are used for practical designs.

11.2.1.2 Safety and Resistance Factors

Table 11.3 gives factors of safety as utilized for ASD. The calculated strength of the part is divided by these factors. This allowable stress must be less than the stress calculated using the total load applied to the part. In LRFD, the calculated strength of the part is multiplied by the resistance factors given in Table 11.4. This calculated stress must be less than that calculated using factored loads. Equations for determining the factored loads are given in the appropriate specifications discussed previously.

11.2.1.3 Buckling Curves for Alloys

The equations for the behavior of aluminum components apply to all thicknesses of material and to all aluminum alloys. Equations for buckling in the elastic and inelastic range are provided. Figure 11.2

TABLE 11.3 Factors of Safety for Allowable Stress Design

Component	Failure mode	Buildings and similar-type structures	Bridges and similar-type structures
Tension	Yielding	1.95	2.20
	Ultimate strength	1.65	1.85
Columns	Yielding (short column)	1.65	1.85
	Buckling	1.95	2.20
Beams	Tensile yielding	1.65	1.85
	Tensile ultimate	1.95	2.20
	Compressive yielding	1.65	1.85
	Lateral buckling	1.65	1.85
Thin plates in compression	Ultimate in columns	1.95	2.20
	Ultimate in beams	1.65	1.85
Stiffened flat webs in shear	Shear yield	1.65	1.85
	Shear buckling	1.20	1.35
Mechanically fastened joints	Bearing yield	1.65	1.85
	Bearing ultimate	2.34	2.64
	Shear str./rivets, bolts	2.34	2.64
Welded joints	Shear str./fillet welds	2.34	2.64
	Tensile str./butt welds	1.95	2.20
	Tensile yield/butt welds	1.65	1.85

Source: The Aluminum Association, *Structural Design Manual,* 2000.

TABLE 11.4 Resistance Factors for LRFD

Component	Limit state	Buildings	Bridges
Tension	Yielding	0.95	0.90
	Ultimate strength	0.85	0.75
Columns	Buckling	Varies with slenderness ratio	Varies with slenderness ratio
Beams	Tensile yielding	0.95	0.90
	Tensile ultimate	0.85	0.80
	Compressive yielding	0.95	0.90
	Lateral buckling	0.85	0.80
Thin plates in compression	Yielding	0.95	0.90
	Ultimate strength	0.85	0.80
Stiffened flat webs in shear	Yielding	0.95	0.90
	Buckling	0.90	0.80

Sources: Buildings — The Aluminum Association, *Structural Design Manual*, 2000; Bridges — American Association of State Highway and Transportation Officials, *AASHTO LRFD Bridge Design Specifications*, 2002.

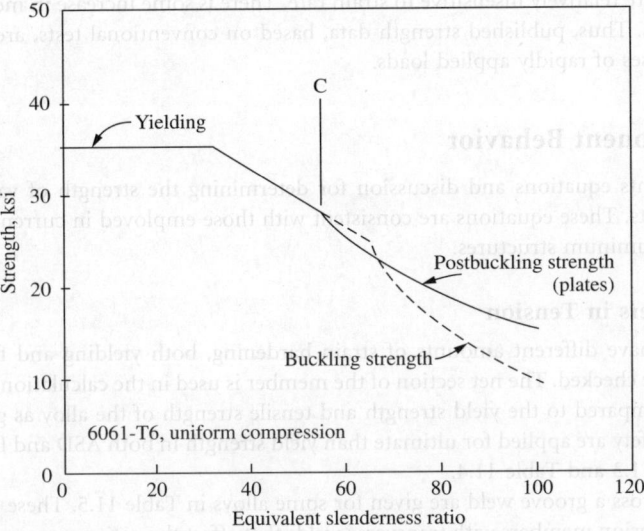

FIGURE 11.2 Buckling of components.

shows the format generally used for both component and element behavior. The strength of the component is normally considered to be limited by the yield strength of the material. For buckling behavior, coefficients are defined for two classes of alloys, those that are heat treated with temper designations -T5 or higher and those that are not heat treated or are heat treated with temper designations -T4 or lower. Different coefficients are needed because of the differences in the shapes of the stress–strain curves for the two classes of alloys.

11.2.1.4 Effects of Welding

In most applications, some efficiency is obtained by using alloys that have been thermally treated or strain hardened to achieve higher strength. The alloys are readily welded. However, welding partially anneals a narrow band of material (about 1.0 in. on either side of the weld) and thus this *heat-affected* material has a lower strength than the rest of the member. The lower strength is accounted for in the design equations presented in Equation 11.1.

   If the strength of the heat-affected material is less than the yield strength of the parent material, the plastic deformation of the component at failure loads will be confined to that of the narrow band of lower-strength material. In this case, the component fails with only a small total deformation, thus exhibiting low structural toughness. For good structural toughness the strength of the heat-affected material should be well above the yield strength of the parent material. In the case of liquid natural gas containers an annealed temper of the plate, 5083-0, has been employed to achieve maximum toughness. The strength of the welded material is the same as that of the parent material and there is essentially no effect of welding on structural behavior.

11.2.1.5 Effects of Temperature

All of the properties important to structural behavior, static strength, elongation, fracture toughness, and fatigue strength, increase with decrease in temperature. Elongation increases but static and fatigue strengths decrease at elevated temperatures. Alloys behave differently but significant changes in mechanical properties can occur at temperatures over 300°F.

11.2.1.6 Effects of Strain Rate

Aluminum alloys are relatively insensitive to strain rate. There is some increase in mechanical properties at high strain rates. Thus, published strength data, based on conventional tests, are normally used for calculations for cases of rapidly applied loads.

11.2.2 Component Behavior

This section presents equations and discussion for determining the strength of various types of aluminum components. These equations are consistent with those employed in current specifications and publications for aluminum structures.

11.2.2.1 Members in Tension

As various alloys have different amounts of strain hardening, both yielding and fracture strength of members should be checked. The net section of the member is used in the calculation. The calculated net section stress is compared to the yield strength and tensile strength of the alloy as given in Table 11.2. Larger factors of safety are applied for ultimate than yield strength in both ASD and LRFD specifications as noted in Table 11.3 and Table 11.4.

   The strengths across a groove weld are given for some alloys in Table 11.5. These properties are used for the design of tension members with transverse welds that affect the entire cross-section. For members

TABLE 11.5 Minimum Strengths of Groove Welds

Parent material	Filler metal	Tension TS[a]	Tension YS[b]	Compression YS[b]	Shear US
3003-H14	1100	14	7	7	10
5456-H116	5556	42	26	24	25
6061-T6	5356	24	20	20	15
6061-T6	4043	24	15	15	15
6063-T5, -T6	4043	17	11	11	11

[a] ASME weld-qualification values. The design strength is considered to be 90% of these values.
[b] Corresponds to a 0.2% set on a 10-in. gage length.
   Notes: All strengths are in ksi. TS is the tensile strength, YS is the yield strength, and US is the ultimate strength.
   Source: The Aluminum Association, *Structural Design Manual,* 2000.

with longitudinal welds in which only part of the cross-section is affected by welds the tensile or yield strength may be calculated using the following equation:

$$F_{pw} = F_n - \frac{A_w}{A}(F_n - F_w) \tag{11.1}$$

where F_{pw} is the strength of the member with a portion of cross-section affected by welding, F_n is the strength of the unaffected parent metal, F_w is the strength of the material affected by welding, A is the area of cross-section, and A_w is the area that lies within 1 in. of a weld.

11.2.2.2 Columns Under Flexural Buckling

The Euler column formula is employed for the elastic region and straight line equations in the inelastic region. The straight line equations are a close approximation to the tangent modulus column curve. The equations for column strength are as follows:

$$F_c = B_c - D_c \frac{KL}{r}, \quad \frac{KL}{r} \le C_c \tag{11.2}$$

$$F_c = \frac{\pi^2 E}{(KL/r)^2}, \quad \frac{KL}{r} > C_c \tag{11.3}$$

where F_c is the column strength (ksi), L is the unsupported length of column (in.), r is the radius of gyration (in.), K is the effective-length factor, E is the modulus of elasticity (ksi), and B_c, D_c, C_c are constants depending on mechanical properties (see Equations 11.4 to 11.9).

For wrought products with tempers starting with, -O, -H, -T1, -T2, -T3, and -T4, and cast products

$$B_c = F_{cy}\left[1 + \left(\frac{F_{cy}}{1000}\right)^{1/2}\right] \tag{11.4}$$

$$D_c = \frac{B_c}{20}\left(\frac{6B_c}{E}\right)^{1/2} \tag{11.5}$$

$$C_c = \frac{2B_c}{3D_c} \tag{11.6}$$

For wrought products with tempers starting with -T5, -T6, -T7, -T8, and -T9

$$B_c = F_{cy}\left[1 + \left(\frac{F_{cy}}{2250}\right)^{1/2}\right] \tag{11.7}$$

$$D_c = \frac{B_c}{10}\left(\frac{B_c}{E}\right)^{1/2} \tag{11.8}$$

$$C_c = 0.41\frac{B_c}{D_c} \tag{11.9}$$

where F_{cy} is the compressive yield strength (ksi).

The column strength of a welded member is generally less than that of a member with the same cross-section but without welds. If the welds are longitudinal and affect part of the cross-section, the column strength is given by Equation 11.1. The strengths in this case are column buckling values assuming all parent metal and all heat-affected metal. If the member has transverse welds that affect the entire cross-section, and occur away from the ends, the strength of the column is calculated assuming that the entire column is a heat-affected material. Note that the constants for the heat-affected materials are given by Equations 11.4 to 11.6. If transverse welds occur only at the ends, the equations for parent metal are used but the strength is limited to the yield strength across the groove weld.

11.2.2.3 Columns Under Flexural–Torsional Buckling

Thin, open sections that are unsymmetrical about one or both principal axes may fail by combined torsion and flexure. This strength may be estimated using a previously developed equation that relates the combined effects to pure flexural and pure torsional buckling of the section. Equation 11.10 is in the form of effective and equivalent slenderness ratios and is in good agreement with test data (Sharp 1993). The equation must be solved by trial for the general case

$$\left[1 - \left(\frac{\lambda_c}{\lambda_y}\right)^2\right]\left[1 - \left(\frac{\lambda_c}{\lambda_x}\right)^2\right]\left[1 - \left(\frac{\lambda_c}{\lambda_\phi}\right)^2\right] - \left(\frac{y_0}{r_0}\right)^2\left[1 - \left(\frac{\lambda_c}{\lambda_x}\right)^2\right] - \left(\frac{x_0}{r_0}\right)^2\left[1 - \left(\frac{\lambda_c}{\lambda_y}\right)^2\right] = 0 \qquad (11.10)$$

where λ_c is the equivalent slenderness ratio for flexural–torsional buckling; λ_x, λ_y are the slenderness ratios for flexural buckling in the x and y directions, respectively; x_0, y_0 are the distances between centroid and shear center, parallel to principal axes, $r_0 = [(I_{xo}I_{yo})/A]^{1/2}$; I_{xo}, I_{yo} are the moments of inertia about axes through shear center; and λ_ϕ is the equivalent slenderness ratio for torsional buckling

$$\lambda_\phi = \sqrt{\frac{I_x + I_y}{(3J/8\pi^2) + (C_w/(K_\phi L)^2)}} \qquad (11.11)$$

where J is the torsion constant, C_w is the warping constant, K_ϕ is the effective length coefficient for torsional buckling, L is the length of the column, and I_x, I_y are the moments of inertia about the centroid (principal axes).

11.2.2.4 Beams

Beams that are supported against lateral–torsional buckling fail by excessive yielding or fracture of the tension flange at bending strengths above that corresponding to stresses reaching the tensile or yield strength at the extreme fiber. This additional strength may be accounted for by applying a shape factor to the tensile or yield strength of the alloy. Nominal shape factors for some aluminum shapes are given in Table 11.6. These factors vary slightly with alloy because they are affected by the shape of the stress–strain curve but the values shown are reasonable for all aluminum alloys.

This higher bending strength can be developed provided that the cross-section is compact enough so that local buckling does not occur at a lower stress. Limitations on various types of elements are given in Table 11.7. The bending moment for compact sections is as follows:

$$M = ZSF \qquad (11.12)$$

where M is the moment corresponding to yield or ultimate strength of the beam, S is the section modulus of the section, F is the yield or tensile strength of the alloy, and Z is the shape factor.

TABLE 11.6 Shape Factors for Aluminum Beams

Cross-section	Yielding, K_y	Ultimate, K_u
I and channel (major axis)	1.07	1.16
I (minor axis)	1.30	1.42
Rectangular tube	1.10	1.22
Round tube	1.17	1.24
Solid rectangle	1.30	1.42
Solid round	1.42	1.70

Sources: Gaylord, Gaylord, and Stallmeyer, *Structural Engineering Handbook*, McGraw-Hill, 1997 and The Aluminum Association, *Structural Design Manual*, 2000.

TABLE 11.7 Limiting Ratios of Elements for Plastic Bending

Element	Limiting ratio
Outstanding flange of I or channel	$b/t \leq 0.30 \ (E/F_{cy})^{1/2}$
Lateral buckling of I or channel	
Uniform moment	$L/r_y \leq 1.2 \ (E/F_{cy})^{1/2}$
Moment gradient	$L/r_y \leq 2.2 \ (E/F_{cy})^{1/2}$
Web of I or rectangular tube	$b/t \leq 0.45 \ (E/F_{cy})^{1/2}$
Flange of rectangular tube	$b/t \leq 1.13 \ (E/F_{cy})^{1/2}$
Round tube	$D/t \leq 2.0 \ (E/F_{cy})^{1/2}$

Source: Gaylord, Gaylord, and Stallmeyer, *Structural Engineering Handbook*, McGraw-Hill, 1997.

11.2.2.5 Effects of Joining

If there are holes in the tension flange the net section should be used for calculating the section modulus. Welding affects beam strength in the same way as it does tensile strength. The groove weld strength is used when the entire cross-section is affected by welds. Beams may not develop the bending strength as given by Equation 11.12 at the locations of the transverse welds. In these locations it is reasonable to use a shape factor equal to 1.0. If only part of the section is affected by welds, Equation 11.1 is used to calculate strength and compact sections can develop the moment as given by Equation 11.12. In the calculation the flange is considered to be the area that lies farther than two-thirds of the distance between the neutral axis and the extreme fiber.

11.2.2.6 Lateral Buckling

Beams that do not have continuous support for the compression flange may fail by lateral buckling. For aluminum beams an equivalent slenderness is defined and substituted in column formulas in place of KL/r. The slenderness ratios for buckling of I-sections, WF-shapes, and channels are as follows:

For beams with end moments only or transverse loads applied at the neutral axis:

$$\lambda_b = 1.4 \frac{L_b}{\sqrt{\dfrac{I_y d C_b}{S_c} \sqrt{1.0 + 0.152 \dfrac{J}{I_y} \left(\dfrac{L_b}{d}\right)^2}}} \tag{11.13}$$

For beams with loads applied to top and bottom flanges where the load is free to move laterally with the beam:

$$\lambda_b = 1.4 \frac{L_b}{\sqrt{\dfrac{I_y d C_b}{S_c}} \left[\pm 0.5 + \sqrt{1.25 + 0.152 \dfrac{J}{I_y} \left(\dfrac{L_b}{d}\right)^2} \right]} \tag{11.14}$$

where

λ_b = equivalent slenderness ratio for beam buckling (to be used in place of KL/r in the column formula)

C_b = coefficient depending on loading and beam supports, which is equal to

  $12.5M_{max}/(2.5M_{max} + 3M_A + 4M_B + 3M_C)$ for simple supports, wherein

  M_{max} = absolute value of maximum moment in the unbraced beam segment

  M_A = absolute value of moment at quarter-point of the unbraced beam segment

  M_B = absolute value of moment at midpoint of the unbraced beam segment

  M_C = absolute value of moment at three-quarter point of the unbraced beam segment

d = depth of beam

I_y = moment of inertia about the axis parallel to the web
S_c = section modulus for compression flange
J = torsion constant

A plus sign is to be used in Equation 11.14 if the load acts on the bottom (tension) flange and a minus sign if it acts on the top (compression) flange.

Equations 11.13 and 11.14 may also be used for cantilever beams of the specified cross-section by the use of the appropriate factor C_b. For a concentrated load at the end the factor is 1.28, and for a uniform lateral load the factor is 2.04.

These equations also may be applied to I-sections in which the tension and compression flanges are of somewhat different sizes. In this case the beam properties are calculated as though the tension flange is of the same size as that of the compression flange. The depth of the section is maintained.

Lateral buckling strengths of welded beams are affected similarly to that of flexural buckling of columns. For cases in which part of the compression flange has a heat-affected material, Equation 11.1 is used. The total flange area is that farther than two-thirds the distance from the neutral axis to the extreme fiber. If the beam has transverse welds away from the supports, the strength of the beam is calculated as though the entire beam is of a heat-affected material.

For other types of cross-sections and loadings not provided for above, and for cases in which the loads cause torsional stresses in the beam, other equations and analysis are needed. Some cases are covered elsewhere (Sharp 1993; Aluminum Association 2000).

11.2.2.7 Members Under Combined Bending and Axial Loads

The same interaction equations may be used for aluminum as for steel members. The following equations are for bending in one direction. Both formulas must be checked:

$$\frac{f_a}{F_{ao}} + \frac{f_b}{F_b} \leq 1.0 \tag{11.15}$$

$$\frac{f_a}{F_a} + \frac{C_b f_b}{F_b(1.0 - f_a/F_e)} \leq 1.0 \tag{11.16}$$

where f_a is the average compressive stress from the axial load, f_b is the maximum compressive bending stress, F_a is the strength of the member as a column, F_b is the strength of the member as a beam, F_{ao} is the strength of the member as a short column, and $F_e = \pi^2 E/(KL/r)^2$.

11.2.2.8 Buckling of Thin, Flat Elements of Columns and Beams Under Uniform Compression

The elastic buckling of plates is calculated using classical plate buckling theory. For inelastic stresses straight line formulas that approximate a secant–tangent modulus combination are used. These straight line formulas give higher stresses than those for columns that use tangent modulus, and they are in close agreement with test data. An equivalent slenderness ratio, $K_p b/t$, is utilized in the equations

$$F_p = B_p - \frac{D_p K_p b}{t}, \quad K_p \frac{b}{t} \leq C_p \tag{11.17}$$

$$F_p = \frac{\pi^2 E}{(K_p b/t)^2}, \quad K_p \frac{b}{t} > C_p \tag{11.18}$$

where F_p is the buckling stress of the plate (ksi), b is the clear width of the plate, t is the thickness of the plate, K_p is the coefficient depending on conditions of edge restraint of the plate (see Table 11.8), and B_p, D_p, C_p are the alloy constants defined in Equations 11.19 to 11.24.

For wrought products with tempers starting with -O, -H, -T1, -T2, -T3, and -T4, and cast products

$$B_p = F_{cy}\left[1 + \frac{(F_{cy})^{1/3}}{7.6}\right] \tag{11.19}$$

TABLE 11.8 Values of K_p for Plate Elements

Type of member	Stress distribution	Edge support	K_p
Column	Uniform compression	One edge free, one edge supported	5.1
		Both edges supported	1.6
Beam (flange)	Uniform compression	One edge free, one edge supported	5.1
		Both edges supported	1.6
Beam (web)	Varying from compression on one edge to tension on the other edge	Compression edge free, tension edge with partial restraint	3.5
		Both edges supported	0.67

Source: Gaylord, Gaylord, and Stallmeyer, *Structural Engineering Handbook*, McGraw-Hill, 1997.

$$D_p = \frac{B_p}{20}\left(\frac{6B_p}{E}\right)^{1/2} \tag{11.20}$$

$$C_p = \frac{2B_p}{3D_p} \tag{11.21}$$

For wrought products with tempers starting with -T5, -T6, -T7, -T8, and -T9

$$B_p = F_{cy}\left[1 + \frac{(F_{cy})^{1/3}}{11.4}\right] \tag{11.22}$$

$$D_p = \frac{B_p}{10}\left(\frac{B_p}{E}\right)^{1/2} \tag{11.23}$$

$$C_p = 0.41\frac{B_p}{D_p} \tag{11.24}$$

11.2.2.9 Buckling of Thin, Flat Elements of Beams Under Bending

For webs under bending loads, Equations 11.17 and 11.18 apply for buckling in the inelastic and elastic ranges. C_p is given by Equation 11.21. However, the values of B_p and D_p are higher than those of elements under uniform compression because they include a shape factor effect, the same as that defined for beams. The constants for the straight-line equation are as follows. They apply to all alloys and tempers

$$B_p = 1.3F_{cy}\left[1 + \frac{(F_{cy})^{1/3}}{7}\right] \tag{11.25}$$

$$D_p = \frac{B_p}{20}\left(\frac{6B_p}{E}\right)^{1/2} \tag{11.26}$$

11.2.2.10 Postbuckling Strength of Thin Elements of Columns and Beams

Most thin elements can develop strengths much higher than the elastic buckling strength as given by Equation 11.18. This higher strength is used in design. Elements of angle, cruciform, and channel (flexural buckling about the weak axis) columns may not develop postbuckling strength. Thus, the buckling strength should be used for these cases. For other cases the postbuckling strength (in the elastic buckling region) is given as follows.

$$F_{cr} = k_2\frac{\sqrt{B_pE}}{K_pb/t} \quad \text{for } \frac{b}{t} > \frac{k_1B_p}{K_pD_p} \tag{11.27}$$

where F_{cr} is the ultimate strength of plate in compression (ksi), B_p, D_p are the coefficients defined in Equations 11.19 to 11.26, and k_1, k_2 are the coefficients ($k_1 = 0.5$ and $k_2 = 2.04$ for wrought products whose temper starts with -O, -H, -T1, -T2, -T3, and -T4, and castings; $k_1 = 0.35$ and $k_2 = 2.27$ for wrought products whose temper starts with -T5, -T6, -T7, -T8 and -T9).

11.2.2.11 Weighted Average Strength of Thin Sections

In many cases a component will have elements with different calculated buckling strengths. An estimate of the component strength is obtained by equating the ultimate strength of the section multiplied by the total area to the sum of the strength of each element times its area, and solving for the ultimate strength. This weighted average approach gives a close estimate of strength for columns and for beam flanges.

11.2.2.12 Effect of Local Buckling on Column and Beam Strength

If local buckling occurs at a stress below that for overall buckling of a column or beam, the strength of the component will be reduced. Thus, the elements should be proportioned such that they are stable at column or beam buckling strengths. There are methods for taking into account local buckling on column or beam strength provided elsewhere (Sharp 1993; Aluminum Association 2000).

11.2.2.13 Shear Buckling of Plates

The same equations apply to stiffened and unstiffened webs. Equivalent slenderness ratios are defined for each case. Straight line equations are employed in the inelastic range and the Euler formula in the elastic range. The equations are as follows:

$$F_s = B_s - D_s \lambda_s, \quad \lambda_s \leq C_s \tag{11.28}$$

$$F_s = \frac{\pi^2 E}{\lambda_s^2}, \quad \lambda_s > C_s \tag{11.29}$$

For tempers -O, -H, -T1, -T2, -T3, -T4

$$B_s = F_{sy}\left(1 + \frac{F_{sy}^{1/3}}{6.2}\right) \tag{11.30}$$

$$D_s = \frac{B_s}{20}\left(\frac{6 B_s}{E}\right)^{1/2} \tag{11.31}$$

$$C_s = \frac{2}{3}\frac{B_s}{D_s} \tag{11.32}$$

For tempers -T5, -T6, -T7, -T8, -T9

$$B_s = F_{sy}\left(1 + \frac{F_{sy}^{1/3}}{9.3}\right) \tag{11.33}$$

$$D_s = \frac{B_s}{10}\left(\frac{B_s}{E}\right)^{1/2} \tag{11.34}$$

$$C_s = 0.41\frac{B_s}{D_s} \tag{11.35}$$

where F_{sy} is the shear yield strength (ksi), λ_s is equal to $1.25h/t$ for unstiffened webs, $= 1.25a_1/t[1 + 0.7(a_1/a_2)^2]^{1/2}$ for stiffened webs, h is the clear depth of the web, t is the web thickness, a_1 is the smallest dimension of the shear panel, and a_2 is the largest dimension of the shear panel.

11.2.2.14 Web Crushing

One of the design limitations of formed sheet members in bending and thin-webbed beams is local failure of the web under concentrated loads. There also is interaction between the effects of the concentrated load and the bending strength of the web.

For interior loads

$$P = \frac{t^2(N + 5.4)(\sin\Theta)(0.46F_{cy} + 0.02\sqrt{EF_{cy}})}{0.4 + r(1 - \cos\Theta)} \quad (11.36)$$

where P is the maximum load on one web (kip), N is the length of the load (in.), F_{cy} is the compressive yield strength (ksi), E is the modulus of elasticity (ksi), r is the radius between web and top flanges (in.), t is the thickness (in.), and Θ is the angle between the plane of the web and the plane of the loading (flange).

For loads at the end of the beam

$$P = \frac{1.2t^2(N + 1.3)(\sin\Theta)(0.46F_{cy} + 0.02\sqrt{EF_{cy}})}{0.4 + r(1 - \cos\Theta)} \quad (11.37)$$

If there is significant bending stresses at the point of concentrated load, the interaction may be calculated using the following equation:

$$\left(\frac{M}{M_u}\right)^{1.5} + \left(\frac{P}{P_u}\right)^{1.5} \leq 1.0 \quad (11.38)$$

where M is the applied moment (in.-kip), P is the applied concentrated load (kip), M_u is the maximum moment in bending (in.-kip), and P_u is the web crippling load (kip).

11.2.2.15 Stiffeners for Flat Plates

The addition of stiffeners to thin elements greatly improves the efficiency of a material. This is especially important for aluminum components because there usually is no need for a minimum thickness based on corrosion, and thus parts can be thin compared to those of steel, for example.

11.2.2.16 Stiffening Lips for Flanges

The buckling strength of the combined lip and flange is calculated using the equations for column buckling given previously and the following equivalent slenderness ratio that replaces the effective slenderness ratio. Element buckling of the flange plate and lip must also be considered:

$$\lambda = \pi\sqrt{\frac{I_p}{\frac{3}{8}J + 2\sqrt{C_w K_\phi/E}}} \quad (11.39)$$

where

λ = equivalent slenderness ratio (to be used in the column buckling equations)
$I_p = I_{xo} + I_{yo}$ = polar moment of inertia of lip and flange about center of rotation, in.[4]
$\quad I_{xo}, I_{yo}$ = moments of inertia of lip and flange about center of rotation, in.[4]

K_ϕ = elastic restraint factor (the torsional restraint against rotation as calculated from the application of unit outward forces at the centroid of the combined lip and flange of a one unit long strip of the section), in. lb/in.

J = torsion constant, in.[4]

$C_w = b^2(I_{yc} - bt^3/12)$ = warping term for lipped flange about center of rotation, in.[6]

I_{yc} = moment of inertia of flange and lip about their combined centroidal axis (the flange is considered to be parallel to the y-axis), in.[4]

b = flange width, in.

t = flange thickness, in.

E = modulus of elasticity, ksi

11.2.2.17 Intermediate Stiffeners for Plates in Compression

Longitudinal stiffeners (oriented parallel to the direction of the compressive stress) are often used to stabilize the compression flanges of formed sheet products and can be effective for any thin element of a column or the compression flange of a beam. The buckling strength of a plate supported on both edges with intermediate stiffeners is calculated using the column buckling equations and an equivalent slenderness ratio. The strength of the individual elements, plate between stiffeners, and stiffener elements also must be evaluated

$$\lambda = \frac{4Nb}{\sqrt{3}t} \sqrt{\frac{1 + A_s/bt}{1 + \sqrt{1 + 32I_e/3t^3b}}} \tag{11.40}$$

where λ is the equivalent slenderness ratio for the stiffener that replaces the effective slenderness ratio in the column formulas, N is the total number of panels into which the longitudinal stiffeners divide the plate (in.), b is the stiffener spacing (in.), t is the thickness of the plate (in.), I_e is the moment of inertia of the plate-stiffener combination about the neutral axis using an effective width of plate equal to b (in.[4]), A_s is the area of the stiffener (not including any of the plates (in.[2]).

11.2.2.18 Intermediate Stiffeners for Plates in Shear (Girder Webs)

Transverse stiffeners on girder webs must be stiff enough so that they remain straight during the buckling of the plate between stiffeners. The following equations are proposed for design:

$$I_s = \frac{0.46Vh^2}{E}(s/h), \quad \text{for } s/h \leq 0.4 \tag{11.41}$$

$$I_s = \frac{0.073Vh^2}{E}(h/s), \quad \text{for } s/h > 0.4, \tag{11.42}$$

where I_s is the moment of inertia of the stiffener (about face of web plate for stiffeners on one side of web only) (in.[4]), s is the stiffener spacing (in.), h is the clear height of the web (in.), V is the shear force on web at stiffener location (kip), and E is the modulus of elasticity (ksi).

11.2.2.19 Corrugated Webs

Corrugated sheet is highly efficient in carrying shear loads. Webs of girders and roofs and side walls of buildings are practical applications. The behavior of these panels, particularly the shear stiffness, is dependent not only on the type and size of corrugation but also on the manner in which it is attached to edge members. Test information and design suggestions are published elsewhere (Sharp 1993). Some of the failure modes to consider are as follows:

1. *Overall shear buckling.* This primarily is a function of the size of corrugations, the length of the panel parallel to the corrugations, the attachment and the alloy.

2. *Local buckling.* Individual flat or curved elements of the corrugations must be checked for buckling strength.
3. *Failure of corrugations and/or fastening at the attachment to the edge framing.* If not completely attached at the ends, the corrugation may roll or collapse at the supports, or fastening may fail.
4. *Excessive deformation.* This characteristic is difficult to calculate and the best guidelines are based on test data. The shear deformation can be many times that of flat webs particularly for those cases in which the fastening at the supports is not continuous.

11.2.2.20 Local Buckling of Tubes and Curved Panels

The strength of these members for each type of loading is defined by an equation for elastic buckling that applies to all alloys and two equations for the inelastic region that are dependent on alloy and temper. The members also need to be checked for overall buckling.

11.2.2.21 Round Tubes Under Uniform Compression

The local buckling strength is given by the following equations:

$$F_t = B_t - D_t \sqrt{\frac{R}{t}}, \quad \frac{R}{t} \le C_t \tag{11.43}$$

$$F_t = \frac{\pi^2 E}{16(R/t)\left(1 + \dfrac{\sqrt{R/t}}{35}\right)^2}, \quad \frac{R}{t} > C_t \tag{11.44}$$

where F_t is the buckling stress for a round tube in end compression (ksi), R is the mean radius of the tube (in.), t is the thickness of the tube (in.), and C_t is the intersection of equations for elastic and inelastic buckling (determined by charting or trial and error).

The values of the constants, B_t and D_t are given by the following formulas:

For wrought products with tempers starting with -O, -H, -T1, -T2, -T3, and -T4, and cast products

$$B_t = F_{cy}\left(1 + \frac{F_{cy}^{1/5}}{5.8}\right) \tag{11.45}$$

$$D_t = \frac{B_t}{3.7}\left(\frac{B_t}{E}\right)^{1/3} \tag{11.46}$$

For wrought products with tempers starting with -T5, -T6, -T7, -T8 and -T9

$$B_t = F_{cy}\left(1 + \frac{F_{cy}^{1/5}}{8.7}\right) \tag{11.47}$$

$$D_t = \frac{B_t}{4.5}\left(\frac{B_t}{E}\right)^{1/3} \tag{11.48}$$

where F_{cy} is the compressive yield strength, ksi.

For welded tubes, Equations 11.45 and 11.46 are used along with the yield strength for welded material. The accuracy of these equations has been verified for tubes with circumferential welds and R/t ratios equal to or less than 20. For tubes with much thinner walls, limited tests show that much lower buckling strengths may occur (Sharp 1993).

11.2.2.22 Round Tubes and Curved Panels Under Bending

For curved elements of panels under bending, such as corrugated sheet, the local buckling strength of the compression flange may be determined using the same equations as given in the preceding section for tubes under uniform compression.

In the case of round tubes under bending a higher compressive buckling strength is available for low R/t ratios due to the shape factor effect. (Tests have indicated that this higher strength is not developed in curved panels.) The equations for tubes in bending for low R/t are given in Equation 11.49. Note that the buckling of tubes in bending is provided by two equations in the inelastic region, that defined in Equation 11.49 and that for intermediate R/t ratios, which is the same as that for uniform compression, and the equation for elastic behavior, which also is the same as that for tubes under uniform compression

$$F_{tb} = B_{tb} - D_{tb}\sqrt{R/t}, \quad R/t \leq C_{tb} \tag{11.49}$$

where F_{tb} is the buckling stress for round tube in bending (ksi), R is the mean radius of the tube (in.), t is the thickness of the tube (in.), and C_{tb} is equal to $[(B_{tb} - B_t)/(D_{tb} - D_t)]^2$, the intersection of curves, Equations 11.49 and 11.43.

The values of the constants B_{tb} and D_{tb} are given by the following formulae:

For wrought products with tempers starting with -O, -H, -T1, -T2, -T3, and -T4, and cast products

$$B_{tb} = 1.5F_y\left(1 + \frac{F_y^{1/5}}{5.8}\right) \tag{11.50}$$

$$D_{tb} = \frac{B_{tb}}{2.7}\left(\frac{B_{tb}}{E}\right)^{1/3} \tag{11.51}$$

For wrought products with tempers starting with -T5, -T6, -T7, -T8, and -T9

$$B_{tb} = 1.5F_y\left(1 + \frac{F_y^{1/5}}{8.7}\right) \tag{11.52}$$

$$D_{tb} = \frac{B_{tb}}{2.7}\left(\frac{B_{tb}}{E}\right)^{1/3} \tag{11.53}$$

where F_y is the tensile or compressive yield strength, whichever is lower, ksi.

11.2.2.23 Round Tubes and Curved Panels Under Torsion and Shear

Thin walled curved members can buckle under torsion. Long tubes are covered in specifications of the Aluminum Association (2000) and provisions for stiffened and unstiffened cases are provided elsewhere (Sharp 1993).

11.2.3 Joints

11.2.3.1 Mechanical Connections

Aluminum components are joined by aluminum rivets, aluminum and steel (galvanized, aluminized, or stainless) bolts, and clinches. The joints are normally designed as bearing-type connections, because the as-received surfaces of aluminum products have a low coefficient of friction and slip often occurs at working loads. Some information has been developed for the amount of roughening of the surfaces and the limiting thicknesses of material for designing a friction-type joint, although current US specifications do not cover this type of design.

Table 11.9 presents strength data for a few of the rivet and bolt alloys available. Rivets are not recommended for applications that introduce large tensile forces on the fastener. The joints are proportioned based on the shear strength of the fastener and the bearing strength of the elements being joined. The bearing strengths apply to edge distances equal to at least twice the fastener diameter, otherwise reduced values apply. Steel bolts are often employed in aluminum structures. They are generally stronger than the aluminum bolts, and may be required for pulling together parts during assembly. They also have high fatigue strength, which is important in applications in which the fastener is subject to cyclic tension. The steel bolts must be properly coated or be of the 300 series stainless to avoid galvanic corrosion between the aluminum elements and the fastener.

Thin aluminum roofing and siding products are commonly used in the building industry. One failure mode is the pulling of the sheathing off the fastener due to uplift forces from wind. The pull-through strength is a function of the strength of the sheet, the geometry of the product, the location of the fastener, the hole diameter, and the size of the head of the fastener (Sharp 1993).

11.2.3.2 Welded Connections

The aluminum alloys employed in most nonaerospace applications are readily welded, and many structures are fabricated with this method of joining. Transverse groove weld strengths and appropriate filler alloys for a few of the alloys are given in Table 11.5. Because the weld strengths are usually less than those of the base material the design of aluminum welded structures is somewhat different from that of steel structures. Techniques for designing aluminum components with longitudinal and transverse welds are provided in the preceding section. If the welds are inclined to the direction of stress, either purely longitudinal or transverse, the strength of the connection more closely approximates that of the transversely welded case.

Fillet weld strengths are given in Table 11.10. Two categories are defined, longitudinal and transverse. These strengths are based on tests of specimens in which the welds were symmetrically placed and had no

TABLE 11.9 Strengths of Aluminum Bolts and Rivets

	Minimum expected strength, ksi	
Alloy and temper	Shear	Tension on net area
	Rivets	
6053-T61	20	—
6061-T6	25	—
	Bolts	
2024-T4	37	62
6061-T6	25	42
7075-T73	41	68

Source: The Aluminum Association, *Structural Design Manual*, 2000.

TABLE 11.10 Minimum Shear Strengths of Fillet Welds

	Shear strength, ksi	
Filler alloy	Longitudinal	Transverse
4043[a]	11.5	15
5356	17	26
5554	17	23
5556	20	30

[a] Naturally aged (2 to 3 months).

Source: Sharp, *Behavior and Design of Aluminum Structures*, McGraw-Hill, 1993.

large bending component. In the case of longitudinal fillets the welds were subjected to primarily shear stresses. The transverse fillet welds carried part of the load in tension. The difference in stress states accounts for the higher strengths for transverse welds. Aluminum specifications utilize the values for longitudinal welds for all orientations of welds, because many types of transverse fillet welds cannot develop the strengths shown because they have a more severe stress state than the test specimens, for example, more bending stress. Proportioning of complex fillet weld configurations is done using structural analysis techniques appropriate for steel and other metals.

11.2.3.3 Adhesive Bonded Connections

Adhesive bonding is not used as the only joining method for main structural components of non-aerospace applications. It is employed in combination with other joining methods and for secondary members. Although there are many potential advantages in the performance of adhesive joints compared to those for mechanical and welded joints, particularly in fatigue, there are too many uncertainties in design to use them in primary structures. Some of the problems in design are as follows:

1. There are no specific adhesives identified for general structures. The designer needs to work with adhesive experts to select the proper one for the application.
2. In order to achieve long-term durability proper pretreatment of the metal is required. There are little data available for long-term behavior, so the designer should supplement the design with durability tests.
3. There is no way to inspect the quality of the joint. Proper quality control of the joining process should result in good joints. However, a mistake can result in very low strengths, and the bad joint cannot be detected by inspection.
4. There are calculation procedures for proportioning simple joints in thin materials. Techniques for designing complex joints of thicker elements are under development, but are not adequate for design at this time.

11.2.4 Fatigue

Fatigue is a major design consideration for many aluminum applications, for example, aircraft, cars, trucks, railcars, bridges, and bridge decks. Most field failures of metal structures are by fatigue. The current design method used for all specifications, aluminum and steel structures, is to define categories of details that have essentially the same fatigue strength and fatigue curves for each of these categories. Smooth components, bolted and riveted joints, and welded joints are covered in the categories. For a new detail the designer must select the category that has a similar local stress. Chapter 34 of this handbook provides details of this method of design.

Many of the unique characteristics related to the fatigue behavior of aluminum components have been summarized (Sharp et al. 1996). Some general comments from this reference follow:

1. Some cyclic loads, such as wind-induced vibration and dynamic effects in forced vibration, are nearly impossible to design for because stresses are high and the number of cycles build up quickly. These loads must be reduced or eliminated by design.
2. Good practice to eliminate known features of structures causing fatigue, such as sharp notches and high local stresses due to concentrated loads, should be employed in all cases. In some applications the load spectrum is not known, for example, light poles, and fatigue-resistant joints must be employed.
3. The fatigue strength of aluminum parts is higher at low temperature and lower at elevated temperature compared to that at room temperature.
4. Corrosion generally does not have a large effect on the fatigue strength of welded and mechanically fastened joints but considerably lowers that of smooth components. Protective measures such as paint improves fatigue strength in most cases.

5. Many of the joints for aluminum structures are unusual in that they are quite different from those of the fatigue categories provided in the specifications. Stress analysis to define the critical local stress is useful in these cases. Test verification is desirable if practical.

11.3   Design

Aluminum should be considered for applications in which life cycle costs are favorable compared to competing materials. The costs include

1. Acquisition, refining, and manufacture of the metal.
2. Fabrication of the metal into a useful configuration.
3. Assembly and erection of the components in the final structure.
4. Maintenance and operation of the structure over its useful life.
5. Disposal after the useful life.

The present markets for aluminum have developed because of life cycle considerations. Transportation vehicles, one of the largest markets, with aerospace applications, aircraft, trucks, cars, and railcars, are light weight, thus saving fuel costs and are corrosion resistant thus minimizing maintenance costs. Packaging, another large market, makes use of close loop recycling that returns used cans to rolling mills that produce sheet for new cans. Building and infrastructure uses developed because of the durability of aluminum in the atmosphere without the need for painting, thus saving maintenance costs.

11.3.1   General Considerations

11.3.1.1   Product Selection

Most aluminum structures are constructed of flat-rolled products, sheet and plate, and extrusions because they provide the least cost solution. The properties and quality of these products are guaranteed by producers. The flat-rolled products may be bent or formed into shapes and joined to make the final structure. Extrusions should be considered for all applications requiring constant section members. Most extruders can supply shapes whose cross-section fits within a 10-in. circle. Larger shapes are made by a more limited number of manufacturers and are more expensive. Extrusions are attractive for use because the designer can incorporate special features to facilitate joining, place material in the section to optimize efficiency, and consolidate number of parts (compared with fabricated sheet parts). Because die costs are low the designer should develop unique shapes for most applications.

Forgings are generally more expensive than extrusions and plate, and are employed in aerospace applications and wheels, where the three-dimensional shape and high performance and quality are essential. Castings are also used for three-dimensional shapes, but the designer must work with the supplier for design assistance.

11.3.1.2   Alloy Selection

For extrusions alloy 6061 is best for higher strength applications and 6063 is preferred if the strength requirements are less. 5XXX alloys have been extruded and have higher as-welded strength and ductility in structures but they are generally much more expensive to manufacture, compared to the 6XXX alloys.

6061 sheet and plate are also available and are used for many applications. For the highest as-welded strength the 5XXX alloys are employed.

Table 11.11 shows alloys that have been employed in some applications. Choice of specific alloy depends on cost, strength, formability, weldability, and finishing characteristics.

11.3.1.3   Corrosion Resistance

Alloys shown in Table 11.11, 3XXX, 5XXX, and 6XXX, have high resistance to general atmospheric corrosion and can be employed without painting. Tests of small, thin specimens of these alloys in

TABLE 11.11   Selection of Alloy

Application	Specific use	Alloys
Architecture		
Sheet	Curtain walls, roofing and siding, mobile homes	3003, 3004, 3105
Extrusions	Window frames, railings, building frames	6061, 6063
Highway		
Plate	Signs, bridge decks	5086, 5456, 6061
Extrusions	Sign supports, lighting standards, bridge railings	6061, 6063
Industrial		
Plate	Tanks, pressure vessels, pipe	3003, 3004, 5083, 5086, 5456, 6061
Transportation		
Sheet/plate	Automobiles, trailers, railcars, shipping containers, boats	5052, 5083, 5086, 6061, 6009
Extrusions	Stiffeners/framing	6061
Miscellaneous extrusions	Scaffolding, towers, ladders	6061, 6063

Source: Gaylord, Gaylord, and Stallmeyer, *Structural Engineering Handbook*, McGraw-Hill, 1997.

a seacoast or industrial environment for over 50 years of exposure have shown that the depth of attack is small and self-limiting. A hard oxide layer forms on the surface of the component which prevents significant additional corrosion.

If aluminum components are attached to steel components protective measures must be employed to prevent galvanic corrosion. These measures include painting the steel components and placing a sealant in the joint. Stainless steel or galvanized fasteners are also required.

Some of the 5XXX alloys with magnesium content over about 3% may be sensitized by sustained elevated temperatures and lose their resistance to corrosion. For these applications alloys 5052 and 5454 may be used.

11.3.1.4 Metal Working

All of the usual fabrication processes can be used with aluminum. Forming capabilities vary with alloy. Special alloys are available for automotive applications in which high formability is required. Aluminum parts may be machined, cut, or drilled and the operations are much easier to accomplish compared to steel parts.

11.3.1.5 Finishing

Aluminum structures may be painted or anodized to achieve a color of choice. These finishes have excellent long-term durability. Bright surfaces also may be accomplished by mechanical polishing and buffing.

11.3.2 Design Studies

Some specific design examples follow in which product form, alloy selection, and joining method are discussed. The Aluminum Association Specifications are used for calculations of component strength.

EXAMPLE 11.1

Lighting standard

Design requirements

1. Withstand wind loads for area.
2. Fatigue and vibration resistant.

3. Heat treatable after welding to achieve higher strength.
4. Base that breaks away under vehicle impact.

Alloy and product

Round, extruded tubes of 6063-T4 are selected for the shaft. This alloy is easily extruded and has low cost and excellent corrosion resistance. The -T4 temper is required so that the pole can be tapered by a spinning operation so that the structure can be heat treated and aged after welding. A permanent mold casting of 356-T6 is selected for the base. The shaft extends through the base. This base may be acceptable for break away characteristics. If not, a break away device must be employed.

Joining

MIG circumferential welds are made at the top and bottom of the base using filler alloy 4043. This filler alloy must be employed because of the heat treat operation after welding. The corrosion resistance of a 5XXX filler alloy may be lowered by the heat treatment and aging.

Design considerations

Wind-induced vibration can be a problem occasionally. The vibration involves both the standard and luminare. There is currently no accurate way to predict whether or not these structures will vibrate. Light pole manufacturers have dampers that they can use if necessary.

Calculation example — bending of welded tube

Determine the bending strength of a 8 in. diameter (outside) × 0.313 in. wall tube of 6063-T4, heat treated and aged after welding using ASD. Factors of safety corresponding to building-type structures apply.

For this special case of fabrication the specifications allow the use of allowable stresses for the welded construction equal to 0.85 times those for 6063-T6. Also, the allowable stresses can be increased one third for wind loading:

1. The allowable tensile stress (tensile properties are given in Table 11.2, shape factors in Table 11.6, and factors of safety in Table 11.3) is as follows:

    tensile strength: $F_{tu} = 0.85(1.24)(1.33)(30)/(1.95) = 21.6$ ksi

    yield strength: $F_{ty} = 0.85(1.17)(1.33)(25)/(1.65) = 20.0$ ksi

2. The allowable compressive strength is given by Equations 11.43 to 11.53: $R/t = (4.0 - 0.313)/(0.313) = 11.8$. Equation 11.49 applies because R/t is less than $69.6(C_{tb})$. Constants are determined from Equations 11.52 and 11.53:

$$F_{tb} = (0.85)(1.33)\left(45.7 - 2.8\sqrt{(R/t)}\right)\Big/1.65 = 24.7 \text{ ksi}$$

3. The lower of the three values, 20.0 ksi, is used for design. This bending stress must be less than that calculated from all loads.

EXAMPLE 11.2

Overhead sign truss

Design requirements

1. Withstand wind loads for locality (signs and truss are considered).
2. Prevent wind induced vibration of truss and members.
3. Provide structure that does not need painting.

Alloy and product

Extruded tubes of 6061-T651 are selected for the truss and end supports. This alloy is readily welded and has excellent corrosion resistance. It also is one of the lower-cost extrusion alloys.

Joining

The individual members will be machined at the ends to fit closely with other parts and welded together using the MIG process. 5356 filler wire is specified to provide higher fillet weld strength compared to that for 4043 filler.

Design considerations

Wind-induced vibration must be prevented in these structures. The trusses are particularly susceptible to the wind when they do not have signs installed. Vibration of the entire truss can be controlled by the addition of a suitable damper (at midspan) and individual members must be designed to prevent vibration by limiting their slenderness ratio. (Sharp 1993).

Calculation example — buckling of a tubular column with welds at ends

The diagonal member of the truss is a 4-in. diameter tube (outside diameter) of 6061-T651 with a wall thickness of 0.125 in. The radius of gyration is 1.37. Its length is 48 in. and it is welded at each end to chords using filler 5356. Use ASD factors of safety corresponding to bridge structures (Table 11.3). Assume that the effective length factor is 1.0. Allow one third increase in stress because of wind loading

$$KL/r = (1.0)(48)/1.37 = 35.0$$

For column buckling, Equation 11.2 applies ($KL/r \leq C_c$). The constants are calculated from Equations 11.7 to 11.9 and parent metal properties (Table 11.2)

$$F_c = (1.33)(39.4 - 0.246(35.0))/2.20 = 18.6 \text{ ksi}$$

For yielding at the welds (the entire cross-section is affected at the ends), the properties in Table 11.5 are employed

$$F_c = (1.33)(20)/(1.85) = 14.4 \text{ ksi}$$

The allowable stress is the lower of these values, 14.4 ksi.

Calculation example — tubular column with welds at ends and midlength

This is the same construction as described in the previous paragraph, except that the designer has specified that a bracket be circumferentially welded to the tube at midlength. This weld lowers the column buckling strength. The column is now designed as though all the material is heat affected. Equation 11.2 still applies but constants are now calculated using Equations 11.4 to 11.6 and properties from Table 11.5:

$$F_c = (1.33)[22.8 - (1.0)(35.0)(0.133)]/2.20 = 11.0 \text{ ksi}$$

This stress is less than that calculated previously for yielding (14.4 ksi) and now governs. This stress must be higher than that calculated using the total load on the structure.

EXAMPLE 11.3

Built-up highway girder

Design requirements

The loads to be used for static and fatigue strength calculations are provided in AASHTO specifications. Long time maintenance-free construction is also specified. The size of the girder is larger than the largest extrudable section.

Riveted construction

Alloy 6061-T6 is selected for the web plate, flanges, and web stiffeners. This alloy has excellent corrosion resistance, is readily available, and has the highest strength for mechanical joining. Rivets of 6061-T6 are used for the joining because they are a good match for the parts of the girder from strength and corrosion considerations. The extrusions for the flanges and stiffeners are special sections designed to facilitate fabrication and to achieve maximum efficiency of material. A sealant is placed in the faying surfaces to enhance fatigue strength and to prevent ingress of detrimental substances.

Welded construction

Alloy 5456-H116 is selected for the web plate and flange plate. This alloy has high as-welded strength compared to 6061-T6 and excellent corrosion resistance. Alloys 5083 and 5086 would also be satisfactory selections. The filler wire selected for MIG welding is 5556, to have high fillet weld strength.

Calculation example — strength of riveted joint

The 6061-T6 parts are assembled as received from the supplier, so that the joint must be designed as a bearing connection. Use ASD for design. The thickness of the web plate is $\frac{1}{2}$ in. and it is attached to the legs of angle flanges (two angles) that are $\frac{3}{4}$ in. thick. One inch diameter rivets (area is 0.785 in.^{2}) are used.

Allowable bearing load on the web (bearing area is $0.50 \times 1.0 = 0.50$ in.^{2}) for one fastener is (see Table 11.2 and Table 11.3):

$$\text{Based on yielding:} \quad P = (58)(0.50)/1.85 = 15.7 \text{ kip}$$
$$\text{Based on ultimate:} \quad P = (88)(0.50)/2.64 = 16.7 \text{ kip}$$

Allowable shear load on one rivet with double shear:

Based on ultimate (see Table 11.3 and Table 11.9): $P = (2)(0.785)(25)/2.64 = 14.9$ kip

The allowable load per rivet is the smaller of the three values or 14.9 kip.

Calculation example — fatigue life of welded girder with longitudinal fillet welds

The allowable tensile strength for a 5456-H116 girder is (see Table 11.2 and Table 11.3):

$$\text{Based on yield:} \quad F = 33/1.85 = 17.8 \text{ ksi (governs)}$$
$$\text{Based on ultimate:} \quad F = 46/2.2 = 20.9 \text{ ksi}$$

Calculate the number of cycles that the girder can sustain at a stress range corresponding to a stress of half the static design value (8.9 ksi).

Category B applies to a connection with the fillet weld parallel to the direction of stress. For this category the fatigue strength is

$$\text{The stress range:} \quad S = 130N^{-0.207} = 8.9 \text{ ksi}$$
$$\text{The number of cycles:} \quad N = 423{,}000 \text{ cycles}$$

(fatigue equations from Aluminum Association 2000).

Calculation example — intermediate stiffeners

Stiffeners on girder webs must be of sufficient size to remain straight when the web buckles. Stiffener sizes are given by Equations 11.41 and 11.42.

EXAMPLE 11.4

Roofing or siding for a building

Design requirements

1. Withstand wind loads (uplift as well as downward pressure).
2. Withstand concentrated loads from foot pressure or from reactions at supports.
3. Corrosion resistant so that painting is not needed.

Alloy and product

Sheet of alloy 3004-H14 is selected. This alloy and temper has sufficient formability to roll-form the trapezoidal shape desired. It also has excellent corrosion and reasonable strength. Other 3XXX alloys would also be good choices.

Design considerations

Attachment of the sheet panels to the supporting structure must be strong enough to resist uplift forces. The pull through strength of the sheet product as well as the fastener strength are considered. Sufficient overlap of panels and fasteners at laps are needed for watertightness.

Calculation example — web crushing load at an intermediate support

Consider the shape shown in Figure 11.3. The bearing length is 2 in. (width of flange of support). Use LRFD specifications for buildings. The material properties for 3004-H14 are given in Table 11.2 and the resistance factors are in Table 11.4. For an interior load use Equation 11.36. $\phi = 0.90$ for this case, the same as that for web buckling.

$$\phi P = \frac{(0.90)(0.032)^2(2.0 + 5.4)(0.866)(0.46 \times 14.0 + 0.02(10100 \times 14.0)^{1/2})}{0.4 + 0.032(1 - 0.5)}$$

$$= 0.198 \text{ kip per web}$$

This load must be higher than that calculated using the factored loads. (Equations for factored loads are given in the Aluminum Association Specifications.)

Calculation example — bending strength of section

To calculate the section strength, the strength of the flange under uniform compression and the strength of the web under bending are calculated separately, and then combined using a weighted average calculation. The area of the web used in the calculation is that area beyond two thirds of the distance from the neutral axis. The resistance factor for the strength calculations from Table 11.4 is 0.85. The radii

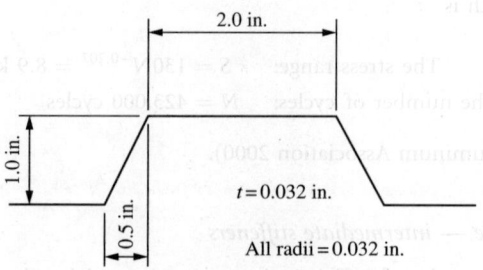

FIGURE 11.3   Example 11.4.

are neglected in subsequent calculations so that plate widths are to the intersection point of elements. The width to points of tangency of the corner radii is more accurate.

Strength of flange

Equation 11.27 governs because the b/t ratio (62.5) is larger than the b/t limit given for that equation. Values for B_p and D_p are given by Equations 11.19 and 11.20; the value of K_p is in Table 11.8

$$\phi F_{cr} = (0.85)(2.04)(18.4 \times 10100)^{1/2}/(1.6)(62.5) = 7.5 \text{ ksi}$$

Strength of web

The web is in bending and has a h/t ratio of 35 so Equation 11.17 governs. Values for B_p and D_p are given by Equations 11.25 and 11.26, and the value of K_p is in Table 11.8

$$\phi F_p = (0.85)(24.5 - (0.67)(0.147)(35)) = 17.9 \text{ ksi}$$

Strength of section

The bending strength of the section is between that calculated for the flange and the web. An accurate estimate of the strength is obtained from a weighted average calculation, which depends on the areas of the elements and the strength of each element. The area of the webs is that portion further than two thirds of the distance from the neutral axis

$$\phi F = [(2.0)t(7.5) + (2)(1.12)t(0.187)(17.9)]/(2.0 + 0.374)t = 9.5 \text{ ksi}$$

This stress must be higher than that calculated using factored loads.

Calculation example — intermediate stiffener

The bending strength of the section in Figure 11.3 can be increased significantly, with a small increase in material, by the addition of a formed stiffener at midwidth as illustrated in Figure 11.4. The strength of the stiffened panel is calculated using an equivalent slenderness ratio as given by Equation 11.40 and column buckling equations. The addition of a few percent more material as illustrated in Figure 11.4 can increase section strength by over 25%.

Calculation example — combined bending and concentration loads

The formed sheet product can experience high longitudinal compressive stresses and a high normal concentrated load at the same location such as an intermediate support. These stresses interact and must be limited as defined by Equation 11.38.

EXAMPLE 11.5

Orthotropic bridge deck

Design requirements

1. Withstand the static and impact loads as provided in an appropriate bridge design specification.
2. Withstand the cyclic loads provided in specifications.

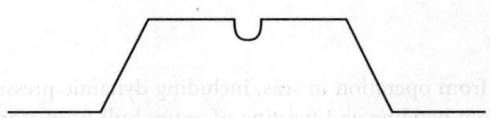

FIGURE 11.4 Example 11.4.

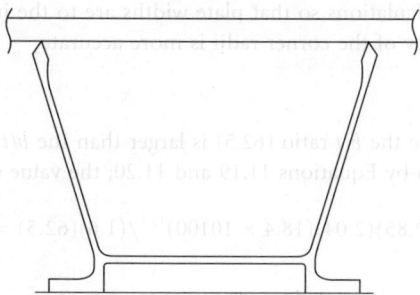

FIGURE 11.5   Example 11.5.

3. Fabricated by the use of welding.
4. Corrosion resistant so that painting is not needed.
5. Large prefabricated panels to shorten erection time.

Alloy and product

The selection depends on the type of construction desired. Figure 11.5 is a plate reinforced by an extruded closed stiffener. This construction has been used successfully. The plate is 5456-H116, chosen because of its high as-welded strength. The extrusion is 6061-T651, which has high strength and reasonable cost. The extrusion is designed to accommodate welding and attachment to supports to minimize fabrication costs. Both alloys have excellent corrosion resistance and will not need to be painted.

All extruded decks with segments either bolted or welded together, to achieve a shape similar to that in Figure 11.5, have also been used. 6061-T651 extrusions for all the segments are the choice in this case.

Joining

MIG welding with filler alloy 5556 is selected for attaching the extrusions to the plate. Fixturing is required to control the final shape of the panel.

Design considerations

Large panels, 11 × 28 ft or larger, complete with wearing surface have been fabricated. The panels must be attached tightly to the supporting structure to avoid fatigue failures of the fasteners.

Galvanized A356 bolts are suggested for the attachment, to obtain high static and fatigue strengths.

Calculation example — bending stresses in plate and section

Fatigue is the major design concern in a metal bridge deck. Wheel pressures cause bending stresses in the deck plate transverse to the direction of the stiffeners. These loads also cause longitudinal bending stresses in the stiffened panel. Fatigue evaluations are needed for both stresses. Deflection and static strength requirements of specifications also must be met.

EXAMPLE 11.6

Ship hull

Design requirements

1. Withstand pressures from operation in seas, including dynamic pressures from storms.
2. Withstand stresses from bending and twisting of entire hull from storm conditions.
3. The hull is of welded construction.

4. The hull must plastically deform without fracture when impacting with another object.
5. Employ joints that are proven to be fatigue resistant in other metal ship structures.
6. Corrosion resistant so that painting is not required even for salt water exposure.

Alloy and product

The hull is constructed of stiffened plate. Alloy 5456-H116 plate and 5456-H111 extruded stiffeners are selected. Main girders are fabricated using 5456-H116 plate. Other lower-strength 5XXX alloys are also suitable choices. This alloy is readily welded and has high as-welded strength. The 5456-H111 extrusions are more expensive than those of 6061-T6 but the welded 5XXX construction is much tougher using 5XXX stiffening, and thus would better accommodate damage without failure. This alloy has excellent resistance to corrosion in a salt water environment.

Joining

MIG welding using 5556 filler is specified. This is a high-strength filler and appropriate for joining parts of this high-strength alloy.

Design considerations

Loadings for hull or component design are difficult to obtain. The American Bureau of Shipping has requirements for the size of some of the components. Fireproofing is required in some areas.

Calculation example — buckling of stiffened panel

For a longitudinally framed vessel the hull plate and stiffeners will be under compression from bending of the ship. The stiffened panel must be checked for column buckling between major transverse members using Equations 11.2 and 11.3. For hull construction that is subjected to normal pressures, the stiffened panel will have lateral bending as well as longitudinal compression. Equations 11.15 and 11.16 are needed in this case.

Elements of the stiffened panel must be checked for strength under the compression loads. In addition, an angle or Tee stiffener can fail by a torsion about an enforced axis of rotation, the point of attachment of the stiffener to the plate. Equation 11.39 may be used for the calculation.

EXAMPLE 11.7

Latticed tower or space frame

Design requirements

1. Withstand wind, earthquake, and other imposed loads.
2. Corrosion resistant so that painting is not required.
3. Prevent wind-induced vibration.

Alloy and product

Extrusions of 6061-T651 are selected for the members because of their corrosion resistance, strength, and economy. 6063 extrusions can be more economical if the higher strength of 6061 is not needed. The designer should make full use of the extrusion process by designing features in the cross-section that will facilitate joining and erection, and that will result in optimal use of the material.

Joining

Mechanical fasteners are selected. Galvanized A325 or stainless steel fasteners are best for major structures because of their higher strength compared to those of aluminum.

Design considerations

Overall buckling of the system as well as the buckling of components must be considered. The manner in which the members are attached at their ends can affect both component and overall strength.

Special extrusions in the form of angles, Y-sections, and hat-sections have been used in these structures. Some of these sections can fail by flexural–torsional buckling under compressive loads. Equation 11.10 covers this case. These sections, because they are relatively flexible in torsion, can vibrate in the wind in torsion as well as flexure.

11.4   Economics of Design

There are two considerations that can affect the economy of aluminum structures: efficiency of design and life cycle costs. These considerations will be summarized briefly here.

Most structural designers are schooled in and are comfortable with design in steel. Although the design of aluminum structures is very similar to that in steel there are differences in their basic characteristics that should be recognized:

1. The density of aluminum alloys is about one third that of steel. Efficiently designed aluminum structures will weigh about one third to half those of efficiently designed steel structures, depending on failure mode. The lighter structures are governed by tensile or yield strength of the material and the heavier ones by fatigue, deflection, or buckling.
2. Modulus of elasticity of aluminum alloys is about one third that of steel. The size/shape of efficient aluminum components will need to be larger for aluminum structures as compared to those of steel for the same performance.
3. The fatigue strength of a joint of aluminum is one third to half that of steel, with identical geometry. The size/shape of the aluminum component will need to be larger than that of steel to have the same performance.
4. The resistance to corrosion from the atmosphere of aluminum is much higher than that of steel. The thickness of aluminum parts can be much thinner than those of steel and painting is not needed for most aluminum structures.
5. Extrusions are used for aluminum shapes of nonrolled sections as used for steel. The designer has much flexibility in the design to: (a) consolidate parts, (b) include features for welding to eliminate machining, (c) include features to snap together parts or to accommodate mechanical fasteners, and (d) to include stiffeners, nonuniform thickness, and other features to provide the most efficient placement of metal. Because die costs are low most extrusions are uniquely designed for the application.

Aluminum applications are economical generally because of life cycle considerations. In some cases, for example, castings, aluminum can be competitive on a first cost basis compared to steel. Light weight and corrosion resistance are important in transportation applications. In this case the higher initial cost of the aluminum structure is more than offset by lower fuel costs and higher pay loads. Closed loop recycling is possible for aluminum and scrap has high value. The used beverage can is converted into sheet to make additional cans with no deterioration of properties.

Glossary

Alloy — Aluminum in which a small percentage of one or more of other elements have been added primarily to improve strength.

Foil — Flat-rolled product that is less than 0.006 in. thick.

Heat-affected zone — Reduced strength material from welding measured 1 in. from centerline of groove weld or 1 in. from toe or heel of fillet weld.

Plate — Flat-rolled product that is greater than 0.25 in. thickness.
Sheet — Flat-rolled product between 0.006 and 0.25 in. thickness.
Temper — The measure of the characteristic of the alloy as established by the fabrication process.

References

Aluminum Association, *Aluminum Design Manual*, Washington DC, 2000.

AASHTO, American Association of State Highway and Transportation Officials, *AASHTO LRFD Bridge Design Specifications*, Washington DC, 2002.

Sharp, M.L., *Behavior and Design of Aluminum Structures*, McGraw-Hill, New York, 1993.

Sharp, M.L., Development of Aluminum Structural Technology in the United States, *Proceedings, 50th Anniversary Conference*, Structural Stability Research Council, Bethlehem, Pennsylvania, 21–22 June 1994.

Sharp, M.L., Nordmark, G.E., and Menzemer, C.C., *Fatigue Design of Aluminum Components and Structures*, McGraw-Hill, New York, 1996.

Further Reading

Aluminum Association, *The Aluminum Extrusion Manual*, Washington DC, 1998.

Aluminum Association, website for book store, www.aluminum.org

American Bureau of Shipping, *Rules for Building and Classing Aluminum Vessels*, New York, 1975.

American Welding Society, *ANSI/AWS D1.2/D1.2M:2003 Structural Welding Code Aluminum*, Miami, Florida, 2003.

Gaylord, E.H., Gaylord, C.N., and Stallmeyer, J.E., *Structural Engineering Handbook*, McGraw-Hill, New York, 1997.

Kissell, J.R. and Ferry, R.L., *Aluminum Structures, A Guide to Their Specifications and Design*, John Wiley & Sons, Inc., New York, 2002.

12

Reliability-Based Structural Design

Achintya Haldar
Department of Civil Engineering and
Engineering Mechanics,
The University of Arizona,
Tucson, AZ

12.1 Introduction

Structural design consists of proportioning elements of a system to satisfy various criteria of performance, safety, serviceability, and durability under various demands. The presence of uncertainty cannot be avoided in every phase of structural engineering analysis and design, but it is not simple to satisfy design requirements in the presence of uncertainty. After three decades of extensive work in different engineering disciplines, several reliability evaluation procedures of various degrees of complexity are now available. First-generation structural design guidelines and codes are being developed and promoted worldwide using some of these procedures.

In civil engineering in particular, the structures group has been providing leadership in implementing the reliability-based design concept. The reliability-based structural design concept, in the form of the load and resistance factor design (LRFD) code, was formally introduced by the American Institute of Steel Construction (AISC) as early as 1986 [1]. The LRFD code was intended to be an alternative to the allowable stress design (ASD) [2] concept used exclusively before 1986. The third edition of the LRFD code was introduced in 2001 [1]. Similar design guidelines for concrete [3], masonry [4], and wood [5,6] are now available, reflecting the reliability-based design concept. Concrete, masonry, and wood codes do not explicitly state that they are based on the LRFD concept, but the intention is the same. Since this handbook will be used by all segments of the structural engineering profession, it would not be appropriate to use the LRFD concept to imply the reliability-based design concept. However, they are essentially the same and it would not be an exaggeration if the terms LRFD and the reliability-based design concept were used synonymously. Most of the worldwide codes applicable to structural engineering have already been modified or are in the process of being modified to implement the concept [7]. At present almost all civil engineering schools in the United States teach steel design using the LRFD concept. It is expected that reliability-based structural design will be the only design option worldwide in the near future. Under the sponsorship of the International Standard Organization (ISO), a standard (ISO 2394) [8] is now being considered for this purpose.

Since AISC's LRFD concept has almost two decades of history, it is used to make some observations here. AISC published the ninth edition of the ASD code in 1989 after the introduction of the LRFD code in 1986, indicating indecisiveness or uneasiness in using the LRFD concept in structural steel design. Even recent graduates who are taught steel design using only the LRFD concept are asked to use the ASD method in everyday practice after graduation. LRFD is calibrated with respect to ASD for a ratio of live load to dead load of 3, that is, the two methods will usually give identical sections. When the ratio is less than 3, as is expected in most design applications, the material savings could be as much as 18%. AISC [9] documented various amounts of savings for office buildings, parking structures, floor and framing systems, and industrial buildings, but the LRFD concept is still being ignored or overlooked by most of the profession. The author believes that the main cause of this nonuniform use of the LRFD code is not its difficulty, but the older generation's lack of familiarity with the concept, new terminologies, and the cost of changing from the old to new guidelines. Since LRFD will save on material cost in most designs, the extra cost of implementing LRFD can be recovered within a short period of time. In the near future, it is expected that the reliability-based design concept, similar to the LRFD concept, may not be an option but a requirement in structural analysis for steel, concrete, wood, and masonry structures.

It can be observed that steel design using the LRFD concept is not that different from concrete design using the ACI's ultimate strength design concept [10], and is not more difficult than using AISC's old ASD method. The ACI recently published strength design guidelines similar to the LRFD concept [3]. Design aids like figures, tables, and charts are very similar in the LRFD and ASD guidelines for steel. It is not difficult for a person familiar with one design concept to use the other with a minor effort. The use of ASD by recent graduates trained in LRFD attests to this statement. Unlike in Europe, a code is not a government document in the United States. It is developed by the profession, and its acceptance is voted by the users and developers. Since all major professional groups are advocating reliability-based design and following the worldwide trend, it would be very appropriate to use only the reliability-based approach in future designs.

12.2   Available Structural Design Concepts

Before introducing the reliability-based design concept in structural engineering, it may be informative to study different deterministic structural design concepts used in the recent past. The fundamental concept behind any structural design is that the resistance of a structural element, joint, or the structure as a whole should be greater than the load or combinations of loads that may act during its lifetime with some conservatism or safety factor built in. The level of conservatism is introduced in the design in several ways depending on the basic design concept being used. In the ASD approach, the basic concept is that the allowable stresses should be greater than the unfactored nominal loads or load combinations expected during its lifetime. The allowable stresses are calculated using a safety factor. In other words, the nominal resistance R_n is divided by a safety factor to compute the allowable resistance R_a, and safe design requires that the nominal load effect S_n is less than R_a. In the ultimate strength design method [10], the loads are multiplied by certain load factors to determine the ultimate load effects and the members are required to resist various design combinations of the ultimate load. In this case, the safety factors are used in the loads and load combinations.

It is well known that the loads that may act on a structure during its lifetime are very unpredictable and different levels of unpredictability or uncertainty exist for each load. The uncertainty associated with predicting dead load is expected to be lower than that of live, wind, or seismic load. Since different assumptions are made in developing the beam and column theories, the theoretical prediction of the resistance or strength of beams and columns is expected to have different levels of uncertainty. In the ASD approach, the safety factor is introduced in predicting resistance, and the loads are assumed at their nominal values ignoring the different levels of uncertainty in predicting them. In the ultimate strength design concept, conservatism is introduced by using different load factors. Conceptually, the use of load- and resistance-related safety factors may not assure a uniform underlying risk for different structural elements, for example, beams, columns, and slabs, or under different loading conditions (e.g., dead, live, wind, or seismic loads). In the LRFD or reliability-based design concept, conservatism is introduced by using both load and resistance factors and satisfying an underlying risk, combining the desirable features of both the ASD and ultimate design concepts. Essentially, the LRFD approach uses safety factors to estimate both the resistance and load under the constraint of an underlying risk. Since satisfying an underlying risk is the main objective of the reliability-based design concept, this assures uniform risk for the structure and may produce a more economical design than other deterministic design concepts.

12.3 Introduction of the Reliability-Based Structural Design Concept

Like the LRFD format, the reliability-based design concept can be represented in its basic form as

$$\sum \gamma_i Q_i \leq \phi R_n \tag{12.1}$$

where Q_i is the ith nominal load effect, γ_i is the load factor corresponding to Q_i, R_n is the nominal resistance, and ϕ is the resistance factor corresponding to R_n. The load factor γ_i and the resistance factor ϕ account for the uncertainties in the parameters related to the loads and resistance. The mathematical expressions for these factors will be derived in Section 12.4. These factors are derived based on reliability analysis of simple "standard" structures such as simple beams, centrally loaded columns, tension members, high-strength bolts, and fillet welds [11], and are calibrated to achieve levels of reliability similar to conventional ASD procedures such as the 1989 AISC specification [2].

One of the most important objectives of the reliability-based design approach is that it provides a reasonable platform to compare different design alternatives by considering the realistic behavior of structures and the uncertainty in the design variables satisfying some underlying design criteria. It tries to reduce the scatter in the underlying risk of different members designed according to the ASD concept. Thus, the risk-consistent load factors used in LRFD help to economize the design. Since dead load has less uncertainty than live load, it should have a smaller load factor. But in the ASD procedure, the dead and live loads have the same load factor of 1. In evaluating the strength or resistance, theoretically predicted values generally do not match experimental results. The strength of a beam is more predictable than column strength. The support conditions and the buckling behavior of a column make its strength prediction more uncertain, indicating that beams and columns should have different capacity reduction factors to satisfy the same underlying risk. This is common-sense logic, and reliability or risk-based design is expected to be superior to ASD.

ASD is essentially a deterministic design concept. A casual evaluation of the LRFD code indicates that it is also deterministic in nature; only the prescribed load and resistance factors are based on reliability analysis. This information may not be of any practical significance to a typical practicing engineer.

The design aids in AISC's LRFD and ASD guidelines are very similar. In both approaches, the isolated member approach was used to develop the design guidelines [11]. There are several advantages to the isolated-member approach: (1) in deterministic design methods that use safety factors, it is not practical to prepare detailed requirements for each structural configuration; (2) the characteristics of the individual members and connections are independent of the framework; and (3) most research has been devoted to the study of such elements, and theoretical and experimental verification of their performance is readily available. Nevertheless, the performance of a member is directly dependent on its location in a structural configuration and on its relationship or connection with the other members in the framework. Such dependence is not restricted to the computation of load effects through a deterministic analysis of the structure, but extends to the probabilistic variation of the load effect as well, which is influenced by the probabilistic characteristics of all the parameters of the structure [12]. Only a probabilistic structural analysis of the entire structure can account for this influence and accordingly determine the risk or reliability of any individual member, enabling an improved approach to reliability-based design. However, this approach could be very complicated [13] and may not be practical for everyday structural design.

The analytical procedures used to estimate the load effects are identical in ASD and LRFD approaches. Thus, the analytical procedures, computer programs, and other analysis aids used for ASD are also applicable for LRFD. Only the treatments of load effects are different in the two concepts. The nominal resistance is also calculated deterministically using a codified approach. The capacity reduction factor is used to address the level of uncertainty in estimating it. As pointed out earlier, for wider applications, LRFD or reliability-based design guidelines were calibrated with respect to time-tested ASD procedure. Thus, in many cases, the final selection of a structural member will be identical for ASD and the reliability-based design concept; however, the design procedures will be very different as outlined below. Using the reliability-based design concept, structural engineers will be more empowered to manage risk in a typical design.

12.4 Fundamental Concept of Reliability-Based Structural Design

It is assumed in the following sections that the readers are familiar with the basic concept of uncertainty analysis. If not, they are urged to refer to a recent book authored by Haldar and Mahadevan [14].

In general, R and S can be used to represent the resistance and load effect as random variables since they are functions of many other random variables. R is a function of material properties and the geometric properties of a structural element including cross-sectional properties. S is a function of the load effect that can be expected during the lifetime of the structural element. The uncertainty in R and S can be completely defined by their corresponding probability density functions (PDFs) denoted as $f_R(r)$ and $f_S(s)$, respectively. Then, the probability of failure of the structural element can be defined as the probability of the resistance being less than the load effect or simply $P(R > S)$. Mathematically, it can be expressed as [14]

$$P(\text{failure}) = P(R < S) = \int_0^\infty \left[\int_0^s f_R(r)\,\mathrm{d}r \right] f_S(s)\,\mathrm{d}s = \int_0^\infty F_R(s)f_S(s)\,\mathrm{d}s \qquad (12.2)$$

where $F_R(s)$ is the cumulative distribution function (CDF) of R evaluated at s. Conceptually, Equation 12.2 states that for a particular value of the random variable $S = s$, $F_R(s)$ is the probability of failure. However, since S is also a random variable, the integration needs to be carried out for all possible values of S, with their respective likelihood represented by the corresponding PDF. Equation 12.2 can be considered as the fundamental equation of the reliability-based design concept.

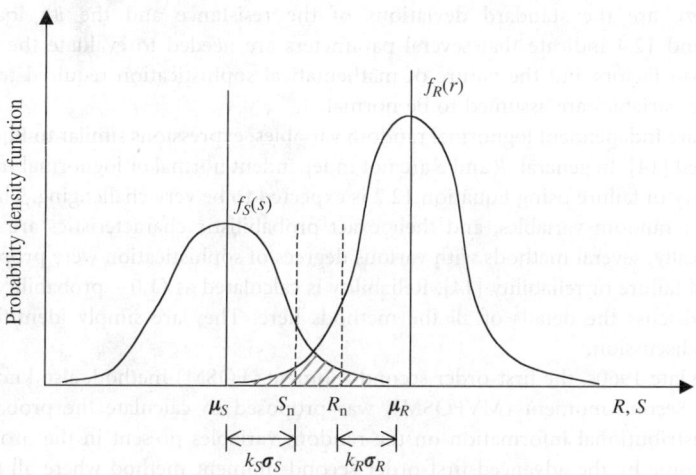

FIGURE 12.1 Reliability-based design concept. (Adopted from *Probability, Reliability and Statistical Methods in Engineering Design*, by Haldar and Mahadevan, 2000, with permission from John Wiley & Sons, Inc.)

The concept is shown in Figure 12.1. In Figure 12.1, the nominal values of resistance and load effect, denoted as R_n and S_n, respectively, and the corresponding PDFs of R and S are shown. The overlapped (dashed area) between the two PDFs provides a *qualitative measure* of the probability of failure. Controlling the size of the overlapped area is essentially the idea behind reliability-based design. Haldar and Mahadevan [14] pointed out that the area could be controlled by changing the relative locations of the two PDFs by separating the mean values of R and S (μ_R and μ_S), the uncertainty expressed in terms of their standard deviations (σ_R and σ_S), and the shape of the PDFs [$f_R(r)$ and $f_S(s)$].

Although conceptually simple, the evaluation of Equation 12.2 may not be easy except for some special cases. Consider $S = S_1 + S_2 + \cdots + S_n$, representing n statistically independent load effects (dead, live, wind loads, etc.), and R and S are independent normal random variables. Under these assumptions, the resistance and the ith load factors in Equation 12.1 can be shown to be [14]

$$\phi = \frac{1 - \varepsilon\beta\delta_R}{1 - k_R\delta_R} \tag{12.3}$$

and

$$\gamma_i = \frac{1 + \varepsilon\varepsilon_{nn}\beta\delta_{S_i}}{1 + k_{S_i}\delta_{S_i}} \tag{12.4}$$

where β is the reliability index, a measure of probability of failure, δ_R and δ_S are the coefficients of variation (COV) of R and S (a measure of uncertainty), k_R is the number of standard deviations below the mean resistance in selecting the nominal value of R (a measure of underestimation of the resistance), k_S is the number of standard deviations above the mean load in selecting the nominal value of the ith load (a measure of overestimation in the load), and

$$\varepsilon = \frac{\sqrt{\sigma_R^2 + \sigma_S^2}}{\sigma_R + \sigma_S} \tag{12.5}$$

$$\varepsilon_{nn} = \frac{\sqrt{\sigma_{S_1}^2 + \sigma_{S_2}^2 + \cdots + \sigma_{S_n}^2}}{\sigma_{S_1} + \sigma_{S_2} + \cdots + \sigma_{S_n}} \tag{12.6}$$

where σ_R and σ_{S_i} are the standard deviations of the resistance and the ith load, respectively. Equations 12.3 and 12.4 indicate that several parameters are needed to evaluate the reliability-based resistance and load factors and the nature of mathematical sophistication required to evaluate them, even when all the variables are assumed to be normal.

When R and S are independent lognormal random variables, expressions similar to Equations 12.3 and 12.4 can be derived [14]. In general, R and S are not independent normal or lognormal random variables, and the probability of failure using Equation 12.2 is expected to be very challenging. They are functions of many different random variables, and their exact probabilistic characteristics are very difficult to evaluate. Historically, several methods with various degrees of sophistication were proposed to evaluate the probability of failure or reliability [14]. Reliability is calculated as (1.0 − probability of failure). It is not possible to discuss the details of all the methods here. They are simply identified here for the completeness of discussion.

Initially, in the late 1960s, the first-order second-moment (FOSM) method, also known as the mean value first-order second-moment (MVFOSM), was proposed to calculate the probability of failure neglecting the distributional information on the random variables present in the problem. This deficiency was overcome by the advanced first-order second-moment method where all the variables are assumed to be normal and independent as proposed by Hasofer and Lind [15]. Rackwitz [16] proposed a more general formulation applicable to different types of distributions. Currently, it is the most widely used reliability evaluation technique. Using this concept, the probability of failure has been estimated using two types of approximations to the limit state at the design point (defined in the following section): first order (leading to the name first-order reliability method or FORM) and second order (leading to the name second-order reliability method or SORM). Since FORM is a commonly used reliability evaluation technique, it is discussed in more detail below. A person without a sophisticated background in probability and statistics can use simulation to evaluate the underlying risk or reliability, as discussed in Section 12.7.

12.4.1 First-Order Reliability Method

The basic idea behind reliability-based structural design is to design a structural member satisfying several performance criteria and considering the uncertainties in the relevant load- and resistance-related random variables, called the basic variables X_i. Since the R and S random variables in Equation 12.2 are functions of many other load- and resistance-related random variables, they are generally treated as basic random variables. The relationship between the basic random variables and the performance criterion, known as the performance or limit state function, can be mathematically represented as

$$Z = g(X_1, X_2, \ldots, X_n) \tag{12.7}$$

The failure surface or the limit state of interest can then be defined as $Z = 0$. The limit state equation plays an important role in evaluating reliability using FORM. It represents the boundary between the safe and unsafe regions and a state beyond which a structure can no longer fulfill the function for which it was designed. Assuming R and S are the two basic random variables, the limit state equation, and the safe and unsafe regions are shown in Figure 12.2. A limit state equation can be an explicit or implicit function of the basic random variables and can be linear or nonlinear. Reliability estimation using explicit limit state functions is discussed here. Haldar and Mahadevan [13] discussed reliability evaluation techniques for implicit limit state functions.

Two types of performance functions are generally used in structural engineering: strength and serviceability. Strength performance functions relate to the safety of the structures and serviceability performance functions are related to the serviceability (deflection, vibration, etc.) of the structure. The reliabilities underlying the strength and serviceability performance functions are expected to be different.

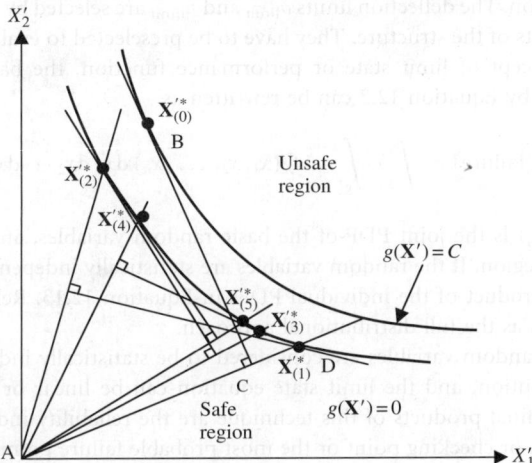

FIGURE 12.2 Limit state concept. (Adopted from *Probability, Reliability and Statistical Methods in Engineering Design*, by Haldar and Mahadevan, 2000, with permission from John Wiley & Sons, Inc.)
Note: A number in parenthesis indicates iteration number.

The limit state equation for the strength limit state in pure bending can be expressed as

$$g(\mathbf{X}) = 1.0 - \frac{M_u}{M_n} \tag{12.8}$$

where M_u is the unfactored applied moment and M_n is the nominal flexural strength of the member.

Using the AISC's LRFD design criteria, the limit state equation for the strength limit state for a beam–column element can be expressed as

$$g(\mathbf{X}) = 1.0 - \left(\frac{P_u}{P_n} + \frac{8}{9}\frac{M_u}{M_n}\right), \quad \text{if } \frac{P_u}{\phi P_n} \geq 0.2 \tag{12.9}$$

$$g(\mathbf{X}) = 1.0 - \left(\frac{P_u}{2P_n} + \frac{M_u}{M_n}\right), \quad \text{if } \frac{P_u}{\phi P_n} < 0.2 \tag{12.10}$$

where P_u is the unfactored tensile and compressive load effects, P_n is the nominal tensile and compressive strength, M_u is the unfactored flexural load effect, and M_n is the nominal flexural strength. Limit state equations can similarly be defined for other strength design criteria like tension, compression, shear, etc. All the strength-related parameters in Equations 12.8 to 12.10 are functions of the geometry, material, and cross-sectional properties of the members under consideration. The code also suggests analytical procedures to evaluate the load effects in some cases. The important point is that the loads effects in Equations 12.8 to 12.10 are to be unfactored in evaluating the reliability index.

For the serviceability limit state, the midspan deflection of beams under live load and the side sway (interstory drift and lateral deflection) of frames are commonly used. They can be represented as

$$g(\mathbf{X}) = 1.0 - \frac{\delta}{\delta_{\text{limit}}} \tag{12.11}$$

$$g(\mathbf{X}) = 1.0 - \frac{u}{u_{\text{limit}}} \tag{12.12}$$

where δ is the vertical deflection of a beam under unfactored live load, δ_{limit} is the allowable or prescribed vertical deflection, u is the lateral deflection under unfactored loads, and u_{limit} is the allowable or prescribed lateral deflection. The deflection limits δ_{limit} and u_{limit} are selected by the designer based on the performance requirements of the structure. They have to be preselected to evaluate the reliability index.

Incorporating the concept of limit state or performance function, the basic reliability evaluation formulation represented by Equation 12.2 can be rewritten as

$$P(\text{failure}) = \int \cdots \int_{g()<0} f_X(x_1, x_2, \ldots, x_n)\, dx_1\, dx_2 \cdots dx_n \qquad (12.13)$$

in which $f_X(x_1, x_2, \ldots, x_n)$ is the joint PDF of the basic random variables, and the integration is performed over the failure region. If the random variables are statistically independent, then the joint PDF can be replaced by the product of the individual PDFs in Equation 12.13. Reliability evaluation using Equation 12.13 is known as the full distributional approach.

Initially, all the basic random variables are considered to be statistically independent; they can have different types of distribution, and the limit state equation can be linear or nonlinear. FORM is an iterative technique. The final products of this technique are the reliability index β, the corresponding coordinates of the design or checking point or the most probable failure point $(x_1^*, x_2^*, \ldots, x_n^*)$, and the sensitivity indexes indicating the influence of the individual random variables on the reliability index. In the context of FROM, the reliability index β has a physical interpretation. It is the shortest distance from the origin to the limit state function at the checking point in the reduced standard normal variable space as shown in Figure 12.2 and Figure 12.3. As will be discussed further later, an optimization technique is used to estimate it iteratively. Once the information on β is available, the probability of failure can be obtained as

$$P(\text{failure}) = \Phi(-\beta) = 1.0 - \Phi(\beta) \qquad (12.14)$$

where Φ is the CDF of the standard normal variable. If β is large, the probability of failure will be small. The coordinates on the limit state surface where the iteration converges represents the worst combination of the random variables that would cause failure and is appropriately named the design point or the most probable failure point $(x_1^*, x_2^*, \ldots, x_n^*)$. All these aspects of FORM are discussed in the following section.

The estimation of probability of failure using Equation 12.13 is expected to be complicated. As mentioned earlier, Rackwitz [16] suggested a solution strategy that can be used to evaluate the

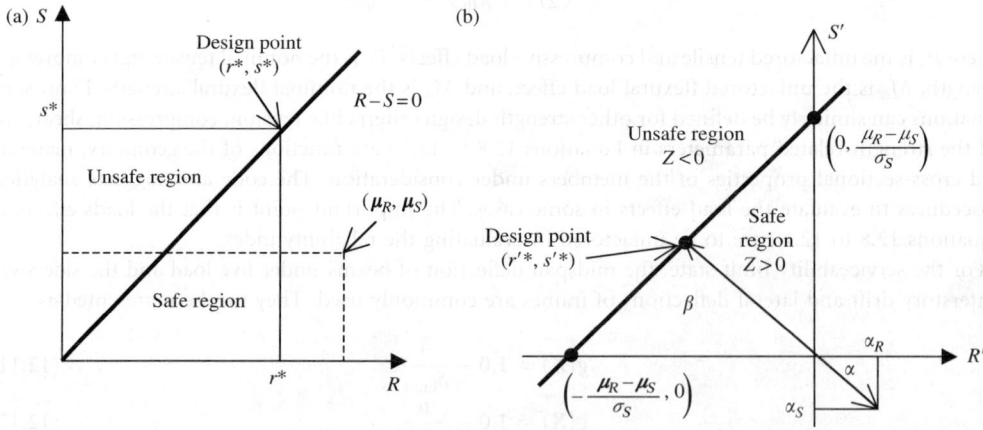

FIGURE 12.3 Reliability evaluation for linear performance function: (a) original coordinates and (b) reduced coordinates. (Adopted from *Probability, Reliability and Statistical Methods in Engineering Design*, by Haldar and Mahadevan, 2000, with permission from John Wiley & Sons, Inc.)

probability of failure for most problems of practical interest. The method requires that the limit state function be available in explicit form, and is discussed next. Haldar and Mahadevan [14] discussed a FORM method where the limit state function is implicit, but it will not be discussed here. Rackwitz's method [16] and some improvements suggested by Ayyub and Haldar [17] are discussed next. For ease of discussion, the method is presented in the form of eight tasks:

- *Task 1.* The limit state equation must be available or defined in terms of the basic random variables and performance criterion.
- *Task 2.* Assume an intelligent initial value of the reliability index β. An initial β value can be 3.0.
- *Task 3.* Assume the coordinates of the initial design or checking point in terms of all the basic random variables. The mean values of the random variables are generally assumed to be the coordinates of the checking point in the first iteration.
- *Task 4.* Not all the random variables in a limit state function are expected to be normal random variables. A nonnormal variable cannot be modified to a normal random variable for all possible values without compromising its underlying uncertainty. It is known that a normal random variable can be uniquely defined in terms of two parameters: its mean and standard deviation. For a nonnormal random variable, the two parameters of an equivalent normal distribution can be evaluated by using two constraints. The two constraints commonly used for this purpose are that the CDF and the PDF of the actual nonnormal variable be equal to the equivalent normal variable at the checking point $(x_1^*, x_2^*, \ldots, x_n^*)$. The coordinates of the checking point are either assumed or known at the beginning of each iteration, and thus the evaluation of two equivalent parameters is not expected to be complicated. Denoting the equivalent normal mean and standard deviation for the ith random variable as $\mu_{X_i}^N$ and $\sigma_{X_i}^N$, respectively, Haldar and Mahadevan [14] have shown that they can be estimated as

$$\mu_{X_i}^N = x_i^* - \Phi^{-1}[F_{X_i}(x_i^*)]\sigma_{X_i}^N \tag{12.15}$$

and

$$\sigma_{X_i}^N = \frac{\phi\{\Phi^{-1}[F_{X_i}(x_i^*)]\}}{f_{X_i}(x_i^*)} \tag{12.16}$$

where $F_{X_i}(x_i^*)$ and $f_{X_i}(x_i^*)$ are the CDF and the PDF of the nonnormal random variable evaluated at the checking point, and $\Phi^{-1}(\ )$ and $\phi(\ )$ are the inverse of the CDF and the PDF of the standard normal distribution.

- *Task 5.* The checking point coordinate for the ith random variable can be shown to be

$$x_i^* = \mu_{X_i}^N - \alpha_{X_i}\beta\sigma_{X_i}^N \tag{12.17}$$

where α_{X_i} is the direction cosine and can be evaluated as

$$\alpha_{X_i} = \frac{(\partial g/\partial X_i)^* \sigma_{X_i}^N}{\sqrt{\sum_{i=1}^{n}(\partial g/\partial X_i\sigma_{X_i}^N)^{*2}}} \tag{12.18}$$

Partial derivatives in Equation 12.18 are evaluated at the checking point x_i^*. If the random variables are normal, then their standard deviations can be used directly; otherwise, for non-normal random variables, the equivalent standard deviations at the checking point need to be used.

- *Task 6.* Using Equation 12.17, the coordinates of the checking point can now be updated. These coordinates are expected to be different from the mean values assumed to start the iteration process. If necessary, Tasks 4 to 6 need to be repeated until the direction cosines α_{X_i}s converge with a predetermined tolerance. A tolerance level between 0.001 and 0.005 is common. Once the direction cosines converge, the coordinates of the checking point can be updated keeping the reliability index β as the unknown parameter.

- *Task 7.* Since the checking point must be on the limit state equation, an updated value for β can be obtained by substituting the updated coordinates of the checking point in terms of β.
- *Task 8.* Recalculate the coordinates of the checking point with the updated value for β as in Task 3. Tasks 3 to 7 need to be repeated until β converges to a predetermined tolerance level.

For hand calculation, a tolerance level of 0.01 is adequate. For computer applications, a tolerance level of 0.001 can be used. The algorithm converges in 5 to 10 cycles in most cases. A small computer program can be written to carry out the calculations.

The procedure for linear limit state equations is shown in Figure 12.3, and for nonlinear limit state equations consisting of two variables in Figure 12.2. In Figure 12.3a, the limit state equation, and the safe and unsafe regions are shown. As mentioned earlier, the limit state equation needs to be presented in the reduced coordinates system for the physical interpretation of the reliability index. The following transformation can be used for this purpose

$$X' = \frac{X_i - \mu_{X_i}}{\sigma_{X_i}}, \quad (i = 1, 2, \ldots, n) \tag{12.19}$$

The limit state equation, the safe and the unsafe regions, and the reliability index are shown in the reduced coordinates in Figure 12.3b. For linear limit state equations, no iteration is necessary and the reliability index can be obtained in one step. However, when the limit state is nonlinear as in Figure 12.2, an iterative procedure is necessary. The reliability index can be evaluated iteratively using the eight tasks discussed earlier. Since FORM is the most commonly used reliability evaluation technique at present, the eight steps are elaborated further with the help of an example.

EXAMPLE

A simply supported beam of span $L = 9.144$ m is loaded by a uniformly distributed load w in kN/m and a concentrated load P (in kN) applied at the midspan. The maximum deflection of the beam at the midspan can be calculated as

$$\delta_{\max} = \frac{5}{384} \frac{wL^4}{EI} + \frac{1}{48} \frac{PL^3}{EI} \tag{12.20}$$

where E is the Young's modulus and I is the moment of inertia of the cross-section of the beam. A beam with $EI = 182{,}262$ kN m^2 is selected to carry the load. Suppose w is a normal random variable with a mean of 35.03 kN/m and a standard deviation of 5.25 kN/m and P is a lognormal random variable with mean of 111.2 kN and a standard deviation of 11.12 kN. Further assume w and P are statistically independent. The allowable deflection, δ_a, is considered to be 38.1 mm. The task is to calculate the probability of failure of the beam in deflection satisfying the allowable deflection criterion.

Solution

This example is a problem of estimating the reliability index and the corresponding probability of failure when the limit state function contains independent nonnormal variables. This type of problem is very common in practice. The procedure is explained using the 8 tasks identified earlier.

Task 1 — Define the limit state function. Substituting deterministic values of all the parameters in Equation 12.20, the limit state function for the problem can be defined as

$$g(\,) = 0.0381 - (4.99444 \times 10^{-4}w + 8.73919 \times 10^{-5}P) \tag{12.21}$$

where w is a normal random variable with specified mean and standard deviation values, that is, $w \sim N(35.03 \text{ kN/m}, 5.25 \text{ kN/m})$ and P is a lognormal random variable with parameters λ_P and ζ_P, that is, $P \sim \ln(\lambda_P, \zeta_P)$. These parameters can be calculated as [14]

$$\delta_P = \frac{11.12}{111.2} = 0.1 \cong \zeta_P \quad \text{and} \quad \lambda_P = \ln 111.2 - \frac{1}{2} \times 0.1^2 = 4.706$$

The probability of failure or the reliability index of the beam in deflection corresponding to Equation 12.21 is evaluated using the eight tasks identified earlier. The results are summarized in Table 12.1. For further clarification, the detail calculations for the third and the final interations are given below.

Third iteration

Task 2. As shown in Table 12.1, an initial value of β of 3.0 is assumed to start the iteration.

Task 3. Using Equation 12.17, the coordinates of the new checking point for w and P can be shown to be

$$w^* = 35.03 + 0.926 \times 3 \times 5.25 = 49.615$$
$$p^* = 110.02 + 0.377 \times 3 \times 12.22 = 123.841$$

Task 4. Since P is a lognormal random variable, the equivalent normal mean and standard deviation at the checking point are calculated using Equations 12.15 and 12.16

$$f_P(123.841) = \frac{1}{\sqrt{2\Pi} \times 0.1 \times 123.841} \exp\left[-\frac{1}{2} \left(\frac{\ln 123.841 - 4.706}{0.1} \right)^2 \right] = 0.0170$$

$$F_P(123.841) = P(P \le 123.841) = \Phi\left(\frac{\ln 123.841 - 4.706}{0.1} \right) = \Phi(1.13)$$

$$\Phi^{-1}[F_P(123.841)] = 1.13$$

$$\phi\{\Phi^{-1}[F_P(123.841)]\} = \frac{1}{\sqrt{2\Pi}} \exp\left[-\frac{1}{2} \times 1.13^2 \right] = 0.21069$$

$$\sigma_P^N = \frac{0.21069}{0.0170} = 12.38$$

$$\mu_P^N = 123.841 - 1.13 \times 12.38 = 109.85$$

TABLE 12.1 Reliability Evaluation Using FORM with Uncorrelated Variables

Step 1	$g() = 0.0381 - (4.99444 \times 10^{-4} w + 8.73919 \times 10^{-5} P)$				
Step 2	β	3			3.88
Step 3	w^*	35.03	49.803	49.615	53.852	53.77
	p^*	111.2	122.216	123.841	128.151	128.813
Step 4	μ_W^N	35.03	35.03	35.03	35.03	35.03
	σ_W^N	5.25	5.25	5.25	5.25	5.25
	μ_P^N	110.61	110.02	109.85	109.28	109.19
	σ_P^N	11.12	12.22	12.38	12.81	12.88
Step 5	$(\partial g / \partial W)^*$	-4.99444×10^{-4}	-4.99444×10^{-4}	-4.99444×10^{-4}	-4.99444×10^{-4}	-4.99444×10^{-4}
	$(\partial g / \partial P)^*$	-8.73919×10^{-5}	-8.73919×10^{-5}	-8.73919×10^{-5}	-8.73919×10^{-5}	-8.73919×10^{-5}
Step 6	α_W	-0.938	-0.926	-0.924	-0.920	-0.919
	α_P	-0.348	-0.377	-0.381	-0.393	-0.394
Step 7				3.88		3.88
Step 8						3.88

Tasks 5 and 6

$$\alpha_w = \frac{-4.99444 \times 10^{-4} \times 5.25}{\sqrt{(-4.99444 \times 10^{-4} \times 5.25)^2 + (-0.873919 \times 10^{-4} \times 12.38)^2}} = \frac{-26.22081}{28.365193} = -0.924$$

$$\alpha_p = \frac{-10.819117}{28.365193} = -0.381$$

The direction cosines converged with a tolerance level of 0.005.

Task 7

$$w^* = 35.03 + 0.924 \times \beta \times 5.25 = 35.03 + 4.8510\beta$$

$$p^* = 109.85 + 0.381 \times \beta \times 12.38 = 109.85 + 4.71678\beta$$

$$0.0381 - 4.99444 \times 10^{-4} \times (35.03 + 4.8510\beta) - 0.873919 \times 10^{-4} \times (109.85 + 4.71678\beta) = 0$$

or

$$\beta = 3.88$$

Task 8. The reliability index did not converge with a tolerance level of 0.005. Go back to Task 3.

Final iteration

Task 3

$$w^* = 53.03 + 0.920 \times 3.88 \times 5.25 = 53.77$$

$$p^* = 109.28 + 0.393 \times 3.88 \times 12.31 = 128.813$$

Task 4. The equivalent normal mean and standard deviation for P at the checking point of 128.813 can be shown to be 109.19 and 12.88, respectively.

Tasks 5 and 6

$$\alpha_w = \frac{-4.9444 \times 10^{-4} \times 5.25}{\sqrt{(-4.9444 \times 10^{-4} \times 5.25)^2 + (-0.873919 \times 10^{-4} \times 12.88)^2}} = \frac{-26.22081}{28.534718} = -0.919$$

$$\alpha_p = \frac{-0.873919 \times 12.88}{28.534718} = -0.394$$

The direction cosines converged.

Task 7

$$w^* = 35.03 + 0.919 \times \beta \times 5.25 = 35.03 + 4.82475\beta$$
$$p^* = 109.19 + 0.394 \times \beta \times 12.88 = 109.19 + 5.07472\beta$$

$$0.0381 - 4.99444 \times 10^{-4} \times (35.03 + 4.82475\beta) - 0.873919 \times 10^{-4} \times (109.19 + 5.07472\beta) = 0$$

or

$$\beta = 3.88$$

Task 8. The reliability index converges with a tolerance level of 0.005.

This simple example outlines the calculation of the underlying risk of a specific design. The design of a structural member is discussed next if the acceptable risk is specified *a priori* for a specific design criterion or limit state.

12.5 Reliability-Based Structural Design Using FORM

In the previous section, procedures were presented to evaluate the reliablity of an already designed structural element using FORM. The next stage is to design a structural element using the reliability-based design procedure or FORM. The steps involved in reliability-based structural design are illustrated with the help of an example.

EXAMPLE

Suppose a simply supported steel beam of span 9.114 m needs to be designed. The beams are spaced 3.048 m apart and are subjected to nominal uniform dead and live loads of 4788.03 and 2394.02 Pa, respectively. Assume the beams are continuously laterally supported by the concrete slab, that is, the unbraced length, L_b, of the beam is zero. Suppose the same beam needs to be designed using the reliability-based design concept. The acceptable risk in terms of the reliability index β for strength, that is, for bending moment is suggested as 3.0.

In the United States, steel members are designed using British units, that is, pound, feet, and second units, and AISC's LRFD design aids were developed using the same units. For this example, all calculations were initially conducted using British units and then converted to SI units to satisfy the requirements of the book.

Solution

The beam is first designed using AISC's LRFD design criteria. The factored load can be calculated as

$$W = 1.2D + 1.6L = 1.2 \times 4788 + 1.6 \times 2394 = 9576 \text{ Pa}$$

The applied design bending moment for the beam can be shown to be

$$M_A = \frac{9576 \times 3.048 \times 9.114^2}{8 \times 1000} = 303 \text{ kN m}$$

Using AISC's LRFD design criteria [1] and Grade 50 steel, the most economical section of W18 $\times$ 35 is selected for the beam. The section has about 10% more plastic section modulus than required by the LRFD guidelines.

As mentioned earlier, in standard codified design the calculations are made using nominal values. However, for reliability-based design using FORM, the statistical descriptions of all the random variables must be known in terms of the underlying distributions and parameters to define them uniquely. For this example, suppose all the random variables are normal. A normal variable can be uniquely defined if the mean and the standard deviation values are known. Denoting μ_D, μ_L, and μ_R as the mean values of the dead and live loads and the resistance of the W section, and the corresponding nominal values as D_n, L_n, and R_n, respectively, Ellingwood et al. [18] showed that it might be reasonable to assume $D_n/\mu_D = 1.05$, $L_n/\mu_L = 1.4$, and $R_n/\mu_R = 0.9$. The uncertainty in the dead load, live load and resistance in terms of COV is generally considered to be 0.13, 0.37, and 0.13, respectively.

To start the design process, the limit state function in strength for the beam is

$$g(\ ) = R - D - L \tag{12.22}$$

where R, D, and L are the resistance, and dead and live loads effects, respectively. From the information given in the problem, the mean values of the dead and live loads are $4788.0/1.05 = 4560.0$ Pa and $2394.02/1.4 = 1710.0$ Pa and the corresponding COVs are 0.13 and 0.37, respectively. The mean value of the moment caused by the applied dead load can be calculated as

$$\mu_{M_D} = \frac{4560.0 \times 3.048 \times 9.114^2}{8 \times 1000} = 144.3 \text{ kN m}$$

The standard deviation of the bending moment due to the dead load M_D is

$$\sigma_{M_D} = 0.13 \times 144.3 = 18.76 \text{ kN m}$$

Similarly, the mean value of the moment caused by the applied live load is

$$\mu_{M_L} = \frac{1710.0 \times 3.048 \times 9.114^2}{8 \times 1000} = 54.1 \text{ kN m}$$

The corresponding standard deviation of the live load moment M_L is

$$\sigma_{M_L} = 0.37 \times 54.1 = 20.02 \text{ kN m}$$

Equation 12.17 can be used to find the checking points for R, D, and L for $\beta = 3$ as

$$r^* = \mu_R - \alpha_R \times \beta \times (0.13 \times \mu_R) = \mu_R - 0.13 \times \beta \times \alpha_R \times \mu_R$$
$$d^* = 144.3 - \alpha_D \times \beta \times 18.76$$

and

$$l^* = 54.1 - \alpha_L \times \beta \times 20.02$$

To calculate the direction cosines α_R, α_D, and α_L using Equation 12.18, the partial derivatives of the performance function with respect to R, D, and L evaluated at the checking point are

$$\frac{\partial g()}{\partial R} = 1, \quad \frac{\partial g()}{\partial D} = -1, \quad \frac{\partial g()}{\partial D} = -1$$

Using Equation 12.18, the corresponding direction cosines are

$$\alpha_R = \frac{\sigma_R}{\sqrt{\sigma_R^2 + \sigma_D^2 + \sigma_L^2}}$$

$$\alpha_D = \frac{\sigma_D}{\sqrt{\sigma_R^2 + \sigma_D^2 + \sigma_L^2}}$$

and

$$\alpha_L = \frac{\sigma_L}{\sqrt{\sigma_R^2 + \sigma_D^2 + \sigma_L^2}}$$

The checking point must satisfy the performance function represented by Equation 12.22; that is

$$g() = 0 = (\mu_R - 0.13 \times \beta \times \alpha_R \times \mu_R) - (144.3 - 18.76 \times \beta \times \alpha_D) - (54.1 - 20.02 \times \beta \times \alpha_L)$$

Substituting the direction cosine values in the preceding equation and simplifying will result in

$$\frac{\mu_R - 144.3 - 54.1}{\sqrt{(0.13 \times \mu_R)^2 + 18.76^2 + 20.02^2}} = \beta = 3$$

This is a quadratic equation in terms of the mean value of R. Solving the equation gives $\mu_R = 361.7$ kN m. To select a section, the nominal value of the resisting bending moment, R_n, is necessary and can be shown to be $0.9 \times 361.7 = 325.5$ kN m. A W18 $\times$ 35 of Grade 50 steel will satisfy this requirement. The same section was obtained using the codified LRFD approach.

On the other hand, suppose the beam needs to be designed for a reliability index of 4, implying that the underlying risk has to be much smaller than before or that the beam needs to be designed more conservatively. Following the same procedure discussed above, it can be shown that a larger size member of W21 $\times$ 44 of Grade 50 steel will be required. This result is expected.

This example clearly demonstrates the advantages of the reliability-based design procedure. It will not only suggest a section but also give the underlying risk in selecting the section. Thus, using the reliability-based design procedure, engineers are empowered to design a structure considering an appropriate acceptable risk different than that considered in the codified approach for a particular structure.

12.6 Reliability Evaluation with Nonnormal Correlated Random Variables

The previous section's discussion of reliability evaluation using FORM implicitly assumes that all the random variables in the performance function are uncorrelated. Considering the practical aspect of structural engineering problems, some of the random variables are expected to be correlated. Thus, the reliability evelution of a structure using FORM for correlated random variables is of considerable interest. Although this is considered to be an advanced topic, it is discussed very briefly below. More detailed information can be found elsewhere [14].

The correlation characteristics of random variables are generally presented in the form of the covariance matrix as

$$[\mathbf{C}] = \begin{bmatrix} \sigma^2_{X_1} & \mathrm{cov}(X_1, X_2) & \cdots & \mathrm{cov}(X_1, X_n) \\ \mathrm{cov}(X_1, X_2) & \sigma^2_{X_2} & \cdots & \mathrm{cov}(X_2, X_n) \\ \vdots & & \vdots & \\ \mathrm{cov}(X_n, X_2) & \mathrm{cov}(X_n, X_2) & \cdots & \sigma^2_{X_n} \end{bmatrix} \quad (12.23)$$

The corresponding correlation matrix can be shown to be

$$[\mathbf{C}'] = \begin{bmatrix} 1 & \rho_{X_1, X_2} & \cdots & \rho_{X_1, X_n} \\ \rho_{X_2, X_1} & 1 & \cdots & \rho_{X_2, X_n} \\ \vdots & \vdots & & \vdots \\ \rho_{X_n, X_1} & \rho_{X_n, X_2} & \cdots & 1 \end{bmatrix} \quad (12.24)$$

where ρ_{X_i, X_j} is the correlation coefficient of the X_i and X_j variables.

Reliability evaluation for correlated nonnormal variables **X** requires the original limit state equation to be rewritten in terms of the uncorrelated equivalent normal variables **Y**. Haldar and Mahadevan [14] showed that this can be done using the following equation:

$$[\mathbf{X}] = [\sigma^N_\mathbf{X}][\mathbf{T}][\mathbf{Y}] + \{\mu^N_\mathbf{X}\} \quad (12.25)$$

where $\mu^N_{X_i}$ and $\sigma^N_{X_i}$ are the equivalent normal mean and standard deviation of X, respectively, evaluated at the checking point using Equations 12.15 and 12.16, and **T** is a transformation matrix. Note that the matrix containing the equivalent normal standard deviation in Equation 12.25 is a diagonal matrix. The matrix **T** can be shown to be

$$[\mathbf{T}] = \begin{bmatrix} \theta^{(1)}_1 & \theta^{(2)}_1 & \cdots & \theta^{(n)}_1 \\ \theta^{(1)}_2 & \theta^{(2)}_2 & \cdots & \theta^{(n)}_2 \\ \vdots & \vdots & & \vdots \\ \theta^{(1)}_n & \theta^{(2)}_n & \cdots & \theta^{(n)}_n \end{bmatrix} \quad (12.26)$$

where $\{\theta^{(i)}\}$ is the normalized eigenvector of the ith mode of the correlation matric $[\mathbf{C}']$ and $\theta^{(i)}_1, \theta^{(i)}_2, \ldots, \theta^{(i)}_n$ are the components of the ith eigenvector. Eigenvalues are the variances of **Y**. **Y** will have zero means [14]. Using Equation 12.25, the correlated **X** variables can be transformed into uncorrelated **Y** variables. Then it is straightforward to rewrite the performance function in terms of the **Y** variables. FORM can then be used to evaluate the corresponding risk and reliability. The additional steps required to evaluate the reliability corresponding to a performance function containing correlated random variables are elaborated further with the help of an example. The example considered in the previous section for uncorrelated variables is modfied for this purpose.

EXAMPLE

Consider the limit state function represented by Equation 12.21 in the previous example. w is considered to be a normal random variable and P is considered to be a lognormal variable with the same means and standard deviation. However, unlike the previous example, assume w and P are correlated. The correlation coefficient between them is 0.7. The task is to evaluate the risk or reliability index of the beam for this situation.

Solution

This is a problem on correlated nonnormal variables. The limit state equation for the problem is given by Equation 12.21. As before, the uniformly distributed load w is normal, that is, $w \sim N(35.03 \text{ kN/m}, 5.25 \text{ kN/m})$, and P is lognormal with $\lambda_P = 4.706$ and $\zeta_P = 0.1$. However, they are now considered to be correlated. Assume that the the correlation coefficient $\rho_{w,P}$ is 0.7. The reliability index calculation is expected to be very complicated. As mentioned earlier, the limit state equation needs to be expressed in terms of the uncorrelated $\mathbf{Y}$ variables. However, since the equivalent normal mean and standard deviation need to be evaluated for P at the checking point and the coordinates of the checking point are expected to be different at each iteration, the limit state function will change for each iteration. Hand calculation is not recommended for this type of problem; a computer program is necessary. However, some important steps are discussed below for ease of comprehension.

The correlation matrix $[\mathbf{C}']$ given by Equation 12.24 for the problem is

$$[\mathbf{C}'] = \begin{bmatrix} 1 & 0.7 \\ 0.7 & 1 \end{bmatrix}$$

The two eigenvalues for the correlation matrix can be shown to be 0.3 and 1.7 [14]. The corresponding normalized eigenvectors can be evaluated and the tranformation matrix $[\mathbf{T}]$ given by Equation 12.26 can be shown to be

$$[\mathbf{T}] = \begin{bmatrix} 0.707 & 0.707 \\ -0.707 & 0.707 \end{bmatrix}$$

The results using hand calculations are summarized in Table 12.2. Calculations for the first and the final iterations are shown in the following sections.

TABLE 12.2 Reliability Evaluation Using FORM for Correlated Nonnormal Variables

Step 1	$g(\ ) = 0.0381 - (4.99444 \times 10^{-4}w + 8.73919 \times 10^{-5}P)$ (original with uncorrelated variables)					
	$g(\ ) = 0.010938 - 1.166536 \times 10^{-3}Y_1 - 2.540336 \times 10^{-3}Y_2$ (first iteration)					
	$g(\ ) = 0.011356 - 9.594176 \times 10^{-4}Y_1 - 2.74745 \times 10^{-3}Y_2$ (final iteration)					
Step 2	β	3.0				3.136
Step 3	w^*	35.03	50.438	50.306	50.132	50.957
	p^*	111.20	138.352	142.733	146.887	144.704
Step 4	μ_W^N	35.03	35.03	35.03	35.03	35.03
	σ_W^N	5.25	5.25	5.25	5.25	5.25
	μ_P^N	110.61	107.389	106.339	105.221	105.823
	σ_P^N	11.12	13.835	14.273	14.688	14.470
Step 5	$(\partial g/\partial Y_1)^*$	-1.166536×10^{-3}	-9.986566×10^{-4}	-9.715649×10^{-4}	-9.459591×10^{-4}	-9.594176×10^{-4}
	$(\partial g/\partial Y_2)^*$	-2.540336×10^{-3}	-2.708217×10^{-3}	-2.735308×10^{-3}	-2.760914×10^{-3}	-2.74745×10^{-3}
Step 6	α_{Y_1}	-0.1894	-0.1530	-0.1491	-0.1425	-0.1451
	α_{Y_2}	-0.9819	-0.9882	-0.9890	-0.9848	-0.9894
Step 7				3.136		3.137
Step 8	β					3.137

First iteration

Task 1. To define the appropriate limit state function, the following steps are followed. Since P is a lognormal random variable, the equivalent normal mean and standard deviation at the checking point are calculated using Equations 12.15 and 12.16 as

$$\mu_P^N = 110.61, \quad \sigma_P^N = 11.12$$

Using Equation 12.23, it can be shown that

$$\begin{Bmatrix} w \\ p \end{Bmatrix} = \begin{bmatrix} 5.25 & 0 \\ 0 & 11.12 \end{bmatrix} \begin{bmatrix} 0.707 & 0.707 \\ -0.707 & 0.707 \end{bmatrix} \begin{Bmatrix} Y_1 \\ Y_2 \end{Bmatrix} + \begin{Bmatrix} 35.03 \\ 110.61 \end{Bmatrix}$$

or

$$w = 3.711Y_1 + 3.711Y_2 + 35.03 \tag{12.27}$$

$$P = -7.86Y_1 + 7.86Y_2 + 110.61 \tag{12.28}$$

Thus, the modified limit state function for the first iteration can be written in terms of the uncorrelated normal **Y** variables as

$$g( ) = 0.0381 - 4.99444 \times 10^{-4}[3.711(Y_1 + Y_2) + 35.03] - 8.73919 \times 10^{-5}[7.86(Y_2 - Y_1) + 110.61]$$

or

$$g( ) = 0.010938 - 1.166536 \times 10^{-3} Y_1 - 2.540336 \times 10^{-3} Y_2 \tag{12.29}$$

Task 2. Assume $\beta = 3.0$.

Task 3. $w^* = 35.03$ and $p^* = 111.20$.

Task 4. The applicable mean and standard deviation values for w and P are given in Table 12.2.

Tasks 5 and 6

$$\frac{\partial g( )}{\partial Y_1} = -1.166536 \times 10^{-3} \quad \text{and} \quad \frac{\partial g( )}{\partial Y_2} = -2.540336 \times 10^{-3}$$

$$\alpha_{Y_1} = \frac{-1.166536 \times 10^{-3} \times \sqrt{0.3}}{\sqrt{(-1.166536 \times 10^{-3})^2 \times 0.3 + (-2.540336 \times 10^{-3})^2 \times 1.7}}$$

$$= \frac{-6.38938 \times 10^{-4}}{3.373257 \times 10^{-3}} = -0.1894$$

$$\alpha_{Y_2} = \frac{-2.540336 \times 10^{-3} \times \sqrt{1.7}}{3.373257 \times 10^{-3}} = -0.9819$$

During the first iteration, no comment can be made about the convergence of the direction cosines. As shown in Table 12.2, the direction cosines converged during the third iteration with a tolerance level of 0.005. Then, Task 7 can be carried out as shown below.

Task 7

$$y_1^* = 0.1491 \times \sqrt{0.3}\beta = 0.081665\beta$$

$$y_2^* = 0.9890 \times \sqrt{1.7}\beta = 1.289498\beta$$

The applicable limit state function, similar to Equation 12.29, can be shown to be

$$g( ) = 0.01131131 - 9.715649 \times 10^{-4} Y_1 - 2.7353082 \times 10^{-3} Y_2 \tag{12.30}$$

Thus

$$0.01131131 - 9.715649 \times 10^{-4} \times 0.081665\beta - 2.7353082 \times 10^{-3} \times 1.289498\beta = 0$$

or

$$\beta = 3.136$$

Final iteration

As shown in Table 12.2, the operating value for β is 3.136 and the direction cosines values are -0.1425 and -0.9848.

Task 3

$$y_1^* = 0.1425 \times \sqrt{0.3} \times 3.136 = 0.24476$$
$$y_2^* = 0.9898 \times \sqrt{1.7} \times 3.136 = 4.0471$$

and

$$w^* = 3.711(0.24476 + 4.0471) + 35.03 = 50.957$$
$$p^* = 10.384(-0.24476 + 4.0471) + 105.221 = 144.704$$

Task 4. The applicable mean and standard deviation values for w and P are given in Table 12.2.

Tasks 5 and 6. The applicable limit state function, similar to Equation 12.30, can be shown to be

$$g(\ ) = 0.0113564 - 9.594176 \times 10^{-4} Y_1 - 2.74745 \times 10^{-3} Y_2 \tag{12.31}$$

$$\frac{\partial g(\ )}{\partial Y_1} = -9.594176 \times 10^{-4} \quad \text{and} \quad \frac{\partial g}{\partial Y_2} = -2.74745 \times 10^{-3}$$

$$\alpha_{Y_1} = \frac{-9.594176 \times 10^{-4} \times \sqrt{0.3}}{\sqrt{(-9.594176 \times 10^{-4})^2 \times 0.3 + (-2.74745 \times 10^{-3})^2 \times 1.7}}$$
$$= \frac{-5.2549466 \times 10^{-4}}{3.62057498 \times 10^{-3}} = -0.1451$$
$$\alpha_{Y_2} = \frac{-2.74745 \times 10^{-3} \times \sqrt{1.7}}{3.62057498 \times 10^{-3}} = -0.9894$$

The direction cosines converged with a tolerance level of 0.005.

Task 7

$$y_1^* = 0.1451 \times \sqrt{0.3}\beta = 0.07947\beta$$
$$y_2^* = 0.9894 \times \sqrt{1.7}\beta = 1.29\beta$$

Using Equation 12.31, it can be shown that

$$0.0113564 - 9.594176 \times 10^{-4} \times 0.07947\beta - 2.74745 \times 10^{-3} \times 1.29\beta = 0$$

or

$$\beta = 3.137$$

The reliability index converges with a tolerance level of 0.005. The reliability index for the correlated random variables case is considerably different than that observed for the uncorrelated case.

12.7　Reliability Evaluation Using Simulation

Reliability evaluation using sophisticated probability and statistical theories may not be practical for many practicing structural engineers. But a simple simulation technique makes it possible to calculate the risk or probability of failure without knowing the analytical techniques and with only a little background in probability and statistics. The advancement in computing power makes simulation an attractive option for risk evaluation at the present time.

International experts agree that simulation can be an alternative for implementing the reliability-based design concept in practical design [19]. Lewis and Orav [20] wrote, "Simulation is essentially a controlled statistical sampling technique that, with a model, is used to obtain approximate answer for questions about complex, multi-factor probabilistic problems." They added, "It is this interaction of experience, applied mathematics, statistics, and computing science that makes simulation such a stimulating subject, but at the same time a subject that is difficult to teach and write about."

Theoretical simulation is usually performed numerically with the help of computers, allowing a more elaborate representation of a complicated engineering system than can be achieved by physical experiments, and it is often less expensive than physical models. It allows a designer to know the uncertainty characteristics being considered in a particular design, to use judgment to quantify randomness beyond what is considered in a typical codified design, to evaluate the nature of implicit or explicit performance functions, and to have control of the deterministic algorithm used to study the realistic structural behavior at the system level.

The method commonly used for this purpose is called the Monte Carlo simulation technique. In the simplest form of the basic simulation, each random variable in a problem is sampled several times to represent the underlying probabilistic characteristics. Solving the problem deterministically for each realization is known as a simulation cycle, trial, or run. Using many simulation cycles will give the probabilistic characteristics of the problem, particularly when the number of cycles tends to infinity. Using computer simulation to study the presence of uncertainty in the problem is an inexpensive experiment compared to laboratory testing. It also helps evaluate different design alternatives in the presence of uncertainty, with the goal of identifying the optimal solution.

12.7.1 Steps in Simulation

The Monte Carlo simulation technique has six essential elements [14]: (1) defining the problem in terms of all the random variables; (2) quantifying the probabilistic characteristics of all the random variables in terms of their PDFs and the corresponding parameters; (3) generating values of these random variables; (4) evaluating the problem deterministically for each set of realizations of all the random variables; (5) extracting probabilistic information from N such realizations; and (6) determining the accuracy and efficiency of the simulation. The success of implementing the Monte Carlo simulation in design will depend on how accurately each element is addressed. All these steps are discussed briefly in the following sections.

Step 1: Defining the problem in terms of all the random variables. The function that needs to be simulated must be defined in terms of all the random variables present in the formulation. For example, if the uncertainty in the applied bending moment, M_a, at the midspan of a simply supported beam of span L loaded with a uniform load w per unit length and a concentrated load P at the midspan needs to be evaluated, the problem can be represented as

$$M_a = wL^2/8 + PL/4 \qquad (12.32)$$

In this equation, if the span is assumed to be a known constant but w and P are random variables with specified statistical characteristics, then the applied moment is also a random variable. Its probabilistic characteristics can be evaluated using simulation.

On the other hand, if the probability of failure of the same beam is of interest and M_R is denoted as its bending moment capacity, the corresponding function to be simulated is

$$g(\ ) = M_R - (wL^2/8 + PL/4) \qquad (12.33)$$

In this case, M_R is expected to be a random variable in addition to w and P. The probability of failure of the beam can be evaluated by studying cases where $g(\ )$ will be negative or where the applied moment is greater than the resisting moment.

Step 2: Quantifying the probabilistic characteristics of all the random variables. The uncertainties associated with most of the random variables used in structural engineering have already been quantified by their underlying distributions and the parameters needed to define them uniquely. The subject has been discussed in detail by Haldar and Mahadevan [14] and will not be discussed further here.

Step 3: Generating random numbers for all the variables. The generation of random numbers according to a specific distribution is the heart of Monte Carlo simulation. All modern computers have the capability to generate uniformly distributed random numbers between 0 and 1. The computer will produce the required number of uniform random numbers corresponding to an arbitrary seed value between 0 and 1. In most cases, these are known as pseudorandom numbers and provide a platform for all engineering simulations.

Since most random variables are not expected to be uniform between 0 and 1, it is necessary to transform a uniform random number u_i between 0 and 1 to another random number with the appropriate statistical characteristics. The inverse transformation technique [14] is commonly used for this purpose. In this approach, the CDF of a random variable X, $F_X(x_i)$ is equated to the generated random number u_i. Thus

$$F_X(x_i) = u_i \tag{12.34}$$

or

$$x_i = F_X^{-1}(u_i) \tag{12.35}$$

If x_i is a uniform random variable between a and b, and u_i is a uniform random number between 0 and 1, then it can be shown that

$$u_i = \frac{x_i - a}{b - a} \tag{12.36}$$

or

$$x_i = a + (b - a)u_i \tag{12.37}$$

when $a = 0$ and $b = 1$, $x_i = u_i$, which is obvious.

If X is a normal random variable with a mean of μ_X and a standard deviation of σ_X, then a normal random number x_i corresponding to a uniform number u_i between 0 and 1 can be shown to be

$$x_i = \mu_X + \sigma_X \Phi^{-1}(u_i) \tag{12.38}$$

where Φ^{-1} is the inverse of the CDF of a standard normal variable. Similarly, if X is a lognormal random variable with parameters λ_X and ς_X, then x_i can be generated according to the lognormal distribution as

$$x_i = \exp[\lambda_X + \varsigma_X \Phi^{-1}(u_i)] \tag{12.39}$$

Most computers will generate random numbers for commonly used distributions. If not, the above procedure can be used to generate random numbers for a specific distribution.

Step 4: Evaluating the problem deterministically for each set of realizations of all the random variables. N random numbers for each of the random variables present in the problem will give N sets of random numbers, each set representing a realization of the problem. Thus, deterministically solving the problem defined in Step 1 N times will give N sample points. The generated information will provide the uncertainty in the response variable. Using N sample points and standard procedures, all the necessary statistical information can be collected, as briefly discussed next.

Step 5: Extracting probabilistic information from N such realizations. Simulation can be used to evaluate the uncertainty in the response variable like M_a in Equation 12.32. However, if the objective is only to estimate the probability of failure, the following procedure can be used.

If the value of $g(\ )$ in Equation 12.33 is negative, it indicates failure. Let N_f be the number of simulation cycles when $g(\ )$ is negative and let N be the total number of simulation cycles. The probability of failure can be expressed as

$$p_f = \frac{N_f}{N} \tag{12.40}$$

Step 6: Determining the accuracy and efficiency of the simulation. The probability of failure using Equation 12.40 is a major concern. The estimated probability of failure will reach the true value when N approaches infinity. When p_f and/or N are small, a considerable amount of error is expected in the estimated value of p_f. Haldar and Mahadevan [14] discussed the related issues in great detail. The following recommendation can be followed. In many structural engineering problems, the probability of failure could be smaller than 10^{-5}, that is, on average only 1 out of 100,000 simulations would show a failure. At least 100,000 simulation cycles are required to predict this behavior. For a reasonable estimate, at least 10 times this minimum, that is, 1 million simulation cycles, is usually recommended to estimate the probability of failure of 10^{-5}. Thus, if n random variables are present in a formulation to be simulated, $n \times 10^6$ random numbers are required. Simulation could be cumbersome or tedious for structural reliability evaluation. However, simulation is routinely used to verify a new theoretical method.

12.7.2 Variance Reduction Techniques

The discussion of simulation will not be complete without discussing variance reduction techniques (VRTs). The concept behind simulation presented in the previous section is relatively simple. However, its application to structural engineering reliability analysis depends on the efficiency of the simulation. The attractiveness of the simulation method can be greatly improved if the probability of failure can be estimated with a reduced number of simulation cycles. This led to the development of many VRTs. The efficiency of simulation can be improved by using VRTs, which can be grouped in several ways [14]. One approach is to consider whether the variance reduction method alters the experiment by altering the input scheme, by altering the model, or by special analysis of the output. The VRTs can also be grouped according to description or purpose (i.e., sampling method, correlation methods, and special methods).

The sampling methods either constrain the sample to be representative or distort the sample to emphasize the important aspects of the function being estimated. Some of the sampling methods are systematic sampling, importance sampling, stratified sampling, Latin hypercube sampling, adaptive sampling, randomization sampling, and conditional expectation. The correlation methods employ strategies to achieve correlation between functions or different simulations to improve the efficiency. Some of the VRTs in correlation methods are common random numbers, antithetic variates, and control variates. Other special VRTs include partition of the region, random quadratic method, biased estimator, and indirect estimator. The VRTs can also be combined to further increase the efficiency of the simulation. The details of these VRTs cannot be presented here but can be found in Haldar and Mahadevan [14].

The type of VRT that can be used depends on the problem under consideration. It is usually impossible to know beforehand how much efficiency can be improved using a given technique. In most cases, VRTs increase the efficiency in the reliability estimation by using a smaller number of simulation cycles. Haldar and Mahadevan [14] noted that VRTs increase the computational difficulty for each simulation, and a considerable amount of expertise may be necessary to implement them. The most desirable feature of simulation, its basic simplicity, is thus lost.

12.7.3 Simulation in Structural Design

As mentioned earlier, simulation can be an attractive alternative to estimate the reliability of a structural system. Simulation will enable reliability estimation considering realistic nonlinear structural behavior, the location of a structural element in a complicated structural system, correlation characteristics of random variables, etc. Reliability evaluation using a classical method like FORM essentially evaluates the reliability at the element level. Thus, simulation has many attractive features. It also has some deficiencies. Like other reliability methods, if the reference or allowable values are not known, it will be unable to estimate the reliability. The outcome of the simulation could be different depending on the number of simulation cycles and the characteristics of the computer-generated random numbers. One fundamental drawback is the time or cost of simulation. Huh and Haldar [21] reported that simulating

100,000 cycles in a supercomputer (SGI Origin 2000) to estimate the reliability of a one-bay two-story steel frame subjected to only 5 s of an earthquake loading may take more than 23 h. Using an ordinary computer, it may take several years.

It is clear that the simulation approach provides a reasonable alternative to the commonly used codified approach. However, there are still some issues that need to be addressed before it can be adopted in structural design. Further evaluation is needed of issues related to the efficiency and accuracy of the deterministic algorithm to be used in simulations, appropriate quantification of randomness, defining the statistical characteristics and performance functions, the selection of reference or allowable values, evaluating the correlation characteristics of random variables in complex systems, simulation of random variables versus random field, simulation of multi-variate random variables, system reliability, the effect of load combinations, time-dependent reliability, available software to implement the simulation-based concept, etc. The documentation of case studies will help in this endeavor.

EXAMPLE

Since each computer is expected to give different sets of random numbers, it may not be practical to give a detailed example to demonstrate the application of the simulation method in reliability analysis. However, a simple example is given to illustrate its many desirable features.

Suppose the probability of failure of a steel beam needs to be evaluated. The limit state function can be defined as

$$g(\ ) = F_y Z - M_a \tag{12.41}$$

where F_y is the yield stress, Z is the plastic section modulus, and M_a is the applied bending moment. The statistical characteristics of these variables are given in Table 12.3.

The probability of failure of the beam can be calculated in several ways [22] and the results are summarized in Table 12.4. If distributional information on the three random variables is ignored, then according to MVFOSM, the probability of failure is found to be 0.007183. However, if FORM is used, the corresponding probability of failure is found to be 0.023270. These probabilities of failure are quite different. Simulation can be used to establish which number is correct.

In the third column of Table 12.4, the probabilities of failure for several simulation cycles are given for direct Monte Carlo simulation. The results indicate that if the simulation cycles are relatively small, the

TABLE 12.3   Statistical Characteristics of Random Variables

Variables	Mean value	COV	Probability distribution
F_y	262.0 MPa	0.10	Normal
Z	$8.19 \times 10^{-4}\,\mathrm{m}^3$	0.05	Lognormal
M_a	113.0 kN m	0.30	Type II

TABLE 12.4   Probability of Failure Evaluations Using Different Methods

		Monte Carlo simulation		
MVFOSM	FORM	Direct	Conditional expectation VRT	Conditional expectation plus antithetic variates VRT
$\beta = 2.448$	$\beta = 1.990$	$N = 10$ 0.000000[a]	0.021626	0.023266
$P(\text{failure}) = 0.007183$	$P(\text{failure}) = 0.023270$	$N = 50$ 0.000000[a]	0.022071	0.024660
		$N = 100$ 0.000000[a]	0.021425	0.024322
		$N = 500$ 0.018000	0.024250	0.024233
		$N = 1000$ 0.021000	0.025025	0.024489

[a] $N_f = 0$, for these cases.

direct Monte Carlo simulation method cannot accurately predict the probability of failure. However, as the number of simulation cycles increases, the probability of failure approaches the value obtained using FORM. The important conclusion is that MVFOSM should not be used for structural reliability evaluation.

Columns 4 and 5 give the probabilities of failure using the conditional expectation VRT and conditional expectation plus antithetic variates VRT. The results indicate the power of the VRTs.

In both VRT schemes, only ten simulations are necessary to predict the underlying probability of failure. This simple example indicates that a considerable amount of additional desirable information can be collected by intelligently solving the problem ten times deterministically instead of solving it only once.

12.8 Future Directions in Reliability-Based Structural Design

The discussion of reliability-based structural design would be incomplete without commenting on future trends. Some important directions are discussed below.

12.8.1 A New Hybrid Reliability Evaluation Method

As pointed out by Haldar and Marek [23], the available reliability evaluation techniques are not capable of estimating the risk of realistic structures. Simulation is an attractive alternative but it can be very inefficient. To address these concerns and combining the desirable features of theoretical and simulation-based techniques, Huh and Haldar [21,24] proposed a hybrid approach. The algorithm intelligently integrates the concepts of the response surface method, the finite element method, FORM, and an iterative linear interpolation scheme. In this algorithm, a real structure is represented as realistically as possible by finite elements. The behavior of the structure is traced considering all major sources of nonlinearity and uncertainty under static and dynamic loading conditions. The coordinates of the most probable failure point or design point are evaluated by FORM, and then a response surface is generated around this point by multiple deterministic analyses of the structure. Conceptually, this part of the algorithm is similar to the Monte Carlo simulation technique; however, the simulation is conducted around highly selective experimental design points required for the response surface method, removing the inefficiency in the algorithm.

Huh and Haldar have shown that instead of conducting 100,000 cycles on Monte Carlo simulation, the probability of failure of a realistic nonlinear steel frame can be obtained with only 50 runs without compromising the accuracy of the estimation. The unique feature of the algorithm is that actual dynamic loading including earthquake loading can be applied in the time domain. This enables the most realistic representation of the dynamic loading condition and provides an alternative to the classical random vibration approach. The method is expected to play a major role in the reliability analysis of real structures under static and dynamic loading conditions in the future.

12.8.2 Education

Lack of education could be a major reason for the profession's avoidance of the reliability-based structural design concept. However, there are some recent major developments in this area. In the Czech Republic, CSN 7314 01-1998 (Appendix A) [25] is one of the pilot codes that allows the Monte Carlo simulation as a design tool. It was pointed out [19] that in Canada, a 14 km long bridge with a span length of 250 m was recently built. The code did not cover this design, and simulation was used. Reliability-based design is very common for offshore structures. In Europe, highway and railway companies are using simulation for assessment purposes. In the United States, the general feeling is that we are safe if we design according to the design code but this is not entirely true. According to a judge, designers should use all available means to satisfy performance requirements. The automotive industry satisfied the code requirements in one case, but a judge ruled that they should have used simulation to address the problem more comprehensively.

Some of the developments in risk-based design using simulation are very encouraging. Simulation could be used in design in some countries, but it is also necessary to look at its legal ramifications. In some countries, code guidelines must be followed to the letter, and other countries permit alternative methods if they are better [19]. In Europe, two tendencies currently exist: Anglo-Saxon (more or less free to do anything) and middle-European (fixed or obligatory requirements). Current Euro-code is obligatory. We need to change the mentality and laws to implement simulation or the reliability-based design concept in addressing real problems.

In the context of education of future structural engineers, the presence of uncertainty must be identified in design courses. Reliability assessment methods can contribute to the transition from deterministic to a probabilistic way of thinking for students as well as designers. In the United States, the Accreditation Board of Engineering and Technology now requires that all civil engineering undergraduate students demonstrate knowledge of the application of probability and statistics to engineering problems, indicating its importance in civil engineering education.

Most of the risk-based design codes are the by-product of education and research at the graduate level. A pilot international project titled TERECO (TEaching REliability COncepts), sponsored by the Leonardo da Vinci Agency in Europe [26], was very successful.

In summary, the profession is moving gradually toward accepting the reliability-based design concept, and the structures group is providing leadership in this regard.

12.8.3 Computer Programs

As mentioned earlier, many theoretical procedures with various degrees of complexity have been developed over the last few decades; however, they are not popular with practicing engineers. One issue could be the lack of user-friendly software. Two types of issues need to be addressed. Reliability-based computer software should be developed for direct applications or the reliability-based design feature should be added to commercially available deterministic software. Some of the commercially available reliability based computer software are briefly discussed next.

NESSUS (Numerical Evaluation of Stochastic Structures Under Stress) was developed by the Southwest Research Institute [27,28] under the sponsorship of the NASA Lewis Research Center. It combines probabilistic analysis with a general-purpose finite element/boundary element code. The probabilistic analysis features an advanced mean value technique. The program also includes techniques such as fast convolution and curvature-based adaptive importance sampling.

PROBAN (PRObability ANalysis) was developed at Det Norske Veritas, Norway, through A.S. Veritas Research [29]. PROBAN was designed to be a general-purpose probabilistic analysis tool. It is capable of estimating the probability of failure using FORM and SORM for a single event, unions, intersections, and unions of intersections. It has a library of standard probability distributions. The approximate FORM/SORM results can be updated through an importance sampling simulation scheme. The probability of general events can be computed using Monte Carlo simulation and directional sampling.

CALREL (CAL-RELiability) is a general-purpose structural reliability analysis program designed to compute probability integrals in the form given by Equation 12.13. CALREL was developed at the University of California at Berkeley by Liu et al. [30]. It incorporates four general techniques for computing the probability of failure: FORM, SORM, directional simulation with exact or approximate surfaces, and Monte Carlo simulation. It has a library of probability distributions of independent and dependent random variables. Additional distributions can be included through a user-defined subroutine.

Under the sponsorship of the Pacific Earthquake Engineering Research (PEER) Center [31], a multiuniversity team developed a general-purpose finite element reliability code within the framework of OpenSees. A web address is given in the references for further information on the program [31].

Structural engineers without a formal education in reliability-based design may not be able to use these computer programs. They can be retrained with very little effort. They may be very knowledgeable using existing deterministic analysis software including commercially available finite element packages.

This expertise needs to be integrated with the reliability-based design concept. Thus, probabilistic features may need to be added to the deterministic finite element packages. Proppe et al. [32] discussed the subject in great detail. For proper interface with deterministic software, they advocated a graphical user interface, a communication interface that must be flexible enough to cope with different application programming interfaces and data formats, and the reduction of problem sizes before undertaking reliability analysis. COSSAN [33] software attempted to implement the concept.

The list of computer programs given here is not exhaustive. However, these programs are being developed and are expected to play a major role in implementing reliability-based structural analysis and design in the near future.

12.9   Concluding Remarks

The state of the art in reliability-based structural analysis and design has been presented in brief in this chapter. Some of the theoretical methods currently available were discussed. The use of the simulation technique is advocated as an alternative to theoretical models. A hybrid method proposed by the author combining the desirable features of theoretical methods and simulation technique was discussed.

All major structural design codes for concrete, masonry, steel, and wood are now using the LRFD concept, which is essentially a reliability-based design concept. In spite of significant developments in the reliability-based structural design concept, it is not popular with practicing structural engineers. Issues related to this were discussed. Since reliability-based structural design may be the only design option worldwide in the near future, it is hoped that this discussion will help readers stay ahead of the curve.

References

[1]   American Institute of Steel Construction (AISC), AISC, *Manual of Steel Construction Load and Resistance Factor Design*, 1st, 2nd, and 3rd Editions, Chicago, IL, 1986, 1994, 2001.

[2]   American Institute of Steel Construction (AISC), *Manual of Steel Construction Allowable Stress Design*, 9th Edition, Chicago, IL, 1989.

[3]   American Concrete Institute (ACI), *Building Code Requirements for Structural Concrete (318-02)*, Farmington Hills, MI, 2002.

[4]   American Concrete Institute (ACI), *Building Code Requirements for Masonry Structures and Specification for Masonry Structures — 2002*, ACI 530-02/ASCE 5-02/TMS 402-02, reported by the Masonry Standards Joint Committee, 2002.

[5]   American Society of Civil Engineers (ASCE), *Standard for Load and Resistance Factor Design (LRFD) for Engineered Wood Construction*, ASCE 16–95, 1995.

[6]   American Wood Council, *The Load and Resistance Factor Design (LRFD) Manual for Engineered Wood Construction*, Washington, DC, 1996.

[7]   American Society of Civil Engineers (ASCE), A panel session on "Past, present and future of reliability-based structural engineering worldwide," *Civil Engineering Conference and Exposition*, Washington, DC, 2002.

[8]   International Standard Organization (ISO), *ISO 2394 — General Principles on Reliability for Structures*, 2nd Edition, 1998-06-01.

[9]   American Institute of Steel Construction (AISC), *Economy in Steel ASD vs. LRFD*, AISC Lecture Series, 1988/1989.

[10] American Concrete Institute (ACI), *Building Code Requirements for Structural Concrete (318-99)*, Farmington Hills, MI, 1999.

[11] Bjorhovde, R., Galambos, T.V., and Ravindra, M.K., LRFD criteria for steel beam-columns, *J. Struct. Eng.*, ASCE, 104(9), 1943, 1982.

[12] Mahadevan, S. and Haldar, A., Stochastic FEM-based validation of LRFD, *J. Struct. Eng.*, ASCE, 117(5), 1393, 1991.

[13] Haldar, A. and Mahadevan, S., *Reliability Assessment Using Stochastic Finite Element Analysis*, John Wiley & Sons, New York, NY, 2000.

[14] Haldar, A. and Mahadevan, S., *Probability, Reliability, and Statistical Methods in Engineering Design*, John Wiley & Sons, New York, NY, 2000.

[15] Hasofar, A.M. and Lind, N.C., Exact and invariant second moment code format, *J. Eng. Mech., ASCE*, 100(EM1), 111, 1974.

[16] Rackwitz, R., Practical probabilistic approach to design, *Bulletin No. 112*, Comite European du Beton, Paris, France, 1976.

[17] Ayyub, B.M. and Haldar, A., Practical structural reliability techniques, *J. Struct. Eng., ASCE*, 110(8), 1707, 1984.

[18] Ellingwood, B., Galambos, T.V., MacGregor, J.G., and Cornell, C.A., Development of a probability based load criterion for American standard A58: Building Code Requirements for Minimum Design Loads in Buildings and Other Structure, *Special Publication 577*, National Bureau of Standards, Washington, DC, 1980.

[19] Marek, P., Haldar, A., Guštar, M., and Tikalsky, P., Editors, *Euro-SiBRAM 2002 Colloquium Proceedings*, ITAM Academy of Sciences of Czech Republic, Prosecka 76, 19000 Prague 9, Czech Republic, 2002.

[20] Lewis, P.A.W. and Orav, E.J., *Simulation Methodology for Statisticians, Operations Analysts, and Engineers*, Vol. 1, Wadsworth & Brooks/Cole Advanced Books & Software, Pacific Grove, CA, 1989.

[21] Huh, J. and Haldar, A., Stochastic finite element-based seismic risk evaluation for nonlinear structures, *J. Struct. Eng., ASCE*, 127(3), 323, 2001.

[22] Ayyub, B.M. and Haldar, A., Improved simulation techniques as structural reliability models, in *Proc. 4th Int. Conf. on Structural Safety and Reliability*, Konishi, I., Ang, A.H.-S., and Shinozuka, M., Eds., IASSR, Japan, 1985, I-17.

[23] Haldar, A., and Marek, P., Role of simulation in engineering design, in *Proc. 9th Int. Conf. on Applications of Statistics and Probability (ICASP9-2003)*, 2, 945, 2003.

[24] Huh, J. and Haldar, A., Seismic reliability of nonlinear frames with PR connections using systematic RSM, *Probab. Eng. Mech.*, 17(2), 177, 2002.

[25] Czech Institute of Standards, *CSN 73 1401-1998 Design of Steel Structures*, Prague, Czech Republic, 1998.

[26] Marek, P., Brozzetti, J., and Gustar, M., *Probabilistic Assessment of Structures using Monte Carlo Simulation*, Academy of Sciences of the Czech Republic, Praha, Czech Republic, 2001.

[27] Cruse, T.A., Burnside, O.H., Wu, Y.-T., Polch, E.Z., and Dias, J.B., Probabilistic structural analysis methods for select space propulsion system structural components (PSAM), *Comput. Struct.*, 29(5), 891, 1988.

[28] Southwest Research Institute, *NEUSS*, San Antonio, Texas, 1991.

[29] Veritas Sesam Systems, *PROBAN*, Houston, Texas, 1991.

[30] Liu, P.-L., Lin, H.-Z., and Der Kiureghian, A., *CALREL*, University of California, Berkeley, CA, 1989.

[31] McKenna, F., Fenves, G.L., and Scott, M.H., *Open System for Earthquake Engineering Simulation*, http://opensees.berkeley.edu/, Pacific Earthquake Engineering Research Center, Berkeley, CA, 2002.

[32] Proppe, C., Pradlwarter, H.J., and Schueller, G.I., Software for stochastic structural analysis — needs and requirements, in *Proc. 4th Int. Conf. on Structural Safety and Reliability*, Corotis, R.B., Schueller, G.I., and Shinizuka, M., Eds., 2001.

[33] COSSAN (Computational Stochastic Structural Analysis) — Stand-Alone Toolbox, *User's Manual*, IfM-Nr: A, Institute of Engineering Mechanics, Leopold-Franzens University, Innsbruck, Austria, 1996.

13

Structure Configuration Based on Wind Engineering

Yoshinobu Kubo
Department of Civil Engineering,
Kyushu Institute of Technology,
Tobata, Kitakyushu,
Japan

13.1 Introduction

When a structure is immersed in an air stream, aerodynamic forces are induced in the structures by the air stream and the aerodynamic forces apply on the structure as wind load generating deflection and vibration. Flexible structures like high-rise buildings and long-span bridges are susceptible to deflection and vibration under wind action. In order to reduce the deflection or to control the vibration of the structure, the effect of aerodynamic/aeroelastic forces should be reduced. The best method for the reduction of the aerodynamic force effect is to adapt a structural configuration that can reduce the aerodynamic/aeroelastic force effect. For this purpose, when designing a flexible structure susceptible to vibration under wind action, it is necessary and useful for structural engineers to understand the mechanism of the process generating aerodynamic forces and the relationship between structural configuration and aerodynamic forces.

The following steps are an outline of the design for the structure susceptible to vibration under wind action:

1. Decide the design wind speed for the structure by using data measured previously at meteorological observatories close to the construction site.
2. Estimate the wind load applying on the structure under the design wind speed.
3. When the wind load exceeds the design wind load, improve the proposed structural configuration so as to possess a smaller wind load based on the data previously measured or the results of wind tunnel tests conducted to find a better configuration.
4. Check the occurrence of aeroelastic vibrations, after confirmation that the wind load is smaller than the design wind load. If the occurrence of aeroelastic vibrations is predicted, the structural configuration should be further improved to be the vibration amplitude less than the allowable value so as not to induce the aeroelastic vibrations.

This chapter deals with the mechanism of the process generating aerodynamic forces and the relationship between the structural configuration and wind load or aeroelastic vibration induced in the structure.

13.2 Effects of Wind Load

Aerodynamic forces are drag force, lift force, and aerodynamic moment. The drag force is a force parallel to the wind direction, the lift force is a force perpendicular to the wind direction, and the aerodynamic moment is a rotating force around a specified point. When the terminology of wind load is used, the wind load usually indicates the drag force.

13.2.1 Mechanism of Wind Load

A cross-sectional shape of structural member is usually a nonstreamline shape, which is called a "bluff body." The representative shapes of a bluff body used in a structure are circular and rectangular cross-sections.

Figure 13.1 shows flow visualization around a circular cross-section structure [1]. The approaching flow separates at an angle θ of about 80°, which is measured at the center of the circular section from the stagnation point to downstream direction along the surface. The separated flows on upper and lower sides roll up from both separation points and vortex streets are generated in a wake of the bluff body. The vortex vibrates with frequency linearly proportional to the wind velocity as discovered by Strouhal [2]. Figure 13.2 shows the mean pressure distribution around the circular structure [3]. Positive pressure is induced on the upstream surface and negative pressure on the downstream surface and on both upper and lower side surfaces. The drag force is calculated by subtracting pressure on downstream surfaces from pressure on upstream surface of the structure. That is, the drag force is generated by the pressure difference between upstream and downstream surfaces. The mechanism generating drag force is simple for a single bluff body as mentioned above, but complicated for multiple bluff bodies. In this section, the drag force coefficients for a single body and multiple bodies are introduced for structural engineers.

Coefficients for drag and lift forces and aerodynamic moment are C_D, C_L, and C_M, respectively, and defined as follows:

$$C_D = \frac{F_D}{\frac{1}{2}\rho U^2 A}, \quad C_L = \frac{F_L}{\frac{1}{2}\rho U^2 A}, \quad C_M = \frac{F_M}{\frac{1}{2}\rho U^2 BA} \qquad (13.1)$$

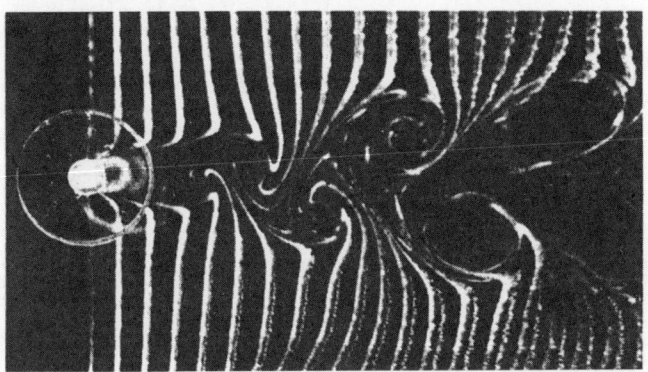

FIGURE 13.1 Flow visualization of the wake of a circular structure [1].

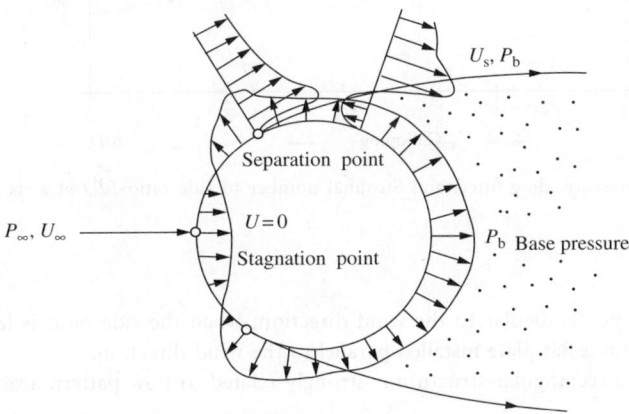

FIGURE 13.2 Rough sketch of the pressure distribution of a circular structure [3].

where ρ, U, A, and B are air density, wind velocity, representative area (usually projected area perpendicular to wind direction), and representative width, respectively. F_D, F_L, and F_M are the drag and the lift forces and the aerodynamic moment, respectively.

Strouhal number St is useful to predict the onset wind velocity of vortex-excited vibration. The definition of Strouhal number is

$$St = \frac{f_v D}{U} \qquad (13.2)$$

where f_v, D, and U are frequency of vortex in the wake, representative length (usually height projected perpendicularly to wind direction), and wind velocity, respectively.

In some following figures, Strouhal number is indicated with drag force coefficients.

13.2.2 Configuration Effect for Single Bluff Body

13.2.2.1 Side Ratio Effect of Rectangular Cross-Section Structure

The drag force of a rectangular cross-section structure (rectangular structure) changes with a variety of side ratios as shown in Figure 13.3 [3]. When side ratio $B/D = 0$, the rectangular structure is

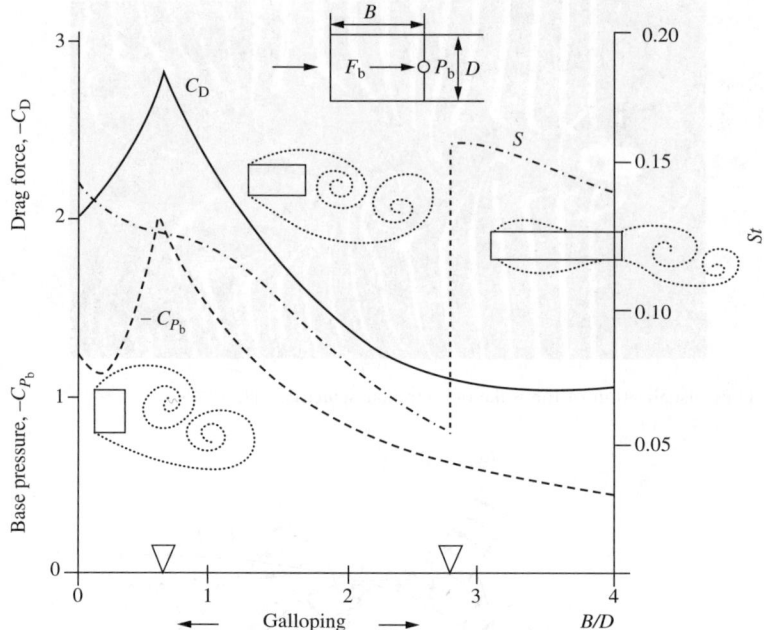

FIGURE 13.3   Base pressure, drag force, and Strouhal number to side ratio B/D of a rectangular cross-section structure [3].

a flat plate installed perpendicular to the wind direction; when the side ratio is large to infinity, the rectangular structure is a flat plate installed parallel to the wind direction.

The drag force of a rectangular structure is strongly related to flow pattern around the rectangular structure:

1. In the case of $B/D < 0.6$, separation flow is completely separated. The separated shear layers from both upstream edges flow down without reattaching to the side surfaces and roll up alternately in the wake to construct Karman vortex streets. In this region, the distance between the rolling up position of the separated flow and the downstream surface of the structure becomes shorter with increment of B/D and the base pressure on the downstream surface decreases remarkably. As a result, the drag force increases with increment of B/D. At about $B/D = 0.6$, the rolling up position becomes nearest to the downstream surface of the structure to be a peak value of the drag force in variety of B/D.

2. In the case of $0.6 < B/D < 2.8$, the approaching flow is also completely separated. In this region, the separated flow interferes with the downstream corners of the structure to increase the distance between the rolling up position and the downstream surface of the body and increases the pressure on the downstream surface (which is called pressure recovery). As a result, the drag force decreases with increment of B/D. At the same time, the interval of vortex occurrence increases with increment of B/D and Strouhal number decreases as shown in Figure 13.3.

3. In the case of $B/D > 2.8$, the separation flow steadily reattaches to the side surface and separates again at a further downstream position than the reattachment position on the side surface to construct the Karman vortex streets in the wake of the body. The separation flow periodically reattaches to the surfaces and constructs vortex as on the surfaces. This region is called the complete reattachment region of separation flow.

13.2.2.2 Drag Reduction Method

13.2.2.2.1 Corner Shape Effect

The drag force of the rectangular cross-section structure can be reduced by making corners round or by cutting corners. Figure 13.4 shows one example for the reduction of the drag force of a square structure [4]. The horizontal axis indicates the ratio of radius of rounded corner to side length and the vertical axis indicates the drag force coefficient. Type 2 is the square structure with a rounded corner at upstream corners, Type 3 has four rounded corners, and Type 4 has a rounded corner at downstream corners. The drag force of Type 2 decreases with increase in the radius of the rounded corner. The drag force of Type 4 takes a little larger value than the value of a square structure without rounded corners. Type 3 takes the value between Type 2 and Type 4. Referring to the results, it is clear that it is very important to reduce the drag force to make upstream corners round or of cutting shape. On the other hand, making downstream corners round has no effect in the reduction of drag force.

13.2.2.2.2 Roughness Effect

Figure 13.5 shows the drag force of a circular cross-section structure to Reynolds number [5]. Reynolds number (Re) is defined as the ratio of inertial force to viscous force of fluid:

$$Re = \frac{UD}{v} \tag{13.3}$$

where U, D, v are wind velocity, representative length, and kinetic viscosity, respectively.

The circular cross-section structure is susceptible to the influence of Re as shown in Figure 13.5. The drag force coefficient of a circular cross-section structure with a smooth surface takes a constant value of 1.2 in region of $Re < 1.5 \times 10^5$ and the drag force coefficient decreases with increment of Re up to $Re = 3.5 \times 10^5$ (which is the critical Reynolds number) and increases gradually with increment of Re. In the region larger than the critical Reynolds number (which is the supercritical region), the drag force is about less than half of the value in the subcritical region. If the same condition of fluid around the circular cross-section structure is realized in the subcritical region as in the supercritical region by using some method, the drag force can be reduced to half the value of the drag force of a cylindrical member in a subcritical region. In the supercritical region, the fluid around the structure is in the state of turbulent flow. By attaching artificial roughness on the surface of the cable, the fluid around the structure can be made turbulent and the drag force reduced. The result in Figure 13.6 was obtained from the work to

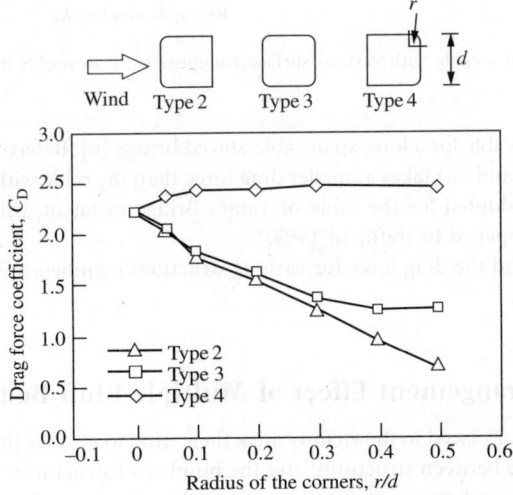

FIGURE 13.4 Drag force coefficient of a square cylinder with a rounded corner [4].

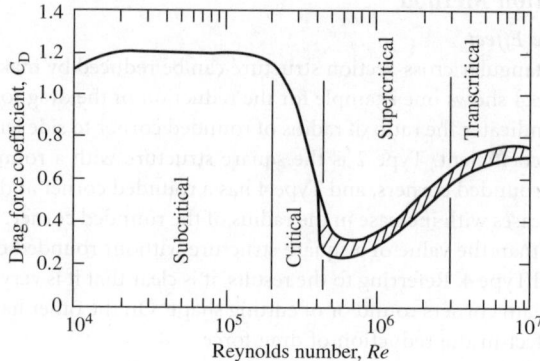

FIGURE 13.5   Drag force coefficient of a circular structure to Reynolds number [5].

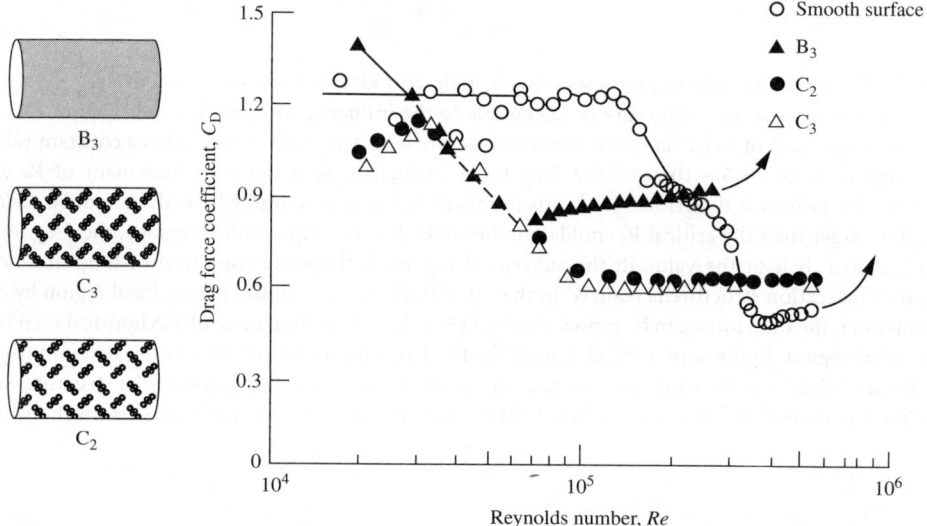

FIGURE 13.6   Drag force of a cable with various surface roughnesses to Reynolds number [6].

reduce the drag force of a cable for a long-span cable-stayed bridge [6]. Referring to the results, the cable with indented surface (C_2 and C_3) takes a smaller drag force than the cable with uniform roughness (B_3). The indented cable was adopted for the cable of Tatara Bridge in Japan, which is the world's longest cable-stayed bridge, and opened to traffic in 1999.

The Strouhal number and the drag force for various structural members are shown in Table 13.1 and Table 13.2 in Section 13.4.

13.2.3   Vicinity Arrangement Effect of Multiple Bluff Bodies [48]

When structures are built or placed in the vicinity area, their wind loads vary through the arrangement of the structures, the distance between structures, and the number of structures. The flow condition varies based on the relative distance between structures and wind load takes various values corresponding to the vicinity arrangement condition.

13.2.3.1 Side by Side Arrangement

The definition of wind load is indicated in Figure 13.7 for a side by side arrangement. In the case of side by side arrangement, since the fluid around a body flows in an asymmetric flow pattern, the lift force is also induced along with the drag force. The wakes of both structures interfere and the lift and drag forces take various values corresponding to the distance between the two structures because of bistable flow running in the gap between both structures. In the following explanations, interval parameter T/D is used. T and D are spacing distance between two structures and representative length, respectively.

Figure 13.8 shows the lift and drag force coefficients of circular structures of side by side arrangement [7]. It is seen that both coefficients take two values against the interval parameter T/D and this fact

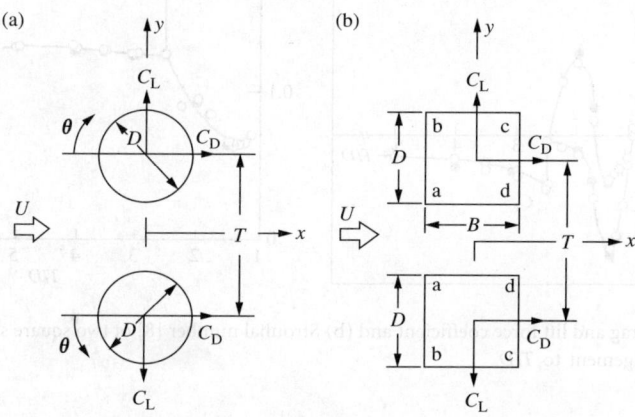

FIGURE 13.7 Side by side arrangement of two structures with: (a) circular and (b) rectangular cross-sections.

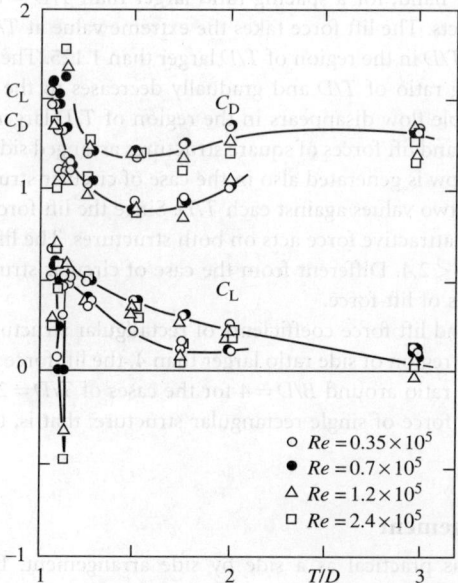

FIGURE 13.8 Lift and drag force coefficients for the side by side arrangement to T/D [7].

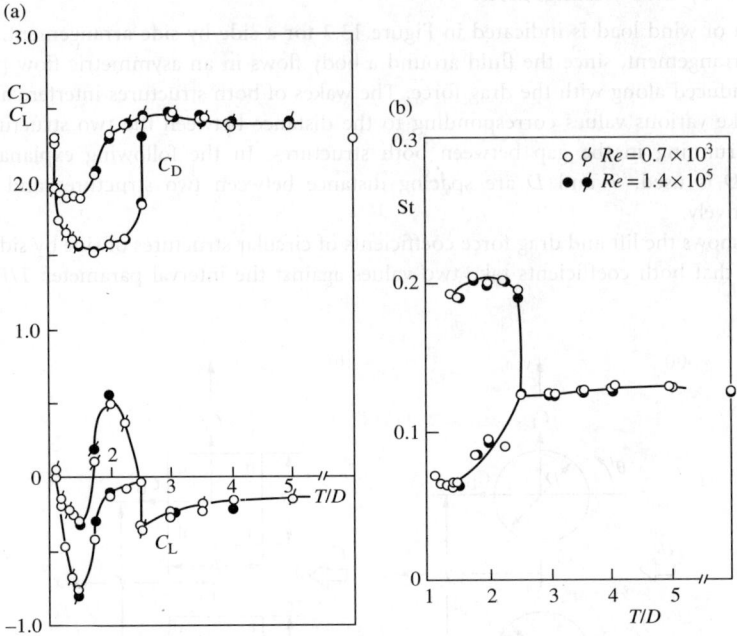

FIGURE 13.9   (a) Drag and lift force coefficient and (b) Strouhal number [8] of two square structures ($B/D = 1$) in the side by side arrangement to T/D.

supports the idea that the flow around the structures is in a bistable state. For a very small value of the spacing ratio $T/D = 1.125$, one structure shows negative lift, that is, the attractive force acts between the two structures. On the other hand, for a spacing ratio larger than $T/D = 1.125$, the repulsive lift force between the two structures acts. The lift force takes the extreme value at T/D just larger than 1.125 and decreases with increment of T/D in the region of T/D larger than 1.125. The drag force takes the extreme value at a very small spacing ratio of T/D and gradually decreases to the value of a single body with increment of T/D. The bistable flow disappears in the region of T/D larger than 3.

Figure 13.9 shows the drag and lift forces of square structures arranged side by side [8]. In the region of T/D less than 3, a bi-stable flow is generated also in the case of circular structures arranged side by side. The drag and lift forces take two values against each T/D. Since the lift force takes a negative value over the whole region of T/D, the attractive force acts on both structures. The life force takes a positive value only in the region $1.8 < T/D < 2.4$. Different from the case of circular structures, the square structures generally take negative values of lift force.

Figure 13.10 shows drag and lift force coefficients of rectangular structures arranged side by side for various side ratios [8]. In the region of side ratio larger than 4, the lift force is positive. Therefore, the lift force changes its sign at side ratio around $B/D = 4$ for the cases of $T/D = 2$ and 3. The drag force has a similar tendency as the drag force of single rectangular structure, that is, the drag force decreases with increment of side ratio.

13.2.3.2   Tandem Arrangement

A tandem arrangement is as practical as a side by side arrangement. In tandem arrangement, the downstream structure is immersed in the wake of the upstream structure. Interference of flow between structures is closely connected with the properties and behavior of the wake of an upstream structure.

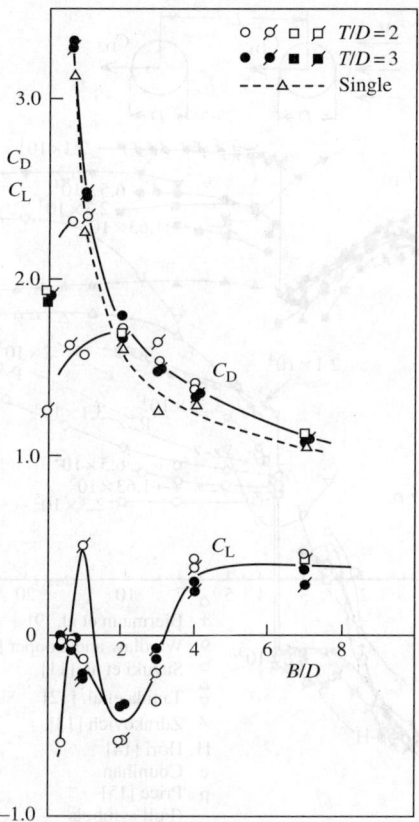

FIGURE 13.10 Variation of drag and lift force coefficients to the side ratio B/D of rectangular structures for $T/D = 2$ and 3 [8].

Figure 13.11 shows the drag force coefficients to L/D (spacing ratio, L is the spacing distance between the two structures) for tandem arrangement of circular structures [9–18]. In the figure, C_{D1} and C_{D2} indicate drag force coefficients of upstream and downstream structures, respectively. C_{D1} is positive over the whole spacing ratio and C_{D2} is negative in the spacing ratio less than 3.6. In the gap between two structures, negative pressure is induced by the wake of the upstream structure. Therefore, focussing on the upstream structure, positive pressure acts on the upstream surface and negative pressure is induced on the downstream surface by the structure's own wake. As a result, the drag force of the upstream structure takes a positive value. On the downstream structure, in case of space distance less than 3.6, negative pressure acts on the upstream surface, induced by the wake of the upstream structure, and negative pressure on the downstream surface. Since the absolute value of pressure for the upstream surface is larger than for the downstream surface, the drag force becomes negative for the downstream structure. The direction of the drag force of the downstream structure is the upstream direction. As shown in Figure 13.11, the drag force coefficients of both structures jump at around $L/D = 3.6$. This is called "critical spacing distance." In the spacing distance larger than the critical spacing distance, the effect of wake interference is weakened and both drag force coefficients take positive values. The flow pattern for a tandem arrangement of circular structures is shown in Figure 13.12 [19]. This figure provides useful information to understand the relationship between spacing distance and flow patterns or wake interference.

Figure 13.13 shows drag force coefficients of a tandem arrangement of rectangular structures with various side ratios [20–22]. The jumps of drag force coefficients are seen in rectangular structures

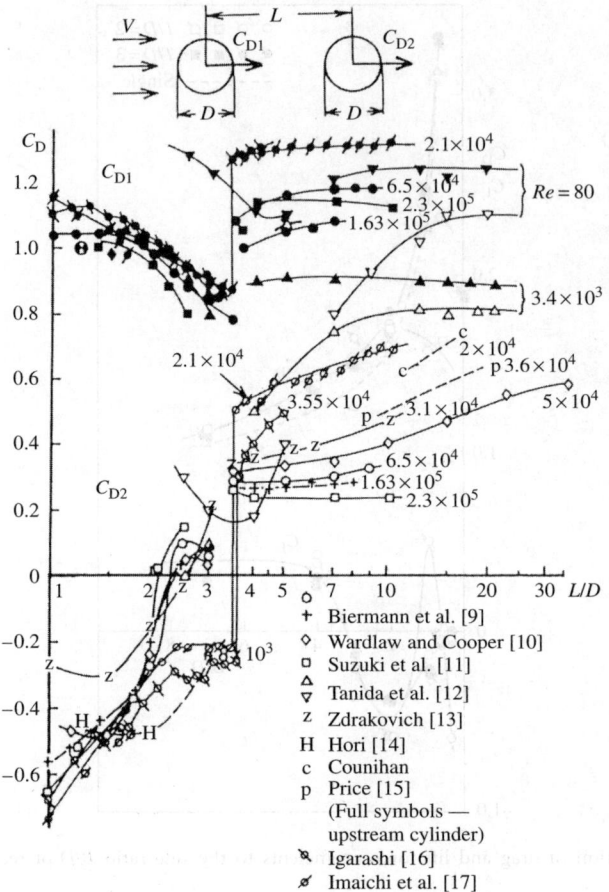

FIGURE 13.11   Drag force coefficient of two tandem circular cylinders [12,16–18,48].

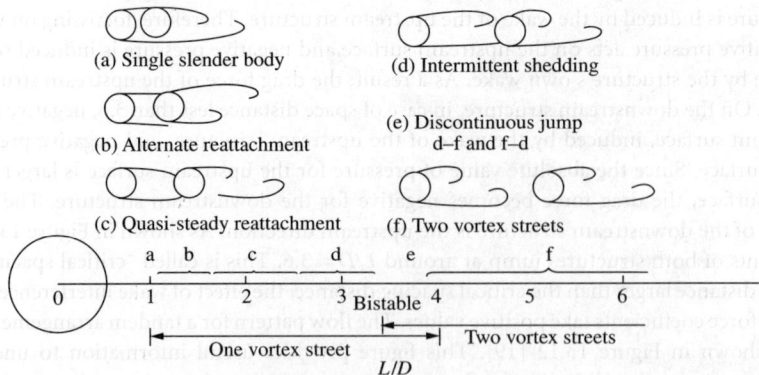

FIGURE 13.12   Classification of flow regimes in a tandem arrangement. (Zdrakovich 1985. Reprinted with permission from Elsevier.)

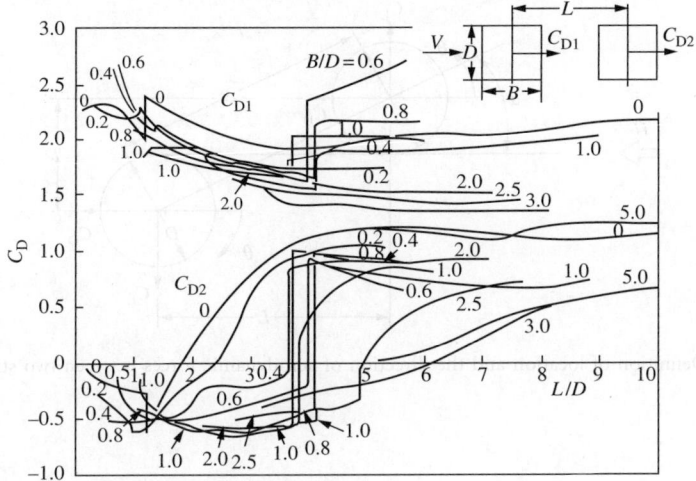

FIGURE 13.13 Drag force coefficients of two tandem rectangular cylinders with various side ratios, $B/D = 0$, 0.2, 0.4, 0.6, 0.8, 1.2, 2.5, 3, and 5 [20–22,48].

with side ratio between 0.4 and 2.0 at a spacing distance of around $L/D = 4$. As seen in the tandem arrangement of circular structures, C_{D2} changes its own sign from negative to positive against increment of L/D. Therefore, the wake interference is supposed to be weakened in the region of spacing distance over $L/D = 4$. Both drag force coefficient curves of the rectangular structures show only small changes to a variety of spacing distances for the cases except a side ratio between 0.4 and 2.0. In the single rectangular structure, it should be pointed out that the side ratio of 0.6 is a critical value. But in the tandem arrangement of rectangular structures the critical value is not found.

13.2.3.3 Staggered Arrangement

When practical structures are considered, the wind direction cannot be fixed. Therefore, even if two structures stand side by side or in tandem arrangement, the arrangement of structures becomes a staggered arrangement depending on the wind direction. Although it is very difficult to find the universality about the behavior of aerodynamic forces for staggered arrangement of the structures, which is induced by the complicated flow conditions, it will be useful for structural engineers to understand the behavior of aerodynamic forces of the staggered arrangement. But the amount of experimental data is restricted and the data presented here are only for circular structures and viaducts.

13.2.3.3.1 Circular Structures

Figure 13.14 shows the definition of location and the direction of aerodynamic forces between two structures. Figure 13.15 and Figure 13.16 show the plot of constant lift and drag force coefficient curves of the downstream structure, respectively [13]. Figure 13.15 shows a positive repulsive force in the vicinity side by side arrangement and negative lift force directed toward the wake axis of the upstream structure in the remainder of the staggered arrangement. The most remarkable feature is that there are two different lines of maximum lift force, shown with a chain-dotted line. One is inside the upstream structure wake for L/D up to 3. The other is nearer to the wake boundary for L/D larger than 2.7.

When the side surface of the downstream structure approaches closely to the axis of the wake of the upstream structure, the gap flow between the two structures generates the negative lift force, and its absolute value increases with decrement of T/D. The gap flow with high velocity induces a very low

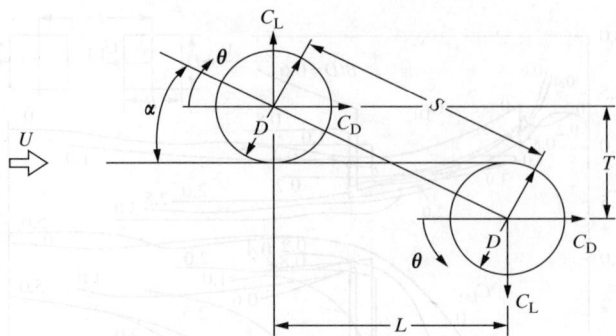

FIGURE 13.14   Definition of location and the direction of aerodynamic forces between two structures.

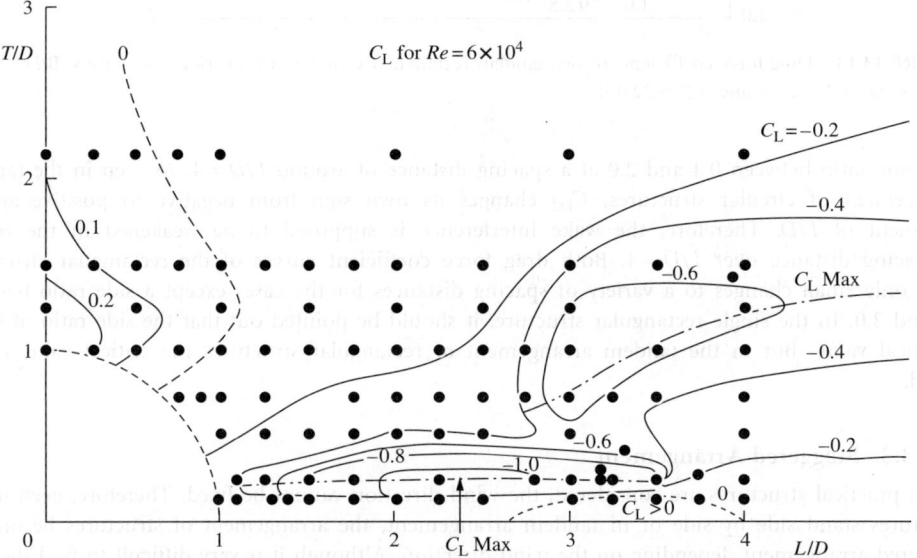

FIGURE 13.15   Lift force coefficient of a downstream structure to L/D. (Zdrakovich and Pridden 1977. Reprinted with permission from Elsevier.)

pressure region around both the inner and the lower sides of the downstream structure as shown in Figure 13.17 [13]. Such a pressure distribution of the downstream structure causes a very higher lift force.

On the other hand, Figure 13.15 and Figure 13.16 show that the slight asymmetry of the flow around the downstream structure has an immediate effect on the lift force, but very little effect on the drag force. A very high lift force is thus generated, but the negative drag force in a tandem arrangement is generated by a no flow condition in the gap.

13.2.3.3.2 Viaducts in Vicinity Arrangement

Figure 13.18 shows the model arrangement to measure the drag and lift forces of staggered arrangement of the viaducts [23]. Wind tunnel tests were conducted to obtain data for wind load to design viaducts with a noise barrier in the staggered arrangement. Figure 13.19 and Figure 13.20 show the drag forces of the model with a relative arrangement. On comparing drag force coefficients with

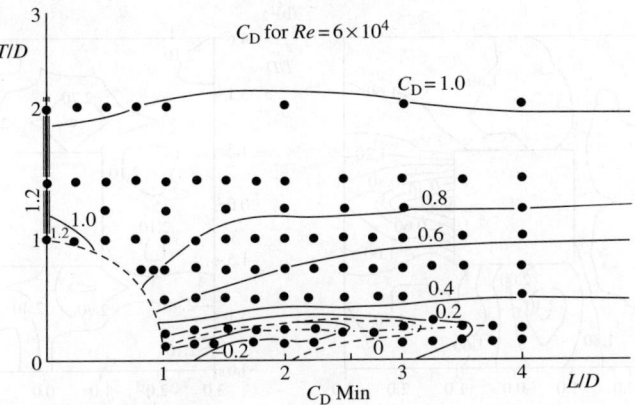

FIGURE 13.16 Drag force coefficient of a downstream structure to L/D. (Zdrakovich and Pridden 1977. Reprinted with permission from Elsevier.)

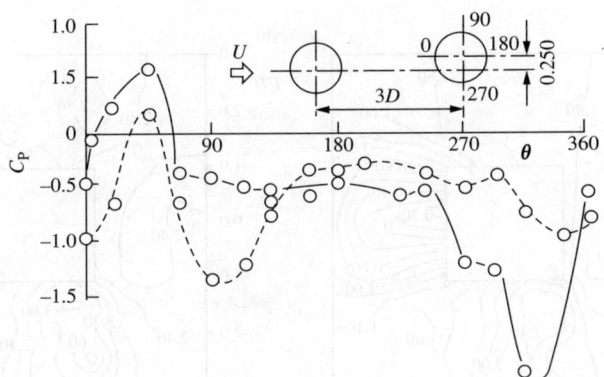

FIGURE 13.17 Pressure distribution around the downstream structure in the staggered arrangement with $L/D = 3$ and $T/D = 0.25$. (Zdrakovich and Pridden 1977. Reprinted with permission from Elsevier.)

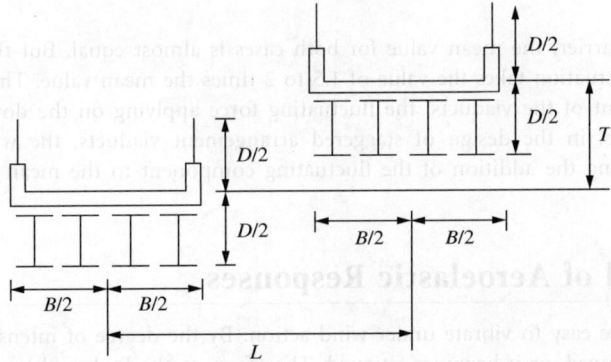

FIGURE 13.18 Definition of location of the model for measurement to the reference model in the staggered arrangement of viaducts.

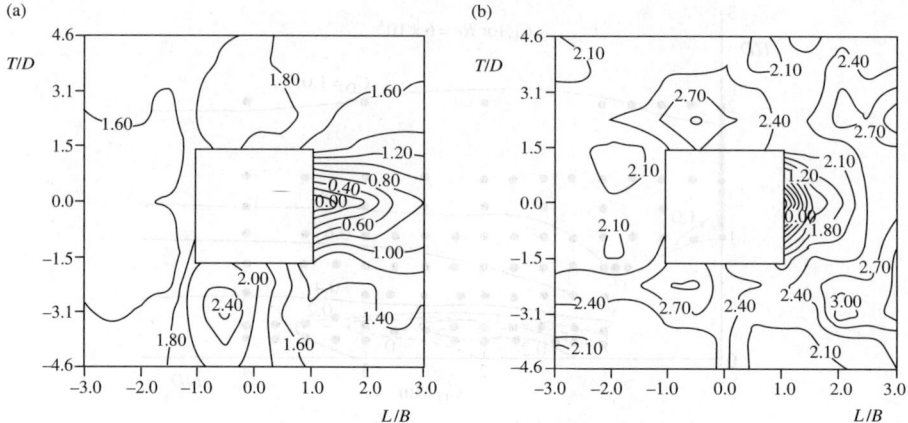

FIGURE 13.19   Drag force coefficient distribution of the model for measurement in the staggered arrangement of viaducts without noise barrier: (a) mean value and (b) mean value plus fluctuation (root mean square).

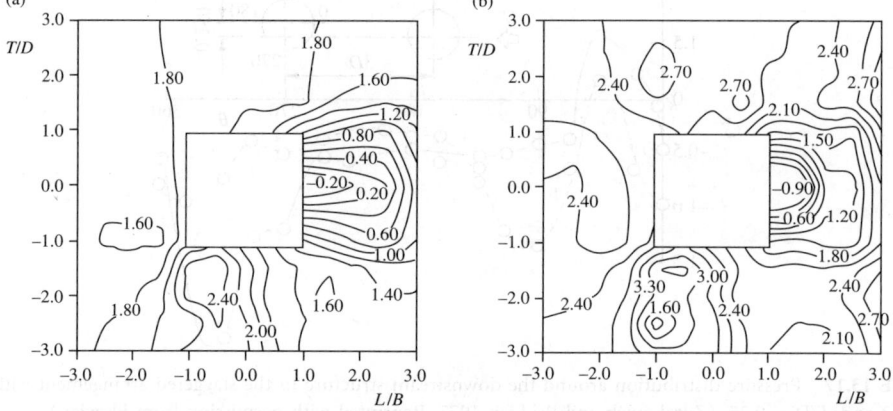

FIGURE 13.20   Drag force coefficient distribution of the model for measurement in the staggered arrangement of viaducts without noise barrier: (a) mean value and (b) mean value plus fluctuation (root mean square).

and without noise barrier, the mean value for both cases is almost equal. But the drag force in the case of including fluctuation takes the value of 1.5 to 2 times the mean value. This means that in the staggered arrangement of the viaducts, the fluctuating force applying on the downstream viaduct is very large. Therefore, in the design of staggered arrangement viaducts, the wind load should be decided by considering the addition of the fluctuating component to the mean value.

13.3 Control of Aeroelastic Responses

Flexible structures are easy to vibrate under wind action. By the degree of intensity of the vibration, the structure is destroyed, or it becomes fatigued. Therefore, methods should be developed to reduce the degree of intensity of the vibration and to keep it safe. One of the methods is to adopt a structural configuration able to reduce the aeroelastic forces. It is, however, necessary to understand the mechanism of aeroelastic vibrations before investigating the method.

13.3.1 Mechanism of Aeroelastic Vibration of Structures

Aeroelastic vibrations of flexible structures are classified into four types of vibrations (Figure 13.21): (1) vortex-excited vibration, (2) self-excited vibrations including galloping and flutter, (3) buffeting in natural wind, and (4) aerodynamic interference vibration. Vertical and horizontal axes indicate amplitude of vibration and wind velocity, respectively. In the following sections, each vibration will be discussed.

13.3.1.1 Vortex-Excited Vibration

The vortex alternately sheds from one side to another in the wake of the bluff body as shown in Figure 13.1. The shedding frequency is linearly proportional to the wind velocity. The pressure fluctuation occurs on the surface of the bluff body corresponding to the shedding frequency. When the shedding frequency is equal to the natural frequency of the structure, the force generated by the fluctuation of the surface pressure becomes resonant with the vibration of the structure. Another definition is that the vortex-excited vibration is the vibration whose frequency coincides with the vortex shedding frequency. The vortex-excited vibration occurs usually at a relatively lower wind velocity as shown in Figure 13.21. As already mentioned, each bluff body possesses an intrinsic value of Strouhal number. Incidentally, $St = 0.2$ for a circular structure, $St = 0.12$ for a square structure.

Figure 13.22 shows a typical example of vortex-excited vibration of a circular structure elastically supported with springs as indicated in the figure. The frequency of the wake of the circular structure in stationary state increases with increment of wind velocity as an inclined line in the figure. But the frequency of the wake of the circular structure supported with elastic springs takes a constant value almost equal to the natural frequency of the vibration system during the vibration of the structure. The region is called "locking-in region" or "locked-in region," which is a kind of synchronization phenomenon that the wake synchronizes with the natural frequency of vibration system. This is called "vortex-excited vibration." In the condition of vortex-excited vibration, the frequencies of both wake and bluff body coincide with each other. Since the amplitude and the region of wind velocity for vortex-excited vibration are limited, this vibration has another name, "limited vibration."

Onset wind velocity of vortex-excited vibration can be estimated from the Strouhal number. In the equation defining Strouhal number (Equation 13.2), all that is needed is to replace f_v as the frequency of the wake with the natural frequency f_0 of the vibration system. The onset wind velocity U_{cr} (critical wind velocity) can be estimated by the following equation:

$$U_{cr} = \frac{f_0 d}{St} \tag{13.4}$$

The expression by reduced wind velocity is as follows:

$$U_{r,cr} = \frac{U_{cr}}{f_0 d} = \frac{1}{St} \tag{13.5}$$

Therefore, Strouhal number directly gives the onset wind velocity by the reduced wind velocity expression for the vortex-excited vibration. $U_{r,cr} = 5$ is for a circular structure and $U_{r,cr} = 8.33$ is for a square structure. Other expressions for the onset wind velocity are $U_{cr} = 5f_0 d$ for the circular structure and $U_{cr} = 8.33f_0 d$ for the square structure.

13.3.1.2 Self-Excited Vibration

Although almost aeroelastic vibration has characteristics of self-excited vibration, galloping and flutter are regarded to be representative self-excited vibrations. The term "galloping" is usually used for

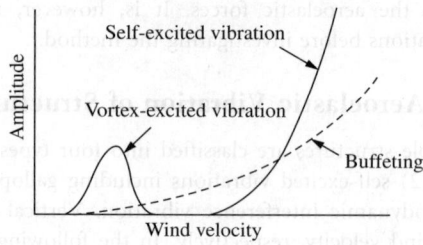

FIGURE 13.21 Types of aeroelastic response.

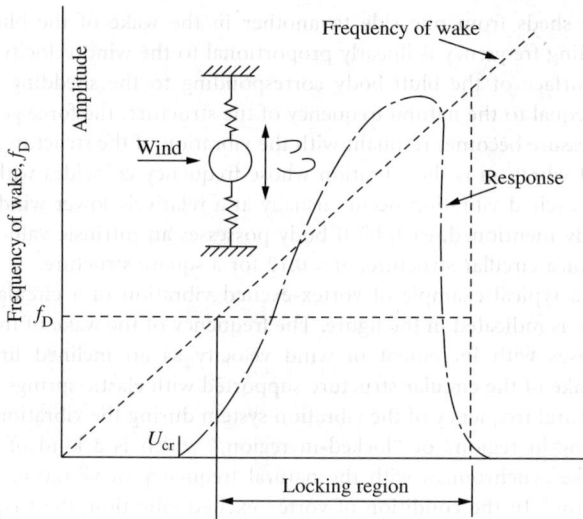

FIGURE 13.22 Vortex-excited vibration of a circular cross-section structure elastically supported with springs.

self-excited vibration including heaving vibration mode, and the term "flutter" is used for self-excited vibration including torsional vibration mode. Self-excited vibration is a vibration that increases the vibration amplitude by adding force induced by the motion of the body in the air as shown in Figure 13.23. Therefore, once a vibration occurs, the vibration grows to a large amplitude to collapse the structure.

The explanation for self-excited vibration by the equation of motion is as follows: m is the mass of the vibration system, c is the structural damping constant, k is the spring constant, y is the displacement, F_0 is the fluctuation amplitude of the aeroelastic force, ω is the frequency of the aeroelastic force, and β is the phase lag of the aeroelastic force to displacement. The equation of motion is

$$m\ddot{y} + c\dot{y} + ky = F_0 \sin(\omega t + \beta)$$
$$= F_0 \cos\beta \sin\omega t + F_0 \sin\beta \cos\omega t$$

by putting $y = y_0 \sin\omega t$, $\dot{y} = \omega y_0 \cos\omega t$

$$= F_0 \cos\beta \frac{y}{y_0} + F_0 \sin\beta \frac{\dot{y}}{\omega y_0} \tag{13.6}$$

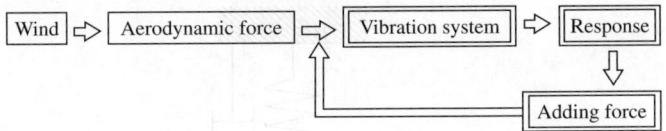

FIGURE 13.23 Explanation of self-excited vibration.

rewriting the equation by moving the right-hand side to the left-hand side

$$m\ddot{y} + \left(c - \frac{F_0 \sin \beta}{\omega y_0}\right)\dot{y} + \left(k - \frac{F_0 \cos \beta}{y_0}\right)y = 0 \tag{13.7}$$

The required condition for vibration growth is that a term proportional to velocity of displacement is negative. This is the condition for the occurrence of self-excited vibration

$$\left(c - \frac{F_0 \sin \beta}{\omega y_0}\right) \leq 0 \tag{13.8}$$

when $c = 0$, $\sin \beta \geq 0$. That is, when $\beta \geq 0$, self-excited vibration occurs. The meaning of $\beta \geq 0$ is that external force F_0 works before displacement increases the amplitude.

An interesting expression is used to explain vibration. Figuratively speaking, a horse beaten to run is forced vibration. A horse is chasing a carrot to eat which is hung down from a bar fixed on her back. The horse will increase its running speed with the intention of bringing the carrot into its mouth. This is literally self-excited phenomenon.

The following phenomena are classified into self-excited vibration.

13.3.1.2.1 Galloping

A square section and D-type section induce galloping. Figure 13.24 shows a spring-mass vibration system under wind action. When the mass goes downward, the mass receives the upward wind. Considering the moment when the mass is going down, the aerodynamic forces act on the mass as shown in Figure 13.25. The upward force F_y from drag and lift forces can be regarded as a driving force to heave the mass up and down. F_y can be expressed as follows:

$$F_y = L \cos \alpha + D \sin \alpha \tag{13.9}$$

By using Taylor's expansion at around $\alpha = 0$ and expressed in first order

$$F_y = \left(\frac{dL}{d\alpha} + D\right)\bigg|_{\alpha=0} \alpha = \frac{1}{2}\rho U^2 A \left(\frac{dC_L}{d\alpha} + C_D\right)\bigg|_{\alpha=0} \alpha \tag{13.10}$$

Assuming that α is very small and satisfies $\tan \alpha \approx \alpha$, $\alpha \approx -\dot{y}/U$ can be assumed. The equation of motion is

$$m\ddot{y} + c\dot{y} + ky = -\frac{1}{2}\rho U^2 A \left(\frac{dC_L}{d\alpha} + C_D\right)\bigg|_{\alpha=0} \frac{\dot{y}}{U} \tag{13.11}$$

by putting $\omega^2 = k/m$, $2\zeta\omega = c/m$, and rearranging the equation

$$\ddot{y} + \left\{2\zeta\omega + \frac{\rho U A}{2m}\left(\frac{dC_L}{d\alpha} + C_D\right)\bigg|_{\alpha=0}\right\}\dot{y} + \omega^2 y = 0 \tag{13.12}$$

The condition toward oscillatory instability is that the second term is negative. If $\zeta = 0$

$$\left(\frac{dC_L}{d\alpha} + C_D\right)\bigg|_{\alpha=0} < 0 \tag{13.13}$$

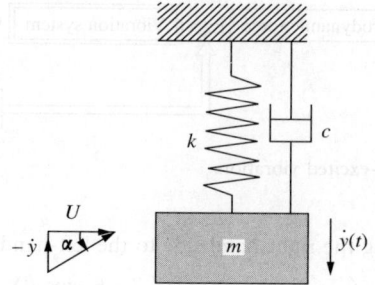

FIGURE 13.24 Spring-mass system subjected to wind action.

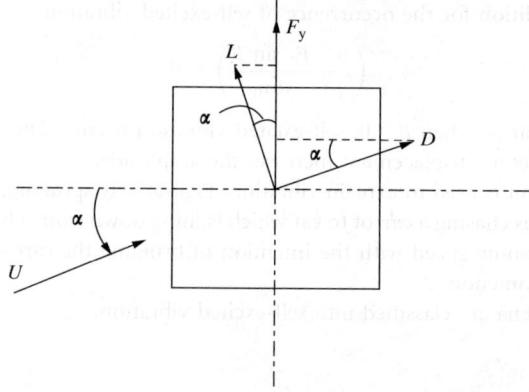

FIGURE 13.25 Heaving force by drag and lift forces and relative wind velocity.

becomes the condition for oscillatory instability. This is the well known Den Hartog criterion. The criterion teaches us that it is possible to estimate the occurrence of galloping from the lift forces to the angle of attack.

Figure 13.26 shows the lift force coefficient to the angle of attack of a square structure [4]. The meaning of the negative slope at zero angle of attack is as follows. When the square structure is going down, the square structure receives upward wind with the positive angle of attack and the downward force F_y is generated through the negative lift force. Therefore, F_y acts to help increase the downward displacement, that is, the force acting on the structure has the same direction as the movement of the structure, and the amplitude of vibration increases infinitely in linear vibration system. This is the mechanism of galloping. The method to estimate the dynamic behavior by using static parameters is called as the quasisteady theory. The galloping of a square structure is a representative example to which the quasisteady theory is applicable.

13.3.1.2.2 Stall Flutter
Shallow rectangular sections and shallow H-sections induce stall flutter, which is the torsional type self-excited vibration with single degree of freedom. Since this type of flutter is induced by the separation flow from the leading edge of the shallow body, it is also called "separation flow flutter."

13.3.1.2.3 Classical Flutter
A flat plate induces classical flutter, which is the self-excited vibration with two degrees of freedom. Heaving and torsional vibrations are coupled and this is the most intense self-excited vibration.

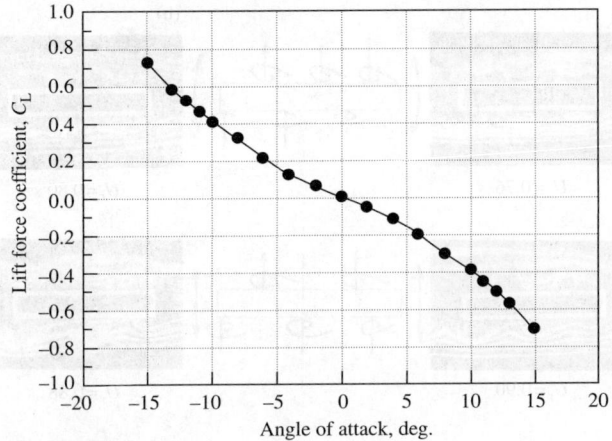

FIGURE 13.26 Lift force coefficient to angle of attack of a square structure [4].

Other names are "coupled flutter" and "potential flutter." In this flutter, the flow around the flat plate is the potential flow without separation from the leading edge.

13.3.1.3 Buffeting and Aerodynamic Interference Vibration

13.3.1.3.1 Buffeting
When the approaching flow to a structure is turbulence, the structure is forced to vibrate randomly due to the random external force induced by the turbulent flow. Buffeting is inherently a random vibration due to the aerodynamic random force induced by upstream turbulence which is generated by the existence of the upstream structure.

13.3.1.3.2 Aerodynamic Interference Vibration
When the structures or structural members are arranged in the vicinity area, an upstream structure or member affects the approaching flow to a downstream structure and the behavior of the downstream structure. For example, in bundle cable of cable-stayed bridges, the downstream cable intensely vibrates, affected by the wake of the upstream cable. As the structure for city life becomes complicated, the number of buildings and viaducts built in the vicinity area increases and the problems associated with aerodynamic interference tend to increase. Since the phenomena are complicated and various, as introduced later, it is very difficult to solve the problem by the deterministic method.

13.3.1.4 Mechanism of aeroelastic vibration of shallow section

Figure 13.27 shows the vortex arrangement on the surface in the motion of torsional and heaving vibrations [24]. Left and right rows show the flow situation around the shallow rectangular structure in torsional and heaving vibration, respectively, the middle row shows sketches of vortex arrangement on the surface, and the arrow indicates force direction acting on the surface. According to the figure, the number of vortices on the surface of the shallow rectangular structure is the same for the identical reduced wind velocity regardless of kinds of motion.

Do inherent differences exist between the vortex-excited vibration and self-excited vibration? According to the research associated with a shallow rectangular structure, there is no inherent difference between them. Figure 13.28 is a summarized expression concerning the mechanism of aeroelastic vibration. The fundamental concept is that aeroelastic vibration is controlled by the vortex

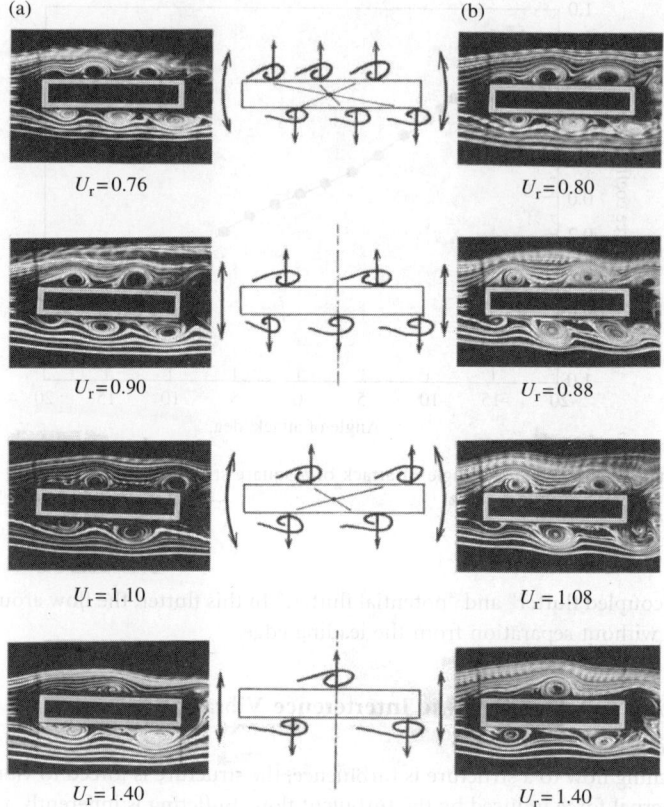

(a)   (b)

$U_r = 0.76$ $U_r = 0.80$

$U_r = 0.90$ $U_r = 0.88$

$U_r = 1.10$ $U_r = 1.08$

$U_r = 1.40$ $U_r = 1.40$

FIGURE 13.27 Vortex arrangement at various wind velocities [24]: (a) torsional vibration and (b) heaving vibration.

arrangement on both upper and lower surfaces. The assumption of the fundamental concept is that the separation flow takes a certain time length to form a vortex on the surface and the formed vortex flows down with the speed equal to wind velocity. Figure 13.28 also shows aeroelastic vibrations to reduced wind velocity for torsional and heaving vibrations. The top panel indicates that torsional vibration and heaving vibration appear alternately to increment of wind velocity, if both vibration modes have the same natural frequency. The torsional vibration mode appears under the condition that the number of vortices on the upper surface is equal to lower surface and the heaving vibration occurs under the condition that there is difference of one vortex between the number of vortices on upper and lower surfaces. From the point of view mentioned above, flutter is regarded as torsional vibration III. In this sense, inherently there is no difference between vortex-excited vibration and stall flutter.

13.3.2 Control of Aeroelastic Vibration of a Single Bluff Body

Since the vortex-excited vibration of a single bluff body is seen at a relatively lower wind velocity, the occurrence frequency is highest among the various aeroelastic vibrations. Various methods have been developed to suppress the vortex-excited vibration of tower-like structures and bridge girders. In the following, methods to control aeroelastic vibrations are introduced as countermeasures for tower-type or bridge girder-type structures.

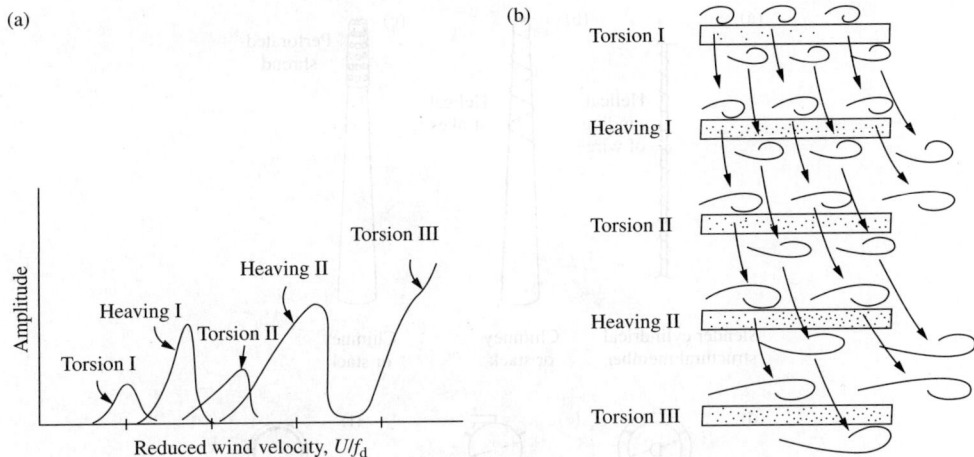

(a)

Amplitude

Torsion I

Heaving I

Torsion II

Heaving II

Torsion III

Reduced wind velocity, U/f_d

(b)

Torsion I

Heaving I

Torsion II

Heaving II

Torsion III

FIGURE 13.28 Summarized expression for aeroelastic vibration and the corresponding vortex arrangement [24]: (a) sketch of aeroelastic response of a shallow rectangular structure and (b) expected vortex arrangement.

13.3.2.1 Tower-Type Structure

The tower-type structures are usually composed of typical bluff bodies that are circular and square structures or shapes nearer to these. A method to suppress the vortex-excited vibrations of circular structures has been developed. The concept of the method is to reduce the space correlation of surface pressure in the axial direction of the bluff body. Another method is to remove the simultaneous fluctuation of surface pressure along the longitudinal axis, or to remove the simultaneous separation of the flow to generate vortices in axial direction of the bluff bodies.

The methods for structures with circular cross-section are helical winding of wire, helical strake, and perforated shroud as shown in Figure 13.29 [25,26]. The wind tunnel test result is shown in Figure 13.30 for a chimney with helical strake [27]. Although the vortex-excited vibration occurs in the case of chimney without helical strakes, it is completely suppressed by attaching the helical strakes.

Since octagonal and square shape cross-sections have the negative slope of lift force around zero degrees of angle of attack as shown in Figure 13.26 and Figure 13.31 [28], they have the possibility of galloping as mentioned in the previous section. It is necessary to develop a method for suppressing the galloping of both shape cross-sections. The methods invented up to now are shown in Figure 13.32. They are corner cut-outs, corner vanes, gap, and protuberant plates. The concept for corner vanes is to suppress the separation from the corner by inducing the approaching flow along the surface. For the corner cuts, corner cut-outs, and protuberant plates, having two corners in close proximity is the concept to control the separation flow from the upstream separation point by contacting it to the downstream separation point. And for gap the concept is to reduce the pressure difference between front and back surfaces. Figure 13.33 and Figure 13.34 show the examples of wind tunnel results for corner vanes and protuberant plates [29,30]. In each figure, the results are compared for both the cases with and without the countermeasures. It is understood that both methods are very useful for suppressing the galloping. Especially interesting is that the angle between two separation points affects the performance of suppression of galloping and that there is an optimum angle around 30° (corresponding to $p/H = 0.3$). The result is obtained from the wind tunnel tests for protuberant plates. These countermeasures were used for steel towers of cable-stayed and suspension bridges. In these bridges, before the cable is installed, the steel tower is in a free standing state with low structural damping and is easy to induce vortex-excited vibration and galloping.

There is another aeroelastic problem on a tower of a cable-stayed bridge. When the tower is in the A-shape, the in-plane vibration is sometimes induced by the wind with bridge axis direction. Since the

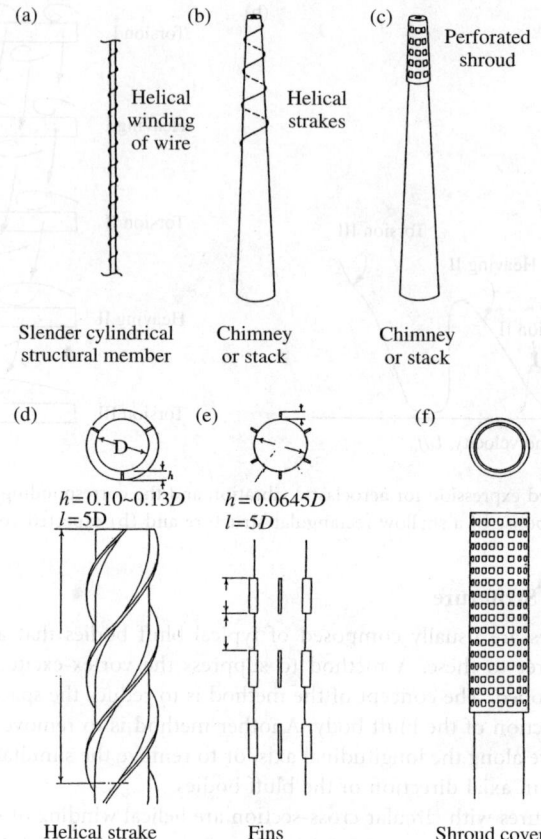

FIGURE 13.29   Examples of countermeasure for vortex-excited vibration of a circular cylinder [25,26].

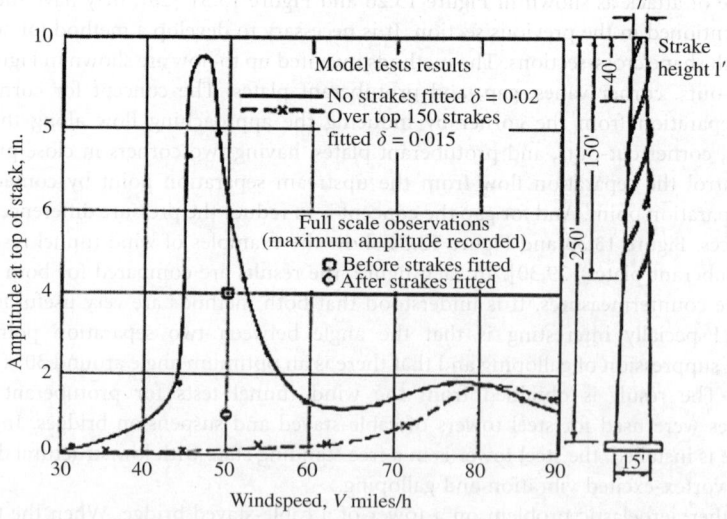

FIGURE 13.30   Wind-induced vibration of chimney improved by helical strake [27].

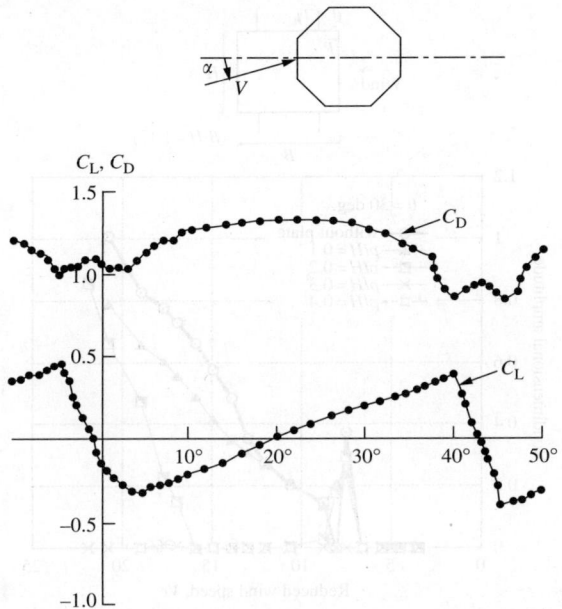

FIGURE 13.31 Drag and lift force coefficients for octagonal cylinder at $Re = 1.2 \times 10^6$ [28].

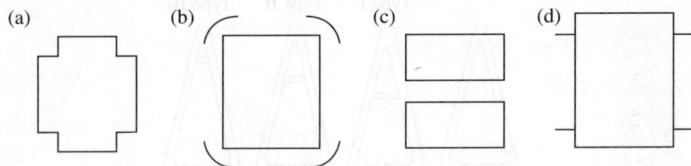

FIGURE 13.32 Types of countermeasure for bridge tower: (a) corner cut-outs; (b) corner vanes; (c) gap; and (d) protuberant plates.

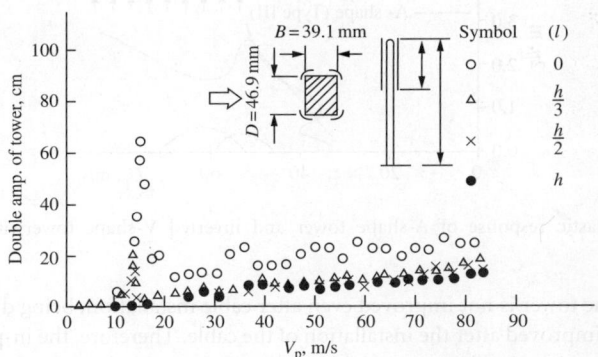

FIGURE 13.33 Effect of corner vanes for bridge tower on reduction of aeroelastic vibration amplitude [29].

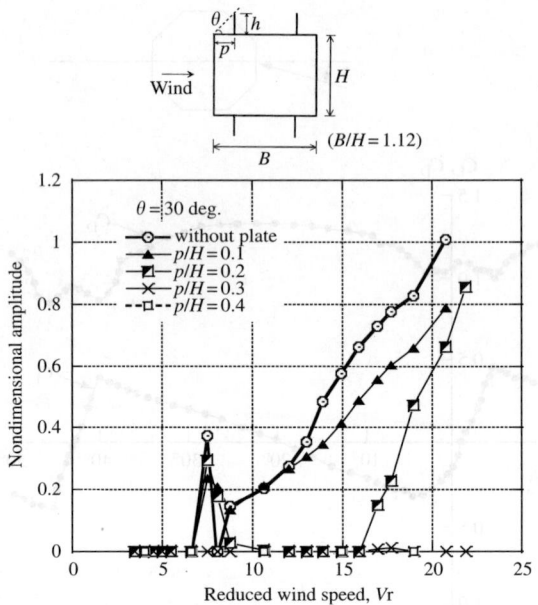

FIGURE 13.34   Aeroelastic response of protuberant plates with various plate locations ($H = 50$ mm, $h = 9$ mm) [30].

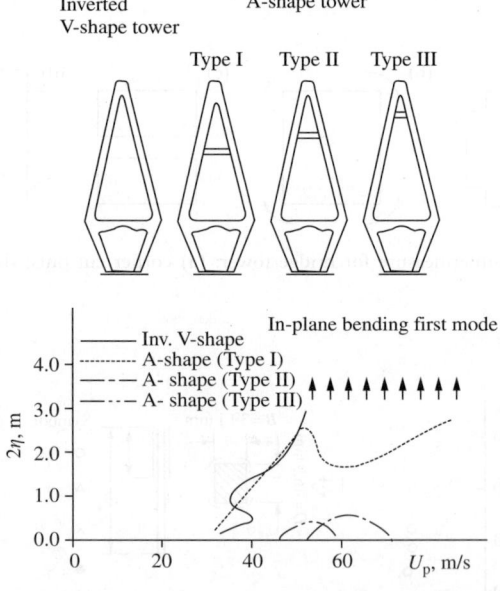

FIGURE 13.35   Aeroelastic response of A-shape tower and inverted V-shape towerwith various locations of horizontal beam [31].

in-plane stiffness of the tower is not improved even after cable installation, being different from the out-of-plane rigidity, it is improved after the installation of the cable. Therefore, the in-plane vibration by the wind in bridge axis direction is still a severe problem even after completion for a cable-stayed bridge with an A-shape tower. Figure 13.35 shows the results of wind tunnel tests for the A- and the inverted V-shape

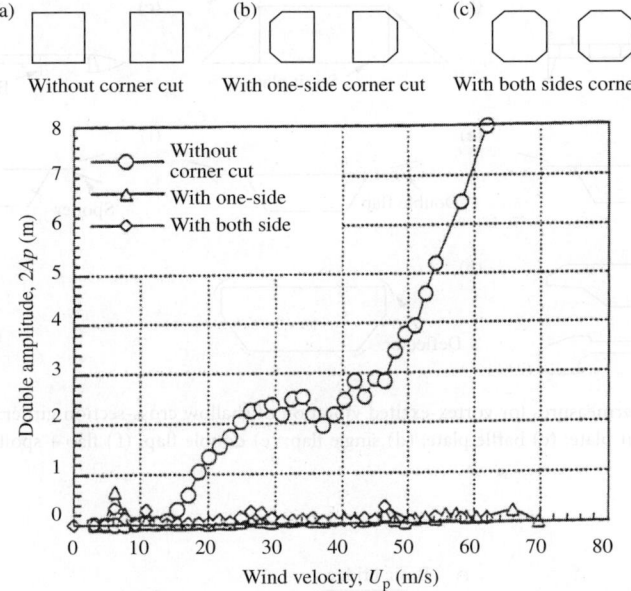

FIGURE 13.36   Aeroelastic response of A-shape tower with/without corner cut column [32].

towers in a free standing state [31,32]. The location of a horizontal beam remarkably affects the aeroelastic vibration in the out-of-plane direction. The key point to suppress the aeroelastic vibration is the selection of mounting position for horizontal beam of the tower.

Figure 13.36 shows the effect of corner-cut for the tower to the wind with perpendicular direction to the bridge axis [32]. When the corner-cuts are conducted at up- and downstream corners, the galloping is not suppressed; however, when the corner-cuts are conducted at only upstream corners, the aeroelastic performance is remarkably improved.

13.3.2.2 Bridge Girder Structures

The relatively shallower cross-section is usually used for bridge girders. Especially since a cable-stayed bridge and a suspension bridge as the representative of longer span bridge and girder bridge with longer span than 80 m have generally low frequency and small structural damping, they are susceptible to vibration under wind action. Various countermeasures have been developed to suppress aeroelastic vibrations induced in bridge girders.

13.3.2.2.1 Vortex-Excited Vibration

Figure 13.37 shows typical examples of countermeasure for vortex-excited vibration of bridge girders. The concept to suppress the vibration is to control the separation flow from the leading edge of the girders. In the figure, (a) horizontal plates were mounted to project outboard from the lower chord of each girder. It is expected for the plate to suppress the separation of flow from the under side of the plate girder. (b) The open girder induced vortex-excited vibration with large amplitude. Figure 13.38 shows the aeroelastic responses during process up to finding the final solution [33]. The figure indicates that the amplitude of vortex-excited vibration is reduced in cases when a soffit plate is added and when a soffit plate with wider fairing width is added. For an open deck, it is useful to suppress the vortex-excited vibration by closing the open side by the soffit plate. The more effective means besides adding the soffit plate is to mount the fairing with wider width. (c) Baffle plates are mounted in the inner part of the open deck. The role of the baffle plates is to prevent the separation flow of the under side of the girder from rolling into the open space of the girder. (d) Flaps are also useful for suppression of vortex-excited vibration. The role of the flap is to suppress the separation at the tip of the upper surface of the deck and

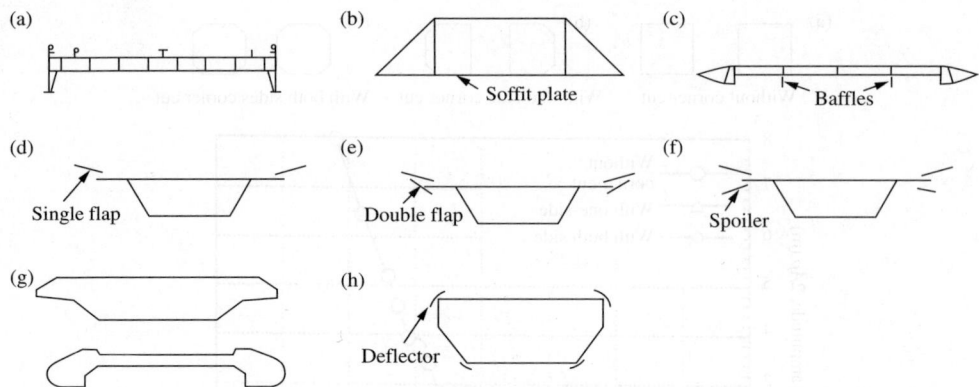

FIGURE 13.37 Countermeasures for vortex-excited vibration of shallow cross-section girder: (a) horizontal plate at lower flange; (b) soffit plate; (c) baffle plate; (d) single flap; (e) double flap; (f) flap + spoiler; (g) blunt fairing; and (h) deflector.

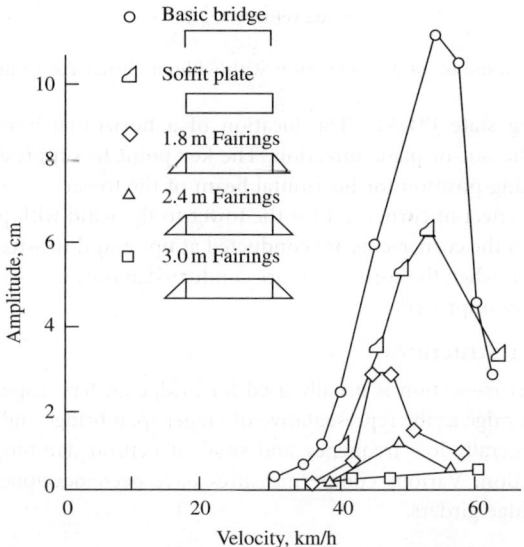

FIGURE 13.38 Effect of soffit plate and fairing for suppressing vortex-excited vibration of open girder [33].

to prevent the flow from rolling on the upper surface of the deck. (e) Deflector has a similar role as the corner vanes for the square-shaped tower introduced in the previous section. The deflector has the role of making the fluid at the leading corner run along the girder shape. (f) Blunt fairing was developed as the fairing for a cable-stayed bridge made of prestressed concrete. At the present time, the blunt fairing is also used in a steel bridge. The concept for the blunt fairing is as follows [34]. The required conditions for a concrete bridge is to eliminate a sharp edge corner, because it is difficult to make the sharp edge corner and the tip of the sharp edge corner is easy to break. After studying aeroelastic performance of various blunt fairings, the upper and lower slopes of the blunt fairing is decided in the shape as shown in Figure 13.39. Figure 13.40 shows the results during the process to decide angle of the lower slope. The lower angle as shown in the figure was changed from 20° to 50° in every 5°. Referring to the figure, the aeroelastic response is remarkably influenced by the lower angles of the blunt fairing. After comparison

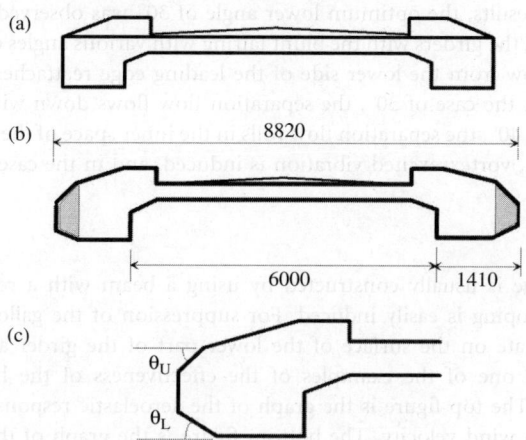

FIGURE 13.39 Open deck girder with blunt fairing made of prestressed concrete: (a) open deck girder without fairing; (b) fundamental cross-section for open deck girder; and (c) definition of θ_U and θ_L. (Kubo et al. 1993. Reprinted with permission from Elsevier.)

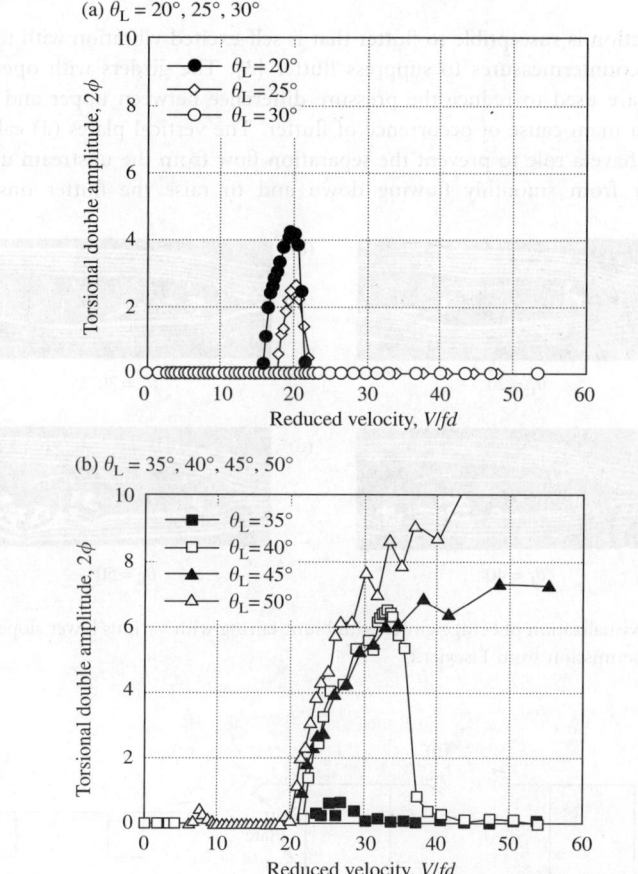

FIGURE 13.40 Aeroelastic response of bridge girder with blunt fairing with various lower slope angles θ_L. (Kubo et al. 1993. Reprinted with permission from Elsevier.)

of the wind tunnel test results, the optimum lower angle of 30° was observed. Figure 13.41 shows the flow visualization around the girders with the blunt fairing with various angles of lower slope. In the case of 20°, the separation flow from the lower side of the leading edge reattaches on the lower surface of the downstream edge. In the case of 30°, the separation flow flows down without reattachment. And in the cases of larger than 40°, the separation flow rolls in the inner space of the open deck. As a result, in the cases of less than 30°, vortex-excited vibration is induced, and in the cases of larger than 40°, stall flutter is induced.

13.3.2.2.2 Galloping
Since a box-girder bridge is usually constructed by using a beam with a rectangular section with a small side ratio, the galloping is easily induced. For suppression of the galloping, it is very useful to mount the horizontal plate on the surface of the lower part of the girder as shown in Figure 13.42 [3]. Figure 13.43 shows one of the examples of the effectiveness of the horizontal plate for suppressing galloping [35]. The top figure is the graph of the aeroelastic response of the original section. Galloping occurs at high wind velocity. The bottom figure is the graph of the aeroelastic response of the improved section by attaching a horizontal plate on the surface of the lower part of the the girder. The figure shows that the galloping which appears in original section is suppressed in the improved section.

13.3.2.2.3 Flutter
The shallow cross-section is susceptible to flutter that is self-excited vibration with torsional vibration. Figure 13.44 shows countermeasures to suppress flutter [3]. The girders with open grating (a) and central opening (b) are used to reduce the pressure difference between upper and lower surfaces of the girder, which is a main cause of occurrence of flutter. The vertical plates (a) called as "stabilizer" and "center barrier" have a role to prevent the separation flow from the upstream upper chord of the truss-stiffened girder from smoothly flowing down and to raise the flutter onset wind velocity.

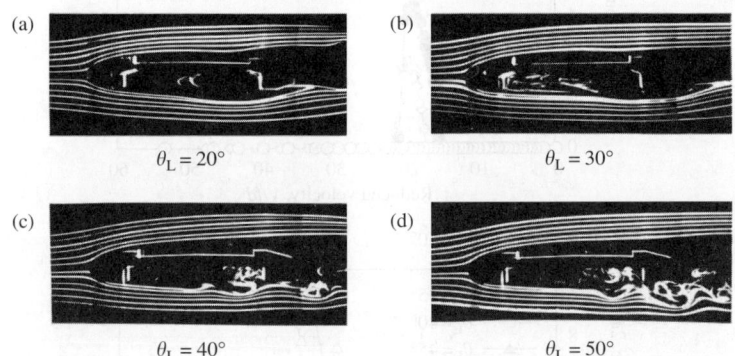

FIGURE 13.41 Flow visualization of bridge girder with blunt fairing with various lower slope angles. (Kubo et al. 1993. Reprinted with permission from Elsevier.)

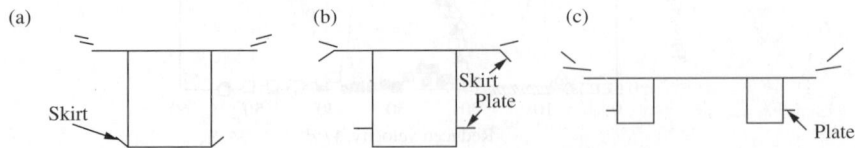

FIGURE 13.42 Countermeasures for galloping of box-girder [3]: (a) double flap + skirt; (b) flap + skirt + plate; and (c) double flap + plate.

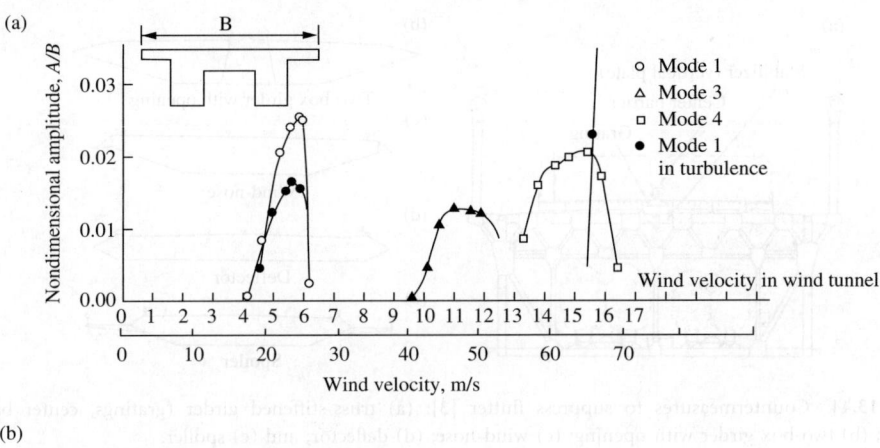

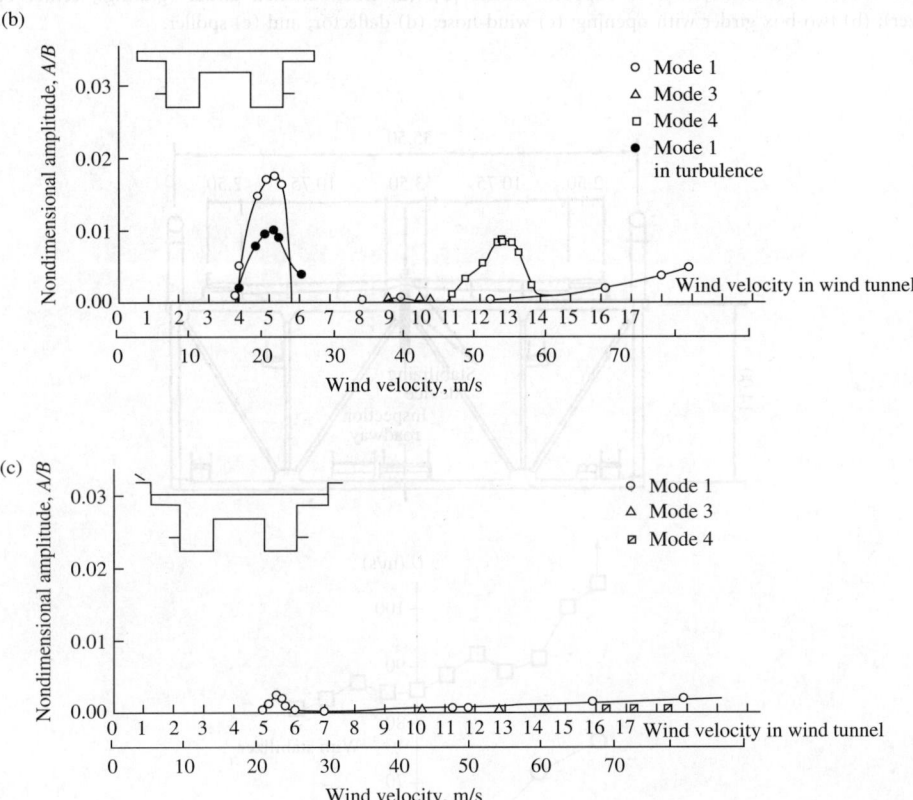

FIGURE 13.43 Effect of horizontal plates and double flap to suppress galloping and vortex-excited vibration [35]: (a) aeroelastic response of the original steel box-girder bridge; (b) aeroelastic response improved by horizontal plates at lower portion of the girder (galloping is suppressed); and (c) aeroelastic response improved by horizontal plate and double flap (galloping and vortex-excited vibration are suppressed).

Figure 13.45 shows the stabilizer for the truss-stiffened girder and the results of sectional model tests [36]. The flutter onset wind velocity of the truss-stiffened girder is remarkably raised by using a stabilizer. Wind-nose (c), deflector (d), and spoiler (e) help make the approaching flow run along the girder shape without separating the flow at the tip of the girder.

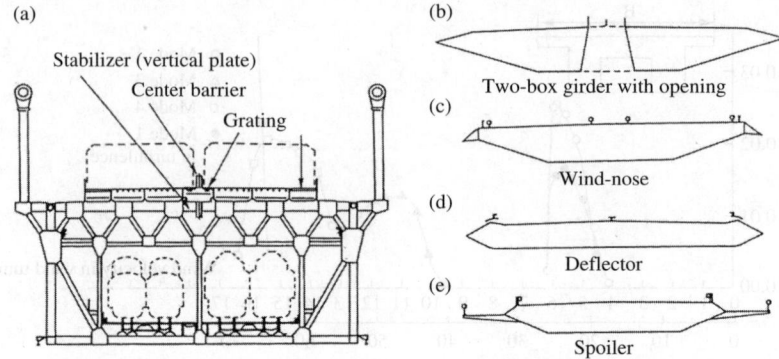

FIGURE 13.44 Countermeasures to suppress flutter [3]: (a) truss-stiffened girder (gratings, center barrier, stabilizer); (b) two-box girder with opening; (c) wind-nose; (d) deflector; and (e) spoiler.

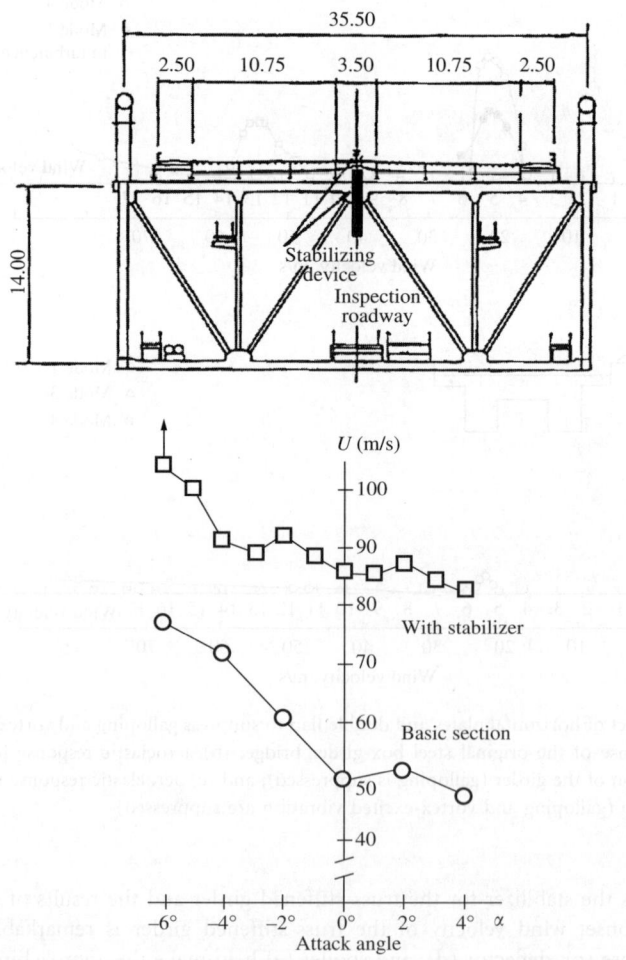

FIGURE 13.45 Flutter onset wind velocity of a truss-stiffened girder with/without stabilizer [36].

The ultimate countermeasure for suppressing aeroelastic vibration should be enabled by using not additive devices as deflector and fairing but only structural members. Figure 13.46 shows a bridge deck with two plate girders as one of the examples for trying to improve aeroelastic performance by only structural members [37]. In order to suppress the aeroelastic vibrations of the bridge girder with two plate girders, investigations were done to find how the locations of the plate girders under the deck influence the aeroelastic performance in both torsional and heaving vibration modes. Figure 13.47 shows the results of wind tunnel tests [37]. Referring to the experimental results, the inner location of the plate girder gives a better performance with regards to aeroelastic behavior. Before deciding on the structural details of a bridge, if the aeroelastic behavior is investigated, more economical and reasonable bridges can be realized without using additive devices.

13.3.2.3 Cables

Transmission lines, cables for cable-stayed bridges, and hanger for suspension bridge induce vortex-excited vibration, rain-wind-induced vibration, and galloping by adhering ice and snow. In order to suppress the vortex-excited vibration, helical winding of wire has been used in towers with circular cross-sections.

For rain-wind-induced vibration, axial grooves and indent shapes have been applied on the surface of the cable for a long-span cable-stayed bridge as shown in Figure 13.6 and Figure 13.48 [2,6]. The axial grooves and indent shapes prevent water rivulets from moving freely on the cable surface to simply induce the aeroelastic vibration. A diagonal hanger of a suspension bridge is susceptible to inducing aeroelastic vibration, explained as the axial flow along the diagonal hanger inducing the vibration. Cable fin is proposed as the countermeasure to prevent the occurrence of axial flow.

It is very difficult to suppress the galloping caused by adhering ice and snow by means of adopting an adequate configuration, because the form by ice and snow is not determined.

13.3.3 Control of Aerodynamic Interference Caused by Multiple Structure

The aerodynamic interference is induced by multiple structure or by multiple structural members and the degree of the interference is different by the difference among the arrangements. An intense vibration is induced by the wake of the upstream structure, which is called as "wake galloping" and "wake flutter."

13.3.3.1 Bundle Cables

In the cable-stayed bridge of prestressed concrete, multicables are sometimes used at one anchorage point from an economical viewpoint. When two cables in a bundle cable are placed closely in tandem arrangement against the wind direction, wake galloping is induced in the downstream cable. In the wake galloping, the downstream cable is vibrated by the wake generated by the upstream cable [38].

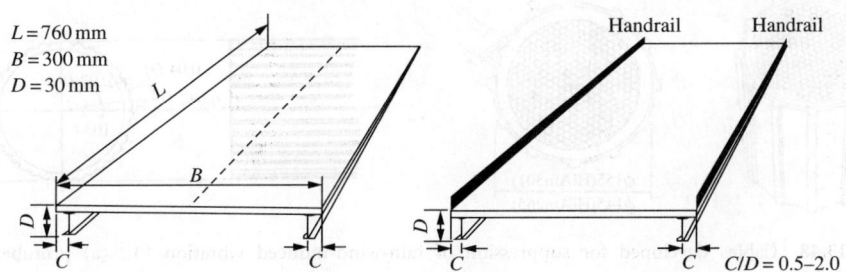

FIGURE 13.46 Bridge girder composed of two plate girders [37].

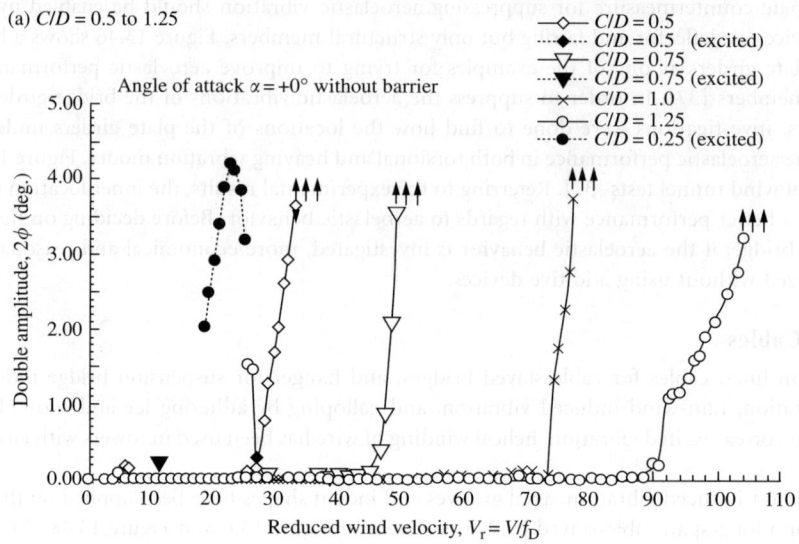

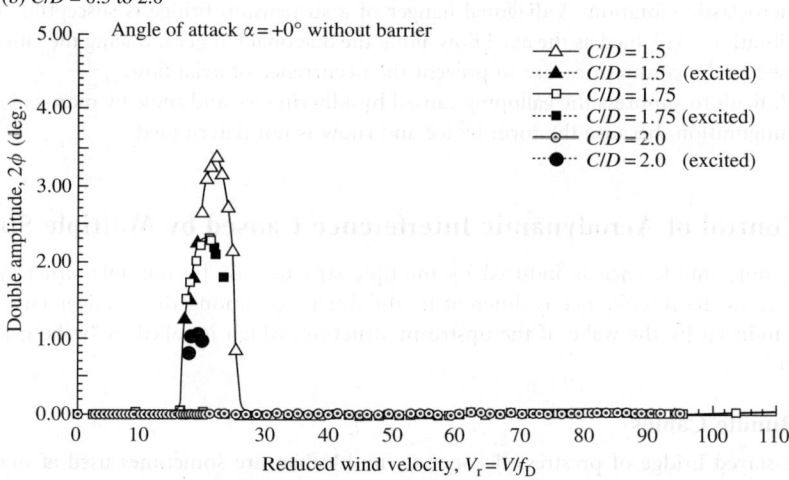

FIGURE 13.47 Torsional response of the bridge girder with various locations of two plate girders [37].

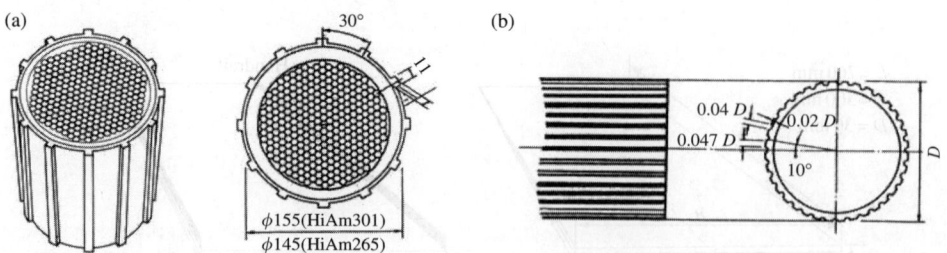

FIGURE 13.48 Cables developed for suppression of rain-wind-induced vibration [3]: (a) protuberant arial groove and (b) U-groove.

The response behavior varies according to the spacing distance between both cables as shown in Figure 13.49. The response behavior drastically changes as the boundary with respect to the critical spacing distance $L/D = 3.6$, which was introduced in Section 13.2.

Aeroelastic performance of three and four bundle cables were also investigated as shown in Figure 13.50 [39]. In the figure, the circle with oblique lines indicates the measured cable. In any of the cases, the downstream cable is vibrated with large amplitude by the wake of the upstream cables. Wake galloping is difficult to be generated in a lateral triangle arrangement, because any cable does not exist in the wake of other cables.

Wake flutter was observed in hanger cables of the Akashi Strait Bridge with large spacing distance of $10D$ (ten times of hanger diameter D) [40]. The wake galloping occurs intensively in tandem arrangement and the direction of vibration of downstream cable is almost perpendicular to the wind direction; the wake flutter, however, occurs intensively in a slightly deviated position from tandem arrangement and the locus of vibration of the downstream cable is an ellipse. Although the mechanism is still not clear, the helical winding of wire was proposed as a countermeasure to suppress the wake flutter. The experimental data are shown in Figure 13.51.

13.3.3.2 Parallel Bridges

A new bridge is sometimes constructed parallel to an old bridge to reduce traffic congestion in a large city. This fact causes a new problem concerning the aerodynamic interference caused by constructing the new bridge close to and along the old bridge.

Figure 13.52 shows the example of the aerodynamic interference caused by the parallel arrangement of a box three-span box-girder bridge [41]. The figure shows clearly that the relative distance among the bridges gives a remarkable effect for the aeroelastic performance of the bridge. The aeroelastic forces applied on the downstream bridge are induced by the wake of upstream bridge and generate the intense vibration with larger amplitude than 3% of deck height. It is very difficult to suppress the vibration induced by aerodynamic interference by the aerodynamic countermeasure by selection of a bridge cross-section providing better aeroelastic performance. Finding the solution is the future subject in the field for suppression of aeroelastic vibrations. One of the solutions is to construct a new bridge in a place away from old bridge where both bridges do not interfere with each other with respect to aerodynamic performance. It is very difficult to find the appropriate place in almost all cases from the viewpoint of effective land use. If possible, it should be investigated at the urban planning stage.

13.4 Wind Design Data

In this section, data for design of structures subjected to wind load and the method to read the experimental data are introduced.

13.4.1 Estimation of Wind Load and Onset Wind Velocity

Table 13.1 and Table 13.2 show Strouhal numbers and aerodynamic coefficients for various cross-sections [5]. The wind load is calculated by Equation 13.1. The stress intensity caused by wind load is compared with the allowable stress intensity of the material used for the structure and confirmed to be less than the allowable stress intensity.

The onset wind velocity of vortex-excited vibration is estimated by Equation 13.5. The following is a summary about the estimation of the onset wind velocity of a bridge girder in *The Wind-Resistant Design Handbook in Japan* [42]. The unit of onset wind velocity is m/s. Onset wind velocity should be larger than the design wind velocity as the design criterion.

(a) $S_H = 2d, 2.5d, 3d$

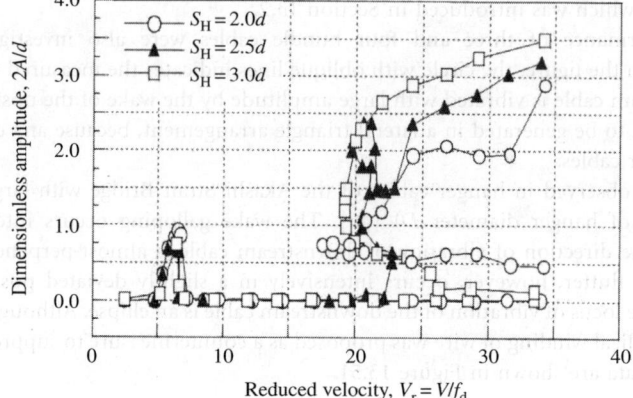

(b) $S_H = 3.5d, 4.0d, 5.0d$

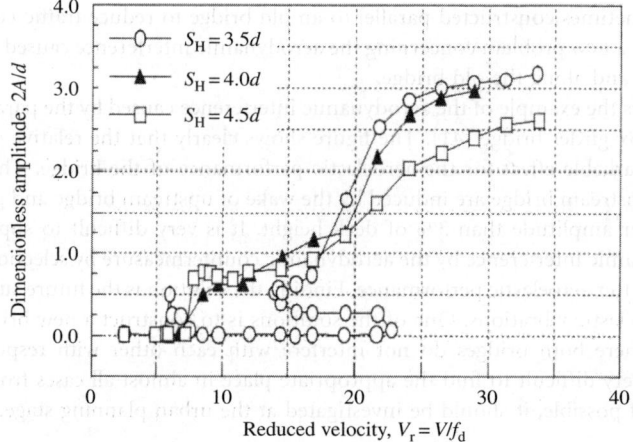

(c) $S_H = 6.0d, 7.0d, 8.0d$

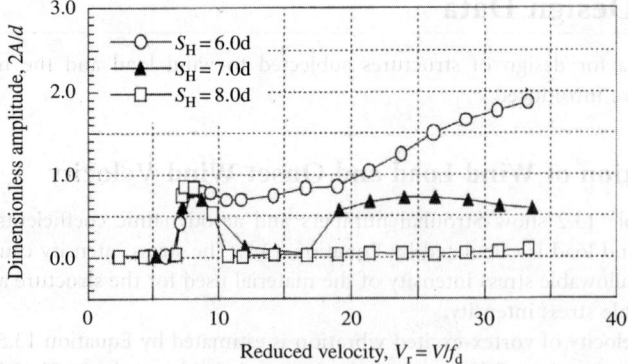

FIGURE 13.49 Aeroelastic response of a downstream cable of a bundle cable in wake-galloping [38].

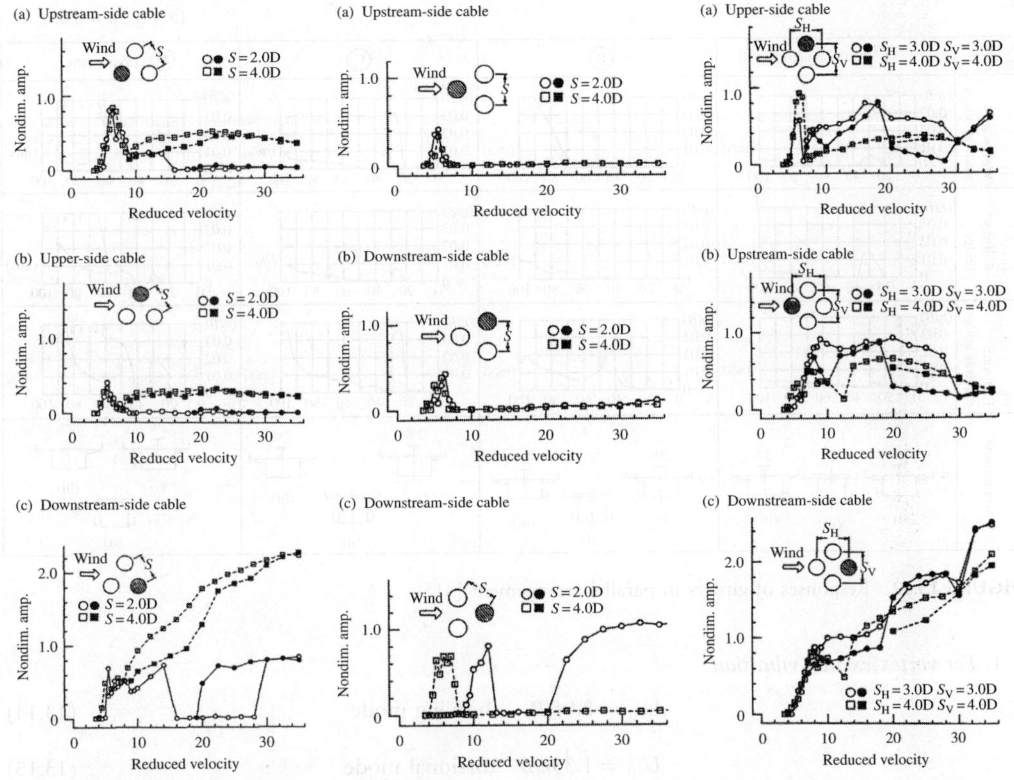

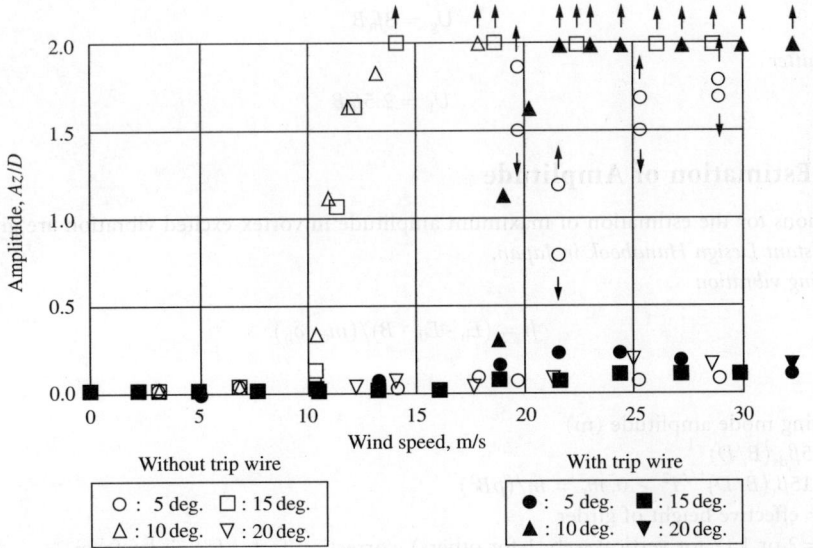

FIGURE 13.50 Aeroelastic responses of a marked cable in multiple cables with various arrangements. (Kubo et al. 1995. Reprinted with permission from Elsevier.)

FIGURE 13.51 Aeroelastic response of downstream cable with/without trip wire (helical wire) in wake flutter against various wind directions. In the figure, deg. means wind direction from tandem arrangement of the cable [40].

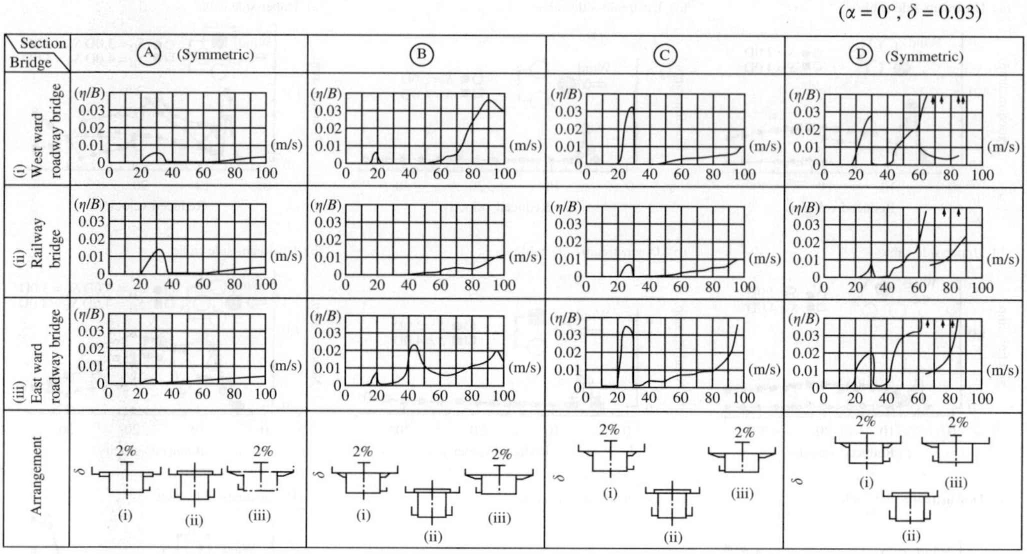

$(\alpha = 0°, \delta = 0.03)$

FIGURE 13.52 Responses of girders in parallel arrangement [41].

1. *For vortex-excited vibration*

$$U_{vh} = 2.0f_h B \quad \text{heaving mode} \tag{13.14}$$

$$U_{v\theta} = 1.33f_\theta B \quad \text{torsional mode} \tag{13.15}$$

where f_h and f_θ are natural frequencies (Hz) of heaving and torsional motion, respectively, and B is the projected width of girder in unit of meter.

2. *For galloping*

$$U_g = 8f_h B \tag{13.16}$$

3. *For flutter*

$$U_f = 2.5f_\theta B \tag{13.17}$$

13.4.2 Estimation of Amplitude

The equations for the estimation of maximum amplitude in vortex-excited vibration are given in *The Wind-Resistant Design Handbook in Japan*.

1. *Heaving vibration*

$$h = (E_h \cdot E_{th} \cdot B)/(m_r \cdot \delta_h) \tag{13.18}$$

where

h = heaving mode amplitude (m)
$E_h = 0.065\beta_{ds}(B/D)^{-1}$
$E_{th} = 1 - 15\beta_t(B/D)^{1/2}I_u^2 \geq 0, m_r = m/(\rho B^2)$
    D = effective height of girder
    β_{ds} = 2 or 1 (2 for vertical web, 1 for others), correction factor for girder form

TABLE 13.1 Strouhal Number of Various Cross-Sections of Structural Members [5]

Wind	Profile dimensions, in mm	Value of $\mathscr{S}$	Wind	Profile dimensions, in mm	Value of $\mathscr{S}$
→	$t=2.0$; 50; 50 (t)	0.120	↓	$t=1.0$; 12.5; 12.5; 25; 50	0.147
↓		0.137			
→	$t=0.5$; 25; 25	0.120	↓	$t=1.0$; 12.5; 12.5; 12.5; 50	0.150
↓	$t=1.0$; 25; 50	0.144	←	$t=1.0$; 50; 50	0.145
			↑		0.142
			↙		0.147
↓	$t=1.5$; 12.5; 50	0.145	←	$t=1.0$; 25; 25	0.131
			↑		0.134
			↙		0.137
↓	$t=1.0$; 25; 50	0.140	→	$t=1.0$; 25; 25; 25; 25	0.121
↑		0.153	↓		0.143
↓	$t=1.0$; 12.5; 50	0.145	→	$t=1.0$; 25; 25; 25; 12.5	0.135
↑		0.168			
→	$t=1.5$; 50	0.156	→	$t=1.0$; 50; 100	0.160
↓		0.145			
Cylinder $11{,}800 < \mathscr{R}e < 91{,}100$	25	0.200	→	$t=1.0$; 25; 50	0.114
			↑		0.115

TABLE 13.2　Drag and Lift Force Coefficients of Structural Members with Various Cross-Sections

Profile and wind direction	C_D	C_L
→ ‖	2.01	0
→ I	2.04	0
→ ⊢⊣	1.81	0
→ L	2.0	0.3
→ ⌐	1.83	2.07
→ L	1.99	−0.09
→ ⌐	1.62	−0.48
→ ⊩	2.01	0
→ ⊤	1.99	−1.19
→ ⊟	2.19	0

β_t　− 1 or 0 (0 for hexagonal cross-section, 1 for others), correction factor for influence of turbulence

I_u　= intensity of turbulence

m　= mass of girder per unit length (kg fs^2/m^2)

ρ　= air density

δ_h　= logarithmic damping decrement of the structure in heaving vibration mode

2. *Torsional vibration*

$$\theta = (E_\theta \cdot E_{t\theta} \cdot B)/(I_{pr} \cdot \delta_\theta) \tag{13.19}$$

where

θ　= torsional mode amplitude (degree)

E_θ　= $17.16\beta_{ds}(B/D)^{-3}$

$E_{t\theta}$　= $1 - 20\beta_i(B/D)^{1/2}I_u^2 \geq 0$

I_{pr}　= $I_p/(\rho B^4)$

　　D = effective height of girder

　　I_p = polar inertia moment of mass of girder per unit length (kg fs^2)

δ_θ　= logarithmic damping decrement of the structure in torsional vibration mode

3. *Amplitude estimation from experimental data:* As indicated in Equation 13.8, the occurrence of self-excited vibration is assessed as the sum of structural damping and aerodynamic damping. In wind tunnel tests, at specified wind velocity, aerodynamic during experiment from measured total damping. The aerodynamic damping usually is as shown in Figure 13.53. That is called the "A–δ curve."

When A–δ curve is obtained, the amplitude of the vibration at specified wind velocity can be estimated for the structure with any structural damping. In the case where the structural damping is δ_s, setting the value of $-\delta_s$ on the vertical axis and obtaining the intersection point on the curve, the corresponding amplitude on the horizontal axis is the estimated amplitude A_s of the steady-state vibration at the specified wind velocity.

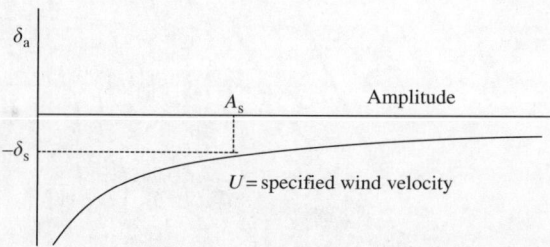

FIGURE 13.53   Aerodynamic damping to amplitude at specified wind velocity (A–δ curve).

13.5   Examples of Real Bridges

In this section, some bridges that had aeroelastic vibration problems are introduced. It is considered that it is very important for bridge engineers to know the damage caused to the bridge by wind action, which would be helpful when they design new bridges.

13.5.1   Collapse of Tacoma Narrows Bridge [43]

The collapse of Tacoma Narrows Bridge in Figure 13.54 is very famous in the bridge engineering field as the accident was caused by a wind action. The accident occurred at a wind speed of 19 m/s in 1940. The Tacoma Narrows Bridge was constructed by utilizing the most advanced analytical method of suspension bridge in those days. Although the effect of wind load as the drag force of the girder was taken into account in the design of the bridge, the occurrence of aeroelastic vibration of the girder was not taken into consideration. The girder configuration was similar as shallow H-shape section shown in Figure 13.55. This type of girder section induces vortex-excited vibration in heaving and torsional modes and the flutter as the torsional self-excited vibration as explained in Section 13.4. As is well known, the bridge collapsed in the torsional vibration with large amplitude caused by the flutter. It was reported that the Tacoma Narrows Bridge was vibrating from the beginning of the erection and investigation was done to control the vibration by attaching fairings on the girder ends so as not to generate the separation flow from the leading edge of the girder as shown in Figure 13.56. Unfortunately, the bridge collapsed before the fairing could be attached. After the collapse, the cause was investigated from various points of view. The girder configuration chosen from various investigations was a truss-stiffened girder reducing the wind load and the possible occurrence of aeroelastic vibrations. The truss-stiffened girder developed in the investigation of the collapse of the Tacoma Narrows Bridge has been used mainly in United States and Japan as the girder of long-span suspension bridges.

13.5.2   Vortex-Excited Vibration of the Great-Belt Bridge [44]

The Great-Belt Bridge (Figure 13.57) was constructed as the second long-span suspension bridge in 1998. The aeroelastic performance was investigated in the wind tunnel by using a full span $\frac{1}{100}$-scaled model. The occurrence of vortex-excited vibration in the smooth flow was predicted through the wind tunnel tests. The vibration, however, was not predicted in the turbulent flow. Therefore, at the beginning of erection, the girder was constructed without attaching a device to suppress the vortex-excited vibrations. Just before being opened to traffic, however, the bridge girder vibrated in the vortex-excited vibration. It was considered that the device to suppress the vortex-excited vibration should have been attached on the bridge girder, when the vortex-excited vibration was observed. After the observation of the vibration, a vane-type device was prepared and attached on the lower corners of the bridge girder as shown in Figure 13.58. As a result, the vortex-excited vibration was suppressed.

FIGURE 13.54 Collapse of the Tacoma Narrows Bridge [43].

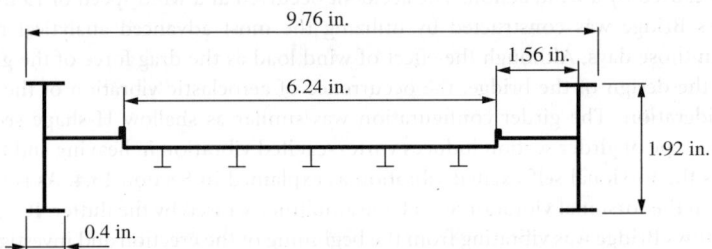

FIGURE 13.55 Cross-section of the girder of the Tacoma Narrows Bridge [43].

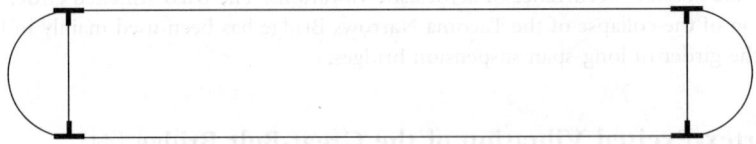

FIGURE 13.56 Fairing planned to be attached on the Tacoma Narrows Bridge before the collapse [43].

13.5.3 Vortex-Excited Vibration of the Box Girder Bridge in Trans-Tokyo Bay Highway [45,46]

This is an example of the vortex-excited vibration occurring in a continuous steel box girder bridge. The outline of the bridge is shown in Figure 13.59. The longest span of 240 m vibrated in the wind with direction perpendicular to the bridge axis under a wind velocity of 20 m/s. Up to this example, it was considered that the vortex-excited vibration is difficult to be induced in a box girder bridge. During the design of the bridge, the wind tunnel test was conducted to check the aeroelastic performance of the bridge.

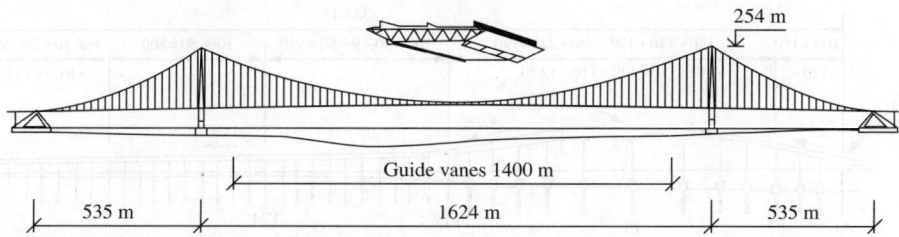

254 m

Guide vanes 1400 m

535 m 1624 m 535 m

FIGURE 13.57 The Great-Belt Bridge [44].

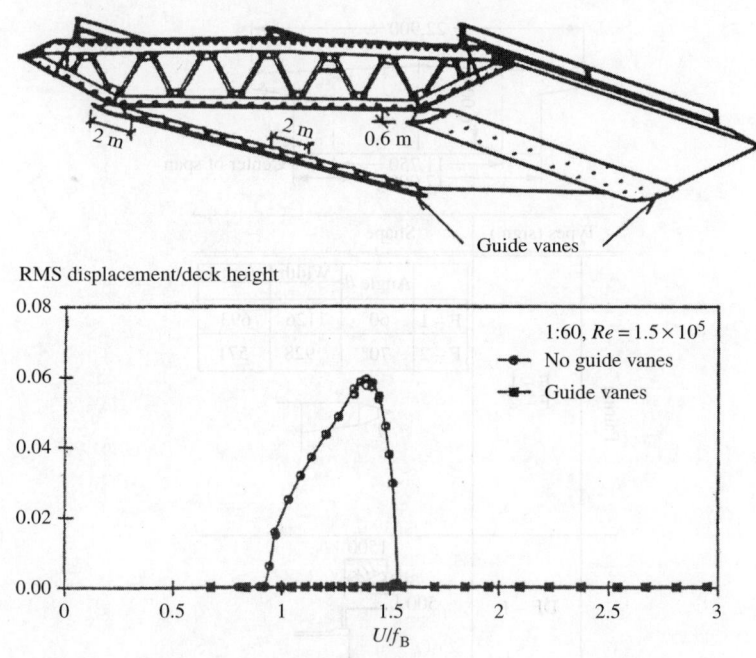

2 m   2 m   0.6 m

Guide vanes

RMS displacement/deck height

1:60, $Re = 1.5 \times 10^5$
— ● — No guide vanes
— ■ — Guide vanes

U/f_B

FIGURE 13.58 Corner vane to suppress vortex-excited vibration of the Great-Belt Bridge and comparison between the aeroelastic response with/without corner vane [44].

The result was that the vortex-excited vibration would be induced in both smooth and turbulent flows. Aerodynamic countermeasure of fairing, double flap, and skirt (as shown in Figure 13.60) was examined to suppress the vortex-excited vibration. The vortex-excited vibration could not be suppressed (as shown in Figure 13.61). This means that it is very difficult to suppress the aeroelastic vibration of a steel box girder bridge by aerodynamic countermeasure through changing the structure configuration. Therefore, in this example, a tuned mass damper (TMD) was used for suppression of vortex-excited vibration.

Figure 13.62 shows the comparison of aerodynamic responses of first mode in both the wind tunnel tests and the field measurement before being opened to the public. In the figure, solid lines show the results of wind tunnel tests for two structural damping decrements of $\delta = 0.028$ and 0.044. The other points are the field measurement results measured in the wind for bridge axis right angle within $\pm 20°$ and the turbulent intensity of 4 to 10%. These results prove that the accuracy of the wind tunnel test is high. The maximum amplitude was measured around a wind velocity of 16 m/s and the result was

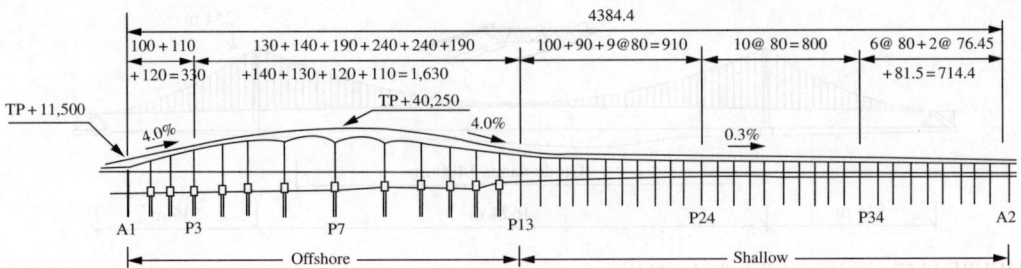

FIGURE 13.59   Outline of the Trans-Tokyo Bay Highway [45].

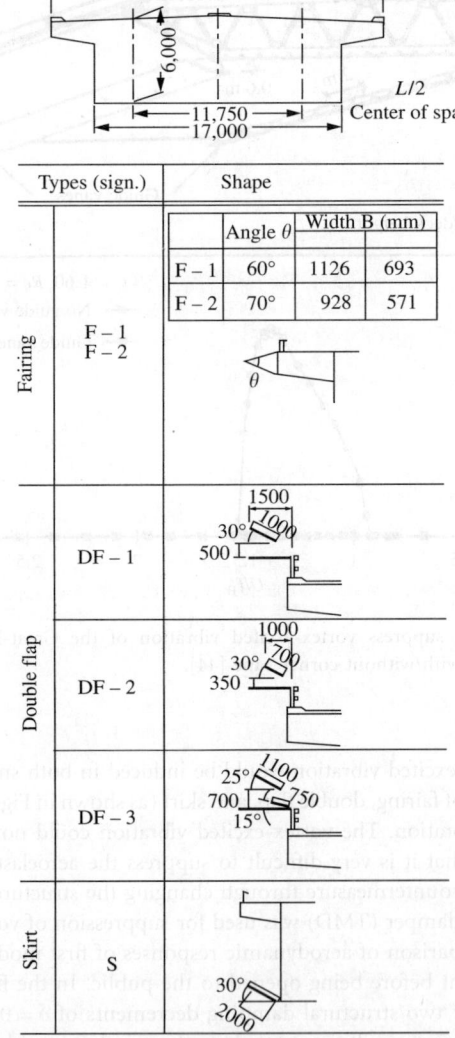

FIGURE 13.60   Bridge girder section and aerodynamic countermeasures tested for the bridge of the Trans-Tokyo Bay Highway to suppress the vortex-excited vibration [45].

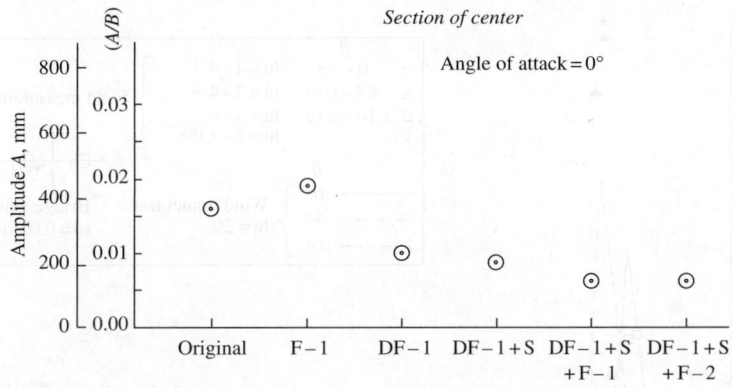

FIGURE 13.61 Comparison of estimated amplitudes of vortex-excited vibration of the box-girder with various aerodynamic countermeasures [45].

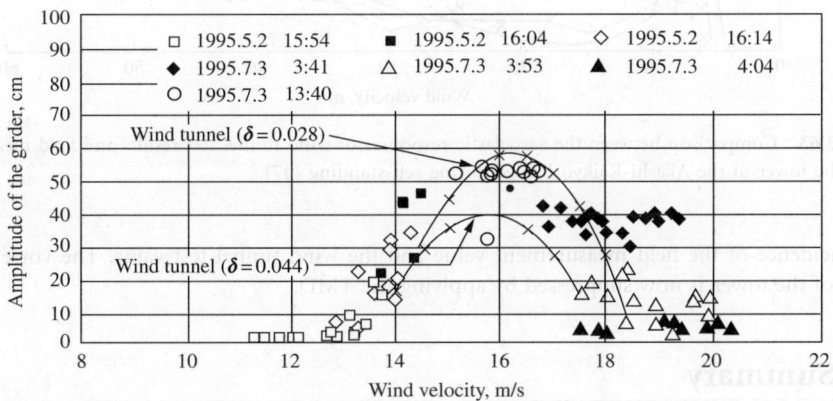

FIGURE 13.62 Comparison between the wind tunnel test results and the field observation results for the continuous steel box girder bridge in the Trans-Tokyo Bay Highway [45].

60 cm; that is, 10% of girder height at the center of the longest span. The vortex-excited vibration was suppressed by TMD and the bridge is now open to traffic.

13.5.4 Vortex-Excited Vibration of the Tower of the Akashi-Strait Bridge During Self-Standing State [47]

During the erection of a suspension bridge and a cable-stayed bridge, there is a period in which the tower is self-supporting. In this period, the tower is in the aeroelastically unstable state, because the structural damping is very small ($\delta = 0.025$) and is susceptible to the vibration under the wind action. The height of the tower is 280 m. When the tower vibrates, a large bending moment occurs at the foundation of the tower. The corner cut-outs introduced in 3–2–1 was chosen as the tower configuration to suppress the galloping of the tower. Even this tower configuration induced the vortex-excited vibration of bending mode in out of plane. Figure 13.63 shows the comparison of results of the field measurement and the wind tunnel tests. The solid and chain lines indicate the results of the wind tunnel tests in three wind directions of 0, 5, and 10° from the right-angled direction at the bridge axis. The dots by the symbol are the measuring results corresponding to the three wind directions. These results show the

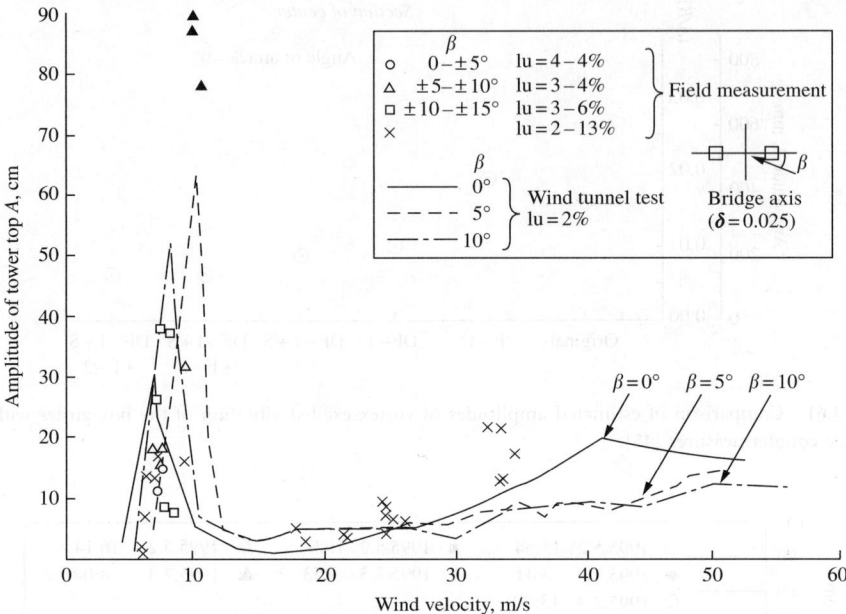

FIGURE 13.63   Comparison between the aeroelastic responses of wind tunnel test results and field measurement results of the tower of the Akashi-Kaikyo Bridge during self-standing [47].

good coincidence of the field measurement value and the wind tunnel test value. The vortex-excited vibration of the tower is now suppressed by applying the TMD.

13.6   Summary

It has been explained that structure configuration is very important in wind engineering. The structure configuration controls the separation flow around the structure. That is, the phenomenon occurring in the structure under wind action is induced by the behavior of the separation flow from the leading edge of the structure and the reattachment of the separation flow on the structure surface. Both the deflection of the structure and the aeroelastic vibration of the structure are induced by the behavior of the separation flow around the structure.

The deflection under wind action is cause by wind load. The wind load is proportional to the square of wind velocity. The reduction of wind load is one of the themes in wind engineering. It was explained that the reduction in the wind load is achieved to some extent by changing the corner configuration of the structure. Another theme is how the aeroelastic vibration of the structure is suppressed. The aeroelastic vibration problem is a peculiar problem in the flexible structure exposed to wind. The suppression of vibration has been achieved by the control of the separation flow and the structural method of adding a damping device including TMD. From the viewpoints of maintenance of the structure with long life, aerodynamic control by structural configuration is rather preferable than the structural method. Because the structural method has movable parts in the damper system, the movable parts should be replaced periodically with new parts.

Until now, it was usual to carry out the examination of the wind-resistant design after the structural design ends. When there was a problem on aeroelastic vibration, the aeroelastic performance was improved by attaching some aerodynamic devices, for example, fairing, deflector, and guide vane so on. The ideal form of the bridge design is the form to fuse the structural design with the wind-resistant design. If it is so done, the ideal bridge cross-section is realized without attaching aerodynamic devices on the bridge girder. In the bridge design, when the lowest natural frequency of the bridge is under 1 Hz, there is the possibility of occurrence of aeroelastic vibration under the design wind velocity. The following equation is very useful to roughly estimate the lowest natural frequency of the girder bridge:

$$f = 100/\text{span length (m)}, \quad \text{in Hz}$$

Therefore, in the case the lowest frequency is under 1 Hz, the design should be proceeded with respect to structural and wind-resistant designs. Then the ideal bridge section will be found also from the viewpoint of both aesthetic and economical reasons.

For the bridge engineers who are interested in wind engineering of structures, some reference books are introduced in the Further reading section.

References

[1] Y. Kubo and K. Kato, "The role of end plates in two dimensional tests," *Proc. JSCE*, vol. 368/I-5, pp. 179–186.

[2] V. Strouhal, "Uber eine besondere Art der Tonerregung," *Ann. Phys.*, vol. 5, pp. 216–250, 1878.

[3] Japan Society of Steel Construction, "Wind Engineering on Structures" (in Japanese), Tokyo Denki University Press, Tokyo, 1997.

[4] Y. Kubo, E. Yamaguchi, S. Kawamura, K. Tou, and K. Hayashida, "Aerodynamic Characteristics of Square Prism with Rounded Corners," Proccedings of the 14th National Symposium on Wind Engineering in Japan, pp. 281–286, 1996 (in Japanese).

[5] Y. Nakamura, "Aerodynamic characteristics of fundamental structures (part 1)," *J. Wind Eng., JAWE*, no. 36, pp. 49–84; E. Simiu and R.H. Scanlan, "*Wind Effects on Structures (Third Edition),*" John Wiley & Sons, Inc., New York, 1996.

[6] T. Miyata, H. Yamada, and T. Hojo, "Aerodynamic Response of PE Stay Cables with Pattern-Indented Surface," Proceedings of Cable-Stayed and Suspension Bridges (Deauville), 1994.

[7] A. Okajima, K. Sugitani, and T. Mizota, "Flow around a pair of circular cylinders arranged side by side at high reynolds numbers," *Trans. JSME*, vol. 52, no. 480, pp. 2844–2850, 1986 (in Japanese).

[8] A. Okajima, K. Sugitani, and T. Mizota, "Flow around two rectangular cylinders in the side-by-side arrangement," *Trans. JSME*, vol. 51, no. 472, pp. 3877–3886, 1985 (in Japanese).

[9] D. Biermann and W.H. Herrnstein, "The Interference Between Struts in Various Combinations," National Advisory Committee for Aeronautics, Tech. Rep. 468, pp. 515–524, 1933.

[10] R.L. Wardlaw and K.R. Kooper, "A Wind Tunnel Investigation of the Steady Aerodynamic Forces on Smooth and Standard Twin Bundled Power Conductors for the Aluminum Company of America," National Aeronautical Establishment, Canada, LTR-LA-117, 1973.

[11] N. Suzuki, H. Sato, M. Iuchi, and S. Yamamoto, "Aerodynamic Forces Acting on Circular Cylinders Arranged in a Longitudinal Row," Proceedings of the 3rd International Symposium on Wind Effects on Buildings and Structures, Tokyo, pp. 377–387, 1971.

[12] Y. Tanida, A. Okajima, and Y. Watanabe, "Stability of circular cylinder oscillating in uniform or in a wake," *J. Fluid Mech.*, vol. 61, pp. 769–784, 1973.

[13] M.M. Zdrakovich and D.L. Pridden, "Interference between two circular cylinders; series of unexpected discontinuities," *J. Ind. Aero.*, vol. 2, pp. 255–270, 1977.

[14] E. Hori, "Experiments on Flow around a Pair of Parallel Circular Cylinders," Proceedings of the 9th Japan National Congress for Applied Mechanics, Tokyo, pp. 231–234, 1959.

[15] S.J. Price and M.P. Paidoussis, "The aerodynamic forces acting on groups of two and three circular cylinders when subject to a cross-flow," *J. Wind Eng. Ind. Aero.*, vol. 17, pp. 329–347, 1984.

[16] T. Igarashi, "Characteristics of the flow around two circular cylinders arranged in tandem (1st Report), _Bull. JSME_, vol. 24, no. 188, pp. 323–331, 1981.

[17] K. Imaichi, Preprint for JSME, no. 734–5, pp. 104–106, 1973 (in Japanese).

[18] M.M. Zdrakovich, "Review of flow interference between two circular cylinders in various arrangement," _Trans. ASME, J. Fluids Eng., Ser. I_, vol. 99, no. 4, pp. 618–633, 1977.

[19] M.M. Zdrakovich, "Flow-induced oscillations of two interfering circular cylinders," _J. Sound Vibration_, vol. 101, no. 4, pp. 511–521, 1985.

[20] A. Takano, I. Arai, and M. Matsuzaka, "An experiment on the flow around a group of square prisms," _Trans. JSME_, vol. 47, no. 417, pp. 982–991, 1981 (in Japanese).

[21] K. Hirano, H. Ohsako, and A. Kawashima, "Interference between two normal flat plates in the various staggered arrangements (1st, Drag and Vortex Shedding Frequency)," _Trans. JSME_, vol. 49, no. 2363–2370, 1983.

[22] A. Okajima, Preprint for JSME, no. 7867–1, pp. 15–16, 1986 (in Japanese).

[23] Y. Kubo, E. Yamaguchi, S. Kawamura, K. Tou, and K. Hayashida, "Aerodynamic Characteristics of Square Prism with Rounded Corners," Proceedings of the 16th National Symposium on Wind Engineering in Japan, 2000 (in Japanese).

[24] Y. Kubo, K. Hirata, and K. Mikawa, "Mechanism of aerodynamic vibration of shallow bridge girder sections," _J. Wind Eng. Ind. Aerodyn._, vol. 41–44, pp. 1297–1308, 1992.

[25] D.E. Walshe and L.R. Wootton, "Preventing Wind-Induced Oscillation of Structures of Circular Section," Proc. ICE, Paper 7289, 1970.

[26] H. Liu, _Wind Engineering, A Hand Book for Structural Engineers_, Prentice Hall, New York p. 125, 1991.

[27] C. Scruton and A.R. Flint, "Wind-Excited Oscillations of Structures," Proc. I.C.E., paper no. 6758, 1964.

[28] R.H. Scanlan and R.L. Wardlaw, "Reduction of Flow-Induced Structural Vibrations," Isolation of Mechanical Vibration, Impact, and Noise, AMD, vol. 1, sec. 2, ASME, 1973.

[29] S. Ohno, S. Sano, and C. Morimoto, "Study of aerodynamic stability of cable-stayed curved bridge and its countermeasures against wind-induced vibration," _J. Wind Eng._, no. 37, pp. 567–579, 1988.

[30] Y. Kubo, A. Koishi, K. Tasaki, and H. Nakagiri, "Mechanism of Separation Interference on Bridge Sections," Proceedings of the 13th National Symposium on Wind Engineering in Japan, pp. 353–358, 1994 (in Japanese).

[31] K. Ogawa, "On 3-D aerodynamic characteristics of bridge tower," _J. Wind Eng._, JAWE, no. 59, pp. 53–54, 1994.

[32] Y. Kubo, K. Kato, E. Yamaguchi, H. Nakagiri, and K. Okumura, "Suppression Mechanism of Aerodynamic Vibration of A-Shape Tower due to Horizontal Beam," Proccedings of the 14th National Symposium on Wind Engineering in Japan, pp. 593–598, 1996 (in Japanese).

[33] R.L. Wardlaw, "Improvement of Aerodynamic Performance," Aerodynamics of Large Bridges edited by A. Larsen, A.A. Balkema, pp. 59–70, 1992.

[34] Y. Kubo, K. Honda, K. Tasaki, and K. Kato, "Improvement of aerodynamic instability of cable-stayed bridge deck by separated flows mutual interference method," _J. Wind Eng. Ind. Aerodyn._, vol. 49, pp. 553–564, 1993.

[35] K. Nagai, M. Oyadomari, and R. Inamuro, "Aeroelastic Response of 3 Span Double Steel Box Girder Bridge," Proceedings of the 40th Annual Conference on the Japan Society of Civil Engineers, 1, pp. 481–482, 1985.

[36] T. Miyata, K. Yokoyam, M. Yasuda, and Y. Hikamai, "Akashi Kaikyo Bridge: Wind Effects and Full Model Wind Tunnel Tests," Aerodynamics of Large Bridges edited by A. Larsen, A.A. Balkema, pp. 217–236, 1992.

[37] K. Sadashima, Y. Kubo, T. Koga, Y. Okamoto, E. Yamaguchi, and K. Kato, "Aeroelastic responses of 2-edge girder for cable-stayed bridges," _J. Struct. Eng._, JSCE, vol. 46A, pp. 1073–1078, 2000 (in Japanese).

[38] Y. Kubo, K. Kato, H. Maeda, K. Oikawa, and T. Takeda, "New Concept on Mechanism and Suppression of Wake-Galloping of Cable-Stayed Bridges," Proceedings of the Cable-stayed and Suspension Bridges, pp. 491–498, 1994.

[39] Y. Kubo, T. Nakahara, and K. Kato, "Aerodynamic behavior of multiple elastic circular cylinders with vicinity arrangement," *J. Wind Eng. Ind. Aerodyn.*, vol. 54/55, pp. 227–237, 1995.

[40] R. Toriumi, N. Furuya, M. Takeguchi, M. Miyazaki, and Y. Saito, "A Study on Wind-Induced Vibration of Parallel Suspenders Observed at the Akashi-Kaikyo Bridge," Proceedings of the 3rd International Symposium on Cable Dynamics, pp. 177–182, 1999.

[41] A. Honda, N. Shiraishi, and S. Motoyama, "Aerodynamic instability of Kansai international airport access bridge," *J. Wind Eng., JAWE*, no. 37, pp. 521–528, 1988.

[42] *The Wind-Resistant Design Handbook for Road Bridges*, Japan Society for Roads, 1991.

[43] F.B. Farquarson (ed.), "Aerodynamic Stability of Suspension Bridges," Parts I–V, Bulletin No. 116, University of Washington Engineering Experiment Station, Seattle, 1949–1954.

[44] A. Larsen, S. Esdahl, J.E. Andersen, and T. Vejrum, "Vortex Shedding Excitation of the Great Belt Suspension Bridge," Proceedings of the 10th ICWE, pp. 947–954, 1999.

[45] Y. Yoshida, Y. Fujino, H. Tokita, and A. Honda, "Wind tunnel study and field measurement of vortex-induced vibration of continuous steel box girder in Trans-Tokyo Bay highway," *J. Struc. Mech. Earthquake Eng., JSCE*, no. 633, I-49, pp. 103–117, 1999.

[46] Y. Yoshida, Y. Fujino, H. Sato, H. Tokita, and S. Miura, "Control of vortex-induced vibration of continuous steel box girder in Trans-Tokyo Bay highway," *J. Struc. Mech. Earthquake Eng., JSCE*, no. 633, I-49, pp. 119–134, 1999.

[47] H. Shimodoi, T. Kanesaki, K. Hata, and N. Sasaki, "Aeroelastic Response of Tower of the Akashi Kaikyo Bridge during Self Standing State," Proceedings of the 40th Annual Conference on the Japan Society of Civil Engineers, 1, pp. 976–977, 1994.

[48] N.P. Cheremisnoff, "Encyclopedia of Fluid Mechanics: Aerodynamics and Compressible Flows," Chapter 10: Wake Interference and Vortex Shedding, Gulf Publishing, 1989.

Further Reading

R.D. Blevins, "Flow-Induced Vibration (Second Edition)," Krieger Publishing Company, New York, 1994.

C. Dyrbye and S.O. Hansen, "Wind Loads on Structures," John Wiley & Sons, Inc., New York, 1996.

Japan Society of Steel Construction, "Wind Engineering on Structures" (in Japanese), Tokyo Denki University Press, Tokyo, 1997.

H. Liu, "Wind Engineering, A Hand Book for Structural Engineers," Prentice Hall, New York, 1991.

P. Sachs, "Wind Forces in Engineering," Pergamon Press, New York, 1972.

E. Simiu and R.H. Scanlan, "Wind Effects on Structures (Third Edition)," John Wiley & Sons, Inc., New York, 1996.

III

Structural Design Using High-Performance Materials

III

14

High-Performance Steel

Eric M. Lui
Department of Civil and Environmental Engineering, Syracuse University, Syracuse, NY

14.1 Introduction

High-performance steels (HPS) were developed through a cooperative research program between the US Navy, the Federal Highway Administration (FHWA), and the American Iron and Steel Institute (AISI). At the time of this writing, two grades of HPS are available and have successfully been used in bridge construction. Their ASTM designations are HPS50W and HPS70W (the equivalent metric designations are HPS345W and HPS485W, respectively). A primary objective of developing HPS is to take advantage of the higher strength offered by high-strength steels without compromising on weldability. Conventional high-strength steels have a relatively high carbon content (approximately 19% by weight), and so special attention is needed to obtain high quality welds. This typically requires preheating of the welded parts, controlling the interpass temperature, controlling the energy input during passes, using special electrodes, careful handling of the welding consumables, and if necessary applying postweld treatment such as controlled cooling and feathering of the welds. When all operations are performed correctly and all prescribed procedures are followed carefully, one can often obtain good quality welds in conventional high-strength steels. Unfortunately, while these conditions can usually be met in a controlled environment such as in a shop, difficulties may arise when such welding has to be done on the field. This is particularly the case for bridge construction, when welding under less than optimal condition is often the norm. HPS have a carbon content that is very comparable to low carbon steel, and they can be welded under a variety of conditions without requiring the rather time consuming and often expensive pre- and postweld treatments. In addition, it has been demonstrated experimentally that HPS exhibit good toughness and satisfactory ductility values for use as an effective construction material.

0-8493-1569-7/05/$0.00+$1.50
© 2005 by CRC Press

14.2 Chemical Compositions

The chemical compositions for HPS50W (HPS345W) and HPS70W (HPS485W) are given in Table 14.1. For purpose of comparison, the chemical compositions of a conventional high-strength steel commonly used for bridge construction, A709-70W (A709-485W), are also given.

It can be seen from Table 14.1 that the carbon level is much lower for HPS, thus enhancing weldability and toughness. There is also a noticeable decrease in the amount of phosphorus, sulfur, and nickel, but with the addition of molybdenum, aluminum, and nitrogen.

14.3 Mechanical Properties

HPS50W steel is produced by conventional hot-rolling or controlled rolling. Plates up to 4 in. (10.2 cm) thick, similar to those of conventional Grade 50W steel, are available. The minimum yield and tensile strengths of HPS50W (HPS345W) grade steel are 50 ksi (345 MPa) and 70 ksi (485 MPa), respectively.

HPS70W steel is produced by either quenching and tempering (Q&T) or thermal–mechanical controlled processing (TMCP). Plates produced using Q&T are available up to 4 in. (10.2 cm) thick and 50 ft (15.25 m) long. Those produced using TMCP are available up to 2 in. (5.1 cm) thick and 125 ft (38 m) long. The minimum yield and tensile strengths of HPS70W (HPS485W) grade steel are 70 ksi (485 MPa) and 85 ksi (585 MPa), respectively.

In addition to strength, another important property of steel is fracture toughness. To avoid brittle failure above an anticipated service temperature, steel use in bridge construction must satisfy an AASHTO Charpy V-Notched (CVN) toughness requirements for a given temperature zone. There are three temperature zones as shown in Table 14.2, with Zone 3 being the most severe.

It has been shown experimentally (Fisher and Wright 2000; Wright et al. 2001) that the two HPS possess toughness not only higher than that of conventional bridge steel but exceeds the AASHTO CVN Zone 3 requirements. Table 14.3 shows a comparison of the *minimum* CVN values between HPS70W and the conventional Grade 70W steels. The enhanced fracture toughness of HPS is apparent.

TABLE 14.1 Comparison of Chemical Compositions (in Weight%) Between HPS and Conventional High-Strength Steels

	C	Mn	P	S	Si	Cu	Ni	Cr	Mo	V	Al	N
HPS50W HPS70W	0.11 max.	1.10–1.35	0.02 max.	0.006 max.	0.30–0.50	0.25–0.40	0.25–0.40	0.45–0.70	0.02–0.08	0.04–0.08	0.01–0.04	0.015 max.
A709-70W	0.19 max.	0.80–1.35	0.035 max.	0.04 max.	0.25–0.65	0.20–0.40	0.50 max.	0.40–0.70	—	0.02–0.10	—	—

TABLE 14.2 Temperature Zones for CVN Requirements

Temperature zone	Temperature range
1	0°F and above (−18°C and above)
2	−30°F to 0°F (−34°C to −18°C)
3	−60°F to −30°F (−51°C to −34°C)

TABLE 14.3 Comparison of CVN Values Between HPS and Conventional High-Strength Steels

Zone	HPS70W	Grade 70W
1	Non-FCM > 2.5 in., 25 ft lb at −10°F	Non-FCM > 2.5 in., 25 ft lb at +50°F
	FCM > 2.5 in., 35 ft lb at −10°F	FCM > 2.5 in., 35 ft lb at +50°F
2	Non-FCM > 2.5 in., 25 ft lb at −10°F	Non-FCM > 2.5 in., 25 ft lb at +20°F
	FCM > 2.5 in., 35 ft lb at −10°F	FCM > 2.5 in., 35 ft lb at +20°F
3	Non-FCM > 2.5 in., 25 ft lb at −10°F	Non-FCM > 2.5 in., 25 ft lb at −10°F
	FCM > 2.5 in., 35 ft lb at −10°F	FCM > 2.5 in., 35 ft lb at −10°F

Notes: FCM, fracture critical member.
1 in. = 2.54 cm, 1 ft lb = 1.356 N m, Temp°F = 1.8 × Temp°C + 32.

The fatigue resistance of HPS, like conventional steels, is more of a function of the weld details and the operating stress range. Tests on HPS have demonstrated that the detail categories for load-induced fatigue given in AASHTO LRFD Section 6.6.1 (AASHTO 1998) for ordinary steels also apply to HPS.

14.4 Ductility

Ductility is the ability of a material to undergo large inelastic deformation before failure. It is an important material property because ductility allows for early warnings of incipient failure and permits force redistribution to occur in statically indeterminate structures. One of the drawbacks of high-strength steels is the reduction in ductility when compared to lower-strength steels, such as A36, A572, and A992 structural steels. However, it has been reported that all HPS steels can achieve uniaxial elongations in the range of 18–30% (Bjorhovde 2004), which is more than adequate for engineering applications. Moreover, tests on HPS wide-plate (8 × 0.75 in., or 20 × 2 cm) specimens with holes (Dexter et al. 2002) under tension have shown that the tensile ductility capacity of HPS70W steels is well within the required range for structural steel. The study also indicates that adequate ductility is achieved for members under tension when

$$\left(\frac{A_n}{A_g}\right)\left(\frac{F_u}{F_y}\right) \geq 1.0 \qquad (14.1)$$

where A_n is the net sectional area, A_g is the gross sectional area, F_y is the yield strength, and F_u is the tensile strength.

Statistical analysis of the mill report tensile properties of more than 2400 heats of HPS70W has shown that $F_y/F_u \approx 0.84$, with a coefficient of variation of 0.06. This means in accordance with Equation 14.1, adequate ductility is expected if $A_n/A_g \geq 0.84$.

Because the current American Institute of Steel and Concrete (AISC) and American Association for State Highway and Transportation Officials (AASHTO) provisions for tension members with holes require that the member be proportioned so that

$$\left(\frac{A_n}{A_g}\right)\left(\frac{F_u}{F_y}\right) \geq 1.2 \qquad (14.2)$$

the condition for adequate ductility expressed in Equation 14.1 is automatically satisfied if the member is designed according to code provisions.

Tests on full-sized girders made of HPS have also been performed (Dexter et al. 2002; Yakel et al. 2002). One objective of these tests was to determine the rotation ductility R of these girders under flexure. R is defined by the equation

$$R = \frac{\theta_{p2} - \theta_{p1}}{\theta_{p1}} \qquad (14.3)$$

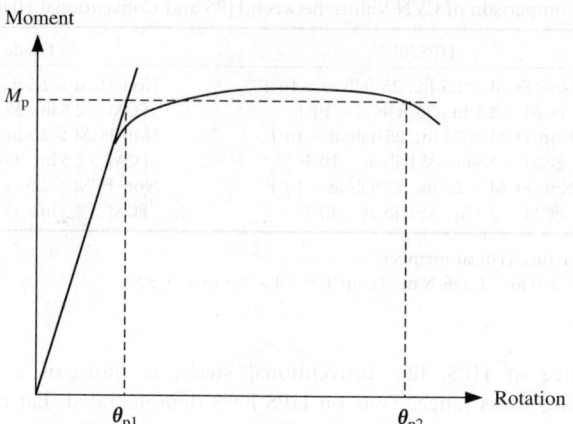

FIGURE 14.1   Moment rotation curve.

where, in reference to Figure 14.1, $\theta_{p1} = M_p/EI$, and θ_{p2} is the maximum value of beam end rotation when a simply supported beam is sustaining M_p. M_p is the plastic moment capacity of the cross-section and EI is the flexure rigidity. To allow for moment redistribution, AASHTO implies an R value of 3. While the test of a simply supported hybrid girder (Dexter et al. 2002) with tension flange and web made of HPS70W (HPS345W) steel and compression flange made of A572 Grade 50 under a three-point bending test has indicated that an R value exceeding 3 was achieved, an R value less than 3 was reported for tests of compact and (web) noncompact girders (Yakel et al. 2002) made entirely of HPS70W (HPS345W) steel. In addition, numerical studies on HPS girders under negative moment (Barth et al. 2000) have shown that while HPS girders are able to develop their plastic moment capacity, they are not able to provide the necessary rotation ductility for moment redistribution. As a result, the 10% reduction in negative moments that is allowed for continuous girders to account for moment redistribution should not be applied to HPS girders.

14.5   Weldability and Weathering Characteristics

14.5.1   Weldability

One advantage of HPS over conventional high-strength steels is its enhanced weldability due to the lower carbon content in HPS. Because HPS can be welded under less stringent conditions when compared to conventional high-strength steels, the recommended procedures for welding high-strength steels outlined in Americal Welding Society (AWS) D1.5 *Bridge Welding Code* (AWS 2002) is also applicable to welding HPS. In addition, an addendum is available (AASHTO 2000a) to provide guidance to bridge engineers interested in using HPS in bridge design and construction.

One of the most common problems for bridge welds is hydrogen induced (cold) cracking in the heat affected and fusion zones of the welds. Because hydrogen can come from many sources such as atmosphere, water/moisture, grease, oxides, and contaminants, it is not an easy task to eliminate all hydrogen sources during the welding process. To reduce the likelihood of hydrogen induced cracking, one needs to select the material carefully, exercise proper handling techniques for the consumables, and follow established guidelines for welding procedures (AWS 2002). Specifying minimum preheat and interpass temperature for welding also helps to minimize diffusible hydrogen during production welding of HPS. High preheat temperature allows more time for hydrogen to diffuse from the welds and it reduces the chance for the formation of brittle microstructures where cracks can initiate. Table 14.4 shows the minimum preheat and interpass temperature for HPS70W using the two recommended welding processes for HPS — submerged arc welding (SAW) and shield metal arc

TABLE 14.4 Minimum Preheat and Interpass Temperature for HPS70W (HPS485W)

		Thickness t of the thickest part at point of welding			
Welding process	Max. H_d	$t \leq \frac{3}{4}$ in. ($t \leq 19$ mm)	$\frac{3}{4} < t \leq 1\frac{1}{2}$ in. ($19 < t \leq 38$ mm)	$1\frac{1}{2} < t \leq 2\frac{1}{2}$ in. ($38 < t \leq 63.5$ mm)	$t > 2\frac{1}{2}$ in. ($t > 63.5$ mm)
SAW[a]/SMAW[b]	4 ml/100 g	50°F (10°C)	70°F (21°C)	70°F (21°C)	125°F (52°C)
SMAW[b]	8 ml/100 g	50°F (10°C)	125°F (52°C)	175°F (79.5°C)	225°F (107°C)

[a] Using LA85 electrode with Mil800HPNi flux (both by Lincoln Electric Company).
[b] Using E9018MR for matching and E7018MR for undermatching electrodes (moisture resistant coating is required for all SMAW electrodes used for welding HPS70W steels).

welding (SMAW) — with consumables having a diffusible hydrogen H_d measured in terms of milliliter per 100 grams of deposited weld material not exceeding the values shown. If the diffusible hydrogen level exceeds the indicated values, or if the consumables are not as specified in Table 14.4, or for HPS50W, the minimum preheat and interpass temperatures should follow those in AWS D1.5 (AWS 2002), which are usually higher than those shown in Table 14.4. Note that the interpass temperatures shown are the minimum values required based on test specimens. If satisfactory results are not achieved, temperatures higher than those shown may be required. In any case, the maximum interpass temperature for welding HPS70W steel should not exceed 450°F (232°C).

In welding HPS, consumables with matching weld strength are recommended for SAW complete penetration groove welding of HPS70W plates. Consumables with undermatched weld strength are strongly recommended for all fillet welding of HPS70W plates because of reduced likelihood of hydrogen cracking. In any event, careful control of heat input is of particular importance in minimizing base metal microstructure effects and thus ensuring the soundness of the welds. The AWS recommended range of heat input is 40 kJ/in. (1.6 kJ/mm) to 90 kJ/in. (3.5 kJ/mm).

For hybrid designs, the minimum preheat and interpass temperature should be based on the higher value required by AWS D1.5 (AWS 2002) and those shown in Table 14.4. Also, when Grade 50W and HPS70W steels are joined, consumables satisfactory for Grade 50W base metals as listed in AWS D1.5 should be used. To minimize the potential for hydrogen induced cracking, it is recommended that consumables be conformed to the diffusible hydrogen requirements of the AWS Filler Metal Specification with optional supplementary moisture-resistance designator of H4 (i.e., $H_d = 4$ ml/100 g) for SAW or H8 (i.e., $H_d = 8$ ml/100 g) for SMAW.

14.5.2 Weathering Characteristics

Used in an engineering context, a material with good weathering characteristics is one that has an improved resistance to physical breakdown and chemical alteration over time when subjected to normal atmospheric conditions. For steel bridges, corrosion of the load carrying members is a primary cause of structural deterioration. The service life of a bridge can be prolonged if a more corrosion resistant material is used and a more vigorous maintenance program is followed. When compared to conventional high-strength steels, HPS have slightly better corrosion resistance. Corrosion tests conducted in accordance with ASTM G101 have indicated that the atmospheric corrosion resistance index (CI) calculated using the equation

$$CI = 1.20(\%Cr) + 26.01(\%Cu) + 3.88(\%Ni) + 17.28(\%P) + 1.49(\%Si)$$
$$- 33.39(\%Cu)^2 - 7.29(\%Cu)(\%Ni) - 9.01(\%Ni)(\%P)$$

(14.4)

is 6.5 for HPS70W steel, as compared to a CI of 6.0 for conventional Grade 70W high-strength steel. A CI of 6.5 means HPS70W possesses the chemical compositions to be classified as a high-performance weathering steel. A bridge designer has the discretion of using HPS70W in an unpainted condition.

The elimination of painting in selected areas may help reduce the life-cycle maintenance cost of the bridge.

Like conventional steels, HPS can also be painted. The paint can act as a protective coat and serve as a barrier to retard atmospheric corrosion. In exposed areas or when used in unpainted conditions, the same guidelines and detailing practice as outlined for conventional weathering grade steels (FHWA 1989) should be followed to assure the successful application and durability of HPS.

14.6   Current Usage

The first bridge that uses HPS70W steel in design and construction was the Snyder South Bridge on State Rt. 79 in Dodge County, Nebraska (Figure 14.2). It is a 150-ft (45.7-m) long simple span bridge that provides a 35.5 ft (10.8 m) clear roadway on a concrete deck supported by five 4′6″ (137 cm) deep welded plate girders. The bridge was opened to traffic in October 1997. Since then, the number of bridges designed and constructed using some HPS components is on the rise. As of 2003, some 35 States in the United States have over 160 HPS bridges in service, or under design, fabrication, and construction. The bridges vary in length from under 100 ft (Rte 3 over Rte 110, MA) to over 5000 ft (I-55/I-64/I-70/US-40 Poplar Street Complex, IL), differ in the number of spans, and represent various bridge types although Plate or I-girder bridges are the most common. HPS have been used on these bridges as truss gusset plates, to floor systems and main girders. When used as main girders, a design that employs HPS70W steel in the negative moment regions, or a hybrid design that uses HPS70W steel for all bottom flanges and negative moment top flanges but uses conventional Grade 50W or HPS50W in other areas is often employed (see Figure 14.3 for a specific design of a uniformly loaded girder) because it represents a good compromise for strength, weight savings, and economy. Although the unit cost of HPS is higher than conventional steels, savings can often be achieved (Barker and Schrage 2000; Horton et al. 2000, 2003; Price and Cassity 2000; Van Ooyen 2002; Clingenpeel and Barth 2003) because the use of HPS generally results in smaller/shallower members and fewer girders (since greater girder spacing can be used due to the higher strength offered by the high-strength steels), and thus fewer diaphragms, stiffeners, and welds are needed. The resulting structure is also lighter, so savings in the construction of the piers and abutments may be realized. However, in using HPS or other high-strength steels for bridge construction, the designers need to pay more attention to the overall deformations of the structure, global instability of the

FIGURE 14.2   Snyder South Bridge (courtesy of Nebraska DOT and FHWA).

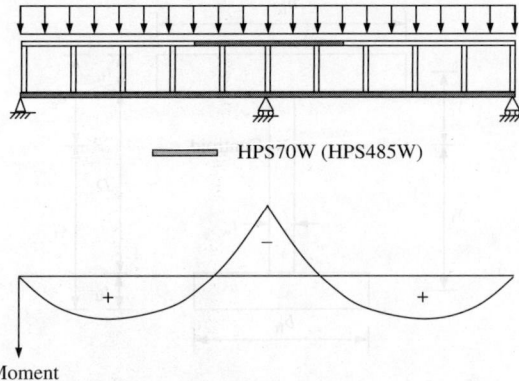

HPS70W (HPS485W)

Moment

FIGURE 14.3 A hybrid girder under a uniformly distributed load.

girders, and local buckling of the component elements. These design aspects will be briefly addressed in the following section.

14.7 Design Considerations

14.7.1 Overall Deformation

Because of the potential for using fewer girder lines or smaller-sized girders when HPS70W or other high-strength steels are used in the design, the overall deformation of the bridge is likely to be higher. For instance, the strong axis moment of inertia calculated using the following equation for an I-shaped cross-section shown in Figure 14.4 will be smaller than that of a similar strength girder made of lower-strength steels:

$$I_x = \frac{1}{12}\left(b_{fc}t_{fc}^3 + b_{ft}t_{ft}^3 + t_w D^3\right) + \left(b_{fc}t_{fc}\right)\left(\frac{t_{fc}}{2} + t_{ft} + D - c_t\right)^2$$

$$+ \left(b_{ft}t_{ft}\right)\left(c_t - \frac{t_{ft}}{2}\right)^2 + \left(c_t - t_{ft} - \frac{D}{2}\right)^2 \tag{14.5}$$

where

$$c_t = \frac{1}{2}\left[\frac{b_{fc}t_{fc}^2 + b_{ft}t_{ft}^2 + t_w D^2 + 2b_{fc}t_{fc}(t_{ft} + D) + 2Dt_w t_{ft}}{b_{fc}t_{fc} + b_{ft}t_{ft} + Dt_w}\right] \tag{14.6}$$

A smaller moment of inertia means the deflection of the girder will be larger. Consequently, if the optional live load deflection criteria of Span/800 (for general vehicular load) or Span/1000 (for vehicular and pedestrian loads) as per AASHTO 2.5.2.6.2 is imposed, there is a possibility that deflection may control the design of the bridge.

14.7.2 Global Instability

The smaller sized cross-section that results in a HPS70W design also means the girder is more susceptible to lateral torsional instability. To prevent lateral torsional instability from occurring prematurely, sufficient compression flange bracing should be provided as per AASHTO (AASHTO 1998)

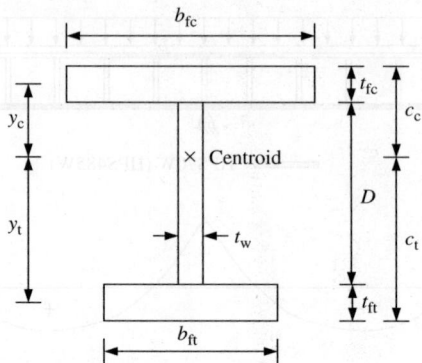

FIGURE 14.4   Nomenclature of an I-shaped cross-section.

Section 6.10.4.1.7 (for a compact section) and Section 6.10.4.1.9 (for a noncompact section). The equations for the maximum unbraced girder length L_b are reproduced as follows:

For compact sections:

$$L_b \leq \left[0.124 - 0.0759\left(\frac{M_l}{M_p}\right)\right]\left(\frac{r_y E}{F_{yc}}\right) \qquad (14.7)$$

where M_l is the the smaller of the two end moments of the unbraced length of the girder due to factored loading, M_p is the plastic moment of the cross-section, r_y is the minimum radius of gyration of the cross-section, E is the elastic modulus, and F_{yc} is the specified minimum compression flange yield stress.

For noncompact sections:

$$L_b \leq 1.76 r_t \sqrt{\frac{E}{F_{yc}}} \qquad (14.8)$$

where E and F_{yc} are as defined in Equation 14.7, r_t is the radius of gyration of a notional cross-section composed of the entire compression flange and one third of the depth of the compression portion of the web, taken about the vertical axis of the section.

14.7.3 Local Buckling

Local buckling of component elements is likely to occur if the slenderness of the flange or web exceeds certain limits. These limits differ depending on whether the section is compact or noncompact:

For compact sections:

$$\text{Flange slenderness limit:} \quad \frac{b_{fc}}{2t_{fc}} \leq 0.382\sqrt{\frac{E}{F_{yc}}} \qquad (14.9)$$

$$\text{Web slenderness limit:} \quad \frac{2D_{cp}}{t_w} \leq 3.76\sqrt{\frac{E}{F_{yc}}} \qquad (14.10)$$

where b_{fc} is the compression flange width, t_{fc} is the compression flange thickness, D_{cp} is the depth of the compression portion of the web at the theoretical plastic moment, t_w is the web thickness, and E and F_{yc} are as defined in Equation 14.7.

For compact sections, local buckling should not occur before the cross-section attains its plastic moment capacity M_p and undergoes an inelastic rotation that corresponds to an inelastic rotational ductility R, as defined in Equation 14.3, of 3. For a noncomposite I-shaped section bent about its major axis with the plastic neutral axis in the web, M_p is given by

$$M_p = A_{fc} F_{yc} \left(D_{cp} + \frac{t_{fc}}{2} \right) + A_{ft} F_{yt} \left(D - D_{cp} + \frac{t_{ft}}{2} \right) + A_w F_{yw} \left(\frac{D}{2} - D_{cp} + \frac{D_{cp}^2}{D} \right) \tag{14.11}$$

where

$$D_{cp} = \frac{1}{2} \left[\frac{1}{t_w} \left(\frac{A_{ft}}{m_t} - \frac{A_{fc}}{m_c} \right) + D \right] \tag{14.12}$$

and

A_{fc} = compression flange area
A_{ft} = tension flange area
D = web depth, D_{cp} is the depth of the compression portion of the web at M_p
F_{yc} = specified minimum compression flange yield stress
F_{yt} = specified minimum tension flange yield stress
F_{yw} = specified minimum web yield stress
m_c = ratio of web to compression flange yield stress = F_w/F_{yc}
m_t = ratio of web to tension flange yield stress = F_w/F_{yt}
t_{fc} = compression flange thickness
t_{ft} = tension flange thickness
t_w = web thickness

If D_{cp} calculated from Equation 14.12 is negative, the plastic neutral axis does not lie in the web. Rather, it lies in the compression flange. In this case, M_p is given by

$$M_p = A_{fc} F_{yc} t_{fc} \left(\frac{1}{2} - \gamma_c + \gamma_c^2 \right) + A_{ft} F_{ft} \left[(1 - \gamma_c) t_{fc} + \frac{t_{ft}}{2} + D \right] + A_w F_{yw} \left[(1 - \gamma_c) t_{fc} + \frac{D}{2} \right] \tag{14.13}$$

where

$$\gamma_c = \frac{1}{2} \left(1 + \frac{A_{ft} F_{yt}}{A_{fc} F_{yc}} + \frac{A_w F_{yw}}{A_{fc} F_{fc}} \right) \tag{14.14}$$

and the other terms are as defined above for Equation 14.11.

On the other hand, if D_{cp} calculated from Equation 14.11 is larger than D, the plastic neutral axis lies in the tension flange. In this case, M_p is given by

$$M_p = A_{ft} F_{yt} t_{ft} \left(\frac{1}{2} - \gamma_t + \gamma_t^2 \right) + A_{fc} F_{fc} \left[(1 - \gamma_t) t_{ft} + \frac{t_{fc}}{2} + D \right] + A_w F_{yw} \left[(1 - \gamma_t) t_{ft} + \frac{D}{2} \right] \tag{14.15}$$

where

$$\gamma_t = \frac{1}{2} \left(1 + \frac{A_{fc} F_{yc}}{A_{ft} F_{yt}} + \frac{A_w F_{yw}}{A_{ft} F_{ft}} \right) \tag{14.16}$$

For noncompact sections:

$$\text{Flange slenderness limit:} \quad \frac{b_{fc}}{2t_{fc}} \leq 12.0 \tag{14.17}$$

where b_{fc} and t_{fc} are as defined in Equation 14.7.

For noncompact sections, local buckling should not occur before the cross-section attains its moment capacity at first yield M_y. For a noncomposite I-shaped section bent about its major axis, M_y is given by

$$M_y = \text{smaller of } \begin{cases} S_c F_{yc} \\ S_t F_{yt} \end{cases} \tag{14.18}$$

where F_{yc} is the specified minimum compression flange yield stress, F_{yt} is the specified minimum tension flange yield stress, S_c is the section modulus with respect to the compression flange, that is equal to I_x/c_c, and S_t is the section modulus with respect to the tension flange, that is equal to I_x/c_t, in which I_x and c_t are given in Equations 14.5 and 14.6, respectively, and c_c is equal to $t_{fc}+t_{ft}+D-c_t$.

Tests on noncomposite bridge girders made of HPS70W steel (Yakel et al. 2002) have demonstrated that while noncompact and compact plate girders could provide the first yield and plastic moment capacities, respectively, the compact plate girders were not able to provide an inelastic rotational ductility of 3, as implied by the AASHTO Specification for compact sections. As a result, the 10% moment redistribution for continuous girders as per Section 6.10.4.4 of the AASHTO Specification (AASHTO 1998) should not be applied to girders made of 70 ksi (485 MPa) steels (AASHTO 2000b).

14.7.4 Load-Induced Fatigue

According to a parametric study reported by Horton et al. (2000, 2003), the design of welded plate girders made entirely from HPS70W or hybrid HPS70W and conventional (Grade 50W) steel is more likely to be controlled by load-induced fatigue when compared to those made entirely of conventional steel (Grade 50W). Using detail category C′ and C (AASHTO 2000b), and assuming full composite action in both positive and negative bending regions as well as a lower bound (conservative) fatigue limit equal to one-half the constant amplitude fatigue threshold value for all designs, Horton et al. concluded that load-induced fatigue often governed the design of HPS welded girders that were less than 150 ft (45.8 m) in length. For span length equal to or exceeding 200 ft (61 m), the effect of load-induced fatigue becomes negligible. The reason load-induced fatigue governs the design of shorter span but not longer span girders is because dead-load to total stress ratio increases with span length. As the dead-load to total stress ratio increases, the corresponding live-load to total stress ratio will decrease, thus rendering a lower stress range for live-load stress and less pronounced fatigue effects. It should be pointed out that the study reported by Horton et al. was conducted using a rather conservative fatigue limit that ignored the effect of average daily truck traffic on the value of the fatigue limit. The actual volume of truck traffic used for the design of a particular bridge may result in an increase in the allowable fatigue limit and thus reduce or even eliminate the effect of load-induced fatigue on the design of the bridge.

14.7.5 Tension Field Action

Tension field action is a design consideration that allows a designer to make use of the postbuckling web strength of a plate girder in computing its shear capacity when certain conditions are met. These conditions are: the girder is homogeneous (i.e., the component elements of the cross-section are made from the same material), the panel in which tension field action is considered in the design is an interior web panel, there is a sufficient number of transverse stiffeners along the length of the girder to act as anchors for the tension stress band (Basler 1961a) after web buckling, and the web panel under consideration must not be subjected to simultaneous high shear and moment (Basler 1961b). As of this writing, the use of tension field action for hybrid girders is *not* allowed. The reason for this is that the web of a hybrid girder is usually fabricated from a lower-strength steel when compared to the flanges. As a result, the web is likely to experience yielding when the nominal girder flexural strength is reached. It is believed that this yielding would compromise the web's ability to act effectively as a tension strut in the Pratt truss model used in the derivation of the postbuckling panel shear strength. Based on Basler's work, Hurst (2000), Aydemir (2000), and Barker et al. (2002) have developed theoretical models that take into consideration this web yielding in calculating the postbuckling panel shear strength of hybrid girders. Modified shear strength and moment–shear interaction equations have also been proposed. The validity of these equations will be checked against numerical and experimental studies to determine if they are appropriate for use in design. The allowance for tension field action in the design of hybrid girders is likely to reduce benefits such as shallower girder depth, fewer transverse stiffeners, lower fabrication cost, lighter structure, etc.

14.7.6 Heat Curving

Tests conducted in a FHWA demonstration project have shown that localized short term application of heat up to 1250°F (677°C) for purposes of heat curving, camber/sweep correction has no noticeable adverse effect on the ultimate strength, yield strength, ductility, and CVN toughness of HPS70W steel. However, as a conservative measure, the Guide (AASHTO 2000a) only allows limited application of heat up to only 1100°F (593°C).

14.7.7 Ultrasonic Testing of HPS Welds

When tested according to current codes and recommended procedures, no problems have been reported in regard to qualifying HPS welds using ultrasonic testing. Like any tests, the accuracy of the test is only as good as the instrumentation used and the person who conducts the test. It is therefore recommended that the test should only be performed by certified engineers/technicians using established procedures (Birks et al. 1991; AWS 2002).

14.8 Summary

High-performance steels offer many advantages over conventional steels in bridge design and construction. Some of the major benefits are

- Improved weldability and lower preheat temperature.
- Enhanced weathering characteristics and durability.
- Higher fracture toughness and crack tolerance.
- Wider girder spacing may help lower fabrication and construction cost.
- Shallower members may help solve or alleviate vertical clearance problem.
- Increased span lengths may help reduce the number of piers needed to support the superstructure.
- Lighter structure.
- Lower initial, or perhaps even life-cycle, costs.

Despite these advantages, the use of smaller/shallower and fewer girders in a HPS bridge means the possibility of excessive deformations, global/local buckling, and increased load-induced fatigue. Care must be exercised to ensure that both strength and serviceability limit states are satisfied. In addition, it has been reported that difficulties were encountered in the drilling (Wasserman et al. 1998) and reaming of HPS70W plates. The use of a larger quantity of lubricant in the holes and on the drill bits and reamers may be needed. Moreover, there have been reports that show HPS70W steel has a tendency to leave tiny pitting residue after blast cleaning, and mill scale removal by descaler is not particularly effective. Finally, if HPS70W steel plates fabricated using the Q&T process are used, its limited length (50 ft or 15 m) means more splice plates are needed to build a long girder. The increase in the number of beam splices, the more expensive flux, and the higher material cost mean the unit cost of a girder made of HPS70W steel is likely to be higher. However, a higher unit cost does not necessarily translate to a higher overall cost because of the benefits outlined above for using HPS70W steel in bridge construction.

The AASHTO load and resistance factor design (LRFD) Specifications (AASHTO 1998) have been updated (AASHTO 2000a, 2000b) to allow for the use of HPS70W steel in bridge design. Studies have shown that these specifications are adequate for use with HPS70W steel provided that the issues of girder deformations, global and local buckling, and tension field action are properly addressed in the design.

References

AASHTO. 1998. *AASHTO LRFD Bridge Design Specifications*, American Association for State Highway and Transportation Officials, Washington, DC.

AASHTO. 2000a. *Guide Specifications for Highway Bridge Fabrication with HPS70W Steel. An Addendum to ANSI/AASHTO/AWS D1.5*, American Association for State Highway and Transportation Officials, Washington, DC.

AASHTO. 2000b. *Interim AASHTO LRFD Bridge Design Specifications*, American Association for State Highway and Transportation Officials, Washington, DC.

Aydemir, M. 2000. *Moment Shear Interaction in HPS Hybrid Plate Girders*, M.S. Thesis, Georgia Institute of Technology, Atlanta, GA.

AWS. 2002. AWS D1.5:2002 *Bridge Welding Code*, American Welding Society, Miami, FL.

Barker, M.G. and Schrage, S.D. 2000. High Performance Steel: Design and Cost Comparisons, *Bridge Crossings-Practical Information for the Bridge Industry*, No. 16, Reprinted from Modern Steel Construction, AISC.

Barker, M.G., Hurst, A.M., and White, D.W. 2002. Tension Field Action in Hybrid Steel Girders, *Eng. J.*, AISC, First Quarter, 52–62.

Barth, K.E., White, D.W., and Bobb, B. 2000. Negative Bending Resistance of HPS70W Girders. *J. Construct. Steel Res.*, 53(1), 1–31.

Basler, K. 1961a. Strength of Plate Girders in Shear, *J. Struct. Div.*, ASCE, 87(7), 151–180.

Basler, K. 1961b. Strength of Plate Girders Under Combined Bending and Shear, *J. Struct. Div.*, ASCE, 87(7), 181–197.

Birks, A.S., Green, R.E., and McIntire, P. 1991. *Nondestructive Testing Handbook, Vol. 7 — Ultrasonic Testing*, American Society of Nondestructive Testing, Columbus, OH.

Bjorhovde, R. 2004. Development and Use of High Performance Steel, *J. Construct. Steel Res.*, 60, 393–400.

Clingenpeel, B.F. and Barth, K.E. 2003. Design Optimization Study of a Three-Span Continuous Bridge Using HPS70W, *Eng. J.*, AISC, Third Quarter, 149–158.

Dexter, R.J., Alttstadt, S.A., and Gardner, C.A. 2002. Strength and Ductility of HPS70W Tension Members and Tension Flanges with Holes, Technical Report to Federal Highway Administration Turner-Fairbank Highway Research Center, March.

FHWA. 1989. *Technical Advisory T 5140.22 — Uncoated Weathering Steel in Structures*, US Department of Transportation, Federal Highway Administration, Washington, DC.

Fisher, J.W. and Wright, W.J. 2000. High Toughness of HPS: Can it Help You in Fatigue Design. *Proceedings, Steel Bridge Design and Construction for the New Millennium with Emphasis on High Performance Steel*, November 30–December 1.

Horton, R., Power, E., Van Ooyen, K., and Azizinamini, A. 2000. High Performance Steel Cost Comparison Study, *Proceedings, Steel Bridge Design and Construction for the New Millennium with Emphasis on High Performance Steel*, November 30–December 1.

Horton, R., Power, E., Van Ooyen, K., and Azizinamini, A. 2003. Can Less be More? *Civil Eng.*, ASCE, November, 70–77.

Hurst, A.M. 2000. *Tension Field Action in HPS Hybrid Plate Girder*, M.S. Thesis, Department of Civil Engineering, University of Missouri-Columbia, 124pp.

Lwin, M.M. 2002. *High Performance Steel Designers' Guide*, 2nd edition, US Department of Transportation, Federal Highway Administration, Western Resource Center, San Francisco, CA.

Price, K.D. and Cassity, P.A. 2000. High Performance Hybrid, *Civil Eng.*, ASCE, 70(6), 66–69.

Van Ooyen, K. 2002. HPS Success, *Modern Steel Construction*, AISC, September, pp. 36–38.

Wasserman, E., Azizinamini, A., Pate, H., and Greer, W. 1998. Making the Grade, *Civil Eng.*, ASCE, 68(4), 69–71.

Wright, W.J., Tjiang, H., and Albrecht, P. 2001. Fracture Toughness of Structural Steels, Technical Report to Federal Highway Administration Turner-Fairbank Highway Research Center, January.

Yakel, A.J., Mans, P., and Azizinamini, A. 2002. Flexural Capacity and Ductility of HPS-70W Bridge Girders, *Eng. J.*, AISC, 39(1), 38–51.

Relevant Websites

www.asnt.org
www.astm.org
www.fhwa.dot.gov/bridges
www.lincolnelectric.com
www.nabro.unl.edu
www.steel.org/infrastructure/bridges

15

High-Performance Concrete

Zongjin Li
Department of Civil Engineering,
Hong Kong University of Science and
Technology,
Kowloon, Hong Kong

Yunsheng Zhang
Department of Materials Science
and Engineering,
Southeast University,
Nanjing, China

15.1 Introduction

15.1.1 Historical Development

Since the 1990s, the expression *high-performance concrete* and the acronym *HPC* have become very popular and very fashionable in civil engineering. At a first glance, it sounds like advertising a new product but, in most respects, HPC is not fundamentally different from the concrete that we have been using all along, as it does not contain any new ingredient and does not involve new practices on site.

No single person invented HPC, and no single country pioneered its use. The development of HPC materials in use today was an incremental and combined effort involving many individuals, companies, government agencies, and countries, particularly in Asia, Europe, and North America.

The full development of HPC took quite a long time. Initially, HPC was mainly related to high-strength concrete. Specified concrete strength for buildings steadily increased from 35 MPa in the 1950s to 100 MPa by the end of the 1980s. During this period, the term "high-strength concrete" was frequently used. However, when the high range water reducer or *superplastizer* was invented and began to be used to decrease the *water/cement* (w/c) or *water/binder* (w/b) ratios rather than being exclusively used as fluid modifiers for normal-strength concretes, it was found that in addition to improvement in strength, concretes with very low w/c or w/b ratios also demonstrated other improved characteristics, such as higher fluidity, higher elastic modulus, higher flexural strength, lower permeability, improved abrasion resistance, and better durability. Today, the definition of HPC has expanded to encompass both durability and strength.

The term "high-performance concrete" originally comes from French. It was coined in 1980 by Roger Lacroix and Yves Malier [1]. In 1986, the French project "New Ways for Concrete" brought together 36 researchers from France, Switzerland, and Canada. The research, findings, and field applications of all the members of this group formed the contents of the first book published that was solely devoted to HPC [2].

By the end of 1988, Pierre-Claude Aïtcin, assisted by Denis Mitchell and Michael Collins, wrote the successful proposal for the Network of Centres of Excellence on High Performance Concrete, funded under the Federal Government "Centres of Excellence Programme." This research program started in 1990, and in its second phase, starting in 1994, the Network became known as Concrete Canada. The researchers who comprised Concrete Canada were not the only Canadians researching and using HPC; however, they were the preeminent and most active group in this field. By virtue of many publications in scientific journals, a Newsletter sent to 7000 persons world-wide, the organization of technology transfer days and seminars, and the construction of demonstration projects, Concrete Canada played the major role in establishing HPC as a widely accepted construction material in Canada.

In the United States, starting from 1989, a 4-year investigation on the mechanical behavior of HPC was initiated by the researchers at North Carolina State University under the support of the Strategic Highway Research Program (SHRP) [3]. The research complied most of the publications in the HPC area and provided very useful information for HPC studies.

Since the publication of these two documents, there has been a phenomenal increase in the development and use of HPC. So a need exists to update the earlier documents and summarize the significant developments during the past several years. This volume is a sequel to the earlier state-of-the-art report and covers the 6-year period from 1989 to 1994. A second annotated bibliography containing 776 references for the 6-year period has also been compiled by the authors as a separate document [3].

In the decade 1990–2000, numerous research programs have been carried out in many countries in Europe, Asia, Australasia, Japan, and North America. Meanwhile, thousands of papers on HPC have also been published [4]. At present, the use of HPC has spread throughout the world.

15.1.2 Definition of HPC

Until now, it is not possible to provide a unique definition of HPC without considering the performance requirements of the intended use of the concrete. Generally, HPC can be defined as a concrete composed of the same raw materials as normal concrete with additional admixtures and engineered to achieve enhanced workability, durability, or strength characteristics to meet the specific demands of a construction project. Hence, any concrete that satisfies certain criteria proposed to overcome limitations of conventional concretes may be called HPC.

In addition to various definitions given by different concrete researchers, many very respected institutitons have regulated the definitions according to their experiences and requirements. Some good examples are given in the following list:

1. *Definition from Strategic Highway Research Program.* Based on the results of SHRP C-205, Zia et al. [5] summarized the requirements for HPC as shown in Table 15.1. It should be noted that the requirements include strength, w/b ratio, and durability factor.

TABLE 15.1   Definition of HPC According to SHRP C-205

Category of HPC	Minimum compressive strength	Maximum water/cement ratio	Minimum frost durability factor
Very early strength (VES)			
Option A (with Type III cement)	2,000 psi (14 MPa) in 6 h	0.40	80%
Option B (with PBC-XT cement)	2,500 psi (17.5 MPa) in 4 h	0.29	80%
High early strength (HES) (with Type III cement)	5,000 psi (17.5 MPa) in 24 h	0.35	80%
Very high strength (VHS) (with Type I cement)	10,000 psi (70 MPa) in 28 h	0.35	80%

Source: Zia et al. Mechanical Behavior of high-programme concreter, vol. 1, National Research Council, Washington DC, 1993.

2. *Definition from FHWA.* According to Federal Highway Administration (FHWA), HPC is defined as "a concrete that has been designed to be more durable and, if necessary, stronger than conventional concrete." Concrete so designated should meet significantly more stringent criteria than those required for normal structural concrete. It should give optimized performance characteristics and should have high workability, very high fluidity, and minimum or negligible permeability. Serviceability as determined by crack control and deflection control, as well as long-term environmental effects, is equally important as durability parameters. In 1996, the FHWA proposed additional criteria for four different performance grades of HPC as shown in Table 15.2.

3. *Definition from ACI.* In 1993, the American Concrete Institute (ACI) published a broad definition for HPC:

> High-performance concrete (HPC) is defined as concrete which meets special performance and uniformity requirements that cannot always be achieved by using only the conventional materials and mixing, placing and curing practices. The performance requirements may involve enhancements of placement and compaction without segregation, long-term mechanical properties, early age strength, toughness, volume stability, or service life in severe environments.

Based on the HPC definitions, it can be seen that HPC performance characteristics consist of two categories: durability and strength. The durability and strength parameters divide field conditions into three categories: climate, exposure effects, and load. To achieve the strength and durability requirements, a lower w/b ratio is necessary. For most application of HPCs, the w/b ratio is less than 0.4. Furthermore, some admixtures (both chemical and mineral) are usually needed to produce HPC.

In general, HPC is not one product but includes a range of materials with special properties beyond conventional concrete and routine construction methods at that time. It only can be made by appropriate materials, suitable mix design and properly mixed, transported, placed, consolidated, and cured so that the resulting concrete will give excellent performance in the structure in which it will be placed, in the environment to which it will be exposed, and with loads to which it will be subjected to during its design life. Presently, the primary characteristics of HPC can be summarized as easy placement, high early age strength, toughness, superior long-term mechanical properties, and prolonged service life in severe environment. Of course, some HPC applied in special fields and occasions possess some other chacteristics, and these characteristics of HPC will change with time.

Herein, an important point must be emphasized. HPC is different from high-strength concrete: a high-strength concrete is usually a HPC, but a HPC is not always a high-strength concrete because HPC stresses concrete durability behavior, not excessively showing indication of high strength. A high-strength concrete does not ensure that a durable concrete will be achieved. Given that the required durability characteristics are more difficult to define than strength characteristics, specifications often use a combination of performance and prescriptive requirements, such as permeability and a maximum water–cementitious material ratio to achieve a durable concrete. The end result may

TABLE 15.2 Definition of HPC According to Federal Highway Administration [6]

Performance characteristics	Standard test method	FHWA HPC performance grade			
		1	2	3	4
Freeze–thaw durability (X = relative dynamic modulus of elasticity after 300 cycles)	AASHTO T 161 ASTM C 666 Procedure A	60% < X < 80%	80% < X		
Scaling resistance (X = visual rating of the surface after 50 cycles)	ASTM C 672	X = 4, 5	X = 2, 3	X = 0, 1	
Abrasion resistance (X = avg. depth of wear in mm)	ASTM C 944	2.0 > X > 1.0	1.0 > X > 0.5	0.5 > X	
Chloride penetration (X = coulombs)	AASHTO T 277 ASTM C 1202	3000 > X > 2000	2000 > X > 800	800 > X	
Strength (X = compressive strength)	AASHTO T 2 ASTM C 39	41 < X < 55 MPa (6 < X < 8 ksi)	55 < X < 69 MPa (8 < X < 10 ksi)	69 < X < 97 MPa (10 < X < 14 ksi)	97 MPa < X (14 ksi < X)
Elasticity (X = modulus)	ASTM C 469	28 < X < 40 Gpa ($4 < X < 6 \times 10^6$ psi)	40 < X < 50 GPa ($6 < X < 7.5 \times 10^6$ psi)	50 GPa < X < (7.5×10^6 psi < X)	
Shrinkage (X = microstrain)	ASTM C 157	800 > X > 600	600 > X > 400	400 > X	
Specific creep (X = microstrain per MPa)	ASTM C 512	75 > X > 60/MPa (0.52 > X > 0.41/psi)	60 > X > 45/MPa (0.41 > X > 0.31/psi)	45 > X > 30/MPa (0.31 > X > 0.21/psi)	30/MPa > X (0.21/psi > X)

be a high-strength concrete, but this only comes as a by-product in the process of preparing a durable concrete.

15.1.3 Types of HPC

Fiber-reinforced concrete (FRC) — Concrete with sufficient fiber reinforcement to provide ductility or toughness equal to at least five times the area under the stress–strain curve for the same concrete mixture without fiber reinforcement. FRC is normally associated with toughness, that is, the ability to absorb energy. This energy absorption occurs primarily after the ultimate strength of the concrete has been attained.

High-durability concrete — Concrete with a minimum durability factor (freezing and thawing) of 80%, as measured by AASHTO T 161 (Method A) (or ASTM C 666), and a w/c ratio of 0.35 or less. A maximum w/c of 0.35 will provide a paste with a discontinuous capillary system after a relatively short curing period (normally about a day). This provides improved resistance to moisture penetration and chemical attack from the environment.

High-strength lightweight concrete — concrete produced by using lightweight aggregates, such as expanded clay, shale, and slate aggregates, so as to reduce the mass from 20 to 25% below that of conventional concrete. Some lightweight concrete can attain compressive strengths greater than 69 MPa (10,000 psi). This type of HPC is desirable in applications where reduction of dead load is a significant consideration.

Self-consolidation concrete — It is also called self-compaction concrete. In fresh stage, such a concrete has the ability to fill formwork and encapsulate reinforcing bars only through the action of gravity, and with maintained homogeneity.

15.1.4 Benefits to Construction

Many advantages have been brought to construction with the application of HPC. The following highlights the primary construction-related benefits:

1. *High strength.* HPC, generally speaking, has high compressive strength. The high-strength property, on the one hand, can increase the load carrying capacity of the structural members such as steel bar reinforced concrete columns and shells, resulting in less reinforcing steel and labor consumption, and materials costs saving. On the other hand, by keeping the same load, the cross-section size of the primary structural members can be considerably diminished. The benefit to flexural strength of construction components can also be improved due to high strength. Although high strength cannot enhance the flexural strength too much, the compressed region depth can be greatly decreased for the bended members and the corresponding ductility is markedly improved. Meanwhile, higher steel bar ratio can be allowed for, which further increases the flexural strength or reduces the cross-section size.
2. *Improved workability.* Due to the addition of water reducing admixtures, especially super-plasticizer, and the increased cement content in HPC, the workability of HPC, including fluidity and cohesiveness, is generally improved. For self-compacting concrete, a type of HPC, the concrete can flow in mold without compaction.
3. *Low volume change.* The lower volume deformation of HPC brings about an increase in stiffness, which is very beneficial to those beams and plates whose cross-section sizes are controlled by deformation. As with prestressed enforcement concrete members, three advances can be obtained: higher prestress gain, earlier applied prestress, and less loss in prestress.
4. HPC has a special significance for these constructions whose loads primarily come from the self-weight. There can be a great improvement in reducing the bearing load of foundations by using HPC. Furthermore, more space can be effectively utilized for housing constructions caused by the smaller cross-section size using HPC. For bridge constructions, more net space below the bridge decks and the lower roadbed can be obtained. For underground engineering, the use of HPC may reduce the excavation mass. In addition, the construction modules can be adjusted into identical sizes, resulting in a lower production cost of modules.

5. *Enhanced durabilities.* Greatly enhanced durability properties of HPC allow the overall life of the structure to be significantly extended. The period of time until maintenance is needed is much longer than conventional concrete, therefore, there is less disruption to the public due to repairs and less maintenance costs. In addition, HPC is less permeable than conventional concrete, which makes it more resistant to salts in seawater, freeze–thaw cycles, and damage caused by chemical deicing, which is very attractive in environments where ordinary concrete would not suffice such as marine, high rate running water, or easily collision engineering.

15.1.5 Future Development

Significant evolutions relating to HPC have been taking place in the past two decades. These evolutions have been facilitated considerably by increased knowledge of the atomic and molecular structure, studies of long-term durability, development of more powerful instruments and monitoring techniques, and the need for stronger, higher-performance materials suitable for larger structures, long span, and more ductility. The last two decades can be described as the decades of HPC.

Despite these tremendous advances in concrete technology and development of HPC, there is still much scope for new mineral and chemical admixtures, microstructure studies, better material selection proportioning, placement techniques, long-term performance, and new structural design formulations compatible with the behavior and potential of HPC.

The present state of the art in concrete materials research has demonstrated that the use of mineral admixtures as cementitious materials used together with Portland cement in concrete not only contributes to high-performance gain of HPC but also to energy saving and a solution to the disposal of industrial by-products. Recent progress in the use of fly ash, slag, silica fume, and natural pozzolans are noteworthy. The development of new cements such as macrodefect-free cement and densified small particle (DSP) cement, and the advances in HPC-based composites such as slurry-infiltrated fiber concrete or slurry-infiltrated fiber mesh concrete, and reactive powder concrete (RPC) have all reached compressive strength levels of 300 MPa, particularly, for RPC, 800 MPa compressive strength can be obtained. Since self-consolidation concrete has the advantages of high performance in both fresh and hardened states, cost competitive, improved working and living environments, and enhancement towards automation of construction process, its applications will become broader and broader in the whole world.

Deterioration of the infrastructure governs replacement or rehabilitation. Rehabilitation often requires new materials and techniques compatible with the parent system. In all these, concrete in various forms and compositions is the only suitable material. Adding the factor of rapid population growth, particularly in the underdeveloped parts of the world, and the continuous increased urbanization and industrialization, it is clear that the 21st century will still be the golden age of HPC.

15.2 Materials

15.2.1 Portland Cement

HPCs have been produced successfully using Types I, II, and III Portland cements that meet the ASTM Standard Specification C 150. Unfortunately, ASTM C 150 is very imprecise in its chemical and physical requirements, and so cements that meet these rather loose specifications can vary quite widely in their fineness and chemical composition. Consequently, the same type of cements have quite different rheological and mechanical characteristics, particularly when used in combination with chemical admixtures and mineral admixtures. Therefore, when choosing Portland cements for producing HPC three requirements should be satisfied: (1) it can develop the appropriate strength at given ages; (2) it exhibits good rheological behavior, particularly when chemical admixtures are incorporated; and (3) it should not generate too much hydration heat.

The strength of the cement has a strong influence on the properties of HPC. If the cement cannot develop reasonable strength at given ages, it is impossible to achieve the mechanical and durability requirements of HPC.

TABLE 15.3 Properties of Cement Compounds

Component	Rate of reaction	Heat liberated	Ultimate cementing value
Tricalcium silicate, C_3S	Medium	Medium	Good
Dicalcium silicate, C_2S	Slow	Small	Good
Tricalcium aluminate, C_3A	Fast	Large	Poor
Tetracalcium aluminoferrate, C_4AF	Slow	Small	Poor

The strength of cement paste is the result of the process of hydration. The hydration process results in the formation of crystal phases such as $Ca(OH)_2$, AFt/AFm, and calcium silicate hydrate (C–S–H) gel phase. The crystal phases interlock each other and construct the bone of the harden paste. For C–S–H gels, it fills in the cavities among the bone. Table 15.3 shows the relative contribution of each component of the cement toward the rate of gain in strength. The early strength of portland cement is mainly controlled by the percentages of C_3S. C_3S contributes most to the strength developed during the first day after placing the concrete.

In general, no special compressive strength criterion is specified for cement used for the manufacture of HPC; however, Portland cement is preferable with 52.5 MPa or greater compressive strength at 28 days.

Increasing the fineness of the Portland cement will, on the one hand, increase the early strength of the concrete. On the other hand, higher fineness will lead to a more rapid release of hydration heat and a big rheological problem, which are attributed to the rapid reaction and the great amount of ettringite formation at early ages. Early work by Perenchio [7] indicated that fine cements produced higher early strengths, though at later ages the strength differences of different fineness cement concrete were not significant. Most cements now used to produce HPC have Blaine finenesses that are in the range of 300 to 400 m²/kg.

The work of Perenchio [7] indicates that cements with higher C_3A contents lead to higher strengths. However, subsequent work [8] has shown that high C_3A contents generally leads to rapid loss of flow for fresh concrete. Therefore, cements with high C_3A contents should be avoided for the production of HPC. Aïtcin [9] has shown that C_3A should be primarily in its cubic, rather than its orthorhombic, form. Further, Aïtcin suggests that attention must also be paid not only to the total amount of SO_3 in the cement but also to the amount of soluble sulfates. Thus, the degree of sulfurization of the clinker is an important parameter. In addition to commercially available cements conforming to ASTM Types I, II, and III, a number of cements have been formulated specifically for HPC.

Generally, HPC typically contains higher cement content than normal concrete in order to develop good workability, high strength, low permeability, and excellent durability. Thus, much attention should be paid on hydration heat for HPC, particularly for massive HPC structures.

Since different types of cement generate different amounts of hydration heat at different rates, the type of structure governs the type of cement to be used. The bulkier and the heavier in cross-section the structure is, the less generation of heat of hydration is desired. In massive HPC structures such as transferring floor, type IV cement is a good candidate to use. In addition, the reduction in hydration heat can also be accomplished in the presence of mineral admixture, ASTM type II cements, or moderate-heat blended cements. From this discussion it is seen that the type of structure, the weather, and other conditions under which the structure is built and that will exist during its life span are the governing factors in the choice of the type of cement that should be used.

15.2.2 Aggregate

The aggregate, including coarse and fine aggregates, consists of about 70 to 80% of the total concrete by volume; thus, it is very important in the production of HPC. The main factors affecting the development of high performance of concrete are: particle shape, particle size distribution, strength of the aggregate, and possible chemical reaction between the aggregate and the paste. Unlike their use in normal concrete, where we rarely consider the strength of the coarse aggregate, in high-strength concrete the aggregates may well become one important strength limiting factor. Also, since it is necessary to maintain a low w/c ratio to achieve high strength, the shape and grading of the aggregate, including fine and coarse aggregates, must also be very tightly controlled.

15.2.2.1 Coarse Aggregate

For HPC, the coarse aggregate particles themselves must be strong. A number of different rock types used to make HPC include: limestone, dolomite, granite, andesite, and diabase. It has been suggested that in most cases the bond strength between the cement matrix and the aggregate is not usually the limiting factor in HPC; rather, it is the aggregate strength itself. In addition, aggregates that may be susceptible to alkali–aggregate reaction or D-cracking, should be avoided if possible, even though the low w/c ratios used will tend to reduce the severity of these types of reaction.

From both concrete strength and fresh concrete mixture rheological considerations, the coarse aggregate particles should be roughly equidimensional: either crushed rock or natural gravels are suitable. Flat or elongated particles must be avoided at all costs. They are inherently weak, and lead to harsh mixes. In addition, it is important to ensure that the aggregate is clean, since a layer of silt or clay may reduce the cement–aggregate bond strength and increase the water demand. Thus, the coarse aggregate should be washed if there is too much fine powder. Finally, the aggregates should not be highly polished (as is sometimes the case with river-run gravels), because this will reduce the cement–aggregate bond.

Not much work has been carried out on the effects of aggregate mineralogy on the properties of HPC. However, a detailed study by Aïtcin and Methta [10], involving four hard strong aggregates (diabase, limestone, granite, natural siliceous gravel), revealed that the granite and the gravel yielded much lower strengths and *E*-values than the other two aggregates. This phenomenon appeared to be related both to aggregate strength and to the strength of the cement–aggregate transition zone. Cook [11] had also pointed out the effect of the elasticity modulus of the coarse aggregate on that of the concrete. However, much work remains to be done to establish the relationship between the mechanical and mineralogical properties of the aggregate and those of the resulting HPC.

It is commonly assumed that a smaller maximum size of coarse aggregate will lead to a more homogenous concrete and higher strengths. The maximum particle size of 20 mm coarse aggregate is commonly recommended for the production of HPC.

15.2.2.2 Fine Aggregate

The fine aggregate should consist of smooth rounded particles to improve the workability of fresh concrete. Normally, the fine aggregate grading should conform to the limits established by the American Concrete Institute [12] for normal-strength concrete. However, for HPC the gradings of fine aggregate should lie on the coarser side of these limits. A fineness modulus of 3.0 or greater is recommended [13,14] for HPC to decrease the water demand and to improve the workability of these paste-rich mixes. Of course, the sand too must be free of silt or clay particles.

15.2.3 Mineral Admixtures (Silica Fume, Fly Ash, Slag, Metakaoline)

As indicated above, most modern HPCs contain at least one type of mineral admixture: silica fume, fly ash, blast-furnace slag, or metakaoline. Very often, the fly ash or slag is used in conjunction with silica fume or metakaoline. In the United States, fly ash is specified in ASTM C 618.1 and blast-furnace slag in ASTM C 989. There is, as yet, no US standard for silica fume and metakaoline. These materials are described in detail in Supplementary Cementing Materials for Concrete.

15.2.3.1 Silica Fume

Silica fume (see Figure 15.1) is a by-product of the induction arc furnaces in the silicon metal and ferro-silicon alloy industries. Reduction of quartz to silicon at temperature up to 2000°C produces SiO vapors, which oxidize and condense in the low-temperature zone to tiny spherical particles consisting of non-crystalline silica. The material is removed by filtering the outgoing gases in bag filters. A size distribution of

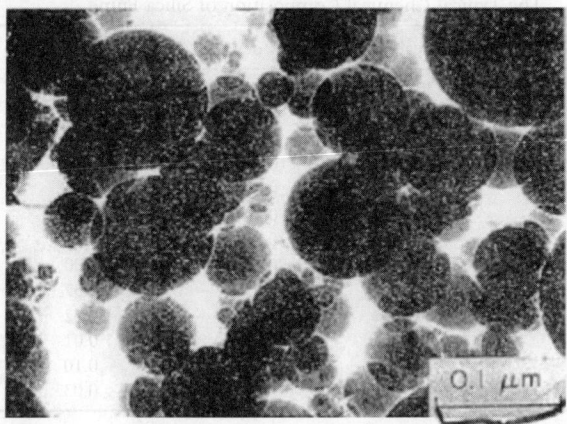

FIGURE 15.1 Silica fume particle.

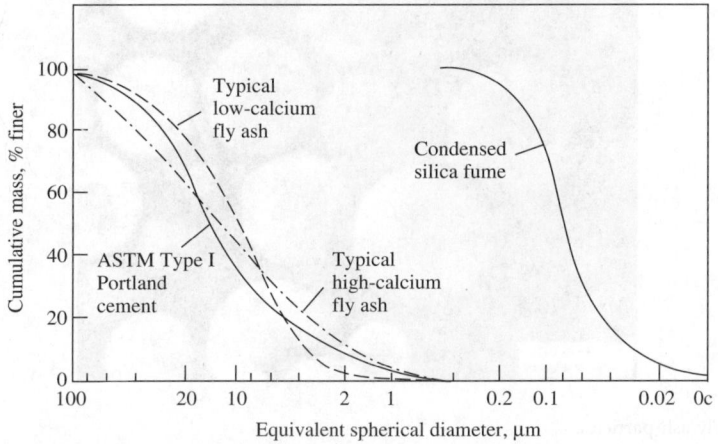

FIGURE 15.2 A comparison of size distribution of silica fume to other materials.

silica fume is shown in Figure 15.2. More accurately, the size distribution of a typical silica fume product is provided as

20% below 0.05 μm
70% below 0.10 μm
95% below 0.20 μm
99% below 0.50 μm

Microsilica has surface area around 20 m²/g and its average bulk density is 586 kg/m³. Compared with normal Portland cement and typical fly ashes, silica fume sizes are two orders of magnitude finer. Thus, the material is highly pozzolanic and reacts very quickly with calcium hydroxide caused by hydration of cement to form high-strength calcium silicate hydrates (C–S–H). However, it creates problems of handling and increases the water requirement in concrete appreciably unless superplasticizer is used.

The typical chemical composition of silica fume is shown in Table 15.4. It can be seen that silica is dominant (>92%) in the material.

TABLE 15.4   The Typical Chemical Composition of Silica Fume

	Typical	St. dev.	Min	Max
Moisture%	0.30	0.09	0.09	0.50
LOI%	1.18	0.26	0.79	0.73
SiO_2%	92.9	0.60	92.0	94.0
Al_2O_3%	0.69	0.10	0.52	0.86
Fe_2O_3%	1.25	0.46	0.74	2.39
CaO%	0.40	0.09	0.28	0.74
MgO%	1.73	0.31	0.23	2.24
K_2O%	1.19	0.15	1.00	1.53
Na_2O%	0.43	0.03	0.37	0.49
C%	0.88	0.19	0.62	1.30
Cl%	0.02	0.01	0.01	0.03
S%	0.20		0.10	0.30
P%	0.07		0.03	0.12

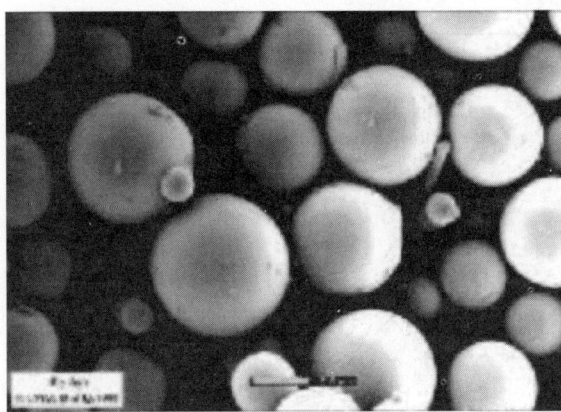

FIGURE 15.3   Fly ash particles.

It is possible to make HPC with compressive strengths of up to about 98 MPa without silica fume. Beyond that strength level, however, silica fume becomes essential, and even at lower strengths of 63 to 98 MPa. It is easier to make HPC with silica fume than without it. Thus, when it is available at a reasonable price, it should generally be a component of the HPC mix.

15.2.3.2 Fly Ash

Fly ash (pulverized fuel ash; see Figure 15.3) is a by-product of electricity generating plants using coal as fuel. During combustion of powdered coal in modern power plants, as coal passes through the high-temperature zone in the furnace, the volatile matter and carbon are burned off, whereas most of the mineral impurities, such as clays, quartz, and feldspar, will melt at the high temperature. The fused matter is quickly transported to lower-temperature zones, where it solidifies as spherical particles of glass. Some of the mineral matter agglomerates forming bottom ash, but most of it flies out with the flue gas stream and thus is called fly ash. This ash is subsequently removed from the gas by electrostatic precipitators.

Fly ash can be divided into two categories according to the calcium content. The ash containing less than 10% CaO (from bituminous coal) is called low-calcium fly ash (Class F) and the ash typically containing 15 to 30% of CaO (from lignite coal) is called high-calcium fly ash (Class C). Usually, the high-calcium fly ash is more reactive because it contains most of the calcium in the form of reactive crystalline compounds, such as C_3A and CS.

Micrographic evidence presented in Figure 15.3 shows that most of the particles in fly ash occurs as solid spheres of glass, but sometimes a small number of hollow spheres, which are called cenospheres (completely empty), and plerospheres (packed with numerous small spheres), present. The size distribution of fly ash is slightly smaller than those of Portland cement with more than 50% under 20 μm.

It was found from the researches carried out in past 40 years or so that the incorporation of fly ash into concrete has certain advantages and some disadvantages. The major advantages of fly ash concrete are low cost, low energy demand, and low hydration heat. The disadvantages of fly ash concrete are low early-age strength, longer initial setting time, and the variability in its physical and chemical characteristics. Thus, for high strength–high performance concrete, silica fume must be used in conjunction with the fly ash, though this practice has not been common in the past.

In general, for high strength–high performance concrete applications, fly ash is used at dosage rates of about 15% of the cement content. Because of the variability of the fly ash produced even from a single plant, however, quality control is particularly important. This involves determinations of the Blaine specific surface area, as well as the chemical composition. And, as with silica fume, it is important to check the degree of crystallinity. The more glassy the fly ash, the better.

15.2.3.3 Blast-Furnace Slag

Blast-furnace slag (see Figure 15.4) is a glassy material that is made by rapidly quenching molten blast-furnace slag and grinding the resulting material into a fine powder. It is composed essentially of silicates and aluminosilicates of calcium. When it is ground to cement fineness, it is referred to as ground granulated blast-furnace slag (GGBFS), and it is commonly used in HPC mixtures. GGBFS is classified by ASTM C 989 (AASHTO M 302) according to its level of reactivity. Depending on the desired properties, the amount of GGBFS can be as high as 80% of the total cementitious materials content.

The use of GGBFS reduces the permeability of the mature concrete. It is believed that this improvement is a result of the reaction of the GGBFS with the calcium hydroxide and alkalis released during hydration of the Portland cement. The reaction products fill the pore spaces in the paste and result in a denser microstructure. In addition to reducing the permeability of concrete, GGBFS also improves resistance to sulfate attack because of the low calcium hydroxide content. Like fly ash, GGBFS is also used to reduce the temperature rise in mass concrete. GGBFS improves the workability of fresh concrete. It is believed that the smooth, dense surfaces of the slag particles (Figure 15.4) result in very little water absorption during the mixing process.

FIGURE 15.4 Ground granulated blast-furnace slag particles.

In North America, slag is not as widely available as in Europe and Asia, and hence there is not much information available as to its performance in high-strength concrete. However, the indications are that, as with fly ash, slags that perform well in ordinary concrete are suitable for use in high-strength concrete, at dosage rates between 15 and 30%. The lower dosage rates should be used in the winter, so that the concrete develops strength rapidly enough for efficient form removal. For very high strengths in excess of 96 MPa, it will likely be necessary to use the slag in conjunction with silica fume.

The chemical composition of slag does not generally vary very much. Therefore, routine quality control is generally confined to Blaine specific surface area tests, and x-ray diffraction studies to check on the degree of crystallinity (which should be low).

15.2.3.4 Metakaoline

Metakaoline, generally called "calcined clay," is a reactive aluminosilicate pozzolan produced by heating kaolinite at a specific temperature regime. It may react with calcium hydroxide to form calcium silicate and calcium aluminate hydrates. The type of mineral admixture conforms to ASTM C618 class N pozzolan specifications. Metakaoline (Figure 15.5) typically has an average particle size of about 1.5 μm in diameter, which is between silica fume (0.1 to 0.12 μm) and Portland cement (15 to 20 μm).

As a new mineral admixture for producing HPC, metakaolinite can produce HPC of compressive strengths in excess of 110 MPa through replacing part of the cement in the mixture. Tests have shown that use of 5 to 10% by mass of metakaoline can produce HPC with performance characteristics comparable to those of silica fume concretes with respect to strength development, chloride ion penetration, drying shrinkage, and resistance to freeze–thaw cycles and scaling. Further tests are required to confirm these findings.

15.2.4 Chemical Admixtures (Superplasticizer)

Chemical admixtures are important and necessary components for the preparation of HPC. The concrete properties, both in fresh and hardened states, can be significantly modified or improved by chemical admixtures. In some countries, 70 to 80% of concrete contains one or more admixtures (88% in Canada, 85% in Australia, and 71% in the United States). It is thus important for civil engineers to be familiar with commonly used admixtures. Presently, many types of chemical admixtures are being used in HPC. The following will only discuss the most commonly used chemical admixture — superplasticizer.

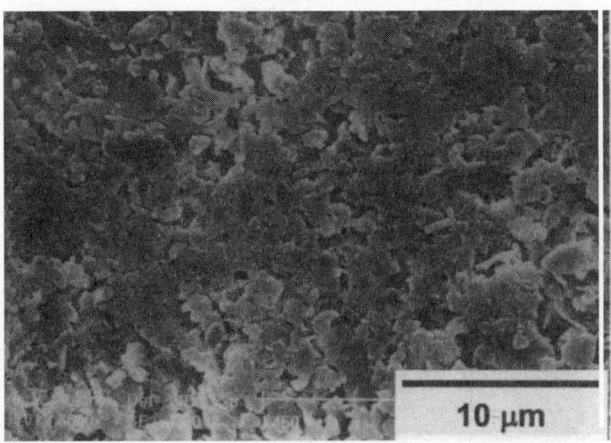

FIGURE 15.5 Metakaoline particles.

15.2.4.1 Superplasticizers

In modern concrete practice, it is almost impossible to make HPC with adequate workability in the field without superplasticizers. Unfortunately, different superplasticizers would behave quite differently with different cements (even cements of nominally the same type). This is due in part to the variability in the minor components of the cement (which are not generally specified), and in part to the fact that the acceptance standards for superplasticizers themselves are not very rigorous. Thus, some cements will simply be found to be incompatible with certain superplasticizers.

There are, basically, three common types of superplasticizer: (1) lignosulfonate based; (2) poly-condensale of formaldehyde and melamine sulfonate (often referred to simply as melamine sulfonate); and (3) polycondensate of formaldehyde and naphthalene sulfonate (often referred to as naphthalene sulfonate).

In addition, a variety of other molecules might be mixed in with these basic formulations. It may thus be very difficult to determine the precise chemical composition of most superplasticizers; certainly manufacturers try to keep their formulations as closely guarded secrets.

It should be noted that much of what we know about superplasticizers comes from tests carried out on normal strength concretes, at relatively low superplasticizer contents. This does not necessarily reflect their performance at very low w/c ratios and very high superplasticizer dosage.

15.2.4.1.1 Lignosulfonate-Based Superplasticizers

In high strength–high performance concrete, lignosulfonate superplasticizers are generally used in conjunction with either melamine or naphthalene superplasticizers. This tends not to be efficient enough for the economic production of very high-strength concretes on their own. Sometimes, lignosulfonates are used for initial slump control, with the melamines or naphthalenes used subsequently for slump control in the field.

15.2.4.1.2 Melamine Sulfonate Superplasticizers

Until recently, only one melamine superplasticizer was available (trade name Melment), but now other melamine-based superplasticizers are likely to become commercially available shortly.

Melamine superplasticizers are clear liquids, containing about 22% solid particles; they are generally in the form of their sodium salt. These superplasticizers have been used for many years now with good results, and so they remain popular with high-strength concrete producers.

15.2.4.1.3 Naphthalene Sulfonate Superplasticizers

Naphthalene superplasticizers have been in use longer than any of the others, and are available under a greater number of brand names. They are available as both a powder and a brown liquid; in the liquid form they typically have a solids content of about 40%. They are generally available as either calcium salts, or more commonly, sodium salts. (Calcium salts should be used in the case where a potentially alkali-reactive aggregate is used.)

The particular advantages of naphthalene superplasticizers, apart from their being slightly less expensive than the other types, appears to be that they make it easier to control the rheological properties of high-strength concrete because of their slight retarding action.

For a superplasticizer, great concern should be paid on the superplasticizer–cement compatibility, apart from the commonly considering water-reducing ratio. As is well known, the production of modern HPC typically requires a low w/b ratio. This means that high dosages of superplasticizers have to be incorporated to improve workability. In conjunction with the relatively low water contents of modern HPC, typically in the range of 125 to 145 L/m^3, this may lead to superplasticizer–cement compatibility problems. In particular, the rapid rate of water consumption, which occurs during the formation of ettringite ($C_3A \cdot 3CaSO_4 \cdot 32H_2O$), drastically reduces the already small amount of mixing water. This can result in insufficient water available to provide proper workability, particularly when there is no optimum content of calcium sulfate [15]. Recent work by Mokhtarzadeh and French [16] has highlighted the care that must be taken in choosing an appropriate superplasticizer for any particular HPC mix; the choice must be made by considering the superplasticizer effects on both workability and long-term strength.

15.2.5 Polymers

The incorporation of a polymer into concrete can produce excellent HPC with compressive strength of 100 MPa or higher and tensile strength of 10 MPa or higher. The commonly used polymers for producing HPC include: epoxy, methyl methacrylate, styrene, unsaturated polyster resins, vinyl esters, and latex.

These polymers are incorporated into Portland cement pastes commonly in the form of water-soluble or emulsified solutions. As the Portland cement hardens, the polymer simultaneously hardens forming a continuous matrix of polymer that penetrates throughout the whole concrete specimen, thus facilitating a rapid decrease in porosity of cement concrete.

15.3   Different Types of High-Performance Concrete

15.3.1 High-Strength Concrete

15.3.1.1 Definition

High-strength concrete is defined as the concrete made with a normal-weight aggregate with a compressive strength higher than 50 MPa.

15.3.1.2 Differences from Normal-Strength Concrete

To make high-strength concrete, more stringent quality control and more care in the selection of materials are needed. The microstructure and failure modes in compression of high-strength concrete are quite different from the normal-strength concrete.

From a microstructure point of view, the behavior of high-strength concrete is more like a homogeneous material. The extent of porosity in the transition zone is greatly reduced and thus almost eliminates the existence of the transition zone. Also, the amount of microcracking in high-strength concrete associated with shrinkage, short-term loading, and sustained loading is significantly less than in normal-strength concrete. As for the failure mode, high-strength concrete usually has a vertical crack going through the aggregate and shows a more brittle mode of fracture and less volumetric dilation.

15.3.1.3 Materials for High-Strength Concrete

Usually, the materials used for normal-strength concrete can be used to make high-strength concrete. However, for the aggregate to be used, it is better to choose a high-strength concrete if possible. The composition of high-strength concrete is different from normal-strength concrete. The maximum size of the aggregate is usually limited to 20 mm. The cement content is high (400 to 600 kg/m^3). The limitation on the maximum aggregate size is to reduce the strength of the transition zone and to get a more homogeneous material. The higher cement content results from the limited maximum aggregate size and the need for workability under the smaller w/c ratio condition.

15.3.1.4 Water/Cement Ratio and Abrams' Law

Water/cement ratio is the most important factor influencing various kinds of concrete properties. The basic measure usually taken in making high-strength concrete is to reduce the w/c ratio from the values for normal-strength concrete of about 0.5 or higher to about 0.3 or less. This result can be inferred from Abrams' law:

$$f_c = \frac{A}{B^{1.5(w/c)}}$$

where f_c is the compressive strength, A is an empirical constant (usually 14,000 psi), and B is a constant depending mostly on the properties of the cement (usually 4).

One of the major drawbacks of a lower w/c ratio is the loss of workability. However, this disadvantage can be overcome by the addition of admixtures. To recover the necessary workability, superplasticizers are usually added to the mix.

15.3.1.5 Silica Fume and Microengineering

Two admixtures are commonly used in high-strength concretes. One is superplasticizer, which is for improvement of workability. Another admixture that has been frequently used in high-strength concrete is silica fume. Silica fume is a waste product from the ferrosilicon manufacturing process. Due to its small size, sometimes it is called microsilica (MS). MS is a highly reactive amorphous silica that is highly pozzolanic.

The reasons that concrete strength can be significantly increased by adding silica fume are

- The size of silica fume particles is smaller than that of cement particles (less than 1 µm) and thus through the so-called packing effect the silica fume can make a more dense concrete.
- The silica fume can react with calcium hydroxide (CH) and the products of this reaction are very efficient in filling up a large capillary space, thus improving the strength and permeability. It begins to react as soon as water is added. It has been shown that this reaction changes: (1) the amount of CH; (2) the composition of C–S–H; and (3) the pore size distribution.

15.3.2 Ultrahigh-Strength Concrete

15.3.2.1 Definition

Ultrahigh-strength concrete is defined as concrete that has compressive strength greater than 200 MPa. In general, high-strength concrete can also be more durable. The comparison of conventional, high-strength, and ultrahigh-strength concretes is given in Table 15.5.

15.3.2.2 Composition of Ultrahigh-Strength Concrete

To illustrate the composition of ultrahigh-strength concrete, two examples are given in Table 15.6. The key characteristics of ultrahigh-strength concrete can be summarized as

- A low water–binder ratio.
- A large quantity of silica fume (or other fine mineral powder).
- Aggregates containing only fine sand.
- A high dosage of superplasticizers. Depending on the level of compressive strength, postset heat treatment, and application of pressure before or during setting may be necessary.

TABLE 15.5 Characteristics of Ultrahigh-Strength Concrete

	Conventional concrete	High-strength concrete	Ultrahigh-strength concrete
Compressive strength (MPa)	<50	~100	>200
Water–binder ratio	>0.5	−0.30	<0.2
Chemical admixture	Not necessary	WRA/HRWRA necessary	HRWRA essential
Mineral admixture	Not necessary	Fly ash (and/or) silica fume commonly used	Silica fume (and/or) fine powder essential
Fibers	Beneficial	Beneficial	Essential
Air entrainment	Necessary	Necessary	Not necessary
Processing	Conventional	Conventional	Heat treatment and pressure
Steady-state chloride diffusion ($\times 10^{-12}\,m^2/s$)	1.0	0.6	0.02

TABLE 15.6 The Two Mix Proportions of Ultrahigh-Strength Concrete

	Cement	Water	Superplasticizers	Silica fume	Fine sand	Quartz flour
No. 1	1	0.28	0.06	0.33	1.43	0.3
No. 2	1	0.15	0.044	0.25	1.1	—

15.3.2.3 Microstructure of Ultrahigh-Strength Concrete

Based on the observations made by optical, SEM, TEM, x-ray diffraction, and mercury intrusion porosimetry, the microstructure of ultrahigh-strength concrete can be concluded to have the following properties:

1. On a millimeter scale, the material is more homogeneous.
2. Absence of a pronounced transition zone between sand and the paste.
3. Low or negligible presence of portlandite.
4. Low porosity (1 to 3%).
5. Proportional lower volume of capillary porosity (10% of total porosity).
6. Strong component of pores with diameter 2.5 nm.

The above-mentioned microstructural features lead to low water absorption, gas permeability, and chloride diffusivity.

15.3.2.4 Limitations (Brittleness)

Ultrahigh-strength concrete is considerably more brittle than conventional concrete. This can be seen by comparing the uniaxial compressive stress–strain curve of conventional concrete with that of a ultrahigh-strength concrete, as shown in Figure 15.6. These figures were obtained with a digitally controlled, closed-loop test system. It can be observed that the postpeak response of ultrahigh-strength concrete is considerably steeper than that of conventional concrete.

15.3.2.5 Applications

Ultrahigh-strength concrete can be used for prestressed concrete structures and nuclear waste storage. A pedestrian bridge has been constructed using ultrahigh-strength concrete of 200 MPa in Canada. The footbridge was built with post-tensioned, precast elements made with FRC (fiber amount: 200 kg/m^3, 12 mm long, and 0.2 mm in diameter). Each element was 6.8 m long, 4.2 m wide, and 3.5 m high. The web of the bridge was designed with concrete without fibers but confined in a steel tube. The cost of concrete with fibers is about US\$1242 per cubic meter, whereas without fibers, the cost is US\$332 per cubic meter. The advantages of using ultrahigh-strength concrete in constructing nuclear waste storage are its low porosity and the elimination of the need for transporting nuclear substances.

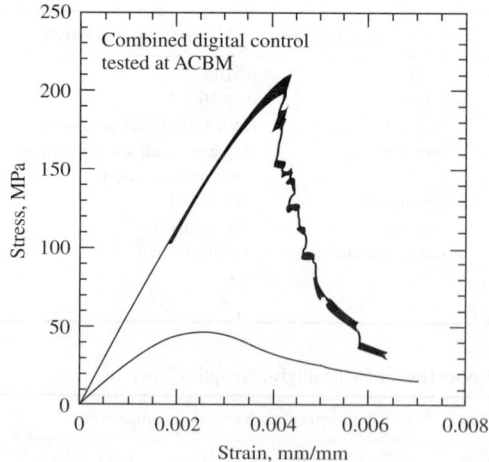

FIGURE 15.6 Comparison of stress–strain curves of reactive powder concrete to that of high-strength concrete.

15.3.3 Self-Consolidation Concrete

Self-consolidation concrete (SCC) or self-compaction concrete means a concrete that has a high fluidity and can be easily placed by itself in formwork (even highly congested ones, without external consolidation by vibration). It is characterized by its high filling capacity caused by high viscoplastic deformability and resistance to segregation.

The mix proportion of SCC has some characteristics comparable to conventional concrete. The unit paste is much more elevated, supplemented by a decrease in the volume of the coarse aggregate, and the water powder ratio of the paste is lowered to assure the required viscosity. Hence, the powder content is elevated up to, in some cases, $600 \, kg/m^3$ or more, and the water content is restrained to an adequately low level. Various mineral admixtures are utilized to increase the powder content. High fluidity is produced by a high-performance superplasticizer, and a viscosity modifying agent may be added to restrain segregation.

Thus, the characteristics of mix proportions of SCC are somewhat near to those of high-strength concrete. Hence, high-strength concrete can be easily transformed to SCC, and vice versa.

Despite reports on self-levelling concrete that appeared in Europe in the 1970s, the concept of "self-compacting" was first systematized by Prof. Okamura [17] in 1988 and he named it "high-performance concrete." This SCC/HPC was initially conceived with the aim of assuring the durability of structures by eliminating consolidation defects. However, its major purposes have shifted toward the simplification of placing and improvement of efficiency. The effects of self-compaction, however, are not limited to these advantages. It enables concrete to be placed into structures that would otherwise be impossible to be constructed with concrete, and to have dramatic improvements in construction efficiency of extra large scale structures.

15.3.4 Microsilica Concrete [18]

15.3.4.1 Definition

MS concrete means the concrete containing microsilica. There are some commercially available MS concrete in the United States. The microstructure of MS concrete has the following characteristics.

15.3.4.2 Microstructure

15.3.4.2.1 Calcium Hydroxide

In MS concretes the range of silica addition is 5 to 15%, which is not enough to completely eliminate CH since theoretically an amount in the range of 15 to 20% of weight of cement is most probably needed for theoretical removal of all CH (depending on the composition of cement).

Reduction in the amount of CH has been documented by some research. Furthermore, large crystals of CH will form early in hydration (1 to 2 days) before the pozzolanic reaction is fully underway (3 to 10 days). Once formed, CH crystals will persist, particularly when moist curing begins.

15.3.4.2.2 Composition of C–S–H

Studies have shown that the composition of C–S–H formed in the presence of reactive pozzolans is slightly different from that formed during normal hydration. In particular, the C/S ratio is reduced from −1.7 to −1.4, but the silicate structure is not significantly different.

It is likely that both types, "normal" (C/S = 1.7) and "pozzolanic" (C/S = 1.4), coexist, giving rise to a mean value of around 1.5.

15.3.4.2.3 Pore Size Distribution

Marked changes in pore size distribution has been shown by using mercury intrusion porosimetry. The addition of MS at quite high water/powder ratios virtually eliminates the macroporosity (>100 nm in diameter) that dominates permeability, after only 7 days of moist curing. Hence, very impermeable concretes should be possibly obtained.

15.3.4.2.4 Cement–Aggregate Bond

There is now accumulated evidence that the microstructure of the interfacial zone between the cement paste and the aggregate is not so typical of the bulk paste. It is characterized by a more porous structure and oriented CH crystals and as a result is more prone to cracking, particularly under stress concentration that may occur at the interface. In the presence of MS, the interfacial zone has a denser, more uniform structure that does not appear to differ significantly from the bulk paste. The small particle size of the microstructure inhibits the development of segregated water films, which is probably the cause of the porous zone and inhibits the growth of oriented CH crystals. Similarly, the bond between the paste and other embedded materials such as fibers will also be improved.

15.3.4.3 Durability

Durability is defined as the capability of concrete to resist chemical and physical attacks from environmental factors or weather.

15.3.4.3.1 Permeability

Predictions made from considerations of pore size distribution are in fact correct. The effectiveness of MS in reducing the permeability of concrete was demonstrated a long time ago at the University of Trondheim. Typical figures are given in Table 15.7.

The data in Table 15.7 show that even with low cement contents and high w/c ratios, low permeability coefficients are attained. The effects are greatest at about 400 lb/cu. yd and fall off at higher cement contents. The efficiency factor is −10 for a 10% addition of silica, that is, the replacement of one part of cement for one part of MS is equivalent to adding ten parts of cement. In other words, the reduction in permeability with 10% of MS is approximately equivalent to doubling the cement content. This assessment is based on well-cured concretes. Inadequate curing will have a greater detrimental effect on MS concrete than on conventional concrete, because the pozzolanic reaction will be inhibited.

15.3.4.3.2 Chemical Resistance

The fine pore structure increases the chemical resistance of concrete (along with a reduction in ionic concentrations) and reduces the access of oxygen. Lowered alkalinity caused by the pozzolanic reaction is not sufficient to destroy the passive layer. Diamond has measured a pH of >12 after 145 days hydration in the presence of 30% MS.

TABLE 15.7 The Effectiveness of Microsilica in Reducing the Permeability of Concrete

Cement content (lb/yd^3)	WDRA^a (wt%)	Microsilica (wt%)	w/p	Permeability coefficient K (m/s)
170	0	0	2.38	120×10^{-10}
170	1	0	2.09	1160×10^{-10}
170	0	10	2.32	10×10^{-10}
170	1	10	2.10	4×10^{-10}
170	2	20	2.02	0.6×10^{-10}
420	0	0	0.89	0.5×10^{-10}
420	1	0	0.81	620×10^{-15}
420	0	10	0.97	95×10^{-15}
420	1	10	0.82	18×10^{-15}
420	2	20	0.79	21×10^{-15}
675	0	0	0.52	7×10^{-15}
675	0	10	0.56	136×10^{-15}
675	1	10	0.47	40×10^{-15}
675	1	10	0.44	8×10^{-15}
845	0	0	0.43	14×10^{-15}
845	1	0	0.49	41×10^{-15}

^{a} WDRA — Lignosulfonate-based water reducing agent.

If properly cured MS concrete is no more susceptible to carbonation than conventional concrete of similar w/c ratio, carbonation could destroy the passive layer only. MS concrete has been exposed to salt spray in Kristiansand, Norway, for over 10 years without any outward sign of corrosion.

When chlorides are added to MS concrete, the chloride ion does not appear to be bound by the hydration products to the same extent as in conventional concretes. Thus, depassivation can occur at lower dosages of added chlorides. However, the ingress of external chloride ion is significantly reduced by at least a factor of 10 ($\sim$1 × 10 cm/s).

The rate of corrosion in the presence of equal amounts of chloride ion is the same regardless of the MS content of the concrete; that is, there is no negative influence of MS to nullify or outweigh the beneficial effects of a refined pore structure. Reduced rates of corrosion have also been reported for DSP materials.

Some years ago, it was reported by Greening and Bhatty that C–S–H made with lower C/S ratios were capable of taking up greater quantities of alkali ions. It was suggested that this might be a major reason for the success of pozzolans in controlling the alkaline–aggregate reaction. The effectiveness of MS in controlling the alkali–aggregate reaction is now well established in several independent studies. In Iceland, 7.5% MS is routinely added to all cement for this purpose. Recently, Diamond et al. have shown that the addition of MS does reduce the concentration of alkali (K^+ and OH^-) in the pore solution. The low permeability and reduction in CH content should also improve resistance to attack sulfates and acids. It has been demonstrated that MS performs better than other pozzolans at equivalent additions. Field trials under severe conditions in Oslo, Norway, have shown good performance over a 20-year period.

15.3.5 Densified with Small Particles (DSP) Concrete [18]

15.3.5.1 Definition

It would be reasonable to expect that the behavior of DSP materials can be estimated from MS concrete properties. MS concrete and DSP materials are quite similar in terms of raw material composition and the microstructure, except that DSP materials usually have much lower w/c ratio ($\sim$0.20) and a higher MS content (>20% of the weight of cement).

15.3.5.2 Microstructure

Very low w/p values are attained by the particle packing attributes of MS, whereby very fine silica particles pack between the much coarser cement grains. By using large quantities of a dispersing agent (usually a superplasticizer), a low water demand is achieved. The particle packing can be further optimized to get a more ideal packing.

A microsilica/cement ratio of 0.45 is needed for a complete reaction. Initially CH is formed as a reaction product before the pozzolanic reaction has advanced but is reduced to very low levels. Large massive crystals of CH are not observed, the microstructure is very uniform in contrast to regular cement paste. Only about 50% of the cement is able to hydrate because of the low w/c used. The pore structure is also quite different. Mercury intrusion porosimetry curves shown in Figure 15.7 indicate a very low pore volume accessible to mercury. This is a good reason to believe that no capillary pores are present in DSP.

15.3.5.3 Strength

For DSP, strength development is very rapid after a delayed setting time of 15 to 20 h, so in 2 days a strength of 100 MPa is obtained. However, strengths continue to increase with extended moist curing. The compressive strength of 345 MPa is the highest reported value for DSP. This may be due to the elimination of macroporosity during the hot curing cycle. Using a convoluted stainless steel aggregate contributes to the strength of DSP.

It is important that a complete dispersion of the MS can be achieved. The use of a chemical dispersing agent is not sufficient if low shear mixing is used. High shear mixes will allow compressive strengths to be significantly increased. It is possible to achieve a paste strength of over 200 MPa by optimizing processing.

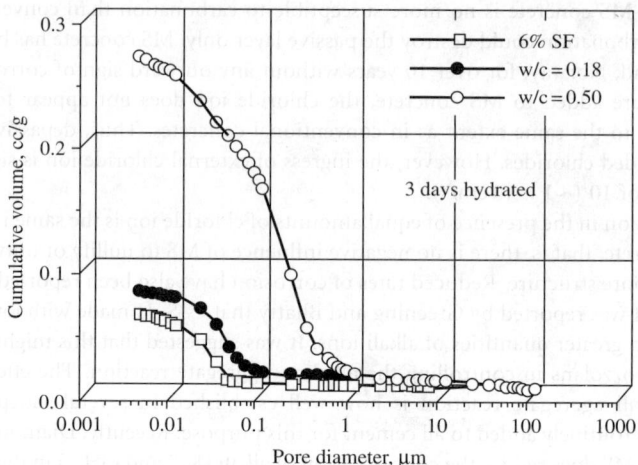

FIGURE 15.7   Pore size distribution of DSP cement paste compared to conventional cement paste.

15.3.5.4 Brittle Behavior

DSP materials are extremely brittle. The ratio of compressive to flexural strength is in the range of 10–12:1 for paste and 7–12:1 for mortar. Thus, reinforcement will be needed to improve ductility.

15.3.5.5 Durability

15.3.5.5.1 Abrasion Resistance

The abrasion resistance of DSP is excellent. A screw feeder for fly ash made from DSP has reported a lifetime of 1250 h compared with 250 h of one made of steel with a wear-resistant carbide coating.

15.3.5.5.2 Permeability

The permeability of DSP would be expected to be extremely low because of both the low w/p and the high MS content. Some DSP products (e.g., DASH 47) have an air permeability even lower than that of cast aluminum. Another DSP product, Densit formulations, is found to be a good corrosion resistant material.

15.3.5.5.3 Temperature Resistance

DSP has excellent ability to withstand temperature cycling without loss of its properties. Hence, its applications include providing a more stable microstructure. This suggests that specialized curing regimes will be necessary for optimum performance. However, 200 to 250°C is the upper limit for this system. As a solution to overcome this limit, polymer impregnation with a high-temperature polymer (Novolac epoxy resin) has been used. It is verified by experiment that vacuum impregnation can significantly improve flexural strength.

15.3.6 Fiber-Reinforced Concrete

For making HPC, it is essential to select proper raw materials, evaluate their properties, and have knowledge of the interaction of different materials for optimum usage.

15.3.6.1 Definition

Fiber-reinforced cementitious composite is a concrete that has fibers in the matrix.

15.3.6.2 Property Improvements Compared with Plain Concrete

Quasibrittle materials such as concrete have two major deficiencies, that is, a rather low tensile strength and a rather low energy consumption capacity or toughness. One way to overcome these weaknesses is to incorporate high-strength small-diameter fibers. The effect of the incorporation of the fiber can be largely attributed to the contribution of the bond between the fiber and matrix. The role of the interface can be viewed in two different aspects. First, in conventional applications of fiber reinforced cementitious composites, which use a low volume fraction of fibers (e.g., 0.5 to 1.5%, 25 mm long steel fibers with a diameter of 0.25 mm), the role of fibers is apparent only after the cracks have formed in the material. Although there is still only one major crack formed and the overall behavior of composites is still characterized by strain softening after the peak load like that of a plain matrix behavior, the incorporation of fibers leads to a significant increase in the total energy consumption and overall toughness of the composites. In such cases, assuming there is no fiber fracture, the fiber debonding and pull-out process plays an important role. Second, with an increase in the volume concentration of the fibers, it is possible that microcracks that are formed in the matrix are stabilized due to the interaction with fibers. If the cement-based matrix is incorporated with a relatively large volume (ranging up to 15%) of small diameter (in the micrometer range) steel, glass, or synthetic fibers, the fiber/matrix bond can lead to strain hardening and multiple cracking behavior. As a result, not only the composites' toughness, but also the matrix tensile strength can be significantly improved. One of the mechanisms in slowing down the growth of a transverse crack in unidirectional fiber composites can be attributed to the development of longitudinal cylindrical shear microcracks located at the boundary between fiber and bulk matrix, allowing the fibers to debond while transferring the force across the faces of the main crack.

The enhancements achieved by adding fibers are listed in Table 15.8.

Here, it is important to recognize that, in general, fiber reinforcement is not a substitute for conventional reinforcement. Fibers and rebars have different roles to play in modern concrete technology, and there are many applications in which both are used.

15.3.6.3 Factors influencing properties

Fiber-reinforced cement-based composites have been developed to improve the toughness and the tensile properties of concrete. The following parameters can influence the fiber reinforced composite properties.

15.3.6.3.1 Fiber Type

The fibers that are commonly used in the fiber-reinforced cement-based composites are asbestos, carbon, glass (borosilicate and alkali resistant), polymers (acrylic, aramid, nylon, polyester, polyethylene, polypropylene, and polyvinyl alcohol), natural materials (wood cellulose, sisal, coir or coconut, bamboo, jute, akwara, and elephant grass), and steel (high tensile and stainless). Some commonly used fibers are shown in Figure 15.8.

In order to get multiple cracking and strain hardening response, two fundamental requirements have to be satisfied. First, the fibers should be strong enough to carry the total load at the position of the first matrix transverse crack. Second, the bond in the fiber–cement interface should be strong enough to transfer the forces from the fiber to the matrix and thus build up the tensile stress in the matrix. According to previous studies, PVA fiber shows a very promising potential in improving the interfacial bond and in achieving multiple cracking.

Different fibers will have different aspect ratios, which is defined as the ratio of fiber length to the fiber diameter. Aspect ratio plays an important role in the bonding between the fiber and the matrix, usually, the larger the aspect ratio, the better the bond.

TABLE 15.8   The Mechanical Enhancing Effect Due to the Addition of Fiber

Compressive strength	Almost no effect for low V_f; small for high V_f
Tensile strength	Can be improved to 50% and above
Impact and flexural strength	Can be improved to 80% and above
Toughness (energy absorption)	Can be improved to 200% and above

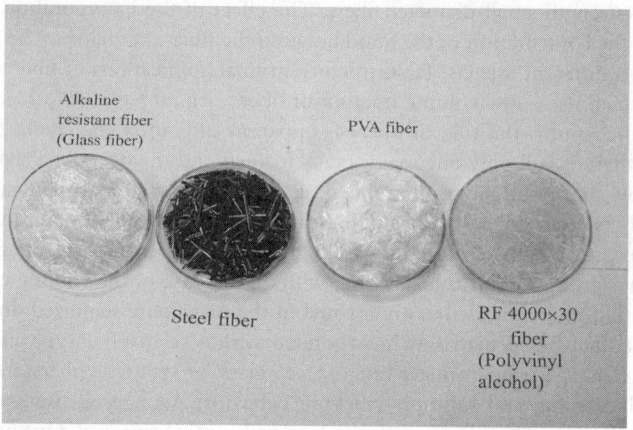

FIGURE 15.8   Different types of fiber.

15.3.6.3.2 Fiber Volume Ratio

Another important factor that greatly influences FRC properties is the fiber fraction ratio. At low fiber volume ratio, the addition of fiber mainly contributes to the energy consuming property. At higher fiber volume ratio, the tensile strength of the matrix can be enhanced and the failure mode can be changed.

According to the rule of mixture, a composite reaches its ultimate strength when the matrix fails at its cracking strength

$$\sigma_{cu} = V_m\sigma_{mu} + V_f\sigma_f = V_m E_m \varepsilon_{mu} + V_f E_f \varepsilon_{mu} = V_m E_m \varepsilon_{mu} + V_f \frac{E_f}{E_m}\sigma_{mu}$$

where σ_{cu} is the ultimate stress in the reinforced concrete, σ_{mu} is the ultimate stress in the matrix, σ_f is the stress in the fiber, E_m is Young's modulus, ε_{mu} is the ultimate strain, V_m is the matrix volume ratio, and V_f is the fiber volume ratio.

This formula is used for a low fiber volume fraction case. For a high fiber volume fraction, the composites ultimate strength is essentially the fiber of σ_{fu}. Hence

$$\sigma_{cu} = V_f\sigma_{fu}$$

where σ_f is the ultimate stress in the fiber.

The intersection of the above two equations separates the single fracture and multiple cracking. Thus, by equating the above two equations, the minimum fiber volume fraction for multiple cracking can be obtained as

$$V_f^{minimum} = \frac{\sigma_{mu}}{\sigma_{fu} + (1 - (E_f/E_m))\sigma_{mu}}$$

It can be seen from the above equation that the critical fiber volume fraction only depends on the fiber and the matrix ultimate strength, in accordance with the theory applied here.

15.3.6.3.3 Matrix Variation

The purpose of changing the matrix composition is to increase the bond properties and improve the matrix toughness.

15.3.6.3.4 Processing Methods

The commonly used processing methods in industries include hatschek process, reticem process, continuous surface reinforced process, wellcrete process, normal casting, pultrution process, and extrusion process. The extrusion technology is an economical mass-production method. The advantage of introducing extrusion into processing of cement products is that the materials are formed under high shear and high compressive forces. With properly designed dyes and properly controlled material mixes and viscosity, the fibers can be aligned in a load-bearing direction. The matrix and fiber packings can be densified to achieve a low porosity, and the interface bond between the fiber and matrix can thus be improved. Owing to these factors, it is expected that the extruded specimens will achieve a better performance than the cast specimens provided other conditions are similar.

15.3.6.4 Mechanical Properties

15.3.6.4.1 Tension

Usually, there are two categories of FRCs according to their tensile response, strain softening, and strain hardening. The strain softening type of fiber-reinforced cement-based composites is usually reinforced with a low volume of short fibers. This kind of composites containing about 1% fiber is typically used for bulk field applications involving a massive volume of concrete. Figure 15.9 illustrates the tensile response of such low volume composites. It shows stress deformation curves for a plain concrete specimen, steel, and a polypropylene fiber reinforced concrete specimen. The length, diameter, and volume fraction of steel fiber are 48 mm, 0.5 mm and 0.5%, respectively, whereas the length and volume fraction of fibrillated polypropylene fiber are 50 mm and 0.5%, respectively. It can be seen from the figure that although all the specimens display a softening behavior after peak load, the area under the load deformation curve is different and it is larger for fiber-reinforced specimens than that for plain concrete specimens. This implies that the fiber-reinforced specimens require more energy to fracture than the plain concrete specimens.

For a strain softening type of failure, the fracture process of the material can be divided into three stages, namely, randomly distributed damage during the loading stage to about 80% of the prepeak load, microcrack localization during the loading period between 80% of prepeak load and 80% of postpeak load, and the major crack propagation during the period after 80% of postpeak load.

A failure mode of multiple matrix cracking can be obtained when the quasibrittle matrix is reinforced with either a high volume fraction of short fibers or with aligned continuous fibers. Such high volume fiber composites have been used in thin sheets (e.g., curtain walls). A typical strain-hardening curve for steel-fiber-reinforced cementitious composites is shown in Figure 15.10. There is a special point (marked B in the curve) called the "bend over point" (BOP) at which the matrix's contribution to the tension capacity reaches a maximum. The stress–strain curve shown in the figure can be roughly divided into four stages:

- Stage 1 is characterized by the elastic behavior of the composite before the initiation of the first matrix cracks. In this stage, a few microcracks are initiated and are randomly dispersed.
- Stage 2 is defined as the period from the initiation to the complete propagation of the first major crack across the specimen's cross-section. The BOP corresponds to the end of stage 2.
- Stage 3 can be called multiple cracking stage. In this stage, the incremental loading of the fibers (at the position of the crack) is transferred back to the matrix through the bond and this process builds up the tensile stresses in the matrix. This can lead to the next matrix crack when the stress

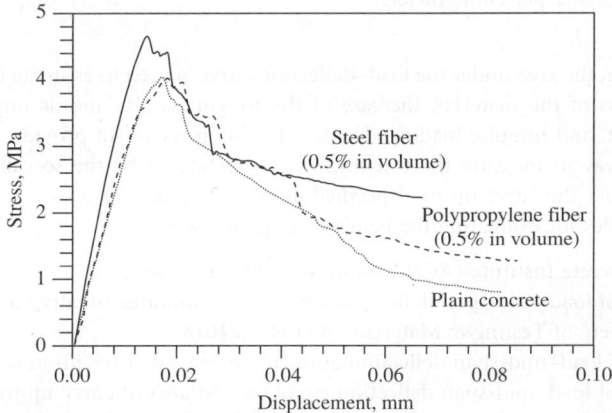

FIGURE 15.9 Stress deformation curves for three different specimens.

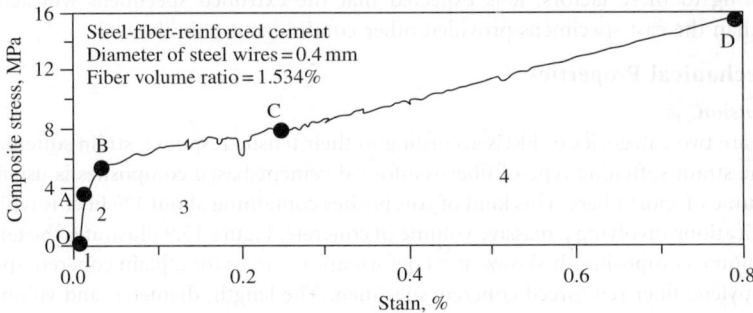

FIGURE 15.10   Strain hardening phenomenon of fiber-reinforced concrete.

    reaches its tensile strength. The multiple cracking process eventually leads to homogenization of matrix cracks, that is, the end of multiple cracking at which the minimum crack spacing, determined by the interfacial bond properties, is reached.

- In stage 4, no further matrix crack is expected and the additional load is only sustained by the fibers.

To study the multiple cracking and associated interfacial bond behavior, experiments have been conducted on the cementitious composites reinforced with continuous glass, polypropylene, and steel fibers. Three recently developed techniques, laser holographic interferometry, quantitative optical microscopy, and moire interferometry, are being employed in the investigation.

Recent studies have shown that the response of FRCs also depends on the methods of processing. Along this direction, successful applications of pultrusion and extrusion techniques in manufacturing cement composite products are two good examples. Pultrusion is employed to incorporate continuous fibres into the cement matrix. Composites with fiber volume ratio of more than 10% could be produced. Strain hardening type of response is the direct result of a tremendous enhancement in tensile strength. Extrusion can also lead to a strain hardening type of response. It is obvious that special processing compacts matrix with fibers to a low porosity, controls fiber direction and distribution, and improves the interfacial bond between fiber and matrix, which in turn leads to a class of high-performance FRCs with strain hardening.

So far, the application of FRCs in structural components are used for inhibiting cracking, for improving resistance to impact or dynamic loading, and for resisting material disintegration such as airport runway and pavement. This is partly due to high cost. For example, for SFRC, 1% by volume will increase the cost by US$52 per cubic meter.

15.3.6.4.2 Bending

In the case of bending, the area under the load–deflection curve is used to estimate the energy-absorbing capacity or toughness of the material. Increase of the toughness also means improved performance under fatigue, impact, and impulse loading. The toughening mechanism provides ductility.

Another popular way to measure the toughness is characterized by the so-called toughness index, which is the area under the curve up to a specified deflection value.

Three common codes for evaluating the bending toughness are

- American Concrete Institute (ACI) Committee 544 Toughness:
  Index area of load–mid-span deflection curve to 1.9 mm/area of curve up to first cracking.
- American Society of Testing & Materials (ASTM) C 1018:
  I5 = Area of load–mid-span deflection curve to 3δ/area of curve up to 8 (first cracking).
  I10 = Area of load–mid-span deflection curve to 5.5δ/area of curve up to 8 (first cracking).
  I20 = Area of load–mid-span deflection curve to 15.5δ/area of curve up to 8 (first cracking).
- Japan Toughness index = Area of load–mid-span deflection curve to 1/150 span.

15.4 Quality Control of High-Performance Concrete

The production of HPC is quite different from that of normal concrete. The fundamental difference is the extreme care needed for HPC in selection of the aggregates and the knowhow needed in the proportioning of the mixture, trial batching, testing, and site delivery. As mentioned previously, HPC has a considerably smaller margin of error than that of normal concrete. A small variation from specified requirements can result in major deficiencies in quality or test results. Thus, quality control for HPC is more critical.

15.4.1 Materials Quality Control

For making HPC, it is essential to select proper raw materials, evaluate their properties, and have knowledge of the interaction of different materials for optimum usage.

15.4.1.1 Cement

The choice of Portland cement for HPC is extremely important. Different brands of cements will have different strength development characteristics because of the variations in chemical composition and fineness that are permitted by ASTM C 150.

Initially, silo test certificates should be obtained from potential suppliers for the previous 6 to 12 months. Not only will this give an indication of strength characteristics from the ASTM C 109 mortar cube test, but also, more importantly, it will provide an indication of cement uniformity. The cement supplier should be required to report uniformity in accordance with ASTM C 917. If the tricalcium silicate content varies by more than 4%, the ignition loss by more than 0.5%, or the fineness by more than $375 \, \text{cm}^2/\text{g}$ (Blaine), then problems in maintaining a uniform high strength may result.

Special concern should be given to the contents of C_3A and SO_3 in the cement, because slump loss in concrete is usually associated with the formation of ettringite caused by the two components. When early setting and hardening properties are desired, it is preferable to use more finely ground Type II Portland cement than Type I and III cements, which generally contain more than 8% C_3A. On the other hand, cements with a very low or negligible C_3A content, such as ASTM Type V, are generally high in C_4AF and their setting and hardening rates may be unusually slower than desired. Portland pozzolan (ASTM Type IP) and portland-slag (ASTM Type IS) cements should also be seriously considered for use in HPC mixtures because of certain desirable physical–chemical effects associated with fine particles of a pozzolan or slag. The same effects can generally be achieved by using a high-quality pozzolan or slag as a mineral admixture in Portland cement concrete mixtures.

In addition, alkali content should also be strictly controlled below the limitation level of ASTM standard specifications in order to prevent the potential alkali–aggregate reaction.

A further consideration is the optimization of the cement–admixture system. The exact effect of a water-reducing agent on water requirement will depend on the cement characteristics. Mechanical and durability performance developments will depend on both cement characteristics and cement content.

15.4.1.2 Chemical Admixtures

Selection of type, brand, and dosage of chemical admixtures should be based on performance relative to other materials being considered or selected for use in the project. Significant increases in compressive strength, control of rate of hardening, accelerated strength gain, unproved workability, and durability are contributions that can be expected from the chemical admixtures. Reliable performance on previous work should be considered during the selection process.

Superplasticizer plays the key role in the development of high performance for HPC. Specifications for superplasticizer are covered in ASTM Types F and G. Matching the admixture to the cement, both in type and dosage, is important. The slump loss characteristics of a superplasticizer will determine whether it should be added at the plant, at the site, or a combination of both.

Use of a superplasticizer in HPC may result in different effects according to specific conditions: to increase strength by keeping the same slump with reference concrete by reducing the w/b ratio or to increase slump by keeping the same w/b ratio with reference concrete. The method of addition should distribute the admixture throughout the concrete. Adequate mixing is critical to uniform performance. Supervision is important for the successful use of a superplasticizer. The use of superplasticizers is discussed further in ACI SP-68.

Combinations of superplasticizer with normal-setting water reducers or retarders have become common to achieve optimum performance at low cost. Improvements in strength gain and control of setting times and workability are possible with optimized combinations.

When using a combination of admixtures, they should be dispensed individually in a manner approved by the manufacturer. Air-entraining admixtures should, if used, be dispensed separately from water-reducing admixtures.

15.4.1.3 Mineral Admixtures

Mineral admixtures commonly used for HPC consist mainly of fly ash, silica fume, and slag.

15.4.1.3.1 Fly Ash

Fly ash for HPC is classified into two classes. Class F Gy ash is normally produced from burning anthracite or bituminous coal and has pozzolanic properties, but little or no cementitious properties. Class C fly ash is normally produced from burning lignite or subbituminous coal, and in addition to having pozzolanic properties, has some autogenous cementitious properties.

Methods for sampling and testing can be found in ASTM C 311. Although this specification permits a higher loss on ignition, an ignition loss of 3% or less is desirable.

Variations in physical or chemical properties of mineral admixtures, although within the tolerances of these specifications, may cause appreciable variations in properties of HPC. Such variations can be minimized by appropriate testing of shipments and increasing the frequency of sampling. ACI 212.2R provides guidelines for the use of admixtures in concrete. It is extremely important that mineral admixtures be tested for acceptance and uniformity and carefully investigated for strength-producing properties and compatibility with the other materials in the HPC mixture before they are used in the work.

15.4.1.3.2 Silica Fume

Silica fume has been widely used in HPC for structural purposes and for surface application, and as repair materials in situations where abrasion resistance and low permeability are advantageous.

When silica fume is used, the following specifications are to be considered carefully.

SO_3 content should be controlled below 4.0%. SiO_2 content directly determines the reactive ability of silica fume, so its content should be controlled over 70% for ensuring the development of high performance of HPC with silica fume added. The loss of ignite of silica fume is closely related to the water demand for HPC, thus the less the content, the better. According to ATSTM C 618, loss of ignite must be no more than 10.0%. In addition, pozzolanic activity indicator of silica fume has a great impact on its reactive properties. At least 75% should be ensured.

The use of silica fume to produce HPC increased dramatically in the 1990s. Both laboratory and field experiences indicate that concrete incorporating silica fume has an increased tendency to develop plastic shrinkage cracks. Thus, it is necessary to quickly cover the surfaces of freshly placed silica fume concrete to prevent rapid water evaporation.

15.4.1.3.3 Slag

Quality control for slag is usually limited to Blaine specific surface area tests and x-ray diffraction studies to check on the degree of crystallinity, which should be low. Detailed specifications for ground granulated blast-furnace slag are given in ASTM C 989.

15.4.1.4 Aggregates

Both fine and coarse aggregates used for making HPC should, as a minimum criterion, meet the requirements of ASTM C 33; however, the following exceptions should also be taken care of and controlled.

15.4.1.4.1 Coarse Aggregate

Some important properties like crushing strength, the maximum size, flake, and elongation indices should be carefully controlled while selecting the aggregate for HPC. As the tensile strength of concrete is of prime importance, it is necessary to reduce the flaky and elongated particles to reduce the weaker zones in the concrete. This is achieved by screening the aggregate through specially designed sieves.

Strict control of coarse aggregate strength is important for HPC. HPC possesses good interfacial bonds between the coarse aggregate and the cement matrix, resulting in the fracture crack usually passing through the aggregate, rather than the interfacial zone under the loading. The aggregate strength typically determines the final strength of HPC. Thus, when making HPC, the aggregate compressive strength should not be less than the required compressive strength of HPC.

Many studies have shown that crushed stone was preferable in HPC than rounded gravel. The most likely reason for this is the greater mechanical bond that can develop with angular particles. The ideal aggregate should be clean, cubical, angular, 100% crushed aggregate with a minimum of flat and elongated particles.

Equidimensional particles from crushing of either dense limestones or igneous rocks, of plutonic type (viz., granite, syenite, diorite, gabbro, diabase) are generally satisfactory as coarse aggregate.

15.4.1.4.2 Fine Aggregate

For fine aggregates, a rounded particle shape and smooth texture have been found to require less mixing water in concrete and for this reason are preferable in HPC. The optimum gradation of fine aggregate for HPC is determined more by its effect on water requirement than on physical packing. A sand with a fineness modulus (FM) below 2.5 gives the concrete a sticky consistency, making it difficult to compact. Sand with an FM of about 3.0 gives the best workability and compressive strength. Thus, it is helpful to increase the fineness modulus. The amounts passing the No. 50 and 100 sieves should be kept low, but still within the requirements of ASTM C 33, and mica or clay contaminants should be avoided.

15.4.2 Production Procedures

The procedures for producing HPC involve the following steps.

15.4.2.1 Mixing

The effect of mixing is to blend all the constituent materials into a homogeneous concrete mixture. This is affected greatly by the sequence used to load the materials into the mixer and the efficiency of the mixer to blend the materials. It should be noted, however, that total automation of concrete production and delivery always improves consistency as a measure of workability in placement. Admixtures seem to be most effective if they are introduced after the cement is wetted. In the case of truck mixing, the temperature of the concrete as delivered should not exceed 42°C. If water demand becomes excessive or slump loss rapid, the dosage of chemical retarders or superplasticizer has to be increased. If the problem persists, such as in hot summers and in hot-weather concreting, ice or chilled aggregates have to be used to lower the temperature.

15.4.2.2 Transportation

Transportation of the concrete from the mixer to the site has to be accomplished without any significant change in the slump, w/b ratio, air content, consistency, and temperature. Quality control and quality assurance personnel have to be cognizant of prolonged mixing causes slump loss, and hence lower workability. Consequently, adequate job control must be established so as to prevent delays in delivery. Where practical, withholding some of the superplasticizer until the truck arrives at the site is desirable, provided that proper mixing is exercised. Particularly in the case of high-strength concrete, all mixer trucks should be inspected and certified regularly to comply with NRMCA inspection requirements. In addition, it is vital that water used to wash mixers between loads should be discharged prior to batching for subsequent loads. In lieu of that, provisions can be made to measure and account for the remaining wash water batched for the next load.

15.4.2.3 Placement

Care has to be exercised to ensure that there is no segregation of the coarse aggregate. This is particularly essential in the case of high flowable HPC, since it is typically produced with slumps in excess of 200 mm. Despite the fluid appearance of high flowable HPC, the mixture requires complete consolidation that should be achieved quickly. Vibratory equipment is recommended, with at least one standard vibrator for every three vibrators. Additionally, the ACI 309 guidelines should be used for proper consolidation.

15.4.2.4 Finishing

The finishing operations have to make sure that the concrete surface is produced with minimum manipulation. Overworking of the surface can result in a reduction of the surface air content; hence, susceptibility to deicing freezing and thawing damage. The concrete should not be troweled while water is present on the surface. Neither should water be applied to aid in the finishing process. Doing so would increase the water/cementitious material ratio, and considerably reduce the compressive strength, durability, and long-term performance of the top layer.

15.4.3 Early Age Curing

Early age curing is a key step to achieve high performance for HPC. If early age curing is not carried out, even the best HPC that is made with excellent materials and proper compaction will not perform well. While this statement is true for all concretes, it is particularly important for HPC; ordinary concretes are rather more forgiving to the usual inadequate curing. This is because HPC does not usually exhibit much bleeding. Without protection from loss of surface moisture, plastic shrinkage cracks may form on the exposed HPC surface.

A great deal of research on the best curing techniques for HPC has been carried out at the Université de Sherbrooke. This work is based on the premise that curing of HPC is carried out not only to maximize the amount of cement hydration, but also to minimize the amount of shrinkage. The basic conclusion of these studies is that, especially for HPC, adequate curing must begin as soon as the concrete is placed, and must be continued for as long as possible. In practice, this means that while the concrete is still in the plastic state (before initial set), when plastic shrinkage might occur, it should either be fog misted or covered with a curing membrane, to prevent water from evaporating. Once the concrete has set, when self-desiccation might occur (leading to cracking), the HPC should either be fog misted or water cured, to prevent the formation of menisci in the capillary pores. Finally, the concrete should then be coated with an impervious film, as it dries, to prevent further desiccation.

Unfortunately, the concrete construction industry is not accustomed to curing even ordinary concretes as well as they should. The steps outlined above will inevitably add some construction complexity and some costs to a project. However, there is little point in going through the trouble of designing and placing HPC mixes if their properties, in terms of both strength and durability, are ultimately going to be degraded by inadequate curing.

15.4.4 Early Age Property Monitoring

Compared to normal concrete, early age property develops very rapidly for HPC. A great amount of hydration products are produced in a relatively short period. After that only a little change can be observed. Thus, monitoring of early age property is very important to understand the setting and hardening processes of HPC, which has a great impact on the final workability, mechanical properties, and durability.

Development of early age properties and strength gain in HPC are intimately related to the hydration of Portland cement. At present, some methods and techniques such as ultrasonic through-transmission measurements [19–22], microwaves monitoring methods [23,24], impact-echo method [25], ultrasonic reflection [26,27], and electrodeless resistivity measurements [28] have been developed for following the property changes of Portland cement at an early stage.

Ultrasonic reflection and electrodatless resistivity measurements are two most promising methods amongst these techniques for *in situ* monitoring early age properties of HPC. A detailed introduction to the two methods will be given below.

15.4.4.1 Ultrasonic Reflection Method

An ultrasonic reflection technique that continuously monitors the setting and hardening of concrete has been recently been developed by Kolluru V. Subramaniam, J.P. Mohsen, C.K. Shaw, and S.P. Shah from Northwest University, USA. The experimental procedure is based on high-frequency ultrasonic measurements and consists of monitoring the wave reflection factor (WRF) at the interface between a steel surface and concrete. The basic principle, experimental setup, and procedure are described as follows.

The ultrasonic reflection method requires access to only one face or side of the structure. Freshly mixed HPC is placed in a mold. Transducers that excite and receive ultrasonic sound wave pulses are attached to the outside of a steel plate that is placed in the HPC. The top surface of the steel plate is flush with the surface of concrete. The reflection of ultrasonic waves at the steel–concrete interface is recorded during the setting and hardening of HPC. A schematic diagram of the test setup is shown in Figure 15.11.

The experimental procedure consists of collecting the first and second reflected echoes from the steel–air interface prior to placing the concrete. The first and second reflections are then collected at regular intervals of time at the steel–concrete interface as shown in Figure 15.11b. These time domain signals captured during the setting process are collected and digitized. Then, the collected time domain signals are transformed into the frequency domain using the FFT subroutine, as shown in Figure 15.11c.

When concrete is placed in the mold, it is in a plastic state that resembles a fluid. According to wave mechanics, a shear wave traveling through metal that is incident upon a steel–fresh concrete interface is entirely reflected. Thus, at early ages, most of the wave energy is reflected and the amplitude of the received wave is almost of the same magnitude as the input signal. As the concrete stiffens, more of the wave energy is transmitted through the concrete and less is reflected at the interface. The process of wave reflection can be quantified using the ultrasonic reflection measurement that defines the ratio of the amount of incident wave energy that is reflected from an interface between two materials. The ultrasonic reflection method as a function of time, determined under controlled laboratory conditions, is shown in Figure 15.12.

Many laboratory- and field-scale experiments have been conducted by using ultrasonic reflection measurements. The results show that the observed trends in the ultrasonic reflection measurements are very similar to those observed in the strength gain obtained from the same samples. Furthermore, the percentage change in the WRF and strength with time is very similar.

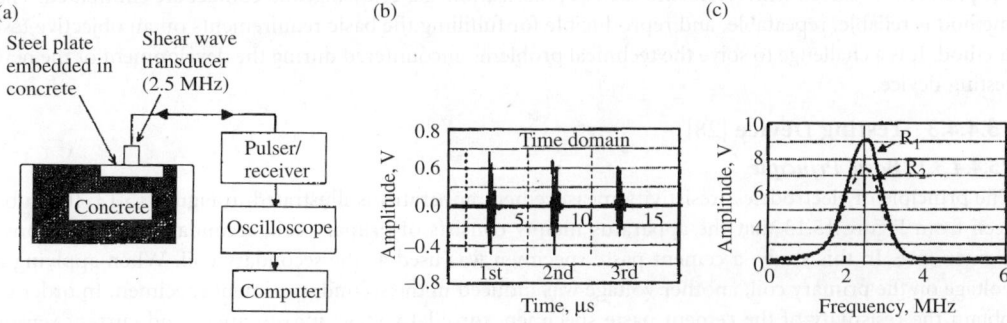

FIGURE 15.11 (a) Schematic representation of experimental setup for ultrasonic reflection measurements; (b) First, second, and third reflections of steel–concrete in time domain; and (c) First and second reflections of steel–concrete in frequency domain.

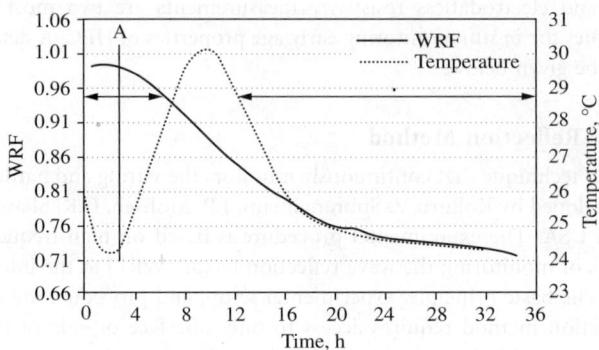

FIGURE 15.12   Typical WRF response during first 36 h after casting.

15.4.4.2 Electrodeless Resistivity Measurement

The electrical resistivity change with time is regarded as a fingerprint of the hydration process of cement-based materials. Through monitoring the electrical resistivity of fresh cement and concrete, different periods in the hydration process can be identified. Thus, the measurement of electrical resistivity can be used to guide the construction process. Moreover, it can be used as a quality index for cement-based materials at an early age.

Previous researches on the electrical property of cement-based materials can be divided into two stages. In the first stage, two electrodes that are connected to a cement-based specimen for electrical resistivity measurement. Since any application of direct current to a cement and concrete specimen produces a polarization effect, this method cannot provide accurate results for the electric resistivity of cement and concrete.

In second stage, to overcome the disadvantage of direct current, high-frequency (1000 Hz) alternating current is applied for measuring the electrical resistivity. In this case, the polarization effect is eliminated. But the alternating current cannot solve the contact problem and gas releasing effect. To solve the contact problem, sometimes the researchers use an external force to fasten the electrodes and concrete specimen from two ends [29].

For these above reasons, the electrodeless resistivity apparatus is invented and used to investigate the hydration process of cement. The present invention overcomes the problems with the prior procedure by providing an apparatus and method for the electrical resistivity of cementitious materials that does not require the use of electrodes in contact with a specimen, but instead uses a contactless transformer-based method in which a specimen serves as the secondary coil of the transformer. As there are no electrodes, the problems involved with electrodes such as polarization, gas releasing, and contact are eliminated. The method is reliable, repeatable, and reproducible for fulfilling the basic requirements on an objective test method. It is a challenge to solve the technical problems encountered during the development of the new testing device.

15.4.4.3 Testing Device [28]

15.4.4.3.1 Basic Principle

The principle of electrodeless resistivity measurement apparatus is illustrated in Figure 15.13. It can be seen from Figure 15.13 that the apparatus mainly consists of primary coil, secondary coil, and transformer core. In this study, a cement paste specimen was used as the secondary coil. When applying a voltage on the primary coil, another voltage was induced in the secondary cement specimen. In order to obtain the resistivity of the cement paste specimen, toroidal voltage measurement and current sensor were employed to monitor the inductive voltage and current of the cement specimen. It should be noted that toroidal voltage and current measurement did not touch cement specimen. As a result, the resistance R_t was calculated based on Ohm's law. It is of great concern for getting the resistivity rather

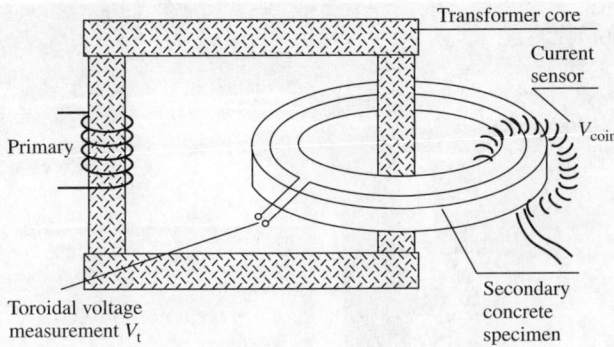

FIGURE 15.13   Schematic of resistance measurement.

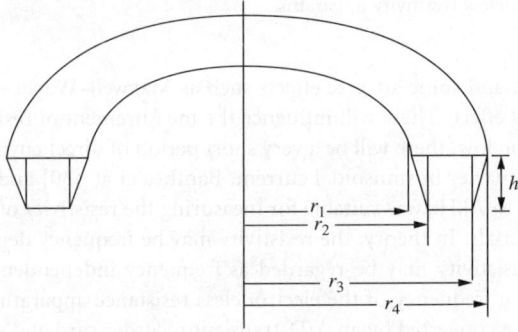

FIGURE 15.14   Cross-sectional schematic of the ladder-shaped polycarbonate container.

TABLE 15.9   The Dimensions of the Ladder-Shaped
Polycarbonate Container (Unit: cm)

r_1	r_2	r_3	r_4	h
9.9	10.4	14.1	14.6	5.4

than resistance, because resistance is related to the specimen shape and size. It is easy to obtain the resistivity of the specimen (ρ) by the following equation:

$$\rho = \frac{R_t S}{L}$$

where ρ is the electrical resistivity, R_t is the electrical resistance, S is the cross-section area of the specimen, and L is the length of specimen.

15.4.4.3.2 Containers
Fresh cement paste has to be poured and compacted in a container. Figure 15.14 shows the dimensional size of the container. In this study, a ladder-shaped polycarbonate (PC) container was designed to easily remove the harden cement paste. The dimensions of the PC container are listed in Table 15.9.

15.4.4.3.3 Frequency and Data Acquisition
When using alternating current to measure the resistivity, the determination of the frequency is of great importance for accurately obtaining the resistivity of the specimen. If the frequency is too high, it will

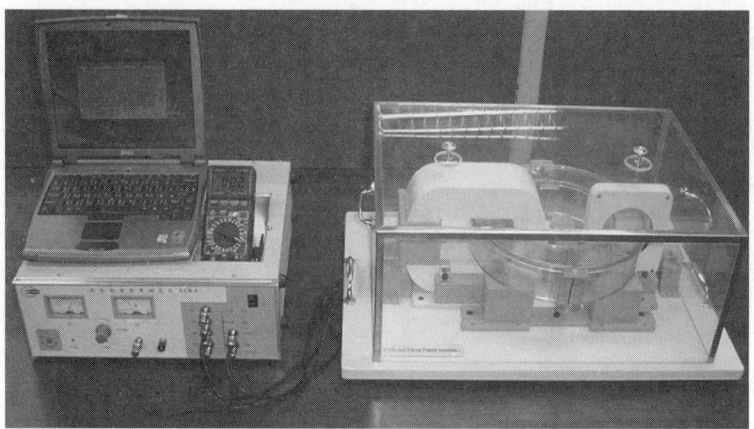

FIGURE 15.15   The electrodeless resistivity apparatus.

lead to electrical radiation and some adverse effects such as Maxwell–Wager effect, viscous conduction effect, and their combined effect. These will influence the measurement of resistivity [29]. On the other hand, if the frequence is too low, there will be a very short period of direct current at that moment of the upper quarter and down quarter in sinusoidal current. Banthea et al. [30] and Lakshminarayanan et al. [31] recommended that 1 to 7 kHz was suitable for measuring the resistivity of high-resistance materials such as cementitious materials. In theory, the resistivity may be frequency dependent, but in a range of around 1 to 2 kHz the resistivity may be regarded as frequency independent. As a result, 1 kHz was employed as the measuring frequency of the electrodeless resistance apparatus.

The signals received were converted by an A/D-transient recorder card and stored on the hard disk of a computer. The whole electrodeless resistivity apparatus is illustrated in Figure 15.15.

15.4.4.3.4 Reliability and Reproducibility

In order to check the reliability of the apparatus, the experimental results of 0.1 N KCl standard solution obtained by the electrodeless resistivity apparatus was compared with the standard resistivity value provided by the international Chemistry and Physics Society [32]. The international standard value of 0.1 N KCl solution, electrical resistivity is 78.3699 Ω cm at 24.5°C. The measured value using the electrodeless resistivity apparatus is 78.6782 Ω cm at 24.5°C. It can be seen that the experimental result is in good agreement with the international standard value. The difference is only 0.3934%. It indicates that the electrodeless resistivity measurement apparatus provides an accurate method.

Figure 15.16 plots the resistivity against time curves for three specimens with an identical formula. The similarity of the three curves showed that the measurement method had very good reproducibility.

15.4.4.4 Applicability

The electrical resistivity developments over time have been measured by using the electrodeless resistivity apparatus for hundreds of HPC mixtures. The typical electrical resistivity curves are shown in Figure 15.17 and Figure 15.18.

Based on the measured results, four periods with their own characteristics can be clearly defined as shown in Figure 15.17 and Figure 15.18 [33]. The first period is from the mixing time to $t(m)$, corresponding to the M point shown in Figure 15.18. The second period is from $t(m)$ to time $t(l)$ corresponding to the L point marked in Figure 15.18. The third period begins from $t(l)$ to $t(i)$ corresponding to the I point, and the fourth period begins from $t(i)$ onwards. Their characteristics are described in Table 15.10.

It is found that the four characteristic periods defined by the electrical resistivity curves are consistent with those obtained by other methods such as hydration heat test, ultrasonic waves. However, electrodeless resistivity measurements are much easier to operate and the apparatus is so sensitive, accurate, and

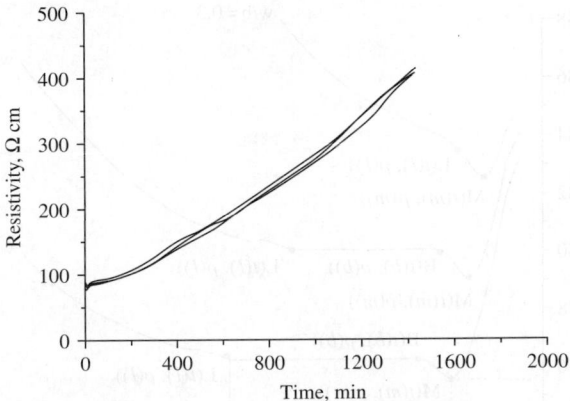

FIGURE 15.16 Reproducibility check on three batches of the same mix.

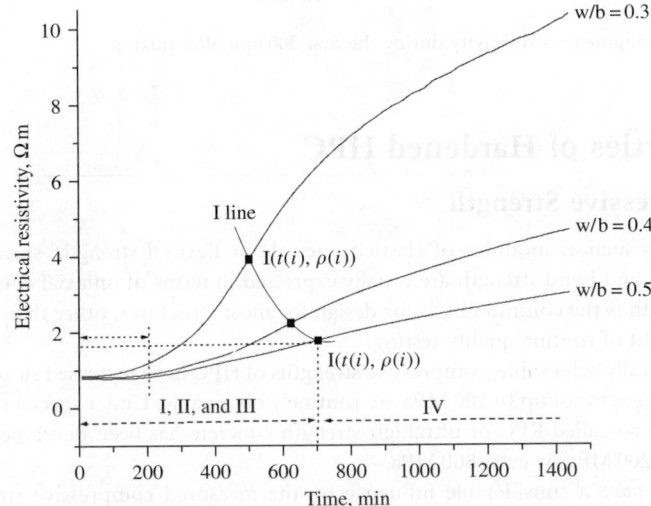

FIGURE 15.17 Development of resistivity during the first 1440 min after mixing [33].

TABLE 15.10 Characteristics of Four Periods of Hydration

Period name	Range	Characteristics
Dissolving period (I)	$t \leq t(m)$	$d\rho/dt \leq 0$, dissolution of cement in water
Competition period (II)	$t(m) < t \leq t(l)$	$d\rho/dt = 0$, competition balance between dissolution and precipitation
Setting period (III)	$t(l) < t \leq t(i)$	$0 < d\rho/dt \leq d\rho(i)/dt$, continuous formation of hydrates
Hardening period (IV)	$t > t(i)$	$0 < d\rho/dt \leq d\rho(i)/dt$, slow and steady formation of hydrates

effective that the effects of the chemical admixtures, mineral admixtures, alkaline content, and temperature on the hydration process of HPC can also be quantitatively recorded. It is because of the above reasons that great attention has been placed on electrodeless resistivity apparatus across the world. Some papers and patents have also been published and the corresponding apparatus commercially available.

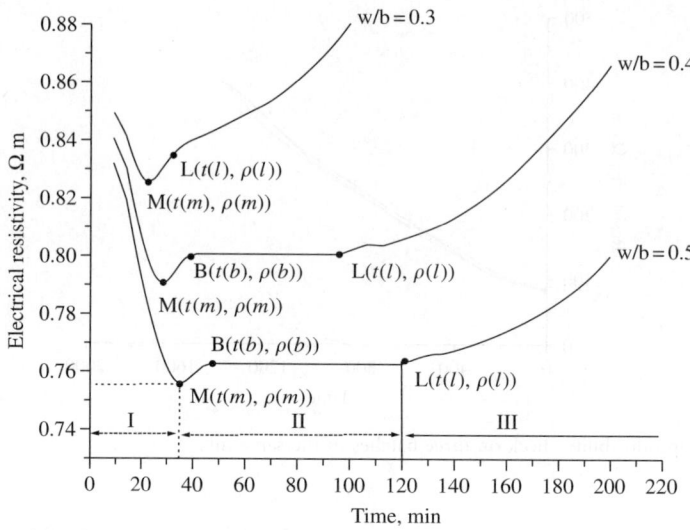

FIGURE 15.18   Development of resistivity during the first 200 min after mixing.

15.5 Properties of Hardened HPC

15.5.1 Compressive Strength

Concrete properties such as modulus of elasticity, tensile or flexural strength, shear strength, stress–strain relationships, and bond strength are usually expressed in terms of uniaxial compressive strength. Compressive strength is the common basis for design for most structures, other than pavements, and is the common method of routine quality testing.

Maximum, practically achievable, compressive strengths of HPC have increased steadily over the years. Presently, 28-day strengths of up to 100 MPa are routinely obtainable. Under special treatment and with a special formula, a so-called RPC or ultrahigh-strength concrete has been developed. Its compressive strength can reach 200 MPa or even 800 MPa.

Testing variables have a considerable influence on the measured compressive strength. The major testing variables are: mold type, specimen size, end conditions, and rate of loading. The sensitivity of measured compressive strength to testing variables varies with the level of compressive strength.

Apart from increasing the compressive strength of HPC, the stress–strain behavior in compression should also be examined in detail. A number of investigations [34–36] have been undertaken to obtain the complete stress–strain curves in compression. Axial stress–strain curves for concretes with compressive strengths up to 200 MPa were obtained by different researchers. It is generally recognized that for HPC with high compressive strength, the shape of the ascending part of the curve becomes more linear and steeper, the strain at maximum stress is slightly high, and the slope of the descending part becomes steeper when compared to normal concrete, as shown in Figure 15.19. It implies that HPC may be more brittle than normal concrete.

15.5.2 Tensile Strength

The tensile strength governs the cracking behavior and affects other properties such as stiffness, damping action, bond to embedded steel, and durability of concrete. It is also of importance with regard to the behavior of concrete under shear loads. The tensile strength is determined either by direct tensile tests or by indirect tensile tests such as flexural or split cylinder tests.

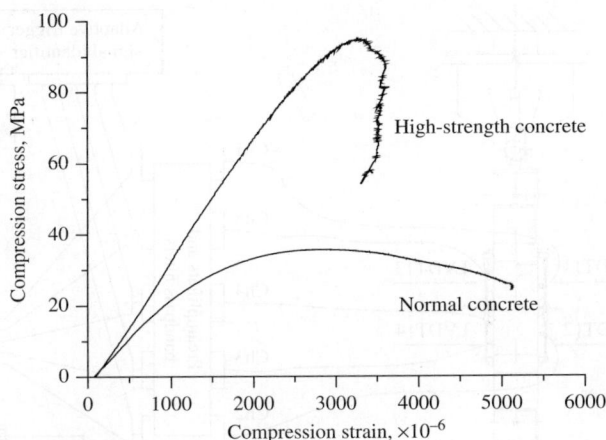

FIGURE 15.19 Comparison of stress–strain curves in compression of HPC and normal concrete.

15.5.2.1 Direct Tensile Strength

The direct tensile test is difficult to conduct. Due to the difficulty in testing, only limited and often conflicting data are available. For normal-strength concrete, it is often assumed that the direct tensile strength of concrete is about 10% of its compressive strength. For HPC, the ratio of tensile strength to compressive strength is greatly reduced to even 5%.

Among more recent studies on the tensile strength of concrete, a new test method for uniaxial tension test of a concrete has been developed by Li et al. [37,38]. Several measures have been taken to ensure the success of the uniaxial tension test. To avoid possible eccentricity or bending of the tension specimen, a set of special loading fixture was developed.

The setup for the uniaxial tension test is shown in Figure 15.20. Two identical loading fixtures were used: one is gripped on the actuator of the MTS machine and the other is connected to the load cell. The loading fixture contains a ball joint that is essential for protecting the tension specimen from possible bending effects. The glued specimen was connected to the loading fixture by using a $\varnothing$ 12 mm pin. Four SCHAEVITZ LVDTs (linear variable differential transformer), with a working range of ± 0.635 mm and a designed gage length of 120 mm, were mounted on the two sides of the specimen for deformation measurement as well as test control. The LVDT holders were specially designed to allow the adjustment of entering and offsetting for the LVDTs.

Tension tests were conducted using an MTS machine with a capacity of 250 kN. The tests were carried out using a newly developed adoptive control method, which ensured the successful and stable measurement of the postpeak response. Figure 15.21 shows the tension stress–deformation curve measured using this method. It can be seen that the postpeak response can be easily recorded even for concrete at early ages [38].

15.5.2.2 Indirect Tensile Strength

The most commonly used tests for estimating the indirect tensile strength of concrete are the splitting tension test (ASTM C 496) and the third-point flexural loading test (ASTM C 78). Both the splitting tensile strength (f_{ct}) and the flexural strength or modulus of rupture (f_r) are related to the compressive strength by the following general expression:

$$f_{ct} \text{ or } f_r = k\sqrt{f_c'}$$

ACI Committee 363 recommends that for concrete strength up to 83 MPa, the coefficient should be taken as 7.4 for tensile strength and as 11.7 for flexural strength. Other investigators have proposed

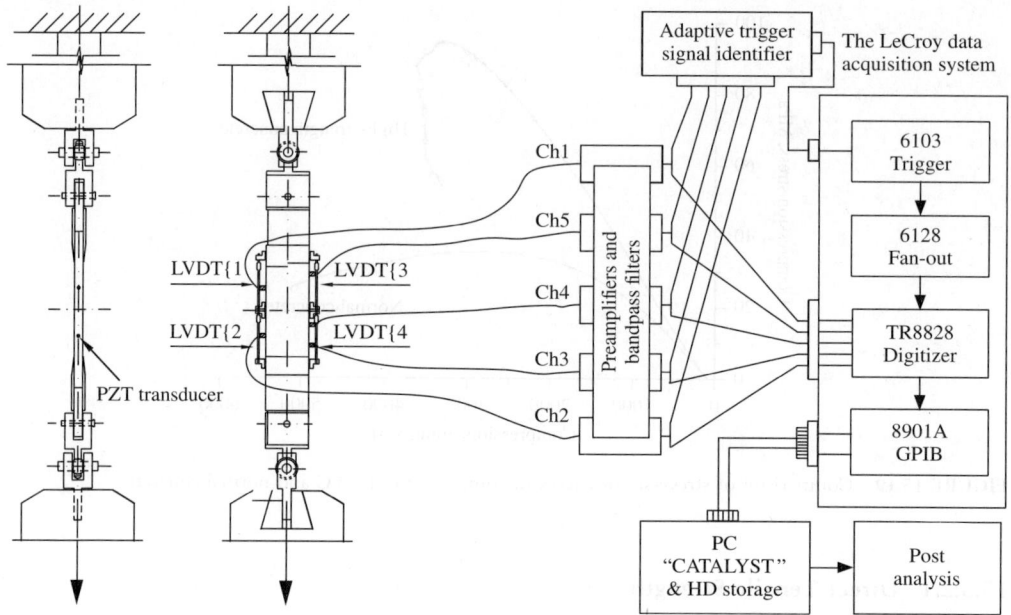

FIGURE 15.20   Uniaxial tension test setup.

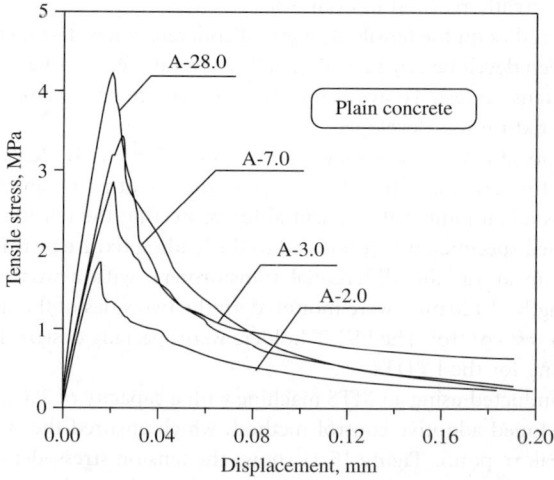

FIGURE 15.21   Tension stress–deformation curves.

slightly different values for the two types of strengths. More details can be found in a previous report [39].

The results of the recent SHRP studies indicated that for the splitting tensile strength the recommendation of ACI Committee 318 is equally acceptable as that of ACI Committee 363. However, for the flexural strength (modulus of rupture), the recommendation of ACI 318 is a better representation than that of ACI Committee 363.

The study by Burg and Ost [40] showed that the average modulus of rupture and splitting tensile strength in comparison with the compressive strength were similar to the recommendation of ACI

Committee 363. The moist cured specimens consistently produced higher strength than air cured specimens.

15.5.3 Flexural Strength

Flexural strength is the strength of concrete under bending. In designing concrete pavements, flexural strength is often used because it better simulates what happens when a slab is loaded by a truck or other vehicle. Most general-use concrete has a flexural strength between 3.5 and 5 MPa.

In general, a relationship exists between the compressive and flexural strengths of concrete. Concrete, which has a higher compressive strength, will have a correspondingly higher flexural strength. This holds true for HPC. However, in many cases, fly ash concrete has demonstrated flexural strength exceeding that of conventional concrete when compressive strengths were roughly equal.

15.5.4 Fracture

In recent years, the fracture behavior of HPC is being studied with great seriousness. HPC with high-strength is nearer to linear theories of fracture and is relatively more brittle. The fracture surfaces are smooth and less tortuous in HPC [41]. It has been reported that the fracture energy decreases and the brittleness index of HPC increases significantly with the incorporation of large size aggregates [42]. A judicious combination of different sizes of coarse aggregates significantly increases the fracture properties. From the knowledge of the variation of load verses CMOD, it can also be noted that the process zone in high-strength concrete has been very small. For HPC, one challenge is whether one can make it relatively more ductile by improving the cohesiveness of cracks.

15.5.5 Shrinkage

Shrinkage and creep are time-dependent deformations that, along with cracking, provide the greatest concern for designers because of the degree of uncertainty associated with their prediction. Concrete exhibits elastic deformations only under loads of short duration, and due to additional deformation with time, the effective behavior is that of an inelastic and time-dependent material.

Shrinkage of HPC may be expected to differ from normal concrete in three broad areas: plastic shrinkage, drying shrinkage, and autogenous shrinkage:

1. Plastic shrinkage occurs during the first few hours after fresh concrete is placed. During this period, moisture may evaporate faster from the concrete surface than the bleeding water from lower layers of the concrete mass. HPC, which is a paste-rich mix, will be more susceptible to plastic shrinkage than normal concretes.
2. Drying shrinkage occurs after the concrete has already attained its final set and a good portion of the chemical hydration process in the cement gel has been accomplished. Drying shrinkage mainly depends on w/b and total amount of water used. For HPC, its w/b ratio is reduced. However, its total water amount may not be very different from normal strength concrete due to increased cement paste. Although HPC's shrinkage is perhaps potentially larger due to higher paste volumes, it does not, in fact, appear to be appreciably larger than normal concretes. This is probably due to the increase in stiffness of the stronger mixes.
3. Autogenous shrinkage due to self-desiccation is one of the typical characteristics for HPC with very low w/b ratio. Of course, autogenous shrinkage also occurs during the hydration process of normal concrete, while, compared with HPC, the autogenous shrinkage of normal concrete is generally negligible. Shrinkage should not be confused with thermal contraction that occurs as concrete loses the heat of hydration.

The shrinkage properties of HPC with higher compressive strengths are summarized in an ACI state-of-the-art report [43]. The basic conclusions were: shrinkage is not only affected by the w/c ratio but also the percentage of water by volume in concrete; laboratory and field studies have shown that shrinkage of higher-strength concrete is similar to that of lower-strength concrete; shrinkage of high-strength concrete containing high-range water reducers is less than for lower-strength concrete; higher-strength–high-performance concrete exhibits relatively higher initial rate of shrinkage, but after drying for 180 days, there is little difference between the shrinkage of high-strength–high-performance concrete and normal-strength concrete made with dolomite or limestone.

Carette et al. [44] reported a study of HPCs with high volume fly ash from sources in the United States. The concretes had low bleeding, satisfactory slump and setting characteristics, and low autogenous temperature rise. These concretes also had excellent mechanical properties at both early and late ages with compressive strength reaching as high as 50 MPa at 91 days and the drying shrinkage of the concretes was relatively low.

Field measurements of surface shrinkage strains on a mock column, fabricated with HPC, after 2 and 4 years were conducted by Sarkar and Aïtcin [45] and the measurements were compared with results on specimens under laboratory conditions. It was shown that the surface shrinkage strains under the field condition were considerably lower than those measured under the laboratory conditions.

In a similar study by Hindy et al. [46], measurements of dry shrinkage were carried out on concrete specimens as well as on instrumented reference columns made with two different ready-mixed HPCs. One had a 91-day compressive strength of 98 MPa and the other had a 91-day compressive strength of 80 MPa. The first contained silica fume but the second did not. The effects of curing time, curing conditions, silica fume content, and w/b ratio were considered. It was found that the longer the curing time the lower the dry shrinkage. The lower the w/b ratio the lower the dry shrinkage. Dry shrinkage of small specimens measured by the conventional laboratory test was found to overestimate shrinkage of the concrete in the real structure. The ACI 209 predictive equation was valid for HPCs only if new values for the parameters were introduced.

15.5.6 Creep

Creep is the time-dependent increase in strain of hardened concrete subjected to sustained stress. It is usually determined by subtracting, from the total measured strain in a loaded specimen, the sum of the initial instantaneous strain (usually considered elastic) due to sustained stress, the shrinkage, and any thermal strain in an identical load-free specimen, subjected to the same history of relative humidity and temperature conditions.

Creep is closely related to shrinkage and both phenomena are related to the hydrated cement paste. As a rule, a concrete that is resistant to shrinkage also has a low creep potential. The principal parameter influencing creep is the load intensity as a function of time; however, creep is also influenced by the composition of the concrete, the environmental conditions, and the size of the specimen.

The composition of concrete can essentially be defined by the w/b ratio, aggregate and cement types, and quantities. Therefore, as with shrinkage, an increase in w/b ratio and in cement content generally results in an increase in creep. Also, as with shrinkage, the aggregate induces a restraining effect so that an increase in aggregate content and stiffness reduces creep.

A widely used predictive equation for creep strain at time t days after loading is that given by the ACI Committee 209. As with shrinkage, use of standard-sized laboratory test specimens for determining creep properties of a concrete mix will require adjustment for size and shape effects to reasonably predict the behavior of a concrete structure.

Earlier studies on creep of concrete containing fly ash, silica fume, and high-range water reducer have been summarized in a previous report [39]. It was observed that, with comparable strength, there was no apparent difference between the specific creep of silica fume concrete and that of Portland cement concrete, or of fly ash concrete. However, creep strains, creep coefficient, and specific creep were all smaller for high-strength–high-performance concrete than for concretes of medium- and low-strengths.

Collins investigated the effect of mix proportions on creep characteristics in a study in which five mixes with 28-day design strengths ranging from 60 to 64 MPa were used. The results indicated that creep was somewhat less for concrete mixtures with lower cement paste and larger aggregate size. It was also shown that the use of a high-range water-reducing admixture did not have a significant effect on the creep deformations.

Carette et al. reported a study of HPCs with high volume fly ash from sources in the United States. The concretes had low bleeding, satisfactory slump and setting characteristics, and low autogenous temperature rise. These concretes also had excellent mechanical properties at both early and late ages with compressive strength reaching a value as high as 50 MPa at 91 days. The creep of the concretes was relatively low.

In a study of long-term deflection of high-strength–high-performance concrete beams, Paulson et al. [47] pointed out that a large body of experimental evidence was available confirming that the creep coefficient of high-strength–high-performance concrete under sustained axial compression was significantly less than that of ordinary concrete. Thus, the ratio of time-dependent deflection to immediate elastic deflection of high-strength–high-performance concrete beams under sustained loads should likewise be lower. With nine beams made of nominal concrete strengths over a range from 40 to 91 MPa and loaded over a 12-month period, they found significant differences between the deflection of high-strength–high-performance concrete beams and that of normal-strength concrete beams. It was suggested that the long-term deflection multipliers of the ACI Code should be modified to account for the reduced creep deflection with high-strength–high-performance concrete.

Since HPC with high strength is increasingly being used in compression members with reinforcement, and creep data are generally obtained with small unreinforced specimens in the laboratory, Yamamoto conducted a creep test of a reinforced concrete column of $25 \times 25 \times 100$ cm containing 2.44% longitudinal reinforcement and 0.79% lateral ties. The concrete strength was 57 MPa at 31 days. When the concrete was 33 days old, a sustained stress of 19.4 MPa was applied on the column. After 170 days of loading, the creep coefficient of the concrete was determined to be 0.57 and the ultimate creep coefficient was estimated to be 1.4, which suggested that the creep deformation of high-performance reinforced concrete columns would be much smaller than that of normal-strength concrete columns (40 MPa or lower).

In the SHRP C-205 studies, Zia et al. evaluated the creep behavior of very high-strength concrete with different aggregates (crushed granite, marine marl, and rounded gravel). Creep strain measurements were made for 90 days in each case. The observed creep strains of the different groups of VHS concrete ranged from 20 to 50% of that of conventional concrete. The creep strains were especially low for concretes with a 28-day strength in excess of 70 MPa.

15.5.7 Durability

When properly designed and carefully produced with good quality control, concrete is inherently a durable material. However, under adverse conditions, concrete is potentially vulnerable to deleterious attacks such as frost, sulfate attack, alkali–aggregate reaction, and corrosion of steel. Each of these processes involves movement of water or other fluids, transporting aggressive agents through the pore structure of concrete. Therefore, permeability is an important property that affects the durability of concrete.

15.5.7.1 Permeability

Compared to that of normal concrete, the main characteristics of HPC are low w/b, and the addition of mineral admixtures, which results in the more uniform and homogeneous microstructure of hardened HPC.

When Portland cement is combined with some fine particle and high reactive mineral admixtures in low w/b ratios, the microstructure of such systems consists mainly of poorly crystalline hydrates forming a denser matrix with low porosity [48,49,50]. This is because that larger amount of CH is transformed into

C–S–H through pozzalanic reaction between CH and mineral admixtures, while the remaining CH tends to form smaller crystals compared to that in pure Portland cement paste. Furthermore, the Ca/Si ratio of the C–S–H gel also decreases with the incorporation of mineral admixture. As a consequence, it appears that the total porosity is reduced and the pore size distribution is also refined to a certain extent, resulting in the very high resistance to permeability of harden cement pastes.

The interfacial zone between cement pastes and aggregates (ITZ) for normal concrete is about 20 to 100 μm in width. It has a very distinct microstructure compared to that of the bulk matrix for normal concrete. The ITZ of normal concrete that is inferior in quality and thus leads to a poorer bond between the aggregates and the cement pastes is typically characterized by the following key elements:

1. The ITZ is richer in CH and ettringite (AFt) than the bulk phase, and the CH is oriented. A rim of massive CH can often be observed around the aggregate.
2. The porosity of the ITZ is greater than that of the bulk phase, and a gradient in porosity can be observed with a declining trend as the distance from the aggregate surface increases.

However, for HPC, the microstructure in the ITZ obviously differs from that of normal concrete due to the adoption of the fine mineral admixture and low w/b. Aïtcin and coworkers [48–50] observed that HPC with silica fume was not as crystallized and porous as normal concrete, and the space in the vicinity of the aggregate was occupied with amorphous and dense C–S–H gel. Also, direct contact was formed between the aggregate and the C–S–H gel rather than with CH crystals as in normal concrete. Scrivener et al. [51] quantified the ITZ and demonstrated that in HPC with silica fume the porosity of the ITZ was practically eliminated and practically no gradient in porosity was observed, in contrast to normal concrete.

The above qualitative and quantitative observations indicate that in HPC the bulk matrix becomes very dense and, typically, the dense matrix extends up to the aggregate surface, in such a way that the inhomogeneity of the ITZ is largely eliminated. It is now well documented that this improved microstructure is closely related to the reduced permeability and improved performance of HPC.

15.5.7.2 Freezing and Thawing

Damage of concrete under repeated cycles of freezing and thawing (frost attack) is a major problem of durability. In a previous state-of-the-art report [39], the mechanism with which freezing and thawing damages concrete has been discussed and some of the earlier researches have been summarized.

For normal concrete, entrained air of 4 to 8% by volume of concrete provides an effective defense against frost damage and the exact amount is dependent on the maximum size of the coarse aggregate, provided that the coarse aggregate itself is frost resistant [5]. The optimum spacing factor of the air voids should be no more than 0.2 mm and the air voids should be small with their diameter being in the range of 0.05 to 1.25 μm to ensure that the required spacing factor is obtained with low air contents. In addition to air entraining agent, a newly developed concrete durability enhancing admixture developed by Li et al. [52] has shown a good capability in improving the freezing and thawing resistance of a concrete [53].

At the present time, there is no clear direction as to whether HPC will require air entrainment and the necessary air-void parameters. Some authors [54–56] argue that with low w/b ratio and mineral admixtures, the amount of freezable water in HPC would be low and its pore size would be decreased to the extent that water in the pore cannot freeze. However, in their study of HPCs they found that 5% entrained air was required to achieve a higher level of frost resistance than that required by the ASTM C 666, Procedure A (i.e., a durability factor of 80 versus 60%).

Okada et al. [57] reported very good frost resistance of non-air-entrained HPC with w/b ratios in the range of 0.25 to 0.35. Similar observations were later on reported by Foy et al. [58] who observed good frost resistance of non-air-entrained concrete with w/b ratios of 0.25 and 0.35, respectively.

Malhotra et al. [59] also tested a number of concretes with different types of cement and w/b ratios of 0.30 and 0.35. They reported, however, that air entrainment was necessary for these concretes to be frost resistant. Hammer and Sellevold [60] tested non-air-entrained concrete with 0 and 10% silica fume and w/b ratios varying from 0.25 to 0.40. Even for the lowest w/b ratios some of the specimens were damaged during testing. These observations were in conflict with low-temperature calorimeter data, which clearly demonstrated very low freezable water content. Based on this, Hammer and Sellevold suggested, therefore, that the observed damage could be due to thermal fatigue caused by too large differences between the thermal expansion coefficients of aggregate and binder rather than ice formation.

15.5.7.3 Chemical Resistance

Chemical attack to concrete can be classified into three types of processes depending on the predominant chemical reaction taking place:

1. *Leaching corrosion.* It is the process where parts or all of the hardened cement paste are removed from the concrete. Normally, this is caused by the action of water of low carbonate hardness or carbonic acid content.
2. *Soluble corrosion.* This process is corrosion by exchange reactions and by removal of readily soluble compounds from the hardened cement paste. This process occurs as a result of a base exchange reaction between the readily soluble compounds of the hardened cement paste and the aggressive solution.
3. *Swelling corrosion.* This process is largely due to the formation of new and aggressive expansive compounds in the hardened cement paste. This process is primarily the result of attack by certain salts (sulfate attack is one of the commonly found chemical attacks). Also, alkali–aggregate reaction causes expansion, where the concrete eventually is destroyed by a swelling pressure.

For all of the above deteriorating processes, the permeability and diffusivity of the concrete are the key factors governing the rate of determination. In addition, CH in cement paste is also an easily soluble constituent and can react with the sulfates to form disruptive expansion products such as gypsum and ettringite. HPC typically possesses quite lower permeability and diffusivity compared to normal concrete. Furthermore, due to the addition of mineral admixtures, the CH products in HPC are effectively reduced. All of the above points make HPC highly resistant to chemical attack.

Some investigations have reported the beneficial effect of mineral admixtures in high-sulfate-containing environments [61,62]. In these investigations, the performance achieved has been equal or better than that obtained by the use of sulfate-resistant cements. Mehta [63] exposed a number of HPCs to solutions of both 5% sodium sulfate and 1% sulfuric and hydrochloric acids for a period of up to 182 days. Although Portland cement contained 7% C_3A, the results showed that w/b ratios of 0.33 to 0.35 gave too low a permeability to cause any deterioration. In more aggressive environments pure Portland cements have shown some deterioration, whereas addition of silica fume has given practically unaffected performance.

The addition of pozzolans such as silica fume can also reduce the alkali–aggregate reaction. Pore water [64,65] analyses of cement paste with silica fume have demonstrated the ability to reduce the alkali concentration in the pore water rapidly, thus making it unavailable for the slower reaction with reactive silica in the aggregate. Also, for HPC the effect of self-desiccation may reduce the moisture content to a level where no alkali–aggregate reaction can take place.

15.5.7.4 Scaling

Scaling is another problem of durability. It is caused by repeated application of deicing salts. Concrete surface damaged by salt scaling becomes roughened and pitted as a result of spalling and flaking of small pieces of mortar near the surface. Even high-quality concrete with adequate air entrainment can still suffer scaling by deicing chemicals.

As far as salt scaling is concerned, there are some conflicting results reported in the literature for HPC. Petersson [66] reported that deterioration of HPC due to salt scaling was small for the first 50 to 100 cycles, but increased very rapidly to total destruction in the following 10 to 20 cycles. Foy et al. [58] however, observed that the resistance to salt scaling of concrete with a w/b ratio of 0.25 was very good even after 150 cycles. Both Hammer and Sellevold [60] have demonstrated that it is possible to produce HPC with high resistance to salt scaling without any air entrainment. These test results included w/b ratios up to 0.37.

In lieu of silica fume, high-reactivity metakaolin has also been used as an effective mineral admixture for high-strength concrete, which proved to have satisfactory performance in scaling resistance [67].

The effect of curing and drying on salt scaling resistance of fly ash concrete was investigated by Bilodeau et al. [68]. Concretes with 20 and 30% fly ash as cement replacement were produced with two types of aggregates, using w/b of 0.35, 0.45, and 0.55. The test results showed that, with few exceptions, concrete with up to 30% fly ash performed well under the scaling test. Extended moist-curing or dry periods did not seem to significantly affect the scaling performance of the reference concrete as well as the fly ash concrete. However, when higher volume of fly ash (55 to 60%) was used in air-entrained concrete, the scaling performance of the concrete was less than satisfactory.

15.5.7.5 Abrasion

Abrasion is the surface wearing of a concrete due to repeated rubbing and friction. For pavements, abrasion results from traffic wear. Adequate abrasion resistance is important for pavements and bridge decks from the standpoint of safety. Excessive abrasion leads to an increase in accidents as the pavement becomes polished and reduces its skid resistance.

Abrasion resistance of concrete is a direct function of its strength, and thus its w/c ratio and constituent materials. High-quality paste and strong aggregates are essential to produce an abrasion-resistant concrete. Considering the use of low w/b ratio and mineral admixtures, HPC made with wear-resistance coarse aggregates typically possesses high abrasion resistance.

Abrasion resistance of HPC containing chemical and mineral admixtures was investigated by de Almeida [70]. Ten concrete mixtures were evaluated for their abrasion resistance according to a Portuguese Standard, which is similar to the Brazilian Standard and the German Standard DIN 52108, using the Dorry apparatus. The compressive strength of the concrete varied from 60 to 110 MPa at 28 days, and the w/b varied from 0.24 to 0.42. The concrete mixtures contained silica fume, fly ash, or natural pozzolan, with or without a superplasticizer to maintain a consistent workability. From the test results, it was concluded that the abrasion resistance of concrete generally varies inversely with the w/b ratio, the porosity, and the cement paste volume in the concrete. Therefore, by using super-plasticizer to substantially reduce the w/b ratio, the abrasion resistance of concrete would be improved considerably. Introducing mineral admixture without using superplasticizer would reduce the abrasion resistance of concrete since more water would be needed to maintain a constant workability. The least abrasion resistant concrete produced in the study resulted in surface wear that was only 17% of normal concrete.

15.6   Design Philosophy [69]

Many of the design provisions that exist in the codes and standards of North America and Europe are based on experimental results obtained with normal concretes. As HPCs have become available, the applicability of the design provisions for HPC has been questioned. The following highlights several areas where designers should give special consideration.

15.6.1   Concrete Compressive Strength Limits

Table 15.11 gives the maximum concrete compressive strength limits for high-strength concrete, and with the shapes and sizes of the corresponding standard compressive test specimens used in different countries [71–75]. Although ACI 318-02 has no upper limit on the compressive strength itself, there are limits on the compressive strength that can be used in some design equations (e.g., development length and shear). CSA A23.3-94 and NZS 3101-95 have reduced limits for the concrete compressive strength

TABLE 15.11 Upper Limits of Specified or Characteristic Concrete Compressive Strengths and Standard Test Specimens

Country/region	Code	Date	Maximum specified or characteristic concrete compressive strengths, MPA	Standard test specimen, mm
Europe	CEB-FIP MC-90[72]	1993	80	Cyl. 150 × 300 Cube 200 × 200
Europe	EC2-02 (Final Draft)[73]	2002	90	Cyl. 150 × 300 Cube 200 × 200
Canada	CSA A23.3-94[74]	1994	80/55[a]	Cyl. 150 × 300 Cyl. 100 × 200
USA	ACI 318-02[71]	2002	No limit[b]	Cyl. 6 × 12 in. Cyl. 152 × 304
New Zealand	NZS 3101-95[75]	1995	100/70[a]	Cyl. 152 × 304 Cyl. 152 × 304

[a] For ductile elements in seismic design.
[b] Concrete strength limited in some design equations.

TABLE 15.12 Load Factors and Resistance Factors for the Different Codes

Code	Load factor		Strength reduction factor			Material strength reduction factor	
	Dead	Live	Flexural	Flexural and axial load	Shear	Concrete, φ_c or $1/\gamma_c$	Steel, φ_s or $1/\gamma_s$
ACI 318-021	1.2	1.6	0.65–0.9	0.65–0.9	0.75	—	—
CSA A23.3-94	1.25	1.5	—	—	—	0.6	0.85
MC-90 VEC2-023	1.35	1.5	—	—	—	1/1.5	1/1.15
NZS 3101-955	1.2	1.6	0.85	0.65–0.85	0.75	—	—

for the seismic design of ductile reinforced concrete members. These more conservative limits were chosen because of the concern over the ductility of high-strength concrete members and the lack of reversed cyclic loading test results at the time the standards were written. In the comparison of the different maximum strengths, it is important to realize that the codes determine the compressive strengths with a number of different standard test specimens (Table 15.11).

15.6.2 Load and Resistance Factors

Table 15.12 provides a comparison of different code approaches for load factors and resistance factors for a HPC structure designer. ACI 318-02 and NZS 3101-95 both use strength reduction factors φ applied to the nominal capacity, which varies with the type of action (e.g., flexure, shear, and axial loads). The other codes in Table 15.12 use a partial safety factor approach with reduction factors applied separately to the stress resultant in the concrete (φ_c or $1/\gamma_c$) and the stress resultant in the steel (φ_s or $1/\gamma_s$). It is interesting to note that the revisions proposed in 1995 to the 1990 CEB-FIP Model Code (MC-90) [76] introduces a high-strength concrete factor γ_{hsc}, to modify γ_c, for concrete having compressive strengths from 50 to 100 MPa.

$$\gamma_{hsc} = \frac{1}{1.1 - (f_{ck}/500)} \tag{15.1}$$

where f_{ck} is the characteristic strength defined in the following. This factor was introduced to account for the brittle behavior associated with spalling, delamination, or failures due to high local stresses.

15.6.3 Modulus of Elasticity

To appreciate the differences in code expressions for the modulus of elasticity of concrete, it is important to realize that both ACI 318-02 and EC2-02 use a secant modulus. The definition of the secant modulus, however, is different in the two codes; and the ACI 318-02 bases the modulus on the specified compressive strength, whereas EC2-02 bases the modulus on the average compressive strength. ACI 318-02 gives an expression for the secant modulus of elasticity — defined as the slope of a line drawn from a stress of zero to a compressive stress equal to $0.45f_c'$ — and given as

$$E_c = w_c^{1.5} 33 \sqrt{f_c'} \text{ psi}$$
$$E_c = 0.043 w_c^{1.5} \sqrt{f_c'} \text{ MPa} \tag{15.2}$$

where w_c is the concrete unit weight varying from 90 to 155 lb/ft³ (1500 to 2500 kg/m³). For normal density concrete, the expression simplifies to

$$E_c = 57,000 \sqrt{f_c'} \text{ psi}$$
$$E_c = 4700 \sqrt{f_c'} \text{ MPa} \tag{15.3}$$

EC2-02 gives the following expression for the secant modulus — defined as the slope of a line drawn from a stress of zero to a compressive stress equal to $0.40 f_{cm}$

$$E_{cm} = 22,000 \left(\frac{f_{cm}}{10} \right)^{0.3} \text{ MPa} \tag{15.4}$$

CSA A23.3-94 and NZS 3101-95 adopted slightly modified versions of an equation based on the research by Carrasquillo et al. [77] and reported by ACI Committee 363 [78]:

$$E_c = \left(3320 \sqrt{f_c'} + 6900 \right) \left(\frac{w_{cm}}{2300} \right)^{1.5} \text{ MPa} \tag{15.5}$$

15.6.4 Concrete Tensile Strength

Tensile strength is an important parameter used to calculate cracking loads and to determine the required minimum amounts of reinforcement. ACI 318-02 uses the modulus of rupture f to determine the flexural cracking moment from the following expressions:

$$f_r = 7.5 \sqrt{f_c'} \text{ psi}$$
$$f_r = 0.6 \sqrt{f_c'} \text{ MPa} \tag{15.6}$$

ACI Committee 363 [78] reported the following relationships for f_r, which was recommended by Carrasquillo et al. [77], for concrete having a compressive strength between 3,000 and 12,000 psi (20 to 80 MPa):

$$f_r = 11.2 \sqrt{f_c'} \text{ psi}$$
$$f_r = 0.94 \sqrt{f_c'} \text{ MPa} \tag{15.7}$$

NZS 3101-95 adopted the following expression for the modulus of rupture:

$$f_r = 0.8 \sqrt{f_c'} \text{ MPa} \tag{15.8}$$

EC2-02 uses the direct tensile strength as the basic parameter to categorize the tensile strength of the concrete. In the absence of direct tensile tests, which are very difficult to perform correctly, EC2-02 provides the following expressions for the mean tensile strength for the concrete f_{ctm}.

$$f_{ctm} = 0.30 f_{ck}^{2/3} \text{ MPa,} \qquad \text{for } f_{ck} \le 50 \text{ MPa}$$
$$f_{ctm} = 2.12 \ln \left(1 + \frac{f_{cm}}{10} \right) \text{ MPa,} \qquad \text{for } f_{ck} > 50 \text{ MPa} \tag{15.9}$$

where f_{cm} is the mean compressive strength taken as $f_{ck} + 8$ MPa. One interesting aspect of EC2-02 is that it makes a clear distinction in applying the different values of tensile strengths. A lower-bound characteristic tensile strength, $\frac{5}{6}$ fractile value ($f_{ctk,0.05} = 0.7f_{ctm}$), is used when accounting for tensile stresses contributing to strength (e.g., shear). An average or mean value is used when computing deflections and crack widths. An upper-bound characteristic tensile strength, 95% fractile value ($f_{ctk,0.95} = 1.3f_{ctm}$), is also provided for the designer and can be used to estimate the upper bound of cracking loads to check minimum reinforcement requirements.

It is a well-known fact in fracture mechanics that for brittle materials, as the member size increases, the tensile strength decreases. Some codes account for this so-called "size effect" when determining the cracking stress. For the data examined in this article for the modulus of rupture, it is concluded that the equation proposed by Carasquillo et al. [77] provides a reasonable estimate of the average modulus of rupture. For this same data set, the ACI 318-02 equation for the modulus of rupture provides a reasonable estimate of the minimum values.

15.6.5 Minimum Reinforcement for Flexure

The primary role of minimum reinforcement is to provide adequate reserve of strength after cracking. ACI 318-02 and NZS 3101-95 code expressions for the minimum reinforcement ratio for rectangular sections are

$$A_{s,min} = \frac{3\sqrt{f_c'}}{f_y} b_w d \geq \frac{200}{f_y} b_w d \text{ in.}^2 \quad \text{(psi and in.)}$$

$$A_{s,min} = \frac{\sqrt{f_c'}}{4f_y} b_w d \geq \frac{1.4}{f_y} b_w d \text{ mm}^2 \quad \text{(MPa and mm)}$$

$$(15.10)$$

where b_w is the web width and d is the effective depth. For a statically determinate T-beam, with the flange in tension, ACI 318-02 requires that b_w be taken as the smaller of $2b_w$ or the flange width.

The approach used by CSA A23.3-94 for determining minimum reinforcement is based on the ACI approach for prestressed concrete members, requiring that the factored flexural strength be at least 1.2 times the cracking moment. The cracking moment is determined with a modulus of rupture $f_r = 0.6\sqrt{f_c'}$ MPa. This philosophy is applied equally to non-prestressed as well as prestressed concrete members. CSA A23.3-94 provides slightly lower values of $A_{s,min}$ than the ACI Code.

$$A_{s,min} = 0.2 \frac{\sqrt{f_c'}}{4f_y} b_w h \text{ mm}^2 \quad \text{(MPa and mm)} \quad (15.11)$$

where b_w is the width of the tension zone and h is the overall depth of the member. ACI 318-02, CSA A23.3-94, and NZS 3101-95 codes allow the minimum reinforcement requirement to be waived if the ultimate or factored flexural strength is at least one-third larger than the factored moment. This represents a trade-off between ductility and over-strength.

EC2-02 requires that the minimum area of flexural reinforcement must satisfy the following expression:

$$A_{s,min} \geq 0.26 \frac{f_{ctm}}{f_{yk}} b_t d \geq 0.0013 b_t d \quad \text{(MPa and mm)} \quad (15.12)$$

where b_t is the mean width of the concrete zone in tension, taken as the web width for a T-beam with the flange in compression, and f_{ctm} is given by Equation 15.9. The 1995 Recommended Extensions to MC-90 require a minimum reinforcement ratio equal to $0.23f_{ctk,max}/f_{yk}$, where $f_{ctk,max} = f_{ctk,0.95}$, which is a function of the maximum tensile strength.

Figure 15.22 compares the minimum flexural reinforcement ratio $\rho_{min} = A_{s,min}/b_w d$, required by ACI 318-02, CSA A23.3-94, and EC2-02, as a function of the concrete compressive strength. It is noted that EC2-02 requires less minimum reinforcement for beams than ACI 318-02 and CSA A23.3-94.

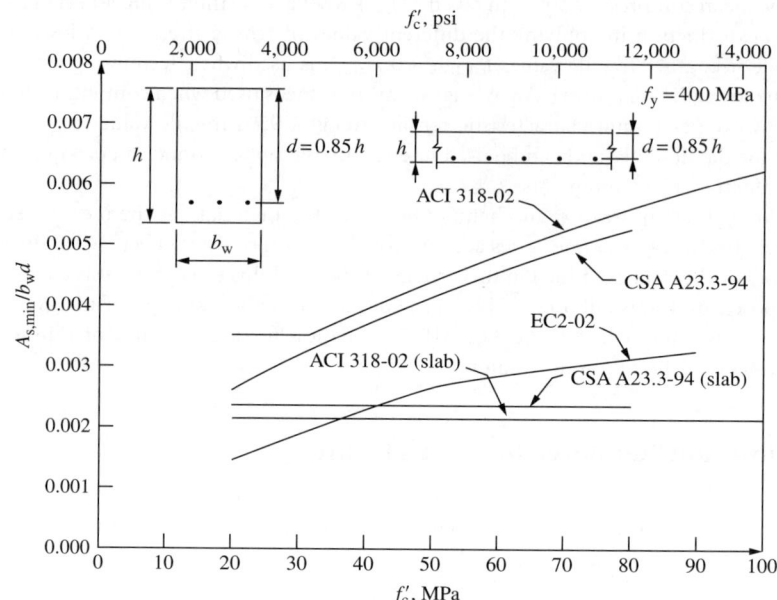

FIGURE 15.22 Comparison of the minimum flexural reinforcement ratio as a function of f_c'.

Unlike the requirements for the EC2-02, the minimum reinforcement requirements for slabs, given by ACI 318-02 and CSA A23.3-94, are not a function of concrete strength.

15.6.6 Flexure and Axial Loads

Figure 15.23 compares the simplified equivalent stress distributions for determining the nominal resistances under flexure and axial load. EC2-02 permits the use of several equivalent stress distributions — the preferred one being the parabola–rectangle stress distribution. In the use of the equivalent rectangular stress distributions prescribed by the different codes, there are significant differences in the treatment of higher strength concretes. CSA A23.3-94 and NZS 3101-95 have departed from the traditional ACI stress block factor α_1 of 0.85 to give α_1 values that are as low as 0.67 for very high-strength concrete. EC2-02 uses stress block factors that are reduced for concrete having a compressive strength greater than 50 MPa. In EC2-02, α_{cc} is a factor accounting for long-term and loading effects, and is recommended to be 1.0. EC2-02, however, indicates that this factor may be changed by amendments in a National Annex when adopted by an individual country. Figure 15.24 illustrates the differences in the axial load bending interaction diagrams for columns containing 1 and 4% of vertical reinforcement as the concrete compressive strength varies from 30 to 80 MPa. In calculating the predictions using the EC2-02 code, the traditional value of 0.85 has been used for α_{cc}. It is noted that the four code approaches, compared in Figure 15.24, give similar nominal strength predictions for the lower-strength concrete columns. Above a concrete strength of 50 MPa, however, the ACI 318-02 approach predicts significantly higher column capacities than the other three codes for "compression controlled" columns. This difference is particularly evident for columns containing the minimum amount of vertical reinforcement (1%) permitted by the codes. In view of observations [79] that there is a tendency for splitting and splitting of the concrete cover in high-strength concrete columns, CSA A23.3-94 requires that, for columns with f_c' greater than 50 MPa, the column ties must have 135°-bend anchorages, and the traditional tie spacing limits must be multiplied by 0.75.

Considering the differences in strength predictions and detailing considerations, it is clear that more research on high-strength concrete column behavior is needed. The behavioral aspects that need to be

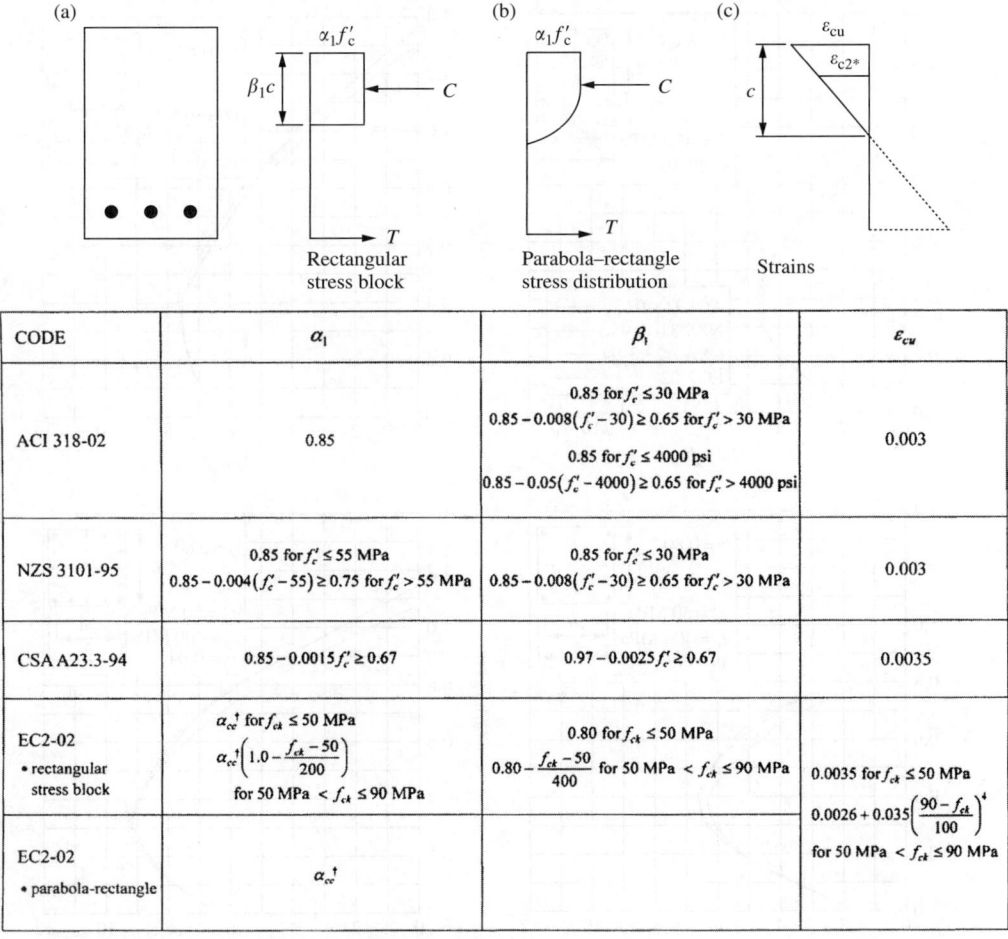

CODE	α_1	β_1	ε_{cu}
ACI 318-02	0.85	0.85 for $f_c' \leq 30$ MPa $0.85 - 0.008(f_c' - 30) \geq 0.65$ for $f_c' > 30$ MPa 0.85 for $f_c' \leq 4000$ psi $0.85 - 0.05(f_c' - 4000) \geq 0.65$ for $f_c' > 4000$ psi	0.003
NZS 3101-95	0.85 for $f_c' \leq 55$ MPa $0.85 - 0.004(f_c' - 55) \geq 0.75$ for $f_c' > 55$ MPa	0.85 for $f_c' \leq 30$ MPa $0.85 - 0.008(f_c' - 30) \geq 0.65$ for $f_c' > 30$ MPa	0.003
CSA A23.3-94	$0.85 - 0.0015 f_c' \geq 0.67$	$0.97 - 0.0025 f_c' \geq 0.67$	0.0035
EC2-02 • rectangular stress block	$\alpha_{cc}^{\dagger}$ for $f_{ck} \leq 50$ MPa $\alpha_{cc}^{\dagger}\left(1.0 - \dfrac{f_{ck} - 50}{200}\right)$ for 50 MPa $< f_{ck} \leq 90$ MPa	0.80 for $f_{ck} \leq 50$ MPa $0.80 - \dfrac{f_{ck} - 50}{400}$ for 50 MPa $< f_{ck} \leq 90$ MPa	0.0035 for $f_{ck} \leq 50$ MPa $0.0026 + 0.035\left(\dfrac{90 - f_{ck}}{100}\right)^4$ for 50 MPa $< f_{ck} \leq 90$ MPa
EC2-02 • parabola-rectangle	$\alpha_{cc}^{\dagger}$		

$^*\varepsilon_{c2} = 0.002$ for $f_{ck} \leq 50$ MPa, $\varepsilon_{c2} = 0.002 + 0.085 \times 10^{-3}(f_{ck} - 50)^{0.53}$ for 50 MPa $< f_{ck} \leq 90$ MPa

$^\dagger\alpha_{cc}$ = Factor accounting for long term and loading effects

FIGURE 15.23 Simplified equivalent stress distributions for determining the nominal resistance under flexure and axial loads.

investigated include the potential for cover splitting, the required confinement reinforcement, detailing requirements for columns, and suitable values for α_1 and β_1 over the full range of practical concrete strengths.

15.6.7 Concrete Contribution to Shear

For normal-density concrete, ACI 318-02 gives the following expression for the nominal shear carried by the concrete V_c:

$$V_c = 2\sqrt{f_c'} b_w d \text{ lb} \quad \text{(psi and in.)}$$

$$V_c = \frac{\sqrt{f_c'}}{6} b_w d \text{ N} \quad \text{(MPa and mm)}$$

(15.13)

where b_w is the web width, d is the effective depth, and f_c' is the concrete compressive strength. ACI 318-02 limits the value of $\sqrt{f_c'}$ to 100 psi (25/3 MPa), unless the required amount of minimum shear reinforcement is provided.

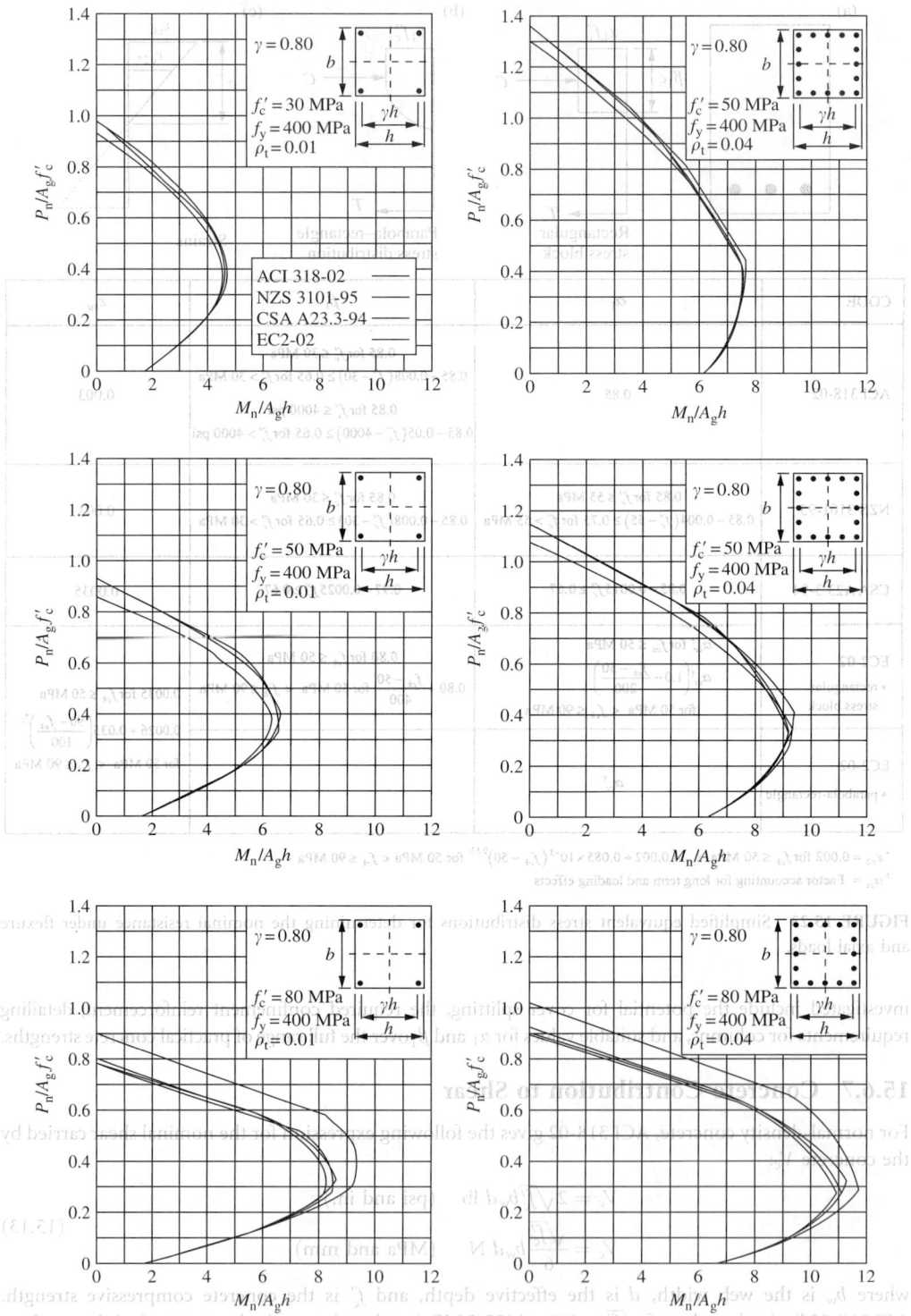

FIGURE 15.24   Comparison of axial load–moment interaction diagrams using different code approaches.

It is noted that the EC2-02 always requires minimum shear reinforcement, except for members where transverse redistribution of loads is possible (e.g., slabs) or for members of minor importance, which do not contribute significantly to the overall resistance and stability of the structure (e.g., short lintels). For members not requiring shear reinforcement and without axial load, EC2-02 gives the following expression for the shear carried by the concrete:

$$V_c = [0.18k(100\rho f_{ck})^{1/3}]b_w d \geq 1.5 v_{min} b_w d \text{ N} \quad \text{(MPa and mm)} \tag{15.14}$$

where $k = 1 + \sqrt{200/d} \leq 2.0$, $\rho = A_s/b_w d$, and $v_{min} = 0.035 k^{3/2} f_{ck}^{1/2}$. This expression includes a size effect factor k and also includes the reinforcement ratio ρ.

CSA A23.3-94 has two approaches for shear design: the simplified method based on the ACI Code and the general method based on modified compression field theory [80]. Although the expression for the factored shear resistance has been adjusted for the low value of φ_c, the nominal shear resistance of the concrete in the simplified method for sections having either: at least minimum amount of transverse reinforcement; or an effective depth $d \leq 300$ mm is the same as the ACI 318-02 expression (Equation 15.13).

For sections with effective depths greater than 300 mm and with no transverse reinforcement, or less than the minimum amount of transverse reinforcement, the nominal shear resistance can be written as

$$V_c = \left(\frac{217}{1000 + d}\right) \sqrt{f_c'} b_w d \geq 0.083 \sqrt{f_c'} b_w d \text{ N} \tag{15.15}$$

It is noted that both EC2-02 and CSA A23.3-94 have a size effect term in the design expressions. Furthermore, EC2-02 accounts for the amount of flexural reinforcement. The general method of the CSA Standard accounts for both the size effect and the amount of longitudinal reinforcement when determining V_c.

Figure 15.25 compares the nominal shear stress carried by the concrete for one-way slabs without shear reinforcement using the expressions from ACI 318-02, EC2-02, and CSA A23.3-94. It is apparent that the shear capacities predicted using the EC2-02 and CSA A23.3-94 expressions for thick slabs are considerably below those predicted using the ACI 318-02 expression (Equation 15.13). It has been shown by tests on large beams without shear reinforcement, particularly those made from high-strength concrete, that failure can occur at loads considerably less than the loads predicted using Equation 15.13 [79,80].

15.6.8 Minimum Shear Reinforcement

The minimum area of shear reinforcement required by ACI 318-02 is given by

$$A_{v,min} = 0.75\sqrt{f_c'}\frac{b_w s}{f_y} \geq 50\frac{b_w s}{f_y} \text{ in.}^2 \quad \text{(psi and in.)}$$

$$A_{v,min} = \frac{1}{16}\sqrt{f_c'}\frac{b_w s}{f_y} \geq 0.33\frac{b_w s}{f_y} \text{ mm}^2 \quad \text{(MPa and mm)} \tag{15.16}$$

This continuous function replaces the more complex minimum shear reinforcement requirements of ACI 318-99 [81].

The minimum amount of shear reinforcement required by EC2-02 is given by

$$A_{v,min} = 0.08\sqrt{f_c'}\frac{b_w s}{f_{yk}} \text{ mm}^2 \quad \text{(MPa and mm)} \tag{15.17}$$

EC2-02 indicates that this minimum area of shear reinforcement is subject to change in a National Annex.

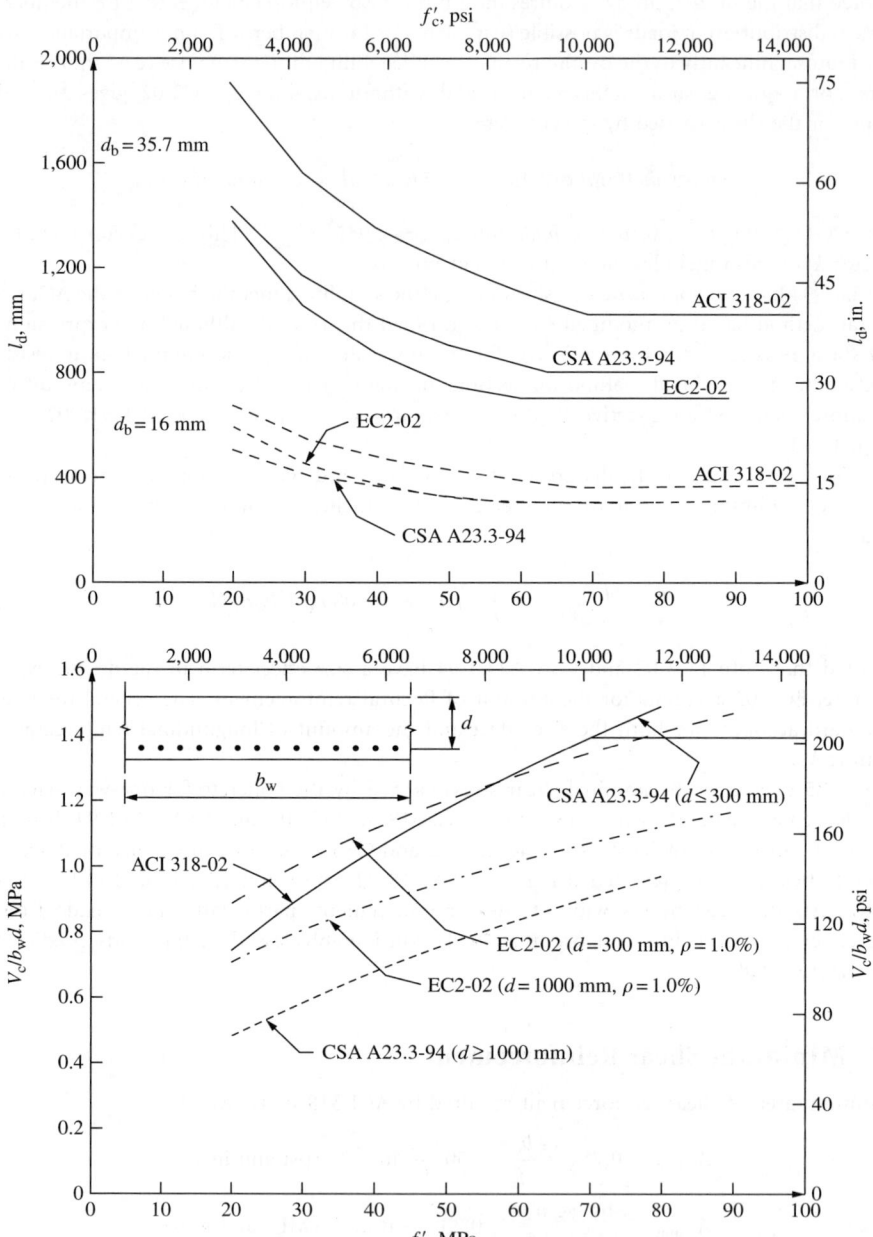

FIGURE 15.25 Comparison of the nominal shear stress carried by the concrete for one-way slabs without shear reinforcement.

The minimum area of shear reinforcement required by CSA A23.3-94 is given by

$$A_{v,\min} = 0.06\sqrt{f'_c}\,\frac{b_w s}{f_{yk}}\ \text{mm}^2 \quad (\text{MPa and mm}) \tag{15.18}$$

Figure 15.26 presents a comparison of the minimum amounts of shear reinforcement. The ACI 318-02 and CSA A23.3-94 expressions give similar amounts of shear reinforcement, while the EC2-02

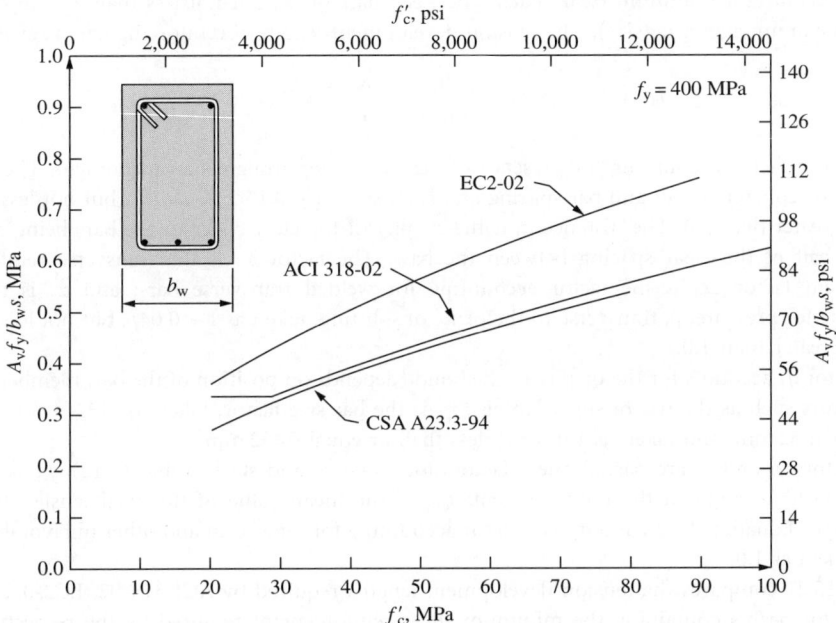

FIGURE 15.26 Minimum shear reinforcement requirements of the different codes as a function of f'_c.

expression gives a larger amount of minimum shear reinforcement for all concrete strengths. Both ACI 318-02 and CSA A23.3-94 require that a minimum amount of shear reinforcement be provided in all reinforced concrete flexural members where the factored shear force exceeds one-half the factored shear resistance of the concrete, except for slabs, footings, joists, and shallow beams meeting geometrical restrictions. In contrast, EC2-02 requires that all beams that contribute significantly to the overall resistance and stability of the structure must contain at least a minimum amount of shear reinforcement.

15.6.9 Development Length for Reinforcing Bars

The ACI 318-02 Code gives the following expression for the tension development length l_d for deformed bars for situations where the clear spacing of the bars being developed and clear cover are not less than d_b, and stirrups or ties throughout l_d are not less than the code minimum.

$$l_d = \frac{1}{20} \alpha \beta \gamma \lambda \frac{f_y}{\sqrt{f'_c}} d_b \quad (\text{psi and in.})$$

$$l_d = \frac{3}{5} \alpha \beta \gamma \lambda \frac{f_y}{\sqrt{f'_c}} d_b \quad (\text{MPa and mm})$$

(15.19)

where α is the reinforcement location factor, β is the coating factor, γ is the reinforcement size factor (0.8 for No. 6 or 19 mm and smaller bars and 1.0 for larger bars), and λ is the lightweight aggregate concrete factor.

The development length requirements in CSA A23.3-94 are based on the ACI Code provisions, but have been modified to account for the larger required minimum spacing between bars of $1.4d_b$, but not less than 30 mm. This results in shorter development lengths with the factor $\frac{3}{5}$ in the metric expression of Equation 15.19 being replaced by 0.45.

EC2-02 requires a minimum clear spacing between bars of d_b, but not less than 25 mm (assuming 20 mm maximum aggregate size). The tension development length of reinforcing bars is given by

$$l_d = \alpha_1 \alpha_2 \alpha_3 \alpha_4 \alpha_5 \frac{1}{0.7\alpha_{ct} f_{ctm}/\gamma_c} \frac{d_b}{4} \quad \text{(MPa and mm)} \tag{15.20}$$

where α_1 is a factor accounting for presence of bends (1.0 for straight bar anchorage). The factor α_2 accounts for concrete cover and bar spacing and is taken as $1 - 0.15(c_d - d_b)/d_b$, but not less than 0.7, and not greater than 1.0. The distance c_d is the smaller of the clear cover to the bars being developed and one half of the clear spacing between the bars. The factor α_3 is the transverse reinforcement confinement factor; α_4 is the factor accounting for welded transverse bars; and α_5 is the factor accounting for pressure, p, transverse to the plane of splitting, taken as $1 - 0.04p$, but not less than 0.7, and not greater than 1.0.

The factor η_1 accounts for the quality of the bond (depends on position of the bar, member size, and other factors such as the use of slip forming); η_2 is the bar size factor, taken as $(132 - d_b)/100$ for d_b greater than 32 mm, and taken as 1.0 for d_b less than or equal to 32 mm.

The factors γ_c and γ_s are partial safety factors for concrete and steel (Table 15.12); f_{yk} is the characteristic yield strength of the reinforcement; f_{ctm} is the mean value of the axial tensile strength of concrete (see Equation 15.9); and σ_{ct} is a factor accounting for long-term and other unfavorable effects, usually taken as 1.0.

Figure 15.27 compares the tension development lengths required by ACI 318-02, EC2-02, and CSA A23.3-94 for beams containing the minimum shear reinforcement required by the respective codes. This comparison is made for two bar sizes, having diameters of 16 mm (0.6 in.) and 35.7 mm (1.4 in.). In making this comparison it has been assumed that the factors in the ACI and CSA code expressions

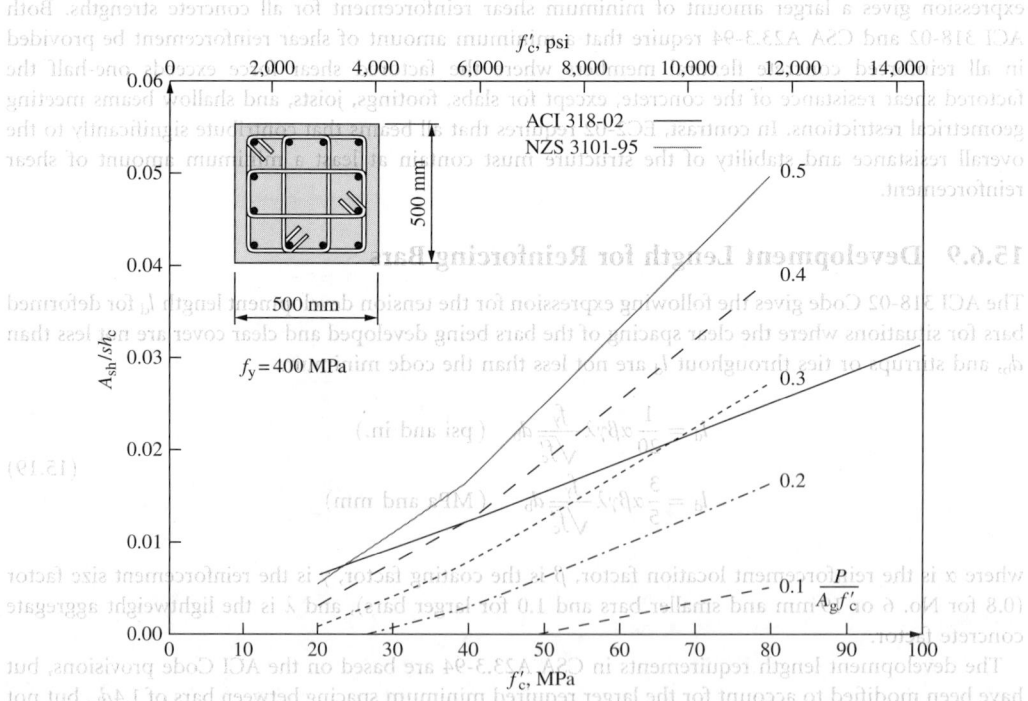

FIGURE 15.27 Comparison of the tension development lengths required by ACI 318-02, EC2-02, and CSA A23.3-94 for beams containing the minimum shear reinforcement, as a function of f_c'.

for reinforcement location, bar coating, and concrete density are all 1.0. It was also assumed that all of the α factors, η_1, and α_{ct} in Equation 15.20 are 1.0. In addition, it was assumed that the minimum concrete cover to the bars being developed was 40 mm (1.5 in.) and that spacing between the bars was between $1.4d_b$ and $2d_b$. Both the EC2-02 and CSA A23.3-94 expressions give shorter development lengths than the ACI 318-02 expression. This is due to the different requirements in the three codes regarding minimum shear reinforcement and minimum bar spacing. On the one hand, EC2-02 requires larger amounts of minimum shear reinforcement (Figure 15.25), but has the same minimum clear bar spacing as ACI 318-02. On the other hand, CSA A23.3-94 requires similar amounts of shear reinforcement as ACI 318-02 (Figure 15.25), but requires a larger minimum clear spacing between bars.

ACI 318-02 and CSA A23.3-94 limit the value of $\sqrt{f_c'}$ to 100 psi (25/3 MPa) and 8 MPa, respectively, in applying the expressions for determining the development length. EC2-02 cautions designers that due to the increasing brittleness of higher-strength concrete, the tensile strength of concrete required in calculating the bond strength should be limited to that corresponding to concrete with a compressive strength of 60 MPa, unless it can be verified that the average bond strength increases above this limit. It is noted that tests [82] indicate that bond failures in high-strength concrete members are more brittle than those in normal-strength concrete members unless adequate transverse reinforcement is provided. More research is needed to determine the relationship between the amount of confinement reinforcement and the required development length for higher-strength concrete.

15.6.10 Column Confinement for Seismic Design

ACI 318-02 requires that the total cross-sectional area of rectangular hoop reinforcement in ductile columns be not less than that given by the following expressions:

$$A_{sh} = 0.3 s h_c \frac{f_c'}{f_{yh}} \left(\frac{A_g}{A_{ch}} - 1 \right) \text{ in.}^2 \qquad (15.21)$$

$$A_{sh} = 0.09 s h_c \frac{f_c'}{f_{yh}} \left(\frac{A_g}{A_{ch}} - 1 \right) \text{ mm}^2 \qquad (15.22)$$

where A_{sh} is the total cross-sectional area of transverse reinforcement within spacing s, s is the hoop spacing, h_c is the cross-sectional dimension of the column core measured center-to-center of confining reinforcement, A_g is the gross cross-sectional area, A_{ch} is the cross-sectional area of the column measured out-to-out of transverse reinforcement, and f_{yh} is the specified yield strength of the transverse reinforcement.

NZS 3101-95 includes the influence of axial load level in determining the required A_{sh} as follows:

$$A_{sh} = \frac{(1.3 - \rho_g m) s h''}{3.3} \frac{A_g}{A_{ch}} \frac{f_c'}{f_{yh}} \frac{N^*}{\phi A_g f_c'} - 0.006 s h'' \qquad (15.23)$$

where ρ_g is the ratio of total longitudinal reinforcement area to cross-sectional area of column; $m = f_y/(0.85 f_c')$; h'' is the dimension of the concrete core measured perpendicular to the direction of the hoop bars to outside of peripheral hoop; and N^* is the design axial load at ultimate limit state.

Figure 15.28 compares the confinement requirements of ACI 318-02 and NZS 3101-95 as a function of concrete compressive strength. Unlike the ACI 318-02 requirements, the NZS 3101-95 requirements are strongly affected by the axial load. For low axial loads and low concrete compressive strengths, the

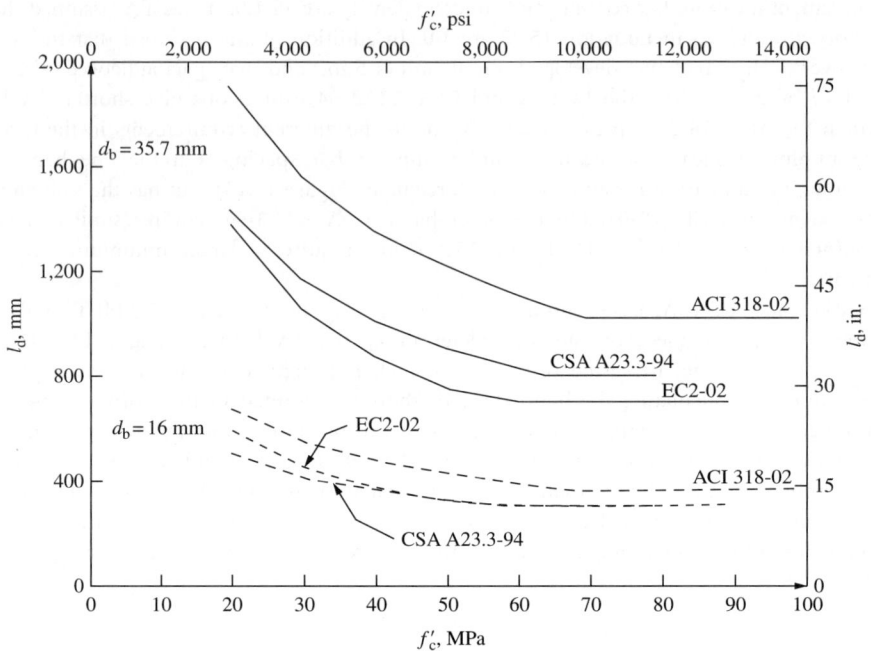

FIGURE 15.28   Transverse confinement reinforcement requirements of ACI 318-02 and NZS 3101-95 for rectangular columns as a function of f_c'.

New Zealand confinement equation results in considerably less confinement reinforcement than the ACI Code. For high axial loads and high-strength concrete, the NZS 3101-95 requirements are more stringent than those in ACI 318-02.

References

[1] Aïtcin, P.-C. 1998. "High-Performance Concrete," Modern Concrete Technology 5, E & FN SPON, London and New York, 591pp.

[2] Malier, Y. 1990. "Les béton à hautes performances," — Du matériaux à l'ouvrage, Presses de l'Ecole Nationale des Ponts et Chaussées, Paris, France, 1990, 507pp. (English edition: High Performance Concrete — from Materials to Structure, Ed. by Yves Malier, E & FN SPON, 1992, 541pp.)

[3] Zia, P., S.H., Ahmad, and Leming, M.L. 1996. "High-Performance Concretes: An Annotated Bibliography (1989–1994)," Federal Highway Administration, McLean, VA, Publication No. FHWA-RD-96-112, iv, 337pp.

[4] Zia, P. 1997. "State-of-the-Art of HPC: An International Perspective," Proceedings PCI/FHWA International Symposium on High Performance Concrete, New Orleans, LA, October 1997, pp. 49–59.

[5] Zia, P., Leming, M.L., Ahmad, S.H., Schemmel, J.J., Elliott, R.P., and Naaman, A.E. 1993. "Mechanical Behavior of High-Performance Concretes, Volume 1: Summary Report," SHRP-C-361, Strategic Highway Research Program, National Research Council, Washington, DC, xi, 98pp.

[6] Goodspeed, C.H., Vanikar, S., and Cook, R.A. 1996. High-performance concrete defined for highway structures, *Concrete Int.*, 18(2), 62–67.

[7] Perenchio, W.F. 1973. An evaluation of some of the factors involved in producing very high-strength concrete, *Res. Dev. Bull.*, No. RD014-01T, Portland Cement Association, Skokie, IL, 7pp.

[8] Mehta, P.K. and Aïtcin, P.C. 1990. "Microstructural basis of selection of materials and mix proportions for high strength concrete," in Second International Symposium on High Strength Concrete, SP-121, ACI, Detroit, MI, pp. 265–286.

[9] Aïtcin, P.-C. 1992. Private communication.

[10] Aïtcin, P.-C. and Mehta, P.K. 1990. Effect of coarse aggregate type or mechanical properties of high strength concrete. *ACI Mater. J.,* ACI, Detroit, MI, 87(2), 103–107.

[11] Cook, J.E. 1989. 10000 psi Concrete. *Concrete Int.,* 11(10), 67–75.

[12] ACI Standard 211.1. 1989. "Recommended practice for selecting proportions for normal weight concrete," ACI, Detroit, MI.

[13] SHPR-C/FR-91-103. 1991. "High Performance Concretes, a State of the Art Report," Strategic Highway Research Program, National Research Council, Washington, DC.

[14] Mehta, P.K. and Aïtcin, P.C. 1990. "Microstructural basis of selection of materials and mix proportions for high strength concrete," in Second International Symposium on High Strength Concrete, SP-121, ACI, Detroit, MI, pp. 265–286.

[15] Tagnit-Hamou, A. and Aïtcin, P.-C. 1993. Cement and superplasticizer compatibility, *World Cement,* 24(8), 38–42.

[16] Mokhtarzadeh, A. and French, C. 2000. Mechanical properties of high-strength concrete with consideration for precast applications, *ACI Mater. J.,* 97(2), 136–147.

[17] Okamura H. and Ouchi M. 2000. "Self-compactability of fresh concrete — liquid and solid," in Proceedings of High Performance Concrete — Workability, Strength and Durability, December 10–15, Ed. by Leung, Li and Ding, Hoang Kong & Shenzhen, China, pp. 161–170.

[18] "Teaching the materials science, engineering, and field aspects of concrete", Faculty Enhancement Workshop, July 25–30, 1993, Ed. by Cohen, Shah and Young, NSF-ACBM Center.

[19] Boutin, C. and Arnaud, L. 1995. Mechanical characterization of heterogeneous materials during setting. *Eur. J. Mech., A/Solids,* 14(4), 633–656.

[20] Keating, J., Hannant, D.J., and Hippert, A.P. 1989. Comparison of shear modulus and pulse velocity techniques to measure buildup structure in fresh cement pastes used in oil-well cementing. *Cement Concrete Res.,* 19, 554–566.

[21] Keating, J., Hannant, D.J., and Hippert, A.P. 1989. Correlation between cube strength, ultrasonic pulse-velocity and volume change for oil-well cement slurries. *Cement Concrete Res.,* 19, 715–726.

[22] Niyogi, S.K., Das Roy, P.K., and Roychaudhari, M. 1990. Acousto-ultrasonic study on hydration of Portland cement. *Ceramic Tran.,* 16, 137–145.

[23] Moukwa, M., Brodwin, M., Christo, S., Chang, J., and Shah, S.P. 1991. The influence of hydration process upon microwave properties of cements. *Cement Concrete Res.,* 21, 863–872.

[24] Beek, A.V. 2000. Dielectric properties of young concrete, no-destructive dielectric sensor for monitoring the strength development of young concrete. Ph.D. thesis, Delft University, The Netherlands.

[25] Pessiki, S.P. and Carino, N.J. 1998. Setting time and strength of concrete using impact-echo method. *ACI Mater. J.,* 95(5), 539–540.

[26] Subramaniam, K.V., Mohsen, J.P., Shaw, C.K., and Shah, S.P. 2002. Ultrasonic technique for monitoring concrete strength gain at early age. *ACI Mater. J.,* 99(5), 458–462.

[27] Rapoport, J., Popovics, J.S., Subramaniam, K.V., and Shah, S.P. 2000. Using ultrasound to monitor stiffening process of concrete with admixtures. *ACI Mater. J.,* 97(6), 675–683.

[28] Zongjin, Li and Li, Wenlai 2001. "No-contacting method for resistivity measurement of concrete specimen." US patent, in process.

[29] Whittington, H.W., McCarter, J., and Forde, M.C. The conduction of electricity through concrete. *Mag. Concrete Res.,* 33(114), 48–60.

[30] Banthea, N., Djeridane, S., and Pigeon, M. 1992. Electrical resistivity of carbon and steel micro-fiber reinforced cements. *Cement Concrete Res.,* 22, 804–814.

[31] Lakshminarayanan, V., Ramesh, P.S., and Rajagopalan, S.R. 1992. A new technique for the measurement of the electrical resistivity of concrete. *Mag. Concrete Res.,* 44(158), 47–52.

[32] Lide, D.R. (ed.). 1913–1915: "Handbook of Chemistry and Physics," 5-86, 75th edition, CRC Press, Boca Raton, FL.

[33] Li, Z.J., Wei, X.S., and Li, W.L. 2003. Preliminary interpretation of hydration process for cement-based materials at early ages. *Journal of ACI Materials*, 100(3), 253–254.

[34] Ahmad, S.H. 1981. Properties of confined concrete subjected to static and dynamic loading. Ph.D. thesis, University of Illinois, Chicago.

[35] Kaar, P.H., Hanson, N.W., and Capell, H.T. 1977. Stress-strain characteristics of high strength concrete. *Res. Dev. Bull.*, RD05101D, Portland Cement Association, Skokie, IL, 11pp.

[36] Nilson, A.H. and Slate, F.O. 1979. "Structural Design Properties of Very High Strength Concrete," Second Progress Report, NSF Grant ENG 7805124, School of Civil and Environmental Engineering, Cornell University, Ithaca, NY.

[37] Zongjin, Li, Kulkarni, S.M., and Shah, S.P. 1993. New test method for determining softening response of unnotched concrete specimen under uniaxial tension, *Exp. Mech.*, 33(3), 181–188.

[38] Xianyu Jin and Zongjin Li 2000. Investigation on mechanical properties of young concrete, *Materials and Structures*, RELIM, 33, 627–633.

[39] Zia, P., Leming, M.L., and Ahmad, S.H. 1991. "High-Performance Concrete: A State-of-the-Art Report," Strategic Highway Research Program, National Research Council, Washington, DC, 251pp. (SHRP-C/FR-91-103; PB92-130087).

[40] Burg, R.G. and Ost, B.W. 1992. Engineering properties of commercially available high-strength concretes. *Res. and Dev. Bull.*, No. RD104.01T, Portland Cement Association, Skokie, IL, 55pp.

[41] Sabir, B.B., Wild, S., and Asili, M. 1997. On the tortuosity of the fracture surface in concrete. *Cement Concrete Res.*, 27(5), 785–795.

[42] Tasdemir, C., Tasdemir, M.A., Lydon, F.D., and Barr, B.I.G. 1996. Effects of silica fume and aggregate size on the brittleness of concrete. *Cement Concrete Res.*, 26(1), 63–68.

[43] ACI Committee 209. 1993. "Prediction of Creep, Shrinkage, and Temperature Effects in Concrete Structures (ACI 209R-92)," ACI Manual of Concrete Practice, American Concrete Institute, Detroit, MI, Part 1, 47pp.

[44] Carette, G.G., Bilodeau, A., Chevrier, R.L., and Malhotra, V.M. 1993. Mechanical properties of concrete incorporating high volumes of fly ash from sources in the U.S. *ACI Mater. J.*, 90(6), 535–544.

[45] Sarkar, S.L. and Aïtcin, P.-C. 1990. "Importance of Petrological, Petrographical and Mineralogical Characteristics of Aggregates in Very High Strength Concrete," ASTM Special Technical Publication, No. 1061, pp. 129–144.

[46] Hindy, E.E., Miao, B., Chaallal, O., and Aïtcin, P.-C. 1994. Drying shrinkage of ready-mixed high-performance concrete, *ACI Mater. J.*, 91(3), 300–305.

[47] Paulson K.A., Nilson A.H., and Hover K.C. 1991. Long-term deflection of high-strength concrete beams, *ACI Materila*, 88(2), 197–266.

[48] Regourd, M., Mortureux, B., Aïtcin, P.-C., and Pinsonneault, P. 1983. "Microstructure of field concrete containing silica fume," in Proceedings, 4th International Symposium on Cement Microscopy, Nevada, pp. 249–60.

[49] Sarkar, S.L. and Aïtcin, P.-C. 1987. Comparative study of the microstructure of very high strength concretes. *Cement, Concrete Aggregates*, 9(2), 57–64.

[50] Aïtcin, P.-C. 1989. "From gigapascals to nanometers," in Proceedings, Engineering Foundation Conference on Advances in Cement Manufacture and Use, Potosi, Missouri, USA, The Engineering Foundation, pp. 105–130.

[51] Scrivener, K.L., Bentur, A., and Pratt, P.L. 1988. Quantitative characterization of the transition zone in high strength concretes, *Adv. Cement Res.*, 1(2), 230–237.

[52] Zongjin, Li, Chau, C.K., Ma, Baoguo, and Li, Faming 2000. "New natural polymer based durability enhancement admixture and corresponding concretes made with the admixture", US Patent No. 6,153,006, November.

[53] Zongjin Li, and Chau, C.K. 2000. Frost resistance of concrete incorporated with a natural polymer-based admixture, *Magazine of Concrete Research*, 53(2), 73–84.

[54] Zia, P., Ahmad, S.H., Leming, M.L., Schemmel, J.J., and Elliott, R.P. 1993. "Mechanical Behavior of High Performance Concretes, Volume 3: Very Early Strength Concrete." Strategic Highway Research Program, National Research Council, Washington, DC, xi, 116pp. (SHRP-C-363).

[55] Zia, P., Ahmad, S.H., Leming, M.L., Schemmel, J.J., and Elliott, R.P. 1993. "Mechanical Behavior of High Performance Concretes, Volume 4: High Early Strength Concrete." Strategic Highway Research Program, National Research Council, Washington, DC, xi, 179pp. (SHRP-C-364).

[56] Zia, P., Ahmad, S.H., Leming, M.L., Schemmel, J.J., and Elliott, R.P. 1993. "Mechanical Behavior of High Performance Concretes, Volume 5: Very High Strength Concrete," Strategic Highway Research Program, National Research Council, Washington, DC, xi, 101pp. (SHRP-C-365).

[57] Okada, E., Hisaka, M., Kazama, Y., and Hattori, K. 1981 "Freeze-Thaw Resistance of Super-plasticized Concretes. Developments in the Use of Superplasticizer," ACI SP-68, pp. 269–282.

[58] Foy, C., Pigeon, M., and Bauthia, N. 1988. Freeze-thaw durability and deicer salt scaling resistance of a 0.25 water-cement ratio concrete. *Cement Concrete Res.*, 18, 604–614.

[59] Malhotra, V.M., Painter, K., and Bilodeau, A. 1987. "Mechanical properties and freezing and thawing resistance of high-strength concrete incorporating silica fume," in Proceedings, CANMET-ACI International Workshop on Condensed Silica Fume in Concrete, Montreal, Canada, p 25.

[60] Hammer, T.A. and Sellevold, E.J. 1990. "Frost Resistance of High Strength Concrete," ACI SP-121, pp. 457–487.

[61] Mather, K. 1982. "Current Research in Sulfate Resistance at the Waterways Experiment Station," George Verbeck Symposium on Sulfate Resistance of Concrete, ACI SP-77, Detroit, USA, pp. 63–74.

[62] Cohen, M.D., and Bentur, A. 1988. Durability of Portland cement-silica fume pastes in magnesium and sodium sulfate solutions. *ACI Mater. J.* 85, 148–157.

[63] Gjørv, O.E. 1983. "Durability of Concrete Containing Condensed Silica Fume," ACI SP-79. Vol. II, pp. 695–708.

[64] Davis, G. and Oberholster, R.E. 1987. Use of the NMRI accelerated test to evaluate the effectiveness of mineral admixtures in preventing the alkali-silica reaction. *Cement Concrete Res.* 16(2), 97–107.

[65] Hooton, R. 1987. "Some aspects of durability with condensed silica fume in concretes," in Proceedings, CANMET-ACI International Workshop on Silica Fume in Concrete, Montréal, Canada.

[66] Petersson, P.E. 1984. "Inverkan av salthaltiga miljøer på betongens frostbestandighet," Technical Report SP-RAPP 1984: 34 ISSN 0280-2503, National Testing Institute, Borås, Sweden.

[67] Caldarone, M.A., Gruber, K.A., and Burg, R.G. 1994. High-reactivity metakaolin: a new generation of mineral admixture. *Concrete Int.* 16(11), pp. 37–40.

[68] Bilodeau, A. Carette, G.G., Malhotra, V.M., and Langley, W.S. 1991. Influence of curing and drying on salt scaling resistance of fly ash concrete. Durability of concrete, in Second International Concrete, 1991, Montreal, Canada, Ed. by V.M. Malhotra, American Concrete Institute, Detroit, MI, Vol. 1, pp. 201–228 (ACI SP-126).

[69] Paultre, P. and Michell, D. 2003, Code provisions for high-strength concrete — an international perspective, *Concrete International*, 25(5), 76–90.

[70] de Almeida, I.R. 1994. "Abrasion resistance of high strength concrete with chemical and mineral admixtures." Durability of concrete in Proceedings of the Third International Conference, May 22–28, 1994, Nice, France, Ed. by V.M. Malhotra, American Concrete Institute, Detroit, MI, pp. 1099–1113 (ACI SP-145).

[71] ACI Committee 318, 2002. "Building Code Requirements for Structural Concrete (ACI 318-02) and Commentary (318R-02)," and "Metric Building Code Requirements for Structural Concrete (ACI 318M-02) and Commentary (318RM-02)," American Concrete Institute, Farmington Hills, MI, 443pp. and 443pp. respectively.

[72] Comite Euro-International du Beton. 1993. "CEB-FIP Model Code 1990," Bulletin d'information No. 213/214, Thomas Telford, London, 437pp.

[73] Comite Europeen de Normalisation (CEN). 2002. "Eurocode 2: Design of Concrete Structures. Part 1 — General Rules and Rules for Buildings," prEN 1992-1, 211pp.

[74] Canadian Standards Association. 1994. "Design of Concrete Structures, CSA A23.3 1994," Rexdale, Ontario, Canada, 199pp.

[75] Standards Association of New Zealand. 1995. "Concrete Design Standard, NZS 3101:1995, Part 1," and "Commentary on the Concrete Design Standard, NZS 3101:1995, Part 2," Wellington, New Zealand, 256pp. and 264pp. respectively.

[76] CEB-FIP Working Group on High-Strength/High-Performance Concrete. 1995. "High-Performance Concrete — Recommended Extension to the Model Code 90," CEB Bulletin d'information No. 228, Comite International du Beton, Lausanne, Switzerland, 33pp.

[77] Carrasquillo, R.L., Nilson, A.H., and Slate, F.O. 1981. Properties of high-strength concrete subjected to short-term loads. *ACI J. Proc.*, 78(3), 171–178.

[78] ACI Committee 363. 1984. State-of-the-art report on high-strength concrete. *ACI J. Proc.*, August 81(4), 364–411.

[79] Collins, M.P., Mitchell, D., and MacGregor, J.G. 1993. Structural design considerations for high-strength concrete. *Concrete Int.*, 15(5), 27–34.

[80] Collins, M.P. and Mitchell, D. 1990. "Prestressed Concrete Structures," Prentice Hall, Englewood Cliffs, NJ, 766pp.

[81] ACI Committee 318. 2002. "Building Code Requirements for Structural Concrete (ACI 318-99) and Commentary (318R-99)," and "Metric Building Code Requirements for Structural Concrete (ACI 318M-99) and Commentary (318RM-99)," American Concrete Institute, Farmington Hills, MI, 391pp. and 391pp. respectively.

[82] Azizinamini, A., Stark, A. Roller, J.J. and Ghosh, S.K. 1993. Bond performance of reinforcing bars embedded in high-strength concrete, *ACI Struct. J.*, 90(5), pp. 554–561.

16

Fiber-Reinforced Polymer Composites

Lawrence C. Bank
*Department of Civil and
Environmental Engineering,
University of Wisconsin,
Madison, WI*

16.1   Analysis and Design of FRP Pultruded Structures

16.1.1   Introduction

Truss structures and braced framed structures have been designed and constructed using thin-walled composite material fiber-reinforced polymer (FRP, commonly referred to as fiber-reinforced plastic) pultruded members for over 30 years. Pultruded structural members (referred to as pultrusions or pultruded shapes or pultruded profiles) have been used in a significant number of structures to-date, such as pedestrian bridges, vehicular bridges, building frames, stair towers, cooling towers, and walkways and platforms (Bakis et al. 2002). Pultruded members are often the materials of choice where significant corrosion and chemical resistance is required (food processing, cooling towers, offshore platforms), where electromagnetic transparency is required (electronics manufacturing, radomes), or where accessibility is limited and lightweight skeletal structures are assembled on-site (pedestrian bridges in parklands).

Pultrusion is a continuous, and highly cost-effective, manufacturing technology for producing constant cross-sectional FRP structural shapes (profiles) (Meyer 1985). Pultruded materials consist of fiber reinforcements (typically, glass fiber or carbon fiber) and thermosetting resins (typically, polyester, vinylesyter, and epoxy polymers). The fiber architecture within a thin "panel" or "plate" in a pultruded shape (such as a web or a flange) typically consists of longitudinal continuous fiber bundles (called

---

rovings or tows) and continuous filament fiber mats (CFM). In specialized custom pultrusions bidirectional stitched fabrics are also used to improve mechanical or physical properties. The volume fraction of the fiber reinforcement in a pultruded shape is typically between 30 and 50%. In addition to the base polymer resin, pultruded materials typically contain inorganic fillers, chemical catalysts and promoters, release agents, ultraviolet retardants, fire retardants, pigments, and surfacing veils. Commercially produced pultruded materials made with glass fibers are sold in the $2 to $3 per pound range. Specialized pultruded products utilizing higher-performance carbon fibers, sophisticated fabrics, and higher-performance resin systems may be considerably more costly.

Currently, no building codes or specifications provide consensus guidelines for the design of composite material framed structures using either "standard" or "custom" pultruded structural shapes. Two design manuals that are currently available provide much of the fundamental information that is needed for the analysis and the design of pultruded structures. These are the *Structural Plastics Design Manual* (SPDM 1984) and the *Eurocomp Design Code and Handbook* (EDCH 1996).

While there are no text books currently available that specifically cover the design of FRP pultruded structures for structural engineering (as there are for steel, concrete, and wood structures), there are many excellent classic texts on the mechanics (and, to a lesser extent, design) of composite materials for aerospace and mechanical engineering structures, such as Tsai and Hahn (1980), Vinson and Sierakowski (1986), Agarwal and Broutman (1990), and Jones (1999). Books by Hollaway (1993), Barbero (1998), and Hollaway and Head (2001) bridge the gap between the mechanics of composite materials and composite structures and lay more emphasis on applications to civil engineering.

Pultruded structural shapes are produced by a number of commercial companies throughout the world (e.g., Strongwell [United States], Creative Pultrusions [United States], Bedford Reinforced Plastics [United States], Fiberline [Denmark], TopGlass [Italy], Pacific Composites [Australia, Asia, United Kingdom]). There is a commonality (similarity) in the types of shapes and the properties of these shapes, and although there is no official standard for these shapes and their properties, they are referred to commonly as standard pultruded structural shapes. Company-specific literature provides details of the geometries and properties of their common shapes (e.g., Fiberline 1995; Creative Pultrusions 2000; Strongwell 2002). Commonly produced standard pultruded shapes that are usually available from stock inventories include I section, wide-flange (WF), square tube, rectangular tube, channels, and angles, and plate materials. Standard sections range from 2 in. in height and width to approximately 12 in. in height and width and have pultruded material thicknesses of $\frac{1}{4}$ to $\frac{1}{2}$ in. Company design guides also provide a design basis and analytical methods used for designing with their products. Design equations in company design guides are often empirical and are based on fits to test data.

Pultrusion companies also produce custom pultruded shapes. Companies generally do not make geometry and property data available for their custom shapes. A number of industry groups represent and loosely coordinate the activities of pultrusion manufacturers. Leading groups are the Pultrusion Industry Council of the American Composites Manufacturers Association (ACMA) and the European Pultrusion Technology Association (EPTA).

Thin-walled shapes can be produced by many other processes, such as filament winding, resin-transfer-molding, and centrifugal casting. No commonality exists in these products (except perhaps for composite poles and pipes). They are generally not intended for use in framed structures and are not discussed in what follows.

Experimental testing has been conducted in earnest since the mid-1980s to characterize the behavior and performance of pultruded structures and structural members. Much effort has been devoted to investigating whether the design equations presented in design manuals are in fact suitable for standard pultruded shapes and whether or not design approaches used for conventional thin-walled metallic members are applicable to pultruded structures. Less work has been done on the applicability of the equations to custom shapes. Based on these studies, there is sufficient evidence that many of the design equations presented in the design manuals are in fact suitable for use when designing framed structures with standard pultruded shapes.

Although the suitability of using specific analytical equations to predict the behavior of pultruded members and structures has been confirmed, there is significantly less consensus on the established safety factors (for allowable-stress design [ASD]) and the material resistance factors (for load and resistance factor design) than exists for other materials, such as steel and reinforced concrete.

16.1.2 Properties of Pultruded Shapes

Standard FRP pultruded shapes are constructed of "specially" orthotropic thin plate elements. The internal architecture of the plate elements (the "lamination"), which may consist of roving bundles, fabrics, and mats, is such that the laminate is both balanced and symmetric to prevent extensional–shear coupling and extensional–flexural coupling (Tsai and Hahn 1980) in the plate and in the section. Specially orthotropic means that the orthotropic axes of the plate elements coincide with the structural axes of the pultruded section.

The in-plane (or extensional) stiffness properties of an orthotropic plate of pultruded material can be defined by five independent in-plane "effective" engineering constants:

E_L = longitudinal elastic modulus

E_T = transverse elastic modulus

G_{LT} = in-plane shear modulus

ν_L = in-plane longitudinal (major) Poisson ratio

ν_T = in-plane transverse (minor) Poisson ratio = $\nu_L E_T / E_L$

It is typically assumed that the stiffness properties are the same in tension and in compression; however, manufacturers generally report tensile and compressive moduli separately. In this case the subscript t is used for tension and the subscript c is used for compression. The longitudinal and the transverse directions are also commonly referred to as lengthwise and the crosswise directions, respectively.

The in-plane extensional strength properties of an orthotropic plate of pultruded material can be defined by five independent constants:

$\sigma_{L,t}$ = longitudinal tensile strength

$\sigma_{L,c}$ = longitudinal compressive strength

$\sigma_{T,t}$ = transverse tensile strength

$\sigma_{T,c}$ = transverse compressive strength

τ_{LT} = in-plane shear strength

The out-of-plane (through-the-thickness [TT]) properties of an orthotropic plate of pultruded material are generally not measured and are not used in design. The exception to this is the interlamina shear strength (ILSS), also known as the short beam shear (SBS) strength, which is routinely measured. It is a measure of the transverse shear strength of the laminate and is denoted by

τ_{TT} = interlamina shear strength

In addition, for design purposes a "quasi-structural" bearing strength, either in the longitudinal or in the transverse direction, is also measured and reported as

$\sigma_{L,br}$ = longitudinal bearing strength

$\sigma_{T,br}$ = transverse bearing strength

Since a plate of pultruded material is not homogenous through its thickness (i.e., it consists of layers or laminae [or plies] of fiber-reinforced materials having different in-plane properties), the in-plane flexural properties of pultruded materials are not the same as the in-plane extensional properties of the material due to the dependence on the location of the layers through the thickness of the material (Tsai and Hahn 1980). Consequently, there exists a set of effective flexural properties for pultruded material that is analogous to the set of effective extensional properties described earlier in this section. These properties are often represented by the superscript b to distinguish them from the extensional properties (often represented by the superscript 0). For example, E_L^b and E_L^0 are the in-plane longitudinal flexural modulus and the in-plane longitudinal extensional modulus, respectively.

The in-plane extensional and flexural engineering properties may be obtained theoretically from classical lamination theory (CLT) in which the laminated plate is characterized by in-plane extensional stiffness coefficients (A_{ij} with $i, j = 1, 2, 6$) and in-plane flexural stiffness coefficients (D_{ij} with $i, j = 1, 2, 6$) (Tsai and Hahn 1980). For global (member) stress calculations the in-plane extensional properties (the A matrix coefficients) are used to determine the engineering extensional properties. For local (plate) stress calculations the flexural properties (the D matrix coefficients) are used to determine the flexural engineering properties. In this approach the laminate is assumed to be inhomogenous on the ply level but nevertheless able to be represented by laminate properties for the purposes of stress-resultant calculations.

Alternatively, the in-plane engineering stiffness properties may be obtained from standard tests on coupons extracted from the pultruded section. In this approach the laminate is assumed to be homogenous. Standard ASTM tests (ASTM 2003) that are recommended for determining the properties of pultruded materials can be found in Bank et al. (2003) and in manufacturers' design guides (Creative Pultrusions 2000; Strongwell 2002) and are shown in Table 16.1.

If the orthotropic plates are assumed to be homogenous, then the plate flexural properties can be calculated from the in-plane extensional engineering properties (either obtained from test data or from

TABLE 16.1   Recommended ASTM Test Methods for Pultruded Materials in Shapes

Measured property	Recommended ASTM test methods
Mechanical properties	
Strength	
Longitudinal tensile strength	D 3039, D 5083, D 638
Longitudinal compressive strength	D 3410, D 695
Longitudinal short beam shear strength	D 2344
Longitudinal bearing strength	D 5961, D 953
In-plane shear strength	D 5379, D3846
Impact resistance	D 256
Transverse tensile strength	D 3039, D 5083, D 638
Transverse compressive strength	D 3410, D 695
Transverse short beam shear strength	D 2344
Transverse bearing strength	D 5961, D 953
Stiffness	
Longitudinal tensile modulus	D 3039, D 5083, D 638
Longitudinal compressive modulus	D 3410, D 695
Major (longitudinal) Poisson ratio	D 3039, D 5083, D 638
In-plane shear modulus	D 5379
Transverse tensile modulus	D 3039, D 5083, D 638
Transverse compressive modulus	D 3410, D 695
Physical properties	
Fiber volume fraction	D 3171, D 2584
Density	D 792
Barcol hardness	D 2583
Glass transition temperature/heat distortion temperature	E 1356, E 1640, D 648, E 2092
Water absorbed when substantially saturated	D 570
Longitudinal coefficient of thermal expansion	E 831, D 696
Transverse coefficient of thermal expansion	E 831, D 696
Flash ignition temperature	D 1929

the in-plane extensional matrix). This assumption is frequently made in the analysis of pultruded structures. The orthotropic plate flexural stiffnesses are then given as

$$D_{\mathrm{L}} = \frac{E_{\mathrm{L}} t_{\mathrm{p}}^3}{12(1 - \nu_{\mathrm{L}} \nu_{\mathrm{T}})} \tag{16.1}$$

$$D_{\mathrm{T}} = \frac{E_{\mathrm{T}} t_{\mathrm{p}}^3}{12(1 - \nu_{\mathrm{L}} \nu_{\mathrm{T}})} \tag{16.2}$$

$$D_{\mathrm{LT}} = \frac{\nu_{\mathrm{T}} E_{\mathrm{L}} t_{\mathrm{p}}^3}{12(1 - \nu_{\mathrm{L}} \nu_{\mathrm{T}})} = \frac{\nu_{\mathrm{L}} E_{\mathrm{T}} t_{\mathrm{p}}^3}{12(1 - \nu_{\mathrm{L}} \nu_{\mathrm{T}})} \tag{16.3}$$

$$D_{\mathrm{SS}} = \frac{G_{\mathrm{LT}} t_{\mathrm{p}}^3}{12} \tag{16.4}$$

Alternatively, the plate flexural properties (the Ds above) may be obtained from the in-plane flexural properties, however, this is seldom done in practice for commercially produced pultruded materials consisting of layers of unidirectional roving and continuous filament mats. In this case, for example, the longitudinal flexural stiffness, D_{L}, is written as (Barbero 1998)

$$D_{\mathrm{L}} = \frac{E_{\mathrm{L}}^{\mathrm{b}} t_{\mathrm{p}}^3}{12(1 - \nu_{\mathrm{L}}^{\mathrm{b}} \nu_{T}^{\mathrm{b}})} \tag{16.5}$$

The in-plane strength properties may be obtained from theoretical calculations or from testing of coupons taken from the laminate. Where theoretical predictions are used, the first-ply-failure (FPF) is assumed to represent the strength of the laminate. Coupon testing is highly recommended for obtaining the strength properties.

Property data provided in pultrusion company manuals and design guides are obtained from tests on coupons of pultruded materials taken from sections. Data reported are typically applicable to a broad range of section sizes and types. The specific fiber architectures (volume fractions of roving, mats, and fabrics) in the different section sizes and types varies. Fiber architecture may also vary within the section (e.g., web reinforcement architecture may be different from flange reinforcement architecture). The properties given by manufacturers can be assumed to be lower bounds for the sections indicated in the manuals. No data are provided to determine the statistics of these properties for use in probabilistic-based design. In addition, the property data are not related to the capacity of members subjected to specific loading conditions (e.g., axial load, flexure) and are based on coupon testing. Representative mechanical property data for pultruded materials used in commercially available pultruded shapes, flat sheets, and rods are shown in Table 16.2. Properties of pultruded carbon-reinforced epoxy strips for structural strengthening of concrete are shown for comparison purposes. Selected physical and electrical property data reported by manufacturers are shown in Table 16.3.

It is assumed that the pultruded material behaves in a linear elastic manner in tension, compression, and shear in both the longitudinal and the transverse directions and that failure is brittle. This assumption is reasonable in the service range ($\sim$20% ultimate) but is not reasonable, especially in the shear and transverse directions, at higher loads where the stress–strain behavior is highly nonlinear.

All orthotropic plates of pultruded material in pultruded shapes are assumed to be thin (i.e., the out-of-plane thickness is an order of magnitude less than the in-plane length and breadth of the plate).

Standard pultruded structural shapes are often assumed to be homogenous on the section level (i.e., the shape consists of plates all having the same properties). However, it should be noted that some manufacturers (e.g., Creative Pultrusions) have optimized their shapes and provide properties separately for the webs and the flanges of many of their shapes.

It is assumed that the properties of the junctions between the plates (e.g., the web–flange junctions in an I beam) are the same as those of the plates themselves.

The ratio of the longitudinal modulus to the shear modulus, $E_{\mathrm{L}}/G_{\mathrm{LT}}$, for pultruded orthotropic material plates can be much larger than that for isotropic material plates. As a result of this, shear deformation

TABLE 16.2 Mechanical Properties of Typical Commercially Produced FRP Pultruded Materials

	Glass-reinforced vinylester shapes (WF)[a,b] (0.25 to 0.5 in. thick)	Glass-reinforced vinylester flat sheet[a,b] ($\frac{3}{8}$ to 1 in. thick)	Glass-reinforced vinylester rods[a,b] (0.25 to 2 in. diameter)	Carbon-reinforced epoxy strip[c] (0.047 in. thick)
Fiber volume (est)[d]	25–40	20–25	50–60	65
Fiber architecture	Roving and CFM	Roving and CFM	Roving only	Tow only
Strength ($\times 10^3$ psi)				
Tensile, longitudinal	30–46	20	100	406
Tensile, transverse	7–12	10	NR[e]	NR
Compressive, longitudinal	30–52	24	65	NR
Compressive, transverse	16–20	16	NR	NR
Shear, in-plane	4.5–7	NR	NR	NR
Shear, out-of-plane	3.9–4.5	6	8	NR
Flexural, longitudinal	30–49	30–35	100	NR
Flexural, transverse	10–19	15–18	NR	NR
Bearing, longitudinal	30–39	32	NR	NR
Bearing, transverse	26–34	32	NR	NR
Stiffness ($\times 10^6$ psi)				
Tensile, longitudinal	2.6–4.1	1.8	6.0	23.9
Tensile, transverse	0.8–1.4	1.4	NR	NR
Compressive, longitudinal	2.6–3.8	1.8	NR	NR
Compressive, transverse	1.0–1.9	1.0	NR	NR
Shear, in-plane	0.43–0.50	NR	NR	NR
Flexural, longitudinal	1.6–2.0	2.0	6.0	NR
Flexural, transverse	0.8–1.7	1.1–1.4	NR	NR
Poisson ratio, longitudinal	0.33–0.35	0.32	NR	NR

[a] Strongwell (2002).
[b] Creative Pultrusions (2000).
[c] Sika (2000).
[d] Estimated.
[e] NR, not reported by manufacturers.

TABLE 16.3 Physical and Electrical Properties of Typical Commercially Produced FRP Pultruded Materials

	Glass-reinforced vinylester shapes (WF)[a,b] (0.25 to 0.5 in. thick)	Glass-reinforced vinylester flat sheet[a,b] ($\frac{3}{8}$ to 1 in. thick)	Glass-reinforced vinylester rods[a,b] (0.25 to 2 in. diameter)	Carbon-reinforced epoxy strips[c] (0.047 in. thick)
Fiber volume (est)[d]	25–40	20–25	50–60	65
Fiber architecture	Roving and CFM	Roving and CFM	Roving only	Tow only
CTE longitudinal ($10^{-6}/°F$)	4.4	8.0	3.0	NR[e]
CTE transverse ($10^{-6}/°F$)	NR	NR	NR	NR
Barcol hardness	45	40	50–55	NR
24-h water absorption (% max.)	0.6	0.6	0.25	NR
Dielectric strength — longitudinal (kV/in.)	35–40	35–40	40	NR
Density (lb/in.³)	0.060–0.070	0.060–0.070	0.072–0.076	NR

[a] Strongwell (2002).
[b] Creative Pultrusions (2000).
[c] Sika (2000).
[d] Estimated.
[e] NR, not reported by manufacturers.

plays a significant role in the analysis of thin-walled pultruded shapes. The effects of shear deformation should be accounted for in deflection calculations as well as in stability calculations wherever possible.

The properties of pultruded shapes are affected by the environment in which they are used and the conditions under which they are used. These "environmental-use" effects can include time, temperature, radiation, solvents, fire, impact, abrasion, and fatigue. The way in which these effects are accounted for in design is not yet clear. In some cases (e.g., temperature, creep) some guidance is available. The safety factors recommended by pultrusion companies for use in design are intended to account for some of these effects.

In the case of elevated temperature service, pultrusion companies specifically recommend reductions in strength and stiffness properties as a function of temperature for different resin systems (usually isophthalic polyester or vinylester). At 65°C (150°F) the strength of isophthalic polyester resin pultrusions is 50% of the room temperature strength and the modulus is 85% of the room temperature modulus. For vinylester resin pultrusions at 65°C (150°F) the strength and modulus are 80 and 90% of their room temperatrue values, respectively. At temperatures higher than 65°C (150°F) the use of isophthalic polyester resin pultrusions is not recommended and vinylester resin pultrusions are recommended to a maximum use temperature of 93°C (200°F) with further strength and stiffness reductions (strength = 50%, moduli = 85%).

Stiffness and strength properties used in design are typically obtained from short-term testing. Long-term changes in stiffness and strength can be accounted for using different models. For creep deflections the models proposed by the SPDM (1984) developed by Findley can be used to determine effective *viscoelastic* moduli for use in predicting long-term deflections of pultruded structures.

Long-term degradation due to corrosive environments is typically determined by use of *corrosion resistance guides* that provide recommendations on the use of different pultruded materials in a variety of different chemical environments and temperatures. These guides are based on coupon test data. Fundamental studies to develop models to predict the long-term degradation of FRP composites in different service environments are ongoing but cannot yet be used for reliable lifetime prediction.

16.1.3 Design Basis for Pultruded Structures

A basis for design of FRP pultruded structures is needed. In the absence of an approved design code for FRP pultruded structures a *design basis* must be agreed upon by all parties (owner, designer, contractor, and local authorities). Four types of design basis approaches are typically used for FRP pultruded structures:

1. Allowable-stress design (ASD).
2. Limit states design (LSD).
3. Load and resistance factor design (LRFD).
4. Performance-based design (PBD).

16.1.3.1 Allowable-Stress Design

ASD requires that the calculated design stresses (the required demand, σ_{reqd}) obtained from nominal service loads be less than the ultimate strengths (the capacity, σ_{ult}) divided by an appropriate factor of safety (SF). The ultimate strength divided by the safety factor is termed the *allowable* stress (σ_{allow}). Safety factors have traditionally been determined from industry practice and are not based on probabilistic methods. They do not provide a measurable reliability index

$$\sigma_{reqd} \leq \frac{\sigma_{ult}}{SF} \qquad (16.6)$$

ASD is recommended by most pultrusion companies (e.g., Creative Pultrusions 2000; Strongwell 2002) in their design manuals. For ASD the safety factors typically recommended by pultrusion companies for their common structural shapes are

Flexural members (beams) = 2.5
Compression members (columns) = 3.0
Shear = 3.0
Connections = 4.0

These factors are applied to strength and stability. For deformation calculations the safety factors recommended are

Longitudinal and transverse moduli $= 1.0$
Shear moduli $\qquad = 1.0$

These factors are applied to the material properties and design equations provided in the company literature. These design equations are often empirically derived from test data. Material properties are provided in the company literature for different grades of pultruded material. Designs are based on unfactored nominal service loads.

ASD is also recommended by AASHTO (2001) in the *Standard Specifications for Structural Supports for Highway Signs, Luminaires and Traffic Signals*. The design equations provided in this specification are based on theoretical equations for orthotropic plates, beams, and columns of linear elastic materials made of glass reinforced polyester. Structural members may be manufactured by pultrusion, centrifugal casting, or filament winding. The following minimum safety factors are recommended:

Bending strength $\quad = 2.5$
Tensile strength $\quad = 2.0$
Compressive strength $= 3.0$
Shear strength $\quad = 3.0$
Moduli $\qquad = 1.0$

Material properties to be used in the design equations are obtained from American Standard for Testing and Materials (ASTM) standard test methods. Designs are based on loads provided in the American Association for State Highway and Transportation Officials (AASHTO) specification. Full-scale testing according to ASTM 4923 is provided as a PBD method as an alternative to the analytical design. Where the analytical ASD method is used "design calculations provided shall be verified by documented test results on similar structures." (AASHTO 2001, p. 8-5).

ASD is also commonly used by structural designers in conjunction with design equations based on theoretical equations for orthotropic plates, beams, and columns. Equations provided in the SPDM (1984) are typically used in calculations. Material properties are often obtained from standard ASTM tests. Minimum properties are usually specified in a *design basis document* (often referred to as a "Special Provision to the Specifications"). Safety factors recommended by the pultrusion companies are usually used in this approach. Designs are based on unfactored nominal service loads.

16.1.3.2 Limit States Design

LSD requires that the design requirement (or actions) of the loads be less than the design resistance (or capacity) of the structure or structural members. The design loads and capacity are both multiplied by appropriate factors. The two primary limit states considered are the ultimate limit state (ULS) and the serviceability limit state (SLS). The ULS is associated with collapse, instability, and failure of the structure or structural members. The SLS is associated with states in which excessive deflection, vibration, and degradation make the structure or structural members unserviceable (EDCH 1996).

LSD is recommended by the *Structural Plastics Design Manual*. The service loads are factored to account for variations in applied loads, eccentricities, and differences between analytical and real behavior. Specific recommendations for load factors are not provided; however, example values are suggested for some cases. Capacity reduction factors are recommended for design of different structural members to account for long-term load effects and manufacturing variations. Example values are suggested. Design equations are based on the analysis of orthotropic plates, beams, and columns. In many cases equations for isotropic metallic structures are suggested where design equations are not available for FRP members. Material properties are obtained from manufacturers or from ASTM standard tests.

For the design of a pultruded column the following load and capacity reduction factors are given as an example (SPDM 1984, p. 717):

Load factors
    All service loads $= 2.5$
Capacity reduction factors
    Ultimate compressive strength $= 0.5$
    Elastic moduli $= 0.7$

For the design of a pultruded beam the following load and capacity reduction factors are given as an example (SPDM 1984, p. 741):

Load factors
    All service loads $= 2.0$
Capacity reduction factors
    Ultimate compressive strength $= 0.5$
    Ultimate tensile strength $= 0.4$
    Ultimate shear strength $= 0.3$
    Elastic moduli $= 0.7$

The load and capacity reduction factors suggested are based on standard industry practice and were not determined by probabilistic methods. They do not provide a measurable reliability index.

LSD is also recommended by *Eurocomp Design Code and Handbook* (EDCH). In the Eurocomp approach load factors are provided by the pertinent European Codes. Material partial safety factors, γ_m, are used to obtain appropriate capacity reduction factors. Material properties are divided by partial safety factors. According to the Eurocode the partial safety factors of FRP composites for structural applications account for (1) the method in which the material property data were obtained ($\gamma_{m,1}$), (2) the material manufacturing process ($\gamma_{m,2}$), and (3) the effects of environment and the duration of loading ($\gamma_{m,3}$). The material partial safety factor is given as the product of the three coefficients, $\gamma_m = \gamma_{m,1} \times \gamma_{m,2} \times \gamma_{m,3}$, and may not be less than 1.5 for ULSs and not less than 1.3 for SLSs. The following are examples of recommended values of the coefficients:

Material property data derived from theory $\gamma_{m,1} = 1.5$
Material property data derived from testing $\gamma_{m,1} = 1.15$
Fully cured pultruded material $\gamma_{m,2} = 1.1$
Nonfully cured pultruded material $\gamma_{m,2} = 1.7$
Hand-layup fully cured material $\gamma_{m,2} = 1.4$
Hand-layup non-fully cured material $\gamma_{m,2} = 2.0$
Long-term loading for $0 < T < 25°C$ and $T_g > 55°C$ $\gamma_{m,3} = 2.5$
Short-term loading for $25 < T < 50°C$ and $55 < T_g < 80°C$ $\gamma_{m,3} = 1.2$
Short-term loading for $25 < T < 50°C$ and $T_g > 90°C$ $\gamma_{m,3} = 1.0$

where T_g is the glass transition temperature (or the heat distortion temperature) of the FRP composite material. The material manufacturer is required to provide proof that the material is fully cured before the designer can use the fully cured partial safety coefficients.

Additionally, the design resistance is multiplied by an *analytical uncertainty factor* that ranges from 0.5 to 1.0 depending on the sophistication of the analytical method used in the design procedure. The material partial safety factors and the analytical uncertainly factors provided are based on standard industry practice and were not determined by probabilistic methods. They do not provide a measurable reliability index.

16.1.3.3 Load and Resistance Factor Design

A probability-based LRFD guide for FRP composite structures does not exist at present. Conceptually, the use of the LRFD approach for the design of FRP pultruded structures is quite possible and a prestandard outline has been prepared by ASCE (Chambers 1997). Fundamental issues associated with developing an LRFD procedure for pultruded FRP structural sections, with measurable reliability indices, have been discussed by Ellingwood (2003).

The difficulties associated with an LRFD reliability approach are primarily associated with the determination of the statistical material design properties for common pultruded sections. Since a standard material specification does not exist at this time, the statistical properties of pultruded materials are not well defined. A material specification for FRP composite materials for civil engineering applications has been proposed (Bank et al. 2002).

An LRFD procedure for the design of doubly symmetric and singly symmetric pultruded columns that accounts for global flexural buckling, global torsional buckling, flexural–torsional buckling, and material compression has been presented by Zureick and Scott (1997) and Zureick and Steffen (2000). According to these studies, load factors provided by ASCE 7-02 (2002) are used in conjunction with resistance factors that have been determined based on detailed tests of pultruded materials and pultruded column buckling experiments. Equations are presented for determining the capacity of the column with appropriate resistance factors as follows:

Doubly symmetric sections
  Flexural buckling $\phi = 0.85$
  Axial shortening $\phi = 0.80$
Singly symmetric angles
  Flexural buckling $\phi = 0.65$
  Flexural–torsional buckling $\phi = 0.85$
  Compressive failure $\phi = 0.50$

For the singly symmetric angles Zureick and Steffen (2000) have targeted reliability indices, β, of 3, 3, and 4 for the flexural buckling, flexural–torsional buckling, and material compressive failure limit states. Design values for the material properties were based on the 95% lower confidence limit on the fifth percentile of the population, obtained from tests conducted on the materials. For doubly symmetric sections (Zureick and Scott 1997) a rigorous statistical analysis was not performed and the resistance factors were obtained from lower bounds of comparisons between experimental data and theoretical predictions. Reliability indices were therefore not provided for these resistance factors.

Based on a recent analysis, Ellingwood (2003) has concluded that as more data become available from experiments on FRP composites for structural applications, LRFD design procedures for FRP pultruded structural members "might be expected to fall in the following ranges" (Ellingwood 2003):

Compression $\phi = 0.7$–0.8
Flexure $\phi = 0.8$–0.9
Shear $\phi = 0.7$–0.8
Tension $\phi = 0.7$–0.8
Fasteners $\phi = 0.6$–0.7
Adhesives $\phi = 0.6$–0.7

16.1.3.4 Performance-Based Design

A design can be based on a performance specification. In this approach the entire structure or a portion of the structure is required to meet certain performance requirements. The performance requirements are typically applied to both local and global deformations and capacities. Full-scale testing of the structure or a portion of the structure is usually required to meet the performance specification. Both proof testing of the actual structure and failure testing of a full-size mock-up of the structure (or parts thereof) are used.

In many cases for FRP composite structures a prescriptive set of material specifications is provided along with the structural performance requirements. The material specification can stipulate the manufacturing method, the limiting mechanical and physical properties of the materials produced, and the requirements for quality assurance testing.

The factor of safety for a PBD is defined in the design basis documents. ASTM and other standard test methods are usually used to conduct the performance testing. Special performance testing is defined in the design basis documents that are incorporated into the construction specifications.

16.1.4 Design of Pultruded Structural Members

Fundamentals needed for the design of transversely loaded members (beams), axially loaded compression members (columns), and axially loaded tension members (ties, truss members) are presented in the following. Connections for beams and columns are treated in the Section 16.1.5. The limit states discussed here for the members assume that the connections are appropriately designed to transfer all loads into the members. For each type of member various limit states are identified and discussed.

Analytically derived equations are presented for analyzing strength, stability, and deformation limit states. While the equations may be appropriately used with any of the four design basis approaches discussed previously, they are only valid if appropriate factors can be determined for the conditions for which they are used. The equations are primarily those presented in the SPDM (which are in most cases identical to those in the EDCH) and can therefore reasonably be assumed to be suitable for use in an LSD approach or in an ASD approach. Many of the equations presented may also be found in pultrusion manufacturer design guides, which are a combination of analytical and empirical equations. It can be assumed that the safety factors recommended by pultrusion manufacturers can be used with confidence with the analytical equations presented in this section. In addition, material properties provided by pultrusion manufactures can similarly be used in an ASD approach with the equations provided herein.

16.1.4.1 Transversely Loaded Members

Members considered are symmetric with respect to the flexural plane and consist of horizontal plates (flanges) and vertical plates (webs). Where singly symmetric sections are used it is assumed that they are loaded in the plane of symmetry through the shear center. Typical common pultruded shapes that fall in this definition, and that are routinely used as beams, include I, WF, and tubular "box" sections. Channel sections and angles are often used in pairs to form built-up doubly symmetric or singly symmetric sections.

The member is designed to resist stress-resultants that are due to the applied transverse forces (loads). These stress-resultants are the bending moment (M) and the transverse shear force (V). Due to the stress-resultant M the member is assumed to develop an axial (flexural) stress,

$$\sigma_z = \frac{M_x y}{I_x} \tag{16.7}$$

Due to the stress-resultant V the member is assumed to develop a transverse shear stress,

$$\tau = \frac{V_y Q_x}{I_x t} \tag{16.8}$$

Due to local concentrated loads P and reactions R at supports in the plane of the web at the point of load application or at the support, the member is assumed to develop a transverse compressive stress given by

$$\sigma_y = \frac{P}{A_{\text{eff}}} \quad \text{or} \quad \sigma_y = \frac{R}{A_{\text{eff}}} \tag{16.9}$$

where A_{eff} is the effective area over which the concentrated load or reaction is applied and can be taken as

$$A_{\text{eff}} = n(t_{\text{w}} + 2t_{\text{f}} + 2t_{\text{bp}})L_{\text{eff}} \tag{16.10}$$

where n is the number of webs, and L_{eff} is the effective bearing length along the beam and is taken as the width of the support or the length over which the concentrated load is applied. t_{w}, t_{f}, and t_{bp}, are the web thickness, the flange thickness, and the thickness of the bearing plate under the flange (if applicable), respectively.

Due to the stress-resultants M and V it is assumed that the member will deflect in its plane of loading. The elastic curve that describes the deflected shape is a function of the flexural rigidity of the member (EI) and the transverse shear rigidity of the member (KAG). It can be obtained from the Timoshenko (shear deformation) beam theory as

$$\delta(z) = \frac{f_1(z)}{EI} + \frac{f_2(z)}{KAG} \qquad (16.11)$$

where the functions $f_1(z)$ and $f_2(z)$ depend on the loading and boundary conditions and can be found in standard engineering texts.

For flexural members SLSs (deflection) and ULSs (strength and stability) are considered. Procedures are presented to determine the short-term and long-term member deflections as a function of the material properties and member geometric properties. The calculated design stresses (or design force resultants) must be less than the critical strengths and critical buckling stresses (the critical resistances or capacities) according to the design basis selected. The calculated deflections (serviceability design deflections) must be less than the building code stipulated deflections. The EDCH (1996) recommends limiting total deflections to the range of $L/400$ to $L/250$ for FRP frame structures.

Since pultruded members have low stiffness to strength ratios relative to conventional structural materials, the stiffness controlled limit states are checked first (i.e., deflection and stability). Thereafter, the ultimate strength limit states are checked. The order of the limit states presented follows this order. It may be noted that this is typically not the order in which limit states are considered for design with conventional materials.

16.1.4.1.1 Transverse Deflection

Shear deformation beam theory is used to determine the deflection of the beam. This is the procedure recommended by the SPDM (1984) and the EDCH (1996). The use of shear deformation beam theory is especially important in FRP pultruded beams due to the relatively low longitudinal modulus (leading to beams with short spans) and the relatively high E/G ratios (and low shear moduli) (see Bank 1989a,b; Mottram 1992). To calculate the design deflection, appropriate choices of the flexural rigidity, EI, and the transverse shear rigidity, KAG, are required. For homogenous (i.e., having the same properties in the flanges and webs of the section) pultruded beams the following are recommended:

$$EI = E_L I_x \qquad (16.12)$$

and

$$KAG = k_{\text{tim}} AG_{\text{LT}} \qquad (16.13)$$

where k_{tim} is the (Timoshenko) shear coefficient. It can be found for both homogenous and non-homogenous thin-walled composite beam sections as described by Bank (1987). The flexural rigidity is primarily a function of the longitudinal moduli of the flanges and the webs. For nonhomogenous sections having different longitudinal moduli in the flanges and the webs the conventional mechanics of *composite* sections or the "transformed-section" method can be used to find the effective flexural rigidity of the section. The section transverse shear rigidity is a function of the *shear flow* in the flanges and the webs of the section (Bank 1987). However, for common pultruded I, WF, and box sections it has been shown that this term can be reasonably replaced by either the full-section shear rigidity defined as AG_b or the area of the web multiplied by the in-plane shear modulus of the web, $A_w G_{\text{LT}}$ (Bank 1989a,b). The full-section shear modulus is found by tests on full-section pultruded beams (Bank 1989a,b; Roberts 2002).

The short-term longitudinal and shear moduli presented in the deflection equation will both decrease as a function of time due to the viscoelastic nature of the FRP composite. To predict the long-term deflection, time-dependent *viscoelastic* moduli are substituted for the short-term moduli that are typically measured by coupon tests. Both the longitudinal modulus and the in-plane shear modulus will be time dependent; however, their time dependency will be different. In general, the shear modulus will be more time dependent than the longitudinal modulus since it is primarily a resin matrix-dependent property. The viscoelastic longitudinal modulus, $E_L = E_v$, and the viscoelastic in-plane shear modulus, $G_{LT} = G_v$, are given as (Bank and Mosallam 1992).

$$E_v(t) = \frac{E_0 E_t}{E_t + E_0 t^{n_e}} \tag{16.14}$$

and

$$G_v(t) = \frac{G_0 G_t}{G_t + G_0 t^{n_g}} \tag{16.15}$$

where t is the time in hours, E_0 and G_0 are the short-term time-independent moduli and E_t and G_t are the (constant) time-dependent moduli determined from the sustained loading (or from long-term creep tests), and n_e and n_g are the empirical constants obtained from curve fitting (from creep tests) according to the linearized version of Findley's theory (SPDM 1984; Bank and Mosallam 1992). For common commercially produced pultruded FRP materials the constants obtained by Mosallam and Bank (1991) based on 2,000 and 10,000-h creep tests are $n_e = n_g = 0.30$, $E_t = 180 \times 10^6$ psi and $G_t = 30 \times 10^6$ psi. Coefficients for a variety of plastics and other nonpultruded composites are reported in the SPDM (1984). In typical pultruded structures the long-term sustained stress is between 10 and 20% of the ultimate strength of the material.

16.1.4.1.2 Global Lateral–Torsional Buckling

Lateral–torsional buckling of symmetric pultruded beams has been studied by a number of researchers. It appears that the well-known equation that is used for isotropic beam sections can be used for standard pultruded I and WF sections provided the appropriate values of E and G are used in the equations (Mottram 1992). Other analytical approaches have been suggested (Barbero and Raftoyiannis 1994; Davalos and Qiao 1997); however, simple design equations are not presented in these works. Inclusion of the effects of shear deformation is discussed by Roberts (2002). The equation to determine the critical lateral–torsional buckling stress for an I or WF section loaded through its centroid (not including the effects of shear deformation) is

$$\sigma_{cr} = \frac{C_1}{S_x} \sqrt{\frac{\pi^2 E_L I_y G_{LT} J}{(k_f L)^2} + \frac{\pi^4 E_L^2 I_y I_\omega}{(k_f L)^2 (k_\omega L)^2}} \tag{16.16}$$

where C_1 is a coefficient that accounts for moment variation along the beam, $S_x = I_x/c$ is the section modulus about the strong (bending) axis, J is the torsional constant, I_y is the second moment about the weak (vertical) axis, I_ω is the warping constant, k_f is an end restraint coefficient for flexural buckling, k_ω is an end restraint coefficient for torsional buckling, and L is the unbraced length of the member. For WF sections $I_\omega = I_y(d^2/4)$, where d is the section depth.

For closed cross-sections such as rectangular tubes the warping torsional resistance is large and lateral buckling is not a critical condition. For singly symmetric sections such as channels and angles little experimental data is available for pultruded members. Use of appropriate equations for isotropic metallic members is recommended at this time with substitution of the isotropic material properties with those of the orthotropic material properties.

16.1.4.1.3 Local Compressive Flange Buckling

Transversely loaded thin-walled sections can fail due to local buckling of the compressive flange (Barbero et al. 1991; Bank et al. 1994b, 1995). The critical buckling load (or stress) in the flange is a function of the boundary conditions on the longitudinal edges of the flange. In I and WF sections one edge is free, while the other edge is elastically restrained. There is no exact closed-form solution for this case. For the characteristic equations and computations see Bank and Yin (1996). For I and WF beams both the SPDM (1984) and the EDCH (1996) recommend that the elastically restrained edge be assumed to be simply supported. This is a conservative assumption. Test results show clearly that the local buckling stress is higher than that predicted by this assumption; however, it is lower than that predicted by assuming that the edge is fixed (see Bank et al. 1995). The Equations presented below are taken from the SPDM (1984) and are attributed to Haaijer (1957). They are shown in terms of the effective engineering properties of the section in terms of the flexural stiffness coefficients (D_{ij}) in the SPDM.

Assuming the restrained edge to be simply supported, the local buckling stress for flanges in compression is calculated as follows:

$$\sigma_{cr} = \frac{\pi^2}{t_f (b_f/2)^2} \left[\frac{E_L t_f^3}{12(1 - \nu_L \nu_T)} \left(\frac{b_f/2}{a} \right)^2 + \frac{G_{LT} t_f^3}{\pi^2} \right] \tag{16.17}$$

where t_f is the flange thickness, b_f is the (entire) flange width, and a is the unbraced length of the flange. For long unbraced flanges the first term in the square brackets is negligible and the critical stress can be found from

$$\sigma_{cr} = 4 G_{LT} \left(\frac{t_f}{b_f} \right)^2 \tag{16.18}$$

If the restrained edge is assumed to be fixed (i.e., built-in or clamped), then the local buckling stress is calculated as follows:

$$\sigma_{cr} = \frac{\pi^2}{t_f (b_f/2)^2} \left[0.935 \left(\frac{t_f^3 \sqrt{E_L E_T}}{12(1 - \nu_L \nu_T)} \right) - 0.656 \left(\frac{t_f^3 \nu_T E_L}{12(1 - \nu_L \nu_T)} \right) + 2.082 \left(\frac{t_f^3 G_{LT}}{12} \right) \right] \tag{16.19}$$

and the length of the half-buckle wavelength is given by

$$a = 1.46 \left(\frac{b_f}{2} \right) \sqrt[4]{\frac{E_L}{E_T}} \tag{16.20}$$

A reasonable approximation based on test data obtained by Bank et al. (1994b, 1995) for common pultruded WF beams is to assume that the critical local buckling stress for the compression flange is equal to the average of the simply supported and the fixed conditions given in Equations 16.17 and 16.19.

For compression flanges where both longitudinal edges are restrained, such as in box beams or hat sections, the flange is assumed to be simply supported on both of its longitudinal edges and the critical buckling stress is given as

$$\sigma_{cr} = \frac{2\pi^2 t_f^2}{b_f^2} \left(\frac{\sqrt{E_L E_T}}{12(1 - \nu_L \nu_T)} + \frac{\nu_T E_L}{12(1 - \nu_L \nu_T)} + \frac{G_{LT}}{6} \right) \tag{16.21}$$

and the length of the half-buckle wavelength is given by

$$a = b_f \sqrt[4]{\frac{E_L}{E_T}} \tag{16.22}$$

The equation for the case of flange fixed (built-in) at both longitudinal edges is given as

$$\sigma_{cr} = \frac{4.52\pi^2 t_f^2}{b_f^2} \left(\frac{\sqrt{E_L E_T}}{12(1 - \nu_L \nu_T)} + 0.543 \frac{\nu_T E_L}{12(1 - \nu_L \nu_T)} + 0.543 \frac{G_{LT}}{6} \right) \tag{16.23}$$

and the length of the half-buckle wavelength is given by

$$a = 0.67 b_f \sqrt[4]{\frac{E_L}{E_T}} \tag{16.24}$$

16.1.4.1.4 Web Flexural Buckling

Deep webs may be susceptible to flexural buckling in the plane of the web due to the linearly distributed axial (flexural) stress along the beam web. Experimental evidence of this buckling mode in common pultruded beams has not been reported in the literature. The critical buckling stress for this case is given as

$$\sigma_{cr} = \frac{k_{LL}\pi^2 E_L t_w^2}{12(1 - \nu_L \nu_T)d_w^2} \tag{16.25}$$

where k_{LL} is a coefficient that depends on the degree of web restraint and the orthotropy ratio of the web. It is given in charts (SPDM 1984, p. 680). In typical pultruded beams where the orthotropy ratio of the web $E_T/E_L < 0.5$, $k_{LL} = 20$ is recommended.

16.1.4.1.5 Web Shear Buckling

The vertical web of a WF or box beam can buckle in shear at locations of high shear forces (typically near supports). The equation for the buckling of an orthotropic plate in pure shear is a function of the restraint provided by the flanges, the aspect ratio, and the orthotropy ratio (E_L/E_T). Experimental evidence of this buckling mode in common pultruded beams has not been reported in the literature. The critical shear stress for an orthotropic web simply supported on its edges is given as

$$\tau_{cr} = \frac{4k_{LT}}{t_w d_w^2} \sqrt[4]{\frac{E_L t_w^3}{12(1 - \nu_L \nu_T)}} \sqrt[4]{\left(\frac{E_T t_w^3}{12(1 - \nu_L \nu_T)}\right)^3} \tag{16.26}$$

where d_w is the depth of the web and k_{LT} is a coefficient given in charts (SPDM 1984, p. 682). A minimum value of $k_{LT} = 8$ is recommended unless the charts are used to obtain a more exact value.

16.1.4.1.6 Transverse Web Crushing and Buckling

Webs of pultruded beams are particularly susceptible to local failure at the location of concentrated loads at applied load points and at seated supports, due to the relatively low transverse compressive strength and stiffness of the web. This is often referred to as "web resistance to transverse forces." Concentrated loads and seated supports for single-web members such as I or WF beams should always be placed directly over (or under) the web of the section and not on outstanding cantilevered flanges. Due to processing, pultruded sections tend to have slightly concave flanges (i.e., "curled" inward) such that the load is transferred directly to the web or webs in a concentrated fashion (Mottram 1992). The critical crushing strength is assumed to be equal to the transverse compressive strength of the material in the web

$$\sigma_{y,\text{crit}} = \sigma_{T,c} \tag{16.27}$$

In addition to the web being susceptible to crushing at locations of local concentrated forces, it may also buckle in the vertical plane. In this case it is assumed that the web acts as a plate simply supported on all four sides loaded in the transverse direction with the load applied over an effective width, b_{eff}. The equation is identical to that of the compression flange simply supported on its two longitudinal edges presented previously. In terms of the effective width (along the beam length) it is written as

$$\sigma_{cr} = \frac{2\pi^2 t_w^2}{b_{\text{eff}}^2} \left(\frac{\sqrt{E_L E_T}}{12(1 - \nu_L \nu_T)} + \frac{\nu_T E_L}{12(1 - \nu_L \nu_T)} + \frac{G_{LT}}{6} \right) \tag{16.28}$$

and

$$a = b_{\text{eff}} \sqrt[4]{\frac{E_T}{E_L}} \tag{16.29}$$

where b_{eff} is assumed to be equal to the web depth (d_w) or the distance between vertical web stiffeners, whichever is smaller. Note that in this case multiple half-wavelengths occur in the vertical direction and the single half-wavelength occurs in the longitudinal direction of the web.

16.1.4.1.7 Longitudinal Flange and Web Crushing or Tensile Rupture

For members where the b/t ratio of the flange or the d/t ratio of the web is small or local buckling is prevented by multiple longitudinal and transverse stiffeners the flexural member may fail due to compressive crushing or tensile rupture of the pultruded material. In this case the critical compressive strength is equal to the longitudinal compressive strength of the pultruded material in the web or the flange and the critical tensile strength is equal to the longitudinal tensile strength of the material,

$$\sigma_{z,\text{crit}} = \sigma_{\text{L},t} \quad \text{or} \quad \sigma_{\text{L},c} \tag{16.30}$$

16.1.4.1.8 Shear Failure in the Flange and Web

The critical shear strength in the flanges and webs of a pultruded thin-walled section is typically taken as the in-plane shear strength of the pultruded material in the web:

$$\tau_{\text{crit}} = \tau_{\text{LT}} \tag{16.31}$$

The interlamina shear strength may also be a critical shear strength in flanges. In this case it is prudent to calculate the out-of-plane shear stress and compare this to the interlamina shear strength. For this case the critical shear stress is

$$\tau_{\text{crit}} = \tau_{\text{TT}} \tag{16.32}$$

16.1.4.2 Axially Loaded Compression Members

Members considered are symmetric with respect to both axes and consist of horizontal plates (flanges) and vertical plates (webs). They are loaded by an axial load that is applied at the centroid of the cross-section. Typical standard pultruded shapes that fall into this definition, and that are routinely used as columns, include I, WF, and tubular box sections. The member is designed to resist a single stress-resultant that is due to the applied axial forces (loads). This stress-resultant is the axial force P. Since the force is applied at the centroid of the section, no eccentricity is considered (i.e., beam–columns are not considered). Due to the stress-resultant P_z the member is assumed to develop a uniform axial stress:

$$\sigma = \frac{P_z}{A_z} \tag{16.33}$$

where A_z is the cross-sectional area of the member.

Due to the stress-resultant P it is assumed that the member will undergo axial deformation (shortening):

$$\delta = \frac{P_z L}{A_z E} \tag{16.34}$$

where $A_z E$ is the axial stiffness of the member in compression.

For compression members serviceability (deflection) and ultimate (strength and stability) design states are considered. The calculated design stresses must be less than the critical strengths and critical buckling stresses according to the design basis selected.

16.1.4.2.1 Axial Shortening

For calculation of axial deformation the axial stiffness is taken as

$$A_z E = A_z E_{\text{L}} \tag{16.35}$$

where E_{L} is the longitudinal modulus of the material. Note that it is often assumed that the tensile and compressive moduli of the FRP material are the same. If experimental tests indicate that the compressive modulus is not the same as the tensile modulus, then it should be used to determine axial shortening of columns. In the case of a nonhomogenous section the effective (composite) axial stiffness is used. To determine long-term axial shortening, the viscoelastic longitudinal modulus should be used.

16.1.4.2.2 Global Flexural Buckling

Global flexural (Euler) buckling of common pultruded columns has been studied in some detail (Barbero and Tomblin 1993; Zureick and Scott 1997; Zureick and Steffen 2000). It has been shown that the well-known Euler equation can be used. The classical equation, modified to account for shear deformation effects, is recommended (Zureick and Scott 1997; Zureick and Steffen 2000). The critical buckling stress including the effects of shear deformation is given as

$$\sigma_{cr} = \frac{\pi^2 E_L}{(kL/r)^2_{MAX}} \left[\frac{1}{1 + (1/k_{tim} A_z G_{LT})(\pi^2 E_L/(kL/r)^2_{MAX})} \right] \tag{16.36}$$

where $(kL/r)_{MAX}$ is the maximum slenderness ratio for bending about the critical axis, k is the end restraint coefficient for the axis under consideration, L is the unbraced length of the column, and r is the radius of gyration for the axis under consideration. k_{tim} is the Timoshenko shear coefficient and A is the cross-sectional area of the section. To neglect the effects of shear deformation, the term in the square brackets is set to unity. The Timoshenko shear coefficient used in this equation is the same as that used in the shear deformation beam theory.

16.1.4.2.3 Local Compressive Flange Buckling
The local buckling of unsupported flanges in columns is identical to the local buckling of compression flanges in beams and has been discussed in Section 16.1.4.1.3.

16.1.4.2.4 Local Compressive Web Buckling
Local buckling of column webs that are supported on both longitudinal edges is identical to that of the flanges supported on two longitudinal edges discussed previously. It is conservatively assumed that the web is simply supported at the longitudinal edges. The equation for determining this buckling stress is

$$\sigma_{cr} = \frac{2\pi^2 t_w^2}{d_w^2} \left(\frac{\sqrt{E_L E_T}}{12(1 - v_L v_T)} + \frac{v_T E_L}{12(1 - v_L v_T)} + \frac{G_{LT}}{6} \right) \tag{16.37}$$

and the length of the half-buckle wavelength is given by

$$a = d_w \sqrt[4]{\frac{E_L}{E_T}} \tag{16.38}$$

The equation for the case of the web fixed (built-in) at both edges is (for comparison purposes)

$$\sigma_{cr} = \frac{4.52\pi^2 t_w^2}{d_w^2} \left(\frac{\sqrt{E_L E_T}}{12(1 - v_L v_T)} + 0.543 \frac{v_T E_L}{12(1 - v_L v_T)} + 0.543 \frac{G_{LT}}{6} \right) \tag{16.39}$$

and the length of the half-buckle wavelength is given by

$$a = 0.67 d_w \sqrt[4]{\frac{E_L}{E_T}} \tag{16.40}$$

16.1.4.2.5 Global Torsional Buckling
Open-section axially loaded compression members can buckle in a pure torsional mode. For doubly symmetric sections such as I and WF sections the critical torsional buckling stress (not including effects of shear deformation) is given as (Roberts 2002)

$$\sigma_{cr} = \frac{A_z}{I_x} \left[\frac{\pi^2 E_L I_\omega}{(k_\omega L)^2} + G_{LT} J \right] \tag{16.41}$$

where A_z is the cross-sectional area, I_x is the second moment of area about the strong axis, J is the torsional constant, I_ω is the warping constant, k_ω is an end restraint coefficient for torsional buckling, and L is the unbraced length of the member. For a wide-flange section $I_\omega = I_y(d^2/4)$, where d is the section depth.

In the case of singly symmetric sections such as angle sections, members may also buckle axially in a flexural–torsional mode consisting of Euler buckling about the strong axis of the cross-section and torsional buckling about the longitudinal axis of the member. Equations for this special case are presented by Zureick and Steffen (2000). It is interesting to note that for the special case of single angles the pure torsional buckling stress is identical to the local flange buckling stress.

16.1.4.2.6 *Longitudinal Flange and Web Crushing*

For members where the b/t ratio of the flange or the d/t ratio of the web is small or local and global buckling is prevented the axial member may fail due to compressive crushing of the pultruded material. In this case the critical compressive strength is equal to the longitudinal compressive strength of the pultruded material in the web or the flange:

$$\sigma_{z,\text{crit}} = \sigma_{\text{L,c}} \tag{16.42}$$

16.1.4.3 Axially Loaded Tension Members

Members considered are symmetric with respect to both axes and consist of horizontal plates (flanges) and vertical plates (webs). They are loaded by an axial load that is applied at the centroid of the cross-section and in the longitudinal direction of the material. Typical standard shapes that fall into this definition, and that are routinely used as tension members in trusses or as bracing members in frames, include I, WF, and tubular box sections. The member is designed to resist a single stress-resultant that is due to the applied axial forces (loads). This stress-resultant is the axial force P. Since the force is applied at the centroid of the section, no eccentricity is considered. Due to the stress-resultant P_z the member is assumed to develop a uniform axial tensile stress:

$$\sigma = \frac{P_z}{A_z} \tag{16.43}$$

where A_z is the cross-sectional area of the member.

Due to the stress-resultant P it is assumed that the member will undergo axial deformation (stretching):

$$\delta = \frac{P_z L}{A_z E} \tag{16.44}$$

where $A_z E$ is the axial stiffness of the member in tension.

For tension members serviceability (deflection) and ultimate (gross and net section strength) design states are considered. The calculated design stresses must be less than the critical strengths according to the design basis selected.

16.1.4.3.1 *Axial Extension*

For calculation of axial deformation the axial stiffness is taken as

$$A_z E = A_z E_{\text{L}} \tag{16.45}$$

where E_{L} is the longitudinal modulus of the material. In the case of a nonhomogenous section the effective (composite) axial stiffness is used.

16.1.4.3.2 *Longitudinal Rupture of the Gross Section*

The critical tensile strength on the gross section of the member in tension is equal to the longitudinal tensile strength of the pultruded material:

$$\sigma_{z,\text{gross}} = \sigma_{\text{L,t}} \tag{16.46}$$

and the design stress is calculated over the gross section, A_{g}.

16.1.4.3.3 *Longitudinal Rupture of the Net Section*

Where the cross-section of a tensile member is reduced due to the presence of holes (typically for bolted connections), the tensile strength is reduced due to the effect of stress concentrations due to the holes. The critical tensile strength for the net section can be taken as (ECDH 1996)

$$\sigma_{z,\text{net}} = 0.9\sigma_{\text{L,t}} \tag{16.47}$$

and the design stress is calculated over the net section, A_{net}.

16.1.4.4 Combined Axial and Flexural Members

Members subjected to combined bending (flexure) and compression (beam–columns) or combined flexure and tension may be designed by assuming a linear interaction between the flexural and axial load effects according to the SPDM (1984), EDCH (1996), and AASHTO (2001). For the ASD basis the interaction formula is given as

$$\frac{\sigma_A}{\sigma_{A,allow}} + \frac{\sigma_F}{\alpha\, \sigma_{F,allow}} \leq 1 \tag{16.48}$$

where the subscript A stands for axial and the subscript F stands for flexural. The allowable axial and flexural stresses in the equation are the critical axial and flexural stresses divided by their appropriate safety factors, respectively. The parameter α represents the effect of moment magnification, due to second-order P–δ effects, for members subjected to combined flexure and compression. It is given as

$$\alpha = \left(1 - \frac{\sigma_A}{\sigma_{cr}^{Euler}}\right) \tag{16.49}$$

where σ_{cr}^{Euler} is the critical Euler buckling load (see Section 16.1.4.2.2) for the axis about which the beam–column is subjected to flexure. Members subjected to combined flexure and compression should also be checked for the case of $\alpha = 1$ at locations where lateral displacement is restrained in the member. For members subjected to combined flexure and tension, $\alpha = 1$.

For LSD or LRFD design bases the equations take an analogous form having a linear interaction between the limiting states. Few experimental data have been presented for pultruded members subjected to combined loads and further research is needed in this area to validate the interaction equation presented previously (16.48).

16.1.5 Design of Pultruded Structural Connections

Pultruded FRP sections can be used in trusses and braced frame structures. As with the design of FRP structural sections there is no design guide or code for connections for pultruded structures. Connections are currently designed based on the first principles of mechanics and tests of full-sized connections.

For common FRP pultruded sections mechanical fasteners (bolts, rivets, screws) and adhesive bonding (epoxy, polyester) are most commonly used. Both steel and FRP bolts are available; however, most structures utilize metallic fasteners (usually galvanized or stainless steel fasteners).

For future routine design of FRP pultruded structures standard connection analysis and design procedures are needed. Connections include both simple single- and double-shear lap joints as well as frame connections consisting of multiple parts (members themselves and gusset plates, inserts, clip angles, and stiffeners) that are typically used to connect beams, columns, and bracing members in structural braced frames and trusses made of I, WF, and tubular FRP pultruded members.

Unique specialty connections can be designed for custom pultruded structures. These include slotted, clipping, and expansion connections and can be designed for custom pultrusions to reduce the number of parts needed in the connection and also to facilitate rapid prefabricated construction.

This discussion will focus on connections for existing standard pultruded sections for use in braced frame and truss structures. Pultruded frames should be designed as simple frames (Type 2). Due to the low stiffness of the materials used in the members themselves and the parts of the connection it is not possible to develop significant moment resistance in the connections (Bank et al. 1990, 1994a, 1996; Bass and Mottram 1994; Mottram and Zheng 1996, 1999a,b; Lopez-Anido et al. 1999; Smith et al. 1999).

Pultruded frames may be designed as semirigid frames (Type 3) provided sufficient information on the moment–rotation characteristics of the frame connection can be obtained. At the present time semirigid analysis and design for pultruded frames is not recommended (Turvey 2001) unless accompanied by a rigorous experimental investigation to determine connection failure modes and moment–rotation characteristics. A state-of-the-art report on semirigid frame connections for FRP structures (primarily, pultruded materials) was published in 1998 (Mottram and Turvey 1998).

In recent years test data have been reported for single and double lap single- and multibolted joints in commercially produced pultruded materials (Abd-el-Naby and Holloway 1993a,b; Cooper and Turvey 1995; Erki 1995; Rosner and Rizkalla 1995; Hassan et al. 1997; Turvey 1998; Wang 2002). It is important to note that most of the data reported have been from tests on commercially produced pultruded plate materials, which may be different from the pultruded material used in structural sections (profile shapes). Test data have also been reported for simple shear and semirigid frame connections (Bank et al. 1990, 1994b, 1996; Bass and Mottram 1994; Mottram and Zheng 1996, 1999a,b; Lopez-Anido et al. 1999; Smith et al. 1999). There is nevertheless very little design guidance that has been developed from these test data for routine design of connections in pultruded structures, and full-scale testing of typical pultruded frame or truss connections is recommended at this time (Turvey 2001).

16.1.5.1 Connections in Existing Pultruded Structures

Typically "standard" braced frame connections are detailed using equal leg or unequal leg angles and pultruded gusset plates together with steel fasteners or FRP threaded rods and nuts. Connections of this type mimic steel Type 2 simple framing connections and have been used in numerous pultruded frame structures to-date. Connection parts are usually bonded with an epoxy adhesive as well as fastened with bolts. Details for standard clip angle shear connections are provided in manufacturers design guides (Creative Pultrusions 2000; Strongwell 2002). Base plate clip angle details are also provided.

Typical standard truss connections are detailed using tubular members with tubular "telescoping" inserts (either hollow or solid) of different sizes with steel bolts. Standard pultruded square tubular members in the 2 by 2 to 4 by 4 sizes are used in this manner to construct lattice-like truss structures usually used for footbridges spanning up to approximately 80 ft (Johansen et al. 1999) and "stick-built" cooling towers. Larger gusset plates may also be used for truss connections with tubular members.

Often, custom framed connections are designed for custom pultruded sections for use in prefabricated FRP structures such as cooling towers, walkways, and platforms. Pultruded and molded grating or plank-type flooring systems are often integrated into these building systems to develop lateral stability.

16.1.5.2 Bolted Connections

Advantages of bolted connections for FRP pultruded structures include amenability to field or shop assembly, simplicity of fabrication, ease of inspection, unsophisticated tooling, familiarity to steel workers, relatively low cost, and speed of assembly (as opposed to bonded connections). Disadvantages of bolted connections for FRP pultruded structures include the lack of high-strength FRP fasteners, stress concentrations in the base pultruded materials, and their structural inefficiency.

16.1.5.2.1 Design Considerations for Bolted FRP Pultruded Connections

The following material properties of pultruded materials should be considered when designing a bolted connection for a pultruded structure: anisotropy, inhomogeneity, viscoelasticity, environmental degradation, and potentially brittle failure. The following geometric parameters should be considered: bolt diameter, d_b, hole diameter, d, edge (side) distance, s, end distance, e, pitch, p (distance between "rows" of fasteners parallel to the load direction), and gage, g (distance between "columns" of fasteners perpendicular to the load direction). The following fabrication parameters should be considered: FRP or steel fasteners, bolt torque, clearance, washer size, and the fact that there are limited sizes of members and connection parts.

16.1.5.2.2 Failure Modes of Bolted Connections

FRP pultruded connections can fail due to failure of the fasteners in the following modes: bolt shear, bolt tension, bolt thread shear, and nut thread shear. The base pultruded material can fail in the plane of loading in the following modes: net tension, bearing, shear-out, cleavage, splitting, or combinations of the above. The base pultruded material can fail out of the plane of the loading due to punching or crushing.

16.1.5.3 Design Recommendations for Bolted Connections

Models are available for predicting failure modes for base pultruded materials used in connections. Hassan et al. (1997) have presented models for single- and multihole double-lap tension connections. Models are based on the Hart-Smith (1978) semiempirical approach and require testing of specific materials to obtain material constants. Models predict failure modes (net tension, bearing and cleavage) and have been calibrated for a typical pultruded plate material. Simplified "strength-of-materials" models based typically on bearing, shear-out, or net-tension failure modes have been discussed by many authors (Mottram 2001).

A review of design recommendations for spacing and edge distances for lap joints in pultruded materials loaded in tension has revealed that there is reasonable agreement between research data and manufacturers' recommendations (Mottram 2001). Based on the data presented by Mottram (2001) the following design geometric parameters are suggested for in-plane tensile loads on connection elements:

1. Edge distance to bolt diameter, $e/d_b \geq 4$.
2. Element (member) width to bolt diameter, $w/d_b = 5$.
3. Side distance to bolt diameter, $s/d_b \geq 2$.
4. Pitch to bolt diameter, $p/d_b \geq 4$.
5. Gage to bolt diameter, $g/d_b \geq 4$.
6. Bolt diameter to plate thickness, $d_b/t_{pl} \geq 1$.
7. Washer diameter to bolt diameter, $d_w/d_b \geq 2$.

It is important to note that the geometric design recommendations presented above are for single- and double-lap joints loaded in tension and are intended to cause bearing failure in the material at the locations of the fasteners and to avoid net tension, shear-out, and combined failure modes (e.g., block shear) that are regarded as more brittle than bearing failures. It is also important to note that the recommendations provided are based on the tensile load being applied in the longitudinal material direction. Little guidance is available for shear loaded (in-plane) joints used in typical simple beam framing connections. These recommendations are often also used for joints loaded in shear.

For constructibility bolt holes should be oversized by $\frac{1}{16}$ in. However, if possible, holes should not be oversized. It is possible to not oversize holes and still fit the connection together (provided holes are accurately drilled) since the pultruded material is relatively soft and bolts (especially when steel bolts are used) can be inserted with light tapping with a rubber mallet.

It is assumed that FRP nuts are tightened to the manufacturer's recommended torque to achieve a measure of clamping pressure on the connection. It has been shown that clamping pressure can increase the bearing strength of the base material (Cooper and Turvey 1995). In addition, clamping pressure is needed where bonded and bolted connections are used in order to achieve a good bond line. Clamping torque depends on the nominal diameter of the FRP bolt and ranges from 15 to 110 ft lb for FRP bolts $\frac{1}{2}$ to 1 in. in diameter, respectively. If steel bolts are used, care must be taken not to overtighten the nuts so as to cause out-of-plane crushing of the base material. Finger tight plus one-half-turn of the nut is recommended. Even though some tightening of the nut is used in bolted connections for pultruded structures, all connections are designed to be bearing-type connections. Friction ("slip-critical") connections can not be achieved in pultruded structures due to the low through-the-thickness stiffness of the pultruded material.

The safety factor is taken as 4.0 for all connection parts when the design is based on the ASD. Partial safety factors for LSD are provided in the EDCH. Tentative resistance factors for LRFD are provided by Prabhakaran et al. (1996).

Pultruded connections are currently designed only for the ultimate strength limit state. Bearing strengths for pultruded materials used are the ultimate bearing strengths (i.e., the load required to cause local failure at the hole) and not the serviceability bearing strength (often defined as the load causing an axial deformation equal to 4% of the hole diameter per ASTM 953). All joints are assumed to be pinned for braced frame analysis or truss analysis for the purposes of serviceability analysis of the structure (and the members). No specific requirements are available for the local deformation of the pultruded connections.

16.1.5.3.1 Design Stress States

For a joint in tension the average bearing design stress at the hole in the base pultruded material is given as

$$\sigma_{br} = \frac{P}{d_b t_{pl}}$$ (16.50)

where P is the far field tensile force, d_b is the bolt diameter, and t_{pl} is the thickness of the base pultruded material.

For a joint in tension the net-tension design stress at the location of the hole in the pultruded material is given as

$$\sigma_{net} = \frac{P}{A_{net}}$$ (16.51)

where

$$A_{net} = t_{pl}(W - nd)$$ (16.52)

for a row of n fasteners where holes are not staggered. W is the plate width perpendicular to the load direction. This assumes that the critical section will be through the row of holes and perpendicular to the longitudinal direction of the material. Due to the orthotropy of the pultruded material, however, the critical section may be "staggered" though multiple rows (Prabhakaran et al. 1996). At this time the empirical formula used for staggered rows for joints in steel members is recommended as a first approximation for orthotropic pultruded materials:

$$A_{net} = t_{pl}\left(W - nd + \sum \frac{p^2}{4g} \right)$$ (16.53)

where the summation is over all diagonal distances in an assumed failure "path." This equation assumes that the strength of the material is the same along all paths, which is not the case for orthotropic materials. Where elements of profile sections are connected in tension (such as flanges of angles and I-sections), it is recommended that only the area of the outstanding element be considered as being effective (and not the area of the entire section). For bolts in rows perpendicular to the load direction the load may not be distributed evenly among the bolt rows. The "forward" row carries a proportionally greater share that depends on fiber orientation. Recommendations are given in the ECDH for load distribution to rows of bolts. For preliminary calculations it can be assumed that load is distributed evenly among all bolts.

For a joint in tension the shear-out design stress at the bolt location at the material edge in the direction of the tensile load is given as

$$\tau_{shear-out} = \frac{P}{2t_{pl}e}$$ (16.54)

The design shear stress on a fastener is given as

$$\tau_f = \frac{V_f}{A_f}$$ (16.55)

where V_f is the shear force on the fastener (accounting for single- of double-shear configurations) and A_f is the cross-sectional area of the fastener shank.

The shear stress in the parts of the connection (angles) is calculated as

$$\tau_{pl} = \frac{V}{A_{pl}}$$ (16.56)

where V is the design shear force on the element and A_{pl} is the effective shear area of the element. Where clip angles are used to transfer shear at the beam web in beam-to-column connections, the shear stress is typically calculated at the "heel" of the angle and is taken as the length of the angle multiplied by the angle thickness.

Combinations of shear and tensile stresses in parts of the connection or the connected members must be checked to prevent "block-shear" failure of the pultruded material through the fastener holes. Where flanges of beams are coped to detail the connection, this is often a governing failure mode.

Even in shear-type connections, tensile loads (due to prying action) may develop in the members and the parts of a connection due to eccentricities in the connection. Such eccentricities are neglected in the design of steel shear connections. However, in pultruded structures these eccentricities should be considered. The local tensile stress on the member or the element should be determined since such loads typically are applied transverse to the longitudinal direction of the pultruded material. The appropriate tensile load and the effective area are difficult to determine precisely due to complex stress states in the connecting elements, and simple approximations based on the geometry of the connection are usually made for design calculations.

16.1.5.3.2 Critical Connection Limit States

Bearing failure in the base pultruded material is the recommended failure mode for pultruded connections. This failure mode is generally ductile and leads to a progressive shear-out failure of the material when loads are in the longitudinal direction of the pultruded material and sufficient transverse reinforcement is provided (in the form of mats and fabrics). (In the transverse direction bearing failure may not be ductile and may lead to brittle net-tension transverse failure of the base material.) The bearing strength provided by pultrusion manufacturers can be used in calculations. If bearing strength is not available from test data, it can be conservatively approximated as the material compressive strength (for bearing in the longitudinal direction) and the transverse tensile strength (for bearing in the transverse direction). Either the longitudinal bearing strength or the transverse bearing strength is used depending on the direction of the member or element relative to the load.

$$\sigma_{cr} = \sigma_{L,br} \quad \text{or} \quad \sigma_{T,br} \tag{16.57}$$

Tensile failure in the elements of the connection is a function of the tensile strength of the pultruded material either in the longitudinal or in the transverse direction. For tensile stress on the gross area

$$\sigma_{cr} = \sigma_{L,t} \quad \text{or} \quad \sigma_{T,t} \tag{16.58}$$

For tensile stress on the net area a reduced ultimate strength is recommended to account for stress concentrations (as with members in tension):

$$\sigma_{cr} = 0.9\sigma_{L,t} \quad \text{or} \quad 0.9\sigma_{T,t} \tag{16.59}$$

Shear failure (and shear-out failure) of the elements of the connection is a function of the in-plane shear strength for the pultruded material:

$$\tau_{cr} = \tau_{LT} \tag{16.60}$$

Shear failure of the fasteners is a function of the shear strength of the fasteners. For steel fasteners this is typically not a likely ULS. For FRP fasteners shear in the fasteners can govern the design of the connection:

$$\tau_{cr} = \tau_{ult} \tag{16.61}$$

where τ_{ult} is the ultimate shear strength of the fastener.

Where FRP threaded rods or bolts are used in connections, the threads of the bolt can shear off longitudinally. This limit state is a function of the longitudinal shear strength of the rods or bolts used. It should be checked when FRP bolts and nuts are used. Ultimate thread shear strength is given in manufacturers' design guides.

In semirigid connections in which top and bottom seats made of pultruded angles are used, extreme caution must be used to ensure that the angles do not fail locally in through-the-thickness radial tension or compression in either an "opening" or a "closing mode" (Bass and Mottram 1994; Smith et al. 1999).

In addition, the flange–web junction behind the flange in the column section to which the top and bottom angles are connected must be checked for transverse tensile local failure (Bank et al. 1994a). As noted previously, semirigid connection design should only be used where test data are available for a full-sized connection as detailed in the structure.

16.2 Analysis and Design of FRP Reinforcements for Concrete

16.2.1 Introduction

FRP reinforcing bars and grids have been produced for reinforcing concrete structures for over 20 years (Nanni 1993; ACI 440R-96 1996). FRP reinforcing bars have been developed for prestressed and non-prestressed (conventional) concrete reinforcement. This section will consider only non-prestressed reinforcement for concrete structures. A review of design recommendations for FRP reinforcement for prestressed concrete structures can be found in Gilstrap et al. (1997) and ACI 440.4R-04 (2004).

Current FRP reinforcing bars and grids (henceforth referred to as FRP rebars) are commercially produced using thermosetting polymer resins (commonly vinylester and epoxy) and glass, carbon or aramid reinforcing fibers. The bars are primarily longitudinally reinforced with volume fractions of fibers in the 50 to 65% range. FRP rebars with thermoplastic polymer resins, which may allow them to be bent in the field, are in the developmental stages. FRP reinforcing bars are usually produced by a process similar to pultrusion and have a surface deformation or texture to develop bond to concrete (see ACI 440R-96 1996 for photographs of typical FRP rebars).

A number of design guides and national standards are currently published to provide recommendations for the analysis, design, and construction of concrete structures reinforced with FRP rebars (JSCE 1997; Sonobe et al. 1997; Bakht et al. 2000; CSA 2002; ACI 440.1R-03 2003). This section will provide a brief review of the ACI 440.1R-03 guidelines. A number of industry groups coordinate the activities of FRP rebar producers in the United States, such as, the ACMA. Activities of these organizations are closely coordinated with the American Concrete Institute (ACI) Technical Committee 440 — FRP Reinforcements.

Research in the use of FRP reinforcements in concrete structures has been the focus of intense international research activity since the late 1980s. A biannual series of symposia entitled "Fiber Reinforced Plastics in Reinforced Concrete Structures" or "FRPRCS" has been the leading venue for reporting and disseminating research results. The most recent symposium, the sixth in the series dating back to 1993, was held in Singapore in July 2003 (Tan 2003).

16.2.2 Properties of FRP Reinforcing Bars

Glass fiber reinforced vinylester bars are the most commonly commercially produced FRP rebars. They are available from a number of manufacturers. Bars are typically produced in sizes ranging from $\frac{3}{8}$ in. to 1 in. in diameter (i.e., #3 to #8 bars). Bars have a sand coated external layer, a molded deformation layer, or a spiral wind layer to create a nonsmooth surface.

The properties of FRP rebars are reported by manufacturers in accordance with ACI 440.1R-03 recommendations for guaranteed tensile strength and longitudinal modulus. It is important to note that the reported strength of FRP rebars decreases with the diameter of the bar. This is attributed to the relatively low in-plane shear strength of the FRP rebars. Designers need to consult manufacturers' recommended properties to be used for design. Typical properties are shown in Table 16.4 for glass fiber FRP rebars and carbon fiber FRP bars. It should be noted that the carbon fiber bars are typically used as prestressing tendons or "near-surface-mounted" (NSM) strengthening and not as conventional reinforcing bars due to cost considerations.

FRP rebars are considered to be transversely isotropic from a mechanics perspective (Bank 1993). Theoretical equations are available to predict the mechanical and physical properties of the FRP rebars from the properties of the fiber and resin constituents. Theoretical methods are not available to predict the bond properties and the long-term durability characteristics of FRP rebars. A specification has been

TABLE 16.4 Properties of Typical Commercially Produced FRP Reinforcing Bars

	Glass-reinforced vinylester bar[a,b] (0.5 in. diameter)	Glass-reinforced vinylester bar[a] (1 in. diameter)	Carbon-reinforced vinylester[a] bar (0.5 in. diameter)	Carbon-reinforced epoxy bar[c] (0.5 in. diameter)
Fiber volume (est)[d]	50–60	50–60	50–60	50–60
Fiber architecture	Unidirectional	Unidirectional	Unidirectional	Unidirectional
Strength ($\times 10^3$ psi)				
Tensile, longitudinal	90–100	80	300	327
Compressive, longitudinal	NR[e]	NR	NR	NR
Bond strength	1.7	1.7	1.3	NR
Shear out-of-plane	22–27	22	NR	NR
Stiffness ($\times 10^6$ psi)				
Tensile, longitudinal	5.9–6.1	5.9	18	21.3
Compressive, longitudinal	NR	NR	NR	NR
CTE longitudinal ($10^{-6}/°F$)	3.7–4.9	3.7	−4.0–0	0.38
CTE transverse ($10^{-6}/°F$)	12.2–18.7	18.7	41–58	NR
Barcol hardness	60	60	48–55	NR
24 h water absorption (% max.)	NR	NR	NR	NR
Density (lb/in.^{3})	0.072	0.072	NR	0.058

[a] Aslan (Hughes Brothers).
[b] Isorod (Pultrall).
[c] Leadline (Mitsubishi).
[d] Estimated.
[e] NR, not reported by manufacturers.

proposed for determining long-term properties for FRP rebars based on the Arrhenius accelerated aging procedure (Bank et al. 2003). Test methods for determining the properties of FRP rebars can be found in JSCE (1997) and in ACI 440.3R-04 (2004).

FRP rebars should only be used at service temperatures below the glass transition temperature of the polymer resin system used in the bar. For typical vinylester polymers this is around 200°F. The bond properties have been shown to be highly dependent on the glass transition temperature of the polymer (Katz et al. 1999). In addition, it is important to note that the coefficients of thermal expansion of FRP rebars are not the same in the transverse (radial) direction as in the longitudinal direction. The coefficient of thermal expansion may be close to an order of magnitude higher in the transverse direction of the bar due to its anisotropic properties (see typical properties in Table 16.4). This may cause longitudinal splitting in the concrete at elevated temperatures if insufficient cover is not provided.

FRP rebars containing glass fibers can fail catastrophically under sustained load at stresses significantly lower than their tensile strengths, a phenomenon known as creep rupture or static fatigue. The amount of sustained load on FRP rebars is therefore limited by design guides.

FRP reinforcing bars made of thermosetting polymers (e.g., vinylester, epoxy) cannot be bent in the field and must be produced by the FRP rebar manufacturer with "bends" for anchorages or stirrups. The strength of the FRP rebar at the bend is substantially reduced and must be considered in the design.

FRP rebars should not be used for carrying compressive stress in concrete members (i.e., compression reinforcement) where they are used in the compression zone they should be suitably confined to prevent local instability.

16.2.3 Design Basis for FRP Reinforced Concrete

A design basis for FRP reinforced concrete has been recommended by a number of standards and professional organizations. The LRFD basis is recommended by the ACI 440.1R-03; however, at this time resistance factors are not probabilistically based. Load factors are those recommended for all concrete structures by ACI 318-95 (1995).

For the design of flexural members reinforced with FRP rebars the ACI 440.1R-03 recommends the following resistance factors:

Flexural capacity (tensile reinforcement only):
    $\phi = 0.5$ for an underreinforced beam section ($\rho_f < \rho_{fb}$)
    $\phi = 0.7$ for a substantially overreinforced beam section ($\rho_f > 1.4\rho_{fb}$)
    $\phi = 0.5\rho_f/\rho_{fb}$ for a lightly overreinforced beam section ($\rho_{fb} < \rho < 1.4\rho_{fb}$)
Shear capacity (stirrups):
    $\phi = 0.85$ per ACI 318-95.

where ρ_f is the FRP reinforcement ratio and ρ_{fb} is the balanced FRP reinforcement ratio.

Characteristic strength (also called the guaranteed strength) and strain to failure of FRP rebars are defined as the mean minus three standard deviations of a minimum of 25 test samples. The design strength, f_{fu}, and design failure strain, ε_{fu}, are obtained from the characteristic strength and failure strain by multiplying them by an environmental reduction factor, C_E, which depends on the fiber type in the bar and the type of intended service of the structure. For example, for weather exposed concrete with glass FRP rebars, C_E is 0.7 (ACI 440.1R-03 2003).

Since FRP rebars typically have a lower modulus than steel rebars, the serviceability limit state (deflections and crack widths) can often control the design of FRP reinforced concrete sections. The ACI 440.1R-03 provides procedures for calculating deflections and crack widths in FRP reinforced members.

16.2.4 Design of Flexural Members with FRP Reinforcing Bars

16.2.4.1 Moment Capacity

Design of flexural members follows the *strength design method* of the ACI limit states version of an LRFD design basis. In this approach the nominal factored moment resistance (ϕM_n) of the member must be greater than the factored ultimate moment required (M_u).

The nominal moment capacity of FRP reinforced concrete members is determined in a similar fashion to that of a steel reinforced section; however, since FRP rebars do not yield (i.e., are linear elastic to failure), the ultimate strength of the bar replaces the yield strength of the steel rebar in the traditional concrete beam design formula based on strain compatibility (assuming plane sections remain plane and bars are perfectly bonded to the concrete) and equilibrium of forces (ACI 440.1R-03 2003). Both underreinforced section design and overreinforced section design are permitted; however, due to serviceability limits (deflections and crack widths) most glass FRP reinforced flexural members will be substantially overreinforced.

The flexural reinforcement ratio for an FRP reinforced rectangular beam section (where the subscript f is used to indicate FRP reinforcement to distinguish it from conventional reinforcement) is given as

$$\rho_f = \frac{A_f}{bd} \tag{16.62}$$

and the balanced FRP reinforcement ratio is given as

$$\rho_{fb} = 0.85\beta_1 \frac{f_c'}{f_{fu}} \frac{E_f\varepsilon_{cu}}{E_f\varepsilon_{cu} + f_{fu}} \tag{16.63}$$

where A_f is the area of FRP reinforcement, b is the beam width, d is the effective depth, β_1 is a factor that depends on concrete strength (0.85 for 4000 psi concrete), f_c' is the cylinder compressive strength of the concrete, E_f is the longitudinal modulus of the FRP rebar, ε_{cu} is the ultimate nominal compressive strain in the concrete (taken as 0.003), and f_{fu} is the longitudinal design strength of the FRP rebar.

When $\rho_f > \rho_{fb}$ the section will fail due to concrete crushing and the nominal moment capacity is given in a similar fashion to that for a section reinforced with steel rebar (where the rebar has not reached its yield stress). The stress in the rebar needs to be calculated to determine the capacity. Design is usually done by trial and error. The nominal moment capacity is given as

$$M_n = A_f f_f \left(d - \frac{a}{2} \right)$$ (16.64)

where

$$a = \frac{A_f f_f}{0.85 f_c' b}$$ (16.65)

$$f_f = \left(\sqrt{\frac{(E_f \varepsilon_{cu})^2}{4} + \frac{0.85 \beta_1 f_c'}{\rho_f} E_f \varepsilon_{cu}} - 0.5 E_f \varepsilon_{cu} \right)$$ (16.66)

where f_f is the stress in the FRP rebar at concrete compressive failure and a is the depth of the equivalent rectangular (Whitney) stress block in the concrete.

When $\rho_f < \rho_{fb}$ the section will fail due to rupture of the FRP rebars in tension and nominal moment capacity is given in a similar fashion to that for a section reinforced with steel rebar if the rebar does not reach its yield stress before tensile failure (which will not happen for a conventional steel bar but will always happen for a FRP rebar). Since the reinforcement has not yielded in this case failure cannot be assumed to occur due to concrete compression failure following yielding of the rebar and the stress in the concrete at the time of rebar failure needs to be calculated. This calculation requires knowledge of the nonlinear stress–strain characteristics of the concrete and requires an incremental solution procedure, which is not suited to design calculations. To overcome this situation the ACI 440.1R-03 guide recommends computing the approximate (and conservative) nominal flexural capacity as follows:

$$M_n = 0.8 A_f f_{fu} \left(d - \frac{\beta_1 c_b}{2} \right)$$ (16.67)

where the depth of the neutral axis, c_b, is given as

$$c_b = \left(\frac{\varepsilon_{cu}}{\varepsilon_{cu} + \varepsilon_{fu}} \right) d$$ (16.68)

16.2.4.2 Shear Capacity

The nominal shear capacity of an FRP reinforced concrete member in flexure is determined in a similar fashion to that of a steel rebar reinforced section; however, due to the lower modulus of FRP rebars, which leads to a shallower compression zone and larger deflections at flexural failure of FRP reinforced beams, the shear capacity of the concrete is reduced from its conventional value. In addition, the strain in the FRP stirrups is limited to prevent large shear cracks from developing in the FRP reinforced member. The strength of the FRP stirrup is also limited by its strength at its bent portion. The nominal shear capacity, V_n, is given as

$$V_n = V_{c,f} + V_f$$ (16.69)

where $V_{c,f}$ is the nominal shear capacity of the concrete with FRP rebars used as main tension reinforcement, and can be taken as

$$V_{c,f} = \frac{\rho_f E_f}{90 \beta_1 f_c'} V_c$$ (16.70)

and V_f (in Equation 16.69) is the nominal shear capacity of the FRP stirrups. For vertical shear stirrups

$$V_f = \frac{A_{fv} f_{fv} d}{s}$$ (16.71)

where the strength of the FRP stirrup, f_{fv}, is limited by the smaller of

$$f_{fv} = 0.004 E_f \tag{16.72}$$

and the strength of the FRP rebar at its bend,

$$f_{fb} = \left(0.05 \frac{r_b}{d_b} + 0.3 \right) f_{fu} \tag{16.73}$$

where r_b is the inside radius of the bend and d_b is the diameter of the FRP rebar. Standard bend radii are reported by manufacturers and range from 4.25 to 6 in. for typical FRP rebars.

For the SLS both crack width and deflections must be checked according to the procedures detailed in ACI 440.1R-03. The designer is required to check stresses under sustained service loads against creep rupture stress limits and fatigue stress limits. These limits depend on the type of fiber in the FRP rebar and are the most restrictive for glass FRP rebars where the limiting stress is only 20% of the design strength (i.e., $0.2 f_{fu}$). Limits on minimum FRP reinforcement ratios for flexural and shear reinforcement are also provided. For further details see ACI 440.1R-03 (2003).

16.2.5 Detailing of FRP Reinforcements

The development length of an FRP rebar is different from the development length of a conventional steel rebar. The properties of hooks in FRP rebars are also different from those in steel rebars due to the decrease in strength at the bends in FRP rebars. The ACI 440.1R-03 guide provides recommendations for detailing FRP rebar anchorages and splices so that the FRP rebar can develop its tensile strength at member ends, at hooks, and at locations where bars are overlapped or terminated. It should be noted that current recommendations are based on limited, and sometimes conflicting, data.

Failure due to insufficient development length is called a bond failure and may be due to splitting of the concrete surrounding the bar or due to pull-out of the bar itself. Two equations are provided by ACI 440.1R-03, one for splitting controlled failure and one for pull-out controlled failure. For splitting failure

$$l_{bf} = K_2 \frac{d_b^2 f_{fu}}{\sqrt{f_c'}} \tag{16.74}$$

and for pull-out failure

$$l_{bf} = \frac{d_b f_{fu}}{2700} \tag{16.75}$$

where l_{bf} is the development length and K_2 ranges from $\frac{1}{21.3}$ to $\frac{1}{5.6}$ depending on FRP rebar type and the source of the test data (ACI 440.1R-03 2003). Manufacturers of FRP rebars also provide data on development lengths for the bars they produce.

For hooked bars the development length is given as a function of the FRP rebar design strength. For FRP rebars with design strengths in the range of 75 to 150 ksi (typical of glass FRP rebars) the length of a hooked bar, l_{bfh}, is given as

$$l_{bfh} = \frac{f_{fu}}{37.5} \frac{d_b}{\sqrt{f_c'}} \tag{16.76}$$

The tail length should not be less than $12 d_b$ or 9 in.

Tension lap splices for FRP rebars are based on recommendations for steel rebars and limited test data. For Class A and Class B lap splices the recommended development lengths are $1.3 l_{df}$ and $1.6 l_{df}$, respectively.

FRP stirrups can be spaced at a maximum of $d/2$ (or 24 in.) and should have a minimum r_b/d_b ratio of 3. The tail length of 90° hooks in the stirrups must be at least $12 d_b$.

Minimum concrete cover for FRP rebars is d_b.

16.3 Analysis and Design of FRP Strengthening for Concrete

16.3.1 Introduction

FRP reinforcing systems for strengthening structurally deficient concrete structural members and for repairing damaged or deteriorated concrete structures have been used since the mid-1980s. The first applications involved beams strengthened to increase their flexural capacity using high-strength, lightweight fiber reinforced epoxy laminates that were bonded to the soffits of the beams (e.g., Meier and Kaiser 1991; Saadatmanesh and Ehsani 1991; Meier 1995). The method is a modification of the one where epoxy bonded steel plates are used to strengthen concrete beams, which has been in use since the mid-1960s. The FRP systems were shown to provide significant benefits in constructibility and durability over the steel plates. Thereafter, significant work was conducted on strengthening of concrete columns to enhance their axial capacity, shear capacity, and ductility, primarily for seismic loadings (e.g., Seible et al. 1997). This method is a modification of the one using steel jackets to strengthen concrete columns. This was followed closely by work on shear strengthening of beams (e.g., Triantafillou 1998). A review of the state-of-the-art on the subject can be found in Teng et al. (2001) and Bakis et al. (2002). The method has also been used to strengthen masonry and timber structures; however, applications of this type are not discussed in this section.

Current FRP strengthening systems for concrete fall into two popular types. One type consists of factory manufactured (typically unidirectional pultruded) laminates (also known as strips or plates) of carbon- or glass-reinforced thermosetting polymers (epoxy or vinylester) that are bonded to the surface of the concrete using an epoxy adhesive. The manufactured laminates typically have a volume fraction of fibers in the range of 55 to 65% and are cured at high temperatures (typically >300°F), but are bonded in the field at ambient temperatures. The other type consists of layers (or plies) of unidirectional sheets or woven or stitched fabrics of dry fibers (glass, carbon, or aramid) that are saturated in the field with a thermosetting polymer (epoxy or vinylester) that simultaneously bonds the FRP laminate (thus formed) to the concrete. These "formed-in-place" or "layed-up" FRP systems typically have a fiber volume fraction of between 20 and 30% and are cured at ambient temperatures in the field.

A number of design guides and national standards are currently published that provide recommendations for the analysis, design, and construction of concrete structures strengthened with FRP materials (AC125 1997; TR55 2000; FIB 2001; JSCE 2001; ACI 440.2R-02 2002). This section will provide a brief review of the ACI guidelines.

Manufacturers of FRP strengthening systems for concrete typically provide their own design and installation guides for their proprietary systems. Since the performance of the FRP strengthening system is highly dependent on the adhesive or saturating polymer used, the preparation of the concrete surface prior to application of the FRP strengthening system, and the field installation and construction procedures, manufacturers typically certify "approved contractors" to ensure that their systems are designed and installed correctly. In addition, code guidance is provided to ensure that FRP strengthening systems are appropriately installed (AC187 2001; ACI 440.2R-02 2002; TR57 2003).

Research in the use of FRP strengthening systems for concrete structures has been the focus of intense international research activity since the early 1990s. A biannual series of symposia entitled *Fiber Reinforced Plastics in Reinforced Concrete Structures* or "FRPRCS" has been the leading venue for reporting and disseminating research results. The most recent symposium, the sixth in the series dating back to 1993, was held in Singapore in July 2003 (Tan 2003).

16.3.2 Properties of FRP Strengthening Systems

Carbon fiber-reinforced epoxy laminates (or strips) are the most commonly used in the adhesively bonded type of products. Depending on the type of carbon fiber used in the strip, different longitudinal strengths and stiffnesses are produced. Strips are typically thin (less than 0.100 in.) and are available in a variety of widths (typically 2 to 4 in.). Since the strips are reinforced with unidirectional fibers they are highly orthotropic with very low properties in the transverse and through the thickness directions. Manufacturer's typically only report properties in the longitudinal directions and report very little data on physical properties. The strips are bonded to the concrete with a compatible adhesive that is supplied by the strip manufacturer. Typical properties of strips are shown in Table 16.5. It is important to note that the properties shown for the strips are properties of the FRP composite and not the properties of the fibers alone.

In the type of products for FRP strengthening that consist of dry fibers in sheet or fabric form with compatible polymer saturating resins there is a greater array of available products that depend on fiber type and sheet or fabric architecture. In this group of products a unidirectional, highly orthotropic carbon fiber tow-sheet is produced by a number of manufacturers and is often used in strengthening applications. The individual carbon tows in the sheet are held together by a very light polymeric binder (or a light stitching) and it is supplied on a wax paper backing. Sheets are typically 10 to 40 in. wide and are often applied in multiple layers with different orientations. Other common fabric materials in this group are woven glass fiber materials typically consisting of 12 to 32 Oz/yd^2 materials with a variety of weaves, which can give the fabric properties from highly orthotropic to square symmetric. Carbon fiber tow fabrics and hybrid fabrics (with more than one fiber type) are also available. Fabrics are typically much thicker than tow-sheets and are also used in multiple layers. Because of the wide variety of products available and their different thicknesses it is not easy to compare their properties directly. In addition, the fibers must be used with a compatible resin system applied with a controlled volume fraction to achieve an FRP composite with measurable properties. In the case of sheet and fabric materials, manufacturers typically report the mechanical properties of the dry fibers and the thickness (or area) of the fibers. It is important to note that when reported in this fashion the properties are not the properties of the FRP composite. Properties of some commonly available fiber sheet materials are listed in Table 16.6.

The performance of the FRP strengthening system is highly influenced by the properties of the adhesive layer in the case of the bonded strip and by the properties of the saturating polymeric resin in the case of the sheets and fabrics. The interface between the FRP composite and the concrete substrate transfers the loads from the concrete to the FRP composite. In the case of flexural, shear, or axial tensile strengthening this load transfer is primarily in shear and the properties and the quality of the interface bond between the FRP composite and the concrete effects load transfer into the FRP strengthening

TABLE 16.5   Properties of Typical Commercially Produced FRP Strengthening Strips

	Standard modulus carbon-reinforced epoxy strip[a–c]	High modulus carbon-reinforced epoxy strip[a]	Glass-reinforced epoxy strip[b]	Carbon-reinforced vinylester strip[d]
Fiber volume (est)[e]	65–70	65–70	65–70	60
Fiber architecture	Unidirectional	Unidirectional	Unidirectional	Unidirectional
Nominal thickness (in.)	0.047–0.075	0.047	0.055–0.075	0.079
Width (in.)	2–4	2–4	2–4	0.63
Strength ($\times 10^3$ psi)				
Tensile, longitudinal	390–406	188	130	300
Rupture strain (%)				
Tensile, longitudinal	1.8	NR[f]	2.2	1.7
Stiffness ($\times 10^6$ psi)				
Tensile, longitudinal	22.5–23.9	43.5	6.0	19.0
CTE longitudinal ($10^{-6}/^\circ$F)	NR	NR	NR	−4.0–0
CTE transverse ($10^{-6}/^\circ$F)	NR	NR	NR	41–58
Barcol hardness	NR	NR	NR	48–55

[a] Carbodur (Sika).
[b] Tyfo (Fyfe).
[c] Mbrace (Wabo/S&P).
[d] Aslan (Hughes Brothers).
[e] Estimated.
[f] NR, not reported by manufacturers. All strips must be bonded with manufacturer-supplied compatible adhesives.

TABLE 16.6 Properties of Typical Commercially Produced FRP Sheet Strengthening Materials

	Standard modulus carbon fiber tow sheet[a–c]	High-modulus carbon fiber tow sheet[a–c]	Glass fiber roving sheet[a,b]
Thickness (in.)	0.0065–0.013	0.0065	0.014
Typical width (in.)	24	24	24
Fiber architecture	Unidirectional	Unidirectional	Unidirectional
Strength ($\times 10^3$ psi)			
Fiber tensile, longitudinal	550	510	220–470
Rupture strain (%)			
Fiber tensile, longitudinal	1.67–1.7	0.94	2.1–4.5
Stiffness ($\times 10^6$ psi)			
Fiber tensile, longitudinal	33.0–33.4	54.0	10.5

[a] Mbrace (Wabo).
[b] Tyfo (FyFe).
[c] Replark (Mitsubishi).

system. Such applications are termed "bond-critical" applications. In the case of axial compressive strengthening where the role of the strengthening system is to confine the lateral expansion of the cracked concrete the interface bond is not as critical as long as the FRP strengthening system is in intimate contact with the concrete and is wrapped around the concrete continuously so as to provide a confining pressure with appropriate hoop stiffness and strength. Such applications are termed "contact-critical." It is important to note, however, that in either type of system the polymer resin still plays a critical role in the FRP composite in transferring load to the fibers and protecting the fibers. A poorly applied sheet or fabric having a low fiber volume fraction (<20%) and a relatively thick (uneven thickness) polymer layer with large void content (>3%) will not provide a durable FRP strengthening system even though the amount of fiber placed in the wrap may be according to design requirements.

Since most FRP strengthening systems depend on the curing of polymer adhesives or resins at ambient temperature in the field, the glass transition temperatures of these components may be quite low (120 to 180°F). Since the effectiveness of the strengthening system depends to a great degree on the stiffness of the adhesive or resin used to bond the FRP composite, to the concrete, an FRP strengthening system can be severely degraded at even moderately high temperatures (~120°F). In certain cases where the fibers are sufficiently anchored away from the region subjected to the high temperature the FRP strengthening system may still be effective (Meier 1995). Designers should always be aware of the transition temperature of the FRP composite or adhesive they are using in a design. In the event of fire the integrity of an FRP strengthening system may be severely compromised in a short time. Test methods for FRP strips and sheets are provided in JSCE (2001) and ACI 440.3R-04 (2004).

16.3.3 Design Basis for FRP Strengthening Systems

A design basis for FRP strengthening systems for concrete structures has been recommended by a number of standards and professional organizations. The LRFD basis is recommended by the ACI 440.2R-02 (2002); however, at this time resistance factors are not probabilistically based. Load factors are those recommended for all concrete structures by ACI 318-99.

For the design of concrete members with FRP strengthening systems the ACI recommends the following resistance factors:

Flexural capacity (tensile strengthening)
  $\phi = 0.9$ for ductile failure of the member following steel yielding ($\varepsilon_s > 0.005$)
  $\phi = 0.7$ for a brittle failure when the member fails prior to steel yielding ($\varepsilon_s < \varepsilon_{sy}$)
  $\phi = 0.7$–0.9 for an intermediate region ($\varepsilon_{sy} < \varepsilon_s < 0.005$)
  $\psi_f = 0.85$ for FRP bond-critical strengths (in addition to ϕ factors)

Shear capacity (shear strengthening)

$\phi = 0.85$ per ACI 318-99

$\psi_f = 0.85$ for FRP bond-critical strengths (in addition to ϕ factor)

$\psi_f = 0.95$ for FRP contact-critical strengths (in addition to ϕ factor)

Axial compressive capacity (confinement)

$\phi = 0.75$ per ACI 318-99 for steel spiral reinforcement

$\phi = 0.70$ per ACI 318-99 for steel tied reinforcement

$\psi_f = 0.95$ for FRP contact-critical strengths (in addition to ϕ factors)

Characteristic strength (also called the guaranteed strength) and strain to failure of FRP composite materials for strengthening are defined as the mean minus three standard deviations of a minimum of 20 test samples. The design strength, f_{fu}, and design failure strain, ε_{fu}, are obtained from the characteristic strength and failure strain by multiplying them by an environmental reduction factor, C_E, which depends on the fiber type in the FRP strengthening system and the type of intended service of the structure. For example, for weather exposed concrete with a glass-reinforced epoxy FRP strengthening system, C_E is 0.65 (ACI 440.2R-02 2002).

While flexural strengthening over 100% of the original strength is achievable, the ACI limits the amount of strengthening to prevent catastrophic failure of the concrete member in the event of loss of, or damage to, the strengthening system (due to vandalism or environmental degradation). The ACI recommends that the strengthened member have sufficient original factored capacity (i.e., discounting the additional strengthening system) to resist a substantial portion of the factored load on the strengthened member, given as

$$1.2D + 0.85L$$

To account for fire, additional restrictions are placed on the factored capacity of the FRP strengthened structured (ACI 440.2R-02 2002).

When a concrete member is strengthened to increase its capacity in a selected mode (e.g., flexure) the member must be checked to ensure that the capacities in other failure modes (e.g., shear) are not exceeded. If this is the case the strengthening should be decreased or the secondary capacity needs to be enhanced with its own strengthening system.

At this time the ACI 440.2R-02 guide does not provide recommendations for the determination of serviceability criteria (such as deflections or crack widths) for FRP strengthened members. The ACI 440.2R-02 guide does, however, limit the stress in the steel at service loads to 80% of the steel yield stress and limits the "sustained plus cyclic" stress in the FRP strengthening system to account for creep rupture and fatigue depending on the fiber system. For carbon FRP strengthening systems this limit is 55% of the ultimate strength. For deflections in flexural members where stresses are in the service range the contribution of the FRP strengthening system is typically small. Flexural deflections in the service range can be estimated by the use of an effective "composite-section" second moment-of-area (I_e) analysis where the tensile contribution of the FRP is added to the contribution of the steel reinforcing. In the inelastic range (after the primary reinforcing steel has yielded) the contribution of the FRP strengthening to the postyield stiffness can be quite considerable and should be accounted for, in inelastic analysis.

It is extremely important to note that the method of determining the tensile force resultant in an FRP strengthening system depends on the type of system used. In the bonded strip (or plate) system the ultimate force is obtained from the strength of the FRP composite (see Table 16.5) and the gross cross-sectional area of the strip. In the dry fiber systems the ultimate force is obtained from the strength of the fibers and the thickness of the net area of the fibers (see Table 16.6). The designer must know if the reported strength (and stiffness) for an FRP strengthening system is for the FRP composite (gross cross-section) or for the fibers alone (net fiber cross-section). Both methods of calculation are permitted by the ACI 440.2R-02 guide at this time.

16.3.4 Design of FRP Flexural Strengthening Systems

Flexural strengthening is achieved by attaching an FRP strengthening system (bonded strip or saturated dry fabric) to the underside (or soffit) of a flexural member to increase the effective tensile force resultant in the member and thereby increase the moment capacity of the member. This is analogous to adding steel strengthening strips (or plates) to the soffit of a member. However, two fundamental differences exist. First, the FRP strengthening system behaves in a linear elastic fashion and does not yield, and second, the FRP strengthening system is more susceptible to detachment (debonding or delamination) failures than steel plate systems. Epoxy bonded steel plates are typically anchored with steel bolts at their ends in addition to the epoxy bonding. Since the steel plates themselves will yield at a similar strain to the internal steel reinforcing, the stress level in the steel strengthening system is limited. In the case of FRP strengthening with FRP systems having ultimate tensile strengths exceeding 300 ksi (see Table 16.5 and Table 16.6) the stress level in the FRP can be significantly higher than that in steel strengthening systems. In the event that the internal steel reinforcing yields before the FRP strengthening system fails (the desired failure mode) the concrete member will undergo large deflections and cracking. All these factors lead to the greater likelihood that the FRP strengthening system will detach from the concrete long before it achieves its ultimate tensile capacity.

Strengthening of members in flexure can only be achieved if there is sufficient additional compressive capacity in the concrete to allow for the increase in internal moment. Therefore, flexural strengthening is most suitable for concrete members that are lightly to moderately reinforced, having steel reinforcement in the range of 20 to 40% of the balanced ratio. This is not uncommon in reinforced concrete members, especially in older structures.

The existing tensile strain in the concrete at the location of the applied FRP strengthening system due to sustained loads when the FRP strengthening system is applied should be accounted for in design calculations if a shoring system is not used.

The key to flexural strengthening with FRP strengthening systems is to understand the failure modes of the system. These include rupture of the FRP strengthening system, detachment of the FRP strengthening system (due to a variety of delamination or debonding modes), or compressive failure of the concrete. All of these modes can occur either before or after the internal steel has yielded. The desired mode of failure is concrete compressive failure after the internal steel has yielded with the FRP strengthening system still attached. This is often difficult to achieve and the mode of FRP detachment (or less frequently rupture) at large deflections after the internal steel has yielded is often achieved.

The FRP strengthening system can detach in a number of modes. The FRP system can delaminate from the concrete substrate (due to failure in the concrete, the adhesive layer, or the FRP laminate itself) either at the ends (due to high peeling and shear stresses) or in the interior of the beam due to flexural and shear cracks in the beam at large deflections. For a detailed discussion on detachment failure modes see Teng et al. (2001). Analytical methods to predict the various detachment failure modes are still not fully developed and the ACI 440.2R-02 guide limits the tensile strain level in the FRP strengthening system to prevent delamination failure by the use of an empirically obtained bond-dependent coefficient, κ_m, which is a function of the unit stiffness of the FRP system and is defined as

$$
\kappa_m = \begin{cases}
\dfrac{1}{60\varepsilon_{fu}}\left(1 - \dfrac{nE_f t_f}{2,000,000}\right) \le 0.90 & \text{for } nE_f t_f \le 1,000,000 \text{ lb/in.} \\[3mm]
\dfrac{1}{60\varepsilon_{fu}}\left(\dfrac{500,000}{nE_f t_f}\right) \le 0.90 & \text{for } nE_f t_f > 1,000,000 \text{ lb/in.}
\end{cases}
\tag{16.77}
$$

where ε_{fu} is the ultimate strain in the FRP, n is the number of layers (or plies) of FRP strips or sheets or fabrics, E_f is the longitudinal tensile modulus of the FRP composite in the case of strips or the longitudinal modulus of the fibers in the strengthening direction in the case of sheets or fabrics, and t_f is the thickness of an individual strip in the case of FRP strips or the net thickness of the fibers in a single sheet or fabric in the case of sheets or fabrics.

The strain level in the FRP strengthening system is limited by the strain in the concrete or the ultimate strain in the FRP system and is given as

$$\varepsilon_{fe} = \varepsilon_{cu}\left(\frac{h-c}{c}\right) - \varepsilon_{bi} \le \kappa_m \varepsilon_{fu} \tag{16.78}$$

where ε_{fe} is the effective ultimate strain in the FRP at failure, ε_{cu} is the ultimate compressive strain in the concrete (0.003), c is the depth of the neutral axis, h is the depth of the section, and ε_{bi} is the existing tensile strain in the concrete substrate at the location of the FRP strengthening system. The effective stress, f_{fe}, in the FRP is the ultimate strength of the FRP that can be achieved at failure and is linearly related to the ultimate strain as

$$f_{fe} = E_f \varepsilon_{fe} \tag{16.79}$$

The nominal moment capacity, M_n, of the strengthened section (with an existing layer of tensile steel reinforcement only) is given as

$$M_n = A_s f_s \left(d - \frac{\beta_1 c}{2}\right) + \psi_f A_f f_{fe} \left(h - \frac{\beta_1 c}{2}\right) \tag{16.80}$$

with

$$c = \frac{A_s f_s + A_f f_{fe}}{\gamma f_c' \beta_1 b} \tag{16.81}$$

and

$$f_s = E_s \varepsilon_s = E_s(\varepsilon_{fe} + \varepsilon_{bi})\left(\frac{d-c}{h-c}\right) \le f_y \tag{16.82}$$

where A_s is the area of the tensile steel, f_s is the stress in the steel at failure, d is the depth of the steel reinforcing, β_1 is the depth ratio of the equivalent Whitney stress block, A_f is the area of the FRP strip or the fibers in a dry fiber system, γ is the concrete stress resultant factor (0.85 when concrete compressive failure governs), b is the width of the section, and f_y is the yield stress in the reinforcing steel. The solution to the above equations is typically found by a trail and error method by assuming a number of plies (or layers) of strengthening and calculating the resulting nominal moment. The desirable failure mode is achieved when the internal steel has yielded. Therefore, the current stress state in the steel can be assumed to be at yield and then checked. If the steel has not yielded before FRP rupture or detachment occurs, a new design should be attempted with a different strengthening system or a different number of plies.

The stresses in the steel and the FRP strengthening system at service loads should be determined using an elastic cracked section and checked against appropriate stress limits for sustained loads on FRP strengthened structures (ACI 440.2R-02 2002).

Mechanical anchorages or FRP wraps can be used to enhance the attachment of the FRP strengthening system to the concrete beam, especially at the ends of the FRP strengthening system. Design guidance is not provided by the ACI 440.2R-02 for this at the present time.

16.3.5 Design of FRP Shear Strengthening Systems

FRP strengthening systems can be used to increase the shear capacity of concrete beams and columns. FRP strengthening systems are applied to the webs of beams (or columns) and function in an analogous fashion to internal steel shear reinforcement. Because FRP shear strengthening systems are applied to concrete members that are often monolithic with other continuous members (such as floors and walls), it is not always possible "wrap" the FRP strengthening system completely around the member (which is the desirable condition). The FRP strengthening system must therefore be terminated at the top of the web (a three-sided "U-wrap") or both at the top and at the bottom of the web (a two-sided system).

The non-fully wrapped systems are susceptible to detachment failures (similar to flexural strengthening) and their strains are limited by a shear bond-reduction coefficient, κ_{v}.

The nominal shear capacity of an FRP strengthened concrete member with existing steel shear reinforcing is determined by adding the contribution of the FRP strengthening system to the existing shear capacity, given as

$$V_{\mathrm{n}} = V_{\mathrm{c}} + V_{\mathrm{s}} + \psi_{\mathrm{f}} V_{\mathrm{f}} \tag{16.83}$$

where

$$V_{\mathrm{f}} = \frac{A_{\mathrm{fv}} f_{\mathrm{fe}} (\sin\alpha + \cos\alpha) d_{\mathrm{f}}}{s_{\mathrm{f}}} \tag{16.84}$$

and

$$A_{\mathrm{fv}} = 2 n t_{\mathrm{f}} w_{\mathrm{f}} \tag{16.85}$$

where V_{c} is the existing shear capacity of the concrete, V_{s} is the shear capacity of the existing steel shear reinforcement, V_{f} is the shear capacity of the FRP strengthening system, $f_{\mathrm{fe}} = E_{\mathrm{f}} \varepsilon_{\mathrm{fu}}$ is the effective tensile stress in the FRP at ultimate, α is the inclination of the fiber in the FRP strengthening system to transverse axis (the horizontal axis in a beam or the vertical axis in a column), d_{f} is the effective depth of the FRP strengthening system, s_{f} is the center-to-center spacing of the FRP shear strengthening strips, and w_{f} is the width of the FRP shear strengthening strip. (For a continuous FRP shear strengthening sheet or fabric $s_{\mathrm{f}} = w_{\mathrm{f}}$.) The FRP capacity reduction factor, ψ_{f}, is taken as 0.95 for completely wrapped sections (contact-critical) and as 0.85 for two- or three-sided wrapped sections (bond-critical).

The effective strain in the FRP shear strengthening system is limited to prevent detachment failures and also to maintain the integrity of the concrete aggregate interlock in the concrete member. For completely wrapped FRP shear strengthening systems the maximum effective strain in the FRP strengthening system at failure is limited to

$$\varepsilon_{\mathrm{fe}} = 0.004 \leq 0.75 \varepsilon_{\mathrm{fu}} \tag{16.86}$$

For three-sided or two-sided shear strengthening the effective shear strain in the FRP strengthening system at failure is limited to

$$\varepsilon_{\mathrm{fe}} = \kappa_{\mathrm{v}} \varepsilon_{\mathrm{fu}} \leq 0.004 \tag{16.87}$$

where the shear bond-reduction coefficient is a function of the concrete strength, the wrapping type used, and the stiffness of the FRP strengthening system. Empirical equations for determining this coefficient are given in the ACI 440.2R-02 guide.

Mechanical anchorages can be used to anchor two- or three-sided wraps in the compression zone of the web; however, design guidance is not provided by the ACI 440.2R-02 for this at the present time.

Limits on spacing of discreet FRP shear strengthening strips and the maximum shear strength enhancement that can be obtained using FRP strengthening are provided in the ACI 440.2R-02 guide.

16.3.6 Design of FRP Axial Strengthening Systems

Concrete compression members can be strengthened to increase their axial load carrying capacity, their shear capacity, their steel rebar lap splice capacity, and their lateral load carrying deformation capacity (ductility). FRP strengthening of columns is most effective when applied to circular columns and always consists of complete wrapping to obtain confinement of the concrete. FRP strengthening systems for confinement of columns are classified as contact-critical applications. It is also important to note that FRP axial strengthening systems are regarded as "passive" systems. That is, they are not effective (or active) until the concrete reaches its transverse cracking strain in compression and begins to dilate, thus placing hoop stress on the FRP wrap. This is in contrast to the FRP flexural and shear strengthening systems, which must be active at all load levels.

For a nonslender, nonprestressed, normal-weight concrete column reinforced with spiral reinforcement and steel ties the nominal axial capacity is given as

$$P_n = 0.85\left(0.85\psi_f f'_{cc}\left(A_g - A_{st}\right) + f_y A_{st}\right) \tag{16.88}$$

and

$$P_n = 0.80\left(0.85\psi_f f'_{cc}\left(A_g - A_{st}\right) + f_y A_{st}\right) \tag{16.89}$$

respectively, with the confined compressive strength, f'_{cc}, given as

$$f'_{cc} = f'_c\left(2.25\sqrt{1 + 7.9\frac{f_l}{f'_c}} - 2\frac{f_l}{f'_c} - 1.25\right) \tag{16.90}$$

in terms of the confining pressure provide by the FRP wrap (or jacket), f_l, given as

$$f_l = \frac{\kappa_a \rho_f E_f \varepsilon_{fe}}{2} \tag{16.91}$$

where A_g is the gross area of the concrete, A_{st} is the area of the longitudinal steel, κ_a is an efficiency factor that depends on the shape of the column, and ρ_f is the reinforcement ratio of the FRP strengthening system. For circular columns, κ_a is 1.0 and the reinforcement ratio is given as

$$\rho_f = \frac{A_f}{A_g} = \frac{4nt_f}{h} \tag{16.92}$$

where h is the diameter of the circular column. It is important to note that the fiber layers must all be oriented in the hoop direction around the column. If layers are also oriented in the longitudinal direction, these layers should not be considered to contribute to axial strengthening. For noncircular columns FRP strengthening for increasing axial capacity is much less effective due to stress concentrations at the corners (even when rounded) and the nonuniform confining pressure developed by the wrap.

Limits are placed on axial strengthening to ensure that the concrete does not approach its transverse cracking strain, or the steel its yield strain, in the service range (ACI 440.2R-02 2002).

The lateral displacement capacity (or ductility) of a concrete column can also be increased by confining it with FRP strengthening wraps. The maximum concrete compressive strain for an FRP wrapped circular column is given as

$$\varepsilon'_{cc} = \frac{1.71\left(5f'_{cc} - 4f'_c\right)}{E_c} \tag{16.93}$$

where E_c is the elastic modulus of the concrete and f'_{cc} is as defined previously (Equation 16.90).

References

AASHTO, 2001, Standard Specifications for Structural Supports for Highway Signs, Luminaires and Traffic Signals, 4th Ed., AASHTO, Washington, DC.

Abd-el-Naby, S.F.M. and Hollaway, L., 1993a, "The Experimental Behavior of Bolted Joints in Pultruded Glass/Polyester Material. Part 1: Single Bolt Joints," *Composites*, Vol. 24, pp. 531–538.

Abd-el-Naby, S.F.M. and Hollaway, L., 1993b, "The Experimental Behavior of Bolted Joints in Pultruded Glass/Polyester Material. Part 2: Two-Bolt Joints," *Composites*, Vol. 24, pp. 539–546.

AC125, 1997, Interim Criteria for Concrete and Reinforced and Unreinforced Masonry Strengthening Using Fiber-Reinforced Polymer (FFP) Composite Systems, ICC Evaluation Service, California.

AC187, 2001, Interim Criteria for Inspection and Verification of Concrete and Reinforced and Unreinforced Masonry Strengthening Using Fiber-Reinforced Polymer (FFP) Composite Systems, ICC Evaluation Service, California.

ACI 318-95, 1995, Building Code Requirements for Structural Concrete and Commentary, ACI 318-95, American Concrete Institute, Michigan.

ACI 440R-96, 1996, State-of-the-Art Report on Fiber Reinforced Plastic (FRP) Reinforcement for Concrete Structures (440R-96), American Concrete Institute, Michigan.

ACI 318-99, 1999, Building Code Requirements for Structural Concrete and Commentary, ACI 318-99, American Concrete Institute, Michigan.

ACI 440.2R-02, 2002, Guide to the Design and Construction of Externally Bonded FRP Systems for Strengthening Concrete Structures, ACI 440.2R-02, American Concrete Institute, Michigan.

ACI 440.1R-03, 2003, Guide for the Design and Construction of Concrete Reinforced with FRP Bars (440.1R-03), American Concrete Institute, Michigan.

ACI 440.3R-04, 2004, Guide Test Methods for Fiber Reinforced Polymers (FRP) for Reinforcing or Strengthening Concrete Structures (440.3R-04), American Concrete Institute, Michigan.

ACI 440.4R-04, 2004, Prestressing Concrete Structures with FRP Tendons (440.4R-04), American Concrete Institute Michigan.

Agarwal, D.H. and Broutman, L.J., 1990, Analysis and Performance of Fiber Composites, 2nd Ed., Wiley, New York.

ASCE 7-02, 2002, Minimum Design Loads for Buildings and Other Structures, American Society of Civil Engineers, Reston, VA.

ASTM, 2003, ASTM 2003 Standards, American Society for Testing and Materials, Pennsylvania.

Bakht, B. et al., 2000, "Canadian Bridge Design Code Provisions for Fiber-Reinforced Structures," *J. Composites Constr.*, Vol. 4, pp. 3–15.

Bakis, C.E., Bank, L.C., Brown, V.L., Cosenza, E., Davalos, J.F., Lesko, J.J., Machida, A., Rizkalla, S.H., and Triantifilliou, T.C., 2002, "Fiber-Reinforced Polymer Composites for Construction — State-of-the-Art Review," *J. Composites Constr.*, Vol. 6, pp. 73–87.

Bank, L.C., 1987, "Shear Coefficients for Thin-Walled Composite Beams," *Composite Struct.*, Vol. 8, pp. 47–61.

Bank, L.C., 1989a, "Flexural and Shear Moduli of Full-Section Fiber Reinforced Plastic (FRP) Pultruded Beams," *J. Test. Eval.*, Vol. 17, No. 1, pp. 40–45.

Bank, L.C., 1989b, "Properties of Pultruded Fiber Reinforced Plastic (FRP) Structural Members," Transportation Research Record 1223, National Research Council, pp. 117–124.

Bank, L.C., 1993, "FRP Reinforcements for Concrete," in Fiber-Reinforced Plastic (FRP) for Concrete Structures: Properties and Applications (ed. A. Nanni), Elsevier Science Publishers, New York, pp. 59–86.

Bank, L.C., Gentry, T.R., and Nadipelli, M., 1996, "Local Buckling of Pultruded FRP Beams — Analysis and Design," *J. Reinforced Plast. Composites*, Vol. 15, pp. 283–294.

Bank, L.C., Gentry, T.R., Thompson, B.P., and Russell, J.S., 2003, "A Model Specification for Composites for Civil Engineering Structures," *Construction and Building Materials*, Vol. 17, pp. 405–437.

Bank, L.C. and Mosallam, A.S., 1992, "Creep and Failure of a Full-Size Fiber Reinforced Plastic Pultruded Frame," *Composites Eng.*, Vol. 2, pp. 213–227.

Bank, L.C., Mosallam, A.S., and Gonsior, H.E., 1990, "Beam-to-Column Connection for Pultruded Frame Structures," in Proceedings of the 1st ASCE Materials Congress (ed. B. Suprenant), Denver, CO, pp. 804–813.

Bank, L.C., Mosallam, A.S., and McCoy, G.T., 1994a, "Design and Performance of Connections for Pultruded Frame Structures," *J. Reinforced Plast. Composites*, Vol. 13, pp. 199–212.

Bank, L.C., Nadipelli, M., and Gentry, T.R., 1994b, "Local Buckling and Failure of Pultruded Fiber-Reinforced Plastic Beams," *J. Eng. Mater. Technol.*, Vol. 116, pp. 233–237.

Bank, L.C., Yin, J., and Nadipelli, M., 1995, "Local Buckling of Pultruded Beams — Nonlinearity, Anisotropy and Inhomogeneity," *Construction and Building Materials*, Vol. 9, pp. 325–331.

Bank, L.C. and Yin, J., 1996, "Buckling of Orthotropic Plates with Free and Rotationally Restrained Unloaded Edges," *Thin-Walled Struct.*, Vol. 24, pp. 83–96.

Bank, L.C. and Yin, J., 1999, "Analysis of Progressive Failure of the Web-Flange Junction in Post-Buckled Pultruded I-Beams," *J. Composites Constr.*, Vol. 3, pp. 177–184.

Bank, L.C., Yin, J., Moore, L.E., Evans, D., and Allison, R., 1996, "Experimental and Numerical Evaluation of Beam-to-Column Connections for Pultruded Structures," *J. Reinforced Plast. Composites*, Vol. 15, pp. 1052–1067.

Barbero, E. and Tomblin, J., 1993, "Euler Buckling of Thin-Walled Composite Columns," *Thin-Walled Struct.*, Vol. 17, pp. 237–258.

Barbero, E.J. and Raftoyiannis, I.G., 1994, "Lateral and Distortional Buckling of Pultruded I-beams," *Composite Structures*, Vol. 27, pp. 261–268.

Barbero, E.J., 1998, Introduction to Composite Materials Design, Taylor & Francis, Philadelphia, PA.

Barbero, E.J., Fu, S.-H., and Raftoyiannis, I., 1991, "Ultimate Bending Strength of Composite Beams," *J. Mater. Civil Eng.*, Vol. 3, pp. 292–306.

Bass, A.J. and Mottram, J.T., 1994, "Behaviour of Connections in Frames of Fibre-Reinforced-Polymer Section," *Struct. Engineer*, Vol. 72, pp. 280–285.

Chambers, R.E., 1997, "ASCE Design Standard for Pultruded Fiber-Reinforced Plastic (FRP) Structures," *J. Composites Constr.*, Vol. 1, pp. 26–38.

Cooper, C., and Turvey, G.J., 1995, "Effects of Joint Geometry and Bolt Torque on the Structural Performance of Single Bolt Tension Joins in Pultruded GRP Sheet Material," *Composite Struct.*, Vol. 32, pp. 217–226.

Creative Pultrusions, 2000, The Pultex Pultrusion Global Design Manual, Creative Pultrusions, Alum Bank, PA. Also at www.creativepultrusions.com.

CSA, 2002, Design and Construction of Building Components with Fibre-Reinforced Polymers, S806–02, Canadian Standards Association (CSA), Toronto, Canada.

Davalos, J.F. and Qiao, P., 1997, "Analytical and Experimental Study of Lateral and Distortional Buckling of FRP Wide-Flange Beams," *J. Composites Constr.*, Vol. 1, pp. 150–159.

Ellingwood, B.R., 2003, "Toward Load and Resistance Factor Design for Fiber-Reinforced Polymer Composite Structures," *J. Struct. Eng.*, Vol. 129, pp. 449–458.

Erki, M.A., 1995, "Bolted Glass-Fibre-Reinforced Plastic Joints," *Can. J. Civil Eng.*, Vol. 22, pp. 736–744.

Eurocomp Design Code and Handbook (EDCH), 1996, Structural Design of Polymer Composites (ed. J. Clarke), E&F Spon, London.

FIB, 2001, Externally Bonded FRP Reinforcement for RC Structures, International Federation for Structural Concrete (fib), Switzerland.

Fiberline, 1995, Fiberline Design Manual, Fiberline, Denmark. Also at www.fiberline.com.

Gilstrap, J.M., Burke, C.R., Dowden, D.M., and Dolan, C.W., 1997, "Development of FRP Reinforcement Guidelines for Prestressed Concrete Structures," *J. Composites Constr.*, Vol. 1, pp. 131–139.

Haaijer, G., 1957, "Plate Buckling in the Strain Hardening Range," *J. Eng. Mech.*, Vol. 83, No. 1212, pp. 1–47.

Hart-Smith, L.J., 1978, "Mechanically Fastened Joints for Advanced Composites — Phenomenological Considerations and Simple Analyses," in Proceedings of the 4th Conference on Fibrous Composites in Structural Design (San Diego, CA, eds. E.M. Lenoe, D.W. Oplinger, and J.W. Burke), Nov. 14–17, 1978, Plenum Publishers, New York, pp. 543–574.

Hassan, N.H., Mohamedien, M.A., and Rizkalla, S.H., 1997a, "Multibolted Joints for GFRP Structural Members," *J. Composites Constr.*, Vol. 1, pp. 3–9.

Hassan, N.H., Mohamedien, M.A., and Rizkalla, S.H., 1997b, "Rational Model for Multibolted Connections for GFRP Members," *J. Composites Constr.*, Vol. 1, pp. 71–78.

Hollaway, L., 1993, Polymer Composites for Civil and Structural Engineering, Chapman and Hall, New York.

Hollaway, L.C. and Head, P.R., 2001, Advanced Polymer Composites and Polymers in the Civil Infrastructure, Elsevier, Oxford, UK.

Johansen, G.E., Wilson, R.J., Roll, F., Levin, B., Chung, D., and Poplawski, E., 1999, "Testing and Evaluation of FRP Truss Connections," in Materials and Construction: Exploring the Connection (ed. L.C. Bank), ASCE, Reston, VA, pp. 68–75.

Johnson, A.F., 1985, "Simplified Buckling Analysis for R.P. Beams and Columns," in Proceedings of ICCM-1, Bordeaux, France, pp. 541–549.

Jones, R.M., 1999, Mechanics of Composite Materials, 2nd Ed., Taylor & Francis, Philadelphia, PA.

JSCE, 1997, Recommendation for Design and Construction of Concrete Structures Using Continuous Fiber Reinforcing Materials, Japan Society of Civil Engineers (JSCE), Tokyo, Japan.

JSCE, 2001, Recommendation for Upgrading of Concrete Structures with Use of Continuous Fiber Sheets, Japan Society of Civil Engineers (JSCE), Tokyo, Japan.

Katz, A., Berman, N., and Bank, L.C., 1999, "Effect of High Temperature on the Bond Strength of FRP Rebars," *ASCE J. Composites Constr.*, Vol. 3, pp. 73–81.

Lopez-Anido, R., Falker, E., Mittlestadt, B., and Troutman, D., 1999, "Shear Tests on Pultruded Beam-to-Column Connections with Clip Angles," in Materials and Construction: Exploring the Connection (ed. L.C. Bank), ASCE, Reston, VA, pp. 92–99.

Meier, U., 1995, "Strengthening of Concrete Structures Using Carbon Fibre/Epoxy Composites," *Constr. Building Mater.*, Vol. 9, pp. 341–351.

Meier, U. and Kaiser, H., 1991, "Strengthening Structures with CFRP Laminates," in Advanced Composites in Civil Engineering Structures (eds S.L. Iyer and R. Sen), ASCE, Reston, VA, pp. 224–232.

Meyer, R.W., 1985, Handbook of Pultrusion Technology, Chapman and Hall, London.

Mosallam, A.S. and Bank, L.C., 1991, "Creep and Recovery of a Pultruded FRP Frame," in Advanced Composite Materials for Civil Engineering Structures," (eds S.L. Iyer and R. Sen), ASCE, Reston, VA, pp. 24–35.

Mosallam, A.S. and Bank, L.C., 1992, "Short-Term Behavior of Pultruded Fiber-Reinforced Plastic Frame," *J. Struct. Eng.*, Vol. 118, pp. 1937–1954.

Mottram, J.T., 1991, "Evaluation of Design Analysis for Pultruded Fibre-Reinforced Polymeric Box Beams," *Struct. Engineer*, Vol. 69, pp. 211–220.

Mottram, J.T., 1992, "Lateral Torsional Buckling of a Pultruded I Beam," *Composites*, Vol. 23, pp. 81–92.

Mottram, J.T., 2001, "Analysis and Design of Connections for Pultruded FRP Structures," in Composites in Construction: A Reality (eds E. Cosenza, G. Manfredi, and A. Nanni), ASCE, Reston, VA, pp. 250–257.

Mottram, J.T. and Turvey, G.J. (eds), 1998, "State-of-the-Art Review on Design, Testing, Analysis and Applications of Polymeric Composite Connections," Report EUR 18172 EN, European Commission, Luxembourg.

Mottram, J.T. and Zheng, Y., 1996, "State-of-the-Art Review on the Design of Beam-to-Column Connections for Pultruded Structures," *Composite Struct.*, Vol. 35, pp. 387–401.

Mottram, J.T. and Zheng, Y., 1999a, "Further Tests on Beam-to-Column Connections for Pultruded Frames: Web-Cleated," *J. Composites Constr.*, Vol. 3, pp. 3–11.

Mottram, J.T. and Zheng, Y., 1999b, "Further Tests on Beam-to-Column Connections for Pultruded Frames: Flange-Cleated," *J. Composites Constr.*, Vol. 3, pp. 108–116.

Nagaraj, V. and GangaoRao, H.V.S., 1997, "Static Behavior of Pultruded GFRP Beams," *J. Composites Constr.*," Vol. 1, pp. 120–129.

Nanni, A. (ed.), 1993, Fiber-Reinforced Plastic (FRP) for Concrete Structures: Properties and Applications, Elsevier Science Publishers, New York.

Prabhakaran, R., Razzaq, Z., and Devera, S., 1996, "Load and Resistance Factor Design (LRFD) Approach to Bolted Joints in Pultruded Structures," *Composites Part B*, Vol. 27B, pp. 351–360.

Roberts, T.M., 2002, "Influence of Shear Deformation on Buckling of Pultruded Fiber Reinforced Plastic Profiles," *J. Composites Constr.*, Vol. 6, pp. 241–248.

Roberts, T.M. and Al-Ubaidi, H., 2002, "Flexural and Torsional Proerties of Pultruded Fiber Reinforced Plastic I-Profiles," *J. Composites Constr.*, Vol. 6, pp. 28–34.

Rosner, C.N. and Rizkalla, S.H., 1995a, "Bolted Connections for Fiber-Reinforced Composite Structural Members: Experimental Program," *J. Mater. Civil Eng.*, Vol. 7, pp. 223–231.

Rosner, C.N. and Rizkalla, S.H., 1995b, "Bolted Connections for Fiber-Reinforced Composite Structural Members: Analytical Model and Design Recommendations," *J. Mater. Civil Eng.*, Vol. 7, pp. 232–238.

Saadatmanesh, H. and Ehsani, M.R., 1991, "RC Beams Strengthened with FRP Plates: Experimental Study," *J. Struct. Eng.*, Vol. 117, pp. 3417–3433.

Scott, D.W., Zureick, A.-H., and Lai, J.S., 1995, "Creep Behavior of Fiber Reinforced Polymeric Composites." *J. Reinforced Plast. Composites*, Vol. 14, pp. 590–617.

Seible, F., Priestly, M.J.N., Hegemier, G.A., and Innamorato, D., 1997, "Seismic Retrofit of RC Columns with Continuous Carbon Fiber Jackets," *J. Composites Constr.*, Vol. 1, pp. 52–62.

Smith, S.J., Parsons, I.D., and Hjelmstad, K.D., 1999, "Experimental Comparison of Connections for GFRP Pultruded Frames," *J. Composites Constr.*, Vol. 3, pp. 20–26.

Sonobe, Y. et al., 1997, "Design Guidelines of FRP Reinforced Concrete Structures," *J. Composites Constr.*, Vol. 1, pp. 90–115.

Strongwell, 2002, Strongwell Design Manual, CD Rom, Strongwell, Bristol, VA. Also available at www.strongwell.com.

Structural Plastics Design Manual (SPDM), 1984, ASCE.

Tan, K.H. (ed.), 2003, Fibre-Reinforced Polymer Reinforcement for Concrete Structures, in Proceedings of the 6th International Symposium on FRP Reinforced Concrete Structures (FRPRCS6), Singapore, World Scientific Publishing, Singapore.

Teng, J.G., Chen, J.F., Smith, S.T., and Lam, L., 2001, "FRP Strengthened RC Structures," Wiley, New York.

TR55, 2000, Design Guidance for Strengthening Concrete Structures Using Fibre Composite Materials, The Concrete Society, U.K.

TR57, 2003, Strengthening Concrete Structures with Fibre Composite Materials: Acceptance, Inspection and Monitoring, The Concrete Society, UK.

Triantafillou, T., 1998, "Shear Strengthening of Reinforced Concrete Beams Using Epoxy-Bonded FRP Composites," *ACI Struct. J.*, Vol. 95, pp. 107–115.

Tsai, S.W. and Hahn, H.T., 1980, Introduction to Composite Materials, Technomic, Lancaster, PA.

Turvey, G.J., 1998, "Single-Bolt Tension Tests on Pultruded GRP Plate — Effects of Tension Direction Relative to Pultrusion Direction," *Composite Struct.*, Vol. 42, pp. 341–351.

Turvey, G., 2001, "Pultruded GRP Frames: Simple (Conservative) Approach to Design, a Rational Alternative and Research Needs for Improved Design," in Composites for Construction: A Reality (eds E. Cosenza, G. Manfredi, and A. Nanni), ASCE, Reston, VA, pp. 258–266.

Vinson, J.R. and Sierakowski, R.L., 1986, The Behavior of Structures Composed of Composite Materials, Martinus Nijhoff, The Netherlands.

Wang, Y., 2002, "Bearing Behavior of Joints in Pultruded Composites," *J. Composite Mater.* Vol. 36, pp. 2199–2216.

Zureick, A. and Scott, D., 1997, "Short-Term Behavior and Design of Fiber-Reinforced Polymeric Slender Members under Axial Compression," *J. Composites Constr.*, Vol. 1, pp. 140–149.

Zureick, A. and Steffen, R., 2000, "Behavior and Design of Concentrically Loaded Pultruded Angles Struts," *J. Struct. Eng.*, Vol. 126, pp. 406–416.

IV

Earthquake Engineering and Design

VI

Earthquake
Engineering and
Design

17

Fundamentals of Earthquake Engineering

Charles Scawthorn
Department of Urban Management,
Kyoto University,
Kyoto, Japan

17.1 Introduction

This chapter provides a basic understanding of earthquakes, by first discussing the causes of earthquakes, then defining commonly used terms, explaining how earthquakes are measured, discussing the distribution of seismicity, and, finally, explaining how seismicity can be characterized.

Earthquakes are broad-banded vibratory ground motions, resulting from a number of causes including tectonic ground motions, volcanism, landslides, rockbursts, and man-made explosions. Of these, naturally occurring tectonic-related earthquakes are the largest and most important. These are caused by the fracture and sliding of rock along *faults* within the Earth's crust. A fault is a zone of the earth's crust within which the two sides have moved — faults may be hundreds of miles long, from one to over one hundred miles deep, and are sometimes not readily apparent on the ground surface. Earthquakes initiate a number of phenomena or agents, termed *seismic hazards*, which can cause significant damage to the built environment — these include fault rupture, vibratory ground motion (i.e., shaking), inundation (e.g., tsunami, seiche, dam failure), various kinds of permanent ground failure (e.g., liquefaction), fire, or hazardous materials release. In a particular earthquake event, any particular hazard can dominate, and historically each has caused major damage and great loss of life in particular earthquakes.

For most earthquakes, shaking is the dominant and most widespread agent of damage. Shaking near the actual earthquake rupture lasts only during the time when the fault ruptures, a process that takes seconds or at most a few minutes. The seismic waves generated by the rupture propagate long after the movement on the fault has stopped, however, spanning the globe in about 20 min. Typically, earthquake ground motions are powerful enough to cause damage only in the near field (i.e., within a few tens of kilometers from the causative fault) — in a few instances, long period motions have caused significant damage at great distances, to selected lightly damped structures. A prime example of this was the 1985 Mexico City Earthquake, where numerous collapses of mid- and high-rise buildings were due to a magnitude 8.1 Earthquake occurring at a distance of approximately 400 km from Mexico City.

17.2 Causes of Earthquakes and Faulting

In a global sense, tectonic earthquakes result from motion between a number of large plates comprising the earth's crust or lithosphere (about 15 large plates, in total), Figure 17.1.

These plates are driven by the convective motion of the material in the earth's mantle, which in turn is driven by the heat generated at the earth's core. Relative plate motion at the fault interface is constrained by friction and/or *asperities* (areas of interlocking due to protrusions in the fault surfaces). However, strain energy accumulates in the plates, eventually overcomes any resistance, and causes slip between the two sides of the fault. This sudden slip, termed *elastic rebound* by Reid (1910) based on his studies of regional deformation following the 1906 San Francisco Earthquake, releases large amounts of energy, which constitutes or *is* the earthquake. The location of initial radiation of seismic waves (i.e., the first location of dynamic rupture) is termed the *hypocenter*, while the projection on the surface of the earth directly above the hypocenter is termed the *epicenter*. Other terminology includes *near-field*[1] (within one source dimension of the epicenter, where source dimension refers to the width or length of faulting, whichever is shorter), *far-field* (beyond near-field), and *meizoseismal* (the area of strong shaking and damage). Energy is radiated over a broad spectrum of frequencies through the earth, in *body waves* and *surface waves* (Bolt 1993). Body waves are of two types: P waves (transmitting energy via push–pull motion) and slower S waves (transmitting energy via shear action at right angles to the direction of motion). Surface waves are also of two types: horizontally oscillating *Love waves* (analogous to S body waves) and vertically oscillating *Rayleigh waves*.

While the accumulation of strain energy within the plate can cause motion (and consequent release of energy) at faults at any location, earthquakes occur with greatest frequency at the boundaries of the tectonic plates. The boundary of the Pacific plate is the source of nearly half of the world's great earthquakes. Stretching 40,000 km (24,000 miles) around the circumference of the Pacific Ocean, it includes Japan, the west coast of North America, and other highly populated areas, and is aptly termed the *Ring of Fire*. The interiors of plates, such as ocean basins and continental shields, are areas of low seismicity but are not inactive — the largest earthquakes known to have occurred in North America, for example, occurred in 1811–1812 in the New Madrid area, far from a plate boundary. Tectonic plates move relatively slowly (5 cm per year is relatively fast) and irregularly, with relatively frequent small and only occasional large earthquakes. Forces may build up for decades or centuries at plate interfaces until a large movement occurs all at once. These sudden, violent motions produce the shaking that is felt as an earthquake. The shaking can cause direct damage to buildings, roads, bridges, and other man-made structures as well as triggering landslides, fires, tidal waves (tsunamis), and other damaging phenomena.

Faults are the physical expression of the boundaries between adjacent tectonic plates and thus may be hundreds of miles long. In addition, there may be thousands of shorter faults parallel to or branching out from a main fault zone. Generally, the longer a fault the larger the earthquake it can generate. Beyond the main tectonic plates, there are many smaller subplates, "platelets," and simple blocks of crust that occasionally move and shift due to the "jostling" of their neighbors and the major plates. The existence of these many subplates means that smaller but still damaging earthquakes are possible almost anywhere, although often with less likelihood.

---

[1]Not to be confused with *near-source* as used in the 1997 Uniform Building Code, which can be as much as 15 km, depending on type of faulting.

(a)

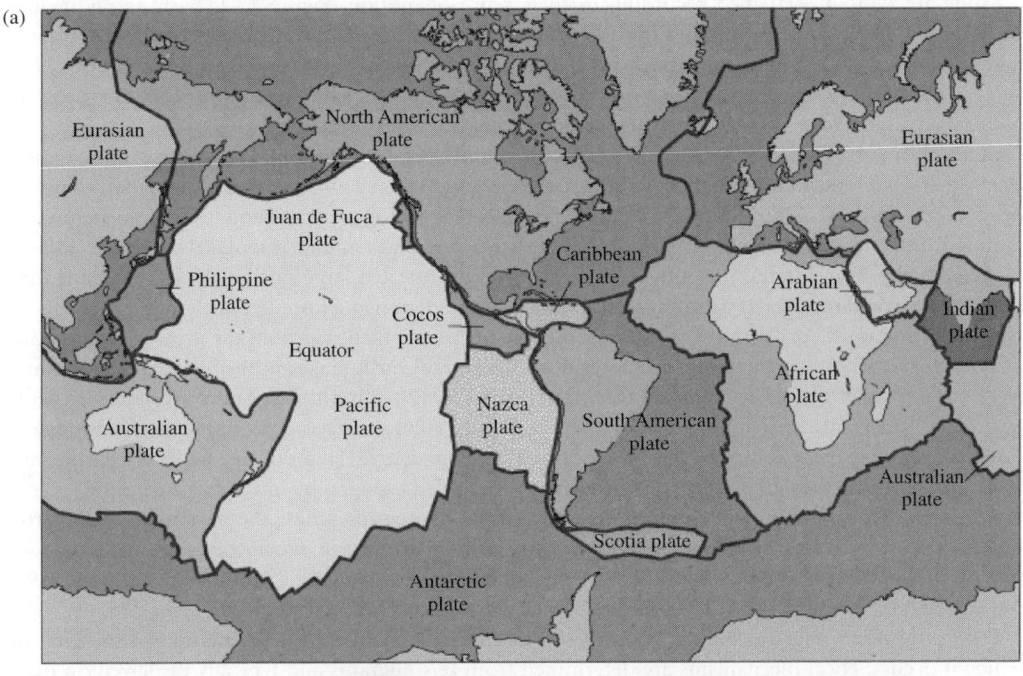

(b)

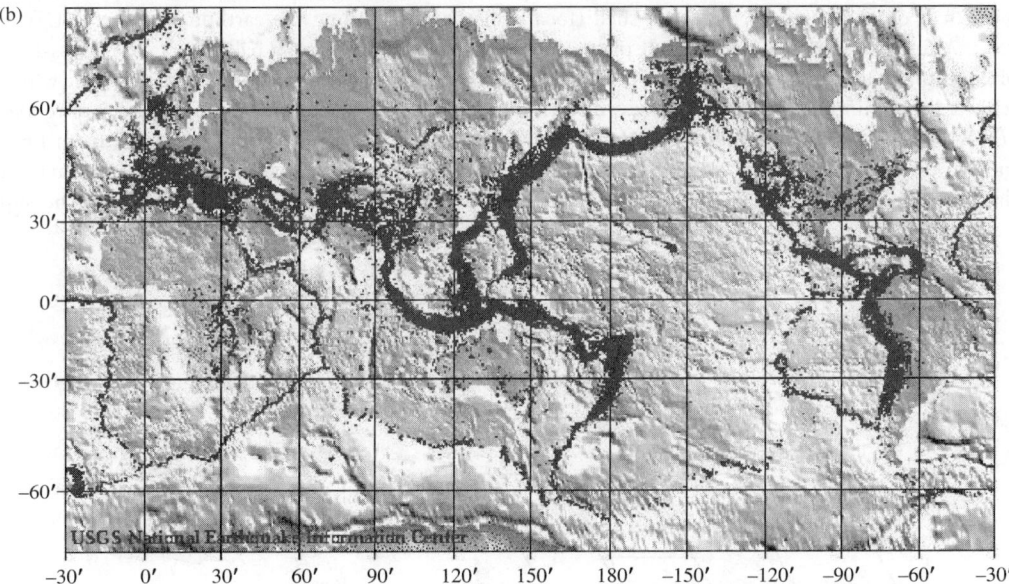

FIGURE 17.1   (a) Global tectonic plate boundaries. (b) Global seismicity 1975–1995 (*from:* U.S. Geological Survey [USGS]).

Faults are typically classified according to their sense of motion, Figure 17.2. Basic terms include *transform* or *strike slip* (relative fault motion occurs in the horizontal plane, parallel to the *strike* of the fault), *dip-slip* (motion at right angles to the strike, up- or down-slip), *normal* (dip-slip motion, two sides in tension, move away from each other), *reverse* (dip-slip, two sides in compression, move toward each other), and *thrust* (low-angle reverse faulting).

Generally, earthquakes will be concentrated in the vicinity of faults, faults that are moving more rapidly than others will tend to have higher rates of seismicity, and larger faults are more likely than others to produce a large event. Many faults are identified on regional geological maps, and useful information on fault location and displacement history is available from local and national geological surveys in areas of high seismicity. Considering this information, areas of an expected large earthquake in the near future (usually measured in years or decades) can, and have, been identified. However, earthquakes continue to occur on "unknown" or "inactive" faults. An important development has been the growing recognition of *blind thrust faults*, which emerged as a result of the several earthquakes in the 1980s, none of which were accompanied by surface faulting (Stein and Yeats 1989). Blind thrust faults are faults at depth occurring under anticlinal folds — since they have only subtle surface expression, their seismogenic potential can only be evaluated by indirect means (Greenwood 1995). Blind thrust faults are particularly worrisome because they are hidden, are associated with folded topography in general, including areas of lower and infrequent seismicity, and, therefore, result in a situation where the potential for an earthquake exists in any area of anticlinal geology, even if there are few or no earthquakes in the historic record. Recent major earthquakes of this type have included the 1980 M_W 7.3 El Asnam (Algeria), 1988 M_W 6.8 Spitak (Armenia), and 1994 M_W 6.7 Northridge (California) events.

Focal mechanism refers to the direction of slip in an earthquake and the orientation of the fault on which it occurs. Focal mechanisms are determined from seismograms and typically displayed on maps as a black and white "beach ball" symbol. This symbol is the projection on a horizontal plane of the lower half of an imaginary, spherical shell (focal sphere) surrounding the earthquake source (USGS, n.d.). A line is scribed where the fault plane intersects the shell. The beach ball depicts the stress-field orientation at the time of rupture such that the black quadrants contain the tension axis (T), which reflects the minimum compressive stress direction, and the white quadrants contain the pressure axis (P), which reflects the maximum compressive stress direction. For mechanisms calculated from first-motion directions (as well as some other methods), more than one focal mechanism solution may fit the data equally well, so that there is an ambiguity in identifying the fault plane on which the slip

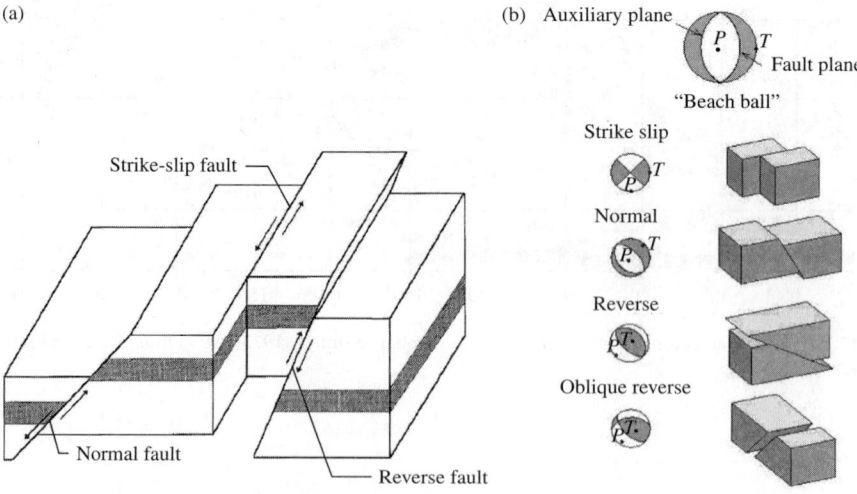

FIGURE 17.2   (a) Types of faulting and (b) focal mechanisms (after U.S. Geological Survey).

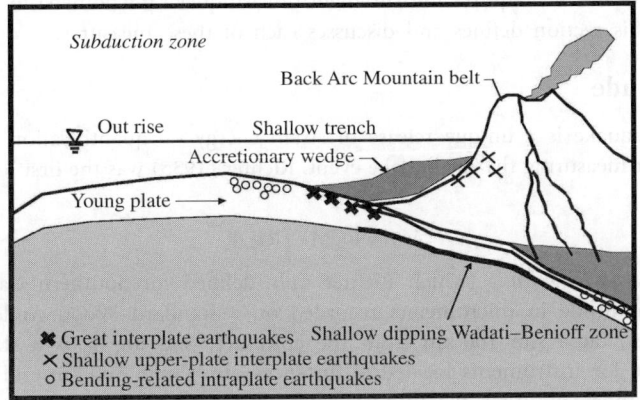

FIGURE 17.3   Schematic diagram of subduction zone, typical of west coast of South America, Pacific Northwest of United States or Japan.

occurred, from the orthogonal, mathematically equivalent, auxiliary plane. The ambiguity may sometimes be resolved by comparing the two fault-plane orientations to the alignment of small earthquakes and aftershocks. The first three examples describe fault motion that is purely horizontal (strike slip) or vertical (normal or reverse). The oblique-reverse mechanism illustrates that slip may also have components of horizontal and vertical motion.

Subduction refers to the plunging of one plate (e.g., the Pacific) beneath another, into the mantle, due to convergent motion, as shown in Figure 17.3. Subduction zones are typically characterized by volcanism, as a portion of the plate (melting in the lower mantle) re-emerges as volcanic lava. Four types of earthquakes are associated with subduction zones: (1) shallow crustal events, in the accretionary wedge; (2) intraplate events, due to plate bending; (3) large interplate events, associated with slippage of one plate past the other; and (4) deep Benioff zone events. Subduction occurs along the west coast of South America at the boundary of the Nazca and South American plate, in Central America (boundary of the Cocos and Caribbean plates), in Taiwan and Japan (boundary of the Philippine and Eurasian plates), and in the North American Pacific Northwest (boundary of the Juan de Fuca and North American plates), among other places.

Probabilistic methods can be usefully employed to quantify the likelihood of an earthquake's occurrence. However, the earthquake generating process is not understood well enough to reliably predict the times, sizes, and locations of earthquakes with precision. In general, therefore, communities must be prepared for an earthquake to occur at any time.

17.3   Measurement of Earthquakes

Earthquakes are complex multidimensional phenomena, the scientific analysis of which requires measurement. Prior to the invention of modern scientific instruments, earthquakes were qualitatively measured by their effect or *intensity*, which differed from point to point. With the deployment of seismometers, an instrumental quantification of the entire earthquake event — the unique *magnitude* of the event — became possible. These are still the two most widely used measures of an earthquake, and a number of different scales for each have been developed, which are sometimes confused.[2]

---

[2]Earthquake magnitude and intensity are analogous to a lightbulb and the light it emits. A particular lightbulb has only one energy level, or wattage (e.g., 100 W, analogous to an earthquake's magnitude). Near the lightbulb, the light intensity is very bright (perhaps 100 ft-candles, analogous to MMI IX), while farther away the intensity decreases (e.g., 10 ft-candles, MMI V). A particular earthquake has only one magnitude value, whereas it has many intensity values.

Engineering design, however, requires measurement of earthquake phenomena in units such as force or displacement. This section defines and discusses each of these measures.

17.3.1 Magnitude

An individual earthquake is a unique release of strain energy — quantification of this energy has formed the basis for measuring the earthquake event. Richter (1935) was the first to define earthquake magnitude, as

$$M_L = \log A - \log A_0 \qquad (17.1)$$

where M_L is the *local magnitude* (which Richter only defined for Southern California), A is the maximum trace amplitude in micrometers recorded on a standard Wood–Anderson short-period torsion seismometer,[3] at a site 100 km from the epicenter, and $\log A_0$ is a standard value as a function of distance for instruments located at distances other than 100 km and less than 600 km. Subsequently, a number of other magnitudes have been defined, the most important of which are surface wave magnitude M_S, body wave magnitude m_b, and moment magnitude M_W. Due to the fact that M_L was only locally defined for California (i.e., for events within about 600 km of the observing stations), surface wave magnitude M_S was defined analogously to M_L, using teleseismic observations of surface waves of 20 s period (Richter 1935). Magnitude, which is defined on the basis of the amplitude of ground displacements, can be related to the total energy in the expanding wave front generated by an earthquake, and thus to the total energy release — an empirical relation by Richter is

$$\log_{10} E_S = 11.8 + 1.5 M_S \qquad (17.2)$$

where E_S is the total energy in ergs.[4] Note that $10^{1.5} = 31.6$, so that an increase of one magnitude unit is equivalent to 31.6 times more energy release, two magnitude units increase equivalent to $998.6 \cong 1000$ times more energy, etc. Subsequently, due to the observation that deep-focus earthquakes commonly do not register measurable surface waves with periods near 20 s, a body wave magnitude m_b was defined (Gutenberg and Richter 1954), which can be related to M_S (Darragh et al. 1994):

$$m_b = 2.5 + 0.63 M_S \qquad (17.3)$$

Body wave magnitudes are more commonly used in eastern North America, due to the deeper earthquakes there. A number of other magnitude scales have been developed, most of which tend to *saturate* — that is, asymptote to an upper bound due to larger earthquakes radiating significant amounts of energy at periods longer than used for determining the magnitude (e.g., for M_S, defined by measuring 20 s surface waves, saturation occurs at about $M_S > 7.5$). More recently, *seismic moment* has been employed to define a *moment magnitude* M_W (Hanks and Kanamori 1979; also denoted as boldface **M**), which is finding increased and widespread use

$$\log M_0 = 1.5 M_W + 16.0 \qquad (17.4)$$

where seismic moment M_0 (dyne cm) is defined as (Lomnitz 1974)

$$M_0 = \mu A \bar{u} \qquad (17.5)$$

where μ is the material shear modulus, A is the area of fault plane rupture, and $\bar{u}$ is the mean relative displacement between the two sides of the fault (the averaged fault slip). Comparatively, M_W and M_S are

---

[3]The instrument has a natural period of 0.8 s, critical damping ration 0.8, magnification 2800.

[4]Richter (1958) gives 11.4 for the constant term, rather than 11.8, which is based on subsequent work — the uncertainty in the data makes this difference inconsequential.

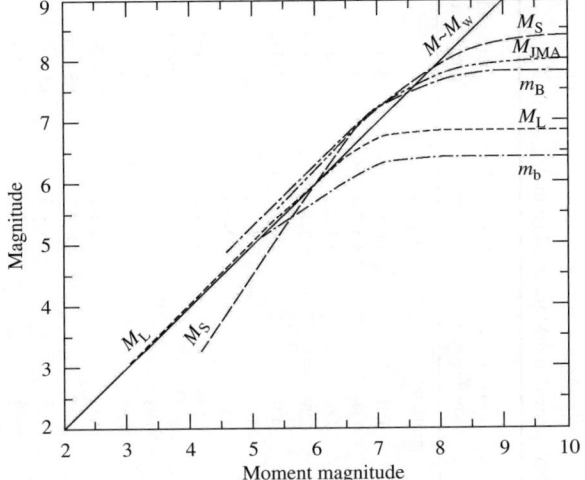

FIGURE 17.4 Relationship between moment magnitude and various magnitude scales (Campbell, K.W. 1985).

numerically almost identical up to magnitude 7.5. Figure 17.4 indicates the relationship between moment magnitude and various magnitude scales.

For lay communications, it is sometimes customary to speak of great earthquakes, large earthquakes, etc. There is no standard definition for these, but the following is an approximate categorization:

Earthquake	Micro	Small	Moderate	Large	Great
Magnitude[a]	Not felt	<5	5–6.5	6.5–8	>8

[a] Not specifically defined.

From the foregoing discussion, it can be seen that magnitude and energy are related to fault rupture length and slip. Slemmons (1977) and Bonilla et al. (1984) have determined statistical relations between these parameters for worldwide and regional data sets, aggregated and segregated by type of faulting (normal, reverse, strike-slip). Bonilla et al.'s worldwide results for all types of faults are

$$M_S = 6.04 + 0.708 \log_{10} L, \qquad s = 0.306 \tag{17.6}$$

$$\log_{10} L = -2.77 + 0.619 M_S, \qquad s = 0.286 \tag{17.7}$$

$$M_S = 6.95 - 0.723 \log_{10} d, \qquad s = 0.323 \tag{17.8}$$

$$\log_{10} d = -3.58 + 0.550 M_S, \qquad s = 0.282 \tag{17.9}$$

which indicates, for example, that, for $M_S = 7$, the average fault rupture length is about 36 km (and the average displacement is about 1.86 m), and s indicates standard deviation. Conversely, a fault of 100 km length is capable of about an $M_S = 7.5$ event[5]. More recently, Wells and Coppersmith (1994) have performed an extensive analysis of a dataset of 421 earthquakes — their results are presented in Table 17.1.

---

[5]Note that $L = g(M_S)$ should not be inverted to solve for $M_S = f(L)$, as a regression for $y = f(x)$ is different than a regression for $x = g(y)$.

TABLE 17.1   Regressions of (a) Rupture Length, Rupture Width, Rupture Area, and Moment Magnitude and (b) Displacement and Moment Magnitude

Equation[a]	Slip type[b]	Number of events	Coefficients and standard errors a(sa)	b(sb)	Standard deviation s	Correlation coefficient r	Magnitude range	Length/width range (km)
(a) Regressions of rupture length, rupture width, rupture area, and moment magnitude								
M = a + b·log(SRL)	SS	43	5.16(0.13)	1.12(0.08)	0.28	0.91	5.6 to 8.1	1.3–432
	R	19	5.00(0.22)	1.22(0.16)	0.28	0.88	5.4–7.4	3.3–85
	N	13	4.86(0.34)	1.32(0.25)	0.34	0.81	5.2–7.3	2.5–41
	All	77	5.08(0.10)	1.16(0.07)	0.28	0.89	5.2–8.1	1.3–432
log(SRL) = a + b·M	SS	43	−3.55(0.37)	0.74(0.05)	0.23	0.91	5.6–8.1	1.3–432
	R	19	−2.86(0.55)	0.63(0.08)	0.20	0.88	5.4–7.4	3.3–85
	N	15	−2.01(0.65)	0.50(0.10)	0.21	0.81	5.2–7.3	2.5–41
	All	77	−3.22(0.27)	0.69(0.04)	0.22	0.89	5.2–8.1	1.3–432
M = a + b × log(RLD)	SS	93	4.33(0.06)	1.49(0.05)	0.24	0.96	4.8–8.1	1.5–350
	R	50	4.49(0.11)	1.49(0.09)	0.26	0.93	4.8–7.6	1.1–80
	N	24	4.34(0.23)	1.54(0.18)	0.31	0.88	5.2–7.3	3.8–63
	All	167	4.38(0.06)	1.49(0.04)	0.26	0.94	4.8–8.1	1.1–350
log(RLD) = a + b·M	SS	93	−2.57(0.12)	0.62(0.02)	0.15	0.96	4.8–8.1	1.5–350
	R	50	−2.42(0.21)	0.58(0.03)	0.16	0.93	4.8–7.6	1.1–80
	N	24	−1.88(0.37)	0.50(0.06)	0.17	0.88	5.2–7.3	3.8–63
	All	167	−2.44(0.11)	0.59(0.02)	0.16	0.94	4.8–8.1	1.1–350
M = a + b × log(RW)	SS	87	3.80(0.17)	2.59(0.18)	0.45	0.84	4.8–8.1	1.5–350
	R	43	4.37(0.16)	1.95(0.15)	0.32	0.90	4.8–8.1	1.1–80
	N	23	4.04(0.29)	2.11(0.28)	0.31	0.86	5.2–7.3	3.8–63
	All	153	4.06(0.11)	2.25(0.12)	0.41	0.84	4.8–8.1	1.5–350
log(RW) = a + b·M	SS	87	−0.76(0.12)	0.27(0.02)	0.14	0.84	4.8–8.1	1.5–350
	R	43	−1.61(0.20)	0.41(0.03)	0.15	0.90	4.8–7.6	1.1–80
	N	23	−1.14(0.28)	0.35(0.05)	0.12	0.86	5.2–7.3	3.8–63
	All	153	−1.01(0.10)	0.32(0.02)	0.15	0.84	4.8–8.1	1.1–350

Equation	Slip Type[b]	N	a(SE)	b(SE)	r	s	Magnitude Range	Displacement or Area Range
M = a + b × log(RA)	SS	83	3.98(0.07)	1.02(0.03)	0.96	0.23	4.8–7.9	3–5184
	R	43	4.33(0.12)	0.90(0.05)	0.94	0.25	4.8–7.6	2.2–2400
	N	22	3.93(0.23)	1.02(0.10)	0.92	0.25	5.2–7.3	19–900
	All	148	4.07(0.06)	0.98(0.03)	0.95	0.24	4.8–7.9	2.2–5184
log(RA) = a + b = M	SS	83	–3.42(0.18)	0.90(0.03)	0.96	0.22	4.8–7.9	3–5184
	R	43	–3.99(0.36)	0.98(0.06)	0.94	0.26	4.8–7.6	2.2–2100
	N	22	–2.87(0.50)	0.82(0.08)	0.92	0.22	5.2–7.3	19–900

(b) Regressions of displacement and moment magnitude

Equation	Slip Type[b]	N	a(SE)	b(SE)	r	s	Magnitude Range	Displacement Range
M = a + b × log(MD)	SS	43	6.81(0.05)	0.78(0.06)	0.90	0.29	5.6–8.1	0.01–14.6
	{R[c]	*21*	*6.52(0.11)*	*0.44(0.26)*	*0.36*	*0.52*	*5.4–7.4*	*0.11–6.51}*
	N	16	6.61(0.09)	0.71(0.15)	0.80	0.34	5.2–7.3	0.06–6.1
	All	80	6.69(0.04)	0.74(0.07)	0.78	0.40	5.2–8.1	0.01–14.6
log(MD) = a + b × M	SS	43	–7.03(0.55)	1.03(0.08)	0.90	0.34	5.6–8.1	0.01–14.6
	{R	*21*	*–1.84(1.14)*	*0.29(0.17)*	*0.36*	*0.42*	*5.4–7.4*	*0.11–6.51}*
	N	16	–5.90(1.18)	0.89(0.18)	0.80	0.38	5.2–7.3	0.06–6.1
	All	80	–5.46(0.51)	0.82(0.08)	0.78	0.42	5.2–8.1	0.0–14.6
M = a + b × log(AD)	SS	29	7.04(0.05)	0.89(0.09)	0.89	0.28	5.6–8.1	0.05–8.0
	{R	*15*	*6.64(0.16)*	*0.13(0.36)*	*0.10*	*0.50*	*5.8–7.4*	*0.06–1.51}*
	N	12	6.78(0.12)	0.65(0.25)	0.64	0.33	6.0–7.3	0.08–2.1
	All	56	6.93(0.05)	0.82(0.10)	0.75	0.39	5.6–8.1	0.05–8.0
log(AD) = a + b × M	SS	29	–6.32(0.61)	0.90(0.09)	0.89	0.28	5.6–8.1	0.05–8.0
	{R	*15*	*–0.74(1.40)*	*0.08(0.21)*	*0.10*	*0.38*	*5.8–7.4*	*0.06–1.51}*
	N	12	–4.45(1.59)	0.63(0.24)	0.64	0.33	6.0–7.3	0.08–2.1
	All	56	–4.80(0.57)	0.69(0.08)	0.75	0.36	5.6–8.1	0.05–8.0

[a] SRL—surface rupture length (km); RLD — subsurface rupture length (km); RW — downdip rupture width (km); RA — rupture area (km^2); MD — maximum displacement (m); AD — average displacement (m).

[b] SS— strike slip; R — reverse; N — normal.

[c] Regressions for reverse-slip relationships shown in italics and brackets are not significant at a 95% probability level.

Source: From Wells, D.L. and Coopersmith, K.J. (1994). Empirical Relationships Among Magnitude, Rupture Length, Rupture Width, Rupture Area and Surface Displacements, *Bull. Seismol. Soc. Am.*, 84 (4), 974–1002. With permission.

17.3.2 Intensity

In general, seismic intensity is a metric of the effect, or the strength, of an earthquake hazard at a specific location. While the term can be generically applied to engineering measures such as peak ground acceleration (PGA), it is usually reserved for qualitative measures of location-specific earthquake effects, based on observed human behavior and structural damage. Numerous intensity scales developed in preinstrumental times — the most common in use today are the modified Mercalli (MMI) (Wood and Neumann 1931), Rossi–Forel (R–F), Medvedev–Sponheur–Karnik (MSK-64 1981; Grunthal 1998) and its successor the European Macroseismic Scale (EMS-98 1998), and Japan Meteorological Agency (JMA) (Kanai 1983) scales.

Modified Mercalli Intensity (MMI) is a subjective scale defining the level of shaking at specific sites on a scale of I to XII. (MMI is expressed in Roman numerals, to connote its approximate nature.) For example, moderate shaking that causes few instances of fallen plaster or cracks in chimneys constitutes MMI VI. It is difficult to find a reliable relationship between magnitude, which is a description of the earthquake's total energy level, and intensity, which is a subjective description of the level of shaking of the earthquake at specific sites, because shaking severity can vary with building type, design and construction practices, soil type, and distance from the event (Table 17.2).

Note that MMI X is the maximum considered physically possible due to "mere" shaking, and that MMI XI and XII are considered due more to permanent ground deformations and other geologic effects than to shaking.

TABLE 17.2   Modified Mercalli Intensity Scale of 1931 (after Wood and Neumann 1931)

I	Not felt except by a very few under especially favorable circumstances
II	Felt only by a few persons at rest, especially on upper floors of buildings. Delicately suspended objects may swing
III	Felt quite noticeably indoors, especially on upper floors of buildings, but many people do not recognize it as an earthquake. Standing motor cars may rock slightly. Vibration like passing track. Duration estimated
IV	During the day felt indoors by many, outdoors by few. At night some awakened. Dishes, windows, and doors disturbed; walls make creaking sound. Sensation like heavy truck striking building. Standing motorcars rock noticeably
V	Felt by nearly everyone; many awakened. Some dishes, windows, etc., broken; a few instances of cracked plaster; unstable objects overturned. Disturbance of trees, poles, and other tall objects sometimes noticed. Pendulum clocks may stop
VI	Felt by all; many frightened and run outdoors. Some heavy furniture moved; a few instances of fallen plaster or damaged chimneys. Damage slight
VII	Everybody runs outdoors. Damage negligible in buildings of good design and construction slight to moderate in well-built ordinary structures; considerable in poorly built or badly designed structures. Some chimneys broken. Noticed by persons driving motor cars
VIII	Damage slight in specially designed structures; considerable in ordinary substantial buildings, with partial collapse; great in poorly built structures. Panel walls thrown out of frame structures. Fall of chimneys, factory stacks, columns, monuments, walls. Heavy furniture overturned. Sand and mud ejected in small amounts. Changes in well water. Persons driving motor cars disturbed
IX	Damage considerable in specially designed structures; well designed frame structures thrown out of plumb; great in substantial buildings, with partial collapse. Buildings shifted off foundations. Ground cracked conspicuously. Underground pipes broken
X	Some well-built wooden structures destroyed; most masonry and frame structures destroyed with foundations; ground badly cracked. Rails bent. Landslides considerable from river banks and steep slopes. Shifted sand and mud. Water splashed over banks
XI	Few, if any (masonry), structures remain standing. Bridges destroyed. Broad fissures in ground. Underground pipelines completely out of service. Earth slumps and land slips in soft ground. Rails bent greatly
XII	Damage total. Waves seen on ground surfaces. Lines of sight and level distorted. Objects thrown upward into the air

TABLE 17.3 Comparison of Modified Mercalli (MMI) and Other Intensity Scales

a, gals	MMI, Modified Mercalli	R–F, Rossi–Forel	MSK, Medvedev–Sponheur–Karnik	JMA, Japan Meteorological Agency
0.7	I	I	I	0
1.5	II	I–II	II	I
3	III	III	III	II
7	IV	IV–V	IV	II–III
15	V	V–VI	V	III
32	VI	VI–VII	VI	IV
68	VII	VIII–	VII	IV–V
147	VIII	VIII+ to IX–	VIII	V
316	IX	IX+	IX	V–VI
681	X	X	X	VI
(1468)*	XI	—	XI	VII
(3162)*	XII	—	XII	

* a values provided for reference only. MMI > X are due more to geologic effects.

Other intensity scales are defined analogously, Table 17.3, which also contains an approximate conversion from MMI to acceleration a (PGA, in cm/s^2 or gals). The conversion is due to Richter (1935) (other conversions are also available: Trifunac and Brady 1975; Murphy and O'Brien 1977)

$$\log a = \text{MMI}/3 - 1/2 \qquad (17.10)$$

Intensity maps are produced as a result of detailed investigation of the type of effects tabulated in Table 17.2, as shown in Figure 17.5 for the 1994 M_W 6.7 Northridge Earthquake. Correlations have been developed between the area of various MMI intensities and earthquake magnitude, which are of value for seismological and planning purposes).

Figure 17.6, for example, correlates A_{felt} versus M_W. For preinstrumental historical earthquakes, A_{felt} can be estimated from newspapers and other reports, which then can be used to estimate the event magnitude, thus supplementing the seismicity catalog. This technique has been especially useful in regions with a long historical record (Ambrayses and Melville 1982; Woo and Muirwood 1984).

17.3.3 Time History

Sensitive strong motion seismometers have been available since the 1930s, and record actual ground motions specific to their location, Figure 17.7. Typically, the ground motion records, termed *seismographs* or *time histories*, have recorded acceleration (these records are termed *accelerograms*) for many years in analog form on photographic film and, more recently, digitally. Analog records required considerable effort for correction due to instrumental drift, before they could be used.

Time histories theoretically contain complete information about the motion at the instrumental location, recording three *traces* or orthogonal records (two horizontal and one vertical). Time histories (i.e., the earthquake motion at the site) can differ dramatically in duration, frequency content, and amplitude. The maximum amplitude of recorded acceleration is termed the peak ground acceleration, PGA (also termed the ZPA, or zero period acceleration) — peak ground velocity (PGV) and peak

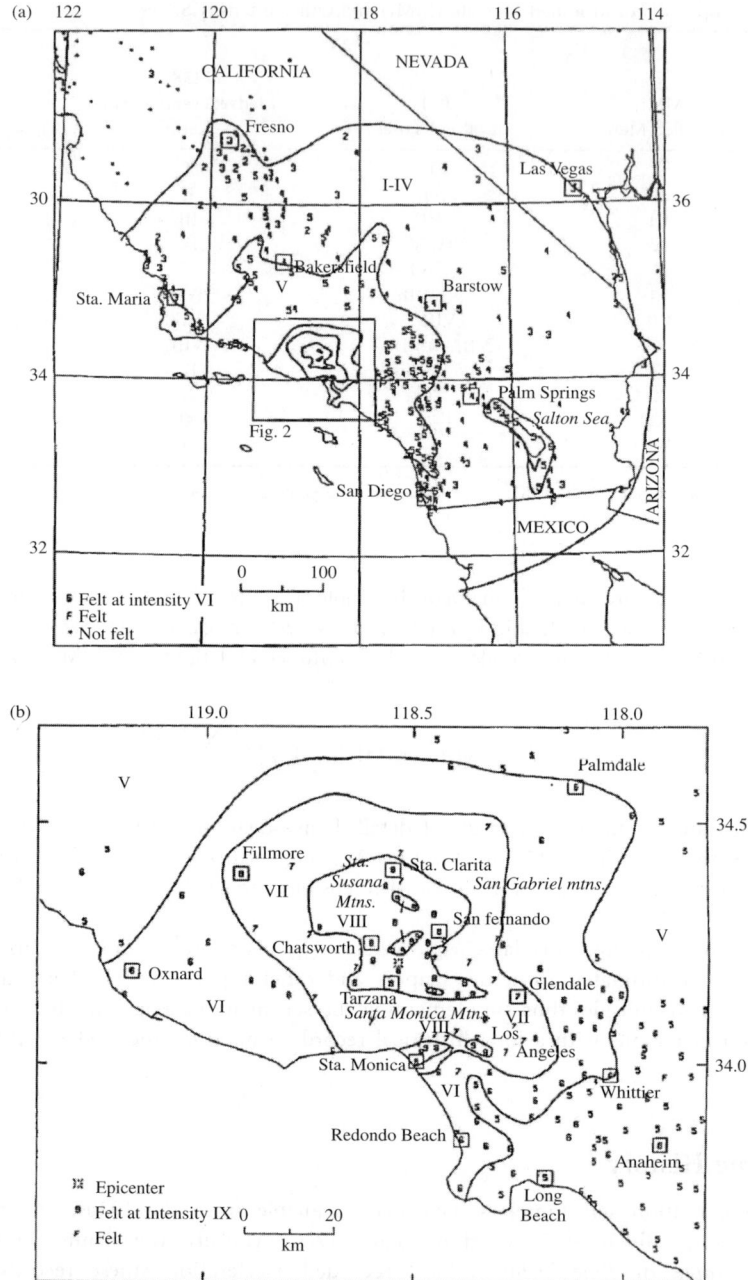

FIGURE 17.5   MMI maps, 1994 M_W 6.7 Northridge Earthquake (courtesy Dewey et al. 1995).

ground displacement (PGD) are the maximum respective amplitudes of velocity and displacement. Acceleration is normally recorded, with velocity and displacement being determined by integration; however, actual velocity and displacement meters are also deployed, to a lesser extent. Acceleration can be expressed in units of cm/s^2 (termed gals), but is often also expressed in terms of the fraction or

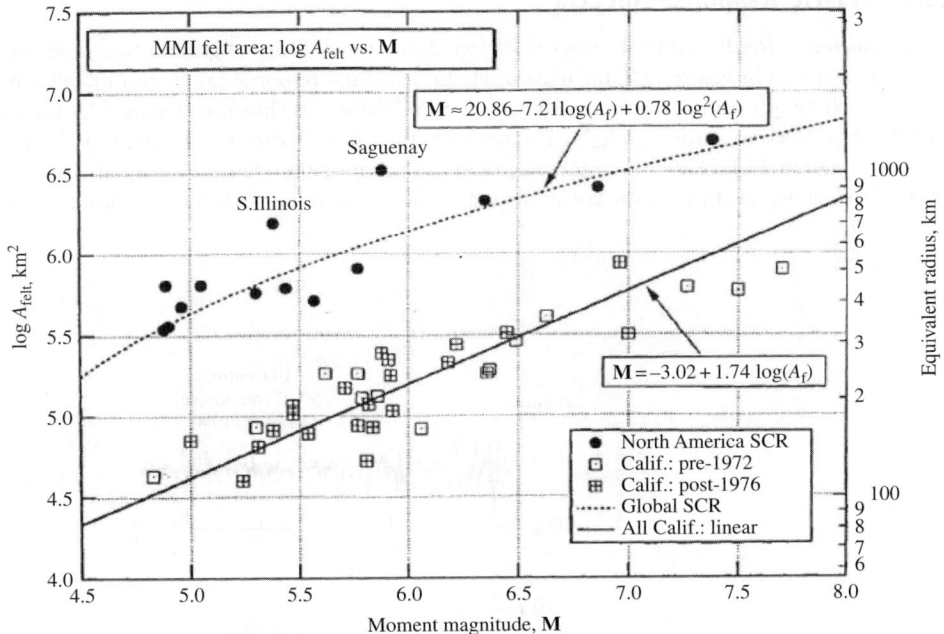

FIGURE 17.6 log A_{felt} (km^2) versus M_W (courtesy Hanks, T.C. and Kanamori, H. 1992).

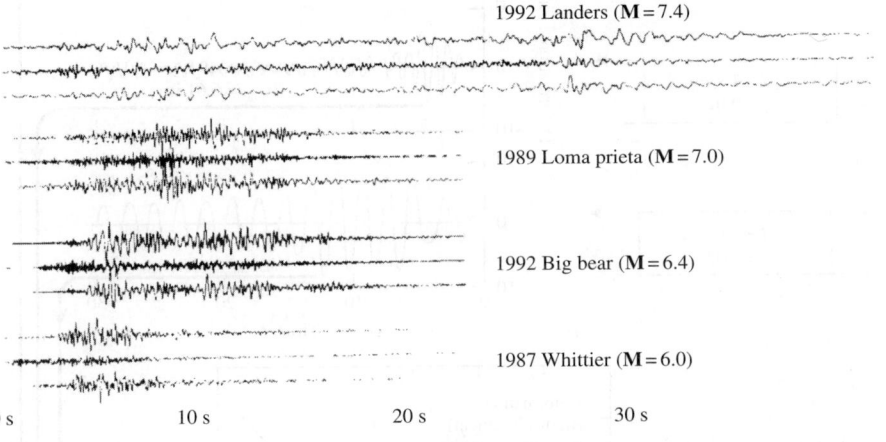

FIGURE 17.7 Typical earthquake accelerograms (courtesy Darragh, R.B., Huang, M.J., and Shakal, A.F. 1994).

percentage of the acceleration of gravity (980.66 gals, termed 1g). Velocity is expressed in cm/s (termed kine). Recent earthquakes (1994 Northridge, M_W 6.7 and 1995 Hanshin [Kobe] M_W 6.9) have recorded PGAs of about 0.8g and PGVs of about 100 kine — almost 2g was recorded in the 1992 Cape Mendocino Earthquake.[6]

---

[6] While almost 2g was recorded in the Cape Mendocino event, the portion of the record was a very narrow spike and while considered genuine, is not considered to be a significant acceleration for structures.

17.3.4 Elastic Response Spectra

If a single degree-of-freedom (SDOF) mass is subjected to a time history of ground (i.e., base) motion similar to that shown in Figure 17.7, the mass or elastic structural response can be readily calculated as a function of time, generating a structural response time history, as shown in Figure 17.8 for several oscillators with differing natural periods. The response time history can be calculated by direct integration of Equation 17.1 in the time domain, or by solution of the Duhamel integral (Clough and Penzien 1975). However, this is time consuming, and the elastic response is more typically calculated

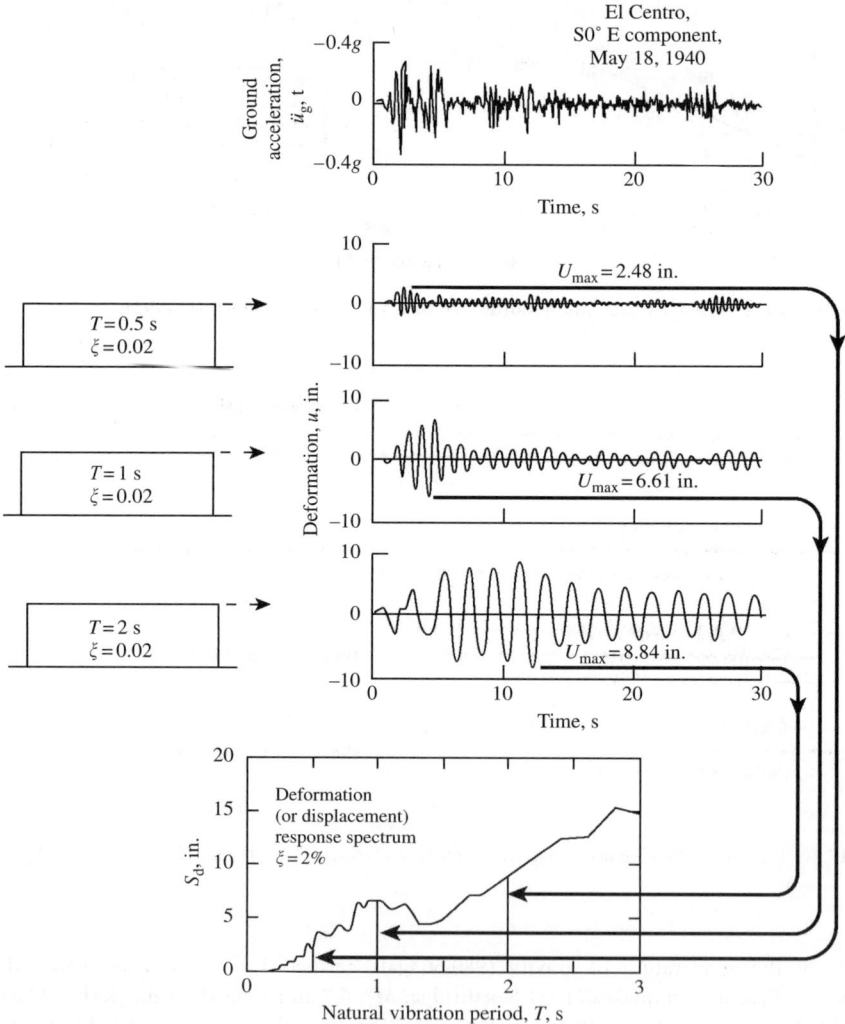

FIGURE 17.8   Computation of deformation (or displacement) response spectrum (Chopra, A.K. 1981).

in the frequency domain

$$v(t) = \frac{1}{2\pi} \int_{\varpi=-\infty}^{\infty} H(\varpi)c(\varpi) \exp(i\varpi t) \, d\varpi \tag{17.11}$$

where $v(t)$ is the elastic structural displacement response time history, ϖ is the frequency,

$$H(\varpi) = \frac{1}{-\varpi^2 m + ic + k}$$

is the complex frequency response function, and

$$c(\varpi) = \int_{\varpi=-\infty}^{\infty} p(t) \exp(-i\varpi t) \, dt$$

is the Fourier transform of the input motion (i.e., the Fourier transform of the ground motion time history), which takes advantage of computational efficiency using the fast fourier transform (Clough and Penzien 1975).

For design purposes, it is often sufficient to know only the maximum amplitude of the response time history. If the natural period of the SDOF is varied across a spectrum of engineering interest (typically, for natural periods from 0.03 to 3 or more seconds, or frequencies of 0.3 to 30+ Hz), then the plot of these maximum amplitudes is termed a response spectrum. Figure 17.8 illustrates this process, resulting in S_d, the *displacement response spectrum*, while Figure 17.9 shows (a) the S_d, displacement response spectrum, (b) S_v, the *velocity response spectrum* (also denoted PSV, the pseudospectral velocity, pseudo to emphasize that this spectrum is not exactly the same as the relative velocity response spectrum; Hudson, 1979), and (c) S_a, the acceleration *response spectrum*. Note that

$$S_v = \frac{2\pi}{T} S_d = \varpi S_d \tag{17.12}$$

and

$$S_a = \frac{2\pi}{T} S_v = \varpi S_v = \left(\frac{2\pi}{T}\right)^2 S_d = \varpi^2 S_d \tag{17.13}$$

Response spectra form the basis for much modern earthquake engineering structural analysis and design. They are readily calculated *if* the ground motion is known. For design purposes, however, response spectra must be estimated — this process is discussed in another chapter. Response spectra may be plotted in any of several ways, as shown in Figure 17.9 with arithmetic axes, and in Figure 17.10, where the velocity response spectrum is plotted on tripartite logarithmic axes, which equally enables reading of displacement and acceleration response. Response spectra are most normally presented for 5% of critical damping.

While actual response spectra are irregular in shape, they generally have a concave-down arch or trapezoidal shape, when plotted on tripartite log paper. Newmark observed that response spectra tend to be characterized by three regions: (1) a region of constant acceleration, in the high frequency portion of the spectra; (2) constant displacement, at low frequencies; and (3) constant velocity, at intermediate frequencies, as shown in Figure 17.11. If a *spectrum amplification factor* is defined as the ratio of the spectral parameter to the ground motion parameter (where parameter indicates acceleration, velocity, or displacement), then response spectra can be estimated from the data in Table 17.4, provided estimates of the ground motion parameters are available. An example spectrum using these data is given in Figure 17.11.

A standardized response spectrum is provided in the Uniform Building Code (UBC 1997). The spectrum is a smoothed average of a normalized 5% damped spectrum obtained from actual ground motion records grouped by subsurface soil conditions at the location of the recording instrument, and are applicable for earthquakes characteristic of those that occur in California (SEAOC 1988). This normalized shape may be employed to determine a response spectra, appropriate for the soil conditions. Note that the maximum amplification factor is 2.5, over a period range approximately 0.15 s to 0.4–0.9 s, depending on the soil conditions.

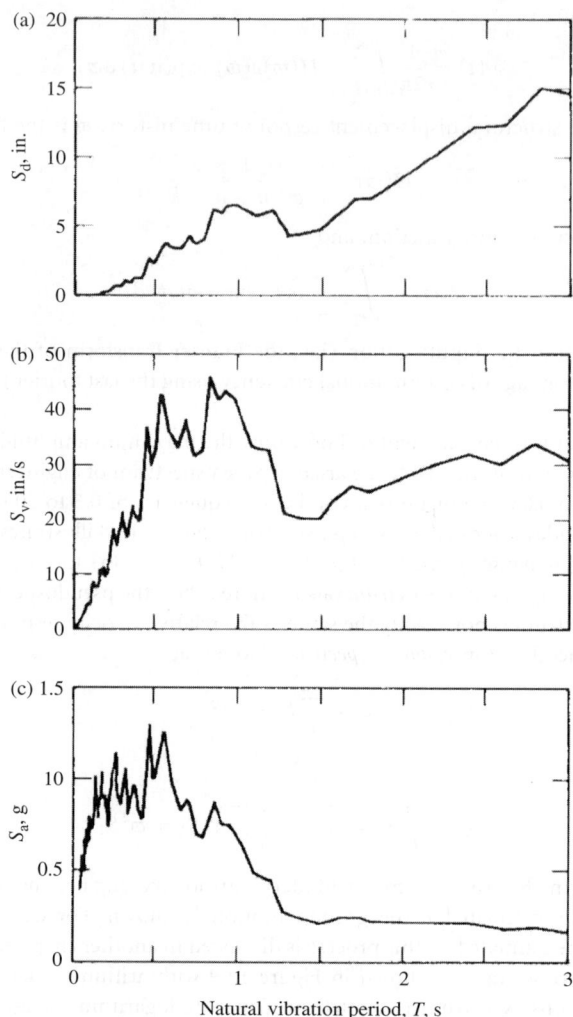

FIGURE 17.9   Response spectra (Chopra, A.K. 1981).

17.3.5 Inelastic Response Spectra

While the foregoing discussion has been for elastic response spectra, most structures are not expected, or even designed, to remain elastic under strong ground motions. Rather, structures are expected to enter the *inelastic* region — the extent to which they behave inelastically can be defined by the ductility factor, μ:

$$\mu = \frac{u_m}{u_y} \tag{17.14}$$

where u_m is the actual displacement of the mass under actual ground motions and u_y is the displacement at yield (i.e., that displacement which defines the extreme of elastic behavior). Inelastic response spectra can be calculated in the time domain by direct integration, analogous to elastic response spectra but with the structural stiffness as a nonlinear function of displacement, $k = k(u)$. If elastoplastic

Response spectrum

Imperial Valley Earthquake
May 18, 1940—2037 PST

III A001 40.001.0 El Centro site
Imperial Valley Irrigation District Comp S0° E
Damping Values are 0, 2, 5, 10, and 20% of critical

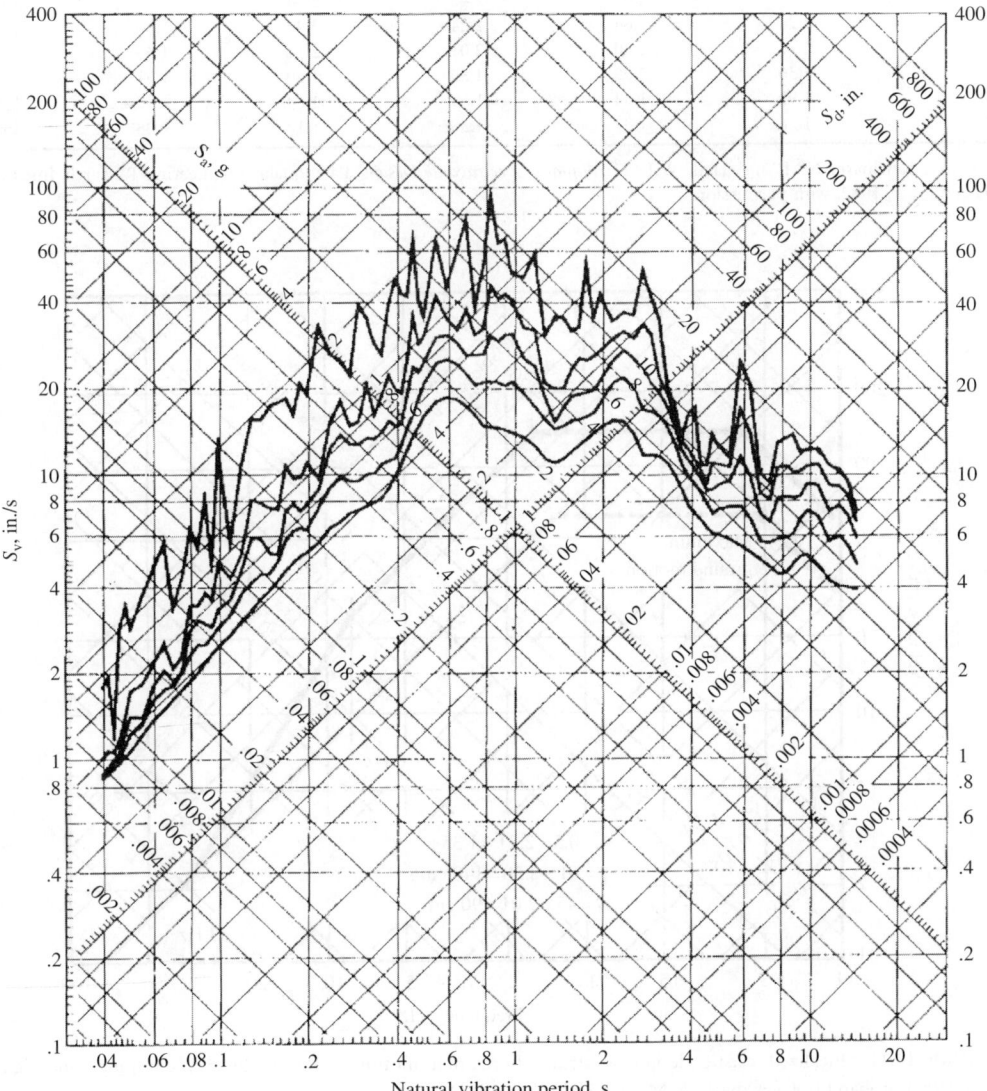

FIGURE 17.10   Response spectra, tripartite plot (El Centro S0° E component) (Chopra, A.K. 1981).

TABLE 17.4   Spectrum Amplification Factors of Horizontal Elastic Response

Damping % critical	One sigma (84.1%)			Median (50%)		
	A	V	D	A	V	D
0.5	5.10	3.84	3.04	3.68	2.59	2.01
1	4.38	3.38	2.73	3.21	2.31	1.82
2	3.66	2.92	2.42	2.74	2.03	1.63
3	3.24	2.64	2.24	2.46	1.86	1.52
5	2.71	2.30	2.01	2.12	1.65	1.39
7	2.36	2.08	1.85	1.89	1.51	1.29
10	1.99	1.84	1.69	1.64	1.37	1.20
20	1.26	1.37	1.38	1.17	1.08	1.01

Source: Newmark, N.M. and Hall, W.J., *Earthquake Spectra and Design*, Earthquake Engineering Research Institute, Oakland, CA, 1982, with permission.

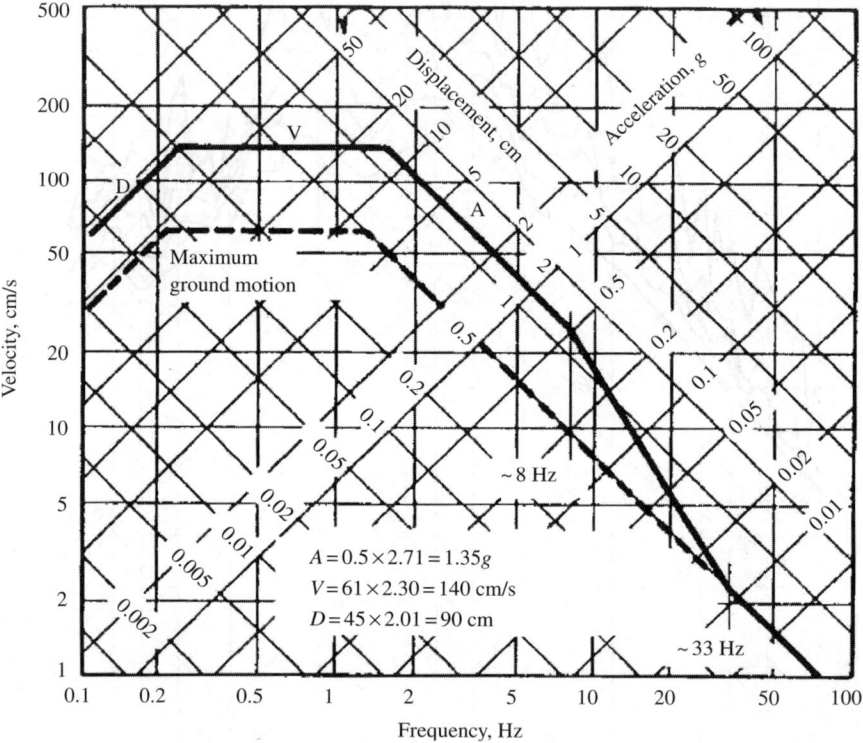

$$A = 0.5 \times 2.71 = 1.35g$$
$$V = 61 \times 2.30 = 140 \text{ cm/s}$$
$$D = 45 \times 2.01 = 90 \text{ cm}$$

FIGURE 17.11   Idealized elastic design spectrum, horizontal motion (ZPA = 0.5g, 5% damping, one sigma cumulative probability (Newmark, N.M. and Hall, W.J. 1982).

behavior is assumed, then elastic response spectra can be readily modified to reflect inelastic behavior (Newmark and Hall 1982), on the basis that (1) at low frequencies (<0.3 Hz) displacements are the same, (2) at high frequencies (>33 Hz), accelerations are equal, and (3) at intermediate frequencies, the absorbed energy is preserved. Actual construction of inelastic response spectra on this basis is shown in Figure 17.13, where $DVAA_0$ is the elastic spectrum, which is reduced to D' and V' by the ratio of $1/\mu$ for frequencies less than 2 Hz, and by the ratio of $1/(2\mu - 1)^{1/2}$ between 2 and 8 Hz. Above 33 Hz there is no reduction. The result is the inelastic acceleration spectrum ($D'V'A'A_0$), while $A''A_0'$ is the inelastic

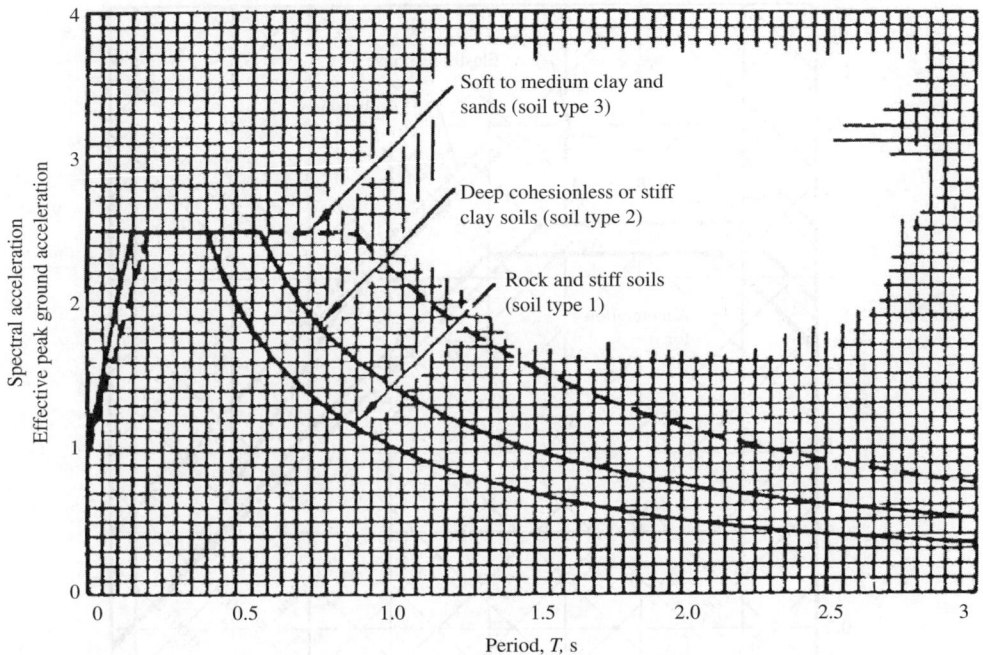

FIGURE 17.12 Normalized response spectra shapes (UBC, 1994).

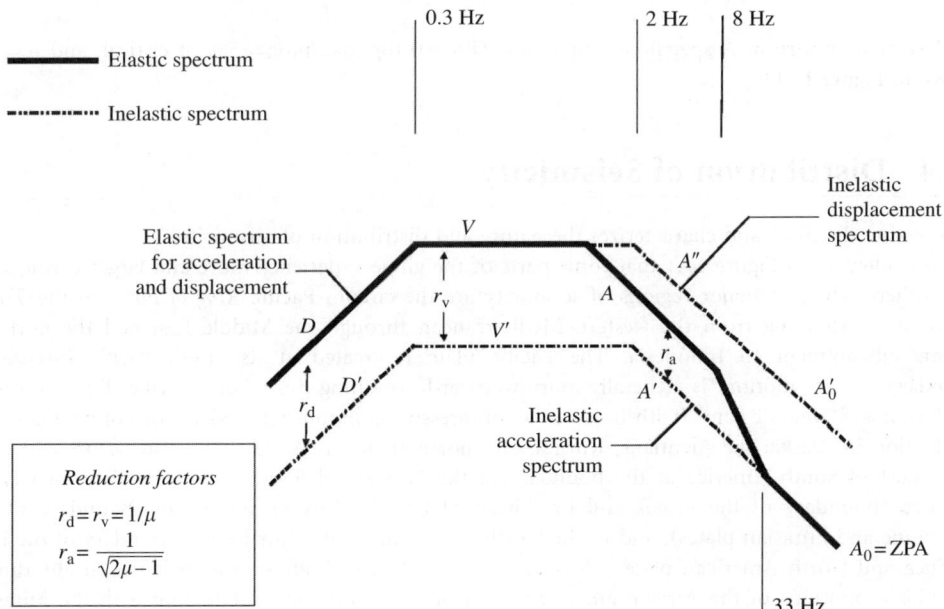

FIGURE 17.13 Inelastic response spectra for earthquakes (Newmark, N.M. and Hall, W.J. 1982).

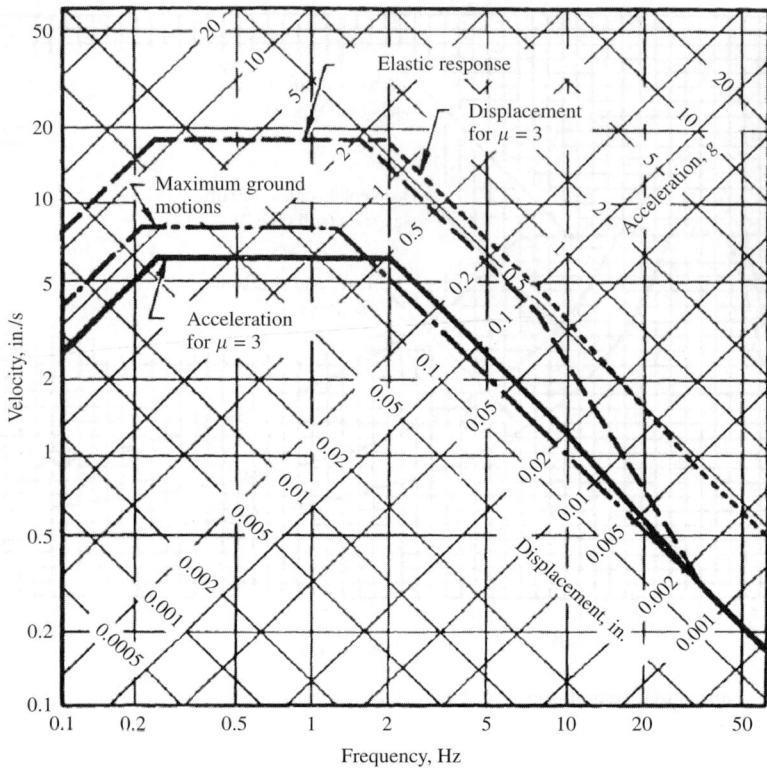

FIGURE 17.14   Example inelastic response spectra (Newmark, N.M. and Hall, W.J. 1982).

displacement spectrum. A specific example, for ZPA = 0.16g, damping = 5% of critical, and $\mu = 3$ is shown in Figure 17.14.

17.4   Distribution of Seismicity

This section discusses and characterizes the nature and distribution of seismicity.

It is evident from Figure 17.1 that some parts of the globe experience more and larger earthquakes than others. The two major regions of seismicity are the circum-Pacific *Ring of Fire* and the *Trans-Alpide belt*, extending from the western Mediterranean through the Middle East and the northern Indian subcontinent to Indonesia. The Pacific plate is created at its South Pacific extensional boundary — its motion is generally northwestward, resulting in relative strike-slip motion in California and New Zealand (with however a compressive component), and major compression and subduction in Alaska, the Aleutians, Kuriles, and northern Japan. Subduction also occurs along the west coast of South America at the boundary of the Nazca and South American plate, in Central America (boundary of the Cocos and Caribbean plates), in Taiwan and Japan (boundary of the Philippine and Eurasian plates), and in the North American Pacific Northwest (boundary of the Juan de Fuca and North American plates). Seismicity in the Trans-Alpide seismic belt is basically due to the relative motions of the African and Australia plates colliding and subducting with the Eurasian plate. The reader is referred to Chen and Scawthorn (2002) for a more extended discussion of global seismicity.

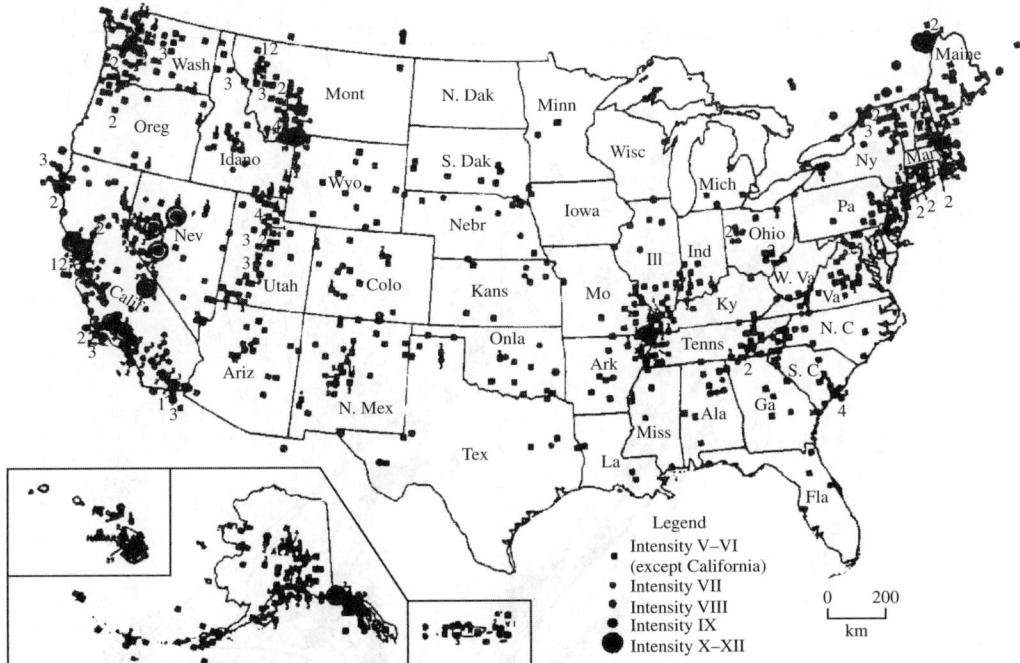

FIGURE 17.15   U.S. seismicity (Algermissen, S.T. 1983; after Coffman et al. 1980).

Regarding U.S. seismicity, the San Andreas fault system in California and the Aleutian Trench off the coast of Alaska are part of the boundary between the North American and Pacific tectonic plates, and are associated with the majority of U.S. seismicity, Figure 17.15. There are many other smaller fault zones throughout the western United States that are also helping to release the stress that is built up as the tectonic plates move past one another, Figure 17.16.

While California has had numerous destructive earthquakes, there is also clear evidence that the potential exists for great earthquakes in the Pacific Northwest (Atwater et al. 1995). On February 28, 2001, the M_W 6.8 Nisqually struck the Puget Sound area, a very similar earthquake to the M_W 6.5 1965 event. Fortunately, the Nisqually event was relatively deep (~50 km), and caused relatively few casualties, although still about $1 billion in damage.

On the east coast of the United States, the cause of earthquakes is less well understood. There is no plate boundary and very few locations of active faults are known so that it is more difficult to assess where earthquakes are most likely to occur. Several significant historical earthquakes have occurred, such as in Charleston, South Carolina, in 1886, and New Madrid, Missouri, in 1811 and 1812, indicating that there is potential for very large and destructive earthquakes (Wheeler et al. 1994; Harlan and Lindbergh 1988). However, most earthquakes in the eastern United States are smaller magnitude events. Because of regional geologic differences, eastern and central U.S. earthquakes are felt at much greater distances than those in the western United States, sometimes up to a thousand miles away (Hopper 1985).

17.5 Strong Motion Attenuation and Duration

The rate at which earthquake ground motion decreases with distance, termed *attenuation*, is a function of the regional geology and inherent characteristics of the earthquake and its source. Three

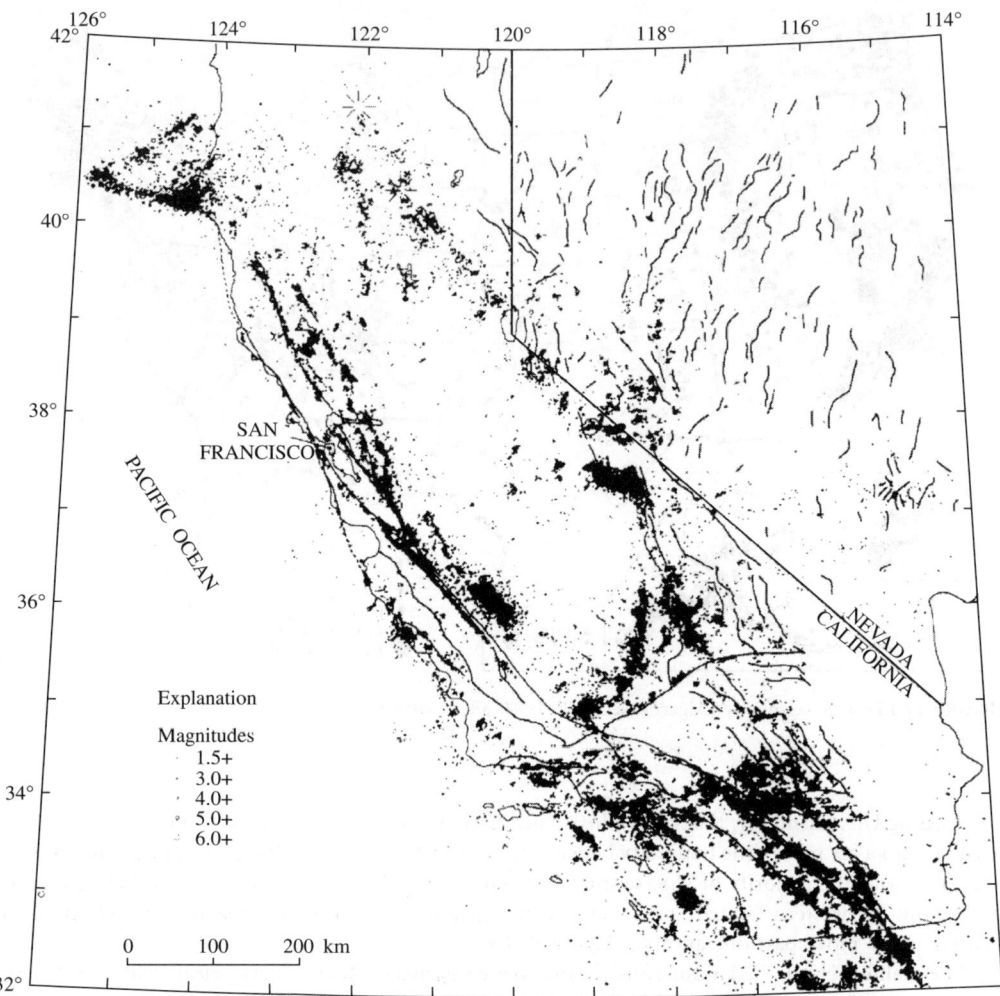

FIGURE 17.16   Seismicity for California and Nevada, 1980–86 **M** > 1.5 (courtesy Jennings, C.W. 1994).

major factors affect the severity of ground shaking at a site: (1) *source* — the size and type of the earthquake; (2) *path* — the distance from the source of the earthquake to the site, and the geologic characteristics of the media earthquake waves pass through; and (3) *site-specific effects* — type of soil at the site. In the simplest of models, if the seismogenic source is regarded as a point then, from considering the relation of energy and earthquake magnitude, and the fact that the volume of a hemisphere is proportion to R^3 (where R represents radius), it can be seen that energy per unit volume is proportional to $C10^{aM} R^{-3}$, where C is a constant or constants dependent on the earth's crustal properties. The constant C will vary regionally — for example, it has long been observed that attenuation in eastern North America (ENA) varies significantly from that in western North America (WNA) — earthquakes in ENA are felt at far greater distances. Therefore, attenuation relations are regionally dependent. Another regional aspect of attenuation is the definition of terms, especially magnitude, where various relations are developed using magnitudes defined by local observatories.

A very important aspect of attenuation is the definition of the distance parameter — since attenuation is the change of ground motion with location, this is clearly important. Many investigators use differing definitions — as study has progressed, several definitions have emerged: (1) hypocentral distance (i.e., straight line distance from point of interest to hypocenter, where hypocentral distance may be arbitrary, or based on regression rather than observation; (2) epicentral distance; (3) closest distance to the causative fault; and (4) closest horizontal distance from the station to the point on the earth's surface that lies directly above the seismogenic source. In using attenuation relations, it is critical that the correct definition of distance is consistently employed.

An extensive discussion of attenuation is beyond the scope of this chapter, and the reader is referred to Chen and Scawthorn for an extended discussion. However, for completeness, we present one attenuation relation, that of Campbell and Bozorgnia (2003 from which the following is excerpted), which can be represented by the expression:

$$\ln Y = c_1 + f_1(M_W) + c_4 \ln \sqrt{f_2(M_W, r_{seis}, S)} + f_3(F) + f_4(S) + f_5(HW, M_W, r_{seis}) + \varepsilon \qquad (17.15)$$

where the magnitude scaling characteristics are given by

$$f_1(M_W) = c_2 M_W + c_3(8.5 - M_W)^2 \qquad (17.16)$$

The distance scaling characteristics are given by

$$f_2(M_W, r_{seis}, S) = r_{seis}^2 + g(S)^2 \left(\exp\left[c_8 M_W + c_9(8.5 - M_W)^2\right]\right)^2 \qquad (17.17)$$

in which the near-source effect of local site conditions is given by

$$g(S) = c_5 + c_6(S_{VFS} + S_{SR})c_7 S_{FR} \qquad (17.18)$$

The effect of faulting mechanism is given by

$$f_3(F) = (c_{10} F_{RV} + c_{11} F_{TH}) \qquad (17.19)$$

The far-source effect of local site conditions is given by

$$f_4(S) = c_{12} S_{VFS} + c_{13} S_{SR} + c_{14} S_{FR} \qquad (17.20)$$

and the effect of the hanging wall (HW) is given by

$$f_5(HW, F, M_W, r_{seis}) = HW f_3(F) f_{HW}(M_W) f_{HW}(r_{seis}) \qquad (17.21)$$

where

$$HW = \begin{cases} 0 & \text{for } r_{jb} \geq 5 \text{ km} \quad \text{or} \quad \delta > 70° \\ (S_{VFS} + S_{SR} + S_{FR})(5 - r_{jb})/5 & \text{for } r_{jb} < 5 \text{ km} \quad \text{and} \quad \delta \leq 70° \end{cases} \qquad (17.22)$$

$$f_{HW}(M_W) = \begin{cases} 0 & \text{for } M_W < 5.5 \\ M_W - 5.5 & \text{for } 5.5 \leq M_W \leq 6.5 \\ 1 & \text{for } M_W > 6.5 \end{cases} \qquad (17.23)$$

and

$$f_{HW}(r_{seis}) = \begin{cases} c_{15}(r_{seis}/8) & \text{for } r_{seis} < 8 \text{ km} \\ c_{15} & \text{for } r_{seis} \geq 8 \text{ km} \end{cases} \qquad (17.24)$$

The parameter HW quantifies the effect of the hanging wall and will always evaluate to zero for firm soil and for a horizontal distance of 5 km or greater from the rupture plane. The standard deviation of $\ln Y$ is defined as a function of magnitude according to the expression:

$$\sigma_{\ln Y} = \begin{cases} c_{16} - 0.07 M_W & \text{for } M_W < 7.4 \\ c_{16} - 0.518 & \text{for } M_W \geq 7.4 \end{cases} \qquad (17.25)$$

or as a function of PGA according to the expression:

$$\sigma_{\ln Y} = \begin{cases} c_{17} + 0.351 & \text{for PGA} \leq 0.07g \\ c_{17} - 0.132 \ln(PGA) & \text{for } 0.07g < PGA < 0.25g \\ c_{17} + 0.183 & \text{for PGA} \geq 0.25g \end{cases} \qquad (17.26)$$

TABLE 17.5   Coefficients for Campbell and Bozorgina Attenuation Relation: Horizontal Component

T_n (s)	c_1	c_2	c_3	c_4	c_5	c_6	c_7	c_8	c_9	c_{10}	c_{11}	c_{12}	c_{13}	c_{14}	c_{15}	c_{16}	c_{17}
Unc PGA	−2.896	0.812	0.000	−1.318	0.187	−0.029	−0.064	0.616	0	0.179	0.307	−0.062	−0.195	−0.320	0.370	0.964	0.263
Cor PGA	−4.033	0.812	0.036	−1.061	0.041	−0.005	−0.018	0.766	0.034	0.343	0.351	−0.123	−0.138	−0.289	0.370	0.920	0.219
0.05	−3.740	0.812	0.036	−1.121	0.058	−0.004	−0.028	0.724	0.032	0.302	0.362	−0.140	−0.158	−0.205	0.370	0.940	0.239
0.075	−3.076	0.812	0.050	−1.252	0.121	−0.005	−0.051	0.648	0.040	0.243	0.333	−0.150	−0.196	−0.208	0.370	0.952	0.251
0.10	−2.661	0.812	0.060	−1.308	0.166	−0.009	−0.068	0.621	0.046	0.224	0.313	−0.146	−0.253	−0.258	0.370	0.958	0.257
0.15	−2.270	0.182	0.041	−1.324	0.212	−0.033	−0.081	0.613	0.031	0.318	0.344	−0.176	−0.267	−0.284	0.370	0.974	0.273
0.20	−2.771	0.812	0.030	−1.153	0.098	−0.014	−0.038	0.704	0.026	0.296	0.342	−0.148	−0.183	−0.359	0.370	0.981	0.280
0.30	−2.999	0.812	0.007	−1.080	0.059	−0.007	−0.022	0.752	0.007	0.359	0.385	−0.162	−0.157	−0.585	0.370	0.984	0.283
0.40	−3.511	0.812	−0.015	−0.964	0.024	−0.002	−0.005	0.842	−0.016	0.379	0.438	−0.078	−0.129	−0.557	0.370	0.987	0.286
0.50	−3.556	0.812	−0.035	−0.964	0.023	−0.002	−0.004	0.842	−0.036	0.406	0.479	−0.122	−0.130	−0.701	0.370	0.990	0.289
0.75	−3.709	0.812	−0.071	−0.964	0.021	−0.002	−0.002	0.842	−0.074	0.347	0.419	−0.108	−0.124	−0.796	0.331	1.021	0.320
1.0	−3.867	0.812	−0.101	−0.964	0.019	0	0	0.842	−0.105	0.329	0.338	−0.073	−0.072	−0.858	0.281	1.021	0.320
1.5	−4.093	0.812	−0.150	−0.964	0.019	0	0	0.842	−0.155	0.217	0.188	−0.079	−0.056	−0.954	0.210	1.021	0.320
2.0	−4.311	0.812	−0.180	−0.964	0.019	0	0	0.842	−0.187	0.060	0.064	−0.124	−0.116	−0.916	0.160	1.021	0.320
3.0	−4.817	0.812	−0.193	−0.964	0.019	0	0	0.842	−0.200	−0.079	0.021	−0.154	−0.117	−0.873	0.089	1.021	0.320
4.0	−5.211	0.812	−0.202	−0.964	0.019	0	0	0.842	−0.209	−0.061	0.057	−0.054	−0.261	−0.889	0.039	1.021	0.320

Note: Uncorrected PGA is to be used only when in estimate of PGA is required. Corrected PGA is to be used when an estimate of PGA compatible with PSA is required.

Source: Adapted from Campbell, K.W. and Bozorgnia, Y. (2003) Updated Near-Source Ground Motion (Attenuation) Relations for the Horizontal and Vertical Components of Peak Ground Acceleration and Acceleration Response Spectra, *Bull. Sesimol. Soc. Am.*

TABLE 17.6 Coefficients for Campbell and Bozorgnia Attenuation Relation: Vertical Component

T_n (s)	c_1	c_2	c_3	c_4	c_5	c_6	c_7	c_8	c_9	c_{10}	c_{11}	c_{12}	c_{13}	c_{14}	c_{15}	c_{16}	c_{17}
Unc PGA	−2.807	0.756	0	−1.391	0.191	0.044	−0.014	0.544	0	0.091	0.223	−0.096	−0.212	−0.199	0.630	1.003	0.320
Cor PGA	−3.108	0.756	0	−1.287	0.142	0.046	−0.040	0.587	0	0.253	0.173	−0.135	−0.138	−0.256	0.630	0.975	0.274
0.05	−1.918	0.756	0	−1.517	0.309	0.069	−0.023	0.498	0	0.058	0.100	−0.195	−0.274	−0.219	0.630	1.031	0.330
0.075	−1.504	0.756	0	−1.551	0.343	0.083	0.000	0.487	0	0.135	0.182	−0.224	−0.303	−0.263	0.630	1.031	0.330
0.10	−1.672	0.756	0	−1.473	0.282	0.062	0.001	0.513	0	0.168	0.210	−0.198	−0.275	−0.252	0.630	1.031	0.330
0.15	−2.323	0.756	0	−1.280	0.171	0.045	0.008	0.591	0	0.223	0.238	−0.170	−0.175	−0.270	0.630	1.031	0.330
0.20	−2.998	0.756	0	−1.131	0.089	0.028	0.004	0.668	0	0.234	0.256	−0.098	−0.041	−0.311	0.571	1.031	0.330
0.30	−3.721	0.756	0.007	−1.028	0.050	0.010	0.004	0.736	0.007	0.249	0.328	−0.026	0.082	−0.265	0.488	1.031	0.330
0.40	−4.536	0.756	−0.015	−0.812	0.012	0	0	0.931	−0.018	0.299	0.317	−0.017	0.022	−0.257	0.428	1.031	0.330
0.50	−4.651	0.756	−0.035	−0.812	0.012	0	0	0.931	−0.043	0.243	0.354	−0.020	0.092	−0.293	0.383	1.031	0.330
0.75	−4.903	0.756	−0.071	−0.812	0.012	0	0	0.931	−0.087	0.295	0.418	0.078	0.091	−0.349	0.299	1.031	0.330
1.0	−4.950	0.756	−0.101	−0.812	0.012	0	0	0.931	−0.124	0.266	0.315	0.043	0.101	−0.481	0.240	1.031	0.330
1.5	−5.073	0.756	−0.150	−0.812	0.012	0	0	0.931	−0.184	0.171	0.211	−0.038	−0.018	−0.518	0.240	1.031	0.330
2.0	−5.292	0.756	−0.180	−0.812	0.012	0	0	0.931	−0.222	0.114	0.115	0.033	−0.022	−0.503	0.240	1.031	0.330
3.0	−5.748	0.756	−0.193	−0.812	0.012	0	0	0.931	−0.238	0.179	0.159	−0.010	−0.047	−0.539	0.240	1.031	0.330
4.0	−6.042	0.756	−0.202	−0.812	0.012	0	0	0.931	−0.248	0.237	0.134	−0.059	−0.267	−0.606	0.240	1.031	0.330

Note: Uncorrected PGA is to be used only when an estimate of PGA is required. Corrected PGA is to be used when an estimate of PGA compatible with PSA is required.

Source: Adapted from Campbell, K.W. and Bozorgnia, Y. (2003) Updated Near-Source Ground Motion (Attenuation) Relations for the Horizontal and Vertical Components of Peak Ground Acceleration and Acceleration Response Spectra, *Bull. Seismol. Soc. Am.*

where PGA is either uncorrected PGA or corrected PGA, depending on the application (see footnote to Table 17.5). The regression coefficients are listed in Table 17.5 and Table 17.6. The relation is considered valid for $MW \geq 4.7$ and $r_{seis} \leq 60$ km.

The relation predicts ground motion for firm soil, equivalent to the condition $SFS = 1$, unless one of the site parameters in $g(S)$ and $f_4(S)$ is set to one, in which case it predicts ground motion for either very firm soil, soft rock, or firm rock. The relationship between the faulting mechanism parameters FRV (reverse faulting with dip greater than 45°) and FTH (thrust faulting with dip less than or equal to 45°) and the rake angle λ is[7] (a) strike slip: $F = 0$, rake angle $\lambda = 0$–22.5, 177.5–202.5, 337.5–360; (b) normal: $F = 0$, rake angle $\lambda = 202.5$–337.5; (c) reverse (FRV $= 1$) $F = 1.0$, rake angle $\lambda = 22.5$–157.5 ($\delta > 45$); and (d) thrust (FTH $= 1$) $F = 1.0$, rake angle $\lambda = 22.5$–157.5 ($\delta \leq 45$).

Sediment depth D was evaluated and found to be important, but it was not included as a parameter, since it is rarely used in engineering practice. If desired, sediment depth can be included in an estimate of ground motion by using the attenuation relation developed by Campbell (1997, 2000, 2001).

17.6 Characterization of Seismicity

The previous section described the global distribution of seismicity, in qualitative terms. This section describes how that seismicity may be mathematically characterized, in terms of magnitude–frequency and other relations.

The term magnitude–frequency relation was first characterized by Gutenberg and Richter (1954) as

$$\log N(m) = a_N - b_N m \tag{17.27}$$

where $N(m)$ is the number of earthquake events equal to or greater than magnitude m occurring on a seismic source per unit time, and a_N and b_N are regional constants ($10a_N$ is equal to the total number of earthquakes with magnitude >0, and b_N is the rate of seismicity; b_N is typically 1 ± 0.3). Gutenberg and Richter's examination of the seismicity record for many portions of the earth indicated this relation was valid for selected magnitude ranges. The Gutenberg–Richter relation can be normalized to

$$F(m) = 1 - \exp[-B_M(m - M_0)] \tag{17.28}$$

where $F(m)$ is the cumulative distribution function (CDF) of magnitude, B_M is a regional constant, and M_0 is a small enough magnitude such that lesser events can be ignored. Combining this with a Poisson distribution to model large earthquake occurrence (Esteva 1976) leads to the CDF of earthquake magnitude per unit time:

$$F(m) = \exp[-\exp\{-a_M(m - \mu_M)\}] \tag{17.29}$$

which has the form of a Gumbel (1958) extreme value type I (largest values) distribution (denoted $EX_{I,L}$), which is an unbounded distribution (i.e., the variate can assume any value). The parameters a_M and μ_M can be evaluated by a least squares regression on historical seismicity data, although the probability of very large earthquakes tends to be overestimated. Several attempts have been made to account for this (e.g., Cornell and Merz 1973). Yegulalp and Kuo (1974) have used Gumbel's Type III (largest value, denoted $EX_{III,L}$) to successfully account for this deficiency. This distribution

$$F(m) = \exp\left[-\left(\frac{w - m}{w - u}\right)^k\right] \tag{17.30}$$

has the advantage that w is the largest possible value of the variate (i.e., earthquake magnitude), thus permitting (when w, u, and k are estimated by regression on historical data) an estimate of the source's largest possible magnitude. It can be shown (Yegulalp and Kuo 1974) that estimators of w, u, and k can

---

[7]Rake is a continuous variable representing the angle between the direction of slip on the fault plane and the **strike** or the orientation of the fault on the Earth's surface.

be obtained by satisfying Kuhn–Tucker conditions although, if the data are too incomplete, the $EX_{III,L}$ parameters approach those of the $EX_{I,L}$:

$$u \longrightarrow \mu_M, \quad k/(w - u) \longrightarrow a_M$$

Determination of these parameters requires careful analysis of historical seismicity data (which is highly complex and something of an art; Donovan and Bornstein 1978), and the merging of the resulting statistics with estimates of maximum magnitude and seismicity made on the basis of geological evidence (i.e., as discussed above, maximum magnitude can be estimated from fault length, fault displacement data, time since last event and other evidence, and seismicity can be estimated from fault slippage rates combined with time since last event, see Schwartz, 1988, for an excellent discussion of these aspects). In a full probabilistic seismic hazard analysis, many of these aspects are treated fully or partially probabilistically, including the attenuation, magnitude–frequency relation, upper and lower bound magnitudes for each source zone, geographical bounds of source zones, fault rupture length, and many other aspects. The full treatment requires complex specialized computer codes, which incorporate uncertainty via use of multiple alternative source zonations, attenuation relations, and other parameters (EPRI 1986; Bernreuter et al. 1989) often using a logic tree format. A number of codes have been developed using the public domain FRISK (Fault Risk) code first developed by McGuire (1978).

Several topics are worth noting briefly:

- While analysis of the seismicity of a number of regions indicates that the Gutenberg–Richter relation $\log N(M) = a - bM$ is a good overall model for the magnitude–frequency or probability of occurrence relation, studies of late Quaternary faults during the 1980s indicated that the exponential model is not appropriate for expressing earthquake recurrence on individual faults or fault segments (Schwartz 1988). Rather, it was found that many individual faults tend to generate essentially the same size or *characteristic earthquake* (Schwartz and Coppersmith 1984), having a relatively narrow range of magnitudes at or near the maximum that can be produced by the geometry, mechanical properties, and state of stress of the fault. This implies that, relative to the Gutenberg–Richter magnitude–frequency relation, faults exhibiting characteristic earthquake behavior will have relatively less seismicity (i.e., higher b value) at low and moderate magnitudes, and more near the characteristic earthquake magnitude (i.e., lower b value).
- Most probabilistic seismic hazard analysis models assume the Gutenberg–Richter exponential distribution of earthquake magnitude, and that earthquakes follow a Poisson process, occurring on a seismic source zone randomly in time and space. This implies that the time between earthquake occurrences is exponentially distributed and that the time of occurrence of the next earthquake is independent of the elapsed time since the prior earthquake.[8] The CDF for the exponential distribution is

$$F(t) = 1 - \exp(-\lambda t) \tag{17.31}$$

Note that this forms the basis for many modern building codes, in that the probabilistic seismic hazard analysis results are selected such that the seismic hazard parameter (e.g., PGA) has a "10% probability of exceedance in 50 years" (UBC 1994) — if $t = 50$ years and $F(t) = 0.1$ (i.e., only 10% probability that the event has occurred in t years), then $\lambda = 0.0021$ per year, or 1 per 475 years. A number of more sophisticated models of earthquake occurrence have been investigated, including time-predictable models (Anagnos and Kiremidjian 1984), renewal models (Kameda and Takagi 1981; Nishenko and Buland 1987), and time-dependent models (Ellsworth et al. 1999). The latter have formed the basis for state-of-the-art estimation of seismic hazard for the San Francisco Bay Area by the U.S. Geological Survey, but can only be used when sufficient data are available.

---

[8]For this aspect, the Poisson model is often termed a *memoryless* model.

- Construction of response spectra is usually performed in one of two ways:
  A. Using probabilistic seismic hazard analysis to obtain an estimate of the PGA, and using this to scale a normalized response spectral shape. Alternatively, estimating PGA and PSV (also perhaps PSD) and using these to fit a normalized response spectral shape, for each portion of the spectrum. Since probabilistic response spectra are a composite of the contributions of varying earthquake magnitudes at varying distances, the ground motions of which attenuate differently at different periods, this method has the drawback that the resulting spectra have varying (and unknown) probabilities of exceedance at different periods. Because of this drawback, this method is less favored at present, but still offers the advantage of economy of effort.
  B. An alternative method results in the development of *uniform hazard spectra* (Anderson and Trifunac 1977), and consists of performing the probabilistic seismic hazard analysis for a number of different periods, with attenuation equations appropriate for each period (e.g., those of Boore, Joyner, and Fumal). This method is currently preferred, as the additional effort is not prohibitive, and the resulting response spectra has the attribute that the probability of exceedance is independent of frequency.

The reader is referred to Chen and Scawthorn (2002) for a more extensive discussion of this topic.

Glossary

Attenuation — The rate at which earthquake ground motion decreases with distance.

Benioff zone — A narrow zone, defined by earthquake foci, that is tens of kilometers thick dipping from the surface under the earth's crust to depths of 700 km (also termed Wadat–Benioff zone).

Body waves — Vibrational waves transmitted through the body of the earth, and are of two types: (1) P waves (transmitting energy via dilatational or push-pull motion) and (2) slower S waves (transmitting energy via shear action at right angles to the direction of motion).

Characteristic, earthquake — A relatively narrow range of magnitudes at or near the maximum that can be produced by the geometry, mechanical properties, and state of stress of a fault (Schwartz and Coppersmith 1984).

Completeness — Homogeneity of the seismicity record.

Corner frequency, f_0 — The frequency above which earthquake radiation spectra vary with ϖ^{-3} below f_0, the spectra are proportional to seismic moment.

Cripple wall — A carpenter's term indicating a wood frame wall of less than full height T, usually built without bracing.

Critical damping — The value of damping such that free vibration of a structure will cease after one cycle ($c_{crit} = 2m\omega$). Damping represents the force or energy lost in the process of material deformation (damping coefficient $c = $ force per velocity).

Dip — The angle between a plane, such as a fault, and the earth's surface.

Dip slip — Motion at right angles to the strike, up- or down-slip.

Ductile detailing — Special requirements such as, for reinforced concrete and masonry, close spacing of lateral reinforcement to attain confinement of a concrete core, appropriate relative dimensioning of beams and columns, 135° hooks on lateral reinforcement, hooks on main beam reinforcement within the column, etc.

Ductile frames — Frames required to furnish satisfactory load-carrying performance under large deflections (i.e., ductility). In reinforced concrete and masonry this is achieved by ductile detailing.

Ductility factor — The ratio of the total displacement (elastic plus inelastic) to the elastic (i.e., yield) displacement.

Epicenter — The projection on the surface of the earth directly above the hypocenter.

Far-field — Beyond near-field, also termed teleseismic.

Fault — A zone of the earth's crust within which the two sides have moved — faults may be hundreds of miles long, from one to over one hundred miles deep, and not readily apparent on the ground surface.

Focal mechanism — Refers to the direction of slip in an earthquake, and the orientation of the fault on which it occurs.

Fragility — The probability of having a specific level of damage given a specified level of hazard.

Hypocenter — The location of initial radiation of seismic waves (i.e., the first location of dynamic rupture).

Intensity — A metric of the effect, or the strength, of an earthquake hazard at a specific location, commonly measured on qualitative scales such as MMI, MSK, and JMA.

Lateral force resisting system — A structural system for resisting horizontal forces, due, for example, to earthquake or wind (as opposed to the vertical force resisting system, which provides support against gravity).

Liquefaction — A process resulting in a soil's loss of shear strength, due to a transient excess of pore water pressure.

Magnitude — A unique measure of an individual earthquake's release of strain energy, measured on a variety of scales, of which the moment magnitude M_W (derived from seismic moment) is preferred.

Magnitude–frequency relation — The probability of occurrence of a selected magnitude — the commonest is $\log_{10} n(m) = a - bm$ (Gutenberg and Richter 1954).

Meizoseismal — The area of strong shaking and damage.

Near-field — Within one source dimension of the epicenter, where source dimension refers to the length or width of faulting, whichever is less.

Nonductile frames — Frames lacking ducility or energy absorption capacity due to lack of ductile detailing — ultimate load is sustained over a smaller deflection (relative to ductile frames), and for fewer cycles.

Normal fault — A fault that exhibits dip-slip motion, where the two sides are in tension and move away from each other.

Peak ground acceleration (PGA) — The maximum amplitude of recorded acceleration (also termed the ZPA, or zero period acceleration).

Pounding — The collision of adjacent buildings during an earthquake due to insufficient lateral clearance.

Response spectrum — A plot of maximum amplitudes (acceleration, velocity, or displacement) of a single-degree-of-freedom (SDOF) oscillator as the natural period of the SDOF is varied across a spectrum of engineering interest (typically, for natural periods from 0.03 to 3 or more seconds or frequencies of 0.3 to 30+ Hz).

Reverse fault — A fault that exhibits dip-slip motion, where the two sides are in compression and move away toward each other.

Ring of fire — A zone of major global seismicity due to the interaction (collision and subduction) of the Pacific plate with several other plates.

Sand boils or mud volcanoes — Ejecta of solids (i.e., sand, silt) carried to the surface by water, due to liquefaction.

Seismic gap — A portion of a fault or seismogenic zone that can be deduced to be likely to rupture in the near term, based on patterns of seismicity and geological evidence.

Seismic hazards — The phenomena and/or expectation of an earthquake-related agent of damage, such as fault rupture, vibratory ground motion (i.e., shaking), inundation (e.g., tsunami, seiche, dam failure), various kinds of permanent ground failure (e.g., liquefaction), fire, or hazardous materials release.

Seismic moment — The moment generated by the forces generated on an earthquake fault during slip.

Seismic risk — The product of the hazard and the vulnerability (i.e., the expected damage or loss, or the full probability distribution).

Seismotectonic model — A mathematical model representing the seismicity, attenuation, and related environment.

Soft story — A story of a building signifiantly less stiff than adjacent stories (i.e., the lateral stiffness is 70% or less than that in the story above, or less than 80% of the average stiffness of the three stories above (BSSC 1194).

Spectrum amplification factor — The ratio of a response spectral parameter to the ground motion parameter (where parameter indicates acceleration, velocity, or displacement).

Strike — The intersection of a fault and the surface of the earth, usually measured from north (e.g., the fault strike is N60° W).

Subduction — Refers to the plunging of a tectonic plate (e.g., the Pacific) beneath another (e.g., the North American) down into the mantle, due to convergent motion.

Surface waves — Vibrational waves transmitted within the surficial layer of the earth, and are of two types: horizontally oscillating Love waves (analogous to S body waves) and vertically oscillating Rayleigh waves.

Tectonic — Relating to, causing, or resulting from structural deformation of the earth's crust, (from Greek *tektonikos*, from *tektn*, builder).

Thrust fault — Low-angle reverse faulting (blind thrust faults are faults at depth occurring under anticlinal folds — they have only subtle surface expression).

Trans-alpide belt — A zone of major global seismicity, extending from the Mediterranean through the Middle East, Himalayas, and Indonesian archipelago, resulting from the collision of several major tectonic plates.

Transform or strike-slip fault — A fault where relative fault motion occurs in the horizontal plane, parallel to the strike of the fault.

Uniform hazard spectra — Response spectra with the attribute that the probability of exceedance is independent of frequency.

Vulnerability — The expected damage given a specified value of a hazard parameter.

References

Algermissen, S.T. (1983) *An Introduction to the Seismicity of the United States*, Earthquake Engineering Research Institute, Oakland, CA.

Ambrayses, N.N. and Finkel, C.F. (n.d.) *The Seismicity of Turkey and Adjacent Areas, A Historical Review, 1500–1800*, EREN, Istanbul.

Ambrayses, N.N. and Melville, C.P. (1982) *A History of Persian Earthquakes*, Cambridge University Press, Cambridge.

Anagnos, T. and Kiremidjian, A.S. (1984) Temporal Dependence in Earthquake Occurrence, in *Proc. Eighth World Conf. Earthquake Eng.*, v. 1, Earthquake Engineering Research Institute, Oakland, CA, pp. 255–262.

Anderson, J.G., Trifunac, M.D. (1977) Uniform Risk Absolute Acceleration Spectra, *Advances in Civil Engineering Through Engineering Mechanics: Proceedings of the Second Annual Engineering Mechanics Division Specialty Conference*, Raleigh, NC, May 23–25, 1977; American Society of Civil Engineers, New York, pp. 332–335.

Bernreuter, D.L. et al. (1989) *Seismic Hazard Characterization of 69 Nuclear Power Plant Sites East of the Rocky Mountains*, U.S. Nuclear Regulatory Commission, NUREG/CR-5250.

Bolt, B.A. (1993) *Earthquakes*, W.H. Freeman and Co., New York.

Bonilla, M.G. et al. (1984) Statistical Relations Among Earthquake Magnitude, Surface Rupture Length, And Surface Fault Displacement, *Bull. Seis. Soc. Am.*, 74 (6), 2379–2411.

Campbell, K.W. (1985) Strong Ground Motion Attenuation Relations: A Ten-Year Perspective, *Earthquake Spectra*, 1 (4), 759–804.

Campbell, K.W. (1997) Empirical Near-Source Attenuation Relationships for Horizontal and Vertical Components of Peak Ground Acceleration, Peak Ground Velocity, and Pseudo-Absolute Acceleration Response Spectra, *Seismol. Res. Lett.*, 68, 154–179.

Campbell, K.W. (2000) Erratum: Empirical Near-Source Attenuation Relationships for Horizontal and Vertical Components of Peak Ground Acceleration, Peak Ground Velocity, and Pseudo-Absolute Acceleration Response Spectra, *Seismol. Res. Lett.*, 71, 353–355.

Campbell, K.W. (2001) Erratum: Empirical Near-Source Attenuation Relationships for Horizontal and Vertical Components of Peak Ground Acceleration, Peak Ground Velocity, and Pseudo-Absolute Acceleration Response Spectra, *Seismol. Res. Lett.*, 72, 474.

Campbell, K.W. and Bozorgnia, Y. (2003). Updated Near-Source Ground Motion (Attenuation) Relations for the Horizontal and Vertical Components of Peak Ground Acceleration and Acceleration Response Spectra, *Bull. Seismol. Soc. Am.*, 93 (1), 314–331.

Chen, W.F. and Scawthorn, C. (2002) *Earthquake Engineering Handbook*, CRC Press, Boca Raton.

Chopra, A.K. (1981) *Dynamics of Structures, A Primer*, Earthquake Engineering Research Institute, Oakland, CA.

Clough, R.W. and Penzien, J. (1975) *Dynamics of Structures*, McGraw-Hill, New York.

Coffman, J.L., von Hake, C.A., and Stover, C.W. (1980) *Earthquake History of the United States*, U.S. Dept. of Commerce, NOAA, Pub. 41-1, Washington.

Cornell, C.A. (1968) Engineering Seismic Risk Analysis, *Bull. Seis. Soc. Am.*, 58 (5), 1583–1606.

Cornell, C.A. and Merz, H.A. (1973). Seismic Risk Analysis Based on a Quadratic Magnitude Frequency Law, *Bull. Seis. Soc. Am.*, 63 (6), 1992–2006.

Darragh, R.B., Huang, M.J., and Shakal, A.F. (1994) Earthquake Engineering Aspects of Strong Motion Data from Recent California Earthquakes, *Proc. Fifth U.S. National Conf. Earthquake Engineering*, v. III, Earthquake Engineering Research Institute, Oakland, CA, 99–108.

Dewey, J.W. and Suárez, G. (1991) Seismotectonics of Middle America, in Slemmons, D.B., Engdahl, E.R., Zoback, M.D., and Blackwell, D.B., eds., *Neotectonics of North America*, GSA DNAG Vol., pp. 309–321.

Dewey, J.W. et al. (1995). Spatial Variations of Intensity in the Northridge Earthquake, in Woods, M.C. and Seiple, W.R., eds., *The Northridge California Earthquake of 17 January 1994*, California Department of Conservation, Division of Mines and Geology, Special Publ. 116, Sacramento, pp. 39–46.

Donovan, N.C. and Bornstein, A.E. (1978) Uncertainties in Seismic Risk Procedures, *J. Geotech. Div.*, ASCE 104 (GT7), 869–887.

Earthquake of 17 January (1994) Special Publ. 116, California Department of Conservation, Division of Mines and Geology, Sacramento, pp. 39–46.

Electric Power Research Institute (1986) *Seismic Hazard Methodology for the Central and Eastern United States*, EPRI NP-4726, Menlo Park, CA.

Ellsworth, W.L. et al. (1999) *A Physically-Based Earthquake Recurrence Model for Estimation of Long-Term Earthquake Probabilities*, Workshop on Earthquake Recurrence: State of the Art and Directions for the Future, Istituto Nazionale de Geofisica, Rome, Italy, February.

Esteva, L. (1976) Seismicity, in Lomnitz, C. and Rosenblueth, E., eds., *Seismic Risk and Engineering Decisions*, Elsevier, New York.

European Seismological Commission (1998) European Macroseismic Scale 1998, EMS-98, Grunthal, G. editor, Subcommission on Engineering Seismology, Working Group Macroseismic Scales, Geo ForschungsZentrum Potsdam, Germany, http: www.gfz-potsdam.de/pb1/pg2/ems_new/INDEX.HTM

Greenwood, R.B. (1995) Characterizing blind thrust fault sources — an overview, in Woods, M.C. and W.R. Seiple., eds., *The Northridge California Earthquake of 17 January 1994*, Calif. Dept. Conservation, Div. Mines and Geology, Special Publ. 116, pp. 279–287.

Grunthal G. (1998) European Macroseismic Scale, Cahiers du Centre Europeen de Geodynamique et de Seismologie, pp. 1–99.

GSHAP North Andes (1998) *Global Seismic Hazard Assessment Program, North Andean Region Final Report. Index Final Report*, Observatorio de San Calixto, Bolivia Instituto de Investigaciones en Geociencias, Minería y Química (INGEOMINAS), Colombia Escuela Politécnica de Quito (EPN), Ecuador Instituto Geofísico del Perú (IGP), Fundación Venezolana de Investigaciones

Sismológica (FUNVISIS), Venezuela GeoforschungsZentrum (GFZ), Germany Institute of Geophysics, ETH, Switzerland Istituto Nazionale di Geofisica (ING), Italy; http://seismo.ethz.ch/gshap/piloto/report.html

Gumbel, E.J. (1958) *Statistics of Extremes*, Columbia University Press, New York.

Gutenberg, B. and Richter, C.F. (1954) *Seismicity of the Earth and Associated Phenom'na*, Princeton University Press, Princeton.

Hanks, T.C. and Johnston, A.C. (1992) Common Features of the Excitation and Propagation of Strong Ground Motion for North American Earthquakes, *Bull. Seis. Soc. Am.*, 82 (1), 1–23.

Hanks, T.C. and Kanamori, H. (1979) A Moment Magnitude Scale, *J. Geophys. Res.*, 84, 2348–2350.

Harlan, M.R. and Lindbergh, C. (1988) An Earthquake Vulnerability Analysis of the Charleston, South Carolina, Area, Rept. No. CE-88-1, Dept. of Civil Engng, The Citadel, Charleston, SC.

Hopper, M.G. (1985) Estimation of Earthquake Effects associated with Large Earthquakes in the New Madrid Seismic Zone, U.S.G.S. Open File Report 85-457, Washington.

Hudson, D.E. (1979) *Reading and Interpreting Strong Motion Accelerograms*, Earthquake Engineering Research Institute, Oakland, CA.

IAEE (1992) *Earthquake Resistant Regulations: A World List-1992*. Rev. ed. Prepared by the International Association for Earthquake Engineering. Tokyo: International Association for Earthquake Engineering, 1992. Approximately 1100 pages. Distributed by Gakujutsu Bunken Fukyu-Kai (Association for Science Documents Information) Oh-Okayama, 2-12-1, Meguroku, Tokyo, 152, Japan.

Jennings, C.W. (1994) *Fault Activity Map of California and Adjacent Areas*, Dept. of Conservation, Div. Mines and Geology, Sacramento.

Kameda, H. and Takagi, H. (1981) *Seismic Hazard Estimation Based on Non-Poisson Earthquake Occurrences*, Mem. Fac. Engng, Kyoto Univ., v. XLIII, Pt. 3, July, Kyoto.

Kanai, K. (1983) *Engineering Seismology*, University of Tokyo Press, Tokyo.

Lomnitz, C. (1974). *Global Tectonics and Earthquake Risk*, Elsevier, New York.

McGuire, R.K. (1978) *FRISK: Computer Program for Seismic Risk Analysis Using Faults as Earthquake Sources*, US Geological Survey, Reports, United States Geological Survey Open file 78-1007, 71 pp.

McGuire, R.K., ed. (1993) *Practice of Earthquake Hazard Assessment.* International Association of Seismology and Physics of the Earth's Interior, 284 pp.

MSK-64 (1981) Meeting on Up-dating of MSK-64. Report on the Ad-hoc Panel Meeting of Experts on Up-dating of the MSK-64 Seismic Intensity Scale, Jene, 10–14 March 1980, Gerlands Beitr. Geophys., Leipzeig 90, 3, 261–268.

Murphy J.R. and O'Brien, L.J. (1977) The Correlation of Peak Ground Acceleration Amplitude with Seismic Intensity and Other Physical Parameters, *Bull. Seis. Soc. Am.*, 67 (3), 877–915.

Newmark, N.M. and Hall, W.J. (1982) *Earthquake Spectra and Design*, Earthquake Engineering Research Institute, Oakland, CA.

Nishenko, S.P. and Buland, R. (1987) A Generic Recurrence Interval Distribution For Earthquake Forecasting, *Bull. Seis. Soc. Am.*, 77, 1382–1399.

Reid, H.F. (1910) The Mechanics of the Earthquake, The California Earthquake of April 18, 1906, Report of the State Investigation Committee, v. 2, Carnegie Institution of Washington, Washington, D.C.

Richter, C.F. (1935) An Instrumental Earthquake Scale, *Bull. Seis. Soc. Am.*, 25, 1–32.

Richter, C.F. (1958) *Elementary Seismology*, W.H. Freeman, San Francisco.

Scholz, C.H. (1990) *The Mechanics of Earthquakes and Faulting*, Cambridge University Press, New York.

Schwartz, D.P. (1988) Geologic Characterization of Seismic Sources: Moving into the 1990s, in J.L. v. Thun, ed., *Earthquake Engineering and Soil Dynamics II — Recent Advances in Ground-Motion Evaluation*, Geotechnical Spec. Publ. No. 20., American Soc. Civil Engrs., New York.

Schwartz, D.P. and Coppersmith, K.J. (1984) Fault Behavior and Characteristic Earthquakes: Examples from the Wasatch and San Andreas Faults, *J. Geophys. Res.*, 89, 5681–5698.

SEAOC (1980) *Recommended Lateral Force Requirements and Commentary*, Seismology Committee, Structural Engineers of California, San Francisco, CA.

Slemmons, D.B. (1977) State-of-the-Art for Assessing Earthquake Hazards in the United States, Report 6: Faults and Earthquake Magnitude, U.S. Army Corps of Engineers, Waterways Experiment Station, Misc. Paper s-73-1, 129 pp.

Stein, R.S. and Yeats, R.S. (1989) Hidden Earthquakes, *Sci. Am.*, June, 260, 48–57.

Structural Engineers Association of California (1988) *Recommended Lateral Force Requirements and Tentative Commentary*, Structural Engineers Association of California, San Francisco, CA.

Trifunac, M.D. and Brady, A.G. (1975) A Study on the Duration of Strong Earthquake Ground Motion, *Bull. Seis. Soc. Am.*, 65, 581–626.

Uniform Building Code (1994). *Volume 2, Structural Engineering Design Provisions*, Intl. Conf. Building Officials, Whittier.

Uniform Building Code (1997) *Volume 2, Structural Engineering Design Provisions*, Intl. Conf. Building Officials, Whittier.

Wells, D.L. and Coppersmith, K.J. (1994) Empirical Relationships among Magnitude, Rupture Length, Rupture Width, Rupture Area and Surface Displacement, *Bull. Seis. Soc. Am.*, 84 (4), 974–1002.

Wheeler, R.L. et al. (1994) Elements of Infrastructure and Seismic Hazard in the Central United States, U.S.G.S. Prof. Paper 1538-M, Washington.

Woo, G., Wood, H.O., and Muir R. (1984) British Seismicity and Seismic Hazard, in *Proceedings of the Eighth World Conference on Earthquake Engineering*, v. I, Earthquake Engineering Research Institute, Oakland, CA, pp. 39–44.

Wood, H.O. and Neumann, Fr. (1931) Modified Mercalli Intensity Scale of 1931, *Bull. Seis. Soc. Am.*, 21, 277–283.

Working Group on California Earthquake Probabilities (1999) Earthquake Probabilities in the San Francisco Bay Region: 2000 to 2030 — A Summary of Findings, Open-File Report 99-517, US Geological Survey, Washington.

Yegulalp, T.M. and Kuo, J.T. (1974) Statistical Prediction of the Occurrence of Maximum Magnitude Earthquakes, *Bull. Seis. Soc. Am.*, 64 (2), 393–414.

Youngs, R.R. and Coppersmith, K.J. (1989) Attenuation Relationships for Evaluation of Seismic Hazards from Large Subduction Zone Earthquakes. *Proceedings of Conference XLVIII: 3rd Annual Workshop on Earthquake Hazards in the Puget Sound, Portland Area*, March 28–30, Portland, Oregon; Hays-Walter-W, Ed. US Geological Survey, Reston, VA, 1989, pp. 42–49.

Youngs, R.R. and Coppersmith, K.J. (1987) Implication of Fault Slip Rates and Earthquake Recurrence Models to Probabilistic Seismic Hazard Estimates, *Bull. Seis. Soc. Am.*, 75, 939–964.

Further Reading

There is a plethora of good references on earthquakes. Chen and Scawthorn (2002) provides an extensive reference. The reader is recommended to *Earthquakes* by B.A. Bolt (1993, Freeman, San Francisco) for an excellent and readable introduction to the subject; to *The Mechanics of Earthquakes and Faulting* by C.A. Scholz (1990, Cambridge University Press, New York) for an erudite treatment of seismogenesis; to *The Geology of Earthquakes* by R.S. Yeats, K. Sieh, and C.R. Allen (1997, Oxford University Press, New York) for an exhaustive review of faulting around the world; and to *Modern Global Seismology* by T. Lay and T.C. Wallace (1995, Academic Press, New York) for a readable theoretical text on seismology (a very rare thing).

18

Earthquake Damage to Structures

Mark Yashinsky
*Division of Structures Design,
California Department of
Transportation,
Sacramento, CA*

18.1 Introduction

18.1.1 Earthquakes

Most earthquakes occur due to the movement of faults. Faults slowly build up stresses that are suddenly released during an earthquake. We measure the size of earthquakes using moment magnitude as defined in Equation 18.1:

$$M = \left(\frac{2}{3}\right)[\log(M_0) - 16.05] \qquad (18.1)$$

where M_0 is the seismic moment as defined in Equation 18.2:

$$M_0 = GAD \quad \text{(in dyne cm)} \qquad (18.2)$$

where G is the shear modulus of the rock (dyne/cm^2), A is the area of the fault (cm^2), and D is the amount of slip or movement of the fault (cm).

The largest-magnitude earthquake that can occur on a particular fault is the product of the fault length times its depth (A), the average slip rate times the recurrence interval of the earthquake (D), and the hardness of the rock (G).

For instance the northern half of the Hayward Fault (in the San Francisco Bay Area) has an annual slip rate of 9 mm/year (Figure 18.1). It has an earthquake recurrence interval of 200 years. It is 50 km long and 14 km deep. G is taken as 3×10^{11} dyne/cm^2.

$$M_0 = (0.9 \times 200)(5 \times 10^6)(1.4 \times 10^6)(3 \times 10^{11}) = 3.78 \times 10^{26}$$

$$M = (2/3)[\log 3.78 \times 10^{26} - 16.05] = 7.01$$

Therefore, about a magnitude 7.0 earthquake is the maximum event that can occur on the northern section of the Hayward Fault. Since G is a constant, the average slip is usually a few meters, and

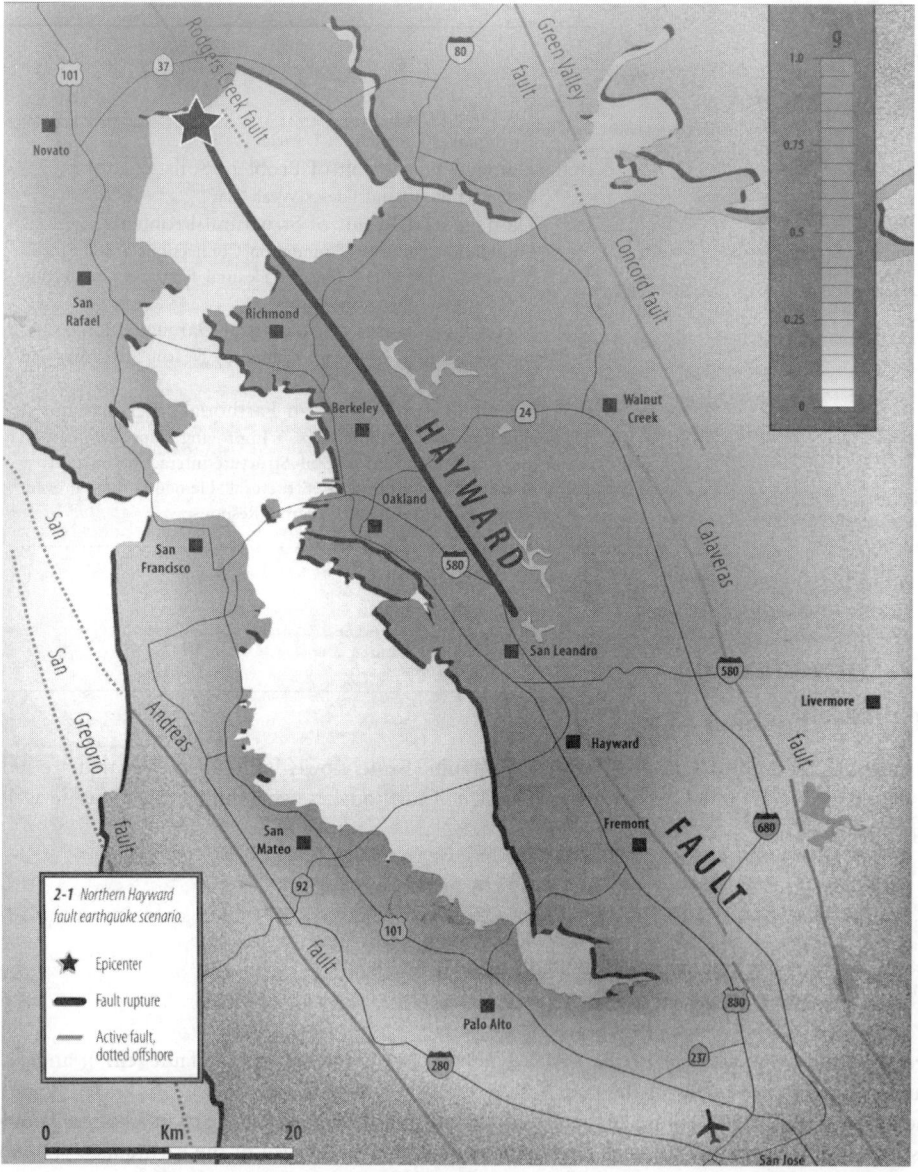

FIGURE 18.1   Map of Hayward Fault (courtesy of EERI, Earthquake Engineering Research Institute, HF-96, The Institute, Oakland, CA, 1996).

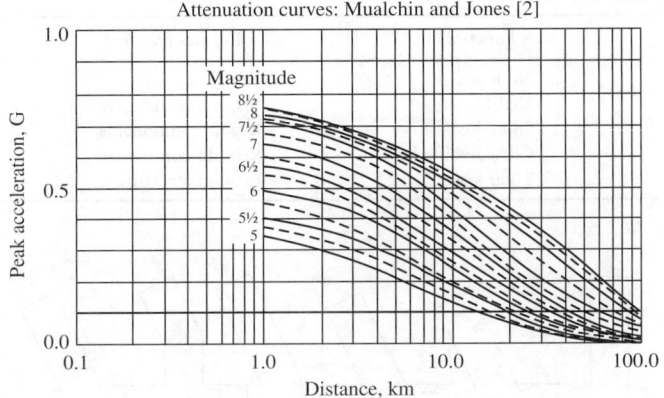

FIGURE 18.2   Attenuation curve developed by Mualchin and Jones [2].

the depth of the crust is fairly constant, the size of the earthquake is usually controlled by the length of the fault.

Magnitude is not particularly revealing to the structural engineer. Engineers design structures for the peak accelerations and displacements at the site. After every earthquake, seismologists assemble the recordings of acceleration versus distance to create attenuation curves that relate the peak ground acceleration (PGA) to the magnitude of earthquakes with distance from the fault rupture (Figure 18.2).

All of the data available on active faults is assembled to create a seismic hazard map. The map has contour lines that provide the peak acceleration based on attenuation curves that provide the reduction in acceleration due to the distance from a fault. The map is based on deterministic derived earthquakes or on earthquakes with the same return period.

18.1.2   Structural Damage

Every day, regions of high seismicity experience many small earthquakes. However, structural damage does not usually occur until the magnitude approaches 5.0. Most structural damage during earthquakes is caused by the failure of the surrounding soil or from strong shaking. Damage also results from surface ruptures, the failure of nearby lifelines, or the collapse of more vulnerable structures. We consider these effects secondary because they are not always present during an earthquake. However, when there is a long surface rupture such as that which accompanied the 1999 Ji Ji, Taiwan, earthquake, secondary effects can dominate.

Since damage can mean anything from minor cracks to total collapse, categories of damage have been developed as shown in Table 18.1. These levels of damage give engineers a choice for the performance of their structure during earthquakes. Most engineered structures are designed only to prevent collapse. This is not only to save money, but also because as a structure becomes stronger it attracts larger forces. Thus, most structures are designed to have sufficient ductility to survive an earthquake. This means that elements will yield and deform but they will be strong in shear and continue to support their load during and after the earthquake.

As shown in Table 18.1, the time that is required to repair damaged structures is an important parameter that weighs heavily on the decision making process. When a structure must be quickly repaired or must remain in service, a different damage state should be chosen.

During large earthquakes the ground is jerked back and forth, causing damage to the element whose capacity is furthest below the earthquake demand. Figure 18.3 shows that the cause may be the supporting soil, the foundation, weak flexural or shear elements, or secondary hazards such as surface

TABLE 18.1   Categories of Structural Damage

Damage state	Functionality	Repairs required	Expected outage
None (preyield) (1)	No loss	None	None
Minor/slight (2)	Slight loss	Inspect, adjust, patch	<3 days
Moderate (3)	Some loss	Repair components	<3 weeks
Major/extensive (4)	Considerable loss	Rebuild components	<3 months
Complete/collapse (5)	Total loss	Rebuild structure	>3 months

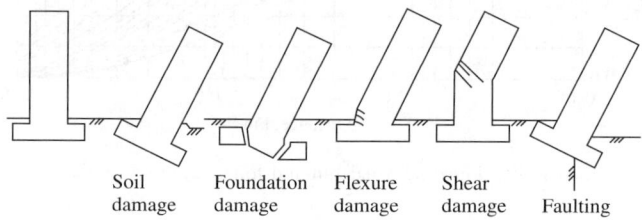

| Soil | Foundation | Flexure | Shear | |
| damage | damage | damage | damage | Faulting |

FIGURE 18.3   Common types of damage during large earthquakes.

faulting or failure of a nearby structure. Damage also frequently occurs due to the failure of connections, from large torsional moments, from tension and compression, buckling, pounding, etc.

In this chapter structural damage as a result of soil problems, structural shaking, and secondary causes will be discussed. These types of damage illustrate the most common structural hazards that have been seen during recent earthquakes.

18.2   Damage as a Result of Problem Soils

18.2.1   Liquefaction

One of the most common causes of damage to structures is the result of liquefaction to the surrounding soil. When loose saturated sands, silts, or gravel are shaken, the material consolidates, reducing the porosity and increasing pore water pressure. The ground settles, often unevenly, tilting and toppling structures that were formerly supported by the soil. During the 1955 Niigata, Japan, earthquake, several four-story apartment buildings toppled over due to liquefaction (Figure 18.4).

These buildings fell as the liquefied soil lost its ability to support them. As can be seen clearly in Figure 18.5, there was little damage to these buildings and it was reported that their collapse took place over several hours.

Partial liquefaction of the soil in Adapazari during the 1999 Kocaeli, Turkey, earthquake caused several buildings to settle or fall over. Figure 18.6 shows a building that settled as pore water was pushed to the surface, reducing the bearing capacity of the soil. Note that the weight of the building squeezed the weakened soil under the adjacent roadway. Another problem during liquefaction is that the increased pore pressure pushes quay walls, riverbanks, and the piers of bridges toward adjacent bodies of water, often dropping the end spans in the process.

The Shukugawa Bridge is a three-span, continuous, steel box girder superstructure with a concrete deck. The end spans are 87.5 m and the center span is 135 m. The superstructure is supported by steel, multicolumn bents with dropped-bent caps. It is part of a long elevated viaduct, and has expansion joints at Pier 131 and Pier 134. The columns are supported by steel piles embedded in reclaimed land along Osaka Bay.

During the 1995 Kobe, Japan, earthquake, increased pore pressure pushed the quay wall near the west end of the bridge toward the river, allowing the soil and westernmost pier (Pier 134) to move 1 m

FIGURE 18.4 Liquefaction caused building failure in Niigata, Japan. (Photo by Joseph Penzien; photo courtesy of Steinbrugge Collection, Earthquake Engineering Research Center, University of California, Berkeley.)

FIGURE 18.5 Liquefaction caused building failure in Niigata, Japan. (Photo by Joseph Penzien; photo courtesy of Steinbrugge Collection, Earthquake Engineering Research Center, University of California, Berkeley.)

eastward (Figure 18.7). This resulted in the girders falling off their bearings, damaging the expansion joint devices and making the bridge inaccessible. The easternmost pier (Pier 131) moved half a meter toward the river. It appears that the restrainers were the only thing that kept the superstructure at the expansion joint above Pier 134 together, preventing the collapse of the west span. The expansion joint had a 0.6-m vertical offset, and excavation showed that the piles at Pier 134 were also damaged due to the longitudinal movement.

Structures supported on liquefied soil topple, structures that retain liquefied soil are pushed forward, and structures buried in liquefied soil (like culverts and tunnels) float to the surface in the newly buoyant medium.

FIGURE 18.6   Settlement of building due to loss of bearing during the 1999 Kocaeli Earthquake.

FIGURE 18.7   Liquefaction caused bridge damage during the Kobe Earthquake.

The Webster and Posey Street Tube Crossings are 4500-ft-long tubes carrying two lanes of traffic under the Oakland, California, Estuary. The Posey Street Tube was built in the 1920s (Figure 18.8) while the Webster Street Tube was built in the 1960s (Figure 18.9). They are reinforced concrete (RC) tubes with a bituminous coating for waterproofing. The ground was excavated and each tube section was joined to the previously laid section. The tube descends to 70 ft below sea level.

During the 1989 Loma Prieta, California, earthquake, the soil surrounding the Webster and Posey Tubes (that carry traffic through the Oakland Estuary) liquefied. The tunnels began to float to the surface, breaking the joints between sections and slowly filling with water (Figure 18.10 and Figure 18.11).

18.2.2 Landslides

When a steeply inclined mass of soil is suddenly shaken, a slip-plane can form and the material slides downhill. During a landslide, structures sitting on the slide move downward and structures below the slide are hit by falling debris (Figure 18.12).

Landslides frequently occur in canyons, along cliffs, on mountains, and anywhere else where unstable soil exists. Landslides can occur without earthquakes (they often occur during heavy rains that increase the weight and reduce the friction of the soil) but the number of landslides is greatly increased wherever large earthquakes occur. Landslides can move a few inches or hundreds of feet. They can be the result of liquefaction, weak clays, erosion, subsidence, ground shaking, etc.

During the 1999 Ji Ji, Taiwan, earthquake, many of the mountain slopes were denuded by slides, which continued to be a hazard for people traveling on mountain roads in the weeks following the earthquake. The many RC gravity retaining walls that supported the road embankments in the mountainous terrain were all damaged: either from being pushed downhill by the slide (Figure 18.13) or in some cases broken when the retaining wall was restrained from moving downhill (Figure 18.14).

One of the more interesting retaining wall failures during the Ji Ji Earthquake was in a geogrid fabric and mechanically stabilized earth (MSE) wall at the entrance to the Southern International University (Figure 18.15). This wall was quite long and tall and its failure was a surprise since MSE walls have a good performance record during earthquakes. It was speculated that the geogrid retaining system had insufficient embedment into the soil and also it was unclear why an MSE wall would be used in a cut roadway section.

One of the best-known and largest landslides occurred at Turnagain Heights in Anchorage during the 1964 Great Alaska Earthquake. The area of the slide was about 8500 ft wide by 1200 ft long. The average drop was about 35 ft. This slide was complex, but the main cause was the failure of the weak clay layer and the unhindered movement of the ground down the wet mud flats to the sea. Figure 18.16 and Figure 18.17 provide a section and plan view of the slide.

The soil failed due to the intense shaking, and the whole neighborhood of houses, schools, and other buildings slid hundreds of yards downhill, many remaining intact during the fall (Figure 18.18).

Bridges are also severely damaged by landslides. During the 1999 Ji Ji, Taiwan, earthquake, landslides caused the collapse of two bridges. The Tsu Wei Bridges were two parallel three-span structures that crossed a tributary of the Dajia River near the city of Juolan. The superstructure was simply supported "T" girders on hammerhead single-column bents with "drum"-type footings and seat-type abutments. The girders sat on elastomeric pads between transverse shear keys. The spans were about 80 ft long by 46 ft wide, and had a 30° skew. The head scarp was clearly visible on the hillside above the bridge. During the earthquake, the south abutment was pushed forward by the landslide, the first spans fell off the bent caps on the (far) north side, and the second span of the left bridge also fell off of the far bent cap (Figure 18.19).

Also, the tops of the columns at Bent 2 had rotated away from the (south) Abutment 1. Therefore, it appears that both the top of Abutment 1 and the top of Bent 2 had moved away from the slide, while the remaining spans, restrained by Bent 3 and Abutment 4, had remained in place. Perhaps the landslide originally had pushed against Bent 2, rotating the columns forward, and the debris had

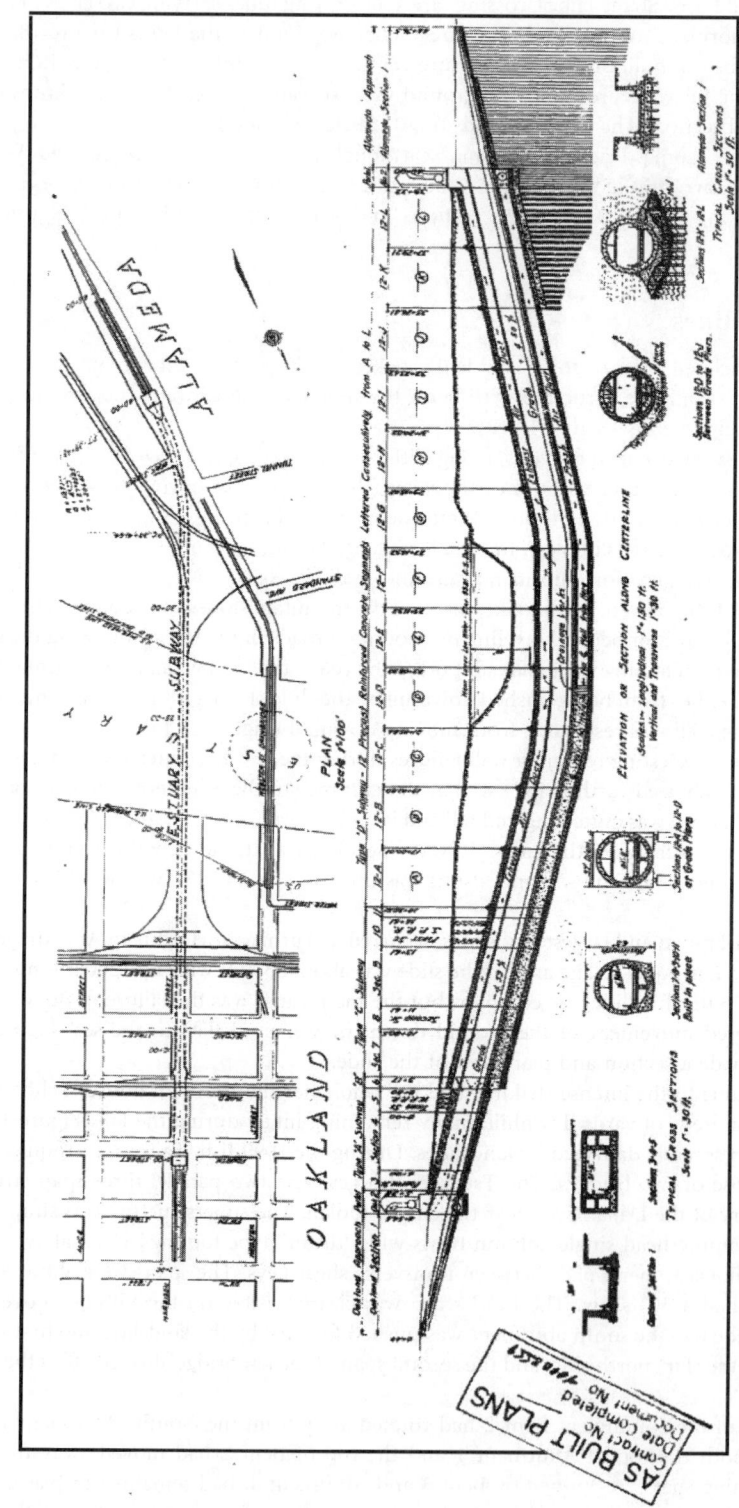

FIGURE 18.8 Elevation view of the Posey Street Tube.

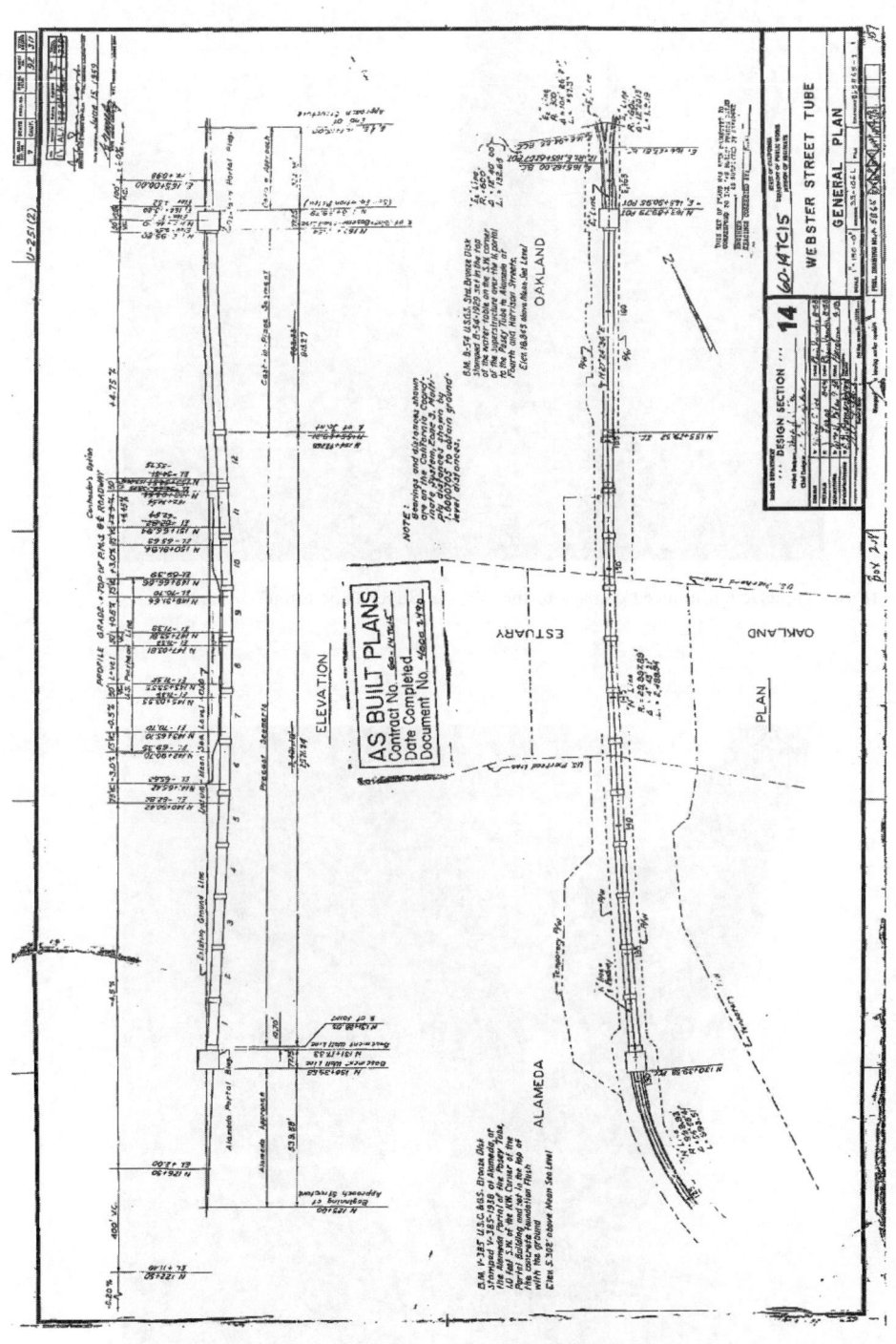

FIGURE 18.9 Elevation view of the Webster Street Tube.

FIGURE 18.10   Liquefaction induced damage to the Webster Street Tube tunnel.

FIGURE 18.11   Liquefaction induced damage to the Webster Street Tube tunnel.

Before landslide

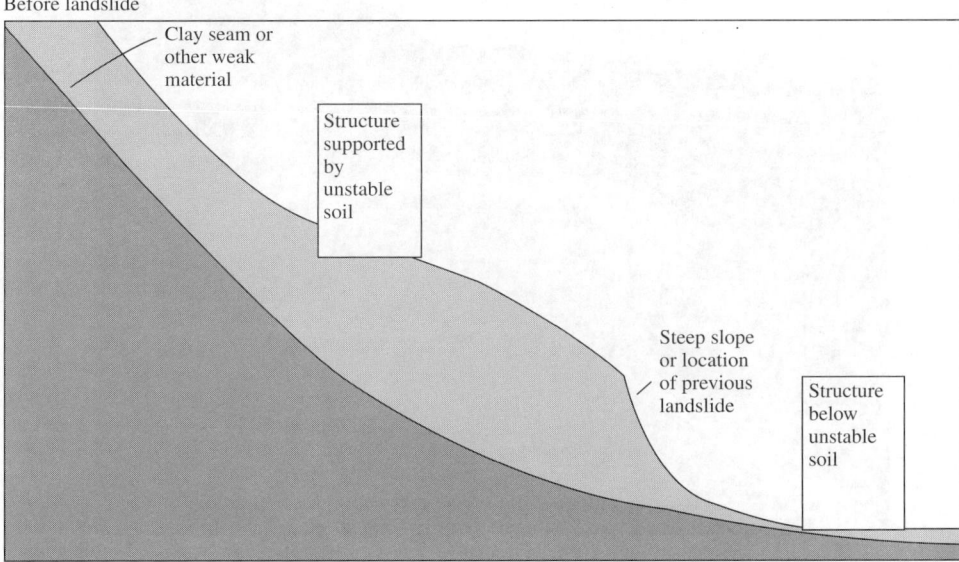

After landslide

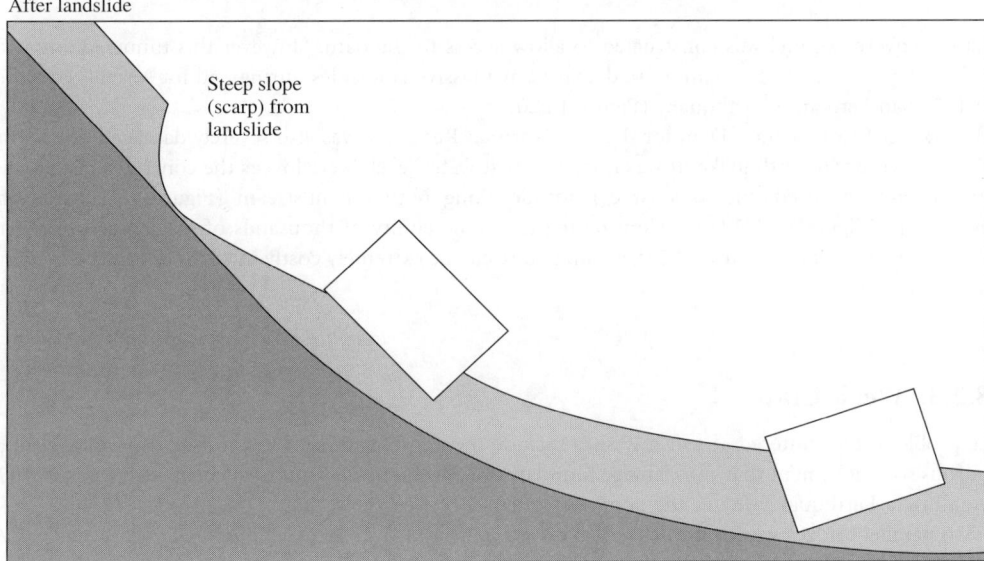

FIGURE 18.12 Diagram showing typical features of landslides.

since been removed by the current or by a construction crew. Perhaps the skew had rotated the spans to the right as they fell, pushing them against the shear keys at Bent 2, which rotated the top of the columns forward and eventually pushed the spans off the tops of Bent 2 and Bent 3. Or perhaps there was an element of strong shaking that combined with the landslide to create the column rotation and fallen spans.

Dams are particularly vulnerable to landslides since they are frequently built to hold back the water in canyons and mountain streams. Moreover, inspection of the dam after an earthquake is often difficult when slides block the roads leading to the dam. When the Pacoima concrete arch dam was built in the

FIGURE 18.13   Gravity retaining wall pushed outward by landslide.

1920s, a covered tunnel was constructed to allow access to the dam. However this tunnel, along with roads and a tramway to the dam, were damaged by massive landslides during and for several days after the 1971 San Fernando Earthquake (Figure 18.20).

The Lower San Fernando Dam for the Van Norman Reservoir was also severely damaged during the 1971 San Fernando Earthquake. It was fortunate that water levels were low as the concrete crest on this earthen dam collapsed due to a large landslide along both the upstream (Figure 18.21) and the downstream (Figure 18.22) faces. Considering the vulnerability of thousands of residences in the San Fernando Valley below (Figure 18.23), a dam failure can be extremely costly in terms of human lives and property damage.

18.2.3 Weak Clay

The problems encountered at soft clay sites include the amplification of the ground motion as well as vigorous soil movement that can damage foundations. Several bridges suffered collapse during the 1989 Loma Prieta Earthquake due to the poor performance of weak clay.

Two parallel bridges were built in 1965 to carry Highway 1 over Struve Slough near Watsonville, CA. Each bridge was 800 ft long with spans ranging from 80 to 120 ft. The superstructures were continuous for several spans with transverse hinges located in spans 6, 11, and 17 on the right bridge and in spans 6, 11, and 16 on the left bridge (they are both 21-span structures). Each bent was composed of four 14-in.-diameter concrete piles extending above the ground into a cap beam acting as an end diaphragm for the superstructure. The surrounding soil was a very soft clay (Figure 18.24). The bridges were retrofit in 1984 by adding cable restrainers to tie the structure together at the transverse hinges.

During the earthquake the soft saturated soil in Struve Slough was violently shaken. The soil pushed against the piles, breaking their connection to the superstructure (Figure 18.25), and pushing them away from the cap beam so that they punctured the bridge deck (Figure 18.26). Investigators arriving at the bridge found shear damage at the top of the piles, indicating that the soil limited the point of fixity of the piles to near the surface. They also found long, oblong holes in the soil, indicating that the piles were dragged from their initial position during the earthquake. It was believed that the damage at Struve

FIGURE 18.14 Gravity retaining wall with shear damage from landslide.

Slough was the result of vertical acceleration, but the structure's vertical period of 0.20 s was too short to be excited by the ground motion at this site.

Similarly, The Cypress Street Viaduct collapsed only at those locations that were underlain by weak Bay mud. This was a very long, two-level structure with a cast-in-place, RC, box girder superstructure with spans of 68 to 90 ft. The substructure was multicolumn bents with many different configurations including some prestressed top bent caps. Most of the bents had pins (shear keys) at the top or bottom of the top columns and all the bents were pinned above the pile caps as well. There was a superstructure hinge at every third span on both superstructures. Design began on the Cypress Viaduct in 1949, and

FIGURE 18.15   Fabric retaining wall damaged during the 1999 Ji Ji, Taiwan, earthquake.

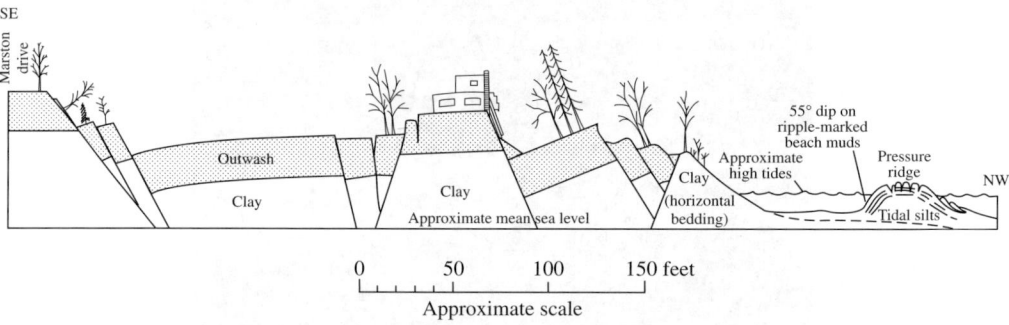

FIGURE 18.16   Section through eastern part of Turnagain Heights Slide. (Drawing courtesy of the National Academy of Sciences. National Research Council (U.S.) no. 1603, National Academy of Sciences, Washington, DC, 1968–1973, 8 v. in 10.)

construction was completed in 1957. The pins and hinges were used to simplify the analysis for this long, complicated structure. The northern two thirds of the Cypress Viaduct was on Bay Mud with 50-ft-long piles while the southern one third was on Merritt sand with 20-ft piles (Figure 18.27).

During the 1989 Loma Prieta Earthquake, the upper deck of the Cypress Viaduct collapsed from Bent 63 in the south all the way to Bent 112 in the north. Only Bents 96 and 97 remained standing. This collapse was the result of the weak pin connections at the base of the columns of the upper frame (Figure 18.28). There was inadequate confinement around the four #10 bars to restrain them during the earthquake. However, the soft Bay mud also played a role in the collapse. The southern portion of the bridge with the same vulnerable details but supported on sand remained standing (Figure 18.29). The northern portion was supported by soft Bay mud that was sensitive to the long-period motion and caused large movements that overstressed the pinned connections.

Buildings on weak clay also are susceptible to earthquake damage. Mexico City was located 350 km from the epicenter of the magnitude 8.1, 1985 Mexico Earthquake, but the city is underlain by an

FIGURE 18.17   Aerial view of Turnagain Slide. (Photograph courtesy of Steinbrugge Collection, Earthquake Engineering Research Center, University of California, Berkeley.)

FIGURE 18.18   About 75 homes were damaged as a result of the Turnagain Heights Slide. (Photograph courtesy of Steinbrugge Collection, Earthquake Engineering Research Center, University of California, Berkeley.)

old lakebed composed of soft silts and clays (Figure 18.30). This material was extremely sensitive to the long-period (about 2 s) ground motion coming from the distant but high-magnitude (8.1) source, as were the many medium-height (10 to 14 story) buildings that were damaged or had collapsed during the earthquake (Figure 18.31). Many much taller and shorter buildings were undamaged due to the difference in their fundamental period of vibration.

FIGURE 18.19   Collapse of Tsu Wei Bridge due to landslide during the Ji Ji, Taiwan, earthquake.

FIGURE 18.20   Landslides at Pacoima Dam following the 1971 San Fernando Earthquake. (Photograph courtesy of Steinbrugge Collection, Earthquake Engineering Research Center, University of California, Berkeley.)

18.3   Damage as a Result of Structural Problems

18.3.1   Foundation Failure

Usually, it is the connection to the foundation or an adjacent member rather than the foundation itself that is damaged during a large earthquake. However, materials that cannot resist lateral forces, such as hollow masonry blocks, make a poor foundation and their use should be avoided (Figure 18.32).

Engineers will occasionally design foundations to rock during earthquakes as a way of dissipating energy and of reducing the demand on the structure. However, when the foundation is too small, it can become unstable and rock over. During the magnitude 7.6, 1999 Ji Ji, Taiwan, earthquake, a local

FIGURE 18.21 Damage to the Lower San Fernando Dam. (Photograph courtesy of Steinbrugge Collection, Earthquake Engineering Research Center, University of California, Berkeley.)

FIGURE 18.22 Closer view of damage to the Lower San Fernando Dam. (Photograph courtesy of Steinbrugge Collection, Earthquake Engineering Research Center, University of California, Berkeley.)

three-span bridge rocked over transversely due to small, drum-shaped footings that provided little lateral stability (Figure 18.33).

We have already seen pile damage as a result of weak clay on the Struve Slough bridges during the 1989 Loma Prieta Earthquake. Similar damage occurred during the 1964 Great Alaska Earthquake. After the 1971 San Fernando and 1995 Kobe Earthquakes, an inspection was made of bridge foundations, but only a little damage to the tops of piles was found. As long as the foundation is embedded in good material, it usually has ample strength and ductility to survive large earthquakes. Usually, it is the more vulnerable

FIGURE 18.23   Aerial view of Lower San Fernando Dam and San Fernando Valley. (Photograph courtesy of Steinbrugge Collection, Earthquake Engineering Research Center, University of California, Berkeley.)

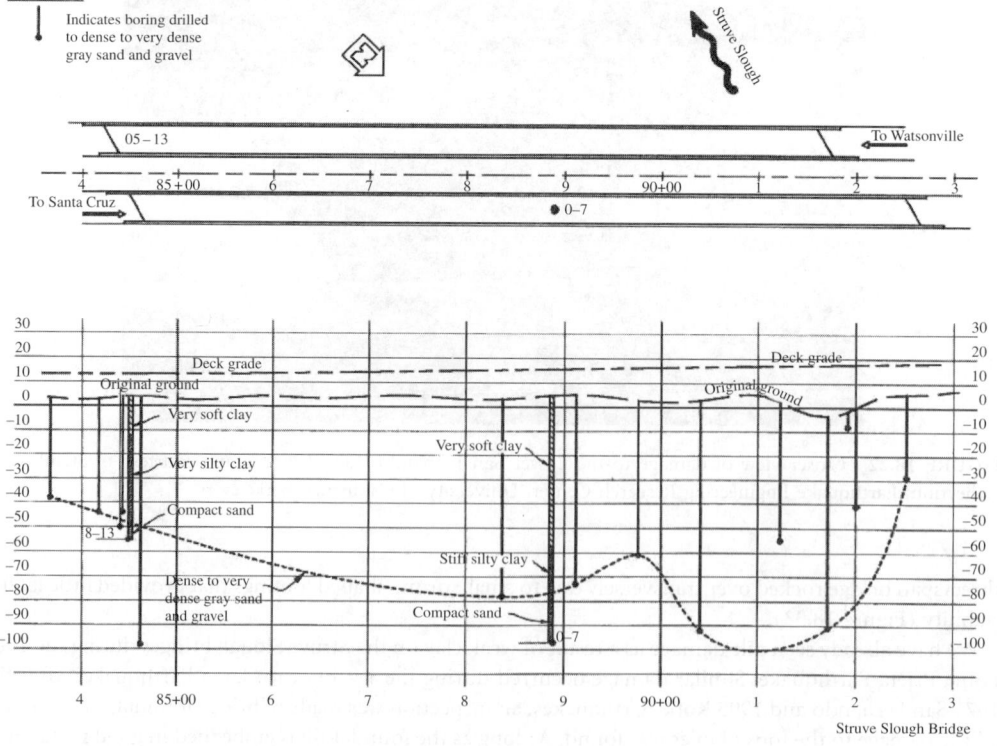

FIGURE 18.24   Soil profile for Struve Slough bridges.

FIGURE 18.25   Broken piles under bridge.

FIGURE 18.26   Piles penetrating bridge deck.

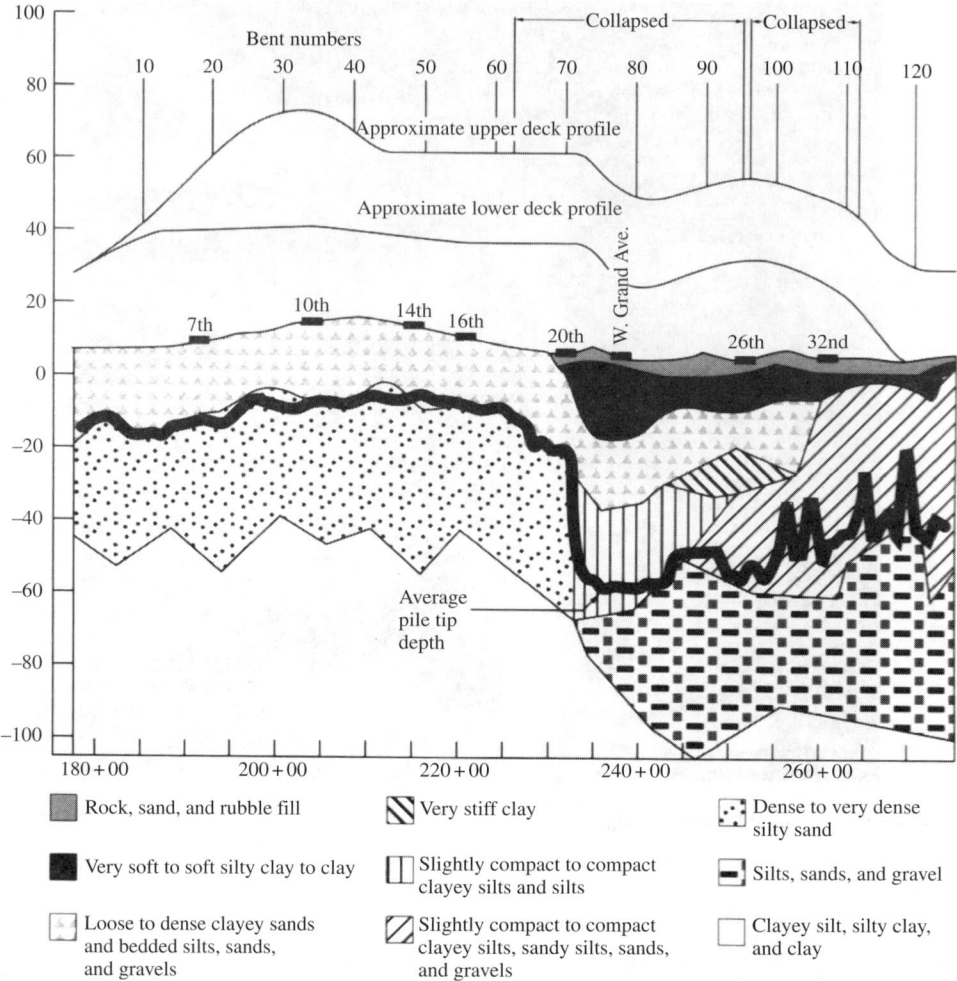

FIGURE 18.27 Geology of Cypress Freeway Viaduct site. (Housner et al., Report to the Governor George Deukmejian, State of California Office of General Services May 1990.)

elements above the foundation that can fail or become damaged during earthquakes. Still, as structures are designed to resist larger and larger earthquakes, we may begin to see more foundation damage.

18.3.2 Foundation Connections

The major cause of damage to electrical transformers, storage bins, and a variety of other structures and lifeline facilities during earthquakes is the lack of a secure connection to the foundation. Houses need to be anchored to the foundation with hold-downs connected to the stud walls and anchor bolts connected to the sill plates. Otherwise, the house will fall off its foundation as shown in Figure 18.34.

The connections to bridge foundations also need to be carefully designed. Route 210/5 Separation and Overhead was a seven-span RC box girder bridge with a hinge at Span 3 and seat-type abutments. The superstructure was 770 ft long on a 800-ft-radius curve. The piers were 4 ft by 6 ft single-column bents. Piers 2 and 3 were on piles, while Piers 4 to 7 were supported by 6-ft-diameter drilled shafts. This interchange was built on consolidated sand.

FIGURE 18.28   Damage to Cypress Street Viaduct.

FIGURE 18.29   Aerial view of Cypress Viaduct showing collapse where the structure crossed over Bay mud.

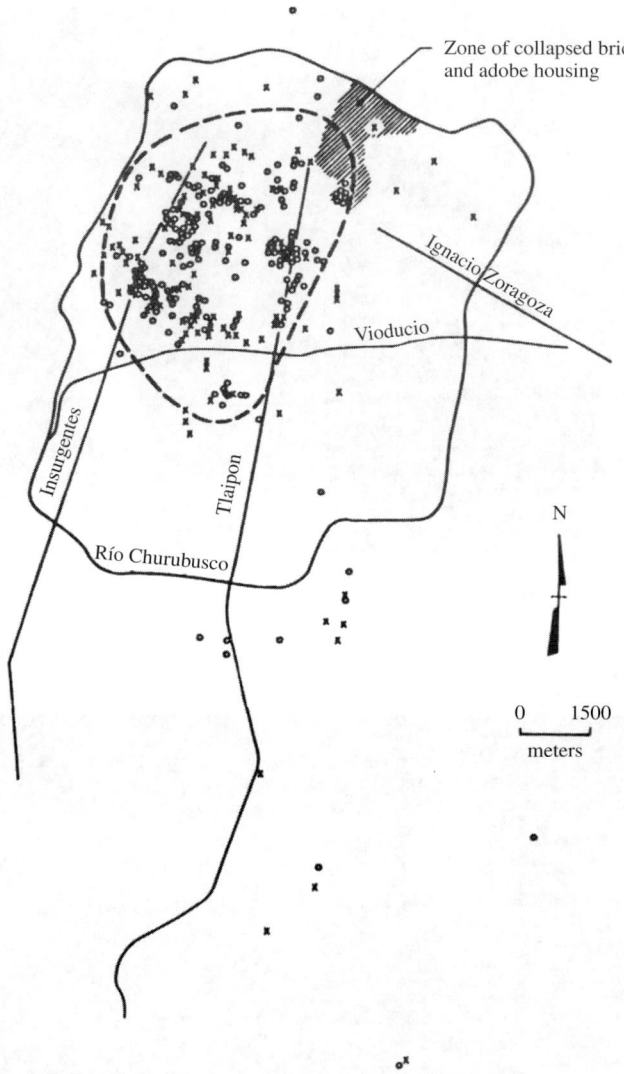

FIGURE 18.30   Locations of building damage at old lake bed in Mexico City (courtesy of EERI; EERI, *Earthquake Spectra*, 4, 3, 569–589, 1988).

During the 1971 San Fernando Earthquake, this structure collapsed onto its west (outer) side, breaking into several pieces and causing considerable damage to two lower-level bridges. A close examination of the fallen structure revealed that the collapse was due to pull-out of the column reinforcement from the foundations.

There was no top mat of reinforcement (and no ties) in the pile caps at Piers 2 and 3. The column longitudinal reinforcement (22 #18 bars) was placed in the footing with 12″ 90° bends at the bottom of the reinforcement. Transverse reinforcement was #4 bars at 12 in. around the longitudinal reinforcement. During the earthquake, the longitudinal reinforcement did not have sufficient development length to transfer the force to the footings. Insufficient confinement reinforcement in the footings and

FIGURE 18.31   Damaged 10-story building between the Plaza de la Constitution and Zona Rosa in Mexico City. (Photo by Karl Steinbrugge, from the EERC NISEE Photo Library.)

FIGURE 18.32   Failed hollow concrete block foundation during the magnitude 6.0, 1987 Whittier, California, earthquake. (Photo by Karl Steinbrugge. Photo courtesy of Steinbrugge Collection, Earthquake Engineering Research Center, University of California, Berkeley.)

columns, and the lack of a top mat of reinforcement resulted in the rebar (and columns) pulling out of the footing (Figure 18.35). Piers 5 to 7 had straight #18 bars embedded 6 ft into pile shafts, and they also pulled cleanly out during the earthquake (Figure 18.36).

After the San Fernando Earthquake, the development length of large-diameter bars was increased, splices to longitudinal rebars were no longer allowed in the plastic hinge area, and more confinement steel was provided in footings and columns.

FIGURE 18.33   Three-span bridge rocked over during the 1999 Ji Ji, Taiwan, earthquake.

FIGURE 18.34   House that fell from its foundation during the 1971 San Fernando Earthquake. (Photograph courtesy of Steinbrugge Collection, Earthquake Engineering Research Center, University of California, Berkeley.)

18.3.3 Soft Story

During the 1989 Loma Prieta Earthquake, many houses and apartment buildings in the San Francisco area had severe damage on the ground floor. These structures had less lateral support on the ground floor to allow room for cars to park under the structure. The remaining supports could not support the movement of the upper stories and dropped the top stories onto the ground (Figure 18.37).

However, a soft story does not always occur on the bottom floor. During the 1995 Kobe, Japan, earthquake, many tall buildings had damage at the midstory, often due to designing the upper floors for a reduced seismic load.

FIGURE 18.35   Failure of column to pile shaft connection.

FIGURE 18.36   Failure of column to footing connection.

Most buildings in Japan are either built of RC or of steel and reinforced concrete (SRC). These SRC buildings, when correctly designed, provide a great deal of ductility and more fire protection during large earthquakes. However, the design practice in Japan was to discontinue either the RC or the SRC above a certain floor. Figure 18.38 shows typical details used in SRC buildings.

Figure 18.39 shows a ten-story SRC building where the third story collapsed during the 1995 Kobe Earthquake.

18.3.4 Torsional Moments

Curved, skewed, and eccentrically supported structures often experience a torsional moment during earthquakes.

FIGURE 18.37   Soft story collapse in San Francisco during the 1989 Loma Prieta Earthquake. (Photo courtesy of the USGS; The Loma Prieta, USGS, 1998.)

A nine-story building in Kobe, Japan, consisted of shear walls along three sides and a moment-resisting frame on the fourth (east) side (Figure 18.40). Shaking during the 1995 Kobe Earthquake caused a torsional moment in the building. The first-story columns on the east side failed in shear, the building leaned to the east (Figure 18.41), and it eventually collapsed (Figure 18.42).

Since rivers, railroad tracks, and other obstacles do not usually cross perpendicularly under bridge alignments, columns and abutments must be built on a skew to accommodate them. These skewed bridges are vulnerable to torsion.

The Gavin Canyon Undercrossing consisted of two bridges over 70 ft tall, with a 67° skew, and was composed of three frames. An integral abutment and a two-column bent supported each end-frame. The center frame was supported by two 2-column bents while supporting the cantilevered end-frames. The superstructure was RC box girders at the end-frames and posttensioned concrete box girders at the center-frame. Each column was a 6 ft by 10 ft rectangular section, fixed at the top and bottom, with a flare at the top. The bridges were retrofitted in 1974 with cable restrainer units at transverse in-span hinges with an 8-in. seat width that connected the frames.

During the 1994 Northridge Earthquake, the superstructures were unseated due to the following factors. The tall, center frame had a large, long-period motion that was out of phase with the stiff end-frames. The end-frame center of stiffness was near the abutment, while its mass was near the bent, causing the end-frame to twist about the abutment. The sharp skew allowed the acute corners to slide off the narrow seats. The cable restrainers, being among the first in the country, were grouted in the ducts, making them too brittle and prone to failure. Both bridges failed as shown in Figure 18.43.

Another interesting example of torsional damage occurred at the Ji Lu Bridge during the 1999 Ji Ji, Taiwan, earthquake. This is a cable-stayed bridge with a single tower and cast-in-place, 102-m-long, box girder spans sitting on two-column end bents that connect the structure to precast "I" girder approach spans (Figure 18.44). The tower is 58 m from the top of the deck to the top of the tower, and 20 m from the top of the footing to the soffit. All the foundations are supported on driven piles. Construction was almost completed on this bridge at the time of the earthquake. All of the cables had been tensioned and all but one had been permanently socketed into the tower. The false work had been pretty much removed except for a few final pours for the portion of the superstructure where it connects to the tower.

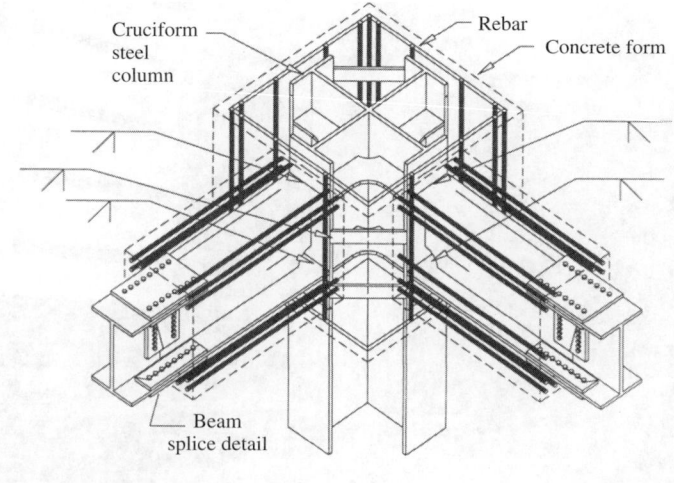

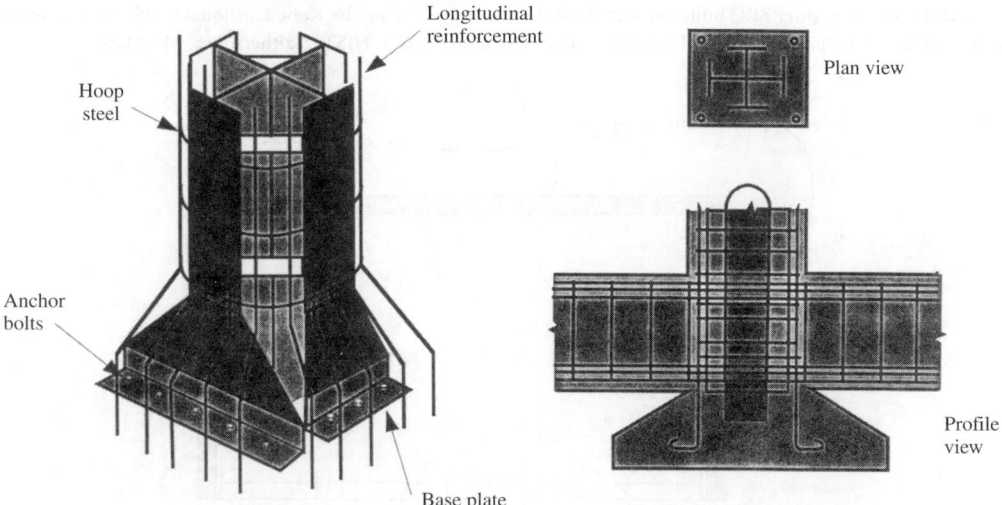

FIGURE 18.38 Examples of SRC construction. (Drawing courtesy of NIST; NIST, The January 17, 1995 Hyogoken (Kobe) Earthquake, U.S. NIST, Gaithersburg, MD, 1969.)

The dominant mode of shaking for this structure was twisting of the tower as the two cantilever spans moved back and forth. Looking at Figure 18.45 we can see that the key at the end of the spans walked up and down the bent seat almost to the end of the support. T.Y. Lin engineers explained that this was because the final pour around the tower had not been completed, making this structure extremely flexible in this direction.

Similar damage occurred to the center piers of curved ramps to the Minatogawa Interchange during the Kobe Earthquake. In this case, the superstructure swung off its end supports and the center column suffered severe torsional damage (Figure 18.46).

As one member of a bridge goes into flexure it can create large torsional moments in adjacent members. For instance, flexure in the columns of outrigger bents causes large torsional moments in the bent cap (Figure 18.47). All of these examples reinforce the idea that most structures require consideration of all three translations and rotations.

FIGURE 18.39 Ten-story SRC building with third floor collapse during the Kobe Earthquake. (Photo courtesy of NIST; NIST, The January 17, 1995 Hyogoken (Kobe) Earthquake, U.S. NIST, Gaithersburg, MD, 1969.)

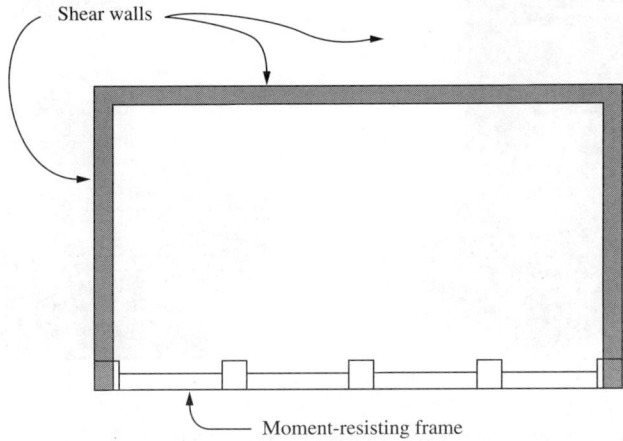

FIGURE 18.40 Plan view of nine-story SRC building in Kobe.

18.3.5 Shear

Most building structures use shear walls or moment-resisting frames to resist lateral forces during earthquakes. Damage to these systems varies from minor cracks to complete collapse. Figure 18.48 is a photo of the Mt. McKinley Apartments after the 1964 Great Alaska Earthquake. It was a 14-story RC building composed of narrow exterior shear walls and spandrel beams (Figure 18.48), as well as interior and exterior columns and a central tower. During the 1964 earthquake, this structure suffered major structural damage to most of the load bearing members.

The most serious damage was to a shear wall on the north side of the building (Figure 18.49). A very wide shear crack split the wall in two directly under a horizontal beam. This crack was because there was not enough transverse reinforcement to hold the wall together as it moved transversely and also due to a cold joint in the concrete at that location. The spandrel beams between the walls had large "X" cracks

FIGURE 18.41 Nine-story SRC building immediately after the 1995 Kobe Earthquake. (Photo courtesy of NIST; NIST, The January 17, 1995 Hyogoken-Nanbu (Kobe) Earthquake, U.S. NIST, Gaithersburg, MD, 1996.)

FIGURE 18.42 Eventual collapse of nine-story SRC building after the Kobe Earthquake. (Photo courtesy of NIST; NIST, The January 17, 1995 Hyogoken-Nanbu (Kobe) Earthquake, U.S. NIST, Gaithersburg, MD, 1996.)

FIGURE 18.43   Damage to Gavin Canyon UC during the 1995 Northridge Earthquake.

FIGURE 18.44   The Ji Lu cable-stayed bridge after the 1999 Ji Ji, Taiwan, earthquake.

FIGURE 18.45   Damage at end-supports to Ji Lu cable-stayed bridge.

FIGURE 18.46 Column damage to the Minatogawa interchange during the Kobe Earthquake.

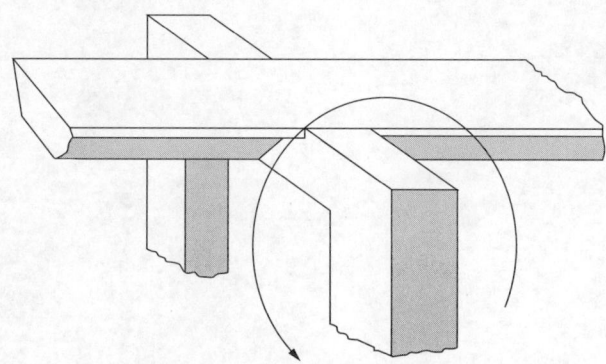

FIGURE 18.47 Column flexure causing torsion in the bent cap.

associated with shear damage as the building moved back and forth. These cracks decreased in size on the upper floors.

There was also shear damage to many of the columns. Figure 18.50 is one of the exterior columns on the south side of the building with a diagonal shear crack. Again, the problem was insufficient transverse reinforcement to resist the large shear forces that occurred during the earthquake. The central tower was also damaged. However, the Mt. McKinley Apartment can be viewed as a success since there was no collapse and no lives lost during this extremely large earthquake.

Bridges are equally susceptible to shear damage. For instance, there was considerable shear damage to piers on elevated Route 3 during the Kobe Earthquake. The superstructure is mostly steel girders and the substructure is RC single-column bents. Between Piers 148 and 150, the superstructure is a three-span, continuous, double-steel box with span lengths of 45 m, 75 m (between Piers 149 and 150), and 45 m. Pier 149 is a 10-m-tall by 3.5-m square RC single-column bent supported on a 12-m square pile cap.

FIGURE 18.48   West elevation of the Mt. McKinley Apartment building after the 1964 Great Alaska Earthquake. (Photograph by Karl Steinbrugge, Earthquake Engineering Research Center, University of California, Berkeley.)

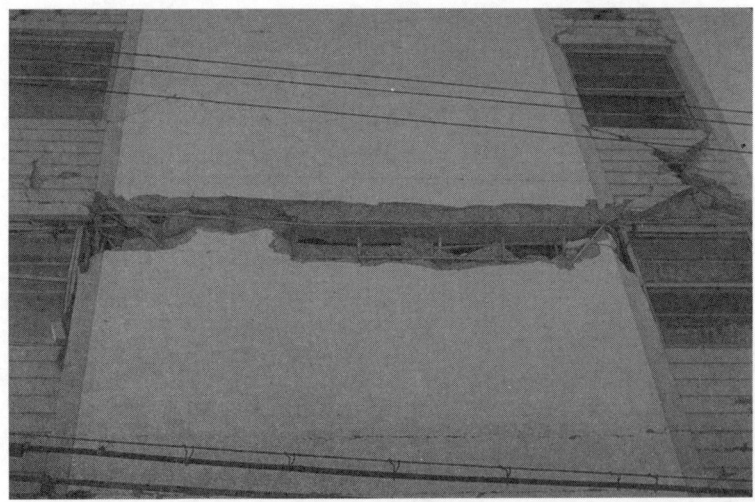

FIGURE 18.49   Damage to north side of Mt. McKinley Apartments. (Photograph by Karl Steinbrugge, Earthquake Engineering Research Center, University of California, Berkeley.)

Pier 150 is a 9.1-m-tall by 3.5-m square RC single-column bent supported on a 14.5-m by 12-m rectangular pile cap.

Figure 18.51 shows the shear failure at Pier 150. This damage was the result of insufficient transverse reinforcement and poor details. During the large initial jolt (amplified by near-field directivity effects) the transverse reinforcement came apart and resulted in the column failing in shear. Pier 149 was also severely damaged, but because it was taller (and more flexible), most of the force went to stiff Pier 150. The three-span continuous superstructure survived the collapse of Pier 150 with minor damage.

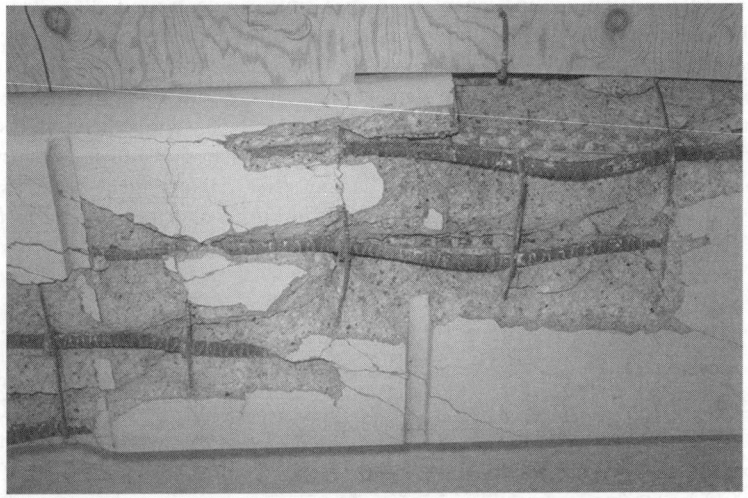

FIGURE 18.50 Damage to the south side of Mt. McKinley Apartments. (Photograph by Karl Steinbrugge, Earthquake Engineering Research Center, University of California, Berkeley.)

FIGURE 18.51 Shear failure of Pier 150 on Kobe Route 3.

18.3.6 Flexural Failure

Flexural members are often designed to form plastic hinges during large earthquakes. A plastic hinge allows a member to yield and deform while continuing to support its load. However, when there is insufficient confinement for RC members (and insufficient *b/t* ratios for SRC members), a flexural failure will occur instead. Often, flexural damage is accompanied by compression or shear damage as the capacity of the damaged area has been lowered.

The Dakkai subway station in Japan is a two-story underground RC structure. It was constructed by removing the ground, building the structure, and then covering it (the-cut-and-cover method). During the 1995 Kobe Earthquake, the center columns on both levels suffered a combination of flexural and compression damage that caused both roofs to collapse along with a roadway that ran above the station (Figure 18.52). Figure 18.53 shows the rather slender center columns at the lower level after the earthquake. The columns had insufficient transverse reinforcement at the location of maximum moment. The transverse reinforcement broke as the columns were displaced, allowing the longitudinal reinforcement to buckle and the concrete to fall out of the column.

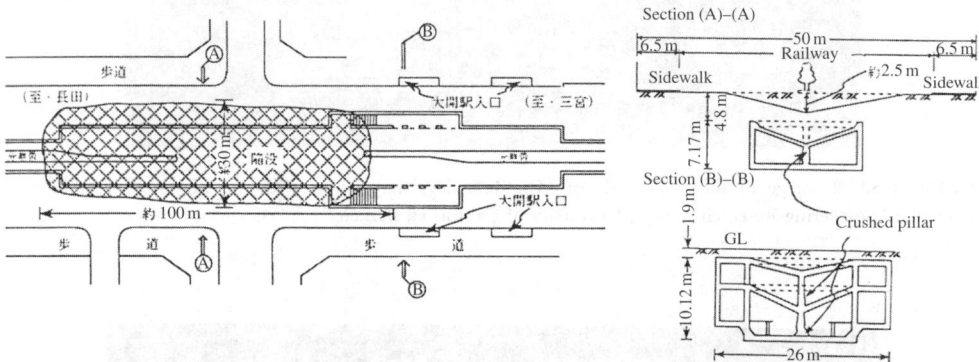

FIGURE 18.52 Plan and section drawings of the Dakkai subway station after the Kobe Earthquake. (Drawing courtesy of JSCE, JSCE, Preliminary report on the Great Hanshin Earthquake, January 17, 1995, JSCE, 1995.)

FIGURE 18.53 Flexural damage to columns at lower level of Dakkai subway during the 1995 Kobe Earthquake. (Photo courtesy of JSCE, JSCE, Preliminary report on the Great Hanshin earthquake, January 17, 1995, JSCE, 1995.)

Steel columns experience similar damage when the flexural demand exceeds the capacity. In downtown Kobe, in the Nagata District, Route 3 splits into two parallel structures with the super-structure composed of 50-m-long simple spans with steel girders and a three-span, continuous, steel box girder section between Piers 585 and 588. The substructure consists of 2.2-m-diameter, 14-m-tall steel hammerhead single columns. The steel columns are bolted onto 4-m-diameter hollow, concrete 20-m-long shafts. The column bottoms are filled with concrete to protect against vehicular impact. During the Kobe Earthquake, these steel columns had damage varying from local buckles to a complete section buckle, and at a few locations, the steel shell had torn, splitting the column. Most of the buckling occurred in a thinner section of the column. In some cases (Figure 18.54), the column underwent an excursion in only one direction and consequently had a buckle on one side of the column.

In some cases the buckled face tore in the tension cycle (Figure 18.55). The tears occurred in low-ductility welds. After the earthquake, the columns were tilting dangerously to the side. Buckling occurred before a plastic hinge was formed. A few columns remained undamaged as a result of failed bearings. Although local buckling cannot be completely eliminated, its spread can be prevented by maintaining smaller *b/t* ratios. This is accomplished with thicker sections, more frequent stiffeners, and diaphragms, or by filling the steel shells with lightweight concrete.

FIGURE 18.54 Pier 585 on Kobe Route 3 during the 1995 Kobe Earthquake. (Photo courtesy of Hanshin Expressway Public Corporation.)

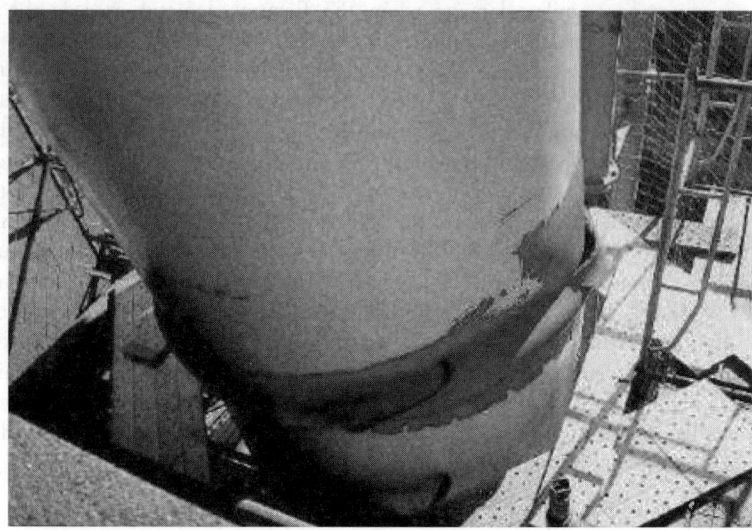

FIGURE 18.55   Torn buckle of steel column on Route 3 after the 1995 Kobe Earthquake. (Photo courtesy of Hanshin Expressway Public Corporation.)

18.3.7   Connection Problems

The most catastrophic type of structural damage is the failure of connections. When a bridge superstructure moves off its expansion joint or when the connection between building columns and beams fail, the result is too often the collapse of the structure.

Between Piers 39 and 43 on Kobe Route 3, the superstructure was a series of 52-m-long simple-span steel boxes supporting a concrete deck. Steel web plate restrainers with oversized holes at one end connected the girders together over each pier support. Fixed pin bearings and movable roller bearings supported each span. The substructure was hammerhead piers with 3.5-m-diameter circular or 3.5-m square concrete columns over 10 m in height. Each column was supported on a rectangular footing supported by 18- to 1.0-m-diameter by 16.5-m-long piles.

During the earthquake, the piers moved back and forth longitudinally, sustaining cracks or shear damage at the column bases. At the same time, the girder spans were moving west (toward Pier 43) as a result of the impact with the five-span, continuous, steel girder bridge east of Pier 38. The relative displacement between the piers and the simple spans exceeded the 0.8-m seat width at Piers 40 and 41 and the expansion end of Spans 40 and 41 fell off the piers (Figure 18.56). Most of the web plate restrainers failed in tension (Figure 18.57) and almost all of the fixed and expansion bearings were damaged, mostly due to the top shoe pulling out of the bearing (Figure 18.58).

RC structures must be carefully designed to allow the shear transfer at joints. A common kind of residence in Turkey was four- to eight-story buildings composed of concrete columns supporting concrete slabs and infill walls of unreinforced masonry blocks. During the 1999 Kocaeli Earthquake, these buildings collapsed at very low levels of acceleration, killing thousands of people. The large inertia force from these heavy structures had to be carried by the slender columns and by the inadequately reinforced connections (Figure 18.59). Sufficient concrete and reinforcement must be provided to resist the large tension and compression forces in moment-resisting joints during earthquakes. Joints of moment-resisting RC frames should be stronger than the elements that join them.

The connections of steel moment-resisting frames, with detailing recommended by design codes prior to 1994, suffered considerable damage during the Northridge (and Kobe) Earthquakes. During the Northridge Earthquake, over 100 steel moment-resisting frame buildings had some damage to the

FIGURE 18.56 Superstructure collapse at Spans 40 and 41 on Kobe Route 3.

FIGURE 18.57 Broken restrainer at Pier 39. (Photo courtesy of Hanshin Expressway Public Corporation.)

column to beam connections. As previously mentioned, flexural elements are supposed to form plastic hinges during large earthquakes. However, during the Northridge Earthquake the connection fractured instead. In this connection the beam flanges are welded to the column with a column flange stiffener placed along the column web to provide continuity to the joint. The beam web is bolted to the column with a shear connection plate. As shown in Figure 18.60, a large crack fractured the flange and web of some columns. Although this damage was unexpected and looked serious, no building during the Northridge Earthquake was reported to have collapsed as a result of the failure of this connection. Still, a few more cycles of motion (from a much larger earthquake than the magnitude 6.7 event) could easily

FIGURE 18.58   Broken bearings at Pier 40. (Photo courtesy of Hanshin Expressway Public Corporation.)

FIGURE 18.59   Collapsed reinforced concrete building during the 1999 Kocaeli Earthquake.

tear apart this fractured column. Consequently, new connection details have been developed as well as retrofit details for the many buildings with this connection.

18.3.8 Problem Structures

Some types of structures have performed particularly poorly during previous earthquakes. Usually, this is the result of vulnerabilities such as weak connections, improper detailing, and eccentric loads that predispose these structures to severe damage and collapse during earthquakes.

Unreinforced masonry should never be used to resist lateral forces since it is very weak in tension and very heavy, resulting in walls that immediately fall over, seriously injuring anyone nearby (Figure 18.61). However, reinforced masonry walls have performed very well when thoughtfully designed.

FIGURE 18.60 Steel moment-resisting frame connection after the 1994 Northridge Earthquake. (Photo courtesy of EQE International; EQE, *EQE Int. Rev.*, Fall, 1–6, 1994.)

FIGURE 18.61 Rear view of damaged unreinforced masonry (URM) building facing Pacific Garden Mall, Santa Cruz Area, California. (Photo by James R. Blacklock, Loma Prieta Collection, Earthquake Engineering Research Center, University of California, Berkeley.)

Tilt-up buildings have also performed poorly during past earthquakes because the walls pull away from the roof diaphragm, because of the discontinuity at the vertical joints between panels, because of poor connections to the roof joists, etc. Similarly, precast prestressed bridges tend to fall apart due to inadequate connections between members.

Tanks, bins, silos, grain elevators, concrete mix plants, etc. are the most commonly damaged structures during earthquakes. They are tall, heavy, and too often designed with weak supports and inadequate anchors (Figure 18.62).

Construction sites are particularly dangerous places to be on during an earthquake. Too often, little thought is given to providing lateral strength to partially constructed structures. This can result in millions of dollars in damage as well as fatalities. For instance a new expressway was being built during the 1999 Ji Ji, Taiwan, earthquake. This included two simple-span, precast, and prestressed "I" girder bridges on two-column bents near the epicenter. The girders were 84 in. tall by 24 in. wide and sat on

FIGURE 18.62   California Water Service 150,000-gal tank on six legs; the typical failure of tanks is upside down, with riser on top (Bakersfield, California, photo by Karl V. Steinbrugge, Steinbrugge Collection, slide/image No. S64, photo date July 25, 1952, Kern County, California, earthquake, July 21, 1952, magnitude 7.69; courtesy of Steinbrugge Collection, Earthquake Engineering Research Center, University of California, Berkeley).

FIGURE 18.63   Collapsed precast, prestressed "I" girders during construction of new expressway in Taiwan due to the 1999 Ji Ji, Taiwan, earthquake.

24 in. by 16 in. by 5 in.-tall elastomeric pads. The bents were at least 30 ft tall from the ground to the top of the bent caps. The bridge had 13 approximately 150-ft-long spans. At the time of the earthquake, the girders had been placed on the bents and the seat-type abutments, and the intermediate and end-diaphragms were just beginning to be cast between the girders. Wherever the intermediate diaphragms had not been placed, the girders had been shaken off of their supports and had fallen to the ground, breaking into pieces (Figure 18.63). This resulted in the loss of about 100 girders. The use of temporary supports before the casting of diaphragms could have saved about a million dollars.

All of these structures can be designed to perform adequately when they are provided with sufficient strength and ductility and a good understanding of their structural behavior during earthquakes.

18.4   Secondary Causes of Structural Damage

18.4.1   Surface Faulting

Few structures are designed to accommodate an offset of several meters laterally or vertically at any location. Yet, that is what is required near active faults. When the fault reaches the surface, the ground can be pushed together, pulled apart, raised, or dropped 5, 6, and even 10 m. Some of the most spectacular structural damage from faulting occurred during the 1999 Ji Ji, Taiwan, earthquake. The Shih Kang Dam was one of several structures across the ill-fated Dajia River north of Route 3 (Figure 18.64). During the earthquake, the north end of the dam dropped 9 m (Figure 18.65).

The Dajia River Bridge was a 14-span, RC bridge with a "T" girder superstructure, single-column bents with hammerhead bent caps, and drum-shaped footings. The girders sat on elastomeric pads and between transverse shear keys. It was one of several bridges a little north of Route 3 across the Dajia River near the town of Shargang (Figure 18.64).

This bridge had spans 30 m long by 10 m wide and columns about 10 m in height at the center of the bridge. During the earthquake, the first three southerly spans collapsed along with Pier 3 (Figure 18.66). The south abutment and first two piers moved 6.5 m vertically and 3 m to the west. The column

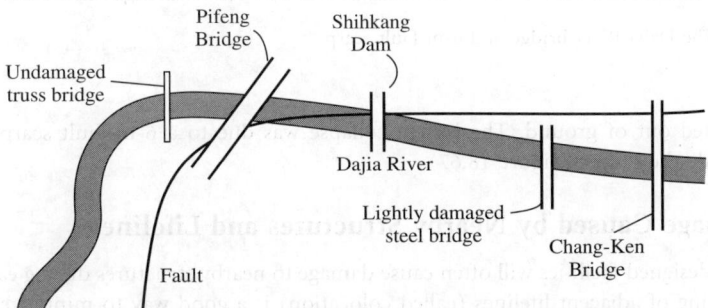

FIGURE 18.64   Location map of damaged structures along the Dajia river after the 1999 Ji Ji, Taiwan, earthquake.

FIGURE 18.65   North end of Shih Kang Dam after the Ji Ji, Taiwan, earthquake.

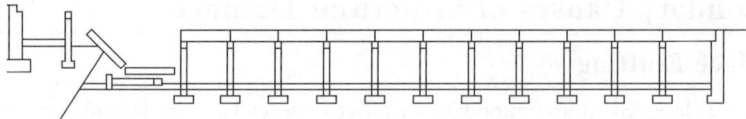

FIGURE 18.66   Elevation drawing of Dajia Bridge collapse.

FIGURE 18.67   The Dajia River bridge and 6-m fault scarp.

foundation rotated out of ground. The bridge collapse was due to a 6-m fault scarp that created a waterfall alongside the bridge (Figure 18.67).

18.4.2 Damage Caused by Nearby Structures and Lifelines

Weak or poorly designed facilities will often cause damage to nearby structures during earthquakes. The intelligent planning of adjacent lifelines (called colocation) is a good way to minimize disruptions to lifelines during earthquakes and other disasters.

When buildings, bridges, and other structures are built too closely together, one problem is pounding. This is particularly a problem for connectors on highway interchanges that are sometimes in such close proximity that the column of a higher structure will go through the lower bridge deck. Buildings are often built too closely together causing pounding and moderate structural damage during earthquakes. Pounding can cause serious damage since impact loads of very short duration can carry a very large force. However, pounding may also be beneficial when it prevents resonance.

More serious damage occurs when a poorly designed structure collapses onto an otherwise seismically resistant structure. There are no benefits from this interaction.

Piers 352 and 353 on the Route 3 Expressway in Kobe, Japan, supported the superstructure on steel hammerhead piers. The columns were 2 m by 1.5 m by 8.8 m in height, supported on pile foundations, and built using the 1964 specifications. The piers were 8.8 m tall and on pile foundations. The superstructure consisted of 30-m-long, steel simply supported plate girders, one end fixed and the other end on elastomeric expansion bearings. During the 1995 Kobe Earthquake, the bridge was damaged when a collapsing building fell against the expressway. The building (shown in Figure 18.68) applied a lateral load to the superstructure. This resulted in buckling of the vertical stiffeners on the girder webs, bearing damage, and also some damage to the bottom of Pier 353. The main damage was separation of a metal plate that was installed to protect the steel column from traffic collisions. However, because there was a collapsed building leaning against the bridge, it was felt to be expedient to provide shoring to the piers after the earthquake. The ability of the superstructure to move laterally over the piers helped to prevent more serious damage from occurring to the substructure.

FIGURE 18.68 Collapsed building leaning against the Kobe Route 3 Expressway. (Photo courtesy of Hanshin Expressway Public Corporation.)

FIGURE 18.69 Washed-out abutment fill caused by water pipe break.

A lifeline can often cause problems to other nearby lifelines during earthquakes. The Balboa Boulevard Overcrossing is a three-span (actually two spans with a 22-ft-long bin-type abutment between Abutment 1 and Bent 2) cast-in-place prestressed concrete bridge (except for a concrete slab for the bin-type abutment) on Highway 118 near Northridge, CA. The superstructure is 283 ft long and 117 ft wide. Bent 3 is a three-column bent on a spread footing. The columns are 5-ft octagonal sections with a large flare. The abutments and wingwalls are on piles. During the 1994 Northridge Earthquake the Balboa Boulevard Overcrossing experienced minor spalling of the concrete cover at the top of the columns during the earthquake. The shaking broke a small-diameter water line under Abutment 1 that washed out the backfill behind the abutment. Figure 18.69 shows the collapsed slab for the bin-type abutment and the cast-in-drilled-hole (CIDH) pile supporting Abutment 1. The grade beam that supports the wingwall is on the left.

These are just a few of the many secondary causes of damage during large earthquakes. What typifies this kind of damage is that its prevention may not rely on structural issues. In fact, the best solution is to find a better site during the planning of the structure. There is little the engineer can do once the structure straddles an active fault or is within a few feet of a vulnerable structure.

18.5   Recent Improvements in Earthquake Performance

Much of the damage that has occurred during recent earthquakes was the result of the failure of the surrounding soil or structural elements. Not surprisingly, most of the recent improvements to the earthquake resistance of structures have focused on methods of improving poor soil, on soil–structure interaction, and on structural elements and systems that increase strength, stability, ductility, etc.

There has also been some work to prevent secondary earthquake damage. For instance, California has begun to retrofit bridges that cross over active faults. On the Colton Interchange, a bridge was provided with additional supports to catch the superstructure if it is pulled off its piers. Another California bridge was provided with large galleries at the abutments to accommodate large movements. A couple of bridges have even been provided with gates that close the bridge when the ground begins to shake.

Some effort has also been made to improve lifeline performance, most notably in Wellington, New Zealand, where facilities are carefully planned and located to prevent disruptions in service during large earthquakes.

18.5.1   Soil Remediation Procedures

Because much of Japan is covered with either weak clay or saturated, loosely consolidated material, the development of soil remediation procedures has flourished in that country. After the 1994 Kobe Earthquake researchers did a careful study of soil-remediation sites and found that these locations performed much better than the surrounding area [10].

The following are a few examples of soil improvement methods commonly used in Japan.

18.5.1.1   Gravel Drains: Ariake Quay-Wall Improvement Project

Ariake is a man-made island in Tokyo Bay. The weak material of the island is supported by a quay-wall made of timber piles and steel sheet piles. The goal of this project was to protect the quay-wall from damage during an earthquake (Figure 18.70). The wharf is composed of loose sand that could liquefy during an earthquake. Increased pore water pressure would then push over the quay-wall. Gravel drains were installed to reduce pore water pressure during an earthquake. The advantage of this method is that it is quick, inexpensive, and free of vibration. The disadvantage is that it does not prevent settlement of

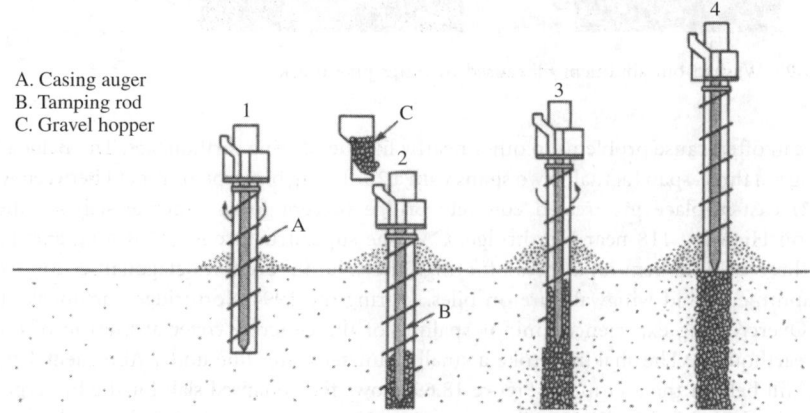

FIGURE 18.70   Gravel drain construction. (Drawing courtesy of Japan's Public Works Research Institute.)

the soil during an earthquake. Since the goal is protection of the quay-wall, this is not considered a problem. The procedure is to drill a hole in the ground using a casing auger. Gravel is carried to the auger by a front-end loader. It is dropped into a hopper, lifted to the top of the auger, and poured into the casing. The casing is then removed from the ground leaving a sand drain. The gravel is fairly uniform. The casing has replaceable steel teeth to help it cut through soil and push away rocks. The gravel drains are placed close enough together to form a grid that will effectively drain out all the water. This project consisted of 3997 gravel drain piles in an area of 2770 m^2 for 1.4 piles/m^2 (the drains are 0.8 m apart). The drains are 0.5 m in diameter and 17 m long and cost about 50,000 Yen ($500) per pile or about 200 million Yen for the project. It takes less than an hour to make a gravel drain and there were four augers at the job site.

18.5.1.2 Deep Mixing Method: Kawaguchi City

The area along the Arakawa River has poorly consolidated soil. This project used deep soil mixing to improve the unconfined compressive strength of the ground (Figure 18.71). Without the DMM the soil would be expected to settle 1.5 m after the embankment is placed. With the DMM the settlement should be less than 0.03 m.

On this project, a modified pile-driving machine rotates a pair of rods with stirring wings for mixing the soil. The distance between piles can be changed by moving the rods closer together or further apart. The pile diameter is 1 m on this project. First, the machine pushes the stirring wings down 30 m,

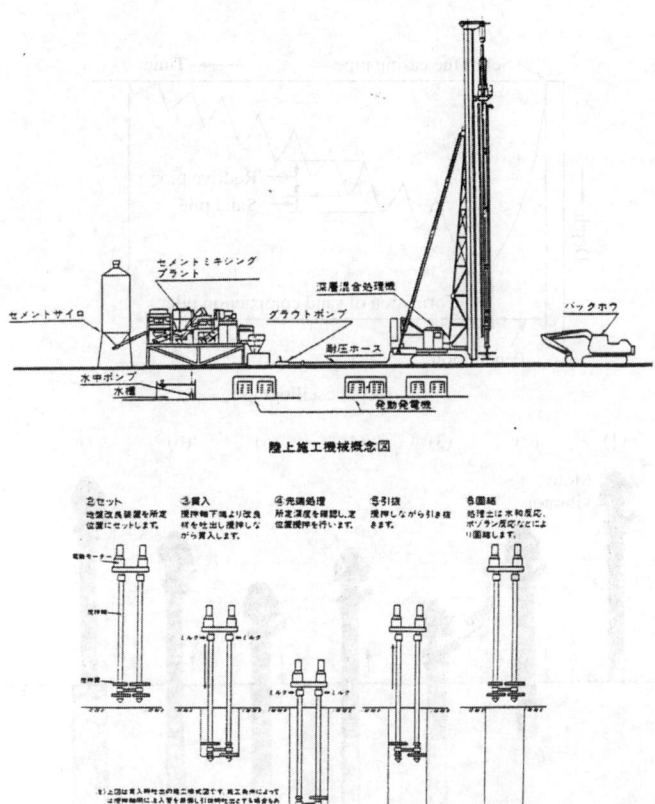

FIGURE 18.71 Soil deep mixing construction. (Drawing courtesy of Japan's Public Works Research Institute.)

breaking up the soil and making it soft and permeable. The w.ngs rotate 20 cycles per minute as they descend into the soil. Then, the stirring wings are pulled up while injecting cement milk into the soil. The milk is composed of an equal weight of cement and water. About 100 to 150 kg of cement milk is injected per cubic yard of soil. The wings rotate at 40 cycles per minute as they ascend. Loose soil comes up during this procedure and is removed with an excavator. The stirring wings descend and ascend at 1 m/min. There is a cement plant for each machine and a separate pump for each rod. The cement milk flows through flexible hoses from the cement plant to the top of the rods where it is injected into the soil. When the piles are completed the machine crawls to the next location on a plywood mat and drills another pile. Display panels in the machine cab give information about depth, rotation, and the amount of cement milk being pumped.

The deep mixing method is also very effective in preventing soil liquefaction. In fact, this method was used to prevent liquefaction for some flood prevention works next to this site. In this case the piles are placed in a lattice pattern to contain the liquefiable soil.

The deep mixing method was found to be the most effective method for creating strong, highly ductile ground that does not liquefy or settle. It is also the most expensive method, but the lack of noise and vibration makes it ideal for city environments.

18.5.1.3 Sand Compaction Pile Method: Ohgishima Island, Tokyo

Sand compaction piles are a popular way of preventing liquefaction of loose alluvium. However, the noise and vibration make it unacceptable at some locations.

This project was on a man-made island in Tokyo Bay (Figure 18.72). It is the location of an LNG tank farm. Soil remediation was required at the toe of an embankment that covers these tanks.

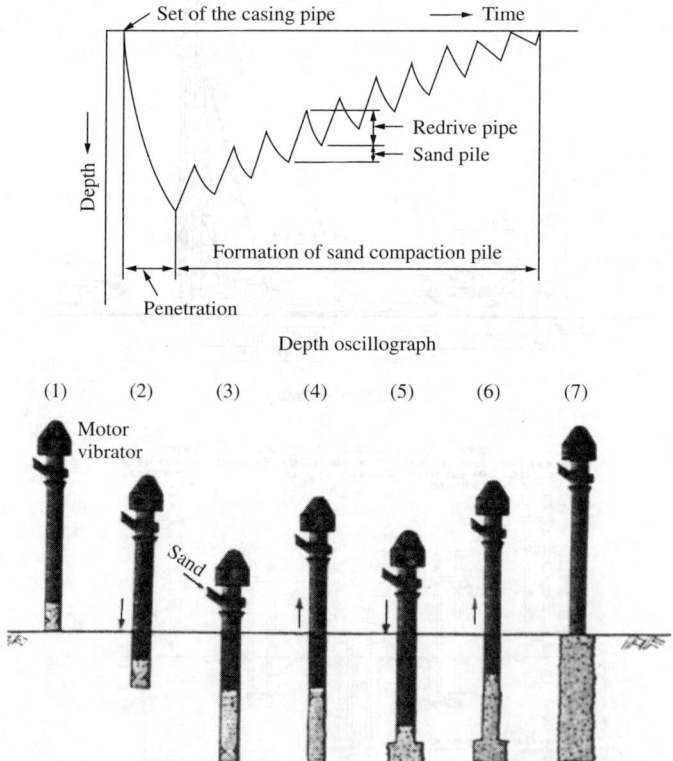

FIGURE 18.72   Sand compaction pile method. (Drawing courtesy of Japan's Public Works Research Institute.)

Because it was a remote site, the loud noise and vibrations were not a problem. The sand compaction pile method uses a modified pile-driving machine to vibrate a steel pipe into the ground. When the penetration reaches the proper depth, sand is carried to the top by a hopper and forced to the bottom of the pipe with compressed air. Then, the pipe is raised and lowered in the hole as sand is repeatedly shot to the bottom of the hole. The result is a pile of compacted sand and an area between the piles of compacted soil. The equipment is similar to the gravel drain method. The pile driver has a steel casing with a lid that can be opened and closed at the bottom. Sand is brought in by dump truck to the site. A front-end loader pours the sand into the hopper. The hopper carries the sand to the top of the pipe where it is poured into the air compression chamber. The steel casing with the lid closed is vibrated to the required depth. In this project the depth was 17 m and the pile diameter was 0.7 m. When the pile is completed, the panels the truck sits on are moved back and the pile driver is moved back to drive a new pile. The pile is driven at about 2 min/m and a pile is completed in about 30 min. Before this project began, the soil had an average N value of 10 to 12. The areas that have been completed now have an N value of 15 to 20.

18.5.2 Improving Slope Stability and Preventing Landslides

The failure of many retaining walls during the 1999 Ji Ji, Taiwan, earthquake was surprising. Well-designed cantilever and MSE walls have performed well during recent earthquakes. For instance, there was an MSE wall that continued supporting the embankment adjacent to the Arifiye Bridge that collapsed during the 1999 Kocaeli, Turkey, earthquake (Figure 18.73). A cribwall in the Santa Cruz Mountains was severely distressed during the 1989 Loma Prieta Earthquake but continued to support the steeply sloping ground (Figure 18.74). There is no record of a cantilever retaining wall failure during a California Earthquake. Therefore, well-designed retaining walls continue to be a good method for supporting most soils.

Other methods of preventing soil movement include planting trees and other vegetation, the use of geotextiles, installing piles, etc. The foundations of structures may also be strengthened to prevent their movement. Many of the methods used for preventing liquefaction can also be used to stabilize soils. However, when an area has a long history of landslides, it may be more prudent to build elsewhere.

FIGURE 18.73 Mechanically stabilized earth wall continues to support the embankment behind the collapsed Arifiye Overcrossing after the 1999 Kocaeli, Turkey, earthquake.

FIGURE 18.74   Crib wall continues to support steeply sloping soil in Santa Cruz after the 1989 Loma Prieta Earthquake. (Photo courtesy of EERI.)

18.5.3 Soil–Structure Interaction to Improve Earthquake Response

Soil–structure interaction modifies the ground-input motion at the foundation. By taking advantage of soil–structure interaction, structures can be protected during earthquakes. Frank Lloyd Wright's 1915 Imperial Tokyo Hotel survived the 1923 Great Kanto Earthquake because he built his foundation in 70 ft of weak clay. During the earthquake, the ground underneath moved violently while his hotel (effectively isolated by this plastic material) remained relatively immobile.

Similarly, a common retrofit technique for bridges is to allow the foundation to rock during large earthquakes. As the foundation rocks, the period is lengthened and damping is increased, all of which lowers the demand on the structure. To insure stability, the foundation may be connected to flexible anchors or an outer perimeter of piles may be placed under (but not connected to) a widened foundation. The piles will provide support for the foundation as it rocks back and forth.

Special foundations are sometimes used to improve seismic response. For instance, a popular bridge foundation in California is the large-diameter drilled pile shaft. These foundations are very flexible, replacing potentially damaging seismic forces with large displacements. Moreover, when the pile shaft is allowed to yield, a large plastic hinge forms, which provides more ductility for the structure.

In contrast to California's efforts to provide more flexibility and ductility in its foundations, Japan has been developing stiffer and more massive foundations for bridges and other structures. Besides the advantage of handling very large forces elastically, many of these new foundations use advanced automation techniques to simplify their construction. *The open caisson construction method* pushes a precast hollow cylindrical caisson into the ground while excavating the ground beneath with a grab bucket. Additional sections are attached to the top until it bears on good material. The caisson segments can be constructed up to 4 m in diameter and are match-cast for a tight fit. After they are completely assembled,

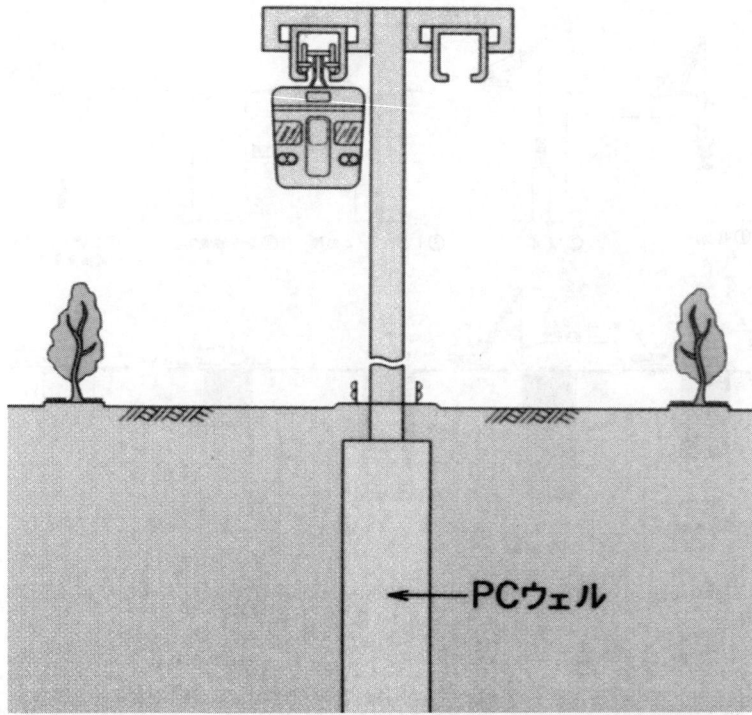

FIGURE 18.75 Precast caissons were used for the Chiba City, Japan, monorail. (Drawing Courtesy of Japan's Public Works Research Institute.)

they are filled with soil, and a steel assembly is attached to mount the substructure. This method was used for the Chiba City Monorail System. The open caisson construction method allowed them to build large caissons often within a few feet of an existing building (Figure 18.75).

Another innovative foundation in Japan is the *continuous diaphragm wall*. This method involves excavating wall-type ditches and casting wall elements in them (Figure 18.76). These walls are connected together with joint elements. Then, an upper slab connects the top of the walls to the pier. The walls typically vary in thickness from 1 to 2.8 m. Continuous diaphragm walls come in a variety of shapes. Circles, rectangles, and grids are all very common. These are used for deep, stiff foundations as well as enormous shafts and storage tanks.

18.5.4 Structural Elements that Prevent Damage and Improve Dynamic Response

Improved structural performance during large earthquakes depends on a balanced structural system. Elements that share the same displacement during large earthquakes must be designed to have about the same stiffness. Otherwise, the stiffer elements will be forced to resist most of the earthquake force. Elements that share the same force are often provided with a fuse to limit the force and protect adjacent members.

Providing great strength to resist earthquakes is usually self-defeating. As the elements are made stronger, they attract larger earthquake forces. If an element along the load path cannot resist these forces, it will break, sometimes with disastrous consequences. Elements like shear walls are often used to limit displacements in buildings, but they are nonductile and will often shatter when unexpectedly large earthquakes occur.

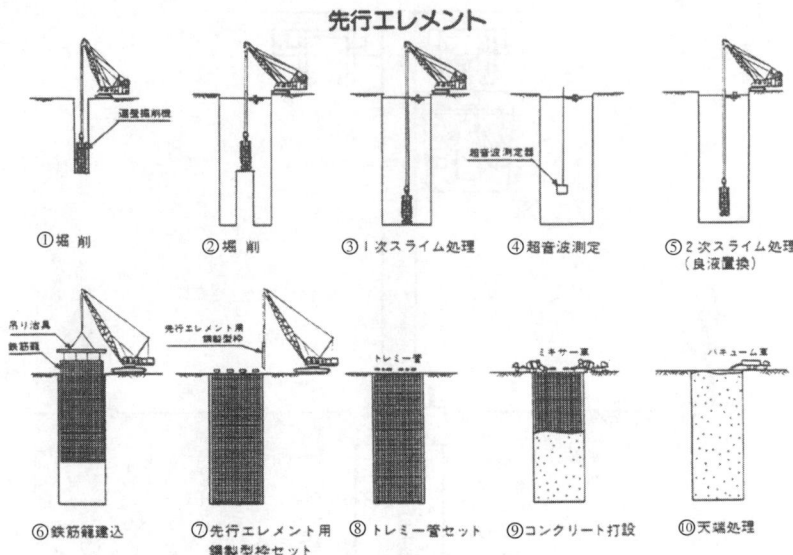

FIGURE 18.76 Construction sequence for diaphragm foundations. (Drawing courtesy of Japan's Public Works Research Institute.)

FIGURE 18.77 Construction of the new San Francisco Airport International Terminal. (Photo courtesy of Earthquake Protection Systems Inc.)

The use of isolation and damping devices has gained popularity because they can protect a structure during several large earthquakes without suffering damage or requiring replacement. Moreover, these devices have proven to be effective for new construction as well as for retrofitting vulnerable existing structures.

The new San Francisco Airport International Terminal is a large steel frame building with a truss roof (Figure 18.77). It is the largest base-isolated building in the world. To ensure that it would remain in service after a very large (magnitude 8) earthquake, 267 friction pendulum seismic isolation bearings were placed between the steel columns and the foundations (Figure 18.78 and Figure 18.79). Each bearing can move 20 in. while supporting 6000 kip. The bearings increased the building's fundamental period to 3 s and reduced the earthquake force by 70%.

These interesting devices are surprisingly simple in principle. For service level loads, static friction restrains the system. During an earthquake, the system operates like a pendulum (Figure 18.80) with the amount of damping controlled by dynamic friction and with the period

FIGURE 18.78 Layout of foundations and friction pendulum devices for the new San Francisco Airport International Terminal. (Photo courtesy of Earthquake Protection Systems Inc.)

FIGURE 18.79 One of the 267 friction pendulum devices for the new San Francisco Airport International Terminal. (Photo courtesy of Earthquake Protection Systems Inc.)

and stiffness as shown below (where R is the radius, W is the weight, and g is the acceleration due to gravity).

$$T = 2\pi \sqrt{\frac{R}{g}} \tag{18.3}$$

$$K = \frac{W}{R} \tag{18.4}$$

The period of the structure is increased by flattening the bearing's concave surface and the force–displacement relationship is modified by changing the friction or the dead load. Similarly, the design of the pyramid-shaped "Money Store" in West Sacramento (Figure 18.81) included fluid viscous dampers (FVD) to absorb energy and allow the steel moment-resisting frame to remain elastic during large

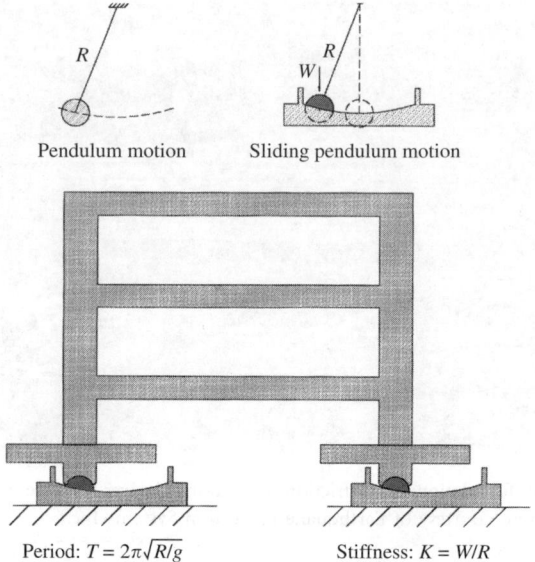

FIGURE 18.80 The structure's period is controlled by the radius of the curved bearing. (Drawing courtesy of Earthquake Protection Systems Inc. Zayas et al., The FPS earthquake resisting system, University of California, Berkeley, CA, 1987.)

FIGURE 18.81 The uniquely shaped "Money Store" was designed with fluid viscous dampers to improve its seismic behavior. (Photo courtesy of Marr-Shaffer & Miyamoto, Structural Engineers.)

earthquakes. The FVDs are cylinders filled with liquid silicon. When a load is applied, a piston pushes through the viscous fluid in the cylinder, absorbing energy (Figure 18.82). Concentrically braced frames were built with FVDs between the diagonal braces and the columns to increase the damping ratio of the building to 15% and reduce the story drift ratio to 0.005 (Figure 18.83).

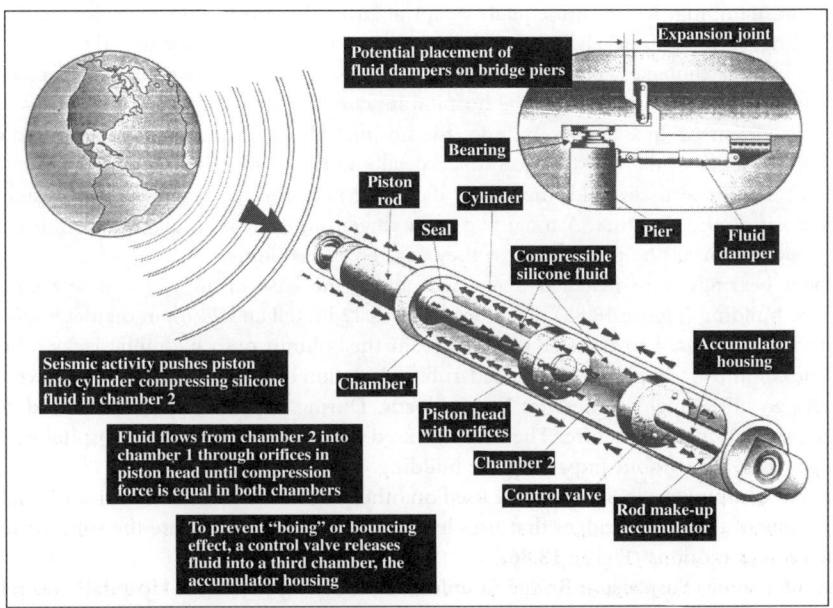

FIGURE 18.82 Schematic drawing of a fluid viscous damper. (Drawing courtesy of Taylor Devices Inc.)

FIGURE 18.83 The steel frames included fluid viscous dampers to reduce the story drift. (Photo courtesy of Marr-Shaffer & Miyamoto, Structural Engineers.)

Isolation and damping devices are equally adept at providing protection for seismically vulnerable existing buildings. The Long Beach VA Hospital is a concrete shear-wall structure that was found to be vulnerable to large earthquakes (Figure 18.84). Many retrofit strategies were studied before isolation was finally accepted as the best way to keep the hospital functioning after a large earthquake on the nearby Pales Verdes and San Andreas Faults. Because the hospital also had to remain in service during construction, the sequence of the construction was crucial and the contractor had to be responsive to any problems that developed that could impact the daily operations of the hospital. Most challenging was completely isolating the structure. A moat had to be dug around the building and flexible connections had to be designed for all the utility lines as they entered the building.

Lead-rubber bearings were installed a few feet above the base of all 150 concrete columns that supported the building (Figure 18.85). Each bearing was 22 in. tall and 24 in. in diameter. First, friction gripping devices were used to transfer the load from the column onto hydraulic jacks. Then, a 20-ft section of the column was replaced with a lead-rubber isolation bearing. These bearings have a lead core that provides an initial high stiffness for service loads. During large earthquakes the lead core yields, providing damping for the structure. The bearings used on the Long Beach VA Hospital will allow the ground to move 16 in. without impacting the building.

Isolation and damping devices can also be used on other structures. The All American Canal Bridge in California is one of the many bridges that uses lead-rubber bearings to isolate the superstructure from earthquake ground motions (Figure 18.86).

The Vincent Thomas Suspension Bridge (a mile from the Long Beach VA Hospital) was retrofit with 80 FVDs to absorb energy and prevent the bridge deck from pounding against the towers and cable bents (Figure 18.87). The connection between the towers and the truss was modified to allow very large relative movements. The truss section on each side of the towers was replaced with a new unit that includes a deck section with 26-ft-long finger joints and large viscous dampers to absorb energy and prevent the truss from pounding against the towers (Figure 18.88).

In general, isolation and damping devices have performed very well during large earthquakes. However, isolated structures have usually been too far away to experience really large accelerations. The one occurrence where isolation and damping devices were very close to a fault rupture was on the Bolu

FIGURE 18.84 The existing Long Beach VA Hospital was seismically retrofit with lead-rubber bearings. (Photo courtesy of Dynamic Isolation Systems, Inc.)

FIGURE 18.85 The concrete columns were cut at midheight and lead-rubber isolation devices were installed. (Photo courtesy of Dynamic Isolation Systems Inc.)

FIGURE 18.86 Lead-rubber isolation bearings on the All American Canal Bridge in California.

FIGURE 18.87 The Vincent Thomas Bridge is a three-span suspension structure built over the port of Los Angeles.

FIGURE 18.88   The Vincent Thomas Bridge retrofit provided gaps between the decks and the towers and also fluid viscous dampers to dissipate energy. (Photo courtesy of Enidine Inc.)

Viaduct during the 1999 Duzce Earthquake. However, in this installation the bearings were eccentric to the dampers, resulting in out-of-plane motions that locked up the dampers and caused significant damage to the bridge. It will probably take several earthquakes to work out all the bugs and come up with installations that work most effectively. Because of the increased use of isolation and damping devices on structures in highly seismic areas, there should be many more opportunities to study their behavior.

When isolation and damping devices are not used, a sacrificial element will often limit the force and increase the damping in a structure. New RC beams and columns are provided with welded hoops and spirals that allow these members to form plastic hinges during large earthquakes. Existing concrete members are sometimes wrapped in steel casings, fiberglass, carbon fiber, and many other materials to increase their shear capacity and allow for the formation of a plastic hinge (Figure 18.89).

Steel structures have undergone similar improvements. Beginning in the late 1970s *eccentrically braced frames* (EBFs) were developed that provide greater stiffness and ductility during earthquakes. The EBF has a ductile link between the connections that is specially designed to act as an energy dissipater. This concept has been expanded to include a variety of different configurations (Figure 18.90).

The poor performance at the connections of moment-resisting frames during the Northridge and Kobe Earthquakes has resulted in a great deal of research and testing. There are now a variety of welded beam–column joints that ensure ductile behavior in the members rather than brittle fracture of the joint. One popular new connection is the "dog bone" (Figure 18.91). Testing has shown that plastic hinging will occur at the reduced section, protecting the connection.

There are a number of other devices and structural elements that improve structural response during large earthquakes. Restrainers, shear keys, catchers, and seat extenders are used to prevent bridge superstructures from falling from their supports. A variety of materials are wrapped around weak columns to increase their ductility. However, isolation and damping devices show the greatest potential

FIGURE 18.89 Steel shell being wrapped around a concrete bridge column to increase its ductility and shear strength.

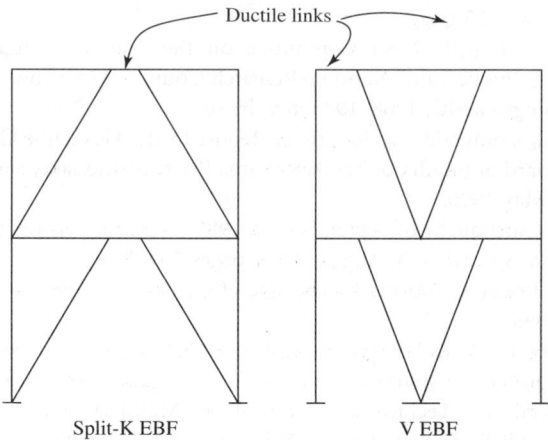

Ductile links

Split-K EBF   V EBF

FIGURE 18.90 Two of the many configurations that have been developed for eccentrically braced connections.

to control the amount of damage on a structure, particularly when the goal is to keep the structure in service after a very large earthquake.

By ensuring that the supporting soil remains undamaged, by avoiding sites near active faults or other secondary hazards, and by the effective use of isolation and damping devices, most serious earthquake damage can be avoided. However, every structure should also be provided with abundant ductility and

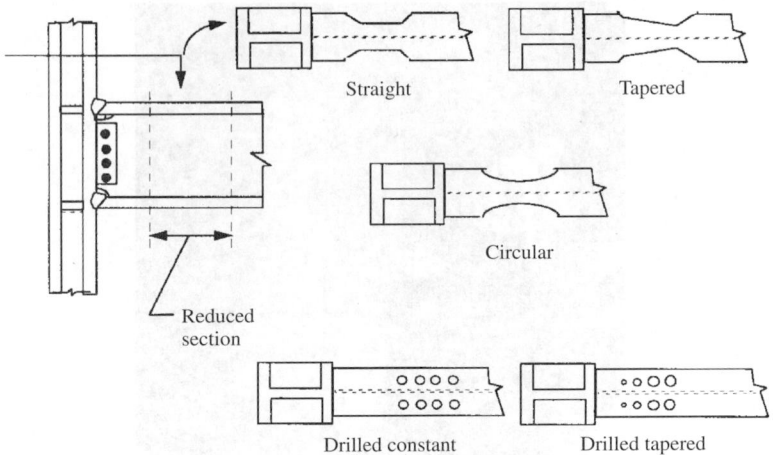

FIGURE 18.91   Examples of beams in moment-resisting frames that use a reduced flange section to prevent fracturing of the connection.

large seats in case an expectedly large earthquake were to occur. Moreover, all structures must be carefully designed to be relatively uniform and without eccentric loads.

References

[1] EERI, Scenario for a magnitude 7.0 earthquake on the Hayward fault, Earthquake Engineering Research Institute, HF-96, The Institute, Oakland, CA, 1996, 109 pages.

[2] Mualchin, L. and Jones, A.L., Peak acceleration from maximum credible earthquakes in California (rock and stiff-soil sites), DMG Open-file Report, 92-1, California Division of Mines and Geology, Sacramento, CA, 1992, 53 pages.

[3] National Research Council (U.S.) Committee on the Alaska Earthquake, The great Alaska earthquake of 1964, Publication (National Research Council [U.S.]) no. 1603, National Academy of Sciences, Washington, DC, 1968–1973, 8 v. in 10.

[4] Housner, G., et al., Competing against time, Report to the Governor George Deukmejian from The Governors' Board of Inquiry on the 1989 Loma Prieta earthquake, State of California Office of General Services, May 1990.

[5] EERI, The Mexico earthquake of September 19, 1985 — on the seismic response of the Valley of Mexico, *Earthquake Spectra*, 4, 3, August 1988, pages 569–589.

[6] USGS, The Loma Prieta, California Earthquake of October 17, 1989, United States Government Printing Office, 1998.

[7] NIST, The January 17, 1995 Hyogoken-Nanbu (Kobe) earthquake: performance of structures, lifelines, and fire protection systems, NIST Special Publication 901 (ICSSC TR18), U.S. National Institute of Standards and Technology, Gaithersburg, MD, July 1996, 538 pages.

[8] Japan Society of Civil Engineers (JSCE), Preliminary report on the Great Hanshin earthquake, January 17, 1995, Japan Society of Civil Engineers, 1995.

[9] EQE, Steel's performance in the Northridge earthquake, *EQE Int. Rev.*, Fall 1994, pages 1–6.

[10] Mitchell, J.K., Baxter, C.D.P., and Munson, T.C., Performance of improved ground during earthquakes, Soil Improvement for Earthquake Hazard Mitigation, American Society of Civil Engineers, New York, 1995, pages 1–36.

[11] Zayas, V.A. Low, S.S., and Mahin, S.A., The FPS earthquake resisting system: experimental report, UCB/EERC-87/01, Earthquake Engineering Research Center, University of California, Berkeley, CA, June 1987, 98 pages.

19

Seismic Design of Buildings

Ronald O. Hamburger
Simpson Gumpertz & Heger, Inc.,
San Francisco, CA

Charles Scawthorn
Department of Urban Management,
Kyoto University,
Kyoto, Japan

19.1 Introduction

Seismic design involves two distinct steps — determining (or estimating) the forces that will act on a structure and designing the structure so as to both resist these forces and keep deflections within prescribed limits. This chapter provides a basic explanation of the seismic design of buildings, by first discussing how earthquake forces are caused in buildings and the systems that have been developed to deal with these forces, then reviewing the most common types of buildings and their typical seismic performance, then discussing selected key aspects of seismic design, and, finally, discussing the force-determination aspect currently used in building codes.

19.2 Earthquakes — Their Cause and Effect

As discussed in Chapter 17, displacement of a fault results in the propagation of time-varying displacements throughout the earth — this vibratory ground motion is what is termed an *earthquake*. From a structural viewpoint, the essence of earthquakes is the dynamic displacement of the ground supporting a building, resulting in lateral and vertical forces on the building. In this chapter, emphasis will be on the lateral forces on a building due to shaking — vertical forces due to shaking are usually of somewhat less significance, but not always, and building codes and good practice require that vertical forces also be considered in the seismic design of buildings. While earthquakes also result in other effects on buildings,

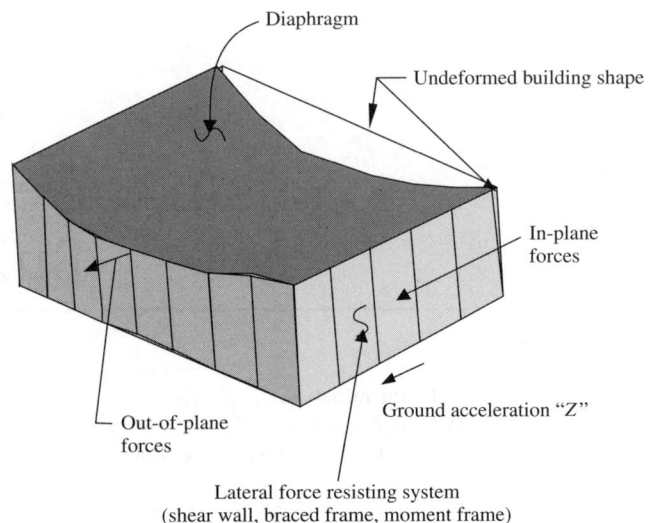

FIGURE 19.1   Effect of ground displacements on a building.

such as partial or complete ground failure, fire, and tsunami, design for these effects is more specialized and will not be covered here.

For a structure connected to the earth (and very few buildings are not connected to the earth through their base or foundation), as the base moves or displaces, the inertia and flexibility of the structure result in a time lag before the rest of the structure can displace in response to its base's displacement. The interaction of these displacements and the response of the structure result in time-varying displacements and strains within the structure, which the structure must be designed to sustain. For a structure responding to a moving base there is an equivalent system, in which the base is fixed and the structure is acted upon by forces (called inertia forces) that cause the same displacements as are occurring in the moving base system. In seismic design it is customary to visualize the structure as a fixed base system acted upon by inertia forces. Figure 19.1 shows such a system, in which the base has moved (due to the ground accelerations) and the roof is deflected. The roof's deflection is relative to its base, and is due to it and its supporting walls not moving or displacing as quickly as the base has displaced underneath it. This is exactly equivalent as if a force had been applied to the roof, which deflected the roof and its supporting walls the same amount, relative to its base.

19.3   Lateral Force Resisting Systems

The parts of the structure that connect the structure's mass to the ground and resist or otherwise accommodate these displacements or equivalent forces are termed the *lateral force resisting system* (LFRS). An LFRS is usually capable of resisting only forces that result from ground motions parallel to them. However, the combined action of LFRS along the width and length of a building can typically resist earthquake motion from any direction. LFRS differ from building to building because the type of system is controlled to some extent by the basic layout and structural elements of the building. Figure 19.2 illustrates the basic elements that may be used in LFRS, which consist of axial (tension and/or compression bracing) elements, shear (wall) elements, and/or bending resistant (frame) elements. Horizontal load distributing elements are termed diaphragms and are most often floor or roof slabs, but can be horizontally braced (i.e., truss) elements. Few buildings would use all of these LFRS elements, but most buildings would use one or more.

The earthquake resisting systems in modern steel buildings take many forms, Figure 19.3. Many types of bracing configurations have been used (diagonal, "X," "V," "K," etc.). Moment resisting steel frames

Diaphragm
Wall
(masonry or reinforced
concrete)
Frame
Bracing

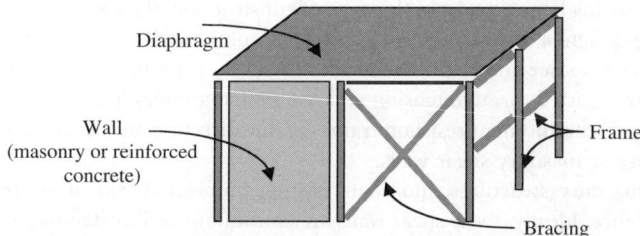

FIGURE 19.2 Types of elements comprising an LFRS.

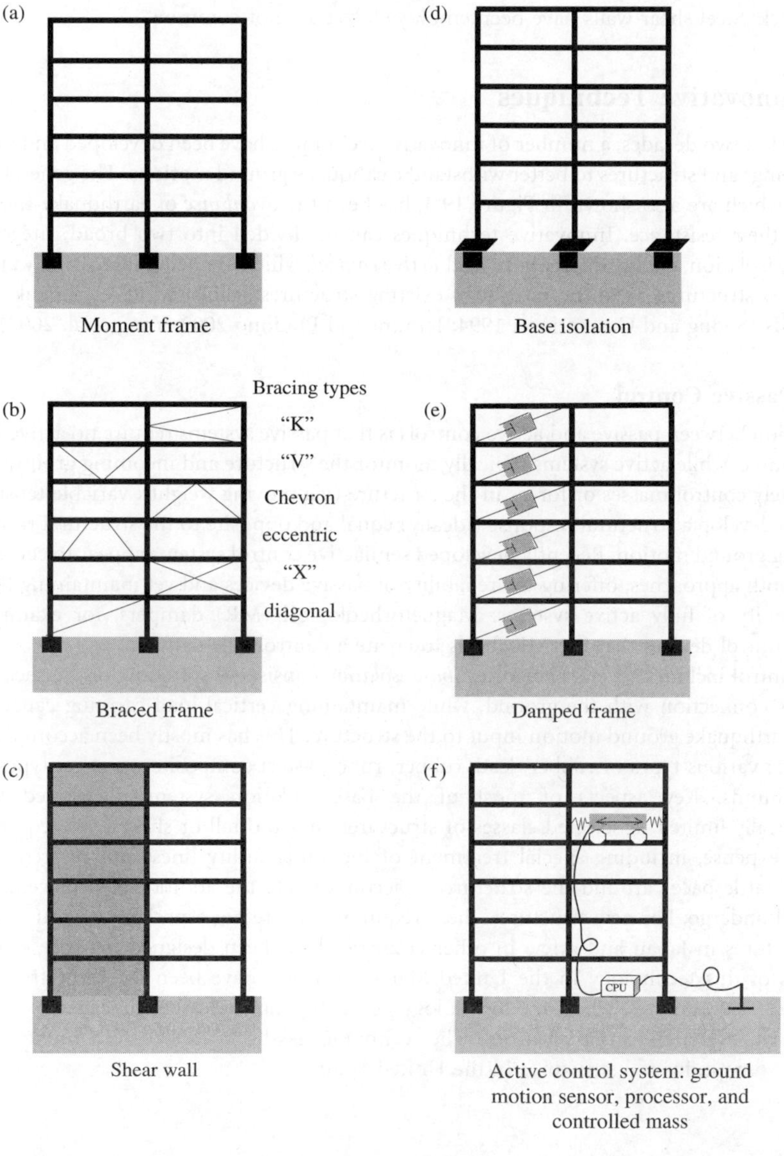

(a)
Moment frame
(d)
Base isolation
(b)
Bracing types
"K"
"V"
Chevron
eccentric
"X"
diagonal
Braced frame
(e)
Damped frame
(c)
Shear wall
(f)
CPU
Active control system: ground
motion sensor, processor, and
controlled mass

FIGURE 19.3 (a)–(c) Traditional earthquake force resisting systems. (d)–(f) Emerging technologies for earthquake force reducing systems.

are also capable of resisting lateral loads. In this type of construction, the connections between the beams and the columns are designed to resist the rotation of the column relative to the beam. Thus, the beam and the column work together and resist lateral movement by bending. This is contrary to the braced frame, where loads are resisted through tension and compression forces in the braces. Steel buildings are sometimes constructed with moment resistant frames in one direction and braced frames in the other, or with integral concrete or masonry shear walls.

In concrete and masonry structures, moment-resisting frames (MRFs) or *shear walls*[1] are used to provide lateral resistance. Ideally, these shear walls are continuous walls extending from the foundation to the roof of the building and can be exterior or interior walls. They are interconnected with the rest of the concrete frame and thus resist the motion of one floor relative to another. Shear walls can be constructed of cast-in-place reinforced concrete, precast concrete, reinforced brick, or reinforced hollow concrete block. Steel shear walls have been employed, but are not common.

19.3.1 Innovative Techniques

In about the last two decades, a number of innovative techniques have been developed and introduced to enable buildings and structures to better withstand earthquake ground motions. The general aim of these techniques, which are also shown in Figure 19.3, has been the *avoidance* of earthquake-induced forces, rather than their resistance. Innovative techniques can be divided into two broad categories, *passive control* (base isolation, energy dissipation) and *active control*, which are being increasingly applied to the design of new structures or to the retrofit of existing structures against wind, earthquakes, and other external loads (Soong and Costantinou 1994; Iemura and Pradono 2002; Yang et al. 2002).

19.3.1.1 Passive Control

The distinction between passive and active controls is that passive systems require no active intervention or energy source, while active systems typically monitor the structure and incoming ground motion and seek to actively control masses or forces in the structure (via moving weights, variable tension tendons, etc.) so as to develop a structural response (ideally) equal and opposite to the structural response due to the incoming ground motion. Recently developed semiactive control systems appear to combine the best features of both approaches, offering the reliability of passive devices and yet maintaining the versatility and adaptability of fully active systems. Magnetorheological (MR) dampers, for example, are new semiactive control devices that use MR fluids to create a controllable damper.

Passive control includes several categories: *base isolation* consists of softening of the shear capacity of a structure's connection with the ground, while maintaining vertical load-carrying capacity, so as to reduce the earthquake ground motion input to the structure. This has mostly been accomplished to date via the use of various types of rubber, lead–rubber, rubber–steel composite, or other types of bearings beneath columns. Key aspects of most of the base isolation systems developed to date are (a) economically limited to selected classes of structures (not too tall or short); (b) require additional foundation expense, including special treatment of incoming utility lines; and (c) require a certain amount of "rattlespace" around the structure to accommodate the additional displacements that the bearings will undergo. For new structures, these requirements are not especially onerous and a number of new structures in Japan and a few in other countries have been designed for base isolation. Most applications of this technology in the United States, however, have been for the retrofit of existing (usually historic) structures, where the technology permitted increased seismic capacity without major modification to the architectural features. The technique has been applied to a number of highway bridges and some industrial structures in the United States.

---

[1]Termed shear walls because the depth to width ratio is so large that deformation is primarily due to shear rather than bending.

19.3.1.1.1 Supplemental Damping

If damping can be significantly increased, then structural responses (and therefore forces and displacements) are greatly reduced (Hanson et al. 1993). Supplemental damping systems include *friction* systems (e.g., Sumitomo, pall, and friction-slip) based on Coulomb friction, *self-centering* friction resistance that is proportional to displacement (e.g., Fluor–Daniel energy dissipating restraint), or various *energy dissipation* mechanisms: ADAS (added damping and stiffness) elements, which utilize the yielding of mild-steel X-plates; viscoelastic shear dampers using a 3M acrylic copolymer as the dissipative element; or nickel–titanium alloy shape-memory devices that take advantage of reversible, stress-induced phase changes in the alloy to dissipate energy (Aiken et al. 1993). These systems are generally still in the developmental stage although there are no special obstacles for the implementation of the ADAS system, which has seen one application to date in the United States (Perry et al. 1993) and more in other countries.

19.3.1.2 Active Control

Active control depends on actively modifying a structure's mass, stiffness, or geometric properties during its dynamic response in such a manner so as to counteract and reduce excessive displacements (Iemura and Pradono 2002). Tuned mass dampers, reliance on liquid sloshing (Lou et al. 1994), and active tensioning of tendons are methods currently under investigation. Most methods of active control are real time, relying on measurement of structural response, rapid structural computation, and fast-acting energy sources. A number of issues of reliability remain to be resolved (Spencer et al. 1994).

19.4 Types of Buildings and Typical Earthquake Performance

There are many different types of buildings, with varying kinds of earthquake performance and seismic design needs. This section discusses general earthquake performance of buildings, with the emphasis more toward those buildings typically built in the western United States. Specific aspects of structural analysis and design of buildings, other structures, steel, concrete, wood, masonry, and other topics are discussed in other chapters.

In buildings, earthquake performance can be divided into two categories: structural and non-structural, both of which when unsatisfactory can be hazardous to building occupants — when *damage* occurs. Structural damage means degradation of the building's structural support systems (i.e., vertical and lateral force resisting systems), such as the building frames and walls. Nonstructural damage refers to any damage that does not affect the integrity of the structural support system. Examples of nonstructural damage are a chimney collapsing, windows breaking, ceilings falling, piping damage, and disruption of pumps, control panels, telecommunications equipment, etc. Nonstructural damage can still be life threatening and costly. The type of damage to be expected is a complex issue that depends on the structural type and age of the building, its configuration, construction materials, the site conditions, the proximity of the building to neighboring buildings, and the type of nonstructural elements.

The typical earthquake performances of different types of common building structural systems are described in this section to provide insights into seismic design for buildings.

19.4.1 Wood Frame

Wood-frame structures tend to be mostly low rise (one to three stories, occasionally four stories). Vertical framing may be of several types, for example, stud wall, braced post and beam, or timber pole:

- Stud walls are typically constructed of 2 in. by 4 in. wood members vertically set about 16 in. apart — multiple story buildings may have 2×6 or larger studs. These walls are braced by plywood sheathing or by diagonals made of wood or steel. Most detached single and low-rise multiple family residences in the United States are of stud wall wood frame construction, Figure 19.4.

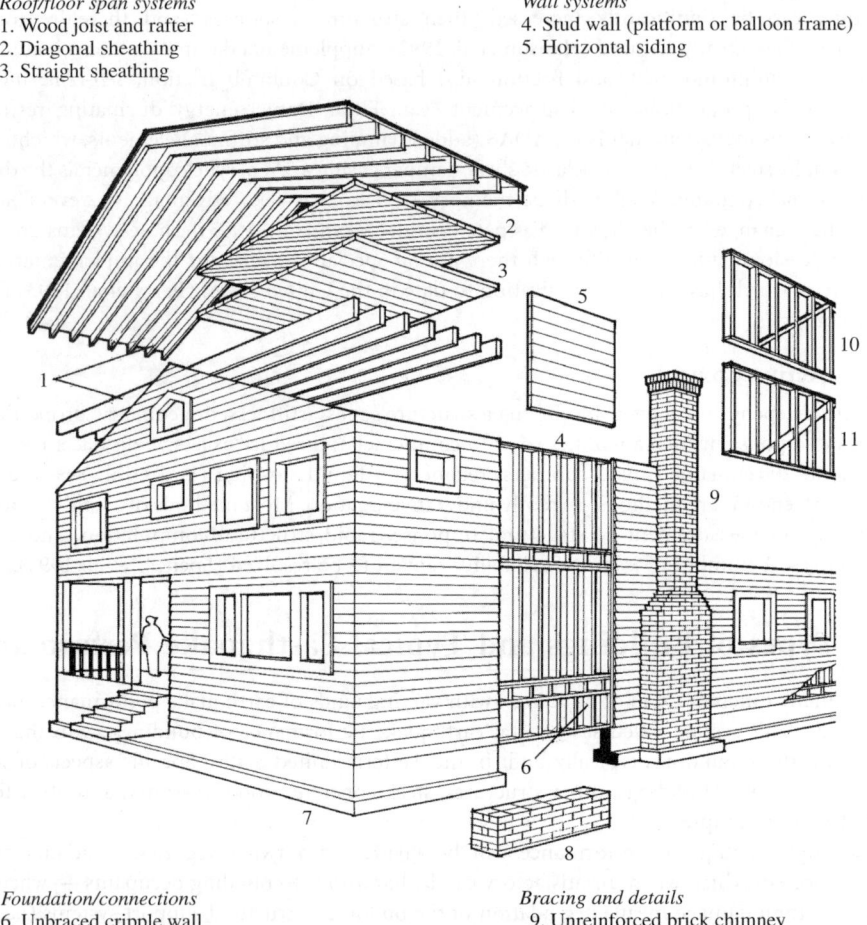

Roof/floor span systems
1. Wood joist and rafter
2. Diagonal sheathing
3. Straight sheathing

Wall systems
4. Stud wall (platform or balloon frame)
5. Horizontal siding

Foundation/connections
6. Unbraced cripple wall
7. Concrete foundation
8. Brick foundation

Bracing and details
9. Unreinforced brick chimney
10. Diagonal blocking
11. Let-in brace (only in vintage)

FIGURE 19.4 Schematic of wood light-frame construction. (From Federal Emergency Management Agency. 1988. *Rapid Visual Screening of Buildings for Potential Seismic Hazards: A Handbook*, FEMA 154, FEMA, Washington, DC.)

- Post and beam construction is not very common in the United States, although it is the basis of the traditional housing in other countries (e.g., Europe, Japan); in the United States, it is found mostly in older housing and larger buildings (i.e., warehouses, mills, churches, and theaters). This type of construction consists of larger rectangular (6 in. by 6 in. and larger) or sometimes round wood columns framed together with large wood beams or trusses.
- Timber pole buildings are a less common form of construction found mostly in suburban/rural areas. Generally adequate seismically when first built, they are more often subject to wood deterioration due to the exposure of the columns, particularly near the ground surface. Together with an often-found "soft story" in this building type, this deterioration may contribute to unsatisfactory seismic performance.

In wood frame stud-wall buildings, the resistance to lateral loads is typically provided by (a) for older buildings, especially houses, wood diagonal "let-in" bracing and (b) for newer (primarily post-World War II) buildings, plywood siding "shear walls." Without the extra strength provided by the bracing or plywood, walls would distort excessively or "rack," resulting in broken windows, stuck doors, cracked plaster, and, in extreme cases, collapse.

Stud-wall buildings have performed very well in past U.S. earthquakes for ground motions of about 0.5*g* or less, due to inherent qualities of the structural system and because they are lightweight and low rise. Cracking in plaster and stucco may occur and these act to degrade the strength of the building to some extent (i.e., the plaster and stucco may in fact form part of the LFRS, sometimes by design) — this is usually classified as nonstructural damage but, in fact, dissipates a lot of the earthquake-induced energy. However, the most common type of structural damage in older wood-frame buildings results from a lack of connection between the superstructure and the foundation — the so-called "cripple wall" construction. This kind of construction is common in the milder climes of the west, where full basements are not required, and consists of an air space (typically 2–3 ft) left under the house — the short stud walls under the first floor (termed by carpenters a cripple wall because of their less than full height) were usually built without bracing so that their is no adequate LFRS for this short height. Plywood sheathing nailed to the cripple studs may have been used to strengthen the cripple walls. Additionally, the mud sill in these older (typically pre-World War II) housing may not be bolted to the foundation. As a result, houses can slide off their foundations when not properly bolted to the foundation, resulting in major damage to the building as well as to plumbing and electrical connections. Overturning of the entire structure is usually not a problem because of the low-rise geometry. In many municipalities, modern codes require wood structures to be bolted to their foundations. However, the year that this practice was adopted will differ from community to community and should be checked.

Garages often have a very large door opening in one wall with little or no bracing. This wall has almost no resistance to lateral forces, which is a problem if a heavy load such as a second story sits on top of the garage (the so-called house over garage, or HOGs). Homes built over garages have sustained significant amounts of damage in past earthquakes, with many collapses. Therefore, the HOG configuration, which is found commonly in low-rise apartment complexes and some newer suburban detached dwellings, should be examined more carefully and perhaps strengthened.

Unreinforced masonry (URM) chimneys also present a life-safety problem. They are often inadequately tied to the building and therefore fall when strongly shaken. On the other hand, chimneys of reinforced masonry generally perform well.

Some wood-frame structures, especially older buildings in the eastern United States, have masonry veneers that may represent another hazard. The veneer usually consists of one wythe of brick (a wythe is a term denoting the width of one brick) attached to the stud wall. In older buildings, the veneer is either insufficiently attached or has poor quality mortar, which often results in peeling off of the veneer during moderate and large earthquakes.

Post and beam buildings tend to perform well in earthquakes if adequately braced. However, walls often do not have sufficient bracing to resist horizontal motion and thus they may deform excessively.

The 1994 M_W 6.7 Northridge earthquake was the largest earthquake to occur directly within an urbanized area since the 1971 San Fernando earthquake — ground motions were as high as 0.9*g* and substantial numbers of modern wood-frame dwellings sustained significant damage, including major cracking of veneers, gypsum board walls, and splitting of wood wall studs. It may be inferred from this, as well as the performance observed in the more sparsely populated epicentral regions of the 1989 M_W 7.1 Loma Prieta Earthquake, that U.S. single family dwelling design begins to sustain substantial nonstructural and structural damage for peak ground acceleration in excess of about 0.5*g*.

19.4.2 Steel-Frame Buildings

Steel-frame buildings generally may be classified as MRFs, braced frames, or mixed construction (e.g., steel frame for vertical forces and reinforced concrete shear wall for the LFRS) based on their LFRSs. In concentric braced frames the lateral forces or loads are resisted by the tensile and compressive strength of the bracing, which can assume a number of different configurations including diagonal, "V," inverted "V" also termed chevron, "K," etc. A recent development in seismic bracing is the eccentric brace frame. Here, the bracing is slightly offset from the main beam to column connection, and the short section of the beam is expected to deform significantly under major seismic forces and thereby dissipate a considerable portion of the energy. MRFs resist lateral loads and deformations by the bending stiffness of the beams and columns (there is no bracing), Figure 19.5.

Steel-frame buildings have tended to perform satisfactory in earthquakes with ground motions less than about $0.5g$ because of their strength, flexibility, and lightness. Collapse in earthquakes has been very rare, although steel-frame buildings did collapse, for example, in the 1985 Mexico City Earthquake. More recently, following the 1994 M_W 6.7 Northridge Earthquake, a number of MRFs were found to have sustained serious cracking in the beam column connection; see Figure 19.6, which shows one of a number of different types of cracking that were found following the Northridge Earthquake. The cracking typically initiated at the lower beam flange location and propagated upward into the shear panel. Similar cracking was also observed following the 1995 M_W 6.9 Hanshin (Kobe) Earthquake, which experienced similar levels of ground motion as Northridge. More worrisome is that, as of this writing, some steel buildings in the San Francisco Bay Area have been found to have similar cracking, presumably as a result of the 1989 M_W 7.1 Loma Prieta Earthquake. As a result, there is an ongoing effort by a consortium of research organizations, termed SAC (funded by the Federal Emergency Management Agency) to better understand and develop solutions for this problem.

Light-gage steel buildings are used for agricultural structures, industrial factories, and warehouses. They are typically one story in height, sometimes without interior columns, and often enclose a large floor area, Figure 19.7. Construction is typically of steel frames spanning the short dimension of the building and resisting lateral forces as moment frames. Forces in the long direction are usually resisted by diagonal steel rod bracing. These buildings are usually clad with lightweight metal or asbestos reinforced concrete siding, often corrugated. Because these buildings are low rise, lightweight, and constructed of steel members, they usually perform relatively well in earthquakes. Collapses do not usually occur. Some typical problems are (a) insufficient capacity of tension braces can lead to their elongation and, in turn, building damage and (b) inadequate connection to the foundation can allow the building columns to slide.

19.4.3 Concrete Buildings

Several construction subtypes fall under this category: (a) MRFs (nonductile or ductile); (b) shear wall structures; and (c) precast, including tilt-up structures. The most prevalent of these is nonductile reinforced concrete frame structures with or without infill walls built in the United States between about 1920 and (in the western United States) 1972. In many other portions of the United States this type of construction continues to the present. This group includes large multistory commercial, institutional, and residential buildings constructed using flat slab frames, waffle slab frames, and the standard girder-column-type frames. These structures generally are more massive than steel frame buildings, are underreinforced (i.e., have insufficient reinforcing steel embedded in the concrete), and display low ductility. Some typical problems are (a) large tie spacings in columns can lead to a lack of concrete confinement and/or shear failure; (b) placement of inadequate rebar splices at the same location can lead to column failure; (c) insufficient shear strength in columns can lead to shear failure prior to the development of moment hinge capacity; (d) insufficient shear tie anchorage can prevent the column from developing its full shear capacity; (e) lack of continuous beam reinforcement can result in hinge formation during load reversal; (f) inadequate reinforcing of beam–column joints or location of beam bar splices at columns can lead to failures; and (g) the relatively low stiffness of the frame can lead to substantial nonstructural damage.

Ductile reinforced concrete frames where special reinforcing details are required in order to furnish satisfactory load-carrying performance under large deflections (termed ductility) have usually only been required in the highly seismic portions of the United States since the mid-1970s. ACI-318 (1995)

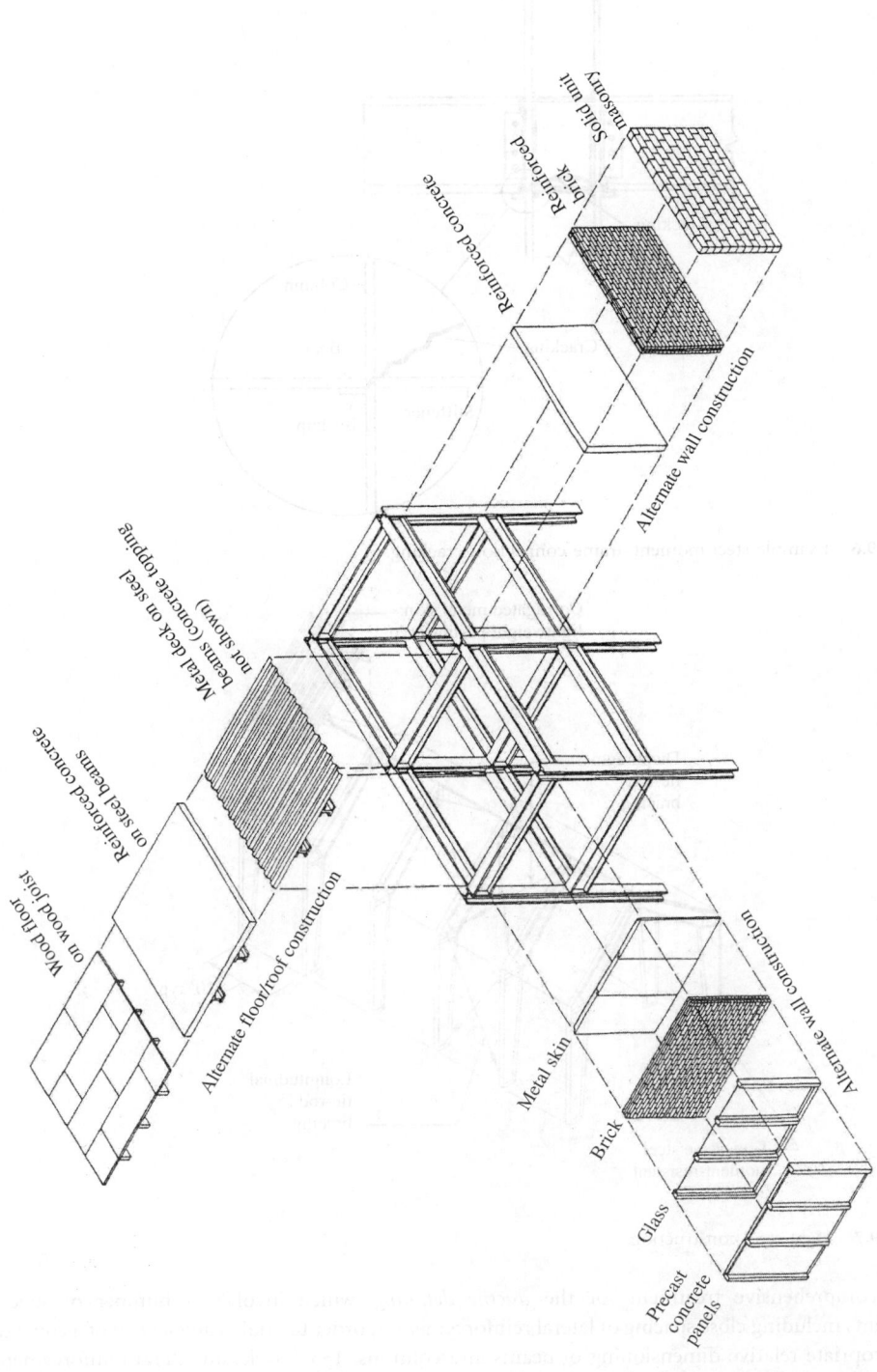

FIGURE 19.5 Steel moment resisting frame construction. (From Federal Emergency Management Agency. 1988. *Rapid Visual Screening of Buildings for Potential Seismic Hazards: A Handbook*, FEMA 154, FEMA, Washington, DC.)

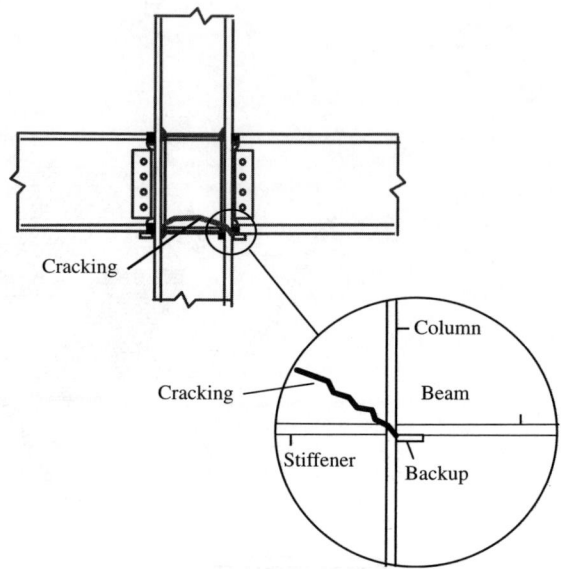

FIGURE 19.6 Example steel moment–frame connection cracking.

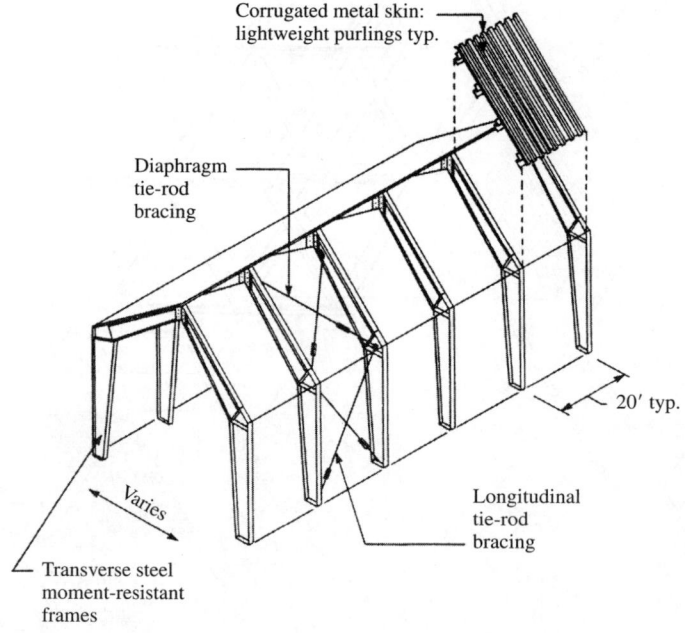

FIGURE 19.7 Light steel construction.

provides comprehensive treatment for the *ductile detailing*, which involves a number of special requirements including close spacing of lateral reinforcement in order to attain *confinement* of a concrete core, appropriate relative dimensioning of beams and columns, 135° hooks on lateral reinforcement, hooks on main beam reinforcement within the column, etc.

Concrete shear wall buildings consist of a concrete box or frame structural system with walls constituting the main LFRS, Figure 19.8. The entire structure, along with the usual concrete diaphragm, is typically cast in place. Shear walls in buildings can be located along the perimeter, as interior walls, or around the service or elevator core. This building type generally tends to perform better than concrete frame buildings. They are heavier than steel-frame buildings but they are also rigid due to the shear walls. Some types of damage commonly observed in taller buildings are caused by vertical discontinuities, pounding, and/or irregular configuration. Other damages specific to this building type are (a) shear cracking and distress can occur around openings in concrete shear walls during large seismic events; (b) shear failure can occur at wall construction joints usually at a load level below the expected capacity; and (c) bending failures can result from insufficient chord steel lap lengths.

Tilt-up buildings are a common type of construction in the western United States and consist of concrete wall panels cast on the ground and then tilted upward into their final positions. More recently, wall panels are fabricated off-site and trucked in. The wall panels are welded together at embedments or held in place by cast-in-place columns or steel columns, depending on the region. The floor and roof beams are often glue-laminated wood or steel open webbed joists that are attached to the tilt-up wall panels; these panels may be load bearing or nonload bearing, depending on the region. These buildings tend to be low-rise industrial or office buildings. Before 1973 in the western United States, many tilt-up buildings did not have sufficiently strong connections or anchors between the walls and the roof and floor diaphragms. During an earthquake, weak anchors pull out of the walls, causing the floors or roofs

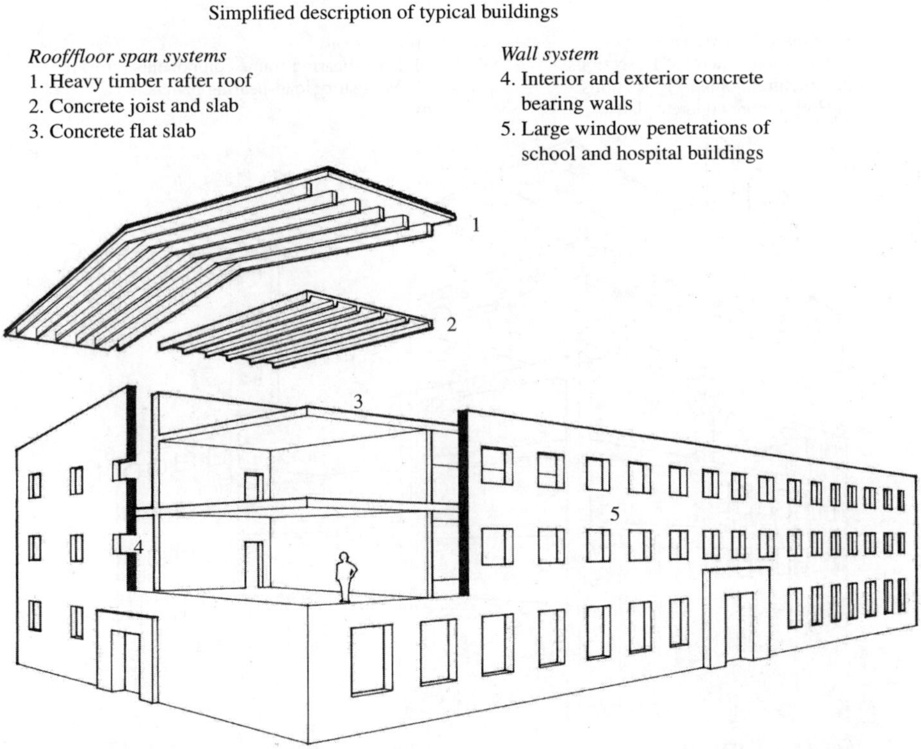

Simplified description of typical buildings

Roof/floor span systems
1. Heavy timber rafter roof
2. Concrete joist and slab
3. Concrete flat slab

Wall system
4. Interior and exterior concrete bearing walls
5. Large window penetrations of school and hospital buildings

FIGURE 19.8 Reinforced concrete shear wall construction.

to collapse. The connections between concrete panels are also vulnerable to failure. Without these, the building loses much of its lateral force resisting capacity. For these reasons, many tilt-up buildings were damaged in the 1971 San Fernando Earthquake. Since 1973, tilt-up construction practices have changed in California and other high-seismicity regions, requiring positive wall–diaphragm connection and prohibiting cross-grain bending in wall ledgers. (Such requirements may not have yet been made in other regions of the country.) However, a large number of these older, pre-1970s vintage tilt-up buildings still exist and have not been retrofitted to correct this wall-anchor defect. These buildings are a prime source of seismic hazards. In areas of low or moderate seismicity, inadequate wall anchor details continue to be employed. Damage to tilt-up buildings was observed again in the 1994 M_W 6.7 Northridge earthquake, where the primary problems were poor wall anchorage into the concrete and excessive forces due to flexible roof diaphragms amplifying ground motion to a greater extent than anticipated in the code.

Precast concrete frame construction, first developed in the 1930s, was not widely used until the 1960s. The precast frame is essentially a post and beam system in concrete where columns, beams, and slabs are prefabricated and assembled on site, Figure 19.9. Various types of members are used: vertical load-carrying elements may be Ts, cross-shapes, or arches and are often more than one story in height. Beams are often Ts and double Ts or rectangular sections. Prestressing of the members, including pretensioning and posttensioning, is often employed. The LFRS is often concrete cast-in-place shear walls. The earthquake performance of this structural type varies greatly and is sometimes poor. This type of building can perform well if the details used to connect the structural elements have sufficient strength and ductility (toughness). Because structures of this type often employ cast-in-place concrete shear walls for lateral load resistance, they experience the same types of damage as other shear wall building types. Some of the problem areas specific to precast frames are (a) poorly designed connections between

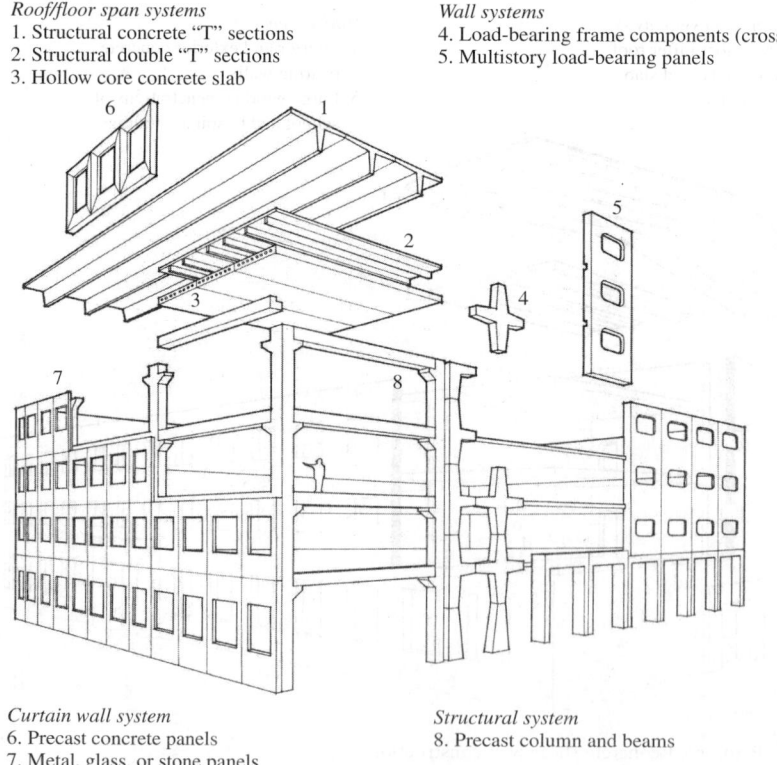

Roof/floor span systems
1. Structural concrete "T" sections
2. Structural double "T" sections
3. Hollow core concrete slab

Wall systems
4. Load-bearing frame components (cross)
5. Multistory load-bearing panels

Curtain wall system
6. Precast concrete panels
7. Metal, glass, or stone panels

Structural system
8. Precast column and beams

FIGURE 19.9   Precast concrete construction.

prefabricated elements can fail; (b) accumulated stresses can result due to shrinkage and creep and due to stresses incurred in transportation; (c) loss of vertical support can occur due to inadequate bearing area and/or insufficient connection between floor elements and columns; and (d) corrosion of metal connectors between prefabricated elements can occur. A number of precast parking garages failed in the 1994 M_W 6.7 Northridge Earthquake, including a large structure at the Cal State Northridge campus that sustained a progressive failure. This structure had a perimeter precast MRF and interior non-ductile columns — the MRF sustained large but tolerable deflections; however, interior nonductile columns failed under these deflections, resulting in an interior collapse, which then pulled the exterior MRFs over.

19.4.4 Masonry Buildings

Reinforced masonry buildings are mostly low-rise perimeter bearing wall structures, often with wood diaphragms although precast concrete is sometimes used. Floor and roof assemblies usually consist of timber joists and beams, glue-laminated beams, or light steel joists. The bearing walls consist of grouted and reinforced hollow or solid masonry units. Interior supports, if any, are often wood or steel columns, wood stud frames, or masonry walls. Generally, they are less than five stories in height although many mid-rise masonry buildings exist. Reinforced masonry buildings can perform well in moderate earthquakes if they are adequately reinforced and grouted and if sufficient diaphragm anchorage exists.

Most URM bearing wall structures in the western United States were built before 1934, although this construction type was permitted in some jurisdictions having moderate or high seismicity until the late 1940s or early 1950s (in low-seismicity jurisdictions URM may still be a common type of construction, even today). These buildings usually range from one to six stories in height and typically construction varies according to the type of use, although wood floor and roof diaphragms are common. Smaller commercial and residential buildings usually have light wood floor/roof joists supported on the typical perimeter URM wall and interior wood load-bearing partitions. Larger buildings, such as industrial warehouses, have heavier floors and interior columns, usually of wood. The bearing walls of these industrial buildings tend to be thick, often as much as 24 in. or more at the base. Wall thicknesses of residential buildings range from 9 in. at upper floors to 18 in. at lower floors. URM structures are recognized as perhaps the most hazardous structural type. They have been observed to fail in many modes during past earthquakes. Typical problems are

1. *Insufficient anchorage.* Because the walls, parapets, and cornices are not positively anchored to the floors, they tend to fall out. The collapse of bearing walls can lead to major building collapses. Some of these buildings have anchors as a part of the original construction or as a retrofit. These older anchors exhibit questionable performance.
2. *Excessive diaphragm deflection.* Because most of the floor diaphragms are constructed of wood sheathing, they are very flexible and permit large out-of-plane deflection at the wall transverse to the direction of the force. The large drift, occurring at the roof line, can cause the masonry wall to collapse under its own weight.
3. *Low shear resistance.* The mortar used in these older buildings is often made of lime and sand, with little or no cement, and has very little shear strength. The bearing walls will be heavily damaged and collapse under large loads.
4. *Wall slenderness.* Some of these buildings have tall story heights and thin walls. This condition, especially in nonload-bearing walls, will result in buckling out-of-plane under severe lateral load. Failure of a nonload-bearing wall represents a falling hazard, whereas the collapse of a load-bearing wall will lead to partial or total collapse of the structure.

19.4.5 Configuration, Irregularities, and Pounding

Certain problems in earthquake performance are common to many building types and include issues of configuration, irregularities, and pounding.

19.4.5.1 Configuration and Irregularities

Configuration, or the general vertical and/or horizontal shape of buildings, is an important factor in earthquake performance and damage. Buildings that have simple, regular, symmetric configurations generally display the best performance in earthquakes. The reasons for this are (a) nonsymmetric buildings tend to have twist (i.e., have significant torsional modes) in addition to shaking laterally and (b) the various "wings" of a building tend to act independently, resulting in differential movements, cracking, and other damage. Rotational motion introduces additional damage, especially at re-entrant or "internal" corners of the building. The term "configuration" also refers to the geometry of lateral load resisting systems as well as the geometry of the building. Asymmetry can exist in the placement of bracing systems, shear walls, or MRFs that are used to provide earthquake resistance in a building. This type of asymmetry, of the LFRS, can result in twisting or differential motion, with the same consequences as asymmetry in the building plan. An important aspect of configuration is *soft story*, which is a story of a building signifiantly less stiff than adjacent stories (i.e., a story in which the lateral stiffness is 70% or less than that in the story above or less than 80% of the average stiffness of the three stories above; BSSC 2001). Soft stories often (but not always) occur on the ground floor, where commercial or other reasons require a greater story height, and large windows or openings for ingress or commercial display (e.g., the building might have masonry curtain walls for the full height, except at the ground floor, where these are replaced with large windows for a store's display). Due to inadequate stiffness, a disproportionate amount of the entire building's drift is concentrated at the soft story, resulting in nonstructural and potential structural damage. Many older buildings with soft stories but built prior to recognition of this aspect collapse due to excessive ductility demands at the soft story.

The National Earthquake Hazard Reduction Program (NEHRP) provisions for the design of new buildings (BSSC 2001) have defined when a building's configuration is "irregular," and provided required strength increase factors or other approaches to deal with these irregularities. The NEHRP definitions for plan and vertical irregularities are illustrated in Table 19.1 and Table 19.2.

19.4.5.2 Pounding

Pounding is the collision of adjacent buildings during an earthquake due to insufficient lateral clearance. Such collision can induce very high and unforeseen accelerations and story shears in the overall structure. Additionally, if adjacent buildings have varying story heights, a relatively rigid floor or roof diaphragm may impact adjacent buildings at or near mid-column height, causing bending or shear failure in the columns and subsequent story collapse. Under earthquake lateral loading, buildings deflect significantly — these deflections or drift are limited by code — and adjacent buildings must be separated by a seismic gap equal to the sum of their actual calculated drifts (i.e., ideally, each building set back from its property line by the drift). Pounding has been the cause of a number of mid-rise building collapses, most notably in the 1985 Mexico City Earthquake.

19.5 2000 NEHRP Recommended Provisions

19.5.1 Overview

The *2000 NEHRP Recommended Provisions for Seismic Regulation for Buildings and Other Structures* (NEHRP Provisions) represents the current state of the art in prescriptive, as opposed to performance-based, provisions for seismic-resistant design. Its provisions form the basis for earthquake design specifications contained in the 2001 edition of *ASCE-7, Minimum Design Loads for Buildings and other Structures*, either through reference or direct incorporation, the seismic regulations in the 2003 edition of the *International Building Code*, and also the 2002 edition of the *NFPA 5000 Building Code* (NFPA, n.d.). As such, it will form the basis for most earthquake-resistant design in the United States, as well as other nations that base their codes on U.S. practices, throughout much of the first decade of the twenty-first century.

TABLE 19.1   NEHRP 2000 Plan Structural Irregularities

1: Torsional irregularity — when diaphragms are not flexible

Torisonal irregularity exists (1a) when the maximum story drift, computed including accidental torsion, at one end of the structure transverse to an axis is more than 1.2 times the average of the story drift at the two ends of the structure. (1b) Extreme torisonal irregularity exists when ratio >1.4

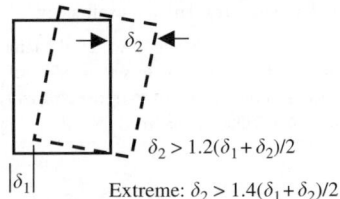

$\delta_2 > 1.2(\delta_1 + \delta_2)/2$

Extreme: $\delta_2 > 1.4(\delta_1 + \delta_2)/2$

2: Re-entrant corners

Plan configurations of a structure and its lateral-force-resisting system contain re-entrant corners where both projections of the structure beyond a re-entrant corner are greater than 15% of the plan dimension of the structure in the given direction

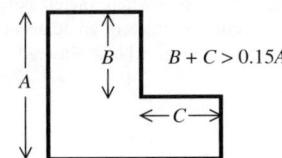

$B + C > 0.15A$

3: Diaphragam discontinuity

Diaphragms with abrupt discontinuities or variations in stiffness including those having cutout or open areas greater than 50% of the gross enclosed diaphragm area or changes in effective diaphragm stiffness of more than 50% from one story to the next

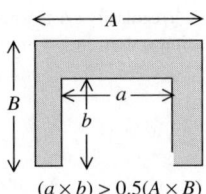

$(a \times b) > 0.5(A \times B)$

4: Out-of-plane offsets

Discontinuities in a lateral-force-resistance path such as out-of-plane offsets of the vertical elements

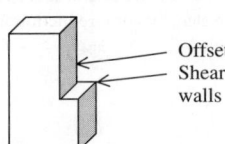

Offset
Shear
walls

5: Nonparallel systems

The vertical lateral-force-resisting elements are not parallel to or symmetric about the major orthogonal axes of the lateral-force-resisting system

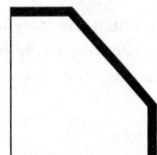

The *NEHRP Provisions* assume significant amounts of nonlinear behavior will occur under design level events. The extent of nonlinear behavior that may occur is dependent on the structural systems employed in resisting earthquake forces, the configuration of these systems, and the extent that the structural systems are detailed for ductile behavior under large cyclic inelastic deformation. The *NEHRP Provisions* may therefore be thought to consist of two component parts:

- One part relates to specification of the required design strength and stiffness of the structural system.
- The second part relates to issues of structural detailing.

TABLE 19.2   NEHRP 2000 Vertical Structural Irregularities

1: Stiffness irregularity — soft story

(1a) A soft *story* is one in which the lateral stiffness is less than 70% of that in the *story* above or less than 80% of the average stiffness of the three stories above. (1b) An extreme soft *story* is for ratios of less than 60% or less than 70%, respectively

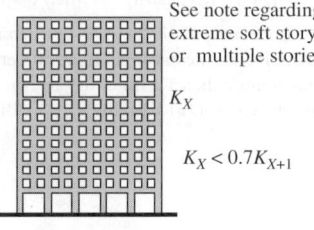

See note regarding extreme soft story, or multiple stories

K_X

$K_X < 0.7K_{X+1}$

2: Weight (mass) irregularity

Mass irregularity exists where the effective mass of any *story* is more than 150% of the effective mass of an adjacent *story*. A roof that is lighter than the floor below need not be considered

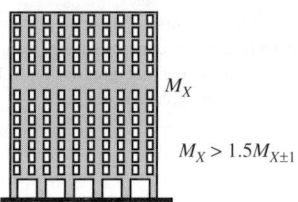

M_X

$M_X > 1.5M_{X\pm1}$

3: Vertical geometric irregularity

Vertical geometric irregularity exists where the horizontal dimension of the lateral-force-resisting system in any *story* is more than 130% of that in an adjacent *story*

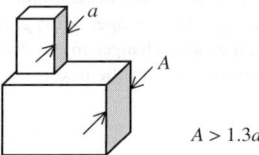

$A > 1.3a$

4: In-plane discontinuity in vertical lateral-force-resisting elements

An in-plane offset of the lateral-force-resisting elements greater than the length of those elements or a reduction in stiffness of the resisting elements in the story below

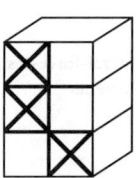

5: Discontinuity in capacity — weak story

A weak *story* is one in which the *story* lateral *strength* is less than 80% of that in the *story* above. The *story strength* is the total *strength* of all seismic-resisting elements sharing the *story* shear for the direction under consideration

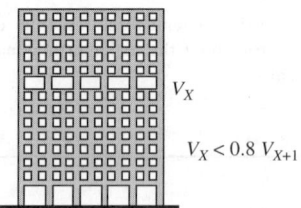

V_X

$V_X < 0.8 V_{X+1}$

For this second part, the *NEHRP Provisions* adopt, with modification, design standards and specifications developed by industry groups such as the American Concrete Institute or the American Institute of Steel Construction. This second part of the *NEHRP Provisions* is not discussed in this chapter, but is covered in detail in each of the following chapters, which treat the individual structural materials.

Instead, this article focuses primarily on the manner in which the *NEHRP Provisions* regulate the required strength and stiffness of structures.

19.5.2 Performance Intent and Objectives

The *NEHRP Provisions* are intended to provide a tiered series of performance capabilities for structures, depending on their intended occupancy and use. Under the *NEHRP Provisions*, each structure must be assigned to a seismic use group (SUG). Three SUGs are defined and are respectively labeled I, II, and III:

- SUG-I encompasses most ordinary occupancy buildings including typical commercial, residential, and industrial structures. For these facilities the basic intent of the *NEHRP Provisions*, just as with earlier codes, is to provide a low probability of earthquake-induced life safety endangerment.
- SUG-II includes facilities that house large numbers of persons, persons who are mobility impaired, or large quantities of materials that if released could pose substantial hazards to the surrounding community. Examples of such facilities include large assembly facilities, housing several thousand persons, daycare centers, and manufacturing facilities containing large quantities of toxic or explosive materials. The performance intent for these facilities is to provide a lower probability of life endangerment, relative to SUG-I structures, and a low probability of damage that would result in release of stored materials.
- SUG-III includes those facilities such as hospitals and emergency operations and communications centers deemed essential to disaster response and recovery operations. The basic performance intent of the *NEHRP Provisions* with regard to these structures is to provide a low probability of earthquake-induced loss of functionality and operability.

In reality, the probability of damage resulting in life endangerment, release of hazardous materials, or loss of function should be calculated using structural reliability methods as the total probability of such damage over a period of time (Ravindra 1994). Mathematically, this is equal to the integral, over all possible levels of ground motion intensity, of the conditional probability of excessive damage given that a ground motion intensity is experienced and the probability that such ground motion intensity will be experienced in the desired period of time. Although such an approach would be mathematically and conceptually correct, it is currently regarded as too complex for practical application in the design office.

Instead, the *NEHRP Provisions* design for desired limiting levels of nonlinear behavior for a single design earthquake intensity level, termed *maximum considered earthquake* (MCE) ground shaking. In most regions of the United States, the MCE is defined as that intensity of ground shaking having a 2% probability of exceedance in 50 years. In certain regions, proximate to major active faults, this probabilistic definition of MCE motion is limited by a conservative deterministic estimate of the ground motion intensity anticipated to result from an earthquake of characteristic magnitude on these faults. The MCE is thought to represent the most severe level of shaking ever likely to be experienced by a structure, though it is recognized that there is some limited possibility of more severe motion occurring. Structures categorized as SUG-I are designed with the expectation that MCE shaking would result in severe damage to both structural and nonstructural elements, with damage perhaps being so severe that following the earthquake the structure would be on the verge of collapse. This damage state has come to be termed *collapse prevention*, because the structure is thought to be at a state of incipient but not actual collapse. Theoretically, SUG-I structures behaving in this manner would be total or near total financial losses, in the event that MCE shaking was experienced. To the extent that shaking experienced by the structure exceeds the MCE level, the structure could actually experience partial or total collapse.

SUG-III structures are designed with the intent that when subjected to MCE shaking they would experience both structural and nonstructural damages; however, the structures would retain significant residual structural resistance or margin against collapse. It is anticipated that when experiencing MCE shaking such structures may be damaged to an extent that they would no longer be suitable for occupancy, until repair work had been instituted, but that repair would be technically and economically feasible. This superior performance relative to SUG-I structures is accomplished through specification that SUG-III structures be designed with 50% greater strength and more stiffness than their SUG-I counterparts. SUG-II structures are designed for performance intermediate to that for SUG-I and SUG-III with strengths and stiffness that are 25% greater than those required for SUG-I structures.

19.5.3 Seismic Hazard Maps and Ground Motion Parameters

The *NEHRP Provisions* incorporate a series of national seismic hazard maps for the United States and territories, developed by the United States Geologic Survey (USGS), specifically for this purpose (available at http://geohazards.cr.usfs.gov/eq/index.html). Two sets of maps are presented. One set presents contours of MCE, 5% damped, elastic spectral response acceleration at a period of 0.2 s, termed S_S. The second set presents contours of MCE, 5% damped, elastic spectral response acceleration at a period of 1.0 s, termed S_1. In both cases, the spectral response acceleration values are representative of sites with subsurface conditions bordering between firm soil or soft rock. Contours are presented in increments of 0.02g in areas of low seismicity and 0.05g in areas of high seismicity. By locating a site on the maps and interpolating between the values presented for contours adjacent to the site, it is possible to rapidly estimate the MCE level shaking parameters for the site, given that it has a soft rock or firm soil profile Figure 19.10 shows, for a portion of the western United States, contours of the 0.2 s spectral acceleration with a 90% probability of not being exceeded in 50 years. As indicated in the figure, in zones of high seismicity these contours are quite closely spaced, making use of the maps difficult. Therefore, the USGS has furnished software, available both over the internet (at the URL indicated above) and on a CD-ROM, which permits determination of the MCE spectral response acceleration parameters based on longitude and latitude.

Since many sites are located neither on soft rock nor on firm soil sites, it is necessary to correct the mapped values of spectral response acceleration to account for site amplification and de-amplification effects. To facilitate this process, a site is categorized into one of six site class groups, labeled A through F. Table 19.3 summarizes the various site class categories.

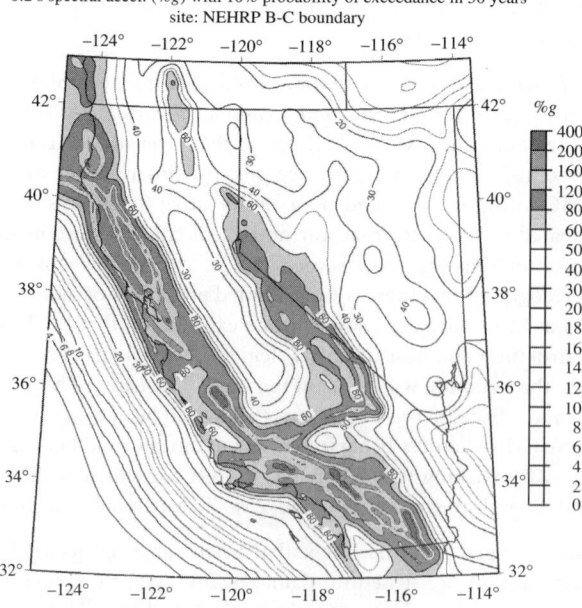

FIGURE 19.10   MCE Seismic Hazard Map (0.2 s spectral response acceleration) for western United States.

TABLE 19.3   Site Categories

Site class	Description	Shear wave velocity, $\bar{v}_s$	Penetration resistance, $\bar{N}$	Unconfined shear strength, $\bar{s}_u$
A	Hard rock	>5000 ft/s		
B	Rock	2500 ft/s < $\bar{v}_s$ ≤ 5000 ft/s		
C	Very firm soil or soft rock	1200 ft/s < $\bar{v}_s$ ≤ 2500 ft/s	>50	>2000 psf
D	Stiff soil	600 ft/s < $\bar{v}_s$ ≤ 1200 ft/s	15–50	1000–2000 psf
E	Soil	$\bar{v}_s$ < 600 ft/s	<15	<1000 psf
F	Special soils	Soils requiring site-specific evaluations: 1. Soils vulnerable to potential failure or collapse under seismic loading such as liquefiable soils, quick and highly sensitive clays, collapsible weakly cemented soils 2. Peats and/or highly organic clays (H > 10 ft of peat and/or highly organic clay where H = thickness of soil) 3. Very high plasticity clays (H > 25 ft [8 m] with PI > 75) 4. Very thick soft/medium stiff clays (H > 120 ft [36 m])		

Note: $\bar{v}_s$, $\bar{N}$, $\bar{s}_u$ represent the average value of the parameter over the top 30 m (100 ft) of soil.

TABLE 19.4   Coefficient F_a as a Function of Site Class and Mapped Spectral Response Acceleration

Site class	Mapped maximum considered earthquake spectral response acceleration at short periods				
	$S_S = 0.25$	$S_S = 0.50$	$S_S = 0.75$	$S_S = 1.00$	$S_S = 1.25$
A	0.8	0.8	0.8	0.8	0.8
B	1.0	1.0	1.0	1.0	1.0
C	1.2	1.2	1.1	1.0	1.0
D	1.6	1.4	1.2	1.1	1.0
E	2.5	1.7	1.2	0.9	a
F	a	a	a	a	a

Note: "a" indicates site-specific evaluation required.

Once a site has been categorized within a site class, a series of coefficients are provided that are used to adjust the mapped values of spectral response acceleration for site response effects. These coefficients were developed based on observed site response characteristics in ground motion recordings from past earthquakes. Two coefficients are provided:

- The F_a coefficient is used to account for site response effects on short period ground shaking intensity.
- The F_v coefficient is used to account for site response effects of longer period motions.

Table 19.4 and Table 19.5 indicate the values of these coefficients as a function of site class and mapped MCE ground shaking acceleration values. Site-adjusted values of the MCE spectral response acceleration parameters at 0.2 and 1 s, respectively, are found from the following equations:

$$S_{MS} = F_a S_s \qquad (19.1)$$

$$S_{M1} = F_v S_1 \qquad (19.2)$$

The two site-adjusted spectral response acceleration parameters, S_{MS} and S_{M1}, permit a 5% damped, maximum considered earthquake ground shaking response spectrum to be constructed for the building site. This spectrum is constructed as indicated in Figure 19.11 and consists of a constant response acceleration range, between periods of T_0 and T_s, a constant response velocity range for periods in excess of T_s, and a short period range that ramps between an estimated zero period acceleration given by $S_{MS}/2.5$ and S_{MS}. Site-specific spectra can also be used. Regardless of whether site-specific spectra or

TABLE 19.5   Coefficient F_v as a Function of Site Class and Mapped Spectral Response Acceleration

Site class	Mapped maximum considered earthquake spectral response acceleration at 1 s periods				
	$S_1 = 0.1$	$S_1 = 0.2$	$S_1 = 0.3$	$S_1 = 0.4$	$S_1 = 0.5$
A	0.8	0.8	0.8	0.8	0.8
B	1.0	1.0	1.0	1.0	1.0
C	1.7	1.6	1.5	1.4	1.3
D	2.4	2.0	1.8	1.6	1.5
E	3.5	3.2	2.8	2.4	a
F	a	a	a	a	a

Note: "a" indicates site-specific evaluation required.

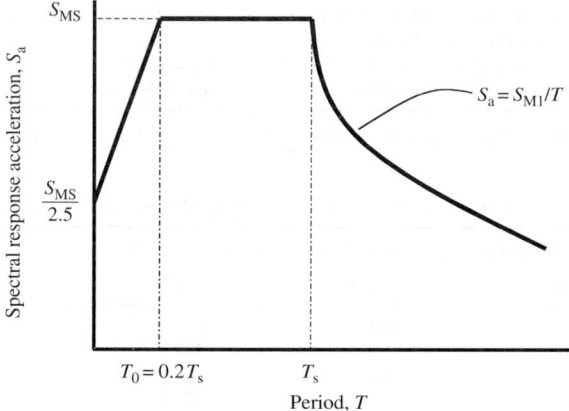

FIGURE 19.11   Maximum considered earthquake response spectrum.

spectra based on mapped values are used, the actual design values are taken as two thirds of the MCE values. The resulting design parameters are, respectively, labeled S_{DS} and S_{D1} and the design spectrum is identical to the MCE spectrum, except that the ordinates are taken as two thirds of the MCE values. The reason for using design values that are two thirds of the maximum considered values is that the design procedures, described in later sections, are believed to provide a minimum margin against collapse of 150%. Therefore, if design is conducted for two thirds of the MCE ground shaking, it is anticipated that buildings experiencing MCE ground shaking would be at incipient collapse, the desired performance objective for SUG-I structures.

19.5.4   Seismic Design Categories

The seismicity of the United States, and indeed the world, varies widely. It encompasses zones of very high seismicity in which highly destructive levels of ground shaking are anticipated to occur every 50 to 100 years and zones of much lower seismicity in which only moderate levels of ground shaking are ever anticipated. The *NEHRP Provisions* recognize that it is neither technically necessary nor economically appropriate to require the same levels of seismic protection for all buildings across these various regions of seismicity. Instead, the *NEHRP Provisions* assign each structure to a seismic design category (SDC) based on the level of seismicity at the building site, as represented by mapped shaking parameters, and the SUG.

Six SDCs, labeled A through F, are defined. SDC A represents the least severe seismic design condition and includes structures of ordinary occupancy located on sites anticipated to experience only very limited levels of ground shaking. SDC F represents the most severe design condition and includes

TABLE 19.6 Categorization of Structures into Seismic Design Category, Based on Design Short Period Spectral Response Acceleration, S_{DS} and Seismic Use Group

Value of S_{DS}	Seismic Use Group		
	I	II	III
$S_{DS} < 0.167g$	A	A	A
$0.167g \leq S_{DS} < 0.33g$	B	B	C
$0.33g \leq S_{DS} < 0.50g$	C	C	D
$0.50g \leq S_{DS}$	D[a]	D[a]	D[a]

[a] *Seismic Use Group* I and II *structures* located on sites with mapped *maximum considered earthquake* spectral response acceleration at 1 s period, S_1, equal to or greater than 0.75g shall be assigned to *Seismic Design Category* E and *Seismic Use Group* III *structures* located on such sites shall be assigned to *Seismic Design Category* F.

TABLE 19.7 Categorization of Structures into Seismic Design Category, Based on Design One-Second Period Spectral Response Acceleration, S_{D1} and Seismic Use Group

Value of S_{D1}	Seismic Use Group		
	I	II	III
$S_{D1} < 0.067g$	A	A	A
$0.067g \leq S_{D1} < 0.133g$	B	B	C
$0.133g \leq S_{D1} < 0.20g$	C	C	D
$0.20g \leq S_{D1}$	D[a]	D[a]	D[a]

[a] See footnote to Table 19.6.

structures assigned to SUG-III and located within a few kilometers of major, active faults, anticipated to produce very intense ground shaking. A designer determines to which SDC a structure should be assigned by reference to a pair of tables, reproduced as Table 19.6 and Table 19.7. A structure is assigned to the most severe category indicated by either table.

Nearly all aspects of the seismic design process are affected by the SDC that a structure is assigned to. This includes designation of the permissible structural systems, specification of required detailing, limitation on permissible heights and configuration, the types of analyses that may be used to determine the required lateral strength and stiffness, and the requirements for bracing and anchorage of non-structural components.

19.5.5 Permissible Structural Systems

The *NEHRP Provisions* define more than 70 individual seismic-force-resisting system types. These systems may be broadly categorized into five basic groups that include: bearing wall systems, building frame systems, moment-resisting frame systems, dual systems, and special systems:

- Bearing wall systems include those structures in which the vertical elements of the LFRS comprise either shear walls or braced frames in which the shear-resisting elements (walls or braces) are required to provide support for gravity (dead and live) loads in addition to providing lateral resistance. This is similar to the "box system" contained in earlier codes.
- Building frame systems include those structures in which the vertical elements of the LFRS comprise shear walls or braces, but in which the shear-resisting elements are not also required to provide support for gravity loads.
- Moment-resisting frame systems are those structures in which the lateral-force resistance is provided by the flexural rigidity and strength of beams and columns, which are interconnected in such a manner that stress is induced in the frame by lateral displacements.

- Dual systems rely on a combination of MRFs and either braced frames or shear walls. In dual systems, the braced frames or shear walls provide the primary lateral resistance and the MRF is provided as a back-up or redundant system, to provide supplemental lateral resistance in the event that earthquake response damages the primary lateral-force-resisting elements to an extent that they lose effectiveness.
- Special systems include unique structures, such as those that rely on the rigidity of cantilevered columns for their lateral resistance.

Within these broad categories, structural systems are further classified in accordance with the quality of detailing provided and the resulting ability of the structure to withstand earthquake-induced inelastic, cyclic demands. Structures that are provided with detailing believed capable of withstanding large cyclic inelastic demands are typically termed "special" systems. Structures that are provided with relatively little detailing and therefore, incapable of withstanding significant inelastic demands are termed "ordinary." Structures with limited levels of detailing and inelastic response capabilities are termed "intermediate." Thus, within a type of structure, for example, moment-resisting steel frames or reinforced concrete bearing walls, it is possible to have "special" MRFs or bearing walls, "intermediate" MRFs or shear walls, and "ordinary" MRFs or shear walls. The various combinations of such systems and construction materials results in a wide selection of structural systems to choose from. The use of "ordinary" and "intermediate" systems, regarded as having limited capacity to withstand cyclic inelastic demands, is generally limited to SDC A, B, and C and to certain low-rise structures in SDC D.

19.5.6 Design Coefficients

Under the *NEHRP Provisions*, required seismic design forces and, therefore, required lateral strength is typically determined by elastic methods of analysis, based on the elastic dynamic response of structures to design ground shaking. However, since most structures are anticipated to exhibit inelastic behavior when responding to the design ground motions, it is recognized that linear response analysis does not provide an accurate portrayal of the actual earthquake demands. Therefore, when linear analysis methods are employed, a series of design coefficients are used to adjust the computed elastic response values to suitable design values that consider probable inelastic response modification. Specifically, these coefficients are the response modification factor, R, the overstrength factor, Ω_0, and the deflection amplification coefficient, C_d. Tabulated values of these factors are assigned to a structure based on the selected structural system and the level of detailing employed in that structural system:

- The response modification coefficient, R, is used to reduce the required lateral strength of a structure, from that which would be required to resist the design ground motion in a linear manner to that required to limit inelastic behavior to acceptable levels, considering the characteristics of the selected structural system. Structural systems deemed capable of withstanding extensive inelastic behavior are assigned relatively high R values, as large as eight, permitting minimum design strengths that are only $\frac{1}{8}$ that required for elastic response to the design motion. Systems deemed to be incapable of providing reliable inelastic behavior are assigned low R values, approaching unity, requiring sufficient strength to resist design motion in a nearly elastic manner.
- The deflection amplification coefficient, C_d, is used to estimate the total elastic and inelastic lateral deformations of the structure when subjected to design earthquake ground motion. Specifically, lateral deflections calculated for elastic response of the structure to the design ground motion, reduced by the response modification coefficient R, are amplified by the factor C_d to obtain this estimate. The C_d coefficient accounts for the effects of viscous and hysteretic damping on structural response, as well as the effects of inelastic period lengthening. Structural systems that are deemed capable of developing significant amounts of viscous and hysteretic damping are assigned C_d values somewhat less than the value of the R coefficient. This results in an estimate of total lateral deformation that is somewhat lower than would be anticipated for a pure elastic response.

For structural systems with relatively poor capability to develop viscous and/or hysteretic damping, the C_d value may exceed R, resulting in estimates of lateral drift that exceed that calculated for elastic response.

- The overstrength coefficient, Ω_0, is used to provide an estimate of the maximum force likely to be delivered to an element in the structure, considering that due to effects of system and material overstrength this may be larger than the force calculated by elastic analysis of the structure's response to design ground motion, reduced by the response modification coefficient R. This overstrength factor is used to compute the required strength to resist behavioral modes that have limited capacity for inelastic response, such as column buckling or connection failure in braced frames.

Figure 19.12 illustrates the basic concepts behind these design coefficients. The figure contains an elastic design response spectrum, an elastic response line, and an inelastic response curve for an arbitrary structure, all plotted in lateral inertial force (base shear) versus lateral roof displacement coordinates. Response spectra are more familiarly plotted in coordinates of spectral response acceleration (S_a) versus structural period (T). It is possible to convert a spectrum plotted in that form to the spectrum shown in the figure through a two-step process. The first step consists of converting the response spectrum for S_a versus T coordinates to S_a versus spectral response displacement (S_d) coordinates. This is performed using the following relationship between S_a, S_d, and T:

$$S_d = \frac{T^2}{4\pi^2} S_a \tag{19.3}$$

Then the response spectrum is converted to the form shown in the figure by recognizing that for a structure responding in a given mode of excitation the base shear is equal to the product of the mass participation factor for that mode, the structure's mass, and the spectral response acceleration, S_a, at that period. Similarly, the lateral roof displacement for a structure responding in that mode is equal to the spectral response displacement times the modal participation factor. For a single degree of freedom structure, the mass participation factor and modal participation factor are both unity and the lateral base shear, V, is equal to the product of the spectral response acceleration at the mode of response and the mass of the structure, while the lateral roof displacement is equal to the spectral response displacement.

The dashed diagonal line in the figure represents the elastic response of the arbitrary structure. It is a straight line because a structure responding in an elastic manner will have constant stiffness and, therefore, a constant proportional relationship between the applied lateral force and resulting

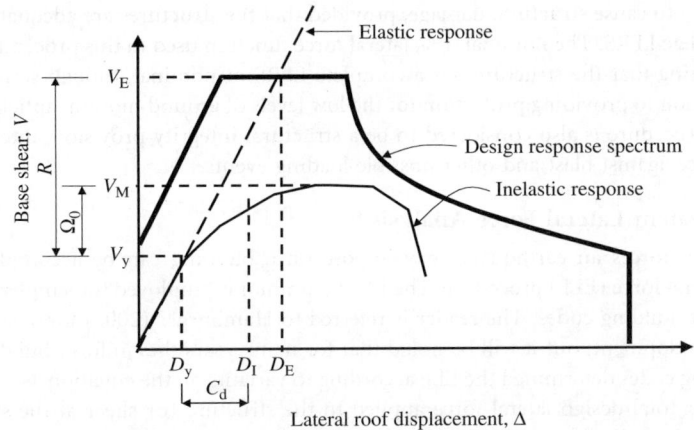

Figure 19.12 Schematic illustration of design coefficients.

displacement. The intersection of this diagonal line with the design response spectrum indicates the maximum total lateral base shear, V_E, and roof displacement, D_E, of the structure would develop if it responded to the design ground motion in an elastic manner. The third plot in the figure represents the inelastic response characteristics of this arbitrary structure, sometimes called a pushover curve. The pushover curve has an initial elastic region having the same stiffness as the elastic response line. The point V_y, D_y on the pushover curve represents the end of this region of elastic behavior. Beyond V_y, D_y, the curve is represented by a series of segments, with sequentially reduced stiffness, representing the effects of inelastic softening of the structure. The lateral base shear force, V_M, at the peak of the pushover curve, represents the maximum lateral force that the structure is capable of developing at full yield.

The response modification coefficient, R, is used in the provisions to set the minimum acceptable strength at which the structure will develop its first significant yielding, V_y. This is given by the simple relationship

$$V_y = \frac{V_E}{R} \tag{19.4}$$

The coefficient Ω_0 is used to approximate the full yield strength of the structure through the relationship

$$V_M = \Omega_0 V_y \tag{19.5}$$

The maximum total drift of the structure, D_I, is obtained from the relationship

$$D_I = C_d D_y \tag{19.6}$$

19.5.7 Analysis Procedures

The *NEHRP Provisions* permit the use of five different analytical procedures to determine the required lateral strength of a structure and to confirm that the structure has adequate stiffness to control lateral drift. The procedures permitted for a specific structure are dependent on the structure's SDC and its regularity.

19.5.7.1 Index Force Procedure

The index force procedure is permitted only for structures in SDC A. In this procedure, the structure must be designed to have sufficient strength to resist a static lateral force equal to 1% of the weight of the structure, applied simultaneously to each level. The forces must be applied independently in two orthogonal directions. Structures in SDC A are not anticipated ever to experience ground shaking of sufficient intensity to cause structural damage, provided that the structures are adequately tied together and have a complete LFRS. The nominal, 1%, lateral force function used in this procedure is intended as a means of ensuring that the structure has a complete LFRS of nominal, though somewhat arbitrary, strength. In addition to providing protection for the low levels of ground motion, anticipated for SDC A structures, this procedure is also considered to be a structural integrity provision, intended to provide nominal resistance against blast and other possible loading events.

19.5.7.2 Equivalent Lateral Force Analysis

The estimation of forces an earthquake may impose on a building can be accomplished by use of an equivalent lateral force (ELF) procedure. The ELF is commonly employed for simpler buildings and is provided in most building codes. The reader is referred to Hamburger (2002) for a review of building code and ELF development, but it will be noted that for many years the uniform building code (UBC) and other building codes determined the ELF according to variants on the equation $V = ZICKSW$, where V was the ELF or total design lateral force applied to the structure (or shear at the structure's base), determined from Z, a seismic zone factor varying between 0.075 (Zone 1, low-seismicity areas) and 0.40 (Zone 4, high-seismicity areas); I, an importance factor varying between 1.0 and 1.5; C, a function of the structure's fundamental period T, which effectively defined a response spectral shape; K, a factor varying by type of structure that accounted for the ductility of the structure and its material; S, a factor to account for site soil conditions; and W, the total seismic dead load.

In the 2000 *NEHRP Provisions*, ELF analysis may be used for any structure in SDC B and C, for any structure of light-frame construction, and for all regular structures, with a calculated structural period, T, not greater than $3.5T_s$, where T_s is as previously defined in Figure 19.11. ELF analysis consists of a simple approximation to modal response spectrum analysis. It only considers the first mode of a structure's lateral response and presumes that the mode shape for this first mode of response is represented by that of a simple shear beam. For structures having sufficiently low periods of first mode response ($T < 3.5T_s$) and regular vertical and horizontal distribution of stiffness and mass, this procedure approximates modal response spectrum analysis well. However, for longer period structures, higher mode response becomes significant and neglecting these higher modes results in significant errors in the estimation of structural response. Also, as the distribution of mass and stiffness in a structure becomes irregular, for example, the presence of torsional conditions or soft story conditions, the assumptions inherent in the procedure with regard to mode shape also become quite approximate, leading to errors. In SDCs D, E, and F, this method is permitted only for those structures where these inaccuracies are unlikely to be significant. The procedure is permitted for more general use in other SDCs because it is felt that the severity of design ground motion is low enough that inaccuracies in analysis of lateral response is unlikely to result in unacceptable structural performance and also because it is felt that designers in these regions of low seismicity may not be able to implement the more sophisticated and accurate methods properly.

As with the index force analysis procedure, the ELF consists of the simultaneous application of a series of static lateral forces to each level of the structure in each of the two independent orthogonal directions. In each direction, the total lateral force, known as the base shear, is given by the formula

$$V = \frac{S_{DS}}{R/I} W \tag{19.7}$$

This formula gives the maximum lateral inertial force that acts on an elastic, single degree of freedom structure with a period that falls within the constant response acceleration (periods shorter than T_s) portion of the design spectrum, reduced by the term R/I. In this formula, S_{DS} is the design spectral response acceleration at short periods, W is the dead weight of the structure and a portion of the supported live load, R is the response modification coefficient, and I is an occupancy importance factor, assigned based on the structure's SUG. For SUG-1 structures, I is assigned a value of unity. For SUG II and III structures, I is assigned values of 1.25 and 1.5, respectively. The effect of I is to reduce the permissible response modification factor, R, for structures in higher SUGs, requiring that the structures have greater strength, thereby limiting the permissible inelasticity and damage in these structures.

The base shear force given by Equation 19.7 need never exceed the following:

$$V = \frac{S_{D1}}{(R/I)T} W \tag{19.8}$$

Equation 19.8 represents the maximum lateral inertial force that acts on an elastic, single degree of freedom structure with period T that falls within the constant response velocity portion of the design spectrum (periods longer than T_s), reduced by the response modification coefficient, R and the occupancy importance factor, I. In this equation, all terms are as previously defined except that S_{D1} is the design spectral response acceleration at 1 s. For short period structures, Equation 19.7 will control. For structures with periods in excess of T_s, Equation 19.8 will control.

The shape of the design response spectrum shown in Figure 19.11 is not representative of the dynamic characteristics of ground motion found close to the fault rupture zone. Such motions are often dominated by a large velocity pulse and very large spectral displacement demands. Therefore, for structures in SDCs E and F, the seismic design categories for structures located close to major active faults, the base shear may not be taken less than the value given by Equation 19.9. Equation 19.9 approximates the effects of the additional long period displacements that have been recorded in some near field ground motion records

$$V = \frac{0.5S_1}{R/I} \qquad (19.9)$$

The total, lateral base shear force given by Equations 19.7–19.9 must be distributed vertically for application to the various mass or diaphragm levels of the structure. For a structure with n levels, the force at diaphragm level x is given by the equation

$$F_x = C_{vx}V \qquad (19.10)$$

where

$$C_{vx} = \frac{w_x h_x}{\sum_{i=1}^{n} w_i h_i} \qquad (19.11)$$

h_x and h_i, respectively, are the heights of levels x and i above the structure's base. These formula are based on the assumption that the structure is responding in its first mode, in pure sinusoidal motion, and that the mode shape is linear. That is, it is assumed that at any instant of time, the displacement at level x of the structure is

$$\delta_x = \frac{h_x}{h_n}\delta_n \qquad (19.12)$$

where δ_x and δ_n are the lateral displacements at level x and the roof of the structure, respectively, and h_n is the total height of the structure. For a structure responding in pure sinusoidal motion, the displacement δ_x, velocity v_x, and acceleration a_x, of level x at any instant of time, t, is given by the following equations:

$$\delta_x = \delta_{x\,\text{max}} \sin\left(\frac{2\pi}{T}t\right) \qquad (19.13)$$

$$v_x = \delta_{x\,\text{max}} \frac{2\pi}{T}\cos\left(\frac{2\pi}{T}t\right) \qquad (19.14)$$

$$a_x = -\delta_{x\,\text{max}} \frac{4\pi}{T^2}\sin\left(\frac{2\pi}{T}t\right) \qquad (19.15)$$

Since acceleration at level x is directly proportional to the displacement at level x, the acceleration at level x in a structure responding in pure sinusoidal motion is given by the equation

$$a_x = \frac{h_n}{h_x}a_n \qquad (19.16)$$

where a_n is the acceleration at the roof level. Since the inertial force at level x is equal to the product of mass at level x and the acceleration at level x, Equation 19.11 can be seen to be an accurate distribution of lateral inertial forces in a structure responding in a linear mode shape.

The lateral forces given by Equation 19.10 are applied to a structural model of the building and the resulting member forces and building interstory drifts are determined. The analysis must consider the relative rigidity of both the horizontal and vertical elements of the LFRS, and when torsional effects are significant, must consider three-dimensional distributions of stiffness, centers of mass, and rigidity. The structure must then satisfy two basic criteria. First, the elements of the LFRS must have sufficient strength to resist the calculated member forces in combination with other loads, and second, the structure must have sufficient strength to maintain computed interstory drifts within acceptable levels. The specific load combinations that must be used to evaluate member strength and the permissible interstory drifts are described in succeeding sections.

In recognition of the fact that higher mode participation can result in significantly larger forces at individual diaphragm levels, than is predicted by Equation 19.11, forces on diaphragms are computed using an alternative equation, as follows:

$$F_{px} = \frac{\sum_{i=x}^{n} F_i}{\sum_{i=x}^{n} w_i} w_{px} \qquad (19.17)$$

where F_{px} is the design force applied to diaphragm level x, F_i is the force computed from Equation 19.11 at level i, w_{px} is the effective seismic weight, at level x, and w_i is the effective weight at level i.

19.5.7.3 Response Spectrum Analysis

Response spectrum analysis is permitted to be used for the design of any structure. The procedure contained in the *NEHRP Provisions* uses standard methods of elastic modal dynamic analysis, which are not described here, but are well documented in the literature, for example, by Chopra (1981). The analysis must include sufficient modes of vibration to capture participation of at least 90% of the structure's mass in each of the two orthogonal directions. The response spectrum used to characterize the loading on the structure may be either the generalized design spectrum for the site, shown in Figure 19.11, or a site-specific spectrum developed considering the regional seismic sources and site characteristics.

Regardless of the spectrum used, the ground motion is scaled by the factor (I/R), just as in the ELF technique. The *NEHRP Provisions* require that the member forces determined by response spectrum analysis be scaled so that the total applied lateral force in any direction be not less than 80% of the base shear calculated using the ELF method for regular structures nor 100% for irregular structures. This scaling requirement was introduced to ensure that assumptions used in building the analytical model does not result in excessively flexible representation of the structure and, consequently, an underestimate of the required strength.

19.5.7.4 Response History Analysis

Response history analysis is also permitted to be used for the design of any structure but, due to the added complexity, is seldom employed in practice except for special structures incorporating special base isolation or energy dissipation technologies. Either linear or nonlinear response history analysis is permitted to be used. When response history analysis is performed, input ground motion must consist of a suite of at least three pairs of orthogonal horizontal ground motion components, obtained from records of similar magnitude, source, distance, and site characteristics as the event controlling the hazard for the building's site. Each pair of orthogonal records must be scaled such that with a period range approximating the fundamental period of response of the structure, the square root of the sum of the squares of the orthogonal component ordinates envelopes 140% of the design response spectrum. Simple amplitude, rather than frequency domain scaling, is recommended. Actual records are preferred, though simulations may be used if a sufficient number of actual records representative of the design earthquake motion are not available. If a suite of less than seven records is used as input ground motion, the maximum of the response parameters (element forces and deformations) obtained from any of the records is used for design. If seven or more records are used, the mean values of the response parameters obtained from the suite of records may be used as design values. This requirement was introduced with the understanding that the individual characteristics of a ground motion record can produce significantly different results for some response quantities. It was hoped that this provision would encourage engineers to use larger suites of records and obtain an understanding of the variability associated with possible structural response.

When linear response history analyses are performed, the ground motion records, scaled as previously described, are further scaled by the quantity (I/R). The resulting member forces are combined with other loads, just as they would be if the ELF or response spectrum methods of analysis were performed.

When nonlinear response history analyses are performed, they must be used without further scaling. Rather than evaluating the strength of members using the standard load combinations considered with other analysis techniques, the engineer is required to demonstrate acceptable performance capability of the structure, given the predicted strength and deformation demands. The intention is that laboratory and other relevant data be used to demonstrate adequate behavior. This is a rudimentary introduction of performance-based design concepts, which will likely have significantly greater influence in future building codes.

19.5.8 Load Combinations and Strength Requirements

Structures must be proportioned with adequate strength to resist the forces predicted by the lateral seismic analysis together with forces produced by response to vertical components of ground shaking as well as dead and live loads. Unless nonlinear response history analysis is performed using ground motion records that include a vertical component of motion, the effects of vertical earthquake shaking are accounted for by the equation

$$E = Q_E \pm 0.2 S_{DS} D \tag{19.18}$$

where Q_E are the element forces predicted by the lateral seismic analysis, S_{DS} is the design spectral response acceleration at a 0.2 s response period, and D are the forces produced in the element by the structure's dead weight. The term $0.2 S_{DS} D$ represents the effect of vertical ground shaking response. For structures in zones of high seismicity, the term S_{DS} has a value approximating $1.0g$ and, therefore, the vertical earthquake effects are taken as approximately 20% increase or decrease in the dead load stress demands on each element. In fact, there are very few cases on record where structural collapse has been ascribed to the vertical response of a structure. This is probably because design criteria for vertical load resistance incorporate substantial factors of safety and also because most structures carry only a small fraction of their rated design live loads when they are subjected to earthquake effects. Therefore, most structures inherently have substantial reserve capacity to resist additional loading induced by vertical ground motion components. In recognition of this, most earlier codes neglected vertical earthquake effects. However, during the formulation of *ATC3.06*, it was felt to be important to acknowledge that ground shaking includes three orthogonal components. The resulting expression, which was somewhat arbitrary, ties vertical seismic forces to the short period design spectral response acceleration, as most structures are stiff vertically and have very short periods of structural response for vertical modes.

The earthquake forces on structural elements derived from Equation 19.18 are combined with dead and live loads in accordance with the standard strength level load combinations of *ASCE-7*. The pertinent load combinations are

$$Q = 1.4D \pm E \tag{19.19}$$

$$Q = 1.4D + 0.75(L + E) \tag{19.20}$$

where D, L, and E are the dead, live, and earthquake forces, respectively. Elements must then be designed to have adequate strength to resist these combined forces. The reduction factor of 0.75 on the combination of earthquake and live loads accounts for the low likelihood that a structure will be supporting full live load at the same time that it experiences full design earthquake shaking. An alternative set of load combinations is also available for use with design specifications that utilize allowable stress design formulations. These are essentially the same as Equations 19.19 and 19.20, except that the earthquake loads are further reduced by a factor of 1.4.

The *NEHRP Provisions* recognize that it is undesirable to allow some elements to experience inelastic behavior as they may be subject to brittle failure and in doing so compromise the ability of the structure to develop its intended inelastic response. The connections of braces to braced frames are an example of such elements. The *Provisions* also recognize that inelastic behavior in some elements, such as columns supporting discontinuous shear walls, could trigger progressive collapse of the structure. For these elements, the earthquake force *E* that must be used in the load combination (Equations 19.19 and 19.20) is given by the formula:

$$E = \Omega_0 Q_E \pm 0.2 S_{DS} D \tag{19.21}$$

where the term $0.2S_{DS}D$ continues to represent the effects of vertical ground shaking response and the term $\Omega_0 Q_E$ represents an estimate of the maximum force likely to be developed in the element as a result of lateral earthquake response, considering the inelastic response characteristics of the entire structural system. In Equation 19.20, the term $\Omega_0 Q_E$ need never be taken larger than the predicted force on the element derived from a nonlinear analysis or plastic mechanism analysis.

19.5.9 Drift Limitations

It is important to control lateral drift in structures because excessive drift can result in extensive damage to cladding and other nonstructural building components. In addition, excessive lateral drift can result in the development of *P*–Δ instability and collapse.

Lateral drift is evaluated on a story by story basis. Story drift, δ, is computed as the difference in lateral deflection at the top of a story and that at the bottom of the story, as predicted by the lateral analysis. If the lateral analysis was other than a nonlinear response history analysis, design story drift, Δ, is obtained from the computed story drift, δ, by the equation:

$$\Delta = C_d \delta \tag{19.22}$$

where C_d is the design coefficient previously discussed. The design interstory drift computed from Equation 19.22 must be less than a permissible amount, dependent on the SUG and structural system as given in Table 19.8.

The provisions require evaluation of potential *P*–Δ instability through consideration of the quantity θ given by the equation:

$$\theta = \frac{P_x \Delta}{V_x h_x C_d} \tag{19.23}$$

In this equation, P_x is the dead weight of the structure above story *x*, Δ is the design story drift, computed from Equation 19.22, V_x is the design story shear obtained from the lateral force analysis, h_x is the story height, and C_d is the coefficient previously discussed. If the quantity θ computed by this equation is found to be less than 0.1, *P*–Δ effects may be neglected. If the quantity θ is greater than 0.1, *P*–Δ effects must be directly considered in performing the LFA. If the quantity θ exceeds 0.3, the structure should be considered potentially unstable and must be redesigned.

This approach to *P*–Δ evaluation has remained essentially unchanged since its initial introduction in *ATC3.06*. It was introduced in that document as a placeholder, pending the development of a more

TABLE 19.8 Permissible Drift Limits[a]

Structure	Seismic use group		
	I	II	III
Structures other than masonry *shear wall* or masonry *wall*-frame *structures*, four stories or less in height with interior *walls*, *partitions*, ceilings, and exterior *wall* systems that have been designed to accommodate the *story* drifts	$0.025h_{sx}$[b]	$0.020h_{sx}$	$0.015h_{sx}$
Masonry cantilever *shear wall* structures[c]	$0.010h_{sx}$	$0.010h_{sx}$	$0.010h_{sx}$
Other masonry *shear wall* structures	$0.007h_{sx}$	$0.007h_{sx}$	$0.007h_{sx}$
Masonry *wall*-frame *structures*	$0.013h_{sx}$	$0.013h_{sx}$	$0.010h_{sx}$
All other *structures*	$0.020h_{sx}$	$0.015h_{sx}$	$0.010h_{sx}$

[a] There shall be no drift limit for single-story structures with interior walls, partitions, ceilings, and exterior wall systems that have been designed to accommodate the story drifts.
[b] h_{sx} is the story height below Level *x*.
[c] Structures in which the basic structural system consists of masonry shear walls designed as vertical elements cantilevered from their base or foundation support which are so constructed that moment transfer between shear walls (coupling) is negligible.

accurate method for evaluating drift-induced instability. Obvious deficiencies in this current approach include the fact that it evaluates drift effects at the somewhat artificial design-base shear levels. A more realistic evaluation would consider the actual expected lateral deformations of the structure as well as the yield level shear capacity of the structure at each story. As contained in the current provisions, evaluation of P–Δ effects seldom controls a structure's design.

19.5.10 Structural Detailing

Structural detailing is a critical feature of seismic-resistant design but is not generally specified by the *NEHRP Provisions*. Rather, the Provisions adopt detailing requirements contained in standard design specifications developed by the various materials industry associations including the American Institute of Steel Construction, the American Concrete Institute, the American Forest Products Association, and the Masonry Society. Other chapters in this handbook present the requirements of these various design standards.

Glossary

Attenuation — The rate at which earthquake ground motion decreases with distance.

Base shear — The total lateral force for which a structure is designed using equivalent lateral force techniques.

Characteristic earthquake — A relatively narrow range of magnitudes at or near the maximum that can be produced by the geometry, mechanical properties, and state of stress of a fault (Schwartz and Coppersmith 1987).

Completeness — Homogeneity of the seismicity record.

Cripple wall — A carpenter's term indicating a wood frame wall of less than full height T, usually built without bracing.

Critical damping — The value of damping such that free vibration of a structure will cease after one cycle (ccrit $= 2\ m\omega$).

Damage — Permanent, cracking, yielding, or buckling of a structural element or structural assemblage.

Damping — Energy dissipation that occurs in a dynamically deforming structure, either as a result of frictional forces or structural yielding. Increased damping tends to reduce the amount that a structure responds to ground shaking.

Degradation — A behavioral mode in which structural stiffness or strength is reduced as a result of inelastic behavior.

Design (basis) earthquake — The earthquake (as defined by various parameters, such as PGA, response spectra, etc.) for which the structure will be, or was, designed.

Ductile detailing — Special requirements, such as for reinforced concrete and masonry, close spacing of lateral reinforcement to attain confinement of a concrete core, appropriate relative dimensioning of beams and columns, 135° hooks on lateral reinforcement, hooks on main beam reinforcement within the column, etc.

Ductile frames — Frames required to furnish satisfactory load-carrying performance under large deflections (i.e., ductility). In reinforced concrete and masonry this is achieved by ductile detailing.

Ductility factor — The ratio of the total displacement (elastic plus inelastic) to the elastic (i.e., yield) displacement.

Elastic — A mode of structural behavior in which a structure displaced by a force will return to its original state upon release of the force.

Fault — A zone of the earth's crust within which the two sides have moved — faults may be hundreds of miles long — from one to over one hundred miles deep, and not readily apparent on the ground surface.

Ground shaking — A random, rapid cyclic motion of the ground produced by an earthquake.

Hysteresis — A form of energy dissipation that is related to inelastic deformation of a structure.

Inelastic — A mode of structural behavior in which a structure, displaced by a force, exhibits permanent unrecoverable deformation.

Lateral force resisting system (LFRS) — A structural system for resisting horizontal forces due, for example, to earthquake or wind (as opposed to the vertical force resisting system, which provides support against gravity).

Liquefaction — A process resulting in a soil's loss of shear strength due to a transient excess of pore water pressure.

Magnitude — A unique measure of an individual earthquake's release of strain energy, measured on a variety of scales, of which the moment magnitude M_w (derived from seismic moment) is preferred.

Mass participation — That portion of total mass of a multidegree of freedom structure that is effective in a given mode of response.

MCE — Maximum considered earthquake — the earthquake intensity forming the basis for design in the *NEHRP Provisions*.

Mode shape — A deformed shape in which a structure can oscillate freely when displaced.

Natural mode — A characteristic dynamic property of a structure in which it will oscillate freely.

Nonductile frames — Frames lacking ducility or energy absorption capacity due to lack of ductile detailing — ultimate load is sustained over a smaller deflection (relative to ductile frames) and for fewer cycles.

Participation factor — A mathematical relationship between the maximum displacement of a multi-degree of freedom structure and a single degree of freedom structure.

Peak ground acceleration (PGA) — The maximum amplitude of recorded acceleration (also termed the ZPA, or zero period acceleration).

Period — The amount of time it takes a structure that has been displaced in a particular natural mode and then released to undergo one complete cycle of motion.

Pounding — The collision of adjacent buildings during an earthquake due to insufficient lateral clearance.

Response spectrum — A plot of maximum amplitudes (acceleration, velocity, or displacement) of a single degree of freedom oscillator (sdof), as the natural period of the sdof is varied across a spectrum of engineering interest (typically, for natural periods from 0.03 to 3 or more seconds, or frequencies of 0.3 to 30+ Hz).

Reverse fault — A fault that exhibits dip-slip motion, where the two sides are in compression and move away toward each other.

Seismic risk — The product of the hazard and the vulnerability (i.e., the expected damage or loss, or the full probability distribution).

Soft story — A story of a building signifiantly less stiff than adjacent stories (i.e., the lateral stiffness is 70% or less than that in the story above, or less than 80% of the average stiffness of the three stories above; BSSC 1994).

Spectral acceleration — The maximum response acceleration that a structure of given period will experience when subjected to a specific ground motion.

Spectral displacement — The maximum response displacement that a structure of given period will experience when subjected to a specific ground motion.

Spectral velocity — The maximum response velocity that a structure of given period will experience when subjected to a specific ground motion.

Spectrum amplification factor — The ratio of a response spectral parameter to the ground motion parameter (where parameter indicates acceleration, velocity, or displacement).

Viscous — A form of energy dissipation that is proportional to velocity.

Yielding — A behavioral mode in which a structural displacement increases under application of constant load.

References

Aiken, I.D., Nims, D.K., Whittaker, A.S., and Kelly, J.M. 1993. Testing of Passive Energy Dissipation Systems. *Earthquake Spectra*, Volume 9, number 3, August, pp. 336–370.

Building Seismic Safety Council. 1997. *NEHRP Recommended Provisions for Seismic Regulations for Buildings and Other Structures*. Report No. FEMA 302/303. Federal Emergency Management Agency, Washington, DC.

Building Seismic Safety Council. 2001. *NEHRP Recommended Provisions for Seismic Regulations for Buildings and Other Structures, 2000 Edition*. Report no. FEMA 368. Federal Emergency Management Agency, Washington, DC.

Chen, W.F. and Scawthorn, C. (eds). 2002. *Earthquake Engineering Handbook*, CRC Press, Boca Raton, FL.

Dowrick, D.J. 1987. *Earthquake Resistant Design*, 2nd edition, John Wiley & Sons, New York.

Earthquake Resistant Design Codes in Japan, January 2000. Japan Society of Civil Engineers, Tokyo.

Federal Emergency Management Agency. 1988. *Rapid Visual Screening of Buildings for Potential Seismic Hazards: A Handbook*, FEMA 154, FEMA, Washington, DC.

Hamburger, R.O. 2002. *Building Code Provisions for Seismic Resistance*, chapter in Chen and Scawthorn (2002).

Hanson, R.D., Aiken, I.D., Nims, D.K., Richter, P.J., and Bachman, R.E. 1993. *State-of-the-Art and State-of-the-Practice in Seismic Energy Dissipation*. Proceedings of ATC-17-1 Seminar on Seismic Isolation, Passive Energy Dissipation, and Active Control; San Francisco, California, March 11–12, 1993, Volume 2: Passive Energy Dissipation, Active Control, and Hybrid Control Systems. Applied Technology Council, Redwood City, CA, pp. 449–471.

IAEE. 1996. *Regulations for Seismic Design: A World List, 1996 (RSD)*. Rev. ed. (update of Earthquake Resistant Regulations: A World List, 1992). Prepared by the International Association for Earthquake Engineering (IAEE), Tokyo, 1996; and its Supplement 2000: Additions to Regulations for Seismic Design: A World List, 1996. Available from Gakujutsu Bunken Fukyu-Kai (Association for Science Documents Information), c/o Tokyo Institute of Technology, 2-12-1 Oh-Okayama, Meguro-Ku, Tokyo, Japan 152-8550 (telephone: +81-3-3726-3117; fax: +81-3-3726-3118; e-mail: gakujyutu-bunken@mvd.biglobe.ne.jp).

ICC. 2000. *International Building Code 2000*, International Code Council, published by International Conference of Building Officials, Whittier, CA, and others.

Iemura, H. and Pradono, M.H. 2002. *Structural Control*, chapter in Chen and Scawthorn (2002).

Lou, J.Y.K., Lutes, L.D., and Li, J.J. 1994. *Active Tuned Liquid Damper for Structural Control*. Proceedings [of the] First World Conference on Structural Control: International Association for Structural Control; Los Angeles, Augusts 3–5, 1994; Housner, G.-W. et al., eds; International Association for Structural Control, Los Angeles, Volume 2, pp. TP1-70–TP1-79.

Naeim, F. (ed.). 2001. *Seismic Design Handbook*, 2nd edition, Kluwer, New York.

NFPA. n.d. *NFPA 5000 Building Code*, National Fire Protection Association, Cambridge, MA (publication pending).

Paz, M. (ed.). 1994. *International Handbook of Earthquake Engineering: Codes, Programs, and Examples*, Chapman & Hall, New York.

Perry, C.L., Fierro, E.A., Sedarat, H., and Scholl, R.E. 1993. Seismic Upgrade in San Francisco Using Energy Dissipation Devices. *Earthquake Spectra*, Volume 9, number 3, August, pp. 559–580.

Soong, T.T. and Costantinou, M.C. (eds.). 1994. *Passive and Active Structural Vibration Control in Civil Engineering*, State University of New York, Buffalo, NY (CISM International Centre for Mechanical Sciences, Vol. 345), VIII, 380 pp. Springer-Verlag, Vienna, New York.

Spencer Jr., B.F., Sain, M.K., Won, C.-H. Kaspari Jr., D.C., and Sain, P.M. 1994. Reliability-Based Measures of Structural Control Robustness, *Struct. Safety*, Volume 15, pp. 111–129.

Structural Engineers Association of California. 1996. *Vision 2000, A Framework for Performance-based Seismic Design*, Structural Engineers Association of California, Sacramento, CA.

Structural Engineers Association of California. 1999. Seismology Committee. *Recommended Lateral Force Requirements and Commentary*. Sacramento, CA.

U.S. Army. 1992. *Seismic Design for Buildings* (Army: TM 5-809-10; Navy: NAVFAC P-355; USAF: AFM-88-3, Chapter 13). Washington, DC: Departments of the Army, Navy, and Air Force, 1992. 407 pages. Available from the National Technical Information Service (NTIS), Military Publications Division, Washington.

Uniform Building Code. 1997. *Volume 2, Structural Engineering Design Provisions*, International Conference on Building Officials, Whittier.

Yang, Y.-B., Chang, K.-C., and Yau, J.-D. 2002. *Base Isolation*, chapter in Chen and Scawthorn (2002).

Further Reading

Chen and Scawthorn (2002) provide an extensive reference on earthquake engineering, while Naiem (2001) provides an excellent resource on seismic design. Both references have individual chapters on design of steel, wood, reinforced concrete, reinforced masonry, and precast structures, and also on nonstructural elements. SEAOC (1999), BSSC (1997), and SEAOC (1996) provide an excellent overview of the current state of seismic design requirements. U.S. Army (1992) and Dowrick (1987) are also useful, although a bit older. Some useful sources on seismic code provisions in countries other than the U.S. include *Earthquake Resistant Design Codes in Japan* (2000), Paz (1995), and IAEE (1996). The last is a comprehensive compendium of seismic regulations for over 40 countries, including *Eurocode 8* (the European Union's seismic provisions).

20

Seismic Design of Bridges*

Lian Duan
Division of Engineering Services, California Department of Transportation, Sacramento, CA

Mark Reno
Quincy Engineering, Sacramento, CA

Wai-Fah Chen
University of Hawaii at Manoa, Honolulu, HI

Shigeki Unjoh
Ministry of Construction, Public Works Research Institute, Tsukuba, Ibaraki, Japan

20.1 Introduction

Bridges are very important elements in the modern transportation system. Recent earthquakes, particularly the 1989 Loma Prieta and the 1994 Northridge Earthquakes in California, the 1995 Hyogo-Ken Nanbu Earthquake in Japan, the 1999 JiJi Earthquake in Taiwan, and the 1999 Kocaeli Earthquake in Turkey, have caused collapse of, or severe damage to, a considerable number of major bridges [1,2]. Since the 1989 Loma Prieta Earthquake in California [3], extensive research [4–18] has been conducted on seismic design and retrofit of bridges in Japan and the United States, especially in California.

---

*Much of the material of this chapter was taken from Duan, L. and Chen, W.F., Chapter 19: Bridges, in Earthquake Engineering Handbook, Chen, W.F. and Scawthorn, C., Ed., CRC Press, Boca Raton, FL, 2002.

0-8493-1569-7/05/$0.00+$1.50
© 2005 by CRC Press

This chapter first addresses the seismic bridge design philosophies and conceptual design in general, then discusses mainly the U.S. seismic design practice to illustrate the process, and finally presents briefly seismic design practice in Japan.

20.2 Earthquake Damages to Bridges

Past earthquakes have shown that the damage induced in bridges can take many forms depending on the ground motion, site conditions, structural configuration, and specific details of the bridge [1]. Damage within the superstructure has rarely been the primary cause of collapse. Most of the severe damage to bridges has taken one of the following forms [1]:

- *Unseating of the superstructure at in-span hinges or simple supports due to inadequate seat lengths or restraint.* A skewed, curved, or complex superstructure framing configuration further increases the vulnerability. Figure 20.1 shows the collapsed upper and lower decks of the eastern portion of the San Francisco-Oakland Bay Bridge (SFOBB) in the 1989 Loma Prieta Earthquake, which can be attributed to anchor bolt failures allowing the span to move. Figure 20.2 shows collapsed I-5 Gavin Canyon Undercrossing, California, in the 1994 Northridge Earthquake, which can be attributed to geometric complexities arising from 66° skew angle abutments, in-span expansion joints, as well as the inadequate 300 mm seat width. For simply supported bridges, these failures are most likely when ground failure induces relative motion between the spans and their supports.
- *Column brittle failure due to deficiencies in shear capacity and inadequate ductility.* In reinforced concrete columns, the shear capacity and ductility concerns usually stem from inadequate lateral and confinement reinforcement. Figure 20.3 shows the collapsed 600 m of Hanshin Expressway in the 1995 Hyogo-Ken Nanbu Earthquake in Japan where failure was attributed to deficiencies in shear design and poor ductility. In steel columns, the inadequate ductility usually stems from progressive local buckling, fracture, and global buckling leading to collapse.
- *Unique failures in complex structures.* Figure 20.4 shows the collapsed Cypress Street Viaduct, California, in the 1989 Loma Prieta Earthquake where the unique vulnerability was the inadequately reinforced pedestal above the first level. In outrigger column bents, the vulnerability may be in the cross-beam or the beam–column joint.

FIGURE 20.1   Collapse of eastern portion of San Francisco-Oakland Bay Bridge in the Loma Prieta Earthquake (courtesy of California Department of Transportation).

FIGURE 20.2   Collapse of I-5 Gavin Canyon Undercrossing, California, in the 1994 Northridge Earthquake.

FIGURE 20.3   Collapse of Hanshin Expressway in the 1995 Hyogo-Ken Nanbu Earthquake in Japan (courtesy of Mark Yashinsky).

20.3   Seismic Design Philosophies

20.3.1   Design Evolution

Seismic bridge design has been improving and advancing, based on research findings and lessons learned from past earthquakes. In the United States, prior to the 1971 San Fernando Earthquake, the seismic design of highway bridges was partially based on lateral force requirements for buildings.

FIGURE 20.4   Collapse of Cypress Viaduct, California, in the 1989 Loma Prieta Earthquake.

Lateral loads were considered as levels of 2 to 6% of dead loads. In 1973, California Department of Transportation (Caltrans) developed new seismic design criteria (SDC) related to site, seismic response of the soils at the site, and dynamic characteristics of bridges. The American Association of State Highway and Transportation Officials (AASHTO) modified the Caltrans 1973 Provisions slightly and adopted Interim Specifications. Applied Technology Council (ATC) developed guidelines ATC-6 [19] for seismic design of bridges in 1981. AASHTO adopted ATC-6 as the Guide Specifications in 1983 and in 1991 incorporated it into the Standard Specifications for Highway Bridges [20].

Prior to the 1989 Loma Prieta Earthquake, bridges in California were typically designed using a single-level force-based design approach based on a "no-collapse" design philosophy. Seismic loads were determined based on a set of soil conditions and a suite of four site-based standard acceleration response spectra (ARS). Structures were analyzed using the three-dimensional elastic dynamic multimodal response spectrum analysis method. Structural components were designed to resist forces from the response spectrum analysis that were modified with a "Z-factor." The "Z-factor" was based on the individual structural element ductility and degree of risk. Minimum transverse reinforcement levels were required to meet confinement criteria [21].

Since 1989, the design criteria specified in Caltrans design manuals [21–27] have been updated continuously to reflect recent research findings and development in the field of seismic bridge design. Caltrans has been shifting toward a displacement-based design approach emphasizing element and system capacity design. For important bridges in California the performance-based project-specific design criteria [28–30] have been developed since 1989.

FHWA updated its *Seismic Design and Retrofit Manual for Highway Bridges* in 1995 [31,32]. ATC published the improved SDC recommendations for California bridges [33] in 1996 and for U.S. bridges and highway structures [34] in 1997. Caltrans published the performance and displacement-based Seismic Design Criteria Version 1.3 [35] in 2004, which focuses mainly on concrete bridges, and the Guide Specifications for Seismic Design of Steel Bridge (*Guide*) [36] in 2001. Most recently, the NCHRP 12-49 team developed a new set of LRFD Guidelines (*Guidelines*) for the seismic design of highway bridges [37], compatible with the AASHTO-LFRD Bridge Design Specifications [38]. Significant advances in earthquake engineering were made during the last decade of the twentieth century.

20.3.2 No-Collapse-Based Design

The basic design philosophy is to prevent bridges from collapsing during severe earthquakes [24–27,35–37] that have only a small probability of occurring during the useful life of the bridge. To prevent collapse, two alternative approaches are commonly used in design. The first is a conventional force-based approach where the adjustment factor Z for ductility and risk assessment [21], or the response modification factor R [20,37], is applied to elastic member forces obtained from a response spectra analysis or an equivalent static analysis. The second approach is a more recent displacement-based approach [24,35] where displacements are a major consideration in design. For more detailed information, references can be made to comprehensive discussions in *Seismic Design and Retrofit of Bridges* by Prietley et al. [16], *Bridge Engineering Handbook* by Chen and Duan [39], and Refs [40,41].

20.3.3 Performance-Based Design

Following the 1989 Loma Prieta Earthquake, bridge engineers [3] have faced three essential challenges:

- Ensure that earthquake risks posed by new construction are acceptable.
- Identify and correct unacceptable seismic safety conditions in existing structures.
- Develop and implement the rapid, effective, and economic response mechanism for the recovering structural integrity after damaging earthquakes.

Performance-based project-specific criteria [28–30] and design memoranda [24] have been developed and implemented for the design and retrofitting of important bridges by California bridge engineers. These performance-based criteria included guidelines for development of site-specific ground motion estimates, ductile design details to preclude brittle failure modes, rational procedures for concrete joint shear design, and the definition of limit states for various performance objectives [15]. The performance-based criteria usually require a two-level design approach. The first level of design is to ensure the performance (service) of a bridge in small-magnitude earthquake events that may occur several times during the life of the bridge. The second level of design is to achieve the performance (no collapse) of a bridge under severe earthquakes that have only a small probability of occurring during the useful life of the bridge. Figure 20.5 shows a flowchart for development of performance-based SDC.

20.4 Seismic Conceptual Design

Bridge design is a complex engineering process involving consideration of numerous important factors, such as bridge structural systems, materials, dimensions, foundations, esthetics, and local landscape and surrounding environment [16,42]. Selecting an appropriate earthquake-resisting system (ERS) to resolve the potential conflicts between the configuration and seismic performance should be completed as early as possible in the design effort. For a desirable seismic resistant design, the following guidelines may be useful:

- Bridge type, component and member dimensions, and esthetics shall be investigated to reduce the seismic demands to the greatest extent possible. Esthetics should not be the primary reason for producing undesirable frame and component geometry.
- Bridges should ideally be as straight as possible. Curved bridges complicate and potentially magnify seismic responses.
- Superstructures should be continuous with as few joints as possible. Necessary restrainers and sufficient seat width shall be provided between adjacent frames at all expansion joints, and at the seat-type abutments to eliminate the possibility of unseating during a seismic event.
- Support skew angles should be as small as possible, that is, abutments and piers should be oriented as close to perpendicular to the bridge longitudinal axis as practical (within 20 to 25°) even at the expense of slightly increasing the bridge length. Highly skewed abutments and piers are vulnerable to damage from undesired rotational response and increased seismic displacement demands.

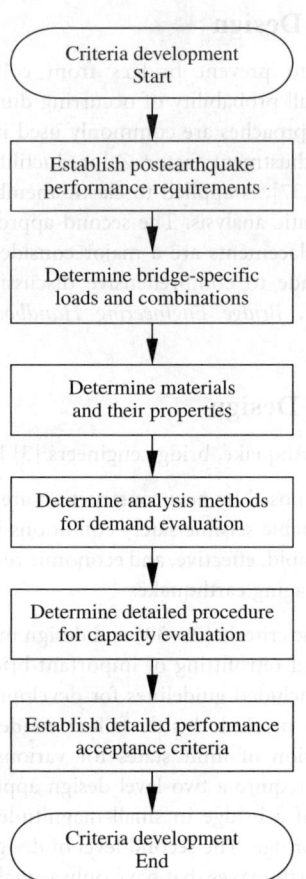

FIGURE 20.5   Development of performance-based seismic design criteria.

- Adjacent frames or piers should be proportioned to minimize the differences in the fundamental periods and skew angles, and to avoid drastic changes in stiffness and strength in both the longitudinal and transverse directions. Dramatic changes in stiffness can result in damage to the stiffer frames or piers. It is strongly recommended [35] that the effective stiffness between any two bents within a frame, or between any two columns within a bent, not vary by a factor of more than two. Similarly, it is highly recommended that the ratio of the shorter fundamental period to the longer fundamental period for adjacent frame in the longitudinal and transverse directions be larger than 0.7.
- Structural configurations that cannot accommodate the recommendations must be capable of accommodating the associated large relative displacement without compromising structural integrity. Each frame shall provide a well-defined load path with predetermined plastic hinge locations and utilize redundancy whenever possible. Balanced mass and stiffness distribution (Figure 20.6) in a frame results in a structure response that is more predictable and is more likely to respond in its fundamental mode of vibration. Simple analysis tools can then be used to predict the structures response with relative accuracy, whereas irregularities in geometry increase the likelihood of complex nonlinear response that is difficult to accurately predict by elastic modeling or plane frame inelastic static analysis. The following various techniques may be used to achieve balanced geometry to create a uniform and more predictable structure response [43]:
  a. Adjust foundation rotational and translation stiffness (e.g., use oversized pile shafts).
  b. Adjust effective column lengths (e.g., lower footings, provide isolation casings).

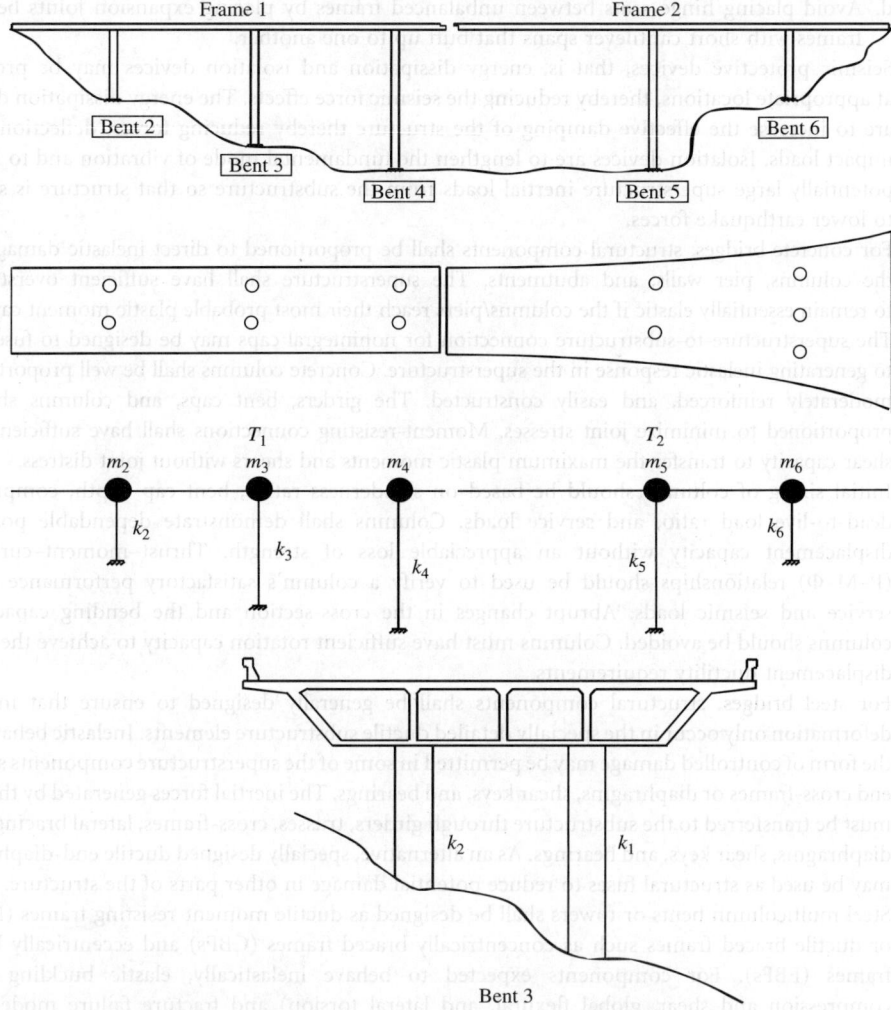

FIGURE 20.6 Frame stiffness [35].

c. Modify end fixities.

d. Reduce/redistribute superstructure mass.

e. Vary the column cross-section and longitudinal reinforcement ratios.

f. Add or relocate columns/piers.

g. Modify the hinge/expansion joint layout.

In the event of other constraints preventing the designer from achieving balance between frames of a bridge, the following recommendations may be considered [35]:

a. Isolate adjacent frames longitudinally by providing a large expansion gap to reduce the likelihood of pounding.

b. Provide adequate seat width to prevent unseating at hinges. Seat extenders may be used; however, they should be isolated transversely to avoid transmitting large lateral shear forces between frames.

c. Limit the transverse shear capacity between frames to prevent large lateral forces from being transferred to the stiffer frame.

d. Avoid placing hinge seats between unbalanced frames by placing expansion joints between frames with short cantilever spans that butt up to one another.

- Seismic protective devices, that is, energy dissipation and isolation devices may be provided at appropriate locations, thereby reducing the seismic force effects. The energy dissipation devices are to increase the effective damping of the structure thereby reducing forces, deflections, and impact loads. Isolation devices are to lengthen the fundamental mode of vibration and to isolate potentially large superstructure inertial loads from the substructure so that structure is subject to lower earthquake forces.

- For concrete bridges, structural components shall be proportioned to direct inelastic damage into the columns, pier walls, and abutments. The superstructure shall have sufficient overstrength to remain essentially elastic if the columns/piers reach their most probable plastic moment capacity. The superstructure-to-substructure connection for nonintegral caps may be designed to fuse prior to generating inelastic response in the superstructure. Concrete columns shall be well proportioned, moderately reinforced, and easily constructed. The girders, bent caps, and columns shall be proportioned to minimize joint stresses. Moment-resisting connections shall have sufficient joint shear capacity to transfer the maximum plastic moments and shears without joint distress.

- Initial sizing of columns should be based on slenderness ratios, bent cap depth, compressive dead-to-live load ratio, and service loads. Columns shall demonstrate dependable postyield displacement capacity without an appreciable loss of strength. Thrust–moment–curvature (P–M–Φ) relationships should be used to verify a column's satisfactory performance under service and seismic loads. Abrupt changes in the cross-section and the bending capacity of columns should be avoided. Columns must have sufficient rotation capacity to achieve the target displacement ductility requirements.

- For steel bridges, structural components shall be generally designed to ensure that inelastic deformation only occur in the specially detailed ductile substructure elements. Inelastic behavior in the form of controlled damage may be permitted in some of the superstructure components such as end cross-frames or diaphragms, shear keys, and bearings. The inertial forces generated by the deck must be transferred to the substructure through girders, trusses, cross-frames, lateral bracings, end diaphragms, shear keys, and bearings. As an alternative, specially designed ductile end-diaphragms may be used as structural fuses to reduce potential damage in other parts of the structure.

- Steel multicolumn bents or towers shall be designed as ductile moment-resisting frames (MRFs) or ductile braced frames such as concentrically braced frames (CBFs) and eccentrically braced frames (EBFs). For components expected to behave inelastically, elastic buckling (local compression and shear, global flexural, and lateral torsion) and fracture failure modes shall be avoided. All connections and joints should preferably be designed to remain essentially elastic. For MRFs, the primary inelastic deformation shall preferably be columns. For CBFs, diagonal members shall be designed to yield when members are in tension and to buckle inelastically when they are in compression. For EBFs, a short beam segment designated as a "link" shall be well designed and detailed to provide ductile structural behavior.

- The ATC/MCEER recommended LRFD *Guidelines* [37] classify the ERS into permissible and not recommended categories (Figure 20.7 to Figure 20.11) based on consideration of the most desirable seismic performance ensuring wherever possible postearthquake serviceability. Figure 20.12 shows design approaches for the permissible ERS.

20.5 Seismic Performance Criteria

20.5.1 ATC/MCEER Guidelines

Table 20.1 gives the seismic performance criteria for highway bridges specified in the proposed ATC/MCEER *Guidelines* [37]. As a minimum, a bridge shall be designed for the life safety level of

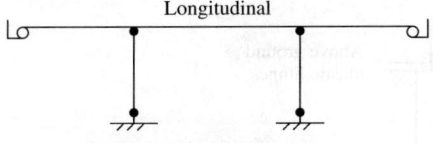

Abutment resistance not
required as part of ERS

Plastic hinges in inspectable locations
or elastic design of columns

Knock-off backwalls permissible

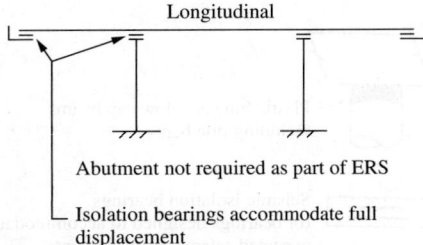

Abutment not required as part of ERS

Isolation bearings accommodate full
displacement

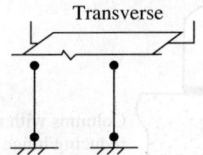

Abutment not required in ERS, breakaway
shear keys permissible

Plastic hinges in inspectable locations
or elastic design of columns

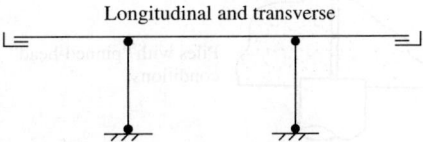

Isolation bearings with significant energy
dissipation capacity or energy dissipators are
used at the abutment to limit overall
displacements

Plastic hinges in inspectable locations
or elastic design of columns

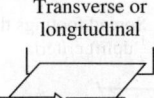

Abutment resistance required, but abutment
able to resist 3% in 75-year earthquake elastically
and passive soil pressure in longitudinal direction
is less than 0.70 × presumptive value given in
Article 7.5.2 in *Guidelines*

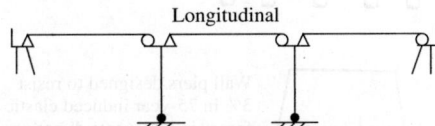

Multiple simply supported spans with adequate
seat widths. Plastic hinges in inspectable locations
or elastic design of columns

FIGURE 20.7 Permissible earthquake resisting systems [37].

performance. Higher level of performance may be required depending upon the bridge's importance and owner's requirements.

The seismic performance criteria shown in Table 20.1 shall be achieved by the following design objectives:

- *Columns as primary energy dissipation mechanism.* The main objective is to force the inelastic deformations to occur primarily in the columns in order that the earthquake damage can be easily inspected and readily repaired after an earthquake. The amount of longitudinal steel in the reinforced columns should be minimized to reduce foundation and connection costs.
- *Abutments as an additional energy dissipation mechanism.* The objective is to expect the inelastic deformations to occur in the columns as well as the abutments in order to either minimize column size and reduce ductility demand on the column.
- *Isolation bearings as main energy dissipation mechanism.* The objective is to lengthen the period of a relatively stiff bridge and which results in a lower design seismic force. Energy dissipation will occur in the isolation bearings and columns are usually then expected to perform elastically.
- *Structural components between deck and columns/abutments as energy dissipation mechanism.* The objective is to design ductile components that do not result in reduced design force but will reduce the ductility demands on the columns in order to minimize the energy that is dissipated in the plastic hinge zone of columns.

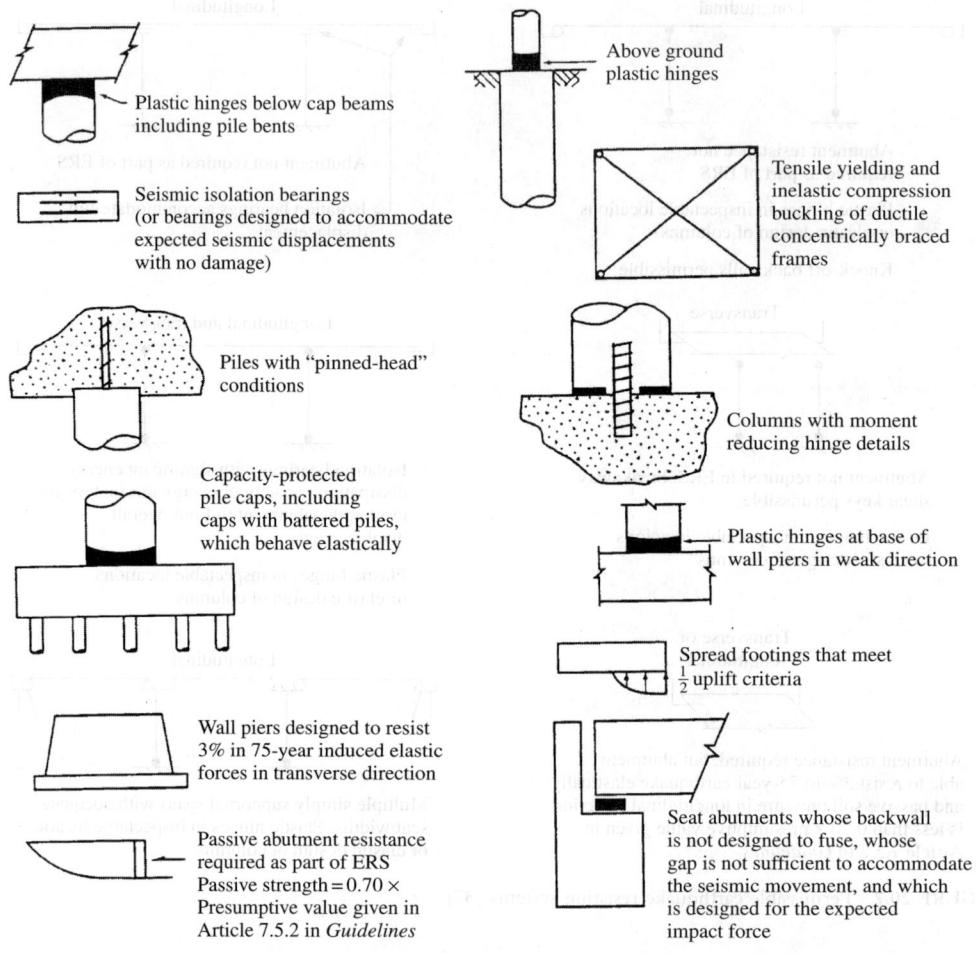

FIGURE 20.8   Permissible earthquake resisting elements [37].

- *Replaceable/renewable sacrificial plastic hinge elements as energy dissipation mechanism.* The objective is to control damage and permit significant inelastic deformation to occur at a specially designed replaceable/renewable sacrificial plastic hinge elements in the plastic hinge zone of a column. The concept is similar to the conventional ductile design concept that permits significant inelastic deformation in the plastic hinge zone of a column. The difference compared with the conventional ductile design is that construction details in the plastic hinge zone of concrete columns provide a replaceable/renewable sacrificial plastic hinge elements. The concept has been extensively tested [44] but has not been used in practice.

20.5.2 Caltrans

Table 20.2 outlines Caltrans seismic performance criteria [24] including the bridge classification, the service, and damage levels established in 1994 [15]. A bridge is categorized as "Important" or "Ordinary."
For Standard "Ordinary" bridges, the displacement-based one-level safety-evaluation design ("no-collapse" design) is only required in the Caltrans SDC [35]. Nonstandard "Ordinary" bridges feature irregular geometry and framing (multilevel, variable width, bifurcating, or highly horizontally

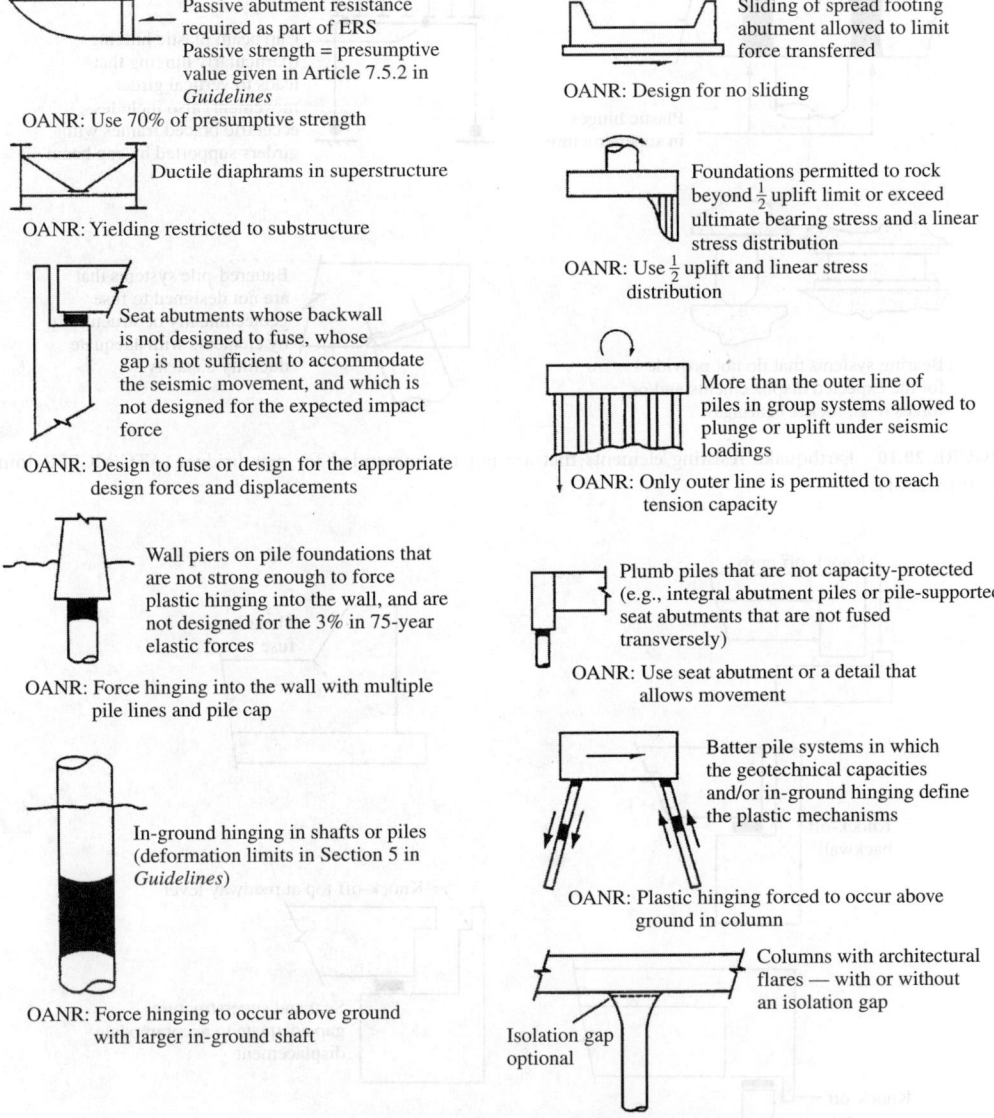

Passive abutment resistance required as part of ERS
Passive strength = presumptive value given in Article 7.5.2 in *Guidelines*

OANR: Use 70% of presumptive strength

Ductile diaphrams in superstructure

OANR: Yielding restricted to substructure

Seat abutments whose backwall is not designed to fuse, whose gap is not sufficient to accommodate the seismic movement, and which is not designed for the expected impact force

OANR: Design to fuse or design for the appropriate design forces and displacements

Wall piers on pile foundations that are not strong enough to force plastic hinging into the wall, and are not designed for the 3% in 75-year elastic forces

OANR: Force hinging into the wall with multiple pile lines and pile cap

In-ground hinging in shafts or piles (deformation limits in Section 5 in *Guidelines*)

OANR: Force hinging to occur above ground with larger in-ground shaft

Sliding of spread footing abutment allowed to limit force transferred

OANR: Design for no sliding

Foundations permitted to rock beyond $\frac{1}{2}$ uplift limit or exceed ultimate bearing stress and a linear stress distribution

OANR: Use $\frac{1}{2}$ uplift and linear stress distribution

More than the outer line of piles in group systems allowed to plunge or uplift under seismic loadings

OANR: Only outer line is permitted to reach tension capacity

Plumb piles that are not capacity-protected (e.g., integral abutment piles or pile-supported seat abutments that are not fused transversely)

OANR: Use seat abutment or a detail that allows movement

Batter pile systems in which the geotechnical capacities and/or in-ground hinging define the plastic mechanisms

OANR: Plastic hinging forced to occur above ground in column

Columns with architectural flares — with or without an isolation gap

Isolation gap optional

OANR: Remove flare

Note: OANR means a design alternate where owners approval is not required and a higher level of analysis (pushover in SDAP E) can be avoided.

FIGURE 20.9 Permissible earthquake resisting elements that require owner's approval [37].

curved superstructures, varying structure types, outriggers, unbalanced mass and stiffness, high skew) and unusual geologic conditions (soft soil, moderate to high liquefaction potential and proximity to an earthquake fault). In this case, project-specific criteria need to be developed to address their nonstandard features.

For important bridges such as the San Francisco-Oakland Bay Bridge and the Benicia-Martinez Bridge or structures in a designated lifeline route, performance-based project-specific two-level seismic design criteria [28–30] are required.

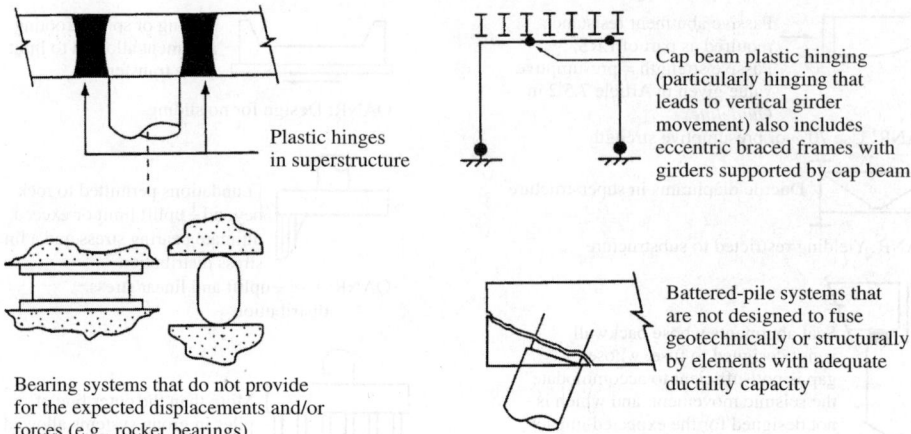

FIGURE 20.10   Earthquake resisting elements that are not recommended for new bridges (ATC/MCEER Joint Venture, 2001).

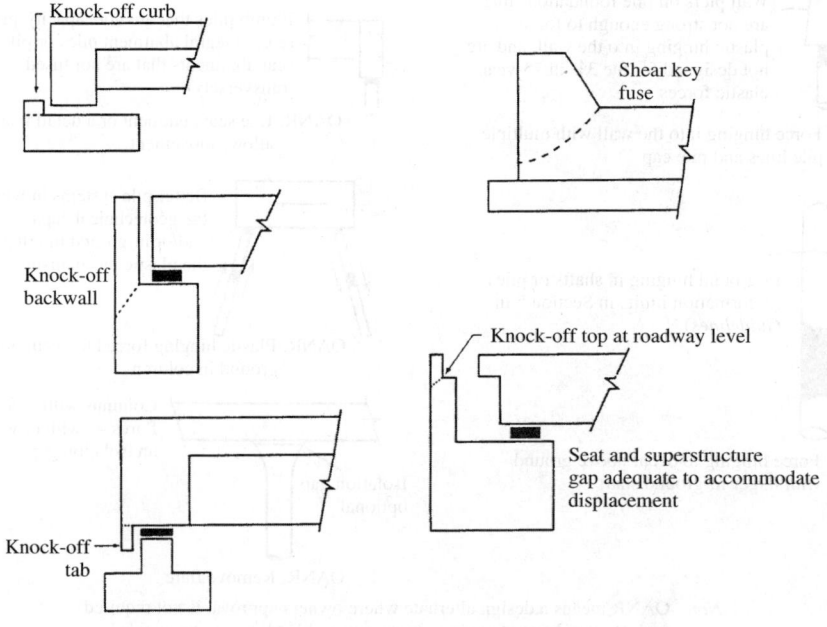

FIGURE 20.11   Methods of minimizing damage to abutment foundation [37].

The SDC shown in Table 20.2 shall be achieved by the following design objectives:

- All bridges shall be designed to withstand deformation imposed by the design earthquake.
- All structure components have sufficient strength and ductility to ensure collapse will not take place during the maximum credible earthquake (MCE). Ductile behavior can be provided by inelastic actions either through selected structural members and through protective systems — seismic isolations and energy dissipation devices.
- Inelastic behavior shall be limited to the preidentified locations, that is, ductile components explicitly designed for ductile performance, within the bridge that are easily inspected and

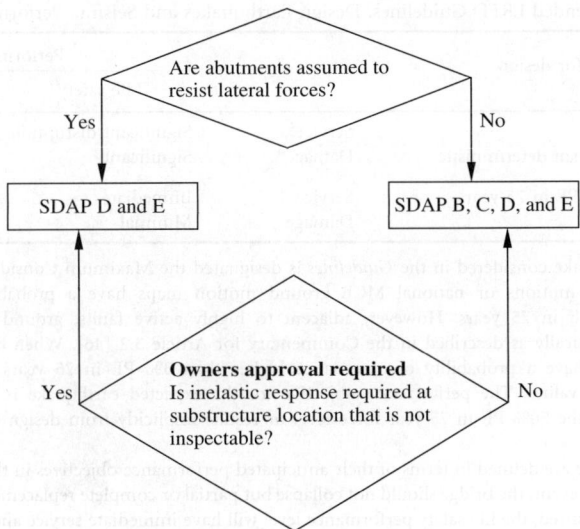

FIGURE 20.12 Design approaches for permissible earthquake resisting systems [37].

repaired following an earthquake. Because inelastic response of a concrete superstructure is difficult to inspect and repair and superstructure damage may cause a bridge to be unserviceable after an earthquake, inelastic behavior on most concrete bridges should be preferably be located in columns, pier walls, backwalls, and wingwalls.

- Structural components not explicitly designed for ductile performance, that is, capacity-protected components shall be designed to remain essentially elastic. That is, (1) response in concrete components shall be limited to minor cracking or limited to force demands not exceeding the nominal strength capacity determined by current Caltrans SDC [35] and (2) response in steel components shall be limited to force demands not exceeding the nominal strength capacity determined by the current Caltrans *Guide* [36]. To assure the yielding mechanism occurs in the desired location, the capacity design principle is used by providing overstrength for those capacity-protected components.

20.6 Seismic Design Approaches

20.6.1 ATC/MCEER Guidelines [37]

20.6.1.1 Seismic Loads

Seismic loads are represented by the design response spectrum curve for a damping ratio of 5% (Figure 20.13):

$$S_a = \begin{cases} 0.6\dfrac{S_{DS}}{T_0}T + 0.4S_{DS} & \text{for } T \leq T_0 = 0.2T_S \\[2mm] S_{DS} & \text{for } T_0 < T \leq T_S = \dfrac{S_{D1}}{S_{DS}} \\[2mm] \dfrac{S_{D1}}{T} & \text{for } T > T_S \end{cases} \tag{20.1}$$

$$S_{DS} = F_a S_s \tag{20.2}$$

$$S_{D1} = F_v S_1 \tag{20.3}$$

TABLE 20.1 Recommended LRFD Guidelines. Design Earthquakes and Seismic Performance Objectives [37]

Probability of exceedance for design earthquake ground motions[a]		Performance level[b]	
		Life safety	Operational
Rare earthquake (MCE)	Service[c]	Significant disruption	Immediate
3% PE in 75 years/1.5 mean deterministic	Damage[d]	Significant	Minimal
Expected earthquake 50% PE in 75 years	Service	Immediate	Immediate
	Damage	Minimal	Minimal to none

[a] The upper-level earthquake considered in the *Guidelines* is designated the Maximum Considered Earthquake or MCE. In general the ground motions or national MCE ground motion maps have a probability of exceedance (PE) of approximately 3% PE in 75 years. However, adjacent to highly active faults, ground motions on MCE maps are bounded deterministically as described in the Commentary for Article 3.2 [46]. When bounded deterministically, MCE ground motions have a probability of exceedance higher than 3% PE in 75 years not to exceed 1.5 times the mean deterministic values. The performance objective for the expected earthquake is either explicitly included as an elastic design for the 50% PE in 75-year force level or results implicitly from design for the 3% PE in 75-year force level.

[b] *Performance levels:* These are defined in terms of their anticipated performance objectives in the upper level earthquake. Life safety — In an MCE event, the bridge should not collapse but partial or complete replacement may be required. Since a dual level design is required, the life safety performance level will have immediate service and minimal damage for the expected design earthquake; Operational — For both rare and expected earthquakes, the bridge should be immediate service and minimal damage.

[c] *Service levels:* Immediate — Full access to normal traffic shall be available to traffic following an inspection of the bridge; Significant disruption — Limited access (reduced lanes, light emergency traffic) may be possible after shoring; however, the bridge may need to be replaced.

[d] *Damage levels:* None — Evidence of movement may be present but no notable damage; Minimal — Some visible signs of damage. Minor inelastic response may occur, but postearthquake damage is limited to narrow flexural cracking in concrete and the onset of yielding in steel. Permanent deformations are not apparent, and any repairs could be made under nonemergency conditions with the exception of superstructure joints; Significant — Although there is no collapse, permanent offsets may be occur and damage consisting of cracking, reinforcement yield, and major spalling of concrete and extensive yielding and local buckling of steel columns, global and local buckling of steel braces, and cracking in the bridge deck slabs at shear studs on the seismic load path is possible. These conditions may require closure to repair the damage. Partial or complete replacement of columns may be required in some cases. For sites with lateral flow due to liquefaction, significant inelastic deformation is permitted in the piles, whereas for all other sites the foundations are capacity-protected and no damage is permitted in the pile. Partial or complete replacement of the columns and piles may be necessary if significant lateral flow occurs. If replacement of columns or other components is to be avoided, the design approaches producing minimal or moderate damage such as seismic isolation or the control and repairability design concept should be assessed.

where S_{DS} is the design earthquake response spectral acceleration at short periods, S_{D1} is the design earthquake response spectral acceleration at 1-s period, S_s is the 0.2-s period spectral acceleration on Class B rock from national ground motion maps [45], S_1 is the 1-s period spectral acceleration on Class B rock from national ground motion maps [45], F_a is the site coefficient (Table 20.3) for the short-period portion of the design response spectrum curve, and F_v is the site coefficient (Table 20.4) for the long-period portion of the design response spectrum curve.

For Site Class F, which is not included in the Table 20.5, such as soils vulnerable to potential failure or collapse under seismic loading, such as liquefiable soils, quick and highly sensitive clays, and collapsible weakly cemented soils, site-specific geotechnical investigation and dynamic site response analyses shall be performed [37].

The effects of vertical ground motion may be ignored if the bridge site is located more than 50 km from an active fault. If the bridge is located within 10 km of an active fault, site-specific response spectra and acceleration time histories including vertical ground motions shall be considered.

TABLE 20.2   Caltrans Seismic Performance Criteria [24]

Ground motions at the site	Level of damage and postearthquake service	
	Ordinary bridge	Important bridge
Functional — evaluation ground motion	Service: immediate Damage: repairable	Service: immediate Damage: minimal
Safety — evaluation ground motion	Service: limited Damage: significant	Service: immediate Damage: repairable

Definition

Important bridge: A bridge meets one or more of the following requirements:
  • Required to provide postearthquake life safety; such as access to emergency facilities.
  • Time for restoration of functionality after closure would create a major economic impact.
  • Formally designed as critical by a local emergency plan.
Ordinary bridge: Any bridge not classified as an important bridge.

Functional — evaluation ground motion (FEGM): This ground motion may be assessed either deterministically or
  probabilistically. The determination of this event is to be reviewed by a Caltrans-approved consensus group.
Safety — evaluation ground motion (SEGM): This ground motion may be assessed either deterministically
  or probabilistically. The deterministic assessment corresponds to the maximum credible earthquake (MCE).
  The probabilistic ground motion for the safety evaluation typically has a long return period (approximately 1000
  to 2000 years).

MCE: The largest earthquake that is capable of occurring along an earthquake fault, based on current geologic information
  as defined in the 1996 Caltrans Seismic Hazard Map.

Service levels
  • *Immediate:* Full access to normal traffic is available almost immediately following the earthquake.
  • *Limited:* Limited access (e.g., reduced lanes, light emergency traffic) is possible with days of the earthquake. Full service
    is restorable within months.

Damage levels
  • *Minimal:* Essentially elastic performance.
  • *Repairable:* Damage that can be repaired with a minimum risk of losing functionality.
  • *Significant:* A minimum risk of collapse, but damage that would require closure to repair.

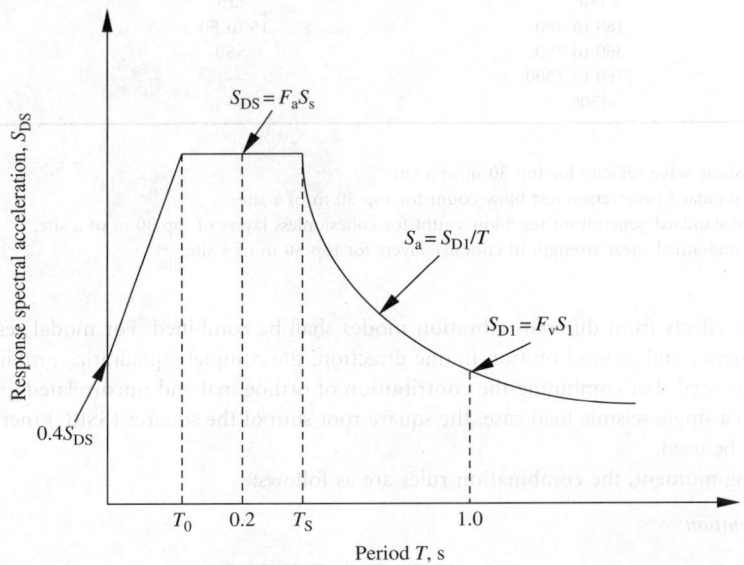

FIGURE 20.13   Design response spectrum curve [37].

TABLE 20.3   Recommended LRFD *Guidelines* — Site Coefficient F_a [37]

Site class	Mapped spectral response acceleration at short periods				
	$S_s \leq 0.25g$	$S_s = 0.5g$	$S_s = 0.75g$	$S_s = 1.00g$	$S_s = 1.25g$
A	0.8	0.8	0.8	0.8	0.8
B	1.0	1.0	1.0	1.0	1.0
C	1.2	1.2	1.1	1.0	1.0
D	1.6	1.4	1.2	1.1	1.0
E	2.5	1.7	1.2	0.9	0.9
F	a.0	a.0	a.0	a.0	a.0

[a] Site-specific geotechnical investigation and dynamic site response analysis shall be performed. For purpose of defining seismic hazard levels, Type E values may be used for Type F soils.

Note: Use straight line interpolation for intermediate values of S_s.

TABLE 20.4   Recommended LRFD *Guidelines* — Site Coefficient F_v [37]

Site class	Mapped spectral response acceleration at 1-s period				
	$S_s \leq 0.1g$	$S_s = 0.2g$	$S_s = 0.3g$	$S_s = 0.4g$	$S_s \geq 0.5g$
A	0.8	0.8	0.8	0.8	0.8
B	1.0	1.0	1.0	1.0	1.0
C	1.7	1.6	1.5	1.4	1.3
D	2.4	2.0	1.8	1.6	1.5
E	3.5	3.2	2.8	2.4	2.4
F	a.0	a.0	a.0	a.0	a.0

[a] Site-specific geotechnical investigation and dynamic site response analysis shall be performed. For purpose of defining seismic hazard levels, Type E values may be used for Type F soils.

Note: Use straight line interpolation for intermediate values of S_1.

TABLE 20.5   Recommended LRFD *Guidelines* — Site Classification [37]

Site class	$\bar{v}_s$ (m/s)	$\bar{N}$ or $\bar{N}_{ch}$ (blows/0.3 m)	$\bar{s}_u$ (kPa)
E	<180	<15	<50
D	180 to 360	15 to 50	50 to 100
C	360 to 760	>50	>100
B	760 to 1500	—	—
A	>1500	—	—

Notes:

$\bar{v}_s$ is the average shear wave velocity for top 30 m of a site.

$\bar{N}$ is the average standard penetration test blow count for top 30 m of a site.

$\bar{N}_{ch}$ is the average standard penetration test blow count for cohesionless layers of top 30 m of a site.

$\bar{s}_u$ is the average undrained shear strength of cohesive layers for top 30 m of a site.

Seismic force effects from different vibration modes shall be combined. For modal response closely spaced in frequency and ground motion in one direction, the complete quadratic combination (CQC) method shall be used. For combining the contribution of orthogonal and uncorrelated ground motion components to a single seismic load case, the square root sum of the squares (SSRC) method or 100 to 40% rule shall be used.

For a bending moment, the combination rules are as follows:

SSRC combination:

$$M_x = \sqrt{\left(M_x^T\right)^2 + \left(M_x^L\right)^2 + \left(M_x^V\right)^2} \tag{20.4}$$

100 to 40% combination:

$$M_x^{LC1} = 1.0M_x^T + 0.4M_x^L + 0.4M_x^V \qquad (20.5)$$

$$M_x^{LC2} = 0.4M_x^T + 1.0M_x^L + 0.4M_x^V \qquad (20.6)$$

$$M_x^{LC3} = 0.4M_x^T + 0.4M_x^L + 1.0M_x^V \qquad (20.7)$$

For circular columns, the vector moments and axial forces shall be obtained for biaxial design:

SSRC combination:

- For bridges with skew angle less than $10°$, the maximum of $\sqrt{M_x^2 + (0.4M_y)^2}$ and $\sqrt{M_y^2 + (0.4M_x)^2}$ with the maximum axial load $\pm P$
- For bridges with skew angle greater than $10°$, $\sqrt{M_x^2 + M_y^2}$ with the maximum axial load $\pm P$

100 to 40% combination:

- The maximum of $\sqrt{(M_x^{LC1})^2 + (M_y^{LC1})^2}$ and $\sqrt{(M_x^{LC2})^2 + (M_y^{LC2})^2}$ and $\sqrt{(M_x^{LC3})^2 + (M_y^{LC3})^2}$ with the maximum axial load $\pm P$

where subscripts x and y represent two horizontal axes, x–x and y–y, respectively; superscripts L, T, and V indicate the longitudinal, transverse, and vertical directions, respectively; and superscripts LC1, LC2, and LC3 are load cases 1, 2, and 3, respectively.

20.6.1.2 Seismic Design and Analysis Procedures (SDAP)

Depending on the seismic hazard levels specified in Table 20.6, each bridge shall be designed, analyzed, and detailed in accordance with Table 20.6.

20.6.1.2.1 Single-Span Bridges

Single-span bridges need not be analyzed for seismic loads and design requirements are limited to the minimum seat widths and connection forces, which shall not be less than the product of $F_aS_S/2.5$ and the tributary permanent load.

20.6.1.2.2 SDAP A1 and A2

For low seismicity areas, only minimum seat widths and connection design forces for bearings and minimum shear reinforcement in concrete columns and piles in the seismic design requirement (SDR) 2 are deemed necessary for the life safety performance objective. The primary purpose is to ensure that the connections between the superstructure and its supporting substructures remain intact during the design earthquake. SDAP A1 and A2 require that the horizontal design connection forces in the restrained directions shall not be taken to be less than 0.1 and 0.25 times the vertical reactions due to tributary permanent loads and assumed existing live loads, respectively.

TABLE 20.6 Recommended LRFD *Guidelines* — SHL, SDAP, and SDR [37]

Seismic hazard level (SHL)	S_{D1} (F_vS_1)	S_{DS} (F_aS_s)	Seismic design and analysis procedure (SDAP) and seismic design requirements (SDR)			
			Life safety		Operational	
			SDAP	SDR	SDAP	SDR
I	$0.15 < S_{D1} \leq 0.15$	$0.15 < S_{DS} \leq 0.15$	A1	1	A2	2
II	$0.15 < S_{D1} \leq 0.25$	$0.15 < S_{DS} \leq 0.35$	A2	2	C/D/E	3
III	$0.25 < S_{D1} \leq 0.40$	$0.35 < S_{DS} \leq 0.60$	B/C/D/E	3	C/D/E	5
IV	$0.40 < S_{D1}$	$0.60 < S_{DS}$	C/D/E	4	C/D/E	6

20.6.1.2.3 SDAP B — No-Analysis Approach

The no-analysis approach allows for the bridge to be designed for all nonseismic requirements without a seismic demand analysis and the capacity design principle is used for all components connected to columns. For geotechnical design of the foundations, the moment overstrength capacity of columns that frame into the foundation, $M_{po} = 1.0M_n$, where M_n is nominal moment capacity of a column. SDAP B applies only to regular bridges meeting the following restrictions:

- The maximum span length is less than both 80 m and 1.5 times the average span length.
- The maximum skew angle is less than 30°.
- The ratio of the maximum interior bent stiffness to the average bent stiffness of the bridge is less than 2.
- The subtended angle in the horizontally curved bridges is less than 30°.
- For frames in which the superstructure is continuous over the bents and some bents do not participate in the ERS, $F_v S_1 (N_{bent}/N_{ers}) < 0.4 \cos \alpha_{skew}$, where N_{bent} and N_{ers} are total number of bents in the frame and number of bents participating in the ERS in the longitudinal direction, respectively; and α_{skew} is the skew angle of the bridge.
- $F_v S_1 < 0.4 \cos \alpha_{skew}$.
- The bridge site has a low potential for liquefaction and the piers are not seated on spread footing.
- For concrete column and pile bents, column axial load, $P_e < 0.15 f'_c A_g$ (f'_c is specified minimum concrete compression strength and A_g is gross cross-sectional area of column); longitudinal reinforcement ratio, $\rho_l > 0.008$; column transverse dimension, $D > 300$ mm; and maximum column moment–shear ratio, $M/VD < 6$.
- For concrete wall piers with low volumes of longitudinal steel, $P_e < 0.1 f'_c A_g$; $\rho_l > 0.0025$; wall thickness or smallest cross-sectional dimension, $t > 300$ mm; and $M/Vt < 10$.
- For steel pile bents framing into concrete caps, $P_e < 0.15 P_y$ (P_y is axial yield force of steel pile); pile dimension about the weak axis bending at the ground level, $D_p > 250$ mm; and $L/b < 10$ (L is the length from point of maximum moment to the inflexion point of the pile when subjected to a pure transverse load and b is the flange width; for a cantilever column in a pile bent configuration, L is equal to the length above ground to the top of the bent plus 3 pile diameters).
- For timber piles framing into concrete caps or steel moment–frame columns, $P_e < 0.1 P_c$ (and P_c is axial compression capacity of the pile or the column); $D_p > 250$ mm; and $M/VD_p < 10$.

20.6.1.2.4 SDAP C — Capacity Spectrum Design Method

The capacity spectrum design method combines a demand and capacity analysis and is conceptually the same as the Caltrans displacement-based design method. The primary difference is that the SDAP C begins with nonseismic design and then assesses the adequacy of the displacement. The key equations is the relationship between the seismic coefficient C_s and displacement, Δ:

$$C_s \Delta = \left(\frac{F_v S_1}{2\pi B_L} \right)^2 g \tag{20.8}$$

$$C_s = \frac{F_a S_s}{B_S} \tag{20.9}$$

where B_L is the response reduction factor for long period structures as specified in Table 20.7, B_S is the response reduction factor for short period structures as specified in Table 20.7, g is the acceleration due to gravity (9.8 m/s^2), and Δ is the lateral displacement of the pier; taken as 1.3 times the yield displacement of the pier when the long period equation governs.

The lesser of Equations 20.8 and 20.9 shall be used to assess C_s for two-level earthquakes. The required lateral strength of the bridge is $V_n = C_s W$ where W is the weight of the bridge responding

TABLE 20.7 Recommended LRFD *Guidelines* — Capacity Spectrum Response Reduction
Factors for Bridge with Ductile Piers [37]

Earthquake	Performance level	B_s	B_L
50% PE in 75-year	Life safety	1	1
	Operational	1	1
3% PE in 75-year/1.5 mean deterministic	Life safety	2.3	1.6
	Operational	1	1

to earthquake ground motion. The procedure applies only to bridges that behave essentially as a single
degree-of-freedom system and very regular bridges satisfying the following requirements:

- The number of spans per frame does not exceed six.
- The number of spans per frame is at least three, unless seismic isolation bearings are utilized
  at the abutments.
- The maximum span length is less than both 60 m and 1.5 times the average span length in a frame.
- The maximum skew angle is less than 30° and skew of piers or bents differs by less than 5° in the
  same direction.
- The subtended angle in the horizontally curved bridges is less than 20°.
- The ratio of the maximum bent or pier stiffness to the average bent stiffness is less than 2,
  including the effects of foundation.
- The ratio of the maximum lateral strength (or seismic coefficient) to the average bent strength is
  less than 1.5.
- Abutment shall not be assumed to resist the significant forces in both the transverse and
  longitudinal directions. Pier wall substructures must have bearings to permit transverse
  movement.
- For concrete column and pile bents, $P_e \leq 0.2 f'_c A_g$; $\rho_l > 0.008$; and $D > 300$ mm.
- Piers and bents must have pile foundations when the bridge site has a potential for liquefaction.

20.6.1.2.5 SDAP D — Elastic Response Spectrum Method

The elastic response spectrum method uses either the uniform load or multimode method of analysis
by considering cracked section properties. The analysis shall be performed for the governing design
earthquakes, either the 50% PE (probability of exceedence) in 75-year or the 3% PE in 75-year/1.5 mean
deterministic earthquake. Elastic forces obtained from analyses shall be modified using the response
modification factor R.

20.6.1.2.6 SDAP E — Elastic Response Spectrum Method with Displacement Capacity Verification

SDAP E is a two-step design procedure. The first step is the same as SDAP D and the second step is
to perform a two-dimensional nonlinear static (push over) analysis to verify substructure displacement
capacity.

20.6.1.3 Response Modification Factor R

It is generally recognized that it is uneconomical, sometimes impractical, and impossible to design
a bridge to resist large earthquake elastically. Columns are assumed to deform inelastically where seismic
forces exceed their design levels, which is established by dividing the elastically computed force effects
by the response modification factor R. The R-factors specified in the following were based on an
evaluation of existing test data, engineering judgement, and the equal displacement principle as shown
in Figure 20.14. It is used in principle for the conventional ductile design.
   For substructures

$$R = 1 + (R_B - 1)\frac{T}{T^*} \leq R_B \qquad (20.10)$$

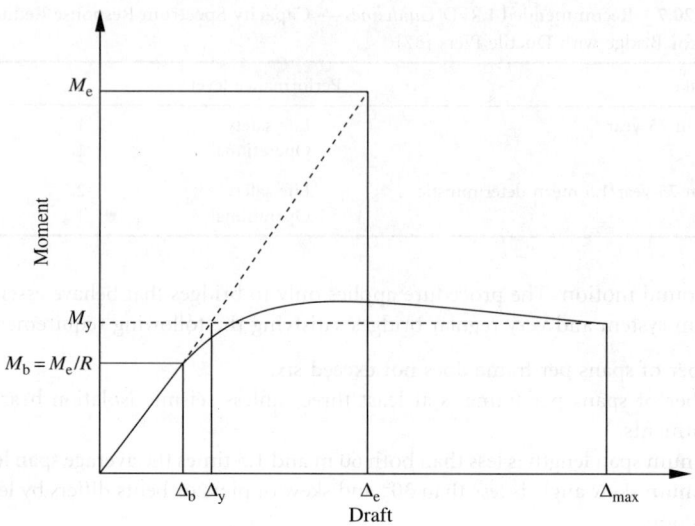

FIGURE 20.14   Basis for conventional ductile design [37].

TABLE 20.8   Recommended LRFD *Guidelines* — Base Response Modification Factor for Substructures, R_B (ATC/MCEER Joint Venture, 2001)

	Performance objective			
	Life safety		Operational	
Substructure element	SDAP D	SDAP E	SDAP D	SDAP E
Wall-type pier — large dimension	2.0	3.0	1.0	1.5
Columns — single and multiple	4.0	6.0	1.5	2.5
Pile bents and drilled shafts — vertical piles — above ground	4.0	6.0	1.5	2.5
Pile bents and drilled shafts — vertical piles — 2 diameters below ground level — no owners approval required	1.0	1.5	1.0	1.0
Pile bents and drilled shafts — vertical piles — in ground — owners approval required	N/A	2.5	N/A	1.5
Pile bents with batter piles	N/A	2.5	N/A	1.5
Seismically isolated structures	1.5	1.5	1.0	1.5
Steel-braced frame — ductile components	3.0	4.5	1.0	1.5
Steel-braced frame — normally ductile components	1.5	2.0	1.0	1.0
All elements for expected earthquake	1.3	1.3	0.9	0.9

Notes:

1. The substructure design forces resulting from the elastic analysis divided by the appropriate *R*-factor for SDAP E can not be reduced below 70% at the *R*-factored reduced forces or the 50% PE in 75-year design forces as part of the pushover analysis.
2. There may be design situation (e.g., architecturally oversized column) where a designer opts to design the column for an $R = 1$ (i.e., elastic design). In concrete columns the associate elastic design shear may be obtained from the elastic analysis forces using an *R*-factor of 0.67 or by calculating the design shear by the capacity design procedures using a flexural overstrength factor of 1.0. In steel-braced frame if an $R = 1.0$ is used the connection design forces shall be obtained using an $R = 0.67$. If an $R = 1.0$ is used in any design the foundation shall be designed for elastic forces plus the SDR detailing requirements are required for concrete piles (i.e., minimum shear requirements).
3. Unless specifically stated, the *R*-factor applies to both steel and concrete.
4. N/A means that owners approval is required and thus the use of SDAP E is required.

where R_B is the base response modification factor specified in Table 20.8, T is the natural period of the structure, $T^* = 1.25T_s$ where T_s is as defined in Section 20.6.1.1.

For connections (superstructure to abutment; expansion joints within a span of the superstructure; columns, piers, or pile bents to cap beam or superstructure; and column or piers to foundations), an R-factor of 0.8 shall be used for those cases where capacity design principles are not used to develop the design forces to design the connections. It is assumed that if the $R < 1.5$, columns should remain essentially elastic for design earthquake; if $1.5 < R > 3.0$, columns should be repairable; if $R > 3.0$ significant plastic hinging may occur and the column may not be repairable; however, collapse is still prevented.

20.6.1.4 Capacity Design Principle

The main objective of the capacity design principle is to ensure the desirable mechanisms can dissipate significant amounts of energy and inelastic deformation (plastic hinging) occurs at expected locations (at top and bottom of columns) where they can be readily inspected and repaired. To achieve this objective, the overstrength force effects developed from the plastic hinges in columns shall be dependably resisted by column shear and adjoining elements such as cap beams, spread footing, pile cap, and foundations. The moment overstrength capacity (M_{po}) can be assessed using one of the following approaches:

- $$M_{po} = \begin{cases} 1.5M_n & \text{for concrete column} \\ 1.2M_n & \text{for steel column, } M_n \text{ based on expected yield strength} \\ 1.3M_n & \text{for concrete filled steel tubes} \\ 1.5M_n & \text{for steel piles in weak axis bending and for steel members} \\ & \quad \text{in shear (e.g., eccentrically braced frames)} \\ 1.0M_n & \text{for geotechnical design force in SDR3} \end{cases} \tag{20.11}$$

- For reinforce concrete columns [46]

$$M_{po} = M_{bo}\left[1 - \left(\frac{P_e - P_b}{P_{to} - P_b}\right)^2\right] \tag{20.12}$$

where

P_e = axial compression load based on gravity load and seismic (framing) action
P_b = axial compression capacity at the maximum nominal (balanced) moment on the section
$\quad = 0.425\,\beta_1 f_c' A_g$
$\quad \beta_1$ = compression stress block factor ≤ 0.85
P_{to} = axial tensile capacity of the column $= -A_{st}f_{su}$
$\quad A_{st}$ = area of longitudinal reinforcement
$\quad f_{su}$ = ultimate tensile strength of the longitudinal reinforcement

$$M_{bo} = K_{shape}A_{st}f_{su}D' + P_bD\left(\frac{1 - \kappa_0}{2}\right) \tag{20.13}$$

D' = pitch circular diameter of the reinforcement in a circular section or the out-to-out dimension of the reinforcement in a rectangular section, this generally may be assumed as $= 0.8D$

$$K_{shape} = \begin{cases} 0.32 & \text{for circular sections} \\ 0.375 & \text{for square sections with 25\% of the longitudinal} \\ & \quad \text{reinforcement placed in each face} \\ 0.25 & \text{for walls with strong axis bending} \\ 0.5 & \text{for walls with weak axis bending} \end{cases} \tag{20.14}$$

κ_0 = a factor related to the centroid of compression stress block and should be taken as 0.6 and 0.5 for circular and rectangular sections, respectively.

It should be pointed out that Equations 20.12 and 20.13 are rearranged in the simpler format given above.

- For reinforced concrete column, a moment–curvature section analysis taking into account the expected strength, confined concrete properties, and strain hardening effects of longitudinal reinforcement.
- For a steel column, nominal flexural resistance (M_n) shall be determined either in accordance with the AASHTO-LRFD [38] or

$$M_n = 1.18 M_{px} \left[1 - \frac{P_u}{A_g F_{ye}} \right] \leq M_{px} \tag{20.15}$$

A_g = gross cross-sectional area of a steel column
F_{ye} = expected specified minimum yield strength of steel
M_{px} = plastic moment under pure bending calculated using F_{ye}
P_u = factored axial compression load

20.6.1.5 Plastic Hinge Zones

The plastic hinge zones (L_p) for typical concrete and steel columns, pile bents, and drilled shaft and zones of a columns above a footing or above an oversized in-ground drilled shaft shall be the maximum of the following:

For reinforced concrete columns

$$L_p = \text{maximum of} \begin{cases} D_{max} \\[4pt] \dfrac{L}{6} \\[4pt] D\left(\cot \theta + \dfrac{\tan \theta}{2} \right) \\[4pt] 1.5\left(0.08 \dfrac{M}{V} + 4000\, \varepsilon_y\, d_b \right) \\[4pt] \dfrac{M}{V}\left(1 - \dfrac{M_y}{M_{po}} \right) \\[4pt] 450 \text{ mm} \end{cases} \tag{20.16}$$

where

D = transverse column dimension in the direction of bending
D_{max} = maximum cross-sectional dimension of a column
d_b = diameter of longitudinal reinforcement
L = clear height of a column
M = maximum column moment
V = maximum column shear
M_y = column yield moment
ε_y = yield strain of longitudinal reinforcement
θ = principal crack angle
    = $\tan^{-1}((1.6/\Lambda)(\rho_v/\rho_t)(A_v/A_g))^{0.25}$ with $\theta \geq 25°$ and $\theta \geq \tan^{-1}(D'/L)$
A_v = shear area of concrete which may be taken as $0.8A_g$ and $b_w d$ for a circular section and a rectangular section, respectively
ρ_v = ratio of transverse reinforcement
ρ_t = volumetric ratio of longitudinal reinforcement
Λ = fixity factor taken as 1 for fixed-pinned and 2 for fixed-fixed ends

For steel columns:

$$L_p = \text{maximum of} \begin{cases} \dfrac{L}{8} \\[4pt] 450 \text{ mm} \end{cases} \tag{20.17}$$

For a flared column, the plastic hinge zone shall be extended from the top of column to a distance equal to the maximum of the above criteria below the bottom of the flare. The areas within the plastic hinge zones shall be detailed to assure ductile element behavior (e.g., providing confinement reinforcement in concrete columns or inelastic local buckling behavior in steel columns).

20.6.2 Caltrans Seismic Design Criteria

20.6.2.1 Seismic Loads

For ordinary bridges, safety-evaluation ground motion is based on deterministic assessment corresponding to the MCE, the largest earthquake, which is capable of occurring based on current geologic information. A set of ARS curves developed by ATC-32 are adopted as standard horizontal ARS curves in conjunction with the peak rock acceleration from the Caltrans Seismic Hazard Map 1996 to determine the horizontal earthquake forces. Figure 20.15 shows typical ARS curves. Vertical acceleration shall be considered for bridges with nonstandard structural components, unusual site conditions, and/or close

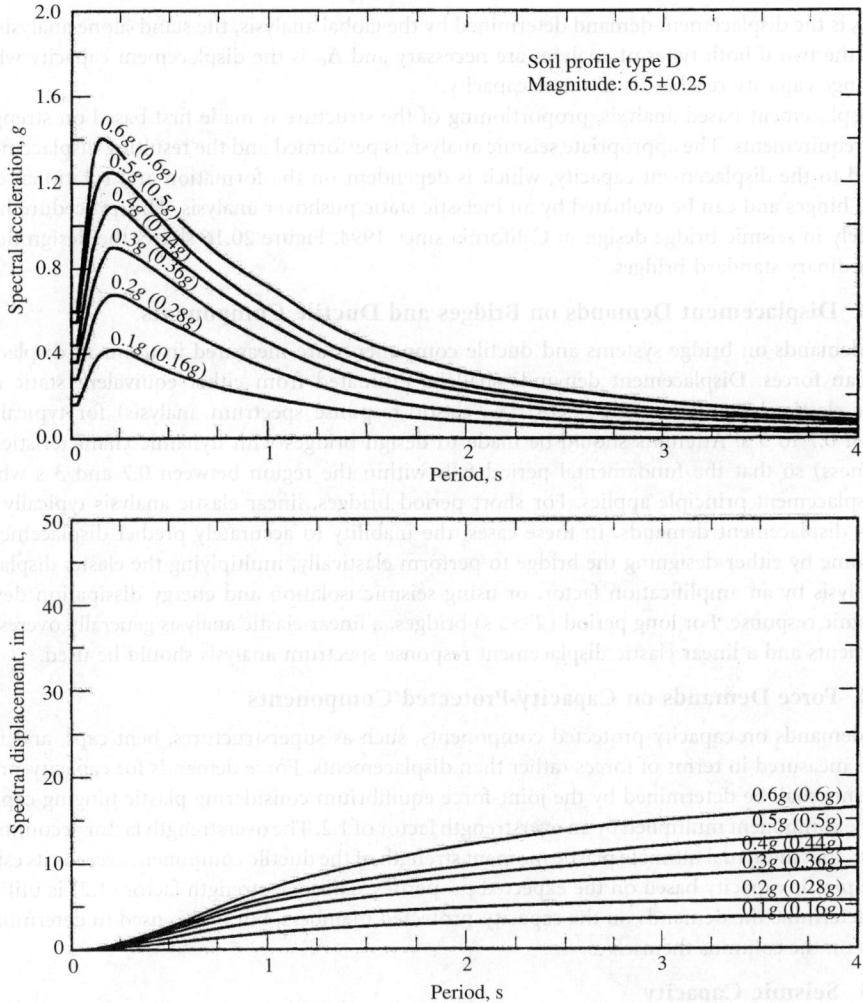

FIGURE 20.15 Typical Caltrans ARS curves [35].

Note: Peak ground acceleration values not in parentheses are for rock (soil profile type B) and peak ground acceleration values in parentheses are for soil profile type D.

proximity to earthquake faults and can be approximated by an equivalent static vertical force applied to the superstructure.

For structures within 15 km from an active fault, the spectral ordinates of the appropriate standard ARS curves shall be increased by 20%. For long period structures ($T \geq 1.5$ s) on deep soil sites (depth of alluvium ≥ 75 m) the spectral ordinates of the appropriate standard ARS curves shall be increased by 20% and the increase applies to the portion of the curves with periods greater than 1.5 s.

20.6.2.2 Design Approaches

The displacement-based design approach is used to ensure that the structural system and its individual components have enough displacement capacities to withstand the deformation imposed by the design earthquake. Using displacements rather than forces as a measurement of earthquake damages allows a structure to fulfill the required functions.

The displacements of the global system and the local ductile system shall satisfy the following requirement:

$$\Delta_D \leq \Delta_C \tag{20.18}$$

where Δ_D is the displacement demand determined by the global analysis, the stand-alone analysis, or the larger of the two if both types of analyses are necessary and Δ_C is the displacement capacity when any plastic hinge capacity reaches its ultimate capacity.

In a displacement-based analysis, proportioning of the structure is made first based on strength and stiffness requirements. The appropriate seismic analysis is performed and the resulting displacements are compared to the displacement capacity, which is dependent on the formation and rotational capacity of plastic hinges and can be evaluated by an inelastic static pushover analysis. This procedure has been used widely in seismic bridge design in California since 1994. Figure 20.16 shows the design flowchart for the ordinary standard bridges.

20.6.2.3 Displacement Demands on Bridges and Ductile Components

Seismic demands on bridge systems and ductile components are measured in terms of displacements rather than forces. Displacement demands shall be estimated from either equivalent static analysis (ESA) or elastic dynamic analysis (EDA, i.e., elastic response spectrum analysis) for typical bridge periods of 0.7 to 3 s. Attempts should be made to design bridges with dynamic characteristics (mass and stiffness) so that the fundamental period falls within the region between 0.7 and 3 s where the equal displacement principle applies. For short period bridges, linear elastic analysis typically underestimates displacement demands. In these cases, the inability to accurately predict displacements can be overcome by either designing the bridge to perform elastically, multiplying the elastic displacement from analysis by an amplification factor, or using seismic isolation and energy dissipation devices to limit seismic response. For long period ($T > 3$ s) bridges, a linear elastic analysis generally overestimates displacements and a linear elastic displacement response spectrum analysis should be used.

20.6.2.4 Force Demands on Capacity-Protected Components

Seismic demands on capacity-protected components, such as superstructures, bent caps, and foundations, are measured in terms of forces rather than displacements. Force demands for capacity-protected components shall be determined by the joint-force equilibrium considering plastic hinging capacity of the ductile component multiplied by an overstrength factor of 1.2. The overstrength factor accounts for the possibility that the actual ultimate plastic moment strength of the ductile component exceeds its estimated idealized plastic capacity based on the expected properties. This overstrength factor (1.2) is utilized not only to determine the demands on the capacity-protected members, but is also used to determine shear demands on the columns themselves.

20.6.2.5 Seismic Capacity

20.6.2.5.1 General

Strength and displacement capacities of a ductile flexural element shall be evaluated by moment–curvature analysis based on the expected material properties and anticipated damages. The impact the second-order

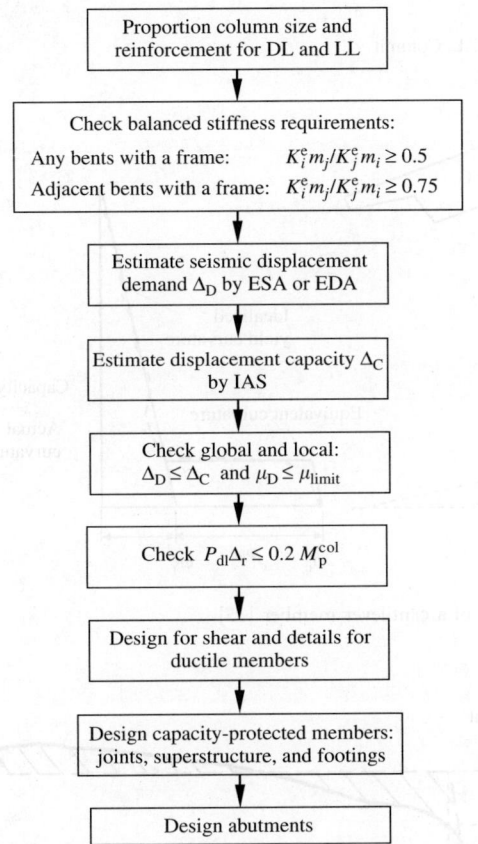

FIGURE 20.16 Caltrans seismic procedure for ordinary standard bridges.

P–Δ effect on the strength and displacement capacities of all members subjected to combined bending and compression shall be considered. Components may require redesign if the P–Δ effect is significant.

20.6.2.5.2 Displacement Capacity

The displacement capacity of a bridge system shall be evaluated by an inelastic static analysis (i.e., a static push over analysis). The rotational capacity of all plastic hinges shall be limited to a safe performance level. The plastic hinge regions shall be designed and detailed to perform with minimal strength degradation under cyclic loading.

The displacement capacity of a local member can be evaluated by its rotational capacity. The displacement capacity of a prismatic cantilever member (Figure 20.17) can be calculated as

$$\Delta_c = \Delta_Y^{col} + \Delta_p \tag{20.19}$$

$$\Delta_Y^{col} = L^2/3 \times \phi_Y \tag{20.20}$$

$$\Delta_p = \theta_p \times \left(L - \frac{L_p}{2}\right) \tag{20.21}$$

$$\theta_p = L_p \times \phi_p \tag{20.22}$$

$$\phi_p = \phi_u - \phi_Y \tag{20.23}$$

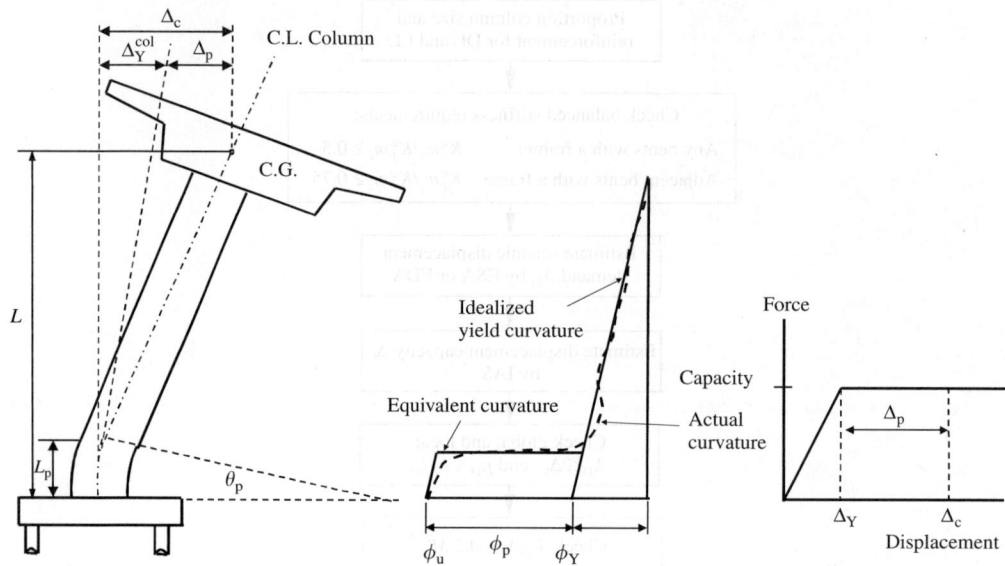

FIGURE 20.17   Displacement of a cantilever member [35].

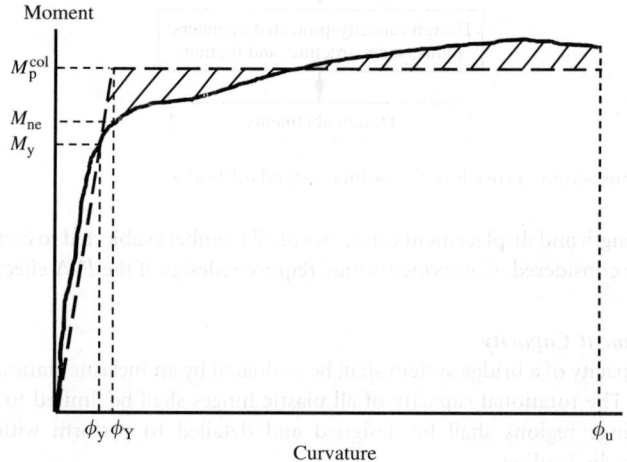

FIGURE 20.18   Idealized moment–curvature curve [35].

where

L = distance from the point of maximum moment to the point of contraflexure

L_p = equivalent analytical plastic hinge length as defined in Section 20.8.2.6

Δ_p = idealized plastic displacement capacity due to rotation of the plastic hinge

Δ_Y^{col} = idealized yield displacement of the column at the formation of the plastic hinge

ϕ_Y = idealized yield curvature defined by an elastic–perfectly plastic representation of the cross-section's M–ϕ curve, see Figure 20.18

ϕ_p = idealized plastic curvature capacity (assumed constant over L_p)

ϕ_u = curvature capacity at the failure limit state, defined as the concrete strain reaching ε_{cu} or the main column reinforcing steel reaching the reduced ultimate strain ε_{su}^R

θ_p = plastic rotation capacity of an equivalent plastic hinge

However, it should be pointed out that Equation 20.19 might overestimate the displacement capacity for a reinforced concrete column [47]. Column slenderness, high compression axial loads, and a low percentage of reinforcement all may contribute to the overestimating of the displacement capacity. Special attention, therefore, should be paid to the estimation of displacement capacity. It was recommended [47] that the P–Δ effect should be taken into account in calculating lateral load-carrying capacity and the displacement capacity, especially for medium-long and long columns. The lateral displacement capacity Δ_c can be chosen as the displacement that corresponds to the condition when lateral load carrying capacity degrades to a certain acceptable level, say a minimum of 80% of the peak resistance (Figure 20.19) or peak load [47–49].

20.6.2.5.3 Shear Capacity

Shear capacity of concrete members shall be calculated using nominal material properties as

$$\phi V_n = \phi(V_c + V_s) \tag{20.24}$$

$$\phi = 0.85 \tag{20.25}$$

$$V_c = v_c(0.8A_g) \tag{20.26}$$

Concrete shear capacity is influenced by flexural and axial loads and is calculated separately for regions within the plastic hinge zone and regions outside this zone. In the plastic hinge zone, concrete shear capacity is modified based on the level of confinement and the displacement ductility demand (Figure 20.20).

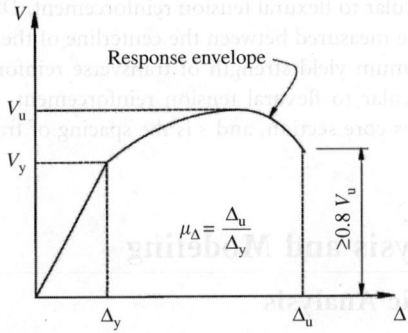

FIGURE 20.19 Lateral load–displacement curve [36].

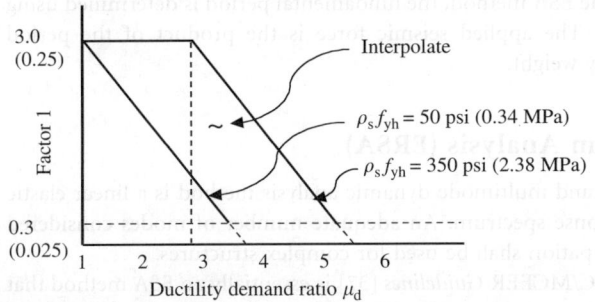
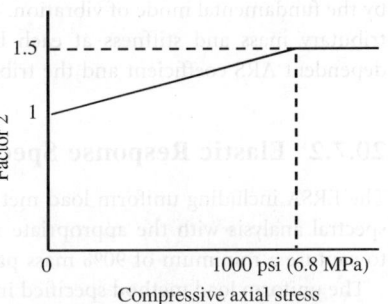

FIGURE 20.20 Shear factors [35].

- Inside the plastic hinge zone

$$v_c = \text{Factor } 1 \times \text{Factor } 2 \times \sqrt{f_c'} \le 0.33\sqrt{f_c'} \quad (\text{MPa}) \tag{20.27}$$

- Outside the plastic hinge zone

$$v_c = 0.25 \times \text{Factor } 2 \times \sqrt{f_c'} \le 0.33\sqrt{f_c'} \quad (\text{MPa}) \tag{20.28}$$

$$\text{Factor } 1 = 0.025 \le \frac{\rho_s f_{yh}}{12.5} + 0.305 - 0.083\mu_d < 0.25 \tag{20.29}$$

$$\text{Factor } 2 = 1 + \frac{P_c}{13.8\,A_g} \le 1.5 \tag{20.30}$$

To ensure reliable capacity in the plastic hinge regions, all column lateral reinforcements are required to be butt welded or spliced hoops capable of resisting the ultimate capacity of the reinforcing steel.

- For confined circular or interlocking core sections:

$$V_s = \frac{\pi}{2}\frac{nA_b f_{yh} D'}{s} \tag{20.31}$$

- For pier wall in weak direction

$$V_s = \frac{A_v f_{yh} d}{s} \tag{20.32}$$

where A_b is the area of an individual interlock spiral or hoop bar, A_v is the total area of shear reinforcement perpendicular to flexural tension reinforcement, D' is the cross-section dimension of confined concrete core measured between the centerline of the peripheral hoop or spiral bars, f_{yh} is the specified minimum yield strength of transverse reinforcement, d is the area of shear reinforcement perpendicular to flexural tension reinforcement, n is the number of individual interlock spirals or hoops core section, and s is the spacing of transverse reinforcement.

20.7 Seismic Analysis and Modeling

20.7.1 Equivalent Static Analysis

The ESA method specified in Caltrans SDC [35] can be used for simpler structures, which have balanced spans and similar bent stiffness. Low skew and seismic response are primarily captured by the fundamental mode of vibration. In the ESA method, the fundamental period is determined using tributary mass and stiffness at each bent. The applied seismic force is the product of the period dependent ARS coefficient and the tributary weight.

20.7.2 Elastic Response Spectrum Analysis (ERSA)

The ERSA including uniform load method and multimode dynamic analysis method is a linear elastic spectral analysis with the appropriate response spectrum. An adequate number of modes considered to capture a minimum of 90% mass participation shall be used for complex structures.

The uniform load method specified in ATC/MCEER *Guidelines* [37] is essentially an ESA method that uses a uniform lateral load distribution to approximate the effect of seismic loads. It may be used for both transverse and longitudinal directions if structures satisfy the requirements in Table 20.9.

TABLE 20.9 Recommended LRFD *Guidelines* — Requirement for Uniform Load Method [37]

Parameter	Value				
Number of spans	2	3	4	5	6
Maximum subtended angle for a horizontally curved bridge	20°	20°	30°	30°	30°
Maximum span length ratio from span to span	3	2	2	1.5	1.5
Maximum bent/pier stiffness ratio from span to span, excluding abutments	—	4	4	4	2

In both ESA and ERSA analyses, "effective" stiffness of the components shall be used in order to obtain realistic evaluation for the structure's period and displacement demands. The effective stiffness of ductile components shall represent the component's actual secant stiffness near first yield of rebar. The effective stiffness shall include the effects of concrete cracking, reinforcement, and axial load for concrete components; residual stresses, out-of-straightness, and axial load for steel components; and the restraints of the surrounding soil for piles. For ductile concrete column members, effective moments of inertia, I_{eff}, shall be based on cracked section properties and can be determined from the initial slope of the M–ϕ curve between the origin and the point designating first yield of the main column reinforcement. The torsional moment of inertia of concrete column J_{eff} may be taken as 0.2 times J_{gross}. For capacity-protected concrete members, I_{eff} shall be based on their level of cracking. For a conventionally reinforced concrete box girder superstructure, I_{eff} can be estimated between 0.5 and 0.75 times I_{gross}, the moment of inertia of a gross section. For prestressed concrete superstructures, I_{eff} is assumed 1.0 times I_{gross}.

The following are major considerations in seismic analysis and design practice:

- A beam–element model with three or more lumped masses for each member is usually used [25,26].
- Larger cap stiffness is often used to simulate a stiff deck.
- Compression and tension models are used to simulate the behavior of expansion joints. In the tension model, superstructure joints including abutments are released longitudinally but the restrainers are modeled as truss elements. In the compression model, all restrainers are considered to be inactive and all joints are locked longitudinally.
- Simplistic analysis models should be used for initial assessment of structural behavior. The results of more sophisticated models shall be compared for reasonableness with the results obtained from the simplistic models. The rotational and translational stiffness of abutments and foundations modeled in the seismic analysis must be compatible with their structural and geotechnical capacity. The energy dissipation capacity of the abutments should be considered for bridges whose response is dominated by the abutments [50].
- For elastic response spectrum analysis, the viscous damping ratio inherent in the specified ground spectra is usually 5%.
- For time history analysis, in lieu of measurements, a damping ratio of 5% for both concrete and timber constructions and 2% for welded and bolted steel construction may be used.
- For one- or two-span continuous bridges with abutment designed to activate significant passive pressure in the longitudinal direction, a damping ratio of up to 10% may be used in longitudinal analysis.
- Soil-spring elements should be used to the soil–foundation–structure interaction. Adjustments are often made to meet force–displacement compatibility, particularly for abutments. The maximum capacity of the soil behind abutments with heights larger than 2.5 m may be taken as 370 kPa and will be linearly reduced for the backwall height less than 2.5 m.

- Pile footing with pile cap and spread footing with soil types A and B [37] may be modeled as rigid. If footing flexibility contributes more than 20% to pier displacement, foundation springs shall be considered.
- For pile bent/drilled shaft, estimated depth to fixity or soil-spring based on idealized p–y curves should be used.
- Force–deformation behavior of a seismic isolator can be idealized as a bilinear relationship with two key variables: second slope stiffness and characteristic strength. For design, the force–deformation relationship can be represented by an effective stiffness based on the secant stiffness and a damping coefficient. For more detailed information, references can be made ATC/MCEER *Guidelines* and AASHTO Guide Specifications [37,51] and a comprehensive chapter by Zhang [52].

20.7.3 Nonlinear Dynamic Analysis

The nonlinear dynamic analysis (NDA) procedure is normally used for the 3% PE in 75-year earthquake. A minimum of three ground motions including two horizontal components and one vertical component shall be used and the maximum actions for those three motions shall be used for design. If more than seven ground motions are used, the design action may be taken as the mean action of ground motions. The result of an NDA should be compared with an ERSA as a check for a reasonableness of the nonlinear model.

20.7.4 Global and Stand-Alone Analysis

The global analysis specified in Caltrans SDC [35] is an EDA considering the entire bridge modeled from abutment to abutment. It is often used to determine displacement demands on multiframe structures. The stand-alone analysis is an elastic dynamic analysis considering only one individual frame. To avoid having individual frames dependent on the strength and stiffness of adjacent frames, the separate stand-alone model for each frame must meet all requirements of the SDC.

20.7.5 Inelastic Static Analysis — Push Over Analysis

Inelastic Static Analysis (ISA), commonly referred to as the "push over analysis," shall be used to determine the displacement capacity of a bridge system. IAS shall be performed using expected material properties for modeled members. ISA can be categorized into three types of analysis: (1) elastic–plastic hinge, (2) refined plastic hinge, and (3) distributed plasticity.

The simplest method, elastic–plastic hinge analysis, may be used to obtain an upper bound solution. The most accurate method, distributed plasticity analysis, can be used to obtain a better (more refined) solution. Refined plastic hinge analysis is an alternative that can reasonably achieve both computational efficiency and accuracy.

In an elastic–plastic hinge (lumped plasticity) analysis, material inelasticity is taken into account using concentrated "zero-length" plastic hinges, which maintain plastic moment capacities and rotate freely. When the section reaches its plastic capacity, a plastic hinge is formed and element stiffness is adjusted [53,54]. For regions in a framed member away from the plastic hinge, elastic behavior is assumed. It does not, however, accurately represent the distributed plasticity and associated P–δ effects. This analysis predicts an upper bound solution.

In the refined plastic hinge analysis [55], a two-surface yield model considers the reduction of plastic moment capacity at the plastic hinge due to the presence of axial force and an effective tangent modulus accounts for the stiffness degradation due to distributed plasticity along a frame member. This analysis is similar to the elastic–plastic hinge analysis in efficiency and simplicity and also accounts for distributed plasticity.

Distributed plasticity analysis models the spread of inelasticity through the cross-sections and along the length of the members. This is also referred to as plastic zone analysis, spread-of-plasticity analysis, and elasto-plastic analysis by various researchers. In this analysis, a member needs to be subdivided into several elements along its length to model the inelastic behavior more accurately. Two main approaches have been successfully used to model plastification of members in a second-order distributed plasticity analysis:

- Cross-sectional behavior is described as an input for the analysis by means of moment–thrust–curvature (M–P–ϕ) and moment–thrust–axial strain (M–P–ε) relations, which may be obtained separately from a moment–curvature analysis or approximated by closed-form expressions [56].
- Cross-sections are subdivided into elemental areas and the state of stresses and strains are traced explicitly using the proper stress–strain relations for all elements during the analysis.

20.7.6 Moment–Curvature Analysis

The main purpose of moment–curvature analysis is to study the section behavior. The following assumptions are usually made:

- Plane section before bending remains plane after bending.
- Shear and torsional deformations are negligible.
- Stress–strain relationships for concrete and steel are given [35].
- For reinforced concrete, a prefect bond between concrete and steel rebar exists.

The mathematical formulas used in the section analysis are (Figure 20.21)

Compatibility equations

$$\phi_x = \varepsilon/y \tag{20.33}$$

$$\phi_y = \varepsilon/x \tag{20.34}$$

Equilibrium equations

$$P = \int_A \sigma \, dA = \sum_{i=1}^{n} \sigma_i A_i \tag{20.35}$$

$$M_x = \int_A \sigma y \, dA = \sum_{i=1}^{n} \sigma_i y_i A_i \tag{20.36}$$

$$M_y = \int_A \sigma x \, dA = \sum_{i=1}^{n} \sigma_i x_i A_i \tag{20.37}$$

For a reinforced concrete member, the cross-section is divided into an appropriate number of equivalent concrete and steel elements or filaments representing the concrete and reinforcing steel as shown in Figure 20.21. Each concrete and steel layer or filament is assigned its corresponding stress–strain relationships. Confined and unconfined stress–strain relationships are used for the core concrete and for the cover concrete, respectively.

For a structural steel member, the section is divided into steel layers or filaments and a typical steel stress–strain relationship is used for tension and compact compression elements, and an equivalent stress–strain relationship with reduced yield stress and strain can be used for a noncompact compression element.

The analysis process starts by selecting a strain for the extreme concrete (or steel) fiber. Using this selected strain and assuming a section neutral axis (NA) location, a linear strain profile is constructed and the corresponding section stresses and forces are computed. Section force equilibrium is then checked for the given axial load. By changing the location of the NA, the process is repeated until

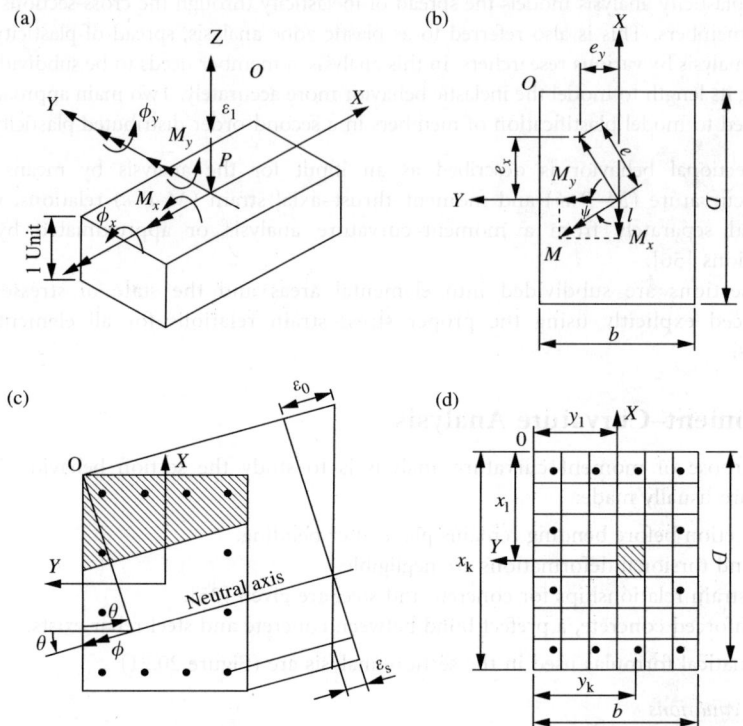

FIGURE 20.21   Section analysis modeling.

equilibrium is satisfied. Once the equilibrium is satisfied, and for the assumed strain and the given axial load, the corresponding section moment and curvature are computed by Equations 20.36 and 20.37.

A moment–curvature (M–ϕ) diagram for a given axial load is constructed by incrementing the extreme fiber strain and finding the corresponding moment and the associated curvature. An interaction diagram (M–P) relating axial load and ultimate moment is constructed by incrementing the axial load and finding the corresponding ultimate moment using the above procedure.

For a reinforced concrete section, the yield moment is usually defined as the section moment at onset of yielding of the tension reinforcing steel. The ultimate moment is defined as the moment at peak moment capacity. The ultimate curvature is usually defined as the curvature when the extreme concrete fiber strain reaches ultimate strain or when the reinforcing rebar reaches its ultimate (rupture) strain (whichever take place first). Figure 20.22a shows typical M–P–ϕ curves for a reinforced concrete section.

For a simple steel section, such as rectangular, circular-solid, and thin-walled circular section, a closed-form of M–P–ϕ can be obtained using the elastic–perfectly plastic stress–strain relations [56]. For all other commonly used steel section, numerical iteration techniques are used to obtain M–P–ϕ curves. Figure 20.22b shows typical M–P–ϕ curves for a wide-flange section.

20.7.7 Random Vibration Approach

The random vibration approach is a well-recognized and advanced seismic-response analysis method for linear multisupport-structural systems and long-span structures [57,58]. This approach provides a statistical measure of the response, which is not controlled by an arbitrary choice of the input motions, and also significantly reduces the response evaluation to that of a series of linear one-degree systems in

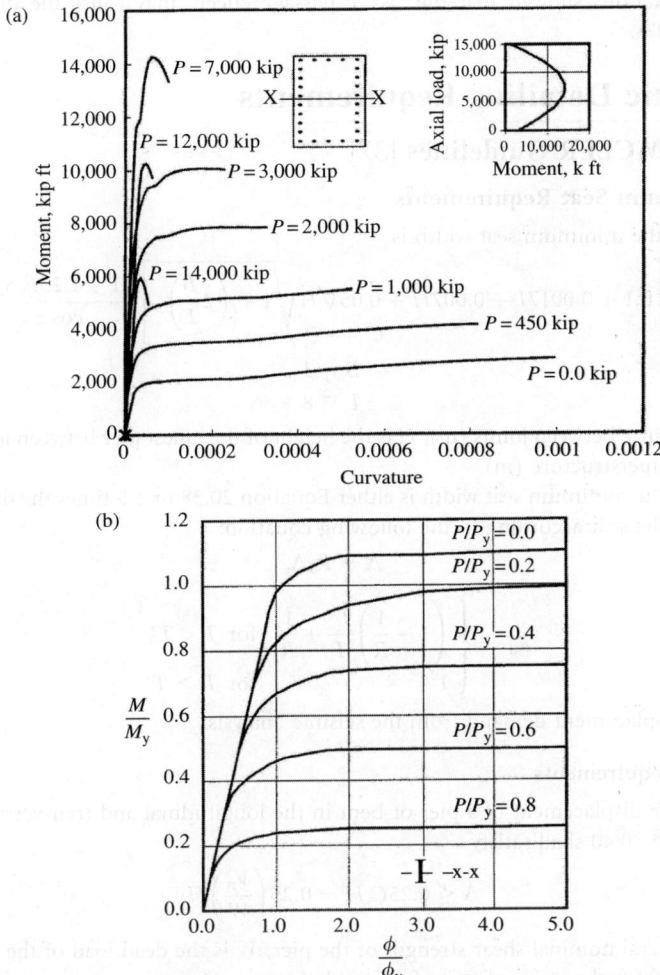

FIGURE 20.22 Typical moment–thrust–curvature curves: (a) reinforced concrete section and (b) steel section.

a way that fully accounts for the multiple-support input and the space–time correlation structure of the ground motion [58].

Although the random vibration approach has been adopted by the Eurocode [59] and widely used by Chinese engineers, it has not been accepted as a practical method of analysis for complex long-span structures by U.S. practicing engineers due to computation difficulties [57].

During 1990s, Lin and coworkers [60–64] have developed a new series of algorithms, that is, the pseudo-excitation method (PEM) series, on structural stationary/nonstationary random response analysis. The PEM is an accurate and extremely efficient method to solve complicated random vibration problems. The cross-correlation terms between all participating modes and between all excitations are all included in the responses. Most recently, the PEM has been used to analyze several long-span bridges in China. Cheng [65] analyzed the Hunan Yue-Yang Cable-stayed Bridge with a total length of 5700 m and a main span of 880 m, in which 2700 degrees of freedom, 15 multisupport ground motions, and 200 modes were used. Fan et al. [66] used PEM to analyze the Second Yangtze River Bridge at Nancha, a cable stay bridge with a total length of 1238 m and a main span of 628 m, in which 300 modes in the fast CQC (i.e., PEM) analysis and 12 multisupport ground motions were used. The computations showed that the "wave passage effect" may cause the differences of some demands up to 40%.

20.8 Seismic Detailing Requirements

20.8.1 ATC/MCEER Guidelines [37]

20.8.1.1 Minimum Seat Requirements

In SDRs 1 and 2, the minimum seat width is

$$N_{min} = \left[0.1 + 0.0017L + 0.007H + 0.05\sqrt{H} \sqrt{1 + \left(2\frac{B}{L}\right)^2} \right] \frac{(1 + 1.25F_v S_1)}{\cos\alpha} \tag{20.38}$$

$$\frac{B}{L} \leq \frac{3}{8} \tag{20.39}$$

where L is the distance between joints (m), H is the height of the tallest pier between joints (m), and B is the width of the superstructure (m).

In SDRs 3 to 6, the minimum seat width is either Equation 20.38 or 1.5 times the displacement of the superstructure at the seat according to the following equation:

$$\Delta = R_d \Delta_e \tag{20.40}$$

$$R_d = \begin{cases} \left(1 - \frac{1}{R}\right)\frac{T^*}{T} + \frac{1}{R} & \text{for } T < T^* \\ 1 & \text{for } T \geq T^* \end{cases} \tag{20.41}$$

where Δ_e is the displacement demand from the seismic analysis.

20.8.1.2 P–Δ Requirements

In SDRs 3 to 6, the displacement of a pier or bent in the longitudinal and transverse directions determined by Equation 20.40 shall satisfy

$$\Delta \leq 0.25C_s H = 0.25\left(\frac{V_n}{W}\right)H \tag{20.42}$$

where V_n is the lateral nominal shear strength of the pier, W is the dead load of the pier, and H is the height of the pier from the point of fixity for foundation.

The basis of this requirement is that maximum displacement is such that the reduction in resisting force is limited to a 25% reduction from the lateral strength assuming no postyield stiffness. The inequality in Equation 20.42 is to keep the bridge pier from being significantly affected by P–Δ moments.

20.8.1.3 Minimum Displacement Capacity Requirements

For SDAP E, the following equation shall be satisfied:

$$\Delta_c \geq 1.5\Delta \tag{20.43}$$

where Δ_c is lateral displacement capacity of the pier or bent. It is defined as the displacement at which the first component reaches its maximum deformation. The factor of 1.5 on displacement demand recognizes the approximation of the modeling and analysis.

20.8.1.4 Structural Steel Design Requirements

20.8.1.4.1 Limiting Width-to-Thickness Ratios

In SDRs 2 to 6, the width-to-thickness ratio of compression elements of the columns in ductile MRF and single-column structures shall satisfy the limiting ratios specified in Table 20.10. Full penetration flange and web welds are required at beam-to-column connections.

TABLE 20.10 Recommended LRFD *Guidelines* — Limiting Width-to-Thickness Ratios [37]

Description of element	Width-to-thickness ratio, b/t	Limiting width-to-thickness ratio, λ_p	Limiting ratio, k
Flanges of I-shaped sections and channels in compression	$b_f/2t_f$	$\dfrac{135}{\sqrt{F_y}}$	0.3
Webs in combined flexural and axial compression	h_c/t_w	For $\dfrac{P_u}{\phi_b P_y} \le 0.125$ $\dfrac{1365}{\sqrt{F_y}}\left(1 - \dfrac{1.54 P_u}{\phi_b P_y}\right)$	For $\dfrac{P_u}{\phi_b P_y} \le 0.125$ $3.05\left(1 - \dfrac{1.54 P_u}{\phi_b P_y}\right)$
		For $\dfrac{P_u}{\phi_b P_y} \le 0.125$ $\dfrac{500}{\sqrt{F_y}}\left(2.33 - \dfrac{P_u}{\phi_b P_y}\right) \ge \dfrac{665}{\sqrt{F_y}}$	For $\dfrac{P_u}{\phi_b P_y} \le 0.125$ $1.12\left(2.33 - \dfrac{P_u}{\phi_b P_y}\right) \ge 1.48$
Hollow circular sections (pipes)	D/t	$\dfrac{8950}{F_y}$	$\dfrac{200}{\sqrt{F_y}}$
Unstiffened rectangular tubes	b/t	$\dfrac{300}{\sqrt{F_y}}$	0.67
Longitudinally stiffened plates in compression	b/t	$\dfrac{145}{\sqrt{F_y}}$	0.32

Notes:
1. b_f and t_f are the width and thickness of an I-shaped section and h_c is the depth of that section and t_w is the thickness of its web.
2. Limits λ_p is for format $b/t \le \lambda_p$.
3. Limits k is for format $b/t \le k\sqrt{E/F_y}$.

TABLE 20.11 Recommended LRFD *Guidelines* — Limiting Slenderness Ratio [37]

Description of members	Limiting slenderness ratio Kl/r	Limiting length, L (m)
Unsupported distance for potential plastic hinge zone of columns		$\dfrac{17250 r_y}{F_y}$
Ductile compression bracing members	$\dfrac{2600}{\sqrt{F_y}}$	
Nominally ductile bracing members	$\dfrac{3750}{\sqrt{F_y}}$ limit is waived if members designed as tension-only bracing	

20.8.1.4.2 Limiting Slenderness Ratio

In SDRs 2 to 6, Table 20.11 summarizes the limiting slenderness ratio (KL/r) for various steel members. Recent studies found that more stringent requirement for slenderness ratios may be unnecessary, provided that connections are capable of conveying at least the member's tension capacity. The ratios shown in Table 20.11 reflect those relaxed limits.

20.8.1.4.3 *Limiting Axial Load Ratio*

High axial load in a column usually results in the early deterioration of strength and ductility.

The ratio of factored axial compression due to seismic load and permanent loads to yield strength ($A_g E_y$) for columns in ductile MRFs and single-column structures shall not exceed 0.4 for SDR 2 and SDRs 3 to 6, respectively.

20.8.1.4.4 *Plastic Rotation Capacities*

In SDRs 3 to 6, the plastic rotational capacity shall be based on the appropriate performance level and may be determined from tests and a rational analysis. The maximum plastic rotational capacity θ_p should be conservatively limited to 0.035, 0.005, and 0.01 radians for life safety, operational performance, and in ground hinges and piles, respectively.

20.8.1.5 Concrete Design Requirements

20.8.1.5.1 *Limiting Longitudinal Reinforcement Ratios*

The ratio of longitudinal reinforcement to the gross cross-section shall be not less than 0.008 and not more than 0.04.

20.8.1.5.2 *Shear Reinforcement*

The shear strength shall be determined by either an implicit approach or an explicit approach. In the end regions, the explicit approach assumes that the shear-resisting mechanism is provided by the strut-tie model (explicit approach) such that

$$\phi V_s \geq V_u - (V_p + V_c) \tag{20.44}$$

$$V_p = \frac{\Lambda}{2} P_e \frac{D'}{L} \tag{20.45}$$

$$V_c = \begin{cases} 0.05\sqrt{f_c'}b_w d & \text{for plastic hinge zone} \\ 0.17\sqrt{f_c'}b_w d & \text{for outside plastic hinge zone} \end{cases} \tag{20.46}$$

$$V_s = \begin{cases} \dfrac{\pi}{2}\dfrac{A_{bh}}{s}f_{yh}D''\cot\theta & \text{for circular section} \\[2mm] \dfrac{A_v}{s}f_{yh}D''\cot\theta & \text{for rectangular section} \end{cases} \tag{20.47}$$

where

P_e = compressive axial force including seismic effects
D' = pitch circle diameter of the longitudinal reinforcement in a circular column, or the distance between the outermost layers of bars in a rectangular column
L = column length
Λ = fixity factor defined in Section 20.6.1.5
b_w = web width of the section
d = effective depth of the section
A_{bh} = area of one circular hoop/spiral rebar
A_{sh} = total area of transverse reinforcement in one layer in the direction of the shear force
D'' = centerline section diameter/width of the perimeter spiral/hoops
θ = principal crack angle defined in Section 20.6.1.5

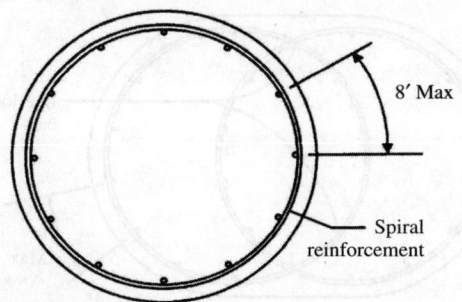

FIGURE 20.23   Column single spiral details [37].

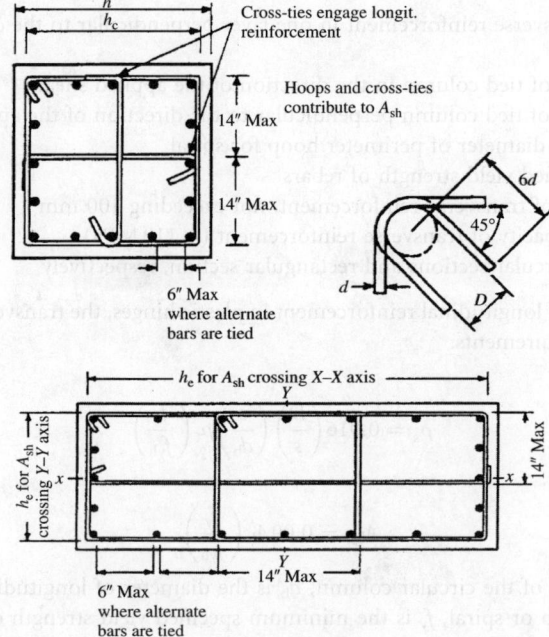

FIGURE 20.24   Column tie details [37].

20.8.1.5.3 Transverse Reinforcement in Plastic Hinge Zones

Figures 20.23 to 20.25 illustrate the typical transverse reinforcement. For confinement in plastic hinge zones, the ratio of transverse reinforcement, ρ_s, shall not be less than

$$\rho_{s\,min} = 0.008 \frac{f_c'}{U_{sf}} \left[\alpha_{shape} \left(\frac{P_e}{f_c' A_g} + \rho_t \frac{f_y}{f_c'} \right) \left(\frac{A_g}{A_{cc}} \right)^2 - 1 \right] \qquad (20.48)$$

$$\rho_s = \begin{cases} \dfrac{4A_{bh}}{D's} & \text{for circular sections} \\[2mm] \dfrac{A_{sh}}{sB'} + \dfrac{A_{sh}'}{sD''} & \text{for rectangular sections} \end{cases} \qquad (20.49)$$

where

A_{cc} = area of column core concrete, measured to the centerline of the perimeter hoop or spiral
A_{sh} = total area of transverse reinforcement in one layer into the direction of the applied shear

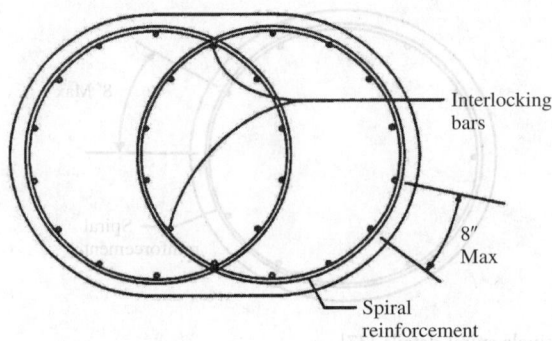

FIGURE 20.25   Column interlocking spiral details [37].

A'_{sh} = total area of transverse reinforcement in one layer perpendicular to the direction of the applied
        shear
B'' = core dimension of tied column in the direction of the applied shear
D'' = core dimension of tied column perpendicular to the direction of the applied shear
D' = center-to-center diameter of perimeter hoop for spiral
f_y = minimum specified yield strength of rebars
s = vertical spacing of transverse reinforcement, not exceeding 100 mm
U_{sf} = strain energy capacity of transverse reinforcement ($= 110$ MPa)
α_{shape} = 12 and 15 for circular sections and rectangular section, respectively

   To restrain buckling of longitudinal reinforcement in plastic hinges, the transverse reinforcement shall
satisfy the following requirements:

For circular section

$$\rho_s = 0.016 \left(\frac{D}{s}\right)\left(\frac{s}{d_b}\right)\rho_t\left(\frac{f_y}{f_{yh}}\right) \tag{20.50}$$

For rectangular section

$$A_{bh} = 0.09 A_b \left(\frac{f_y}{f_{yh}}\right) \tag{20.51}$$

where D is the diameter of the circular column, d_b is the diameter of longitudinal reinforcement bars
being restrained by hoop or spiral, f_y is the minimum specified yield strength of longitudinal reinforcement bars, f_{yh} is the minimum specified yield strength of transverse reinforcement bars, A_b is the area
of longitudinal reinforcement bars being restrained by hoop or spiral, A_{bh} is the area of hoops or spiral or
ties restraining the longitudinal steel, and s is the vertical spacing of transverse reinforcement restraining
the longitudinal steel.
   The transverse reinforcement shall be provided in the plastic hinge zones defined in Section 20.6.1.5.
The spacing the transverse reinforcement shall meet

$$s \leq \begin{cases} 6d_b \\ 0.25 \text{ (minimum member dimension)} \\ 150 \text{ mm} \end{cases} \tag{20.52}$$

20.8.1.5.4 *Joint Reinforcement*

   - Moment-resisting integral connections shall be designed to resist the maximum plastic moment.
   - The principal tension stress p_t and compression stress p_c shall be calculated by

$$\frac{p_t}{p_c} = \frac{f_h + f_v}{2} \mp \sqrt{\left(\frac{f_h + f_v}{2}\right)^2 + v_{hv}^2} \tag{20.53}$$

where f_h is the average axial stress in the horizontal direction within the plans of the connection under consideration (positive is compressive stress) (MPa), f_h is the average axial stress in the vertical direction within the plans of the connection under consideration (positive is compressive stress) (MPa), and v_{hv} is the average shear stress within the plans of the connection (MPa).

- The principal compression stress p_c shall not exceed $0.25f'_c$.
- When the principal tension stress $p_t \leq 0.29\sqrt{f'_c}$ for circular columns or columns with intersecting spirals, the volumetric ratio of transverse reinforcement in the form of spirals or hoops to be continued into the cap or footing ρ_s shall not be less than $0.29\sqrt{f'_c}/f_{yh}$ where f_{yh} is yield stress of horizontal hoop/tie reinforcement in the joint.
- When the principal tension stress $p_t > 0.29\sqrt{f'_c}$ the additional reinforcement shown in Figure 20.26 shall be provided as follows:
  - On each side column, vertical stirrups: $A_{jv} = 0.16A_{st}$.
  - Inside joint: the required vertical tie: $A_{jt} = 0.08A_{st}$.
  - Longitudinal reinforcement: $A_{jt} = 0.08A_{st}$.
  - Column hoop or spiral reinforcement into the cap: $\rho_s \geq 0.4A_{st}/l_{ac}^2$.
    where A_{st} is the total area of longitudinal steel anchored in the joint and l_{ac} is the length of column reinforcement embedded into the joint.

20.8.1.5.5 Plastic Rotation Capacities

In SDRs 3 to 6, the plastic rotational capacity shall be based on the appropriate performance level and may be determined from tests and a rational analysis. The maximum plastic rotational capacity θ_p should be conservatively limited to 0.035 (0.05), 0.01, and 0.02 radians for life safety (for liquifiable pile foundation), operational performance, and in ground hinges, respectively. For the life safety performance, the plastic hinge of a column shall be calculated by

$$\theta_p = 0.11\frac{L_p}{D'}(N_f)^{-0.5} \tag{20.54}$$

$$2 \leq N_f = 3.5(T_n)^{-1/3} \leq 10 \tag{20.55}$$

where

N_f = number of cycles of loading expected at maximum displacement amplitude for liquefiable soil, take $N_f = 2$

L_p = effective plastic hinge length

$$L_p = \begin{cases} 0.08\dfrac{M}{V} + 4400\varepsilon_y d_b & \text{for common columns} \\ L_g + 8800\varepsilon_y d_b & \text{for columns with isolation gap} \end{cases} \tag{20.56}$$

ε_y = yield strain of longitudinal reinforcement
L_g = gap length between the flare and adjacent element
D' = distance between outer layers of the longitudinal reinforcement
d_b = diameter of the main longitudinal reinforcement bars

20.8.2 Caltrans SDC

20.8.2.1 Minimum Seat Width Requirements

To prevent unseating of superstructures at hinges, piers, and abutments, the seat width shall be available to accommodate the anticipated thermal movement, prestressing shortening, concrete creep and

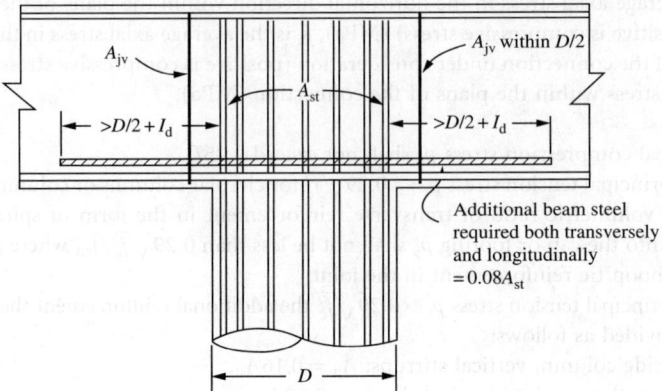

FIGURE 20.26 Additional cap beam bottom reinforcement for joint force transfer [37].
　　Note: I_d = development length.

shrinkage, and the relative longitudinal earthquake displacement, and shall not exceed the following minimum seat requirements [35]:

For seat width at hinges

$$N_{min} = \text{larger of} \begin{cases} \Delta_{ps} + \Delta_{cr+sh} + \Delta_{temp} + \Delta_{eq} + 100 \\ 600 \text{ mm} \end{cases} \qquad (20.57)$$

where Δ_{eq} is total relative displacement between frames.

For seat width at abutments

$$N_{min} = \text{larger of} \begin{cases} \Delta_{ps} + \Delta_{cr+sh} + \Delta_{temp} + \Delta_{eq} + 100 \\ 760 \text{ mm} \end{cases} \qquad (20.58)$$

where Δ_{ps}, Δ_{cr+sh}, Δ_{temp}, and Δ_{eq} are relative displacement due to prestressing, concrete creep and shrinkage, and earthquake, respectively (mm).

20.8.2.2 P–Δ Effects

The P–Δ effects tend to increase the displacement and decrease the lateral load-carrying capacity of a bridge column. These effects can typically be ignored if the moment ($P_{DL}\Delta$) is less than or equal to 20% of the column plastic moment, that is, $P_{DL}\Delta \leq 0.2 M_p^{col}$.

20.8.2.3 Minimum Displacement Ductility Capacity

To ensure the dependable ductile behavior of all columns regardless of seismic demand, a minimum local displacement ductility capacity of $\mu_c = \Delta_c/\Delta_Y \geq 3$ is required and target local ductility of $\mu_c \geq 4$ is recommended. The local displacement ductility capacity shall be calculated for an equivalent member that approximates a fixed base cantilever element.

20.8.2.4 Maximum (Target) Displacement Ductility Demand

The engineers are encouraged to limit displacement ductility demands, defined as $\mu_D = \Delta_D/\Delta_Y$ to the values shown in Table 20.12.

TABLE 20.12　Caltrans Limiting Displacement Ductility Demand Values [35]

Item	Limiting $\mu_D = \Delta_D/\Delta_Y$
Single column bents supported on fixed foundations	4
Multicolumn bents supported on fixed or pinned footings	5
Pier walls (weak direction) supported on fixed or pinned footings	5
Pier walls (strong direction) supported on fixed or pinned footings	1

20.8.2.5 Minimum Lateral Strength

Although providing ductile detailing is essential for achieving the expected performance requirements, each column shall be designed to have a minimum lateral flexural to resist a lateral force of $0.1g$ or 0.1 times P_{dl}.

20.8.2.6 Structural Steel Design Requirements (Guide) [36]

20.8.2.6.1 Limiting Width-to-Thickness Ratios

For capacity-protected components, width–thickness ratios of compression elements shall not exceed the limiting value λ_r as specified in Table 20.13. For ductile components, width–thickness ratios shall not exceed the λ_p as specified in Table 20.13. Welds located in the expected inelastic region of ductile components are preferably complete penetration welds. Partial penetration groove welds are not recommended in these regions. If the fillet welds are only practical solution for an inelastic region, quality control and quality assurance inspection procedures for the fracture critical members shall be followed (Figure 20.27).

20.8.2.6.2 Limiting Slenderness Ratio

The slenderness parameter λ_c for compression members and λ_b for flexural members shall not exceed the limiting values, λ_{cp} and λ_{bp}, as specified in Table 20.14, respectively.

20.8.2.6.3 Limiting Axial Load Ratio

High axial load in a column usually results in the early deterioration of strength and ductility. The ratio of factored axial compression due to seismic load and permanent loads to yield strength $(A_g F_y)$ for columns in ductile moment-resisting frames and single-column structures shall not exceed 0.3.

20.8.2.6.4 Shear Connectors

Shear connectors shall be provided on the flanges of girders, end cross-frames, or diaphragms to transfer seismic loads from the concrete deck to the abutments or pier supports. The cross-frames or diaphragms at the end of each span are the main components to transfer the lateral seismic loads from the deck down to the bearing locations. Recent tests on a 0.4 scale experimental steel girder bridge (18.3 m long) conducted by University of Nevada, Reno [69] indicated that too few shear connectors between the girders and deck at the bridge end did not allow the end cross-frame to reach its ultimate capacity.

20.8.2.7 Concrete Design Requirements

20.8.2.7.1 Limiting Longitudinal Reinforcement Ratios

The ratio of longitudinal reinforcement to the gross cross-section shall not be less than 0.01 and 0.005 for columns and pier walls, respectively, and not more than 0.04.

20.8.2.7.2 Transverse Reinforcement in Plastic Hinge Zones

For confinement in plastic hinge zones (larger of 1.5 times cross-sectional dimension in the direction of bending and the regions of column where the moment exceeds 75% of overstrength plastic moment, M_p^{col}), transverse reinforcement shall not be less than [27]

For spiral and hoops $(\rho_s = 4A_{bh}/D's)$

$$\rho_{s,\min} = \begin{cases} 0.45\dfrac{f_c'}{f_y}\left(\dfrac{A_g}{A_c} - 1\right)\left(0.5 + \dfrac{1.25P_e}{f_c'A_g}\right) & \text{for } D \le 0.9\,\text{m} \\[4mm] 0.12\dfrac{f_c'}{f_y}\left(0.5 + \dfrac{1.25P_e}{f_c'A_g}\right) & \text{for } D > 0.9\,\text{m} \end{cases} \tag{20.59}$$

For ties

$$A_{sh,\min} = \text{larger of} \begin{cases} 0.3 s h_c\dfrac{f_c'}{f_y}\left(\dfrac{A_g}{A_c} - 1\right)\left(0.5 + \dfrac{1.25P_e}{f_c'A_g}\right) \\[4mm] 0.12 s h_c\dfrac{f_c'}{f_y}\left(0.5 + \dfrac{1.25P_e}{f_c'A_g}\right) \end{cases} \tag{20.60}$$

TABLE 20.13 Caltrans Limiting Width–Thickness Ratios [36]

No.	Description of elements	Examples	Width–thickness ratios	λ_r	λ_p
Unstiffened elements					
1	Flanges of I-shaped rolled beams and channels in flexure	Figure 20.27a Figure 20.27c	b/t	$\dfrac{370}{\sqrt{F_y}-69}$	$\dfrac{137}{\sqrt{F_y}}$
2	Outstanding legs of pairs of angles in continuous contact; flanges of channels in axial compression; angles and plates projecting from beams or compression members	Figure 20.27d Figure 22.27e	b/t	$\dfrac{250}{\sqrt{F_y}}$	$\dfrac{137}{\sqrt{F_y}}$
Stiffened elements					
3	Flanges of square and rectangular boxes and hollow structural section of uniform thickness subject to bending or compression; flange cover plates and diaphragm plates between lines of fasteners or welds	Figure 20.27b	b/t	$\dfrac{625}{\sqrt{F_y}}$	$\dfrac{290}{\sqrt{F_y}}$ (tubes) $\dfrac{400}{\sqrt{F_y}}$ (others)
4	Unsupported width of cover plates perforated with a succession of access holes	Figure 20.27d	b/t	$\dfrac{830}{\sqrt{F_y}}$	$\dfrac{400}{\sqrt{F_y}}$
5	All other uniformly compressed stiffened elements, i.e., supported along two edges	Figures 20.27a, c, d, f	b/t h/t_w	$\dfrac{665}{\sqrt{F_y}}$	$\dfrac{290}{\sqrt{F_y}}$ (w/lacing) $\dfrac{400}{\sqrt{F_y}}$ (others)
6	Webs in flexural compression	Figures 20.27a, c, d, f	h/t_w	$\dfrac{2550}{\sqrt{F_y}}$	$\dfrac{1365}{\sqrt{F_y}}$
7	Webs in combined flexural and axial compression	Figures 20.27a, c, d, f	h/t_w	$\dfrac{2550}{\sqrt{F_y}}\times\left(1-\dfrac{0.74P}{\phi_b P_y}\right)$	For $P_u \le 0.125\phi_b P_y$ $\times\dfrac{1365}{\sqrt{F_y}}\left(1-\dfrac{1.54P}{\phi_b P_y}\right)$ For $P_u > 0.125\phi_b P_y$ $\times\dfrac{500}{\sqrt{F_y}}\left(2.33-\dfrac{P}{\phi_b P_y}\right)$ $\ge\dfrac{665}{\sqrt{F_y}}$
8	Longitudinally stiffened plates in compression	Figure 20.27e	b/t	$\dfrac{297\sqrt{k}}{\sqrt{F_y}}$	$\dfrac{197\sqrt{k}}{\sqrt{F_y}}$
9	Round HSS in axial compression or flexure		D/t	$\dfrac{7930}{F_y}$	$\dfrac{8950}{F_y}$

n = number of equally spaced longitudinal compression flange stiffeners.

I_s = moment of inertia of a longitudinal stiffener about an axis parallel to the bottom flange and taken at the base of the stiffener.

Notes:

1. Width–thickness ratios shown in **Bold** are from AISC-LRFD (1999) [67] and AISC-Seismic Provisions (1997) [68]. F_y is MPa.

2. k = buckling coefficient specified by Article 6.11.2.1.3a of AASHTO-LRFD [38]. For $n = 1$, $k = (8I_s/bt^3)^{1/3} \le 4.0$; for $n = 2, 3, 4,$ and 5, $k = (14.3I_s/bt^3 n^4)^{1/3} \le 4.0$.

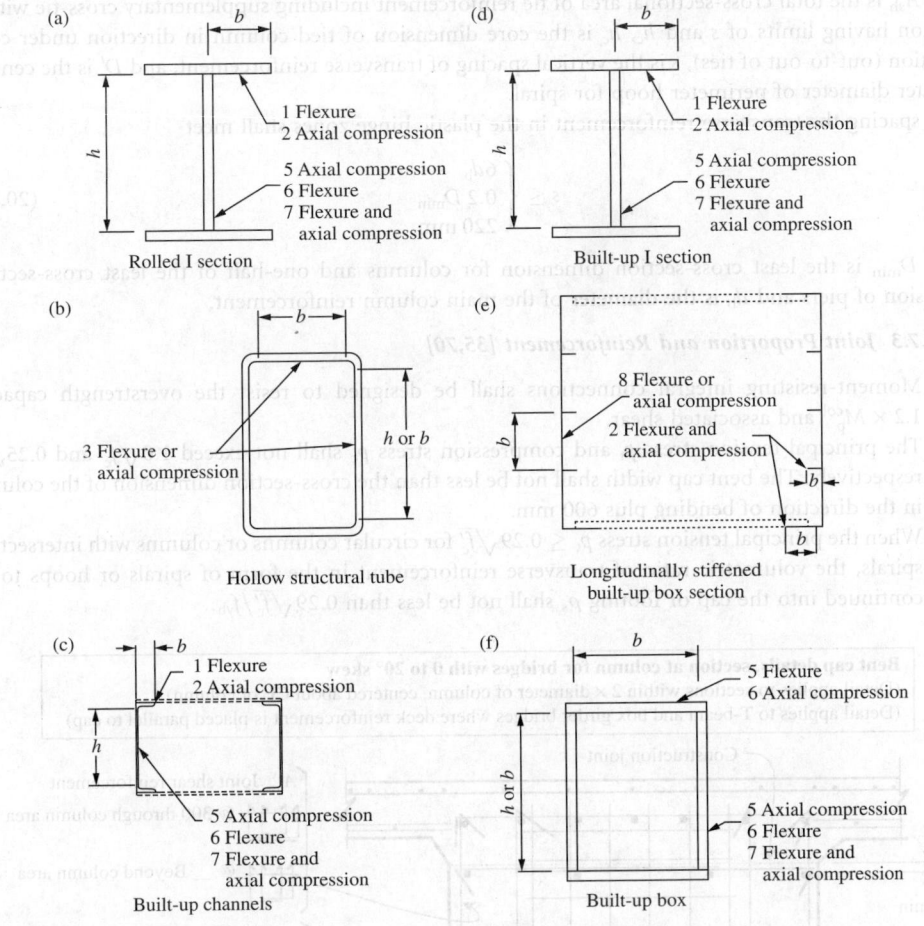

FIGURE 20.27 Selected steel cross-sections [36].

TABLE 20.14 Caltrans Limiting Slenderness Parameters [36]

Member classification		Limiting slenderness parameters
Ductile		
Compression member	λ_{cp}	0.75
Flexural member	λ_{bp}	$17240/F_y$
Capacity protected		
Compression member	λ_{cp}	1.5
Flexural member	λ_{bp}	$1970/\sqrt{F_y}$

Notes:

$\lambda_c = (KL/r\pi)\sqrt{(F_y/E)}$ (slenderness parameter for compression members).

$\lambda_b = L/r_y$ (slenderness parameter for flexural members).

λ_{cp} = limiting slenderness parameter for compression members.

λ_{bp} = limiting slenderness parameter for flexural members.

K = effective length factor of a member.

L = unsupported length of a member (mm).

r = radius of gyration (mm).

r_y = radius of gyration about the minor axis (mm).

F_y = specified minimum yield strength of steel (MPa).

E = modulus of elasticity of steel (200,000 MPa).

where A_{sh} is the total cross-sectional area of tie reinforcement including supplementary cross-tie within a section having limits of s and h_c, h_c is the core dimension of tied column in direction under consideration (out-to-out of ties), s is the vertical spacing of transverse reinforcement, and D' is the center-to-center diameter of perimeter hoop for spiral.

The spacing the transverse reinforcement in the plastic hinge zones shall meet

$$s \leq \begin{cases} 6d_b \\ 0.2\,D_{min} \\ 220\ \mathrm{mm} \end{cases} \tag{20.61}$$

where D_{min} is the least cross-section dimension for columns and one-half of the least cross-section dimension of piers and d_b is the diameter of the main column reinforcement.

20.8.2.7.3 Joint Proportion and Reinforcement [35,70]

- Moment-resisting integral connections shall be designed to resist the overstrength capacity $1.2 \times M_p^{col}$ and associated shear.
- The principal tension stress p_t and compression stress p_c shall not exceed $1.0\sqrt{f_c'}$ and 0.25, f_c', respectively. The bent cap width shall not be less than the cross-section dimension of the column in the direction of bending plus 600 mm.
- When the principal tension stress $p_t \leq 0.29\sqrt{f_c'}$ for circular columns or columns with intersecting spirals, the volumetric ratio of transverse reinforcement in the form of spirals or hoops to be continued into the cap or footing ρ_s shall not be less than $0.29\sqrt{f_c'}/f_{yh}$.

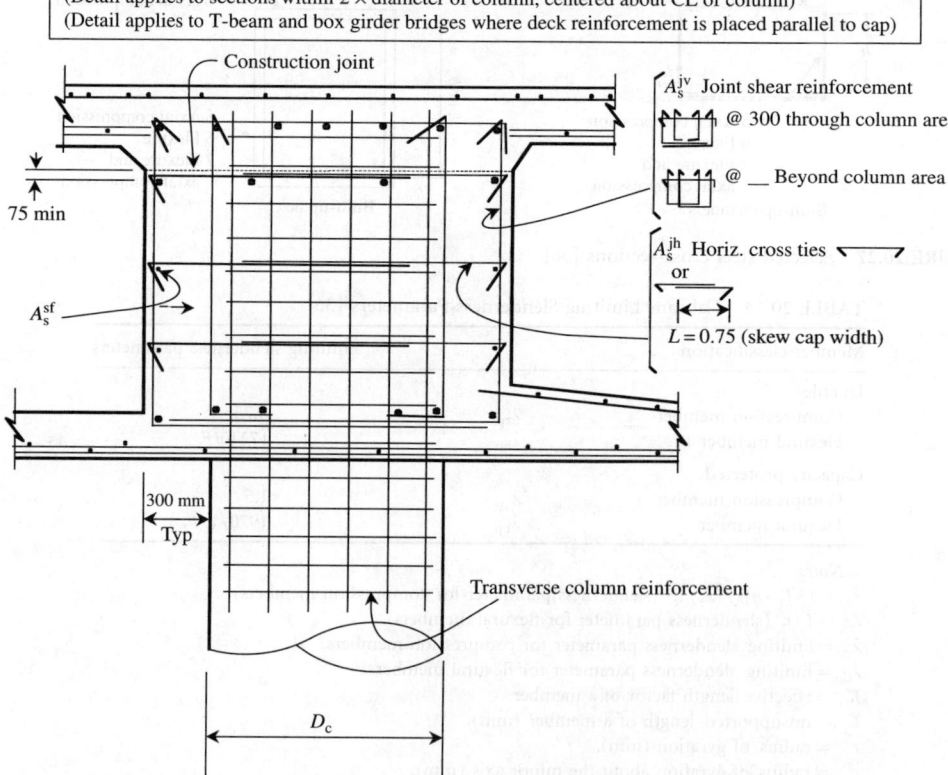

FIGURE 20.28   Example cap joint shear reinforcement — skews 0 to 20° [35].

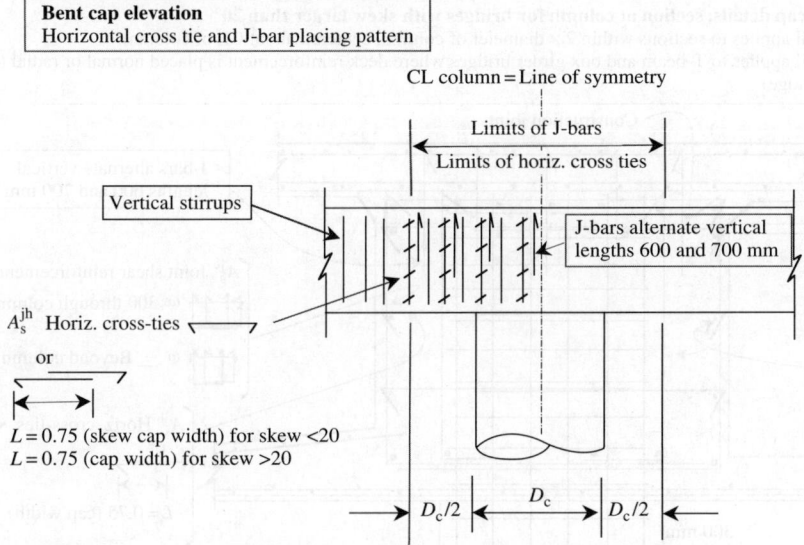

FIGURE 20.29 Location of horizontal joint shear reinforcement [35].

- When the principal tension stress $p_t > 0.29\sqrt{f_c'}$ the additional reinforcement as shown in Figure 20.28 and Figure 20.29 shall be provided and well distributed within $D/2$ from the face of the column. All joint shear reinforcement shall be as follows:
- On each side column, vertical stirrups: $A_s^{jv} = 0.2A_{st}$.
- Horizontal stirrups or tiles: $A_s^{jh} = 0.1A_{st}$.
- Longitudinal side reinforcement: $A_s^{sf} \geq 0.1A_{cap}$, where A_{cap} is area of bent cap top or bottom flexural steel.
- Column hoop or spiral reinforcement into the cap: $\rho_s \geq 0.4A_{st}/l_{ac}^2$.
- Horizontal reinforcement shall be stitched across the cap in two or more intermediate layers. The reinforcement shall be shaped as hairpins, spaced vertically at not more than 460 mm. The hairpins shall be 10% of column reinforcement. Spacing shall be denser outside the column than that used within the column.
- For bent caps skewed greater than 20°, the vertical J-bars hooked around longitudinal deck and bent cap steel shall be 8% of column steel (see Figure 20.30). The J-bars shall be alternatively 600 and 750 mm long and placed within a width of column dimension either side of the column centerline.
- All vertical column bars shall be extended as high as practically possible without interfering with the main cap bars.

20.8.2.7.4 Effective Plastic Hinge Length
The effective plastic hinge length is used to evaluate the plastic rotation of columns.

$$L_p = \begin{cases} 0.08\,L + 0.022\,f_{ye}\,d_b \geq 0.044\,f_{ye}\,d_b & \text{for columns and Type II shafts} \\ G + 0.044\,f_{ye}\,d_b & \text{for horizontally isolated flared columns} \\ D^* + 0.06\,H' & \text{for noncased Type I shafts} \end{cases} \quad (20.62)$$

where f_{ye} is the expected yield stress of longitudinal reinforcement, L is the member length from the maximum moment to the point of contraflexure, H' is the length of pile shaft/column from the ground surface to point of contraflexure, above ground, and D^* is the lesser cross-section dimension of shafts.

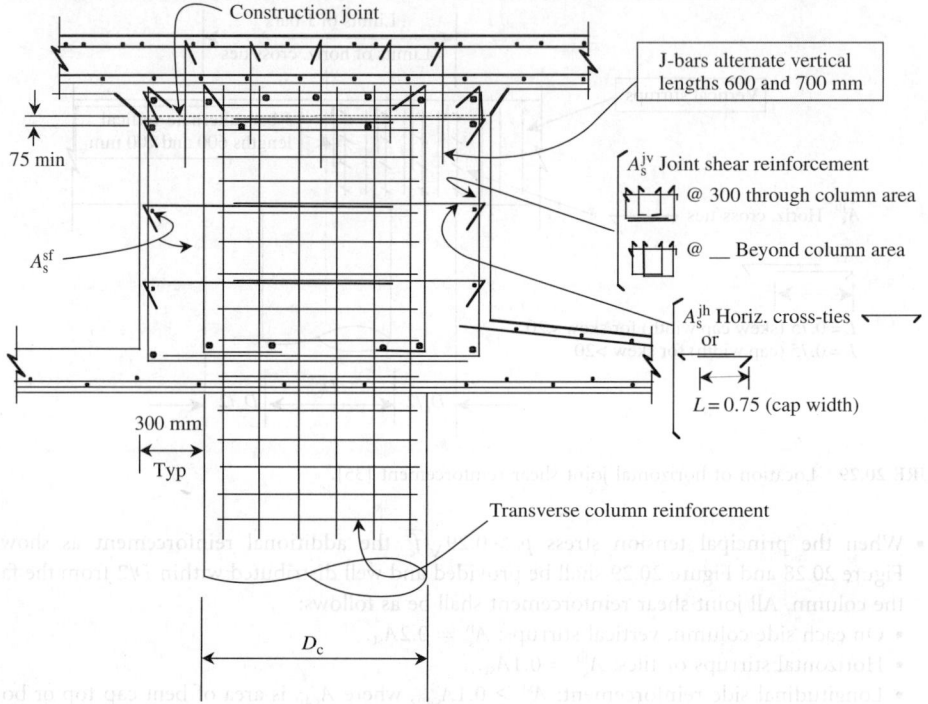

Bent cap details, section at column for bridges with skew larger than 20°
(Detail applies to sections within 2 × diameter of column, centered about CL of column)
(Detail applies to T-beam and box girder bridges where deck reinforcement is placed normal or radial to CL bridge)

Construction joint

J-bars alternate vertical lengths 600 and 700 mm

75 min

A_s^{jv} Joint shear reinforcement

@ 300 through column area

@ __ Beyond column area

A_s^{sf}

A_s^{jh} Horiz. cross-ties
or

$L = 0.75$ (cap width)

300 mm
Typ

Transverse column reinforcement

D_c

FIGURE 20.30   Example cap joint shear reinforcement — Skews > 20° [35].

20.9 Seismic Design Practice in Japan

20.9.1 Introduction

Seismic design methods for highway bridges in Japan have been developed and improved based on the lessons learned from the various past bitter experiences after the Kanto Earthquake (M7.9) in 1923. By introducing various provisions such as soil liquefaction considerations and unseating prevention devices to prevent bridges from serious damage, only a few highway bridges suffered complete collapse of superstructures in the recent past earthquakes.

However, the Hyogo-ken-Nanbu (Kobe) Earthquake (M7.3) of January 17, 1995, caused destructive damage to highway bridges. Collapse and near-collapse of superstructures occurred at nine sites and other destructive damages occurred at 16 sites [71,72]. The earthquake revealed that there were a number of critical design issues that should be reevaluated and revised in the seismic design and seismic strengthening of bridges.

Just after the earthquake, the "Committee for Investigation on the Damage of Highway Bridges Caused by the Hyogo-ken-Nanbu Earthquake" was established in the Ministry of Construction (currently Ministry of Land, Infrastructure and Transport (MLIT)) to investigate the damage and to clarify the factors that affected the damage. The committee published the intermediate investigation report in March 1995, and the final one in December 1995 [71]. Besides the investigation of damage of highway bridges, the Committee approved the "Guide Specifications for Reconstruction and Repair of Highway Bridges which Suffered Damage due to the Hyogo-ken-Nanbe Earthquake" on February 27,

1995, and the Ministry of Construction adopted on the same day that the reconstruction and repair of highway bridges that suffered damage during the Hyogo-ken-Nanbu Earthquake should be made according to the "Guide Specifications." Then, the Ministry of Construction decided on May 25, 1995, that the "Guide Specifications" should be tentatively used in all sections of Japan as emergency measures for seismic design of new highway bridges and seismic strengthening of existing highway bridges until the Design Specifications of Highway Bridges were revised.

Based on the lessons learned from the Hyogo-ken-Nanbu Earthquake through the various investigations, the seismic design specifications for highway bridges were significantly revised in 1996 [73–75]. The intensive earthquake motion with a short distance from the inland earthquakes with Magnitude 7 class as the Hyogo-ken-Nanbu Earthquake has been considered in the design.

After that, the revision works of the design specifications of highway bridges have been continuously made. The revised specifications were targeted to use the performance-based design concept and to enhance the durability of bridge structures for a long-term use, as well as the inclusion of the improved knowledge on the bridge design and construction methods after the 1996 specifications. The 2002 design specifications of highway bridges were issued by the Ministry of Land, Infrastructure and Transport on December 27, 2001, and were published with commentary from the Japan Road Association (JRA) in March 2003 [76,77].

20.9.2 Performance-Based Design Specifications

The 2002 JRA design specifications [76] are based on the performance concept for the purpose to respond the international harmonization of design codes and the flexible applications of new structures and new construction methods. The performance-based design concept is that the necessary performance requirements and the verification policies are clearly specified. The JRA specifications employ the style to specify both the requirements and the acceptable solutions including the detailed performance verification methods which are design methods and details in the 1996 design specifications [73]. For example, the analysis method to evaluate the response against the loads is placed as one of the verification methods or acceptable solutions. Therefore, designers can propose new ideas or select other design methods with the necessary verification.

The code structure of the JRA Seismic Design Specifications is as shown in Figure 20.31. The static and dynamic verification methods of the seismic performance as well as the evaluation methods of the

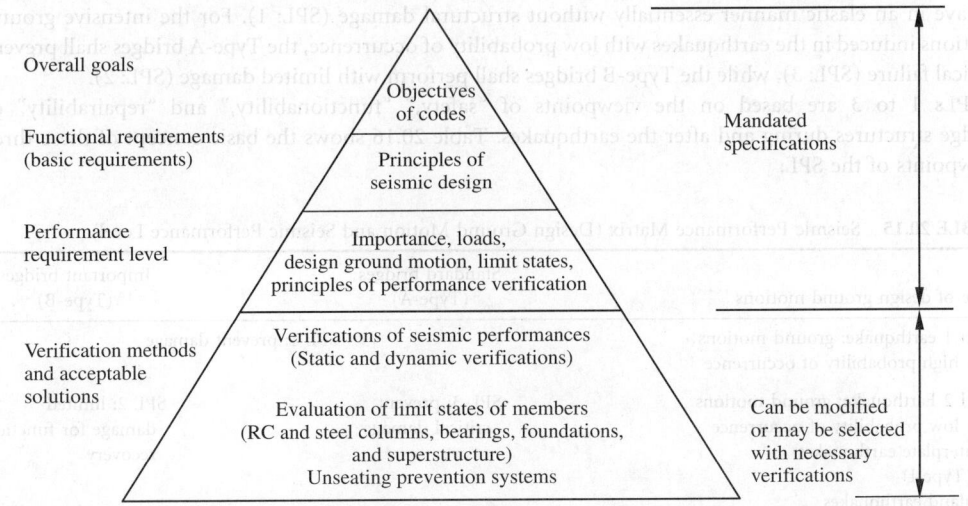

FIGURE 20.31 Code structure of JRA seismic design specification.

strength and ductility capacity of the bridge members are placed as the verification methods and the acceptable solutions, which can be modified by the designers with the necessary verifications.

20.9.3 Basic Principles of Seismic Design

Table 20.15 shows the performance matrix including the design earthquake ground motion and the seismic performance level (SPL) provided in the JRA Seismic Design Specifications. The two-level ground motions are the moderate ground motions induced in the earthquakes with high probability to occur (Level 1 earthquake) and the intensive ground motions induced in the earthquakes with low probability to occur (Level 2 earthquake).

The Level 1 earthquake provides the moderate earthquake ground motions considered in the conventional elastic design method. For the Level 2 earthquake, two types of intensive ground motions are considered. The first is the ground motions that are induced in the interplate-type earthquakes with a magnitude of around 8. The ground motion at Tokyo in the 1923 Kanto Earthquake is a typical target of this type of ground motion. The second is the ground motion developed in earthquakes with magnitude of around 7 at a very short distance. The ground motion at Kobe during the Hyogo-ken-Nanbu Earthquake is a typical target of this type of ground motion. The first and the second ground motions are named as Type-I and Type-II ground motions, respectively.

Figure 20.32 shows the acceleration response spectra of the design ground motions. It is specified that the site-specific design ground motions shall be considered if the ground motion can be appropriately estimated based on the information on the earthquake including past history and the location and detailed condition of the active faults, ground conditions including the condition from the faults to the construction sites. The site-specific design ground motion shall be developed by utilizing necessary and accurate information on the earthquake and ground conditions as well as the verified evaluation methodology of the fault-induced ground motions. However, such detailed information in the regions of Japan is very limited so far. Therefore, continuous investigation and research on this issue as well as the reflection on the practical design of highway bridges is expected.

20.9.4 Ground Motion and Seismic Performance Level

The seismic design of bridges is according to the performance matrix as shown in Table 20.15. The bridges are categorized into two groups depending on their importance: standard bridges (Type-A bridges) and important bridges (Type-B bridges). The SPL depends on the importance of bridges. For the moderate ground motions induced in the earthquakes with high probability of occurrence, both A and B bridges shall behave in an elastic manner essentially without structural damage (SPL: 1). For the intensive ground motions induced in the earthquakes with low probability of occurrence, the Type-A bridges shall prevent critical failure (SPL: 3), while the Type-B bridges shall perform with limited damage (SPL: 2).

SPLs 1 to 3 are based on the viewpoints of "safety," "functionability," and "repairability" of bridge structures during and after the earthquakes. Table 20.16 shows the basic concept of these three viewpoints of the SPL.

TABLE 20.15 Seismic Performance Matrix (Design Ground Motion and Seismic Performance Level)

Type of design ground motions	Standard bridges (Type-A)	Important bridges (Type-B)
Level 1 earthquake: ground motions with high probability of occurrence	SPL 1: prevent damage	
Level 2 Earthquake: ground motions with low probability of occurrence Interplate earthquakes (Type-I) Inland earthquakes (Type-II)	SPL 3: prevent critical damage	SPL 2: limited damage for function recovery

Note: SPL, seismic performance level.

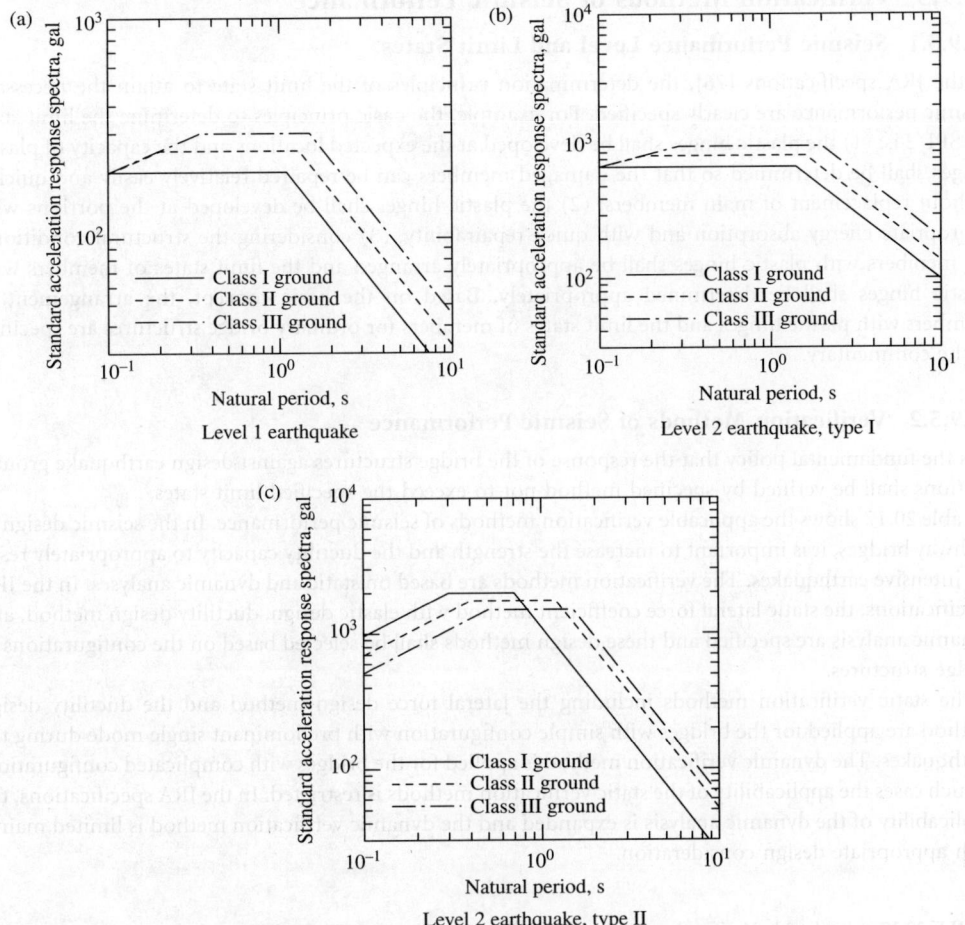

FIGURE 20.32 Design acceleration spectrum. (Class I ground: stiff ground, Class II ground: medium ground, Class III ground: soft ground.)

TABLE 20.16 Key Issues of Seismic Performance

SPL	Safety	Functionability	Repairability	
			Short term	Long term
SPL 1, Prevent damage	Safety against unseating of superstructure	Same function as before earthquake	No need of repair for function recovery	Simple repair
SPL 2, Limited damage for function recovery	Safety against unseating of superstructure	Early function recovery can be made	Function recovery can be made by temporary repair	Relatively easy permanent repair work can be made
SPL 3, Prevent critical damage	Safety against unseating of superstructure	—	—	—

20.9.5 Verification Methods of Seismic Performance

20.9.5.1 Seismic Performance Level and Limit States

In the JRA specifications [76], the determination principles of the limit state to attain the necessary seismic performance are clearly specified. For example, the basic principles to determine the limit state for SPL 2 is: (1) the plastic hinges shall be developed at the expected locations and the capacity of plastic hinges shall be determined so that the damaged members can be repaired relatively easily and quickly without replacement of main members; (2) the plastic hinges shall be developed at the portions with appropriate energy absorption and with quick repairability; (3) considering the structural conditions, the members with plastic hinges shall be appropriately arranged and the limit states of members with plastic hinges shall be determined appropriately. Based on the basic concept, the arrangement of members with plastic hinges and the limit states of members for ordinary bridge structures are specified in the commentary.

20.9.5.2 Verification Methods of Seismic Performance

It is the fundamental policy that the response of the bridge structures against design earthquake ground motions shall be verified by specified method not to exceed the specified limit states.

Table 20.17 shows the applicable verification methods of seismic performance. In the seismic design of highway bridges, it is important to increase the strength and the ductility capacity to appropriately resist the intensive earthquakes. The verification methods are based on static and dynamic analyses. In the JRA specifications, the static lateral force coefficient method with elastic design, ductility design method, and dynamic analysis are specified and these design methods shall be selected based on the configurations of bridge structures.

The static verification methods including the lateral force design method and the ductility design method are applied for the bridges with simple configuration with predominant single mode during the earthquakes. The dynamic verification method is applied for the bridges with complicated configuration; in such cases the applicability of the static verification methods is restricted. In the JRA specifications, the applicability of the dynamic analysis is expanded and the dynamic verification method is limited mainly with appropriate design consideration.

TABLE 20.17   Applicable Verification Methods of Seismic Performance Depending on Earthquake Response Characteristics of Bridge Structures

Dynamic characteristics; SPL to be verified	Bridges with simple configuration	Bridges with multi-plastic hinges and without verification of applicability of energy constant rule	Bridges with limited application of static analysis	
			With multimode response	Bridges with complicated configuration
SPL 1	Static verification	Static verification	Dynamic verification	Dynamic verification
SPL 2/SPL 3	Static verification	Dynamic verification	Dynamic verification	Dynamic verification
Example of bridges	Other bridges	(1) Bridges with rubber bearings to distribute inertia force of superstructures (2) Seismically isolated bridges (3) Rigid frame bridges (4) Bridges with steel columns	(1) Bridges with long natural period (2) Bridge with high piers	(1) Cable-stayed bridges, suspension bridges (2) Arch bridges (3) Curved bridges

Glossary

Capacity-protected component — A component expected to experience minimum damage and to behave essentially elastic during the design earthquakes.

Concentrically braced frame (CBF) — A diagonally braced frame in which all members of the bracing system are subjected primarily to axial forces.

Connections — A combination of joints used to transmit forces between two or more members.

Design earthquake — Earthquake loads represented by acceleration response spectrum (ARS) curves specified in design specifications or codes.

Displacement ductility — Ratio of ultimate-to-yield displacement.

Ductile component — A component expected to experience repairable damage during the FEE and significant damage but without failure during the SEE.

Ductility — Ratio of ultimate-to-yield deformation.

Eccentrically braced frame (EBF) — A diagonally braced frame that has at least one end of each bracing member connected to a link.

Expected nominal strength — Nominal strength of a component based on its expected yield strength.

Functional evaluation earthquake (FEE) — A lower level design earthquake that has relatively small magnitude but may occur several times during the life of the bridge. It may be assessed either deterministically or probabilistically. The determination of this event is to be reviewed by a Caltrans-approved consensus group.

Joint — An area where member ends, surfaces, or edges are attached.

Link — In EBF, the segment of a beam that is located between the ends of two diagonal braces or between the end of a diagonal brace and a column. Under lateral loading, the link deforms plastically in shear thereby absorbing energy. The length of the link is defined as the clear distance between the ends of two diagonal braces or between the diagonal brace and the column face.

Liquefaction — Seismically induced loss of shear strength in loose, cohesionless soil that results from a build up of pour pressure as the soil tries to consolidate when exposed to seismic vibrations.

Maximum credible earthquake (MCE) — The largest earthquake that is capable of occurring along an earthquake fault, based on current geologic information as defined by the 1996 Caltrans Seismic Hazard Map.

Moment-resisting frame (MRF) — A frame system in which seismic forces are resisted by shear and flexure in members and connections in the frame.

Nominal strength — The capacity of a component to resist the effects of loads, as determined by computations using specified material strength, dimensions, and formulas derived form acceptable principles of structural mechanics or by field tests or laboratory test of scaled models, allowing for modeling effects, and differences between laboratory and field conditions.

Overstrength capacity — The maximum possible strength capacity of a ductile component considering actual strength variation between the component and adjacent components. It is estimated by an overstrength factor of 1.2 times expected nominal strength.

Plastic hinge — A concentrated "zero length" hinge that maintains its plastic moment capacity and rotates freely.

Plastic hinge zone — A region of structural components that are subject to potential plastification and thus must be detailed accordingly.

Seismic performance criteria — The levels of performance in terms of postearthquake service and damage that are expected to result from specified earthquake loadings.

Safety evaluation earthquake (SEE) — An upper level design earthquake that has only a small probability of occurring during the life of the bridge. It may be assessed either deterministically or probabilistically. The deterministic assessment corresponds to the maximum credible earthquake. The probabilistically assessed earthquake typically has a long return period (approximately 1000 to 2000 years).

Ultimate displacement — The lateral displacement of a component or a frame corresponding to the expected damage level, not to exceed the displacement when the lateral resistance degrades to a minimum of 80% of the peak resistance.

Upper bound solution — A solution calculated on the basis of an assumed mechanism that is always at best equal to or greater than the true ultimate load.

Yield displacement — The lateral displacement of a component or a frame at the onset of forming the first plastic hinge.

References

[1] Moehle, J.P. and Eberhard, M.O. Chapter 34: Earthquake Damage to Bridges, *Bridge Engineering Handbook*, ed., Chen, W.F. and Duan, L., CRC Press, Boca Raton, FL, 2000.

[2] Yashinsky, M. Chapter 29a: Earthquake Damage to Structures, *Structural Engineering Handbook CRCnetBase 2000*, ed., Chen, W.F. and Duan, L., CRC Press, Boca Raton, FL, 2000.

[3] Housner, G.W. *Competing Against Time*, Report to Governor George Deuknejian from The Governor's Broad of Inquiry on the 1989 Loma Prieta Earthquake, Sacramento, CA, 1990.

[4] Caltrans, *The First Annual Seismic Research Workshop*, Division of Structures, California Department of Transportation, Sacramento, CA, 1991.

[5] Caltrans, *The Second Annual Seismic Research Workshop*, Division of Structures, California Department of Transportation, Sacramento, CA, 1993.

[6] Caltrans, *The Third Annual Seismic Research Workshop*, Division of Structures, California Department of Transportation, Sacramento, CA, 1994.

[7] Caltrans, *The Fourth Caltrans Seismic Research Workshop*, Engineering Service Center, California Department of Transportation, Sacramento, CA, 1996.

[8] Caltrans, *The Fifth Caltrans Seismic Research Workshop*, Engineering Service Center, California Department of Transportation, Sacramento, CA, 1998.

[9] FHWA and Caltrans, *The Proceedings of First National Seismic Conference on Bridges and Highways*, San Diego, CA, 1995.

[10] FHWA and Caltrans, *The Proceedings of Second National Seismic Conference on Bridges and Highways*, Sacramento, CA, 1997.

[11] Kawashima, K. and Unjoh, S., The Damage of Highway Bridges in the 1995 Hyogo-Ken Naubu Earthquake and its Impact on Japanese Seismic Design, *J. Earthquake Eng.*, 1(2), 1997, 505.

[12] Park, R., Ed., Seismic Design and Retrofitting of Reinforced Concrete Bridges, *Proceedings of the Second International Workshop*, Queenstown, New Zealand, August, 1994.

[13] Astaneh-Asl, A. and Roberts, J., Ed., Seismic Design, Evaluation and Retrofit of Steel Bridges, *Proceedings of the First U.S. Seminar*, San Francisco, CA, 1993.

[14] Astaneh-Asl, A. and Roberts, J., Ed., Seismic Design, Evaluation and Retrofit of Steel Bridges, *Proceedings of the Second U.S. Seminar*, San Francisco, CA, 1997.

[15] Housner, G.W. *The Continuing Challenge — The Northridge Earthquake of January 17, 1994*, Report to Director, California Department of Transportation, Sacramento, CA, 1994.

[16] Priestley, M.J.N., Seible, F., and Calvi, G.M., *Seismic Design and Retrofit of Bridges*, John Wiley and Sons, New York, NY, 1996.

[17] Caltrans, *The Sixth Caltrans Seismic Research Workshop*, Divisions of Engineering Services, California Department of Transportation, Sacramento, CA, June 12–13, 2001.

[18] FHWA-NSF, *Proceedings of 16th US–Japan Bridge Engineering Workshop*, Lake Tahoe, Nevada, October 2–4, 2000.

[19] ATC, *Seismic Design Guidelines for Highway Bridges*, Report No. ATC-6, Applied Technology Council, Redwood City, CA, 1981.

[20] AASHTO, *Standard Specifications for Highway Bridges*, 17th ed., American Association of State Highway and Transportation Officials, Washington, DC, 2002.

[21] Caltrans, *Bridge Design Specifications*, California Department of Transportation, Sacramento, CA, 1990.

[22] Caltrans, *Bridge Design Specifications, LFD Version*, California Department of Transportation, Sacramento, CA, 2000.

[23] Caltrans, *Bridge Memo to Designers (20-1) — Seismic Design Methodology*, California Department of Transportation, Sacramento, CA, January 1999.

[24] Caltrans, *Bridge Memo to Designers (20-4)*, California Department of Transportation, Sacramento, CA, 1995.

[25] Caltrans, *Bridge Memo to Designers (20-11) — Establishing Bridge Seismic Design Criteria*, California Department of Transportation, Sacramento, CA, January 1999.

[26] Caltrans, *Bridge Design Aids*, California Department of Transportation, Sacramento, CA, 1995.

[27] Caltrans, *Bridge Design Specifications — LFD Version*, California Department of Transportation, Sacramento, CA, April 2000.

[28] Caltrans, *San Francisco — Oakland Bay Bridge West Spans Seismic Retrofit Design Criteria*, Prepared by Reno, M. and Duan, L. Edited by Duan, L., California Department of Transportation, Sacramento, CA, 1997.

[29] Caltrans, *San Francisco-Oakland Bay Bridge East Span Seismic Safety Project Design Criteria, Version 7*, Prepared by T.Y. Lin, Moffatt & Nichol Engineers, California Department of Transportation, Sacramento, CA, July 6, 1999.

[30] IAI, *Benicia-Martinez Bridge Seismic Retrofit — Main Truss Spans Final Retrofit Strategy Report*, Imbsen and Association, Inc., Sacramento, CA, 1995.

[31] FHWA, *Seismic Design and Retrofit Manual for Highway Bridges*, Report No. FHWA-IP-87-6, Federal Highway Administration, Washington, DC, 1987.

[32] FHWA, *Seismic Retrofitting Manual for Highway Bridges*, Publication No. FHWA-RD-94-052, Federal Highway Administration, Washington, DC, 1995.

[33] ATC, *Improved Seismic Design Criteria for California Bridges: Provisional Recommendations*, Report No. ATC-32, Applied Technology Council, Redwood City, CA, 1996.

[34] Rojahn, C. et al. *Seismic Design Criteria for Bridges and Other Highway Structures*, Report NCEER-97-0002, National Center for Earthquake Engineering Research, State University of New York at Buffalo, Buffalo, NY, 1997. Also refer as ATC-18, Applied Technology Council, Redwood City, CA, 1997.

[35] Caltrans, *Seismic Design Criteria, Version 1.3*, California Department of Transportation, Sacramento, CA, February 2004.

[36] Caltrans, *Guide Specifications for Seismic Design of Steel Bridges*, 1st ed., California Department of Transportation, Sacramento, CA, December 2001.

[37] ATC/MCEER Joint Venture, *Recommended LRFD Guidelines for the Seismic Design of Highway Bridges, Part I: Specifications; Part II: Commentary and Appendixes, Preliminary Report*, ATC Report Nos. ATC-49a and ATC-49b, Applied Technology Council, Redwood City, California and MCEER Technical Report No. MCEER-02-SP01, Multidisciplinary Center for Earthquake Engineering Research, State University of New York at Buffalo, Buffalo, NY, November 2001.

[38] AASHTO, *LRFD Bridge Design Specifications*, 3rd ed., American Association of State Highway and Transportation Officials, Washington, DC, 2004.

[39] Chen, W.F. and Duan, L., Ed., *Bridge Engineering Handbook*, CRC Press, Boca Raton, FL, 2000.

[40] Priestley, N., Myths and Fallacies in Earthquake Engineering — Conflicts Between Design and Reality, *Proceedings of Tom Paulay Symposium — "Recent Development in Lateral Force Transfer in Buildings,"* University of California, San Diego, CA, 1993.

[41] Kowalsky, M.J., Priestley, M.J.N., and MacRae, G.A., *Displacement-Based Design*, Report No. SSRP-94/16, University of California, San Diego, CA, 1994.

[42] Troitsky, M.S. Chapter 1: Conceptual Bridge Design, *Bridge Engineering Handbook*, ed., Chen, W.F. and Duan, L., CRC Press, Boca Raton, FL, 2000.

[43] Keever, M.D., Caltrans Seismic Design Criteria, *Proceedings of 16th U.S.–Japan Bridge Engineering Workshop*, Lake Tahoe, Nevada, October 2–4, 2000.

[44] Cheng, C.T. and Mander, J.B. *Seismic Design of Bridge Columns Based on Control and Repairability of Damage*, NCEER-97-0013, State University of New York at Buffalo, Buffalo, NY, 1997.

[45] USGS, http://www.geohazards.cr.usgs.gov/eq/, 2001.

[46] Mander, J.B. and Cheng, C.T. *Seismic Resistance of Bridge Piers Based on Damage Avoidance Design*, NCEER-97-0014, State University of New York at Buffalo, Buffalo, NY, 1997.

[47] Duan, L. and Cooper, T.R. Displacement Ductility Capacity of Reinforced Concrete Columns, *ACI Concrete Int.*, 17(11), 1995, 61–65.

[48] Park, R., and Paulay, T. *Reinforced Concrete Structures*, John Wiley and Sons, New York, NY, 1975.

[49] Akkari, M. and Duan, L., Chapter 36: Nonlinear Analysis of Bridge Structures, *Bridge Engineering Handbook*, ed., Chen, W.F. and Duan, L., CRC Press, Boca Raton, FL, 2000.

[50] Caltrans, *Seismic Design of Abutments for Ordinary Standard Bridges*, Division of Structure Design, California Department of Transportation, Sacramento, CA, March 20, 2001.

[51] AASHTO, *Guide Specifications for Seismic Isolation Design*, American Association of State Highway and Transportation Officials, Washington, DC, 2000.

[52] Zhang, R., Chapter 41: Seismic Isolation and Supplemental Energy Dissipation, *Bridge Engineering Handbook*, ed., Chen, W.F. and Duan, L., CRC Press, Boca Raton, FL, 2000.

[53] King, W.S., White, D.W., and Chen, W.F. Second-Order Inelastic Analysis Methods for Steel-Frame Design, *J. Struct. Eng.*, ASCE, 118(2), 1992, 408–428.

[54] Levy, R., Joseph, F., and Spillers, W.R. Member Stiffness with Offset Hinges, *J. Struct. Eng.*, ASCE, 123(4), 1997, 527–529.

[55] Chen, W.F. and Toma, S. *Advanced Analysis of Steel Frames*, CRC Press, Boca Raton, FL, 1994.

[56] Chen, W.F. and Atsuta, T. *Theory of Beam-Columns*, Vols 1 and 2, McGraw-Hill Inc., New York, NY, 1977.

[57] Kiureghian, A.D. and Neuenhofer, A. Response Spectrum Method for Multi-Support Seismic Excitations. *Earthquake Eng. Struct. Dyn.*, 21, 1992, 713–740.

[58] Heredia-Zavoni, E. and Vanmarcke, E.H. Seismic Random-Vibration Analysis of Multisupport-Structural Systems, *J. Eng. Mech.*, ASCE, 120(5), 1994, 1107–1128.

[59] European Committee for Standardization, *Eurocode 8: Structures in Seismic Regions — Design Part 2: Bridges*, 1995.

[60] Lin, J.H., Zhang, W.S., and Williams, F.W. Pseudo-Excitation Algorithm for Nonstationary Random Seismic Responses, *Eng. Struct.*, 16, 1994, 270–276.

[61] Lin, J.H., Zhang, W.S., and Li, J.J. Structural Responses to Arbitrarily Coherent Stationary Random Excitations, *Comput. Struct.*, 50, 1994, 629–633.

[62] Lin, J.H., Li, J.J., Zhang, W.S., and Williams, F.W. Non-stationary Random Seismic Responses of Multi-support Structures in Evolutionary Inhomogeneous Random Fields, *Earthquake Eng. Struct. Dyn.*, 26, 1997, 135–145.

[63] Lin, J.H., Zhao, Y., and Zhang, Y.H., Accurate and Highly Efficient Algorithms for Structural Stationary/Non-stationary Random Responses, *Comput. Meth. Appl. Mech. Eng.*, 191(1–2), 2001, 103–111.

[64] Lin, J.H., Zhang, Y.H., and Zhao, Y., High Efficiency Algorithm Series of Random Vibration and Their Applications, *Proceedings of the 8th International Conference on Enhancement and Promotion of Computational Methods in Engineering and Science*, Shanghai, China, 2001, pp. 120–131.

[65] Cheng, W., Spectrum Simulation Models for Random Ground Motions and Analysis of Long-Span Bridges under Random Earthquake Excitations, *Doctoral Dissertation*, Hunan University, Changsha, China, 2000.

[66] Fan, L.C., Wang, J.J., and Chen, W., Response Characteristics of Long-Span Cable-Stayed Bridges under Non-uniform Seismic Action, Dept of Bridge, *Chinese J. Comput. Mech.*, 18(3), 2001, 358–363.

[67] AISC, *Load and Resistance Factor Design Specification for Structural Steel Buildings*, 3rd ed., American Institute of Steel Construction, Chicago, IL, 1999.

[68] AISC, *Seismic Provisions for Structural Steel Buildings (1997), Supplement No. 1* (1999) and *Supplement No. 2* (2000), American Institute of Steel Construction, Chicago, IL, 1997.

[69] Carden, L., Garcia-Alvarez, S., Itani, A., and Buckle, I., Cyclic Response of Steel Plate Girder Bridges in the Transverse Direction, *The Sixth Caltrans Seismic Research Workshop*, Divisions of Engineering Services, California Department of Transportation, Sacramento, CA, June 12–13, 2001.

[70] Zelinski, R. *Seismic Design Momo Various Topics Preliminary Guidelines*, California Department of Transportation, Sacramento, CA, 1994.

[71] Ministry of Construction. *Report on the Damage of Highway Bridges by the Hyogo-ken Nanbu Earthquake*, Committee for Investigation on the Damage of Highway Bridges Caused by the Hyogo-ken Nanbu Earthquake, 1995 (in English).

[72] Kawashima, K. *Impact of Hanshin/Awaji Earthquake on Seismic Design and Seismic Strengthening of Highway Bridges*, Report No. TIT/EERG 95-2, Tokyo Institute of Technology, 1995 (in English).

[73] Japan Road Association, *Design Specifications of Highway Bridges, Part I Common Part, Part II Steel Bridges, Part III Concrete Bridges, Part IV Foundations, and Part V Seismic Design*, 1996 (in Japanese, Part. V: English version, July 1998).

[74] Kawashima, K. et al. *1996 Design Specifications for Highway Bridges*, 29th UJNR Joint Panel Meeting, May 1996 (in English).

[75] Unjoh, S. Chapter 44: Seismic Design Practice in Japan, *Bridge Engineering Handbook*, ed., Chen, W.F. and Duan, L., CRC Press, Boca Raton, FL, 2000.

[76] Japan Road Association, *Design Specifications of Highway Bridges, Part I Common Part, Part II Steel Bridges, Part III Concrete Bridges, Part IV Foundations, and Part V Seismic Design*, 2002 (in Japanese, Part. V: English version, June 2003).

[77] Unjoh, S. et al. *Design Specifications for Highway Bridges*, 34th UJNR Joint Panel Meeting, May 2002.

21

Performance-Based Seismic Design and Evaluation of Building Structures

Sashi K. Kunnath
Department of Civil Engineering
University of California,
Davis, CA

Performance-based design (PBD) has emerged as the new paradigm in seismic engineering. The concept of PBD is not limited to the field of seismic design, though the material covered in this chapter focuses primarily on recent developments in earthquake engineering. Performance-based seismic engineering (PBSE) is still an evolving methodology; hence, the information presented in this chapter should not be interpreted as an existing design specification or an adopted standard. Rather, the contents of this chapter should be viewed as an introduction to an emerging concept for seismic design of building systems.

There are several completed and ongoing efforts to develop performance-based seismic design methodologies. Published documents such as ATC-40 (1996), FEMA-350 (2000), and FEMA-356 (2000) embody key aspects of PBD; however, they were each developed with limited objectives. FEMA-350 applies to new steel moment frames only. ATC-40 and FEMA-356 contain guidelines for the evaluation and rehabilitation of existing buildings; the former is limited to the evaluation of existing reinforced concrete buildings, while the latter covers all building types and is, therefore, referred to as a "prestandard" suggesting that it may resemble the model of a PBD code.

The purpose of this chapter is to provide readers with an overview of existing procedures that attempt to incorporate performance-based concepts in seismic design of new structures or evaluation of existing structures. Several potential approaches to accomplishing the goals of PBD will be discussed and an illustrative example that applies each conceptual framework in the evaluation of a hypothetical design will be presented so as to offer a basis for comparing the relative merits and potential drawbacks of each approach. Much of the emphasis will be placed on buildings since they constitute the largest stock of structures in the built environment, though the procedures outlined here can be extended with minimal effort to other structural systems.

21.1   Some Issues in Current Seismic Design

The first question that comes to mind when introducing a new methodology is the obvious one: why do we need a new procedure? What is inherently wrong or inadequate in the existing provisions for design that warrants a new look at the entire process? Hence, the task of introducing PBSE is more easily accomplished by highlighting the limitations and drawbacks of existing seismic design procedures.

Since the early development of seismic design codes, global response modification factors (or R-factors) have remained at the core of seismic force formulas. The main purpose of the force-reduction factors used in seismic design is to simplify the analysis process so that elastic methods can be used to approximately predict the expected inelastic demands in a structure subjected to the design loads. They account for reductions in seismic force values due to a variety of factors including system inherent ductility, overstrength, and redundancy. Of these, only the ductility component of the R-factor is generally implied in the design provisions because systems with larger expected ductility have the lowest reduction factors. Current codes also specify a displacement amplification factor C_d that quantifies the expected inelastic displacement of the system. Both R and C_d factors are global response measures that do not provide an assessment of structural performance at the component level. There is growing awareness that force-based design using R and C_d factors has serious shortcomings. For instance, these factors are independent of the building period and ground motion characteristics. Additionally, the same R-factor is used for moment-resisting reinforced concrete (RC), steel, and braced frames. It is clear that a single global response modifier cannot capture the progressive distribution of nonlinearities between various structural elements, the resulting redistribution of seismic demands inside the structure, and the changes that occur during the course of the seismic motion. In addition to differences in seismic demands and failure mechanisms, the damage distribution is also likely to vary from one structure to the other even though they have all been designed for the same R-value. A coherent description of the meaning and basis of establishing force-reduction factors is outlined in two papers by Uang (1991) and Uang and Bertero (1991). Figure 21.1 shows the base-shear versus roof displacement response of a typical building structure. The vertical axis in Figure 21.1 shows the base shear coefficient, which is the total shear normalized by the seismic weight of the building (V/W). The design base shear coefficient of the building is C_s, while the corresponding elastic strength is C_e. As is evident from the response of the structure, the yield strength of the building is C_y (assuming a bilinear idealization as shown). First yielding in a member in the system should typically commence at C_s, though material overstrength and member sizing

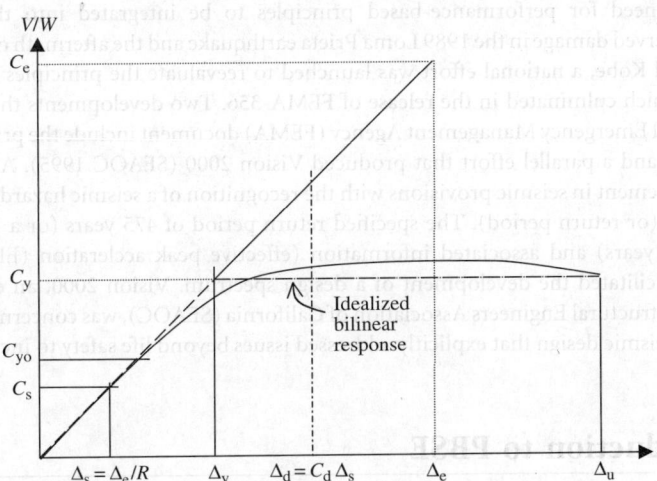

FIGURE 21.1 Conceptual basis behind use of *R*-factors.

may delay initial yielding beyond this value. Hence, the actual force-reduction factor or response modification factor as defined in IBC (2000) can be defined as follows:

$$R = \frac{C_e}{C_s} = \frac{C_e}{C_y} \cdot \frac{C_y}{C_s} \tag{21.1}$$

The ratio C_e/C_y can be viewed as the response modification factor related to ductility (R_μ), while the ratio C_y/C_s contains two components: overstrength (R_Ω) and redundancy (R_R). Hence, the *R*-factor can be broken down into three main components as indicated in Equation 21.2:

$$R = R_\mu R_\Omega R_R \tag{21.2}$$

Numerous factors influence each of the modifiers that appear in Equation (21.2). For example, the postyield strain (or stiffness) can affect both R_μ and R_Ω. The use of higher material strengths than those specified in the design, satisfying minimum code requirements for detailing, etc., the presence of nonstructural components, and the oversizing of members have an impact primarily on R_Ω and R_R. The product $R_\Omega R_R$ is likely to have greater variability for RC than for steel structures. It is clearly difficult to isolate the different components of the reduction factor and thereby provide engineers with an understanding of the demands imposed not only on the overall system but also on individual components in the system and the margin of safety against failure.

Since the total lateral force or base shear is the primary design parameter, the current code format is regarded as a "strength-based" design procedure. The idea of distributing strength throughout the structure rather than relying on a single base-shear parameter is recognized in the concept of capacity design (Park and Paulay 1976), which forms the basis of the New Zealand building code. The realization that displacements are more critical than forces initiated the move toward displacement-based design (Moehle 1992, 1996; Priestley and Calvi 1997; Chopra and Goel 2001). Procedures to enhance displacement-based methods yet retain the simplicity of a response spectrum to characterize the hazard were developed by Fajfar and Gasperic (1996). Other interesting research on the shortcomings of existing procedures and the promise of new ideas to advance seismic design methodologies was presented at a workshop in Slovenia (Fajfar and Krawinkler 1997).

While displacement (or deformation) based criteria still remain at the core of ongoing developments to improve the state of practice, the need to correlate measures of performance with measures of demand has ushered the need for performance-based principles to be integrated into the design process. Following the observed damage in the 1989 Loma Prieta earthquake and the aftermath of the seismic events in Northridge and Kobe, a national effort was launched to reevaluate the principles governing modern seismic design, which culminated in the release of FEMA-356. Two developments that laid the foundation for the Federal Emergency Management Agency (FEMA) document include the prior effort that led to ATC 3-06 (1978) and a parallel effort that produced Vision 2000 (SEAOC 1995). ATC-3-06 marked a significant improvement in seismic provisions with the recognition of a seismic hazard in terms of a mean annual frequency (or return period). The specified return period of 475 years (or a 10% probability of exceedance in 50 years) and associated information (effective peak accleration (EPA), effective peak velocity (EPV)) facilitated the development of a design spectrum. Vision 2000, an effort initiated and supported by the Structural Engineers Association of California (SEAOC), was concerned with developing a framework for seismic design that explicitly addressed issues beyond life safety to include damageability and functionality.

21.2 Introduction to PBSE

The concepts of hazard and performance level are implied in the provisions of standard building codes such as International Building Code (IBC). The *SEAOC Blue Book,* regarded as the commentary for the Uniform Building Code (now IBC) provisions, clearly identifies the following performance objectives:

- No damage for minor levels of ground motion.
- No structural damage but some nonstructural damage for moderate ground motions.
- Structural and nonstructural damage but no collapse for major levels of ground motion (which include the strongest credible earthquake at the site).

Despite the recognition of three damage states implied in modern codes, life safety has historically been the principle concern in seismic design. Furthermore, there are no explicit procedures or processes that allow an engineer to evaluate the expected performance of the final design or assess the margin of safety provided by satisfying code requirements. Recent experience has shown that property damage and related losses resulting from minor to moderate earthquakes is significant. Nonstructural considerations and business losses can dominate the cost–benefit ratio in a life-cycle cost analysis (Krawinkler 1997). Hence, emphasis on nonstructural issues and business interruption losses must become part of the equation in a holistic treatment of seismic design. This and other concerns discussed in the previous section have led researchers and engineers to rethink the principles and process governing modern seismic design. The various alternative strategies proposed by numerous individual and group research efforts have collectively led to the evolution of performance-based concepts that allow the design of buildings (or the rehabilitation of existing buildings) with due consideration to different design objectives and varying levels of risk and loss. While current design methodology relies considerably on empirical formulations and prescriptive procedures that require only minimal design checks (such as drift limits following the sizing of sections), PBD methodology calls for a detailed demand evaluation of the building model under simulated earthquake loads to assess if the design objective has been achieved. The assessment is intended to project the potential damage and associated losses, which in turn allows all stakeholders (from the building owner to insurers and the building occupants) to specify desired levels of performance, which then becomes the basis of design for the structural engineer. While this entails a more significant responsibility on the part of the design professional, it also means that greater flexibility is possible in the design — in terms of new design alternatives and techniques (the use of seismic protection devices is one example) that can be utilized to achieve the expected performance objective.

21.2.1 Elements of a Typical Performance-Based Methodology

The first definitive document that laid the basis for PBD in seismic engineering is FEMA-356 (2000), which essentially synthesizes two earlier reports, FEMA-273 and FEMA-274, and deals with seismic rehabilitation of existing buildings. A parallel effort that resulted in the publication of ATC-40 (1996) is limited to RC buildings but is more comprehensive in its treatment. At least one other guideline that builds on FEMA-273 is the result of the FEMA-sponsored SAC project that produced FEMA-350 (2000). Other notable research that addresses issues in performance-based engineering includes ongoing efforts at the Mid-America Earthquake (MAE) and Pacific Earthquake Engineering Research (PEER) centers.

The contents of this chapter are based on the concepts and procedures outlined in the various documents and guidelines cited above. All performance-based methodologies share common elements. Though the details of implementation may vary from one methodology to the next, the basic objectives of the each approach are essentially the same. An attempt is made in the following sections to classify the key components of a typical PBD methodology in three collective steps:

Step 1. a. Define a performance objective that incorporates a description of both the hazard and the expected level of performance.

      b. Select a trial design.

Step 2. Determine seismic demands on the system and its components through an analysis of a mathematical model of the structure.

Step 3. Evaluate performance (at the system and component levels) to verify if the performance objective defined in Step 1 has been met. If performance levels are not satisfactory, revise design and return to Step 2.

A brief overview of the three-step process is outlined in the following sections.

21.2.1.1 Quantifying Performance Objectives

Stated simply, a performance objective specifies the desired seismic performance target for the structure. A performance objective consists of two parts: a level of performance that is typically expressed in terms of a *damage state* or *decision variable* and a *hazard level* that describes the expected seismic load at the site. Some examples of performance levels include: collapse prevention, life safety (preservation of human life), and immediate occupancy of the building following the design seismic event. Alternatively, a performance level can be expressed in terms of its economic impact (dollar loss or downtime). A commonly used description of hazard is the probability of the event in a given duration, for example, 10% probability of the design seismic loads being exceeded in 50 years. The intent of this notion of a performance objective is conveyed in Figure 21.2, which quantifies the allowable damage or loss as

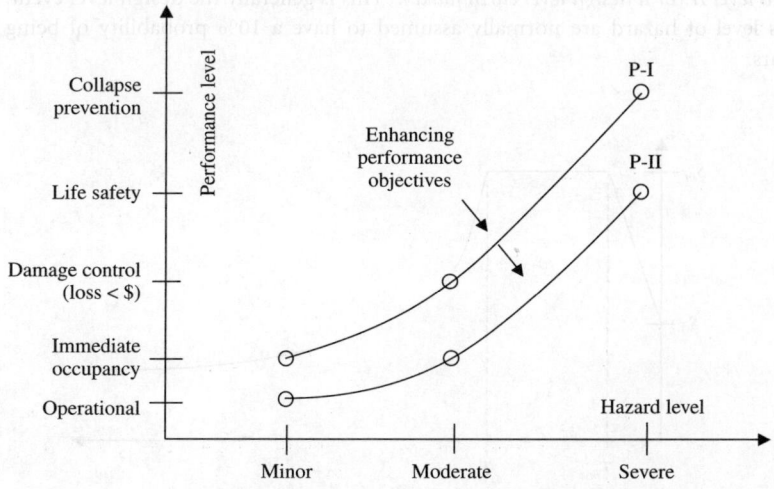

FIGURE 21.2 Defining performance objectives.

a function of the intensity of the earthquake load. As suggested earlier, a hazard level denotes the severity of the earthquake. A minor hazard level would represent an event with a higher probability of being exceeded during the life of the structure, while an increased hazard level representing a more severe event would indicate a reduced probability for the same duration. Damage states may be quantified in numerous ways, but two of the most common approaches are physical damage states based on the degree of inelastic response of members and expected losses including economic and indirect losses.

Since a structural system is composed of both primary and secondary elements, the specification of a performance level can be distinguished depending on the structural function. Hence, the structural performance level can be different from the nonstructural performance level for a given performance objective.

21.2.1.1.1 Description of Hazard

Earthquake ground motions, or the characteristics of the expected ground motions, are generally used in conjunction with a performance objective. Given the uncertainty in predicting ground motions, the task of establishing a site-specific design event is a difficult but critical part of any PBD methodology. The description of the hazard is meant to represent the seismic threat given knowledge of the potential earthquake sources at the site. This is typically accomplished through a probabilistic seismic hazard assessment that results in a hazard curve describing mean annual frequency of exceeding a certain spectral acceleration magnitude.

The specification of ground shaking is inevitably linked to two factors: (1) site geology and related site characteristics, which may be classified by soil profile or other soil parameters such as shear wave velocity, undrained shear strength, etc. and (2) site seismicity, which includes information on the seismic source. Ultimately, the design earthquake needs to be quantified in terms of a response spectrum as shown in Figure 21.3. The spectrum is characterized by three critical points: the peak ground acceleration (PGA) S_0, the region of maximum spectral demands (S_m between T_m and T_n), and finally a description of the spectral demands beyond T_n.

To allow for a range of performance objectives, it is necessary to define more than a single hazard level. The prevalent thinking in most PBD guidelines is to consider at least three levels:

- *Hazard level I (or a basic service level earthquake)*. Earthquakes representing this level of hazard are expected to occur more frequently, and therefore, the likelihood of such an event during the life of the structure is high. Probabilistically speaking, this event is typically equated with a 50% probability of being exceeded in 50 years or a mean return period of 72 years.
- *Hazard level II (or a design level earthquake)*. This is generally the design level event. Earthquakes at this level of hazard are normally assumed to have a 10% probability of being exceeded in 50 years.

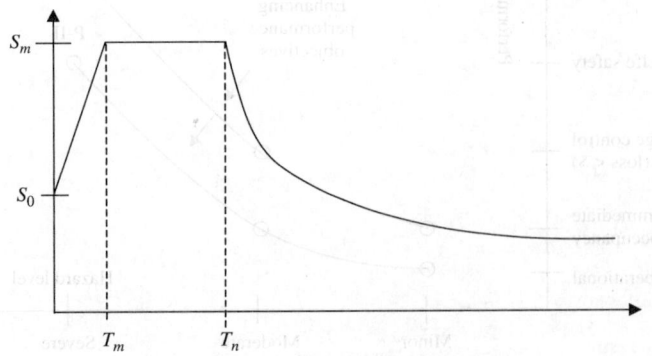

FIGURE 21.3 Key parameters in design response spectrum.

- *Hazard level III (or a maximum credible level earthquake).* Finally, it will be necessary to define a maximum credible event at the site. Such an event can be used to enable higher performance objectives (particularly for critical facilities). Here, the event is assumed to have only a 2% probability of being exceeded in 50 years.

21.2.1.2 Estimation of Demand

Once a performance objective has been defined, the next task is to determine if the design of the structure meets the stated performance objective. This means that a mathematical model of the structure is subjected to the seismic design loads (as specified by the hazard at the site in the previous step) and the demands on the system and its components are evaluated. The task of estimating seismic demand involves the determination of deformations and forces in both structural and nonstructural elements in the structure.

The prediction of deformation demands is a crucial and challenging aspect of a PBD approach. Currently, the only requirement in design that examines demand is the check on drift limitations, which is accomplished by linear elastic methods. Though some of the proposed PBD approaches suggest the use of linear methods to determine demands, there is concern that the introduction of a simplified technique into an otherwise advanced and comprehensive process will prove to be the undoing of the methodology. Errors in estimating the demand can propagate through the process and lead to misleading conclusions on the performance of the structure. Hence, for PBD methodology to become acceptable to engineers and simultaneously pass the scrutiny of researchers, there has to be consensus on an analytical approach (linear or nonlinear and static or dynamic) to evaluate seismic demands.

The process of estimating seismic demands can be subdivided into two primary tasks: developing a model of the structure wherein all structural components and essential nonstructural components are adequately represented and carrying out an analysis of the model subjected to loads that characterize the hazard level. The model may be simple or complex and is dependent on the computational tool (software) that will be used in the analysis.

The deformation demands in a structure and its elements are dependent on a variety of factors:

- *The characteristics of the seismic input.* A response spectrum is valid for static methods including modal analysis, while ground motions (recorded or simulated) are required for time history analyses.
- *Modeling of elements.*
- *Consideration of nonstructural components.*
- *Accurate estimation of material properties.*
- *Modeling of material behavior.* This relates to modeling the nonlinear response characteristics of the elements of the structure, particularly the nature of energy dissipation (hysteresis) of members. Consideration of cyclic degradation is critical to the assessment of performance and must be incorporated in the analytical model.
- *Method of analysis.* This refers to the four possible methods that can be used to estimate forces and deformations in the structural model. Static methods can be used if the seismic input is specified in terms of a response spectrum while transient methods employing a time-integration scheme must be used for ground motion input. In either case, the system of equations may be solved using a constant stiffness matrix resulting in a linear solution, or advanced techniques to incorporate material or geometric nonlinearities can be utilized.

The seismic input will already be defined in the first step in terms of either a response spectra or a set of actual ground motions that satisfy certain criteria. The modeling of structural elements that includes the specification of material-dependent force–deformation behavior is limited by the analytical tool being used to evaluate the model. Most PBD approaches utilize an element-by-element discretization of the building and specify modeling characteristics based on moment and rotation at the ends of the element with due consideration to interaction with axial and shear forces. Assuming an acceptable mathematical model of the system is developed, the next step is to carry out an analysis of the model. This brings us to a crucial phase in the process. What kind of analysis is appropriate to accurately estimate the imposed seismic demands? Since the motivation for moving away from traditional seismic design is to abandon linear force based methods, it can be argued that the analysis has to be nonlinear. In this context, pushover methods or nonlinear static approaches have gained considerable popularity since they avoid the difficulties and uncertainties associated with time-history analyses. However, it must not be over-looked that an earthquake is a dynamic event and the reliability of using static methods to predict dynamic behavior must be carefully evaluated before prescribing its use in a performance-based eva-luation. FEMA-356 prescribes the use of both linear and nonlinear methods, though certain restrictions are placed on linear and static approaches. ATC-40 restricts the analytical solution to nonlinear methods only. All four methods of analysis are discussed later in this chapter, including limitations and potential issues on their application in PBSE.

21.2.1.3 Assessment of Performance

This can be regarded as the third and final phase of a performance-based evaluation. If the estimation of seismic demands is a critical phase in a PBD methodology that must be carried out cautiously with attention to detail, then the next stage in a performance-based evaluation wherein the estimated demands are transformed into performance measures is probably the most contentious. In this phase of the evaluation, demand values such as displacement, drift, and plastic rotation need to be interpreted on a damage scale D $(0 < D < 1)$ that encompasses the complete range from elastic undamaged response to near or total collapse.

The question that needs to be addressed here is, what is the most convenient yet rational manner to translate response quantities from a structural analysis into qualitative measures of performance? Much of the work that has been conducted in the area of damage modeling becomes relevant in this context. Currently, only the PEER methodology (to be discussed later in this chapter) uses damage measures explicitly in its formulation. The approach that is favored by engineers (if FEMA-356 is used as a frame of reference) is the use of so-called acceptance criteria that demystify the demand-to-damage trans-formation through simple subjective but quantitative guidelines. The need to improve and fine-tune these guidelines is well recognized, but the process is well defined and deterministic.

Response quantities estimated from the analysis of the building model in Step 2 need to be compared to acceptable or allowable limits that are termed "acceptance criteria." These limits can be specified both at the global level in terms of interstory drift limits and at the local level in terms of component demand limits. The limits are a function of performance levels. Hence, to achieve a higher performance level for a given hazard level, the acceptability criteria will be more stringent.

The introduction of acceptance criteria for different performance levels represents a significant departure from traditional seismic design and is obviously the singular step in the overall process that qualifies the term "performance based" in the terminology of structural design. To develop acceptance criteria for a range of element types, it is necessary to have access to information on physically observed damage states during laboratory testing of components and subassemblies. FEMA-356 and ATC-40 provide an initial compilation of such data based on the input of experts and must be regarded as a preliminary guideline for use in the evaluation of existing structures. What is important is that the framework now exists to establish new acceptance limits and engineers have the flexibility to develop their own data using the results of experimental research.

An alternative approach to assessing performance is to describe it in a probabilistic sense given the fact that it is difficult to establish quantitative or deterministic estimates of the consequences of demand limits or damage states. For this reason, FEMA-350 proposes "confidence levels" on the demand esti-mates that result in a probabilistic description of the performance level. Ongoing work in PEER and MAE attempts to further expand this description using concepts in probability theory. The result is the evolution of cumulative distribution functions or fragility curves that characterize the conditional probability of a decision, damage, or demand variable as a function of the seismic hazard.

To wrap up the introduction to the three-step methodology, it is necessary to return to one of the fundamental objectives that advanced the art of PBSE, namely, the consideration of losses from structural and nonstructural damage. The basic framework of FEMA-356 or ATC-40 does not explicitly link acceptance criteria to expected losses, but it is possible to generate estimates of potential losses given a known damage state. Some of the ongoing research in PBSE (Miranda and Aslani 2003; Krawinkler and Miranda 2004) attempts to address this issue. For example, the PEER methodology explicitly includes decision variables (such as dollar loss, downtime, or other quantifiable economic impact) in its performance-based formulation and will be briefly discussed later in this chapter.

21.3 Performance-Based Methodologies: Deterministic Approach

The precursors to performance-based design in seismic engineering, as suggested earlier in this chapter, can be found in two separate published reports: ATC-40 and FEMA-356. However, both documents arose from the urgent need to develop systematic guidelines for seismic rehabilitation of buildings. The effort to produce ATC-40 was prompted by the stock of pre-1970s cast-in-place concrete buildings in California. The lack of guidelines to assist engineers to rehabilitate these relatively nonductile buildings resulted in the ATC-40 project. Given the narrow focus of ATC-40, it provides a fairly detailed and in-depth look at different perspectives in performance-based design and evaluation. FEMA-356, on the other hand, was a much larger effort and covers all building types: steel, concrete, masonry, wood, and structures incorporating isolation or energy dissipating devices. Though both documents employ a similar format for describing performance objectives, only ATC-40 endorses the capacity spectrum method (CSM) as a basis for evaluating structural performance.

We begin with the common thematic elements in ATC-40 and FEMA-356, namely, the classification of performance levels leading to the specification of performance objectives.

21.3.1 Classifying Performance Levels

Structural and nonstructural performance levels are generally combined to yield the overall building performance level. At the structural level, the following major performance levels are used:

- Immediate occupancy (IO) level requires the building to be safe for unlimited egress, ingress, and occupancy.
- Damage control is described as a performance range from IO to life safety level rather than a specific performance level.
- Life safety (LS) level of performance suggests structural damage without partial or total collapse of structural members, which might pose a risk to life.
- Limited safety, like damage control, is another performance range that lies between LS and the structural stability limit.
- Structural stability level can be viewed as an alternative definition of collapse prevention (CP) wherein the structure is still capable of maintaining gravity loads though structural damage is severe and the risk of falling hazards is high.

Each structural performance level is associated with a damage state that can be observed or quantified. For example, FEMA-356 describes the expected damage at LS performance level for concrete frames as "extensive damage to beams," "hinge formation" in ductile secondary elements with crack widths less than $\frac{1}{8}$ in., and permanent drifts up to 1%. At CP performance level, concrete frames could experience permanent drifts up to 4%, while steel frames may see drift (transient or permanent) magnitudes of 5%. Both FEMA-356 and ATC-40 (which applies to concrete structures only) contain detailed descriptions of damage for different element types (columns, beams, structural walls, etc). It must be reiterated that the damage measures are based on the expected performance of existing buildings. The magnitudes of

TABLE 21.1 Major Performance Levels Recommended in ATC-40 and FEMA-356

	Structural performance levels		
Nonstructural performance levels	I: immediate occupancy	III: life safety	V: structural stability (collapse prevention)
A: operational	I-A: operational	NR	NR
B: immediate occupancy	I-B: immediate occupancy	III-B	NR
C: life safety	I-C	III-C: life safety	V-C
E: not considered	NR	NR	V-E: structural stability

Note: NR, not recommended.

these measures can be adjusted if experimental data are available to support alternative performance ranges.

Nonstructural performance levels include the performance of components such as elevators, piping, fire sprinkler systems, and heating, ventilation, and air conditioning (HVAC) equipment and building contents such as computers, bookshelves, and art objects. Nonstructural performance levels can be defined as follows:

- Operational performance level requires the postearthquake damage state of all nonstructural elements to remain functional and in operation.
- IO performance level allows for minor disruption due to shifting and damage of components, but all nonstructural components are generally in place and functional.
- LS performance level allows for damage to components but does not include failure of items heavy enough to pose a risk of severe injuries or secondary hazards from damage to high-pressure toxic and fire-suppressing piping.
- Reduced hazard is a postearthquake damage state that considers risk to groups of people from falling heavy objects such as cladding and heavy ceilings.
- Finally, it is also possible to include what is called a "not considered" performance level. This performance level is provided to cover those nonstructural elements that have not been evaluated as having an impact on the overall structural response. In fact, it is common for computer models evaluating seismic demands to ignore the presence of nonstructural elements.

When structural and nonstructural performance levels are combined, the resulting building performance level is established. Table 21.1 shows how such a performance level is determined. Note that only a few selected performance levels are displayed in Table 21.1. In the context of Table 21.1, if a building performance level of III-C were selected, this would require LS performance level at both structural and nonstructural elements.

21.3.2 Defining Performance Objectives

To prescribe a set of performance objectives, ATC-40 defines the following three hazard levels: (1) a serviceable earthquake (SE) with a 50% probability of being exceeded in 50 years, (2) a design earthquake (DE) with a 10% probability of being exceeded in 50 years, and (3) a maximum earthquake (ME) with a 5% chance of being exceeded in 50 years. FEMA-356, on the other hand, specifies four hazard levels in which the earthquake event has the following probabilities: 50% in 50 years, 20% in 50 years, 10% in 50 years, and 2% in 50 years.

A performance objective is now defined by selecting a building performance objective (Table 21.1) for a given hazard level. A possible set of combinations is displayed in Table 21.2 for three hazard levels and three performance levels, making a total of nine possible combinations. For example, a possible performance objective would be to achieve an overall building performance level of III-C for a DE event which translates into LS performance for a 10%/50-year earthquake (or *b* in Table 21.2). It is also

TABLE 21.2   Defining Performance Objectives

	Target building performance levels		
Hazard level	Immediate occupancy	Life safety	Structural stability or collapse prevention
10%/50 years	a	b	c
5%/50 years	d	e	f
2%/50 years	g	h	i

possible to create dual or multiple performance objectives by selecting different performance levels and different hazard levels. An example of this would be achieving I-B performance level for a DE event but satisfying III-C for an ME earthquake (this would correspond to **a** and **e** in Table 21.2).

Both ATC-40 and FEMA-356 prescribe a basic safety objective (BSO), which comprises a dual-level performance objective. In ATC-40, this would entail satisfying III-C (LS performance level) for a DE and also achieving V-E (structural stability) for an ME event. In FEMA-356, the BSO criterion requires LS performance for a 10% in 50-year event and CP performance level for a 2% in 50-year earthquake. With reference to Table 21.2, the BSO in ATC-40 is **b** + **f** and in FEMA-356 it is **b** + **i**.

21.3.3   Design Response Spectra and Ground Motions

The selection of a performance objective, as just described, involves the specification of a hazard level. Unless ground motion time histories are used in a dynamic time-history analysis, it is customary to specify the hazard in terms of a response spectrum. The generation of the ground motion hazard spectrum is a function of several parameters, most of which pertain to site characteristics.

Site-specific hazard analysis is recommended for very soft soils that are vulnerable to failure under seismic action such as liquefiable soil, highly sensitive clays, high-plasticity clays with depths exceeding 25 ft and $PI > 75$, and soft- to medium-stiff clays with depths exceeding 120 ft. Soil properties are based on average values for the top 100 ft (30 m) of soil profile. Site-specific studies are also required for special and critical structures located near an active fault. Even if a site-specific hazard spectrum is developed, both FEMA-356 and ATC-40 place limits on the values of site response coefficients C_A and C_V.

In all other cases, an elastic response spectrum as shown in Figure 21.4 can be constructed from known values of C_A and C_V. Typical soil profile types are described in Table 21.3.

21.3.3.1   Generating the Design Spectrum Using ATC-40 Provisions

The site response coefficient C_A is simply the effective peak acceleration (EPA) at the site while the coefficient C_V when divided by the period defines the acceleration in the constant velocity domain. ATC-40 provides the following three options when developing the elastic design spectra:

1. Site specific hazard analysis.
2. Spectral contour maps developed by U.S. Geological Survey (USGS).
3. Site seismic coefficients given in Table 21.4.

If the USGS contour maps (developed for rock sites) are used, then the following modifications are prescribed for buildings on S_B soil sites: $C_A = 0.4S_{MS}$, $C_V = S_{M1}$, where S_{MS} is the spectral acceleration at 0.3 s (short-period range) and S_{M1} is the spectral acceleration at 1.0 s (velocity-sensitive range).

To use Table 21.4, it is first necessary to determine the shaking intensity, which is defined as the product of three quantities: $Z * E * N =$ zone factor × earthquake hazard level × near source factor. Using the shaking intensity value, the corresponding site response coefficients C_A and C_V are obtained for a known soil profile. The response spectrum can now be easily generated as indicated in Figure 21.4.

The potential for liquefaction and landsliding at a site are also important considerations when evaluating a structure for earthquake hazards. ATC-40 provides some guidelines for determining when

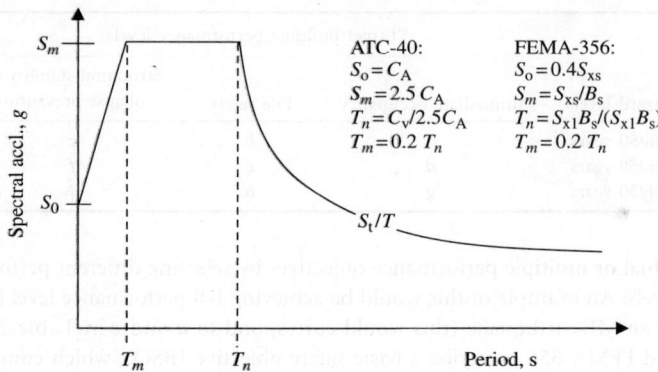

FIGURE 21.4 ATC-40 and FEMA-356 representation of the 5% damped design response spectrum.

TABLE 21.3 Average Soil Properties Used to Establish Soil Profile

Soil profile type	Soil profile name	Shear wave velocity (m/s)	Standard penetration test (blows/30 cm)	Undrained shear strength (kPa)
S_A	Hard rock	>1520	n/a	n/a
S_B	Rock	760–1520	n/a	n/a
S_C	Very dense soil	365–760	>50	>2
S_D	Stiff soil	180–365	15–150	1–2

Note: n/a, not applicable.

TABLE 21.4 ATC-40 Specifications for Developing Hazard Spectrum

Soil profile		Shaking intensity, ZEN					
		0.075	0.150	0.200	0.300	0.400	>0.400
S_B	C_A	0.08	0.15	0.20	0.30	0.40	$1.0 * ZEN$
	C_V	0.08	0.15	0.20	0.30	0.40	$1.0 * ZEN$
S_C	C_A	0.09	0.18	0.24	0.33	0.40	$1.0 * ZEN$
	C_V	0.13	0.25	0.32	0.45	0.56	$1.4 * ZEN$
S_D	C_A	0.12	0.22	0.28	0.36	0.44	$1.1 * ZEN$
	C_V	0.18	0.32	0.40	0.54	0.64	$1.6 * ZEN$

Notes: Z = zone factor (IBC-2000: e.g., Zone 4 = 0.4g); E = 0.5 (SE), 1.0 (DE), and 1.25 (Zone 4) and 1.50 (Zone 3) (ME); N = near source factor (typically, N = 1 for faults not capable of producing events with maximum moment magnitude M > 6.5 and faults with slip rates less than 2 mm/year). Linear interpolation is permitted for intermediate values.

a more detailed study of these hazards is warranted. In both cases, the expected EPA is used to establish the likelihood of ground failure given the liquefaction potential or the geologic group at the site.

21.3.3.2 FEMA-356 Provisions for Generating the Design Spectrum

National Earthquake Hazards Reduction Program (NEHRP) seismic maps (also available at http://eqhazmaps.usgs.gov) provide peak acceleration plots for 5% damped response spectrum at two characteristic periods (short duration: 0.2 s, long duration: 1 s). These maps are referenced to site class B and must be adjusted for the applicable site class through the following equations:

$$S_{xs} = F_a S_s \tag{21.3}$$

$$S_{x1} = F_v S_1 \tag{21.4}$$

where S_{xs} is the design short-period spectral response parameter, S_{x1} is the design spectral response acceleration parameter at 1 s, F_a and F_v are the site coefficients (provided in FEMA-356), S_s is the short-period spectral response parameter, and S_1 is the spectral response acceleration parameter at 1 s.

The site-specific response spectrum is then generated using the following equations:

$$S_a = (S_{xs}/B_s)(0.4 + 3T/T_n) \quad \text{for } 0 < T \leq T_m \tag{21.5}$$

$$S_a = (S_{xs}/B_s) \qquad\qquad \text{for } T_m < T \leq T_n \tag{21.6}$$

$$S_a = (S_{x1}/B_1 T) \qquad\qquad \text{for } T > T_n \tag{21.7}$$

where

$$T_n = (S_{x1}B_s)/(S_{xs}B_1) \tag{21.8}$$

and B_s and B_1 are damping coefficients, quantified in FEMA-356 as a function of effective damping in the system. The resulting response spectrum is conceptually similar to ATC-40 and is shown in Figure 21.4.

21.3.4 Ground Motions

Both ATC-40 and FEMA-356 contain provisions for using acceleration time histories in conjunction with transient response analyses. A minimum of three ground motion sets must be used. The selected ground motions are expected to have magnitude, fault distances, and source mechanisms that are comparable to the ground-shaking hazard at the building site. FEMA-356 requires that the data sets be scaled in such a manner that the average value of the square root of sum of squares (SRSS) spectra (that are generated for each data set) does not fall below 1.4 times the 5% damped spectrum for the DE for periods between $0.2T$ and $1.5T$ (where T is the fundamental period of the structure).

When only three time history sets (one horizontal and one vertical component for two-dimensional [2D] systems and all three components for a 3D model) are used in the evaluation process, the maximum value of each response parameter should be considered. If seven or more records are used, then both FEMA-356 and ATC-40 permit mean response estimates to be considered to assess design acceptability.

ATC-40 actually provides a list of ten earthquake data sets for two scenarios: sites that are farther than 10 km from the source and sites that are within 5 km from the source to be considered near-fault events. The ground motions listed in ATC-40 are free-field motions recorded on firm soil sites with magnitude greater than 5.5 and a PGA exceeding 0.2g. A subset of these records is used in the design example presented later in this chapter. It is expected that the listed ground motions be scaled such that the average value of the spectra matches the site response spectrum in the period range of interest (an effective period range near the performance point of the structure — see ensuing discussion on CSM for a definition of the performance point).

21.4 Evaluating Seismic Demands

The estimation of seismic demands requires the development of a mathematical model of the building. The model should incorporate all components that influence the mass, stiffness, and strength of the building, particularly in the inelastic regime of the response. Hence, issues like soil–structure interaction must be carefully evaluated before a decision is made to include or exclude the soil–foundation system in the final model. The model should properly account for gravity loads that comprise dead loads and other permanent fixtures. Consideration of live loads is also necessary if the presence of additional gravity loads is likely to create a situation resulting in adverse seismic response or the shifting of the location of the plastic hinge.

A structural model also includes specification of the expected behavior of all of the elements used to develop the building model. A linear elastic analysis requires only the estimation of the effective stiffness of each element, whereas a nonlinear analysis demands a more concerted effort to establish the expected local behavior of every element in the overall structural model. The guidelines summarized in the next section are partly based on recommendations collectively provided in ATC-40, FEMA-356, and FEMA-350.

21.4.1 Modeling Guidelines

The development of analytical models must account for all possible aspects of behavior while still working within the limitations of the analytical tools being used to carry out the evaluation of seismic demands. Both commercial and noncommercial software is available to assist engineers in estimating seismic demands, and the modeling requirements vary from one tool to the next. Most programs employ simple hinge-based inelastic models with multilinear characterization of force–deformation behavior. In general, it is necessary to model three commonly recognized aspects of cyclic behavior: stiffness degradation, strength deterioration, and pinching response. Hysteretic models that incorporate some or all of these behavioral aspects have been proposed by numerous researchers (Kunnath et al. 1997; Sivaselan and Reinhorn 2000). There is also a large body of work on degrading force–deformation models developed by Japanese researchers, many of which are summarized in a report by Umemura and Takizawa (1982).

A general hysteretic model that incorporates all three elements is shown in Figure 21.5. This basic model, which is implemented in the inelastic damage analysis of RC structures (IDARC) (Kunnath et al. 1992; Kunnath 2004) series of programs, uses several control parameters to establish the rules under which inelastic loading reversals take place. A variety of hysteretic properties can be achieved through the combination of a nonsymmetric trilinear curve and certain control parameters that characterize the shape of the force–deformation loops. For example, α, which can be expressed as a function of the deformation, controls the amount of stiffness loss; ϕ and χ control the initiation and degree of pinching respectively; and the slope s and the change in expected peak strength (M to M^*) control the softening due to system deterioration. A sample simulation of observed behavior using this model is shown in Figure 21.6. The hysteresis curves in this case were obtained from tests of a precast concrete connection with a hybrid combination of mild steel and posttensioning steel (Cheok et al. 1998). However, the specification of hysteretic rules in an actual analysis is rather empirical and should be based on available experimental data. Even so, a parametric study to evaluate the sensitivity of these parameters is necessary.

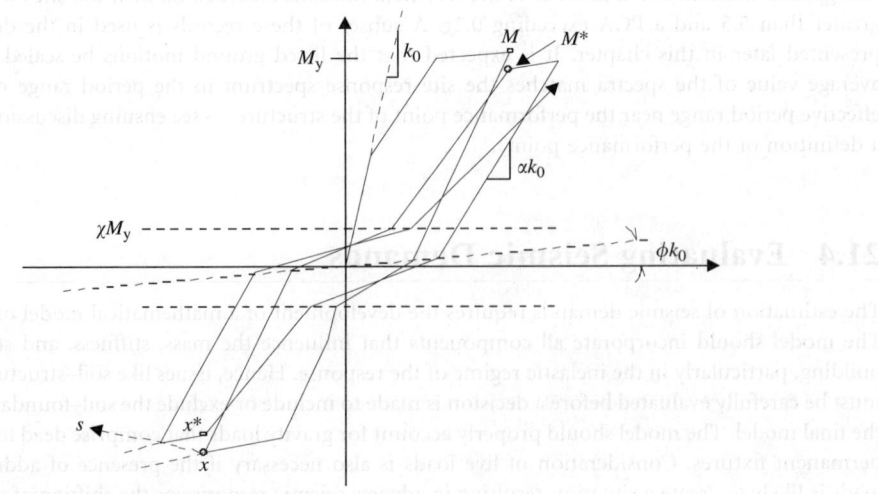

FIGURE 21.5 General-purpose hysteretic model.

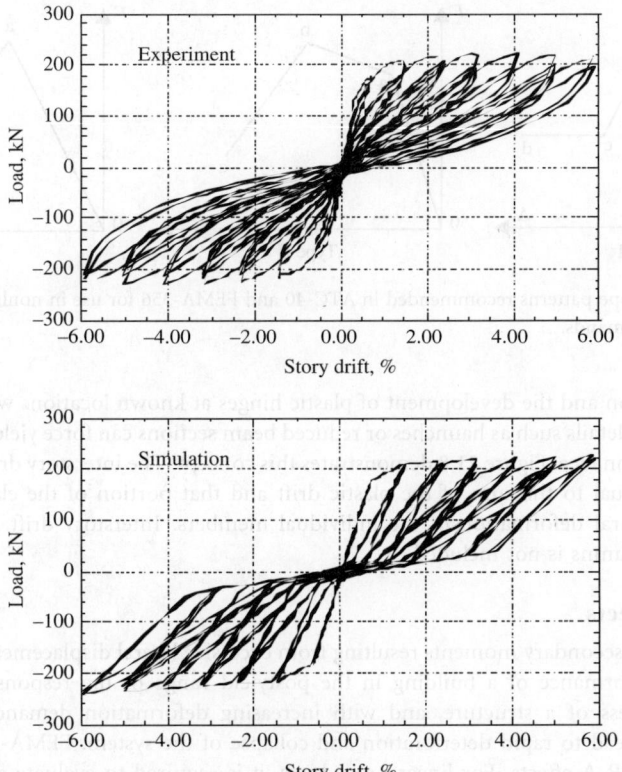

FIGURE 21.6   Simulation of observed hysteretic behavior.

The degree of degeneration is a function of the connection detail. Welded steel moment connections or well-detailed RC members may display stable hysteresis with little degradation of strength in the performance range of the element. Bolted connections or concrete sections with inadequate details may exhibit pinching behavior. Shear dominated members will experience rapid decay in strength after yielding. In general, it is necessary to know in advance the expected behavior of an element when using force–deformation type models. Complete cyclic description of behavior, as illustrated in Figure 21.5 and Figure 21.6, is essential for nonlinear time-history methods only. For nonlinear static approaches, only the force–deformation envelope is needed. FEMA-356 and ATC-40 specify three types of force–deformation envelopes for use with nonlinear static analyses. These are displayed in Figure 21.7. A Type I envelope is meant to represent the expected behavior of a well-detailed ductile connection with a well-defined postyield range (a to b), a region of gradual decay (b to c), and a nonnegligible residual strength that is capable of supporting gravity loads in the regime (c to d). A Type II curve is similar to Type I with the exception that an element modeled as Type II cannot be relied upon to support gravity loads beyond "b." An important characterization in FEMA is the classification of members as being force controlled or deformation controlled. Members modeled as Type I are generally deformation-controlled members while Type II elements can be classified as deformation controlled only if the ductility at peak strength (b/a) is greater than 2.0. A Type III curve is representative of brittle failure and is, therefore, always a force-controlled element.

An additional detail that is outlined in FEMA-350 is the identification and specification of the interstory drift angle and the associated plastic hinge locations. The idea is that frames should be detailed in such a manner that the required interstory drift angle can be accommodated through a combination

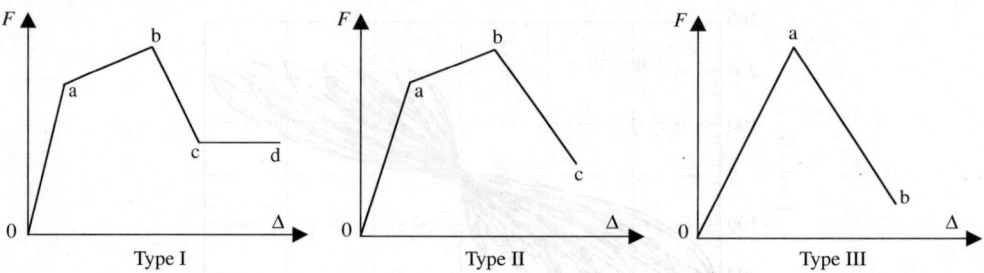

FIGURE 21.7   Envelope patterns recommended in ATC-40 and FEMA-356 for use in nonlinear analysis methods to estimate seismic demands.

of elastic deformation and the development of plastic hinges at known locations within the frame. For example, the use of details such as haunches or reduced beam sections can force yielding in the beam but away from the column face. Figure 21.8 demonstrates this concept. The interstory drift angle, as specified in FEMA-350, is equal to the sum of the plastic drift and that portion of the elastic interstory drift resulting from flexural deformation of the individual members. Interstory drift resulting from axial deformations of columns is not included.

21.4.1.1   *P*–Δ Effects

The consequence of secondary moments resulting from excessive lateral displacements can have a major impact on the performance of a building in the postyield range of the response. *P*–Δ effects alter the postyield stiffness of a structure, and with increasing deformation demands the net effect of *P*–Δ softening can lead to rapid deterioration and collapse of the system. FEMA-356 recognizes both static and dynamic *P*–Δ effects. For linear procedures, it is required to evaluate a stability coefficient given by

$$\theta_i = \frac{P_i \delta_i}{V_i h_i} \tag{21.9}$$

where P_i is the weight of structure acting on story level i, V_i is the total lateral shear force acting on story level i, h_i is the height of story level i, and δ_i is the lateral drift of story i.

If the stability coefficient is less than 0.1 in all stories, then *P*–Δ effects can be ignored in the analysis. For values of the coefficient between 0.1 and 0.33, the seismic effects are to be magnified by a factor $1/(1 - \theta_i)$, and for values in excess of 0.33, the system is considered unstable and should be redesigned.

Even if nonlinear procedures are used, a magnification factor (C_3) is used in FEMA-356 for static methods because dynamic *P*–Δ effects are recognized as being different from static effects. Static *P*–Δ should be incorporated in pushover analysis, while both static and dynamic *P*–Δ can be automatically incorporated in nonlinear dynamic procedures.

21.4.1.2   Other Effects and Considerations

In general, it is expected that a building be modeled as a 3D assembly of elements. However, most structures tend to be fairly regular with assumptions of symmetry commonly employed in routine analysis. Some modeling issues that must be resolved before finalizing a building model for use in demand prediction include the following:

- The assumption of rigid diaphragms, which allows axial deformations in floor beams to be ignored and simplifies the resulting system of equations that need to be solved.
- Any out-of-plane offsets in those vertical elements that resist lateral forces must be accounted for when estimating demands on floor diaphragms.

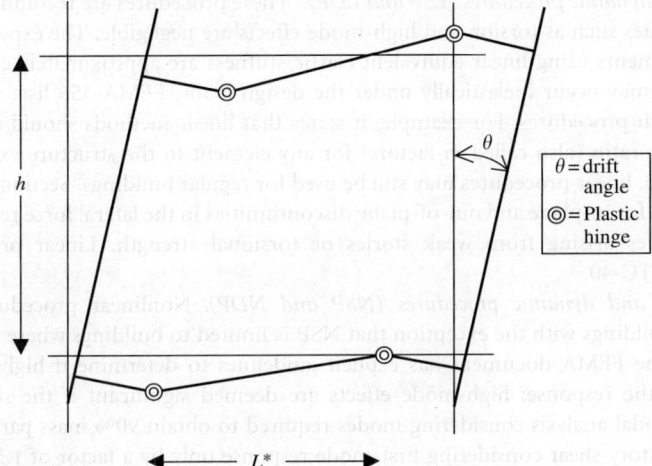

FIGURE 21.8　Definition of interstory drift angle.

- Consideration of actual and accidental horizontal torsion: FEMA-356 lists specific criteria for considering the effects of horizontal torsion.

Normally, it is adequate to consider seismic input in each principal direction of the building only. However, for irregular buildings, concurrent seismic effects in two orthogonal directions must be accounted for by combining 100% of the design effects in one direction with 30% of the design effects in the orthogonal direction. Similarly, the need to consider vertical seismic input arises only in special circumstances such as cantilevered or prestressed elements and those components in which gravity load demands are excessive.

Other considerations not discussed in this section include

- Foundation and SFSI interaction effects.
- Variability in ground motion.
- Modeling of secondary and nonstructural components.
- Introduction and modeling of seismic protection devices such as isolators, dampers, and other energy dissipating devices.

A description of the details that take into consideration the consequence of the above effects is beyond the scope of this chapter.

21.4.2　Methods of Analysis

The estimation of demands can be accomplished using a variety of available procedures. In this section, discussion will be limited to a generic set of evaluation methods that best lend themselves to an assessment within a performance-based framework. The primary objective is to determine forces and deformations both at the global and at the local level when the structure is subjected to seismic loads that characterize the hazard at the building site. As such, there are four possible methods to analyze a mathematical model of a building structure. They may be classified into two broad categories depending on the treatment of the response or the treatment of the loads. The former category results in the distinction between linear and nonlinear methods of analysis, while the latter distinguishes static and dynamic application of the seismic loads.

Linear static and dynamic procedures (LSP and LDP). These procedures are recommended for regular buildings where issues such as torsion and high-mode effects are negligible. The expectation is that the computed displacements using linear equivalent elastic stiffness are approximately equal to the actual displacements that may occur inelastically under the design loads. FEMA-356 lists specific criteria to limit the use of such procedures. For example, it states that linear methods should not be used if the demand to capacity ratio (also called *m*-factors) for any element in the structure exceeds 2.0. Even if *m*-factors exceed 2.0, linear procedures may still be used for regular buildings. Section 2.4.1.1 of FEMA-356 outlines criteria for in-plane and out-of-plane discontinuities in the lateral force resisting system and also for irregularities arising from weak stories or torsional strength. Linear procedures are not recommended in ATC-40.

Nonlinear static and dynamic procedures (NSP and NDP). Nonlinear procedures are generally applicable for all buildings with the exception that NSP is limited to buildings where high-mode effects are small. Again, the FEMA document has explicit guidelines to determine if higher modes play an important role in the response: high-mode effects are deemed significant if the shear in any story resulting from a modal analysis considering modes required to obtain 90% mass participation exceeds the corresponding story shear considering first-mode response only by a factor of 1.3.

The LSP is the simplest procedure described in FEMA-356. LSP is an equivalent lateral load analysis procedure that attempts to represent the seismic loading as a static force. The reason for retaining a linear static approach in a performance-based guideline is debatable, but the motivation stems from the simplicity of the method and the current familiarity among engineers with force-based design. It is also felt that such an approach may still be valid for a large group of regular structures. The total lateral load applied to the structure is calculated from

$$V = C_1 C_2 C_3 C_m S_a W \tag{21.10}$$

where V is the total base shear, C_1, C_2, and C_3 are modification factors to account for the effects of inelasticity, system degradation, and P–Δ effects, respectively, C_m is an effective mass factor to account for high-mode participation, and S_a is the spectral acceleration at the fundamental period of the structure. The seismic weight of the structure, W, includes 25% of live load in addition to dead loads. The fundamental period, T, of the structure can be computed using an eigenvalue analysis or through an empirical equation, given by the following expression:

$$T = C_t (H_n)^\beta \tag{21.11}$$

where H_n is the height of the building measured from base to roof, and coefficient C_t and power β depend on the structural system. The fundamental period is a critical parameter in the evaluation procedure for both static procedures (LSP and NSP), since they can change the applied lateral force in the linear procedure or alter the target displacement in the nonlinear procedure (discussed in Section 21.5.2). The total shear computed in Equation 21.10 is distributed over the building height as follows:

$$F_x = \frac{w_x h_x^k}{\sum w_i h_i^k} V \tag{21.12}$$

where w_i is the portion of weight at floor level i, w_x is the portion of weight at floor level x, H_i is the height from base to floor level i, H_x is the height from base to floor level x, and $k = 1.0$ for $T < 0.5$ s and 2.0 for $T \geq 2.5$ s.

The summation in the denominator of Equation 21.12 is carried over all stories, and the coefficient k in the period range from 0.5 and 2.5 s is established through linear interpolation. Note that a reduction factor is not applied to the computed base shear. Hence, the demands will exceed the strength capacity of those components that are expected to yield under the design earthquake loads. The resulting demand-to-capacity ratios are evaluated at the component level.

The LDP involves a linear elastic dynamic analysis using a response spectrum approach or direct time-history evaluation to determine the building response. Linear elastic stiffness properties of elements and equivalent viscous damping at or near the yield level are to be used in the analysis. An unmodified elastic spectrum is utilized in a response spectrum analysis. Peak forces and deformations are estimated using an SRSS or complete quadratic combination (CQC) of modal quantities. For time-history methods, if less than seven earthquake records are used, the maximum response quantities are to be used in the evaluation procedure. For seven or more records, an average value of the response is adequate. As mentioned previously, FEMA-356 requires that each record be scaled such that the average value of the SRSS of the 5% damped spectrum does not fall below 1.4 times the design spectrum value (also constructed for 5% damping) between $1.2T$ and $1.5T$ (where T is the fundamental building period).

The NSP or pushover analysis takes into consideration material nonlinearities. The concept gained prominence after its introduction in the CSM by Freeman (1978). The nonlinear static method is also called the displacement coefficient method in that the expected demand (target displacement) of a single degree-of-freedom (SDOF) system is modified by a set of coefficients to approximate the multiple DOF response. The procedure involves selecting a control node (typically the center of mass at the roof level of the building) and subjecting the structural model to monotonically increasing lateral loads till a target displacement is reached. The lateral load applied to the building to achieve the control node target displacement can be distributed in several ways. The choice of a lateral load pattern can have a significant influence on the calculated response. FEMA-356 suggests that at least two patterns be used: a uniform pattern that results in lateral forces proportional to the total mass of each floor level and a modal pattern, which should be selected from the following two options: (1) when more than 75% of the total mass participates in the fundamental mode, a lateral load pattern represented by Equation 21.12 is used and (2) a lateral load pattern that is proportional to the story shear distribution, calculated by combining modal responses from a response spectrum analysis wherein at least 90% modal mass participation is incorporated, is used. The target displacement is established from the following equation:

$$\delta_t = C_0 C_1 C_2 C_3 S_a \left[\frac{T_e^2}{4\pi^2} \right] g \qquad (21.13)$$

where δ_t is the target displacement (note that response quantities are determined when the roof displacement reaches this target displacement), C_0 is a modification factor relating spectral displacement estimated for an equivalent SDOF system to the likely roof displacement of a multistory structure, and the remaining coefficients C_1 to C_3 have the same meaning as previously defined for linear procedures. The effective fundamental period T_e is determined from

$$T_e = T_i (K_i / K_e)^{0.5} \qquad (21.14)$$

where T_i is the elastic fundamental period, K_i is the elastic lateral stiffness, and K_e is the secant stiffness at 60% of the yield strength of the building.

Pushover methods have been the subject of numerous studies following numerous questions about the validity of using static procedures to predict dynamic demands. Modified and advanced pushover techniques are available in the literature and are discussed later in this chapter following the design example.

The NDP refers to a complete nonlinear dynamic time history analysis and is generally regarded as the most accurate method of demand evaluation. A complete nonlinear description, in terms of force–deformation behavior at a cross-section (either using hysteretic models or from direct integration of stresses derived from nonlinear constitutive relationships), of all elements expected to experience inelastic deformations should be developed. No fewer than three ground motion sets (comprising of two horizontal components and a vertical component, if necessary) need to be considered. The selected time histories should have magnitude, fault distance, and source mechanisms that are equivalent to the design ground motion at the site. More detailed criteria are specified in FEMA-356 if simulated motions are used. Similar to the linear dynamic procedure, mean response quantities can be used if seven or more records are utilized in the simulations, and peak response quantities should be used if fewer than seven ground motions are used in the evaluation.

21.5 Assessing Performance: The Development of Acceptance Criteria

The third and final step in a general PBD process is to determine if the demands computed in step 2 using one of the four methods described in the previous section are within the acceptance criteria that satisfy the performance level defined as part of the overall performance objective in step 1 of the procedure. The current thinking in FEMA-356 is that the performance of a component in the system is critical to the overall performance of the building. Consequently, acceptance criteria are specified at the element level. The resulting responses from the analysis of the structural model of the building system are classified as force-controlled (zero or limited ductility) or deformation-controlled actions. It is also necessary to distinguish primary and secondary components.

21.5.1 Linear Procedures

For linear procedures (LSP and LDP), deformation-controlled design actions should include the combined effects of earthquake and gravity loads as follows:

$$Q_{UD} = Q_G \pm Q_E \tag{21.15}$$

where Q_{UD} is the deformation-controlled design action, Q_G is the action due to gravity loads, and Q_E is the action due to earthquake forces.

In the case of force-controlled actions (defined as actions in components where nonlinear deformations are not permitted or are limited), the following expression is used in FEMA-356:

$$Q_{UF} = Q_G \pm \frac{Q_E}{C_1 C_2 C_3 J} \tag{21.16}$$

where Q_{UF} is the force-controlled design action, Q_G is the action due to gravity loads, Q_E is the action due to earthquake forces, C_i are the amplification coefficients introduced in the previous section, and J is the smallest demand-to-capacity ratio of components delivering forces to the component (≥ 1); alternatively, $J = 2$ in high seismic zones, 1.5 in moderate seismic zones, and 1.0 in zones, of low seismicity.

Note that both the positive and the negative signs that appear in Equations 21.15 and 21.16 should be considered when determining the design actions, and the worst-case scenario should be used in the evaluation. Following the analysis using either an LSP or an LDP, the demand-to-capacity ratio (or m-factors) for each component is established as follows:

$$m = \frac{Q_{UD}}{\kappa Q_{CE}} \tag{21.17}$$

where Q_{CE} is the expected capacity of the component (typically the yield capacity of the section considering possible interaction with other actions) and κ is the knowledge factor that accounts for the uncertainty in estimating the material properties of the section (this factor is unity for new designs).

The demand-to-capacity ratio for force-controlled elements is limited to unity: elements with limited ductility (less than 2.0) are also classified as force-controlled elements within the scope of FEMA-356 guidelines. For deformation-controlled elements (those elements where inelastic action is expected and permitted), a set of acceptance (or allowable) limits need to be developed for different components. Typically, such values come from an evaluation of experimental data supplemented by analytical studies. FEMA-356 provides an initial set of numbers that are recommended for existing buildings. However, the FEMA tables also contain recommended values for well-detailed sections that can form the basis of

TABLE 21.5　Typical Acceptance Criteria for Primary Components in a Fully Restrained Steel Frame Building

	m-Factors for linear procedures			Plastic rotation angle for nonlinear methods		
Component	IO	LS	CP	IO	LS	CP
Beams — flexure						
$\dfrac{b_f}{2t_f} \le \dfrac{52}{\sqrt{f_{ye}}}$ and $\dfrac{h}{t_w} \le \dfrac{418}{\sqrt{f_{ye}}}$	2	4	6	θ_y	$6\theta_y$	$8\theta_y$
Columns — flexure $P/P_{CL} < 0.20$						
$\dfrac{b_f}{2t_f} \le \dfrac{52}{\sqrt{f_{ye}}}$ and $\dfrac{h}{t_w} \le \dfrac{300}{\sqrt{f_{ye}}}$	2	4	6	θ_y	$6\theta_y$	$8\theta_y$

TABLE 21.6　Typical Acceptance Criteria for Primary Components in a Reinforced Concrete Frame Building

			m-factors			Plastic rotation angle		
			IO	LS	CP	IO	LS	CP
Beams (flexure controlled)								
$\dfrac{\rho - \rho'}{\rho_{bal}}$	Transverse reinforcement	$\dfrac{V}{b_w d \sqrt{f_c'}}$						
≤ 0.0	C	≤ 3	3	6	7	0.010	0.020	0.025
≤ 0.0	NC	≤ 3	2	3	4	0.005	0.010	0.020
Columns (flexure controlled)								
$\dfrac{P}{A_g f_c'}$	Transverse reinforcement	$\dfrac{V}{b_w d \sqrt{f_c'}}$						
≤ 0.1	C	≤ 3	2	3	4	0.005	0.015	0.020

acceptance criteria for new design. A sample of typical *m*-factors for both steel and concrete members is shown in Table 21.5 and Table 21.6.

21.5.2　Nonlinear Procedures

Acceptance criteria for nonlinear procedures, both in ATC-40 and in FEMA-356 are based on peak demands without explicit consideration of cumulative effects resulting from cyclic effects. To some extent, cyclic degrading effects can be implicitly included in the specification of the backbone envelope. To process the results of a nonlinear analysis, it is essential to define response measures that form the basis of acceptance criteria. Currently, the response measure of choice is the rotation demand at the plastic hinge. If a concentrated hinge model is used in the analysis, this is a relatively direct response quantity. However, if a spread-plastic model or a fiber-section model is used to represent material nonlinearities in the element, then the computation of plastic rotations is not straightforward. Either a plastic hinge length needs to be defined to integrate curvature estimates or alternative procedures need to be developed to estimate the plastic rotation demand. FEMA-356 and FEMA-350 define deformations in terms of chord rotations as shown in Figure 21.9.

If the FEMA definition of element rotation is used, then two possibilities exist for defining the plastic rotation. The first approach is to track the moment–rotation behavior at every connection and then compute the difference between the peak rotation and the recoverable rotation (see Figure 21.10). In a nonlinear time-history analysis, the recoverable rotation is more completely defined because member behavior is explicitly defined using constitutive material models or hysteresis models. For nonlinear static procedures, the recoverable rotation may be estimated using the initial stiffness path as the unloading path. Alternatively, the yield rotation can be predetermined using conventional concepts in structural mechanics. For example, FEMA suggests the following expression for steel frame members:

$$\theta_y = \frac{Zf_{ye}L}{6EI}\left(1 - \frac{P}{P_{ye}}\right) \tag{21.18}$$

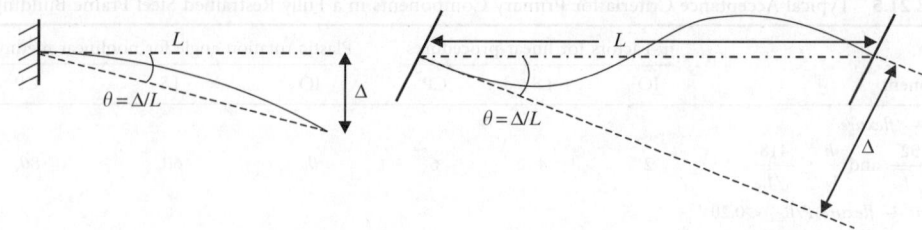

FIGURE 21.9 Definition of chord rotations to be used in calculating plastic rotation demands.

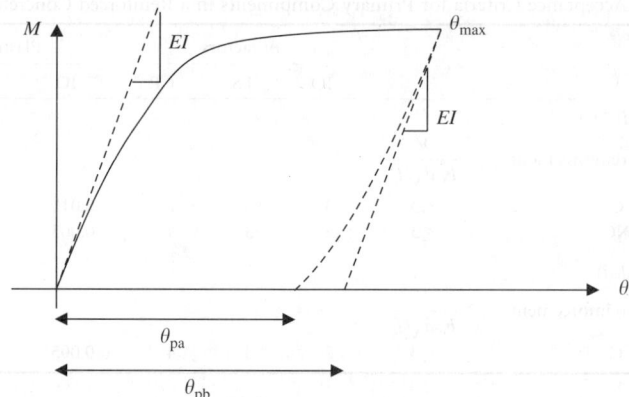

FIGURE 21.10 Defining plastic rotations for seismic loading.

where Z is the plastic section modulus, f_{ye} is the yield stress of the material, L is the length of the member, EI is the flexural rigidity, P is the axial force in member (may be taken as zero for beams), and P_{ye} is the expected axial capacity ($A_g f_{ye}$).

However, Equation 21.18 is derived on the assumption that the inflection point occurs at the midpoint of the element. If this is not the case, then the yield rotation should be computed using the following expression:

$$\theta_y = \frac{M_y^2 \ell}{3(M_y + M_2)EI} \tag{21.19}$$

where M_2 is the moment at one end of the member ($M_2 < M_y$), ℓ is the length of member, E is the elastic modulus, I is the moment of inertia, and M_y is the yield moment of the section. In FEMA-356, M_2 is assumed to be equal to M_y. This assumption is reasonably true for beams in moment frame structures but is generally not valid for columns.

In the case of RC section, the determination of yield rotation presents significantly greater challenges. For purely flexural elements, it may be adequate to determine the yield moment and then assume an equivalent flexural rigidity so as to establish the yield curvature. The yield rotation is then calculated by assuming a plastic hinge length. There are widely accepted guidelines for assigning effective stiffness values to RC components. FEMA-356, for example, recommends the following: $0.5EI_g$ for beams and columns with gravity compressive loads $<0.3A_g f_c'$, $0.7EI_g$ for columns with gravity compressive loads $>0.5A_g f_c'$, $0.8\ EI_g$ for uncracked shear walls, and $0.5EI_g$ for walls with visible cracking. The interaction of flexure and axial loads must be considered in developing nonlinear modeling parameters. The interaction of shear and flexure (in the presence of axial forces) is an issue that is still the subject of ongoing research.

Typical limits on plastic rotation demands for different performance levels, as specified in FEMA-356, are also shown in Table 21.5 and Table 21.6 for steel and RC frame buildings.

Ultimately, the intent of the above discussion is to make readers aware of the possible complexities in modeling and evaluating the expected demands at various connections and elements in the structural system. While simple representations of nonlinear behavior lead to more comprehensible results, it is necessary to gain confidence in the computed demands. The desired confidence can be achieved with sensitivity studies on the impact of uncertain variables on the computed demand estimates. For this reason, it is not surprising that FEMA-350 takes a probabilistic approach to PBD. Similarly, ongoing work at PEER evolves around a probabilistic format. Both these methodologies are discussed later in this chapter.

21.5.3 ATC-40 and Capacity Spectrum Procedures

ATC-40 advocates the use of the CSM to evaluate the overall adequacy of design of a structural system. The term "capacity spectrum" refers to an altered form of the capacity or pushover curve for the building. As discussed earlier, a pushover curve provides a representation of both the displacement and the force capacity of a building in terms of roof drift and base shear, respectively. The quantities computed to develop the pushover curve should be recast into spectral displacements and spectral accelerations using simple concepts in structural dynamics. Such a format is conventionally referred to as the acceleration–displacement response spectrum or (ADRS) format (Mahaney et al. 1993). The ATC-40 methodology involves a simple conceptual procedure wherein the reformatted capacity curve is compared to the seismic demand curve, which is also expressed in a similar format. In the context of ATC-40 and capacity spectrum procedures, it should be noted that Reinhorn (1997) proposes an interesting variation on CSM to evaluate the seismic response of buildings. Recently, Fajfar (1999) advanced CSM to incorporate the so-called N2 method into the formulation. These and other related developments on CSM are beyond the scope of this chapter.

21.5.3.1 Determining Capacity

The capacity of a structure is a function of the capacity of its individual components and the interaction of its elements. The preferred method of choice in ATC-40 to establish system capacity is the nonlinear static or pushover procedure in which a mathematical model of the structure is subjected to a lateral force incrementally or iteratively till a capacity curve is obtained. Unlike the pushover procedure in FEMA-356 wherein the lateral load is applied until a target displacement is obtained, in the CSM, the model is pushed till a stability limit is reached. The objective in CSM is to determine the "performance" point of the structure that identifies the demand corresponding to the hazard at the site specified in terms of a response spectrum.

The capacity curve is essentially a pushover curve, and consequently, the NSP procedure of FEMA-356 can be used to develop the curve. If the nonlinear behavior of each element is modeled explicitly, then the resulting base shear versus roof displacement plot is the required capacity curve. However, with the realization that many engineers may only have access to a linear analysis program, ATC-40 also recommends a simplified procedure. In the simplified approach, an elastic analysis is used to approximate the inelastic behavior of the building. The procedure to accomplish this is as follows: a lateral load is applied on the structure, and the magnitude of the lateral load is increased till some element (or a group of elements) reaches approximately 90% of its capacity. At this stage, the analysis is stopped, and the stiffness properties of these elements are set to a very small value or the elements are removed from the model. The lateral loads are reapplied incrementally as before till another set of elements approach their strength capacities. The base shear and roof displacement are recorded at the end of each phase of the analysis, and cumulative values represent points on the capacity curve. The process continues till a stability limit is reached. Figure 21.11a displays the process of combining the results of each phase of analysis. Figure 21.11b shows the same capacity curve, identified in the diagram as Curve #1.

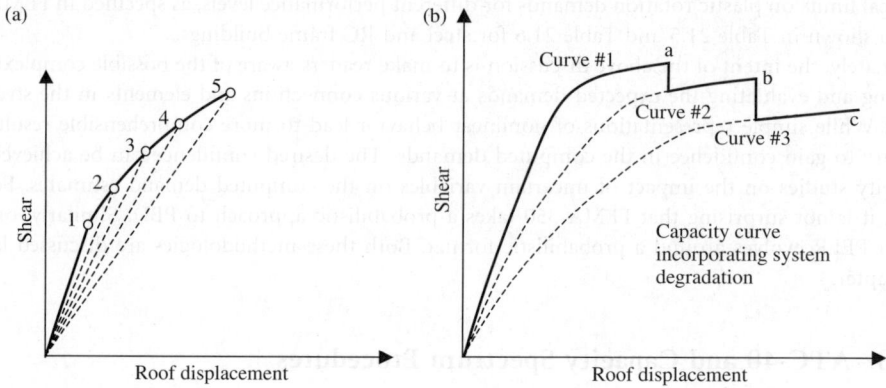

FIGURE 21.11   Generating the capacity curve using linear solution methods (ATC-40).

ATC-40 also recommends changing the lateral load pattern to reflect the displaced shape of the building at the end of each phase of the analysis.

It is also possible to consider degrading effects in the simplified procedure described above by rein-itializing the analysis at the end of each phase and plotting a new capacity curve. Each new curve begins with a degraded system that accounts for the state of elements at the end of each phase. Figure 21.11b provides a conceptual view of three such capacity curves. The final capacity curve is drawn by connecting the final point on the previous curve to a new point on the next capacity curve at the next displacement increment so as to produce a saw-tooth response as shown in Figure 21.11b.

21.5.3.1.1 Conversion to ADRS Format

The capacity curve obtained either by the simplified process described in the previous section or by a detailed nonlinear analysis technique is transformed into ADRS format. The following relationships provide the conversion to ADRS format:

$$S_{a,n} = \alpha_n(V_n/W) \text{ (expressed in units of } g)$$ (21.20)

$$S_{a,n} = \frac{\Delta_{c,n}}{\beta_n \Phi_{c,n}}$$ (21.21)

$$\alpha_n = \frac{\sum_{i=1}^{N} w_i \Phi_{i,n}^2}{\left(\sum_{i=1}^{N} w_i \Phi_{i,n}\right)^2}$$ (21.22)

$$\beta_n = \frac{\sum_{i=1}^{N} w_i \Phi_{i,n}}{\sum_{i=1}^{N} w_i \Phi_{i,n}^2}$$ (21.23)

where $S_{a,n}$ is the spectral acceleration for mode n, $S_{d,n}$ is the spectral displacement for mode n, V_n is the base shear for mode n, W is the seismic weight of the building, w_i is the seismic weight of the floor at level i, $\phi_{i,n}$ is the modal amplitude at level i for mode n, and $\Delta_{c,n}$ and $\Phi_{c,n}$ are the displacement and modal amplitude of control node for mode n.

The theoretical basis for making the conversion is outlined in the Appendix A.

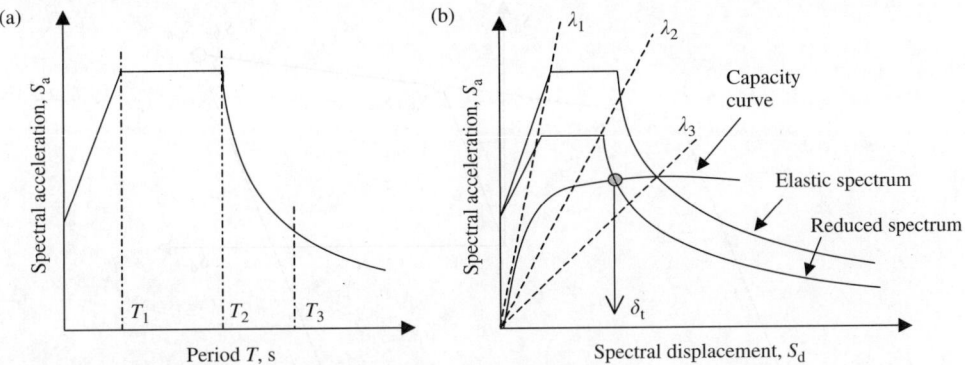

FIGURE 21.12 Generating the ADRS demand and capacity spectrum.

21.5.3.2 Determining Demand

The next step in CSM is to convert the response spectrum (which represents the demand side of the equation) into ADRS format as well, thereby permitting a comparison of demand versus capacity. The design response spectrum is usually expressed in terms of spectral acceleration and period as shown in Figure 21.12a. Since a response spectrum results from the analysis of an SDOF system, the following relationships between pseudospectral acceleration and spectral displacement can be used:

$$S_d = \frac{S_a T^2}{(2\pi)^2} \tag{21.24}$$

$$\frac{S_a}{S_d} = \lambda = \left(\frac{2\pi}{T}\right)^2 \tag{21.25}$$

Hence, it is possible to convert the spectral coordinates for every value of the period T_i into spectral displacements. The resulting curve is in ADRS format and the radial lines shown in Figure 21.12b represent λ values for corresponding T_i values. The capacity curve obtained in the previous step can now be superimposed on the demand curve. If the elastic design spectrum is used to create the demand spectrum, the overlay is valid only if the structural response is also elastic. Hence, the next step in the process is to reduce the elastic response spectrum to an inelastic spectrum using the concept of equivalent damping. Using the fundamental principles of structural mechanics, the equivalent damping ζ_d associated with dissipated energy during inelastic response is given by

$$\zeta_d = \frac{1}{(\Omega/\omega)} \frac{1}{4\pi} \frac{E_D}{E_S} \tag{21.26}$$

where (Ω/ω) is the ratio of the forcing frequency to the natural frequency of the system, E_D is the energy dissipated through hysteretic behavior, and E_S is the strain energy at the maximum displacement. If it is assumed that the peak response is associated with the resonant frequency, then the ratio $\Omega/\omega = 1.0$.

The ATC-40 methodology for estimating the equivalent viscous damping is derived for a bilinear capacity curve, therefore, it is necessary to transform the capacity curve into bilinear form. Figure 21.13 shows a bilinear capacity curve and the energy dissipated in a single cycle given the maximum displacement demand and the corresponding spectral acceleration (S_{dm}, S_{am}). The equivalent damping corresponding to the dissipated energy can be computed using Equation 21.26 and reduced to the following form:

$$\zeta_d = \frac{0.637(S_{ay} S_{dm} - S_{dy} S_{am})}{S_{am} S_{dm}} \tag{21.27}$$

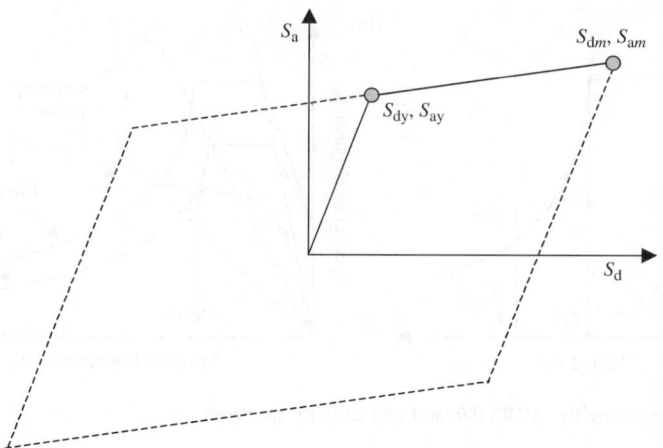

FIGURE 21.13   Bilinear representation of capacity curve and peak demand parameters.

Note that the elastic design spectrum already incorporates 5% damping, hence the equivalent damping (Equation 21.27) from inelastic behavior must be added to the elastic viscous damping. For behavior other than bilinear hysteresis, a modification factor κ is introduced. The final damping value incorporating elastic damping, equivalent inelastic damping, and general hysteretic behavior is given by

$$\zeta_{eq} = \kappa\zeta_d + 0.05 \tag{21.28}$$

Finally, the elastic spectrum is transformed into a reduced spectrum for the damping ratio given by Equation 21.28. ATC-40 provides the following spectral reduction factors, which are derived using the well-known Newmark–Hall relationships:

$$SR_A = \frac{3.21 - 0.68\,\ln(100\zeta_{eq})}{2.12} \tag{21.29}$$

$$SR_V = \frac{2.31 - 0.41\,\ln(100\zeta_{eq})}{1.65} \tag{21.30}$$

where SR_A is the reduction factor in the constant acceleration region of the spectrum and SR_V is the reduction factor in the constant velocity region. There are imposed limits on the above reduction factors depending on the expected shape of the hysteresis loops.

21.5.3.2.1 Performance Point
The CSM methodology attempts to predict the expected peak displacement given a demand response spectrum. A trial displacement value δ_t (Figure 21.12b) and the corresponding spectral acceleration is selected on the capacity curve. The equivalent viscous damping associated with these spectral magnitudes can now be estimated using Equation 21.27. The 5% damped elastic spectrum is transformed into a reduced inelastic spectrum using Equations 21.20 and 21.21. The demand and capacity curves in ADRS format are overlaid. If the trial displacement is within 5% of the displacement at the intersection of the demand and capacity curves, the performance point has been located. Figure 21.12b shows a case where the trial displacement is the performance point.

21.5.3.3 Performance Assessment
Once the performance point has been established, the acceptability of the design can be assessed by comparing the demand to the acceptance criteria. ATC-40 global acceptance criteria for different performance levels are listed in Table 21.7. Component acceptance criteria are essentially similar to FEMA-356 specifications.

TABLE 21.7 Global Acceptance Criteria for Use with ATC-40 CSM

	Performance level		
Parameter	IO	Damage control	LS
Maximum interstory drift	0.010	0.01–0.02	0.02
Maximum inelastic drift	0.005	0.005–0.015	No limit

21.6 Performance-Based Methodologies: Probabilistic Approaches

21.6.1 The SAC-FEMA Project and FEMA-350

The discovery of numerous brittle fractures of beam-to-column connections in fully restrained steel moment frames following the 1994 Northridge Earthquake dramatically altered a widely held conviction that welded-steel moment-frame buildings are among the most ductile systems contained in the building code. The observed damage was not limited to any particular class of steel structures: varying degrees of brittle fracture were reported in buildings ranging from 1 to 26 stories in height and in a range of ages from 30-year-old buildings to structures being constructed at the time of the earthquake.

The need to reevaluate both existing code provisions and construction practice in welded-steel frames led to the organization of the SAC Joint Venture in 1994, which eventually produced FEMA-350 (2000), a performance-based guideline for the design of new steel structures. The basic procedure described in FEMA-350 is similar to that in FEMA-356 with the major exception that the eventual acceptance criteria assume a probabilistic format. Additionally, it also provides specific qualification criteria for various types of connections that are beyond the scope of this chapter.

The conceptual basis of the FEMA-350 evaluation methodology, outlined in a paper by Cornell et al. (2002), can be expressed in probabilistic terms as follows:

$$P(D > PL) = \int P_{D > PL}(x) h(x) \, dx \qquad (21.31)$$

$P(D > PL)$ is the probability that the damage exceeds a specified performance level within a given duration (say, t years). $P_{D > PL}(x)$ is the probability that the damage exceeds a specified performance level, as a function of x, given that the ground motion intensity is level x. $h(x) \, dx$ is the probability of experiencing a ground motion of level x to $(x + dx)$ in a duration of t years.

The steps involved in evaluating the performance of a building using the above concept are summarized here:

1. As with any performance-based process, the performance objective consisting of a hazard level and a performance level is first defined. FEMA-350 lists only two primary hazard levels based on current specifications in FEMA-302: a maximum considered earthquake (MCE), which corresponds to a 2% probability of being exceeded in 50 years, and a DE, which is specified as having an intensity equal to two thirds of MCE. Recall that the DE in ATC-40 was defined as an event with a 10% chance of being exceeded in 50 years. More importantly, only two performance levels are considered: IO and CP. To this extent, FEMA-350 is a limited application of PBD concepts. However, this is an acknowledgment of the lack of available data to classify additional performance limit states.
2. A hazard spectrum or a set of ground motions that meet the criteria specified in the performance objective is developed. The procedure for generating a site-specific response spectrum (similar to

Figure 21.3) is identical to FEMA-356. For probabilistic-based assessment, the ground motion intensity is expressed in terms of the 5% spectral acceleration (S_a) magnitude at the fundamental period of the building. Procedures for determining S_a for other exceedance probabilities are outlined in FEMA-356.

3. The next step in the process is to establish seismic demands that are established from a static or dynamic analysis of a mathematical model of the building — a process not different from that described earlier in this chapter. However, FEMA-350 clearly defines the scope of each analytical method to carry out the task of demand estimation. A table listing the applicability of each method (LSP, NSP, LDP, and NDP) is provided for guidance. For example, if IO performance level is being evaluated and the fundamental period (T) of the structure is less than 3.5 times the characteristic period of the design response spectrum (T_n), then any of the four methods of analysis is valid, but if $T > 3.5T_n$, then static methods are not permitted. Similarly, if CP performance level is being considered for a regular structure with $T < 3.5T_n$ and strong column conditions (sum of column moment capacity is greater than sum of beam moment capacity at connection) exist throughout the frame, then all four methods of analysis are permitted while only nonlinear methods are allowed for irregular structures. Additionally, the demand values of interest are somewhat different from those used in FEMA-356. The following comprise the demand quantities of interest in FEMA-350: interstory drift angles, column compression, and column splice tensile demands.

4. Following the determination of the global and local demand measures, median estimates of structural capacity are established. These capacity parameters include system level measures such as interstory drift capacity and component level measures such as column compressive capacity. FEMA-350 recommends using an incremental dynamic analysis (IDA) to estimate interstory drift capacity, while statistical analyses of available experimental data, as described later in this section, can be utilized for component measures of capacity.

5. Finally, a demand and resistance factor design format is used to determine the confidence level associated with the probability that a building will have less than a specified exceedance probability for a desired performance level. The confidence index is determined through evaluation of the factored-demand-to-capacity ratio given by the equation:

$$\lambda = \frac{\gamma \gamma_a D}{\phi C} \tag{21.32}$$

where C is the estimated capacity of the structure for the primary demand measures, namely, interstory drift, column compression, and column splice tensile demand; D is the estimated demand for the structure, obtained from the structural analysis of a building model for the corresponding demand measure; γ is a factor that accounts for the variability inherent in the prediction of demand related to structural modeling assumptions and specification of ground motion parameters; γ_a is a factor that accounts for the bias and uncertainty inherent in the method of analysis used to estimate demands; and ϕ is a resistance factor that accounts for the uncertainty and variability inherent in the prediction of structural capacity as a function of ground-shaking intensity.

The confidence level of the performance evaluation is then back-calculated from the following equation:

$$\lambda_L = \exp(-b\beta_{UT}[K_x - 0.5k\beta_{UT}]) \tag{21.33}$$

where b is a coefficient relating an incremental change in the seismic demand measure to an incremental change in the intensity measure of the ground motion, determined from the demand hazard curve; β_{UT} is a parameter to account for uncertainties in demand and capacity calculated from $\sqrt{(\sum \beta_{ui}^2)}$, where β_{ui} are the standard deviations of the natural logarithms of the variations in demand and capacity, outlined in subsequent sections; k is the slope of the hazard curve (in ln–ln coordinates) at the hazard level of interest; and

K_x is the standard Gaussian variate associated with probability x of not being exceeded as a function of standard deviation measures (obtained readily from standard probability tables).

FEMA-350 includes a table that provides the solution of the above equation for various values of k, λ, and β_{UT}.

The overall confidence level is controlled by the lowest λ value for each of the demand measures (interstory drift, column compression, and column splice tensile demand). The various parameters needed to evaluate the confidence measure are summarized in the following.

21.6.1.1 Slope of the Hazard Curve

A plot of the probability of exceedance (or annual rate of exceedance) of a spectral amplitude versus the spectral value (at a period typically corresponding to the fundamental period of the structure), plotted on a log–log scale, is represented by the following functional form in the hazard range of interest:

$$v(S_a) = k_0 (S_a)^{-k} \tag{21.34}$$

where k is the slope of the hazard curve and $v(S_a)$ is the probability of an event having a spectral amplitude greater than S_a. Hazard curves can be developed from seismic hazard maps provided by the USGS and are assessable at http://eqhazmaps.usgs.gov.

21.6.1.2 Demand Variability Factor (γ)

This factor accounts for the variability in the prediction of demand measures such as interstory drift ratio and column splice tension demands. This is determined from the following expression:

$$\gamma = \exp\left(\frac{k}{2b}\beta_{DR}^2\right) \tag{21.35}$$

where k and b are the same parameters described in Equation 21.33, and β_{DR} is the standard deviation of the natural logarithms of the selected response measure developed from nonlinear time-history analyses of a mathematical model of the building subjected to a suite of ground motions, all of which are scaled to match the 5% damped spectral response acceleration of the hazard spectrum.

21.6.1.3 Analysis Uncertainty Factor (γ_a)

This represents the bias associated with the analytical method (LSP, LDP, or NSP) used to determine the demand measure. It is computed from the following equation:

$$\gamma_a = C_B \exp\left(\frac{k}{2b}\beta_{DU}^2\right) \tag{21.36}$$

where C_B is the bias factor given by the ratio of the demand predicted by nonlinear time-history analysis to the demand predicted by the chosen analytical method. In this case, β_{DU} represents the uncertainty that results from variability in the model parameters such as material strength, assumed damping, hysteretic behavior, and soil–foundation interaction modeling. A series of analyses with different structural models (that account for critical model uncertainties) are carried out using a single ground motion, and statistical measures of dispersion are obtained. β_{DU} is the standard deviation of the natural logarithm of the selected response parameter.

21.6.1.4 Resistance Factor (ϕ)

This factor accounts for the uncertainty in the estimation of the system or component capacity. The resistance factor will have the following general form:

$$\phi = \phi_R \phi_U = \phi_R \exp\left(\frac{k}{2b}\beta^2\right) \tag{21.37}$$

where ϕ_U is the factor accounting for uncertainty in the relationship between laboratory findings and behavior in real buildings (when considering component effects) or the uncertainty in analytical predictions (when considering system response measures such as global drift capacity). FEMA-350 suggests a value of $\beta = 0.2$ for component response measures and a period-dependent value for system response measures. ϕ_R is the randomness inherent in the computation of the resistance factor and is expressed in a form identical to ϕ_U. When computing the resistance factor for component response, β corresponds to the logarithmic standard deviation of the capacity measure based on observed data from laboratory testing. When estimating the resistance factor at the system level, β corresponds to the logarithmic standard deviation of the global stability limit determined from an IDA, as described later.

21.6.2 Simplified Evaluation Method (FEMA-350)

The overall procedure outlined in the Section 21.6.1 is the detailed method that requires numerous simulations of response data and careful statistical analysis of capacity measures. To facilitate routine engineering assessment of regular buildings, FEMA-350 also provides a set of tables to approximately evaluate the uncertainty parameters needed to establish the confidence index parameter. These uncertainty and variability factors are approximate estimates derived from a limited subset of parametric studies based on typical characteristics of regular buildings. Hence, they incorporate a certain degree of conservativeness. Whenever possible, the detailed method described earlier in this section should be used to improve the reliability of the performance evaluation.

A summary of a selected subset of uncertainty factors for special moment frames is shown in Table 21.8 and Table 21.9. Once the uncertainty factors are known, the factored demand-to-capacity index λ can be established for a particular demand measure. Finally, the confidence level of the computed "factored" demand-to-capacity ratio is determined using another table provided in FEMA-350. Table 21.10 shows a slice of the FEMA-350 listing: here, the values tabulated are valid only for midrise frames (4 to 12 stories) at CP performance level. To assess the adequacy of the design, the calculated confidence levels are compared to minimum acceptance criteria. Table 21.11 lists the FEMA criteria for global interstory drift and column compression.

TABLE 21.8 Typical Uncertainty Factors Recommended in FEMA-350 to be Used in Assessing Global Interstory Drift in Special Moment Frames

		Low-rise (<4 stories)	Mid-rise (4–12 stories)	High-rise (>12 stories)
γ_a				
LSP	IO	0.94	1.15	1.12
	CP	0.70	0.97	1.21
NSP	IO	1.13	1.45	1.36
	CP	0.89	0.99	0.95
γ	IO	1.50	1.40	1.40
	CP	1.30	1.20	1.50
ϕ	IO	1.00	1.00	1.00
	CP	0.90	0.85	0.75

TABLE 21.9 Global Interstory Drift Capacity (C) and Associated Uncertainty

		Low-rise (<4 stories)	Mid-rise (4–12 stories)	High-rise (>12 stories)
IO	C	0.02	0.02	0.02
	β_{UT}	0.20	0.20	0.20
CP	C	0.10	0.10	0.085
	β_{UT}	0.30	0.40	0.50

TABLE 21.10   Typical Confidence Levels Listed in FEMA-350 for Computed Factored Demand-to-Capacity Index

Confidence level	10	20	30	40	50	60	70	80	90	95	99
$\beta_{UT} = 0.4$											
λ		2.12	1.79	1.57	1.40	1.27	1.15	1.03	0.90	0.76	0.66	0.51

TABLE 21.11   Sample FEMA-350 Criteria for Assessing Adequacy of Performance

	Performance level	
	Immediate occupancy (IO) (%)	Collapse prevention (CP) (%)
Global behavior limited by interstory drift	50	90
Column compression behavior	50	90

21.6.3   Incremental Dynamic Analysis

The global stability limit corresponds to the CP limit state and is determined using an IDA as proposed by Vamvatsikos and Cornell (2002). An IDA curve is developed from the following procedure:

1. A suite of accelerograms (a minimum of ten records is recommended) is selected representative of the site and hazard level for which the CP level (or global stability limit) is desired. Artificially synthesized records may be used, but it is preferable to choose recorded ground motions that represent the expected hazard at the site in terms of source mechanism, fault distance, and geological characteristics.
2. For each earthquake record, conduct a series of time-history analyses by uniformly scaling the accelerogram each time so as to induce increasing levels of inelasticity in the structure and thereby characterize the behavior of the system from its elastic limit to its failure state.
3. Plot the magnitude of the spectral acceleration of the 5% damped response spectrum of the earthquake record at the fundamental period $S_a(T_0)$ versus the control node (typically the roof) displacement (Δ). Note that the spectral values will scale linearly during the scaling process. The resulting plot is called the IDA curve. Repeat the process for each of the selected ground motions thereby constructing a set of IDA curves.
4. The scale factors used to generate the IDA curve must be carefully selected to capture the elastic, postyield, and collapse limit states. The number of analyses required to develop an IDA curve will vary from one structure to the next; however, Vamvatsikos and Cornell (2002) suggest that the global drift capacity be defined at a point where the slope of the S_a–Δ curve drops below 20% of the elastic slope. Additionally, FEMA-350 places a limit of 10% on the global drift capacity.
5. Establish the distribution of spectral magnitudes at the global drift capacity to facilitate the computation of the logarithmic standard deviation of the drift capacity, β. The global resistance factor is then determined using

$$\phi_R = \exp\left(\frac{k}{2b}\beta^2\right) \tag{21.38}$$

where k and b are the same parameters described in previous expressions.

21.6.4   The PEER PBD Methodology

The methodology in development at the PEER center may be regarded as a holistic approach to PBD in that it involves a measure of performance that is relevant to stakeholders.

A probabilistic framework is used to assess the performance of the structure. This framework is based on a similar format used in the SAC Steel Frame project (Cornell et al. 2002) described in the previous section. While the original formulation by Cornell and coworkers was cast in demand versus capacity format and expressed the performance objective function as the probability of exceeding a certain performance level, the methodology developed for PEER extends this concept to include decision variables and damage measures. A conceptual description of the PEER methodology, using the total probability theorem, is expressed as follows:

$$\nu(\text{DV}) = \iint G(\text{DV}|\text{DM})\, \mathrm{d}G(\text{DM}|\text{EDP})\, \mathrm{d}G(\text{EDP}|\text{IM})\, \mathrm{d}\lambda(\text{IM}) \qquad (21.39)$$

where $\nu(\text{DV})$ is the probabilistic description of the decision variable (expressed in terms of the mean annual probability of repair or replacement cost, net dollar loss, downtime, etc.), DM represents the damage measure (an index value that expresses the state of damage resulting from a given demand), EDP represents the engineering demand parameter (drift, plastic rotation, etc.), and IM represents the intensity measure (characterizing the hazard). The expression of the form $P(A|B)$ is essentially a cumulative distribution function or the conditional probability that A exceeds a specified limit for a given value of B. The term of the form $\mathrm{d}P(A|B)$ is the derivative with respect to A of the conditional probability $(A|B)$.

One of the objectives of the PEER methodology is to incorporate all significant sources of uncertainty that arise in the specification of the ground motion, the material properties, and the modeling and evaluation process. The methodology also insists on a rigorous evaluation utilizing state-of-the-art tools for both the prediction of demand and the probabilistic assessment of each random variable. Hence, it would generally be expected that the evaluation be based on nonlinear time-history analyses using a suite of ground motions since it offers more options for the treatment of uncertainties. However, it does not preclude any particular method of analysis as long as the inherent bias and system uncertainties are quantified and probabilistic descriptions of the demand measures are established.

A brief description is now provided of the variables that appear in Equation 21.39.

21.6.4.1 Hazard Description (IM)

This refers to the hazard curve that expresses the annual frequency or exceedance probability of the design seismic event in a specified duration. In developing the hazard model, a rigorous seismic hazard analysis must be carried out that takes into consideration the seismic source, recurrence, and attenuation relations. Once a hazard model is developed, a set of ground motions must be synthesized (artificially) or selected from available records that satisfy the magnitude–distance combinations and geologic and fault characteristics at the site. An intensity measure must then be selected that quantifies the hazard. The idea of introducing an intensity measure to characterize the hazard provides a framework to develop fragility functions. A simple choice of intensity measure would be the PGA. Another option would be the 5% damped spectral acceleration at the characteristic period of the structure ($S_a(T_0)$) since it introduces an additional parameter in the description. There could potentially be numerous such measures with increasing complexity. The goal in selecting a measure of earthquake intensity is to minimize the variability in the performance assessment arising from ground motion parameters.

21.6.4.2 Engineering Demand Parameter (EDP)

The nonlinear analysis of a mathematical model of the structural system produces a wealth of information. The process of selecting one or more demand measures to characterize the performance of the system is not an easy task. The PEER methodology does not identify any specific demand measure since the choice of an appropriate demand parameter is a function of both the system and the desired performance objectives. A performance objective, for example, that attempts to minimize losses in a moderate earthquake might depend on the damage to nonstructural elements. In this case, the demand measure of interest may be interstory drift or peak floor acceleration. On the other hand, if the performance objective were linked to global measures such as LS or CP, it would be prudent to select demand measures that provide a more rational evaluation of the damage states of components. The treatment of modeling uncertainties is important in the process of establishing demands.

21.6.4.3 Damage Measure (DM)

Relevant measures of damage are identified for both structural and nonstructural components. However, unlike deterministic approaches where performance limits are specified for different damage states, the damage description in a probabilistic framework takes the form of a fragility function. The likelihood of a certain damage state is conditioned on a structural response measure or EDP. Similar to the development of acceptance criteria in FEMA-356, a damage model in the PEER framework is expected to rely on damage data from field observations and laboratory testing. Since the database of experimental information is limited, analytical simulations wherein calibrated models can be utilized to generate fragility curves for different variables can aid the development of damage fragilities. Again, this approach provides a means to incorporate material and model uncertainties into the evaluation process.

21.6.4.4 Decision Variable (DV)

The final phase of the PEER methodology involves loss modeling. The terminology that is introduced in this context is a decision variable that is expected to measure system performance in terms of cost estimates, such as mean annual dollar loss (repair costs), or the probability that the damage resulting from the seismic event will result in a specified downtime (loss of operability) and similar economic measures.

The PEER methodology offers distinct advantages over existing PBD frameworks. First, it recognizes the need to incorporate uncertainties in every phase of the process. This enables an assessment of the sensitivity of critical decision variables to uncertain input parameters. Approaches to evaluating system uncertainties and reducing the number of variable parameters that need to be considered have been investigated by Haukaas and Der Kiureghian (2002) and Porter (2002). Second, it provides an explicit probabilistic expression of the system performance in terms of economic measures that are more intelligible to stakeholders. Finally, the methodology is a generic representation of input and output that permits maximum flexibility to the user to define model parameters.

21.7   PBD Application: An Illustrative Example

The basic procedures outlined in this chapter are now applied to a sample six-story steel building to illustrate the application of performance-based concepts in routine seismic evaluation and design. The intention here is to merely demonstrate the process rather than establish a valid design that conforms to a specific performance-based format. As reiterated in several instances in this chapter, PBD is an evolving methodology that still lacks final details on acceptance criteria and loss estimates associated with different damage states. In fact, this example will also serve to illustrate potential unresolved issues in the current state-of-the-art.

For the purpose of this illustration, it will be assumed that the site of the building is located in an active seismic zone in Southern California. The building has a regular and symmetric plan 120 ft^2. The primary lateral load resisting system is a moment frame around the perimeter of the building. Moment continuity of each of the perimeter frames is interrupted at the ends where a simple shear connection is used to connect to the weak column axis. The plan view of the building and the elevation of a typical perimeter frame are shown in Figure 21.14. The interior frames were designed as gravity frames and consist of simple shear connections only. It will be further assumed that the columns of the perimeter frames are supported on foundation beams and rigid pile groups that permit the structure to be modeled as fully restrained at the supports.

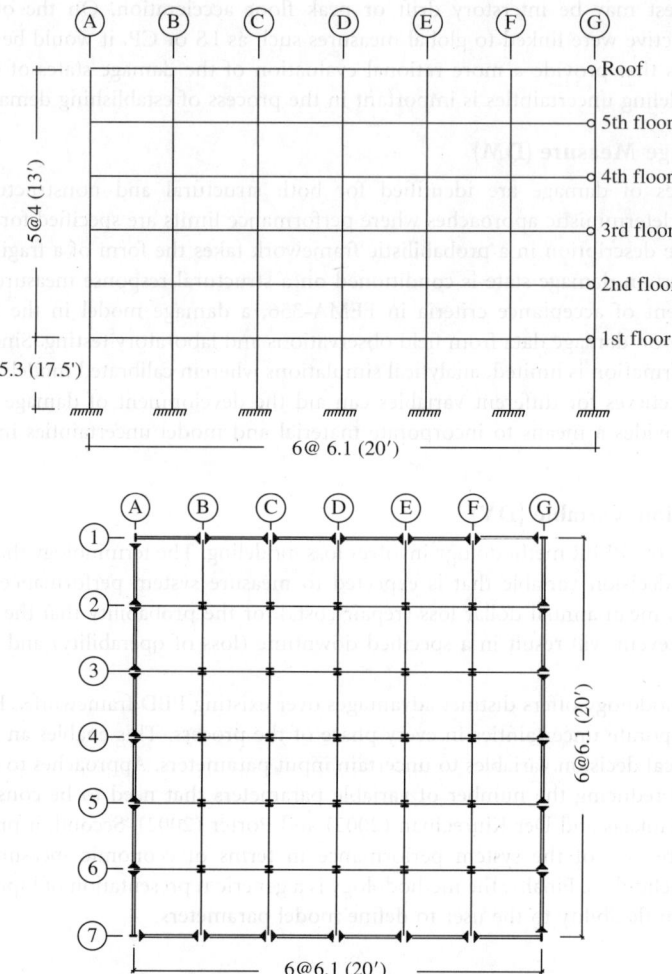

FIGURE 21.14 Building plan and elevation of typical moment frame considered in sample problem.

21.7.1 Performance Objective

The building will be designed for an event with a 10% probability of being exceeded in 50 years. This is referred to as the DE in ATC-40 or a BSE-1 hazard level in FEMA-356. In the present example, a performance level corresponding to LS for all primary structural elements is selected. Secondary or nonstructural performance is not considered in this example. Hence, the performance objective for the trial design is to achieve LS performance in all components when the building is subjected to a DE with the specified probability of occurrence (10% in 50 years).

It is now necessary to define the design event. Since both static and dynamic methods will be used to estimate demands, it is necessary to define a response spectrum and to develop a set of acceleration time histories. To construct the hazard spectrum, the soil profile, the geological characteristics, and source mechanism at the site need to be established. The following values will be assumed to generate the spectrum: seismic zone 4, firm soil with a shear wave velocity exceeding 1000 m/s and the closest distance to a seismic source capable of producing large-magnitude events to be 10 km. The spectral acceleration at a 1.0-s period for site class B is taken as 0.48g. Note that both the short-period and the long-period spectral acceleration values can be determined from USGS seismic maps. This results in the following parameters:

ATC-40:
$$C_A = 0.40g$$
$$C_V = 0.48g$$
$$T_n = C_V/2.5C_A = 0.48 \text{ s}$$
$$T_m = 0.2T_n = 0.096 \text{ s}$$
$$S_t = C_V = 0.48g$$

FEMA-356:
$$S_{xs} = 1.0g$$
$$S_{x1} = S_{m1} = 0.48g$$
$$S_t = S_{x1} = 0.48g$$

Using the parameters defined above, the hazard spectrum for the building site corresponding to a design event with a 10% chance of being exceeded in 50 years is developed and is shown in Figure 21.15.

21.7.2 PBD and Assessment

The next step in the process is to select preliminary sections for the building elements. This stage of the design is similar to current practice wherein preliminary sizing of members must precede the analysis of the building under the imposed seismic loads. In the present example, we will focus on the performance-based design and evaluation of a typical perimeter frame which is designated to carry half the lateral load in each direction. The initial selection of members is shown in Table 21.12. Section properties are

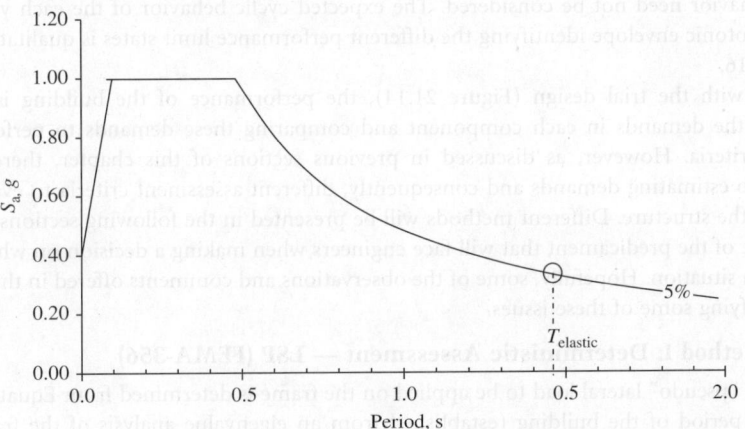

FIGURE 21.15 Design spectrum for building site.

TABLE 21.12 Section Properties of Frame Elements Used in Design Example

Level	Beam section	Column section
6	W24 × 69	W14 × 90
5	W24 × 84	W14 × 90
3–4	W24 × 84	W14 × 132
2	W27 × 102	W14 × 176
1	W30 × 116	W14 × 176

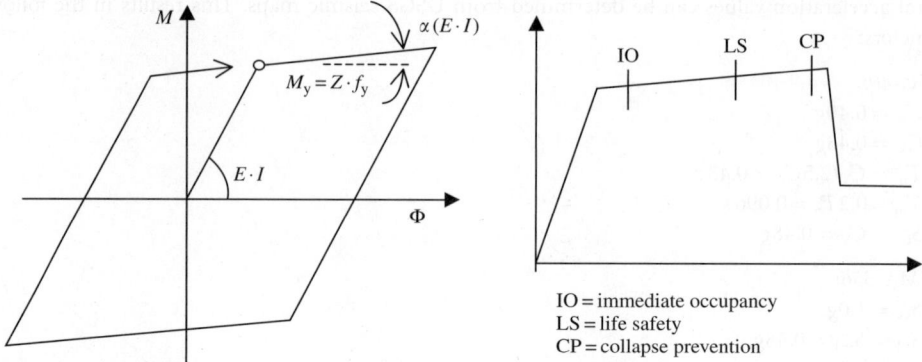

IO = immediate occupancy
LS = life safety
CP = collapse prevention

FIGURE 21.16   Prescribed force–deformation behavior of members and monotonic behavior identifying different performance limit states.

computed assuming a yield stress of 414 MPa (60 ksi). Using nominal dimensions for the floor slab and standard weights for nonstructural elements, the floor weights are estimated as 6227 kN (1400 kip) at the first level, 5338 kN (1200 kip) at the second through fifth level, and 7562 kN (1700 kip) at the roof level, resulting in a total building weight of approximately 33,805 kN (7600 kip). A 2D model of the frame was developed to estimate seismic demands. The computer program OpenSees (2003) was used for the analytical simulations since it provided the capability to carry out nonlinear static and dynamic analyses.

Nonlinearity in members is modeled using a bilinear force–deformation behavior in view of the fact that fully restrained connections are used and no brittle fracture is expected. The performance regime of the components will, therefore, lie in the ascending slope of the force–deformation envelope, and softening behavior need not be considered. The expected cyclic behavior of the each yielding section and the monotonic envelope identifying the different performance limit states is qualitatively displayed in Figure 21.16.

Beginning with the trial design (Figure 21.14), the performance of the building is evaluated by determining the demands in each component and comparing these demands to performance-based acceptance criteria. However, as discussed in previous sections of this chapter, there are different approaches to estimating demands and consequently, different assessment criteria to evaluate the performance of the structure. Different methods will be presented in the following sections so as to make readers aware of the predicament that will face engineers when making a decision on which method to use in a given situation. Hopefully, some of the observations and comments offered in this example will assist in clarifying some of these issues.

27.7.2.1   Method I: Deterministic Assessment — LSP (FEMA-356)

The so-called "pseudo" lateral load to be applied on the frame is determined from Equation 21.10. The fundamental period of the building (established from an eigenvalue analysis of the frame model) is 1.45 s. The coefficients needed to evaluate the total lateral load V are

$C_1 = 1.0$ (since $T > T_n$)
$C_2 = 1.0$ for linear procedures
$C_3 = 1.0$ (assuming P–Δ effects can be ignored)
$C_m = 1.0$ (since $T > 1$ s)
$S_a = 0.48/1.45 = 0.33g$

This results in a total lateral force of $V = 6672$ kN (1500 kip) to be applied on each perimeter frame in each direction. The lateral load is distributed over the building height using the expression given in Equation 21.12. The resulting moments on the structural components are recorded, and the peak m-factors are evaluated. Recall that the m-factor is the ratio of the peak moment to the moment capacity (defined here as the yield moment and computed by taking the product of the yield stress of the material and the plastic section modulus) of the beam. A typical summary of such an evaluation is shown in Table 21.13 for all the beam elements on the first level of the building.

This process is carried through all the beams and columns in the structure. The resulting *m*-factors for the entire frame are exhibited in Figure 21.17.

These *m*-factors are then compared to the acceptance criteria listed in FEMA-356 to check if the desired performance level has been satisfied. The limits on *m*-factors for different performance levels are a function of certain section characteristics as identified previously in Table 21.5. For the beams on the first level for which *m*-factor computations were listed in Table 21.13, the following section characteristics are established:

$$b_f = 267 \text{ mm} (10.5 \text{ in.}), \quad t_f = 21.6 \text{ mm} (0.85 \text{ in.}), \quad t_w = 14.5 \text{ mm} (0.57 \text{ in.}), \quad h = 762 \text{ mm} (30 \text{ in.})$$

$$\frac{b_f}{2t_f} = 61.8, \quad \frac{52}{\sqrt{f_{ye}}} = 6.71, \quad \frac{h}{t_w} = 50.6, \quad \frac{418}{\sqrt{f_{ye}}} = 54$$

The above numbers, using Table 21.5, require that for IO performance the *m*-factors be less than 2.0. Since this is true for all beams in the first-story level, they satisfy IO criteria. This performance limit is summarized in Figure 21.18 along with the performance limit states for all elements in the building. As displayed in Figure 21.18, several columns fail LS performance level at the third and fifth floor levels. The notation CP indicates that the element meets the criteria for CP but fails LS performance levels. Since the objective of the design was to ensure LS performance, the result of this evaluation indicates that the

TABLE 21.13 Peak Demands in Beams on First Floor Using LSP

Beam no.	Left	Right	Z	M_y	m-factor Left	m-factor Right	Max, m
101	23,670	19,059	378	22,680	1.39	1.12	1.39
102	17,050	17,414			1.00	1.02	1.02
103	17,546	17,507			1.03	1.03	1.03
104	17,437	17,361			1.03	1.02	1.03
105	17,808	19,277			1.05	1.13	1.13
106	11,963	0			0.70	0.00	0.70

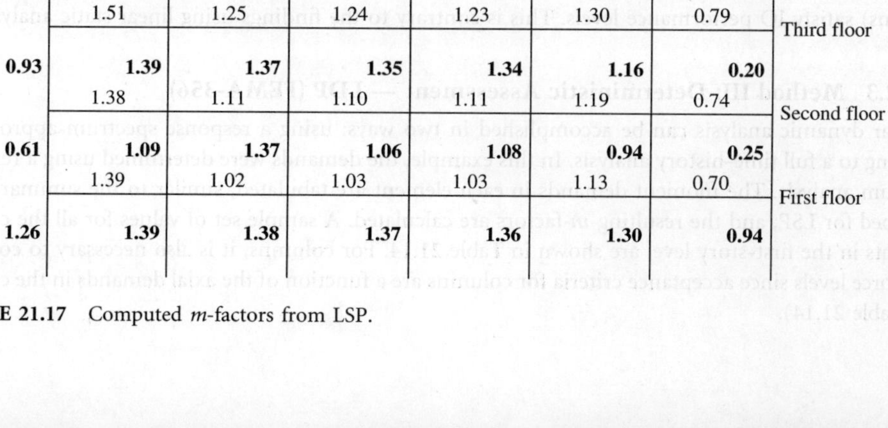

								Sixth floor
0.47	0.77	0.76	0.74	0.72	0.64	0.22		
	0.83	0.68	0.66	0.66	0.67	0.46		Fifth floor
0.85	1.26	1.26	1.25	1.23	1.10	0.22		
	1.24	1.02	1.01	1.01	1.06	0.67		Fourth floor
0.71	1.19	1.17	1.17	1.17	1.01	0.08		
	1.51	1.25	1.24	1.23	1.30	0.79		Third floor
0.93	1.39	1.37	1.35	1.34	1.16	0.20		
	1.38	1.11	1.10	1.11	1.19	0.74		Second floor
0.61	1.09	1.37	1.06	1.08	0.94	0.25		
	1.39	1.02	1.03	1.03	1.13	0.70		First floor
1.26	1.39	1.38	1.37	1.36	1.30	0.94		

FIGURE 21.17 Computed *m*-factors from LSP.

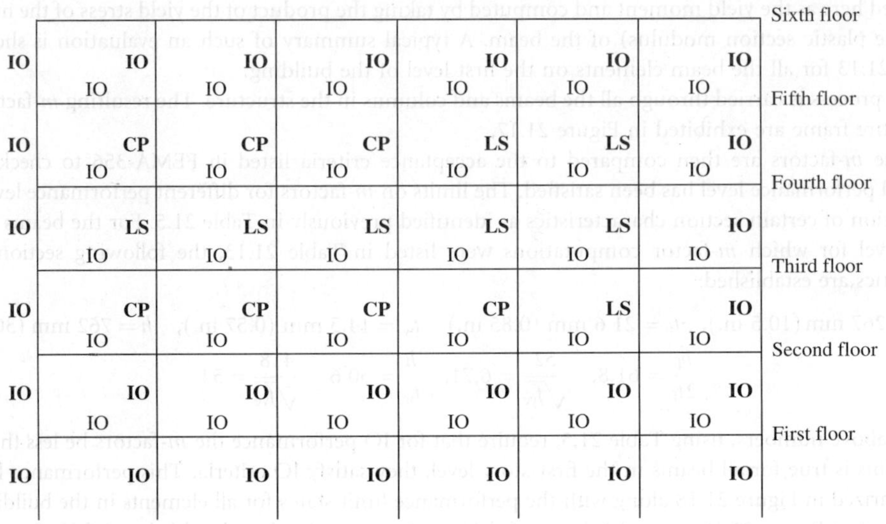

IO	IO	IO	IO	IO	IO	IO	Sixth floor
IO	IO	IO	IO	IO	IO	IO	
IO	CP	CP	CP	LS	LS	IO	Fifth floor
IO	IO	IO	IO	IO	IO	IO	
IO	LS	LS	LS	LS	LS	IO	Fourth floor
IO	IO	IO	IO	IO	IO	IO	
IO	CP	CP	CP	CP	LS	IO	Third floor
IO	IO	IO	IO	IO	IO	IO	
IO	IO	IO	IO	IO	IO	IO	Second floor
IO	IO	IO	IO	IO	IO	IO	
IO	IO	IO	IO	IO	IO	IO	First floor

FIGURE 21.18 Performance assessment of frame based on LSP evaluation.

Note: IO = element passing immediate occupancy criteria; LS = component failing IO but passing life safety criteria; CP = component failing LS but passing collapse prevention criteria.

desired performance objective has not been met. It is, therefore, necessary to redesign the frame and go through an iterative process till the design objective is satisfied.

21.7.2.2 Method II: Deterministic Assessment — NSP (FEMA-356)

The same frame will now be evaluated using a pushover analysis to determine expected demands. The building model is subjected to a monotonically increasing inverted triangular lateral load pattern till the roof drift reaches the target value. The target roof displacement is estimated using the following coefficients:

C_0: 1.42 (From table 3.2 of FEMA-356)
C_1: 1.0 since $T_e > T_S$ (from Hazard Spectra)
C_2: 1.0 (from table 3.3 of FEMA-356)
C_3: 1.0 for positive postyield stiffness

The spectral acceleration at the fundamental period T: 1.45 s is 0.33g. This results in a target displacement of 260 mm (10.22 in.). When the building model is pushed using an inverted triangular lateral load distribution till the roof displacement reaches this value, none of the elements reach their yield value. This means that the plastic rotation demands are zero and, consequently, all elements (beams and columns) satisfy IO performance levels. This is contrary to the findings using linear static analysis.

21.7.2.3 Method III: Deterministic Assessment — LDP (FEMA-356)

A linear dynamic analysis can be accomplished in two ways: using a response spectrum approach or resorting to a full time-history analysis. In this example, the demands were determined using a response spectrum analysis. The moment demands in each element are tabulated, similar to the summary table described for LSP, and the resulting m-factors are calculated. A sample set of values for all the column elements in the first-story level are shown in Table 21.14. For columns, it is also necessary to compute axial force levels since acceptance criteria for columns are a function of the axial demands in the column (see Table 21.14).

TABLE 21.14 Peak Demands in Columns on First Story Using LDP

Column no.	P	A_g	P_{cl}	P/P_{cl}	Bottom	Top	Z	M_y	Co-factor Left	Co-factor Right	Max, m
11	471	56.8	3408	0.14	15,081	10,726	355	21,300	0.94	0.67	0.94
12	157			0.05	16,665	14,123			1.04	0.88	1.04
13	108			0.03	16,524	13,803			1.03	0.86	1.03
14	108			0.03	16,492	13,778			1.03	0.86	1.03
15	119			0.03	16,432	13,749			1.03	0.86	1.03
16	359			0.11	15,777	12,486			0.99	0.78	0.99
17	176			0.05	11,542	3,509			0.72	0.22	0.72

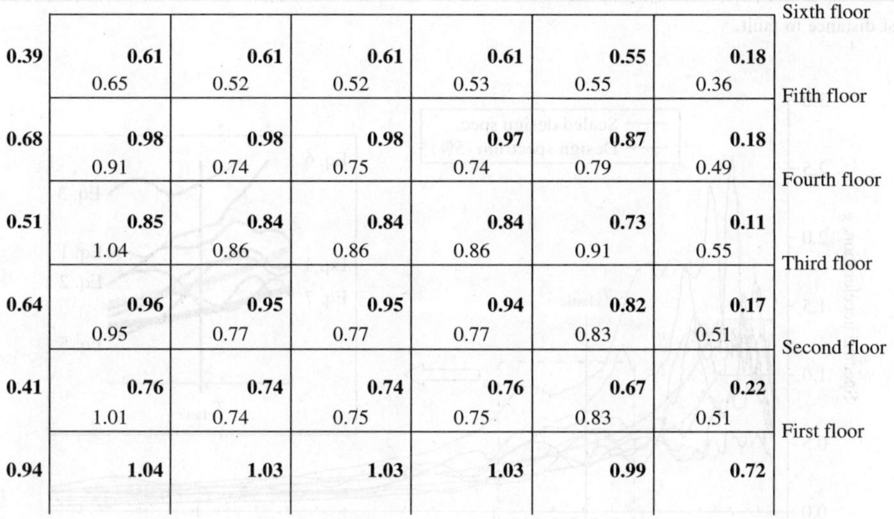

Sixth floor

0.39	**0.61**	**0.61**	**0.61**	**0.61**	**0.55**	**0.18**
	0.65	0.52	0.52	0.53	0.55	0.36

Fifth floor

0.68	**0.98**	**0.98**	**0.98**	**0.97**	**0.87**	**0.18**
	0.91	0.74	0.75	0.74	0.79	0.49

Fourth floor

0.51	**0.85**	**0.84**	**0.84**	**0.84**	**0.73**	**0.11**
	1.04	0.86	0.86	0.86	0.91	0.55

Third floor

0.64	**0.96**	**0.95**	**0.95**	**0.94**	**0.82**	**0.17**
	0.95	0.77	0.77	0.77	0.83	0.51

Second floor

0.41	**0.76**	**0.74**	**0.74**	**0.76**	**0.67**	**0.22**
	1.01	0.74	0.75	0.75	0.83	0.51

First floor

0.94	**1.04**	**1.03**	**1.03**	**1.03**	**0.99**	**0.72**

FIGURE 21.19 Computed *m*-factors using LDP.

The *m*-factors determined in the table are shown in Figure 21.19 along with the demand-to-capacity values for the remaining elements. Using the criteria listed in FEMA-356, the acceptance levels corresponding to these *m*-factors are established. As can be seen from the *m*-factors, most values are below unity, which means that the demands are generally below the yield capacity of the corresponding components. The demand-to-capacity ratios for a few columns are marginally greater than 1.0. Overall, all elements in the structure pass the criteria for IO. This is consistent with the findings using the NSP.

21.7.2.4 Method IV: Deterministic Assessment — NDP (FEMA-356)

The most rigorous procedure to estimate demands is to carry out a complete nonlinear time-history analysis using a set of ground motions. To accomplish this, a subset of seven earthquake ground motions recommended in ATC-40 is used. Basic seismic parameters of the selected record set are listed in Table 21.15.

As required in FEMA-356, the ground motions are scaled such that the average value of the SRSS spectra does not fall below 1.4 times the 5% damped spectrum for the DE for periods between $0.2T$ and $1.5T$. The spectra for the scaled records along with the original design spectra and the scaled design spectra (wherein the ordinates of the original spectrum are scaled by a factor of 1.4) are shown in Figure 21.20.

TABLE 21.15 Characteristics of Earthquake Records Used in NDP Evaluation

Earthquake no.	Magnitude	Year	Earthquake	Recording station	PGA(g) Horizontal component 1	Horizontal component 2	Vertical component	Distance (km)[a]
1	6.6	1971	San Fernando	Station 241	0.25	0.13	0.17	16.5
2	6.6	1971	San Fernando	Station 458	0.12	0.11	0.11	18.3
3	7.1	1989	Loma Prieta	Hollister, South & Pine	0.18	0.37	0.20	17.2
4	• 7.1	1989	Loma Prieta	Gilroy #2	0.32	0.35	0.28	4.5
5	7.5	1992	Landers	Yermo	0.15	0.24	0.14	31.0
6	7.5	1992	Landers	Joshua Park	0.28	0.27	0.18	10.0
7	6.7	1994	Northridge	Century City LACC North	0.26	0.22	0.12	23.7

[a] Closest distance to fault.

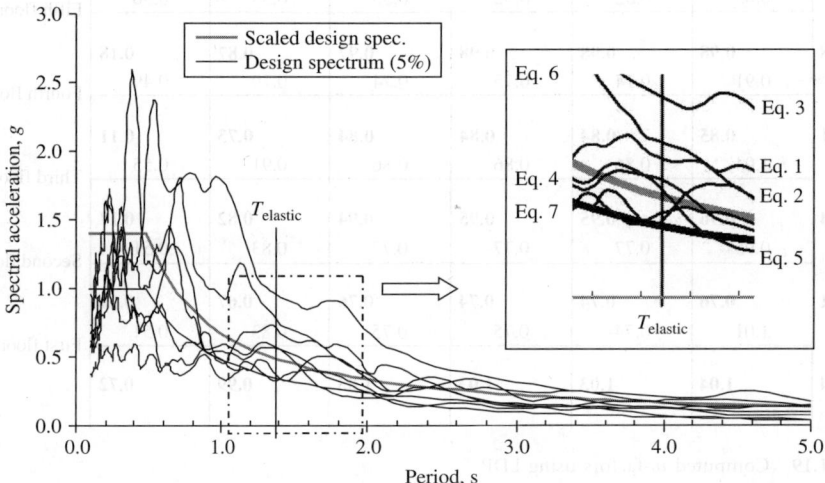

FIGURE 21.20 Original design spectrum, scaled spectrum, and spectra of selected ground motions after scaling.

Component demands from each time-history analysis are recorded, and the maximum values from each earthquake are used to estimate plastic rotation demands. Since seven records were used in the simulation, both FEMA-356 and ATC-40 stipulate that mean values of the peak demands are an adequate measure of the seismic demand. Table 21.16 summarizes the peak and average plastic rotation demands on the columns in the first story for all seven earthquakes. Table 21.16 also shows the maximum demands. This observation is crucial to our understanding of the process governing the choice of earthquake records for an NDP evaluation. If, for example, only three records were used in the evaluation, and Earthquake no. 3 was among the three records, then the seismic demands on the structure would be controlled by the peak values rather than the mean estimates. The maximum demands are considerably different from the mean values, suggesting that engineers should use caution when opting to select mean demands when evaluating performance limits.

In the present case, since all seven data points are used, performance limit states are going to be evaluated on the basis of mean demand estimates. The resulting performance of the building is shown in Figure 21.21. Though only the mean values were used, the NDP results indicate that several columns do not pass IO criteria but do pass LS criteria. Since LS performance is achieved for all elements, the trial design is acceptable.

TABLE 21.16 Peak and Mean Plastic Rotation Demands on First-Story Columns Using NDP

Column no.	Earthquake no.							Mean	Max, m
	1	2	3	4	5	6	7		
11	0.14	0.00	1.15	0.00	0.00	1.08	0.00	0.34	1.15
12	0.41	0.00	1.78	0.14	0.23	1.56	0.00	0.59	1.78
13	0.39	0.00	1.71	0.12	0.20	1.59	0.00	0.57	1.71
14	0.38	0.00	1.70	0.11	0.19	1.58	0.00	0.57	1.70
15	0.36	0.00	1.65	0.11	0.18	1.54	0.00	0.55	1.65
16	0.24	0.00	1.74	0.00	0.19	1.30	0.00	0.50	1.74
17	0.00	0.00	0.00	0.00	0.00	0.00	0.00	0.00	0.00

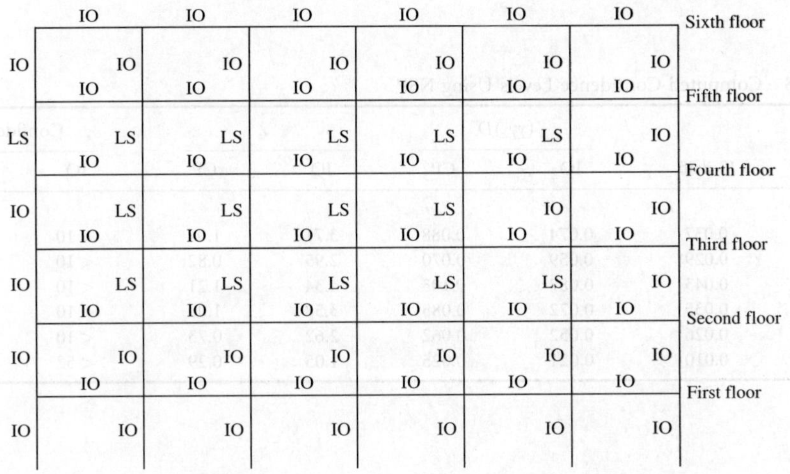

FIGURE 21.21 Performance assessment of frame based on LDP evaluation (see notation in Figure 21.18).

21.7.2.5 Method V: Probabilistic Assessment — LSP (FEMA-350)

The previous four methods provided a deterministic assessment of the building performance. The same design will now be evaluated in a probabilistic manner using the criteria in FEMA-350. The process itself is not probabilistic, though the final evaluation results in a nondeterministic evaluation of the computed demands. We begin with the LSP. The steps involved in predicting demands are similar to method I. The total lateral force is applied as an inverted triangular load using the distribution given by Equation 21.11, and the resulting moment demands in each component are tabulated. These demands are then factored by the uncertainty factors (see Table 21.8) to establish the factored demand value $\gamma\gamma_a D$. Acceptable capacity values are then determined (see typical capacity limits given in Table 21.9) and modified by a capacity reduction factor (shown in Table 21.8), to account for the inherent uncertainty in establishing such limits. The factored demand-to-capacity ratio, λ, is used to establish a confidence level. The computed confidence levels for global interstory drift are tabulated in Table 21.17. The acceptability of the estimated confidence level is specified in FEMA-350 as follows:

Recommended Minimum Confidence Levels (%)

	IO	CP
Global behavior limited by interstory drift	50	90
Local connection behavior limited by interstory drift	50	50
Column compression behavior	50	90
Column splice tension behavior	50	50

TABLE 21.17   Confidence Levels from Factored Demand and Capacity Estimates (LSP)

Story level	IS drift	$(\gamma\gamma_a)D$		λ		Confidence level	
		IO	CP	IO	CP	IO	CP
0							
1	0.014	0.023	0.027	1.15	0.31	35	99
2	0.014	0.022	0.026	1.11	0.31	47	99
3	0.017	0.028	0.033	1.40	0.38	10	99
4	0.016	0.026	0.030	1.29	0.35	20	99
5	0.015	0.024	0.028	1.19	0.33	30	99
6	0.009	0.014	0.017	0.72	0.20	98	99

TABLE 21.18   Computed Confidence Levels Using NSP

Story level	IS drift	$(\gamma\gamma_a)D$		λ		Confidence level	
		IO	CP	IO	CP	IO	CP
0							
1	0.037	0.074	0.088	3.71	1.04	< 10	69
2	0.029	0.059	0.070	2.95	0.82	< 10	87
3	0.043	0.087	0.103	4.34	1.21	< 10	55
4	0.035	0.072	0.085	3.58	1.00	< 10	72
5	0.026	0.052	0.062	2.62	0.73	< 10	92
6	0.010	0.021	0.025	1.05	0.29	< 52	99

Based on the recommended acceptance criteria, it is deduced that the confidence level associated with preventing building collapse at the computed drift values is extremely high (actually, this value is at its upper limit). However, the confidence level on IO performance is unacceptable. The minimum recommended value in FEMA-350 is 50%, while the confidence levels based on computed interstory drift are well below these limits for the first five story levels. Unfortunately, FEMA-350 does not specify confidence levels for LS performance. It will be necessary to use the detailed approach to establish the confidence level for this performance state. Consequently, it will not be feasible to evaluate the performance objective of this example using FEMA-350.

21.7.2.6 Method VI: Probabilistic Assessment — NSP (FEMA-350)

Though we have seen that FEMA-350 cannot be used to evaluate LS performance using the approximate method, the evaluation is extended in this section to include NSPs. The intent here is to provide a comparison of the performance evaluation using the same guideline (in this case FEMA-350) but employing different analytical methods to estimate seismic demands. It was shown earlier that the FEMA-356 evaluation using LSP resulted in unacceptable performance, while all the remaining procedures indicated satisfactory performance for the stated design objective.

Results of the evaluation using a pushover analysis (or NSP) are given in Table 21.18. The confidence levels are unacceptable for IO performance at all levels and for CP performance at the third story-level. It is observed that the confidence levels are generally lower than those estimated using LSP. This is interesting because the drift values using NSP are identical for both FEMA-356 and FEMA-350. However, FEMA-356 determines acceptance criteria using local component demands only while FEMA-350 uses a multilevel acceptance criterion. In the present case, global interstory drift was used to evaluate the building performance.

21.7.2.7 Method VII: Capacity Spectrum Based Evaluation (ATC-40)

The next evaluation methodology that will be discussed is the CSM outlined in ATC-40. The first step is to establish the capacity curve, which in this case is simply the pushover curve for the building. However, unlike the process discussed in Method II, no target displacement is used. Instead, the lateral loads are applied incrementally until a stability point is reached (impending P–Δ collapse). Since a nonlinear computer program is employed for the analysis, the approximate technique (Figure 21.11) discussed earlier does not apply. The pushover curve, which is obtained in terms of base shear and roof displacement, is converted into ADRS format (Equations 21.20 and 21.21). The next step is to select a trial performance point. Before doing this, it is instructive to overlay the 5% damped spectrum, also transformed into ADRS format, with the capacity curve to get a sense of the demand region. Figure 21.22 shows the original demand spectrum and the pushover capacity spectrum. The intersection of the demand and capacity curves occurs at approximately 180 mm (7.1 in.). At this level of displacement, the system is still elastic and the effective period is still 1.45 s. This implies that the intersection point is indeed the performance point since introducing additional damping will only reduce the demand. This result is consistent with the finding using Method II, which also concluded that all members will remain in the elastic range when the structure is subjected to design lateral forces.

21.7.2.8 Method VIII: PEER Probabilistic Framework for Performance Assessment

The final evaluation method to be considered is the PEER probabilistic approach. The first step in the PEER methodology calls for a site-specific analysis to develop the probabilistic hazard curve. As outlined in the detailed FEMA-350 procedure, it will be necessary to determine the slope of the hazard curve to carry out the integration implied in the PEER methodology (Equation 21.39). This means that we need to define at least two hazard levels. In this case, the following hazard levels are selected: events with a 10% probability of being exceeded in 50 years and events with a 2% probability of being exceeded in 50 years. To create a new hazard level beyond that specified in the design objective, the design spectrum for the 10%/50 hazard is scaled by a factor of 1.5 to produce the 2%/50 hazard spectrum. Earlier, a set of seven records corresponding to a 10%/50 hazard level was used in the FEMA-356 NDP evaluation. The same seven records are scaled in a manner consistent with the criteria specified in ATC-40 and FEMA-356 to produce records for the evaluation of the building model for 2%/50 hazard. The design spectral curves and the mean and median of the seven records for both hazard levels are displayed in Figure 21.23.

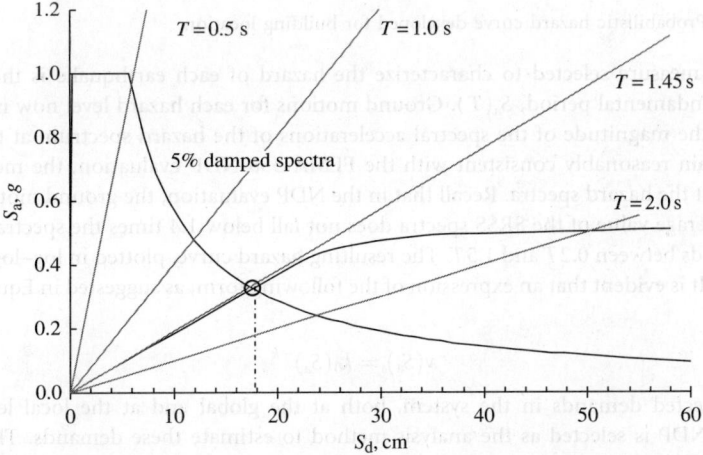

FIGURE 21.22 Capacity spectrum analysis of the building using ATC-40.

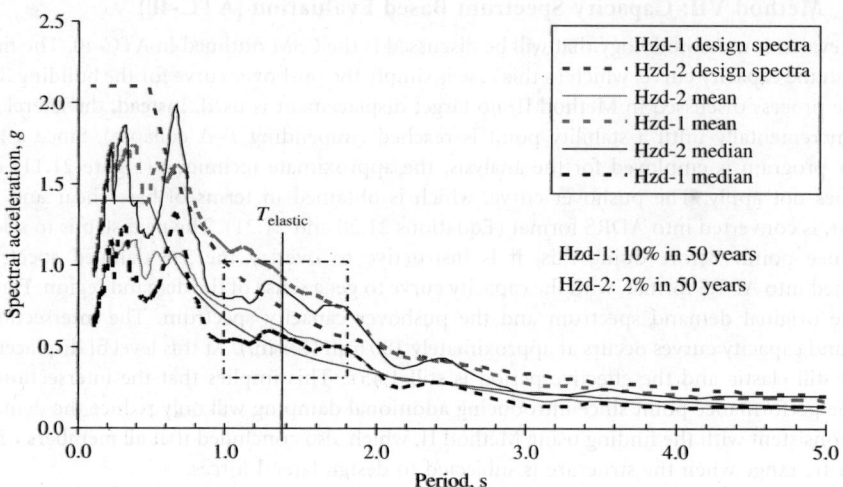

FIGURE 21.23   Mean and median spectra of scaled ground motions for both hazard levels.

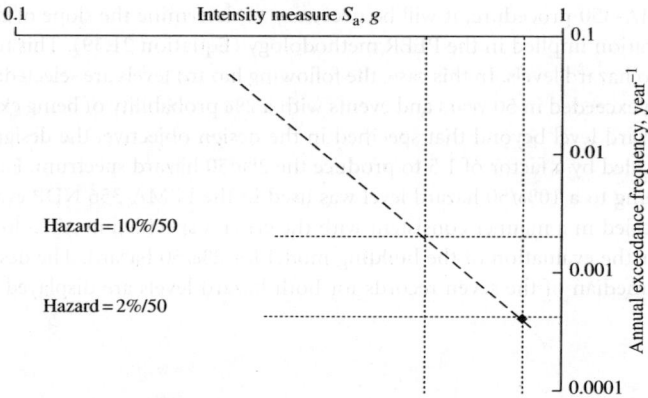

FIGURE 21.24   Probabilistic hazard curve developed for building location.

The intensity measure selected to characterize the hazard of each earthquake is the spectral acceleration at the fundamental period, $S_a(T)$. Ground motions for each hazard level now need to be scaled so as to match the magnitude of the spectral accelerations of the hazard spectrum at the fundamental period. To remain reasonably consistent with the FEMA-356 NDP evaluation, the median curves are used to represent the hazard spectra. Recall that in the NDP evaluation, the ground motions were scaled such that the average value of the SRSS spectra does not fall below 1.4 times the spectral magnitudes of the DE for periods between $0.2T$ and $1.5T$. The resulting hazard curve, plotted in log–log scale, is shown in Figure 21.24. It is evident that an expression of the following form, as suggested in Equation 21.34, can be developed:

$$v(S_a) = k_0(S_a)^{-k}$$

Next, the expected demands in the system, both at the global and at the local level, need to be estimated. The NDP is selected as the analysis method to estimate these demands. The model of the building is subjected to a total of 14 earthquake records, seven for each hazard level. The peak interstory drifts resulting from the nonlinear dynamic analyses are summarized in Figure 21.25.

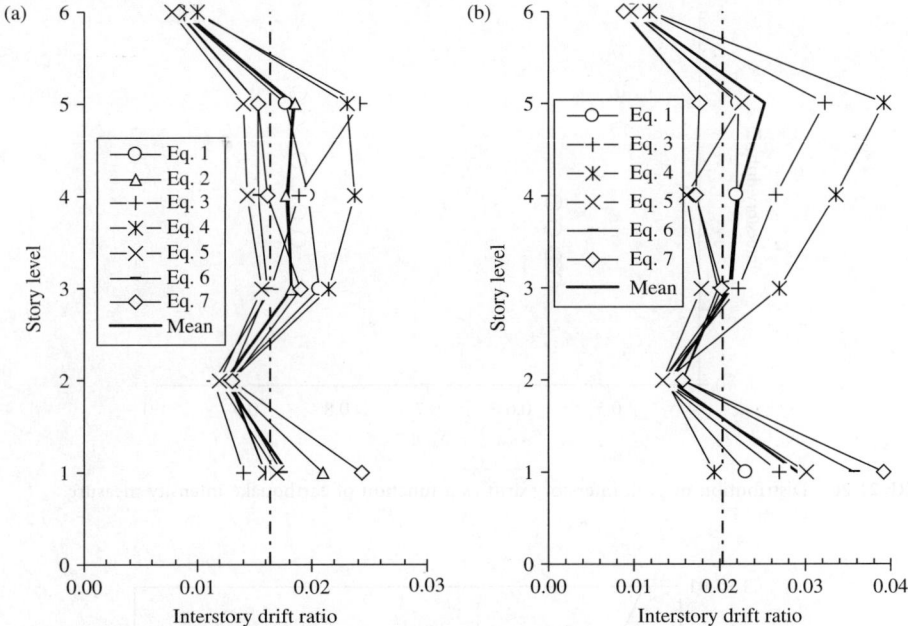

FIGURE 21.25 Interstory drift across building height for both hazard levels.

For the FEMA NDP evaluation, only the peak plastic rotation demands were needed to assess the performance of the system. In the PEER approach, the choice of a demand measure is left to the discretion of the user. For example, we could choose to use peak plastic rotations in any component at a given story level. However, in this case, the peak interstory drift ratios (IDR) are utilized to evaluate performance. In general, it would be advisable to evaluate system performance using several demand measures ranging from global measures such as drift to local measures such as plastic rotation demand. The distribution of the IDR as a function of the intensity measure is shown in Figure 21.26. Only the peak IDRs for each analysis are used to generate this plot. It is now possible to develop a drift hazard curve of the following form:

$$e^{\mu_{\ln[x|S_a]}} = a(S_a)^b \tag{21.40}$$

where $\mu_{\ln[x|S_a]}$ is the mean value of the natural logarithm of the IDRs associated with the given intensity measure and a and b are constants to be evaluated through curve-fitting the data.

The objective in this phase of the evaluation is to estimate the probability of occurrence of a peak IDR conditioned on a given spectral ordinate, as follows:

$$P(x|\text{IM} = S_a) = \Phi\left(\frac{\ln(x) - \mu_{\ln[x]}}{\sigma_{\ln[x|S_a]}}\right) \tag{21.41}$$

where x is the peak IDR, IM is the intensity measure which has been selected as $S_a(T)$, $\mu_{\ln[x]}$ is as defined previously, and $\sigma_{\ln[x|S_a]}$ is the standard deviation of the natural logarithm of the IDRs at the same intensity measure. If the drift demand is considered to be a random variable, and the drift hazard curve is established as described in Equation 21.40, Luco and Cornell (1998) show that the annual exceedance probability can be estimated from

$$v(x) = k_0 \left[\left(\frac{x}{a}\right)^{1/b}\right]^{-k} \exp\left[0.5k^2\left(\frac{\sigma_{\ln(x|S_a)}}{b}\right)^2\right] \tag{21.42}$$

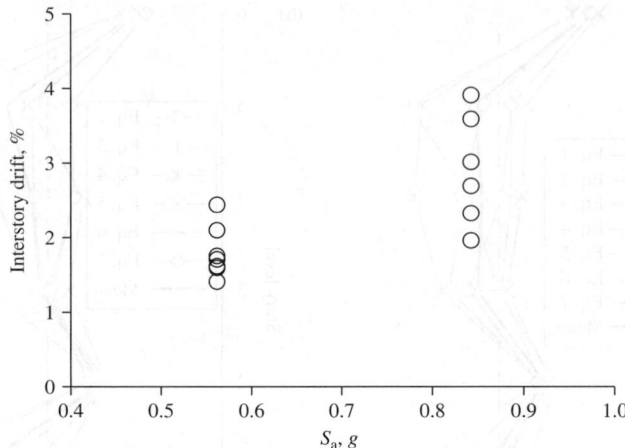

FIGURE 21.26   Distribution of peak interstory drift as a function of earthquake intensity measure.

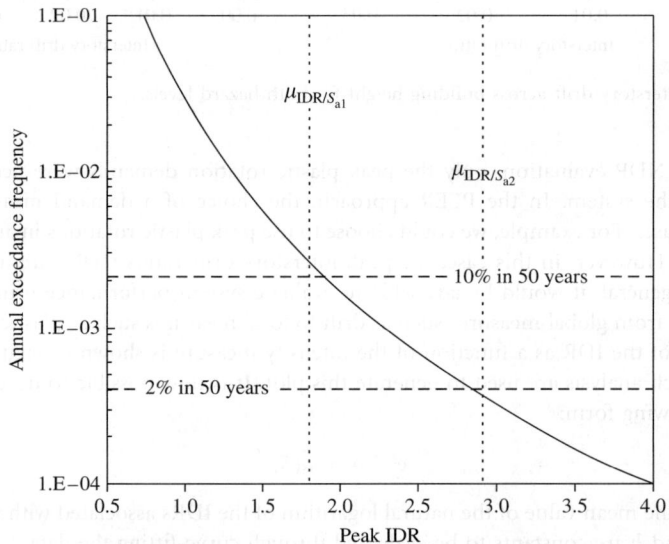

FIGURE 21.27   Interstory drift hazard.

All terms in the above expression have been previously defined. Since we have two distributions of the IDRs for two hazard levels, separate standard deviations are available for each hazard. However, for the design example under consideration, the dispersion associated with the 10%/50 year hazard is utilized to construct the plots of exceedance probability shown in Figure 21.27.

The information presented in Figure 21.27 is only an intermediate step in the PEER methodology. Next, we need to establish a damage definition given a demand value. There are obviously many ways to do this. For the purpose of this study, we will examine the use of IDA to identify critical limit states and use these limits to define a measure of damage. For example, IDA curves are recommended in FEMA-350 to determine the collapse limit state of the building. In this example, the IDA curve is used to specify several possible limit states. In the IDA analysis, each of the seven earthquake records is scaled repeatedly to produce a range of demand values from the elastic state to near collapse. The collapse limit state in most applications is a stability limit state. When using a nonlinear computer program with ability to incorporate P–Δ effects, this limit state corresponds to numerical instability (or inability to converge within the specified tolerance) arising from P–Δ softening.

Results of the IDA evaluation are displayed in Figure 21.28. For each analysis, both the roof drift and the peak interstory drift values are recorded. In many cases, failure may be triggered by local soft stories, and damage states are more discernable when examining story drift measures rather than global response measures. An interesting feature of IDA is that it provides a sense of the complexity of inelastic dynamic response. Scaling the accelerogram does not necessarily result in increased demands. Each earthquake produces a different pattern of demands. Hence, it is necessary to consider a statistically significant sample of earthquake records to extract meaningful information from an IDA.

Since the primary objective of this example is to demonstrate the methodology rather than offer a complete evaluation of the building, the IDA curves presented in Figure 21.28 will now be evaluated to define performance limit states. The response of the system is essentially in the elastic range up to about 1% peak interstory drift (at any level). However, damage to the contents in the building may occur at lower drift limits. A deviation from the initial linear portion of the curve is evident around 2% for most records. Earthquake No. 2 produces considerable softening in the system beyond 2% interstory drift. Based on these qualitative observations, the following limit states are defined:

- *Immediate occupancy limit state.* IDR of 1.0%.
- *Life safety limit state.* IDR of 3.0%.

Assuming that the above IDR values represent the mean estimates and using the dispersion (standard deviation) in the IDR data at the above two limit states, the damage probabilities conditioned on the IDRs are determined and displayed in Figure 21.29a. It is important to reitrate that the above limit states are rather arbitrary and correspond to global response measures only. Detailed evaluation of demands at the component (and nonstructural component) level is necessary to develop more rational limit states in an actual building evaluation. For the present example, therefore, if the above limit states are defined, it is possible to assess the building performance from a probabilistic standpoint. If the PEER methodology is carried through as suggested in Equation 21.33, it is also necessary to define a decision variable. A careful examination of the different variables that appear in the

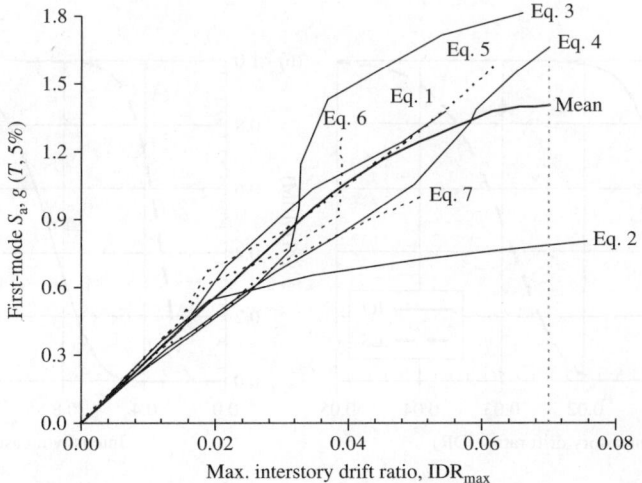

FIGURE 21.28 IDA curves for the six-story frame using the records listed in Table 21.15 (peak IDR at stability limit is identified as 0.07).

PEER equation indicates that it may not be necessary to integrate across three separate variables. For example, a demand measure may be translatable directly into a decision variable since damage parameters and decision variables are closely interlinked. In this illustration, the evaluation will not be carried through to the level of a decision variable (meaning a quantified economic loss measure) as intended in the PEER methodology primarily because the task of defining and calibrating such variables is beyond the scope of this chapter. Instead, the damage fragility will be established by considering the individual probabilities of damage and demand in the following manner:

$$P(\text{DM}|S_a) = \int P(\text{DM}|x)P(x\,|\,S_a)\,\mathrm{d}x \qquad (21.43)$$

The variable x in the above expression is the IDR. $P(\text{DM}|x)$ is plotted in Figure 21.29a. Combining this with $P(x\,|\,\text{IM})$ and carrying out the assessment indicated by Equation 21.43 results in the fragility curves shown in Figure 21.29b. The results indicate that the likelihood of exceeding 1% drift (which we have defined to correspond to IO performance criteria) is almost certain if the spectral acceleration of the ground motion (based on constructing a 5% damped elastic spectrum of the accelerogram) exceeds about 0.6g at a fundamental period of approximately 1.39 s. The required magnitude of the spectral acceleration at the fundamental period of the building for LS performance is about twice as much.

21.7.3 Notes and Observations

The primary purpose of the illustrative exercises presented previous subsections was to highlight issues in the current state-of-the-art in performance-based seismic design. There are clearly many unresolved issues, particularly pertaining to acceptance criteria and method of analysis. For example, in the sample evaluations discussed here, the findings were inconsistent for different methods. Using the same design spectrum, the demands from different methods produced different performance levels. With reference to FEMA-356, it is observed that in this case the linear static method was the most conservative. The remaining three methods indicated satisfactory behavior with the performance objective being achieved. However, the results from LDP and NSP indicated an essentially elastic response, while demands from NDP showed inelastic behavior in some elements though the demands were within LS levels. Analysis using ATC-40 and CSMs was similar to FEMA-356 NSP. In the case of FEMA-350, which is written in a format conforming to modern code specifications, no criteria are specified for LS performance.

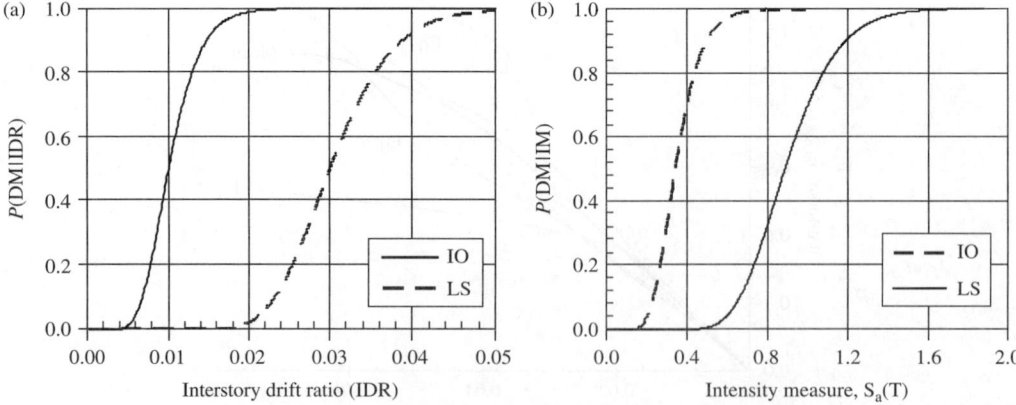

FIGURE 21.29 Fragility functions showing damage probabilities for two performance states: (a) damage probability conditioned on IDR and (b) damage probability conditioned on intensity measure.

Note: IO and LS limit states correspond to 1% and 3% IDR, respectively.

However, an interesting observation here is that an evaluation using NSP was less conservative than that with LSP, which is contrary to a similar comparison using FEMA-356. The results of the evaluation using the PEER methodology cannot be directly compared to the results of the previous procedures. Yet, it is possible to arrive at a nondeterministic understanding of the response: if we were to use the spectral acceleration of 0.58*g* corresponding to the 10%/50 year hazard, then the probability of achieving IO performance level is low (in other words, there is a high probability that the 0.01 IDR threshold for IO performance will be exceeded). This is based on the conclusion from the IDA curve that suggests a 0.01 peak IDR to correlate with IO performance level (or likely elastic limit). However, the likelihood of more severe damage states decreases dramatically (for example, Life Safety performance at the 10%/50 year hazard level is easily achieved since the probability of exceeding 3% drift corresponding to LS performance is nearly zero). Ideally, it should also be possible to develop loss estimates associated with different damage states. The reader is referred to the work of Miranda and Aslani (2003) and Krawinkler and Miranda (2004) for a more in-depth treatment of this subject.

21.8   Concluding Remarks

While much progress has been made in the last decade to develop processes and procedures that incorporate performance-based concepts in design and evaluation of structural systems, the findings and observations presented in the previous section raise important issues that point to the need for additional research and the development and validation of methodologies that minimize uncertainty and maximize confidence. The ongoing work at PEER is clearly a major advancement that deserves attention. Another ongoing effort is the FEMA-sponsored ATC-58 project, which has been charged with developing the next-generation performance-based seismic design procedures and guidelines. The objective of the ATC endeavor is to "express performance directly in terms of the quantified risks" in a format that is comprehensible to the decision maker. The project is expected to be conducted in two phases: in the first phase a set of procedures will be developed using probabilistic methods to assist an engineer to evaluate either an existing building or the proposed design of a new building utilizing terminology defined by decision makers; in the second phase a complimentary set of guidelines for use by stakeholders will be developed.

Since PBEE seeks to define a range of damage states, considerable work remains to be done to develop advanced computational tools that predict seismic demands of a complete soil–foundation–structure system reliably. Precise modeling guidelines should accompany such tools. Work is needed to quantify damage states and to develop models that transform damage states into loss estimates. Methods of analysis need to be validated with more rigor, particularly simplified methods that can be used in lieu of fully nonlinear time-history evaluations. Pushover procedures, which work well for low-rise, regular structures must be enhanced so that they can be used for a wider range of structural configurations. The work of Gupta and Kunnath (2000) and Chopra and Goel (2002) are important developments in this area. But more importantly, the propagation of uncertainties must be incorporated through the demand and assessment process. The degree of sophistication or level of detail will depend on the application, on whether the structure is classified as being ordinary or critical, and numerous other factors.

The success of a new seismic design methodology lies in the success of its implementation in building practice. Based on patterns in other earthquake innovations (e.g., base isolation and load and resistance factor design (LRFD)) May (2000) suggests that it may take at least two decades for ideas to move from the initial onset of preliminary guidelines (such as FEMA-356 and FEMA-350) to widespread adoption. Some of the key obstacles to advancing and implementing performance-based earthquake engineering (PBEE) cited by May include uncertainties in the methodology and its benefits, costs associated with adopting the methodology, complexity of the methodology in comparison to current code format, and issues related to validating and facilitating the adoption of the methodology. Another factor that can impact the development of PBEE is the interest of stakeholders in the success of the methodology. Steps to mitigate these barriers are essential to hasten the adoption of PBEE.

In conclusion, it must be reiterated that the intent of this chapter is to provide readers with a glimpse of evolving views on PBSE so that they can be better prepared as the first performance-based code for seismic design becomes a reality in the future. At a time when performance-based concepts were still being debated, Krawinkler (1997) noted that PBEE "appears to promise engineered structures whose performance can be quantified and conform to the owner's desires. If rigorously held to this promise, performance-based engineering will be a losing cause." This prediction can be regarded as ominous or enlightening depending on the perspective of the reader. Rather than offer an unpromising indictment, Krawinkler's purpose was to underscore the myriad of uncertainties associated with the overall process. In fact, he continues by saying that "significant improvements beyond the status quo will not be achieved without a new and idealistic target to shoot for. We need to set this target high and strive to come close to its accomplishment ... Performance based seismic engineering is the best target available and we need to focus on it." The real challenge in PBSE is not simply to better predict seismic demands or improve our loss-assessment methodology, but it is "in contributing effectively to the reduction of losses and the improvement of safety." (Cornell and Krawinkler 2000).

Appendix A

Conceptual Basis of Pushover Analysis and CSM

In this section, a few conceptual details on some of the processes and methods discussed in this chapter are outlined. These concepts should aid the reader in comprehending the basis of both pushover analysis and CSMs. Figure 21.A1 shows an idealized model of a four-story frame. In a standard computer analysis of the 2D frame model, each node will be assigned three DOFs corresponding to axial, shear, and bending deformations. However, the primary degree of interest in pushover analysis and CSMs is the lateral floor DOF since it is generally assumed that floors are rigid and a SDOF is adequate to represent the drift at a story level. When constructing the system stiffness matrix, all three DOFs are used; however, only the floor DOFs are retained (the remaining DOFs are condensed, not eliminated or ignored).

The idea behind static pushover methods is to characterize the nature of seismic loads acting on the frame. The response of the frame to any dynamic load is a combination of the dynamic modes of vibration of the system. Hence, it is instructive to begin with fundamental concepts in modal analysis of structures. As is well known, any vector of order n can be expressed by a set of n independent vectors. In this case, the independent eigenvectors resulting from the solution of the eigenvalue problem serve as appropriate vectors to express the floor displacements of a multistory building. The variable n refers to the number of DOFs which in this case is the number of floor levels for the reasons cited:

$$\{u_i\}_m = \sum_{m=1}^{N} \Phi_m q_m = [\Phi]\{q\} \tag{21.A1}$$

where $\{u_i\}$ is the displacement vector, $\{q\}$ is the normal or modal coordinates, $[\phi]$ is the matrix of eigen vectors, and m and i are the mode number and floor level, respectively.

Let us now consider the equilibrium relationship for a multi-DOF system:

$$[m]\{\ddot{u}\} + [c]\{\dot{u}\} + [k]\{u\} = -[m]\{\imath\}\ddot{u}_g(t) \tag{21.A2}$$

where $[m]$, $[c]$, and $[k]$ are the mass, damping, and stiffness matrices, respectively; $\{u\}$, $\{\dot{u}\}$, and $\{\ddot{u}\}$ are the displacement, velocity, and acceleration vectors, respectively; $\{\imath\}$ is a vector of unit values; and $\ddot{u}_g(t)$ is the imposed ground motion acceleration.

Utilizing the modal decomposition given in Equation 21.A1 and applying orthogonality relationships, the equilibrium expression reduces to the following form:

$$\ddot{q}_n + 2\zeta\omega_n\dot{q}_n + \omega_n^2 = -\Gamma_n\ddot{u}_g(t) \tag{21.A3}$$

where $\Gamma_n = ([\Phi]^{\mathrm{T}}[m]\{\imath\})/M_n$, in which $M_n = [\Phi]^{\mathrm{T}}[m][\Phi]$.

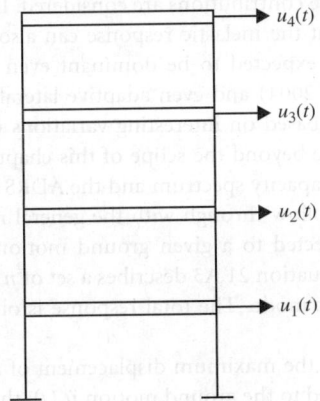

FIGURE 21.A1 Typical multistory frame model with reduced DOFs.

Actually, a more informative way of expressing the right-hand side of Equation 21.A2 is to consider independent modal contributions as presented by Chopra (2001):

$$[m]\{\ddot{u}\} + [c]\{\dot{u}\} + [k]\{u\} = -\sum_{n=1}^{N} R_n \ddot{u}_g \qquad (21.A4)$$

Comparing Equation 21.A4 to Equation 21.A2 and following through with the modal transformation that results in Equation 21.A3, it can be shown that

$$\{R\} = \sum R_n = \Gamma_n m \Phi_n \qquad (21.A5)$$

Each term in the above expansion contains the modal contribution of the respective mode. Another way of visualizing Equation 21.A5 is to consider the load vector on the right-hand side of Equation 21.A2 as follows:

$$[m]\{\imath\}\ddot{u}_g = \{R\}f(t) \qquad (21.A6)$$

where $\{R\}$ is a load distribution vector. For a general loading function $\{p(t)\} = \{r\}f(t)$, the vector $\{r\}$ represents a displacement transformation vector resulting from a unit support displacement. For earthquake loading, this simply becomes a vector with unit values. The external loading can obviously vary as a function of time in terms of both amplitude and spatial distribution. The objective of deriving an expression of the form given by Equation 21.A6 is to separate the spatial distribution from the time-varying amplitude function. This concept is not new and has been discussed in standard textbooks in dynamics (Clough and Penzien 1993).

The next step is to introduce features of the earthquake loading. Since the procedure being developed is a static procedure, the most appropriate form of earthquake loading that can be considered is a response spectrum. The spatial distribution of lateral forces to be used in conjunction with a pushover analysis is approximated in terms of the peak modal contributions as follows:

$$\{f_n\} = \Gamma_n [m] \{\Phi_n\} S_a(\zeta_n, T_n) \qquad (21.A7)$$

where S_a is the spectral acceleration for the given earthquake loading at a frequency corresponding to the period T and damping ratio ζ for mode n.

The modal forces computed using Equation 21.A7 will represent the contributions to mode n only. Equation 21.A7 represents the most general form of the lateral force vector to be used in a pushover analysis. If $n = 1$, only the first mode contributions are considered. In the approach proposed by Chopra and Goel (2002), it is assumed that the inelastic response can also be approximated by modal super-position because the nth mode is expected to be dominant even for inelastic systems. Other modal combination techniques (Kunnath 2004) and even adaptive lateral force distributions have been proposed (Gupta and Kunnath 2000) based on interesting variations in applying the load represented by Equation 21.A7. These concepts are beyond the scope of this chapter.

To understand the concept of a capacity spectrum and the ADRS conversion, it is necessary to return to Equations 21.A1 to 21.A3 and follow through with the general modal analysis procedure. The peak response of an SDOF system subjected to a given ground motion can be obtained from a response spectrum of the ground motion. Equation 21.A3 describes a set of n SDOF systems with each expression providing the solution to a specific mode. The total response is obtained through the transformation given by Equation 21.A1.

Assuming $S_d(\zeta_n, \omega_n)$ to represent the maximum displacement of an SDOF system with frequency ω_n and damping ratio ζ_n when subjected to the ground motion $\ddot{u}_g(t)$, the peak displacement response of the system represented in Equation 21.A3 is given by

$$\{q_n\}_{\max} = \Gamma_n S_d(\zeta_n, \omega_n) \tag{21.A8}$$

The peak story displacements can now be determined using Equation 21.A1 as follows:

$$\left\{ \begin{array}{c} u_1 \\ u_2 \\ \vdots \\ u_n \end{array} \right\}_{\max} = \Gamma_1 S_d(\zeta_1, \omega_1) \left\{ \begin{array}{c} \Phi_{11} \\ \Phi_{21} \\ \vdots \\ \Phi_{n1} \end{array} \right\} + \Gamma_2 S_d(\zeta_2, \omega_2) \left\{ \begin{array}{c} \Phi_{12} \\ \Phi_{22} \\ \vdots \\ \Phi_{n2} \end{array} \right\} + \cdots + \Gamma_n S_d(\zeta_n, \omega_n) \left\{ \begin{array}{c} \Phi_{1n} \\ \Phi_{2n} \\ \vdots \\ \Phi_{nn} \end{array} \right\} \tag{21.A9}$$

The above expression contains the contributions of all modes. Let us assume that only the peak displacement at a particular DOF is needed. For example, if DOF n represents the roof level, and only the first-mode contribution is considered, then the following expression is obtained:

$$u_{n,\max} = \Gamma_1 S_d(\zeta_1, \omega_1) \Phi_{n1} \tag{21.A10}$$

This equation is used to convert the roof displacement resulting from a pushover analysis to the first-mode spectral displacement in the capacity spectrum procedure.

To establish the equivalent first-mode spectral acceleration from the base shear, the peak displacement

$$\{f_n\}_{\max} = \omega_n^2 [m] \{u_n\}_{\max} \tag{21.A11}$$

$$\{f_n\}_{\max} = \omega_n^2 [m] \Gamma_n S_d(\zeta_n, \omega_n)[\Phi] \tag{21.A12}$$

$$\{f_n\}_{\max} = \Gamma_n S_a(\zeta_n, \omega_n)[m][\Phi]$$

$$= \left\langle \Gamma_1 S_a(\zeta_1, \omega_1) \left\{ \begin{array}{c} m_1 \Phi_{11} \\ m_2 \Phi_{21} \\ \cdots \end{array} \right\} + \Gamma_2 S_a(\zeta_2, \omega_2) \left\{ \begin{array}{c} m_1 \Phi_{12} \\ m_2 \Phi_{22} \\ \cdots \end{array} \right\} + \cdots \right\rangle \tag{21.A13}$$

If only the first-mode contribution is considered

$$\{f_n\}_{\max} = \Gamma_1 S_a(\zeta_1, \omega_1) \left\{ \begin{array}{c} m_1 \Phi_{11} \\ m_2 \Phi_{21} \\ \cdots \end{array} \right\} \tag{21.A14}$$

The base shear is the sum of story forces, hence the first-mode contribution to the base shear is given by

$$V = \Gamma_1 S_a(\zeta_1, \omega_1) \sum_{i=1}^{n} m_i \Phi_{i1} \qquad (21.A15)$$

Acknowledgments

I would like to acknowledge the assistance and contributions of Erol Kalkan in preparing the design example presented in this chapter. Conversations with Helmut Krawinkler on performance-based engineering and with Eduardo Miranda on the PEER methodology, provided new perspectives that have aided my thought process in the preparation of this chapter.

References

ATC 3-06 (1978). *Tentative Provisions for the Development of Seismic Regulations for Buildings*. Applied Technology Council, Redwood City, CA.

ATC-40 (1996). *Seismic Evaluation and Retrofit of Concrete Buildings*. Report SSC 96-01, California Seismic Safety Commission, Applied Technology Council, Redwood City, CA.

Cheok, G., Stone, W., and Kunnath, S.K. (1998). Seismic response of precast concrete frames with hybrid connections. *ACI Struct. J.* 95(5), 527–539.

Chopra, A. (2001). *Dynamics of Structures: Theory and Applications to Earthquake Engineering*. Prentice Hall, New York.

Chopra, A. and Goel, R. (2001). Direct displacement-based design: use of inelastic vs. elastic design spectra. *Earthquake Spectra* 17(1), 47–64.

Chopra, A. and Goel, R. (2002). A modal pushover analysis procedure for estimating seismic demands for buildings. *Earthquake Engineering and Structural Dynamics* 31 (3), 561–582.

Clough, R. and Penzien, J. (1993). *Dynamics of Structures*. McGraw-Hill, New York.

Cornell, C.A. (1996). Calculating building seismic performance reliability: a basis for multilevel design norms. *11th World Conference on Earthquake Engineering*, Paper No. 2122. Elsevier Science Ltd., Amsterdam.

Cornell, C.A. and Krawinkler, H. (2000). Progress and challenges in seismic performance assessment. *PEER Center News* 3(2), 1.

Cornell, C.A., Jalayer, F., Hamburger, R.O., and Foutch, D.A. (2002). Probabilistic basis for 200 sac federal emergency management agency steel moment frame guidelines. *ASCE J. Struct. Eng.* 128(4), 526–533.

Fajfar, P. (1999). Capacity spectrum method based on inelastic demand spectra. In *Earthquake Engineering and Structural Dynamics*. Vol. 28, pp. 979–993.

Fajfar, P. and Gasperic, P. (1996). The N2 method for the seismic damage analysis of RC buildings. *Earthquake Engineering and Structural Dynamics*. 28(1), 31–46.

Fajfar, P. and Krawinkler, H., eds (1997). *Seismic Design Methodologies for the Next Generation of Codes*. Balkema Publishers, Rotterdam.

FEMA-350 (2000). *Recommended Seismic Design Criteria for New Steel Moment-Frame Buildings*. Developed by the SAC Joint Venture for the Federal Emergency Management Agency, Washington, DC.

FEMA-356 (2000). *Prestandard and Commentary for the Seismic Rehabilitation of Buildings*. Federal Emergency Management Agency, Washington, DC.

Freeman, S.A. (1978). Prediction of response of concrete buildings to severe earthquake motion. *Douglas McHenry International Symposium on Concrete and Concrete Structures*, ACI SP-55. American Concrete Institute, Detroit MI. pp. 589–605.

Gupta, B. and Kunnath, S.K. (2000). Adaptive spectra-based pushover procedure for seismic evaluation of structures. *Earthquake Spectra* 16(2), 367–392.

Haukaas, T. and Der Kiureghian, A. (2003). Finite element reliability and sensitivity analysis in performance-based engineering. *Proceedings*, ASCE Structures Congress, Seattle, WA.

IBC (2000). *International Building Code.* International Code Council, ICBO, Whittier, CA.

Krawinkler, H. (1997). Research issues in performance based seismic engineering. In *Seismic Design Methodologies for the Next Generation of Codes* (P. Fajfar and H. Krawinkler, eds). Balkema Publishers, Rotterdam.

Krawinkler, H. and Miranda, E. (2004). Performance-based earthquake engineering. In *Earthquake Engineering: From Engineering Seismology to Performance-Based Engineering* (Y. Bozorgnia and V.V. Bertero, eds). CRC Press, Boca Raton, FL.

Kunnath, S.K. (2004). IDASS: inelastic dynamic analysis of structural systems. http://cee.engr.ucdavis.edu/faculty/kunnath/idass.htm.

Kunnath, S.K. (2004). Identification of modal combinations for nonlinear static analysis of building structures. *Comput. Aided Civil Infrastruct. Eng.* 19, 282–295.

Kunnath, S.K., Mander, J.B., and Lee, F. (1997). Parameter identification for degrading and pinched hysteretic structural concrete systems. *Eng. Struct.* 19(3), 224–232.

Kunnath, S.K., Reinhorn, A.M., and Lobo, R.F. (1992). *IDARC Version 3.0 — A Program for Inelastic Damage Analysis of RC Structures.* Technical Report NCEER-92-0022, National Center for Earthquake Engineering Research, SUNY, Buffalo, NY.

Luco, N. and Cornell, A. (1998). Effects of random connection fractures on the demands and reliability for a 3-story pre-Northridge SMRF structure. *Proceedings of the 6th National Conference on Earthquake Engineering*, Seattle, WA.

Mahaney, J.A., Paret, T.F., Kehoe, B.E., and Freeman, S. (1993). The capacity spectrum method for evaluating structural response during the Loma Prieta earthquake. *Proceedings of the National Earthquake Conference*, Memphis, TN.

May, P.J. (2002). *Barriers to Adoption and Implementation of PBEE Innovations.* Technical Report PEER 2002/20, Pacific Earthquake Engineering Research Center, University of California, Berkeley, CA.

Miranda, E. and Aslani, H. (2003). *Probabilistic Response Assessment for Building Specific Loss Estimation.* Report PEER 2003/03, Pacific Earthquake Engineering Research Center, University of California, Berkeley, CA.

Moehle, J.P. (1992) Displacement-based design of RC structures subjected to earthquakes. *EERI Spectra*, 8(3), 403–428.

Moehle, J.P. (1996). Displacement based seismic design criteria. *Proceedings of the 11th World Conference on Earthquake Engineering*, Acapulco, Mexico.

OpenSees (2003). Open system for earthquake engineering simulation. http://opensees.berkeley.edu.

Park, R. and Paulay, T. (1976). *Reinforced Concrete Structures.* John Wiley & Sons, New York.

Porter, K.A. (2002). An overview of PEER's performance-based earthquake engineering methodology. *Proceedings, Conference on Applications of Statistics and Probability in Civil Engineering (ICASP9)*, July 6–9, 2003, San Francisco, CA.

Priestley, M.J.N. and Calvi, G.M. (1997). Concepts and procedures for direct-displacement based design. In *Seismic Design Methodologies for the Next Generation of Codes* (P. Fajfar and H. Krawinkler eds). Balkema Publishers, Rotterdam, pp. 171–181.

Reinhorn, A.M. (1997). Inelastic analysis techniques in seismic evaluations. In *Seismic Design Methodologies for the Next Generation of Codes* (P. Fajfar and H. Krawinkler, eds). Balkema Publishers, Rotterdam. pp. 277–287.

SEAOC (1995). *Vision 2000: Performance Based Seismic Engineering of Buildings.* Structural Engineers Association of California (SEAOC), Sacramento, CA.

Sivaselvan, M.V. and Reinhorn, A.M. (2000). Hysteretic models for deteriorating inelastic structures. *J. Eng. Mech.* 126(6), 633–640.

Uang, C.-M. (1991). Establishing R (or R_w) and C_d factors for building seismic provisions. *ASCE J. Struct. Eng.* 117(1), 19–28.

Uang, C.-M. and Bertero, V.V. (1991). UBC seismic serviceability regulations: critical review. *ASCE J. Struct. Eng.* 117(7), 2055–2068.

Umemura, H. and Takizawa, H. (1982). *Dynamic Response of Reinforced Concrete Buildings*. Structural Engineering Documents 2, International Association for Bridge and Structural Engineering (IABSE), Switzerland.

Vamvatsikos, D. and Cornell, C.A. (2002). Incremental dynamic analysis. *Earthquake Engineering and Structural Dynamics* 31(3), 491–514.

V

Special Structures

V

Special Structures

22

Multistory Frame Structures

J. Y. Richard Liew
Department of Civil Engineering,
National University of Singapore,
Singapore

T. Balendra
Department of Civil Engineering,
National University of Singapore,
Singapore

22.1 Modern Techniques in Steel Frame Construction

22.1.1 Constructability

One of the main considerations in planning a building project is to have the building ready and occupied as early as possible. In order to reduce the time over which the investment is tied up in construction and maximize the return of investment through the use of the building, the design needs to consider the constructability aspects of the construction.

Speed in construction is achieved through a number of factors, some of which are listed below:

- Simple building design to avoid complicated site works.
- Design for minimum delay in construction.
- Maximize use of prefabricated and precast elements to avoid delays on site.

22-1

- Reduce the number of operations on the critical path.
- Complete all the designs before starting work on site.

Complicated geometry and building layout design should be avoided where possible. This is especially critical in crowded city sites where access and storage of materials may be a problem. Repetition of work means that the work can be done in a much faster process. The more repetition in elements, the quicker the site team goes through the process of familiarization.

In a steel framed building the positioning of services may need careful consideration at the design stage to allocate service zones. Hence, conflict of interests between various professions can be avoided.

Steel offers the best framing material for prefabrication. With the use of metal decks, the concept of fast track construction is introduced. The metal decking can be placed easily and used as the slab reinforcement. Trough deck stud welding for composite action reduces beams weights and depths. It also helps ensure that the floor slab can be used as a diaphragm to transfer lateral loads to bracing frames or stiff cores. Lightweight fire protection can be applied at a later stage, taking it off the critical path.

22.1.2 Prefabrication and Ease of Construction

Steel members and plates can be shop-fabricated using computer-controlled machinery, which have less chance of mistakes. On site the assembly is mainly carried out by a bolting procedure. Lateral-load-resisting systems should be located at the lifts, stair towers, etc., to provide stability throughout, rather than a rigid unbraced frame where temporary bracing may be required during erection.

Structural members delivered to site should be lifted directly and fixed into position, avoiding storage on site. Steel stairs, erected along with the frame, give immediate access for quicker and safer erection.

Metal decking may be lifted in bundles and no further craning is required as it is laid by hand and fixed by welded studs. This gives both a working and safety platform against accidents.

Secondary beams should be placed close enough to suit the deck, so that temporary propping can be avoided, and the deck could be concreted immediately.

22.1.3 Steel–Concrete Composite Design

Considerable benefit is gained by composition of the slab with the steel beam with possible weight savings of up to 30%. An effective width of slab is assumed to carry the compressive stresses leaving virtually the whole of the steel beam in tension creating a T-beam effect. Interaction between the slab and the beam is generated by "through deck" stud welding on to the beam flange.

22.1.4 Deflection and Cambering

Where the floors are unpropped, the deflection due to wet construction requires consideration to avoid the problem of ponding.

Dead load deflections exceeding 15 to 20 mm can be easily offset by cambering, which is best achieved by cold rolling the beam. This is a specialized operation as the camber is permanent because of stress redistribution due to controlled yielding. This will depend upon a number of factors and the advice of the specialist should be sought.

Because cambering can add 10 to 20% to the basic steel cost, this should be compared with the cost of a deeper and stiffer beam section provided that the increase in building height does not compromise additional cladding costs.

22.1.5 Fire Resistance

In the event of fire the metal deck unit would cease to function effectively due to loss of strength. However, additional strength can be provided by the added wire mesh for up to 1 h fire rating.

For higher period of fire resistance or for exceptionally high imposed loads, heavier reinforcement in the form of bars placed within the deck through can be used. Up to 4 h fire rating can be obtained using this method based upon fire engineering calculations with the deck units serving only as a permanent formwork.

For beams and columns, fire resistance may be provided by lightweight systems, which are quick to apply and economic. Normally cement-based spays are applied to beams and boards around columns. For tall buildings, steel columns may be encased or circular steel columns infilled with high-strength concrete to enhance resistance against compression and fire.

22.2 Classification of Multistory Frames

It is useful to define various frame systems to simplify the modeling of multistory frames. For more complicated three-dimensional structures involving the interaction of different structural systems, simple models are useful for preliminary design and for checking computer results. These models should capture the behavior of individual subframes and their effects on the overall structures.

This section describes what a framed system represents, defines when a framed system can be considered to be braced by another system, defines what is meant by a bracing system, and provides the difference between sway and nonsway frames. Various structural schemes for multistory building construction are also given.

22.2.1 Moment Frames

A moment frame derives its lateral stiffness mainly from the bending rigidity of frame members interconnected by rigid joints. The joints shall be designed in such a manner that they have enough strength and stiffness and negligible deformation. The deformation must be small enough to have any significant influence on the distribution of internal forces and moments in the structure or on the overall frame deformation.

An unbraced rigid frame should be capable of resisting lateral loads without relying on an additional bracing system for stability. The frame, by itself, has to resist the design forces, including gravity and lateral forces. At the same time, it should have adequate lateral stiffness against sidesway when it is subjected to horizontal wind or earthquake loads. Even though the detailing of the rigid connections results in a less economic structure, rigid unbraced frame systems perform better in load reversal situation or in earthquakes. From the architectural and functional points of view, it can be advantageous not to have any triangulated bracing systems or solid wall systems in the building.

22.2.2 Simple Frames

A simple frame refers to a structural system in which the beams and columns are pin connected and the system is not capable of resisting any lateral loads. The stability of the entire structure must be provided by attaching the simple frame to some forms of bracing systems. The lateral loads are resisted by the bracing systems while the gravity loads are resisted by both the simple frame and the bracing system.

In most cases, the lateral load response of the bracing system is sufficiently small such that second-order effects may be neglected for the design of the frames. Thus, the simple frames that are attached to the bracing system may be classified as nonsway frames. Figure 22.1 shows the principal components — simply frame and bracing system — of such a structure.

There are several reasons for adopting pinned connections in the design of steel multistory frames:

1. Pin-jointed frames are easier to fabricate and erect. For steel structures, it is more convenient to join the webs of the members without connecting the flanges.
2. Bolted connections are preferred compared to welded connections, which normally require weld inspection, weather protection, and surface preparation.
3. It is easier to design and analyze a building structure that can be separated into system resisting vertical loads and system resisting horizontal loads. For example, if all the girders are simply

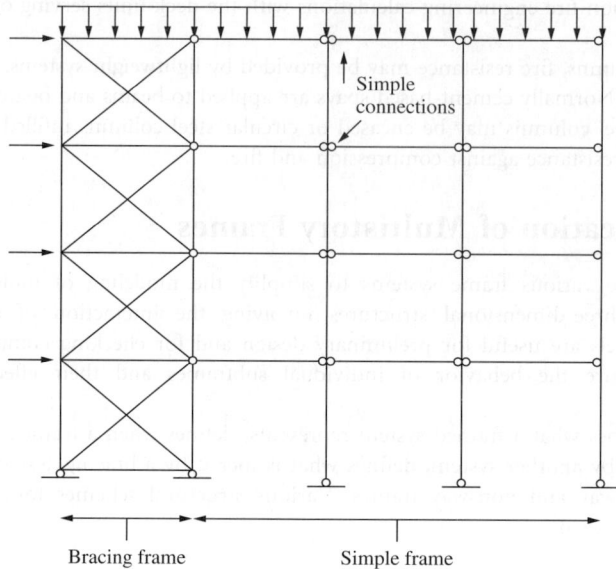

Bracing frame   Simple frame

FIGURE 22.1   Simple braced frame.

supported between the columns, the sizing of the simply supported girders and the columns
is a straightforward task.

4. It is more cost-effective to reduce the horizontal drift by means of bracing systems added to the
   simple framing than to use unbraced frame systems with rigid connections.

Actual connections in structures do not always fall within the categories of pinned or rigid connections. Practical connections are semirigid in nature and therefore the pinned and rigid conditions are
only idealizations. Modern design codes allow the design of semirigid frames using the concept of wind
moment design (type 2 connections). In wind moment design, the connection is assumed to be capable
of transmitting only part of the bending moments (those due to the wind only). Recent development in
the analysis and design of semirigid frames can be obtained from Chen et al. (1996). Design guidance is
given in Eurocode 3 (EC3) (1992a,b).

22.2.3   Bracing Systems

Bracing systems provide lateral stability to the overall framework. It may be in the form of triangulated
frames, shear wall/cores, or rigid-jointed frames. It is common to find bracing systems represented as
shown in Figure 22.2. They are normally located in buildings to accommodate lift shafts and staircases.

In steel structures, it is common to have triangulated vertical truss to provide bracing (see
Figure 22.2a). Unlike concrete structures where all the joints are naturally continuous, the most direct
way of making connections between steel members is to hinge one member to the other. For a very stiff
structure, shear wall or core wall is often used (Figure 22.2b). The efficiency of a building to resist lateral
forces depends on the location and the types of the bracing systems employed and the presence or
otherwise of shear walls, and cores around lift shafts and stair wells.

22.2.4   Braced versus Unbraced Frames

Building frame systems can be separated into vertical load-resistance and horizontal load-resistance
systems. The main function of a bracing system is to resist lateral forces. In some cases the vertical load-
resistance system also has some capability to resist horizontal forces. It is necessary, therefore, to identify

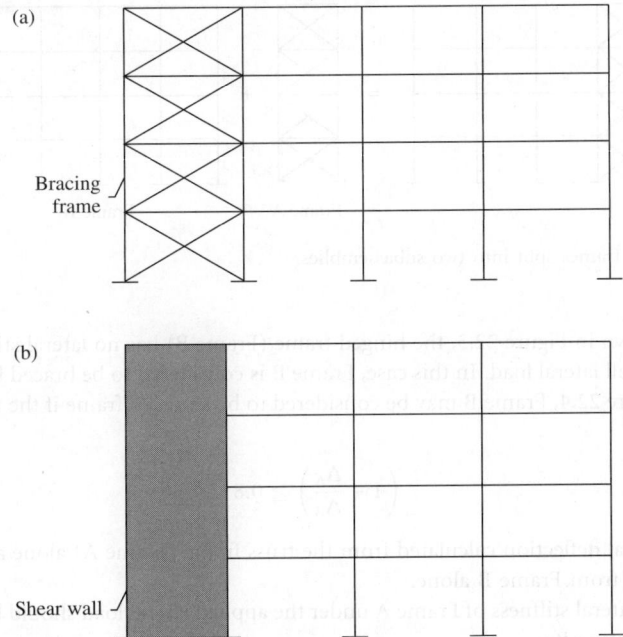

FIGURE 22.2   Common bracing systems: (a) vertical truss system and (b) shear wall.

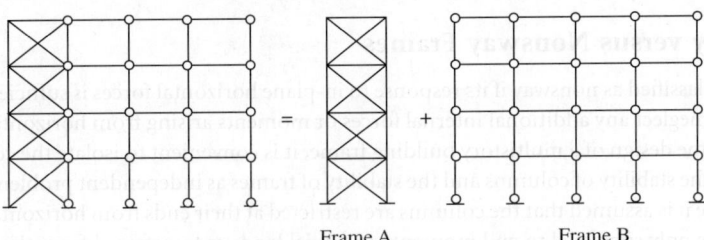

Frame A   Frame B

FIGURE 22.3   Frames split into two subassemblies.

the sources of resistance and to compare their behavior with respect to the horizontal actions. However, this identification is not that obvious since the bracing is integral within the structure. Some assumptions need to be made in order to define the two structures for the purpose of comparison.

Figure 22.3 and Figure 22.4 represent the structures that are easy to define, within one system, two subassemblies identifying the bracing system and the system to be braced. For the structure shown in Figure 22.3, there is a clear separation of functions in which the gravity loads are resist by the hinged subassembly (Frame B) and the horizontal loads are resisted by the braced assembly (Frame A). In contrast, for the structure in Figure 22.4, since the second subassembly (Frame B) is able to resist horizontal actions as well as vertical actions, it is necessary to assume that practically all the horizontal actions are carried by the first subassembly (Frame A) in order to define this system as braced.

According to EC3 (1992a,b), a frame may be classified as braced if its sway resistance is supplied by a bracing system in which its response to lateral loads is sufficiently stiff for it to be acceptably accurate to assume all horizontal loads are resisted by the bracing system. The frame can be classified as braced if the bracing system reduces its horizontal displacement by at least 80%.

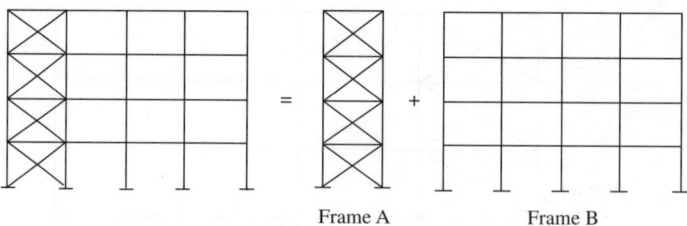

FIGURE 22.4   Mixed frames split into two subassemblies.

For the frame shown in Figure 22.3, the hinged frame (Frame B) has no lateral stiffness and Frame A (truss frame) resists all lateral load. In this case, Frame B is considered to be braced by Frame A. For the frame shown in Figure 22.4, Frame B may be considered to be a braced frame if the following deflection criterion is satisfied:

$$\left(1 - \frac{\Delta_A}{\Delta_B}\right) \geq 0.8 \qquad (22.1)$$

where Δ_A is the lateral deflection calculated from the truss frame (Frame A) alone and Δ_B is the lateral deflection calculated from Frame B alone.

Alternatively, the lateral stiffness of Frame A under the applied lateral load should be at least five times larger than that of Frame B:

$$K_A \geq 5K_B \qquad (22.2)$$

where K_A is the lateral stiffness of Frame A and K_B is the lateral stiffness of Frame B.

22.2.5 Sway versus Nonsway Frames

A frame can be classified as nonsway if its response to in-plane horizontal forces is sufficiently stiff for it to be acceptable to neglect any additional internal forces or moments arising from horizontal displacements of the frame. In the design of a multistory building frame, it is convenient to isolate the columns from the frame and treat the stability of columns and the stability of frames as independent problems. For a column in a braced frame it is assumed that the columns are restricted at their ends from horizontal displacements and therefore are only subjected to end moments and axial loads as transferred from the frame. It is then assumed that the frame, possibly by means of a bracing system, satisfies global stability checks and that the global stability of the frame does not affect the column behavior. This gives the commonly assumed *nonsway frame*. The design of columns in a nonsway frame follows the conventional beam–column capacity check approach and the column effective length may be evaluated based on the column end restraint conditions. Interaction equations for various cross-section shapes have been developed through years of research in the field of beam–column design (Chen and Atsuta 1976).

Another reason for defining "sway" and "nonsway frames" is the need to adopt conventional analysis in which all the internal forces are computed on the basis of the undeformed geometry of the structure. This assumption is valid if second-order effects are negligible. When there is an interaction between overall frame stability and column stability, it is not possible to isolate the column. The column and the frame have to act interactively in a "sway" mode. The design of sway frames has to consider the frame sub-assemblage or the structure as a whole. Moreover, the presence of "inelasticity" in the columns will render some doubts on the use of the familiar concept of "elastic effective length" (Liew et al. 1991, 1992).

On the basis of the above considerations, a definition can be established for sway and nonsway frames as

A frame can be classified as nonsway if its response to in-plane horizontal forces is sufficiently stiff for it to be acceptably accurate to neglect any additional internal forces or moments arising from horizontal displacements of its nodes.

British Code: BS5950: Part 1 (1990) provides a procedure to distinguish between sway and nonsway frames as follows:

1. Apply a set of notional horizontal loads to the frame. These notional forces are to be taken as 0.5% of the factored dead plus vertical imposed loads and are applied in isolation, that is, without the simultaneous application of actual vertical or horizontal loading.
2. Perform a first-order linear elastic analysis and evaluate the individual relative sway deflection δ for each story.
3. If the actual frame is uncladed, the frame may be considered to be nonsway if the interstory deflection satisfies the following limit:

$$\delta < \frac{h}{4000}$$

   where h is the story height for every story.
4. If the actual frame is claded but the analysis is carried out on the bare frame, then in recognition of the fact that the cladding will substantially reduce deflections, the condition is reflected and the frame may be considered to be nonsway if

$$\delta < \frac{h}{2000}$$

   where h is the story height for every story.
5. All frames not complying with the criteria in (3) or (4) are considered to be sway frames.

EC3 (1992a,b) also provides some guidelines to distinguish between sway and nonsway frames. It states that a frame may be classified as nonsway for a given load case if $P_{cr}/P \geq 10$ for that load case, where P_{cr} is the elastic critical buckling value for sway buckling and P is the design value of the total vertical load. When the system buckling load factor is ten times more than the design load factor, the frame is said to be stiff enough to resist lateral load and it is unlikely to be sensitive to sidesway deflections. AISC LRFD (1993) does not give specific guidance on frame classification. However, for frames to be classified as nonsway in AISC LRFD format, the moment amplification factor, B_2, has to be small (a possible range is $B_2 \leq 1.10$) so that sway deflection would have negligible influence on the final value obtained from the beam–column capacity check.

22.2.6 Classification of Multistory Buildings

The selection of appropriate structural systems for tall buildings must satisfy both the strength and stiffness requirements. The structural system must be adequate to resist lateral and gravity loads that cause horizontal shear deformation and overturning deformation. Other important issues that must be considered in planning the structural schemes and layout are the requirements for architectural details, building services, vertical transportation, and fire safety, among others. The efficiency of a structural system is measured in terms of their ability to resist higher lateral load, which increases with the height of the frame (Iyengar et al. 1992). A building can be considered as tall when the effect of lateral loads is reflected in the design. Lateral deflections of tall buildings should be limited to prevent damage to both structural and nonstructural elements. The accelerations at the top of the building during frequent windstorms should be kept within acceptable limits to minimize discomfort to the occupants (see Section 22.5).

Figure 22.5 shows a chart which defines, in general, the limits to which a particular framing system can be used efficiently for multistory building projects. The various structural systems in Figure 22.5 can be broadly classified into two main types: (1) medium-height buildings with shear-type deformation predominant and (2) high-rise cantilever structures such as framed tubes, diagonal tubes, and braced trusses. This classification of system forms is based primarily on their relative effectiveness in resisting lateral loads. At one end of the spectrum in Figure 22.5 is the moment resisting frames, which are efficient for buildings of 20 to 30 stories, and at the other end is the tubular systems with high cantilever

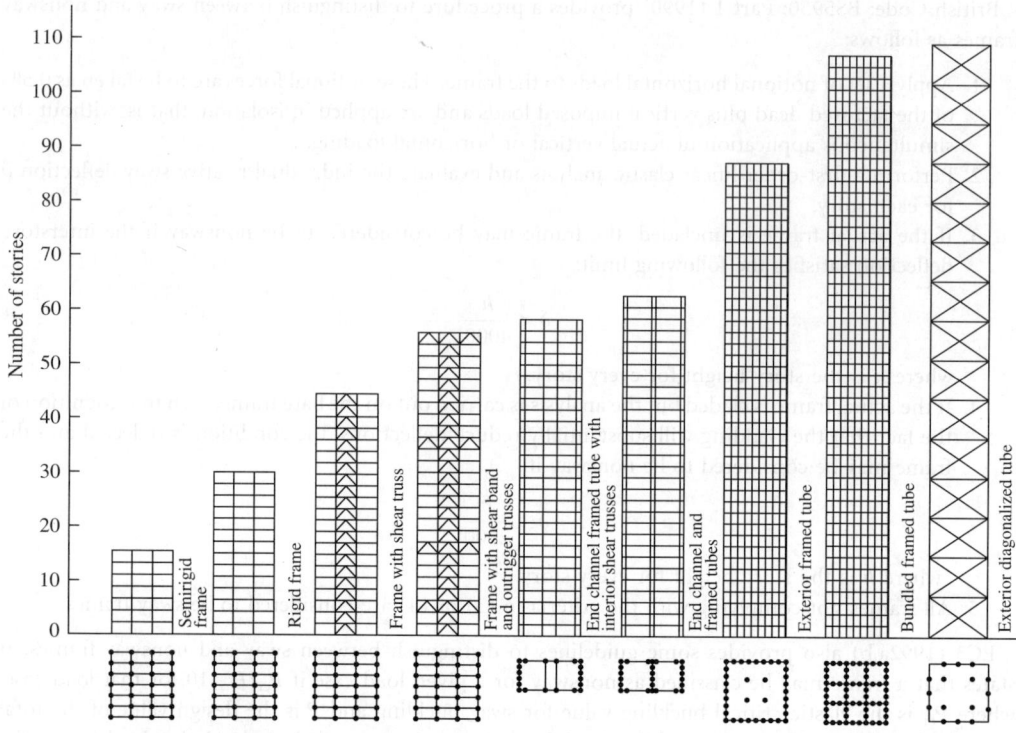

FIGURE 22.5   Categorization of tall building systems.

efficiency. Other systems are placed with the idea that the application of any particular form is economical only over a limited range of building heights.

An attempt has been made to develop a rigorous methodology for the cataloguing of tall buildings with respect to their structural systems (Council on Tall Buildings and Urban Habitat 1995). The classification scheme involves four levels of framing division: primary framing system, bracing sub-system, floor framing, and configuration and load transfer. While any cataloguing scheme must address the preeminent focus on lateral load resistance, the load-carrying function of the tall building subsystems is rarely independent. An efficient high-rise system must engage vertical gravity load-resisting elements in the lateral load subsystem in order to reduce the overall structural premium for resisting lateral loads. Further readings on design concepts and structural schemes for steel multistory buildings can be found in Liew (2001) and the design calculations and procedure for building frame structures using the AISC LRFD procedure are given in ASCE (1987).

Some degree of independence can be distinguished between the floor framing systems and the lateral-load-resisting systems, but the integration of these subassemblies into the overall structural scheme is crucial. Section 22.3 provides some advice for selecting composite floor systems to achieve the required stiffness and strength and also highlights the ways where building services can be accommodated within normal floor zones. Several practical options for long-span construction are discussed and their advantages and limitations are compared and contrasted. Design considerations for floor diaphragm are discussed. Section 22.4 provides some advice on the general principles to be applied when preparing a structural scheme for multistory steel and composite frames. The design procedure and construction considerations that are specific to steel gravity frames, braced frames, moment-resisting frames, and the design approaches to be adopted for sizing multistory building frames are given. The potential use of steel–concrete composite material for high-rise construction is presented. Section 22.5 deals with the issues related to wind-induced effects on multistory frames. Dynamic effects due to along wind, across wind, and torsional response are considered with examples.

22.3 Floor Systems

22.3.1 Design Consideration

Floor systems in tall buildings generally do not differ substantially from those in low-rise buildings. However, the following aspects need to be considered in design:

1. Weight to be minimized
2. Self-supporting during construction
3. Mechanical services to be integrated in the floor zone
4. Adequate fire resistance
5. Buildability
6. Long spanning capability
7. Adequate floor diaphragm

Modern office buildings require large floor span in order to create greater space flexibility for the accommodation of greater variety of tenant floor plans. For building design, it is necessary to reduce the weight of the floors so as to reduce the size of columns and foundations and thus permit the use of larger space. Floors are required to resist vertical loads and they are usually supported by secondary beams. The spacing of the supporting beams must be compatible with the resistance of the floor slabs.

The floor systems can be made buildable using prefabricated or precasted elements of steel and reinforced concrete in various combinations. Floor slabs can be precast concrete slab or composite slabs with metal decking. Typical precast slabs are 4 to 7 m, thus avoiding the need of secondary beams. For composite slabs, metal deck spans ranging from 2 to 7 m may be used depending on the depth and shape of the deck profile. However, the permissible spans for steel decking are influenced by the method of construction; in particular, it depends on whether temporary propping is provided. Propping is best avoided as the speed of construction is otherwise diminished for the construction of tall buildings.

Sometimes openings in the webs of beams are required to permit passage of horizontal services, such as pipes (for water and gas), cables (for electricity, telecommunications, and electronic communication), and ducts (air conditioning), etc.

In addition to strength, floor spanning systems must provide adequate stiffness to avoid large deflections due to live load, which could lead to damage of plaster and slab finishers. Where the deflection limit is too severe, precambering with an appropriate initial deformation equal and opposite to that due to the permanent loads can be employed to offset part of the deflection. In steel construction, steel members can be partially or fully encased in concrete for fire protection. For longer period of fire resistance, additional reinforcement bars may be required. Various long-span flooring systems in Section 22.3.4 offer solutions to integrate building service into the structural depth leading to potential savings in weight and cladding cost.

22.3.2 Composite Floor Systems

Composite floor systems typically involve structural steel beams, joists, girders, or trusses linked via shear connectors with a concrete floor slab to form an effective T-beam flexural member resisting primarily gravity loads. The versatility of the system results from the inherent strength of the concrete floor component in compression and the tensile strength of the steel member. The main advantages of combining the use of steel and concrete materials for building construction are:

- Steel and concrete may be arranged to produce an ideal combination of strength, with concrete efficient in compression and steel in tension.

- Composite flooring is lighter in weight than pure concrete.
- The construction time is reduced since casting of additional floors may proceed without having to wait for the previously cast floors to gain strength. The steel decking system provides positive-moment reinforcement for the composite floor and requires only a small amount of reinforcement to control cracking and for fire resistance.
- The construction of a composite floor does not require highly skilled labor. The steel decking acts as a permanent formwork. Composite beams and slabs can accommodate raceways for electrification, communication, and air distribution systems. The slab serves as a ceiling surface to provide easy attachment of a suspended ceiling.
- The composite floor system produces a rigid horizontal diaphragm, providing stability to the overall building system while distributing wind and seismic shears to the lateral-load-resisting systems.
- Concrete provides thermal protection to steel at elevated temperature. Composite slabs of 2 h fire rating can be easily achieved for most building requirements.

The floor slab can be formed by the following methods:

- a flat-soffit reinforced concrete slab (Figure 22.6a);
- precast concrete planks with cast *in situ* concrete topping (Figure 22.6b);
- precast concrete slab with *in situ* grouting at the joints (Figure 22.6c);
- a metal steel deck tops with concrete, either composite or noncomposite (Figure 22.6d).

The composite action of the metal deck results from side embossments incorporated into the steel sheet profile.

22.3.3 Composite Beams and Girders

Steel and concrete composite beams may be formed by shear connectors connecting the concrete floor to the top flange of the steel member. Concrete encasement will provide fire resistance to the steel member.

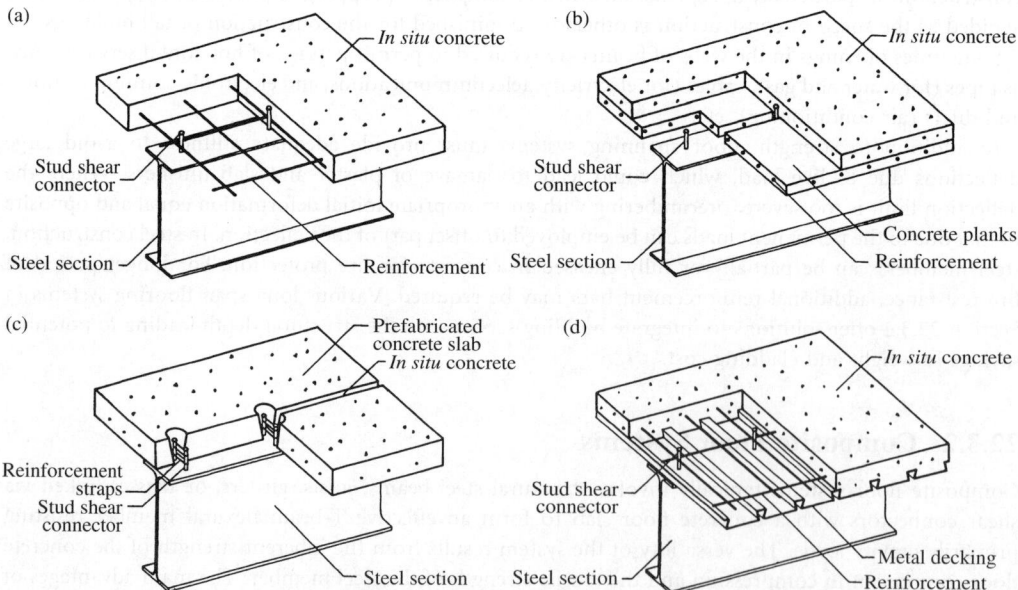

FIGURE 22.6 Composite beams with: (a) flat-soffit reinforced concrete slab, (b) precast concrete planks and cast *in situ* concrete topping, (c) precast concrete slab and *in situ* concrete at the joints, and (d) metal steel deck supporting concrete slab.

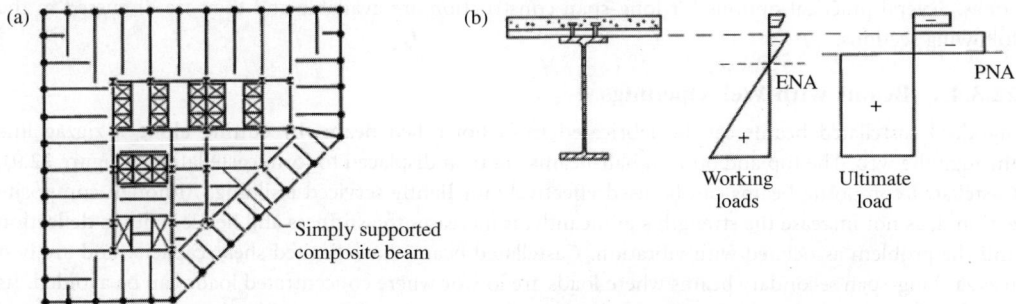

FIGURE 22.7 (a) Composite floor plan. (b) Stress distribution in a composite cross-section.

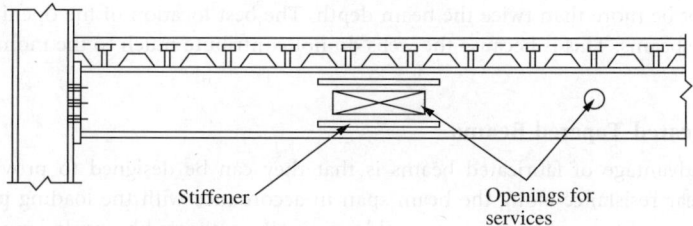

FIGURE 22.8 Web opening with horizontal reinforcements.

Alternatively, direct sprayed-on cementitious and board-type fireproofing materials may be used economically to replace the concrete insulation on the steel members. The most common arrangement found in composite floor systems is a rolled or built-up steel beam connected to a formed steel deck and concrete slab (Figure 22.6d). The metal deck typically spans unsupported between steel members while also providing a working platform for concreting work.

Figure 22.7a shows a typical building floor plan using composite steel beams. The stress distribution at working loads in a composite section is shown schematically in Figure 22.7b. The neutral axis is normally located very near to the top flange of the steel section. Therefore, the top flange is lightly stressed. From a construction point of view, a relatively wide and thick top flange must be provided for proper installation of shear stud and metal decking. However, the increased fabrication costs must be evaluated, which tend to offset the saving from material efficiency.

A number of composite girder forms allowing passage of mechanical ducts and related services through the depth of the girder (Figure 22.8). Successful composite beam design requires the consideration of various serviceability issues such as long-term (creep) deflections and floor vibrations. Of particular concern is the occupant-induced floor vibrations. The relatively high flexural stiffness of most composite floor framing systems results in relatively low vibration amplitudes and therefore is effective in reducing perceptibility. Studies have shown that short to medium span (6 to 12 m) composite floor beams perform quite well and are rarely found to transmit annoying vibrations to the occupants. Particular care is required for long-span beams more than 12 m in range.

22.3.4 Long-Span Flooring Systems

Long spans impose a burden on the beam design in terms of larger required flexural stiffness for serviceability design. Besides satisfying both serviceability and ultimate strength limit states, the proposed system must also accommodate the incorporation of mechanical services within normal floor zones. Several practical options for long-span construction are available and they are discussed in the following sections.

22.3.4.1 Beams with Web Openings

Standard castellated beams can be fabricated from hot-rolled beams by cutting along a zigzag line through the web. The top and bottom half-beams are then displaced to form castellations (Figure 22.9). Castellated composite beams can be used effectively for lightly serviced building. Although composite action does not increase the strength significantly, it increases the stiffness and hence reduces deflection and the problem associated with vibration. Castellated beams have limited shear capacity and are best used as long-span secondary beams where loads are low or where concentrated loads can be avoided. Its use may be limited due to the increased fabrication cost and the fact that the standard castellated openings are not large enough to accommodate the large mechanical ductwork common in modern high-rise buildings.

   Horizontal stiffeners may be required to strengthen the web opening and they are welded above and below the opening. The height of the opening should not be more than 70% of the beam depth and the length should not be more than twice the beam depth. The best location of the openings is in the low shear zone of the beams. This is because the webs do not contribute much to the moment resistance of the beam.

22.3.4.2 Fabricated Tapered Beams

The economic advantage of fabricated beams is that they can be designed to provide the required moment and shear resistance along the beam span in accordance with the loading pattern along the beam. Several forms of tapered beams are possible. A simply supported beam design with a maximum bending moment at the mid-span would require that they all effectively taper to a minimum at both ends (Figure 22.10). A rigidly connected beam would have minimum depth toward the mid-span. To make best use of this system, services should be placed toward the smaller depth of the beam cross-sections. The spaces created by the tapered web can be used for running services of modestly size (Figure 22.10).

   A hybrid girder can be formed with the top flange made of lower-strength steel in comparison with the steel grade for the bottom flange. The web plate can be welded to the flanges by double-sided fillet welds. Web stiffeners may be required at the change of section when taper slope exceeds approximately 6°. Stiffeners are also required to enhance the shear resistance of the web, especially when the web slenderness ratio is too high. A tapered beam is found to be economical for spans up to 20 m. Further information on the design of fabricated beams with tapered webs can be found in Owens (1989).

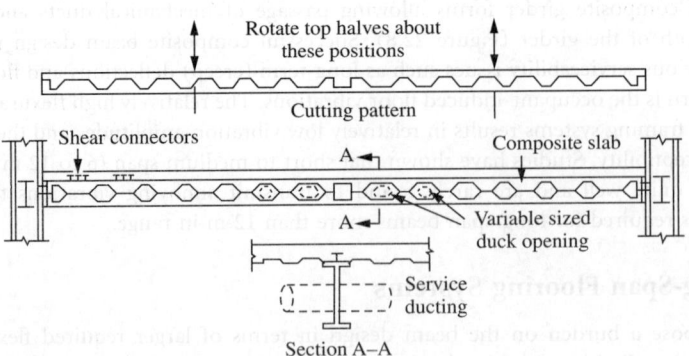

FIGURE 22.9 Composite castellated beams.

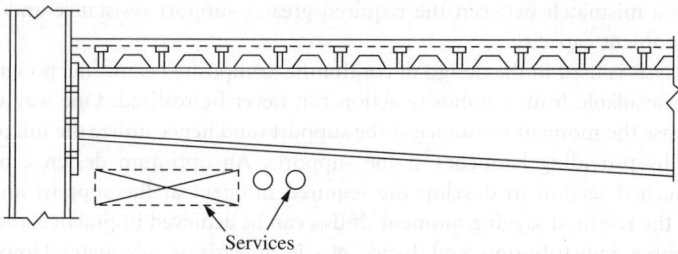

FIGURE 22.10 Tapered composite beam.

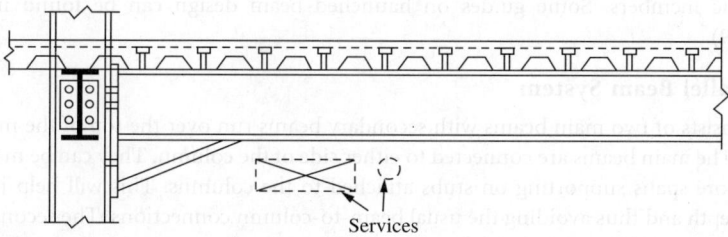

FIGURE 22.11 Haunched composite beam.

22.3.4.3 Haunched Beams

Haunched beams are designed by forming a rigid moment connection between the beams and columns. The haunch connections offer restraints to beam and it helps to reduce mid-span moment and deflection. The beams are designed in a manner similar to continuous beams. Considerable economy can be gained in sizing the beams using continuous design that may lead to a reduction in beam depth up to 30% and deflection up to 50%.

The haunch may be designed to develop the required moment that is larger than the plastic moment resistance of the beam. In this case, the critical section is shifted to the tip of the haunch. The depth of the haunch is selected based on the required moment at the beam-to-column connections. The length of the haunch is typically 5 to 7% the span length for nonsway frames or 7 to 15% for sway frames. Service ducts can pass below the beams (Figure 22.11).

Haunched composite beams are usually used in the case where the beams frame directly into the major axis of the columns. This means that the columns must be designed to resist the moment transferred from the beam to the column. Thus, a heavier column and a more complex connection would be required in comparison with a structure designed based on the assumption that the connections are pinned. The rigid frame action derived from the haunched connections can resist lateral loads due to wind without the need of vertical bracing. Haunched beams offer higher strength and stiffness during the steel erection stage thus making this type of system particular attractive for long-span construction. However, haunched connections behave differently under positive and negative moments, as the connection configuration is not symmetrical about the bending axis.

The rationale of using the haunched beam approach is explained as follows. In continuous beam design, the moment distribution of a continuous beam would show that the support moment is generally larger than the mid-span moment up to the ratio of 1.8. The effective cross-sections of typical steel–concrete composite beams under hogging and sagging moment can be determined according to the usual stress block method of design. It can be observed that the hogging moment capacity of the composite section at the support is smaller than the sagging moment capacity near the mid-span.

Therefore, there is a mismatch between the required greater support resistance and the much larger available sagging moment capacity.

When elastic analysis is used in the design of continuous composite beams, the potential large sagging moment capacities available from composite action can never be realized. One way to overcome this problem is to increase the moment resistance at the support (and hence utilize the full potential of larger sagging moment) by providing haunches at the supports. An optimum design can be achieved by designing the haunched section to develop the required moment at the support and the composite section to develop the required sagging moment. If this can be achieved in practice, the design does not require inelastic force redistribution and hence elastic analysis is adequate. However, analysis of haunched composite beams is more complicated because the member is nonprismatic (i.e., cross-section property varies along the length). The analysis of such beams requires the evaluation of section properties such as beam's stiffness (*EI*) at different cross-sections. The analysis/design process is more involved because it requires the evaluation of serviceability deflection and ultimate strength limit state of nonprismatic members. Some guides on haunched beam design can be found in Lawson and Rackham (1989).

22.3.4.4 Parallel Beam System

The system consists of two main beams with secondary beams run over the top of the main beams (see Figure 22.12). The main beams are connected to either side of the column. They can be made continuous over two or more spans supporting on stubs attached to the columns. This will help in reducing the construction depth and thus avoiding the usual beam-to-column connections. The secondary beams are designed to act compositely with the slab and may also be made to span continuously over the main

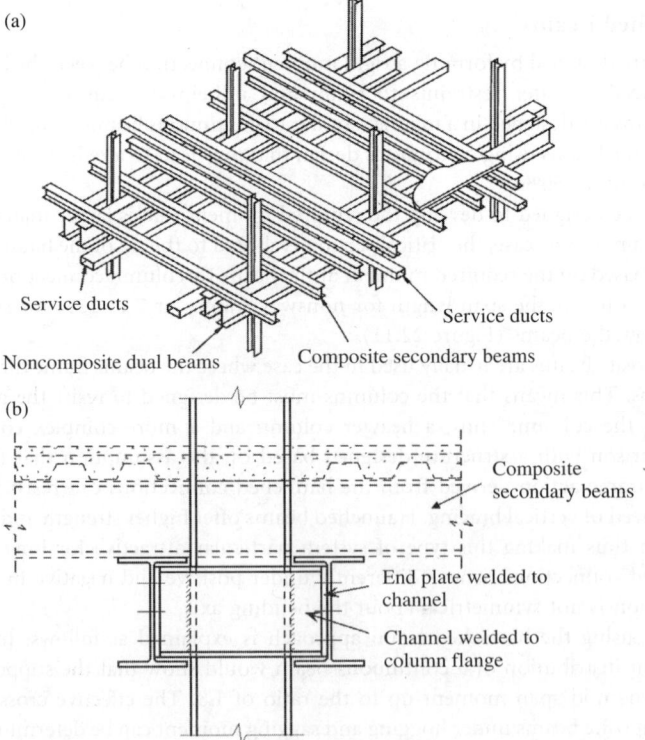

FIGURE 22.12   (a) Parallel composite beam system. (b) Parallel beam connection.

beams. The need to cut the secondary beams at every junction is thus avoided. The parallel beam system is ideally suited for accommodating large service ducts in orthogonal directions (Figure 22.12). A small saving in steel weight is expected from the continuous construction because the primary beams are noncomposite. However, the main beam can be made composite with the slab by welding beam stubs to the top flange of the main beam and connected to the concrete slab through the use of shear studs. The simplicity of connections and ease of fabrication make this long-span beam option particularly attractive. Competitive pricing can be obtained from the fabricator. Further details on parallel beam approach can be found in Brett and Rushton (1990).

22.3.4.5 Composite Trusses

Trusses are frequently used in multistory buildings for very long-span supports. The openings created in the truss braces can be used to accommodate large services. Although the cost of fabrication is higher in relation to the material cost, truss construction can be cost-effective for a very long span when compared to other structural schemes. One disadvantage of the truss configuration is that fire protection is labor intensive and sprayed protection systems cause a substantial mess to the services that pass through the web opening (see Figure 22.13).

Several forms of truss arrangement are possible. The three most common web framing configurations in floor truss and joist designs are (1) Warren truss, (2) modified Warren truss, and (3) Pratt truss, as shown in Figure 22.14. The efficiency of various web members in resisting vertical shear forces may be affected by the choice of a web-framing configuration. For example, the selection of a Pratt web over a Warren web may effectively shorten compression diagonals resulting in more efficient use of these members.

Experience has shown that both Pratt and Warren configurations of web framing are suitable for short-span trusses with shallow depths. For truss with spans greater than 10 m, or effective depths larger than 700 mm, a modified Warren configuration is generally preferred. The Warren and modified Warren trusses are more popular for building construction since they offer larger web openings for services between bracing members.

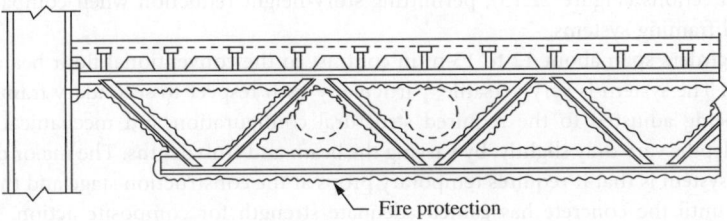

FIGURE 22.13 Composite truss.

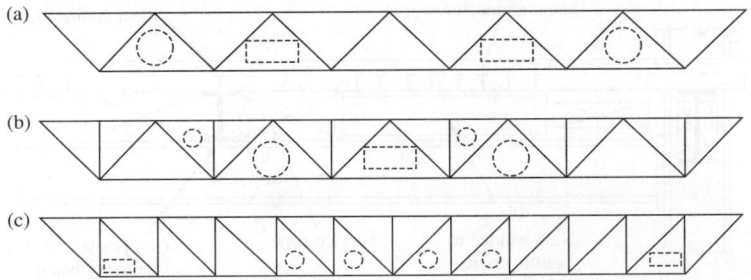

FIGURE 22.14 Truss configurations: (a) Warren truss, (b) modified Warren truss, and (c) Pratt truss.

The resistance of a composite truss is governed by (1) yielding of the bottom chord, (2) crushing of the concrete slab, (3) failure of the shear connectors, (4) buckling of top chord during construction, (5) buckling of web members, and (6) instability occurring during and after construction. To avoid brittle failures, ductile yielding of the bottom chord is the preferred failure mechanism. Thus, the bottom chord should be designed to yield prior to crushing of concrete slab. The shear connectors should have sufficient capacity to transfer the horizontal shear between the top chord and the slab. During construction, adequate plan bracing should be provided to prevent top chord buckling. When considering composite action, the top steel chord is assumed not to participate in the moment resistance of the truss, since it is located very near to the neutral axis of the composite truss and, thus, contributes very little to the flexural capacity. However, the top chord has two functions: (1) it provides an attachment surface for the shear connectors and (2) it resists the forces in the end panel without reliance on composite action unless shear connectors are placed over the seat or along a top chord extension. Thus, the top chord must be designed to resist the compressive force equilibrating the horizontal force component of the first web member. In addition, the top chord also transfers the factored shear force to the support and thus must be designed accordingly.

The bottom chord shall be continuous and may be designed as an axially loaded tension member. The bottom chord shall be proportioned to yield before the concrete slab, web members, or the shear connectors fail.

The shear capacity of the steel top and bottom chords and concrete slab can be ignored in the evaluation of the shear resistance of a composite truss. The web members should be designed to resist vertical shear. Further references on composite trusses can be found in ASCE Task Committee (1996) and Neals and Johnson (1992).

22.3.4.6 Stub Girder System

The stub girder system involves the use of short beam stubs that are welded to the top flange of a continuous, heavier bottom girder member, and connected to the concrete slab through the use of shear studs. Continuous transverse secondary beams and ducts can pass through the openings formed by the beam stub. The natural openings in the stub girder system allow the integration of structural and service zones in two directions (Figure 22.15), permitting story-height reduction when compared with some other structural framing systems.

Ideally, stub girders span about 12 to 15 m in contrast to the conventional floor beams, which span about 6 to 9 m. The system is very versatile, particularly with respect to secondary framing spans with beam depths being adjusted to the required structural configuration and mechanical requirements. Overall girder depths vary only slightly, by varying the beam and stub depths. The major disadvantage of the stub girder system is that it requires temporary props at the construction stage and these props have to be retained until the concrete has gained adequate strength for composite action. However, it is possible to introduce additional steel top chord, such as a T-section, which acts in compression to

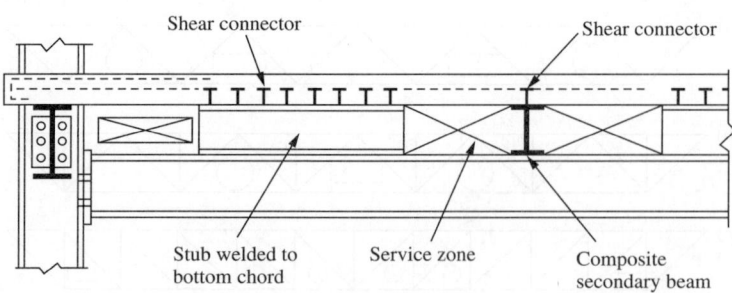

FIGURE 22.15 Stub girder system.

develop the required bending strength during construction. For span length greater than 15 m, stub girders become impractical, because the slab design becomes critical.

In the stub girder system, the floor beams are continuous over the main girders and splice at the locations near the points of inflection. The sagging moment regions of the floor beams are usually designed compositely with the deck–slab system, to produce savings in structural steel as well as to provide stiffness. The floor beams are bolted to the top flange of the steel bottom chord of the stub girder and two shear studs are usually specified on each floor beam, over the beam–girder connection for anchorage to the deck–slab system. The stub girder may be analyzed as a Vierendeel girder, with the deck–slab acting as a compression top chord, the full length steel girder as a tensile bottom chord, and the steel stubs as vertical web members or shear panels.

22.3.4.7 Prestressed Composite Beams

Prestressing of the steel girders is carried out such that the concrete slab remains uncracked under the working loads and the steel is utilized fully in terms of stress in the tension zone of the girder.

Prestressing of steel beam can be carried out using a precambering technique as depicted in Figure 22.16. First, a steel girder member is prebent (Figure 22.16a) and is then subjected to preloading in the direction against the bending curvature until the required steel strength is reached (Figure 22.16b). Second, the lower flange of the steel member, which is under tension, is encased in a reinforced concrete chord (Figure 22.16c). The composite action between the steel beam and the concrete slab is developed by providing adequate shear connectors at the interface. When the concrete gains adequate strength, the steel girder is prestressed by stress relieving the precompressed tension chord (Figure 22.16d). Further composite action can be achieved by supplementing the girder with *in situ* or prefabricated reinforcement concrete slabs and this will produce a double composite girder (Figure 22.16e).

The major advantages of this system are that the steel girders are encased in concrete on all sides, no corrosion and fire protection are required on the sections. The entire process of precambering and prestressing can be performed and automated in a factory. During construction, the lower concrete chord cast in the works can act as the formwork. If the distance between two girders is large, precast planks can be supported by the lower concrete chord as permanent formwork.

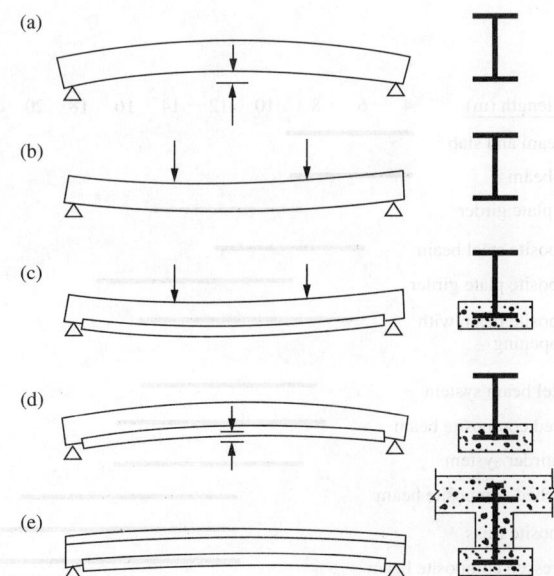

FIGURE 22.16 Process of prestressing using precambering technique.

Prestressing can also be achieved by using tendons that can be attached to the bottom chord of a steel composite truss or the lower flange of a composite girder to enhance the load-carrying capacity and stiffness of long-span structures (Figure 22.17). This technique has been found to be popular for bridge construction in Europe and the United States, although less common for building construction.

22.3.5 Comparison of Floor Spanning Systems

The conventional composite beams are the most common forms of floor construction for a large number of building projects. Typically, they are highly efficient and economic with bay sizes in the range of 6 to 12 m. There is, however, much demand for larger column free areas where, with a traditional composite approach, the beams tend to become excessively deep, with unnecessary increases in the overall building height and the consequent increases in cladding costs. Spans exceeding 12 m are generally achieved by choosing an appropriate structural form that integrates the services within the floor structure, thereby reducing the overall floor zone depths. Although a long-span solution may entail a small increase in structural costs, the advantages of greater flexibility and adaptability in service and the creation of column-free space often represent the most economic option over the design life of the building. Figure 22.18 compares the various structural options of a typical range of span lengths used in practice.

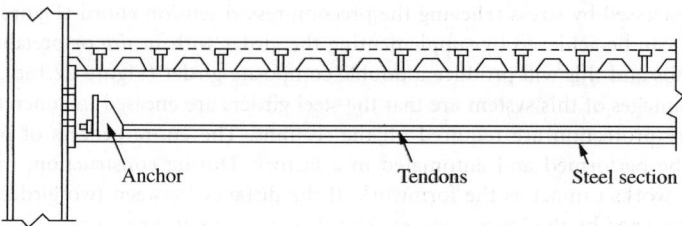

Anchor   Tendons   Steel section

FIGURE 22.17   Prestressing of composite steel girders with tendons.

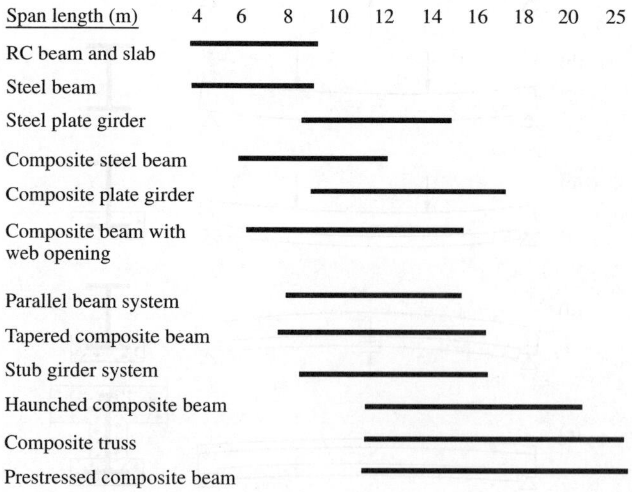

Span length (m)	4	6	8	10	12	14	16	18	20	25
RC beam and slab										
Steel beam										
Steel plate girder										
Composite steel beam										
Composite plate girder										
Composite beam with web opening										
Parallel beam system										
Tapered composite beam										
Stub girder system										
Haunched composite beam										
Composite truss										
Prestressed composite beam										

FIGURE 22.18   Comparison of composite floor systems.

22.3.6 Floor Diaphragms

Typically, beams and columns rigidly connected for moment resistance are placed in orthogonal directions to resist lateral loads. Each plane frame would assume to resist a portion of the overall wind shear, which is determined from the individual frame stiffness in proportion to the overall stiffness of all frames in that direction. This is based on the assumption that the lateral loads are distributed to the various frames by the floor diaphragm. In order to develop proper diaphragm action, the floor slab must be attached to all columns and beams that participate in lateral-force resistance. For building relying on bracing systems to resist all lateral loads, the stability of a building depends on rigid floor diaphragm to transfer wind shears from their point of application to the bracing systems such as lattice frames, shear walls, or core walls.

The use of composite floor diaphragms in place of in-plane steel bracing has become an accepted practice. The connection between slab and beams are often through shear studs that are welded directly through the metal deck to the beam flange. The connection between beams of adjacent deck panels is crucial and often through interlocking of panels overlapping each other. The diaphragm stresses are generally low and can be resisted by floor slabs that have adequate thickness for most buildings.

Plan bracing is necessary when the diaphragm action is not adequate. Figure 22.19a shows a triangulated plan bracing system that resists lateral load on one side and spans between the vertical walls. Figure 22.19b illustrated the case where the floor slab has adequate thickness and it can act as diaphragm resisting lateral loads and transmitting the forces to the vertical walls. However, if there is an abrupt

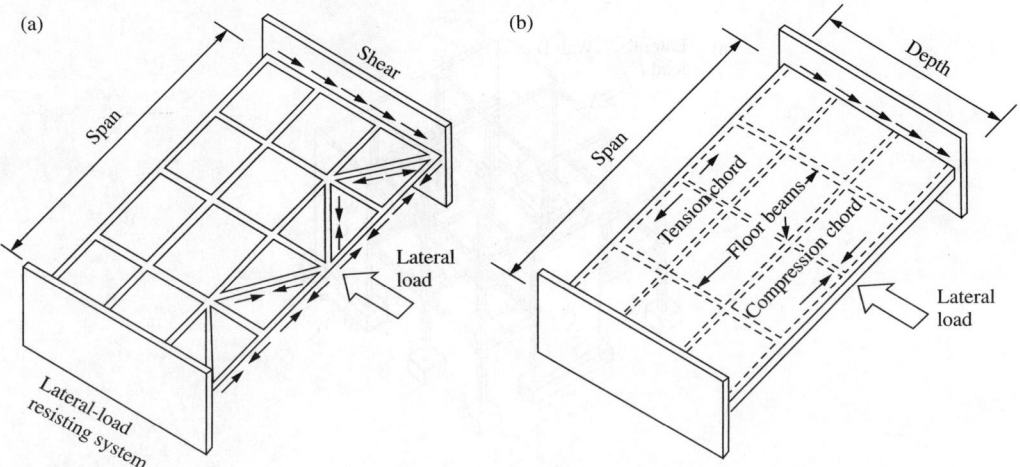

FIGURE 22.19 (a) Triangulated plan bracing system. (b) Concrete floor diaphragm.

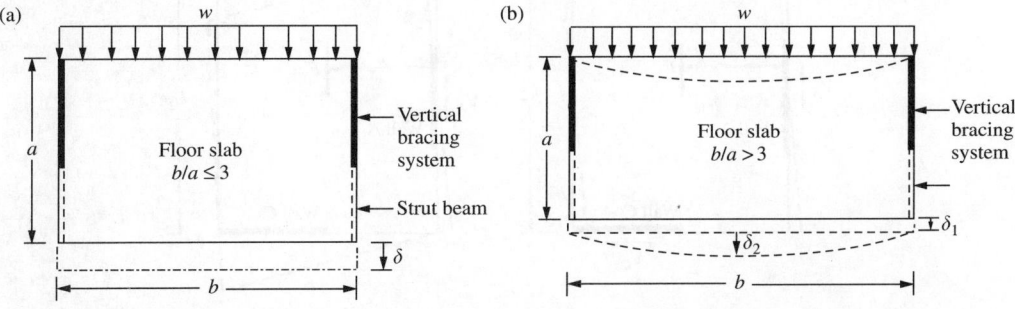

FIGURE 22.20 Diaphragm rigidity: (a) plan aspect ratio ≤ 3 and (b) plan aspect ratio > 3.

change in lateral stiffness or where the shear must be transferred from one frame to the other due to the termination of lateral bracing system at certain height, large diaphragm stresses may be encountered and they must be accounted for through proper detailing of slab reinforcement. Also, diaphragm stresses may be high where there are large openings in the floor, in particular at the corners of the openings.

The rigid diaphragm assumption is generally valid for most high-rise buildings (Figure 22.20a); however, as the plan aspect ratio (b/a) of the diaphragm linking two lateral systems exceeds 3 in 1 (see the illustration in Figure 22.20b), the diaphragm may become semirigid or flexible. For such cases, the wind shears must be allocated to the parallel shear frames according to the attributed area rather than relative stiffness of the frames.

From the analysis point of view, a diaphragm is analogous to a deep beam with the slab forming the web and the peripheral members serving as the flanges as shown in Figure 22.20b. It is stressed principally in shear, but tension and compression forces must be accounted for in design.

A rigid diaphragm is useful to transmit torsional forces to the lateral-load-resistance systems to maintain lateral stability. Figure 22.21a shows a building frame consisting of three shear walls resisting lateral forces acting in the direction of Wall A. The lateral load is assumed to act as a concentrated load with a magnitude F on each story. Figure 22.21b and c shows the building plan having dimensions of L_1 and L_2. The lateral-load-resisting system is represented in the plan by solid lines that represent Wall A, Wall B, and Wall C. Since there is only one lateral resistance system (Wall A) in the direction of

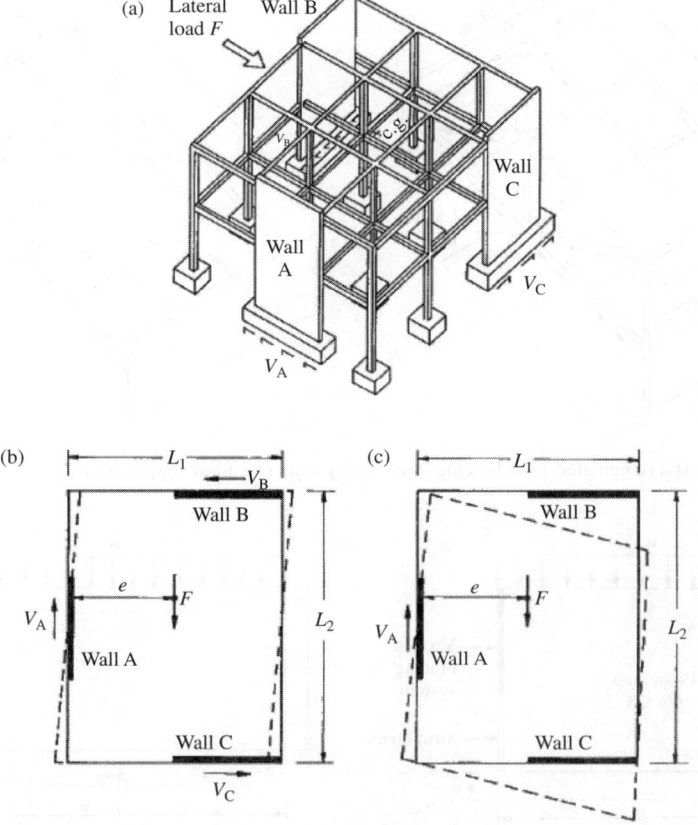

FIGURE 22.21   (a) Lateral-force-resisting system in a building. (b) Rigid diaphragm. (c) Flexible diaphragm.

TABLE 22.1 Details of Typical Flooring Systems and their Relative Merits

Floor system	Typical span length (m)	Typical depth (mm)	Construction time	Degree of lateral restraint to beams	Degree of diaphragm action	Usage
In situ concrete	3–6	150–250	Medium	Very good	Very good	All categories but not often used in multistory buildings
Steel deck with *in situ* concrete	2.5–3.6 unshore; >3.6 shore	110–150	Fast	Very good	Very good	All categories especially in multistory office buildings
Precast concrete	3–6	110–200	Fast	Fair–good	Fair–good	All categories with cranage requirements
Prestressed concrete	6–9	110–200	Medium	Fair–good	Fair–good	Multistory buildings and bridges

the applied load, the loading condition creates a torsion (*Fe*) and the diaphragm tends to rotate as shown by the dashed lines in Figure 22.21b. The lateral-load-resistance systems in Wall B and Wall C will provide the resistance forces to stabilize the torsional force by generating a couple of shear resistance as

$$V_B = V_C = \frac{Fe}{L_2}$$

Figure 22.21c illustrates the same condition except that a flexible diaphragm is used. The same torsional tendency exists, but the flexible diaphragm is unable to generate a resisting couple in Wall B and Wall C and the structure will collapse as shown by the dashed lines. To maintain stability, a minimum of two vertical bracings in the direction of the applied force is required to eliminate the possibility of any torional effects.

The adequacy of the floor to act as a diaphragm depends very much on its type. Precast concrete floor planks without any prestressing offer limited resistance to the racking effects of diaphragm action. In such cases, supplementary bracing systems in plan, such as those shown in Figure 22.19a, are required for resistance of lateral forces. Where precast concrete floor units are employed, sufficient diaphragm action can be achieved by using a reinforced structural concrete topping, so that all individual floor planks are combined to form a single floor diaphragm. Composite concrete floors, incorporating permanent metal decking, provide excellent diaphragm action provided that the connections between the diaphragm and the peripheral members are adequate. When composite beams or girders are used, shear connectors will usually serve as boundary connectors and intermediate diaphragm-to-beam connectors. By fixing the metal decking to the floor beams, an adequate floor diaphragm can be achieved during the construction stage.

It is essential at the start of the design of structural steelworks to consider the details of the flooring system to be used, since these have a significant effect on the design of the structure. Table 22.1 summarizes the salient features of the various types of flooring systems in terms of their diaphragm actions.

22.4 Design Concepts and Structural Schemes

22.4.1 Introduction

Multistory steel frames consist of columns and beams interconnected to form a three-dimensional structure. A building frame can be stabilized either by some forms of bracing systems (braced frames) or by itself (unbraced frames). All building frames must be designed to resist lateral load to ensure overall stability. A common approach is to provide a gravity framing system with one or more lateral bracing systems attached to it. This type of framing system, which is generally referred to as simple braced frames, is found to be cost-effective for multistory buildings of moderate height (up to 20 stories).

For gravity frames, the beams and columns are pinned connected and the frames are not capable of resisting any lateral loads. The stability of the entire structure is provided by attaching the gravity frames to some forms of bracing systems. The lateral loads are resisted mainly by the bracing systems while the gravity loads are resisted by both the gravity frame and the bracing system. For buildings of moderate height, the bracing system's response to lateral forces is sufficiently stiff such that second-order effects may be neglected for the design of such frames.

In moment-resisting frames, the beams and columns are rigidly connected to provide moment resistance at joints, which may be used to resist lateral forces in the absence of any bracing systems. However, moment joints are rather costly to fabricate. In addition, it takes longer time to erect a moment frame than a gravity frame.

A cost-effective framing system for multistory buildings can be achieved by minimizing the number of moment joints, replacing field welding by field bolting, and combining various framing schemes with appropriate bracing systems to minimize frame drift. A multistory structure is most economical and efficient when it can transmit the applied loads to the foundation by the shortest and most direct routes. For ease of construction, the structural schemes should be simple enough, which implies repetition of member and joints, adoption of standard structural details, straightforward temporary works, and minimal requirements for an inter-related erection procedure to achieve the intended behavior of the completed structure. Sizing of structural members should be based on the longest spans and largest attributed roof and floor areas. The same sections should be used for similar but less onerous cases.

Scheme drawings for multistory building design should include the following:

1. General arrangement of the structure including column and beam layout, bracing frames, and floor systems.
2. Critical and typical member sizes.
3. Typical cladding and bracing details.
4. Typical and unusual connection details.
5. Proposals for fire and corrosion protection.

This section offers advice on the general principles to be applied when preparing a structural scheme for multistory steel and composite frames. The aim is to establish several structural schemes that are practicable, sensibly economic, and functional to the changes that are likely to be encountered as the overall design develops. The section begins by examining the design procedure and construction considerations that are specific to steel gravity frames, braced frames, and moment-resisting frames and the design approaches to be adopted for sizing tall building frames. The potential use of steel–concrete composite material for high-rise construction is then presented. Finally, the design issues related to braced and unbraced composite frames are discussed.

22.4.2 Frame and Member Stability

There are many parameters and behavioral effects that influence the stability of steel-framed structures. The extent to which these factors are modeled in analysis will affect the criteria that one applies in design of the frame, its members, and connections. Three basic aspects of behavior, geometric nonlinearities, imperfection effects, and member limit states, will ultimately govern frame deformations under applied loads and the resulting internal load effects.

22.4.2.1 Geometric Nonlinearities and Imperfection Effects

Modern stability design provisions are based on the premise that the member forces are calculated by second-order elastic analyses, where equilibrium is satisfied on the deformed structure. When stability effects are significant, consideration must be given to initial geometric imperfections. Many code provisions for calibrating the stability requirements are based on initial geometric imperfections conservatively assumed as equal to the maximum fabrication and erection tolerances permitted by the Code. For columns and frames, this would normally imply a member out-of-straightness equal to $0.001L$, where L is the member length between brace or framing points, and a frame out-of-plumb equal to $0.002H$, where H is the story height.

Many of the code's specified analysis/design approaches are calibrated against inelastic distributed-plasticity analyses that account for yielding through the member cross-section and along the member length. Thermal residual stresses in I- or H- (wide flange) shape members are assumed to have maximum values of $0.3F_y$, which are linearly varying across the flanges and uniform tension in the web.

22.4.2.2 Member Limit States

Member strength may be controlled by any one of the following limit states: cross-section yielding, local buckling, flexural buckling, and torsional–flexural buckling. Most of the structural analyses envisioned for routine design use do not model limit states associated with local buckling of cross-section or torsional–flexural buckling of member. Therefore, these limits must be considered in separate member design checks. For inelastic analyses, the effect of cross-section yielding is incorporated directly in the analysis; for elastic analyses, the cross-section strength can be checked by an interaction equation that approximates the $P–M$ yield surface. Whether or not the analysis captures in-plane flexural buckling depends on the extent to which the maximum moments are affected by distributed plasticity and member imperfections. Concerns as to whether the analysis captures these effects suggest the need to apply a member check for in-plane flexural buckling, even when an accurate second-order analysis is used. A major consideration for the in-plane flexural buckling check relates to the assumed buckling length used in calculating the design compression strength.

22.4.2.3 Second-Order Analysis

In practice, there are alternative approaches one can employ for conducting second-order analyses, some of which are more rigorous than others. The difference between simplified and rigorous analyses depends on the extent to which $P–\delta$ effects due to curvature of the member relative to its chord are modeled and whether the problem is "linearized" to expedite the solution. Rigorous second-order analyses are those that accurately model all significant second-order effects such as to include solution of the governing differential equation, either through stability functions or computer frame analysis programs that model these effects.

Not all modern commercial computer programs are capable of rigorous analyses. Methods that modify first-order analysis results through second-order amplifiers are in some cases accurate enough to constitute a rigorous analysis, but this depends on the magnitude of second-order effects and other characteristics of the problem. A common type of approximate analyses are those that only capture $P–\Delta$ due to member end translations (e.g., interstory drift) but fail to capture $P–\delta$ effects. Where $P–\delta$ effects are significant, errors arise in approximate methods that do not accurately account for the effect of $P–\delta$ moments on amplification of both local member moments and the calculated global displacements.

22.4.2.4 Direct Analysis Approach

The direct analysis approach is developed with the goal to accurately model frame stability effects and thereby eliminate the need for calculating effective buckling lengths for column designs. The new provisions in the 2005 AISC Standard involve reducing the nominal elastic stiffness and applying a notional load to the frame. The notional load approach can be found in steel standards such as Eurocode, but many aspects of the proposed provisions given in the new 2005 AISC Standard address known shortcomings of conventional notional load approaches in other standards.

In the reduced stiffness approach, the reduction on the flexural stiffness EI can be applied by modifying E in the analysis. For computer programs that perform semiautomated design checks, one should be sure that the elastic modulus should not be reduced in design equations, which involve E to evaluate the design strength. The reduced stiffness and notional load requirements only pertain to analysis of the strength limit state and they do not apply to analysis of other serviceability conditions.

Notional load is applied to represent the destabilizing effect of geometric imperfections and other effects such as yielding, nonidealized boundary, and loading conditions. Notional loads are applied as lateral loads at each floor level and specified in terms of the gravity loads applied at that floor level. The notional load magnitude of 0.002 corresponds to a frame out-of-plumb equal to $0.002H$ (where H is the story height). Notional loads shall be applied in the direction that adds to the destabilizing effects under the specified strength load combination.

22.4.3 Gravity Frames

Gravity frames refer to structures that are designed to resist only gravity loads. The bases for designing gravity frames are as follows:

1. The beam and girder connections transfer only vertical shear reactions without developing bending moment that will adversely affect the members and the structure as a whole.
2. The beams may be designed as a simply supported member.
3. Columns must be fully continuous. The columns are designed to carry axial loads only. Some codes of practice (e.g., BS5950 1990) require the column to carry nominal moments due to the reaction force at the beam end, applied at an appropriate eccentricity.
4. Lateral forces are resisted entirely by bracing frames or by shear walls, lift, or staircase closures through floor diaphragm action.

22.4.3.1 General Guides

The following points should be observed in the design of gravity frames:

1. Provide lateral stability to gravity framing by arranging suitable braced bays or core walls deployed symmetrically in orthogonal directions, or wherever possible, to resist lateral forces.
2. Adopt a simple arrangement of slabs, beams, and columns so that loads can be transmitted to the foundations by the shortest and most direct load paths.
3. Tie all the columns effectively in orthogonal directions at every story. This may be achieved by the provision of beams or ties that are placed as close as practicable to the columns.
4. Select a flooring scheme that provides adequate lateral restraint to the beams and adequate diaphragm action to transfer the lateral load to the bracing system.
5. For tall building construction, choose a profiled-steel-decking composite floor construction if uninterrupted floor space is required and/or height is at a premium. As a guide, limit the span of the floor slab to 2.5 to 3.6 m, the span of the secondary beams to 6 to 12 m, and the span of the primary beams to 5 to 7 m. Otherwise, choose a precast or an *in situ* reinforced concrete floor, limiting their span to 5 to 6 m and the span of the beams to 6 to 8 m approximately.

22.4.3.2 Structural Layout

In building construction, greater economy can be achieved through a repetition of similarly fabricated components. A regular column grid is less expensive than a nonregular grid for a given floor area. In addition, greater economies can be achieved when the column grids in plan are rectangular in which the secondary beams should span in the longer direction and the primary beams in the shorter direction as shown in Figure 22.22a and b. This arrangement reduces the number of beam-to-beam connections and the number of individual members per unit area of supported floor (Owens and Knowles 1992).

In gravity frames, the beams are assumed to be simply supported between columns. The effective beam span to depth ratio (L/D) is about 12 to 15 for steel beams and 18 to 22 for simply supported composite beams. The design of the beam is often dependent on the applied load, the type of beam system employed, and the restrictions on structural floor depth. The floor-to-floor height in a multistory building is influenced by the restrictions on overall building height and the requirements for services above and below the floor slab. Naturally, flooring systems involving the use of structural steel members that act compositely with the concrete slab achieve the longest spans.

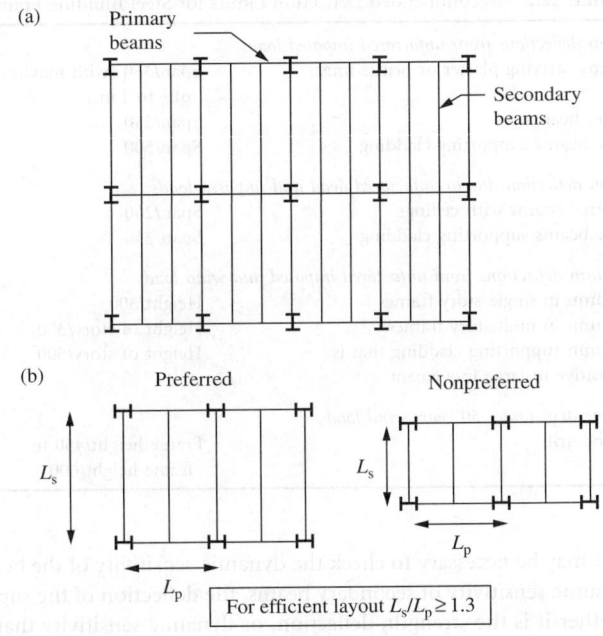

FIGURE 22.22   (a) Rectangular grid layout. (b) Preferred and nonpreferred grid layouts.

22.4.3.3 Analysis and Design

The analysis and design of a simple braced frame must recognize the following points:

1. The members intersecting at a joint are pin connected.
2. The columns are not subjected to any direct moment transferred through the connection, but nominal moments due to eccentricity of the beam reaction forces should be considered. The design axial force in the column is predominately governed by floor loading and the tributary areas.
3. The structure is statically determinate. The internal forces and moments are therefore determined from a consideration of statics.
4. Gravity frames must be attached to a bracing system so as to provide lateral stability to the part of the structure resisting gravity load. The frame can be designed as a nonsway frame and the second-order moments associated with frame drift can be ignored.
5. The leaning column effects due to column sidesway must be considered in the design of the frames that participate in sidesway resistance.

Since the beams are designed as simply supported beams between their supports, the bending moments and shear forces are independent of beam size. Therefore, initial sizing of beams is a straightforward task. Beam or girder members supporting more than 40 m^2 of floor at one story should be designed for a reduced live load in accordance with ASCE (1990).

Most conventional types of floor slab construction will provide adequate lateral restraint to the compression flange of the beam. Consequently, the beams may be designed as laterally restrained beams without the moment resistance being reduced by lateral–torsional buckling.

Under service loading, the total central deflection of the beam or the deflection of the beam due to unfactored live load (with proper precambering for dead load) should satisfy the deflection limits as given in Table 22.2.

TABLE 22.2 Recommended Deflection Limits for Steel Building Frames

Beam deflections from unfactored imposed loads	
Beams carrying plaster or brittle finish	Span/360 (with maximum of $\frac{1}{4}$ to 1 in.)
Other beams	Span/240
Edge beams supporting cladding	Span/500
Beam deflection due to unfactored dead and imposed loads	
Internal beams with ceilings	Span/200
Edge beams supporting cladding	Span/350
Column deflections from unfactored imposed and wind loads	
Column in single-story frames	Height/300
Column in multistory frames	Height of story/300
Column supporting cladding that is sensitive to large movement	Height of story/500
Frame drift under 50 years wind load	
Frame drift	Frame height/450 to frame height/600

In some occasions, it may be necessary to check the dynamic sensitivity of the beams. When assessing the deflection and dynamic sensitivity of secondary beams, the deflection of the supporting beams must also be included. Whether it is the strength, deflection, or dynamic sensitivity that controls the design will depend on the span-to-depth ratio of the beam. Figure 22.18 gives typical span ranges for beams in office buildings for which the design would be optimized for strength and serviceability. For beams with their span lengths exceeding those shown in Figure 22.18, serviceability limits due to deflection and vibration will most likely be the governing criteria for design.

The required axial forces in the columns can be derived from the cumulative reaction forces from those beams that frame into the columns. Live load reduction should be considered in the design of columns in a multistory frame (ASCE 1990). If the frame is braced against sidesway, the column node points are prevent from lateral translation. A conservative estimate of column effective length, KL, for buckling considerations is $1.0L$, where L is the story height. However, in cases where the columns above and below the story under consideration are underutilized in terms of load resistance, the restraining effects offered by these members may result in an effective length of less than $1.0L$ for the column under consideration. Such a situation arises where the column is continuous through the restraint points and the columns above and below the restraint points are of different length.

An example of such cases is the continuous column shown in Figure 22.23 in which Column AB is longer than Column BC and hence Column AB is restrained by Column BC at the restraint point B. A buckling analysis shows that the critical buckling load for the continuous column is $P_{cr} = 5.89EI/L^2$, which gives rise to an effective length factor of $K = 0.862$ for Column AB and $K = 1.294$ for Column BC. Column BC has a larger effective length factor because it provides restraint to Column AB, whereas Column AB has a smaller effective length factor because it is restrained by column BC during buckling. Figure 22.24 summarizes the reductions in effective length that may be considered for columns in a frame with different story heights having various values of a/L ratios (Owens and Knowles 1992).

22.4.3.4 Simple Connections

Simple connections should be designed and detailed to allow free rotation and to prevent excessive transfer of moment between the beams and columns. Such connections should comply with the classification requirement for a "nominally pinned connection" in terms of both strength and stiffness. Computer program for connection classification has been made available in a book by Chen et al. (1996); their design implications for semirigid frames are discussed in Liew et al. (1993a).

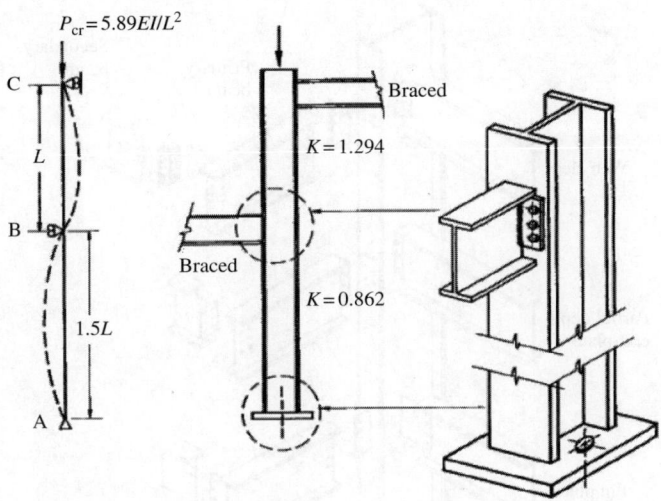

FIGURE 22.23 Buckling of a continuous column with intermediate restraint.

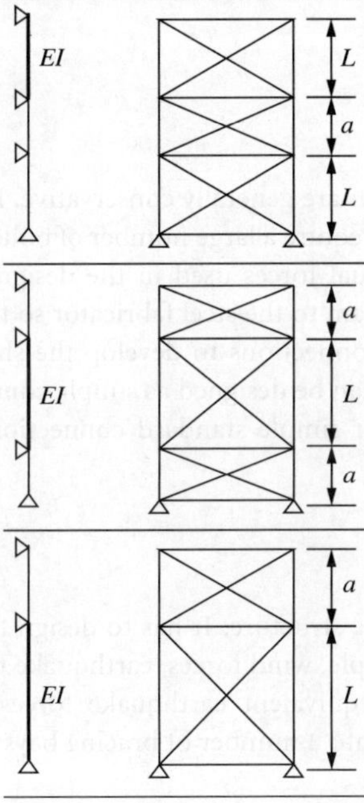

Column	Frame		a/L				
			0.2	0.4	0.6	0.8	1.0
EI		L	0.76	0.82	0.88	0.94	1.0
		a					
		L					
		a					
EI		L	0.57	0.65	0.75	0.87	1.0
		a					
EI		a	0.74	0.79	0.84	0.91	1.0
		L					

FIGURE 22.24 Effective length factors of continuous braced columns.

Simple connections are designed to resist vertical shear at the beam end. Depending on the connection details adopted, it may also be necessary to consider an additional bending moment resulting from the eccentricity of the bolt line from the supporting face. Often the fabricator is told to design connections based on the beam end reaction for one-half uniformed distributed load (UDL). Unless the concentrated

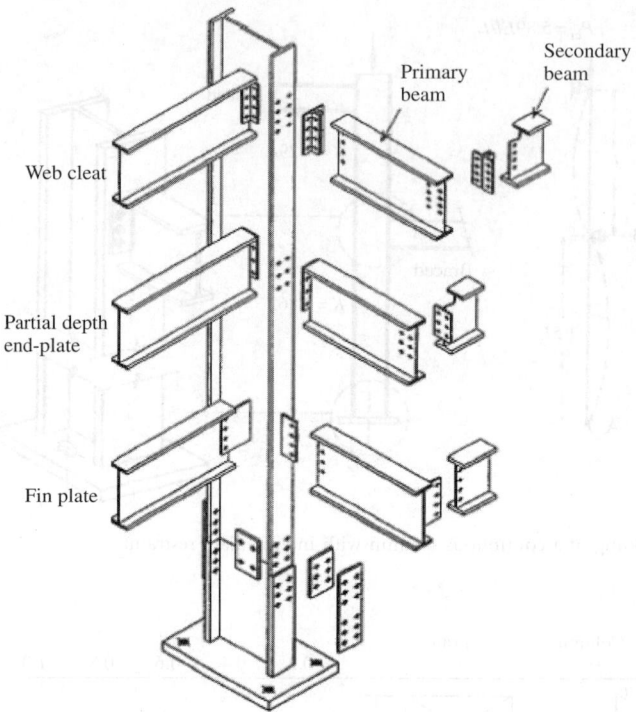

FIGURE 22.25 Simple beam-to-column connections.

load is located very near to the beam end, UDL reactions are generally conservative. Because of the large reaction, the connection becomes very strong and may require a large number of bolts. Thus, it would be a good practice to design the connections for the actual forces used in the design of the beam. The engineer should give the design shear force for every beam to the steel fabricator so that a more realistic connection can be designed, instead of requiring all connections to develop the shear capacity of the beam. Figure 22.25 shows the typical connections that can be designed as simple connections. When the beam reaction is known, capacity tables developed for simple standard connections can be used for detailing such connections (AISC 1990).

22.4.4 Bracing Systems

Bracing frames provide the lateral stability to the entire structure. It has to design to resist all possible kinds of lateral loading due to external forces, for example, wind forces, earthquake forces, and "leaning forces" from the gravity frames. The wind or the equivalent earthquake forces on the structure, whichever are greater, should be assessed and divided into a number of bracing bays resisting the lateral forces in each direction.

22.4.4.1 Structural Forms

Steel braced systems are often in a form of vertical truss that behaves like cantilever elements under lateral loads developing tension and compression in the column chords. Shear forces are resisted by the bracing members. The truss diagonalization may take various forms, as shown in Figure 22.26. The design of such structures must take into account the manner in which the frames are erected, the distribution of lateral forces, and their side sway resistance.

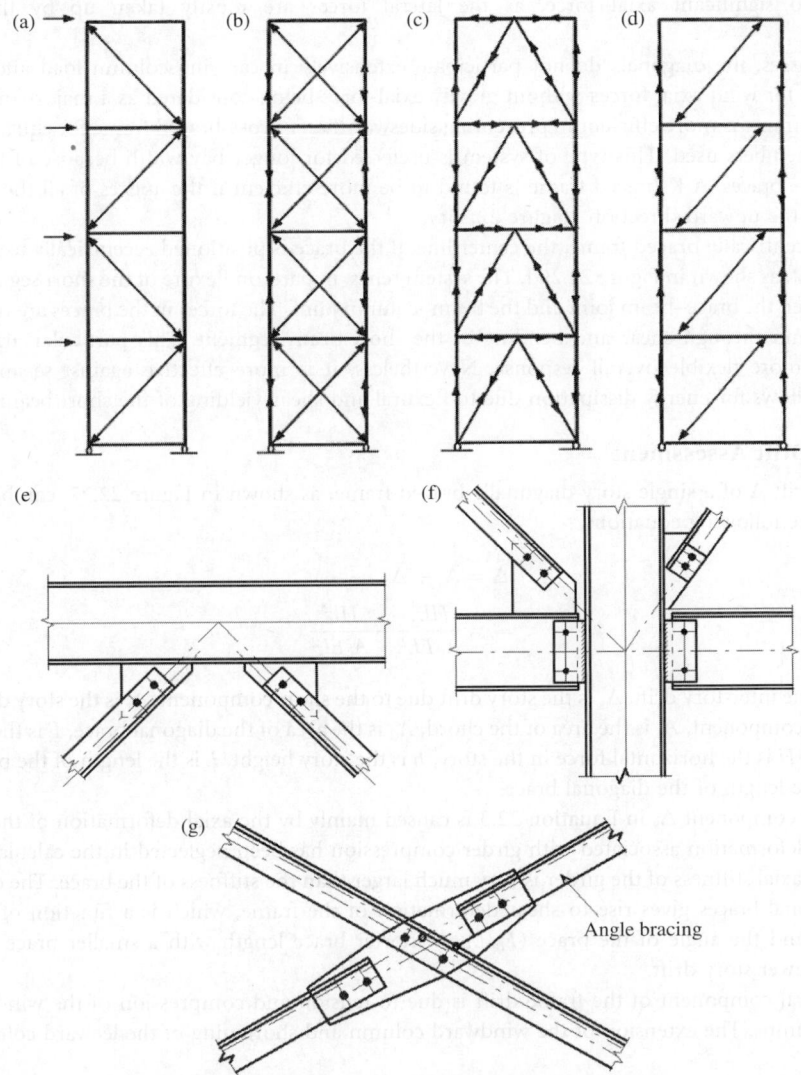

FIGURE 22.26 (a) Diagonal bracing, (b) cross-bracing, (c) K-bracing, (d) eccentric bracing, (e) braces connected to beam, (f) bracing connection to either beam or column, and (g) X-bracing connections.

In the single braced forms, where a single diagonal brace is used (Figure 22.26a), it must be capable of resisting both tensile and compressive axial forces caused by the alternate wind load. Hollow sections may be used for the diagonal braces as they are stronger in compression. In the design of diagonal braces, gravity forces may tend to dominate the axial forces in the members and due consideration must be given in the design of such members. It is recommended that the slenderness ratio of the bracing member (L/r) must not be greater than 200 to prevent the self-weight deflection of the brace limiting its compressive resistance.

In a cross-braced system (Figure 22.26b), the brace members are usually designed to resist tension only. Consequently, light sections such as structural angles and channels or tie rods can be used to provide a very stiff bracing. The advantage of the cross-braced system is that the beams are not subjected to significant axial force, as the lateral forces are mostly taken up by the bracing members.

For K-trusses, the diagonals do not participate extensively in carrying column load and can thus be designed for wind axial forces without gravity axial force being considered as a major contribution. A K-braced frame is more efficient in preventing sidesway than a cross-braced frame for equal steel areas of braced members used. This type of system is preferred for longer bay width because of the shorter length of the braces. A K-braced frame is found to be more efficient if the apexes of all the braces are pointing in the upward direction (Figure 22.26c).

For an eccentrically braced frame, the center line of the brace is positioned eccentrically to the beam–column joint, as shown in Figure 22.26d. The system relies, in part, on flexure of the short segment of the beam between the brace–beam joint and the beam–column joint. The forces in the braces are transmitted to the column through shear and bending of the short beam segment. This particular arrangement provides a more flexible overall response. Nevertheless, it is more effective against seismic loading because it allows for energy dissipation due to flexural and shear yielding of the short beam segment.

22.4.4.2 Drift Assessment

The story drift Δ of a single story diagonally braced frame, as shown in Figure 22.27, can be approximated by the following equation:

$$\Delta = \Delta_s + \Delta_f$$
$$= \frac{HL_d^3}{A_d EL^2} + \frac{Hh^3}{A_c EL^2} \tag{22.3}$$

where Δ is the interstory drift, Δ_s is the story drift due to the shear component, Δ_f is the story drift due to the flexural component, A_c is the area of the chord, A_d is the area of the diagonal brace, E is the modulus of elasticity, H is the horizontal force in the story, h is the story height, L is the length of the braced bay, and L_d is the length of the diagonal brace.

The shear component Δ_s in Equation 22.3 is caused mainly by the axial deformation of the diagonal brace. The deformation associated with girder compression has been neglected in the calculation of Δ_s because the axial stiffness of the girder is very much larger than the stiffness of the brace. The elongation of the diagonal braces gives rise to shear deformation of the frame, which is a function of the brace length, L_d, and the angle of the brace (L_d/L). A shorter brace length with a smaller brace angle will produce a lower story drift.

The flexural component of the frame drift is due to tension and compression of the windward and leeward columns. The extension of the windward column and shortening of the leeward column cause

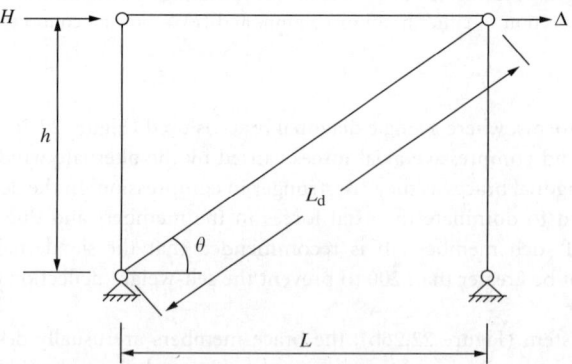

FIGURE 22.27 Lateral displacement of a diagonally braced frame.

flexural deformation of the frame, which is a function of the area of the column and the ratio of the height to bay length (h/L). For a slender bracing frame with large h/L ratio, the flexural component can contribute significantly to the overall story drift.

Low-rise braced frame deflects predominantly in shear mode while high-rise braced frames tend to deflect more in flexural mode.

22.4.4.3 Design Considerations

Frames with braces connecting columns may obstruct locations of access openings such as windows and doors, and so they should be placed where such access is not required, for example, around elevators, service, and stair wells. The location of the bracing systems within the structure will influence the efficiency with which the lateral forces can be resisted. The most appropriate position for the bracing systems is at the periphery of the building (Figure 22.28a) since this arrangement provides greater

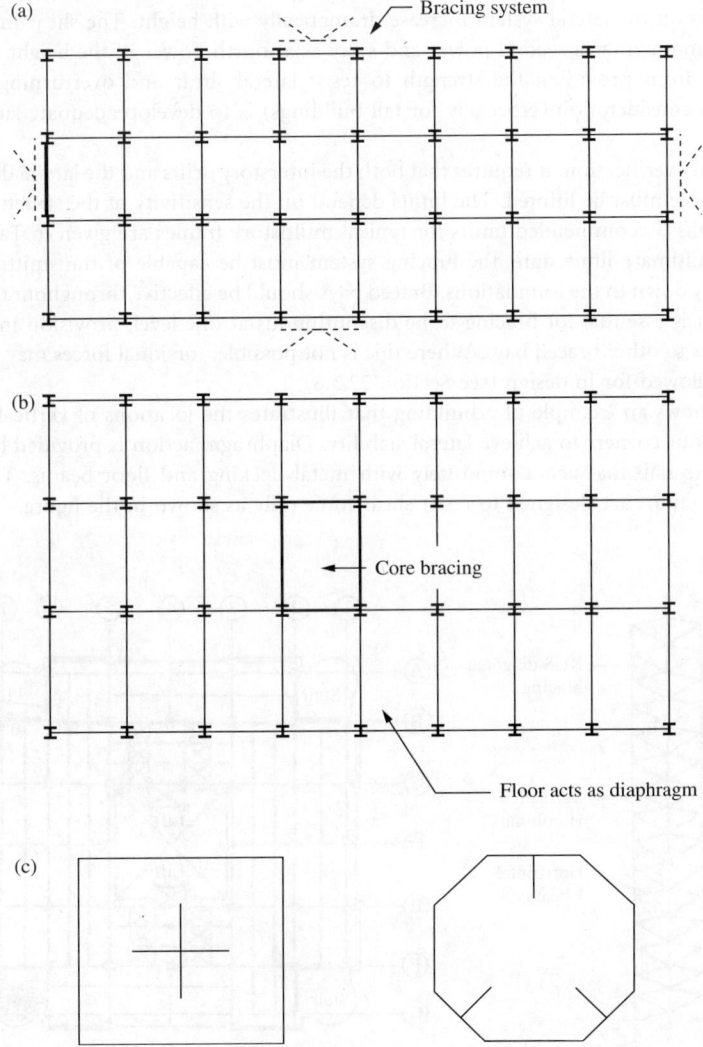

FIGURE 22.28 Location of bracing systems: (a) exterior braced frames, (b) internal braced core, and (c) bracing arrangements to be avoided.

torsional resistance. Bracing frames should be situated where the center of lateral resistance is approximately equal to the center of shear resultant on plan. Where this is not possible, torsional forces will be induced and they must be considered when calculating the load carried by each braced system.

When core braced systems are used, they are normally located in the center of the building (Figure 22.28b). The torsional stability is then provided by the torsional rigidity of the core brace. For tall building frames, a minimum of three braced bents are required to provide transitional and torsional stability. These bents should be carefully arranged so that their planes of action do not meet at one point so as to form a center of rotation. The bracing arrangement shown in Figure 22.28c should be avoided.

The flexibility of different bracing systems must be taken into account in the analysis, since the stiffer braces will attract a larger share of the applied lateral load. For tall and slender frames, the bracing system itself can be a sway frame and a second-order analysis is required to evaluate the required forces for ultimate strength and serviceability checks.

Lateral loads produce transverse shears, over turning moments and sidesway. The stiffness and strength demands on the lateral system increase dramatically with height. The shear increases linearly, the overturning moment as a second power and sway as a fourth power of the height of the building. Therefore, apart from providing the strength to resist lateral shear and overturning moments, the dominant design consideration (especially for tall buildings) is to develop adequate lateral stiffness to control sway.

For serviceability verification, it requires that both the interstory drifts and the lateral deflections of the structure as a whole must be limited. The limits depend on the sensitivity of the structural elements to shear deformations. Recommended limits for typical multistory frames are given in Table 22.2. When considering the ultimate limit state, the bracing system must be capable of transmitting the factored lateral loads safely down to the foundations. Braced bays should be effective throughout the full height of the building. If it is essential for bracing to be discontinuous at one level, provision must be made to transfer the forces to other braced bays. Where this is not possible, torsional forces may be induced and they should be allowed for in design (see Section 22.3.6).

Figure 22.29 shows an example of a building that illustrates the locations of vertical braced trusses provided at the four corners to achieve lateral stability. Diaphragm action is provided by a lightweight aggregate concrete slab that acts compositely with metal decking and floor beams. The floor beam-to-column connections are designed to resist shear force only as shown in the figure.

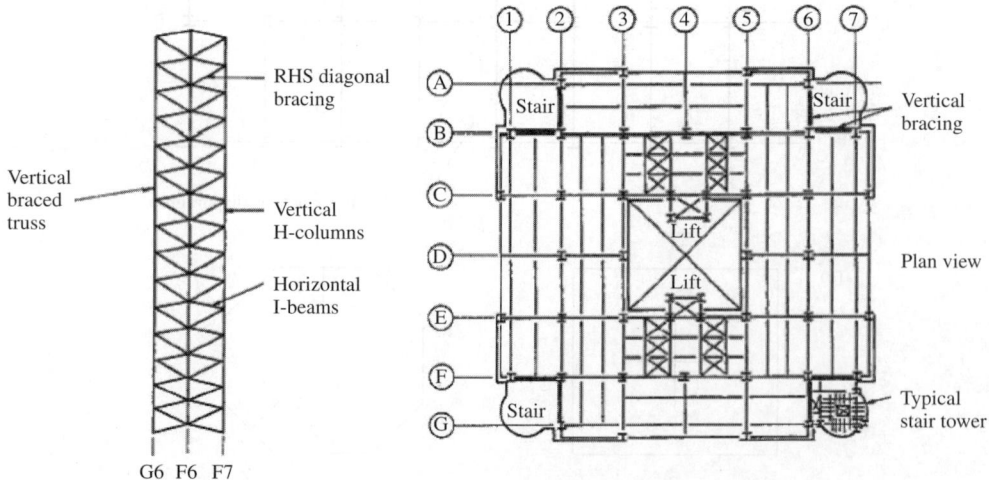

FIGURE 22.29 Simple building frame with vertical braced trusses located at the corners.

22.4.5 Moment-Resisting Frames

In cases where bracing systems would disturb the functioning of the building, rigidly jointed moment-resisting frames can be used to provide lateral stability to the building, as illustrated in Figure 22.30a. The efficiency of development of lateral stiffness depends on bay span, number of bays in the frame, number of frames, and the available depth in the floors for the frame girders. For buildings with heights not more than three times the plan dimension, the moment frame system is an efficient form. Bay dimensions in the range of 6 to 9 m and structural height up to 20 to 30 stories are commonly used. However, as the building height increases, deeper girders are required to control drift, thus the design becomes uneconomical.

When a rigid unbraced frame is subjected to lateral load, the horizontal shear in a story is resisted predominantly by the bending of columns and beams. These deformations cause the frame to deform in a shear mode. The design of these frames is controlled therefore by the bending stiffness of individual members. The deeper the member, the more efficiently the bending stiffness can be developed. A small part of the frame sidesway is caused by the overturning of the entire frame resulting in shortening and elongation of the columns at opposite sides of the frame. For unbraced rigid frames up to 20 to 30 stories, the overturning moment contributes for about 10 to 20% of the total sway, whereas shear racking accounts for the remaining 80 to 90% (Figure 22.30b). However, the story drift due to overall bending tends to increase with height, while that due to shear racking tends to decrease.

22.4.5.1 Drift Assessment

Since shear racking accounts for most of the lateral sway, the design of such frames should be directed toward minimizing the sidesway due to shear. The shear displacement Δ in a typical story in a multistory frame, as shown in Figure 22.31, can be approximated by the equation

$$\Delta_i = \frac{V_i h_i^2}{12E}\left(\frac{1}{\sum(I_{ci}/h_i)} + \frac{1}{\sum(I_{gi}/L_i)}\right) \tag{22.4}$$

where Δ_i is the shear deflection of the *i*th story, E is the modulus of elasticity, I_c, I_g are second moments of area for columns and girders, respectively, h_i is the height of the *i*th story, L_i is the length of girder in the *i*th story, V_i is the total horizontal shear force in the *i*th story, $\sum(I_{ci}/h_i)$ is the sum of the column stiffness in the *i*th story, and $\sum(I_{gi}/L_i)$ is the sum of the girder stiffness in the *i*th story.

Examination of Equation 22.4 shows that sidesway deflection caused by story shear is influenced by the sum of column and beam stiffnesses in a story. Since for multistory construction span lengths are generally larger than the story height, the moment of inertia of the girders needs to be larger to match the

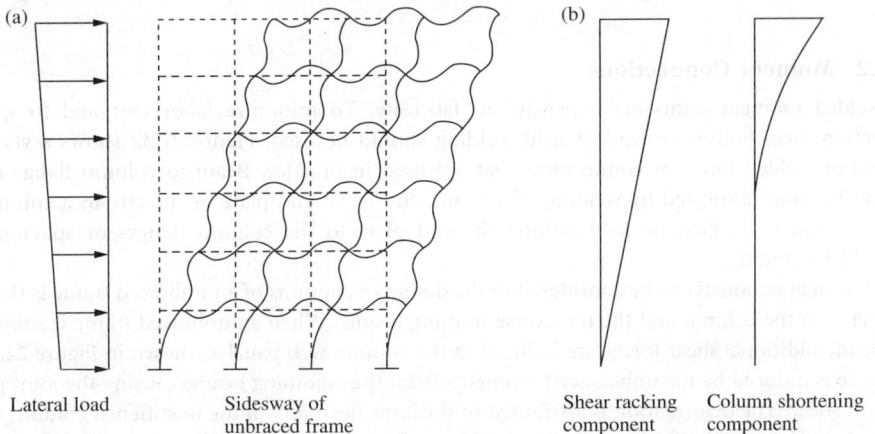

(a)

Lateral load   Sidesway of unbraced frame

(b)

Shear racking component   Column shortening component

FIGURE 22.30 Sidesway resistance of a rigid unbraced frame.

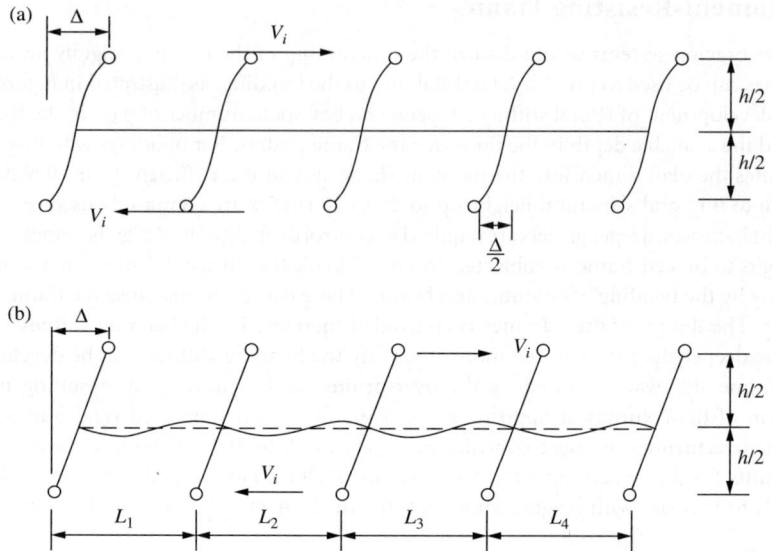

FIGURE 22.31 Story drift due to (a) bending of columns and (b) bending of girders.

column stiffness, as both of these members contribute equally to the story drift. As the beam span increases, considerably deeper beam sections will be required to control frame drift.

Since the gravity forces in columns are cumulative, larger column sizes are needed in lower stories as the frame height increases. Similarly, story shear forces are cumulative and, therefore, larger beam properties in lower stories are required to control lateral drift. Because of limitations in available depth, heavier beam members will need to be provided at lower floors. This is the major shortcoming of unbraced frames because considerable premium for steel weight is required to control lateral drift as building height increases.

Apart from the beam span, height-to-width ratios of the building play an important role in the design of such structures. Wider building frames allow a larger number of bays (i.e., larger values for story summation terms $\sum(I_{ci}/h_i)$ and $\sum(I_{gi}/L_i)$ in Equation 22.4 with consequent reduction in frame drift. Moment frames with closed spaced columns that are connected by deep beams are very effective in resisting sidesway. This kind of framing system is suitable for use in the exterior planes of the building.

22.4.5.2 Moment Connections

Fully welded moment joints are expensive to fabricate. To minimize labor cost and to speed up site erection, field bolting instead of field welding should be used. Figure 22.32 shows several types of bolted or welded moment connections that are used in practice. Beam-to-column flange connections can be shop-fabricated by welding of a beam stub to an end plate or directly to a column. The beam can then be erected by field bolting the end plate to the column flanges or splicing beams (Figure 22.32c and d).

An additional parameter to be considered in the design of columns of an unbraced frame is the "panel zone" between the column and the transverse framing beams. When an unbraced frame is subjected to lateral load, additional shear forces are induced in the column web panel as shown in Figure 22.33. The shear force is induced by the unbalanced moments from the adjoining beams causing the joint panel to deform in shear. The deformation is attributed to the large flexibility of the unstiffened column web. To prevent shear deformation so as to maintain the moment joint assumption as assumed in the global analysis, it may be necessary to stiffen the panel zone using either a doubler plate or a diagonal stiffener

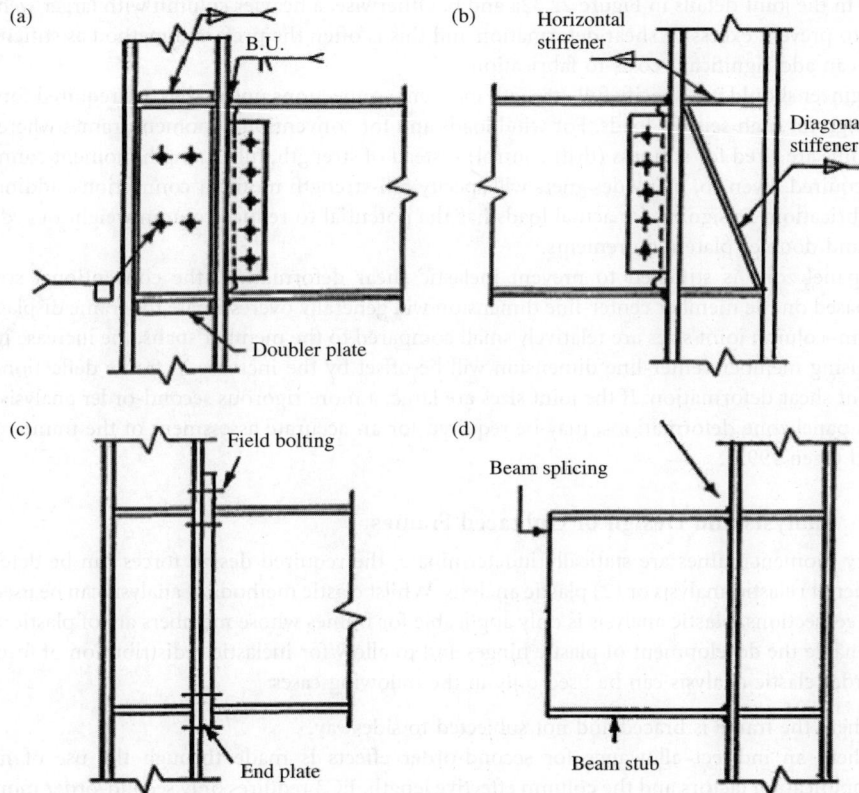

FIGURE 22.32 Rigid connections: (a) bolted and welded connection with doubler plate; (b) bolted and welded connection with diagonal stiffener; (c) bolted end-plate connection; and (d) beam-stub welded to column.

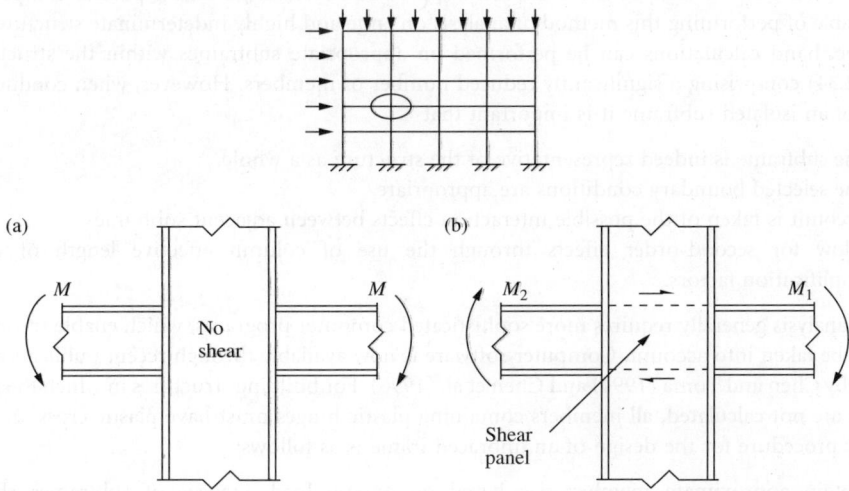

FIGURE 22.33 Force acting on a panel joint: (a) balanced moment due to gravity load and (b) unbalanced moment due to lateral load.

as shown in the joint details in Figure 22.32a and b. Otherwise, a heavier column with larger web area is required to prevent excessive shear deformation and this is often the preferred method as stiffeners and doublers can add significant costs to fabrication.

The engineer should not specify full-strength moment connections unless they are required for ductile frame design for high seismic loads. For wind loads and for conventional moment frames where beams and columns are sized for stiffness (drift control) instead of strength, full-strength moment connections are not required. Even so, many designers will specify full-strength moment connections, adding to the cost of fabrication. Designing for actual loads has the potential to reduce column weight or reduce the stiffener and doubler plate requirements.

If the panel zone is stiffened to prevent inelastic shear deformation, the conventional structural analysis based on the member center-line dimension will generally overestimate the frame displacement. If the beam–column joint sizes are relatively small compared to the member spans, the increase in frame stiffness using member center-line dimension will be offset by the increase in frame deflection due to panel–joint shear deformation. If the joint sizes are large, a more rigorous second-order analysis, which considers panel zone deformations, may be required for an accurate assessment of the frame response (Liew and Chen 1995).

22.4.5.3 Analysis and Design of Unbraced Frames

Multistory moment frames are statically indeterminate, the required design forces can be determined using either (1) elastic analysis or (2) plastic analysis. Whilst elastic methods of analysis can be used for all kind of steel sections, plastic analysis is only applicable for frames whose members are of plastic sections so as to enable the development of plastic hinges and to allow for inelastic redistribution of forces.

First-order elastic analysis can be used only in the following cases:

1. Where the frame is braced and not subjected to sidesway.
2. Where an indirect allowance for second-order effects is made through the use of moment amplification factors and the column effective length. EC3 requires only second-order moment or effective length factor to be used in the beam–column capacity checks. However, column and frame imperfections need to be modeled explicitly in the analysis. In AISC LRFD (1993), both factors need to be computed for checking the member strength and stability and the analysis is based on structures without initial imperfections.

The first-order elastic analysis is a convenient approach. Most design offices possess computer software capable of performing this method of analysis on large and highly indeterminate structures. As an alternative, hand calculations can be performed on appropriate subframes within the structure (see Figure 22.34) comprising a significantly reduced number of members. However, when conducting the analysis of an isolated subframe it is important that

1. The subframe is indeed representative of the structure as a whole.
2. The selected boundary conditions are appropriate.
3. Account is taken of the possible interaction effects between adjacent subframes.
4. Allow for second-order effects through the use of column effective length or moment amplification factors.

Plastic analysis generally requires more sophisticated computer programs, which enable second-order effects to be taken into account. Computer software is now available through recent publications made available by Chen and Toma (1994) and Chen et al. (1996). For building structures in which the required rotations are not calculated, all members containing plastic hinges must have plastic cross-sections.

A basic procedure for the design of an unbraced frame is as follows:

1. Obtain approximate member size based on gravity load analysis of subframes shown in Figure 22.34. If sidesway deflection is likely to control (e.g., slender frames) use Equation 22.4 to estimate the member sizes.

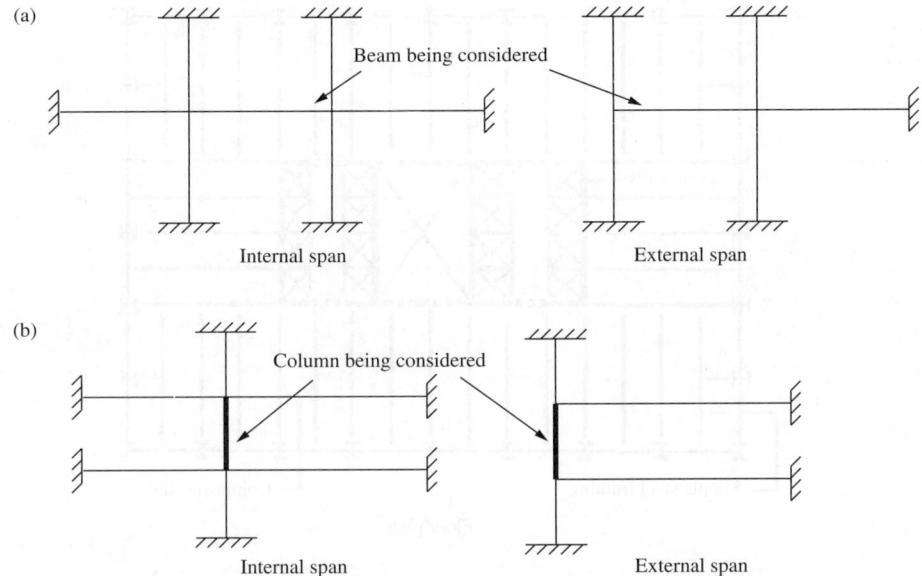

FIGURE 22.34 Subframe analysis for gravity loads: (a) subframe for beam and (b) subframe for column design.

2. Determine wind moments from the analysis of the entire frame subjected to lateral load. A simple portal wind analysis may be used in lieu of the computer analysis.
3. Check member capacity for the combined effects of factored lateral load plus gravity loads.
4. Check beam deflection and frame drift.
5. Redesign the members and perform final analysis/design check (a second-order elastic analysis is preferable at the final stage).

The need to repeat the analysis to correspond to changed section sizes is unavoidable for highly redundant frames. Iteration of Steps 1 to 5 gives results that will converge to an economical design satisfying the various design constraints imposed on the analysis.

22.4.6 Multistory Building Framing Systems

The following subsections discuss four classical systems that have been adopted for tall building constructions, namely (1) core braced system, (2) moment–truss system, (3) outriggle and belt system, and (4) tube system. Tall frames that utilize cantilever action will have higher efficiencies, but the overall structural efficiency depends on the height-to-width ratio. Interactive systems involving moment frame and vertical truss or core are effective up to 40 stories and represent most building forms for tall structures. Outrigger truss and belt truss help to further enhance the lateral stiffness by engaging the exterior frames with the core braces to develop cantilever actions. Exterior framed tube systems with closely spaced exterior columns connected by deep girders mobilize the three-dimensional action to resist lateral and torsional forces. Bundled tubes improve the efficiency of exterior frame tubes by providing internal stiffening to the exterior tube concept. Finally, by providing diagonal braces to the exterior framework, a superframe is formed and can be used for ultratall megastructures.

22.4.6.1 Core Braced Systems

This type of structural system relies entirely on the internal core for lateral load resistance. The basic concept is to provide an internal shear wall core to resist the lateral forces (Figure 22.35). The surround

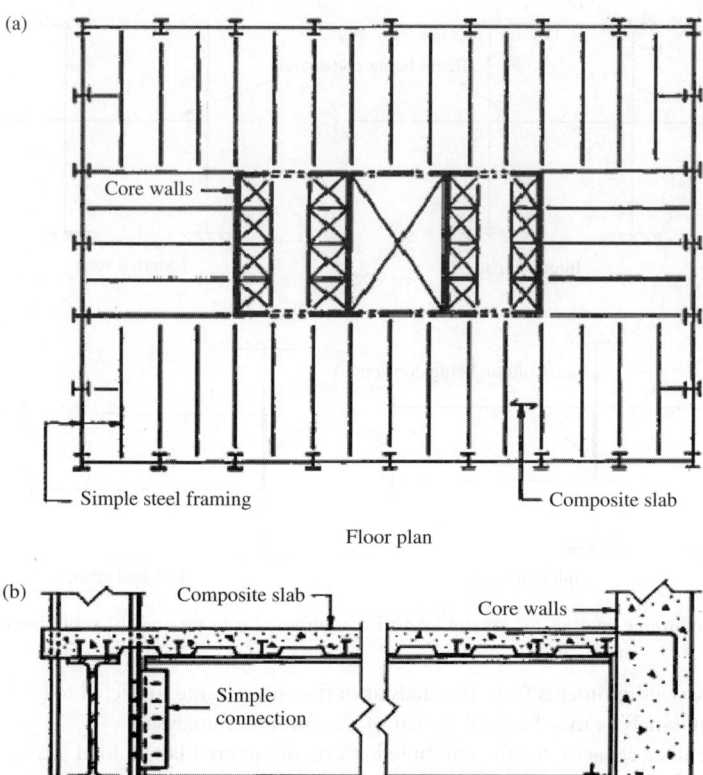

FIGURE 22.35   Core braced frame: (a) internal core walls with simple exterior framing and (b) beam-to-wall and beam-to-exterior column connections.

steel framing is designed to carry gravity load only if simple framing is adopted. Otherwise, a rigid framing surrounding the core will enhance the overall lateral-force resistance of the structure. The steel beams can be simply connected to the core walls using a typical corbel detail, by bearing in a wall pocket, or by a shear plate embedded in the core wall through studs. If a rigid connection is required, the steel beams should be rigidly connected to steel columns embedded in the core wall. Rigid framing surrounding the cores are particularly useful in high seismic areas and for very tall buildings that tend to attract stronger wind loads. They act as moment frames and provide resistance to some part of the lateral loads by engaging the core walls in the building.

The core generally provides all torsional and flexural rigidity and strength with no participation from the steel system. Conceptually, the core system should be treated as a cantilever wall system with punched openings for access. The floor framing should be arranged in such a way that it distributes enough gravity loads to the core walls so that their design is controlled by compressive stresses even under wind loads. The geometric location of the core should be selected so as to minimize eccentricities for lateral load. The core walls need to have adequate torsional resistance for possible asymmetry of the core system where the center of the resultant shear load is acting at an eccentricity from the center of the lateral-force resistance.

(a)    (b)

FIGURE 22.36   A steel building braced by two external core walls with large open lobby at the bottom: (a) during construction and (b) after construction.

A simple cantilever model should be adequate to analyze a core-wall structure. However, if the structural form is a tube with openings for access, it may be necessary to perform a more accurate analysis to include the effect of openings. The walls can be analyzed by a finite element analysis using thin-walled plate elements. An analysis of this type may also be required to evaluate torsional stresses when the vertical profile of the core-wall assembly is asymmetrical.

The concrete core walls can be constructed using slip-form techniques, where the core walls could be advanced several floors (typically four to six stories) ahead of the steel framing (see Figure 22.36a). A core-wall system represents an efficient type of structural system up to certain height premium because of its cantilever action. However, when it is used alone, the massiveness of the wall structure increases with height, thereby inhabiting the free planning of interior spaces, especially in the core. The space occupied by the shear walls leads to loss of overall floor area efficiency, as compared to the tube system which could otherwise be used.

In commercial buildings where floor space is valuable, the large area taken up by a concrete column can be reduced by the use of an embedded steel column to resist the extreme loads encountered in tall building. Sometimes, particularly at the bottom open floors of a high-rise structure where large open lobbies or atriums are utilized as part of the architectural design (see Figure 22.36a and b), a heavy embedded steel section as part of a composite column is necessary to resist high load and due to the long unbraced length. A heavy steel section in a composite column is often utilized where the column size is restricted architecturally and where reinforcing steel percentages would otherwise exceed the maximum code allowed values for the design of reinforced concrete columns.

22.4.6.2 Frame–Truss Systems

Vertical shear trusses located around the inner core in one or both directions can be combined with perimeter moment-resisting frames in the facade of a building to form an efficient structure for lateral load resistance. An example of a building consisting of moment frames with shear trusses located at the center of the building is shown in Figure 22.37a. For the vertical trusses arranged in the north–south direction, either the K- or X-form of bracing is acceptable since access to lift shafts is not required. However, K-trusses are often preferred because in the case of X or single brace form bracings the influence of gravity loads is rather significant. In the east–west direction, only the knee bracing is effective in resisting lateral load.

In some cases internal bracing can be provided using concrete shear walls as shown in Figure 22.37b. The internal core walls substitute the steel trusses in K, X, or single brace form, which may interfere with openings that provide access to, for example, elevators.

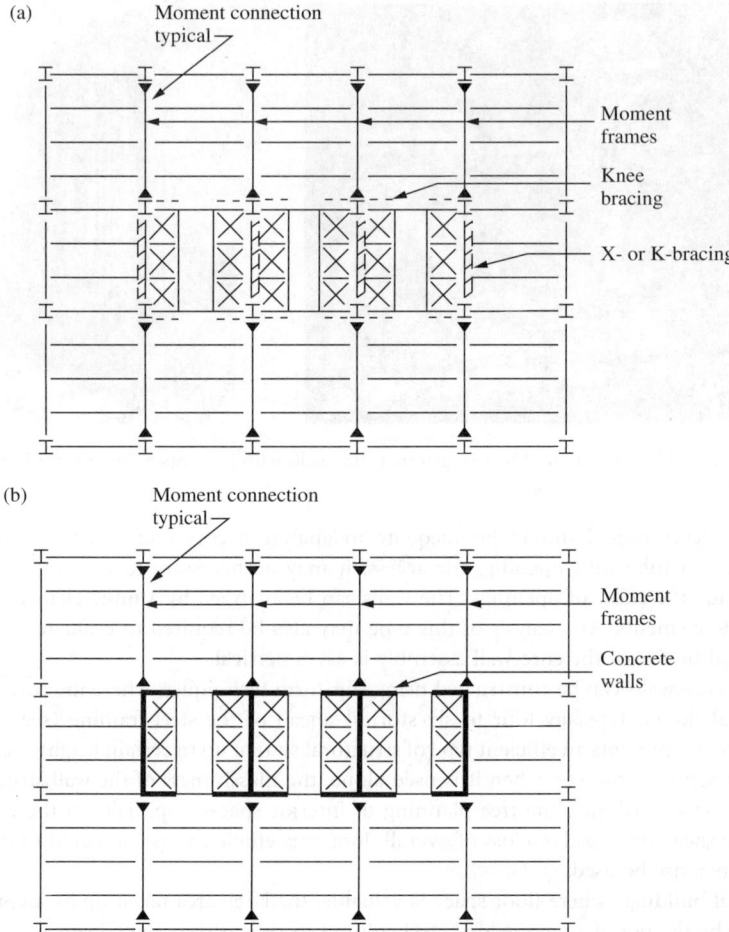

FIGURE 22.37   (a) Moment frames with internal braced trusses. (b) Moment frames with internal core walls.

The interaction of shear frames and vertical trusses produces a combination of two deflection curves with the effect of more efficient stiffness. These moment frame–truss interacting systems are considered to be the most economical steel systems for buildings up to 40 stories. Figure 22.38 compares the sway characteristic of a 20-story steel frame subjected to same lateral forces, but with different structural schemes namely (1) unbraced moment frame, (2) simple-truss frame, and (3) moment–truss frame. The simple-truss frame helps to control lateral drift at the lower stories, but the overall frame drift increases toward the top of the frame. The moment frame, on the other hand, shows an opposite characteristic for sidesway in comparison with the simple braced frame. The combination of moment frame and truss frame provides overall improvement in reducing frame drift; the benefit becomes more pronounce toward the top of the frame. The braced truss is restrained by the moment frame at the upper part of the building, while at the lower part the moment frame is restrained by the truss frame. This is because the slope of frame sway displacement is relatively smaller than that of the truss at the top while the proportion is reversed at the bottom. The interacting forces between the truss frame and moment frame, as shown in Figure 22.39, enhance the combined moment–truss frame stiffness to a level larger than the summation of individual moment frame and truss stiffnesses.

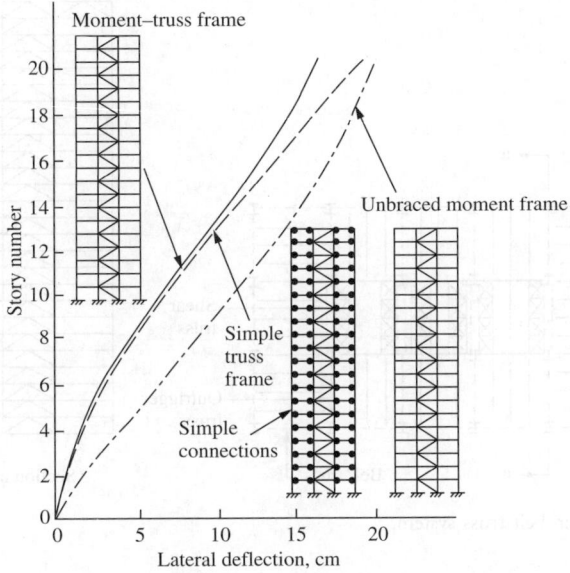

FIGURE 22.38 Sway characteristics of rigid braced frame, simple braced frame, and rigid unbraced frame.

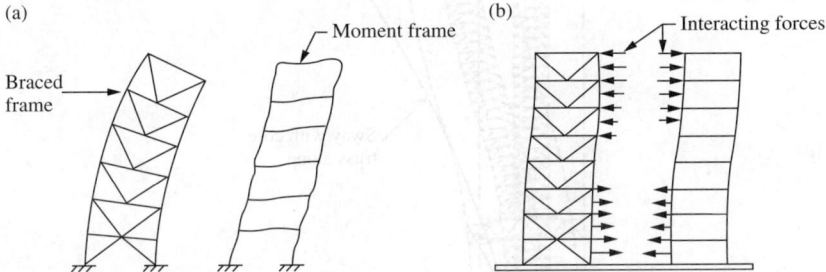

FIGURE 22.39 Behavior of frames subjected to lateral load: (a) independent behavior and (b) interactive behavior.

22.4.6.3 Outrigger and Belt Truss Systems

Another significant improvement of lateral stiffness can be obtained if the vertical truss and the perimeter shear frame are connected on one or more levels by a system of outrigger and belt trusses. Figure 22.40 shows a typical example of such a system. The outrigger truss leads the wind forces of the core truss to the exterior columns providing cantilever behavior of the total frame system. The belt truss in the facade improves the cantilever participation of the exterior frame and creates a three-dimensional frame behavior.

Figure 22.41 shows a schematic diagram that demonstrates the sway characteristic of the overall building under lateral load. Deflection is significantly reduced by the introduction of the outrigger–belt trusses. Two kinds of stiffening effects can be observed: one is related to the participation of the external columns together with the internal core to act in a cantilever mode and the other is related to the stiffening of the external facade frame by the belt truss to act as a three-dimensional tube. The overall stiffness can be increased up to 25% as compared to the shear truss and frame system without such outrigger–belt trusses.

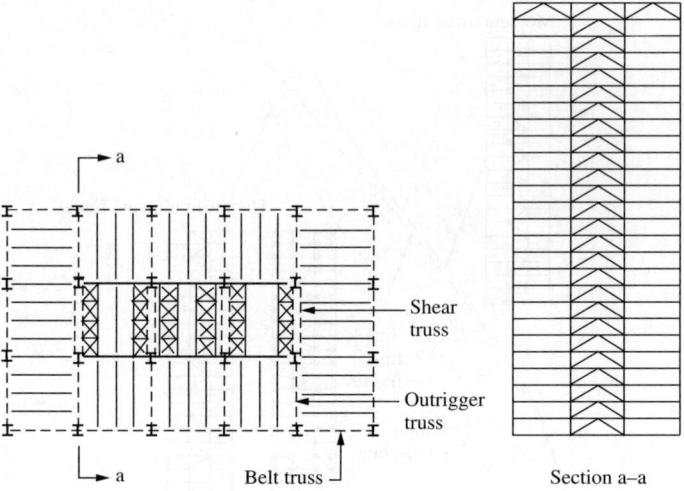

FIGURE 22.40   Outrigger–belt truss system.

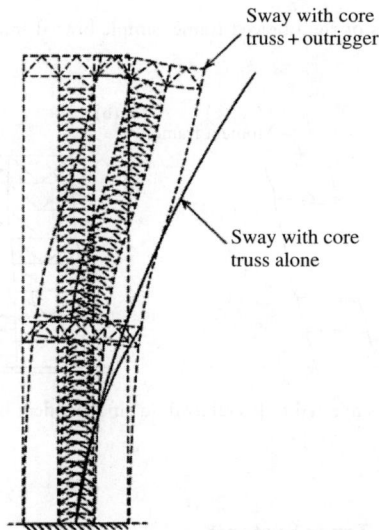

FIGURE 22.41   Improvement of lateral stiffness using outrigger–belt truss system.

The efficiency of this system is related to the number of trussed levels and the depth of the truss. In some cases the outrigger and belt trusses have a depth of two or more floors. They are located in services floors where there are no requirements for wide open spaces. These trusses are often pleasingly integrated into the architectural conception of the facade.

22.4.6.4 Frame Tube Systems

Figure 22.42 shows a typical frame tube system, which consists of a frame tube at the exterior of the building and gravity steel framing at the interior. The framed tube is constructed from wide columns placed at close centers connected by deep beams creating a punched wall appearance. The exterior frame tube structure resists all lateral loads of wind or earthquake whereas the gravity steel framing in the interior resists only its share of gravity loads. The behavior of the exterior frame tube is similar to a

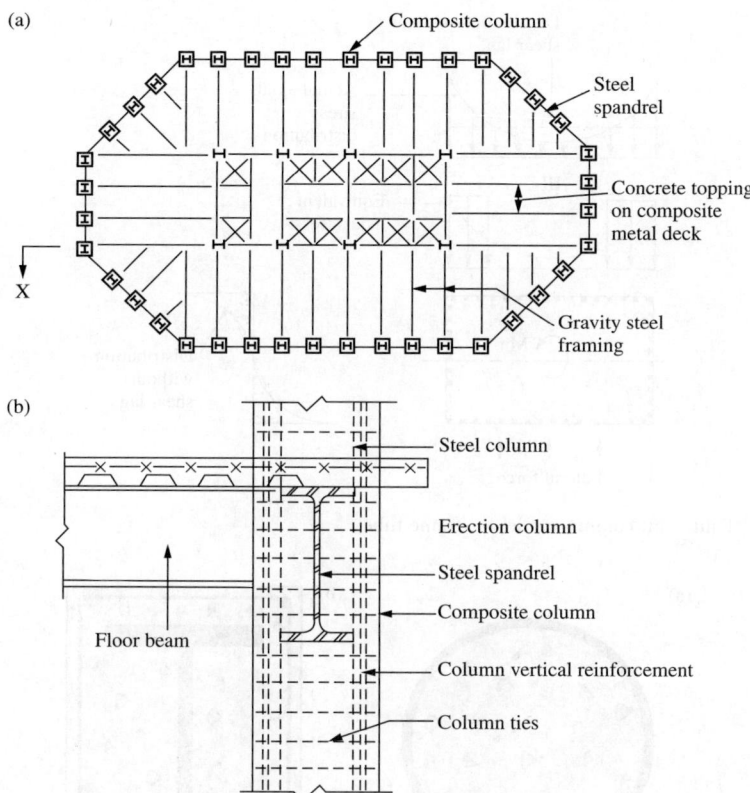

FIGURE 22.42 Composite tubular system: (a) plan view and (b) beam-to-column connection.

hollow perforated tube. The overturning moment under the action of lateral load is resisted by compression and tension of the leeward and windward columns, which are called the flange columns. The shear is resisted by bending of the columns and beams at the two sides of the building parallel to the direction of the lateral load, which are called the web frames.

Deepening on the shear rigidity of the frame tube, there may exist a shear lag across the windward and leeward sides of the tube. As a result of this, not all the flange columns resist the same amount of axial force. An approximate approach is to assume an equivalent column model as shown in Figure 22.43. In the calculation of the lateral deflection of the frame tube it is assumed that only the equivalent flange columns on the windward and leeward sides of the tube and the web frames would contribute to the moment of inertia of the tube.

The use of exterior framed tube has the following distinct advantages: (1) It develops high rigidity and strength for torsional and lateral-load resistance, since the structural components are effectively placed at the exterior of the building forming a three-dimensional closed section. (2) Massiveness of the frame tube system eliminates potential uplift difficulties and produces better dynamic behavior. (3) The use of gravity steel framing in the interior has the advantages of flexibility and enables rapid construction. If a composite floor with metal decking is used, electrical and mechanical services can be incorporated in the floor zone.

Composite columns are frequently used in the perimeter of the building where the closely spaced columns work in conjunction with the spandrel beam (either steel or concrete) to form a

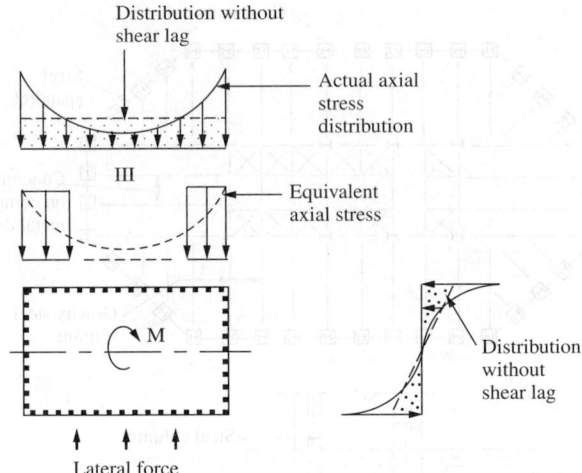

FIGURE 22.43   Equivalent column model for frame tube.

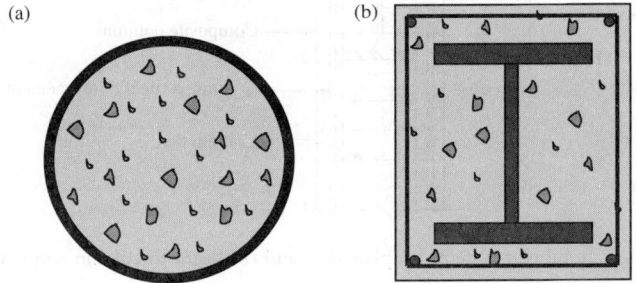

FIGURE 22.44   (a) Encased composite section. (b) Concrete-filled steel column.

three-dimensional cantilever tube rather than an assembly of two-dimensional plane frames. The exterior frame tube significantly enhances the structural efficiency in resisting lateral loads and thus reduces the shear wall requirements. However, in cases where a higher magnitude of lateral stiffness is required (such as for vary tall buildings), internal wall cores and interior columns with floor framing can be added to transform the system into a tube-in-tube system. The concrete core may be strategically located to recapture elevator space and to provide transmission of mechanical ducts from shafts and mechanical rooms.

22.4.7   Steel–Concrete Composite Systems

Steel–concrete composite construction has gained wide acceptance as an alternative to pure steel and pure concrete construction. Composite building systems can be broadly categorized into two forms: one utilizes the core braced system by means of interior shear walls and the other utilizes exterior framing to form a tube for lateral load resistance. Combining these two structural forms will enable taller buildings to be constructed.

22.4.7.1   Composite Column

Composite columns have been used for many decades, with steel-encased sections similar that shown in Figure 22.44a being incorporated in multistory buildings in the United States during the late 19th century. The initial application of composite columns was for fire rating requirements of the steel section. Later developments saw the composite action fully utilized for strength and stability. Composite action in columns utilizes the favorable tensile and compressive characteristics of the steel and concrete, respectively. These types of columns are still in use today where steel sections are used as erection columns, with reinforced concrete cast around them. One major benefit of this system has been the ability to achieve higher steel percentages than conventional reinforced concrete structures, and the steel erection column allows rapid construction of steel floor systems in steel-framed buildings.

Concrete-filled steel columns, as illustrated in Figure 22.44b, were developed much later during the last century but are still based on the fundamental principle that steel and concrete are most effective in tension and compression, respectively. The major benefits also include constructibility issues, whereby the steel section acts as a permanent and integral formwork for the concrete. These sections were essentially expensive, as the steel section was designed to be hollow, thus requiring large steel plate thickness. This lack of constructional economy has seen the use of concrete-filled steel columns limited in their application through the world. Furthermore, restrictive cross-section sizes have rendered them unsuitable for application in tall buildings where demand on axial strength is high.

22.4.7.2 Composite Connection

For composite frames resisting gravity load only, the beam-to-column connections behave as pinned before the placement of concrete. During construction, the beam is designed to resist concrete dead load and the construction load (to be treated as temporary live load). At the composite stage, the composite strength and stiffness of the beam should be utilized to resist the full design loads. For gravity frames consisting of bare steel columns and composite beams, there is now sufficient knowledge available for the designer to use composite action in the structural element as well as the semirigid composite joints to increase design choices, leading to more economical solutions (Leon and Ammerman 1990; Eurocode 4 1994; Leon 1994).

Figure 22.45a and b shows the typical beam-to-column connections, one using flushed end-plate bolted to the column flange and the other using bottom angle with double web cleats. Composite action in the joint is developed based on the tensile forces developed in the rebars acting with the balancing compression forces transmitted by the lower portion of the steel section that bear against the column flange to form a couple. Properly designed and detailed composite connections are capable of providing moment resistance up to the hogging resistance of the connecting members.

In designing the connections, slab reinforcements placed within a horizontal distance of six times the slab depth are assumed to be effective in resisting the hogging moment. Reinforcement steels that fall outside this width should not be considered in calculating the resisting moment of the connection (see Figure 22.46). The connections to edge columns should be carefully detailed to ensure adequate anchorage of rebars. Otherwise they shall be designed and detailed as simply supported. In a braced

(a)   (b)

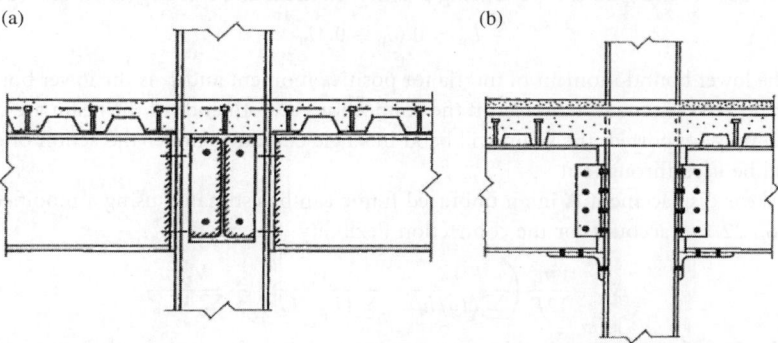

FIGURE 22.45 Composite beam-to-column connections with: (a) flush end plate and (b) seat and double web angles.

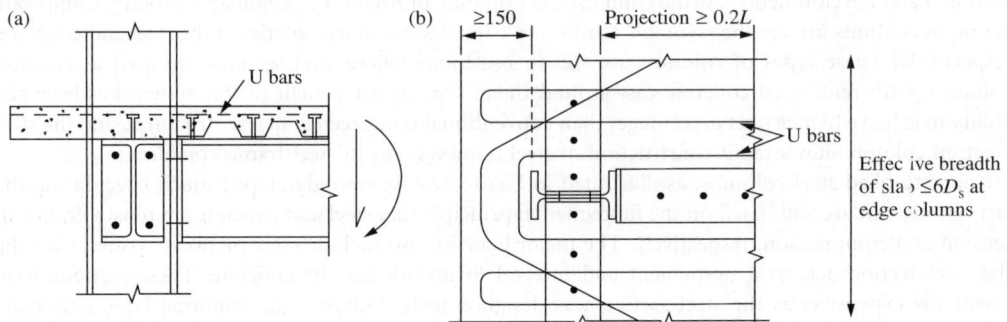

FIGURE 22.46 Moment transfer through reinforcement at perimeter columns: (a) connection details and (b) reinforcement detail.

frame a moment connection to the exterior column will increase the moments in the column, resulting in an increase of column size. Although the moment connections restrain the column from buckling by reducing the effective length, this is generally not adequate to offset the strength required to resist this moment.

22.4.7.3 Unbraced Composite Frames

For unbraced frame subjected to gravity and lateral loads, the beam typically bends in double curvature with negative moment at one end of the beam and positive moment on the other end. The concrete is assumed to be ineffective in tension; therefore, only the steel beam stiffness on the negative moment region and the composite stiffness on the positive moment region can be utilized for frame action. The frame analysis can be performed with variable moment of inertia for the beams (see Figure 22.47). Further research is still needed in order to provide tangible guidance for design.

If semirigid composite joints are used in unbraced frames, the flexibility of the connections will contribute to additional drift over that of a fully rigid frame. In general, semirigid connections do not require the column size to be increased significantly over an equivalent rigid frame. This is because the design of frames with semirigid composite joints takes advantage of the additional stiffness in the beams provided by the composite action. The increase in beam stiffness would partially offset the additional flexibility introduced by the semirigid connections.

The moment of inertia of the composite beam I_{cp} may be estimated using a weighted average of moment of inertia in the positive moment region (I_p) and negative moment regions (I_n). For interior spans, approximately 60% of the span is experiencing positive moment and it is suggested that (Leon 1990)

$$I_{cp} = 0.6I_p + 0.4I_n \tag{22.5}$$

where I_p is the lower bound moment of inertia for positive moment and I_n is the lower bound moment of inertia for negative moment. However, if the connections at both ends of the beam are designed and detailed for simply supported, the beam will bend in single curvature under the action of gravity loads and I_p should be used throughout.

The story shear displacement Δ in an unbraced frame can be estimated using a modified expression from Equation 22.4 to account for the connection flexibility

$$\Delta_i = \frac{V_i h_i^2}{12E}\left(\frac{1}{\sum(I_{ci}/h_i)} + \frac{1}{\sum(I_{cpi}/L_i)}\right) + \frac{V_i h_i^2}{\sum K_{con}} \tag{22.6}$$

where Δ_i is the shear deflection of the ith story, E is the modulus of elasticity, I_c is the moment of inertia for columns, I_g is the moment of inertial of a composite girder based on the weighted average method, h_i is the height of the ith story, L_i is the length of girder in the ith story, V_i is the total horizontal shear force

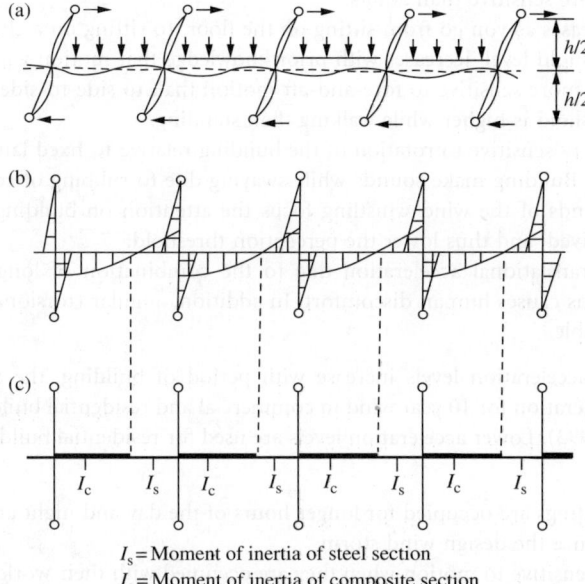

I_s = Moment of inertia of steel section
I_c = Moment of inertia of composite section

FIGURE 22.47 Composite unbraced frames: (a) story loads and idealization, (b) bending moment diagrams, and (c) composite beam stiffness.

in the ith story, $\sum(I_{ci}/h_i)$ is the sum of the column stiffnesses in the ith story, $\sum(I_{gi}/L_i)$ is the sum of the girder stiffnesses in the ith story, and $\sum K_{con}$ is the sum of the connection rotational stiffness in the ith story.

Further research is required to assess the performance of various types of composite connections used in building construction. Issues related to accurate modeling of effective stiffness of composite members and joints in unbraced frames for the computation of second-order effects and drifts need to be addressed.

22.5 Wind Effects on Buildings

22.5.1 Introduction

With the development of lightweight high-strength materials, the recent trend is to build tall and slender buildings. The design of such buildings in nonseismic areas is often governed by the need to limit the wind-induced accelerations and drift to acceptable levels for human comfort and integrity of nonstructural components, respectively. Thus, to check for serviceability of tall buildings, the peak resultant horizontal acceleration and displacement due to the combination of along wind, across wind, and torsional loads are required. As an approximate estimation, the peak effects due to along wind, across wind, and torsional responses may be determined individually and then combined vectorally. A reduction factor of 0.8 may be used on the combined value to account for the fact that in general the individual peaks do not occur simultaneously. If the calculated combined effect is less than any of the individual effects, then the latter should be considered for the design.

The effects of acceleration on human comfort are given in Table 22.3. The factors affecting the human response are

1. Period of building — tolerence to acceleration tends to increase with period.
2. Women are more sensitive than men.

3. Children are more sensitive than adults.
4. Perception increases as you go from sitting on the floor, to sitting on a chair, to standing.
5. Perception threshold level decreases with prior knowledge that motion will occur.
6. Human body is more sensitive to fore-and-aft motion than to side-to-side motion.
7. Perception threshold is higher while walking than standing.
8. Visual cue — very sensitive to rotation of the building relative to fixed landmarks outside.
9. Acoustic cue — Building make sounds while swaying due to rubbing of contact surfaces. These sounds and sounds of the wind whistling focus the attention on building motion even before motion is perceived, and thus lower the perception threshold.
10. The resultant translational acceleration due to the combination of longitudinal, lateral, and torsional motions causes human discomfort. In addition, angular (torsional) motion appears to be more noticeable.

Since the tolerable acceleration levels increase with period of building, the recommended design standard for peak acceleration for 10 year wind in commercial and residential buildings is as depicted in Figure 22.48 (Griffis 1993). Lower acceleration levels are used for residential buildings for the following reasons:

1. Residential buildings are occupied for longer hours of the day and night and are therefore more likely to experience the design wind storm.
2. People are less sensitive to motion when they are occupied with their work than when they relax at home.
3. People are more tolerant of their work environment than of their home environment.
4. Occupancy turnover rates are higher in commercial buildings than in residential buildings.
5. People can be easily evacuated from commercial buildings than residential buildings in the event of a peak storm.

TABLE 22.3 Acceleration Limits for Different Perception Levels

Perception	Acceleration limits
Imperceptible	$a < 0.005g$
Perceptible	$0.005g < a < 0.015g$
Annoying	$0.015g < a < 0.05g$
Very annoying	$0.05g < a < 0.15g$
Intolerable	$a > 0.15g$

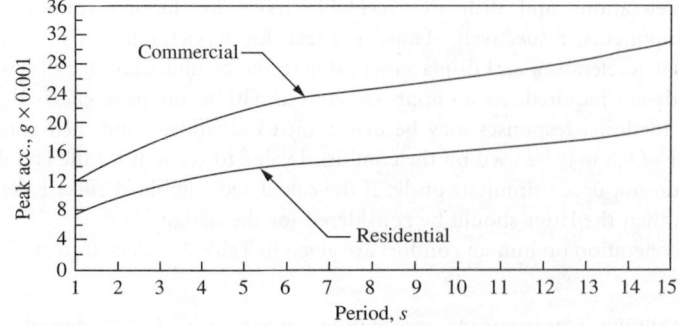

FIGURE 22.48 Design standard on peak acceleration for a 10-year return period.

TABLE 22.4 Serviceability Problems at Various Deflection or Drift Indices

Deformation as a fraction of span or height	Visibility of deformation	Typical behavior
$\frac{1}{500}$	Not visible	Cracking of partition walls
$\frac{1}{300}$	Visible	General architectural damage
		Cracking in reinforced walls
		Cracking in secondary members
		Damage to ceiling and flooring
		Facade damage
		Cladding leakage
		Visual annoyance
$\frac{1}{200}$ $\frac{1}{300}$	Visible	Improper drainage
$\frac{1}{100}$ $\frac{1}{200}$	Visible	Damage to lightweight partitions, windows, finishes
		Impaired operation of removable components such as doors, windows, sliding partitions

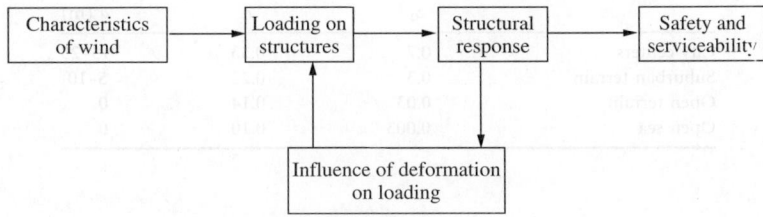

FIGURE 22.49 Schematic diagram for wind-resistant design of structures.

The effects of excessive deflection on building components is described in Table 22.4. Thus, the allowable drift, defined as the resultant peak displacement at the top of the building divided by the height of the building, is generally taken to be in the range $\frac{1}{450}$ to $\frac{1}{600}$.

Figure 22.49 depicts schematically the procedure of estimating the wind-induced accelerations and displacements in a building. The steps involved in this design procedure are described below with numerical examples for situations where the motion of the building does not affect the loads acting on the building. Finally, the situations when a wind tunnel studies is required is listed at the end of this section.

22.5.2 Characteristics of Wind

22.5.2.1 Mean Wind Speed

The velocity of wind (wind speed) at great heights above the ground is constant and is called the gradient wind speed. As shown in Figure 22.50, closer to the ground surface, the wind speed is affected by frictional forces caused by the terrain and thus there is a boundary layer within which the wind speed varies from zero to the gradient wind speed. The thickness of the boundary layer (gradient height) depends on the ground roughness. For example, the gradient height is 457 m for large cities, 366 m for suburbs, 274 m for open terrain, and 213 m for open sea.

The velocity of wind averaged over 1 h is called the hourly mean wind speed. The mean wind velocity profile within the atmospheric boundary layer is described by a power law:

$$\bar{U}(z) = \bar{U}(z_{\text{ref}})\left(\frac{z}{z_{\text{ref}}}\right)^{\alpha} \tag{22.7}$$

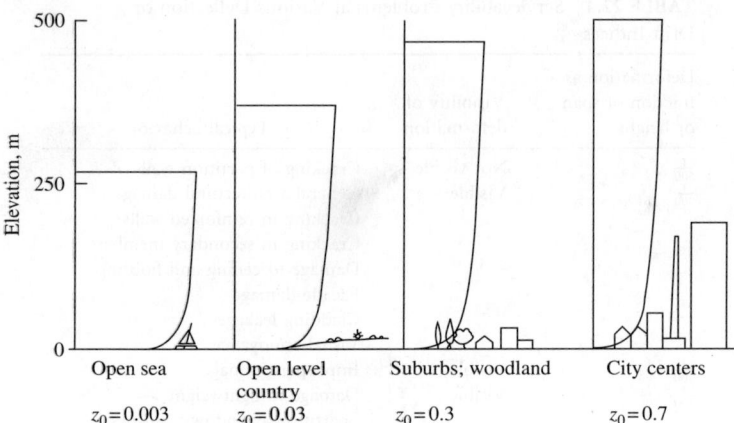

FIGURE 22.50   Mean wind profiles for different terrains.

TABLE 22.5   Typical Values of Terrain

	z_0	a	d (m)
City centers	0.7	0.33	15–25
Suburban terrain	0.3	0.22	5–10
Open terrain	0.03	0.14	0
Open sea	0.003	0.10	0

in which $\bar{U}(z)$ is the mean wind speed at height z above the ground, z_{ref} is the reference height normally taken to be 10 m, and α is the power law exponent.

An alternative description of the mean wind velocity is by the logarithmic law:

$$\bar{U}(z) = \frac{1}{k} u_* \ln\left(\frac{z - d}{z_0}\right) \tag{22.8}$$

in which u_* is the friction velocity, k is von Karmon's constant equal to 0.4, z_0 is the roughness length, and d is the height of zero-plane above the ground where the velocity is zero. Generally, zero plane is about 1 or 2 m below the average height of buildings and trees providing the roughness. Typical values of α, z_0, and d are given in Table 22.5 (ANSI 1982; ESDU 1985).

The roughness affects both the thickness of the boundary layer and the power law exponent. The thickness of the boundary layer and the power law exponent increase with the roughness of the surface. Consequently, the velocity at any height decreases as the surface roughness increases. However, the gradient velocity will be the same for all surfaces. Thus, if the velocity of wind for a particular terrain is known, using Equation 22.7 and Table 22.5 the velocity at some other terrain can be computed.

22.5.2.2 Turbulence

The variation of wind velocity with time is shown in Figure 22.51. The eddies generated by the action of wind blowing over obstacles cause the turbulence. In general, the velocity of wind may be represented in a vector form as

$$U(z, t) = \bar{U}(z)\underline{i} + u(z, t)\underline{i} + v(z, t)\underline{j} + w(z, t)\underline{k} \tag{22.9}$$

where u, v, and w are the fluctuating components of the gust in x, y, z (longitudinal, lateral, and vertical axes) as shown in Figure 22.52 and $\bar{U}(z)$ is the mean wind along the x-axis. The fluctuating component

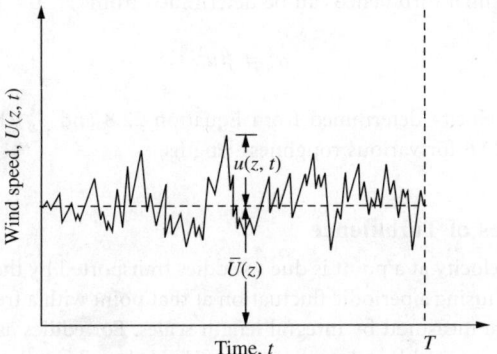

FIGURE 22.51   Variation of longitudinal component of tubular wind with time.

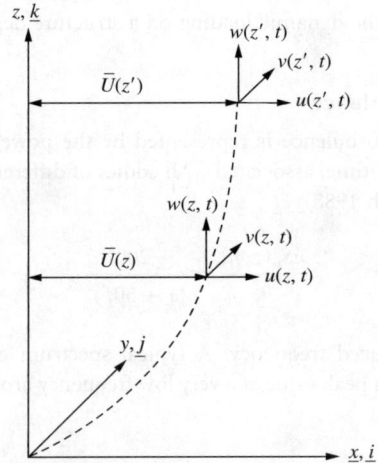

FIGURE 22.52   Velocity components of turbulent wind.

along the mean wind direction, u, is the largest and is therefore the most important for the vertical structures such as tall buildings, which are flexible along the wind direction. The vertical component w is important for horizontal structures, which are flexible vertically such as long-span bridges.

An overall measure of the intensity of turbulence is given by the root mean square value. Thus, for the longitudinal component of the turbulence

$$\sigma_u(z) = \left[\frac{1}{T_0} \int_0^{T_0} \{u(z, t)^2\} \, dt \right]^{1/2} \tag{22.10}$$

where T_0 is the averaging period. For the statistical properties of the wind to be independent on which part of the record is being used, T_0 is taken to be 1 h. Thus, the fluctuating wind is a stationary random function, over 1 h.

The value of $\sigma_u(z)$ divided by the mean velocity $\bar{U}(z)$ is called the turbulence intensity

$$I_u(z) = \frac{\sigma_u(z)}{\bar{U}(z)} \tag{22.11}$$

which increases with ground roughness and decreases with height.

The variance of longitudinal turbulence can be determined from

$$\sigma_u^2 = \beta u_*^2 \tag{22.12}$$

where u_* is the friction velocity determined from Equation 22.8 and β, which is independent of the height, is given in Table 22.6 for various roughness lengths.

22.5.2.3 Integral Scales of Turbulence

The fluctuation of wind velocity at a point is due to eddies transported by the mean wind $\bar{U}$. Each eddy may be considered to be causing a periodic fluctuation at that point with a frequency n. The average size of the turbulent eddies are measured by integral length scales. For eddies associated with longitudinal velocity fluctuation u, the integral length scales are L_u^x, L_u^y, L_u^z describing the size of the eddies in longitudinal, lateral, and vertical directions, respectively. If L_u^y and L_u^z are comparable to the dimension of the structure normal to the wind, then the eddies will envelope the structure and give rise to well-correlated pressures and thus the effect is significant. On the other hand, if L_u^y and L_u^z are small, then the eddies produce uncorrelated pressures at various parts of the structure and the overall effect of the longitudinal turbulence will be small. Thus, the dynamic loading on a structure depends on the size of eddies.

22.5.2.4 Spectrum of Turbulence

The frequency content of the turbulence is represented by the power spectrum, which indicates the power or kinetic energy per unit time, associated with eddies of different frequencies. An expression for the power spectrum is (Taranath 1988)

$$\frac{nS_u(z, n)}{u_*^2} = \frac{200f}{(1 + 50f)^{5/3}} \tag{22.13}$$

where $f = nz/\bar{U}(z)$ is the reduced frequency. A typical spectrum of wind turbulence is shown in Figure 22.53. The spectrum has a peak value at a very low frequency around 0.04 Hz. As the typical range

TABLE 22.6 Values of β for Various Roughness Lengths

z_0 (m)	0.005	0.7	0.30	1.0	2.5
b	6.5	6.0	5.25	4.85	4.0

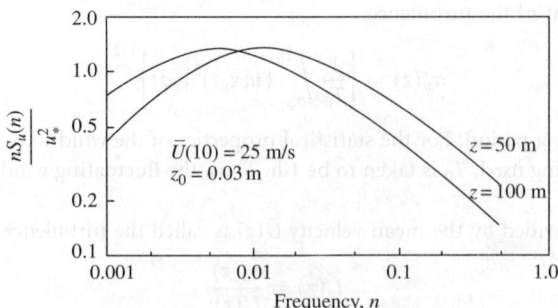

FIGURE 22.53 Power spectrum of longitudinal turbulence.

for the fundamental frequency of a tall building is 0.1 to 1 Hz, the buildings are affected by high-frequency small eddies characterizing the descending part of the power spectrum.

22.5.2.5 Cross-Spectrum of Turbulence

The cross-spectrum of two continuous records is a measure of the degree to which the two records are correlated. If the records are taken at two points M_1 and M_2 separated by a distance r, then the cross-spectrum of longitudinal turbulent component is defined as

$$S_{u_1 u_2}(r, n) = S^c_{u_1 u_2}(r, n) + i S^q_{u_1 u_2}(r, n) \tag{22.14}$$

where the real and imaginary parts of the cross-spectrum are known as the cospectrum and the quadrature spectrum, respectively. However, the latter is small enough to be neglected. Thus, the cospectrum may be expressed nondimensionally as the coherence and is given by

$$\gamma^2(r, n) = \frac{[S_{u_1 u_2}(r, n)]^2}{S_{u_1}(n) S_{u_2}(n)} \tag{22.15}$$

where $S_{u_1}(n)$ and $S_{u_2}(n)$ are the longitudinal velocity spectra at M_1 and M_2 respectively.

The square root of the coherence is given by the following expression (Davenport 1968):

$$\gamma(r, n) = e^{-\hat{f}} \tag{22.16}$$

where

$$\hat{f} = \frac{n[c_z^2(z_1 - z_2)^2 + c_y^2(y_1 - y_2)^2]^{1/2}}{\left(\frac{1}{2}\right)[\bar{U}(z_1) + \bar{U}(z_2)]} \tag{22.17}$$

in which y_1, z_1 and y_2, z_2 are the coordinates of points M_1 and M_2. The line joining M_1 and M_2 is assumed to be perpendicular to the direction of the mean wind. The suggested values of c_y and c_z for engineering calculation are 16 and 10, respectively (Vickery 1970).

22.5.3 Wind-Induced Dynamic Forces

22.5.3.1 Forces Due to Uniform Flow

A bluff body in a two-dimensional flow, as shown in Figure 22.54, is subjected to a nett force in the direction of flow (drag force) and a force perpendicular to the flow (lift force). Furthermore, when the

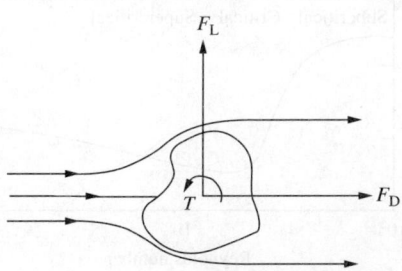

FIGURE 22.54 Drag and lift forces and torsional moment on a bluff body.

resultant force is eccentric to the elastic center, the body will be subjected to torsional moment. For uniform flow these forces and moment per unit height of the object are determined from

$$F_D = \tfrac{1}{2}\rho C_D B \bar{U}^2 \qquad\qquad (22.18)$$

$$F_L = \tfrac{1}{2}\rho C_L B \bar{U}^2 \qquad\qquad (22.19)$$

$$T = \tfrac{1}{2}\rho C_T B^2 \bar{U}^2 \qquad\qquad (22.20)$$

where $\bar{U}$ is the mean velocity of the wind, ρ is the density of air, C_D and C_L are the drag and lift coefficients, C_T is the moment coefficient, and B is the characteristic length of the object such as the projected length normal to the flow.

The drag coefficient for a rectangular building in plan is shown in Figure 22.55 for various depth-to-breadth ratios (ASCE 1987). The shear layers originating from the separation points at the windward corners surround a region known as the wake. For elongated sections, the stream lines that separate at the windward corners reattach to the body to form a narrower wake. This attributes to the reduction in the drag for larger aspect ratios. For cylindrical buildings in plan the drag coefficient is dependent on the Reynolds number as indicated in Figure 22.56.

Unlike the drag force, the lift force and torsional moment do not have a mean value for a symmetric object with a symmetric flow around it, as the symmetrical distribution of mean forces acting in the across wind direction cannot cause a nett force. If the direction of wind is not parallel to the axes of

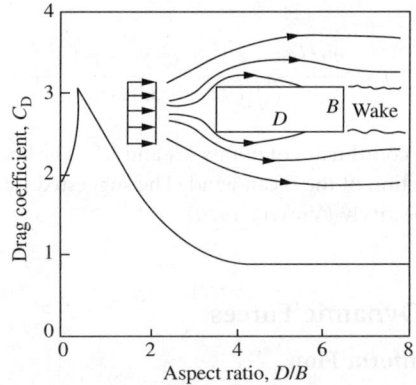

FIGURE 22.55   Drag coefficient for a rectangular section with different aspect ratios.

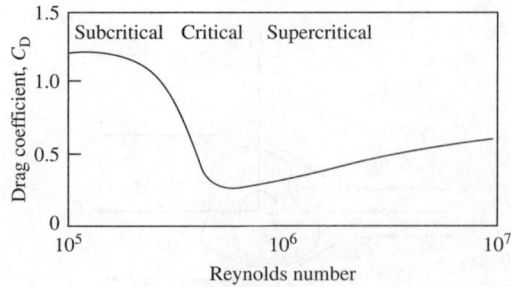

FIGURE 22.56   Effects of Reynolds number on drag coefficient of a circular cylinder.

symmetries or if the object is asymmetrical, then there will be a mean lift force and torsional moment. However, due to vortex shedding, fluctuating lift force and torsional moment will be present in both symmetric and nonsymmetric structures. Figure 22.57 shows the mechanism of vortex shedding. Near the separation zones, strong shear stresses impart rotational motions to the fluid particles. Thus, discrete vortices are produced in the separation layers. These vortices are shed alternatively from the sides of the object. The asymmetric pressure distribution created by the vortices around the cross-section leads to an alternating transverse force (lift force) on the object. The vortex shedding frequency in Hz, n_s, is related to a nondimensional parameter called the Strouhal number, S, defined as

$$S = \frac{n_s B}{\bar{U}} \qquad (22.21)$$

where $\bar{U}$ is the mean wind speed and B is the width of the object normal to the wind. For objects with rounded profiles such as circular cylinders, the Strouhal number varies with Reynolds number "Re" defined as

$$Re = \frac{\rho \bar{U} B}{\mu} \qquad (22.22)$$

where ρ is the density of air and μ is the dynamic viscosity of the air. The vortex shedding becomes random in the transition region of $4 \times 10^5 < Re < 3 \times 10^6$ where the boundary layer at the surface of the cylinder changes from laminar to turbulent. Outside this transition range, the vortex shedding is regular producing a periodic lift force. For cross-sections with sharp corners, the Strouhal number is independent of the Reynolds number. The variation of Strouhal number with length-to-breadth ratio of a rectangular cross-section is shown in Figure 22.58.

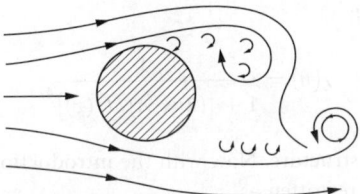

FIGURE 22.57 Vortex shedding in the wake of a bluff body.

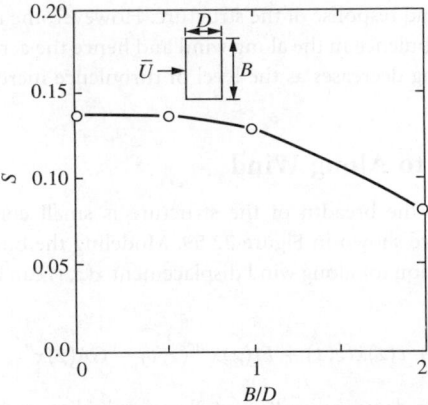

FIGURE 22.58 Strouhal number for a rectangular section.

22.5.3.2 Forces Due to Turbulent Flow

If the wind is turbulent, then the velocity of the wind along the wind direction is described as follows:

$$U(t) = \bar{U} + u(t) \tag{22.23}$$

where $\bar{U}$ is the mean wind and $u(t)$ is the turbulent component along the wind direction. The time dependent drag force per unit height is obtained from Equation 22.18 by replacing $\bar{U}$ by $U(t)$. As the ratio $u(t)/\bar{U}$ is small, the time-dependent drag force can be expressed as

$$f_D(t) = \bar{f}_D + f_D'(t) \tag{22.24}$$

where $\bar{f}_D$ and f_D' are the mean and the fluctuating parts of the drag force per unit height, which are given by

$$\bar{f}_D = \tfrac{1}{2}\rho \bar{U}^2 C_D B \tag{22.25}$$

$$f_D' = \rho \bar{U} u C_D B \tag{22.26}$$

The spectral density of the fluctuating part of the drag force is obtained from the Fourier transformation of the autocorrelation function as

$$S_{fD}(n) = \rho^2 \bar{U}^2 B^2 C_D^2 S_u(n) \tag{22.27}$$

where $S_u(n)$ is the spectral density of the turbulent velocity and may be obtained from Equation 22.13.

In practice, the presence of the structure distorts the turbulent flow, particularly the small high-frequency eddies. A correction factor known as the aerodynamic admittance function $\chi(n)$ may be introduced (Davenport 1961) to account for these effects. The following empirical formula has been suggested for $\chi(n)$ (Vickery 1970):

$$\chi(n) = \frac{1}{1 + \left[(2n\sqrt{A})/\bar{U}(z)\right]^{4/3}} \tag{22.28}$$

where A is the frontal area of the structure. Now with the introduction of the aerodynamic admittance function, Equation 22.27 may be rewritten as

$$S_{fD}(n) = \rho^2 \bar{U}^2 B^2 C_D^2 \chi^2(n) S_u(n) \tag{22.29}$$

It is evident from Equation 22.26 that the fluctuating drag force varies linearly with the turbulence. Thus, large integral length scale and high turbulent intensities will cause strong buffeting and consequently increases the along wind response of the structure. However, the regularity of vortex shedding is affected by the presence of turbulence in the along wind and hence the across wind motion and torsional motion due to vortex shedding decreases as the level of turbulence increases.

22.5.4 Response Due to Along Wind

Tall slender buildings, where the breadth of the structure is small compared to the height, can be idealized as a line-like structure shown in Figure 22.59. Modeling the building as a continuous system, the governing equation of motion for along wind displacement $x(z, t)$ can be written as (Heidebrecht and Smith 1973):

$$m(z)\ddot{x}(z, t) + c(z)\dot{x}(z, t) + EI(z)x''''(z, t) - GA(z)x''(z, t) = f(z, t) \tag{22.30}$$

where m, c, EI, GA are the mass, damping coefficient, flexural rigidity, and shear rigidity, per unit height, respectively. Furthermore, $f(z, t)$ is the fluctuating wind load per unit height given in Equation 22.26.

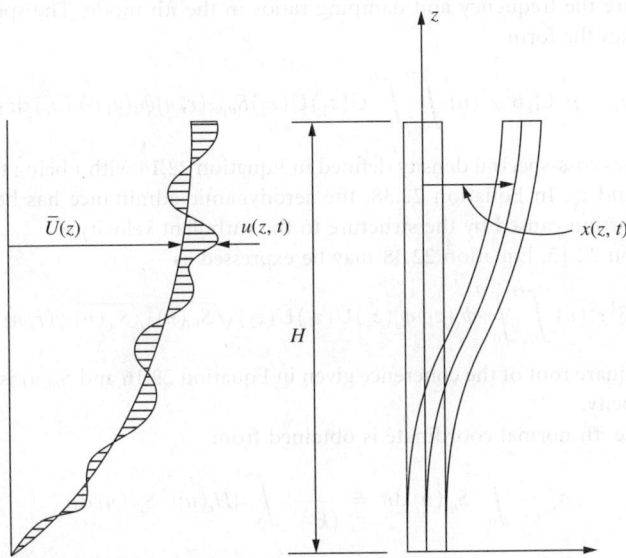

FIGURE 22.59 Typical deflection mode of a shear wall-frame building.

Expressing the displacement in terms of the normal coordinates

$$x(z, t) = \sum_{i=1}^{N} \phi_i(z) q_i(t) \tag{22.31}$$

where ϕ_i is the ith vibration mode shape and q_i is the ith normal coordinate. Using the orthogonality conditions of mode shapes, Equation 22.30 can be expressed as (Balendra 1993)

$$m_i^* \ddot{q}_i + c_i^* \dot{q}_i + k_i^* q_i = p_i^*, \quad i = 1 \text{ to } N \tag{22.32}$$

where m_i^*, c_i^*, k_i^*, and p_i^* are the generalized mass, damping, stiffness, and force in the ith mode of vibration. The generalized mass and force are determined from

$$m_i^* = \int_0^H m(z) \phi_i^2(z) \, dz \tag{22.33}$$

$$p_i^* = \int_0^H f(z, t) \phi_i(z) \, dz$$

$$= \rho C_D B \int_0^H \bar{U}(z) u(z, t) \phi_i(z) \, dz \tag{22.34}$$

Equation 22.32 consists of a set of uncoupled equations, each representing a single degree of freedom system. Using the random vibration theory (Robson 1963), the power spectrum of the response in each normal coordinate is given by

$$S_{q_i}(n) = |H_i(n)|^2 S_{p_i^*}(n) \frac{1}{(k_i^*)^2} \tag{22.35}$$

where

$$|H_i(n)| = \frac{1}{\left(\left[1 - (n/n_i)^2 \right]^2 + 4\zeta_i^2 (n/n_i)^2 \right)^{1/2}} \tag{22.36}$$

and

$$k_i^* = 4\pi^2 n_i^2 m_i^* \tag{22.37}$$

in which n_i and ζ_i are the frequency and damping ratios in the ith mode. The spectral density of the generalized force takes the form

$$S_{p_i^*}(n) = \rho^2 C_D^2 B^2 \chi^2(n) \int_0^H \int_0^H \bar{U}(z_1)\bar{U}(z_2) S_{u_1 u_2}(r, n) \phi_i(z_1)\phi_i(z_2) \, dz_1 \, dz_2 \qquad (22.38)$$

where $S_{u_1 u_2}(r, n)$ is the cross-spectral density defined in Equation 22.14 with r being the distance between the coordinates z_1 and z_2. In Equation 22.38, the aerodynamic admittance has been incorporated to account for the distortion caused by the structure to the turbulent velocity.

In view of Equation 22.15, Equation 22.38 may be expressed as

$$S_{p_i^*}(n) = \rho^2 C_D^2 B^2 \chi^2(n) \int_0^H \int_0^H \phi_i(z_1)\phi_i(z_2)\bar{U}(z_1)\bar{U}(z_2)\sqrt{S_{u_1}(n)}\sqrt{S_{u_2}(n)} \, \gamma(r, n) \, dz_1 \, dz_2 \qquad (22.39)$$

where $\gamma(r, n)$ is the square root of the coherence given in Equation 22.16 and $S_u(n)$ is the spectral density of the turbulent velocity.

The variance of the ith normal coordinate is obtained from

$$\sigma_{q_i}^2 = \int_0^\infty S_{q_i}(n) \, dn = \frac{1}{(k_i^*)^2} \int_0^\infty |H_i(n)|^2 S_{p_i^*}(n) \, dn \qquad (22.40)$$

The calculation of the above integral is very much simplified by observing the plot of the two components of the integrant shown in Figure 22.60. The mechanical admittance function is either 1.0 or zero for most of the frequency range. However, over a relatively small range of frequencies around the natural frequency of the system, it attains very high values if the damping is small. As a result, the integrant takes the shape shown in Figure 22.60c. It has a sharp spike around the natural frequency of the system.

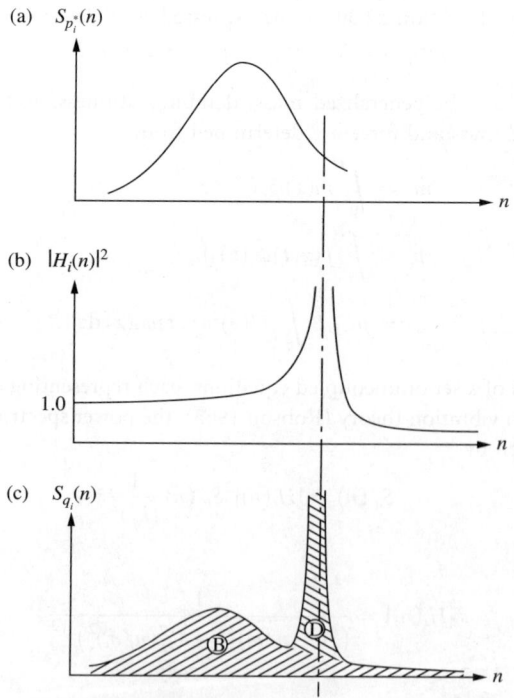

FIGURE 22.60 Schematic diagram for computation of response.

The broad hump is governed by the shape of the turbulent velocity spectrum which is modified slightly by the aerodynamic admittance function. The area under the broad hump is the broad band or nonresonant response, whereas the area in the vicinity of the natural frequency gives the narrow band or resonant response. Thus, Equation 22.40 can be rewritten as

$$\sigma_{q_i}^2 = \frac{1}{\left(k_i^*\right)^2}\left[\int_0^{n_i-\Delta n} S_{P_i^*}(n)\,\mathrm{d}n + \frac{\pi n_i}{4\zeta_i} S_{P_i^*}(n_i)\right] = \sigma_{Bq_i}^2 + \sigma_{Dq_i}^2 \tag{22.41}$$

in which σ_{Bq_i} and σ_{Dq_i} are the nonresonating and resonating root mean square response of the ith normal coordinate. As the response due to various modes of vibration are statistically uncorrelated, the response of the system is given by

$$\sigma_x^2(z) = \sum_{i=1}^N \phi_i^2(z)\sigma_{Bq_i}^2 + \sum_{i=1}^N \phi_i^2(z)\sigma_{Dq_i}^2 \tag{22.42}$$

which gives the variance and hence the root mean square displacement at various heights.

The total displacement is obtained by including the static deflection due to the mean drag load, which is determined conveniently as follows:

In view of Equation 22.25, the mean generalized force is given by

$$\bar{f}_i = \int_0^H \frac{1}{2}\rho C_\mathrm{D}\bar{U}^2(z)B\phi_i(z)\,\mathrm{d}z = \frac{1}{2}\rho C_\mathrm{D}B\int_0^H \bar{U}^2(z)\phi_i(z)\,\mathrm{d}z \tag{22.43}$$

Then, the mean displacement is determined from

$$\bar{x}(z) = \sum_{i=1}^N \phi_i(z)\left[\frac{\bar{f}_i}{\left(2\pi n_i\right)^2 m_i^*}\right] \tag{22.44}$$

The root mean square acceleration is obtained from

$$\sigma_{\ddot{x}}(z) = \left[\sum_{i=1}^N \left(2\pi n_i\right)^4 \phi_i^2(z)\sigma_{Dq_i}^2\right]^{1/2} \tag{22.45}$$

The dynamic shear and bending moment at any height is obtained from the vibratory inertia forces in each mode and then by summing the modal contributions.

The probability of the response exceeding certain magnitude is determined using a peak factor on the root mean square response, Davenport (1964) recommended the following expression for 50% probability of exceedence:

$$g_\mathrm{D} = \sqrt{[2\,\ln(\nu T_0)]} + \frac{0.577}{\sqrt{[2\,\ln(\nu T_0)]}} \tag{22.46}$$

where g_D is the peak factor, ν is the expected frequency at which the fluctuating response crosses the zero axis with positive slope, and T_0 is the period (usually 3600 s) during which the peak response is assumed to occur.

For resonant response, ν is equal to the natural frequency and thus the peak factor for the resonant response g_D is obtained by setting $\nu = n$. For the nonresonating or broadband response the peak factor g_B has been evaluated to be 3.5 (ESDU 1976).

Using these peak factors, the most probable maximum value of the load effect, E, such as displacement, shear, bending moment, are determined as follows:

$$E_{\max} = \bar{E} + [(g_\mathrm{B}\sigma_\mathrm{BE})^2 + (g_\mathrm{D}\sigma_\mathrm{DE})^2]^{1/2} \tag{22.47}$$

where σ_BE and σ_DE are the nonresonating and resonating components of the load effect and $\bar{E}$ is the load effect due to mean wind.

EXAMPLE 22.1

A rectangular building of height $H = 194$ m is situated in a suburban terrain. The breadth B and width D of the building are 56 m and 32 m, respectively. The period of the building corresponding to the fundamental sway mode is 5.15 s. The values of the mode shape at various heights are given below:

H(m)	0	20	40	75	95	135	150	170	194
ϕ	0	0.032	0.096	0.248	0.365	0.611	0.746	0.849	1.0

The generalized mass and damping ratio corresponding to this mode are 18×10^6 kg and 2%, respectively.

Assuming that the mean wind profile follows the power law with a power law coefficient $\alpha = 0.22$, determine the maximum drift for a 50-year wind storm of 21 m/s at 10 m height, blowing normal to the breadth of the building. Given that the friction velocity is 2.96 m/s, the drag coefficient C_D is 1.3 and density of air $\rho = 1.2$ kg/m^3.

Solution

The mean height of the building $\bar{H} = 97$ m

$$\bar{U}(97) = \bar{U}(10)\left(\frac{97}{10}\right)^{0.22} = 21\left(\frac{97}{10}\right)^{0.22} = 34.6 \text{ m/s}$$

At mid-height, the reduced frequency

$$f = \frac{n\bar{H}}{\bar{U}(\bar{H})} = \frac{97n}{34.6} = 2.8n$$

From Equation 22.13, the spectrum of turbulent wind is given by

$$S_u(\bar{H}, n) = \frac{2.96^2 \times 200 \times 2.8}{(1 + 50 \times 2.8n)^{5/3}} = \frac{4906}{(1 + 140n)^{5/3}}$$

Resonant displacement

$$n_1 = \frac{1}{5.15} = 0.194 \text{ Hz,}$$

$$S_u(\bar{H}, n_1) = \frac{4906}{(1 + 140 \times 0.194)^{5/3}} = 18.8 \text{ m}^2/\text{s}$$

The admittance function, from Equation 22.28, becomes

$$\chi(n) = \frac{1}{1 + ((2n\sqrt{56 \times 194})/34.6)^{4/3}} = \frac{1}{1 + 10.96n^{4/3}}$$

$$\chi(n_1) = 0.45$$

From Equation 22.39

$$S_{P_1^*}(n) = \rho^2 C_D^2 B^2 \chi^2(n) S_u(\bar{H}, n) \frac{[\bar{U}(\bar{H})]^2}{\bar{H}^{2\alpha}} \int_0^H \int_0^H \phi_1(z_1)\phi_1(z_2) z_1^\alpha z_2^\alpha \gamma(z_1, z_2, n) \, dz_1 \, dz_2$$

The square root of coherence γ is determined from Equation 22.16, considering only the vertical correlation. Thus

$$S_{P_1^*}(n_1) = 1.2^2 \times 1.3^2 \times 56^2 \times (0.45)^2 \times 18.8 \times \frac{(34.6)^2}{97^{0.44}} \times 16{,}900$$

$$= 7.85 \times 10^{10} \ N^2/Hz$$

From Equations 22.41 and 22.42, the variance of resonant displacement at the top of the building is obtained as

$$\sigma_D^2 = \phi_1^2(H)\sigma_{Dq_1}^2 = \frac{1}{(k_1^*)^2}\frac{\pi n_1}{4\zeta_1}S_{P_1^*}(n_1)$$

$$\sigma_D^2 = \left(\frac{1}{26.8 \times 10^6}\right)^2 \left(\frac{\pi(0.194)}{4(0.02)}\right)(7.85 \times 10^{10})10^6 \ mm^2$$

$$\sigma_D = 28.9 \ mm$$

Nonresonant displacement
The variance of nonresonant displacement at the top of the building is determined from Equations 22.41 and 22.42 as

$$\sigma_B^2 = \phi_1^2(H)\sigma_{Bq_1}^2 = \frac{1}{(k_1^*)^2}\int_0^{n_1-\Delta n} S_{P_1^*}(n) \, dn$$

$$= \frac{565 \times 10^9}{(26.8 \times 10^6)^2} \times 10^6 \ mm$$

$$\sigma_B = 28 \ mm$$

Response to mean wind
From Equation 22.43, the mean generalized force

$$\bar{f}_1 = \frac{1}{2}\rho C_D B \int_0^H \bar{U}^2(z)\phi_1(z) \, dz$$

$$= \frac{1}{2}\rho C_D B[\bar{U}(H)]^2 \left(\frac{1}{H}\right)^{2\alpha}\int_0^H z^{2\alpha}\phi_1(z) \, dz$$

$$= \frac{1}{2} \times 1.2 \times 1.3 \times 56 \times (34.6)^2 \left(\frac{1}{97}\right)^{0.44} \times 684$$

$$= 4.8 \times 10^6 \ N$$

The generalized stiffness

$$k_1^* = \left(\frac{2\pi}{5.15}\right)^2 \times 18 \times 10^6$$

$$= 26.8 \times 10^6 \ N/m$$

Thus, the mean displacement

$$\bar{X} = \frac{4.8 \times 10^6}{26.8 \times 10^6} \times 10^3 = 179 \ mm$$

The peak factor g_D for resonant response is determined from Equation 22.46 as 3.78. Using a peak factor of 3.5 for nonresonant response, the most probable maximum displacement is

$$
\begin{aligned}
X_{\max} &= \overline{X} + \sqrt{(g_B \sigma_B)^2 + (g_D \sigma_D)^2} \\
&= 179 + \sqrt{(3.5 \times 28)^2 + (3.78 \times 28.9)^2} \\
&= 326 \, \text{mm}
\end{aligned}
$$

The most probable maximum drift would be

$$
= \frac{326}{194{,}000} = \frac{1}{595}
$$

The peak acceleration would be

$$
\begin{aligned}
&= 0.0289 \times 3.78 \times (2\pi)^2 \times (0.194)^2 \\
&= 0.16 \, \text{m/s}^2 \ (1.6\% \ \text{g})
\end{aligned}
$$

22.5.5 Response Due to Across Wind

For most modern tall buildings, the across wind response is more significant than the along wind response. Across wind vibration of building is caused by the combination of forces from three sources: (1) buffeting by the turbulence in the across wind direction, (2) wake excitation due to vortex shedding, and (3) lock-in, a displacement-dependent excitation.

The across wind force due to lateral turbulence in the approaching flow is generally small compared to the effects due to other mechanisms. Lock-in is the term used to describe large amplitude across wind motion that occurs when the vortex shedding frequency is close to the natural frequency. If the across wind response exceeds a certain critical value, the across wind response causes an increase in the excitation force, which in turn increases the response. The vortex shedding frequency tends to couple with the natural frequency of the structure for a range of wind velocities and the large amplitude response will persists. Lock-in is likely to occur only in the case of structures with relatively low stiffness and low damping, operating near the critical wind velocity given by

$$
\overline{U}_{\text{crit}} = \frac{n_0 B}{S} \tag{22.48}
$$

in which $\overline{U}_{\text{crit}}$ is the critical wind speed, B is the breadth of the structure normal to the wind stream, n_0 (in Hz) is the fundamental natural frequency of the structure in the across wind direction, and S is the Strouhal number.

Buildings should.be designed so that lock-in effects do not occur during their anticipated life. If the root mean square displacement at the top of the structure is less than a certain critical value, then lock-in will not occur. For square tall buildings, the critical root mean square displacements σ_{yc} expressed as a ratio with respect to the breadth (σ_{yc}/B) are approximately 0.015, 0.025, and 0.045, respectively (Rosati 1968) for open terrain ($z_0 = 0.07$ m), suburban terrain ($z_0 = 1.0$ m), and city centers ($z_0 = 2.5$ m). For circular sections with diameter D, the value of σ_{yc}/D is approximately 0.006 for suburban terrain.

Thus for buildings the most common cause for across wind motion is the wake excitation. Although the turbulence in the atmospheric boundary layer affects the regularity of vortex shedding, the shed vortices have a predominant period that could be determined from an appropriate Strouhal number. Because the vortex shedding is random, the fluctuating across wind force is effectively broadband as shown in Figure 22.61. The band width and the energy concentration near the vortex shedding frequency depend on the geometry of the building and the characteristics of the approach flow.

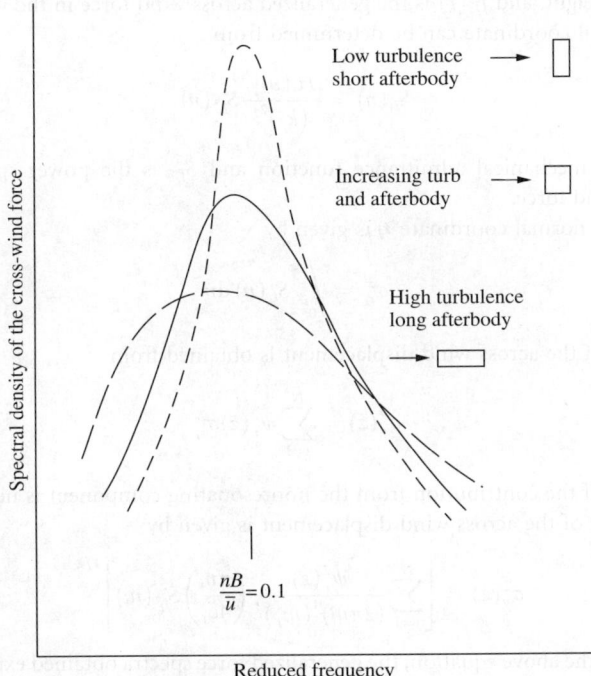

FIGURE 22.61 Effects of turbulence intensity and afterbody length on across wind force spectra.

The response due to this across wind random excitation can be determined using the random vibration theory. Idealizing the tall building as a line-like structure, the across wind displacement $y(z, t)$ may be expressed in terms of the normal coordinates $r_i(t)$ as

$$y(z, t) = \sum_{i=1}^{N} \psi_i(z) r_i(t) \qquad (22.49)$$

where $\psi_i(z)$ is the ith vibration mode in the across wind direction and N is the total number of modes considered to be significant. The governing equation of motion in terms of generalized mass, m_i^*, generalized damping, c_i^*, and generalized stiffness, k_i^*, takes the form

$$m_i^* \ddot{r}_i + c_i^* \dot{r}_I + k_i^* r_i = f_i^*(t), \quad i = 1 \text{ to } N \qquad (22.50)$$

in which

$$m_i^* = \int_0^H m(z) \psi_i^2(z) \, \mathrm{d}z$$

$$k_i^* = (2\pi n_i)^2 m_i^*$$

$$c_i^* = 2\zeta_i \sqrt{m_i^* k_i^*} \qquad (22.51)$$

$$f_i^*(t) = \int_0^H f(z, t) \psi_i(z) \, \mathrm{d}z$$

where H is the height of the building, $m(z)$ is the mass per unit length, n_i is the frequency of the ith mode in the across wind direction, ζ_i is the damping ratio in the ith mode, $f(z, t)$ is the across wind force per unit height, and $f_i^*(t)$ is the generalized across wind force in the ith mode. The spectral density of each normal coordinate can be determined from

$$S_{r_i}(n) = \frac{|H_i(n)|^2}{(k_i^*)^2} S_{f_i^*}(n) \tag{22.52}$$

where $|H_i(n)|$ is the mechanical admittance function and $S_{f_i^*}$ is the power spectral density of the generalized across wind force.

The variance of the normal coordinate r_i is given by

$$\sigma_{r_i}^2 = \int_0^\infty S_{r_i}(n)\,dn \tag{22.53}$$

Hence, the variance of the across wind displacement is obtained from

$$\sigma_y^2(z) = \sum_{i=1}^{N} \psi_i^2(z)\sigma_{r_i}^2 \tag{22.54}$$

In Equation 22.53, if the contribution from the nonresonating component is neglected then the root mean square response of the across wind displacement is given by

$$\sigma_y^2(z) = \left[\sum_{i=1}^{N} \frac{\psi_i^2(z)}{(2\pi n_i)^4 (m_i^*)^2} \left(\frac{\pi n_i}{4\zeta_i}\right) S_{f_i^*}(n_i) \right]^{1/2} \tag{22.55}$$

For convenient use of the above equation, the generalized force spectra obtained experimentally by Kwok and Melbourne (1981) and Saunders and Melbourne (1975) are presented in Figure 22.62 for various aspect ratios of square and rectangular buildings deflecting in a linear mode.

EXAMPLE 22.2

Consider the building of Example 22.1. If the period of vibration in the across wind direction is 5.2 s, assuming a linear mode determine the acceleration in the across wind direction for a 10-year wind storm of 14 m/s at 10 m height. Given that the generalized mass corresponding to the linear mode is 17.5×10^6 kg and the damping in this mode of oscillation is 2%.

Solution

The building is rectangular with an aspect ratio of

$$H{:}B{:}D = 6{:}1.75{:}1$$

Since the building is in a suburban terrain, the generalized cross wind force can be determined from Figure 22.62e.

The wind speed at the tip of the building

$$\bar{U}(H) = \bar{U}(10) \times \left(\frac{194}{10}\right)^{0.22} = 26.9 \text{ m/s}$$

The reduced frequency

$$\frac{n_1 B}{\bar{U}(H)} = \frac{0.192 \times 56}{26.9} = 0.40$$

then from Figure 22.62e

$$S_{f_1^*}(n_1) = \left(\frac{0.00018}{0.192}\right)\left(\frac{1}{2} \times 1.2 \times 26.9^2 \times 56 \times 194\right)^2$$

$$= 2.09 \times 10^{10} \text{ N}^2/\text{Hz}$$

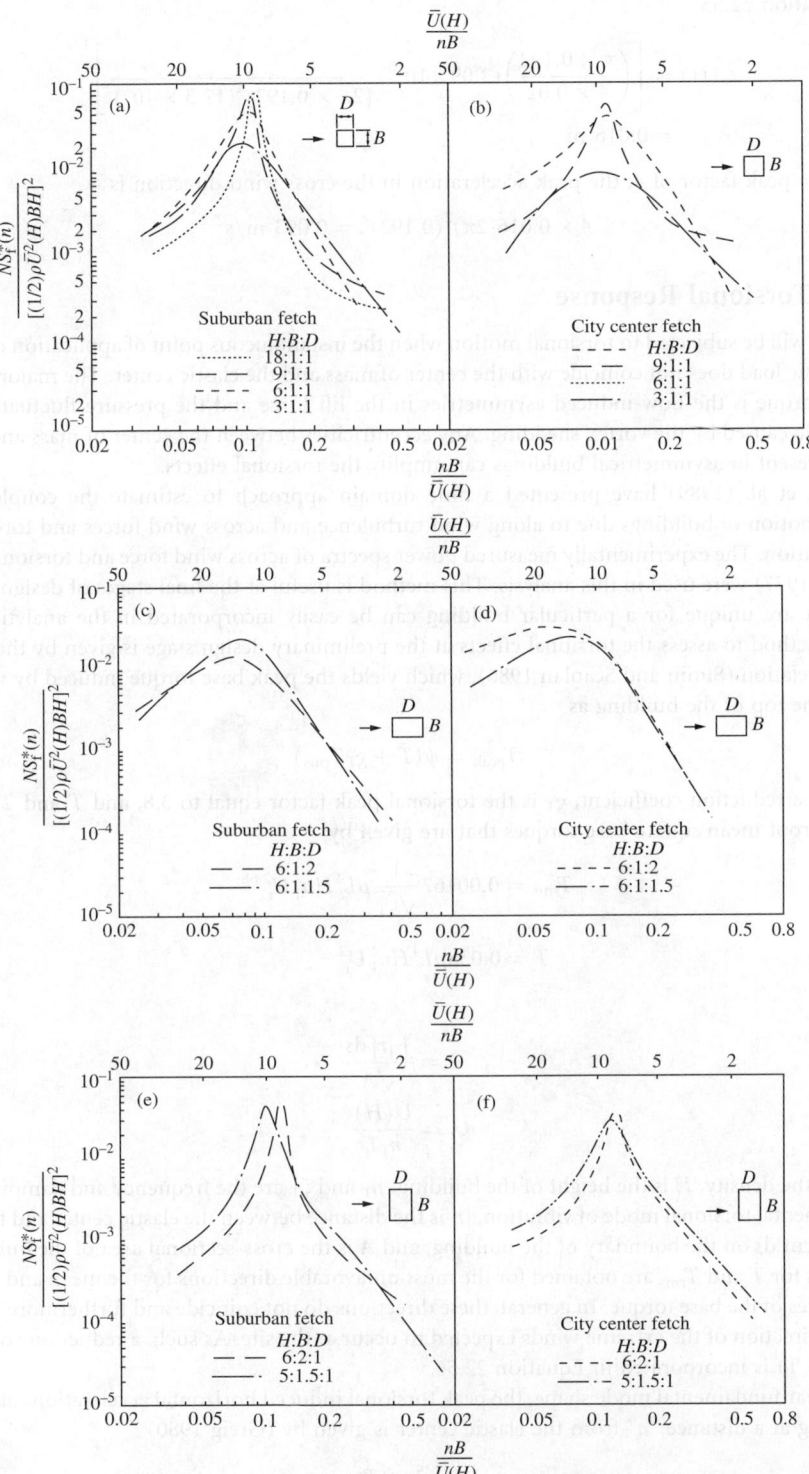

FIGURE 22.62 Generalized force spectra for a square and a rectangular building in suburban and city center fetch.

From Equation 22.55

$$\sigma_y(H) = \left[\left(\frac{\pi \times 0.192}{4 \times 0.02} \right) (2.09 \times 10^{10}) \frac{1}{(2\pi \times 0.192)^4 (17.5 \times 10^6)^2} \right]^{1/2}$$

$$= 0.016 \text{ m}$$

Assuming a peak factor of 4, the peak acceleration in the cross wind direction is

$$4 \times 0.016(2\pi)^2 (0.192)^2 = 0.093 \text{ m/s}^2$$

22.5.6 Torsional Response

A building will be subjected to torsional motion when the instantaneous point of application of resultant aerodynamic load does not coincide with the center of mass and the elastic center. The major source for dynamic torque is the flow-induced asymmetries in the lift force and the pressure fluctuation on the leeward side caused by the vortex shedding. Any eccentricities between the center of mass and center of stiffness present in asymmetrical buildings can amplify the torsional effects.

Balendra et al. (1989) have presented a time domain approach to estimate the coupled lateral–torsional motion of buildings due to along wind turbulence and across wind forces and torque due to wake excitation. The experimentally measured power spectra of across wind force and torsional moment (Reinhold 1977) were used in this analysis. This method is useful at the final stages of design as specific details that are unique for a particular building can be easily incorporated in the analytical model. A useful method to assess the torsional effects at the preliminary design stage is given by the following empirical relation (Simiu and Scanlan 1986), which yields the peak base torque induced by wind speed $\bar{U}(H)$ at the top of the building as

$$T_{peak} = \psi(\bar{T} + g_T T_{rms}) \tag{22.56}$$

where ψ is a reduction coefficient, g_T is the torsional peak factor equal to 3.8, and $\bar{T}$ and T_{rms} are the mean and root mean square base torques that are given by

$$T_{rms} = 0.00167 \frac{1}{\sqrt{\zeta_T}} \rho L^4 H n_T^2 U_r^{2.68} \tag{22.57}$$

$$\bar{T} = 0.038 \rho L^4 H n_T^2 U_r^2 \tag{22.58}$$

in which

$$L = \frac{\int |r| \, ds}{\sqrt{A}} \tag{22.59}$$

$$U_r = \frac{\bar{U}(H)}{n_T L} \tag{22.60}$$

where ρ is the density, H is the height of the building, n_T and ζ_T are the frequency and damping ratio in the fundamental torsional mode of vibration, $|r|$ is the distance between the elastic center and the normal to an element ds on the boundary of the building, and A is the cross-sectional area of the building. The expressions for $\bar{T}$ and T_{rms} are obtained for the most unfavorable directions for the mean and root mean square values of the base torque. In general, these directions do not coincide and furthermore will not be along the direction of the extreme winds expected to occur at the site. As such, a reduction coefficient ψ ($0.75 < \psi \leq 1$) is incorporated in Equation 22.56.

For a linear fundamental mode shape, the peak torsional induced horizontal accelerations at the top of the building at a distance "a" from the elastic center is given by (Greig 1980)

$$a\ddot{\theta} = \frac{2ag_T T_{rms}}{\rho_b BDH r_m^2} \tag{22.61}$$

where $\ddot{\theta}$ is the peak angular acceleration, ρ_b is the mass density of the building, B and D are the breadth and depth of the building, and r_m is the radius of gyration. For a rectangular building with uniform mass density

$$r_m^2 = \tfrac{1}{12}(B^2 + D^2) \tag{22.62}$$

EXAMPLE 22.3

If the torsional frequency of the building in Example 22.2 is 0.8 Hz, assuming a linear mode and 2% damping ratio, determine the peak acceleration at the corner of the building due to torsional motion for a 10-year wind storm of 14 m/s. Given that the center of rigidity is at the geometric center of the building.

Solution

For a rectangular building

$$\int |r|\, ds = \frac{1}{2}(B^2 + D^2)$$

Thus, from Equations 22.59 and 22.60

$$L = \frac{1}{\sqrt{BD}}(B^2 + D^2)\frac{1}{2} = 49.1 \text{ m}$$

$$U_r = \frac{\bar{U}(H)}{n_T L} = \frac{26.9}{0.8 \times 49.1} = 0.685$$

From Equation 22.58

$$T_{rms} = 0.00167\left(\frac{1}{\sqrt{0.02}}\right)(1.2)(49.1)^4(194)(0.8)^2(0.685)^{2.68}$$

$$= 3.71 \times 10^6 \text{ N m}$$

The average density of the building is determined as

$$\rho_b = \frac{3m_1^*}{AH} = \frac{3 \times 17.5 \times 10^6}{56 \times 32 \times 194} = 151 \text{ kg/m}^3$$

Thus, the peak torsional acceleration of the corner for which $a = 32.2$ m is

$$a\ddot{\theta} = \frac{2 \times 32.2 \times 3.8 \times 3.71 \times 10^6}{151 \times 56 \times 32 \times 194 \times 346.7} = 0.05 \text{ m/s}^2$$

22.5.7 Response by Wind Tunnel Tests

There are many situations where analytical methods cannot be used to estimate certain type of wind loads and associated structural response. For example, the aerodynamic shape of the building is rather uncommon or the building is very flexible so that its motion affects the aerodynamic forces acting on the building. In such situations, a more accurate estimate of wind effects on buildings are obtained through aeroelastic model tests in a boundary-layer wind tunnel (Balendra 1993).

The aeroelastic model studies would provide the overall mean and dynamic loads, displacements, rotations, and accelerations. The aeroelastic model studies may be required under the following situations:

1. When the height-to-width ratio exceeds 5.
2. When the structure is light with a density in the order of 1.5 kN/m³.

3. The fundamental period is long in the order of 5 to 10 s.
4. When the natural frequency of the building in the cross wind direction is in the neighborhood of the shedding frequency.
5. When the building is torsionally flexible.
6. When the building is expected to execute strongly coupled lateral–torsional motion.

Glossary

Aeroelastic model — The model which simulates the dynamic properties of buildings to capture the motion-dependent loads.
Along wind response — Response in the direction of wind.
Boundary layer — The layer within which the velocity varies because of ground roughness.
Bracing frames — Frames that provide lateral stability to the overall framework.
Composite beams — Steel beam acting compositely with part of the concrete slab through shear connectors.
Cross wind response — Response perpendicular to the direction of wind.
Drag force — Force in the direction of wind.
Frequency — Number of cycles per second.
Gradient height — Thickness of the boundary layer.
Gradient wind — Wind velocity above the boundary layer.
Generalized force — Force associated with a particular mode of vibration.
Generalized mass — Participating mass in a particular mode of vibration.
Integral length scale — A measure of average size of the eddies.
Lift force — Force perpendicular to the flow.
Lock-in — Situation where the vortex shedding frequency tends to couple with the frequency of the structure.
Long-span systems — Structural systems that span a long distance. The design is likely to be governed by serviceability limit states.
Mode shapes — Free vibration deflection configurations in each frequency of the structure.
Nonresonating response — Response due to eddies whose frequencies are remote from the structural frequency.
Normal coordinates — Coordinates associated with modes of vibration.
Peak factor — Ratio between the peak and root mean square values.
Period — Duration of one complete cycle.
Power spectral density — Kinetic energy per unit time associated with eddies of different frequencies.
Resonant response — Response due to eddies whose frequencies are in the neighborhood of structural frequency.
Rigid frames — Frames resisting lateral load by bending of members that are rigidly connected.
Simple frames — Frames that have no lateral resistance and whose members are pinned connected.
Stiffness — Force required to produce unit displacement.
Sway frames — Frames in which the second-order effects due to gravity load acting on the deformed geometry can influence the force distribution in the structure.
Torsional response — Response causing twisting motion.
Turbulent intensity — Overall measure of intensity of turbulence.
Wake — Region surrounded by the shear layers originating from separation points.
Wake excitation — Excitation caused by the vortices in the wake.

References

AISC. 1989. *Allowable Stress Design and Plastic Design Specifications for Structural Steel Buildings*, 9th ed., American Institute of Steel Construction, Chicago, IL.
AISC. 1990. *LRFD — Simple Shear Connections*, American Institute of Steel Construction, Chicago, IL.

AISC. 1993. *Load and Resistance Factor Design Specification for Structural Steel Buildings*, 2nd ed., American Institute of Steel Construction, Chicago, IL.

ANSI. 1982. *American National Standard Building Code Requirements for Minimum Design Loads in Buildings and Other Structures*, A 58.1, New York.

ASCE. 1987. Wind loading and wind induced structural response, *State-of-the-Art Report*, Committee on Wind Effects, New York.

ASCE. 1990. Minimum design loads for buildings and other structures, ASCE Standard, *ASCE 7-88*, American Society of Civil Engineers.

ASCE Task Committee. 1996. Proposed specification and commentary for composite joints and composite trusses, ASCE Task Committee on Design Criteria for Composite in Steel and Concrete, *J. Struct. Eng., ASCE*, April, 122(4), 350–358.

Balendra, T., Nathan, G.K., and Kang, K.H. 1989. Deterministic model for wind induced oscillations of buildings, *J. Eng. Mech., ASCE*, 115, 179–199.

Balendra, T. 1993. *Vibration of Buildings to Wind and Earthquake Loads*, Springer-Verlag, Berlin.

Brett, P. and Rushton, J. 1990. *Parallel Beam Approach — A Design Guide*, The Steel Construction Institute, U.K.

BS5950:Part 1. 1990. *Structural Use of Steelwork in Building. Part 1: Code of Practice for Design in Simple and Continuous Construction: Hot Rolled Section*. British Standards Institution, London.

Chen, W.F. and Atsuta, T. 1976. *Theory of Beam-Column, Vol. 1, In-Plane Behavior and Design*, McGraw-Hill, New York.

Chen, W.F. and Lui, E.M. 1991. *Stability Design of Steel Frames*, CRC Press, Boca Raton, FL.

Chen, W.F. and Toma, S. 1994. *Advanced Analysis in Steel Frames: Theory, Software and Applications*, CRC Press, Boca Raton, FL.

Chen, W.F., Goto, Y., and Liew, J.Y.R. 1996. *Stability Design of Semi-Rigid Frames*, John Wiley & Sons, New York.

Council on Tall Buildings and Urban Habitat. 1995. *Architecture of Tall Buildings*, Armstrong, P.J., Ed., McGraw-Hill, New York.

Davenport, A.G. 1961. The application of statistical concepts to the wind loading of structures. *Proc. Inst. Civil Eng.*, 19, 449–472.

Davenport, A.G. 1964. Note on the distribution of the largest value of a random function with application to gust loading. *Proc. Inst. Civil Eng.*, 28, 187–196.

Davenport, A.G. 1968. The dependence of wind load upon meteorological parameters. *Proc. Intl. Res. Sem. Wind Effects on Buildings and Structures*, University of Toronto Press, Toronto, pp. 19–82.

ESDU. 1976. *The Response of Flexible Structures to Atmospheric Turbulence*. Item 76001, Engineering Sciences Data Unit, London.

ESDU. 1985. *Characteristics of Atmospheric Turbulence Near the Ground, Part II: Single Point Data for Strong Winds* (Neutral Atmosphere). Item 85020, Engineering Sciences Data Unit, London.

Eurocode 3. 1992a. *Design of Steel Structures: Part 1.1 — General Rules and Rules for Buildings*, National Application Document for use in the U.K. with ENV1993-1-1:1991, Draft for Development.

Eurocode 3. 1992b. *Design of Steel Structures: Part 1.1 — General Rules and Rules for Buildings*, National Application Document for use in the U.K. with ENV1993-1-1:1991, Draft for Development.

Eurocode 4. 1994. *Design of Composite Steel and Concrete Structures: General Rules for Buildings*, preENV 1994-1-1, European Committee for Standardization.

Fishers, J.M. and West, M.A. 1990. *Serviceability Design Considerations for Low-Rise Buildings*, American Institute of Steel Construction, Chicago, IL.

Geschwindner, L.F., Disque, R.O., and Bjorhovde, R. 1994. *Load and Resistance Factored Design of Steel Structures*, Prentice Hall, Englewood Cliffs, NJ.

Greig, L. 1980. *Toward an Estimate of Wind Induced Dynamic Torque on Tall Buildings*, MSc thesis, Department of Engineering, University of Western Ontario, London, Ontario.

Griffis, L.G. 1993. Serviceability limit states under wind load. *Eng. J., AISC*, Vol. 1, 1–16.

Heidebrecht, A.C. and Smith, B.S. 1973. Approximate analysis of tall wall-frame structures. *J. Struct. Div., ASCE*, 99, 199–221.

Iyengar, S.H., Baker, W.F., and Sinn, R. 1992. Multi-Story Buildings, in *Constructional Steel Design, An International Guide*, Dowling, P.J. et al., Eds., Elsevier, London, pp. 645–670, Chap. 6.2.

Knowles, P.R. 1985. *Design of Castellated Beams*, The Steel Construction Institute, U.K.

Kwok, K.C.S. and Melbourne, W.H. 1981. Wind induced lock-in excitation of tall structures. *J. Struct. Div., ASCE*, 107, 57–72.

Lawson, R.M. 1987. *Design for Openings in Webs of Composite Beams CIRIA*, The Steel Construction Institute, U.K.

Lawson, R.M. 1993. *Comparative Structure Cost of Modern Commercial Buildings*, The Steel Construction Institute, U.K.

Lawson, R.M. and McConnel, R.E. 1993. *Design of Stub Girders*, The Steel Construction Institute, U.K.

Lawson, R.M. and Rackham, J.W. 1989. *Design of Haunched Composite Beams in Buildings*, The Steel Construction Institute, U.K.

Leon, R.T. 1990. Semi-rigid composite construction. *J. Construct. Steel Res.*, 15(1&2), 99–120.

Leon, R.T. 1994. *Composite Semi-Rigid Construction, Steel Design: An International Guide*, R. Bjorhovde, Harding, J. and Dowling, P., Eds., Elsevier, Amsterdam, pp. 501–522.

Leon, R.T. and Ammerman, D.J. 1990. Semi-rigid composite connections for gravity loads. *Eng. J., AISC*, 1st Qrt., 1–11.

Leon, R. T., Hoffman, J.J., and Staeger, T. 1996. Partially restrained composite connections. *AISC Steel Design Guide Series 8*, AISC.

Liew, J.Y.R. 1995. Design concepts and structural schemes for steel multi-story buildings. *J. Singapore Struct. Steel Soc., Steel Struct.*, 6(1), 45–59.

Liew, J.Y.R. 2001, *A Resource Book for Structural Steel Design and Construction*, SSSS/BCA Joint Publication, Singapore Structural Steel Society, 83 p.

Liew, J.Y.R. and Chen, W.F. 1994. Implications of using refined plastic hinge analysis for load and resistance factor design, *J. Thin-Walled Struct.*, 20(1–4), 17–47.

Liew, J.Y.R. and Chen, W.F. 1995. Analysis and design of steel frames considering panel joint deformations. *J. Struct. Eng., ASCE*, 121(10), 1531–1540.

Liew, J.Y.R. and Chen, W.F. 1997. LRFD — limit design of frames, in *Steel Design Handbook*, Tamboli, A., Ed., McGraw-Hill, New York, Chap. 6.

Liew, J.Y.R., White, D.W., and Chen, W.F. 1991. Beam-column design in steel frameworks — insight on current methods and trends. *J. Construct. Steel Res.*, 18, 259–308.

Liew, J.Y.R., White, D.W., and Chen, W.F. 1992. Beam-columns, in *Constructional Steel Design, An International Guide*, Dowling, P.J. et al., Eds., Elsevier, London, pp. 105–132, Chap. 5.1.

Liew, J.Y.R., White, D.W., and Chen, W.F. 1993a. Limit-states design of semi-rigid frames using advanced analysis. Part 1: Connection modeling and classification. Part II: Analysis and design. *J. Construct. Steel Res.*, 26(1), 1–57.

Liew, J.Y.R., White, D.W., and Chen, W.F. 1993b. Second-order refined plastic hinge analysis for frame design: Parts 1 & 2. *J. Struct. Eng., ASCE*, 119(11), 3196–3237.

Liew, J.Y.R., White, D.W., and Chen, W.F. 1994. Notional load plastic hinge method for frame design. *J. Struct. Eng., ASCE*, 120(5), 1434–1454.

Neals, S. and Johnson, R. 1992. *Design of Composite Trusses*, The Steel Construction Institute, U.K.

Owens, G. 1989. *Design of Fabricated Composite Beams in Buildings*, The Steel Construction Institute, U.K.

Owens, G.W. and Knowles, P.R. 1992. *Steel Designers' Manual*, 5th ed., Blackwell Scientific Publications, London.

Reinhold, T.A. 1977. *Measurements of Simultaneous Fluctuating Loads at Multiple Levels on a Model of Tall Building in a Simulated Urban Boundary Layer*, PhD thesis, Department of Civil Engineering, Virginia Polytechnic Institute and State University.

Robson, J.D. 1963. *An Introduction to Random Vibration*, Edinburgh University Press, Scotland.

Rosati, P.A. 1968. *An Experimental Study of the Response of a Square Prism to Wind Load*, Faculty of Graduate Studies, BLWT II-68, University of Western Ontario, London, Ontario, Canada.

Saunders, J.W. and Melbourne, W.H. 1975. Tall rectangular building response to cross-wind excitation, *Proceedings of the 4th International Conference on Wind Effects on Building Structures*, Cambridge University Press, Cambridge.

SCI. 1995. *Plastic Design of Single-Story Pitched-Roof Portal Frames to Eurocode 3*, Technical Report, SCI Publication 147, The Steel Construction Institute, U.K.

Simiu, E. 1974. Wind spectra and dynamic along wind response. *J. Struct. Div., ASCE*, 100, 1897–1910.

Simiu, E. and Scanlan, R.H. 1986. *Wind Effects on Structures*, 2nd ed., John Wiley & Sons, New York.

Taranath, B.S. 1988. Structural *Analysis and Design of Tall Buildings*, McGraw-Hill, New York.

Vickery, B.J. 1965. *On the Flow Behind a Coarse Grid and its Use as a Model of Atmospheric Turbulence in Studies Related to Wind Loads on Buildings*, Nat. Phys. Lab. Aero. Report 1143.

Vickery, B.J. 1970. On the reliability of gust loading factors, *Proc. Tech. Meet. Concerning Wind Loads on Buildings and Structures*, National Bureau of Standards, Building Science Series 30, Washington, D.C.

Further Reading

ASCE. 1997. *Effective Length and Notional Load Approaches for Assessing Frame Stability*, ASCE, New York, NY.

Chen, W.F. and Kim, S.E. 1997. *LRFD Steel Design Using Advanced Analysis*, CRC Press, Boca Raton, FL.

Chen, W.F. and Sohal, I. 1995. *Plastic Design and Second-Order Analysis of Steel Frames*, Springer-Verlag, New York.

Council of Tall Buildings and Urban Habitat. 1995. *Structural Systems for Tall Buildings*, McGraw-Hill, Singapore.

Deierlein, G.G., and White, D.W. 1988. Chapter 16 — Frame stability, in *Guide to Stability Design Criteria for Metal Structures*, 5th ed., Galambos, T.V., Ed., John Wiley & Sons, New York.

Lawson, T.V. 1980. *Wind Effects on Buildings*, Applied Science Publishers, London.

Smith, J.W. 1988. *Vibration of Structures — Application in Civil Engineering Design*, Chapman & Hall, London.

Smith, B.S. and Coull, A. 1991. *Tall Building Structures*, John Wiley & Sons, New York.

23

Semirigid Frame Structures

Lei Xu
Department of Civil Engineering,
University of Waterloo,
Waterloo, Ontario, Canada

23.1 Introduction

The analysis and design of building steel frames in practice has long been based on simplified assumptions concerning the behavior of beam-to-column connections as two idealized models, fully rigid and ideally pinned, even though the semirigid concept was introduced many years ago. The first model implies displacement and slope continuity between the column and the beam, together with the full transfer of bending moments. The latter one, on the other hand, implies that rotation continuity is nonexistent and, consequently, no bending moment may be transmitted to the column by the beam. Although adoption of such idealized models leads to great simplification in structural analysis and design, it by no means represents the actual behavior of the connections. Consequently, the predicted structural responses may be inaccurate as a result of disregarding the true behavior of beam-to-column connections. To date, numerous experimental investigations on connection behavior have clearly demonstrated that a pinned connection possesses a certain amount of rotational stiffness, while a rigid connection possesses some degree of rotational flexibility. Most beam-to-column connections commonly used in practice actually exhibit semirigid behavior characterized by the moment–rotational relationship as shown in Figure 23.1.

The importance of the semirigid behavior of connections and the semirigid construction of steel framing systems has been recognized in different structural steel design standards in both North American and European communities for a period of time. In general, the process of designing semirigid steel frames involves the following three tasks: assessing the behavior of beam-to-column connections,

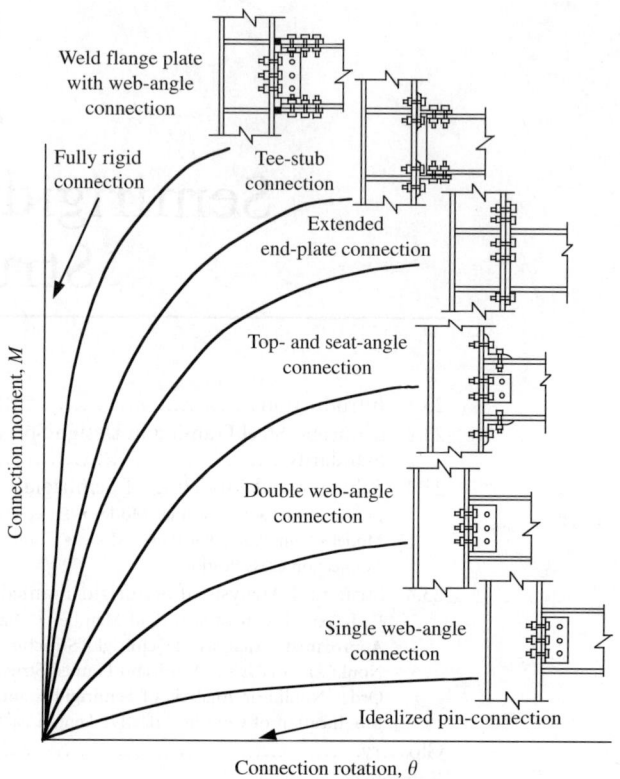

FIGURE 23.1   Moment–rotation behavior for various beam-to-column connections.

incorporating the connection behavior into the structural analysis of the frames, and designing the frames in accordance with the pertained specifications. In this chapter, together with a brief review of different design specifications on semirigid construction, a discussion on each of the three foregoing aspects is presented.

23.2   Semirigid Steel Framing in Design Specifications and Standards

In the United States of America, the *Specification for Allowable Stress Design for Structural Steel Buildings* of the American Institute of Steel Construction (ASD) [1] provides three connection types for construction and associated analysis assumptions: Type 1, commonly designated as "rigid framing or continuous framing"; Type 2, commonly designated as "simple framing"; and Type 3, commonly designated as "semirigid framing." The rigid frame assumes that beam-to-column connections have sufficient rigidity to maintain the original angles between intersecting members virtually unchanged. The simple framing assumes that the ends of beams are connected for shear only and are free to rotate under load. The semirigid framing assumes that the connections of beams and girders possess a dependable and known moment capacity intermediate in degree between the complete rigidity of Type 1 and complete flexibility of Type 2.

The *Load and Resistance Factored Design Specification for Structural Steel Building* of the American Institute of Steel Construction (LRFD) [2] classifies two types of constructions: Type FR (fully restrained) and Type PR (partially restrained). FR constructions are basically the same as the Type 1 construction in ASD. PR constructions are those where the connections have insufficient rigidity to maintain the original angles between the beams and the columns. Type PR comprises two cases depending on whether connection restraint is ignored or not, which correspond to Types 2 and 3 of ASD, respectively. When connection restraint is considered, it is required that the strength, stiffness, and ductility characteristics of the connections be incorporated into the analysis and design. These characteristics shall be as documented in technical literature or established by analytical or experimental means.

The Canadian standard [3] on *Limit States Design of Steel Structures*, CAN/CSA-S16, stipulates three types of steel framing constructions; rigidly connected and continuous construction, simple construction, and semirigid (partially restrained) construction. The design and construction of semirigid frames is required to comply with the following:

1. The positive and negative moment–rotation relation up to the maximum capacity for a connection is to be established by testing and be either as is published in technical literature or be available from a reputable testing facility.
2. The design of the frame shall be based on either linear analysis using the secant stiffness of the connections at ultimate load or on nonlinear analysis using a nonlinear moment–rotation relationship for connections.
3. The design should consider effects of repeated loading, load reversals, incremental strain in connections, and low-cycle fatigue.

The European standard [4] on *Design of Steel Structures*, commonly called "Eurocode 3," defines three framing types: continuous, simple, and semicontinuous. The standard precisely distinguishes between these three types of connections and also recognizes that the extent of semirigid action depends to a large extent on the type of structure, such as braced or unbraced frame.

23.3 Behavior and Modeling of Semirigid Connections

A beam-to-column connection is generally subjected to axial and shear forces and bending and torsional moments. The effect of torsion is usually neglected for planar frames. Furthermore, axial and shearing deformations are also neglected since they are small compared to the bending deformation of most connections. Thus, only the rotational deformation of a connection alone is considered in semirigid framing for practical purposes. As shown in Figure 23.2, a beam-to-column connection rotates through an angle θ when a moment M is applied. The angle θ corresponds to the relative rotation of the beam and column at the connection. Therefore, the dominant in-plane behavior of a semirigid connection is represented by its moment–rotation (M–θ) relationship.

Shown in Figure 23.1 is the semirigid moment–rotation (M–θ) behavior of some common beam-to-column connections used in practice. It is obvious from the figure that the behavior of all types of connections falls in between the two extremes of ideally pinned and fully rigid connections.

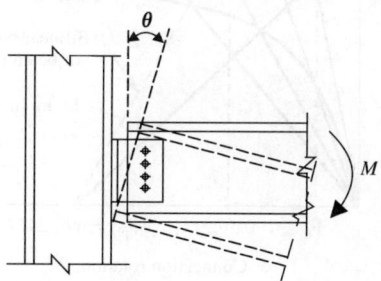

FIGURE 23.2 Semirigid beam-to-column connections.

The single-angle connection and the weld flange plate with web-angle connection represent very flexible and rather rigid connections, respectively. Also, as shown in Figure 23.1, experimental tests have clearly demonstrated that there is a nonlinear relationship between M and θ for almost all types of connections over the entire loading range. The nonlinearity is mainly because a connection is an assemblage of several components that interact differently at different levels of applied loads. Material discontinuity of the connection subassemblage, local yielding of some components, local buckling of a plate element, and so on, all contribute to the nonlinearity. One of the primary obstacles to implementing semirigid framing in design practice is that the nonlinear M–θ relationship of semirigid connections complicates the evaluation of the structural response of semirigid frames.

The modeling of the M–θ relationship is a fundamental requirement for any consideration of the interaction of connection and member behavior. The most accurate and reliable knowledge of the behavior of a beam-to-column connection is obtained through experimental testing, but this technique is too expensive for design practice and is usually reserved for research purposes only. With the aim to incorporate connection behavior into structural analysis, connection behavior is generally simulated by means of a mathematical representation of the M–θ relationship. It is difficult to model this nonlinear relationship by rigorous and exact mathematical procedures and the connection behavior modeling in practical design is usually approximate in nature with drastic simplifications. Tests of prototype connections are commonly carried out to obtain actual moment–rotation behavior, which is then modeled approximately by means of different mathematical representation. The degree of the refinement of the mathematical representations of the M–θ relationship depends somewhat on the computational capabilities of available design aids.

23.3.1 Linear Model

The simplest connection model is the single-stiffness linear model that is expressed as follows:

$$M = R\theta \tag{23.1}$$

where M is the connection moment and R and θ are the stiffness and rotation of the connection, respectively.

The connection stiffness R in Equation 23.1 can be the initial connection rotational stiffness, R_i, as shown in Figure 23.3. Alternatively, a secant stiffness of the connection, R_s, which corresponds to a rotation of θ_0 ($= M_u/R_i$), can be determined from the initial stiffness R_i, the ultimate moment capacity M_u, and the moment–rotation curve of the connection. R_s is recommended instead of R_i as a representative connection

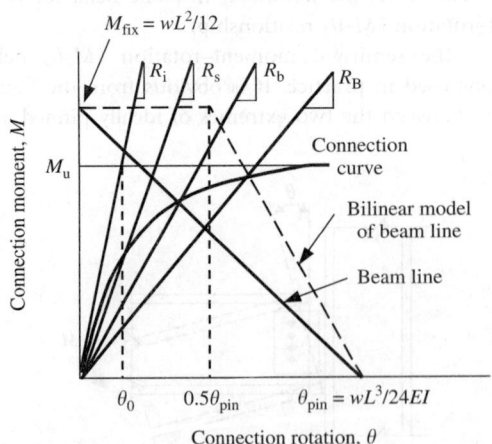

FIGURE 23.3   Various linear models for semirigid connection.

stiffness value because the initial stiffness of the connection does not represent connection response adequately.

To recognize the degradation of connection stiffness under increased load, the secant stiffness, R_b, determined in accordance with the so-called *beam line* method can be used in Equation 23.1. As shown in Figure 23.3, the beam line is defined as the straight line connecting two extreme connection conditions for a uniformly loaded beam, the end restrained moment when the connection is fully rigid and the end rotation when the connection is ideally pinned; that is, the rotation is zero for fully rigid end-connections with a resulting end-restrained moment of $wL^2/12$ and the moment is zero when the ends of the beam are pinned with resulting end rotation of $wL^3/24EI$. The connection stiffness R_b is then determined by the intersection of the beam line and moment–rotation curve of the connection. However, it is evident that the beam line does not represent the attainment of yield stress for any condition other than for a simply supported beam and a fixed ended beam, that is, the two end points of the line. To overcome this shortcoming, the bilinear model of the beam line [5] shown as the dashed line in Figure 23.3 can be used establish a connection stiffness R_B, which corresponds to an end rotation midway between pinned and fully fixed.

23.3.2 Polynomial Model

While the linear model in Equation 23.1 is simple and easy to use, it does not represent the true behavior of the connection. Bilinear and piecewise linear models have been proposed to provide a closer approximation of connection behavior. With the aim of providing a more accurate representation of nonlinear behavior of semirigid connections, nonlinear empirical models based on curve-fitting experimental results are widely employed to replicate and predict the nonlinear M–θ relationship of various types of beam-to-column connections. Shown in Equation 23.2 is an odd-power polynomial model [6]:

$$\theta = C_1(KM) + C_2(KM)^3 + C_3(KM)^5 \tag{23.2}$$

where θ is the connection rotation and M is the moment acting on the connection. Parameter K is a standardization factor determined by the connection type and geometry and C_1, C_2, and C_3 are curve-fitting constants obtained by using the method of least squares. The standardization factor and curve-fitting constants for various connections are shown in Table 23.1.

23.3.3 Three-Parameter Power Model

Chen and Kishi [7] adopted the power model developed by Richard and Abbott [8] to represent the moment–rotation characteristics of steel beam-to-column connections. The model is composed of three parameters: initial stiffness R_{ki}, ultimate moment capacity M_u, and a shape parameter n, and can be expressed as in Equation 23.3, taking the shape shown in Figure 23.4.

$$M = \frac{R_{ki}\theta}{\{1 + [\theta/\theta_0]^n\}^{1/n}} \tag{23.3}$$

where θ_0 is equal to M_u/R_{ki} is the reference plastic rotation. From Figure 23.4, it is recognized that the larger the shape parameter n the steeper the curve, which represents more rigid connections. The shape parameter n can be determined by using the least squares method applied to the differences between the predicted moments and the experimental test moments. The tangent connection stiffness R_{kt} and connection rotation can be determined from Equation 23.3 as follows:

$$R_{kt} = \frac{dM}{d\theta} = \frac{R_{ki}}{(1 + [\theta/\theta_0]^n)^{(n+1)/n}} \tag{23.4}$$

$$\theta = \frac{M}{R_{ki}(1 - [M/M_u]^n)^{1/n}} \tag{23.5}$$

TABLE 23.1 Curve-Fitting Constants and Standardization Constants for Polynomial Model (All Size Parameters are in Centimeters)

Connection types	Curve-fitting and standardization constants
Single web-angle connection	$C_1 = 1.67 \times 10^0$ $C_2 = 8.56 \times 10^{-2}$ $C_3 = 1.35 \times 10^{-2}$ $K = d_a^{-2.4} t_c^{-1.81} g^{0.15}$
Double web-angle connection	$C_1 = 1.43 \times 10^{-1}$ $C_2 = 6.79 \times 10^1$ $C_3 = 4.09 \times 10^5$ $K = d_a^{-2.4} t_c^{-1.81} g^{0.15}$
Header plate connection	$C_1 = 6.14 \times 10^{-3}$ $C_2 = 1.08 \times 10^{-3}$ $C_3 = 6.05 \times 10^{-3}$ $K = t_p^{-1.6} g^{1.6} d_p^{-2.3} t_w^{-0.5}$
Top- and seat-angle connections	$C_1 = 2.59 \times 10^{-1}$ $C_2 = 2.88 \times 10^3$ $C_3 = 3.31 \times 10^4$ $K = d^{-1.5} t_a^{-0.5} l_a^{-0.7} d_b^{-1.1}$
Top- and seat-angle with double web-angle connection	$C_1 = 1.50 \times 10^{-3}$ $C_2 = 5.60 \times 10^{-3}$ $C_3 = 4.35 \times 10^{-3}$ $K = d^{-1.287} t_a^{-1.128} t_c^{-0.415} l_a^{-0.694} (g - 0.5 d_b)^{1.35}$
Extend end-plate connection without column stiffeners	$C_1 = 8.91 \times 10^{-1}$ $C_2 = -1.20 \times 10^4$ $C_3 = 1.75 \times 10^8$ $K = d_g^{-2.4} t_p^{-0.4} t_f^{-1.5}$
Extended end-plate connection with column stiffeners	$C_1 = 2.60 \times 10^{-1}$ $C_2 = 5.36 \times 10^2$ $C_3 = 1.31 \times 10^7$ $K = d_g^{-2.4} t_p^{-0.6}$
T-stub connection	$C_1 = 6.42 \times 10^{-2}$ $C_2 = 1.77 \times 10^2$ $C_3 = -2.03 \times 10^4$ $K = d^{-1.5} t^{-0.5} l_t^{-0.7} d_b^{-1.1}$

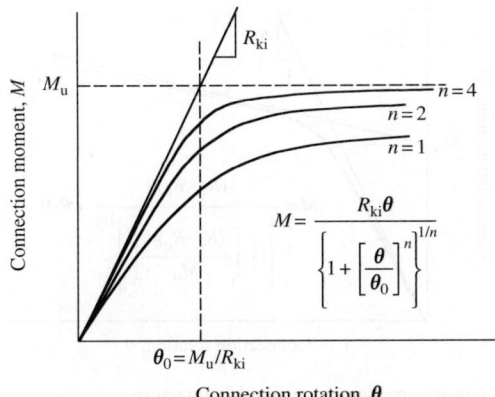

FIGURE 23.4 Three-parameter power model of semirigid connection.

Kishi and Chen [9] applied the model to four types of connections: single web-angle connections, double web-angle connections, top- and seat-angle connections, and top- and seat-angle with double web-angle connections. The detailed procedure to evaluate the three parameters R_{ki}, M_u, and n for the four types of connections is described by Chen and Lui [10].

23.3.4 Four-Parameter Power Model

As experimental tests have shown that many semirigid connections do not exhibit the plateau in their M–θ relationship shown in Figure 23.4 even at large rotation, the four-parameter power model shown in Equation 23.6 and Figure 23.5 has been adopted by a number of researchers [11–13] to simulate the nonlinear moment–rotation relationship of such connections:

$$M = \frac{(R_e - R_p)\theta}{\{1 + [((R_e - R_p)\theta)/M_0]^n\}^{1/n}} + R_p\theta \tag{23.6}$$

where, as shown in Figure 23.5, R_e is the initial stiffness of the connection, R_p is the plastic stiffness of the connection, M_0 is a reference moment, and n is a shape parameter. One of the advantages of this model is that it can easily represent simpler models, such as the linear model by setting $R_e = R_p$ and the three-parameter power model by setting $R_p = 0$. The model also approaches a bilinear model when n becomes large.

23.3.5 Connection Database

The connection database considered herein is a collection of experimental test results for different types of steel beam-to-column connections. The data usually collected are the corresponding details and dimensions of the connection, fastener, and connected beam and column. The purpose of the connection database is to provide the necessary information for the determination of the moment–rotation characteristics of beam-to-column connections, which are often required by design standards in facilitating the analysis and design of semirigid framed structures.

As extensive tests had been previously conducted to investigate the moment–rotation characteristics of various types of connections, efforts to establish connection databases were initiated in 1980s. Goverhan [14] collected a total of 230 experimental moment–rotation curves and digitized them to form the corresponding database of connection behavior. Nethercot [15] conducted an extensive literature survey of more than 800 individual tests from over 70 studies. Kishi and Chen [16] expanded and extended Goverdhan's collection to a total 303 tests, which consisted of experimental data published from 1936 to 1985. After that, Kishi and his colleagues revised the database and appended 93 sets of experimental data published from 1986 to 1998.

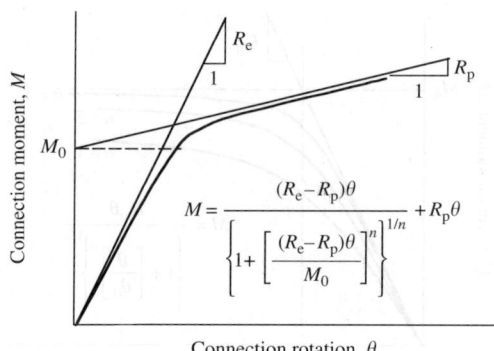

FIGURE 23.5 Four-parameter power model of semirigid connection.

Three prediction moment–rotation equations were incorporated with Kishi and Chen's database: the polynomial model shown in Equation 23.2, a modified exponential model [17], and the three-parameter power model shown in Equation 23.3. The validity of either the three-parameter power model or the modified experimental model for practical use was examined by Kishi et al. [18], who concluded that the three-parameter power model adequately describes the experimental test data for modeling the connection moment–rotation behavior for practical use.

23.3.6 Connection Classification

In practice, steel framed structures are to be designed in accordance with one of the three types of steel framing constructions: continuous construction, simple construction, and semirigid construction. Therefore, designers must be knowledgeable as to when the connections can be assumed to be rigid, semirigid, or flexible. The purpose of connection classification is to select a suitable basis on which to conduct the frame analysis and design. Based on the stiffness and strength of a connection, the connection can be classified as follows:

- *Stiffness criterion.* A connection is categorized as rigid, semirigid, and flexible based on the ratio of its rotational stiffness (R) to the stiffness of the beam (EI/L) it connects, RL/EI, where L and EI are the length and bending rigidity of the beam, respectively.
- *Strength criterion.* The most important aspect of the strength of a connection is its relationship to the strength of the connected beam. Based on comparison of the moment resistance (M_n) of a connection to the plastic moment resistance (M_p) of the beam it connects, the connection can be classified as a full strength, partial strength, or flexible connection.

Table 23.2 summarizes three classifications of connections presented by Bjorhovde et al. [19], Eurocode 3 [4], and LRFD [2] based on the foregoing criteria. The initial stiffness of the connection (R_i) was adopted in classifications by Bjorhovde et al. and Eurocode 3, while the secant stiffness of the connection (R_s) was used in the LRFD classification. As shown in Table 23.2, the secant stiffness, R_s, is defined on the basis of either the moment, M_s, or the rotation, θ_s, which would occur under the applied loads. The LRFD classification suggests that two distinct values of secant stiffness should be adopted to characterize the connection behavior under the two limit states of serviceability and ultimate load. The moment resistance of the connection, M_n, can be determined on the basis of an ultimate limit-state model of the connection or from tests. When a semirigid connection does not exhibit a plateau in its moment–rotation relationship, the moment resistance of the connection is defined at the rotation $\theta_n = 0.02$ radians [2,12,13,20].

In addition to the strength and stiffness of the connection, a third important characteristic that should be considered in connection classification is the ductility of the connection. Connection ductility

TABLE 23.2 Summary of Connection Classification Systems

Bjorhovde classification [19]	Eurocode 3 classification [4]	AISC LRFD classification [3]

Stiffness criterion

Bjorhovde classification [19]	Eurocode 3 classification [4]	AISC LRFD classification [3]
Rigid:	Rigid:	Rigid:
$R_i \geq EI/2d$	$R_i L/EI \geq 25$ (unbraced frame)	$R_s L/EI \geq 20$
	$R_i L/EI \geq 8$ (braced frame)	
Semirigid:	Semirigid:	Semirigid:
$EI/10d < R_i < EI/2d$	$2 < R_i L/EI < 20$ (unbraced frame)	$2 < R_s L/EI < 20$
	$2 < R_i L/EI < 8$ (braced frame)	
Flexible:	Flexible:	Flexible:
$EI/10d \geq R_i$	$R_i L/EI \leq 2$	$R_s L/EI \leq 2$

Strength criterion

Bjorhovde classification [19]	Eurocode 3 classification [4]	AISC LRFD classification [3]
Rigid:	Full strength:	Full strength:
$M_n \geq 0.7 M_p$	$M_n \geq M_p$	$M_n \geq M_p$
Semirigid:	Partial strength:	Partial strength:
$0.2 M_p < M_n < 0.7 M_p$	$0.25 M_p < M_n < M_p$	$0.2 M_p < M_{n\_\theta=0.02};\ M_n < M_p$
Flexible:	Flexible:	Flexible:
$M_n \leq 0.7 M_p$	$M_n \leq 0.25 M_p$	$M_{n\_\theta=0.02} \leq 0.2 M_p$
$M' = M/M_p;\ \theta' = \theta/\theta_p$	$M^* = M/M_p;\ \theta^* = \theta/\theta_p$	$M_{n_\theta=0.02} = $ moment resistance of the
$\theta_p = M_p/(EI/5d)$	$\theta_p = M_p/(EI/L)$	connection at a rotation of
$d =$ depth of beam		0.02 radians

is a key parameter when joint deformations are concentrated in the connection elements, as is the typical case for partial strength connections where the strength of the connected beam exceeds that of the connection. The connection ductility can be represented by its rotational capacity, θ_u, which is defined in the LRFD classification as the value at which the moment resistance of the connection has dropped to $0.8 M_n$ or the connection has deformed beyond 0.03 radians. That is, a connection is classified as ductile if its rotation capacity is equal or greater than 0.03 radians, otherwise the connection is considered to be nonductile.

The difference between the LRFD and Eurocode 3 classifications is that the latter one acknowledges the fact that the extent of semirigid action is largely associated with the type of structure, such as a braced or unbraced frame. Therefore, in Eurocode 3 the connections are classified separately with regard to braced and unbraced frames. The Eurocode 3 classification considers the load-carrying capacity of frames, which is more rational. For example, the stiffness boundaries between the rigid and semirigid zones are chosen such that the drop in load carrying capacity, due to semirigid behavior, is not more than 5% (in terms of the Euler buckling load). However, the ductility demand is not considered in Eurocode 3.

As the purpose of connection classification is to facilitate frame design for different limit states, it is necessary that a classification system be able to reflect the connection assumptions used in the frame analysis and design at each limit state. To that purpose, the connections should be classified corresponding not only to the ultimate limit state but also to the serviceability limit state, as a connection may respond differently at the two different limit states.

23.4   Structural Analysis of Semirigid Framed Structures

The structural analysis procedures for semirigid frames presented in this section are based on the following assumptions and idealizations:

1. All members are prismatic and straight.
2. Only the moment–rotation behavior of connections is considered (i.e., axial and shear deformations in a connection are ignored).
3. Members display linear-elastic or second-order elastic behavior, while the connections display nonlinear moment–rotation behavior.
4. Connection dimensions are assumed to be negligible compared to the lengths of the beams and columns; that is, the rotational deformation of a connection is assumed to be concentrated at a point at the end of the semirigid member.
5. The effects of eccentricity at joints are neglected.

23.4.1   End-Fixity Factor of Semirigid Member

Shown in Figure 23.6 is a semirigid member comprising a finite-length beam–column member with a zero-length rotational spring at each end (the symbol @ represents the spring). The connection flexibilities are modeled through rotational springs of stiffness R_1 and R_2 at the two ends of the beam. The relative stiffness of the beam–column member and the rotational end-spring connection is measured by an end-fixity factor defined as follows [21]:

$$r_j = \frac{1}{1 + 3EI/R_jL} \quad (j = 1, 2) \tag{23.7}$$

where R_j is the end-connection spring stiffness and EI/L is the flexural stiffness of the attached member. The fixity factor r_j defines the rotational stiffness of each end-connection relative to that of the attached member. For a pinned connection, the rotational stiffness of the connection is idealized as zero and thus the value of the end-fixity factor is zero ($r_j = 0$). For a rigid connection, the connection rotational stiffness is taken to be infinite and the end-fixity factor has a value of unity ($r_j = 1$). Therefore, a semirigid connection has an end-fixity factor between zero and one ($0 < r_j < 1$).

Upon the introduction of the end-fixity factor, different member-end restraint conditions, such as rigid–pinned, rigid–semirigid, pinned–semirigid, are readily modeled simply by setting the end-fixity factors at the two ends of the member to appropriate values. This comprehensive model of member end-rotational conditions can be employed for the analysis of frames with any combination of pinned, rigid, and semirigid connections.

The end-fixity factor also simplifies the analysis procedure for semirigid framed structures. Figure 23.7 and Figure 23.8 illustrate the variation of mid-span and end moments and mid-span deflection of a uniformly loaded semirigid beam versus the end-fixity factor. Comprehensive formulations for calculation of both end-reactions and span deflections in terms of end-fixity factors for a semirigid member under different types of loads are listed in Table 23.3 and Table 23.4. As will be

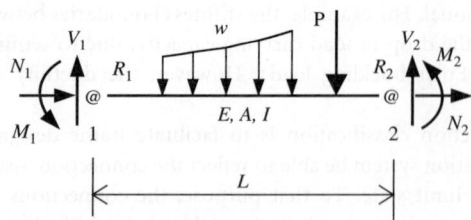

FIGURE 23.6   Model of semirigid member.

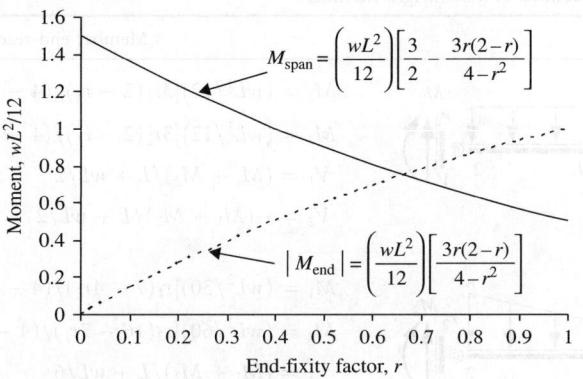

FIGURE 23.7 Moments of a uniformly loaded beam versus end-fixity factor.

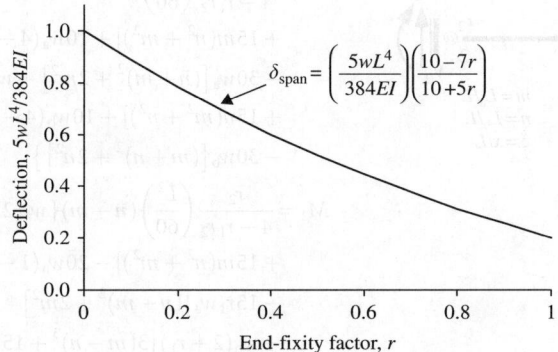

FIGURE 23.8 Deflection of a uniformly loaded beam versus end-fixity factor.

demonstrated subsequently, the formulations of stiffness matrices and effective-length factors of beam-columns can all be expressed in terms of the end-fixity factors. The end-fixity factor has further value in design because it provides a physical interpretation of the extent of rigidity available in a connection. It also provides designers with a convenient way to compare the structural responses of a member with semirigid connections to that with rigid or pinned connections.

By Equation 23.7, the relationship between the end-fixity factor and the connection stiffness is nonlinear, as shown in Figure 23.9. It is observed from the figure that when the connection stiffness is large, very significant changes in stiffness produce only very small changes in the end-fixity factor. Consequently, from Figure 23.7 and Figure 23.8, such changes have a negligible influence on both moments and deflection of the beam. Conversely, from Figure 23.9, with low values of connection stiffness, small increases in the stiffness result in appreciable increases in the end-fixity factor. In this case, as Figure 23.7 and Figure 23.8 show, there is a considerable effect on the bending moments and the deflection. Thus, in practice, when a real pinned connection has some stiffness, a considerable restraining moment may develop to the benefit of the structure. At the other extreme, attempting to achieve further increase in connection stiffness beyond that of a nearly rigid connection is not efficient and economical because it involves only a small change in the end-fixity factor and, consequently, has little effect on the response of structure.

TABLE 23.3 End-Reactions of a Semirigid Member

Member span loads	Member end-reactions
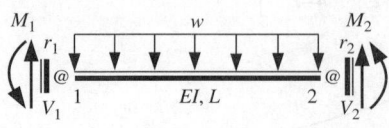	$M_1 = (wL^2/12)[3r_1(2-r_2)/(4-r_1r_2)]$ $M_2 = (wL^2/12)[3r_2(2-r_1)/(4-r_1r_2)]$ $V_1 = (M_1+M_2)/L + wL/2$ $V_2 = -(M_1+M_2)/L + wL/2$
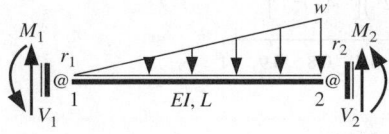	$M_1 = (wL^2/30)[r_1(7-4r_2)/(4-r_1r_2)]$ $M_2 = (wL^2/60)[r_2(16-7r_1)/(4-r_1r_2)]$ $V_1 = (M_1+M_2)/L + wL/6$ $V_2 = -(M_1+M_2)/L + wL/3$

$$M_1 = \frac{r_1}{4-r_1r_2}\left(\frac{L^2}{60}\right)(n-m)\{w_a(2+r_2)[3(n-m)^3$$
$$+15m(n^2+m^2)]+10w_a(4-r_2)(n+2m)$$
$$-30w_a[(n+m)^2+2m^2]+w_b(2+r_2)[3(m-n)^3$$
$$+15n(m^2+n^2)]+10w_b(4-r_2)(m+2n)$$
$$-30w_b[(m+n)^2+2n^2]\}$$

$$M_2 = \frac{r_2}{4-r_1r_2}\left(\frac{L^2}{60}\right)(n-m)\{w_a(2+r_1)[3(n-m)^3$$
$$+15m(n^2+m^2)]-20w_a(1-r_1)(n+2m)$$
$$-15r_1w_a[(n+m)^2+2m^2]$$
$$+w_b(2+r_1)[3(m-n)^3+15n(m^2+n^2)]$$
$$-20w_b(1-r_1)(m+2n)-15r_1w_b[(m+n)^2+2n^2]\}$$

$$V_1 = \frac{(M_1+M_2)}{L}+\frac{(n-m)L}{6}[w_a(3-2m-n)$$
$$+w_b(3-m-2n)]$$

$$V_2 = -\frac{(M_1+M_2)}{L}+\frac{(n-m)L}{6}[w_a(2m+n)+w_b(m+2n)]$$

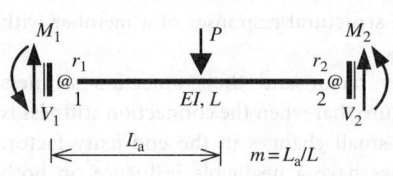	$M_1 = \frac{r_1}{4-r_1r_2}m(1-m)[2(2-m)-r_2(1+m)]PL$ $M_2 = -\frac{r_2}{4-r_1r_2}m(1-m)[2(1+m)-r_1(2-m)]PL$ $V_1 = (M_1+M_2)/L + P(1-m)$ $V_2 = -(M_1+M_2)/L + Pm$
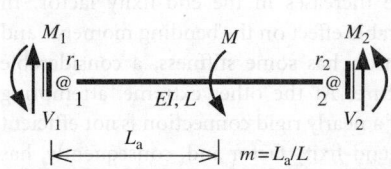	$M_1 = -\frac{r_1}{4-r_1r_2}[6(1-m)^2-2+r_2(3m^2-1)]M$ $M_2 = -\frac{r_2}{4-r_1r_2}[2(3m^2-1)-3r_1(1-m)^2-r_1]M$ $V_1 = -V_2 = (M+M_1+M_2)/L$

TABLE 23.4 Span Deflection of a Semirigid Member

Member span loads	Member span deflection
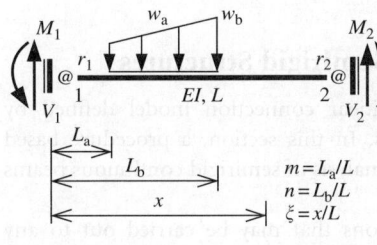	$\delta(\xi) = \delta_M(\xi) - \dfrac{wL^4\xi}{24EI}(\xi^3 - 2\xi^2 + 1) \quad (0 \le \xi \le 1)$ $\delta_M(\xi) = \dfrac{M_1 L^2 \xi}{6EI}(1-\xi)(2-\xi) - \dfrac{M_2 L^2 \xi}{6EI}(1-\xi^2)$

$$\delta(\xi) = \delta_M(\xi) - \frac{wL^4\xi}{360EI}(3\xi^4 - 10\xi^2 + 7) \quad (0 \le \xi \le 1)$$

$$\delta_M(\xi) = \frac{M_1 L^2 \xi}{6EI}(1-\xi)(2-\xi) - \frac{M_2 L^2 \xi}{6EI}(1-\xi^2)$$

$$\delta_1(\xi) = \delta_M(\xi) + \frac{L^4\xi}{36EI}(n-m)\{[\xi^2 - n^2(3-2n)]$$
$$\times [w_a(3-2m-n) + w_b(3-m-2n)]$$
$$- 2(1-n)^3 [w_a(2m+n) + w_b(m+2n)]\}$$
$$+ \frac{L^4\xi}{120EI}(n-m)^3[5(1-n)(w_b + 3w_a)$$
$$+ (n-m)(w_b + 4w_a)]$$

$$\delta_2(\xi) = \delta_1(\xi) - \frac{L^4}{120EI}(\xi - m)^4 \left[5w_a + \frac{w_b - w_a}{n-m}(\xi-m)\right]$$

$$\delta_3(\xi) = \delta_M(\xi) + \frac{L^4(1-\xi)}{36EI}(n-m)\{[(1-\xi^2)$$
$$- (1-n^2)(1+2n)] \times [w_a(2m+n) + w_b(m+2n)]$$
$$- 2n^3[w_a(3-2m-n) + w_b(3-m-2n)]\}$$
$$- \frac{L^4(1-\xi)}{120EI}(n-m)^3$$
$$\times [(n-m)(w_b + 4w_a) + 5n(w_b + 3w_a)]$$

$$\delta_M(\xi) = \frac{M_1 L^2 \xi}{6EI}(1-\xi)(2-\xi) - \frac{M_2 L^2 \xi}{6EI}(1-\xi^2)$$

$$\delta(\xi) = \delta_1(\xi), \quad 0 \le \xi \le m$$
$$\delta(\xi) = \delta_2(\xi), \quad n \le \xi \le m$$
$$\delta(\xi) = \delta_3(\xi), \quad m \le \xi \le n$$

Member span loads	Member span deflection
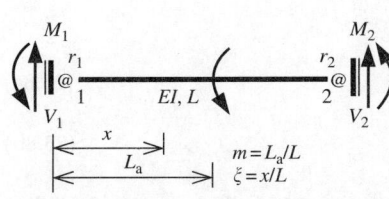	$\delta_1(\xi) = \delta_M(\xi) - \dfrac{PL^3\xi}{6EI}(1-m)[1 - \xi^2 - (1-m)^2] \quad (0 \le \xi \le m)$ $\delta_2(\xi) = \delta_M(\xi) - \dfrac{PL^3(1-\xi)}{6EI}m(2\xi - \xi^2 - m^2) \quad (m < \xi \le 1)$ $\delta_M(\xi) = \dfrac{M_1 L^2 \xi}{6EI}(1-\xi)(2-\xi) - \dfrac{M_2 L^2 \xi}{6EI}(1-\xi^2)$

$$\delta_1(\xi) = \delta_M(\xi) + \frac{ML^2\xi}{6EI}(\xi^2 - 6m + 3m^2 + 2) \quad (0 \le \xi \le m)$$

$$\delta_2(\xi) = \delta_M(\xi) - \frac{ML^2(1-\xi)}{6EI}[(1-\xi)^2 + 3m^2 - 1] \quad (m < \xi \le 1)$$

$$\delta_M(\xi) = \frac{M_1 L^2 \xi}{6EI}(1-\xi)(2-\xi) - \frac{M_2 L^2 \xi}{6EI}(1-\xi^2)$$

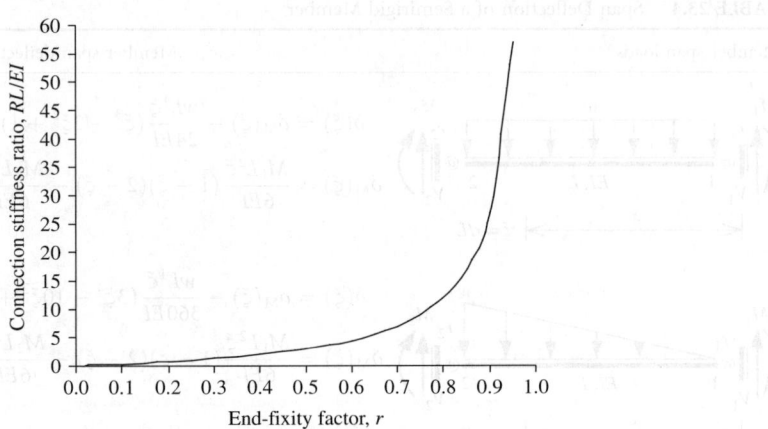

FIGURE 23.9   Relationship between connection stiffness and end-fixity factor.

23.4.2 Manual-Based Approximate Analysis of Semirigid Structures

For simple semirigid framed structures, analysis using the linear connection model defined by
Equation 23.1, can be conducted manually as for rigid frames. In this section, a procedure based
on a modified moment distribution method is presented for the analysis of semirigid continuous beams
and braced frames.

Moment distribution is a method of successive approximations that may be carried out to any
desired degree of accuracy. The convergence of the method is guaranteed for stable linear elastic
structures under small deformations and displacements. Its primary advantage over other hand-
calculation structural analysis procedures, such as the method of consistent deformations and the slope
deflection method, is that it eliminates the requirement to solve a set of simultaneous equations to
determine the member end moments for statically indeterminate beams or nonsway plane frames in
which the joints are restrained against translation.

Upon the introduction of the end-fixity factor defined in Equation 23.7 to take into account semirigid
connection behavior based on Equation 23.1, the analysis of semirigid continuous beams and frames can
be conducted through minor modification of the conventional moment distribution method for rigid
continuous beams and frames, as described in the following:

- *Sign convention.* All loads acting on the members in the mathematical model of a semirigid
  continuous beam or a plane frame are referenced to the individual local member coordinate
  system. As indicated in Figure 23.6, positive load acts downward and positive end moment and
  shear act counter-clockwise and upward, respectively.
- *Member end-reaction moments.* The end moments of a semirigid member due to different types of
  applied loads are listed in Table 23.3.
- *Member rotational stiffness factor K_{ij}.* For a semirigid member i, with different end-connection
  stiffnesses R_1 and R_2 determined by Equation 23.1, let the end-fixity factors defined in
  Equation 23.7 for the ends 1 and 2 be r_1 and r_2, respectively. The end-stiffness factors K_{i1} and K_{i2}
  for the member can be expressed as

$$K_{i1} = \frac{3r_1}{4 - r_1 r_2}\frac{4EI}{L} \qquad\qquad (23.8a)$$

$$K_{i2} = \frac{3r_2}{4 - r_1 r_2}\frac{4EI}{L} \qquad\qquad (23.8b)$$

in which $3r_1/(4 - r_1 r_2)$ and $3r_2/(4 - r_1 r_2)$ are factors accounting for the end-connection flexibility.

- *Distribution factor D_{ij}.* The distribution factor is defined as the stiffness factor of the member divided by the sum of the stiffness factors of all members connecting at the joint,

$$D_{ij} = \frac{K_{ij}}{\sum_{i=1}^{n} K_{ij}} \quad (j = 1, 2) \tag{23.9}$$

This factor corresponds to the fraction of the total applied moments at a joint that will be resisted by any member i as the joint rotates through an angle ϕ, where n is the number of members connected at the joint.

- *Carry-over factor C_{ij}.* When a moment is applied to the near end of any member whose far end is restrained against rotation during the distribution process, a moment must be applied to the far end by the joint to resist rotation of the member end. This moment is called the carry-over moment and the ratio of the far-end moment to the near-end moment is known as the carry-over factor. The carry-over factors for a semirigid member i are

$$C_{i1} = 0.5r_2 \tag{23.10a}$$
$$C_{i2} = 0.5r_1 \tag{23.10b}$$

Implementation of the mathematical model of the moment distribution method for semirigid continuous beams, or semirigid braced frames in which the joints are restrained against translation, consists of the following steps:

1. Evaluate end-fixity factors r_1 and r_2 according to Equation 23.7 for each member for the given values of connection stiffness R_j ($j = 1, 2$). For cases of pinned and fixed-end connections, assign the end-fixity factor to zero and unity, respectively.
2. Calculate end-stiffness factors K_{ij} ($j = 1, 2$) according to Equations 23.8 for each member i and determine moment distribution factors D_{ij} from Equation 23.9.
3. Compute the end-reaction moments $M_{ij}^{(k)}$ ($j = 1, 2$) due to the applied loads for each member through Table 23.3, assuming that all joints are locked by fictitious rotation restraints, and set index $k = 0$.
4. Sum the member end moments at each joint to determine the unbalanced moment $M^{(k)} = -\sum M_{ij}^{(k)}$ there.
5. Release or "unlock" the joints by removing the fictitious rotation restraint at each joint in turn, assuming that the other joints are still restrained against rotation, and distribute the unbalanced moment at the joint to the near end of each connecting member as $M_{ij}^{(k+1)} = D_{ij} \times M^{(k)}$.
6. Carry the near-end moment of each member over to its other end by multiplying each moment with the carry-over factor, that is, $M_{i1}^{(k+1)} = C_{i2} \times M_{i2}^{(k+1)}$ or $M_{i2}^{(k+1)} = C_{i1} \times M_{i1}^{(k+1)}$.
7. If the unbalanced moment $M^{(k)}$ is small enough to be considered insignificant, then terminate the iteration and proceed to step 8; otherwise, set $k = k + 1$ and return to step 4.
8. Sum the end moments of each member. Calculate the axial forces (for nonsway frames) and the shear forces of the members satisfying equilibrium and determine the reaction forces and moments at the supports based on the joint equilibrium conditions.

EXAMPLE 23.1

For the three-span continuous beam shown in Figure 23.10, all of the end-connections are semi-rigid except for the connection at joint D, which is rigid ($r = 1$). The secant rotational stiffnesses of the semirigid connections associated with the serviceability and ultimate load states are 1.15×10^{10} and 5.75×10^9 N mm/rad, respectively. The beam is subjected to uniformly distributed dead and live loads, 3.2 and 12.0 kN/m, respectively. The beam is laterally braced and has a W410 × 39 section with cross-sectional properties $I_x = 126 \times 10^6$ mm^4 and $Z_x = 730 \times 10^3$ mm^3. The yield stress F_y and elastic modulus E of the steel are 345 MPa and 2.0×10^5 MPa, respectively. It is required to employ the modified moment distribution method for semirigid continuous beams to determine the bending moment distribution

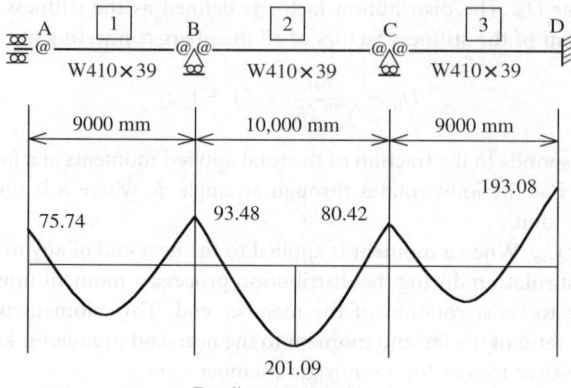

FIGURE 23.10   Example 1: three-span semirigid continuous beam.

of the structure and to then check if the flexural strength of the beam is adequate based on LRFD [2]. The mid-span deflection of the beam due to the service load is required to be less than or equal to $L/240$.

For the ultimate load state and connection stiffness 5.75×10^9 N mm/rad, the values of the corresponding end-fixity factor for beams 1 and 3 of length $L = 9000$ mm are

$$r_{1A} = r_{1B} = r_{3C} = \frac{1}{1 + 3EI/RL} = \frac{1}{1 + (3 \times 200 \times 10^3 \times 126 \times 10^6)/(5.75 \times 10^9 \times 9000)} = 0.406$$

and $r_{3D} = 1.0$.

For beam 2, $L = 10{,}000$ mm

$$r_{2B} = r_{2C} = \frac{1}{1 + 3EI/RL} = \frac{1}{1 + (3 \times 200 \times 10^3 \times 126 \times 10^6)/(5.75 \times 10^9 \times 10{,}000)} = 0.432$$

Carry-over factors for the beams are

$$C_{1A} = 0.406/2 = 0.203, \quad C_{2B} = C_{2C} = 0.432/2 = 0.216, \quad C_{3D} = 1.0/2 = 0.5$$

Stiffness factors K_{1B} and K_{3C} for beams 1 and 3 are

$$K_{1B} = \frac{3r_{1B}}{4 - r_{1A}r_{1B}} \cdot \frac{4EI}{L} = \frac{3 \times 0.406}{4 - 0.406^2} \cdot \frac{4EI}{9000} = 1.411 \times 10^{-4} EI/\text{mm}$$

$$K_{3C} = \frac{3r_{3C}}{4 - r_{3C}r_{3D}} \cdot \frac{4EI}{L} = \frac{3 \times 0.406}{4 - 0.406 \times 1.0} \cdot \frac{4EI}{9000} = 1.506 \times 10^{-4} EI/\text{mm}$$

Stiffness factors K_{2B} and K_{2C} for beam 2 are

$$K_{2B} = K_{2C} = \frac{3r_{2B}}{4 - r_{2B}r_{2C}} \cdot \frac{4EI}{L} = \frac{3 \times 0.432}{4 - 0.432^2} \cdot \frac{4EI}{10{,}000} = 1.359 \times 10^{-4} EI/\text{mm}$$

At joint B, the distribution factors for beams 1 and 2 are

$$DF_{1B} = \frac{K_{1B}}{K_{1B} + K_{2B}} = \frac{1.411 \times 10^{-4} EI/\text{mm}}{1.413 \times 10^{-4} EI/\text{mm} + 1.359 \times 10^{-4} EI/\text{mm}} = 0.509$$

$$DF_{2B} = 1 - DF_{1B} = 1 - 0.509 = 0.491.$$

At joint C, the distribution factors for beams 2 and 3 are

$$DF_{2C} = \frac{K_{2C}}{K_{2C} + K_{3C}} = \frac{1.359 \times 10^{-4} EI/\text{mm}}{1.359 \times 10^{-4} EI/\text{mm} + 1.506 \times 10^{-4} EI/\text{mm}} = 0.474$$

$$DF_{3C} = 1 - DF_{2C} = 1 - 0.474 = 0.526.$$

The uniformly distributed factored load of the beam is

$$w_f = 1.2 \text{ dead load} + 1.6 \text{ live load} = 1.2 \times 3.2 \text{ kN/m} + 1.6 \times 12 \text{ kN/m} = 23.04 \text{ kN/m}$$

From Table 23.3, the end-reaction moments of the beams associated with the factored load are

$$M_{1A}^{(0)} = -M_{1B}^{(0)} = \frac{wL^2}{12} \cdot \frac{3r_{1A}(2 - r_{1B})}{4 - r_{1A}r_{1B}} = \frac{23.04(9)^2}{12} \cdot \frac{3 \times 0.406(2 - 0.406)}{4 - 0.406^2} = 78.73 \text{ kN m}$$

End-reaction moments of beam 2 due to the load are

$$M_{2B}^{(0)} = -M_{2C}^{(0)} = \frac{wL^2}{12} \cdot \frac{3r_{2B}(2 - r_{2C})}{4 + r_{2B}r_{2C}} = \frac{23.04(10)^2}{12} \cdot \frac{3 \times 0.432(2 - 0.432)}{4 - 0.432^2} = 102.32 \text{ kN m}$$

End-reaction moments of beam 3 due to the load are

$$M_{3C}^{(0)} = \frac{wL^2}{12} \cdot \frac{3r_{3C}(2 - r_{3D})}{4 + r_{3C}r_{3D}} = \frac{23.04(9)^2}{12} \cdot \frac{3 \times 0.406(2 - 1)}{4 - 0.406 \times 1} = 52.71 \text{ kN m}$$

$$M_{3D}^{(0)} = \frac{wL^2}{12} \cdot \frac{3r_{3D}(2 - r_{3C})}{4 + r_{3C}r_{3D}} = \frac{23.04(9)^2}{12} \cdot \frac{3 \times 1(2 - 0.406)}{4 - 0.406 \times 1} = -206.93 \text{ kN m}$$

The modified moment distribution is conducted as shown in Table 23.5 and the final moment distribution of the beam is shown in Figure 23.10. The maximum span moment of Beam 2 is 201.09 kN m, which is located 5.06 m from joint B. The maximum negative moment at joint D is 193.08 kN m. Since the beam is laterally braced and W410 × 39 is a compact section, the flexural strength of the beam based on LRFD [2] is

$$M_r = \phi_b M_p = 0.9 Z_x F_y = 0.9 \times 730 \times 10^3 \text{ mm}^3 \times 345 \text{ MPa} \times 10^{-6} = 226.67 \text{ kN m} > 201.09 \text{ kN m}$$

Therefore, the flexural strength of the beam is adequate.

To calculate the mid-span deflection of Beam 2 due to the service load $w = 3.2 \text{ kN/m} + 12.0 \text{ kN/m} = 15.2 \text{ kN/m}$, and the connection rotational stiffness $R = 1.15 \times 10^{10} \text{ N mm/rad}$, the corresponding end-fixity factors of the beam are

$$r_{1A} = r_{1B} = r_{3C} = 0.578, \quad r_{3D} = 1.0, \quad \text{and} \quad r_{2B} = r_{2D} = 0.603$$

Following the foregoing procedure, the final moments obtained by the modified moment distribution at joints B and C are $M_{2B} = 81.67 \text{ kN m}$ and $M_{2B} = -73.04 \text{ kN m}$.

TABLE 23.5 Moment Distribution of Three-Span Semirigid Continuous Beam

Joint	A	B		C		D
Member end	1A	1B	2B	2C	3C	3D
DF		0.509	0.491	0.474	0.526	
FEM	78.73	−78.73	102.32	−102.32	52.71	−206.93
Distribution		−12.01	−11.58	23.52	26.09	
Carry-over	−2.44		5.08	−2.50		13.05
Distribution		−2.59	−2.49	1.19	1.32	
Carry-over	−0.52		0.26	−0.54		0.66
Distribution		−0.13	−0.13	0.26	0.28	
Carry-over	−0.03		0.06	−0.03		0.14
Distribution		−0.03	−0.03	0.01	0.01	
Final moment (kN m)	75.74	−93.48	93.48	−80.42	80.42	−193.08

Therefore, the mid-span deflection of beam 2 can be obtained from the deflection equation shown in Table 23.4 as

$$|\delta(\xi)| = \left| \left[\frac{M_1 L^2 \xi}{6EI}(1-\xi)(2-\xi) - \frac{M_2 L^2 \xi}{6EI}(1-\xi^2) - \frac{wL^4 \xi}{24EI}(\xi^3 - 2\xi^2 + 1) \right]_{\xi=0.5} \right|$$

$$= \left| \frac{81.67 \times 10^6 \times 10{,}000^2 \times 0.5}{6 \times 200 \times 10^3 \times 126 \times 10^6} \times (1-0.5)(2-0.5) - \frac{-73.04 \times 10^6 \times 10{,}000^2 \times 0.5}{6 \times 200 \times 10^3 \times 126 \times 10^6}(1-0.5^2) \right.$$

$$\left. - \frac{15.2 \times 10{,}000^4 \times 0.5}{24 \times 200 \times 10^3 \times 126 \times 10^6}(0.5^3 - 2 \times 0.5^2 + 1) \right|$$

$$= |-40.2 \text{ mm}| = 40.2 \text{ mm} < \frac{L}{240} = \frac{10{,}000}{240} = 41.7 \text{ mm}$$

EXAMPLE 23.2

A one-storey two-bay semirigid braced frame of a low-rise building is shown in Figure 23.11. The spacing between the frames is 10 m. The roof dead load is 1.25 kN/m² and the snow load is 1.82 kN/m². The secant stiffness of the beam-to-column connections obtained from experimental tests associated with the connection rotation $\theta = 0.02$ radians are 2.8×10^{10} N mm/rad and 1.7×10^{10} N mm/rad for beams BC and CE, respectively. The rotational stiffnesses of the column bases are 3.0×10^9 N mm/rad for columns AB and DC and 1.75×10^9 N mm/rad for column EF. The diagonal bracing is to be designed as a tension-only member with pinned ends. It is required to find the bending moment distribution of the frame and axial loads of the columns for the load case of dead load plus live load by using the modified moment distribution method.

Load calculation:

The uniformly distributed nominal dead load is 10 m × 1.25 kN/m² = 12.5 kN/m
The uniformly distributed nominal snow load is 10 m × 1.82 kN/m² = 18.2 kN/m
Gravity load case (dead load + snow load):

$$w = 1.2D + 1.6L = 1.2 \times 12.5 + 1.6 \times 18.2 = 44.12 \text{ kN/m}$$

Properties of cross-sections:
Beam: W530 × 85, $I_{bx} = 485 \times 10^6$ mm⁴; Column: W250 × 33, $I_{cx} = 48.9 \times 10^6$ mm⁴

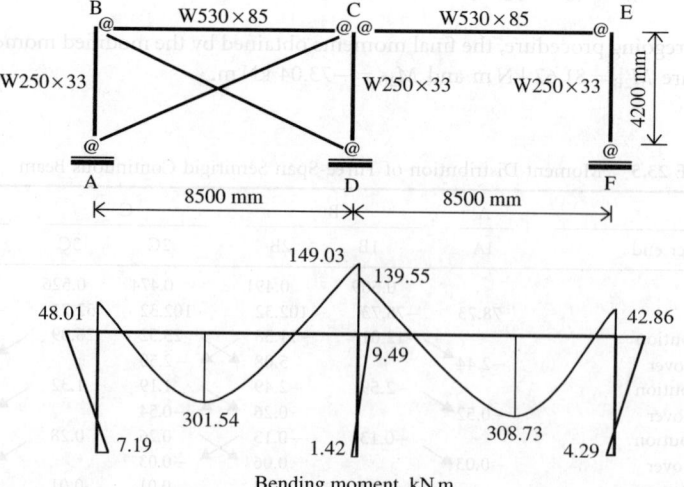

FIGURE 23.11 Example 2: two-bay semirigid braced frame.

End-fixity factor r of beams:

$$r_{BC} = r_{CB} = \frac{1}{1 + (3EI_{bx}/RL)} = \frac{1}{1 + \dfrac{3(200 \times 10^3)485 \times 10^6}{(2.8 \times 10^{10})8500}} = 0.45$$

$$r_{CE} = r_{EC} = \frac{1}{1 + (3EI_{bx}/RL)} = \frac{1}{1 + \dfrac{3(200 \times 10^3)485 \times 10^6}{(1.7 \times 10^{10})8500}} = 0.332$$

Carry-over factor for beams:

$$C_{BC} = C_{CB} = r_{BC}/2 = 0.45/2 = 0.225 \quad \text{and} \quad C_{CE} = C_{EC} = r_{CE}/2 = 0.332/2 = 0.166$$

End-fixity factor r of columns:

$$r_{AB} = r_{DC} = \frac{1}{1 + 3EI_{cx}/RL} = \frac{1}{1 + \dfrac{3(200 \times 10^3)48.9 \times 10^6}{(3 \times 10^9)4200}} = 0.30$$

$$r_{FE} = \frac{1}{1 + 3EI_{cx}/RL} = \frac{1}{1 + \dfrac{3(200 \times 10^3)48.9 \times 10^6}{(1.75 \times 10^9)4200}} = 0.20$$

$$r_{BA} = r_{CD} = r_{EF} = 1.0$$

Carry over factor for columns:

$$C_{BA} = C_{CD} = r_{AB}/2 = 0.30/2 = 0.15, \; C_{EF} = r_{FE}/2 = 0.20/2 = 0.10$$

Member stiffness factor:
Beams:

$$K_{BC} = K_{CB} = \frac{3r_{BC}}{4 - r_{BC}r_{CB}} \times \frac{4EI}{L} = \frac{3 \times 0.45}{4 - 0.45^2} \times \frac{4E(485 \times 10^6)}{8500} = 81{,}137E \times \text{mm}^3$$

$$K_{CE} = K_{EC} = \frac{3r_{CE}}{4 - r_{CE}r_{DE}} \times \frac{4EI}{L} = \frac{3 \times 0.332}{4 - 0.332^2} \times \frac{4E(485 \times 10^6)}{8500} = 58{,}441E \times \text{mm}^3$$

Columns:

$$K_{AB} = K_{DC} = \frac{3r_{AB}}{4 - r_{AB}r_{BA}} \times \frac{4EI}{L} = \frac{3 \times 0.3}{4 - 1 \times 0.3} \times \frac{4E(485 \times 10^6)}{4200} = 11{,}328E \times \text{mm}^3$$

$$K_{BA} = K_{CD} = \frac{3r_{BA}}{4 - r_{BA}r_{AB}} \times \frac{4EI}{L} = \frac{3 \times 1.0}{4 - 1.0 \times 0.30} \times \frac{4E(48.9 \times 10^6)}{4200} = 37{,}761E \times \text{mm}^3$$

$$K_{EF} = \frac{3r_{EF}}{4 - r_{EF}r_{FE}} \times \frac{4EI}{L} = \frac{3 \times 1.0}{4 - 1.0 \times 0.20} \times \frac{4E(48.9 \times 10^6)}{4200} = 36{,}767E \times \text{mm}^3$$

$$K_{FE} = \frac{3r_{FE}}{4 - r_{FE}r_{EF}} \times \frac{4EI}{L} = \frac{3 \times 0.2}{4 - 1.0 \times 0.20} \times \frac{4E(48.9 \times 10^6)}{4200} = 7353.4E \times \text{mm}^3$$

Distribution factor:
Joint B:

$$DF_{BA} = \frac{K_{BA}}{K_{BA} + K_{BC}} = \frac{37{,}761E \times \text{mm}^3}{37{,}761E \times \text{mm}^3 + 81{,}137E \times \text{mm}^3} = 0.318 \quad \text{and} \quad DF_{BC} = 1 - 0.318 = 0.682$$

Joint C:

$$DF_{CB} = \frac{K_{CB}}{K_{CB} + K_{CD} + K_{CE}} = \frac{81{,}136E \times \text{mm}^3}{81{,}136E \times \text{mm}^3 + 37{,}761E \times \text{mm}^3 + 58{,}441E \times \text{mm}^3} = 0.458$$

$$DF_{CE} = \frac{K_{CE}}{K_{CB} + K_{CD} + K_{CE}} = \frac{58{,}441E \times \text{mm}^3}{81{,}136E \times \text{mm}^3 + 37{,}761E \times \text{mm}^3 + 58{,}441E \times \text{mm}^3} = 0.330$$

$$DF_{CD} = 1 - 0.457 - 0.330 = 0.212$$

TABLE 23.6   Moment Distribution of the Two-Bay Semirigid Braced Frame

Joint	A	B			C		D	E		F
Member	AB	BA	BC	CB	CD	CE	DC	EC	EF	FE
DF		0.318	0.682	0.458	0.212	0.330		0.614	0.386	
FEM	0	0	146.37	−146.37	0	113.46	0	−113.46	0	0
Distribution		−46.55	−99.82	15.08	6.98	10.86		69.66	43.79	
Carry-over	−6.98		3.39	−22.46		11.56	1.05	1.80	0	4.38
Distribution		−1.08	−2.31	4.99	2.31	3.60		−1.11	−0.70	
Carry-over	−1.6		1.12	−0.52		−0.18	0.35	0.60	0	−0.07
Distribution		−0.36	−0.76	0.32	0.15	0.23		−0.37	−0.23	
Carry-over	−0.05		0.07	−0.17		−0.06	0.02	0.04	0	−0.01
Distribution		−0.02	−0.05	0.11	0.05	0.08		−0.02	−0.01	
Final moment (kN M)	−7.19	−48.01	48.01	−149.03	9.49	139.55	1.43	−42.86	42.86	4.29

Joint E:

$$DF_{EC} = \frac{K_{EC}}{K_{EC} + K_{EF}} = \frac{58441E \times mm^3}{58441E \times mm^3 + 36767E \times mm^3} = 0.614$$

$$DF_{EF} = 1 - 0.614 = 0.386$$

End moments of the beams under the gravity load:

$$M_{BC}^{(0)} = -M_{CB}^{(0)} = \frac{wL^2}{12} \cdot \frac{3r_{BC}(2 - r_{CB})}{4 - r_{BC}r_{CB}} = \frac{44.12(8.5)^2}{12} \cdot \frac{3 \times 0.45(2 - 0.45)}{4 - 0.45^2} = 146.37 \text{ kN m}$$

$$M_{CE}^{(0)} = -M_{EC}^{(0)} = \frac{wL^2}{12} \cdot \frac{3r_{CE}(2 - r_{EC})}{4 - r_{CE}r_{EC}} = \frac{44.12(8.5)^2}{12} \cdot \frac{3 \times 0.332(2 - 0.332)}{4 - 0.332^2} = 113.46 \text{ kN m}$$

Following the procedure for the modified moment distribution method, the detailed moment distribution process is shown in Table 23.6 and the final moment distribution of the frame under the gravity loads is shown in Figure 23.11. It can be seen from Figure 23.11 that the bending moment is not distributed symmetrically due to the differences in connection stiffness between the two bays. The maximum span moment for beam BC is located 4.01 m from joint B with a value of 296.25 kN m, while the maximum span for beam CE is located 4.51 m from joint C with a value of 309.10 kN m. The axial loads in the columns are found as

$$P_{f_{AB}} = \frac{wL}{2} + \frac{M_{BC} + M_{CB}}{L} = \frac{44.12 \times 8.5}{2} + \frac{48.01 - 149.03}{8.5} = 175.63 \text{ kN}$$

$$P_{f_{DC}} = wL - \frac{M_{BC} + M_{CB}}{L} + \frac{M_{CE} + M_{EC}}{L} = 44.12 \times 8.5 - \frac{48.01 - 149.03}{8.5} + \frac{139.55 - 42.86}{8.5}$$
$$= 398.28 \text{ kN}$$

$$P_{f_{DC}} = \frac{wL}{2} - \frac{M_{CE} + M_{EC}}{L} = \frac{44.12 \times 8.5}{2} - \frac{139.55 - 42.86}{8.5} = 176.13 \text{ kN}$$

23.4.3 First-Order Nonlinear Analysis of Semirigid Framed Structures

Computer-based analysis of semirigid framed structures can be generally achieved by modifying existing programs of rigid frame analysis in one of two ways. The first method introduces additional "connection elements" that model the beam-to-column connections, while the second method modifies the member stiffness matrix to account for the connection flexibility. The application of the first approach is limited to only computer-based analysis and design and is inconvenient for practical use because the beam–column members are separated from the attached end-connections. Therefore, the second approach is considered to be more general as the analysis and design of such structures can also be facilitated manually. Most research and engineering applications have adopted the second approach as the means to incorporate semirigid connection behavior into frame analysis and design.

The implementation of the concept of end-fixity factor into frame analysis is straightforward. For an existing rigid frame analysis computer program, only minor modifications to the member stiffness matrix and the evaluation of member end-reactions due to applied member loads are required. The elastic stiffness matrix of a member i with two semirigid end-connections having stiffness moduli R_1 and R_2, as shown in Figure 23.6, can be represented by the stiffness matrix for the member taken to have rigid end-connections modified by a correction matrix [21]; that is

$$K_i^{SR} = S_i \cdot C_{e-i} \tag{23.11}$$

where K_i^{SR} is the stiffness matrix of member i with semirigid end-connections taken into account, S_i is the stiffness matrix of the member taken to have rigid ends, and C_{e-i} is the required correction matrix. For a planar beam–column element with six degrees of freedom, the matrices S_i and C_{e-i} have the following form:

$$S_i = \begin{bmatrix} \dfrac{EA}{L} & 0 & 0 & \dfrac{-EA}{L} & 0 & 0 \\[2ex] & \dfrac{12EI}{L^3} & \dfrac{6EI}{L^2} & 0 & \dfrac{-12EI}{L^3} & \dfrac{6EI}{L^2} \\[2ex] & & \dfrac{4EI}{L} & 0 & \dfrac{-6EI}{L^2} & \dfrac{2EI}{L} \\[2ex] & & & \dfrac{EA}{L} & 0 & 0 \\[2ex] & \text{sym} & & & \dfrac{12EI}{L^3} & \dfrac{-6EI}{L^2} \\[2ex] & & & & & \dfrac{4EI}{L} \end{bmatrix} \tag{23.12}$$

$$C_{e-i} = \begin{bmatrix} 1 & 0 & 0 & 0 & 0 & 0 \\[2ex] 0 & \dfrac{4r_2 - 2r_1 + r_1 r_2}{4 - r_1 r_2} & \dfrac{-2Lr_1(1 - r_2)}{4 - r_1 r_2} & 0 & 0 & 0 \\[2ex] 0 & \dfrac{6(r_1 - r_2)}{L(4 - r_1 r_2)} & \dfrac{3r_1(2 - r_2)}{4 - r_1 r_2} & 0 & 0 & 0 \\[2ex] 0 & 0 & 0 & 1 & 0 & 0 \\[2ex] 0 & 0 & 0 & 0 & \dfrac{4r_1 - 2r_2 + r_1 r_2}{4 - r_1 r_2} & \dfrac{2Lr_2(1 - r_1)}{4 - r_1 r_2} \\[2ex] 0 & 0 & 0 & 0 & \dfrac{6(r_1 - r_2)}{L(4 - r_1 r_2)} & \dfrac{3r_2(2 - r_1)}{4 - r_1 r_2} \end{bmatrix} \tag{23.13}$$

where E is elastic modulus and L, A, and I are the length, cross-sectional area, and moment of inertia of the member, respectively. The end-fixity factors r_1 and r_2 in Equation 23.13 are defined by Equation 23.7. Knowing the semirigid beam–column member stiffness matrix from Equation 23.11 for specified values of end-fixity factors reflecting connection stiffness, the analysis of frames with semirigid connections can then be carried out directly using the conventional displacement method.

To take into account the nonlinear behavior of semirigid connections, an iterative procedure is applied to obtain the solution. In each iteration, the member stiffness matrix K_i^{SR} is modified using the correction matrix C_{e-i} with updated end-fixity factors r. In addition, when there are member loads on semirigid members, the equivalent joint loads that facilitate matrix analysis of the structure are updated since member end-reactions are also functions of the end-fixity factors. The secant-stiffness-based iterative procedure for the first-order analysis of semirigid frameworks accounting for the nonlinear behavior of connections defined by Equation 23.6 is described in the following [where subscript j ($j = 1, 2$) denotes the two ends of semirigid member i]:

Step 1. Input data, including the nonlinear parameters and initial stiffness R_e defining the specified connections in the connection database. Set the iteration index $k = 0$ and assign connection stiffness $R_{ij}^{(k)} = R_e$.

Step 2. For each beam–column element i, update the end-fixity factors r_{ij} through Equation 23.7 and compute the corresponding equivalent joint loads. Generate the correction matrix C_{e-i} and member stiffness matrix K_i^{SR} through Equations 23.13 and 23.11, respectively.

Step 3. Assemble the structure stiffness matrix and solve for the nodal displacements, member forces, and connection moments $M_{ij}^{(k)}$.

Step 4. Calculate connection rotations $\theta_{ij}^{(k)} = M_{ij}^{(k)}/R_{ij}^{(k)}$ and obtain the corresponding moments $M(\theta_{ij}^{(k)})$ from Equation 23.6.

Step 5. Check convergence by comparing the connection moments $M_{ij}^{(k)}$ obtained through analysis with the moments $M(\theta_{ij}^{(k)})$ calculated in Step 4; if $|M_{ij}^{(k)} - M(\theta_{ij}^{(k)})| \leq \varepsilon$ (a predefined tolerance), then stop; otherwise, go to Step 6.

Step 6. Update connection secant stiffnesses $R_{ij}^{(k)} = M(\theta_{ij}^{(k)})/\theta_{ij}^{(k)}$ and iteration index $k = k + 1$, and return to Step 2.

23.4.4 Second-Order Nonlinear Analysis of Semirigid Framed Structures

The stability of framed structures has long been a concern in the design of such structures. In current design practice, this concern can be addressed by performing the so-called second-order P–Δ analysis for frames. Unlike the first-order elastic analysis, in which the equilibrium and kinematic relationships of a frame are established with respect to the undeformed geometry of the structure, the second-order elastic analysis is associated with the interaction between forces and deformation of the structure. For that reason, the second-order elastic analysis is referred to as geometrical nonlinear analysis. By inducing additional stresses in the members and by causing the frame members to deform more, the second-order effects (P–Δ and P–δ), have a destabilizing effect on the structure. For rigid frames, the computer-based second-order elastic analysis is often pursued as an iterative procedure and the stiffness matrix of each member i with rigid ends is expressed as

$$K_i = S_i + G_i \tag{23.14}$$

where S_i is the elastic stiffness matrix and is as shown in Equation 23.12 and G_i is the geometrical stiffness matrix that accounts for the second-order effects and is expressed as

$$G_i = \frac{N}{L} \begin{bmatrix} 0 & 0 & 0 & 0 & 0 & 0 \\ & \dfrac{6}{5} & \dfrac{L}{10} & 0 & -\dfrac{6}{5} & \dfrac{L}{10} \\ & & \dfrac{2L^2}{15} & 0 & -\dfrac{L}{10} & -\dfrac{L^2}{30} \\ & & & 0 & 0 & 0 \\ & \text{sym.} & & & \dfrac{6}{5} & -\dfrac{L}{10} \\ & & & & & \dfrac{2L^2}{15} \end{bmatrix} \tag{23.15}$$

where N is the axial force in the member (positive when the member is in tension).

Knowing that semirigid frames are more flexible than conventional rigid frames, the stability of semirigid frames becomes even more a concern in the design of such structures. Together with consideration of the nonlinear behavior of connections, the analysis of semirigid frames requires both geometrical nonlinearity (second-order) and material nonlinearity (semirigid connection behavior) effects to be taken into account. The semirigid member stiffness matrices for second-order elastic analysis can be obtained by modifying the corresponding stiffness matrices of rigid members as follows [22]:

$$K_i^{SR} = S_i C_{e-i} + G_i C_{g-i} \tag{23.16}$$

where the matrices S_i, C_{e-i}, and G_i are as defined in Equations 23.12, 23.13, and 23.15, respectively, and C_{g-i} is the correction matrix for the geometrical stiffness of the semirigid member and is expressed as

$$C_{g-i} = \begin{bmatrix} 0 & 0 & 0 & 0 & 0 & 0 \\ 0 & 1 & 0 & 0 & 0 & 0 \\ 0 & G_{32} & G_{33} & 0 & G_{35} & G_{36} \\ 0 & 0 & 0 & 0 & 0 & 0 \\ 0 & 0 & 0 & 0 & 1 & 0 \\ 0 & G_{62} & G_{63} & 0 & G_{65} & G_{66} \end{bmatrix} \tag{23.17}$$

where

$$G_{32} = -G_{35} = \frac{-4}{5L(4 - r_1 r_2)^2} \left(8r_1^2 r_2 - 13r_1 r_2^2 - 32r_1^2 - 8r_2^2 + 25r_1 r_2 + 20 \right)$$

$$G_{33} = \frac{r_1}{5(4 - r_1 r_2)^2} \left(16r_2^2 + 25r_1 r_2^2 - 96r_1 r_2 + 128r_1 - 28r_2 \right)$$

$$G_{36} = \frac{4r_2}{5(4 - r_1 r_2)^2} \left(16r_1^2 - 5r_1^2 r_2 + 9r_1 r_2 - 28r_1 + 8r_2 \right)$$

$$G_{62} = -G_{65} = \frac{-4}{5L(4 - r_1 r_2)^2} \left(8r_1 r_2^2 - 13r_1^2 r_2 - 32r_2^2 - 8r_1^2 + 25r_1 r_2 + 20 \right)$$

$$G_{63} = \frac{4r_1}{5(4 - r_1 r_2)^2} \left(16r_2^2 - 5r_1 r_2^2 + 9r_1 r_2 + 8r_1 - 28r_2 \right)$$

$$G_{66} = \frac{r_2}{5(4 - r_1 r_2)^2} \left(16r_1^2 + 25r_1^2 r_2 - 96r_1 r_2 - 28r_1 + 128r_2 \right)$$

in which r_1 and r_2 are end-fixity factors of the member as defined by Equation 23.7 and L is the length of the member.

Since the correction matrices C_{e-i} and C_{g-i} that take into account the semirigid connection behavior are in terms of end-fixity factors instead of connection stiffness, the stiffness matrix K_i^{SR} applies to a beam–column member with any combination of pinned, semirigid, and rigid connections.

The secant-stiffness-based iterative procedure for the second-order analysis of semirigid frames, with nonlinear behavior of connections taken into account, is readily developed through modification of the first-order analysis procedure as described in the following:

Step 1. Input the data, including the parameters for the specified connections from Equation 23.6. Set the iteration index $k = 0$, assign connection stiffnesses $R_{ij}^{(k)} = R_e$, and set member axial forces $N_i^{(k)} = 0$. Generate elastic stiffness matrix from Equation 23.12 for each member.

Step 2. Set iteration index $k = k + 1$, update the end-fixity factors r_{ij} using Equation 23.7 for each semirigid member i; evaluate member equivalent joint loads; calculate the correction matrices C_{e-i} and C_{g-i} by Equations 23.13 and 23.17, respectively. Evaluate geometrical stiffness matrix G_i based on Equation 23.15 and the axial force $N_i^{(k-1)}$ from the previous iteration. Generate the member stiffness matrix K_i^{SR} according to Equation 23.16 for each semirigid member.

Step 3. Assemble the structure stiffness matrix and solve for nodal displacements, member axial forces $N_i^{(k)}$, and connection moments $M_{ij}^{(k)}$.

Step 4. Calculate connection rotations $\theta_{ij}^{(k)} = M_{ij}^{(k)} / R_{ij}^{(k)}$ and obtain the corresponding moments $M(\theta_{ij}^{(k)})$ from Equation 23.6.

Step 5. Check for convergence for specified tolerances ε_1 and ε_2 by comparing the connection moments $M_{ij}^{(k)}$ obtained through analysis with the moments $M(\theta_{ij}^{(k)})$ calculated in Step 23.4 and by also comparing the member axial forces $N_i^{(k)}$ with the forces $N_i^{(k-1)}$ from the previous iteration; if $|M_{ij}^{(k)} - M(\theta_{ij}^{(k)})| \le \varepsilon_1$ and $|N_{ij}^{(k)} - N_{ij}^{(k-1)}| \le \varepsilon_2$, then stop; otherwise, go to Step 6.

Step 6. Update connection secant stiffness $R_{ij}^{(k)} = M(\theta_{ij}^{(k)})/\theta_{ij}^{(k)}$ and go back to Step 2.

With the adoption of the end-fixity factor defined in Equation 23.7 and the member stiffness matrix Equation 23.16 accounting for both second-order effects and semirigid connection behavior, it is a straightforward matter to modify linear elastic analysis computer programs for rigid frames to perform second-order analysis of semirigid frames. Figure 23.12 illustrates a program flow chart for the second-order analysis of semirigid frames.

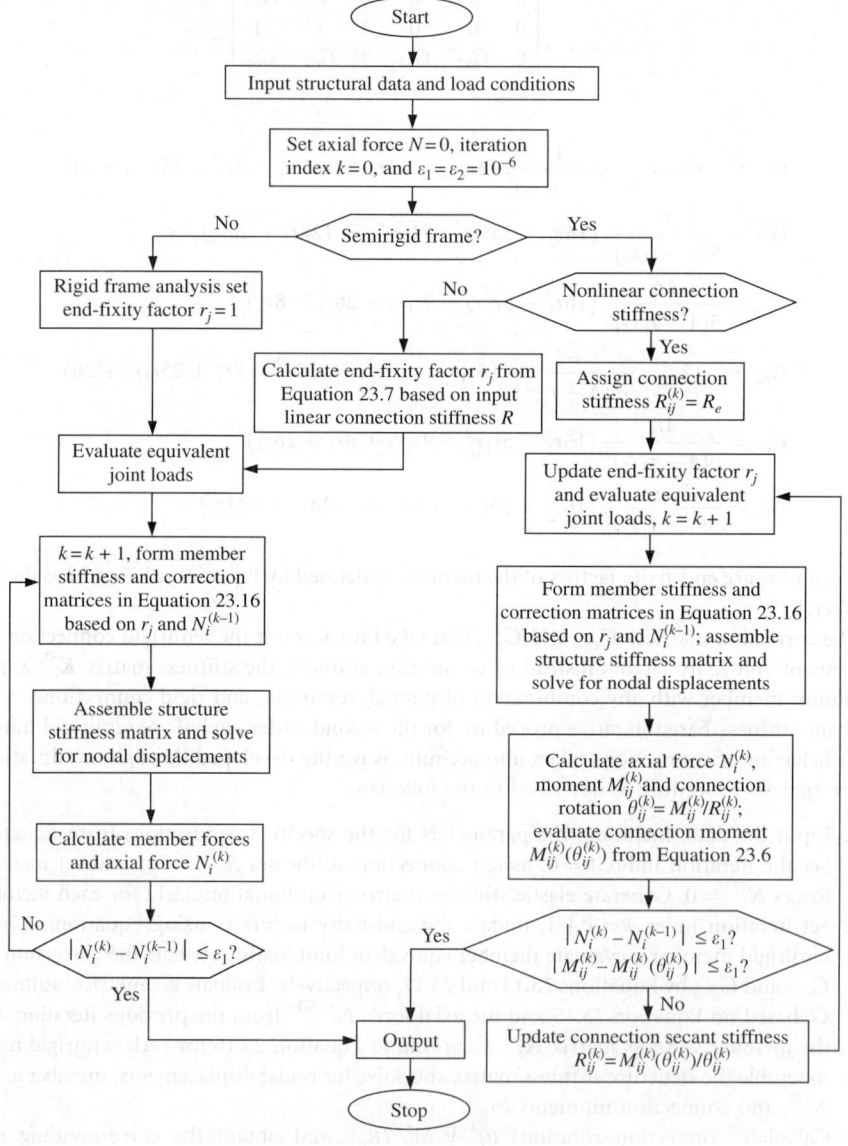

FIGURE 23.12   Flowchart of second-order semirigid frame elastic analysis.

23.4.5 Modification of Column Effective-Length Factor

The determination of column strength is of primary importance in the design of framed structures. The strength and stability of a frame and its columns are interrelated. The effective-length concept for evaluating column strength in rigid frames used to be the primary method to evaluate the stability of columns of frames. According to this concept, the strength of a column is evaluated from that of an equivalent pin-ended member of length KL, subject to axial load only, by means of a so-called K factor found as

$$K = \sqrt{\frac{P_e}{P_{cr}}} \qquad (23.18)$$

where $P_e = \pi^2 EI/L^2$ is the Euler buckling load of the pinned-end column, in which L is the length of the column and I is the moment of inertia of the column section about the axis of bending, and P_{cr} is the buckling load of the end-restrained column. For a laterally braced semirigid column, the corresponding effective-length factor based on Newmark's approximation [23] can be expressed in terms of column end-fixity factors r_1 and r_2 defined in Equation 23.7 as

$$K = \sqrt{\frac{[\pi^2(1 - r_1) + 6r_1] \times [\pi^2(1 - r_2) + 6r_2]}{[\pi^2(1 - r_1) + 12r_1] \times [\pi^2(1 - r_2) + 12r_2]}} \qquad (23.19)$$

Equation 23.19 yields K factor values having a maximum error of less than 4% compared to corresponding theoretically correct values.

The effective-length factor concept is considered to be an essential part of many analysis procedures, and a corresponding alignment chart approach is recommended by almost all current design standards [1–4]. The idealizations and assumptions associated with the approach are as follows:

1. All beams and columns are purely elastic.
2. All members have constant cross-section.
3. All joints are rigid.
4. For braced (sidesway inhibited or sway prevented) frames, rotations at opposite ends of the restraining beams are equal in magnitude, producing single-curvature bending.
5. For unbraced (sidesway uninhibited or sway permitted) frames, rotations at opposite ends of the restraining beams are equal in magnitude, producing reverse-curvature bending.
6. The stiffness parameters $L\sqrt{P/EI}$ of all columns are equal.
7. Joint restraint is distributed to the column above and below the joint in proportion to the I/L values for the two columns.
8. All columns buckle simultaneously.
9. No significant axial compression force exists in the beams.

Based on the foregoing assumptions, the effective-length factor K of a column in a braced frame is determined from the following equation:

$$\frac{G_U G_L}{4}\left(\frac{\pi}{K}\right)^2 + \frac{G_U + G_L}{2}\left(1 - \frac{\pi/K}{\tan \pi/K}\right) + 2\left(\frac{\tan \pi/K}{\pi/K}\right) = 1 \qquad (23.20)$$

For an unbraced frame, the current Canadian standard CSA S16 [3] takes sway effects into account directly by using $K = 1.0$ for columns and performing a second-order elastic analysis under actual and notional loads (i.e., in lieu of using the traditional effective-length factor to account for frame stability, as is permitted by the American design specifications [1,2]). Following the alignment chart approach, the K factor for a column in an unbraced frame is determined from the following equation:

$$\frac{G_U G_L (\pi/K)^2 - 36}{6(G_U + G_L)} = \left(\frac{\pi/K}{\tan \pi/K}\right) \qquad (23.21)$$

In Equations 23.20 and 23.21, G_U and G_L are the stiffness factors for the upper and lower ends of the column, respectively. A stiffness factor G is defined as

$$G = \frac{\sum(I_c/L_c)}{\sum(I_b/L_b)} \tag{23.22}$$

where the summation is taken over all members connected to the joint, I_c is the moment of inertia of the column section corresponding to the plane of buckling, L_c is the unsupported length of the column, I_b is the moment of inertia of the beam/girder section corresponding to the plane of buckling, and L_b is the unsupported length of the beam/girder. Having the stiffness factors G_U and G_L evaluated in accordance with Equation 23.22, the effective-length factor K of a column in a braced or unbraced frame can be obtained from the applicable alignment charts developed from Equations 23.20 and 23.21.

The evaluation of the effective-length factor K for columns in semirigid frames can be conveniently achieved by pursuing the foregoing approach for rigid frames with a necessary modification to account for connection flexibility when calculating a stiffness factor as

$$G = \frac{\sum(I_c/L_c)}{\sum(\alpha_b I_b/L_b)} \tag{23.23}$$

where α_b is a modification factor applied to the moment of inertia I_b of the restraining beams that can be expressed as a function of the end-fixity factors associated with each restraining beam depending on whether the frame is of braced or unbraced frames. For braced semirigid frames

$$\alpha_b = \frac{3r_1(2 - r_2)}{4 - r_1 r_2} \tag{23.24}$$

while for unbraced semirigid frames

$$\alpha_b = \frac{r_1(2 + r_2)}{4 - r_1 r_2} \tag{23.25}$$

where the end-fixity factors r_1 and r_2 in Equations 23.24 and 23.25 correspond to the so called "near end" and "far end," respectively, of the beam. In the case where the "far end" of the restraining beam is connected to a rotation-restrained support instead of a column, the corresponding modification factors for braced and unbraced semirigid frames are

$$\alpha_b = \frac{6r_1}{4 - r_1 r_2} \tag{23.26}$$

$$\alpha_b = \frac{2r_1}{4 - r_1 r_2} \tag{23.27}$$

As the modification factor α_b is expressed by the end-fixity factors, beams with pinned, semirigid, and rigid connections can all be considered. For instance, Equation 23.24 will yield a value of 1.5 for the modification factor α_b for a beam in a braced frame with a rigid connection at the "near end" $(r_1 = 1)$ and a pinned connection at the "far-end" $(r_2 = 0)$. If the "near end" connection is a pinned connection $(r_1 = 0)$, the corresponding modification factor α_b becomes zero, which indicates the beam does not provide any rotational restraint to the connected column.

Having the modification factor value α_b based on the end-fixity of the restraining beams, the stiffness factors for the upper and lower ends of the column can be calculated according to Equation 23.23. The corresponding column effective-length factor is then obtained from the applicable alignment chart. However, unlike that for the rigid frame case, in which the effective-length factor of a column is evaluated regardless of the applied loads, the rotational stiffnesses of beam-to-column connections in semirigid frames are interrelated to the loads and, hence, so are the end-fixity factors r_j $(j = 1, 2)$ and modification factors α_b. As a frame is loaded, the connection stiffness and, thus, the restraint provided to the columns gradually decreases, causing the column effective-length factor to increase. Consequently, the effective-length factor of the column in a semirigid frame must be evaluated for each applied load case.

EXAMPLE 23.3

The 11-storey unbraced frame shown in Figure 23.13 is analyzed using the second-order elastic analysis approach presented earlier, with the intention to illustrate frame behavior due to the interaction of second-order effects and the semirigid behavior of the beam-to-column connections. All of the beam-to-column connections of the frame are assumed to be identical.

In order to trace the effect of semirigid behavior on geometric second-order effects, nine (9) analyses were conducted for the frame with the end-fixity factor decreasing in decrements of 0.1 from that for a rigid connection ($r = 1.0$) to that for a flexible connection ($r = 0.2$). The frame was analyzed with and without considering second-order effects for each case. The lateral deflection obtained at the roof level for each case is presented in Table 23.7 along with the values of the end-fixity factor, the corresponding

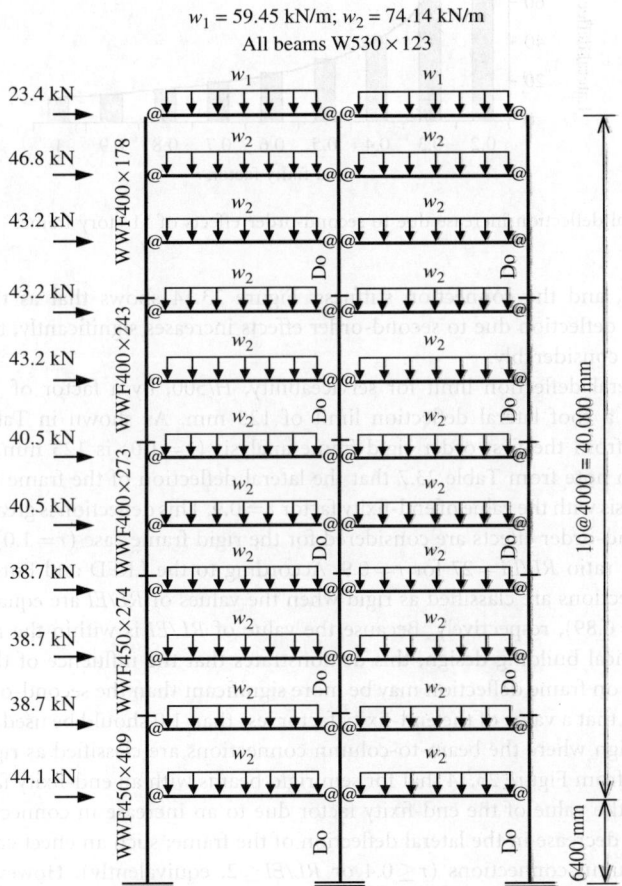

FIGURE 23.13 Example 3: two-bay 11-story semirigid steel frame.

TABLE 23.7   Lateral Deflection of Two-Bay 11-Story Steel Frame

End-fixity factor (r)	Stiffness ratio (RL/EI)	Connection stiffness (N/mm/rad)	Lateral deflection Δ_1, first-order analysis (mm)	Lateral deflection Δ_2, second-order analysis (mm)	$\dfrac{\Delta_2}{\Delta_1}$
1.0	∞	∞	125	140	1.12
0.9	27.00	4.723×10^{11}	145	165	1.14
0.8	12.00	2.099×10^{11}	168	196	1.17
0.7	7.00	1.225×10^{11}	197	237	1.20
0.6	4.50	7.872×10^{10}	235	293	1.25
0.5	3.00	5.248×10^{10}	286	377	1.32
0.4	2.00	3.499×10^{10}	358	514	1.44
0.3	1.29	2.249×10^{10}	478	780	1.66
0.2	0.75	1.312×10^{10}	675	1531	2.27

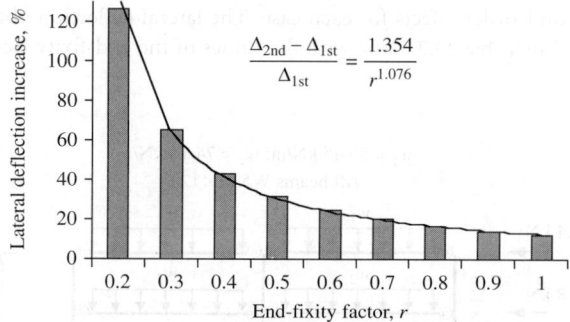

$$\frac{\Delta_{2nd} - \Delta_{1st}}{\Delta_{1st}} = \frac{1.354}{r^{1.076}}$$

FIGURE 23.14   Lateral deflection increase due to second-order effects of 11-story frame.

stiffness ratio *RL/EI*, and the connection stiffness. Figure 23.14 shows that as the end-fixity factor decreases, the lateral deflection due to second-order effects increases significantly, thus intensifying the second-order effects considerably.

Increasing the lateral deflection limit for serviceability, *H*/500, by a factor of 1.35 to account for factored loads gives a roof lateral deflection limit of 125 mm. As shown in Table 23.7, the lateral deflection obtained from the first-order rigid frame analysis ($r = 1.0$) is 125 mm, which satisfies the deflection limit. Also note from Table 23.7 that the lateral deflection of the frame is 145 mm for first-order semirigid analysis with the value of end-fixity factor $r = 0.9$. This deflection is greater than the 140 mm obtained when second-order effects are considered for the rigid frame case ($r = 1.0$). Also note that the value of the stiffness ratio $RL/EI = 27$ for $r = 0.9$. According to the LRFD and Eurocode 3 connection classifications, connections are classified as rigid when the values of *RL/EI* are equal or greater than 20 ($r = 0.87$) or 25 ($r = 0.89$), respectively. Because the value of *RL/EI* is within the range for rigid connections used in typical building design, this demonstrates that the influence of the beam-to-column connection behavior on frame deflection may be more significant than the second-order effects. In turn, this perhaps suggests that a value of the end-fixity factor less than 1.0 should be used in mid- to high-rise rigid steel frame design where the beam-to-column connections are classified as rigid connections.

It is also observed from Figure 23.14 that for semirigid beams with an end-fixity factor value less than 0.75, an increase in the value of the end-fixity factor due to an increase in connection stiffness would result in appreciable decrease in the lateral deflection of the frame; such an effect can be substantial for flexible beam-to-column connections ($r \leq 0.4$ or $RL/EI \leq 2$, equivalently). However, for nearly rigid connections ($r \geq 0.75$), further increase in the end-fixity factor has a trivial effect on the deflection. A similar phenomenon is also observed for the strength of columns in the frame. Figure 23.15 illustrates

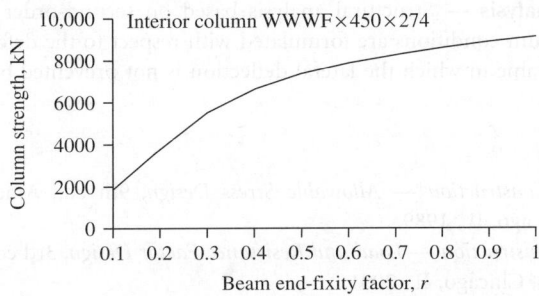

FIGURE 23.15 Interior column strength versus end-fixity factor.

the influence of beam end-fixity on the strength of the interior column (WWF400 × 274) between the third and fourth floors, in which the strength of the column is calculated based on LRFD with steel yield stress 350 MPa. Considering the relationship between the end-fixity factor and connection stiffness shown in Figure 23.9, it is concluded that it is important and economical to consider the semi-rigid behavior of connections in the design of unbraced steel frames having flexible beam-to-column connections.

This study also finds that the buckling load capacity of the frame increases considerably as the end-fixity factors increase from zero to 0.4 for beam-to-column connections. For an unbraced frame having flexible connections having a low value of connection stiffness, even a small increase in the stiffness would result in an appreciable increase in the end-fixity factor, which consequently causes a significant increase in the buckling load capacity. In practice, pinned connections always have some rotational stiffness, which considerably benefits the load capacity of the frame. Conversely, reducing the connection stiffness from rigid connection with full fixity (e.g., from $r = 1.0$ to 0.75) will result in an insignificant decrease in the critical buckling loads. This is because large reductions in connection stiffness from rigid frames are shown to have little effect on the frame stability in that a large change to the stiffness of rigid connections will result in only a small change in the end-fixity factor, as shown in Figure 23.9. Consequently, potential savings in connection cost may be achieved from the replacement of rigid connections with semirigid connections. Therefore, considering the behavior of semirigid connections in frame stability can produce more appropriate and economical designs.

Glossary

ASD — Allowable stress design.

LRFD — Load and resistance factor design.

Beam-to-column connection — A connection that connects a beam to a column.

Braced frame — A frame to which lateral translation is prevented by a system of bracing.

Connection rotation capacity — The angular rotation that a connection can undergo prior to local failure.

Effective-length factor — The ratio between the effective length and the unbraced length of a member.

End-fixity factor — The factor that defines the rotational stiffness of an end-connection relative to that of the attached member.

First-order elastic analysis — Structural analysis based on linear elastic deformation, in which equilibrium conditions are formulated with respect to the undeformed structure.

M–θ relationship — The relationship between moment and rotation of a connection.

P–Δ effect — Secondary effect of axial loads and lateral deflection on the moments in members.

Rigid frame — A structure in which connections maintain the angular relationship between beam and column members under load.

Second-order elastic analysis — Structural analysis based on second-order elastic deformation, in which equilibrium conditions are formulated with respect to the deformed structure.
Unbraced frame — A frame in which the lateral deflection is not prevented by a system of bracing.

References

[1] *Manual of Steel Construction — Allowable Stress Design*, 9th ed., American Institute of Steel Construction, Chicago, IL, 1989.

[2] *Manual of Steel Construction — Load and Resistance Factor Design*, 3rd ed., American Institute of Steel Construction, Chicago, IL, 2001.

[3] *Limit States Design of Steel Structures*, CSA Standard CAN/CSA S16-01, Canadian Standards Association, Toronto, Ontario, 2001.

[4] *Design of Steel Structures*, Part 1.1, Eurocode 3, European Committee for Standardization, CEN, Brussels, 1992.

[5] Kennedy, D.J.L., Moment–rotation characteristics of shear connections, *AISC Eng. J.*, 6(4), 105, 1969.

[6] Frye, M.J. and Morris, G.A., Analysis of frames with flexible connected steel frames, *Can. J. Civil Eng.*, 2–3, 280, 1975.

[7] Chen, W.F. and Kishi, N. Moment–rotation relation of top- and seat-angle connections, Report CE-STR-87-4, School of Civil Eng., Purdue Univ., W. Lafayette, IN, 1987.

[8] Richard, R.M. and Abbott, B.J., Versatile elastic–plastic stress–strain formula, *J. Eng. Mech. Div.*, ASCE, 101(4), 511, 1975.

[9] Kishi, N. and Chen, W.F., Moment–rotation relations of semi-rigid connections with angles, *J. Struct. Eng.*, ASCE, 116(7), 1813, 1990.

[10] Chen, W.F. and Lui, E.M., Behaviour and modeling of semi-rigid connections, in *Stability Design of Steel Frames*, CRC Press, Boca Raton, FL, 1991, chap. 5.

[11] Yee, Y.L. and Melchers, R.E., Moment–rotation curves for bolted connections. *J. Struct. Eng.*, ASCE, 112(3), 615, 1986.

[12] Hsieh, S.H. and Deierlein, G.G., Nonlinear analysis of three-dimensional steel frames with semi-rigid connections, *Comput. Struct.*, 41(5), 995, 1991.

[13] Xu, L., Sherbourne, A.N., and Grierson, D.E., Optimal cost design of semi-rigid low-rise industrial frames, *Eng. J.*, AISC, 32(3), 87, 1995.

[14] Goverdhan, A.V., A collection of experimental moment rotation curves and the evaluation of predicting equations for semi-rigid connections, Master's thesis, Vanderbilt University, Nashville, TN, 1983.

[15] Nethercot, D.A., Steel beam to column connections — a review of test data and their applicability to the evaluation of joint behaviour of the performance of steel frames, CIRIA, London, 1985.

[16] Kishi, N. and Chen, W.F., Data base of steel beam-to-column connections, Report, CE-STR-86-26, School of Civil Eng., Purdue University, W. Lafayette, IN, 1986.

[17] Kishi, N. and Chen, W.F., Steel connection data bank program, Report, CE-STR-86-18, School of Civil Eng., Purdue University, W. Lafayette, IN, 1986.

[18] Kishi, N. et al. Design aid of semi-rigid connections for frame analysis, *AISC Eng. J.*, 30(3), 90, 1993.

[19] Bjorhovde, R., Colson, A., and Brozzetti, J., Classification system for beam-to-column connections, *J. Struct. Eng.*, ASCE, 116(11), 3059, 1990.

[20] Leon, R.T., Horffinman, J., and Staeger, T. *Design of Partially-Restrained Composite Connections*, Steel Design Guide Series No. 9, AISC, Chicago, IL, 1996.

[21] Monforton, G.R. and Wu, T.S., Matrix analysis of semi-rigidly connected frames, *J. Struct. Eng.*, ASCE, 89(6), 13, 1963.

[22] Xu, L., Second-order analysis for semi-rigid steel frame design, *Can. J. Civ. Eng.*, 28, 59, 2001.

[23] Newmark, N.M., A simple approximate formula for effective fixity of columns, *J. Aeronaut. Sci.*, 16(2), 116, 1949.

Further Reading

The following publications provide additional sources of information for semi-rigid frame structures:

Chan, S.L. and Chui, P.P.T., *Static and Cyclic Analysis of Semi-Rigid Steel Frames*, Elsevier Science, London, 2000.

Chen, W.F., Goto, Y., and Liew, J.Y.R., *Stability Design of Semi-rigid Frames*, John Wiley & Sons, New York, 1996.

Chen, W.F. and Lui, E.M., *Stability Design of Steel Frames*, CRC Press, Boca Raton, FL, 1991.

Faella, C., Piluso, V., and Rizzano, G., *Structural Steel Semirigid Connections: Theory, Design and Software*, CRC Press, Boca Raton, FL, 2000.

Practical Analysis for Semi-Rigid Frame Design, Chen, W.F., Ed., World Scientific, Singapore, 2000.

Semi-Rigid Construction in Structural Steelwork, Ivany, M. and Baniotopoulos, C. Eds., Springer-Verlag, Wien, New York, 2000.

24

Space Frame Structures

Tien T. Lan
Institute of Building Structures,
Chinese Academy of
Building Research,
Beijing, China

24.1 Introduction to Space Frame Structures

24.1.1 General Introduction

A growing interest in space frame structures has been witnessed worldwide over the last half-century. The search for new structural forms to accommodate large unobstructed areas has always been the main objective of architects and engineers. With the advent of new building techniques and construction materials, space frames frequently provide the right answer and satisfy the requirements for lightness, economy, and speedy construction. Significant progress has been made in the process of development of space frame. A large amount of theoretical and experimental research programs were carried out by many universities and research institutions in various countries. As a result, a great deal of useful information has been disseminated, and fruitful results have been put into practice.

In the past few decades, the proliferation of space frame was mainly due to its great structural potential and visual beauty. New and imaginative applications of space frames are being demonstrated in the total range of building types, such as sports arenas, exhibition pavilions, assembly halls, transportation terminals, airplane hangars, workshops, and warehouses. They have been used not only on long-span roofs, but also on mid- and short-span enclosures as roofs, floors, exterior walls, and canopies. Many interesting projects have been designed and constructed all over the world using a variety of configurations.

Some important factors that influence the rapid development of space frame can be cited as follows. First of all, the search for large indoor space has always been the focus of human activities. Consequently, sports tournaments, cultural performances, mass assemblies, and exhibitions can be held under one roof. The modern production and the needs of greater operational efficiency also created demand for large space with minimum interference from internal supports. Space frame provides the benefit that the interior space can be used in a variety of ways and thus is ideally suited for such requirements.

Space frames are highly statically indeterminate, and their analysis leads to extremely tedious computation if done by hand. The difficulty of the complicated analysis of such a system has contributed to its limited use. The introduction of the electronic computer has radically changed the whole approach to the analysis of space frames. By using computer programs, it is possible to analyze very complex space structures with great accuracy and less time involved.

Last, the space frame has also the problem of connecting a large number of members (sometimes up to 20) in space through different angles at a single point. The emergence of several connecting methods of proprietary systems has made great improvement in the construction of space frame, which offers simple and efficient means for making connection of members. The exact tolerances required by these jointing systems can be achieved in the fabrication of the members and joints.

24.1.2 Definition of Space Frame

If one looks at technical literature on structural engineering, one finds that the meaning of the *space frame* has been very diverse or even confusing. In a very broad sense, space frame is literally a three-dimensional structure. However, in a more restricted sense, space frame means some type of special structural action in three dimensions. Sometimes, structural engineers and architects fail to convey what they really mean by the term. Thus it is appropriate to define here the term space frame as understood throughout this section. It is best to quote a definition given by a Working Group on Spatial Steel Structures of the International Association on Shell and Spatial Structures [1]:

> A space frame is a structure system assembled of linear elements so arranged that forces are transferred in a three-dimensional manner. In some cases, the constituent element may be two-dimensional. Macroscopically a space frame often takes the form of a flat or curved surface.

It should be noted that virtually the same structure defined as space frame here is referred to as *latticed structures* in a state-of-the-art report prepared by the Task Committee on Latticed Structures [2], which states

> A latticed structure is a structure system in the form of a network of elements (as opposed to a continuous surface). Rolled, extruded or fabricated sections comprise the member elements. Another characteristic of latticed structural system is that their load-carrying mechanism is three-dimensional in nature.

The American Society of Civil Engineers (ASCE) report also specifies that the three-dimensional character includes flat surfaces with loading perpendicular to the plane as well as curved surfaces. The Report excludes structural systems such as common trusses on building frames, which can appropriately be divided into a series of planar frameworks with loading in the plane of the framework. In this section the terms "space frames" and "latticed structures" are considered synonymous.

A space frame is usually arranged in an array of single, double, or multiple layers of intersecting members. Some authors define space frames only as double-layer grids. A single-layer space frame that has the form of a curved surface is termed as *braced vault*, *braced dome*, or *latticed shell*.

Occasionally, the term *space truss* appears in the technical literature. According to the structural analysis approach, a space frame is analyzed by assuming rigid joints that cause internal torsions and moments in the members, whereas a space truss is assumed as hinged joints and therefore has no internal member moments. The choice between space frame and space truss action is mainly determined by the joint-connection detailing and the member geometry is no different for both. However, in engineering practice, there are no absolutely rigid or hinged joints. For example, a double-layer flat surface space frame is usually analyzed as hinged connections, while a single-layer curved surface space frame may be analyzed either as hinged or as rigid connections. The term "space frame" will be used to refer to both space frames and space trusses.

24.1.3 Basic Concepts

The space frame can be formed on either a flat or a curved surface. The earliest form of space frame structure is single-layer grid. By adding intermediate grids and including rigid connection to the joist and girder framing system, the single-layer grid is formed. The major characteristic of grid construction is the omnidirectional spread of the load as opposed to the linear transfer of the load in an ordinary framing system. Since such load transfer is mainly by bending, for larger spans the bending stiffness is increased most efficiently by changing to a double-layer system. The load transfer mechanism of a curved surface space frame is essentially different from the grid system that is primarily membrane-like action. The concept of space frame can be best explained by the following example.

It is necessary to design a roof structure for a square building. Figure 24.1a and b shows two different ways of roof framing. The roof system shown in Figure 24.1a is a complex roof composed of planar latticed trusses. Each truss resists the load acting on it independently and transfers the load to the columns on each end. To ensure the integrity of the roof system, usually purlins and bracings are used between trusses. In Figure 24.1b latticed trusses are laid orthogonally to form a system of space latticed grids that will resist the roof load through its integrated action as a whole and transfer the loads to the columns along the perimeters. Since the loads can be taken by the members in three dimensions, the corresponding forces in space latticed grids are usually less than that in planar trusses and hence the depth can be decreased in a space frame.

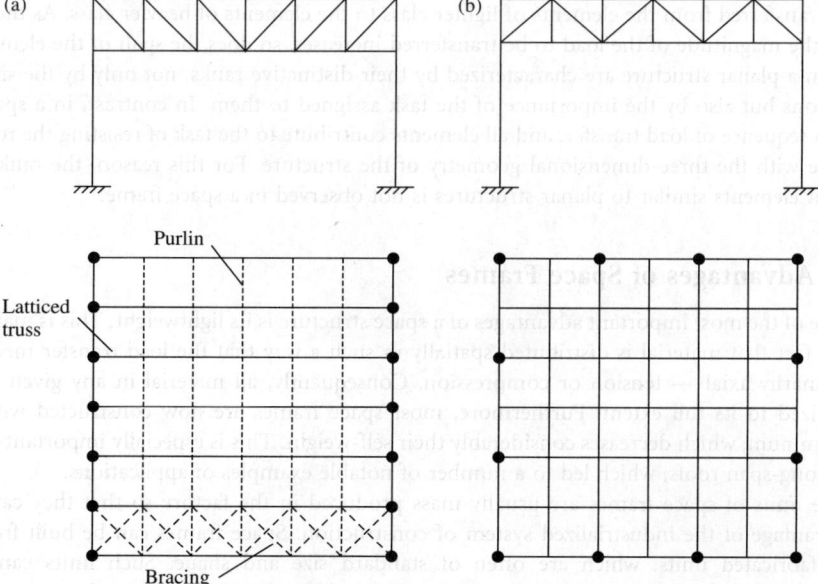

FIGURE 24.1 Roof framing for a square plan.

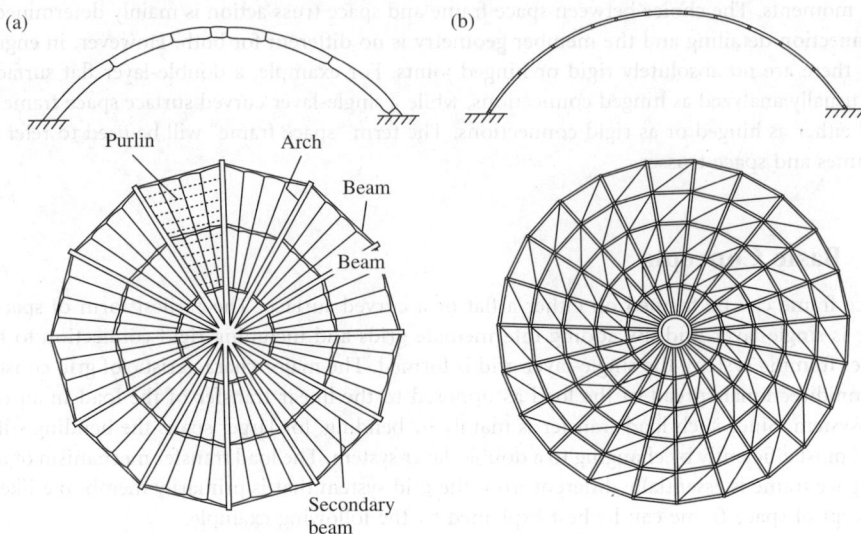

FIGURE 24.2   Roof framing for a Circular Dome.

The same concept can be observed in the design of a circular dome. Again, there are two different ways of framing a dome. The dome shown in Figure 24.2a is a complex of elements like arches, primary and secondary beams, and purlins, which all lie in a plane. Each of these elements constitutes a system that is stable by itself. In contrast, the dome shown in Figure 24.2b is an assembly of a series of longitudinal, meridional, and diagonal members, a form of latticed shell. It is a system whose resisting capacity is ensured only through its integral action as a whole.

The difference between planar structures and space frames can be understood also by examining the sequence of flow of forces. In a planar system, the force due to the roof load is transferred successively through the secondary elements, the primary elements, and then finally to the foundation. In each case, loads are transferred from the elements of lighter class to the elements of heavier class. As the sequence proceeds, the magnitude of the load to be transferred increases, so does the span of the element. Thus, elements in a planar structure are characterized by their distinctive ranks, not only by the size of their cross-sections but also by the importance of the task assigned to them. In contrast, in a space system there is no sequence of load transfer, and all elements contribute to the task of resisting the roof load in accordance with the three-dimensional geometry of the structure. For this reason, the ranking of the constituent elements similar to planar structures is not observed in a space frame.

24.1.4 Advantages of Space Frames

1. One of the most important advantages of a space structure is its lightweight. This is mainly due to the fact that material is distributed spatially in such a way that the load transfer mechanism is primarily axial — tension or compression. Consequently, all material in any given element is utilized to its full extent. Furthermore, most space frames are now constructed with steel or aluminum, which decreases considerably their self-weight. This is especially important in the case of long-span roofs, which led to a number of notable examples of applications.

2. The units of space frames are usually mass produced in the factory so that they can take full advantage of the industrialized system of construction. Space frames can be built from simple prefabricated units, which are often of standard size and shape. Such units can be easily transported and rapidly assembled on site by semi-skilled labor. Consequently, space frames can be built at a lower cost.

3. A space frame is usually sufficiently stiff in spite of its lightness. This is due to its three-dimensional character and to the full participation of its constituent elements. Engineers appreciate the inherent rigidity and great stiffness of space frames and their exceptional ability to resist unsymmetrical or heavy concentrated load. Possessing greater rigidity, the space frames allow also greater flexibility in layout and positioning of columns.

4. Space frames possess a versatility of shape and form and can utilize a standard module to generate various flat space grids, latticed shell, or even free-form shapes. Architects appreciate the visual beauty and the impressive simplicity of lines in space frames. A trend is very noticeable in which the structural members are left exposed as a part of the architectural expression. Desire for openness for both visual impact as well as the ability to accommodate variable space requirements always calls for space frames as the most favorable solution.

24.1.5 Preliminary Planning Guidelines

In the preliminary stage of planning a space frame to cover a specific building, a number of factors should be studied and evaluated before proceeding to structural analysis and design. These include not only structural adequacy and functional requirements but also the esthetic effect desired.

1. In its initial phase, structural design consists of choosing the general form of the building and the type of space frame appropriate to this form. Since a space frame is assembled of straight, linear elements connected at nodes, the geometrical arrangement of the elements — surface shape, number of layers, grid pattern, etc. — needs to be studied carefully in the light of various pertinent requirements.

2. The geometry of the space frame is an important factor to be planned, which will influence both the bearing capacity and the weight of the structure. The *module* size is developed from the overall building dimensions, while the *depth* of the grid (in the case of double-layer), the size of cladding, and the position of the supports will also have a pronounced effect upon it. For curved surface, the geometry is also related to the curvature, or more specifically to the rise of the span. A compromise between these various aspects usually has to be made to achieve a satisfactory solution.

3. In a space frame, connecting joints play an important role, both functional and esthetic, which derives from their rationality during construction and after completion. Since joints have a decisive effect on the strength and stiffness of the structure and compose around 20 to 30% of the total weight, joint design is critical to space frame economy and safety. These are quite a few proprietary systems that are used for space frame structures. They should be selected on the basis of quality, cost, and erection efficiency. In addition, custom-designed space frames have been developed, especially for long-span roofs. Regardless of the type of space frame, the essence of any system is the jointing system.

4. At the preliminary stage of design, the choosing of the type of space frames has to be closely related with the constructional technology. The space frames do not have such a sequential order of erection for planar structures and require special consideration on the method of construction. Usually, a complete falsework has to be provided so that the structure can be assembled in the high position. Alternatively, the structure can be assembled on the ground, and a certain technique can be adopted to lift the whole structure, or its major part, to the final position.

24.2 Double-Layer Grids

24.2.1 Types and Geometry

Double-layer grids, or flat surface space frames, consist of two planar networks of members forming the top and bottom-layers parallel to each other and interconnected by vertical and inclined web members. Double-layer grids are characterized by hinged joints with no moment or torsional resistance; therefore,

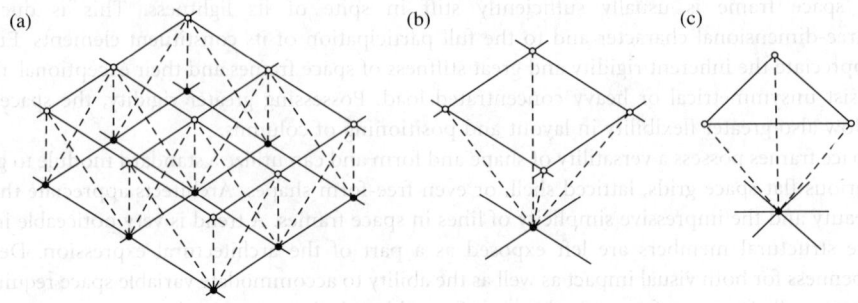

FIGURE 24.3   Basic elements of double-layer grids.

all members can only resist tension or compression. Even in the case of connection by comparatively rigid joints, the influence of bending or torsional moment is insignificant.

Double-layer grids are usually composed of basic elements such as

1. Planar latticed truss.
2. A pyramid with a square base that is essentially a part of an octahedron.
3. A pyramid with a triangular base (tetrahedron).

The basic elements used for various types of double-layer grids are shown in Figure 24.3.

Several types of double-layer grids can be formed by these basic elements. They are developed by varying the direction of the top and bottom-layers with respect to each other and also by the positioning of the top-layer nodal points with respect to the bottom-layer nodal points. Additional variations can be introduced by changing the size of the top-layer grid with respect to the bottom-layer grid. Thus, internal openings can be formed by omitting every second element in a normal configuration. According to the form of basic elements, double-layer grids can be divided in two groups, *latticed grids* and *space grids*. The latticed grids consist of intersecting vertical latticed trusses and form a regular grid. Two parallel grids are similar in design, with one layer directly over the top of another. Both top and bottom grids are directionally the same. The space grids consist of a combination of square or triangular pyramids. This group covers the so-called offset grids, which consist of parallel grids having an identical layout with one grid offset from the other in plane but remaining directionally the same, as well as the so-called "differential" grids in which two parallel top and bottom grids are of a different layout but are chosen to coordinate and form a regular pattern [3].

The type of double-layer grid can be chosen from the following most commonly used framing systems, which are shown in Figure 24.4a through j. In Figure 24.4, top chord members are depicted with heavy solid lines, bottom chords are depicted with light solid lines, and web members are depicted with dashed lines, while the upper joints are depicted by hollow circles and the bottom joints by solid circles. Different types of double-layer grids are grouped and named according to their composition, and the names in the parentheses indicate those suggested by other authors.

- Group 1: composed of latticed trusses
    A. *Two-way orthogonal latticed grids (square on square) (Figure 24.4a).* This type of latticed grid has the advantage of simplicity in configuration and in joint detail. All chord members are of the same length and lie in two planes that intersect at 90° to each other. Because of its weak torsional strength, horizontal bracings are usually established along the perimeters.
    B. *Two-way diagonal latticed grids (Figure 24.4b).* The layout of the latticed grid is exactly the same as in type 1, except that it is offset by 45° from the edges. The latticed trusses have different spans along two directions at each intersecting joint. Since the depth is all the same, the stiffness of each latticed truss varies according to its span. The latticed trusses of shorter span may be considered as a kind of support for latticed trusses of longer span, hence more spatial action is obtained.

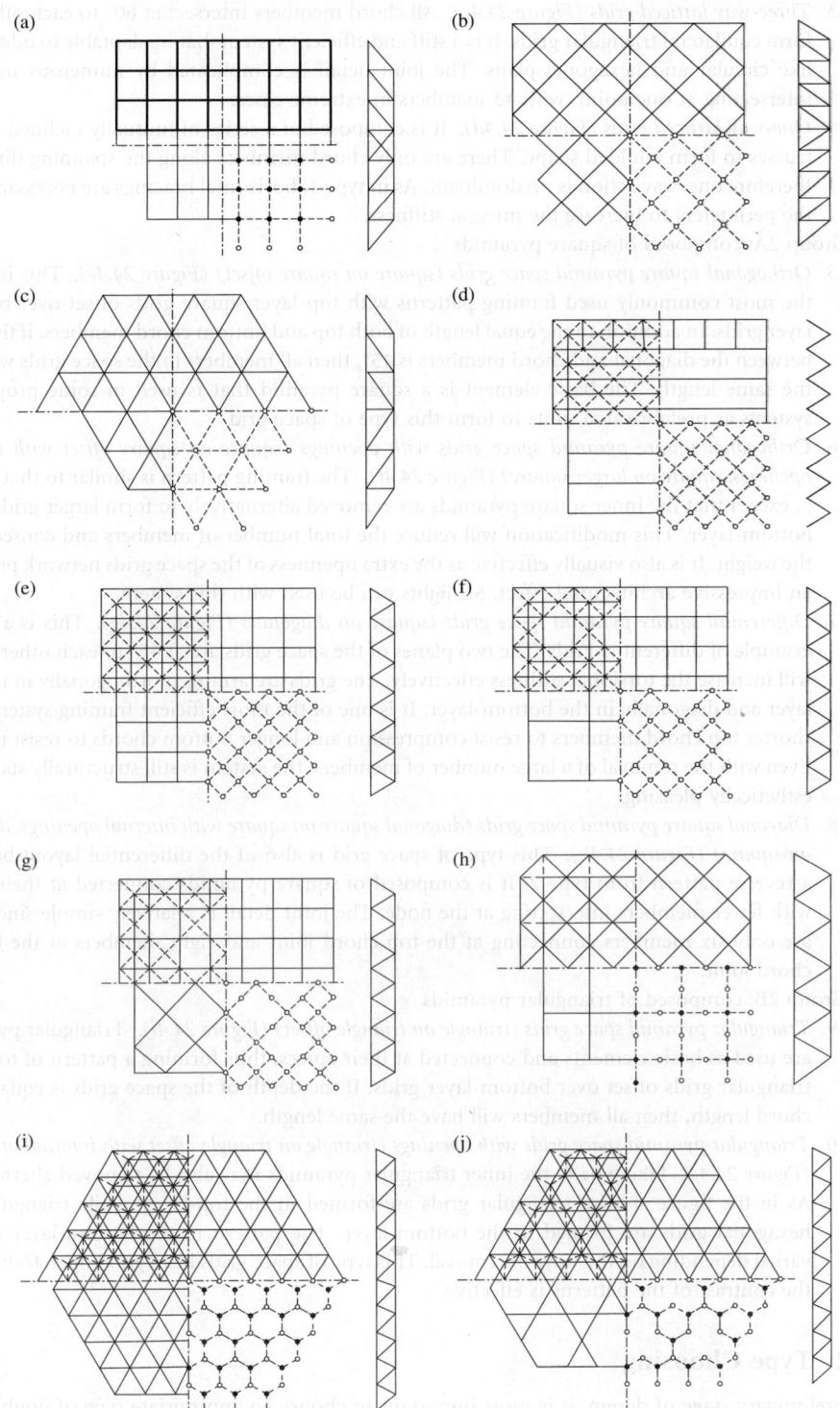

FIGURE 24.4 Framing system of double-layer grids.

3. *Three-way latticed grids (Figure 24.4c).* All chord members intersect at 60° to each other and form equilateral triangular grids. It is a stiff and efficient system that is adaptable to odd shapes like circular and hexagonal plans. The joint detail is complicated by numerous members intersecting at one point, with 13 members in extreme cases.

4. *One-way latticed grids (Figure 24.4d).* It is composed of a series of mutually inclined latticed trusses to form a folded shape. There are only chord members along the spanning direction, therefore one-way action is predominant. As in type 1, horizontal bracings are necessary along the perimeters to increase the integral stiffness.

- Group 2A: composed of square pyramids

5. *Orthogonal square pyramid space grids (square on square offset) (Figure 24.4e).* This is one of the most commonly used framing patterns with top-layer square grids offset over bottom-layer grids. In addition to the equal length of both top and bottom chord members, if the angle between the diagonal and chord members is 45°, then all members in the space grids will have the same length. The basic element is a square pyramid that is used in some proprietary systems as prefabricated units to form this type of space grid.

6. *Orthogonal square pyramid space grids with openings (square on square offset with internal openings, square on larger square) (Figure 24.4f).* The framing pattern is similar to that of type 5, except that the inner square pyramids are removed alternatively to form larger grids in the bottom-layer. This modification will reduce the total number of members and consequently the weight. It is also visually effective as the extra openness of the space grids network produces an impressive architectural effect. Skylights can be used with this system.

7. *Differential square pyramid space grids (square on diagonal) (Figure 24.4g).* This is a typical example of differential grids. The two planes of the space grids are at 45° to each other, which will increase the torsional stiffness effectively. The grids are arranged orthogonally in the top-layer and diagonally in the bottom-layer. It is one of the most efficient framing systems with shorter top chord members to resist compression and longer bottom chords to resist tension. Even with the removal of a large number of members, the system is still structurally stable and esthetically pleasing.

8. *Diagonal square pyramid space grids (diagonal square on square with internal openings, diagonal on square) (Figure 24.4h).* This type of space grid is also of the differential layout but with a reverse pattern from type 7. It is composed of square pyramids connected at their apices with fewer members intersecting at the node. The joint detail is relatively simple since there are only six members connecting at the top chord joint and eight members at the bottom chord joint.

- Group 2B: composed of triangular pyramids

9. *Triangular pyramid space grids (triangle on triangle offset) (Figure 24.4i).* Triangular pyramids are used as basic elements and connected at their apices, thus forming a pattern of top-layer triangular grids offset over bottom-layer grids. If the depth of the space grids is equal to $\sqrt{\frac{2}{3}}$ chord length, then all members will have the same length.

10. *Triangular pyramid space grids with openings (triangle on triangle offset with internal openings) (Figure 24.4j).* Like type 6, the inner triangular pyramids may also be removed alternatively. As in the figure shown, triangular grids are formed in the top-layer, while triangular and hexagonal grids are formed in the bottom-layer. The pattern in the bottom-layer may be varied depending on the ways of removal. This type of space grids has a good open feeling, and the contrast of the patterns is effective.

24.2.2 Type Choosing

In the preliminary stage of design, it is most important to choose an appropriate type of double-layer grid that will have direct influence on the overall cost and speed of construction. It should be determined comprehensively by considering the shape of building plan, size of span, supporting conditions, magnitude of loading, roof construction, and architectural requirements. In general, the system should be chosen so that the space grid is built of relatively long tension members and short compression members.

In choosing the type, the steel weight is one of the important factors for comparison. If possible, the cost of the structure should also be taken into account, which is complicated by the different costs of joint and member. By comparing the steel consumption of various types of double-layer grids with rectangular plan and supported along perimeters, it was found that the *aspect ratio* of the plan, defined here as the ratio of longer span to shorter span, has more influence than the span of the double-layer grids. When the plan is square or nearly square (aspect ratio = 1 to 1.5), two-way latticed grids and all space grids of group 2A, that is, types 1, 2, and 5 through 8, could be chosen. Of these types, the diagonal square pyramid space grid or differential square pyramid space grid has the minimum steel weight. When the plan is comparatively narrow (aspect ratio = 1.5 to 2), then those double-layer grids with the orthogonal grid system in the top-layer will consume less steel than those with the diagonal grid system. Therefore, two-way orthogonal latticed grids, orthogonal square pyramid space grids, and also those with openings and differential square pyramid space grids, that is, types 1, 5, 6, and 7, could be chosen. When the plan is long and narrow, the one-way latticed grid is the only choice. For square or rectangular double-layer grids supported along perimeters on three sides and free on the other side, the selection of the appropriate types for different cases is essentially the same. The boundary along the free side should be strengthened by increasing either the depth or the number of layers. An individual supporting structure like truss or girder along the free side is not necessary.

In case the double-layer grids are supported on intermediate columns, one could choose from two-way orthogonal latticed grids, orthogonal square pyramid space grids, and also those with openings, that is, types 1, 5, and 6. If the supports for multispan double-layer grids are combined with those along perimeters, then two-way diagonal latticed grids and diagonal square pyramid space grids, that is, types 2 and 8, could also be used.

For double-layer grids with circular, triangular, hexagonal, and other odd shapes supporting along perimeters, types with triangular grids in the top-layer, that is, types 3, 9, and 10, are appropriate for use.

The recommended types of double-layer grids are summarized in Table 24.1 according to the shape of the plan and their supporting conditions.

24.2.3 Method of Support

Ideal double-layer grids would be square, circular, or other polygonal shapes with overhanging and continuous supports along perimeters. This approaches more of a plate type of design, which minimizes the maximum bending moment. However, the configuration of building has a great number of varieties, and the support of the double-layer grids can take the following locations:

1. *Support along perimeters.* This is the most commonly used support location. The supports of double-layer grids may directly rest on the columns or on ring beams connecting the columns or

TABLE 24.1 Type Choosing for Double-Layer Grids

Shape of the plan	Supporting condition	Recommended types
Square, rectangular, aspect ratio = 1–1.5	Along perimeters	1, 2, 5–8
Rectangular, aspect ratio = 1.5–2	Along perimeters	1, 5–7
Long strip, aspect ratio > 2	Along perimeters	4
Square, rectangular	Intermediate support	1, 5, 6
Square, rectangular	Intermediate support combined with support along perimeters	1, 2, 5, 6, 8
Circular, triangular, hexagonal, and other odd shapes	Along perimeters	3, 9, 10

exterior walls. Care should be taken that the module size of grids should match the column spacing.

2. *Multicolumn supports.* For single-span buildings, like sports hall, double-layer grids can be supported on four intermediate columns as shown in Figure 24.5a. For buildings like workshops, usually multispan columns in the form of grids as shown in Figure 24.5b are used. Sometimes the column grids are used in combination with supports along perimeters as shown in Figure 24.5c. Overhangs should be employed where possible in order to provide some amount of stress reversal to reduce the interior chord forces and deflections. For those double-layer grids supported on intermediate columns, it is best to design with overhangs, which are taken as quarter to one third of the midspan. Corner supports should be avoided if possible, since this cause large forces in the edge chords. If only four supports are to be provided, then it is more desirable to locate them in the middle of the sides rather at the corners of the building.

3. *Support along perimeters on three sides and free on the other side.* For buildings of rectangular shape, it is necessary to have one side open, such as in the case of airplane hangar or for future extension. Instead of establishing the supporting girder or truss on the free side, triple-layer grids can be formed by simply adding another layer of several module widths (Figure 24.6). For shorter spans, this can also be solved by increasing the depth of the double-layer grids. The sectional area of the members along the free side will increase accordingly.

The columns for double-layer grids must support gravity loads and possible lateral forces. Typical types of support on multicolumns are shown in Figure 24.7. Usually, the member forces around the support will be excessively large, and some means of transferring the loads to column is necessary. It may carry the space grids down to the column top by an inverted pyramid as shown in Figure 24.7a or by triple-layer grids as shown in Figure 24.7b, which can be employed to carry skylights. If necessary, the inverted pyramids may be extended down to the ground level as shown in Figure 24.7c. The spreading out of the concentrated column reaction on the space grids reduces the maximum chord and web member forces adjacent to the column supports and reduces the effective spans. The use of a vertical strut on column tops as shown in Figure 24.7d enables the space grids to be supported on top chords, but the vertical strut and the connecting joint have to be very strong. The use of crosshead beams on

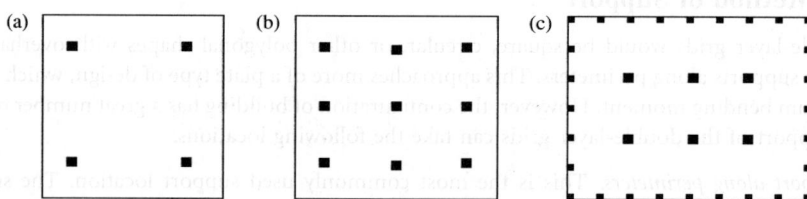

FIGURE 24.5 Multicolumn supports.

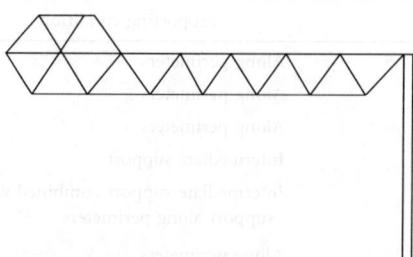

FIGURE 24.6 Triple-layer grids on the free side.

column tops as shown in Figure 24.7e produces the same effect as the inverted pyramid but usually costs more in material and special fabrication.

24.2.4 Design Parameters

Before any work can proceed on the analysis of a double-layer grid, it is necessary to determine the depth and the module size. The depth is the distance between the top and bottom-layers and the module is the distance between two joints in the layer of the grid (see Figure 24.8). Although these two parameters seem simple enough to determine, yet they will play an important role in the economy of the roof design. There are many factors influencing these parameters, such as the type of double-layer grid, the span between the supports, the roof cladding, and the proprietary system used. In fact, the depth and module size are mutually dependent, which is related by the permissible angle between the center line of web members and the plane of the top and bottom chord members. This should be less than 30°, or the forces in the web members and the length will be relatively excessive, or greater than 60°, or the density of the web members in the grid will become too high. For some of the proprietary systems, the depth and module are standardized.

The depth and module size of double-layer grids are usually determined by practical experience. In some of the papers and handbooks, figures on these parameters are recommended, and one may find that the difference is quite large. For example, the span–depth ratio varies from 12.5 to 25, or even more. It is usually considered that the depth of space frame can be relatively small when compared with more conventional structures. This is generally true because double-layer grids produce smaller deflections under load. However, depths that are small in relation to span will tend to use smaller modules, and hence a heavier structure will result. In the design, almost unlimited possibilities exist in practice for the choice of geometry. It is best to determine these parameters through structural optimization.

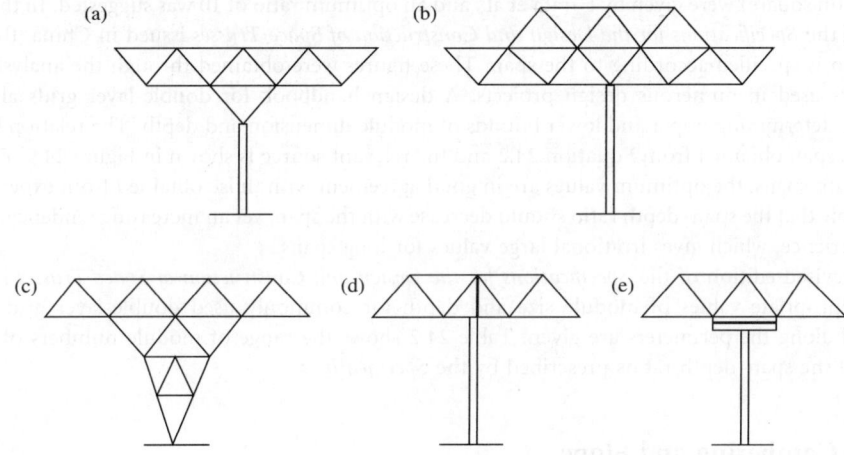

FIGURE 24.7 Supporting columns.

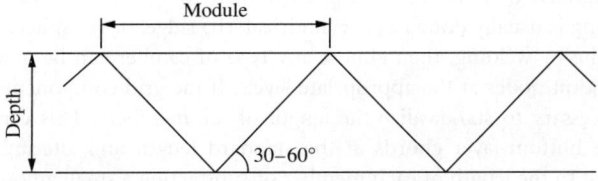

FIGURE 24.8 Depth and module.

Work has been done on the optimum design of double-layer grids supported along perimeters. In an investigation by Lan [4], seven types of double-layer grids were studied. The module dimension and depth of the space frame are chosen as the design variables. The total cost is taken as the objective function, which includes the cost of members and joints as well as of roofing systems and enclosing walls. Such assumption makes the results realistic to a practical design. A series of double-layer grids of different types spanning from 24 to 72 m were analyzed by optimization. It was found that the optimum design parameters were different for different types of roof system. The module number generally increases with the span, and the steel purlin roofing system allows larger module sizes than that of reinforced concrete. The optimum depth is less dependent on the span and a smaller depth can be used for steel purlin roofing system. It should be observed that a smaller member density will lead to a grid with relatively few nodal points and thus the least possible production costs for nodes, erection expense, etc.

Through regression analysis of the calculated values by optimization method where the costs are within the 3% optimum, the following empirical formulas for optimum span–depth ratio are obtained. It was found that the optimum depths are distributed in a belt and all the span–depth ratios within this range will have optimum effect in construction.

For roofing system composed of reinforced concrete slabs

$$L/d = 12 \pm 2 \tag{24.1}$$

and for roofing system composed of steel purlins and metal decks

$$L/d = (510 - L)/34 \pm 2 \tag{24.2}$$

where L is the short span and d is the depth of the double-layer grids.

Few data could be obtained from the past works. Regarding the optimum depth for steel purlin roofing systems, Geiger suggested the span–depth ratio to be varied from 10 to 20 with less than 10% variation in cost. Motro recommended a span–depth ratio of 15. Curves for diagonal square pyramid space grids (diagonal on square) were given by Hirata et al., and an optimum ratio of 10 was suggested. In the earlier edition of the *Specifications for the Design and Construction of Space Trusses* issued in China, the span–depth ratio is specified according to the span. These figures were obtained through the analysis of the parameters used in numerous design projects. A design handbook for double-layer grids also gives graphs for determining upper and lower bounds of module dimension and depth. The relation between depth and span obtained from Equation 24.2 and the relevant source is shown in Figure 24.9. For short and medium spans, the optimum values are in good agreement with those obtained from experience. It is noticeable that the span–depth ratio should decrease with the span, yet an increasing tendency is found from experience, which gives irrational large values for long spans.

In the revised edition of the *Specifications for the Design and Construction of Space Trusses* issued in China, appropriate values of module size and depth for commonly used double-layer grids simply supported along the perimeters are given. Table 24.2 shows the range of module numbers of the top chord and the span–depth ratios prescribed by the *Specifications*.

24.2.5 Cambering and Slope

Most double-layer grids are sufficiently stiff so that camber is often not required. Cambering is considered when the structure under load appears to be sagging and the deflection might be visually undesirable. It is suggested that the cambering be limited to $\frac{1}{300}$ of the shorter span. As shown in Figure 24.10, cambering is usually done in (a) cylindrical, (b) ridge, or (c) spherical shape. If the grid is being fabricated on site by welding, then almost any type of camber can be obtained as this is just a matter of setting the joint nodes at the appropriate levels. If the grid components are fabricated in the factory, then it is necessary to standardize the length of the members. This can be done by keeping either the top or the bottom-layer chords at the standard length and altering the others either by adding a small amount to the length of each member or subtracting a small amount from it to generate the camber required.

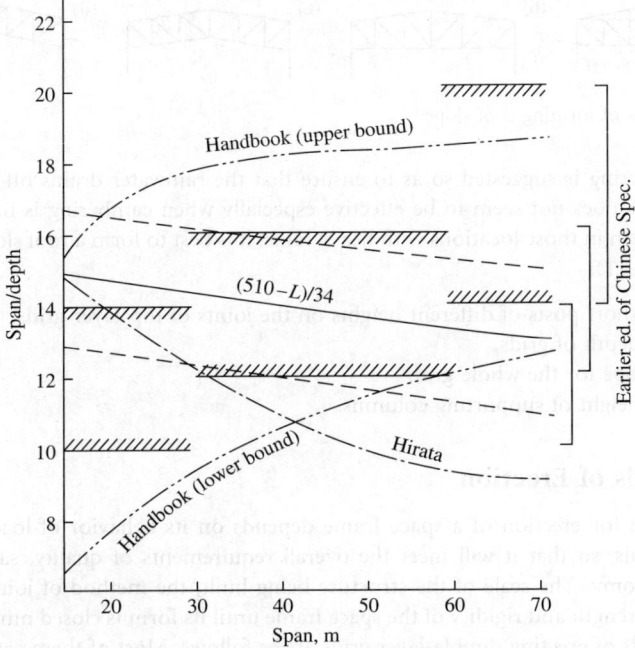

FIGURE 24.9 Relation between depth and span of double-layer grids.

TABLE 24.2 Module Number and Span–Depth Ratio

Type of double-layer grids	Reinforced concrete slab roofing system		Steel purlin roofing system	
	Module number	Span–depth ratio	Module number	Span–depth ratio
1, 5, 6	$(2–4) + 0.2L$	10–14	$(6–8) + 0.07L$	$(13–17) − 0.03L$
2, 7, 8	$(6–8) + 0.08L$			

Notes:
1. L denotes the shorter span in meters.
2. When the span is less than 18 m, the number of modules may be decreased.

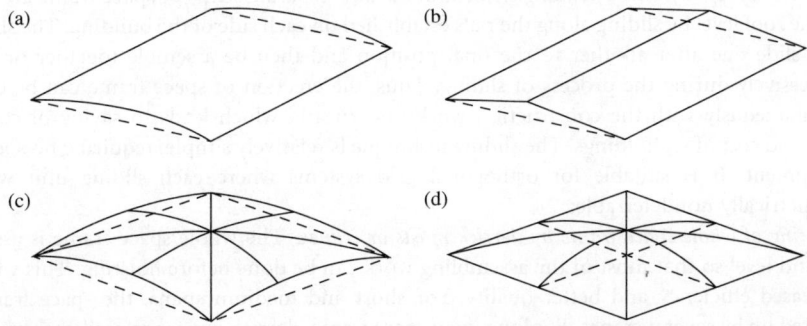

FIGURE 24.10 Ways of cambering.

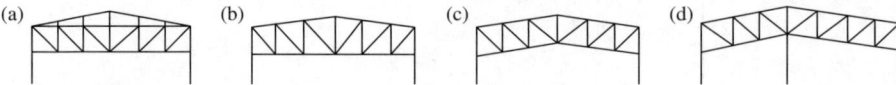

FIGURE 24.11　Ways of forming roof slope.

Sometimes cambering is suggested so as to ensure that the rainwater drains off the roof quickly to avoid ponding. This does not seem to be effective especially when cambering is limited. To solve the water run-off problem in those locations with heavy rains, it is best to form a roof slope by the following methods (Figure 24.11):

1. Establishing short posts of different heights on the joints of top-layer grids.
2. Varying the depth of grids.
3. Forming a slope for the whole grid.
4. Varying the height of supporting columns.

24.2.6　Methods of Erection

The method chosen for erection of a space frame depends on its behavior of load transmission and constructional details, so that it will meet the overall requirements of quality, safety, speed of construction, and economy. The scale of the structure being built, the method of jointing the individual elements, and the strength and rigidity of the space frame until its form is closed must all be considered. The general methods of erecting double-layer grids are as follows. Most of them can also be applied to the construction of latticed shells.

1. *Assembly of space frame elements in the air.* Members and joints or prefabricated subassembly elements are assembled directly on their final position. Full scaffoldings are usually required for this type of erection. Sometimes only partial scaffoldings are used if cantilever erection of space frame can be executed. The elements are fabricated at the shop and transported to the construction site, and no heavy lifting equipment is required. It is suitable for all types of space frame with bolted connections.
2. *Erection of space frame by strips or blocks.* The space frame is divided on its plan into individual strips or blocks. These units are fabricated on the ground level, then hoisted up into its final position and assembled on the temporary supports. With more work being done on the ground, the amount of assembling work at high elevation is reduced. This method is suitable for those double-layer grids where the stiffness and load-resisting behavior will not change considerably after dividing into strips or blocks, such as two-way orthogonal latticed grids, orthogonal square pyramid space grids, and the those with openings. The size of each unit will depend on the hoisting capacity available.
3. *Assembly of space frame by sliding element in the air.* Separate strips of space frame are assembled on the roof level by sliding along the rails established on each side of the building. The sliding units may slide one after another to the final position and then be assemble together or assembled successively during the process of sliding. Thus, the erection of space frame can be carried out simultaneously with the construction work underneath, which leads to saving of construction time and cost of scaffoldings. The sliding technique is relatively simple, requiring no special lifting equipment. It is suitable for orthogonal grid systems where each sliding unit will remain geometrically nondeferrable.
4. *Hoisting of whole space frame by derrick masts or cranes.* The whole space frame is assembled at ground level so that most of the assembling work can be done before hoisting. This will result in increased efficiency and better quality. For short and medium spans, the space frame can be hoisted up by several cranes. For long-span space frame, derrick masts are used as the support and electric winches as the lifting power. The whole space frame can be translated or rotated in the air

and then seated on its final position. This method can be employed to all types of double-layer grids.

5. *Lifting up the whole space frame.* This method also has the benefit of assembling space frame at ground level, but the structure cannot move horizontally during lifting. Conventional equipment used is hydraulic jacks or lifting machines for lift-slab construction. An innovative method has been developed by using the center hole hydraulic jacks for slipforming. The space frame is lifted up simultaneously with the slipforms for reinforced concrete columns or walls. This lifting method is suitable for double-layer grids supported along perimeters or on multipoint supports.

6. *Jacking-up the whole space frame.* Heavy hydraulic jacks are established on the position of columns that are used as supports for jacking-up. Occasionally, roof claddings, ceilings, and mechanical installations are also completed with the space frame at the ground level. They are appropriate for use in space frame with multipoint supports, the number of which is usually limited.

24.3 Latticed Shells

24.3.1 Form and Layer

The main difference between double-layer grids and latticed shells is the form. For double-layer grid, it is simply a flat surface. For latticed shell, the variety of forms is almost unlimited. A common approach to the design of latticed shells is to start with the consideration of the form — a surface curved in space. The geometry of basic surfaces can be identified, according to the method of generation, as surface of translation and surface of rotation. A number of variations of form can be obtained by taking segments of the basic surfaces or by combining or adding them. In general, the geometry of surface has a decisive influence on essentially all characteristics of the structure: the manner in which it transfers loads, its strength and stiffness, the economy of construction, and the esthetic quality of the completed project.

Latticed shells can be divided into three distinct groups forming singly curved, synclastic, and anticlastic surfaces. A barrel vault (cylindrical shell) represents a typical developable surface, having a zero curvature in the direction of generatrices. A spherical or elliptical dome (spheroid or elliptic paraboloid) is a typical example of synclastic shell. A hyperbolic paraboloid is a typical example of anticlastic shell.

Besides the mathematical generation of surface systems, there are other methods for finding shapes of latticed shell. Mathematically, the surface can be defined by a high-degree polynomial with the unknown coefficients determined from the known shape of the boundary and the known position of certain points at the interior required by the functional and architectural properties of the space. Experimentally, the shape can be obtained by loading a net of chain wires or a rubber membrane or a soap membrane in the desired manner. In each case the membrane is supported along a predetermined contour and at pre-determined points. The resulting shape will produce a minimal surface, which is characterized by a least surface area for a given boundary and also constant skin stress. Such an experimental model helps to develop a first understanding about the nature of structural forms.

The inherent curvature in latticed shell will give the structure greater stiffness. Hence, latticed shell can be built in single layer, which is a major difference from the double-layer grid. Of course, latticed shell may also be built in double-layer. Although single- and double-layer latticed shells are similar in shape, the structural analysis and connecting detail are quite different. The single-layer latticed shell is a structural system with rigid joints, while the double-layer latticed shell has hinged joints. In practice, single-layer latticed shells of short span with lightweight roofing may also be built with hinged joints. The members and connecting joints in a single-layer shell of large span will resist not only axial forces as in a double-layer shell but also the internal moments and torsions. Since the single-layer latticed shells are easily liable to buckling, the span should not be too large. There is no distinct limit between single and double layers, which will depend on the type of shell, the geometry and size of the framework, and the section of members.

24.3.2 Braced Barrel Vaults

The braced barrel vault is composed of member elements arranged on a cylindrical surface. The basic curve is a circular segment; occasionally, a parabola, ellipse or funicular line may also be used. Figure 24.12 shows the typical arrangement of a braced barrel vault. Its structural behavior depends mainly on the type and location of supports, which can be expressed as L/R, where L is the distance between the supports in longitudinal direction and R is the radius of curvature of the transverse curve.

If the distance between the supports is long and usually edge beams are used in the longitudinal direction (Figure 24.12a), the primary response will be beam action. For $1.67 < L/R < 5$, the barrel vaults are called long shells, which can be visualized as beams with curvilinear cross-sections. The beam theory with the assumption of linear stress distribution may be applied to barrel vaults that are of symmetrical cross-section and under uniform loading if $L/R > 3$. This class of barrel vault will have longitudinal compressive stresses near the crown of the vault, longitudinal tensile stresses toward the free edges, and stresses toward the supports.

As the distance between transverse supports becomes closer, or as the dimension of the longitudinal span becomes smaller than the dimension of the shell width, such that $0.25 < L/R < 1.67$, the primary response will be arch action in the transverse direction (Figure 24.12b). The barrel vaults are called short shells; their structural behavior is rather complex and dependent on their geometrical proportions. The force distribution in the longitudinal direction is not linear anymore, but in a curvilinear manner, trusses or arches are usually used as the transverse supports.

When a single braced barrel vault is supported continuously along its longitudinal edges on foundation blocks, or the ratio of L/R becomes very small, that is, < 0.25 (Figure 24.12c), the forces are carried directly in the transverse direction to the edge supports. Its behavior may be visualized as the response of parallel arches. Displacement in the radial direction is resisted by circumferential bending stiffness. This type of barrel vault can be applied to buildings like airplane hangars or gymnasia where the wall and roof are combined together.

There are several possible types of bracing that have been used in the construction of single-layer braced barrel vaults. Figure 24.13 shows five principle types:

1. Orthogonal grid with single bracing of Warren truss (a).
2. Orthogonal grid with single bracing of Pratt truss (b).
3. Orthogonal grid with double bracing (c).
4. Lamella (d).
5. Three way (e).

The first three types of braced barrel vaults can be formed by composing latticed trusses with difference in the arrangement of bracings (Figure 24.13a–c). In fact, the original barrel vault was introduced by Foppl. It consists of several latticed trusses, spanning the length of the barrel and supported on the gables. After connection of the longitudinal booms of the latticed trusses, they became a part of the braced barrel vault of the single-layer type.

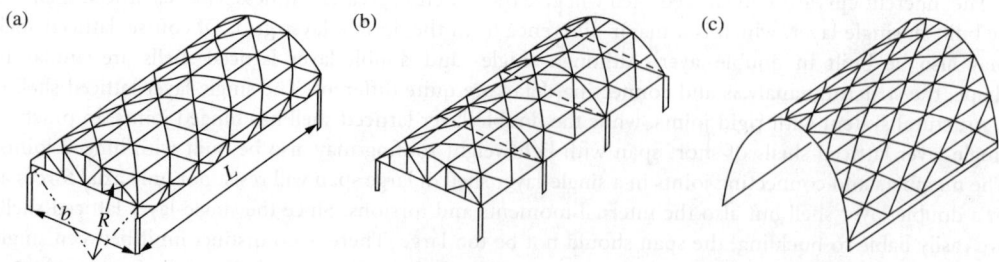

FIGURE 24.12   Braced barrel vaults.

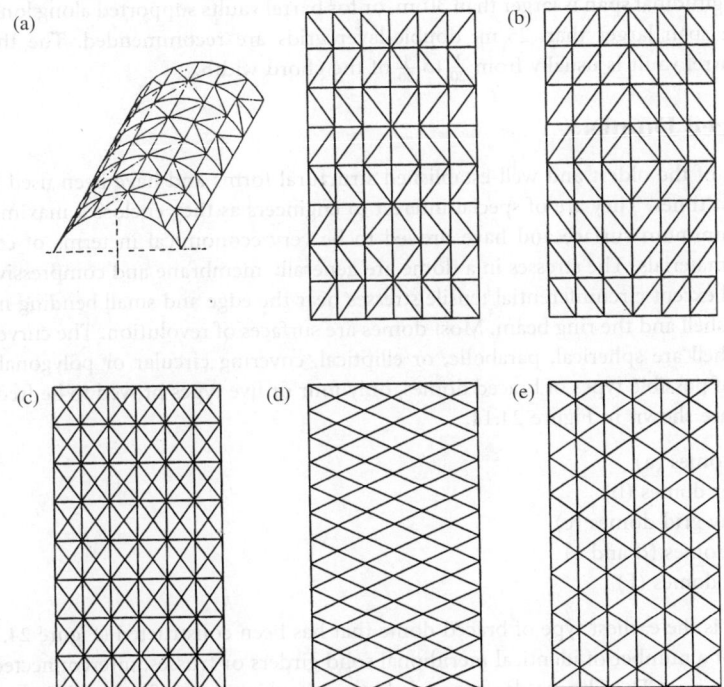

FIGURE 24.13   Types of bracing for braced barrel vaults.

The popular diamond-patterned lamella type of braced barrel vault consists of a number of inter-connected modular units forming a rhombus-shaped grid pattern (Figure 24.13d). Each unit, which is twice the length of the side of a diamond, is called a *lamella*. Lamella roofs proved ideal for prefabricated construction as all the units are of standard size. They were originally constructed of timber, but with the increase of span, steel soon became the most frequently used material.

To increase the stability of the structure and to reduce the deflections under unsymmetrical loads, purlins were employed for large-span lamella barrel vaults. This created the three-way grid type of bracing and became very popular (Figure 24.13e). The three-way grid enables the construction of such a system using equilateral triangles composed of modular units, which are of identical length and can be connected with simple nodes.

Research investigations have been carried out on braced barrel vaults. One aspect of this research referred to the influence of different types of bracing on the resulting stress distribution. The experimental tests on the models proved that there are significant differences in the behavior of the structures, and the type of bracing has a fundamental influence on the strength and load-carrying capacity of the braced barrel vaults. The three-way single-layer barrel vaults exhibited a very uniform stress distribution under uniformly distributed load and much smaller deflections in the case of unsymmetrical loading than for any of the other type of bracing. The experiments also showed that large-span single-layer braced barrel vaults are prone to instability, especially under the action of heavy unsymmetrical loads, and that the rigidity of joints can exert an important influence on the overall stability of the structure.

For double-layer braced barrel vaults, if two- or three-way latticed trusses are used to form the top and bottom-layers of the latticed shell, the grid pattern is identical as shown in Figure 24.13 for single-layer shells. If square or triangular pyramids are used, either the top-layer grid or the bottom-layer grid may follow the same pattern as shown in Figure 24.13.

The usual height to width ratio for long shells varies from $\frac{1}{3}$ to $\frac{1}{6}$. When the barrel vault is supported along the longitudinal edges, the height can be increased to half to one fifth of the chord width. For long shells, if the longitudinal span is larger than 30 m, or for barrel vaults supported along longitudinal edges with transverse span larger than 25 m, double-layer grids are recommended. The thickness of the double-layer barrel vault is usually from $\frac{1}{20}$ to $\frac{1}{50}$ of the chord width.

24.3.3 Braced Domes

Domes are one of the oldest and well-established structural forms and have been used in architecture since the earliest times. They are of special interest to engineers as they enclose a maximum amount of space with a minimum surface and have proved to be very economical in terms of consumption of constructional materials. The stresses in a dome are generally membrane and compressive in the major part of the shell except circumferential tensile stresses near the edge and small bending moments at the junction of the shell and the ring beam. Most domes are surfaces of revolution. The curves used to form the synclastic shell are spherical, parabolic, or elliptical, covering circular or polygonal areas. Out of a large variety of possible types of braced domes, only four or five types proved to be frequently used in practice. They are shown in Figure 24.14.

1. Ribbed domes (a)
2. Schwedler domes (b)
3. Three-way grid domes (c)
4. Lamella domes (d and e)
5. Geodesic domes (f)

Ribbed dome is the earliest type of braced dome that has been constructed (Figure 24.14a). A ribbed dome consists of a number of identical meridional solid girders or trusses, interconnected at the crown by a compression ring. The ribs are also connected by concentric rings to form grids in trapezium shape. The ribbed dome is usually stiffened by a steel or reinforced concrete tension ring at its base.

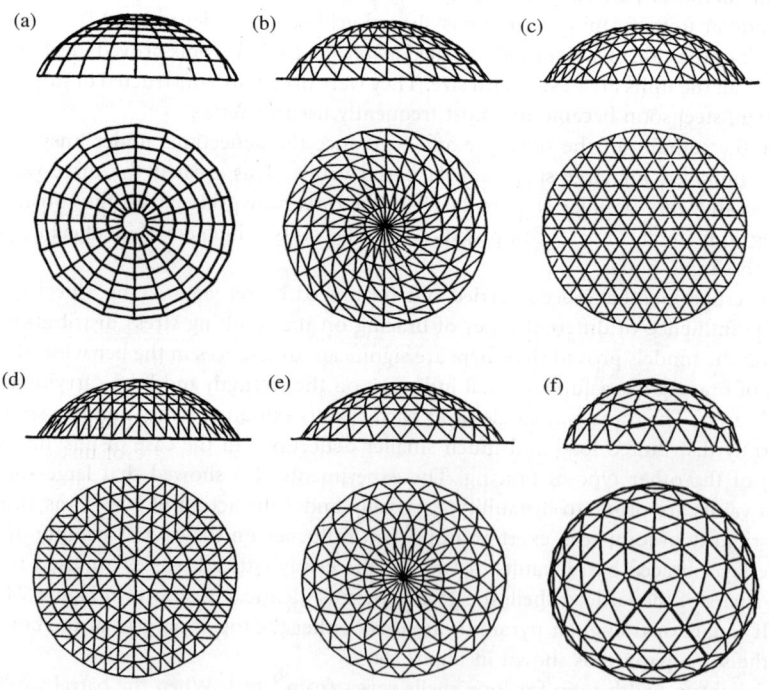

FIGURE 24.14   Braced domes.

A Schwedler dome also consists of meridional ribs connected together to a number of horizontal polygonal rings to stiffen the resulting structure so that it will be able to take unsymmetrical loads (Figure 24.14b). Each trapezium formed by intersecting meridional ribs with horizontal rings is sub-divided into two triangles by a diagonal member. Sometimes the trapezium may also be subdivided by two cross-diagonal members. This type of dome was introduced by a German engineer J.W. Schwedler in 1863. The great popularity of Schwedler domes is due to the fact that, on the assumption of pin-connected joints, the structure can be analyzed as statically determinate. In practice, in addition to axial forces, all the members are also under the action of bending and torsional moments. Many attempts have been made in the past to simplify their analysis, but precise methods of analysis using computers have finally been applied to find the actual stress distribution.

The construction of three-way grid dome is self-explanatory, which may be imagined as a curved form of three-way double-layer grids (Figure 24.14c). It can also be constructed in single layer for the dome. The Japanese "diamond dome" system by Tomoegumi Iron Works belongs to this category. The theoretical analysis of three-way grid domes shows that even under unsymmetrical loading the forces in this configuration are very evenly distributed leading to economy in material consumption.

A lamella dome is formed by intersecting two-way ribs diagonally to form a rhombus-shaped grid pattern. As in lamella braced barrel vault, each lamella element has a length that is twice the length of the side of a diamond. The lamella dome can further be distinguished into parallel and curved domes. For parallel lamella as shown in Figure 24.14d, the circular plan is divided into several sectors (usually six or eight), and each sector is subdivided by parallel ribs into rhombus grids of the same size. This type of lamella dome is very popular in the United States. It is sometimes called a Kiewitt dome, after its developer. For curved lamella as shown in Figure 24.14e, rhombus grids of different sizes, gradually increasing from the center of the dome, are formed by diagonal ribs along the radial lines. Sometimes, for the purpose of establishing purlins for roof decks, concentric rings are introduced, and a triangular network is generated.

The geodesic dome was developed by the American designer Buckminster Fuller, who turned architects' attention to the advantages of braced domes in which the elements forming the framework of the structure are lying on the great circle of a sphere. This is where the name "geodesic" came from (Figure 24.14f). The framework of these intersecting elements forms a three-way grid comprising virtually equilateral spherical triangles. In Fuller's original geodesic domes, he used an icosahedron as the basis for the geodesic subdivision of a sphere; then, the spherical surface is divided into 20 equilateral triangles as shown in Figure 24.15a. This is the maximum number of equilateral triangles into which a sphere can be divided. For domes of larger span, each of these triangles can be subdivided into six triangles by drawing medians and bisecting the sides of each triangle. It is therefore possible to form 15 complete great circles regularly arranged on the surface of a sphere (see Figure 24.15b). Practice shows that the primary type of bracing, which is truly geodesic, is not sufficient since it would lead to an excessive length for members in geodesic dome, therefore a secondary bracing has to be introduced. To obtain a more or less regular network of the bracing bars, the edges of the basic triangle are divided modularly. The number of modules into which each edge of the spherical icosahedron is divided depends mainly on the size of the dome, its span, and the type of roof cladding. This subdivision is usually referred to as "frequency," as depicted in Figure 24.15c. It must be pointed out that during such a subdivision the resulting triangles are no longer equilateral. The members forming the skeleton of the dome show slight variation in their length. As the frequency of the subdivision increases, the member length reduces, the number of components as well as the number of types of connecting joints increase. Consequently, this reflects in the increase of the final price of geodesic dome and is one of the reasons why geodesic domes, in spite of their undoubted advantages for smaller spans, do not compare equally well with other types of braced domes for larger span.

The rise of a braced dome can be as flat as one seventh of the diameter or as high as three fourths of the diameter, which will constitute the greater part of a sphere. For diameters larger than 60 m, double-layer grids are recommended. The ratio of the thickness to the diameter of double-layer braced dome is in the range of $\frac{1}{30}$ to $\frac{1}{60}$; for long spans the thickness can be as small as $\frac{1}{100}$ of the diameter.

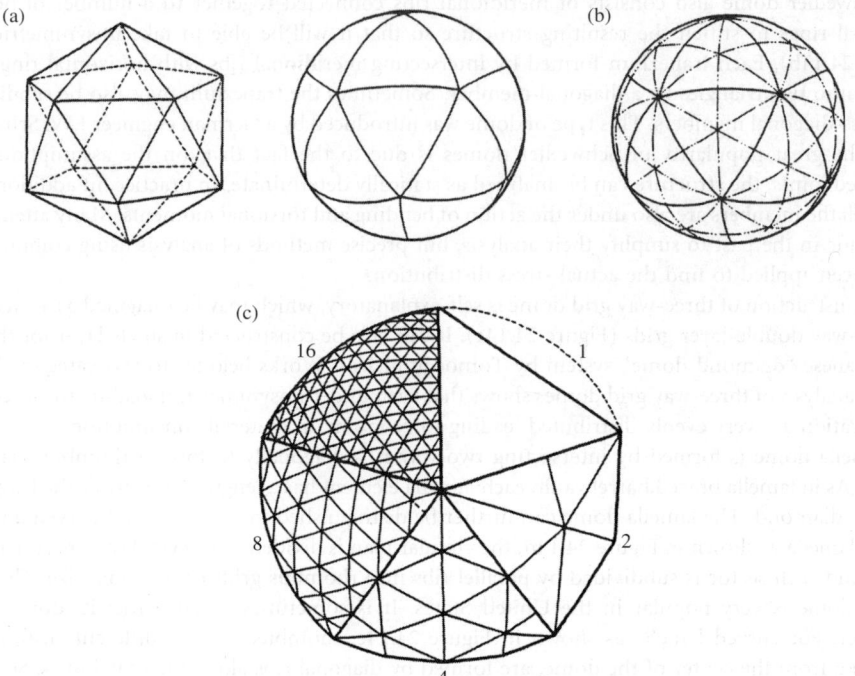

FIGURE 24.15   Geodesic subdivision.

The subdivision of the surface of the braced dome can also be carried out using the following three methods. The first method is based on the surface of revolution, and the first set of lines of division are drawn as the meridional lines from the apex. Next, circumferential rings are added. This results in a ribbed dome and further a Schwedler dome. Alternately, the initial set may be taken as a series of spiral arcs, resulting in dividing the surface into triangular units as uniform as possible. This is achieved by drawing great circles in three directions as shown in the case of grid dome. A noteworthy type of division of a braced dome is the parallel lamella dome, which is obtained by combining the first and second methods described above. The third type of subdivision results from projecting the edges of in-polyhedra onto the spherical surface and then inscribing a triangular network of random frequency into this basic grid. Geodesic dome represents an application of this method, with the basic field derived from the isosahedron further subdivided into equilateral triangles.

24.3.4 Hyperbolic Paraboloid Shells

The hyperbolic paraboloid or hypar is a translational surface formed by sliding a concave paraboloid called generatrix parallel to itself along a convex parabola called directrix, which is perpendicular to the generatrix (Figure 24.16a). By cutting the surface vertically, parabolas can be obtained, and cutting horizontally will give hyperbolas. Such a surface can also be formed by sliding a straight line along two other straight lined skewed with respect to each other (Figure 24.16b). The hyperbolic paraboloid is a doubly ruled surface; it can be defined by two families of intersecting straight lines, which form in plan projection a rhombic grid. This is one of the main advantages of a hyperbolic paraboloid shell. Although it has a double curvature anticlastic surface, it can be built by only using linear structural members. Thus, single-layer hypar shells can be fabricated from straight beams and double-layer hypar shells from linear latticed trusses. The single hypar unit shown in Figure 24.16 is suitable for use in the building of square, rectangular, or elliptic plan. In practice, there exist an infinite number of ways of combining hypar units to enclose a given building space.

A shallow hyperbolic paraboloid under uniform loading acts primarily as a shear system, where the shear forces, in turn, cause diagonal tension and compression. The behavior of the surface can be visualized as thin compression arches in one direction and tension cables in the perpendicular direction. In reality, additional shear and bending may occur along the vicinity of the edges.

24.3.5 Intersection and Combination

The basic forms of latticed shells are single-curvature cylinders, as well as double-curvature sphere and hyperbolic paraboloids. Many interesting new shapes can be generated by intersecting and combining of these basic forms. The art of intersection and combination is one of the important tools in the design of latticed shells. To fulfill the architectural and functional requirements, the load-resisting behavior of the structure as a whole and also its relation to the supporting structure should be taken into consideration.

For cylindrical shells, a simply way is to intersect through the diagonal as shown in Figure 24.17a. Two types of groined vaults on square plan can be formed by combining the corresponding intersected curve surfaces shown in Figure 24.17b and c. Likewise, combination of curved surfaces intersected from a cylinder produces a latticed shell on a hexagonal plan as shown in Figure 24.17d.

For spherical shells, segments of the surface are used to cover planes other than circular, such as triangular, square, and polygonal as shown in Figure 24.18a–c, respectively. Figure 24.18d shows a latticed shell on a square plan by combining the intersected curved surfaces from a sphere.

It is usual to combine a segment of cylindrical shell with hemispherical shells at two ends as shown in Figure 24.19. This form of latticed shell is an ideal plan for indoor track fields and ice skating rink.

Different solutions for assembling single hyperbolic paraboloid units to cover a square plan are shown in Figure 24.20. The combination of four equal hypar units produces different types of latticed shell supported on a central column as well as two or four columns along the outside perimeter. These basic blocks, in turn, can be added in various ways to form the multibay buildings.

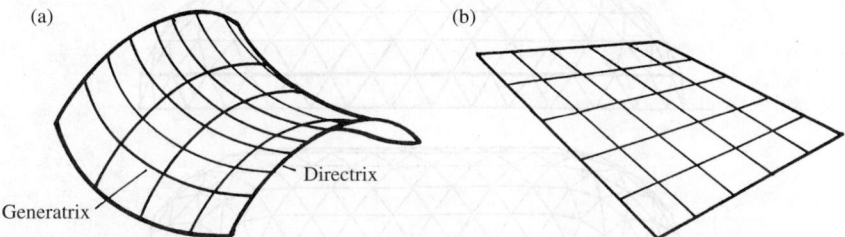

FIGURE 24.16 Hyperbolic paraboloid shells.

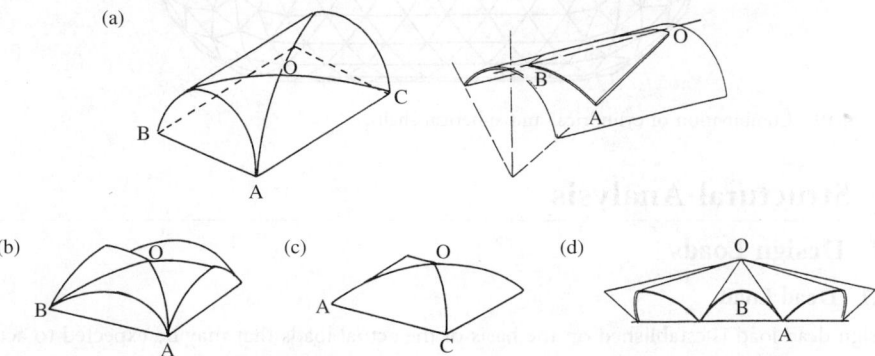

FIGURE 24.17 Intersection and combination of cylindrical shells.

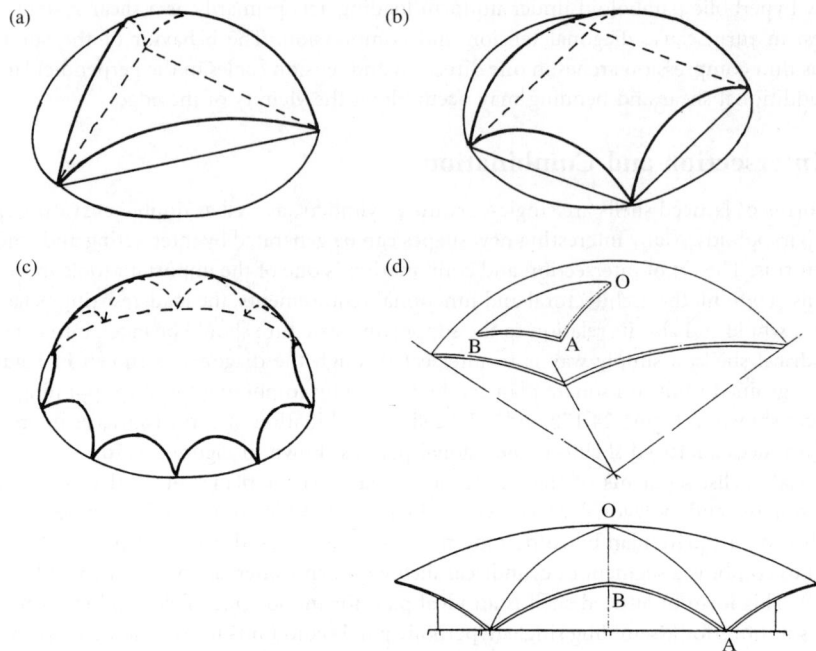

FIGURE 24.18   Intersection and combination of spherical shells.

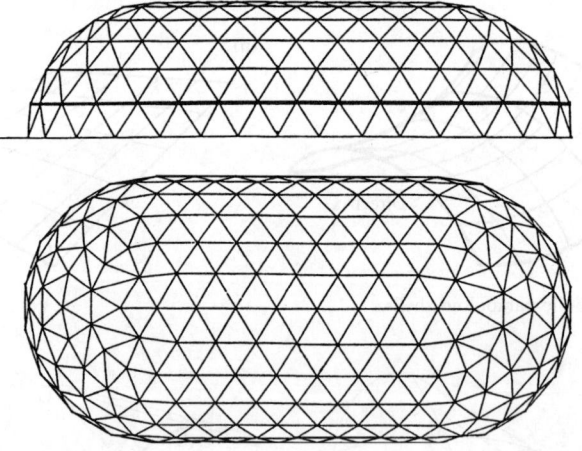

FIGURE 24.19   Combination of cylindrical and spherical shells.

24.4   Structural Analysis

24.4.1   Design Loads

24.4.1.1   Dead Load

The design dead load is established on the basis of the actual loads that may be expected to act on the structure of constant magnitude. The weight of various accessories — cladding, supported lighting, heat and ventilation equipment — and the weight of space frame comprise the total dead load. An empirical

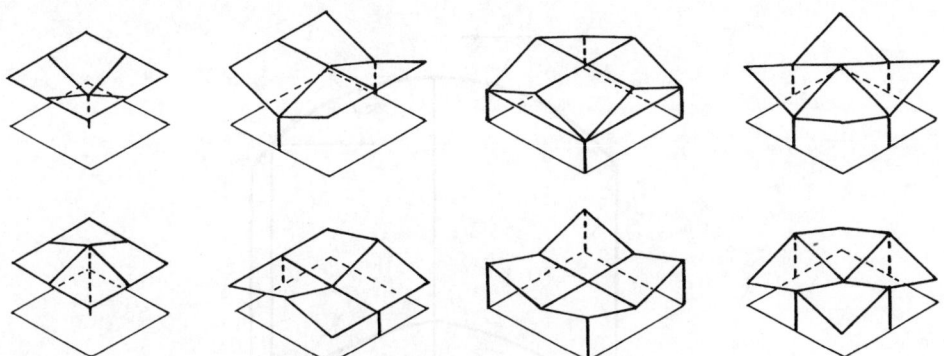

FIGURE 24.20 Combination of hyperbolic paraboloids.

formula is suggested to estimate the dead weight g of double-layer grids:

$$g = \tfrac{1}{200}(\zeta\sqrt{q_w}L)\ \text{kN/m}^2 \tag{24.3}$$

where q_w indicates all dead and live loads acting on double-layer grid except its self-weight (in kN/m^2); L is the shorter span (in m); and ζ is a coefficient, 1.0 for steel tubes and 1.2 for mill sections.

24.4.1.2 Live Load, Snow, or Rain Load

Live load is specified by the local building code and compared with the possible snow or rain load; the larger one should be used as the design load. Each space frame is designed with uniformly distributed snow load and further allowed for drifting depending on the shape and slope of the structure. Often more than one assumed distribution of snow load is considered. It was recommended by ISO for the determination of snow loads on simple curved roofs, pointed arches, and domes. The intensity of snow load as specified in *Bases for Design of Structures: Determination of Snow Loads on Roofs* [5] is reproduced as Figure 24.21. For domes of circular plan form, an axially symmetrical balanced load may be given as the corresponding balanced arch load. The drift load may likewise be given by the corresponding arch drift load along the plan diameter being parallel to the wind direction multiplied by a reduction factor $(1-a/r)$ where r is the plan radius and a is the horizontal distance from the wind direction diameter to any parallel plan chord. The snow loads can be calculated by the following formulas:

Windward side

$$s = s_b \tag{24.4a}$$

Leeward side

$$s = s_b + s_d \tag{24.4b}$$

Balanced load part

$$s_b = s_0 C_e C_t \mu_b$$
$$\mu_b = \sqrt{\cos(C_m 1.5\beta)} \quad \text{for } (C_m 1.5\beta) \le 90° \tag{24.5}$$
$$\mu_b = 0 \qquad\qquad\qquad \text{for } (C_m 1.5\beta) > 90°$$

Drifted load part

$$s_b = s_0 C_e C_t (\mu_b \mu_d)$$
$$\mu_d = (2.2C_e - 2.1C_e^2)\sin(3\beta) \quad \text{for } \beta \le 60° \tag{24.6}$$
$$\mu_d = 0 \qquad\qquad\qquad\qquad\qquad \text{for } \beta > 60°$$

where s_0 is the characteristic snow load on the ground (in kN/m^2), μ_b is the slope reduction coefficient, μ_d is the drift load coefficient, C_e is the exposure reduction coefficient, C_t is the thermal reduction coefficient, and C_m is the surface material coefficient.

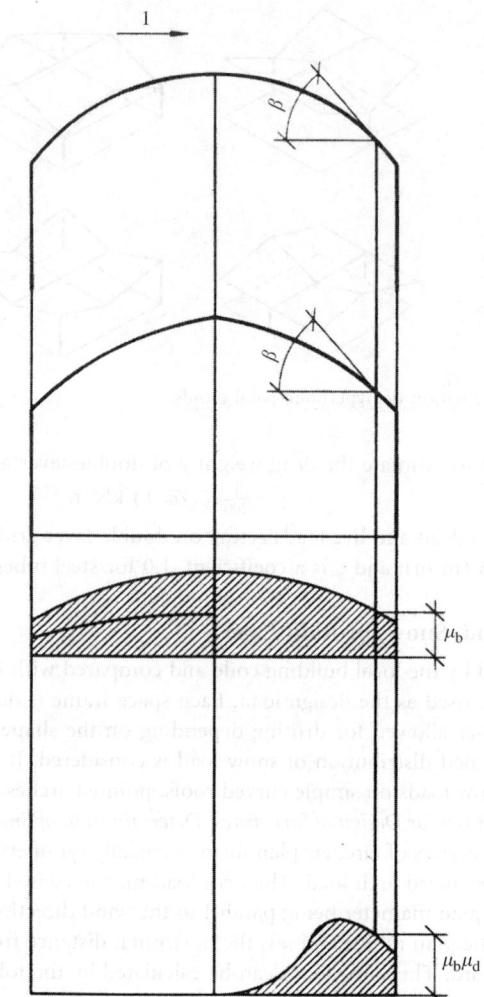

Key
1 Wind direction

FIGURE 24.21 Snow loads on simple curved roofs and domes.

The exposure coefficient, C_e, defines the balanced load on a flat horizontal roof of a cold building as a fraction of the characteristic load on the ground. For regions where there are not sufficient sinter climatological data available, it is recommended to set $C_e = 0.8$. However, the designer should always assess whether calm weather conditions (i.e., $C_e = 10$) during the snowfall season might yield more severe conditions for the structure. The thermal coefficient, C_t, is introduced to account for the reduction of snow load on roofs with high thermal transmittance, in particular glass-covered roofs, from melting caused by heat loss through the roof. For such cases C_t may take values less than unity. For all other cases, $C_t = 1.0$ applies. The surface material coefficient, C_m, defines a reduction of the snow load on roofs made of surface material with low surface roughness. It varies between 1.333, for slippery, unobstructed surfaces, to 1.0, for other surfaces. Detailed methods for the determination of these coefficients are given in the annexes of the Standard.

Rain load may be important in the tropical climate especially if the drainage provisions are insufficient. Ponding results when water on a double-layer grid flat roof accumulates faster than it runs off, thus causing excessive load on the roof.

24.4.1.3 Wind Load

The wind loads usually represent a significant proportion of the overall forces acting on barrel vaults and domes. A detailed comparison of the available codes concerning wind loads has revealed quite a large difference between the practices adopted by various countries. Pressure coefficients for arched roof springing from ground surface that can be used for barrel vault design are shown in Figure 24.22 and Table 24.3. For arched roof resting on an elevated structure like enclosure walls, the pressure coefficients are shown in Table 24.4. It should be noticed that ANSI is no longer used in practice in the United States. The ASCE has issued *Minimum Design Loads for Buildings and Other Structures* (SEI/ASCE 7-02). In Chapter C6 "Wind Loads," detailed provisions for determining wind loads are given.

The wind pressure distribution on buildings is also recommended by the European Committee for Standardization as European Prestandard [6]. The pressure coefficients for arched roof and spherical domes, either resting on the ground or on elevated structure, are presented in graphical forms in Figure 24.23 and Figure 24.24 respectively.

It can be seen that significant variations in pressure coefficients from different codes of practice exist for three-dimensional curved space frames. This is due to the fact that these coefficients are highly dependent on Reynolds number, surface roughness, wind velocity profile, and turbulence. It may be concluded that the codes of practice are only suitable for preliminary design purposes, especially for the important long-span space structures and those lattice shells with peculiar shapes. It is therefore necessary to undertake further wind tunnel tests in an attempt to establish more accurately the pressure distribution over the roof surface. For such tests, it is essential to simulate the velocity profile and turbulence of the natural wind and the Reynolds number effects associated with curved surface.

24.4.1.4 Temperature Effect

Most space frames are subject to thermal expansion and contraction due to changes in temperature and thus may be subject to axial loads if restrained. Potential temperature effect must be considered in the design especially when the span is comparatively large. The choice of support locations — perimeter, intermediate columns, etc. — and types of support — fixed, slid, or free rotation and translation — as

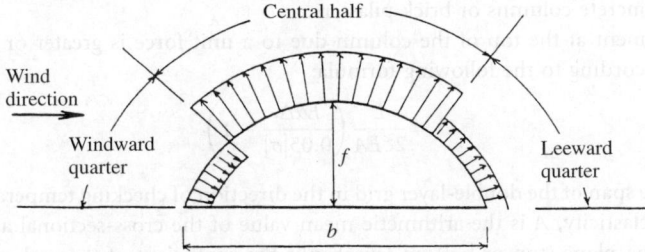

FIGURE 24.22 Wind pressure on an arched roof.

TABLE 24.3 Pressure Coefficient for Arched Roof on Ground

Country code	Windward quarter	Central half	Leeward quarter	Rise/span, r
United States: ANSI A 58.1-1982	$1.4r$	$-0.7 - r$	-0.5	$0 < r < 0.6$
USSR: BC&R 2.01.07-85	0.1	-0.8	-0.4	0.1
	0.3	-0.9	-0.4	0.2
	0.4	-1.0	-0.4	0.3
	0.6	-1.1	-0.4	0.4
	0.7	-1.2	-0.4	0.5
China: GB 50009-2001	0.1	-0.8	-0.5	0.1
	0.2	-0.8	-0.5	0.2
	0.6	-0.8	-0.5	0.5

TABLE 24.4   Pressure Coefficient for Arched Roof on Elevated Structure

Country code	Windward quarter	Central half	Leeward quarter	Rise/span, r
United States: ANSI A 58.1-1982	−0.9	−0.7 − r	−0.5	0 < r < 0.2
	1.5r − 0.3[a]	−0.7 − r	−0.5	0.2 < r < 0.3
	2.75r − 0.7	−0.7 − r	−0.5	0.3 ≤ r ≤ 0.6
USSR: BC&R 2.01.07-85[b]				
$h_e/b = 0.2$	−0.2	−0.8	−0.4	0.1
	−0.1	−0.9	−0.4	0.2
	0.2	−1.0	−0.4	0.3
	0.5	−1.1	−0.4	0.4
	0.7	−1.2	−0.4	0.5
$h_e/b > 1$	−0.8	−0.8	−0.4	0.1
	−0.7	−0.9	−0.4	0.2
	−0.3	−1.0	−0.4	0.3
	0.3	−1.1	−0.4	0.4
	0.7	−1.2	−0.4	0.5
China: GB50009-2001	−0.8	−0.8	−0.5	0.1
	0	−0.8	−0.5	0.2
	0.6	−0.8	−0.5	0.5

[a] Alternate coefficient 6r − 2.1 can also be used.
[b] h_e, height of the elevated structure.

well as the geometry of members adjacent to the support, all contribute to minimizing the effect of thermal expansion. The temperature effect of a space frame may be calculated by the ordinary matrix displacement method of analysis, and most computer programs provided such a function.

For double-layer grid, if it satisfies one of the following requirements, the calculation for temperature effect may be exempted:

1. The joints on supports allow the double-layer grid to move horizontally.
2. Double-layer grids of less than 40 m span are supported along perimeters by independent reinforced concrete columns or brick pilasters.
3. The displacement at the top of the column due to a unit force is greater or equal to the value calculated according to the following formula:

$$\delta = \frac{L}{2\xi EA}\left(\frac{E\alpha\Delta_t}{0.05[\sigma]} - 1\right) \tag{24.7}$$

where L is the span of the double-layer grid in the direction of checking temperature effect, E is the modulus of elasticity, A is the arithmetic mean value of the cross-sectional area of members in the supporting plane (top or bottom-layer), α is the coefficient of thermal expansion, Δ_t is the temperature difference, $[\sigma]$ is the allowable stress of steel, and ξ is a coefficient: when the chords in the supporting plane are arranged in orthogonal grids, $\xi = 1$, in diagonal grids, $\xi = 2$, and in three-way grids, $\xi = 2$.

24.4.1.5 Construction Loads

During construction, structures may be subjected to loads different from the design loads after completion, depending on the sequence of construction and method of scaffoldings. For example, a space frame may be lifted up at points different from the final supports or it may be constructed in blocks or strips. Therefore, the whole structure, or a portion of it, should be checked during various stages of construction.

24.4.2 Static Analysis

There are generally two different approaches in use for the analysis of space frames. In the first approach the structure is analyzed directly as a general assembly of discrete members, that is, *discrete method*.

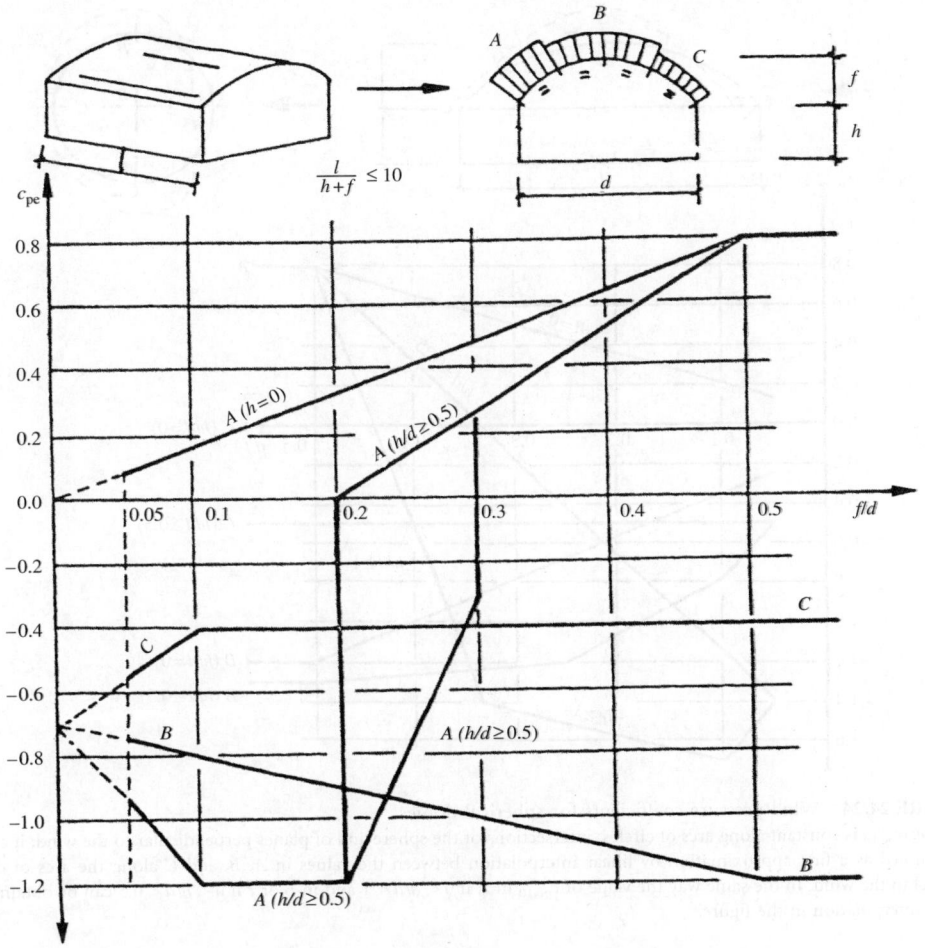

FIGURE 24.23 Wind pressure coefficients for arched roof.

Notes: (a) For $0 \leq h/d \leq 0.5$, $c_{pe,10}$ is obtained by linear interpolation; (b) for $0.2 \leq f/d \leq 0.3$ and $h/d \geq 0.5$, two values of $c_{pe,10}$ have to be considered; and (c) the diagram is not applicable for slat roofs.

In the second approach the structure is represented by an equivalent continuum like a plate or shell, that is, *continuum analogy method.*

The advent of computers has radically changed the whole approach to the analysis and design of space frames. It has also been realized that matrix methods of analysis provide an extremely efficient means for rapid and accurate treatment of many types of space structures. In the matrix analysis, a structure is represented as a discrete system, and all the usual equations of structural mechanics are written conveniently in matrix form. Thus, matrix analysis is particularly suitable to computer formulation, with an automatic sequence of operations. A number of general purpose computer programs like STRUDL, SAP, and ANSYS have been developed and are available to designers.

The two common formulations of the matrix analysis are the stiffness method and the flexibility method. The stiffness method is also referred to as displacement method, since the displacements of the redundant members are treated as unknowns. The flexibility method (or force method) treats the forces in the members as unknowns. Of these two methods, the displacement method is widely used in most computer programs.

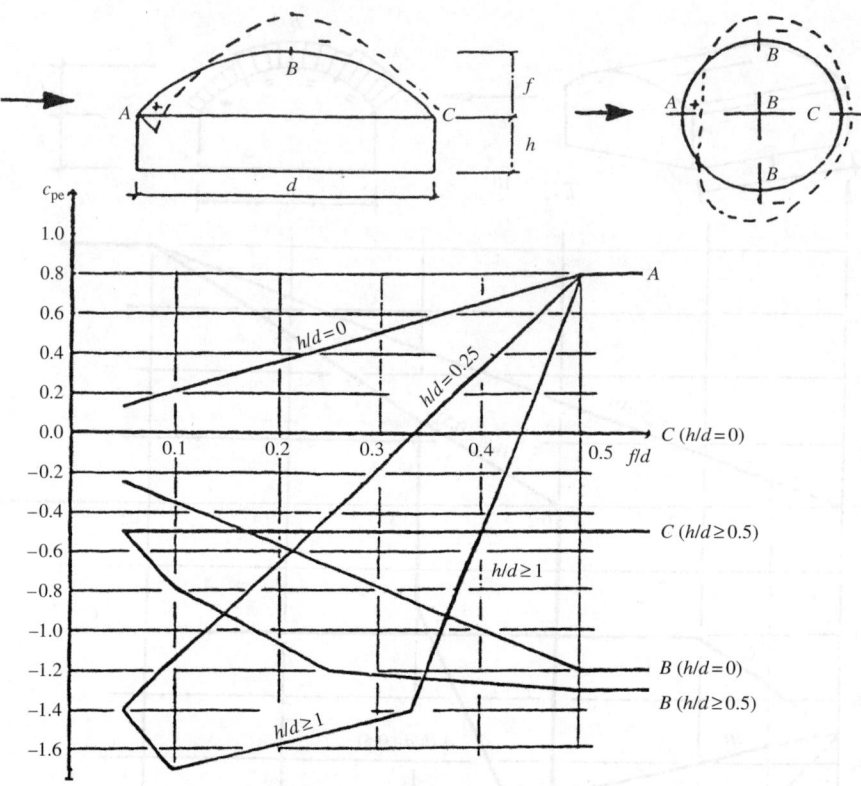

FIGURE 24.24 Wind pressure coefficients for spherical domes.

Note: $c_{pe,10}$ is constant along arcs of circles, intersections of the sphere and of planes perpendicular to the wind; it can be determined as a first approximation by linear interpolation between the values in A, B, and C along the arcs of circles parallel to the wind. In the same way the value of $c_{pe,10}$ in A if $0 < h/d < 1$ and in B or c if $0 < h/d < 0.5$ can be obtained by linear interpolation in the figure.

In the *displacement method*, the stiffness matrix of the whole structure is obtained by adding appropriately the stiffness matrices of the individual elements. Supports are then introduced since the displacements at these points are known. A set of simultaneous equations are solved for displacements. From the joint displacements the member elongation can be found and, hence, the member forces and reaction at supports.

The matrix displacement method is by far the most accurate method for the analysis of space frames. It can be used without any limit on the type and shape of the structure, the loadings, the supporting conditions, or the variation of stiffness. The effect of temperature or uneven settlement of supports can be also analyzed conveniently by this method. For design work, a special purpose computer program for space frames is preferred, otherwise the input of generating nodal coordinates and member connectivity plus loading information will be a tremendous amount of work. Some sophisticated computer programs provide the functions of automatic design, optimization, and drafting.

Double-layer grids can be analyzed as pin connected, and rigidity of the joints does not change the stress by more than 10 to 15%. In the displacement method, bar elements are used with three unknown displacements in the x, y, and z directions at each end. For single-layer reticulated shells with rigid joints, bar elements are used and the unknowns are doubled, that is, three displacements and three rotations. Under specific conditions, single-layer braced domes may be analyzed as pin connection joints with reasonable accuracy if the rise of the dome is comparatively large and under symmetric loading.

When using a computer, the engineer must know the assumptions on which the program is based, the particular conditions for its use (boundary conditions for example), and the manner of introducing the input data. In the static analysis of space frames, care should be taken on the following issues.

24.4.2.1 Support Conditions

A fixed support (bolted or welded) in construction should not be treated as a completely fixed node in analysis literally. As a matter of fact, most space frames are supported on columns or walls that have a lateral flexibility. Upon the acting of external loads, there will be lateral displacements on the top of columns. Therefore, it is more reasonable to assume the support as horizontally movable rather than fixed, or as an elastic support by considering the stiffness of the supporting column.

24.4.2.2 Criteria for the Number of Reanalyses

Usually, a set of sectional areas are assumed for members, and the computer will proceed to analyze the structure to obtain a set of member forces. Then, the members are checked to see if the assumed areas are appropriate. If not, the structure should be reanalyzed until the forces and stiffness will completely match with each other. However, such extended reanalyses by the stiffness method will induce a high concentration of stiffness and, hence, a great difference of member sections, which is unacceptable for practical use. Therefore, it is necessary to limit the number of reanalyses. In practice, certain criteria are specified such that the reanalyses will terminate automatically. One of the criteria suggested is the number of modified members should be less than 5% of the total number of members. Usually, three or four runs will produce a satisfactory result.

24.4.2.3 Checking of Computer Output

It is dangerous for an engineer to rely on the computer output as being infallible. Always try to estimate and anticipate results approximately. A simple manual calculation by the approximate method and comparing it with the computer output will be beneficial. By doing so, at least one can get an order of magnitude for the results. In this operation, intuition also plays an important role. At the same time, simple checks should be done to test the reliability of the computer program, such as the equilibrium of forces at nodes and equilibrium of total loading with the summation of reactions. A check on the deflections along certain axes of the structure would also be helpful. The size and location of any large deflection should be noted. All deflections should be scanned to look for possible bad solutions caused due to improper modeling of the structure. This check is made easily if the program has the ability to produce a deformed geometry plot.

A *continuum analogy method* may also be used for the static analysis of space frames. This is to replace a latticed structure by an equivalent continuum that exhibits equivalent behavior with respect to strength and stiffness. The equivalent rigidity is used for the stress and displacement analysis in the elastic range, particularly for stability and dynamic analysis. It is also useful for providing an understanding of the overall behavior of the structure. By using equivalent rigidity, the thickness, elastic moduli, and Poisson's ratio are determined for the equivalent continuum, and the fundamental equations that govern the behavior of the equivalent continuum are established as in the usual continuum theory. Therefore, the methods of solution and the results of the theory of plates and shells are directly applicable. Thus, certain types of latticed shells and double-layer grids can be analyzed by treating them as a continuum and applying the shell or plate analogy. This method has been found to be satisfactory where the loading is uniform and the load transfer is predominantly through membrane action.

Some difficulties may occur in the application of continuum analogy method. The boundary conditions of the continuum cannot be entirely analogous to the boundary condition of the discrete prototype. Also, some of the effects that are relatively unimportant in the case of continua may be significant in the case of space frames. Two of these merit mention. The effect of shear deformation in elastic plates and shells is essentially negligible, whereas the contributions of web members connecting the layers of a space frame can be significant to the total deformation. Similarly, the correct continuum model of a rigidly connected space frame must allow for the possibility of rotation of joints independent of the rotations of normal sections. Such models are more complex than the usual ones, and few solutions of the governing equations exist.

It is useful to compare the discrete method and continuum analogy method. The continuum analogy method can only be applied to regular structures, while the discrete method can handle arbitrary structural configuration. The computational time is much less for an equivalent shell analysis than for a stiffness method analysis. The work involved in a continuum analogy method includes calculating the equivalent rigidity, calculating the forces in the equivalent continuum and finally calculating the forces in the members. This will go through a discrete–continuum–discrete process and hence involves further approximation. To summarize, the continuum analogy method is most valuable at the stage of conceptual and preliminary design, while the discrete method should be used for a working design.

24.4.3 Earthquake Resistance

One of the important issues that must be taken into consideration in the analysis and design of space frames is the earthquake excitation in case the structure is located in a seismic area. The response of the structure to earthquake excitation is dynamic in nature, and usually a dynamic analysis is necessary. The analysis is complicated due to the fact that the amplitude of ground accelerations, velocities, and motions is not clearly determined. Furthermore, the stiffness, mass distribution, and damping characteristics of the structure will have a profound effect on its response: the magnitude of internal forces and deformations.

The dynamic behavior of a space frame can be studied first through the vibration characteristics of the structure, which is represented by its natural frequencies. The earthquake effects can be reflected in response amplification through interaction with the natural dynamic characteristics of the structures. Thus, double-layer grids can be treated as a pin-connected space truss system, and its free vibration is formulated as an equation of motion for a freely vibrating undamped multi-degree-of-freedom system. By solving the generalized eigenvalue problems, the frequencies and vibration modes are obtained.

A series of double-layer grids of different types and spans were taken for dynamic analysis [7]. The calculation results show some interesting features of the free vibration characteristics of the space frames. The difference between the frequencies of the first ten vibration modes is so small that the frequency spectrums of space frames are rather concentrated. The variation of any design parameter will lead to the change of frequency. For instance, the boundary restraint has a significant influence on the fundamental period of the space frame: the stronger the restraint, the smaller the fundamental period.

The fundamental periods of most double-layer grids range from 0.37 to 0.62 s which is less than that of planar latticed trusses of comparable size. This fact shows clearly that the space frames have a relatively higher stiffness. Investigating into the relationship of fundamental periods of different types of double-layer grids with span, it is found that fundamental period increases with the span, that is, the space frames will be more flexible for longer span. The response of space frame with shorter span will be stronger.

The vibration modes of double-layer grids could be classified mainly as vertical modes and horizontal modes that appear alternately. In most cases, the first vibration modes are vertical. The vertical modes of different types of double-layer grids demonstrate essentially the same shape, and the vertical frequencies for different space frames of equal span are very close to each other. It was found that the forces in the space frame due to vertical earthquake are mainly contributed by the first three symmetrical vertical modes. Certain relations could be established between the first three frequencies of the vertical mode as follows:

$$\omega_{v2} = (2 - 3.5)\omega_{v1} \tag{24.8}$$

$$\omega_{v3} = (4 - 4.6)\omega_{v2} \tag{24.9}$$

where ω_{v1}, ω_{v2}, and ω_{v3} are the first, second, and third vertical frequencies, respectively.

The simplest way to estimate the earthquake effect is a quasi-static model, in which the dynamic action of the ground motion is simulated by a static action of equivalent loads. The manner in which the equivalent static loads are established is introduced in many seismic design codes of different countries. In the region where the maximum vertical acceleration is 0.05g, usually the earthquake effect is not the governing factor in design and it is not necessary to check the forces induced by vertical or horizontal earthquake. In the area where the maximum vertical acceleration is 0.1g or greater, a factor of 0.08 to 0.2, depending on different codes, is used to multiply the gravitational loads to represent the equivalent vertical earthquake load. It should be noticed that in certain seismic codes, the live load that forms a part of gravitational loads is reduced by 50%. The values of vertical seismic forces in the members of double-layer grids are higher near the central region and decrease gradually toward the perimeters. Thus, the ratio between the forces in each member due to vertical earthquake and static load is not constant over the whole structure. The method of employing equivalent static load serves only as an estimation of the vertical earthquake effect and provides an adequate level of safety.

Due to the inherent horizontal stiffness of double-layer grids, the forces induced by horizontal earthquake can be resisted effectively. In the region of 0.1g maximum acceleration, if the space grids are supported along perimeters with short or medium span, it is not required to check the horizontal earthquake. However, for double-layer grids of longer span or if the supporting structure underneath is rather flexible, seismic analysis in the horizontal direction should be done. In the case of latticed shells with curved surface, the response to horizontal earthquake is much stronger than in double-layer grids depending on the shape and supporting condition. Even in the region with maximum vertical acceleration of 0.05g, the horizontal earthquake effect on latticed shells should be analyzed. In such an analysis, the coordinating action of space frame and the supporting structure should be considered. A simple way of coordination is to include the elastic effect of the supporting structure. This is represented by the elastic stiffness provided by the support in the direction of restraint. The space frame is analyzed as if the supports have elastic restraints horizontally. For more accurate analysis, the supporting columns are taken as member with bending and axial stiffness and analyzed together with the space frame. In the analysis, it is also important to include the inertial effect of the supporting structure, which has influence on the horizontal earthquake response of the space frame.

In the case of more complex structures or large span, dynamic analysis such as response spectrum method for modal analysis should be used. Such a method gives a good estimate of the maximum response during which the structure behaves elastically. For space frames, the vertical seismic action should be considered. However, few recorded data on the behavior of such a structure under vertical earthquake exist. In some seismic design codes, the magnitude of the vertical component may be taken as 50 to 65% of the horizontal motions. It is recommended to use 10 to 20 vibration modes for the space frames when applying the response spectrum method.

For space frames with irregular and complicated configuration or important long-span structures, time-history analysis method should be used. The number of acceleration records or synthesized acceleration curves for time-history analysis is determined according to intensity, near or far earthquake, and site category. In usual practice, at least three records are used for comparison. This method is an effective tool to calculate the earthquake response when large, inelastic deformations are expected.

The behavior of latticed structures under dynamic loads or, more specifically, the performance of latticed structures due to earthquake was the main concern of structural engineers. An ASCE Task Committee on Latticed Structures under Extreme Dynamic Loads was formed to investigate this problem. One of the objectives was to determine if dynamic conditions have historically been the critical factor in the failure of lattice structures. A short report on "Dynamic Considerations in Latticed Structures" [8] was submitted by the Task Committee in 1984. Eight major failures of latticed roof structures were reported, but notably none of them was due to earthquake.

Since the ASCE report was published, valuable information on the behavior of space frame during earthquake has been obtained through two seismic events. In 1995, the Hanshin area of Japan suffered a strong earthquake, and many structures were heavily damaged or destroyed. However, comparing with other types of structures, most of the damage to space frames located in that area was relatively minor [9]. It is worthwhile to mention that two long-span sports arenas of space frame construction were built on an artificial island in Kobe, and no major structural damage was found. On the other hand, serious damage to a latticed shell was found on the roof structure of a hippodrome stand where many members were buckled. The cause of the damage is not due to the strength of the space frame itself but the failure of the supporting structure. Another example of serious damage to double-layer grids in the roof structure of a theater occurred in 1985, when a strong earthquake struck the Kashigor District of Sinkiang Uygur Autonomous Region in China [10]. Failure was caused by a flaw in the design as the elastic stiffness and inertial effect of the supporting structures were completely ignored. Behavior of space frame structures under strong earthquake has generally been satisfactory from a strength point of view. Experience gained from strong earthquakes shows that the space frames demonstrate an effective spatial action and consequently a reasonably good earthquake resistant behavior.

24.4.4 Stability

Although a great amount of research work has been carried out to determine the buckling load of latticed shells, the available solutions are not satisfactory for practical use. The problem is complicated by the effect of geometric nonlinearity of the structure and also the influence of the joint system according to which the members can be considered as pin connected, partially or completely restrained at the nodes. The following points are important in the buckling analysis of latticed shells [11]:

1. Decision on which kind of nonlinearity is to be used — only geometrical nonlinearity with the elastic analysis or geometrical and material nonlinearities with the elastic–plastic analysis.
2. Choosing the physical model — equivalent continuum or discrete structure.
3. Choosing the computer model and numerical procedure for tracing the nonlinear response for precritical behavior, collapse range, and postcritical behavior.
4. Study of factors that influence load-carrying capacity — buckling modes, density of network, geometrical and mechanical imperfections, plastic deformations, rigidity of joints, load distributions, etc.
5. Experimental investigations to provide data for analysis (rigidity of joints, postbuckling behavior of individual members, etc.) and confirmation of theoretical values.

Generally speaking, there are three types of buckling that may occur in latticed shells:

1. Member buckling (Figure 24.25a)
2. Local or dimple buckling at a joint (Figure 24.25b)
3. General or overall buckling of the whole structure (Figure 24.25c)

Member buckling occurs when an individual member becomes unstable, while the rest of the space frame (members and nodes) remains unaffected. The buckling load P_{cr} of a straight prismatic bar under axial compression is given by

$$P_{cr} = \alpha \frac{\pi^2 E_e I}{l^2}, \quad \alpha = \alpha(c_i, c_j, w_0, e, m) \tag{24.10}$$

where E_e is the effective modulus of elasticity that coincides with Young's modulus in the elastic range, I is the moment of inertia of member, and l is the length of the member. The coefficient α takes different values depending on the parameter in the parentheses. The quantities c_i and c_j characterize the rotational stiffness of the joints, w_0 is the initial imperfection, e is the eccentricity of the end compressive forces, and m indicates the end shear forces and moments. A reduced length l_0 should be used in place of l when the ratio of the joint diameter to member length is relatively large. On the basis of Equation 24.10, the design code for steel structures in different countries provides methods for estimating member buckling, usually by introducing the slenderness ratio $\lambda = l/r$, where r is the radius of gyration of the member's section.

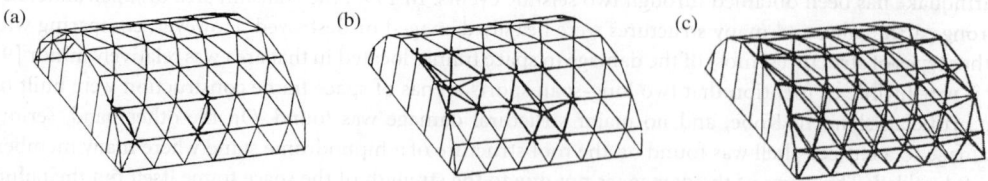

(a)   (b)   (c)

FIGURE 24.25 Different types of buckling.

The *local buckling* of space frame consists of a snap-through buckling, which takes place at one joint. Snap-through buckling is characterized by a strong geometrical nonlinearity. Local buckling is apt to occur when the ratio of t/R (where t is the equivalent shell thickness and R is the radius of curvature) is small. Similarly, local buckling of a space frame is likely to occur in single layer latticed shells.

Local buckling is greatly affected by the stiffness of and the loads on the adjacent members. Consider the pin-connected structure shown in Figure 24.26. Buckling load q_{cr} in terms of uniform normal load per unit area can be expressed as

$$\frac{AEl}{12R^3} \leq q_{cr} \leq \frac{AEl}{6R^3} \tag{24.11}$$

where A is the cross-sectional area of the member, E is the modulus of elasticity, and R is the radius of an equivalent spherical shell through points B–A–B.

In practice, different types of joints used in the design will possess different flexural strengths; thus, the actual behavior of the joint and member assembly should be incorporated in determining the local buckling load. An approximate formula was proposed by Lind [12] and is applicable to triangular networks having all elements of the same cross-sections. For the uniform load, the critical load is

$$Q_{cr} = \frac{E_t}{1 + \alpha^2/8\pi^2}\left(0.47\frac{Al^3}{R^3} + 3\frac{BI}{lR}\right), \quad \alpha = l^2/rR \tag{24.12}$$

where E_t is the tangent modulus of elasticity, R is the radius of curvature of the framework midsurface, r is the radius of gyration, and B is the nondimensional bending stiffness of the grid given in Table 24.5. For the concentrated load, the following two formulas are presented:

$$W_{cr} = \frac{3EAh^3}{l^3}\left[\frac{8B}{\alpha^2} + 0.241\left(1 - 5.95\frac{8B}{\alpha^2}\right)\right] \tag{24.13}$$

which is valid for $\alpha > 9$, and

$$W_{cr} = 0.0905EA\left(\frac{l}{R}\right)^3 \tag{24.14}$$

for a regular pin-jointed triangular network.

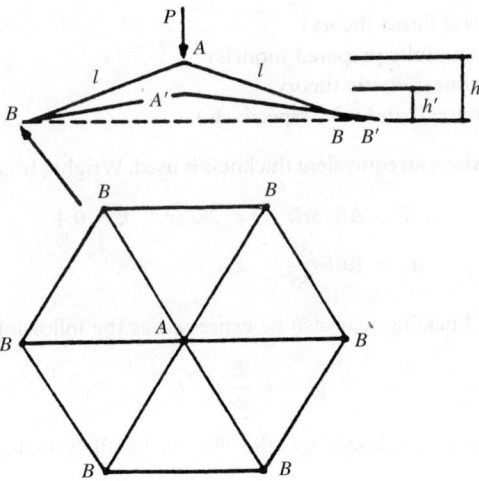

FIGURE 24.26 Local buckling of a pin-connected structure.

TABLE 24.5 Equivalent Bending Stiffness B

α	$\frac{1}{32}$	$\frac{1}{16}$	$\frac{1}{8}$	$\frac{1}{4}$	$\frac{1}{2}$	1	2	4	8	16	32	64
B	0.868	0.873	0.886	0.950	1.176	1.85	3.15	4.83	6.48	7.35	7.80	7.90

The *overall buckling* occurs when a relatively large area of the space frame becomes unstable and a relatively large number of joints are involved in the buckle. For most cases, in overall buckling of a space frame, the wave length is significantly greater than the member length. Local buckling often plays the role of a trigger for overall buckling.

The type of buckling collapse of a space frame is greatly influenced by the following factors: its Gaussian curvature, whether it is single- or double-layer system, the degree of statical indeterminacy, and the manner of supporting and loading. Generally speaking, a shallow shell of positive Gaussian curvature, like a dome, is more prone to overall buckling than a cylindrical shell of zero Gaussian curvature. Recent research reveals that hyperbolic paraboloid shell is less vulnerable to overall buckling and the arrangement of the grids has a considerable influence on the stability and stiffness of the shell. It is best to arrange the members along the direction of compressive forces. Single-layer space frame exhibits greater sensitivity to buckling than the double-layer structure. Moreover, various types of buckling behavior may take place simultaneously in a complicated relation. For double-layer grids, in most cases, it is sufficient to examine the member collapse that may occur in the compressive chord members.

The theoretical analysis of buckling behavior may be approached by two methods: continuum analogy analysis and discrete analysis. Since almost all space frames are constructed from nearly identical units arranged in a regular pattern, it is generally accepted that the analysis on the basis of the equivalent continuum serves as an important tool in the investigation of the buckling behavior of space frames. Numerous analytical and experimental studies on the buckling of continuous shells have been performed, and the results can be applied to the latticed shells.

The buckling formula for a spherical shell subjected to a uniformly distributed load normal to the middle surface can be expressed as

$$q_{cr} = kE\left(\frac{t}{R}\right)^2 \tag{24.15}$$

where t is the thickness and R is the radius of the shell.

Different values of the coefficient k were obtained by various investigators:

1.21 ([13], based on classical linear theory)
0.7 (experiments on very carefully prepared models)
0.366 ([14], based on nonlinear elastic theory)
0.228 and 0.246 ([15], for $\mu = 0$ and 0.3, respectively).

For a triangulated dome, where an equivalent thickness is used, Wright [16] derived the formula by using

$$E = AE/3rl, \quad t = 2\sqrt{3}r, \quad k = 0.4$$
$$q_{cr} = 1.6E\frac{Ar}{lR^2} \tag{24.16}$$

The critical load for overall buckling may also be expressed as the following formula for comparison:

$$q_{cr} = k'\frac{E}{R^2}t_m^{1/2}t_b^{3/2} \tag{24.17}$$

where t_m is the effective in-plane thickness, t_b is the effective bending thickness, and the values of k' are

0.377 [16]
0.365 [17]
0.247 [15]
0.294 [18]

Discrete Analysis is a more powerful tool to study the whole process of instability for space frames. As shown in Figure 24.27, a structure may lose its stability when it has reached a "limit point," where

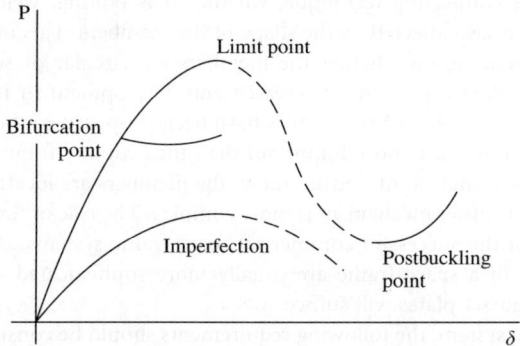

FIGURE 24.27   Instability points.

the stiffness is lost completely. On the other hand, a structure, such as a dome, may lose its stability by a sudden buckling into a mode of deformation before the limit point, which occurs at a distinct *critical point* — the "bifurcation point" on the load path. It should be noted that the initial imperfection of the structure will greatly reduce the value of critical load, and certain types of space frame are very sensitive to the presence of imperfection.

In the stability analysis, usually the characteristic at a certain special state is investigated, that is, the stability mode and critical load are analyzed as an eigenvalue problem. The researchers are now more interested to study the whole process of nonlinear stability. As a result of the development of the computer matrix method, numerical analyses of the large system have become straightforward. Therefore, the discrete analysis of a space frame, itself a discrete structure, is very suitable for the study of stability problems. Major problems encountered in the nonlinear stability process are mathematical and mechanical modeling of the structure, numerical technique for solving nonlinear equations, and tracing method for the nonlinear equilibrium path. A great many research works have been carried out in the above area.

The Newton–Raphson method or the modified Newton–Raphson method is the fundamental method for solving the nonlinear equilibrium equations and has proved to be one of the most effective methods. The purpose of tracing the nonlinear equilibrium path is as follows: (a) to provide equilibrium analysis for the prebuckling state, (b) to determine the critical point, such as the limit point or bifurcation point, on the load path and its critical load, and (c) to trace the postbuckling response. On the basis of the increment-iteration process for finite element method, techniques for the analysis of nonlinear equilibrium path and its tracing tactics have made significant progress in recent years. Numerical methods used for the construction of equilibrium path associated with nonlinear problems, like load incremental method, constant arc-length method, displacement control method, etc., were developed by different authors. Since each technique has its advantages and disadvantages in the derivation of fundamental equations, accuracy of solution, computing time, etc., the selection of an appropriate method has a profound influence on the efficiency of computation. In the present stage of development, complicated equilibrium path can be traced with the aid of the above technique. Computer programs have been developed for the whole process of nonlinear stability and can be used for the design of various types of latticed shells.

24.5   Jointing Systems

24.5.1   General Description

The jointing system is an extremely important part of a space frame design. An effective solution of this problem may be said to be fundamental to successful design and construction. The type of jointing depends primarily on the connecting technique, whether it is bolting, welding, or applying special mechanical connectors. It is also affected by the shape of the members. This usually involves a different connecting technique depending on whether the members are circular or square hollow sections or rolled steel sections. The effort expended on research and development of jointing systems has been enormous, and many different types of connectors have been proposed in the past decades.

The joints for the space frame are more important than the ordinary framing systems because more members are connected to a single joint. Furthermore, the members are located in a three-dimensional space, and hence the force transfer mechanism is more complex. The role of the joints in a space frame is so significant that most of the successful commercial space frame systems utilize proprietary jointing systems. Thus, the joints in a space frame are usually more sophisticated than the joints in planar structures, where simple gusset plates will suffice.

In designing the jointing system, the following requirements should be considered. The joints must be strong and stiff, simple structurally and mechanically, and easy to fabricate without recourse to more advanced technology. The eccentricity at a joint should be kept to a minimum, and yet the joint detailing should provide for the necessary tolerances that may be required during the construction. Finally, joints of space frames must be designed to allow for easy and effective maintenance.

The cost of the production of joints is one of the most important factors affecting the final economy of the finished structure. Usually, the steel consumption of the connectors will constitute 15 to 30% of the total. Therefore, a successful prefabricated system requires joints that must be repetitive, mass produced, simple to fabricate, and able to transmit all the forces in the members interconnected at the node.

All connectors can be divided into two main categories: the purpose-made joint and the proprietary joint used in the industrialized system of construction. The purpose-made joints are usually used for long-span structure where the application of standard proprietary joints is limited. An example of such a joint is the cruciform gusset plate for connecting rolled steel sections shown in Figure 24.28.

A survey around the world will reveal that there are over 250 different types of jointing system suggested or used in practice, and there are some 50 commercial firms trying to specialize in the manufacture of proprietary jointing systems for space frames. Unfortunately, many of these systems have not proved to be very successful mainly because of the complexity of the connecting method. Table 24.6, Table 24.7, and Table 24.8 give a comprehensive survey of the jointing systems all over the world. All the connection techniques can be divided into three main groups: (a) with a node, (b) without a node, and (c) with prefabricated units.

24.5.2 Proprietary System

Some of the most successful prefabricated jointing systems are summarized in Table 24.9. This is followed by further description of each system.

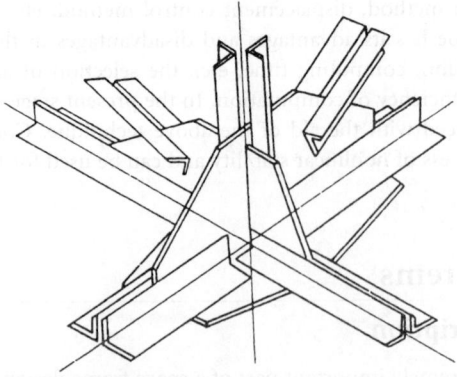

FIGURE 24.28   Connecting joint with cruciform gusset plate.

TABLE 24.6   Connection Types with a Node

Node		Connector	Member	Cross-section	Examples
Sphere	Solid			◯	Mero KK, Germany; Montal, Germany; Uzay, Italy; Zublin, Germany
				◯	Steve Baer, United States; Van Tiel, Netherlands; KT space truss, Japan
				•	Mero MT, Germany
	Hollow			◯	Spherobat, France
				◯	NS space trusses, Japan; Tubal, Netherlands; Orbik, United Kingdom
				I	NS space trusses, Japan; Tubal, Netherlands; Orbik, United Kingdom
	Hollow			◯	SDC, France
	Hollow			◯	Oktaplatta, Germany
				◯	WHSJ, China
	Hollow			◯	Vestrut, Italy
Cylinder	Solid			◯	Triodetic, Canada; Nameless, East Germany
				◯	Octatube Plus, Netherlands; Nameless, Singapore
				◯	Pieter Huybers, Netherlands
	Hollow			▢	Nameless system, United Kingdom
Disc	Flat			□	Palc, Spain
				□	Power strut, United States
				◯	Pieter Huybers, Netherlands
				○	Tridimatec, France
				□ □	Moduspan (Unistrut), United States; Space-frame system VI (Unistrut), United States
	Welded			◯	Boyd Auger, United States; Octatube, Netherlands
				◯	Piramodul large span, Netherlands
				◯ □	Nodus, United Kingdom
Prism	Solid			□	Montal, Germany
				▯	Mero BK, Germany
				▮	Mero TK and ZK, Germany
				▯	Mero NK, Germany
	Hollow			◯	Satterwhite, United States

Source: Gerrits [19].

TABLE 24.7   Connection Types without a Node

Node		Connector Member	Cross-section	Examples
Form of member	Forming			Buckminster Fuller
			○	Nonadome, Netherlands
	Flattened and bending		○ Σ □	Radial, Australia
			○ Σ	Harley, Australia
Addition of member	Plate(s)		□ ○ ○	Mai Sky, United States
	Member end		○	Pieter Huybers, Netherlands
			○	Pierce, United States
			○	Buckminster Fuller

Source: Gerrits [19].

TABLE 24.8   Connection Types with Prefabricated Units

Node	Prefabricated unit	Member cross-section top / bracing / bottom	Example
Geometrical solid		⌐ ○ ○	Space deck, United Kingdom
		[○ ○	Mero DE, Germany
		⌐[] ○ ○	Unistrut, France
		⊓ ⌐ ⊔	Nameless system, Italy
2D components		□ □ □	Ruter, Germany
		[○ [	Nameless system, Italy
3D components		I □ I	Cubic, United Kingdom

Source: Gerrits [19].

TABLE 24.9   Commonly Used Proprietary Systems

Name	Country	Period of development	Material	Connecting method
MERO	Germany	1940–1950	Steel Aluminum	Bolting
Space Deck	United Kingdom	1950–1960	Steel	Bolting
Triodetic	Canada	1950–1960	Aluminum Steel	Inserting member ends into hub
Unistrut (Moduspan)	United States	1950–1960	Steel	Bolting
Oktaplatte	Germany	1950–1960	Steel	Welding
Unibat	France	1960–1970	Steel	Bolting
Nodus	United Kingdom	1960–1970	Steel	Bolting and using pins
NS	Japan	1970–1980	Steel	Bolting

24.5.2.1 Mero

The Mero connector, introduced some 50 years ago by Dr. Mengeringhausen, proved to be extremely popular and has been used for numerous temporary and permanent buildings. Its joint consists of a node that is a spherical hot-pressed steel forging with flat facets and tapped holes. Members are circular hollow sections with cone-shaped steel forgings welded at the ends, which accommodate connecting bolts. Bolts are tightened by means of a hexagonal sleeve and dowel pin arrangement, resulting in a completed joint such as that shown in Figure 24.29. Up to 18 members can be connected at a joint with no eccentricity. The manufacturer can produce nodes of different sizes with diameters ranging from 46.5 to

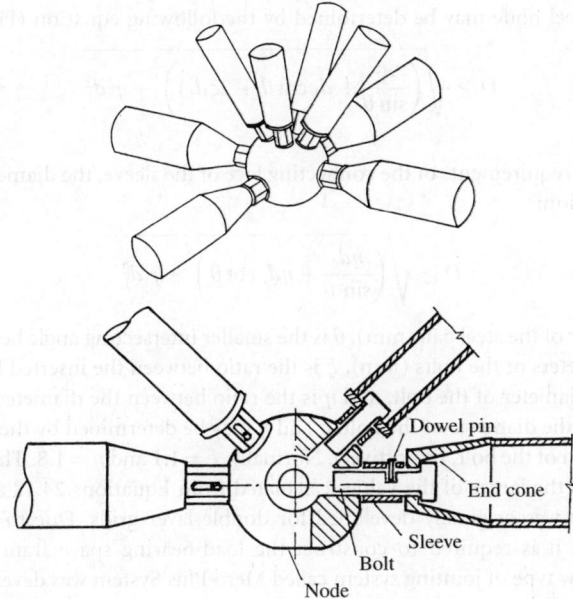

Dowel pin

End cone

Sleeve

Bolt

Node

FIGURE 24.29 Mero system.

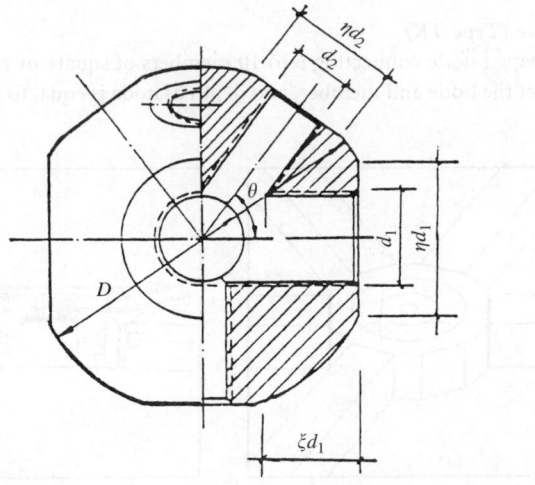

FIGURE 24.30 Dimensions of spherical node.

350 mm, the corresponding bolts ranging from M12 to M64 with a maximum permissible force of 1413 KN. A typical space module of the Mero system is a square pyramid (half an octahedron) with both chord and diagonal members of the same length, 'a'; angles extended are 90 or 60°. Thus, the depth of the space module is $a/\sqrt{2}$, and the vertical angle between diagonal and chord member is 54.7°.

The Mero connector has the advantage that the axes of all members pass through the center of the node, eliminating eccentricity loading at the joint. Thus, the joint is only under the axial forces. Then, tensile forces are carried along the longitudinal axis of the bolts and resisted by the tube members through the end cones. The compressive forces do not produce any stresses in the bolts; they are distributed to the node through the hexagonal sleeves. The size of the connecting bolt of compression members based on the diameter calculated from its internal forces may be reduced by 6 to 9 mm.

The diameter of a steel node may be determined by the following equation (Figure 24.30):

$$D \geq \sqrt{\left(\frac{d_2}{\sin\theta} + d_1 \cot\theta + 2\xi d_1\right)^2 + \eta^2 d_1^2} \tag{24.18}$$

However, to satisfy the requirements of the connecting face of the sleeve, the diameter should be checked by the following equation:

$$D \geq \sqrt{\left(\frac{\eta d_2}{\sin\theta} + \eta d_1 \cot\theta\right)^2 + \eta^2 d_1^2} \tag{24.19}$$

where D is the diameter of the steel ball (mm), θ is the smaller intersecting angle between two bolts (rad), d_1 and d_2 are the diameters of the bolts (mm), ξ is the ratio between the inserted length of the bolt into the steel ball and the diameter of the bolt, and η is the ratio between the diameter of the circumscribed circle of the sleeve and the diameter of the bolt. ξ and η may be determined by the design tension values or compression strength of the bolt, respectively. Normally, $\xi = 1.1$ and $\eta = 1.8$. The diameter of the steel ball should be taken as the larger of the values calculated from Equations 24.18 and 24.19.

The Mero connector was originally developed for double-layer grids. Due to the increasing use of nonplanar roof forms, it is required to construct the load-bearing space frame integrated with the cladding element. A new type of jointing system called Mero Plus System was developed so that a variety of curved and folded structures are possible. Square or rectangular hollow sections are used to match the particular requirements of the cladding so that a flush transition from member to connecting node can be executed. The connector can transmit shear force, resist torsion and, in special cases, resist bending moment. There are four groups in this system, which are described as follows.

24.5.2.1.1 Disk Node (Type TK)
This a planar ring-shaped node connecting 5 to 10 members of square or rectangular sections. A single bolt is used to connect the node and member, and depth of node is equal to member section depth. Such

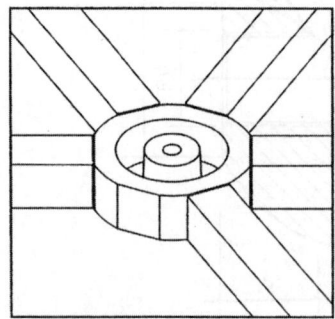 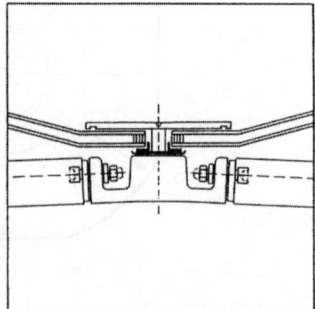

FIGURE 24.31   Disk node (type TK).

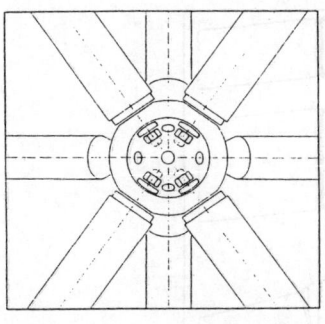
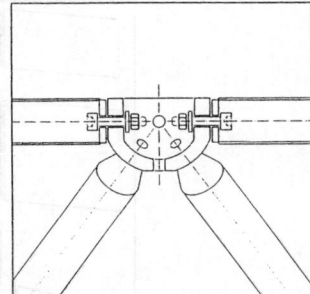

FIGURE 24.32 Bowl node (type NK).

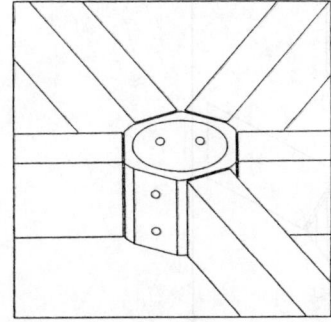
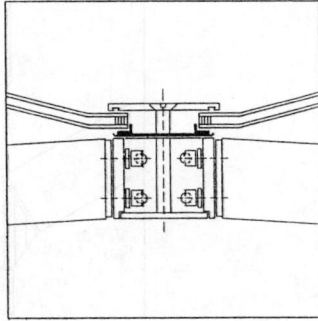

FIGURE 24.33 Cylinder node (type ZK).

a jointing system can transmit shear force and resist rotation (Figure 24.31). In the following discussion the U-angle is designated as the angle between two members connected to the same node. Also, the V-angle is the angle between the member axis and the normal in the plane of the node, which is a measure of curvature. For disk node, the U-angle varies from 30 to 80°, and the V-angle varies from 0 to 10°. This type of jointing system is essentially pin-jointed connections and be suitable for latticed shells made of triangular meshes.

24.5.2.1.2 Bowl Node (Type NK)
This is a hemispherical node connecting top chord and diagonal members. A single bolted connection from node to member is used. The top chord members of square or rectangular sections can be loaded in shear and are fitted flush to the nodes. Bowl nodes are used for double-layer planar and curved surfaces, in particular buildings irregular in plan or pyramid in shape. The diagonals and lower chords are constructed in ordinary Mero system with circular tubes and spherical nodes (Figure 24.32).

24.5.2.1.3 Cylinder Node (Type ZK)
This is a cylindrical node with multiple bolted connection, which can transmit bending moment (Figure 24.33). Usually, the node can connect 5 to 10 square or rectangular sections that can take transverse loading. Connection angle varies from 30 to 100° for U-angle and from 0 to 10° for V-angle. Cylinder nodes are used in singly or doubly curved surface of latticed shells with trapezoidal meshes where flexural rigid connections are required.

24.5.2.1.4 Block Node (Type BK)
This is a block- or prism-shaped solid node connecting members of square or rectangular sections. The U-angle varies from 70 to 120°, and the V-angle varies from 0 to 10°. It can be used for singly or doubly curved surfaces with pin-jointed or rigid connections where the number of members is small. The structure is of simple geometry and small dimensions (Figure 24.34).

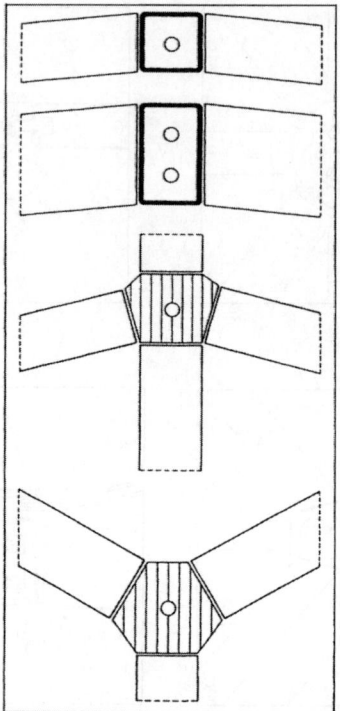

FIGURE 24.34   Block node (type BK).

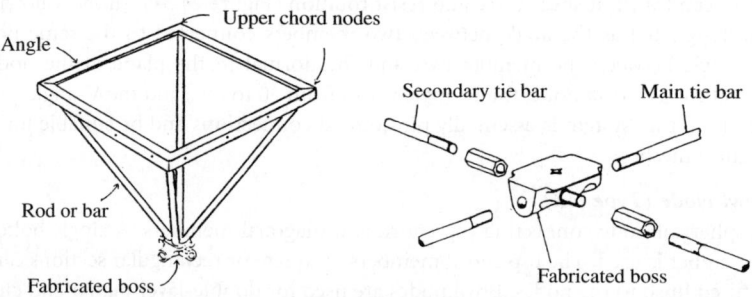

FIGURE 24.35   Space deck system.

24.5.2.2 Space Deck

The Space Deck System, introduced in England in the early 1950s, utilizes pyramidal units that are fabricated in the shop, as shown in Figure 24.35. The four diagonals made of rods or bars are welded to the corners of the angle frame and joined to a fabricated boss at the apex. It is based on square pyramid units that form a configuration of square on square offset double-layer space grids. The units are field-bolted together through the angle frames. The apexes of the units are connected in the field by using tie bars made from high-tensile steel bars. Camber can be achieved by adjusting the tie bar lengths, since right- and left-hand threading is provided in the boss. The Space Deck System is usually used for buildings of span less than 40 m with a standard module and depth, both of 1.2 m; a minimum structural depth of 0.75 m is also provided. For higher design loading and larger spans, alternative production modules of 1.5 and 2.0 m with the same depth as the module are also available.

24.5.2.3 Triodetic

The joint for the Triodetic System, developed in Canada, consists of an extruded aluminum connector hub with serrated keyways. Each member end is pressed to form a coined edge that fits into the hub keyway. The joint is completed when the members are inserted into the hub, washers are placed at each end of the hub, and a screw bolt is passed through the center of the hub, as shown in Figure 24.36. The triodetic connector can be used for any type of three-dimensional space frame. Originally, only aluminum structures were built in this system, but now, space frames are erected using galvanized steel tubes and aluminum hubs. Triodetic double-layer grids have been used up to 33-m clear span. The basic module can be almost any size up to approximately 2.7 m^2. The depth is usually 70% of the module size.

24.5.2.4 Unistrut

The Unistrut System was developed in the United States in the early 1950s. Its joint consists of a connector plate that is press-formed from steel plate. The members are channel-shaped, cold-formed sections and are fastened to the connector plate by using a single bolt at each end. The connectors for the top and bottom-layers are identical, and therefore the Unistrut double-layer grids consist only of four components, that is, the connector plate, the strut, the bolt, and the nut (see Figure 24.37). The maximum span for this system is approximately 40 m with standard modules of 1.2 and 1.5 m. The name of Moduspan has also been used for this system.

24.5.2.5 Oktaplatte

The Oktaplatte System utilizes hollow steel spheres and circular tube members that are connected by welding. The node is formed by welding two hemispherical shells together, which are made from steel plates either by hot or cold pressing. The hollow sphere may be reinforced with an annular diaphragm. This type of node was popular at the early stage of development of space frames. It is also useful for the long-span structures where other proprietary systems are limited by their bearing capacity. Hollow spheres with diameter up to 500 mm are used. It can be applied to single-layer latticed shells as the joint can be considered as semi- or fully rigid. The whole jointing system and the hollow sphere with its parts are shown in Figure 24.38.

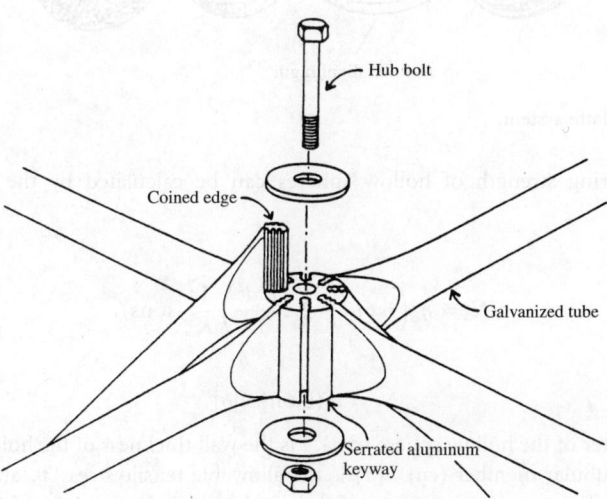

FIGURE 24.36 Triodetic system.

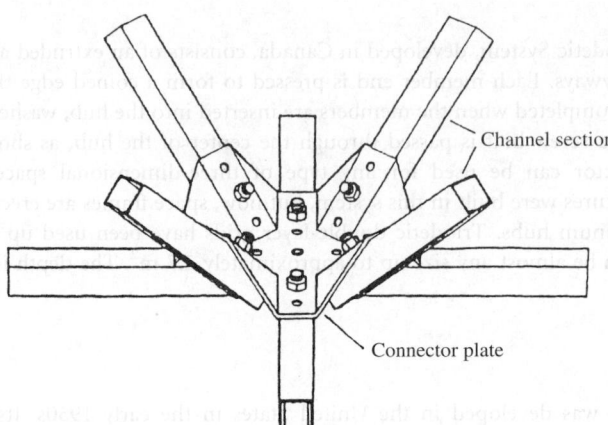

Channel section

Connector plate

FIGURE 24.37 Unistrut system.

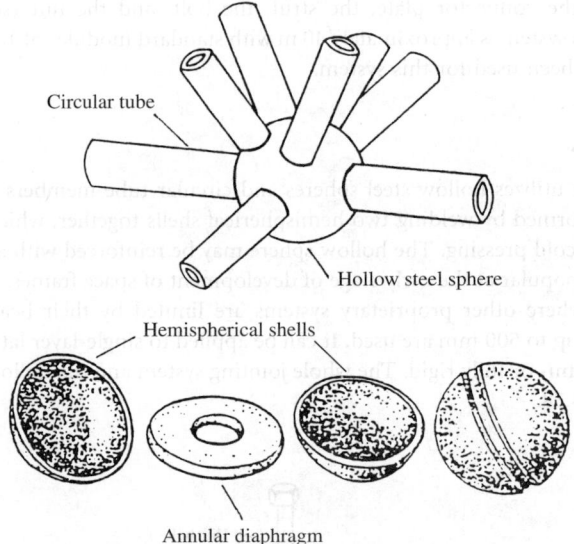

Circular tube

Hollow steel sphere

Hemispherical shells

Annular diaphragm

FIGURE 24.38 Oktaplatte system.

The allowable bearing strength of hollow spheres can be calculated by the following empirical formulas:

Under compression

$$N_c = \eta_c \left(6.6td - 2.2 \frac{t^2 d^2}{D} \right) \frac{1}{K} \text{ (tons)} \tag{24.20}$$

Under tension

$$N_t = \eta_t (0.6td\pi)[\sigma] \tag{24.21}$$

where D is the diameter of the hollow sphere (cm), t is the wall thickness of the hollow sphere (cm), d is the diameter of the tubular member (cm), $[\sigma]$ is the allowable tensile stress, η_c and η_t are the amplification factors due to the strengthening effect of the diaphragm, taken as 1.4 and 1.1, respectively, and K is a factor of safety.

24.5.2.6 Unibat

The Unibat System, developed in France, consists of pyramidal units by arranging the top-layer set on a diagonal grid relative to the bottom-layer. The short length of the top chord members results in less material being required in these members to resist the applied compressive and bending stresses. The standard units are connected to the adjacent units by means of a single high-tensile bolt at each upper corner. The bottom-layer is formed by a two-way grid of circular hollow sections, which are interconnected with the apex of pyramidal units by a single vertical bolt (Figure 24.39). Numerous multistory buildings, as well as large-span roofs over sports buildings, have been built using the Unibat System since 1970.

24.5.2.7 Nodus

The Nodus System was developed in England in the early 1970s. Its joint consists of half-casing, which is made of cast steel and has machined grooves and drilled holes, as shown in Figure 24.40. The chord connections are made of forged steel, have machined teeth, and are full-strength welded to the member ends. The teeth and grooves have an irregular pitch to ensure proper engagement. The forked connectors are made of cast steel and are welded to the diagonal members. In the completed joint, the centroidal axes of the diagonals intersect at a point that generally does not coincide with the corresponding intersecting points of the chord members. This eccentricity produces some amount of local bending in the chord members and the joint components. Destructive load tests performed on typical joints usually result in failures due to bending of the teeth in the main half-casing. The main feature of the Nodus jointing system is that all fabrication is carried out in the workshop so that only the simplest erection techniques are necessary for the assembly of the structure on-site.

24.5.2.8 NS Space Truss

The NS Space Truss System was introduced around 1970 by the Nippon Steel Corporation. It originated from the space truss technology developed for the construction of the huge roof at the symbol zone for Expo '70 in Japan. The NS Space Truss System has a joint consisting of thick spherical steel shell connectors open at the bottom for bolt insertion. The structural members are steel hollow sections having specially shaped end cones welded to both ends of the tube. End cones have threaded bolt holes. Special high-strength bolts are used to join the tubular members to the spherical shell connector. The NS nodes enable several members to be connected to one node from any direction without any eccentricity of internal forces. The NS Space Truss System has been used successfully for many large-span double- and triple-layer grids, domes, and other space structures. The connection detail of the NS node is shown in Figure 24.41.

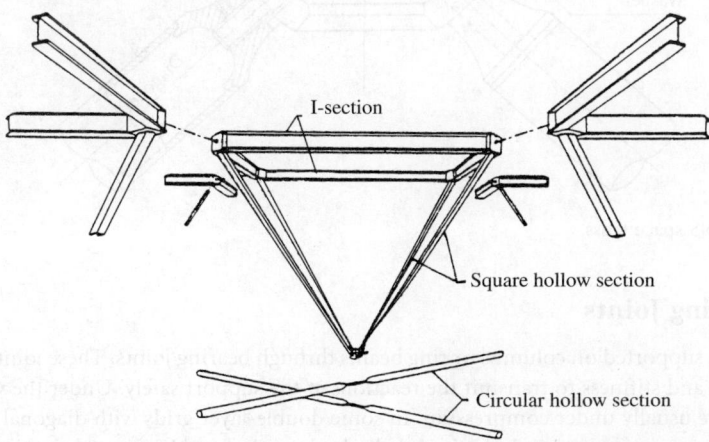

I-section

Square hollow section

Circular hollow section

FIGURE 24.39 Unibat system.

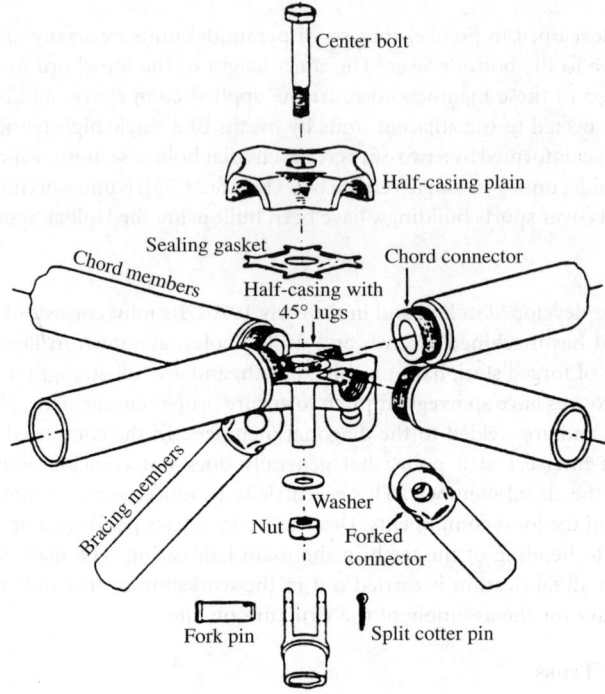

FIGURE 24.40 Nodus system.

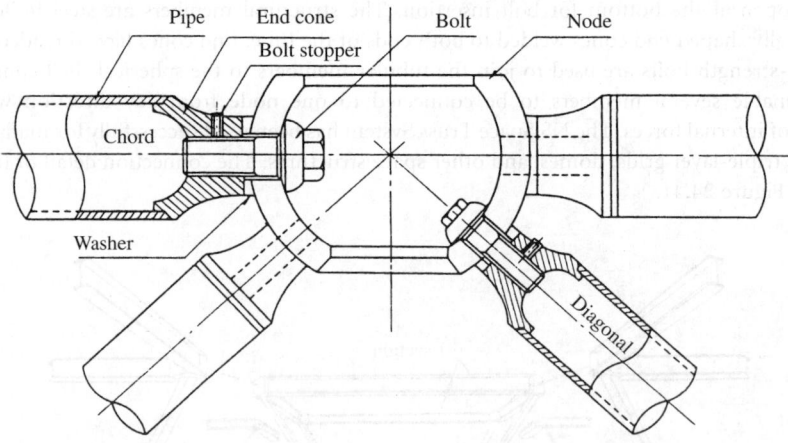

FIGURE 24.41 NS space truss.

24.5.3 Bearing Joints

Space frames are supported on columns or ring beams through bearing joints. These joints should posses enough strength and stiffness to transmit the reactions at the support safely. Under the vertical loading, bearing joints are usually under compression. In some double-layer grids with diagonal layout, bearing joints at corners may resist tension. In latticed shells, both vertical and horizontal reactions are acting on the bearing joints.

The restraint of the bearing joint has a distinct influence on the joint displacement and member forces. The construction detail of bearing support should conform to the restraint assumed in the design as near as possible. If this requirement is not satisfied, the magnitude or even the sign of the member forces may be changed.

The axes of all connecting members and the reaction should be intersected at one point at the support where a hinged joint is used. This will allow free rotation of the joint. From an engineering standpoint, the space frame may be fixed in the vertical direction, while in the horizontal direction, it may be fixed either tangential or normal to the boundary or both. The way the space frame is fixed often depends on the temperature effect. If the bearing support can allow a horizontal motion normal to the boundary, then the member forces due to the temperature variation can be neglected. In such a case, the bearing should be constructed so that it can slide horizontally. For those space frames with large-span or complicated configuration, especially curved surface structures supported on a sloped base, care should be exercised to ensure a reliable bearing support.

Typical details for bearing joints are shown in Figure 24.42. The simplest form of bearing is to establish the joint on a flat plate and anchored by bolts as shown in Figure 24.42a and b. This joint seems to be fixed at the support, but in structural analysis it has to be incorporated with the supporting structure, such as columns or walls that have a lateral flexibility. Figure 24.42c shows the joint resting on a curved bearing block, which allows rotation along the curved surface. This type of construction is considered as a hinged joint. If a laminated elastomeric pad is used under the joint as shown in Figure 24.42d, a new type of bearing joint is formed. Due to the shear deformation of the elastomeric pad, the joint can produce both rotation and horizontal movements. It is very effective to accommodate the horizontal deformation caused by temperature variation or earthquake action.

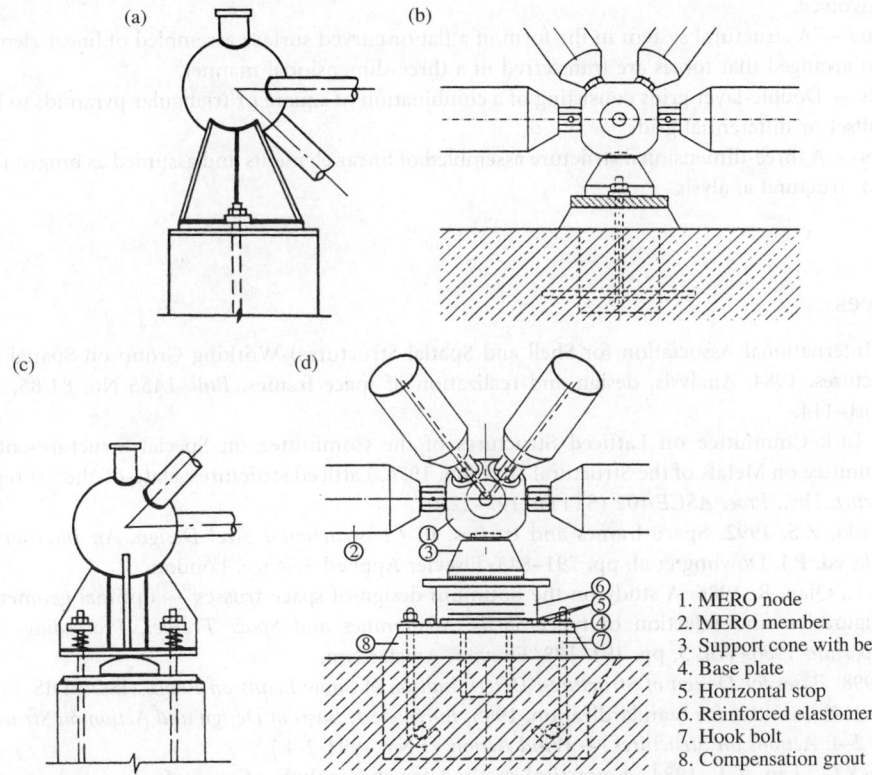

(a)   (b)

(c)   (d)

1. MERO node
2. MERO member
3. Support cone with bearing
4. Base plate
5. Horizontal stop
6. Reinforced elastomer
7. Hook bolt
8. Compensation grout

FIGURE 24.42 Bearing joints.

Glossary

Aspect ratio — Ratio of longer span to shorter span of a rectangular space frame.

Braced (barrel) vault — A space frame composed of member elements arranged on a cylindrical surface.

Braced dome — A space frame composed of member elements arranged on a spherical surface.

Continuum analogy method — A method for the analysis of space frame where the structure is analyzed by assuming it as an equivalent continuum.

Depth — Distance between the top and bottom-layers of a double-layer space frame.

Discrete method — A method for the analysis of space frame where the structure is analyzed directly as a general assembly of discrete members.

Double-layer grids — A space frame consisting of two planar networks of members forming the top and bottom-layers parallel to each other and interconnected by vertical and inclined members.

Geodesic dome — A braced dome in which the elements forming the network lie on the great circle of a sphere.

Lamella — A unit used to form diamond-shaped grids, the size being twice the length of the side of the diamond.

Latticed grids — Double-layer grids consist of intersecting vertical latticed trusses to form regular grids.

Latticed shell — A space frame consisting of curved networks of members built either in single or in double layer.

Latticed structure — A structural system in the form of a network of elements whose load-carrying mechanism is three-dimensional in nature.

Local buckling — A snap-through buckling that takes place at one point.

Module — Distance between two joints in the layer of grid.

Overall buckling — Buckling that takes place in a relatively large area where a large number of joints are involved.

Space frame — A structural system in the form of a flat or curved surface assembled of linear elements so arranged that forces are transferred in a three-dimensional manner.

Space grids — Double-layer grids consisting of a combination of square or triangular pyramids to form offset or differential grids.

Space truss — A three-dimensional structure assembled of linear elements and assumed as hinged joints in structural analysis.

References

[1] IASS (International Association for Shell and Spatial Structures) Working Group on Spatial Steel Structures. 1984. Analysis, design and realization of space frames. *Bull. IASS* No. 84/85, XXV (1/2):1–114.

[2] ASCE Task Committee on Latticed Structures of the Committee on Special Structures of the Committee on Metals of the Structural Division. 1976. Latticed structures: state-of-the-art report. *J. Struct. Div., Proc. ASCE* 102 (ST11):2197–2230.

[3] Makowski, Z.S. 1992. Space frames and trusses. In *Constructional Steel Design. An International Guide*, ed. P.J. Dowling et al. pp. 791–843. Elsevier Applied Science, London.

[4] Lan, T.T., Qian, R. 1986. A study on the optimum design of space trusses — optimal geometrical configuration and selection of type. *Shells, Membranes and Space Frames. Proceedings IASS Symposium 1986.* Vol. 3, pp. 191–198. Elsevier, Amsterdam.

[5] ISO. 1998. *Bases for Design of Structures: Determination of Snow Loads on Roofs.* (ISO/FDIS 4355.)

[6] European Committee for Standardization. 1995. *Eurocode 1: Basis of Design and Action on Structures Part 2–4: Actions on Structures — Wind Action.* (ENV 1991-2-4.)

[7] Zhang, Y.G., Lan, T.T. 1984. A practical method for the analysis of space frames under vertical earthquake loads. *Final Report, IABSE 12th Congress 1984.* pp. 169–176. IABSE, Zurich.

[8] ASCE Task Committee on Latticed Structures under Extreme Dynamic Loads of the Committee on Special Structures of the Committee on Metals of the Structural Division. 1984. Dynamic considerations in latticed structures. *J. Struct. Eng.* 110(10):2547–2550.

[9] Kato, S., Kawaguchi, K., Saka, T. 1995. Preliminary report on Hanshin earthquake. *Spatial Structures: Heritage, Present and Future. Proceedings of the IASS Symposium 1995.* Vol. 2, pp. 1059–1066. S. G. Editoriali, Padova, Italy.

[10] Lan, T.T. 1998. Investigation of the failure of a space truss subjected to earthquake. *Adv. Struct. Eng.* 1(4):307–311.

[11] Gioncu, V. 1995. Buckling of reticulated shells: state-of-the-art. *Int. J. Space Struct.* 10(1):1–46.

[12] Lind, N.C. 1969. Local instability analysis of triangulated dome framework. *Struct. Eng.* 47(8).

[13] Zoelly, R. 1915. Über ein knickungsproblem an der kugelschale. *Dissertation,* Zürich, Switzerland.

[14] von Kármán, T. and Tsien, H.-S. 1939. The buckling of spherical shells by external pressure. *J. Aero. Sci.* 7(2):43–50.

[15] Del Pozo, M. 1979. Scattering and active acoustic control from a submerged spherical shell. *J. Acoust. Soc. of Am.,* 111:1–15.

[16] Wright, D.T. 1965. Membrane forces and buckling in reticulated shells. *J. Struct. Div. Proc. ASCE* 91(ST1):173–201.

[17] Buchert, K.P. 1976. Shells and shell-like structures. In *Guide to Stability Design Criteria for Metal Structures,* ed. B.G. Johnston, 3rd edition, Chapter 18. John Wiley and Sons, New York.

[18] Hangai, Y., Tsuboi, Y. 1985. Buckling loads of reticulated single-layer space frames. *Theory and Experimental Investigation of Spatial Structures. Proceedings IASS Congress 1985,* Moscow.

[19] Gerrits, J.M. 1996. The architectural impact of space frame systems. *Proceedings of Asia-Pacific Conference on Shell and Spatial Structures 1996.* China Civil Engineering Society, Beijing.

Further Reading

An introduction to the practical design of space structures is presented in *Horizontal-span Building Structures* by W. Schueller. It covers a wide range of topics, including the development, structural behavior, simplified analysis, and application of different types of space structures.

For further study of continuum analogy method of space frames, *Analysis and Design of Space Frames by the Continuum Method* by L. Kollar and I. Hegedus provides a good reference.

The quarterly journal *International Journal of Space Structures* reports advances in the theory and practice of space structures. Special issues treating individual topics of interest were published, such as *Stability of Space Structures* (ed. V. Gioncu) Vol. 7, No. 4, 1992 and *Prefabricated Spatial Frame Systems* by A. Hanaor, Vol. 10, No. 3, 1995.

Conferences and symposiums are organized by IASS (International Association for Shell and Spatial Structures) annually. The proceedings document the latest development in this field and provide a wealth of information on theoretical and practical aspects of space structures. The proceedings of conferences recently held are as follows:

1. *Spatial Structures at the Turn of Millennium,* IASS Symposium 1991, Copenhagen.
2. *Innovative Large Span Structures,* IASS-CSCE International Congress 1992, Toronto.
3. *Public Assembly Structures from Antiquity to the Present,* IASS Symposium 1993, Istanbul.
4. *Nonlinear Analysis and Design for Shell and Spatial Structures,* Seiken-IASS Symposium 1993, Tokyo.
5. *Spatial, Lattice and Tension Structures,* IASS-ASCE International Symposium 1994, Atlanta.
6. *Spatial Structures: Heritage, Present and Future,* IASS Symposium 1995, Milan.
7. *Conceptual Design of Structures,* IASS International Symposium 1996, Stuttgart.
8. *Design, Construction, Performance and Economics,* IASS International Symposium 1997, Singapore.

9. *Lightweight Structures in Architecture and Engineering*, IASS Symposium 1998, Sydney.
10. *Shell and Spatial Structures: from Recent Past to the Next Millennium*, 40th Anniversary Congress of IASS 1999, Madrid.
11. *Bridging Large Spans, from Antiquity to the Present*, IASS-MSU Symposium 2000, Istanbul.
12. *Theory, Design and Realization of Shell and Spatial Structures*, IASS Symposium 2001, Nagoya.
13. *Lightweight Structures in Civil Engineering*, IASS Symposium 2002, Warsaw.

Journal of IASS is published three times a year and it covers design, analysis, construction, and other aspects of technology of all types of shell and spatial structures.

Additional information can be found in the following:

ASCE Subcommittee on Latticed Structures of the Task Commitee on Special Structures of the Committee on Metals of the Structural Division. 1972. Bibliography on latticed structures. *J. Struct. Div., Proc. ASCE* 98(ST7):1545–1566.

Chinese Academy of Building Research. 1981. *Specifications for the Design and Construction of Space Trusses (JGJ 7-80)*. China Building Industry Press, Beijing. (English translation: *Int. J. Space Struct.* 16[3].)

Chinese Academy of Building Research. 2003. *Technical Specification for Latticed Shells (JGJ 61-2003)*. China Building Industry Press, Beijing.

Heki, K. 1993. Buckling of lattice domes — state of the art report. *Nonlinear Analysis and Design for Shell and Spatial Structures. Proceedings of the Seiken-IASS Symposium 1993*, Tokyo. pp. 159–166.

Iffland, J.S.B. 1987. Preliminary design of space trusses and frames. In *Building Structural Design Handbook*, ed. R.N. White, C.G. Salmon. pp. 403–423. Wiley, New York.

Makowski, Z.S. (ed.) 1981. *Analysis, Design and Construction of Double-Layer Grids*. Applied Science, London.

Makowski, Z.S. (ed.) 1984. *Analysis, Design and Construction of Braced Domes*. Granada, London.

Makowski, Z.S. (ed.) 1985. *Analysis, Design and Construction of Braced Barrel Vaults*. Elsevier Applied Science, London.

Saitoh, M., Hangai, Y., Todu, I., Okuhara, T. 1987. Design procedure for stability of reticulated single-layer domes. *Building Structures. Proceedings of Structure Congress 1987*. pp. 368–376. ASCE, New York.

25

Bridge Structures

Shouji Toma
*Department of Civil Engineering,
Hokkai-Gakuen University,
Sapporo, Japan*

Lian Duan
*Division of Engineering Services,
California Department of
Transportation,
Sacramento, CA*

Wai-Fah Chen
*College of Engineering,
University of Hawaii at Manoa,
Honolulu, HI*

25.1 Introduction

A bridge is a structure that crosses over a river, bay, or other obstruction and permits the smooth and safe passage of vehicles, trains, and pedestrians. An elevation view of a typical bridge is shown in Figure 25.1. A bridge structure can be divided into an upper part (the superstructure), which consists of the deck,

0-8493-1569-7/05/$0.00+$1.50
© 2005 by CRC Press

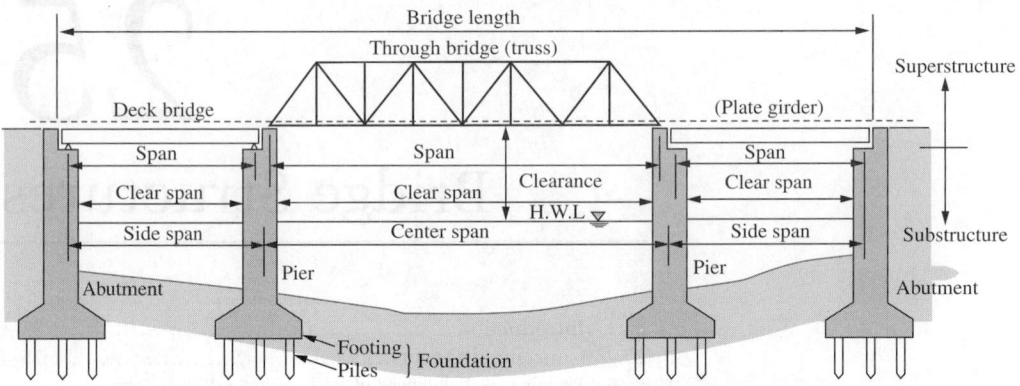

FIGURE 25.1   Elevation view of a typical bridge.

the floor system, and the main trusses or girders, and a lower part (the substructure), which consists of piers, columns, footings, piles, and abutments. The superstructure provides horizontal spans and carries traffic loads directly. The substructure supports the horizontal spans elevating above the ground surface. The bridges can be classified into the following categories:

1. *Classification by materials (superstructures)*
   Steel bridges. Bridges using a wide variety of structural steel components and systems: decks, girders, trusses, arches, stayed, and suspension cables.
   Concrete bridges. Bridges using reinforced and prestressed concrete.
   Timber bridges. Bridges using wood when the span is relatively short.
   Metal alloy bridges. Bridges using metal alloys such as aluminum alloy and stainless steel.
   Advanced composite bridges. Bridges using advanced composite materials or fiber reinforced plastics composites.
   Stone bridges. Bridges using stone; in ancient times stone was the most common material used to construct magnificent arch bridges.
2. *Classification by objectives*
   Highway bridges. Bridges carrying vehicle traffic.
   Railway bridges. Bridges carrying trains.
   Combined bridges. Bridges carrying vehicles and trains.
   Pedestrian bridges. Bridges carrying pedestrian traffic.
   Aqueduct bridges. Bridges supporting pipelines with channeled waterflow.
   Bridges can alternatively be classified into movable for ships to pass the river, or fixed and permanent, or temporary categories.
3. *Classification by structural systems (superstructures)*
   I-girder (beam) bridges. The main girders consist of either plate girders (plate assemblages of top and bottom flanges and a web) or rolled I-shapers. I-sections can effectively resist bending and shear.
   Box-girder bridges. The main girders consist of a single or multiple box beams fabricated from steel plates and formed from concrete, which resist not only bending and shear but also torsion effectively.
   T-beam bridges. A number of reinforced concrete T-beams are placed side by side to support the live load.
   Composite girder bridges. The concrete deck slab works in conjunction with the steel girders to support loads as a united beam. The steel girder takes mainly tension, while the concrete slab takes the compression component of the bending moment.

Grillage girder bridges. The main girders are connected transversely by floor beams to form a grid pattern, which shares the loads with the main girders.

Othotropic deck bridges. The deck consists of a steel deck plate and rib stiffeners.

Truss bridges. The bar members are arranged in a continuous pattern based on structural rigidity of triangles. Truss members are theoretically considered to be connected with pins at their ends. Each member resists an axial force, either in compression or tension. Figure 25.1 shows a Warren truss bridge with vertical members, which is a "through bridge," that is, the deck slab is placed through the lower part of the bridge.

Arch bridges. The arch is a vertically curved structure that resists load mainly in axial compression.

Cable stayed bridges. The girders are supported by highly strengthened cables slopping directly from one or more towers. These are most suited to bridge long distances.

Suspension bridges. The girders are supported by vertical or near-vertical hangers, which are, in turn, supported by the main suspension cables extending over towers from anchorage to anchorage. The load is transmitted mainly by tension in cables. This design is suitable for longest bridges.

4. *Classification by support conditions*

Figure 25.2 shows three different support conditions for girder and truss bridges.

Simply supported bridges. The main girders or trusses are supported by a movable hinge at one end and a fixed hinge at the other end.

Continuously supported bridges. Girders or trusses are supported continuously by more than three hinges, resulting in a structurally indeterminate system. These tend to be more economical since the fewer expansion joints will have less service and maintenance problems. Settlements at the supports shall be avoided.

Gerber bridges (cantilever bridges). A continuous bridge is rendered determinate by placing intermediate hinges between the supports.

Rigid frame bridges. The girders are rigidly connected to the substructures.

In this chapter, main structural features of common types of steel and concrete bridges for short and mid-long spans are discussed. Cable-supported long-span bridges are addressed in Chapter 26. The recently published Bridge Engineering Handbook (Chen and Duan 2000) provides a unique, comprehensive, state-of-the-art, and state-of-the-practice reference work and resource book covering the major area of bridge engineering. For a more detailed look at design procedures, references should be made to the books of Troitsky (1994), Xanthakos (1994, 1995), Tonias (1995), and Barker and Puckett (1997).

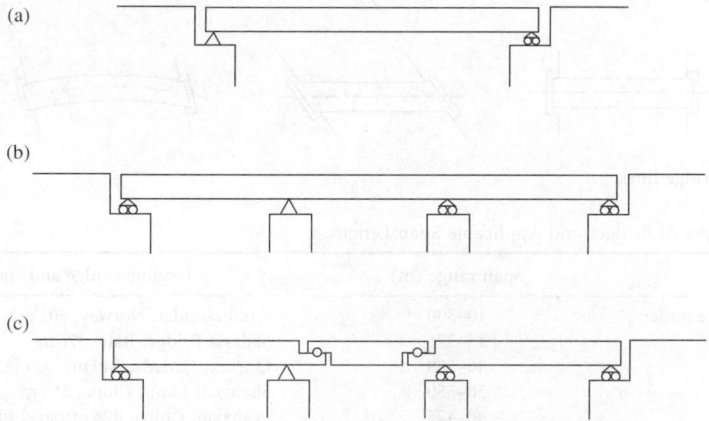

FIGURE 25.2 Supporting conditions of girder bridges: (a) simple girder; (b) continuous girder; and (c) Gerber girder.

25.2   Conceptual Design

Bridge design is a combination of art–creation, science–natural laws, and technology–engineering reality. The conceptual design is the first step, which the designers must to take, to visualize and imagine the bridge in order to determine its fundamental function and performance before any theoretical analysis and detailing design can be proceeded. The design process includes considerations of important factors, such as selection of bridge systems, materials, proportions, dimensions, foundations, esthetics, and surrounding landscape and environment.

25.2.1   Planning — Fulfillment of its Function

The first principle is planning a bridge to optimally meet the specified functions. A bridge project will start with planning the fundamental design conditions. To meet the specific purposes, a bridge may have different alignments: straight, skewed, and horizontally curved (Figure 25.3). A straight bridge is easy to design and construct but often needs longer spans. A skewed bridge or a horizontally curved bridge is commonly required for expressways or railroads where the road line must be kept straight or horizontally curved, even at the cost of a more difficult design. The width of a bridge is dependent of the traffic requirements. For a highway bridge, its width is usually determined by the width of the traffic lanes and the sidewalk width, and often the same dimension as that of the approaching road.

25.2.2   Bridge Types — Challenging Spans

The type of bridges is usually determined by factors such as design loads, surrounding geographical features, soil and foundations, passing line and its width, the length and span of the bridge, esthetics, the requirement for clearance below the bridge, transportation of the construction materials, erection procedures, construction cost and period. Table 25.1 shows the span lengths appropriate to each type of bridge.

25.2.3   Proportioning — Golden Mean and Image of Human Body

The geometry proportioning of a bridge structure is to establish the relationships between horizontal superstructures and supporting substructures, between the depth and span of the beams, between the height, length, and width of the opening, and between sizes of individual members. The ideal bridge is structurally straightforward and elegant. The golden mean provides the key to squaring the circle. The image of a beautiful human body provides the harmonious proportion — all measurements and their relationship are found all numerical relationships (Leonhardt 2000).

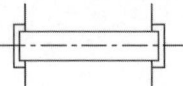 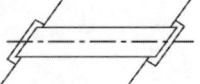 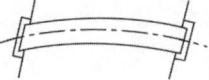

FIGURE 25.3   Bridge lines.

TABLE 25.1   Types of Bridges and Applicable Span Length

Bridge type	Span range (m)	Leading bridge and span length
Prestressed concrete girder	10–300	Stolmasundet, Norway, 301 m
Steel I/box girder	15–376	Sfalassa Bridge, Italy, 376 m
Steel truss	40–550	Quebec, Canada, 549 m
Steel arch	50–550	Shanghai Lupu, China, 550 m
Concrete arch	40–425	Wanxian, China, 425 m (steel tube filled concrete)
Cable stayed	110–1100	Sutong, China, 1088 m
Suspension	150–2000	Akaski-Kaikyo, Japan, 1991 m

FIGURE 25.4 Firth of Forth Bridge and bridge model in U.K. (1980).

A structure designed in harmony with one other not only appeals to our hereditary sense of beauty, but also behaves a smooth load transfer path. Figure 25.4 shows the Firth of Forth Bridge built in Scotland, U.K., in 1890 and the bridge model. This bridge is a gain steel railway bridge with a main span of 521 m. The bridge designer Baker stated (Collins 2001) that

> When a load is put on the central girder by a person setting on it, the men's arms and the anchor ropes come into the tension, and the men's bodies and the sticks come into compression. The chairs are representative of the circular granite piers. Imagine of chairs one third of a mile apart and the men's head as high as the cross of St. Paul's, their arms represented by the huge lattice steel girders and the sticks by the tubes 12 ft (3.6 m) in diameter at the base, and a very good notion of the structure is obtained.

25.2.4 Esthetics — Harmonizing Surroundings

A bridge is required not only to fulfill its function as a thoroughfare, but must use its structure and form to blend, harmonize, and enhance its surroundings. Although there are different views regarding esthetic practice in bridge engineering, the following guidelines presented by Svensson (1998) may be useful:

- "Choice of a clean and simple structural system," like a beam, a frame, an arch, or a suspended structure, the bridge must look trustworthy and stable.
- "Good proportion in all three dimensions" between the structural members or between length and depth of bridges openings.

- "Good order of all the lines of edges of a structures," which determine the appearance. One should limit the number of directions that cause unrest, confusion, and worried feelings. For the transition from straight lines to curved lines, the curvature should steadily increase like a second-order parabola.
- The compatible integration of a structure into its environment, into the landscape of city. This is specially important with regard to the scale of the structure compared to the scale of the surroundings.
- The choice of the materials has considerable influence on the esthetic effects.
- Simplicity and restriction to the pure structural shape is important.
- Pleasing appearance can be enhanced by color.
- The space above the bridge should be shaped in such a way that the driver experiences the bridge and gets a comfortable feeling.
- A structure must be designed that the flow of forces is evident to the causal observer.
- Moderate esthetic lighting can enhance the appearance of a bridge at night.

Figure 25.5 shows a conceptual design of Ruck-a-Chucky Bridge crossing American River about 17 km from the Auburn Dam in California. This horizontally curved cable-stayed bridge spanning 396 m was designed to anchor cables in the hillsides. Although the bridge was never built, the design fits the topography, the surrounding environment, and is a real well-conceived design (Lin 2001; Ruck-a-Chucky 2003). It was awarded by *Progressive Architecture* magazine in 1979. For more detailed discussion about esthetics, references are made to Leonhardt (1984, 2000), Billington (1983, 2000), and TRB (1991).

25.2.5 Bridge-Type Selection — Comprehensive Decision

The selection of bridge types is a complex task to achieve the owner's objectives. Table 25.2 shows a sample evaluation matrix, which may be used to select the bridge types. For the items listed in Table 25.2, a priority factor may be assigned from 1 to 5 (1 = low, 2 = standard, 3 = high, 4 = very high, and 5 = extremely high). The quality rating can be on a scale of 1 to 5 (1 = poor, 2 = fair, 3 = good, 4 = very good, and 5 = excellent). The weighted rating is obtained by multiplying the priority factor with the quality-rating

FIGURE 25.5   Conceptual design of Ruck-a-Chucky Bridge (courtesy of T.Y. Lin).

TABLE 25.2 Bridge-Type Evaluation Form

Item (1)	Bridge type		
	Priority (2)	Quality (3)	Weighted rating (2) × (3)
Structural			
Traffic			
Constructibility			
Maintenance and inspection			
Construction schedule impact			
Esthetics			
Environmental			
Future expansion			
Cost			
Total rating			

FIGURE 25.6 San Francisco-Oakland Bay Bridge — east span.

factor and summed for each bridge alternative. The bridge type with the highest total score shall be the best candidate.

The collapse of a portion of upper and lower decks of the eastern portion (Figure 25.6) of the San Francisco-Oakland Bay Bridge (SFOBB) in the 1989 Loma Prieta Earthquake, California, demonstrated the critical need for seismic safety on bridges in San Francisco Bay Area. California Department of Transportation (SFOBB 2003) determined that it is more cost-effective to replace the existing east span with a new bridge than it would be to seismically retrofit the existing structure (Figure 25.6). Figure 25.7 to Figure 25.10 show four structure alternatives for the new SFOBB east signature span across the shipping channel. The Engineering and Design Advisory Panel (SFOBB 2003) for the new SFOBB east span adopted the following guidelines:

- The bridge should integrate into the site and surrounding environment by reflecting the grand scale of San Francisco Bay, by harmonizing with the existing western span of the bridge, and by landing gracefully on the Oakland shore and Yerba Buena Island.

FIGURE 25.7   San Francisco-Oakland Bay Bridge east span — cable-stayed alternative 1.

FIGURE 25.8   San Francisco-Oakland Bay Bridge east span — cable-stayed alternative 2.

- The replacement bridge should, by contrast or similarity, complement the existing Bay Bridge suspension span. One bridge should not diminish the visual quality or importance of the other.
- The new bridge should be visually memorable and convey a sense of gateway to Oakland.
- Views from the bridge when traveling toward Oakland should consider Oakland's central business district and waterfront.
- The bridge should provide a measure of visual continuity for motorists.
- The girders, piers, and rails should generally appear slender and provide for views by motorists on the bridge.
- Guardrails and handrails should be designed for maximum transparency to maintain views while meeting safety criteria.
- Night lighting on the bridge is an important consideration.

The single-tower self-anchored asymmetrical suspension span (180 m + 385 m) with a steel tower and a steel orthotropic deck (Figure 25.10) was the final selection from the above four alternatives (Nader et al. 2001). The bridge is under construction and expected to open to traffic in 2007.

FIGURE 25.9 San Francisco-Oakland Bay Bridge east span — suspension alternative 1.

FIGURE 25.10 San Francisco-Oakland Bay Bridge east span — suspension alternative 2 (selected).

Figure 25.11 shows a comparison of the four design alternatives evaluated for Minato Oh-Hashi in Osaka, Japan. The truss frame design was selected.

25.3 Final Design

The final design is to produce a set of constructible, biddable, and cost-effective construction documents. The design process involves structural analysis, member and detailing design, and preparation of a set of construction documents including plans, drawings, and special provisions. Structural analysis shall consider the appropriate material properties, member joints, and boundary conditions. The members and joints shall be proportioned to carry various possible loads (dead loads, live loads, wind loads, earthquake loads, etc.) in accordance with the requirements of design standards and codes, such as AASHTO (2004) for highway bridges and AREMA (2003) for railway bridges in the United States or JRA (2002) in Japan, or project-specific design criteria. For major bridges, project-specific design criteria are usually needed and shall be developed (Caltrans 2002). A well-prepared construction document shall be biddable and constructable and cost-effective.

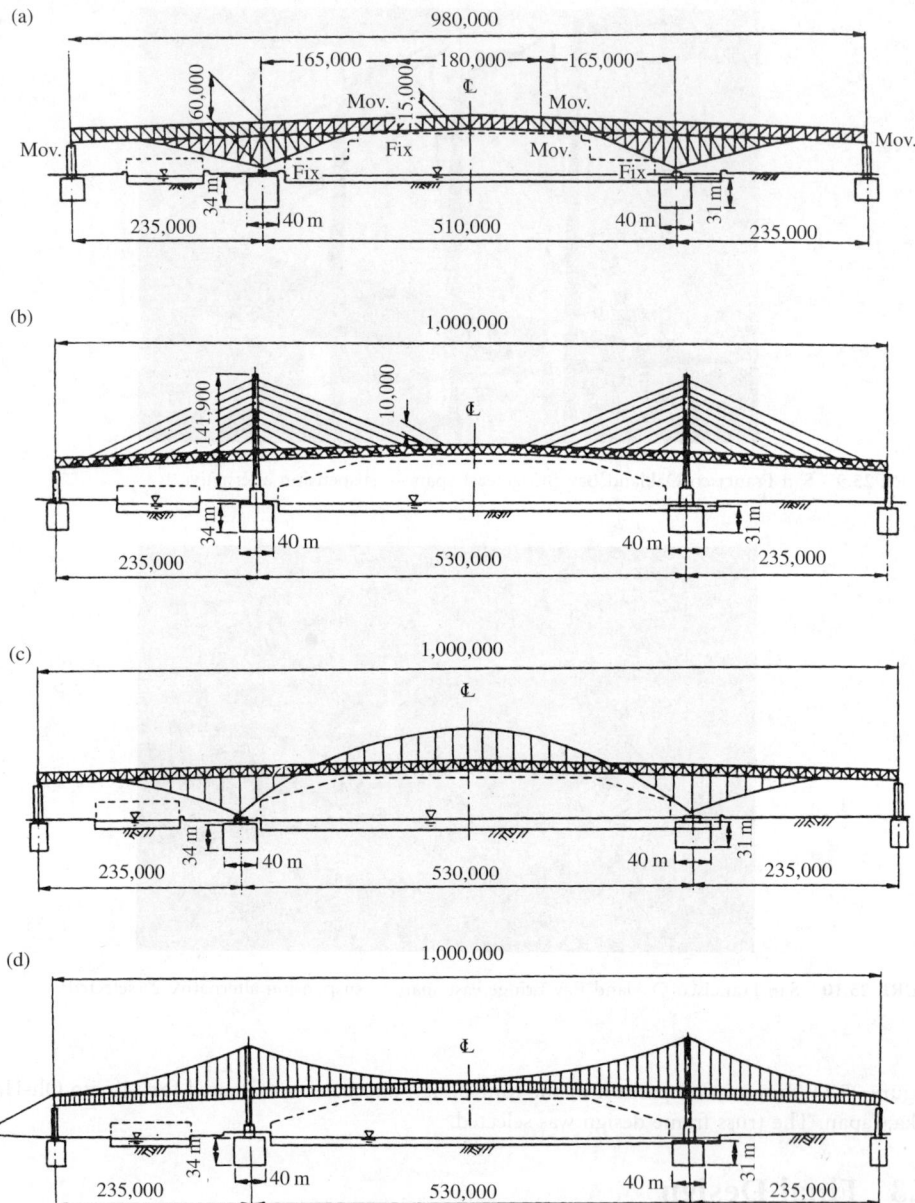

FIGURE 25.11 Design comparison of Minato Oh-Hashi, Japan: (a) K-truss bridge (this was selected); (b) cable-stayed bridge; (c) arch bridge; and (d) suspension bridge (HEPC 1975).

25.4 Steel Girder Bridges

25.4.1 Introduction

Steel has higher strength, ductility, and toughness than many other structural materials such as concrete or wood, and thus makes a vital material for bridge structures. In addition to the conventional steel, there are many types of high-performance steel (HPS) developed recently for the bridge application.

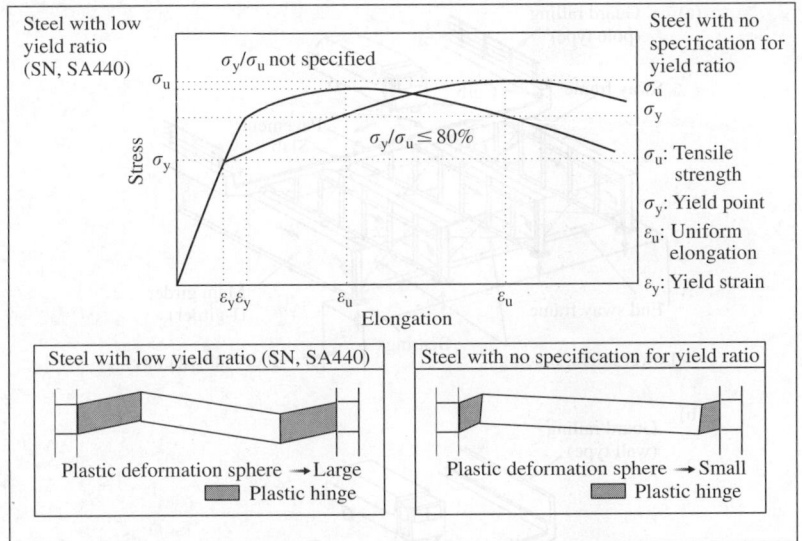

FIGURE 25.12 High-performance steel with low yield ratio (from Kozai Club, Japan, with permission).

HPS can be defined as having an optimized balance of their properties, such as strength, weldability, toughness, ductility, corrosion resistance, and formability, to give maximum performance in bridge structures while remaining cost-effective (Mistry 2002). The two main differences compared to conventional weathering steels are improved weldability and toughness. Other properties such as corrosion resistance and ductility will be essentially the same. There are two HPS grades, ASTM A 709 HPS 70W and HPS 50W, available for bridge structures in the United States. There are 91 HPS bridges built in the United States so far (FHWA 2003). HPS has very high strength, with constant yield point, narrow range of yield point, low yield ratio, excellent ductility, and ultrathickness. Figure 25.12 shows the performance mechanism of the steel with low yield ratio that provides higher ductility at plastic hinges.

Girder bridges are structurally the simplest and the most common. They consist of a floor slab, girders, and bearings, which support and transmit gravity loads to the substructure. Girders resist bending moments and shear forces and are used to span short distances. Steel girders are classified as plate and box girders.

Figure 25.13 shows the structural composition of plate and box girder bridges and the load transfer path. In plate girder bridges, the live load is directly supported by the slab and then by the main girders. In box girder bridges, the forces are taken first by the slab, then supported by the stringers and floor beams in conjunction with the main box girders, and finally taken to the substructure and foundation through the bearings.

Girders are classified as noncomposite or composite, that is, whether the steel girders act in tandem with the concrete slab (using shear connectors) or not. Since composite girders make use of best properties of both steel and concrete, they are often the rational and economic choice. Less frequently noncomposite I-shapes are used for short-span noncomposite bridges.

25.4.2 Plate Girder (Noncomposite)

The plate girder is the most economical shape designed to resist bending and shear and the moment of inertia is the greatest for a relatively low weight per unit length. Figure 25.14 shows a plan of a typical plate girder bridge with four main girders spanning 30 m and a width of 8.5 m.

The gravity loads are supported by several main plate girders, each manufactured by welding three plates: top and bottom flanges and a web. Figure 25.15 shows a piece of plate girder and its fabrication

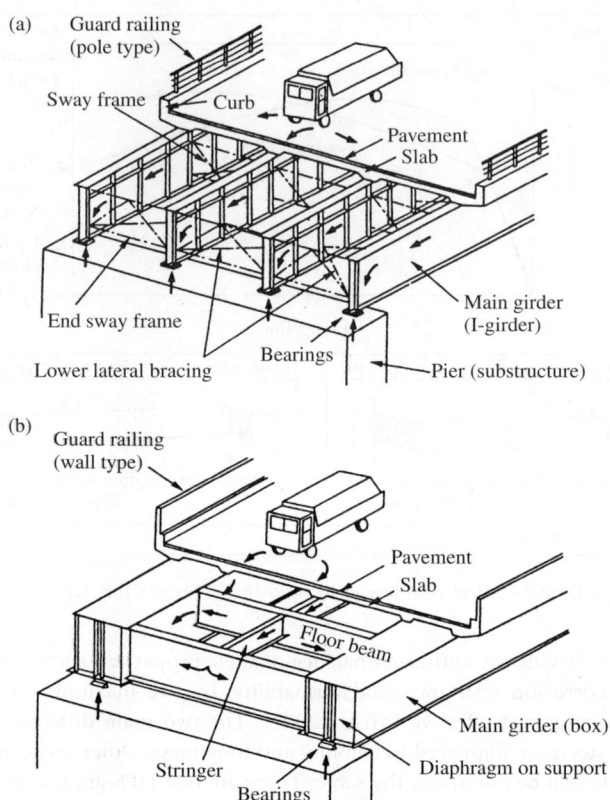

FIGURE 25.13 Steel girder bridges: (a) plate girder bridge and (b) box girder bridge (Nagai 1994).

process. The web and the flanges are cut from the steel plate and welded. The piece is fabricated in shop and transported to the construction site for erection.

The design procedure for plate girders, primarily the sizing of the three plates, is discussed as follows:

1. *Web height.* The web height is the fundamental design factor affecting the weight and cost of the bridge. If the height is too small, the flanges need to be large and the dead weight increases. The height (D) is determined empirically by dividing the span length (L) by a "reasonable" factor. Common ratios are $D/L = \frac{1}{18}$ to $\frac{1}{20}$ for highway bridges and a little smaller for railway bridges. The web height also influences the flexural stiffness of bridge. Greater heights generally produce greater stiffness. However, if the height is too large, the web may becomes unstable and must have its thickness supplemented or stiffeners added. These measures increase the weight and the cost. In addition, plate girders with excessively deep web and small flanges are vulnerable to buckle laterally.

2. *Web thickness.* The web primarily resists shear forces that are not usually significant when the web height is properly designed. The shear force is generally assumed to be distributed uniformly across the web instead of using the exact equation of beam theory. The web thickness (t) is determined such that thinner is better as long as buckling is prevented. Since the web does not contribute much to the bending resistance, thin webs are most economical but the possibility of buckling increases. Therefore, a noncompact web thickness is usually selected and stiffened by transverse (vertical) and longitudinal (horizontal) stiffeners. It is not primarily strength but rather stiffness that controls the design of webs.

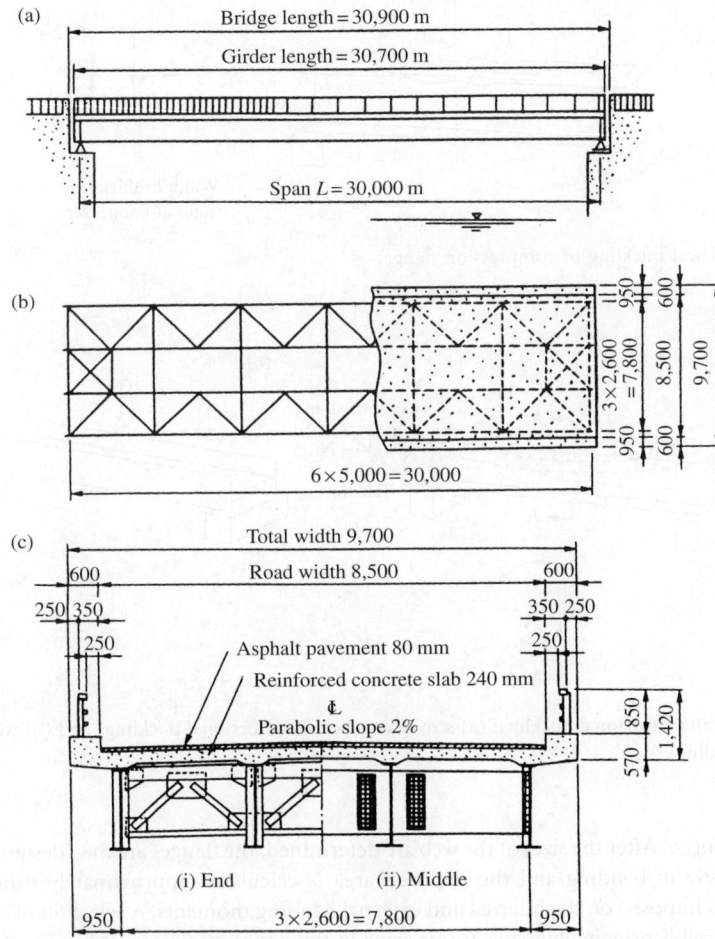

FIGURE 25.14 General plans of a typical plate girder bridge: (a) side view; (b) plan; and (c) cross-section (Tachibana and Nakai 1996).

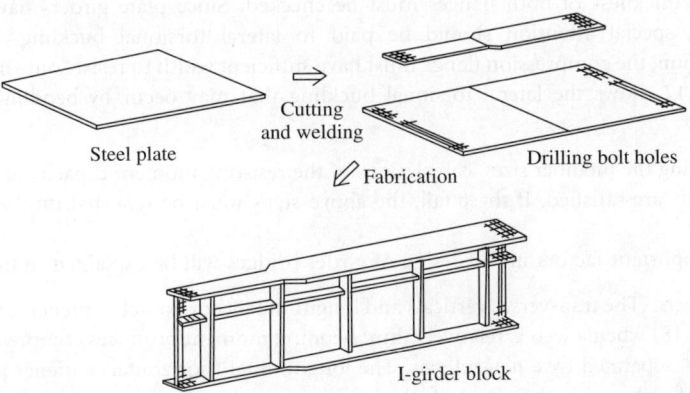

FIGURE 25.15 Fabrication of plate girder piece.

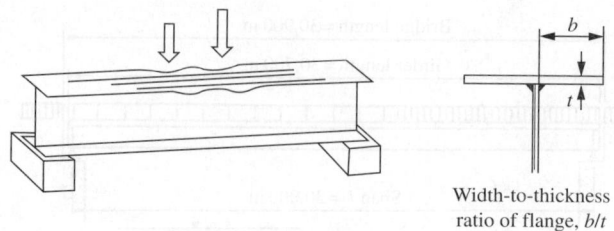

FIGURE 25.16   Local buckling of compression flange.

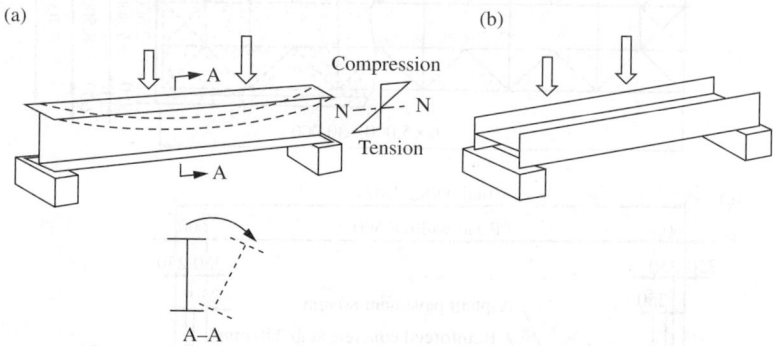

FIGURE 25.17   Lateral torsional buckling: (a) strong axis bending (torsional buckling) and (b) weak axis bending (no torsional buckling).

3. *Area of flanges.* After the sizes of the web are determined, the flanges are then designed. The flanges work mostly in bending and the required area is calculated approximately using equilibrium conditions imposed on the internal and external bending moments. A selection of strength for the steel material is principally made at this stage in the design process.

4. *Width and thickness of flanges.* The width and thickness can be determined by ensuring that the area of the flanges falls under the limiting width-to-thickness ratio, b/t (Figure 25.16), as specified in design codes. If the flanges are too thin (i.e., the width-to-thickness ratio is too large), the compression flange may buckle or the tension flange may be distorted by heat of welding. Thus, the thickness of both flanges must be checked. Since plate girders have little torsional resistance, special attention should be paid to lateral torsional buckling. To prevent this phenomenon, the compression flange must have sufficient width to resist "out-of-plane" bending. Figure 25.17 shows the lateral torsional buckling that may occur by bending with respect to strong axis.

After determining the member sizes, calculations of the resisting moment capacity are made to ensure code requirements are satisfied. If these fail, the above steps must be repeated until the specifications are met.

A few other important factors in the design of girder bridges will be explained in the following:

1. *Web stiffeners.* The transverse (vertical) and longitudinal (horizontal) stiffeners are usually needed (Figure 25.18) when a web is relatively thin. Bending moment produces compression and tension in the web, separated by a neutral axis. The longitudinal (horizontal) stiffener prevents bending buckling of web and is therefore attached to the compression side of the web (the top half for a simply supported girder). Since the bending moment is the largest near the midspan for a simply

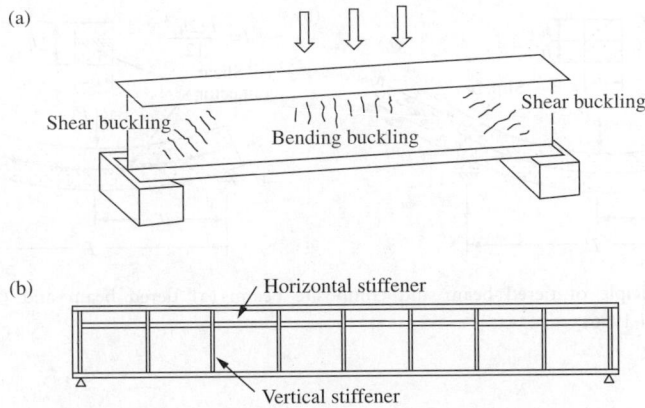

(a)

Shear buckling

Bending buckling

Shear buckling

(b)

Horizontal stiffener

Vertical stiffener

FIGURE 25.18 Stiffeners of web: (a) buckling of web and (b) stiffeners of web.

supported girder, the longitudinal (horizontal) stiffeners are usually located here. The longitudinal stiffeners are not recommended due to its poor fatigue resistance. Transverse (vertical) stiffeners, on the other hand, prevent shear buckling and provide postelastic shear buckling capacity by the tension field action. The most transverse (vertical) stiffeners are placed near the support since the largest shear force occurs at those locations. Bearing stiffeners are also required at the supports to combat large reaction forces, which are designed independently just as any other compression member would be. Buckling patterns of a web are shown in Figure 25.18. If the web is not too deep nor its thickness is too small, no stiffeners are necessary and fabrication costs are reduced.

2. *Variable sections.* The variable cross-sections may be used to save material and cost where the bending moment is smaller, that is, near the end of the span (see Figure 25.15). However, this reduction increases the manpower required for welding and fabrication. The cost of manpower and material must be balanced and traded off. In today's industrial climate, manpower is more important and costly than the material. Therefore, the change of girder section may not economic. Likewise, thick plates are often specified to eliminate the number of stiffeners needed, thus to reduce the necessary manpower.

25.4.3 Composite Girder

If two beams are simply laid one upon the other as shown in Figure 25.19a, they act separately and only share the load depending on their relative flexural stiffness. In this case, slip occurs along the boundary between the beams. However, if the two beams are connected and slip prevented as shown in Figure 25.19b, they act as a unit, that is, a composite girder. For composite plate girder bridges, the steel girder and the concrete slab are joined by shear connectors. In this way, the concrete slab becomes integral with the girder and usually takes most of the compression component of the bending moment while the steel plate girder takes the tension. Composite girders are much more effective than the simply tiered girder.

Let us consider the two cases shown in Figure 25.19 and note the difference between tiered beams and composite beams. Both have the same cross-sections and are subjected to a concentrated load at mid-span. The moment of inertia for the composite beam is four times that of the tiered beams, thus the resulting vertical deflection is one fourth. The maximum bending stress in the outer (top or bottom) fiber is a half that of the tiered beam configuration.

The corresponding stress distributions are shown in Figure 25.20. Points "S" and "V" are the center of area of the steel section and the composite section, respectively. According to beam theory, the strain distribution is linear but the stress distribution has a step change at the boundary between the steel and concrete.

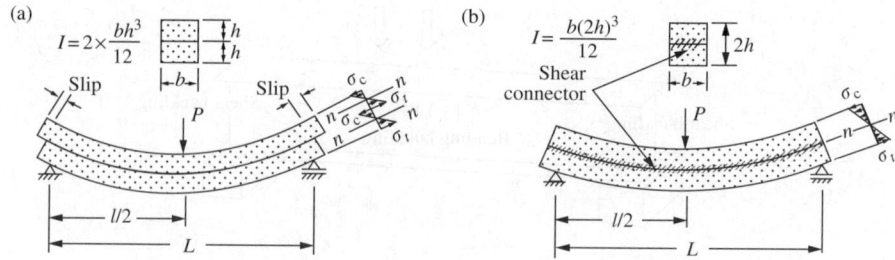

FIGURE 25.19 Principle of tiered beam and composite beam: (a) tiered beam and (b) composite beam (Tachibana and Nakai 1996).

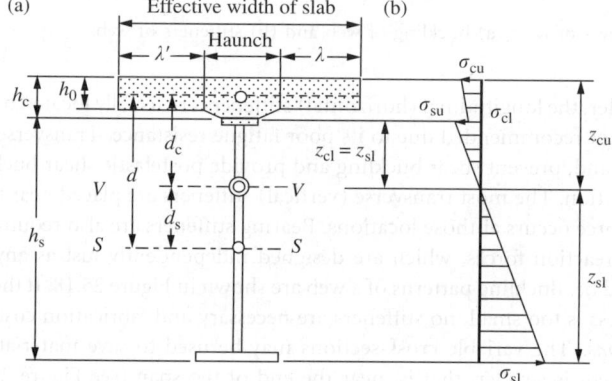

FIGURE 25.20 Section of composite girder: (a) composite girder section and (b) stress distribution (Tachibana and Nakai 1996).

Three types of shear connectors, studs, horse shoes, and steel blocks, are shown in Figure 25.21. Studs are most commonly used since they are easily welded to the compression flange by the electric resistance welding, but the weld inspection is a cumbersome task. If the weld on a certain stud is poor, the stud may shear off and trigger a totally unforeseen failure mode. Other types are considered to maintain more reliability.

Shear connectors are needed most near the ends of the span where the shear force is largest. This region is illustrated in Figure 25.19a, which shows the maximum shift due to slip occurs at ends of tiered beams. It is this slip that is restrained by the shear connectors.

25.4.4 Grillage Girder

When girders are placed in a row and connected transversely by floor beams, the truck loads are distributed by the floor beams to the girders. This system is called as "grillage girder." If the main girders are plate girders, no stiffness in torsion is considered. On the other hand, box girders and concrete girders can be analyzed assuming stiffness is available to resist torsion. Floor beams increase the torsional resistance of the whole structural system of bridge.

Let us consider the structural system shown in Figure 25.22a to observe the load distribution in a grillage system. This grillage has three girders with one floor beam at midspan. In this case, there are three nodal forces at the intersections of the girders and the floor beam but only two equilibrium equations ($V = 0$ and $M = 0$). Thus, it becomes one degree statically indeterminate. If we disconnect the intersection

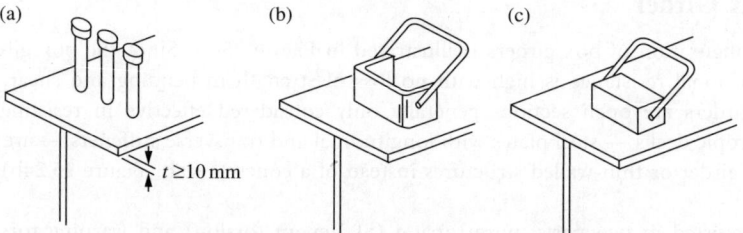

FIGURE 25.21 Type of shear connectors: (a) stud; (b) horse shoe; and (c) steel block (Nagai 1994).

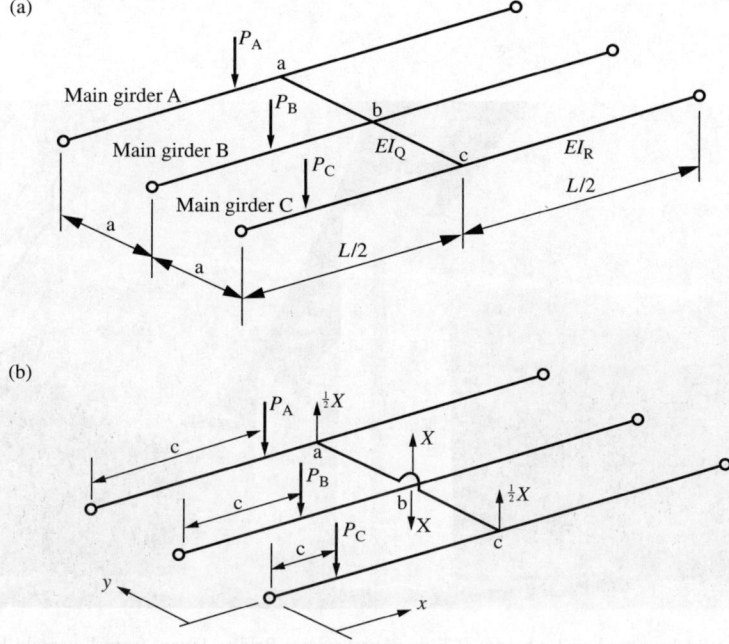

FIGURE 25.22 Grillage girders: (a) one-degree indeterminate system and (b) statically determinate system (Tachibana and Nakai 1996).

between the main girder B and the floor beam and apply a pair of indeterminate forces "X" at point "b" as shown in Figure 25.22b, X can be obtained using the compatibility condition at point "b." Once the force X is found, the sectional forces in the girders can be calculated. This structural system is commonly applied to the practical design of plate girder bridges.

25.4.5 Widely Spaced Plate Girder

To compete with concrete bridges, a new design concept of steel bridge has been developed by mini-mizing the number of girders and pieces of fabrication blocks, thus reducing the construction cost. The space between girders is taken wide and the lateral bracing is neglected. An example of the bridge is shown in Figure 25.23, which has only two girders with the space 5.7 m and the prestressed concrete slab deck thickness of 320 mm.

25.4.6 Box Girder

Structural configuration of box girders is illustrated in Figure 25.24. Since the box girder is a closed section, its torsional resistance is high with no loss of strength in bending and shear. On the other hand, plate girders are open sections generally only considered effective in resisting bending and shear. Orthotropic decks — steel plates with longitudinal and transverse stiffeners — are often used for decks on box girder or thin-walled structures instead of a concrete slab (Figure 25.24b) for long-span bridges.

Torsion is resisted in two parts: pure torsion (St Venant torsion) and warping torsion. The pure torsional resistance of I-plate girders is negligible. However, for closed sections such as a box girder, the pure torsional resistance is considerable, making them particularly suited for horizontally curved bridges or long-span bridges. On the other hand, the warping torsion for box section is negligible. The I-section girder has some warping resistance but it is not so large compared to the pure torsion of closed sections.

FIGURE 25.23   Widely spaced girder bridge (Chidorinosawagawa Bridge, Japan, from Kawasaki Heavy Ind., Ltd., with permission).

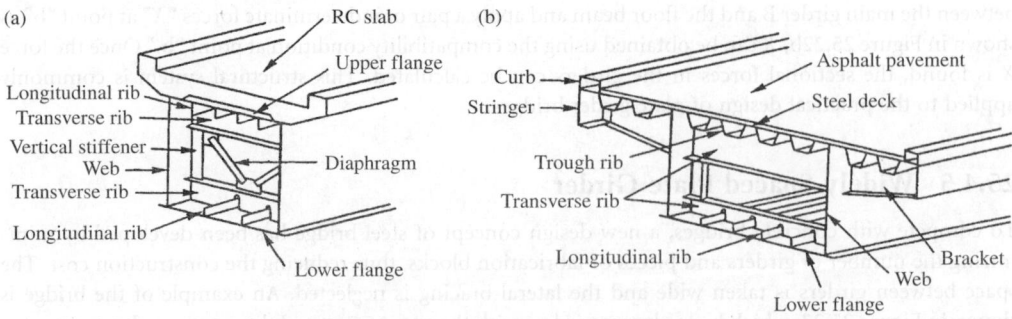

FIGURE 25.24   Box girder: (a) with reinforced concrete deck and (b) with steel deck (Nagai 1994).

25.4.7 Fabrication

Gas flame cutting is generally used to cut steel plates to designated dimensions. Fabrication by welding is conducted in shop where the bridge components are prepared before being assembled (usually bolted) on the construction site.

25.4.7.1 Welding

Welding is the most effective means of connecting steel plates. The properties of steel change when heated and this change is usually for the worse. Molten steel must be shielded from the air to prevent oxidization. Welding can be categorized by the method of heating and the shielding procedure. Shielded metal arc welding (SMAW), submerged arc welding (SAW), CO_2 gas metal arc welding (GMAW), tungsten arc inert gas welding (TIG), metal arc inert gas welding (MIG), electric beam welding, laser beam welding, and friction welding are common methods.

The first two welding procedures mentioned above, SMAW and SAW, are used extensively in bridge construction due to their high efficiency. These both use an electric arc that is generally considered the most efficient method of applying heat. SMAW is done by hand and is suitable for welding complicated joints but less efficient than SAW. SAW is generally automated and can be very effective for welding simple parts such as the connection between flange and web of plate girders. A typical placement of these welding methods is shown in Figure 25.25. TIG and MIG use an electric arc for heat source and inert gas for shielding.

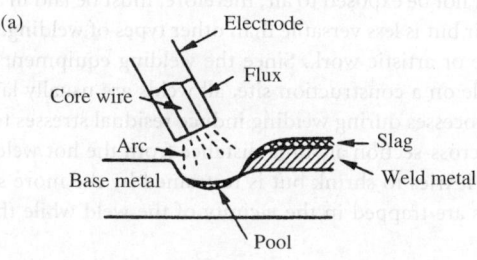

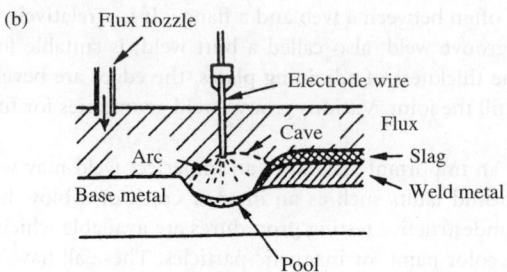

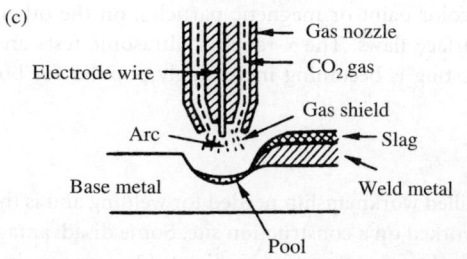

FIGURE 25.25 Welding methods: (a) shielded metal arc welding; (b) submerged arc welding; and (c) gas metal arc welding (Nagai 1994).

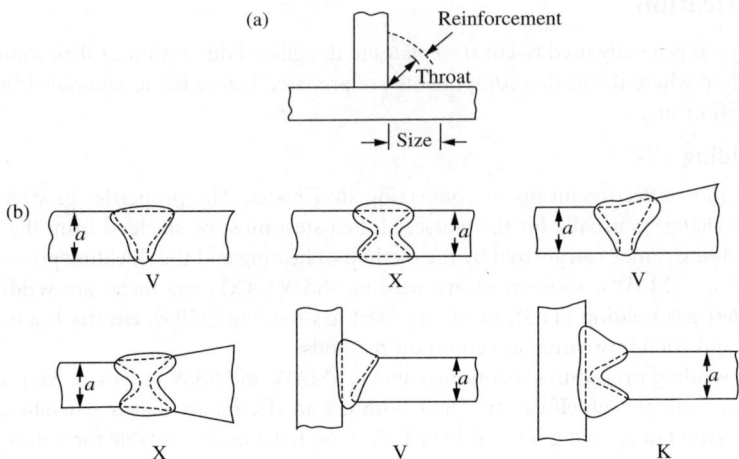

FIGURE 25.26   Types of welding joints: (a) fillet weld and (b) groove weld (Tachibana 1996).

An electric beam weld must not be exposed to air; therefore, must be laid in a vacuum chamber. A laser beam weld can be placed in air but is less versatile than other types of welding. It cannot be used on thick plates but is ideal for minute or artistic work. Since the welding equipment necessary for heating and shielding is not easy to handle on a construction site, all welds are usually laid in the fabrication shop.

The heating and cooling processes during welding induce residual stresses to the connected parts. The steel surfaces or parts of the cross-section at some distance from the hot weld cool first. When the area close to the weld then cools, it tries to shrink but is restrained by the more solidified and cooler parts. Thus, tensile residual stresses are trapped in the vicinity of the weld while the outer parts are put into compression.

There are two types of welded joints: fillet and groove welds (Figure 25.26). The fillet weld is placed at the junction of two plates, often between a web and a flange. It is a relatively simple procedure with no machining required. The groove weld, also called a butt weld, is suitable for joints requiring greater strength. Depending on the thickness of adjoining plates, the edges are beveled in preparation for the weld to allow the metal to fill the joint. Various groove weld geometries for full penetration welding are shown in Figure 25.26b.

Inspection of welding is an important task since an imperfect weld may well have catastrophic consequences. It is difficult to find faults such as an interior crack or a blow hole by observing only the surface of a weld. Many nondestructive testing procedures are available which use various devices, such as x-ray, ultrasonic waves, color paint, or magnetic particles. These all have their own advantages and disadvantages. For example, the x-ray and the ultrasonic tests are suitable for interior faults but require expensive equipment. Use of color paint or magnetic particles, on the other hand, is a less-expensive alternative but only detects surface flaws. The x-ray and ultrasonic tests are used in common bridge construction, but ultrasonic testing is becoming increasingly popular for both its "high tech" and its economical features.

25.4.7.2 Bolting

Bolting does not require the skilled workmanship needed for welding and is thus a simpler alternative. It is applied to the connections worked on a construction site. Some disadvantages, however, are incurred: (1) splice plates are needed and the force transfer is indirect; (2) screwing-in of the bolts creates noise; and (3) esthetically bolts are less appealing. In special cases that need to avoid these disadvantages, the welding may be used even for site connections.

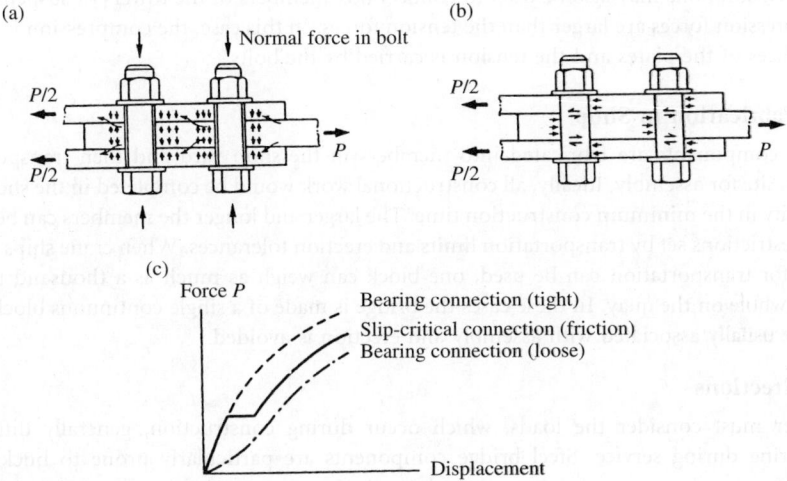

FIGURE 25.27 Slip-critical and bearing-type connections: (a) slip-critical connection; (b) bearing-type connection; and (c) behavior of bolt connections (Nagai 1994).

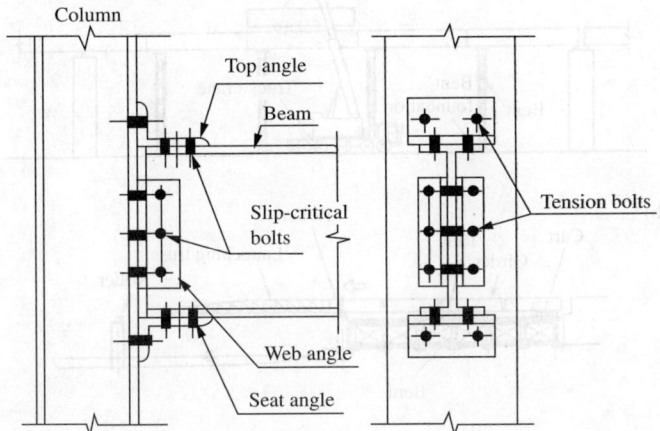

FIGURE 25.28 Tension-type connection.

There are three types of high-tensile-strength bolted connections: the slip-critical connection, the bearing-type connection (Figure 25.27), and the tensile connection (Figure 25.28). The slip-critical (friction) connection is most commonly used in bridge construction as well as other steel structures because it is simpler than a bearing-type connection and more reliable than a tension connection. The force is transferred by the friction generated between the base plates and the splice plates. The friction resistance is induced by the axial compression force in the bolts.

The bearing-type bolt transfers the force by bearing against the plate as well as making some use of friction. The bearing-type bolt can transfer larger force than the friction bolts but are less forgiving with respect to the clearance space often existing between the bolt and the plate. These require that precise holes be drilled and at exact spacings. The force transfer mechanism for these connections is shown in Figure 25.27. In the beam-to-column connection shown in Figure 25.26, the bolts attached to the column are tension bolts while the bolts on the beam are slip-critical bolts.

The tension bolt transfers force in the direction of bolt axis. Tension type of bolt connection is easy to connect on site but difficulties arise in distributing forces equally to each bolt, resulting in reduced reliability. Tension bolts may also be used to connect box members of the towers of suspension bridges where compression forces are larger than the tension forces. In this case, the compression is shared with butting surfaces of the plates and the tension is carried by the bolts.

25.4.7.3 Fabrication in Shop

Steel bridge components are fabricated into members in the shop yard and then transported to the construction site for assembly. Ideally, all constructional work would be completed in the shop to get the highest quality in the minimum construction time. The larger and longer the members can be, the better, within the restrictions set by transportation limits and erection tolerances. When crane ships for erection and barges for transportation can be used, one block can weigh as much as a thousand tons and be erected as a whole on the quay. In these cases the bridge is made of a single continuous block and much of the hassle usually associated with assembly and erection is avoided.

25.4.7.4 Erections

The designer must consider the loads, which occur during construction, generally different from those occurring during service. Steel bridge components are particularly prone to buckling during

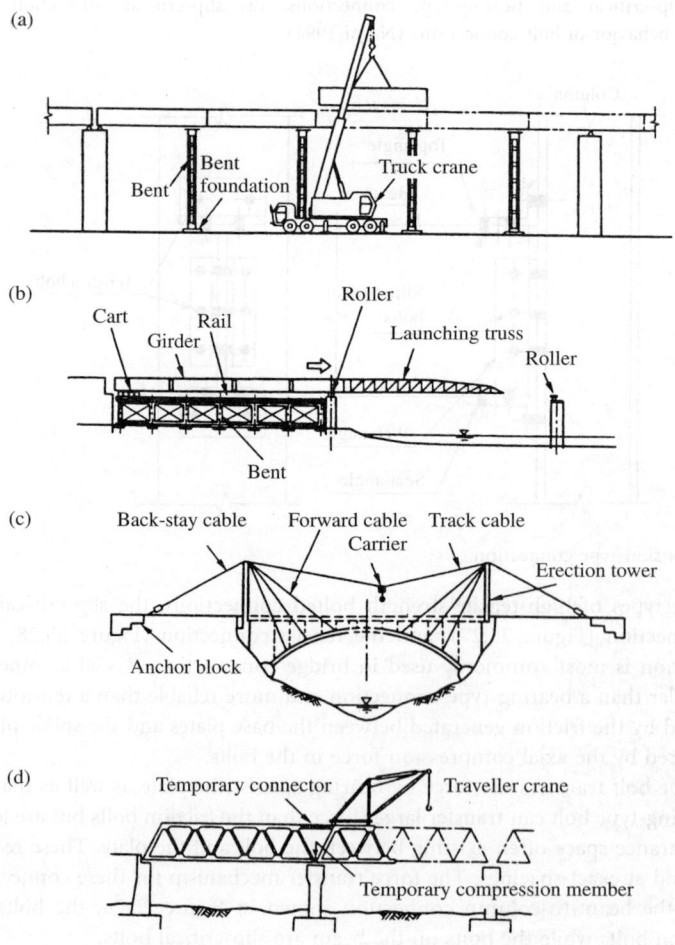

FIGURE 25.29 Erection methods: (a) truck crane and bent erection; (b) launching erection; (c) cable erection; and (d) cantilever erection (JMCA 1991).

construction. The erection plan must be made prior to the main design and must be checked for strength and stability for every possible load case that may arise during erection. Truck crane and bent erection (or staging erection), launching erection, cable erection, cantilever erection, and large block erection (or floating crane erection) are several techniques (see Figure 25.29). An example of the large block erection is shown in Figure 25.46, in which a 186 m–4500 ton center block is transported by barge and lifted.

25.4.7.5 Painting

Steel must be painted to protect it from rusting. There is a wide variety of paints and the life of a steel structure is largely influenced by its quality. In areas near the sea, the salty air is particularly harmful to exposed steel. The cost of painting is high but is essential to the continued good condition of the bridge. The color of the paint is also an important consideration in terms of its public appeal or esthetic quality.

25.5 Concrete Girder Bridges

25.5.1 Introduction

For modern bridges, both structural concrete and steel give satisfactory performance. The choice between the two materials depends mainly upon the cost of construction and maintenance. Generally, concrete structures require less maintenance than steel structures, but since the relative cost of steel and concrete is different from country to country, and may even vary throughout different parts of the same country, it is impossible to put one definitively above the other in terms of "economy."

In this section, the main features of common types of concrete bridge superstructures are briefly discussed. Concrete bridge substructures will be discussed in Section 25.6.

25.5.2 Reinforced Concrete Bridges

Figure 25.30 shows the typical reinforced concrete sections commonly used in highway bridge superstructures.

25.5.2.1 Slab

A reinforced concrete slab (Figure 25.30a) is the most economical bridge superstructure for spans of up to approximately 40 ft (12.2 m). The slab has simple details, standard formwork, and is neat, simple, and pleasing in appearance. Common spans range from 16 to 44 ft (4.9 to 13.4 m) with structural-depth-to-span ratios of 0.06 for simple spans and 0.045 for continuous spans.

25.5.2.2 T-Beam (Deck Girder)

The T-beams (Figure 25.30b) are generally economic for spans of 40 to 60 ft (12.2 to 18.3 m), but does require complicated formwork, particularly for skewed bridges. Structural-depth-to-span ratios are 0.07 for simple spans and 0.065 for continuous spans. The spacing of girders in a T-beam bridge depends on the overall width of the bridge, the slab thickness, and the cost of the formwork and may be taken as 1.5 times the structural depth. The most commonly used spacings are between 6 and 10 ft (1.8 to 3.1 m).

25.5.2.3 Cast-In-Place Box Girder

Box girders like the one shown in Figure 25.30c, are often used to span 50 to 120 ft (15.2 to 36.6 m). Its formwork for skewed structures is simpler than that required for the T-beam. Due to excessive dead load deflections, the use of reinforced concrete box girders over simple spans of 100 ft (30.5 m) or more may uneconomical. The depth-to-span ratios are typically 0.06 for simple spans and 0.055 for continuous spans with the girders spaced at 1.5 times the structural depth. The high torsional resistance of the box girder makes it particularly suitable for curved alignments, such as the ramps onto freeways. Its smooth flowing lines are appealing in metropolitan cities.

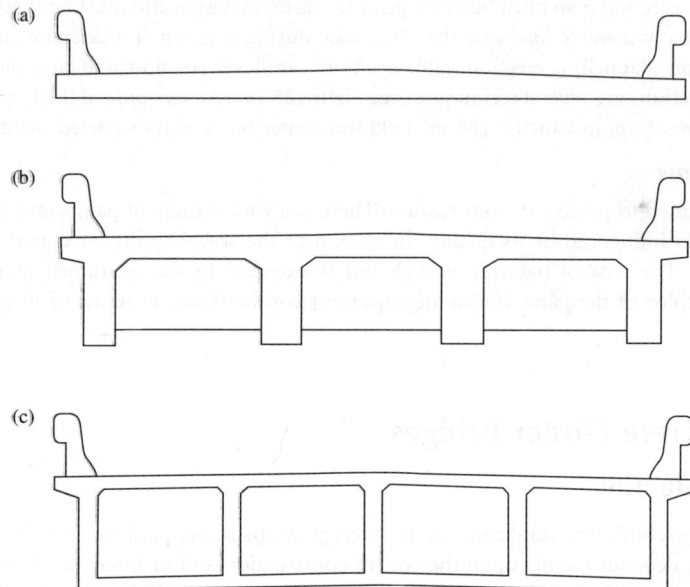

FIGURE 25.30 Typical reinforced concrete sections in bridge superstructures: (a) solid slab; (b) T-beam; and (c) box girder.

25.5.2.4 Design Consideration

A reinforced concrete highway bridge should be designed to satisfy the specification or code requirements, such as the AASHTO-LRFD (2004) requirements for all appropriate service, fatigue, strength, and extreme event limit states. In the AASHTO-LRFD (2004), service limit states include cracking and deformation effects, and strength limit states consider the strength and stability of a structure. A bridge structure is usually designed for the strength limit states and is then checked against the appropriate service and extreme event limit states.

25.5.3 Prestressed Concrete Bridges

Prestressed concrete, using high-strength materials, makes it an attractive alternative for long-span bridges. It has been widely used in bridge structures since the 1950s.

25.5.3.1 Slab

Figure 25.31 shows FHWA (1990) standard types of precast prestressed voided slabs and their sectional properties. While a cast-in-place prestressed slab is more expensive than reinforced concrete slab, a precast prestressed slab is economical when many spans are involved. Common spans range from 20 to 50 ft (6.1 to 15.2 m). Structural-depth-to-span ratios are 0.03 for both simple and continuous spans.

25.5.3.2 Precast I-Girder

Figure 25.32 shows AASHTO (FHWA 1990) standard types of I-beams. Figure 25.33 and Figure 25.34 show Caltrans standard "I"-girder "Bulb-Tee" girders (Caltrans 2001), respectively. These compete with steel girders and generally cost more than reinforced concrete with the same depth-to-span ratios. The formwork is complicated, particularly for skewed structures. These sections are applicable to spans 30 to 120 ft (9.1 to 36.6 m). Structural-depth-to-span ratios are 0.055 for simple spans and 0.05 for continuous spans.

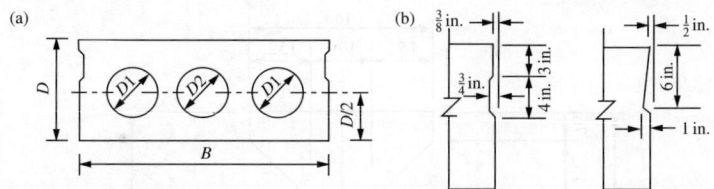

Span range, ft (m)	Section dimensions				Section properties		
	Width B, in. (mm)	Depth D, in. (mm)	$D1$, in. (mm)	$D2$, in. (mm)	A, in.^{2} (mm^2 10^6)	I_x, in.^{4} (mm^4 10^9)	S_x, in.^{3} (mm^3 10^6)
25 (7.6)	48 (1,219)	12 (305)	0 (0)	0 (0)	576 (0.372)	6,912 (2.877)	1,152 (18.878)
30–35 (9.1–10.7)	48 (1,219)	15 (381)	8 (203)	8 (203)	569 (0.362)	12,897 (5.368)	1,720 (28.185)
40–45 (12.2–13.7)	48 (1,219)	18 (457)	10 (254)	10 (254)	628 (0.405)	21,855 (9.097)	2,428 (39.788)
50 (15.2)	48 (1,219)	21 (533)	12 (305)	10 (254)	703 (0.454)	34,517 (1.437)	3,287 (53.864)

FIGURE 25.31 FHWA precast prestressed voided slab sections: (a) typical section and (b) alternative shear key (FHWA 1990).

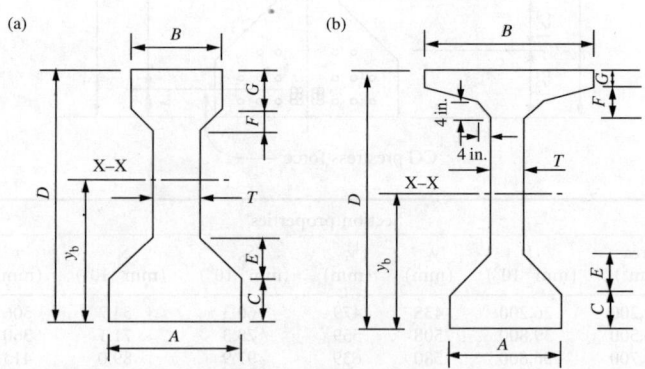

AASHTO beam type	Section dimensions in (mm)							
	Depth D	Bot width A	Web width T	Top width B	C	E	F	G
II	36 (914)	18 (457)	6 (152)	12 (305)	6 (152)	6 (152)	3 (76)	6 (152)
III	45 (1,143)	22 (559)	7 (178)	16 (406)	7 (178)	7.5 (191)	4.5 (114)	7 (178)
IV	54 (1,372)	26 (660)	8 (203)	20 (508)	8 (203)	9 (229)	6 (152)	8 (203)
V	65 (1,651)	28 (711)	8 (203)	42 (1,067)	8 (203)	10 (254)	3 (76)	5 (127)
VI	72 (1,829)	28 (711)	8 (203)	42 (1,067)	8 (203)	10 (254)	3 (76)	5 (127)

AASHTO beam type	Section properties				
	A, in.^{2} (mm^2 10^6)	Y_b, in. (mm)	I_x, in.^{4} (mm^4 10^9)	S_b, in.^{3} (mm^3 10^6)	S_t, in.^{3} (mm^3 10^6)	Span ranges, ft (m)
II	369 (0.2381)	15.83 (402.1)	50,980 (21.22)	3,220 (52.77)	2,528 (41.43)	40–45 (12.2–13.7)
III	560 (0.3613)	20.27 (514.9)	125,390 (52.19)	6,186 (101.38)	5,070 (83.08)	50–65 (15.2–19.8)
IV	789 (0.5090)	24.73 (628.1)	260,730 (108.52)	10,543 (172.77)	8,908 (145.98)	70–80 (21.4–24.4)
V	1013 (0.6535)	31.96 (811.8)	521,180 (216.93)	16,307 (267.22)	16,791 (275.16)	90–100 (27.4–30.5)
VI	1085 (0.7000)	36.38 (924.1)	733,340 (305.24)	20,158 (330.33)	20,588 (337.38)	110–120 (33.5–36.6)

FIGURE 25.32 Precast AASHTO I-beam sections: (a) AASHTO beam types II, III, and IV and (b) AASHTO beam types V and VI (FHWA 1990).

25.5.3.3 Box Girder

Figure 25.35 shows FHWA (1990) standard types of precast box sections. Figure 25.36 shows Caltrans standard precast "Bathtub" girder (Caltrans 2001). The shape of a cast-in-place prestressed concrete box girder is similar to the conventional reinforced concrete box girder (Figure 25.31c). The spacing of the

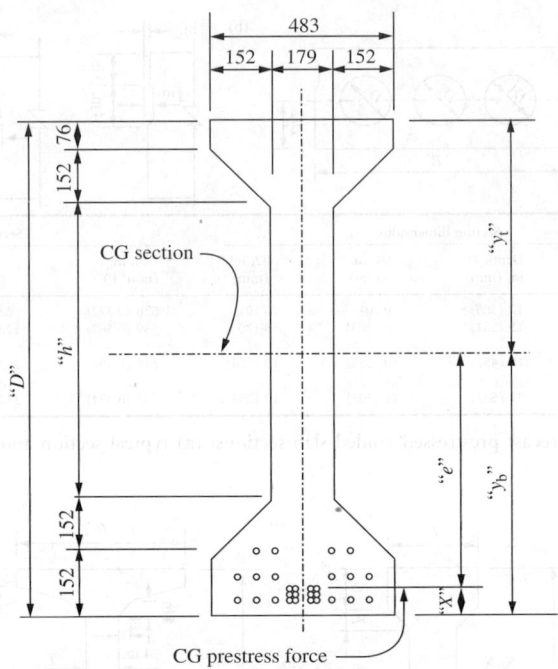

Section properties										
"D" (mm)	"h" (mm)	Area (mm²)	I (mm⁴ 10⁶)	y_b (mm)	y_t (mm)	S_b (mm³ 10⁶)	S_t (mm³ 10⁶)	r (mm)	Force (N/m)	Mass (kg/m)
914	382	279,200	26,200	435	479	60.3	54.7	306	6,850	671
1,067	535	306,500	39,800	508	559	78.3	71.1	360	7,223	736
1,219	687	333,700	56,800	580	639	97.9	89.0	413	7,866	802
1,372	840	361,000	77,800	654	718	118.9	108.4	464	8,509	867
1,524	992	388,300	102,900	728	796	141.4	129.3	515	9,152	933
1,676	1,144	415,600	132,600	802	874	165.3	151.6	565	9,795	998

FIGURE 25.33 Caltrans precast standard "I"-girder (Caltrans 2001).

girders can be taken as twice the structural depth. It is used mostly for spans of 100 to 600 ft (30.5 to 182.9 m). Structural depth-to-span ratios are 0.045 for simple spans and 0.04 for continuous spans. These sections are used frequently for simple spans of over 100 ft (30.5 m) and are particularly suitable for widening in order to control deflections. About 70 to 80% of California's highway bridge system is composed of prestressed concrete box girder bridges.

25.5.3.4 Segmental Concrete Bridge

The segmentally constructed concrete bridges have been successfully developed by combining the concepts of prestressing, box girder, and the cantilever construction (Podolny and Muller 1982; AASHTO 1999). The first prestrssed segmental box girder bridge was built in West Europe in 1950. California's Pine Valley Bridge as shown in Figure 25.37 and Figure 25.38 (composed of three spans of 340 ft (103.6 m), 450 ft (137.2 m), and 380 ft (115.8 m) with the pier height of 340 ft (103.6 m)) is the first cast-in-place segmental bridge built in the United States in 1974.

The prestressed segmental bridges with precast or cast-in-place segmental can be classified by the construction methods (1) balanced cantilever, (2) span-by-span, (3) incremental launching,

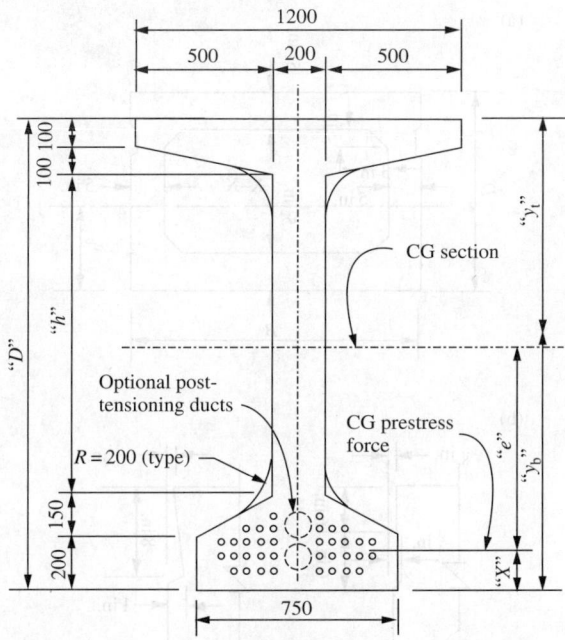

FIGURE 25.34 Caltrans precast standard "Bulb-Tee" girder (Caltrans 2001).

"D" (mm)	"h" (mm)	Area (mm²)	I (mm⁴ 10⁶)	yb (mm)	yt (mm)	Sb (mm³ 10⁶)	St (mm³ 10⁶)	r (mm)	Force (N/m)	Mass (kg/m)
1,400	850	596,500	155,400	721	679	215.5	228.8	510	14,060	1,433
1,550	1,000	626,500	201,200	795	755	253.0	266.5	567	14,770	1,505
1,700	1,150	656,700	254,200	870	830	292.1	306.2	622	15,480	1,577
1,850	1,300	686,100	314,500	945	905	333.0	347.5	677	16,180	1,650
2,000	1,450	716,300	382,600	1,019	981	375.4	390.0	731	16,890	1,722
2,150	1,600	746,600	458,800	1,094	1,056	419.4	434.5	784	17,600	1,794

Section properties

and (4) progressive placement. The selection between cast-in-place and precast segmental and among various construction methods is dependent of project features, site conditions, environmental and public constraints, construction time for the project, and equipment available. Table 25.3 lists the range of the application of segmental bridges by the span lengths (Podolny and Muller 1982).

25.5.3.5 Design Consideration

Compared to reinforced concrete, the main design features of prestressed concrete are that stresses for concrete and prestressing steel and deformation of structures at each stage, that is, during construction, stressing, handling, transportation, and erection as well as during the service life, and stress concentrations need to be investigated in accordance with the design specifications and codes.

25.5.4 Corrugated Steel Web Bridges

The corrugated steel web bridge is used in prestressed concrete to reduce the weight and increase the span length. The corrugated web also has the advantages that it does not take the axial force by accordion effect, thus introducing prestressing force in the concrete more effectively. An example of the bridge is shown in Figure 25.39.

(a)

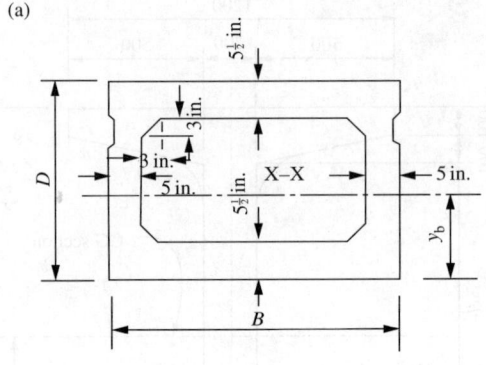

(b)

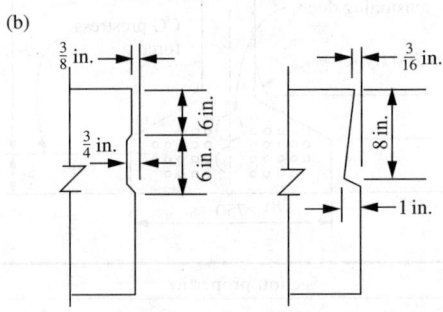

	Section dimensions		Section properties				
Span, ft (m)	Width B, in. (mm)	Depth D, in (mm)	A, in.^{2} (mm^2 10^6)	Y_b, in. (mm)	I_x, in.^{4} (mm^4 10^9)	S_b, in.^{3} (mm^3 10^6)	S_t, in.^{3} (mm^3 10^6)
50	48	27	693	13.37	65,941	4,932	4,838
(15.2)	(1,219)	(686)	(0.4471)	(339.6)	(27.447)	(80.821)	(79.281)
60	48	33	753	16.33	110,499	6,767	6,629
(18.3)	(1,219)	(838)	(0.4858)	(414.8)	(45.993)	(110.891)	(108.630)
70	48	39	813	19.29	168,367	8,728	8,524
(21.4)	(1,219)	(991)	(0.5245)	(490.0)	(70.080)	(143.026)	(139.683)
80	48	42	843	20.78	203,088	9,773	9,571
(24.4)	(1,219)	(1,067)	(0.5439)	(527.8)	(84.532)	(160.151)	(156.841)

FIGURE 25.35   FHWA precast pretensioned box sections: (a) typical section and (b) alternative shear key (FHWA 1990).

25.6 Concrete Substructures

25.6.1 Introduction

Bridge substructures transfer traffic loads from the superstructure to the footings and foundations. Vertical intermediate support — piers or bents and end supports — abutments, are included.

25.6.2 Bents and Piers

25.6.2.1 Pile Bents

Pile extension as shown in Figure 25.40a is used for slab and T-beam bridges. It is usually used to cross streams when debris is not a problem.

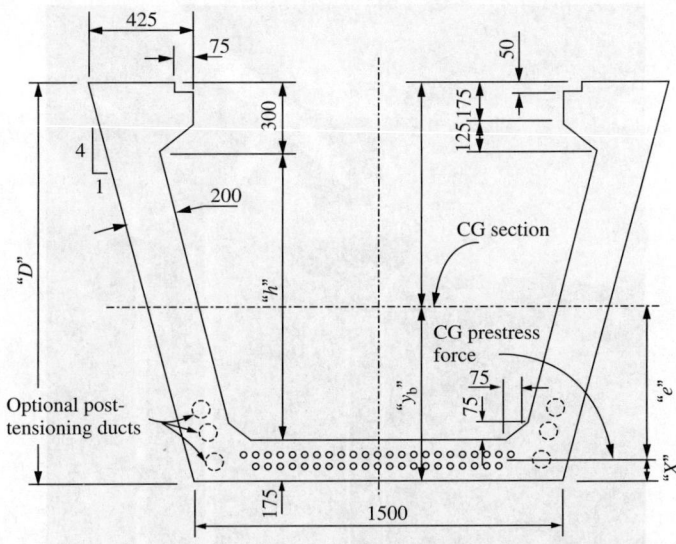

FIGURE 25.36 Caltrans precast standard "bathtub" girder (Caltrans 2001).

Section properties										
"D" (mm)	"h" (mm)	Area (mm^2)	I (mm^4 10^6)	y_b (mm)	y_t (mm)	S_b (mm^3 10^6)	S_t (mm^3 10^6)	r (mm)	Force (N/m)	Mass (kg/m)
1,400	925	864,000	191,500	612	788	313.0	243.1	471	20,370	2,076
1,550	1,075	925,900	251,500	683	867	368.2	290.0	521	21,820	2,225
1,700	1,255	987,700	321,800	755	945	426.2	340.5	571	23,280	2,374
1,850	1,375	1,049,600	403,200	827	1,023	487.5	394.1	620	24,740	2,522
2,000	1,525	1,111,400	496,400	899	1,101	552.2	450.9	668	26,200	2,671
2,150	1,675	1,173,300	602,100	972	1,178	619.4	511.1	716	27,660	2,819

25.6.2.2 Solid Piers

Figure 25.40b shows a typical solid pier used mostly when stream debris or fast currents are present. These are used for long spans and can be supported by spread footings or pile foundations.

25.6.2.3 Column Bents

Column bents (Figure 25.40c) are generally used on the dry land structures and supported by spread footings or pile foundations. Multicolumn bents are desirable for bridges in seismic zones. The single-column bent, such as a T-bent (Figure 25.40d), modified T-bent, C-bent (Figure 25.40e), or outrigger bent (Figure 25.40f) may be used when the location of the columns is restricted and changes of the alignment are impossible. To achieve a pleasing appearance at minimum cost using standard column shapes, Caltrans (1990) developed "Standard Architectural Columns" (Figure 25.41). Prismatic sections of Column Types 1 and 1W, with one-way flares of Column Types 2 and 2W, and with two-way flares of Column Types 3 and 3W may be used for various highway bridges.

25.6.3 Abutments

Abutments are the end supports of a bridge. Figure 25.42 shows the typical abutments used for highway bridges. The seven types of abutments can be divided into two categories: open and closed ends. Selection of an abutment type depends on the requirements for structural support, movement, drainage, road approach, and earthquakes.

FIGURE 25.37   Pine Valley Bridge — construction stage, California.

FIGURE 25.38   Pine Valley Bridge — construction completed, California.

TABLE 25.3   Range of Application of Segmental Bridge Type by Span Length (Podolny and Muller 1982)

Span, ft (m)	Bridge types
0–150 (0–45.7)	I-type pretensioned girder
100–300 (30.5–91.4)	Cast-in-place posttensioned box girder
100–300 (30.5–91.4)	Precast-balanced cantilever segmental, constant depth
200–600 (61.0–182.9)	Precast-balanced cantilever segmental, variable depth
200–1000 (61.0–304.8)	Cast-in-place cantilever segmental
800–1500 (243.8–457.2)	Cable-stay with balanced cantilever segmental

FIGURE 25.39   Prestressed concrete bridge with corrugated steel web (Hondani Birdge, Japan, from P.S. Mitsubishi Construction Co. Ltd., with permission).

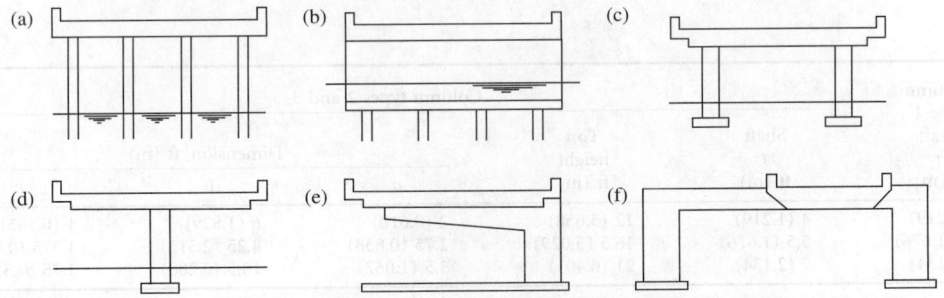

FIGURE 25.40   Bridge substructures — piers and bents: (a) pile bents; (b) solid pier; (c) column bent; (d) "T" bent; (e) "C" bent; and (f) outrigger bent (Caltrans 1990).

25.6.3.1 Open End Abutments

Open end abutments include diaphragm abutments and short seat abutments. These are the most frequently used abutments and are usually the most economical, adaptable, and attractive. The basic structural difference between the two types is that seat abutments permit the superstructure to move

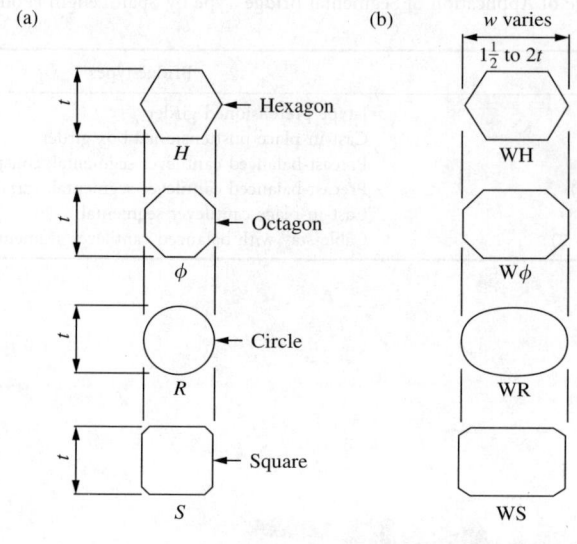

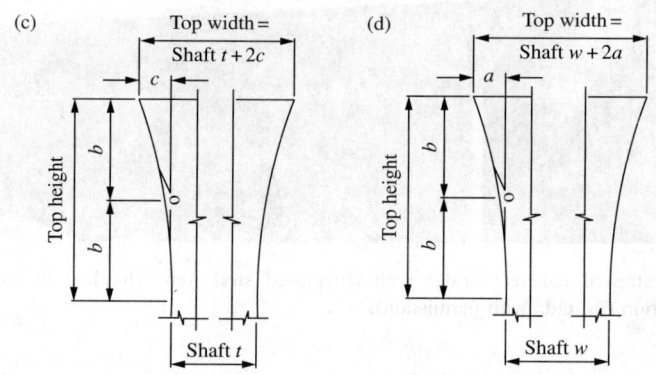

Column type 1	Column types 2 and 3				
Shaft t ft (m)	Shaft t ft (m)	Top height ft (m)	Dimension, ft (m)		
			a	b	c
4 (1.219)	4 (1.219)	12 (3.658)	2 (0.610)	6 (1.829)	1 (0.305)
5.5 (1.676)	5.5 (1.676)	16.5 (5.029)	2.75 (0.838)	8.25 (2.515)	1.375 (0.419)
7 (2.134)	7 (2.134)	21 (6.401)	3.5 (1.067)	10.5 (3.200)	1.75 (0.533)

Notes:
1. Square shape S section is only for Column type 1.
2. Flare curve is parabolic and dimension C is only for Column type 3.

FIGURE 25.41 Caltrans standard architectural columns: (a) Column types 1, 2, 3; (b) Column types 1W, 2W, 3W; (c) side view; and (d) front view (Caltrans 1990).

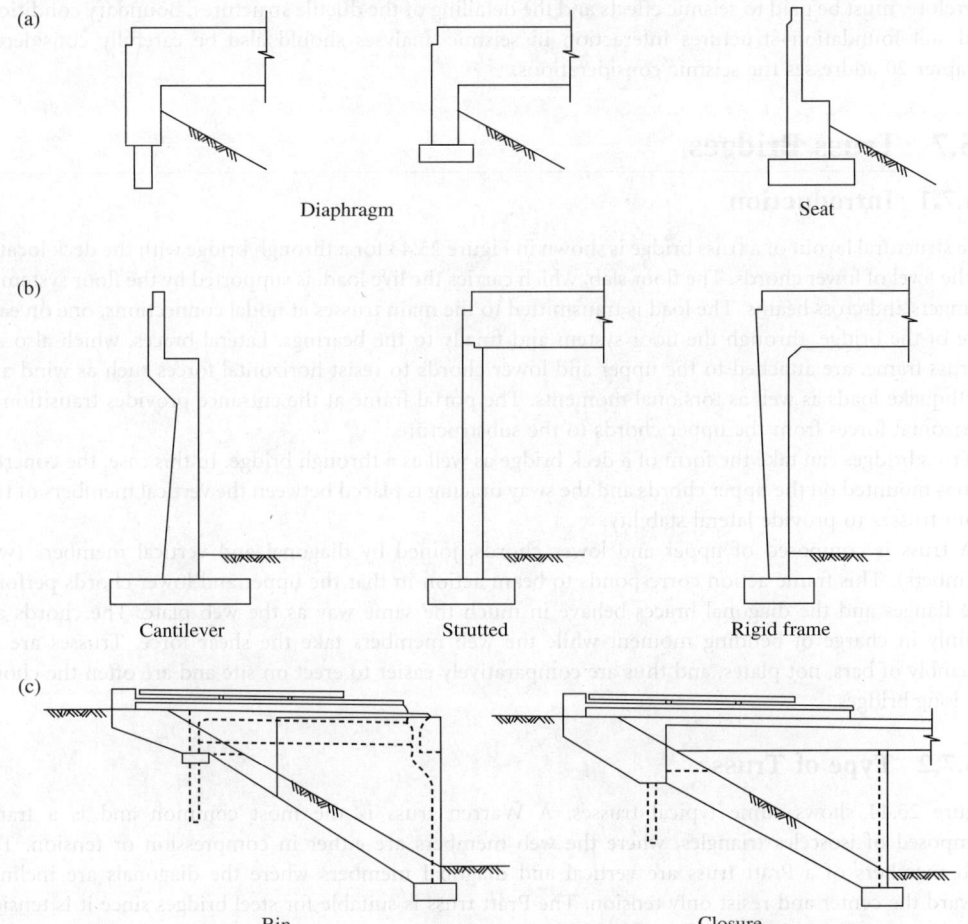

FIGURE 25.42 Typical types of abutments: (a) open end; (b) closed end — backfilled; and (c) closed end — cellular (Caltrans 1990).

independently from the abutment while the diaphragm abutment does not. Since they have lower height of abutment walls, there is less settlement in the road approaches than that experienced by higher backfilled closed abutments. These also provide more economical widening than closed abutments.

25.6.3.2 Closed End Abutments

Closed end abutments include cantilever, strutted, rigid frame, bin, and closure abutments. These are less commonly used, but are used for bridge widenings of the same kind, unusual sites, or in tightly constrained urban locations. Rigid frame abutments are generally used with tunnel-type single-span connectors and overhead structures that permit passage through a roadway embankment. Because the structural supports are adjacent to traffic these have a high initial cost and present a closed appearance to approaching traffic.

25.6.4 Design Consideration

During the recent 1989 Loma Prieta and the 1994 Northridge Earthquakes in the United States and the 1995 Kobe Earthquake in Japan, major damages have been found in substructures. Special attention, therefore, must be paid to seismic effects and the detailing of the ductile structures. Boundary conditions and soil–foundation–structures interaction in seismic analyses should also be carefully considered. Chapter 20 addresses the seismic considerations.

25.7 Truss Bridges

25.7.1 Introduction

The structural layout of a truss bridge is shown in Figure 25.43 for a through bridge with the deck located at the level of lower chords. The floor slab, which carries the live load, is supported by the floor system of stringers and cross beams. The load is transmitted to the main trusses at nodal connections, one on each side of the bridge, through the floor system and finally to the bearings. Lateral braces, which also are a truss frame, are attached to the upper and lower chords to resist horizontal forces such as wind and earthquake loads as well as torsional moments. The portal frame at the entrance provides transition of horizontal forces from the upper chords to the substructure.

Truss bridges can take the form of a deck bridge as well as a through bridge. In this case, the concrete slab is mounted on the upper chords and the sway bracing is placed between the vertical members of two main trusses to provide lateral stability.

A truss is composed of upper and lower chords, joined by diagonal and vertical members (web members). This frame action corresponds to beam action in that the upper and lower chords perform like flanges and the diagonal braces behave in much the same way as the web plate. The chords are mainly in charge of bending moment while the web members take the shear force. Trusses are an assembly of bars, not plates, and thus are comparatively easier to erect on site and are often the choice for long bridges.

25.7.2 Type of Truss

Figure 25.44 shows some typical trusses. A Warren truss is the most common and is a frame composed of isosceles triangles, where the web members are either in compression or tension. The web members of a Pratt truss are vertical and diagonal members where the diagonals are inclined toward the center and resist only tension. The Pratt truss is suitable for steel bridges since it is tension that is most effectively resisted. It should be noted, however, that vertical members of Pratt truss are in compression. A Howe truss is similar to the Pratt except that the diagonals are inclined toward the ends, leading to axial compression forces, and the vertical members resist tension. Wooden bridges

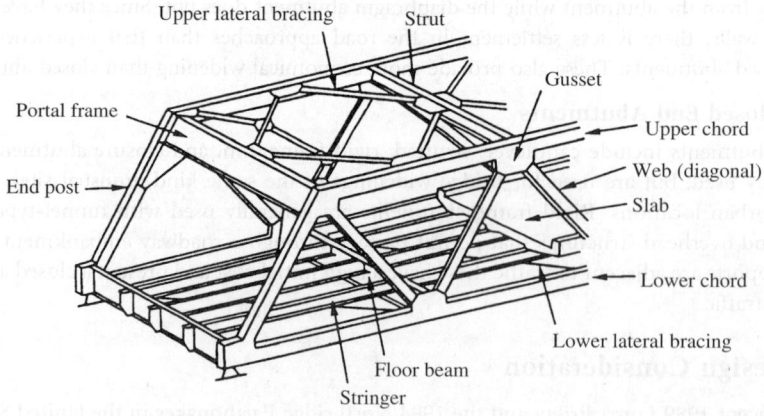

FIGURE 25.43   Truss bridge (Nagai 1994).

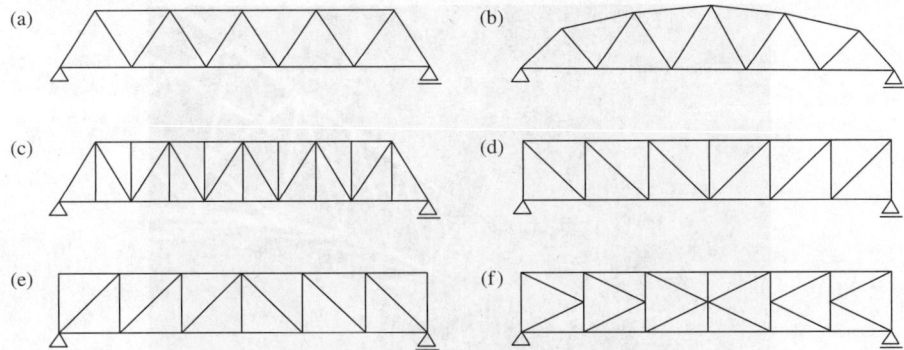

FIGURE 25.44 Types of trusses: (a) Warren truss (straight upper chord); (b) Warren truss (curved upper chord); (c) Warren truss with vertical members; (d) Prutt truss; (e) Howe truss; and (f) K-truss.

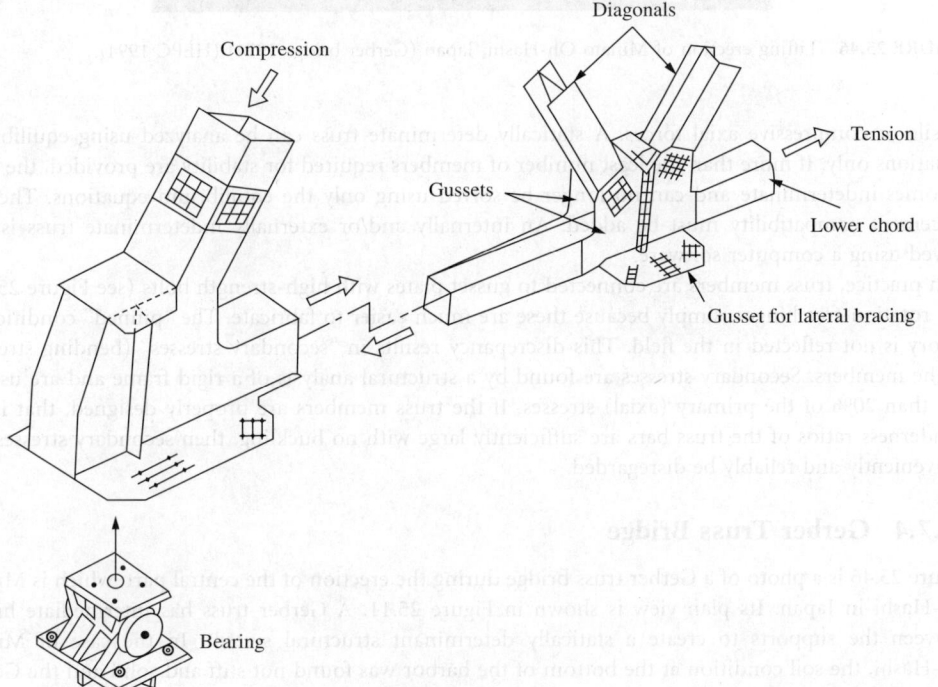

FIGURE 25.45 Nodal joints of truss bridge (JASBC 1985).

often make use of the Howe truss since the connections of the diagonals in wood prefer compression. A K-truss, so named since the web members form a "K," is most economical in large bridges because the short member lengths reduce the risk of buckling.

25.7.3 Structural Analysis and Secondary Stress

The truss is a framed structure of bars, theoretically assumed to be connected by hinges, forming stable triangles. Trusses contain triangle framed units to keep it stable. Its members are assumed to resist only

FIGURE 25.46   Lifting erection of Minato Oh-Hashi, Japan (Gerber bridge 1974) (HEPC 1994).

tensile or compressive axial forces. A statically determinate truss can be analyzed using equilibrium equations only. If more than the least number of members required for stability are provided, the truss becomes indeterminate and can no longer be solved using only the equilibrium equations. The displacement compatibility must be added. An internally and/or externally indeterminate truss is best solved using a computer software.

In practice, truss members are connected to gusset plates with high-strength bolts (see Figure 25.45), not rotation-free hinges, simply because these are much easier to fabricate. The "pinned" condition of theory is not reflected in the field. This discrepancy results in "secondary stresses" (bending stresses) in the members. Secondary stresses are found by a structural analysis of a rigid frame and are usually less than 20% of the primary (axial) stresses. If the truss members are properly designed, that is the slenderness ratios of the truss bars are sufficiently large with no buckling, then secondary stresses can conveniently and reliably be disregarded.

25.7.4 Gerber Truss Bridge

Figure 25.46 is a photo of a Gerber truss bridge during the erection of the central part, which is Minato Oh-Hashi in Japan. Its plan view is shown in Figure 25.11. A Gerber truss has intermediate hinges between the supports to create a statically determinant structural system. In the case of Minato Oh-Hashi, the soil condition at the bottom of the harbor was found not stiff and solid and the Gerber truss proved the wisest choice.

25.8 Rigid Frame Bridges (Rahmen Bridges)

25.8.1 Introduction

The members are rigidly connected in "Rahmen" structures or "rigid frames." Unlike the truss and the arch bridges that will be discussed in the following subsection, all the members are subjected to both axial force and bending moments. Figure 25.47 shows various types of Rahmen bridges.

The members of a rigid frame bridge are much larger than those in a typical building. Consequently, stress concentrations occur at the junctions of beams and columns that must be carefully designed using

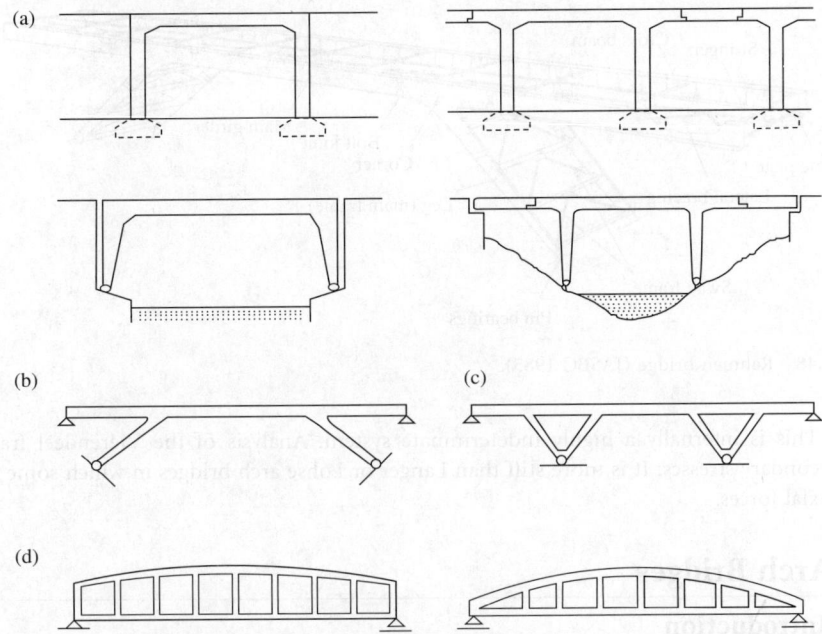

FIGURE 25.47 Type of Rahmen bridges: (a) portal frame; (b) π-Rahmen; (c) V-leg Rahmen; and (d) Vierendeel Rahmen.

finite element analyses or experimental verification. The supports of Rahmen bridges are either hinged or fixed, making it an externally indeterminate structure, and it is therefore not suitable when the foundation is likely to sink. The reactions at supports are horizontal and vertical forces at hinges, and with the addition of a bending moment at a fixed base.

25.8.2 Portal Frame

A portal frame is the simplest design (Figure 25.47a) and is widely used for the piers of elevated highway bridges because the space underneath can be effectively used for other roads or parking lots. These piers were proved in the 1995 Kobe Earthquake in Japan to be more resilient, that is, to retain more strength and absorb more energy than single-column piers.

25.8.3 π-Rahmen (Strutted Beam Bridge)

The π-Rahmen design is usually used for bridges in mountainous regions where the foundation is firm and it can pass over a deep valley with a relatively long span or for bridges crossing over expressways (Figure 25.47b). As shown in the structural layout of π-Rahmen in Figure 25.48, the two legs support the main girders, inducing axial compression in the center span of the girder. Live load on the deck is transmitted to the main girders through the floor system. Intermediate hinges may be inserted in the girders to make Gerber girders. A V-leg Rahmen bridge is similar to a π-Rahmen bridge but can span longer distances with no axial force in center span of the girder (Figure 25.47c).

25.8.4 Vierendeel Bridge

The Vierendeel bridge is a rigid frame whose upper and lower chords are connected rigidly to the vertical members (Figure 25.47d). All the members are subjected to axial and shear forces as well as bending

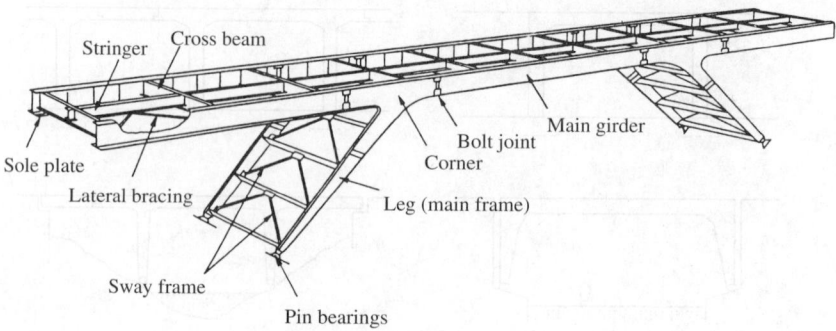

FIGURE 25.48 Rahmen bridge (JASBC 1985).

moments. This is internally a highly indeterminate system. Analysis of the Vierendeel frame must consider secondary stresses. It is more stiff than Langer or Lohse arch bridges in which some members take only axial forces.

25.9 Arch Bridges

25.9.1 Introduction

An arch rib acts like a circular beam restrained not only vertically but also horizontally at both ends and thus results in vertical and horizontal reactions at the supports. The horizontal reaction causes axial compression in addition to bending moments in the arch rib. The bending moments caused by the horizontal force balances those due to gravity loads. Comparing with the axial force, the effect of the bending moment is usually small. That is why the arch is often made of materials that have high compressive strength such as concrete, stone, or brick.

25.9.2 Type of Arch

An arch bridge includes the road deck and the supporting arch. Various types of arches are shown in Figure 25.49. In the figure, the thick line represents the members carrying bending moment, shear, and axial forces. The thin line represents members taking axial forces only. Arch bridges are classified into the deck and the through-deck types according to the location of the road surface as shown in Figure 25.49. Since the deck in both type of bridges is sustained by either vertical columns or hangers to the arch, structurally the same axial force action, either compression or tension, is in effect in the members. The difference is that the vertical members of deck bridges take compressive forces and the hangers of through-deck bridges take tension. The live load acts on the arch only indirectly.

A basic structural type for an arch is a two-hinged arch (see Figure 25.49a). The two-hinge arch has one degree of indeterminacy externally because there are four end reactions. If one hinge is added at the crown of the arch, creating a three-hinge arch, it is rendered determinate. If the ends are clamped, turning it into a fixed arch, it becomes indeterminate to the third degree. The tied arch is subtended by two hinges by a tie and simply supported (Figure 25.49b). The tied arch is externally determinate but internally has one degree of indeterminacy. The floor structures hang from the arch and are isolated from the tie. Other type of arch bridges will be discussed later in more detail.

25.9.3 Structural Analysis

Almost all bridge design analyses, in this age of super computing power, use finite element methods. The analysis of an arch is basically the same as that for a frame. The web members are analyzed as truss bars

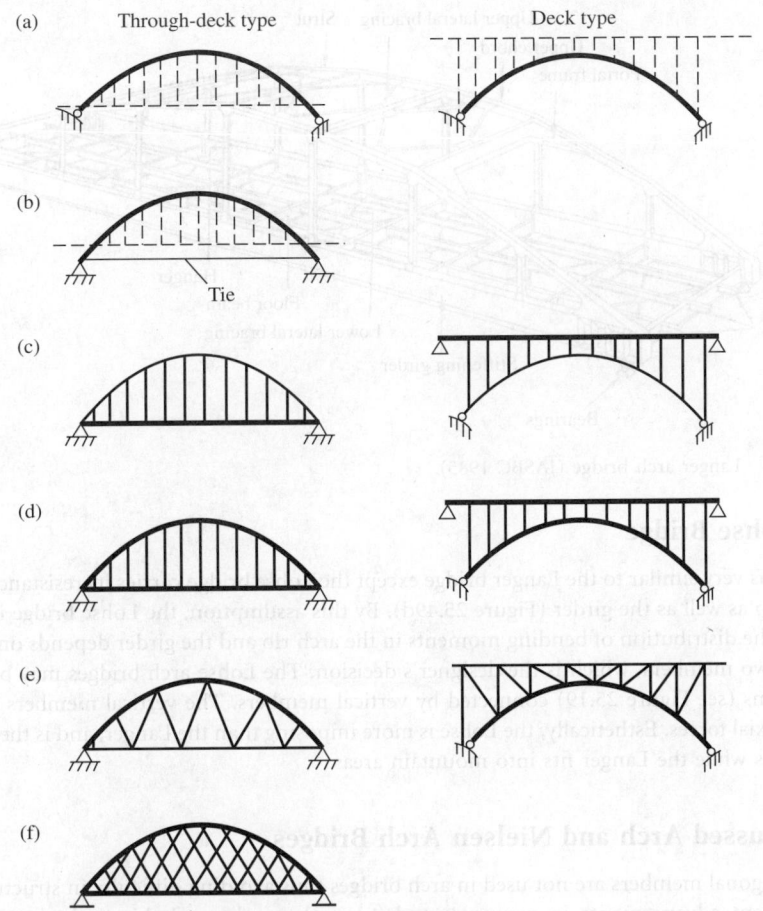

FIGURE 25.49   Type of arch bridges: (a) two-hinge arch; (b) tied arch; (c) Langer arch; (d) Lohse arch; (e) trussed arch; and (f) Nielsen arch (Shimada 1991).

that take only axial forces. The arch rib and the girders are analyzed as either trusses or beam-columns depending on the type of arch considered. Beam-columns take axial and shear forces and bending moments. An arch rib is usually made up of straight piecewise components, not curved segments, and it is so analyzed.

25.9.4 Langer Bridge

The Langer arch is analyzed by assuming that the arch rib takes only axial compression (Figure 25.49c). The arch rib is thin, but the girders are deep and resist moment and shear as well as axial tension. The girders of the Langer bridge are regarded as being strengthened by the arch rib. Figure 25.50 shows structural components of a Langer bridge.

If diagonals are used in the web, it is called a trussed Langer. The difference of the trussed Langer against a standard truss is that the lower chord is a girder instead of just a bar. The Langer bridge is also determinate externally and indeterminate internally. The deck-type bridge of the Langer is often called a "reversed" Langer.

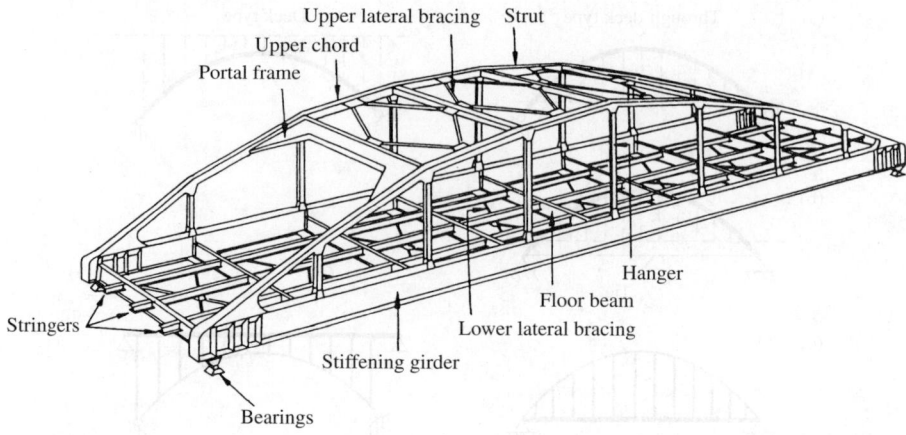

FIGURE 25.50   Langer arch bridge (JASBC 1985).

25.9.5 Lohse Bridge

Lohse bridge is very similar to the Langer bridge except the Lohse bridge carries its resistance to bending in the arch rib as well as the girder (Figure 25.49d). By this assumption, the Lohse bridge is stiffer than the Langer. The distribution of bending moments in the arch rib and the girder depends on the stiffness ratio of the two members, which is the designer's decision. The Lohse arch bridges may be thought of as tiered beams (see Figure 25.19) connected by vertical members. The vertical members are assumed to take only axial forces. Esthetically, the Lohse is more imposing than the Langer, and is therefore suited to urban areas while the Langer fits into mountain areas.

25.9.6 Trussed Arch and Nielsen Arch Bridges

Generally, diagonal members are not used in arch bridges thus avoiding difficulty in structural analysis. However, recent advancements in computer technology have changed this outlook. New types of arch bridges, such as the trussed arch in which diagonal truss bars are used instead of vertical members or the Nielsen Lohse design in which tension rods are used for diagonals, have now been introduced (see Figure 25.49e and f). Diagonal web members increase the stiffness of a bridge more than the vertical members.

   All the members of the truss bridge take only axial forces. On the other hand, the trussed arch bridge, however, may resist bending in the arch rib, the girder, or both. Since the diagonals of the Nielsen Lohse bridge carry only axial tension, they are prestressed by the dead load to compensate for the compression force due to the live load.

25.10 Floor System

25.10.1 Introduction

The floor system of a bridge usually consists of a deck carried by girders. The deck directly supports the live load. Floor beams as well as stringers shown in Figure 25.51 form a grillage and transmit the load from the deck to the main girders. The floor beams and stringers are used for framed bridges, that is, truss, Rahmen, and arch bridges (see Figure 25.43, Figure 25.48, and Figure 25.50), in which the spacing of the main girders or trusses is large. In an upper deck type of plate girder bridge the deck is directly supported by the main girders and often there is no floor system because the main girders run in parallel and close together.

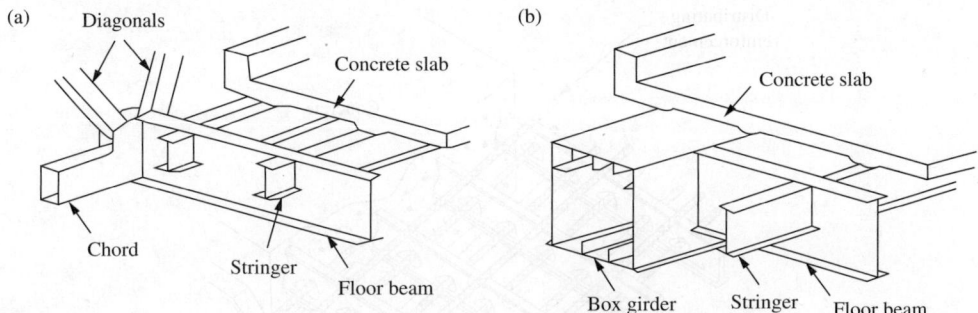

FIGURE 25.51   Floor system: (a) truss bridge and (b) box girder bridge (Nagai 1994).

The floor system is classified as suitable for either highway or railroad bridges. The deck of a highway bridge is designed for the wheel loads of trucks using plate bending theory in two dimensions. Often in design practice, however, this plate theory is reduced to equivalent one-dimensional beam theory. The materials used are also classified into concrete, steel, or wood.

The recent influx of traffic flow has severely fatigued existing floor systems. Cracks in concrete decks and connections of floor system are often found in old bridges that have been in service for many years.

25.10.2   Decks

25.10.2.1   Concrete Deck

A reinforced concrete deck slab is most commonly used in highway bridges. It is the deck that is most susceptible to damage caused by the flow of traffic, which continues to increase over the years. Urban highways are exposed to heavy traffic and forced to be repaired frequently.

Recently, a composite deck slab was developed to increase the strength, ductility, and durability of decks without increasing their weight or affecting the cost and duration of construction. In a composite slab, the bottom steel plate serves both as a part of the slab and the form work for pouring the concrete. There are many ways of combining the steel plate and the reinforcement. A typical example is shown in Figure 25.52. This slab is prefabricated in the yard and then the concrete is poured on site after girders have been placed. A precast prestressed deck may reduce the time required to complete construction.

25.10.2.2   Orthotropic Deck

For long spans, the othotropic deck is used to minimize the weight of the deck. The orthotropic deck is a steel deck plate stiffened with longitudinal and transverse ribs as shown in Figure 25.53. The steel deck also works as the upper flange of the supporting girders. The pavement or wearing surfacing on the steel deck should be carefully finished to prevent water from penetrating through the pavement and causing the steel deck to rust.

25.10.3   Pavement

The pavement on the deck provides both a smooth driving surface and prevents rain water from seeping into the reinforcing bars and steel deck below. A layer of waterproofing may be inserted between the pavement and the deck. Asphalt is most commonly used to pave highway bridges. Its thickness is usually 5 to 10 cm on highways and 2 to 3 cm on pedestrian bridges.

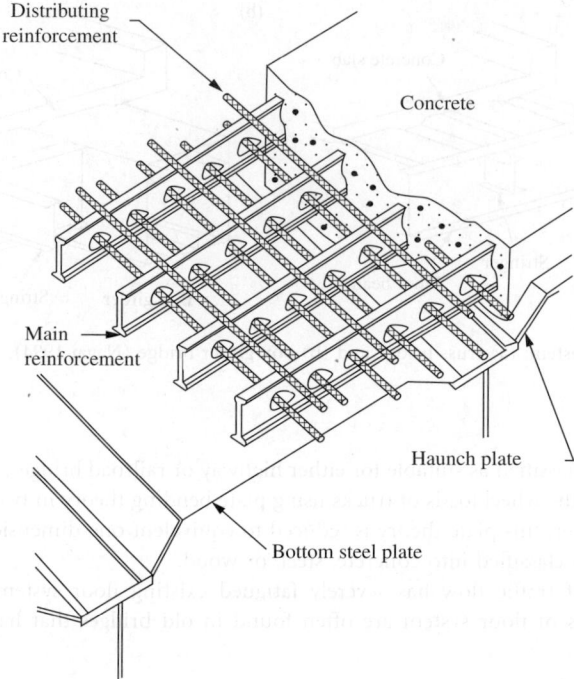

FIGURE 25.52 Composite deck (JASBC 1988).

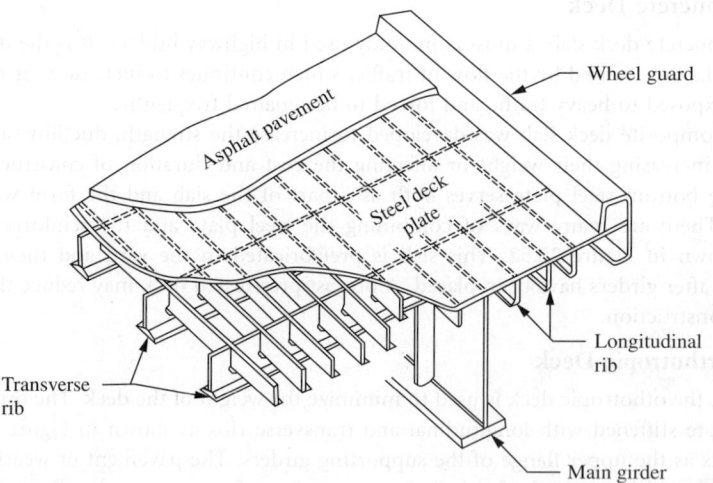

FIGURE 25.53 Orthotropic deck (JASBC 1985).

25.10.4 Stringers

The stringers support the deck directly and transmit the loads to floor beams as can be seen in Figure 25.51. They are placed in the longitudinal direction just like the main girders are in a plate girder bridge and thus provide much the same kind of support.

The stringers must be sufficiently stiff in bending to prevent cracks forming in the deck or on the pavement surface. The design codes usually limit the vertical displacement caused by the weight of a truck.

25.10.5 Floor Beams

The floor beams are placed in the transverse direction and connected by high-strength bolts to the truss frame or arch as shown in Figure 25.51. The floor beams support the stringers and transmit the loads to main girders, trusses, or arches. In other words, the main truss or arch receives the loads indirectly via the floor beams. The floor beams also provide transverse stiffness to bridges and thus improve the overall torsional resistance.

25.11 Bearings, Expansion Joints, and Railings

25.12.1 Introduction

Aside from the main components such as the girders or the floor structure, some other parts such as bearings (shoes), expansion joints, guard railings, drainage paths, lighting, and sound-proofing walls also make up the structure of a bridge. Each plays a minor part but provides an essential function. Drains flush rain water off and wash away dust. Guard railings and lights add to the esthetic quality of the design as well as providing their obvious original functions. A sound-proofing wall may take away from the beauty of the structure but might be required by law in urban areas to isolate the sound of traffic from the surrounding residents. In the following section, bearings, expansion joints, and guard railings are discussed.

25.11.2 Bearings

Bearings (shoes) support the superstructure (the main girders, trusses, or arches) and transmit the loads to the substructure (abutments or piers). The bearings connect the upper and lower structures and carry the whole weight of the superstructure. The bearings are designed to resist these reaction forces by providing support conditions that are fixed or hinged. The hinged bearings may be movable or immovable; that is, horizontal movement is restrained or unrestrained (horizontal reaction is produced or not). The amount of the horizontal movement is determined by calculating the elongation due to a temperature change.

During the 1995 Kobe earthquake in Japan, many bearings were found to have sustained extensive damage due to stress concentrations, which are the weak spots along the bridge. The bearings may play the role of a fuse to keep damage from occurring at vital sections of the bridge, but the risk of the superstructure falling down goes up. The girder-to-girder or girder-to-abutment connections prevent the girders from collapsing during strong earthquakes.

Many types of bearings are available and some are shown in Figure 25.54 and briefly explained in the following:

1. *Line bearings.* The contacting line between the upper plate and the bottom round surface provides rotational capability as well as sliding. These are used in small bridges.
2. *Plate bearings.* The bearing plate has plane surface on the top side which allows sliding and a spherical surface on the bottom allowing rotation. The plate is placed between the upper and lower shoes.
3. *Hinged bearings (pin bearings).* A pin is inserted between the upper and lower shoes allowing rotation but no translation in the longitudinal direction.
4. *Roller bearings.* Lateral translation is unrestrained by using single or multiple rollers for hinged bearings or spherical bearings.

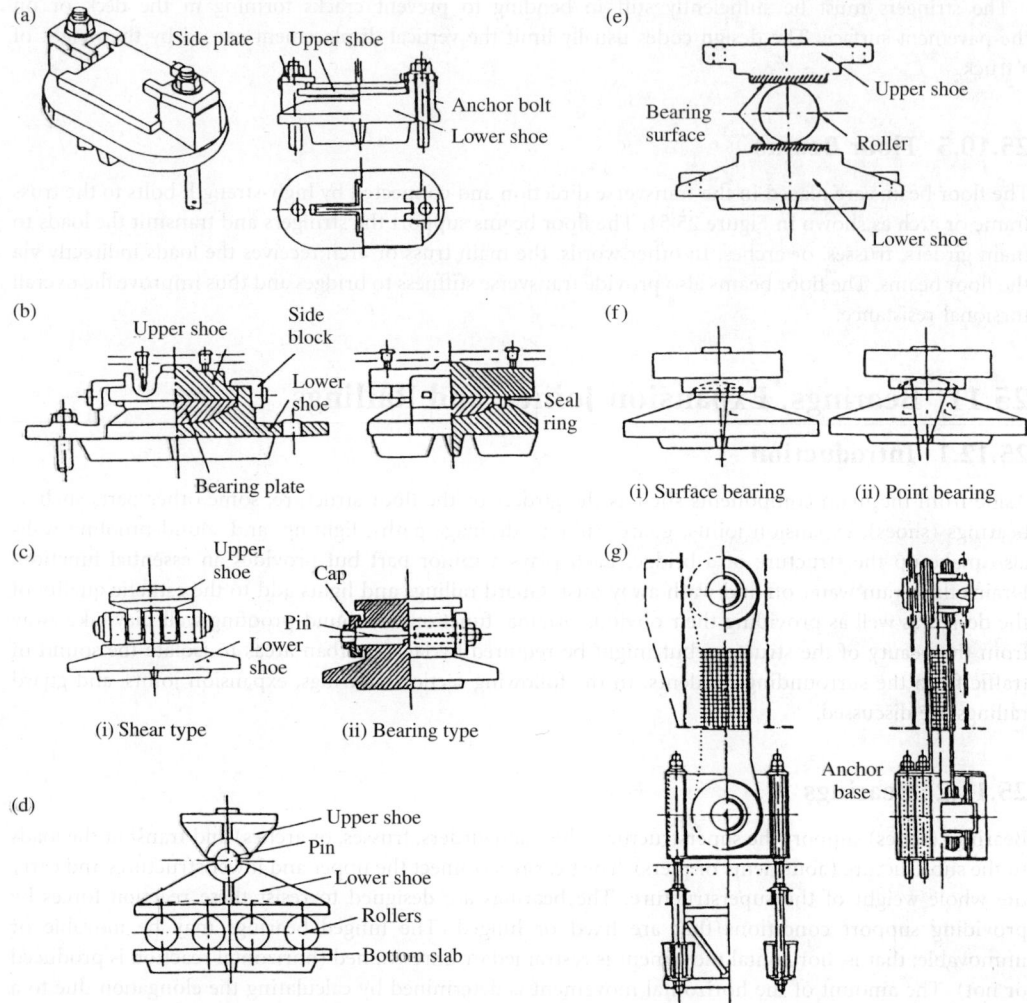

FIGURE 25.54 Types of bearings: (a) line bearing; (b) plate bearing; (c) hinged bearing; (d) multiple roller bearing; (e) single roller bearing; (f) spherical bearing; and (g) pendel bearing (JASBC 1984).

5. *Spherical bearings (pivot bearings).* Convex and concave spherical surfaces allow rotation in all directions and no lateral movement. The two types are: a point contact for large differences in the radii of each sphere and a surface contact for small differences in their radii.

6. *Pendel bearings.* An eye bar connects the superstructure and the substructure by a pin at each end. Longitudinal movement is permitted by inclining the eye bar; therefore, the distance of the pins at ends should be properly determined. These are used to provide a negative reaction in cable-stayed bridges. There is no resistance in transverse direction.

7. *Wind bearings.* This type of bearings provides transverse resistance for wind and is often used with pendel bearings.

8. *Elastomeric bearings.* The flexibility of elastomeric or lead rubber bearings is that they allow both rotation and horizontal movement. Figure 25.55 explains a principle of rubber-layered bearings by

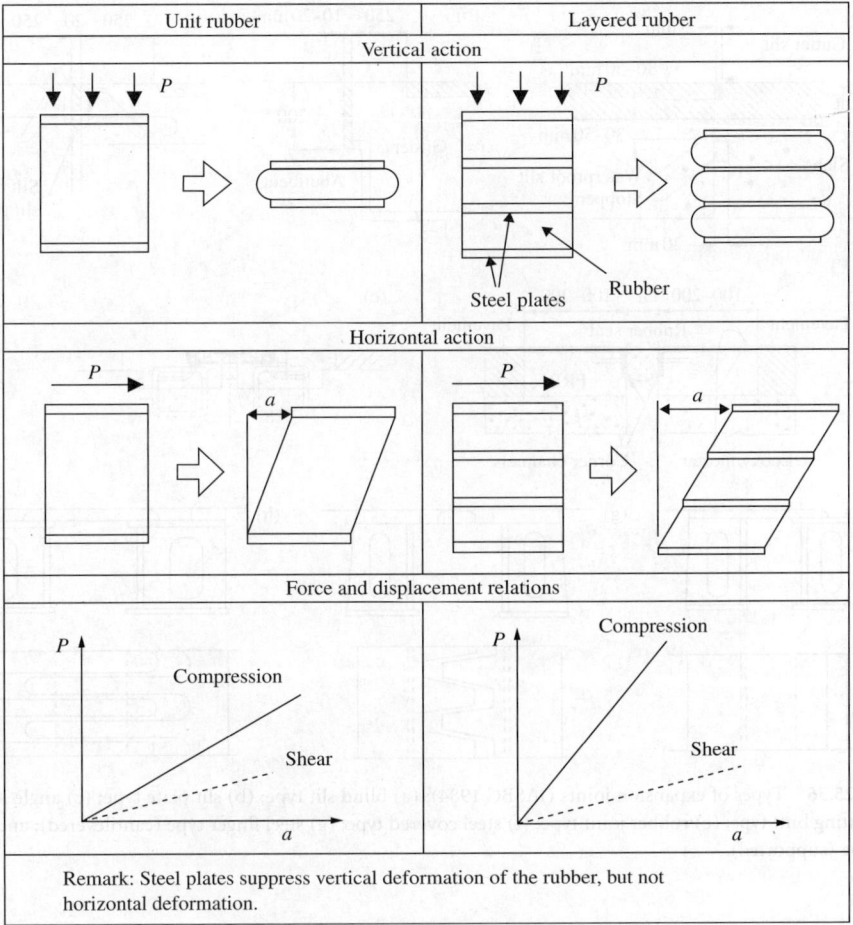

FIGURE 25.55 Properties of elastomeric bearings (Bridgestone Co.).

comparing with a unit rubber. A layered rubber is stiff, unlike a unit rubber, for vertical compression since the steel plates placed between the rubber restrain the vertical deformation of the rubber, but flexible for horizontal shear force like a unit rubber. The flexibility absorbs horizontal seismic energy and are ideally suited to resist earthquake actions. After the disaster of the 1995 Kobe Earthquake in Japan the elastomeric rubber bearings are becoming more and more popular, but whether they effectively sustain severe vertical actions without damage is not certified.

9. *Seismic isolation bearings.* Many different types of seismic isolation bearings are available, such as elastomeric isolators and sliding isolators. When installed on the bridge piers and abutments, the isolation bearings serve both vertical bearing devices for gravity loads and lateral isolation devices for seismic load. The basic purpose of isolation devices is to change the fundamental mode of vibration so that the structure is subjected to lower earthquake force. However, reduction in force may be accompanied by an increase in displacement demand that shall be accommodated within the isolation system and any adjacent structures.

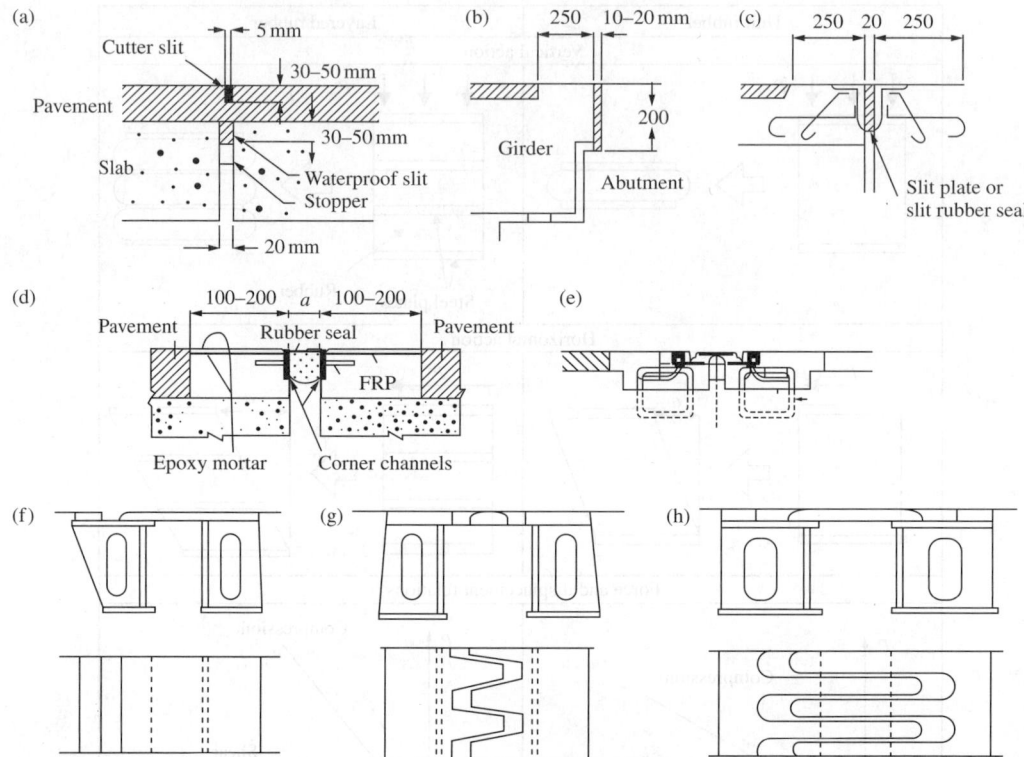

FIGURE 25.56 Types of expansion joints (JASBC 1984): (a) blind slit type; (b) slit plate type; (c) angle joint type; (d) postfitting butt type; (e) rubber joint type; (f) steel covered type; (g) steel finger type (cantilevered); and (h) steel finger type (supported).

A selection from these types of bearings is made according to the size of bridge and the magnitude of predicted downward or upward reaction forces.

25.11.3 Expansion Joints

Expansion joints are provided to allow a bridge to adjust its length under changes in temperature or deformation by external loads. They are designed according to expanding length and material as classified in Figure 25.56. Steel expansion joints are most commonly used. A defect is often found at the boundary between the steel and the concrete slab where the disturbing jolt is given to drivers as they pass over the junction. To solve this problem, rubber joints are used on the road surface to provide a smooth transition for modern bridge construction (see Figure 25.56e) or continuous girders are more commonly adopted than simple girders.

25.11.4 Railings

Guard railings are provided to ensure vehicles and pedestrians do not fall off from the bridge. They may be a hand rail for pedestrians, a heavier guard for vehicles, or a common railing for both. These are made from materials such as concrete, steel, or aluminum. The guard railings are located prominently and are thus open to the critical eye of the public. It is important that they do not only keep traffic within boundaries but also add to the esthetic appeal of the whole bridge (Figure 25.57).

FIGURE 25.57 Pedestrian railing (JASBC 1985): (a) railing images waves and (b) Kanazawa Kakkei Oh-Hashi, Kanagawa, Japan (1982).

Acknowledgments

Many of the figures in this section are copied from other textbooks and journals. The authors would like to express sincere gratitude to these original authors. Special thanks go to Professor N. Nagai of Nagaoka Institute of Science and Technology, Professors Y. Tachibana and H. Nakai of Osaka City University, Japan Association of Steel Bridge Construction, American Association of State Highway and Transportation Officials, and California Department of Transportation for their generosity.

Glossary

Abutment — An end support for a bridge structure.

Arch bridge — A bridge that includes the road deck and the supporting arch.

Bridge — A structure that crosses over a river, bay, or other obstruction, permitting the smooth and safe passage of vehicles, trains, and pedestrians.

Cable-stayed bridge — A bridge in which the superstructure is hanging from the diagonal cables that are tensioned from the tower.

Cast-in-place concrete — Concrete placed in its final position in the structure while still in a plastic state.

Composite girder — A steel girder connected to a concrete deck so that they respond to force effects as a unit.

Deck (slab) — A component, with or without wearing surface, directly supporting wheel loads.

Floor system — A superstructure, in which the deck is integral with its supporting components, such as floor beams and stringers.

Girder — A structural component whose primary function is to resist loads in flexure and shear. Generally, this term is used for fabricated section.

Girder bridge — A bridge superstructure that consist of a floor slab, girders, and bearings.

Influence line — A continuous or discretized function over a section of girder whose value at a point, multiplied by a load acting normal to the girder at that point, yields the force effect being sought.

Lever rule — The static summation of moments about one point to calculate the reaction at the second point.

LRFD (load and resistance factor design) — A method of proportioning structural components (members, connectors, connecting elements, and assemblages) such that no applicable limit state is exceeded when the structure is subjected to all appropriate load combinations.

Precast member — Concrete element cast in a location other than its final position.

Prestressed concrete — Concrete components in which the stresses and deformations are introduced by application of prestressing forces.

Rigid frame bridge — A bridge in which the superstructure and substructure members are rigidly connected.

Segmental bridge — A bridge in which primary load-supporting members are composed of individual members called segments post-tensioned together to act as a monolithic unit under loads.

Substructure — Structural parts of the bridge that provide the horizontal span.

Superstructure — Structural parts of the bridge that support the horizontal span.

Suspension bridge — A bridge in which the superstructure is suspended by two main cables and anchored to end blocks.

Truss bridge — A bridge superstructure that consist of a floor system and main trusses.

References

AASHTO. 1999. Guide Specifications for Design and Construction of Segmental Concrete Bridges, 2nd Edition, American Association of State Highway and Transportation Officials, Washington, DC.

AASHTO. 2004. *AASHTO LRFD Bridge Design Specifications*, 3rd Edition, American Association of State Highway and Transportation Officials, Washington, DC.

AREMA. 2003. *Manual for Railway Engineering*, The American Railway Engineering and Maintenance-of-Way Association, Landover, MD.

Barker, R.M. and Puckett, J.A. 1997. *Design of Highway Bridges*, John Wiley & Sons, Inc., New York, NY.

Billington, D. 1983. *The Tower and the Bridge — The New Art of Structural Engineering*, Basic Books, New York, NY.

Billington, D. 2000. "Chapter 3: Bridge Aesthetics — Structural Art," *Bridge Engineering Handbook*, Ed. Chen, W.F. and Duan, L., CRC Press, Boca Raton, FL.

Caltrans. 1990. *Bridge Design Details Manual*, California Department of Transportation, Sacramento, CA.

Caltrans. 1994. *Bridge Design Aids Manual*, California Department of Transportation, Sacramento, CA.

Caltrans. 2001. *Bridge Design Aids 6-1 Precast Prestressed Girders*, California Department of Transportation, Sacramento, CA.

Caltrans. 2002. *San Francisco-Oakland Bay Bridge East Span Seismic Safety Project — Design Criteria*, Prepared by T.Y. Lin International/Moffatt & Nichol Engineers, a Joint Venture, California Department of Transportation, Sacramento, CA.

Chen, W.F. and Duan, L., Ed., 2000. *Bridge Engineering Handbook*, CRC Press, Boca Raton, FL.

Collins, M.P. 2001. "In search of Elegance: The Evolution of the Art of Structural Engineering in the Western World," *ACI Concrete Int.*, Vol. 23, No. 7, pp. 57–72.

Duan, L. and Chen, W.F. 2002. "Chapter 18: Bridges," *Earthquake Engineering Handbook*, Ed. Chen, W.F. and Scawthron, C., CRC Press, Boca Raton, FL.

FHWA. 1990. *Standard Plans for Highway Bridges, Vol. I, Concrete Superstructures*, U.S. Department of Transportation, Federal Highway Administration, Washington, DC.

FHWA. 2003. http://www.fhwa.dot.gov/bridge/hps.htm

HEPC. 1975. *Construction Records of Minato Oh-Hashi*, Hanshin Expressway Public Corporation, Japan Society of Civil Engineers, Tokyo, Japan (in Japanese).

HEPC. 1994. *Techno Gallery*, Hanshin Expressway Public Corporation, Osaka, Japan.

JASBC. 1981. *Manual Design Data Book*, Japan Association of Steel Bridge Construction, Tokyo, Japan (in Japanese).

JASBC. 1984. *A Guide Book of Bearing Design for Steel Bridges*, Japan Association of Steel Bridge Construction, Tokyo, Japan (in Japanese).

JASBC. 1984. *A Guide Book of Expansion Joint Design for Steel Bridges*, Japan Association of Steel Bridge Construction, Tokyo, Japan (in Japanese).

JASBC. 1985. *Outline of Steel Bridges*, Japan Association of Steel Bridge Construction, Tokyo, Japan (in Japanese).

JASBC. 1988. *Planning of Steel Bridges*, Japan Association of Steel Bridge Construction, Tokyo, Japan (in Japanese).

JCMA. 1991. *Cost Estimation of Bridge Erection*, Japan Construction Mechanization Association, Tokyo, Japan (in Japanese).

JRA. 2002. *Specifications for Highway Bridges, Part I Common Provisions, Part II Steel Bridges and Part III Concrete Bridges*, Japan Road Association, Tokyo, Japan (in Japanese).

Leonhardt, F. 1984. *Bridges: Aesthetics and Design*, The MIT Press, Cambridge, MA.

Leonhardt, F. 2000. "Chapter 2: Bridge Aesthetics — Basics," *Bridge Engineering Handbook*, Ed. Chen, W.F. and Duan, L., CRC Press, Boca Raton, FL.

Lin, T.Y. 2001. *"The Father of Prestressed Concrete" Teaching Engineers, Bridging Rivers and Borders, 1931 to 1999*, Regents of the University of California, UC Berkeley, CA.

Mistry, V.C. 2002. "High Performance Steel For Highway Bridges," *The Proceedings, 2002 FHWA Steel Bridge Conference for the Western United States*, Ed. Azizinamini, A., December 12–13, Salt Lake City, UT.

Nader, M., Manzanarez, R., Baker, G., and Toan, V. 2001. "San Francisco-Oakland Bay Bridge Self Anchored Suspension Span Steel Design Challenges and Detailing Solutions," *The Proceedings of 2001 World Steel Bridge Symposium*, Chicago, IL.

Nagai, N. 1994. *Bridge Engineering*, Kyoritsu Pub. Co., Tokyo, Japan (in Japanese).

Podolny, W. and Muller, J.M. 1982. *Construction and Design of Prestressed Concrete Segmental Bridges*, John Wiley & Sons, Inc., New York, NY.

Ruck-a-Chucky. 2003. http://www.ketchum.org/ruckachucky/index.html

SFOBB. 2003. Internet References: http://www.mtc.ca.gov/projects/bay_bridge/bbmain.htm; http://www.dot.ca.gov/dist4/eastspans/index.html

Shimada, S. 1991. "Basic Theory of Arch Structures," *J. Bridge Foundation Eng., Kensetsu-Tosho*, Vol. 25, No. 8 (in Japanese).

Svensson, H.S. 1998. "Aesthetics of Steel Bridges," *The Proceedings of 1998 World Steel Bridge Symposium*, Chicago, IL.

Tonias, D.E. 1995. *Bridge Engineering*, McGraw-Hill, Inc., New York, NY.

Tachibana, Y. and Nakai, H. 1996. *Bridge Engineering*, Kyoritsu Pub. Co., Tokyo, Japan (in Japanese).

TRB. 1991. *Bridge Aesthetics Around the World*, Committee on General Structures, Subcommittee on Bridge Aesthetics, Transportation Research Board, National Research Council, Washington, DC.

Troitsky, M.S. 1994. *Planning and Design of Bridges*, John Wiley & Sons, Inc., New York, NY.

Xanthakos, P.P. 1994. *Theory and Design of Bridges*, John Wiley & Sons, Inc., New York, NY.

Xanthakos, P.P. 1995. *Bridge Substructure and Foundation Design*, Prentice Hall, Inc., Upper Saddle River, NJ.

26

Cable-Supported Bridges

Manabu Ito
University of Tokyo,
Tokyo, Japan

26.1 Introduction

26.1.1 Features of Cable-Supported Bridges

Cable-supported bridges or cable-suspended bridges are defined as bridges whose decks are supported by flexible cables. They are, in principle, classified into a suspension type where the bridge deck is continuously supported by stretched catenary cable(s), a cable-stayed type where the deck is discretely and directly suspended by straight stay cables, and their combined type. Suspension and stayed structures are applied to roof and buildings too.

Although the load bearing mechanisms are different, suspension and cable-stayed bridges have in common the following features:

1. They generally consist of cables, bridge deck incorporated with solid-web girder or truss, and towers.
2. They are advantageous for spanning long distances as seen in Figure 26.1 because cables are subject to only tension and steel wires consisting of a cable have very high tensile strength, although they may also be used economically on short- and medium-span pedestrian bridges.
3. The entire structure is much more flexible than other types of structures having equivalent span length.
4. The complete structure can be mostly erected without intermediate staging from the ground.
5. The main structure is elegant and neatly expresses its function owing to its transparent appearance.

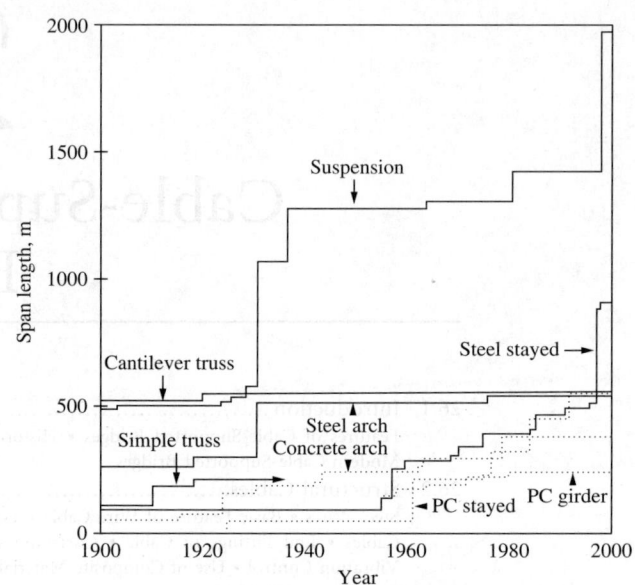

FIGURE 26.1   Transition of maximum span length of bridges in the 20th century.

Since the cable-supported bridges are mostly used outside the span range of the standard bridge specifications and are quite flexible compared with other structural types, their design specifications are often provided peculiarly.

26.1.2 Historical Sketch of Modern Cable-Supported Bridges

The history of cable-supported bridges is very long because the concept of suspending a bridge deck by cables might be easily thought of. However, construction of reliable structures of these sorts had to wait until the age of the industrial revolution when strong and homogeneous wrought iron, and later steels, were manufactured. Since then, techniques of suspension bridge construction have made steady progress despite frequent accidents, whereas early attempts to construct bridges as cable-stayed systems were not successful probably due to lack of technical and analytical understanding.

Roeblings' Brooklyn Bridge (486 m span) built in 1883 marked the beginning of the golden age of modern suspension bridges in the United States spanning the next half a century. With the successive construction of the Williamsburg (1903; 488 m), Benjamin Franklin (1926; 533 m), Ambassador (1929; 564 m), George Washington (1931; 1067 m), Golden Gate (1937; 1280 m), and Verrazano Narrows (1965; 1298 m) bridges, the United States stood at the vanguard of long-span suspension bridge technology. In the mid-1950s, 15 of the longest span suspension bridges in the world were in the United States. Among them, the Manhattan Bridge built in 1909 is featured as the first suspension bridge designed on the basis of the deflection theory. It provides a contrast to the neighboring Williamsburg Bridge, built only 6 years earlier by using the elastic theory, when slenderness of stiffening truss is compared. The George Washington Bridge was epoch-making because its main span length exceeded 1 km for the first time, almost doubling the previous record. Not to speak of the technological excellence, the Golden Gate Bridge is featured by its Art Deco style towers and iron oxide red color that harmonize so well with the magnificent site. In 1940, the Tacoma Narrows Bridge, then the world's third-longest suspension bridge, collapsed due to dramatic flutter oscillation caused by a gale. The lesson from this tragedy was reflected in the design of the Mackinac Strait and the Verrazano Narrows bridges afterwards in different styles.

From around 1960, a wave of long-span suspension bridges moved to western Europe. In particular, the unprecedented ideas of streamlined box girder and inclined hangers adopted first for the design of the Severn Bridge in the United Kingdom, completed in 1966, were further applied to the Humber Bridge, the world's longest span bridge at the time of completion in 1981, and the first Bosporus bridge (Kemal Ataturk Bridge), designed by the same group. Later, in the last quarter of the 20th century, the construction of long-span suspension bridges boomed in Japan and the Scandinavian countries, and then in China. Such chronological trends as mentioned above may be recognized from Table 26.1a.

Among the recent suspension bridges in the Far East, the Seto Bridges in Japan (1988) and the Tsin Ma Bridge in Hong Kong (1997) are featured as long-span suspension bridges carrying both road and substantial rail traffics. Particularly in the former, new techniques such as innovative track structures and new design and fabrication provisions against fatigue were first developed. The longest span of a suspension bridge

TABLE 26.1 Span Length Ranking of (a) Suspension Bridges (Span >1000 m) and (b) Cable-Stayed Bridges (Span > 500 m)

Name	Span (m)	Country	Year	Girder
		(a) Suspension bridges		
Akashi	1991	Japan	1998	Truss
Great Belt	1624	Denmark	1998	Box
Humber	1410	U.K.	1981	Box
Jiangyin	1385	China	1999	Box
Tsin Ma	1377	Hong Kong, China	1997	Road/rail
Verrazano	1298	U.S.A.	1964	Double deck
Golden Gate	1280	U.S.A.	1937	Truss
Hoga Kusten	1210	Sweden	1998	Box
Mackinac	1158	U.S.A.	1957	Truss
South Bisan	1100	Japan	1988	Road/rail
F.S. Mehmet	1090	Turkey	1988	Box
Kemal Ataturk	1074	Turkey	1973	Box
G.Washington	1067	U.S.A.	1931	Double deck
Kurushima III	1030	Japan	1999	Box
Kurushima II	1020	Japan	1999	Box
April 25th	1013	Portugal	1966	Road/rail
Forth Road	1006	U.K.	1964	Truss
Runyang Yangtze	1490	China	(2005)	Box
Tsing Lung	1418	Hong Kong, China	u.d.	Twin box
		(b) Cable-Stayed Bridges		
Tatara	0890	Japan	1999	Steel + PC[a]
Normandy	0856	France	1995	Steel + PC
Nanjing Yangtze II	0628	China	2001	Steel
Baishazhou Yangtze	0618	China	2000	Steel + PC
Qingzou	0605	China	2001	Steel
Yanpu	0602	China	1995	Composite
Xupu	0590	China	1997	Composite + PC
Meiko Central	0590	Japan	1998	Steel
Skarnsundet	0530	Norway	1991	PC
Jueshi	0518	China	1998	Steel + PC
Tsurumi	0510	Japan	1995	Steel
Jingzhou Yangutze	0500	China	2002	PC
Nanjing Yangtze III	0648	China	(2006)	Steel
Stonecutters	1017	Hong Kong, China	(2008)	Steel
Sutong Yangtze	1088	China	(2008)	Steel

[a] PC: prestressed concrete.

has reached nearly 2 km in the 20th century, while the Messina Strait Bridge in Italy, which will also carry both road and rail and is awaiting construction, is expected to have the formidable span length of 3300 m.

Although the concept of cable-stayed bridge was known since ancient times, the first outstanding application of stay cables was realized by Roebling in his Niagara suspension bridge in the mid-1850s, carrying rail and road. The same system was used for the Brooklyn Bridge also though these stays were not effective structural components. More than a half century later, F. Dischinger proposed the combined system of suspension and stay cables in 1938 but did not use it for actual construction.

The construction of modern cable-stayed bridges was led by German engineers after the Second World War. In particular, three large bridges, Nord, Knie, and Oberkassel, built from 1956 to 1972 for the city of Duesseldorf crossing the River Rhine, were a dramatic display of the new form. Following these early German designs, cable-stayed bridges spread rapidly throughout the world, marked especially by a variety of forms chosen for esthetic and technical reasons. As seen in Figure 26.1, the development of record span length of this type has been very rapid. The longest span exceeds 800 m in the late 1990s, and now is expected to exceed 1000 m by 2010 (see Table 26.1b). As far as the number of long-span cable-stayed bridges is concerned, Japan and China have taken lead; in 2010, China will have eight among the top ten longest bridges.

Since the tower and girder of a cable-stayed bridge are mostly subject to bending and compression, concrete structures have been widely used from the early stage, even for spans as long as 500 m or more. On the other hand, its application to short-span pedestrian bridges has been also very widespread, being appreciated for its esthetic value and economical advantage.

26.2 Structural Cables

26.2.1 Steel Wires

Cables are naturally the most important element of a cable-supported bridge. Although eye-bar chain was used in the early days and now the use of new composite materials is going to increase for small-span bridges, the material most frequently employed in modern bridge cables is cold-drawn high-strength steel wires, which have a diameter of 3 to 7 mm and ultimate tensile strength of 1.5 to 1.9 GPa.

As shown in Figure 26.2, the tensile strength of the steel wires used in suspension bridge cables increased up to 1.8 GPa in constructing the world's longest bridge, Akashi Kaikyo, which was completed in 1998 by adding an appropriate amount of silicon. On the other hand, the wire element of the seven-wire strand as used extensively for prestressed concrete, consisting of a straight core wire surrounded by a single layer of long-pitch six wires, has already had a higher tensile strength between 1.8 and 1.9 GPa.

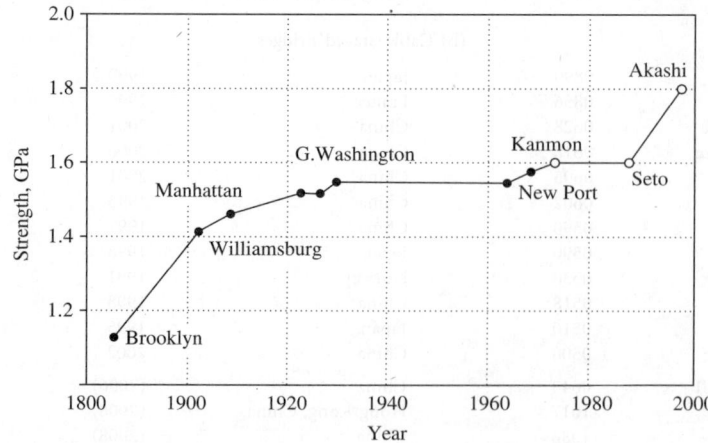

FIGURE 26.2 Tensile strength of galvanized steel wires.

26.2.2 Basic Feature of Wire Cable

A wire cable has to be composed of a large number of single wires to meet the strength requirement. A certain number of wires is often shop-assembled to form a prefabricated strand, subsequently at the site as basic elements for the final cable by being bundled into generally a circular cross-section.

Cable strands are classified into the wire rope formed by helically twisted wires and the parallel-wire strand (PWS), as shown in Table 26.2. A strand rope is manufactured by further twisting spiral rope strands around a single straight core strand. Mechanical properties of a cable go down in order of the PWS, the spiral rope, and the strand rope (see Table 26.3), while order of ease of handling is the reverse. Due to the inferiority of mechanical properties, the strand rope is rarely used in bridge cables. Stay cables used in cable-stayed bridges have more diversity as will be mentioned in the next section.

The cables in a cable-stayed bridge are all inclined. The actual stiffness of such an inclined cable is lower than that of the cable material itself because of sag due to its own weight. The equivalent Young's modulus of an inclined cable is expressed by

$$E_{\text{eff}} = \frac{E_0}{1 + \gamma^2 l^2 E_0/(12\sigma^3)} \tag{26.1}$$

where E_0 is Young's modulus of the straight cable, γ is the weight of the unit length cable, l is the horizontally projected length of the cable, and σ is the tensile stress in the cable [1]. Although this

TABLE 26.2 Types of Suspension Bridge Cables

Name	Shape of section	Structure
Parallel wire strand		Wires are hexagonally bundled in parallel
Strand rope		Six strands made of several wires are closed around a core strand
Spiral rope		Wires are stranded in several layers mainly in opposite lay directions
Locked coil rope		Deformed wires are used for the outside layers of spiral rope

Source: Okukawa, A., Suzuki, S., and Harazaki, I., in *Bridge Engineering Handbook*, CRC Press, Boca Raton, FL, 2000, with permission.

TABLE 26.3 Mechanical Properties of Steel Wire Cable

Type	Void ratio (%)	P/NS[a]	Elastic modulus (MPa)
Strand rope	35–42b	0.80–0.85	1.35×10^5
Spiral rope	23–25b	0.9	1.55×10^5
Locked-coil rope	10–14[b]	0.9	1.55×10^5
Parallel wire strand	11–14[b]	0.95–0.98	1.95×10^5

[a] P: rupture load of strand, S: tensile strength of a wire, N: number of wire.
[b] Value for a circumscribed hexagon.

reduction of the stiffness is practically negligible in most cases, it must be properly considered for a very long cable or during some construction stage.

The sectional area of cables is determined on the basis of the maximum cable tension. The safety factor of structural cables has been normally 2.0 to 2.5 for the guaranteed tensile strength, while that of main cables of long-span suspension bridges may be a little lower. This depends on the ratio of the dead load stress to the total stress, the nature of secondary stress, and the existence of fatigue stress.

26.2.3 Types of Bridge Cables

26.2.3.1 Spiral Rope

Spiral ropes have been used for main cables of many short- and medium-span suspension bridges, hanger ropes of almost all suspension bridges, and stay cables of some cable-stayed bridges mainly in the United Kingdom. When a large sectional area is required as in the case of main cables for a medium-span suspension bridge, a certain number of spiral rope strands are arranged in parallel and usually compacted in one cable. In order to remove nonelastic elongation of the spiral rope strand caused by the compaction of the strand, prestressing is performed before erection. Because spiral ropes are self-compacting, it is not required to wrap or provide with bands around the cable if a strand is separately arranged.

26.2.3.2 Locked-Coil Rope

Locked-coil ropes (LCRs) are a family of spiral ropes, but noncircular or deformed wires are arranged in the outside layers of the strand as seen in Table 26.2. The strength of the wires in an LCR is lower (1370 to 1570 MPa), while a nominal modulus of elasticity of an LCR strand is slightly higher than that of a normal spiral rope. The Z-shaped wires of the outer layers make the LCR strand more compact owing to a very small void ratio inside the strand, more water-tight, and less sensitive to side pressures at saddles and anchorages, whereas an LCR is not easy to handle when compared to normal spiral ropes due to its bending stiffness.

The LCR has been widely used in steel cable-supported bridges in Europe, and even in the main cables of suspension bridges with a main span of up to 850 m. Its maximum size so far is approximately 1250 m in length and 180 mm in diameter.

26.2.3.3 Parallel-Wire Cable

A parallel-wire main cable of a large suspension bridge is made either by the air-spinning method where the total cable section is assembled on site from individual wires or by the prefabricated parallel-wire strand (PPWS) method where the PPWS fabricated at the shop is coiled around the reel and transported to the site. The diameter of the individual wire used in suspension bridge main cables is about 5 mm.

A PPWS with covering sheath is widely used in cable-stayed bridges. A parallel-wire bundle of prestressing wires with a diameter of about 5 or 7 mm is incorporated as a stay cable with a polyethylene pipe filled by cement grout as corrosion protection and with HiAm anchor sockets as the end fittings. These parallel-wire cables have been widely used in both concrete and steel cable-stayed bridges.

26.2.3.4 Ultra-Long Lay Cable

The idea of an ultra-long lay cable strand was initiated in the 1980s as the improved variant of PWS parallel-wire cables. Twisting the wires up to 3 to 4° enables the wire bundle to ease reeling and make the strand self-compacting under axial tension without spoiling the mechanical properties. These cable strands were designated as "New PWS" in Japan and as "HiAm-SPWC" in Europe. The New PWS is also featured by extruding high-density polyethylene (HDFE) cover directly onto the wire bundle so that no void will exist between the outer wires and the surrounding cover. The strand is assembled by

7 mm wires and the thickest one comprises 421 wires. The longest stay cable of this type is 460 m long with an outer diameter of 165 mm, as used in the Tatara Bridge.

26.2.3.5 Parallel-Strand Cable

The seven-wire strand that has been extensively used as tendon for a prestressed concrete structure is the simplest and most prevalent stay cable of prestressed concrete cable-stayed bridges. As the pitch of twisted wires is relatively long, the stiffness of the strand is close to that of a straight wire strand and its breaking strength is even higher. For cables, the strand is normally made from 5 mm wires and its nominal diameter is 12.7 or 15.2 mm and these strands are arranged in parallel to form a stay cable. The number of the seven-wire strands varies from 7 to 127 depending on the required design force.

There are a number of strand or cable systems using the seven-wire strands according to corrosion protection, assembling method, and end-fitting techniques, for example, Freyssinet, Dywidag, VSL, Stronghold, SEEE, ASP (at-site prefabricated cable system), and so on. The parallel-strand cables are either shop-fabricated or site-fabricated, and sometimes their combination like SEEE. Cost saving is attained by the site-fabrication of stay cables with individual strands pushed through a preinstalled sheath. The corrosion protection will be illustrated later in Section 26.2.5. These types of cables have also been applied recently to steel or steel/concrete hybrid cable-stayed bridges.

26.2.3.6 Bar Stay Cable

A bar stay cable consists of round steel bars with a diameter of 26 to 36 mm, being covered by a steel pipe, the inside of which is filled with cement grout. The external steel tube is considered in the cable cross-section when live load is applied. Since the length of a bar cannot be long, coupling is normally needed. This type of stay member is scarcely used, particularly for large cable-stayed bridges.

26.2.4 End Fittings of Cable

Bridge cables are connected and anchored to the anchoring substructures or other structural members such as the girder and towers through the end fittings.

26.2.4.1 Strand Shoes

For parallel-wire cables erected by the air-spinning method, the anchoring is generally made by looping the wires around a strand shoe disc that is semicircular or circular and is fixed to the anchor frame through anchoring rods.

26.2.4.2 Metallic Alloy Sockets

For prefabricated strands, either parallel-wire cables or wire ropes, both ends of the strands are fitted by metallic alloy sockets. The socket consists of a thick-walled steel cylinder with a conical cavity, in which the disentangled end of the strand is inserted and the molten metallic alloy is filled. The wedge action of the hardened alloy serves the force transmission. The sockets are supported by the anchor fittings or other anchoring members. The metallic alloy should be provided with large bonding force for wires, low melting temperature, superplasticity, and favorable creep properties. Zn–Cu alloy with 2% Cu is usually used for suspension bridge main cables.

26.2.4.3 Sockets for Stay Cables

The Zn–Cu alloy used in suspension bridges is also popular in the socket of the stay cables, although Zn–Al alloy has been employed in Germany. In the case of cables of the cable-stayed bridges, the fatigue under repeated stress due to vehicle loading and the fretting corrosion due to the abrasion of metals, which occur at the entrance of the socket, are the problems to be considered. As exemplified in

the following, a variety of high-fatigue-resistant sockets have been developed for the stay cables
(Figure 26.3):

1. *HiAm anchor socket.* The anchoring mechanism is mainly arch action of wires and steel balls in
   the conical hole of the steel socket.

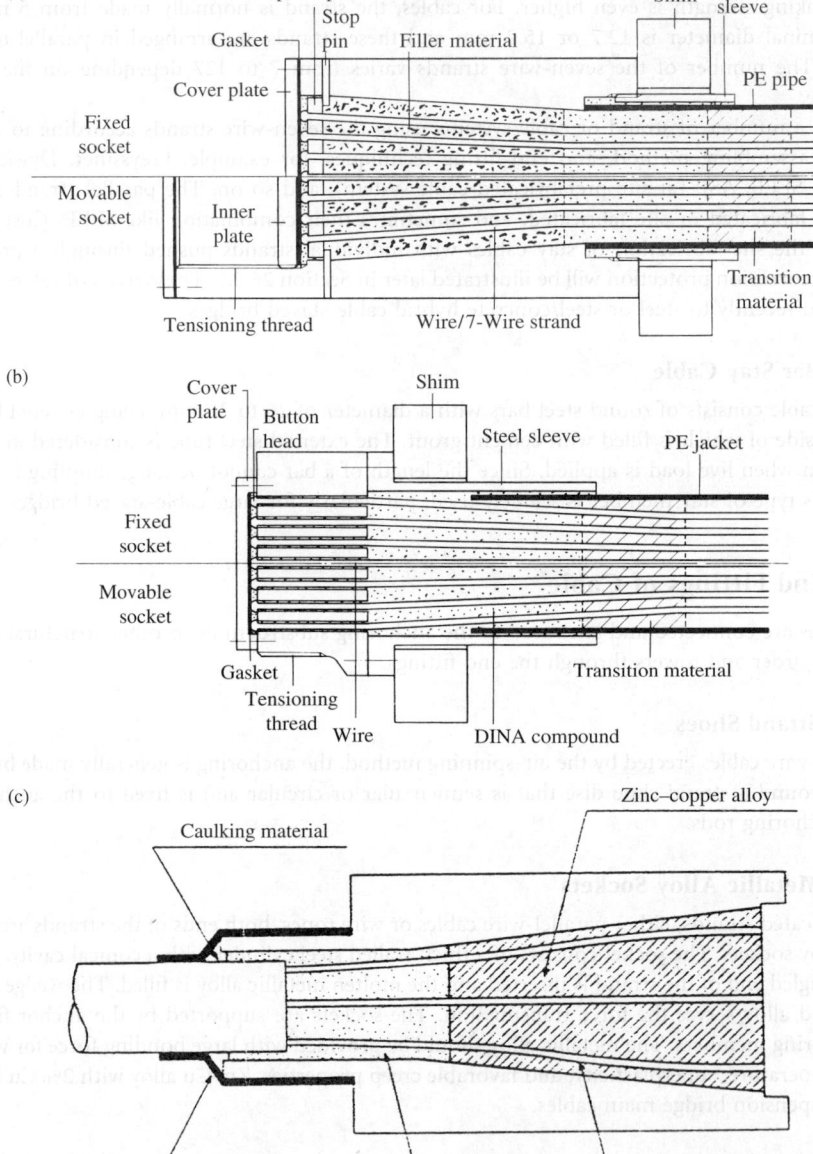

FIGURE 26.3 Examples of high-fatigue-resistant socket: (a) HiAm anchor socket; (b) DINA anchor head; and
(c) NS anchor socket. (Adapted from M. Ohashi, Cables for Cable-Stayed Bridges, in *Cable-Stayed Bridges — Recent
Developments and their Futures,* M. Ito et al., Eds., Elsevier, Amsterdam, 125, 1991, with permission.)

2. *NS socket.* This consists of a hot poured Zn–Cu alloy cone and epoxy resin at the mouth of the socket. The epoxy resin prevents fretting corrosion, buffers stress concentration, and reduce unfavorable thermal effect of the molten alloy.

3. *DINA anchor head.* It consists of button heads of wires, a steel anchor head, and epoxy resin. One of its advantages is the short length that can be accommodated into a small space.

Furthermore, various stay-cable anchorages such as Freyssinet, Dywidag, VSL, Stronghold, and SEEE, evolved from the post-tension systems, have been applied to the cable-stayed bridges.

26.2.5 Corrosion Protection

Cable corrosion is caused by water and ion invasion from outside and by dew resulting from the alternating dry and humid conditions inside the cable void. The corrosion protection system usually consists of at least two barriers: the internal barrier immediately adjacent to the main tension element and the external barrier or covering that is exposed to the outside environment and often the blocking compound that is the corrosion inhibiting or water repelling material used to fill the voids inside the covering [2,3]. The internal barrier for steel wires is usually a zinc galvanized barrier, but mainly in North America, epoxy coating of individual wires or seven-wire strands has been widely used on stay cables.

26.2.5.1 Parallel-Wire Main Cables of Suspension Bridges

The conventional method of corrosion protection for the main cables of most long-span suspension bridges has been a covering over the zinc-galvanized parallel wires by layers of paste, wrapping with galvanized soft wires, and painting, all of which aim at preventing water from permeating into the cable, as seen in Figure 26.4a. The paste materials have usually been red lead, polymer organic lead, zinc dust, or the recent lead-acid calcium, while the paint has been futal acid, rubber chloride, or polyurethane.

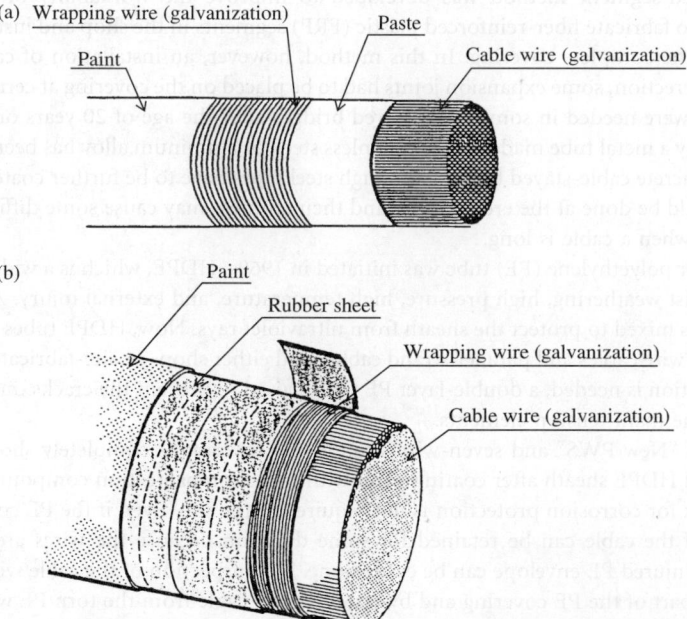

FIGURE 26.4 Corrosion protection of suspension-bridge cable: (a) conventional system and (b) new system for the Akashi Kaikyo Bridge.

However, the above-mentioned system was found unsatisfactory under highly humid environments because the conventional types of paste are not effective in protecting against water and corrosion inside the cable occurs irrespective of the effectiveness of the paste if moisture exists in the cable. Accordingly, new corrosion protection technologies have been introduced in the recent Japanese suspension bridges. One is the use of a combination of aluminum triphosphate and organic lead paste and wrapping with S-shaped deformed steel wires adopted in the Hakucho Bridge (1998), and the other is the system consisting of, in addition to the conventional wire wrapping, the wrapping with neoprene rubber sheet instead of paste, as shown in Figure 26.4b; the adoption of a dehumidified air-injection device has also been used, which was used in the Akashi Kaikyo (1998) and Kurushima (1999) bridges. The rubber sheet may be replaced by fiberglass acrylic sheet. In the latter, double layers of butane-contained rubber and denatured silicone were used at the cable band portion.

26.2.5.2 Wire Ropes

The wires for spiral ropes are zinc galvanized and the voids are filled with a sealing compound such as metalcoat, which is a suspension of aluminum flakes incorporated into a hydrocarbon resin carrier suitably diluted with a solvent for ease of application.

LCRs used in Japanese cable-stayed bridges are manufactured from galvanized wires, applying a minimum amount of lubricating oil during rope closing to avoid any concern about future stains of the surface. The outer surfaces are usually painted after the dead load has been fully applied. In European practice, the inner voids of the zinc-galvanized wires are filled by polyurethane with zinc dust or linseed oil with red lead, and outer surface of the rope is coated with polyurethane. Metalcoat is also sometimes applied as the second barrier during its fabrication.

26.2.5.3 Covering of Stay Cables

Covering the strand or cable as the external barrier has been common to other stay cables than helical wire ropes. One of the methods was to wrap a foamed polyethylene tape with glass fiber reinforced plastic covering over the PWS. In the early 1970s this was executed by hand-lay-up method on site, but later prefabricated-segment method was developed to improve the workability of the hand-lay-up method. This is to fabricate fiber-reinforced plastic (FRP) segments in the shop and just to connect them on site to form the complete covering. In this method, however, an installation of catwalk was indispensable for the erection, some expansion joints had to be placed on the covering at certain intervals, and repairing works were needed in some cable-stayed bridges after the age of 20 years or so.

The covering by a metal tube made of steel, stainless steel, or aluminum alloy has been often applied to stay cables of concrete cable-stayed bridges although steel pipes have to be further coated. Installation of metal pipes should be done at the erection site and their stiffness may cause some difficulty in handling during erection when a cable is long.

Use of a FRP or polyethylene (PE) tube was initiated in 1960s. HDPE, which is a widely used material, is selected to resist weathering, high pressure, high temperature, and external injury. At the same time, 2 to 3% carbon is mixed to protect the sheath from ultraviolet rays. Now, HDPE tubes are most popular for both parallel-wire cables and parallel-strand cables, and either shop- or site-fabricated. When heavier corrosion protection is needed, a double-layer PE tube is used to prevent the cracks on the outer surface from reaching the main tension elements.

In the cases of "New PWS" and seven-wire strands, the covering is completely shop-fabricated by a directly extruded HDPE sheath after coating wires with corrosion protection compound. In the former, any further work for corrosion protection is not required at the site. Even if the PE covering is injured, the durability of the cable can be retained for some duration because the wires are galvanized. The repairing of the injured PE envelope can be easily done. The inspection of the cable will also be possible by tearing off a part of the PE covering and by watching the cable from the torn PE window. Although the original color of the PE covering is black due to the mixed carbon, cable coloring techniques have been developed. One is to extrude a colored thin fluoro-polymer on the black PE layer; another method is a paint coating system that consists of an application of primer made from adhesive components for

PE envelope and for the finish coat, and a baking of the primer with a far-infrared ray. The finish coat is usually done by fluoro-olefin paint. The light color is preferred not only for good looks but also for reducing the temperature effect. Supplementary wrapping with colored Tedler tapes is an alternative.

26.2.5.4 Blocking Compound

Cement grout has been the most popular blocking compound for its alkaline properties providing an active corrosion protection to the steel wires. Cement mortar is injected after the stay cables are erected on the bridge when the cable is under full dead load stress to suppress the formation of cracks in the grout, because the presence of cracks may be associated with the potential for fretting corrosion of steel wires. However, cracks may still occur due to shrinkage of cement mortar and stress repetition under cyclic live loading.

Another problem of cement grout combined with galvanized steel wires is a fear of hydrogen brittlement caused by reaction of zinc and cement milk. In order to avoid it, the nongalvanized wires or the galvanized wires coated with polyester to isolate zinc from cement milk have been used in this case. Another measure is to substitute normal Portland cement by polymer cement, the advantages of which are that it is far more ductile, does not shrink after grouting nor bleed during placing, does not require special technique and equipment, and that it can be used in combination with galvanized wire without a fear of chemical reaction between the zinc layer and the cement. On the other hand, its disadvantages are relatively high materials cost and temperature-dependent viscosity and hardening. Cement grout plasticized with polyurethane was used in some bridges.

An example of alternatives to cementitious grout once used by the Honshu Shikoku Bridge Authority in Japan was synthetic resin material based on polybutadiene. This two-component material has very low viscosity during pouring, is very flexible after hardening, and has such low density as about a half of that of cement grout. But the material and execution costs were high and it is highly temperature dependent and flammable. Epoxy resin is specified in the American Society for Testing and Materials provisions as a filling material into the interstices of epoxy-coated seven-wire prestressing strands.

Other blocking compounds are grease used on PWS or prestressing strands and wax for parallel-strand cables. The petroleum is injected in a liquid state at temperatures of 85 to 105°C, which solidifies upon cooling. However, it shrinks during the cooling process and cracks may develop. A soft petroleum base wax that can be applied at ambient temperature on the monostrand system seems promising [4]. It has a melting point over 260°C and displaces any moisture on the surface of steel. In any case, grease and wax are to be used in combination with other corrosion protection measures. It is noted that the nongrout type PWS, which is completely fabricated in the shop, is now available.

26.2.6 Vibration Control

Wind-induced vibrations of stay cables have revealed themselves after the introduction of multistay systems with thin and long cables covered by PE sheath having a smooth surface. The long hangers of a suspension bridge using similar cables are also vulnerable to vibrations under wind. Among several kinds of wind-induced vibrations considered on cables, those where special care should be taken are vortex excitation, wake galloping, and rain/wind-induced vibration. Although tentative or approximate stability criteria for typical vibratory phenomena have been proposed [5], these wind-induced vibrations are dependent on the local terrain features and the peculiar conditions of the respective structural elements.

The following measures are generally considered to suppress the wind-induced vibrations of cables when the responses are not acceptable:

1. Increase of structural stiffness and natural frequency.
2. Increase of structural damping.
3. Modification of the cable surface.

Points (1) and (2) are categorized as mechanical or structural means, while point (3) is the aero-dynamic means to weaken the exciting mechanisms by disturbing or reducing wind-induced dynamic forces acting on the cable. However, because the exciting mechanisms of different vibratory phenomena differ, the countermeasures shall fit for the phenomenon concerned.

Occurrence of wake galloping depends on the spacing of neighboring parallel cables. Very small spacing or quite wide spacing, more than six times the cable diameter, can remarkably moderate the response. If these conditions cannot be satisfied for other design reasons or when undesirable wake galloping is observed after erection, the cable vibration can be suppressed by connecting both cables by a few spacers or small mechanical dampers.

One of the common countermeasures that has been often adopted is to connect the cables with secondary thin cables which may terminate at a cable or at the deck (tie-down cable). Even with a few and small stabilizing ropes, cable movement can be restrained. The natural frequencies of each cable and thus the resistance to dynamic excitation can be raised by shortening the effective free length of the primary cables. Such elements may be also sources of additional damping. But even if the size of stabilizing ropes is significantly smaller than that of the primary cables, they may affect the appearance of the structure to some extent. Furthermore, the rupture of these interconnecting ropes or the fatigue failure of the connection fittings have been reported in several bridges. Viscoelastic bushings or special clamps with damping devices in the joint of the both cables can reduce fatigue and provide additional damping.

Increase of structural damping is effective in suppressing the amplitude of buffeting, vortex excitation, and rain/wind-induced vibration and in raising the critical wind speed for the onset of galloping. It is first recommended to place such damping material as neoprene ring or high-damping rubber between the cable and the steel exit pipe at the pylon and deck anchorage. Use of damping material results in the additional benefit of reduced local bending moment in the cable. Further additional damping, if necessary, can be provided by mechanical damping devices. Very simple and small tuned mass dampers represented by the classical Stockbridge damper that has been prevailing on transmission power lines were applied to stay cables or diagonal hangers of some European cable-supported bridges, but is not very popular now for esthetic reasons.

When rather high additional damping is required, the most prevalent method in cable-stayed bridges is to install a dash-pot type viscous damper between the stay cable and the bridge deck. Shock absorbers similar to those used in automobiles or hydraulic oil dampers are the examples. These dampers shall not spoil the appearance of the bridge. The mechanical dampers utilizing shear-viscous material can be more compact. The attached position of these dampers influences the mode shape and the dampening effect of the cables. It is not easy to make a compromise between the requirement to lower the damper position and the efficacy of the damper when the cable is long.

The aerodynamic countermeasures for the round cables are to modify the sheath surface. The idea of helical fins often used on circular stacks to prevent vortex excitation will be also effective against rain vibration, but care shall be taken on the appearance. The idea of axial protuberances in the form of longitudinal ribs on the HDPE tube surface was developed in the Higashi Kobe Bridge in Japan (Figure 26.5a), aiming at preventing the rain vibration. A similar idea was later seen in the more simpler HDPE sheaths with fine grooves. Further on the Tatara Bridge, HDPE sheaths are provided a pattern-indented surface with roughness of 1% applied disorderly in a convex or a concave pattern (Figure 26.5b). The effect of these surface modifications is linked to influence the water rivulets. However, it should be noted that these means are not necessarily effective in suppressing vortex excitation and that the drag coefficient may be increased. In case of the Tatara Bridge, however, the drag coefficient in the supercritical Reynolds number range could be reduced to 0.6.

26.2.7 Use of Composite Materials

A fiber-reinforced polymer composite or a FRP is a matrix of polymeric material reinforced by fibers or other reinforcement that has a discernible aspect ratio of length to thickness. Reinforcements are fibers made from glass, carbon, or aramid, and the polymer resin can be either thermoset or thermoplastic. Thermosets are typically used for construction applications and can be polyester, vinyl ester, phenolic, or epoxy. FRP composites have been used in a variety of engineering products.

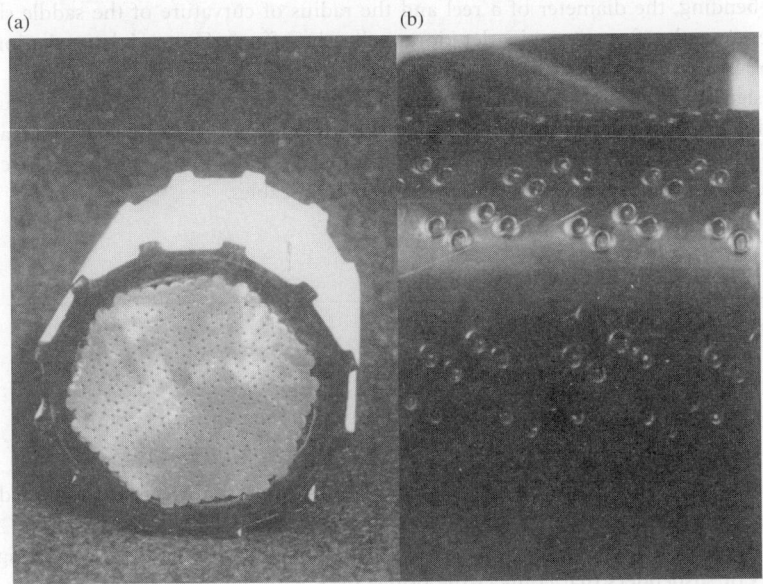

FIGURE 26.5 High-density polyethylene sheath of (a) the Higashi Kobe and (b) the Tatara bridges.

TABLE 26.4 Mechanical Properties of Different Cable Materials

Material	Specific gravity	Tensile strength (MPa)	Young's modulus (MPa)	Failure strain
Steel	7.85	1750–1900	200	6
Aramid fiber	1.35	1400–1800	50–70	2–4
Glass fiber	1.85	600–900	30	2
Carbon fiber	1.55	1900–2300	120–400	0.6–1.9

FRP composites have many advantages as structural materials such as high strength, relatively high elastic modulus, light weight, long-term durability, high dielectric strength, low axial thermal expansion, low maintenance, design flexibility, tailored esthetic appearance, and low tooling and installed costs. On the other hand, their demerits are the rupture without yielding, low elastic modulus of some kinds of the materials, anisotropic properties, low shear and impact resistance, and high material cost that may be gradually improved by the progress of manufacturing technology and widespread use.

Although the use of FRP composites is now becoming popular in structural engineering, carbon fiber-reinforced plastic (CFRP) as a cable material seems the most ideal application. Actually, CFRP stay cables have already been employed on several cable-stayed bridges for pedestrian use. In the far future, CFRP or some other FRPs may be used for the main cables of suspension bridges and greatly increase the limiting span length. The mechanical properties of different cable materials are shown in Table 26.4.

CFRP cables are produced as assemblies of the wires made up of carbon fibers and an epoxy resin matrix. The design considerations for cables made of unidirectional CFRP wires are similar to those for steel cables, with a few exceptions due to the highly anisotropic nature of the material. So far, CFRP stay cables have been made by twisting CFRP wires. Although twisted fiber rope makes coiling possible, resulting in ease in handling, care must be taken on the strength loss of the wires in a bundle and friction between wires. Since the decrease in cable strength is also brought by side pressure and bending, the diameter of a reel and the radius of curvature of the saddle should not be small. As the strength of epoxy resin deteriorates by the effect of ultraviolet radiation, the CFRP cables shall be protected by covering.

The trial design of a super-long-span suspension bridge for which CFRP main cables are used indicates that the bending moment and vertical deflection of the stiffening girder significantly increase due to the light weight of the structure [6], while the dimensions of towers and foundations can be greatly reduced. It is recommended to reduce the sag/span ratio of the main cables.

26.3 Suspension Bridges

26.3.1 Structural System

26.3.1.1 Components of a Suspension Bridge

A suspension bridge typically consists of the following components (Figure 26.6):

1. Main cables that suspend the bridge deck.
2. Main towers that support the main cables. Sometimes lower subtowers are positioned between the main tower and the cable anchorage to lead the cable to the anchorage.
3. Stiffening girder, either solid-web girder or truss, being incorporated with the bridge deck.
4. Hangers or suspenders that connect the bridge deck with the main cables.
5. Anchorages that anchor the main cable. They are usually massive concrete blocks, in which the anchor frame is embedded.

In very special situations occasionally encountered in short-span bridges, some components other than the main cables and the bridge deck may be omitted.

26.3.1.2 Classification

The structural system of suspension bridges may be classified by the following factors:

Number of spans. A suspension bridge may be single-span, two-span, three-span, or multispan. The number of main towers is one for two-span, two for single- and three-span, and more than two for multispan bridges (Figure 26.7). Two-span suspension bridges have been rare because they are less efficient as recognized from Equation 26.2. Single-span suspension bridges have straight backstays. Three-span suspension bridges are the most popular, in particular for long-span bridges and the ratio of side spans to main span is mostly 0.2 to 0.5. Although multispan suspension bridges have been rare because of great flexibility, their applicability is being studied for the future straits crossings [7]. The major concerns about multispan suspension bridges are the design of intermediate towers and cable erection.

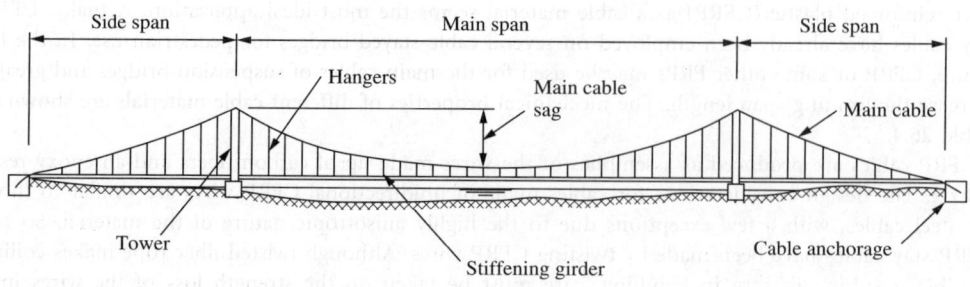

FIGURE 26.6 Suspension bridge.

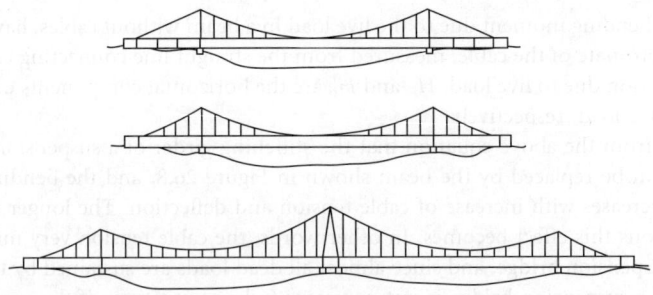

FIGURE 26.7 Single-, three-, and multispan suspension bridge (from top).

Continuity of stiffening girder. The stiffening girders are typically either simply supported on each span or continuous over two or more spans. The former is called two-hinged and commonly used for road bridges. Although continuous girder with intermediate supports is not economical as also recognized from Equation 26.2, it is advantageous for a rail bridge to improve runnability of trains. Furthermore, in the Great Belt Bridge, which is a road bridge with a center span of 1624 m, the girder is continuously supported by the cable system through three spans. The economical design was attained in this case by omitting the vertical support of the girder at the towers and fixing the main cables to the stiffening girder by clamps at the midspan.

Arrangement of hangers. Hangers are either vertical or diagonal. The latter makes the entire suspended structure a kind of truss incorporated with the main cables and the bridge deck.

Method of cable anchoring. The main cables of a suspension bridge are either externally anchored to the anchor blocks or self-anchored to the stiffening girder. Although the latter is rare, a more detailed description will be made in the following.

26.3.1.3 Self-Anchored Suspension Bridge [8]

Self-anchored suspension bridges do not require massive end anchor blocks. Instead, the main cables are anchored to the stiffening girder. Accordingly, the girder is subject to large axial force in addition to bending and shear, and the girder must be placed before the main cables are erected. Therefore, this type of structure is limited to moderate spans. Recent examples of the self-anchored suspension bridge are the Konohana Bridge in Japan (1990, three spans), the Yongjong Bridge in Korea (2000, three-span double-deck truss), and the replaced San Francisco Oakland Bay East Bridge (two spans). The first two bridges have 300 m center span, whereas the span of the Bay Bridge will be 385 m long. The Konohana Bridge is a monocable type, while two main cables of the other two bridges are converged at the peaked tower top.

26.3.2 Analysis of Suspension Bridges

Because a suspension bridge is very flexible and the bridge deck is suspended from the main cables by thin hangers, a different approach from other types of bridge structures should be taken in its analysis.

26.3.2.1 Behavior under Vertical Loads

The overall behavior of a suspension bridge is given by the deflection theory where the vertical deflection of cables due to live load is considered in the equilibrium of the structure. The premised assumptions in this case are (1) the cable is completely flexible, (2) the stiffening girder is horizontal and straight, (3) the original form of the main cable is a parabola, (4) all dead loads are sustained by the main cable, and (5) the hangers are inextensible and closely spaced. Thus, the bending moment $M(x)$ of the stiffening girder under live load is given by

$$M(x) = M_0(x) - H_\mathrm{p} y(x) - (H_\mathrm{w} + H_\mathrm{p}) v(x) \tag{26.2}$$

where $M_0(x)$ is the bending moment due to the live load in a beam without cables, having the same span length, $y(x)$ is the ordinate of the cable, measured from the straight line connecting cable supports, $v(x)$ is the vertical deflection due to live load, H_w and H_p are the horizontal components of cable tension due to dead load and live load, respectively.

It is understood from the above equation that the stiffening girder of a suspension bridge subject to vertical live load can be replaced by the beam shown in Figure 26.8, and the bending moment in the stiffening girder decreases with increase of cable tension and deflection. The longer the span length is, the more conspicuous this effect becomes. In other words, the cable tension very much contributes to the stiffness of a suspension bridge, and since almost all dead loads are sustained by the main cable, the stiffening girder of a suspension bridge is not necessary to be so stiff even if the span is long.

26.3.2.2 Behavior under Horizontal Load

Lateral forces caused by wind or earthquake in horizontal direction are to be transmitted from the stiffening girder to the main cables through flexible hangers because the deformation of the girder is larger than that of the main cables due to the difference of the horizontal loads and their stiffness. This effect is more pronounced as seen in Figure 26.9 in the center region of the span owing to large inclination of hangers. Consequently, the lateral bending moment induced in the girder is not so large in the region of the mid-span.

26.3.2.3 Design Analysis

Equation 26.2 indicates the nonlinearity between live load and the bridge responses. In long-span suspension bridges, however, $H_w + H_p$ can be assumed nearly constant because H_w is much larger than H_p, and therefore the analysis becomes quasilinear and the influence line analysis can be used on the conservative side (the linearized deflection theory) [9]. This simplified treatment is useful in a very early stage of preliminary design.

More accurate analysis shall be performed at the subsequent design stages including detailed design. With the progress of computer application nowadays, the finite displacement analysis for the entire

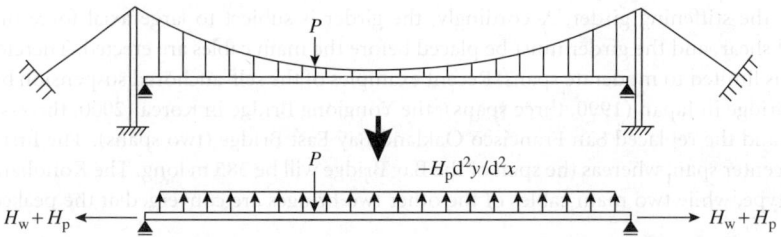

FIGURE 26.8 The equivalent beam of a suspension bridge.

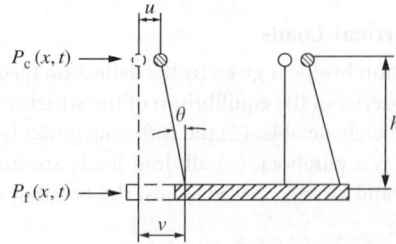

FIGURE 26.9 Transverse deformation of a suspension bridge under wind load.

suspension bridge structure is available taking into account the discreteness and elongation of hangers, deformation of towers, and so on.

26.3.3 Design Procedure

After the type of structural system and the span arrangement are decided, the design of suspension bridge superstructure is performed according to the procedure shown in Figure 26.10 [10].

26.3.3.1 Design Loads

Accurate evaluation of dead load is important in suspension bridge design because its contribution to the main components of the bridge is large. Design live load can be reduced from a probabilistic viewpoint with increasing span length and road width. The dynamic magnification due to running vehicles is usually negligible for design of the main cables, towers, and stiffening girder of a large suspension bridge owing to its long span and low natural frequencies.

The dynamic analysis is inevitable in seismic and wind-resistant design of flexible suspension bridges. Although the seismic forces induced in the superstructure are relatively small due to its long natural period, attention should be paid on possible large displacement of the stiffening girder and great forces transferred to the supports under seismic loading. Aerodynamic stability often dominates the design of a suspension bridge and some appropriate cures may be needed (cf. Chapter 9 of Section II) [11].

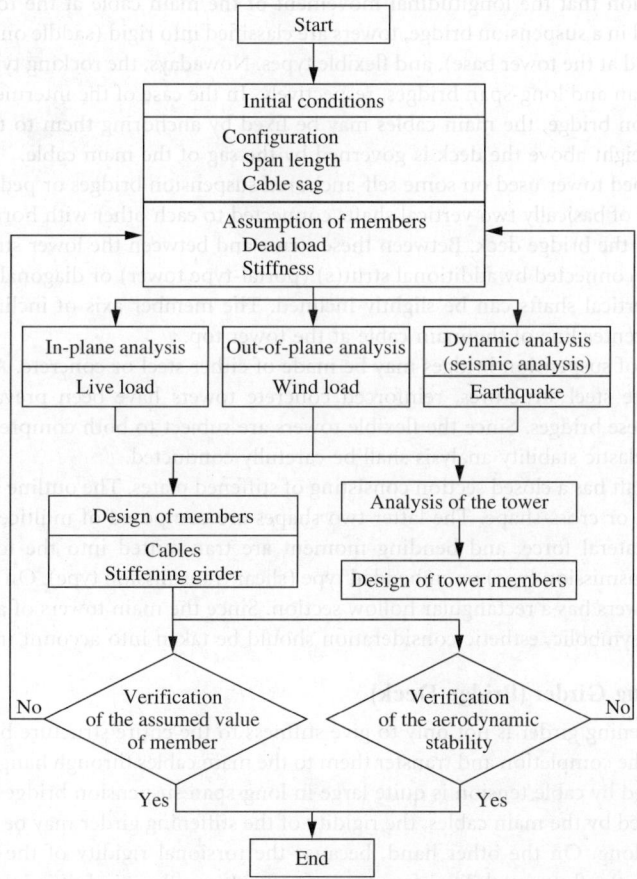

FIGURE 26.10 Design process of a suspension bridge superstructure.

26.3.3.2 Main Cables

The final form of the main cable of a suspension bridge is a parabola. Sag/span ratio is determined to optimize construction cost of the entire bridge, and sometimes in consideration of the aerodynamic stability as it affects the dynamic characteristics of the entire structure. In general, the sag/span ratio is around $\frac{1}{10}$ irrespective of span length.

As mentioned in Section 26.2.2, the parallel-wire cable is exclusively used in very long-span bridges, whereas the spiral ropes or LCRs have been sometimes adopted in moderate- or short-span suspension bridges.

26.3.3.3 Hangers

Hanger cables are arranged to be either vertical or diagonal (forming a truss). The latter aimed at increasing the damping properties and stiffness of the structure, and was first adopted in the Severn Bridge in the United Kingdom in 1966, followed by two large suspension bridges in Europe. But it was abandoned afterwards due to the fatigue injuries in the Severn Bridge.

The spiral or strand ropes have been widely used on the hangers of suspension bridges. Exceptions are the LCRs as in the Little Belt Bridge in Denmark and the PWS in the latest Japanese suspension bridges (Akashi and Kurushima). The hangers are connected to the main cable either by pin or by laying around the cable.

26.3.3.4 Main Towers

To meet the condition that the longitudinal movement of the main cable at the tower top should be theoretically allowed in a suspension bridge, towers are classified into rigid (saddle on rollers at the tower top), rocking (hinged at the tower base), and flexible types. Nowadays, the rocking type and flexible type are used in short-span and long-span bridges, respectively. In the case of the intermediate tower(s) of a multispan suspension bridge, the main cables may be fixed by anchoring them to the top of the rigid tower. The tower height above the deck is governed by the sag of the main cable.

Except the A-shaped tower used on some self-anchored suspension bridges or pedestrian bridges, the main towers consist of basically two vertical shafts connected to each other with horizontal struts at the top and underneath the bridge deck. Between these struts and between the lower strut and tower base, two shafts are often connected by additional strut(s) (portal-type tower) or diagonal members (braced-type tower). The vertical shafts can be slightly inclined. The member axis of inclined shafts typically coincides with the center line of the main cable at the tower top.

The main towers of suspension bridges may be made of either steel or concrete. Although almost all stiffening girders are steel structures, reinforced concrete towers have been prevalent in the recent European and Chinese bridges. Since the flexible towers are subject to both compression and bending, strength as well as elastic stability analysis shall be carefully conducted.

The steel tower shaft has a closed section consisting of stiffened plates. The outline of the cross-section is a rectangular, T-, or cross-shape. The latter two shapes are composed of multicells. The tower base where axial force, lateral force, and bending moment are transmitted into the foundation is either grillage (bearing transmission type) or embedded type (shear transmission type). On the other hand, the shaft of concrete towers has a rectangular hollow section. Since the main towers of a suspension bridge is outstanding and symbolic, esthetic consideration should be taken into account in its configuration.

26.3.3.5 Stiffening Girder (Bridge Deck)

The role of the stiffening girder is not only to give stiffness to the entire structure but to distribute the loads applied after the completion and transfer them to the main cables through hangers. However, since the stiffness provided by cable tension is quite large in long-span suspension bridges and almost all the dead load is sustained by the main cables, the rigidity of the stiffening girder may be not so large even if the span length is long. On the other hand, because the torsional rigidity of the stiffening girder is significant to ensure the flutter stability of a suspension bridge under wind, it is intended to constitute closed box effects structurally even if the girder is a trussed structure, by providing with both upper and lower lateral bracings. The ratio of fundamental natural frequencies in torsion and in bending is recommended to be 2.0 or higher.

Although short-span bridges are often provided with I-girders, the stiffening girder of a long-span suspension bridge is usually either a truss girder or a box girder. Its design is mainly governed by the function such as traffic requirement, the necessary stiffness, and aerodynamic stability, as well as the site conditions of erection and maintenance. Selecting an aerodynamically stable cross-section has usually priority. Use of either streamlined box section or the section with openings such as a truss of small solidity factor and steel open-grating floor will meet this requirement. Three typical examples of the stiffening girder are shown in Figure 26.11. Although a truss girder has been used on double-deck

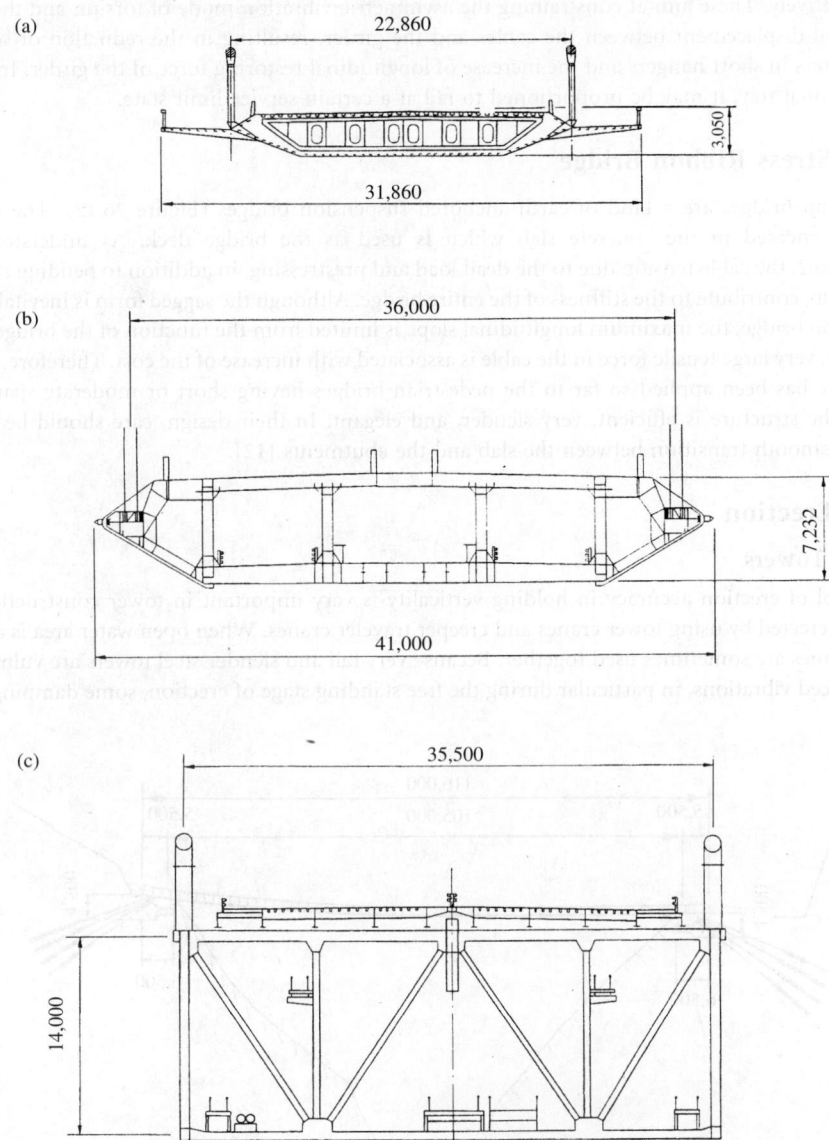

FIGURE 26.11   Examples of stiffening girder of modern suspension bridge: (a) Severn Bridge; (b) Tsing Ma Bridge and (c) Akashi Kaikyo Bridge.

bridges, adoption of big fairings on both sides of the cross-section and longitudinal openings on upper and lower decks features the Tsing Ma Bridge in Hong Kong (Figure 26.11b).

Even if the stiffening girder is simply supported, both supports are designed as movable shoes in long-span suspension bridges; that is, the stiffening girder is suspended by tower links at the tower and end links at the abutment. The wind bearings preventing transverse displacement of the girder are installed on the horizontal strut of the tower and on the abutment. When no vertical support is provided at the tower, the lateral support at the tower can be accomplished by applying vertical sliding bearings between the girder and the inner surfaces of the tower shafts as in the case of the Great Belt Bridge.

In many long-span suspension bridges, the stiffening girder is fixed to the main cables at the center of main span by either rigid clamps or diagonal stays, which are called the center tie and the center diagonal stay, respectively. These aim at constraining the asymmetric vibration mode of torsion and the relative longitudinal displacement between the cables and the girder, resulting in the reduction of secondary bending stress in short hangers and the increase of longitudinal restoring force of the girder. In the case of the diagonal stay, it may be proportioned to fail at a certain service limit state.

26.3.4 Stress Ribbon Bridge

Stress ribbon bridges are a kind of earth-anchored suspension bridges (Figure 26.12). The stretched cables are encased in the concrete slab which is used as the bridge deck. As understood from Equation 26.2, the cable tension due to the dead load and prestressing, in addition to bending rigidity of the deck slab, contribute to the stiffness of the entire bridge. Although the sagged form is inevitable in the stress ribbon bridge, the maximum longitudinal slope is limited from the function of the bridge. On the other hand, very large tensile force in the cable is associated with increase of the cost. Therefore, this type of structure has been applied so far to the pedestrian bridges having short or moderate span length, although the structure is efficient, very slender, and elegant. In their design, care should be taken in forming a smooth transition between the slab and the abutments [12].

26.3.5 Erection

26.3.5.1 Towers

The control of erection accuracy in holding verticality is very important in tower construction. Steel towers are erected by using tower cranes and creeper traveler cranes. When open water area is available, floating cranes are sometimes used together. Because very tall and slender steel towers are vulnerable to wind-induced vibrations, in particular during the free standing stage of erection, some damping devices

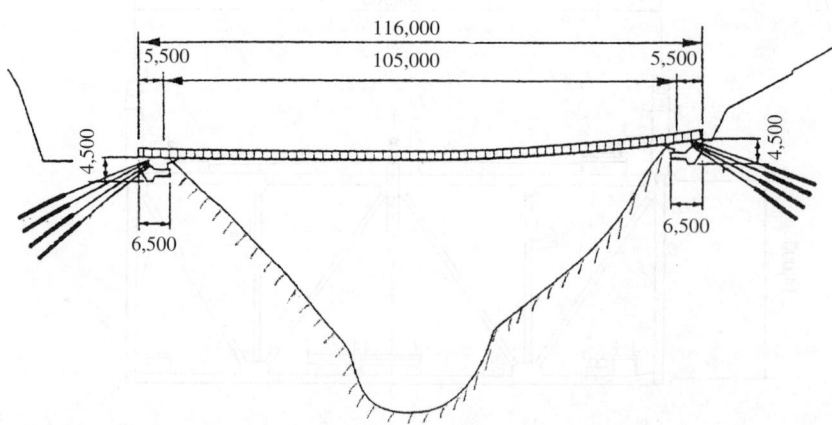

FIGURE 26.12   An example of stress ribbon bridge.

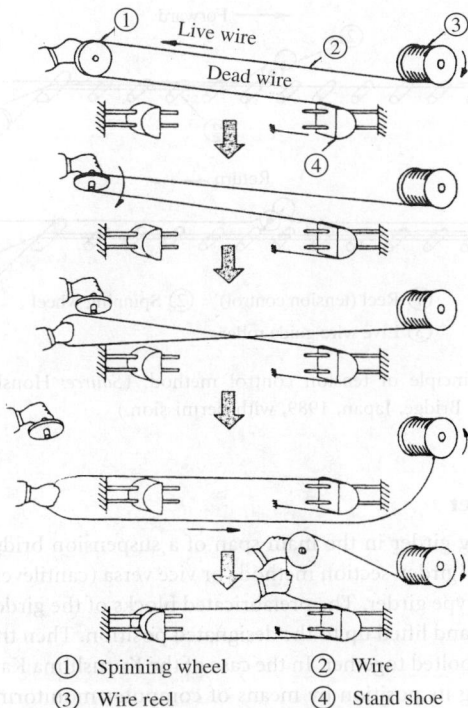

FIGURE 26.13 Operating principle of aerial spinning. (*Source:* Honshu-Shikoku Bridge Authority, Technology of the Seto-Ohashi Bridge, Japan, 1989, with permission.)

are often installed. In the case of the Akashi Kaikyo Bridge, a set of tuned mass dampers are left as permanent fixtures inside the shafts. In constructing concrete towers, the towers are slip-formed in a continuous operation, employing self-climbing forms, tower cranes, and concrete-pumping buckets.

26.3.5.2 Main Cables

The erection of parallel-wire main cables of a suspension bridge is executed by either aerial spinning (AS) method or prefabricated strand (PPWS) method. In the former, the total cable section is assembled on site from individual 5 mm wires pulled across an anchorage to the other anchorage over the tower saddles. The conventional sag-control method spans individual wires in free-hang condition and the sag of each wire is individually adjusted to ensure the wires to be of equal length (Figure 26.13). The problem of this method was the sensitiveness of the quality of the cables and the erection period to such site working conditions as wind environment and spinning equipment.

A new method that is called the tension-control method was developed in Japan (Figure 26.14) and has been applied not only to the Japanese bridges but also to the Second Bosporus and Great Belt bridges. Although adjustment of individual strands is still required, the idea of this method is to keep the tension in the wire constant during cable spinning to obtain uniform wire lengths.

On the other hand, the PPWS method was introduced to reduce the labor and the weather sensitivity and consequently to speed up erection work. Although the limitation of this method may lie in the weight of the strand and the size of the reel, it was successfully applied to the world's longest Akashi Kaikyo Bridge, requiring strands with a length of 4000 m and a strand plus reel weight of 95 tons.

Now that the AS and PPWS methods have been improved step by step and experience has been gained, the method undertaken depends on the cost and period of construction at the specific site or area.

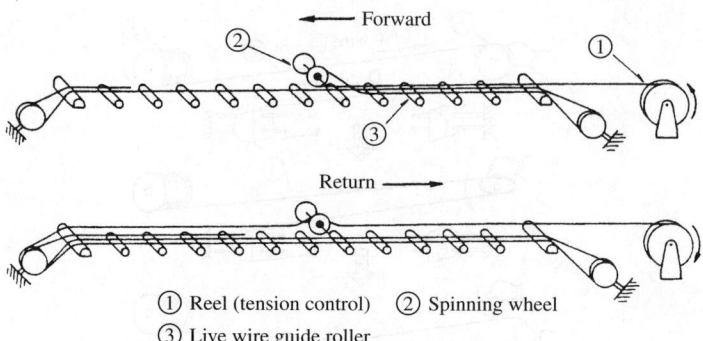

FIGURE 26.14 Operating principle of tension control method. (*Source:* Honshu-Shikoku Bridge Authority, Technology of the Seto-Ohashi Bridge, Japan, 1989, with permission.)

26.3.5.3 Stiffening Girder

The erection of the stiffening girder in the main span of a suspension bridge proceeds either from the mid-span to the both towers (girder-section method) or vice versa (cantilevering method). The former is typically applied to the box-type girder. The prefabricated blocks of the girder are transported to the site by deck barges and tugboats and lifted up to the designated position. Then the blocks are connected with hinges and finally welded or bolted together. In the case of the Kurushima Kaikyo bridges in Japan, a self-controlled barge maintaining its position by means of computer monitoring and a quick-joint system shortening the erection time were developed and used to cope with the fast and complex tidal currents at the site.

On the other hand, the cantilevering method is typically adopted when the open sea area cannot be utilized for erection or when a trussed stiffening girder is to be erected. Preassembled panels of the truss girder are erected by cantilevering them from the towers and anchorage blocks. If the situation allows, some large blocks are erected by floating cranes.

The connections of the girder section during erection is temporarily hinged or rigidly connected. In the former case, the joints are loosely connected until all sections are positioned. It enables simple and easy analysis of the behavior of the structure under erection and usually temporary reinforcement of the members is not needed, while the countermeasures against wind instability should be taken if necessary. In the rigid-connection method, full-splice joints are immediately completed as each girder block is erected into place. This keeps the girder smooth and rigid, resulting in high construction accuracy. But temporary reinforcement of the girders and hangers to resist transient excessive stresses or controlled operation to avoid overstress will be sometimes required.

26.4 Cable-Stayed Bridges

26.4.1 Layout of Structural System

A cable-stayed bridge is considered a girder elastically supported at the anchoring points of stay cables, but the girder is subjected to not only bending and shear but also axial forces due to the horizontal component of tension in stay cables (Figure 26.15). The degree of statical indeterminacy increases with increasing number of stay cables. The jacking devices arranged at the cable anchorages can not only adjust for relaxation in the cables or errors in the cable length but also modify the cable forces, and hence the prestressing by cables can improve the dead load stress distribution in the girder.

The wide design possibility is available in cable-stayed bridges because of a variety of alternatives for configuration, structural system, and relative stiffness of each element. This is the reason why

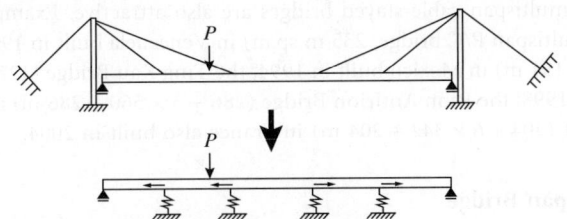

FIGURE 26.15 The equivalent beam of a cable-stayed bridge.

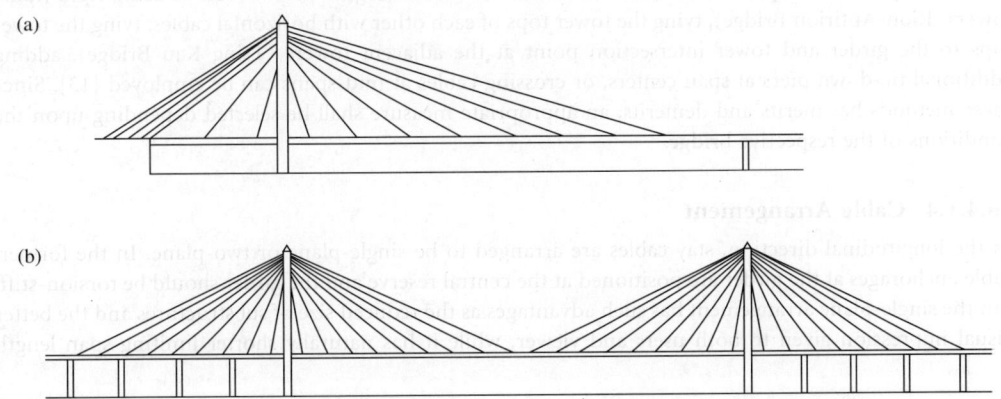

FIGURE 26.16 Typical cable-stayed bridges: (a) partially earth-anchored two-span bridge and (b) three-span bridge with intermediate piers in the side span.

cable-stayed bridges can be applied to not only very long-span bridges but also to short-span pedestrian bridges. In contrast to a suspension bridge, a cable-stayed bridge is typically a closed structural system; in other words, mostly a self-anchored system. Since a cable-stayed bridge can be built usually without massive anchor block and temporary staging, it is particularly advantageous in areas where the soil condition is not so good.

As compared with a suspension bridge, the stiffness of a cable-stayed bridge is greater because the cables are straight though the limit span length of the former may be longer than the latter.

26.4.1.1 Alignment of Bridge

Although most bridges are straight or skewed, a cable-stayed bridge can be designed as a curved bridge. The Katsushika Harp Bridge in Tokyo has the bridge axis forming an S-shape curve because of the complex site conditions from the adjacent rivers and road alignment. The anchors of single-plane stay cables were positioned symmetrically with respect to the tower, so as to minimize the transverse bending moment in the tower. The girder of a curved bridge should be torsion-stiff.

26.4.1.2 Span Allocation

Although the three-span structure is most prevalent, two-span layouts are widely acceptable in a cable-stayed bridge. Even the asymmetric span allocation can be designed economically and be esthetically pleasing. When the side spans are very short, all or some of the stay cables may be earth-anchored (Figure 26.16a). The earth-anchored stay cables make the whole structure stiffer and hence more advantageous in planning super-long-span cable-stayed bridges.

If the situation allows, it is advantageous to provide the intermediate piers in the side spans (Figure 26.16b) or to extend the suspended side spans continuously toward further one or a few spans in order to increase the stiffness of the entire structure.

On the other hand, multispan cable-stayed bridges are also attractive. Examples are the Maracaibo Bridge (cantilevered multispan P/C bridge, 235 m span) in Venezuela built in 1962, the Mezcara Bridge (167 + 311.4 + 299.5 + 191 m) in Mexico built in 1994, the Ting Kau Bridge (127 + 448 + 475 + 127 m) in Hong Kong built in 1998, the Rion Antirion Bridge (286 + 3 × 560 + 286 m) in Greece built in 2004, and the Millau Viaduct (204 + 6 × 342 + 204 m) in France also built in 2004.

26.4.1.3 Multiple-Span Bridge

When a cable-stayed bridge has more than three spans, the bending moment in the intermediate towers may be very large or the structure will be subject to large deformations under the action of live load. In order to cope with the problem, various alternative methods (Figure 26.17) such as using rigid frame towers (Rion-Antirion Bridge), tying the tower tops of each other with horizontal cables, tying the tower tops to the girder and tower intersection point at the adjacent towers (Tsing Kau Bridge), adding additional tie-down piers at span centers, or crossing cables at mid-spans can be employed [13]. Since these methods has merits and demerits, an appropriate measure shall be selected depending upon the conditions of the respective bridge.

26.4.1.4 Cable Arrangement

In the longitudinal direction, stay cables are arranged to be single-plane or two-plane. In the former, cable anchorages at the girder are positioned at the central reserve and the girder should be torsion-stiff. But the single-plane arrangement has such advantages as the reduced size of substructures and the better visual impression given to both users and viewer, while it has naturally shorter limiting span length

FIGURE 26.17 Alternate solutions for multispan cable-stayed bridges (from top: rigid intermediate pylons; head-cables connecting tower tops; cables connecting tower top and an adjacent pylon at the deck level; overlapped stays in the mid-span area). (Adapted from Virlogeux, M., *Struct. Eng. Int.*, IABSE, 1, 61, 2001, with permission.)

compared with a two-plane cable system: currently the Tsurumi Bridge in Yokohama, Japan, has the maximum span length of 510 m. On the other hand, the double-plane system may be formed either as two vertical planes or as twin inclined planes connected from the edge of the deck to either an A frame or inverted Y frame tower. Inclined stays increase the stiffness and stability, particularly in torsion, of the structure and are advantageous for very long spans.

Viewing from the transverse direction, the cable arrangement can be classified into harp, fan, and radial types (Figure 26.18). Although the fan-type system is popular because of the advantages in proportioning bridge components, the harp-type cable system is sometimes preferred for either pleasing appearance or more constraint upon longitudinal displacement of the bridge deck.

The use of a small number of stay cable(s) with large diameter has shifted to a multistay system since the end of the 1960s. The closely spaced multistay system has such advantages that additional bending moment in the girder is reduced, a stay cable is individually replaceable, and the anchorage details become more compact and simple.

26.4.1.5 Extradosed Prestressed Concrete Bridge

In an extradosed prestressed concrete bridge, the concept of which was first proposed by a French engineer U. Mathivat, the cables behave as external prestressing with a large eccentricity and the girder provides a significant part of the global load carrying capacity of the superstructure, whereas a typical cable-stayed bridge has a more flexible deck and relies on the stay cables to provide the global vertical load carrying capacity for the structure. In appearance, however, it is difficult to distinguish an extradosed bridge (e.g., Figure 26.19) from a cable-stayed bridge. Its apparent feature is lower towers and thick deck as compared with a cable-stayed bridge.

Extradosed bridges are used to span normally 100 to 200 m, but the Kiso River Bridge (160 + 3 × 275 + 160 m) and the Ibi River Bridge (154 + 4 × 271.5 + 157 m) near Nagoya, Japan, could span about 275 m by using the steel box girders in the middle part of main span.

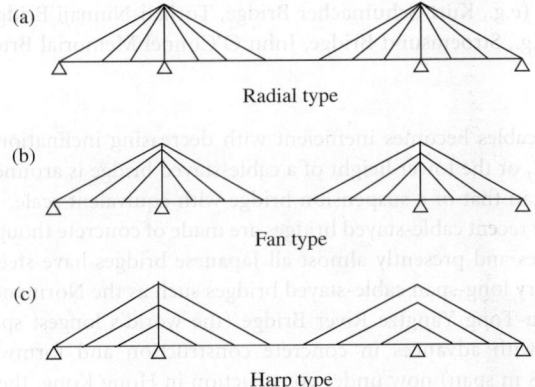

FIGURE 26.18 Stay-cable layout.

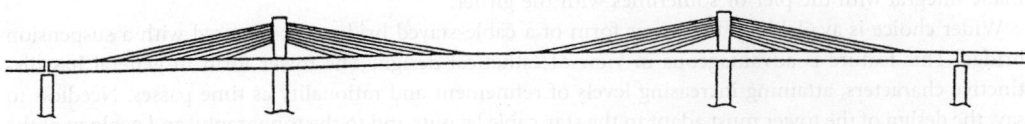

FIGURE 26.19 Extradosed bridge.

26.4.2   Configuration and Design of Structural Components

26.4.2.1   Stay Cables

Since the types, corrosion protection, and vibration control of cables were already mentioned in Section 26.2, just the anchoring techniques of cables in cable-stayed steel bridges will be reviewed herein.

The structural details to anchor stay cables at the main girder or the tower shall be designed so as to smoothly transmit cable tension to the main structural component, to be as simple and compact as possible, to be easy in fabrication and maintenance, and to have good appearance. Care shall be also taken on the fatigue and secondary stresses of both cables and anchorages. Although the types of structural details are manifold, the typical ones are given in the Sections 26.4.2.1.1 and 26.4.2.1.2.

26.4.2.1.1   The Stay Anchor at the Girders

1. Spray saddle/anchor girder type — The combination of spray saddle and anchor girder has been used for thick cables, but is not prevailing now.
2. Anchor girder (block) type — The anchor girder or the anchor block is inserted in the main girder by welding or bolting.
3. Bracket type — The bracket that anchors the stay cable is projected outside the main girder.
4. Pipe anchor type — The pipe that anchors the stay cable is incorporated with the web of a main girder.
5. Gusset type — The cable socket is connected to the gusset extended from the web of a main girder.

26.4.2.1.2   The Stay Anchor at the Tower

In the earlier bridges having a small number of thick stay cables, the hinged or fixed saddle over which cables are continuous was used. But nowadays, use of thin stay cables is prevailing and the typical types of anchorage are as follows:

1. Saddle type (e.g., Zarate-Brazo Largo Bridge, Yokohama Bay Bridge).
2. Anchor girder type (e.g., Speyer Bridge, Tempozan Bridge).
3. Bearing plate type (e.g., Kurt-Schumacher Bridge, Torikai-Ninnaji Bridge).
4. Pin-socket type (e.g., Stroemsund Bridge, John O'Connel Memorial Bridge).

26.4.2.2   Towers

Since the action of stay cables becomes inefficient with decreasing inclination, the stay inclination is usually taken as 25 to 65°, or the tower height of a cable-stayed bridge is around $\frac{1}{4}$ to $\frac{1}{5}$ of the main span and hence much taller than that of a suspension bridge with equivalent scale.

Almost all towers of the recent cable-stayed bridges are made of concrete though steel towers were used on the earlier steel bridges and presently almost all Japanese bridges have steel girders. Now concrete towers can be used for very long-span cable-stayed bridges such as the Normandy Bridge over the Seine (850 m span) and the Su-Tong Yangtze River Bridge (the world's longest span 1088 m) now under construction in China, with advances in concrete construction and formwork technology. In the Stonecutters Bridge (1018 m span) now under construction in Hong Kong, the top one third of 290 m high single-pylon tower will be a composite steel/concrete structural section. The steel skin will be fabricated from stainless steel for reasons of appearance and durability.

Development of bending moment in pylons was prevented in the earlier bridges by the use of rocker or sliding saddles and pinned tower feet, but those of recent bridges except for very short spans are mostly made integral with the pier or sometimes with the girder.

Wider choice is available in the tower form of a cable-stayed bridge as compared with a suspension bridge. This feature is advantageous in view of esthetical design. The tower form in general has distinctive characters, attaining increasing levels of refinement and rationality as time passes. Needless to say, the design of the tower must adapt to the stay cable layouts and to the topography and geology of the bridge site, and carry the forces economically.

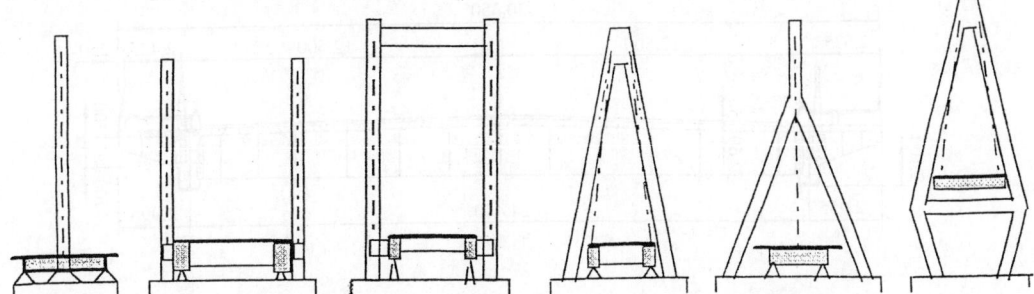

FIGURE 26.20 Various forms of cable-stayed bridge tower.

The typical forms of cable-stayed bridge towers are shown in Figure 26.20. The single pylon adapting to a single-plane cable system has to be usually situated within the central median of the roadway and hence an additional width of deck is required for the necessary clearance to traffic. The inverted Y-shape tower is used for both single-plane and double-plane cable systems, while A- or inverse V-shape towers adapt to a double-plane cable system. The H-shape tower is the most logical form structurally for a two-plane cable system. In the case of the A or inverted Y towers, the spacing of two shafts under the deck is often narrowed to reduce the size of the foundation. These diamond-shape towers have been used for long-span cable-stayed bridges such as the Tatara (Japan), Yangpu (China), and Baytown (United States), but the visual aspect of securing the strength of the portion under the deck should be carefully pursued.

26.4.2.3 Bridge Deck

In cable-stayed bridges, the solid-web girders precede the truss girders except for double-deck designs for both structural and esthetic reasons. These girders or bridge deck are subjected to not only bending and torsional moments as well as shear forces but also axial forces. The closer spacing of cables in a multistay system allows a more slender bridge deck. Although the stability of the girder as a beam-column is generally not yet a serious problem owing to elastic and almost continuous support by stay cables so far, the overall stability of the structure must be checked by using a nonlinear, second-order analysis for super-long-span cable-stayed bridges [14].

As for the cross-section of the steel girder in a cable-stayed bridge, there seems two ways at the moment. One is a single or double box as seen in many long-span cases. Orthotropic steel plate deck is incorporated with the steel girder and the circumscribed shape of the cross-section is usually trapezoidal or hexagonal, mainly for aerodynamic reason. Another way is the use of shallow-plated edge girders connected with cross-girders. For wider bridges and longer span bridges than 500 m, Leonhardt and Zeller [15] suggest the use of similar cross-sections of all steel structures with an orthotropic steel deck, and that no box girder is needed and simple edge beams are sufficient. However, a box girder with high torsional rigidity should be used when the stay cables are arranged in a single plane along the center line of the bridge deck.

The use of a composite steel/concrete girder on cable-stayed bridges may be categorized into the latter. Prefabricated or *in situ* concrete slabs are connected with longitudinal edge girders and cross-beams in steel by stud shear connectors. The compressive forces from the stay cables can be sustained by the concrete slab. The Yangpu Bridge in Shanghai with a long main span of 602 m is provided with double-web edge girders. Figure 26.21 shows two typical examples of steel girders used in large cable-stayed bridges.

One trapezoidal or two rectangular box designs are most popular in prestressed concrete cable-stayed bridges. But as in the case of the composite girder mentioned above, the longitudinal edge beams

(a)

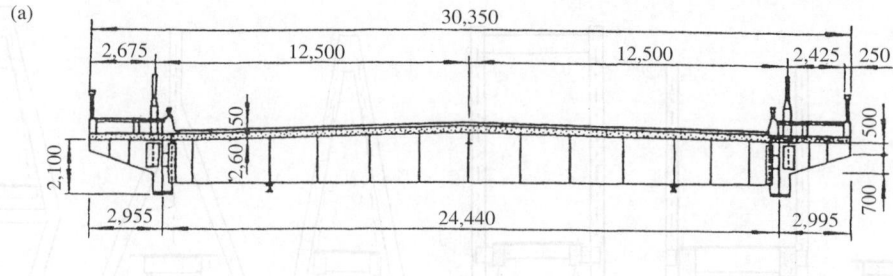

(b)

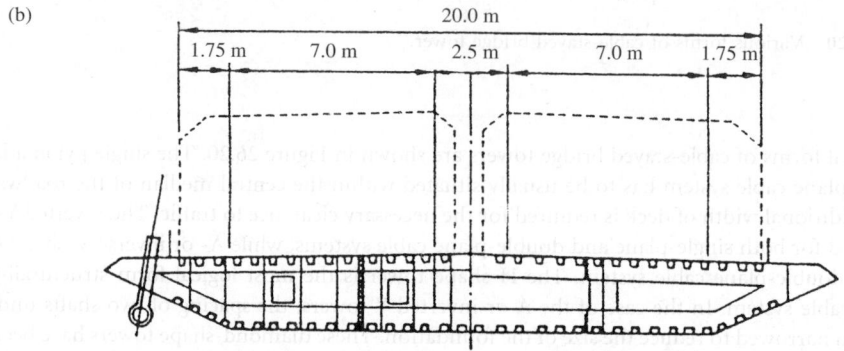

FIGURE 26.21 Examples of steel girder of cable-stayed bridge: (a) composite girder of the Nanpu Bridge and (b) steel box girder of the Tatara Bridge.

connected by the transverse beams and incorporated with the stay anchors are sometimes employed. Even in the prestressed concrete girders, wind nose may be added to the cross-section when the span is long.

When the ratio of the side span to main span length is small, the use of steel girders in the main span and continuously extended concrete girders in the side spans is one of the solutions to attain the rational and economical design. If the situation allows, it is advantageous to provide with intermediate supports for the side-span girders in this case.

Most of the truss girders for cable-stayed bridges are employed in double-deck cases, as exemplified in several Japanese designs. Among them, the Hitsuishijima and Iwakurojima bridges (Figure 26.22) were designed to carry four lanes of roadway traffic on the upper deck and ordinary type railway tracks as well as double Shinkansen rail tracks on the lower deck. Yokohama Bay Bridge, which carries six lanes of roadway traffic on the respective decks, is unique in that the upper chord of its truss is a shallow steel box section stiffening the girder system and concealing power and communication cables. The truss girder of the Higashi-Kobe Bridge is shallow and of Warren type without vertical members, so it appears slender for a double-deck road bridge.

26.4.2.4 Vertical Support of Girder

A variety of supporting conditions have been adopted in modern cable-stayed bridges [16] because the selection of the supporting conditions for longitudinal movement is rather adaptable owing to the existence of stay cables and flexible towers. Less constraint in the bridge axis direction yields longer natural periods of the corresponding motion and thus reduces the seismic inertia forces. However, less constraint on the longitudinal movement of the girder may cause large bending moment in the towers and larger displacement of the girder. The effect of temperature change should be also taken into account. Figure 26.23 illustrates the potential combinations of supporting conditions for three span cases that are most prevalent in long span cable-stayed bridges.

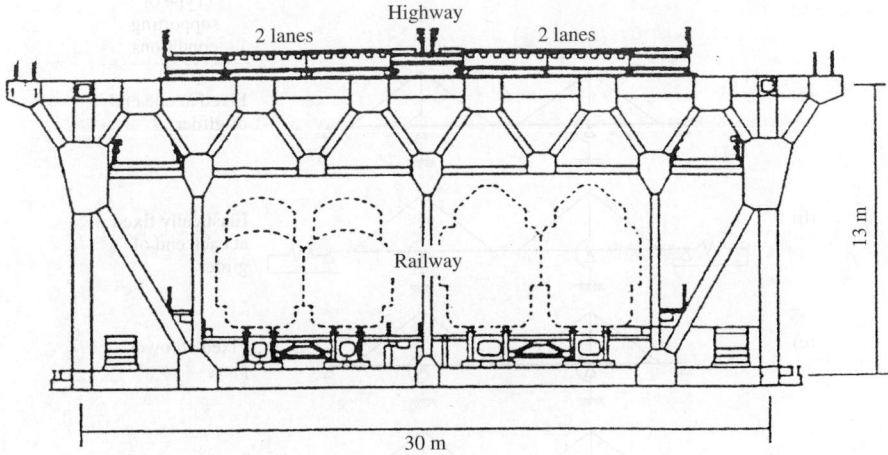

FIGURE 26.22 Typical cross-section of the Seto Bridge.

Type (a) or (e) in the figure, where one support is made as a fixed hinge and all other supports are longitudinally movable, have been widely used for bridges with medium or short span length because temperature effects are released and the seismic force applied to the substructure is relatively small. With increasing span length, design and construction of the pier fixed to the girder become difficult due to the increase of seismic reaction. The supporting condition (e) has been preferred to (a) in Japanese bridges because the expansion of the girder end under temperature change is smaller and the size of the tower foundation is relatively large for reasons other than earthquake effect. The supporting condition (a) may, however, be preferable when the height of the tower below the bridge deck is large and the end support is on the abutment.

When large clearance height is required, the bridge deck is often provided with fixed hinges at both flexible towers as in (c) or is rigidly connected to these towers. The stresses due to temperature change can be released owing to the flexibility of the towers and seismic reactions can be sustained by both tower piers. Some difficulties may arise, however, in giving the towers both the flexibility to absorb deformation of the girder due to temperature change and the stiffness to cope with seismic effects. The additional thrust induced in the girder should be also borne in mind in this case.

Prevalent in long-span cable-stayed steel bridges in Japan are the various devices connecting the girder elastically with towers or abutments ((b), (d), or (h) in Figure 26.23). In the Meikoh-West bridge and some other bridges, elastic restraint was provided by attaching horizontal cables between the girder and the tower, while large belleville springs were fitted to the rocker bearings at each end pier of the Hitsuishijima and Iwakurojima bridges. The purpose of these devices was to reduce and distribute seismic forces, to control the longitudinal movement of the girder, and also to find a compromise with temperature effect.

The advantage of all movable support types, (f) and (g) in Figure 26.23, is to reduce the seismic inertial force of the girder by attaining very long natural periods of longitudinal sway motion. The so-called floating type (g), where the vertical support at the tower is omitted, leads to a noticeable reduction of the bending moment in the girder at the towers. However, care should be given to excessive displacement of the girder and proneness of instability of the towers. In the Higashi-Kobe Bridge, newly developed vane-type dampers were installed on the end piers as stopping devices against unexpectedly severe earthquakes. The use of short tower links in the Yokohama Bay Bridge or thick rubber shoes on the side span of the Ikuchi Bridge was also aimed at optimizing seismic design under the given conditions.

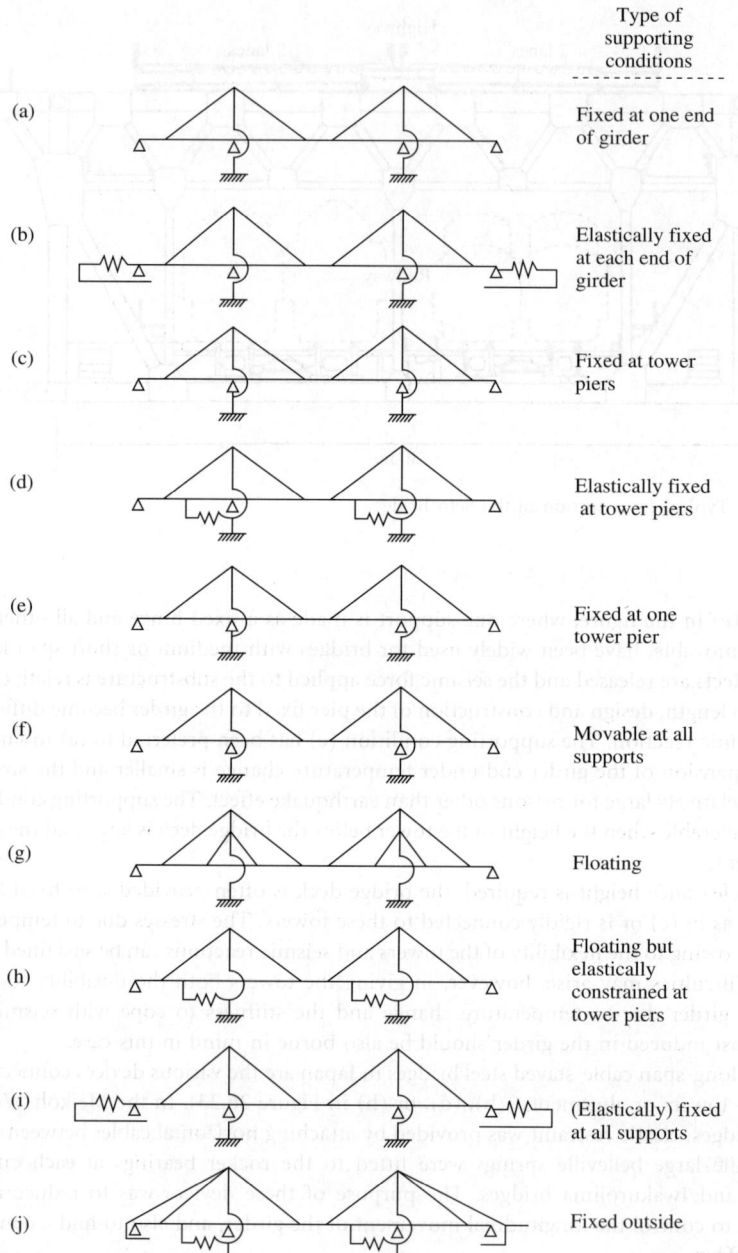

Type of
supporting
conditions
- - - - - - - - - - - - - - -

(a) Fixed at one end
     of girder

(b) Elastically fixed
     at each end of
     girder

(c) Fixed at tower
     piers

(d) Elastically fixed
     at tower piers

(e) Fixed at one
     tower pier

(f) Movable at all
     supports

(g) Floating

(h) Floating but
     elastically
     constrained at
     tower piers

(i) (Elastically) fixed
     at all supports

(j) Fixed outside

FIGURE 26.23   Different supporting systems of cable-stayed bridge.

26.4.3 Erection of Cable-Stayed Bridges

Different from an earth-anchored suspension bridge, the erection of which proceeds from anchorages and towers, to cable, and then stiffening girder, the different main components of the self-anchored cable-stayed bridges may be erected in parallel. During cantilever erection of the girder from the tower toward mid-span, the stay cables will have to be installed and tensioned every time an anchor point at the girder is reached. In some cases, even the construction of the pylons will proceed in parallel with the erection of the girder and stay cables. More detailed description on the erection of cable-stayed bridges are given in the book by Gimsing [17].

The construction methods used on the towers of cable-stayed bridges are typically not different from those of suspension bridges. For smaller or moderate size of steel towers, the erection may be carried out by traveler crane or floating crane with tall boom.

References

[1] Ernst, H.J., Der E-Modul von Seilen unter Beruecksichitigung des Durchhanges, *Bauingenieur*, No. 2, 52, 1965.

[2] Podolny, Jr., W., Current corrosion proyection methods for stay cables, in *IABSE Report* 83-2, *Symp. Extending the Life-Span of Structures*, 855, 1995.

[3] Ito, M., Tada, K., and Kitagawa, M., Cable corrosion protection system for cable-supported bridges in Japan, op. cit. [2], 873.

[4] Lapsley, R.D. and Granz, H.R., Experience, developments and trends for improved durability of stay cables, op. cit. [2], 879.

[5] *Recommendations for Stay Cable Design, Testing and Installation*, 4th ed., Post-Te.isioning Institute, 2001, sec. 5.2.

[6] Maeda, K. et al., Applicability of CFRP cables to ultra long span suspension bridges, *IABSE Report* 84, *Conf. Cable-Supported Bridges* (in CD-ROM), 2001.

[7] Lin, T.Y. and Chow, P., Gibraltar Strait crossing — a challenge to bridge and structural engineering, *Struct. Eng. Int., IABSE*, No. 2, 53, 1991.

[8] Ochsendorf, J.A. and Billington, D.P., Self-anchored suspension bridges, *J. Bridge Eng., ASCE*, 4–3, 151, August 1999.

[9] Peery, D.J., An influence line analysis for suspension bridges, *Proc. ASCE* (ST), 80-581, 558, 1954.

[10] Okukawa, A., Suzuki, S., and Harazaki, I., Suspension bridges, in *Bridge Engineering Handbook*, Chen, W.-F. and Duan, L., Eds., CRC Press, Boca Raton, FL, 2000, chapter 18.

[11] Ito, M., Suppression of wind-induced vibrations of structures, in *A State of the Art in Wind Engineering*, IAWE, Wiley Eastern, New Delhi, 281, 1995.

[12] Schlaich, J. and Engelsmann, S., Stress ribbon concrete bridges, *Struct. Eng. Int., IABSE*, 4, 271, 1996.

[13] Virlogeux, M., Bridges with multiple cable-stayed spans, *Struct. Eng. Int., IABSE*, 1, 61, 2001.

[14] Tang, M.-C., Buckling of cable-stayed girder bridges, *Proc. ASCE*, 102-ST9, 1695, 1976.

[15] Leonhardt, F. and Zeller, W., Past, present and future of cable-stayed bridges, in *Cable-Stayed Bridges — Recent Developments and their Future*, Ito, M. et al., Eds., Elsevier, Amsterdam, 1, 1991.

[16] Ito, M., Supporting devices of long span cable-stayed bridge girder, in *Innovative Large Span Structures*, Srivastava, N.K. et al., Eds., Canadian Soc. Civil Engineers, Montreal, 2, 255, 1992.

[17] Gimsing, N.J., *Cable Supported Bridges — Concept & Design*, 2nd ed., Wiley, New York, 1998, chapter 7.

27

Cooling Tower Structures

Phillip L. Gould
Department of Civil Engineering,
Washington University,
St. Louis, MO

Wilfried B. Krätzig
Department of Civil Engineering,
Ruhr-University Bochum,
Bochum, Germany

27.1 Introduction

Hyperbolic cooling towers are large, thin shell reinforced concrete structures that contribute to environmental protection and to power generation efficiency and reliability. As shown in Figure 27.1, they may dominate the landscape but they possess a certain esthetic eloquence due to their doubly curved form. The operation of a cooling tower is illustrated in Figure 27.2. In a thermal power station, heated steam drives the turbogenerator that produces electric energy. To create an efficient heat sink at the end of this process, the steam is condensed and recycled into the boiler. This requires a large amount of cooling water, whose temperature is raised and then recooled in the tower.

In a so-called "wet" natural draft cooling tower, the heated water is distributed evenly through channels and pipes above the fill. As the water flows and drops through the fill sheets, it comes into contact with the rising cooler air. Evaporative cooling occurs and the cooled water is then collected in the water basin to be recycled into the condenser. The difference in density of the warm air inside and the colder air outside creates the natural draft in the interior. This upward flow of warm air, which leads to a continuous stream of fresh air through the air inlets into the tower, is protected against atmospheric turbulence by the reinforced concrete shell. The cooling tower shell is supported by a truss or framework of columns bridging the air inlet to the tower foundation.

There are also "dry" cooling towers that operate simply on the basis of convective cooling. In this case, the water distribution, the fill, and the water basin are replaced by a closed piping system around the air inlet, resembling, in fact, a gigantic automobile radiator. While dry cooling towers are doubtless superior

FIGURE 27.1   Computer vision of the lignite power plant at Niederaussem.

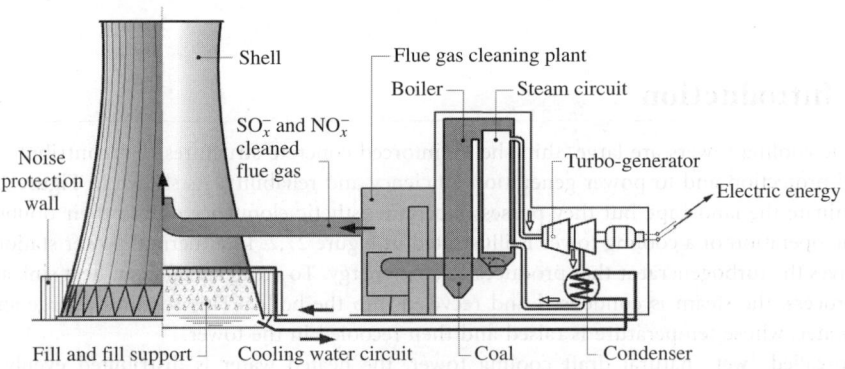

FIGURE 27.2   Thermal power plant with cleaned flue gas injection.

from the point of view of environmental protection, their thermal efficiency is only about 30% of comparable wet towers. If the flue gas is cleaned by a washing technology, it is frequently discharged into the atmosphere by the cooling tower upward flow. This saves reheating of the cleaned flue gas and the construction of a smoke stack (see Figure 27.2).

Figure 27.3 summarizes the historical development of natural draft cooling towers. Technical cooling devices first came into use at the end of the 19th century. The well-known hyperbolic shape of cooling towers was introduced by two Dutch engineers, Van Iterson and Kuyper, who in 1914 constructed the first hyperbolic towers, which were 35 m high. Soon, capacities and heights increased until around 1930,

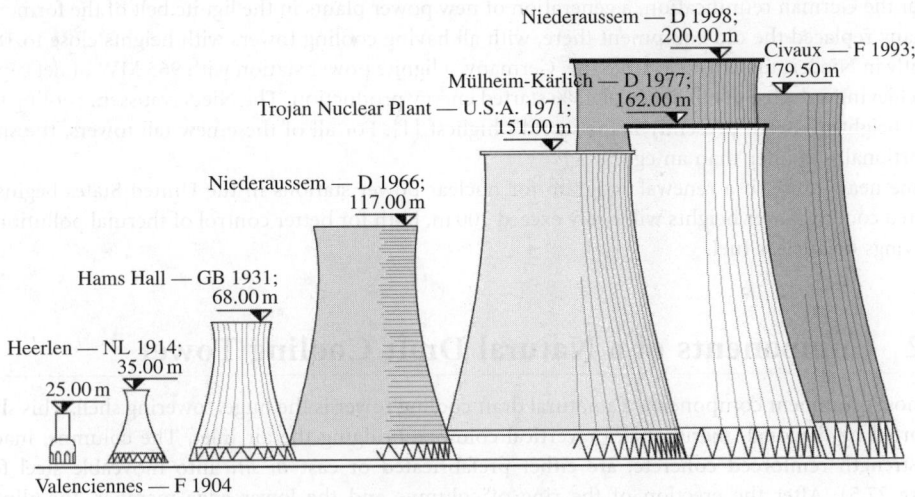

Niederaussem — D 1998;
200.00 m

Civaux — F 1993;
179.50 m

Mülheim-Kärlich — D 1977;
162.00 m

Trojan Nuclear Plant — U.S.A. 1971;
151.00 m

Niederaussem — D 1966;
117.00 m

Hams Hall — GB 1931;
68.00 m

Heerlen — NL 1914;
35.00 m

25.00 m

Valenciennes — F 1904

FIGURE 27.3   Historical development of natural draft cooling towers.

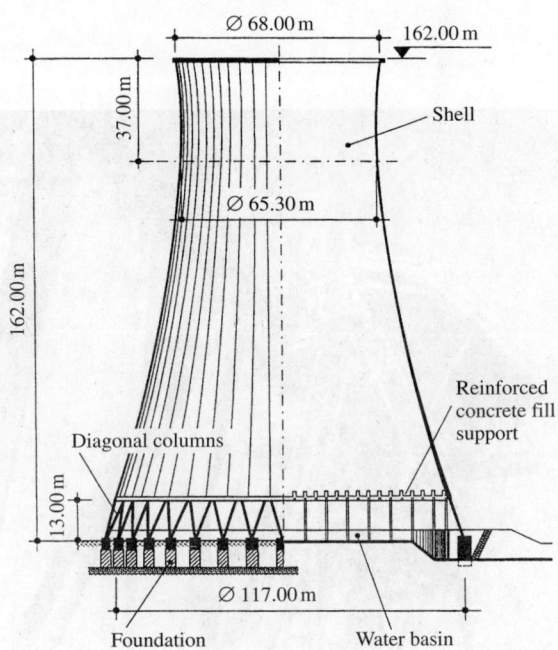

Ø 68.00 m   162.00 m

37.00 m

Shell

Ø 65.30 m

162.00 m

Reinforced
concrete fill
support

Diagonal columns

13.00 m

Ø 117.00 m

Foundation   Water basin

FIGURE 27.4   Cooling tower, Gundremmingen, Germany.

when tower heights of 65 m were achieved. The first such structures to reach higher than 100 m were the towers of the High Marnham Power Station in Britain.

In the 1970s, cooling towers for nuclear power plants exceeded elevations of 160 m, having key dimensions as given in Figure 27.4. The increase of efficiency at French nuclear stations led to tower heights close to 190 m, demonstrated by the highest cooling tower for nearly one decade in Civaux, France.

After the German reunification, a generation of new power plants in the lignite belt of the former East Germany replaced the old equipment there, with all having cooling towers with heights close to 180 m. Recently in Niederraussem, near Cologne, Germany, a lignite power station with 965 MW of net capacity and achieving a degree of efficiency of 43% started energy production. The Niederraussem cooling tower with a height of 200 m presently is the world's highest [1]. For all of these new tall towers, the shell is proportionally thinner than an egg.

In the near future, if a renewal program for nuclear power stations in the United States begins, the expected cooling tower heights will easily exceed 200 m, both for better control of thermal pollution and for savings of nuclear fuel.

27.2　Components of a Natural Draft Cooling Tower

The most prominent component of a natural draft cooling tower is the huge, towering shell. This shell is supported by diagonal, meridional, or vertical columns bridging the air inlet. The columns, made of high-strength reinforced concrete, are either prefabricated or cast *in situ* into moveable steel forms (Figure 27.5). After the erection of the ring of columns and the lower edge member, the climbing formwork is assembled and the stepwise climbing construction of the cooling tower shell wall or veil begins (Figure 27.6). Fresh concrete and reinforcement steel are supplied to the working site by a central

FIGURE 27.5　Fabrication of supporting columns.

FIGURE 27.6 Climbing construction of the shell.

crane anchored to the completed parts of the shell and are placed in lifts up to 2 m high (Figure 27.7). After sufficient strength has been gained, the complete forms are raised for the next lift.

To enhance the durability of the concrete and to provide sufficient cover for the reinforcement, the cooling tower shell thickness should not be less than 16 to 18 cm. The shell itself should be sufficiently stiffened by upper and lower edge members. In order to achieve sufficient resistance against instability, large cooling tower shells may be stiffened by additional internal or external rings. These stiffeners may also be used for repair or rehabilitation of deficient shells.

Wet cooling towers have a water basin with a cold water outlet at the base. These are both large engineered structures, able to handle up to 50 m^3/s of water circulation, as indicated in Figure 27.8. The fill construction inside the tower is a conventional frame structure and is always prefabricated. It carries the water through a large piping system, the spray nozzles, and the fill package. Often, dripping traps are applied on the upper surfaces of the fill to keep water losses through the uplift stream under 1%. Finally, noise protection elements around the inlet decrease the noise caused by the continuously dripping water, as illustrated in Figure 27.2.

27.3 Damage, Failures, and Loss of Durability

Today's natural draft cooling towers are safe and durable structures if properly designed and constructed. Nevertheless, it should be recognized that this high-quality level has been achieved only after the lessons learned from a series of collapsed or heavily damaged towers have been incorporated into the relevant body of engineering knowledge.

FIGURE 27.7   Formwork and steel reinforcement of shell wall.

FIGURE 27.8   Water basin.

While cooling towers have been the largest existing shell structures for many decades, their design and construction were formerly carried out simply by following the existing "recognized rules of crafts-manship," which had never envisaged constructions of this type and scale. This changed radically, however, in the wake of the Ferrybridge failures in 1965 [2]. On November 1, 1965, three of eight 114 m high cooling towers collapsed during a Beaufort 12 gale in an obviously identical manner (Figure 27.9).

FIGURE 27.9 Collapse of Ferrybridge Power Station shell.

Within a few years of this spectacular accident, the response phenomena of cooling towers had been studied in detail, and safety concepts with improved design rules were developed. These international research activities gained further momentum after the occurrence of failures in Ardeer (Britain) in 1973, Bouchain (France) in 1979, and Fiddler's Ferry (Britain) in 1984, the last case clearly displaying the influence of dynamic and stability effects.

In surveying these failures, one can recognize at least four common circumstances:

1. The maximum design wind speed was often underestimated, so that the safety margin for the wind load was insufficient.
2. Group effects leading to higher wind speeds and increased vortex shedding influence on downstream towers were neglected.
3. Large regions of the shell were reinforced only in one central layer (in two orthogonal directions) or the double layer reinforcement was insufficient.
4. The towers had no upper edge members or the existing members were too weak to stiffen the structure against dynamic wind actions.

Two towers in the United States, at Willow Island, West Virginia, and at Port Gibson, Mississippi, were heavily damaged during their construction stage, the latter by a tornado-initiated crane impact. The Port Gibson tower repair included the addition of ring stiffeners [3]. Another tower in Poland collapsed without any definitive explanation having been published up to now, but probably because of large imperfections.

In addition to these cases, cracking of many cooling towers has been observed, often due to ground motions following underground coal mining or just because of faulty design and construction.

One of the main European design mistakes was in the amount of circumferential reinforcement, too low to resist shell bending due to wind dynamics, thermal, and hydrated effects. Such deficiencies can lead to extensive, even progressive vertical crack-damage on the outer shell faces [4]. Obviously, any visible crack in a cooling tower shell is an indication of deterioration of its safety, reliability, and durability. Corrosion of concrete with low resistance to SO_x and NO_x contents in the air around fossil fuel plants adds severely to this degradation. It is thus imperative to conform to a design and construction concept that guarantees sufficiently safe and reliable structures over a predetermined lifetime.

Although power plant construction over much of the industrialized west has slowed in the last decade, research and development on the structural aspects of hyperbolic cooling towers has continued [5,6] and a new wave of construction for these impressive structures seems to be approaching. Engineers face this challenge with confidence in their improved analytical tools, in their ability to employ improved materials, and in their valuable experience in construction.

27.4 Geometry

The main elements of a cooling tower shell in the form of a hyperboloid of revolution are shown in Figure 27.10. This form falls into the class of structures known as thin shells. The cross-section as shown depicts the ideal profile of a shell generated by rotating the hyperboloid $R = f(Z)$ about the vertical (Z) axis. The coordinate Z is measured from the throat while z is measured from the base. All dimensions in the R–Z plane are specified on a reference surface, theoretically the middle surface of the shell but possibly the inner or outer surface. Dimensions through the thickness $h(Z)$ are then referred to this surface. There are several variations possible on this idealized geometry such as a cone-toroid with an upper and lower cone connected by a toroidal segment, two hyperboloids with different curves meeting at the throat, and an offset of the curve describing the shell wall from the axis of rotation. Classically, the columns are arranged as V- or X-frameworks, the latter for very high air inlets; vertical (meridional) columns now popular and were used for the world's highest tower shown later in Figure 27.17.

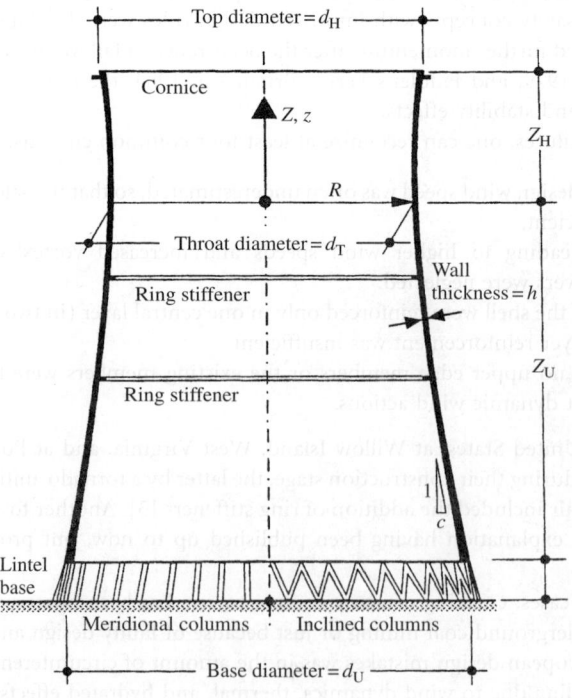

FIGURE 27.10   Hyperbolic cooling tower.

Referring to Figure 27.10, the equation of the generating curve is given by

$$4R^2/d_T^2 - Z^2/b^2 = 1 \tag{27.1}$$

where d_T is the diameter at the throat and b is a characteristic dimension of the shell that may be evaluated from the diameters d_H and d_U by

$$b = d_T Z_H / \sqrt{(d_H^2 - d_T^2)} \tag{27.2}$$

or from

$$b = d_T Z_U / \sqrt{(d_U^2 - d_T^2)} \tag{27.3}$$

if the upper and lower curves are different. The dimension b is related to the slope of the asymptote of the generating hyperbola (see Figure 27.10) by

$$b = 2d_T/c \tag{27.4}$$

For a cooling tower shell where the axis of the generating hyperboloid is offset from the axis of the cooling tower shell middle surface by a constant radial distance R_o, Equations 27.1–27.4 are modified slightly. However, the former equations may be used directly if the dimension b in Equations 27.2 and 27.3 is calculated using the values of d_H and d_T referred to the axis of the hyperboloid. Then, R_o is added to the value of R calculated by Equation 27.1 to determine the total radial distance from the axis of the cooling tower to the middle surface of the shell at any vertical coordinate Z. Such a shell is known in Figure 27.3, where the value of $R_o = -1.0370$ m for the part of the shell below the throat.

27.5 Loading

Hyperbolic cooling towers may be subjected to a variety of loading conditions. Most commonly, these are dead load (D), wind load (W), earthquake load (E), temperature variations (T), construction loads (C), and settlement (S). For the proportioning of the elements of the cooling tower, the effects of the various loading conditions should be factored and combined in accordance with the applicable codes or standards. If no other codes or standards specifically apply, the factors and combinations given in ASCE 7 [7] are appropriate.

Dead load consists of the self-weight of the shell wall and the ribs and the superimposed load from attachments and equipment. Wind loading is extremely important in cooling tower design for several reasons. First, the amount of reinforcement, beyond a prescribed minimum level, is often controlled by the *net difference* between the *tension* due to wind loading and the dead load *compression*, and is therefore especially sensitive to variations in the wind-produced tension. Second, the quasistatic velocity pressure on the shell wall is sensitive to the vertical variation of the wind, as it is for most structures, and also to the circumferential variation of the wind around the tower, which is peculiar to cylindrical bodies. While the vertical variation is largely a function of the regional climatic conditions and the ground surface irregularities, the circumferential variation is strongly dependent on the roughness properties of the shell wall surface. There are also additional wind effects such as internal suction, dynamic amplification, and group configuration.

The external wind pressure acting at any point on the shell surface is computed as [5,8]

$$q(z,\theta) = q(z)H(\theta)(1 + g) \tag{27.5}$$

where $q(z)$ is the effective velocity pressure at a height z (Figure 27.10) above the ground level, $H(\theta)$ is the coefficient for circumferential distribution of external wind pressure, and $1 + g$ is the gust response factor, and g is the peak factor.

As mentioned above, $q(z)$ should be obtained from applicable codes or standards such as Ref. [7].

The circumferential distribution of the wind pressure is denoted by $H(\theta)$ and is shown in Figure 27.11. The key regions are the windward meridian, $\theta = 0°$, the maximum side suction, $\theta \approx 70°$, and the back suction, $\theta \geq 90°$. These curves were determined by laboratory and field measurements as a function of

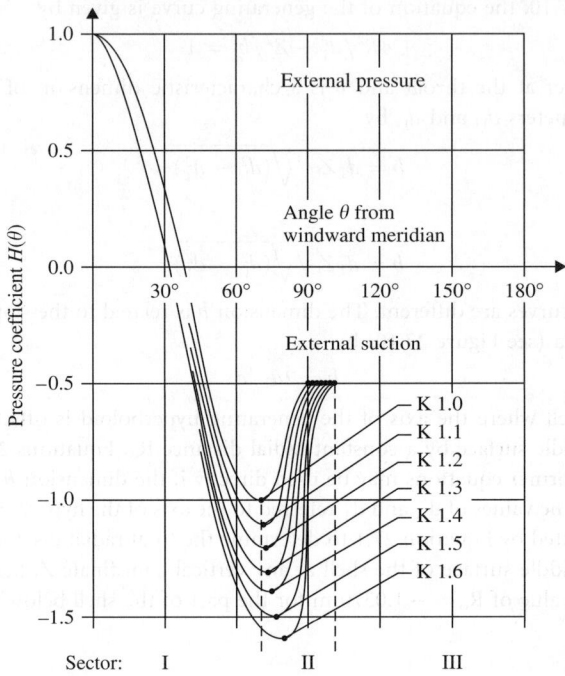

FIGURE 27.11 Types of circumferential pressure distribution.

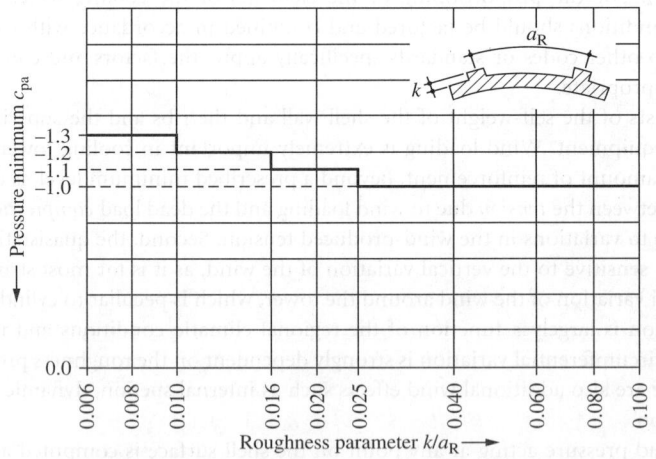

Execution	Roughness parameter k/a_R	Pressure min c_{pa}	Pressure distribution
With ribs	0.025–0.100	−1.0	K 1.0
	0.016–0.025	−1.1	K 1.1
	0.010–0.016	−1.2	K 1.2
	0.006–0.010	−1.3	K 1.3
Without ribs	Smooth	−1.5	K 1.5
		−1.6	K 1.6

FIGURE 27.12 Surface roughness and type of pressure distribution.

the roughness parameter k/a_R as shown in Figure 27.12, in which k is the height of the rib and a_R is the mean distance between the ribs measured at about one-third of the height of the tower. Note that the coefficient along the windward meridian $H(\theta)$ reflects the so-called *stagnation pressure* while the side suction is, remarkably, significantly affected by the surface roughness k/a_R. As will be discussed in Section 27.6.2, the meridional forces in the shell wall and hence the required reinforcing steel are very sensitive to $H(\theta)$. In turn, the costs of construction are affected. Thus, the design of the ribs, or of alternative roughness elements, is an important consideration. For quantitative purposes, the equations of the various curves are given in Table 27.1 and tabulated values at 5° intervals are available [4].

The circumferential distribution of the external wind pressure may be presented in another manner that accents the importance of the asymmetry. If the distribution $H(\theta)$ is represented in a Fourier cosine series of the form

$$H(\theta) = \sum_{n=0}^{n=\infty} A_n \cos n\theta \qquad (27.6)$$

the Fourier coefficients A_n for a distribution most similar to the curve for K 1.4 are as follows [5]:

n	A_n
0	−0.3922
1	0.2602
2	0.6024
3	0.5046
4	0.1064
5	−0.0948
6	−0.0186
7	0.0468

Representative modes are shown in Figure 27.13. The $n=0$ mode represents uniform expansion and contraction of the circumference, while $n=1$ corresponds to beam-like bending about a diametrical axis resulting in translation of the cross-section. The higher modes $n>1$ are peculiar to shells in that they produce undulating deformations around the cross-section with no net translation. The relatively large Fourier coefficients associated with $n=2, 3, 4, 5$ indicate that a significant portion of the loading will cause shell deformations in these modes. In turn, the corresponding local forces are significantly higher than a beam-like response would produce.

To account for the internal conditions in the tower during operation, it is common practice to add an axisymmetric internal suction coefficient $H=0.5$ to the external pressure coefficients $H(\theta)$. In terms of the Fourier series representation, this would increase A_0 to −0.8922.

TABLE 27.1 Equations for $H(\theta)$

Curve	Minimum pressure	Sector I	II	III	c_W
K 1.0	−1.0	$1 - 2.0\left(\sin\dfrac{90}{71}\theta\right)^{2.267}$	$-1.0 + 0.5\left(\sin\left(\dfrac{90}{21}(\theta - 70)\right)\right)^{2.395}$	−0.5	0.66
K 1.1	−1.1	$1 - 2.1\left(\sin\dfrac{90}{71}\theta\right)^{2.239}$	$-1.1 + 0.6\left(\sin\left(\dfrac{90}{22}(\theta - 71)\right)\right)^{2.395}$	−0.5	0.64
K 1.2	−1.2	$1 - 2.2\left(\sin\dfrac{90}{71}\theta\right)^{2.205}$	$-1.2 + 0.7\left(\sin\left(\dfrac{90}{23}(\theta - 72)\right)\right)^{2.395}$	−0.5	0.60
K 1.3	−1.3	$1 - 2.3\left(\sin\dfrac{90}{73}\theta\right)^{2.166}$	$-1.3 + 0.8\left(\sin\left(\dfrac{90}{24}(\theta - 73)\right)\right)^{2.395}$	−0.5	0.56
K 1.5	−1.5	$1 - 2.5\left(\sin\dfrac{90}{75}\theta\right)^{2.104}$	$-1.5 + 1.0\left(\sin\left(\dfrac{90}{27}(\theta - 75)\right)\right)^{2.395}$	−0.5	0.49
K 1.6	−1.6	$1 - 2.6\left(\sin\dfrac{90}{76}\theta\right)^{2.085}$	$-1.6 + 1.1\left(\sin\left(\dfrac{90}{28}(\theta - 76)\right)\right)^{2.395}$	−0.5	0.46

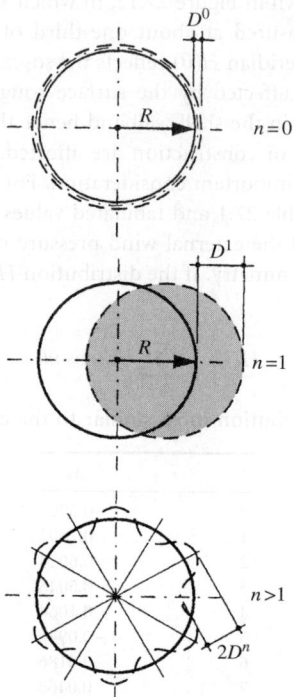

FIGURE 27.13   Harmonic components of the radial displacement.

The dynamic amplification of the effective velocity pressure is represented by the parameter g in Equation 27.5. This parameter reflects the resonant part of the response of the structure and may be as much as 0.2, depending on the dynamic characteristics of the structure. However, when the basis of $q(z)$ includes some dynamic portion such as the fastest-mile-of-wind, $(1 + g)$ is commonly taken as 1.0.

Cooling towers are often constructed in groups and close to other structures, such as chimneys or boiler houses, which may be higher than the tower itself. When the spacing of towers is closer than 1.5 times the base diameter or two times the throat diameter, or when other tall structures are nearby, the wind pressure on any single tower may be altered in shape and intensity. Such effects should be studied carefully in boundary-layer wind tunnels in order not to overlook dramatic increases in the wind loading.

Earthquake loading on hyperbolic cooling towers is produced by ground motions transmitted from the foundation through the supporting columns and the lintel into the shell. If the base motion is assumed to be uniform vertically and horizontally, the circumferential effects are axisymmetrical ($n = 0$) and anti-symmetrical ($n = 1$), respectively (see Figure 27.13). In the meridional direction, the magnitude and dis-tribution of the earthquake-induced forces is a function of the mass of the tower and the dynamic properties of the structure (natural frequencies and damping) as well as the acceleration produced by the earthquake at the base of the structure. The most appropriate technique for determining the loads applied by a design earthquake to the shell and components is the response spectrum method, which, in turn, requires a free vibration analysis to evaluate the natural frequencies [8–10]. It is common to use elastic spectra with 5% of critical damping. The supporting columns and foundation are critical for this loading condition and should be modeled in appropriate detail [9,10]. Just as for the wind load case, earthquake loading may produce tensile forces opposite to the dead load compression, so the net effect may be quite sensitive.

Temperature variations on cooling towers arise from two sources: operating conditions and sunshine on one side. Typical operating conditions for Central European and North American climatic conditions are

FIGURE 27.14 Central crane during construction attached to shell wall.

an external temperature of $-15°C$ and internal temperature of $+30°C$. This is an axisymmetrical effect, $n = 0$ in Figure 27.13. For sunshine, a temperature gradient of $25°C$ constant over the height and distributed as a half-wave around one-half of the circumference is appropriate. This loading would require a Fourier expansion in the form of Equation 27.6 and higher harmonic components, $n > 1$, to be considered.

For power plants operating as the base supply (e.g., nuclear plants) and situated in cold winter regions, the inner cooling tower faces may be wetted by condensed vapor for several months. Such hydrated actions may cause swelling of the inner concrete, the intensity of which is dependent on the properties of gravel and cement. Without detailed concrete investigations, an additional thermal gradient of $\sim 10°C$ (warmer at the inner face) may account for this deteriorating effect.

Construction loads are generally caused by the fixing devices of climbing formwork, by tower crane anchors, and by attachments for material transport equipment as shown in Figure 27.14. These loads must be considered on the portion of the shell extant at the phase of construction.

Nonuniform settlement due to varying subsoil stiffness may be a consideration. Such effects should be modeled considering the interaction of the foundation and the soil.

27.6 Methods of Analysis

27.6.1 Response Behavior

Thin shells may resist external loading through forces acting parallel to the shell surface, forces acting perpendicular to the shell surface, and moments. While the analysis of such shells may be formulated within the three-dimensional theory of elasticity, there are reduced theories that are two-dimensional and are expressed in terms of force and moment *intensities*. These intensities are traditionally based on a reference surface, generally the middle surface, and are forces and moments per unit length of the middle surface element upon which they act: they are called *stress resultants* and *stress couples*, respectively, and are associated with the three directions — circumferential, θ^1; meridional, θ^2; and normal, θ^3. In Figure 27.15, the extensional stress resultants, n_{11} and n_{22}, the in-plane shearing stress resultants, $n_{12} = n_{21}$, and the transverse shear stress resultants, $q_{12} = q_{21}$, are shown in the left diagram along with the components of the applied loading in the circumferential, meridional, and normal directions, p_1, p_2, and p_3, respectively. The bending stress couples, m_{11} and m_{22}, and the twisting stress couples, $m_{12} = m_{21}$, are shown in the right diagram along with the displacements v_1, v_2, and v_3 in the respective directions.

Historically, doubly curved thin shells have been designed to resist applied loading primarily through the extensional and shearing forces in the "plane" of the shell surface, as opposed to the transverse shears and bending and twisting moments, which predominate in flat plates loaded normally to their surface. This is known as *membrane* action, as opposed to *bending* action, and is consistent with an accompanying theory and calculation methodology having the advantage of being statically determinate. This methodology was well suited for the precomputer age and enabled many large thin shells, including cooling towers, to be rationally designed and economically constructed [5]. Because the conditions that must be provided at the shell boundaries in order to insure membrane action are not always achievable, shell bending should be taken into account, even for shells designed by membrane theory. Remarkably, the accompanying bending often is confined to narrow regions in the vicinity of the boundaries and other discontinuities and may have only a minor effect on the shell design, such as local thickening and additional reinforcement. Many clever and insightful techniques have been developed over the years to approximate the effects of local bending in shells designed by the membrane theory.

As we have passed into and advanced in the computer age, it is no longer appropriate to use the membrane theory to analyze such extraordinary thin and massive shells, except perhaps for preliminary design purposes. The finite element method is widely accepted as the standard contemporary analysis technique and the attention shifts to the level of sophistication required. Usually, the more sophisticated the analysis, more data are required. Consequently, a model may evolve through several stages, starting with a relatively simple version that enables the structure to be sized, to the most complex version that may depict such phenomena as the sequence of progressive collapse of the as-built shell under various static and dynamic loading scenarios, the incremental effects of the progressive stages of construction, the influence of the operating environment, aging and deterioration on the structure, etc. The techniques described in the following paragraphs form a hierarchical progression from the relatively simple to the very complex, depending on the objective of a particular analysis.

In modeling cooling tower shells using the finite element method, there are a number of options. For the shell wall, a variety of ring elements, triangular elements, and quadrilateral elements have been used. Early on, flat elements adapted from two-dimensional elasticity and plate formulations were used to

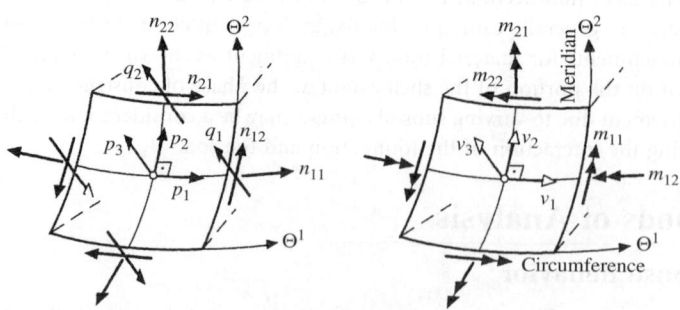

FIGURE 27.15 Surface loads, stress resultants, stress couples, and displacements.

approximate the doubly curved surface. Such elements present a number of theoretical and computational problems and are *not* recommended for the analysis of shells. Currently, shell elements degenerated from three-dimensional solid elements are very popular. These elements have been utilized in both ring and quadrilateral forms.

The column region at the base of the shell presents a special modeling challenge. For static analysis, the lower boundary is often idealized as a uniform support at the lintel level. Then, a portion of the lower shell and the columns is considered in a subsequent analysis to account for the concentrated actions of the columns, which may penetrate only a relatively short distance into the shell wall. For dynamic analysis, it is important to include the column region along with the veil in the model. An equivalent shell element has proved useful in this regard if ring elements are used to model the shell [9,10]. It may also be desirable to include some of the foundation elements, such as a ring beam at the base and even the supporting piles in a dynamic or settlement model. Figure 27.16 shows such a finite element model for the final tower design with approximately 51,000 degrees of freedom, shown here in a five times coarser resolution.

27.6.2 Linear Elastic Analysis

The linear static analysis method is based on the classical bending theory of thin shells. While this theory has been formulated for many years, solutions for doubly curved shells have not been readily achievable until the development of computer-based numerical methods, most notably the finite element method. The outputs of such an analysis are the stress resultants and couples, defined in Figure 27.15, over the entire shell surface and the accompanying displacements. The analysis is based on the initial geometry, linear elastic material behavior, and a linear kinematic law. Representative results of such analyses for a

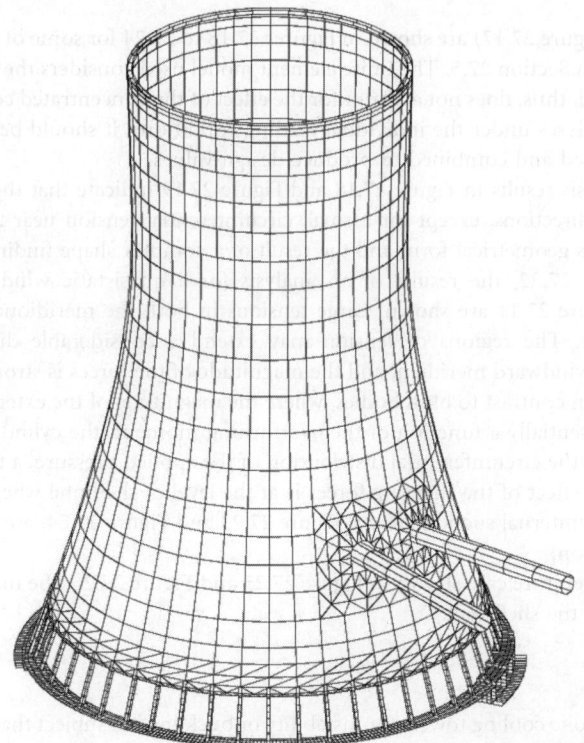

FIGURE 27.16 Finite element mesh for tower design including shell, supports, and foundation.

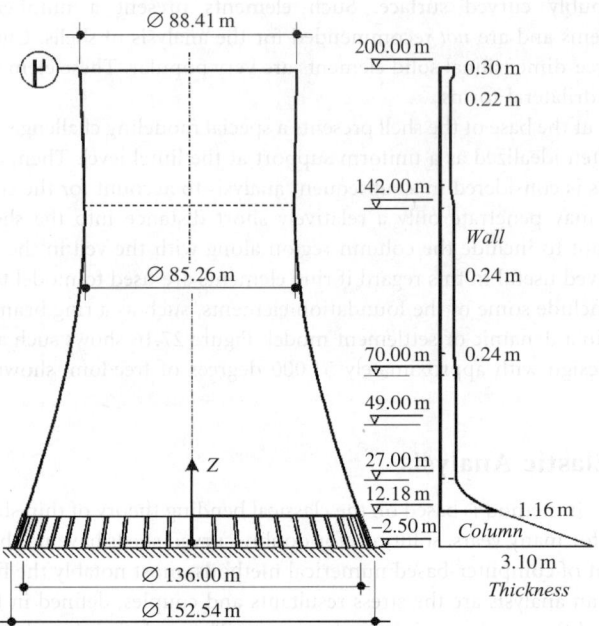

Ø 88.41 m

200.00 m

0.30 m
0.22 m

142.00 m

Wall

Ø 85.26 m

0.24 m

70.00 m   0.24 m

49.00 m

Z

27.00 m
12.18 m
−2.50 m   *Column*

1.16 m

3.10 m
Thickness

Ø 136.00 m

Ø 152.54 m

FIGURE 27.17   Prototype of cooling tower at Niederaussem.

large cooling tower (Figure 27.17) are shown in Figures 27.18 to 27.24 for some of the important loading conditions discussed in Section 27.5. The finite element model used considers the shell to be fixed at the top of the columns and, thus, does not account for the effect of the concentrated column reactions. Also, in considering the analyses under the individual loading conditions, it should be remembered that the effects are to be factored and combined to produce design values.

The dead load analysis results in Figure 27.18 and Figure 27.19 indicate that the shell is always under compression in both directions, except for a small circumferential tension near the top. This is a very desirable feature of this geometrical form and the result of a complex shape finding process [6,11].

In Figures 27.20 to 27.22, the results of an analysis for a quasistatic wind load using the K1.0 distribution from Figure 27.11 are shown. Large tensions in both the meridional and circumferential directions are present. The regions of tension may extend a considerable distance along the circumference from the windward meridian, and the magnitude of the forces is strongly dependent on the distribution selected. In contrast to bluff bodies, where the magnitude of the extensional force along the meridian would be essentially a function of the overturning moment, the cylindrical-type body is also strongly influenced by the circumferential distribution of the applied pressure, a function of the surface roughness. The major effect of the shearing forces is at the level of the lintel where they are transferred into the columns. The internal suction effects, Figure 27.23 and Figure 27.24, are significant only in the circumferential direction.

For the service temperature case shown in Figure 27.25 and Figure 27.26, the main effects are bending in the lower region of the shell wall.

27.6.3 Stability

The analysis of hyperbolic cooling towers for instability or buckling is a subject that has been investigated for several decades [12]. Shell buckling is a complex topic to treat analytically in any case, due to the influence of imperfections; for reinforced concrete, it is even more difficult. While the governing

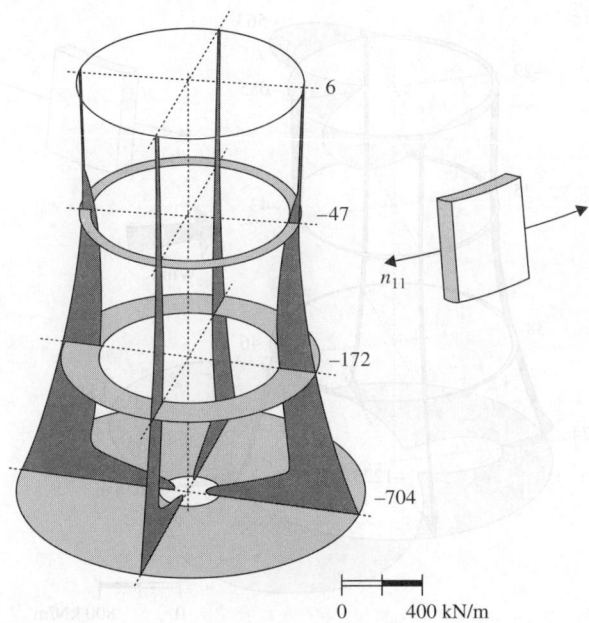

FIGURE 27.18 Circumferential forces n_{11D} under deadweight.

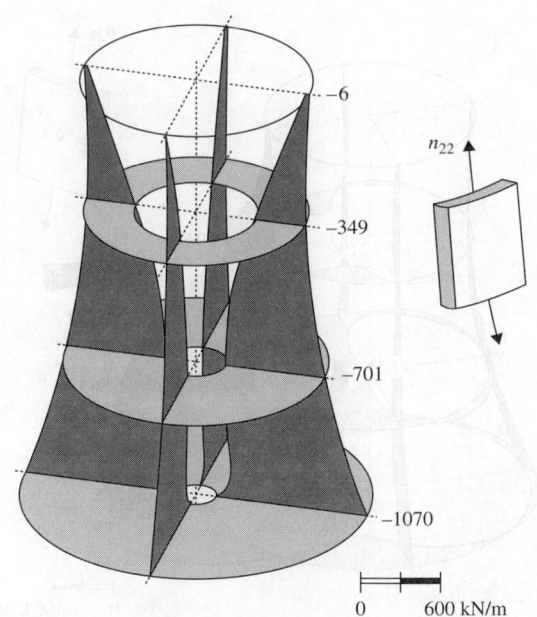

FIGURE 27.19 Meridional forces n_{22D} under deadweight.

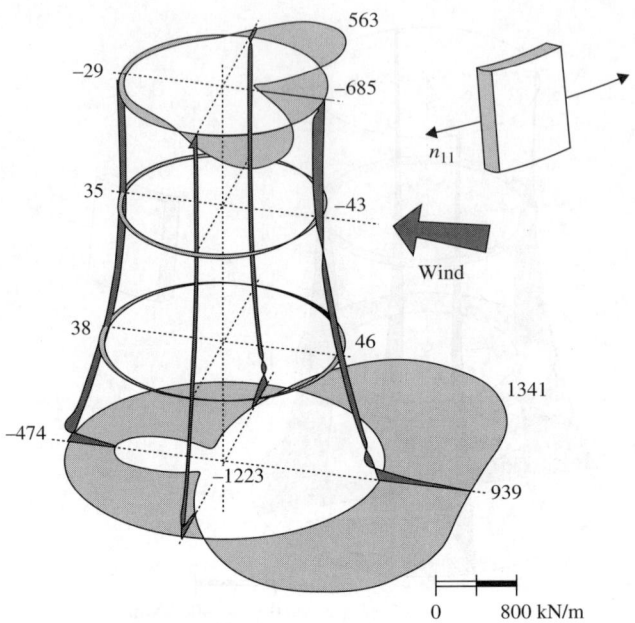

FIGURE 27.20   Circumferential forces n_{11W} under wind load.

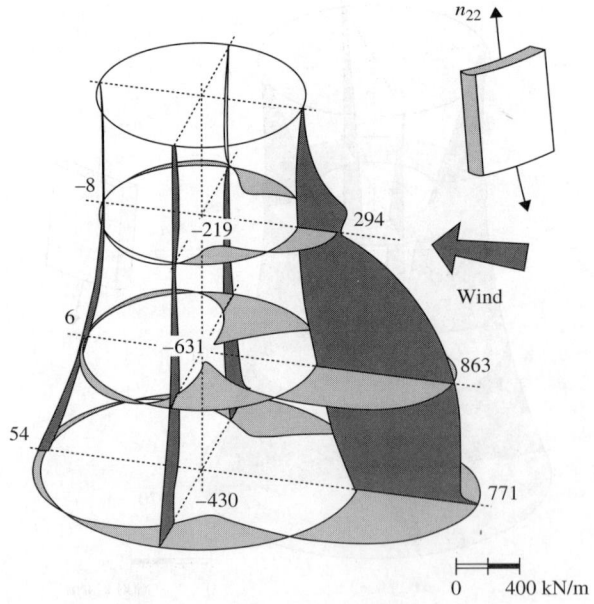

FIGURE 27.21   Meridional forces n_{22W} under wind load.

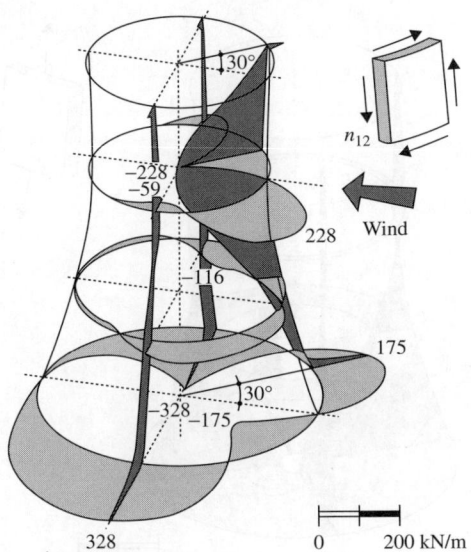

FIGURE 27.22   Shear forces n_{12W} under wind load.

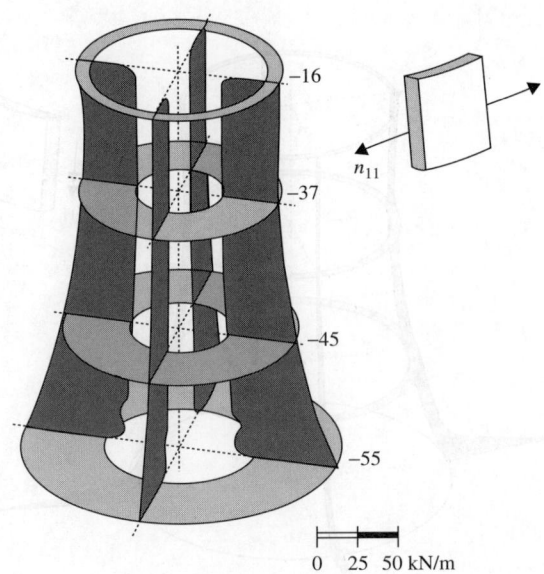

FIGURE 27.23   Circumferential forces n_{11S} under internal suction.

equations may be generalized to treat instability by using nonlinear strain–displacement relations and thereby introducing the geometric stiffness matrix, the correlation between the resulting analytical solutions and the possible failure of a reinforced concrete cooling tower is questionable.

Nevertheless, it has been common to analyze cooling tower shells for instability under an unfactored combination of dead load plus wind load plus internal suction. The corresponding instability buckling pattern is shown in Figure 27.27.

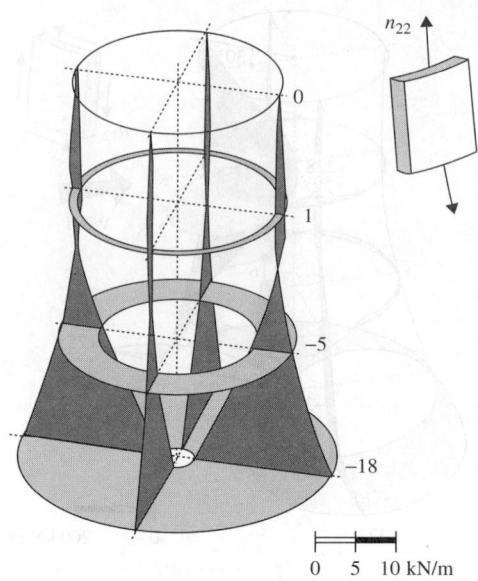

FIGURE 27.24 Meridional forces n_{22S} under internal suction.

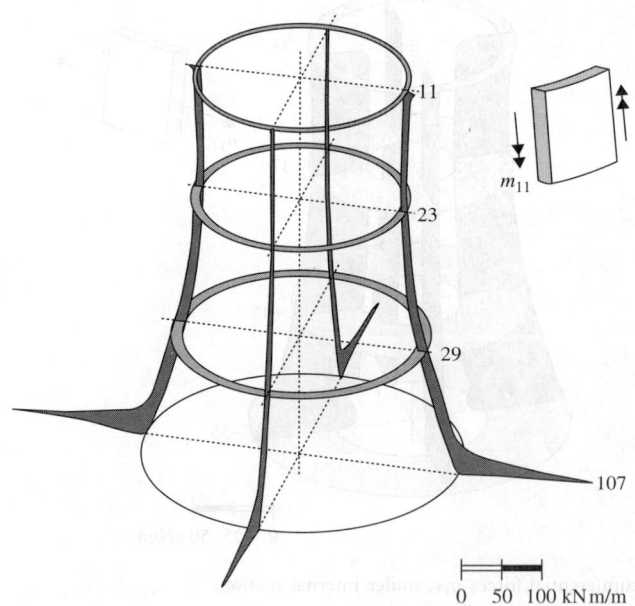

FIGURE 27.25 Circumferential bending moments m_{11T} under service temperature.

Interaction diagrams calibrated from experimental studies based on bifurcation buckling are also available [4,5,13]. Additionally, there are empirical methods based on wind tunnel tests that consider a snap-through buckle at the upper edge at each stage of construction [4]. These formulas are proportional to h/R and are convenient for establishing an appropriate shell thickness profile. If buckling safety is evaluated based on such a linear buckling analysis or an experimental investigation, the buckling safety

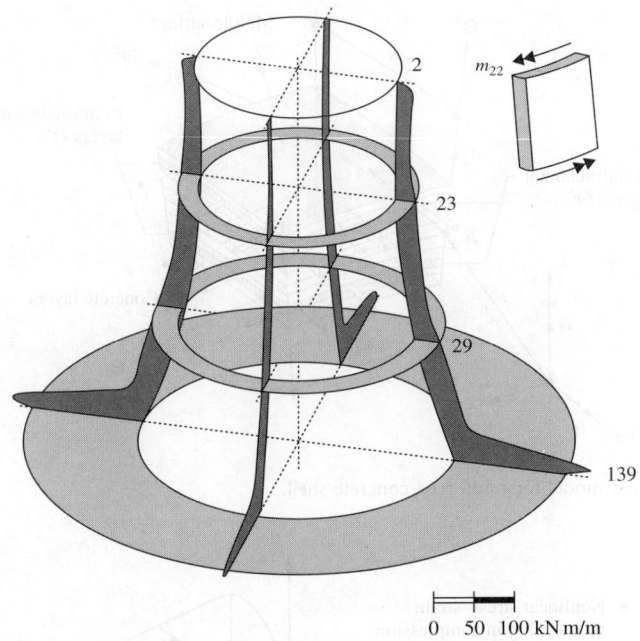

FIGURE 27.26 Meridional bending moments m_{22T} under service temperature.

FIGURE 27.27 Buckling pattern of tower shell with upper ring beam, load combination $D + W + S$.

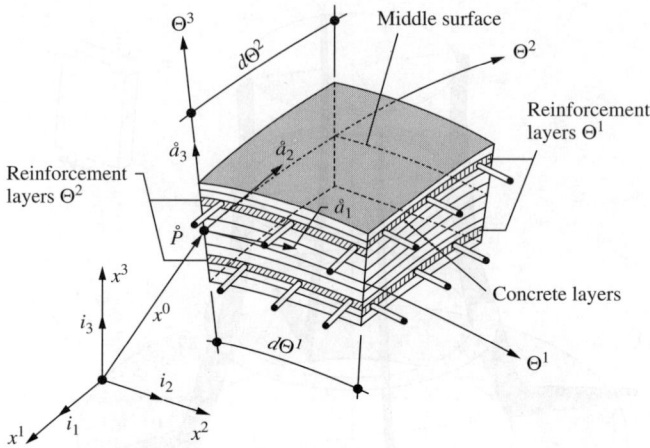

FIGURE 27.28 Layered model for reinforced concrete shell.

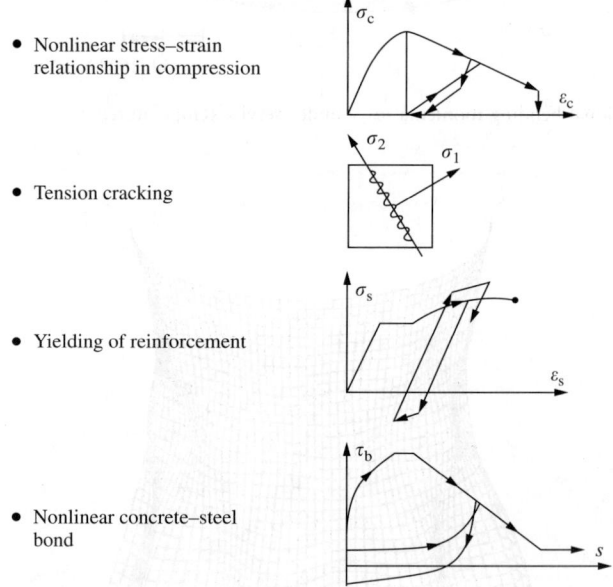

- Nonlinear stress–strain relationship in compression

- Tension cracking

- Yielding of reinforcement

- Nonlinear concrete–steel bond

FIGURE 27.29 Nonlinear material properties of reinforced concrete.

factor for realistic material parameters should exceed 5.0. Presently, however, the use of bifurcation buckling analyses should be confined to preliminary proportioning and to wall strengthening around shell openings, since more rational procedures based on nonlinear analysis have been developed to predict the collapse of reinforced concrete shells, as discussed in the following section.

27.6.4 Nonlinear Analysis

Advances in the analyses of reinforced concrete have produced the capability to analyze shells taking into account the layered composition of the cross-section as shown in Figure 27.28. Using realistic material properties for steel and for concrete, including tension stiffening in the form shown in Figures 27.29 to 27.32,

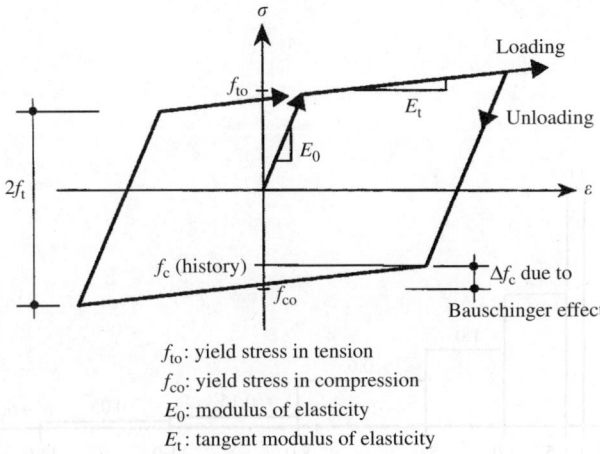

f_{to}: yield stress in tension
f_{co}: yield stress in compression
E_0: modulus of elasticity
E_t: tangent modulus of elasticity

FIGURE 27.30 Elastic–plastic material law for steel.

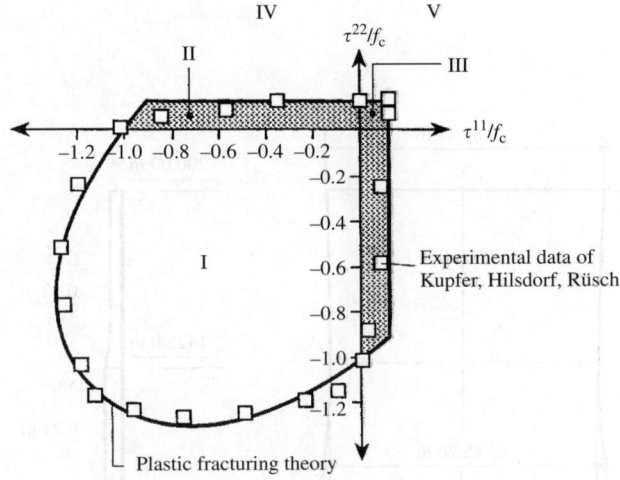

FIGURE 27.31 Biaxial failure envelope of Kupfer/Hilsdorf/Rüsch.

load–deflection relationships may be constructed for appropriate load combinations. These relationships progress from the linear elastic phase to initial cracking of the concrete through spreading of the cracks until collapse.

Results from a nonlinear study of a shell using a shell geometry given in Figure 27.33, a finite element mesh shown in Figure 27.34, and a reinforcement as specified in Figure 27.35 are presented in Figures 27.36 to 27.41. The wind load factor λ is plotted against the maximum lateral displacement at different heights on the shell in Figure 27.36 and the deformed shape for the collapse load is shown in Figure 27.37. Also, the pattern of cracking corresponding to the initial yielding of the reinforcement is indicated in Figure 27.38. For reinforced concrete shells, this type of analysis represents the state-of-the-art and provides a realistic evaluation of the capacity of such shells against extreme loading [13]. Also, durability assessments can be performed by this concept, from which particularly weak and crack-endangered regions of the shell can be identified and further reinforced [6].

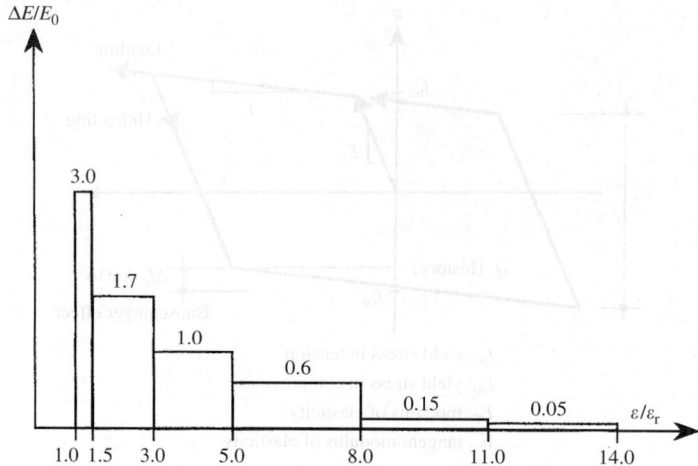

FIGURE 27.32 Additional modulus of elasticity due to tension stiffening.

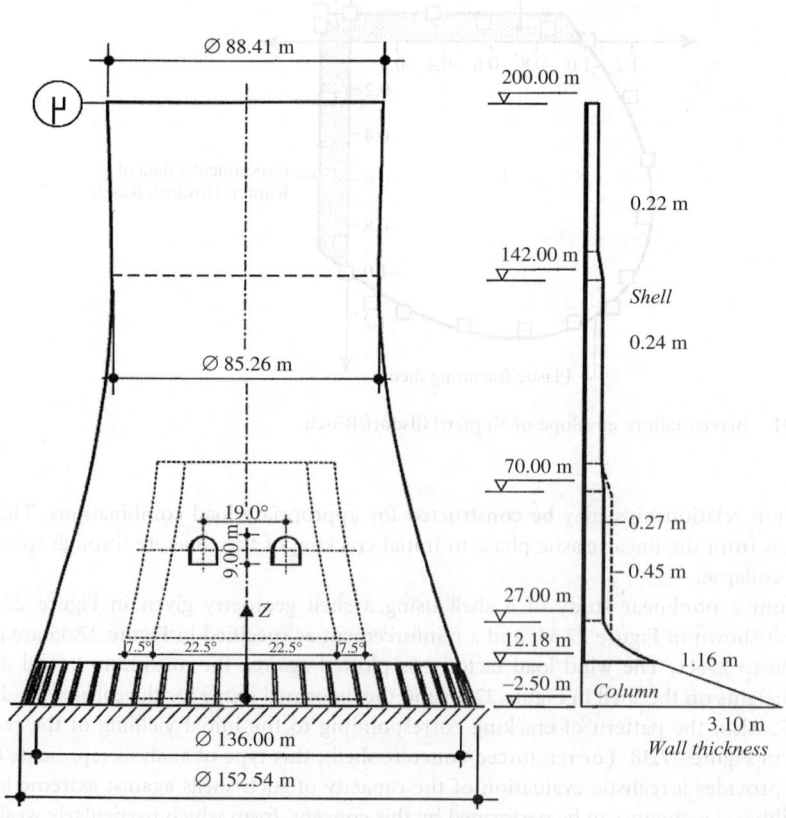

FIGURE 27.33 Cooling tower geometry with injection holes and shell thickness.

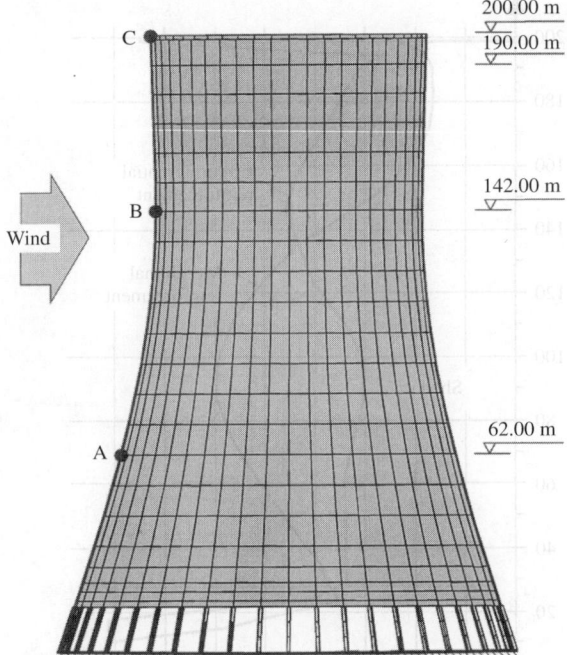

FIGURE 27.34   Finite element mesh for prototype cooling tower Niederaussem.

Figure 27.39 and Figure 27.40 show the additional weakening and cracking of the shell by wind effects *W* acting on the tower during a cold winter night ($\Delta T = 45°C$).

It is possible to obtain an estimate of the wind load factor, λ, from the results of a linear elastic analysis, even from a calculation based on membrane theory. This estimate is computed as cracking load for the shell under a combination of $D + \lambda W$ and is predicated on the notion that the reinforcement may add only a modest amount of capacity to the tower beyond the cracking load [15]. The amount of reinforcement in the wall is often controlled by a specified minimum percentage, augmented by that required to resist the net tension due to the factored load combinations. The steel provided is often less than the capacity of the concrete in tension, which is presumed to be lost when the concrete cracks. Therefore, the cracking load represents most of the ultimate capacity of the tower.

The maximum meridional tension location under the wind loading is identified, for example, as the value of $n_{22} = 863$ kN/m in Figure 27.21. The dead load at this location is obtained from Figure 27.19 as -701 kN/m. Taking the concrete tensile capacity as 2400 kN/m² and the wall thickness as 16 cm, the tensile strength is 384 kN/m. Therefore, we have

$$-701 + \lambda 863 = 384 \tag{27.7}$$

giving $\lambda = 1.26$ as the *lower bound* on the ultimate strength of the tower. Note that this value characterizes the end of the elastic response phase of the tower. What follows then is the long redistribution phase by secondary concrete cracking and yielding of the reinforcement, which nearly might double λ up to failure. Or, λ might not increase very much, depending on the amount of reinforcement in the critical regions.

27.6.5 Dynamic Analysis

The dynamic analysis of cooling towers is usually associated with design for earthquake-induced forces. The most efficient approach is the response spectrum method, but a time-history analysis may be appropriate if nonlinearities are to be included [2,8]. For large shells the dynamic response due to wind

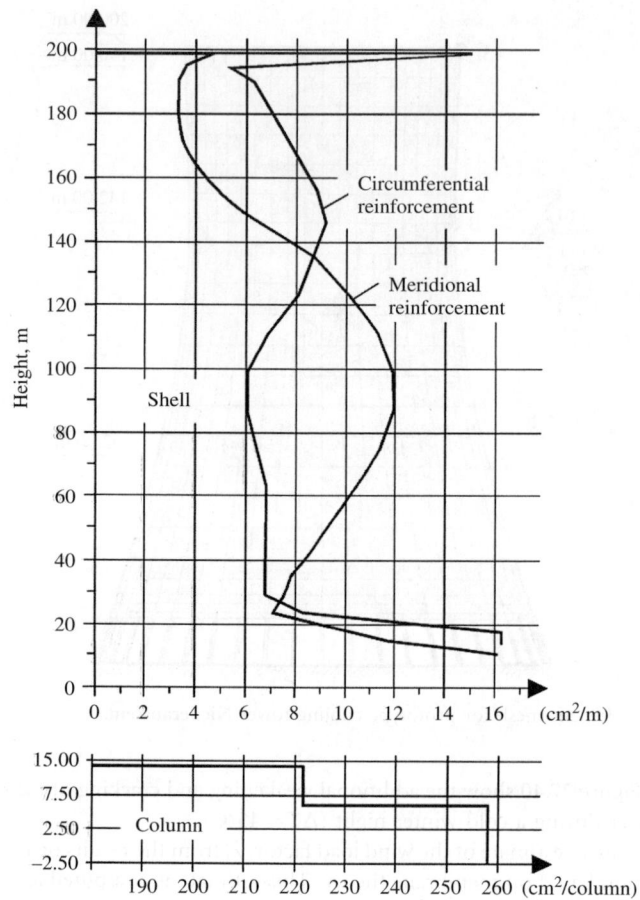

FIGURE 27.35 Reinforcement in shell and supports for prototype tower.

is often investigated, at least to determine the positions of the nodal lines and areas of particularly intensive vibrations. In any case the first step is to carry out a free vibration analysis. This analysis represents the modes of free vibration associated with each natural frequency, f, or its inverse the natural period T, as the product of a circumferential mode proportional to $\sin n\theta$ or $\cos n\theta$ and a longitudinal mode along the z-axis [9,10]. Some results are shown in Figure 27.41, which demonstrate the lowest vibration modes of a tower without and then with holes for flue gas inlets. Such inlets require great strengthening of the injection area. Further representative results are shown in Figure 27.42 and Figure 27.43, as discussed below.

As an illustration, the cooling tower from Figure 27.4 is again considered. Some key circumferential and longitudinal modes for a fixed-base boundary condition are shown in Figure 27.42. Also, the effects of different cornice stiffnesses are demonstrated. This model may be regarded as preliminary in that the relatively soft column supports are not properly represented, but it illustrates the salient characteristics of the modes of vibration. Most interesting are the frequency curves in Figure 27.43 for the first ten harmonics. Note that the natural frequencies *decrease* with increasing n until a minimum is reached whereupon they increase, a characteristic behavior for cylindrical-type shells. Also, the stiffening of the cornice tends to raise the minimum frequency, which is desirable for resistance to dynamic wind. Longitudinally, the cornice stiffness effect is significant for odd modes only, as shown in Figure 27.42.

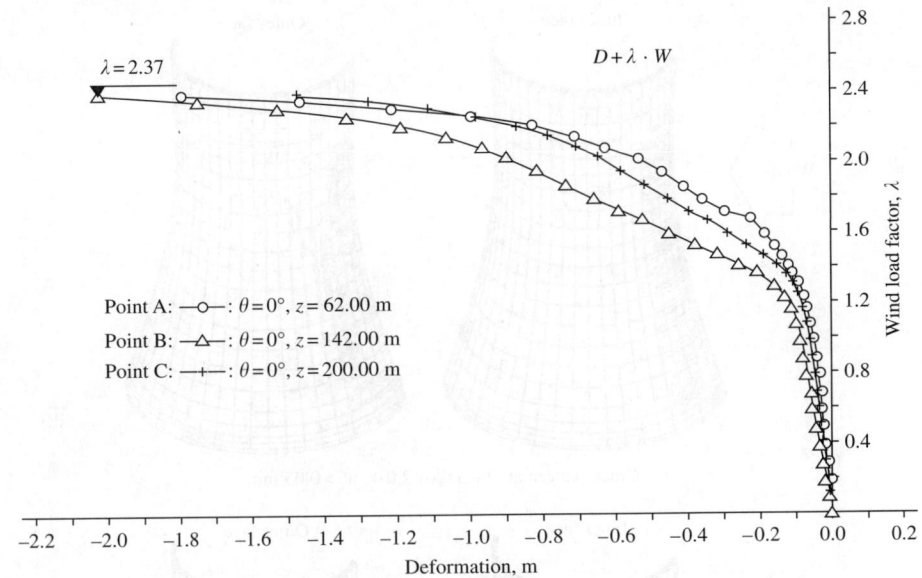

FIGURE 27.36 Load–displacement diagrams for load combination $D + \lambda W$.

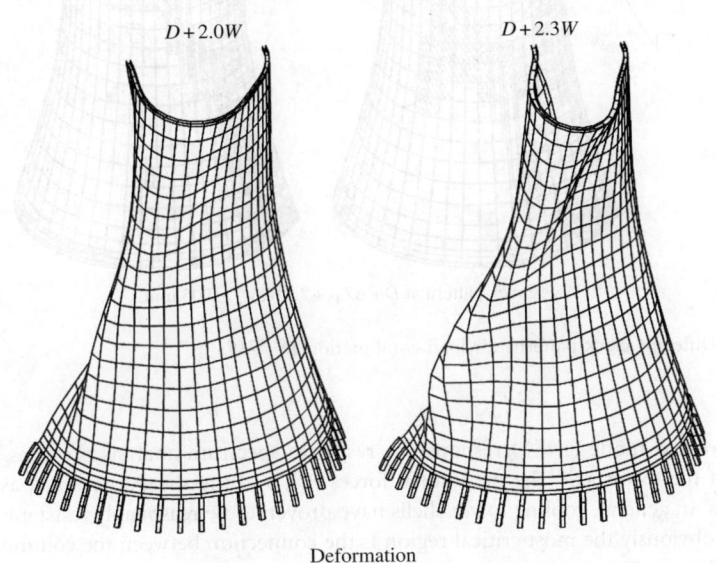

FIGURE 27.37 Displacement plot for load combination $D + \lambda W$ (exaggeration: 40).

Specifically for earthquake effects and other coherent excitations, only the first mode participates in a linear analysis for uniform horizontal base motion and the respective values for $n = 1$ should be entered into the design response spectrum.

Results from a seismic analysis of a cooling tower are presented in Figures 27.44 to 27.47. The cooling tower of Figure 27.4 is subjected to a horizontal base excitation based on Figure 27.44, leading to

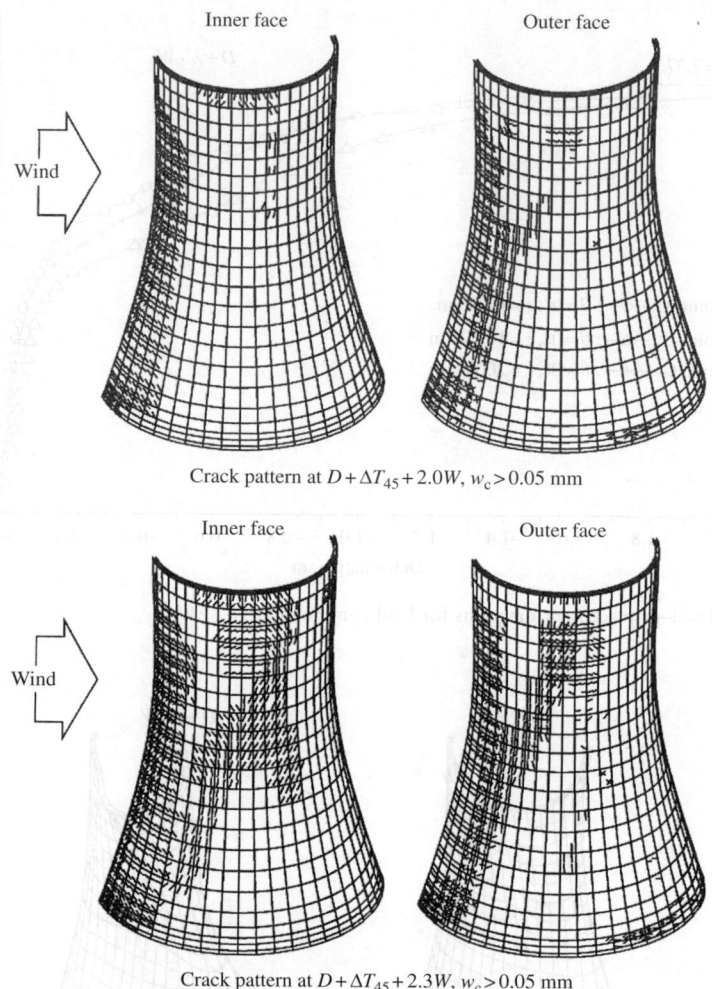

Crack pattern at $D + \Delta T_{45} + 2.0W$, $w_c > 0.05$ mm

Crack pattern at $D + \Delta T_{45} + 2.3W$, $w_c > 0.05$ mm

FIGURE 27.38 Different crack patterns for load combination $D + \lambda W$.

a first circumferential mode ($n = 1$) response. A response spectrum analysis provides the lateral displacements w of the tower axis, the meridional forces n_{22}, and the shear forces n_{12}, as shown on the indicated figures. In general, cooling tower shells have proven to be reasonably resistant against seismic excitations, but obviously the most critical region is the connection between the columns and the lintel as portrayed in Figure 27.48.

27.7 Design and Detailing of Components

The dimensioning of the tower components should follow the requirements of the applicable codes of practice. In Ref. [5], limit states of serviceability and of failure are presented to provide sufficient safety. In the limit states of serviceability, a total safety factor concept is applied and the following load combinations are recommended:

$$D + W, \quad D + W + T, \quad D + W/3 + T + E, \quad D + 0.70W + T$$

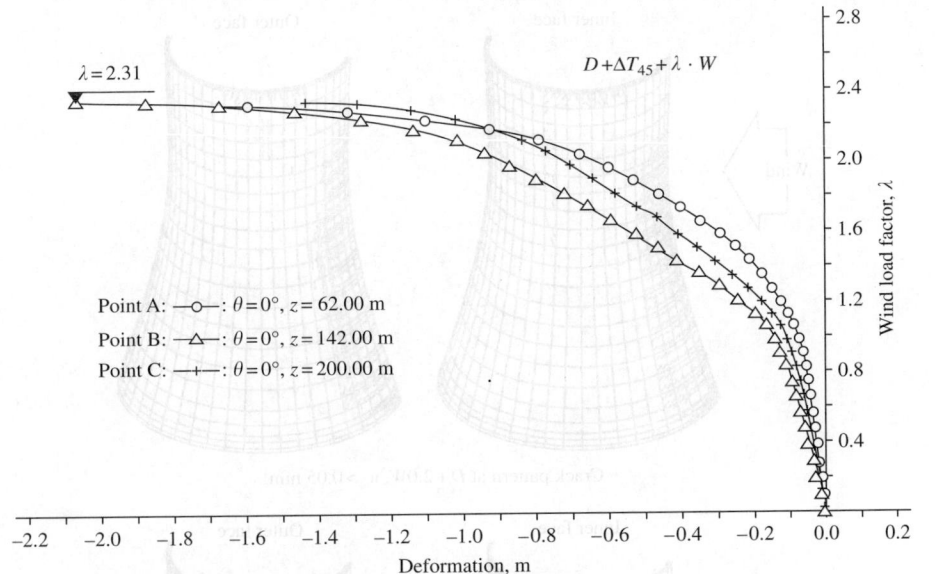

FIGURE 27.39 Load–displacement diagrams for load combination $D + \Delta T_{45} + \lambda W$.

the latter for checking crack widths. Here, T is taken for standard temperature service conditions. In the failure limit state design, partial safety factors are used for the following load combinations:

$$\gamma_D G + \gamma_W W, \qquad \gamma_D G + \gamma_W W + \gamma_T T$$

The structural elements of the tower should be constructed with a suitable grade of concrete following the provisions of applicable codes and standards. The design of the mixture should reflect the site conditions for placement of the concrete and the external and internal environment of the tower.

The shell wall should be of a thickness that will permit two layers of reinforcement in two perpendicular directions to be covered by a minimum of 3 cm of concrete, and should be no less than 16 cm thick [2,4,8]. The buckling considerations mentioned in Section 27.6.3 have proven to be convenient and evidently acceptable criteria for setting the minimum wall thickness, subject to a nonlinear analysis. The formula

$$q_c = 0.052E(h/R)^{2.3} \tag{27.8}$$

where E is the modulus of elasticity has been used to estimate the critical shell buckling pressure, q_c [4,12]. Then, $h(z)$ is selected to provide a factor of safety of at least 5.0 with respect to the maximum velocity pressure along the windward meridian, $q(z)(1 + g)$. Also, the cornice should have a minimum stiffness of

$$I_x/d_H = 0.0015 \text{ m}^3 \tag{27.9}$$

where I_x is the moment of inertia of the uncracked cross-section about the vertical axis and d_H is shown in Figure 27.10 [4]. Some typical forms of the cornice cross-section are shown in Figure 27.49.

The elements of the cooling tower should be reinforced with deformed steel bars so as to provide for the tensile forces and moments arising from the controlling combination of factored loading cases. The shell walls may be proportioned as rectangular cross-sections subjected to axial forces and bending. As mentioned above, a mesh of two orthogonal layers of reinforcement should be provided in the shell walls, generally in the meridional and circumferential directions [8]. In each direction, the inner and outer layers should generally be the same, except near the edges where the bending may require an

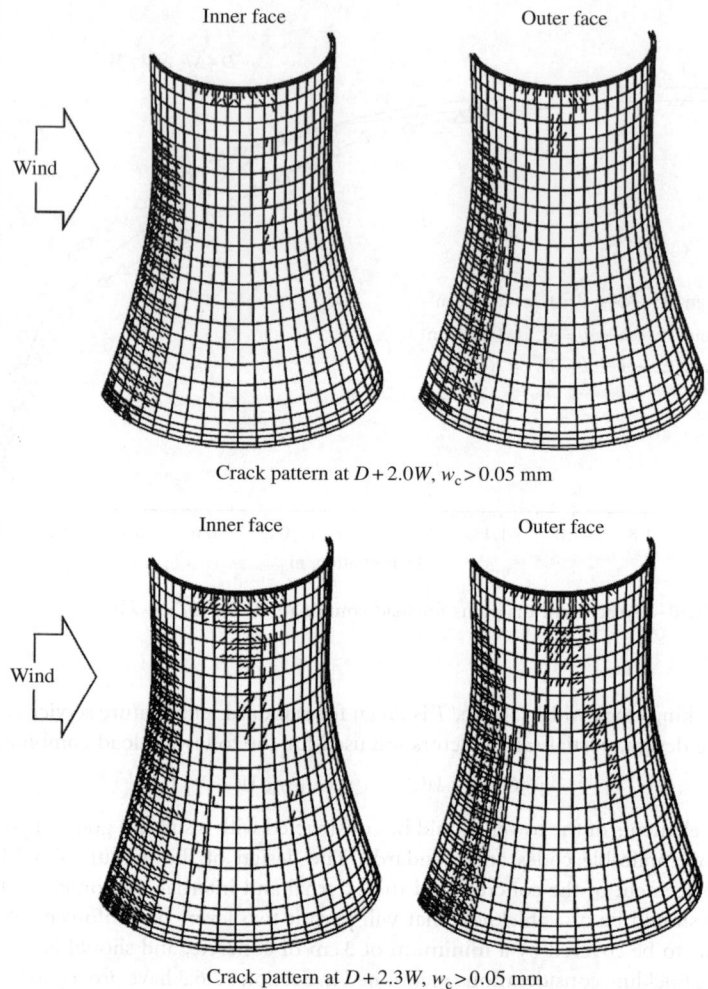

Crack pattern at $D + 2.0W$, $w_c > 0.05$ mm

Crack pattern at $D + 2.3W$, $w_c > 0.05$ mm

FIGURE 27.40   Different crack patterns for load combination $D + \Delta T_{45} + \lambda W$.

unsymmetrical mesh. It is preferable to locate the circumferential reinforcement outside of the meridional reinforcement except near the lintel, where the meridional reinforcement should be on the outside to stabilize the circumferential bars [16]. A typical heavily reinforced segment of the lintel, also showing the anchorage of the column reinforcement into the shell, is depicted in Figure 27.50.

A summary of the most important minimum construction tolerance values for the shell wall reinforcement is given in Figure 27.51 [2,4,8]. The bars should not be smaller than 8 mm diameter, ϕ, and, for meridional bars, not smaller than 10 mm. Further, a minimum of 0.35 to 0.45%, depending on the admissible cracking, should be used in each direction. The minimum cover, as mentioned previously, should be 3 cm, the maximum spacing of the bars, e, should be 20 cm, and the splices should be staggered as specified for the construction of walls in the applicable codes or standards. Particular attention should be given to splices in tensile zones.

The supporting columns should ideally be proportioned for the forces and moments computed from an analysis in which they are represented as discrete members, using the appropriate factored loading

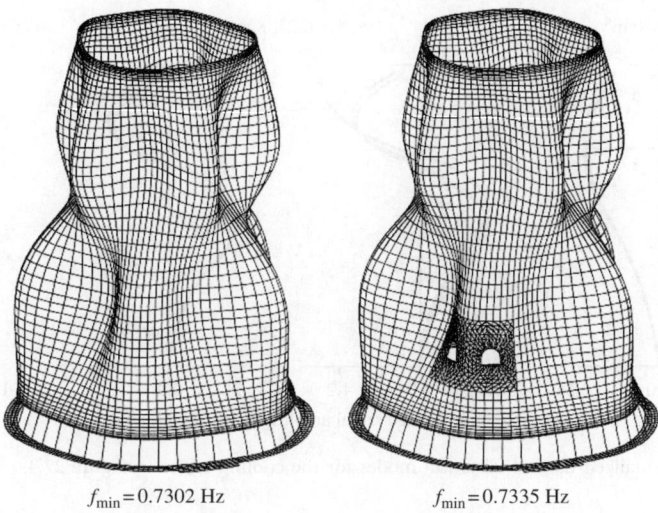

$$f_{min} = 0.7302 \text{ Hz} \qquad f_{min} = 0.7335 \text{ Hz}$$

FIGURE 27.41 Lowest natural vibration modes with/without flue gas inlets.

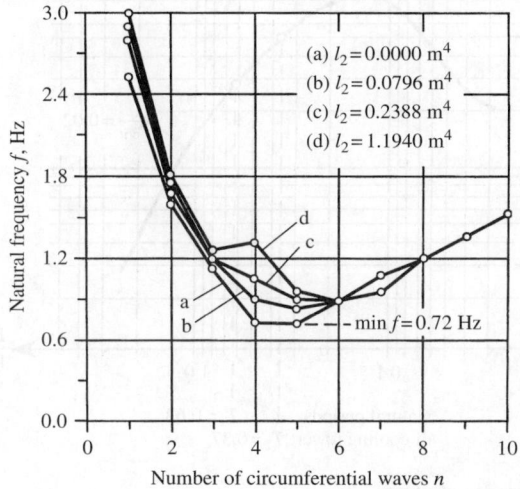

FIGURE 27.42 Natural frequencies for different cornice stiffnesses for the cooling tower in Figure 27.4.

combinations [9]. If the column region has not been modeled discretely, but rather by a continuum approximation, the columns may be proportioned to resist the tributary factored forces and moments at the interface with the lintel, as computed from the shell analysis. The effective length may be taken as unity. Particular attention should be directed toward splices of the column bars when net tension is present. Since large bars will be involved, welded or mechanical splices capable of transmitting tension forces are recommended in such regions.

It is possible to add discrete circumferential stiffeners to the shell to increase the stability or to restore capacity that may have been lost due to cracking or other deterioration [3] (see Figure 27.10). Such stiffeners can generally be included in a finite element model of the shell wall and should be

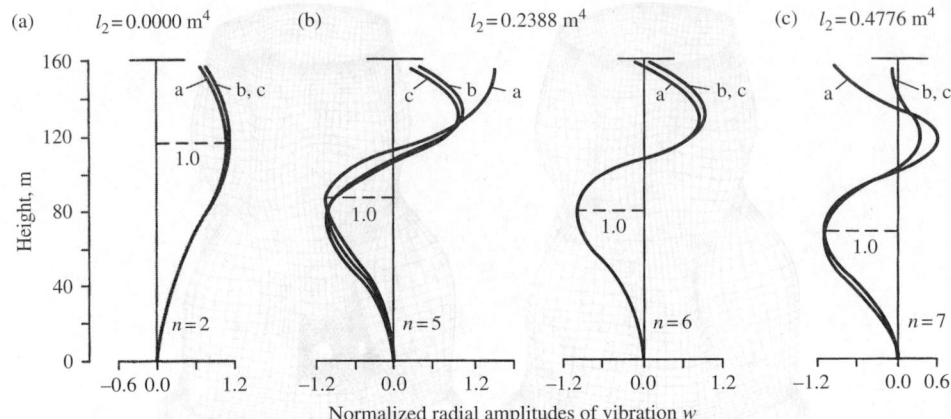

FIGURE 27.43 Normalized natural vibration modes for the cooling tower in Figure 27.4.

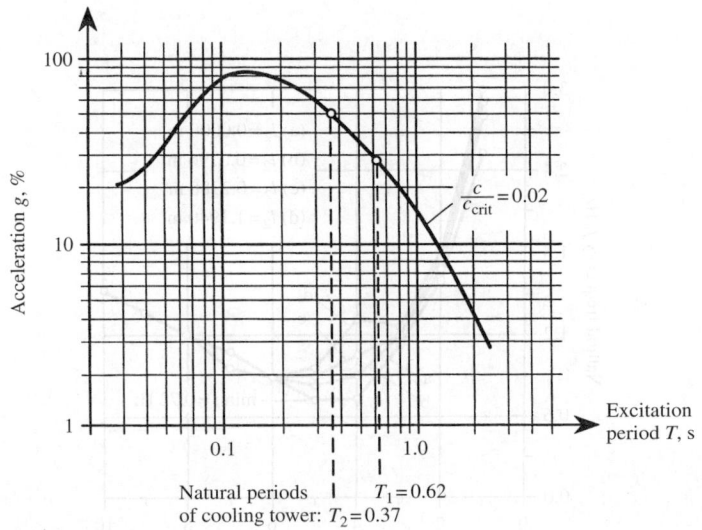

FIGURE 27.44 Seismic response spectrum.

proportioned for the forces computed from such an analysis. The eccentricity of the stiffeners with respect to the circumferential axis should be considered when the stiffeners are only on one side of the shell. This technique is discussed further in Section 27.9 and some applications are given there.

The foundations should be proportioned for the factored forces induced by the column reactions or from the computed forces if the foundation is included in the model with the shell and columns. Reinforcement detailing and cover should be in accordance with the applicable codes or standards. Several improved forms for cooling tower foundations have been suggested. Figure 27.52 shows a flat ring footing suitable for uniform soil conditions, while Figure 27.53 portrays a stiff ring beam foundation appropriate for soil conditions that are nonuniform around the circumference. An example of an individual pier on bedrock is given in Figure 27.54.

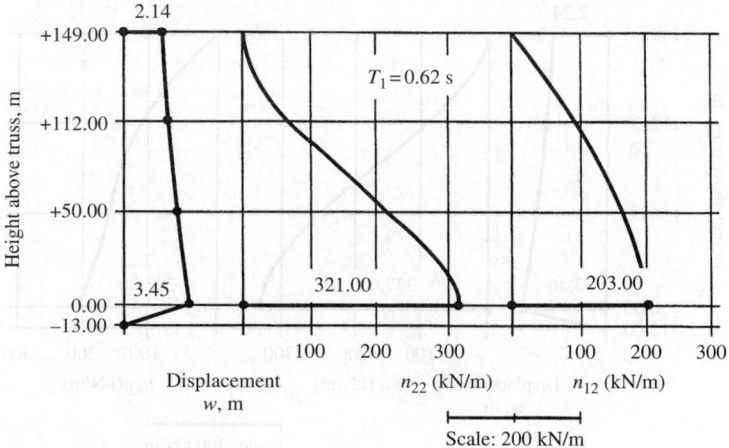

FIGURE 27.45 First axial mode seismic response.

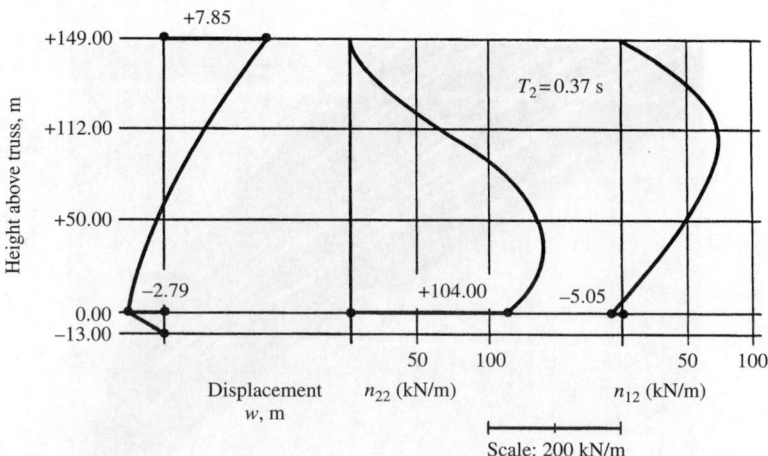

FIGURE 27.46 Second axial mode seismic response.

27.8 Construction

Tolerances for tall concrete cooling tower shells have been debated for many years and reasonable values should take into consideration what is achievable and what is measurable. It should be noted that state-of-the-art finite element models are capable of analyzing the as-built shell as well as the design configuration, so that the effects of those irregularities arising during construction, or even those discovered later, may be quantitatively studied and sometimes corrected.

It is recommended that the actual wall thickness be no less than the design thickness and exceed this thickness by not more than 10%. The imperfections of the shell wall middle surface should not exceed one-half of the wall thickness or 10 cm. Deviations from the design geometry occurring during the construction should be corrected gradually, limiting the angular change in either direction to 1.5%. The column heads should be within 0.005 times the column height or ±6.0 cm of the design position, and

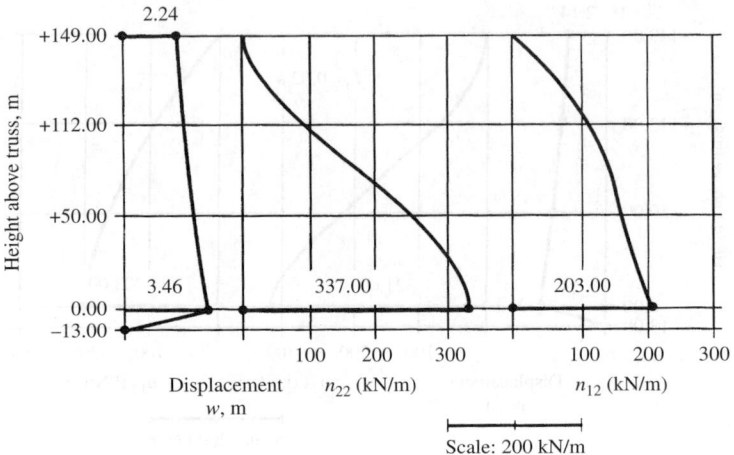

FIGURE 27.47 SRSS superposition of first and second axial modes.

FIGURE 27.48 Column to lintel connection.

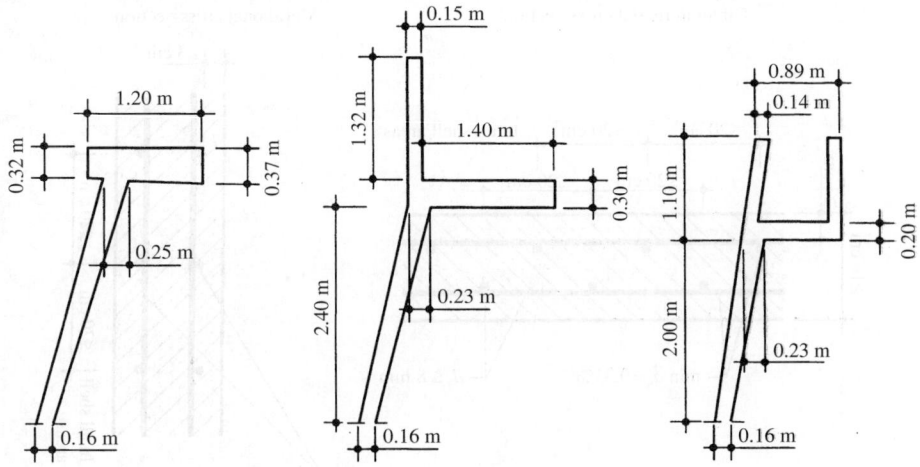

FIGURE 27.49 Suitable forms of the cornice.

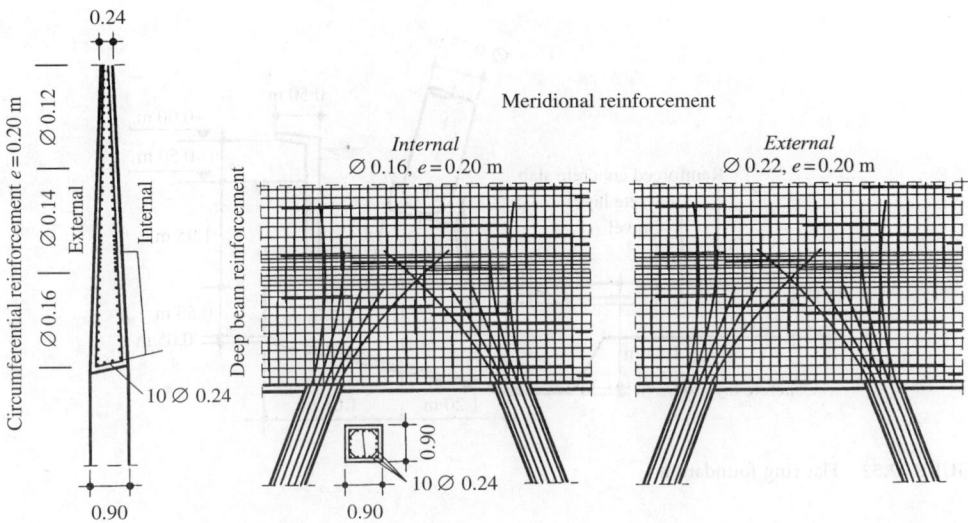

FIGURE 27.50 Lintel reinforcement.

foundation structures should also be within ±6.0 cm of the design location [2,4,8]. Tolerances for the reinforcement are discussed in Section 27.7.

Formwork and scaffolding systems are generally proprietary and are provided by the constructor. Nevertheless, their influence on the shell quality is of utmost importance and diligent attention of the engineer is required. In general, the system should be designed to provide safety to operating personnel and to produce a sound structure. The working platforms should be designed for realistic loading and scaffolding systems used for continuous material transport should be designed and built taking into account the resulting loads.

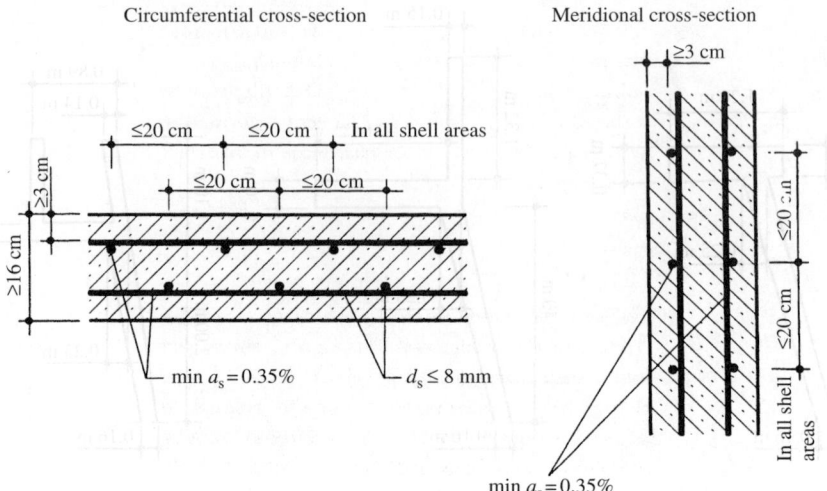

FIGURE 27.51   Important minimum construction values.

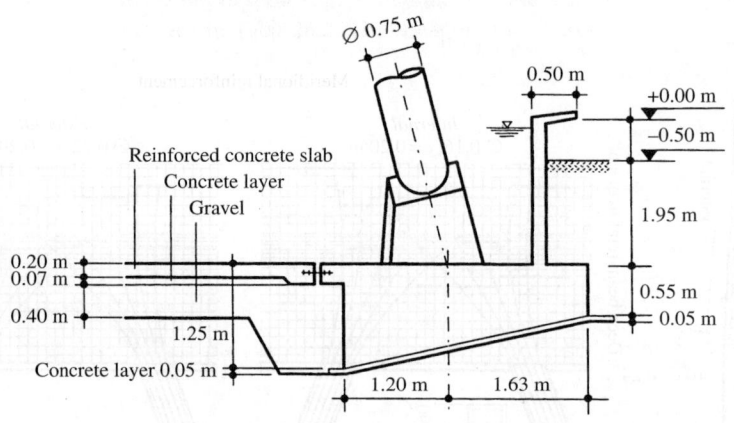

FIGURE 27.52   Flat ring foundation.

The connections and joints between individual scaffolding units should be designed and built to act independently in case of collapse, so that the loss of one unit would not affect the adjacent units. Furthermore, at least two independent safety devices should be in place to prevent collapse of the scaffolding.

The shell wall should be designed to resist the anchor loads of the scaffolding, based on the actual strength of the concrete that is expected to be available when the anchors are loaded. Continuous monitoring of the concrete strength during the climbing process is essential.

27.9 Durability

Cooling tower shells, particularly in fossil fuel power plants, are subjected to a relatively severe environment over their lifetime, which may span several decades, and special care must be taken in order to provide a durable structure. The tower is subjected to the physical loads produced by wind, temperature,

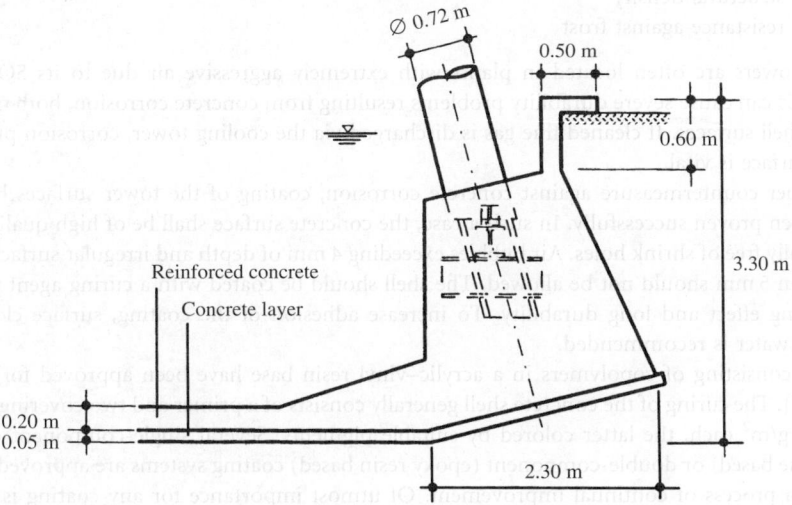

FIGURE 27.53 Ring beam foundation.

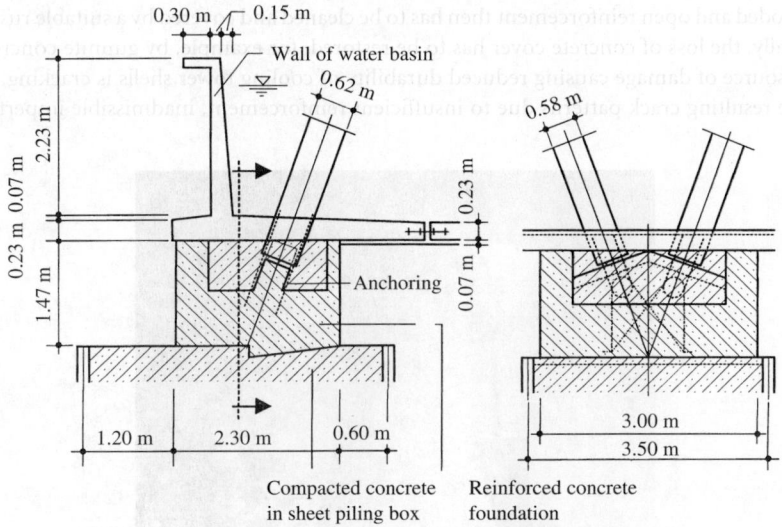

FIGURE 27.54 Individual reinforced concrete foundation on concrete base.

and moisture acting on concrete which may still be drying and hardening. Over the lifetime of the structure, it may be exposed to severe frost action in a saturated state, chemical attacks due to noxious substances in the atmosphere and in the water and water vapor, biological attacks due to microorganisms, and possibly additional chemical attacks due to reintroduced cleaned flue gases.

The concrete should be of high-quality approved materials including fly ash. It should have the following properties:

- High resistance against chemical attacks
- High early strength

- High structural density
- High resistance against frost

Cooling towers are often located in plants with extremely aggressive air due to its SO_x and NO_x content. This can cause severe durability problems resulting from concrete corrosion, both on the outer and inner shell surfaces. If cleaned flue gas is discharged via the cooling tower, corrosion protection of the inner surface is vital.

As a proper countermeasure against concrete corrosion, coating of the tower surfaces by synthetic resin has been proven successfully. In such a case, the concrete surface shall be of high quality, smooth, and essentially free of shrink holes. Air bubbles exceeding 4 mm of depth and irregular surface elevations of more than 5 mm should not be allowed. The shell should be coated with a curing agent providing a high blocking effect and long durability. To increase adhesion of the coating, surface cleaning with pressurized water is recommended.

Coatings consisting of copolymers in a acrylic–vinyl resin base have been approved for more than 20 years [17]. The curing of the concrete shell generally consists of a primer and two covering coatings of at least $300 \, g/m^2$ each, the latter colored by suitable pigments. Several single-component (acrylate or polyurethene based) or double-component (epoxy resin based) coating systems are approved worldwide and are in a process of continual improvement. Of utmost importance for any coating is the homogeneity of the applied film, between $\geq 200 \, \mu m$ for single- and $\geq 300 \, \mu m$ for double-component systems, since the durability of the complete coating is determined by the thinnest film spots. An example of a recently built coated cooling tower is shown in Figure 27.55.

If the shell concrete is already severely corroded, all loose particles have to be removed carefully by water jetting. Corroded and open reinforcement then has to be cleaned and covered by a suitable rust preventing coating. Finally, the loss of concrete cover has to be restored, for example, by gunnite concreting.

A second source of damage causing reduced durability of cooling tower shells is cracking, as demonstrated by the resulting crack patterns due to insufficient reinforcement, inadmissible imperfections, soil

FIGURE 27.55   Coated cooling tower shell at Herne Power Station, Germany.

settlements, or even accidents [18]. All these cracks, if exceeding a width of ~0.5 mm, should be carefully repaired, preferably by pressurized injection techniques, to maintain sufficient durability.

In order to prevent local reopening of repaired cracks, they can be covered by carbon-fiber bands glued to the concrete with resin. If the shell requires more global stiffening, the application of additional ring stiffeners is an approved measure [11]. This procedure has been discussed in Section 27.7.

Figure 27.56 shows a comparison of the stiffening effect from two different configurations, related to classical stability limits. For the two 165 m high tower variants of identical thickness, the stability limited is increased from $\lambda = 4.029$ (unstiffened except on boundaries) to $\lambda = 6.379$ (two additional stiffeners), both for the load combination $D + 0.6W$. Figure 27.57 shows the cross-section of such a ring stiffener,

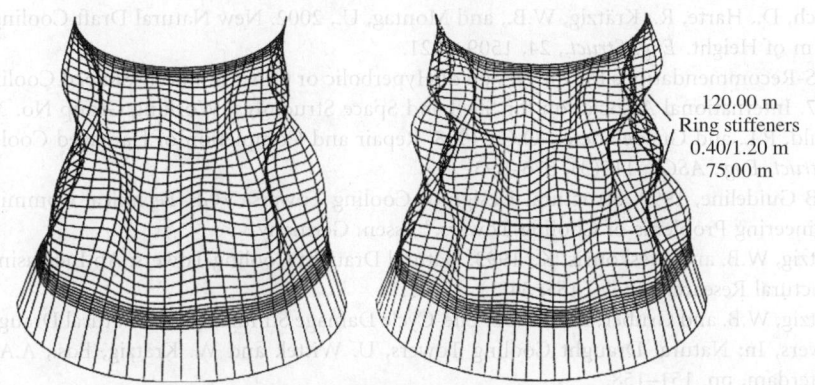

120.00 m
Ring stiffeners
0.40/1.20 m
75.00 m

FIGURE 27.56 Buckling patterns of 165 m high tower with upper edge member and with/without additional stiffening rings.

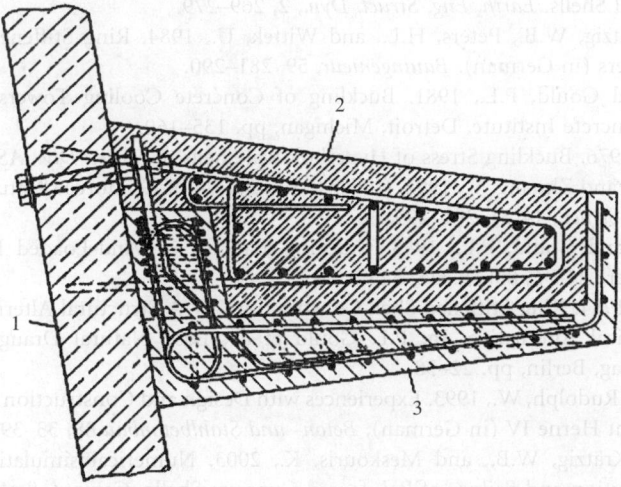

FIGURE 27.57 Cross-section of ring stiffeners at Mississippi Tower: 1, original shell wall; 2, *in situ*; and 3, prefabricated.

composed of combined prefabricated and *in situ* cast reinforced concrete parts, which was attached to the Mississippi Tower after it had been hit and damaged by the collapsing central crane during a tornado [3,11].

In the early 1980s, several ring stiffened large cooling towers were erected in Germany and in Belgium [16,19]. Initial use of circumferential ring stiffeners may save concrete in the tower shell, but increases the amount of reinforcement required. Since the ring construction has to be separated from the continuing climbing process of the shell, their erection is the same as for the rehabilitation applications on existing towers. After crane capacities increased, new high towers were again designed without stiffening rings. Reference [20] reports on the ring stiffening of several cooling towers in South Africa that were heavily cracked by soil settlements due to underground mining, starting in 1985.

References

[1] Busch, D., Harte, R., Krätzig, W.B., and Montag, U., 2002. New Natural Draft Cooling Tower of 200 m of Height. *Eng. Struct.*, 24, 1509–1521.

[2] IASS-Recommendations for the Design of Hyperbolic or Other Similarly Shaped Cooling Towers, 1977. International Association for Shell and Space Structures, Working Group No. 3, Brussels.

[3] Gould, P.L. and Guedelhoefer, O.C., 1988. Repair and Completion of Damaged Cooling Tower, *J. Struct. Eng., ASCE*, 115(3), 576–593.

[4] VGB Guideline, 1990. Structural Design of Cooling Towers, VGB-Technical Committee, "Civil Engineering Problems of Cooling Towers," Essen, Germany.

[5] Krätzig, W.B. and Meskouris, K., 1993. Natural Draught Cooling Towers: An Increasing Need for Structural Research, *Bull. IASS*, 34(1), 37–51.

[6] Krätzig, W.B. and Gruber, K.P., 1996. Life-Cycle Damage Simulations of Natural Draught Cooling Towers, In: Natural Draught Cooling Towers, U. Wittek and W. Krätzig, Eds., A.A. Balkema, Rotterdam, pp. 151–158.

[7] Minimum Design Loads for Buildings and Other Structures, 2002. SEI/ASCE 7-02, ASCE, New York.

[8] ACI-ASCE Committee 334, 1977. Recommended Practice for the Design and Construction of Reinforced Concrete Cooling Towers, *ACI J.*, 74(1), 22–31.

[9] Gould, P.L, 1985. Finite Element Analysis of Shells of Revolution, Pittman, London.

[10] Gould, P.L., Suryoutomo, H., and Sen, S.K., 1974. Dynamic Analysis of Column-Supported Hyperboloidal Shells. *Earth. Eng. Struct. Dyn.*, 2, 269–279.

[11] Form, J., Krätzig, W.B., Peters, H.L., and Wittek, U., 1984. Ring Stiffened Natural Draft RC Cooling Towers (in German). *Bauingenieur*, 59, 281–290.

[12] Abel, J.F. and Gould, P.L., 1981. Buckling of Concrete Cooling Towers Shells, ACI SP-67, American Concrete Institute, Detroit, Michigan, pp. 135–160.

[13] Mungan, I., 1976. Buckling Stress of Hyperboloidal Shells, *J. Struct. Div., ASCE*, 102, 2005–2020.

[14] Krätzig, W.B. and Zhuang, Y., 1992. Collapse Simulation of Reinforced Natural Draught Cooling Towers, *Eng. Struct.*, 14(5), 291–299.

[15] Hayashi, K. and Gould, P.L., 1983. Cracking load for a Wind-Loaded Reinforced Concrete Cooling Tower, *ACI J.*, 80(4), 318–325.

[16] Peters, H.L., 1984. Ring-Stiffened Shell Constructions — A Structural Alternative or a Technical and Economical Alternative. In: P.L. Gould et al., Eds., Natural Draught Cooling Towers, Springer-Verlag, Berlin, pp. 22–38.

[17] Harte, R. and Rudolph, W., 1993. Experiences with Design and Construction of the Cooling tower at Power-Plant Herne IV (in German), *Beton- und Stahlbetonbau* 88, 33–39.

[18] Noh, S.-Y., Krätzig, W.B., and Meskouris, K., 2003. Numerical simulation of serviceability, Damage Evolution and Failure of Reinforced Concrete Shells, *Comput. Struct.*, 81, 843–857.

[19] Eckstein, U., Eller, C., Harte, R., Sanal, Z., Krätzig, W.B., and Wittek, U., 1984. Improvement of the Structural Behavior of Cooling Tower Shells by Ring-Stiffeners. In: P.L. Gould et al., Eds., Natural Draught Cooling Towers, Springer-Verlag, Berlin, pp. 61–76.

[20] Bosman, P.B., Strickland, I.G., and Prukl, R.P., 1996. Strengthening of Natural Draught Cooling Tower Shell with Stiffening Rings. In: U. Wittek, W.B. Krätzig, Eds., Natural Draught Cooling Towers, A.A. Balkema, Rotterdam, pp. 293–301.

Further Reading

Busch, D., Harte, R., and Niemann, H.-J., 1998. Study of a Proposed 200 m High Natural Draught Cooling Tower at Power Plant Niederaussem/Germany, *Eng. Struct.*, 19, 920–927.

Gould, P.L., Krätzig, W.B., Mungan, I., and Wittek, U., Eds., 1984. Natural Draught Cooling Towers. Proceedings of the Second International Symposium on Natural Draught Cooling Towers, Springer-Verlag, Heidelberg.

Proceedings (First) International Symposium on Very Tall Reinforced Concrete Cooling Towers, 1978. I.A.S.S., E.D.F., Paris, France, November.

28

Tunnel Structures

Christian Ingerslev
Parsons Brinckerhoff, Inc.,
New York, NY

Brian Brenner
Department of Civil and
Environmental Engineering,
Tufts University,
Medford, MA

Jaw-Nan Wang
Parsons Brinckerhoff, Inc.,
New York, NY

Phil Rice
Parsons Brinckerhoff, Inc.,
New York, NY

Birger Schmidt[*]
Parsons Brinckerhoff, Inc.,
New York, NY

28.1 Introduction

This chapter deals with structural design and the selection of structural systems for several types of tunnel structure. Methods of construction are important in the selection of structural systems and their details, but space does not permit this chapter to do justice to every complex method of tunnel construction possible in the great variety of geologic environments that exist. The chapter includes methods for seismic analysis of underground structures.

---

[*]Deceased.

28.1.1 What is a Tunnel?

Tunnels are located either below ground or below a body of water. Almost all tunnels serve as part of the infrastructure of cities and countries, conveying utilities, water, or sewage, or serving as transportation arteries. Some tunnels serve as storage, and some caverns are used as stadiums, swimming pools, or even for cleaning sewage. In cities, tunnels are often preferred because they can be built without taking up precious surface space. In the open country, tunnels are most likely to be found where it is necessary to traverse obstacles such as mountains, rivers, lakes, and fjords. The depth, cross-section, and alignment of tunnels are determined by the functional requirements of the facility, by geographical and environmental constraints, and by issues of constructibility under the prevailing geotechnical conditions.

The three principal types and methods of tunnel construction are

- *Cut-and-cover tunnels*, built by excavating a trench, constructing the concrete structure in the trench, and covering it up. Methods used include top-down and bottom-up.
- *Bored or mined tunnels*, excavated through rock either by blasting or by tunnel boring machines, or excavated through soil using shields, mechanized to a greater or lesser extent.
- *Immersed or floating tunnels*, made from very large precast concrete or concrete-filled steel elements, floated out, and then either lowered into a prepared trench and covered over (immersed), or anchored or supported below the water surface but not buried (floating).

In this chapter a short section on *tunnel jacking* is included, a procedure used recently in Boston and which is on occasion used in Europe and elsewhere. It is used only for very shallow tunnels when it is vital not to interrupt surface facilities and if other methods are not suitable.

At shallow depth and for short water crossings, the cut-and-cover tunnels are the most economical. Per unit length, immersed or floating tunnels can be more costly than bored or mined tunnels, but because they can often be made shallower and therefore shorter, and because they may serve their function better than other choices, the overall project cost can be less.

28.1.2 Fundamental Approach to Underground Design

Underground design must achieve functionality, stability, and safety of the underground openings during construction and for the lifetime of the structure thereafter. There is no recognized U.S. standard, practice, or code for the design of most underground structures. Designers may apply codes such as ACI Codes and Practices for concrete design, but these were developed for structures above ground, not for underground structures, and only parts of these codes apply to underground structures such as cut-and-cover structures.

There are five basic steps in the design of underground structures:

1. Define functional requirements, including design life and durability requirements.
2. Acquire and analyze geologic, geotechnical, and cultural data relevant to the design and construction of the underground facilities.
3. Determine plausible and possible modes of failure, including construction events, unsatisfactory long-term performance, failure to meet environmental requirements, and flooding; determine means of analyzing these modes of failure and acquire the necessary data.
4. Establish appropriate method(s) of construction considering the geologic and groundwater information, constructibility, and economy.
5. Establish and design appropriate initial and final ground support and lining systems, considering both ground conditions and the method(s) of construction.

28.1.3 Cut-and-Cover Tunnels

28.1.3.1 Introduction

Cut-and-cover tunnels are by far the most common of all the types of tunnel. They are constructed by one of two main methods. With the "bottom-up" approach, a trench is excavated from the surface, either supporting the sides of the excavation or sloping the sides. The complete structure is then constructed within the excavation; temporary decking may be installed across the top of the excavation to permit early reinstatement of traffic. With "top-down" construction, the walls (in narrow trenches) and the tunnel roof are built first and the surface reinstated; thereafter, the tunnel section is excavated down from the underside of the already constructed roof structure, and the tunnel is completed. For both methods, the tunnel structure is usually either cast-in-place concrete or erected steel structural elements with infill concrete. It is unusual for prefabricated elements to be used for cut-and-cover tunnel structures.

The cut-and-cover method is most attractive for relatively shallow tunnel sections. Cut-and-cover construction, particularly when the bottom-up method is used, can be disruptive to both surface traffic and existing facilities even when temporary decking is used over the cut.

28.1.3.2 Trench Excavation

The tunnel trench may be constructed by

- Sloping back the excavation, with little or no excavation support.
- Using the temporary support of excavation walls, such as soldier piles and lagging or steel sheet piles that do not form part of the permanent structure.
- Using excavation support walls that are designed to be part of the permanent structure.

The first option can be the most economical, but it is often impractical due to adjacent structures and facilities or due to depth. Comparing the second and third options, increased cost and staging difficulties for building permanent excavation support walls must be balanced against the simplicity and ease of building a permanent structure within temporary walls.

28.1.3.3 Top-Down Construction

As a cost-saving measure, this method typically uses the excavation support walls as part of the final structure (for e.g., concrete diaphragm walls cast in slurry-filled trenches, known as slurry walls, or precast wall sections lowered into slurry-filled trenches). After the walls are constructed downward from the surface, excavation proceeds down to the underside of roof level, and the roof is installed. The surface is then restored to its final condition, and excavation proceeds below the finished roof.

Top-down construction requires a different set of analysis assumptions and staging approaches than bottom-up construction, whether or not it has temporary decking. For example, the staging must lay out specific glory hole locations for all stages, being holes that provide access for construction, materials, and removal of excavated soil and rock. This type of staging planning can be more complex than cut-and-cover work with temporary decking or an open cut.

28.1.3.4 Groundwater Impacts

It is important to evaluate groundwater impacts for cut-and-cover construction. The evaluation includes

- Analysis to determine impacts on the water table beyond the zone of excavation.
- Analysis of the stability of the cut during construction.
- Evaluation of buoyancy under final conditions.

In all stages of cut-and-cover excavation and construction, the weight of the structure and backfill will tend to be less than the weight of the soil removed. The staging evaluation can be more complicated in areas with high groundwater levels.

28.1.4 Bored and Mined Tunnels in Soil or Rock

Transportation tunnels in built-up areas are often constructed by tunneling methods such as machine boring or hand mining. Tunneling can cause much less disruption at the surface, especially of local utilities, and is less expensive for deep tunnels. It is common practice to provide at least one diameter of soil or rock cover, though shallower cover is possible.

Depending on functional requirements, bored or mined tunnels come in almost all sizes, ranging from some 2 m in diameter for medium-sized water and sewage conveyance up to about 20 m in diameter for rail stations and some hydropower plant caverns. While rock tunnels excavated by blasting methods can be built to almost any size and shape, tunnels in soil are usually built using a shield, currently to a maximum circular diameter of about 15 m, although larger-diameter machines are being designed. Most rapid transit tunnels are single-track tunnels with a diameter of about 6 m, but a few have been built as double-track tunnels with a diameter of about 10 m. Even smaller tunnels (microtunnels) are used for utilities, but they may be constructed differently.

Functional and environmental requirements determine basic tunnel support and lining requirements such as shape and size, smoothness, water tightness, and durability. Some tunnels through high-quality rock are unlined or only furnished with nominal support consisting of rock bolts, wire mesh, or shotcrete. At the other extreme, transportation tunnels through soft ground below the groundwater table are furnished with durable, watertight linings, generally designed at present as moment resistant reinforced concrete linings with waterproofing membranes. Many sewer tunnels are furnished with corrosion resistant internal linings.

28.1.5 Immersed and Floating Tunnels

28.1.5.1 Immersed Tunnels

Whenever there is a need to cross water, an immersed tunnel should be considered. Besides their use as road and rail tunnels, many have been built for utilities or pedestrian-only use. Immersed tunnels may appear complex to those unfamiliar with the technology because of the operations in and over water that they entail (Figure 28.1). In reality though, the technique is often less risky than bored tunneling, and tunnel element manufacture can be better controlled since it is precast in the dry. As a result, immersed tunnels are nearly always much more watertight and therefore drier than bored tunnels.

Immersed tunnels offer a number of special advantages:

- They do not have to have circular cross-sections unless the external pressure is particularly high. Almost any cross-section can be accommodated, making immersed tunnels particularly attractive for wide highways and combined road or rail tunnels.
- In contrast to a bored tunnel, they can be placed immediately beneath a waterway with a nominal cover of only 1.5 to 2.0 m of backfill to provide protection against dropping anchors or sinking ships. Because they are shallower, this allows immersed tunnel approaches to be shorter and gradients to be flatter, which is an advantage for all tunnels, but especially so for rail tunnels.
- They can be constructed in ground conditions that would make bored tunneling difficult or expensive, such as the soft alluvial deposits characteristic of large river estuaries.

Immersed tunnels have been located in both freshwater and seawater, in hot and cold climates, in busy waterways, and in earthquake zones. Alignments do not need to be straight, so that the tunnels can be designed to suit design speeds, existing land use, topography, and connections to the existing road or rail system. The ideal alignment for an immersed tunnel may not coincide with the ideal alignment for a bridge. Soft ground is ideal for immersed tunnels, but if excavation into rock is required, the high cost of rock excavation under water could make other tunneling methods more cost-effective.

Immersed tunnels are not suitable for every situation. However, if there is water to cross or water can be used to float-in a precast structure, they usually present a feasible alternative to bored tunnels at a comparable or better price. A catalog of existing immersed tunnels is given in Ref. [1].

28.1.5.2 Floating Tunnels

As water depth increases, particularly if the waterway is narrow, tunnels may become excessively long when compared to the width of the crossing. Using shallow alignments, a new option has been developed in which the tunnel is exposed within the water column, resulting in much shorter

FIGURE 28.1 Lowering an immersed tunnel element into the excavated trench (© Parsons Brinckerhoff, Inc. 1967, with permission).

tunnels. Such tunnels are known as *floating* tunnels and appear to be viable using available technology; designs for several locations are well advanced, although none has yet been constructed. While very short *floating* tunnels might not need intermediate supports, longer nonbuoyant tunnels might be supported by columns up from the bed (like an underwater bridge) or suspended from pontoons on the surface. Conversely, anchors might hold down a permanently buoyant tunnel (similar to tension-leg oil platform technology). Further information on floating tunnels can be found in Ref. [1].

Although most immersed tunnels are built for water depths between 5 and 20 m, schemes have been postulated for depths up to 100 m. Such great depths could be avoided by using *floating* tunnel technology, though if submarines ply the waters, the risk of collision might preclude the use of a *floating* tunnel or at the very least will complicate the design.

28.1.6 Jacked Tunnels

Tunnel jacking is a specialized technique for constructing a concrete tunnel, typically rectangular, under surface areas that are being used for facilities considered critical to the degree that interrupting them would result in serious impacts, not the least of which are significant costs. Examples of such critical surface facilities include operating railway tracks, major roadways, and airport runways. The technique is generally applied in soft ground for relatively short tunnel sections and shallow depth of cover. The method has been employed to construct tunnels of varying uses, including pedestrian passageways, flood control structures, and roadways.

The tunnel jacking method is costly and is not normally considered the preferred technique for constructing an underground crossing. Conventional mined tunneling would be the method of choice if the depth of cover were sufficient to limit the influence of excavation on the surface facilities to acceptable or manageable levels. For shallow tunnels where mined tunneling would be too risky, cut-and-cover methods would be preferred. However, when the costs of shutting down critical surface facilities completely or rerouting or relocating them or providing substitute services for them for the duration of the construction period are taken into account, tunnel jacking may become the most economical method.

28.2 Cut-and-Cover Tunnels

28.2.1 Introduction

For structural evaluation of cut-and-cover tunnel structures, a spectrum can be considered where cut-and-cover tunnels fall between immersed and bored tunnels. Immersed tunnels are largely made of prefabricated structural sections, and their structural analysis resembles the more traditional approach of loads applied to a structure; construction staging must be carefully considered as the tunnel is floated and immersed. Bored tunnels can be thought of as reinforced openings in a competent subgrade material; their structural analysis is usually not a problem of loads applied to a structure, but one of considering how the soil behaves once a gap is constructed and then lined. Analysis of cut-and-cover tunnels will typically consider both of the following features:

- Soil–structure interaction.
- The traditional approach of loads applied to a structural model.

Computer modeling is now able to handle sophisticated soil–structure interaction analysis of cut-and-cover tunnel construction. As the software becomes easier to use and more flexible, more sophisticated analyses of tunnel sections are being performed, as discussed in Section 28.2.4.

An initial distinction can be made between tunnels for which support of excavation (SOE) walls are temporary and those for which cofferdam walls are incorporated into the final structure. In the past, construction walls such as sheet piling and lagging were "throw-away" temporary walls. More recently, however, some engineers have specified that the construction excavation walls be used as part of the final tunnel section. Examples include conventional concrete slurry walls, soldier pile tremie concrete walls, and tangent and secant pile walls. There are important differences in terms of construction issues between incorporating the excavation support or not, with the focus here being on structural analysis and design characteristics. Figure 28.2 shows a schematic of the two different types of section.

When the SOE walls are temporary, the tunnel box is assumed to experience mostly final loading conditions. However, when the SOE walls are included as a part of the final structure, the design loading condition are much more time dependent. The designer will need to consider not just the final loading conditions but also all of the states of stress experienced by the cofferdam walls as they are constructed, braced or tied back, and finally connected to the tunnel as a whole. Many of these stresses are locked-in stresses to be superimposed on later conditions.

The evaluation of impacts on existing facilities is another complication for cut-and-cover tunnel analysis that is not as much of a concern for bored and immersed tunnels. Of the three tunnel types, cut-and-cover tunnels are the most likely to be built adjacent to existing buildings, utilities, and other facilities, so the nature of the tunneling process can lead to significant impacts on existing structures.

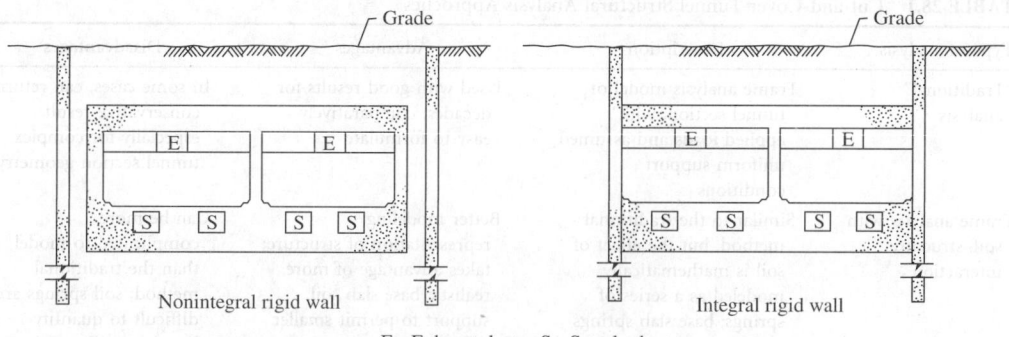

FIGURE 28.2 Typical sections for cut-and-cover tunnels.

28.2.2 Structural Analysis

Cut-and-cover tunnel sections are most often modeled as structural frames. In general, however, there are three ways to analyze cut-and-cover tunnel sections:

- *Traditional frame analysis.* A stick figure is drawn to represent the tunnel section. Soil–structure interaction is treated as an applied load to the frame.
- *Frame analysis with a more rigorous soil–structure interaction.* The soil is not just an applied load; its properties are modeled along with the tunnel. An example is treating the subgrade material as a series of springs, the "beam on elastic foundation" (BOEF) approach.
- *Finite element/finite difference analysis.* The continuous soil and the tunnel structural material together are modeled as a continuum.

Table 28.1 describes the attributes, advantages, and disadvantages of the three analysis approaches. Two-dimensional (2D) sectional analysis remains as good practice for most tunnel conditions, even though the most accurate tunnel model is 3D. Structural analysis of tunnels in three dimensions has been performed, but is very complex. A good case can also be made for the fact that a complex computer analysis exceeds the accuracy of its input for this type of design. Given the uncertainty about the way soil behaves, its inherent lack of homogeneity, and difficulties in modeling three-dimensional soil–structure interaction, the 3D structural analysis and design of cut-and-cover tunnels is reserved mostly for special analysis conditions, such as boxes with rigid end diaphragm walls for rail transit stations.

28.2.3 Methods of Framing

A difference between the tunnels with temporary support walls and those with permanent support walls as part of the final tunnel section is related to the framing model for the structure. When SOE walls are temporary, the tunnel section is almost always treated as a box with fixed joints. However, when rigid cofferdam walls become part of the final structure, it can become difficult to construct fixed connections in the field. Therefore, designers may specify partially fixed or shear connections at the roof to wall joints or base slab to wall joints. The number of different ways to model the tunnel frame increases in this case.

Considering the shape of the frame, most cut-and-cover tunnel sections are rectangular, unlike immersed and bored tunnels, which are often round or oval. A rectangular cross-section is the best-fitting shape for most uses of the tunnel, and it is the one requiring the least excavation for the relatively shallow cut-and-cover tunnel and has the simplest formwork. However, the box shape is also a less efficient structural system for carrying compression loading than circular bores. This structural inefficiency is often mitigated by inserting corner haunches (Figure 28.3).

TABLE 28.1   Cut-and-Cover Tunnel Structural Analysis Approches

Type of analysis	Description	Advantages	Disadvantages
"Traditional" analysis	Frame analysis model of tunnel section, with applied loads and assumed uniform support conditions	Used with good results for decades; comparatively easy to formulate	In some cases, can return conservative results, especially for complex tunnel section geometry
Frame analysis with soil–structure interaction	Similar to the traditional method, but the effect of soil is mathematically modeled as a series of springs; base slab springs use the beam on elastic foundation method; lateral springs can also be applied, partly in place of assumed soil loads	Better modeling representation of structure; takes advantage of more realistic base slab soil support to permit smaller slabs in design	Can be more complicated to model than the traditional method; soil springs are difficult to quantify based on soil test data
Finite element analysis	The section soil and structure are modeled as a grid of geometric elements; this category includes many different types of analyses, depending on the mathematical model assumed for the soil type. Structrual elements are usually treated as linear elastic	Closest modeling representation to "reality"; better numerical resolution for conditions of difficult tunnel geometry and framing details	Usually complex to set up and run; results require careful interpretation; precision of analysis method can be many times better than the data that form the input, especially for soil model

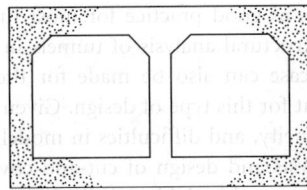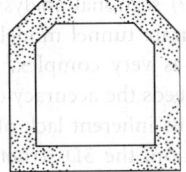

FIGURE 28.3   Haunched section.

Most cut-and-cover tunnels are constructed from reinforced concrete. When using temporary SOE walls, the tunnel section is formed and cast within them, often with a layer of waterproofing completely enveloping the section. The construction procedure for the tunnel using the SOE walls as part of the structure will involve placement of keys in the cofferdam wall into which the roof and base slab can then frame. Another method is to cast an inside wall against the face of the SOE wall. In this case, the base and roof slabs are supported by the inside wall. The SOE wall may also be structurally connected by a composite design with the inside wall.

Concrete framing is usual, except for very deep or very wide tunnels that require excessively thick concrete slabs. In these cases, one structural type that has been used frequently is steel composite concrete construction. For the roof, a relatively thin concrete slab can be cast with a steel beam, compositely connected using shear studs. Wall members may be designed as steel columns without composite connections to the enveloping concrete, or composite action can be specified. The base slab is the least likely location these days for composite steel–concrete members, although some tunnel structures are constructed with base, walls, and roof consisting of steel framing with concrete infill. Taking advantage of the comparatively thinner composite beams and columns in the roof and walls, designers raise the tunnel vertical profile and take advantage of its fit in the plan. However, space constraints are typically least severe on the underside where a concrete base slab helps to resist uplift through its additional weight.

28.2.4   Analysis in Section: Typical Frame and BOEF Methods of Analysis

Although there are many types of cut-and-cover tunnel sections, a typical section may have the following features:

- Construction and excavation via temporary, flexible SOE walls like soldier piles and lagging. A mud mat is constructed at the bottom of the excavation to facilitate casting of the base slab. For some types of soil, a geotextile fabric can be placed to help stabilize the excavation.
- Reinforced concrete base slab, roof slab, and walls.
- A layer of waterproofing beneath the base slab, on the outside surfaces of the walls, and above the roof slab. The waterproofing is protected by a layer of lean concrete before backfilling.

A typical structural analysis of this section would use a frame model. Different loadings would be calculated and applied to the model. Analysis would be performed using an indeterminate method like moment distribution or with the assistance of a frame analysis program like STAAD-III or STRUDL.

Based on the results of the analysis, the components of the cut-and-cover tunnel would be sized. Concrete slabs would be designed typically as bending members. There could be some amount of iteration to the process: the engineer, following sizing of the structure, may need to repeat the analysis using the new member sizes.

In the analysis, each member must be assigned initial trial analysis properties such as sectional moment of inertia and cross-sectional area. For a reinforced concrete tunnel, a 1-ft-long length of tunnel is typical for modeling, and the corresponding member properties are calculated based on the assumed structural depth. Similarly, material properties must be calculated and assigned to the members. For example, each member in the frame analysis must be given a modulus of elasticity and its material unit weight.

Assumptions for supports to the frame analysis model are typically handled in one of two ways (Figure 28.4):

- *Traditional frame analysis.* For the simple tunnel box shown, two imaginary support points are placed in the model. An upward distributed vertical pressure, usually uniform, is assumed to act upward below the base slab such that all vertical loads are balanced and the resulting reactions at the imaginary support points are zero.
- *BOEF method.* Spring supports are specified. The value of the spring constant is determined from consideration of the underlying strata and its modulus of subgrade reaction. As shown in Figure 28.4, this method results in a nonuniform base slab pressure, in contrast to the conventional design method of balancing the vertical loads. The resulting soil pressure depends on the width of the tunnel box and the relative stiffness of the soil. If the soil were infinitely stiff, all loads would be transferred to the soil directly beneath the walls of the box. If the soil behaved as a fluid, the load would be uniformly distributed below the base slab.

While much of the current process for cut-and-cover tunnel analysis is a refinement of tried and true methods that have been used for decades, computerized design has brought some new issues to the surface. One such concern is lateral load balancing. In the past, tunnels were analyzed structurally and designed by drawing a frame model, applying some assumed loads, performing frame analysis, and sizing the members. With the increased power of computer frame analysis comes realization of a curious problem: the frames tend to distort laterally.

Unless the assumed lateral loads are perfectly balanced and the frame geometry is perfectly symmetric, the frame models will tend to lean to one side. Studies that use hand calculations would show the same

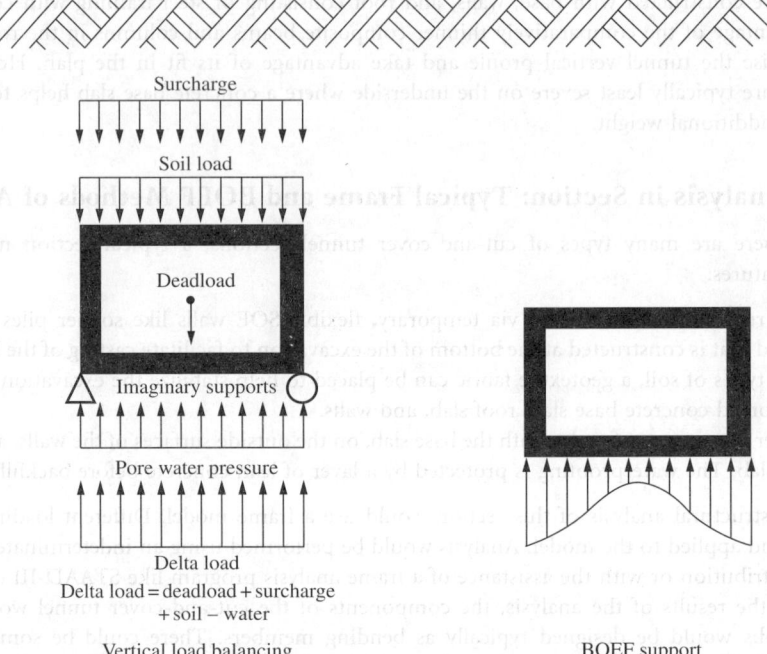

FIGURE 28.4   Base slab pressure: traditional and beam on elastic foundation (BOEF) analysis.

results were engineers to have the ability to display the deformed shape of the tunnel model in a matter of seconds, as is done using computers.

In reality, the tunnel model would not sway. The "sway" side would be held back by a higher soil pressure than that assumed by the preliminary analysis. For frame models, the issue can be approached in two ways:

- *Add soil springs to wall members.* This is a complex treatment, because of difficulties in calculating properties for lateral springs and in calculating applied loads for the springs to react against.
- *Use lateral load balancing.* Lateral loads are estimated and additional, triangular "balancing" loads are applied to the sway side. The reasoning is that the sway side would be held back via passive pressure from the soil. The amount of passive pressure used in the analysis must be limited to ensure that the structure is not induced to sway in the opposite direction. While it is difficult to specify the resulting stress distribution exactly, a triangular loading approximation is reasonable.

Other conditions in the frame model that need to be considered, which differ from the "typical" section described earlier, include the following:

- *Use of haunches.* For the cast-in-place tunnel section, the designer may place haunches to reduce design spans and bending moments (Figure 28.3). This is an attempt to develop a tunnel shape more in compression, as achieved by a circular bore, and less in bending, which is a disadvantage of the straight, rectangular sections.
- *Integral, rigid SOE walls.* For this condition, a rigid, impervious SOE wall, such as a concrete slurry wall or an SPTC wall, is incorporated into the final structure. For the purpose of structural modeling, the designer has to deal with the added concerns of complex tunnel shapes, loadings on wall segments not directly a part of the box, complicated construction sequence loadings, and others.
- *Different material types and framing methods.* These can include steel composite concrete construction and prestressed concrete elements.

- *Different modeling methods.* All-concrete box tunnels are usually designed with moment connections at the joints. This is not difficult to detail and construct when the box is cast-in-place within temporary cofferdam walls. Other structural models may be more appropriate for some of the other tunnel types and construction methods.
- *Different construction staging approaches (top-down).* The order of construction will influence loading and assumptions. For example, in top-down construction, permanent support of excavation walls used as part of the final structure will be more heavily loaded vertically, because the roof is placed and loaded before the base slab is constructed. The base slab acts as a mat for supporting vertical loads, but it is not available until toward the end of construction of the section.
- *Framing of ducts.* In highway tunnels, ventilation supply ducts may be placed in the base slab, in the roof slab, or as side ducts adjacent to the walls. The ducts introduce additional framing conditions and considerations. For example, ducts formed as part of the base slab can be modeled as a type of Vierendeel truss in a stick frame model.

28.2.5 Loading

Cut-and-cover tunnel loads include the following (note that hydrostatic loads always act normal to the exterior surface of the tunnel):

- *Vertical loads, roof slab.* Soil pressure on the roof slab from backfill, own weight of the roof, a live load surcharge typically uniquely specified for each project, and water below the water table; often, a future air-rights development load.
- *Vertical loads, base slab.* Hydrostatic pressure (acts up), upward soil pressure to balance resultant weight of tunnel and all other imposed loads, own weight of base slab (acts down).
- *Lateral loads.* Hydrostatic pressure, effective soil pressure, lateral soil load due to surcharge loading, special loading conditions due to nearby foundations, and other underground conditions.

Vehicle loading (transit or highway) needs to be considered but may not contribute to the most severe tunnel loading case. The dominant loading direction on a tunnel base slab is up from hydrostatic pressure and soil pressure, so including the downward acting vehicle loads would reduce the overall loads on the base slab. Therefore, the controlling load case is often for an empty tunnel.

In congested urban areas, where many cut-and-cover tunnels are located, the tunnel design may need to consider temporary loads from underpinning existing structures and permanent loads from construction of future structures above the new tunnel. A good example of both concerns is the tunnel design for Boston's Central Artery. The depressed artery is constructed directly beneath the original overhead viaduct, so bridge loads from underpinning of the elevated expressway needed to be included in the tunnel design. In addition, once construction is complete, acres of developable land in downtown Boston will become available, so the tunnel design had to incorporate provisions for future building loads directly on the tunnel boxes.

28.2.6 Finite Element Analysis

The preceding discussion focused on the use of frame models for structural analysis of cut-and-cover tunnel sections. As introduced in Section 28.2.2, frame models with load balancing and frame models with soil–structure interaction assumptions represent the first two of the three analysis methods. These are the traditional approaches for this type of design and have been used since before the advent of computers. However, ways in which frame models form an inaccurate representation of the analysis problem include the following:

- Frame models feature line elements, but in reality, elements of the tunnel section have thickness. For example, base slabs with air vents are thick and massive, and the line elements of the frame are a coarse representation of the structure. Steps can be taken to modify the frame models to account for the mass of the tunnel.

- Frame models treat the surrounding soil as an equivalent applied load. In reality, the soil and structure behave together in a complex soil–structure interaction. Assumed soil loads, used for decades in tunnel design, constitute a conservative but imprecise representation of the analysis problem.

Another concern is the issue of construction staging. Underground structures experience several different states of stress. These are time dependent and depend on the methods of excavation and stabilizing the sides and on the order in which the tunnel components are installed.

For the purposes of this discussion, the finite element method and the finite difference method will be considered the same. Although they differ mathematically, the general approach of the two is similar. They are both used to model cut-and-cover tunnel analysis by modeling the continuous soil as discrete space and having the computer calculate most applied loads by assumptions about staging, soil properties, and boundary conditions.

In general, the finite element method is an analytical tool for solving partial differential equations by first making the equations discrete in the solution domain. The solution domain refers to the physical definition of a problem. For a cut-and-cover tunnel section, the solution domain includes the structural elements of the tunnel, the backfill, the soil behind the walls, and soil beneath the excavation. Transforming the solution domain into discrete elements is done by dividing it into small regions of simple but arbitrary shape — the finite elements. The locations of elements are defined using nodal points (nodes). To solve the equations over the entire domain, the equations for the finite elements are summed, resulting in global matrix equations. The finite element method is usually performed by computer because of the large number of equations and difficulties in incorporating boundary conditions.

A finite element solution can be run for each construction stage. The output from the previous stage — one line of data for each node and element plus loading and material property data — is used as the input for the subsequent stage. Initial conditions are established by applying self-weight. The soil (finite elements) is then excavated (elements removed), and supports are installed for each stage.

The finite element analysis loads the structural model based on the input soil properties and constitutive relationships. Unlike the traditional approach, no assumed applied soil loads are calculated. Although frame models have been used for cut-and-cover tunnel sections traditionally, finite element analysis provides a more rigorous analysis of the problem. A few words of caution are in order, however

- Available information for soil–structure interaction problems will often be less sophisticated than for the analysis model. The finite element study assumes that the soil mass behaves in an elastic or other mathematically quantifiable way. The interpreted soil layer itself, based on investigations that can only approximate actual conditions in place, is taken as a group of homogenous layers. At best, a computer model may need to include certain simplifying assumptions about soil behavior. To complicate matters further, the stress–strain relationship at the soil–structure interface is highly indeterminate, time dependent, and not well understood. Research has focused on ways to extrapolate laboratory conditions to mathematical models, but it is difficult to quantify conditions such as a tunnel wall against several different soil layers.
- In reality, a model with better accuracy would include 3D geometry and loading. However, whatever uncertainties are present for 2D finite element studies are magnified for 3D work.
- Soil–structure finite element analysis is invariably more complicated and time-consuming than frame analysis. Interpretation of results requires much greater care. A highly automated approach can be difficult to run for this type of analysis because the results need to be carefully inspected. Fortunately, software and interpretation methods continue to improve with time.
- While the behavior of the soil mass is very sensitive to the way it is modeled, the ultimate design of the tunnel structure is not. Discounting a few exceptions, the finite element and frame modeling will result in a tunnel design of about the same structural dimensions. Differences that are more significant will be apparent in the calculation of temporary support of excavation structures like struts and tiebacks and the estimation of soil movement in a temporary construction.

Overall, frame analysis of cut-and-cover tunnels is still a viable, appropriate design tool. A suggested approach for structural analysis and design of cut-and-cover tunnels is as follows:

- Prepare sections, taken at a suitable interval, and run frame analysis and design. The frame can be modified to reflect conditions like special structural framing, vents, cofferdam walls incorporated in the final structure, and other features.
- Run a few soil–structure interaction computer analyses for the study of special problems. These can include construction staging, stress analysis of thick, vented members, predictions of soil movement during excavation, estimates of the effects of lateral cross walls (e.g., in subway stations), and other problems where the improved resolution of the finite elements outweighs the increased effort in their application.
- Check the results against frame analysis for a "reality check" of the finite element runs.
- Prepare a few computer frame analyses to study longitudinal conditions. While analysis in section largely governs the tunnel design, special longitudinal conditions should be studied as well. These can include framing of transverse ducts to vent buildings, lateral diaphragm walls at the ends of transit stations, and modeling of changes in foundation material in the longitudinal direction.

28.2.7 Buoyancy

In areas with a high water table, buoyancy will be a concern. The weight of soil taken out for the cut-and-cover tunnel can be less than the weight of the tunnel structure. In addition to the procedures above, buoyancy calculations involve a stability analysis. For the final condition, the weight of the tunnel section must be greater than the upward net hydrostatic pressure. It is conservative not to take advantage of the side friction acting on the tunnel walls, although in certain cases if construction methods and the undisturbed soil are considered, it might be acceptable to use side friction in the calculations. For tunnel sections where the excavation support wall is used as the final structure, the weight of the full wall above and below the tunnel section may be used to resist buoyancy.

Calculations for buoyancy must be done for temporary construction stages as well as for the final condition. The calculations must be consistent with the specified requirements for dewatering for each stage of the excavation and construction.

28.2.8 Evaluation of Construction Impact and Mitigation

An important issue for cut-and-cover tunnel analysis and design is the evaluation and mitigation of construction impacts. By the nature of the methods used, cut-and-cover construction can be more disruptive than bored tunnels. It is important for structural engineers to be familiar with the analytical aspects of evaluating soil movement and the impacts it can have on existing buildings and utilities at the construction site. Soil movement can be due to

- *Inward movement of excavation support walls.* Walls will deflect into cut-and-cover excavation prior to installation of each level of struts or tiebacks supporting the wall. The movement is not recoverable. At each level of excavation support, a certain amount of inward movement will have been recorded, and this is the starting deflection for the next stage of excavation. Therefore, whether using a frame model with assumed applied loads or a finite element analysis, the analysis is nonlinear.
- *Consolidation due to dewatering.* In excavations where the water table is high, it is often necessary to dewater inside to avoid instability. Dewatering inside the cut leads to a drop in the hydrostatic pressure outside the cut. Depending on the soil strata, this can lead to consolidation.

Existing buildings and facilities must be evaluated for estimated soil movement during excavation. This evaluation depends on the type of existing structure, its distance and orientation from the excavation, and other parameters. The analysis is site-specific, and it can be very complex. Empirical methods and screening tools are available to characterize impacts more generally. These approaches are a good way to get analytically a handle on this complex problem.

Many tools are available to mitigate, measure, and control adverse impacts from soil movement, including

- Design of stiffer and watertight excavation support walls.
- More closely spaced and stiffer excavation support braces and tiebacks.
- Pre-excavation soil improvement.
- Specification of limiting values for allowable movements of excavation support walls, soil profiles, and allowable impacts of depressing the water table during construction.
- Measurement of these impacts by a program of geotechnical and structural instrumentation during excavation.
- Requirement for mitigation plans to react to situations where movement measurements approach allowable limits.

28.3   Tunnel Linings for Bored and Mined Tunnels

This section describes basic principles for the selection and design of lining systems for bored and mined tunnels. For a greater overview of these principles, see Kuesel [3]. For detailed application in the realm of tunnels in hard ground, see U.S. Army Corps of Engineers Manual *Tunnels and Shafts in Rock* [4].

28.3.1   Introduction

The design of bored and mined tunnels differs in principle from the design of most other types of structures. Two important aspects largely govern the design of these tunnels:

- More than for other structures, the design details of the tunnel structure, often even its basic shape, depend on the method and details of construction.
- The principal load-carrying component of the tunnel structure is the soil or the rock mass surrounding the tunnel opening, and the structural lining serves largely to maintain the integrity of the surrounding ground mass.

For these reasons, the method of construction of a bored or mined tunnel must be established before its design can be contemplated, and the design must include analyses or assumptions concerning the interaction between the tunnel lining and the ground.

Principles and details of tunnel linings depend very much on the functional requirements of the finished structure. The lining concept can vary from an unlined and leaky tunnel in rock, supported only by occasional initial ground support such as rock bolts, to a full concrete or steel lining designed to carry the entire rock and groundwater load and to be watertight. The lining concept may be no lining, one-pass lining, or two-pass lining. The lining concept is selected based on the following basic criteria:

- Functional requirements.
- Ground conditions.
- Constructibility.
- Economy.

One of the most important decisions in the development of the lining concept is what to do with the groundwater, both during construction and in the permanent structure.

Highway tunnels and most railway tunnels today are designed to eliminate water inflow through the crown and sidewalls of the tunnels. If the tunnels are above the groundwater table, making the top of the tunnel watertight prevents infiltration of occasional seepage water. If the groundwater is not far above the tunnel, this method will still work, provided adequate drainage is placed outside the water-proofing and through pipes in the tunnel invert. This is a so-called drained tunnel. If the groundwater table is high above the tunnel and the ground is very pervious, then a drained tunnel is not feasible, and the tunnel must be waterproofed all around, and the lining must be designed for the full groundwater pressure.

Tunnels conveying water for power, irrigation, drainage, or other purposes may be pressurized structures or operated with open-channel flow. These tunnels must be designed for the highest operating and transient internal pressures, compatible with the hydraulic grade-line and the hydraulic controls, in addition to the external loads. Air relief must be provided as required. The tunnel finish must match the roughness (Manning's coefficient) assumed for the hydraulic analyses.

For the design of pressurized tunnels, the external load and the strength of the surrounding rock mass may be considered if the rock is competent, but often the lining must be designed for the full internal pressure. When the internal water pressure is relatively low, less than 100 to 150 psi, it is possible to design steel reinforcement in the concrete lining to accept the hoop tension. With higher pressures, it is usually necessary to use a steel lining. For the design of pressurized tunnels, see the penstock literature or, for example, the U.S. Army Corps of Engineers Manual, *Tunnels and Shafts in Rock* [4].

When the groundwater is aggressive or contains chemicals that cannot be permitted to flow into the potable water in a water supply tunnel, the tunnel should be made watertight. Some groundwater contains chemicals that will clog up drainage facilities. Under those conditions, the drainage paths must be serviceable or the tunnel made watertight. Sewage tunnels with open-channel flow in warm climates will typically require protection against corrosion resulting from generation of hydrogen sulfide. Frequently this protection takes the form of an internal polyvinyl chloride (PVC) or high-density polyethylene (HDPE) membrane cast into the internal tunnel lining.

Another important functional requirement deals with the tolerance of the installation of the tunnel facility. A pressurized-water tunnel can tolerate a wide placement tolerance in all directions without serious impairment of function. A tunnel with gravity flow must satisfy the vertical flow requirement but can tolerate wide horizontal tolerances. The interior of traffic tunnels (rail as well as highway or street) must meet the civil requirements of the guideway and the dynamic envelope of the vehicles.

Setting a concrete form precisely within an excavated tunnel or cavern is usually much easier than producing a precise excavation in the first place. Therefore, a wide tolerance can be applied to the initial excavation and still permit proper placement of the concrete. On the other hand, the one-pass lining is also the final lining and must be built precisely. One-pass linings are often designed with several inches of additional construction tolerance just to make sure the finished structure meets the functional requirements.

28.3.2 Mechanized Tunneling through Soil

Modern tunnel excavation through soil is usually done with a shield, most often a mechanized shield. A shield is usually a circular steel cylinder furnished with a cutting edge. The shield is jacked through the soil as soil is excavated at the face, with the soil being removed through the back. A tunnel lining is erected as the shield advances, against which the propulsion jacks of the shield push. The shield provides the initial ground support. A fully mechanized shield is much like a factory, producing a tunnel in the fashion of an assembly line. Details of the mechanized shield are selected primarily to meet expected ground and groundwater conditions.

28.3.2.1 Dry Soil

In dry ground above the groundwater table, it is usually possible to excavate with an open face or with partial face support. Hand tools or, more commonly, large hydraulic scrapers or hoes are used to loosen the soil and scoop it onto a conveyor belt for loading into muck cars.

If the ground is firm, an initial lining consisting of steel ribs and timber lagging, or unbolted concrete segments, may be expanded against the soil behind the advancing shield tail. On the other hand, a cohesionless sand or silt material will tend to collapse behind the shield tail, and it may be necessary to erect the initial lining inside the shield tail. In this case, the lining will usually be erected and bolted together to a diameter smaller than the diameter cut by the shield, and the resulting tail void will need to be filled, often with pea gravel and grout.

28.3.2.2 Wet Soil

In soft clay, the ground will tend to be overstressed in the face of an open shield, and ground movements can be unacceptably large. Thus, continuous support is required. This is also the case in most soils below the groundwater table. Without continuous support, the ground will tend to flow uncontrollably into the shield, driven by the seepage forces in the ground. In these cases, a shield with face pressure control is often used today. The front end of the shield is sealed off with a sturdy bulkhead, and sufficient pressure is maintained in the front to keep the soil and the groundwater stable. There are two types of face-pressure shields:

- Slurry tunnel boring machine (slurry TBM).
- Earth-pressure-balance (EPB) TBM.

28.3.2.2.1 Slurry TBM

With a slurry TBM, the sealed front compartment is filled with clay slurry, usually a bentonite slurry, pressurized to balance the groundwater pressure plus at least the active earth pressure acting in front of the face. The pressure is maintained constant either through delicate flow controls or through a compressed-air buffer chamber. The earth is loosened by a rotary cutterhead and kept in suspension in the slurry. The muck-laden slurry is pumped to the ground surface, where the muck is separated and the slurry reconditioned before returning to the tunnel face.

In recent years, the slurry TBM has lost much of its market in favor of the EPB TBM, which does not require an expensive slurry separation and treatment plant. However, the required rotary torque for a large-diameter EPB cutterhead becomes excessive, and all machines over 10 m in diameter to date have been slurry TBMs. The slurry TBM also offers advantages of better face control, in particular in mixed-face and bouldery ground. It also offers the opportunity to place a boulder-crushing unit at the bottom of the face to reduce large boulders to manageable sizes. Figure 28.5 shows a typical slurry TBM.

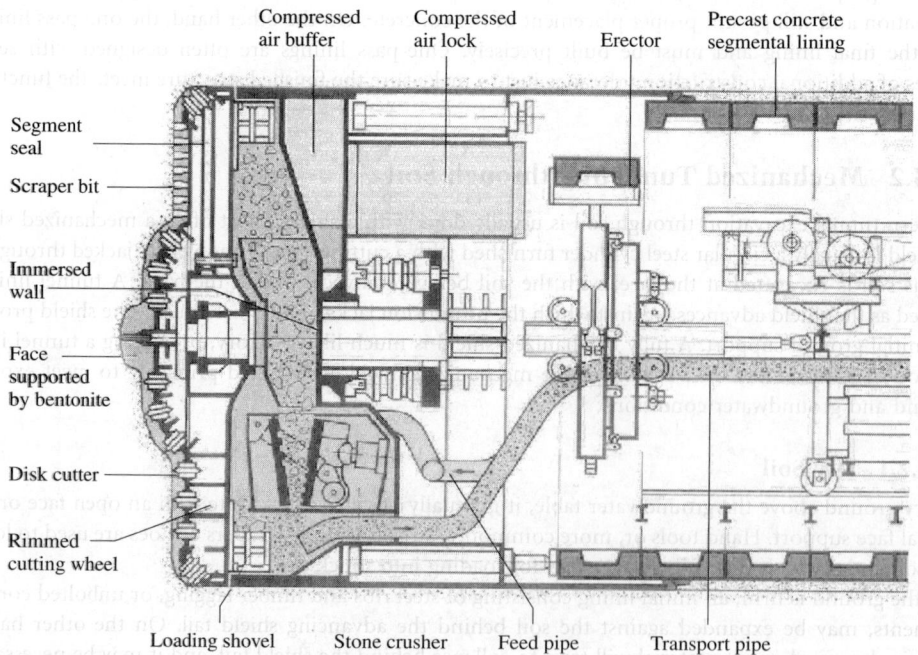

FIGURE 28.5 Schematic of slurry TBM (courtesy of Parsons Brinckerhoff, Inc.).

28.3.2.2.2 EPB TBM

With the EPB TBM, the front compartment is kept full of soil under sufficient pressure to maintain the stability of the face and minimize ground movement ahead of the face. The soil excavated by the rotary cutterhead is removed from the bottom of the compartment through one or two screw conveyors in series, discharging onto a conveyor belt or directly into muck cars. Groundwater is prevented from flashing through the screw conveyor system by foam, sometimes supplemented by a polymer, or clay slurry added to the soil in the front compartment. The additives must provide sufficient cohesion to counter the groundwater pressure along the length of the screw conveyor. The additives also help lubricate the soil mass, reduce the required torque for turning the cutterhead, and reduce wear on the components.

28.3.3 Linings for Tunnels in Soil

28.3.3.1 One-Pass Lining

Both types of face-pressure TBMs usually require the installation of a bolted, gasketed tunnel lining usually made of concrete segments. In the past, cast iron and fabricated steel segments have been popular but are now much too expensive except in special circumstances. The lining must be installed under the protection of the shield tail, and the tail void must be filled. To prevent excessive movement of soft or loose soil into the tail void, it is usually necessary to fill the tail void as the shield advances and keep the tail void under pressure equal to the ground and groundwater pressure.

After the installation of the segmental lining in the tail of the shield, usually, no additional internal lining is required. This type of lining is usually required when the ground conditions are sufficiently poor to require a face-pressure TBM. The one-pass lining must be fabricated and erected with great precision since once in place, misalignment can be fixed only with great difficulty. The one-pass lining must be designed for all transport and construction loads, including the forces from the shield propulsion jacks. Its adequacy to resist ground loads must also be checked.

28.3.3.2 Two-Pass Lining

When an initial lining can be installed behind the tail of the shield, the advance rate of tunnel excavation is usually much faster and the cost of tunneling cheaper, even if a second lining must be placed inside. The initial lining may be steel ribs with lagging or unbolted concrete segments, just adequate to permit installation of the final lining. The final lining is cast-in-place concrete with one, sometimes two, layer of reinforcement. A waterproofing membrane is usually installed before placing the internal concrete. With a two-pass lining, the initial lining is often considered part of the permanent structure, participating in carrying the loads; however, the inside final lining usually must be designed at least for the groundwater pressure.

28.3.4 Bored Tunnels in Rock

28.3.4.1 Hard-Rock TBM

Long tunnels in rock are now mostly excavated using hard-rock TBMs, which excavate a tunnel of circular cross-section. The rotary cutterhead of a hard-rock TBM is fitted with disk-shaped rock cutters that roll over the face of the tunnel under great thrust, breaking the rock into slivers and advancing the tunnel by a fraction of an inch per rotation, up to tens of feet per hour. In excellent rock with few discontinuities, no immediate ground support is required. More commonly, the rock mass is fractured, and initial ground support in the form of rock bolts or dowels, or even steel sets or lattice girder support, is required.

Propulsion of the advancing TBM, as well as the thrust required for breaking the rock, is usually derived from large pads jacked firmly against the sides of the tunnel. Every 4 to 6 ft or so, the jacks are retracted from the walls, advanced forward toward the face, and re-engaged against the sidewall.

In poor rock, or even in good rock where other parts of the tunnel are advanced through soil-like materials, the TBM may be outfitted with a shield much like a soil TBM. A one-pass or two-pass lining system can be erected in the tail of the shield, as in a soil tunnel. Figure 28.6 shows the cutterhead of a hard-rock TBM, holing through.

28.3.4.2 Roadheader

Another type of rock excavator is the roadheader, which is a smaller excavator or ripper mounted on a slewing and elevating arm. The excavator is equipped with ripper or point-attack teeth that rotate around the axis of the arm or around an axis at right angles with the arm. The roadheader cannot excavate as strong a rock mass as the hard-rock TBM, but it is able to excavate a rectangular or horseshoe-shaped opening. Figure 28.7 shows a typical roadheader.

FIGURE 28.6   Rock TBM holing through (© David Sailors 1988, with permission).

FIGURE 28.7   Roadheader with double rotating cutterhead and gathering apron (© Sandvik 2000, with permission).

28.3.5   Sequential Excavation and Support for Rock Tunnels

Shorter tunnels in rock, and rock tunnels of large cross-sections, are sometimes excavated by roadheader, if the rock is soft enough, but more often by blasting. Blasting is a cyclic operation, which typically includes the following activities in sequence:

- Drilling blast holes in a predesigned pattern and loading them with blasting agents.
- Setting off the blast using millisecond delays so that there is a series of closely timed blasts (reduces vibrations and facilitates the blasting process).
- Removing the loosened rock, loading it into trucks or rail cars, and bringing it out.
- Scaling loosened rock from the crown and the side walls.
- Installing initial ground support, rock bolts, shotcrete, steel ribs, or other components as required.
- Repeating the cycle.

With small cross-sections in good rock, the full face can be blasted at one time. Often a top heading is excavated first, followed by excavation of the bench. In very poor rock, it may be necessary to excavate and support multiple headings.

The final lining is usually made of reinforced concrete cast in place. Below the groundwater table, a waterproofing membrane is placed to cover the initial ground support before placing the cast-in-place lining. As discussed earlier, the tunnel lining may be designed to be undrained, with the final lining accepting the full groundwater pressure. Alternatively, drainage may be provided outside the waterproofing membrane, in which case the final lining may be designed for only a portion of the full groundwater pressure.

28.3.6   Selection of Lining System in a Rock Tunnel

For most transportation tunnels, a substantial final lining is installed. For many tunnels, however, there is a wide selection of tunnel ground support. In some cases, a different ground support system will be selected for different parts of the tunnel depending on geologic conditions and local variations in construction methodology. For example, a steel lining may be required for a portion of a pressure tunnel with high pressure and poor rock, while other parts may require no lining at all. A watertight tunnel may be required through a permeable shattered-rock zone, or through seams of gypsum or anhydrite, which may expand or deteriorate, but it may not be required elsewhere. If a TBM and a substantial initial lining are required in part of the tunnel and if the lining is used to provide part of the reaction for the TBM propulsion, then a contractor may choose to line the entire tunnel even if a lining is not required everywhere for ground support.

28.3.6.1   Unlined Tunnel

In an unlined water tunnel, the water has direct access to the rock, and leakage will occur into and out of the tunnel. Changes in pressure can cause fluctuations and flows in and out of fissures, washing out fines and eventually lead to instability. For an unlined tunnel to be feasible, the rock must be inert to water, free of significant filled joints and faults, able to withstand the internal pressure in the tunnel without hydraulic fracture, and sufficiently tight so that leakage rates are acceptable. Unlined tunnels should be furnished with an invert pavement to provide a suitable surface for vehicles and help to control erosion. Norwegian hydropower tunnels in good crystalline rock are often unlined for most of their length.

28.3.6.2   Shotcrete Lining

A shotcrete lining will provide ground support and may improve leakage and hydraulic characteristics of the tunnel. It also protects the rock against erosion and the deleterious effects of water. To protect water-sensitive ground, the shotcrete should be continuous and crack free and reinforced with wire fabric or fibers. As with unlined tunnels, shotcrete-lined tunnels should be furnished with an invert pavement to provide a suitable surface for vehicles and help control erosion.

28.3.6.3 Unreinforced Concrete Lining

An unreinforced concrete lining is placed to protect the rock and provide a smooth interior surface. Most shafts and tunnels that are not subject to internal pressure are lined with unreinforced concrete. This type of tunnel or shaft lining is acceptable if the rock is in equilibrium before concrete placement, and loads are expected to be uniform and radial. Shrinkage and temperature cracks are expected to occur and can cause leakage. If the groundwater is corrosive to concrete, a tighter lining may be required to prevent corrosion by the seepage water. An unreinforced lining is generally not acceptable through soil overburden or through badly squeezing ground, which can exert nonuniform displacement loads.

The obvious attraction of the unreinforced lining is the substantially lower expense of the lining when no reinforcing steel cage is needed, even if a waterproofing membrane might be set or fibers used to minimize cracks. A considerable amount of information on the use of plain concrete in tunnels is available in the AFTES (French Tunneling Association) *Recommendations in Respect of the Use of Plain Concrete in Tunnels* (AFTES c/o SNCF, 17 Rue d'Amsterdam, F75008 Paris, France) [5].

28.3.6.4 Reinforced Concrete Lining

The reinforcement in linings with only one layer of steel should preferably be placed close to the inside face to resist temperature stresses and shrinkage. Depending on the moment distribution calculated in the lining based on exterior loads, the single steel layer may move from inside to middle to outside around the ring. A lining of this type can remain undamaged at distortions of 0.5% or more, measured as relative diameter change, and can remain functional even at greater distortions.

Multiple layers of reinforcement may be required due to large internal pressures or in a squeezing or swelling ground to resist potential nonuniform distortions. Nonuniform pressures requiring a second layer of steel can also occur through faults and shear zones, near the ground surface, and at connections to adjacent structures.

28.3.6.5 Pipe in Tunnel

This method is often used in water and sewer tunnels up to a size of pipe manageable on the job or on the road, the maximum usually about 14 ft (4.2 m) in diameter. The tunnel is driven and provided with initial ground support, and a steel or concrete pipe is installed. The concrete pipe may be a standard pipe or prestressed pipe with steel core. The pipe may be furnished with interior corrosion protection made of PVC or HDPE to resist sulfuric acids generated by hydrogen sulfide in an anerobic condition. Plastic pipes as well as glass-reinforced plastic or plastic mortar pipes have also been used.

After placing the pipe, aligning it, and securing it against flotation, the void between the pipe and the rock is filled with flowable mass concrete or, more economically, cellular concrete.

28.3.6.6 Steel Lining

Where the internal pressure in the tunnel exceeds the external (confining) ground and groundwater pressure, a steel lining is usually required to prevent hydro-jacking of the rock mass. With only steel reinforcement (two layers), the ability of the tunnel lining to withstand internal pressure is limited. The guideline concerning confinement is that the weight of the rock mass above the tunnel must exceed the internal operating water pressure, with a suitable safety factor, usually 1.2. This is generally conservative with a level ground surface but not for tunnels running in a valley wall, where the side cover may be smaller. Here, more precise *in situ* stress analyses must be made or *in situ* stress tests performed.

In addition to the interior pressure, the steel lining must also be designed for the exterior pressure. Assuming that pipe leakage occurs over the long-term, the external water pressure is often assumed equal to the interior pressure. The critical design condition is buckling, and external stiffeners are often required. Methods of analyses of both internal and external water pressures for tunnels and shafts are found in the U.S. Army Corps of Engineers Manual, *Tunnels and Shafts in Rock* [4].

28.3.6.7 Concrete Segmental Lining

The one-pass concrete segmental lining is usually used for lining of shield or TBM-driven tunnels through soil. Where portions of a tunnel are in soil and others in rock, it may be convenient to build the entire length of tunnel using concrete segmental lining. Particulars on the design of concrete segmental linings are found in Section 28.3.8.

28.3.7 Structural Design of Permanent Concrete Linings in Rock

28.3.7.1 Rock–Lining Interaction

The most important material for the stability of a tunnel in rock is the rock mass, which accepts most or all of the distress associated with the excavation of the rock opening by redistributing stresses around the opening. The initial ground support and the final lining often serve primarily a confining function, in addition to providing the appropriate interior finish. A lining placed in a rock opening that has reached stability will experience no stresses other than self-weight and internal loads. On the other hand, a lining placed inside an opening that has experienced, say, 70% of its elastic (or elasto-plastic) latent displacement will experience stresses from the release of the remaining 30% of displacement. The actual stresses and distortions will depend on the modulus of the lining and of the rock mass. If the rock modulus or the *in situ* stresses are anisotropic, then the lining will distort as a complex equilibrium is sought.

External loads on a tunnel lining may be following or nonfollowing loads. Typical following loads are the contributions from groundwater pressure, which are independent of the displacement of the lining resulting from the loading. Other types of following loads come from squeezing or swelling ground, where only very large distortions may reduce the load intensity. The typical soil or rock load, however, is to a large measure nonfollowing. As the crown of the tunnel lining yields away from high loads at the top, the loads are relaxed. At the same time, the lining distortion will expand the horizontal diameter of the tunnel, increasing reaction loads on the sides, thus rendering the lining loads more nearly uniform.

28.3.7.2 Cracking of Linings

With nonuniform loads or distortions, moments will occur both in circular and noncircular tunnel linings. Thus, tension cracks can form, the extent of which depends on the combinations of moments and thrusts. Tension cracks in a tunnel lining usually do not form a failure mode. There is usually a compressive component so that a tension crack does not penetrate through the entire section. In fact, separated blocks in compression form a stable arch. Without through-going cracks, water tightness also is not significantly impaired. Calculated extension cracks at the exterior of the lining, facing the rock, may be fictitious because the rock outside the lining is usually in compression and shear bond between lining and rock will prevent a tension crack from actually opening.

As a rule of design, the calculated crack length is usually permitted to reach half the section thickness, whether the lining is reinforced or not. Depending on the end-use requirements of the tunnel, this can be relaxed.

Cracking will also occur in linings due to shrinkage during curing, temperature variations, and the like. Because a blasted rock surface is uneven and the lining thickness, therefore, variable, such cracks can vary in dimension and location and often are concentrated into a few large cracks at the locations of the lesser thicknesses rather than many small cracks. If steel ribs are used for initial ground support, cracks will occur typically at the locations of the ribs. Incorporation of expansion joints is usually not effective in controlling these types of cracks because the concrete is bonded firmly to the irregular face of the rock and initial ground support.

These types of shrinkage cracks may go through the entire cross-section and may be undesirable. To control them, reinforcing steel is often placed close to the interior surface up to a percentage of 0.28 (up to 0.40% in a corrosive environment) of the maximum concrete section thickness. Alternatively, polypropylene, olefin, or steel fibers have been used with success. Construction joints between concrete pours are appropriately furnished with waterstops.

28.3.7.3 Lining Loads for Design

It would be misleading to present an exhaustive treatment of the subject of lining loads for tunnels in rock in this brief treatise. The design loads on the final lining will vary with the rock quality as well as the proportion of the rock load assumed to be carried by the initial ground support (if any). They will also include the external water pressure, if the tunnel is watertight, or at least a part of the external water pressure if the tunnel is drained. Additional nonuniform loads can be derived from squeezing layers of rock and from irregularities in the excavated surfaces. Different assumptions are usually made for circular and noncircular tunnels and for tunnels driven by TBM as opposed to tunnels excavated by blasting, which tends to disturb the rock and add loads.

Guidance to estimate the design rock loads can be found in the rock mechanics literature and in publications such as the U.S. Army Corps of Engineers Manual cited previously. Experienced geologists and rock tunnel engineers should be consulted when establishing these design loads. Conditions in the field should be observed to verify the estimated loads.

28.3.7.4 Methods of Analysis

Closed-form solutions are available for the simple conditions of a circular elastic tunnel lining embedded in a homogenous elastic ground, where the induced stresses are the assumed *in situ* stresses in the ground. The closed-form solution presented in the following is based on the following simplifying assumptions:

- Plain strain, elastic radial lining pressures are equal to *in situ* stresses, or a proportion thereof.
- It includes tangential bond between lining and ground.
- Lining distortion and compression are resisted or relieved by ground reactions.
- Vertical and lateral *in situ* soil or rock stresses are σ_v and $K_0 \sigma_v$.
- Modulus and Poisson's ratio of the soil or rock mass are E_r and v_r, respectively.
- Concrete lining modulus, moment of inertia, and area are E_c, I, and A, respectively, and mean radius is R.

Maximum and minimum bending moments can be determined by the following equation:

$$M = \pm \frac{\sigma_v(1 - K_0)R^2}{4 + (3 - 2v_r)/[3(1 + v_r)(3 - 4v_r)](E_r R^3/E_c I)} \tag{28.1a}$$

Maximum or minimum hoop force can be determined by the following equation:

$$N = \frac{\sigma_v(1 + K_0)R}{2 + (1 - K_0)2(1 - v_r)/[(1 - 2v_r)(1 + v_r)](E_r R/E_c A)}$$
$$\pm \frac{\sigma_v(1 - K_0)R}{2 + 4v_r E_r R^3/[(3 - 4v_r)(12(1 + v_r)E_c I + E_r R^3)]} \tag{28.1b}$$

Maximum or minimum radial displacement equation:

$$\frac{u}{R} = \frac{\sigma_v(1 + K_0)R^3}{[2/(1 + v_r)]E_r R^3 + 2E_c A R^2 + 2E_c I} \pm \frac{\sigma_v(1 - K_0)R^3}{(3 - 2v_r)/[(1 + v_r)(3 - 4v_r)]E_r R^3 + 12E_c I} \tag{28.1c}$$

These simplifying assumptions are hardly ever met in real life and presume essentially the lining "wished in place" into the ground. Nonetheless, the equations are useful in examining the effects of variations of important parameters. The maximum moment is controlled by the flexibility ratio:

$$\alpha = \frac{E_r R^3}{E_c I} \tag{28.2}$$

For a large value of the flexibility ratio (large rock mass modulus), the moment becomes very small. Conversely, for a small value (very rigid lining), the moment is large. If the rock mass modulus is zero, the rock does not restrain the movement of the lining, and the maximum moment in the lining is

$$M = 0.25\,\sigma_v(1 - K_0)R^2 \tag{28.3}$$

With $K_0 = 1$ (uniform stress field) the moment is zero and with $K_0 = 0$ (simple vertical loading of vertical ring) the moment is maximum.

More commonly, structural finite element programs are used for analysis, where the lining ring is modeled as beam elements and the ground as springs representing the modulus of the ground. From these analyses the sectional forces, moment, thrust, and shear are deduced. The structural analysis offers little difficulty, but the loading assumptions and the elastic constants assumed are dependent on both the nature of the ground and the method of construction.

Continuum analyses using finite element or finite difference methods, where the ground is represented by a grid of elements, can analyze more accurately the variations in the ground stratification as well as steps in the construction process. Thrusts and moments in the lining elements can be derived directly.

The concrete cross-section is designed for the moment and thrust using the standard moment thrust or capacity interaction diagram, described in concrete design texts. The typical analyses may be modified to fit just one layer or reinforcement, fiber reinforcement, or none at all. Sometimes a larger theoretical zone of tension in the concrete section may be accepted than strictly according to code.

28.3.8 Design of Segmental Concrete Linings

28.3.8.1 Components of the Lining

The number of segments in a lining ring is decided as a compromise. A small number of segments will give the contractor fewer pieces to handle, for quicker erection, and a smaller number of segment forms. On the other hand, fewer segments are heavier and more difficult to handle. With a wide-open shield, as few as three segments per ring can be handled, but this is rare. More typically, there are five to eight segments per ring, sometimes more for a very large tunnel. The last segment placed is usually wedge-shaped for ease of insertion. Each ring is between 1 and 1.5 m wide, matching the stroke of the shield propulsion jacks, with the narrower rings better suited for tunnels with short-radius curves. Segment rings for curved tunnels are tapered (with ring widths on one side of the tunnel slightly wider than on the other) to allow placement as secants to the curve of the tunnel. Segment joints are usually staggered from ring to ring to provide greater rigidity and redundancy. Figure 28.8 shows the installation of a one-pass liner ring.

The most important parts of the lining are the joints. A one-pass lining must be firmly fixed in a circular form. This usually requires bolts to be installed between rings and between segments in a ring. Sometimes dowels are used between rings. Joints in a one-pass lining must also usually be made watertight, which is usually accomplished by installing a recessed rubber or neoprene gasket around each segment.

The rubber gasket is vulcanized and made to fit the segment precisely. The two opposing gaskets in the joint will be compressed sufficiently so that water leakage through the joint is prevented. The gasket recess and the size and geometry of the gasket are designed to meet the leakage requirements, even if the segments are joined out of tolerance. Laboratory testing of gaskets in adverse installation configuration is often performed.

Another type of gasket is the hydro-swelling gasket, made with rubber and bentonite or similar materials. When exposed to seepage water this gasket will swell three to ten times and fill the available space for a watertight fit. Some important linings are fitted with a hydro-swelling gasket toward the exterior and a rubber gasket toward the interior.

Several types of bolt systems are employed. The most common type consists of straight bolts and nuts with washers and grommets, requiring bolt holes and pockets cast into the segment for their insertion. Curved bolts are easier to install and require smaller bolt pockets. More common today are the straight

FIGURE 28.8   Installation of one-pass liner (© Bernie Martin, with permission).

bolts installed at an angle into an insert cast into the segment. Some designers assume that bolts are for erection only and do not require to be considered for the permanent structure, except at transitions and other critical locations. Some contractors will retrieve most of the bolts after reuse once the lining is in place and the surrounding grout fully cured.

28.3.8.2 Design Conditions

The segments must be analyzed and designed for both the ultimate functional design and the lifetime criteria. They must also withstand all other events during the life of the segments, casting, deforming at low strength, storing, transport, erection, bolting/doweling, shield jacking, forces from out-of-tolerance placement, gasket forces, etc.

Reinforcement is placed not only to help withstand bending moments around the ring but also other forces such as

- Splitting or tension across the section from the force between the segments in a ring.
- Eccentric jacking force tending to break off a corner of the joint.
- Forces from tightening the bolts, in particular when curved bolts are used.
- Torsion across the segment resulting from only partial contact between adjacent ring segments, when segment joints are staggered.

From the point of view of fabrication, it is an advantage if the reinforcing cage is manufactured in part from welded-wire fabric. Recently steel-fiber reinforced concrete without other reinforcing steel has been used successfully for the manufacture of tunnel lining segments.

In the long-term, elements of the lining can corrode if exposed to corrosive environmental agents. For example, sulfates in the soil and groundwater can attack the concrete. Salt in the groundwater will tend to seep through the concrete and concentrate at the inside face as the water evaporates, eventually causing reinforcing steel to corrode. Remedies to improve long-term durability include at least the following:

- Use of high-density, low-permeability concrete with pozzolanic admixtures and microsilica.
- Coating the exterior concrete face with impervious bitumen.
- Epoxy coating of reinforcing steel after fabrication of cages.
- Placement of thick grout or mortar backfill around the lining, designed for water tightness.
- Use of corrosion-resistant bolts and nuts.

28.4 Immersed and Floating Tunnels

28.4.1 Introduction

Although written for immersed tunnels, most concepts in this chapter apply equally to floating tunnels, except for the comments on foundations and backfilling. The primary difference between the design of immersed and floating tunnels is that floating tunnels have to be designed for dynamic loads due to the presence of moving water throughout their life, whereas immersed tunnels need only to consider dynamic loads that occur until finally placed.

28.4.2 Sizing of Tunnel Sections

The needs of the client, the projected traffic volume, the projected vehicle types, and the profiles of the tunnel will dictate the number of traffic lanes required. If uphill grades are long or steep, climbing lanes may be required for heavy vehicles, though it is usual to avoid climbing lanes within the immersed tunnel elements themselves. Due to cost, it is unusual to have full-width emergency lanes or shoulders within tunnels, nominal widths usually sufficing. Very long tunnels may have a short extra lane at intervals to permit emergency stopping.

Unmountable barriers, such as Jersey barriers, are generally used to protect the walls from traffic impact. When the heights of such barriers exceed 600 mm, traffic tends to shy away, making the lane width seem narrower and slowing down motorists.

Walkways within road tunnels should be discouraged, although emergency access through a central wall into an adjacent tunnel should be available. This would require an emergency walkway at least 0.6 m wide on the side adjacent to emergency doors at the level of the top of the barrier and so arranged that vehicle overhangs do not endanger users of the walkway. Such emergency "cross-passages" may need to be provided at intervals of, say, 100 m. Passenger rail tunnels today require an emergency walkway and a place of refuge, often the adjacent tunnel. In immersed rail tunnels, emergency cross-passage doors may be set closer at 50-m intervals, and the emergency walkway may need to be 1.2 m wide to comply with National Fire Protection Association, Inc. (NFPA) 130 escape time limits.

In some cases, it may be desirable to provide an escape duct, usually having an air pressure slightly above that of the highway ducts. Utility ducts may be provided with space in excess of that needed to service the tunnel so that space may be rented out to others. Space above the minimum vertical clearance for road traffic may be needed for lighting and signs. The clearance height for permitted vehicles is usually held as low as possible. Permitted classes of vehicles in any tunnel may be restricted by legislation or owner requirements.

Having thus determined the width and height of internal spaces, preliminary calculations can be made regarding proposed methods of ventilation, jet fan sizes, or duct areas. Although normal ventilation may be designed for stalled traffic conditions, the critical case that defines ventilation may be that of a fuel fire [6]. Space within the tunnel envelope yet outside the traffic envelope may be required to accommodate jet fans, or alternatively, ventilation ducts may be needed.

Air requirements must also be determined. Most immersed tunnel elements are designed only just to float when completed (Figure 28.9), yet be heavy enough to stay submerged by the addition of permanent ballast. A few have been provided with temporary external buoyancy to avoid adding the ballast later. Because of this very delicate balance of being able to float yet later to stay submerged, it is very important that the internal air volume is kept to a minimum, consistent with the demand for internal space. Tolerances to allow for construction variations, fireproofing material on the walls and ceiling, and possible misalignment of elements are additional to other space requirements. In contrast, floating tunnels that rely on buoyancy must have sufficient compartmentalized buoyancy to stay afloat in case of accidental damage, usually considered to be to two adjacent compartments (damaged on the joint).

Once minimum air, traffic, electrical, mechanical, and utility duct requirements have been evaluated, temporary end bulkheads can be sized and calculations made to ensure that tunnel elements float. Initial

FIGURE 28.9   Floating element — airport rail tunnel, Hong Kong (© Christian Ingerslev).

estimates of slab and wall thicknesses, concrete and water densities and steel content (in the form of either structural steel or reinforcement) will be needed to do this. To prevent the incomplete tunnel from floating after removal of the bulkheads, space for additional ballast to be added later (or other means) will need to be provided. If subsequent structural analysis requires any increase in thickness, it may have to be achieved at the expense of adding extra internal space to compensate for the additional weight. Haunches may be necessary to alleviate shear problems at the edges of larger spans. If shear does require reinforcement, it is better to use bent-up bars and avoid shear links in concrete tunnels. Allowance must be made for the addition, if needed, of more concrete or permanent ballast to meet the requirement for an immersed tunnel not to float under final conditions.

28.4.3   Types of Immersed Tunnel

Two main types of tunnel have emerged, known as steel and concrete. Steel tunnels use structural steel, usually in the form of stiffened plate, working compositely with the interior concrete, whereas concrete tunnels do not, relying on steel reinforcing bars or prestressing cables. The number of concrete tunnels is almost twice that of steel tunnels. Steel tunnels can be designed like a ship to have an initial draft of as little as about 2.5 m, as long as the remaining concrete is placed afloat at an outfitting site with greater draft, whereas concrete tunnels have a draft of almost their full depth. Tunnel cross-sections may have flat sides or curved sides.

Historically, the predominant shape of concrete tunnels has been rectangular, which is particularly attractive for wide highways and combined road–rail tunnels, but they have also been circular or curved with a flat bottom. In Europe, Southeast Asia, and Australia, virtually all immersed tunnels are of concrete. In Japan, steel and concrete tunnels are in approximately equal numbers. Most tunnels in North America are steel tunnels.

Steel tunnels have been circular, curved with a flat bottom, and rectangular (particularly in Japan). However, the predominant shape in the past in the United States has been a circular shell within an octagonal shape, with ventilation supply ducts above and exhaust below the roadway. Some of these tunnels consist of a single octagonal tube, while others have a binocular arrangement with two adjacent octagonal tubes forming each tunnel element. This arrangement of ventilation ducts may change, since current techniques permit the use of longitudinal ventilation in much longer tunnels, often obviating the need for separate ventilation ducts. Most or all of the concrete in steel tunnels is placed while the steel shell is afloat, in direct contrast to concrete tunnels that are virtually complete before being floated out. The order in which concrete is placed for a steel tunnel is tightly controlled to minimize deformations and the resulting locked-in moments. Steel tunnels can be categorized into three subtypes:

- Single shell, where the external structural shell plate works compositely with the interior reinforced concrete and the shell plate requires corrosion protection. The BART tunnel is typical of this type.
- The so-called double-shell tunnel, where the interior structural shell plate works compositely with the reinforced concrete within it and is protected by external concrete placed within nonstructural steel form plates (the second "shell"). This shape has also been used in pairs for several tunnels, the most recent being the Ted Williams in Boston, MA. Double-shell tunnels are found only in the United States. The composite action of the structural steelwork with the concrete infill in this arrangement generally requires a circular tube (Figure 28.10 and Figure 28.11) for the greatest efficiency.
- Sandwich, where internal steel diaphragms connect structural steel plates, both internal and external, and the space between them is subsequently filled with unreinforced nonshrink self-compacting concrete. This type of tunnel is a recent development in Japan. Osaka South Port and Kobe Port tunnels are constructed by this method, the latter being the only steel tunnel so far to carry three lanes per duct.

Concrete can be cast into any shape, so economics can influence whether rectangular or circular sections are used. For the shallowest profiles, rectangular sections (Figure 28.12) may provide the best alternative, side vented if required. Where more than two lanes per tube are needed, concrete rectangular tunnels are usually more economical than steel tunnels. Construction joints in the concrete are areas of weakness in all tunnels and can be vulnerable to leakage, particularly if the external membrane is punctured.

28.4.4 Principles of Design

Elements are designed to resist dead loads, live loads, exceptional loads, and extreme loads [7]. These should be applied in accordance with the relevant codes only where they consider the particular conditions of an immersed tunnel. It is important that in considering the design of any one element, all aspects of the design of that element meet the same code. Because of the requirements during construction to float and later not float, actual loads and densities are important, and incorrect

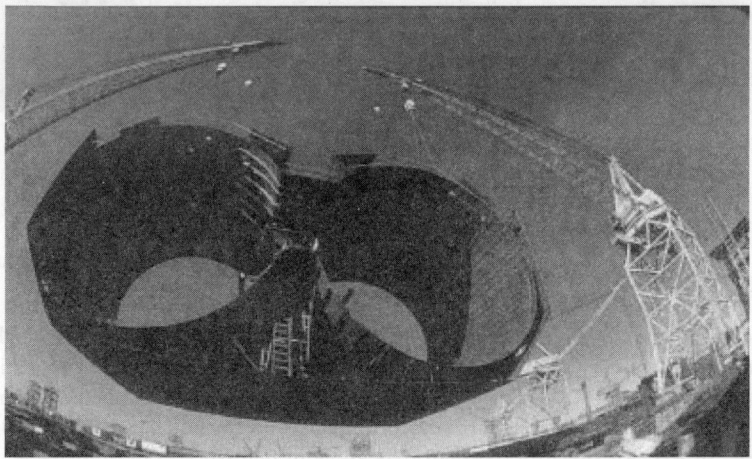

FIGURE 28.10 Steel immersed tunnel — Ted Williams Tunnel, Boston, MA (© David Sailors 1993, with permission).

FIGURE 28.11   Steel immersed tunnel — Cross Harbour Tunnel, Hong Kong (© Parsons Brinckerhoff, Inc. 1971, with permission).

FIGURE 28.12   Concrete immersed tunnel — Western Harbour Crossing, Hong Kong (© Christian Ingerslev).

assumptions can lead to serious errors. In the absence of relevant codes, the following assumptions may be made:

- Dead load includes all long-term loads and mean water level, that is, the self-weight of the basic structure and secondary elements supported and the pressure due to soil and static water level.
- Live load includes creep, shrinkage, prestress, temperature, backfill, seabed erosion and siltation, traffic, variations in water level, current, storm loads, and earthquakes, each with a return period of 5 years or less. Air pressure transient may apply in high-speed rail tunnels.
- Exceptional loads include loss of support (subsidence) below the tunnel or to one side and storms and extreme water levels with a probability of being exceeded once during the design life.

- Extreme loads include falling anchors, sunken or stranding ships, ship collision, water-filled tunnel, vehicular accidents or derailments with collision, explosion (e.g., vehicular), fire, the design seismic event predicted for the location, and the resulting movement of soils. Some of these loads may be affected by categories of dangerous goods permitted through the tunnel.
- Construction loads.
- Life safety; the tunnel should be checked at ultimate limit state for the safety evaluation earthquake (SEE) using load factors of unity. Life safety is directly linked to the survivability of the structure under the most severe seismic event considered at the location.

Load combinations should be selected with regard to simultaneous probability. For example, extreme seismic events may be assumed to occur without storm loads. Structures should be designed to accommodate expected movements due to deformation of foundations without limiting normal operations. Soil pressures should take account of the soil surface profile as well as the geometry of the structure. Geotechnical considerations should also include effects due to seepage, erosion, a change from drained to undrained conditions, liquefaction, and cuts in soft clays. Differential settlements can be expected at interfaces between types of construction, at locations where immersed tunnels extend into the shoreline, and during construction. Tunnels may require fire protection where spalling and loss of section could cause catastrophic failure and inundation.

At least the following conditions should be considered during analysis and design, with levels of damage and resulting performance levels agreed on with the client:

- Normal operating conditions, with full performance level.
- Abnormal conditions are the conditions immediately after the tunnel has experienced extreme or exceptional loads when no loads except the sunken ship are acting, whereas soil and other conditions may have changed. As a result of applying factored dead, live, and ship loads during abnormal conditions, the tunnel should remain operational under the after-effects of any of the extreme or exceptional loads (but not flooded) and settlement (if applicable). Under the worst possible combination of events, the tunnel should sustain no more than light damage requiring minor repairs, public traffic, perhaps immediately, minor leaks, and effectively full or limited performance level. With operations ceased and closed to traffic, the tunnel should survive some loss of support beneath or to one side.
- As a result of applying factored extreme loads, excluding earthquake, full performance level should result with no more than light damage, minor repairs, brief interruption to public traffic, minor leaks, and effectively no significant loss of service under the most adverse combination of events. For extreme actions, including earthquake, limited or minimum performance level may result.
- Construction conditions, including temporary structures (e.g., sheet piles) and loads due to handling, launching, transporting (including expected wave conditions), and placing, combined with environmental and seismic loads appropriate to the season, duration of use, and location. Abnormal and extreme conditions may be inappropriate.

28.4.5 Analysis

28.4.5.1 General

Methods used for analysis may include the usual frame and finite element methods, but where subsoils are nonuniform or where sudden changes in loading on top of the tunnel occur, such at the banks of a river, effects may require soil–structure interaction analysis to correctly model soil behavior. Both ultimate and serviceability analysis, as appropriate, should be investigated. Effects to be considered include adequate safety against failure of complete structures and of their components, static equilibrium, buckling, water tightness, fatigue, durability, vibration, cracking, and deformations.

In the absence of other data for immersed tunnels, methods may use the load factors (appropriate load combination multipliers) given in Table 28.2. Where alternate factors are given, the most adverse combinations should be used.

TABLE 28.2 Load Combination Multipliers for Immersed Tunnels

Load case	Normal	Abnormal	Extreme	Construction
Loading				
Dead load and road ballast	1.4/0.9	1.2/0.9	1.05	1.1
Traffic load				
Normal	1.6/0	1.2/0	—	—
Exceptional vehicle	1.4/0	1.2/0	—	—
Centrifugal	1.6/0	1.3/0	—	—
Braking load				
Normal	1.6/0	1.3/0	—	—
Exceptional vehicle	1.4/0	1.2/0	—	—
Wind and waves	—	1.2	1.05	1.2
Current	1.4	1.2	1.05	1.2
Water pressure allowing for tidal variation	1.4/0.9	1.2/0.9	1.05	1.2
Additional water pressure due to surge	—	1.2/0.9	1.05	1.2
Backfill pressure				
Horizontal	0.9/1.6	0.9/1.4	1.05	1.3/0
Vertical	1.6/0.9	1.4/0.9	1.05	1.3/0
Creep and shrinkage	1.3	1.3	1.05	1.1/0.9
Temperature effects and prestressing	1.3	1.3	1.05	1.1
Sunken ship	—	1.2/0	1.05	—
Earthquake	—	1.2	1.05	—
Other extreme loads	—	—	1.05	—
Imposed loads during construction	—	—	—	1.3

28.4.5.2 Analysis of Earthquake Effects

All immersed tunnels should be designed for seismic events appropriate to their location [1, Chapter 8]. Seismic events during the construction phase should also be considered. Liquefaction of soils around an immersed tunnel should be avoided, perhaps requiring special measures to be taken. An appropriate level of risk should be agreed on with the client because cost implications may be significant.

It may be appropriate to consider three magnitudes of earthquake loading:

- The functional evaluation earthquake (FEE), also known as the design basis earthquake (DBE), should be used first to design the structure for either limited or full performance. A magnitude corresponding to a return period of one to three times the design life is appropriate.
- The SEE should be checked to ensure compliance with minimum performance for life safety and survivability of a design made to FEE. The SEE is the most severe seismic event considered at the location. A performance level agreed on with the client may or may not assume progressive collapse under SEE, but ductility of the structure must be ensured to prevent sudden fracture. A return period of 1000 years or more should be used. As an alternative, however, selecting the maximum credible earthquake (MCE) (i.e., the maximum foreseeable earthquake) or that of large earthquakes that occur at a lesser frequency may sometimes be appropriate.
- A smaller serviceability limit state earthquake, corresponding to a 5- to 10-year return period, may also be included as an ordinary static live load to be combined with other live loads.

For each of these magnitudes, an acceptable structural response to or performance with these loads, including the extent of cracking, movement, damage, formation of plastic hinges, etc., needs to be defined and agreed on with the owner so that the corresponding allowable design stresses and displacements can be determined. (Structural response resulting in collapse or catastrophic inundation is not acceptable.) Typical acceptable performance criteria are

- Minimum performance level
  - Significant damage, repairable or perhaps not.
  - May require full closure or replacement of tunnel.

- Emergency vehicles and slow-moving traffic can still pass any flooding on roadway.
- Limited lighting and ventilation.
- Limited performance level
  - Intermediate damage, repairable over 12 months.
  - Limited emergency and public traffic possible within hours.
  - Limited leakage.
  - Full lighting and ventilation.
- Full performance level
  - Light damage requiring minor repairs.
  - Public traffic immediately.
  - Minor leaks.
  - No significant loss of service.

Not all combinations of earthquake and performance level may be useful to consider. Those that should be used for design would depend on the strategic importance of the tunnel route, the availability of alternative routes, the risks that the owner is prepared to carry, and the cost. Dynamic structural response analysis and assessment of displacements using soil–structure interaction may be necessary and may need to include the effective mass of water. An effective soil drainage system can reduce soil pore pressures.

28.4.6 Methods of Constructing Elements: Concrete and Steel

28.4.6.1 Concrete

An element is a length of tunnel that is floated and immersed as a single rigid unit. The rigidity may be temporary and later released. In other words, elements either may be monolithic or may consist of a number of discrete segments stressed temporarily together longitudinally for ease of transportation and placing. After placing, the stressing may be removed so that each segment may act as a miniele-ment that is free to move at each segment joint. Some Dutch tunnels and the Øresund tunnel between Denmark and Sweden consist of such released segments. The ability to use discrete segments can depend on subsurface conditions, acceptable displacements, and sufficient capacity to resist seismic effects.

28.4.6.1.1 Monolithic Elements

Monolithic elements are cast in bays (equivalent to segments), typically with the floor slab cast first, and followed by the walls and roof in either one or two operations (Figure 28.12). Special efforts to reduce cracking adjacent to previously cast concrete could include low-heat concrete mixes and cooling pipes embedded in the concrete. Reinforcement is continuous across construction joints.

28.4.6.1.2 Discrete Segments

Discrete segments lend themselves to being cast in a single continuous operation, either horizontally (Øresund Tunnel) or vertically and then rotated (Tuas Bay Tunnel). Either way has the advantage of eliminating horizontal construction joints in the walls and the associated thermal cracking. If assembly can be achieved reasonably quickly, a basin or dock sized for a single element can be used. Operations can be tailored to obviate the need to store completed elements before immersion.

External walls and slabs are usually 1 m or more in thickness. These must therefore be considered thick concrete, so special precautions must be taken to avoid cracking during casting. It is particularly important to avoid cracking caused by heat of hydration because such cracks will leak.

28.4.6.2 Steel

A steel tunnel is usually designed to be able to float initially with little or none of the internal concrete having been placed. The bulk or all of the concrete is then placed after launching, either close to the fabrication site or more usually at an outfitting site close to the immersion site. If a graving dock is used (Figure 28.12), a compromise must be reached between the size, the number of reuses, and the schedule.

Testing and repair of any cracks and leaks is needed before submerging the elements. Figure 28.11 shows the Cross Harbour Tunnel on the quayside in Hong Kong almost ready for side launching.

Ease of construction can be achieved by a high degree of mechanization and line fabrication. Prefabricated forms, work shelters (heated in winter), shop welding of any steel plate, and prefabrication of reinforcement cages and prestressing cables can all be used to advantage.

28.4.6.3 Waterproofing

The need for waterproofing concrete tunnels is still a hotly debated issue. A barrier between concrete and salt water, warmer waters, or corrosive waters would appear particularly beneficial. Waterproofing will also reduce the amount of water penetrating the remaining cracks, particularly if the waterproofing adheres to the concrete. In case of fire, concrete that is not saturated with water is less likely to spall since the formation of steam is less likely.

Steel tunnels do not need waterproofing since the structural steel shell serves this purpose well, yielding and not cracking when overloaded, although measures to inhibit corrosion of the steel may be required.

28.4.7 Tunnel Joints

28.4.7.1 Final Joint

Some tunnels are constructed progressively from one end to the other, after which the landside structures are completed. Others may require the last element to be inserted rather than appended to the end of the previous element. To achieve this, a small final gap will have to remain. This closure or final joint corresponds to a short length of tunnel that will need to be cast in place, inserted like a wedge, or jacked out like a piston. Methods used include tremie concrete to seal the joint and dewatering to complete the joint in the dry from the inside.

28.4.7.1.1 Construction Joints

These are horizontal or vertical connections between monolithic parts of a structure. Usually, a waterstop is placed in such a joint. Typically, this would be one of the traditional types of waterstops, although hydrophilic and groutable waterstops are sometimes used. Special moveable watertight joints are required between discrete segments of a tunnel element, where used, designed as expansion joints and perhaps also with shear capacity.

28.4.7.1.2 Immersion Joints

These joints are the ones that are dewatered between tunnel elements subsequent to the immersing and joining operations at the seabed. The joints may remain flexible, or they can be made rigid, as has been common with many steel tunnels. Flexible joints are generally sealed with a temporary immersion gasket or soft-nosed gasket in compression. The use of a secondary independent flexible seal, capable of being replaced from within the tunnel, is common practice (often an omega seal). Each seal should be capable of resisting the external hydrostatic pressure and should allow for expected future movements. Protection should also be provided against damage to seals from within the tunnel, such as impact damage or airborne contaminant damage.

28.4.7.1.3 Seismic Joints

A seismic joint, which can be an immersion joint of special design, may be required to accommodate large differential movements in any direction due to a seismic event. Such a joint would most likely be located at significant changes in cross-section. Figure 28.13 shows a typical seismic joint of BART in San Francisco. Semirigid or flexible joints between elements may also need to be strengthened to carry seismic loads and to prevent catastrophic inundation, typically by using stressed or unstressed prestressing components across the joints or by using bearings and shear keys.

28.4.7.1.4 Terminal Joints (Land Connections)

The terminal joints between the shore ends of an immersed tunnel and the land portions may also be immersion joints. Direct joint connections may be made to land-based structures such as cut-and-cover

FIGURE 28.13   Seismic joint of BART, San Francisco, CA (© Parsons Brinckerhoff, Inc. 1967, with permission).

tunnels or ventilation buildings. These structures may be constructed either before or after placing the immersed tunnel, depending on schedule constraints and local conditions. For a bored or mined tunnel connection, the backfill around the end of the immersed tunnel would first need to be made relatively impermeable, such as by grouting, to allow boring to continue into the end of the immersed tunnel.

Steel immersed tunnels, because of their shallow draft capabilities, may eliminate the need for cut-and-cover tunnel construction in poor ground. After backfilling, the end of the tunnel can be exposed and open depressed highway sections constructed against it. The end of the immersed tunnel was exposed at the Second Downtown Elizabeth River Tunnel in Virginia (Figure 28.14).

Wherever the cross-section changes significantly, seismic actions may generate significant differential movements and the design must accommodate these. Drains, sumps, and pumping stations are required at both portals to remove rainwater that falls within the open sections, and at the tunnel low points to remove wash and leakage water.

Ventilation buildings, if needed, should preferably be sited on land, where they cannot be hit by shipping. The end faces of the buildings may also make suitable interfaces between immersed tunnel elements and on-shore techniques.

28.4.8 Construction Aspects

Dimensional and density checking of the concrete is necessary at all stages of construction to ensure that the design weight is not exceeded, since the tunnel might then not float: it is easier to add more weight than to remove it, and weight can always be added externally.

Temporary bulkheads are needed at the ends of each element, just visible in Figure 28.9 and clearly visible in Figure 28.14. They need to be watertight and yet reasonably simple to remove later. Bulkheads, if made of steel, are often designed for reuse.

Typical foundation layers are 600 mm to 1 m thick above the bottom of the predredged trench. They are mostly formed either by a screeded mattress of stone placed before the tunnel element is immersed or by jetted sand after the element is set on temporary supports, conditions to be allowed for in design. Such supports at the free end of the element are made adjustable, while the other end is first guided and then held by the previously laid element. Other types of design including injection of concrete are sometimes used. Before immersing (Figure 28.1), elements are usually held by temporary lifting hooks while ballast is

FIGURE 28.14   Second Downtown Elizabeth River Tunnel, Virginia (© Parsons Brinckerhoff, Inc. 1983, with permission).

added to provide the necessary negative buoyancy. Survey towers or similar devices are attached to the elements to enable monitoring of position after lowering starts, while bollards and other towing equipment are removed. Calculations to ensure stability at all stages of placement are necessary. Back-filling must be carried out in such a manner that unbalanced lateral forces do not move the element.

28.4.9 Protection against Ship Traffic and Currents

Immersed tunnels should be protected against falling anchors by a layer of either graded material or sacrificial concrete (see Chapter 6 on "Hazard Analysis" in Ref. [1]). The same reference also provides guidance on expected loads from sinking ships, although this is very much dependent on the cargo that the ships would be carrying. Typical results for actual ships passing a particular location, assuming the worst cargo, can be plotted as shown in Figure 28.15. These results may vary further, depending on

- Partially sunken vessels, which may increase these values, particularly if the vessel has a rigid stern post that can produce point loading.
- Critical water depths over which the vessel must pass.
- The water depth at the tunnel location.

If storms cause vessels to drag their anchors, it may be desirable to protect the tunnel further with rock berms over which anchors would be forced. Once clear of the tunnel, the anchors could re-engage.

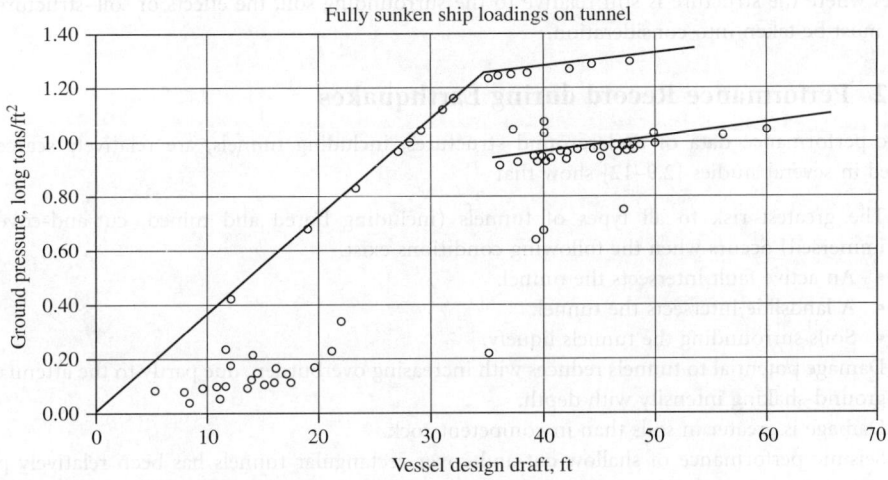

FIGURE 28.15 Ground pressures from fully sunken ships (courtesy of Christian Ingerslev).

At the proposed site of the tunnel, the depth to the structure below the clearance envelope required for the shipping channel will vary according to the amount of overdredge expected during maintenance and the amount of protective backfill required over the tunnel. Protection may be required to the surface of the bed near the tunnel against scour, not only from currents but also from propellers. At some vessel terminals where propellers are kept turning to maintain position, significant local scour can occur. Overdredge of 1 m is not unusual, depending on the method used for dredging. Anchor loads are usually small compared with structural capacity but may cause local surface damage if insufficient backfill is provided [1]. The type and grading of backfill layers are selected so that they do not damage the tunnel structure and waterproofing, if used, and so that material does not get washed away under anticipated currents.

It may not be necessary to completely bury a tunnel and restore the original bed levels (a *floating* tunnel is an extreme case). In such cases, the effects of hydraulic intrusion of a tunnel into an existing waterway regime may require study since water flow is obstructed. For a floating tunnel, tidal and current effects may also cause dynamic and fatigue effects. In some instances all of these measures have been implemented on a single project.

28.5 Seismic Analysis and Design

28.5.1 Introduction

Tunnels, in general, have performed better during earthquakes than have above ground structures such as bridges and buildings. Tunnel structures are constrained by the surrounding ground and, in general, cannot be excited independent of the ground or be subjected to strong vibratory amplification, such as the inertial response of a bridge structure during earthquakes. Adequate design and construction of seismic resistant tunnel structures, however, should never be overlooked. Their seismic performance could be vital, particularly when they comprise important components of a critical transportation system (e.g., a transit system) in which little redundancy exists.

The general procedure for seismic design and analysis of tunnel structures should be based primarily on the ground deformation approach, that is, the structures should be designed to accommodate the deformations imposed by the ground. The analysis of the structure response can be conducted first by ignoring the stiffness of the structure, leading to a conservative estimate of the ground deformations. This simplified procedure is generally applicable for structures embedded in rock or stiff or dense soil.

In cases where the structure is stiff relative to the surrounding soil, the effects of soil–structure interaction must be taken into consideration.

28.5.2 Performance Record during Earthquakes

Seismic performance data of underground structures, including tunnels, are relatively scarce. Data reported in several studies [2,9–12] show that

- The greatest risk to all types of tunnels (including bored and mined, cut-and-cover, and immersed) occurs when the following conditions exist.
  - An active fault intersects the tunnel.
  - A landslide intersects the tunnel.
  - Soils surrounding the tunnels liquefy.
- Damage potential to tunnels reduces with increasing overburden, due partly to the attenuation of ground-shaking intensity with depth.
- Damage is greater in soils than in competent rock.
- Seismic performance of shallow cut-and-cover rectangular tunnels has been relatively poor in comparison to that of bored tunnels, as evidenced during the 1995 Kobe, Japan, earthquake [13,14]. These rectangular box-type structures are particularly vulnerable at the joints connecting the slabs with the walls or columns when subjected to cyclic racking deformations imposed by the ground.
- Immersed tunnels are susceptible to permanent ground deformations resulting from liquefaction induced settlements, lateral spread, and uplift (flotation), and slope instability (landslides) in soft cohesive soils. Joints connecting tube segments are particularly vulnerable to the relative movements between two adjacent segments during shaking, noting that water tightness is one of the critical performance requirements of immersed tunnels.
- Damage potential to bored tunnels due to ground-shaking effects (excluding permanent displacements due to faulting, landslides, and liquefaction) increases with ground-shaking intensity and decreases with better tunnel lining and support system. Figure 28.16 presents performance data of bored tunnels under the effects of seismic ground shaking alone [10]. The figure suggests that
  - Ground shaking caused little damage in tunnels for peak ground acceleration (PGA) less than about 0.2g, where g is the gravity of acceleration.
  - Tunnels with ductile lining (reinforced concrete or steel) tend to have performed better, with damage observed only when the PGA exceeds 0.5g.

28.5.3 Design and Analysis Approach for Ground-Shaking Effects

28.5.3.1 General

Underground tunnel structures undergo three primary modes of deformation during seismic shaking: ovaling or racking, axial, and curvature deformations (see Figure 28.17 and Figure 28.18) [2,12]. The ovaling or racking deformation is caused primarily by seismic waves propagating perpendicular to the tunnel longitudinal axis, causing deformations in the plane of the tunnel cross-section. Vertically propagating shear waves are generally considered the most critical type of waves for this mode of deformation. The axial and curvature deformations are induced by components of seismic waves that propagate along the longitudinal axis.

28.5.3.2 Evaluation of Axial and Curvature Deformations

28.5.3.2.1 Free-Field Deformation Procedure

This procedure assumes that the tunnel lining conforms to the axial and curvature deformations of the ground in the free field (i.e., without the presence of the tunnel). While conservative, this assumption provides a reasonable evaluation because, in most cases, the tunnel lining stiffness is considered relatively

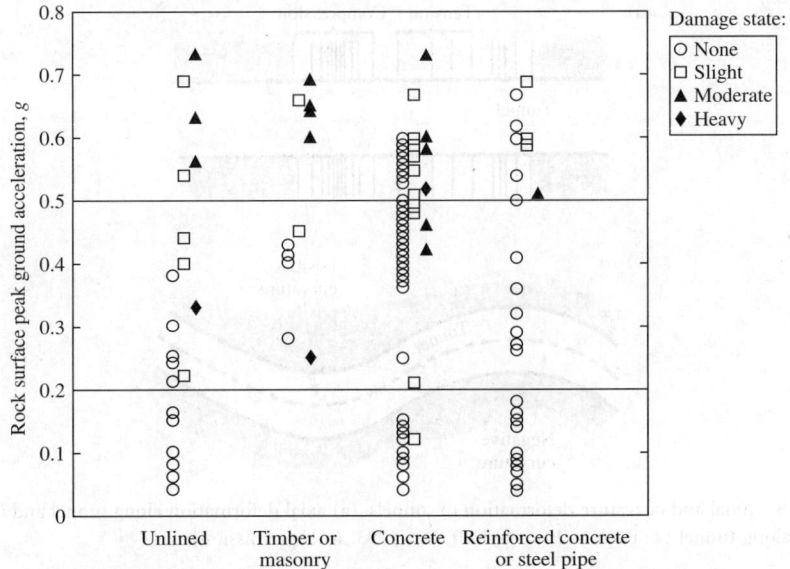

FIGURE 28.16 Rock surface peak ground acceleration for various tunnels (courtesy of M.S. Power, *Summary and Evaluation of Seismic Design of Tunnels*, Draft report submitted to Multidisciplinary Center for Earthquake Engineering Research (MCEER) 1998).

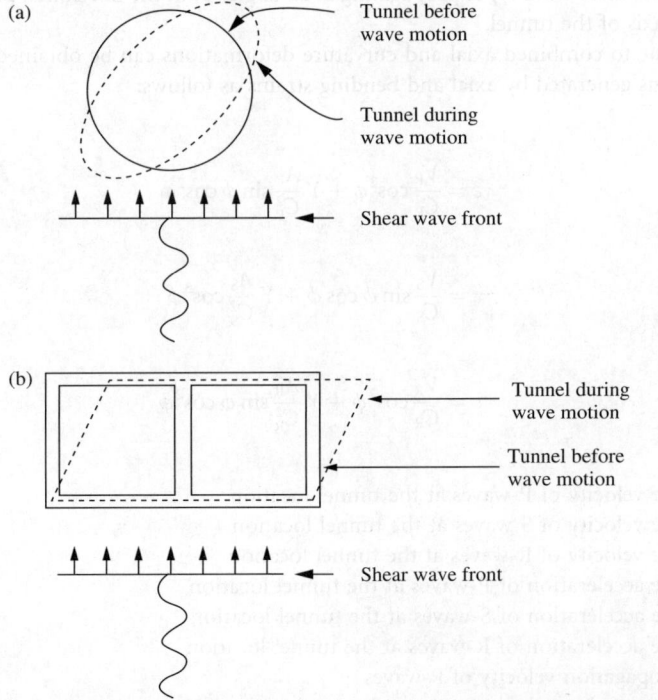

FIGURE 28.17 Ovaling and racking deformation of tunnels: (a) Ovaling deformation of a circular cross-section and (b) racking deformation of a rectangular cross-section (© Parsons Brinckerhoff, Inc. 1993, with permission).

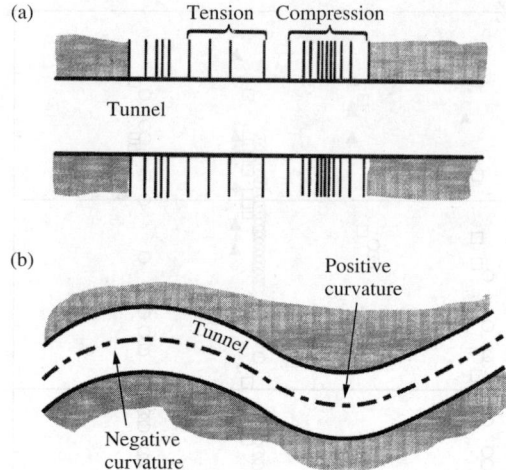

FIGURE 28.18 Axial and curvature deformation of tunnels: (a) axial deformation along tunnel and (b) curvature deformation along tunnel (© Parsons Brinckerhoff, Inc. 1993, with permission).

flexible to the ground. This procedure requires minimum input, making it useful as an initial design tool and as a method of design verification.

The lining will develop axial and bending strains to accommodate the axial and curvature deformations imposed by the surrounding ground. St. John and Zahran [15] developed solutions for these strains due to compression P-waves, shear S-waves, and Rayleigh R-waves. Table 28.3 presents the solutions due to the three wave types propagating at an angle, ϕ, in the horizontal plane with respect to the longitudinal axis of the tunnel.

The strains ε due to combined axial and curvature deformations can be obtained by combining the longitudinal strains generated by axial and bending strains as follows:

For P-waves

$$\varepsilon = \frac{V_P}{C_P}\cos^2\phi + Y\frac{A_P}{C_P^2}\sin\phi\cos^2\phi \qquad (28.4a)$$

For S-waves

$$\varepsilon = \frac{V_S}{C_S}\sin\phi\cos\phi + Y\frac{A_S}{C_S^2}\cos^3\phi \qquad (28.4b)$$

For R-waves

$$\varepsilon = \frac{V_R}{C_R}\cos^2\phi + Y\frac{A_R}{C_R^2}\sin\phi\cos^2\phi \qquad (28.4c)$$

where

V_P = peak particle velocity of P-waves at the tunnel location
V_S = peak particle velocity of S-waves at the tunnel location
V_R = peak particle velocity of R-waves at the tunnel location
A_P = peak particle acceleration of P-waves at the tunnel location
A_S = peak particle acceleration of S-waves at the tunnel location
A_R = peak particle acceleration of R-waves at the tunnel location
C_P = apparent propagation velocity of P-waves
C_S = apparent propagation velocity of S-waves
C_R = apparent propagation velocity of R-waves, and
Y = distance from the neutral axis of the tunnel cross-section to the extreme fiber of the lining

TABLE 28.3 Strain and Curvature Due to Three Wave Types

Wave type	Longitudinal strain		Curvature	
P-wave	$\varepsilon = \left(\dfrac{V_P}{C_P}\right)\cos^2\phi$		$\left(\dfrac{1}{\rho}\right) = \left(\dfrac{a_P}{C_P^2}\right)\sin\phi\cos^2\phi$	
	$\varepsilon_{max} = \dfrac{V_P}{C_P}$ for $\phi = 0°$		$\left(\dfrac{1}{\rho_{max}}\right) = 0.385\left(\dfrac{a_P}{C_P^2}\right)$ for $\phi = 35.27°$	
S-wave	$\varepsilon = \left(\dfrac{V_s}{C_s}\right)\sin\phi\cos\phi$		$\left(\dfrac{1}{\rho}\right) = \left(\dfrac{a_s}{C_s^2}\right)\cos^3\phi$	
	$\varepsilon_{max} = \left(\dfrac{V_s}{2C_s}\right)$ for $\phi = 45°$		$\left(\dfrac{1}{\rho_{max}}\right) = \left(\dfrac{a_s}{C_s^2}\right)$ for $\phi = 0°$	
R-wave Compression component	$\varepsilon = \left(\dfrac{V_R}{C_R}\right)\cos^2\phi$		$\left(\dfrac{1}{\rho}\right) = \left(\dfrac{a_R}{C_R^2}\right)\sin\phi\cos^2\phi$	
	$\varepsilon_{max} = \left(\dfrac{V_R}{C_R}\right)$ for $\phi = 0°$		$\left(\dfrac{1}{\rho_{max}}\right) = 0.385\left(\dfrac{a_R}{C_R^2}\right)$ for $\phi = 35.27°$	
Shear component			$\left(\dfrac{1}{\rho}\right) = \dfrac{a_R}{C_R^2}\cos^2\phi$	
			$\left(\dfrac{1}{\rho_{max}}\right) = \left(\dfrac{a_R}{C_R^2}\right)$ for $\phi = 0°$	

Note:
V_P = soil particle velocity caused by P-waves
a_P = soil particle acceleration caused by P-waves
C_P = apparent propagation velocity of P-waves
V_s = soil particle velocity caused by S-waves
a_s = soil particle acceleration caused by S-waves
C_s = apparent propagation velocity of S-waves
V_R = soil particle velocity caused by R-waves
a_R = soil particle acceleration caused by R-waves
C_R = propagation velocity of R-waves
$1/\rho$ = curvature
Source: St. John and Zahrah 1987.

It should be noted that

- S-waves generally cause the largest strains and are the governing wave type.
- The angle of wave propagation, ϕ, should be the one that maximizes the combined axial strains.

The horizontal propagation S-wave velocity, C_S, in general, reflects the seismic shear wave propagation through the deeper rocks rather than that of the shallower soils where the tunnel is located. In general, this velocity value varies from about 2 to 4 km/s. Similarly, the P-wave propagation velocities, C_P, generally vary between 4 and 8 km/s. The designer should consult with experienced geologists and seismologists for determining C_S and C_P.

When the tunnel is located at a site underlain by deep deposits of soil sediments, the induced strains may be governed by the R-waves. In such deposits, detailed geological and seismological analyses should be performed to derive a reliable estimate of the apparent R-wave propagation velocity, C_R.

The combined strains calculated from Equations 28.4a, 28.4b and 28.4c represent the seismic loading effect only. To evaluate the adequacy of the structure under the seismic loading condition, the seismic loading component has to be added to the static loading components using the appropriated loading combination criteria developed for the structures. The resulting combined strains are then compared against the allowable strain limits, which should be developed based on the performance goal established for the structures (e.g., the required service level and acceptable damage level).

28.5.3.2.2 Procedure Accounting for Soil–Structure Interaction Effects

If a very stiff tunnel is embedded in a soft soil deposit, significant soil–structure interaction effects exist, and the free-field deformation procedure presented in the previous subsection may lead to an overly conservative design. In this case, a simplified BOEF procedure should be used to account for the soil–structure interaction effects. According to St. John and Zahran [15], the effects of soil–structure interaction can be accounted for by applying reduction factors to the free-field axial strains and the free-field curvature strains, as follows:

For axial strains

$$R = 1 + \frac{E_{l}A_{l}}{K_{a}}\left(\frac{2\pi}{L}\right)^{2}\cos^{2}\phi \qquad (28.5a)$$

For bending strains

$$R = 1 + \frac{E_{l}I_{l}}{K_{h}}\left(\frac{2\pi}{L}\right)^{4}\cos^{4}\phi \qquad (28.5b)$$

where E_{l} is Young's modulus of tunnel lining, A_{l} is the cross-sectional area of the lining, K_{h} is the transverse soil spring constant, K_{a} is the longitudinal soil spring constant, L is the wave length of the P-, S-, or R-waves, and I_{l} is the moment of inertia of the lining cross-section.

It should be noted that the axial strain calculated from Equation 28.5a should not exceed the value that could be developed using the maximum frictional forces, Q_{max}, between the lining and the surrounding soils. Q_{max} can be estimated using the following expression:

$$Q_{max} = \frac{fL}{4} \qquad (28.6)$$

where f is the maximum frictional force per unit length of the tunnel.

28.5.3.3 Evaluation of Ovaling Deformations of Bored or Mined Circular Tunnels

The seismic ovaling effect on the lining of bored or mined circular tunnels is best defined in terms of change of tunnel diameter, ΔD_{EQ}. For practical purposes, the ovaling deformations can be assumed to be caused primarily by the vertically propagating shear waves. ΔD_{EQ} can be considered as seismic ovaling deformation demand for the lining. The procedure for determining ΔD_{EQ} and the corresponding lining strains is outlined as follows [12].

28.5.3.3.1 Step 1

Estimate the expected free-field ground strains caused by the vertically propagating shear waves of the design earthquakes. The free-field ground strains can be estimated using the following formula:

$$\gamma_{max} = \frac{V_{S}}{C_{SE}} \qquad (28.7)$$

where γ_{max} is the maximum free-field shear strain at the elevation of the tunnel, V_{S} is the S-wave peak particle velocity at the tunnel elevation, and C_{SE} is the effective shear wave velocity of the medium surrounding the tunnel.

Alternatively, the maximum free-field shear strain can be estimated by a more refined free-field site response analysis (such as in Ref. [16]).

The effective shear wave velocity of the vertically propagating shear wave, C_{SE}, must be compatible with the level of the shear strain that may develop in the ground at the elevation of the tunnel under the design earthquake shaking. A rough estimate of the ratio C_{SE}/C_{SS} (where C_{SS} is the low-strain shear wave velocity of the surrounding medium) can be made as follows:

- For rock, $C_{SE}/C_{SS} \cong 1.0$.
- For stiff to very stiff soil, C_{SE}/C_{SS} may range from 0.7 to 0.9 for low to moderate earthquake shaking and from 0.5 to 0.7 for strong shaking.

Alternatively, site-specific response analyses can be performed for estimating C_{SE}. Site-specific response analyses should be performed on soft soil sites.

The values of the low-strain shear wave velocity, C_{SS}, can be determined using geophysical testing techniques in the field such as P-S logger, cross-hole, and seismic cone penetration methods or estimated from empirical correlation.

28.5.3.3.2 Step 2
By ignoring the stiffness of the tunnel, which is applicable for tunnels in rock or in stiff or dense soils, the lining can be reasonably assumed to conform to the surrounding ground with the presence of a cavity due to the excavation of the tunnel (but without the presence of the lining). The resulting diameter change of the tunnel is

$$\Delta D_{EQ} = \pm 2\gamma_{max}(1 - v_m)D \tag{28.8}$$

where v_m is Poisson's ratio of the surrounding ground and D is the diameter of the tunnel.

28.5.3.3.3 Step 3
If the structure is stiff relative to the surrounding soil, then the effects of soil–structure interaction should be taken into consideration. The relative stiffness of the lining is measured by the flexibility ratio, F, defined as follows:

$$F = \frac{E_m(1 - v_l^2)R_l^3}{6E_l I_{l,1}(1 + v_m)} \tag{28.9}$$

where E_m is the strain-compatible elastic modulus of the surrounding ground, R_l is the nominal radius of the tunnel lining, V_l is Poisson's ratio of the tunnel lining, and $I_{l,1}$ is the moment of inertia of the lining per unit width of tunnel along the tunnel axis.

The strain-compatible elastic modulus of the surrounding ground, E_m, should be derived using the strain-compatible shear modulus, G_m, corresponding to the effective shear wave propagating velocity, C_{SE}.

The moment of inertia of the tunnel lining per unit width, $I_{l,1}$, should be determined based on the expected behavior of the selected lining under the combined seismic and static loads, accounting for cracking and joints between segments and between rings as appropriate.

28.5.3.3.4 Step 4
The diameter change, ΔD_{EQ}, accounting for the soil–structure interaction effects can then be estimated using the following equations:

$$\Delta D_{EQ} = \pm \frac{1}{3} k_1 F \gamma_{max} D \tag{28.10}$$

$$k_1 = \frac{12(1 - v_m)}{2F + 5 - 6v_m} \tag{28.11}$$

where k_1 is the Seismic ovaling coefficient.

The seismic ovaling coefficient curves plotted as a function of F and v_m are presented in Figure 28.19.

The resulting bending moment induced maximum fiber strain, ε_m, and the axial force (i.e., thrust) induced strain, ε_T, in the lining can be derived as follows:

$$\varepsilon_m = \frac{1}{6} k_1 \frac{E_m}{(1 + v_m)} R_l^2 \frac{\gamma_{max} t_l}{2E_l I_{l,1}} \tag{28.12}$$

$$\varepsilon_T = \frac{1}{6} k_1 \frac{E_m}{(1 + v_m)} R_l \frac{\gamma_{max}}{E_l t_l} \tag{28.13}$$

where t_l is the thickness of the lining.

The solutions presented in Equations 28.10 through 28.13 assume that a full-slippage condition exists along the soil–lining interface, which allows normal stresses (without normal separation) but no tangential shear force. The full-slippage assumption yields slightly more conservative results in estimating

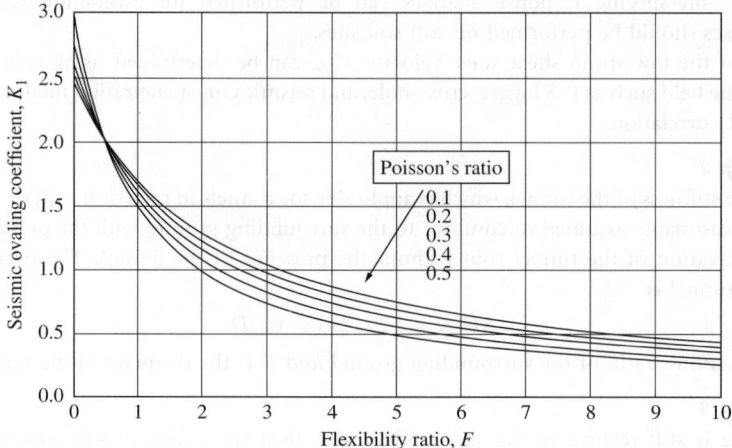

FIGURE 28.19 Seismic ovaling coefficient curves (© Parsons Brinckerhoff, Inc. 1993, with permission).

the diameter change and bending strain but significantly lower values of thrust-induced strain than the no-slippage condition. Therefore, Equation 28.13 should not be used unless a full-slippage mechanism is incorporated in the design. Instead, the no-slippage condition should be assumed in deriving the thrust-induced strain as follows [12]:

$$\varepsilon_T = k_2 \frac{E_m}{2(1+v_m)} R_l \frac{\gamma_{max}}{E_l t_l} \tag{28.14}$$

$$k_2 = 1 + \frac{F[(3-2v_m)-(1-2v_m)C]-\frac{1}{2}(1-2v_m)^2+2}{F[(3-2v_m)+(1-2v_m)C]+C[\frac{5}{2}-8v_m+6v_m^2]+6-8v_m} \tag{28.15}$$

$$C = \frac{E_m(1-v_l^2)R_l}{E_l t(1+v_m)(1-2v_m)} \tag{28.16}$$

where C is the compressibility ratio.

The seismically induced strains due to the ovaling effect need to be combined with the strains resulting from nonseismic loading and then checked against the allowable strain limits consistent with the performance goal established for the design of the tunnel lining.

28.5.3.4 Evaluation of Racking Deformations of Rectangular Tunnels

Racking deformations are defined as the differential sideways movements between the top and bottom elevations of the rectangular structures, shown as Δ_s in Figure 28.20. The resulting material strains in the lining associated with the seismic racking deformation, Δ_s, can be derived by imposing the differential deformation on the structure in a structural frame analysis. The procedure for determining, Δ_s taking into account the soil–structure interaction effects, is presented below [12].

28.5.3.4.1 Step 1
Estimate the free-field ground strains γ_{max} (at the structure elevation) caused by the vertically propagating shear waves of the design earthquakes (see Section 28.5.3.3). Determine $\Delta_{\text{free-field}}$, the differential free-field relative displacements corresponding to the top and the bottom elevations of the rectangular structure (see Figure 28.20), by

$$\Delta_{\text{free-field}} = h\gamma_{max} \tag{28.17}$$

where h is the height of the structure.

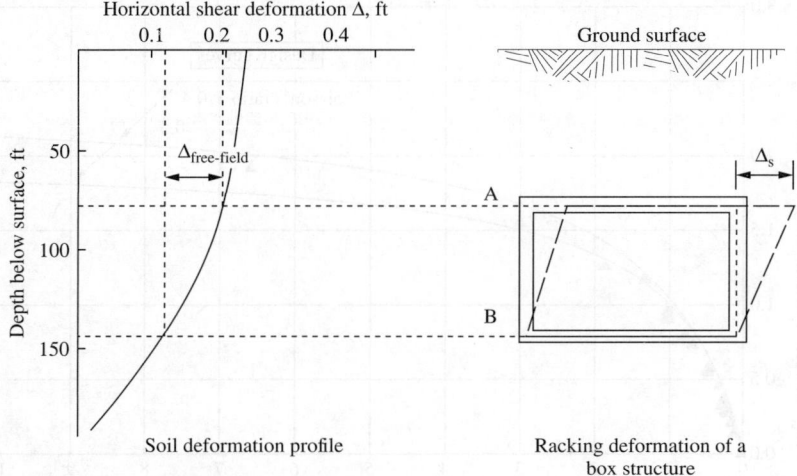

FIGURE 28.20 Racking deformation for a box structure (© Parsons Brinckerhoff, Inc. 1993, with permission).

28.5.3.4.2 Step 2

Determine the racking stiffness, K_s, of the structure from a structural frame analysis. For practical purposes, the racking stiffness can be obtained by applying a unit lateral force at the roof level, while the base of the structure is restrained against translation, but with the joints free to rotate. The structural racking stiffness is defined as the ratio of the applied force to the resulting lateral displacement. In performing the structural frame analysis, it is important to use the appropriate moment of inertia, taking into account the potential development of the cracked section, particularly for the vertical walls.

28.5.3.4.3 Step 3

Determine the flexibility ratio, F_{rec}, of the proposed design of the structure using the following equation:

$$F_{rec} = \frac{G_m}{K_s} \frac{w}{h} \tag{28.18}$$

where w is the width of the structure and G_m is the average strain-compatible shear modulus of the surrounding ground.

The flexibility ratio is a measure of the relative racking stiffness of the surrounding ground to the racking stiffness of the structure.

28.5.3.4.4 Step 4

Based on the flexibility ratio obtained from Step 3, determine the racking reduction ratio, R_{rec}, for the structure using Figure 28.21 or the following expression [17]:

$$R_{rec} = \frac{4(1 - \nu_m)}{((3 - 4\nu_m)/F_{rec}) + 1} \tag{28.19}$$

The triangular points in Figure 28.21 correspond to published results [12]. The data in [12] were generated by performing a series of dynamic finite element analyses on a number of cases with varying soil and structural properties, structural configurations, and ground motion characteristics. As indicated in the figure, if $F_{rec} = 1$, the structure is considered to have the same racking stiffness as the surrounding ground, and therefore the racking distortion of the structure is about the same as that of the ground in the free field. When F_{rec} approaches zero, representing a perfectly rigid structure, the structure does not rack regardless of the distortion of the ground in the free field. For $F_{rec} > 1.0$ the structure becomes flexible relative to the ground, and the racking distortion is magnified in comparison to the shear distortion of the ground in

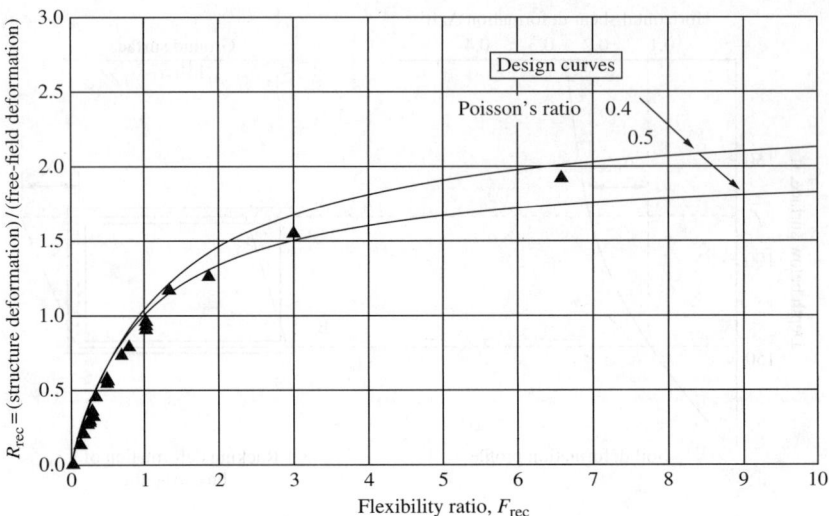

FIGURE 28.21 Racking reduction ratio (© Parsons Brinckerhoff, Inc. 1993, with permission).

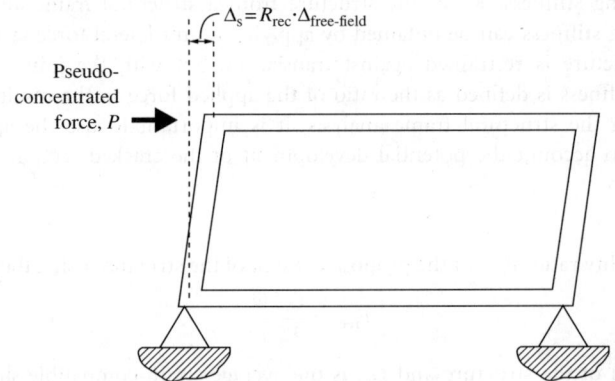

FIGURE 28.22 Frame analysis modeling of racking deformations (© Parsons Brinckerhoff, Inc. 1993, with permission).

the free field. This magnification effect is not caused by the effect of dynamic amplification. Rather it is attributed to the fact that the ground has a cavity in it as opposed to the free-field condition.

28.5.3.4.5 Step 5

Determine the racking deformation of the structure, Δ_s, using the following relationship:

$$\Delta_s = R_{rec}\Delta_{\text{free-field}} \tag{28.20}$$

28.5.3.4.6 Step 6

The seismic demand in terms of internal forces as well as material strains can be calculated by imposing Δ_s upon the structure in a frame analysis as depicted in Figure 28.22.

28.5.3.5 Loads Due to Vertical Seismic Motions

The effects of vertical seismic motions can be accounted for by applying a vertical pseudostatic loading, equivalent to the product of the vertical seismic coefficient and the combined dead and design overburden loads used in static design. The vertical seismic coefficient can be reasonably assumed to be two thirds of the design peak horizontal acceleration divided by gravity. This vertical pseudostatic loading should be applied by considering both up and down directions of motion. Whichever results in a more critical load case should govern.

If there is a potential for a seismically loosened zone to develop above the crown of the tunnel, then the inertial effects of this loosened zone should also be included in the design, unless appropriate stabilization measures are taken to prevent it from occurring.

28.5.4 Tunnel Subjected to Large Displacements

As indicated in Section 28.5.2, the greatest risk to tunnel structures is the potential for large ground movements as a result of unstable ground conditions (e.g., liquefaction and landslides) or fault displacements. In general, it is not feasible to design a tunnel structure to withstand large ground displacements. The proper design measures in dealing with unstable ground conditions may consist of

- Ground stabilization.
- Removal and replacement of the problem soils.
- Rerouting or deep burial to bypass the problem zone.

With regard to the fault displacements, the best strategy is to avoid any potential crossing of active faults. If this is not possible, then the general design philosophy is to accept and accommodate the displacements by either employing an oversized excavation, perhaps backfilled with compressible or collapsible material, or using ductile lining to minimize the instability potential of the lining. In cases where the magnitude of the fault displacement is limited or the width of the sheared fault zone is considerable such that the displacement is dissipated gradually over a distance, design of a strong lining to resist the displacement may be technically feasible. The structures, however, may be subjected to large axial, shear, and bending forces. Many factors need to be considered in the evaluation, including the stiffness of the lining and the ground, the angle of the fault plane intersecting the tunnel, the width of the fault, the magnitude as well as orientation of the fault movement, among others. Analytical procedures are generally used for evaluating the effects of fault displacement on lining response. Some of these procedures were originally developed for buried pipelines [18]. Continuum finite element or finite difference methods have also been used effectively for evaluating the tunnel–ground–faulting interaction effects.

28.5.5 Shaft Structures and Interface Joints

The seismic considerations for the design of vertical shaft structures are similar to those for the lined circular tunnel structures, except that ovaling and axial deformations in general do not govern the design. Considerations should be given to the curvature strains and shear forces of the lining resulting from vertically propagating shear waves. Force and deformation demands may be considerable in cases where shafts are embedded in deep, soft deposits. In addition, potential stress concentrations at the following critical locations along the shaft need to be properly assessed and designed for

- Abrupt change of the stiffness between two adjoining geologic layers.
- Shaft–tunnel or shaft–station interfaces.
- Shaft–surface building interfaces.

Flexible connections to accommodate the potential differential movements are the recommended design strategy between any two structures with drastically different stiffness or mass in poor ground conditions.

28.6 Tunnel Jacking

Tunnel jacking evolved from pipe jacking, retaining many of the same features. Technological advancement from pipe jacking to tunnel jacking occurred in the 1960s, when circular pipe jacked sections were found to be either too small or inefficient for their intended purpose, so they were replaced by rectangular sections. Over the past four decades, the technique has developed significantly, with tunnel sections ranging from 3 m² to 25 m wide × 12 m high and larger, with lengths exceeding 110 m. The method has been used in many parts of the world, including Europe (particularly in the United Kingdom and Germany), Australia, India, South Africa, Canada, and Japan. In the United States, however, tunnel jacking is a relatively new technique, with only one major application completed to date, consisting of three tunnels constructed under an operating rail network for a major highway extension in Boston, MA. A view of one of the jacking operations for that project is shown in Figure 28.23.

Features of a typical tunnel jacking operation are shown in Figure 28.24. The tunnel box structure is constructed in one or more sections immediately adjacent to the planned alignment, either in a pit for an application that is completely below existing grade or in an open space for the case where the jacking is being performed through an embankment. As with conventional pipe jacking, the frictional resistance that needs to be overcome to jack the complete tunnel length may exceed the total capacity of the jacking system that can be installed in the space available at the rear of the tunnel structure. In that case, the tunnel structure will be divided into sections, with intermediate jacking stations established between them.

The tunnel structure is open at the front end and fitted with a shield, which may be constructed of either steel or reinforced concrete. Depending on the characteristics of the ground through which the

FIGURE 28.23 Tunnel jacking operation in Boston, MA, for a major highway project.

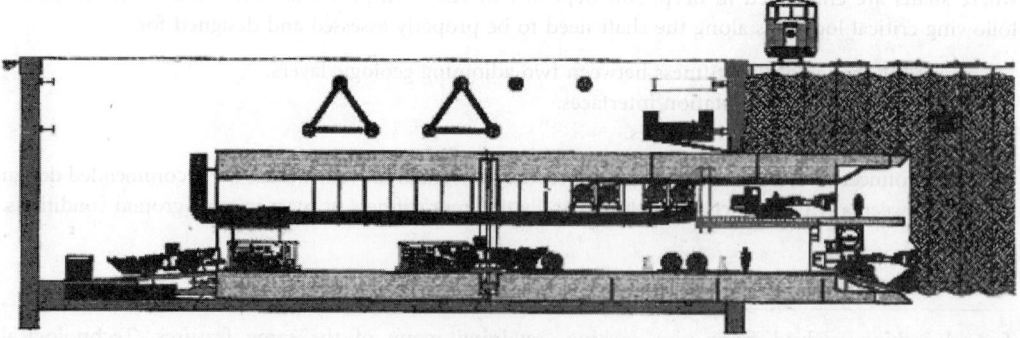

FIGURE 28.24 Cross-section of a typical tunnel jacking operation.

jacking is being done, and on the size of the tunnel cross-section, the shield may be compartmentalized and the individual cells outfitted with breasting doors to minimize the area of unsupported face exposed at the tunnel heading. In some cases, ground improvement techniques such as grouting or even freezing may be implemented in advance of the jacking operation to increase the stand-up time of the ground.

Excavation is typically performed in small increments to avoid creating an unstable unsupported span beyond the leading edge of the shield. Overcutting along the excavation perimeter is carefully controlled to limit the potential for settlement of the ground surface above. Once an excavation cycle is complete, the tunnel structure is jacked ahead either in one step from a rear jacking station or in stages if intermediate jacking stations are being used.

The tunnel structure is advanced with a series of hydraulic jacks, with stroke lengths coordinated with the excavation increments. Depending on the particular system design, the jacks may be situated only along the base of the concrete section or around the entire section perimeter. Required operating capacities and pressures vary with the size of the tunnel being jacked; in larger applications, these parameters can be as much as 500 tons and greater than 40 MPa, respectively. Alignment control can be accomplished by varying the thrust applied in individual or clusters of jacks and through the use of guide walls constructed in the launch area outside the tunnel headwall. Reaction for the jacking force is provided by either a structural frame or thrust blocks at the rear of the tunnel, with spacer blocks or pipe sections or a series of anchorage points used as necessary to maintain continuity of the reaction structure as the tunnel advances into the ground and progressively farther from its starting point.

When the depth of cover is small relative to the width of the tunnel structure, there is a tendency for the overlying soil to be dragged along with the tunnel. To prevent this from occurring, a separation layer is typically introduced between the top surface of the tunnel and the adjacent ground to reduce the frictional resistance. This can be accomplished in a number of ways, with common methods including a series of steel sheets or an array of closely spaced wire ropes. Side friction is controlled by designing a small overcut into the shield structure and if necessary, by the injection of lubricating agents into the interface through multiple ports in the walls of the tunnel structure.

To maintain and protect critical surface facilities during tunnel jacking, tunnel jacking projects nearly always include some form of surface survey program to monitor the impact of the underground excavation, with predetermined corrective action plans already in place, ready to be implemented quickly in the event that ground deformations exceed tolerable thresholds.

Glossary

Advance rate — The amount of progress that a tunnel-boring machine makes — the distance it covers in a day, usually measured in feet or meters; also used for tunneling by other means.

Backfill — Any material such as sand, gravel or crushed stone, grout or concrete, used to fill the remainder of an excavation after a tunnel or other underground structure has been constructed within the excavation.

Ballast — A stabilizing weight (stones or concrete), either temporary or permanent, often added to immersed tunnel elements while floating, being immersed, or in their final position.

Bentonite — A clay mineral that can absorb large amounts of water. When mixed with water, it forms a slurry used to support deep trenches or bore holes against collapse until they can be filled with concrete.

Bottom-up — A method of construction in which a tunnel is built within an excavated trench in a conventional way or sequence: base slab first, then the walls, and finally, the roof. The completed structure is then backfilled and the ground surface reinstated.

Cofferdam — A temporary structure, constructed in water and pumped dry, used to keep water out so that work can be performed in dry conditions.

Controlled blasting — Use of patterned drilling and optimum amounts of explosives and detonating devices to control blasting damage.

Cross-passage — A short passage or tunnel connecting two adjacent tunnels, often used to allow people to escape from one tunnel to another in an emergency.

Crown — The top of a tunnel.

Cut-off wall — An underground wall, either temporary or permanent, used in tunnel excavations to prevent the passage of ground water.

Decking — A plank cover over a work area that serves as a temporary surface for pedestrian and vehicular traffic (including construction equipment), usually made of wood, concrete, or steel.

Dewatering — The removal of ground water from the area to be excavated.

Drill-and-blast — A method of mining in which small-diameter holes are drilled into the rock and then loaded with explosives. The blast from the explosives fragments and breaks the rock away from the face so the rock can be removed. Repeated drilling and blasting advances the underground opening.

Drawdown — A lowering of the normal groundwater level (water table) as the result of dewatering.

Face — The soil being excavated directly in front of the TBM or the surface at the head of a tunnel excavation. A mixed-face is a condition with more than one type of material, such as clay, sand, gravel, cobbles, or rock.

Ground reinforcement — Structural elements installed in the ground by drilling and insertion, including rock dowels or anchors, often grouted in place.

Ground support — Installation of any type of engineering structure around or inside the excavation, such as steel sets, wood cribs, timbers, or lining.

Heading — A smaller tunnel used when a larger tunnel is excavated in several stages. Smaller headings are used when full tunnel excavation at once would not be prudent. Sometimes multiple headings are used simultaneously to increase the number of excavation faces and compensate for slower excavation rates.

Invert — The bottom of a tunnel.

Lining — A temporary or permanent structure, made of concrete or other materials, to secure and finish the tunnel interior or to support an excavation.

Mole — Slang term for a tunnel engineer; also used to describe a TBM.

Open cut — A method of construction in which the excavated trench is left uncovered while the tunnel is constructed.

Overburden — The soil between the ground surface and the roof of a tunnel.

Pilot tunnel — An exploratory tunnel, usually smaller and driven ahead of the main tunnel.

Portal — The entrance to a tunnel.

Sandhog — Slang term for a tunnel worker.

Shaft — A vertical excavation, often used to provide access to a tunnel from the surface.

Shield — A structure used in soft ground to provide support at the face of the tunnel for the soil above the tunnel, to provide space for erecting supports, and to protect the workers excavating and erecting supports.

Shotcrete — Concrete pneumatically projected at high velocity onto a surface; pneumatic method of applying a lining of concrete.

Soft ground — Soils and weak rock that can be easily removed.

TBM — Tunnel boring machine, a machine that excavates a tunnel by drilling out the heading to full size in one operation.

Top-down — A method of construction in which the tunnel walls are built first, using special machinery, within a narrow trench. Next, the roof is built in a shallow excavation, and the ground surface is reinstated. The rest of the tunnel is then excavated and constructed underneath the roof.

Underpinning — The installation of new supports beneath the foundation of a building or other structure to protect it from settlement caused by adjacent tunneling or other construction; sometimes used in place of an old foundation that was removed.

References

[1] International Tunnelling Association Immersed and Floating Tunnels Working Group, *State-of-the-Art Report*, second edition, Pergamon Press, Oxford, 1997.

[2] Owen, G.N. and Scholl, R.E., *Earthquake Engineering of Large Underground Structures*, prepared for the Federal Highway Administration, FHWA/RD-80/195, 1981.

[3] Kuesel, T.R., *Tunnel Stabilization and Lining*, in *Tunnel Engineering Handbook*, Bickel, J.O., Kuesel, T.R., and King, E.H., editors, Chapman & Hall, New York, 1996.

[4] U.S. Army Corps of Engineers Manual, *Tunnels and Shafts in Rock*, EM 1110-2-2901, 1997.

[5] AFTES (French Tunneling Association), *Recommendations in Respect of the Use of Plain Concrete in Tunnels*, AFTES Paris, 1998 (translated into English, 1999).

[6] Bechtel/Parsons Brinckerhoff, *Memorial Tunnel Fire Ventilation Test Program, Comprehensive Test Report*, prepared for Massachusetts Highway Department, November 1995.

[7] Ingerslev, L.C.F., *Developments in Immersed Tunnels*, in *Options for Tunnelling, 1993*, Burger, H., editor, Elsevier Science Publishers B.V., Amsterdam, pp. 79–88, 1993.

[8] Ingerslev, L.C.F., *Concrete Immersed Tunnels: The Design Process*, in *Immersed Tunnel Techniques*, The Institution of Civil Engineers, Telford, United Kingdom, 1989.

[9] Dowding, C.H. and Rozen, A., *Damage to Rock Tunnels from Earthquake Shaking*, J. Geotech. Eng. Div, ASCE, Vol. 104, No. GT2, 1978.

[10] Power, M.S. and Rosidi, D., *Seismic Vulnerability of Tunnels and Underground Structures Revisited*, North American Tunneling '98, 1998.

[11] Sharma, S. and Judd, W.R., *Underground Opening Damage from Earthquakes*, Eng. Geol., Vol. 30, 1991.

[12] Wang, J., *Seismic Design of Tunnels — A Simple State-of-the-Art Design Approach*, Parsons Brinckerhoff Monograph No. 7, 1993.

[13] O'Rourke, T.D. and Shiba, Y., *Seismic Performance and Design of Tunnels*, Annual Report, NCEER Highway Project, sponsored by U.S. Department of Transportation and Federal Highway Administration, 1997.

[14] Nakamura, S., Yoshida, N., and Iwatate, Y., *Damage to Daikai Subway Station during the 1995 Hyogoken-Nambu Earthquake and Its Investigation*, Japan Society of Civil Engineers, Committee of Earthquake Engineering, 1996.

[15] St. John, C.M. and Zahrah, T.F., *Aseismic Design of Underground Structures*, Tunneling Underground Space Technol., Vol. 2, No. 2, 1987.

[16] Idriss, I.M. and Sun, J.I., *SHAKE91 — A Computer Program for Conducting Equivalent Linear Seismic Response Analyses of Horizontally Layered Soil Deposits*, Center for Geotechnical Modeling, Department of Civil and Environmental Engineering, University of California at Davis, 1992.

[17] Penzien, J., ASCE Committee on Gas and Liquid Fuel Lifelines, *Guidelines for the Seismic Design of Oil and Gas Pipeline Systems*, Technical Council on Lifeline Earthquake Engineering, ASCE, New York, 1984 (personal communication).

[18] International Tunnelling Association Immersed and Floating Tunnels Working Group, *State-of-the-Art Report*, first edition, Pergamon Press, Oxford, 1993.

Further Reading

AFTES (French Tunneling Association), *The Design, Sizing and Construction of Precast Concrete Segments Installed at the Rear of a Tunnel Boring Machine (TBM)*, AFTES, Paris, France, 1997 (translated into English, 1999).

American Society of Civil Engineers, *Steel Penstocks, Manual on Engineering Practice No. 79*, 1993.

Hoek, E. and Brown, E.T., *Underground Excavations in Rock*, Institution of Mining and Metallurgy, London, 527 pp., 1980.

Peck, R.B., *Deep Excavations and Tunneling in Soft Ground, State-of-the-Art Report,* 7th International
Conference on Soil Mechanics and Fundamental Engineering Mexico City, pp. 225–290, 1969.

Rice, P.M., Manville, P.A., Taylor, S., and Powderham, A.J. *Development of Design and Construction
Concepts for Jacked Tunnel Sections of I-93/I-90 Interchange, Central Artery/Tunnel Project,* Boston,
MA, ASCE Geotechnical Special Publication Number 90, June 1999.

Taylor, S. and Winsor, D., *Developments in Tunnel Jacking,* ASCE Special Technical Publication
Number 87, October 1998.

van Dijk, P., Almeraris, G., and Rice, P. *Construction of I-90 Highway Tunnels under Boston's South
Station Rail Yard by Box Jacking,* Proceedings, Rapid Excavation and Tunneling Conference, SME/
ASCE, San Diego, CA, June 2001.

29

Glass Structures

A. K. W. So
*Research Engineering Development
Façade and Fire Testing
Consultants Ltd.,
Yuen Long, Hong Kong*

Andy Lee
*Ove Arup & Partners
Hong Kong Ltd.,
Kowloon, Hong Kong*

Siu-Lai Chan
*Department of Civil and Structural
Engineering,
Hong Kong Polytechnic University,
Kowloon, Hong Kong*

29.1 Introduction

Glass is a brittle material that is weak in tension because of its noncrystalline molecular structure. When glass is stressed beyond its strength limit, breakage occurs immediately without warning, unlike steel and aluminum where plastic mechanism can be formed. Stress or moment redistribution does not occur in glass, and local and then consequential global failure is very common. Testing has shown that glass strength is statistical in nature.

The main constituent of glass is silica sand. Zachariasen and Warren [1] suggested that glass is made up of network formers and modifiers (Figure 29.1). Silicon and oxygen ions bonded together (formers) to form the basic three-dimensional network structure in which ions of sodium, potassium, calcium, and magnesium (modifiers) are bonded in the holes inside the silicon–oxygen former network. Glass is one of the most durable building materials. An extremely important property of glass is its resistance to corrosion attack by water and acid.

There are three basic types of glass: float glass, plate glass, and sheet glass. Float glass is produced by pouring continuously from a furnace onto a large shallow bath of molten tin. In the flow chamber the atmosphere is controlled to prevent oxidation. The second type of glass is plate glass, which is produced by grinding and polishing rough glass. The third type is sheet glass, which is produced by continuously drawing molten glass from a bath through an annealing lehr. A simplified diagrammatic presentation of the production process is illustrated in Figure 29.2. Nowadays, more than 90% of glass is produced by the float process. The float glass is available in a number of modified forms: reflective coated glass, heat-absorbing glass, tempered glass, insulating glass, acoustical glass, etc.

Glass as a building material has been widely used in curtain wall and glass wall systems, which generally provide an esthetic appearance to the complete building. Large glass panels of size in

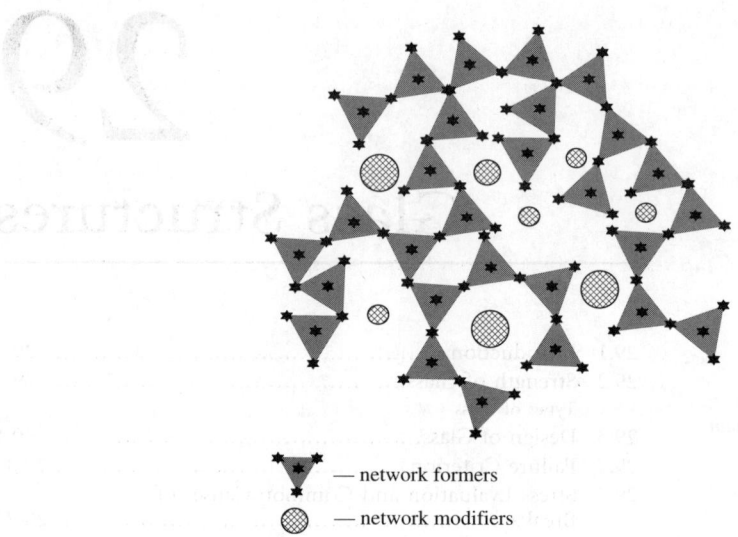

— network formers
— network modifiers

FIGURE 29.1   Simplified two-dimensional representation of a glass network.

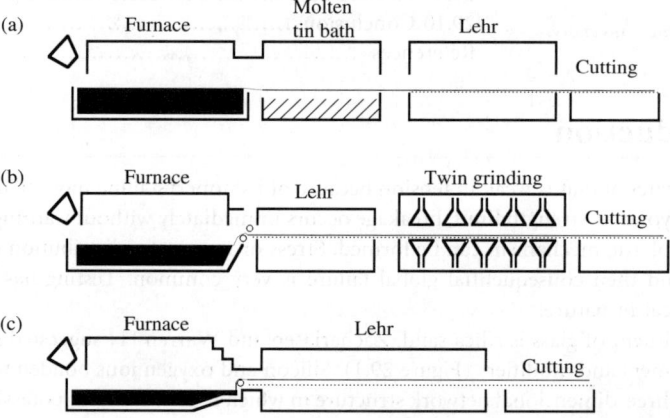

FIGURE 29.2   Manufacturing processes of glass: (a) float glass, (b) plate glass, and (c) sheet glass.

excess of 1.5 m² are commonly used in commercial buildings to date. In practice, they are structurally glazed with structural sealants, on the four sides or on two sides with the other two edges clamped mechanically along the transoms. For shop fronts and entrances of prestigious buildings, an unobstructed view and architectural appearance can normally be provided by using a glass wall system. Generally speaking, their esthetic appearance is more appealing than that of to other finishes.

When compared to other building materials such as concrete, steel, or even timber, glass receives relatively less attention from the researcher and the engineer. The probable reason for the lack of research in glass may be the perceptively moderate tendency for collapse when compared with other materials. However, as glass structures have no allowance for plastic deformation, overloading will not be shed to other parts of a structure, and their breakage normally is without warning due to their brittleness. The breakage will lead to casualties when debris falls onto the street from a high-rise building. In the past two decades or so, it has been noted that glass structures are commonly constructed in areas of high human exposure such as shopping arcades and city malls. The failure of the structure may be catastrophic and cannot therefore be overlooked. Although the uses of laminated and tempered glass can lower the chance of harmful damage, they may not be preferred as they reduce the vision quality of glass.

The major reason for special care in the design of glass is that it has no ductility to allow moment or force redistribution like steel and concrete frames. Further, the overdesign is costly. In Hong Kong, the facade system normally takes a share from 15 to 20% of the total construction cost in a commercial building. Obviously, the resources spent on research in glass are far less than for other materials like concrete and steel. Although glass manufacturers provide design manuals for glass panels, many of these are based on the linear theory [2], which is of inadequate accuracy under high wind pressure. The American [3] and the Canadian [4] design codes of practice for glass require the consideration of nonlinear effect when the glass plate deflection is large and of a magnitude more than three-fourths of its thickness, which is very common in practice. In a general design of glass structures, the glass panel exhibits considerable change in geometry, and an accurate analysis should allow for the geometrically nonlinear effects in accordance with these design codes. Figure 29.3 shows the damage of buildings after a typhoon attack.

Studies have shown that breakage of annealed glass is due to the tensile stress on the hairy cracks on the surface of the panel, resulting in a serious stress concentration. Due to the difficulty in estimating the density and the extent of these hairy cracks in all glass panels, the failure probability instead of direct specification of failure load for a glass panel is usually used as a reference for safety of glass structures. Generally speaking, the probability of failure (POF) of 8/1000 is acceptable for most purposes. In congested areas, the POF should be further reduced.

In recent years, the extensive construction of high-rise buildings with curtain wall envelops in many cities in China and Hong Kong has further highlighted the importance of conducting more research on

FIGURE 29.3 Damage of buildings after a typhoon attack.

the safety of these structures. In fact, at the time of writing this chapter, use of glass curtain walling is heavily criticized in China as a "hanging bomb."

29.2 Strength of Glass

In most structural applications of glass it is necessary for the components to sustain mechanical stress. When a material is stressed, it deforms, and strains are created. At a low level of stress, most materials obey Hook's law, that is, strain is proportional to stress. While the stress level is high, most materials deform plastically. Glass is a brittle material, which cannot accommodate this plastic deformation but breaks without warning. The stress–strain curve in Figure 29.4 shows a perfect linearity from zero strain to failure. The mechanical properties of glass as an engineering material are tabulated in Table 29.1.

Generally speaking, it can be stated that the theoretical strength of a piece of glass is equal to about one tenth of its modulus of elasticity [5]. Glass in compression is extremely strong. The compressive strength can approach 10,000 MPa without breakage. However, glass in tension usually fails at stress levels less than 100 MPa. It has been pointed out that the failure of glass [6] results from a tensile component of stress. Nowadays, it is generally accepted that the failure of glass originates at surface flaws [7] at which stresses are concentrated, as shown in Figure 29.5. Since basically no plastic flow is possible in glass, these flaws lead to high stress concentrations when glass surface is in tension. Because of the random nature of the flaws, a large variability in the strength of individual pieces of glass has been observed and reported [6]. Therefore, the failure strength of glass can only be expressed by means of a statistical analysis. Based on these statistical results, we can only obtain a design value at which the risk of fracture of glass is sufficiently low, but it provides no guarantee that the glass will survive under the design load level.

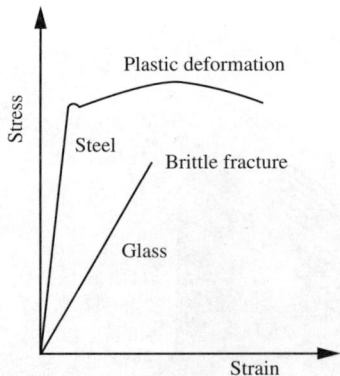

FIGURE 29.4   Stress–Strain diagram.

TABLE 29.1   Mechanical Properties of Glass

E — Young's modulus of elasticity	10.4×10^6 psi or 7.2×10^{10} N/m^2
G — Modulus of rigidity	4.3×10^6 psi or 3.0×10^{10} N/m^2
μ — Poisson's ratio	0.22
α — Coefficient of thermal expansion	$88 \times 10^{-7}/°C$
ρ — Density	157 lb/ft^3 or 2.5 g/cm^3

An important property of glass is that its strength depends on the duration of load [6] application and on the environmental conditions. This concept is not familiar to engineers and architects. Basically, the relationship between stress and time can be expressed as

$$\sigma^n T = \text{constant} \tag{29.1}$$

where σ is the applied stress, while T is the duration of the stress, and n is a constant with a value between 12 and 20. Figure 29.6 illustrates the strength of glass against time. Since the duration of loading is

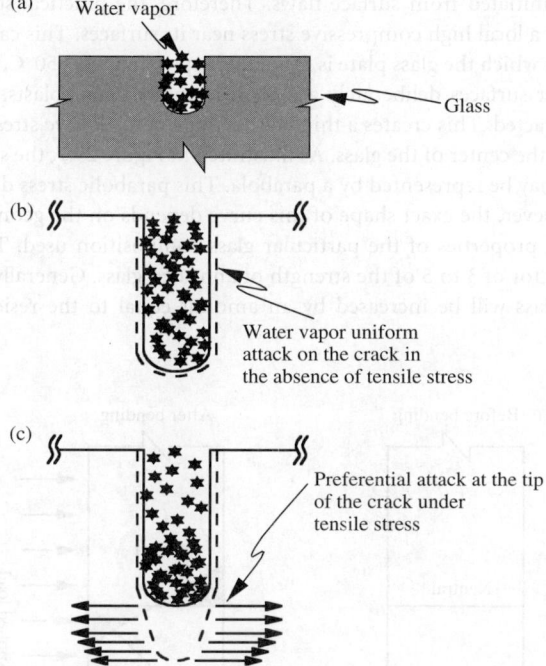

FIGURE 29.5 Surface flaw: (a) two-dimensional model of flaw on glass surface, (b) attack of water vapor on the crack, and (c) attack of water vapor on the crack under tensile stress.

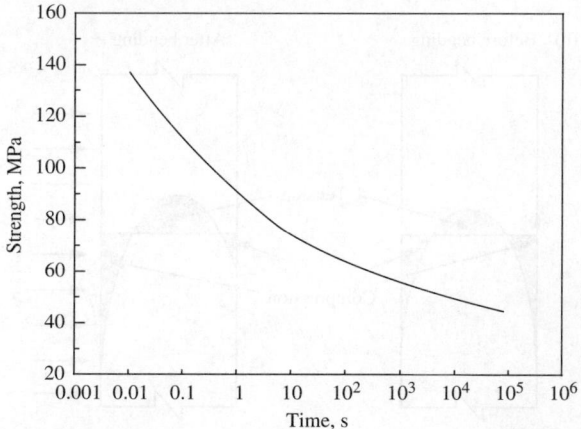

FIGURE 29.6 Glass strength and load duration.

important in determining the failure load for a given glass panel, it is necessary to define the loadings in a time-dependent form.

29.2.1 Types of Glass

From a structural point of view, there are several types of glass used in buildings. Their basic properties are discussed as follows.

29.2.1.1 Tempered (Toughened) and Heat-Strengthened Glass

The fracture of glass is initiated from surface flaws. Therefore, the practical strength of glass may be increased by introducing a local high compressive stress near its surfaces. This can be achieved by means of thermal toughening in which the glass plate is heated to approximately 650°C, at which point it begins to soften. Then, its outer surfaces deliberately are cooled rapidly by air blasts. The exterior layers are quickly cooled and contracted. This creates a thin layer of high compressive stress at the surfaces, with a region of tensile stress at the center of the glass. As illustrated in Figure 29.7, the stress distribution across the thickness of a plate may be represented by a parabola. This parabolic stress distribution must also be in self-equilibrium. However, the exact shape of this curve depends on the geometric shape of the glass section and the physical properties of the particular glass composition used. The bending strength is usually increased by a factor of 3 to 5 of the strength of annealed glass. Generally speaking, the nominal breaking stress of the glass will be increased by an amount equal to the residual compressive stress

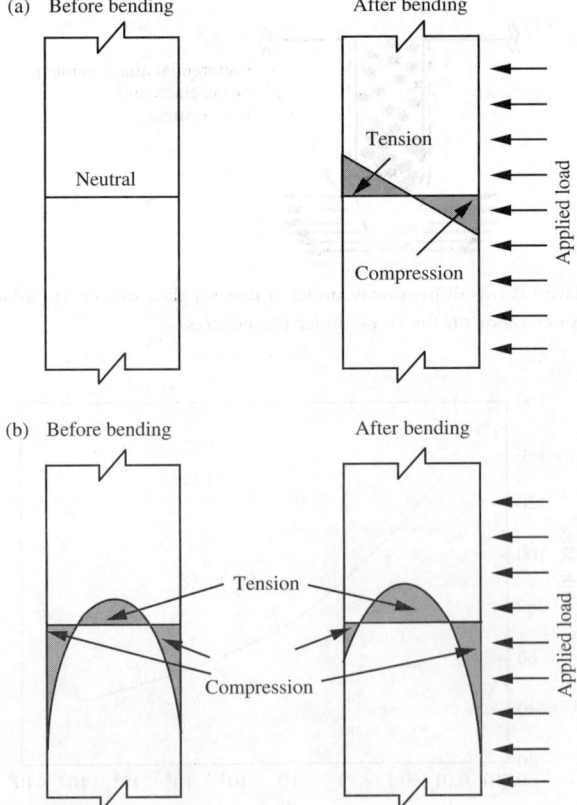

FIGURE 29.7   Stress profiles: (a) annealed glass and (b) toughened glass.

developed at the surface. When the toughened glass is broken, it fractures into small, harmless dice, which result from multiple crack branching due to the release of elastic energy.

29.2.1.2 Annealed Glass

This refers to those glass panels without heat treatment. The permissible stress is taken approximately as 15 N/mm^2. Sometimes we cannot avoid using annealed glass because of manufacturing difficulties such as the glass panels being too large for heat treatment. Due to its small strength, annealed glass is weak in thermal resistance. Partial shading causes annealed glass to fail by thermal stress. Very often, glass fins are annealed.

29.2.1.3 Tinted Glass

Tinted glass or heat-absorbing glass is made by adding colorant to normal clear glass. Light transmittance varies from 14 to 85%, depending on color and thickness. Because of this, the tinted glass is hot, and heat-strengthened glass is normally used in making tinted glass.

29.2.1.4 Coated Glass

Coated glass is manufactured by placing layers of coating onto the glass surfaces. There are two types, the solar control (reflective) and the low-emissivity (low-e) types. They are more related to energy absorption and light transmission and only indirectly affect the structural strength by changing the thermal stress. Because of this, for colored glass to prevent excessive thermal stress, at least heat-strengthened glass should be used.

29.2.1.5 Wired Glass

Wired glass is made by introducing a steel mesh into molten glass during the rolling process. It is weak in resisting thermal stress and therefore has a high rate of breakage due to sunlight, etc. Polished wired glass is generally used for fire rating since after its breakage, it is stuck to the wire mesh and prevents passage of smoke. However, it is weak in resisting thermal stress. Figure 29.8 shows the damaged wired glass panels under sunlight.

FIGURE 29.8 Broken glass panel due to thermal stress.

FIGURE 29.9   Laminated glass when it is broken.

29.2.1.6 Laminated Glass

This is a very common form of glass formed by bonding two or more glass panes by interlayers like polyvinyl butyral (PVB) or resin. The thickness of this interlayer is normally 0.38, 0.76, 1.52 mm, etc. The major problem for laminated glass is the validity of composite action. Can we assume a composite action, that is, an 8- + 6-mm-thick laminated glass is equivalent to a 14-mm-thick glass? If not, does it behave as two separated panes 8 mm and 6 mm thick?

The actual response for a laminated glass is somewhere between these two extremes. For short-term load, the behavior is closer to composite assumption, while for long-term load, it behaves as separated panes because of creeping effect in the interlayer. However, as the actual response is dependent on the property of the interlayer, it may not be overgeneralized. One method is to use a simple test to measure the deflection of the panel under a specific load and then compare this with the deflection calculated by a finite element program. We can then adjust the equivalent thickness in the program to give the same deflection so that we can determine the equivalent thickness of the laminated glass pane and use it for economical and rational design. ASTM C1172 is a relevant standard for further information and testing. Figure 29.9 shows the property of laminated glass when broken.

29.3   Design of Glass

The linear deflection theory, which assumes that deflections are directly proportional to applied load, is of sufficient accuracy for many engineering applications. However, for a thin glass plate simply supported on four sides, the linear theory is invalidated when the deflection is larger than three fourths of its thickness (Canadian code [4]). The typical load versus central deflection curve for a glass panel is shown in Figure 29.10, and it can be seen that the linear theory is only valid in a small loading range before deflection is significant. The use of the linear theory will result in a deviation from the real solution as shown in Figure 29.10 for deflection and Figure 29.11 for stress. In the linear theory, the location of maximum stress is predicted at the plate center. In fact, the

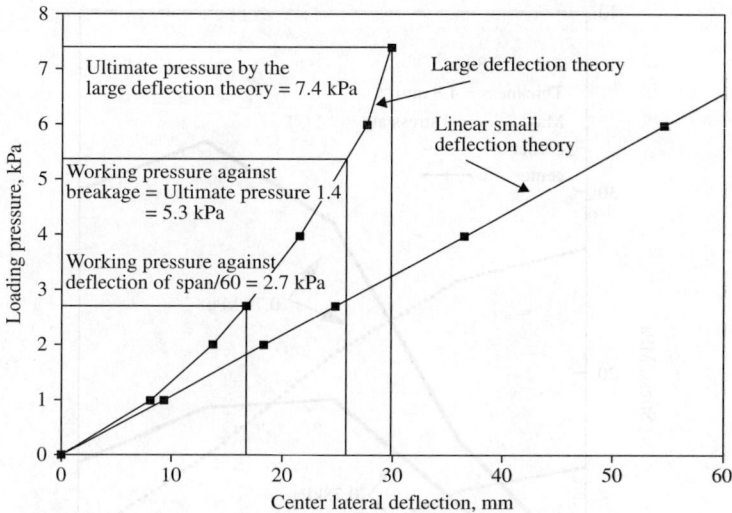

FIGURE 29.10 Load vs. center deflection of a 4-side simply supported glass pane of 2000 mm × 1000 mm × 5.6 mm.

location of maximum stress changes with the load level and the aspect ratio of glass plate. This change is illustrated in Figure 29.11. From the figure, it can be seen that the maximum principal stresses at the corner and the center are more or less the same for aspect ratio equal to 1 and load level equal to 0.1 kPa in the glass plate under consideration. When the load level is increased to 0.76 kPa, the maximum principal stress at the corner increases more rapidly than the stress at the center. Thus, the maximum stress location is at the corner of the plate. On the other hand, the rate of increase of the maximum principal stress at the center is much faster than the rate of increase of the maximum principal stress at the corner for aspect ratio equal to 5. In this case, the maximum stress is located at the center of the plate. As mentioned above, the failure of glass depends on the stress state and surface flaws. Thus, there is a need to develop a numerical procedure to find out the stresses at various locations of the glass plate under different load levels in order to determine its load capacity in terms of probability of failure.

Glass type	Canadian[a]/U.S.[b]	Australian[c]	U.K.[d]	Chinese[e]
Load duration (s)	60	3	3	60
Load factor	1.5	—	—	1.4
Annealed (N/mm²)	20–25 (edge/center, following similar)	20	41 for $t \leq 6$, 34.5 for $t \leq 8$, 28 for $t \leq 10$	28 for $5 < t < 12$ 20 for $15 < t < 19$
Heat-strengthened (N/mm²)	40–50	32	—	—
Tempered (N/mm²)	80–100	50	59	84 for $5 < t < 12$ 59 for $15 < t < 19$

[a] Canadian General Standards Board (1989), "Structural design of glass for buildings," CAN/CGSB-12.20-M89. The first one refers to center stress and the second one to edge stress.
[b] ASTM (1997), "Standard practice for determining minimum thickness and type of glass required to resist a specified load," E1300–97.
[c] Standards Australia (1994), "Glass in buildings — selection and installation."
[d] Pilkington Glass (see IStructE, Structural Use of Glass in Buildings, 1999).
[e] "Technical code for glass curtain wall engineering," JGJ 102-96, 1996, Beijing, China.
   Note: t = thickness of glass plate.

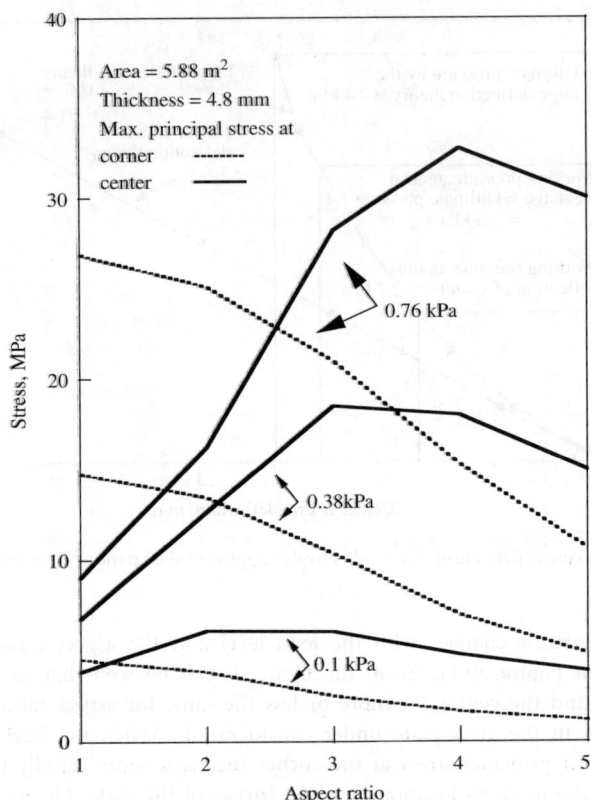

FIGURE 29.11 Stress against aspect ratio at different load levels.

The design load for glass is time-dependent. It is accepted worldwide that it should be based on the 1-min constant and uniform loads. However, for most applications of glass, load duration has little to no effect on the long-term performance of glass. There are two common methods in the industry for determining glass strength. The first is the empirical glass-to-destruction test method, and the second is the analytical nondestructive computer method.

The empirical glass strength curves were developed from destructive test of glass plates to provide factual data on glass strength. At least 25 glass panes each thickness and area were tested to produce a statistical validity for the average breaking pressure under uniform load conditions. A typical design chart produced from the results of testing glass-to-destruction is included in the appendix of the ASTM Standard E300-84. The effect of aspect ratio of glass plate has not been indicated in this design chart.

Nowadays, the advances in computer technology and the dramatic reductions in computer cost make the computer method for determining glass strength more practical than previously. The finite element method has made it possible to determine glass design data with various supporting systems and loading cases effectively. The finite element method is adopted to calculate the magnitudes and orientations of stress and deflections of the glass plate. The computer outputs are then used with a statistical or failure prediction model [8] to find out the glass breakage probabilities under the design condition. Glass breaks when the maximum principal tensile stress reaches the critical value determined by the failure prediction model, which is discussed in the subsequent section. Finite element computer analysis indicates that as the aspect ratio of the glass plate changes, the levels and locations of maximum tension stress are also varied. Thus, the method is more realistic in representing the glass strength.

29.4 Failure Criterion

For commercial glass widely used in curtain wall systems, failure and breakage are due to the stress concentrated at the invisible hairy crack on its surfaces. The failure stress of a piece of glass is more dependent on the density of these hairy cracks than the theoretical breakage stress, which can be as high as 10,000 MPa. Thus, a rational design failure stress is expressed in terms of the duration of load (Weibull's theory [9] for failure of brittle material). Treatment of glass to reduce surface tensile stress and the area of the glass panel is being considered by glass manufacturers. As glass plates are usually thin and undergo large displacements, the use of conventional thin plate linear bending theory will yield erroneous results. Indeed, to accurately compute the maximum stress in a panel for checking of stress against failure, the large deflection theory allowing for membrane stress should be used. In the breakage analysis of glass panels, failure is assumed to occur when the maximum tensile stress is equal to the breaking stress of the glass. For ductile material, the yield strength can be accurately measured and, typically, varies over a narrow range. However, as a brittle material, glass has no observable yield strength as other materials such as steel. Thus, the failure of glass can only be represented by breakage stress, which is obtained from a statistical basis. For tempered glass, the breakage stress is usually taken to be four times the failure stress for clear float glass. For heat-strengthened glass, where the tempering process is lighter than for tempered glass, the strength is twice that of annealed glass.

The Canadian Code has adopted the failure prediction model developed by Beason and Morgan [8]. The failure prediction model is based on the simplified formulation, which has been presented by Brown [10] to model the glass strength with load duration. The resistance to failure of a surface flaw can be expressed as follows:

$$K_f = \int_0^{T_f} [\sigma(T)]^n \, dT \tag{29.2}$$

where T is the load duration and K_f is the resistance to failure of a surface flaw exposed to tensile stress and water vapor. The nominal tensile stress, $\sigma(T)$, at the flaw is expressed as a function of time, and n is a constant of which the value of the best fit, from experimental data, is found to be 16 (Dalgliesh and Taylor [11]). The duration of the loading causing failure is expressed as T_f. The glass plate fails when K_f reaches some critical values which depend on the flaw's characteristics and stress state at the flaw. With Equation 29.2, we can adjust the strength of glass for different load durations. In a computer analysis, we can compute constant pressure causing glass breakage and relate this to failure pressure with different load durations as follows:

$$P_{60} = P_f \left[\frac{T_f}{60} \right]^{1/n} \tag{29.3}$$

where P_{60} is the constant pressure causing failure of the panel in 60 s and P_f is the constant pressure causing failure at a duration of T_f s.

The use of a design factor of 2.5 has been introduced to control the POF to 0.008. The POF can be expressed in terms of Weibull distribution as follows:

$$POF = 1 - e^{-B} \tag{29.4}$$

where B is a function that reflects the risk of failure and is given as

$$B = \left(\frac{A}{A_0}\right)\left(\frac{S_{m,p,r}}{S_0}\right)^m \tag{29.5}$$

where e is a natural number, A_0 and S_0 are the area and characteristic strength of the reference glass panel, respectively and A and $S_{m,p,r}$ are the area and characteristic strength of the glass panel, respectively. $S_{m,p,r}$ is a function of the Weibull parameter, m, pressure, p, and aspect ratio, r. Failure data for in-service glass were collected (see Ref. [12]) and fitted to Equation 29.5. The fitted Weibull parameters $m = 7$ and $S_0 = 32.1$ MPa are adopted in Canadian Code, and the reference area, A_0, is equal to $1\,m^2$.

29.5 Stress Evaluation and Common Causes of Breakages

The failure of glass is assumed when the principal tensile stress is equal to or greater than the characteristic strength calculated in Equation 29.5. The bending stress is assumed to vary linearly across the thickness of the plate, and the membrane stress is constant across the thickness of the plate. The total stress is obtained by superimposing the bending and membrane stresses. The stress components at each of the three nodes of the element are then used to calculate the principal stresses within the element. The nodal stresses are averaged at nodes that are attached to more than one element.

29.5.1 Common Causes of Glass Breakage

The causes of breakage for glass can be due to (*not* in order of importance)

- Excessive stress from wind pressure or other loads.
- Thermal stress due to differential temperature on different parts of the pane (for 33°C, the thermal stress is 20.7 N/mm^2).
- Buckling due to large compression (e.g., glass rod and glass fins).
- Surface or edge damage.
- Deep scratches or gouges.
- Severe weld splatter.
- Windborne missiles (i.e., debris impact).
- Direct contact with metal (e.g., window aluminum frame).
- Impurities like nickel sulfide (NiS).
- Excessive deflection bringing glass in contact with other hard objects.

29.5.2 Impurities

One big disadvantage in using tempered glass is the problem of spontaneous breakage due to impurities like NiS. NiS is formed when nickel-rich contaminants like nichrome wire and stainless steel are unavoidably introduced into the glass melting furnace, and when they are mixed with sulfur, NiS is formed. They are harmless in annealed or heat-strengthened glass since the induced stress cannot break the tensile failure stress of glass but causes instantaneous breakage when they are located at the tension zone of tempered glass and expand with temperature and time. For surface stress less than 52 N/mm^2, NiS is not a problem since its expansion, together with the tensile prestress, cannot generate a breaking stress higher than the failure tension stress of the glass. Therefore, using heat-strengthened glass is a means of solving the problem of NiS. Heat-soaking test is a procedure to break the glass panels

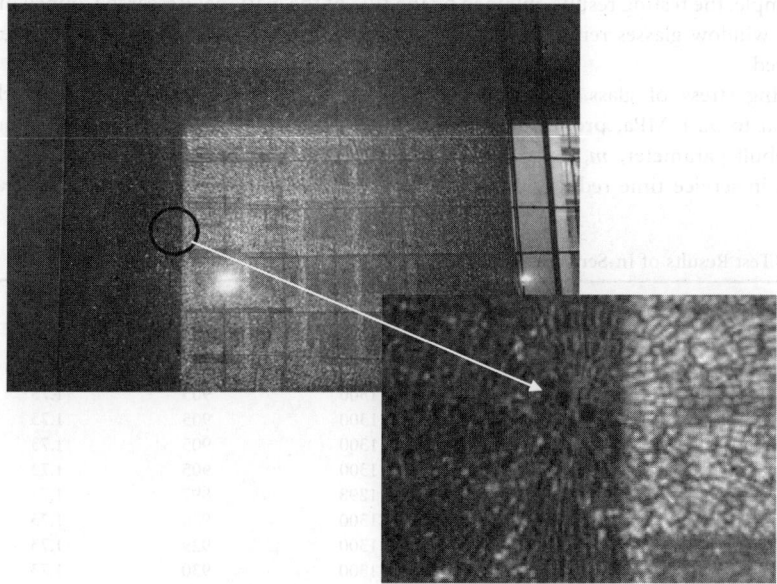

FIGURE 29.12 Glass breakage due to nickel sulfide.

containing NiS in the factory rather than after installation. The time and temperature are important in heat-soaking, and their requirement varies from one country to another. Figure 29.12 shows a picture of glass breakage due to NiS, which is signified by the origin as a pair of butterfly wings.

29.6 Numerical Examples for Breakage Analysis of Glass Structure

Any numerical or analytical method must first be tested and validated before it can be actually used. The limitations and scopes of the method must be clearly investigated and defined. This chapter presents a verification study on the application of the developed finite element method for several nonlinear problems for glass structures. The accuracy of the developed method is then compared with the available solution.

The first example is to compare the results of in-service glass obtained by the Institute for Research in Construction (IRC) [12]. Totally, 47 pieces of in-service glasses obtained from the University of Ottawa's Thompson Residence tested to failure. The second example is the simulation of a curved glass panel under positive and negative wind load. Curved glass panels are frequently used in the construction of observation lift cladding or the exterior staircase of modern prestigious buildings. The third and fourth examples are concerned with glass fin systems that are widely used in shop fronts and entrances of buildings. Two examples of elastic supports are presented that visualize the effects of out-of-plane and in-plane stiffnesses of sealants on the stress distribution within the loaded glass plate. Finally, the results of the simulation of flexible support are compared with a full-scale mock-up test.

29.7 Failure Test of In-Service Glass

The strengths of new glass and in-service glass differ considerably. While the design of glass panels is mostly based on new glass, the actual failure load of in-service glass is of greater interest when one considers safety during the service life of a building.

In this example, the testing results obtained by the IRC of the National Research Council of Canada for 47 in-service window glasses removed from the University of Ottawa's Thompson Residence in 1986 were compared.

The breaking stress of glass is determined by Equations 29.4 and 29.5 with the characteristic strength equal to 32.1 MPa, probability of failure equal to 0.008, reference area, A_0, equal to $1\,\mathrm{m}^2$, and the Weibull parameter, m, equal to 7 as recommended by the Canadian Code. It is generally believed that in-service time reduces the breakage stress of a glass panel due to the increased density

TABLE 29.2 Test Results of In-Service Glass (Mean = 2.51, Standard Deviation = 0.62)

No.	P_{60} (kPa)	Thickness (mm)	X (mm)	Y (mm)	NAShell	Ratio
1	2.84	4.10	1300	905	1.73	1.64
2	4.57	4.00	1300	905	1.73	2.64
3	2.26	4.10	1300	905	1.73	1.30
4	5.27	4.00	1300	905	1.73	3.04
5	4.47	4.05	1300	905	1.73	2.58
6	4.10	4.00	1300	905	1.73	2.37
7	5.62	4.07	1298	897	1.73	3.24
8	4.12	3.90	1300	930	1.73	2.38
9	5.29	4.00	1300	929	1.73	3.05
10	5.01	4.00	1300	930	1.73	2.89
11	4.47	3.93	1300	928	1.73	2.58
12	5.75	3.95	1300	925	1.73	3.32
13	3.66	3.84	1300	925	1.73	2.11
14	5.39	3.90	1300	924	1.73	3.11
15	4.79	3.95	1300	925	1.73	2.76
16	5.70	3.88	1300	930	1.73	3.29
17	5.73	4.04	1300	925	1.73	3.31
18	6.08	3.93	1300	900	1.73	3.51
19	4.58	3.90	1300	900	1.73	2.64
20	4.77	4.09	1300	900	1.73	2.75
21	5.16	4.01	1300	895	1.73	2.98
22	2.92	4.00	1300	900	1.73	1.68
23	3.38	3.97	1300	899	1.73	1.95
24	5.14	3.86	1300	930	1.73	2.97
25	5.54	4.00	1300	900	1.73	3.20
26	6.18	3.96	1300	975	1.70	3.65
27	5.04	4.00	1300	975	1.70	2.97
28	5.03	3.96	1300	975	1.70	2.97
29	4.04	3.96	1300	975	1.70	2.38
30	5.02	3.91	1300	975	1.70	2.96
31	2.19	3.83	1340	916	1.66	1.32
32	2.46	3.77	1340	916	1.66	1.48
33	2.54	3.69	1340	916	1.58	1.61
34	3.75	3.75	1342	916	1.66	2.26
35	2.73	4.00	1357	1300	1.43	1.91
36	3.11	4.05	1356	1300	1.43	2.17
37	3.11	3.93	1356	1300	1.36	2.29
38	4.64	4.01	1358	1300	1.43	3.24
39	4.83	4.81	1358	1300	1.81	2.67
40	2.66	3.80	1374	1342	1.21	2.21
41	1.86	3.83	1374	1342	1.21	1.54
42	3.01	4.05	1300	1062	1.81	1.66
43	3.27	3.94	1300	1062	1.66	1.97
44	4.27	4.04	1300	1065	1.73	2.46
45	4.51	3.86	1300	1065	1.66	2.72
46	3.98	3.87	1300	1065	1.66	2.40
47	3.08	3.74	1300	1065	1.58	1.95

of hairy cracks on glass surfaces in the course of resisting wind loads and also when subjected to natural or man-made scratches.

The equivalent 60-s pressure of the testing results and output by NAShell [13] are tabulated in Table 29.2. The average ratio of failure load to the predicted breaking load by NAShell is 2.51. This ratio is considered to be in a reasonable range because the failure stress used in NAShell has included the probability of failure of 8/1000. Not a single sample has a failure load lowered than the predicted load, indicating the reliability of the suggested method in the design of in-service glass panels.

The standard deviation, however, for the failure loads is quite large and is equal to 0.62. This demonstrates the variability of glass strength in practice and also that the nature and behavior of glass strength can only be represented as a probability of failure.

29.8   Curved Glass Panel

In this example, a curved glass panel with base or projected dimension of 1500 mm × 1500 mm, radius of 1500 mm, Young's modulus of 70,000 MPa, Poisson's ratio of 0.22 and thickness of 8 mm, and under uniform lateral load is analyzed (see Figure 29.13). The longitudinal boundaries are hinged and immovable, while the curved edges are restrained in the longitudinal direction. Due to symmetry, only a quarter of the panel is analyzed with mesh size of 10 × 10. In Figure 29.14, we can see the load–deflection path at the plate center and the failure loads for annealed glass and tempered glass under positive and negative pressure. Failure is assumed when the maximum principal tensile stress reaches the characteristic strength of 14.25 MPa, which is calculated from Equations 29.4 and 29.5. For tempered glass, the failure stress is assumed to be four times the value for annealed glass [12].

From the figures, it can be seen that the failure pressure ratio for annealed and tempered glasses is the same as the ratio of their stresses where the geometrical change is not significant. However, for compressive load case, the failure pressure ratio for annealed to tempered glasses may not be equal to the ratio of their failure stresses. This is due to the large change in geometry resulting in the nonlinearity between the stress and the load.

29.9   Flexible Support for Full-Scale Mock-Up Test

In this example, the results obtained from a full-scale curtain wall test were compared with the numerical results obtained from the computer program NAShell. In this analysis, the mullions and

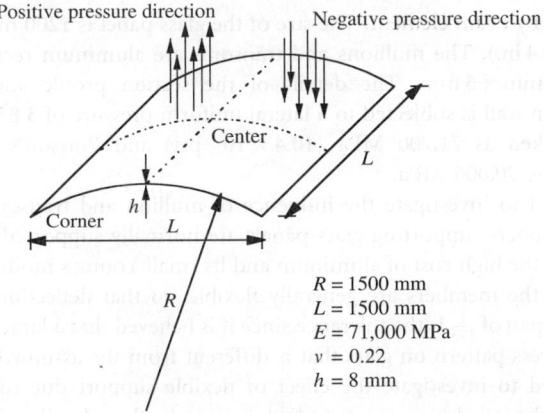

FIGURE 29.13   Layout and properties of curved glass panel.

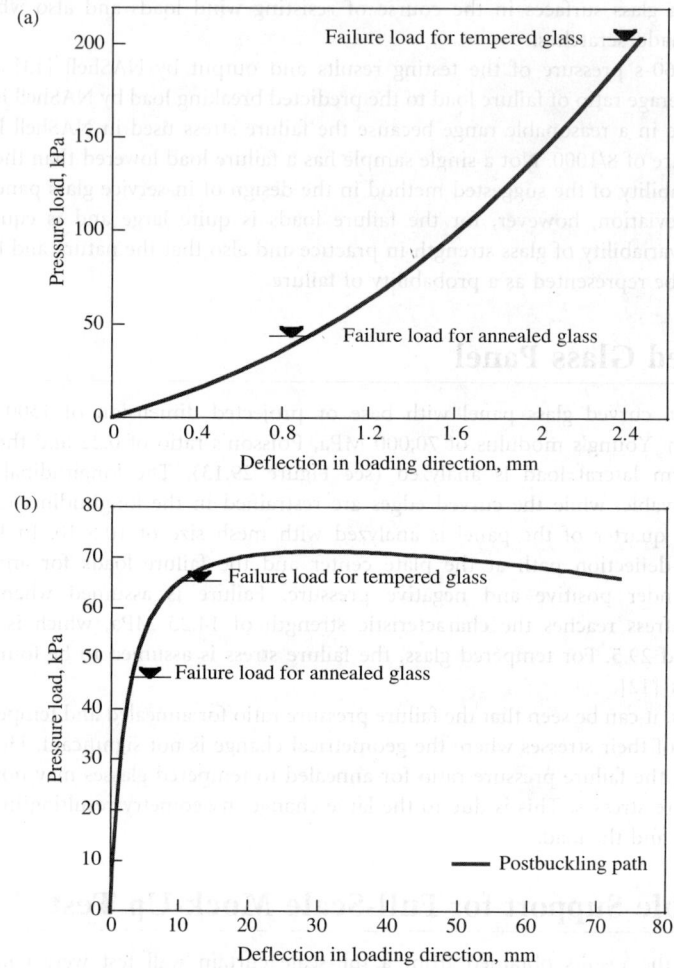

FIGURE 29.14 Load–deflection path of the curved glass at the center: (a) positive pressure and (b) negative pressure.

transoms were modeled by beam element. The size of the glass panel is 1200 mm × 1800 mm × 10 mm (47.25 in. × 70.9 in. × 0.4 in.). The mullions and transoms are aluminum rectangular hollow sections of size 45 mm × 100 mm × 3 mm. The details of the section profile and layout are shown in Figure 29.15. The curtain wall is subjected to a lateral uniform pressure of 3.85 kPa (0.56 psi). Young's modulus of glass is taken as 71,700 MPa (10.4×10^6 psi) and Poisson's ratio, as 0.22. Young's modulus of aluminum is 70,000 MPa.

This problem is aimed to investigate the influence of mullion and transom flexibility to the glass strength. Structural members supporting glass panels are normally supported by brackets to concrete slab or spandrel. Due to the high cost of aluminum and its small Young's modulus of elasticity of about one third that of steel, the members are generally flexible, so that deflection is commonly a design criterion. In practice, a span of $\frac{1}{175}$ is the tolerance since it is believed that a large deflection in mullion or transom will create a stress pattern on glass that is different from the assumed rigid support case.

This example is aimed to investigate the effect of flexible support due to flexibility in structural members, which, to our knowledge, was not studied previously, though a limiting value of $\frac{1}{175}$ of span is recommended in the Canadian Code of Practice [12].

FIGURE 29.15 Layout of the tested full-scale sample.

TABLE 29.3 Flexible Supports

	Central deflection of glass (mm)	Midpoint deflection of mullion (mm)	Midpoint deflection of transom (mm)
Full-scale mock-up test	12.84	7.45	1.43
NAShell	12.14 (94.55%)	7.36 (98.79%)	1.36 (95.10%)
Built-in fixed support[a]	6.15 (47.90%)	N/A	N/A
Roller simply support[b]	8.99 (70.02%)	N/A	N/A

[a] Lateral deflection and rotation are fixed.
[b] Lateral deflection is fixed but free to rotate.
Note: Values in parentheses refer to the ratio of the deflections to the measured deflections in the test.

From Table 29.3, we can see that we have underestimated the glass deflection of the glass plate if we consider that the glass plate is fixed support (in rotation and translation) or even simply support (restrained only in translation). The actual deflection of the glass is about double in the case of fixed support and about 1.4 times in the case of simply support. On the other hand, we observe that the deflections obtained by NAShell are close to those in the mock-up test, and the errors in the prediction of deflections are acceptable in engineering practice. The underestimation of deflection would probably increase the chance of contact with hard objects such as a concrete wall behind the glass and hence increase the failure rate. Deflection limit for serviceability requirement is normally taken as one-sixtyth of shorter span.

29.10 Conclusions

The concept and method for design and analysis of glass panels is described in this chapter. The validity of the finite element formulation has been demonstrated for flat and curved panels with in-plane edge support flexibility. All these problems are related to the practical design of the glass system. Further, a summary of possible causes of breakage for glass panels is presented. It can be seen that, with a proper design and analysis method and methods of installation, glass structures can be designed to meet the safety and serviceability requirements.

References

[1] Zachariasen, W.H. and Warren, B.E., *The atomic arrangement in glass, J. Am. Chem. Soc.*, Vol. 54, pp. 3841–3851, 1932.

[2] Libbey-Owens-Ford Co., *Technical Information — Strength of Glass under Wind Loads*, ATS-109, Toledo, OH, 1980.

[3] ASTM Standard E1300-89, *Standard Practice for Determining the Minimum Thickness of Annealed Glass Required to Resist a Specified Load*, 1989.

[4] National Standard of Canada, *Structural Design of Glass for Buildings*, CAN/CGSB-12.20-M89, Canadian General Standards Board, 1989.

[5] Scholze, H., *Glass — Nature, Structure, and Properties*, Springer-Verlag, New York, 1990 (translated by M.J. Lakin).

[6] Shand, E.B., *Glass Engineering Handbook*, McGraw-Hill, New York, 3rd edition, 1984.

[7] Griffith, A.A., *The phenomena of rupture and flows in solids, Trans. R. Soc., Ser. A*, Vol. 221, pp. 163–198, 1921.

[8] Beason, W.L. and Morgan, J.R., *Glass failure prediction model, J. Struct. Eng.*, ASCE, Vol. 110, No. 2, pp. 197–212, 1984.

[9] Weibull, W., *A Statistical Theory of the Strength of Materials*, Royal Swedish Institute for Engineering Research, Stockholm, Sweden, 1939.

[10] Brown, W.G., *A practicable formulation for the strength of glass and its special application to large plates*, Publication No. NRC 14372, National Research Council of Canada, Ottawa, Ontario, Canada, 1974.

[11] Dalgliesh, W.A. and Taylor, D.A., *The strength and testing of window glass, Can. J. Civ. Eng.*, Vol. 17, pp. 752–762, 1990.

[12] National Standard of Canada, *Structural Design of Glass for Buildings*, CAN/CGSB-12.20-M89, Canadian General Standards Board, 1989.

[13] So, A.K.W. and Chan, S.L., *"NASHELL", Computer program for geometrically nonlinear analysis of glass panels*, User's Manual, 1995.

VI

Special Topics

IV

Special Topics

30

Welded Tubular Connections — CHS Trusses

Peter W. Marshall
MHP Systems Engineering,
Houston, TX

30.1 Introduction

Truss connections in circular hollow sections (CHS) present unique design challenges. This chapter discusses the following elements of the subject: Architecture, Characteristics of Tubular Connections, Nomenclature, Failure Modes, Reserve Strength, Empirical Formulations, Design Charts, Application, and Summary and Conclusions.

30.2 Architecture

Architecture is defined as the art and science of designing and successfully executing structures in accordance with aesthetic considerations and the laws of physics, as well as practical and material considerations. Where tubular structures are exposed for dramatic effect, it is often disappointing to see grand concepts fail in execution due to problems in the structural connections of tubes. Such "failures" range from awkward ugly detailing, to learning curve problems during fabrication, to excessive deflections or even collapse. Such failures are unnecessary, as the art and science of welded tubular connections has been codified in the American Welding Society (AWS) *Structural Welding Code* [1]. The AWS design criteria have also been incorporated into the American Institute of Steel Construction (AISC) Specification for the design of steel hollow structural sections [2].

30-1

A well-engineered structure requires that a number of factors be in reasonable balance. Factors to be considered in relation to economics and risk in the design of welded tubular structures and their connections include (1) static strength, (2) fatigue resistance, (3) fracture control, and (4) weldability. Static strength considerations are so important that they often dictate the very architecture and layout of the structure; certainly, they dominate the design process and are the focus of this chapter. Many of the other factors also require early attention in design, and it is of benefit to set up quality control/quality assurance programs during construction; these are discussed further in sections of the Code dealing with materials, welding technique, qualification, and inspection.

The designer of record is responsible for structural connections and should not attempt to abdicate his responsibility by placing a "Truss Note" on the drawings such as "All welded truss connections shall develop the full axial capacity of each member at a stress of $0.6F_y$." The fabricator might simply provide matching weld sizes and let it go at that.

30.3 Characteristics of Tubular Connections

Tubular members benefit from an efficient distribution of their material, particularly in regard to beam bending or column buckling about multiple axes. However, their resistance to concentrated radial load is more problematic. For architecturally exposed applications, the clean lines of a closed section are esthetically pleasing, and they minimize the amount of surface area for dirt, corrosion, or other fouling. Simple welded tubular joints can extend these clean lines to include the structural connections.

Although many different schemes for stiffening tubular connections have been devised [3], the most practical connection is made by simply welding the branch member to the outside surface of the main member (or chord). Where the main member is relatively compact (D/T less than 15 or 20), the branch member thickness is limited to 50 or 60% of the main member thickness, and a prequalified weld detail is used, the connection can develop the full static capacity of the members joined. Where the foregoing conditions are not met, for example, with large-diameter tubes, a short length of heavier material (or joint can) is inserted into the chord to locally reinforce the connection area. Here, the design problem reduces to one of selecting the right combination of thickness, yield strength, and notch toughness for the chord or joint can. Sometimes the entire chord of a truss is sized to satisfy connection requirements. The detailed considerations involved in this design process are the subject of this chapter.

30.4 Nomenclature

Nondimensional parameters for describing the geometry of a tubular connection are given in the following list. Beta, eta, theta, and zeta describes the surface topology. Gamma and tau are two very important thickness parameters. Alpha (not shown) is an ovalizing parameter, depending on load pattern (it was formerly used for span length in beams loaded via tee connections).

β (beta)	d/D, branch diameter/main diameter
η (eta)	branch footprint length/main diameter
θ (theta)	angle between branch and main member axis
ζ (zeta)	g/D, gap/diameter (between balancing branches of a K connection)
γ (gamma)	R/T, main member radius/thickness ratio
τ (tau)	t/T, branch thickness/main thickness

In AWS D1.1 [1], the term "T-, Y-, and K connection" is used generically to describe simple structural connections or nodes, as opposed to coaxial butt and lap joints. A letter of the alphabet (T, Y, K, X) is used to evoke a picture of what the node subassemblage looks like.

30.5 Failure Modes

A number of unique failure modes are possible in tubular connections. In addition to the usual checks on weld size, provided for in most design codes, the designer must check for the following failure modes, listed together with the relevant AWS and AISC Code sections.

	AWS	AISC
Local failure (punching shear)	2.24.1.1	9.4 ¶2
General collapse	2.24.1.2	9.4 ¶2.6.3
Unzipping (progressive weld failure)	2.24.1.3	Ref. to AWS
Materials problems	2.26	—
Fracture	C4.12.4.4	—
Delamination	C2.6.3	—
Fatigue	2.36.6	—

30.5.1 Local Failure

AWS design criteria for this failure mode have traditionally been formulated in terms of punching shear. The main member acts as a cylindrical shell in resisting the concentrated radial line loads (force/length, e.g., kip/in.) delivered to it at the branch member footprint. Although the resulting localized shell stresses in the main member are quite complex, a simplified, but still quite useful, representation can be given in terms of punching shear stress, v_p:

$$\text{acting } v_p = f_n \tau \sin \theta \qquad (30.1)$$

where f_n is the nominal stress at the end of the branch member, either axial or bending, which is treated separately. Punching shear is the notional stress on the potential failure surface, as illustrated in Figure 30.1. The overriding importance of chord thickness is reflected in tau, while $\sin \theta$ indicates that it is the radial component of load that causes all the mischief.

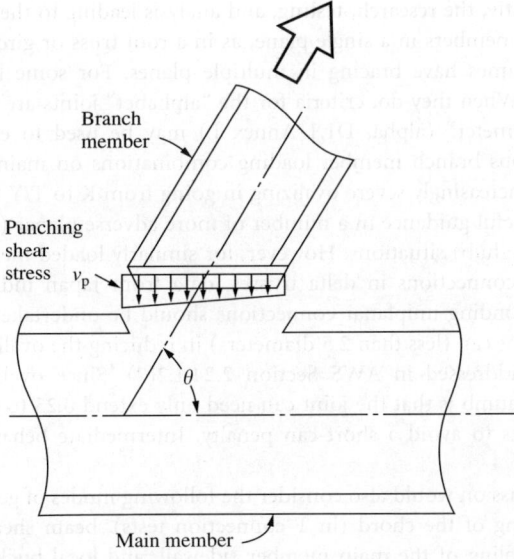

FIGURE 30.1 Local failure mode and punching shear v_p.

The allowable punching shear stress is given in the AWS Code as

$$\text{allowable } v_p = \frac{F_{y0}}{0.6\gamma} \cdot Q_q \cdot Q_f \tag{30.2}$$

We see that the allowable punching shear stress is primarily a function of main member yield strength (F_{y0}) and gamma ratio (main member radius/thickness), with some trailing terms that tend toward unity. The term Q_q reflects the considerable influence of connection type, geometry, and load pattern, while interactions between branch and chord loads are covered by the reduction factor Q_f. Interactions between brace axial load and bending moments are treated analogous to those for a fully plastic section.

Since 1992, the AWS Code also includes tubular connection design criteria in total load ultimate strength format, compatible with a load and resistance factor design (LRFD) design code formulation. This was derived from and is roughly equivalent to the original punching shear criteria and is the format adopted by AISC.

30.5.2 General Collapse

In addition to local failure of the main member in the vicinity of the branch member, a more widespread mode of collapse may occur, for example, general ovalizing plastic failure in the cylindrical shell of the main member. To a large extent, this is now covered by strength criteria that are specialized by connection type and load pattern, as reflected in the Q_q factor.

For balanced K connections, the inward radial loads from one branch member is compensated by outward loads on the other, ovalizing is minimized, and capacity approaches the local punching shear limit. For T and Y connections, the radial load from the single branch member is reacted by beam shear in the main member or chord, and the resulting ovalizing leads to lower capacity. For cross- or X-connections, the load from one branch is reacted by the opposite branch, and the resulting double dose of ovalizing in the main member leads to still further reductions in capacity. The Q_q term also reflects reduced ovalizing and increased capacity, as the branch member diameter approaches that of the main member.

Thus, for design purposes, tubular connections are classified according to their configuration (T, Y, K, X, etc.). For these "alphabet" connections, different design strength formulae are often applied to each different type. Until recently, the research, testing, and analysis leading to these criteria dealt only with connections having their members in a single plane, as in a roof truss or girder.

Many tubular space frames have bracing in multiple planes. For some loading conditions, these different planes interact. When they do, criteria for the "alphabet" joints are no longer satisfactory. In AWS, an "ovalizing parameter" (alpha, D1.1 Annex L) may be used to estimate the beneficial or deleterious effect of various branch member loading combinations on main member ovalizing. This reproduces the trend of increasingly severe ovalizing in going from K to T/Y to X connections and has been shown to provide useful guidance in a number of more adverse planar (e.g., all-tension double-K [4]) and multiplanar (e.g., hub) situations. However, for similarly loaded members in adjacent planes, for example, paired KK connections in delta trusses, data from Japan indicate that no increase in capacity over the corresponding uniplanar connections should be undertaken [5].

The effect of a short joint can (less than 2.5 diameters) in reducing the ovalizing or crushing capacity of cross-connections is addressed in AWS Section 2.24.1.2(2). Since ovalizing is less severe in K connections, the rule of thumb is that the joint can need only extend 0.25 to 0.4 diameters beyond the branch member footprints to avoid a short-can penalty. Intermediate behavior would apply to T/Y connections.

A more exhaustive discussion would also consider the following modes of general collapse in addition to ovalizing: beam bending of the chord (in T-connection tests), beam shear (in the gap of K connections), transverse crippling of the main member sidewall, and local buckling due to uneven load transfer (either brace or chord). These are illustrated in Figure 30.2.

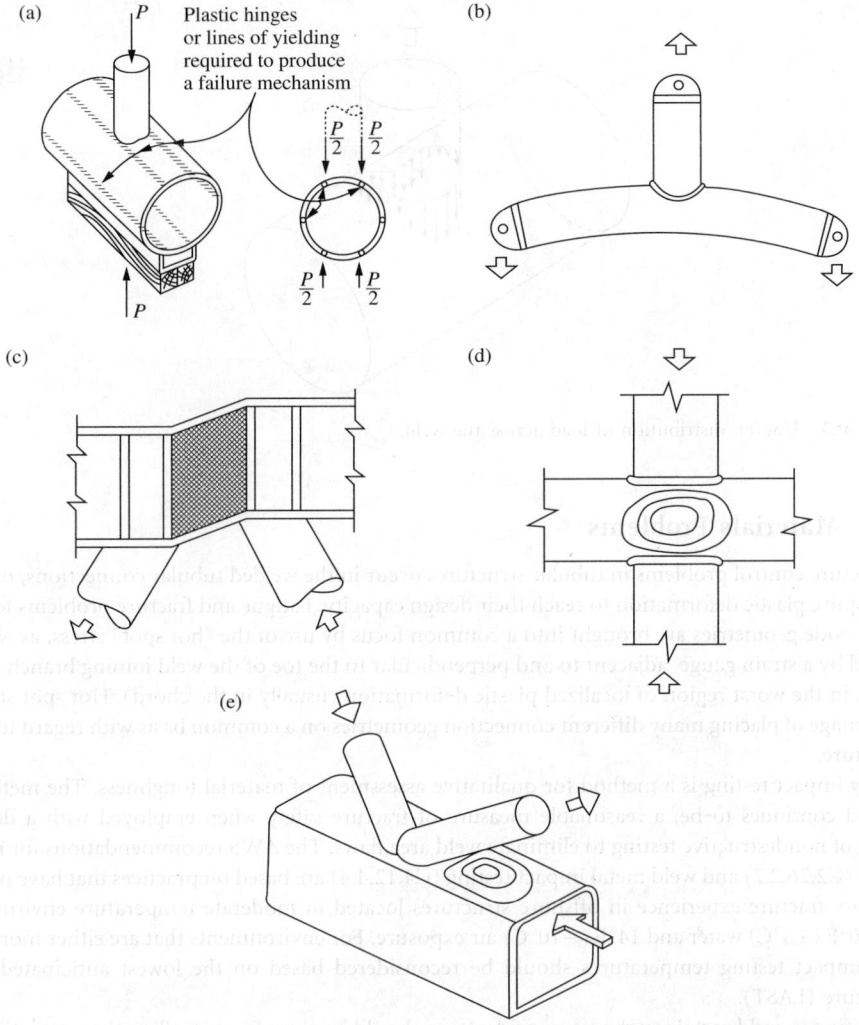

FIGURE 30.2 Failure modes — general collapse: (a) ovalizing, (b) beam bending, (c) beam shear in the gap, (d) sidewall (web) crippling, and (e) local buckling due to uneven distribution of axial load.

30.5.3 Unzipping or Progressive Failure

The initial elastic distribution of load transfer across the weld in a tubular connection is highly nonuniform, as illustrated in Figure 30.3, with the peak line load often being a factor of 2 higher than that indicated on the basis of nominal sections, geometry, and statics. Some local yielding is required for tubular connections to redistribute this and reach their design capacity. If the weld is a weak link in the system, it may "unzip" before this redistribution can happen. Criteria given in the AWS Code are intended to prevent this unzipping, taking advantage of the higher reserve strength in weld allowable stresses than is the norm elsewhere. For traditional mild steel tubes (not multigrade material of higher actual strength) and overmatched E70 weld metal, a weld effective throat as small as 70% of the branch member thickness has been permitted. Other codes and design situations require effective throats of 100 or 110%.

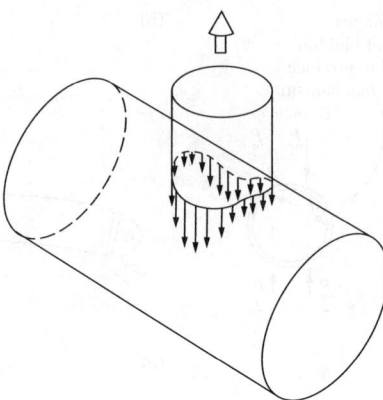

FIGURE 30.3   Uneven distribution of load across the weld.

30.5.4 Materials Problems

Most fracture control problems in tubular structures occur in the welded tubular connections, or nodes. These require plastic deformation to reach their design capacity. Fatigue and fracture problems for many different node geometries are brought into a common focus by use of the "hot spot" stress, as would be measured by a strain gauge, adjacent to and perpendicular to the toe of the weld joining branch to main member, in the worst region of localized plastic deformation (usually in the chord). Hot spot stress has the advantage of placing many different connection geometries on a common basis with regard to fatigue and fracture.

Charpy impact testing is a method for qualitative assessment of material toughness. The method has been, and continues to be, a reasonable measure of fracture safety when employed with a definitive program of nondestructive testing to eliminate weld area flaws. The AWS recommendations for material selection (C2.26.2.2) and weld metal impact testing (C4.12.4.4) are based on practices that have provided satisfactory fracture experience in offshore structures located in moderate temperature environments, that is, 40°F (+5°C) water and 14°F (−10°C) air exposure. For environments that are either more or less hostile, impact testing temperatures should be reconsidered based on the lowest anticipated service temperature (LAST).

In addition to weld metal toughness, consideration should be given to controlling the properties of the heat affected zone (HAZ). Although the heat cycle of welding sometimes improves hot rolled base metals of low toughness, this region will more often have degraded toughness properties. A number of early failures in welded tubular connections involved fractures that either initiated in or propagated through the HAZ, often obscuring the identification of other design deficiencies, for example, inadequate static strength.

A more rigorous approach to fatigue and fracture problems in welded tubular connections has been taken by using fracture mechanics [6]. The crack tip opening displacement (CTOD) test is used to characterize materials that are tough enough to undergo some plasticity before fracture.

Underneath the branch member footprint, the main member is subjected to stresses in the thru-thickness or short transverse direction. Where these stresses are tensile, due to weld shrinkage or applied loading, delamination may occur — either by opening up preexisting laminations or by laminar tearing in which microscopic inclusions link up to give a fracture having a woody appearance, usually in or near the HAZ. These problems are addressed in API joint can steel specifications 2H, 2W, and 2Y. Preexisting laminations are detected with ultrasonic testing. Microscopic inclusions are prevented by restricting sulfur to very low levels (<60 ppm) and by inclusion of shape control metallurgy in the steel-making ladle. As a practical matter, weldments that survive the weld shrinkage phase usually perform satisfactorily in ordinary service.

Joint can steel specifications also seek to enhance weldability with limitations on carbon and other alloying elements, as expressed by carbon equivalent or P_{cm} formulae. Such controls are increasingly important as residual elements accumulate in steel made from scrap. In AWS Appendix XI [1], the preheat required to avoid HAZ cracking is related to the carbon equivalent, base metal thickness, hydrogen level (from welding consumables), and degree of restraint.

30.5.5   Fatigue

This failure mode has been observed in tubular joints in offshore platforms, dragline booms, drilling derricks, radio masts, crane runways, and bridges. The nominal stress, or detail classification approach, used for nontubular structures fails to recognize the wide range of connection efficiencies and stress concentration factors that can occur in tubular structures. Thus, fatigue design criteria based on either punching shear or hot spot stress appear in the AWS Code. The subject is also summarized in recent papers on tubular offshore structures [7,8].

30.6   Reserve Strength

While the elastic behavior of tubular joints is well predicted by shell theory and finite element analysis, there is considerable reserve strength beyond theoretical yielding due to triaxiality, plasticity, large deflection effects, and load redistribution. Practical design criteria make use of this reserve strength, placing considerable demands on the notch toughness of joint can materials. Through joint classification (API) or an ovalizing parameter (AWS), they incorporate elements of general collapse as well as local failure. The resulting criteria may be compared against the supporting database of test results to ferret out bias and uncertainty as measures of structural reliability. Data for K, T/Y, and X joints in compression show a bias on the safe side of 1.35, beyond the nominal safety factor of 1.8, as shown in Figure 30.4. Tension joints appear to show a larger bias of 2.85; however, this reduces to 2.05 for joints over 0.12 in. thick and 1.22 for joints over 0.5 in., suggesting a thickness effect for tests that end in fracture.

For overload analysis of tubular structures (e.g., earthquake), we need not only ultimate strength but also the load–deflection behavior. Early tests showed ultimate deflections of 0.03 to 0.07 chord diameters, giving a typical ductility of 0.10 diameters for a brace with weak joints at both ends. As more different types of joints were tested, a wider variety of load–deflection behaviors emerged, making such generalizations tenuous.

Cyclic overload raises additional considerations. One issue is whether the joint will experience a ratcheting or progressive collapse failure or will achieve stable behavior with plasticity contained at local hotshots, a process called "shakedown" (as in shakedown cruise). Tubular connections have withstood 60 to several hundred repetitions of plastic strain at loads in excess of their nominal capacity. A conservative analytical treatment is to extend the hot spot S–N design curve into the low-cycle range as strain.

When tubular joints and members are incorporated into a space frame, the question arises as to whether computed bending moments are primary (i.e., necessary for structural stability, as in a sideway portal situation, and must be designed for) or secondary (i.e., an unwanted side effect of deflection, which may be safely ignored or reduced). When proportional loading is imposed, with both axial load and bending moment being maintained regardless of deflection, the joint simply fails when it reaches its failure envelope. However, when moments are due to imposed lateral deflection, and then axial load is imposed, the load path skirts along the failure envelope, shedding the moment and sustaining further increases in axial load.

Another area of interaction between joint behavior and frame action is the influence of brace bending or rotation on the strength of gap K connections. If rotation is prevented, bending moments develop that permit the gap region to transfer additional load. If the loads remain strictly axial, brace end rotation occurs in the absence of restraining moments, and a lower joint capacity is found. These problems arise for circular tubes as well as box connections, and a recent trend has been to conduct joint-in-frame tests to achieve a realistic balance between the two limiting conditions. Loads that maintain their original direction

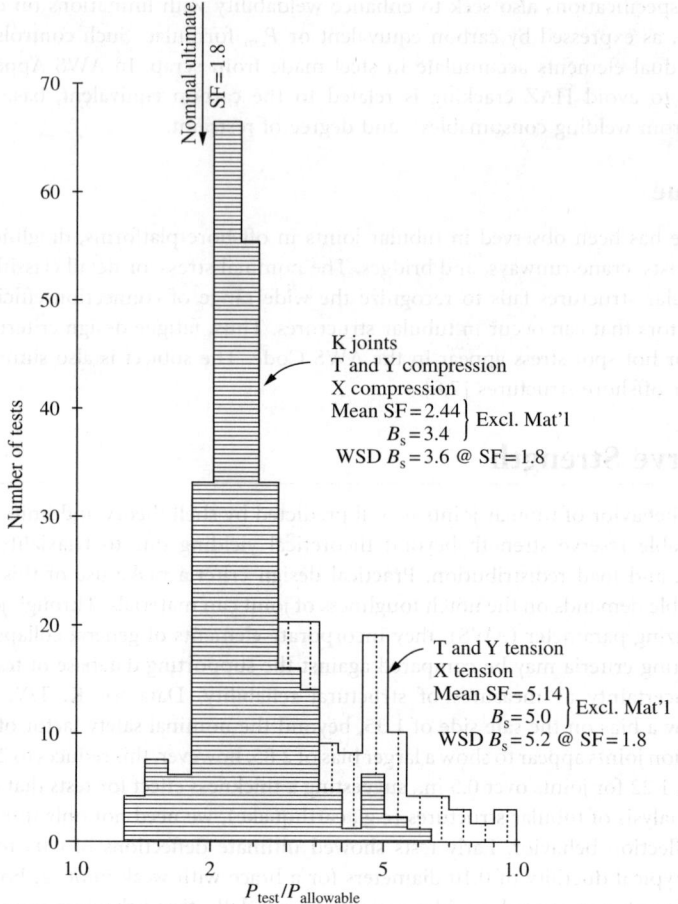

FIGURE 30.4 Comparison of AWS design criteria with the WRC database.

(as in an inelastic finite element analysis) or, worse yet, follow the deflection (as in testing arrangements with a two-hinge jack) result in a plastic instability of the compression brace stub as a column, which grossly understates the actual joint strength. Existing databases may need to be screened for this problem.

30.7 Empirical Formulations

Because of the foregoing reserve strength issues, AWS design criteria have been derived from a database of ultimate strength tubular joint tests. Comparison with the database (Figure 30.4) indicates a safety index of 3.6 against known static loads for the AWS punching shear criteria.

Safety index is the safety margin, including hidden bias, expressed in standard deviations of total uncertainty. Since these criteria are used to select the main member chord or joint can, the choice of safety index is similar to that used for sizing other structural members, rather than the higher safety margins used for workmanship-sensitive connection items such as welds or bolts.

When the ultimate axial load is used in the context of AISC-LRFD, with a resistance factor of 0.8, AWS ultimate strength is nominally equivalent to punching shear allowable stress design (ASD) for structures having 40% dead load and 60% live load. LRFD falls on the safe side of ASD for structures having a lower proportion of dead load. AISC criteria for tension and compression members appear to have made the equivalency trade-off at 25% dead load; thus, the LRFD criteria given by AWS would appear to be conservative for a larger part of the population of structures. In Canada, using these resistance factors with slightly different load factors, a 4.2% difference in overall safety factor results — within calibration accuracy [9].

30.8   Design Charts

Research, testing, and applications have progressed to the point where tubular connections are about as reliable as the other structural elements with which designers deal. One of the principal barriers to more widespread use seems to be unfamiliarity. To alleviate this problem, design charts for connections of circular tubes have been presented by Marshall [10]. Packer's original charts for square and rectangular tubes appear in the AISC Manual.

The capacity of simple, direct, welded, tubular connections are given in terms of punching shear efficiency, E_v, where

$$E_v = \frac{\text{allowable punching shear stress}}{\text{main member allowable tension stress}} \tag{30.3}$$

Charts for punching shear efficiency for axial load, in-plane bending, and out-of-plane bending appear as Figure 30.5 to Figure 30.9. Note that for axial load, separate charts are given for K connections, T/Y connections, and X connections, reflecting their different load patterns and different values of the ovalizing parameter (alpha). Within each connection or load type, punching shear efficiency is a function of the geometry parameters, diameter ratio (beta), and chord radius or thickness (gamma), as defined earlier. For K connections, the gap, g, between braces (of diameter d) is also significant, with the behavior reverting to that of T/Y connections for a very large gap. Punching shear efficiency cannot exceed a value of $0.4F_{yo}/0.6F_y = 0.67$, the material limit for shear.

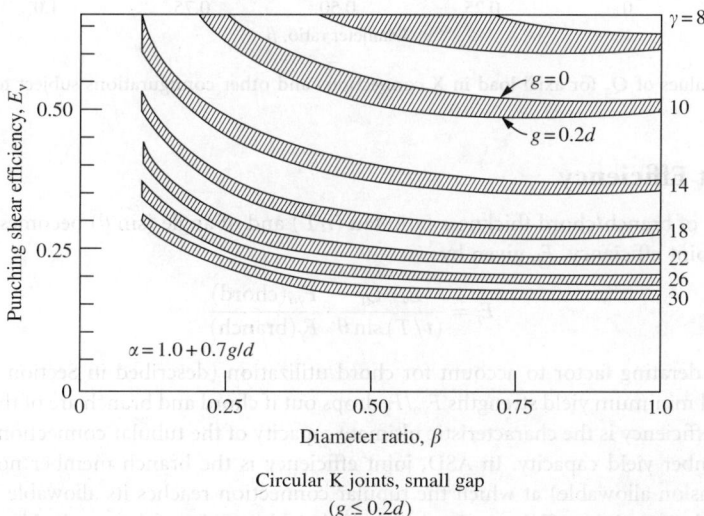

Circular K joints, small gap
$(g \le 0.2d)$

FIGURE 30.5   Values of Q_q for axial load in K connections.

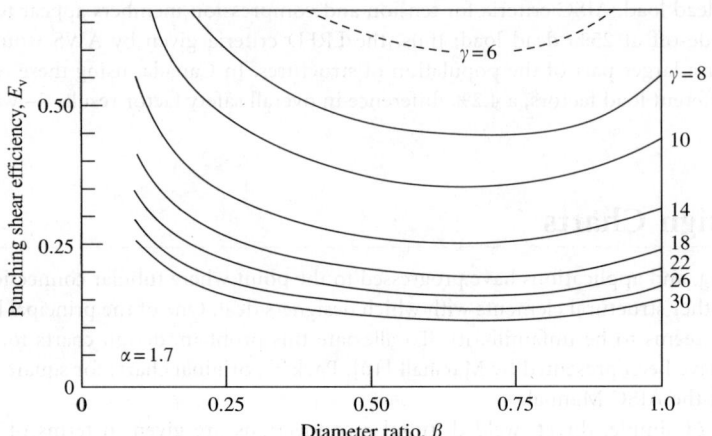

FIGURE 30.6   Values of Q_q for axial load in T and Y connections.

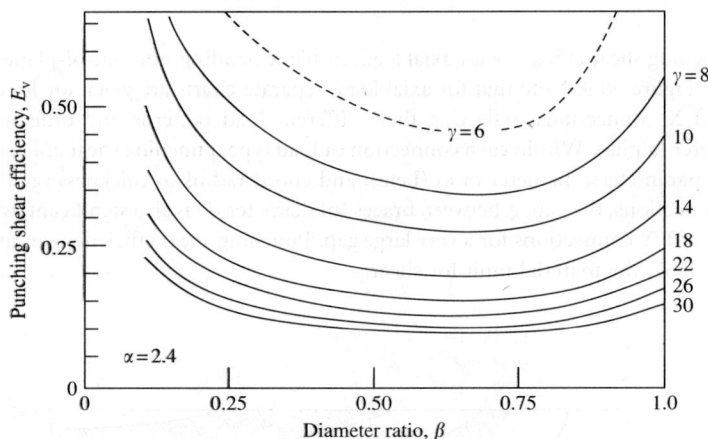

FIGURE 30.7   Values of Q_q for axial load in X connections and other configurations subject to crushing.

30.8.1 Joint Efficiency

The importance of branch/chord thickness ratio tau (t/T) and of angle $(\sin \theta)$ becomes apparent in the expression for joint efficiency, E_j, given by

$$E_j = \frac{E_v \cdot Q_f}{(t/T) \sin \theta} \cdot \frac{F_{yo}(\text{chord})}{F_y(\text{branch})} \qquad (30.4)$$

where Q_f is the derating factor to account for chord utilization (described in Section 30.8.2), and the ratio of specified minimum yield strengths F_{yo}/F_y drops out if chord and branch are of the same material. In LRFD, joint efficiency is the characteristic ultimate capacity of the tubular connection, as a fraction of the branch member yield capacity. In ASD, joint efficiency is the branch member nominal stress (as a fraction of tension allowable) at which the tubular connection reaches its allowable punching shear. Connections with 100% joint efficiency develop the full yield capacity of the attached branch member, in either design format.

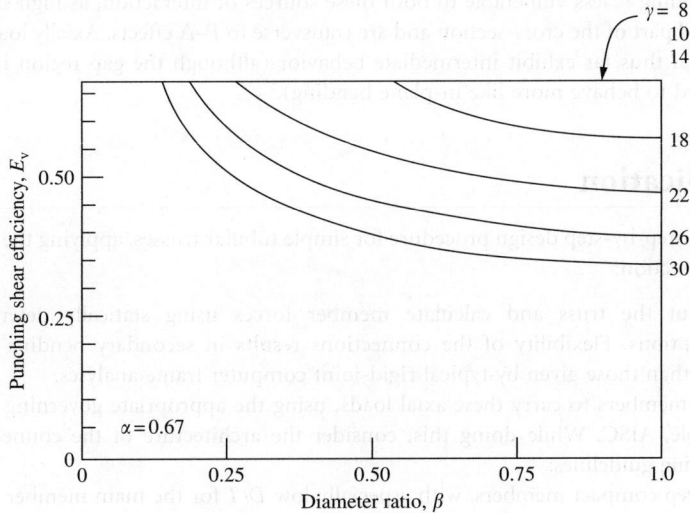

FIGURE 30.8 Values of Q_q for in-plane bending.

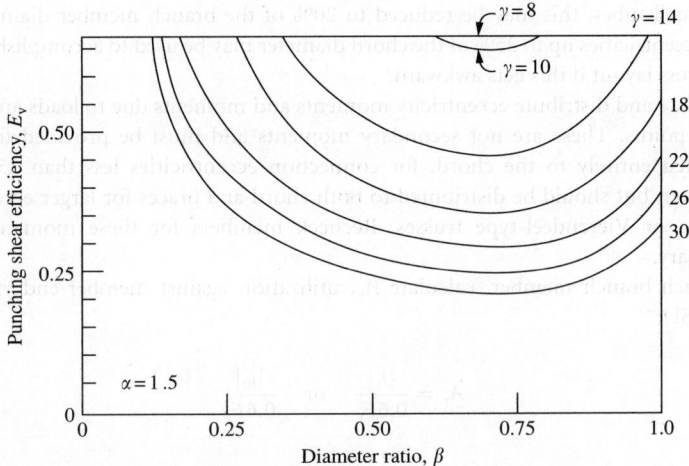

FIGURE 30.9 Values of Q_q for out-of-plane bending.

30.8.2 Derating Factor

In most structures, the main member (chord) at tubular connections must do double duty, carrying loads of its own (axial stress f_a and bending f_b) in addition to the localized loadings (punching shear) imposed by the branch members. Interaction between these two causes a reduction in the punching shear capacity, as reflected in the Q_f derating factor, shown in Figure 30.10.

In-plane bending experiences the most severe interaction, as localized shell bending stresses at the tubular intersection are in the same direction and directly additive to the chord's own nominal stresses over a large part of the cross-section. For chords with very high R/T (gamma) and high nominal compressive stresses, buckling tendencies further reduce the capacity for localized shell stresses.

Out-of-plane bending is less vulnerable to both these sources of interaction, as high shell stresses only occupy a localized part of the cross-section and are transverse to P–Δ effects. Axially loaded connections of the types tested thus far exhibit intermediate behavior (although the gap region in K connections might be expected to behave more like in-plane bending).

30.9 Application

What follows is a step-by-step design procedure for simple tubular trusses, applying the charts presented in the foregoing section:

Step 1. Lay out the truss and calculate member forces using statically determinate pin-end assumptions. Flexibility of the connections results in secondary bending moments being lower than those given by typical rigid-joint computer frame analyses.

Step 2. Select members to carry these axial loads, using the appropriate governing design code, for example, AISC. While doing this, consider the architecture of the connections along the following guidelines:

  A. Keep compact members, with especially low D/T for the main member (chord).

  B. Keep branch/main thickness ratio (tau) less than unity, preferably about 0.5.

  C. Select branch members to aim for large beta (branch/main diameter ratio), subject to avoidance of large eccentricity moments.

  D. In K connections, use a minimum gap of 2 in. between the braces for welding access. For small tubes, this may be reduced to 20% of the branch member diameter. Connection eccentricities up to 25% of the chord diameter may be used to accomplish this. Reconsider truss layout if this gets awkward.

Step 3. Calculate and distribute eccentricity moments and moments due to loads applied in-between panel points. These are not secondary moments and must be provided for. They may be allocated entirely to the chord, for connection eccentricities less than 25% of the chord diameter, but should be distributed to both chord and braces for larger eccentricities, portal frames, or Vierendeel-type trusses. Recheck members for these moments and resize as necessary.

Step 4. For each branch member, calculate A_y, utilization against member-end yield at the joint. For ASD:

$$A_y = \frac{|f_a|}{0.6F_y} \quad \text{or} \quad \frac{|f_b|}{0.6F_y} \tag{30.5}$$

where f_a is the nominal axial stress and f_b is the bending in the branch. Where used, the $\frac{1}{3}$ increase is applicable to the denominator.

Step 5. Also calculate chord utilization, using the formula in Figure 30.10 with chord nominal stresses and specified minimum yield strength. Use the appropriate chart in the figure to determine the derating factor Q_f. At heavily sheared gap K connections and at eccentric bearing shoes, it may (rarely) also be necessary to check beam shear in the main member and its interaction with other chord stresses, for example, using AISC criteria. For circular sections, the effective area for beam shear is half the gross area.

Step 6. For each end of each branch member, calculate the joint efficiency, E_j, using Equation 30.4 and the appropriate charts for punching shear efficiency, E_v. Joint efficiencies less than 0.5 are sometimes considered poor practice, rendering the structure vulnerable to incidental loads that the members could resist but not the weaker joints.

Step 7. For axial loading alone, or bending alone, the connection is satisfactory if member-end utilization is less than joint efficiency, that is, $A_y/E_j \leq 1.0$. For the general case, with

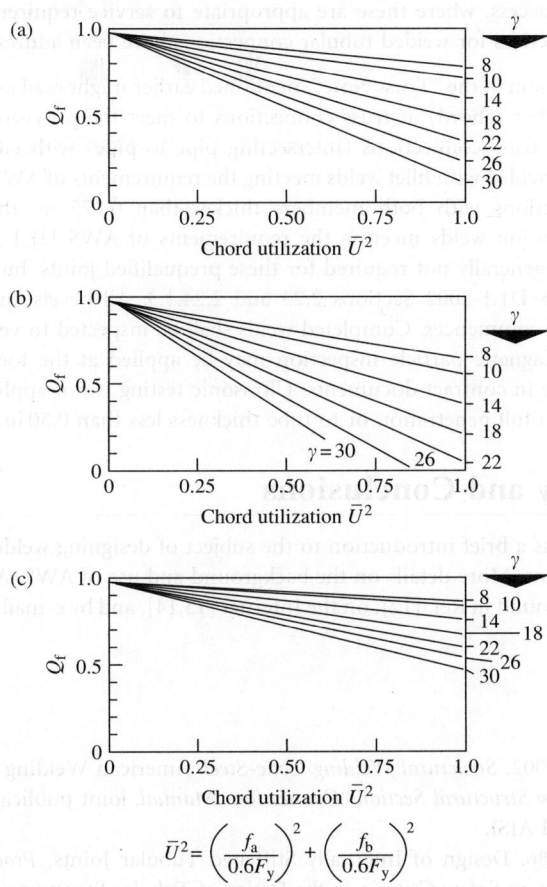

$$\bar{U}^2 = \left(\frac{f_a}{0.6F_y}\right)^2 + \left(\frac{f_b}{0.6F_y}\right)^2$$

FIGURE 30.10　Derating factor Q_q for (a) axial loads in branch, (b) in-plane bending, and (c) out-of-plane bending.

combinations of axial load and bending, the connection must satisfy the following interaction formula:

$$(A_y/E_j)^{1.75}_{\text{axial}} + (A_y/E_j)_{\text{bending}} \leq 1.0 \qquad (30.6)$$

Step 8.　To redesign an unsatisfactory connection, go back to Step 2 and do one of the following:

    A.　Increase the chord thickness.

    B.　Increase the branch diameter.

    C.　Both of the above.

    Consider overlapped connections (AWS Section 2.24.1.6) or stiffened connections only as a last resort. Overlapped connections increase the complexity of fabrication but can result in substantial reductions in the required chord wall thickness.

Step 9.　When the designer thinks he is done, he should talk to potential fabricators and erectors. Their feedback could be valuable for avoiding unnecessary difficult and expensive construction headaches. Also, make sure they are familiar with and prepared to follow AWS Code requirements for special welder qualifications and that they are capable of coping the brace ends with sufficient accuracy to apply AWS prequalified procedures. Considerable savings can be realized by specifying partial joint penetration welds for tubular T, Y, and K connections with no root access, where these are appropriate to service requirements. Fabrication and inspection practices for welded tubular connections have been addressed by Post [11].

A more satisfactory version of the "Truss Note" mentioned earlier might read as follows: "The designer has sized the main member (chord) at truss connections to meet the provisions of AWS D1.1-2002 Section 2.24. All welded truss connections (intersecting pipe-to-pipe) with either member less than 0.375 in. thick shall be provided with fillet welds meeting the requirements of AWS D1.1-2002 Figure 3.2. All welded truss connections with both members thicker than 0.375 in. thick shall be provided with partial joint penetration welds meeting the requirements of AWS D1.1-2002 Figure 3.5. Weld strength calculations are generally not required for these prequalified joints, but (if provided) shall be in accordance with AWS D1.1-2002 Sections 2.23 and 2.24.1.3. All bevels and their fit-up shall be inspected before welding commences. Completed welds shall be inspected to verify the full leg lengths as tabulated in AWS. Magnetic particle inspection may be applied at the toes of completed welds, where specified elsewhere in contract documents. Ultrasonic testing is not applicable to welds that are designated to be less than full penetration or to tube thickness less than 0.50 in."

30.10 Summary and Conclusions

This chapter has served as a brief introduction to the subject of designing welded tubular connections for circular hollow sections. More details on the background and use of AWS, AISC, and international codes in this area can be found in Ref. [12], on the Internet [13,14], and by e-mail to the source 4 of these criteria [15].

References

[1] AWS D1.1-2002. 2002. *Structural Welding Code-Steel*, American Welding Society, Miami, FL.
[2] AISC. 1997. *Hollow Structural Sections Connections Manual*, joint publication of AISC, the Steel Tube Institute, and AISI.
[3] Marshall, P.W. 1986. Design of Internally Stiffened Tubular Joints, *Proceedings IIW/AIJ International Conference on Safety Criteria in the Design of Tubular Structures*, Tokyo.
[4] Marshall, P.W. and Luyties, W.H. 1982. Allowable Stresses for Fatigue Design, *Proceedings of the International Conference on Behavior of Off-Shore Structures*, BOSS-82 at MIT, McGraw-Hill, New York.
[5] Kurobane, Y. 1995. Comparison of AWS vs. International Criteria, ASCE Structures Congress, Atlanta, GA.
[6] Marshall, P.W. 1990. Advanced Fracture Control Procedures for Deepwater Offshore Towers, *Weld. J.* vol. 69, no. 5.
[7] Marshall, P.W. 1993. API Provisions for Scf. S-N and Size-Profile Effects, *Proceedings of the Offshore Technical Conference*, OTC 7155, Houston, TX.
[8] Marshall, P.W. 1996. Offshore Tubular Structures, *Proceedings of the AWS International Conference on Tubular Structures*, Vancouver.
[9] Packer, J.A. et al. 1984. Canadian Implementation of CIDECT Monograph 6, IIW Doc. XV-E-84-072.
[10] Marshall, P.W. 1989. Designing Tubular Connections with AWS D1.1, *Weld. J.* vol. 68, no. 3.
[11] Post, J.W. 1996. Fabrication and Inspection Practices for Welded Tubular Connections, *Proceedings of the AWS International Conference on Tubular Structures*, Vancouver.
[12] Marshall, P.W. 1992. *Design of Welded Tubular Connections: Basis and Use of AWS D1.1*, Elsevier Science Publishers, Amsterdam.
[13] Comite Internationale de l'Etude et Construction Tubulaire, http://www.cidect/org/
[14] Steel Tube Institute of North America, http://www.steeltubeinstitute.org/
[15] Moonshine Hill Proprietary, mhpsyseng@aol.com

31

Effective Length Factors of Compression Members

Lian Duan
*Division of Engineering Services,
California Department of
Transportation,
Sacramento, CA*

Wai-Fah Chen
*College of Engineering,
University of Hawaii at Manoa,
Honolulu, HI*

0-8493-1569-7/05/$0.00+$1.50
© 2005 by CRC Press

31.1 Introduction

The concept of the *effective length factor* has been well established and widely used by practicing engineers and plays an important role in compression member and column design. The essence of the concept is to estimate the interaction effects of the whole frame on an individual compression member. In the development of design interaction equations for beam–columns, much discussion has been focused on the need and validity of using the *effective length factor K* in the equations (Cheong-Siat-Moy 1986; Liew et al. 1991; ASCE 1997; White and Clarke 1997a,b; Schmidt 1999). Although attempts were made to formulate the general interaction equations without *K* factors, it was found that this was almost impossible if the interaction equations were to be versatile enough for a wide range of slenderness ratios and load combinations (Liew et al. 1991). It is well known that the effective length factor approach introduces inaccuracies into the design process; the simplicity of the approach, however, is likely to still make the approach an important part of compression member design in the foreseeable future (Hellesland and Bjorhovde 1996). The most structural design codes, standards, and specifications worldwide have provisions concerning the effective length factor.

The aim of this chapter is to present a state-of-the-art engineering practice of the effective length factor for the design of compression members and columns. In the first part of this chapter, the basic concept of the effective length factor is discussed. Then, the design implementation for individual columns, framed columns, crossing bracing systems, latticed members, tapered columns, crane columns, gable frames, columns in fire, space frames, truss-type highway sign support structures, precast concrete skeletal frames, and steel moment frames is presented. The determination of whether a frame is braced or unbraced is also addressed. Several detailed examples are given to illustrate the determination of the effective length factor for different cases of engineering applications.

31.2 Basic Concept

Mathematically, the effective length factor or the *elastic K-factor* is defined as

$$K = \sqrt{\frac{P_e}{P_{cr}}} = \sqrt{\frac{\pi^2 EI}{L^2 \, P_{cr}}} \tag{31.1}$$

where P_e is the Euler load, the elastic buckling load of a pin-ended compression member or column, P_{cr} is the elastic buckling load of an end-restrained compression member or column, E is the modulus of elasticity of the material, I is the moment of inertia in the flexural buckling plane, and L is the unsupported length of the compression member or column.

Physically, the *K*-factor is a factor that when multiplied by the actual length of the end-restrained column (Figure 31.1a) gives the length of an equivalent pin-ended column (Figure 31.1b) whose buckling load is the same as that of the end-restrained column. It follows that effective length *KL* of an end-restrained column is the length between the adjacent inflection points of its pure flexural buckling shape.

Practically, the design specifications usually provide the resistance equations for pin-ended columns, while the resistance of framed columns can be estimated through the *K*-factor to the pin-ended column strength equation. Theoretical *K*-factor is determined from an elastic eigenvalue analysis of the entire structural system, while practical methods for the *K*-factor are based on an elastic eigenvalue analysis of selected subassemblages. The effective length concept is the only tool currently available for the design of compression members in engineering structures and it is an essential part of analysis and design procedures.

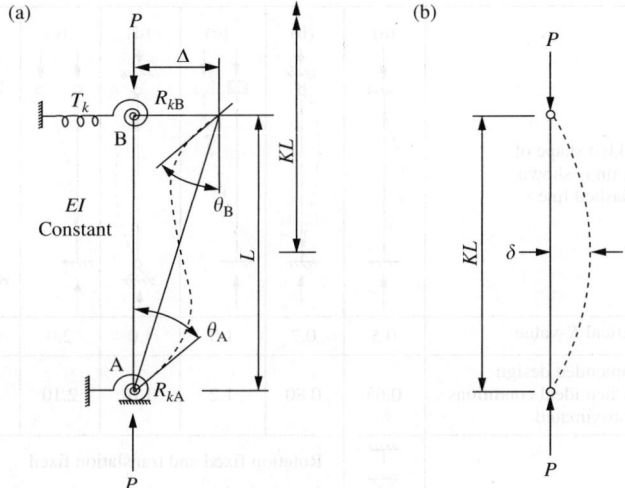

FIGURE 31.1   Individual column: (a) end-restrained columns and (b) pin-ended column.

31.3   Individual Columns

From an eigenvalue analysis, the general K-factor equation of an individual column as shown in Figure 31.1 is obtained as

$$\det \begin{vmatrix} C + \dfrac{R_{kA}L}{EI} & S & -(C+S) \\[2mm] S & C + \dfrac{R_{kB}L}{EI} & -(C+S) \\[2mm] -(C+S) & -(C+S) & 2(C+S) - \left(\dfrac{\pi}{K}\right)^2 + \dfrac{T_k L^3}{EI} \end{vmatrix} = 0 \qquad (31.2)$$

where the stability functions C and S are defined as

$$C = \frac{(\pi/K)\sin(\pi/K) - (\pi/K)^2 \cos(\pi/K)}{2 - 2\cos(\pi/K) - (\pi/K)\sin(\pi/K)} \qquad (31.3)$$

$$S = \frac{(\pi/K)^2 - (\pi/K)\sin(\pi/K)}{2 - 2\cos(\pi/K) - (\pi/K)\sin(\pi/K)} \qquad (31.4)$$

The largest value of K that satisfies Equation 31.2 gives the elastic buckling load of an end-restrained column.

Figure 31.2 (AISC 1989, 1999; AASHTO 2004) summarizes the theoretical K-factors for end-restrained columns with some idealized end conditions. The recommended K-factors are also shown in Figure 31.2 for practical design applications. Since actual column conditions seldom comply fully with idealized end conditions used in buckling analysis, the recommended K-factors are always equal or greater than their theoretical counterparts.

31.4   Framed Columns — Alignment Chart Method

In theory, the effective length factor, K, for any columns in a framed structure can be determined from a stability analysis of the entire structural analysis — eigenvalue analysis. Methods available for stability

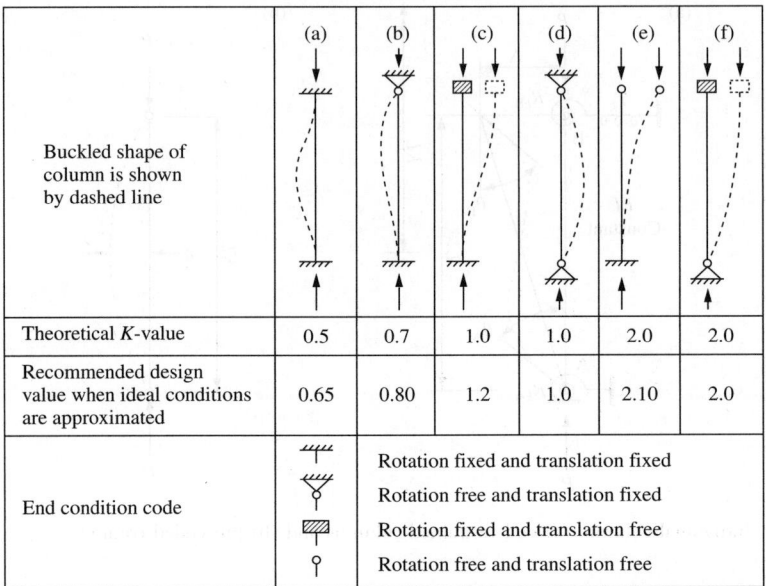

	(a)	(b)	(c)	(d)	(e)	(f)
Buckled shape of column is shown by dashed line						
Theoretical *K*-value	0.5	0.7	1.0	1.0	2.0	2.0
Recommended design value when ideal conditions are approximated	0.65	0.80	1.2	1.0	2.10	2.0
End condition code		Rotation fixed and translation fixed				
		Rotation free and translation fixed				
		Rotation fixed and translation free				
		Rotation free and translation free				

FIGURE 31.2 Theoretical and recommended *K*-factors for individual columns with idealized end conditions (AISC 1999).

analysis include the slope deflection method (Winter et al. 1948; Galambos 1968; Chen and Lui 1991), the three-moment equation method (Bleich 1952), and energy methods (Johnson 1960). In practice, however, such an analysis is not practical, and simple models are often used to determine the effective length factors for framed columns (Kavanagh 1962; Lu 1962; Gurfinkel and Robinson 1965; Wood 1974). One such practical procedure that provides an approximate value of the elastic *K*-factor is the alignment chart method (Julian and Lawrence 1959). This procedure has been adopted by the AISC (1989, 1999), ACI (2002), AASHTO (2004), and CSA (1994) *Specifications*, among others. At present, most engineers use the alignment chart method in lieu of an actual stability analysis.

31.4.1 Alignment Chart Method

The structural models employed for determination of *K*-factor for framed columns in the alignment chart method are shown in Figure 31.3. The assumptions used in these models are (Chen and Lui 1991; AISC 1999)

1. All members have constant cross-section and behave elastically.
2. Axial forces in the girders are negligible.
3. All joints are rigid.
4. For braced frames, the rotations at the near and far ends of the girders are equal in magnitude and opposite in direction (i.e., girders are bent in single curvature).
5. For unbraced frames, the rotations at the near and far ends of the girders are equal in magnitude and direction (i.e., girders are bent in double curvature).
6. The stiffness parameters, $L\sqrt{P/EI}$, of all columns are equal.
7. All columns buckle simultaneously.
8. Joint restraint is distributed to the column above and below the joint in proportion to I/L of the two columns.

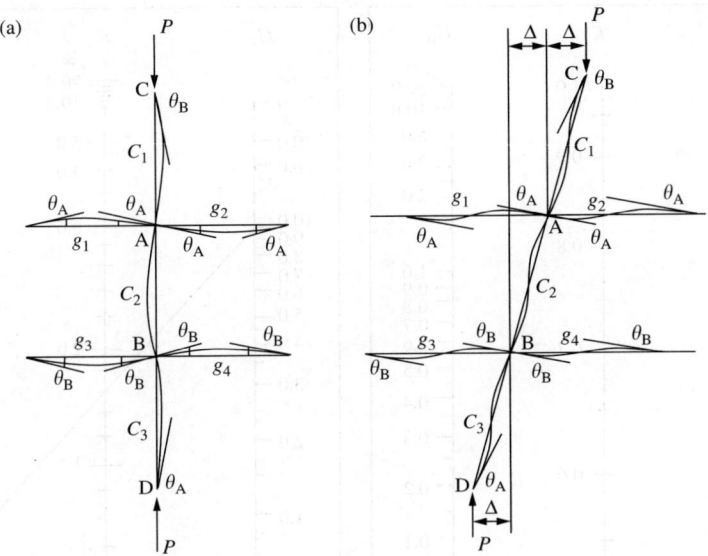

FIGURE 31.3 Subassemblage models for *K*-factors of framed columns: (a) braced frames and (b) unbraced frames.

Using the slope deflection equation method and stability functions, the effective length factor equations of framed columns are obtained as

For columns in braced frames:

$$\frac{G_A G_B}{4}(\pi/K)^2 + \left(\frac{G_A + G_B}{2}\right)\left(1 - \frac{\pi/K}{\tan(\pi/K)}\right) + \frac{2\tan(\pi/2K)}{\pi/K} - 1 = 0 \qquad (31.5)$$

For columns in unbraced frames:

$$\frac{G_A G_B (\pi/K)^2 - 36^2}{6(G_A + G_B)} - \frac{\pi/K}{\tan(\pi/K)} = 0 \qquad (31.6)$$

where G_A and G_B are stiffness ratios of columns and girders at two end joints A and B of the column section being considered, respectively. They are defined by

$$G_A = \frac{\sum_A (E_c I_c / L_c)}{\sum_A (E_g I_g / L_g)} \qquad (31.7)$$

$$G_B = \frac{\sum_B (E_c I_c / L_c)}{\sum_B (E_g I_g / L_g)} \qquad (31.8)$$

where $\sum$ indicates the summation of all members rigidly connected to the joint and lying in the plane in which buckling of column is being considered, and subscripts c and g represent columns and girders, respectively.

Equations 31.5 and 31.6 can be expressed in the form of alignment charts as shown in Figure 31.4. It is noted that for columns in braced frames, the range of *K* is $0.5 \le K \le 1.0$; for columns in unbraced frames, the range is $1.0 \le K \le \infty$. For column ends supported by, but not rigidly connected to, a footing or foundations, *G* is theoretically infinity, but unless actually designed as a true friction-free pin, it may be taken as 10 for practical design. If the column end is rigidly attached to a properly designed footing, *G* may be taken as 1.0 (AISC 1999).

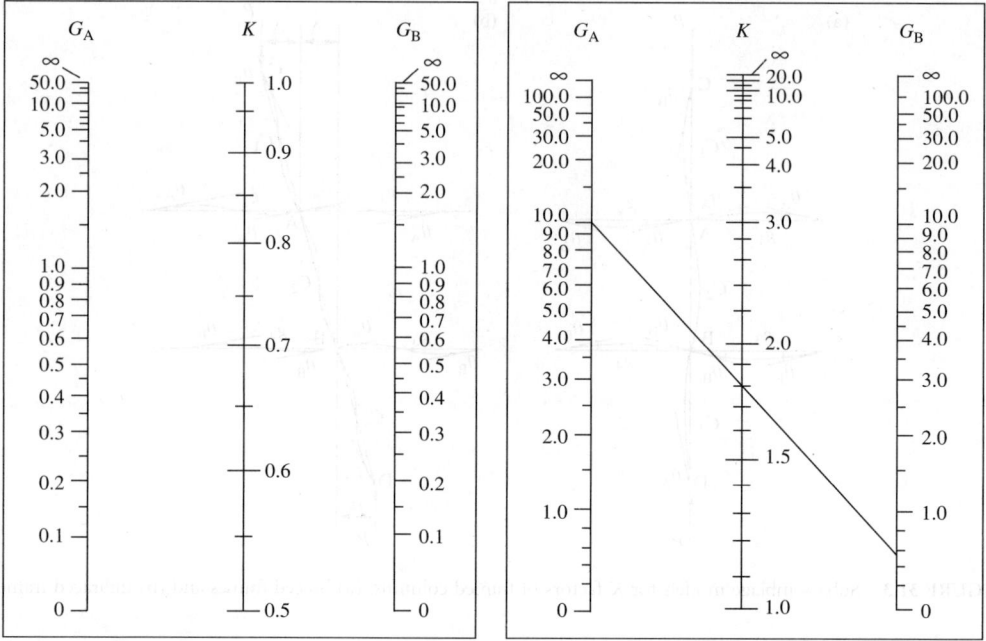

FIGURE 31.4 Alignment charts for effective length factors of framed columns: (a) braced frames and (b) unbraced frames.

EXAMPLE 31.1

Given: A two-story steel frame is shown in Figure 31.5. Using the alignment chart, determine the K-factor for the elastic column DE. $E = 29,000$ ksi (200 GPa) and $F_y = 36$ ksi (248 MPa).

Solution

(1) For the given frame, section properties are

Members	Section	I_x (in.⁴, mm⁴ × 10⁸)	L (in., mm)	I_x/L (in.³, mm³)
AB and GH	W 10 × 22	118, 0.49	180, 4572	0.656, 10,750
BC and HI	W 10 × 22	118, 0.49	144, 3658	0.819, 13,412
DE	W 10 × 45	248, 1.03	180, 4572	1.378, 22,581
EF	W 10 × 45	248, 1.03	144, 3658	1.722, 28,219
BE	W 18 × 50	800, 3.33	300, 7620	2.667, 43,704
EH	W 18 × 86	1530, 6.37	360, 9144	4.250, 69,645
CF	W 16 × 40	518, 2.16	300, 7620	1.727, 28,300
FI	W 16 × 67	954, 3.97	360, 9144	2.650, 43,426

(2) Calculate G-factor for column DE:

$$G_{\mathrm{E}} = \frac{\sum_{\mathrm{E}}(E_c I_c/L_c)}{\sum_{\mathrm{E}}(E_g I_g/L_g)} = \frac{1.378 + 1.722}{2.667 + 4.250} = 0.448$$

$$G_{\mathrm{D}} = 10 \text{ (AISC 1999)}$$

(3) From the alignment chart in Figure 31.4b, $K = 1.8$ is obtained.

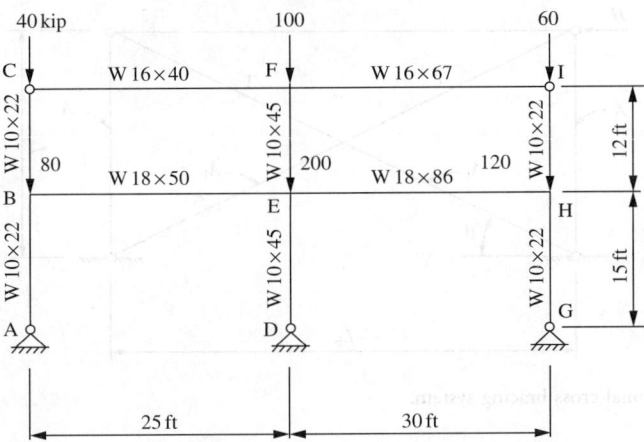

FIGURE 31.5 An unbraced two-story frame.

31.4.2 Requirements for Braced Frames

In stability design, one of the major decisions engineers have to make is the determination of whether a frame is braced or unbraced. In actual structures, a completely braced frame seldom exists. But in practice, some structures can be analyzed as braced frames as long as the lateral stiffness provided by bracing systems such as diagonal bracing, shear walls, or equivalent means is large enough. The following brief discussion may provide engineers with the tools to make engineering decisions regarding the basic requirements for a braced frame.

31.4.2.1 Lateral Stiffness Requirement

Galambos (1964) presented a simple conservative procedure to evaluate the minimum lateral stiffness provided by a bracing system so that the frame is considered braced:

$$\text{Required lateral stiffness} \quad T_k = \frac{\sum P_n}{L_c} \tag{31.9}$$

where $\sum$ represents the summation of all columns in one story, P_n is the nominal axial compression strength of the column using the effective length factor $K = 1$, and L_c is the unsupported length of the column.

31.4.2.2 Bracing Size Requirement

Galambos (1964) applied Equation 31.9 to a diagonal bracing (Figure 31.6) and obtained the minimum requirements of diagonal bracing for a braced frame as

$$A_b = \frac{\left[1 + (L_b/L_c)^2\right]^{3/2} \sum P_n}{(L_b/L_c)^2 E} \tag{31.10}$$

where A_b is the cross-sectional area of diagonal bracing and L_b is the span length of the beam.

A recent study by Aristizabal-Ochoa (1994a) indicates that the size of diagonal bracing required for a totally braced frame is about 4.9 and 5.1% of the column cross-section for "rigid frame" and "simple framing," respectively, and increases with the moment of inertia of the column, the beam span, and the beam to column span ratio L_b/L_c.

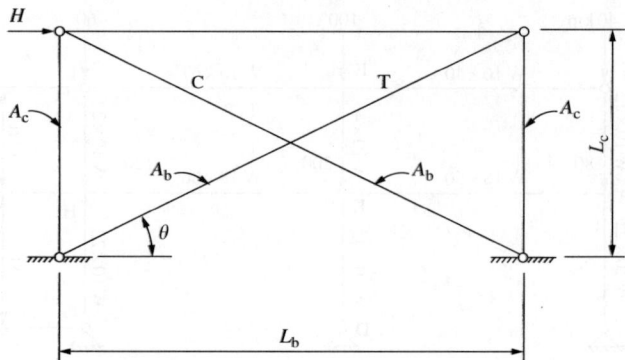

FIGURE 31.6 Diagonal cross bracing system.

31.4.2.3 Threshold Lateral Stiffness of Braces

For a multistory braced frame, Tong and Shi (2001) defined the threshold lateral stiffness as the bracing stiffness at which the critical buckling load in sway buckling mode is equal to the nonsway buckling mode. They found out that most threshold stiffness values are from π^2 to $4\pi^2$ and proposed the following approximate formula (Tong and Shi 2001):

$$T_k = G_{\text{TH}} \left(\frac{EI}{L^3} \right)_{\text{c}} \tag{31.11}$$

$$G_{\text{TH}} = \frac{2}{3} \left(\frac{\pi^2}{K_b^2} - \frac{\pi^2}{K_{\text{ub}}} \right) \left[\left(\frac{G_A}{G_B} \right)^{0.18} + \left(\frac{G_B}{G_A} \right)^{0.18} \right] \le 60 \tag{31.12}$$

where K_b and K_{ub} are the effective length factors for braced and unbraced frames, respectively.

31.4.2.4 Stability Index

ACI 318-02 (ACI 2002) permits us to assume a column in a structure is braced if the increase in the lateral load moment due to second-order (P–Δ) effects does not exceed 5% of the first-order moment (MacGregor and Hage 1977). The American Concrete Institute (ACI) also provides an alternative method based on the stability index for a story. The story within a structure can be assumed as braced if the stability index

$$Q = \frac{\sum P_u \Delta_0}{V_u L_c} \le 0.05 \tag{31.13}$$

where $\sum P_u$ and V_u are the total vertical load and the story shear, respectively, in the story in question; Δ_0 is the first-order relative deflection between the top and bottom of that story due to V_u; and L_c is the length of the compression member in a frame, measured from center to center of the joints in the frame.

Menon (2001) used the "fuzzy logic" concept and proposed that a frame is "adequately braced" if its stability index Q is less than 0.02, unbraced if Q exceeds 0.06 and "partially braced" if Q lies in the intermediate domain {0.02, 0.06}, and the K-factor can be estimated by the following equations:

$$K = K_{\text{ub}} - q_x(K_{\text{ub}} - K_b) \tag{31.14}$$

$$q_x = \begin{cases} 1.0 & \text{for } Q \le 0.02 \\ 1.0 - 0.5(Q/0.02 - 1.0)^2 & \text{for } 0.02 < Q \le 0.04 \\ 0.5(3.0 - Q/0.02)^2 & \text{for } 0.04 < Q \le 0.06 \\ 0.0 & \text{for } Q > 0.06 \end{cases} \tag{31.15}$$

where K_{ub} and K_b are the effective length factors for unbraced and braced frames, respectively.

31.4.2.5 AISC Load and Resistance Factor Design Requirements

Based on the concept of dual criterion for bracing design, strength, and stiffness proposed by Winter (1958, 1960), AISC (1999) provides the stability bracing requirements for braced frames, in which the effective length factor K for compression members is taken as unity, unless structural analysis shows that a smaller value may be used.

For frames:

The required story or panel bracing shear force is

$$P_{br} = 0.004 \sum P_u \tag{31.16}$$

The required story or panel shear stiffness is

$$B_{br} = \frac{2 \sum P_u}{\phi L} \tag{31.17}$$

where ϕ is equal to 0.75, $\sum P_u$ is the summation of the factored column axial loads in the story or panel supported by the bracing, and L is the story height or panel spacing.

For individual columns:

1. For relative bracing

    The required brace strength is

$$P_{br} = 0.004 P_u \tag{31.18}$$

   The required brace stiffness is

$$B_{br} = \frac{2 P_u}{\phi L_b} \tag{31.19}$$

2. For nodal bracing

    The required brace strength is

$$P_{br} = 0.01 P_u \tag{31.20}$$

   The required brace stiffness is

$$B_{br} = \frac{8 P_u}{\phi L_b} \tag{31.21}$$

where ϕ is equal to 0.75, P_u is the factored column axial load, and L_b is the distance between braces. When the actual spacing of braced points is less than the maximum unbraced length for the required column force with K is equal to 1, then L_b in Equations 31.19 and 31.21 is permitted to be taken equal to L_q. For a more detailed discussion, see Yura (1995).

31.4.3 Simplified Equations to Alignment Charts

A graphical alignment chart determination of the K-factor is easy to perform, while solving the chart Equations 31.5 and 31.6 always involves iteration. With today's personal computer, especially spreadsheet applications in design offices, the following simplified equations may serve as the equivalent of the alignment charts.

31.4.3.1 ACI 318-02 Equations

The ACI Building Code (ACI 318-02) recommends the use of alignment charts as the primary design aid for estimating K-factors, with the following two sets of simplified K-factor equations as an alternative.

For braced frames (Cranston 1972)

$$K = 0.7 + 0.05(G_A + G_B) \leq 1.0 \tag{31.22}$$

$$K = 0.85 + 0.05 G_{min} \leq 1.0 \tag{31.23}$$

The smaller of the above two expressions provides an upper bound to the effective length factor for braced compression members.

For unbraced frames (Furlong 1971)

$$K = \frac{20 - G_m}{20} \sqrt{1 + G_m} \quad \text{for } G_m < 2 \tag{31.24}$$

$$K = 0.9\sqrt{1 + G_m} \quad \text{for } G_m \geq 2 \tag{31.25}$$

For columns hinged at one end

$$K = 2.0 + 0.3G \tag{31.26}$$

where G_m is the average of the G values at the two ends of columns.

It is found that ACI simplified Equations 31.22 to 31.26 estimate very conservative K-factors when the difference between the relative stiffness ratios at the two ends of an unbraced column becomes larger and when end restraints of a braced column are large. In general, they may not lead to an economical design (Hu et al. 1993).

31.4.3.2 Duan–King–Chen Equations

To achieve both accuracy and simplicity for design purpose, the following alternative K-factor equations were proposed by Duan et al. (1993).

For braced frames

$$K = 1 - \frac{1}{5 + 9G_A} - \frac{1}{5 + 9G_B} - \frac{1}{10 + G_A G_B} \tag{31.27}$$

For unbraced frames

$$K = 4 - \frac{1}{1 + 0.2G_A} - \frac{1}{1 + 0.2G_B} - \frac{1}{1 + 0.01G_A G_B} \quad \text{for } K < 2 \tag{31.28}$$

$$K = \frac{2\pi a}{0.9 + \sqrt{0.81 + 4ab}} \quad \text{for } K \geq 2 \tag{31.29}$$

where

$$a = \frac{G_A G_B}{G_A + G_B} + 3 \tag{31.30}$$

$$b = \frac{36}{G_A + G_B} + 6 \tag{31.31}$$

Equation 31.28 shall be used to first calculate K. If the value of K calculated by Equation 31.28 is greater than 2, Equation 31.29 shall then be used.

31.4.3.3 French Equations

For braced frames

$$K = \frac{3G_A G_B + 1.4(G_A + G_B) + 0.64}{3G_A G_B + 2.0(G_A + G_B) + 1.28} \tag{31.32}$$

For unbraced frames

$$K = \sqrt{\frac{1.6G_A G_B + 4.0(G_A + G_B) + 7.5}{G_A + G_B + 7.5}} \tag{31.33}$$

Equations 31.32 and 31.33 first appeared in the *French Rules for the Design of Steel Structure* in 1966 (CM 1966) and were later incorporated into the *European Recommendation for Steel Construction* in 1978 (ECCS 1978). Neither French CM 66 Rules nor the European Recommendations give the origin of these two formulas. They provide a good approximation to the alignment charts (Dumonteil 1992). For braced frames, Equation 31.32 underestimates K by not more than 0.5% and overestimates it by 1.5%. For unbraced frames, Equation 31.33 approximates K within 2% (Dumonteil 1999).

31.4.3.4 Donnell Equation for Braced Frames

The Donnell equation, developed in 1949 as reported by Rondal (1988), has the following format:

$$K = \sqrt{\frac{G_A G_B + 0.43(G_A + G_B) + 0.17}{G_A G_B + 0.86(G_A + G_B) + 0.68}} \tag{31.34}$$

This is the formula presented by Dumonteil and Valley (1995) in their discussion paper. Its accuracy ranges from 0.4 to 1.3%.

31.4.3.5 Newmark Equation for Braced Frames

Introducing the *G*-factor, the Newmark (1949) equation has the following form:

$$K = \sqrt{\frac{(G_A + 4/\pi^2)(G_B + 4/\pi^2)}{(G_A + 8/\pi^2)(G_B + 8/\pi^2)}} \tag{31.35}$$

While the accuracy of Equation 31.35 is remarkable, it could still be slightly improved by replacing the term $4/\pi^2$ with 0.41, and the Newmark equation becomes (Dumonteil 1999)

$$K = \sqrt{\frac{(G_A + 0.41)(G_B + 0.41)}{(G_A + 0.82)(G_B + 0.82)}} \tag{31.36}$$

Equation 31.36 underestimates *K* by no more than 0.1% and overestimates it by less than 1.5%.

31.5 Modifications to Alignment Charts

In using the alignment charts in Figure 31.4 and Equations 31.5 and 31.6, engineers must always be aware of the assumptions used in the development of these charts. When actual structural conditions differ from these assumptions, unrealistic design may result (Johnston 1976; Liew et al. 1991; Hajjar and White 1994; AISC 1999; Dafedar et al. 2001). The Structural Stability Research Council (SSRC) Guide (Johnston 1976) provides methods enabling engineers to make simple modifications of the charts for some special conditions, such as unsymmetrical frames, column base conditions, girder far end conditions, and flexible conditions. Procedures that can be used to account for far ends of restraining columns being hinged or fixed were proposed by Duan and Chen (1988, 1989, 1996) and Essa (1997). Consideration of effects of material inelasticity on the *K*-factor was developed originally by Yura (1971) and expanded by Disque (1973). LeMessurier (1977) presented an overview of unbraced frames with or without leaning columns. An approximate procedure was also suggested by AISC (1999). Special attention should also be paid to calculation of the proper *G*-values (Barakat and Chen 1991; Duan 1996; Kishi et al. 1997) when partially restrained (PR) connections are used in frames. Several commonly used modifications are summarized in this section.

31.5.1 Different Restraining Girder End Conditions

When the end conditions of restraining girders are not rigidly joined to columns, the girder stiffness (I_g/L_g) used in the calculation of G_A and G_B in Equations 31.7 and 31.8 should be multiplied by a modification factor α_k as

$$G = \frac{\sum(E_c I_c/L_c)}{\sum \alpha_k(E_g I_g/L_g)} \tag{31.37}$$

where the modification factor α_k developed by Duan (1996) for braced frames and proposed by Kishi et al. (1997) for unbraced frames are given in Table 31.1 and Table 31.2, respectively. In these tables, R_{kN} and R_{kF} are elastic spring constants at the near and far ends of a restraining girder, respectively.

TABLE 31.1   Modification Factor α_k for Braced Frames with Semirigid Connections (Duan 1996)

End conditions of restraining girder		α_k
Near end	Far end	
Rigid	Rigid	1.0
Rigid	Hinged	1.5
Rigid	Semirigid	$\left(1+\dfrac{6E_gI_g}{L_gR_{kF}}\right)\Big/\left(1+\dfrac{4E_gI_g}{L_gR_{kF}}\right)$
Rigid	Fixed	2.0
Semirigid	Rigid	$1\Big/\left(1+\dfrac{4E_gI_g}{L_gR_{kN}}\right)$
Semirigid	Hinged	$1.5\Big/\left(1+\dfrac{3E_gI_g}{L_gR_{kN}}\right)$
Semirigid	Semirigid	$\left(1+\dfrac{6E_gI_g}{L_gR_{kF}}\right)\Big/R^*$
Semirigid	Fixed	$2\Big/\left(1+\dfrac{4E_gI_g}{L_gR_{kN}}\right)$

$$R^* = \left(1+\frac{4E_gI_g}{L_gR_{kN}}\right)\left(1+\frac{4E_gI_g}{L_gR_{kF}}\right) - \left(\frac{E_gI_g}{L_g}\right)^2\frac{4}{R_{kN}R_{kF}}$$

TABLE 31.2   Modification Factor α_k for Unbraced Frames with Semirigid Connections (Kishi, Chen and Goto, 1997)

End conditions of restraining girder		α_k
Near end	Far end	
Rigid	Rigid	1
Rigid	Hinged	0.5
Rigid	Semirigid	$\left(1+\dfrac{2E_gI_g}{L_gR_{kF}}\right)\Big/\left(1+\dfrac{4E_gI_g}{L_gR_{kF}}\right)$
Rigid	Fixed	2/3
Semirigid	Rigid	$1\Big/\left(1+\dfrac{4E_gI_g}{L_gR_{kN}}\right)$
Semirigid	Hinged	$0.5\Big/\left(1+\dfrac{3E_gI_g}{L_gR_{kN}}\right)$
Semirigid	Semirigid	$\left(1+\dfrac{2E_gI_g}{L_gR_{kF}}\right)\Big/R^*$
Semirigid	Fixed	$\left(\dfrac{2}{3}\right)\Big/\left(1+\dfrac{4E_gI_g}{L_gR_{kN}}\right)$

$$R^* = \left(1+\frac{4E_gI_g}{L_gR_{kN}}\right)\left(1+\frac{4E_gI_g}{L_gR_{kF}}\right) - \left(\frac{E_gI_g}{L_g}\right)^2\frac{4}{R_{kN}R_{kF}}$$

R_{kN} and R_{kF} are the tangent stiffnesses of a semirigid connection at buckling. ASCE Task Committee on Effective Length (1997) provides a detailed discussion of frame stability with PR connection.

EXAMPLE 31.2

Given: A steel frame is shown in Figure 31.5. Using the alignment chart with the necessary modifications, determine the K-factor for elastic column EF. $E = 29{,}000$ ksi (200 GPa) and $F_y = 36$ ksi (248 MPa).

Solution

1. Calculate the G-factor with modification for column EF. Since the far end of the restraining girders is hinged, girder stiffness should be multiplied by 0.5 (see Table 31.2). Using the section properties in Example 31.1, we obtain

$$G_F = \frac{\sum(E_c I_c/L_c)}{\sum \alpha_k (E_g I_g/L_g)} = \frac{1.722}{0.5(1.727) + 0.5\,(2.650)} = 0.787$$

$$G_E = 0.448$$

2. From the alignment chart in Figure 32.4b, $K = 1.22$ is obtained.

31.5.2 Different Restraining Column End Conditions

To consider different far end conditions of restraining columns, the general effective length factor equations for column C2 (Figure 31.3) were derived by Duan and Chen (1988, 1989, 1996). By assuming that the far ends of columns C1 and C3 are hinged and using the slope deflection equation approach for the subassemblies shown in Figure 31.3, we obtain the following.

31.5.2.1 For a Braced Frame (Duan and Chen 1988)

$$C^2 - S^2 \left[G_{AC1} + G_{BC3} + G_{AC2}G_{BC2} + \frac{(2G_{BC3}/G_A) + (2G_{AC1}/G_B)}{C} - G_{AC1}G_{BC3}\left(\frac{S}{C}\right)^2 \right]$$

$$+ 2C\left(\frac{1}{G_A} + \frac{1}{G_B}\right) + \frac{4}{G_A G_B} = 0 \tag{31.38}$$

where C and S are stability functions as defined by Equations 31.3 and 31.4; G_A and G_B are defined in Equations 31.7 and 31.8; and G_{AC1}, G_{AC2}, G_{BC2}, and G_{BC3} are the stiffness ratios of columns at the A-th and B-th ends of the columns being considered, respectively. They are defined as

$$G_{Ci} = \frac{E_{ci}I_{ci}/L_{ci}}{\sum(E_{ci}I_{ci}/L_{ci})} \tag{31.39}$$

where $\sum$ indicates the summation of all columns rigidly connected to the joint and lying in the plane in which buckling of column is being considered.

Although Equation 31.38 was derived for the special case in which the far ends of both column C1 and column C3 are hinged, this equation is also applicable if adjustment to G_{Ci} is made as follows: (1) if the far end of column Ci (C1 or C3) is fixed, then take $G_{Ci}=0$ (except for G_{C2}) and (2) if the far end of column Ci (C1 or C3) is rigidly connected, then take $G_{Ci}=0$ and $G_{C2}=1.0$. Therefore, Equation 31.38 can be specialized for the following conditions:

1. If the far ends of both column C1 and column C3 are fixed, we have $G_{AC1}=G_{BC3}=0$ and Equation 31.38 reduces to

$$C^2 - S^2(G_{AC2}G_{BC2}) + 2C\left(\frac{1}{G_A} + \frac{1}{G_B}\right) + \frac{4}{G_A G_B} = 0 \tag{31.40}$$

2. If the far end of column C1 is rigidly connected and the far end of column C3 is fixed, we have $G_{AC2}=1.0$ and $G_{AC1}=G_{BC3}=0$, and Equation 31.38 reduces to

$$C^2 - S^2 + G_{BC2} + 2C\left(\frac{1}{G_A} + \frac{1}{G_B}\right) + \frac{4}{G_A G_B} = 0 \tag{31.41}$$

3. If the far end of column C1 is rigidly connected and the far end of column C3 is hinged, we have $G_{AC1} = 0$ and $G_{AC2} = 1.0$, and Equation 31.38 reduces to

$$C^2 - S^2 \left(G_{BC3} + G_{BC2} + \frac{2G_{BC3}}{G_A C} \right) + 2C \left(\frac{1}{G_A} + \frac{1}{G_B} \right) + \frac{4}{G_A G_B} = 0 \qquad (31.42)$$

4. If the far end of column C1 is hinged and the far end of column C3 is fixed, we have $G_{BC3} = 0$, and Equation 31.38 reduces to

$$C^2 - S^2 \left(G_{AC1} + G_{AC2} G_{BC2} + \frac{2G_{AC1}}{G_B C} \right) + 2C \left(\frac{1}{G_A} + \frac{1}{G_B} \right) + \frac{4}{G_A G_B} = 0 \qquad (31.43)$$

5. If the far ends of both columns C1 and C3 are rigidly connected (i.e., the assumptions used in developing the alignment chart), we have $G_{C2} = 1.0$ and $G_{Ci} = 0$, and Equation 31.38 reduces to

$$C^2 - S^2 + 2C \left(\frac{1}{G_A} + \frac{1}{G_B} \right) + \frac{4}{G_A G_B} = 0 \qquad (31.44)$$

which can be rewritten in the form of Equation 31.5.

31.5.2.2 For an Unbraced Frame (Duan and Chen 1989, 1996)

$$\det \begin{vmatrix} a_{11} & a_{12} & a_{13} \\ a_{21} & a_{22} & a_{23} \\ a_{31} & a_{32} & a_{33} \end{vmatrix} = 0 \qquad (31.45)$$

or

$$a_{11} a_{22} a_{33} + a_{21} a_{32} a_{13} + a_{31} a_{23} a_{12} - a_{31} a_{22} a_{13} - a_{21} a_{12} a_{33} + a_{11} a_{23} a_{32} = 0 \qquad (31.46)$$

where

$$a_{11} = C + \frac{6}{G_A} - G_{AC1} \frac{S^2}{C} \qquad (31.47)$$

$$a_{22} = C + \frac{6}{G_B} - G_{BC3} \frac{S^2}{C} \qquad (31.48)$$

$$a_{33} = -2 \left[C + S - \frac{1}{2} \left(\frac{\pi}{K} \right)^2 \right] \qquad (31.49)$$

$$a_{12} = G_{AC2} S \qquad (31.50)$$

$$a_{21} = G_{BC2} S \qquad (31.51)$$

$$a_{31} = a_{32} = C + S \qquad (31.52)$$

$$a_{13} = -(C + S) + G_{AC1} \left(S + \frac{S^2}{C} \right) \qquad (31.53)$$

$$a_{23} = -(C + S) + G_{BC3} \left(S + \frac{S^2}{C} \right) \qquad (31.54)$$

Although Equation 31.45 was derived for the special case in which the far ends of both column C1 and column C3 are hinged, it can be adjusted to account for the following cases: (1) if the far end of column Ci (C1 or C3) is fixed, then take $G_{Ci} = 0$ (except for G_{C2}) and (2) if the far end of column Ci (C1 or C3) is rigidly connected, then take $G_{Ci} = 0$ and $G_{C2} = 1.0$. Therefore, Equation 31.45 can be used for the following conditions:

1. If the far ends of both column C1 and column C3 are fixed, we take $G_{C1} = G_{C3} = 0$ and obtain from Equations 31.47, 31.48, 31.53, and 31.54

$$a_{11} = C + \frac{6}{G_A} \tag{31.55}$$

$$a_{22} = C + \frac{6}{G_B} \tag{31.56}$$

$$a_{13} = a_{23} = -(C + S) \tag{31.57}$$

2. If the far end of column C1 is rigidly connected and the far end of column C3 is fixed, we take $G_{AC2} = 1.0$ and $G_{AC1} = G_{BC3} = 0$ and obtain from Equations 31.47, 31.48, 31.50, 31.53, and 31.54

$$a_{11} = C + \frac{6}{G_A} \tag{31.58}$$

$$a_{22} = C + \frac{6}{G_B} \tag{31.59}$$

$$a_{12} = S \tag{31.60}$$

$$a_{13} = a_{23} = -(C + S) \tag{31.61}$$

3. If the far end of column C1 is rigidly connected and the far end of column C3 is hinged, we take $G_{AC1} = 0$ and $G_{AC2} = 1.0$ and obtain from Equations 31.47, 31.50, and 31.52

$$a_{11} = C + \frac{6}{G_A} \tag{31.62}$$

$$a_{12} = S \tag{31.63}$$

$$a_{13} = -(C + S) \tag{31.64}$$

4. If the far end of column C1 is hinged and the far end of column C3 is fixed, we have $G_{BC3} = 0.0$ and obtain from Equations 31.48 and 31.54

$$a_{22} = C + \frac{6}{G_B} \tag{31.65}$$

$$a_{23} = -(C + S) \tag{31.66}$$

5. If the far ends of both column C1 and column C3 are rigidly connected (i.e., the assumptions used in developing the alignment chart, that is, $\theta_C = \theta_B$ and $\theta_D = \theta_A$), we take $G_{C2} = 1.0$ and $G_{Ci} = 0$ and obtain from Equations 31.47 to 31.54

$$a_{11} = C + \frac{6}{G_A} \tag{31.67}$$

$$a_{22} = C + \frac{6}{G_B} \tag{31.68}$$

$$a_{12} = a_{21} = S \tag{31.69}$$

$$a_{13} = a_{23} = -(C + S) \tag{31.70}$$

Equation 31.45 is reduced to the form of Equation 31.6.

The procedures to obtain the K-factor directly from the alignment charts without resorting to solve Equations 31.38 and 31.45 were also proposed by Duan and Chen (1988, 1989).

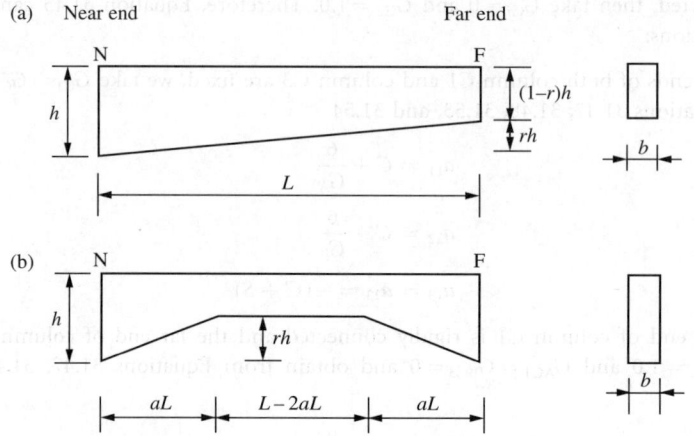

FIGURE 31.7 Tapered rectangular girders: (a) linearly tapered girder and (b) symmetrically tapered girder.

31.5.3 Columns Restrained by Tapered Rectangular Girders

A modification factor α_T was developed by King et al. (1993) for those framed columns restrained by tapered rectangular girders with different far end conditions. The following modified G-factor is introduced in connection with the use of alignment charts:

$$G = \frac{\sum(E_c I_c / L_c)}{\sum \alpha_T (E_g I_g / L_g)} \tag{31.71}$$

where I_g is the moment of inertia of the girder at the near end. Both closed-from and approximate solutions for modification factor α_T were derived. It is found that the following two-parameter power function can describe the closed-from solutions very well:

$$\alpha_T = D(1 - r)^\beta \tag{31.72}$$

where the parameter D is a constant depending on the far end conditions and β is a function of far end conditions and tapering factors a and r as defined in Figure 31.7.

For a braced frame

$$D = \begin{cases} 1.0 & \text{rigid far end} \\ 2.0 & \text{fixed far end} \\ 1.5 & \text{hinged far end} \end{cases} \tag{31.73}$$

For an unbraced frame

$$D = \begin{cases} 1.0 & \text{rigid far end} \\ 2/3 & \text{fixed far end} \\ 0.5 & \text{hinged far end} \end{cases} \tag{31.74}$$

1. *For a linearly tapered rectangular girder* (Figure 31.7a)
   For a braced frame

$$\beta = \begin{cases} 0.02 + 0.4r & \text{rigid far end} \\ 0.75 - 0.1r & \text{fixed far end} \\ 0.75 - 0.1r & \text{hinged far end} \end{cases} \tag{31.75}$$

For an unbraced frame

$$\beta = \begin{cases} 0.95 & \text{rigid far end} \\ 0.70 & \text{fixed far end} \\ 0.70 & \text{hinged far end} \end{cases} \tag{31.76}$$

2. *For a symmetrically tapered rectangular girder* (Figure 31.7b)
   For a braced frame

$$\beta = \begin{cases} 3 - 1.7a^2 - 2a & \text{rigid far end} \\ 3 + 2.5a^2 - 5.55a & \text{fixed far end} \\ 3 - a^2 - 2.7a & \text{hinged far end} \end{cases} \tag{31.77}$$

For an unbraced frame

$$\beta = \begin{cases} 3 + 3.8a^2 - 6.5a & \text{rigid far end} \\ 3 + 2.3a^2 - 5.45a & \text{fixed far end} \\ 3 - 0.3a & \text{hinged far end} \end{cases} \tag{31.78}$$

EXAMPLE 31.3

Given: A one-story frame with a symmetrically tapered rectangular girder is shown in Figure 31.8. Assuming $r = 0.5$, $a = 0.2$, and $I_g = 2I_c = 2I$, determine the K-factor for column AB.

Solution

1. *Using the alignment chart with modification*
   For joint A, since the far end of the girder is rigid, use Equations 31.78 and 31.72:

$$\beta = 3 + 3.8(0.2)^2 - 6.5(0.2) = 1.852$$
$$\alpha_T = (1 - 0.5)^{1.852} = 0.277$$
$$G_A = \frac{\sum E_c I_c / L_c}{\sum \alpha_T E_g I_g / L_g} = \frac{EI/L}{0.277E(2I)/2L} = 3.61$$
$$G_B = 1.0 \text{ (AISC 1999)}$$

From the alignment chart in Figure 31.4b, $K = 1.59$ is obtained.

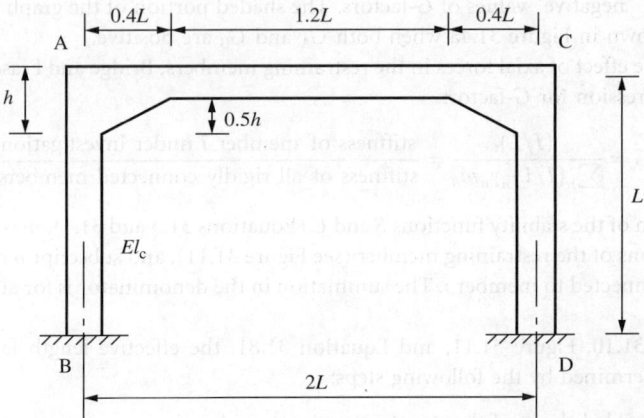

FIGURE 31.8 A simple frame with rectangular sections.

2. *Using the alignment chart without modification*

A direct use of Equations 31.7 and 31.8 with an average section $(0.75h)$ results in

$$I_g = 0.75^3(2I) = 0.844I$$

$$G_A = \frac{EI/L}{0.844EI/2L} = 2.37, \quad G_B = 1.0$$

From the alignment chart in Figure 31.4b, $K = 1.49$ or $(1.49 - 1.59)/1.59 = -6\%$ in error on the less conservative side.

31.5.4 Unsymmetrical Frames

When the column sizes or column loads are not identical, adjustments to the alignment charts are necessary to obtain correct K-factors. Figure 31.9 presents a set of curves for a modification factor β developed by Chu and Chow (1969).

$$K_{\text{adjusted}} = \beta K_{\text{alignment chart}} \tag{31.79}$$

If the K-factor of the column under the load λP is desired, further modifications to K are necessary. Denoting K' as the effective length factor of column with $I'_c = \alpha I_c$ subjected to the axial load $P' = \lambda P$ as shown in Figure 31.9, we have

$$K' = K_{\text{adjusted}} \frac{L}{L'} \sqrt{\frac{\alpha}{\lambda}} \tag{31.80}$$

Equation 31.80 can be used to determine K-factors for columns in adjacent stories with different heights L'.

31.5.5 Effects of Axial Forces in Restraining Members

Compressive axial load in a restraining girder reduces its flexural stiffness and then affects adversely the K-factor of the column (AISC 1999). To account for any compression axial load in a girder, the girder stiffness parameter $(E_g I_g / L_g)$ in Equations 31.7 and 31.8 should be modified by the factor $[1 - (Q/Q_{cr})]$, where Q is the axial compression load in the girder, and Q_{cr} is the in-plane buckling load of the girder based on $K = 1.0$. Tensile axial load in the girder can be ignored when determining the G-factor.

Bridge and Fraser (1987) observed that K-factors of a column in a braced frame may be greater than unity due to "negative" restraining effects. Figure 31.10 shows the solutions obtained by considering the both "positive" and "negative" values of G-factors. The shaded portion of the graph corresponds to the alignment chart shown in Figure 31.4a when both G_A and G_B are positive.

To account for the effect of axial forces in the restraining members, Bridge and Fraser (1987) proposed a more general expression for G-factor:

$$G = \frac{(I/L)}{\sum_n (I/L)_n \gamma_n m_n} = \frac{\text{stiffness of member } i \text{ under investigation}}{\text{stiffness of all rigidly connected members}} \tag{31.81}$$

where γ is a function of the stability functions S and C (Equations 31.3 and 31.4), m is a factor to account for the end conditions of the restraining member (see Figure 31.11), and subscript n represents the other members rigidly connected to member i. The summation in the denominator is for all members meeting at the joint.

By using Figure 31.10, Figure 31.11, and Equation 31.81, the effective length factor K_i for the ith member can be determined by the following steps:

1. Sketch the buckled shape of the structure under consideration.
2. Assume a value of K_i for the member being investigated.

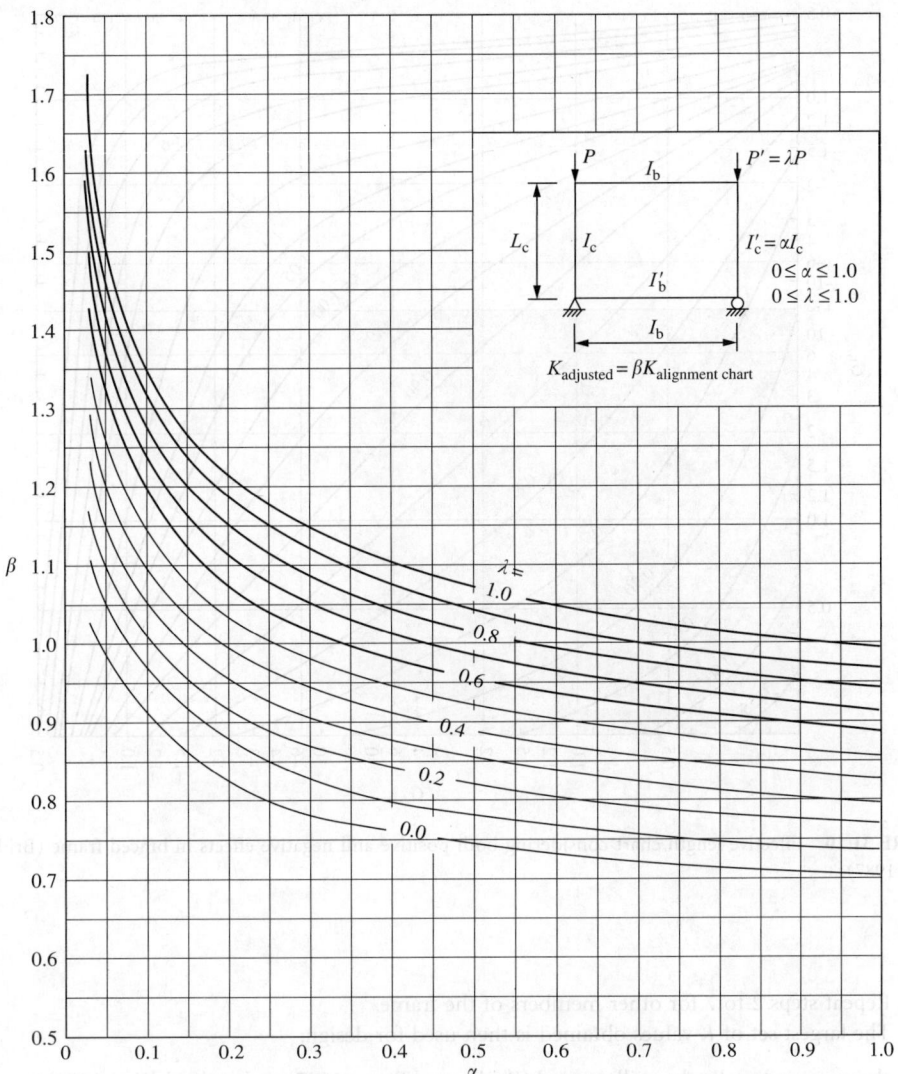

FIGURE 31.9 Chart for the modification factor β in an unsymmetrical frame.

3. Calculate values of K_n for each of the other members that are rigidly connected to the *i*th member using the equation

$$K_n = K_i \frac{L_i}{L_n} \sqrt{\left(\frac{P_i}{P_n}\right)\left(\frac{I_n}{I_i}\right)}$$ (31.82)

4. Calculate γ and obtain m from Figure 31.11 for each member.
5. Calculate G_i for the *i*th member using Equation 31.81.
6. Obtain K_i from Figure 31.10 and compare with the assumed K_i at step 2.
7. Repeat the procedure by using the calculated K_i as the assumed K_i until K_i calculated at the end of the cycle is approximately (say, 10%) equal to the K_i at the beginning of the cycle.

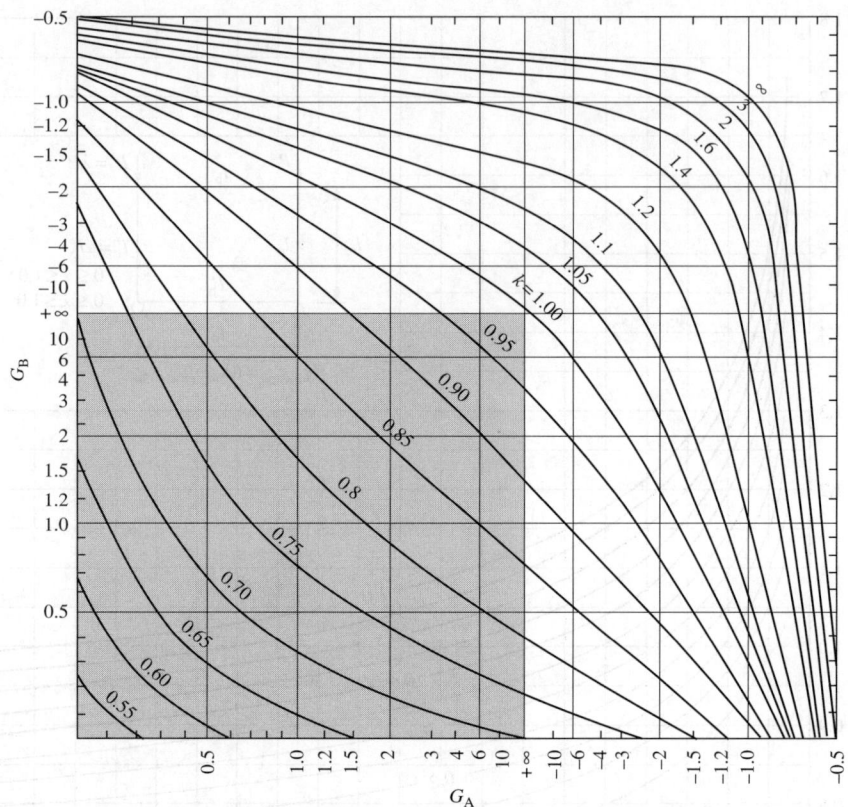

FIGURE 31.10   Effective length chart considering both positive and negative effects in braced frame (Bridge and Fraser 1987).

8. Repeat steps 2 to 7 for other members of the frame.
9. The largest set of K values obtained is then used for design.

The above procedure has been illustrated (Bridge and Fraser 1987) and verified (Koo 1988) to provide a good elastic K-factor of columns in braced frames.

EXAMPLE 31.4

Given: A braced column is shown in Figure 31.12. Consider axial force effects to determine K-factors for columns AB and BC.

Solution

1. Sketch buckled shape as shown in Figure 31.12b.
2. Assume $K_{AB} = 0.94$.
3. Calculate K_{BC} by Equation 31.82:

$$K_{BC} = K_{AB} \frac{L_{AB}}{L_{BC}} \sqrt{\left(\frac{P_{AB}}{P_{BC}}\right)\left(\frac{I_{BC}}{I_{AB}}\right)} = 0.94 \frac{L}{L}\sqrt{\frac{2PI}{P(1.2I)}}$$

$$= 1.22$$

Axial forces	Cases	Exact γ formula	Approximate γ formula	m
Tension	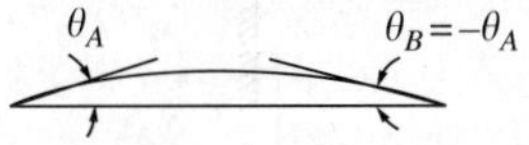 θ_A $\quad$ $\theta_B=-\theta_A$	$\dfrac{C-S}{2}$	$1+\dfrac{1}{1.5\,K_n^2}$	1
	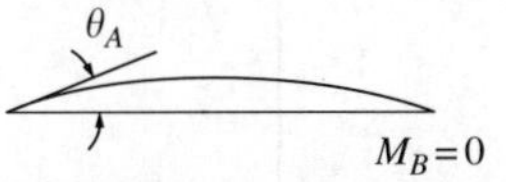 θ_A $\quad$ $M_B=0$	$\dfrac{C-S^2/C}{3}$		1.5
	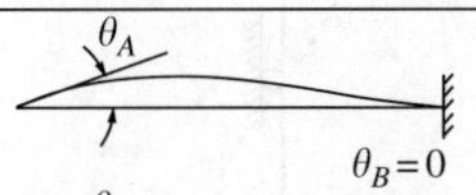 θ_A $\quad$ $\theta_B=0$	$\dfrac{C}{4}$	$1+\dfrac{1}{4\,K_n^2}$	2
	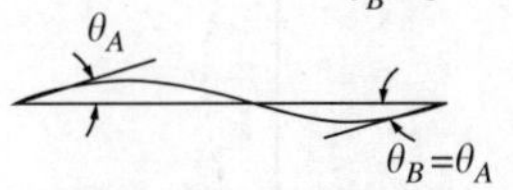 θ_A $\quad$ $\theta_B=\theta_A$	$\dfrac{C+S}{6}$		3
Compression	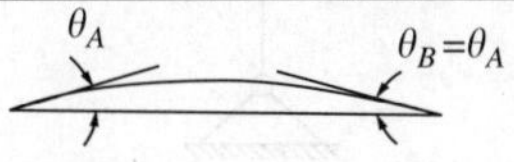 θ_A $\quad$ $\theta_B=\theta_A$	$\dfrac{C-S}{2}$	$\begin{cases} 1-\dfrac{1}{K_n^2} & \text{for } K_n>1.0 \\[2mm] 2-\dfrac{2}{K_n^2} & \text{for } K_n\le 1.0 \end{cases}$	1
	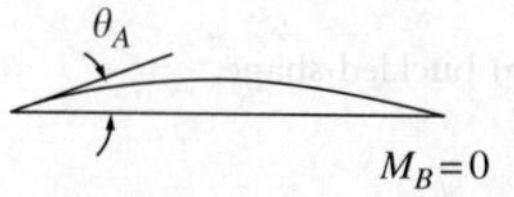 θ_A $\quad$ $M_B=0$	$\dfrac{C-S^2/C}{3}$		1.5
	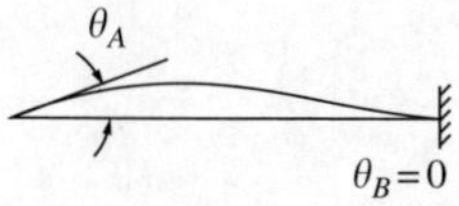 θ_A $\quad$ $\theta_B=0$	$\dfrac{C}{4}$	$\begin{cases} 1-\dfrac{1}{2\,K_n^2} & \text{for } K_n>0.7 \\[2mm] 2-\dfrac{2}{K_n^2} & \text{for } K_n\le 0.7 \end{cases}$	2
	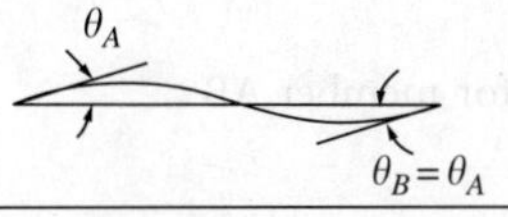 θ_A $\quad$ $\theta_B=\theta_A$	$\dfrac{C+S}{6}$	$1-\dfrac{1}{4\,K_n^2}$	3
Note:	C and S are stability equations			

FIGURE 31.11 Values of γ and m to account for the effect of axial forces in the restraining members.

4. Calculate γ and obtain m from Figure 31.11 for member BC.
   Since $K_{BC} > 1.0$

$$\gamma_{BC} = 1 - \frac{1}{K_{BC}^2} = 1 - \frac{1}{1.22^2} = 0.33$$

   Far end is pined, $m_{BC} = 1.5$
5. Calculate G-factor for member AB using Equation 31.81.

$$G_B = \frac{(I/L)}{\sum_n (I/L)_n \gamma_n m_n} = \frac{(1.2I/L)}{(I/L)(0.33)(1.5)} = 2.42$$

$$G_A = \infty$$

6. From Figure 31.10, $K_{AB} = 0.93$. Comparing with the assumed $K_{AB} = 0.94$ it is alright.

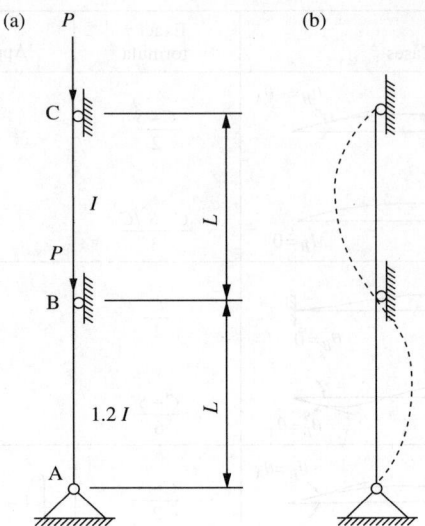

FIGURE 31.12   Braced columns: (a) braced columns and (b) buckled shape.

7. Repeat the above procedure for member BC.
   Assume $K_{BC} = 1.2$
   Calculate K_{AB} by Equation 31.82

$$K_{AB} = K_{BC}\frac{L_{BC}}{L_{AB}}\sqrt{\left(\frac{P_{BC}}{P_{AB}}\right)\left(\frac{I_{AB}}{I_{BC}}\right)} = 1.2\frac{L}{L}\sqrt{\frac{P(1.2I)}{2PI}} = 0.93$$

Calculate γ and obtain m from Figure 31.11 for member AB
Since $K_{AB} < 1.0$

$$\gamma_{AB} = 2 - \frac{2}{K_{AB}^2} = 2 - \frac{2}{0.93^2} = -0.312$$

Far end is pined, $m_{AB} = 1.5$
Calculate G-factor for the member BC using Equation 31.81.

$$G_B = \frac{(I/L)}{\sum_n(I/L)_n\gamma_n m_n} = \frac{(I/L)}{(1.2I/L)(-0.312)(1.5)} = -1.78$$
$$G_c = \infty$$

See Figure 31.10, $K_{BC} = 1.18$
Comparing with the assumed $K_{AB} = 1.20$ it is alright.

8. It is seen that the largest set of K-factors is

$$K_{AB} = 1.22 \quad \text{and} \quad K_{BC} = 0.93$$

31.5.6   Consideration of Partial Column Base Fixity

In computing the effective length factor for monolithic connections, it is important to properly evaluate the degree of fixity in foundation. The following two approaches can be used to account for foundation fixity.

31.5.6.1 Fictitious Restraining Beam Approach

Galambos (1960) proposed that the effect of partial base fixity can be modeled as a fictitious beam. The approximate expression for the stiffness of the fictitious beam accounting for rotation of foundation in the soil has the form

$$\frac{I_s}{L_B} = \frac{qBH^3}{72E_{\text{steel}}} \tag{31.83}$$

where q is the modulus of subgrade reaction (varies from 50 to 400 lb/in.^{3}, 0.014 to 0.109 N/mm^3), B and H are the width and length (in bending plane) of the foundation, respectively, and E_{steel} is the modulus of elasticity of steel.

Based on the studies of Salmon et al. (1957), the approximate expression for the stiffness of the fictitious beam accounting for the rotations between column ends and footing due to deformation of base plate, anchor bolts, and concrete can be written as:

$$\frac{I_s}{L_B} = \frac{bd^2}{72E_{\text{steel}}/E_{\text{concrete}}} \tag{31.84}$$

where b and d are the width and length of the base plate, respectively and subscripts concrete and steel represent concrete and steel, respectively. Galambos (1960) suggested that the smaller stiffness calculated by Equations 31.83 and 31.84 be used in determining K-factors.

31.5.6.2 AASHTO Load and Resistance Factor Design Approach

The following values are suggested by AASHTO (2004):

$G = 1.5$ footing anchored on rock

$G = 3.0$ footing not anchored on rock

$G = 5.0$ footing on soil

$G = 1.0$ footing on multiple rows of end bearing piles

31.5.7 Inelastic K-factor

The effect of material inelasticity and end restrain on the K-factors has been studied during the last two decades (Yura 1971; Disque 1973; Jones et al. 1980, 1982; Chapius and Galambos 1982; Sugimoto and Chen 1982; Vinnakota 1982; Lui and Chen 1983; Razzaq 1983; Bjorhovde 1984; Sohal et al. 1995). The inelastic K-factor developed originally by Yura (1971) and expanded by Disque (1973) makes use of alignment charts with simple modifications. It is conservative to design column on the base of elastic K-factors. It is less conservative to design column on the base of inelastic K-factors. To consider the inelasticity of material, the G-values as defined by Equations 31.7 and 31.8 are replaced by G^* (Disque 1973) as follows:

$$G^* = \text{SRF}(G) = \frac{E_t}{E}G \tag{31.85}$$

where E_t is the tangent modulus of the material. For practical application, the stiffness reduction factor $(\text{SRF}) = (E_t/E)$ and can be taken as the ratio of the inelastic to elastic buckling stress of the column

$$\text{SRF} = \frac{E_t}{E} \approx \frac{(F_{cr})_{\text{inelastic}}}{(F_{cr})_{\text{elastic}}} \approx \frac{(P_u/A_g)}{(F_{cr})_{\text{elastic}}} \tag{31.86}$$

where P_u is the factored axial load and A_g is the gross section area of member. $(F_{cr})_{inelastic}$ and $(F_{cr})_{elastic}$ can be calculated by AISC (1999) column equations:

$$(F_{cr})_{inelastic} = (0.658)^{\lambda_c^2} F_y \tag{31.87}$$

$$(F_{cr})_{elastic} = \left[\frac{0.877}{\lambda_c^2}\right] F_y \tag{31.88}$$

$$\lambda_c = \frac{KL}{r\pi}\sqrt{\frac{F_y}{E}} \tag{31.89}$$

where K is the elastic effective length factor and r is the radius of gyration about the plane of buckling. Table 31.3 gives the SRF values for different stress levels and slenderness parameters. AISC (1999) provides a direct calculation of SRF as follows:

$$SRF = \begin{cases} 1.0 & \text{for } (P_u/P_y) \leq \frac{1}{3} \\ -7.38\left(\frac{P_u}{P_y}\right)\log\left(\frac{(P_u/P_y)}{0.85}\right) & \text{for } (P_u/P_y) > \frac{1}{3} \end{cases} \tag{31.90}$$

EXAMPLE 31.5

Given: A two-story steel frame is shown in Figure 31.5. Using the alignment chart to determine the K-factor for inelastic column DE, $E = 29,000$ ksi (200 GPa) and $F_y = 36$ ksi (248 MPa).

Solution

1. Calculate the axial stress ratio

$$\frac{P_u}{A_g F_y} = \frac{300}{13.3(36)} = 0.63$$

2. Obtain SRF = 0.793 from Table 31.3

TABLE 31.3 Stiffness Reduction Factor (SRF) for G-values

	$(KL/r)_{elastic}$			
	36 ksi	50 ksi		SRF
$P_u/(A_g F_y)$	(248 MPa)	(345 MPa)	λ_c	(Equation 31.86)
1.00	0.0	0.0	0.155	0.000
0.95	31.2	26.5	0.350	0.133
0.90	44.7	38.0	0.502	0.258
0.85	55.6	47.1	0.623	0.376
0.80	65.1	55.2	0.730	0.486
0.75	73.9	62.7	0.829	0.588
0.70	82.3	69.8	0.923	0.680
0.65	90.5	76.8	1.015	0.763
0.60	98.5	83.6	1.105	0.835
0.55	106.6	90.4	1.195	0.896
0.50	114.7	97.4	1.287	0.944
0.45	123.2	104.5	1.381	0.979
0.40	131.9	111.9	1.480	0.998
0.39	133.7	113.5	1.500	1.000

From Equation 31.83b

$$\text{SRF} = -7.38\left(\frac{P_u}{P_y}\right)\log\left(\frac{(P_u/P_y)}{0.85}\right) = -7.38(0.63)\log\left(\frac{0.63}{0.85}\right) = 0.605$$

3. Calculate the modified G-factor

$G_E = 0.448$ (Example 31.1)

$G_E^* = \text{SRF}(G_E) = 0.794(0.448) = 0.355$

$G_D = 10$ (AISC 1999)

4. From the alignment chart in Figure 31.4b, we have

$(K_{DE})_{\text{inelastic}} = 1.75$

31.6 Framed Columns — Alternative Methods

31.6.1 LeMessurier Method

Considering that all columns in a story buckle simultaneously and strong columns will brace weak columns (Figure 31.13), a more accurate approach to calculate K-factors for columns in a sidesway frame was developed by LeMessurier (1977). The K_i value for the ith column in a story can be obtained by the following expression:

$$K_i = \sqrt{\frac{\pi^2 EI_i}{L_i^2 P_i}\left(\frac{\sum P + \sum C_L P}{\sum P_L}\right)} \tag{31.91}$$

where P_i is the axial compressive force for member i, subscript i represents the ith column, and $\sum P$ is the sum of axial force of all columns in a story.

$$P_L = \frac{\beta EI}{L^2} \tag{31.92}$$

$$\beta = \frac{6(G_A + G_B) + 36}{2(G_A + G_B) + G_A G_B + 3} \tag{31.93}$$

$$C_L = \left(\beta\frac{K_0^2}{\pi^2} - 1\right) \tag{31.94}$$

where K_0 is the effective length factor obtained by the alignment chart for unbraced frames and P_L is

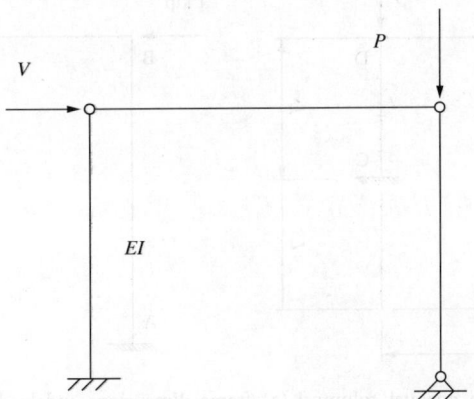

FIGURE 31.13 Subassemblage of LeMessurier method.

only for *rigid columns*, which provide sidesway stiffness. For a cantilever column, $C_L = 0.216$. In multistory structures, C_L may be expediently approximated by 0.2 for all columns except for pin-ended columns, for which $C_L = 0$.

EXAMPLE 31.6

Given: A sway frame with unequal height columns is shown in Figure 31.14a. Determine the elastic *K*-factors for columns by using the LeMessurier method. Member properties are

Member	A (in.², mm²)	I (in.⁴, mm⁴, × 10⁸)	L (in., mm)
AB	21.5, 13,871	620, 2.58	240, 6096
BD	21.5, 13,871	620, 2.58	240, 6096
CD	7.65, 4,935	310, 1.29	120, 3048

Solution

The detailed calculations are listed in Table 31.4.
By using Equation 31.91, we obtain

$$K_{AB} = \sqrt{\frac{\pi^2 EI_{AB}}{L_{AB}^2 P_{AB}} \left(\frac{\sum P + \sum C_L P}{\sum P_L} \right)}$$

$$= \sqrt{\frac{\pi^2 E(620)}{(240)^2(2P)} \left(\frac{3P + 0.495P}{0.271E} \right)} = 0.83$$

$$K_{CD} = \sqrt{\frac{\pi^2 EI_{CD}}{L_{CD}^2 P_{CD}} \left(\frac{\sum P + \sum C_L P}{\sum P_L} \right)}$$

$$= \sqrt{\frac{\pi^2 E(310)}{(120)^2(P)} \left(\frac{3P + 0.495P}{0.271E} \right)} = 1.66$$

31.6.2 Lui Method

A simple and straightforward approach for determining the effective length factors for framed columns without the use of alignment charts and other charts was proposed by Lui (1992). The formula takes into

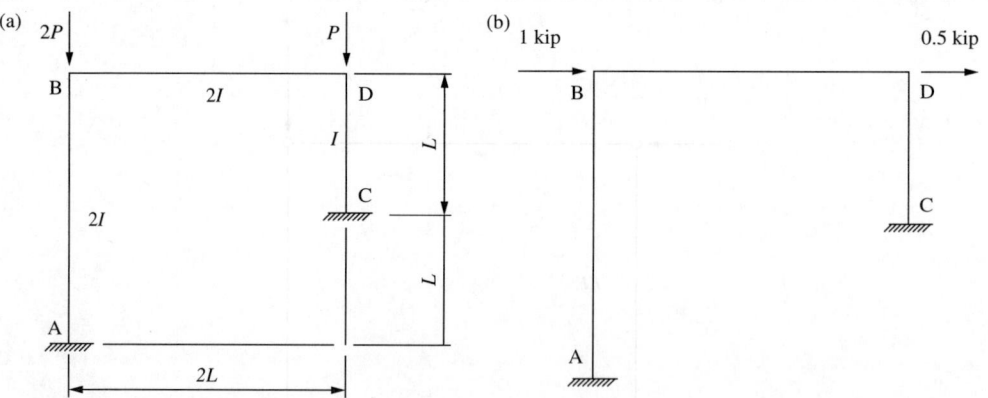

FIGURE 31.14 A frame with unequal columns: (a) frame dimensions and loads and (b) frame subjected to fictitious lateral loads.

TABLE 31.4   Example 31.6 — Detailed Calculation by LeMessurier Method

Members	AB	CD	Sum	Notes
I (in.^{4}, mm$^4 \times 10^8$)	620, 2.58	310, 1.29	—	—
L (in., mm)	240, 6096	120, 3048	—	—
G_{top}	1.0	1.0	—	Equation 31.7
G_{bottom}	0.0	0.0	—	Equation 31.7
β	8.4	8.4	—	Equation 31.93
K_{i0}	1.17	1.17	—	Alignment chart
C_L	0.165	0.165	—	Equation 31.94
P_L	0.09E	0.181E	0.271E	Equation 31.92
P	2P	P	3P	
$C_L P$	0.33P	0.165P	0.495P	

account both the member instability and the frame instability effects explicitly. The K-factor for the ith column in a story was obtained in a simple form:

$$K_i = \sqrt{\left(\frac{\pi^2 EI_i}{P_i L_i^2}\right)\left[\left(\sum \frac{P}{L}\right)\left(\frac{1}{5\sum \eta} + \frac{\Delta_1}{\sum H}\right)\right]} \tag{31.95}$$

where $\sum(P/L)$ represents the sum of axial force to length ratio of all members in a story, $\sum H$ is the story lateral load producing Δ_1, Δ_1 is the first-order interstory deflection, and η is the member stiffness index and can be calculated by

$$\eta = \frac{(3 + 4.8m + 4.2m^2)EI}{L^3} \tag{31.96}$$

where m is ratio of the smaller to larger end moments of the member, it is taken as positive if the member bents in reverse curvature, and negative for single curvature.

It is important to note that the term $\sum H$ used in Equation 31.95 is not the actual applied lateral load. Rather, it is a small disturbing or fictitious force (taken as a fraction of the story gravity loads) to be applied to each story of the frame. This fictitious force is applied in a direction such that the deformed configuration of the frame will resemble its buckled shape.

EXAMPLE 31.7

Given: Determine K-factors by using the Lui method for the frame shown in Figure 31.14a. $E = 29,000$ ksi (200 GPa).

Solution

Apply fictitious lateral forces at B and D (Figure 31.14b) and perform a first-order analysis. Detailed calculation is shown in Table 31.5.

By using Equation 31.95, we obtain

$$K_{\text{AB}} = \sqrt{\left(\frac{\pi EI_{\text{AB}}}{P_{\text{AB}} L_{\text{AB}}^2}\right)\left[\left(\sum \frac{P}{L}\right)\left(\frac{1}{5\sum \eta} + \frac{\Delta_1}{\sum H}\right)\right]}$$

$$= \sqrt{\left(\frac{\pi^2 (29,000)(620)}{(2P)(240)^2}\right)\left[\left(\frac{P}{60}\right)\left(\frac{1}{5(56.24)} + 0.019\right)\right]} = 0.76$$

TABLE 31.5   Example 31.7 — Detailed Calculation by Lui Method

Members	AB	CD	Sum	Notes
I (in.^{4}, mm$^4 \times 10^8$)	620, 2.58	310, 1.29	—	
L (in., mm)	240, 6096	120, 3048	—	
H (kip, kN)	1.0, 4.448	0.5, 2.224	1.5, 6.672	
Δ_1 (in., mm)	0.0286, 0.7264	0.0283, 0.7188	—	
$\Delta_1/\sum H$ (in./kip, mm/kN)	—	—	0.019, 0.108	Average
M_{top} (k in., kN m)	−38.8, 4.38	56.53, 6.39	—	
M_{bottom} (k in., kN m)	−46.2, −5.22	81.18, −9.17	—	
M	0.84	0.69	—	
η (kip/in., kN/mm)	13.00, 2.28	43.24, 7.57	56.24, 9.85	Equation 31.96
P/L (kip/in., kN/mm)	P/120, P/3048	P/120, P/3048	P/60, P/1524	

$$K_{CD} = \sqrt{\left(\frac{\pi^2 EI_{CD}}{P_{CD} L_{CD}^2}\right)\left[\left(\sum\frac{P}{L}\right)\left(\frac{1}{5\sum\eta} + \frac{\Delta_1}{\sum H}\right)\right]}$$

$$= \sqrt{\left(\frac{\pi^2 (29,000)(310)}{P(120)^2}\right)\left[\left(\frac{P}{60}\right)\left(\frac{1}{5(56.24)} + 0.019\right)\right]} = 1.52$$

31.6.3 Essa Method

Considering different story drafts, different stiffness parameters, $L\sqrt{P/EI}$, for columns, and different girder connections for the structural models shown in Figure 31.3, Essa (1998) presented a general effective length factor equation for column C_2 by solving five equations. Those five equations are obtained by considering the equilibrium of joints A and B and story shear for columns C_1, C_2, and C_3.

31.6.4 Cheong-Siat-Moy Method

Considering an individual end-restrained column as shown in Figure 31.1a, Equation 31.2 was presented in detail by Cheong-Siat-Moy (1999). When the beams bend in exact double curvature, R_{kA} and R_{kB} are directly related to G_A and G_B, respectively ($R_k = 6/G$). Equation 31.2 will become Equation 31.5 for braced frames when $T_k = \infty$ and Equation 31.6 for unbraced frames when $T_k = 0$.

T_k is the difference between the final lateral stiffness (T_{kf}) and the initial lateral stiffness (T_{ki}) of a column as follows:

$$T_k = T_{kf} - T_{ki} = \frac{P_i}{\sum P} T_{story} - T_{0i} \tag{31.97}$$

where T_{story} is the first-order lateral story stiffness and T_{0i} is the first-order lateral stiffness of column i.

An improved lateral stiffness T_k associated with γ, P, and h and the first-order stiffness property was proposed by Hellesland (2000) as follows:

$$T_k = T_{kf} - T_{ki} = \frac{(\gamma P/h)_i}{\sum \gamma P/h} T_{story} - T_{0i} \tag{31.98}$$

The term γ is the flexibility parameter and can be expressed as $\gamma = 1 + C_L$. A rather simple, yet fairly accurate expression was developed by Hellesland (1998):

$$\gamma = 1 + 0.11 \frac{1 + [1 - (0.5 G_{max})^p]^3}{(1 + 0.5 G_{min})^2} \tag{31.99}$$

TABLE 31.6 Comparison of *K*-factors for Frame in Figure 31.14a

Columns	Theoretical	Alignment chart	Lui, Equation (31.80)	LeMessurier, Equation (31.76)
AB	0.70	1.17	0.76	0.83
CD	1.40	1.17	1.52	1.67

where $p = 1$ for $G_{max} \leq 2$ and $p = -1$ for $G_{max} > 2$. G_{max} is the larger and G_{min} is the smaller of the *G*-factors at the column ends.

31.6.5 Remarks

For a comparison, Table 31.6 summarizes *K*-factors for the frames shown in Figure 31.14a obtained from the alignment chart, the LeMessurier and Lui methods, as well as an eigenvalue analysis. The methods of both LeMessurier and Lui are based on the story-buckling concept. It is seen that errors in alignment chart results are rather significant in this case. Although the *K*-factors predicted by Lui's and LeMessurier's formulas are identical in most cases, the simplicity and independence of any chart in the case of Lui's formula make it more desirable for design office use (Shanmugam and Chen 1995). Essa's (1998) method overcomes some of the limitations imposed on the development of alignment chart and incorporates effects of inelastic behavior, different column stiffness parameters, and different restraining girder conditions. The Cheong-Siat-Moy (1999) method is dependent on the nondimensionalized lateral stiffness parameter of the column and can be used for partially braced frames. A comprehensive parametric study encouraged engineers to use the story-based *K*-factors for stability assessment (Roddis et al. 1998). Xu and Liu (2002) developed a story-based approach for both unbraced partially and fully restrained frames.

31.7 Unbraced Frames with Leaning Columns

A column framed with simple connections is often called a *leaning column*. It has no lateral stiffness or sidesway resistance. A column framed with the rigid moment-resisting connection is called a *rigid column*. It provides the lateral stiffness or sidesway resistance to the frame. When a frame system (Figure 31.15a) includes leaning columns, the effective length factors of rigid columns must be modified. Several approaches to account for the effect of leaning columns were reported in the literature (Yura 1971; Lim and McNamara 1972; LeMessurier 1977; Cheong-Siat-Moy 1986; Aristizabal-Ochoa 1994). A detailed discussion about leaning columns for practical applications was presented by Geschwindner (1995, 2002).

31.7.1 Rigid Columns

31.7.1.1 Yura Method

Yura (1971) discussed frames with leaning columns and noted the behavior of stronger columns assisting weaker ones in resisting sidesway. He concluded that the alignment chart gives valid sidesway buckling solutions if the columns are in the elastic range, and all columns in a story reach their individual buckling loads simultaneously. For columns that do not satisfy these two conditions, the alignment charts is generally overly conservative. The Yura approach states that

1. The maximum load-carrying capacity of an individual column is limited to the load permitted on that column for the braced case $K = 1.0$.
2. The total gravity loads that produce sidesway are distributed among the columns, which provides lateral stiffness in a story.

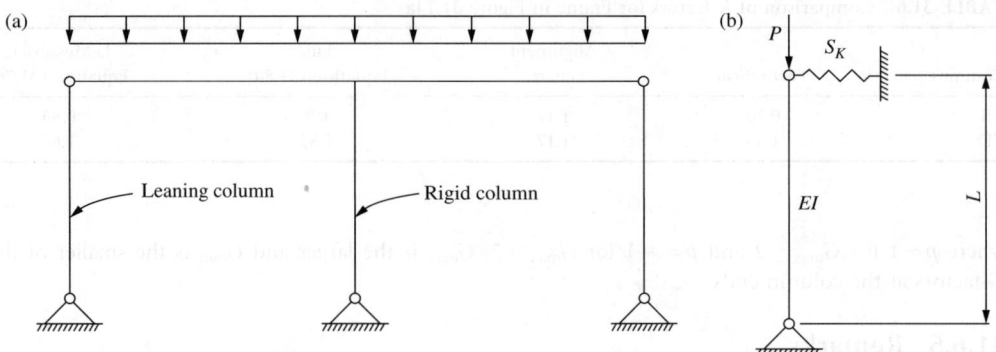

FIGURE 31.15 A frame with leaning columns: (a) a leaning column frame and (b) model for a leaning column.

31.7.1.2 Lim and McNamara Method

Based on the story-buckling concept and using the stability functions, Lim and McNamara (1972) presented the following formula to account for the leaning column effect:

$$K_n = K_0 \sqrt{1 + \frac{\sum Q}{\sum P} \left(\frac{F_0}{F_n} \right)} \tag{31.100}$$

where K_n is the effective length factor accounting for leaning columns; K_0 is the effective length factor determined by the alignment chart (Figure 31.3b) not accounting for the leaning columns; $\sum P$ and $\sum Q$ are the loads on the restraining columns and on the leaning columns in a story, respectively; and F_0 and F_n are the eigenvalue solutions for a frame without and with leaning columns, respectively. For normal column end conditions that fall somewhere between fixed and pinned, $F_0/F_n = 1$ provides a K-factor on the conservative side by less than 2% (Geschwindner 1995). Using $F_0/F_n = 1$, Equation 31.100 becomes

$$K_n = K_0 \sqrt{1 + \frac{\sum Q}{\sum P}} \tag{31.101}$$

Equation 31.93 gives the same K-factor as the modified Yura approach (Geschwindner 1995).

31.7.1.3 LeMessurier and Lui Methods

Equation 31.91 developed by LeMessurier and Equation 31.95 proposed by Lui (1992) can be used for frames both with and without leaning columns. Since the K-factor expressions (Equations 31.91 and 31.95) were derived for an entire story of the frame, they are applicable to frames with and without leaning columns.

31.7.1.4 AISC Load and Resistance Factor Design Method

The current AISC (1999) Commentary adopts the following two modified effective length factors K' for rigidly connected columns.

For story stiffness method

$$K' = \sqrt{\frac{P_e}{0.822 P_u} \sum P_u \left(\frac{\Delta_{0h}}{\sum HL} \right)} \tag{31.102}$$

where

$$P_e = \frac{\pi^2 EI}{L^2} \tag{31.103}$$

P_u is the required axial compressive strength of the column under consideration, $\sum P_u$ is the required axial compressive strength of all columns in a story, Δ_{oh} is the lateral interstory deflection, $\sum H$ is the sum of all story horizontal forces producing Δ_{oh}, and L is the story height. The 0.822 factor is the ratio of the lateral column shear force per radian of drift to the buckling load of a sway permitted column with large end restraint, $G = 0$. This factor will approach 1.0 for more flexible systems with a large percentage of leaner columns.

For the story buckling method

$$K' = \sqrt{\frac{P_e}{P_u}\left(\frac{\sum P_u}{\sum P_{e2}}\right)} \tag{31.104}$$

where $\sum P_{e2}$ is the sum of Euler loads of all columns in a story providing lateral stiffness for the frame based on the effective length factor obtained from the alignment chart for an unbraced frame, P_u is the required axial compressive strength for rigid column, and $\sum P_u$ is the required axial compressive strength of all columns in a story.

EXAMPLE 31.8

Given: A frame with a leaning column is shown in Figure 31.16a (Lui and Sun 1995). Evaluate the K-factor for column AB using various methods. The bottom of column AB is assumed to be ideally pin-ended for comparison purposes. $E = 29,000\,\text{ksi}$ (200 GPa).

Solution

Alignment Chart Method

$$G_A = \infty$$

$$G_B = \frac{\sum E_c I_c / L_c}{\sum \alpha_k E_g I_g / L_g} = \frac{EI/L}{0.5EI/L} = 2.0$$

From Figure 31.3b, we have $K_{AB} = 2.6$.

Lima and McNamara Method
For this frame, $\sum P = \sum Q = P$ and $K_0 = 2.6$. From Equation 31.100, we have

$$K_{AB} = K_0\sqrt{1 + \frac{\sum Q}{\sum P}} = 2.6\sqrt{1 + 1} = 3.68$$

LeMessurier Method
For column AB, $G_A = \infty$ and $G_B = 2.0$; from the alignment chart, $K_0 = 2.6$.
According to Equations 31.91 to 31.94, we have

$$\beta|_{G_A = \infty} = \frac{6(G_A + G_B) + 36}{2(G_A + G_B) + G_A G_B + 3}\bigg|_{G_A = \infty} = \frac{6}{2 + G_B} = \frac{6}{2 + 2} = 1.5$$

$$\sum P_L = (P_L)_{AB} = \frac{\beta EI}{L^2} = 1.5\frac{EI}{L^2}$$

$$C_L = \left(\beta\frac{K_0^2}{\pi^2} - 1\right) = (1.5)\frac{2.6^2}{\pi^2} - 1 = 0.0274$$

$$K_{AB} = \sqrt{\frac{\pi^2 EI_{AB}}{L_{AB}^2 P_{AB}}\left(\frac{\sum P + \sum C_L P}{\sum P_L}\right)} = \sqrt{\frac{\pi^2 EI}{L^2 P}\left(\frac{2P + 0.0274P}{1.5EI/L^2}\right)}$$

$$= \sqrt{13.34} = 3.65$$

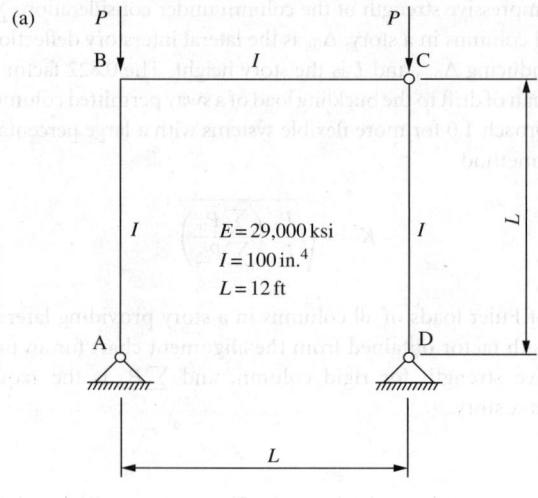

(a)

P

B

I

C

I $E = 29{,}000\,\text{ksi}$ I L
$I = 100\,\text{in.}^4$
$L = 12\,\text{ft}$

A   D

L

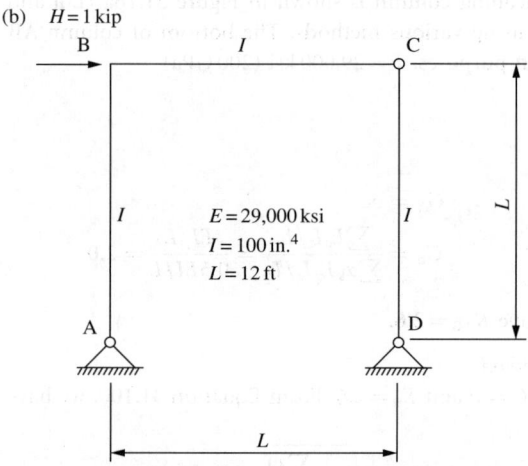

(b) $H = 1\,\text{kip}$

B I C

I $E = 29{,}000\,\text{ksi}$ I L
$I = 100\,\text{in.}^4$
$L = 12\,\text{ft}$

A   D

L

FIGURE 31.16 A leaning column frame: (a) frame dimension and loads and (b) frame subjected to fictitious lateral loads.

AISC Load and Resistance Factor Design Method
Using Equation 31.102 for column AB

$$K_{AB} = \sqrt{\frac{\sum P_u}{P_{AB}} \times \left(\frac{I_{AB}}{\sum (I/K_{i0}^2)} \right)} = K_{i0}\sqrt{2} = 3.68$$

Lui Method

1. Apply a small lateral force $H = 1$ kip as shown in Figure 31.16b.
2. Perform a first-order analysis and find $\Delta_1 = 0.687$ in. (17.45 mm).
3. Calculate η factors from Equation 31.96.
   Since column CD buckles in a single curvature, $m = -1$,

$$\eta_{CD} = \frac{(3 + 4.8m + 4.2m^2)EI}{L^3} = \frac{(3 - 4.8 + 4.2)EI}{L^3} = \frac{2.4EI}{L^3}$$

For column AB, $m = 0$,

$$\eta_{AB} = \frac{(3 + 4.8m + 4.2m^2)EI}{L^3} = \frac{3EI}{L^3}$$

$$\sum \eta = \frac{3EI}{L^3} + \frac{2.4EI}{L^3} = \frac{5.4(29,000)(100)}{(144)^3}$$

$$= 5.245 \text{ kip/in. } (0.918 \text{ kN/mm})$$

4. Calculate the K-factor from Equation 31.95

$$K_{AB} = \sqrt{\left(\frac{\pi^2 EI_{AB}}{P_{AB}L_{AB}^2}\right)\left[\left(\sum \frac{P}{L}\right)\left(\frac{1}{5\sum\eta} + \frac{\Delta_1}{\sum H}\right)\right]}$$

$$= \sqrt{\left(\frac{\pi^2(29,000)(100)}{P(144)^2}\right)\left[\left(\frac{2P}{144}\right)\left(\frac{1}{5(5.245)} + \frac{0.687}{1}\right)\right]} = 3.73$$

From an eigenvalue analysis, $K_{AB} = 3.69$ is obtained. It is seen that a direct use of the alignment chart leads to a significant error for this frame, and other approaches give good results. However, the LeMessurier approach requires the use of the alignment chart, and the Lui approach requires a first-order analysis subjected to a fictitious lateral loading.

31.7.2 Leaning Columns

Recognizing that a leaning column is being braced by rigid columns, a model for the leaning column as shown in Figure 31.15b was proposed by Lui (1992). Rigid columns provide lateral stability to the whole structure and are represented by a translation spring with a spring stiffness S_K. The K-factor for a leaning column can be obtained as

$$K = \text{larger of } \begin{cases} 1 \\ \sqrt{\dfrac{\pi^2 EI}{S_K L^3}} \end{cases} \tag{31.105}$$

For most commonly framed structures, the term $(\pi^2 EI/S_K L^3)$ normally does not exceed unity, and so $K = 1$ often governs. AISC (1999) suggests that leaning columns with $K = 1$ may be used in unbraced frames provided that the lack of lateral stiffness from simple connections to the frame $(K = \infty)$ is included in the design of moment frame columns. Aristizabal-Ochoa (1994b) recommended that (1) the K-factors of leaning columns are identical to the K-factors of the rigid columns when they are subjected to the same magnitude axial loads and are made of the same section and (2) the K-factors of leaning columns must be greater than 1.0 or the K-factor corresponding to the fully braced column with the same supports or boundary conditions.

31.7.3 Remarks

Numerical studies by Geschwindner (1995, 2002) found that the Yura approach gives overly conservative results for some conditions, the Lim and McNamara approach provides sufficiently accurate results for design, and the LeMessurier approach is the most accurate, among the three. The Lim and McNamara approach could be appropriate for preliminary design, while the LeMessurier and Lui approaches would be appropriate for final design.

31.8 Crossing Bracing Systems

Diagonal bracing or X-bracing is commonly used in steel structures to resist horizontal loads. In current practice, the design of these types of bracing system is based on the assumptions that the compression diagonal has negligible capacity and the tension diagonal resists the total load. The assumption that the compression diagonal has a negligible capacity usually results in an overdesign (Picard ar d Beaulieu 1987, 1988).

Picard and Beaulieu (1987, 1988) reported theoretical and experimental studies on double-diagonal cross bracings (Figure 31.6) and found that

1. A general effective length factor equation (Figure 31.17) is given as

$$K = \sqrt{0.523 - \frac{0.428}{C/T}} \geq 0.50 \qquad\qquad (31.106)$$

   where C and T represent the compression and tension forces obtained from an elastic analysis, respectively.
2. When the double diagonals are continuous and attached at the intersection point, the effective length of the compression diagonal is 0.5 times the diagonal length, that is, $K = 0.5$, because the C/T ratio is usually smaller than 1.6.

EL-Tayem and Goel (1986) reported a theoretical and experimental study about the X-bracing system made from single equal-leg angles. They concluded that

1. Design of X-bracing system should be based on an exclusive consideration of one half-diagonal only.
2. For X-bracing systems made from single equal-leg angles, an effective length of 0.85 times the half-diagonal length is reasonable, that is, $K = 0.425$.

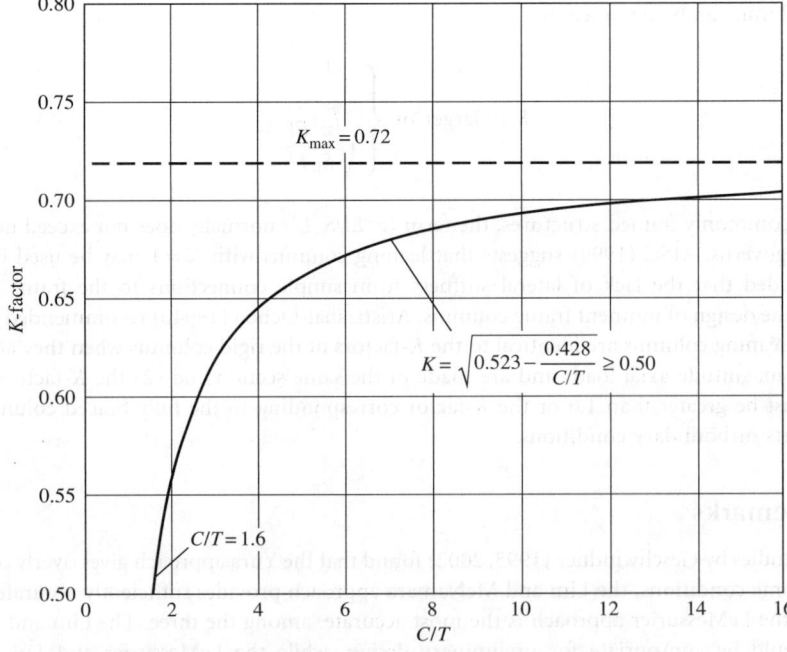

FIGURE 31.17   Effective length factor of compression diagonal (Picard and Beaulieu 1987).

Sabelli and Hohbach (1999) studied the relationship between axial load and end rotational stiffness for cross-braced frames. For in-plane buckling, the lower bound values of K-factor equal 0.422 for pinned-end and 0.245 for fixed-end; the upper bound values of K-factor equal 0.4 for pinned-end and 0.272 for fixed-end. For out-of-plan buckling, the lower bound values of K-factor equal 0.5 for pinned-end and 0.35 for fixed-end.

31.9   Latticed and Built-Up Members

The main difference of behavior between solid-webbed members and latticed members and built-up members is the effect of shear deformation on their buckling strength. For solid-webbed members, shear deformation has a negligible effect on their buckling strength, while for latticed structural members using lacing bars and batten plates, shear deformation has a significant effect on their buckling strength. It is common practice that when the buckling model involves relative deformation produced by shear forces in the connectors, such as lacing bars and batten plates, between individual components, a modified effective length factor K_m is defined as follows:

$$K_{\mathrm{m}} = \alpha_{\mathrm{v}} K \tag{31.107}$$

where K is the usual effective length factor of a latticed member acting as a unit obtained from a structural analysis and α_v is the shear factor to account for shear deformation on the buckling strength, or the modified *effective slenderness ratio* $(KL/r)_m$ should be used in the determination of the compressive strength. Details of the development of the shear factor α_v can be found in textbooks by Bleich (1952) and Timoshenko and Gere (1961). The following section briefly summarizes α_v formulas for various latticed members.

31.9.1   Laced Columns

For laced members as shown in Figure 31.18, by considering shear deformation due to the lengthening of diagonal lacing bars in each panel and assuming hinges at joints, the shear factor α_v has the form

$$\alpha_{\mathrm{v}} = \sqrt{1 + \frac{\pi^2 EI}{(KL)^2} \frac{1}{A_{\mathrm{d}} E_{\mathrm{d}} \sin\phi \cos^2\phi}} \tag{31.108}$$

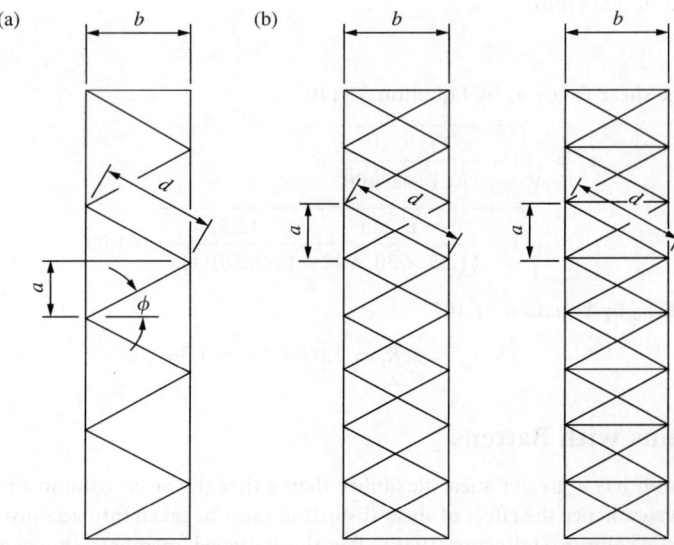

FIGURE 31.18   Typical configurations of laced members: (a) single lacing and (b) double lacing.

where E_d is the modulus of elasticity of materials for lacing bars, A_d is the cross-sectional area of all diagonals in one panel, and ϕ is the angle between the lacing diagonal and the axis that is perpendicular to the member axis.

If the lengths of the lacing bars are given (Figure 31.18), Equation 31.108 can be rewritten as

$$\alpha_v = \sqrt{1 + \frac{\pi^2 EI}{(KL)^2} \frac{d^3}{A_d E_d ab^2}} \tag{31.109}$$

where a, b, and d are height of the panel, depth of the member, and length of the diagonal, respectively.

The SSRC (Galambos 1988) suggested that a conservative estimate of the influence of 60 or 45° lacing, as generally specified in bridge design practice, can be made by modifying the overall effective length factor K by multiplying a factor α_v, originally developed by Bleich (1952) as follows:

$$\alpha_v = \sqrt{1 + 300/(KL/r)^2} \quad \text{for } \frac{KL}{r} > 40 \tag{31.110}$$

$$\alpha_v = 1.1 \quad \text{for } \frac{KL}{r} \leq 40 \tag{31.111}$$

EXAMPLE 31.9

Given: A laced column with angles and cover plates is shown in Figure 31.19. As usual, $K_y = 1.25$, $L = 30$ ft (9144 mm). Determine the modified effective length factor $(K_y)_m$ by considering the shear deformation effect.

Section properties

$I_y = 2259$ in.^{4} $(9.4 \times 10^8 \text{ mm}^4)$

$E = E_d$

$A_d = 1.69$ in.^{2} (1090 mm^2)

$a = 6$ in. (152 mm)

$b = 11$ in. (279 mm)

$d = 12.53$ in. (318 mm)

Solution

1. Calculate the shear factor α_v by Equation 31.110

$$\alpha_v = \sqrt{1 + \frac{\pi^2 EI}{(KL)^2} \frac{d^3}{A_d E_d ab^2}}$$

$$= \sqrt{1 + \frac{\pi^2 E(2259)}{(1.25 \times 30 \times 12)^2} \frac{12.53^3}{1.69E(6)(11)^2}} = 1.09$$

2. Calculate $(K_y)_m$ by Equation 31.107

$$(K_y)_m = \alpha_v K_y = 1.09(1.25) = 1.36$$

31.9.2 Columns with Battens

The battened column has a greater shear flexibility than either the laced column or the column with perforated cover plates, hence the effect of shear distortion must be taken into account in calculating the effective length of a column (Johnston 1976). For the battened members shown in Figure 31.20a, assuming that the points of inflection in the battens are at the batten midpoints and that the points

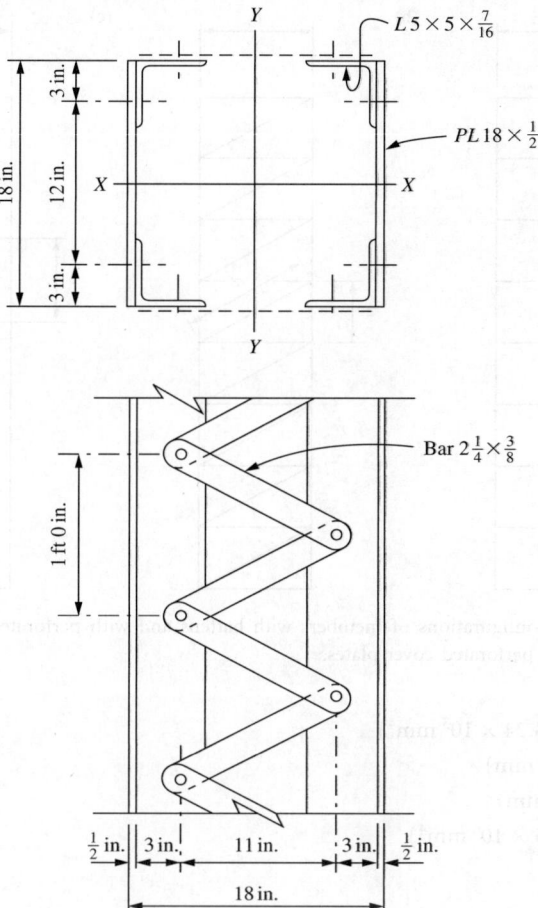

FIGURE 31.19 A laced column.

of inflection in the longitudinal element occur midway between the battens, the shear factor α_v is obtained as

$$\alpha_v = \sqrt{1 + \frac{\pi^2 EI}{(KL)^2}\left(\frac{ab}{12 E_b I_b} + \frac{a^2}{24 E I_f}\right)} \tag{31.112}$$

where E_b is the modulus of elasticity of materials for batten plates, I_b is the moment inertia of all battens in one panel in the buckling plane, and I_f is the moment inertia of one side of the main components taken about the centroid axis of the flange in the buckling plane.

EXAMPLE 31.10

Given: A battened column is shown in Figure 31.21. As usual, $K_y = 0.8$, $L = 30$ ft (9144 mm). Determine the modified effective length factor $(K_y)_m$ by considering the shear deformation effect.

Section properties

$I_y = 144$ in.^{4} $(6.0 \times 10^7$ mm$^4)$

$E = E_b$

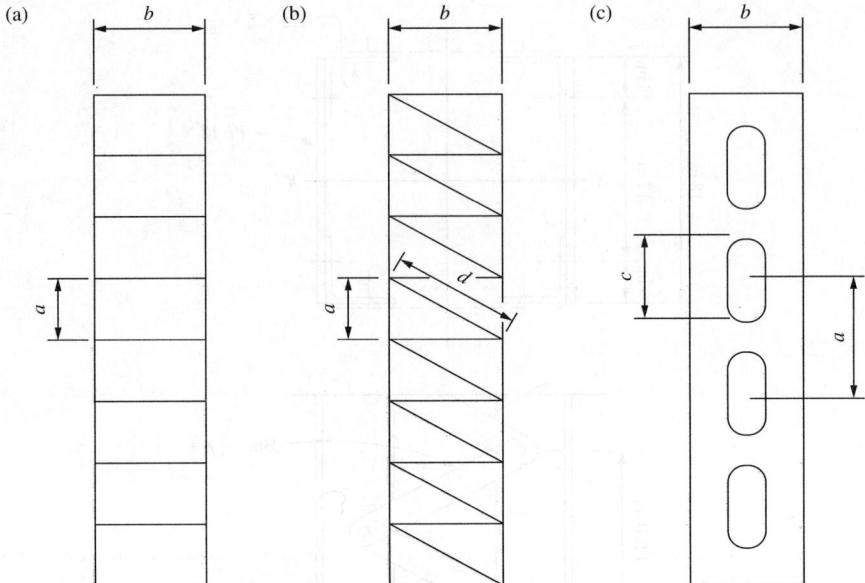

FIGURE 31.20 Typical configurations of members with battens and with perforated cover plates: (a) battens, (b) lacing-battens, and (c) perforated cover plates.

$I_f = 1.98$ in.^{4} $(8.24 \times 10^5$ mm$^4)$

$a = 15$ in. (381 mm)

$b = 9$ in. (229 mm)

$I_b = 9$ in.^{4} $(3.75 \times 10^6$ mm$^4)$

Solution

1. Calculate the shear factor α_v by Equation 31.112

$$\alpha_v = \sqrt{1 + \frac{\pi^2 EI}{(KL)^2}\left(\frac{ab}{12EI_b} + \frac{a^2}{24EI_f}\right)}$$

$$= \sqrt{1 + \frac{\pi^2 E(144)}{(0.8 \times 30 \times 12)^2}\left(\frac{15(9)}{12E(9)} + \frac{15^2}{24E(1.98)}\right)} = 1.05$$

2. Calculate $(K_y)_m$ by Equation 31.107

$$(K_y)_m = \alpha_v K_y = 1.05(0.8) = 0.84$$

31.9.3 Laced-Battened Columns

For the laced-battened columns as shown in Figure 31.20b, considering the shortening of the battens and the lengthening of the diagonal lacing bars in each panel, the shear factor α_v can be expressed as

$$\alpha_v = \sqrt{1 + \frac{\pi^2 EI}{(KL)^2}\left(\frac{d^3}{A_d E_d ab^2} + \frac{b}{aA_b E_b}\right)} \tag{31.113}$$

where E_b is the modulus of elasticity of materials for battens and A_b is the cross-sectional area of all battens in one panel.

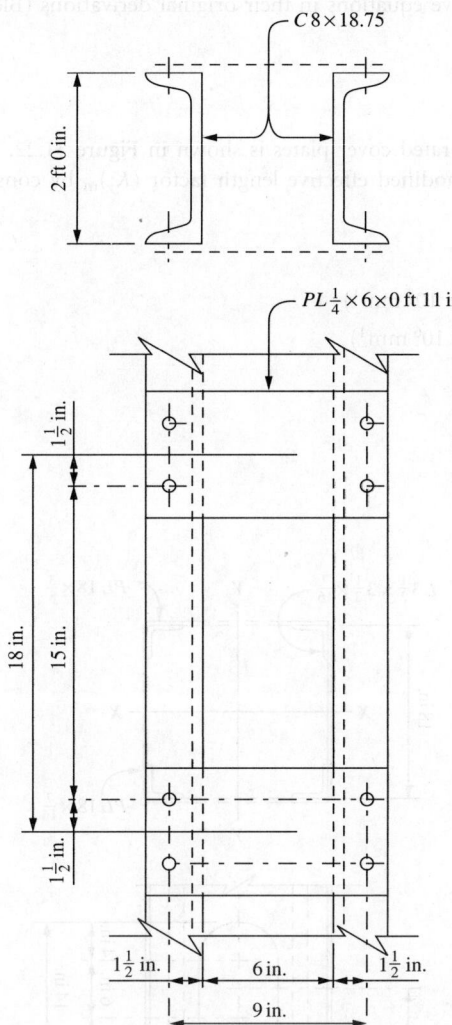

FIGURE 31.21   A battened column.

31.9.4 Columns with Perforated Cover Plates

For members with perforated cover plates shown in Figure 31.20c, considering the horizontal cross-member as infinitely rigid, the shear factor α_v has the form

$$\alpha_v = \sqrt{1 + \frac{\pi^2 EI}{(KL)^2}\left(\frac{9c^3}{64aEI_f}\right)} \tag{31.114}$$

where c is the length of a perforation.

It should be pointed out that the usual K-factor based on a solid member analysis is included in Equations 31.108 to 31.114. However, since the latticed members studied previously have pin-ended conditions, the K-factor of the member in the frame was not included in the second terms of the square root of the above equations in their original derivations (Bleich 1952; Timoshenko and Gere 1961).

EXAMPLE 31.11

Given: A column with perforated cover plates is shown in Figure 31.22. As usual, $K_y = 1.3$, $L = 25$ ft (7620 mm). Determine the modified effective length factor $(K_y)_m$ by considering the shear deformation effect.

Section properties

$$I_y = 2467 \text{ in.}^4 \ (1.03 \times 10^8 \text{ mm}^4)$$
$$I_f = 35.5 \text{ in.}^4 \ (1.48 \times 10^6 \text{ mm}^4)$$
$$a = 30 \text{ in. } (762 \text{ mm})$$
$$c = 14 \text{ in. } (356 \text{ mm})$$

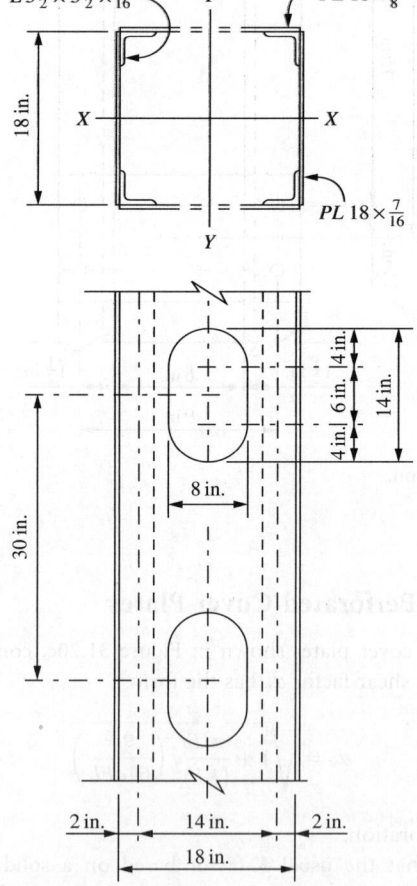

FIGURE 31.22 A column with perforated cover plates.

Solution

1. Calculate the shear factor α_v by Equation 31.114

$$\alpha_v = \sqrt{1 + \frac{\pi^2 EI}{(KL)^2} \left(\frac{9c^3}{64aEI_f} \right)}$$

$$= \sqrt{1 + \frac{\pi^2 E(2467)}{(1.3 \times 25 \times 12)^2} \left(\frac{9(14)^3}{64(30)E(35.5)} \right)} = 1.03$$

2. Calculate $(K_y)_m$ by Equation 31.107

$$(K_y)_m = \alpha_v K_y = 1.03(1.3) = 1.34$$

31.9.5 Built-Up Members with Bolted and Welded Connectors

AISC (1999) specifies that if the buckling of a built-up member produces shear forces in the connectors between individual component members, the usual slenderness ratio KL/r for compression members must be replaced by the modified slenderness ratio $(KL/r)_m$ in determining the compressive strength:

1. *For snug-tight bolted connectors*

$$\left(\frac{KL}{r} \right)_m = \sqrt{\left(\frac{KL}{r} \right)_0^2 + \left(\frac{a}{r_i} \right)^2} \tag{31.115}$$

2. *For welded connectors and for fully tightened bolted connectors*

$$\left(\frac{KL}{r} \right)_m = \sqrt{\left(\frac{KL}{r} \right)_0^2 + 0.82 \frac{\alpha^2}{(1 + \alpha^2)} \left(\frac{a}{r_{ib}} \right)^2} \tag{31.116}$$

where $(KL/r)_0$ is the slenderness ratio of the built-up member acting as a unit, $(KL/r)_m$ is the modified slenderness ratio of the built-up member, a/r_i is the largest slenderness ratio of the individual components, a/r_{ib} is the slenderness ratio of the individual components relative to its centroidal axis parallel to the axis of buckling, a is the distance between connectors, r_i is the minimum radius of gyration of individual components, r_{ib} is the radius of gyration of individual components relative to its centroidal axis parallel to the member axis of buckling, α is the separation ratio $= h/2r_{ib}$, and h is the distance between centroids of individual components perpendicular to the member axis of buckling

Equation 31.115 is the same as that used in the current Italian code as well as in other European specifications, based on test results (Zandonini 1985). In the equation, the bending effect is considered in the first term in square root, and shear force effect is taken into account in the second term. Equation 31.116 was derived from elastic stability theory and was verified by test data (Aslani and Goel 1991). In both cases the end connectors must be welded or slip-critical bolted (Aslani and Goel 1991).

EXAMPLE 31.12

Given: A built-up member with two back-to-back angles is shown in Figure 31.23. Determine the modified slenderness ratio $(KL/r)_m$ in accordance with AISC (1999), Equation 31.108:

$$r_{ib} = 0.735 \text{ in. } (19 \text{ mm})$$
$$a = 48 \text{ in. } (1219 \text{ mm})$$
$$h = 1.603 \text{ in. } (41 \text{ mm})$$
$$(KL/r)_0 = 70$$

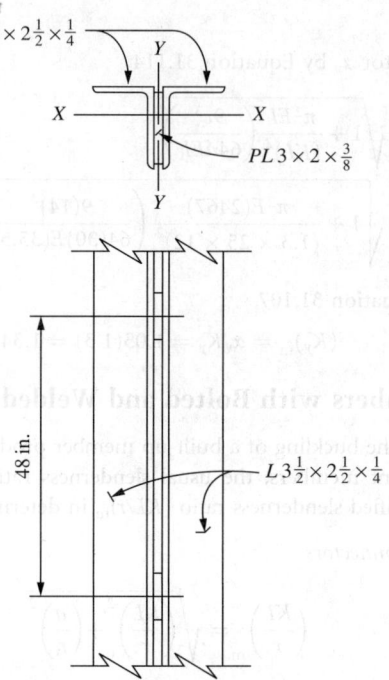

FIGURE 31.23   A built-up member with back-to-back angles.

Solution

1. Calculate the separation factor α

$$\alpha = \frac{h}{2r_{ib}} = \frac{1.603}{2(0.735)} = 1.09$$

2. Calculate the modified slenderness ratio $(KL/r)_m$ by Equation 31.116

$$\left(\frac{KL}{r}\right)_m = \sqrt{\left(\frac{KL}{r}\right)_0^2 + 0.82\frac{\alpha^2}{(1+\alpha^2)}\left(\frac{a}{r_{ib}}\right)^2}$$

$$= \sqrt{(70)^2 + 0.82\frac{1.09^2}{(1+1.09^2)}\left(\frac{48}{0.735}\right)^2}$$

$$= 82.5$$

31.10 Tapered Columns

The state-of-the-art design for tapered structural members was provided in the SSRC guide (Galambos 1988). The charts as shown in Figure 31.24 and Figure 31.25 can be used to evaluate the effective length factors for tapered column restrained by prismatic beams (Galambos 1988). In these figures, I_T and I_B are the moments of inertia of top and bottom beams, respectively; b and L are lengths of beam and column, respectively; and γ is the tapering factor as defined by

$$\gamma = \frac{d_1 - d_0}{d_0} \tag{31.117}$$

where d_0 and d_1 are the section depths of column at the smaller and larger ends, respectively.

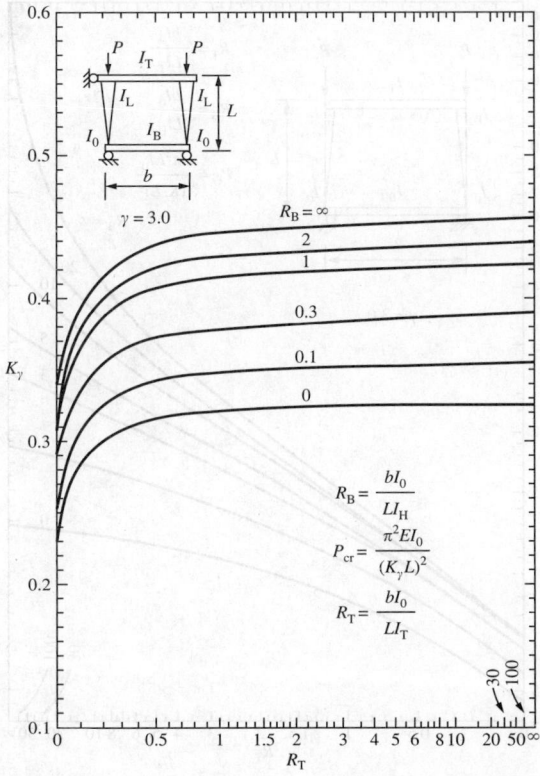

FIGURE 31.24 Effective length factor for tapered columns in braced frames (Galambos 1988).

Effects of shear deformation on the effective length factors of tapered I-section columns in a portal frame Figure 31.26 were studied by Li and Li (2000).

$$n = \frac{I_{c1}}{I_{c0}} \tag{31.118}$$

$$\lambda_c = \frac{L_c}{\sqrt{I_{c0}/A_{c0}}} = \frac{L_c}{r_{c0}} \tag{31.119}$$

When $\lambda_c \geq 36\sqrt{0.02n} + 26$, shear effect can be ignored.

The following shear factor was proposed to account for the tapered columns (Li and Li 2000):

$$\alpha_v = 1.0 + f_1(n)e^{-(\lambda_c - 17.75)/4} + f_2(n)e^{-(\lambda_c - 17.75)/22} \tag{31.120}$$

$$f_1(n) = 0.029031 + 0.0088n - 0.0003416n^2 + 0.000004155n^3 \tag{31.121}$$

$$f_2(n) = 0.06523 + 0.0112n - 0.0001056n^2 \tag{31.122}$$

31.11 Crane Columns

The columns in mill buildings and warehouses are designed to support overhead crane loads. The cross-section of a crane column may be uniform or stepped (see Figure 31.27). Over the past two decades, a number of simplified procedures have been developed for evaluating the K-factors for crane columns (Huang 1968; Anderson and Woodward 1972; Lay 1973; Agarwal and Stafiej 1980; Moore 1986;

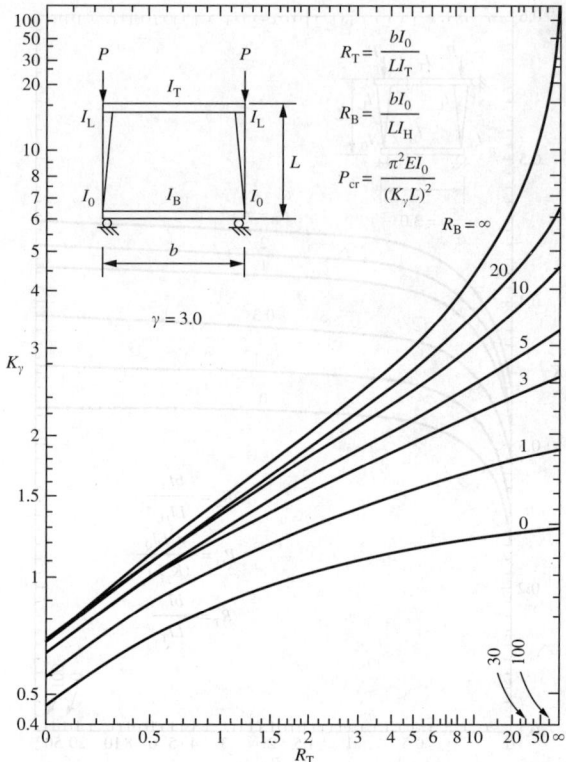

FIGURE 31.25 Effective length factor for tapered columns in unbraced frames (Galambos 1988).

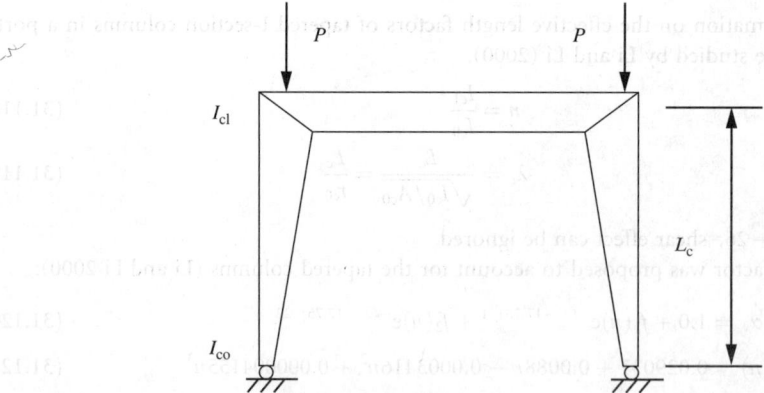

FIGURE 31.26 A tapered portal frame.

Fraser 1989, 1990; AISE 1991; Bendapudi 1994). Those procedures have limitation in terms of column geometry, loading, and boundary conditions. Most importantly, most of these studies ignored the interaction effect between the left and right columns of frames and were based on isolated member analyses (Lui and Sun 1995). Recently, a simple, yet reasonably accurate, procedure for calculating the *K*-factors for crane columns with any value of relative shaft length, moment of inertia, loading, and

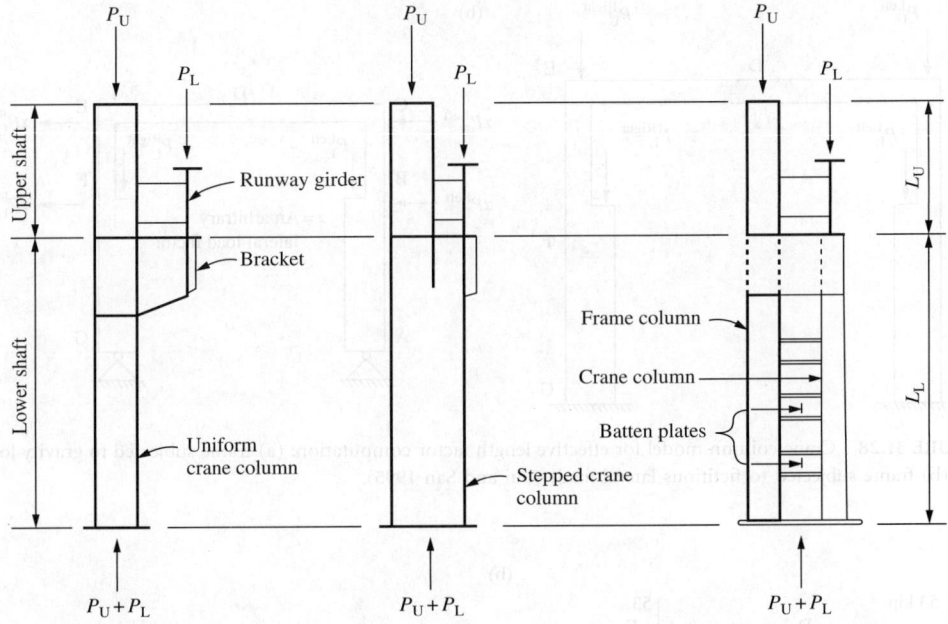

FIGURE 31.27 Typical crane columns.

boundary conditions was developed by Lui and Sun (1995). On the basis of the story stiffness concept and accounting for both member and frame instability effects in the formulation, Lui and Sun (1995) proposed the following procedure (see Figure 31.28):

1. Apply the fictitious lateral loads αP (α is an arbitrary factor, 0.001 may be used) in such a direction as to create a deflected geometry for the frame that closely approximates its actual buckled configuration.
2. Perform a first-order elastic analysis on the frame subjected to the fictitious lateral loads (Figure 31.28b). Calculate $\Delta_1/\sum H$, where Δ_1 is the average lateral deflection at the intermediate load points (i.e., points B and F) of columns and $\sum H$ is the sum of all fictitious lateral loads that act at and above the intermediate load points.
3. Calculate η using the results obtained from a first-order elastic analysis for lower shafts (i.e., segments AB and FG), according to Equation 31.89.
4. Calculate the K-factor for the lower shafts using Equation 31.88.
5. Calculate the K-factor for upper shafts using the following formula:

$$K_U = K_L \left(\frac{L_L}{L_U}\right) \sqrt{\left(\frac{P_L + P_U}{P_U}\right)\left(\frac{I_U}{I_L}\right)} \tag{31.123}$$

where P is the applied load and subscripts U and L represent the upper and lower shafts, respectively.

EXAMPLE 31.13

Given: A stepped crane column is shown in Figure 31.29a. The example is the same frame as that used by Fraser (1990) and Lui and Sun (1995). Determine the effective length factors for all columns using the Lui approach. $E = 29{,}000$ ksi (200 GPa).

$$I_{AB} = I_{FG} = I_L = 30{,}000 \text{ in.}^4 \ (1.25 \times 10^{10} \text{ mm}^4)$$
$$A_{AB} = A_{FG} = A_L = 75 \text{ in.}^2 \ (48{,}387 \text{ mm}^2)$$

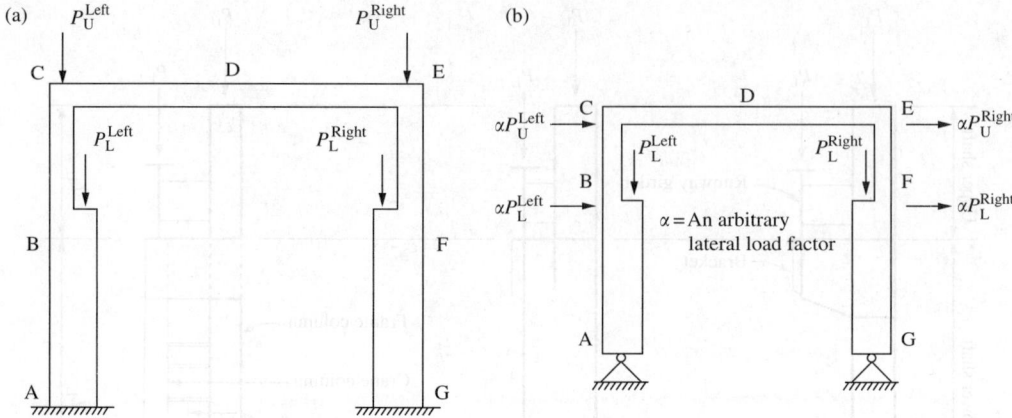

FIGURE 31.28 Crane column model for effective length factor computation: (a) frame subjected to gravity loads and (b) frame subjected to fictitious lateral loads (Lui and San 1995).

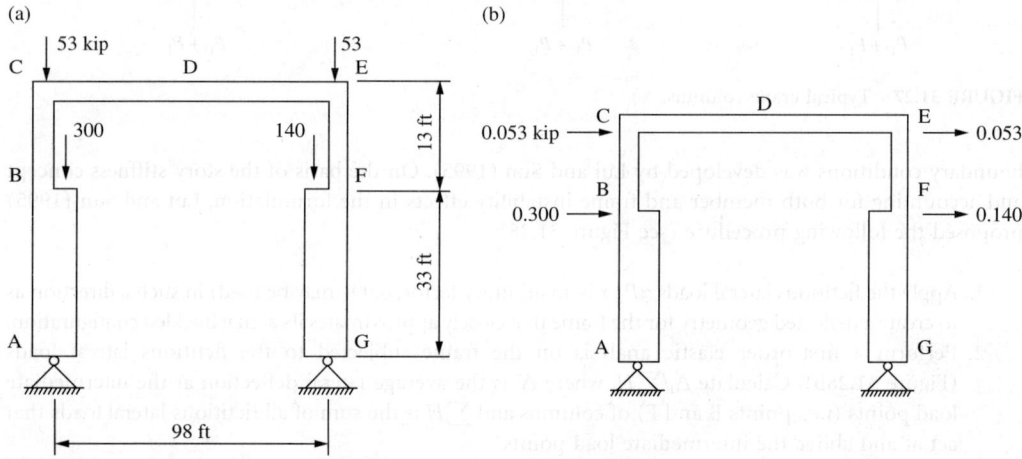

FIGURE 31.29 Pin-based stepped crane columns: (a) frame subjected to gravity loads and (b) frame subjected to fictitious lateral loads.

$$I_{BC} = I_{EF} = I_{CE} = I_U = 5420 \text{ in.}^4 \quad (2.26 \times 10^9 \text{ mm}^4)$$
$$A_{BC} = A_{EF} = A_{CE} = A_U = 34.14 \text{ in.}^2 \quad (22{,}026 \text{ mm}^2)$$

Solution

1. Apply a set of fictitious lateral forces with $\alpha = 0.001$ as shown in Figure 31.29b
2. Perform a first-order analysis and find

$$(\Delta_1)_B = 0.1086 \text{ in. } (2.76 \text{ mm}) \quad \text{and} \quad (\Delta_1)_F = 0.1077 \text{ in. } (2.74 \text{ mm})$$

so,

$$\frac{\Delta_1}{\sum H} = \frac{(0.1086 + 0.1077)/2}{0.053 + 0.3 + 0.053 + 0.14} = 0.198 \text{ in./kip} \quad (1.131 \text{ mm/kN})$$

3. Calculate η factors from Equation 31.89

Since the bottom of columns AB and FG is pin-based, $m = 0$,

$$\eta_{AB} = \eta_{FG} = \frac{(3 + 4.8m + 4.2m^2)EI}{L^3} = \frac{3EI}{L^3}$$

$$= \frac{(3)(29,000)(30,000)}{(396)^3} = 42.03 \text{ kip/in. } (7.36 \text{ mm/kN})$$

$$\sum \eta = 42.03 + 42.03 = 84.06 \text{ kip/in. } (14.72 \text{ mm/kN})$$

4. Calculate the K-factors for columns AB and FG using Equation 31.88

$$K_{AB} = \sqrt{\left(\frac{\pi^2(29,000)(30,000)}{(353)(396)^2}\right)\left[\left(\frac{353 + 193}{396}\right)\left(\frac{1}{5(84.06)} + 0.198\right)\right]}$$

$$= 6.55$$

$$K_{FG} = \sqrt{\left(\frac{\pi^2(29,000)(30,000)}{(193)(396)^2}\right)\left[\left(\frac{353 + 193}{396}\right)\left(\frac{1}{5(84.06)} + 0.198\right)\right]}$$

$$= 8.85$$

5. Calculate the K-factors for columns BC and EF using Equation 31.123

$$K_{BC} = K_{AB}\left(\frac{L_{AB}}{L_{BC}}\right)\sqrt{\left(\frac{P_{AB} + P_{BC}}{P_{BC}}\right)\left(\frac{I_{BC}}{I_{AB}}\right)}$$

$$= 6.55\left(\frac{396}{156}\right)\sqrt{\left(\frac{353}{53}\right)\left(\frac{5,420}{30,000}\right)} = 18.2$$

$$K_{EF} = K_{FG}\left(\frac{L_{FG}}{L_{EF}}\right)\sqrt{\left(\frac{P_{FG} + P_{EF}}{P_{EF}}\right)\left(\frac{I_{EF}}{I_{FG}}\right)}$$

$$= 8.85\left(\frac{396}{156}\right)\sqrt{\left(\frac{193}{53}\right)\left(\frac{5,420}{30,000}\right)} = 18.2$$

The K-factors calculated above are in good agreement with the theoretical values reported by Lui and Sun (1995).

31.12 Columns in Gable Frames

For a pin-based gable frame subjected to a uniformly distributed load on the rafter as shown in Figure 31.30a, Lu (1965) presented a graph (Figure 31.30b) to determine the effective length factors of columns. For frames having different member sizes for rafter and columns having (L/h) of (f/h) ratios not covered in Figure 31.30, an approximate method is available for determining K-factors of columns (Hansell 1964). The method is to find an equivalent portal frame whose span length is equal to twice the rafter length L_r (see Figure 31.30a). The K-factors can be determined from the alignment charts using $G_{top} = (I_c/h)/(I_r/2L_r)$ and the corresponding G_{bottom}.

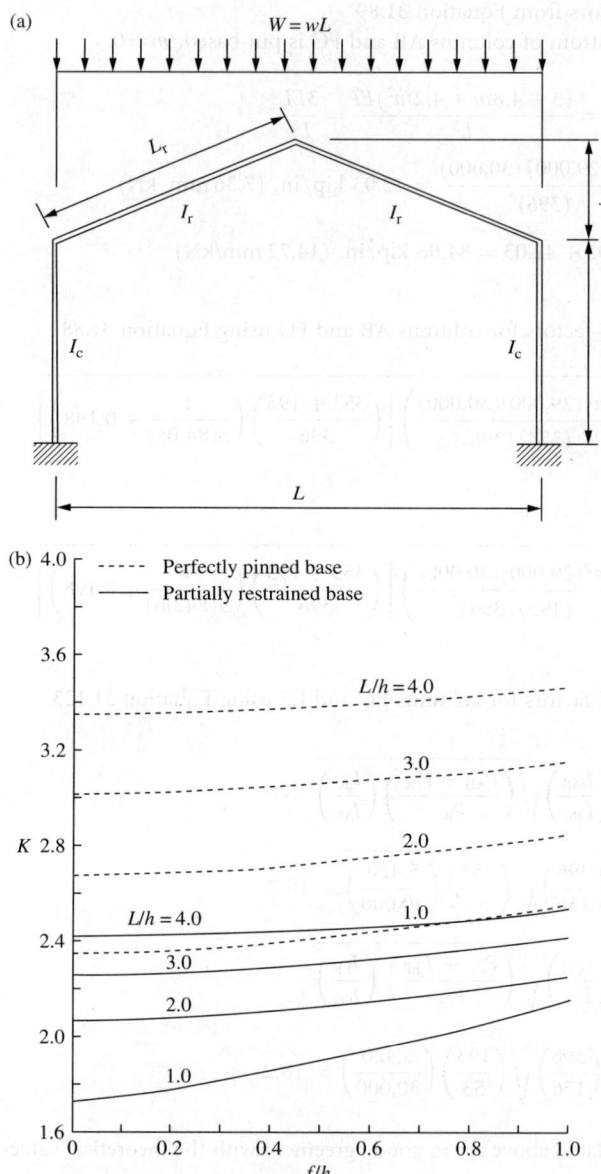

FIGURE 31.30 Effective length factor for columns in pinned-base gable frames: (a) a pinned-base gable frame and (b) effective length factors (Lu 1965).

31.13 Columns in Fire

The effects of structural continuity on fire exposed columns have been investigated over the last two decades. In the fire limit state, unexposed relative cool parts impose restraint on the structure's elements under fire. The Eurocode 4 Part 1.2 (ECS 1994) recommends that the K-factor for braced frames for a column continuous at both ends may be taken as 0.5 and that for a column continuous at one end may be taken as 0.7.

Experimental studies on fixed-ended steel columns in fire performed by Ali (2000) and Ali and O'Connor (2001) gave an average value of 0.56 for highly restrained columns and 0.61 for a low rotational restraint.

Theoretical studies on concrete-filled columns (Wang 1999) found that Eurocode recommendations give very accurate results of fire resistance for unreinforced subframe columns. The fire resistance of reinforced subframe columns were slightly overestimated, but the difference was small. An extensive parametric study on concrete-filled steel square hollow sections (Bailey 2000) concluded that for braced columns continuous at both ends, $K = 0.55$ without considering local buckling at the top of the heated column, and $K = 0.75$ when considering the effects of possible local buckling; for braced columns continuous at one end, $K = 0.8$; for an idealized pinned foundation, $K = 1.0$.

31.14 Members in Space Frames

Space frames with various nodal connection systems are widely used all over the world. The current design method is based on the assumption that the joints are pin-ended. Experimental and theoretical investigations on Oktalok space frame as shown in Figure 31.31 provided the following formula to estimate effective length factor of members in Oktalok space frames (Zhao et al. 2000):

$$K = 1.25 - \frac{L/r}{246} \tag{31.124}$$

where $98 \leq L/r \leq 140$.

31.15 Truss-Type Highway Sign Support Structures

For a typical truss-type highway sign support structure as shown in Figure 31.32, the main vertical support columns are usually assumed as fixed at the base and pinned to the top truss box. $K = 1.0$ is

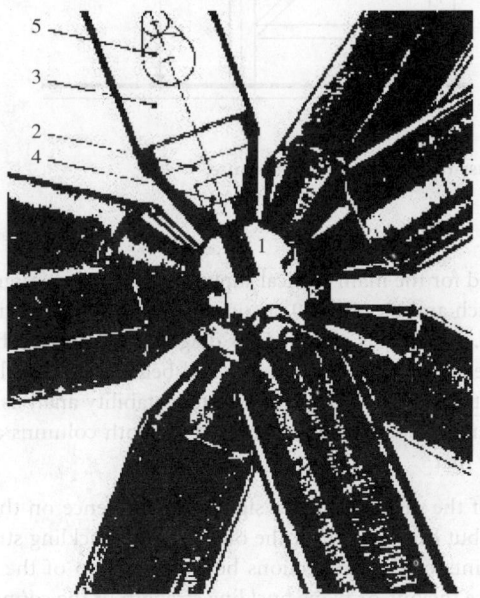

FIGURE 31.31 Oktalok nodal system.

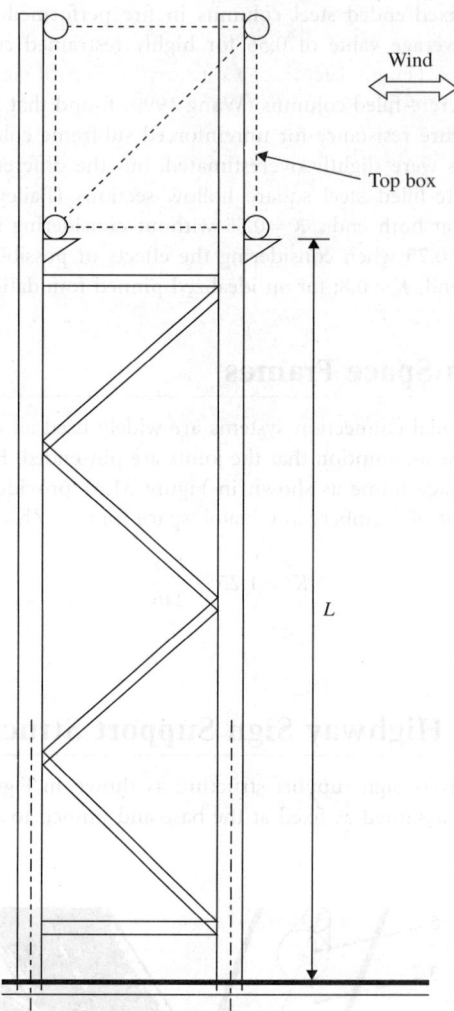

FIGURE 31.32 A Typical truss-type sign support structure.

often conservatively assumed for the main vertical support columns in design. Some engineers may take a more conservative approach and use $K = 2.0$, assuming the frame to be unbraced against sidesway in the perpendicular direction. $K = 0.85$ is assumed for diagonal members. The current overly simplified procedure does not provide a true presentation of actual behavior (DeWolf and Yang 2000) and may lead to excessively conservative design. A structural system stability analysis indicated that a significant reduction in the effective length factors can be achieved for both columns and diagonals. DeWolf and Yang (2000) recommended that

- Changing the sizes of the diagonals has a significant influence on the overall in-plane buckling strength of columns but does not affect the out-of-plane buckling strength.
- Increasing the restraints of the connections between the top of the columns and the sign box structures can increase the out-of-plane buckling strength. If the connection between top box and column is pinned, $K = 1.37$. If it is fully restricted against rotation, $K = 0.82$.

31.16 Precast Concrete Skeletal Frames

Precast skeletal structures are usually designed either as unbraced structures, up to three or four stories, or as fully braced structures, up to 15 to 20 stories in height. The design and analysis of precast skeletal structures is greatly influenced by the behavior of beam-to-column connections. Eillott et al. (1998) conducted a series of experimental studies of precast concrete beam–column connections and proposed an approach to calculate the K-factors.

Precast concrete frames (Figure 31.33) are analyzed either as fully unbraced structures (upper frame) or as partially braced structures (lower frame) where shear walls or cores provide lateral bracing up to a certain level and the frame is unbraced above this point. The following equations were proposed to evaluate the K-factors for the subframes, F1, F2, and F3 (Figure 31.33):

For frame F1

$$K = \begin{cases} 1 + \dfrac{1}{0.2 + 10G_s} + \dfrac{G}{0.3 + 1.8G_s - 0.45G_s^2} & \text{for } 0.1 < G_s \leq 2 \\[4mm] 1.1 + \dfrac{1}{7.4 + 7.4G_s - 0.4G_s^2} + \dfrac{G}{1.6 + 0.3G_s} & \text{for } 2 \leq G_s \leq 10 \end{cases} \qquad (31.125)$$

For frame F2

$$K = \begin{cases} 1 + \dfrac{1}{2.0 + 2G_s + 4G_s^2} + \dfrac{G}{4 + 0.5G_s} & \text{for } 0.1 < G_s \leq 2 \\[4mm] 1 + \dfrac{1}{8.6 + 8.4G_s - 0.4G_s^2} + \dfrac{G}{3.9 + 0.9G_s} & \text{for } 2 \leq G_s \leq 10 \end{cases} \qquad (31.126)$$

For frame F3

$$K = \begin{cases} 1 + \dfrac{1}{1.25 + 2.5G_s + 2.5G_s^2} + \dfrac{G}{2.25 + 0.5G_s} & \text{for } 0.1 < G_s \leq 2 \\[4mm] 1 + \dfrac{1}{6.5 + 5.6G_s - 0.3G_s^2} + \dfrac{G}{2.7 + 0.3G_s} & \text{for } 2 \leq G_s \leq 10 \end{cases} \qquad (31.127)$$

where G is the column end stiffness ratio as defined in Equation 31.7 and G_s is the relative semirigidity of the beam-to-column connection and is defined as joint stiffness J/beam flexural stiffness $4EI/L$.

31.17 Steel Moment Frame

In the proposed American Institute of Steel Construction (AISC) 2005 Specification, attempts are made to use $K = 1.0$ for nominal column strengths, P_n, in the interaction equations, when the required

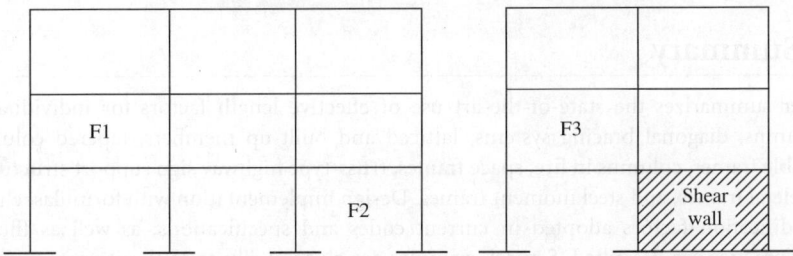

FIGURE 31.33 Types of precast frames.

strengths for members, connections, and other structural elements are obtained from the second-order elastic analysis under load and resistance factor design (LRFD) combinations. Its appendix provides the direct second-order analysis method for steel moment frames.

31.17.1 Second-Order Elastic Analysis

The second-order analysis required for using $K = 1.0$ in the interaction equations should capture both the P–δ and the P–Δ effects. The analysis may be performed either using a direct second-order analysis or by modifying the results of a first-order analysis using the B_1 and B_2 amplification factors (AISC 1999), provided that the B_1 and B_2 factors are based on the reduced flexural stiffness. Approximate P–Δ methods should be permitted when the factored axial loads in all columns are less than 15% of their respective Euler buckling loads. When the second-order displacement based on 20% reduced elastic stiffness is greater than six times the first-order displacement, it is recommended to increase the frame stiffness to limit the amplification to 6. Otherwise, the high nonlinearity would be such that small changes in gravity load or member stiffness will result in large changes in the calculated second-order effects.

31.17.2 Direct Second-Order Analysis

The direct second-order analysis uses the notional load concept and the reduced flexural stiffness principle.

31.17.2.1 Notional Load Concept

Notional loads are lateral loads that are applied at each floor level and are specified in terms of the gravity loads applied at that floor level to account for the effects of geometric imperfections, inelasticity, or both. A notional load applied at floor i, $N_i = 0.002 Y_i$ should be added to the factored lateral load in all LRFD load combinations. Y_i is the gravity load from the load combination acting on floor i and should be equal to or greater than the gravity load associated with the load combination being evaluated. The notional load should be applied in the direction that adds to the destabilizing effects under the specified load combination.

31.17.2.2 Reduced Stiffness Principle

A reduced flexural stiffness, $(EI)^*$, should be used for all members.

$$(EI)^* = \begin{cases} 0.8\tau EI & \text{for columns} \\ 0.8 EI & \text{for other members} \end{cases}$$

$$\tau = \begin{cases} 1.0 & \text{for } P_u/P_y \le 0.5 \\ 4\left[\dfrac{P_u}{P_y}\left(1 - \dfrac{P_u}{P_y}\right)\right] & \text{for } P_u/P_y > 0.5 \end{cases} \tag{31.128}$$

Alternatively, where $P_u > 0.5 P_y$ for any column in the moment frame, an additive notional load of $N_i = 0.001 Y_i$ should be added to the required notional load discussed in Section 31.17.2.1.

31.18 Summary

This chapter summarizes the state-of-the-art use of effective length factors for individual columns, framed columns, diagonal bracing systems, latticed and built-up members, tapered columns, crane columns, gable frames, columns in fire, space frames, truss-type highway sign support structures, precast concrete skeletal frames, and steel moment frames. Design implementation with formulas, charts, tables, various modification factors adopted in current codes and specifications, as well as those used in engineering practice are described. Several examples are given to illustrate the steps of practical applications of various methods.

Glossary

Alignment chart — A nomograph for determining the effective length factor K for some types of compression members.

Braced frame — A frame in which the resistance to lateral load or frame instability is primarily provided by diagonal bracing, shear walls, or equivalent means.

Built-up member — A member made of structural metal elements that are welded, bolted, and riveted together.

Column — A vertical structural member whose primary function is to carry loads parallel to its longitudinal axis.

Compression member — A structural member whose primary function is to carry compression loads parallel to its longitudinal axis.

Crane column — A column that is designed to support overhead crane loads.

Effective length factor K — A factor that when multiplied by actual length of the end-restrained column gives the length of an equivalent pin-ended column whose elastic buckling load is the same as that of the end-restrained column.

Framed column — A column in a framed structure.

Gable frame — A frame with a gabled roof.

Latticed member — A member made of two or more rolled shapes that are connected to one another by means of lacing bars, batten plates, or perforated plates.

Leaning column — A column that is connected to a frame with simple connections and does not provide lateral stiffness or sidesway resistance.

LRFD (load and resistance factor design) — A method of proportioning structural components (members, connectors, connecting elements, and assemblages) such that no applicable limit state is exceeded when the structure is subjected to all appropriate load combinations.

Tapered column — A column that has a continuous reduction in section from top to bottom.

Unbraced frame — A frame in which the resistance to lateral loads is provided by the bending stiffness of frame members and their connections.

References

AASHTO. 2004. *LRFD Bridge Design Specifications*, 3rd ed., American Association of State Highway and Transportation Officials, Washington, DC.

ACI. 2002. *Building Code Requirements for Structural Concrete (ACI 318-02) and Commentary (ACI 318R-02)*, American Concrete Institute, Farmington Hills, MI.

Agarwal, K.M. and Stafiej, A.P. 1980. Calculation of Effective Lengths of Stepped Columns. *AISC Eng. J.*, 15(4): 96–105.

AISC. 1989. *Allowable Stress Design Specification for Structural Steel Buildings*, 9th ed., American Institute of Steel Construction, Chicago, IL.

AISC. 1999. *Load and Resistance Factor Design Specification for Structural Steel Buildings*, 3rd ed., American Institute of Steel Construction, Chicago, IL.

AISE. 1991. *Guide for the Design and Construction of Mill Buildings*, Association of Iron and Steel Engineers, Technical Report, No. 13, Pittsburgh, PA.

Ali, F. 2000. Determining the Effective Length of Fixed End Steel Columns in Fire, *J. Appl. Fire Sci.*, 10(1): 41–44.

Ali, F. and O'Connor, D. 2001. Structural Performance of Rotationally Restrained Steel Columns in Fire, *Fire Saf. J.*, 36: 679–691.

Anderson, J.P. and Woodward, J.H. 1972. Calculation of Effective Lengths and Effective Slenderness Ratios of Stepped Columns, *AISC Eng. J.*, 7(4): 157–166.

Aristizabal-Ochoa, J.D. 1994a. K-Factors for Columns in Any Type of Construction: Nonparadoxical Approach, *J. Struct. Eng.*, ASCE, 120(4): 1272–1290.

Aristizabal-Ochoa, J.D. 1994b. Slenderness K Factors for Leaning Columns, *J. Struct. Eng., ASCE*, 120(10): 2977–2991.

ASCE Task Committee on Effective Length. 1997. *Effective Length and Notional Load Approaches for Assessing Frame Stability: Implications for American Steel Design*, American Society of Civil Engineers, Reston, VA.

Aslani, F. and Goel, S.C. 1991. An Analytical Criteria for Buckling Strength of Built-Up Compression Members, *AISC Eng. J.*, 28(4): 159–168.

Bailey, C. 2000. Effective Lengths of Concrete-Filled Steel Square Hollow Sections in Fire, *Struct. Build.*, 140(2): 167–178.

Barakat, M. and Chen, W.F. 1991. Design Analysis of Semi-Rigid Frames: Evaluation and Implementation, *AISC Eng. J.*, 28(2): 55–64.

Bendapudi, K.V. 1994. Practical Approaches in Mill Building Columns Subjected to Heavy Crane Loads, *AISC Eng. J.*, 31(4): 125–140.

Bjorhovde, R. 1984. Effect of End Restraints on Column Strength — Practical Application, *AISC Eng. J.*, 21(1): 1–13.

Bleich, F. 1952. *Buckling Strength of Metal Structures*, McGraw-Hill, New York.

Bridge, R.Q. and Fraser, D.J. 1987. Improved G-Factor Method for Evaluating Effective Length of Columns, *J. Struct. Eng., ASCE*, 113(6): 1341–1356.

Chapius, J. and Galambos, T.V. 1982. Restrained Crooked Aluminum Columns, *J. Struct. Div., ASCE*, 108(ST12): 511–524.

Cheong-Siat-Moy, F. 1986. K-factor Paradox, *J. Struct. Eng., ASCE*, 112(8): 1647–1760.

Cheong-Siat-Moy, F. 1999. An Improved K-factor Formula, *J. Struct. Eng., ASCE*, 125(2): 169–174.

Chen, W.F. and Lui, E.M. 1991. *Stability Design of Steel Frames*, CRC Press, Boca Raton, FL.

Chu, K.H. and Chow, H.L. 1969. Effective Column Length in Unsymmetrical Frames, *Publ. Int. Assoc. Bridge Struct. Eng.*, 29(1).

CM. 1995. *Rules for Design of Steel Structure — 1996*, Eyrolles, Paris, pp. 154–157, 247–261.

Cranston, W.B. 1972. Analysis and Design of Reinforced Concrete Columns. *Research Report No. 20*, Paper 41.020, Cement and Concrete Association, London.

CSA. 1994. *Limit States Design of Steel Structures*, Standard CAN/CSA S-16.1, Canadian Standards Association, Rexdale, Ontario, Canada.

Dafedar, J.B., Desai, Y.M., and Shiyekar, M.R. 2001. Review of IS Code Provisions for Effective Length of Framed Columns, *Indian Concrete J.*, 75(6): 402–407.

DeWolf, J.T. and Yang, J. 2000. Stability Analysis of Truss Type Highway Sign Support Structures, *Report JHR* 00-280, University of Connecticut, Storrs, CT.

Disque, R.O. 1973. Inelastic K-factor in Design. *AISC Eng. J.*, 10(2): 33–35.

Duan, L. 1996. A Modified G-factor for Columns in Semi-Rigid Frames. *Research Report*, Division of Structures, California Department of Transportation, Sacramento, CA.

Duan, L. and Chen, W.F. 1988. Effective Length Factor for Columns in Braced Frames. *J. Struct. Eng., ASCE*, 114(10): 2357–2370.

Duan, L. and Chen, W.F. 1989. Effective Length Factor for Columns in Unbraced Frames. *J. Struct. Eng., ASCE*, 115(1): 149–165.

Duan, L. and Chen, W.F. 1996. Errata of Paper: Effective Length Factor for Columns in Unbraced Frames. *J. Struct. Eng., ASCE*, 122(1): 224–225.

Duan, L., King, W.S., and Chen, W.F. 1993. K-factor Equation to Alignment Charts for Column Design. *ACI Struct. J.*, 90(3): 242–248.

Dumonteil, P. 1992. Simple Equations for Effective Length Factors. *AISC Eng. J.*, 29(3): 111–115.

Dumonteil, P. 1999. Historical Note on K-Factor Equations. *AISC Eng. J.*, 36(2): 102–103.

Dumonteil, P. and Valley, M. 1995. Discussion of "Novel Design Algorithms for K Factor Calculation and Beam–Column Selection." *J. Struct. Eng., ASCE*, 121(2): 384–385.

ECCS, 1978. *European Recommendations for Steel Construction*, European Convention for Constructional Steelworks, Brussels, pp. 77–81.

ECS 1994. European Committee for Standardization, Eurocode 4, *Design of Composite Steel and Concrete Structure, Part 1.2: Structural Fire Design*, ENV 1994-101, British Standards Institution, London.

Elliott, K., Davies, G., and Gorgun, H. 1998. The Stability of Precast Concrete Skeletal Structures, *PCI J.*, 43(2): 42–60.

El-Tayem, A.A. and Goel, S.C. 1986. Effective Length Factor for the Design of X-Bracing Systems, *AISC Eng. J.*, 23(4): 41–45.

Essa, H.S. 1997. Stability of Columns in Unbraced Frames, *J. Struct. Eng., ASCE*, 123(7): 952–957.

Essa, H.S. 1998. New Stability Equation for Columns in Braced Frames, *Struct. Eng. Mech.*, 6(4): 411–425.

Fraser, D.J. 1989. Uniform Pin-Based Crane Columns, Effective Length, *AISC Eng. J.*, 26(2): 61–65.

Fraser, D.J. 1990. The In-Plane Stability of a Frame Containing Pin-Based Stepped Column, *AISC Eng. J.*, 27(2): 49–53.

Furlong, R.W. 1971. Column Slenderness and Charts for Design, *ACI J., Proc.*, 68(1): 9–18.

Galambos, T.V. 1960. Influence of Partial Base Fixity on Frame Instability, *J. Struct. Div., ASCE*, 86(ST5): 85–108.

Galambos, T.V. 1964. Lateral Support for Tier Building Frames, *AISC Eng. J.*, 1(1): 16–19.

Galambos, T.V. 1968. *Structural Members and Frames*, Prentice Hall International, London.

Galambos, T.V., ed. 1988. *Structural Stability Research Council, Guide to Stability Design Criteria for Metal Structures*, 4th ed., John Wiley & Sons, New York.

Geschwindner, L.F. 1995. A Practical Approach to the "Leaning" Column, *AISC Eng. J.*, 32(2): 63–72.

Geschwindner L.F. 2002. A Practical Look at Frame Analysis, Stability and Leaning Column, *AISC Eng. J.*, 39(4): 167–181.

Gurfinkel, G. and Robinson, A.R. 1965. Buckling of Elasticity Restrained Column, *J. Struct. Div., ASCE*, 91(ST6): 159–183.

Hajjar, J.F. and White, D.W. 1994. The Accuracy of Column Stability Calculations in Unbraced Frames and the Influence of Columns with Effective Length Factors Less Than One, *AISC Eng. J.*, 31(3): 81–97.

Hellesland, J. 1998. Application of the Method of Means to the Stability Analysis of Unbraced Frames, *J. Constr. Steel Res.* 46(1–3): 98.

Hellesland, J. 2000. Discussion of an Improved K-Factor Formula, *J. Struct. Eng., ASCE*, 126(5): 633–635.

Hellesland, J. and Bjorhovde, R. 1996. Restraint Demand Factors and Effective Length of Braced Columns, *J. Struct. Eng., ASCE*, 122(10): 1216–1224.

Hansell, W.C. 1964. Single-Story Rigid Frames, Chapter 20 in *Structural Steel Design*, Ronald Press, New York.

Hu, X.Y., Zhou, R.G., King, W.S., Duan, L., and Chen, W.F. 1993. On Effective Length Factor of Framed Columns in ACI Code, *ACI Struct. J.*, 90(2): 135–143.

Huang, H.C. 1968. Determination of Slenderness Ratios for Design of Heavy Mill Building Stepped Columns, *Iron Steel Eng.*, 45(11): 123.

Johnson, D.E. 1960. Lateral Stability of Frames by Energy Method. *J. Eng. Mech., ASCE*, 95(4): 23–41.

Johnston, B.G., ed. 1976. *Structural Stability Research Council, Guide to Stability Design Criteria for Metal Structures*. 3rd ed., John Wiley & Sons, New York.

Jones, S.W., Kirby, P.A., and Nethercot, D.A. 1980. Effect of Semi-Rigid Connections on Steel Column Strength, *J. Constr. Steel Res.*, 1(1): 38–46.

Jones, S.W., Kirby, P.A., and Nethercot, D.A. 1982. Columns with Semi-Rigid Joints, *J. Struct. Div., ASCE*, 108(ST2): 361–372.

Julian, O.G. and Lawrence, L.S. 1959. *Notes on J and L Nomograms for Determination of Effective Lengths*, Unpublished Report.

Kavanagh, T.C. 1962. Effective Length of Framed Column, *Trans., ASCE*, 127(II): 81–101.

King, W.S., Duan, L., Zhou, R.G., Hu, Y.X., and Chen, W.F. 1993. K-factors of Framed Columns Restrained by Tapered Girders in US Codes, *Eng. Struct.*, 15(5): 369–378.

Kishi, N., Chen, W.F., and Goto, Y. 1997. Effective Length Factor of Columns in Semi-Rigid and Unbraced Frames, *J. Struct. Eng.*, ASCE, 123(3): 313–320.

Koo, B. 1988. Discussion of Paper "Improved G-Factor Method for Evaluating Effective Length of Columns" by Bridge and Fraser, *J. Struct. Eng.*, ASCE, 114(12): 2828–2830.

Lay, M.G. 1973. Effective Length of Crane Columns, *Steel Constr.*, Aust. Inst. Steel Constr., 7(2): 9–19.

LeMessurier, W.J. 1977. A Practical Method of Second Order Analysis Part 2 — Rigid Frames, *AISC Eng. J.*, 14(2): 49–67.

Li, G.Q. and Li, J.J. 2000. Effects of Shear Deformation on the Effective Length of Tapered Columns with I-Section for Steel Portal Frames, *Struct. Eng. Mech.*, 10(5): 479–489.

Liew, J.Y.R., White, D.W., and Chen, W.F. 1991. Beam-Column Design in Steel Frameworks — Insight on Current Methods and Trends, *J. Constr. Steel Res.*, 18: 269–308.

Lim, L.C. and McNamara, R.J. 1972. Stability of Novel Building System, *Structural Design of Tall Steel Buildings, Vol. II-16, Proceedings*, ASCE-IABSE International Conference on the Planning and Design of Tall Buildings, Bethlehem, PA, pp. 499–524.

Lu, L.W. 1962. A Survey of Literature on the Stability of Frames, *Weld. Res. Conc. Bull.*, New York.

Lu, L.W. 1965. Effective Length of Columns in Gable Frame, *AISC Eng. J.*, 2(2): 6–7.

Lui, E.M. 1992. A Novel Approach for K-factor Determination, *AISC Eng. J.*, 29(4): 150–159.

Lui, E.M. and Chen, W.F. 1983. Strength of Columns with Small End Restraints, *J. Inst. Struct. Eng.*, 61B(1): 17–26.

Lui, E.M. and Sun, M.Q. 1995. Effective Length of Uniform and Stepped Crane Columns, *AISC Eng. J.*, 32(2): 98–106.

MacGregor, J.G. and Hage, S.E. 1977. Stability Analysis and Design of Concrete Frames, *J. Struct. Div.*, ASCE, 103(10): 1953–1979.

Maquoi, R. and Jaspart, J.P. 1989. Contribution to the Design of Braced Framed with Semi-Rigid Connections. *Proceedings, 4th International Colloquium, Structural Stability Research Council*, pp. 209–220. Lehigh University, Bethlehem, PA.

Menon, D. 2001. Fuzzy Logic Based Estimation of Effective Lengths of Columns in Partially Braced Multi-Storey Frames, *Struct. Eng. Mech.*, 11(3): 287–299.

Moore, W.E. II. 1986. A Programmable Solutions for Steeped Crane Columns, *AISC Eng. J.*, 21(2): 58–59.

Newmark, N.M. 1949. A Simple Approximate Formula for Effective End-Fixity of Columns, *J. Aero. Sci.*, 16(2).

Picard, A. and Beaulieu, D. 1987. Design of Diagonal Cross Bracings Part 1: Theoretical Study, *AISC Eng. J.*, 24(3): 122–126.

Picard, A. and Beaulieu, D. 1988. Design of Diagonal Cross Bracings Part 2: Experimental Study, *AISC Eng. J.*, 25(4): 156–160.

Roddis, W.M. K., Hamid, H.A., and Guo, C.Q. 1998. K Factors for Unbraced Frames: Alignment Chart Accuracy for Practical Frame Variations, *AISC Eng. J.*, 35(3): 81–93.

Razzaq, Z. 1983. End Restraint Effect of Column Strength, *J. Struct. Div.*, ASCE, 109(ST2): 314–334.

Rondal, J. 1988. Effective Length of Tubular Lattice Girder Members — Statistical Tests, *CIDECT Report*, 3K-88/9, Liege, Belgium.

Sabelli, R. and Hohbach, D. 1999. Design of Cross-Braced Frames for Predictable Buckling Behavior, *J. Struct. Eng.*, ASCE, 125(1): 163–168.

Salmon, C.G., Schenker, L., and Johnston, B.G. 1957. Moment-Rotation Characteristics of Column Anchorage, *Transactions*, ASCE, 122: 132–154.

Schmidt, J.A. 1999. Design of Steel Columns in Unbraced Frames Using Notional Loads, *Pract. Period. on Struct. Des. Constr.*, ASCE, 4(1): 24–28.

Shanmugam, N.E. and Chen, W.F. 1995. An Assessment of K Factor Formulas, *AISC Eng. J.*, 32(3): 3–11.

Sohal, I.S., Yong, Y.K., and Balagura, P.N. 1995. K-factor in Plastic and SOIA for Design of Steel Frames, *Proceeding, International Conference on Stability of Structures*, ICSS 95, June 7–9, Coimbatore, India, pp. 411–421.

Sugimoto, H. and Chen, W.F. 1982. Small End Restraint Effects on Strength of H-Columns, *J. Struct. Div., ASCE* 108(ST3): 661–681.

Timoshenko, S.P. and Gere, J.M. 1961. *Theory of Elastic Stability*, 2nd ed., McGraw-Hill Book Co, New York.

Tong, G.S. and Shi, Z.Y. 2001. The Stability of Weakly Braced Frames, *Adv. Struct. Eng.*, 4(4): 211–215.

Vinnakota, S. 1982. Planar Strength of Restrained Beam Columns. *J. Struct. Div., ASCE*, 108(ST11): 2349–2516.

Wang, Y.C. 1999. The Effects of Structural Continuity on the Fire Resistance of Concrete Filled Columns in Non-Sway Frames, *J. Constr. Steel Res.*, 50: 177–197.

White, D.W. and Clarke, M.J. 1997a. Design of Beam-Columns in Steel Frames. I: Study of Philosophies and Procedures, *J. Struct. Eng., ASCE*, 123(12): 1556–1564.

White, D.W. and Clarke, M.J. 1997b. Design of Beam-Columns in Steel Frames. II: Comparison of Standards, *J. Struct. Eng., ASCE*, 123(12): 1565–1575.

Winter, G. 1958. Lateral Bracing of Columns and Beams, *J. Struct. Div. ASCE*, 84(ST2): 1561-1–1561-22.

Winter, G. 1958. Lateral Bracing of Columns and Beams, *Transactions, ASCE*, 125(1): 809–825.

Winter, G. et al. 1948. Buckling of Trusses and Rigid Frames, *Cornell Univ. Bull.* No. 36, Engineering Experimental Station, Cornell University, Ithaca, NY.

Wood, R.H. 1974. Effective Lengths of Columns in Multi-Storey Buildings, *Struct. Eng.*, 52(7,8,9): 234–244, 295–302, 341–346.

Xu, L. and Liu, Y.X. 2002. Story-Based Effective Length Factor for Unbraced PR Frames, *AISC Eng. J.*, 39(1): 13–29.

Yura, J.A. 1971. The Effective Length of Columns in Unbraced Frames, *AISC Eng. J.*, 8(2): 37–42.

Yura, J.A. 1995. Bracing for Stability-State-of-the-Art, *Proceedings, Structures Congress XIII*, April ASCE, Boston, MA, pp. 88–103.

Zandonini, R. 1985. Stability of Compact Built-Up Struts: Experimental Investigation and Numerical Simulation (in Italian), *Construzioni Metalliche*, No. 4.

Zhao, X.L., Lim, P., Joseph, P., and Pi, Y.L. 2000. Member Capacity of Columns with Semi-Rigid End Conditions is Oktalok Space Frames, *Structural Engineering and Mechanics*, 10(1): 27–36.

Further Reading

Chen, W.F. and Lui, E.M. 1987. *Structural Stability: Theory and Implementation*, Elsevier, New York.

Chen, W.F., Goto, Y., and Liew, J.Y.R. 1996. *Stability Design of Semi-Rigid Frames*, John Wiley & Sons, New York.

Chen, W.F. and Kim, S.E. 1997. *LRFD Steel Design Using Advanced Analysis*, CRC Press, Boca Raton, FL.

32

Structural Bracing

Brian Chen
Wiss, Janney, Elstner
Associates, Inc.,
Irving, TX

Joseph Yura
Department of Civil Engineering,
University of Texas,
Austin, TX

32.1 Introduction

This chapter presents an overview of aspects related to the design of structural bracing used in beams, columns, and frame structures and is intended for practicing civil and structural engineers. Many of the design guidelines presented were incorporated into the 2002 *Load and Resistance Factor Design Manual* published by the American Institute of Steel Construction (AISC). The intended focus is on simplicity and ease of implementation over exact formulations. The basis for the design formulations along with a classification system for bracing systems is first presented. Design formulations are presented with illustrative numerical examples. Finally, common faulty bracing details are presented.

32.2 Types of Bracing

Bracing used in structural systems generally serve two primary functions. They resist secondary loads on structures (e.g., wind bracing) and increase the strength of individual members by resisting deformation in the weakest direction [1]. For the latter case, structural bracing forces higher modes of deformation by providing resistance to lateral and/or rotational displacement. This is achieved through axial, shear, and/ or flexural deformations of the bracing member. Diaphragms, for instance, provide restraint through their shear stiffness while diagonal cross-bracing relies on axial stiffness.

Bracing systems used to control instability fall into four general classifications: relative, nodal, continuous, or lean-on. Common configurations of each type are shown in Figure 32.1. Relative bracing

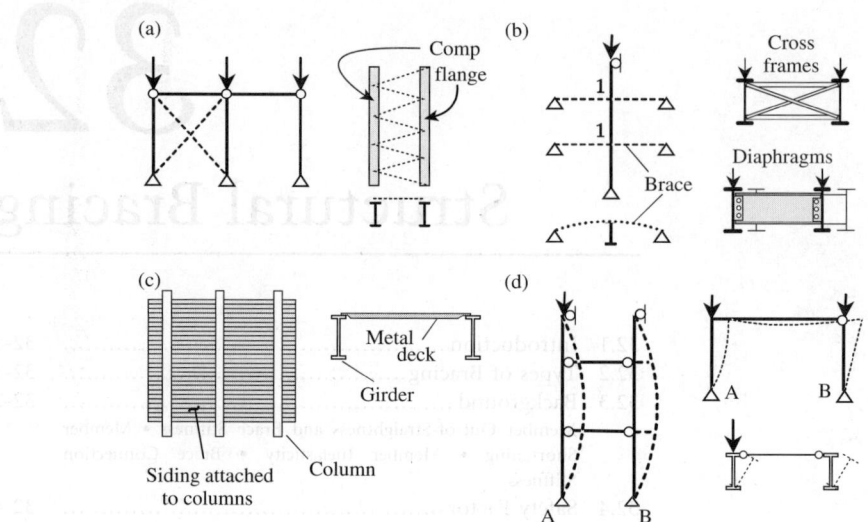

FIGURE 32.1   Types of bracing: (a) relative; (b) nodal; (c) continuous; and (d) lean-on.

systems, such as diagonal bracing or shear walls, prevent the relative lateral movement of adjacent stories or of adjacent points along the length of a member. Relative systems can be readily identified if a cut at any location along the length of a braced member passes through the brace member itself. Nodal systems control the movement only where they attach to the braced member and do not directly interact with adjacent brace points. Cross-frames or diaphragms between two adjacent beams are considered nodal braces. Continuous systems provide uninterrupted support along the entire length of a member, leaving no unbraced length. Shear walls and roof or floor deck are examples of continuous bracing systems. Lean-on systems rely on adjacent structural members to provide support. Lean-on bracing links together adjacent structural members such that buckling of one member requires all members in the system to buckle with the same lateral displacement.

32.3   Background

Structural bracing used to increase the strength of members must possess both sufficient *strength* and *stiffness* [2]. Simple bracing design rules such as designing a brace to resist 2% of the member compressive force address only the strength criterion. The stiffness of the brace along with the out-of-straightness of the member has a direct effect on the magnitude of the brace force [1]. Design recommendations based on perfectly straight members should not be used directly in design since extremely large brace forces and displacements may result [3].

32.3.1   Member Out-of-Straightness and Brace Stiffness

Winter [1,2] developed the concept of a dual *strength* and *stiffness* criterion for the design of bracing used to control instability. The required brace strength cannot be uniquely determined, but depends on both the magnitude of the brace stiffness and member initial out-of-straightness. The relationship between these parameters is illustrated for the relative column brace in Figure 32.2 and can be extended to other types of bracing systems. In order to reach the Euler buckling load, P_e, the brace must possess a minimum stiffness known as the *ideal stiffness*, β_i. Figure 32.3 shows the relationship between the brace stiffness and force. When the ideal stiffness is used ($\beta = \beta_i$), as the column load approaches P_e the sway deflections become very large. Unfortunately, this results in very large brace forces since $P_{br} = \beta\Delta$. At twice the ideal stiffness ($\beta = 2\beta_i$), the brace force equals 0.4% of the column load when $P = P_e$. For practical designs, the

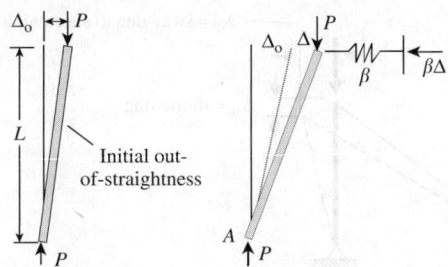

FIGURE 32.2   Relative brace.

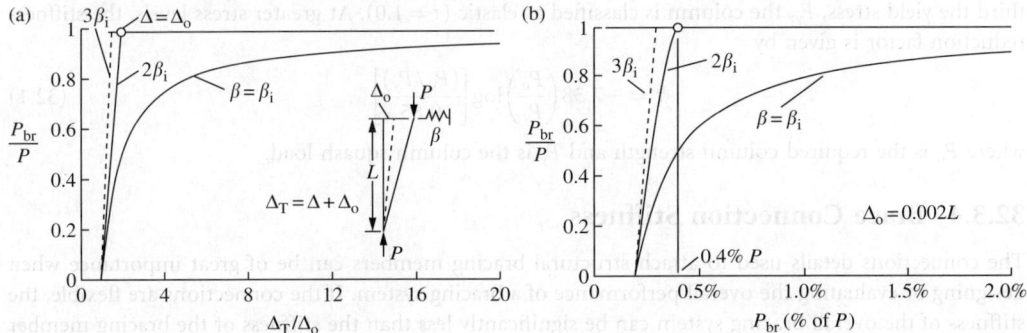

FIGURE 32.3   Effect of initial out-of-plumbness.

deflections and corresponding brace forces are kept small by using brace stiffnesses greater than the ideal stiffness. The plots developed in Figure 32.3 were based on an assumed initial out-of-straightness equal to $0.002L$. Larger out-of-straightness values linearly increase the magnitude of the brace forces.

32.3.2   Member Shortening

Compression elements such as columns or the compression flanges of beams shorten under externally applied compressive forces. For relative tension braces, shortening increases the brace force requirements by causing an apparent increase in the member out-of-straightness, as shown in Figure 32.4. As the member being braced shortens, slack is introduced into the bracing. Lateral movement at the brace points is necessary in order to return the tension brace to its original length prior to member shortening. The increases in brace force due to shortening can be accounted for by adding the out-of-straightness due to shortening to the initial out-of-straightness (see Example 32.1).

32.3.3   Member Inelasticity

The bracing requirements for relative braces are a function of the load on the member and the distance between adjacent brace points and not the elasticity or inelasticity of the member. For a nodal bracing system, there has been some debate about the effects of inelasticity on the bracing requirements. Research, however, has indicated that inelasticity of the main members does not affect the bracing requirements [3]. For continuous and lean-on bracing systems, the bracing requirements, which will be presented later, are based on the elastic and inelastic stiffnesses of the braced members. For these bracing systems, the influence of inelasticity on the buckling solution can be reasonably approximated using the tangent modulus $E_T = \tau E$, where E is the elastic modulus and τ is the inelastic stiffness reduction factor.

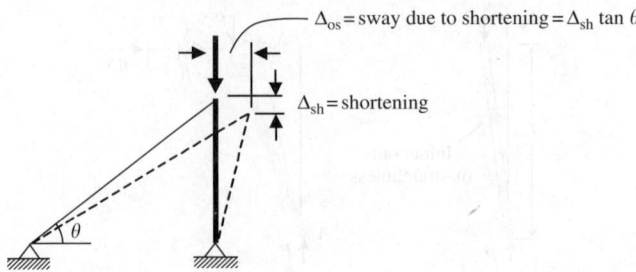

FIGURE 32.4 Effect of member shortening for relative bracing.

In the AISC load and resistance factor design (LRFD) specification, when the axial stress is less than one third the yield stress, F_y, the column is classified as elastic ($\tau = 1.0$). At greater stress levels, the stiffness reduction factor is given by

$$\tau = -7.38 \left(\frac{P_u}{P_y} \right) \log \left[\frac{(P_u/P_y)}{0.85} \right] \tag{32.1}$$

where P_u is the required column strength and P_y is the column squash load.

32.3.4 Brace Connection Stiffness

The connections details used to attach structural bracing members can be of great importance when designing or evaluating the overall performance of a bracing system. If the connections are flexible, the stiffness of the overall bracing system can be significantly less than the stiffness of the bracing member alone. The stiffness of a bracing system can be evaluated as springs in series using

$$\frac{1}{\beta_{sys}} = \frac{1}{\beta_{br}} + \sum \frac{1}{\beta_{conn}} \tag{32.2}$$

The system stiffness, β_{sys}, will always be less than the smaller of the brace member stiffness, β_{br}, and any of the connection stiffnesses, β_{conn}.

32.4 Safety Factors

The bracing design recommendations that will follow are based on ultimate strength. The loads on the members being braces are assumed to be factored. Since both strength and stiffness are essential requirements for adequate bracing, resistance factors for each are necessary. When using an LRFD approach, the required stiffness is divided by a resistance factor of $\phi = 0.75$ to obtain a conservative requirement. The design brace force is based on factored loads and compared to the factored design strength of the brace and its connections.

32.5 Column or Frame Bracing

Column or frame bracing systems can be relative, nodal, continuous, or lean-on. The design recommendations for relative and nodal column bracing are based on an initial out-of-straightness $\Delta_o = 0.002L$, where L is the column length and a brace stiffness equal to twice the ideal stiffness. The initial displacement, Δ_o, is defined as the lateral offset between two adjacent brace points caused by sources other than brace elongations from gravity loads or compressive forces. For example, Δ_o may be a displacement due to wind or other lateral forces, erection tolerances, or column shortening. If Δ_o differs from $0.002L$, the brace force, P_{br}, will change in direct proportion to the actual Δ_o. In frame systems where a story may contain n_0 columns, each having a random out-of-plumbness, an average value for Δ_o can be used [4]

$$\Delta_o = 0.002L\sqrt{n_0} \tag{32.3}$$

32.5.1 Column Buckling and Design Philosophy

For no-sway columns, the effective length factor, K, will be less than 1.0. Most designs conservatively use $K = 1.0$ and in most situations achieving $K < 1.0$ is not economical. For sway permitted columns, the effective length $K \geq 1.0$. Bracing used to prevent sway can reduce the effective length to $K = 1.0$ and achieve significant economic savings.

The bracing design criterion for columns is based on providing sufficient strength and stiffness to allow a sway column to achieve the Euler buckling load corresponding to $K = 1.0$. For columns that possess nonzero end restraint, this *does not* correspond to the no-sway buckling load since for these cases K is theoretically less than unity.

For flexural buckling modes, brace points attached in-line with the centroid of the structural member are most effective. For torsional buckling modes, bracing must prevent twist of the cross-section to be effective.

32.5.2 Relative Systems

AISC LRFD brace requirements for *relative* column bracing are

$$\beta_{br} = \frac{2P_u}{\phi L_b} \tag{32.4}$$

$$P_{br} = 0.004P_u \tag{32.5}$$

where $\phi = 0.75$, P_u is the required compressive strength of the column, and L_b is the required brace spacing.

When the actual brace stiffness provided, β_{act}, differs from the required value given in Equation 32.4, the brace strength requirement can be modified using

$$P_{br} = 0.004 \sum P_u \frac{1}{2 - (\beta_{br}/\beta_{act})} \tag{32.6}$$

EXAMPLE 32.1

Relative Column Brace Design

A typical tension-only X-brace must stabilize three bents. Each bent carries a total factored load of 600 kip $(125 + 300 + 175)$. Assume the floor acts as a rigid diaphragm and all $\Delta_o = 0.002L$ and all columns are W14 × 53 $(A_g = 15.6 \text{ in.}^2)$.

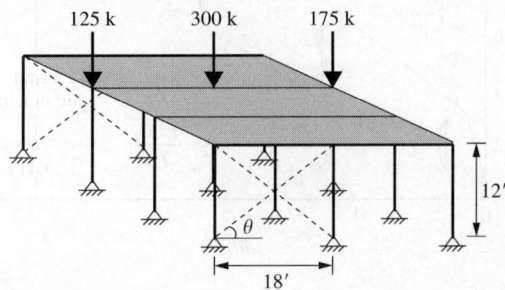

$$\sum P_u = (3\text{ bents})(125 + 300 + 175) = 1800\text{ kip}$$

Bracing shear force:

$$P_{br} = 0.004(1800) = 7.2\text{ kip}$$

Bracing shear stiffness:

$$\beta_{br} = \frac{2(1800)}{0.75(12)} = 400\text{ kip/ft}$$

Design recommendations assume brace shear force and stiffness are perpendicular to column. Therefore, for an inclined threaded rod (A36 steel):

Strength:

$$P_{br} = \frac{7.2\text{ kip}}{\cos\theta} = 8.64\text{ kip} = 0.9(36)A_g \quad (A_g)_{req'd} = 0.27\text{ in.}^2$$

Stiffness:

$$\frac{A_g E}{21.6\text{ ft}}\cos^2\theta = 400\text{ k/ft} \qquad (A_g)_{req'd} = 0.43\text{ in.}^2 \leftarrow \text{Controls}$$

$$\text{Use } \tfrac{3}{4}\text{ in. dia. rod }\left(A_g = 0.44\text{ in.}^2\right)$$

Consider effects of shortening

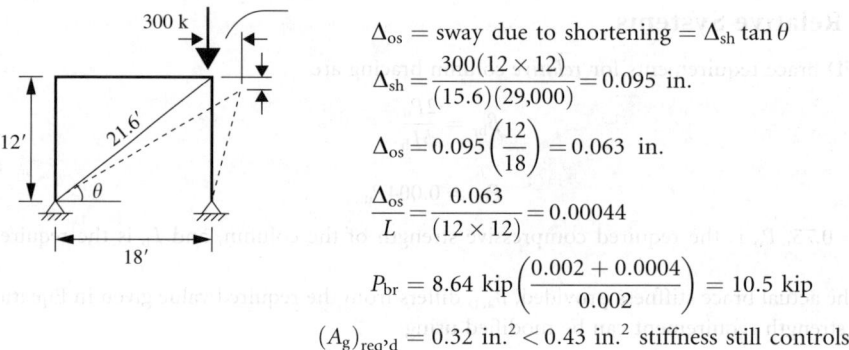

$$\Delta_{os} = \text{sway due to shortening} = \Delta_{sh}\tan\theta$$

$$\Delta_{sh} = \frac{300(12\times12)}{(15.6)(29{,}000)} = 0.095\text{ in.}$$

$$\Delta_{os} = 0.095\left(\frac{12}{18}\right) = 0.063\text{ in.}$$

$$\frac{\Delta_{os}}{L} = \frac{0.063}{(12\times12)} = 0.00044$$

$$P_{br} = 8.64\text{ kip}\left(\frac{0.002 + 0.0004}{0.002}\right) = 10.5\text{ kip}$$

$$(A_g)_{req'd} = 0.32\text{ in.}^2 < 0.43\text{ in.}^2 \text{ stiffness still controls}$$

32.5.3 Nodal Systems

Figure 32.5 shows the relationship between brace stiffness and buckling load for a column with three equally spaced nodal braces. The exact solution taken from Timoshenko and Gere [5] illustrates the increase in buckling load and changes in mode shapes as the brace stiffness increases.

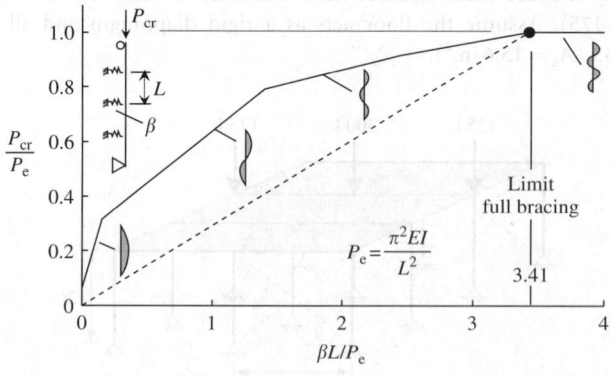

FIGURE 32.5 Discrete bracing.

For nodal column bracing, the ideal brace stiffness is a function of the number of intermediate braces [1,2]. For a single brace at midheight $\beta_i = 2P/L$. For many closely spaced braces the ideal stiffness approaches $\beta_i = 4P/L$. Twice the ideal stiffness for the most severe case was adopted by the AISC LRFD for many braces.

AISC LRFD brace requirements for *nodal* column bracing are

$$\beta_{br} = \frac{8P_u}{\phi L_b} \tag{32.7}$$

$$P_{br} = 0.01 P_u \tag{32.8}$$

where $\phi = 0.75$, P_u is the required compressive strength of the column, and L_b is the required brace spacing.

For n equally spaced braces, the ideal stiffness can be approximated as

$$\beta_i = N_i \frac{P_u}{L_b} \tag{32.9}$$

where $N_i \approx 4 - 2/n$. Using the recommended stiffness equal to twice the ideal stiffness and applying the resistance factor gives

$$\beta_{br} = N_i \frac{2P_u}{\phi L_b} \tag{32.10}$$

Equation 32.10 is based on equally spaced braces and is unconservative for unequal spacings. The required stiffness for unequal brace spacings can be obtained using Winter's rigid bar model [6]. In this model, the column is represented using rigid links with ficticious hinges at brace locations and the braces are represented using simple springs. Under the applied load, displacements are imposed at brace locations and equilibrium is enforced to obtain N_i. This technique is illustrated in Example 32.2. For a single nodal brace at any location along the length of a column, with the longest segment defined as L and the shorter segment as αL, N_i can be conservatively determined using

$$N_i = 1 + \frac{1}{\alpha} \tag{32.11}$$

EXAMPLE 32.2

Nodal Column Brace — Unequal Spacing

1. Introduce hinge at **B** and displace arbitrary distance Δ.
2. Sum forces and moments to obtain reactions at **A** and **C**.
3. Cut and sum moments at **B** on **AB** to obtain

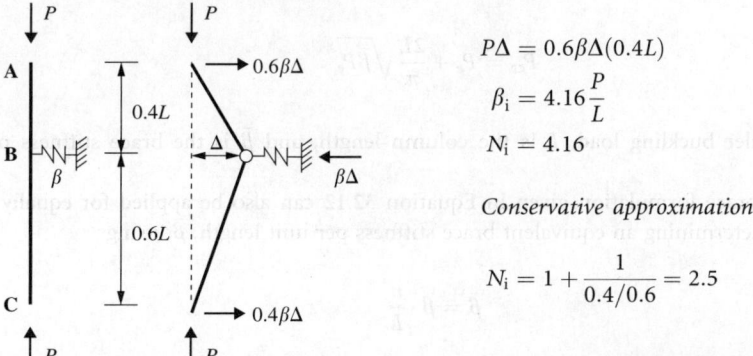

$$P\Delta = 0.6\beta\Delta(0.4L)$$

$$\beta_i = 4.16 \frac{P}{L}$$

$$N_i = 4.16$$

Conservative approximation

$$N_i = 1 + \frac{1}{0.4/0.6} = 2.5$$

The brace stiffness requirements for nodal bracing are inversely proportional to the unbraced length, L_b. Closer-spaced braces require more stiffness because the derivations are based upon allowing the column to reach a load that corresponds to buckling of the most critical unbraced length with a *K*-factor equal to 1.0. In many instances, there are more potential brace points than necessary to support the member forces required. Using the actual unbraced length may result in excessively conservative stiffness requirements. Therefore, the maximum unbraced length that enables the column to reach the required loading, L_q, can be used. For example, say the column shown in Figure 32.5 is supported against weak-axis buckling at three locations giving an unbraced length of *L*. If a single brace at midheight giving an unbraced length of 1.5*L* would be sufficient t﹚ carry the load on the column, then the required stiffness for the three braces could be conservativel﹀ estimated using the permissible unbraced length of 1.5*L* in Equation 32.10 in place of the actual unbraced length of *L* (see Example 32.3).

EXAMPLE 32.3

Nodal Column Brace Design

Two 8 ft cross-members brace a 30 ft WT5 × 19.5 compression member. Buckling about *x–x* axis controls since brace flexural stiffness is much lower than axial stiffness. $F_y = 36$ ksi. Find the required brace strength and stiffness.

$P_u = 120 \, \text{k}$

$$n = 2, \quad N_i = 4 - \frac{2}{2} = 3$$

$$\beta_{br} = 3 \frac{2(200)}{0.75(10 \times 12)} = 13.33 \, \text{k/in.} = \frac{F}{\Delta} = \frac{48EI}{(8 \times 12)^3}$$

$$I_{req'd} = \frac{13.33(96)^3}{48(29,000)} = 8.5 \, \text{in.}^4$$

$$P_{br} = 0.01(200) = 2 \, \text{kip}$$

$$M_{br} = \frac{2(96)}{4} = 48 \, \text{kip in.}$$

32.5.4 Continuous Systems

Figure 32.6 shows the relationship between brace stiffness and buckling load for a continuously braced column. The exact solution can be approximated using the following equation [3]:

$$P_{cr} = P_e + \frac{2L}{\pi} \sqrt{\bar{\beta} P_e} \tag{32.12}$$

where P_e is the Euler buckling load, *L* is the column length, and $\bar{\beta}$ is the brace stiffness per unit length.

The continuous brace formulation given in Equation 32.12 can also be applied for equally spaced discrete braces by determining an equivalent brace stiffness per unit length, $\bar{\beta}$, using

$$\bar{\beta} = \beta \cdot \frac{n}{L} \tag{32.13}$$

where *n* is the number of braces within the column length, *L*. This method is accurate for two or more discrete braces and is illustrated in Example 32.4.

Corrugated metal deck is a common type of continuous lateral bracing and acts like a shear diaphragm with the properties of a relative brace. The stiffness and strength properties of the metal

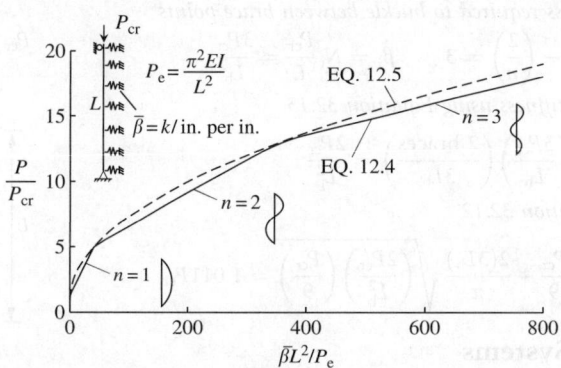

FIGURE 32.6 Continuous bracing.

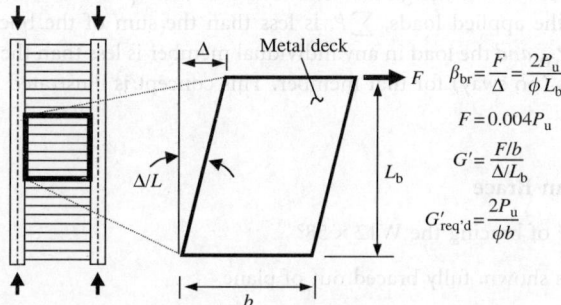

FIGURE 32.7 Continuous metal-deck bracing.

deck are generally defined in a per unit width basis (e.g., shear stiffness $= G'$ kip/rad per ft width). The bracing requirements for the shear diaphragm can be determined from the relative brace requirements presented in Section 32.5.2, as shown in Figure 32.7. Properties for corrugated deck can be obtained from the Steel Deck Institute *Diaphragm Design Manual* [7]. The required shear diaphragm stiffness per unit width is

$$G'_{\text{req'd}} = \frac{2P_u}{\phi b} \qquad (32.14)$$

Dividing the perpendicular brace force requirement by the diaphragm width gives the required shear strength per unit width, S_u

$$S_u = \frac{0.004P_u}{b} \qquad (32.15)$$

It should be noted that the brace force requirements given in Equation 32.13 are in addition to other load demands placed on the diaphragm.

EXAMPLE 32.4

Nodal Column Braces as Effective Continuous Braces

Consider a column of length $3L_b$ with two equally spaced nodal braces giving an unbraced length of L_b. Estimate the critical load of an ideally-braced column by approximating the nodal braces as an equivalent continuous brace.

Ideal nodal brace stiffness required to buckle between brace points

$$N_i \approx 4 - \left(\frac{2}{n}\right) = 3 \qquad \beta_i = N_i\frac{P_{cr}}{L} = \frac{3P_{cr}}{L_b}$$

$$P_{cr} = \frac{\pi^2 EI}{L_b^2}$$

Equivalent continuous stiffness using Equation 32.13

$$\overline{\beta} = \left(\frac{3P_{cr}}{L_b}\right)\left(\frac{2 \text{ braces}}{3L_b}\right) = \frac{2P_{cr}}{L_b^2}$$

Critical load using Equation 32.12

$$P_{cr} = \frac{P_{cr}}{9} + \frac{2(3L_b)}{\pi}\sqrt{\left(\frac{2P_{cr}}{L_b^2}\right)\left(\frac{P_{cr}}{9}\right)} = 1.011 P_{cr}$$

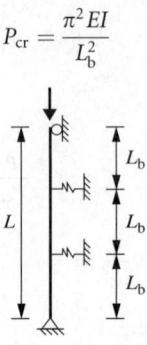

32.5.5 Lean-On Systems

Lean-on bracing systems provide a means for some members in a structural system to rely on other members for stability. One of the most commonly encountered examples of lean-on systems are frames with "leaning" pin-ended columns connected to columns with nonzero end restraint. For these systems, the $\sum P$ concept [8] can be used to design the members. This concept states that a system remains stable so long as the sum of the applied loads, $\sum P$, is less than the sum of the buckling strength of each individual member, $\sum P_{cr}$, and the load in any individual member is less than the load corresponding to buckling between braces (no sway) for that member. This concept is illustrated in Example 32.5.

EXAMPLE 32.5

Lean-On Column Brace

Is the W10 × 33 capable of bracing the W12 × 58?

A36 steel, factored loads shown, fully braced out of plane

W10 × 33   **W12 × 58**

$A_g = 9.71$ in.^{2} $A_g = 17.0$ in.^{2} *No sway capacity* — From AISC Manual ($KL_y = 8$ ft)

$I_x = 171$ in.^{4} $I_y = 107$ in.   W12 × 58: $(\phi P_n = 482 \text{ k}) > (P_u = 450 \text{ k})$ OK

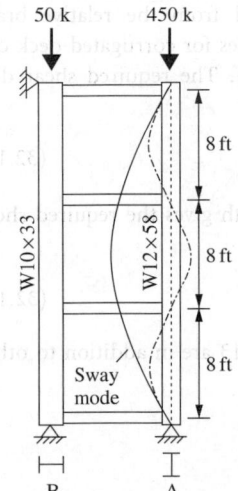

Sway capacity — Using $\sum P$ concept

Column A (W12 × 58)

$$\frac{P_u}{F_y A_g} = \frac{450}{(36)(17.0)} = 0.735 > \frac{1}{3} \quad \therefore \text{ inelastic}$$

$$\tau = -7.38(0.735)\log\left(\frac{0.735}{0.85}\right) = 0.342$$

$$\phi P_n = 0.85(0.342)(0.877)\frac{\pi^2(29,000)(107)}{(288)^2} = 94 \text{ k}$$

Column B (W10 × 33)

$$\frac{P_u}{F_y A_g} = \frac{50}{(36)(9.71)} = 0.143 < \frac{1}{3} \quad \therefore \text{ elastic } \tau = 1.0$$

$$\phi P_n = 0.85(1.0)(0.877)\frac{\pi^2(29,000)(171)}{(288)^2} = 440 \text{ k}$$

Using $\sum P$ concept

$$\left(\sum \phi P_n = 94 + 440 = 534 \text{ k}\right) > \left(\sum P_u = 50 + 450 = 500 \text{ k}\right) \text{ OK}$$

32.5.6 Torsional Bracing

In order for a brace to effectively increase the load-carrying capacity of a member, it must restrain the movement of the lowest buckling mode. Depending on the cross-section, the lowest buckling mode of a compression member may be flexural (lateral), torsional, or a combined flexural–torsional. Bracing against flexural modes must prevent the lateral translation of the member cross-section. Bracing against torsional modes must restrain the twist of the cross-section. Bracing details such as a rod framing into a wide-flange column web (Figure 32.8) resist lateral buckling about the weak axis but do not prevent twist and are ineffective torsional braces.

For doubly symmetric sections such as wide-flange columns, weak-axis flexural buckling controls for an unbraced member. Providing sufficient weak-axis lateral bracing, in some instances, will result in the section being controlled by the torsional buckling mode. If bracing is provided that prevents both translation and twist of a doubly symmetric cross-section, weak-axis buckling will always control.

The torsional buckling load, P_T, for a column restrained about an axis modified by τ is given by the following (see Figure 32.9) [5]:

Axis of restraint along weak axis of column:

$$P_T = \frac{\tau P_{ey}\left[(d^2/4) + y_{br}^2\right] + GJ}{y_{br}^2 + r_x^2 + r_y^2} \tag{32.16}$$

Axis of restraint along strong axis of column:

$$P_T = \frac{\tau P_{ey}\left[(d^2/4) + (I_x/I_y)x_{br}^2\right] + GJ}{x_{br}^2 + r_x^2 + r_y^2} \tag{32.17}$$

where x_{br}, y_{br} are the coordinates of axis of restraint with respect to column centroid, d is the column depth, and P_{ey} is the Euler load based on a column length between points of zero twist.

To compensate for the assumption in the derivation of Equations 32.16 and 32.17 that the brace is infinitely stiff, the maximum factored column load should be limited to $0.90P_T$ [3].

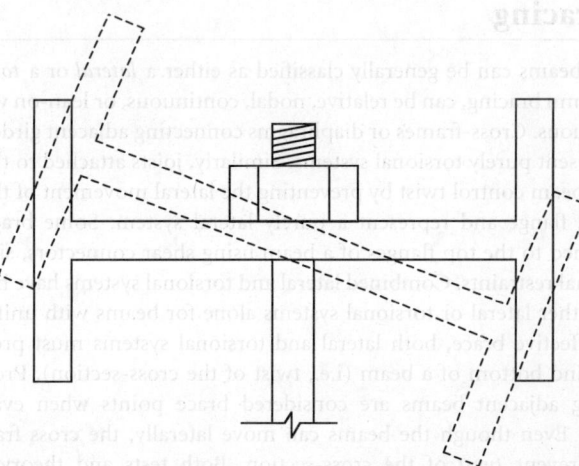

FIGURE 32.8 Ineffective torsional column brace.

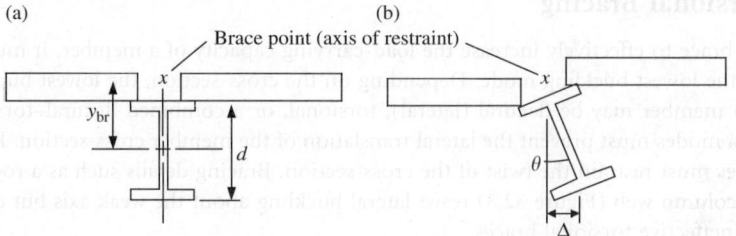

FIGURE 32.9   Buckling of a column about a restrained axis: (a) lateral brace at flange and (b) buckled shape.

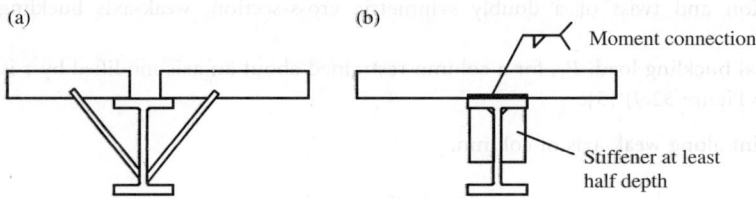

FIGURE 32.10   Typical torsional brace details: (a) control twist with struts and (b) moment connection with stiffener.

If column loads greater than P_T are required, torsional bracing must be provided. Two typical bracing schemes are shown in Figure 32.10. For continuous girts with moment connections, twisting restraint is provided. However, partial depth stiffeners should be used to control web distortion. The design requirements for torsional bracing are based on the nodal requirements presented in Section 32.5.3 and are obtained by introducing equal and opposite brace forces on each flange. The magnitude of these forces is based on the assumption that each flange carries one-half of the total column load. The resulting brace moment, $M_T = 0.5P_{br}d$. Using the angle of twist $\theta = \Delta/d$ as shown in Figure 32.9b, the stiffness requirement $\beta_T = M_T/\theta = 0.5P_{br}d^2/\Delta$ reduces to

$$\beta_T = 0.5\beta_{br}d^2 \tag{32.18}$$

where β_{br} is the nodal brace stiffness requirement from Section 32.5.3.

32.6   Beam Bracing

Structural bracing for beams can be generally classified as either a *lateral* or a *torsional* system. Lateral beam bracing, like column bracing, can be relative, nodal, continuous, or lean-on while torsional systems can be nodal or continuous. Cross-frames or diaphragms connecting adjacent girders control twist of the cross-section and represent purely torsional systems. Similarly, joists attached to the compression flange of a simply supported beam control twist by preventing the lateral movement of the compression flange relative to the tension flange and represent a purely lateral system. Some bracing systems, such as a composite slab attached to the top flanges of a beam using shear connectors, simultaneously provide both lateral and torsional restraints. Combined lateral and torsional systems have been shown to be more effective braces than either lateral or torsional systems alone for beams with uniform moment [9,10].

In order to be an effective brace, both lateral and torsional systems must prevent the relative displacement of the top and bottom of a beam (i.e., twist of the cross-section). Properly designed cross-frames interconnecting adjacent beams are considered brace points when evaluating the buckling strength of the beams. Even though the beams can move laterally, the cross frames are still effective braces because they prevent *twist* of the cross-section. Both tests and theory have confirmed this fact [11,12].

32.6.1 Inflection Points as Brace Points

Inflection points are sometimes mistakenly identified as brace points in restrained beams. This often occurs when the top flange of a beam is laterally braced by a slab or joists all along the span while the bottom flange remains unbraced. Even in these instances, the inflection point *cannot* be considered a brace point. The justification for this can be illustrated by comparing the two beams shown in Figure 32.11. Beam A has a moment on one end with an unbraced length $L_b = L$ while Beam B has equal and opposite end moments resulting in an inflection point at midspan and an unbraced length $L_b = 2L$. The buckling moment for Beam B is 68% of Beam A. If the inflection point were a brace point, the critical moment for both beams would be identical. A plan view of the buckled shape of Beam B illustrates how the top and bottom flanges move in opposite directions at midspan. Even an actual brace on one flange at the inflection point does not provide effective bracing at midspan because twist is not restrained [12].

32.6.2 Lateral Bracing

The primary factors influencing the effectiveness of *lateral* beam bracing are

- Number or spacing of braces
- Vertical position of the braces on the cross-section
- Vertical position of the loads on the cross-section
- Moment gradient on the beam

Increasing the number of braces along the length of a beam can increase the load-carrying capacity by reducing the unbraced length. Like column bracing, however, the brace stiffness requirements increase with the number of braces. This is because the design recommendations assume the beam must reach a load corresponding to buckling between brace points. Similarly, if more brace points are provided than necessary to support the required loads, the maximum permissible unbraced length, L_q, may be used in place of the actual unbraced length, L_b.

The vertical position of a lateral brace along the height of a beam significantly affects the effectiveness of lateral bracing. Lateral bracing is most effective when positioned at the compression flange of

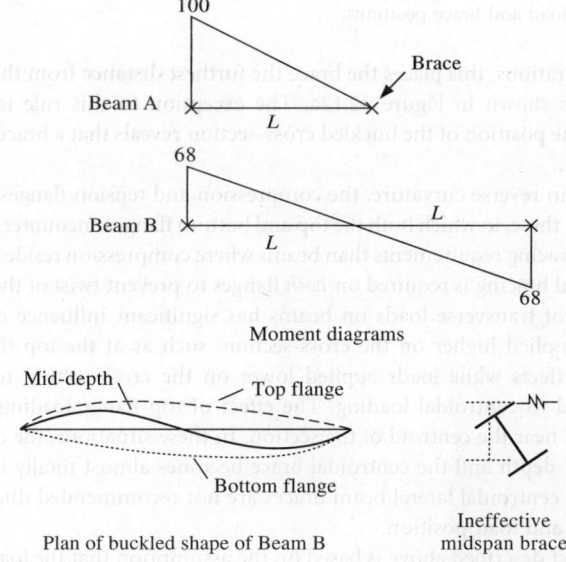

FIGURE 32.11 Beam with inflection point.

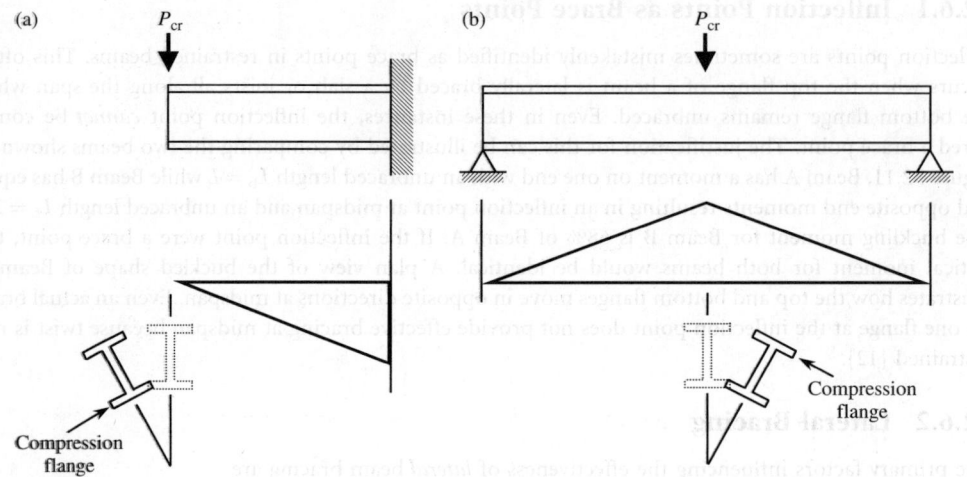

FIGURE 32.12 Lateral buckling of cantilever and simple beams: (a) max moment at max twist and (b) zero moment at max twist.

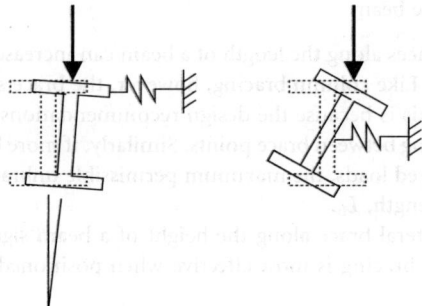

FIGURE 32.13 Effect of load and brace positions.

a beam. In most configurations, this places the brace the furthest distance from the center of twist of the buckled cross-section as shown in Figure 32.12a. The exception to this rule is for cantilever beams (Figure 32.12b) where the position of the buckled cross-section reveals that a brace located at the tension flange is most beneficial.

When a beam is bent in reverse curvature, the compression and tension flanges switch at the point of inflection. Beams such as these, in which both the top and bottom flanges encounter compression along the span, have more severe bracing requirements than beams where compression resides only in one flange [3]. In these situations, lateral bracing is required on *both* flanges to prevent twist of the cross-section.

The vertical position of transverse loads on beams has significant influence on the effectiveness of lateral bracing. Loads applied higher on the cross-section, such as at the top flange, have more pronounced destabilizing effects while loads applied lower on the cross-section tend to provide added stability when compared to centroidal loading. The effect of top-flange loading is even greater when lateral bracing is located near the centroid of the section. In these situations, the center of twist shifts to a position closer to mid-depth and the centroidal brace becomes almost totally ineffective as shown in Figure 32.13. Therefore, centroidal lateral beam braces are not recommended due to the effects of both cross-section distortion and load position.

The load position effect described above is based on the assumption that the load remains vertical and passes through the original point of contact on the member as it buckles. For many structural systems,

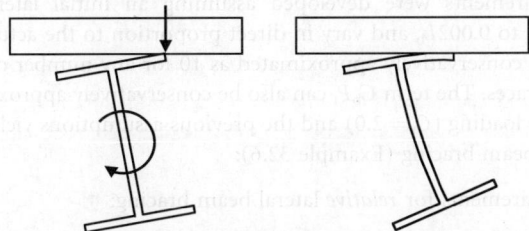

FIGURE 32.14 Tipping effect and cross-section distortion.

the load transferred to beams is applied through secondary members or a floor slab. When loading is through a slab, for example, a restoring torque is created by a tipping effect during buckling as illustrated in Figure 32.14. This tipping effect has been shown to significantly increase the lateral buckling capacity even if the slab is only resting (not positively attached) on the top flange [12]. Unfortunately, the benefits of tipping are severely limited by distortion of the cross-section and are difficult to quantify. As a result, the beneficial effects of tipping are generally neglected.

The AISC LRFD lateral beam brace requirements were based on the following design recommendations developed by Yura [3]:

The brace stiffness requirement for both *relative* and *nodal* beam bracings is

$$\beta_{br} = \frac{2N_i (C_b P_f) C_t C_d}{\phi L_b} \tag{32.19}$$

The brace strength requirement for *relative* bracing is

$$P_{br} = 0.004 \frac{M_u C_t C_d}{h_{oh}} \tag{32.20}$$

and for *nodal* bracing it is

$$P_{br} = 0.01 \frac{M_u C_t C_d}{h_{oh}} \tag{32.21}$$

where

N_i $= 1.0$ for relative bracing
     $= (4 - 2/n)$ for nodal bracing
     $n =$ number of intermediate braces
P_f $=$ beam compressive flange force
     $= \pi^2 E I_{yc}/L_b^2$
     $I_{yc} =$ out-of-plane moment of inertia of the compression flange
C_t $=$ top-flange loading factor
     $= 1.0$ for centroidal loading
     $= 1 + (1.2/n)$ for top-flange loading
C_d $= 1 + (M_S/M_L)^2$ for reverse-curvature bending
     $= 1.0$ for single-curvature bending
     $M_S =$ smallest moment causing compression in each flange
     $M_L =$ largest moment causing compression in each flange
C_b $=$ nonuniform moment modification factor
     $= \dfrac{12.5 M_{max}}{2.5 M_{max} + 3 M_A + 4 M_B + 3 M_C}$
     $M_{max} =$ absolute value of maximum moment in unbraced segment
     M_A $=$ absolute value of moment at quarter point of unbraced segment
     M_B $=$ absolute value of moment at midspan of unbraced segment
     M_C $=$ absolute value of moment at three-quarter point of unbraced segment

The brace force requirements were developed assuming an initial lateral displacement of the compression flange equal to $0.002L_b$ and vary in direct proportion to the actual out-of-straightness.

The term $2N_iC_t$ can be conservatively approximated as 10 for any number of nodal braces and 4 for any number of relative braces. The term C_bP_f can also be conservatively approximated as M_u/h_{oh}. Using the worst-case top-flange loading ($C_t = 2.0$) and the previous assumptions yields the AISC LRFD brace requirements for lateral beam bracing (Example 32.6):

AISC LRFD brace requirements for *relative* lateral beam bracing:

$$\beta_{br} = \frac{4M_u C_d}{\phi L_b h_{oh}} \tag{32.22}$$

$$P_{br} = 0.008 \frac{M_u C_d}{h_{oh}} \tag{32.23}$$

AISC LRFD brace requirements for *nodal* lateral beam bracing:

$$\beta_{br} = \frac{10M_u C_d}{\phi L_b h_{oh}} \tag{32.24}$$

$$P_{br} = 0.02 \frac{M_u C_d}{h_{oh}} \tag{32.25}$$

where

$\phi = 0.75$
M_u = required flexural strength
h_o = distance between flange centroids
$C_d = 1 + (M_S/M_L)^2$ for reverse-curvature bending
$\quad = 1.0$ for single-curvature bending
L_b = distance between braces

EXAMPLE 32.6

Relative Lateral Beam Brace

Design the diagonals of the top flange horizontal truss to stabilize the five simply-supported 90 ft girders. The factored moment at midspan $M_u = 1200$ k ft and $F_y = 36$ ksi for bracing.

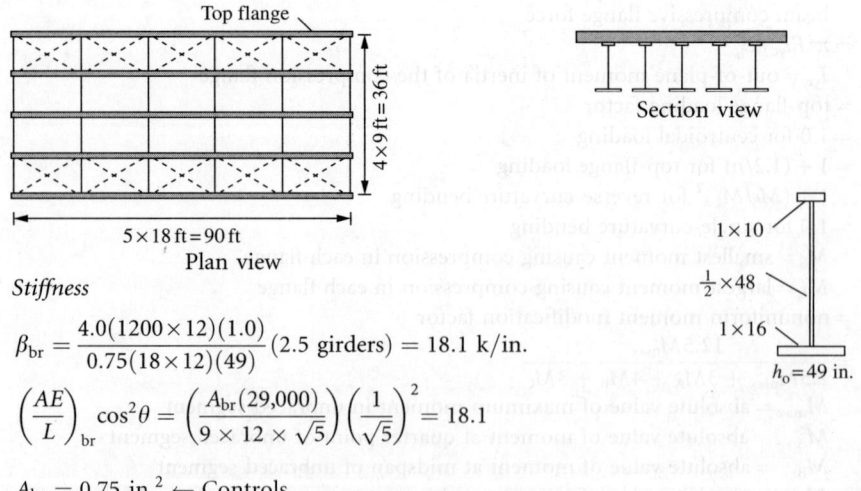

Stiffness

$$\beta_{br} = \frac{4.0(1200 \times 12)(1.0)}{0.75(18 \times 12)(49)} (2.5 \text{ girders}) = 18.1 \text{ k/in.}$$

$$\left(\frac{AE}{L}\right)_{br} \cos^2\theta = \left(\frac{A_{br}(29,000)}{9 \times 12 \times \sqrt{5}}\right)\left(\frac{1}{\sqrt{5}}\right)^2 = 18.1$$

$$A_{br} = 0.75 \text{ in.}^2 \leftarrow \text{Controls}$$

Strength

$$P_{br} = 0.008 \frac{(1200 \times 12)(1.0)}{49} (2.5 \text{ girders}) = 5.88 \text{ k}$$

$$A_{br} = \frac{5.88\sqrt{5}}{(0.9 \times 36)} = 0.41 \text{ in.}^2 \qquad\qquad \text{Use } L2 \times 2 \times \frac{1}{4} \ (A_g = 0.944 \text{ in.}^2)$$

32.6.3 Torsional Bracing

The primary factors influencing the effectiveness of *lateral* beam bracing have relatively little influence on the design of *torsional* beam bracing. Unlike lateral beam bracing, the number of braces, brace location on the cross-section, and load position are relatively unimportant when sizing a torsional brace. Not only is a torsional brace equally effective if attached to the top or bottom flange, but beams in reverse curvature do not alter the torsional brace requirements. The effectiveness of a torsional brace, however, is greatly affected by distortion of the cross-section as illustrated in Figure 32.15. Although the torsional brace prevents twist at the top flange, distortion of the web permits a relative displacement between the two flanges. A web stiffener located at the brace location is often used to control the distortion.

AISC LRFD brace requirements for torsional beam bracing:

$$\beta_{Tb} = \frac{\beta_T}{(1 - (\beta_T/\beta_{sec}))} \tag{32.26}$$

$$M_{br} = \frac{0.024 M_u L}{n C_b L_b} \tag{32.27}$$

where

$$\beta_{sec} = \frac{3.3E}{h_{oh}} \left(\frac{1.5 h_{oh} t_w^3}{12} + \frac{t_s b_s^3}{12} \right) \tag{32.28}$$

$$\beta_T = \frac{2.4 L M_u^2}{\phi n E I_y C_b^2} \tag{32.29}$$

where

$\phi = 0.75$
$L = $ span length
$n = $ number of nodal braced points within span
$E = $ modulus of elasticity
$I_y = $ out-of-plane moment of inertia of beam
$C_b = $ nonuniform moment modification factor
$t_w = $ thickness of beam web

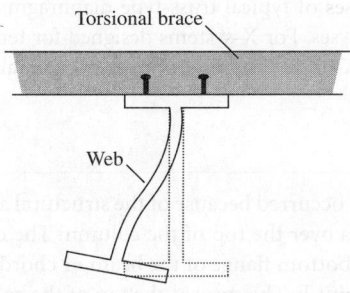

Torsional brace

Web

FIGURE 32.15 Cross-section distortion.

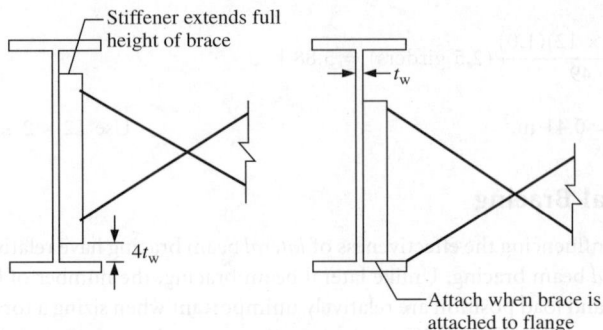

FIGURE 32.16 Web stiffener details for torsional beam bracing.

t_s = thickness of web stiffener(s)

b_s = stiffener width for one-sided stiffeners, which is twice the individual stiffener width for pairs of stiffeners

β_T represents the torsional stiffness of the brace member itself. For flexible connections, Equation 32.2 should be used. β_{sec} is the torsional stiffness associated with the beam web and any transverse web stiffeners that may be present (web distortional stiffness). The required torsional brace stiffness, β_{Tb}, will be negative if $\beta_{sec} < \beta_T$, which indicates the brace will not be effective due to insufficient web distortional stiffness. The coefficient 2.4 in Equation 32.29 comes from using twice the ideal stiffness with a 20% increase to account for top-flange loading. The required strength, M_{br}, assumes an initial twist of $1°$ (0.0175 rad).

When web stiffeners are required, the AISC LRFD specification requires they extend the full depth of the brace as shown in Figure 32.16. When the brace is attached to the beam flange the stiffeners must also be attached to the flange. When the brace is not attached to the beam flange, the stiffener may be terminated at a distance equal to $4t_w$ from the flange.

The continuous torsional beam bracing requirements use the same formulations as the nodal bracing requirements. For continuous bracing, the term L/n in Equations 32.27 and 32.29 is taken equal to 1.0 and the unbraced length, L_b, in Equation 32.27 is taken equal to L_q, the maximum unbraced beam length necessary to carry the required factored loads.

For singly symmetric cross-sections, an effective moment of inertia, I_{eff}, is used in place of I_y as given by

$$I_{eff} = I_{yc} + \frac{y_t}{y_c} I_{yt} \tag{32.30}$$

where I_{yc} is the compression flange out-of-plane moment of inertia, I_{yt} is the tension flange out-of-plane moment of inertia, y_c is the distance from centroid to extreme compression fiber, and y_t is the distance from centroid to extreme tension fiber.

The torsional stiffness of typical beam-type braces is shown in Figure 32.17. Adjacent girders connected on the top flanges with decking, for example, will buckle in the same direction and develop the reverse-curvature stiffness of the brace. The stiffnesses of typical truss-type diaphragm systems are shown in Figure 32.18 and are based on elastic truss analyses. For X-systems designed for tension only, the horizontal members in the truss are not required. For K-brace systems, only one horizontal member is necessary.

32.7 Faulty Details

Numerous structural failures have occurred because of the structural arrangement shown in Figure 32.19. The beam (or truss) is continuous over the top of the column. The critical components are: column in compression, compression in the bottom flange of the beam or chord of the truss, and no bottom flange bracing at point *a* and possibly point *b*. The sway at the top of the column shown in Sect B–B can result in a K-factor much greater than 2.0. The bottom flange of the beam can possibly provide bracing to the

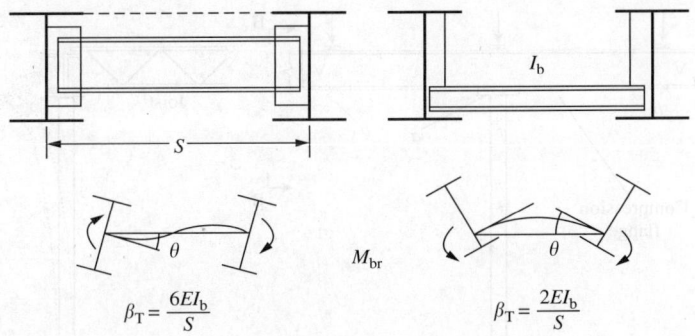

$$\beta_T = \frac{6EI_b}{S} \qquad\qquad \beta_T = \frac{2EI_b}{S}$$

FIGURE 32.17 Stiffness formulas for diaphragms.

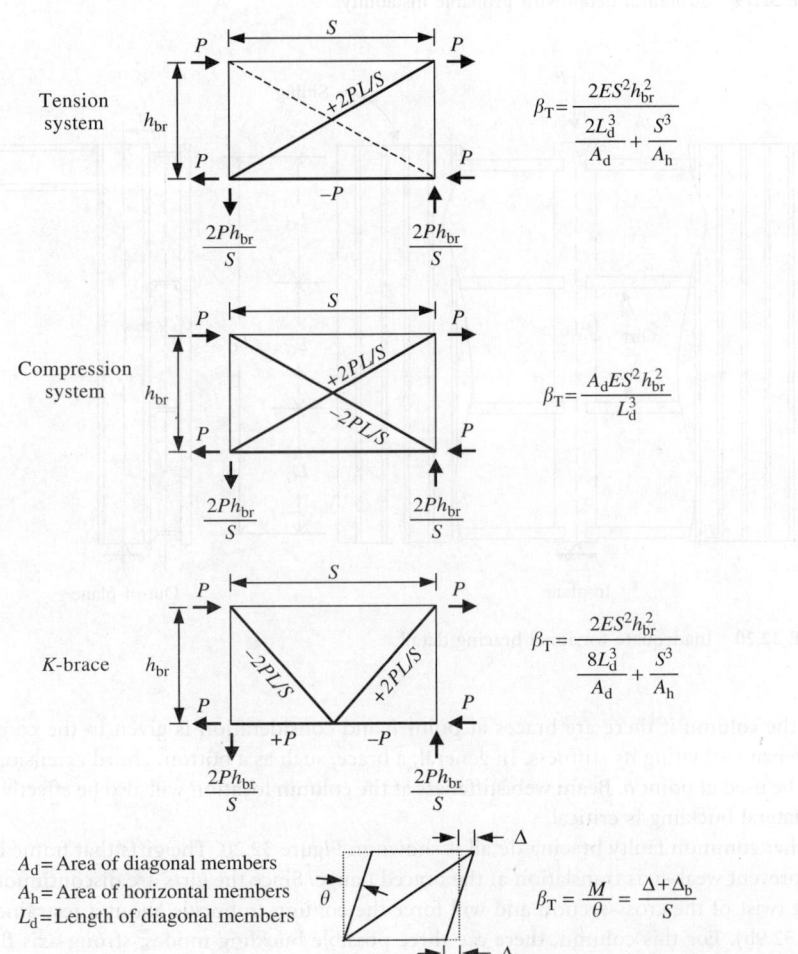

Tension system

$$\beta_T = \frac{2ES^2 h_{br}^2}{\dfrac{2L_d^3}{A_d} + \dfrac{S^3}{A_h}}$$

Compression system

$$\beta_T = \frac{A_d ES^2 h_{br}^2}{L_d^3}$$

K-brace

$$\beta_T = \frac{2ES^2 h_{br}^2}{\dfrac{8L_d^3}{A_d} + \dfrac{S^3}{A_h}}$$

A_d = Area of diagonal members
A_h = Area of horizontal members
L_d = Length of diagonal members

$$\beta_T = \frac{M}{\theta} = \frac{\Delta + \Delta_b}{S}$$

FIGURE 32.18 Stiffness formulas for cross-frames.

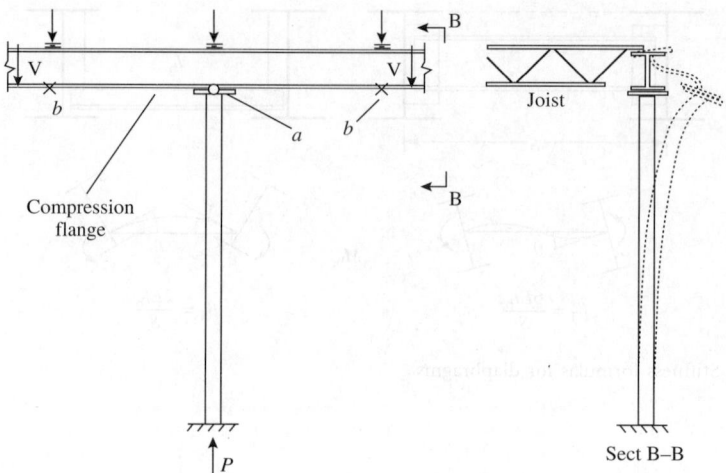

FIGURE 32.19   Structural detail with probable instability.

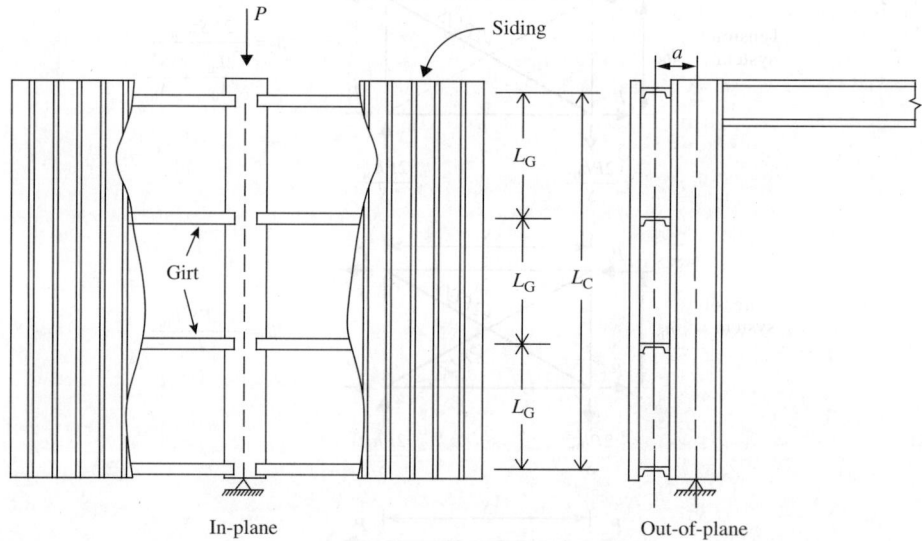

FIGURE 32.20   Inadequate torsional bracing detail.

top of the column if there are braces at point *b* and consideration is given to the compression in the flange when evaluating its stiffness. In general, a brace, such as a bottom chord extension from the joist, should be used at point *a*. Beam web stiffeners at the column location will also be effective unless bottom flange lateral buckling is critical.

Another common faulty bracing detail is shown in Figure 32.20. The girts that frame into the column flange prevent weak-axis translation at the braced flange. Since the girts are discontinuous, they will not prevent twist of the cross-section and will force the column to buckle about a restrained axis (see also Figure 32.9b). For this column, there are three possible buckling modes: strong-axis flexural buckling ($L_b = KL_C$), weak-axis flexural buckling ($L_b = KL_G$), and torsional buckling about a restrained axis ($L_b = KL_C$, $y_{br} = a$ assuming no twist at column ends).

Nomenclature

A	cross-sectional area of primary member	M_S	smallest moment causing compression in each flange along beam length
A_{br}	cross-sectional area of brace member	M_T	torsional brace moment
A_d	area of diagonal member in cross-frame	M_u	required bending strength
A_h	area of horizontal member in cross-frame	N_i	brace stiffness coefficient for nodal braces
A_g	gross cross-sectional area	P_{br}	brace force
C_b	bending coefficient dependent on moment gradient	P_{cr}	member buckling load
		P_e	Euler column buckling load
C_d	reverse-curvature bending factor	P_{ey}	Euler buckling load based on distance between points of zero twist
C_t	top-flange loading factor		
E	modulus of elasticity	P_f	beam compressive flange force
E_T	tangent modulus of elasticity	P_T	torsional buckling load
F_y	specified minimum yield stress	P_u	required compressive strength of column
G	shear modulus of elasticity		
G'	diaphragm shear stiffness per unit width	P_y	column squash load
		S	cross-frame or diaphragm length
I	moment of inertia	S_x	strong-axis section modulus
I_{eff}	effective moment of inertia for singly symmetric beam sections	S_u	required diaphragm shear strength per unit width
$I_{req'd}$	required moment of inertia	V	shear force
I_x, I_y	moment of inertia about strong and weak axes, respectively	b	orthogonal distance between point of restraint and weak axis of member
I_{yc}, I_{yt}	out-of-plane moment of inertia of compression and tension flanges, respectively	b_s	stiffener width for one-sided stiffeners
		d	member depth
J	torsion constant for section	f_b	bending stress
K	column effective length factor	h_{br}	deight of cross frame torsional brace
L	member length		
L_b	required brace spacing or laterally unbraced length; length between points that are either braced against lateral displacement of the compression flange or braced against twist of the cross-section	h_{oh}	distance between flange centroids
		n	rumber of braces within span
		n_0	rumber of columns in a story
		r_x, r_y	radius of gyration about strong and weak axes, respectively
		t_w	thickness of beam web
		t_s	thickness of web stiffener
L_q	maximum unbraced length for required forces	x_{br}, y_{br}	coordinates of axis of restraint with respect to column centroid
M_A	absolute value of moment at quarter point of unbraced beam segment	y_c, y_t	coordinates with respect to centroid of expreme compression and tension fibers, respectively
M_B	absolute value of moment at midspan of unbraced beam segment	$\overline{\beta}$	continuous brace stiffness per unit length
M_C	absolute value of moment at three-quarter point of unbraced beam segment	β_{act}	stiffness provided by brace member
		β_{br}	required lateral brace stiffness
		β_{conn}	stiffness of brace connection
M_L	largest moment causing compression in each flange along beam length	β_{sec}	web distortional stiffness including any transverse web stiffeners if present
		β_{sys}	stiffness of brace system
M_{max}	absolute value of max moment in unbraced beam segment	β_T	nodal torsional brace stiffness

β_{Tb}	required nodal torsional brace stiffness including web distortion	Δ_T	total column sway deflection
Δ	translational displacement	ϕ	resistance factor
Δ_o	column initial out-of-straightness	θ	complementary angle between diagonal brace and axial member or
Δ_{os}	column out-of-straightness due to shortening		twist of member cross-section
Δ_{sh}	shortening or compression element	τ	inelastic stiffness reduction factor

References

[1] Winter, G. (1958), "Lateral Bracing of Columns and Beams," *Trans. ASCE*, Vol. 125, Part 1, pp. 809–825.

[2] Winter, G. (1960), "Lateral Bracing of Columns and Beams," *Proc. ASCE*, Vol. 84 (ST2), pp. 1561-1–1561-22.

[3] Yura, J.A., Bracing, in *Stability Design Criteria for Metal Structures*, 5th Edition, Galambos, T.V. Ed.; John Wiley & Sons, Inc., New York, 1998; Chapter 12.

[4] Chen, S. and Tong, G. (1994), "Design for Stability: Correct Use of Braces," *Steel Struct., J. Singapore Struct. Steel Soc.*, Vol. 5, No. 1, pp. 15–23.

[5] Timoshenko, S. and Gere, J. (1961), *Theory of Elastic Stability*, McGraw-Hill Book Company, New York.

[6] Yura, J.A. (1994), "Winters Bracing Model Revisited," *50th Anniversary Proc., Struc. Stability Research Council*, pp. 375–382.

[7] Luttrell, L.D. (1987), *Diaphragm Design Manual*, 2nd Edition, Steel Deck Institute, Fox River Grove, IL.

[8] Yura, J.A. (1971), "The Effective Length of Columns in Unbraced Frames," *Eng. J. Am. Inst. Steel Const.*, Vol. 8, No. 2, pp. 37–42.

[9] Mutton, B.R. and Trahair, N.S. (1973), "Stiffness Requirements for Lateral Bracing," *ASCE J. Struct. Div.*, Vol. 99, No. ST10, pp. 2167–2182.

[10] Tong, G. and Chen, S. (1988), "Buckling of Laterally and Torsionally Braced Beams," *J. Const. Steel Res.*, Vol. 11, pp. 41–55.

[11] Flint, A.R. (1951), "The Stability of Beams Loaded Through Secondary Members," *Civil Eng. Public Works Rev.*, Vol. 46, No. 537–8 (see also pp. 259–260).

[12] Yura, J.A. (1993), "Fundamentals of Beam Bracing," *Proc. Struc. Stability Research Council Annual Technical Session*, "Is Your Structure Suitably Braced?" Milwaukee, April, 20 pp.

Further Reading

Akay, H.U., Johnson, C.P. and Will, K.M. (1977), "Lateral and Local Buckling of Beams and Frames," *ASCE J. Struct. Div.*, Vol. 103, No. ST9, pp. 1821–1832.

Ales, J.M. and Yura, J.A. (1993), "Bracing Design for Inelastic Structures," *Proc., SSRC Conf. "Is Your Structure Suitably Braced?,"* Milwaukee, April, pp. 29–37.

American Institute of Steel Construction (1995), *Manual of Steel Construction: Load & Resistance Factor Design*, Vol. 1, 2nd Edition, Chicago.

American Society of Civil Engineers (1971), "Commentary on Plastic Design in Steel," *ASCE Manual No. 41*, 2nd Edition, New York.

Essa, H.S. and Kennedy, D.J.L. (1995), "Design of Steel Beams in Cantilever-Suspended-Span Construction," *ASCE J. Struct. Div.*, Vol. 121, No. 11, pp. 1667–1673.

Gil, H. (1966), "Bracing Requirements for Inelastic Steel Members," PhD dissertation, The University of Texas at Austin, May, 156 pp.

Helwig, T.A., Yura, J.A., and Frank, K.H. (1993), "Bracing Forces in Diaphragms and Cross Frames," *Proc., SSRC Conf., "Is Your Structure Suitably Braced?"* Milwaukee, April, pp. 129–140.

Horne, M.R. and Ajmani, J.L. (1971), "Design of Columns Restrained by Side Rails," *Struct. Eng.*, Vol. 49, No. 8, pp. 329–345.

Horne, M.R. and Ajmani, J.L. (1972), "Failure of Columns Laterally Supported on One Flange," *Struct. Eng.*, Vol. 50, No. 9, pp. 355–366.

Lutz, A.L. and Fisher, J. (1985), "A Unified Approach for Stability Bracing Requirements," *Eng. J., Am. Inst. Steel Constr.*, Vol. 22, No. 4, pp. 163–167.

Medland, I.C. and Segedin, C.M. (1979), "Brace Forces in Interbraced Column Structures," *ASCE J. Struct. Div.*, Vol. 105, No. ST7, pp. 1543–1556.

Milner, H.R. and Rao, S.N. (1978), "Strength and Stiffness of Moment Resisting Beam-Purlin Connections," *Civil Eng. Trans.*, Inst. of Engrg, Australia, CE 20(1), pp. 37–42.

Nakamura, T. (1988), "Strength and Deformability of H-Shaped Steel Beams and Lateral Bracing Requirements," *J. Const. Steel Res.*, Vol. 9, 217–228.

Plaut, R.H. (1993), "Requirements for Lateral Bracing of Columns with Two Spans," *ASCE J. Struct. Div.*, Vol. 119, No. 10, pp. 2913–2931.

Pincus, G. (1964), "On the Lateral Support of Inelastic Columns," *Eng. J., AISC*, Vol. 1, No. 4, pp. 113–115.

Salmon, C.S. and Johnson, J.E. (1996), *Steel Structures — Design and Behavior*, 4th Edition, Harper and Row, New York.

Taylor, A.C. and Ojalvo, M. (1966), "Torsional Restraint of Lateral Buckling," *ASCE J. Struct. Div.*, Vol. 92, No. ST2, pp. 115–129.

Timoshenko, S. and Gere, J. (1961), *Theory of Elastic Stability*, McGraw-Hill Book Company, New York.

Tong, G. and Chen, S. (1987), "Design Forces of Horizontal Inter-Column Braces," *J. Const. Steel Res.*, Vol. 7, pp. 363–370.

Tong, G. and Chen, S. (1989), "The Elastic Buckling of Interbraced Girders," *J. Const. Steel Res.*, Vol. 14, pp. 87–105.

Trahair, N.S. and Nethercot, D.A. (1984), "Bracing Requirements in Thin-Walled Structures," *Chapter 3, Developments in Thin-Walled Structures*, Vol. 2, J. Rhodes and A.C. Walker, Eds., Elsevier, Amsterdam, pp. 93–130.

Wang, Y.C. and Nethercot, D.A. (1989), "Ultimate Strength Analysis of Three-Dimensional Braced I-Beams," *Proc. Inst. of Civil Engrg*, Part 2, 87, March, London, pp. 87–112.

Winter, G. (1960), "Lateral Bracing of Columns and Beams," *Trans. ASCE*, Vol. 125, Part 1, pp. 809–825.

Yura, J.A. (1995), "Bracing for Stability-State-of-the-Art," *Proceedings, Structures Congress XIII*, ASCE, Boston, April, pp. 88–103.

Yura, J.A. and Phillips, B.A. (1992), "Bracing Requirements for Elastic Steel Beams," Research Report 1239-1, Center for Transportation Research, Univ. of Texas at Austin, May, 73 pp.

Yura, J.A., Phillips, B., Raju, S., and Webb, S. (1992), "Bracing of Steel Beams in Bridges," Research Report 1239-4F, Center for Transportation Research, Univ. of Texas at Austin, October, 80 pp.

Zuk, W. (1956), "Lateral Bracing Forces on Beams and Columns," *ASCE J. Eng. Mech. Div.*, Vol. 82, No. EM3, pp. 1032-1–1032-11.

33

Stub Girder Floor Systems

Reidar Bjorhovde
The Bjorhovde Group,
Tucson, AZ

33.1 Introduction

The stub girder system was developed in response to a need for new and innovative construction techniques that could be applied to certain parts of multistory steel-framed buildings. Originated in the early 1970s, the system aimed at providing construction economies through the integration of the electrical and mechanical service ducts into the part of the building volume that is occupied by the floor framing system of the building [1,2]. Since the overall height of the floor system at times could become large, increases in the overall height of the structure and hence the steel tonnage for the project could be significant. At other times the height could be reduced, but only at the expense of having sizeable web penetrations for the ductwork to pass through. This solution often entailed reinforced web openings, further increasing the construction cost.

The composite stub girder floor system was developed subsequently. Making extensive use of relatively simple shop fabrication techniques, basic elements with limited fabrication needs, simple connections between the main floor system elements and the structural columns, and composite action between the concrete floor slab and the steel load-carrying members, a floor system of significant strength, stiffness, and ductility was devised. This led to a reduction in the amount of structural steel that traditionally had been needed for the floor framing. When coupled with the use of continuous, composite transverse floor beams and the shorter erection time that was needed for the stub girder system, this yielded attractive cost savings.

Since its introduction, the stub girder floor system has been used for a variety of steel-framed buildings in the United States, Canada, Mexico, and Europe, ranging in height from 2 to 72 stories. Despite this relatively widespread usage, the analysis techniques and the design criteria remain unknown to many designers. This chapter will offer examples of practical uses of the system, together with recommendations for suitable design and performance criteria.

33.2 Description of the Stub Girder Floor System

The main element of the system is a special girder, fabricated from standard hot-rolled wide-flange shapes, that serves as the primary framing element of the floor. Hot-rolled wide-flange shapes are also used as transverse floor beams, running in a direction perpendicular to the main girder. The girder and the beams are usually designed for composite action, although the system does not rely on having composite floor beams, and the latter are normally analyzed as continuous beams. The transverse floor beams normally use a smaller size member as a drop-in span within the positive moment region.

Load and resistance factor design (LRFD) is preferred for stub girders, since it gives lower steel weights and simple connections. The costs of an LRFD-designed stub girder are therefore lower than those associated with allowable stress design (ASD).

Figure 33.1 shows the elevation of a typical stub girder. The girder that is shown uses four stubs, oriented symmetrically with respect to the midspan of the member. The locations of the transverse floor beams are assumed to be the quarter points of the span, and the supports are simple. In practice many variations of this layout are used, to the extent that the girders may utilize any number of stubs. However, three to five stubs is the most common choice. The locations of the stubs may differ significantly from the symmetrical case and the exterior stubs may have been placed at the very ends of the bottom chord. However, this is not difficult to address in the modeling of the girder, and the essential requirements are that the forces that develop as a result of the choice of girder geometry be accounted for in the design of the girder components and the adjacent structure. These actual forces are used in the design of the various elements, as distinguished from the simplified models that are currently used for many structural components.

The choices of elements, etc., are at the discretion of the designer and depend on the service requirements of the building as seen from the architectural, structural, mechanical, and electrical viewpoints. Unique design considerations must be made by the structural engineer, for example, if it is decided to eliminate the exterior openings and connect the stubs to the columns in addition to the chord and the slab.

Figure 33.1 shows the main components of the stub girder, as follows:

1. Bottom chord.
2. Exterior and interior stubs.

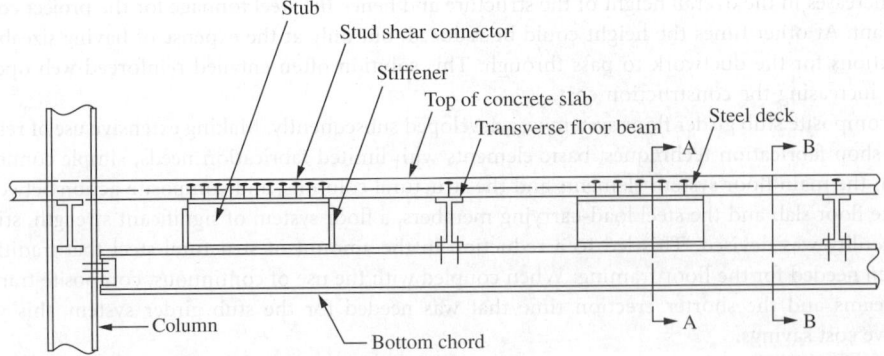

FIGURE 33.1   Elevation of a typical stub girder.

3. Transverse floor beams.
4. Formed steel deck.
5. Concrete slab with longitudinal and transverse reinforcement.
6. Stud shear connectors.
7. Stub stiffeners.
8. Beam-to-column connection.
9. The bottom chord should preferably be a hot-rolled wide-flange shape of column-type proportions, most often in the W12 to W14 series of wide-flange shapes. Other chord cross-sections have been considered; for example, tee shapes and rectangular tubes have certain advantages as far as welded attachments and fire protection are concerned, respectively. However, these other shapes also have significant drawbacks. The rolled tube, for example, cannot accommodate the shear stresses that develop in certain regions of the bottom chord. Rather than using a tee or a tube, therefore, a smaller W-shape (in the W10 series, for example) is most likely the better choice under these conditions.

The steel grade for the bottom chord, in particular, is important, since several of the governing regions of the girder are located within this member, and tension is the primary stress resultant. It is therefore possible to take advantage of higher strength steels, and 50 ksi yield stress steel has typically been the choice.

The floor beams and the stubs are mostly of the same size W-shape and are normally selected from the W16 and W18 series of shapes. This is directly influenced by the size(s) of the heating, ventilating, and air conditioning (HVAC) ducts that are to be used. Although it is not strictly necessary that the floor beams and the stubs use identical shapes, it avoids a number of problems if such a choice is made. These two components of the floor system at least should have the same height.

The concrete slab and the steel deck constitute the top chord of the stub girder. It is either made from lightweight or normal weight concrete, although if the former is available, even at a modest cost premium, it is preferred. The reason is the lower dead load of the floor, especially since the shores that will be used are strongly influenced by the concrete weight. Further, the shores must support several stories before they can be removed. In other words, the stub girders must be designed for shored construction, since the girder requires the slab to complete the system. In addition, the bending rigidity of the girder is substantial and a major fraction is contributed by the bottom chord. The reduction in slab stiffness that is prompted by the lower value of the modulus of elasticity for the lightweight concrete is therefore not as important as it may be for other types of composite bending members.

Concrete strengths of 3000 to 4000 psi are most common, although the choice also depends on the limit state of the stud shear connectors. The concrete does not control the strength of the stub girder, apart from in certain long-span girders, some local regions in the slab, and the desired mode of behavior of the slab-to-stub connection (which limits the maximum f_c'-value that can be used). Consequently, there is little to gain by using high-strength concrete.

The steel deck should be of the composite type, and a number of manufacturers produce suitable types. Normal deck heights are 2 and 3 in., but most floors are designed for the 3 in. deck. The deck ribs are run parallel to the longitudinal axis of the girder, since this gives better deck support on the transverse floor beams. It also increases the top chord area, which lends additional stiffness to a member that can span substantial distances. Finally, the parallel orientation provides a continuous rib trough directly above the girder centerline, improving the composite interaction of the slab and the girder.

Due to fire protection requirements, the thickness of the concrete cover over the top of the deck ribs is either $4\frac{3}{16}$ in. (normal weight concrete) or $3\frac{1}{4}$ in. (lightweight concrete). This eliminates the need for applying fire protective material to the underside of the steel deck.

Stud shear connectors are distributed uniformly along the length of the exterior and interior stubs, as well as on the floor beams. The number of connectors is determined on the basis of the computed shear forces that are developed between the slab and the stubs. This is in contrast to the current design practice for simple composite beams, which is based on the smaller of the ultimate axial load-carrying capacity of the slab and the steel beam [3]. However, the simplified approach of current specifications is not applicable to members where the cross-section varies significantly along the length (nonprismatic beams). The computed shear force design approach also promotes connector economy, in the sense that a much smaller number of shear connectors is required in the interior shear transfer regions of the girder [4–6].

The stubs are welded to the top flange of the bottom chord with fillet welds. In the original uses of the system, the design called for all-around welds [1,2]; subsequent studies demonstrated that the forces that are developed between the stubs and the bottom chord are concentrated toward the end of the stubs [4,5,7]. The welds should therefore be located in these regions.

The type and locations of the stub stiffeners that are indicated for the exterior stubs in Figure 33.1, as well as the lack of stiffeners for the interior stubs, represent one of the major improvements that were made to the original stub girder designs. Based on extensive research [4,5], it was found that simple end-plate stiffeners were as efficient as the traditional fitted ones, and in many cases the stiffeners could be eliminated at no loss in strength and stiffness to the overall girder.

Figure 33.1 shows that a simple (shear) connection is used to attach the bottom chord of the stub girder to the adjacent structure (column, concrete building core, etc.). This is the most common solution, especially when a duct opening needs to be located at the exterior end of the girder. If the support is an exterior column, the slab will rest on an edge member; if it is an interior column, the slab will be continuous past the column and into the adjacent bay. This may or may not present problems in the form of slab cracking, depending on the reinforcing steel placement around the column.

The stub girder has sometimes been used as part of the lateral load-resisting system of steel-framed buildings [8,9]. Although this has certain disadvantages insofar as column moments and the concrete slab reinforcement are concerned, the girder does provide significant lateral stiffness and ductility for the frame. As an example, the maintenance facility for Mexicana Airlines at the Mexico City International Airport, a structure utilizing stub girders in this fashion, survived the 1985 Mexico City Earthquake with no structural damage [9].

Expanding on the details that are shown in Figure 33.1, Figure 33.2 illustrates two cross-sections of a typical stub girder and Figure 33.3 shows a complete girder assembly with lights, ducts, and suspended

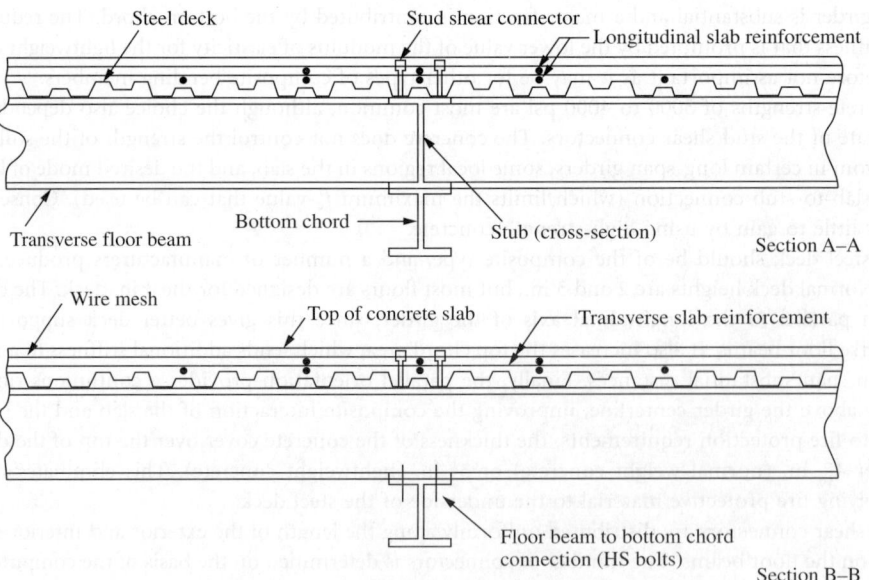

FIGURE 33.2 Cross-sections of a typical stub girder.

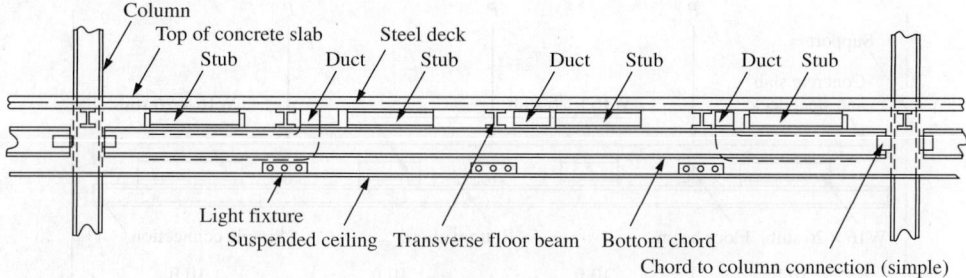

FIGURE 33.3 Stub girder with suspended ceiling, lights, and ducts.

ceiling. Of particular note are the longitudinal reinforcing bars. They add flexural strength as well as ductility and stiffness to the girder. These bars are commonly placed in two layers, with the top one just below the heads of the stud shear connectors. The lower rebars must be raised above the deck proper, using high chairs or other means.

The transverse rebars are important for adding shear strength to the slab, and they also help in the shear transfer from the connectors to the slab. The transverse bars also increase the overall ductility of the stub girder, and placing the bars in a herringbone pattern leads to a small improvement in the effective width of the slab.

Thirty-six or 50 ksi yield stress steel has been the common choice for stub girder floor systems, with a preference for the latter, because of the smaller bottom chord size that can be used. However, all detail materials (stiffeners, connection angles, etc.) are made from 36 ksi steel. Welding is usually done with 70-grade low hydrogen electrodes, using either the Shielded Metal Arc Welding (SMAW), Flux Cored Arc Welding (FCAW), or Gas Metal Arc Welding (GMAW) process, and the stud shear connectors are welded in the normal fashion. All of the work is done in the fabricating shop, except for the shear connectors, which are applied in the field where they are welded directly through the steel deck.

33.3 Methods of Analysis and Modeling

33.3.1 General Observations

In general, any number of methods of analysis may be used to determine the bending moments, shear forces, and axial forces in the components of the stub girder. However, it is essential to bear in mind that the modeling of the girder, or, in other words, how the actual girder is transformed into an idealized structural system, should reflect the relative stiffness of the elements. This means that it is important to establish realistic trial sizes of the components, through an appropriate preliminary design procedure. The subsequent modeling will then lead to stress resultants that are close to the magnitudes that can be expected in actual stub girders. Based on this approach, the design that follows is likely to require few changes, and those that are needed are often so small that they have no practical impact on the overall stiffness distribution and the final member forces. The preliminary design procedure is therefore a very important step in the overall design.

33.3.2 Preliminary Design Procedure

It is not necessary to make any assumptions about the stress distribution over the depth of the girder, other than to adhere to the strength model that was developed for normal composite beams [3,10]. The stress distribution will vary anyway along the span because of the openings. The strength model of Hansell et al. [10] assumes that when the ultimate moment is reached, all or a portion of the slab is failing in compression, with a uniformly distributed stress of $0.85f_c'$. The steel shape is simultaneously yielding in tension. Equilibrium is therefore maintained and the internal stress resultants are determined using first principles. Tests have demonstrated excellent agreement with theoretical analyses that utilize this approach [4–6,10].

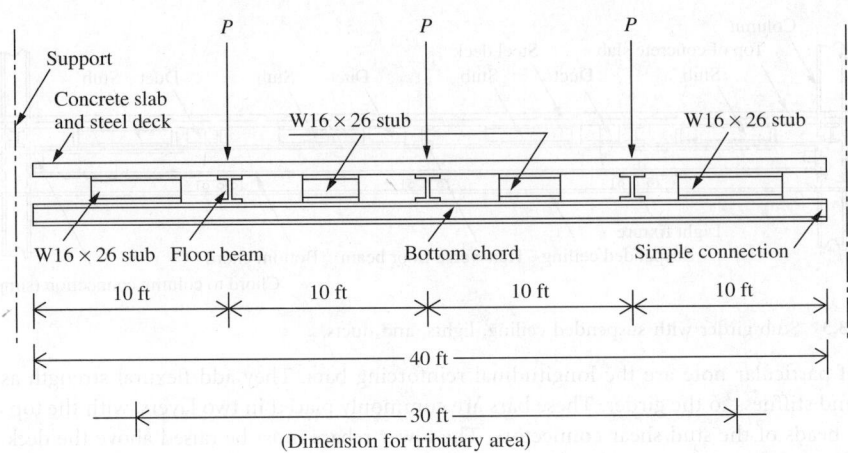

FIGURE 33.4 Stub girder for design example.

The design procedure follows the criteria of the AISC LRFD Specification [3]. The applicable resistance factor is given for the case of gross cross-section yielding. This is because the preliminary design is primarily needed to find the bottom chord size and this component is primarily loaded in tension [4–6,11]. The load factors are given by the ASCE 7 load standard [12] for the combination of dead plus live load.

The load computations follow the choice of the layout of the floor framing plan, whereby girder and floor beam spans are determined. This gives the tributary areas that are needed to calculate the dead and live loads. The load intensities are governed by local building code requirements or by the American Society of Civil Engineers (ASCE) recommendations, in the absence of a local code.

Reduced live loads should be used wherever possible. This is especially advantageous for stub girder floor systems, since spans and tributary areas tend to be large. ASCE 7 [12] makes use of a live load reduction factor, RF, which is significantly simpler to use and also less conservative than that of earlier codes. The standard places some restrictions on the value of RF, to the effect that the reduced live load cannot be less than 50% of the nominal value for structural members that support only one floor. Similarly, it cannot be less than 40% of the nominal live load if two or more floors are involved.

Proceeding with the preliminary design, the stub girder and its floor beam locations determine the magnitudes of the concentrated loads that are to be applied at each of the latter locations. The following illustrative example demonstrates the steps of the solution.

Figure 33.4 shows the layout of the stub girder for which the preliminary sizes are needed. Other computations have already given the sizes of the floor beam, the slab, and the steel deck. The span of the girder is 40 ft, the distance between adjacent girders is 30 ft, and the floor beams are located at the quarter points. The steel grade remains to be chosen; the concrete is lightweight, with $w_c = 120$ pcf and a compressive strength of $f'_c = 4000$ psi.

Loads:

   Estimated dead load: 74 psf

   Nominal live load: 50 psf

Live load reduction factor:

$$\text{RF} = 0.25 + 15/\sqrt{[2 \times (30 \times 30)]} = 0.60$$

Reduced live load:

   RLL $= 0.60 \times 50 = 30$ psf

Load factors (for $D + L$ combination):

> For dead load: 1.2
>
> For live load: 1.6

Factored distributed loads:

> Dead load, $DL = 74 \times 1.2 = 88.8$ psf
>
> Live load, $LL = 30 \times 1.6 = 48.0$ psf
>
> Total $= 136.8$ psf

Concentrated factored load at each floor beam location: Due to the locations of the floor beams and the spacing of the stub girders, the magnitude of each load, P, is

$$P = 136.8 \times 30 \times 10 = 41.0 \text{ kip}$$

Maximum factored midspan moment: The girder is symmetric about midspan and the maximum moment therefore occurs at this location

$$M_{max} = 1.5 \times P \times 20 - P \times 10 = 820 \text{ k ft}$$

Estimated interior moment arm for full stub-girder cross-section at midspan (refer to Figure 33.2 for typical details): The interior moment arm, that is, the distance between the compressive stress resultant in the concrete slab and the tensile stress resultant in the bottom chord is set equal to the distance between the slab centroid and the bottom chord (wide flange shape) centroid. This is simplified and conservative. In the example, the distance is estimated as

$$\text{Interior moment arm, } d = 27.5 \text{ in.}$$

This is based on having a 14-series W-shape for the bottom chord, W16 floor beams and stubs, a 3 in. high steel deck, and $3\frac{1}{4}$ in. of lightweight concrete over the top of the steel deck ribs (this allows the deck to be used without having sprayed-on fire protective material to the underside). These are common sizes for the stub-girder system.

In general, the interior moment arm varies between 24.5 and 29.5 in., depending on the heights of the bottom chord, floor beams/stubs, steel deck, and concrete slab.

Slab and bottom chord axial forces, F (these are the compressive and tensile stress resultants)

$$F = M_{max}/d = (820 \times 12)/27.5 = 357.9 \text{ kip}$$

Required cross-sectional area of bottom chord, A_s: The required cross-sectional area of the bottom chord can now be found. Since the chord is loaded in tension, the φ-value is 0.9.

It is also important to note that in the Vierendeel analysis that is commonly used in the final evaluation of the stub girder, the member forces will be somewhat larger than those determined through the simplified preliminary procedure. It is therefore recommended that an allowance of some magnitude be given for the Vierendeel action. This is done most easily by increasing the area A_s by a certain percentage. Based on experience [6,11], an increase of one third is suitable, and such has been done in the computations that follow.

On the basis of the data that have been developed, the required area of the bottom chord is

$$A_s = \frac{(M_{max}/d)}{\varphi \times F_y} \times \frac{4}{3} = \frac{F}{0.9 \times F_y} \times \frac{4}{3}$$

which gives A_s-values for 36 and 50 ksi steel of

$$A_s = \frac{357.9}{0.9 \times 36} \times \frac{4}{3} = 14.73 \text{ in.}^2 \quad (F_y = 36 \text{ ksi})$$

$$A_s = \frac{357.9}{0.9 \times 50} \times \frac{4}{3} = 10.60 \text{ in.}^2 \quad (F_y = 50 \text{ ksi})$$

Conclusions: If 36 ksi steel is chosen for the bottom chord of the stub girder, the wide-flange shapes W12 × 50 and W14 × 53 will be suitable. If 50 ksi steel is the choice, the sections may be W12 × 40 or W14 × 38.

The final decision is made by the structural engineer. However, since W12-series shapes will save approximately 2 in. in net floor system height, per story of the building, this would mean significant savings if the overall structure is 10 to 15 stories or more. The differences in stub girder strength and stiffness are not likely to play a role [7,12].

33.3.3 Choice of Stub Girder Component Sizes

Some examples were given in the earlier sections for the choices of chord and floor beam sizes, deck height, and slab configuration. These were made primarily on the basis of acceptable geometries, deck size, and fire protection requirements, to mention some examples. However, construction economy is critical and the following guidelines will assist the user. The data that are given are based on actual construction projects.

Economical span lengths for the stub girder range from 30 to 50 ft, although the preferable spans are 35 to 45 ft. Fifty-feet span girders are erectable, but these are close to the limit where the dead load becomes excessive, which has the effect of making the slab govern the overall design. This is usually not an economical solution. Spans shorter than 30 ft are known to have been used successfully; however, this depends on the load level and the type of structure, to mention the key considerations.

Depending on the type and configuration of steel deck that has been selected, the floor beam spacing should generally be maintained between 8 and 12 ft, although larger values have been used. The decisive factor is the ability of the deck to span the distance between the floor beams.

The performance of the stub girder is not particularly sensitive to the stub lengths that are used, as long as these are kept within reasonable limits. In this context, it is important to observe that it is usually the exterior stub that controls the behavior of the stub girder. As a practical guideline, the exterior stubs are normally 5 to 7 ft long; the interior stubs are considerably shorter, normally around 3 ft, but components up to 5 ft long are known to have been used. When the stub lengths are chosen, it is necessary to bear in mind the actual purpose of the stubs and how they carry the loads on the stub girder. That is, the stubs are loaded primarily in shear, which explains why the interior stubs can be kept so much shorter than the exterior ones.

The shear connectors that are welded to the top flange of the stub, the stub web stiffeners, and the welds between the bottom flange of the stub and the top flange of the bottom chord are crucial to the function of the stub girder system. For example, the first application of stub girders utilized fitted stiffeners at the ends and sometimes at midlength of all of the stubs. Subsequent research demonstrated that the midlength stiffener did not perform any useful function and that only the exterior stubs needed stiffeners in order to provide the requisite web stability and shear capacity [5,6]. Regardless of the span of the girder, it was found that the interior stubs could be left unstiffened, even when they were made as short as 3 ft [6].

Similar savings were realized for the welds and the shear connectors. In particular, in lieu of all-around fillet welds for the connection between the stub and the bottom chord, the studies showed that a significantly smaller amount of welding was needed, and often only in the vicinity of the stub ends. However, specific weld details must be based on appropriate analyses of the stub, considering over-turning, weld capacity at the tension end of the stub, and adequate ability to transfer shear from the slab to the bottom chord.

33.3.4 Modeling of the Stub Girder

The original work of Colaco [1,2] utilized a Vierendeel modeling scheme for the stub girder to arrive at a set of stress resultants, which in turn were used to size the various components. Elastic finite element analyses were performed for some of the girders that had been tested, mostly to examine local stress distributions and the correlation between test and theory. However, the finite element solution is not a practical design tool.

Other studies have examined approaches such as nonprismatic beam analysis and variations of the finite element method [5]. The nonprismatic beam solution is relatively simple to apply. On the other hand, it is not as accurate as the Vierendeel approach, since it tends to overlook some important local effects and overstates the service load deflections [4,5].

On the whole, therefore, the Vierendeel modeling of the stub girder has been found to give the most accurate and consistent results, and the correlation with test results is good [1,4,5,7]. Finally, it offers the best physical similarity with actual girders; many designers have found this to be an important advantage.

There are no "simple" methods of analysis that can be used to find the bending moments, shear forces, and axial forces in Vierendeel girders. Once the preliminary sizing has been accomplished, a computer solution is required for the girder. In general, all that is required for the Vierendeel evaluation is a two-dimensional plane frame program for elastic structural analysis. This gives moments, shears, and axial forces, as well as deflections, joint rotations, and other displacement characteristics. The stress resultants are used to size the girder and its elements and connections; the displacements reflect the serviceability of the stub girder.

Once the stress resultants are known, the detailed design of the stub girder can proceed. A final run-through of the girder model should then be done, using the components that were chosen, to ascertain that the performance and strength are sufficient in all respects. Under normal circumstances no alterations are necessary at this stage.

As an illustration of the Vierendeel modeling of a stub girder, the girder itself is shown in Figure 33.5a and the Vierendeel model in Figure 33.5b. The girder is the same as the one used for the preliminary design example. It has four stubs and is symmetrical about midspan; therefore, only one half is illustrated. The boundary conditions are shown in Figure 33.5b.

The bottom chord of the model is assigned a moment of inertia equal to the major axis *I*-value, I_{xo} of the wide-flange shape that was chosen in the preliminary design. However, some analysts believe that

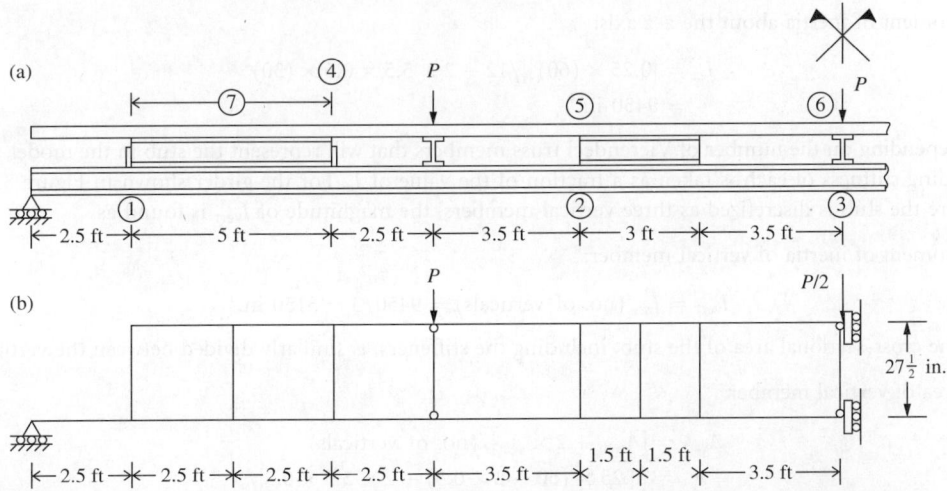

FIGURE 33.5 Vierendeel modeling of a stub girder: (a) stub girder and (b) its Vierendeel model.

since the stub is welded to the bottom chord, a portion of its flexural stiffness should be added to that of the moment of inertia of the wide-flange shape [4,5,7]. This approach is identical to treating the bottom chord W-shape as if it has a cover plate on its top flange. The area of this "cover plate" is the same as the area of the bottom flange of the stub. This should be done only in the areas where the stubs are placed. In the regions of the interior and exterior stubs it is therefore realistic to increase the moment of inertia of the bottom chord by the parallel-axis value of $A_f \times d_f^2$, where A_f designates the area of the bottom flange of the stub and d_f is the distance between the centroids of the flange plate and the W-shape.

The bending stiffness of the top Vierendeel chord equals that of the effective width portion of the slab. This should include the contributions of the steel deck as well as the reinforcing steel bars that are located within this width. In particular, the influence of the deck is important. The effective width is determined from the criteria in the AISC LRFD Specification [3]. These were originally developed on the basis of analyses and tests of prismatic composite beams. The approach has been found to give conservative results [4,5], but should continue to be used until more accurate criteria are available.

In the computations for the slab, the cross-section is conveniently subdivided into simple geometric shapes. The individual areas and moments of inertia are determined on the basis of the usual transformation from concrete to steel, using the modular ratio $n = E/E_c$, where E is the modulus of elasticity of the steel and E_c is that of concrete. The latter must reflect the density of the concrete that is used, and can be computed from [13]

$$E_c = 33 \times w_c^{1.5} \times \sqrt{f_c'} \qquad (33.1)$$

The shear connectors used for the stub are required to develop 100% interaction, since the design is based on the computed shear forces, rather than the axial capacity of the steel beam or the concrete slab, as is used for prismatic beams in the AISC Specification [3]. However, it is neither common nor proper to add the moment of inertia contribution of the top flange of the stub to that of the slab, contrary to what is done for the bottom chord. The reason for this is that dissimilar materials are joined and some local concrete cracking and/or crushing can be expected to take place around the shear connectors.

The discretization of the stubs into vertical Vierendeel girder components is relatively straightforward. Considering the web of the stub and any stiffeners, if applicable (for exterior stubs, most commonly, since interior stubs usually can be left unstiffened), the moment of inertia about an axis that is perpendicular to the plane of the web is calculated. As an example, Figure 33.6 shows the stub and stiffener configuration for a typical case. The stub is a 5 ft long W16 × 26 with $5\frac{1}{2} \times \frac{1}{2}$ in. end-plate stiffeners. The computations give

Moment of inertia about the z–z axis:

$$I_{zz} = [0.25 \times (60)^3]/12 + 2 \times 5.5 \times 0.5 \times (30)^2$$
$$= 9450 \text{ in.}^4$$

Depending on the number of Vierendeel truss members that will represent the stub in the model, the bending stiffness of each is taken as a fraction of the value of I_z. For the girder shown in Figure 33.5, where the stub is discretized as three vertical members, the magnitude of I_{vert} is found as

Moment of inertia of vertical member:

$$I_{vert} = I_{zz}/(\text{no. of verticals}) = 9450/3 = 3150 \text{ in.}^4$$

The cross-sectional area of the stub, including the stiffeners, is similarly divided between the verticals

Area of vertical member:

$$A_{vert} = [A_{web} + 2 \times A_{st}]/(\text{no. of verticals})$$
$$= [0.25 \times (60 - 2 \times 0.5) + 2 \times 5.5 \times 0.5]/3$$
$$= 6.75 \text{ in.}^2$$

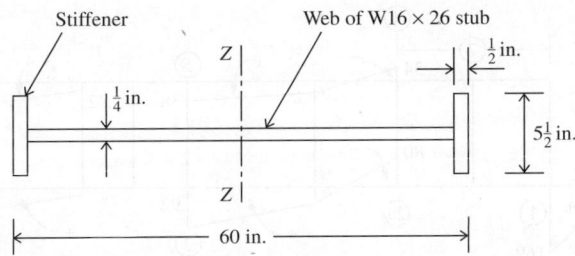

FIGURE 33.6 Plan view of stub with end plate stiffeners.

Several studies have aimed at finding the optimum number of vertical members to use for each stub. However, the strength and stiffness of the stub girder are insignificantly affected by this choice and a number between 3 and 7 is usually chosen. As a rule of thumb, it is advisable to have one vertical per foot length of stub, but this is only a guideline.

The verticals are placed at uniform intervals along the length of the stub, usually with the outside members close to the stub ends. Figure 33.5 illustrates the approach. As for end conditions, these vertical members are assumed to be rigidly connected to the top and bottom chords of the Vierendeel girder.

One vertical member is placed at each of the locations of the floor beams. This member is assumed to be pinned to the top and bottom chords, as shown in Figure 33.5, and its stiffness is conservatively set equal to the moment of inertia of a plate with a thickness equal to that of the web of the floor beam and a length equal to the beam depth. In the example, $t_w = 0.25$ in.; the beam depth is 15.69 in. This gives a moment of inertia of

$$([15.69 \times 0.25^3]/12) = 0.02 \text{ in.}^4$$

and the cross-sectional area is

$$(15.69 \times 0.25) = 3.92 \text{ in.}^2$$

The Vierendeel model shown in Figure 33.5b indicates that the portion of the slab that spans across the opening between the exterior end of the exterior stub and the support for the slab (a column, or a corbel of the core of the structural frame) has been neglected. This is a realistic simplification, considering the low rigidity of the slab in negative bending.

Figure 33.5b also shows the support conditions that are used as input data for the computer analysis. In the example, the symmetrical layout of the girder and its loads make it necessary only to analyze one half of the span. This cannot be done if there is any kind of asymmetry and the entire girder must then be analyzed. For the girder that is shown, it is known that only vertical displacements can take place at midspan; horizontal displacements and end rotations are prevented at this location. At the far ends of the bottom chord only horizontal displacements are permitted and end rotations are free to occur. The reactions that are found are used to size the support elements, including the bottom chord connections and the column.

The structural analysis results are shown in Figure 33.7, in terms of the overall bending moment, shear force and axial force distributions of the Vierendeel model given in Figure 33.5b. Figure 33.7d repeats the layout details of the stub girder, to help identify the locations of the key stress resultant magnitudes with the corresponding regions of the girder.

The design of the stub girder and its various components can now be done. This must also include deflection checks, even though research has demonstrated that the overall design will not be governed by deflection criteria [6]. However, since the girder has to be built in the shored condition, the girder is often fabricated with a camber, approximately equal to the dead load deflection [6,11].

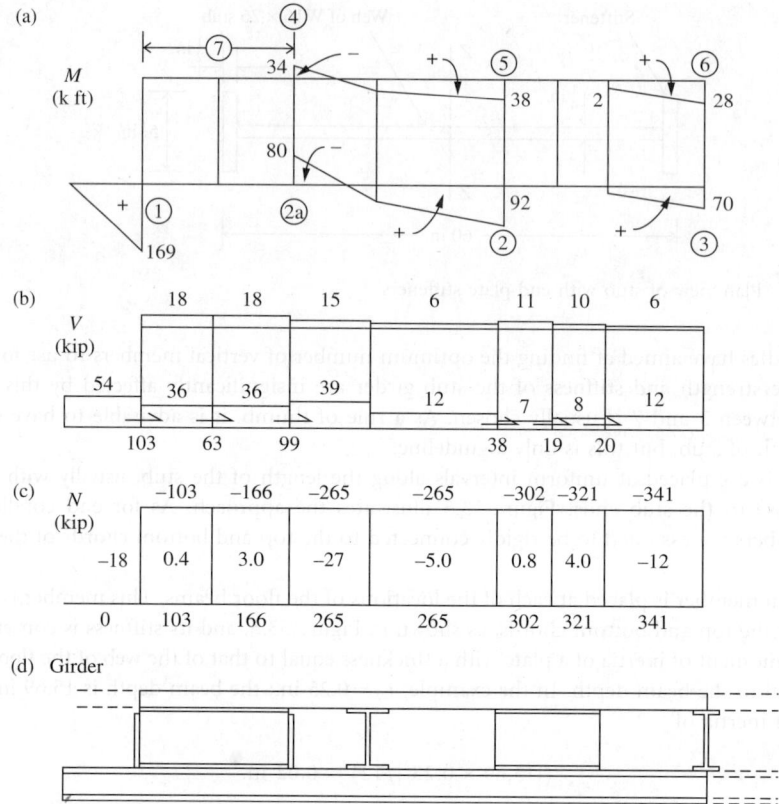

FIGURE 33.7 Bending moment, shear force, and axial force distributions for stub girder and critical points for strength checks.

33.4 Design Criteria for Stub Girders

33.4.1 General Observations

In general, the design of the stub girder and its components must consider overall member strength criteria as well as local checks. For most of these, the AISC Specification [3] gives requirements that address the needs.

In several important areas there are no standardized rules that can be used in the design of the stub girder, and the designer must rely on rational engineering judgment to arrive at satisfactory solutions. This applies to the parts of the girder that have to be designed on the basis of computed forces, such as shear connectors, stiffeners, stub-to-chord welds, and slab reinforcement. The modeling and evaluation of the capacity of the central portion of the concrete slab are also subject to interpretation. However, the design recommendations that are given in the following are based on a wide variety of practical and successful applications.

It is again emphasized that the design throughout is based on the stress resultants that have been determined in the Vierendeel or other analysis, rather than on idealized code criteria. However, the capacities of materials and fasteners, as well as the requirements for the stability and strength of tension and compression members, adhere strictly to the AISC design requirements. Any interpretations that have been made are conservative.

33.4.2 Governing Sections of the Stub Girder

Figure 33.5 and Figure 33.7 show certain circled numbers at various locations throughout the span of the stub girder. These reflect the sections of the girder that are the most important for one reason or another and are the ones that must be examined to determine the required member size, etc. These are the *governing sections of the stub girder* and are itemized as follows:

1. Points 1, 2, and 3 indicate the critical sections for the bottom chord.
2. Points 4, 5, and 6 indicate the critical sections for the concrete slab.
3. Point 7, which is a region rather than a specific point, indicates the critical shear transfer region between the slab and the exterior stub.

The design checks that must be made for each of these areas are discussed in the following.

33.4.3 Design Checks for the Bottom Chord

The size of the bottom chord is almost always governed by the stress resultants at midspan or Point 3 in Figure 33.5 and Figure 33.7. This is also why the preliminary design procedure focused almost entirely on determining the required chord cross-section at this location. As the stress resultant distributions in Figure 33.7 show, the bottom chord is subjected to combined positive bending moment and tensile force at Point 3, and the design check must consider the beam-tension member behavior in this area.

Combined bending and tension must also be evaluated at Point 2, the exterior end of the interior stub. The local bending moment in the chord is generally larger here than at midspan, but the axial force is smaller. Only a computation can confirm whether Point 2 will govern in lieu of Point 3. Further, although the location at the interior end of the exterior stub (Point 2a) is rarely critical, the combination of negative moment and tensile force should be evaluated.

At Point 1 of the bottom chord, which is located at the exterior end of the exterior stub, the axial force is equal to zero. At this location the bottom chord must therefore be checked for pure bending as well as shear.

The preceding applies only to a girder with simple end supports. When it is part of the lateral load-resisting system, axial forces will exist in all parts of the chord. These must be resisted by the adjacent structural members.

33.4.4 Design Checks for the Concrete Slab

The top chord carries varying amounts of bending moment and axial force, as illustrated in Figure 33.7, but the most important areas are indicated as Points 4 to 6. The axial forces are always compressive in the concrete slab; the bending moments are positive at Points 5 and 6 but negative at Point 4. As a result, this location is normally the one that governs the performance of the slab, not the least because the reinforcement in the positive moment region includes the cross-sectional area of the steel deck.

The full effective width of the slab must be analyzed for combined bending and axial force at all of Points 4 through 6. Either the composite beam–column criteria of the AISC LRFD Specification [3] or the criteria of the reinforced concrete structures code of American Concrete Institute (ACI) [13] may be used for this purpose.

33.4.5 Design Checks for the Shear Transfer Regions

Region 7 is the shear transfer region between the concrete slab and the exterior stub, and the combined shear and longitudinal compressive capacity of the slab in this area must be determined. The shear transfer region between the slab and the interior stub always has a smaller shear force.

Region 7 is critical and several studies have shown that the slab in this area will fail in a combination of concrete crushing and shear [4–7]. The shear failure zone usually extends from corner to corner of the steel deck, over the top of the shear connectors, as illustrated in Figure 33.8. This also emphasizes why the placement of the longitudinal reinforcing steel bars in the central flute of the steel deck is important,

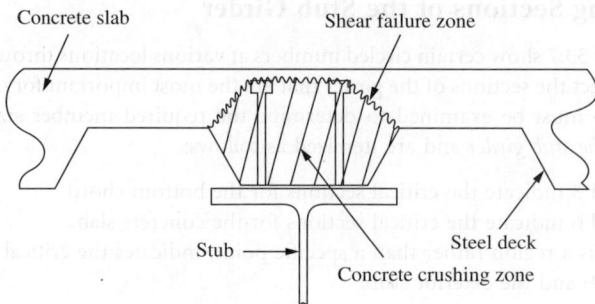

FIGURE 33.8 Shear failure zone in concrete slab above stub.

as well as the location of the transverse bars: both groups should be placed just below the level of the top of the shear connectors (see Figure 33.2). The welded wire mesh reinforcement that is used as a matter of course, mostly to control shrinkage cracking in the slab, also assists in improving the strength and ductility of the slab in this region.

33.4.6 Design of Stubs for Shear and Axial Load

The shear and axial force distributions indicate the governing stress resultants for the stub members. It is important to note that since the Vierendeel members are idealized from the real (i.e., continuous) stubs, bending is not a governing condition. Given the sizes and locations of the individual vertical members that make up the stubs, the design checks are made for axial load and shear. For example, referring to Figure 33.7, it is seen that the shear and axial forces in the exterior and interior stubs, and the axial forces in the verticals that represent the floor beams, are the following:

Exterior stub verticals:

Shear forces: 103 kip   63 kip   99 kip
Axial forces: −18 kip   0.4 kip   3 kip

Interior stub verticals:

Shear forces: 38 kip   19 kip   20 kip
Axial forces: −5 kip   0.8 kip   4 kip

Floor beam verticals:

Exterior: Axial force = −39 kip
Interior: Axial force = −12 kip

Shear forces are zero in these members.

The areas and moments of inertia of the verticals are known from the modeling of the stub girder. Figure 33.7 also shows the shear and axial forces in the bottom and top chords, but the design for these elements has been addressed earlier in this chapter.

The design checks that are made for the stub verticals will also indicate whether there is a need for stiffeners for the stubs, since the evaluations for axial load capacity should always first be made on the assumption that there are no stiffeners. However, experience has shown that the exterior stubs always must be stiffened; the interior stubs, on the other hand, will almost always be satisfactory without stiffeners, although exceptions can occur.

The axial forces that are shown for the stub verticals in the preceding are small but typical, and in all probability only the exterior end of the exterior stub really requires a stiffener. This was examined in one of the stub girder research studies, where it was found that a single stiffener would suffice, although the resulting lack of structural symmetry gave rise to a tensile failure in the unstiffened area of the stub [5].

Although this occurred at a very late stage in the test, the type of failure represents an undesirable mode of behavior, and the use of single stiffeners therefore was discarded. Further, by reason of ease of fabrication and erection, stiffeners should always be provided at both stub ends.

It is essential to bear in mind that if stiffeners are required, the purpose of such elements is to add to the area and moment of inertia of the web, to resist the axial load that is applied. There is no need to provide *bearing* stiffeners, since the load is not transmitted in this fashion. The most economical solution is to make use of end-plate stiffeners of the kind that is shown in Figure 33.1; extensive research evaluations showed that this was the most efficient and economical choice [4,5,7].

The vertical stub members are designed as columns, using the criteria of the AISC Specification [3]. For a conservative solution, an effective length factor of 1.0 may be used. However, it is more realistic to utilize a *K*-value of 0.8 for the verticals of the stubs, recognizing the end restraint that is provided by the connections between the chords and the stubs. The *K*-factor for the floor beam verticals must be 1.0, due to the pinned ends that are assumed in the modeling of these components, as well as the flexibility of the floor beam itself for buckling of the vertical member.

33.4.7 Design of Stud Shear Connectors

The Vierendeel girder shear force diagram gives the shear forces that must be transferred between the slab and the stubs. These are the factored shear force values that are to be resisted by the connectors. The example shown in Figure 33.7 indicates the individual shear forces for the stub verticals, as listed in the preceding section. However, in the design of the overall shear connection, the total shear force that is to be transmitted to the stub is used, and the stud connectors are then distributed uniformly along the stub. The design strength of each connector is determined in accordance with the LRFD Specification [3], including any deck profile reduction factor.

Analyzing the girder whose data are given in Figure 33.7, the following is known:

Exterior stub:

$$\text{Total shear force} = V_{es} = 103 + 63 + 99 = 265 \text{ kip}$$

Interior stub:

$$\text{Total shear force} = V_{is} = 38 + 19 + 20 = 77 \text{ kip}$$

The nominal strength, Q_n, of the stud shear connectors is given by the LRFD Specification; thus

$$Q_n = 0.5 \times A_{sc}\sqrt{f'_c \times E_c} \leq A_{sc} \times F_u \qquad (33.2)$$

where A_{sc} is the cross-sectional area of the stud shear connector, f'_c and E_c are the compressive strength and modulus of elasticity of the concrete, respectively, and F_u is the specified minimum tensile strength of the stud shear connector steel (60 ksi) (ASTM A108).

In the equation for Q_n, the left-hand side reflects the ultimate limit state of shear yield failure of the connector; the right-hand side gives the ultimate limit state of tension fracture of the stud. Although shear almost always governs, and is the desirable mode of behavior, a check has to be made to ensure that tension fracture will not take place. This is achieved by the appropriate value of E_c, setting $F_u = 60$ ksi, and solving for f'_c from the Equation 33.2. The requirement that must be satisfied in order for the stud shear limit state to govern is given by Equation 33.3:

$$f'_c \leq \frac{57,000}{w_c} \qquad (33.3)$$

This gives the limiting values for concrete strength as related to the density (Table 33.1).

For concrete with $w_c = 120$ pcf and $f'_c = 4000$ psi, as used in the design example, $E_c = 2,629,000$ psi. Using $\frac{3}{4}$-in. diameter studs, the nominal shear capacity is

$$Q_n = 0.5[\Pi(0.75)^2/4]\sqrt{(4.2, 629)} \leq [\Pi(0.75)^2/4]60$$

TABLE 33.1   Concrete Strength Limitations for Ductile Shear Connector Failure

Concrete density w_c (pcf)	Maximum concrete strength, f'_c (psi)
145 (= NW)	4000
120	4800
110	5200
100	5700
90	6400

which gives

$$Q_n = 22.7 \text{ kip} < 26.5 \text{ kip}$$

The LRFD Specification [3] does not give a resistance factor for shear connectors, on the premise that the φ-value of 0.85 for the overall design of the composite member incorporates the stud strength variability. This is not satisfactory for composite members such as stub girders and composite trusses. However, a study was carried out to determine the resistance factors for the two ultimate limit states for stud shear connectors [14]. Briefly, on the basis of extensive analyses of test data from a variety of sources, and using the Q_n-equation as the nominal strength expression, the values of the resistance factors that apply to the shear yield and tension fracture limit states, respectively, are

Stud shear connector resistance factors:

  Limit state of shear yielding: $\varphi_{conn} = 0.90$
  Limit state of tension fracture: $\varphi_{conn} = 0.75$

The required number of shear connectors can now be found as follows, using the total stub shear forces, V_{es} and V_{is}, computed earlier in this section:

Exterior stub:

$$n_{es} = V_{es}/(0.9 \times Q_n) = V_{es}/(\varphi_{conn} Q_n)$$
$$= 265/(0.9 \times 22.7) = 13.0$$

That is, use 14 $\frac{3}{4}$-in. diameter stud shear connectors, placed in pairs and distributed uniformly along the length of the top flange of each of the exterior stubs.

Interior stub:

$$n_{is} = V_{is}/(0.9 \times Q_n) = V_{is}/(\varphi_{conn} Q_n)$$
$$= 77/(0.9 \times 22.7) = 3.8$$

That is, use 4 $\frac{3}{4}$-in. diameter stud shear connectors, placed singly and distributed uniformly along the length of the top flange of each of the interior stubs.

Considering the shear forces for the stub girder of Figure 33.5 and Figure 33.7, the number of connectors for the exterior stub is approximately three times that for the interior one. Depending on span, loading, etc., there are instances when it will be difficult to fit the required number of studs on the exterior stub, since typical usage entails a double row, spaced as closely as permitted. Several avenues may be followed to remedy such a problem; the easiest one is most likely to use a higher strength concrete, as long as the limit state requirements for Q_n and Table 33.1 are satisfied. This entails only minor reanalysis of the girder.

33.4.8 Design of Welds Between Stub and Bottom Chord

The welds that are needed to fasten the stubs to the top flange of the bottom chord are primarily governed by the shear forces that are transferred between these components of the stub girder. The shear force distribution gives these stress resultants, which are equal to those that must be transferred between the slab and the stubs. Thus, the factored forces V_{es} and V_{is} that were developed in Section 33.4.7 are used to size the welds.

Axial loads also act between the stubs and the chord; these may be compressive or tensile. In Figure 33.7, it is seen that the only axial force of note occurs in the exterior vertical of the exterior stub (load $= 18$ kip); the other loads are very small compressive or tensile forces. Unless a significant tensile force is found in the analysis, it will be a safe simplification to ignore the presence of the axial forces insofar as the weld design is concerned.

The primary shear forces that have to be taken by the welds are developed in the outer regions of the stubs, although it is noted that in the case of Figure 33.5, the central vertical element in both stubs carry forces of some magnitude (63 and 19 kip, respectively). However, this distribution is a result of the modeling of the stubs; analyses of girders where many more verticals were used have confirmed that the major part of the shear is transferred at the ends [5,6,11]. The reason is that the stub is a full shear *panel*, where the internal moment is developed through stress resultants that act at points toward the ends, in a form of bending action. Tests have also verified this characteristic of the girder behavior [5,7]. Finally, concentrating the welds at the stub ends will have significant economic impact [4–6].

In view of these observations, the most effective placement of the welds between the stubs and the bottom chord is to concentrate them across the ends of the stubs and along a short distance of both sides of the stub flanges. For ease of fabrication and structural symmetry, the same amount of welding should be placed at both ends, although the forces are always smaller at the interior ends of the stubs. Such U-shaped welds were used for a number of the full-size girders that were tested [4,5,7], with only highly localized yielding occurring in the welds. A typical detail is shown in Figure 33.9; this reflects what is recommended for use in practice.

Prior to the research that led to the change of the welded joint design, the stubs were welded with all-around fillet welds for the exterior as well as the interior elements. The improved, U-shaped detail provided for weld metal savings of approximately 75% for interior stubs and around 50% for exterior stubs.

For the sample stub girder, W16 × 26 shapes are used for the stubs. The total forces to be taken by the welds are

Exterior stub: $V_{es} = 265$ kip
Interior stub: $V_{is} = 77$ kip

Using E70XX electrodes and $\frac{5}{16}$-in. fillet welds (the fillet weld size must be smaller than the thickness of the stub flange, which is $\frac{3}{8}$ in. for the W16 × 26), the total weld length for each stub is L_w, given by (refer to Figure 33.9)

$$L_w = 2(b_{fs} + 21)$$

since U-shaped welds of length $(b_{fs} + 21)$ are placed at each stub end. The total weld lengths required for the stub girder in question are therefore

Exterior stub:

$$
\begin{aligned}
(L_w)_{es} &= V_{es}/(0.707 a\varphi_w F_w) \\
&= 265/\left[0.707\left(\tfrac{5}{16}\right) \cdot 0.75(0.6 \cdot 70)\right] = 38.1 \text{ in.}
\end{aligned}
$$

Interior stub:

$$
\begin{aligned}
(L_w)_{is} &= V_{is}/(0.707 a\varphi_w F_w) \\
&= 77/\left[0.707\left(\tfrac{5}{16}\right) \cdot 0.75(0.6 \cdot 70)\right] = 11.1 \text{ in.}
\end{aligned}
$$

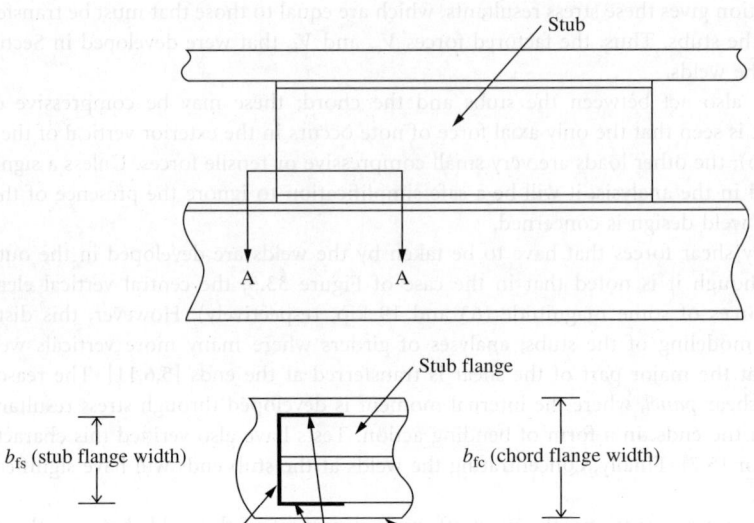

FIGURE 33.9 Welding details for end of stub to bottom chord of stub girder.

In the above expressions, $a = \frac{5}{16}$ in. = fillet weld size, $\varphi_w = 0.75$, and $F_w = 0.6F_{EXX} = 0.6 \times 70 = 42$ ksi for E70XX electrodes. The total U-weld lengths at each stub end are therefore

    Exterior stub: $L_{Ues} = 19.1$ in.
    Interior stub: $L_{Uis} = 5.6$ in.

With a flange width for the W16 × 26 of 5.50 in., the above lengths can be simplified as

$$L_{Ues} = 5.50 + 7.0 + 7.0$$

where L_{es} is chosen as 7.0 in. For the interior stub

$$L_{Uis} = 5.50 + 2.0 + 2.0$$

where L_{is} is chosen as 2.0 in.

The details chosen are a matter of judgment. In the example, the interior stub for all practical purposes requires no weld other than the one across the flange, although at least a minimum weld return of $\frac{1}{2}$ in. should be used.

33.4.9 Floor Beam Connections to Slab and Bottom Chord

In the Vierendeel model, the floor beam is represented as a pinned-end compression member. It is designed using a K-factor of 1.0, and the floor beam web by itself is almost always sufficient to take the axial load. However, the floor beam must be checked for web crippling and web buckling under shoring conditions.

No shear is transferred from the beam to the slab or the bottom chord. In theory, therefore, any attachment device between the floor beam and the other components should not be needed. However, due to construction stability requirements, as well as the fact that the floor beam usually is designed for composite action normal to the girder, fasteners are needed. In practice, these are not actually designed; rather, one or two stud shear connectors are placed on the top flange of the beam and two high strength bolts attach the lower flange to the bottom chord.

33.4.10 Connection of Bottom Chord to Supports

In the traditional use of stub girders, the girder is supported as a simple beam and the bottom chord end connections need to be able to transfer vertical reactions to the supports. The supports may be columns or the girder may rest on corbels or other types of supports that are part of the concrete core of the building. For both of these cases the reactions that are to be carried to the adjacent structure are given by the analysis.

Any shear-type beam connections may be used to connect the bottom chord to a column or a corbel or similar bracket. It is important to ascertain that the chord web shear capacity is sufficient, including block shear.

Some designers prefer to use slotted holes for the connections and to delay the final tightening of the bolts until after the shoring has been removed. This is done on the premise that the procedure will leave the slab essentially stress free from the construction loads, leading to less cracking in the slab during service. Other designers specify additional slab reinforcement to take care of any cracking problem. Experience has shown that both methods are suitable.

The slab may be supported on an edge beam or similar element at the exterior side of the floor system. There is no force transfer ability required of this support. In the interior of the building the slab will be continuously cast across other girders and around columns; this will almost always lead to some cracking, both in the vicinity of the columns as well as along beams and girders. With suitable placement of floor slab joints, this can be minimized, and appropriate transverse reinforcement for the slab will reduce, if not eliminate, the longitudinal cracks.

Data on the effects of various types of cracks in composite floor systems are scarce. Current opinion appears to be that the strength is not significantly influenced. In any case, the mechanics of the short- and long-term service response of composite beams is not well understood. Studies have developed models for the cracking mechanism and the crack propagation [15]; the correlation with a wide variety of laboratory tests is good. However, a comprehensive study of concrete cracking and its implications for structural service and strength needs to be undertaken.

33.4.11 Use of Stub Girder for Lateral Load System

The stub girder was originally conceived only as being part of the vertical load-carrying system of structural frames, and the use of simple connections, as discussed in Section 33.4.9, came from this development. However, recognizing that a deep, long-span member can be very effective as a part of the lateral load-resisting system for a structure, several attempts have been made to incorporate the stub girder into moment frames and similar systems. The projects of Colaco in Houston [8] and Martinez-Romero [9] in Mexico City were successful, although the designers noted that the cost premium could be substantial.

For the Colaco structure, his applications reduced drift, as expected, but gave much more complex beam-to-column connections and reinforcement details in the slab around the columns. Thus, the exterior stubs were moved to the far ends of the girders and moment connections were designed for the full depth. For the Mexico City building, the added ductility was a prime factor in the survival of the structure during the 1985 earthquake.

The advantages of using the stub girders in moment frames are obvious. Some of the disadvantages have been outlined; in addition, it must be recognized that the lack of room for perimeter HVAC ducts may be undesirable. This can only be addressed by the mechanical engineering consultant. As a general rule, a designer who wishes to use stub girders as part of the lateral load-resisting system should examine all structural effects and also incorporate nonstructural considerations such as those prompted by HVAC and electronic communication needs.

33.4.12 Deflection Checks

The service load deflections of the stub girder are needed for several purposes. First, the overall dead load deflection is used to assess the camber requirements. Due to the long spans of typical stub girders, as well as the flexibility of the framing members and the connections during construction, it is important to end up with a floor system that is as level as possible by the time the structure is ready to be occupied. Thus, the girders must be built in the shored condition and the camber should be approximately equal to 75% of the dead load deflection.

Second, it is essential to bear in mind that each girder will be shored against a similar member at the level below the current construction floor. This member, in turn, is similarly shored, albeit against a girder whose stiffness is greater, due to the additional time of curing of the concrete slab. This has a cumulative effect for the structure as a whole and the dead load deflection computations must take this response into account.

In other words, the support for the shores is a flexible one and deflections therefore will occur in the girder as a result of floor system movements of the structure at levels in addition to the one under consideration. Although this is not unique to the stub girder system, the span lengths and the interaction with the frame accentuate the influence on the girder design.

Depending on the structural system, it is also likely that the flexibility of the columns and the connections will add to the vertical displacements of the stub girders. The deflection calculations should incorporate these effects, preferably by utilizing realistic modified E_c-values, and determining displacements as they occur in the frame. Thus, the curing process for the concrete might be considered, since the strength development as a function of time is directly related to the value of E_c [13]. This is a subject that is open for study, although similar criteria have been incorporated in studies of the strength and behavior of composite frames [16,17]. However, detailed evaluations of the influence of time-dependent stiffness still need to be made for a wide variety of floor systems and frames. The cumulative deflection effects can be significant for the construction of the building and consequently also must enter into the contractor's planning. This subject is addressed briefly in Section 33.5.

Third, the live load deflections must be determined to assess the serviceability of the floor system under normal operating conditions. Several studies have demonstrated that such displacements will be significantly smaller than the $L/360$ requirement that is normally associated with live load deflections [5–7,11]. It is therefore rarely possible to design a girder that meets the strength and the deflection criteria simultaneously. In other words, strength governs the overall design.

Finally, although they rarely play a role in the overall response of the stub girder, the deflections and end rotations of the slab across the openings of the girder should also be checked. This is primarily done to assess the potential for local cracking, especially at the stub ends and at the floor beams. However, proper placement of the longitudinal girder reinforcement is usually sufficient to prevent problems of this kind, since the deformations tend to be small.

33.5 Influence of Method of Construction

A number of construction-related considerations have already been addressed in various sections of this chapter. The most important ones relate to the fact that the stub girders must be built in the shored condition. The placement and removal of the shores may have a significant impact on the performance of the member and the structure as a whole. In particular, too early shore removal may lead to excessive deflections in the girders at levels above the one where the shores were located. This is a direct result of the low stiffness of "green" concrete. It can also lead to "ponding" of the concrete slab, producing larger dead loads than accounted for in the original design. Finally, larger girder deflections can be translated into an "inward pulling" effect on the columns of the frame. However, this is clearly a function of the framing system.

On the other hand, the use of high early strength cement and similar products can reduce this effect significantly. Further, since the concrete usually is able to reach about 75% of the 28-day strength after 7 to 10 days, the problem is less severe than originally thought [4,6,11]. In any case, it is important for the structural engineer to interact with the general contractor, in order that the influence of the method of construction on the girders as well as the frame can be quantified, however simplistic the analysis procedure may be.

Due to the larger loads that can be expected for the shores, the latter must either be designed as structural members or at least be evaluated by the structural engineer. The size of the shores is also influenced by the number of floors that are to have these supports left in place. As a general rule, when stub girders are used for multistory frames, the shores should be left in place for at least three floor levels. Some designers prefer a larger number; however, any choices of this kind should be based on computations for sizes and effects. Naturally, the more floors that are specified, the larger the shores will have to be.

References

[1] Colaco, J.P., "A Stub Girder System for High-Rise Buildings," *AISC Eng. J.*, Vol. 9, No. 2, Second Quarter, 1972 (pp. 89–95).

[2] Colaco, J.P., "Partial Tube Concept for Mid-Rise Structures," *AISC Eng. J.*, Vol. 11, No. 4, Fourth Quarter, 1974 (pp. 81–85).

[3] American Institute of Steel Construction (AISC), "*Specification for the Load and Resistance Factor Design, Fabrication, and Erection of Structural Steel for Buildings*," 3rd Edition, AISC, Chicago, IL, 1999.

[4] Bjorhovde, R. and Zimmerman, T.J., "Some Aspects of Stub Girder Design," *AISC Eng. J.*, Vol. 17, No. 3, Third Quarter, 1980 (pp. 54–69).

[5] Zimmerman, T.J. and Bjorhovde, R., "*Analysis and Design of Stub Girders*," Structural Engineering Report No. 90, University of Alberta, Edmonton, Alberta, Canada, 1981.

[6] Bjorhovde, R., "Behavior and Strength of Stub Girder Floor Systems," in *Composite and Mixed Construction*, Special Publication, ASCE, Reston, VA, 1985.

[7] Bjorhovde, R., "*Full-Scale Test of a Stub Girder*," Report submitted to Dominion Bridge Company, Calgary, Alberta, Canada. Department of Civil Engineering, University of Alberta, Edmonton, Alberta, Canada, 1981.

[8] Colaco, J.P. and Banavalkar, P.V., "*Recent Uses of the Stub Girder System*," Proceedings, 1979 National Engineering Conference, American Institute of Steel Construction, Chicago, IL, 1979.

[9] Martinez-Romero, E., "*Continuous Stub Girder Structural System for Floor Decks*," Technical Report, EMRSA, Mexico City, Mexico, 1983.

[10] Hansell, W.C., Galambos, T.V., Ravindra, M.K., and Viest, I.M., "Composite Beam Criteria in LRFD," *J. Struct. Div.*, ASCE, Vol. 104, No. ST9, September 1978 (pp. 1409–1426).

[11] Chien, E.Y.L. and Ritchie, J.K., "*Design and Construction of Composite Floor Systems*," Canadian Institute of Steel Construction (CISC), Willowdale (Toronto), Ontario, Canada, 1984.

[12] American Society of Civil Engineers (ASCE), "*Minimum Design Loads for Buildings and Other Structures*," ASCE/ANSI Standard No. 7-02, ASCE, Reston, VA, 2002.

[13] American Concrete Institute (ACI), "*Building Code Requirements for Reinforced Concrete*," ACI Standard No. 318-02, ACI, Detroit, MI, 2002.

[14] Zeitoun, L.A., "*Development of Resistance Factors for Stud Shear Connectors*," MS Thesis, University of Arizona, Tucson, AZ, 1984.

[15] Morcos, S.S. and Bjorhovde, R., "Fracture Modeling of Concrete and Steel," *J. Struct. Eng.*, ASCE, Vol. 121, No. 7, July 1995 (pp. 1125–1133).

[16] Bjorhovde, R., "*Design Considerations for Composite Frames*," Proceedings, 2nd International and 5th Mexican National Symposium on Steel Structures, IMCA and SMIE, Morelia, Michoacan, Mexico, November 23–24, 1987.

[17] Bjorhovde, R., "Concepts and Issues in Composite Frame Design," *Steel Struct.*, J. Singapore Soc. Steel Struct., Vol. 5, No. 1, December 1994 (pp. 3–14).

34

Fatigue and Fracture

Robert J. Dexter
*Department of Civil Engineering,
University of Minnesota,
Minneapolis, MN*

34.1 Introduction

Fatigue and fracture are two important limit states that need to be checked in the design and evaluation of structures. Fatigue is the formation of a crack due to cyclic loading. The fatigue limit state is defined as the development of a through-thickness crack. This is a serviceability limit state and does not necessarily mean that the structure is in danger of fracture or collapse. Fracture is the rupture in tension or rapid extension of a crack leading to gross deformation, loss of function or serviceability, or complete separation of the component.

This section of the handbook presents an overview of information useful to structural engineers in evaluating the fatigue and fracture limit states of steel, aluminum, and concrete structural components. Topics include materials selection, design, and detailing for new structures, as well as assessment of existing structures. The emphasis of this chapter is on structural steel components, since aluminum and other metal components are not common in the primary load-carrying systems of most civil structures. Fatigue of concrete components is covered only briefly since it is rarely a significant problem. As a practical matter, fracture of concrete is checked by usual strength design calculations and therefore is not covered here. The fracture mechanics of concrete is covered elsewhere [1].

Although most of the examples involve buildings or bridges, the information is equally applicable to similar details in cranes, ships, offshore structures, heavy vehicle frames, etc. The primary difference between various structure types is in the applied loading, whereas the type of structure does not affect the resistance of the details to fatigue or fracture.

Since the scope of this section is limited to practical information, there are many interesting aspects of fatigue and fracture that are not discussed. There are several good texts that can serve as a starting point for more in-depth studies [2–4].

0-8493-1569-7/05/$0.00+$1.50
© 2005 by CRC Press

34.2 Fatigue

Fatigue resistance for a particular detail is the allowable constant-amplitude stress range (S) for a specified number of cycles of loading (N), typically expressed in an S–N curve. S–N curves are based on fatigue tests of full-scale members with welded or bolted details. The assignment of various details to specific categories for S-N curve analysis is discussed below, including bolts, anchor rods, hollow structural sections, concrete members, and cables.

When information about a specific crack is available, a fracture mechanics crack growth rate analysis should be used to calculate the remaining life [3,4]. However, in the design stage, without specific initial crack size data, the fracture mechanics approach is not any more accurate than the S–N curve approach [5]. Therefore, the fracture mechanics crack growth analysis will not be discussed further.

34.2.1 Fatigue Resistance

The approach to designing and assessing structures for fatigue is empirical and is based on tests of full-scale members with welded or bolted details. Such tests indicate that

- The strength and type of steel have only a negligible effect on the fatigue resistance expected for a particular detail [6–10].
- The welding process also does not typically have an effect on the fatigue resistance [11,12].
- The primary effect of constant-amplitude fatigue loading can be accounted for in the live-load stress range [6–9], that is, the mean stress is not significant.

The independence of fatigue resistance from the type of steel greatly simplifies the development of design rules for fatigue since it eliminates the need to generate fatigue data for every type of structural steel. Testing that has been performed on welded details in stainless steel show that the fatigue strength of stainless steel details is comparable to that of ferritic structural steel details [13].

The reason that the dead load has little effect is that, locally, there are very high residual stresses. In details that are not welded, such as anchor rods, there is a mean stress effect [14]. A worst-case conservative assumption, that is, a high tensile mean stress, is made in the testing and in the design of these nonwelded details.

34.2.1.1 Standard S–N Curves

The nominal stress approach is the established approach for fatigue design and assessment of steel and aluminum structures. The nominal stress approach is based on S–N curves, where S is the nominal stress range and N is the number of cycles until the appearance of a visible crack. Details are designed based on the nominal stress range in the connecting members rather than the local "concentrated" stress at the detail. The nominal stress is usually obtained from standard design equations for bending and axial stress and does not include the effect of stress concentrations of welds and attachments. Usually, the nominal stress in the members can be easily calculated without excessive error. However, the proper definition of the nominal stresses may become a problem in regions of high stress gradients.

Figure 34.1 shows the S–N curves for steel that are used throughout North America for a variety of welded structures, including the American Association of State Highway and Transportation Officials (AASHTO) bridge design specifications [15], the American Institute for Steel Construction (AISC) *Manual of Steel Construction* [16], the American Railway Engineering and Maintenance-of-Way Association (AREMA) *Manual for Railway Engineering* [17], the American Welding Society (AWS D1.1) *Structural Welding Code* [18], and the Canadian Standards Association (CSA S16–2001) *Limit States Design of Steel Structures* [19].

AASHTO has seven S–N curves for seven categories (A through E') of weld details. The fatigue design procedure is based on associating the weld detail under consideration with a specific category. The effects of the welds and other stress concentrations, including the typical defects and residual stresses, are reflected in the ordinate of the S–N curves for the various detail categories.

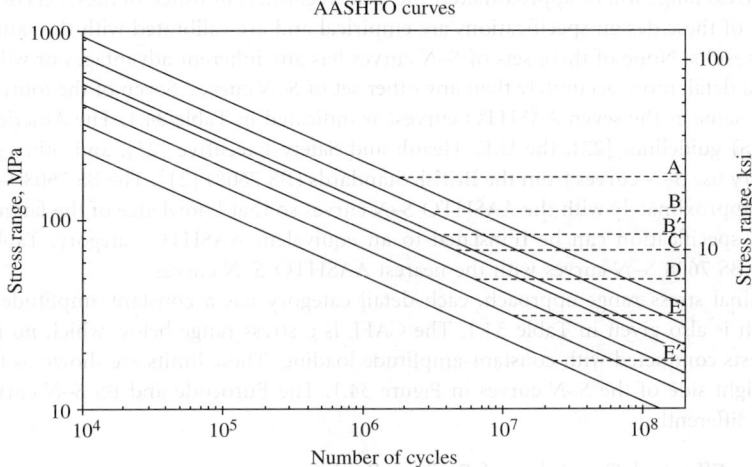

FIGURE 34.1 *S–N* curves for the seven primary fatigue categories from the AASHTO, AREMA, AWS, and AISC specifications; the dotted lines are the constant-amplitude fatigue limits (CAFLs) and indicate the detail category.

TABLE 34.1 Parameters for *S–N* Curves

AASHTO, AISC, and AWS category	Coefficient A for steel (MPa^3)	CAFL^a for steel (MPa)	Equivalent Eurocode 3 category	Nearest BS 7608 category	Coefficient A for aluminum (MPa^3)	Exponent for aluminum	CAFL for aluminum (MPa)
A	$.81.9 \times 10^{11}$	165	160	B	2.18×10^{19}	6.85	70
B	$.39.3 \times 10^{11}$	110	125	C	1.95×10^{14}	4.84	37
B'	$.20.0 \times 10^{11}$	83	100	D	N/A	N/A	N/A
C	$.14.4 \times 10^{11}$	69	90	D	8.88×10^{11}	3.64	28
D	$.7.21 \times 10^{11}$	48	71	F	2.08×10^{11}	3.73	17
E	$.3.61 \times 10^{11}$	31	56	G	3.14×10^{11}	3.45	13
E'	$.1.28 \times 10^{11}$	18	40	W	N/A	N/A	N/A

^{a} CAFL, constant-amplitude fatigue limit.

The inverse slope of the regression line fit to the test data for welded details is typically in the range 2.9 to 3.1 [8]. In the AISC, AASHTO, AREMA, and CSA codes as well as in Eurocode 3 (EU3) [20], the slopes have been standardized at 3.0. Therefore, the *S–N* curves can be represented by the following power–law relationship:

$$N = \frac{A}{S^3} \qquad (34.1)$$

where *N* is the number of stress cycles, *S* is the nominal stress range, and *A* is a constant particular to the detail category, as given in Table 34.1.

AISC and AWS also have a Category F with a slope greater than 3 for checking the shear stress in the throat of welds. AASHTO and CSA S16 require checking the shear stress in the throat of welds according to Category E rather than Category F. However, the minimum weld size requirements and the weld strength requirements generally will ensure sufficient weld throat to avoid fatigue. There are few, if any, documented fatigue cracking cases associated with shear stress through the throat of fillet welds, therefore it is unnecessary to check the shear stress range in the throat of welds for fatigue and this will not be discussed further.

Eurocode 3 (EC3)' [20] and the British Standard 7608 [21] also use a nominal stress approach, but they each have unique sets of *S–N* curves with different category labels. However, the end result of checking the stress range will be approximately the same regardless of which of these sets of *S–N* curves is used, since all of these design specifications are empirical and are calibrated with the same database of full-scale test results. None of these sets of *S–N* curves has any inherent advantages or will estimate the fatigue life of a detail more accurately than any other set of *S–N* curves. Seven of the fourteen Eurocode curves are the same as the seven AASHTO curves, as indicated in Table 34.1. The American Bureau of Shipping (ABS) guidelines [22], the U.K. Health and Safety Executive [23], and other groups in the marine industry use *S–N* curves from the British Standards (BS 7608) [21]. The BS 7608 *S–N* curves can be associated approximately with the AASHTO *S–N* curves so that knowledge of the fatigue strength of details in this specification can be translated to an equivalent AASHTO category. Table 34.1 cross-references the BS 7608 *S–N* curves with the nearest AASHTO *S–N* curves.

In the nominal stress range approach, each detail category has a constant-amplitude fatigue limit (CAFL), which is also given in Table 34.1. The CAFL is a stress range below which no fatigue cracks occurred in tests conducted with constant-amplitude loading. These limits are shown as the horizontal lines on the right side of the *S–N* curves in Figure 34.1. The Eurocode and BS *S–N* curves define the fatigue limits differently.

34.2.1.2 The Effect of Corrosion of Fatigue Resistance

The full-scale fatigue experiments on which the fatigue rules are based were carried out in moist air and therefore reflect some degree of environmental effect or corrosion fatigue. Therefore, the lower-bound *S–N* curves in Figure 34.1 can be used for the design of details with a mildly corrosive environment (such as the environment for bridges, even if salt or other corrosive chemicals are used for deicing) or provided with suitable corrosion protection (galvanizing, other coating, or cathodic protection). Some design codes for offshore structures have reduced fatigue life (by approximately a factor of 2) when details are exposed to seawater [21,23].

At relatively high stress ranges at which most accelerated tests are conducted, the effect of seawater is clearly detrimental. However, there is evidence that the effect of corrosion in seawater is not so severe for long-life variable-amplitude fatigue of welded details. Full-scale fatigue experiments in seawater at realistic service stress ranges do not show significantly lower fatigue lives [24], provided that corrosion is not so severe that it causes pitting or significant section loss. At relatively low stress ranges near the CAFL, typical of service loading, it appears that build-up of corrosion product in the crack may increase crack closure and retard crack growth, at least enough to offset the increase that would otherwise occur due to the environmental effect. The fatigue lives seem to be more significantly affected by the stress concentration at the toe of welds than by the corrosive environment.

Severely corroded members may be evaluated as Category E details [25], regardless of their original category (unless of course they were Category E′ or worse to start with). However, pitting or significant section loss from severe corrosion can lower that fatigue strength [25,26].

34.2.1.3 *S–N* Curves for Aluminum

S–N curves have been proposed by the Aluminum Association for welded aluminum structures [27,28]. The categories for details are similar to those for steel, although the fatigue strength of a detail is approximately 38% of the fatigue strength of the same detail in steel. For aluminum, the exponent in Equation 34.1 is variable and is given in Table 34.1 with the coefficient A and CAFL. The design procedures are similar and the classification of details into categories is approximately the same.

34.2.2 Fatigue Loading

Since fatigue is typically only a serviceability problem, members and connections are designed for fatigue using service loads. Most structures experience what is known as long-life variable-amplitude loading, that is, very large numbers of random-amplitude cycles greater than the number of cycles associated with the CAFL [29]. For example, a structure loaded continuously at an average rate of once per minute (0.016 Hz), would accumulate 10 million cycles in 19 years. With long-life variable-amplitude loading, structures are usually designed so that only a small fraction of cycles, on the order of 0.01%, exceed the CAFL [29,30].

If the percentage of stress ranges exceeding the CAFL is greater than 0.01%, the history of N variable stress ranges can be converted to N cycles of an effective stress range that can then be used just like the constant-amplitude stress range in S–N curve analysis. Typically, Miner's Rule [31] is used to calculate an effective stress range from a histogram of variable stress ranges. Theoretically, this effective constant-amplitude stress range results in approximately the same fatigue damage for a given number of cycles as that for the same number of cycles of the variable-amplitude service history. If the stress ranges are counted in discrete "bins," as in a histogram, the effective stress range, S_{Re} [29] can be calculated as

$$S_{Re} = \left(\sum_i (\alpha_i S_{ri}^3) \right)^{1/3} \tag{34.2}$$

where α_i is the number of stress cycles with stress range in the bin with average value S_{ri} divided by the total number of stress cycles (N).

Variable-amplitude fatigue tests conducted with various sequences in the variable-amplitude loading history have shown that Miner's Rule is reasonably accurate in most cases but can be unconservative with some load histories with unusual sequences. For this reason, some fatigue design specifications for offshore structures put a safety factor of 2.0 on life if Miner's Rule is used [23].

In the AASHTO specifications [15], the stress range from the fatigue design truck represents the effective stress range. No additional safety factor is used for Miner's Rule since it is relatively accurate for truck loading on bridges. For large numbers of cycles, the AASHTO specification has another check that involves comparing the stress range from the fatigue design truck to half of the CAFL. The rationale for this check is that if the effective stress range is less than half the CAFL, most of the stress ranges should be below the CAFL, but occasionally (about once a day) the stress range can exceed the CAFL with no significant effect.

Misalignment at a welded joint is a primary factor in susceptibility to cracking. The misalignment causes eccentric loading, local bending, and stress concentration. The stress concentration factor (SCF) associated with misalignment is

$$SCF = 1.0 + 6e/t \tag{34.3}$$

where e is the eccentricity and t is the smaller of the thicknesses of two opposing loaded members. The nominal stress times the SCF should then be compared to the appropriate category. Generally, such misalignment should be avoided at fatigue critical locations. Equation 34.3 can also be used where e is the distance that the weld is displaced out of plane due to angular distortion. A thorough guide to the SCF for various types of misalignment and distortion, including plates of unequal thickness, can be found in British Standard BS 7910 [32].

34.2.3 Categorization of Details

The following is a brief simplified overview of the categorization of fatigue details. In all cases, the applicable specifications should also be checked. Several reports have been published that show a large number of illustrations of details and their categories in addition to those in AISC and AASHTO specifications [33,34]. Also, the EC3 [20] and the British Standard BS 7608 [21] have more detailed illustrations for their categorization than AISC or AASHTO specifications. A book by Maddox [9] discusses categorization of many details in accordance with BS 7608, from which roughly equivalent AASHTO categories can be inferred (using Table 34.1). In most cases, the fatigue strength recommended in these European standards is similar to the fatigue strength in the AISC and AASHTO specifications. However, there are several cases where the fatigue strength is significantly different; usually, the European specifications are more conservative.

There have been very few, if any, failures that have been attributed to details with fatigue strength greater than Category C. Most structures have many Category C or even more severe details, and these will generally govern the fatigue design. Therefore, only Category C and more severe details will be discussed in depth.

34.2.3.1 Welded Joints

Welded joints are considered transverse if the axis of the weld is perpendicular to the primary stress range. Unless there is a stress concentration from the configuration of the detail, transverse welds are typically Category C details. For example, the transverse welds connecting stiffeners to the web and flange of girders are Category C details.

Full-penetration groove welded butt joints subjected to nondestructive evaluation (NDE) such as ultrasonic testing or radiographic testing are also Category C details. If the reinforcement of these full-penetration butt joints is ground smooth, they are Category B details. Tests show that groove welds that contain large internal discontinuities that were not screened out by NDE had fatigue strength comparable to Category E [5]. BS 7608 [21] and BS 7910 [32] have reduced fatigue strength curves for groove welds with defects that are generally in agreement with these experimental data. Transverse groove welds with a permanent backing bar are reduced to Category D [35,36]. One-sided welds with melt through (without backing bars) are also classified as Category D [35,36].

Continuous longitudinal welds are Category B or B' details. However, the termination of a long-itudinal fillet weld is more severe (Category E). The termination of full-penetration groove longitudinal welds requires a ground transition radius but gives greater fatigue strength, depending on the radius. If longitudinal welds must be terminated, it is better to extend the welds to a location along the structural member where the stress ranges are small or entirely in compression.

Cope holes, cutouts, and snipes should be used for weld access and to avoid intersecting welds. If the thermally cut surfaces have edges conforming to the ANSI smoothness of 1000, these may be considered Category D details, even if welds terminate at these details [35,36]. A rougher cope hole may be treated as a Category E detail. If the steel at the thermally cut edge transforms to martensite, in some cases small cracks may occur that will propagate at even lower stress ranges.

Attachments normal to flanges or plates that do not carry significant load are rated Category C if less than 51 mm long in the direction of the primary stress range, D if between 51 and 101 mm long, and E if greater than 101 mm long. If there is not at least a 10-mm edge distance, then Category E applies for an attachment of any length. Category E', slightly worse than Category E, applies if the attachment plates or the flanges exceed 25 mm in thickness.

The cruciform joint where the load-carrying member is discontinuous is considered a Category C detail because it is assumed that the plate transverse to the load-carrying member does not have any stress range. A special reduction factor for the fatigue strength is provided when the load-carrying plate exceeds 13 mm in thickness. This factor accounts for the possible crack initiation from the unfused area at the root of the fillet welds (as opposed to the typical crack initiation at the weld toe for thinner plates) [37].

In most other types of load-carrying attachments, there is interaction between the stress range in the transverse load-carrying attachment and the stress range in the main member. In practice, each of these stress ranges is checked separately. The attachment is evaluated with respect to the stress range in the main member, and then it is separately evaluated with respect to the transverse stress range. The combined multiaxial effect of the two stress ranges is taken into account by a relative decrease in the fatigue strength, that is, most load-carrying attachments are considered Category E details.

34.2.3.2 Bolted Joints and Anchor Rods

Small holes are considered Category D details. Therefore, rivetted and mechanically fastened joints (other than high-strength bolted joints) loaded in shear are evaluated as Category D in terms of the net-section nominal stress. High-strength A325 and A490 bolted joints that are properly pretensioned and loaded in shear are Category B details in terms of the net-section nominal stress. Pin plates and eyebars are designed as Category E details in terms of the net-section nominal stress.

Bolted joints loaded in direct tension are more complicated. Typically, these provisions are applied to hanger-type or bolted flange connections where the bolts are pretensioned against the plies. In this case, the total fluctuating load is resisted by the area of the precompressed plies, so that the bolts are subjected to only a fraction of the total load [38]. However, the analysis to determine this fraction is difficult. In the AISC Specifications [16] and in BS7608 [21], the designer may assume the load range in the bolts is 20% of the total applied service load (dead plus live load). The total applied service load must include any prying load. Prying is very detrimental to fatigue, so it is best to minimize prying forces by using sufficiently thick plates [16].

In the AISC Specifications [16], the stress range in the bolts is calculated on the tensile stress area, A_t, given by

$$A_t = \frac{\pi}{4}\left(d_b - \frac{0.9743}{n}\right)^2 \qquad (34.4)$$

where d_b is the nominal diameter (the body or shank diameter) and n is threads per inch. (Note that the constant would be different if metric threads are used.)

Test data on bolts in direct tension (with the stress range computed on the tensile stress area) are shown in Figure 34.2 [39,40]. The M20 bolts and B7 rods are similar in strength to A325 high-strength bolts (862-MPa tensile strength). Different size diameters of the B7 bolts were also tested, with the results showing that size did not have an impact on the fatigue limit. Category E seems to be the lower bound for bolts in direct tension.

Anchor rods in concrete cannot be adequately pretensioned and therefore do not behave like hanger-type or bolted flange connections. In the double-nut configuration, they are pretensioned between nuts on either side of the base plate, but the part below the bottom nut is still exposed to the full load range. Fatigue test data are available for anchor rods [14,41,42]. The data for anchor rods with a double-nut configuration tightened one third of a turn past snug tight are shown in Figure 34.3. These data agree well with the bolt data shown in Figure 34.2, that is, the lower-bound fatigue strength is Category E.

These test data were obtained at low enough stress ranges to allow for definition of the CAFL. There was one failure at 69 MPa but below this stress range there were all runouts. Therefore, the CAFL is much greater than the CAFL for other Category E details, which is 31 MPa. Choosing from among the CAFLs for the other categories, the CAFL was taken as that for Category D or 48 MPa.

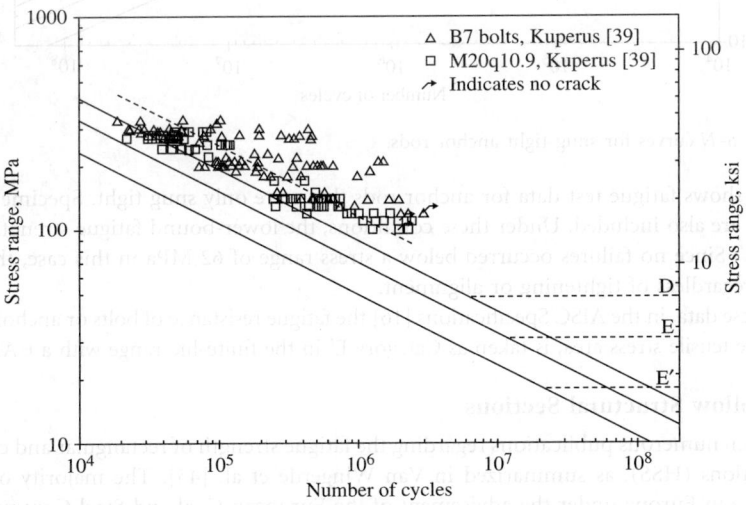

FIGURE 34.2 *S–N* curves for B7 and M20 snug-tight bolts in tension.

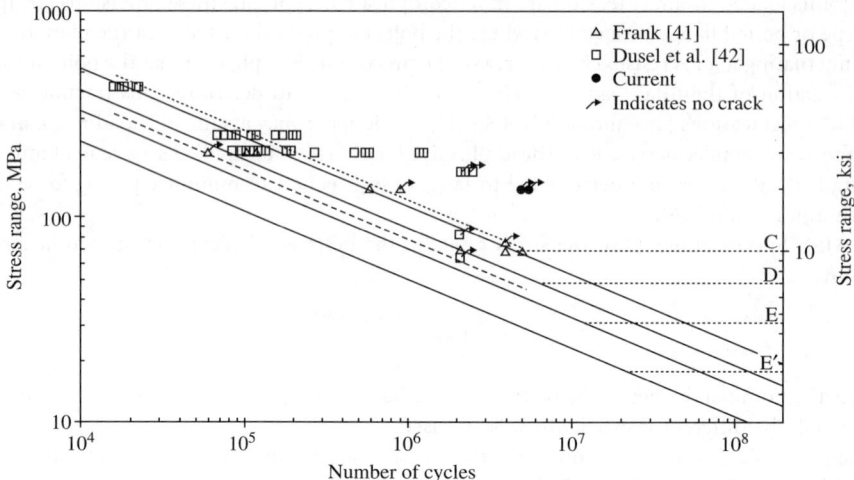

FIGURE 34.3   *S–N* curves for double-nut anchor rods tightened one third of a turn past snug tight.

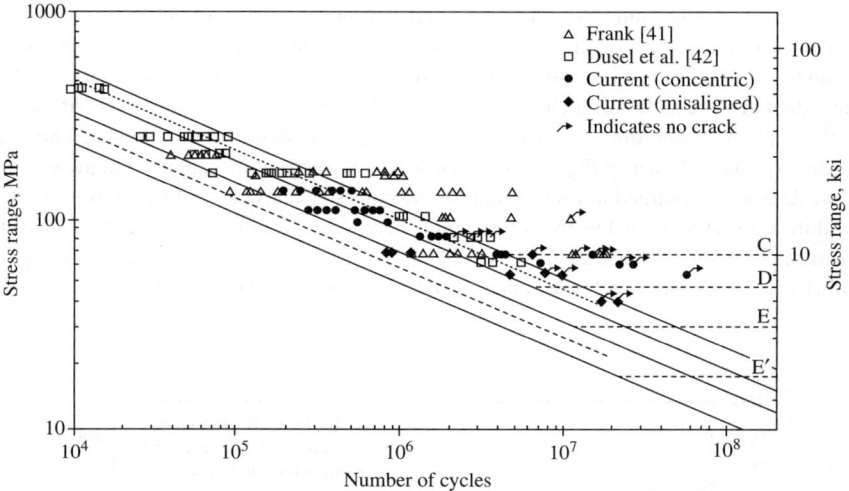

FIGURE 34.4   *S–N* curves for snug-tight anchor rods.

Figure 34.4 shows fatigue test data for anchor rods that were only snug tight. Specimens with a 1:40 misalignment are also included. Under these conditions, the lower-bound fatigue strength corresponds to Category E′. Since no failures occurred below a stress range of 62 MPa in this case, the CAFL at 48 MPa applies regardless of tightening or alignment.

Based on these data, in the AISC Specifications [16] the fatigue resistance of bolts or anchor rods in direct tension, on the tensile stress area, is taken as Category E′ in the finite-life range with a CAFL of 48 MPa.

34.2.3.3 Hollow Structural Sections

There have been numerous publications regarding the fatigue strength of rectangular and circular hollow structural sections (HSS), as summarized in Van Wingerde et al. [43]. The majority of recent work has taken place in Europe under the advisement of the European Coal and Steel Community (ECCS), the British Department of Energy (DEn), now called the Health and Safety Executive, and the Comité

International pour le Developement et l'Etude de la Construction Tubulaire (CIDECT) [44]. The research has resulted in design recommendations that have been adopted by the International Institute of Welding (IIW Document XIII-1804-99 and XV-1035-99) [45] and CIDECT *Design Guide* #8 [46] and have been proposed for inclusion into AWS D1.1 [18] and EC3 [20]. The CIDECT fatigue design guidelines include both the nominal-stress approach and the hot-spot stress approach. AWS D1.1 [18] also has nominal-stress and hot-spot stress *S–N* curves for HSS, although these are not considered as accurate as the CIDECT guidelines. Unlike AWS, the CIDECT guidelines distinguish between circular hollow sections (CHS) and rectangular hollow sections (RHS).

Many sign, signal, and light supports are fabricated from HSS. AASHTO has published a new specification for these types of structures [47] that includes fatigue design using the nominal-stress approach and classification of the common details with respect to the AWS *S–N* curves, based on research in NCHRP Report 412 [14]. In addition, this specification has fatigue design loads for natural wind gusts, galloping, and vortex shedding. Many aspects of this specification could also be applied to wind turbines, communication towers, and other flexible structures exposed to wind.

The hot-spot stress approach is similar to the nominal-stress approach and is also based on full-scale tests, except that the *S–N* curves in this approach are based on the hot-spot stress range [48]. The hot-spot stress is the stress normal to the weld axis, originally defined at some small distance (about 6 mm) from the weld toe. The hot-spot stress can be measured, calculated using finite-element analysis, or determined from parametric formulas derived from the finite-element analyses and measurements.

A distinction is made between the geometric stress concentration, that is, that associated with the arrangement of the members, and the local stress concentration, that is, that associated with the weld toe. In the hot-spot approach, the effect of the geometric stress concentration, expressed as the SCF, is taken out of the fatigue resistance and instead is included in the analysis. This has the advantage of collapsing all the *S–N* curves for different categories into a single baseline *S–N* curve, but it has the disadvantage of increasing the complexity of the analysis. A simple structural analysis is usually performed to obtain the axial and bending forces entering a joint, from which nominal stresses are calculated. The hot-spot stress at a weld is determined by multiplying the appropriate nominal stress by the SCF. The test data and the baseline *S–N* curve still include the effect of the local stress concentration, which is impossible to calculate accurately and therefore must still be treated empirically.

A simple approach that works well is to define the hot-spot stress as the stress measured with a 3-mm strain gage placed as close as practically possible to the weld toe, that is, centered about 6 mm from the weld toe [49]. This is essentially the definition originally used by AWS [48]. The baseline *S–N* curve used with the hot-spot stress defined this way is essentially the same as the nominal-stress *S–N* curve (Category C) for a transverse butt or fillet weld in a nominal membrane stress field, that is, a stress field without any geometric stress concentration [49]. This makes sense since the stress at the weld toe of this detail would include the local stress concentration but would not be affected by geometric stress concentration. In other words, the SCF is equal to 1.0, and the hot-spot stress is equal to the nominal stress in this detail.

The CIDECT research updated parametric equations used for SCF calculations and unified the definition of hot-spot stress. The definition of hot-spot stress is more complex than the simple AWS definition and involves extrapolating the stress from multiple strain gage measurements or analysis points [46]. However, the CIDECT design guidelines contain parametric equations that can be used to calculate SCF based on nondimensional parameters (brace to chord diameter ratio, brace diameter to thickness ratio, brace to chord thickness ratio, and chord length to diameter ratio) for different types of loading. A minimum SCF of 2.0 is recommended.

Obviously, the SCF and the baseline curve are dependent on the definition of the hot-spot stress [50]. The CIDECT baseline hot-spot *S–N* curve is the T′ curve from BS 7608 (equivalent to the Eurocode class 114 curve) for use with 16-mm-thick sections (see Figure 34.5). This is about 25% higher than the baseline *S–N* curve used with the AWS-defined hot-spot stress (6 mm from the weld toe). In hollow structural joints, there is a pronounced thickness effect, with thinner sections having higher fatigue lives. CIDECT included the thickness effect in the definition of the hot-spot *S–N* curves, as shown in Figure 34.5.

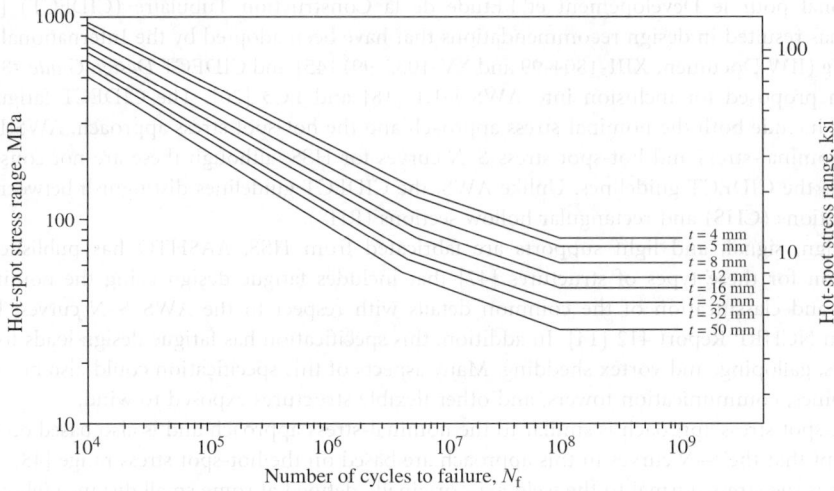

FIGURE 34.5 Hot-spot *S–N* curves for CHS joints (4 mm ≤ *t* ≤ 50 mm) and RHS joints (4 mm ≤ *t* ≤ 16 mm).

34.2.3.4 Reinforced and Prestressed Concrete, Strands, and Cables

Because of the relatively low stresses, concrete structures are typically less sensitive to fatigue than welded steel and aluminum structures. However, fatigue may govern the design when impact loading is involved, such as pavement, bridge decks, and rail ties. Also, in high-strength concrete the applied stress ranges increase, and so should the concern for fatigue.

According to ACI Committee Report 215R-74 in the *Manual of Standard Practice* [51], the fatigue strength of plain concrete at 10 million cycles is approximately 55% of the ultimate strength. However, even if failure does not occur, repeated loading may contribute to premature cracking of the concrete, such as inclined cracking in prestressed beams. This cracking could then lead to localized corrosion and fatigue of the reinforcement [52].

The fatigue strength of straight, unwelded reinforcing bars and prestressing strand can be described (in terms of the categories for steel details described earlier) with the Category B *S–N* curve. The lowest stress range that has been known to cause a fatigue crack in a straight reinforcing bar was 145 MPa, which occurred after more than a million cycles. Mean stress level, yield strength, bar size, geometry, and deformations had minimal effect. ACI Committee 215 suggests that members be designed to limit the stress range in the reinforcing bar to 138 MPa for high levels of mean stress (increases up to 160 MPa as mean stress is reduced).

Fatigue tests show that previously bent bars had only about half the fatigue strength of straight bars and failures have occurred down to 113 MPa [53]. Committee 215 recommends that half of the stress range for straight bars be used, that is, 69 MPa for the worst-case mean stress. Equating this recommendation to the *S–N* curves for steel details, bent reinforcement may be treated as a Category D detail.

Provided the quality is good, butt welds in straight reinforcing bars do not significantly lower the fatigue strength. However, tack welds reduce the fatigue strength of straight bars about 33%, with failures occurring as low as 138 MPa. Fatigue failures have been reported in welded wire fabric and bar mats [54].

Batchelor et al. [55] tested deck panels with a single stationary concentrated load applied at the center. The contact area represented the assumed contact area of the pneumatic tires of large trucks. The only fatigue tests that did not result in failure at the end of 2.5 and 3 million cycles were performed with a loading of less than the 50% of the estimated ultimate punching strength of the deck (P_u). Other research also supports a fatigue limit of $0.5P_u$ at 2 to 3 million cycles for a stationary pulsating load [56,57].

However, the behavior and the fatigue limit are different in the case of rolling loads, as on a bridge. Flexural cracking along the reflection of the longitudinal and transverse reinforcement on the bottom surface was detected in bridges in service, and this is different from cracking patterns in tests with stationary pulsating loads [58]. Tests with moving or rolling loads indicate that the fatigue limit was as low as $0.21P_u$ at 2 to 3 million cycles, comparable to the average flexural cracking load level of about $0.26P_u$ [58–60]. In some of the tests [56,58,59] it was demonstrated that the transverse cracks from the constructional period and the water penetration during service life decreased the ultimate punching shear and fatigue strengths of the reinforced concrete deck. However, these studies have not established any quantitative interaction between the deterioration from the environmental factors and the repetitive axle load.

Test data show that measured stress ranges in the reinforcement at the location of cracks in a highly deteriorated deck under high axle loads are less than 35 MPa, well below the 138-MPa threshold discussed earlier [61]. Therefore, under service load, fatigue does not appear to be a problem for deck reinforcement. This is consistent with the fact that fatigue of reinforced concrete decks is governed by punching failure of the concrete part of the structure.

If prestressed members are designed with sufficient precompression so that the section remains uncracked, there is not likely to be any problem with fatigue. This is because the entire section is resisting the load ranges, and the stress range in the prestressing strand is minimal. Similarly, for unbonded prestessed members, the stress ranges will be very small. However, there is reason to be concerned about bonded prestressing at cracked sections because the stress range increases locally. The concern for cracked sections is even greater if corrosion is involved. The pitting from corrosive attack can dramatically lower the fatigue strength of the reinforcement [52].

Although the fatigue strength of the prestressing strand in air is about equal to Category B, when the anchorages are tested as well, the fatigue strength of the system is as low as half the fatigue strength of the wire alone (i.e., about Category E). When actual beams are tested, the situation is very complex, but it is clear that much lower fatigue strength can be obtained [62,63]. Committee 215 has recommended the following for prestressed beams:

1. The stress range in prestressed reinforcement, determined from an analysis considering the section to be cracked, shall not exceed 6% of the tensile strength of the reinforcement. (Author's note: this is approximately equivalent to Category C.)
2. Without specific experimental data, the fatigue strength of unbonded reinforcement and its anchorages shall be taken as half of the fatigue strength of the prestressing steel. (Author's note: this is approximately equivalent to Category E.) Lesser values shall be used at anchorages with multiple elements.

The Post-Tensioning Institute (PTI) has issued *Recommendations for Stay Cable Design, Testing, and Installation* [64]. The PTI recommends that uncoupled bar stay cables are Category B details, while coupled (glued) bar stay cables are Category D. The fatigue strengths of stay cables are verified through fatigue testing. Two types of tests are performed: (1) fatigue testing of the strand and (2) testing of relatively short lengths of the assembled cable with anchorages. The recommended test of the system is 2 million cycles at a stress range (158 MPa) that is 35 MPa greater than the fatigue allowable for Category B at 2 million cycles. This test should pass with less than 2% wire breaks. A subsequent proof test must achieve 95% of the guaranteed ultimate tensile strength of the tendons.

34.2.4 Low-Cycle Fatigue and Seismic Frame Details

Steel braced frames and moment-resisting frames are expected to withstand cyclic plastic deformation without cracking in a large earthquake. If brittle fracture of these moment frame connections is suppressed by good detailing and minimum notch toughness levels, as discussed in the next section, the connections can be cyclically deformed into the plastic range and will eventually fail by tearing at a location of strain concentration [65]. This limit state is referred to as low-cycle fatigue. Some examples of low-cycle fatigue cracks are shown in Figure 34.6, Figure 34.7, Figure 34.8, and Figure 34.9 from full-scale beam-to-column

(a)

(b)

FIGURE 34.6 Low-cycle fatigue crack developing at the toe of the beam flange weld in a moment–frame connection after (a) 11 cycles and (b) 17 cycles of 4% drift.

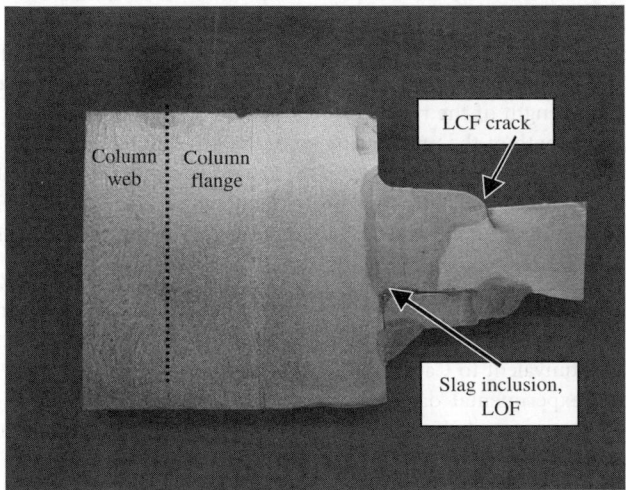

FIGURE 34.7 Cross-section of beam flange weld showing low-cycle fatigue crack developing at the weld toe.

connection tests of the welded-unreinforced flange, welded web (WUF-W) connection, which is one of the FEMA 350 prequalified connections for special moment frames in high seismic regions [66]. Figure 34.6 and Figure 34.7 show low-cycle fatigue cracks forming at the beam flange weld. Figure 34.8 shows a low-cycle fatigue crack forming at the weld of the beam web to the column flange, and Figure 34.9 shows a low-cycle fatigue crack forming at the weld access hole. The detailing in these connections is well balanced since the low-cycle fatigue failure at all of these details occurred between 12 and 16 cycles of the maximum required drift angle in each case.

Most past research on low-cycle fatigue has involved pressure vessels and some other types of mechanical engineering structures. Since low-cycle fatigue is an inelastic phenomenon, the strain range is the key parameter rather than the stress range. The Coffin–Manson rule [67] has been used to relate the strain range in smooth tensile specimens to life. Manson suggested a conservative lower-bound simplification, called Manson's universal slopes equation [68]:

$$\Delta\varepsilon = 3.5\frac{\sigma_u}{E}N^{-0.12} + \varepsilon_f^{0.6}N^{-0.6} \tag{34.5}$$

where $\Delta\varepsilon$ is the total strain range, σ_u is the tensile strength, and ε_f is the elongation at fracture.

FIGURE 34.8 Low-cycle fatigue crack developing at the end of the beam web to column flange weld in a moment–frame connection.

FIGURE 34.9 Low-cycle fatigue crack developing at the weld access hole in a moment–frame connection.

Note that the first term in Equation 34.5 is the elastic part of the total strain range (which is relatively insignificant when there are fewer than 100 cycles), and the second term is the plastic part of the total strain range. Figure 34.10 shows a plot of Manson's universal slopes equation where σ_u is 450 MPa and ε_f is 25%, typical minimum properties for Grade 50 structural steel. Many studies have shown that Manson's universal slopes equation is conservative compared to experimental data from smooth specimens [68,69]. However, because of buckling at greater strain ranges, most of the experimental data are for strain ranges less than 1%, that is, for cycles greater than 1000. Limited data exist at higher strain ranges — some are shown in Figure 34.10 for A36 steel smooth specimens machined from the flanges of wide-flange sections [69].

At present, very little is understood about low-cycle fatigue in welded or bolted structural details. For example, it is a very difficult task just to predict accurately the local strain range at a location of cyclic local flange buckling. However, Krawinkler and Zohrei [70] and Ballio and Castiglioni [71,72] showed that the number of cycles to failure by low-cycle fatigue of welded connections could be predicted by the local strain range in a power law that is analogous to an *S–N* curve. Ballio and Castiglioni [71,72] showed that the power law would have an exponent of 3, just like the elastic *S–N* curves. Krawinkler and Zohrei [70] also showed that Miner's rule could be used to predict the number of variable-amplitude cycles to failure based on constant-amplitude test data.

Therefore, it may be possible to predict and design against low-cycle fatigue using strain-range versus number-of-cycles curves that are extrapolated from the high-cycle fatigue design *S–N* curves. Figure 34.10 shows the AASHTO/AISC *S–N* curves for Categories A and C, converted from stress range to strain range by dividing the stress ranges by the elastic modulus and extrapolated up to one cycle.

There are only limited data to support this approach. Figure 34.10 shows some data from the same full-scale WUF-W beam-to-column connection tests for which the low-cycle fatigue cracks were shown in Figure 34.6, Figure 34.7, Figure 34.8, and Figure 34.9. These tests are subjected to a standard series of cycles of increasing ranges of total drift as shown in Figure 34.11. They are cycled at 4% drift until they

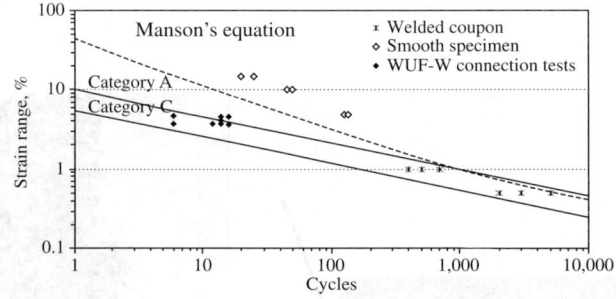

FIGURE 34.10   Comparison of standard *S–N* curves presented in terms of strain range and Manson's universal slopes equation for Grade 50 (350-MPa yield strength) steel to low-cycle fatigue test data.

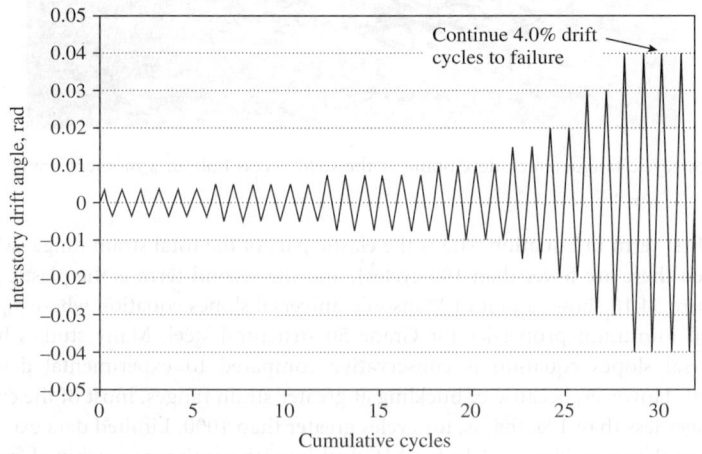

FIGURE 34.11   Standard drift-cycle history for testing moment–frame subassemblies.

fail by low-cycle fatigue. The measured flange strain ranges in these tests varied from 3.7 to 4.8% during the 4% drift cycles. The number of cycles plotted in Figure 34.10 is the equivalent number of cycles at 4% drift, which is the actual number of cycles at 4% plus one additional cycle which, according to Miner's rule, is equivalent in damage to all of the cycles at 3% and less. Since maximum flange strains were used rather than nominal values, this is analogous to a hot-spot approach for high-cycle fatigue. As discussed previously in Section 34.2.3.3, the Category C *S–N* curve is a suitable baseline *S–N* curve for the hot-spot approach. It appears that the Category C *S–N* curve is also a good lower bound to these low-cycle fatigue data. The scatter in the data is substantial, as is also true in high-cycle fatigue.

Also shown in Figure 34.10 are data for smaller coupon-type specimens with transverse butt welds, which would be expected to be Category C details. These are some of the only available data with fewer than 5000 cycles. These coupon data are also in reasonable agreement with the extrapolated Category C curve as a lower bound.

34.3   Fracture

34.3.1   Fracture Resistance

Unlike fatigue, fracture behavior depends strongly on the type and strength level of the steel or filler metal. In fact, the fracture resistance of each type of steel or weld metal varies significantly from heat to heat, or from lot to lot. In fact, in rolled shapes, the fracture resistance may vary significantly within the cross-section. Therefore, design or assessment for fracture resistance will usually involve a measurement of the material's fracture resistance, or fracture toughness. Although fracture toughness can be measured directly in fracture mechanics tests, the usual practice is to characterize the toughness of steel in terms of the impact energy absorbed by a Charpy V-notch (CVN) specimen. Because the Charpy test is relatively easy to perform, it will likely continue to be the measure of toughness used in steel specifications.

Since it is not directly related to the fracture toughness, CVN energy is often referred to as notch toughness. The notch toughness is still very useful, however, since it can often be correlated to the fracture toughness and then used in a fracture mechanics assessment [3,32,73,74]. Figure 34.12 shows a plot of the CVN energy of A588 Grade 50 (350-MPa yield strength) structural steel at varying temperature. These results are typical for ordinary hot-rolled structural steel.

The fracture limit state includes phenomena ranging from brittle fracture of low-toughness materials at service load levels to ductile tensile rupture of a component. The transition between these phenomena

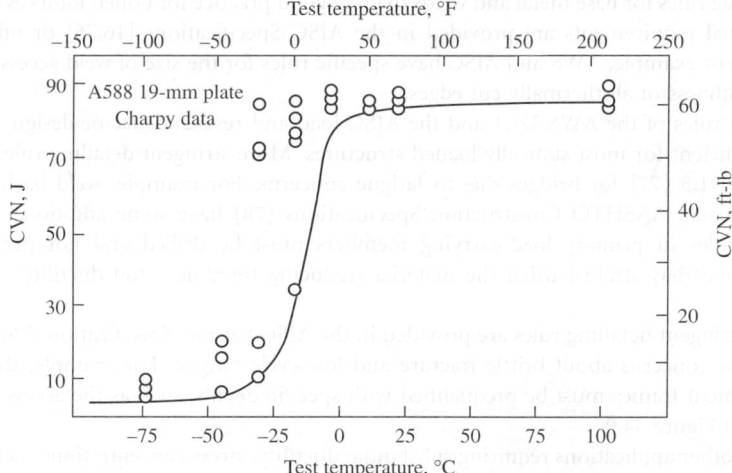

FIGURE 34.12   Charpy transition curve for A588 Grade 50 (350-MPa yield strength) structural steel.

depends on temperature, as reflected by the variation of CVN with temperature as shown in Figure 34.12. The transition is a result of changes in the underlying microstructural fracture mode.

Brittle fracture on the so-called lower shelf in Figure 34.12 is associated with cleavage of individual grains on select crystallographic planes. Brittle fracture may be analyzed with linear-elastic fracture mechanics theory because the plastic zone at the crack tip is very small.

At the high end of the temperature range, the so-called upper shelf, ductile fracture is associated with the initiation, growth, and coalescence of microstructural voids, a process requiring much energy. The net section of plates or shapes fully yields and then ruptures with large slanted shear lips on the fracture surface.

Transition-range fracture occurs at temperatures between the lower and upper shelves and is associated with a mixture of cleavage and shear fracture. Large variability in toughness at constant temperature and large changes with temperature are typical of transition-range fractures.

Brittle fracture can be thought of as an interruption of what would otherwise be ductile plastic deformation in tension, much like buckling can interrupt ductile plastic deformation in compression. For the purposes of structural engineering, it is usually only necessary to make sure that brittle fracture does not occur, so that is the emphasis in this chapter. As long as the behavior is in the transition range or upper shelf, the resulting ductility is generally sufficient. For example, as discussed in the previous section, moment-frame connections will tolerate a certain minimum number of cycles of large rotations governed by the low-cycle fatigue limit state, as long as specific details are used and minimum CVN requirements are assured. If low-toughness weld metal that does not meet minimum requirements is used, brittle fracture may occur before the expected number of cycles [65].

34.3.2 Material Specification and Detailing to Avoid Fracture

Details that have good fatigue resistance, Category C and better, are usually also optimized for resistance to fracture. Detailing rules to avoid fracture are very similar to the common-sense rules to avoid fatigue. For example, intersecting welds should always be avoided due to the probability of defects and excessive constraint. Intersecting welds, or even welds of too close proximity, have caused brittle fractures. For example, in December 2000, fractures occurred in large plate girders of the Hoan Bridge in Milwaukee [75]. These fractures occurred at a shelf plate detail for attachment of the lateral wind bracing to the girders where the shelf plate fit around, and intersected the welds of, a vertical stiffener, as shown in Figure 34.13.

Fractures have also occurred at poorly executed weld access holes, as shown in Figure 34.14. AWS D1.1 [18] has detailing rules for base metal and welds that are good practice for bolted joints as well as welded joints. Additional requirements are provided in the AISC Specifications [16,76] or other applicable specifications. For example, AWS and AISC have specific rules for the size of weld access holes and the radii and smoothness of all thermally cut edges.

The detailing rules of the AWS D1.1 and the AISC load and resistance factor design (LRFD) specification are sufficient for most statically loaded structures. More stringent detailing rules are also provided in AWS D1.5 [77] for bridges due to fatigue concerns. For example, weld backing bars must usually be removed. AASHTO Construction Specifications [78] have some additional requirements, for example, holes in primary load-carrying members must be drilled and not punched. This is because punching may strain-harden the material (reducing toughness and ductility) and introduce microcracks.

Also, more stringent detailing rules are provided in the AISC Seismic Specification [76] and in FEMA 350 [66] due to concerns about brittle fracture and low-cycle fatigue. For example, the connections for special moment frames must be prequalified with specific details, such as the access hole visible in Figure 34.8 and Figure 34.9.

In these and other applications requiring substantial ductility, stress concentrations such as re-entrant corners should be avoided and instead transition radii that are ground smooth should be provided, as shown in Figure 34.8 and Figure 34.9. The ends of butt welds are always a potential location of defects.

FIGURE 34.13   Fractured girder of the Hoan Bridge in Milwaukee and view of critical shelf plate detail featuring intersecting welds.

If possible, it is better to locate butt welds away from such stress concentrations. If not, it is important to use runout tabs and to later grind the ends of the weld to a radius. Fillet weld terminations should not be ground, for this will expose a very thin ligament near the weld root that will tear easily.

The enhanced detailing required by the bridge [77,78] or seismic [66,76] specifications are expensive to implement relative to what would be required by AWS D1.1 [18] and AISC LRFD Specification [16]. Therefore, they should not be specified if extraordinary ductility is not required in a structural component, which is the case for most statically loaded structures.

AASHTO specifications for bridge steel and weld filler metal and ASTM specifications for ship steel (A131) require minimum CVN values at specific temperatures. However, since fatigue is not expected in conventional buildings and therefore the risk of fracture is much lower, CVN is often not explicitly specified for the steel or weld metal for buildings. Rather, the strategy is to allow only specific American Society for Testing and Materials (ASTM) specifications for steel and AWS classifications for weld filler metal that are known to function well in buildings. However, if there is special concern about brittle fracture and (1) high-ductility demand (as in seismic frames), (2) low-temperature exposure,

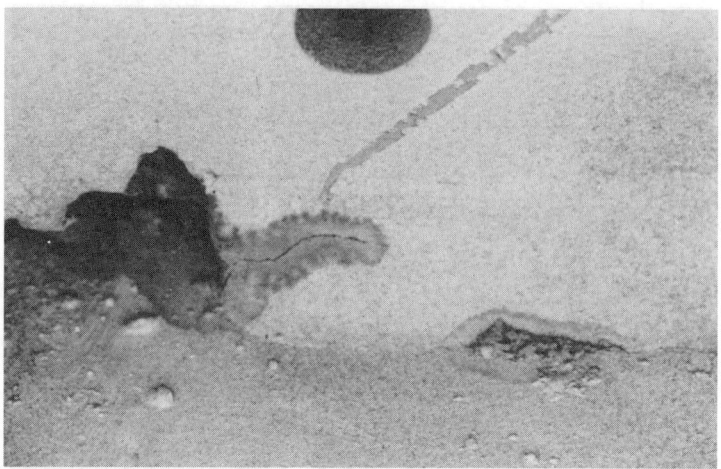

FIGURE 34.14   Fracture emanating from a poorly executed weld access hole.

or (3) fatigue or impact loading, then a supplemental requirement for a minimum CVN at a specific temperature should be considered, along with enhanced detailing. Since there is usually a premium cost, a supplemental CVN requirement should only be specified when necessary.

As shown in Figure 34.12, the typical lower-shelf CVN is about 10 J. Therefore, when a minimum CVN of 20 J or more is specified at some temperature, the most important result of such a specification is that the lower shelf of the Charpy curve will start at a temperature lower than the specified temperature. In fact, this indicates that the lower shelf of a structure loaded statically or at intermediate strain rates such as traffic loading on a bridge is even lower, a phenomenon known as the temperature shift [3]. Because of the temperature shift, the temperature at which the CVN requirement is specified may be greater than the lowest anticipated service temperature.

As long as the material is not on the lower shelf at service temperature, brittle fracture will not occur as long as large cracks do not develop. It almost does not matter what the specified CVN value is as long as it is at least 20 J. Usually, an average from three tests of 34 J (25 ft lbs) or 27 J (20 ft lbs) is specified at a particular temperature. The greater the value of the average CVN requirement, the more certain it is that the material is well above the lower shelf, but there may be a greater premium to be paid with diminishing increases in certainty.

In addition to brittle fracture, there is occasionally a problem with steel or weld metal that has low upper-shelf toughness. Such a material can give a ductile failure mode but without sufficient ductility. To guard against this type of material, a CVN requirement of 54 J (40 ft lbs) is often specified at 21°C (room temperature).

34.3.2.1 Steel Plates and Shapes

AASHTO specifications for Grade 50 (minimum 350-MPa yield strength) bridge steel up to 38-mm thick require a CVN at a temperature that is 38°C *greater* than the minimum service temperature. The temperature shift accounts for the effect of strain rates, which are lower in the service loading of bridges (on the order of 10^{-3}) than in the Charpy test (greater than 10^1). The temperature shift means that when a CVN is specified at some temperature, it ensures that the transition temperature for the structure is at least 38°C less than the specified temperature. For bridge steel, the specified minimum CVN is 20 J for nonfracture critical members and 34 J for fracture critical members. Fracture critical members are defined as those that if fractured would result in collapse of the bridge.

Most of the United States (except Alaska and a small part of the northern tier states) has a lowest anticipated service temperature greater than $-34°C$, the limit for AASHTO Zone II. In Zone II, bridge steel is required to have a minimum of 20 J at 4°C.

ASTM A673 has specifications for the frequency of Charpy testing. The H frequency requires a set of three CVN specimens to be tested from one location for each heat or about 50 tons. These CVN test specimens can be taken from a plate with thickness up to 9 mm different from the product thickness if it is rolled from the same heat. The P frequency requires a set of three specimens to be tested from one end of every plate, or from one shape in every 15 tons of that shape. For bridge steel, the AASHTO code requires CVN tests at the H frequency as a minimum. For fracture critical members, CVN testing at the P frequency is required.

For buildings, A36, A572, A588, or A992 do not have a specified minimum CVN, unless supplemental specifications (CVN testing in accordance with ASTM A6/A6M, Supplementary Requirement S5) should be cited.

One exception is shapes meeting A913, which are produced using a special quenched and self-tempering process that results in good toughness. A913 shapes have a specified minimum CVN (average of three tests) of 54 J at 21°C.

In most cases, ordinary structural steel has sufficient toughness for the required performance in buildings, even under these demanding conditions. For example, Figure 34.15 shows a histogram of CVN data at 4°C from the flanges of shapes for more than 2200 heats of A992 steel from five producers in 1998 [79]. The CVN values were widely dispersed, and the mean values were typically very high, on the order of 160 J or more. Table 34.2 shows some summary statistics for the A992 steel CVN. These data show that 99.9% of the A992 steel meets Zone II, bridge steel requirement of 20 J at 4°C. Therefore, as stated previously, it may not be worth paying the premium for specifying the supplemental CVN requirement since the requirement is almost always met by all structural steel, at least that from the major suppliers (U.S. and European) of the North American market.

Although the toughness of the flanges is generally quite good, the toughness of shapes is usually not homogenous. For example, in rotary straightened W-shapes there is often an area of reduced notch toughness in a limited region of the web immediately adjacent to the flange, referred to as the "k-area," as illustrated in Figure 34.16 [80]. Following the 1994 Northridge earthquake, there was a tendency to specify thicker continuity plates that were groove welded to the web and flange and thicker doubler plates that were often groove welded in the gap between the doubler plate and the flanges. These welds were highly restrained and may have caused cracking during fabrication in some cases [81].

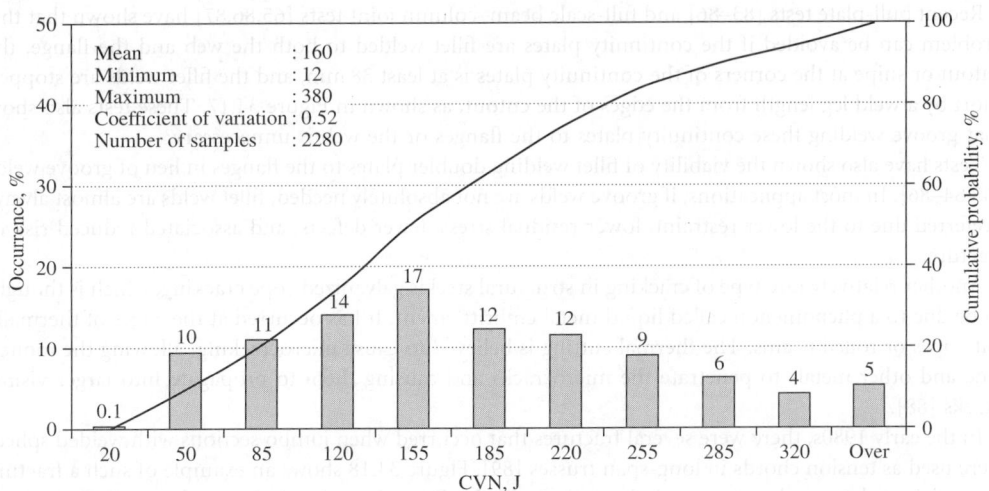

FIGURE 34.15 CVN histogram for A992 steel at 4°C.

TABLE 34.2   Summary of Statistics for CVN for A992 Structural Steel

Temperature	% less than 29 Joules (15 ft-lbs)	First quartile (75% exceedence) Joules (ft-lbs)	Mean Joules (ft-lbs)	Samples
0°C (32°F)	0.25	54 (40)	143 (106)	395
4°C (40°F)	0.13	88 (65)	163 (121)	2280
21°C (70°F)	0.09	104 (77)	175 (160)	1058

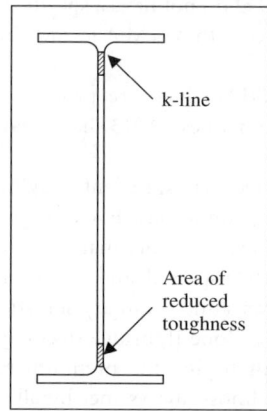

FIGURE 34.16   Location of the k-area affected by rotary straightening.

AISC issued an Advisory in 1997 [82] that recommended that the welds for continuity plates should terminate well away from the k-area. The Advisory defined the k-area as the "region extending from approximately the midpoint of the radius of the fillet into the web approximately 1 to 1.5 in beyond the point of tangency between the fillet and web."

Recent pull-plate tests [83–86] and full-scale beam–column joint tests [65,86,87] have shown that this problem can be avoided if the continuity plates are fillet welded to both the web and the flange, the cutout or snipe at the corners of the continuity plates is at least 38 mm, and the fillet welds are stopped short by a weld leg length from the edges of the cutout, as shown in Figure 34.17. These tests also show that groove welding these continuity plates to the flanges or the web is unnecessary.

Tests have also shown the viability of fillet welding doubler plates to the flanges in lieu of groove welds [65,84–86]. In most applications, if groove welds are not absolutely needed, fillet welds are almost always preferred due to the lower restraint, lower residual stress, fewer defects, and associated reduced risk of fracture.

Another relatively rare type of cracking in structural steel is galvanized cope cracking, which is thought to be due to a phenomenon called liquid metal embrittlement. It has occurred at the edges of thermally cut copes or rolled beams. The thermal cutting is believed to cause microcracking, allowing the molten zinc and other metals to penetrate the microcracks and causing them to propagate into larger visible cracks [88].

In the early 1980s, there were several fractures that occurred when jumbo sections with welded splices were used as tension chords in long-span trusses [89]. Figure 34.18 shows an example of such a fracture that originated at poorly cut access holes at the welded splices where they intersected the web/flange core region of jumbo shapes [89]. This web/flange core region (not the same as the k-area) often had course grain regions with very low toughness. Plates greater than 50 mm thick may also have regions of low toughness at midthickness.

Therefore, AISC specifications now have special detailing requirements for weld access holes and a supplemental Charpy requirement for shapes with flange thickness greater than 38 mm and plates thicker than 51 mm, when these are welded and subject to primary tensile stress from axial load or bending. These jumbo shapes and thick plates must exhibit an average of 27 J at 21°C. In the shapes, the central longitudinal axis of the specimens must be located on a line in a plane one fourth of the way through the thickness from the inside flange surface at the intersection with the web midthickness. The test specimens are taken from the top of each ingot used to produce the product.

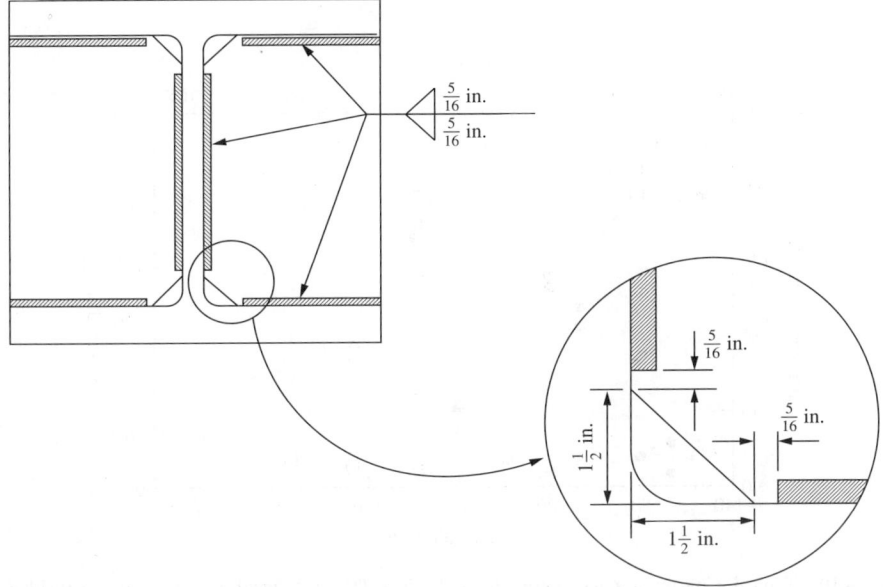

FIGURE 34.17 Recommended placement of continuity plate fillet welds to avoid contact of welds with k-area.

FIGURE 34.18 Jumbo section used as tension chord in a roof truss and fracture in web originating from weld access holes at welded splice.

34.3.2.2 Weld Filler Metal

Prior to the 1994 Northridge Earthquake, the welds in the welded special moment frame (WSMF) connections were commonly made with the self-shielded flux-cored arc welding (FCAW-S) process using an E70T-4 weld wire. Figure 34.19 shows a plot of CVN versus temperature for deposited weld metal using E70T-4, E70TG-K2 (another FCAW-S filler metal), and ordinary E7018 shielded metal arc welding (SMAW) electrodes. The CVN for E70T-4 is less than 20 J for temperatures up to 40°C [74].

Many of these low-toughness WSMF connections fractured in the earthquake [90]. Figure 34.20 shows a fractured beam flange end and a cross-section. Figure 34.21 shows the crack-like notch created by the backing bar. The fracture surfaces, such as those shown in Figure 34.20, indicate that the fractures originate in the root of the weld, typically at a lack-of-fusion defect adjacent to the backing bar notch

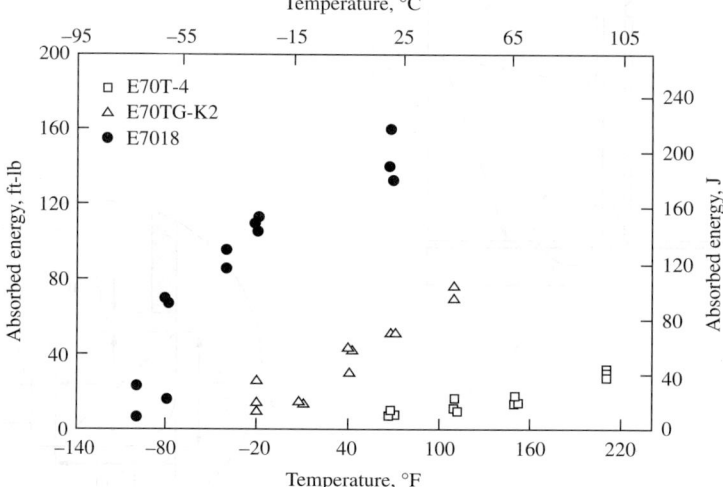

FIGURE 34.19 Typical Charpy impact energy from E70T-4 FCAW-S weld metal from Northridge WSMF connections compared to another FCAWS-S (E70TG-K2) and SMAW (E7018) weld metal.

FIGURE 34.20 Fracture surface and weld cross-section from moment–frame connection that fractured in the Northridge Earthquake showing a typical crack that originated at the backing bar notch.

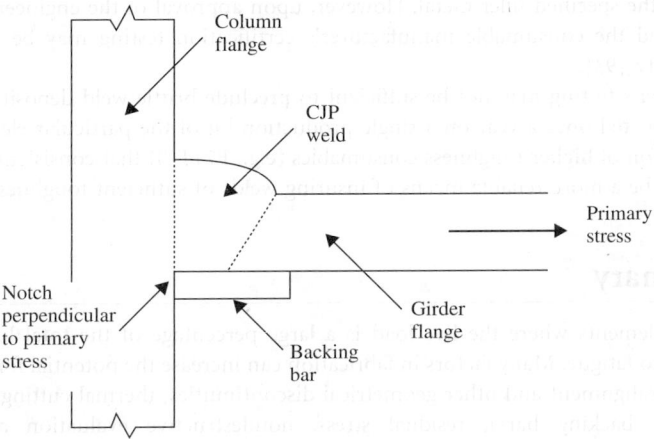

FIGURE 34.21 Schematic cross-section of pre-Northridge groove weld detail with backing bar.

[91,92]. This lack-of-fusion defect is difficult to avoid when the weld must be stopped on one side of the web and started on the other side.

A fracture mechanics assessment shows that the low-toughness E70T-4 weld metal would be predicted to fracture in the presence of the crack-like notch at stress levels well below the yield point [73,74]. In addition to the low toughness and the crack-like notch and defect, there were many other important factors in these fractures. The overall lack of redundancy, that is, the reliance on only one or two massive WSMFs to resist lateral load in each direction, contributes to large forces, increase in the thickness of the members, and the high constraint of the connections.

FEMA 350 has proposed improved connections, such as the WUF-W connection [66]. In addition to specifying notch-tough filler metals, the detailing of the WUF-W connection is also improved relative to the pre-Northridge connection. On the bottom flange weld, the backing bar must be removed, the weld back gouged, and a reinforcing fillet placed underside. On the top flange, the backing bar may be left in place, but the notch is sealed with a fillet weld as shown in Figure 34.7, reducing the adverse effect of the notch.

The FEMA guidelines [66,93] recommend that weld metal be used that meets two CVN test requirements: (1) 54 J at 21°C (to avoid low upper-shelf weld metal) and (2) 27 J at −18°C (to ensure that the service temperature is above the lower shelf). Testing done as part of the FEMA program suggests that another FCAW-S filler metal, E70T-6, can meet these requirements. The AWS classification [94] requires E70T-6 to have 27 J at −29°C. The AISC seismic specifications [76] also require the weld metal for these welds to have a minimum CVN of 27 J at −29°C. However, weld deposits with CVN far less than required have been obtained under some conditions with E70T-6, despite certifications indicating that the filler metal met the AWS classification requirements, leading to brittle fractures in full-scale tests [65].

The FEMA requirements for minimum toughness are adequate, provided they can be consistently met. Toughness is an inherently variable material property, particularly in a nonhomogenous material such as a weld. For this reason, the toughness requirements should be treated as a lower-bound value and not as an average. This can be accomplished either through strict quality assurance or specification of welding consumables that have lower-bound toughness consistently above the FEMA minimum.

The former is the approach required by the FEMA *Recommended Specifications and Quality Assurance Guidelines* [93]. These Recommended Specifications require toughness testing on each production lot of the specified filler metal. However, upon approval of the engineer, this requirement may be waived and the consumable manufacturer's certification testing may be used to verify the material's suitability [93].

The manufacturer's testing may not be sufficient to preclude brittle weld deposits, since the testing need only be conducted once a year on a single production lot of the particular electrode [94]. Alternatively, specification of higher toughness consumables (e.g., E71T-8) that consistently meet minimum requirements may be a more reliable means of insuring welds of sufficient toughness.

34.4 Summary

1. Structural elements where the live load is a large percentage of the total load are potentially susceptible to fatigue. Many factors in fabrication can increase the potential for fatigue, including notches, misalignment and other geometrical discontinuities, thermal cutting, weld joint design (particularly backing bars), residual stress, nondestructive evaluation and weld defects, intersecting welds, and inadequate weld access holes.
2. The fatigue design procedures in the AASHTO and AISC specifications are based on control of the stress range and knowledge of the fatigue strength of the various details. Using these specifications, it is possible to identify and avoid details expected to have low fatigue strength.
3. Low-cycle fatigue is a limit state for members and connections repeatedly cycled in the inelastic range, such as for seismic loading. Low-cycle fatigue can be predicted using strain-range versus cycles curves derived from the stress-based S–N curves for high-cycle fatigue.
4. Welded connections and thermal-cut holes copes, blocks, or cuts are potentially susceptible to brittle fracture. Many interrelated design variables can increase the potential for brittle fracture, including lack of redundancy, large forces and moments with dynamic loading rates, thick members, geometrical discontinuities, and high constraint of the connections. Low temperature can be a factor for exposed structures. The factors mentioned above, which influence the potential for fatigue, have a similar effect on the potential for fracture. In addition, cold work (e.g., from rotary straightening or punching holes), flame straightening, weld heat input, and weld sequence can also affect the potential for fracture.
5. The AASHTO specifications require a minimum CVN notch toughness at a specified temperature for the base metal and the weld metal of members loaded in tension or tension due to bending. Almost two decades of experience with these bridge specifications have proved that they are successful in significantly reducing the number of brittle fractures.
6. Surveys of CVN for wide-flange shapes sold in North America show that 99.9% of this steel meets the AASHTO bridge steel requirements for service down to $-34°C$. Therefore, under most circumstances, there is no need to specify CVN for shapes used in buildings.
7. Achieving the required minimum CVN toughness in the girder flange-to-column flange groove welds is critical for good performance in the prequalified steel moment connections. Lot testing should be considered, or it may be necessary to specify electrodes that typically far exceed the minimum toughness levels. Either step would assure that minimum CVN requirements represent a realistic lower-bound of weld toughness deposited in the field.

References

[1] Bazant, Z. and Planas, J., *Fracture and Size Effect in Concrete and Other Quasibrittle Materials*, CRC Press, Boca Raton, FL, 1997.
[2] Anderson, T.L., *Fracture Mechanics — Fundamentals and Applications*, Second Edition, CRC Press, Boca Raton, FL, 1995.

[3] Barsom, J.M. and Rolfe, S.T., *Fracture and Fatigue Control in Structures: Applications of Fracture Mechanics*, Third Edition, ASTM, West Conshohocken, PA, 1999.

[4] Broek, D., *Elementary Fracture Mechanics*, Fourth Edition, Martinis Nijhoff Publishers, Dordrecht, Netherlands, 1987.

[5] Kober, G.R., Dexter, R.J., Kaufmann, E.J., Yen, B.T., and Fisher, J.W., "The Effect of Welding Discontinuities on the Variability of Fatigue Life," *Fracture Mechanics, Twenty-Fifth Volume*, ASTM STP 1220, F. Erdogan and R.J. Hartranft, Eds., American Society for Testing and Materials, Philadelphia, PA, 1994.

[6] Fisher, J.W., Frank, K.H., Hirt, M.A., and McNamee, B.M., *Effect of Weldments on The Fatigue Strength of Steel Beams*, National Cooperative Highway Research Program (NCHRP) Report 102, Highway Research Board, Washington, DC, 1970.

[7] Fisher, J.W., Albrecht, P.A., Yen, B.T., Klingerman, D.J., and McNamee, B.M., *Fatigue Strength of Steel Beams with Welded Stiffeners and Attachments*, National Cooperative Highway Research Program (NCHRP) Report 147, Transportation Research Board, Washington, DC, 1974.

[8] Keating, P.B. and Fisher, J.W., *Evaluation of Fatigue Tests and Design Criteria on Welded Details*, National Cooperative Highway Research Program (NCHRP) Report 286, Transportation Research Board, Washington, DC, 1986.

[9] Maddox, S.J., *Fatigue Strength of Welded Structures*, Second Edition, Abington Publishing, Cambridge, U.K., 1991.

[10] Dexter, R.J., Fisher, J.W. and Beach, J.E., "Fatigue Behavior of Welded HSLA-80 Members," *Proceedings, 12th International Conference on Offshore Mechanics and Arctic Engineering*, Vol. III, Part A, *Materials Engineering*, M.M. Salama et al., Eds., pp. 493–502, ASME, New York, 1993.

[11] Petershagen, H. and Zwick, W., *Fatigue Strength of Butt Welds Made by Different Welding Processes*, IIW-Document XIII-1048-82, 1982.

[12] Petershagen, H., "The Influence of Undercut on the Fatigue Strength of Welds — A Literature Survey," *Weld. World*, Vol. 28, No. 7/8, pp. 29–36, 1990.

[13] Metrovich, B., Fisher, J.W., Yen, B.T., Kaufmann, E.J., Cheng, X., and Ma, Z., "Fatigue Strength of Welded AL-6XN Superaustenitic Stainless Steel," *Int. J. Fatigue*, Vol. 25, No. 9–11, pp. 1309–1315.

[14] Kaczinski, M.R., Dexter, R.J., and Van Dien, J.P., *Fatigue-Resistant Design of Cantilevered Signal, Sign, and Light Supports*, National Cooperative Highway Research Program, NCHRP Report 412, Transportation Research Board, Washington, DC, 1998.

[15] AASHTO, *AASHTO LRFD Bridge Design Specifications*, Second Edition, The American Association of State Highway and Transportation Officials, Washington, DC, 1998.

[16] AISC, *Load and Resistance Factor Design Specification for Structural Steel Buildings*, Third Edition, American Institute of Steel Construction (AISC), Chicago, 1999.

[17] AREMA, *AREMA Manual for Railway Engineering*, Chapter 15: "Steel Structures," American Railway Engineering and Maintenance of Way Association, 2002.

[18] AWS, *Structural Welding Code — Steel*, ANSI/AWS D1.1-02, American Welding Society, Miami, FL, 2002.

[19] CSA, CSA S16-2001, *Limit States Design of Steel Structures*, Canadian Standards Association, Toronto, Ontario, 2001.

[20] CEN, ENV 1999-1-1, *Eurocode 3: Design of Steel structures — Part 1.1: General Rules and Rules for Buildings*, European Committee for Standardization (CEN), Brussels, May 1998.

[21] BSI, BS 7608, *Specification of Practice for Fatigue Design and Assessment of Steel Structures*, British Standards Institute, London, 1994.

[22] ABS, *Guide for Fatigue Strength Assessment of Tankers*, American Bureau of Shipping, New York, June 1992.

[23] UK Health & Safety Executive (formerly the UK Department of Energy), *Fatigue Design Guidance for Steel Welded Joints in Offshore Structures*, H.M.S.O., London, 1984.

[24] Roberts, R. et al., *Corrosion Fatigue of Bridge Steels*, Vols. 1–3, Reports FHWA/RD-86/165, 166, and 167, Federal Highway Administration, Washington, DC, May 1986.

[25] Outt, J.M.M., Fisher, J.W., and Yen, B.T., *Fatigue Strength of Weathered and Deteriorated Riveted Members*, Report DOT/OST/P-34/85/016, Department of Transportation, Federal Highway Administration, Washington, DC, October 1984.

[26] Albrecht, P. and Shabshab, C., "Fatigue Strength of Weathered Rolled Beam Made of A588 Steel," *J. Mater. Civ. Eng.*, Vol. 6, No. 3, pp. 407–428, 1994.

[27] The Aluminum Association, *The Aluminum Design Manual*, The Aluminum Association, Washington, DC, 1986.

[28] Menzemer, C.C. and Fisher, J.W., "Revisions to the Aluminum Association Fatigue Design Specifications, *6th International Conference on Aluminum Weldments*, Cleveland, OH, April 3–5, 1995, AWS, Miami, FL, pp. 11–23, 1995.

[29] Fisher, J.W., et al., *Resistance of Welded Details under Variable Amplitude Long-Life Fatigue Loading*, National Cooperative Highway Research Program Report 354, Transportation Research Board, Washington, DC, 1993.

[30] Dexter, R.J., Wright, W.J., and Fisher, J.W., "Fatigue and Fracture of Steel Girders," *J. Bridge Eng.*, Vol. 9, Issue 3, May/June 2004, pp. 278–286.

[31] Miner, M.A., "Cumulative Damage in Fatigue," *J. Appl. Mech.*, Vol. 12, A-159, 1945.

[32] BS 7910, *Guide on Methods for Assessing the Acceptability of Flaws in Metallic Structures*, British Standards Institute, London, 1999.

[33] Demers, C. and Fisher, J.W., *Fatigue Cracking of Steel Bridge Structures, Volume I: A Survey of Localized Cracking in Steel Bridges – 1981 to 1988*, Report No. FHWA-RD-89-166, and *Volume II: A Commentary and Guide for Design, Evaluation, and Investigating Cracking*, Report No. FHWA-RD-89-167, FHWA, McLean, VA, March 1990.

[34] Yen, B.T., Huang, T., Lai, L.-Y., and Fisher, J.W., *Manual for Inspecting Bridges for Fatigue Damage Conditions*, Report No. FHWA-PA-89-022 + 85-02, Fritz Engineering Laboratory Report No. 511.1, Pennsylvania Department of Transportation, Harrisburg, PA, January 1990.

[35] Dexter, R.J. and Kelly, B.A., "Research on Repair and Improvement Methods," *International Conference on Performance of Dynamically Loaded Welded Structures*, Proceedings of the IIW 50th Annual Assembly Conference, San Francisco, CA, July 13–19, 1997, Welding Research Council, Inc., New York, pp. 273–285, 1997.

[36] Kelly, B.A. and Dexter, R.J., *Adequacy of Weld Repairs for Ships*, SSC-424, Ship Structure Committee, Washington, DC, 2003.

[37] Frank, K.H. and Fisher, J.W., "Fatigue Strength of Fillet Welded Cruciform Joints," *J. Struct. Div.*, ASCE, Vol. 105, No. ST9, pp. 1727–1740, September 1979.

[38] Kulak, G.L., Fisher, J.W., and Struick, J.H., *Guide to Design Criteria for Bolted and Riveted Joints*, Second Edition, Prentice Hall, Englewood Cliffs, NJ, 1987.

[39] Kuperus, A., *The Fatigue Strength of Tensile Loaded Non-Tightened HSFG Bolts*. Delft University of Technology, Report 6-73-3, June 1973.

[40] Kuperus, A., *The Fatigue Strength of Tensile Loaded Tightened HSFG Bolts*. Delft University of Technology, Report 6-74-4, October 1974.

[41] Frank, K.H., "Fatigue Strength of Anchor Bolts," *J. Struct. Div.*, ASCE, Vol. 106, No. ST, June 1980.

[42] Dusel, J.P., Stoker, J.P. and Travis, R., *Determination of Fatigue Characteristics of Hot-Dipped Galvanized A307 and A449 Anchor Bars and A325 Cap Screws*, California Department of Transportation Reports, February 1984.

[43] Van Wingerde, A., Packer, J., and Wardenier, J., "Criteria for the Fatigue Assessment of Hollow Structural Section Connections," *J. Constr. Steel Res.*, Vol. 35, pp. 71–115, 1995.

[44] Van Wingerde, A., Packer, J., and Wardenier, J., "New Guidelines for Fatigue Design of HSS Connections," *J. Struct. Eng.*, Vol. 122, No. 2, pp. 125–132, February 1996.

[45] Zhao, X. and Packer, J., *Fatigue Design Procedure for Welded Hollow Section Joints*, International Institute of Welding documents XIII-1804-99 and IIW Document XV-1035-99, Abington Publishing, Cambridge, U.K., 2000.

[46] Zhao, X. et al., *Design Guide for Circular and Rectangular Hollow Section Welded Joints under Fatigue Loading,* Comite International pour le Developpement et l'Etude de la Construction Tubulaire, 1999.

[47] AASHTO, *Standard Specifications for Structural Supports for Highway Signs, Luminaires and Traffic Signals,* Fourth Edition, American Association of State Highway and Transportation Officials, Washington, DC, 2001.

[48] Marshall, P., "Welded Tubular Connections — CHS Trusses," *Handbook of Structural Engineering,* W.F. Chen, Ed., CRC Press, Boca Raton, FL, 1999.

[49] Dexter, R., Tarquinio, J., and Fisher, J., "Application of Hot-Spot Stress Fatigue Analysis to Attachments on Flexible Plate," *Proceedings of the 13th International Conference on Offshore Mechanics and Arctic Engineering Conference (OMAE),* M.M. Salama et al., Eds., American Society of Mechanical Engineers, *Materials Engineering,* Vol. III, pp. 85–92, 1994.

[50] Yagi, J., Machida, S., Tomita, Y., Matoba, M., Kawasaki, T., *Definition of Hot-Spot Stress in Welded Plate Type Structure for Fatigue Assessment,* International Institute of Welding, IIW-XIII-1414-91, 1991.

[51] ACI Committee 215, "Considerations for Design of Concrete Structures Subjected to Fatigue Loading," ACI 215R-74 (Revised 1992), *ACI Manual of Standard Practice, Part 1,* 2000.

[52] Hahin, C., *Effects of Corrosion and Fatigue on the Load-Carrying Capacity of Structural Steel and Reinforcing Steel,* Illinois Physical Research Report No. 108, Illinois Department of Transportation, Springfield, IL, March 1994.

[53] Pfister, J.F. and Hognestad, E., "High Strength Bars as Concrete Reinforcement, Part 6, Fatigue Tests," *J. PCA Res. Dev. Lab.,* Vol. 6, No. 1, pp. 65–84, January 1964.

[54] Sternberg, F., *Performance of Continuously Reinforced Concrete Pavement, I-84 Southington,* Connecticut State Highway Department, June 1969.

[55] Batchelor, B., Hewitt B.E., and Csagoly, P., "An Investigation of the Fatigue Strength of Deck Slabs of Composite Steel/Concrete Bridges," *Transportation Research Record 664,* Vol. 1, TRB, National Research Council, Washington, DC, pp. 153–161, 1978.

[56] Azad, A.K., Baluch M.H., Al-Mandil M.Y., and Al-Suwaiyan, "Static and Fatigue Tests of Simulated Bridge Decks," *Experimental Assessment of Performance of Bridges,* L.R. Wang and G.M. Salnis, Eds., Proceedings of ASCE Convention, Boston, MA, pp. 30–41, October 1986.

[57] Youn, S. and Chang S., "Behavior of Composite Bridge Decks Subjected to Static and Fatigue Loading," *ACI Struct. J.,* Vol. 95, No. 3, pp. 249–259, 1998.

[58] Okada, K., Okamura H., and Sonada K., "Fatigue Failure Mechanism of Reinforced Concrete Deck Slabs," *Transportation Research Record 664,* National Research Council, Washington, DC, pp. 136–144, 1978.

[59] Kato, T. and Goto Y., "Effect of Water Infiltration of Penetration Deterioration of Bridge Deck Slabs," *Transportation Research Record 950,* National Research Council, Washington, DC, pp. 202–209, 1978.

[60] Petrou, M., Perdikaris P.C., and Wang, A., "Fatigue Behavior of Non-Composite Reinforced Concrete Bridge Deck Models," *Transportation Research Record 1460,* National Research Council, Washington, DC, pp. 73–80, 1994.

[61] Fang, I.K., Tsui, C.K.T., Burns N.H., and Klinger, R.E., "Fatigue Behavior of Cast-in-Place and Precast Panel Bridge Decks with Isotropic Reinforcement," *PCI J.,* Vol. 35, No. 3, pp. 28–39, May–June 1990.

[62] Rabbat, B.G., et al., "Fatigue Tests of Pretensioned Girders with Blanketed and Draped Strands," *J., Prestressed Concrete Inst.,* Vol. 24, No. 4, pp. 88–115, July/August 1979.

[63] Overnman, T.R., Breen, J.E., and Frank, K.H., *Fatigue Behavior of Pretensioned Concrete Girders,* Research Report 300-2F, Center for Transportation Research, The University of Texas at Austin, November 1984.

[64] Ad-Hoc Committee on Cable-Stayed Bridges, *Recommendations for Stay Cable Design, Testing, and Installation.* Fourth Edition, Post-Tensioning Institute, Phoenix, AZ, March 2001.

[65] Lee, D., Cotton, S., Dexter, R.J., Hajjar, J.F., Ye, Y., and Ojard, S.D., *Column Stiffener Detailing and Panel Zone Behavior of Steel Moment Frame Connections*, Report No. ST-01-3.2, Department of Civil Engineering, University of Minnesota, Minneapolis, MN, 2002.

[66] FEMA, *Recommended Seismic Design Criteria for New Steel Moment-Frame Buildings*, Report No. FEMA 350, Federal Emergency Management Agency, Washington, DC, 2000.

[67] Coffin, L.F., Jr., "A Note on Low Cycle Fatigue Laws," *J. Mater.*, Vol. 6, No. 2, pp. 388-402, 1971.

[68] Itoh, Y.Z. and Kashiwaya, H., "Low-Cycle Fatigue Properties of Steel Sand Their We d Metals," *J. Eng. Mater. Technol.*, Vol. 111, No. 4, pp. 431–437, 1989.

[69] Howdyshell, P., Trovillion, J.C., and Wetterich, J.L., "Low-Cycle Fatigue of Structural Materials," *Materials and Construction: Proceedings of MatCong 5*, The 5th ASCE Materials Engineering Congress, Cincinnati, OH, 10–12 May 1999, L. Bank, Ed., American Society of Civil Engineers, Reston, VA, pp. 148–155.

[70] Krawinkler, H. and Zohrei, M., "Cumulative Damage in Steel Structures Subjected to Earthquake Ground Motion," *Comput. Struct.*, Vol. 16, No. 1–4, pp. 531–541, 1983.

[71] Ballio, G. and Castiglioni, C.A., "A Unified Approach for the Design of Steel Structures under Low and/or High Cycle Fatigue", *J. Constr. Steel Res.*, Vol. 34, No. 1, pp. 75–101, 1995.

[72] Castiglioni, C.A., "Cumulative Damage Assessment in Structural Steel Details," IABSE Symposium, San Francisco, 1995, *Extending the Lifespan of Structures*, pp. 1061–1066, 1995.

[73] Dexter, R.J. and J.W. Fisher, "Fatigue and Fracture," Chapter 8, *Steel Design Handbook, LRFD Method*, A.R. Tamboli, Ed., McGraw-Hill, New York, 1997.

[74] Fisher, J.W., Dexter, R.J., and Kaufmann, E.J., *Fracture Mechanics of Welded Structural Steel Connections*, Report No. SAC 95-09, FEMA-288, March 1997.

[75] Wright, W.J., Fisher, J.W., and Sivakumar, B., *Hoan Bridge Failure Investigation*, Federal Highway Administration, Washington, DC, 2001.

[76] AISC, *Seismic Provisions for Structural Steel Buildings*, Second Edition, American Institute of Steel Construction (AISC), Chicago, 2002.

[77] AWS, *Bridge Welding Code*, AASHTO-AWS-D1.5M-D1.5: 2002, American Welding Society, Miami, FL, 2002.

[78] AASHTO, *LRFD Bridge Construction Specifications*, First Edition, The American Association of State Highway and Transportation Officials, Washington, DC, 1998.

[79] Dexter, R.J., *Structural Shape Material Property Survey*, Final Report to Structural Shape Producer's Council, University of Minnesota, Minneapolis, MN, 2000.

[80] Kaufmann, E.J., Metrovich, B., Pense, A.W., and Fisher, J.W., "Effect of Manufacturing Process on k-Area Properties and Service Performance," *Proceedings of the North American Steel Construction Conference*, Fort Lauderdale, FL, May 9–12, 2001, AISC, Chicago, IL, pp. 17-1–17-24, 2001.

[81] Tide, R.H., "Evaluation of Steel Properties and Cracking in the 'k'-Area of W Shapes," *Eng. Struct.*, Vol. 22, pp. 128–124, 1999.

[82] AISC, "AISC Advisory Statement on Mechanical Properties near the Fillet of Wide Flange Shapes and Interim Recommendations January 10, 1997," *Modern Steel Construction*, p. 18, February 1997.

[83] Dexter, R.J. and Melendrez, M.I., "Through-Thickness Properties of Column Flanges in Welded Moment Connections," *J. Struct. Eng.*, ASCE, Vol. 126, No. 1, pp. 24–31, 2000.

[84] Prochnow, S.D., Ye, Y., Dexter, R.J., Hajjar, J.F., and Cotton, S.C., "Local Flange Bending and Local Web Yielding Limit States in Steel Moment Resisting Connections," *Connections in Steel Structures IV*, Roanoke, VA, October 22–25, 2000, R.T. Leon and W.S. Easterling, Eds., AISC, Chicago, 2002.

[85] Hajjar, J.F., Dexter, R.J., Ojard, S.D., Ye, Y., and Cotton, S.C., "Continuity Plate Detailing for Steel Moment-Resisting Connections," *Eng. J.*, Vol. 40, No. 4, 4th Qrt., pp. 189–211, 2003.

[86] Dexter, R.J., Hajjar, J.F., Prochnow, S.D., Graeser, M.D., Galambos, T.V., and Cotton, S.C., "Evaluation of the Design Requirements for Column Stiffeners and Doublers and the Variation in Properties of A992 Shapes," *2001 North American Steel Construction Conference*, Ft. Lauderdale, FL, May 9–12, 2001, AISC, Chicago, 2001.

[87] Bjorhovde, R., Goland, L.J., and Benac, D.J., "Performance of Steel in High-Demand Full-Scale Connection Tests," *2000 North American Steel Construction Conference*, Las Vegas, NV, May 2000, AISC, pp. 3.1–3.22, February 2000.

[88] Langill, T.J. and Schlafly, T., "Cope Cracking in Structural Steel after Galvanizing," *Mod. Steel Constr.* (USA), Vol. 35, No. 10, pp. 40–43, October 1995.

[89] Fisher, J.W. and Pense, A.W., "Experience with Use of Heavy W Shapes in Tension," *Eng. J., AISC*, Vol. 24, No. 2, 1987.

[90] Northridge Reconnaissance Team (1996). *Northridge Earthquake of January 17, 1994*, Reconnaissance Report (Supplement C-2 to Volume 11), EERI, Oakland, CA.

[91] Kaufmann, E.J., Fisher, J.W., Di Julio, R.M., Jr., and Gross, J.L., *Failure Analysis of Welded Steel Moment Frames Damaged in the Northridge Earthquake*, NISTIR 5944, National Institute of Standards and Technology, Gaithersburg, MD, January 1997.

[92] Tide, R.H.R., Fisher, J.W., and Kaufmann, E.J., "Substandard Welding Quality Exposed: Northridge, California Earthquake, January 17, 1994," *IIW Asian Pacific Welding Congress*, Auckland, New Zealand, February 4–9, 1996.

[93] FEMA, *Recommended Specifications and Quality Assurance Guidelines for Steel Moment-Frame Construction for Seismic Applications*, Report No. FEMA 353, FEMA, Washington, DC, 2000.

[94] AWS, AWS A5.20–95, *Specification for Carbon Steel Electrodes for Flux Cored Arc Welding*, American Welding Society AWS, Miami, FL, 1995.

35

Passive Energy Dissipation and Active Control

T. T. Soong

Department of Civil Engineering, State University of New York, Buffalo, NY

G. F. Dargush

Department of Civil Engineering, State University of New York, Buffalo, NY

35.1   Introduction

In recent years, innovative means of enhancing structural functionality and safety against natural and man-made hazards have been in various stages of research and development. By and large, these approaches can be grouped into three broad areas as shown in Table 35.1: (1) base isolation; (2) passive energy dissipation; and (3) active control. Of the three, base isolation can be considered a more mature technology (e.g., ATC 1993; Skinner et al. 1993) with wider applications as compared with the other two.

Passive energy dissipation systems encompass a range of materials and devices for enhancing damping, stiffness, and strength, and can be used both for natural hazard mitigation and for rehabilitation of aging or deficient structures (e.g., Soong and Dargush 1997; Constantinou et al. 1998; Hanson and Soong 2001). In recent years, serious efforts have been undertaken to develop the concept of energy dissipation, or supplemental damping, into a workable technology, and a number of these devices have been installed in structures throughout the world. In general, such systems are characterized by a capability to enhance energy dissipation in the structural systems in which they are installed. This effect may be achieved either by conversion of kinetic energy to heat or by transferring of energy among vibrating modes. The first method includes devices that operate on principles such as yielding of metals, phase transformation in metals, frictional sliding, deformation of viscoelastic solids or fluids, and fluid orificing. The latter method includes supplemental oscillators that act as dynamic absorbers. A list of such devices, which have found applications, is given in Table 35.1.

TABLE 35.1   Structural Protective Systems

Base isolation	Passive energy dissipation	Active control
Elastomeric bearings	Metallic dampers Friction dampers	Active bracing systems Active mass dampers
Lead rubber bearings	Viscoelastic dampers Viscous fluid dampers	Variable stiffness or damping systems
Sliding friction pendulum	Tuned mass dampers Tuned liquid dampers	

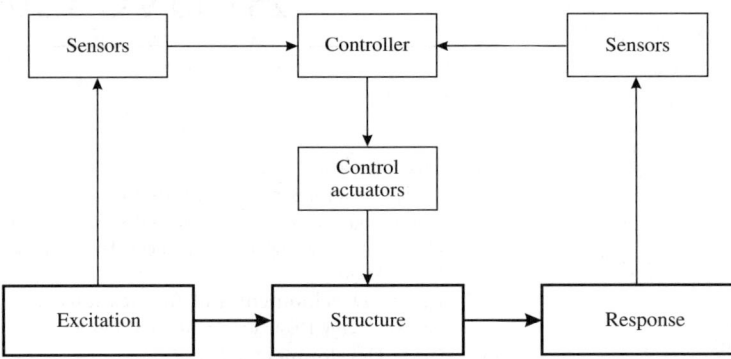

FIGURE 35.1   Block diagram of active structural control.

Among the current passive energy dissipation systems, those based on fluid orificing and on deformation of viscoelastic polymers represent technologies in which the U.S. industry has a worldwide lead. Originally developed for industrial and military applications, these technologies have found recent applications in natural hazard mitigation in the form of either energy dissipation within the primary structural system or as elements of base isolation systems.

The possible use of active control systems and some combinations of passive and active systems, so-called hybrid systems, as a means of structural protection against wind and seismic loads has also received considerable attention in recent years. Active/hybrid control systems are force delivery devices integrated with real-time processing evaluators/controllers and sensors within the structure. These systems must react simultaneously with the hazardous excitation to provide enhanced structural behavior for improved service and safety. Figure 35.1 is a block diagram of the active structural control problem. The basic task is to find a control strategy that uses the measured structural responses to calculate the control signal that is appropriate to send to the actuator. Structural control for civil engineering applications has a number of distinctive features, largely due to implementability issues, that set it apart from the general field of feedback control. First of all, when addressing civil structures, there is considerable uncertainty, including nonlinearity, associated with both physical properties and disturbances such as earthquake and wind. Additionally, the scale of the forces involved is quite large, there are a limited number of sensors and actuators, the dynamics of the actuators can be quite complex, and the systems must be fail-safe (e.g., Soong 1990; Dyke et al. 1994, 1995; Housner et al. 1994; Kobori 1994).

Nonetheless, remarkable progress has been made over the last 25 years in research on using active and hybrid systems as a means of structural protection against wind, earthquakes, and other hazards (e.g., Soong 1990; Soong and Reinhorn 1993; Soong and Constantinou 1994). A number of semiactive systems have also been developed, in which a control algorithm is used to adjust device properties with only minimal power input (e.g., Kobori et al. 1993; Patten et al. 1996; Soong and Spencer 2002). Research to date has reached the stage where active systems, such as those listed in Table 35.1, and semiactive systems have been installed in full-scale structures. Some active systems are also used temporarily in construction of bridges or large span structures (e.g., lifelines, roofs) where no other means can provide adequate protection. Additionally, most of the full-scale systems have been subjected to actual wind forces and ground motions, and their observed performances provide invaluable information in terms of (1) validating analytical and simulation procedures used to predict system performance, (2) verifying complex electronic–digital–servohydraulic systems under actual loading conditions, and (3) verifying the capability of these systems to operate or shutdown under prescribed conditions.

The focus of this section is on passive energy dissipation and active control systems. Their basic operating principles and methods of analysis are given in Section 35.2, followed by a review in Section 35.3 of recent development and applications. Code development is summarized in Section 35.4, and some comments on possible future directions in this emerging technological area are advanced in Section 35.5. In the following sections, we shall use the term *structural protective systems* to represent either passive energy dissipation systems or active control systems.

35.2 Basic Principles and Methods of Analysis

With recent development and implementation of modern structural protective systems, the entire structural engineering discipline is now undergoing a major change. The traditional idealization of a building or bridge as a static entity is no longer adequate. Instead, structures must be analyzed and designed by considering their dynamic behavior. It is with this in mind that we present some basic concepts related to topics that are of primary importance in understanding, analyzing, and designing structures that incorporate structural protective systems.

In what follows, a simple single-degree-of-freedom (SDOF) structural model is discussed. This represents the prototype for dynamic behavior. Particular emphasis is given to the effect of damping. As we shall see, increased damping can significantly reduce system response to time-varying disturbances. While this model is useful for developing an understanding of dynamic behavior, it is not sufficient for representing real structures. We must include more detail. Consequently, a multi-degree-of-freedom (MDOF) model is then introduced and several numerical procedures are outlined for general dynamic analysis. A discussion comparing typical damping characteristics in traditional and control-augmented structures is also included. Finally, a treatment of energy formulations is provided. Essentially, one can envision an environmental disturbance as an injection of energy into a structure. Design then focuses on the management of that energy. As we shall see, these energy concepts are particularly relevant in the discussion of passively or actively damped structures.

35.2.1 Single-Degree-of-Freedom Structural Systems

Consider the lateral motion of the basic SDOF model, shown in Figure 35.2, consisting of a mass m, supported by springs with total linear elastic stiffness k, and a damper with linear viscosity c. This SDOF system is then subjected to an external disturbance characterized by $f(t)$. The excited model responds with a lateral displacement $x(t)$ relative to the ground that satisfies the equation of motion:

$$m\ddot{x} + c\dot{x} + kx = f(t) \tag{35.1}$$

in which a superposed dot represents differentiation with respect to time. For a specified input, $f(t)$, and with known structural parameters, the solution of this equation can be readily obtained.

In the above, $f(t)$ represents an arbitrary environmental disturbance such as wind or an earthquake. In the case of an earthquake load

$$f(t) = -m\ddot{x}_g(t) \tag{35.2}$$

where $\ddot{x}_g(t)$ is ground acceleration.

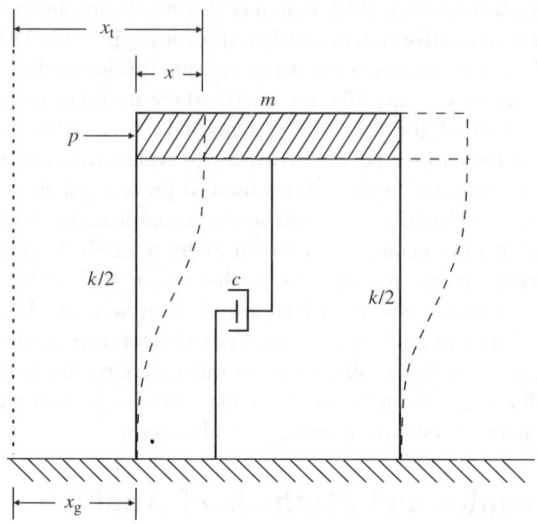

FIGURE 35.2   SDOF structure model.

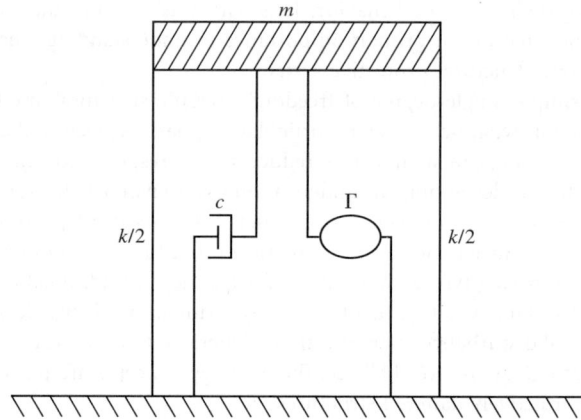

FIGURE 35.3   SDOF structure model with passive damper element.

Consider now the addition of a generic passive or active control element into the SDOF model, as indicated in Figure 35.3. The response of the system is now influenced by this additional element. The symbol Γ in Figure 35.3 represents a generic integrodifferential operator, such that the force corresponding to the control device is written simply as Γx. This permits quite general response characteristics, including displacement-, velocity-, or acceleration-dependent contributions, as well as hereditary effects. The equation of motion for the extended SDOF model then becomes, in the case of an earthquake load,

$$m\ddot{x} + c\dot{x} + kx + \Gamma x = -(m + \bar{m})\ddot{x}_{g} \tag{35.3}$$

with $\bar{m}$ representing the mass of the control element.

The specific form of Γx needs to be specified before Equation 35.3 can be analyzed, which is necessarily highly dependent on the device type. For passive energy dissipation systems, it can be represented by a force–displacement relationship such as the one shown in Figure 35.4, representing a rate-independent elastic–perfectly plastic element. For an active control system, the form of Γx is governed by the control law chosen for a given application. Let us first note that, denoting the control force applied to the structure in Figure 35.1 by $u(t)$, the resulting dynamical behavior of the structure is governed by Equation 35.3 with

$$\Gamma x = -u(t) \tag{35.4}$$

Suppose that a feedback configuration is used in which the control force $u(t)$ is designed to be a linear function of measured displacement $x(t)$ and measured velocity $\dot{x}(t)$. The control force $u(t)$ takes the form

$$u(t) = g_1 x(t) + g_2 \dot{x}(t) \tag{35.5}$$

In view of Equation 35.4, we have

$$\Gamma x = -[g_1 + g_2 d/dt]x \tag{35.6}$$

The control law is, of course, not necessarily linear in $x(t)$ and $\dot{x}(t)$ as given by Equation 35.5. In fact, nonlinear control laws may be more desirable for civil engineering applications (Wu and Soong 1995). Thus, for both passive and active control cases, the resulting Equation 35.3 can be highly nonlinear.

Assume for illustrative purposes that the base structure has a viscous damping ratio $\zeta = 0.05$ and that a simple massless yielding device is added to serve as a passive element. The force–displacement relationship for this element, depicted in Figure 35.4, is defined in terms of an initial stiffness $\bar{k}$ and a yield force $\bar{f}_y$. Consider the case where the passively damped SDOF model is subjected to the 1940 El Centro S00E ground motion as shown in Figure 35.5. The initial stiffness of the elastoplastic passive device is specified as $\bar{k} = k$, while the yield force $\bar{f}_y$ is equal to 20% of the maximum applied ground force. That is,

$$\bar{f}_y = 0.20 \, \text{Max}\{m|\ddot{x}_g|\} \tag{35.7}$$

The resulting relative displacement and total acceleration time histories are presented in Figure 35.6. There is significant reduction in response compared to that of the base structure without the control element, as shown in Figure 35.7. Force–displacement loops for the viscous and passive

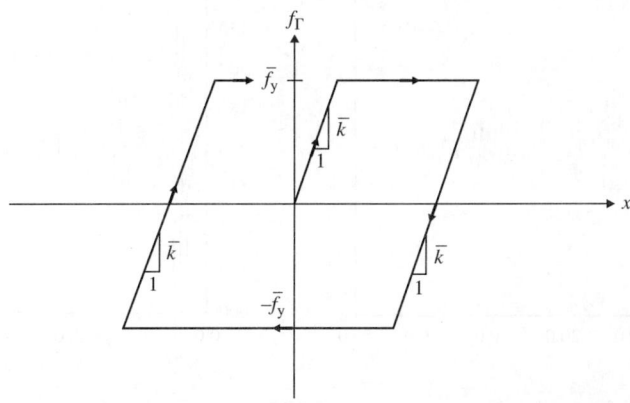

FIGURE 35.4 Force–displacement model for elastic–perfectly plastic passive element.

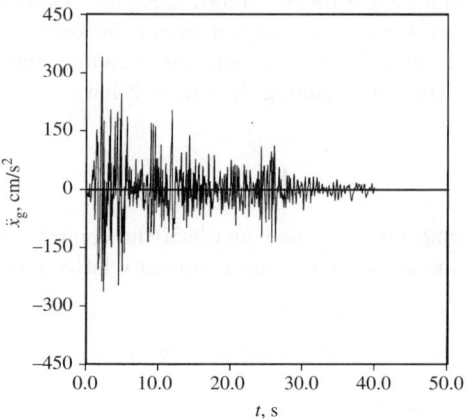

FIGURE 35.5   1940 El Centro S00E accelerogram.

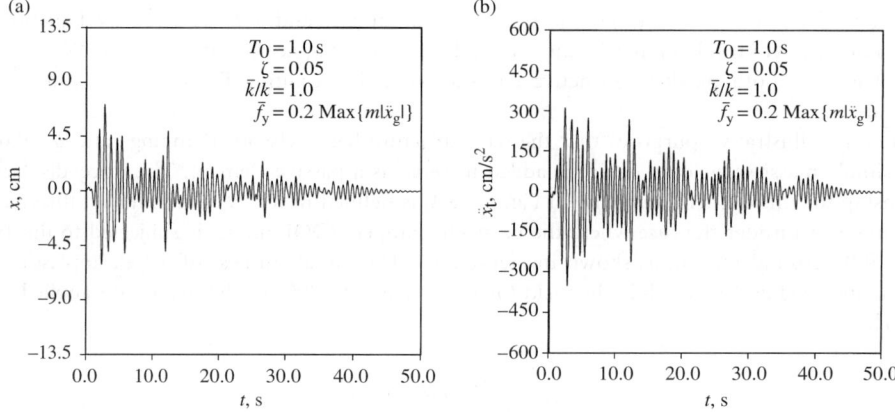

FIGURE 35.6   1940 El Centro time history response for SDOF structure with passive element: (a) displacement and (b) acceleration.

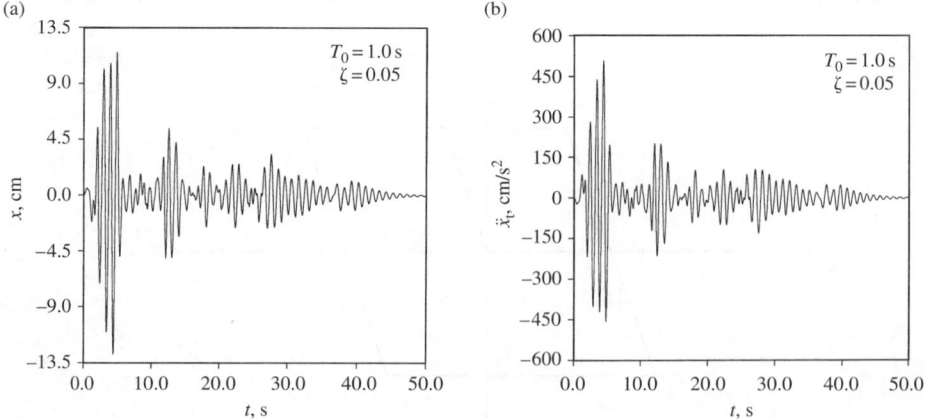

FIGURE 35.7   1940 El Centro time history response for SDOF structure without passive element: (a) displacement and (b) acceleration.

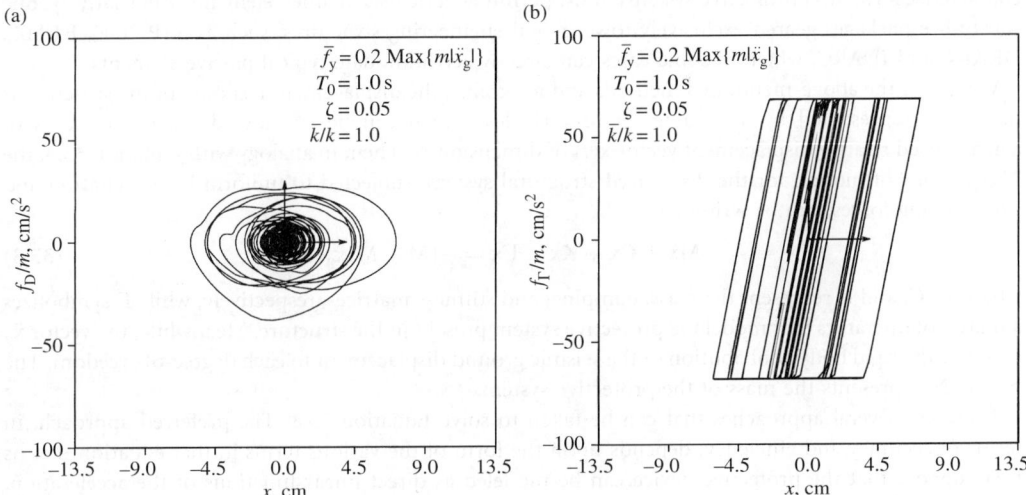

FIGURE 35.8 1940 El Centro force–displacement response for SDOF structure with passive element: (a) viscous element and (b) passive element.

elements are displayed in Figure 35.8. In this case, the size of these loops indicates that a significant portion of the energy is dissipated in the control device. This tends to reduce the forces and displacements in the primary structural elements, which of course is the purpose of adding the control device.

35.2.2 Multi-Degree-of-Freedom Structural Systems

In light of the preceding arguments, it becomes imperative to accurately characterize the behavior of any control device by constructing a suitable model under time-dependent loading. Multiaxial representations may be required. Once that model is established for a device, it must be properly incorporated into a mathematical idealization of the overall structure. Seldom is it sufficient to employ an SDOF idealization for an actual structure. Thus, in the present subsection, the formulation for dynamic analysis is extended to a MDOF representation.

The finite element method (FEM) (e.g., Zienkiewicz and Taylor 1989) currently provides the most suitable basis for this formulation. From a purely physical viewpoint, each individual structural member is represented mathematically by one or more finite elements having the same mass, stiffness, and damping characteristics as the original member. Beams and columns are represented by one-dimensional elements, while shear walls and floor slabs are idealized by employing two-dimensional finite elements. For more complicated or critical structural components, complete three-dimensional models can be developed, and incorporated into the overall structural model in a straightforward manner via substructuring techniques.

The FEM actually was developed largely by civil engineers in the 1960s from this physical perspective. However, during the ensuing decades the method has also been given a rigorous mathematical foundation, thus permitting the calculation of error estimates and the utilization of adaptive solution strategies (e.g., Szabo and Babuska 1991). Additionally, FEM formulations can now be derived from variational principles or Galerkin weighted residual procedures. Details of these formulations is beyond our scope. However, it should be noted that numerous general-purpose finite element software packages currently exist to solve the structural dynamics problem, including ABAQUS, ADINA, ANSYS, and NASTRAN. While none of these programs specifically addresses the special formulations needed to characterize structural protective systems, most permit generic user-defined elements. Alternatively, one can utilize packages geared exclusively toward civil engineering structures, such as SAP 2000, ETABS, DRAIN, and IDARC, which in some cases can already accommodate typical passive elements.

Via any of the above-mentioned methods and programs, the displacement response of the structure is ultimately represented by a discrete set of variables, which can be considered the components of a generalized relative displacement vector $\mathbf{x}(t)$ of dimension N. Then, in analogy with Equation 35.3, the N equations of motion for the discretized structural system, subjected to uniform base excitation and time varying forces, can be written as

$$\mathbf{M}\ddot{\mathbf{x}} + \mathbf{C}\dot{\mathbf{x}} + \mathbf{K}\mathbf{x} + \Gamma\mathbf{x} = -(\mathbf{M} + \bar{\mathbf{M}})\ddot{\mathbf{x}}_g \tag{35.8}$$

where $\mathbf{M}$, $\mathbf{C}$, and $\mathbf{K}$ represent the mass, damping, and stiffness matrices, respectively, while Γ symbolizes a matrix of operators that model the protective system present in the structure. Meanwhile, the vector $\ddot{\mathbf{x}}_g$ contains the rigid body contribution of the seismic ground displacement to each degree-of-freedom. The matrix $\bar{\mathbf{M}}$ represents the mass of the protective system.

There are several approaches that can be taken to solve Equation 35.8. The preferred approach, in terms of accuracy and efficiency, depends upon the form of the various terms in that equation. Let us first suppose that the protective device can be modeled as direct linear functions of the acceleration, velocity, and displacement vectors. That is,

$$\Gamma\mathbf{x} = \bar{\mathbf{M}}\ddot{\mathbf{x}} + \bar{\mathbf{C}}\dot{\mathbf{x}} + \bar{\mathbf{K}}\mathbf{x} \tag{35.9}$$

Then, Equation 35.8 can be rewritten as

$$\hat{\mathbf{M}}\ddot{\mathbf{x}} + \hat{\mathbf{C}}\dot{\mathbf{x}} + \hat{\mathbf{K}}\mathbf{x} = -\hat{\mathbf{M}}\ddot{\mathbf{x}}_g \tag{35.10}$$

in which

$$\hat{\mathbf{M}} = \mathbf{M} + \bar{\mathbf{M}} \tag{35.11a}$$

$$\hat{\mathbf{C}} = \mathbf{C} + \bar{\mathbf{C}} \tag{35.11b}$$

$$\hat{\mathbf{K}} = \mathbf{K} + \bar{\mathbf{K}} \tag{35.11c}$$

Equation 35.10 is now in the form of the classical matrix structural dynamic analysis problem. In the simplest case, which we will now assume, all of the matrix coefficients associated with the primary structure and the passive elements are constant. As a result, Equation 35.10 represents a set of N linear second-order ordinary differential equations with constant coefficients. These equations are, in general, coupled. Thus, depending upon N, the solution of Equation 35.10 throughout the time range of interest could become computationally demanding. This required effort can be reduced considerably if the equation can be uncoupled via a transformation; that is, if $\hat{\mathbf{M}}$, $\hat{\mathbf{C}}$, and $\hat{\mathbf{K}}$ can be diagonalized. Unfortunately, this is not possible for arbitrary matrices $\hat{\mathbf{M}}$, $\hat{\mathbf{C}}$, and $\hat{\mathbf{K}}$. However, with certain restrictions on the damping matrix $\hat{\mathbf{C}}$, the transformation to modal coordinates accomplishes the objective via the modal superposition method (see, e.g., Clough and Penzien 1975).

As mentioned earlier, it is more common having $\Gamma\mathbf{x}$ in Equation 35.9 nonlinear in $\mathbf{x}$ for a variety of passive and active control elements. Consequently, it is important to develop alternative numerical approaches and design methodologies applicable to more generic passively or actively damped structural systems governed by Equation 35.8. Direct time-domain numerical integration algorithms are most useful in that regard. The Newmark beta algorithm, for example, is one of these algorithms and is used extensively in structural dynamics.

35.2.3 Energy Formulations

In Section 35.2.1 and Section 35.2.2, we have considered SDOF and MDOF structural systems. The primary thrust of our analysis procedures has been the determination of displacements, velocities, accelerations, and forces. These are the quantities that, historically, have been of most interest. However, with the advent of innovative concepts for structural design, including structural protective systems, it is important to rethink current analysis and design methodologies. In particular, a focus on energy as a design criterion is conceptually very appealing. With this approach, the engineer is concerned, not so much with the resistance to lateral loads but rather, with the need to dissipate the energy input into the structure from environmental disturbances. Actually, this energy concept is not new. Housner (1956) suggested an energy-based design approach even for more traditional structures several decades ago. The resulting formulation is quite appropriate for a general discussion of energy dissipation in structures equipped with structural protective systems.

In what follows, an energy formulation is developed for an idealized structural system, which may include one or more control devices. The energy concept is ideally suited for application to nontraditional structures employing control elements, since for these systems proper energy management is a key to successful design. To conserve space, only SDOF structural systems are considered. However, these ideas can be easily generalized to MDOF systems.

Consider once again the SDOF oscillator shown in Figure 35.2 and governed by the equation of motion defined in Equation 35.1. An energy representation can be formed by integrating the individual force terms in Equation 35.1 over the entire relative displacement history. The result becomes

$$E_K + E_D + E_S = E_I \tag{35.12}$$

where

$$E_K = \int m\ddot{x}\,dx = \frac{m\dot{x}^2}{2} \tag{35.13a}$$

$$E_D = \int c\dot{x}\,dx = \int c\dot{x}^2\,dt \tag{35.13b}$$

$$E_S = \int kx\,dx = \frac{kx^2}{2} \tag{35.13c}$$

$$E_I = \int f\,dx \tag{35.13d}$$

The individual contributions included on the left-hand side of Equation 35.12 represent the relative kinetic energy of the mass (E_K), the dissipative energy caused by inherent damping within the structure (E_D), and the elastic strain energy (E_S). The summation of these energies must balance the input energy (E_I) imposed on the structure by the external disturbance. Note that each of the energy terms is actually a function of time and that the energy balance is required at each instant throughout the duration of the loading.

Consider aseismic design as a more representative case. It is unrealistic to expect that a traditionally designed structure will remain entirely elastic during a major seismic disturbance. Instead, inherent ductility of structures is relied upon to prevent catastrophic failure, while accepting the fact that some damage may occur. In such a case, the energy input (E_I) from the earthquake simply exceeds the capacity of the structure to store and dissipate energy by the mechanisms specified in Equations 35.13a–c. Once this capacity is surpassed, portions of the structure typically yield or crack. The stiffness is then no longer a constant and the spring force in Equation 35.1 must be replaced by a more general functional relation $g_S(x)$, which will commonly incorporate hysteretic effects. In general, Equation 35.13c is redefined as follows for inelastic response:

$$E_S = \int g_S(x)\,dx = E_{S_e} + E_{S_p} \tag{35.14}$$

in which E_S is assumed separable into additive contributions E_{S_e} and E_{S_p}, representing the fully recoverable elastic strain energy and the dissipative plastic strain energy, respectively.

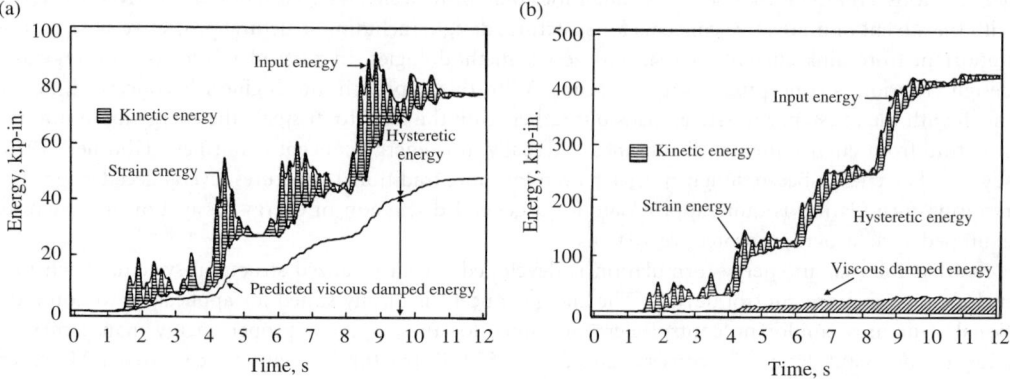

FIGURE 35.9   Energy response of traditional structure: (a) damageability limit state and (b) collapse limit state (Uang and Bertero 1986).

Figure 35.9a provides the energy response of a 0.3-scale, six-story concentrically braced steel structure as measured by Uang and Bertero (1986). The seismic input consisted of the 1978 Miyagi-Ken-Oki Earthquake signal scaled to produce a peak shaking table acceleration of 0.33g, which was deemed to represent the damageability limit state of the model. At this level of loading, a significant portion of the energy input to the structure is dissipated, with both viscous damping and inelastic hysteretic mechanisms having substantial contributions. If the intensity of the signal is elevated, an even greater share of the energy is dissipated via inelastic deformation. Finally, for the collapse limit state of this model structure at 0.65g peak table acceleration, approximately 90% of the energy is consumed by hysteretic phenomena, as shown in Figure 35.9b. Evidently, the consumption of this quantity of energy has destroyed the structure.

From an energy perspective, then, for proper aseismic design, one must attempt to minimize the amount of hysteretic energy dissipated by the structure. There are basically two viable approaches available. The first involves designs that result in a reduction in the amount of energy input to the structure. Base isolation systems and some active control systems, for example, fall into that category. The second approach, as in the passive and semiactive control system cases, focuses on the introduction of additional energy dissipating mechanisms into the structure. These devices are designed to consume a portion of the input energy, thereby reducing damage to the main structure caused by hysteretic dissipation. Naturally, for a large earthquake, the devices must dissipate enormous amounts of energy.

The SDOF system with a control element is displayed in Figure 35.3, while the governing integro-differential equation is provided in Equation 35.3. After integrating with respect to x, an energy balance equation can be written

$$E_K + E_D + E_{S_e} + E_{S_p} + E_C = E_I \tag{35.15}$$

where the energy associated with the control element is

$$E_C = \int \Gamma x \, dx \tag{35.16}$$

and the other terms are as previously defined.

As an example of the effects of control devices on the energy response of a structure, consider the tests of a one-third scale three-story lightly reinforced concrete framed building conducted by Lobo et al. (1993). Figure 35.10a displays the measured response of the structure due to the scaled 1952 Taft

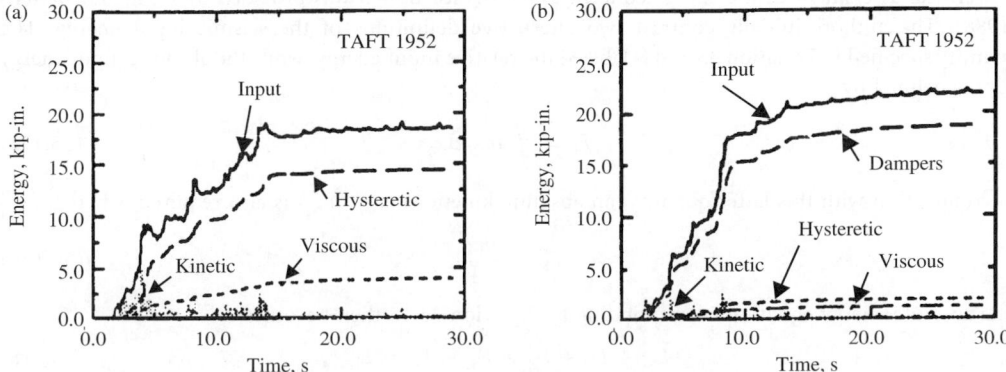

FIGURE 35.10 Energy response of test structure: (a) without passive devices and (b) with passive devices (Lobo et al. 1993).

N21E earthquake signal normalized for peak ground accelerations of 0.20*g*. A considerable portion of the input energy is dissipated via hysteretic mechanisms, which tend to damage the primary structure through cracking and the formation of plastic hinges. On the other hand, damage is minimal with the addition of a set of viscoelastic braced dampers. The energy response of the braced structure, due to the same seismic signal, is shown in Figure 35.10b. Notice that although the input energy has increased slightly, the dampers consume a significant portion of the total, thus protecting the primary structure.

35.2.4 Energy-Based Design

While the energy concept, as outlined briefly above, does not currently provide the basis for aseismic design codes, there is a considerable body of knowledge that has been developed from its application to traditional structures. Housner (1956, 1959) was the first to propose an energy-based philosophy for earthquake resistant design. In particular, he was concerned with limit-design methods aimed toward preventing collapse of structures in seismically active regions. Housner assumed that the energy input calculated for an undamped, elastic idealization of a structure provided a reasonable upper bound to that for the actual inelastic structure.

Berg and Thomaides (1960) examined the energy consumption in SDOF elastoplastic structures via numerical computation and developed energy input spectra for several strong-motion earthquakes. These spectra indicate that the amount of energy E_I imparted to a structure from a given seismic event is quite dependent upon the structure itself. The mass, the natural period of vibration, the critical damping ratio, and yield force level were all found to be important characteristics.

On the other hand, their results did suggest that the establishment of upper bounds for E_I might be possible and thus provided support for the approach introduced by Housner. However, the energy approach was largely ignored for a number of years. Instead, limit-state design methodologies were developed that utilized the concept of displacement ductility to construct inelastic response spectra as proposed initially by Veletsos and Newmark (1960).

More recently, there has been a resurgence of interest in energy-based concepts. For example, Zahrah and Hall (1982) developed a MDOF energy formulation and conducted an extensive parametric study of energy absorption in simple structural frames. Their numerical work included a comparison between energy-based and displacement ductility-based assessments of damage, but the authors stopped short of issuing a general recommendation.

A critical assessment of the energy concept as a basis for design was provided by Uang and Bertero (1988). The authors initially contrast two alternative definitions of the seismic input energy. The quantity specified in Equation 35.13d is labeled the relative input energy, while the absolute input energy (E_{I_a}) is defined by

$$E_{I_a} = \int m\ddot{x}_t \, dx_g \tag{35.17a}$$

In conjunction with this latter quantity, an absolute kinetic energy (E_{K_a}) is also required, where

$$E_{K_a} = \frac{m\dot{x}_t^2}{2} \tag{35.17b}$$

The absolute energy equation corresponding to Equation 35.15 then becomes

$$E_{K_a} + E_D + E_{S_e} + E_{S_p} + E_C = E_{I_a} \tag{35.18}$$

Based upon the development of input energy spectra for an SDOF system, the authors conclude that, while both measures produce approximately equivalent spectra in the intermediate period range, E_{I_a} should be used as a damage index for short period structures and E_I is more suitable for long period structures. Furthermore, an investigation revealed that the assumption of Housner to employ the idealized elastic strain energy, as an estimate of the actual input energy, is not necessarily conservative. Uang and Bertero also studied a MDOF structure and concluded that the input energy spectra for an SDOF can be used to predict the input energy demand for that type of building. In a second portion of the report, an investigation was conducted on the validity of the assumption that energy dissipation capacity can be used as a measure of damage. In testing cantilever steel beams, reinforced concrete shear walls, and composite beams the authors found that damage depends upon the load path.

The last observation should come as no surprise to anyone familiar with classical failure criteria. However, it does highlight a serious shortcoming for the use of the energy concept for limit design of traditional structures. As was noted above, in these structures, a major portion of the input energy must be dissipated via inelastic deformation, but damage to the structure is not determined simply by the magnitude of the dissipated energy. On the other hand, in nontraditional structures incorporating passive damping mechanisms, the energy concept is much more appropriate. The emphasis in design is directly on energy dissipation. Furthermore, since an attempt is made to minimize the damage to the primary structure, the selection of a proper failure criterion is less important.

35.3 Recent Development and Applications

As a result of serious efforts that have been undertaken in recent years to develop and implement the concept of passive energy dissipation and active control, a number of these devices have been installed in structures throughout the world, including Japan, New Zealand, Italy, Mexico, Canada, and the United States. In what follows, advances in terms of their development and applications are summarized.

35.3.1 Passive Energy Dissipation

As alluded to in Section 35.1 and Table 35.1, a number of passive energy dissipation devices have been developed and installed in structures for performance enhancement under wind or earthquake loads. Discussions presented below are centered around some of the more common devices that have found applications in these areas.

35.3.1.1 Metallic Yield Dampers

One of the effective mechanisms available for the dissipation of energy input to a structure from an earthquake is through inelastic deformation of metals. The idea of utilizing added metallic energy dissipators within a structure to absorb a large portion of the seismic energy began with the conceptual and experimental work of Kelly et al. (1972) and Skinner et al. (1975). Several of the devices considered included torsional beams, flexural beams, and U-strip energy dissipators. During the ensuing years, a wide variety of such devices have been studied or tested (Tyler 1985; Bergman and Goel 1987; Whittaker et al. 1991; Tsai et al. 1993). Many of these devices use mild steel plates with triangular or X shapes so that yielding is spread almost uniformly throughout the material. A typical X-shaped plate damper or added damping and stiffness (ADAS) device is shown in Figure 35.11. Other materials, such as lead and shape-memory alloys, have also been evaluated (Aiken and Kelly 1992). Some particularly desirable features of these devices are their stable hysteretic behavior, low-cycle fatigue property, long-term reliability, and relative insensitivity to environmental temperature. Hence, numerous analytical and experimental investigations have been conducted to determine these characteristics of individual devices.

After gaining confidence in their performance based primarily on experimental evidence, implementation of metallic devices in full-scale structures has taken place. The earliest implementations of metallic dampers in structural systems occurred in New Zealand and Japan. A number of these interesting applications are reported in Skinner et al. (1980) and Fujita (1991). More recent applications include the use of ADAS dampers in seismic upgrades of existing buildings in Mexico (Martinez-Romero 1993) and in the United States (Perry et al. 1993). The seismic upgrade project discussed in Perry et al. (1993) involves the retrofit of the Wells Fargo Bank building in San Francisco, CA. The building is a two-story nonductile concrete frame structure originally constructed in 1967 and subsequently damaged in the 1989 Loma Prieta earthquake. The voluntary upgrade by Wells Fargo utilized chevron braces and ADAS damping elements. More conventional retrofit schemes were rejected due to an inability to meet the performance objectives while avoiding foundation work. A plan view of the second floor including upgrade details is provided in Figure 35.12. A total of seven ADAS devices were employed, each with a yield force of 150 kps. Both linear and nonlinear analyses were used in the retrofit design process. Further three-dimensional response spectrum analyses, using an approximate equivalent linear representation for the ADAS elements, furnished a basis for the redesign effort. The final design was verified with DRAIN-2D nonlinear time history analyses. A comparison of computed response before and after the upgrade is contained in Figure 35.13. The numerical results indicated that the revised design was

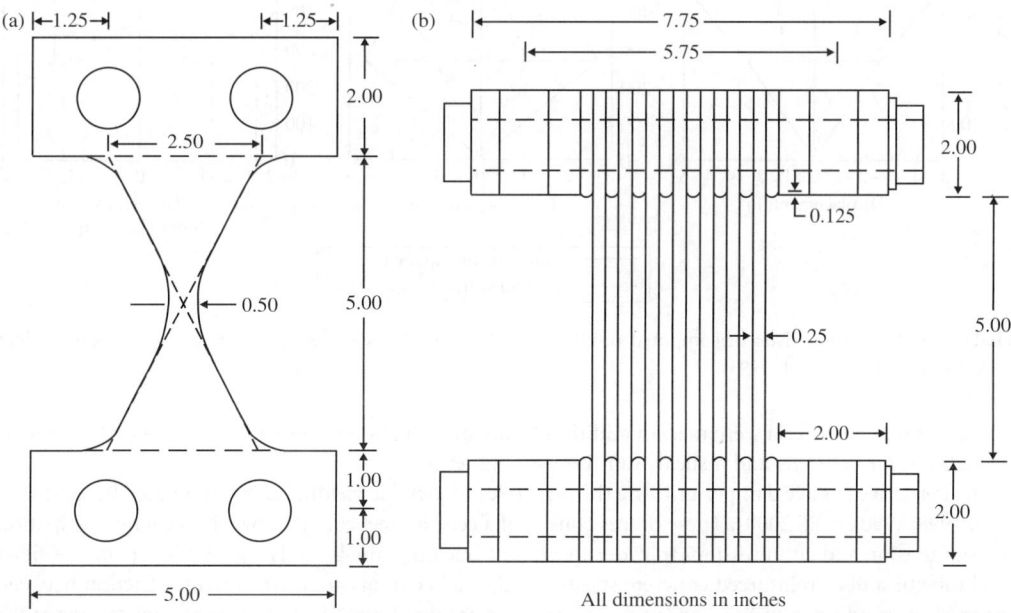

FIGURE 35.11 ADAS device (Whittaker et al. 1991).

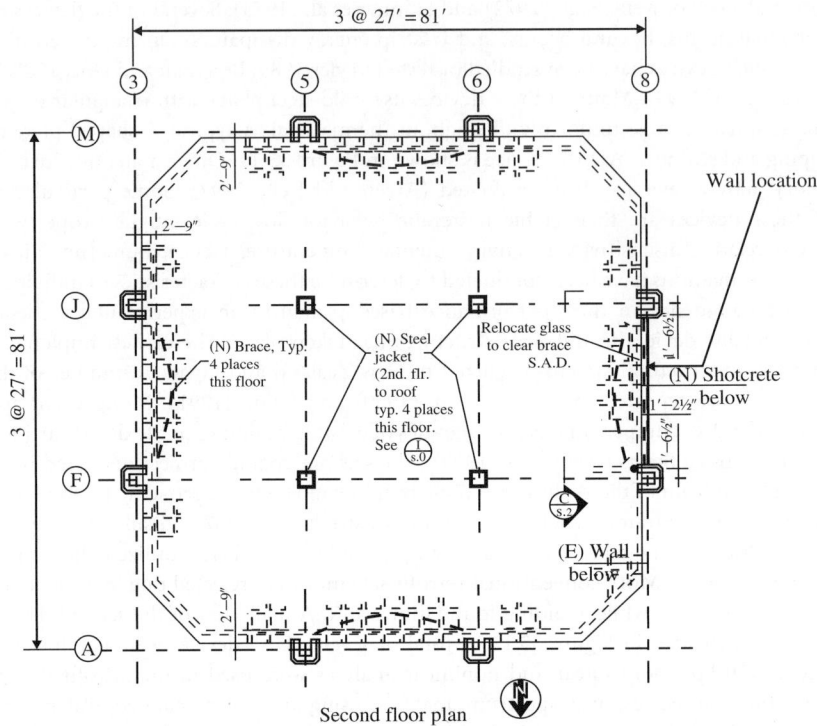

FIGURE 35.12 Wells Fargo Bank building retrofit details (Perry et al. 1993).

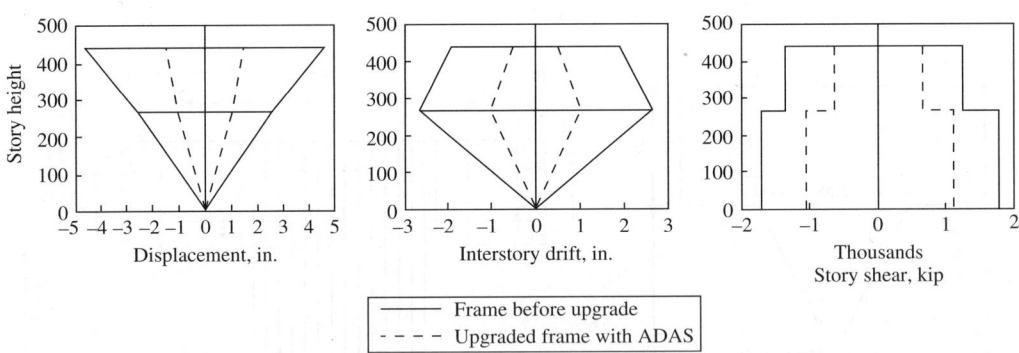

FIGURE 35.13 Comparison of computed results for Wells Fargo Bank Building — envelope of response values in X direction (Perry et al. 1993).

stable and that all criteria were met. In addition to the introduction of the bracing and ADAS dampers, several interior columns and a shear wall were strengthened.

An alternative passive damper utilizing the principles of metallic yielding is the unbonded brace (Shiba et al. 1998; Wada et al. 2000). These braces consist of a central steel core that provides energy dissipation in both tension and compression. In order to prevent buckling, the brace is encased in a concrete-filled steel tube or a fiber-reinforced concrete sheathing. Special coatings are used to reduce friction between the steel core and surrounding concrete. Despite its recent development, a number of applications of the unbonded brace have already appeared in Japan and the United States.

35.3.1.2 Friction Dampers

Friction dampers utilize the mechanism of solid friction that develops between two solid bodies sliding relative to one another to provide the desired energy dissipation. Several types of friction dampers have been developed for the purpose of improving seismic response of structures. In some cases, additional research on the long-term behavior of frictional systems may be appropriate.

A simple brake lining frictional system was studied by Pall et al. (1980); however, a special damper mechanism, devised by Pall and Marsh (1982), and depicted in Figure 35.14, permits much more effective operation. During cyclic loading, the mechanism tends to straighten buckled braces and also enforces slippage in both tensile and compressive directions.

Several alternate friction damper designs have also been proposed in the literature. For example, Roik et al. (1988) discuss the use of three-stage friction-grip elements. A simple conceptual design, the slotted bolted connection (SBC), was investigated by FitzGerald et al. (1989) and Grigorian et al. (1993). Another design of a friction damper is the energy dissipating restraint (EDR) manufactured by Fluor Daniel, Inc. There are several novel aspects of the EDR that combine to produce very different response characteristics. A detailed presentation of the design and its performance is provided in Nims et al. (1993).

In the last decade, there have been several commercial applications of friction dampers aimed at providing enhanced seismic protection of new and retrofitted structures. This activity in North America is primarily associated with the use of Pall friction devices in Canada. For example, the applications of friction dampers to the McConnel Library of the Concordia University in Montreal, Canada, is discussed in Pall and Pall (1993). A total of 143 dampers were employed in this case. Interestingly, the architects chose to expose 60 of the dampers to view due to their esthetic appeal. A series of nonlinear DRAIN-TABS (Guendeman-Israel and Powell 1977) analyses were utilized to establish the optimum slip load for the devices, which ranges from 600 to 700 kN depending upon the location within the structure. For the three-dimensional time-history analyses, artificial aseismic signals were generated with a wide range of frequency contents and a peak ground acceleration scaled to 0.18*g* to represent expected ground motion in Montreal. Under this level of excitation, an estimate of the equivalent damping ratio for the structure with frictional devices is approximately 50%. In addition, for this library complex, the use of the friction dampers resulted in a net savings of 1.5% of the total building cost.

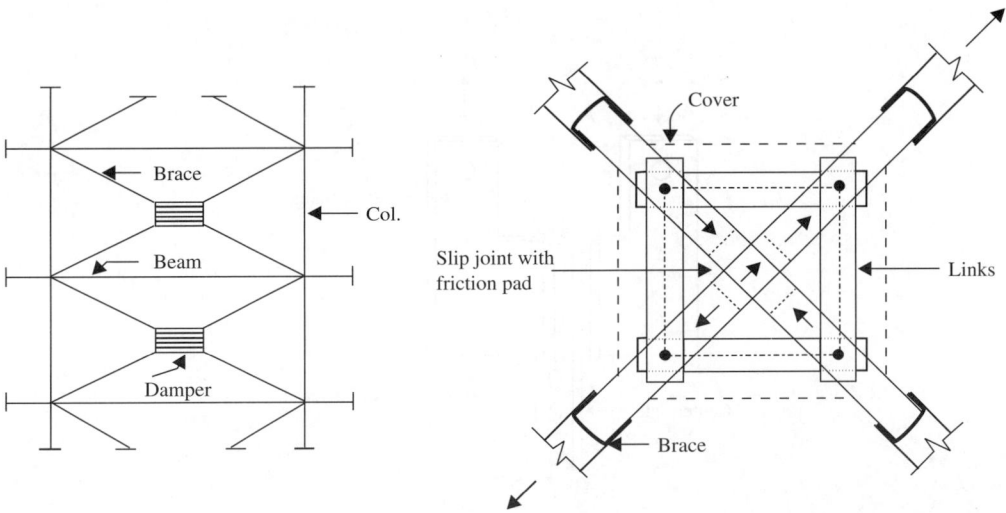

FIGURE 35.14   X braced friction damper (Pall and Marsh 1982).

35.3.1.3 Viscoelastic Dampers

Viscoelastic (VE) materials used in structural applications are usually copolymers or glassy substances that dissipate energy through shear deformation. A typical VE damper, which consists of VE layers bonded with steel plates, is shown in Figure 35.15. When mounted in a structure, shear deformation and hence energy dissipation takes place when structural vibration induces relative motion between the outer steel flanges and the center plates. Significant advances in research and development of VE dampers, particularly for seismic applications, have been made in recent years through analyses and experimental tests (e.g., Chang et al. 1994; Lai et al. 1995; Shen et al. 1995).

The first applications of VE dampers to structures were for reducing acceleration levels, or increasing human comfort, due to wind. In 1969, VE dampers were installed in the twin towers of the World Trade Center in New York, NY, as an integral part of the structural system. There were about 10,000 VE dampers in each tower, evenly distributed throughout the structure from the 10th to the 110th floor. Prior to their collapse, the towers experienced a number of moderate to severe wind storms and the observed performance of the VE dampers has been found to agree well with theoretical values. In 1982, VE dampers were incorporated into the 76-story Columbia SeaFirst Building in Seattle against wind-induced vibrations (Keel and Mahmoodi 1986). To reduce the wind-induced vibration, the design called for 260 dampers to be located alongside the main diagonal members in the building core. The addition of VE dampers to this building was calculated to increase its damping ratio in the fundamental mode from 0.8 to 6.4% for frequent storms and to 3.2% at design wind. Similar applications of VE dampers were made to the Two Union Square Building in Seattle in 1988. In this case, 16 large VE dampers were installed parallel to four columns in one floor.

Seismic applications of VE dampers to structures began more recently. A seismic retrofit project using VE dampers began in 1993 for the 13-story Santa Clara County building in San Jose, CA (Crosby et al. 1994). Situated in a high seismic risk region, the building was built in 1976. It is approximately 64 m in height and nearly square in plan, with 51×51 m on typical upper floors. The exterior cladding consists of full-height glazing on two sides and metal siding on the other two sides. The exterior cladding, however, provides little resistance to structural drift. The equivalent viscous damping in the fundamental mode is less than 1% of critical.

The building has been extensively instrumented, providing invaluable response data obtained during a number of past earthquakes. A plan for seismic upgrade of the building was developed, in part, when the response data indicated large and long-duration response, including torsional coupling, to even moderate earthquakes. The final design called for installation of two dampers per building face per floor

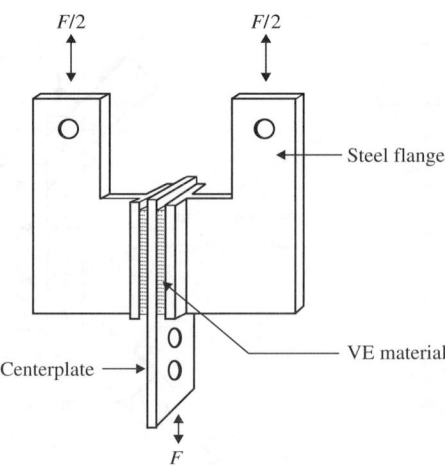

FIGURE 35.15 Typical VE damper configuration.

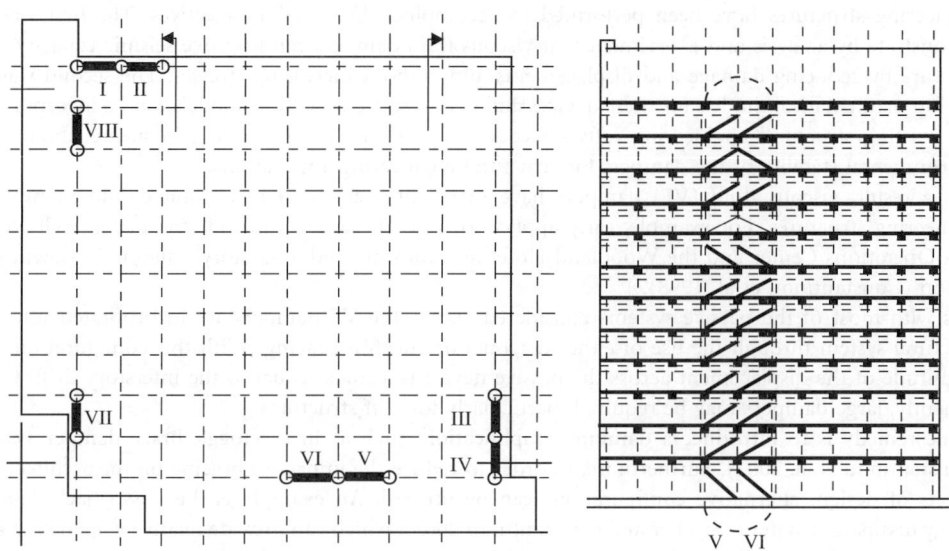

FIGURE 35.16 Location of VE dampers in Santa Clara County building.

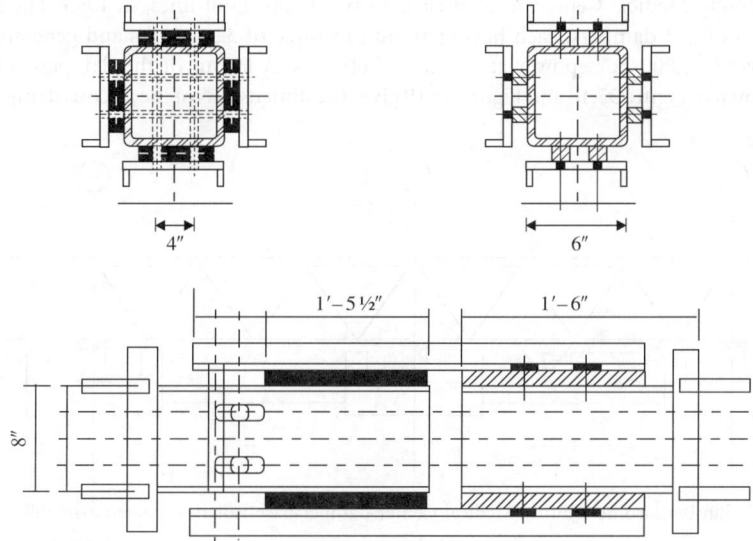

FIGURE 35.17 Santa Clara County building VE damper configuration.

level as shown in Figure 35.16, which would increase the equivalent damping in the fundamental mode of the building to about 17% of critical, providing substantial reductions to building response under all levels of ground shaking. A typical damper configuration is shown in Figure 35.17.

35.3.1.4 Viscous Fluid Dampers

Damping devices based on the operating principle of high-velocity fluid flow through orifices have found numerous applications in shock and vibration isolation of aerospace and defense systems. In the past decade, research and development of viscous fluid dampers for seismic applications to civil engineering structures have been performed to accomplish three major objectives. The first was to demonstrate by analysis and experiment that viscous fluid dampers can improve seismic capacity of a structure by reducing damage and displacements and without increasing stresses. The second was to develop mathematical models for these devices and demonstrate how these models can be incorporated into existing structural engineering software codes. Finally, the third was to evaluate reliability and environmental stability of the dampers for structural engineering applications.

As a result, viscous fluid (VF) dampers have in recent years been incorporated into many civil engineering structures. For example, early applications of VF dampers include the Pacific Bell North Area Operations Center and the Woodland Hotel in California and a 35-story building in downtown Boston (Constantinou et al. 1998).

As with most of the passive systems considered above, the VF dampers are incorporated into the structural system through the use of either diagonal or chevron bracing. With this configuration, the magnitude of the displacement across the passive device is at most equal to the interstory drift. Consequently, large dampers may be required, particularly for stiff structures.

One remedy is to introduce a damping amplification, such as in the toggle-brace damper system (Constantinou et al. 2001). Furthermore, in order to reduce the impact of bracing on the architectural aspects of design, alternative configurations can be utilized. An example is the scissor-jack-damper energy dissipation system (Sigaher and Constantinou 2003), which can provide relatively open bays and damping magnification factors between two and five.

In several other applications, VF dampers were used in combination with seismic isolation systems. For example, VF dampers were incorporated into base isolation systems for five buildings of the new San Bernardino County Medical Center, located close to two major fault lines, in 1995. The five buildings required a total of 233 dampers, each having an output force of 320,000 lbs and generating an energy dissipation level of 3,000 horsepower at a speed of 60 in./s. A layout of the damper-isolation system assembly is shown in Figure 35.18 and Figure 35.19 gives the dimensions of the viscous dampers employed.

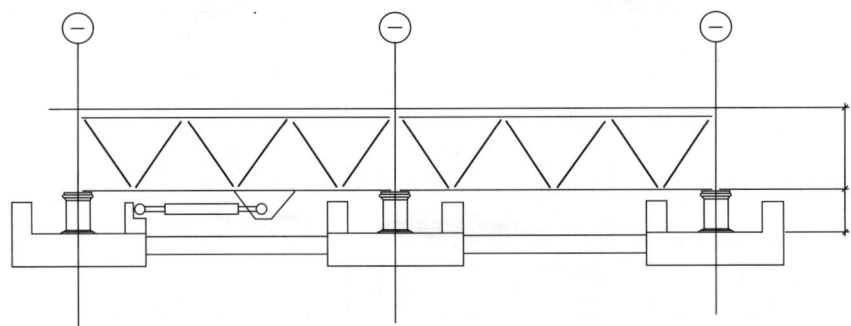

FIGURE 35.18   San Bernardino County Medical Center damper-base isolation system assembly.

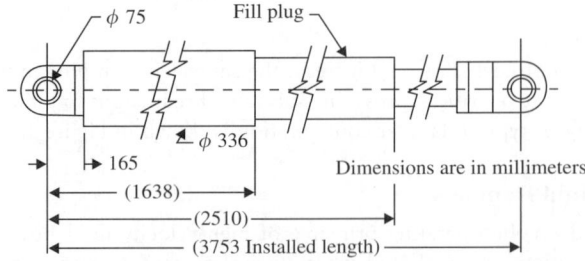

FIGURE 35.19   Dimensions of viscous fluid damper for San Bernardino County Medical Center.

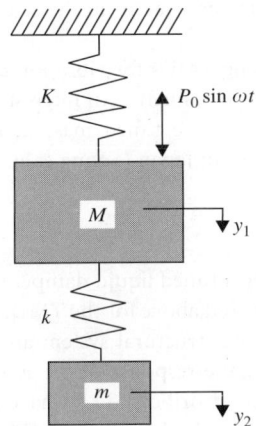

FIGURE 35.20 Undamped absorber and main mass subject to harmonic excitation (Frahm's absorber).

35.3.1.5 Tuned Mass Dampers

The modern concept of tuned mass dampers (TMDs) for structural applications has its roots in dynamic vibration absorbers studied as early as 1909 by Frahm (Den Hartog 1956). A schematic representation of Frahm's absorber is shown in Figure 35.20, which consists of a small mass m and a spring with spring stiffness k attached to the main mass M with spring stiffness K. Under a simple harmonic load, one can show that the main mass M can be kept completely stationary when the natural frequency ($\sqrt{k/m}$) of the attached absorber is chosen to be (or tuned to) the excitation frequency.

As in the case of VE dampers, early applications of TMDs have been directed toward mitigation of wind-induced excitations. It appears that the first structure in which a TMD was installed is the Centerpoint Tower in Sydney, Australia (ENR 1971; Kwok and MacDonald 1987). One of only two buildings in the United States equipped with a TMD is the 960-ft Citicorp Center in New York, NY, in which the TMD is situated on the 63rd floor. At this elevation, the building can be represented by a simple modal mass of approximately 20,000 tons, to which the TMD is attached to form a two-degrees-of-freedom system. Tests and actual observations have shown that the TMD produces an approximate effective damping of 4% as compared to the 1% original structural damping, which can reduce the building acceleration level by about 50% (Petersen 1980, 1981). The same design principles were followed in the development of the TMD for installation in the John Hancock Tower, Boston, MA (ENR 1975). In this case, however, the TMD consists of two 300-ton mass blocks. They move in phase to provide lateral response control and out-of-phase for torsional control.

Numerical and experimental studies have been carried out to examine the effectiveness of TMDs in reducing seismic response of structures. It is noted that a passive TMD can only be tuned to a single structural frequency. While the first-mode response of an MDOF structure with TMD can be substantially reduced, the higher-mode response may in fact increase as the number of stories increases. For earthquake-type excitations, it has been demonstrated that, for shear structures up to 12 floors, the first-mode response contributes more than 80% to the total motion (Wirsching and Yao 1970). However, for a taller building on a firm ground, higher modal response may be a problem that needs further study. In Villaverde (1994), three different structures were studied; the first is a two-dimensional ten-story shear building, the second a three-dimensional one-story frame building, and the third a three-dimensional cable-stayed bridge, by using nine different kinds of real earthquake records. Numerical and experimental results show that the effectiveness of TMDs on reducing the response of the same structure under different earthquakes or of different structures under the same earthquake is significantly different; some cases give good performance and some have little or even no effect. It implies that there is a dependency of the attained reduction in response on the characteristics of the ground motion that excites the structure. This response reduction is large for resonant ground motions and diminishes as the dominant frequency of the ground motion gets further away from the structure's natural frequency to which the TMD is tuned.

It is also noted that the interest in using TMDs for vibration control of structures under earthquake loads has resulted in some innovative developments. An interesting approach is the use of a TMD with active capability, the so-called AMD or active tuned mass damper. Systems of this type have been implemented in a number of tall buildings in Japan (Soong et al. 1994). Some examples of such systems will be discussed in Section 35.3.2.

35.3.1.6 Tuned Liquid Dampers

The basic principles involved in applying a tuned liquid damper (TLD) to reduce the dynamic response of structures is quite similar to that discussed above for the TMD. In effect, a secondary mass in the form of a body of liquid is introduced into the structural system and tuned to act as a dynamic vibration absorber. However, in the case of TLDs, the response of the secondary system is highly nonlinear due either to liquid sloshing or the presence of orifices. TLDs have also been used for suppressing wind-induced vibrations of tall structures. In comparison with TMDs, the advantages associated with TLDs include low initial cost, virtually free maintenance, and ease of frequency tuning.

It appears that TLD applications have taken place primarily in Japan. Examples of TLD-controlled structures include the Nagasaki Airport Tower, installed in 1987, the Yokohama Marine Tower, also installed in 1987, the Shin-Yokohama Prince Hotel, installed in 1992, and the Tokyo International Airport Tower, installed in 1993 (Tamura et al. 1994, 1995). The TLD installed in the 77.6 m Tokyo Airport Tower, for example, consists of about 1400 vessels containing water, floating particles, and a small amount of preservatives. The vessels, shallow circular cylinders 0.6 m in diameter and 0.125 m in height, are stacked in six layers on steel-framed shelves. The total mass of the TLD is approximately 3.5% of the first-mode generalized mass of the tower and its sloshing frequency is optimized at 0.743 Hz. Floating hollow cylindrical polyethylene particles were added in order to optimize energy dissipation through an increase in surface area together with collisions between particles.

The performance of the TLD has been observed during several storm episodes. In one of such episodes with a maximum instantaneous wind speed of 25 m/s, the observed results show that the TLD reduced the acceleration response in the cross-wind direction to about 60% of its value without the TLD.

35.3.2 Active Control

As mentioned in Section 35.1, the development of active, hybrid, and semiactive control systems has reached the stage of full-scale applications to actual structures. Since 1989, over 40 active or hybrid systems have been installed in building structures, primarily in Japan. In addition, a number of bridge towers have employed active systems during erection (Fujino 1994).

In recent years, significant effort has also been directed toward the development of semiactive control devices and systems. These systems permit the variation of device properties (e.g., stiffness, damping) based upon structural response. Often very little power is required to operate these devices and fail-safe passive modes are provided. Examples of semiactive devices include variable stiffness systems (Kobori et al. 1993; Liang et al. 1999a,b), variable fluid dampers (Patten et al. 1996; Symans and Constantinou 1997; Kurata et al. 1999), electrorheological dampers (Gavin et al. 1996a,b; Makris et al. 1996), and magnetorheological dampers (Spencer et al. 1997; Yang et al. 2002).

A complete review of active, hybrid, and semiactive systems will not be attempted here. Instead, described briefly below are two active control applications and their observed performances. The performances of these systems under recent wind and earthquake episodes are summarized in this section. More details of these applications can be found in Soong et al. (1994).

35.3.2.1 Sendagaya INTES Building

An AMD system was installed in the Sendagaya INTES building in Tokyo in 1991. As shown in Figure 35.21, the AMD was installed atop the 11th floor and consists of two masses to control transverse

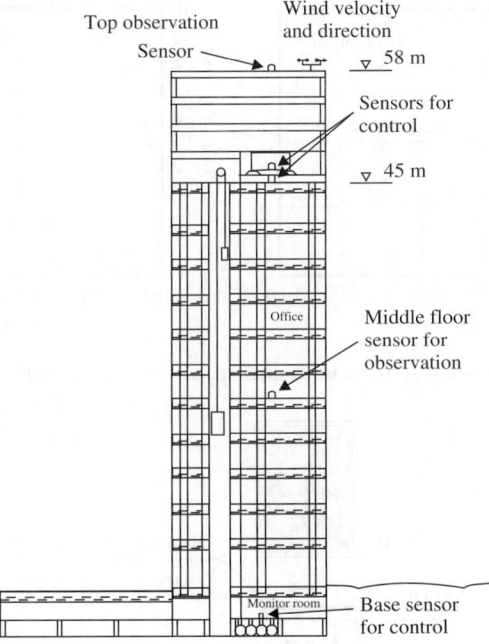

FIGURE 35.21   Sendagaya INTES building.

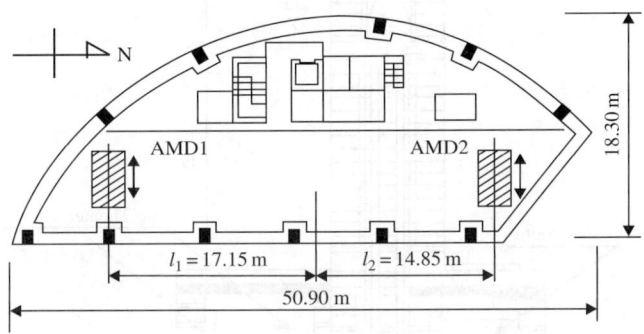

FIGURE 35.22   Top view of AMD in Sendagaya building.

and torsional motions of the structure while hydraulic actuators provide the active control capabilities. The top view of the control system is shown in Figure 35.22 where ice thermal storage tanks are used as mass blocks so that no extra mass is introduced. The masses are supported by multistage rubber bearings intended for reducing the control energy consumed in the AMD and for insuring smooth mass movements (Higashino and Aizawa 1993).

Sufficient data were obtained for evaluation of the AMD performance when the building was subjected to strong wind on March 29, 1993, with peak instantaneous wind speed of 30.6 m/s. An example of the response Fourier spectra using samples of 30-s durations are shown in Figure 35.23, showing good performance in the low-frequency range. The response of the fundamental mode was reduced by 18 and 28% for translation and torsion, respectively. Similar performance characteristics were observed during a series of earthquakes recorded between May 1992 and February 1993.

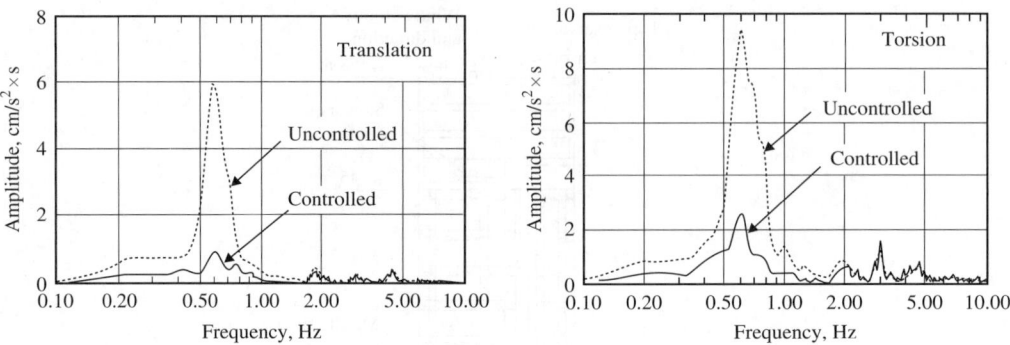

FIGURE 35.23 Sendagaya building — response Fourier spectra (March 29, 1993).

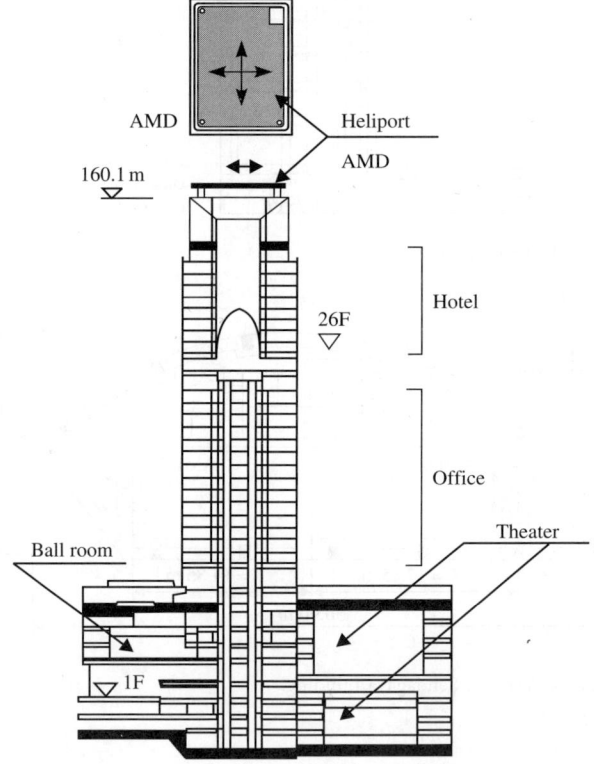

FIGURE 35.24 Hankyu Chayamachi building.

35.3.2.2 Hankyu Chayamachi Building

The 160-m 34-story Hankyu Chayamachi building, as shown in Figure 35.24, is located in Osaka, Japan, where an AMD system was installed in 1992 for the primary purpose of occupant comfort control. In this case, the heliport at the roof top is utilized as the moving mass of the AMD, which weighs 480 tons and is about 3.5% of the weight of the tower portion. The heliport is supported by six multistage rubber bearings. The natural period of rubber and heliport system was set to 3.6 s, slightly lower than that of the building (3.8 s). The AMD mechanism used here has the same architecture as that of Sendagaya INTES, namely, scheme of the digital controller, servomechanism, and the hydraulic design, except that two

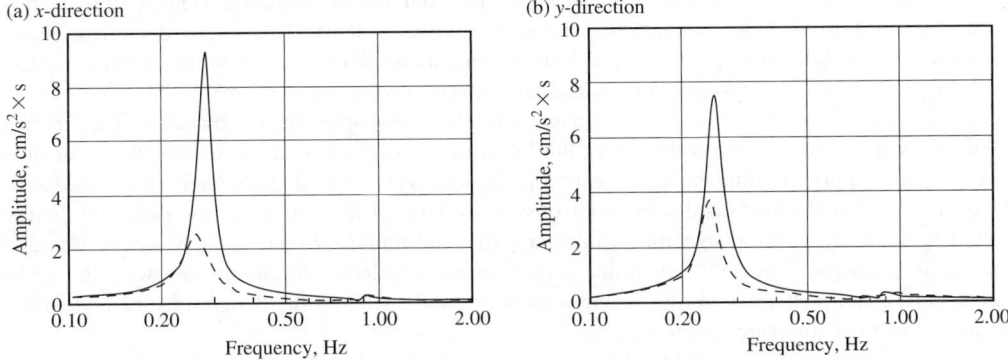

FIGURE 35.25   Hankyu Chayamachi building — acceleration Fourier spectra.

actuators of 5-ton thrusts are attached in horizontal orthogonal directions. Torsional control is not considered here.

Acceleration Fourier spectra during a recent typhoon are shown in Figure 35.25. Since the building in this case oscillated primarily in its fundamental mode, significant reductions in acceleration levels were observed.

An observation to be made in the performance of control systems such as those described above is that efficient active control systems can be implemented with existing technology under practical constraints such as power requirements and stringent demand of reliability. Thus, significant strides have been made considering that serious implementational efforts began less than 10 years prior to construction. On the other hand, the active dampers developed for Sendagaya INTES and Hankyu Chayamachi buildings were designed primarily for response control due to wind and moderate earthquakes. In order to reach the next level in active/hybrid control technology, an outstanding issue that needs to be addressed is whether such systems, with limited control resources and practical constraints such as mass excursions, can be made effective under strong earthquakes.

35.4   Code Development

Extensive efforts in the field of passive energy dissipation and the increased interest of the engineering profession in this area over the past decade has resulted in the development of guidelines and recommended provisions for the design and implementation of passive energy dissipation devices. A brief review of that development in the United States is provided in this section.

Initially, the Energy Dissipation Working Group (EDWG) of the Base Isolation Subcommittee of the Structural Engineers Association of Northern California (SEAONC) developed a document that provided design guidelines applicable to a wide range of system hardware (Whittaker et al. 1993). The scope included metallic, friction, viscoelastic, and viscous devices. On the other hand, TMDs and TLDs were not addressed. The general philosophy of that document was to confine inelastic deformation primarily to the energy dissipators, while the main structure was to remain elastic for the design basis earthquake. Furthermore, since passive energy dissipation technology was still relatively new, a conservative approach was taken on many issues. For example, an experienced independent engineering review panel was required to conduct a review of the energy dissipation system and testing programs.

The 1994 edition of the National Earthquake Hazard Reduction Program (NEHRP) Recommended Provisions for Seismic Regulations for New Buildings (FEMA 1995) then introduced an appendix on passive energy dissipation systems, which is similar to the SEAONC document in both scope and philosophy. Again, a conservative approach was adopted. However, the primary intention was to introduce passive energy dissipation technology to potential users, rather than to provide design requirements.

More recently, NEHRP Guidelines for the Seismic Rehabilitation of Buildings (FEMA 1997a,b) have been developed to include sections on seismic protective systems. This entire document uses a performance-based design approach in which displacements (or damage) control the design. Chapter 9 of these guidelines, and the associated commentary, specifically addresses seismic isolation and passive energy dissipation systems. The passive systems may involve displacement-dependent (e.g., metallic yielding, friction), velocity-dependent (e.g., fluid viscous, viscoelastic solid, viscoelastic fluid), or other (e.g., shape memory, friction-spring re-centering) devices. Very limited guidance is also provided for dynamic vibration absorbers and active control systems. Four distinct analysis procedures are defined, including linear static, linear dynamic, nonlinear static, and nonlinear dynamic approaches. However, the linear procedures are only applicable under certain restrictive conditions. As with the earlier documents, a comprehensive testing program must be conducted and the design must be reviewed by an independent engineering review panel.

For new construction, the FEMA (1995) document mentioned above has now been superceded by FEMA (2001), which includes an appendix to Chapter 13 on structures with damping systems. As stated in the preface to that appendix, it is intended for trial use by design professionals and other interested parties. Both displacement-dependent and velocity-dependent passive devices are considered. Along with non-linear static and dynamic analysis procedures, this appendix includes an equivalent lateral force analysis procedure and a response spectrum analysis procedure. (Further details on the theoretical basis for these two new linear methods can be found in Ramirez et al. 2000.) A prototype test program, quality control plan, and independent review panel are again required elements in these recommended provisions.

35.5 Concluding Remarks

An attempt is made in this section to introduce the basic concepts of passive energy dissipation and active control, and to review current development, structural applications, and code-related activities in this exciting and fast expanding field. While significant strides have been made in terms of implementation of these concepts to structural design and retrofit, it should be emphasized that this entire technology is still evolving. Significant improvements in both hardware and design procedures will certainly continue for a number of years to come.

The acceptance of innovative systems in structural engineering is based on a combination of performance enhancement versus construction costs and long-term effects. Continuing efforts are needed in order to facilitate wider and more timely implementation. These include effective system integration and further development of analytical and experimental techniques by which performances of these systems can be realistically assessed. Structural systems are complex combinations of individual structural components. New innovative devices need to be integrated into these complex systems, with realistic evaluation of their performance and impact on the structural system, as well as verification of their ability of long-term operation. Additionally, innovative ideas for new devices require exploration through experimentation and adequate basic modeling. A series of standardized benchmark structural models, representing large buildings, bridges, towers, lifelines, etc., with standardized realistically scaled-down excitations, representing natural hazards, can be of significant value in helping to provide an experimental and analytical testbed for proof-of-concept of existing and new devices. In the United States, the Network for Earthquake Engineering Simulation also can be expected to have a significant impact on the further development of structural protective systems.

References

Aiken, I.D. and Kelly, J.M. 1992. Comparative Study of Four Passive Energy Dissipation Systems. *Bull. N.Z. Nat. Soc. Earthquake Eng.*, 25(3):175–192.

ATC. 1993. *Proceedings on Seismic Isolation, Passive Energy Dissipation, and Active Control*, ATC-17-1, Appl. Tech. Council, Redwood City, CA.

Berg, G.V. and Thomaides, S.S. 1960. Energy Consumption by Structures in Strong Motion Earthquakes, *Proc. 2nd World Conf. on Earthquake Engrg.*, II:681–697, Tokyo.

Bergman, D.M. and Goel, S.C. 1987. *Evaluation of Cyclic Testing of Steel-Plate Devices for Added Damping and Stiffness*, Report No. UMCE 87-10, The University of Michigan, Ann Arbor, MI.

Chang, K.C., Shen, K.L., Soong, T.T., and Lai, M.L. 1994. Seismic Retrofit of a Concrete Frame with Added Viscoelastic Dampers, *5th Nat. Conf. on Earthquake Eng.*, Chicago, IL.

Clough, R.W. and Penzien, J. 1975. *Dynamics of Structures*, McGraw-Hill, New York.

Constantinou, M.C., Soong, T.T., and Dargush, G.F. 1998. *Passive Energy Dissipation Systems for Structural Design and Retrofit*, Monograph No. 1, Multidisciplinary Center for Earthquake Engineering Research, Buffalo, NY.

Constantinou, M.C., Tsopelas, P., Hammel, W., and Sigaher, A.N. 2001. Toggle-Brace-Damper Seismic Energy Dissipation Systems, *J. Struct. Eng.*, ASCE, 127:105–112.

Crosby, P., Kelly, J.M., and Singh, J. 1994. Utilizing Viscoelastic Dampers in the Seismic Retrofit of a Thirteen Story Steel Frame Building, *Struct. Congress XII*, Atlanta, GA, pp. 1286–1291.

Den Hartog, J.P. 1956. *Mechanical Vibrations*, 4th Edition, McGraw-Hill, New York.

Dyke, S.J., Spencer Jr., B.F., Quast, P., Sain, M.K., Kaspari Jr., D.C., and Soong, T.T. 1994. *Experimental Verification of Acceleration Feedback Control Strategies for an Active Tendon System*, NCEER-94-0024, Buffalo, NY.

Dyke, S.J., Spencer Jr., B.F., Quast, P., and Sain, M.K. 1995. The Role of Control–Structure Interaction in Protective System Design, *J. Eng. Mech.*, ASCE, 121(2):322–338.

ENR. 1971. Tower Cables Handle Wind, Water Tank Dampens It, *Eng. News-Record*, December 9, 23.

ENR. 1975. Hancock Tower Now to Get Dampers, *Eng. News-Record*, October 30, 11.

FEMA. 1995. *1994 NEHRP Recommended Provisions for Seismic Regulations for New Buildings*, Report FEMA 222A, Washington, DC.

FEMA. 1997a. *NEHRP Guidelines for the Seismic Rehabilitation of Buildings*, Report FEMA 273, Building Seismic Safety Council.

FEMA. 1997b. *Commentary to NEHRP Guidelines for the Seismic Rehabilitation of Buildings*, Report FEMA 274, Building Seismic Safety Council.

FEMA. 2001. *NEHRP Recommended Provisions for Seismic Regulations for New Buildings and Other Structures*, 2000 Edition, Report FEMA 356, Building Seismic Safety Council.

FitzGerald, T.F., Anagnos, T., Goodson, M., and Zsutty, T. 1989. Slotted Bolted Connections in Aseismic Design for Concentrically Braced Connections, *Earthquake Spectra*, 5(2):383–391.

Fujino, Y. 1994. Recent Research and Developments on Control of Bridges under Wind and Traffic Excitations in Japan, *Proc. Int. Workshop Struct. Control*, pp. 144–150.

Fujita, T. (Ed.) 1991. Seismic Isolation and Response Control for Nuclear and Non-Nuclear Structures, Special Issue for the Exhibition of *11th Int. Conf. on SMiRT*, Tokyo, Japan.

Gavin, H.P., Hanson, R.D., and Filisko, F.E. 1996a. Electrorheological Dampers I: Analysis and Design, *ASME J. Appl. Mech.*, 63:669–675.

Gavin, H.P., Hanson, R.D., and Filisko, F.E. 1996b. Electrorheological Dampers II: Testing and Modeling, *ASME J. Appl. Mech.*, 63:676–682.

Grigorian, C.E., Yang, T.S., and Popov, E.P. 1993. Slotted Bolted Connection Energy Dissipators, *Earthquake Spectra*, 9(3):491–504.

Guendeman-Israel, R. and Powell, G.H. 1977. *DRAIN-TABS — A Computerized Program for Inelastic Earthquake Response of Three Dimensional Buildings*, Report No. UCB/EERC 77-08, University of California, Berkeley, CA.

Hanson, R.D. and Soong, T.T. 2001. *Seismic Design with Supplemental Energy Dissipation Devices*, EERI Monograph No. 8, Earthquake Engineering Research Institute, Oakland, CA.

Higashino, M. and Aizawa, S. 1993. Application of Active Mass Damper System in Actual Buildings, G.W. Housner and S.F. Masri (Eds.), *Proc. Int. Workshop on Structural Control*, Los Angeles, CA, pp. 194–205.

Housner, G.W. 1956. Limit Design of Structures to Resist Earthquakes, *Proc. 1st World Conf. on Earthquake Eng.*, Earthquake Engrg. Research Center, Berkeley, CA, pp. 5-1–5-13.

Housner, G.W. 1959. Behavior of Structures During Earthquakes, *J. Eng. Mech. Div., ASCE,* 85(EM4):109–129.

Housner, G.W., Soong, T.T., and Masri, S.F. 1994. Second Generation of Active Structural Control in Civil Engineering, *Proc. 1st World Conf. on Struct. Control,* Los Angeles, CA, FA2:3–18.

Keel, C.J. and Mahmoodi, P. 1986. Designing of Viscoelastic Dampers for Columbia Center Building, *Building Motion in Wind,* N. Isyumov and T. Tschanz (Eds.), ASCE, New York, pp. 66–82.

Kelly, J.M., Skinner, R.I., and Heine, A.J. 1972. Mechanisms of Energy Absorption in Special Devices for Use in Earthquake Resistant Structures, *Bull. N.Z. Nat. Soc. Earthquake Eng.,* 5:63–88.

Kobori, T. 1994. Future Direction on Research and Development of Seismic-Response-Controlled Structure, *Proc. 1st World Conf. on Struct. Control,* Los Angeles, CA, Panel:19–31.

Kobori, T., Takahashi, M., Nasu, T., Niwa, N., and Ogasawara, K. 1993. Seismic Response Controlled Structure with Active Variable Stiffness System, *Earthquake Eng. Struct. Dyn.,* 22:925–941.

Kurata, N., Kobori, T., Takahashi, M., Niwa, N., and Midorikawa, H. 1999. Actual Seismic Response Controlled Building with Semi-active Damper System, *Earthquake Eng. Struct. Dyn.,* 28:1427–1448.

Kwok, K.C.S. and MacDonald, P.A. 1987. Wind-induced Response of Sydney Tower, *Proc. 1st National Struct. Eng. Conf.,* pp. 19–24.

Lai, M.L., Chang, K.C., Soong, T.T., Hao, D.S., and Yeh, Y.C. 1995. Full-scale Viscoelastically Damped Steel Frame, *ASCE J. Struct. Eng.,* 121(10):1443–1447.

Liang, Z., Tong, M., and Lee, G.C. 1999a. A Real-time Structural Parameter Modification (RSPM) Approach for Random Vibration Reduction: Part I — Principle, *J. Probabilistic Eng.,* 14: 349–362.

Liang, Z., Tong, M., and Lee, G.C. 1999b. A Real-time Structural Parameter Modification (RSPM) Approach for Random Vibration Reduction: Part II — Experimental Verification, *J. Probabilistic Eng.,* 14:362–385.

Lobo, R.F., Bracci, J.M., Shen, K.L., Reinhorn, A.M., and Soong, T.T. 1993. Inelastic Response of R/C Structures with Viscoelastic Braces, *Earthquake Spectra,* 9(3):419–446.

Makris, N., Burton, S.A., Hilt, D., and Jordan, M. 1996. Analysis and Design of ER Damper for Seismic Protection of Structures, *J. Eng. Mech., ASCE,* 122(10):1003–1011.

Martinez-Romero, E. 1993. Experiences on the Use of Supplemental Energy Dissipators on Building Structures, *Earthquake Spectra,* 9(3):581–624.

Nims, D.K., Richter, P.J., and Bachman, R.E. 1993. The Use of the Energy Dissipating Restraint for Seismic Hazard Mitigation, *Earthquake Spectra,* 9(3):467–489.

Pall, A.S. and Marsh, C. 1982. Response of Friction Damped Braced Frames. *ASCE, J. Struct. Div.,* 1208(ST6):1313–1323.

Pall, A.S. and Pall, R. 1993. Friction-Dampers Used for Seismic Control of New and Existing Building in Canada, *Proc. ATC 17-1 Seminar on Isolation, Energy Dissipation and Active Control,* San Francisco, CA, 2, pp. 675–686.

Pall, A.S., Marsh, C., and Fazio, P. 1980. Friction Joints for Seismic Control of Large Panel Structures, *J. Prestressed Concrete Inst.,* 25(6):38–61.

Patten, W.N., Sack, R.L., and He, Q.W. 1996. Controlled Semiactive Hydraulic Vibration Absorber for Bridges, *J. Structural Eng., ASCE,* 122(2):187–192.

Perry, C.L., Fierro, E.A., Sedarat, H., and Scholl, R.E. 1993. Seismic Upgrade in San Francisco Using Energy Dissipation Devices, *Earthquake Spectra,* 9(3):559–579.

Petersen, N.R. 1980. Design of Large Scale TMD, *Struct. Control,* North Holland, pp. 581–596.

Petersen, N.R. 1981. Using Servohydraulics to Control High-rise Building Motion, *Proc. Nat. Convention on Fluid Power,* Chicago, pp. 209–213.

Ramirez, O.M., Constantinou, M.C., Kircher, C.A., Whittaker, A.S., Johnson, M.W., and Gomez, J.D. 2000. Development and Evaluation of Simplified Procedures for Analysis and Design of Buildings with Passive Energy Dissipation Systems, MCEER-00-0010, Multidisciplinary Center for Earthquake Engineering Research, Buffalo, NY.

Roik, K., Dorka, U., and Dechent, P. 1988. Vibration Control of Structures under Earthquake Loading by Three-Stage Friction-Grip Elements, *Earthquake Eng. Struct. Dyn.*, 16:501–521.

Shen, K.L., Soong, T.T., Chang, K.C., and Lai, M.L. 1995. Seismic Behavior of Reinforced Concrete Frame with Added Viscoelastic Dampers, *Eng. Struct.*, 17(5):372–380.

Shiba, K., Mase, S., Yabe, Y., and Tamura, K. 1998. Active/Passive Vibration Control Systems for Tall Buildings, *Smart Mater. Struct.*, 7(5):588–598.

Sigaher, A.N. and Constantinou, M.C. 2003. Scissor-Jack-Damper Energy Dissipation System, *Earthquake Spectra*, 19:133–158.

Skinner, R.I., Kelly, J.M., and Heine, A.J. 1975. Hysteresis Dampers for Earthquake-Resistant Structures, *Earthquake Eng. Struct. Dyn.*, 3:287–296.

Skinner, R.I., Robinson, W.H., and McVerry, G.H. 1993. *An Introduction to Seismic Isolation*, Wiley, London.

Skinner, R.I., Tyler, R.G., Heine, A.J., and Robinson, W.H. 1980. Hysteretic Dampers for the Protection of Structures from Earthquakes, *Bull. N.Z. Soc. Earthquake Eng.*, 13(1):22–36.

Soong, T.T. 1990. *Active Structural Control: Theory and Practice*, Longman Scientific and Technical, Essex, England, and Wiley, New York, NY.

Soong, T.T. and Constantinou, M.C. (Eds.) 1994. *Passive and Active Structural Vibration Control in Civil Engineering*, Springer-Verlag, Wien and New York.

Soong, T.T. and Dargush, G.F. 1997. *Passive Energy Dissipation Systems in Structural Engineering*, Wiley, London and New York.

Soong, T.T. and Reinhorn, A.M. 1993. An Overview of Active and Hybrid Structural Control Research in the U.S., *J. Struct. Design Tall Bldg.*, 2:193–209.

Soong, T.T. and Spencer Jr., B.F. 2002. Supplemental Energy Dissipation: State-of-the-Art and State-of-the-Practice, *Eng. Struct.*, 24:243–259.

Soong, T.T., Reinhorn, A.M., Aizawa, S., and Higashino, M. 1994. Recent Structural Applications of Active Control Technology, *J. Struct. Control*, 1(2):5–21.

Spencer Jr., B.F., Dyke, S.J., Sain, M.K., and Carlson, J.D. 1997. Phenomenological Model of a Magnetorheological Damper, *J. Eng. Mech.*, ASCE, 123(3):230–238.

Symans, M.D. and Constantinou, M.C. 1997. Seismic Testing of a Building Structure with a Semi-active Fluid Damper Control System, *Earthquake Eng. Struct. Dyn.*, 26(7):759–777.

Szabo, B. and Babuska, I. 1991. *Finite Element Analysis*, Wiley, New York.

Tamura, Y., Shimada, K., Sasaki, A., Kohsaka, R., and Fuji, K. 1994. Variation of Structural Damping Ratios and Natural Frequencies of Tall Buildings During Strong Winds, *Proc. 9th Int. Conf. on Wind Engrg.*, New Delhi, India, 3, pp. 1396–1407.

Tamura, Y., Fujii, K., Ohtsuki, T., Wakahara, T., and Kohsaka, R. 1995. Effectiveness of Tuned Liquid Dampers and Wind Excitations, *Eng. Struct.*, 17(9):609–621.

Tsai, K.C., Chen, H.W., Hong, C.P., and Su, Y.F. 1993. Design of Steel Triangular Plate Energy Absorbers for Seismic-Resistant Construction, *Earthquake Spectra*, 9(3):505–528.

Tyler, R.G. 1985. Test on a Brake Lining Damper for Structures, *Bull. N.Z. Soc. Earthquake Eng.*, 18(3):280–284.

Uang, C.M. and Bertero, V.V. 1986. *Earthquake Simulation Tests and Associated Studies of a 0.3 Scale Model of a Six-Story Concentrically Braced Steel Structure*, Report No. UCB/EERC-86/10, Earthquake Engineering Research Center, Berkeley, CA.

Uang, C.M. and Bertero, V.V. 1988. *Use of Energy as a Design Criterion in Earthquake Resistant Design*, Report No. UCB/EERC-88/18, Earthquake Engineering Research Center, Berkeley, CA.

Veletsos, A.S. and Newmark, N.M. 1960. Effect of Inelastic Behavior on the Response of Simple Systems to Earthquake Motions, *Proc. 2nd World Conf. on Earthquake Eng.*, Tokyo II, pp. 895–912.

Villaverde, R. 1994. Seismic-Control of Structures with Damped Resonant Appendages, *Proc. 1st World Conf. on Struct. Control*, Los Angeles, CA, 1, pp. WP4-113–WP4-122.

Wada, A., Huang, Y.-H., and Iwata, M. 2000. Passive Damping Technology for Buildings in Japan, *Prog. Struct. Eng. Mater.*, 2:335–350.

Whittaker, A.S., Bertero, V.V., Thompson, C.L., and Alonso, L.J. 1991. Seismic Testing of Steel Plate Energy Dissipation Devices, *Earthquake Spectra*, 7(4):563–604.

Whittaker, A.S., Aiken, I., Bergman, P., Clark, J., Cohen, J., Kelly, J.M., and Scholl, R. 1993. Code Requirements for the Design and Implementation of Passive Energy Dissipation Systems, *Proc. ATC 17-1 Seminar on Seismic Isolation, Passive Energy Dissipation and Active Control*, San Francisco, 2, pp. 497–508.

Wirsching, P.H. and Yao, J.T.P. 1970. Modal Response of Structures, *ASCE J. Struct. Div.*, 96(4): 879–883.

Wu, Z. and Soong, T.T. 1995. Nonlinear Feedback Control for Improved Peak Response Reduction, *J. Smart Mater. Struct.*, 4:A140–A147.

Yang, G., Spencer, B.F., Carlson, J.D., and Sain, M.K. 2002. Large-scale MR Fluid Dampers: Modeling and Dynamic Performance Considerations, *Eng. Struct.*, 24(3):309–323.

Zahrah, T.F. and Hall, W.J. 1982. *Seismic Energy Absorption in Simple Structures*, Structural Research Series No. 501, University of Illinois, Urbana, IL.

Zienkiewicz, O.C. and Taylor, R.L. 1989. *The Finite Element Method*, Volumes 1 and 2, 4th Edition, McGraw-Hill, London.

36

Life Cycle Evaluation and Condition Assessment of Structures

Allen C. Estes
*Department of Civil and
Mechanical Engineering,
United States Military Academy,
West Point, NY*

Dan M. Frangopol
*Department of Civil, Environmental,
and Architectural Engineering,
University of Colorado,
Boulder, CO*

36.1 Introduction

Society relies on its engineers to design structures that are safe and perform as intended. The public wants to cross bridges, enter buildings, and live downstream of dams without having to give conscious thought as to whether there is any danger of collapse. The nation's civil engineers have a distinguished history of performing this service. As research has progressed, experience has developed and been codified, and computers have increased computational speed and ability, design methods have become increasingly sophisticated. Traditional design approaches that rely on allowable stress and factor of safety analysis have expanded to incorporate reliability and system-based performance methods. Design optimization methods are beginning to examine the entire life cycle performance of a structure. While this is a new and evolving (and therefore incomplete) field, this chapter explains many of the principles, concepts, and issues involved in examining the performance of a structure over its entire useful life.

Design optimization research often attempts to produce the lowest cost structure that meets specific design criteria. During the 1960s through the 1990s, the objective often involved minimizing the weight of a structure and considered only the costs of design and construction (labor, equipment materials, etc.). Over the past 15 years, research efforts have increasingly demonstrated that these initial costs are often dwarfed by the costs of repairing, inspecting, and maintaining a structure over its useful life. A truly optimum and efficient solution needs to account for these costs. In its simplest form, the difference is illustrated in the following two examples.

EXAMPLE 36.1

The local highway department advertised for bids for a concrete highway bridge. The requirements specified that the concrete must exceed 25 MPa in compressive strength and must have a factor of safety of at least 2.0 with respect to a standard HS-20 truck crossing the bridge at all locations. Company A submitted a bid for $70,000, and Company B submitted a bid for $100,000. Both designs met the specifications, and both bidders were qualified to do the work. Which company should get the job?

Answer

Company A, based on the lower cost.

EXAMPLE 36.2

The local highway department expects the useful life of the highway bridge to be 60 years. The highway department engineer investigates the two proposals in Example 36.1 further and determines that the design submitted by Company A is less expensive because the concrete is not as durable. This engineer estimates that the Company A bridge will require a major rehabilitation every 20 years for a cost of $30,000 and will require a major inspection every 5 years at a cost of $5,000 each time. The Company B bridge will require a major rehabilitation every 30 years for a cost of only $20,000. The inspection of this bridge will also cost $5,000, but it only needs to be done every 10 years. The engineer uses a discount rate of money (which accounts for inflation) of 3% and determines that the preventive maintenance and failure costs of both designs are about equal and, therefore, irrelevant to the decision. Both options require a major rehabilitation at year 60, so there is no difference in salvage value. Which company should get the job?

Answer

The analysis requires that the costs associated with both projects be converted to present value for a valid comparison. Even though the Company A bid appears to now cost $185,000 (i.e., undiscounted total cumulative cost = $70,000 + 2 × $30,000 + 11 × $5,000), over the life of the structure compared to $145,000 (i.e., undiscounted total cumulative cost = $100,000 + $20,000 + 5 × $5,000) for Company B, the timing of the payments is important.

The equation for converting a future value (FV) to a present value (PV) is

$$PV = \frac{FV}{(1+r)^n}$$

where r is the discount rate and n is the number of years.

Company A

$$\text{Total cost} = \$70,000 + \frac{\$30,000}{(1+0.03)^{20}} + \frac{\$30,000}{(1+0.03)^{40}} + \frac{\$5,000}{(1+0.03)^{5}} + \frac{\$5,000}{(1+0.03)^{10}} + \cdots + \frac{\$5,000}{(1+0.03)^{55}}$$

$$= \$70,000 + \$16,610 + \$9,197 + \$4,313 + \$3,720 + \cdots + \$1,141 + \$984 = \$121,022$$

Company B

$$\text{Total cost} = \$100,000 + \frac{\$20,000}{(1+0.03)^{30}} + \frac{\$5,000}{(1+0.03)^{10}} + \frac{\$5,000}{(1+0.03)^{20}} + \frac{\$5,000}{(1+0.03)^{30}}$$

$$+ \frac{\$5,000}{(1+0.03)^{40}} + \frac{\$5,000}{(1+0.03)^{50}}$$

$$= \$100,000 + \$8,240 + \$3,720 + \$2,768 + \$2,060 + \$1,533 + \$1,141 = \$119,462$$

The two options are quite similar, but it now looks like Company B bid would provide the lower life cycle cost and would be the better choice. The result is highly sensitive to the discount rate. If the discount rate had been 4% instead of 3%, Company A would have been a better choice with a life cycle cost of $110,350 versus a cost to Company B of $115,113. Note that (a) the total present value costs decrease as the discount rate rises and (b) the present value costs of expenses late in the life of the structure can become negligible. For this problem the crossover discount rate is about 3.2%, where for any discount rate above that value, Company A will be the preferred alternative. Company B is the best choice for any rate below 3.2% as shown in Figure 36.1a. However, if the service life is reduced to 40 years, Company A is the better choice for any value of the discount rate different from 0% (i.e., at 0% Companies A and B have the same lifetime cost of $135,000, as shown in Figure 36.1b).

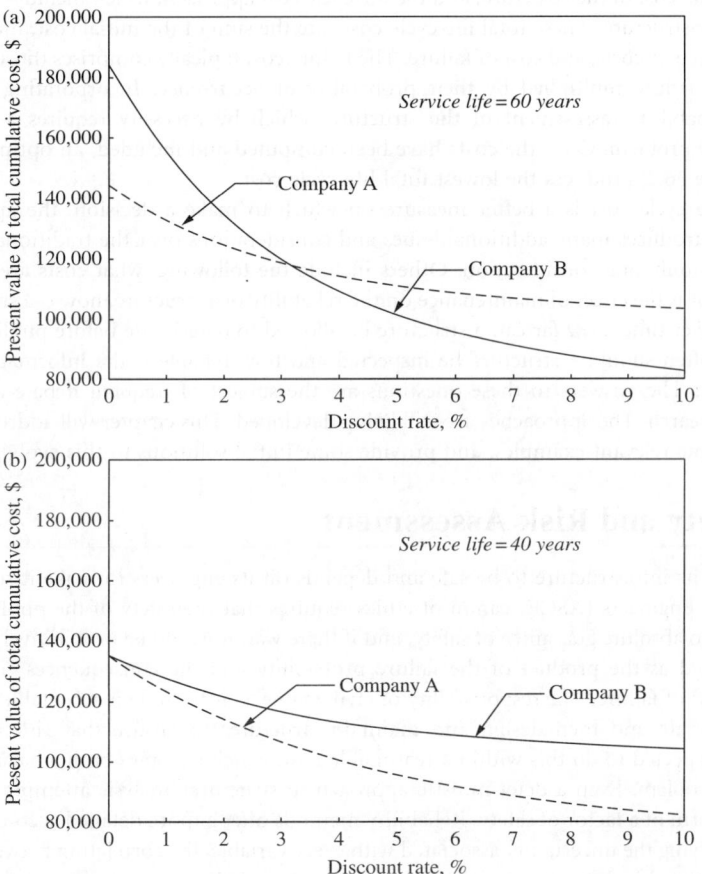

FIGURE 36.1 Present value of total cumulative cost for two competing bids associated with various discount rates for a structure with (a) 60-year service life and (b) 40-year service life.

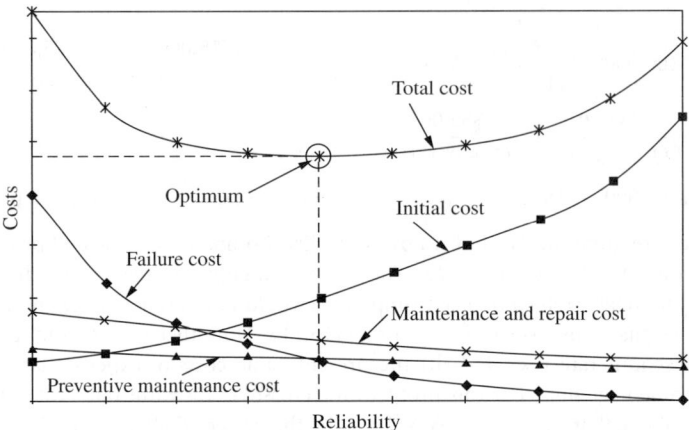

FIGURE 36.2 Optimum reliability of a structure based on total life cycle cost.

These two simple examples illustrate the difference between a traditional design approach, which relies only on the initial cost of the structure, and the life cycle cost approach, which includes all costs over the useful life of the structure. These total life cycle costs are the sum of the initial cost, maintenance costs, inspection and repair costs, and cost of failure. The failure cost typically comprises those costs associated with structural failure multiplied by their probability of occurrence. Incorporating the failure cost requires a probabilistic assessment of the structure, which by necessity requires a reliability-based approach to the problem. Once the costs have been computed and included, an optimum solution, as shown in Figure 36.2, produces the lowest total life cycle cost.

While the life cycle cost is a better measure on which to make a decision, the approach is more complex and introduces many additional issues and considerations over the traditional approach. The discount rate is only one consideration. Others include the following: what costs are included in the failure cost, what is the effect of maintenance on the reliability of a structure, how is a structure expected to deteriorate over time, how far can a structure be allowed to deteriorate before public safety requires a repair, how often should a structure be inspected and how reliable is the information provided by each inspection? The answers to these questions are the subject of frequent debate and the focus of considerable research. The approaches are still being developed. This chapter will address many of these issues, show some relevant examples, and provide some initial solutions to this evolving field.

36.2 Safety and Risk Assessment

Society expects its infrastructure to be safe and depends on its engineers to make it so. The American Society of Civil Engineers (ASCE) canon of ethics requires that the safety of the public is paramount. Sadly, there is no absolute guarantee of safety, and if there was, it would be prohibitively expensive. Risk is usually defined as the product of the failure probability and the consequences (i.e., measured in monetary terms) of failure. The responsibility of civil engineers is to find an acceptable level of risk that society can tolerate and then design and maintain structures to ensure that risk is not exceeded. Engineers are expected to do this within a reasonable cost, which becomes both a balancing act and an optimization problem. Even a deterministic approach to structural analysis attempts to define safety, usually in the form of a factor of safety. Reliability methods offer a more detailed accounting of risk that involves quantifying the uncertainty associated with every variable, the correlation between the variables, the system relationship between structural components, and the consequences of failure. Reliability methods are complex and require significant input data to execute. An elementary example is illustrated as follows.

EXAMPLE 36.3

Two structures are designed to withstand a load with an expected value of 100 kN. The load can vary slightly; in fact, it is estimated to be a normally distributed variable with a standard deviation of 10 kN. Structure A is made of a highly tested material that produces consistent and repeatable results. Its strength is normally distributed with a mean strength of 150 kN and a standard deviation of only 5 kN. Structure B is composed of a stronger material but one that shows greater variability in its characteristics. Its strength is also normally distributed and has a mean of 200 kN but a standard deviation of 30 kN. The load and the strength in both structures are independent — and therefore uncorrelated. Quantify the safety using both a deterministic and a probabilistic approach and decide which structure is safer.

Answer

Using the deterministic approach, the factor of safety (FS) is defined as the ratio of the capacity of the structure to the demand placed on it.

Structure A

$$FS = \frac{\text{Capacity}}{\text{Demand}} = \frac{150}{100} = 1.5$$

Structure B

$$FS = \frac{200}{100} = 2.0$$

Structure B has the larger factor of safety and is, therefore, safer.

Taking a probabilistic approach for two statistically independent, normally distributed random variables, the safety is expressed in terms of a reliability index (β).

Structure A

$$\beta = \frac{\mu_{\text{Capacity}} - \mu_{\text{Demand}}}{\sqrt{\sigma^2_{\text{Capacity}} + \sigma^2_{\text{Demand}}}} = \frac{150 - 100}{\sqrt{10^2 + 5^2}} = 4.47$$

Structure B

$$\beta = \frac{200 - 100}{\sqrt{10^2 + 30^2}} = 3.16$$

where μ_x and σ_x are the mean and standard deviation of the random variable X. The random variables in this problem are capacity and demand.

Under these conditions, the reliability index is translated to a probability of failure as follows: $p_f = \Phi(-\beta)$, where Φ is the distribution function of the standard normal variate. From a Table of Standard Normal Probability, one observes that the probability of failure for Structure A is approximately $p_f = 0.0000038$ and for Structure B, $p_f = 0.00079$. By this analysis, Structure A is the safer structure by a considerable margin. Figure 36.3 illustrates the demand and capacities for the two structures. Failure occurs when demand exceeds capacity. The reliability analysis becomes more complex, as illustrated in Section II, Chapter 12 of the *Handbook*, when the types of probability density distributions vary, the number of random variables increases, and correlation is considered.

While the safety of both structures has been quantified, the question remains as to whether either structure is safe enough to meet societal expectations. Acceptable risk, especially when considering the value of a human life and injuries is difficult to quantify. People die on the highways every day; bridges occasionally collapse; and hurricanes, wildfires, and earthquakes seem to destroy homes and lives every year. For each of these events, there is some level at which the public will demand improvement from their government. The issue of a threshold level of safety is not trivial. Menzies (1996) contends after considerable analysis that society will accept a risk to life of 1 in 1,000,000 per year for the specific example of short-span bridges in the United Kingdom. He further states that acceptable risk levels depend on value judgments and societal norms and cannot be found solely by analysis.

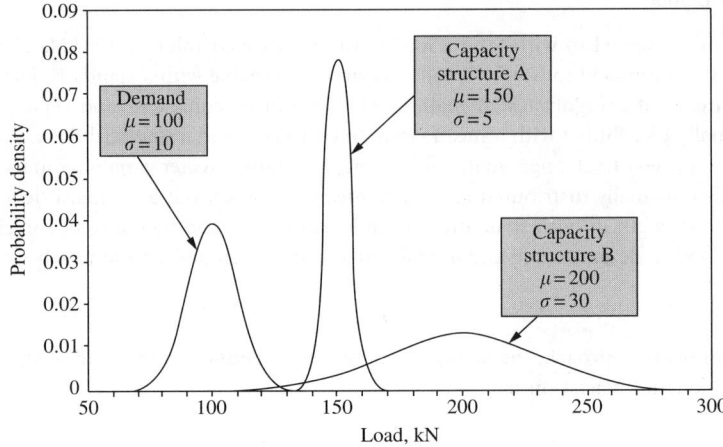

FIGURE 36.3 The capacity and demand for Structures A and B. Failure occurs when demand exceeds capacity.

Fortunately, every engineer does not have to examine societal values and make this determination; those engineers who write the building codes provide that service. As the United States has transitioned to probability-based codes in steel, concrete, and wood, the code committees have wisely calibrated these new codes to those design requirements that have been successful in the past. The American Association of State Highway and Transportation Officials (AASHTO) load and resistance factor design (LRFD) code (AASHTO 1998) calibrates the design of elements of a new bridge in the reliability index range $\beta = 3.0$ to $\beta = 4.0$. The exact value is determined by a load factor modifier η, which accounts for ductility, redundancy, and importance of the structure. If the modifiers, for example, dictated a reliability requirement of $\beta = 3.5$ for Structures A and B in the previous example, it would appear that Structure B ($\beta = 3.16$) was inadequate and Structure A ($\beta = 4.47$) was substantially overdesigned.

While the codes provide a great means to compare different designs, they do not necessarily yet measure society's preferences and risk tolerances. Since there will always be a scarcity of resources to fund everything the public desires, every decision to improve the safety of a bridge or building is a trade-off and a consent to accept additional risk in some other area such as disease prevention or police protection. Risk assessment ultimately needs to be viewed in this broader context. People face risks every day and most of these risks are quantifiable. *Technology Review* (1979) lists smoking 1.4 cigarettes, drinking 0.5 l of wine, traveling 300 miles by car, and flying 1000 miles by airplane as activities that increase the chance of death by one in a million. Even with these statistics, people will not make rational preferences such as the family who drives for three days rather than fly to a location because they perceive it to be safer. Corotis (2003a,b) contends that people's perceptions of risk are affected by whether the risk is objective or subjective, aleatoric or epistemic, familiar or unfamiliar, and voluntary or involuntary. Real world decisions are based on that perception of risk which can include dread, familiarity, number of people exposed, trust, and technological stigma. Corotis further suggests that the public uses several standards to assess risk (cost-benefit analysis, revealed preferences, expressed preferences, and calibrated values) and that future building codes should take all of these into account if society's preferences are to be effectively addressed. This remains a fruitful area for future research.

Even if one concedes that the current building codes provide a reasonable assessment of society's tolerance for risk acceptance, there are additional issues with life cycle cost analysis. Any structure will start deteriorating from the day it is placed in service and, if not maintained, will continually worsen over time. In a design situation, the optimization of a life cycle cost must consider the level to which a structure should be designed and to what level it will be allowed to deteriorate before a repair or replacement is needed. Should the reliability specified in the codes be an average reliability over the lifetime of the structure (β_{lifetime}), the reliability to which a structure is designed (β_{design}),

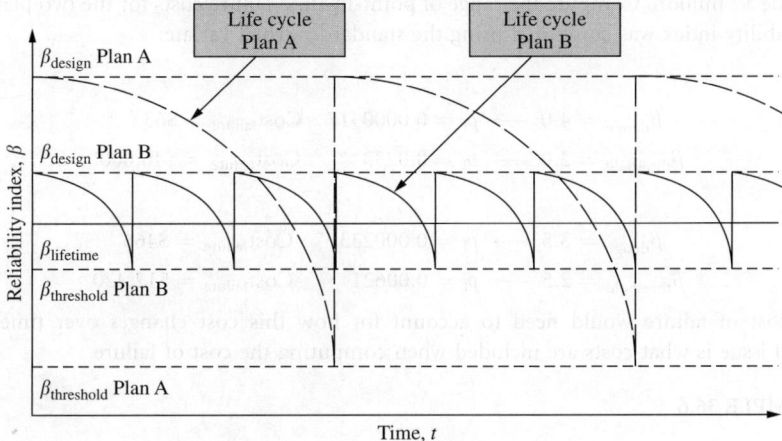

FIGURE 36.4 Alternate life cycle Plans A and B for a designated structure.

or the threshold reliability below which a structure is not allowed to fall ($\beta_{threshold}$), or do multiple reliabilities need to be specified? The answer to this question has not yet been adequately addressed and remains a promising area of research.

EXAMPLE 36.4

A designer proposes two alternate life cycle plans (Plan A and Plan B) for a structure, as illustrated in Figure 36.4. Which plan should the client accept?

Answer

There is not enough information available to decide. If the plans are to be evaluated on the average reliability over the lifetime of the structure, the plans are equivalent since $\beta_{lifetime}$ (i.e., $0.5(\beta_{min} + \beta_{max})$) is the same for both. Plan A initially designs the structure to a higher level of safety, which incurs greater initial costs of construction and allows the structure to deteriorate to a lower level of safety, which increases the likelihood of failure and thus incurs greater failure costs. Plan B, on the other hand, has a narrower band of acceptable reliability and requires more frequent (and probably more costly) maintenance actions throughout the structural life. All estimated costs and the discount rate would be needed to determine the better plan. In addition, the analysis needs to include societal standards to determine a minimum acceptable value of $\beta_{threshold}$. A single value for reliability does not completely define the problem.

36.3 Failure Cost

In most life cycle cost analyses, the failure cost represents the costs incurred by a structural failure multiplied by the probability of failure. What appears to be a straight-forward concept can become quite complex. Costs of failure may be expressed in terms of a point-in-time, annual, or lifetime cost depending on the type of problem. The costs of failure can vary by orders of magnitude over the life of the structure depending on the reliability range the structure is allowed to pass through.

EXAMPLE 36.5

Referring to the life cycle plans in Example 36.4 (see Figure 36.4), assume that for Plan A, $\beta_{design} = 4.0$ and $\beta_{threshold} = 2.0$ and for Plan B, $\beta_{design} = 3.5$ and $\beta_{threshold} = 2.5$. The cost if the structure fails is expected to be $2 million. Compute the range of point-in-time failure costs for the two plans assuming that the reliability index was computed using the standard normal variate:

Plan A

$$\beta_{design} = 4.0 \longrightarrow p_f = 0.0000317 \quad \text{Cost}_{failure} = \$63$$

$$\beta_{threshold} = 2.0 \longrightarrow p_f = 0.0233 \quad \quad \text{Cost}_{failure} = \$46,000$$

Plan B

$$\beta_{design} = 3.5 \longrightarrow p_f = 0.000233 \quad \text{Cost}_{failure} = \$466$$

$$\beta_{threshold} = 2.5 \longrightarrow p_f = 0.00621 \quad \text{Cost}_{failure} = \$12,420$$

A lifetime cost of failure would need to account for how this cost changes over time. The more controversial issue is what costs are included when computing the cost of failure.

EXAMPLE 36.6

The Highway Department commissioned a study to determine the costs associated with the failure of a highway bridge that crosses the river going through the center of town. The study reported a failure cost of $67 million broken down as follows:

Cost of clean-up. $1.2 million
Design cost of new bridge. $2.2 million
Cost of constructing a new bridge. $13.5 million
Expected cost of human injuries. $2.5 million (based on the estimated ten people injured and relying on insurance actuaries to estimate a cost)
Expected cost of additional lawsuits. $5 million
Cost of accident investigation. $0.9 million
Cost of legislation and oversight to prevent similar occurrence. $2.7 million
User costs. $37.0 million (based on the 10,000 people using the bridge every day, who will lose 1 hour on alternate crossing means for 180 days, plus the additional commercial costs associated with transporting goods on alternate routes)
Loss of goodwill and public trust. $2.0 million (almost impossible to evaluate)

What is a realistic cost of failure to use for a life cycle analysis?

Answer

There is no correct answer, which is why this cost is so controversial. The most rigid approach is for the Highway Department to only count those costs that come out of its own budget and that have to be paid directly. In that case, the failure cost would be the $16.9 million associated with the costs of clean-up, establishing an alternate crossing, and providing a new bridge (i.e., $1.2 million + $2.2 million + $13.5 million). In the extreme, a private owner might only consider the amount that insurance premiums would rise as a result of a failure.

A more reasonable approach is to include those costs where money will change hands in some form. The failure cost might then be $28 million, which includes everything except the goodwill and user costs. It is typical for user costs to dwarf all other costs in a life cycle analysis, so a decision on whether or not to use them is critical. Finally, there are others who would include additional costs that are not listed here, and the failure cost might be higher than $67 million.

In most life cycle analyses, the failure cost, however it is computed, is treated as a deterministic quantity. In reality, it is more uncertain and probabilistic than most other quantities in the analysis. Every cost listed above has some uncertainty associated with it. Attempting to put a cost on a human life or injury is a highly emotional, social, and political issue. Predicting the number of accident victims is highly dependent on the circumstances associated with the structural failure. The cost of repercussions will also depend on the details of the failure and whether the failure is an explainable isolated phenomenon or follows on the heels of other structural failures and generates public pressure for greater action. The collapse of the Silver Bridge in 1967 resulted in the monumental bridge inspection and reporting program in place today. If that cost was included in the failure cost of the Silver Bridge, it would add substantially to the amount. If it was, it would also have to be offset by the benefit derived from a mandated biennial bridge inspection program. Again, the point is that failure costs are hotly debated, highly uncertain, and tremendously difficult to compute — another area ripe for future research.

Ayyub and Popescu (2003) have developed software that includes the probability of failure and consequence assessment in deciding among competing projects. Reliability and failure consequences can be considered along with cost, project criticality, use level, and societal preference in making a risk-informed decision on an expenditure. An analytical hierarchy is established, and pairwise comparisons of the various alternatives are made with respect to the relevant criteria. The results are comparison matrices and priority vectors that provide a weighted summation that is used to prioritize the choices. This represents just one attempt to incorporate the failure cost with other relevant criteria to make an optimal life cycle decision.

36.4 Condition Assessment and System Performance

Life cycle cost requires that the condition of a structure be assessed throughout its useful life. While it would be possible to track the factor of safety as a structure deteriorates and experiences different loads over time, reliability methods are much more reasonable — especially if a probability of failure is needed to compute the failure cost. Since the fundamentals of a reliability analysis are covered in Section II, Chapter 12 of the *Handbook*, they are not repeated here. To summarize, if the capacity, R, and the demand, L, are random and the uncertainties can be quantified, then the reliability or probability of safe performance, p_s, can be expressed as

$$p_s = P(R - L > 0) = \iint\limits_{R > L} f_{R,L}(r, l) \, dr \, dl$$

where $f_R(r)$ and $f_L(l)$ are the probability density functions (PDFs) of R and L, respectively and $f_{R,L}(r, l)$ is their joint PDF. A limit state function, $g(\mathbf{X}) = 0$, describes the performance of the system in terms of the vector of basic random variables, $\mathbf{X}$, and defines the failure surface, which separates the survival region from the failure region.

While current design codes focus on designing for specific components and failure modes, structural systems may have multiple components or failure modes. There are many advantages in quantifying the interrelationship between these components and analyzing a structure as an entire system. For example, a system analysis can reveal that some repairs are more important than others. It may also indicate that while each individual component of a structure may have adequate safety, the structure as a whole may still be unsafe.

36.4.1 Series Systems

A series system, also called a weakest-link system, fails when any individual member in the system fails. A chain is only as strong as its weakest link — and that is only true if the failure events are perfectly correlated. If a series system is treated as a series of z elements, the probability of failure of the system p_f is written as the probability of a union of events:

$$p_f = P\left(\bigcup_{a=1}^{z} \{ g_a(\mathbf{X}) \leq 0 \} \right)$$

where $\bigcup_{a=1}^{z} \{ g_a(\mathbf{X}) \leq 0 \}$ is the union of all z failure mode events. Depending on the correlation between the failure modes, the possible range of values for p_f are (Cornell 1967)

$$\max[p_f(a)] \leq p_f \leq 1 - \prod_{a=1}^{z} (1 - p_f(a))$$

The lower bound occurs when the failure modes are perfectly correlated ($\rho = 1.0$) and the upper bound when they are statistically independent ($\rho = 0.0$), where ρ is the correlation coefficient between the safety margins of any pair of failure modes. Ditlevsen (1979) developed tighter bounds using joint-event probabilities, which accounted for failure mode correlation.

EXAMPLE 36.7

A series system consists of three components, where the probability of failure of each individual component is $p_f = 0.01$. What is the probability of failure of the system if the failure modes are independent? If they are perfectly correlated?

Answer

If the three failure modes are independent, so the correlation is $\rho = 0.0$, the failure probability of the system is

$$p_{f\,series} = 1.0 - (1.0 - 0.01)(1.0 - 0.01)(1.0 - 0.01) = 0.0297$$

If the three failure modes are perfectly correlated, so the correlation is $\rho = 1.0$, the failure probability of the system is $p_{f,series} = 0.01$.

36.4.2 Parallel Systems

A parallel system, also called a redundant or fail-safe system, requires every individual member to fail for the system to fail. A parallel system is at least as safe as its most reliable member. The probability of failure of a parallel system is the probability of an intersection of failure events:

$$p_f = P\left(\bigcap_{a=1}^{z} \{g_a(\mathbf{X}) \leq 0\} \right)$$

where $\bigcap_{a=1}^{z} \{g_a(\mathbf{X}) \leq 0\}$ is the intersection of all z failure mode events.

The possible range of values for this system failure probability are (Ang and Tang 1984)

$$\prod_{a=1}^{z} (1 - p_f(a)) \leq p_f \leq \min[p_f(a)]$$

where the lower bound results from mutual independence and the upper bound from perfect correlation. These bounds are often too wide to provide a useful solution. An alternative approach is to reduce all random variables to their equivalent normal distributions and solve the n-dimensional joint standardized distribution integral (Thoft-Christensen and Murotsu 1986):

$$p_f = \int_{\beta_1}^{\infty} \int_{\beta_2}^{\infty} \cdots \int_{\beta_z}^{\infty} \frac{1}{(2\pi)^{z/2} \sqrt{\det[\rho_{sys}]}} e^{-1/2\{\beta\}[\rho_{sys}]^{-1}\{\beta\}^{\mathrm{T}}} \, d\{\beta\}$$

where $\{\beta\} = \{\beta_a, \beta_b, \ldots, \beta_z\}$, ρ_{sys} is the system correlation matrix, and z is the number of members in the parallel system.

EXAMPLE 36.8

A parallel system consists of three components where the probability of failure of each individual component is $p_f = 0.01$. What is the probability of failure of the system if the failure modes are independent? If they are perfectly correlated?

Answer

For a parallel system consisting of three components whose individual probabilities of failure are $p_f = 0.01$, the system failure probability is upper bounded (first-order) by $p_{f,parallel} = \min(p_{f1}, p_{f2}, p_{f3}) = 0.01$ if the three failure modes are perfectly correlated, and it is lower bounded (first-order) by $p_{f,\text{parallel}} = (0.01)(0.01)(0.01) = 0.000001$ if the three failure modes are independent. As indicated in this simplified example, there can be huge errors if correlation is neglected.

36.4.3 General Systems

Many general engineering systems can be modeled as a combination of series and parallel systems. For example, a series of y parallel systems where each parallel system a has z_a components would have a probability of failure expressed as

$$p_f = p\left(\bigcup_{a=1}^{y} \bigcap_{b=1}^{z_a} \{g_{ab}(\mathbf{X}) \leq 0\}\right)$$

where $g_{ab}(\mathbf{X})$ identifies the limit state equation for a specific component in the system model. Most complex systems can be sequentially broken down into simpler equivalent subsystems. The reliabilities of a series subsystem and parallel subsystem are solved individually as described above using the reliabilities and direction cosines at the points of failure of individual components. An example of how a series–parallel system is reduced is shown in Figure 36.5. The system shown contains six components (1–6). Initially, the two parallel subsystems are reduced to equivalent components (7 and 8) that formed part of a series system. The series system is then reduced to a single equivalent component (9). The correlation of the equivalent components is computed using equivalent alpha vectors that are a function of the equivalent direction cosines as described in Estes and Frangopol (1998). Any structural system that can be modeled as a combination of series and parallel components can be analyzed.

EXAMPLE 36.9

The three-bar indeterminate truss shown in Figure 36.6 (Estes 1997; Estes and Frangopol 2001b) is modeled as a series–parallel system where the failure of any two bars will cause failure of the system. Let the event "Bar 2 | 1" indicate failure of bar 2 given that bar 1 has already failed, and let "Bar 1" indicate failure of bar 1. The resistances of the three perfectly ductile bars are assumed random variables as follows:

$$R_{\text{bar1}} \longrightarrow N[15, 1.5], \quad R_{\text{bar2}} \longrightarrow N[15, 1.5], \quad R_{\text{bar3}} \longrightarrow N[10, 1.0]$$

where $N[10, 1.0]$ indicates that the variable is normally distributed with a mean value of 10 kN and a standard deviation of 1.0 kN. The load on the truss P is also a random variable with parameters

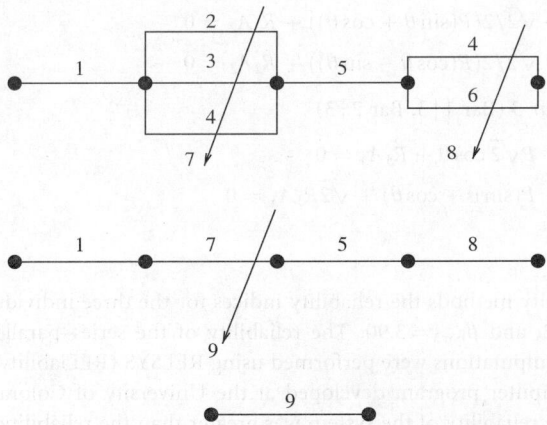

FIGURE 36.5 Reduction of a series–parallel system to a single equivalent component. (Estes and Frangopol 2001b. Reprinted with permission from the Computational Structural Engineering Institute.)

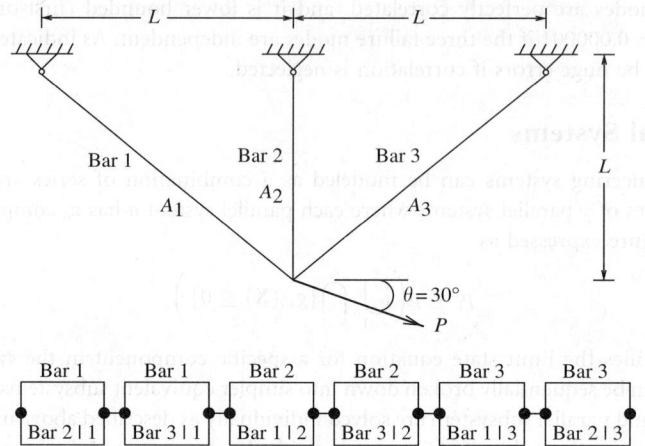

FIGURE 36.6   Three-bar indeterminate truss modeled as a series–parallel system. (Estes and Frangopol 2001b. Reprinted with permission from the Computational Structural Engineering Institute.)

$N[20, 4.0]$, where 20 and 4.0 are the mean and standard deviation (in kN), respectively. The resistances are independent. The limit state equations that describe the components in the series–parallel model are (Estes and Frangopol 2001b)

1. *Prior to any bars failing* (Bar 1, Bar 2, Bar 3)

$$g(1) = R_1(2.0A_1A_3 + \sqrt{2}(A_1A_2 + A_3A_2)) - \sqrt{2}PA_1(\cos\theta + \sin\theta) - 2.0PA_2\cos\theta = 0$$
$$g(2) = R_2(2.0A_1A_2 + \sqrt{2}(A_1A_2 + A_3A_2)) - \sqrt{2}P((A_1 - A_3)\cos\theta + (A_1 + A_3)\sin\theta) = 0$$
$$g(3) = R_3(\sqrt{2}A_1A_3 + A_1A_2 + A_3A_2) + P(A_3\sin\theta - A_3\cos\theta - \sqrt{2}A_2\cos\theta)$$

2. *Given failure of bar 1* (Bar 2 | 1, Bar 3 | 1)

$$g(4) = R_2A_2 - P(\sin\theta + \cos\theta) + \sqrt{2}R_1A_1 = 0$$
$$g(5) = R_3A_3 - P\sqrt{2}\cos\theta + R_1A_1 = 0$$

3. *Given failure of bar 2* (Bar 1 | 2, Bar 3 | 2)

$$g(6) = R_1A_1 - \sqrt{2}/2(P(\sin\theta + \cos\theta)) + R_2A_2 = 0$$
$$g(7) = R_3A_3 - \sqrt{2}/2(P(\cos\theta - \sin\theta)) + R_2A_2 = 0$$

4. *Given failure of bar 3* (Bar 1 | 3, Bar 2 | 3)

$$g(8) = R_1A_1 - P\sqrt{2}\cos\theta + R_3A_3 = 0$$
$$g(9) = R_2A_2 - P(\sin\theta + \cos\theta) + \sqrt{2}R_3A_3 = 0$$

Answer

Using first-order reliability methods the reliability indices for the three individual bars are found to be $\beta_{bar1} = 3.48$, $\beta_{bar2} = 7.42$, and $\beta_{bar3} = 3.90$. The reliability of the series–parallel system $\beta_{system} = 4.19$. These reductions and computations were performed using RELSYS (RELiability of SYStems) (Estes and Frangopol 1998), a computer program developed at the University of Colorado. Due to the parallel nature of the model, the reliability of the system was greater than the reliability of two of the bars. The most critical link in the simplified series system was the one involving bars 1 and 3 in parallel, which is why the system reliability was so much lower than the reliability of bar 2. Correlation between failure modes improves the reliability of a series system and decreases the reliability of a parallel system. Although the resistances are independent, the failure modes are correlated because the same load source P was common to all three bars. Estes and Frangopol (2001b) examine the performance and maintenance of this truss system over time.

In a life cycle cost analysis, the system reliability index accounts for the degree of redundancy and extra safety in a structure. As seen in AASHTO (1998), the importance, ductility, and redundancy of a structure all affect the allowable safety level of a structure. For system performance purposes, Frangopol et al. (1992) classified structural redundancy based on a system redundancy measure (α_{system}) and a redundancy range (γ_{system}), where

$$\alpha_{system} = \beta_{system} - \beta_{weakest\ member}$$

$$\gamma_{system} = \beta_{strongest\ member} - \beta_{weakest\ member}$$

Based on these probabilistic measures, structural systems are classified as follows:

Very redundant systems

$$\alpha_{system} > \gamma_{system} \geq 0$$

Redundant systems

$$\gamma_{system} \geq \alpha_{system} > 0$$

Nonredundant systems

$$\alpha_{system} \leq 0$$

EXAMPLE 36.10

Based on the classification system above, categorize the redundancy of the truss shown in Example 36.9.

Answer

$$\alpha_{system} = 4.19 - 3.48 = 0.71$$
$$\gamma_{system} = 7.42 - 3.48 = 3.94$$

Since $\gamma_{system} \geq \alpha_{system} > 0$, the truss structure is classified as redundant.

36.5 Damage and Deterioration

A life cycle cost analysis for a structure requires a number of assumptions about the loading, resistance, and deterioration. In general, a structure will begin to deteriorate and experience reliability reduction from the day it is placed in service. A predictive model is needed to estimate how the loads and resistances will change over time. The deterioration model is usually derived theoretically, obtained from laboratory data, or extrapolated from the behavior of similar structures under similar conditions. These models predict future performance over several decades and are the basis for optimum life cycle inspection and repair planning. Common deterioration mechanisms include corrosion, fatigue, rotting, and spalling, among others.

EXAMPLE 36.11

A structural engineer is designing a retaining wall where the main structural members are exposed, unpainted steel I-beams located in the downtown area. In performing a life cycle cost analysis of the structure, the designer needs to predict the degree of section loss in the beams over time due to corrosion. Corrosion is a common deterioration mechanism that is difficult to predict since it is dependent on the type of steel, the local environment, the presence of moisture, and even the location of the steel member in the structure. Ideally, the designer would prefer to allow samples of the same type of steel that is used in the project to corrode over time under the same conditions as the proposed structure and use those data to develop a deterioration model. However, it is costly and time consuming to obtain this information. A literature search reveals that Albrecht and Naeemi (1984) developed a commonly referenced corrosion propagation model that predicts the average corrosion penetration $C(t)$ (μm) at any time t (years) as follows:

$$C(t) = At^B$$

where A and B are regression parameters based on the environment and type of steel. For carbon steel in an urban environment, A has a mean value of $\mu_A = 80.2$ and a standard deviation of $\sigma_A = 33.68$, while $\mu_B = 0.593$ and $\sigma_B = 0.24$. The correlation coefficient between A and B is $\rho_{A,B} = 0.68$. What is the estimated section loss after 10 years? After 20 years?

Answer

A probabilistic analysis using Monte Carlo simulation with @RISK software (Palisade 2003) (100,000 simulations) reveals a distribution for $C(t)$ that looks lognormal in shape and has the parameters shown in Table 36.1. In all cases, the uncertainty in the result, as indicated by the extremely large standard deviation, is quite high and increases with time. Column (1) represents the values if the correlation between A and B is not included in the analysis. Column (2) includes the correlation, and Column (3) truncates the distribution for A at 0.0 on the lower end to prevent any results that show a thickness increase due to corrosion. Such results can be quite different depending on the tools that the analyst chooses to use. The information derived from these tools is not exact, and there is no guarantee that the model from the literature matches how the retaining wall will actually behave. Still, the model provides a reasonable predictive tool to make an estimate, but in each case, the analyst needs to understand its limitations.

EXAMPLE 36.12

The support beams in the retaining wall in Example 36.11 are European wide-flange beams HE 180 M having a yield strength of 355 MPa. The initial dimensions of the beam as shown in Figure 36.7 are $b = 186$ mm, $h = 200$ mm, $t_w = 14.5$ mm, and $t_f = 24$ mm. Apply the thickness loss from Example 36.11 to reflect the loss in moment capacity in the I-beam after 10 years. After 20 years.

Answer

The plastic moment capacity of the beam is the product of the yield strength and the plastic section modulus. Because the beam and the assumed corrosion pattern as shown in Figure 36.7 are symmetrical, the neutral axis is located at the centroid of the beam and does not shift. The plastic section modulus (Z) is the first moment of area about the neutral axis:

$$Z = 2(y_1 A_1 + y_2 A_2)$$

TABLE 36.1 Thickness Loss on Retaining Wall Support Beam Due to Corrosion, Based on Albrecht and Naeemi (1984) Model

	Thickness loss $C(t)$					
	$\rho = 0.0$ (1)		$\rho = 0.68$ (2)		$\rho = 0.68$[a] (3)	
t (years)	μ (μm)	σ (μm)	μ (μm)	σ (μm)	μ (μm)	σ (μm)
10	366	283	423	409	426	408
20	614	603	739	908	743	904

[a] Distribution of A is truncated.

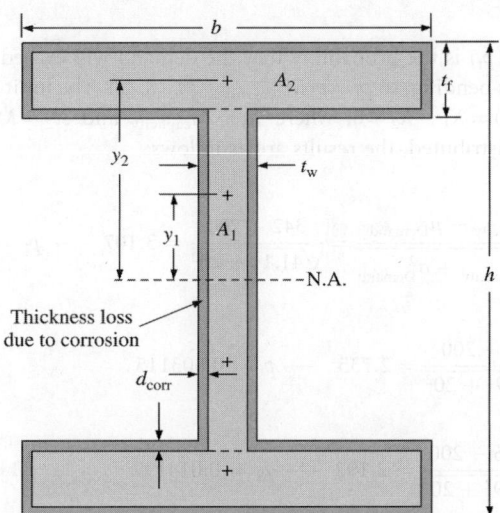

FIGURE 36.7 Dimensions necessary for computing the plastic section modulus (*Z*) on a corroding wide-flange beam.

TABLE 36.2 The Plastic Moment Capacity at Various Points in Time of a Retaining Wall Support Beam that is Deteriorating Due to Corrosion

Time (years)	Plastic section modulus, Z (cm³)		Yield strength, F_y (MPa)		Plastic moment capacity, $M_{capacity}$ (kN m)	
	μ	σ	μ	σ	μ	σ
0	869	0	394	47.2	342	41.0
10	830	37.2	394	47.2	327	41.9
20	802	78.6	394	47.2	316	49.0
40	754	159	394	47.2	297	72.7

where

$A_1 = (0.5\,h - t_f + d_{corr})(t_w - 2d_{corr})$

$y_1 = 0.5(0.5h - t_f + d_{corr})$

$A_2 = (b - 2d_{corr})(t_f - 2d_{corr})$

$y_2 = 0.5h - d_{corr} - 0.5t_f$

$d_{corr} = C(t)/1000$

Applying a bias factor of 1.11 and the coefficient of variation of 0.12 for the yield strength of steel (Nowak and Collins 1995), the yield strength of steel is a normally distributed variable with a mean of 394 MPa and a standard deviation of 47.2 MPa. Simulation using these parameters provides a moment capacity for the beam shown in Table 36.2.

EXAMPLE 36.13

Assuming that the load effect (i.e., maximum moment caused by the load on the member) on the retaining wall beam is normally distributed with a mean value of 250 kN m and a standard deviation of 25 kN m, assess the condition of the support beam from Example 36.12 in terms of probability of failure at years 10, 20, and 40.

Answer

The probability of failure (p_f) is the probability that the demand will exceed the capacity. In this case, the relevant failure mode is bending, so $p_f = p(M_{demand} \geq M_{capacity})$. The limit state equation that defines the failure surface is $g(\mathbf{X}) = X_1 - X_2 = 0$, where $X_1 = M_{capacity}$ and $X_2 = M_{demand}$. Because both are assumed to be normally distributed, the results are as follows:

Year 0

$$\beta = \frac{\mu_{Capacity} - \mu_{Demand}}{\sqrt{\sigma_{Capacity}^2 + \sigma_{Demand}^2}} = \frac{342 - 200}{\sqrt{41.1^2 + 20^2}} = 3.107 \longrightarrow p_f = 0.000946$$

Year 10

$$\beta = \frac{327 - 200}{\sqrt{41.9^2 + 20^2}} = 2.735 \longrightarrow p_f = 0.003115$$

Year 20

$$\beta = \frac{316 - 200}{\sqrt{49^2 + 20^2}} = 2.192 \longrightarrow p_f = 0.014197$$

Year 40

$$\beta = \frac{297 - 200}{\sqrt{72.7^2 + 20^2}} = 1.286 \longrightarrow p_f = 0.099142$$

While the loss in plastic section modulus may not appear extreme, the probability of failure has increased dramatically with time.

This example shows how reliability can be computed at points in time for a deteriorating structure. The example was greatly simplified as it assumed that all distributions were normal and that the load did not change over time. It ignored extreme loading events and other failure modes, all of which must be accounted for in a real world problem.

36.6 Time-Dependent Reliability

The methods described above apply to computing the reliability of a structure at a specific point in time. When attempting to make decisions about a structure over its useful life, time becomes an important variable. If the future load and resistance of the structure can be predicted, the simplest approach is a point-in-time method where the reliability is computed at various specific times in the future. A trend is established, and the structure is planned for a repair when the reliability falls below an acceptable target reliability level. Loads tend to increase over time, and the resistance tends to decrease as the structure deteriorates, so the overall reliability can generally be expected to decrease over time. A weakness of this approach is that it fails to account for previous structural performance.

A more accepted, yet increasingly complex, approach to time-dependent reliability is to compute the probability that a structure will perform safely for a specified period of time (Enright and Frangopol 1998a,b). Whereas reliability is defined as the probability that an element is safe at one particular time, the survivor function $S(t)$ defines the probability that an element is safe at any time t:

$$S(t) = P(T \geq t) = p_s(t)$$

where the random variable T represents time and $t \geq 0$. The probability that a failure, $p_f(t)$, takes place over a time interval Δt is expressed as

$$f(t)\Delta t = P[t_1 < T \leq t_1 + \Delta t]$$

where the PDF $f(t) = -S'(t)$. It is assumed that the derivative $S'(t)$ exists. The probability of failure between times t_a and t_b is

$$P(t_a \leq T \leq t_b) = \int_{t_a}^{t_b} f(t)\,dt$$

The reliability is often expressed in terms of a hazard function, $H(t)$, also called the instantaneous failure rate, the hazard rate, or simply the failure rate. The hazard function expresses the likelihood of failure in the time interval t_1 to $t_1 + dt$ given that the failure has not already occurred prior to t_1 and can be expressed as

$$H(t) = \frac{f(t)}{p_s(t)} = -\frac{S'(t)}{S(t)} = -\frac{1}{p_s(t)} \frac{dp_s(t)}{dt}$$

Thus, the hazard function is the ratio of $f(t)$ to the survivor function $S(t)$. All hazard functions must satisfy the nonnegativity requirement. Their units are typically given in failures per unit time. Large and small values of $H(t)$ indicate great and small risks, respectively (Leemis 1995).

EXAMPLE 36.14

For the deteriorating retaining wall beam in Example 36.13, use the hazard function to determine the probability that the retaining wall will fail in its 20th year if it has not failed earlier. Table 36.3 shows the hazard functions in 2-year increments based on the following computations:

$$p_s(t) = p_s(0) = 1.0 - p_f(0) = 1.0 - 0.000946 = 0.999054$$

$$\frac{dp_s(t)}{dt} = \frac{dp_s(2)}{dt} = \frac{(0.998844 - 0.999054)}{2 - 0} = -0.00011$$

$$H(t) = H(2) = \frac{-1}{p_s(2)} \frac{dp_s(2)}{dt} = \frac{-1(-0.00011)}{0.998844} = 0.000105$$

The graph of the hazard function is shown in Figure 36.8 Because the numerical differentiation causes uneven jumps in the curve, a best fit line through the points, often by fitting a Weibull distribution to the points, provides a smooth curve. Therefore, as shown in Table 36.3, the hazard rate at year 20, provided by best fit, is $H(20) = 0.001951$.

EXAMPLE 36.15

The retaining wall in the previous example has survived and performed well without maintenance for 19 years. In developing his annual budget, the owner is trying to decide whether this is a cost-effective time to replace the support beams in the structure. The replacement cost for new supporting beams is $67,000. The new support beams are expected to have a service life of 20 years before they would need replacing again. The discount rate r is 6%. If the wall fails, there are several possible consequences as follows: (a) There is a 30% chance that the wall will simply bow excessively and there will be long-term

TABLE 36.3 The Hazard Function $H(t)$ for the Support Beams in a Retaining Wall That Is Deteriorating Due to Corrosion

Year	Reliability index (β)	Probability of failure (p_f)	Probability of survival (p_s)	Derivative (dp_s/dt)	Hazard function ($H(t)$)	Best fit $H(t)$
0	3.107	0.000946	0.999054	—	—	—
2	3.047	0.001156	0.998844	−0.00011	0.000105	0.000098
4	2.987	0.001411	0.998589	−0.00013	0.000128	0.000137
6	2.912	0.001798	0.998202	−0.00019	0.000194	0.000191
8	2.834	0.002297	0.997703	−0.00025	0.000250	0.000266
10	2.735	0.003115	0.996885	−0.00041	0.000410	0.000371
12	2.646	0.004073	0.995927	−0.00048	0.000481	0.000517
14	2.550	0.005381	0.994619	−0.00065	0.000657	0.000721
16	2.422	0.007717	0.992283	−0.00117	0.001177	0.001004
18	2.312	0.010399	0.989601	−0.00134	0.001355	0.001400
20	2.192	0.014197	0.985803	−0.0019	0.001926	0.001951

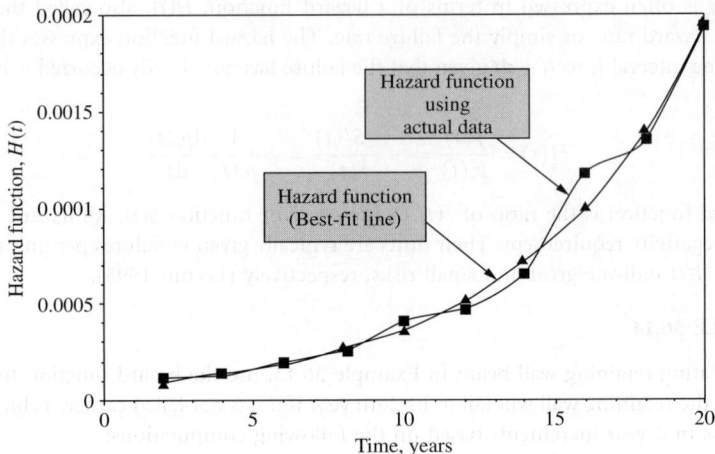

FIGURE 36.8   The hazard function for a retaining wall beam over a 20-year period.

warning that something is wrong. The entire wall would need to be replaced, control measures would be installed, and an accelerated schedule would result for an estimated cost of $200,000. (b) There is a 50% chance that the wall will fail catastrophically, slide down the hill, but not spill onto the roadway below. The cost associated with a clean-up, immediate repair, and delaying of traffic while the project is under way is $450,000. (c) There is a 15% chance that the wall would fail catastrophically, extend onto the highway below, and cause accidents. The estimated cost for this scenario is $1.2 million. (d) There is a 5% chance of resulting injuries and fatalities, which raises the estimated cost to $3.6 million. From a purely financial perspective, should the owner replace the wall this year?

Answer

The probability that the wall will fail this year given that it has not failed already is provided by the hazard function in Table 36.3 as $H(t) = H(20) = 0.001951$. The expected cost of failure is illustrated in the event tree in Figure 36.9 and calculated as

$$E(\text{Cost}_{\text{failure}}) = 0.30(\$200{,}000) + 0.50(\$450{,}000) + 0.15(\$1{,}200{,}000) + 0.05(\$3{,}600{,}000)$$
$$= \$645{,}000$$

$$E(\text{Cost}_{\text{year 20}}) = \$645{,}000(0.001951) + \$0(1 - 0.001951) = \$1{,}242$$

This value represents the annual cost of keeping the retaining wall in service in its present state. In this example, no maintenance is assumed, and thus there is no cost associated with the wall performing as intended.

The annual cost of the support beam replacement is obtained by converting the present value of the $67,000 replacement cost (C_{pv}) to an annual cost (C_{annual}) over the expected 20-year life.

$$C_{\text{annual}} = \frac{C_{pv} r (1 + r)^n}{(1 + r)^n - 1} = \frac{\$67{,}000(0.06)(1 + 0.06)^{20}}{(1 + 0.06)^{20} - 1} = \$5{,}841$$

The beam replacement is not justified at this time based on the comparative annual costs of the alternatives (i.e., $5841 > $1242). This methodology is used by the Army Corps of Engineers to justify major rehabilitation projects on structures (USACE 1996).

EXAMPLE 36.16

Based on the information in Example 36.15, what is the annual probability of failure that would justify a replacement of the retaining wall support beams? When is this expected to occur?

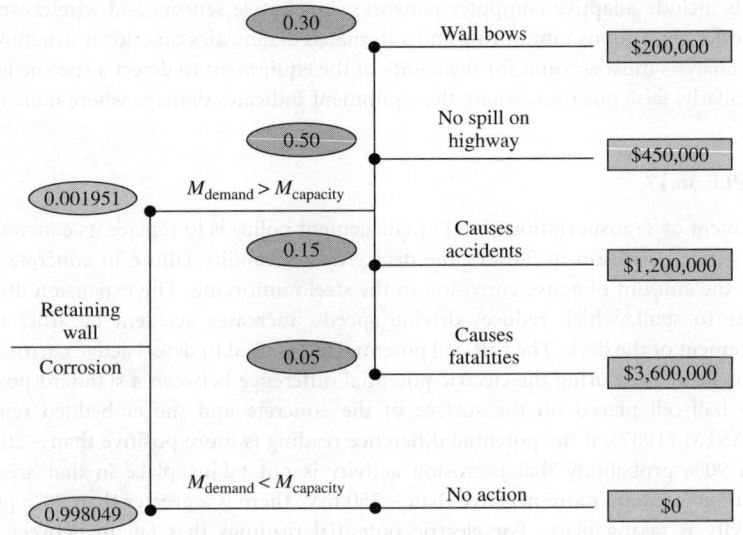

FIGURE 36.9 Event tree that describes the failure consequences and costs associated with the failure of a retaining wall.

The annual probability of failure that would justify the support beam replacement is \$5,841/ \$645,000 = 0.0091. By continuing the analysis in Table 36.3, the best-fit values for $H(28) = 0.0074$ and $H(30) = 0.0103$ indicate that the replacement should be justified by the 30th year of service.

36.7 Inspection

The reliability of a structure that is forecasted over several decades is only as valid as the input data and models that support it. Despite the best intentions, a deterioration model taken from the literature or from the behavior of a similar structure under similar conditions could be grossly inaccurate. A systematic inspection program is needed to verify that the structure is behaving as predicted and to provide new data to modify the prediction models if it is not. Inspections can be expensive, so it becomes important to schedule them to optimize the benefit of the information gathered through inspections.

36.7.1 Nondestructive Evaluation Inspection

A nondestructive evaluation (NDE) inspection that is targeted for specific information and does not damage the structure is usually the preferred type of inspection. The information obtained from these inspections is not readily observable and often correlates with a specific defect such as fatigue cracks in members, corrosion of girders, or deterioration of concrete. These techniques (AASHTO 1994) can include radiography, acoustics travel tomography (Bond et al. 2000; Kepler et al. 2000), ultrasonics, thermography, or electric potential, among others. In some cases, such as the thickness of corrosion on exposed steel, a direct measurement is made. In other cases where a defect is hidden, such as scaling of concrete, acoustics travel tomography, thermography, or radar are useful to infer damage. Special methods are used when there is a high degree of correlation between a test result such as electric potential or an acoustical sounding and the defect being sought such as the presence of corrosion in the steel reinforcement of a concrete structure. Sometimes several techniques may be used in combination to improve the confidence in the structural condition. For example, magnetic particle tests, dye penetrants, and ultrasonics are all useful methods to detect fatigue cracks, and if similar conclusions are reached by all three methods, the credibility of the results is enhanced.

Future trends include adaptive computer networks, innovative sensors and wireless technology that will be able to offer continuous monitoring and automated diagnostics on critical structures. In all cases, a probabilistic analysis must account for the ability of the equipment to detect a specific level of damage or distress. Similarly, false positives, where the equipment indicates damage where none exists, must be considered.

EXAMPLE 36.17

A local Department of Transportation (DOT) management policy is to replace its concrete bridge decks when there is active corrosion in 50% of the deck. A serviceability failure in concrete decks is often determined by the amount of active corrosion in the steel reinforcing. The expansion due to corrosion causes concrete to spall, which reduces driving speeds, increases accident hazards, and eventually requires replacement of the deck. The half-cell potential test is used to detect active corrosion in concrete slab reinforcement by measuring the electric potential difference between a standard portable copper–copper sulfate half-cell placed on the surface of the concrete and the embedded reinforcing steel. According to ASTM (1987), if the potential difference reading is more positive than −200 mV, there is a greater than 90% probability that corrosion activity is not taking place in that area. Similarly, if the potentials of an area are more negative than −350 mV, there is a greater than 90% probability that corrosion activity is taking place. For electric potential readings that fall in between those ranges, corrosion activity is more uncertain.

In an effort to better quantify the uncertainty, Marshall (1996) compiled the half-cell potential data from 89 different bridges. The PDF of half-cell potentials in areas where the deck was known to be undamaged (i.e., there was no reinforcement corrosion) was a normal distribution with a mean value of −207 mV and a standard deviation of 80.4 mV. Similarly, the PDF for half-cell readings in damaged areas of deck was found to be $N[-354 \text{ mV}, 69.7 \text{ mV}]$ as shown in Figure 36.10. The overlap of the two PDFs indicates a significant risk of misinterpretation of actual readings based on half-cell potential mapping.

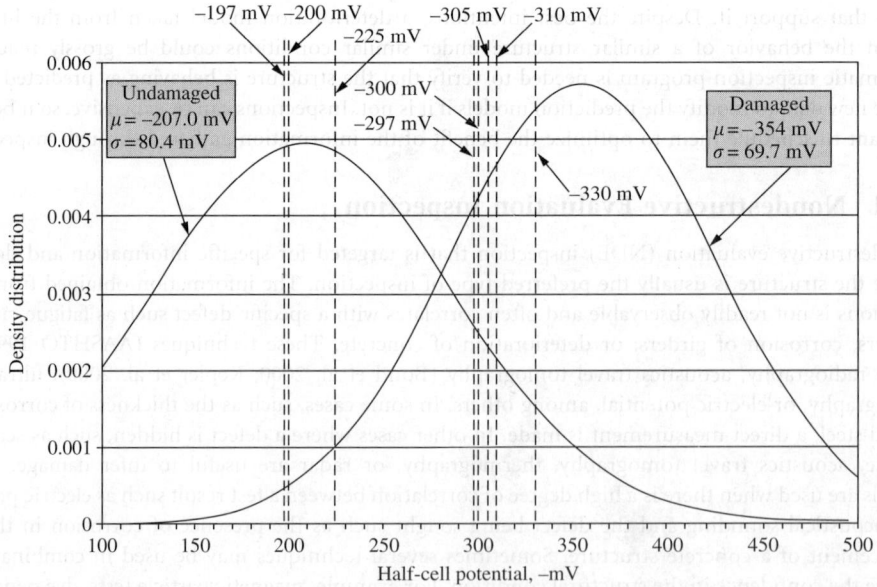

FIGURE 36.10 Probability density functions (PDFs) of half-cell potentials in areas where the deck is known to be damaged and undamaged.

In a half-cell potential test for a bridge deck, readings were taken at eight locations evenly spread throughout the deck. The potential difference readings in (mV) were -200, -225, -297, -300, -305, -310, -197, and -330. What is the probability of active corrosion at each of these locations? Estimate the percentage of damage to the deck. Does it need to be replaced at this time?

Answer

The probability of damage (p_{dam}) at one of the readings can be determined as

$$P_{dam} = \frac{pdf_{bad-x}}{pdf_{bad-x} + pdf_{good-x}}$$

where pdf_{bad-x} and pdf_{good-x} are the values of the density distributions in Figure 36.10 for the damaged and undamaged decks for a half-cell reading of x, respectively. Because the distributions in Figure 36.10 are both normal, the probability of damage when the half-cell reading is -200 mV is obtained as follows:

$$pdf_{good-(-200\,mV)} = \frac{1}{\sigma\sqrt{2\pi}} e^{-(1/2)((x-\mu)/\sigma)^2} = \frac{1}{80.4\sqrt{2\pi}} e^{-(1/2)((200-207)/80.4)^2} = 0.004943$$

$$pdf_{bad-(-200\,mV)} = \frac{1}{69.7\sqrt{2\pi}} e^{-(1/2)((200-354)/69.7)^2} = 0.000432$$

$$P_{dam} = \frac{0.000432}{(0.000432 + 0.004943)} = 0.080$$

The results for the other seven readings are computed in a similar manner and are shown in Table 36.4. The assessed percentage of damage over the deck ($p_{dam\,deck}$) would be the weighted average of the probabilities of damage:

$$P_{dam\,deck} = \frac{\sum_{i=1}^{n} P_{dam\,i}}{n} = \frac{(0.080 + 0.156 + 0.573 + 0.591 + 0.621 + 0.651 + 0.074 + 0.752)}{8} = 43.7\%$$

where i is an individual half-cell reading and n is the total number of readings taken. According to the repair policy, the deck does need to be replaced until $p_{dam\,deck} = 0.50$, but at $p_{dam\,deck} = 0.437$, it is getting close.

There are three main sources of uncertainty that need to be quantified with an NDE inspection: (a) the quality of the inspection equipment and its ability to accurately read whatever measurement is being taken, (b) the correlation between whatever the equipment is measuring and the presence of an actual defect, and (c) the ability to assess the condition of an entire structure based on a finite number of readings. In Example 36.17, the accuracy of the equipment and whether eight evenly spaced readings were sufficient to assess the condition of the entire deck were not addressed. For a probabilistic analysis and an update of a structure's reliability, these uncertainties should all be quantified and considered. Such data are not readily available for most NDE techniques, are often difficult to obtain, and remain a bountiful area for continued research. Estes and Frangopol (2001c) used these half-cell results to optimize maintenance planning for an existing concrete deck.

TABLE 36.4 Probability of Damage for Various Half-Cell Potential Readings on a Reinforced Concrete Deck

Half-cell reading ($-$mV)	Probability density distribution for undamaged deck, pdf_{good-x}	Probability density distribution for damaged deck, pdf_{bad-x}	Probability of damage, P_{dam}
200	0.004943	0.000432	0.080
225	0.004839	0.000895	0.156
297	0.002652	0.003552	0.573
300	0.002542	0.003675	0.591
305	0.002361	0.003876	0.621
310	0.002184	0.004066	0.651
197	0.004924	0.000393	0.074
330	0.001540	0.004676	0.752

36.7.2 Visual Inspection — Condition Ratings

Many structures undergo periodic, formalized visual inspections that assess the overall condition of a structure. The Federal Highway Administration (FHWA 2002), for example, requires that every bridge in the National Bridge Inventory be inspected every 2 years and is given a numerical rating ranging from 9 (best) to 1 (worst). Similarly, the Army Corps of Engineers uses a 100-point condition index system to classify lock and dam structures based on a visual inspection (Stecker et al. 1997).

The inspection results come from inspector observations rather than a series of specific tests. The advantages of visual inspections are that they are easy to conduct, require minimal equipment, are not overly time consuming, and are economical. The primary disadvantage is that a visual inspection only discovers surface effects and the minimum detectable flaws are often a function of lighting, visual acuity, viewing angle, and inspector accessibility to the defect (FHWA 1986). Furthermore, the numerical rating is often based on a vague word description. Without making some assumptions, it would be impossible to translate those condition ratings into useful information for a reliability analysis. In the absence of any other information, these visual inspections can be useful.

EXAMPLE 36.18

A bridge engineer is attempting to update the life cycle maintenance plans for the managed highway bridges. When the bridges were designed, the girders were scheduled for replacement based on the reliability after a specific degree of corrosion deterioration. There is neither time nor money to perform a specific inspection on every bridge, but the records of the mandated biennial visual inspections are available. The State DOT has implemented the PONTIS Bridge Management System (PONTIS 1995), which assigns specialized condition ratings to virtually all bridge components. One of the components is Element 107 (Painted Open Steel Girders) which is classified into one of five condition states as shown in Table 36.5 (CDOT 1995; Estes and Frangopol 2003). The word description is a bit vague, so the bridge manager interviews the expert inspectors and further quantifies the definitions of the condition states (CSs) as shown in the last two columns of Table 36.5.

TABLE 36.5 CDOT (1995) Suggested Condition State (CS) Ratings for Element 107: Painted Open Steel Girders (First Three Columns) plus Necessary Revisions (Last Two Columns) to Update Reliability

CS	Description	Rust code	Section loss[a] (%)	Density distribution[a]
1	No evidence of active corrosion; paint system sound and protecting the girder	—	0–2	Lognormal
2	Slight peeling of the paint, pitting, or surface rust, etc.	Light R1	0–5	Normal[b]
3	Peeling of the paint, pitting, surface rust, etc.	R1	0–10	Normal[b]
4	Flaking, minor section loss (<10% of the original thickness)	R2	10–30	Normal[b]
4	Flaking, swelling, moderate section loss (>10% but <30% of the original thickness); structural analysis not warranted	R3		
5	Flaking, swelling, moderate section loss (>10% but <30% of the original thickness); structural analysis not warranted due to location of corrosion on member	R3		
5	Heavy section loss (>30% of original thickness); may have holes through base metal	R4	>30	Lognormal

[a] Not part of the PONTIS definition — created to quantify the observed corrosion.
[b] The normal distributions are truncated at zero.
 Source: Estes and Frangopol 2003. Reprinted with permission from the American Society of Civil Engineers.

For example, the bridge manager discovers that when an inspector classifies a girder as CS 2, there will be somewhere between 0 and 5% section loss. Due to the training that the inspectors receive, they can be expected to classify the bridge correctly 98% of the time. The bridge manager observes the inspection results for three different girders. The first girder is given a rating of CS = 2. It was rated as CS = 1 during the last inspection and has just made the transition to CS = 2. The second girder has been classified as CS = 2 for the fourth consecutive inspection. The third girder has been classified as CS = 2 for the past eight inspections. The bridge manager uses historical data on many similar bridges to determine that a bridge typically remains in CS = 2 for 10.8 years before transitioning to CS = 3. What values for thickness loss might the bridge manager use to conduct the reliability analysis and update the maintenance plan for this bridge?

Answer

It is assumed that a girder enters a CS for the first time at the mean value of section loss associated with this CS. CS 2 is assumed to be normally distributed and is correctly classified 98% of the time. The girder enters this CS at mean section loss value $\mu = 2.5\%$. Assuming that the incorrect inspector classifications (2%) are evenly divided between 1% too low and 1% too high, the standard deviation of section loss is computed using the cumulative distribution function of the standard normal variate $\Phi(s)$. Therefore, if the girder is classified as CS = 2, the probability that all values of section loss are less than 5% is 0.99. Therefore,

$$p(\text{loss} < 5\%) = 0.99 = \Phi\left(\frac{\text{loss} - \mu}{\sigma}\right) = \Phi\left(\frac{5.0\% - 2.5\%}{\sigma}\right)$$

The standard deviation is therefore

$$\sigma = \frac{5.0\% - 2.5\%}{\Phi^{-1}(0.99)} = \frac{2.5\%}{2.33} = 1.074\%$$

The standard deviation would have been higher if the quality of the inspectors had been lower and vice versa. For the first girder that has just entered the CS, the percent section loss parameters are $N[2.5, 1.074]$. As the girder remains longer in a given CS, it is assumed to deteriorate. Assuming linear transition through the CS, the mean value of the distribution is expected to shift progressively higher until it reaches 5% section loss as follows:

$$\frac{(5.0 - 2.5)\%}{10.8 \text{ years}} = 0.23\% \text{ per year}$$

Girder 2 has been at CS = 2 for 6 years (i.e., four consecutive inspections that represent three inspection intervals, each lasting 2 years). The mean value for the Girder 2 section loss is

$$\mu = 2.5\% + 0.23\% \text{ per year} (6 \text{ years}) = 3.89\%$$

Assuming the same rate of mean section loss for all girders and a constant standard deviation of the loss, the percent section loss distribution for Girder 2 is therefore $N[3.89, 1.074]$. Finally, Girder 3 has remained at CS = 2 for 14 years (seven inspection intervals), which exceeds the expected 10.8-year transition period. It would be unreasonable to assume a mean value of section loss greater than 5% because an inspector would have logically classified such a girder as CS = 3. The percent section loss distribution for Girder 3 is therefore $N[5.0, 1.074]$ and will remain at that rating until an inspector rates the girder into a higher CS.

This example illustrates just one approach to the problem. The bridge manager could have assumed a constant coefficient of variation rather than a constant standard deviation throughout the CS. The object is to make conservative assumptions and determine how a structure is actually behaving in relation to the deterioration model that may have been postulated several decades earlier. While this approach is not a substitute for a specified NDE inspection, it is better than no additional information. At least, it will identify those structures where a more detailed inspection is needed. This approach has been used on an actual highway bridge (Estes and Frangopol 2003) and on the miter gate for a navigational structure (Estes et al. 2002, 2004).

36.7.3 Inspection Optimization

Because targeted NDE inspections are time consuming and expensive, they should be planned and timed for when they will produce the most benefit. Such an inspection will typically trigger a decision on whether or not to make a repair. The timing of the inspection should coincide with when the information is most needed.

EXAMPLE 36.19

A highway bridge is expected to have a service life of 45 years. The concrete bridge deck is not expected to last that long and will potentially have to be replaced several times during the life of the bridge. The reinforcing in the bridge deck is expected to corrode as chlorides from the deicing salts penetrate the deck and reach a critical concentration at the reinforcing steel. The tensile forces resulting from the corrosion expansion cause the bridge deck to spall. Based on the deterioration model provided in Estes and Frangopol (2001c), the corrosion initiation time is a normally distributed variable with a mean value of 19.6 years and a standard deviation of 7.51 years. The bridge deck will be replaced when there is active corrosion over 50% of the bridge deck (see Example 36.17). The bridge manager plans for three lifetime inspections using the half-cell potential test. The present value cost of replacing the bridge deck is $225,600, and the cost of conducting the half-cell test is $1,072. Using a discount rate of 2% along with the assumptions provided in Estes and Frangopol (2001c), what are the optimal inspection times for the half-cell tests? What is the expected total cost of the inspections and deck replacements?

Answer

Every time an inspection is made, there will be a decision to replace the deck or not replace the deck. With three lifetime inspections, there are eight possible repair/no repair paths available as shown in Figure 36.11. The inspection times are determined by optimizing the minimum expected total lifetime cost $E(C_{tot})$. The constraints are that the expected damage at each inspection time (t_1, t_2, and t_3) and at the end of service life ($t_4 = 45$ years) will be less than the maximum allowable damage of 50% active corrosion. Additional constraints were added so that inspections must be at least 2 years apart, but not more than 20 years apart. Using the optimization program ADS (Vanderplaats 1986), Figure 36.12 shows that the optimum inspection times were at $t_1 = 10.05$ years, $t_2 = 19.76$ years, and $t_3 = 35.45$ years (Estes 1997; Frangopol and Estes 1999; Estes and Frangopol 2001c). The expected percentage of damage, which is a weighted average of the eight possible paths in Figure 36.11, is shown for each time, t. The probability of taking any given path is a function of the probability of failure at the time of inspection, as shown in Figure 36.11. The most likely path, for example, was Branch 6, where the likelihood was 58.9%. The path involves a single repair after the second inspection and no repair after the first and third inspections. This reflects that the deck was 10.2% damaged at t_1, 50.9% damaged at t_2, and 30.1% damaged at t_3. Conversely, the least likely paths (0.1%) were Branch 1 and Branch 8, which are associated with repair after every inspection and after none of the inspections, respectively. The expected damage at any point in time $E(\text{Damage})(t)$ is equal to the sum over all possible branches of the event tree given that the particular branch is taken as $E[\text{Damage}(t) \mid \text{Branch}_i]$ multiplied by the probability of taking that branch P_{b_i}.

$$E(\text{Damage})(t) = \sum_{i=1}^{2^m} E[\text{Damage}(t) \mid \text{Branch}_i] \times P_{b_i}$$

where m is the number of inspections.

The minimized expected total cost $E(C_{tot})$ is the sum of the expected repair costs $E(C_{rep})$ and the inspection costs C_{insp}. The inspection costs are the discounted actual costs of inspection:

$$C_{insp} = \sum_{k=1}^{m} \frac{C_{insp_k}}{(1+r)^{t_k}} = \frac{\$1,072}{(1+0.02)^{10.06}} + \frac{\$1,072}{(1.02)^{19.76}} + \frac{\$1,072}{(1.02)^{35.45}} = \$2,045$$

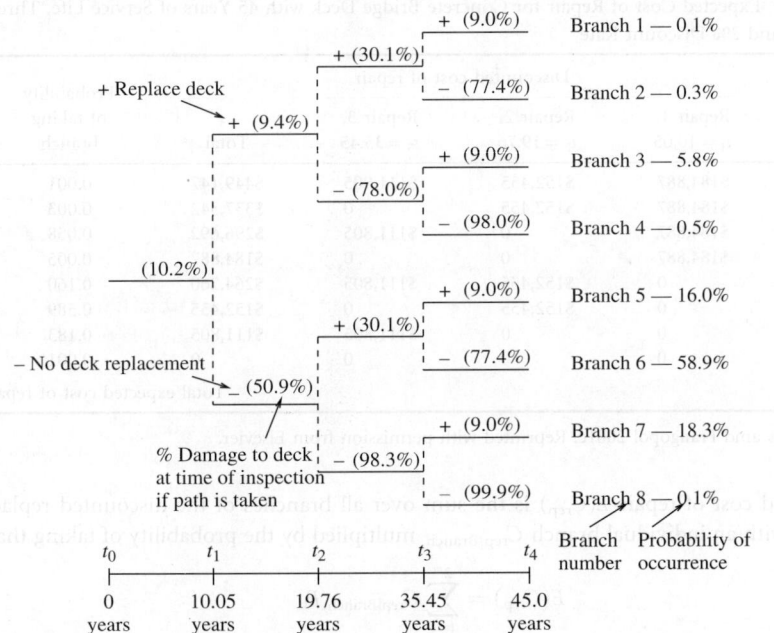

FIGURE 36.11 Event tree for the optimum inspection strategy for a 45-year bridge deck using three lifetime inspections. (Estes and Frangopol 2001c. Reprinted with permission from Elsevier.)

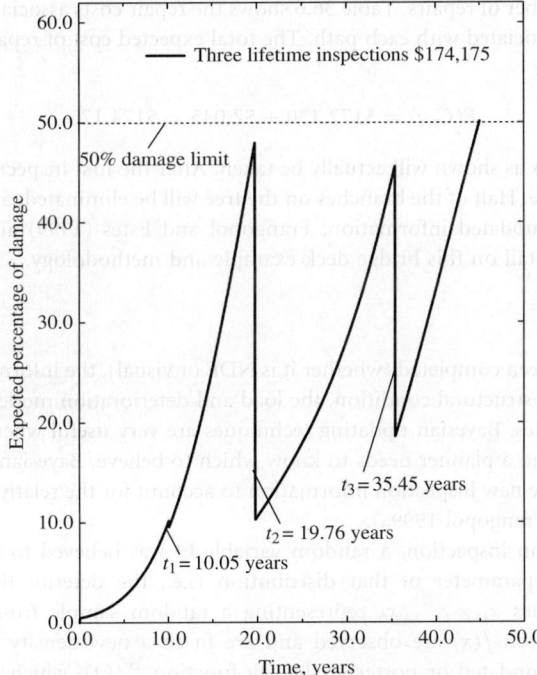

FIGURE 36.12 Optimum inspection strategy and expected damage for a bridge deck with 45 years of service life, three lifetime inspections, and 2% discount rate. (Estes and Frangopol 2001c. Reprinted with permission from Elsevier.)

TABLE 36.6   Expected Cost of Repair for Concrete Bridge Deck with 45 Years of Service Life, Three Lifetime Inspections, and 2% Discount Rate

Event tree	Discounted cost of repair				Probability of taking branch	Expected cost of repair
	Repair 1, $t_1 = 10.05$	Repair 2, $t_2 = 19.76$	Repair 3, $t_3 = 35.45$	Total		
Branch 1	$184,887	$152,455	$111,805	$449,147	0.001	$449
Branch 2	$184,887	$152,455	0	$337,342	0.003	$1,012
Branch 3	$184,887	0	$111,805	$296,692	0.058	$17,208
Branch 4	$184,887	0	0	$184,887	0.005	$924
Branch 5	0	$152,455	$111,805	$264,260	0.160	$42,281
Branch 6	0	$152,455	0	$152,455	0.589	$89,796
Branch 7	0	0	$111,805	$111,805	0.183	$20,460
Branch 8	0	0	0	0	0.001	0
					Total expected cost of repair	$172,130

Source: Estes amd Frangopol 2001c. Reprinted with permission from Elsevier.

The expected cost of repair $E(C_{rep})$ is the sum over all branches of the discounted replacement costs associated with an individual branch $C_{rep|Branch_i}$ multiplied by the probability of taking that branch P_{b_i}.

$$E(C_{rep}) = \sum_{i=1}^{2^m} C_{rep|Branch_i} P_{b_i}$$

$$C_{rep|Branch_i} = \sum_{j=1}^{nr} \frac{C_{rep_j}}{(1+r)^{t_j}} = \sum_{j=1}^{nr} \frac{\$225,600}{(1+r)^{t_j}}$$

where nr is the total number of repairs. Table 36.6 shows the repair costs associated with each inspection and the expected cost associated with each path. The total expected cost of repair is $172,130. The total expected cost is therefore

$$E(C_{tot}) = \$172,130 + \$2,045 = \$174,175$$

None of the eight paths as shown will actually be taken. After the first inspection, a decision to repair or not repair will be made. Half of the branches on the tree will be eliminated, and the optimization will be run again using the updated information. Frangopol and Estes (1999) and Estes and Frangopol (2001c) provide more detail on this bridge deck example and methodology.

36.7.4 Updating

Once an inspection has been completed (whether it is NDE or visual), the information should be used to update the assessment of structural condition, the load and deterioration models, and the future repair and maintenance strategies. Bayesian updating techniques are very useful when faced with two sets of uncertain information and a planner needs to know which to believe. Bayesian updating uses both the prior information and the new inspection information to account for the relative uncertainty associated with each (Enright and Frangopol 1999a).

Assume that prior to an inspection, a random variable Θ was believed to have a density function $f'(\Theta)$, where Θ is the parameter of that distribution (i.e., the deterioration model). During an inspection, a set of values $x_1, x_2, \ldots, x_n$ representing a random sample from a population X with underlying density function $f(x)$ are observed and are fit to a new density function $f(x_i)$ (i.e., the inspection results). The updated or posterior density function $f''(\Theta)$, which uses both sets of information and provides the best use of both, can be expressed as (Ang and Tang 1975)

$$f''(\Theta) = kL(\Theta)f'(\Theta)$$

where $L(\Theta)$ is a likelihood function and k is a normalizing constant. For the case where both $f'(\Theta)$ and $f(x)$ are normally distributed, the posterior function $f''(\Theta)$ is also normally distributed and has the mean value and standard deviation, respectively, as

$$\mu'' = \frac{\mu(\sigma')^2 + \mu'(\sigma)^2}{(\sigma')^2 + (\sigma)^2} \quad \text{and} \quad \sigma'' = \sqrt{\frac{(\sigma')^2(\sigma)^2}{(\sigma')^2 + (\sigma)^2}}$$

where μ, μ', and μ'' represent the mean values of the inspection results, the prior distribution, and the posterior distribution, respectively. The values σ, σ', and σ'' represent the standard deviations of those same distributions.

EXAMPLE 36.20

An engineer is attempting to update the reliability analysis of a miter gate on a navigation lock to justify a major rehabilitation. The miter gate is supported by vertically framed steel beams that are corroding over time. Separate deterioration models were used for the atmospheric zone of the gate, which is only exposed to the air, and the splash zone, which is exposed to raising and lowering of the water in the lock chamber. The predicted values for the mean and standard deviation of the thickness loss (in μm) after 50 years based on the deterioration models for the atmospheric zone are $\mu = 298$ and $\sigma = 197$, respectively. The splash zone is expected to corrode much faster, and its 50-year predicted loss is $\mu = 5,085$ and $\sigma = 1,301$. The miter gate undergoes a periodic visual inspection. With some conservative assumptions, these visual inspections are converted to probabilistic quantities (Estes et al. 2004). The inspection results reveal that the actual section loss is $\mu = 285$ and $\sigma = 151$ and was about the same in both the splash zone and the atmospheric zone. What updated values for section loss will be used in the updated reliability analysis?

Answer

For the atmospheric zone, the posterior distribution is characterized by

$$\mu'' = \frac{285(197)^2 + 298(151)^2}{(197)^2 + (151)^2} = 290$$

$$\sigma'' = \sqrt{\frac{(197)^2(151)^2}{(197)^2 + (151)^2}} = 120$$

For the atmospheric zone, the posterior distribution is characterized by

$$\mu'' = \frac{285(1,301)^2 + 5,085(151)^2}{(1,301)^2 + (151)^2} = 348$$

$$\sigma'' = \sqrt{\frac{(1,301)^2(151)^2}{(1,301)^2 + (151)^2}} = 150$$

Figure 36.13 (Estes et al. 2004) shows the prior, inspection, and posterior distributions for the corrosion loss in both the splash zone and the atmospheric zone. Because of the high degree of corrosion expected in the splash zone and its associated uncertainty, the inspection results dominated the posterior distribution. The projected deterioration and its uncertainty were much less in the atmospheric zone. The deterioration model and the inspection results received about equal weight in computing the posterior distribution. Estes et al. (2002, 2004) provide a complete description of this example.

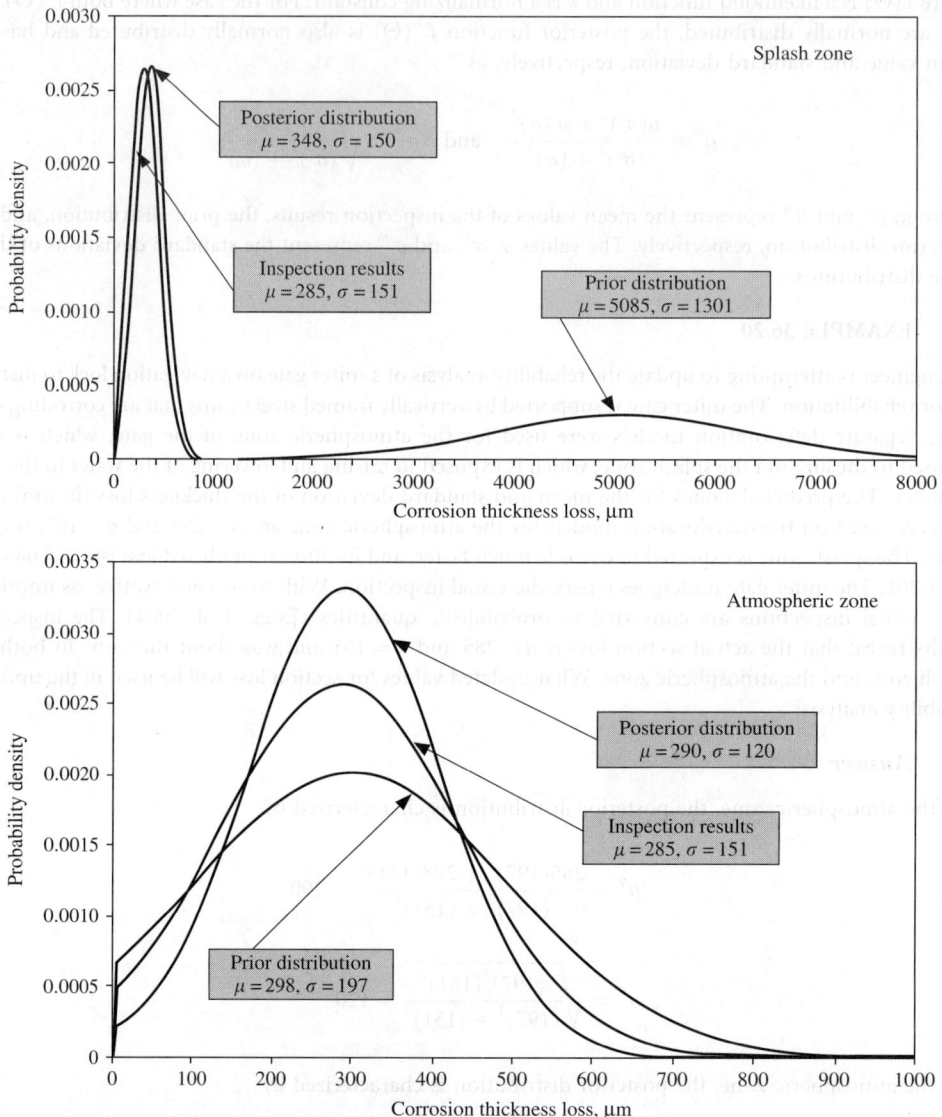

FIGURE 36.13 Bayesian updating based on inspection results for the atmospheric and splash zones on a corroding miter gate. (Estes et al. 2004. Reprinted with permission from Elsevier.)

36.8 Maintenance and Repair

Maintenance and repair are actions taken to slow a structure's rate of deterioration, improve its performance, and lengthen its useful life. The options range from preventive maintenance to partial repairs to replacement of major components. For each option, the analyst needs to know the cost of the remediation and its positive effect on the safety of the structure. Based on the answer, the analyst needs to optimize the timing of the maintenance action and compare its benefits to other available options. The situation will be different for every type of structure and deterioration.

The problem is illustrated conceptually in Figure 36.14 (Frangopol et al. 2001c), where under ideal conditions, a structure would perform as intended without maintenance throughout its useful life.

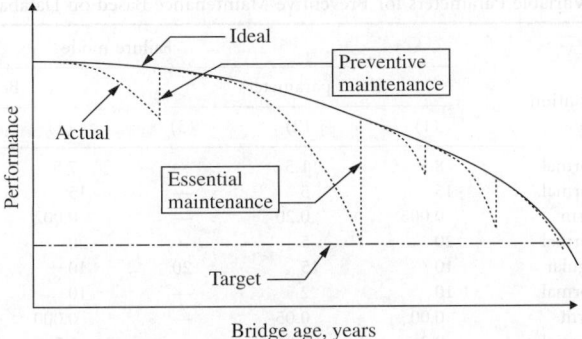

FIGURE 36.14 The effect of various types of maintenance on a typical structure. (Frangopol et al. 2001c. Reprinted with permission from the International Association for Bridge and Structural Engineering.)

TABLE 36.7 Description of the Random Variables Incorporated into an Analysis of the Effects of Preventive Maintenance

Variable	Description
β_0	Initial reliability index
t_I	Time of damage initiation
α	Performance deterioration rate without maintenance
γ	Immediate improvement in reliability index after the application of preventive maintenance
t_R	Rehabilitation time (i.e., time at which the minimum acceptable reliability level is reached without maintenance)
t_{PI}	Time of first application of preventive maintenance
t_P	Time of reapplication of preventive maintenance
t_{PD}	Duration of preventive maintenance effect on bridge reliability
θ	Reliability index deterioration rate during preventive maintenance effect
t_{RP}	Rehabilitation time associated with a preventive maintenance program (i.e., time at which the minimum acceptable reliability level is reached considering preventive maintenance)

Its reliability would decline over time due to a natural aging process, fair wear and tear on the structure, or a gradual increase in the load on the structure and would not fall below an acceptable level during the life of the structure. In reality, structures deteriorate, and various maintenance or repair actions can improve the reliability. It may be preventive maintenance long before a deterioration becomes serious or some form of essential maintenance to prevent the structure from falling below a critical performance level (see Figure 36.14).

36.8.1 Preventive Maintenance

Preventive maintenance may delay the onset of deterioration. A fresh coat of paint will often prevent a steel beam from corroding or a wooden support from rotting. This type of maintenance may also relieve additional stresses on the structure by cleaning an expansion joint or lubricating a bearing. It is often a continuous program with a predictable cost, but the benefits can be difficult to quantify.

Frangopol (1998) and Frangopol et al. (2001b) defined the eight random variable models associated with every aspect of the preventive and essential maintenance processes. This model was successfully applied to individual and groups of bridges in the United Kingdom and the United States (Frangopol et al. 2001b; Frangopol 2003; Frangopol and Neves 2003; Neves and Frangopol 2004; Petcherdchoo et al. 2004). One example of application consisted of maintenance of 713 existing steel/concrete bridges constructed between 1955 and 1998 in the United Kingdom (Frangopol et al. 2001b). Table 36.7 describes all the random variables considered in this example, and Table 36.8 lists their values with

TABLE 36.8 Random Variable Parameters for Preventive Maintenance Based on Database of Existing Bridges

Random variable	Distribution type	Failure mode					
		Shear parameters			Bending parameters		
		(1)	(2)	(3)	(1)	(2)	(3)
β_0	Lognormal	8.5	1.5	—	7.5	1.2	—
t_I (years)	Lognormal	15	5	—	15	5	—
α (years^{-1})	Uniform	0.005	0.20	—	0.002	0.10	—
t_{PI} (years)	Lognormal	20	5	—	20	5	—
t_P (years)	Triangular	10	15	20	10	15	20
t_{PD} (years)	Lognormal	10	2	—	10	2	—
θ (years^{-1})	Uniform	0.00	0.05	—	0.000	0.025	—
γ	Lognormal	0.2	0.04	—	0.2	0.04	—

Note: Distribution parameters — lognormal: (1) mean value and (2) standard deviation; uniform: (1) minimum value and (2) maximum value; triangular: (1) minimum value, (2) mode value, and (3) maximum value.

Source: Frangopol et al. 2001b. Reprinted with permission from the American Society of Civil Engineers.

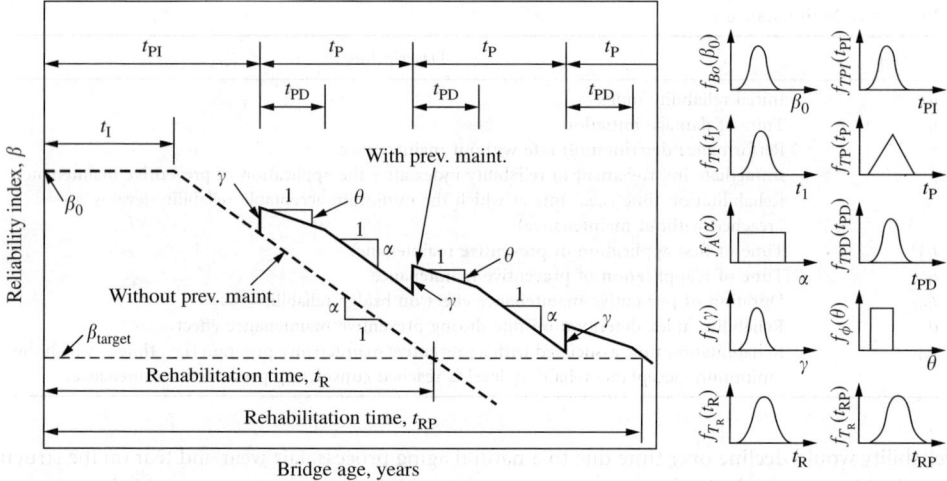

FIGURE 36.15 Random variables affecting the reliability of a group of bridges based on a preventive maintenance program. (Frangopol et al. 2001b. Reprinted with permission from the American Society of Civil Engineers.)

respect to shear and moment failure modes of this group of bridges (Frangopol et al. 2001b). Figure 36.15 illustrates the distributions associated with these random values and a sample life cycle plan that shows three applications of preventive maintenance (Frangopol et al. 2001b). Given these variables, Monte Carlo simulation was used to obtain the distributions of the times when essential maintenance is required (i.e., the so called rehabilitation time) with and without previous preventive maintenance actions. The analysis provides the basis for planning the number and timing of preventive maintenance activities to extend the service life of the group of bridges by the desired time period.

EXAMPLE 36.21

Using the mean values for the parameters in Table 36.8, devise a deterministic preventive maintenance plan for a bridge with respect to shear that will extend its useful life by 10 years. Assume that the minimum acceptable reliability index is $\beta_{target} = 3.0$. If each application of preventive maintenance costs $3,000, what is the present value preventive maintenance cost using a discount rate of 4%?

Answer

For these values, the current life of the bridge structure is

$$t_R = t_1 + \frac{\beta_0 - \beta_{\text{target}}}{\alpha} = 15 + \frac{8.5 - 3.0}{0.1025} = 15 + 53.6 = 68.6 \approx 69 \text{ years}$$

If the bridge life is to be extended 10 years, then a preventive maintenance program will be designed such that $t_{RP} = 79$ years. The preventive maintenance cycle, t_P, lasts 15 years and the change in reliability index during that time is

$$\Delta\beta_{\text{maint}} = t_{PD}\theta + (t_p - t_{PD})\alpha - \gamma = 10(0.025) + (15 - 10)(0.1025) - 0.12 = 0.5625$$

For the equivalent 15-year time period without maintenance, the change in reliability index is

$$\Delta\beta_{\text{no-maint}} = t_p\alpha = 15(0.1025) = 1.5375$$

The benefit of a preventive maintenance application in terms of increment in reliability index is

$$\Delta\beta = \Delta\beta_{\text{no-maint}} - \Delta\beta_{\text{maint}} = 1.5375 - 0.5625 = 0.975$$

To extend the life of the bridge 10 years, $\Delta\beta_{\text{reqd}}$ needs to be

$$\Delta\beta_{\text{reqd}} = 10\alpha = 10(0.1025) = 1.025$$

The number of maintenance applications (n) to achieve this is

$$n = \frac{\Delta\beta_{\text{reqd}}}{\Delta\beta} = \frac{1.025}{0.975} = 1.05$$

Therefore, two preventive maintenance actions must be applied ($n = 2$). One cost-effective strategy is to schedule the maintenance as late in the life of the structure as possible. One approach is to gain as much as possible from the positive effect of the final preventive maintenance action which lasts $t_{PD} = 10$ years. That would suggest applying the second maintenance action at year 69 and the first maintenance action 15 years earlier, at year 54. Figure 36.16 illustrates this strategy, which combines scheduled maintenance late in the life of the structure and maintains the reliability at a high level. The present value of preventive maintenance cost (C_{PM}), discounted at year 0 (i.e., time of construction), is

$$C_{PM} = \frac{\$3,000}{(1 + 0.04)^{54}} + \frac{\$3,000}{(1.04)^{69}} = \$360 + \$200 = \$560$$

The plan should also be checked with respect to the moment failure mode.

In many real world examples, it would not make sense to move preventive maintenance until the end of the life of the structure, but the strategy makes sense in the limited context of the problem. If failure costs were included in the analysis, there may be advantages to keeping the lifetime reliability of the structure higher and thus maintaining the structure sooner.

36.8.2 Repairs

The strategy for making repairs is similar to that for preventive maintenance as different options and their effects must be evaluated. The costs of some structural repairs are functions of the level of deterioration. Automobile body work is an example, where the amount of filling, sanding, and addition of extra material is proportional to the amount of deterioration. The costs of other structural repairs are independent of the level of structural deterioration, as, for example, adding a reinforcing plate or replacing a member, where the cost is the same regardless of the degree of deterioration. Choosing the proper mix of repairs and preventive maintenance and then timing their application in an efficient manner can be a difficult challenge.

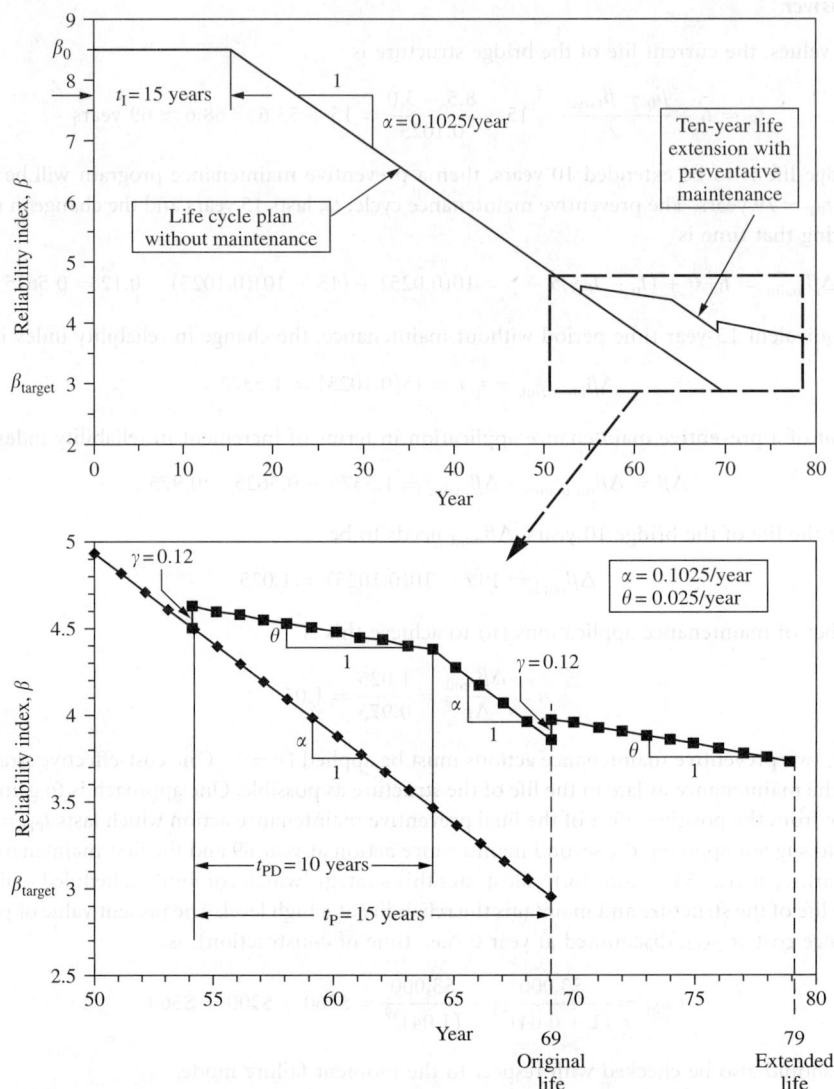

FIGURE 36.16 Preventive maintenance plan to extend the life of a bridge deck by 10 years.

EXAMPLE 36.22

How can one determine the best combination of repair and maintenance options for a given structure or category of structures?

Answer

Kong and Frangopol (2003b) looked at eight different maintenance scenarios and used a program LCADS (Life-Cycle Analysis of Deteriorating Structures) to predict the reliability profiles for an individual bridge associated with each maintenance strategy. The eight scenarios shown in Figure 36.17 include both preventive and essential maintenance. Sample combinations include no maintenance (E0), three applications of essential maintenance (E3), and two applications of essential maintenance with periodic preventive maintenance (E2,P). Figure 36.18 shows the mean system reliability index

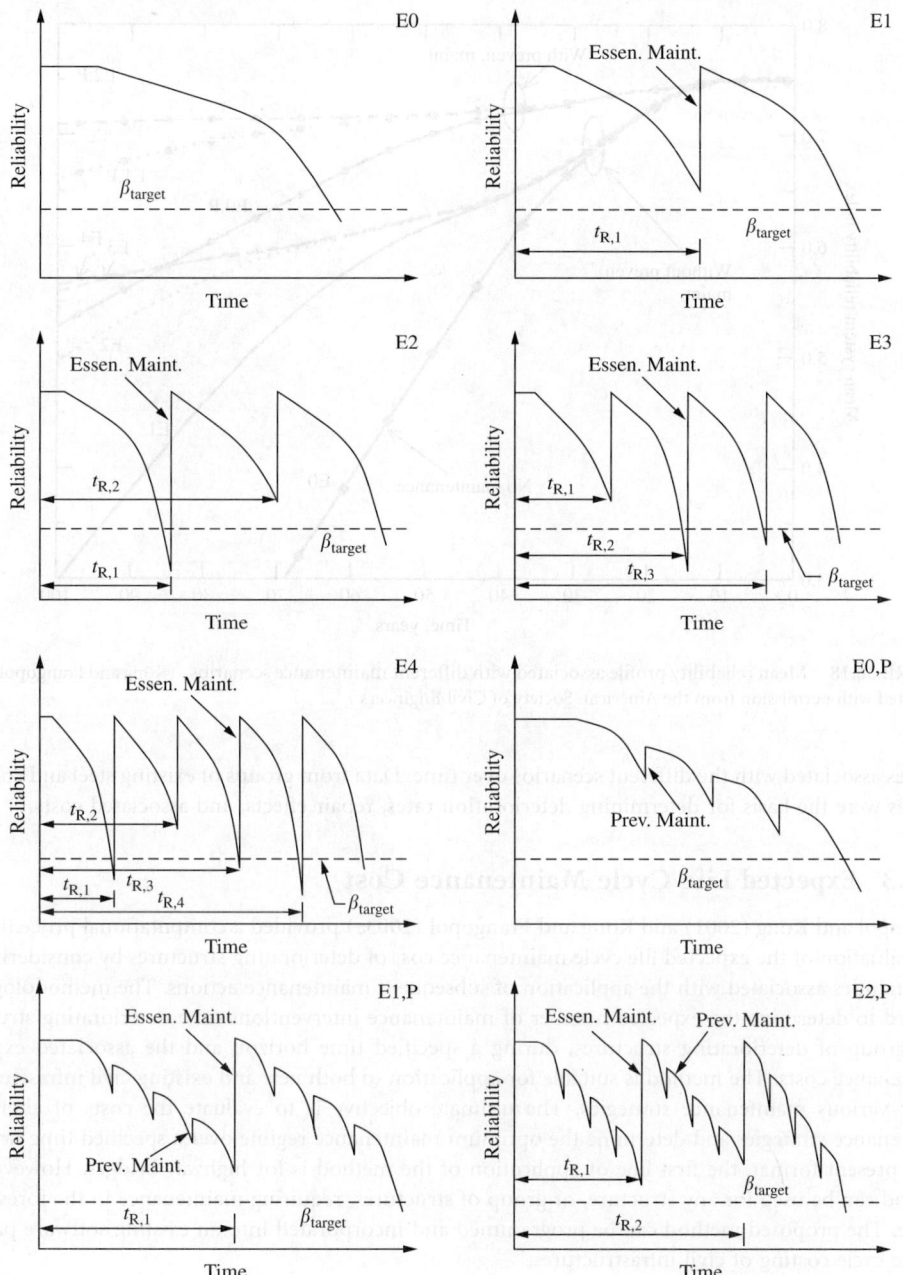

FIGURE 36.17 Eight different maintenance scenarios for a steel/concrete bridge involving various combinations of preventive and essential maintenance. (Kong and Frangopol 2003b. Reprinted with permission from the American Society of Civil Engineers.)

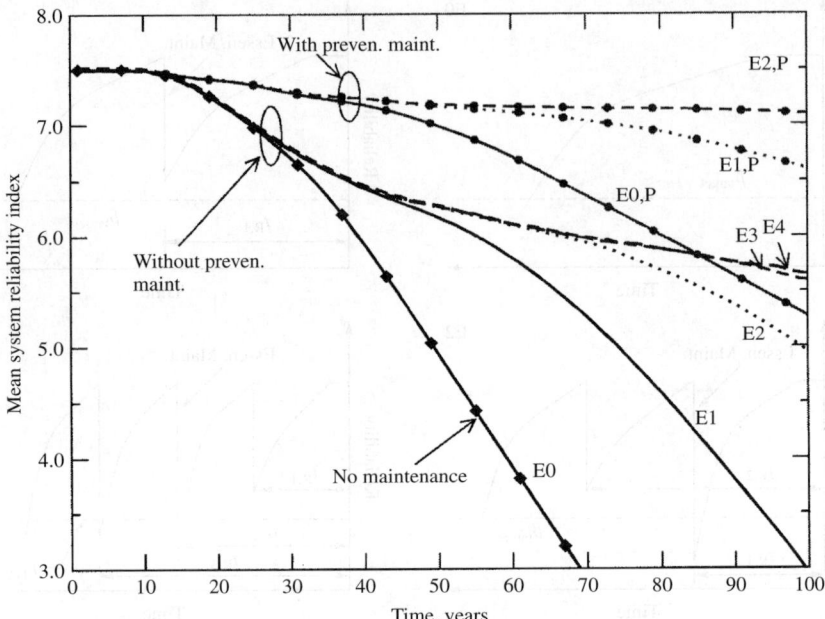

FIGURE 36.18 Mean reliability profile associated with different maintenance scenarios. (Kong and Frangopol 2003b. Reprinted with permission from the American Society of Civil Engineers.)

profiles associated with the different scenarios over time. Data from groups of existing steel and concrete bridges were the basis for determining deterioration rates, repair effects, and associated costs.

36.8.3 Expected Life Cycle Maintenance Cost

Frangopol and Kong (2001) and Kong and Frangopol (2003a) provided a computational procedure for the evaluation of the expected life cycle maintenance cost of deteriorating structures by considering the uncertainties associated with the application of subsequent maintenance actions. The methodology can be used to determine the expected number of maintenance interventions on a deteriorating structure, or a group of deteriorating structures, during a specified time horizon and the associated expected maintenance costs. The method is suitable for application to both new and existing civil infrastructures under various maintenance strategies. The ultimate objective is to evaluate the costs of alternative maintenance strategies and determine the optimum maintenance regime over a specified time horizon. In its present format, the first line of application of the method is for highway bridges. However, the method can be used for any structure, or group of structures, requiring maintenance in the foreseeable future. The proposed method can be programmed and incorporated into an existing software package for life cycle costing of civil infrastructures.

36.9 Discount Rate

As shown in several earlier examples, the choice of repair and the timing of both inspection and repair can be highly dependent on the discount rate of money. With a higher discount rate, expenditures and investments made later in the life of a structure become more attractive. There is considerable debate as to what the correct discount rate should be. Some governments mandate a rate (such as 6%) that must be used on all public civil infrastructure projects, which at least ensures consistency between these

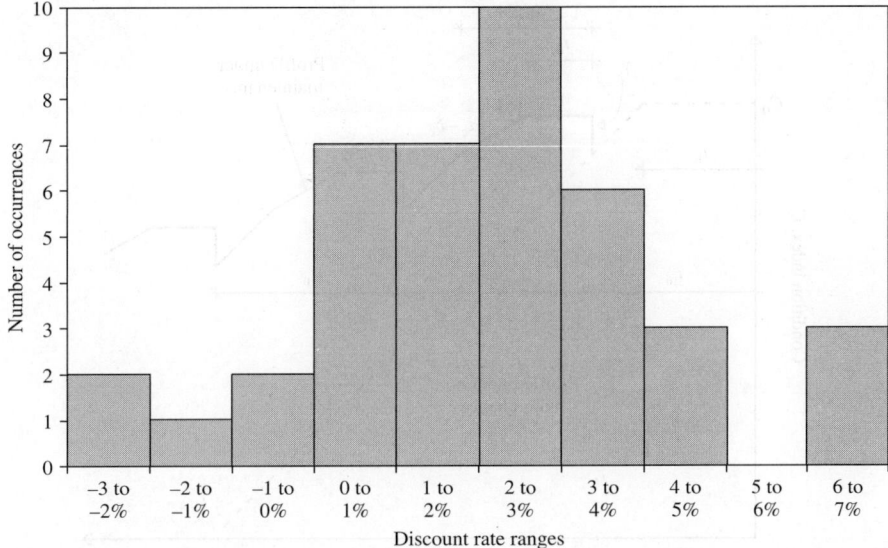

FIGURE 36.19 Histogram showing the historical discount rates from the years 1962 to 2002; the discount rate was defined as the interest rate on 1-year U.S. Treasury bills minus the inflation rate (from Federal Reserve 2003; McMahon 2003).

projects. A more rational approach is to use the difference between the rate of return on a risk-free investment and the inflation rate as the appropriate discount rate, which is about 2 to 3%. Since these rates are well publicized over time, the discount rate could easily become a random variable based on years of past experience. Using this approach, Figure 36.19 shows a histogram reflecting the discount rate in the United States from 1962 to 2002, which reflects the difference between the short-term interest rate offered on 1-year U.S. Treasury bills (Federal Reserve 2003) and the annual inflation rate (McMahon 2003). With these data, the discount rate has a mean value of 2.07% and a standard deviation of 2.11%. The results would change somewhat depending on the years included in the study, the type of investment classified as risk free, and the commodity prices used to determine the inflation rate.

36.10 Condition and Reliability Indices as Joint Performance Indicators

The use of visual inspections and the resulting condition index has been, in the past three decades, the primary source of information on deteriorating bridge structures. As was previously discussed, the reliability index is a consistent measure of structural safety. Therefore, structure management systems should be based on reliability (Frangopol et al. 2000). Frangopol (2002) and Frangopol and Neves (2003) and Neves and Frangopol (2004) proposed a model incorporating condition and reliability indices as joint measures of performance. The condition and reliability indices are considered constant during a period after construction denoted as time of initiation of damage, t_{ic} and t_i, respectively. After this period, the deterioration rates of condition and reliability indices, α_c and α, are considered constant.

Each maintenance action can lead to one, several, or all of the following effects: (a) increase in the condition index or reliability index immediately after application, (b) suppression of the deterioration in condition index or reliability index during a time interval after application, and (c) reduction of the deterioration rate of condition index or reliability index during a time interval after application.

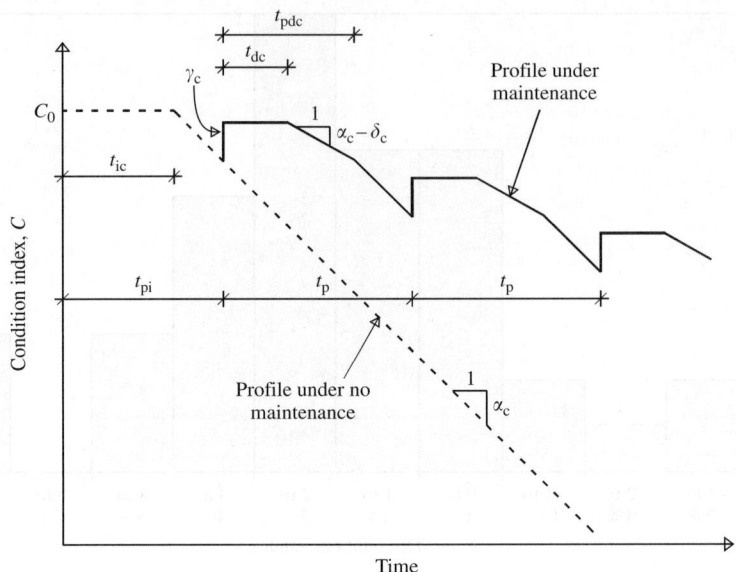

FIGURE 36.20 Condition index profile without or with maintenance. (Neves and Frangopol 2004. Reprinted with permission from Elsevier.)

These effects are modeled through several random variables such as (a) increase in condition index and reliability index immediately after application, γ_c and γ, respectively; (b) time during which the deterioration processes of condition index and reliability index are suppressed, t_{dc} and t_d, respectively; and (c) time during which the deterioration rates in condition index and reliability index are suppressed or reduced, t_{pdc} and t_{pd}, respectively. All variables characterizing the condition profile without or with preventive maintenance are indicated in Figure 36.20. The variables defining the reliability index profiles with or without maintenance are defined in a similar manner (Frangopol and Neves 2003; Neves and Frangopol 2004).

EXAMPLE 36.23

A bridge manager is responsible for maintaining a group of highway bridges. One of these bridges is currently 15 years old and is deteriorating fast. The manager feels that the bridge might become unsafe ($\beta < 3.0$) before reaching its prescribed service life (i.e., 75 years). Analyzing the existing records on the performance of the bridge during the past 15 years the manager realizes that there were no visual defects in the first 3 years of service, after which small cracks were visible. The first signs of corrosion appeared when the bridge was 10 years old. Measuring the loss of reinforcement at the present time (at year 15) the manager estimated that a loss of bending resistance of 4% had occurred. The most economical maintenance action available is the replacement of the concrete cover. This will correct visual defects and delay deterioration. The bending load effect and the bending resistance (both in kN m) at the beginning of service life are characterized by normal distributions $N[85, 12]$ and $N[180, 9]$, respectively.

It is assumed that the replacement of concrete cover improves the condition index C to its initial value (at time $t = 0$). After this concrete replacement, the condition index does not deteriorate for a period of 3 years, as is the case of a new structure. The replacement of the concrete cover does not improve the reliability index but delays any further deterioration until the condition index reaches the level corresponding to initiation of corrosion. Using the above assumptions, estimate the effect of concrete cover replacement on condition and reliability and find the time of application of maintenance such that the target reliability index is not violated during the entire service life.

Answer

The visual deterioration of a bridge can be assessed by the condition index. This index is defined in discrete terms as 0 — no chloride contamination; 1 — onset of corrosion; 2 — onset of cracking; and 3 — loose concrete/significant delamination. According to existing records, the initiation time of condition deterioration is 3 years. Assuming that the coefficient of variation of the resistance is constant over time, the mean resistance reduction of 4% results in a bending resistance after 15 years that is characterized by a normal distribution with a mean and standard deviation as follows:

$$\mu^{R}_{15\,years} = 180 \times 0.96 = 172.8 \text{ kN m} \quad \text{and} \quad \sigma^{R}_{15\,years} = 9 \times 0.96 = 8.64 \text{ kN m}$$

The initial (at time $t = 0$) reliability index and the reliability index at 15 years are

$$\beta(t = 0) = \frac{180 - 85}{\sqrt{12^2 + 9^2}} = 6.333$$

$$\beta(t = 15) = \frac{172.8 - 85}{\sqrt{12^2 + 8.64^2}} = 5.938$$

The time of initiation of deterioration of reliability is equal to the time for corrosion to start (i.e., 10 years). The deterioration rate of the reliability index is given by

$$\alpha = \frac{\beta(t = 10) - \beta(t = 15)}{15 - 10} = \frac{6.333 - 5.938}{5} = 0.079 \text{ per year}$$

The condition index deterioration rate, in absolute value, is

$$\alpha_c = \frac{C(t = 10) - C(t = 3)}{10 - 3} = \frac{1 - 0}{7} = 0.143 \text{ per year}$$

The effect of concrete cover replacement is analyzed at different points in time. To ensure a reliability index above the minimum target during the entire lifetime, maintenance should be applied when the condition index reaches the value 3.0. Figure 36.21 shows the effect of this maintenance action on the condition and reliability indices. Therefore, the times of application of maintenance action such that the target reliability index is not violated during the entire service life are when the bridge is 24, 48, and 72 years old.

EXAMPLE 36.24

Consider the previous example, but assume that uncertainties are associated with the parameters defining the condition and reliability indices as shown in Table 36.9. Compute the mean and standard deviation profiles of the performance indicators of the bridge considering that maintenance is applied when the condition index reaches $C = 3.0$.

Answer

The mean and standard deviation of condition and reliability can be computed using Monte-Carlo simulation. The mean and standard deviation profiles for this scenario obtained by using 50,000 simulations are shown in Figure 36.22. These probabilistic results, which consider the uncertainties associated with both bridge deterioration and the effect of maintenance actions, show that the resulting performance profiles are considerably smoother than those obtained deterministically. Furthermore, the standard deviation profile shows that the uncertainty in performance varies over time. The mean reliability index at $t = 75$ years is close to the value obtained in the previous analysis. However, due to uncertainty in the variables that define the reliability index profile, there exists a probability that the reliability index is less than 3.0.

These results illustrate the importance of probabilistic analysis. The uncertainties associated with the deterioration and effects of maintenance actions can lead to performance much different from that predicted from a deterministic analysis.

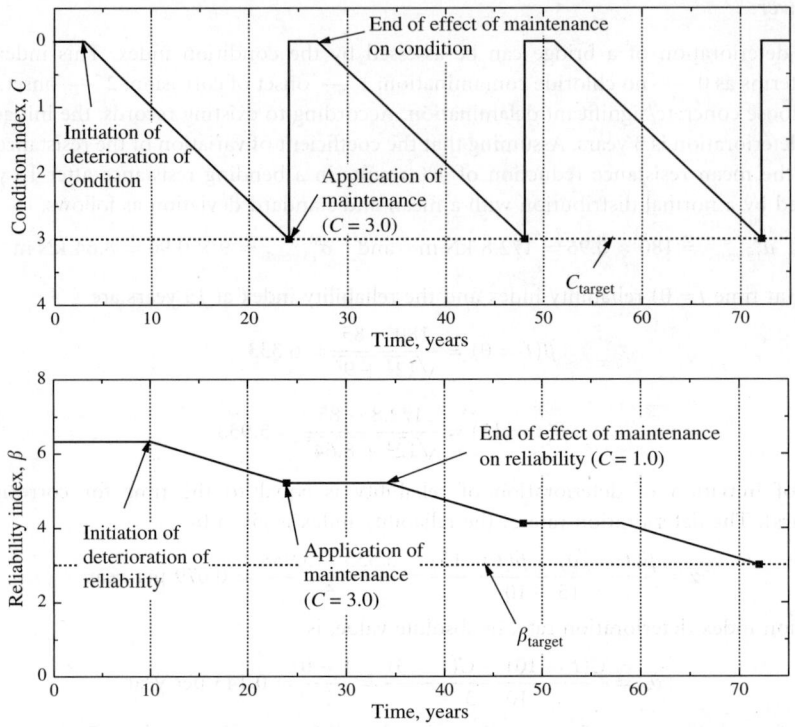

FIGURE 36.21 Condition and reliability index profiles for bridge considering cyclic replacement of concrete cover.

TABLE 36.9 Parameters Defining the Profiles under no Maintenance and Effects of Replacing Concrete Cover

	Variables	Values
Profiles under no maintenance	Initial condition index	0
	Deterioration rate of condition index (year^{-1})	$T(0.09; 0.14; 0.19)$[a]
	Time of initiation of deterioration of condition index (years)	3
	Initial reliability index	6.33
	Deterioration of reliability index (year^{-1})	$T(0.03; 0.079; 0.13)$[a]
	Time of initiation of deterioration of reliability index (years)	10
Effects of replacement of concrete cover	Improvement in condition index	To $C = 0$
	Delay in condition index deterioration (years)	$T(2; 3; 4)$[a]
	Improvement in reliability index	0
	Delay in reliability index deterioration	Until $C = 1$

[a] $T(a, b, c)$ represents a triangular distributed random variable with $a = $ minimum, $b = $ mode, and $c = $ maximum.

36.11 Life Cycle Cost Examples

Since life cycle cost analysis and reliability methods are still relatively new, the methodology and procedures are still being developed. The research is admittedly incomplete. There are many researchers who are suggesting approaches, developing techniques, and applying them to real structures. It is from these studies that standardized procedures and prescribed codes will eventually emerge. This final section briefly examines three such case studies that have attempted to apply many of the concepts discussed in this chapter to actual structures.

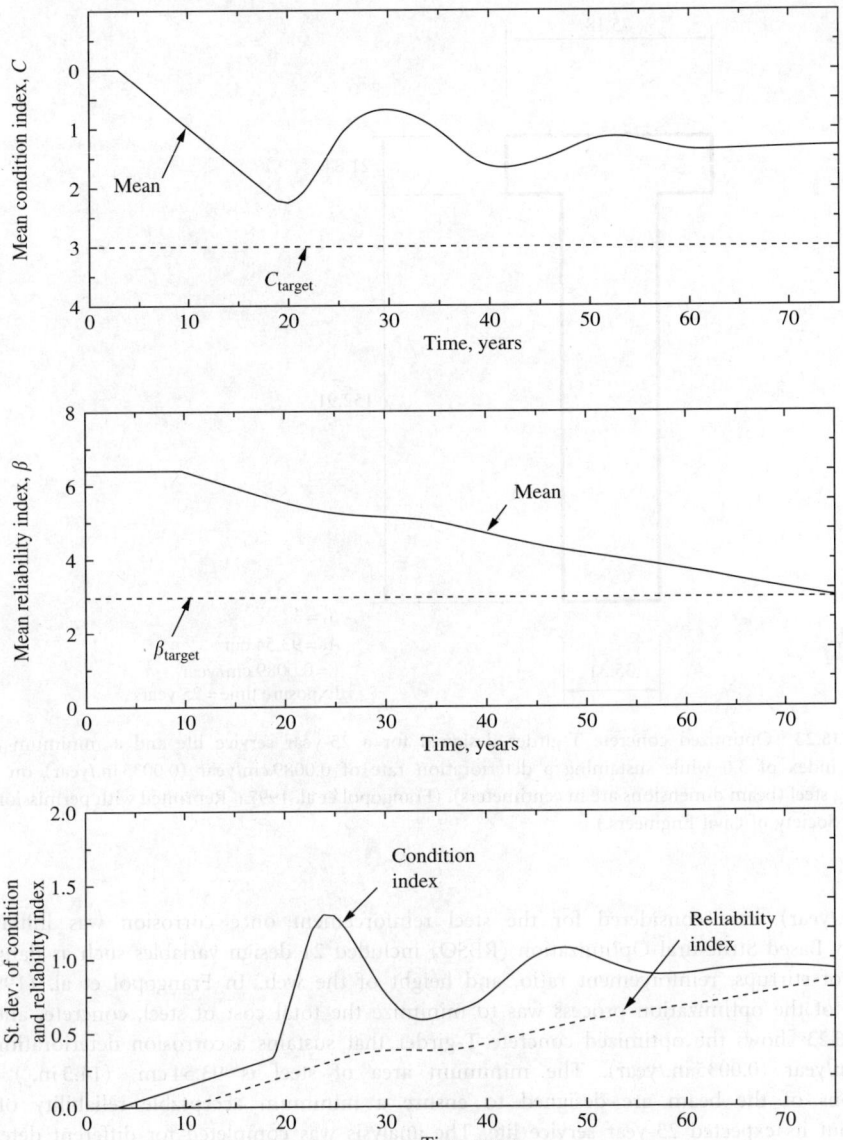

FIGURE 36.22 Mean and standard deviation of bridge condition and reliability index profiles considering cyclic replacement of concrete cover.

Case 36.1

Frangopol et al. (1997a,b) examined the life cycle cost design of reinforced concrete T-girders on highway bridges. The deterioration mechanism was corrosion of the reinforcing steel after sufficient chloride penetration into the concrete. The AASHTO (1992) bridge specification equations for both moment and shear capacities were used to optimize the T-girder for various life spans without the need for maintenance. Corrosion rates v of 0.0064, 0.0089, and 0.0114 cm/year (0.0025, 0.0035, and

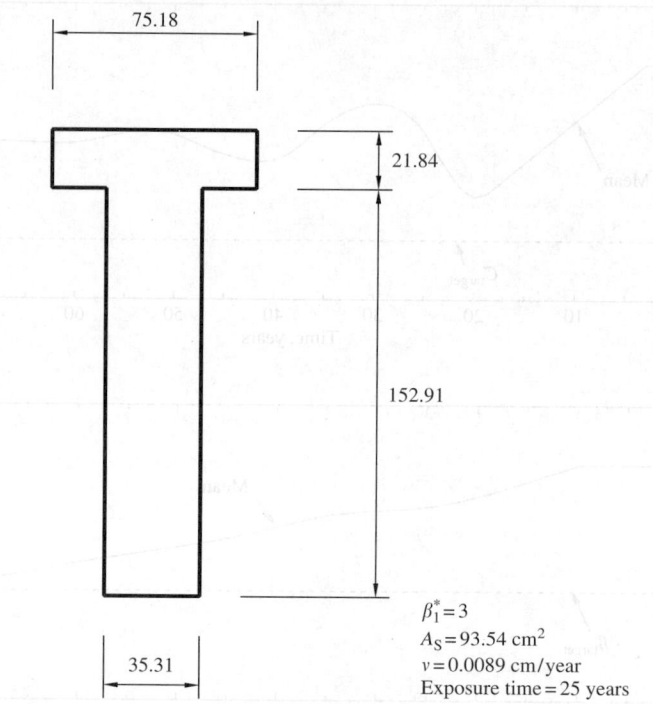

FIGURE 36.23 Optimized concrete T-girder designed for a 25-year service life and a minimum acceptable reliability index of 3.0 while sustaining a deterioration rate of 0.0089 cm/year (0.0035 in./year) on corroding reinforcing steel (beam dimensions are in centimeters). (Frangopol et al. 1997a. Reprinted with permission from the American Society of Civil Engineers.)

0.0045 in./year) were considered for the steel reinforcement once corrosion was initiated. The Reliability Based Structural Optimization (RBSO) included 21 design variables such as flange width, number of stirrups, reinforcement ratio, and height of the web. In Frangopol et al. (1997a), the objective of the optimization process was to minimize the total cost of steel, concrete, and failure. Figure 36.23 shows the optimized concrete T-girder that sustains a corrosion deterioration rate of 0.0089 cm/year (0.0035 in./year). The minimum area of steel is 93.54 cm^2 (14.5 in.^{2}) and the dimensions of the beam are designed to ensure a minimum acceptable reliability of $\beta = 3.0$ throughout its expected 25-year service life. The analysis was completed for different deterioration rates, service lives, and acceptable reliabilities. Figure 36.24 shows that the optimum reliability index is $\beta^* = 2.5$ for a T-girder with an expected 75-year life and a deterioration rate of 0.0064 cm/year (0.0025 in./year) based on the minimum total life cycle cost. The cost of the steel was assumed to be 50 times as great as the cost of the concrete, and the failure cost was 10,000 times the cost of the concrete.

Next, Frangopol et al. (1997b) included preventive maintenance costs, which increased linearly over time. Inspection costs were included, and four inspection methods with varying costs and capabilities were available. The cost of repair was a function of the degree of damage, and an applied aging factor dictated the effect of the postrepair reliability. With multiple lifetime inspections, an event tree accounted for the repair/no repair decisions made after the inspection. The total expected life cycle cost was based on the consequences and probability associated with each branch on the event tree (Frangopol et al. 1997b). Figure 36.25 shows the reliability of a concrete T-girder over a 75-year life using a life cycle strategy that applies four inspections ($m = 4$) and three repairs ($n = 3$). The corrosion rate is

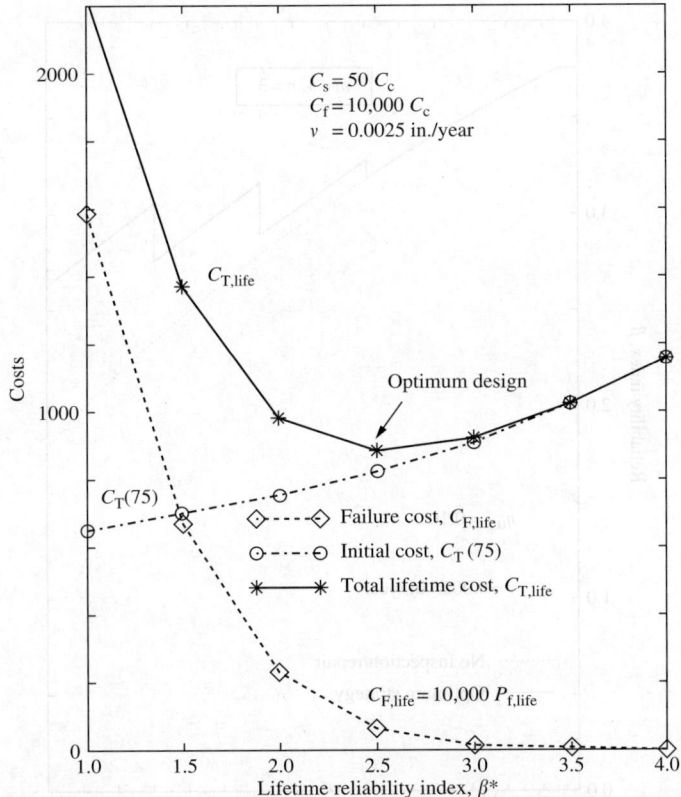

FIGURE 36.24 Optimum design of a concrete T-girder based on minimum total life cycle cost. The corrosion rate is 0.0064 cm/year (0.0025 in./year) due to corrosion of the reinforcement, and the expected service life is 75 years. (Frangopol et al. 1997a. Reprinted with permission from the American Society of Civil Engineers.)

$v = 0.0064$ cm/year (0.0025 in./year), and the inspection technique has a 50% probability of detection when there is 10% damage ($\eta_{0.5} = 0.10$). The number of inspections, inspection technique, and number of repairs were varied until an optimal life cycle cost solution was obtained. The study included both uniform and nonuniform inspection intervals. While a lot of the information was hypothetical, the study was one of the first to incorporate so many key aspects of life cycle analysis and design.

Case 36.2

Estes and Frangopol (1999) treated an existing Colorado highway bridge as a structural system and developed an optimal repair strategy based on minimum life cycle cost. The bridge in Figure 36.26 consisted of three simple spans where each span contained nine girders. The girders were classified as interior, exterior, and interior–exterior based on their load and exposure condition. Using 13 failure modes and 24 random variables, the bridge was modeled as a series–parallel system. After eliminating the nonrelevant failure modes and taking advantage of symmetry and assumed failure mode correlation, the simplified series-parallel model shown in Figure 36.26 was developed assuming that any three adjacent girders must fail for the bridge superstructure to fail. Both shear and moment failure modes were considered for the girders. The girders were corroding over time, and both the concrete slab and the pier cap were deteriorating due to exposure to chlorides from road salts. Meanwhile the live load was increasing over time.

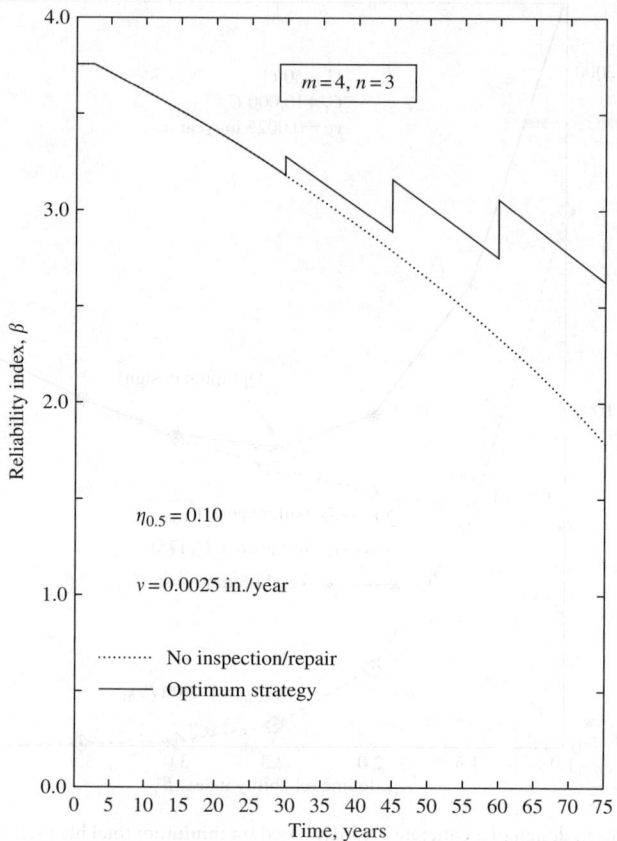

FIGURE 36.25 The reliability of a concrete T-girder over its 75-year life under a life cycle plan that implements four inspections ($m = 4$) and three repairs ($n = 3$). The corrosion rate of the reinforcement is 0.0064 cm/year (0.0025 in./year), and the capability of the inspection technique is a 50% probability of detection when there is 10% damage. (Frangopol et al. 1997b. Reprinted with permission from the American Society of Civil Engineers.)

A repair was required any time the reliability of the bridge system fell below $\beta_{system} = 2.0$. Five repair options that included replacing the deck, replacing exterior girders, replacing exterior girders and deck, replacing superstructure, and replacing the entire bridge were considered along with their associated costs using a discount rate of 2%. Figure 36.27 shows the lifetime reliability of the relevant individual failure modes and the bridge system when the repair strategy is to continually replace the slab. The slab gets replaced at year 50 and year 94 when the system reliability falls below acceptable levels. At year 106, replacement of the slab is no longer sufficient to raise the system reliability above the minimum requirement, and some other repair option is necessary. The optimal solution was obtained by applying various combinations of the discrete repair alternatives.

Figure 36.27 indicates that some failure modes (shear in the exterior and interior–exterior girders) were allowed to fall below the minimum threshold $\beta = 2.0$ due to the parallel nature of the structural model. Due to different deterioration rates of various bridge components and effects of repair on reliability of these components, the most relevant failure mode in the beginning of the life of the structure was not necessarily the most critical mode later on. Different repair strategies were developed for different series–parallel models, different deterioration rates, and different correlations among girder resistances.

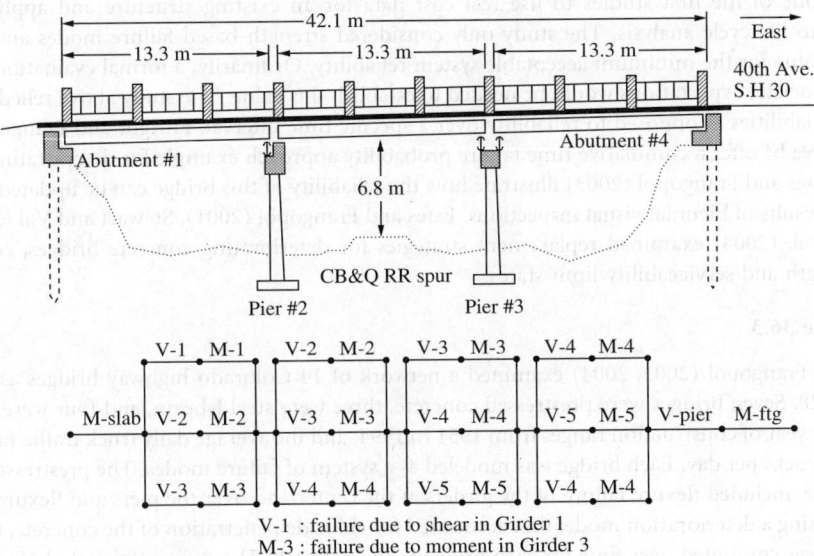

FIGURE 36.26 Colorado highway bridge modeled as a series–parallel system of failure modes. (Estes and Frangopol 1999. Reprinted with permission from the American Society of Civil Engineers.)

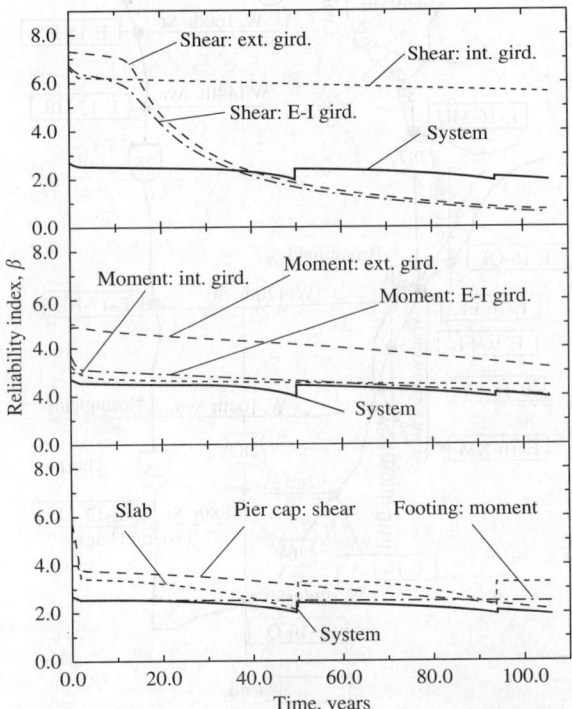

FIGURE 36.27 System reliability over time for deteriorating Colorado highway bridge where deck is replaced when system reliability falls below $\beta_{System} = 2.0$. (Estes and Frangopol 1999. Reprinted with permission from the American Society of Civil Engineers.)

This is one of the first studies to use real cost data for an existing structure and apply a system approach to life cycle analysis. The study only considered strength-based failure modes and chose an arbitrary value for the minimum acceptable system reliability. Ordinarily, a formal evaluation of failure costs and societal expectations would be needed to establish this value. The study above relied on point-in-time reliabilities as opposed to reliability over a specific time interval. Enright and Frangopol (1996, 1998a, 1999a,b) offer a cumulative time failure probability approach example for deteriorating concrete bridges. Estes and Frangopol (2003) illustrate how the reliability of this bridge can be updated over time using the results of biennial visual inspections. Estes and Frangopol (2001), Stewart and Val (2003), and Stewart et al. (2004) examined replacement strategies for deteriorating concrete bridges, considering both strength and serviceability limit states.

Case 36.3

Akgül and Frangopol (2003, 2004) examined a network of 14 Colorado highway bridges as shown in Figure 36.28. Seven bridges were prestressed concrete, three were steel I-beam, and four were steel plate girder. The year of construction ranges from 1951 to 1994, and the average daily truck traffic ranges from 5 to 2955 trucks per day. Each bridge was modeled as a system of failure modes. The prestressed bridges, for example, included flexure failure of the girders at the center and over the piers and flexure failure of the slab. Using a deterioration model that accounted for chloride penetration of the concrete, the system reliability was computed over time for each bridge in the network. The computational platform used in

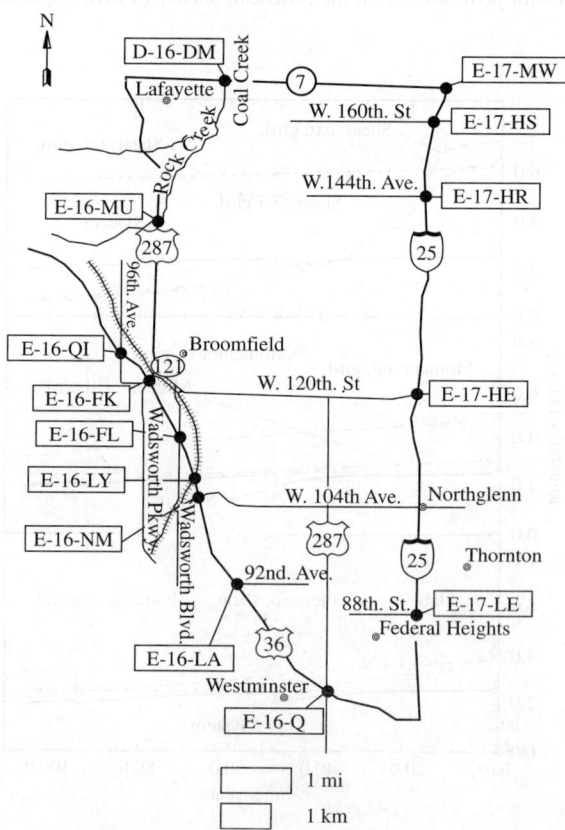

FIGURE 36.28 Bridge network consisting of 14 highway bridges in the Denver–Boulder corridor of Colorado. (Agkül and Frangopol 2003. Reprinted with permission from the American Society of Civil Engineers.)

this study is described in Akgül and Frangopol (2004). A study of networks of bridges (Frangopol et al. 2000a) offers new opportunities for life cycle cost. Akgül and Frangopol (2003) relate the system reliability of the bridges to the load rating for which there are existing data on all bridges. Bridges can be prioritized based on their importance to the overall network and historical inspection data become more useful for maintenance planning of a large category of bridges.

While all three examples in this section involve highway bridges, the concepts are applicable for any structures. Highway bridges are common examples because bridges are critical structures; there exist over 600,000 in the national database, and the Federal Highway Administration and the State DOTs have invested greatly in bridge management systems that provide much of the supporting data needed to support life cycle analyses.

36.12 Conclusions

A life cycle analysis of a structure introduces a number of issues that are not considered in deterministic or probabilistic time invariant solutions. Factors such as deterioration rates, inspection capabilities, repair effects, failure costs, discount rates, and threshold reliabilities present a number of challenges — many of which have not yet been solved. Life cycle analyses are necessarily more complex and require additional input data than those used in conventional time-invariant analyses. It is important to dismiss any misconception that very extensive data are required to perform probabilistic life cycle structural analysis. In fact, probabilistic methods allow the analysis of incomplete data in a much more reliable and consistent manner than conventional deterministic methods. Such analyses also reveal that the costs of maintenance, repair, and inspection may greatly exceed the original cost of construction. In times of scarce and competing resources for an aging infrastructure, these costs cannot be ignored. This chapter has used examples to illustrate the issues involved in condition assessment and life cycle cost analysis and design of deteriorating structures and highlighted some studies that have attempted to address these challenges. There is an urgent need for civil engineers to (a) acquire a good understanding of financial management under uncertainty, (b) have a stronger grasp of reliability-based optimal decision making, and (c) increase their role in the ultimate decisions on life cycle management of our decaying civil infrastructure systems.

Acknowledgments

The partial financial support of the U.S. National Science Foundation through grants CMS-9506435, CMS-9912525, and CMS-0217290, of the U.S. Army Corps of Engineers Construction, Engineering Research Laboratory, of the U.K. Highways Agency and of the Civil Engineering Division of the Netherlands Ministry of Transport, Public Works, and Water Management is gratefully acknowledged. The support of the Colorado Department of Transportation is also gratefully acknowledged. The second author thanks his former and current students Dr. Michael Enright, Dr. Emhaidy Gharaibeh, Dr. Jung Kong, Dr. Seung Yang, Mr. Masaru Miyake, Dr. Ferhat Akgül, Mr. Luis Neves, and Dr. Aruz Petcherdchoo for their cooperation. The opinions and conclusions presented in this chapter are those of the authors and do not necessarily reflect the views of the sponsoring organizations.

Glossary

Bias factor — A factor that relates the nominal deterministic value of a quantity to the mean value of that quantity when it is considered a random variable.

Coefficient of variation — Ratio of the mean value to the standard deviation of a random variable; it is a convenient nondimensional measure of variability (Ang and Tang 1975).

Component (with one failure mode) — A portion of a system represented by a single failure mode and thus a single limit state equation.

Condition rating — A rating assigned by an inspector to identify the structural condition, usually by matching the visual condition of a structure to the most appropriate written description of distress.

Correlation (linear) — A measure of the (linear) relationship between two random variables, usually indicated by a correlation coefficient that ranges from $\rho = 1$ (perfect positive correlation) to $\rho = -1$ (perfect negative correlation), where $\rho = 0$ indicates no correlation.

Deterioration model — A mathematical relationship that describes how a structure is expected to deteriorate over time, usually obtained from laboratory data, theory, or other studies.

Deterministic — No (aleatory and epistemic) uncertainties are present; everything is certain.

Discount rate — The interest rate to be used to convert future monetary sums to present value to allow alternatives to be compared over time.

Ductility — A measure of the ability of a structural member to elongate prior to rupture; a ductile member is still able to carry a load after yielding and allows warning prior to failure, which normally produces a safer structure.

Expected cost — A mean value of the cost; expected cost is used when there is uncertainty associated with the cost and is often a weighted average of the costs associated with various alternatives and their probability of occurrence.

Factor of safety — The ratio of the capacity of a structure (resistance) to the demand placed upon it (load or load effect).

Failure — When a structure no longer performs as intended; using the performance function, failure is defined as $g(\mathbf{X}) < 0$.

Failure cost — Costs incurred by a structural failure multiplied by the probability of failure.

Hazard function — Provides the instantaneous rate of failure; expresses the likelihood of failure in a specific time interval given that the structure has not already failed.

Importance — A measure of the criticality of a structure, usually determined by the degree of consequence if the structure fails.

Independent (statistically) — Two events are independent if the occurrence or (nonoccurrence) of one event does not affect the probability of the other event (Ang and Tang 1975); if two variables are independent, their correlation coefficient is zero.

Initial cost — Those costs associated with the design and construction of a structure (e.g., permits, architect fees, materials, equipment, labor); typically, those costs associated with placing a structure into service.

Life cycle cost — Total costs associated with a structure throughout its entire life; typically includes initial, maintenance, inspection, repair, and failure costs.

Limit state — The boundary between the desired and undesired performance of a structure; the boundary is typically represented by a limit state equation $g(\mathbf{X}) = 0$.

Mean value — The central tendency of a random variable; in the discrete case, the mean value is the average value of the data, in the continuous case where $f_X(X)$ is the density function of a random variable, the mean value is $\mu = \int_{-\infty}^{\infty} x f_x(x)\,\mathrm{d}x$ (Rao 1992).

Nondestructive evaluation — The assessment or inspection of a structure conducted without damaging the structure; the evaluation can be visual or involve testing equipment.

Parallel system — A system where all members must fail for the system to fail.

Preventive maintenance — Precautionary maintenance actions, design to forestall damage and deterioration of a structure.

Probability density function — For a continuous random variable, $f_X(X)$ is a function that defines the probability of detecting X in the infinitesimal interval $(x, x + \mathrm{d}x)$; therefore, $f_X(X)\,\mathrm{d}x = P(x \leq X \leq x + \mathrm{d}x)$.

Random variable — A function that maps events from the sample space into the axis of real numbers; it can be continuous, discrete, or mixed.

Redundancy — Measure of reserve capacity; failure of a single member will not cause failure of a redundant structure.

Reliability — Probability of safe performance of a structure.

Reliability index — A measure of reliability, denoted as β; the shortest distance from the origin to the failure surface in standard normal space; when both load and resistance are normally distributed, $p_f = \Phi(-\beta)$, where Φ is the distribution function of the standard normal variate.

Risk — A function of the probability of occurrence of an adverse event and the consequence of the event; often risk is defined by the probability of failure of the adverse event and its consequence (Ang and De Leon 2005).

Safe — A structure performing as intended.

Series system — A system that fails if any member of the system fails.

Survivor function — Defines the probability that a structure is safe at any particular time.

Visual inspection — A periodic inspection where an inspector observes a structure and classifies its condition in terms of predefined condition ratings.

Notation

A	= cross-sectional area; regression parameter in corrosion model	k	= normalizing constant
		L	= load or load effect
		$L(\Theta)$	= likelihood function
B	= regression parameter in corrosion model	M	= bending moment
		m	= number of inspections
b	= length of flange	$N[7, 1.5]$	= normally distributed variable with a mean value of 7 and standard deviation of 1.5
C	= cost; condition index value		
CS	= condition state	n	= number of time intervals; number of maintenance applications; number of repairs
$C(t)$	= average corrosion penetration at time t (µm)		
d_{corr}	= depth of corrosion penetration	nr	= number of repairs
		p_f	= probability of failure
$E(C)$	= expected value of cost	p_s	= probability of safe performance
$E[\text{Damage}(t) \mid \text{Branch}_i]$	= expected damage at time t given that branch i was taken	pdf$_{bad\text{-}x}$	= value of probability density function for damaged deck
$f_L(l)$	= probability density function of load	PV	= present value of cost
		R	= resistance
FS	= factor of safety	R1	= rust code 1
$f_R(r)$	= probability density function of resistance	r	= discount rate of money
		$S(t)$	= survival function
$f_{R,L}(r,l)$	= joint probability density function of resistance and load	$T(a, b, c)$	= triangular distribution where minimum value is a, mode is b, and maximum value is c
$f(t)$	= probability density function	t	= time
		t_I	= damage initiation time
$f(x_i)$	= probability density function of inspection results	t_f	= thickness of flange
		t_w	= thickness of web
FV	= future value	X	= random variable
F_y	= yield strength	y	= number of parallel systems in the series system; distance from neutral axis to centroid of shape in composite cross-section
$f'(\Theta)$	= prior density function		
$f''(\Theta)$	= posterior density function		
$g(X)$	= performance function		
$H(t)$	= hazard function	Z	= plastic section modulus
h	= depth of column cross-section	z	= number of members in the parallel system

α = system redundancy measure; reliability index deterioration rate without maintenance

α_c = condition index deterioration rate

β = reliability index

Φ = the distribution function of the standard normal variate

γ = redundancy range; immediate improvement in reliability index after application of preventive maintenance

$\eta_{0.5}$ = amount of damage when there is a 50% probability of detection

μ = mean value

ν = corrosion rate

θ = angle; reliability index deterioration rate during preventive maintenance effect

Θ = random variable

ρ = correlation coefficient

σ = standard deviation

References

AASHTO (1992). *AASHTO Standard Specifications for Highways Bridges,* American Association of State Highway and Transportation Officials, 14th Ed., Washington, DC.

AASHTO (1994). *AASHTO Manual for Condition Evaluation of Bridges,* American Association of State Highway and Transportation Officials, Washington, DC.

AASHTO (1998). *LRFD Bridge Design Specification.* 2nd Ed., American Association of State Highway and Transportation Officials, Washington, DC.

Albrecht, P. and Naeemi, A. (1984). *Performance of Weathering Steel in Bridges,* NCHRP Report 272, Washington, DC.

Akgül, F. and Frangopol, D.M. (2003). "Rating and Reliability of Existing Bridges in a Network," *J. Bridge Eng.,* ASCE, **8**(6), 383–393.

Akgül, F. and Frangopol, D.M. (2004). "Computational Platform for Predicting Lifetime System Reliability Profiles for Different Structure Types in a Network," *J. Comput. Civ. Eng.,* ASCE **18**(2), 92–104.

Ang, A.H.-S. and De Leon, D. (2005). "Modeling and Analysis of Uncertainties for Risk-Informed Decisions in Infrastructures Engineering," *Struct. and Infrastruct. Eng.,* Taylor & Francis, **1**(1), 19–31.

Ang, A.H.-S. and Tang, W.H. (1975). *Probability Concepts in Engineering Planning and Design,* **I**, *Basic Principles,* John Wiley & Sons, New York.

Ang, A.H.-S. and Tang, W.H. (1984). *Probability Concepts in Engineering Planning and Design,* **II**, John Wiley & Sons, New York.

ASTM (1987). American Society for Testing and Materials, *Standard Test Method for Half-Cell Potentials of Uncoated Reinforcing Steel in Concrete,* C876-87, Philadelphia, PA.

Ayyub, B.M. and Popescu, C. (2003). "Risk-Based Expenditure Allocation for Infrastructure Improvement," *J. Bridge Eng.,* ASCE, **8**(6), 394–404.

Bond, L.J., Kepler, W.F., and Frangopol, D.M. (2000). "Improved assessment of Mass Concrete Dams Using Acoustic Travel Tomography. Part I — Theory," *Constr. Build. Mater.,* Elsevier, **14**(3), 133–146.

CDOT (1995). Colorado Department of Transportation, *BMS PONTIS Bridge Inspection Manual,* Denver, CO.

Cornell, C.A. (1967). "Bounds on the Reliability of Structural Systems," *J. Struct. Div.,* ASCE, **93**(ST1), 171–200.

Corotis, R.B. (2003a). "Socially Relevant Structural Safety," *Applications of Statistics and Probability in Civil Engineering,* A. Der Kiureghian, S. Madanat, and J.M. Pestana, eds., **1**, 15–24, Millpress, Rotterdam (and on associated CD-ROM).

Corotis, R.B. (2003b). "Risk and Uncertainty," *Applications of Statistics and Probability in Civil Engineering,* A. Der Kiureghian, S. Madanat, and J.M. Pestana, eds., 781–786, Millpress, Rotterdam, (and on associated CD-ROM).

Ditlevsen, O. (1979). "Narrow Reliability Bounds for Structural Systems," *J. Struct. Mech.,* **7**(4), 453–472.

Enright, M.P. and Frangopol, D.M. (1996). "Reliability-Based Analysis of Degrading Reinforced Concrete Bridges," *Structural Reliability in Bridge Engineering*, D.M. Frangopol and G. Hearn, eds., 257–263, McGraw-Hill Book, Inc., New York.

Enright, M.P. and Frangopol, D.M. (1998a). "Service-Life Prediction of Deteriorating Concrete Bridges," *J. Struct. Eng.*, ASCE, **124**(3), 309–317.

Enright, M.P. and Frangopol, D.M. (1998b). "Failure Time Prediction of Deteriorating Fail-Safe Structures," *J. Struct. Eng.*, ASCE, **124**(12), 1448–1457.

Enright, M.P. and Frangopol, D.M. (1999a). "Condition Prediction of Deteriorating Concrete Bridges Using Bayesian Updating," *J. Struct. Eng.*, ASCE, **125**(10), 1118–1124.

Enright, M.P. and Frangopol, D.M. (1999b). "Maintenance Planning for Deteriorating Concrete Bridges," *J. Struct. Eng.*, ASCE, **125**(12), 1407–1414.

Estes, A.C. (1997). *A System Reliability Approach to the Lifetime Optimization of Inspection and Repair of Highway Bridges*, Ph.D. Thesis, Department of Civil, Environmental, and Architectural Engineering, University of Colorado, Boulder, CO.

Estes, A.C. and Frangopol, D.M. (1998). "RELSYS: A Computer Program for Structural System Reliability Analysis," *Struct. Eng. Mech.*, Techno-Press, **6**(8), 901–919.

Estes, A.C. and Frangopol, D.M. (1999). "Repair Optimization of Highway Bridges Using System Reliability Approach," *J. Struct. Eng.*, ASCE, **125**(7), 766–775.

Estes, A.C. and Frangopol, D.M. (2001a). "Bridge Lifetime System Reliability Under Multiple Limit States," *J. Bridge Eng.*, ASCE **6**(6), 523–528.

Estes, A.C. and Frangopol, D.M. (2001b). "Using System Reliability to Evaluate and Maintain Structural Systems," *J. Comput. Struct. Eng.*, Engineering Press **1**(1), 71–80.

Estes, A.C. and Frangopol, D.M. (2001c). "Minimum Expected Cost-Oriented Optimal Maintenance Planning of Deteriorating Structures: Application to Concrete Bridge Decks," *Reliability Eng. System Saf.*, Special Issue on Reliability Oriented Optimal Design, **73**(3), 281–291.

Estes, A.C. and Frangopol, D.M. (2003). "Updating Bridge Reliability Based on Bridge Management System Visual Inspection Results," *J. Bridge Eng.*, ASCE, **8**(6), 374–382.

Estes, A.C., Frangopol, D.M., and Foltz, S.D. (2002). "Using Visual Inspection Results to Update Reliability Analyses of Highway Bridges and River Lock Structures," *Proceedings of the 8th International Conference on Structural Safety and Reliability*, ICOSSAR'01, Newport Beach, CA, June 17–22, 2001, in *ICOSSAR'01, Structural Safety and Reliability*, R.B. Corotis, G.I. Schueller, and M. Shinozuka, eds., Sweets and Zeitlinger Publishers, (8 pages on CD-ROM).

Estes, A.C., Frangopol, D.M., and Foltz, S. (2004). "Updating the Time-Dependent Reliability of Steel Miter Gates Using Visual Condition Index Inspection Results," *Eng. Struct.*, Elsevier, **26**(3), 319–333.

Federal Reserve (2003). *Federal Reserve Statistical Release: H.15 Selected Interest Rates*, http://www.federalreserve.gov/releases/h15/data/a/tcm1y.txt, Board of Governors of the Federal Reserve System, Washington, DC.

FHWA (1986). Federal Highway Administration, *Nondestructive Testing Methods for Steel Bridges*, Participants Training Manual, U.S. Department of Transportation, Washington, DC.

FHWA (2002). Federal Highway Administration, National Bridge Inspection Standards, http://www.fhwa.dot.gov/tndiv/brinsp.htm, U.S. Department of Transportation, Washington, DC.

Frangopol, D.M. (1998). "A Probabilistic Model Based on Eight Random Variables for Preventive Maintenance of Bridges," Presented at the Progress Meeting, *Optimum Maintenance Strategies for Different Bridge Types*, Highways Agency, London, November.

Frangopol, D.M. (2002). "Reliability Deterioration and Lifetime Maintenance Cost Optimization," Keynote lecture in *Proceedings of the First International ASRANet Colloquium on Integrating Structural Reliability Analysis with Advanced Structural Analysis*, Glasgow, Scotland, July 8–10, in *ASRANet: Integrating Structural Reliability Analysis with Advanced Structural Analysis*, Glasgow, Scotland (14 pages on CD-ROM).

Frangopol, D.M. (2003). *Preventive Maintenance Strategies for Bridge Groups — Analysis,* Final Project Report to the Highways Agency, London, March, 139 pages.

Frangopol, D.M., Brühwiler, E., Faber, M.H., and Adey, B., eds. (2004). *Life-Cycle Performance of Deteriorating Structures: Assessment, Design and Management,* ASCE, Reston, Virginia, 456 pages.

Frangopol, D.M. and Estes, A.C. (1999). "Optimum Lifetime Planning of Bridge Inspection and Repair Programs." *Struct. Eng. Int.,,* SEI, **9**(3), 219–223.

Frangopol, D.M., Gharaibeh, E.S., Kong, J.S., and Miyake, M. (2000a). "Optimal Network-Level Bridge Maintenance Planning Based on Minimum Expected Cost," *J. Transp. Res. Bd.,* Transportation Research Record, National Academy Press, **1696**(2), 26–33.

Frangopol, D.M., Gharaibeh, E.S., Kong, J.S., and Miyake, M. (2001a). "Reliability-based evaluation of rehabilitation rates of bridge groups," *Proceedings of the International Conference on Safety, Risk and Reliability — Trends in Engineering,* IABSE, Malta, March 21–23; *Safety, Risk and Reliability — Trends in Engineering,* Conference Report, IABSE-CIB-ECCS-*fib*-RILEM, Malta, 267–272, also on CD-ROM, 1001–1006.

Frangopol, D.M., Gharaibeh, E.S., Kong, J.S., and Miyake, M. (2001c). "Reliability-based evaluation of rehabilitation rates of bridge groups," *Proceedings of the International Conference on Safety, Risk and Reliability — Trends in Engineering,* IABSE, Malta, March 21–23; *Safety, Risk and Reliability — Trends in Engineering,* Conference Report, IABSE-CIB-ECCS-*fib*-RILEM, Malta, 267–272.

Frangopol, D.M., Iizuka, M., and Yoshida, K. (1992). "Redundancy Measures for Design and Evaluation of Structural Systems," *J. Offshore Mech. Arctic Eng.,* ASME, **114**(4), 285–290.

Frangopol, D.M., Kallen, M.-J., and van Noortwijk, J.M. (2004). "Probabilistic Models for Life Cycle-Performance of Deteriorating Structures: Review and Future Directions," *Progress in Structural Engineering and Mechanics,* John Wiley & Sons, **6**(4) (in press).

Frangopol, D.M. and Kong, J.S. (2001). "Expected Maintenance Cost of Deteriorating Civil Infrastructures," Keynote paper in *Life-Cycle Cost Analysis and Design of Civil Infrastructure Systems,* D.M. Frangopol and H. Furuta, eds., 22–47, ASCE, Reston, VA.

Frangopol, D.M., Kong, J.S., and Gharaibeh, E.S. (2000). "Bridge Management Based on Lifetime Reliability and Whole Life Costing: The Next Generation," *Bridge Management 4,* M.J. Ryall, G.A.R. Parke, and J.E. Harding, eds., 392–399, Thomas Telford, London.

Frangopol, D.M., Kong, J.S., and Gharaibeh, E.S. (2001b). "Reliability-Based Life-Cycle Management of Highway Bridges," *J. Comput. Civ. Eng.,* ASCE, **15**(1), 27–34.

Frangopol, D.M., Lin, K.Y., and Estes, A.C. (1997a). "Reliability of Reinforced Concrete Girders Under Corrosion Attack," *J. Struct. Eng.,* ASCE, **123**(3), 286–297.

Frangopol, D.M., Lin, K.Y., and Estes, A.C. (1997b). "Life-Cycle Cost Design of Deteriorating Structures," *J. Struct. Eng.,* ASCE, **123**(10), 1390–1401.

Frangopol, D.M. and Neves, L.C. (2003). "Life-cycle Maintenance Strategies for Deteriorating Structures Based on Multiple Probabilistic Performance Indicators," *Proceedings of the Second International Structural Engineering and Construction Conference,* ISEC-02, Rome, Italy, September 23–26 (keynote paper); *System-based Vision for Strategic and Creative Design,* F. Bontempi, ed., **1**, 3–9, Sweets & Zeitlinger, Lisse (keynote paper).

Kepler, W.F., Bond, L.J., and Frangopol, D.M. (2000). "Improved Assessment of Mass Concrete Dams Using Acoustic Travel Tomography. Part II — Application," *Constr. Build. Mater.,* Elsevier, **14**(3), 147–156.

Kong, J.S. and Frangopol, D.M. (2003a). "Evaluation of Expected Life-Cycle Maintenance Cost of Deteriorating Structures," *J. Struct. Eng.,* ASCE, **129**(5), 682–691.

Kong, J.S. and Frangopol, D.M. (2003b). "Life-Cycle Reliability-Based Maintenance Cost Optimization of Deteriorating Structures with Emphasis on Bridges," *J. Struct. Eng.,* ASCE, **129**(6), 818–828.

Leemis, L.M. (1995). *Reliability: Probabilistic Models and Statistical Methods,* Prentice Hall, Englewood Cliffs, NJ.

Marshall, S.J. (1996). *Evaluation of Instrument — Based, Nondestructive Inspection Methods for Bridges,* M.Sc. Thesis, Department of Civil, Environmental, and Architectural Engineering, University of Colorado, Boulder, CO.

McMahon, T. (2003). Historical U.S. Inflation Data Table, InflationData.com, http://inflationdata.com/ inflation, *Financial Trend Forecaster*, Richmond, VA.

Menzies, J.B. (1996). "Bridge Safety Targets and Needs for Performance Feedback," *Structural Reliability in Bridge Engineering*, D.M. Frangopol and G. Hearn, eds., 156–159. McGraw-Hill, New York.

Neves, L.C. and Frangopol, D.M. (2004). "Condition, Safety and Cost Profiles for Deteriorating Structures with Emphasis on Bridges," *Reliability Eng. System Saf.*, (in press).

Nowak, A.S. and Collins, K.R. (1995). "Calibration of LRFD Bridge Code," *J. Struct. Eng.*, ASCE, **121**(8), 1245–1252.

Palisade (2003). *@RISK Advanced Risk Analysis for Spreadsheets*, Palisade Corporation, Newfield, NY.

Petcherdchoo, A., Neves, L.C., and Frangopol, D.M. (2004). "Combinations of Probabilistic Maintenance Actions for Minimum Life-Cycle Cost of Deteriorating Bridges," *Proceedings of the Second International Conference on Bridge Maintenance, Safety, and Management*, IABMAS '04, Tokyo, October 19–22; Bridge Maintenance, Safety, Management and Cost, E. Watanabe, D.M. Frangopol, and T. Utsunomiya, eds., Balkema (8 pages on CD-ROM).

PONTIS (1995). *Release 3.0 User's Manual*, Cambridge Systematics, Inc., Cambridge, MA.

Rao, S.S. (1992). *Reliability-Based Design*, McGraw-Hill, New York.

Stecker, J.H., Greimann, L.F., Mellema, M., Rens, K., and Foltz, S.D. (1997). *REMR Management Systems — Navigation and Flood Control Structures, Condition Rating Procedures for Lock and Dam Operating Equipment*, Technical Report REMR-OM-19, U.S. Army Corps of Engineers, Washington, DC.

Stewart, M.G., Estes, A.C., and Frangopol, D.M. (2004). "Bridge Deck Replacement for Minimum Expected Cost under Multiple Reliability Constraints," *J. Struct. Eng.*, ASCE **130**(9), 1414–1419.

Stewart, M.G. and Val, D.V. (2003). "Multiple Limit States and Expected Failure Costs for Deteriorating Reinforced Concrete Bridges," *J. Bridge Eng.*, ASCE, **8**(6), 405–415.

Technology Review (1979). February, Cambridge, MA, 45.

Thoft-Christensen, P. and Murotsu, Y. (1986) *Application of Structural Systems to Reliability Theory*, Springer-Verlag, Berlin.

USACE: U.S. Army Corps of Engineers (1996). *Project Operations: Partners and Support (Work Management Guidance and Procedures)*, Engineer Pamphlet EP 1130-2-500, Washington, DC.

Vanderplaats, G.N. (1986). *ADS — A Fortran Program for Automated Design Synthesis*, Version 1.10, Engineering Design Optimization Inc., Santa Barbara, CA.

van Noortwijk, J.M. and Frangopol, D.M. (2004). "Two Probabilistic Life-Cycle Maintenance Models for Deteriorating Civil Infrastructure," *Probabilistic Engineering Mechanics*, Elsevier, **19**(4), 345–359.

37
Structural Design for Fire Safety

Yong C. Wang
School of Aerospace,
Mechanical and Civil Engineering,
The University of Manchester,
Manchester, U.K.

37.1 Introduction

Structural fire safety is one of the three requirements that have to be fulfilled by a fire resistant construction, whose function is to ensure that a fire in a building is contained within the compartment of origin so that occupants in other parts of the building can escape to safety and fire damages do not become excessive. To achieve this, the load bearing structure of a fire resistant construction should not collapse in fire. The other two fire resistance requirements are

- *Insulation.* The unexposed surface of a fire resistant construction should not be heated excessively and cause further ignition. Clearly, whether any material will be ignited or not will not only depend on the temperature of the unexposed surface, but also on its nature and its relative position to the unexposed surface. Nevertheless, at present, regulations worldwide limit the average temperature rise on the unexposed surface to 140°C and the maximum local temperature rise to 180°C.
- *Integrity.* Gaps should not develop in fire resistant construction to spread fire.

The practice of dividing a building into a number of compartments bounded by fire resistant construction is called fire resistant compartmentation. It should be pointed out that fulfillment of the above three fire resistance requirements only applies to those elements of construction that are necessary for the fire resistant compartment to contain fire. Other elements, whose failure to fulfill these requirements does not lead to a failure of the fire resistant compartment, do not require any fire safety design consideration. Before the designer commences detailed structural fire safety design calculations, he should work with the client and the fire service authority to determine the size of the fire resistant construction. This will depend on factors such as fire regulations on the maximum size of fire resistant compartmentation, insurance premium, and fire brigade access and is beyond the scope of this chapter.

Until recently, assessment of fire resistance of a construction is performed experimentally in standard fire resistance test furnaces and under the standard fire condition. Each country has its own fire resistance test standard [e.g., ASTM E-119 (ASTM 1985) in the United States, BS 476 (BSI 1987) in the United Kingdom, and ISO 834 (ISO 1975)], but they are largely similar. The standard fire resistance test has many shortcomings, for example, high cost, time consuming, limitation of specimen size, idealized loading condition, idealized support condition, lack of repeatability, and unrealistic fire exposure. It is now possible to perform some fire resistant design by calculations and the objective of this chapter is to introduce the reader to structural design calculations to ensure stability of load bearing members in the event of a fire.

It should be pointed out that the behavior of a complete structure in fire and that of isolated elements can be different because a complete structure will have characteristics, such as load redistribution, structural interactions, that will not exist in isolated elements. Also, it is worth noting that current calculation methods are based on flexural behavior at small deflections. The behavior of structural elements at large deflections can be vastly different and such a different behavior may be explored to improve structural fire safety design. Behaviors of elements at large deflections and complete structures in fire are beyond the scope of this chapter. Interested readers may consult the book by Wang (2002).

In general, design calculations to check structural safety in fire involves three parts:

1. Assessment of the fire severity to which a structural member is exposed. For structural fire safety design, a fire is usually quantified by a temperature–time relationship of the fire.
2. Evaluation of the temperature field in the structural member under the above fire condition.
3. Calculation of the remaining load carrying capacity of the structural member at elevated temperatures and comparison with the applied load.

This chapter will introduce the reader to all three aspects of structural fire safety design, while emphasizing on the third.

It is understandable that the September 11 tragedy has initiated the interest of many engineers in structural fire safety design. However, it must be mentioned that there had already been great progresses on this topic well before the September 11 event in many parts of the world, particularly in Europe and the United Kingdom. At present, there is a systematic and comprehensive coverage of structural fire safety design methods in Europe, through developments of the so-called Eurocodes. Thus, this chapter will adopt Eurocodes as the basis of its design guidance. However, it is hoped that there will be sufficient explanations of the fundamental engineering principles so that the basis of Eurocode design rules can be similarly adopted in different design environments.

37.2 Fire Severity

Fire severity to structural fire safety design is akin to applied mechanical loads on a structure to structural design at ambient temperature. Fires can occur anywhere; however, for structural fire safety design, they are often assumed to take place in a building enclosure. In such a situation, starting from ignition, a fire can go through a number of stages. In the early stages, combustion is restricted to local areas near the ignition source and temperatures of the combustion gases are low, the safety of a structure exposed to fire attack is very rarely threatened. Later, under certain circumstances, a localized fire can transform very quickly to involve all the combustible materials in the fire enclosure. The transition from a localized fire to a fire engulfing the entire enclosure is called flashover and the fire at this stage is called a postflashover fire. At this stage, the combustion gas temperatures are very high and stability of the building structure may be threatened, the consequence of which can lead to rapid fire spread and loss of life and property. It is this stage of the fire behavior that will be described below. Interested readers should consult some excellent textbooks, such as Drysdale (1999) and Karlsson and Quinterie (2000), to obtain a deep understanding of enclosure fire behavior.

37.2.1 Standard Fire Exposure

There are two ways of dealing with postflashover fires for structural fire safety design. First, the standard fire temperature–time relationship may be adopted. The standard fire equation is similar in different countries of the world. In the European standard (CEN 2000a), the standard fire temperature–time relationship is given by

$$T_{\text{fi}} = T_{\text{a}} + 345 \log(8t + 1) \tag{37.1}$$

where the standard fire exposure time (t) is in minutes and the fire temperature T_{fi} and ambient temperature T_{a} are in degrees celsius.

Equation 37.1 is used for wood based, or cellulosic, fires. For fire resistant design of offshore structures, the standard hydrocarbon fire curve should be used. This fire has a much faster rate of initial increase in temperature. The standard hydrocarbon fire temperature–time relationship is given by (CEN 2000a)

$$T_{\text{fi}} = 1080\left(1 - 0.325e^{-0.167t} - 0.675e^{-2.5t}\right) + T_{\text{a}} \tag{37.2}$$

Figure 37.1 plots the two standard fire curves. It is obvious that the standard cellulosic fire curve gives a monotonically increasing temperature–time relationship that cannot be sustained in any real fire. In order to reflect some reality in the standard fire exposure, a limiting time of fire exposure is specified. This is the familiar standard fire resistance rating. In standard fire resistance design calculations, specifications for the required standard fire resistance rating are based on very broad criteria such as the occupancy type and height of a building. Whilst these criteria give a broad indication of fire load and consequence of fire exposure, they do not consider other important factors that affect the behavior of an enclosure fire such as ventilation condition and construction materials.

37.2.2 Natural Fire Exposure

As a regulatory control tool, the standard fire exposure is simple to use. However, fires cannot be expected to behave according to the standard temperature–time relationship. For more realistic

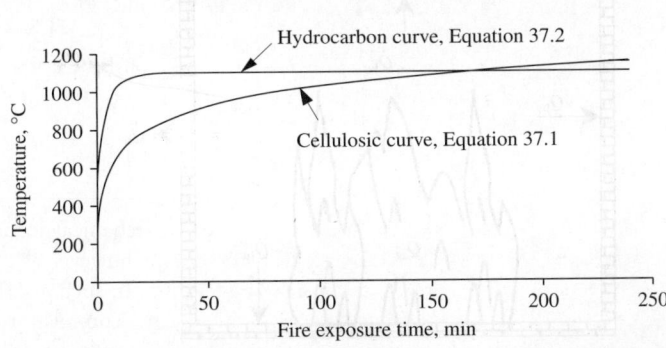

FIGURE 37.1 A comparison of standard cellulosic and hydrocarbon fire temperature–time relationships.

assessment of performance of structures in fire, it is necessary to quantify realistic fire behavior. Assuming that a fire enclosure is at the same temperature during the postflashover phase, the fire temperature–time relationship may be determined by carrying out an energy balance analysis for the fire enclosure; that is

heat input into the fire (heat released from combustion) = heat losses from the fire

Figure 37.2 depicts a postflashover enclosure fire situation. Heat losses from the fire include

1. Heat lost to the outside by hot gases flowing out of the fire compartment through openings ($\dot{Q}_{lc}$).
2. Heat lost to the enclosure lining ($\dot{Q}_{lw}$).
3. Heat lost to the outside environment by radiation through the opening ($\dot{Q}_{lr}$).
4. Heat required to increase the combustion gas temperature ($\dot{Q}_{lg}$).

The rate of heat release is the most important factor. At present, due to the difficulty of numerically modeling fire spread and random distribution of combustible materials in a fire enclosure, it is not possible to accurately calculate the rate of heat release of a fire. Nevertheless, from considerations of the main governing factors of burning, a number of empirical equations have been derived. From basic hydrodynamics, it can be shown that the amount of hot gases flowing out of a fire compartment is related to $A_v \sqrt{h_v}$, the so-called ventilation factor, where A_v is the opening area and h_v is the opening height. In order to sustain burning, cold air should be supplied into the fire compartment to replenish the lost hot gases. Thus, the amount of cold air entering the fire compartment is also related to $A_v \sqrt{h_v}$. It follows that the amount of fresh oxygen supply to the fire is related to $A_v \sqrt{h_v}$. If burning is ventilation controlled, that is, the rate of burning is governed by the amount of fresh oxygen available, the rate of heat release of a fire is a function of $A_v \sqrt{h_v}$. On the other hand, if the opening is large but the burning area is small, a fire can become fuel controlled, that is, the rate of burning is governed by the available surface area of the fuel bed or combustible materials. Also, the amount of available combustible materials, that is, the fuel load, will determine the duration of burning.

Fire development inside a fire enclosure will also be affected by thermal properties of the fire enclosure lining materials, that is, the bounding walls and floors. A material that has a low thermal conductivity, that is, heat is difficult to penetrate the material, will lose a small amount of heat through the material. A material that has a high thermal capacitance, that is, a large amount of heat is required to raise its temperature, will absorb a large amount of heat of the burning fire and vice versa. Combining these two factors, the quantity that is used to describe the thermal properties of fire enclosure lining materials is $k\rho C$ where k and ρC are the thermal conductivity and thermal capacitance of the lining materials, respectively.

Using the three quantities mentioned above, that is, the ventilation factor, the fire enclosure lining material property, and the fuel load, a number of approximate temperature–time relationships of

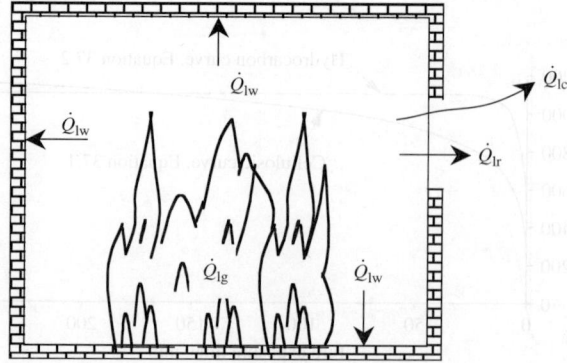

FIGURE 37.2   Fully developed enclosure fire, showing various heat losses.

postflashover enclosure fires have been developed. Among them, the so-called parametric temperature–time curves of Eurocode 1 Part 1.2 (CEN 2000a), based on the results of Pettersson et al. (1976), are widely accepted. As shown in Figure 37.3, a parametric fire curve has an ascending branch and a descending branch. The ascending branch is used to describe the temperature–time relationship of a fire during its growth and steady burning stages, when it is ventilation controlled. The descending branch describes the decay period of the fire. The ascending branch is expressed by

$$T_{\text{fi}} = 1325\left(1 - 0.324e^{-0.2t^*} - 0.204e^{-1.7t^*} - 0.472e^{-19t^*}\right) \tag{37.3}$$

where the modified time t^* (in hours) is related to the real time t (in hours) by

$$t^* = t\Gamma \tag{37.4}$$

in which Γ is a dimensionless parameter, given by

$$\Gamma = \left(\frac{O}{0.04}\right)^2 \left(\frac{1160}{b}\right)^2 \tag{37.5}$$

In Equation 37.5, O is the ventilation factor defined as

$$O = \frac{A_v\sqrt{h_v}}{A_t} \tag{37.6}$$

in which A_t is the total enclosure (including openings) area.

In Equation 37.5, $b = \sqrt{k\rho C}$ [in J/(m^2 s$^{1/2}$ K)] is the overall thermal property of the fire enclosure lining material. For a fire enclosure constructed of a combination of different lining materials, complicated equations have recently been introduced in Eurocode 1 Part 1.2 (CEN 2000a) to find an equivalent value of b.

The ascending branch of the fire temperature–time relationship terminates at time (t_d^*) when the maximum fire temperature is obtained. This time is a function of the fire load in the fire enclosure and is given by

$$t_d^* = 0.00013\frac{q_{t,d}\Gamma}{O} \text{ (in hours)} \tag{37.7}$$

In Equation 37.7, $q_{t,d}$ is the fire load density (in MJ/m^2) related to the total surface area of the fire enclosure A_t. Since fire load density is usually specified with regard to the floor area A_f, the fire load per enclosure area $q_{t,d}$ is related to the fire load per floor area ($q_{f,d}$) using

$$q_{t,d} = q_{f,d}A_f/A_t \tag{37.8}$$

It can be seen that the ascending branch of the fire temperature–time curve is not dependent on the fire load. This is because a fire is assumed to be ventilation controlled and the rate of heat release is the

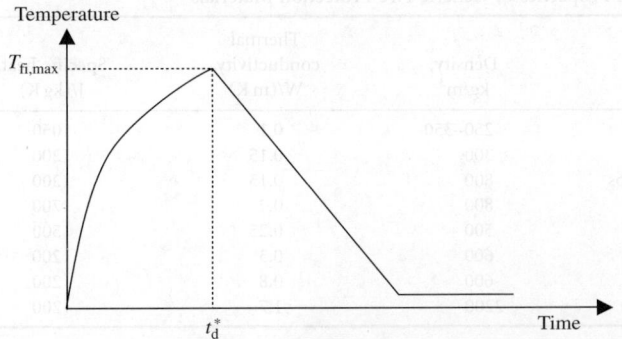

FIGURE 37.3 Parametric time–temperature curve of Eurocode 1 Part 1.2.

same, depending only on the ventilation condition. The effect of fire load is to change the duration of burning t_d^* according to Equation 37.7.

For simplicity, the descending branch is given by a straight line. Since structural behavior is only slightly affected by the descending branch of the fire temperature–time relationship, more complicated equations for the descending branch are not justified. The rate of the descending branch depends on the fire duration. The fire temperature during cooling is given by

$$
\begin{aligned}
T_{fi} &= T_{fi,max} - 625(t^* - t_d^*) & \text{for } t_d^* \leq 0.5 \\
T_{fi} &= T_{fi,max} - 250(3 - t_d^*)(t^* - t_d^*) & \text{for } 0.5 < t_d^* < 2.0 \\
T_{fi} &= T_{fi,max} - 250(t^* - t_d^*) & \text{for } t_d^* \geq 2.0
\end{aligned}
\tag{37.9}
$$

In Equation 37.9, $T_{fi,max}$ is the maximum fire temperature, obtained by substituting the time in Equation 37.7 into Equation 37.3.

In Eurocode 1 Part 1.2 (CEN 2000a), the limit of application of the above fire temperature–time relationship is for fire compartments up to $100\,m^2$ in floor area with the maximum compartment height at $4\,m$. For larger or taller compartments, the effect of nonuniform temperature distribution in the fire enclosure cannot be ignored. Unfortunately, simple methods are not available yet.

From previous discussions, it is clear that the fire temperature–time relationship depends on the amount of combustible materials (or fuel load) in a fire enclosure, the ventilation condition, and thermal properties of the fire enclosure lining material. During a design, the ventilation condition and thermal properties of the fire enclosure lining material may be estimated from construction details, that is, the window size and construction materials. Thermal properties of some enclosure lining materials may be found in Table 37.1.

The design fire load is building specific. However, since the exact type and amount of combustible materials will not be known during the design stage, it is unlikely that the design fire load can be known with any certainty. In fire engineering design calculations, it is common to specify a generic fire load for a type of building, depending on its proposed use. This is similar to specifying a general structural load for structural design at ambient temperature. Values in Table 37.2 may be used as a guide. More detailed information on fire load may be obtained from a Conseil International du Batiment (CIB) report (CIB 1986). It is important to point out that there are many uncertainties about the design fire load. When conducting a fire engineering design, the designer should perform a sensitivity study to investigate the consequence of adopting a range of possible fire loads.

EXAMPLE 37.1

Natural fire exposure

Figure 37.4 shows the dimensions and other design data of a fire enclosure. Evaluate the postflashover fire temperature–time curve inside the enclosure.

TABLE 37.1 Thermal Properties of Generic Fire Protection Materials

Generic material	Density, kg/m^3	Thermal conductivity, $W/(m\,K)$	Specific heat, $J/(kg\,K)$	Moisture content, (% by wt.)
Sprayed mineral fiber	250–350	0.1	1050	1.0
Vermiculite slabs	300	0.15	1200	7.0
Vermiculite/gypsum slabs	800	0.15	1200	15.0
Gypsum plaster	800	0.2	1700	20.0
Mineral fiber sheets	500	0.25	1500	2.0
Aerated concrete	600	0.3	1200	2.5
Lightweight concrete	600	0.8	1200	2.5
Normal weight concrete	2200	1.7	1200	1.5

Source: Lawson, R.M. and Newman, G.M., 1996, *Structural Fire Design to EC3 & EC4, and Comparison With BS 5950*, Technical Report, SCI Publication 159, The Steel Construction Institute.

TABLE 37.2 Design Fuel Load Density

Fuel load density band	Type of construction	Fuel load density, $q_{f,d}$ (MJ/m^2)
I	Assembly, low hazard	350
	Open sided car parks	
II	Assembly, ordinary hazard	500
	Residential	
	Office	
	Industrial, low hazard	
	Storage, low hazard	
III	Assembly, high hazard	750
	Shops and commercial	
IV	Industrial, high hazard	1000
V		1250
VI	Storage, high hazard	1500

Source: British Standards Institution (BSI), 2001, *Draft BS 9999, Code of Practice for Fire Safety in the Design, Construction and Use of Buildings* (London: British Standards Institution).

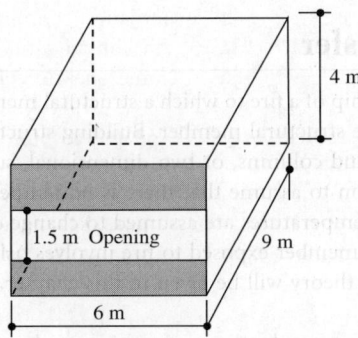

Other design data
Design fire load
720 MJ/m^2 of floor area

Enclosure lining material
lightweight concrete blocks,
density 1600 kg/m^3,
specific heat 1200 J/(kg K)
thermal conductivity 0.8 W/(m K)

FIGURE 37.4 Enclosure information for Example 37.1.

Calculation results

Floor area: $A_f = 54 \, \text{m}^2$

Total enclosure area: $A_t = 228 \, \text{m}^2$

Design fire load density per m^2 enclosure area: $q_{t,d} = 720 * 54/228 = 170.5 \, \text{MJ/m}^2$

Window area: $A_v = 9 \, \text{m}^2$

Opening factor: $O = \dfrac{9 \times \sqrt{1.5}}{228} = 0.04835 \, \text{m}^{1/2}$

Lining property: $b = \sqrt{1600 \times 0.8 \times 1200} = 1239.4 \, \text{J/(m}^2 \, \text{s}^{1/2} \, \text{K)}$

Equation 37.5 gives $\Gamma = \left(\dfrac{0.04835}{0.04}\right)^2 \times \left(\dfrac{1160}{1239.4}\right)^2 = 1.28$

Equation 37.7 gives $t_d^* = 0.00013 \times \dfrac{170.5 \times 1.28}{0.04835} = 0.587 \, \text{h}$

Equation 37.3 gives $T_{fi,max} = 844 + 20 = 864°\text{C}$

The real time at the maximum fire temperature $t_d = 0.587/1.28 = 0.458 \, \text{h}$

For the cooling part, Equation 37.9 will be used, giving the time t^* necessary to reach the ambient temperature as $t^* = 1.937 \, \text{h}$. The real time is $1.937/1.28 = 1.5136 \, \text{h}$.

Figure 37.5 plots the complete fire temperature–time relationship.

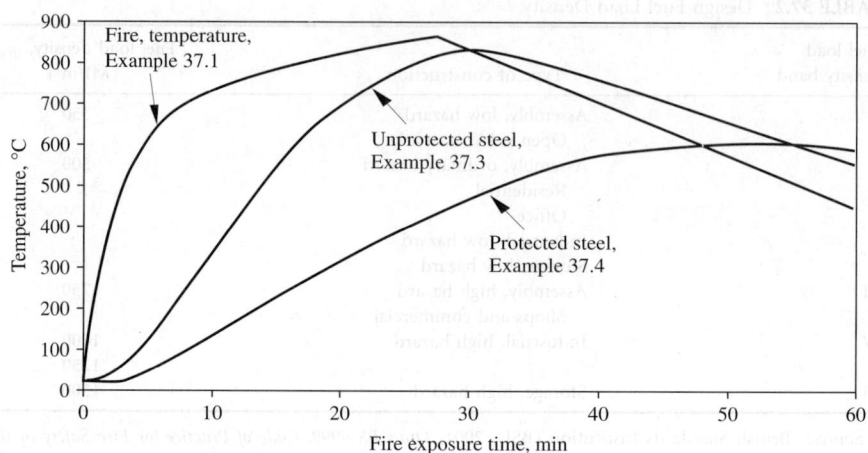

FIGURE 37.5   Temperature–time relationships.

37.3   Introduction to Heat Transfer

Having determined the temperature–time relationship of a fire to which a structural member is exposed, the next step is to calculate the temperatures in the structural member. Building structures are usually treated as either one-dimensional, such as beams and columns, or two-dimensional, such as slabs and walls. For a one-dimensional member, it is common to assume that there is no temperature variation along its length. For a two-dimensional member, temperatures are assumed to change only through its thickness. Calculating temperatures in a structural member exposed to fire involves using heat transfer analysis. Only a brief introduction to heat transfer theory will be given in this chapter, with particular emphasis on applications under fire conditions.

There are three basic mechanisms of heat transfer: conduction, convection, and radiation. In conduction, energy or heat is exchanged in solids on a molecular scale without any movement of macroscopic portions of matter relative to one another. Convection refers to heat transfer at the interface between a fluid and a solid surface. Here, the exchange of heat is due to fluid motion. This motion may be the result of an external force, causing the fluid to flow over the solid surface at speed. This is called forced convection. Convection can also occur due to buoyancy-induced flow when there is a temperature gradient in the fluid, causing a density gradient. This is called natural convection. Radiation is the exchange of energy by electromagnetic waves that, like visible light, can be absorbed, transmitted or reflected at a surface. Unlike conduction and convection, heat transfer by radiation does not require any intervening medium between the heat source and the receiver. Thus, in the context of structural fire safety design, heat conduction describes the heat transfer process inside a structural member and heat convection and radiation describe the thermal boundary condition of the structural member.

37.3.1   Conduction

The basic equation for one-dimensional heat conduction is Fourier's law of heat conduction. It is expressed as

$$\dot{Q} = -k \frac{dT}{dx} \qquad (37.10)$$

where, refering to Figure 37.6, dT is the temperature difference across an infinitesimal thickness dx. $\dot{Q}$ is the rate of heat transfer (heat flux) across the material thickness. The minus sign in Equation 37.10 indicates that heat flows from the higher temperature side to the lower temperature side.

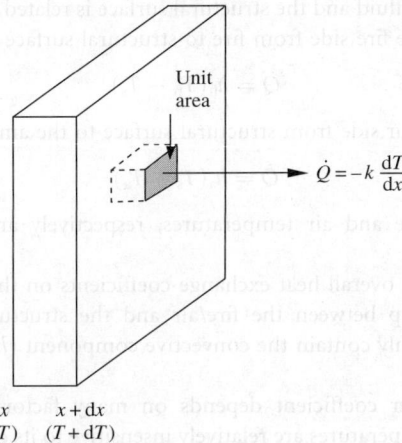

FIGURE 37.6   Heat conduction in one dimension.

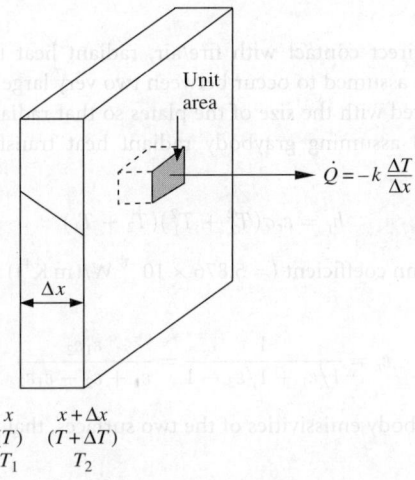

FIGURE 37.7   Temperature distribution with constant thermal conductivity.

The constant of proportionality k is the thermal conductivity of the material. In many practical applications of fire safety engineering, the material thermal conductivity within the relevant temperature range may be approximated as a constant. Thus, Equation 37.10 may be replaced by its finite difference equivalent

$$\dot{Q} = -k\frac{T_2 - T_1}{\Delta x} \quad \text{or} \quad T_1 - T_2 = \dot{Q}\frac{\Delta x}{k} \tag{37.11}$$

where, refering to Figure 37.7, T_1 and T_2 are temperatures at the two sides of a material and Δx is the material thickness. $\Delta x/k$ expresses the thermal resistance of the material.

37.3.2   Convection

Heat convection and radiation are considered at the interface between a structural member and the fire or the ambient temperature air. When applying thermal boundary conditions, it is often assumed that the heat exchange between the fluid and the structural surface is related to the temperature difference at the interface. Therefore, on the fire side from fire to structural surface

$$\dot{Q} = h_{\text{fi}}(T_{\text{fi}} - T_{\text{s}}) \tag{37.12}$$

On the ambient temperature air side from structural surface to the ambient temperature air

$$\dot{Q} = h_{\text{a}}(T_{\text{s}} - T_{\text{a}}) \tag{37.13}$$

where T_{fi} and T_{a} are the fire and air temperatures, respectively and T_{s} is the structural surface temperature.

Quantities h_{fi} and h_{a} are the overall heat exchange coefficients on the fire and air side, respectively. Depending on the relationship between the fire/air and the structural surface, the heat exchange coefficients (h_{fi} and h_{a}) may only contain the convective component (h_{c}), the radiant component (h_{r}), or both.

The convective heat transfer coefficient depends on many factors. However, for structural fire applications, the structural temperatures are relatively insensitive to its exact values. Eurocode 1 Part 1.2 (CEN 2000a) recommends constant convective heat transfer coefficients as follows: on the fire side, $h_{\text{c}} = 25\ \text{W/m}^2$ and on the air side, $h_{\text{c}} = 10\ \text{W/m}^2$.

37.3.3 Radiation

If a structural surface is in direct contact with fire/air, radiant heat transfer between the structural surface and the fire/air may be assumed to occur between two very large parallel plates of area A, whose distance apart is small compared with the size of the plates so that radiation at their edges is negligible. Under this circumstance and assuming graybody radiant heat transfer, the radiant heat exchange coefficient is

$$h_{\text{r}} = \varepsilon_{\text{r}}\sigma(T_2^2 + T_1^2)(T_2 + T_1) \tag{37.14}$$

where σ is the Stefan–Boltzmann coefficient ($= 5.876 \times 10^{-8}\ \text{W/(m\,K}^4))$ and ε_{r} is often referred to as the resultant emissivity given by

$$\varepsilon_{\text{r}} = \frac{1}{1/\varepsilon_1 + 1/\varepsilon_2 - 1} = \frac{\varepsilon_1\varepsilon_2}{\varepsilon_1 + \varepsilon_2 - \varepsilon_1\varepsilon_2} \tag{37.15}$$

in which ε_1 and ε_2 are the graybody emissivities of the two surfaces, that is, that of the structural surface and fire/air, respectively.

37.3.4 Some Simplified Solutions of Heat Transfer

General heat transfer problems are difficult to solve and will usually require the use of numerical heat transfer procedures. However, for two common cases of unprotected and protected steelworks exposed to fire attack, simple analytical solutions have been derived to enable their temperatures to be calculated quickly. These simple analytical solutions have been derived by using the "lumped mass method," that is, the entire steel mass is given the same temperature. For unprotected steelwork

$$\Delta T_{\text{s}} = \frac{h}{\rho_{\text{s}} C_{\text{s}}} \frac{A_{\text{s}}}{V} (T_{\text{fi}} - T_{\text{s}})\, \Delta t \tag{37.16}$$

where V and A_{s} are the volume and exposed surface area of the steel element, respectively, ρ_{s} is the density of steel, and C_{s} is the specific heat of steel. The ratio A_{s}/V in Equation 37.16 is often referred to as the section factor of the steel element. T_{fi} and T_{s} are the fire and steel temperatures, respectively. h is the total heat transfer coefficient between the fire and the steel surface, including both the convective and radiant components. When using Equation 37.16, a step-by-step approach is necessary and the time increment should be small ($\Delta t < 5$ s).

For protected steelwork

$$\Delta T_{\rm s} = \frac{(T_{\rm fi} - T_{\rm s})A_{\rm s}/V}{(t_{\rm p}/k_{\rm p})C_{\rm s}\rho_{\rm s}(1 + (1/3)\phi)}\Delta t - (e^{\phi/10} - 1)\Delta T_{\rm fi}, \quad \text{where } \phi = \frac{C_{\rm p}\rho_{\rm p}}{C_{\rm s}\rho_{\rm s}}t_{\rm p}\frac{A_{\rm s}}{V} \qquad (37.17)$$

Additional symbols in Equation 37.17 include $t_{\rm p}$, the fire protection thickness; $k_{\rm p}$, thermal conductivity of the fire protection; $C_{\rm p}$, specific heat of the fire protection; $\rho_{\rm p}$, density of the fire protection; and $\Delta T_{\rm fi}$, the increment in fire temperature during the time interval Δt. The time increment should not be too large. When using Equation 37.17, the time increment (Δt) should not exceed 30 s.

Because of the second term in Equation 37.17, it is possible that at the early stage of increasing fire temperature, the increase in steel temperature ($\Delta T_{\rm s}$) may be negative. In this case, the steel temperature increase should be taken as zero.

37.3.5 Section Factors

Equations 37.16 and 37.17 clearly indicate that the temperature rise in a steel element is directly related to the section factor $A_{\rm s}/V$, that is, the ratio of the heated surface area to the volume of the steel element. Consider a unit length of a steel element where the end effects are ignored, the section factor may alternatively be expressed as $H_{\rm p}/A$, where $H_{\rm p}$ is the fire exposed perimeter length of the steel cross-section and A is the cross-sectional area of the steel element. Section factors for a few common types of steel sections exposed to fire are given in Table 37.3.

EXAMPLE 37.2

Section factor

Calculate the section factor ($H_{\rm p}/A$) for the two cases shown in Figure 37.8.

Calculation results

Case 1, Figure 37.8a

$$H_{\rm p} = 2 \times 400 + 150 \times 3 - 2 \times 10 = 1230 \, {\rm mm},$$
$$A = 2 \times 15 \times 150 + (400 - 15 \times 2) \times 10 = 8200 \, {\rm mm}^2$$
$$H_{\rm p}/A = 0.15 \, {\rm mm}^{-1} = 150 \, {\rm m}^{-1}$$

Case 2, Figure 37.8b

$$H_{\rm p} = 2\pi R_{\rm o} = 300\pi, \quad A = \pi(R_{\rm o}^2 - R_{\rm i}^2) = 2900\pi$$
$$H_{\rm p}/A = 0.1034 \, {\rm mm}^{-1} = 103.4 \, {\rm m}^{-1}$$

37.3.6 Thermal Properties of Materials

In order to use Equations 37.16 and 37.17, it is necessary to have available information on the thermal properties (thermal conductivity k, density ρ, and specific heat C) of steel and insulation materials.

37.3.6.1 Steel

The thermal properties of steel are known with reasonable accuracy and the following values are given in Eurocode 3 Part 1.2 (CEN 2000b):

Density

$$\rho_{\rm s} = 7850 \, {\rm kg/m}^3$$

Thermal conductivity [W/(m K)]

$$k_{\rm s} = 54 - \frac{T_{\rm s}}{300} \quad \text{for } 20°{\rm C} \le T_{\rm s} \le 800°{\rm C}$$
$$k_{\rm s} = 27.3 \quad \text{for } T_{\rm s} > 800°{\rm C}$$

TABLE 37.3 Section Factors of a Steel Element

Fire exposure situation	A_p/V
Unprotected steel section exposed to fire exposure around all sides	$\dfrac{2(2B - t_w + D)}{A_s}$

Fire exposure on all sides of board protection	$\dfrac{2(B + D)}{A_s}$

Fire protection on three sides: profile protection	$\dfrac{2(B - t_w) + B + 2D}{A_s}$

Fire protection on three sides: board protection	$\dfrac{2D + B}{A_s}$

Note: B = section width, D = steel depth, A_s = cross-sectional area, t_w = web thickness.

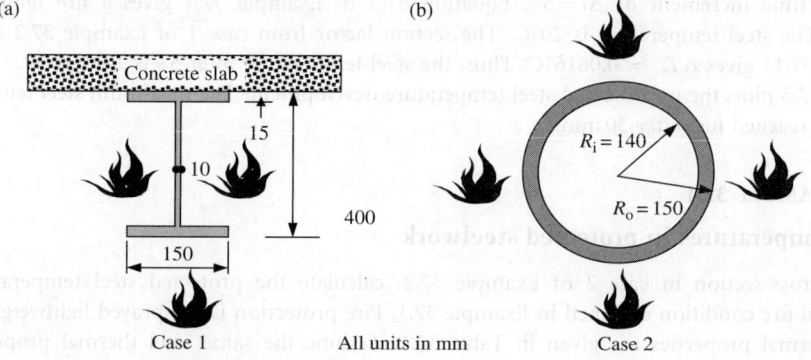

FIGURE 37.8   Dimensions of steel cross-sections.

Specific heat [J/(kg K)]

$$C_s = 425 + 0.773T_s - 0.00169T_s^2 + 2.22 \times 10^{-6}T_s^3 \quad \text{for } 20°C \le T_s \le 600°C$$

$$C_s = 666 - \frac{13,002}{T_s - 738} \quad \text{for } 600°C < T_s \le 735°C$$

$$C_s = 545 - \frac{17,820}{T_s - 731} \quad \text{for } 735°C < T_s \le 900°C$$

$$C_s = 650 \quad \text{for } T_s > 900°C$$

37.3.6.2 Insulation Materials

It is much more difficult to obtain information on the thermal properties of fire protection materials. This is partly due to the specific nature of fire protection materials, which have complicated and variable chemical reactions at high temperatures. An important factor contributing to this lack of information is that most fire protection materials are proprietary systems from different manufacturers and commercial sensitivity prevents publication of this type of information.

Information in Table 37.1 should only be used as a general guide for a few generic types of fire protection material.

Intumescent coatings offer a number of advantages, for example, architectural appearance and the possibility of offsite applications and are increasingly being used as fire protection materials to steel structures. At present, there are no suitable simplified design equations for steel structures protected with intumescent coatings. The designer has to rely on information provided by the intumescent coating manufacturer from standard fire resistance tests.

EXAMPLE 37.3

Temperatures in unprotected steelwork

For the cross-section in Example 37.2, case 1, calculate the unprotected steel temperature under the natural fire condition evaluated in Example 37.1. For steel, assume a constant density of 7850 kg/m³ and a constant specific heat of 650 J/(kg K). Also, assume a constant resultant emissivity of 0.5 and a convective heat transfer coefficient of 25 W/(m² K).

Calculation results

The calculations are performed for intervals of 5 s and results of only the first time increment are shown.
Equation 37.15 gives $h_r = 2.95$ W/(m² K). The total heat transfer coefficient is 27.95 W/(m² K).

After a time increment of $\Delta t = 5\,\mathrm{s}$, Equation 37.3 in Example 37.1 gives a fire temperature of 41.74°C. The steel temperature is 20°C. The section factor from case 1 of Example 37.2 is $150\,\mathrm{m}^{-1}$. Equation 37.17 gives $\Delta T_{\mathrm{s}} = 0.0616$°C. Thus, the steel temperature after 5 s is 20.0616°C.

Figure 37.5 plots the unprotected steel temperature development. The maximum steel temperature is 829.25°C, reached just after 30 min.

EXAMPLE 37.4

Temperatures in protected steelwork

For the cross-section in case 2 of Example 37.2, calculate the protected steel temperature under the natural fire condition obtained in Example 37.1. Fire protection is by sprayed lightweight concrete whose thermal properties are given in Table 37.1. Assume the same steel thermal properties as in Example 37.3.

Calculation results

The calculations are performed for intervals of 5 s and results of only the first time increment are shown.

Equation 37.17 gives $\phi = 1.129$ and $\Delta T_{\mathrm{s}} = -2.54$°C. As pointed out in Section 37.3.4, this negative number should be changed to 0.

Figure 37.5 plots the protected steel temperature development. The maximum steel temperature is 602.58°C, reached at 51.25 min.

37.4 Design of Structural Elements at Elevated Temperatures

37.4.1 General

The remaining sections of this chapter will introduce the reader to design calculations to assess structural stability at elevated temperatures. In Europe, the structural fire safety design of steel elements is covered in ENV 1993-1-2 (CEN 2000b), commonly known as Eurocode 3 Part 1.2 (to be referred to as Eurocode 3 hereafter), composite steel/concrete elements in ENV 1994-1-2 (CEN 2001) or Eurocode 4 Part 1.2 (Eurocode 4), reinforced concrete structures in ENV 1992-1-2 (CEN 1996), or Eurocode 2 Part 1.2, timber structures in ENV 1995-1-2 (CEN 2000c) or Eurocode 5 Part 1.2 and masonry structures in ENV 1996-1-2 (CEN 1997) or Eurocode 6 Part 1.2. At present, design calculations for steel and composite steel/concrete structures are advanced. Design calculations for concrete and timber structures are relatively brief and there is a significant lack of information for fire safety design of masonry structures. Also, there is very little information available to enable fire safety design calculations for structures made of more "specialist" construction materials such as glass, fiber reinforced plastics, aluminum. This chapter will present detailed information to enable fire safety design calculations of steel and composite steel/concrete structural elements. Structures using other materials will be not be dealt with in this chapter due to a significant lack of information.

It should be borne in mind that fire is an accidental event and its coincidence with extreme structural loading is rare. Fire attack is more likely to occur during normal use of a building. Therefore, for structural fire safety design, the design structural loads should be those present during normal service of a building, and further reduced to allow for occupant escape and combustible materials burning off.

Investigations into the collapse of the World Trade Center (FEMA 2002) have raised concern over the reliability of fire protection materials. It is assumed in this chapter that a fire protection material is able to fulfill its intended functions during the life of the protected structure and also during a fire exposure. This may be ensured by making sure that the fire protection material can stick to the protected structure or by limiting deflections of the structure. Also, fire protection materials may get damaged. At present, there are very few studies of this problem and there is insufficient information to help develop a sensible simple design guide to assess the acceptable extent of damage to fire protection materials. In the light of this, the designer has to ensure that any damage to the fire protection material is repaired.

The structural fire safety design criterion is the same as at ambient temperature, that is, the residual load carrying capacity of a structural member should not be lower than the applied load under the fire condition.

37.4.2 Mechanical Properties of Steel and Concrete at Elevated Temperatures

37.4.2.1 Steel

The stress–strain relationships of steel at elevated temperatures depend on whether steady state or transient state testing is employed. In steady state testing, the material temperature is held at a constant value and stress is changed. In transient state testing, stress is applied and the material temperature is then changed. Transient state testing is preferred since it reflects the realistic situation of a structure in fire, where structural loads are applied before fire exposure. In transient state testing, the rate of heating has some influence due to creep strain. But since the steel creep strain is small, mechanical testing of steel at elevated temperatures is usually carried out by using a typical heating rate of about 10°C/min as found in realistic steel structures exposed to fire conditions.

In Eurocode 3, the stress–strain curve of steel consists of a straight line for the initial response, followed by an elliptical relationship and then a plateau. Table 37.4 gives the mathematical descriptions used in Eurocode 3 and Figure 37.9 provides an illustration of this model and shows various parameters to be used in the mathematical model. In order to use this model, the reduced strength and stiffness of steel at elevated temperatures are required as input data and Table 37.5 gives their values, expressed as ratios of the values at elevated temperatures to that at ambient temperature. These ratios are often referred to as retention factors.

The thermal expansion strain (ε_{th}) of steel is given by

$$\varepsilon_{th} = -2.416 \times 10^{-4} + 1.2 \times 10^{-5}T + 0.4 \times 10^{-8}T^2 \quad \text{for } T \leq 750°C$$
$$\varepsilon_{th} = 0.011 \qquad \qquad \qquad \qquad \qquad \qquad \qquad \text{for } 750 < T \leq 860°C$$
$$\varepsilon_{th} = -0.0062 + 2 \times 10^{-5}T \qquad \qquad \qquad \quad \text{for } T > 860°C$$

TABLE 37.4 Mathematical Model of the Stress–Strain Relationship of Steel at Elevated Temperatures

Strain range	Stress σ
$\varepsilon \leq \varepsilon_{p,T}$	εE_T
$\varepsilon_{p,T} < \varepsilon < \varepsilon_{y,T}$	$f_{p,T} - c + \dfrac{b}{a}\sqrt{\left[a^2 - \left(\varepsilon_{y,T} - \varepsilon\right)^2\right]}$
$\varepsilon_{y,T} \leq \varepsilon \leq \varepsilon_{t,T}$	$f_{y,T}$
Parameters	$\varepsilon_{p,T} = \dfrac{f_{p,T}}{E_T},\ \varepsilon_{y,T} = 0.02,\ \varepsilon_{t,T} = 0.15$
Functions	$a^2 = \left(\varepsilon_{y,T} - \varepsilon_{p,T}\right)\left(\varepsilon_{y,T} - \varepsilon_{p,T} + \dfrac{c}{E_T}\right)$
	$b^2 = c\left(\varepsilon_{y,T} - \varepsilon_{p,T}\right)E_T + c^2$
	$c = \dfrac{\left(f_{y,T} - f_{p,T}\right)^2}{\left(\varepsilon_{y,T} - \varepsilon_{p,T}\right)E_T - 2\left(f_{y,T} - f_{p,T}\right)}$

Source: European Committee for Standardisation (CEN), 2000b, *Draft prEN 1993-1-2, Eurocode 3: Design of Steel Structures, Part 1.2: General Rules, Structural Fire Design* (London: British Standards Institution).

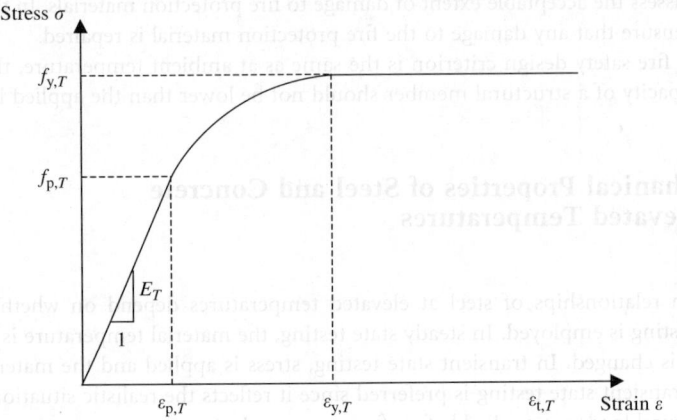

FIGURE 37.9 Stress–strain relationship of hot-rolled steel at elevated temperatures.

TABLE 37.5 Retention Factors of Steel at Elevated Temperatures

Steel temperature, T (°C)	Effective yield strength (relative to f_y at 20°C), $k_{y,T} = f_{y,T}/f_y$	Proportional limit (relative to f_y at 20°C), $k_{p,T} = f_{p,T}/f_y$	Slope of the linear elastic range (relative to E_a at 20°C), $k_{E,T} = E_T/E_a$
20	1	1	1
100	1	1	1
200	1	0.807	0.9
300	1	0.613	0.8
400	1	0.42	0.7
500	0.78	0.36	0.6
600	0.47	0.18	0.31
700	0.23	0.075	0.13
800	0.11	0.050	0.09
900	0.06	0.0375	0.0675
1000	0.04	0.025	0.045
1100	0.02	0.0125	0.0225
1200	0	0	0

Note: The effective yield strength is defined at 2% strain.

Source: European Committee for Standardisation (CEN), 2000b, *Draft prEN 1993-1-2, Eurocode 3: Design of Steel Structures, Part 1.2: General Rules, Structural Fire Design* (London: British Standards Institution).

In simple calculations, the coefficient of thermal expansion of steel may be assumed to be a constant so that the incremental thermal expansion strain is given by

$$\varepsilon_{th} = 14 \times 10^{-5} \, \Delta T$$

37.4.2.2 Concrete

37.4.2.2.1 Thermal Strains

The thermal strain of concrete is complex and is influenced by a number of factors. According to Anderberg and Thelandersson (1976) and Khoury and coworkers (Khoury 1983; Khoury et al. 1986), the thermal strain of concrete may be divided into thermal expansion strain, creep strain, and a stress induced transient thermal strain. Interested readers should refer to the above references for more

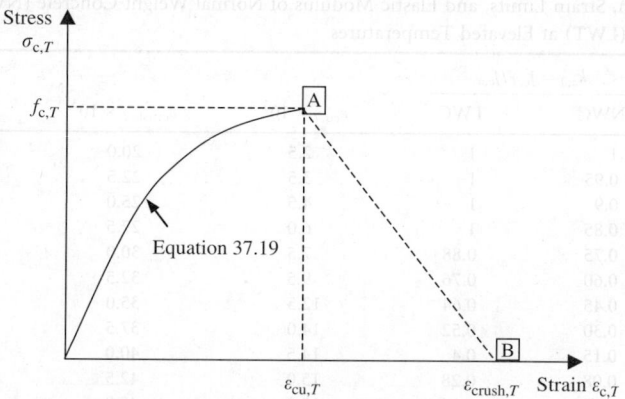

FIGURE 37.10 Stress–strain relationship of concrete at elevated temperatures.

detailed information on how to evaluate these different strain components. Eurocode 4 (CEN 2001) takes a simple approach and gives the coefficient of thermal expansion of concrete as

$$\varepsilon_{th} = -1.8 \times 10^{-4} + 9 \times 10^{-6}T + 2.3 \times 10^{-11}T^3 \quad \text{for } 20°C \leq T \leq 700°C;$$
$$\varepsilon_{th} = 0.0014 \quad \text{for } T > 700°C.$$

37.4.2.2.2 Stress–Strain Relationships

The mechanical properties of concrete are more variable than those of steel. Phan and Carino (1998, 2000) recently carried out a survey of mechanical properties of concrete (including high strength concrete) at elevated temperatures. Khoury (1992) provided an explanation of the variability in concrete mechanical properties at elevated temperatures. Values in Eurocode 4 (CEN 2001) may be regarded as the lower bound values of different test results for normal strength concrete.

Figure 37.10 shows the Eurocode 4 model for the stress–strain relationship of concrete and definitions of various parameters. The stress–strain relationship is divided into two parts: the ascending part and the descending part.

The Eurocode 4 equation for the ascending part is

$$\sigma_{c,T} = f_{c,T}\left\{3\left(\frac{\varepsilon_{c,T}}{\varepsilon_{cu,T}}\right) \bigg/ \left[2 + \left(\frac{\varepsilon_{c,T}}{\varepsilon_{cu,T}}\right)^3\right]\right\} \tag{37.18}$$

where $\sigma_{c,T}$, $\varepsilon_{c,T}$, $f_{c,T}$, $\varepsilon_{cu,T}$, are, respectively, the stress, strain, peak stress, and strain at peak stress for concrete at elevated temperature T.

From Equation 37.18, the initial Young's modulus of concrete may be obtained from

$$E_{c,T} = \frac{3}{2}\frac{f_{c,T}}{\varepsilon_{cu,T}} \tag{37.19}$$

The descending part is a straight line, joining the peak point (A) with the point of concrete crush (B) in Figure 37.10.

Values of $f_{c,T}$, $\varepsilon_{cu,T}$, and $\varepsilon_{crush,T}$, are required to determine the complete stress–strain relationship of concrete at elevated temperatures. Table 37.6 gives their values recommended by Eurocode 4 (CEN 2001). This table also gives the retention factors for modulus of elasticity.

37.4.3 Design of Steel Elements

37.4.3.1 Steel Beams

In general, the load carrying capacity of a steel beam depends on the bending moment capacity of its cross-section and its slenderness if lateral torsional buckling occurs. Unlike structural design at ambient

TABLE 37.6 Strength, Strain Limits, and Elastic Modulus of Normal Weight Concrete (NWC) and Lightweight Concrete (LWT) at Elevated Temperatures

| Temperature | $k_{c,T} = f_{c,T}/f_{c,a}$ | | $\varepsilon_{cu,T} \times 10^3$ | $\varepsilon_{crush,T} \times 10^3$ | $k_{E,T} = E_T/E_a$ | |
	NWC	LWC			NWC	LWC
20	1	1	2.5	20.0	1	1
100	0.95	1	3.5	22.5	0.844	0.889
200	0.9	1	4.5	25.0	0.72	0.8
300	0.85	1	6.0	27.5	0.618	0.727
400	0.75	0.88	7.5	30.0	0.5	0.587
500	0.60	0.76	9.5	32.5	0.369	0.468
600	0.45	0.64	12.5	35.0	0.257	0.366
700	0.30	0.52	14.0	37.5	0.16	0.277
800	0.15	0.4	14.5	40.0	0.075	0.2
900	0.08	0.28	15.0	42.5	0.038	0.132
1000	0.04	0.16	15.0	45.0	0.018	0.071
1100	0.01	0.04	15.0	47.5	0	0.017
1200	0.0	0.0	15.0	50.0	0	0

Source: European Committee for Standardisation (CEN), 2001, *prEN 1994-1-2, Eurocode 4: Design of Composite Steel and Concrete Structures, Part 1.2: Structural Fire Design* (London: British Standards Institution).

temperature, lateral torsional buckling of a steel beam is usually not a problem in fire. This is because steel beams that are required to have fire resistance are floor beams whose compression flanges are restrained by the floor slabs. Steel beams that should be checked for lateral torsional buckling at ambient temperature, for example, roof beams, do not require fire resistance. Therefore, the following design method will only consider the cross-sectional bending resistance of a beam.

Eurocode 3 gives two methods to calculate the plastic bending moment capacity of a steel beam. The first method is the bending moment capacity method that is generally applicable to cross-sections with nonuniform temperature distributions. In this method, the steel cross-section is divided into a number of thin slices of approximately the same temperature. The plastic bending moment capacity of the cross-section is calculated according to the reduced strengths of steel at the temperatures of these slices.

EXAMPLE 37.5

Plastic bending moment capacity of a nonuniformly heated steel beam

An example is given in Table 37.7 to illustrate this method. Input information for this example are shown in Figure 37.11.

As can be seen, the plastic bending moment capacity method requires many calculations. Hence, the only benefit of using this more elaborate method is to explore possible benefits of nonuniform temperature distribution in the cross-section of a beam, particularly to justify the use of unprotected steelwork.

Instead of using the plastic bending moment method, Eurocode 3 also gives an alternative method that is much simpler to use. In this simple method, the plastic bending moment capacity of a beam is given by

$$M_{p,fi} = k_{y,T} M_p / \kappa_1 \qquad (37.20)$$

where M_p is the plastic bending moment capacity of the cross-section at ambient temperature; $k_{y,T}$ is the retention factor for the effective yield strength of steel at the maximum temperature in the lower flange. Thus, $k_{y,T} M_p$ gives the reduced plastic bending moment capacity of the cross-section at a uniform temperature T. The modification factor κ_1 is used to account for nonuniform temperature distribution in the cross-section and $\kappa_1 = 0.7$. For the above example, $k_{y,T} M_p = 137.5$ kN m. Using Equation 37.20 gives a value of 196.5 kN m, which is close to that obtained using the more time-consuming plastic bending moment capacity method.

TABLE 37.7 An Example of Using the Bending Moment Capacity Method

Part i	Temperature, T_i (°C)	Resultant force, $F_i = p_y(T_i) \times A_i$ (kN)	Lever arm from top, d_i (mm)	Bending moment $= F_i \times d_i$ (kN m)
Upper flange	$650 \times 0.8 = 520$	−565.81	8	−4.53
Web 1	528.13	−86.08	39.8	−3.43
Web 2	544.38	−24.10	70.76	−1.71
		55.96	94.56	5.29
Web 3	560.63	73.93	135.0	9.98
Web 4	576.88	67.74	182.6	12.37
Web 5	593.13	61.56	230.2	14.17
Web 6	609.38	55.75	277.8	15.49
Web 7	625.63	50.21	325.4	16.34
Web 8	641.88	44.68	373.0	16.66
Lower flange	650.0	266.16	404.8	107.74
Total	NA	0	NA	188.39

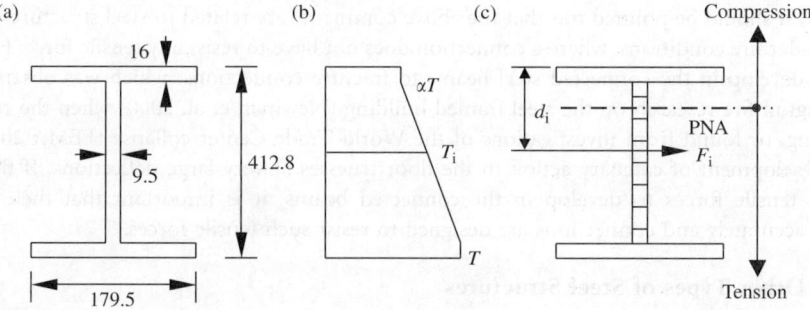

FIGURE 37.11 Input data for calculations in Table 37.7: (a) dimensions (mm), (b) temperature distribution, and (c) cross-section layers.

37.4.3.2 Steel Columns

It is commonly assumed that a steel column is surrounded by fire on all sides so that the steel temperature is uniform. Because the stress–strain relationships of steel at elevated temperatures are non-linear and cannot be assumed to be elastic–perfectly plastic, calculations of the column resistance in fire are slightly different from those at ambient temperature. In Eurocode 3, the column axial compressive resistance is calculated from

$$P_{c,fi} = \chi_{fi} P_{u,fi} \tag{37.21}$$

where $P_{u,fi}$ is the column squash load at elevated temperature T and is calculated from

$$P_{u,fi} = A_s k_{y,T} f_y \tag{37.22}$$

χ_{fi} is the column strength reduction factor to account for the column slenderness effect and is given by

$$\chi_{fi} = \frac{1}{\phi_{fi} + \sqrt{\phi_{fi}^2 - \bar{\lambda}_{fi}^2}}, \quad \text{with } \phi_{fi} = \frac{1}{2}\left(1 + \alpha\bar{\lambda}_{fi} + \bar{\lambda}_{fi}^2\right) \tag{37.23}$$

where α is an imperfection factor, $\alpha = 0.65\sqrt{f_y/235}$.

The column slenderness in fire $\bar{\lambda}_{fi}$ is calculated using

$$\bar{\lambda}_{fi} = \bar{\lambda}\sqrt{\frac{k_{y,T}}{k_{E,T}}} \tag{37.24}$$

where $\bar{\lambda}$ is the column slenderness at ambient temperature. $\bar{\lambda}$ is defined as

$$\bar{\lambda} = \lambda\sqrt{\frac{f_y}{\pi^2 E}}, \quad \text{with } \lambda = \frac{L_e}{r_y} \tag{37.25}$$

where L_e and r_y are the column buckling length and radius of gyration of the column cross-section about the relevant axis of buckling.

37.4.3.3 Steel Connections

The behavior of steel connections in fire is complicated due to complex temperature distributions in connections and connection interactions with the adjacent structure. Fortunately, since a connection is much "bulkier" than the connected members, its temperatures are much lower, and, if designed properly, it is rarely the weak link in the structure. Current design methods require that when fire protection materials are applied to a structure, the thickness of protection applied to a connection should be based on the thickness required for whichever of the members jointed by the connection that has the highest section factor H_p/A.

However, it should be pointed out that the above comments are related to steel structures in flexural bending under fire conditions, where a connection does not have to resist any tensile force. High tensile forces can develop in the connected steel beams to fracture connections, which was observed during the Cardington fire research on the steel framed building (Newman et al. 2000) when the connections were cooling, or found from investigations of the World Trade Center collapse (FEMA 2002) due to possible development of catenary action in the floor trusses at very large deflections. If the designer anticipates tensile forces to develop in the connected beams, it is important that their values are quantified accurately and connections are designed to resist such tensile forces.

37.4.3.4 Other Types of Steel Structures

37.4.3.4.1 Stainless Steel Structures

Due to architectural demand and superior corrosion resistance, stainless steels are becoming more widely used. Although their fire resistance is only a minor factor in determining whether to use stainless steel or not, stainless steel does have superior fire resistance to conventional carbon steels.

Baddoo (1999) gives some information on the strength retention factors of stainless steels at different elevated temperatures. Whilst conventional carbon steel loses about 50% of its strength at a temperature of around 600°C, the temperature that gives the same loss in the strength of stainless steel is much higher, at about 800°C. Also, the surface of stainless steel has a much higher reflectivity, hence the emmissivity (1 − reflectivity) of stainless steel is much lower. Typically, the emmissivity of stainless steel is about 0.3 to 0.4, compared to about 0.8 for carbon steel. This gives a much lower temperature in a stainless steel structure in fire. For realistic loading conditions, it is almost certain that stainless steel structures will be able to be engineered to provide sufficient fire resistance without the need for fire protection.

With suitable modifications to take into consideration the reduced strength and stiffness of stainless steel at elevated temperatures, the same design method for carbon steel may be extended to stainless steel structures (Baddoo and Burgen 1998).

37.4.3.4.2 Portal Frames

Portal frames are usually single storey buildings with a small density of occupants. Escape in the case of fire attack is relatively easy. Therefore, fire safety of a portal frame only becomes a requirement when the portal frame is adjacent to another building and it is necessary to prevent fire spread from the portal frame building to the adjacent building. Fire spread from a portal frame building is usually through collapsed walls; consequently, the structural safety requirement for a portal frame under fire attack is to ensure that the portal frame columns in the walls remain upright so that the walls are stable. The portal frame girders may be allowed to collapse.

The Steel Construction Institute in the United Kingdom has developed a design guide (Newman 1990) for portal frames in fire. Using this guide, portal frames are usually designed without fire protection.

37.4.3.4.3 Water Cooled Structures

The design of water cooled structures (Bond 1975) relies on the principle that boiling water has a very high value of convective heat transfer coefficient, some 3 or 4 orders higher than the convective heat transfer coefficient of air. If boiling water is replenished, heat on a water cooled steel structure is taken away and steel temperatures remain low. Typically, steel temperatures in a water cooled structure do not exceed 150°C. Therefore, structural stability of a water cooled structure is rarely a design issue. Design is mainly concerned with hydraulic calculations to ensure sufficient water supply and circulation in case of fire.

Water cooling a steel structure to achieve fire protection is expensive and because of this it is rarely used solely for the purpose of fire protection. It is usually combined with other functions. An excellent example (Bressington 1997) of recent application of this technique is in the roof truss of the cargo handling facility of Hong Kong Air Cargo Terminals Ltd. The steel structural roof truss is made of circular hollow sections and is used as the water distribution pipe for sprinklers. On operation of the sprinklers in the event of a fire, internal water flow through the steelwork members also provides sufficient cooling.

37.4.3.4.4 External Steelwork

In the case of a building enclosure fire, fire exposure on the external steelwork differs from that on the interior steelwork in two ways:

- The fire temperature to the external steelwork is much lower than that to the interior steelwork.
- The external steelwork may not be directly engulfed in fire.

These two differences ensure that temperatures in the external steelwork are kept lower than their failure temperatures so that fire protection is not necessary.

Law and O'Brien (1989) developed a design method for external steelwork and the design method in Eurocode 3 is based on their work.

37.4.4 Composite Steel/Concrete Members

37.4.4.1 Composite Slabs

Composite slabs are constructed from reinforced concrete slabs in composite action with steel decking underneath. The steel decking acts as support to the concrete during construction and is generally profiled to maximize structural efficiency. Composite slabs usually form the floor of a fire resistant compartment. Hence, they should meet all the requirements of fire resistant construction; that is, in addition to sufficient load bearing resistance, they should also have adequate insulation and maintain their integrity during fire attack.

Composite floor slabs are noncombustible and will not suffer integrity failure by burning through. However, the problem of integrity failure may occur at the junctions between a composite slab and other construction elements. Particular attention should be paid to the slab edges where large cracks may occur due to large rotations. It is important that reinforcement bars should be made continuous over the supports.

To check whether a slab can fulfill the insulation requirement, it is necessary to carry out a heat transfer analysis to determine temperatures on the unexposed surface of the slab. Results of this temperature analysis can then be used to determine the minimum slab thickness above which the unexposed surface temperature is unlikely to exceed the allowed values. For most applications where the required standard fire resistance rating does not exceed 90 min, the required minimum slab thickness will almost certainly be less than that required by other functions such as control of deflections. Hence, the insulation requirement for composite slabs is very rarely a problem in fire resistant design.

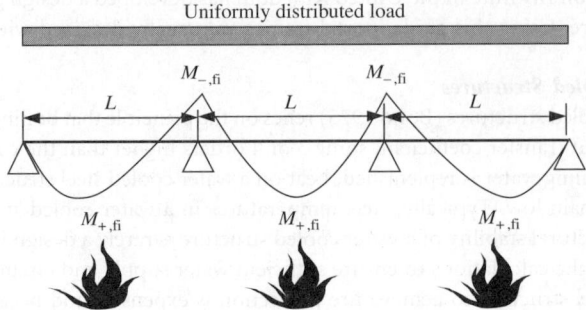

FIGURE 37.12 A continuous slab exposed to fire.

37.4.4.1.1 Load Bearing Capacity of One-Way Spanning Composite Slabs

When calculating the load bearing resistance of a composite slab in fire, it is often assumed that it is one-way spanning, being effective only in the direction of the concrete rib. For a continuous composite slab, the plastic design method may be used, in which plastic hinges are assumed to form at the supports and locations of the maximum bending moment in the span of the slab. For example, for the interior span of a continuous slab under uniformly distributed load, as shown in Figure 37.12,

$$M_{+,\text{fi}} + M_{-,\text{fi}} \geq M_{\text{fi, max}}, \quad \text{i.e., } M_{+,\text{fi}} + M_{-,\text{fi}} \geq \frac{1}{8} w L^2 \qquad (37.26)$$

where $M_{+,\text{fi}}$ and $M_{-,\text{fi}}$ are the sagging and hogging bending moment resistance of the slab, respectively, and $M_{\text{fi,max}}$ is the maximum free bending moment in the slab under fire conditions.

For the end span of a continuous slab with uniformly distributed load, as shown in Figure 37.12,

$$M_{+,\text{fi}} + \frac{1}{2} M_{-,\text{fi}} = \frac{1}{8} w L^2 + \frac{M_{-,\text{fi}}^2}{2 L^2 w} \qquad (37.27)$$

or approximately

$$M_{+,\text{fi}} + 0.45 M_{-,\text{fi}} \geq \frac{1}{8} w L^2 \qquad (37.28)$$

In order to determine the slab load carrying capacity in fire, the sagging bending moment capacity $M_{+,\text{fi}}$ and hogging bending moment capacity $M_{-,\text{fi}}$ should be evaluated.

When calculating the sagging bending moment capacity of a slab, the reinforcement near the fire side is in tension and the compressive concrete is on the unexposed side of the slab. Since the temperature rise on the unexposed side of the slab is required to be below 140°C to fulfill the insulation requirement, the concrete in compression can be assumed to be cold and its cold strength may be used when calculating the sagging bending moment capacity of the slab. Contributions from the steel decking are usually ignored because the decking will be unprotected and may debond because it is under direct fire attack. Figure 37.13 shows the calculation procedure. In Figure 37.13, f_c is the design strength of concrete in bending at ambient temperature, A_r, $p_{y,r}$, and $k_{y,r}(T)$ are the area, design strength at ambient temperature, and strength retention factor of the reinforcement at temperature T, and $k_{y,r}(T)$ may be obtained from Figure 37.14.

Under a hogging bending moment, the compression face of a composite slab is exposed to fire where there is a very steep temperature gradient. When calculating the hogging bending moment capacity, the composite slab, including the concrete in the ribs, should be divided into a number of layers each of approximately constant temperature. The contribution of each layer should be evaluated separately and then integrated to give the total slab resistance.

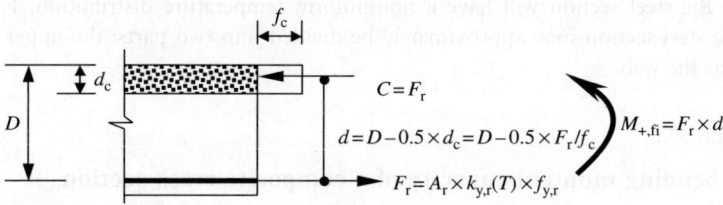

FIGURE 37.13 Calculation method for sagging bending moment capacity.

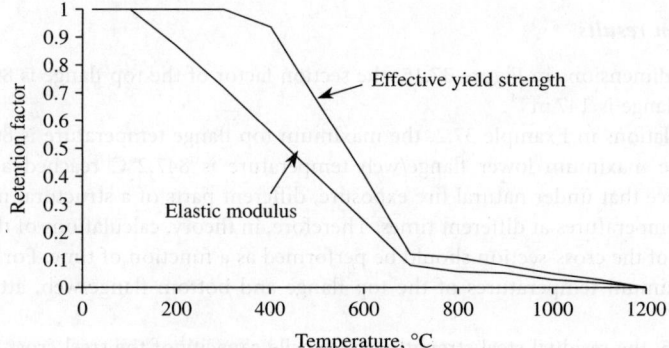

FIGURE 37.14 Retention factors of cold worked reinforcing steel (from European Committee for Standardisation (CEN), 2001, *prEN 1994-1-2, Eurocode 4: Design of Composite steel and Concrete Structures, Part 1.2: Structural Fire Design* (London: British Standards Institution)).

37.4.4.1.2 Load Bearing Capacity of Two-Way Spanning Slabs

The load carrying capacity of a composite slab in one-way spanning is usually sufficient under fire conditions. However, in some cases, it may be necessary to utilize the slab strength in two-way spanning. This is particularly the case when it is necessary to justify the elimination of fire protection from some of the slab-supporting steel beams. The yield line analysis (Johansen 1962) for reinforced concrete slabs may be used to give a safe estimate of the slab load carrying capacity in two-way spanning.

The real benefit of utilizing the strength of a slab in two-way spanning is the possibility of using tensile membrane action in the slab, under which the strength of the slab can be many times higher than that given by yield line analysis. Recently, a design method has been developed to use tensile membrane action in steel framed buildings to eliminate fire protection to some steel beams. For detailed design equations and examples, reference should be made to the two papers by Bailey and Moore (2000a, 2000b) and a publication by the United Kingdom's Steel Construction Institute (Newman et al. 2000).

37.4.4.2 Composite Beams

For a conventional composite beam with concrete slabs on top of the steel section, the sagging bending moment capacity of the composite cross-section may be calculated by the bending moment capacity method, similar to steel beams (Section 37.4.3.1). The concrete in compression may be assumed to be cold and the steel temperatures may be calculated using Equations 37.16 or 37.17. If the steel section is protected, it may be assumed to have a uniform temperature distribution and its section factor A_p/V is that of the entire section, calculated according to Table 37.3. If the steel section is unprotected, the steel section will have a nonuniform temperature distribution. For temperature calculations, the steel section may approximately be divided into two parts: the upper flange and the lower flange plus the web.

EXAMPLE 37.6

Plastic bending moment capacity of a composite cross-section

Figure 37.15 shows a composite cross-section exposed to fire underneath the slab. The steel section is unprotected and fire exposure is according to the temperature–time relationship in Example 37.1. At ambient temperature, the design strength of steel is 275 N/mm^2 and the design compression strength of concrete is 20 N/mm^2. Calculate the minimum sagging bending moment capacity of the composite cross-section in fire.

Calculation results

According to the dimensions in Figure 37.15, the section factor of the top flange is 80 m^{-1} and that of the web/bottom flange is 147 m^{-1}.

Following calculations in Example 37.2, the maximum top flange temperature is 802.8°C reached at 32.25 min and the maximum lower flange/web temperature is 847.2°C reached at 28.75 min. It is interesting to notice that under natural fire exposure, different parts of a structural member will reach their maximum temperatures at different times. Therefore, in theory, calculations of the plastic bending moment capacity of the cross-section should be performed as a function of time. For simplicity, in this example, the maximum temperatures of the top flange and bottom flange/web, attained at different times, are used.

From Table 37.5, the residual steel strengths and tensile capacity of the steel cross-section are

Top flange: $f_y = 0.1086 \times 275 = 29.9$ N/mm^2, $A_s = 2250$ mm^2, $N_{uf} = 67.3$ kN
Bottom flange/web: $f_y = 0.08639 \times 275 = 23.76$ N/mm^2
The lower flange tension resistance: $N_{lf} = 53.5$ kN
The web area is 3700 mm^2, giving the web tension resistance: $N_{wf} = 87.9$ kN
The total tensile resistance of the steel cross-section: $N_s = N_{uf} + N_{wf} + N_{lf} = 208.7$

Assume the top of the concrete slab is cold. The depth of concrete in compression is $208.7 \times 1000/(20 \times 2000) = 5.22$ mm.

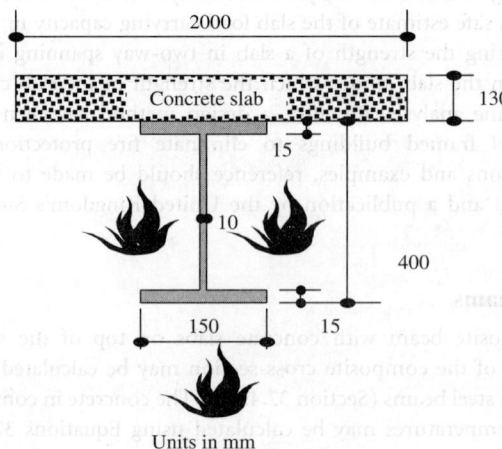

Units in mm

FIGURE 37.15 Composite cross-section dimensions for Example 37.6.

Taking moment about the top surface of the slab, the residual sagging bending moment capacity of the composite cross-section is

$$M_f = [-208.7 \times 5.22/2 + 67.3 \times (130 + 15/2) + 87.9 \times (130 + 400/2)$$
$$+ 53.5 \times (130 + 400 - 15/2)]/1000$$
$$= 65.7 \, \text{kN m}$$

37.4.4.3 Composite Columns

37.4.4.3.1 Resistance to Axial Load According to Eurocode 4

The general equation for calculating the squash load of a composite column is

$$P_{u,fi} = \sum (A_i f_i)_c + \sum (A_i f_i)_s + \sum (A_i f_i)_r \tag{37.29}$$

where f_i is the design strength of the ith layer and subscripts "c," "s," and "r" represent concrete, steel, and reinforcement, respectively. Due to nonuniform temperature distribution, each component of the composite cross-section is divided into a number of layers of approximately the same temperature.

Similarly, the rigidity (EI) of the composite cross-section is calculated using the following equation:

$$(EI)_{fi} = \sum (EI)_{i,s} + \sum 0.8(EI)_{i,c} + \sum (EI)_{i,r} \tag{37.30}$$

where symbols E and I are the initial modulus of elasticity and second moment of area of the appropriate component material about the relevant axis of buckling of the entire composite cross-section, respectively.

The composite column compression resistance in fire is given by

$$P_{c,fi} = \frac{\chi_{fi}}{1.2} P_{u,fi} \tag{37.31}$$

where the compression strength reduction factor in fire is calculated by

$$\chi_{fi} = \frac{1}{\phi_{fi} + \sqrt{\phi_{fi}^2 - \bar{\lambda}_{fi}^2}} \quad \text{with} \quad \phi_{fi} = 0.5 \left[1 + \alpha (\bar{\lambda}_{fi} - 0.2) + \bar{\lambda}_{fi}^2 \right] \tag{37.32}$$

in which the initial imperfection factor α has a value of 0.49.

The column slenderness factor $\bar{\lambda}_{fi}$ in fire is defined by

$$\bar{\lambda}_{fi} = \sqrt{\frac{P_{u,fi}}{P_{cr,fi}}} \tag{37.33}$$

where the Euler buckling load in fire is calculated using

$$P_{cr,fi} = \frac{\pi^2 (EI)_{fi}}{L_e^2} \tag{37.34}$$

in which L_e is the column effective length.

37.4.4.3.2 Simplified Temperature Calculation Method for Unprotected Concrete Filled Columns

The temperature calculation method is based on the method of Lawson and Newman (1996) with modification by Wang (2000). This method assumes that the composite column is unprotected.

In this method, the steel shell temperature is calculated by

$$T_s = C_2 T_{fi} \tag{37.35}$$

where T_{fi} is the standard fire temperature and C_2 is a multiplication factor depending on the fire resistance time. C_2 is given by

$$C_2 = 1 - 0.02t \times \frac{120 - FR}{120}, \quad \text{but } C_2 \leq 1.0 \tag{37.36}$$

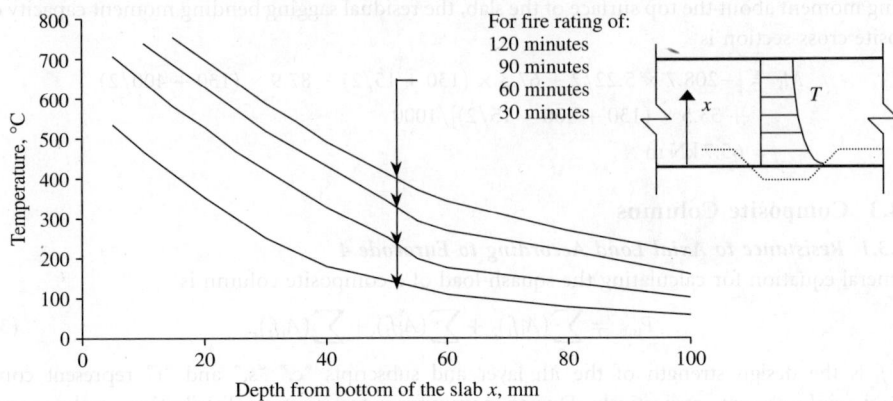

FIGURE 37.16 Temperatures of a concrete slab exposed to the standard fire from underneath (from European Committee for Standardisation (CEN), 2001, *prEN 1994-1-2, Eurocode 4: Design of Composite steel and Concrete Structures, Part 1.2: Structural Fire Design* (London: British Standards Institution)).

TABLE 37.8 Multiplication Factor C_1

Diameter or size of square section (mm)	Distance of center of layer from out surface (mm)				
	10	30	50	70	>70
200	1.08	1.22	1.41	1.60	1.80
300	1.05	1.14	1.22	1.36	1.50
400	1.03	1.09	1.18	1.25	1.35
500	1.02	1.07	1.12	1.18	1.25

Source: Lawson, R.M. and Newman, G.M., 1996, *Structural fire design to EC3 & EC4, and comparison with BS 5950*, Technical Report, SCI Publication 159, The Steel Construction Institute.

where FR is the fire resistance rating in minutes and t is the thickness of the steel shell in mm. The concrete temperature is calculated by

$$T_c = C_1 C_2 T_{\text{slab}} \qquad (37.37)$$

where C_1 is a multiplication factor depending on the composite section size and location of the concrete (and is independent of the standard fire resistance time) and T_{slab} is the temperature in an infinitely wide concrete slab exposed to fire on one side (given in Figure 37.16). Values of C_1 are given in Table 37.8.

If reinforcement is used, the reinforcement temperature should be taken as that of the concrete at the same location. Furthermore, for reinforcement in the corners of a square section, due to heating from two sides of the composite section, the reinforcement temperature should be calculated from an equivalent depth of half the concrete cover depth.

EXAMPLE 37.7

Compression resistance of a composite column

Figure 37.17 shows the dimensions of a concrete filled circular steel section. The effective length of the column is 4 m. Calculate the column compression resistance for a standard fire resistance period of 60 min. At ambient temperature, steel has a design strength of 275 N/mm² and Young's modulus of 205,000 N/mm². Concrete has a design compression strength of 40 N/mm² and modulus of elasticity of 20,000 N/mm².

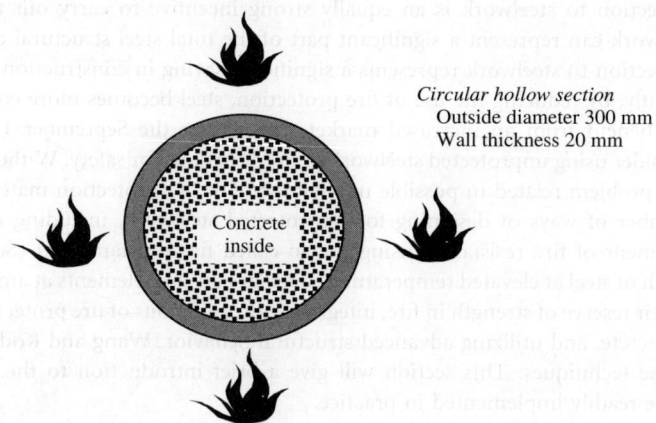

Circular hollow section
Outside diameter 300 mm
Wall thickness 20 mm

Concrete inside

FIGURE 37.17 Column cross-section size for Example 37.7.

TABLE 37.9 Calculation Results for Example 37.7

Zone	Radius (mm)	T_{slab} (°C)	T_{zone} (°C)	A (mm²)	I (cm⁴)	f_i (N/mm²)	E_i (N/mm²)	$(fA)_I$ (kN)	$(EI)_I$ (kN m²)
Steel	130–150	—	756	17,593	17,329	44.77	22,058	787.6	3322.4
Con.	110–130	421	480	15,080	10,933	25.2	7,875	380	861
Con.	90–110	250	305	12,566	6,346	33.84	12,250	425	777.4
Con.	70–90	150	204	10,053	3,267	35.92	14,310	361	467.5
Con.	0–70	130	195	15,394	1,886	36.1	14,512	556	273.7

Calculation results

Divide the composite cross-section into steel tube, three concrete rings of equal thickness of 20 mm, and a concrete core of 70 mm radius. Table 37.9 gives, for each zone of the composite cross-section, temperature (in °C, calculated using Equations 37.35–37.37), area (mm²), second moment of area about a principle axis of the entire cross-section (cm⁴), reduced design strength (in N/mm²), and Young's modulus (in N/mm², from Table 37.5 for steel and Table 37.6 for concrete) at elevated temperatures, compression resistance (in kN) and rigidity (EI, in kN m²).

Equation 37.29 gives $P_{u,fi} = 2509.6$ kN. Equation 37.34 gives the Euler buckling load in fire $P_{cr,fi} = 5226.1$ kN, giving a column slenderness in fire $\bar{\lambda}_{fi} = 0.7885$. Equation 37.32 gives $\chi_{fi} = 0.563$ and Equation 37.31 gives the column compression resistance in fire $P_{c,fi} = 1413$ kN.

37.4.4.3.3 High Strength Concrete Filled Columns
With the introduction of high strength concrete, the load carrying capacity of a concrete filled column can be further enhanced. However, the increase in fire resistance is relatively small because high strength concrete loses its strength at a much lower temperature than normal strength concrete. By adding a small amount of steel fibre to the concrete, the elevated temperature performance of high strength concrete can be much improved and the performance of fibre reinforced high strength concrete filled steel columns is similar to that with normal strength concrete filling (Kodur and Wang 2001). Provided the strength and stiffness retention factors are available, Equations 37.29 and 37.30 can also be used.

37.5 Design for Unprotected Steelwork

The quest for knowledge is one of the main drivers of research to investigate the behavior of steel and composite structures under fire conditions. However, it should be recognized that the desire to reduce or eliminate fire protection to steelwork is an equally strong incentive to carry out these studies. Fire protection to steelwork can represent a significant part of the total steel structural cost and the elimination of fire protection to steelwork represents a significant saving in construction cost to the client. But more importantly, by reducing the use of fire protection, steel becomes more competitive and the steel industry can benefit from an increased market share. After the September 11 event, it is also appropriate to consider using unprotected steelwork in fire situations for safety. Without fire protection, there would be no problem related to possible unreliable use of fire protection materials.

There are a number of ways of designing for unprotected steelwork, including risk assessment to reduce the requirement of fire resistance, using the so-called fire resistant steel (Sakumoto 1998) to increase the strength of steel at elevated temperatures, over-design steel elements at ambient temperature so as to increase their reserve of strength in fire, integrating the functions of fire protection and structural load bearing of concrete, and utilizing advanced structural behavior. Wang and Kodur (2000) provide a summary of these techniques. This section will give a brief introduction to the last two methods because they can be readily implemented in practice.

37.5.1 Integration of Structural Load Bearing and Fire Protection Functions of Concrete

It should be appreciated that it is very rare for steelwork to be used alone. Steel is usually used in combination with other materials, in particular with concrete. Concrete is not only a structural material, it also has good thermal insulation properties. Therefore, by combining these two functions of concrete, composite structures may be constructed to give inherently high fire resistance. The systems that will be described below have made special considerations of fire resistance in their design and construction. The following paragraphs will give a short description of their main features and the inherent standard fire resistance that they can achieve. This should enable the designer to determine quickly a possible structural load bearing system where the main design concern is to use unprotected steelwork. More detailed information may be found in Bailey and Newman (1998).

37.5.1.1 Beams

Three types of construction may be used:

1. Slim floor/asymmetric beam, shown in Figures 37.18a and b. In the slim floor construction, a wide plate is welded to the bottom flange of a universal column section and composite floor slabs are supported on the wide plate. In an asymmetric steel beam, the bottom flange is rolled wider than the top flange. Both systems use the same principle to achieve unprotected steelwork: the

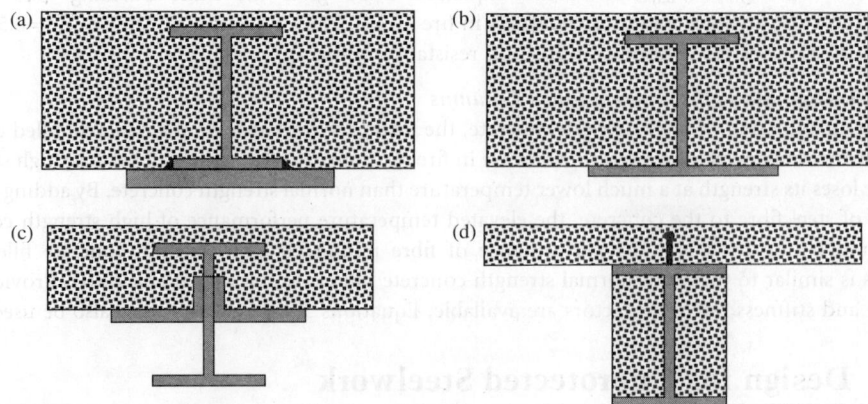

FIGURE 37.18 Steel/composite beams of high fire resistance: (a) slim floor beam, (b) asymmetrical beam, (c) shelf angle beam, and (d) partially encased steel beam.

web of the steel section is protected by the concrete and provides the majority of the bending resistance of the steel beam at elevated temperatures. Only the steel section is assumed to have load carrying capacity, but lateral torsional buckling is prevented by the concrete slabs.

2. Shelf angle beams, shown in Figure 37.18c. In this system, steel angles are welded to the web of a steel beam and these angles are used to support precast concrete floor units. This system is mainly used to reduce the structural depth of the floor. Since the angles, the upper flange, and the upper portion of the web of the steel section are shielded from fire exposure, 60 min of fire resistance can be achieved using this system without fire protection.

3. Partially encased beams, shown in Figure 37.18d. By casting concrete in between the flanges of a regular universal beam section, only the downward side of the lower flange will be exposed to fire. Both the web and the upper flange are shielded from fire exposure and can provide high structural resistance. Composite floor slabs may be connected to the top of the partially encased steel beam via shear connectors to obtain composite action. Since concrete is cast between flanges of the steel section, no temporary formwork is necessary. By using reinforcement, standard fire resistance of up to 3 h can be obtained without fire protection to the steelwork.

Table 37.10 summarizes the standard fire resistance rating that can be achieved by different types of unprotected steel beams.

37.5.1.2 Columns

Three types of unprotected columns may be used:

1. Columns with blocked-in webs as shown in Figure 37.19a. In this construction, lightweight aerated concrete blocks are placed between the flanges of a universal steel section. The aerated concrete blocks not only provide good insulation to the column web, they also reduce the average column flange temperature compared to a bare steel column. A standard fire resistance rating of 30 min can be achieved without additional fire protection.

2. Partially encased steel columns with unreinforced and reinforced concrete as shown in Figure 37.19b. In a column with blocked-in web, the lightweight aerated concrete only provides insulation to the steel section and the system cannot provide 60 min fire resistance. If normal

TABLE 37.10 Standard Fire Resistance Rating of Unprotected Steel Beams

Type of construction	Standard fire resistance time (min)
Bare universal beams	15
Slim floor/asymmetrical beams	60
Shelf angle beams	60
Partially encased beams	>60

Source: Bailey, C.G. and Newman, G.N., 1998, The design of steel framed building without applied fire protection, *The Structural Engineer,* **76**(5), 77–81.

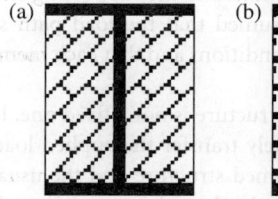
(a)
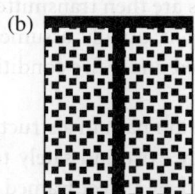
(b)
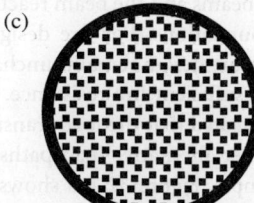
(c)

FIGURE 37.19 Steel/composite columns of high fire resistance: (a) blocked-in web, (b) partially encased, and (c) concrete filled.

TABLE 37.11   Standard Fire Resistance Rating of Unprotected Steel Columns

Type of column	Standard fire resistance time (min)
Universal column	15
Blocked-in column	30
Partially encased with unreinforced concrete	60
Partially encased with reinforced concrete	>60
Concrete filled hollow section without reinforcement	60
Concrete filled hollow section with reinforcement	>60

Source: Bailey, C.G. and Newman, G.N., 1998, The design of steel framed building without applied fire protection, *The Structural Engineer*, **76**(5), 77–81.

strength concrete is used to provide composite action, much higher fire resistance can be obtained. If unreinforced concrete is used, 60 min of fire resistance can be obtained. Reinforcement may be used to give much higher fire resistance.

3. Concrete filled hollow steel sections as shown in Figure 37.19c. Concrete filling of hollow steel sections is a very practical solution to form composite columns. Either unreinforced or reinforced concrete may be used. This type of column has been described in some detail in Section 37.4.4.3. To summarize, unreinforced concrete filled columns can achieve 60 min of fire resistance. If reinforcement is used, much higher fire resistance may be obtained.

To summarize, Table 37.11 gives the standard fire resistance time that can be achieved by different types of unprotected columns.

The usefulness of Table 37.10 and Table 37.11 is to enable readers to reach a decision quickly on the form of construction to achieve the required standard fire resistance without fire protection.

37.5.2   Utilizing Whole Building Performance in Fire

It has long been recognized that the behavior of a whole building in fire is much better than that of its individual members. By achieving a better understanding of whole building behavior, it is possible to achieve the objective of eliminating fire protection in conventionally designed and constructed steel framed buildings. Two methods based on the whole building behavior may be considered.

37.5.2.1   Utilization of Structural Redundancy

Up to now, the design of steel structures for fire safety has generally been based on the assessment of individual structural members, that is, each structural member should achieve the required fire resistance. However, it should be realized that for a building to remain stable under fire conditions, serving the principal need of containing a fire and preventing its spread, it is not absolutely necessary for every individual structural member to remain stable. Fire protection may be eliminated for some steel members if, in the absence of these members, an alternative load path due to structural redundancy can be developed to retain structural stability.

For example, in a multistory steel framed structure, floor loads are transferred by floor slabs to the supporting steel beams and the beam reactions are then transmitted to the supporting steel columns and thence to the foundations. For fire design, it is usually assumed that the load path selected for the ambient temperature design remains unchanged under fire conditions and that each member in this load path has to have sufficient fire resistance.

It is now appreciated that the load transmission path in a structure is not a fixed one. If one load path breaks down, other alternative load paths may exist and safely transfer the applied loads to the foundation. For example, Figure 37.20a shows part of a steel framed structure and the usual load carrying path adopted in the ambient temperature design. However, if the floor slabs are designed to have higher load bearing resistance than required at ambient temperature, it is quite possible for the alternative load carrying sequence in Figure 37.20b to develop in fire.

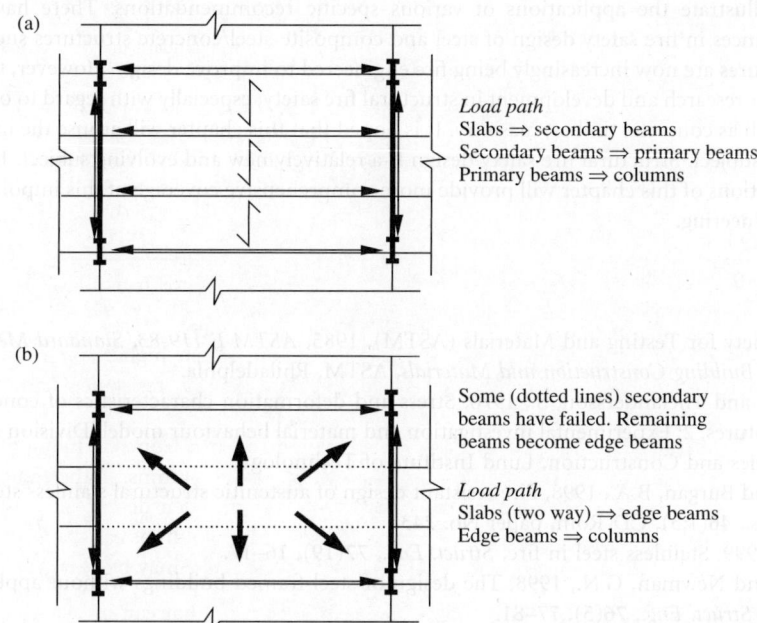

FIGURE 37.20 An example of alternative load paths in a structure: (a) load path of ambient temperature and (b) possible load path in fire.

At ambient temperature (Figure 37.20a), the secondary beams are needed to control excessive slab deflections. Under fire conditions, applied floor loads are reduced and large slab deflections are permissible. Thus, failure of some secondary beams is permissible provided a sufficient network of beams remain available to keep transfer the slab load to the columns. A possible system of this type is shown in Figure 37.20b. In Figure 37.20b, fire protection for the dotted secondary beams is not required. Of course, the design load carrying capacity of the slab can be further increased even to bypass some main steel beams.

37.5.2.2 Developing Better Understanding of Structural Behavior in Fire

Structural fire safety design has been developed based on extending current structural design methods at ambient temperature to elevated temperatures. At ambient temperature, due to the need to avoid large deflections, design methods are based on flexural behavior at small deflections. These current design methods can be very conservative and structural members may have much higher strength if large deflections are allowed. By exploiting this enhanced load bearing capacity, fire protection may be eliminated. An example is the utilization of tensile membrane action in reinforced composite slabs in steel framed buildings (Newman et al. 2000).

37.6 Concluding Remarks

This chapter has given an introduction to different aspects of structural fire safety design, including evaluation of the severity of fire exposure, calculations of temperatures in structures exposed to fire, and assessment of load carrying capacities of structures at elevated temperatures. An important emphasis of this chapter is structural fire safety design from first principles, rather than following the prescriptive route. Materials presented in this chapter are based on developments in Europe. However, the fundamental knowledge should be equally applicable in other situations. A number of examples have been provided to illustrate the applications of various specific recommendations. There have been tremendous advances in fire safety design of steel and composite steel/concrete structures such that these types of structures are now increasingly being fire engineered to improve design. However, there are still large scopes for research and development in structural fire safety, especially with regard to other types of structures, such as concrete, timber, masonry. It is hoped that this chapter will arouse the interest of the reader in this subject. Structural fire safety design is a relatively new and evolving subject. It is expected that future editions of this chapter will provide more comprehensive coverage of this important topic of structural engineering.

References

American Society for Testing and Materials (ASTM), 1985, *ASTM E 119-83, Standard Methods of Fire Tests of Building Construction and Materials*, ASTM, Philadelphia.

Anderberg, Y. and Thelandersson, S., 1976, Stress and deformation characteristics of concrete at high temperatures, 2: Experimental investigation and material behaviour model, Division of Structural Mechanics and Construction, Lund Institute of Technology.

Baddoo, N. and Burgan, B.A., 1998, Fire resistant design of austenitic structural stainless steel, *J. Constr. Steel Res.*, **46**(1:3), CD-Rom, paper No. 243.

Baddoo, N., 1999, Stainless steel in fire, *Struct. Eng.*, **77**(19), 16–17.

Bailey, C.G. and Newman, G.N., 1998, The design of steel framed buildings without applied fire protection, *Struct. Eng.*, **76**(5), 77–81.

Bailey, C.G. and Moore, D.B., 2000a, The structural behaviour of steel frames with composite floor slabs subject to fire, Part 1: Theory, *Struct. Eng.*, **78**(11), 19–27.

Bailey, C.G. and Moore, D.B., 2000b, The structural behaviour of steel frames with composite floor slabs subject to fire, Part 2: Design, *Struct. Eng.*, **78**(11), 28–33.

Bond, G.V.L., 1975, *Fire and Steel Construction: Water Cooled Hollow Columns* (London: Constrado).

Bressington, P., 1997, Integrated active fire protection: two systems for the price of one, *Building Eng.*, June, 10–13.

British Standards Institution (BSI), 1987, *British Standard 476, Fire Tests on Building Materials and Structures, Part 20: Method for Determination of the Fire Resistance of Elements of Construction (General Principles)* (London: British Standards Institution).

British Standards Institution (BSI), 2001, *Draft BS 9999, Code of Practice for Fire Safety in the Design, Construction and Use of Buildings* (London: British Standards Institution).

Conseil International du Batiment (CIB), 1986, Design guide — structural fire safety, *Fire Safety J.*, **10**, 79–137.

Drysdale, D., 1999, *An Introduction to Fire Dynamics* (2nd edition) (Chichester: John Wiley & Sons).

European Committee for Standardisation (CEN), 1996, *ENV 1992-1-2, Eurocode 2: Design of Concrete Structures, Part 1.2: General Rules, Structural Fire Design* (London: British Standards Institution).

European Committee for Standardisation (CEN), 1997, *ENV 1996-1-2, Eurocode 6: Design of Masonry Structures, Part 1.2: General Rules, Structural Fire Design* (London: British Standards Institution).

European Committee for Standardisation (CEN), 2000a, *Draft prEN 1991-1-2, Eurocode 1: Basis of Design and Actions on Structures, Part 1.2: Actions on Structures — Actions on Structures Exposed to Fire* (London: British Standards Institution).

European Committee for Standardisation (CEN), 2000b, *Draft prEN 1993-1-2, Eurocode 3: Design of Steel Structures, Part 1.2: General Rules, Structural Fire Design* (London: British Standards Institution).

European Committee for Standardisation (CEN), 2000c, *ENV 1995-1-2, Eurocode 5: Design of Timber Structures, Part 1.2: General Rules, Structural Fire Design* (London: British Standards Institution).

European Committee for Standardisation (CEN), 2001, *prEN 1994-1-2, Eurocode 4: Design of Composite Steel and Concrete Structures, Part 1.2: Structural Fire Design* (London: British Standards Institution).

FEMA, 2002, *World Trade Center Building Performance Study: Data Collection, Preliminary Observations, and Recommendations*, Federal Emergency Management Agency.

International Standards Organization (ISO), 1975, *ISO 834: Fire Resistance Tests, Elements of Building Construction* (Geneva: International Organization for Standardization).

Johansen, K.W., 1962, *Yield Line Theory* (London: Cement and Concrete Association).

Karlsson, B. and Quantiere, J.G., 2000, *Enclosure Fire Dynamics* (Boca Raton, FL: CRC Press).

Khoury, G.A., 1983, *Transient Thermal Creep of Nuclear Reactor Pressure Vessel Type Concretes*, PhD Thesis, Department of Civil Engineering, Imperial College of Science, Technology and Medicine.

Khoury, G.A., Grainger, B.N., and Sullivan, P.J.E., 1986, Transient thermal strain of concrete during first heating cycle to 600°C, *Mag. Concrete Res.*, **37**(133), 195–215.

Khoury, G.A., 1992, Compressive strength of concrete at high temperatures: a reassessment, *Mag. Concrete Res.*, **44**(161), 291–309.

Kodur, V.K.R. and Wang, Y.C., 2001, Performance of high strength concrete filled steel columns at ambient and elevated temperatures, in *Tubular Structures IX* (ed. Puthli, R. and Herion, S.) (Lisse: Balkema Publishers).

Law, M. and O'Brien, T., 1989, *Fire Safety of Bare External Structural Steel* (Ascot: The Steel Construction Institute).

Lawson, R.M. and Newman, G.M., 1996, *Structural Fire Design to EC3 & EC4, and Comparison with BS 5950*, Technical Report, SCI Publication 159, The Steel Construction Institute.

Newman, G.M., 1990, *Fire and Steel Construction: The Behaviour of Steel Portal Frames in Boundary Conditions* (2nd edition) (Ascot: The Steel Construction Institute).

Newman, G.M., Robinson, J.T., and Bailey, C.G., 2000, *Fire Safety Design: A New Approach to Multi-Storey Steel-Framed Buildings*, SCI Publication P288, The Steel Construction Institute.

Pettersson, O., Magnusson, S.E., and Thor, J., 1976, *Fire Engineering Design of Steel Structures*, Publication 50, Swedish Institute of Steel Construction.

Phan, L.T. and Carino, N.J., 1998, Review of mechanical properties of HSC at elevated temperatures, *ASCE J. Mater. Civil Eng.*, **10**(1), 58–64.

Phan, L.T. and Carino, N.J., 2000, Fire performance of high strength concrete: research needs, *Proceedings of ASCE Congress 2000*, May, Philadelphia, USA.

Sakumoto, Y., 1998, Research on new fire-protection materials and fire-safe design, *ASCE J. Struct. Eng.*, **125**(12), 1415–1422.

Wang, Y.C., 2000, A simple method for calculating the fire resistance of concrete-filled CHS columns, *J. Constr. Steel Res.*, **54**, 365–386.

Wang, Y.C. and Kodur, V.K.R., 2000, Research towards use of unprotected steel structures, *ASCE J. Struct. Eng.*, **126**(12), 1442–1450.

Wang, Y.C., 2002, *Steel and Composite Structures, Behaviour and Design for Fire Safety* (London: Spon Press).

Appendix

WWW Sites

Nowadays the internet can provide a rich source of information on everything. There are many web sites that give information on fire related topics. The following web sites are well-known organizations that have a interest in the topic of this chapter, structural fire engineering. They are divided into three groups: government and research organizations whose main functions are legislation, research, and dissemination; academic institutions whose main functions are research and education; and industrial companies that are involved in research and development to some extent, but whose main interest is in the application of fire engineering in practical projects. It is inevitable that this list is biased toward U.K. organizations.

A: Government and Research Organizations

www.bre.co.uk: Building Research Establishment (BRE), U.K.
www.safety.odpm.gov.uk/fire: Office of the Deputy Prime Minister, U.K.
www.nrc.ca: National Research Council of Canada (NRCC), Canada
www.bfnl.nist.org: Building and Fire Research Lab, National Institution of Science and Technology (NIST), U.S.
www.nfpa.org: National Fire Protection Association (NFPA), U.S.
www.sfpe.org: Society of Fire Protection Engineers (SFPE), U.S.
www.iafss.org: International Association for Fire Safety Science (IAFSS)
www.vtt.fl: VTT Building Technology, Finland
www.sp.se: Swedish National Testing and Research Institute (SP), Sweden
www.sintef.no: Norwegian Fire Research Laboratory (SINTEF), Norway
www.factorymutual.com: Factory Mutual, U.S.
www.cticm.fr: Centre Technique Industriel de la Construction Métallique (CTICM), France
www.tno.bouw.nl: TNO Building and Construction Research, The Netherlands

B: Academic institutions

www.structuralfiresafety.com: One-stop-shop for structural fire safety, set up by Manchester Centre for Civil and Construction Engineering, University of Manchester Institute of Science and Technology (UMIST), U.K.
www.steelinfire.org.uk: Steel In Fire Forum (STIFF), a network coordinated by the University of Sheffield, U.K.
www.ed.ac.uk: University of Edinburgh, U.K.
www.ulst.ac.uk: University of Ulster, U.K.
www.brand.lth.se: Lund University, Sweden
www.wpi.edu: Worcester Polytechnic Institute, U.S.
www.enfp.umd.edu: University of Maryland, U.S.
www.civil.canterbury.ac.nz: University of Canterbury, New Zealand
www.vut.edu.au: Victoria University of Technology, Australia

C: Industrial Companies

www.corusgroup.com: U.K. steel manufacturer
www.arup.com: Ove Arup & Partners, Engineering consultancy, U.K.
www.burohappold.com: Buro Happold Consulting Engineers, Engineering Consultancy, U.K.
www.steel-sci.org: The Steel Construction Institute, U.K.

Index